WIRELESS APPLICATIONS of SPREAD SPECTRUM SYSTEMS

Selected Readings

Edited by

Dr. Sawasd Tantaratana
&
Dr. Kazi M. Ahmed

Printed in the United States of America

Editing, Typography & Layout - Jill R. Cals

Published by the Institute of Electrical and Electronics Engineers, Inc.
445 Hoes Lane, PO Box 1331, Piscataway, NJ 08855-1331.

http://www.ieee.org/eab/

CONTENTS

INTRODUCTION iii

UNIT 1 SPREAD-SPECTRUM TECHNOLOGY

1.1 Tutorials 3

1. Theory of Spread-Spectrum Communications - A Tutorial 5
   R. L. Pickholtz, D. L. Schilling, and L. B. Milstein

2. Pseudo-Random Sequences and Arrays 35
   F. J. MacWilliams and N. J. A. Sloane

3. Spread-Spectrum Technology for Commercial Applications 51
   D. T. Magill, F. D. Natali, and G. P. Edwards

1.2 Multipath and Fading 65

1. Introduction to Spread-Spectrum Antimultipath Techniques and Their Application to Urban Digital Radio 67
   G.L. Turin

2. Rayleigh Fading Channels in Mobile Digital Communication Systems Part I: Characterization 93
   B. Sklar

3. Rayleigh Fading Channels in Mobile Digital Communication Systems Part II: Mitigation 105
   B. Sklar

1.3 Receivers and Their Performances 113

1. Direct-Sequence Spread-Spectrum Multiple-Access Communications Over Nonselective and Frequency-Selective Rician Fading Channels 115
   E. Geraniotis

2. Error Probabilities for Binary Direct-Sequence Spread-Spectrum Communications with Random Signature Sequences 125
   J. W. Lehnert and M. B. Pursley

3. Multi-User Detection for DS-CDMA Communications 137
   S. Moshavi

4. Multiuser Detection for CDMA Systems 151
A. Duel-Hallen, J. Holtzman and Z. Zvonar

5. Minimum Probability of Error for Asynchronous Gaussian Multiple-Access Channels 165
S. Verdu

6. Single-User Detectors for Multiuser Channels 177
H. V. Poor and S. Verdu

1.4 Acquisition and Tracking of PN Sequence 189

1. A Unified Approach to Serial Search Spread-Spectrum Code Acquisition-Part I: General Theory 191
A. Polydoros and C. Weber

2. Direct-Sequence Spread-Spectrum Parallel Acquisition in a Fading Mobile Channel 199
E. A. Sourour and S. C. Gupta

3. Noncoherent Sequential Acquisition of PN Sequences for DS/SS Communications with/without Channel Fading 207
S. Tantaratana, A. W. Lam and P. J. Vincent

4. A Closed-Loop Coherent Acquisition Scheme for PN Sequences Using an Auxiliary Sequence 215
M. Salih and S. Tantaratana

5. Decision-Directed Coherent Delay-Lock Tracking Loop for DS-Spread-Spectrum Signals 223
R. De Gaudenzi and M. Luise

6. A New Tracking Loop for Direct Sequence Spread Spectrum Systems on Frequency-Selective Fading Channels 231
W.H. Sheen and G.L.Stuber

1.5 CDMA System Performance 237

1. Spread Spectrum for Mobile Communications 239
R. L. Pickholtz, L. B. Milstein, and D. L. Schilling

2. Erlang Capacity of a Power Controlled CDMA System 249
A. M. Viterbi and A. J. Viterbi

3. Reverse Link Performance of IS-95 Based Cellular Systems 259
R. Padovani

4. Spread Spectrum Access Methods for Wireless Communications 267
R. Kohno, R. Meidan, and L. B. Milstein

5. Design Study for a CDMA-Based Third-Generation Mobile Radio System 277
A. Baier et al

6. Wireless Communications Going Into 21st Century 289
D.L. Schilling

1.6 Frequency Hopping Spread-Spectrum Systems 297

1. Error Probability of Fast Frequency Hopping Spread Spectrum with BFSK 299
Modulation in Selective Rayleigh and Selective Rician Fading Channels
B. Solaiman, A. Glavieux and A. Hillion

2. Probability of Error in Frequency-Hop Spread-Spectrum Multiple-Access 307
Communication Systems with Noncoherent Reception
K. Cheun and W. E. Stark

3. Optimum Detection of Slow Frequency-Hopped Signals 319
B. K. Levitt, U. Cheng, A. Polydoros and M. K. Simon

UNIT 2 SPREAD-SPECTRUM APPLICATIONS IN CELLULAR MOBILE

2.1 CDMA Overview and Performance 333

1. Overview of Cellular CDMA 335
W.C.Y. Lee

2. On the Capacity of a Cellular CDMA System 347
K.S. Gilhousen, et. al

3. Performance Evaluation for Cellular CDMA 357
L.B. Milstein, et al

4. Effects of Radio Propagation Path Loss on DS-CDMA Cellular 367
Frequency Reuse Efficiency for the Reverse Channel
T. S. Rappaport and L.B. Milstein

5. Multipath Propagation Effects on a CDMA Cellular System 379
N. L. B. Chan

6. Analysis of a Direct-Sequence Spread-Spectrum Cellular Radio System 387
C. Kchao and G. L. Stüber

7. Analysis of a Multiple-Cell Direct-Sequence CDMA Cellular 397
Mobile Radio System
G. L. Stüber and C. Kchao

2.2 CDMA Power Control Issues 409

1. Cellular CDMA Networks Impaired by Rayleigh Fading: System Performance with Power Control 411
   O. K. Tonguz and M. M. Wang

2. Spectral Efficiency of a Power-Controlled CDMA Mobile Personal Communication System 425
   J.H. Gass Jr., et al.

3. Capacity, Throughput, and Delay Analysis of a Cellular DS CDMA System With Imperfect Power Control and Imperfect Sectorization 437
   M. G. Jansen and R. Prasad

4. Performance Analysis of CDMA with Imperfect Power Control 447
   R. Cameron and B. Woerner

5. Effects of Imperfect Power Control and User Mobility on a CDMA Cellular Network 453
   F. D. Priscoli and F. Sestini

6. The Capacity of a Spread Spectrum CDMA System for Cellular Mobile Radio with Consideration of System Imperfections 463
   P. Newson and M.R. Heath

2.3 CDMA Overlay 475

1. CDMA Cellular Engineering Issues 477
   K.I. Kim

2. Overlay of Cellular CDMA on FSM 483
   K.G. Filis and S.C. Gupta

3. Microcellular Engineering in CDMA Cellular Networks 497
   J. Shapira

4. A Micro-Cellular CDMA System Over Slow and Fast Rician Fading Radio Channels with Forward Error Correcting Coding and Diversity 507
   C.A.F.J. Wijffels, et al.

5. CDMA Overlay Situations for Microcellular Mobile Communications 519
   J. Wang and L. B. Milstein

2.4 CDMA Traffic Issues and CDMA Advantages 531

1. SIR-Based Call Admission Control for DS-CDMA Cellular Systems 533
   Z. Liu and M. E. Zarki

2. Congestion Relief on Power-Controlled CDMA Networks 541
J.M. Jacobsmeyer

3. Performance Analysis of Soft Handoff in CDMA Cellular Networks 545
S.L. Su, et al.

4. Advantages of CDMA and Spread Spectrum Techniques over FDMA 553
and TDMA in Cellular Mobile Radio Applications
P. Jung, et al.

UNIT 3 SPREAD-SPECTRUM APPLICATIONS IN MOBILE SATELLITE

1. Increased Capacity Using CDMA for Mobile Satellite Communication 563
K. S. Gilhousen, et. al.

2. Direct-Sequence Spread Spectrum in a Shadowed Rician Fading 575
Land-Mobile Satellite Channel
R.D.J. van Nee, et al.

3. Open-Loop Power Control Error in a Land Mobile Satellite System 583
A. M. Monk and L.B. Milstein

4. Performance of DS-CDMA with Imperfect Power Control Operating 591
Over a Low Earth Orbiting Satellite Link
B.R. Vojcic, et. al.

5. A Performance Comparison of Orthogonal Code Division Multiple-Access 599
Techniques for Mobile Satellite Communications
R. De Gaudenzi, et al.

6. Design Study for a CDMA-Based LEO Satellite Network: Downlink 607
System Level Parameters
S.G. Glisic, et al.

UNIT 4 SPREAD-SPECTRUM APPLICATIONS IN INDOOR WIRELESS

1. Performance of DS-CDMA over Measured Indoor Radio Channels Using 623
Random Orthogonal Codes
M. Chase and K. Pahlavan

2. Performance Evaluation of Direct-Sequence Spread Spectrum Multiple- 631
Access for Indoor Wireless Communication in a Rician Fading Channel
R. Prasad, et al.

3. Performance of BPSK and TCM Using the Exponential Multipath Profile 643
Model for Spread-Spectrum Indoor Radio Channels
J. M. Bargallo and J. A. Roberts

4. Decision Feedback Equalization for CDMA in Indoor Wireless Communications 653
M. Abdulrahman, et al.

5. Effects of Diversity, Power Control, and Bandwidth on the Capacity 663
of Microcellular CDMA Systems
A. Jalali and P. Mermelstein

6. Hybrid DS/SFH Spread-Spectrum Multiple Access with Predetection 673
Diversity and Coding for Indoor Radio
J. Wang and M. Moeneclaey

7. Performance Analysis of a Hybrid DS/SFH CDMA System Using 683
Analytical and Measured Pico Cellular Channels
F. Çakmak, et al.

8. A CDMA-Distributed Antenna System for In-Building Personal 693
Communications Services
H. H. Xia, et al.

9. Traffic Handling Capability of a Broadband Indoor Wireless Network 701
Using CDMA Multiple Access
C.G. Zhang, et al.

UNIT 5 OTHER APPLICATIONS OF SPREAD SPECTRUM

1. The Global Positioning System 713
I. A. Getting

2. Mobile Access to an ATM Network Using a CDMA Air Interface 721
M. J. McTiffin et al.

3. The Application of a Novel Two-Way Mobile Satellite Communications 731
and Vehicle Tracking System to the Transportation Industry
I.M. Jacobs, A. Salmasi, and T.J. Bernard

About the Editors

Sawasd Tantaratana was born in Thailand and received his B.E.E. degree from the University of Minnesota in 1971, his M.S. degree from Stanford University in 1972, and his Ph.D. degree from Princeton University in 1977, all in electrical engineering. He has spent most of his career in teaching in both Thai and U.S. universities, which include King Mongkut's Institute of Technology, Thonburi, Thailand, Auburn University, Alabama, the University of Massachusetts at Amherst, Massachusetts, and Sirindhorn International Institute of Technology of Thammasat University, Thailand, where he is a Professor of Electrical Engineering. From 1980-1981, he was a visiting faculty member at the University of Illinois at Urbana-Champaign, Illinois, and from 1984 to 1986 he was a member of the Technical Staff at AT&T Bell Laboratories in Holmdel, New Jersey. His research interests are signal processing and communications. He has worked in the area of spread spectrum systems for about ten years, and has published approximately 100 technical articles in various journals, conference proceedings, and book chapters, as well as an IEEE self-study course on spread-spectrum systems. He is currently an Associate Editor for the IEEE Transactions on Signal Processing.

Kazi Ahmed, a Bangladesh national, graduated with a Master of Science in Telecommunications from the Leningrad Bonch-Bruevitch Electrical Engineering Institute of Communications in 1978. He obtained his Ph.D. in Electrical Engineering from University of Newcastle, N.S.W., Australia, in 1983, under a University Postgraduate Scholarship. He has vast teaching and research experience as an Assistant and later as Associate Professor since 1983, first at Bangladesh University of Engineering and Technology, Dhaka, and later at Asian Institute of Technology, Bangkok, Thailand. He spent one year in University of Leeds, U.K. as a Commonwealth Academic Staff Fellow doing research in mobile radio propagation. Since 1994, he has been involved in research on CDMA techniques in cellular and satellite systems. He has several conference and journal publications in antenna array processing, radio propagation and CDMA systems.

INTRODUCTION

In conventional communication systems, bandwidth is of major concern and systems are designed to use as little bandwidth as possible. The bandwidth needed to transmit an analog signal source is twice that of the source in double-sideband amplitude-modulated systems. It is also several times the bandwidth of the source in frequency-modulated systems, depending on the modulation index. For a digital signal source, the bandwidth required is in the same order of magnitude as the bit rate of the source. The exact bandwidth required depends on the type of modulation (i.e., BPSK, QPSK, etc.).

In spread-spectrum (SS) communication systems, the bandwidth of a signal is expanded, usually by a few or several orders of magnitude, before transmission. When there is only one user in a SS band, the bandwidth utilization is inefficient. However, in a multiple-user environment, many users can share the same frequency band and the system can become bandwidth efficient while maintaining the advantages of spread-spectrum systems.

The development of SS systems has a long history. For details see Simon et al. [1], Scholtz [2,3], and Price [4]. Since SS signals possess anti-jamming property, as well as being difficult to detect, spectrum spreading was originally used for military communications, where secure communications are essential. However in the last two to three decades, SS techniques have been studied for commercial applications. The introduction of the ISM band (Industrial, scientific, and medical band) and the auction of frequencies in the US in 1996 provided stimulation to the deployment of SS systems for commercial uses. Currently there are several commercial systems that employ SS techniques. More systems are being proposed for use in many new applications, such as Personal Communication Service (PCS), Wireless Local Area Networks (WLAN), Wireless Private Branch Exchanges (WPBX), wireless inventory control systems, and the Global Positioning Satellite (GPS) system.

SS modulation has many attractive features. The important features include:

- resistance to intentional and non-intentional interferences;
- the ability to eliminate or alleviate the effect of multipath propagation;
- the ability to share the same frequency band with other users by using noise-like spreading signal;
- suitabilty for low-power unlicensed SS radios in the ISM bands: 902-928 MHz, 2.4-2.4835 GHz, and 5.725-5.85 GHz;
- offering a certain degree of privacy, due to the use of pseudo-random spreading codes, making it difficult to intercept the signal.

Figure 1 shows a functional block diagram of a typical SS communication system for the terrestrial or satellite configurations. The source may be digital or analog. If the source is analog, it is first digitized by some analog-to-digital (A/D) conversion scheme such as pulse-code modulation (PCM) or delta-modulation (DM) encoding. The data compressor eliminates or reduces the information redundancy in the digital source. The output is then encoded by the error correction encoder, which introduces coding redundancy for detection and correction of the errors that may arise from transmission through the RF channel. The spectrum of the resulting signal is spread out to the desired bandwidth, followed by a modulator, which shifts the spectrum to the assigned frequency range for transmission. The modulated signal is then amplified and sent through the transmission channel, which may be a terrestrial or a satellite channel. The channel introduces some impairments: interference, noise, and attenuation of the signal power. Note that the data compressor/decompressor and the error correction encoder/decoder are optional as far as the SS concept is concerned. They serve to improve the performance of the system. Note also that the locations of the spectrum spreading and the modulation functions may be interchanged. These two functions are usually integrated and implemented as a single unit.

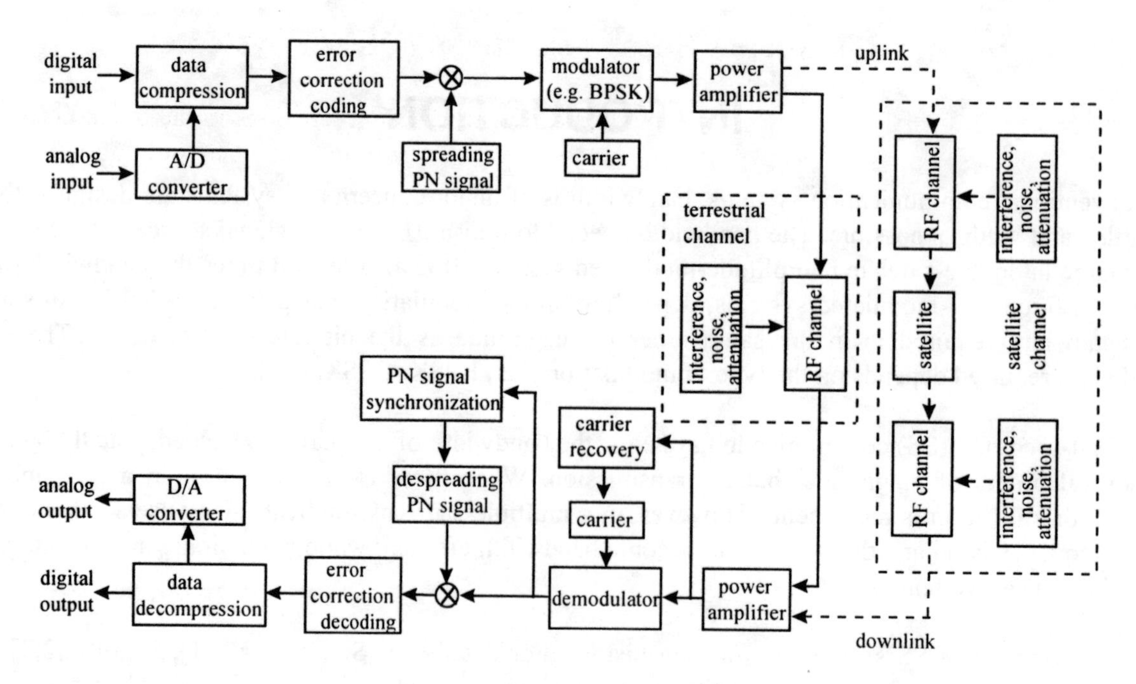

Figure 1: Block diagram of a typical digital communication system with spectrum spreading.

At the receiving end, the receiver attempts to reconstruct the original signal by undoing the processes used in the transmitter; i.e., the received signal is demodulated, spectrum-despread, decoded, and decompressed to obtain a digital signal. If the source is analog, the digital signal is converted to an analog signal using a D/A conversion.

Note that in a conventional (non-SS) system, the spectrum-spreading and spectrum-despreading functions are absent from the block diagram in Figure 1. In fact, those are the only functional differences between a conventional narrowband system and an SS system.

A digital communication system is considered to be a SS system if the transmitted signal occupies a bandwidth larger than the minimum bandwidth required to transmit the information; and the bandwidth spread is accomplished by means of a code which is independent of the data. Note that the definition rules out FM system because its bandwidth depends on the bandwidth of the source.

There are three basic types of SS systems: direct-sequence (DS), frequency-hopping (FH), and time-hopping (TH) systems. Hybrids of such systems can also be obtained. A DS/SS system achieves spectrum spreading by multiplying the source with a pseudo-random signal. A FH/SS system achieves spectrum spreading by hopping its carrier frequency over a (large) set of frequencies in a pseudo-random pattern. In a TH/SS system, a block of data bits is compressed and transmitted in a pseudo-randomly chosen time slot within a frame that consists of a large number of time slots. Pseudo-random code is the key to the operation of SS systems. It helps create a noise-like signal.

In this Selected Readings, the collection of papers introduce the fundamentals of SS techniques, their development, and their applications especially in the wireless environment, as well as various problems and solutions. Many papers are classic reviews, while others are reports of research results.

To provide you with a basic understanding of SS systems, we briefly describe direct-sequence spread-spectrum (DS/SS) systems, frequency-hopped systems, and code-division multiple access (CDMA) in the following sections.

Direct-Sequence Spread-spectrum (DS/SS) Systems

We now consider only the spectrum spreading and carrier modulation functions in the block diagram of Figure 1. A direct-sequence spread-spectrum (DS/SS) system transmitter is depicted in Figure 2, together with examples of the waveforms.

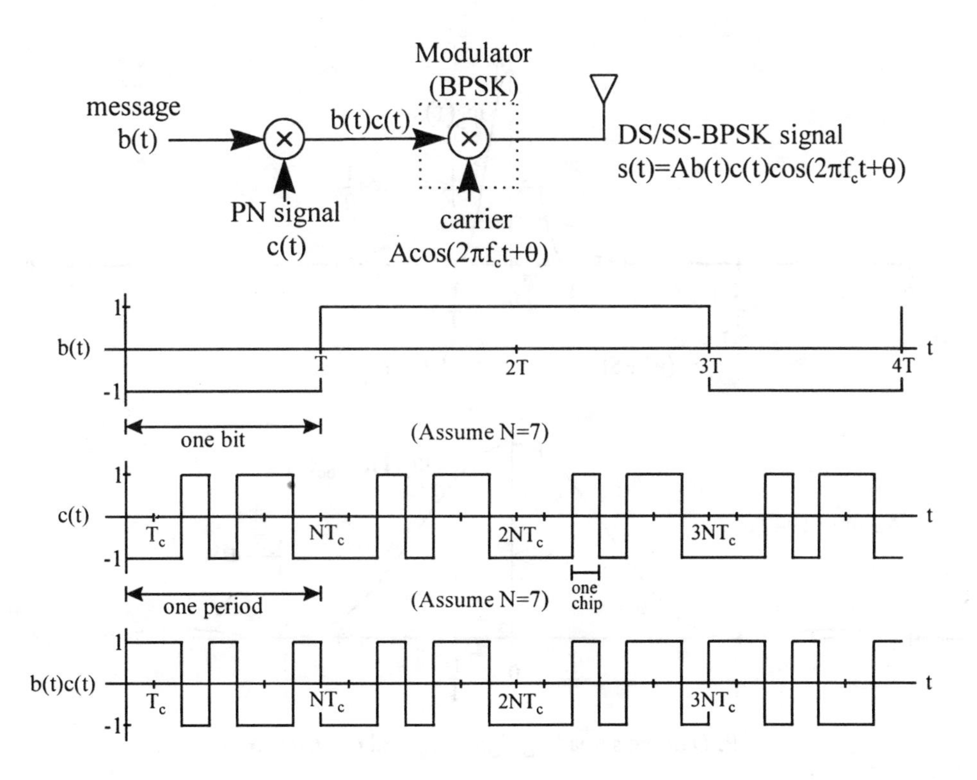

Figure 2: Block diagram of a DS/SS transmitter and waveforms of signals at various points

The binary data $b(t)$ to be transmitted has a bit rate of $1/T$ bps, i.e., one bit duration is T seconds. The pseudo-noise (PN) signal $c(t)$ is also a binary signal with a bit duration of T_c seconds. To distinguish the rate of $c(t)$ from that of $b(t)$, one bit of $c(t)$ is called one chip. Therefore, $c(t)$ has a rate of $1/T_c$ chips/s. Note that a truly random binary signal would not repeat itself. However, in order that the receiver can despread the SS signal, we cannot use a truly random signal. We use pseudo-random signal, which repeats itself with a period of N chips, or NT_c seconds, where N is generally a large number. Here, as in many applications, it is assumed that $T=NT_c$. The product $b(t)c(t)$ has the same rate as $c(t)$, which N times that of $b(t)$. The product signal then modulates the carrier, using a digital modulation scheme, such as a binary phase-shift keying (BPSK), as shown in Figure 2. The resulting signal is

$$s(t) = A\, b(t)c(t) \cos(2\pi f_c t+\theta)$$

as shown in Figure 2, where A is the amplitude, f_c is the carrier frequency, and θ is the carrier phase.

For simplicity, assume that $b(t)$ and $c(t)$ are random binary signals. Then, their power spectral densities (PSD) have the sinc square shape, as shown in Figure 3. Specifically, Figure 3(a) depicts $\varphi_b(f)=T\text{sinc}^2(fT)$ (solid line) and $\varphi_c(f)=T_c\text{sinc}^2(fT_c)$ (dashed line), where $\text{sinc}(x) = \sin(\pi x)/(\pi x)$. To see the sidelobes, $\varphi_c(f)$ is also shown in Figure 3(b). Note that the vertical scale in Figure 3(a) and Figure 3(b) are different, i.e., the magnitude in Figure 3(b) is much smaller than that in Figure 3(a) since $T_c=T/N$. PSD is often plotted in deciBell (dB) to clearly show small values. The first-null bandwidths of Figures 3(a) and 3(b) are $1/T$ and $1/T_c$ Hz, recpectively. The product $b(t)c(t)$ has a PSD of the same form as that of $c(t)$. Therefore, we see that the bandwidth $1/T$ Hz of $b(t)$ has now been spread N folds, to $1/T_c = N/T$ Hz, simply by multiplication with $c(t)$. After carrier modulation, the PSD of the resulting signal $s(t)$ is shifted to a center frequency of by multiplication with $c(t)$. After carrier modulation, the PSD of the resulting signal $s(t)$ is shifted to a center frequency of f_c Hz, as shown in Figure 3(c), with a null-to-null bandwidth of $2N/T = 2/T_c$ Hz. The PSD $\varphi_u(f)$ of the narrowband BPSK signal (unspread signal) $u(t) = b(t)\cos(2\pi f_c t+\theta)$ is depicted in Figure 3(d) with solid line. The bandwidth is $2/T$, which is smaller than the bandwidth of $s(t)$ by a factor of N. Just for comparison, $\varphi_s(f)$ is also plotted in Figure 3(d) with dashed line.

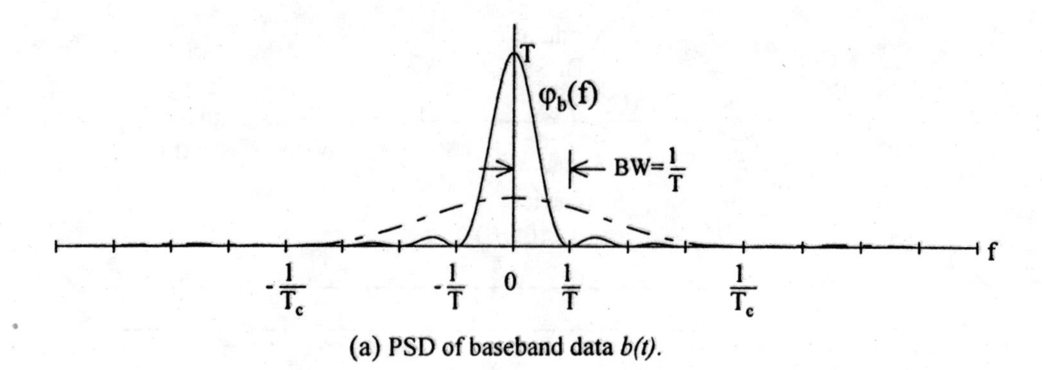

(a) PSD of baseband data $b(t)$.

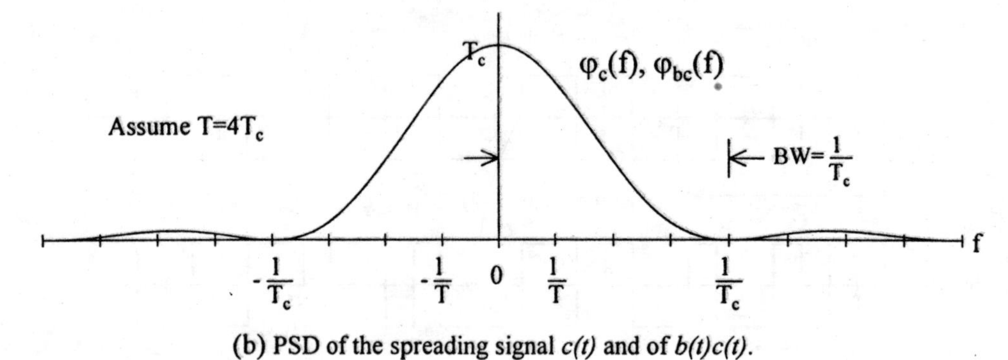

(b) PSD of the spreading signal $c(t)$ and of $b(t)c(t)$.

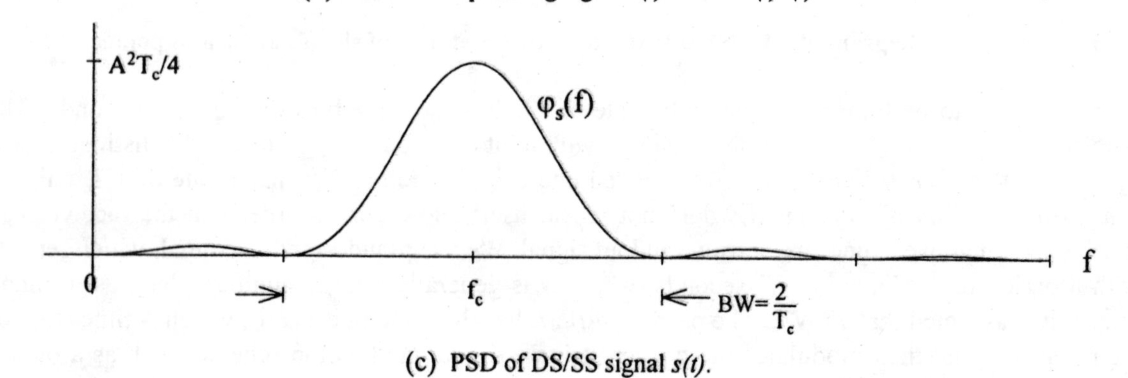

(c) PSD of DS/SS signal $s(t)$.

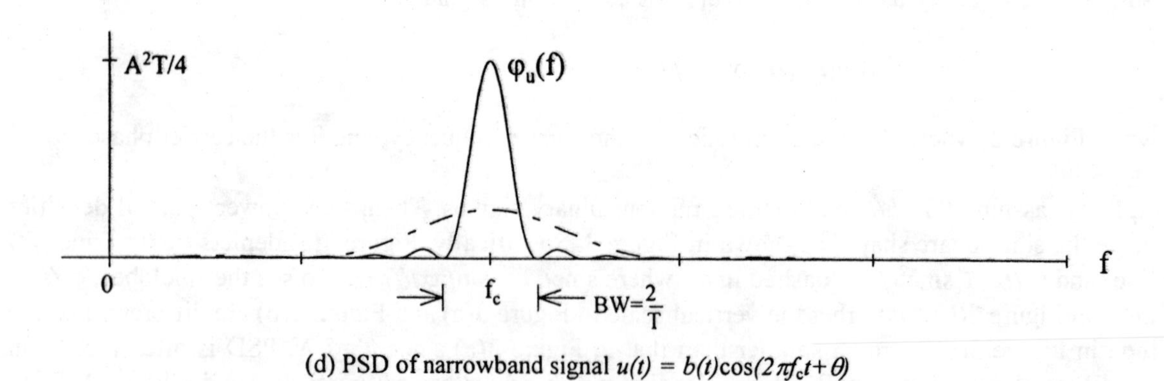

(d) PSD of narrowband signal $u(t) = b(t)\cos(2\pi f_c t+\theta)$

Figure 3: Power spectral densities of various signals in Figure 2.

A receiver for a DS/SS-BPSK system is depicted in Figure 4. The signal arriving at the receiver is $s(t-\tau)$, which is $s(t)$ delayed by τ seconds due to propagation time:

$$s(t-\tau) = A\, c(t-\tau)\, b(t-\tau) \cos(2\pi f_c t+\theta')$$

To despread the bandwidth from $2N/T$ Hz back to $2/T$ Hz, we simply multiply the received signal $s(t-\tau)$ by $c(t-\tau)$, yielding $w(t) = A\, b(t-\tau) \cos(2\pi f_c t+\theta')$ since $c^2(t-\tau)=1$. The signal $w(t)$ is only a BPSK signal, which can be demodulated by a narrowband BPSK demodulator to obtain estimates of the bits b_i of the signal $b(t)$.

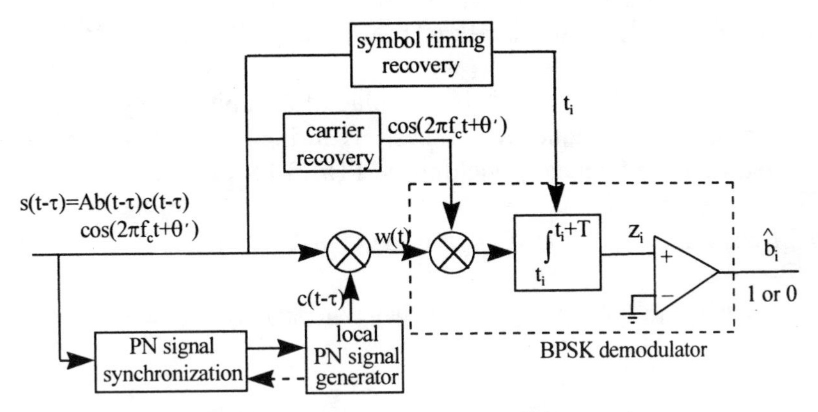

Figure 4: A block diagram of a DS/SS-BPSK receiver.

Note that the signal $c(t-\tau)$ used to despread $s(t-\tau)$ must have the correct phase τ. Otherwise, $w(t)$ is not equal to $A\, b(t-\tau) \cos(2\pi f_c t+\theta')$, i.e., despreading is not accomplished. The process of obtaining τ and generating $c(t-\tau)$ is called PN signal synchronization, a necessary function of the receiver. A local PN signal is generated and its phase is adjusted until it is synchronized to the phase of the incoming PN signal. PN signal synchronization is generally performed in two steps: *(i)* coarse synchronization, which is called acquisition, to bring the phase of the incoming PN signal and that of the locally generated PN signal to within a given range, and *(ii)* fine tuning, which is called tracking.

Besides PN synchronization, the receiver needs the carrier frequency and phase in order to perform BPSK demodulation. They are obtained by the carrier recovery subsystem. Also, the beginning of each data bit must be obtained for the demodulation process. This function is carried out by the symbol timing recovery subsystem, as depicted in Figure 4.

Frequency-Hopped Spread-spectrum (FH/SS) Systems

In a frequency-hopped spread-spectrum (FH/SS) system, the carrier frequency hops among a given set of frequencies. During each hop, the carrier frequency to be used is determined by a PN sequence. If the duration of each hop is smaller than the duration of one data bit, then the system is called a fast frequency-hopped spread-spectrum system (FFH/SS). Therefore, each bit is transmitted using more than one carrier frequency. On the other hand, if the hop duration is larger than the bit duration, the system is called a slow frequency-hopped spread-spectrum system (SFH/SS). In this case, more than one consecutive bits use the same carrier frequency, before jumping to another carrier frequency.

The digital modulation which is often used in conjunction with FH spectrum spreading is the FSK (frequency shift keying) signaling. Figure 5 depicts block diagrams for FH/SS-FSK transmitter and receiver. In the transmitter, the binary data $b(t)$ first modulates a sinusoidal signal $A\cos(2\pi f_0 t+\theta)$ using FSK modulation, yielding

$$d(t) = A\cos(2\pi(f_0+0.5(b_i+1)\Delta f)t+\theta) \qquad \text{for } iT < t < (i+1)T,$$

i.e., during the ith bit duration, where b_i (+1 or -1) is the value of the ith bit of the data, and Δf is the frequency separation. Then, we form the product of this narrowband FSK signal $d(t)$ and the output $g(t)$ of the frequency synthesizer. The frequency synthesizer generates sinusoidal signal whose frequency hops among a set of 2^j frequencies, with the hopping pattern being determined by j bits from the PN sequence generator. Let T_h be the duration of one hop. Assume that $T_h<T$, with $T/T_h=L$, i.e., we have a FFH. The product signal during the mth hop, $mT_h<t<(m+1)T_h$, is given by

$$\begin{aligned} p(t) &= d(t)g(t) = A\cos(2\pi(f_0+0.5(b_k+1)\Delta f)t+\theta)\cos(2\pi f_m t+\alpha) \\ &= 0.5A[\cos(2\pi(f_0+0.5(b_k+1)\Delta f+f_m)t+\theta+\alpha)+\cos(2\pi(f_0+0.5(b_k+1)\Delta f-f_m)t+\theta-\alpha)], \quad mT_h<t<(m+1)T_h \end{aligned}$$

where k is the integer part of m/L, so that b_k is the value of data bit during the mth hop, f_m is the frequency of the signal generated by the frequency synthesizer during the mth hop. The signal $p(t)$ is bandpass filtered to block the difference frequency, yielding a FFH/SS-FSK signal

$$s(t) = 0.5A[\cos(2\pi(f_0+0.5(b_k+1)\Delta f+f_m)t+\theta+\alpha), \quad mT_h<t<(m+1)T_h$$

Therefore, during the mth hop, the transmitted frequency is f_0+f_m if the data bit is -1 or $f_0+f_m+\Delta f$ if the data bit is +1.

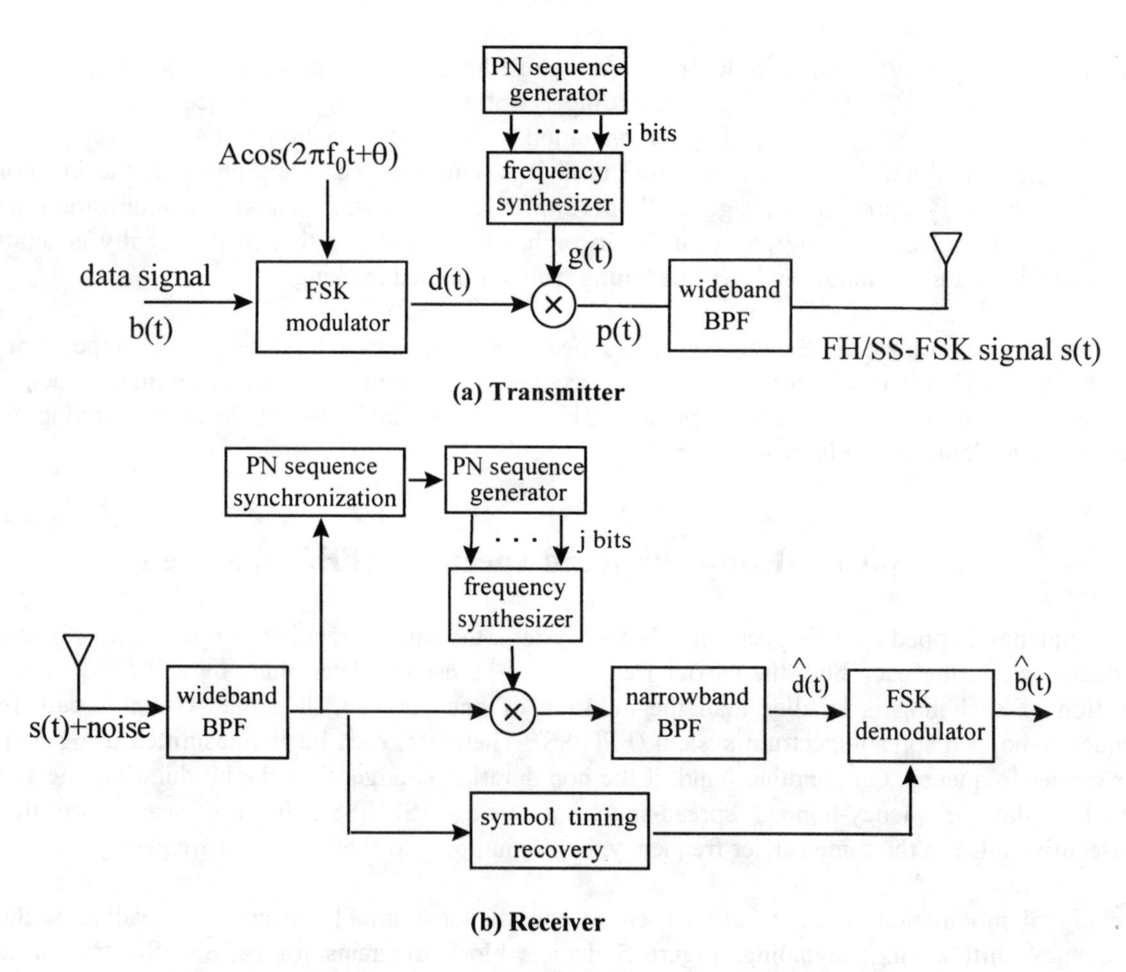

Figure 5: A block diagram of FH/SS-FSK transmitter and receiver.

The frequency of the signal *s(t)* during each hop is depicted in Figure 6, which assumes that $T/T_h=3$. Note that the minimum frequency is f_{min}, which equals f_0+f_l, where f_l is the lowest frequency generated by the frequency synthesizer, and the maximum frequency is $f_{min}+(J-1)\Delta f$, which equals $f_0+f_h+\Delta f$, where f_h is the highest frequency generated by the frequency synthesizer. Note that the diagram in Figure 6 divides the transmitted frequencies into pairs of frequencies, separated by Δf. If the data bit is -1, then the transmitted frequency would be the lower one of a frequency pair. If the data bit is +1, then the transmitted frequency is the upper one of a frequency pair, as depicted in Figure 6. It is clear from Figure 6 that the total bandwidth of the FH/SS-FSK signal is $J\Delta f$ Hz.

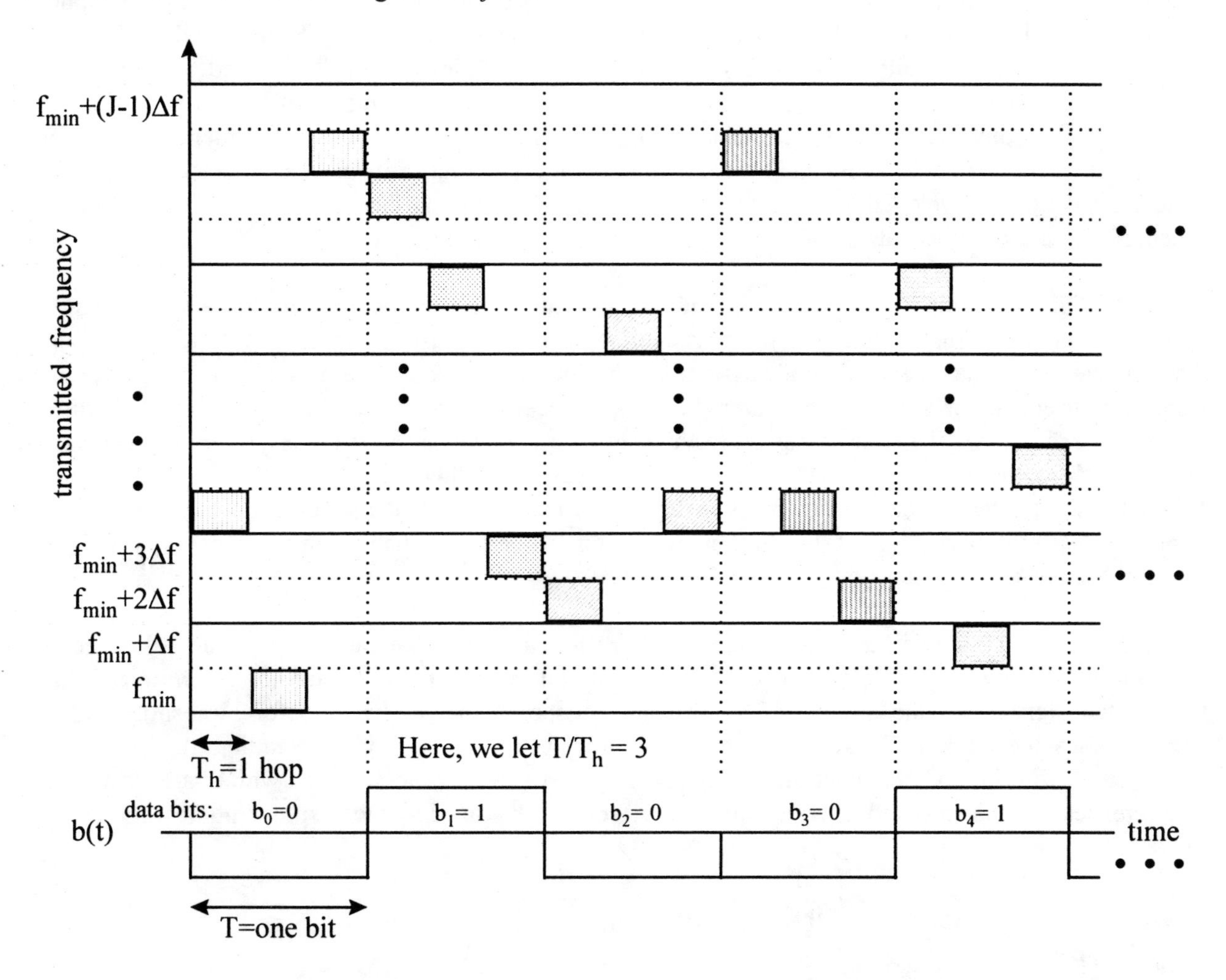

Figure 6: Frequency hopping diagram for a FFH/SS system.

The receiver in Figure 5(b) receives an incoming FH/SS-FSK signal *s(t)* plus noise. The wideband bandpass filter serves to weed out noise and interferences outside the band of frequencies being used by *s(t)*. The local frequency synthesizer generates frequency during each hop in synchronism with the one embedded in the incoming signal. The PN sequence synchronization subsystem acquires and maintains such synchronization. The signal from the local frequency synthesizer mixes with the signal from the wideband BPF, followed by a narrowband BPF, which gives an output corresponding the narrowband FSK signal plus noise. After FSK demodulation, we have an estimate of the data signal *b(t)*.

Code-Division Multiple-Access (CDMA) Communication Systems

In a multiple-access application, there are many users wishing to use the same channel. In a frequency-division multiple-access (FDMA) system, each active user is allocated an individual frequency band for transmission, thus avoiding interference by way of using different frequencies. In a time-division multiple-access (TDMA) system, each active user is allocated an individual time slot in each frame to transmit data, i.e., the users take turns in transmitting. Therefore, interference is avoided. The FDMA system is used in conventional analog cellular phone systems such as the AMPS (advanced mobile phone system), while the TDMA is used in digital cellular phone systems such as the US IS-54 and the European GSM (general system mobile) systems. There is an inherent inefficiency in FDMA and TDMA systems that each user occupies a frequency band or a time slot for the whole duration of his/her call. Normally each party in a conversation speaks less than half of the time (there are periods of silence). Therefore, during the silent periods, the frequency band or the time slot transmits no data, hence it is wasted. Schemes have been developed to make the silent periods available to other users, but the control of such schemes become complicated.

In a code-division multiple-access system, which is also called spread-spectrum multiple access (SSMA) system, all the active users share the same frequency band and transmit at the same time. Each user is distinguished from the others by a code assigned to him/her. If the codes are orthogonal, then the signals from other users can be separated and filtered out at the receiver. Otherwise, the signals from other users become interferences. The codes assigned to the users are PN sequences. It is not possible to generate a large set of orthogonal sequences. Therefore, interferences from other users become a major impairment. When the total interference reaches a level that the performance of the receiver is unacceptable, the number of active users cannot be further increased. Therefore, such multiple-access systems are interference limited.

A DS/SS CDMA system is depicted in Figure 7. Here it is assumed that each user transmits his/her data asynchronously with respect to other users, i.e., the data bits among the users are not synchronized. Each transmitted signal uses the assigned PN sequence to spread the spectrum before transmission. The receiver uses the correct PN sequence to extract the desired signal by despreading it, followed by narrowband filtering. Other undesired signals have low-level wide-spectra, which are low-power interferences after narrowband filtering, and their effects are averaged by the despreading operation.

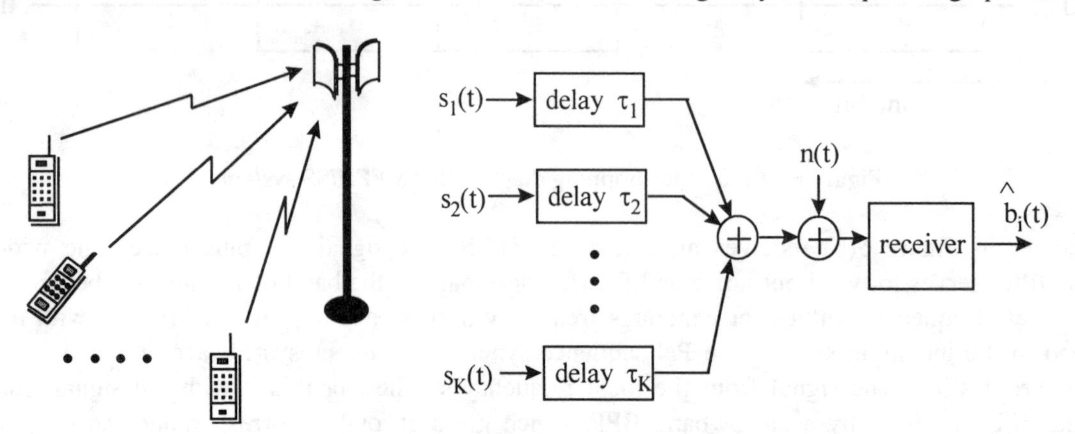

Figure 7: A multiple-access mobile system and a block-diagram model.

In FH/SS systems, each user is assigned a different PN code in such a way that no two transmitters use the same frequency simultaneously, i.e., the transmitters avoid collision with one another. Therefore, FH is an avoidance type of systems, while DS is an averaging type of system. When the number of users is large, total avoidance may be impossible. Therefore, some collisions are allowed. After despreading, the collisions become interferences.

Outline of This Selected Readings

This selected readings collects recent papers on spread spectrum systems, with special emphasis on wireless communications. There were two previous IEEE-published collections of papers on spread spectrum systems, given by references [5] and [6]. Papers already published there are not included in this collection, except for a tutorial paper [7], which is included due to its tutorial value.

Papers are divided into 5 major categories: spread-spectrum technology, which is divided into six sub-sections; applications in cellular mobile, which is divided into four sub-sections; applications in mobile satellite; applications in indoor wireless; and a few papers on other applications. A brief introduction is given for each group of papers.

References

[1] M.K. Simon, J.K. Omura, R.A. Scholtz, and B.K. Levitt, *Spread Spectrum Communications Handbook.* McGraw Hill, 1994.

[2] R.A. Scholtz, "The Origins of Spread-Spectrum Communications," *IEEE Trans. On Communications*, Vol. COM-30, pp. 822-854, May, 1982.

[3] R.A. Scholtz, "Notes on Spread-Spectrum History," *IEEE Trans. On Communications*, Vol. COM-31, pp. 82-84, Jan. 1983.

[4] R. Price, "Further Notes ad Anecdotes on Spread-Spectrum Origins," *IEEE Trans. On Communications*, Vol. COM-31, pp. 85-97, Jan. 1983.

[5] R.C. Dixon, Editor. *Spread Spectrum Techniques*. IEEE Press, 1976.

[6] C.E. Cook, F.W. Ellersick, L.B. Milstein, and D.L. Schilling, Editors. *Spread-Spectrum Communications*. IEEE Press, 1983.

[7] R.L. Pickholtz, D.L. Schilling, and L.B. Milstein, *Theory of Spread-Spectrum Communications - A Tutorial*, IEEE Transactions on Communications, May 1982.

UNIT 1
SPREAD SPECTRUM TECHNOLOGY

In this unit, we have gathered papers that relate to fundamental issues and techniques of spread spectrum. In the last three decades, there have been a large number of papers appearing in transactions and journals addressing various aspects of spread-spectrum techniques. We have selected 27 papers for this unit. We have divided these into six sections, each representing results of these various aspects.

SECTION 1.1
TUTORIALS

In this section three papers of tutorial nature are presented. The first paper, "*Theory of Spread-Spectrum Communications - A Tutorial*" by Pickholtz et al., is a classic tutorial paper on spread-spectrum systems. It presents basic concepts of spectrum spreading, discusses pseudonoise sequences, considers antijamming properties and multipath effects, as well as acquisition and tracking of pseudonoise sequences.

The second paper, "*Pseudo-random Sequences and Arrays*" by MacWilliams and Sloane, provides a tutorial on pseudonoise sequences with an emphasis on maximal-length sequences (m-sequences). Various properties are discussed. Although m-sequences are the most extensively studied pseudorandom sequences, many other types of peudorandom sequences, both real and complex, have also been investigated.

The third paper, "*Spread-Spectrum Technology for Commercial Applications*" by Magill et al., gives an overview of spectrum-spreading methods, applications in satellite mobile and indoor wireless, and compares several personal communication systems.

Theory of Spread-Spectrum Communications—A Tutorial

RAYMOND L. PICKHOLTZ, FELLOW, IEEE, DONALD L. SCHILLING, FELLOW, IEEE, AND LAURENCE B. MILSTEIN, SENIOR MEMBER, IEEE

Abstract—**Spread-spectrum communications, with its inherent interference attenuation capability, has over the years become an increasingly popular technique for use in many different systems. Applications range from antijam systems, to code division multiple access systems, to systems designed to combat multipath. It is the intention of this paper to provide a tutorial treatment of the theory of spread-spectrum communications, including a discussion on the applications referred to above, on the properties of common spreading sequences, and on techniques that can be used for acquisition and tracking.**

I. INTRODUCTION

SPREAD-spectrum systems have been developed since about the mid-1950's. The initial applications have been to military antijamming tactical communications, to guidance systems, to experimental antimultipath systems, and to other applications [1]. A definition of spread spectrum that adequately reflects the characteristics of this technique is as follows:

> "Spread spectrum is a means of transmission in which the signal occupies a bandwidth in excess of the minimum necessary to send the information; the band spread is accomplished by means of a code which is independent of the data, and a synchronized reception with the code at the receiver is used for despreading and subsequent data recovery."

Under this definition, standard modulation schemes such as FM and PCM which also spread the spectrum of an information signal do not qualify as spread spectrum.

There are many reasons for spreading the spectrum, and if done properly, a multiplicity of benefits can accrue simultaneously. Some of these are

- Antijamming
- Antiinterference
- Low probability of intercept
- Multiple user random access communications with selective addressing capability
- High resolution ranging
- Accurate universal timing.

Manuscript received December 22, 1981; revised February 16, 1982.
R. L. Pickholtz is with the Department of Electrical Engineering and Computer Science, George Washington University, Washington, DC 20052.
D. L. Schilling is with the Department of Electrical Engineering, City College of New York, New York, NY 10031.
L. B. Milstein is with the Department of Electrical Engineering and Computer Science, University of California at San Diego, La Jolla, CA 92093.

The means by which the spectrum is spread is crucial. Several of the techniques are "direct-sequence" modulation in which a fast pseudorandomly generated sequence causes phase transitions in the carrier containing data, "frequency hopping," in which the carrier is caused to shift frequency in a pseudorandom way, and "time hopping," wherein bursts of signal are initiated at pseudorandom times. Hybrid combinations of these techniques are frequently used.

Although the current applications for spread spectrum continue to be primarily for military communications, there is a growing interest in the use of this technique for mobile radio networks (radio telephony, packet radio, and amateur radio), timing and positioning systems, some specialized applications in satellites, etc. While the use of spread spectrum naturally means that each transmission utilizes a large amount of spectrum, this may be compensated for by the interference reduction capability inherent in the use of spread-spectrum techniques, so that a considerable number of users might share the same spectral band. There are no easy answers to the question of whether spread spectrum is better or worse than conventional methods for such multiuser channels. However, the one issue that is clear is that spread spectrum affords an opportunity to give a desired signal a power advantage over many types of interference, including most intentional interference (i.e., jamming). In this paper, we confine ourselves to principles related to the design and analysis of various important aspects of a spread-spectrum communications system. The emphasis will be on direct-sequence techniques and frequency-hopping techniques.

The major systems questions associated with the design of a spread-spectrum system are: How is performance measured? What kind of coded sequences are used and what are their properties? How much jamming/interference protection is achievable? What is the performance of any user pair in an environment where there are many spread spectrum users (code division multiple access)? To what extent does spread spectrum reduce the effects of multipath? How is the relative timing of the transmitter-receiver codes established (acquisition) and retained (tracking)?

It is the aim of this tutorial paper to answer some of these questions succinctly, and in the process, offer some insights into this important communications technique. A glossary of the symbols used is provided at the end of the paper.

II. SPREADING AND DIMENSIONALITY—PROCESSING GAIN

A fundamental issue in spread spectrum is how this technique affords protection against interfering signals with

0090-6778/82/0500-0855$00.75 © 1982 IEEE

finite power. The underlying principle is that of distributing a relatively low dimensional (defined below) data signal in a high dimensional environment so that a jammer with a fixed amount of total power (intent on maximum disruption of communications) is obliged to either spread that fixed power over all the coordinates, thereby inducing just a little interference in each coordinate, or else place all of the power into a small subspace, leaving the remainder of the space interference free.

A brief discussion of a classical problem of signal detection in noise should clarify the emphasis on finite interference power. The "standard" problem of digital transmission in the presence of thermal noise is one where both transmitter and receiver know the set of M signaling waveforms $\{S_i(t), 0 \leq t \leq T; 1 \leq i \leq M\}$. The transmitter selects one of the waveforms every T seconds to provide a data rate of $\log_2 M/T$ bits/s. If, for example, $S_j(t)$ is sent, the receiver observes $r(t) = S_j(t) + n_w(t)$ over $[0, T]$ where $n_w(t)$ is additive, white Gaussian noise (AWGN) with (two-sided) power spectral density $\eta_0/2$ W/Hz.

It is well known [3] that the signal set can be completely specified by a linear combination of no more than $D \leq M$ orthonormal basis functions (see below), and that although the white noise, similarly expanded, requires an infinite number of terms, only those within the signal space are "relevant" [3]. We say that the signal set defined above is D-dimensional if the minimum number of orthonormal basis functions required to define all the signals is D. D can be shown to be [3] approximately $2B_DT$ where B_D is the total (approximate) bandwidth occupancy of the signal set. The optimum (minimum probability of error) detector in AWGN consists of a bank of correlators or filters matched to each signal, and the decision as to which was the transmitted signal corresponds to the largest output of the correlators.

Given a specific signal design, the performance of such a system is well known to be a function only of the ratio of the energy per bit to the noise spectral density. Hence, against white noise (which has infinite power and constant energy in every direction), the use of spreading (large $2B_DT$) offers no help. The situation is quite different, however, when the "noise" is a jammer with a fixed *finite* power. In this case, the effect of spreading the signal bandwidth so that the jammer is uncertain as to where in the large space the components are is often to force the jammer to distribute its finite power over many different coordinates of the signal space.

Since the desired signal can be "collapsed" by correlating the signal at the receiver with the known code, the desired signal is protected against a jammer in the sense that it has an effective power advantage relative to the jammer. This power advantage is often proportional to the ratio of the dimensionality of the space of code sequences to that of the data signal. It is necessary, of course, to "hide" the pattern by which the data are spread. This is usually done with a pseudonoise (PN) sequence which has desired randomness properties and which is available to the cooperating transmitter and receiver, but denied to other undesirable users of the common spectrum.

A general model which conveys these ideas, but which uses random (rather than pseudodrandom) sequences, is as follows. Suppose we consider transmission by means of D equiprobable and equienergy orthogonal signals imbedded in an n-dimensional space so that

$$S_i(t) = \sum_{k=1}^{n} S_{ik}\phi_k(t); \qquad 1 \leq i \leq D; \qquad 0 \leq t \leq T$$

where

$$S_{ik} = \int_0^T S_i(t)\phi_k(t)\,dt$$

and where $\{\phi_k(t); 1 \leq k \leq n\}$ is an orthonormal basis spanning the space, i.e.,

$$\int_0^T \phi_l(t)\phi_m(t)\,dt = \delta_{lm} \triangleq \begin{cases} 1 & l = m \\ 0 & l \neq m. \end{cases}$$

The average energy of each signal is

$$\int_0^T \overline{S_i^2(t)}\,dt = \sum_{k=1}^{n} \overline{S_{ik}^2} \triangleq E_s; \qquad 1 \leq i \leq D \tag{1}$$

(the overbar is the expected value over the ensemble).

In order to hide this D-dimensional signal set in the larger n-dimensional space, choose the sequence of coefficients S_{ik} independently (say, by flipping a fair coin if a binary alphabet is used) such that they have zero mean and correlation

$$\overline{S_{ik}S_{il}} \triangleq \frac{E_s}{n}\delta_{kl}; \qquad 1 \leq i \leq D. \tag{2}$$

Thus, the signals, which are also assumed to be known to the receiver (i.e., we assume the receiver had been supplied the sequences S_{ik} before transmission) but denied to the jammer, have their respective energies uniformly distributed over the n basis directions as far as the jammer is concerned.

Consider next a jammer

$$J(t) = \sum_{k=1}^{n} J_k\phi_k(t); \qquad 0 \leq t \leq T \tag{3}$$

with total energy

$$\int_0^T J^2(t)\,dt = \sum_{k=1}^{n} J_k^2 \triangleq E_J \tag{4}$$

which is added to the signal with the intent to disrupt communications. Assume that the jammer's signal is independent of the desired signal. One of the jammer's objectives is to devise a strategy for selecting the components J_k^2 of his fixed total energy E_J so as to minimize the postprocessing signal-to-noise ratio (SNR) at the receiver.

The received signal

$$r(t) = S_i(t) + J(t) \tag{5}$$

is correlated with the (known) signals so that the output of the ith correlator is

$$U_i \triangleq \int_0^T r(t)S_i(t)\,dt = \sum_{k=1}^{n} (S_{ik}^2 + J_k S_{ik}). \tag{6}$$

Hence,

$$E(U_i \mid S_i) = \sum_{k=1}^{n} \overline{S_{ik}^2} = E_s \tag{7}$$

since the second term averages to zero. Then, since the signals are equiprobable,

$$E(U_i) = \frac{E_s}{D}. \tag{8}$$

Similarly, using (1) and (2),

$$\begin{aligned} \operatorname{var}(U_i \mid S_i) &= \sum_{k,l} J_k J_l \overline{S_{ik} S_{il}} \\ &= \sum_{k=1}^{n} J_k^2 \overline{S_{ik}^2} \\ &= \frac{E_s}{n} E_J \end{aligned} \tag{9}$$

and

$$\operatorname{var} U_i = \frac{E_s}{nD} E_J. \tag{10}$$

A measure of performance is the signal-to-noise ratio defined as

$$\mathrm{SNR} = \frac{E^2(U)}{\operatorname{var}(U)} = \frac{E_s}{E_J} \cdot \frac{n}{D}. \tag{11}$$

This result is *independent of the way that the jammer distributes his energy*, i.e., regardless of how J_k is chosen subject to the constraint that $\Sigma_k J_k^2 = E_J$, the postprocessing SNR (11) gives the signal an advantage of n/D over the jammer. This factor n/D is the *processing gain* and it is exactly equal to the ratio of the dimensionality of the *possible* signal space (and therefore the space in which the jammer must seek to operate) to the dimensions needed to actually transmit the signals. Using the result that the (approximate) dimensionality of a signal of duration T and of approximate bandwidth B_D is $2B_DT$, we see the processing gain can be written as

$$G_P = \frac{n}{D} \cong \frac{2B_{ss}T}{2B_DT} = \frac{B_{ss}}{B_D} \tag{12}$$

where B_{ss} is the bandwidth in hertz of the (spread-spectrum) signals $S_i(t)$ and B_D is the minimum bandwidth that would be required to send the information if we did not need to imbed it in the larger bandwidth for protection.

A simple illustration of these ideas using random binary sequences will be used to bring out some of these points. Consider the transmission of a single bit $\pm\sqrt{E_b/T}$ with energy E_b of duration T seconds. This signal is one-dimensional. As shown in Figs. 1 and 2, the transmitter multiplies the data bit $d(t)$ by a binary ± 1 "chipping" sequence $p(t)$ chosen randomly at rate f_c chips/s for a total of f_cT chips/bit. The dimensionality of the signal $d(t)p(t)$ is then $n = f_cT$. The received signal is

$$r(t) = d(t)p(t) + J(t), \qquad 0 \leqslant t \leqslant T, \tag{13}$$

ignoring, for the time being, thermal noise.

The receiver, as shown in Fig. 1, performs the correlation

$$U \triangleq \sqrt{\frac{E_b}{T}} \int_0^T r(t)p(t)\,dt \tag{14}$$

and makes a decision as to whether $\pm\sqrt{E_b/T}$ was sent depending upon $U \gtrless 0$. The integrand can be expanded as

$$r(t)p(t) = d(t)p^2(t) + J(t)p(t) = d(t) + J(t)p(t), \tag{15}$$

and hence the data bit appears in the presence of a code-modulated jammer.

If, for example, $J(t)$ is additive white Gaussian noise with power spectral density $\eta_{0J}/2$ (two-sided), so is $J(t)\,p(t)$, and U is then a Gaussian random variable. Since $d(t) = \pm\sqrt{E_b/T}$, the conditional mean and variance of U, assuming that $\pm\sqrt{E_b/T}$ is transmitted, is given by E_b and $E_b(\eta_{0J}/2)$, respectively, and the probability of error is [3] $Q(\sqrt{2E_b/\eta_{0J}})$ where $Q(x) \triangleq \int_x^\infty (1/\sqrt{2\pi})e^{-y^2/2}\,dy$. Against white noise of unlimited power, spread spectrum serves no useful purpose, and the probability of error is $Q(\sqrt{2E_b/\eta_{0J}})$ regardless of the modulation by the code sequence. White noise occupies all dimensions with power $\eta_{0J}/2$. The situation is different, however, if the jammer power is limited. Then, not having access to the random sequence $p(t)$, the jammer with available energy E_J (power E_J/T) can do better than to apply this energy to one dimension. For example, if $J(t) = \sqrt{E_J/T}$, $0 \leqslant t \leqslant T$, then the receiver output is

$$U = E_b + \sqrt{E_bE_J}\,\frac{1}{n}\sum_{i=1}^{n} X_i \tag{16}$$

where the X_i's are i.i.d.[1] random variables with $P(X_i = +1) = P(X_i = -1) = \frac{1}{2}$. The signal-to-noise ratio (SNR) is

$$\frac{E^2(U)}{\operatorname{var}(U)} = \frac{E_b}{E_J}\,n. \tag{17}$$

Thus, the SNR may be increased by increasing n, the process-

[1] Independent identically distributed.

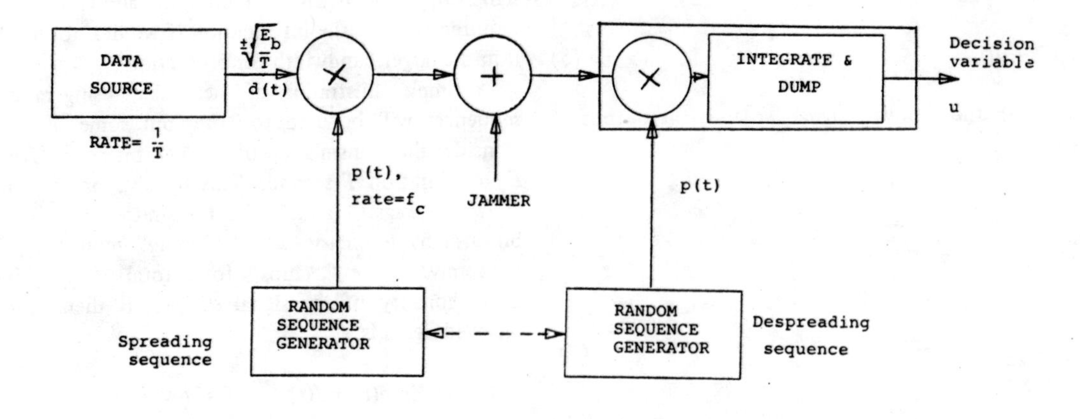

Fig. 1. Direct-sequence spread-spectrum system for transmitting a single binary digit (baseband).

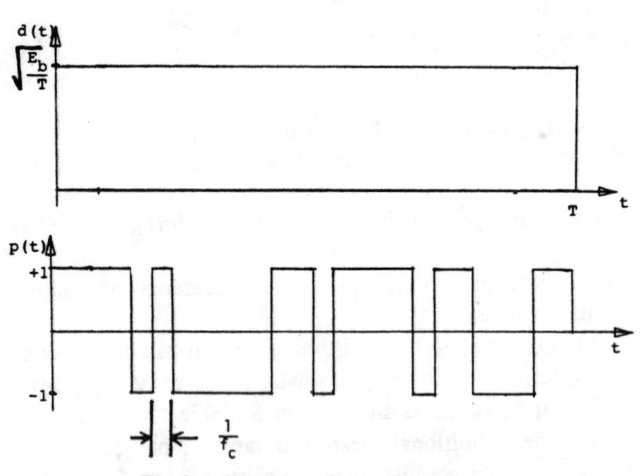

Fig. 2. Data bit and chipping sequence.

ing gain, and it has the form of (11). As a further indication of this parameter, we may compute the probability P_e that the bit is in error from (16). Assuming that a "minus" is transmitted, we have

$$P_e = P(U > 0) = P(Z_n > \alpha n) = \begin{cases} \frac{1}{2^n} \sum_{[\alpha n]}^{n} \binom{n}{k}; & \frac{E_b}{E_J} < 1 \\ 0; & \frac{E_b}{E_J} \geqslant 1 \end{cases} \tag{18}$$

where

$Z_n \triangleq \frac{1}{2} \sum_{i=1}^{n} (1 + X_i)$ is a Bernoulli random variable with mean $\frac{n}{2}$ and variance $\frac{n}{4}$,

$$\alpha \triangleq \frac{1}{2}\left(1 + \sqrt{\frac{E_b}{E_J}}\right),$$

and $[X]$ is defined as the integer portion of X. The partial binomial sum on the right-hand side of (18) may be upper bounded [2] by

$$P_e \leqslant \frac{1}{2^n}\left(\frac{1}{1-\alpha}\right)^n \left(\frac{1-\alpha}{\alpha}\right)^{\alpha n}; \quad \tfrac{1}{2} < \alpha \leqslant 1$$

or

$$P_e \leqslant 2^{-n[1-H(\alpha)]}; \quad \tfrac{1}{2} < \alpha \leqslant 1 \tag{19}$$

where $H(\alpha) \triangleq -\alpha \log_2 \alpha - (1-\alpha) \log_2 (1-\alpha)$ is the binary entropy function. Therefore, for any $\alpha > \frac{1}{2}$ (or $E_b \neq 0$), P_e may be made vanishingly small by increasing n, the processing gain. (The same result is valid even if the jammer uses a chip pattern other than the constant, all-ones used in the example above.) As an example, if $E_J = 9E_b$ (jammer energy 9.5 dB larger than that of the data), then $\alpha = 2/3$ and $P_e \leqslant 2^{-0.085n}$ If $n = 200$ (23 dB processing gain), $P_e < 7.6 \times 10^{-6}$.

An approximation to the same result may be obtained by utilizing a central limit type of argument that says, for large n, U in (16) may be treated as if it were Gaussian. Then

$$P_e = P(U < 0) \cong Q\left(\sqrt{\frac{E_b}{E_J} n}\right) \tag{20}$$

and, if $E_b/E_J = -9.5$ dB and $n = 200$ (23 dB), $P_e \cong Q(\sqrt{22}) \simeq 1.5 \times 10^{-6}$. The processing gain can be seen to be a multiplier of the "signal-to-jamming" ratio E_b/E_J.

A more traditional way of describing the processing gain, which brings in the relative bandwidth of the data signal and that of the spread-spectrum modulation, is to examine the power spectrum of an *infinite sequence* of data, modulated by the rapidly varying random sequence. The spectrum of the random data sequence with rate $R = 1/T$ bits/s is given by

$$S_D(f) = T\left(\frac{\sin \pi f T}{\pi f T}\right)^2 \tag{21}$$

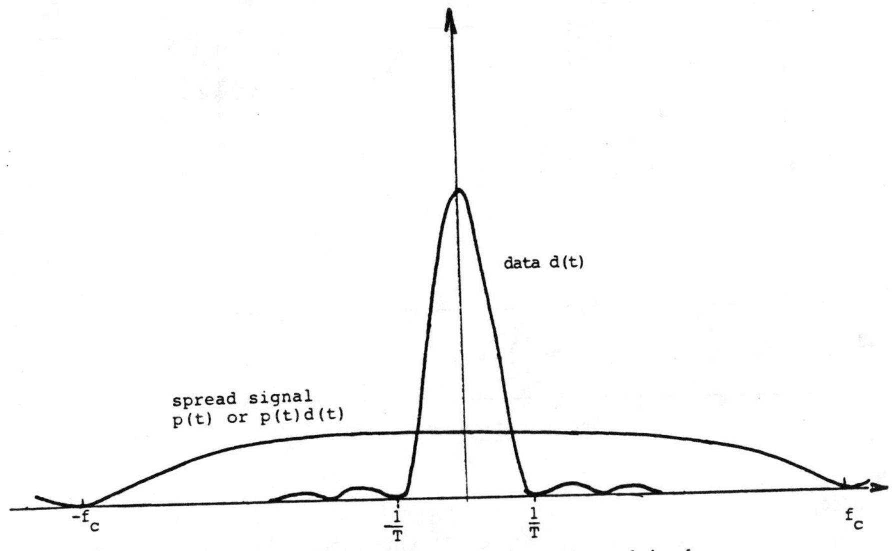

Fig. 3. Power spectrum of data and of spread signal.

and that of the spreading sequence [and also that of the product $d(t)\,p\,(t)$] is given by

$$S_{ss}(f) = \frac{1}{f_c}\left(\frac{\sin \pi f/f_c}{\pi f/f_c}\right)^2. \tag{22}$$

Both are sketched in Fig. 3. It is clear that if the receiver multiplies the received signal $d(t)p(t) + J(t)$ by $p(t)$ giving $d(t) + J(t)p(t)$, the first term may be extracted virtually intact with a filter of bandwidth $1/T \triangleq B_D$ Hz. The second term will be spread over at least f_c Hz as shown in Fig. 3. The fraction of power due to the jammer which can pass through the filter is then roughly $1/f_cT$. Thus, the data have a power advantage of $n = f_cT$, the processing gain. As in (12), the processing gain is frequently expressed as the ratio of the bandwidth of the spread-spectrum waveform to that of the data, i.e.,

$$G_p \triangleq \frac{B_{ss}}{B_D} = f_cT = n. \tag{23}$$

The notion of processing gain as expressed in (23) is simply a power improvement factor which a receiver, possessing a replica of the spreading signal, can achieve by a correlation operation. It must not be automatically extrapolated to anything else. For example, if we use frequency hopping for spread spectrum employing one of N frequencies every T_H seconds, the total bandwidth must be approximately N/T_H (since keeping the frequencies orthogonal requires frequency spacing $\approx 1/T_H$). Then, according to (12), $G_p = (N/T_H)/B_D$. Now if we transmit 1 bit/hop, $T_HB_D \simeq 1$ and $G_P = N$, the number of frequencies used. If $N = 100$, $G_p = 20$ dB, which seems fairly good. But a single spot frequency jammer can cause an average error rate of about 10^{-2}, which is not acceptable. (A more detailed analysis follows in Section IV below.) This effectiveness of "partial band jamming" can be reduced by the use of coding and interleaving. Coding typically precludes the possibility of a small number of frequency slots (e.g., one slot) being jammed causing an unacceptable error rate (i.e., even if the jammer wipes out a few of the code symbols, depending upon the error-correction capability of the code, the data may still be recovered). Interleaving has the effect of randomizing the errors due to the jammer. Finally, an analogous situation occurs in direct sequence spreading when a pulse jammer is present.

In the design of a practical system, the processing gain G_p is not, by itself, a measure of how well the system is capable of performing in a jamming environment. For this purpose, we usually introduce the *jamming margin* in decibels defined as

$$M_J = G_P - \left(\frac{E_b}{\eta_{0J}}\right)_{\min} - L. \tag{24}$$

This is the residual advantage that the system has against a jammer after we subtract both the minimum required energy/bit-to-jamming "noise" power spectral density ratio $(E_b/\eta_{0J})_{\min}$ and implementation and other losses L. The jamming margin can be increased by reducing the $(E_b/\eta_{0J})_{\min}$ through the use of coding gain.

We conclude this section by showing that regardless of the technique used, spectral spreading provides protection against a broad-band jammer with a finite power P_J. Consider a system that transmits R_0 bits/s designed to operate over a bandwidth B_{ss} Hz in white noise with power density η_0 W/Hz. For any bit rate R,

$$\left(\frac{E_b}{\eta_0}\right)_{\text{actual}} = \frac{P_s}{\eta_0 R} = \frac{P_s}{P_N}\frac{B_{ss}}{R} \tag{25}$$

where

$$P_s \triangleq E_bR = \text{signal power}$$

$$P_N \triangleq \eta_0 B_{ss} = \text{noise power}.$$

Then for a specified $(E_b/\eta_0)_{\min}$ necessary to achieve mini-

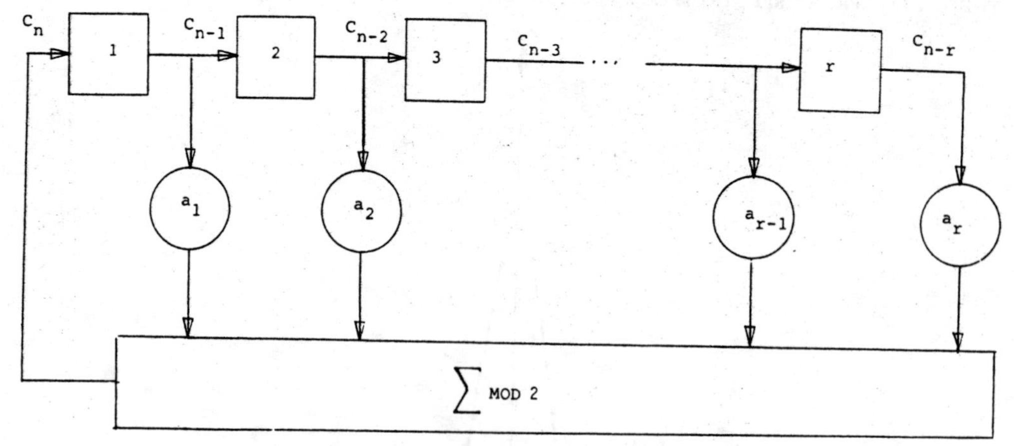

Fig. 4. Simple shift register generator (SSRG).

mum acceptable performance,

$$R \leqslant \frac{P_s}{P_N} \frac{B_{ss}}{(E_b/\eta_0)_{\min}} \triangleq R_0. \tag{26}$$

If a jammer with power P_J now appears, and if we are already transmitting at the maximum rate R_0, then (25) becomes

$$\left(\frac{E_b}{\eta_0}\right)_{\text{actual}} = \frac{P_s}{P_N + P_J} \frac{B_{ss}}{R_0}$$

$$= \left(\frac{E_b}{\eta_0}\right)_{\min} \frac{P_N}{P_N + P_J}$$

or

$$\left(\frac{E_b}{\eta_0}\right)_{\text{actual}} = \left(\frac{E_b}{\eta_0}\right)_{\min} \frac{\eta_0}{\eta_0 + P_J/B_{ss}}. \tag{27}$$

Thus, if we wish to recover from the effects of the jammer, the right-hand side of (27) should be not much less than $(E_b/\eta_0)_{\min}$. This clearly requires that we increase B_{ss}, since for any finite P_J, it is then possible to make the factor $\eta_0/(\eta_0 + P_J/B_{ss})$ approach unity, and thereby retain the performance we had before the jammer appeared.

III. PSEUDORANDOM SEQUENCE GENERATORS

In Section II, we examined how a purely random sequence can be used to spread the signal spectrum. Unfortunately, in order to despread the signal, the receiver needs a replica of the transmitted sequence (in almost perfect time synchronism). In practice, therefore, we generate pseudorandom or pseudonoise (PN) sequences so that the following properties are satisfied. They

1) are easy to generate
2) have randomness properties
3) have long periods
4) are difficult to reconstruct from a short segment.

Linear feedback shift register (LFSR) sequences [4] possess properties 1) and 3), most of property 2), but not property 4). One canonical form of a binary LFSR known as a simple shift register generator (SSRG) is shown in Fig. 4. The shift register consists of binary storage elements (boxes) which transfer their contents to the right after each clock pulse (not shown). The contents of the register are linearly combined with the binary (0, 1) coefficients a_k and are fed back to the first stage. The binary (code) sequence C_n then clearly satisfies the recursion

$$C_n = \sum_{k=1}^{r} a_k C_{n-k} \,(\text{mod } 2); \qquad a_r = 1. \tag{28}$$

The periodic cycle of the states depends on the initial state and on the coefficients (feedback taps) a_k. For example, the four-stage LFSR generator shown in Fig. 5 has four possible cycles as shown. The all-zeros is always a cycle for any LFSR. For spread spectrum, we are looking for *maximal length* cycles, that is, cycles of period $2^r - 1$ (all binary r-tuples except all-zeros). An example is shown for a four-state register in Fig. 6. The sequence output is 100011110101100 ··· (period $2^4 - 1 = 15$) if the initial contents of the register (from right to left) are 1000. It is always possible to choose the feedback coefficients so as to achieve maximal length, as will be discussed below.

If we do have a maximal length sequence, then this sequence will have the following pseudorandomness properties [4].

1) There is an approximate balance of zeros and ones (2^{r-1} ones and $2^{r-1} - 1$ zeros).

2) In any period, half of the runs of consecutive zeros or ones are of length one, one-fourth are of length two, one-eighth are of length three, etc.

3) If we define the ± 1 sequence $C_n' = 1 - 2C_n$, $C_n = 0, 1$, then the autocorrelation function $R_C'(\tau) \triangleq 1/L \sum_{k=1}^{L} C_k' C_{k+\tau}'$ is given by

$$R_{C'}(\tau) = \begin{cases} 1, & \tau = 0, L, 2L \cdots \\ -\dfrac{1}{L}, & \text{otherwise} \end{cases} \tag{29}$$

where $L = 2^r - 1$. If the code *waveform* $p(t)$ is the "square-wave" equivalent of the sequences C_n', if $L \gg 1$, and if we

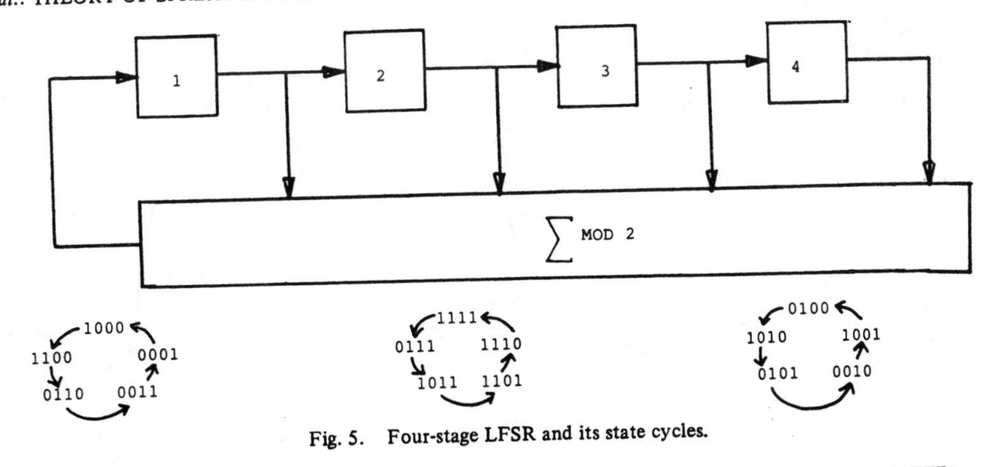

Fig. 5. Four-stage LFSR and its state cycles.

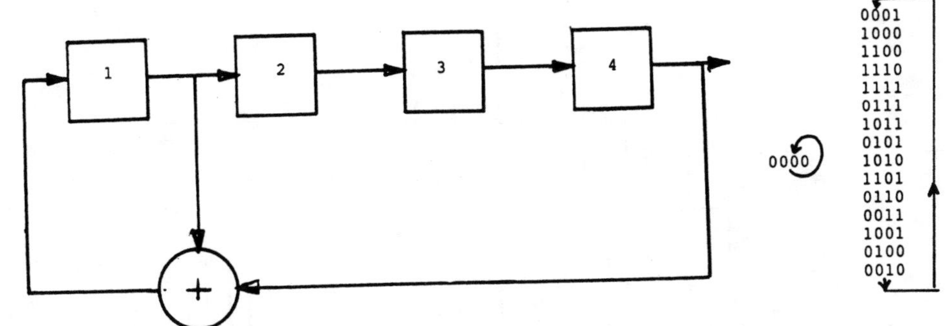

Fig. 6. Four-stage maximal length LFSR and its state cycles.

define

$$q(\tau) \triangleq \begin{cases} 1-|\tau|f_c; & |\tau| \leqslant \dfrac{1}{f_c} \\ 0; & \text{otherwise} \end{cases}$$

then

$$R_p(\tau) \simeq \sum_i q\left(\tau - \frac{iL}{f_c}\right). \tag{30}$$

Equation (29), and therefore (30), follow directly from the "shift-and-add" property of maximal length (ML) LFSR sequences. This property is that the chip-by-chip sum of an MLLFSR sequence C_k and any shift of itself $C_{k+\tau}$, $\tau \neq 0$ is the *same* sequence (except for some shift). This follows directly from (28), since

$$(C_n + C_{n+\tau}) = \sum_{k=1}^{L} a_k(C_{n-k} + C_{n+\tau-k}) \,(\text{mod } 2). \tag{31}$$

The shift-and-add sequence $C_n + C_{n+\tau}$ is seen to satisfy the same recursion as C_n, and if the coefficients a_k yield maximal length, then it must be the same sequence regardless of the initial (nonzero) state. The autocorrelation property (29) then follows from the following isomorphism:

$$(\{0,1\},+) \leftrightarrow (\{1,-1\},\times).$$

Therefore,

$$C_k + C_{k+\tau} \leftrightarrow C_k' C_{k+\tau}'$$

and if C_k' is an MLLFSR $\pm$ 1 sequence, so is $C_k' C_{k+\tau}'$, $\tau \neq 0$. Thus, there are 2^{r-1} 1's and $(2^{r-1}-1)$ -1's in the product and (29) follows. The autocorrelation function is shown in Fig. 7(a).

Property 3) is a most important one for spread spectrum since the autocorrelation function of the code sequence *waveform* $p(t)$ determines the spectrum. Note that because $p(t)$ is *pseudorandom*, it is periodic with period $(2^r-1)\cdot 1/f_c$, and hence so is $R_p(\tau)$. The spectrum shown in Fig. 7(b) is therefore the line spectrum

$$S_p(f) = \left[\sum_{\substack{m=-\infty \\ m\neq 0}}^{\infty} \delta(f - mf_0)\right] \frac{L+1}{L^2} \left(\frac{\sin \pi f/f_c}{\pi f/f_c}\right)^2 + \frac{1}{L^2}\delta(f) \tag{32}$$

where

$$f_0 = \frac{f_c}{2^r - 1}.$$

If $L = 2^r - 1$ is very large, the spectral lines get closer together, and for practical purposes, the spectrum may be viewed as being continuous and similar to that of a purely random binary waveform as shown in Fig. 3. A different, but commonly used implementation of a linear feedback shift register is the modular shift register generator (MSRG) shown in Fig. 8. Additional details on the properties of linear feedback shift registers are provided in the Appendix.

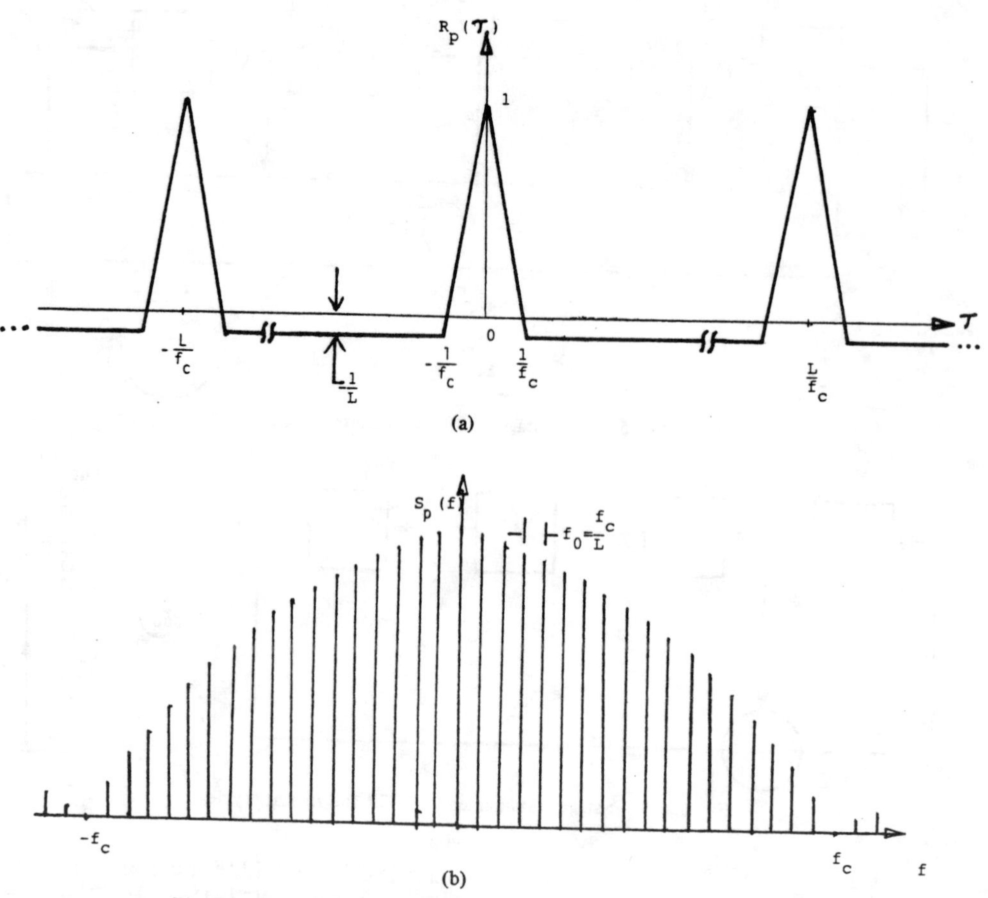

Fig. 7. Autocorrelation function $R_p(\tau)$ and power spectral density of MLLFSR sequence waveform $p(t)$. (a) Autocorrelation function of $p(t)$. (b) Power spectral density of $p(t)$.

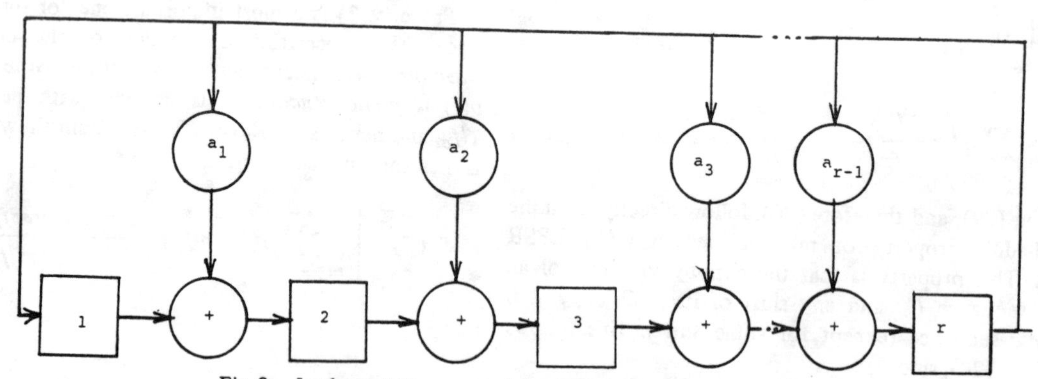

Fig. 8. Implementation as a modular shift register generator (MSRG).

For spread spectrum and other secure communications (cryptography) where one expects an adversary to attempt to recover the code in order to penetrate the system, property 4) cited in the beginning of this section is extremely important. Unfortunately, LFSR sequences do not possess that property. Indeed, using the recursion (28) or (A8) and observing only $2r-2$ consecutive bits in the sequence C_n allows us to solve for the $r-2$ middle coefficients and the r initial bits in the register by linear simultaneous equations. Thus, even if $r = 100$ so that the length of the sequence is $2^{100} - 1 \simeq 10^{30}$, we would be able to construct the entire sequence from 198 bits by solving 198 linear equations (mod 2), which is neither difficult nor that time consuming for a large computer. Moreover, because the sequence C_n satisfies a recursion, a very efficient algorithm is known [7], [8] which solves the equations or which equivalently synthesizes the shortest LFSR which generates a given sequence.

In order to avoid this pitfall, several modifications of the LFSR have been proposed. In Fig. 9(a) the feedback function is replaced by an arbitrary Boolean function of the contents of the register. The Boolean function may be implemented by ROM or random logic, and there are an enormous number of these functions (2^{2^r}). Unfortunately, very little is known [4] in the open literature about the properties of such non-

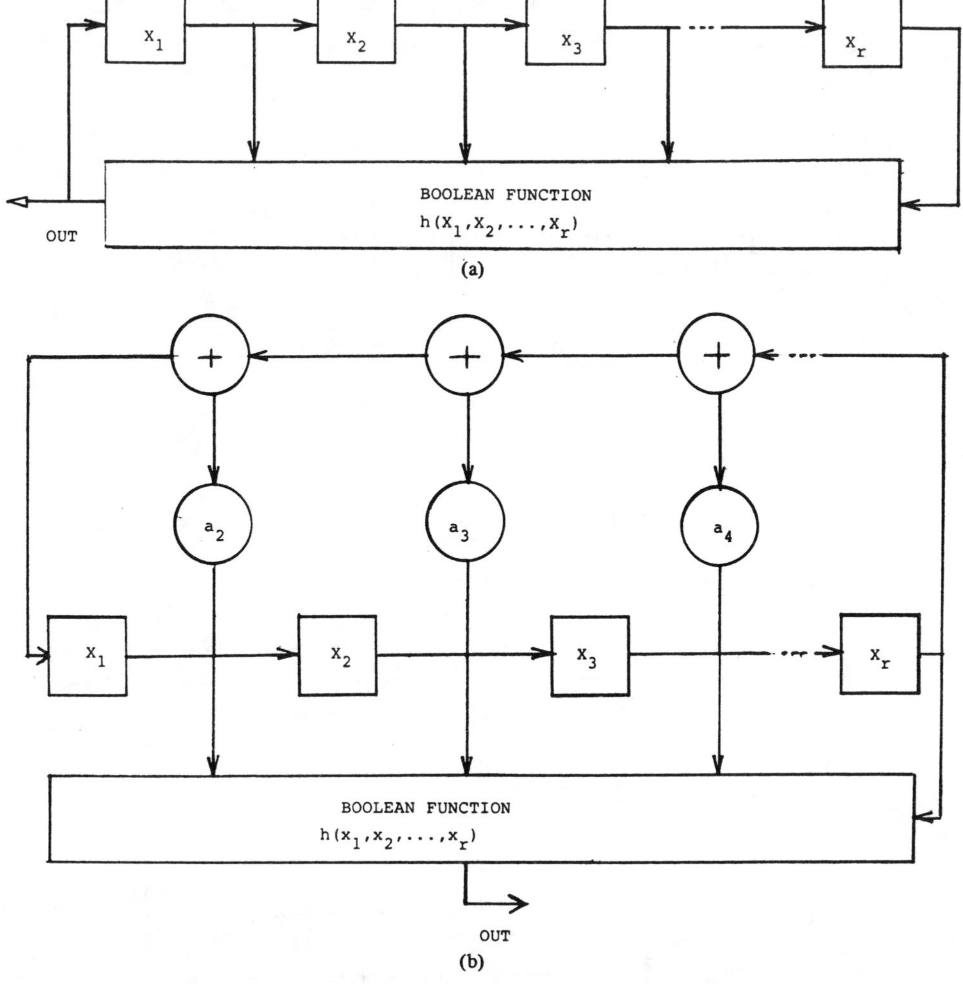

Fig. 9. Nonlinear feedback shift registers. (a) Nonlinear FDBK. Number of Boolean functions = 2^{2^r}. (b) Linear FSR, nonlinear function of state, i.e., nonlinear output logic (NOL).

linear feedback shift registers. Furthermore, some nonlinear FSR's may have no cycles or length > 1 (e.g., they may have only a *transient* that "homes" towards the all-ones state after any initial state). Are there feedback functions that generate only *one* cycle of length 2^r? The answer is yes, and there are exactly $2^{2^{r-1}-r}$ of them [9]. How do we find them? Better yet, how do we find a subset of them with all the "good" randomness properties? These are, and have been, good research problems for quite some time, and unfortunately no general theory on this topic currently exists.

A second, more manageable approach is to use an MLLFSR with nonlinear output logic (NOL) as shown in Fig. 9(b). Some clues about designing the NOL while still retaining "good" randomness properties are available [10]-[12], and a measure for judging how well condition 4) is fulfilled is to ask: What is the degree of the shortest LFSR that would generate the same sequence? A simple example of an LFSR with NOL having three stages is shown in Fig. 10(a). The shortest LFSR which generates the same sequence (of period 7) is shown in Fig. 10(b) and requires six stages.

When using PN sequences in spread-spectrum systems, several additonal requirements must be met.

1) The "partial correlation" of the sequence C_n' over a window w smaller than the full period should be as small as possible, i.e., if

$$\rho(w;j,\tau) \triangleq \sum_{n=j}^{j+w-1} C_n' C_{n+\tau}',$$

$$\rho(w) = \max_{j,\tau} |\rho(w;j,\tau)| \tag{33}$$

should be $\ll L = 2^r - 1$.

2) Different code pairs should have uniformly low cross correlation, i.e.,

$$R_{C'C''}(\tau) \triangleq \frac{1}{L} \sum_{k=1}^{L} C_k' C_{k+\tau}'' \tag{34}$$

should be $\ll 1$ for all values of τ.

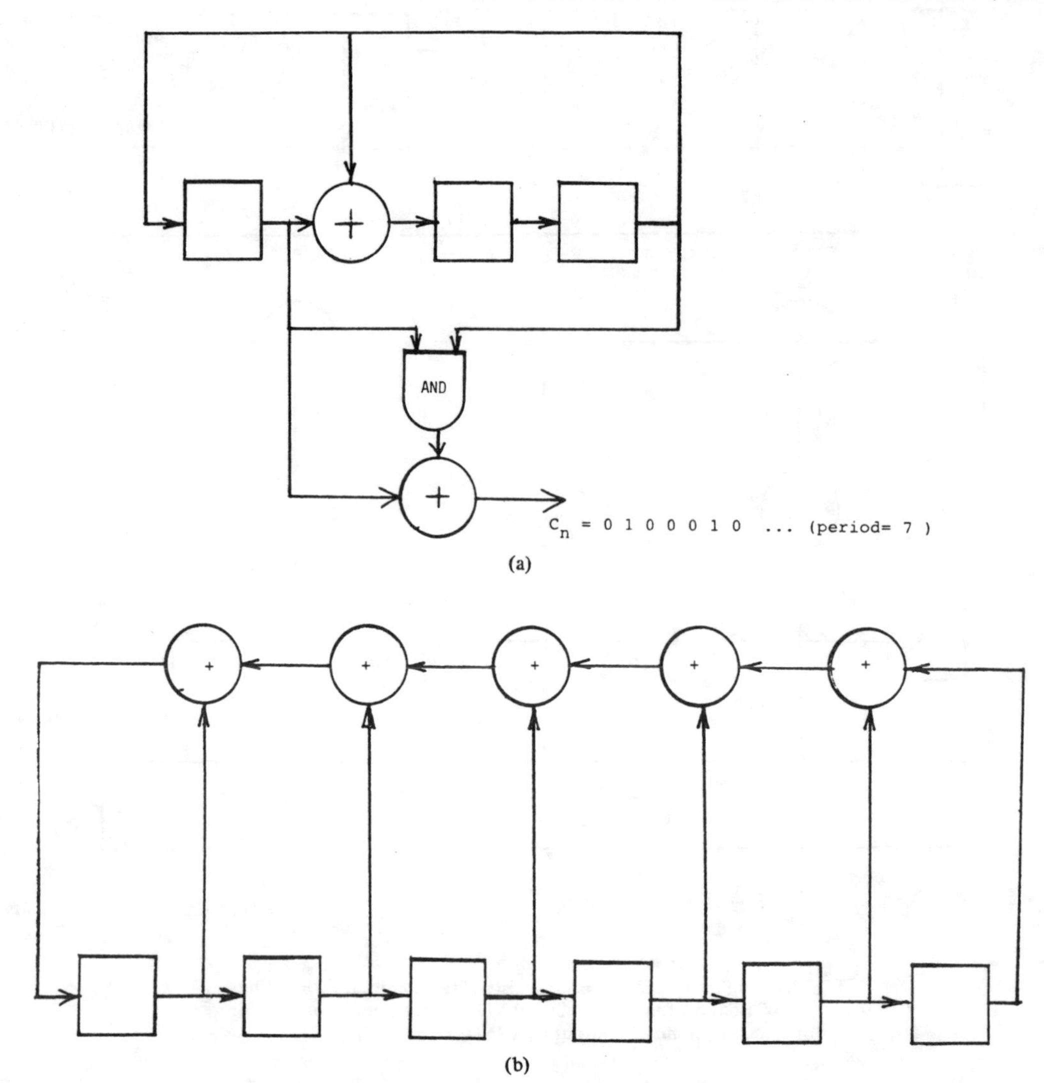

Fig. 10. LFSR with NOL and its shortest linear equivalent. (a) Three-stage LFSR with NOL. (b) LFSR with $f(x) = 1 + x + x^2 + x^3 + x^4 + x^5 + x^6$ which generates the same sequence as that of (a) under the initial state 1 0 0 0 1 0.

3) Since the code sequences are periodic with period L, there are two correlation functions (depending on the relative polarity of one of the sequences in the transition over an initial point τ on the other). If we define the finite-cross-correlation function [13] as

$$f_{C'C''}(\tau) \triangleq \frac{1}{L} \sum_{k=1}^{\tau} C_k' C_{k+\tau}'', \tag{35}$$

then the so-called even and odd cross-correlation functions are, respectively,

$$R_{C'C''}^{(e)}(\tau) = f_{C'C''}(\tau) + f_{C'C''}(L-\tau)$$

and

$$R_{C'C''}^{(o)}(\tau) = f_{C'C''}(\tau) - f_{C'C''}(L-\tau)$$

and we want

$$\max_{\tau} |R_{C'C''}^{e}(\tau)| \quad \text{and} \quad \max_{\tau} |R_{C'C''}^{(o)}(\tau)|$$

to be $\ll 1$.

The reason for 1) is to keep the "self noise" of the system as low as possible since, in practice, the period is very long compared to the integration time per symbol and there will be fluctuation in the sum of any filtered (weighted) subsequence. This is especially worrisome during acquisition where these fluctuations can cause false locking. Bounds on $\rho(w)$ [14] and averages over j of $\rho(w; j, \tau)$ are available in the literature.

Properties 2) and 3) are both of direct interest when using PN sequences for code division multiple access (CDMA) as will be discussed in Section V below. This is to ensure minimal cross interference between any pair of users of the common spectrum. The most commonly used collection of sequences which exhibit property 2) are the Gold codes [15]. These are sequences derived from an MLLFSR, but are *not* of maximal length. A detailed procedure for their construction is given in the Appendix.

Virtually all of the known results about the cross-correlation properties of useful PN sequences are summarized in [16].

As a final comment on the generation of PN sequences for spread spectrum, it is not at all necessary that feedback shift registers be used. *Any* technique which can generate "good" pseudorandom sequences will do. Other techniques are described in [4], [16], [17], for example. Indeed, the generation of good pseudorandom sequences is fundamental to other fields, and in particular, to cryptography [18]. A "good" cryptographic system can be used to generate "good" PN sequences, and vice versa. A possible problem is that the specific additional "good" properties required for an operational spread-spectrum system may not always match those required for secure cryptographic communications.

IV. ANTIJAM CONSIDERATIONS

Probably the single most important application of spread-spectrum techniques is that of resistance to intentional interference or jamming. Both direct-sequence (DS) and frequency-hopping (FH) systems exhibit this tolerance to jamming, although one might perform better than the other given a specific type of jammer.

The two most common types of jamming signals analyzed are single frequency sine waves (tones) and broad-band noise. References [19] and [20] provide performance analyses of DS systems operating in the presence of both tone and noise interference, and [21]-[26] provide analogous results for FH systems.

The simplest case to analyze is that of broad-band noise jamming. If the jamming signal is modeled as a zero-mean wide sense stationary Gaussian noise process with a flat power spectral density over the bandwidth of interest, then for a given fixed power P_J available to the jamming signal, the power spectral density of the jamming signal must be reduced as the bandwidth that the jammer occupies is increased.

For a DS system, if we assume that the jamming signal occupies the total RF bandwidth, typically taken to be twice the chip rate, then the despread jammer will occupy an even greater bandwidth and will appear to the final integrate-and-dump detection filter as approximately a white noise process. If, for example, binary PSK is used as the modulation format, then the average probability of error will be approximately given by

$$P_e = Q\left(\sqrt{\frac{2E_b}{\eta_0 + \eta_{0J}}}\right). \tag{36}$$

Equation (36) is just the classical result for the performance of a coherent binary communication system in additive white Gaussian noise. It differs from the conventional result because an extra term in the denominator of the argument of the $Q(\cdot)$ function has been added to account for the jammer. If P_e from (36) is plotted versus E_b/η_0 for a given value of P_J/P_s, where P_s is the average signal power, curves such as the ones shown in Fig. 11 result.

Expressions similar to (36) are easily derived for other modulation formats (e.g., QPSK), and curves showing the performance for several different formats are presented, for example, in [19]. The interesting thing to note about Fig. 11 is that for a given η_{0J}, the curve "bottoms out" as E_b/η_0 gets larger and larger. That is, the presence of the jammer will cause an irreducible error rate for a given P_J and a given f_c. Keeping P_J fixed, the only way to reduce the error rate is to increase f_c (i.e., increase the amount of spreading in the system). This was also noted at the end of Section II.

For FH systems, it is not always advantageous for a noise jammer to jam the entire RF bandwidth. That is, for a given P_J, the jammer can often increase its effectiveness by jamming only a fraction of the total bandwidth. This is termed *partial-band jamming*. If it is assumed that the jammer divides its power uniformly among K slots, where a slot is the region in frequency that the FH signal occupies on one of its hops, and if there is a total of N slots over which the signal can hop, we have the following possible situations. Assuming that the underlying modulation format is binary FSK (with noncoherent detection at the receiver), and using the terminology MARK and SPACE to represent the two binary data symbols, on any given hop, if

1) $K = 1$, the jammer might jam the MARK only, jam the SPACE only, or jam neither the MARK nor the SPACE;

2) $1 < K < N$, the jammer might jam the MARK only, jam the SPACE only, jam neither the MARK nor the SPACE, or jam both the MARK and the SPACE;

3) $K = N$, the jammer will always jam both the MARK and the SPACE.

To determine the average probability of error of this system, each of the possibilities alluded to above has to be accounted for. If it is assumed that the N slots are disjoint in frequency and that the MARK and SPACE tones are orthogonal (i.e., if a MARK is transmitted, it produces no output from the SPACE bandpass filter (BPF) and vice versa), then the average probability of error of the system can be shown to be given by [23], [24]

$$P_e = \frac{(N-K)(N-K-1)}{N(N-1)} \frac{1}{2} \exp\left(-\tfrac{1}{2}\,\mathrm{SNR}\right) + \frac{K(N-K)}{N(N-1)} \exp\left[-\frac{1}{\dfrac{2}{\mathrm{SNR}} + \dfrac{1}{\mathrm{SJR}}}\right] + \frac{K(K-1)}{N(N-1)} \frac{1}{2} \cdot \exp\left[-\frac{1}{2} \frac{1}{\dfrac{1}{\mathrm{SNR}} + \dfrac{1}{\mathrm{SJR}}}\right] \tag{37}$$

where SNR is the ratio of signal power to thermal noise power at the output of the MARK BPF (assuming that a MARK has been transmitted) and SJR is the ratio of signal

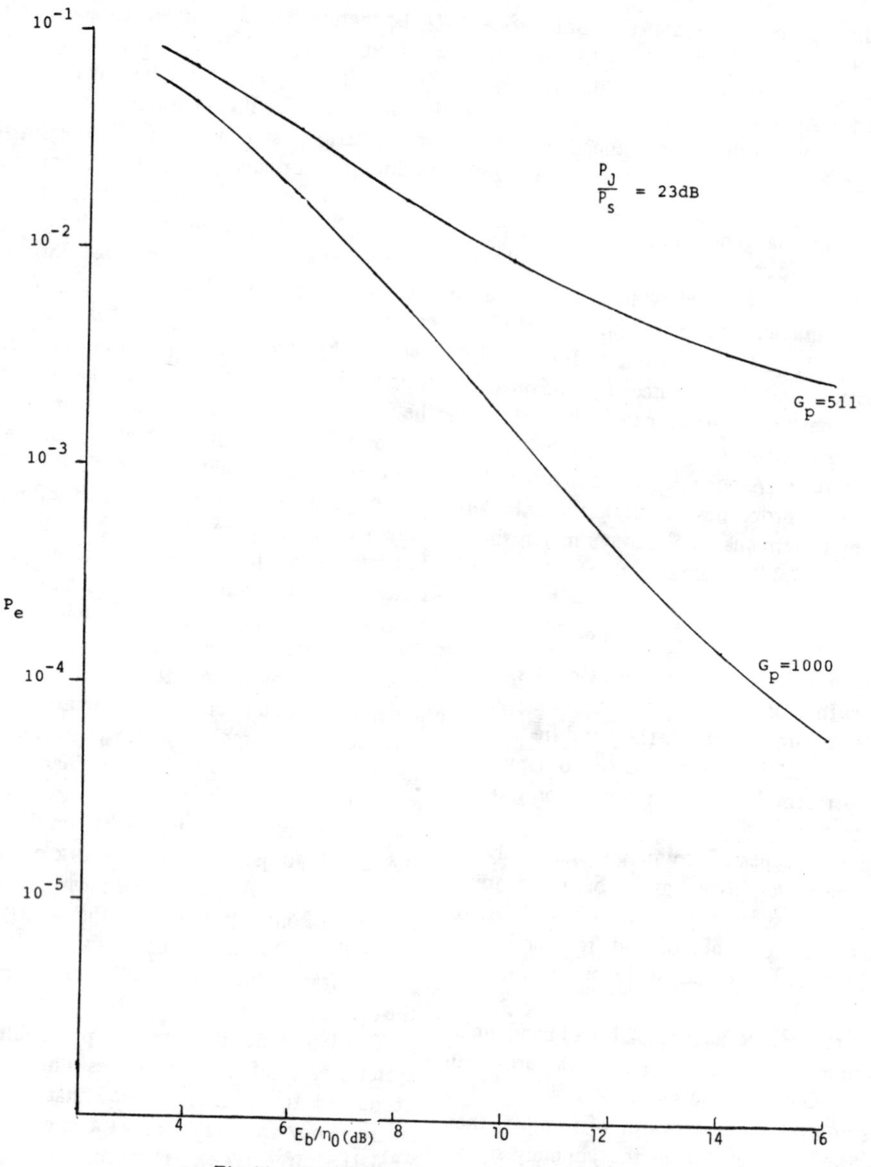

Fig. 11. Probability of error versus E_b/η_0.

power to *jammer power per slot* at the output of the MARK BPF. By jammer power per slot, we mean the total jammer power divided by the number of slots being jammed (i.e., SJR $= P_s/(P_J/K)$).

The coefficients in front of the exponentials in (37) are the probabilities of jamming neither the MARK nor the SPACE, jamming only the MARK or only the SPACE, or jamming both the MARK and the SPACE. For example, the probability of jamming both the MARK and the SPACE is given by $K(K-1)/N(N-1)$. In Fig. 12, the P_e predicted by (37) is plotted versus SNR for $K = 1$ and $K = 100$ for a P_J/P_s of 10 dB. These two curves are labeled "uncoded" on the figure.

Often, a somewhat different model from that used in deriving (37) is considered. This latter model is used in [26], and effectively assumes that either MARK and SPACE are simultaneously jammed or that neither of the two is jammed. For this case, a parameter ρ, where $0 < \rho \leqslant 1$, representing the fraction of the band being jammed, is defined. The resulting average probability of error is then maximized with respect to ρ (i.e., the worst case ρ is found), and it is shown in [26] that

$$P_{e_{max}} > \frac{e^{-1}}{E_b/\eta_0}$$

where e is the base of the natural logarithm. It can be seen that partial band jamming affords the jammer a strategy whereby he can degrade the performance significantly (i.e., P_e can be forced to be inversely proportional to E_b/η_0 rather than exponential).

For tone jamming, the situation becomes somewhat more complicated than it is for noise jamming, especially for DS systems. This is because the system performance depends upon the location of the tone (or tones), and upon whether the period of the spreading sequence is equal to or greater than the duration of a data symbol. Oftentimes the effect

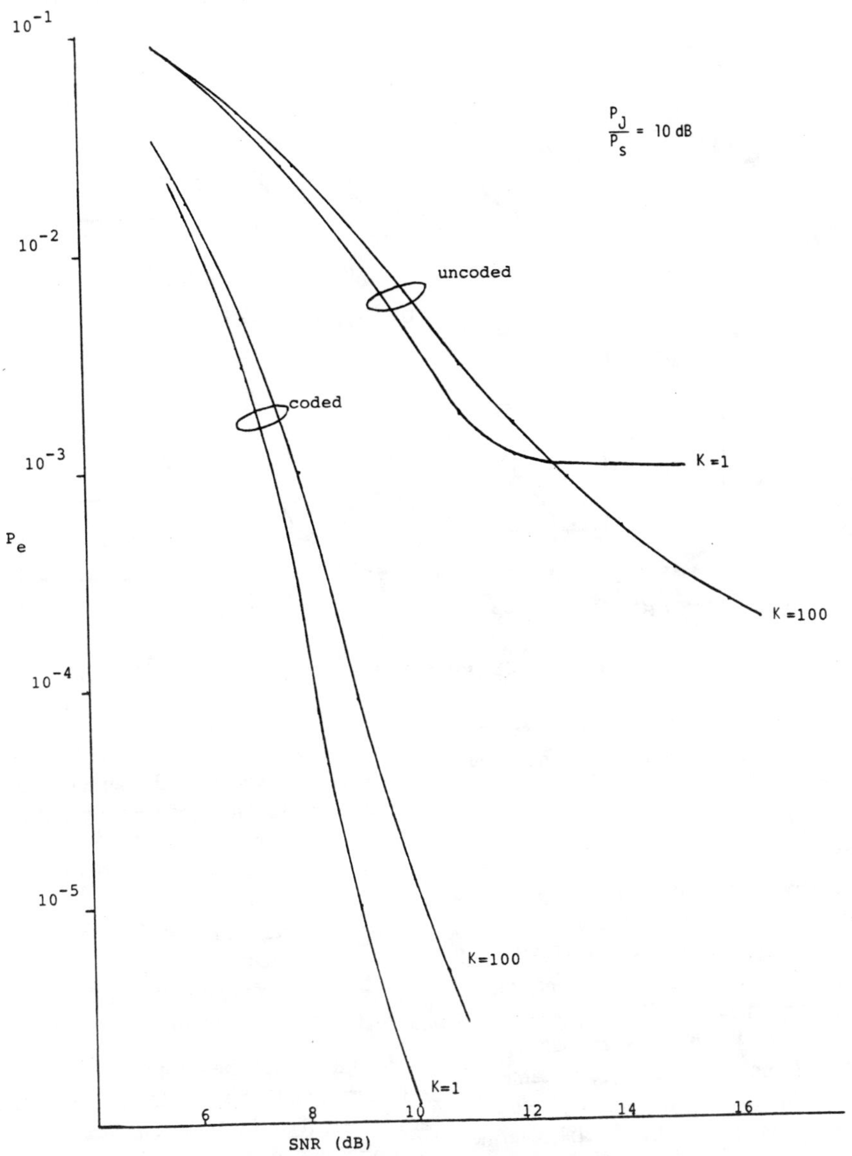

Fig. 12. Probability of error versus SNR.

of a despread tone is approximated as having arisen from an equivalent amount of Gaussian noise. In this case, the results presented above would be appropriate. However, the Gaussian approximation is not always justified, and some conditions for its usage are given in [20] and [27].

The situation is simpler in FH systems operating in the presence of partial-band tone jamming, and as shown, for example, in [24], the performance of a noncoherent FH–FSK system in partial-band tone jamming is often virtually the same as the performance in partial-band noise jamming. One important consideration in FH systems with either noise or tone jamming is the need for error-correction coding. This can be seen very simply by assuming that the jammer is much stronger than the desired signal, and that it chooses to put all of its power in a single slot (i.e., the jammer jams one out of N slots). The $K = 1$ uncoded curve of Fig. 12 corresponds to this situation. Then with no error-correction coding, the system will make an error (with high probability) every time it hops to a MARK frequency when the corresponding SPACE frequency is being jammed or vice versa. This will happen on the average one out of every N hops, so that the probability of error of the system will be approximately $1/N$, independent of signal-to-noise ratio. This is readily seen to be the case in Fig. 12. The use of coding prevents a simple error as caused by a spot jammer from degrading the system performance. To illustrate this point, an error-correcting code (specifically a Golay code [2]) was used in conjunction with the system whose uncoded performance is shown in Fig. 12, and the performance of the coded system is also shown in Fig. 12. The advantage of using error-correction coding is obvious from comparing the corresponding curves.

Finally, there are, of course, many other types of common jamming signals besides broad-band noise or single frequency tones. These include swept-frequency jammers, pulse-burst jammers, and repeat jammers. No further discussion of these

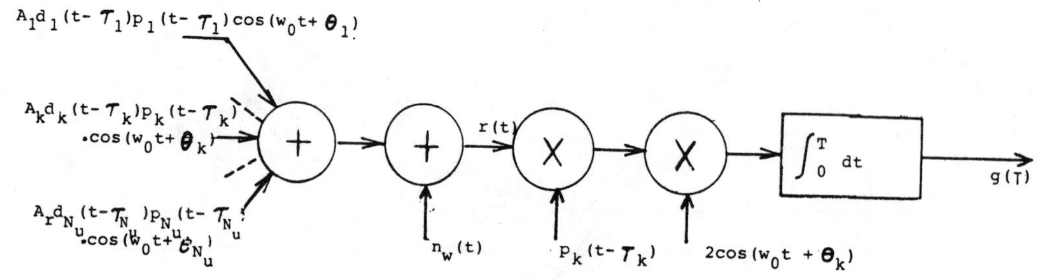

Fig. 13. DS CDMA system.

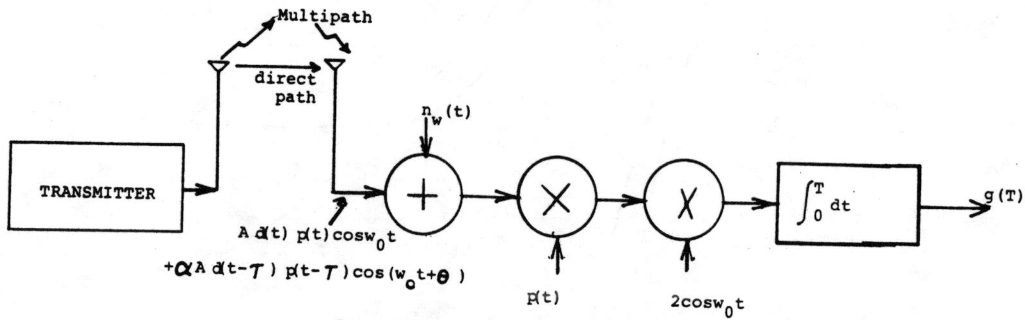

Fig. 14. DS used to combat multipath.

jammers will be presented in this paper, but references such as [28]-[30] provide a reasonable description of how these jammers affect system performance.

V. CODE DIVISION MULTIPLE ACCESS (CDMA)

As is well known, the two most common multiple access techniques are frequency division multiple access (FDMA) and time division multiple access (TDMA). In FDMA, all users transmit simultaneously, but use disjoint frequency bands. In TDMA, all users occupy the same RF bandwidth, but transmit sequentially in time. When users are allowed to transmit simultaneously in time and occupy the same RF bandwidth as well, some other means of separating the signals at the receiver must be available, and CDMA [also termed spread-spectrum multiple access (SSMA)] provides this necessary capability.

In DS CDMA [31]-[33], each user is given its own code, which is approximately orthogonal (i.e., has low cross correlation) with the codes of the other users. However, because CDMA systems typically are asynchronous (i.e., the transition times of the data symbols of the different users do not have to coincide), the design problem is much more complicated than that of having, say, N_u spreading sequences with uniformly low cross correlations such as the Gold codes discussed in Section III and in the Appendix. As will be seen below, the key parameters in a DS CDMA system are both the cross-correlation and the partial-correlation functions, and the design and optimization of code sets with good partial-correlation properties can be found in many references such as [16], [34], and [35].

The system is shown in Fig. 13. The received signal is given by

$$r(t)=\sum_{i=1}^{N_u} A_i d_i(t-\tau_i)p_i(t-\tau_i)\cos(\omega_0 t+\theta_i)+n_w(t) \quad (38)$$

where

$d_i(t)$ = message of ith user and equals ± 1
$p_i(t)$ = spreading sequence waveform of ith user
A_i = amplitude of ith carrier
θ_i = random phase of ith carrier uniformly distributed in $[0, 2\pi]$
τ_i = random time delay of ith user uniformly distributed in $[0, T]$
T = symbol duration
$n_w(t)$ = additive white Gaussian noise.

Assuming that the receiver is correctly synchronized to the kth signal, we can set both τ_k and θ_k to zero without losing any generality. The final test statistic out of the integrate-and-dump receiver of Fig. 14 is given by

$$g(T)=A_k+\frac{1}{T}\sum_{\substack{i=1\\i\neq k}}^{N_u} A_i\int_0^T d_i(t-\tau_i)\cdot p_i(t-\tau_i)p_k(t)\cos(\theta_i)\,dt+\frac{2}{T}\int_0^T n_w(t)p_k(t)\cos(\omega_0 t)\,dt \quad (39)$$

where double frequency terms have been ignored.

Consider the second term on the RHS of (39). It is a sum of N_u-1 terms of the form

$$A_i\cos(\theta_i)\int_0^T d_i(t-\tau_i)p_i(t-\tau_i)p_k(t)\,dt.$$

Notice that, because the ith signal is not, in general, in sync with the kth signal, $d_i(t-\tau_i)$ will change signs somewhere in the interval $[0, T]$ 50 percent of the time. Hence, the above integral will be the sum of two partial correlations of $p_i(t)$ and $p_k(t)$, rather than one total cross correlation. Therefore, (39) can be rewritten

$$g(T) = A_k + \sum_{\substack{i=1 \\ i \neq k}}^{N_u} A_i[\pm\rho_{ik}(\tau_i) \pm \hat{\rho}_{ik}(\tau_i)] \cos(\theta_i) + n(T) \quad (40)$$

where

$$\rho_{ik}(\tau_i) \triangleq \frac{1}{T}\int_0^{\tau_i} p_i(t-\tau_i)p_k(t)\,dt$$

$$\hat{\rho}_{ik}(\tau_i) \triangleq \frac{1}{T}\int_{\tau_i}^{T} p_i(t-\tau_i)p_k(t)\,dt$$

and

$$n(T) \triangleq \frac{2}{T}\int_0^T n_w(t)p_k(t)\cos\omega_0 t\,dt.$$

Notice that the coefficients in front of $\rho_{ik}(\tau_i)$ and $\hat{\rho}_{ik}(\tau_i)$ can independently have a plus or minus sign due to the data sequence of the ith signal. Also notice that $\rho_{ik}(\tau_i) + \hat{\rho}_{ik}(\tau_i)$ is the total cross correlation between the ith and kth spreading sequences. Finally, the continuous correlation functions $\rho_{ik}(\tau) \pm \hat{\rho}_{ik}(\tau)$ can be expressed in terms of the discrete even and odd cross-correlation functions, respectively, that were defined in Section III.

While the code design problem in CDMA is very crucial in determining system performance, of potentially greater importance in DS CDMA is the so-called "near-far problem." Since the N_u users are typically geographically separated, a receiver trying to detect the kth signal might be much closer physically to, say, the ith transmitter rather than the kth transmitter. Therefore, if each user transmits with equal power, the signal from the ith transmitter will arrive at the receiver in question with a larger power than that of the kth signal. This particular problem is often so severe that DS CDMA cannot be used.

An alternative to DS CDMA, of course, is FH CDMA [36]-[40]. If each user is given a different hopping pattern, and if all hopping patterns are orthogonal, the near-far problem will be solved (except for possible spectral spillover from one slot into adjacent slots). However, the hopping patterns are never truly orthogonal. In particular, any time more than one signal uses the same frequency at a given instant of time, interference will result. Events of this type are sometimes referred to as "hits," and these hits become more and more of a problem as the number of users hopping over a fixed bandwidth increases. As is the case when FH is employed as an antijam technique, error-correction coding can be used to significant advantage when combined with FH CDMA.

FH CDMA systems have been considered using one hop per bit, multiple hops per bit (referred to as fast frequency hopping or FFH), and multiple bits per hop (referred to as slow frequency hopping or SFH). Oftentimes the characteristics of the channel over which the multiple users transmit play a significant role in influencing which type of hopping one employs. An example of this is the multipath channel, which is discussed in the next section.

It is clearly of interest to consider the relative capacity of a CDMA system compared to FDMA or TDMA. In a perfectly linear, perfectly synchronous system, the number of orthogonal users for all three systems is the same, since this number only depends upon the dimensionality of the overall signal space. In particular, if a given time-bandwidth product G_P is divided up into, say, G_P disjoint time intervals for TDMA, it can also be "divided" into N binary orthogonal codes (assume that $G_P = 2^m$ for some positive integer m).

The differences between the three multiple-accessing techniques become apparent when various real-world constraints are imposed upon the ideal situation described above. For example, one attractive feature of CDMA is that it does not *require* the network synchronization that TDMA requires (i.e., if one is willing to give up something in performance, CDMA can be (and usually is) operated in an asynchronous manner). Another advantage of CDMA is that it is relatively easy to add additional users to the system. However, probably the dominant reason for considering CDMA is the need, in addition, for some type of external interference rejection capability such as multipath rejection or resistance to intentional jamming.

For an asynchronous system, even ignoring any near-far problem effects, the number of users the system can accommodate is markedly less than G_P. From [31] and [35], a rough rule-of-thumb appears to be that a system with processing gain G_P can support approximately $G_P/10$ users. Indeed, from [31, eq. (17)], the peak signal voltage to rms noise voltage ratio, averaged over all phase shifts, time delays, and data symbols of the multiple users, is approximately given by

$$\overline{\mathrm{SNR}} \simeq \left[\frac{N_u - 1}{3G_P} + \frac{\eta_0}{2E_b}\right]^{-1/2}$$

where the overbar indicates an ensemble average. From this equation, it can be seen that, given a value of E_b/η_0, $(N_u - 1)/G_P$ should be in the vicinity of 0.1 in order not to have a noticeable effect on system performance.

Finally, other factors such as nonlinear receivers influence the performance of a multiple access system, and, for example, the effect of a hard limiter on a CDMA system is treated in [45].

VI. MULTIPATH CHANNELS

Consider a DS binary PSK communication system operating over a channel which has more than one path linking the transmitter to the receiver. These different paths might consist of several discrete paths, each one with a different attenuation and time delay relative to the others, or it might consist of a continuum of paths. The RAKE system described in [1] is an example of a DS system designed to operate effectively in a multipath environment.

For simplicity, assume initially there are just two paths, a direct path and a single multipath. If we assume the time delay the signal incurs in propagating over the direct path is smaller than that incurred in propagating over the single multipath, and if it is assumed that the receiver is synchronized to the time delay and RF phase associated with the direct path, then the system is as shown in Fig. 14. The received signal is given by

$$r(t) = Ad(t)p(t) \cos \omega_0 t + \alpha Ad(t-\tau)p(t-\tau) \cdot \cos(\omega_0 t + \theta) + n_w(t) \tag{41}$$

where τ is the differential time delay associated with the two paths and is assumed to be in the interval $0 \leqslant \tau \leqslant T$, θ is a random phase uniformly distributed in $[0, 2\pi]$, and α is the relative attenuation of the multipath relative to the direct path. The output of the integrate-and-dump detection filter is given by

$$g(T) = A + [\pm \alpha A \rho(\tau) + \alpha A \hat{\rho}(\tau)] \cos \theta \tag{42}$$

where $\rho(\tau)$ and $\hat{\rho}(\tau)$ are partial correlation functions of the spreading sequence $p(t)$ and are given by

$$\rho(\tau) \triangleq \frac{1}{T} \int_0^{\tau} p(t)p(t-\tau)\, dt \tag{43}$$

and

$$\hat{\rho}(\tau) \triangleq \frac{1}{T} \int_{\tau}^{T} p(t)p(t-\tau)\, dt. \tag{44}$$

Notice that the sign in front of the second term on the RHS of (42) can be plus or minus with equal probability because this term arises from the pulse preceding the pulse of interest (i.e., if the ith pulse is being detected, this term arises from the $i-1$th pulse), and this latter pulse will be of the same polarity as the current pulse only 50 percent of the time. If the signs of these two pulses happen to be the same, and if $\tau > T_c$ where T_c is the chip duration, then $\rho(\tau) + \hat{\rho}(\tau)$ equals the autocorrelation function of $p(t)$ (assuming that a full period of $p(t)$ is contained in each T second symbol), and this latter quantity equals $-(1/L)$, where L is the period of $p(t)$. In other words, the power in the undesired component of the received signal has been attenuated by a factor of L^2.

If the sign of the preceding pulse is opposite to that of the current pulse, the attenuation of the undesired signal will be less than L^2, and typically can be much less than L^2. This is analogous, of course, to the partial correlation problem in CDMA discussed in the previous section.

The case of more than two discrete paths (or a continuum of paths) results in qualitatively the same effects in that signals delayed by amounts outside of $\pm T_c$ seconds about a correlation peak in the autocorrelation function of $p(t)$ are attenuated by an amount determined by the processing gain of the system.

If FH is employed instead of DS spreading, improvement in system performance is again possible, but through a different mechanism. As was seen in the two previous sections, FH systems achieve their processing gain through interference avoidance, not interference attenuation (as in DS systems). This same qualitative difference is true again if the interference is multipath. As long as the signal is hopping fast enough relative to the differential time delay between the desired signal and the multipath signal (or signals), all (or most) of the multipath energy will fall in slots that are orthogonal to the slot that the desired signal currently occupies.

Finally, the problems treated in this and the previous two sections are often all present in a given system, and so the use of an appropriate spectrum-spreading technique can alleviate all three problems at once. In [41] and [42], the joint problem of multipath and CDMA is treated, and in [43] and [44], the joint problem of multipath and intentional interference is analyzed. As indicated in Section V, if only multiple accessing capability is needed, there are systems other than CDMA that can be used (e.g., TDMA). However, when multipath is also a problem, the choice of CDMA as the multiple accessing technique is especially appropriate since the same signal design allows both many simultaneous users and improved performance of each user individually relative to the multipath channel.

In the case of signals transmitted over channels degraded by both multipath and intentional interference, either factor by itself suggests the consideration of a spectrum-spreading technique (in particular, of course, the intentional interference), and when all three sources of degradation are present simultaneously, spread spectrum is a virtual necessity.

VII. ACQUISITION

As we have seen in the previous sections, pseudonoise modulation employing direct sequence, frequency hopping, and/or time hopping is used in spread-spectrum systems to achieve bandwidth spreading which is large compared to the bandwidth required by the information signal. These PN modulation techniques are typically characterized by their very low repetition-rate-to-bandwidth ratio and, as a result, synchronization of a receiver to a specified modulation constitutes a major problem in the design and operation of spread-spectrum communications systems [46]-[50].

It is possible, in principle, for spread-spectrum receivers to use matched filter or correlator structures to synchronize to the incoming waveform. Consider, for example, a direct-sequence amplitude modulation synchronization system as shown in Fig. 15(a). In this figure, the locally generated code $p(t)$ is available with delays spaced one-half of a chip ($T_c/2$) apart to ensure correlation. If the region of uncertainty of the code phase is N_c chips, $2N_c$ correlators are employed. If no information is available regarding the chip uncertainty and the PN sequence repeats every, say, 2047 chips, then 4094 correlators are employed. Each correlator is seen to examine λ chips, after which the correlator outputs $V_0, V_1, \cdots, V_{2N_c-1}$ are compared and the largest output is chosen. As λ increases, the probability of making an error in synchronization decreases; however, the acquisition time increases. Thus, λ is usually chosen as a compromise between the probability of a synchronization error and the time to acquire PN phase.

A second example, in which FH synchronization is employed, is shown in Fig. 15(b). Here the spread-spectrum signal

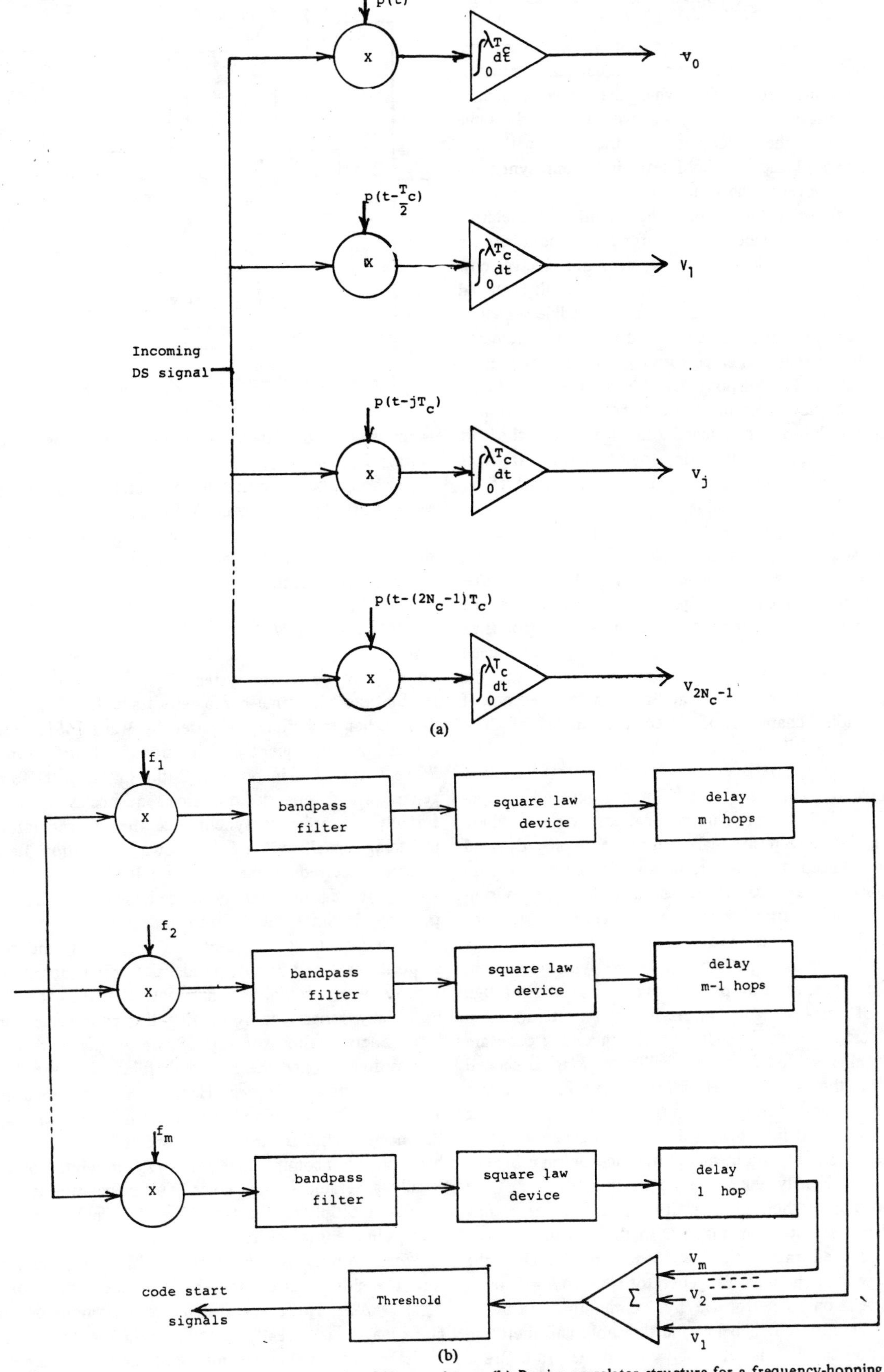

Fig. 15. (a) Direct sequence acquisition using $2N_c$ correlators. (b) Passive correlator structure for a frequency-hopping coarse acquisition scheme.

hops over, for example, $m = 500$ distinct frequencies. Assume that the frequency-hopping sequence is $f_1, f_2, \cdots, f_m$ and then repeats. The correlator then consists of $m = 500$ mixers, each followed by a bandpass filter and square law detector. The delays are inserted so that when the correct sequence appears, the voltages $V_1, V_2, \cdots, V_m$ will occur at the same instant of time at the adder and will, therefore, with high probability, exceed the threshold level indicating synchronization of the receiver to the signal.

While the above technique of using a bank of correlators or matched filters provides a means for rapid acquisition, a considerable reduction in complexity, size, and receiver cost can be achieved by using a single correlator or a single matched filter and repeating the procedure for each possible sequence shift. However, these reductions are paid for by the increased acquisition time needed when performing a serial rather than a parallel operation. One obvious question of interest is therefore the determination of the tradeoff between the number of parallel correlators (or matched filters) used and the cost and time to acquire. It is interesting to note that this tradeoff may become a moot point in several years as a result of the rapidly advancing VLSI technology.

No matter what synchronization technique is employed, the time to acquire depends on the "length" of the correlator. For example, in the system depicted in Fig. 15(a), the integration is performed over λ chips where λ depends on the desired probability of making a synchronization error (i.e., of deciding that a given sequence phase is correct when indeed it is not), the signal-to-thermal noise power ratio, and the signal-to-jammer power ratio. In addition, in the presence of fading, the fading characteristics affect the number of chips and hence the acquisition time.

The importance that one should attribute to acquisition time, complexity, and size depends upon the intended application. In tactical military communications systems, where users are mobile and push-to-talk radios are employed, rapid acquisition is needed. However, in applications where synchronization occurs once, say, each day, the time to synchronize is not a critical parameter. In either case, once acquisition has been achieved and the communication has begun, it is extremely important not to lose synchronization. Thus, while the acquisition process involves a search through the region of time-frequency uncertainty and a determination that the locally generated code and the incoming code are sufficiently aligned, the next step, called *tracking*, is needed to ensure that the close alignment is maintained. Fig. 16 shows the basic synchronization system. In this system, the incoming signal is first locked into the local PN signal generator using the acquisition circuit, and then kept in synchronism using the tracking circuit. Finally, the data are demodulated.

One popular method of acquisition is called the *sliding correlator* and is shown in Fig. 17. In this system, a single correlator is used rather than L correlators. Initially, the output phase k of the local PN generator is set to $k = 0$ and a partial correlation is performed by examining λ chips. If the integrator output falls below the threshold and therefore is deemed too small, k is set to $k = 1$ and the procedure is repeated. The determination that acquisition has taken place

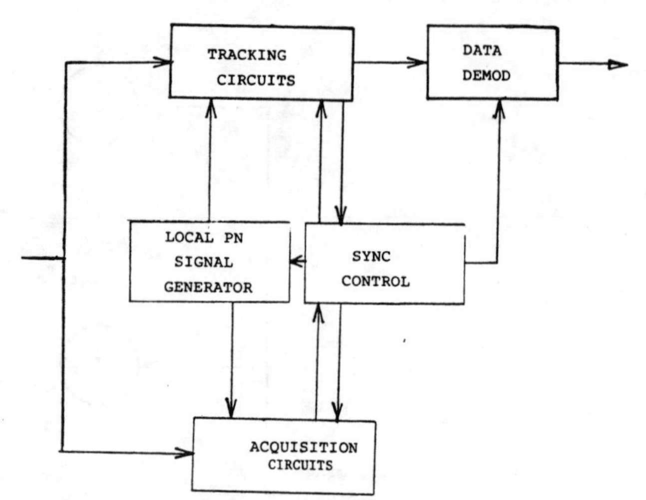

Fig. 16. Functional diagram of synchronization subsystem.

is made when the integrator output V_I exceeds the threshold voltage $V_T(\lambda)$.

It should be clear that in the worst case, we may have to set $k = 0, 1, 2, \cdots$, and $2N_c - 1$ before finding the correct value of k. If, during each correlation, λ chips are examined, the worst case acquisition time (neglecting false-alarm and detection probabilities) is

$$T_{\mathrm{acq,max}} = 2\lambda N_c T_c. \tag{45}$$

In the $2N_c$-correlator system, $T_{\mathrm{acq,max}} = T_c\lambda$, and so we see that there is a time-complexity tradeoff.

Another technique, proposed by Ward [46], called rapid acquisition by sequential estimation, is illustrated in Fig. 18. When switch S is in position 2, the shift register forms a PN generator and generates the same sequence as the input signal. Initially, in order to synchronize the PN generator to the incoming signal, switch S is thrown to position 1. The first N chips received at the input are loaded into the register. When the register is fully loaded, switch S is thrown to position 2. Since the PN sequence generator generates the same sequence as the incoming waveform, the sequences at positions 1 and 2 must be identical. That such is the case is readily seen from Fig. 19 which shows how the code $p(t - jT_C)$ is initially generated. Comparing this code generator to the local generator shown in Fig. 18, we see that with the switch in position 1, once the register is filled, the outputs of both mod 2 adders are *identical*. Hence, the bit stream at positions 1 and 2 are the same and switch S can be thrown to position 2. Once switch S is thrown to position 2, correlation is begun between the incoming code $p(t - jT_c)$ in white noise and the locally generated PN sequence. This correlation is performed by first multiplying the two waveforms and then examining λ chips in the integrator.

When no noise is present, the N chips are correctly loaded into the shift register, and therefore the acquisition time is $T_{\mathrm{acq}} = NT_c$. However, when *noise is present*, one or more chips may be incorrectly loaded into the register. The resulting waveform at 2 will then not be of the same phase as the sequence generated at 1. If the correlator output after λT_c ex-

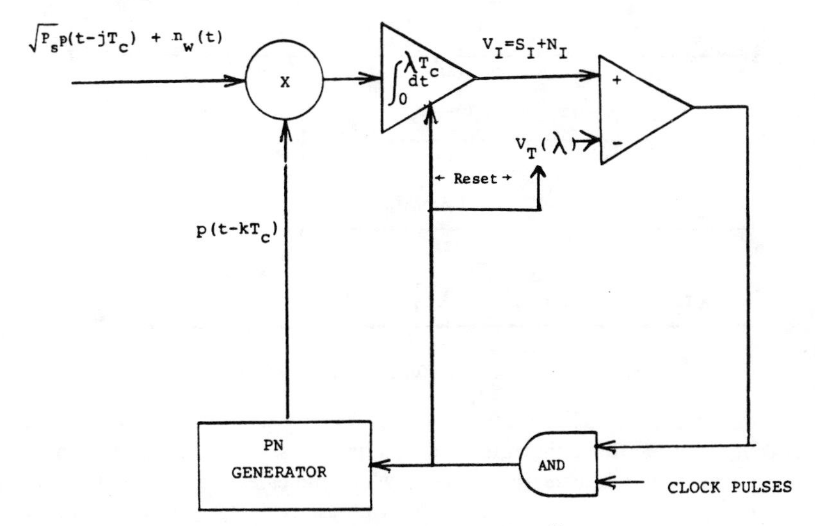

Fig. 17. The "sliding correlator."

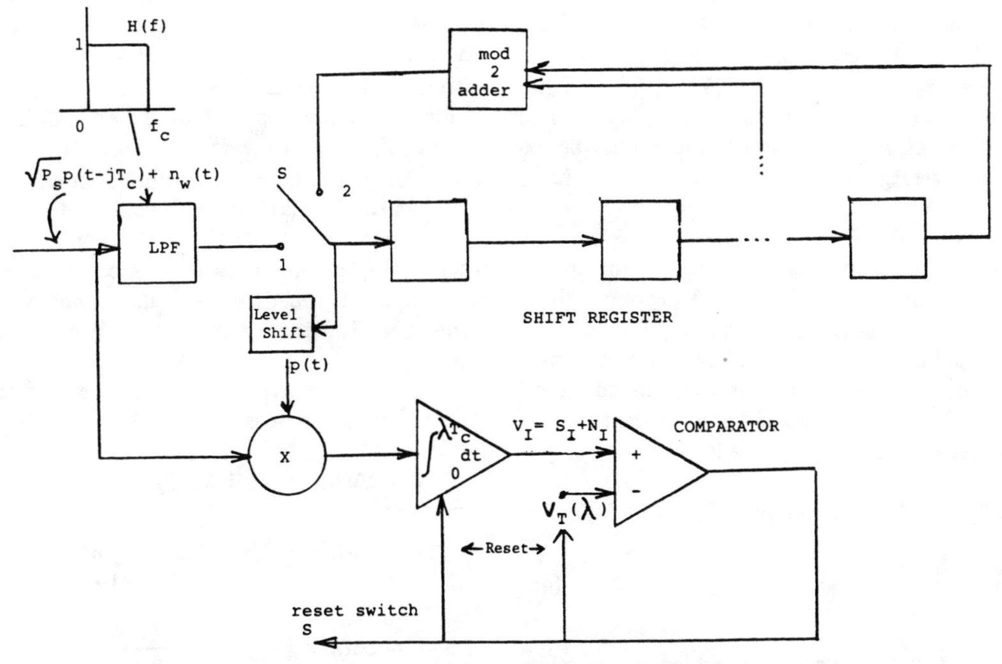

Fig. 18. Shift register acquisition circuit.

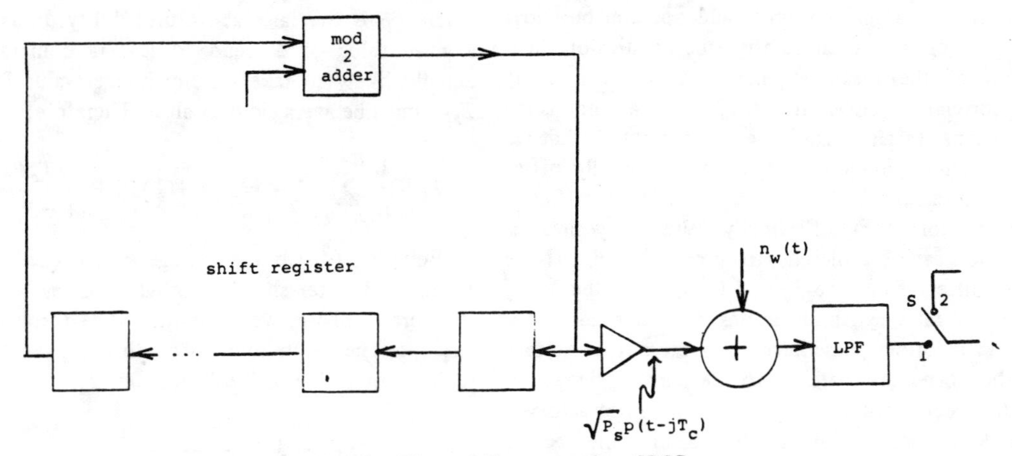

Fig. 19. The equivalent transmitter SRSG.

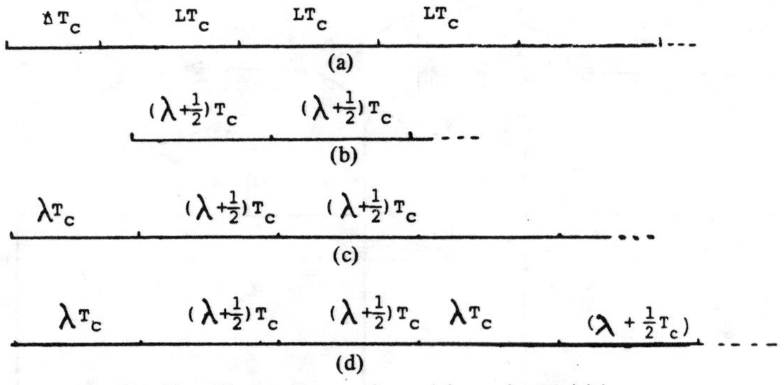

ΔT_c LT_c LT_c LT_c

(a)

$(\lambda+\frac{1}{2})T_c$ $(\lambda+\frac{1}{2})T_c$

(b)

λT_c $(\lambda+\frac{1}{2})T_c$ $(\lambda+\frac{1}{2})T_c$

(c)

λT_c $(\lambda+\frac{1}{2})T_c$ $(\lambda+\frac{1}{2})T_c$ λT_c $(\lambda+\frac{1}{2}T_c)$

(d)

Fig. 20. Timing diagram for serial search acquisition.

ceeds the threshold voltage, we assume that synchronization has occurred. If, however, the output is less than the threshold voltage, switch S is thrown to position 1, the register is reloaded, and the procedure is repeated.

Note that in both Figs. 17 and 18, correlation occurs for a time λT_c before predicting whether or not synchronism has occurred. If, however, the correlator output is examined after a time nT_c and a decision made at each $n \leqslant \lambda$ as to whether 1) synchronism has occurred, 2) synchronism has not occurred, or 3) a decision cannot be made with sufficient confidence and therefore an additional chip should be examined, then the average acquisition time can be reduced substantially.

One can approximately calculate the mean acquisition time of a parallel search acquisition system, such as the system shown in Fig. 15, by noting that after integrating over λ chips, a correct decision will be made with probability P_D where P_D is called the probability of detection. If, however, an incorrect output is chosen, we will, after examining an additional λ chips, again make a determination of the correct output. Thus, on the average, the acquisition time is

$$\bar{T}_{\mathrm{acq}} = \lambda T_c P_D + 2\lambda T_c P_D(1-P_D) + 3\lambda T_c P_D(1-P_D)^2 + \cdots$$
$$= \frac{\lambda T_c}{P_D} \tag{46}$$

where it is assumed that we continue searching every λ chips even after a threshold has been exceeded. This is not, in general, the way an actual system would operate, but does allow a simple approximation to the true acquisition time.

Calculation of the mean acquisition time when using the "sliding correlator" shown in Fig. 17 can be accomplished in a similar manner (again making the approximation that we never stop searching) by noting that we are initially offset by a random number of chips Δ as shown in Fig. 20(a). After the correlator of Fig. 17 finally "slides" by these Δ chips, acquisition can be achieved with probability P_D. (Note that this P_D differs from the P_D of (46), since the latter P_D accounts for false synchronizations due to a correlator matched to an incorrect phase having a larger output voltage than does the correlator matched to the correct phase.) If, due to an incorrect decision, synchronization is not achieved at that time, L additional chips must then be examined before acquistion can be achieved (again with probability P_D).

We first calculate the average time needed to slide by the Δ chips. To see how this time can change, refer to Fig. 20(b) which indicates the time required if we are not synchronized. λ chips are integrated, and if the integrator output $V_I < V_T$ (the threshold voltage), a $\frac{1}{2}$ chip delay is generated, and we then process an additional λ chips, etc. We note that in order to slide Δ chips in $\frac{1}{2}$ chip intervals, this process must occur 2Δ times. Since each repetition takes a time $(\lambda + \frac{1}{2})T_c$, the total elapsed time is $2\Delta(\lambda + \frac{1}{2})T_c$.

Fig. 20(b) assumes that at the end of each examination interval, $V_I < V_T$. However, if a false alarm occurs and $V_I > V_T$, no slide of $T_c/2$ will occur until after an additional λ chips are searched. This is shown in Fig. 20(c). In this case, the total elapsed time is $2\Delta(\lambda + \frac{1}{2})T_c + \lambda T_c$. Fig. 20(d) shows the case where false alarms occurred twice. Clearly, neither the separation between these false alarms nor where they occur is relevant. The total elapsed time is now $2\Delta(\lambda + \frac{1}{2})T_c + 2\lambda T_c$.

In general, the average elapsed time to reach the correct synchronization phase is

$$\bar{T}_{s/\Delta} = 2\Delta(\lambda + \tfrac{1}{2})T_c + \lambda T_c P_F + 2\lambda T_c P_F^2 + \cdots$$
$$= 2\Delta(\lambda + \tfrac{1}{2})T_c + \lambda T_c P_F \sum_{n=1}^{\infty} n P_F^{n-1}$$
$$= 2\Delta(\lambda + \tfrac{1}{2})T_c + \frac{\lambda T_c P_F}{(1-P_F)^2} \tag{47}$$

where P_F is the false alarm probability. Equation (47) is for a given value of Δ. Since Δ is a random variable which is equally likely to take on any integer value from 0 to $L-1$, $\bar{T}_{s/\Delta}$ must be averaged over all Δ. Therefore,

$$\bar{T}_s \triangleq \frac{1}{L}\sum_{\Delta=0}^{L-1} \bar{T}_{s/\Delta} = L(\lambda + \tfrac{1}{2})T_c + \frac{\lambda T_c P_F}{(1-P_F)^2}\,. \tag{48}$$

Equation (48) is the average time needed to slide through Δ chips. If, after sliding through Δ chips, we do not detect the correct phase, we must now slide through an additional L chips. The mean time to do this is given by (47), with Δ replaced by L. We shall call this time $\bar{T}_{s/L}$:

$$\bar{T}_{s/L} = 2L(\lambda + \tfrac{1}{2})T_c + \frac{\lambda T_c P_F}{(1-P_F)^2}\,. \tag{49}$$

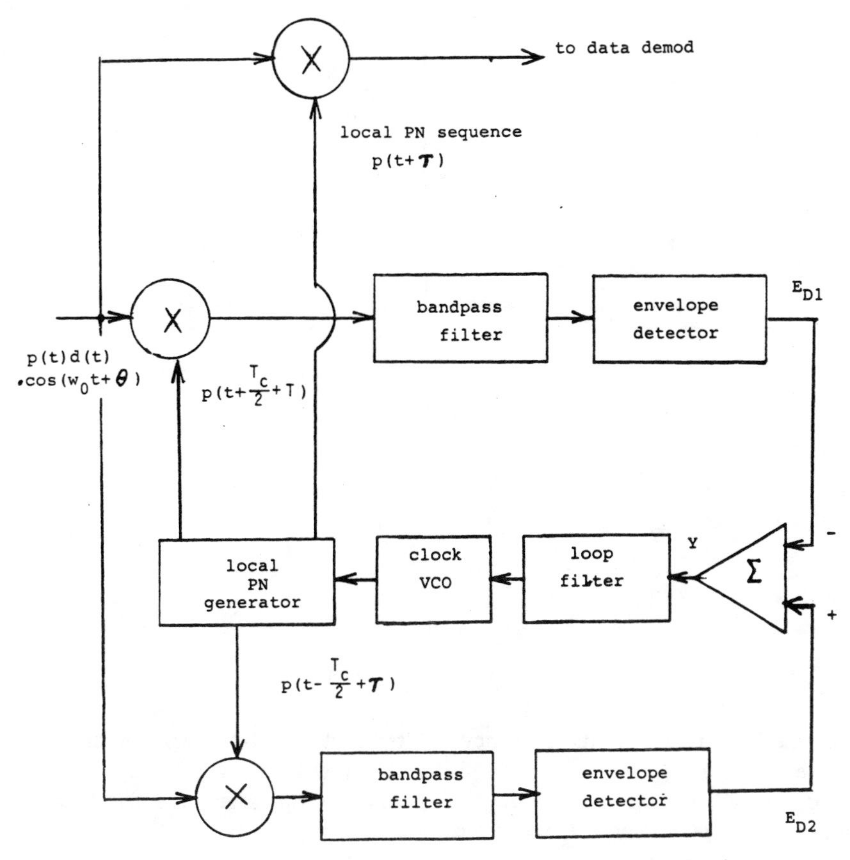

Fig. 21. Delay-locked loop for tracking direct-sequence PN signals.

The mean time to acquire a signal can now be written as

$$\bar{T}_{acq} = \bar{T}_s + \bar{T}_{s/L}[P_D(1-P_D) + 2P_D(1-P_D)^2 + \cdots]$$

$$\bar{T}_s + \frac{1-P_D}{P_D}\bar{T}_{s/L}$$

or

$$\bar{T}_{acq} = \left[L(\lambda + \tfrac{1}{2})T_c + \frac{\lambda T_c P_F}{(1-P_F)^2}\right] + \frac{1-P_D}{P_D}\left[2L(\lambda + \tfrac{1}{2})T_c + \frac{\lambda T_c P_F}{(1-P_F)^2}\right]. \quad (50)$$

VIII. TRACKING

Once acquisition, or coarse synchronization, has been accomplished, tracking, or fine synchronization, takes place. Specifically, this must include chip synchronization and, for coherent systems, carrier phase locking. In many practical systems, no data are transmitted for a specified time, sufficiently long to ensure that acquisition has occurred. During tracking, data are transmitted and detected. Typical references for tracking loops are [51]-[54].

The basic tracking loop for a direct-sequence spread-spectrum system using PSK data transmission is shown in Fig. 21. The incoming carrier at frequency f_0 is amplitude modulated by the product of the data $d(t)$ and the PN sequence $p(t)$. The tracking loop contains a local PN generator which is offset in phase from the incoming sequence $p(t)$ by a time τ which is less than one-half the chip time. To provide "fine" synchronization, the local PN generator generates two sequences, delayed from each other by one chip. The two bandpass filters are designed to have a two-sided bandwidth B equal to twice the data bit rate, i.e.,

$$B = 2R = 2/T. \quad (51)$$

In this way the data are passed, but the product of the two PN sequences $p(t)$ and $p(t \mp T_c/2 + \tau)$ is *averaged*. The envelope detector eliminates the data since $|d(t)| = 1$. As a result, the output of each envelope detector is approximately given by

$$E_{D1,2} \cong \overline{\left|p(t)p\left(t \pm \frac{T_c}{2} + \tau\right)\right|} = \left|R_p\left(\tau + \frac{T_c}{2}\right)\right| \quad (52)$$

where $R_p(x)$ is the autocorrelation function of the PN waveform as shown in Fig. 7(a). [See Section III for a discussion of the characteristics of $R_p(x)$.]

The output of the adder $Y(t)$ is shown in Fig. 22. We see from this figure that, when τ is positive, a positive voltage, proportional to Y, instructs the VCO to increase its frequency, thereby forcing τ to decrease, while when τ is negative, a

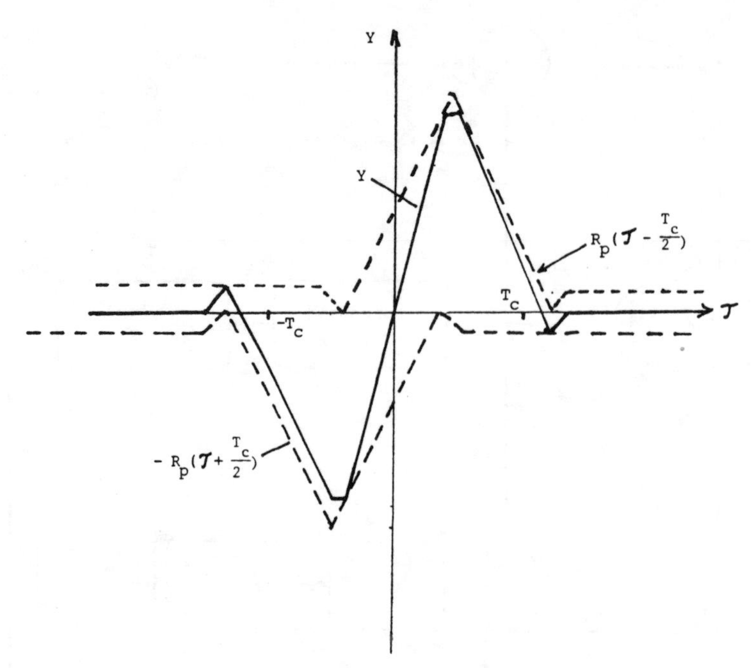

Fig. 22. Variation of Y with τ.

negative voltage instructs the VCO to reduce its frequency, thereby forcing τ to increase toward 0.

When the tracking error τ is made equal to zero, an output of the local PN generator $p(t + \tau) = p(t)$ is correlated with the input signal $p(t) \cdot d(t) \cos(\omega_0 t + \theta)$ to form

$$p^2(t)d(t) \cos(\omega_0 t + \theta) = d(t) \cos(\omega_0 t + \theta).$$

This despread PSK signal is inputted to the data demodulator where the data are detected.

An alternate technique for synchronization of a DS system is to use a tau-dither (TD) loop. This tracking loop is a delay-locked loop with only a single "arm," as shown in Fig. 23(a). The control (or gating) waveforms $g(t)$, $\bar{g}(t)$, and $g'(t)$ are shown in Fig. 23(b), and are used to generate both "arms" of the DLL even though only one arm is present. The TD loop is often used in lieu of the DLL because of its simplicity.

The operation of the loop is explained by observing that the control waveforms generate the signal

$$V_p(t) = g(t)p(t + \tau - T_c/2) + \bar{g}(t)p(t + \tau + T_c/2). \quad (53)$$

Note that either one or the other, but not both, of these waveforms occurs at each instant of time. The voltage $V_p(t)$ then multiplies the incoming signal

$$d(t)p(t) \cos(\omega_0 t + \theta).$$

The output of the bandpass filter is therefore

$$E_f(t) = d(t)g(t)\,|p(t)p(t + \tau + T_c/2)| + d(t)g(t)\,|p(t)p(t + \tau - T_c/2)| \quad (54)$$

where, as before, the average occurs because the bandpass filter is designed to pass the data and control signals, but cuts off well below the chip rate. The data are eliminated by the envelope detector, and (54) then yields

$$E_d(t) = g(t)\,|R_p(\tau + T_c/2)| + g(t)|R_p(\tau - T_c/2)|. \quad (55)$$

The input $Y(t)$ to the loop filter is

$$Y(t) = E_d(t)g'(t) = g(t)\,|R_p(\tau - T_c/2)| - \bar{g}(t)\,|R_p(\tau - T_c/2)| \quad (56)$$

where the "—" sign was introduced by the inversion caused by $g'(t)$.

The narrow-band loop filter now "averages" $Y(t)$. Since each term is zero half of the time, the voltage into the VCO clock is, as before,

$$V_c(t) = |R_p(\tau - T_c/2)| - |R_p(\tau + T_c/2)|. \quad (57)$$

A typical tracking system for an FSK/FH spread-spectrum system is shown in Fig. 24. Waveforms are shown in Fig. 25. Once again, we have assumed that, although acquisition has occurred, there is still an error of τ seconds between transitions of the incoming signal's frequencies and the locally generated frequencies. The bandpass filter BPF is made sufficiently wide to pass the product signal $V_p(t)$ when $V_1(t)$ and $V_2(t)$ are at the same frequency f_i, but sufficiently narrow to reject $V_p(t)$ when $V_1(t)$ and $V_2(t)$ are at different frequencies f_i and f_{i+1}. Thus, the output of the envelope detector $V_d(t)$, shown in Fig. 24, is unity when $V_1(t)$ and $V_2(t)$ are at the same frequency and is zero when $V_1(t)$ and $V_2(t)$ are at different frequencies. From Fig. 25, we see that $V_g(t) = V_d(t)\,V_c(t)$ and is a three-level signal. This three-level signal is filtered to form a dc voltage which, in this case, presents a negative voltage to the VCO.

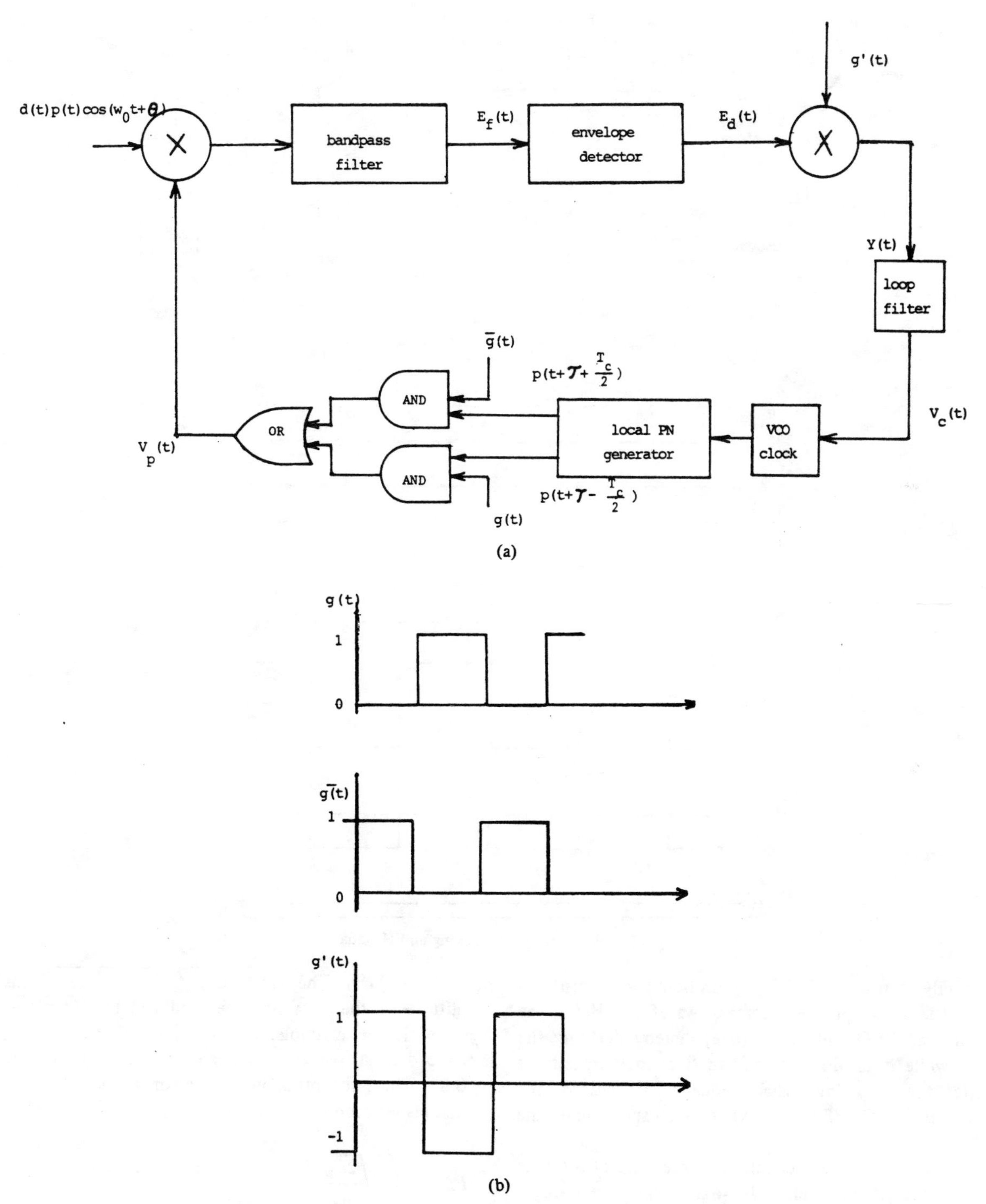

Fig. 23. The tau-dither loop. (a) Block diagram. (b) Control waveforms.

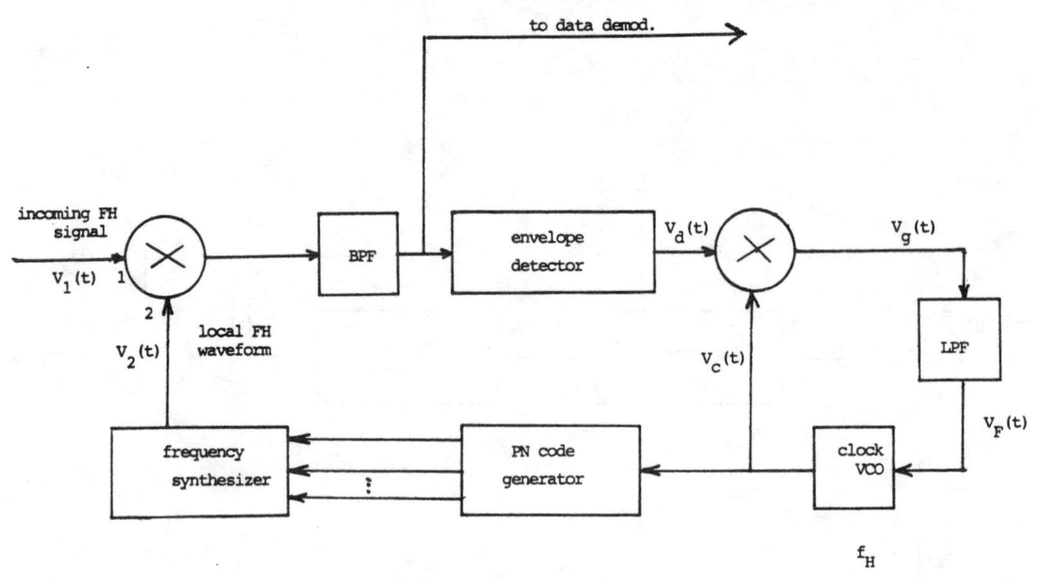

Fig. 24. Tracking loop for FH signals.

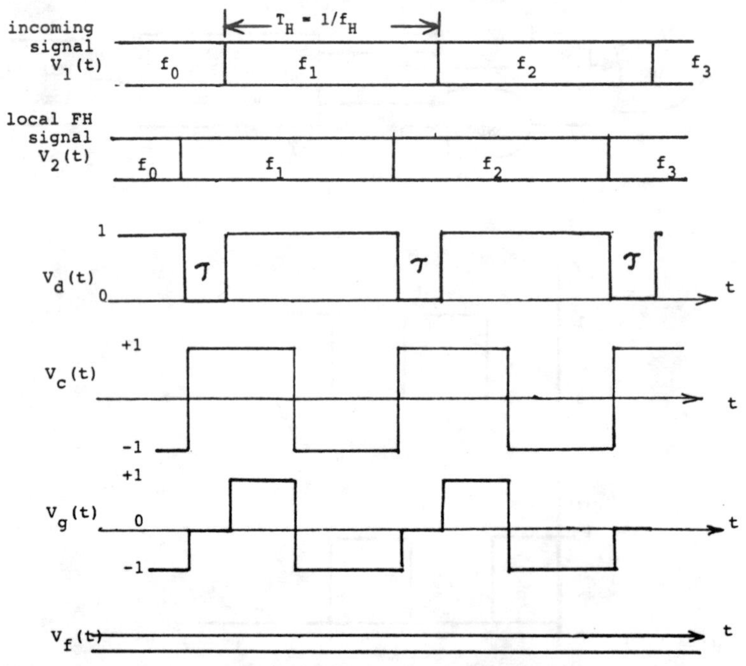

Fig. 25. Waveforms for tracking an FH signal.

It is readily seen that when $V_2(t)$ has frequency transitions which precede those of the incoming waveform $V_1(t)$, the voltage into the VCO will be negative, thereby delaying the transition, while if the local waveform frequency transitions occur after the incoming signal frequency transitions, the voltage into the VCO will be positive, thereby speeding up the transition.

The role of the tracking circuit is to keep the offset time τ small. However, even a relatively small τ can have a major impact on the probability of error of the received data. Referring to the DS system of Fig. 21, we see that if τ is not zero, the input to the data demodulator is $p(t)p(t + \tau)d(t) \cos(\omega_0 t + \theta)$ rather than $p^2(t)d(t) \cos(\omega_0 t + \theta) = d(t) \cos(\omega_0 t + \theta)$. The data demodulator removes the carrier and then averages the remaining signal, which in this case is $p(t)p(t + \tau)d(t)$. The result is $\overline{p(t)p(t + \tau)d(t)}$. Thus, th amplitude of the data has been reduced by $\overline{p(t)p(t + \tau)} = R_p(\tau) \leqslant 1$. For example, if $\tau = T_c/10$, that data amplitud is reduced to 90 percent of its value, and the power is reduce to 0.81. Thus, the probability of error in correctly detectin the data is reduced from

$$P_e = Q\left(\sqrt{\frac{2E_b}{\eta_0}}\right)$$

to

$$P_e(\tau = T_c/10) = Q\left(\sqrt{\frac{1.62E_b}{\eta_0}}\right),$$

and at an E_b/η_0 of 9.6 dB, P_e is increased from 10^{-5} to 10^{-4}.

IX. CONCLUSIONS

This tutorial paper looked at some of the theoretical issues involved in the design of a spread-spectrum communication system. The topics discussed included the characteristics of PN sequences, the resulting processing gain when using either direct-sequence or frequency-hopping antijam considerations, multiple access when using spread spectrum, multipath effects, and acquisition and tracking systems.

No attempt was made to present other than fundamental concepts; indeed, to adequately cover the spread-spectrum system completely is the task for an entire text [55], [56]. Furthermore, to keep this paper reasonably concise, the authors chose to ignore both practical system considerations such as those encountered when operating at, say, HF, VHF, or UHF, and technology considerations, such as the role of surface acoustic wave devices and charge-coupled devices in the design of spread-spectrum systems.

Spread spectrum has for far too long been considered a technique with very limited applicability. Such is not the case. In addition to military applications, spread spectrum is being considered for commercial applications such as mobile telephone and microwave communications in congested areas.

The authors hope that this tutorial will result in more engineers and educators becoming aware of the potential of spread spectrum, the dissemination of this information in the classroom, and the use of spread spectrum (where appropriate) in the design of communication systems.

APPENDIX

ALGEBRAIC PROPERTIES OF LINEAR FEEDBACK SHIFT REGISTER SEQUENCES

In order to fully appreciate the study of shift register sequences, it is desirable to introduce the polynomial representation (or generating function) of a sequence

$$C(x) = \sum_{i=0}^{\infty} C_i x^i \leftrightarrow (C_0, C_1, C_2, \cdots). \tag{A1}$$

If the sequence is periodic with period L, i.e.,

$$C_0, C_1, C_2, \cdots, C_{L-1} C_0 C_1, \cdots, C_{L-1}, C_0, \cdots,$$

then since $x^L C(x) \leftrightarrow (0, 0, \cdots, 0, C_0, C_1, C_2, \cdots)$,

$$C(x)(1 - x^L) = \sum_{i=0}^{L-1} C_i x^i \triangleq R(x) \tag{A2}$$

with $R(x)$ the (finite) polynomial representation of one period.

Thus, for any periodic sequence of period L,

$$C(x) = \frac{R(x)}{1 - x^L}; \qquad \deg R(x) < L. \tag{A3}$$

Next consider the periodic sequence generated by the LFSR recursion. Multiplying each side by x^n and summing gives

$$\sum_{n=0}^{\infty} C_n x^n = \sum_{k=1}^{r} a_k \sum_{n=0}^{\infty} C_{n-k} x^n$$

$$= \sum_{k=1}^{r} a_k \sum_{l=0}^{k-1} C_{l-k} x^l + \sum_{k=1}^{r} a_k x^k \left(\sum_{n=0}^{\infty} C_n x^n \right).$$

The left-hand side is the generating function $C(x)$ of the sequence. The first term on the right is a polynomial of degree $< r$, call it $g(x)$, which depends only on the initial state of the register $C_{-1}, C_{-2}, C_{-3}, \cdots, C_{-r}$. Thus, the basic equation of the register sequence may be written as

$$C(x) = \frac{g(x)}{f(x)}; \qquad \deg g(x) < r \tag{A4}$$

where $f(x) \triangleq 1 - \sum_{k=1}^{r} a_k x^k$ is the characteristic polynomial[2] (or connection polynomial) of the register. Since $C(x)$ is the generating polynomial of a sequence of period $L = 2^r - 1$, it can be shown from (A3) and (A4) that $f(x)$ must divide $1 - x^L$. This is illustrated in the following example.

Example

The three-stage binary maximal length register with $f(x) = 1 + x + x^3$ has period 7. If the initial contents of the register are $C_{-3} = 1$, $C_{-2} = 0$, $C_{-1} = 0$, then $g(x) = a_3 = 1$ and $C(x) = 1/(1 + x + x^3)$. Long division (modulo 2) yields

$$C(x) = 1 + x + x^2 + x^4 + x^7 + x^9 + \cdots$$

which is the generating function of the periodic sequence

1 1 1 0 1 0 0 ⋮ 1 1 1 0 1 ⋯,

and which is precisely the sequence generated by the corresponding recursion

$$C_n = C_{n-1} + C_{n-3} \pmod 2.$$

Observe that

$$(1 + x + x^3)(1 + x + x^2 + x^4) = 1 + x^7$$

so that $f(x)$ divides $1 + x^7$. Also, we may write

$$C(x) = \frac{1}{1 + x + x^3} \cdot \frac{1 + x + x^2 + x^4}{1 + x + x^2 + x^4} = \frac{1 + x + x^2 + x^4}{1 + x^7}$$

which is in the form of (A3).

[2] For binary sequences, all sums are modulo two and minus is the same as plus. The polynomials defining them have 0, 1 coefficients and are said to be polynomials over a finite field with two elements. A field is a set of elements, and two operations, say, + and ·, which obey the usual rules of arithmetic. A finite field with q elements is called a Galois field and is designated as $GF(q)$.

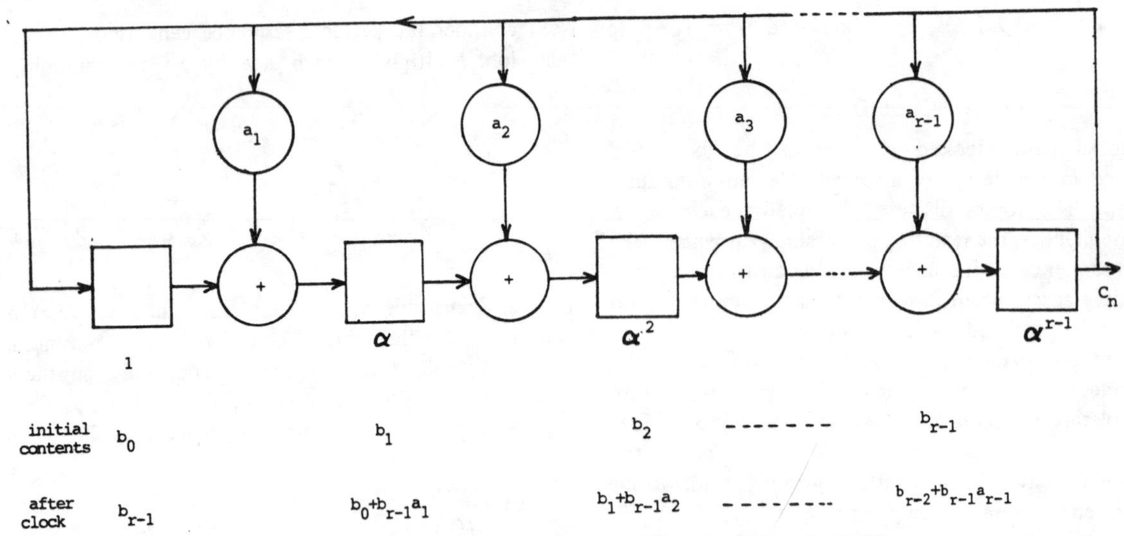

Fig. 26. Binary modular shift register generator with polynomial $f_M(x) = 1 + a_1x + a_2x^2 + \cdots + a_{r-1}x^{r-1} + x^r$.

It is easy to see (by multiplying and equating coefficients of like powers) that if

$$\frac{1}{f(x)} = \frac{1}{1 + a_1x + a_2x^2 + \cdots + a_rx^r}$$

$$= C_0 + C_1x + C_2x^2 + \cdots = C(x)$$

then

$$C_n = \sum_{k=1}^{r} a_k C_{n-k},$$

so that (except for initial conditions) $f(x)$ completely describes the maximal length sequence. Now what properties must $f(x)$ possess to ensure that the sequence is maximal length? Aside from the fact that $f(x)$ must divide $1 + x^L$, it is necessary (but not sufficient) that $f(x)$ be irreducible, i.e., $f(x) \neq f_1(x) \cdot f_2(x)$. Suppose that $f(x) = f_1(x)\, f_2(x)$ with $f_1(x)$ of degree r_1, $f_2(x)$ of degree r_2, and $r_1 + r_2 = r$. Then we can write, by partial fractions,

$$\frac{1}{f(x)} = \frac{\alpha(x)}{f_1(x)} + \frac{\beta(x)}{f_2(x)}; \qquad \deg \alpha(x) < r_1 \quad \deg \beta(x) < r_2.$$

The maximum period of the expansion of the first term is $2^{r_1}-1$ and that of the second term is $2^{r_2}-1$. Hence, the period of $1/f(x) \leqslant$ least common multiple of $(2^{r_1}-1, 2^{r_2}-1) < 2^r-3$. This is a contradiction, since if $f(x)$ were maximal length, the period of $1/f(x)$ would be 2^r-1. Thus, a *necessary* condition that the LFSR is maximal length is that $f(x)$ is irreducible.

A *sufficient* condition is that $f(x)$ is *primitive*. A primitive polynomial of degree r over $GF(2)$ is simply one for which the period of the coefficients of $1/f(x)$ is 2^r-1. However, additional insight can be had by examining the roots of $f(x)$. Since $f(x)$ is irreducible over $GF(2)$, we must imagine that the roots are elements of some larger (extension) field. Suppose that α is such an element and that $f(\alpha) = 0 = \alpha^r + a_{r-1}\alpha^{r-1} + \cdots + a_1\alpha + 1$ or

$$\alpha^r = a_{r-1}\alpha^{r-1} + \cdots + a_1\alpha + 1. \tag{A5}$$

We see that *all* powers of α can be expressed in terms of a linear combination of $\alpha^{r-1}, \alpha^{r-2}, \cdots, \alpha, 1$ since any powers larger than $r-1$ may be reduced using (A5). Specifically, suppose we have some power of α that we represent as

$$\beta \triangleq b_0 + b_1\alpha + b_2\alpha^2 + \cdots + b_{r-1}\alpha^{r-1}. \tag{A6}$$

Then if we multiply this β by α and use (A5), we obtain

$$\beta\alpha = b_{r-1} + (b_0 + b_{r-1}a_1)\alpha + (b_1 + b_{r-1}a_2)\alpha^2 + \cdots + (b_{r-2} + b_{r-1}a_{r-1})\alpha^{r-1} \tag{A7}$$

The observations above may be expressed in another, more physical way with the introduction of an LFSR in modular form [called a modular shift register generator (MSRG)] shown in Fig. 26. The feedback, modulo 2, is between the delay elements. The binary contents of the register at any time are shown as $b_0, b, \cdots, b_{r-1}$. This vector can be identified with β as

$$\beta = b_0 + b_1\alpha + \cdots + b_{r-1}\alpha^{r-1} \leftrightarrow [b_0, b_1, \cdots b_{r-1}],$$

the contents of the first stage being identified with the coefficient of α^0, those of the second stage with the coefficient of α^1, etc. After one clock pulse, it is seen that the register contents correspond to

$$\beta\alpha = b_{r-1} + (b_0 + b_{r-1})\alpha + \cdots + (b_{r-2} + b_{r-1}a_{r-1})\alpha^{r-1} \leftrightarrow [b_{r-1}, \cdots, b_{r-2} + b_{r-1}a_{r-1}].$$

Thus, the MSRG is an α-multiplier. Now if $\alpha^0, \alpha, \alpha^2, \alpha^3, \cdots, \alpha^{L-1}$, $L = 2^r-1$ are all *distinct*, we call α a *primitive* element of $GF(2^r)$. The register in Fig. 26 cycles through all states (starting in any nonzero state), and hence generates a maximal length sequence. Thus, another way of describing that the polynomial $f_M(x)$ is primitive (or maximal length) is that it has a primitive element in $GF(2^r)$ as a root.

There is an intimate relationship between the MSRG shown in Fig. 26 and the SSRG shown in Fig. 4. From Fig. 26 it is easily seen that the output sequence C_n satisfies the recursion

$$C_n = \sum_{k=0}^{r-1} a_k C_{n-r+k}. \tag{A8}$$

Multiplying both sides by x^n and summing yields

$$\begin{aligned} C(x) &\triangleq \sum_{n=-\infty}^{\infty} C_n x^n = \sum_{k=0}^{r-1} a_k \sum_{n=0}^{\infty} C_{n-r+k} x^n \\ &= \sum_{k=0}^{r-1} a_k x^{r-k} \sum_{l=-r+k}^{-1} C_l x^l \\ &\quad + x^r \sum_{k=0}^{r-1} a_k x^{-k} \left(\sum_{n=0}^{\infty} C_n x^n \right) \end{aligned} \tag{A9}$$

or

$$C(x) = g_M(x) + x^r \sum_{k=0}^{r-1} a_k x^{-k} C(x). \tag{A10}$$

$g_M(x)$ is the first term on the right-hand side of (A9) and is a polynomial of degree $< r$ which depends on the initial state. Then we have

$$C(x) = \frac{g_M(x)}{f_M(x)} \tag{A11}$$

where

$$f_M(x) = 1 - \{a_0 x^r + a_1 x^{r-1} + a_2 x^{r-2} + \cdots + a_{r-1} x\}$$

(recall that in $GF(2)$, minus is the same as plus) is the characteristic (or connection) polynomial of the MSRG. Since the sequence C_n [of coefficients of $C(x)$] when $f_M(x)$ is primitive depends *only* on $f_M(x)$ (discounting phase), the relationship between the SSRG and the MSRG which generates the *same* sequence is

$$f(x) = x^r f_M\left(\frac{1}{x}\right). \tag{A12}$$

$f_M(x)$ is called the *reciprocal* polynomial of $f(x)$ and is obtained from $f(x)$ by reversing the order of the coefficients.

There are several good tables of irreducible and primitive polynomials available [2], [5], [6], and although the tables do not list all the primitive polynomials, algorithms exist [7] which allow one to generate all primitive polynomials of a given degree if one of them is known. The number $\lambda(r)$ of primitive polynomials of degree r is [4]

$$\lambda(r) = \frac{\phi(2^r - 1)}{r} \tag{A13}$$

where $\phi(m)$ is the number of integers less than m which are relatively prime to m (Euler totient function). The growth of this number with r is shown in Table I.

TABLE I
THE NUMBER OF MAXIMAL LENGTH LINEAR SRG SEQUENCES OF DEGREE $r = \lambda(r) = \phi(2^r - 1)/r$

r	2^r-1	$\lambda(r)$
1	1	1
2	3	1
3	7	2
4	15	2
5	31	6
6	63	6
7	127	18
8	255	16
9	511	48
10	1,023	60
11	2,047	176
12	4,095	144
13	8,191	630
14	16,383	756
15	32,767	1,800
16	65,535	2,048
17	131,071	7,710
18	262,143	8,064
19	524,287	27,594
20	1,048,575	24,000
21	2,097,151	87,672
22	4,194,303	120,032

The algebra of LFSR's is useful in constructing codes with uniformly low cross correlation known as Gold codes. The underlying principle of these codes is based on the following *theorem* [15].

If $f_1(x)$ is the minimal polynomial of the primitive element $\alpha \in GF(2^r)$ and $f_t(x)$ is the minimal polynomial of α^t, where both $f_1(x)$ and $f_t(x)$ are of degree r and

$$t = \begin{cases} 2^{\frac{r+1}{2}} + 1, & r \text{ odd} \\ 2^{\frac{r+2}{2}} + 1, & r \text{ even}, \end{cases}$$

then the product $f(x) \triangleq f_1(x) f_t(x)$ determines an LFSR which generates $2^r + 1$ different sequences (corresponding to the $2^r + 1$ states in distinct cycles) of period $2^r - 1$, and such that for any pair C' and C'',

$$L\,|R_{C'C''}(\tau)| < t.$$

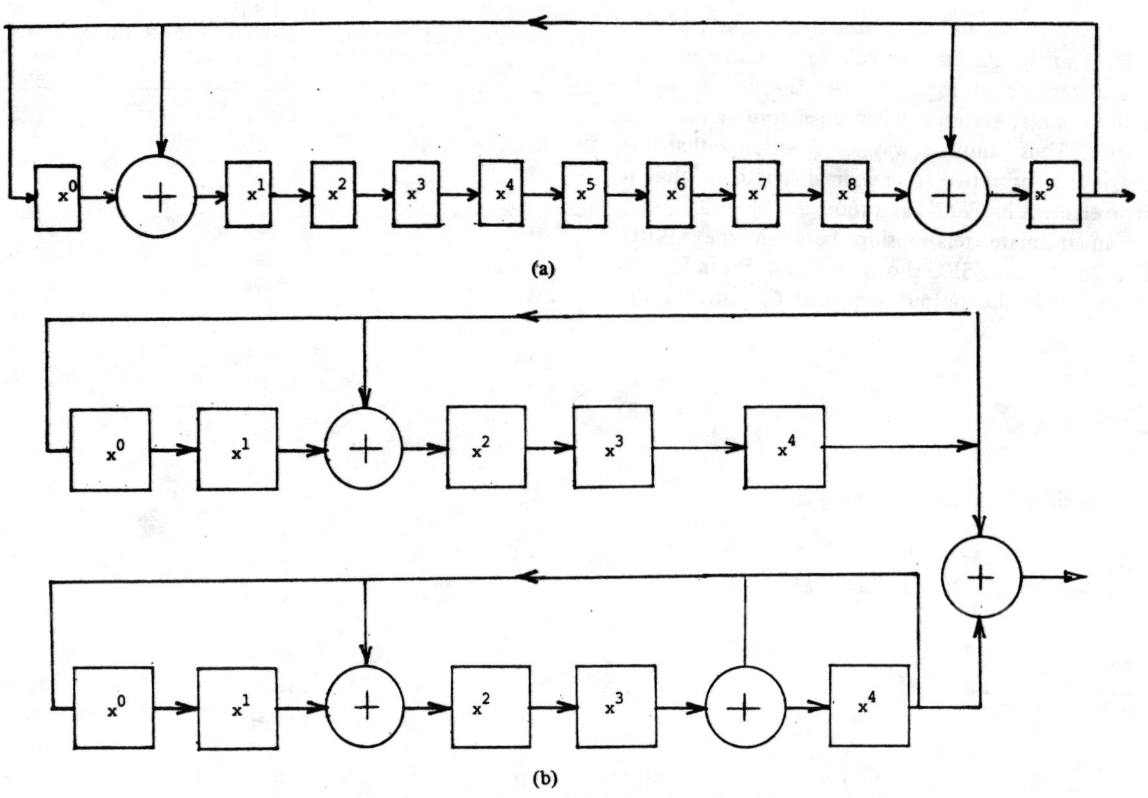

Fig. 27. Two implementations of LFSR which generate Gold codes of length $2^5 - 1 = 31$ with maximum cross correlation $t = 9$. (a) LFSR with $f(x) = 1 + x + x^3 + x^9 + x^{10}$. (b) LFSR which generates sequences corresponding to $f(x) = (1 + x^2 + x^5)\cdot(1 + x^2 + x^4 + x^5) = 1 + x + x^3 + x^9 + x^{10}$.

Futhermore, $R_{C'C''}(\tau)$ is only a three-valued function for any integer τ.

A minimal polynomial of α is simply the smallest degree monic[3] polynomial for which α is a root. With the help of a table of primitive polynomials, we can identify minimal polynomials of powers of α and easily construct Gold codes. For example, if $r = 5$ and $t = 2^3 + 1 = 9$, using [2] we find that $f_1(x) = 1 + x^2 + x^5$ and $f_9(x) = 1 + x^2 + x^4 + x^5$. Then $f(x) = 1 + x + x^3 + x^9 + x^{10}$. The two ways to represent this LFSR (in MSRG form) are shown in Fig. 27. Fig. 27(a) shows one long nonmaximal length register of degree 10 which generates sequences of period $2^5 - 1 = 31$. Since there are $2^{10} - 1$ possible nonzero initial states, the number of initial states that result in distinct cycles is $(2^{10} - 1)/(2^5 - 1) = 2^5 + 1 = 33$. Each of these initial states specifies a different Gold code of length 31. Fig. 27(b) shows how the same result can be obtained by adding the outputs of the two MLFSR's of degree 5 together modulo two. This follows simply from the observation that the sequence(s) generated by $f(x)$ are just the coefficients in the expansion of $1/f(x) = 1/f_1(x)\cdot f_9(x)$. By using partial fractions, one can see that the resulting coefficients are the (modulo two) sum of the coefficients of like powers in the expansion of $1/f_1(x)$ and $1/f_9(x)$. Naturally, the sequence resulting will depend on the relative *phases* of the two degree-5 registers. As before, there are $(2^{10} - 1)/(2^5 - 1) = 2^5 + 1 = 33$ relative phases which result in 33 different sequences satisfying the cross-correlation bound given by the theorem.

[3] A monic polynomial is one whose coefficient of its highest power is unity.

GLOSSARY OF SYMBOLS

a_n	{0, 1} feedback taps for LFSR.
B_D	One-sided bandwidth (Hz) for data signal(s).
B_{ss}	One-sided bandwidth (Hz) of baseband spread-spectrum signal.
$C(x)$	Generating function of C_n; $C(x) = \Sigma_{n=0}^{\infty} C_n x^n$.
C_n	{0, 1} LFSR sequence.
C_n' or C_n''	{1, −1} LFSR sequence.
D	Dimensionality of underlying signal space.
$d(t)$	Data sequence waveform.
Δ	Initial offset, in chips, of incoming signal and locally generated code.
DS	Direct sequence.
E_b	Energy/information bit.
E_J	Jammer energy over the correlation interval.
E_s	Energy/symbol.
$f(x)$	Characteristic (connection) polynomial of an LFSR, $f(x) = 1 + a_1 x + \cdots + a_{r-1} x^{r-1} + x^r$.
f_c	Chip rate; $T_c = 1/f_c$.
FH	Frequency hopping.
G_P	Processing gain.
$J(t)$	Jammer signal waveform.

K Number of frequencies jammed by partial-band jammer.
$L = 2^r - 1$ Period of PN sequence.
L Implementation losses.
λ Number of chips examined during each search in the process of acquisition.
$\lambda(r)$ Number of binary maximal length PN codes of degree r (length $L = 2^r - 1$).
M Signal alphabet size.
M_J Jamming margin.
n Number of chips/bit or number of dimensions of spread signal space.
N Number of frequencies in FH.
N_c Number of chips in uncertainty region at start of acquisition.
N_u Number of users in CDMA system.
η_0 One-sided white noise power spectral density (W/Hz).
η_{0J} $P_J/2f_c$ = power density of jammer.
$n_w(t)$ Additive white Gaussian noise (AWGN).
$p(t)$ Spreading sequence waveform.
P_D Probability of detection.
P_e Probability of error.
P_F Probability of false alarm.
P_J Jammer power.
P_N Noise power.
P_s Signal power.
PN Pseudonoise sequence.
r Number of stages of shift register.
$r(t)$ Received waveform.
R Data rate (bits/s).
$R_p(\tau)$ Autocorrelation function.
$R_{C'C''}(\tau)$ Cross-correlation function of two (periodic) ± 1 sequences C_n', C_n''.
$\rho(\tau)$ $(1/T) \int_0^\tau p(t)\, p(t-\tau)\, dt$ (partial correlation function).
$\hat{\rho}(\tau)$ $(1/T) \int_\tau^T p(t)\, p(t-\tau) dt$.
$S(t)$ Transmitted signal waveform.
$S_p(f)$ Power spectral density of spreading sequence waveform [also denoted $S_{ss}(f)$].
SNR Signal-to-noise power ratio.
SJR Signal-to-jammer power ratio.
T Signal or symbol duration.
T_c Chip duration.
T_H Time to hop one frequency; $1/T_H$ = hopping rate.
V Correlator output voltage.

ACKNOWLEDGMENT

The authors wish to thank the anonymous reviewers for their constructive suggestions in the final preparation of this paper.

REFERENCES

[1] R. A. Scholtz, "The origins of spread-spectrum communications," this issue, pp. 822–854.
[2] W. W. Peterson and E. J. Weldon, Jr., *Error Correcting Codes*, 2nd ed. Cambridge, MA: M.I.T. Press, 1972.
[3] J. M. Wozencraft and I. M. Jacobs, *Principles of Communication Engineering*. New York: Wiley, 1965.
[4] S. W. Golomb, *Shift Register Sequences*. San Francisco, CA: Holden Day, 1967.
[5] R. W. Marsh, *Table of Irreducible Polynomials over GF(2) Through Degree 19*. Washington, DC: NSA, 1957.
[6] W. Stahnke, "Primitive binary polynomials," *Math. Comput.*, vol. 27, pp. 977–980, Oct. 1973.
[7] E. R. Berlekamp, *Algebraic Coding Theory*. New York: McGraw-Hill, 1968.
[8] J. L. Massey, "Shift-register synthesis and BCH decoding," *IEEE Trans. Inform. Theory*, vol. IT-15, pp. 122–127, Jan. 1969.
[9] N. G. deBruijn, "A combinatorial problem," in *Koninklijke Nederlands Akademie Van Wetenschappen Proc.*, 1946, pp. 758–764.
[10] E. J. Groth, "Generation of binary sequences with controllable complexity," *IEEE Trans. Inform. Theory*, vol. IT-17, pp. 288–296, May 1971.
[11] E. L. Key, "An analysis of the structure and complexity of nonlinear binary sequence generators," *IEEE Trans. Inform. Theory*, vol. IT-22, pp. 732–736, Nov. 1976.
[12] H. Beker, "Multiplexed shift register sequences," presented at CRYPTO '81 Workshop, Santa Barbara, CA, 1981.
[13] J. L. Massey and J. J. Uhran, "Sub-baud coding," in *Proc. 13th Annu. Allerton Conf. Circuit and Syst. Theory*, Monticello, IL, Oct. 1975, pp. 539–547.
[14] J. H. Lindholm, "An analysis of the pseudo randomness properties of the subsequences of long *m*-sequences," *IEEE Trans. Inform. Theory*, vol. IT-14, 1968.
[15] R. Gold, "Optimal binary sequences for spread spectrum multiplexing," *IEEE Trans. Inform. Theory*, vol. IT-13, pp. 619–621, 1967.
[16] D. V. Sarwate and M. B. Pursley, "Cross correlation properties of pseudo-random and related sequences," *Proc. IEEE*, vol. 68, pp. 598–619, May 1980.
[17] A. Lempel, M. Cohn, and W. L. Eastman, "A new class of balanced binary sequences with optimal autocorrelation properties," IBM Res. Rep. RC 5632, Sept. 1975.
[18] A. G. Konheim, *Cryptography, A Primer*. New York: Wiley, 1981.
[19] D. L. Schilling, L. B. Milstein, R. L. Pickholtz, and R. Brown, "Optimization of the processing gain of an *M*-ary direct sequence spread spectrum communication system," *IEEE Trans. Commun.*, vol. COM-28, pp. 1389–1398, Aug. 1980.
[20] L. B. Milstein, S. Davidovici, and D. L. Schilling, "The effect of multiple-tone interfering signals on a direct sequence spread spectrum communication system," *IEEE Trans. Commun.*, vol. COM-30, pp. 436–446, Mar. 1982.
[21] S. W. Houston, "Tone and noise jamming performance of a spread spectrum *M*-ary FSK and 2, 4-ary DPSK waveforms," in *Proc. Nat. Aerosp. Electron. Conf.*, June 1975, pp. 51–58.
[22] G. K. Huth, "Optimization of coded spread spectrum systems performance," *IEEE Trans. Commun.*, vol. COM-25, pp. 763–770, Aug. 1977.
[23] R. H. Pettit, "A susceptibility analysis of frequency hopped *M*-ary NCPSK—Partial-band noise on CW tone jamming," presented at the Symp. Syst. Theory, May 1979.
[24] L. B. Milstein, R. L. Pickholtz, D. L. Schilling, "Optimization of the processing gain of an FSK-FH system," *IEEE Trans. Commun.*, vol. COM-28, pp. 1062–1079, July 1980.
[25] M. K. Simon and A. Polydoros, "Coherent detection of frequency-hopped quadrature modulations in the presence of jamming—Part I: QPSK and QASK; Part II: QPR class I modulation," *IEEE Trans. Commun.*, vol. COM-29, pp. 1644–1668, Nov. 1981.
[26] A. J. Viterbi and I. M. Jacobs, "Advances in coding and modulation for noncoherent channels affected by fading, partial band, and multiple access interference," in *Advances in Communication Systems*, vol. 4. New York: Academic, 1975.
[27] J. M. Aein and R. D. Turner, "Effect of co-channel interference on CPSK carriers," *IEEE Trans. Commun.*, vol. COM-21, pp. 783–790, July 1973.
[28] R. H. Pettit, "Error probability for NCFSK with linear FM jamming," *IEEE Trans. Aerosp. Electron. Syst.*, vol. AES-8, pp. 609–614, Sept. 1972.
[29] A. J. Viterbi, "Spread spectrum communications—Myths and realities," *IEEE Commun. Mag.*, pp. 11–18, May 1979.
[30] D. J. Torrieri, *Principles of Military Communication Systems*. Dedham, MA: Artech House, 1981.

[31] M. B. Pursley, "Performance evaluation for phase-coded spread spectrum multiple-access communication—Part I: System analysis," *IEEE Trans. Commun.*, vol. COM-25, pp. 795–799, Aug. 1977.
[32] K. Yao, "Error probability of asynchronous spread-spectrum multiple access communication systems," *IEEE Trans. Commun.*, vol. COM-25, pp. 803–809, Aug. 1977.
[33] C. L. Weber, G. K. Huth, and B. H. Batson, "Performance considerations of code division multiple access systems," *IEEE Trans. Veh. Technol.*, vol. VT-30, pp. 3–10, Feb. 1981.
[34] M. B. Pursley and D. V. Sarwate, "Performance evaluation for phase-coded spread spectrum multiple-access communication—Part II. Code sequence analysis," *IEEE Trans. Commun.*, vol. COM-25, pp. 800–803, Aug. 1977.
[35] M. B. Pursley and H. F. A. Roefs, "Numerical evaluation of correlation parameters for optimal phases of binary shift-register sequences," *IEEE Trans. Commun.*, vol. COM-25, pp. 1597–1604, Aug. 1977.
[36] G. Solomon, "Optimal frequency hopping sequences for multiple access," in *Proc. 1973 Symp. Spread Spectrum Commun.*, vol. 1, AD915852, pp. 33–35.
[37] D. V. Sarwate and M. B. Pursley, "Hopping patterns for frequency-hopped multiple-access communication," in *Proc. 1978 IEEE Int. Conf. Commun.*, vol. 1, pp. 7.4.1–7.4.3.
[38] P. S. Henry, "Spectrum efficiency of a frequency-hopped-DPSK spread spectrum mobile radio system," *IEEE Trans. Veh. Technol.*, vol. VT-28, pp. 327–329, Nov. 1979.
[39] O. C. Yue, "Hard-limited versus linear combining for frequency-hopping multiple-access systems in a Rayleigh fading environment," *IEEE Trans. Veh. Technol.*, vol. VT-30, pp. 10–14, Feb. 1981.
[40] R. W. Nettleton and G. R. Cooper, "Performance of a frequency-hopped differentially modulated spread-spectrum receiver in a Rayleigh fading channel," *IEEE Trans. Veh. Technol.*, vol. VT-30, pp. 14–29, Feb. 1981.
[41] D. E. Borth and M. B. Pursley, "Analysis of direct-sequence spread-spectrum multiple-access communication over Rician fading channels," *IEEE Trans. Commun.*, vol. COM-27, pp. 1566–1577, Oct. 1979.
[42] E. A. Geraniotis and M. B. Pursley, "Error probability bounds for slow frequency-hopped spread-spectrum multiple access communications over fading channels," in *Proc. 1981 Int. Conf. Commun.*
[43] L. B. Milstein and D. L. Schilling, "Performance of a spread spectrum communication system operating over a frequency-selective fading channel in the presence of tone interference," *IEEE Trans. Commun.*, vol. COM-30, pp. 240–247, Jan. 1982.
[44] ——, "The effect of frequency selective fading on a noncoherent FH-FSK system operating with partial-band interference," this issue, pp. 904–912.
[45] J. M. Aein and R. L. Pickholtz, "A simple unified phasor analysis for PN multiple access to limiting repeaters," this issue, pp. 1018–1026.
[46] R. B. Ward, "Acquisition of pseudonoise signals by sequential estimation," *IEEE Trans. Commun. Technol.*, vol. COM-13, pp. 474–483, Dec. 1965.
[47] R. B. Ward and K. P. Yiu, "Acquisition of pseudonoise signals by recursion-aided sequential estimation," *IEEE Trans. Commun.*, vol. COM-25, pp. 784–794, Aug. 1977.
[48] P. M. Hopkins, "A unified analysis of pseudonoise synchronization by envelope correlation," *IEEE Trans. Commun.*, vol. COM-25, pp. 770–778, Aug. 1977.
[49] J. K. Holmes and C. C. Chen, "Acquisition time performance of PN spread-spectrum systems," *IEEE Trans. Commun.*, vol. COM-25, pp. 778–783, Aug. 1977.
[50] S. S. Rappaport, "On practical setting of detection thresholds," *Proc. IEEE*, vol. 57, pp. 1420–1421, Aug. 1969.
[51] J. J. Spilker, Jr., "Delay-lock tracking of binary signals," *IEEE Trans. Space Electron. Telem.*, vol. SET-9, pp. 1–8, Mar. 1963.
[52] P. T. Nielson, "On the acquisition behavior of delay lock loops," *IEEE Trans. Aerosp. Electron. Syst.*, vol. AES-12, pp. 415–523, July 1976.
[53] ——, "On the acquisition behavior of delay lock loops," *IEEE Trans. Aerosp. Electron. Syst.*, vol. AES-11, pp. 415–417, May 1975.
[54] H. P. Hartman, "Analysis of the dithering loop for PN code tracking," *IEEE Trans. Aerosp. Electron. Syst.*, vol. AES-10, pp. 2–9, Jan. 1974.
[55] R. C. Dixon, *Spread Spectrum Systems*. New York: Wiley, 1976.
[56] J. K. Holmes, *Coherent Spread Spectrum Systems*. New York: Wiley, 1982.

Raymond L. Pickholtz (S'54–A'55–M'60–SM'77–F'82) received the B.E.E. and M.S.E.E. degrees from the City College of New York, New York, NY, and the Ph.D. degree from the Polytechnic Institute of Brooklyn, Brooklyn, NY.

He is a Professor and was Chairman of the Department of Electrical Engineering and Computer Science, George Washington University, Washington, DC. He was a Research Engineer at RCA Laboratories and at ITT Laboratories for a period of ten years, working on problems ranging from color television to secure communications and guidance before returning to academia. He was an Associate Professor at the Polytechnic Institute of Brooklyn until 1972 when he joined the faculty at George Washington University. He taught part-time at New York University, New York, and in the Department of Physics, Brooklyn College, Brooklyn. He was a Visiting Professor at the University of Quebec, Quebec, P.Q., Canada. He has been an active consultant in communications to industry and government for many years and has lectured extensivley in this country, Canada, Europe, and South America on various aspects of communications. He is President of Telecommunications Associates, a small consulting and research firm.

Dr. Pickholtz has been very active in the IEEE and in the Communications Society (ComSoc) in particular. He was the first Editor for Computer Communication of the IEEE TRANSACTIONS ON COMMUNICATIONS. He organized the Technical Committee on Computer Communication of ComSoc. He was Guest Editor of the Special TRANSACTIONS Issue on Computer Communications. He was General Chairman of the Third Data Networks Symposium and of the Workshop on Data Networks. In addition, he was Chairman of the New York Chapter of the Information Theory Group and has also contributed his efforts to the Computer Society.

Donald L. Schilling (S'56–M'58–SM'69–F'75), for a photograph and biography, see this issue, p. 821.

Laurence B. Milstein (S'66–M'68–SM'77), for a photograph and biography, see this issue, p. 820.

Pseudo-Random Sequences and Arrays

F. JESSIE MACWILLIAMS AND NEIL J. A. SLOANE, MEMBER, IEEE

Abstract—Binary sequences of length $n = 2^m - 1$ whose autocorrelation function is either 1 or $-1/n$ have been known for a long time, and are called pseudo-random (or PN) sequences, or maximal-length shift-register sequences. Two-dimensional arrays of area $n = 2^{lm} - 1$ with the same property have recently been found by several authors. This paper gives a simple description of such sequences and arrays and their many nice properties.

I. INTRODUCTION

PSEUDO-RANDOM SEQUENCES (which are also called pseudo-noise (PN) sequences, maximal-length shift-register sequences, or m-sequences) are certain binary sequences of length $n = 2^m - 1$ (the construction is given in Section II). They have many useful properties, one of which is that their periodic autocorrelation function is given by

$$\rho(0) = 1 \qquad \rho(i) = -\frac{1}{n}, \qquad \text{for } 1 \leqslant i \leqslant n - 1 \qquad (1)$$

(see Section II-D and especially Fig. 9). These sequences have been known for a long time, are used in range-finding, scrambling, fault detection, modulation, synchronizing, etc., and a considerable body of literature exists. See for example [2], [13], [16], [19], [20], [24], [27], [31], [36], [43b], [55], [58], [67a], [68], [72], [74], [75], and especially Golomb [28], [29], Kautz [42], Selmer [60], Zierler [76], [77], and [44, ch. 14].

Manuscript received June 1, 1976; revised August 19, 1976.
The authors are with Bell Laboratories, Murray Hill, NJ 07974.

Nevertheless they are not as widely known as they should be, especially outside of the area of communication theory—see [61], [62], [66]—and there does not seem to exist a simple, comprehensive account of their properties. The reader will find such an account in Section II of this paper.

Recently, several applications ([37], [38], [43], [59]) have called for two-dimensional arrays whose two-dimensional autocorrelation function should satisfy $\rho(0, 0) = 1$, $\rho(i, j)$ small for $(i, j) \neq (0, 0)$. We shall see in Section III that it is very easy to use pseudo-random sequences to obtain $n_1 \times n_2$ arrays with

$$\rho(0, 0) = 1 \qquad \rho(i, j) = -\frac{1}{n}, \qquad \text{for } 0 \leqslant i < n_1, \quad 0 \leqslant j < n_2, \quad (i, j) \neq (0, 0) \qquad (2)$$

where $n = 2^m - 1 = n_1 n_2$, provided n_1 and n_2 are relatively prime. The construction is a standard one in studying product codes (see [26], [10]). For earlier work on two-dimensional arrays see Reed and Stewart [54], Spann [63], Gordon [30], Calabro and Wolf [12], Nomura *et al.* [48]–[51], Ikai and Kojima [40], and Imai [41]. The construction given in Section III does not seem to be mentioned in these papers, although it can be shown ([44, ch. 18]) to be equivalent to a special case of the constructions given by Nomura *et al.*

One of the problems considered in these papers is the construction of arrays with the *window property*. This means that if a window of prescribed size, say $k_1 \times k_2$, is slid over the array, each of the $2^{k_1 k_2} - 1$ possible nonzero $k_1 \times k_2$ arrays is seen through the window exactly once. (To avoid trouble at the edges, either several copies of the array are placed side by side, or alternatively the array is written on a torus.) We shall see in Section III-B and the Appendix that our arrays have the window property.

It is straightforward to generalize the construction given in Section III to obtain three and higher dimensional arrays with flat autocorrelation function; we leave the details to the reader.

To find sequences (and arrays) whose *aperiodic* autocorrelation function is flat is a different problem altogether—see Turyn [69] and Lindner [43c].

The outline of this paper is as follows. Section II describes pseudo-random arrays and their many nice properties and connections with other parts of mathematics. A particularly useful property is Property XIII, in Section II-K, which says that the pseudo-random sequences of length $2^m - 1$, together with the zero sequence, are isomorphic to a field with 2^m elements. Section III describes pseudo-random arrays and properties. So far everything has been binary, but in Section IV we describe pseudo-random sequences and arrays with elements taken from an alphabet of q symbols, where q is a prime power. A kind of generalized autocorrelation function of pseudo-random arrays, called the transmission function, is studied in Section V. A short summary appears in Section VI.

II. Pseudo-Random Sequences

A. The Shift Register

To construct a pseudo-random sequence of length $n = 2^m - 1$, one needs a primitive polynomial $h(x)$ of degree m. This term is defined below; for the moment we take

$$h(x) = x^4 + x + 1 \tag{3}$$

as an example of degree $m = 4$. This polynomial specifies a *feedback shift register* as shown in Fig. 1. In general this is a shift register consisting of m little boxes, representing memory elements or flip-flops, each containing a 0 or 1. At each time unit the contents of the boxes are shifted one place to the right, and the boxes corresponding to the terms in $h(x)$ are added and fed into the left-hand box. The sum is calculated mod 2, so →⊕→ in the figures represents a mod-2 adder or EXCLUSIVE-OR gate, defined by $0 + 0 = 1 + 1 = 0$, $0 + 1 = 1 + 0 = 1$.

In the above example, if the register contains $a_{i+3}, a_{i+2}, a_{i+1}, a_i$ at time i, then at time $i + 1$ it contains

$$a_{i+4} = a_{i+1} + a_i, \quad a_{i+3}, a_{i+2}, a_{i+1}$$

as shown in Fig. 2. In other words, this feedback shift register generates an infinite sequence $a_0 a_1 a_2 \cdots a_i \cdots$ which satisfies the recurrence

$$a_{i+4} = a_{i+1} + a_i, \quad i = 0, 1, \cdots, \tag{4}$$

where + denotes addition mod 2. The shift register needs to be started up, so we must specify the initial values $a_0, a_1, \cdots, a_{m-1}$.

Here is a more complicated example. When $m = 8$, we take

$$h(x) = x^8 + x^6 + x^5 + x + 1 \tag{5}$$

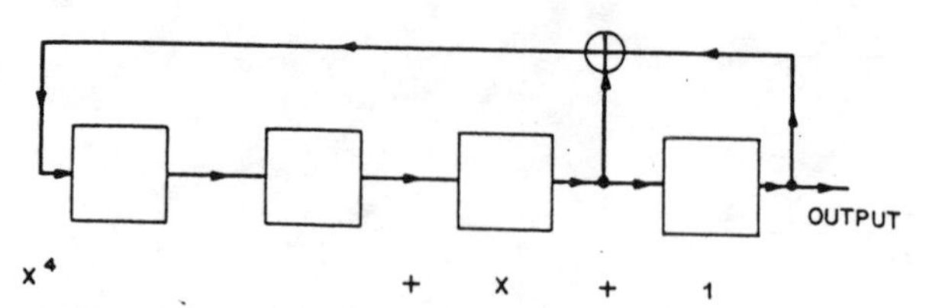

Fig. 1. Feedback shift register corresponding to $x^4 + x + 1$.

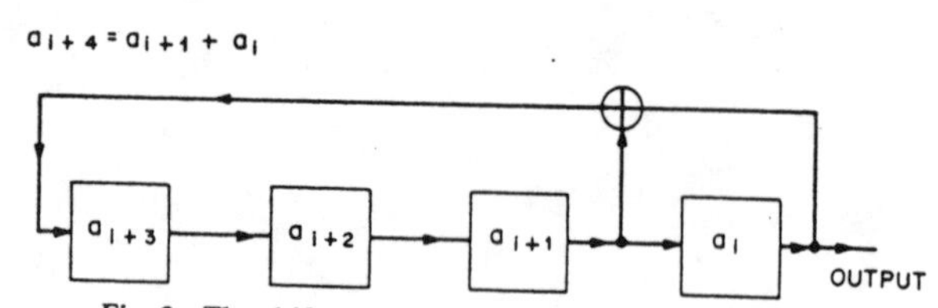

Fig. 2. The shift register specifies a recurrence relation.

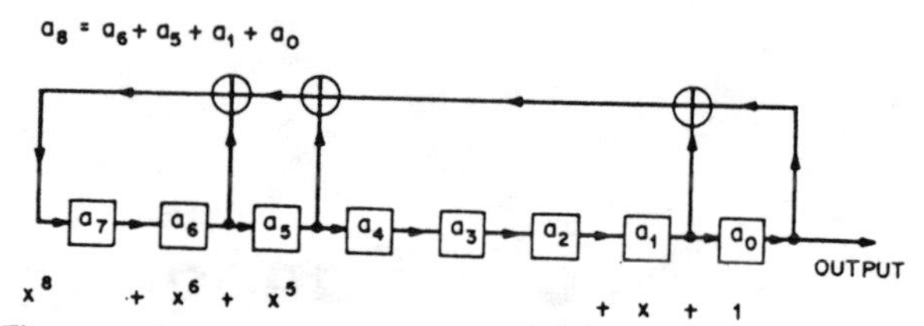

Fig. 3. Feedback shift register corresponding to $x^8 + x^6 + x^5 + x + 1$.

the shift register is shown in Fig. 3, and the output sequence satisfies the recurrence

$$a_{i+8} = a_{i+6} + a_{i+5} + a_{i+1} + a_i, \quad i = 0, 1, \cdots, \tag{6}$$

with initial values $a_0, a_1, \cdots, a_7$.

B. Pseudo-Random Sequences

Since each of the m boxes contains a 0 or 1, there are 2^m possible states for the shift register. Thus the sequence $a_0 a_1 a_2 \cdots$ must be periodic. But the zero state $00 \cdots 0$ can't occur unless the sequence is all zeros. So the maximum possible period is $2^m - 1$.

We can now define a *primitive* polynomial $h(x)$. This is one for which $a_0 a_1 a_2 \cdots$ has period $2^m - 1$ (for some starting state). For example, Fig. 4 shows the successive states and output sequence of Fig. 1 if the initial state is 1000. (Note that the output sequence is the same as the right-hand column of the list of states. The output is equal to the parity of the binary number corresponding to the state.) Since the output sequence has period $15 = 2^4 - 1$, $x^4 + x + 1$ *is* a primitive polynomial. Similarly the output of Fig. 3 has period $255 = 2^8 - 1$.

We ask the reader to accept this fact: there exist primitive polynomials of degree m for every m. (The rather complicated proof can be found for example in [8] or [44, ch. 4].) Fig. 5 gives a table for $m \leqslant 40$, sufficient to generate sequences of period up to $2^{40} - 1 \approx 10^{12}$, enough for most purposes. Fig. 5 is taken from Stahnke [64], who gives a table for $m \leqslant 168$. Primitive polynomials of much higher degree have been found by Zierler and Brillhart [78].

It follows that if $h(x)$ is a primitive polynomial of degree m, the shift register goes through all $2^m - 1$ distinct nonzero states before repeating, and produces an output sequence

$$a_0 a_1 a_2 \cdots \tag{7}$$

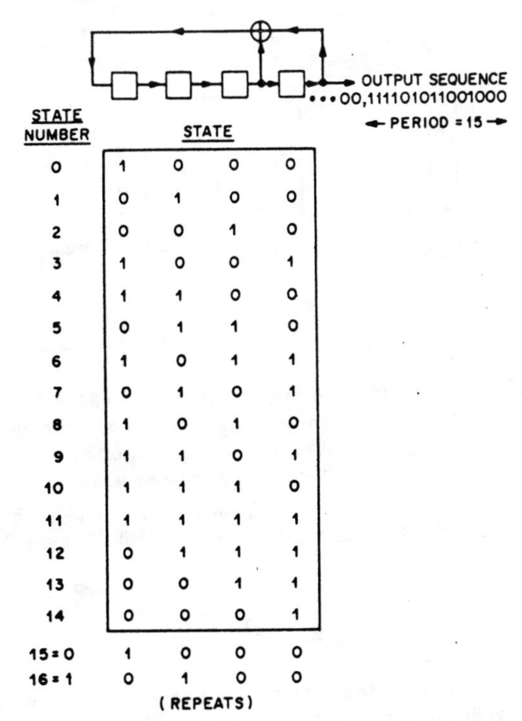

Fig. 4. Successive states and output sequence from shift register.

0 0 0 1 0 0 1 1 0 1 0 1 1 1 1
0 0 1 0 0 1 1 0 1 0 1 1 1 1 0
0 1 0 0 1 1 0 1 0 1 1 1 1 0 0
1 0 0 1 1 0 1 0 1 1 1 1 0 0 0
0 0 1 1 0 1 0 1 1 1 1 0 0 0 1
0 1 1 0 1 0 1 1 1 1 0 0 0 1 0
1 1 0 1 0 1 1 1 1 0 0 0 1 0 0
1 0 1 0 1 1 1 1 0 0 0 1 0 0 1
0 1 0 1 1 1 1 0 0 0 1 0 0 1 1
1 0 1 1 1 1 0 0 0 1 0 0 1 1 0
0 1 1 1 1 0 0 0 1 0 0 1 1 0 1
1 1 1 1 0 0 0 1 0 0 1 1 0 1 0
1 1 1 0 0 0 1 0 0 1 1 0 1 0 1
1 1 0 0 0 1 0 0 1 1 0 1 0 1 1
1 0 0 0 1 0 0 1 1 0 1 0 1 1 1

Fig. 6. The 15 pseudo-random sequences obtained from Fig. 4.

deg m	$h(x)$	deg m	$h(x)$
1	$x+1$	21	$x^{21}+x^2+1$
2	x^2+x+1	22	$x^{22}+x+1$
3	x^3+x+1	23	$x^{23}+x^5+1$
4	x^4+x+1	24	$x^{24}+x^4+x^3+x+1$
5	x^5+x^2+1	25	$x^{25}+x^3+1$
6	x^6+x+1	26	$x^{26}+x^8+x^7+x+1$
7	x^7+x+1	27	$x^{27}+x^8+x^7+x+1$
8	$x^8+x^6+x^5+x+1$	28	$x^{28}+x^3+1$
9	x^9+x^4+1	29	$x^{29}+x^2+1$
10	$x^{10}+x^3+1$	30	$x^{30}+x^{16}+x^{15}+x+1$
11	$x^{11}+x^2+1$	31	$x^{31}+x^3+1$
12	$x^{12}+x^7+x^4+x^3+1$	32	$x^{32}+x^{28}+x^{27}+x+1$
13	$x^{13}+x^4+x^3+x+1$	33	$x^{33}+x^{13}+1$
14	$x^{14}+x^{12}+x^{11}+x+1$	34	$x^{34}+x^{15}+x^{14}+x+1$
15	$x^{15}+x+1$	35	$x^{35}+x^2+1$
16	$x^{16}+x^5+x^3+x^2+1$	36	$x^{36}+x^{11}+1$
17	$x^{17}+x^3+1$	37	$x^{37}+x^{12}+x^{10}+x^2+1$
18	$x^{18}+x^7+1$	38	$x^{38}+x^6+x^5+x+1$
19	$x^{19}+x^6+x^5+x+1$	39	$x^{39}+x^4+1$
20	$x^{20}+x^3+1$	40	$x^{40}+x^{21}+x^{19}+x^2+1$

Fig. 5. Primitive polynomials.

of period $2^m - 1$. We call any segment

$$a_i a_{i+1} \cdots a_{i+2^m-2} \tag{8}$$

of length $2^m - 1$ a *pseudo-random sequence*. There are $2^m - 1$ different pseudo-random sequences [taking $i = 0, 1, \cdots, 2^m - 2$ in (8)]; those corresponding to Fig. 4 are shown in Fig. 6.

Note that if a different nonzero initial state is used, this is still one of the states the shift register goes through, and the new output sequence is just a shift of (7), namely $a_r a_{r+1} a_{r+2} \cdots$ for some r. Therefore the same set of pseudo-random sequences is obtained from any nonzero starting state. (Of course

LENGTH = 3

011

LENGTH = 7

0010111

LENGTH = 15

000100110101111

LENGTH = 31

0000100101100111110001101110101

LENGTH = 63

00000100001100010100111101000111001001011011101100
1101010111111

LENGTH = 127

00000010000011000010100011110010001011001110101001
111101000011100010010011011010110111110110001101001
011101110011001010101111111

LENGTH = 255

00000001011100011101111000101100110110000111100111
00001010111111110010111101001010000110111011011111
01011101000001100101010100011010110001100000100101
101101011001101001111110111001100111101100100001000
00011100100100110001001110101011010001000101001000
11111

Fig. 7. Examples of pseudo-random sequences.

changing $h(x)$ will give a different set of sequences. For example using $x^4 + x^3 + 1$ instead of $x^4 + x + 1$ reverses the sequences in Fig. 6.)

Fig. 7 gives one pseudo-random sequence of length $2^m - 1$ for each m between 2 and 8, obtained from the primitive polynomials of Fig. 5 with initial state $100 \cdots 0$.

Remark

It is possible to attain period 2^m (rather than $2^m - 1$) using a nonlinear shift register. The corresponding output sequence is called a *de Bruijn cycle* ([11a], [19a], [19b], [29], [43a]).

C. Properties of Pseudo-Random Sequences

Let $h(x)$ be a fixed primitive polynomial of degree m, and let $\mathcal{S}_m$ be the set consisting of the pseudo-random sequences obtained from $h(x)$, together with the sequence of $2^m - 1$ zeros (denoted by $\mathbf{0}$). These pseudo-random sequences are the $2^m - 1$ different segments

$$a_i a_{i+1} \cdots a_{i+2^m-2}, \quad i = 0, 1, \cdots, 2^m - 2$$

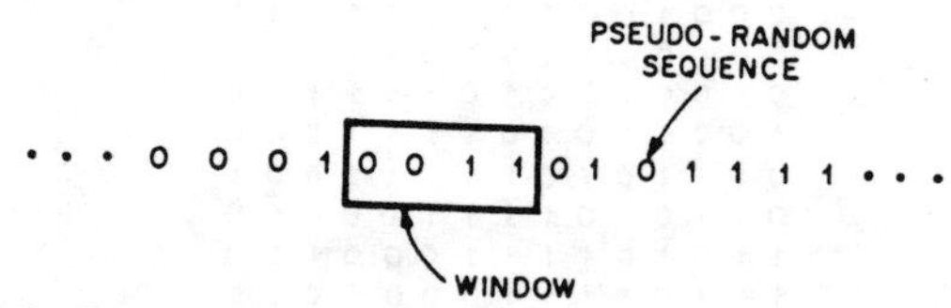

Fig. 8. The window property: Every nonzero 4-tuple is seen once.

of length $2^m - 1$ from the output of the shift register specified by $h(x)$. For example, $\mathcal{S}_4$ consists of **0** and the rows of Fig. 6.

We proceed to give the properties of these sequences. Properties I and II follow from the preceding discussion.

Property I–The Shift Property: If $b = b_0 b_1 \cdots b_{2^m-2}$ is any pseudo-random sequence in $\mathcal{S}_m$, then any cyclic shift of b, say

$$b_j b_{j+1} \cdots b_{2^m-2} b_0 \cdots b_{j-1}$$

is also in $\mathcal{S}_m$.

Property II–The Recurrence: Suppose $h(x) = \sum_{i=0}^{m} h_i x^i$, with $h_0 = h_m = 1$, $h_i = 0$ or 1 for $0 < i < m$. Any pseudo-random sequence $b \in \mathcal{S}_m$ satisfies the recurrence

$$b_{i+m} = h_{m-1} b_{i+m-1} + h_{m-2} b_{i+m-2} + \cdots + h_1 b_{i+1} + b_i \quad (9)$$

for $i = 0, 1, \cdots$. (This generalizes (4) and (6).) Conversely any solution of (9) is in $\mathcal{S}_m$. Using all $2^m - 1$ distinct nonzero initial values $b_0, \cdots, b_{m-1}$ in (9) we obtain the $2^m - 1$ pseudo-random sequences. There are m linearly independent solutions to (9), hence m linearly independent sequences in $\mathcal{S}_m$.

Property III–The Window Property: If a window of width m is slid along a pseudo-random sequence in $\mathcal{S}_m$, each of the $2^m - 1$ nonzero binary m-tuples is seen exactly once—see Fig. 8 for the case $m = 4$. (This follows from the fact that $h(x)$ is a primitive polynomial.)

To avoid difficulties at the ends, either imagine three copies of the sequence are placed next to each other, or alternatively that the sequence is written in a circle.

Property IV–Half 0's and Half 1's: Any pseudo-random sequence in $\mathcal{S}_m$ contains 2^{m-1} 1's and $2^{m-1} - 1$ 0's. (This is because there are 2^{m-1} odd numbers between 1 and $2^m - 1$, with binary representation ending in 1, and $2^{m-1} - 1$ even numbers in the same range, with binary representation ending in 0. The state of the shift register runs through these numbers, and the output is equal to the parity of the state—see Fig. 4.)

Property V–The Addition Property: The sum of two sequences in $\mathcal{S}_m$ (formed componentwise, modulo 2, without carries) is another sequence in $\mathcal{S}_m$. (For the sum of two solutions to (9) is another solution.) E.g. the sum of the first two sequences in Fig. 6 is the fifth sequence.

Property VI–The Shift-and-Add Property: The sum of a pseudo-random sequence and a cyclic shift of itself is another pseudo-random sequence. (From Properties I and V.)

D. Autocorrelation Function

We come now to the most important property, the autocorrelation function. The *autocorrelation function* $\rho(i)$ of a real (or complex) sequence $s_0 s_1 \cdots s_{n-1}$ of length n is defined by

$$\rho(i) = \frac{1}{n} \sum_{j=0}^{n-1} s_j \bar{s}_{i+j}, \qquad i = 0, \pm 1, \pm 2, \cdots \quad (10)$$

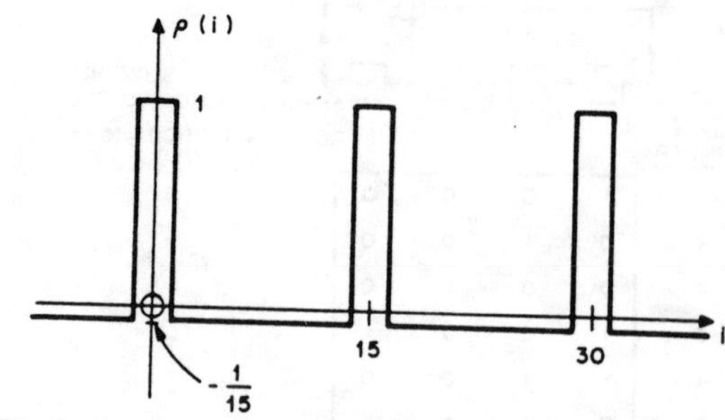

Fig. 9. Autocorrelation function of a pseudo-random sequence.

where subscripts are reduced mod n if they exceed $n - 1$, and the bar denotes complex conjugation. This is a periodic function: $\rho(i) = \rho(i + n)$. The autocorrelation function of a binary sequence $a_0 a_1 \cdots a_{n-1}$ is then defined to be equal to the autocorrelation function of the real sequence $(-1)^{a_0}, (-1)^{a_1}, \cdots, (-1)^{a_{n-1}}$ obtained by replacing 1's by −1's and 0's by +1's. Thus

$$\rho(i) = \frac{1}{n} \sum_{j=0}^{n-1} (-1)^{a_j + a_{i+j}}.$$

Alternatively, let A be the number of places where $a_0 \cdots a_{n-1}$ and the cyclic shift $a_i a_{i+1} \cdots a_{i-1}$ agree, and D the number of places where they disagree (so $A + D = n$). Then

$$\rho(i) = \frac{A - D}{n}. \quad (11)$$

For example, the pseudo-random sequence given in the first row of Fig. 6 has autocorrelation function $\rho(0) = 1$, $\rho(i) = -\frac{1}{15}$ for $1 \leqslant i \leqslant 14$, as shown in Fig. 9.

Property VII–Autocorrelation Function: The autocorrelation function of a pseudo-random sequence of length $n = 2^m - 1$ is given by

$$\rho(0) = 1$$

$$\rho(i) = -1/n, \qquad \text{for } 1 \leqslant i \leqslant 2^m - 2.$$

(For $a + a^{(i)} = a^{(j)}$ for some j by the shift-and-add property. Then D = number of 1's in $a^{(j)} = 2^{m-1}$, by Property IV, $A = n - D = 2^{m-1} - 1$, and the result follows from (11).)

It can be shown ([28, p. 48]) that this is the best possible autocorrelation function of any binary sequence of length $2^m - 1$, in the sense of minimizing $\max_{0<i<n} \rho(i)$.

E. Runs

Define a *run* to be a maximal string of consecutive identical symbols. For example, the first row of Fig. 6 contains runs of four 1's, three 0's, two 1's, two 0's, two runs of a single 1, and two runs of a single 0, for a total of 8 runs.

Property VIII–Runs: In any pseudo-random sequence, one-half of the runs have length 1, one-quarter have length 2, one-eighth have length 3, and so on, as long as these fractions give integral numbers of runs. In each case the number of runs of 0's is equal to the number of runs of 1's. (This follows easily from the window property—see [29, ch. 3].)

F. Random Sequences

Properties III, IV, VII, VIII justify the name pseudo-random sequences, for these are the properties that one would expect from a sequence obtained by tossing a fair coin $2^m - 1$ times. Such properties make pseudo-random sequences very useful in a number of applications, such as range-finding [17a], [28, ch. 6], [52a], synchronizing [28, ch. 8], [65], modulation [28, ch. 5], scrambling [18], [39], [46], [57], etc.

Of course these sequences are not random, and one way this shows up is that the properties we have mentioned hold for *every* pseudo-random sequence, whereas in a coin-tossing experiment there would be some variation from sequence to sequence. For this reason these sequences are unsuitable for serious encryption ([21], [22]).

Property IX–A Test to Distinguish a Pseudo-Random Sequence from a Coin-Tossing Sequence (E. N. Gilbert [23]*):* Let $c_0, \cdots, c_{N-1}$ be N consecutive binary digits from a pseudo-random sequence of length $2^m - 1$, where $N < 2^m - 1$, where $N < 2^m - 1$, and form the matrix

$$M = \begin{bmatrix} c_0 & c_1 & \cdots & c_{N-b} \\ c_1 & c_2 & \cdots & c_{N-b+1} \\ \cdots & \cdots & \cdots & \\ c_{b-1} & c_b & \cdots & c_{N-1} \end{bmatrix}$$

where $m < b < \frac{1}{2}N$. Then the rank of M over $GF(2)$ is less than b. (From Property II, since there are only m linearly independent sequences in $\mathcal{S}_m$.) On the other hand, if $c_0, \cdots, c_{N-1}$ is a segment of a coin-tossing sequence, where each c_i is 0 or 1 with probability $\frac{1}{2}$, then the probability that rank $(M) < b$ can be shown to be at most 2^{2b-N-1}. This is very small if $b << \frac{1}{2}N$.

Thus the question, "is rank $(M) = b$?" is a test on a small number of digits from a pseudo-random sequence which shows a departure from true randomness. For example, if $m = 11$, $2^m - 1 = 2047$, $b = 15$, the test will fail if applied to any $N = 50$ consecutive digits of a pseudo-random sequence, whereas the probability that a coin-tossing sequence fails is at most 2^{-21}.

G. Hadamard Matrices

Recall that a *Hadamard matrix* is a real $n \times n$ matrix H_n of +1's and −1's which satisfies

$$H_n H_n^T = nI \qquad (12)$$

where the T denotes transpose and I is an $n \times n$ unit matrix. (See for example Hall [35], Wallis *et al.* [71].)

Property X–Construction of Hadamard Matrices: Take the array whose rows are the sequences in $\mathcal{S}_m$, change 1's to −1's, and 0's to +1's, and add an initial column of +1's. The resulting $2^m \times 2^m$ array is a Hadamard matrix. (This follows from Property VII and equation (12).)

For example, Fig. 10 shows the 8 X 8 Hadamard matrix obtained in this way from the pseudo-random sequence 0010111.

Note that this Hadamard matrix can be constructed so that, except for the first row and column, it is a circulant matrix. For other Hadamard matrices with this property see [5], [67] and [28, Appendix 2].

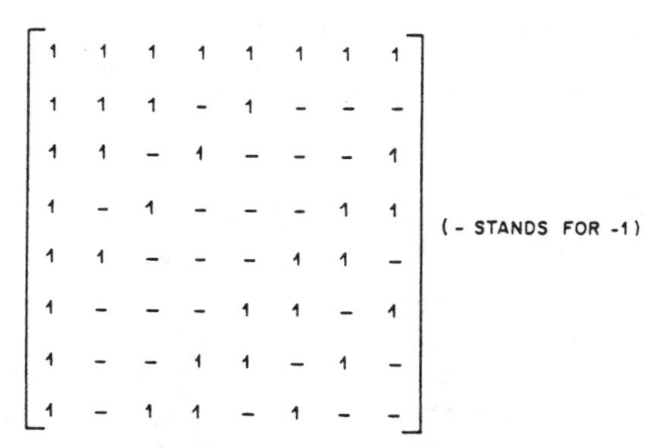

Fig. 10. An 8 X 8 Hadamard matrix.

H. Error-Correcting Codes

Definition: A binary linear *code* of *length n*, *dimension k*, and minimum *distance d* consists of 2^k binary vectors $u_1 \cdots u_n$, $u_i = 0$ or 1, called *codewords*, which: i) form a linear space (i.e. the modulo 2 sum of two codewords is a codeword), and ii) are such that any two codewords differ in at least d places. Such a code can correct $[\frac{1}{2}(d-1)]$ errors, where $[x]$ denotes the greatest integer not exceeding x. See for example [8], [44], or [52].

For example $\mathcal{C}_1 = [000, 011, 101, 110]$ is a code of length 3, dimension 2, and minimum distance 2.

Property XI–Pseudo-Random Sequences form a Simplex Code: The sequences in $\mathcal{S}_m$ form a linear code of length $2^m - 1$, dimension m, and minimum distance 2^{m-1}. All the nonzero codewords are cyclic shifts of any one of them. (From Properties, I, IV, V, and VI.)

For example the rows of Fig. 6 together with **0** form the code $\mathcal{S}_4$ of length 15, dimension 4, and minimum distance 8. Also $\mathcal{C}_1 = \mathcal{S}_2$.

Let the weight of a vector $v = v_1 \cdots v_n$, denoted by $wt(v)$, be the number of nonzero v_i's, and let the Hamming distance between vectors $u = u_1 \cdots u_n$, $v = v_1 \cdots v_n$, denoted by dist (u, v), be the number of i's such that $u_i \neq v_i$. Clearly dist $(u, v) = wt(u - v)$. For example, dist $(1100, 1010) = wt(0110) = 2$. Exactly the same definitions apply to nonbinary vectors. Condition ii) of the definition of a code requires that the codewords have Hamming distance at least d apart.

From Properties VII and IV any two codewords in $\mathcal{S}_m$ have Hamming distance exactly 2^{m-1} apart, and for this reason $\mathcal{S}_m$ is called a *simplex* code. (The codewords are at the vertices of a regular simplex.)

Let G_m be the $(2^m - 1) \times m$ matrix consisting of a list of the states of the shift register defined by $h(x)$, assuming that the initial state is $100 \cdots 0$. For example, G_4 is shown in Fig. 4. The columns of G_m are linearly independent. The set of all 2^m mod-2 linear combinations of these columns are exactly the sequences in $\mathcal{S}_m$. For this reason G_m is called a *generator matrix* for $\mathcal{S}_m$.

Definition: The *dual* code to $\mathcal{S}_m$, denoted by $\mathcal{S}_m^\perp$, consists of all binary vectors u such that $uG_m = 0 \pmod 2$.

Then $\mathcal{S}_m^\perp$ is a code of length $2^m - 1$ and dimension $2^m - 1 - m$. Since the rows of G_m are distinct, u is either zero or has at least three nonzero components. Hence $\mathcal{S}_m^\perp$ has minimum distance 3. In fact $\mathcal{S}_m^\perp$ is a Hamming single-error-correcting code (see [44, ch. 1]).

It is worth mentioning that coding theory provides the answers to the following questions, which sometimes arise in applications (see for example [9], [53], [66]):

i) Given a binary vector v of length $2^m - 1$, which pseudo-random sequence in $\mathcal{S}_m$ is closest to it in Hamming distance? A fast way to find the answer uses a discrete version of the fast Fourier transform–see [33], [34], [44, ch. 14], [53], [73].

ii) How far away can a vector be from the closest pseudo-random sequence in $\mathcal{S}_m$? In other words, what is

$$f(m) = \max_{v} \min_{u \in \mathcal{S}_m} \text{dist}(u, v)?$$

The vectors furthest away from $\mathcal{S}_m$ are called *bent*–see [44, ch. 14], [17], [45], [56].

I. Representation of Pseudo-Random Sequences by Polynomials

The pseudo-random sequences in $\mathcal{S}_m$ have a concise description by polynomials. We may represent any binary sequence $a = a_0 a_1 \cdots a_{n-1}$ of length n by the polynomial

$$a(x) = a_0 + a_1 x + a_2 x^2 + \cdots + a_{n-1} x^{n-1}. \tag{13}$$

For example 000100110101111 is represented by

$$a(x) = x^3 + x^6 + x^7 + x^9 + x^{11} + x^{12} + x^{13} + x^{14}. \tag{14}$$

A cyclic shift of a by one place to the right, $a_{n-1} a_0 a_1 \cdots a_{n-2}$, is represented by $a_{n-1} + a_0 x + \cdots + a_{n-2} x^{n-1}$. If we agree that $x^n = 1$, then this is just $xa(x)$. For

$$xa(x) = a_0 x + a_1 x^2 + \cdots + a_{n-2} x^{n-1} + a_{n-1} x^n$$
$$= a_{n-1} + a_0 x + \cdots + a_{n-2} x^{n-1}.$$

Multiplying by x corresponds to a cyclic shift to the right.

A primitive polynomial $h(x)$ of degree m always divides $x^n + 1$, where $n = 2^m - 1$ and the division is carried out modulo 2 ([44, ch. 4]). Let

$$\tilde{h}(x) = \sum_{k=0}^{m} h_k x^{m-k} = x^m h\left(\frac{1}{x}\right). \tag{15}$$

This is the *reciprocal polynomial* of $h(x)$, obtained by reversing the coefficients. Define

$$g(x) = \frac{x^n + 1}{\tilde{h}(x)}. \tag{16}$$

E.g. if $h(x) = x^4 + x + 1$, $\tilde{h}(x) = x^4 + x^3 + 1$, and

$$g(x) = \frac{x^{15} + 1}{x^4 + x^3 + 1} = 1 + x^3 + x^4 + x^6 + x^8 + x^9 + x^{10} + x^{11}. \tag{17}$$

Notice that (14) is x^3 times (17). This is a special case of

Property XII–Polynomial Representation: The pseudo-random sequences in $\mathcal{S}_m$ consist of the polynomials $x^i b(x)$, $i = 0, \cdots, 2^m - 2$, where $b(x)$ is any polynomial in $\mathcal{S}_m$ (from Properties I or XI). Alternatively, they are the polynomials $t(x)g(x)$ where $t(x)$ is any polynomial of degree less than m.

(For let $b(x) = t(x)g(x) = \sum_{j=0}^{n-1} b_j x^j$. Then $b(x)\tilde{h}(x) = t(x)(x^m + 1)$. Equating coefficients of x^{m+i} in this identity gives

$$\sum_{k=0}^{m} b_{i+k} h_k = 0, \qquad i = 0, \cdots, n - m - 1.$$

But this is the recurrence of Property II, so $b(x) \in \mathcal{S}_m$. Since $g(x), xg(x), \cdots, x^{m-1} g(x)$ represent m linearly independent sequences in $\mathcal{S}_m$, everything in $\mathcal{S}_m$ can be represented by $t(x)g(x)$ where $\deg t(x) < m$.)

Hence $g(x)$ is distinguished by being the polynomial of lowest degree in $\mathcal{S}_m$, and is called the *generator polynomial* of $\mathcal{S}_m$.

J. Finite Fields

A *field* is a set of elements in which it is possible to add, subtract, multiply, and divide. For any α, β, γ in the field we must have

$$\alpha + \beta = \beta + \alpha, \quad \alpha\beta = \beta\alpha$$
$$\alpha + (\beta + \gamma) = (\alpha + \beta) + \gamma, \quad \alpha(\beta\gamma) = (\alpha\beta)\gamma$$
$$\alpha(\beta + \gamma) = \alpha\beta + \alpha\gamma$$

and, furthermore, elements $0, 1, -\alpha$, and (for $\alpha \neq 0$) α^{-1} must exist such that

$$0 + \alpha = \alpha, \quad (-\alpha) + \alpha = 0, \quad 0\alpha = 0$$
$$1\alpha = \alpha, \quad (\alpha^{-1})\alpha = 1.$$

A *finite* field contains a finite number of elements. A field with q elements is called a *Galois field* and is denoted by $GF(q)$.

The simplest fields are the following: Let p be a prime number. Then the integers modulo p form the field $GF(p)$. The elements of $GF(p)$ are $\{0, 1, 2, \cdots, p - 1\}$, and $+, -, \times, \div$ are carried out mod p. For example, $GF(2)$ is the binary field $\{0, 1\}$. $GF(3)$ is the ternary field $\{0, 1, 2\}$, with $1 + 2 = 3 = 0 \pmod 3$, $2 \cdot 2 = 4 = 1 \pmod 3$, $1 - 2 = -1 = 2 \pmod 3$, etc.

A field with p^m elements, when p is a prime and m is any positive integer, can be constructed as follows. The elements of the field are all polynomials in x of degree $\leqslant m - 1$ with coefficients from $GF(p)$. Addition is done in the ordinary way, modulo p.

A polynomial with coefficients from $GF(p)$ which is not the product of two polynomials of lower degree is called *irreducible*. A primitive polynomial is automatically irreducible. Choose a fixed irreducible polynomial $h(x)$ of degree m.

Then the product of two field elements is obtained by multiplying them in the usual way (modulo p), and taking the remainder when divided by $h(x)$. The field obtained in this way contains p^m elements and is denoted by $GF(p^m)$.

As in illustration we take $p = 2$, $m = 4$, $h(x) = x^4 + x + 1$ and construct $GF(2^m)$. The 16 elements are

$$0, 1, x, x + 1, x^2 + 1, \cdots, \quad x^4 + x^3 + x^2 + x + 1.$$

Addition is like this

$$(x^2 + 1) + (x + 1) = x^2 + x$$

and multiplication like this: The ordinary product of $x^3 + 1$ and $x^3 + x + 1$ is $x^6 + x^4 + x + 1$. But

$$x^6 + x^4 + x + 1 = (x^2 + 1)h(x) + x^3 + x^2$$

so in the field

$$(x^3 + 1)(x^3 + x + 1) = x^3 + x^2.$$

It is easy to check that this set has all the properties of a field; the inverses $a(x)^{-1}$ exist because $h(x)$ is irreducible (see [44, chs. 3, 4], [4], [6], [8] or [15] for details).

Just as complex numbers can be written either in rectangular coordinates, as $z = x + iy$, or in polar coordinates, as $z = re^{i\theta}$, so elements of a finite field have two representations. The first is the polynomial form given above, and the second is as a power of a certain fixed field element called a *primitive* element, denoted by ξ. (If $h(x)$ is a primitive polynomial we can take ξ to be a zero of $h(x)$.)

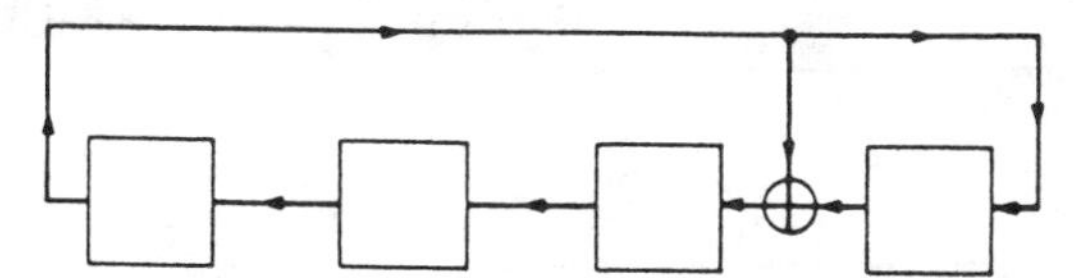

Fig. 11. A shift register generating the elements of $GF(2^4)$.

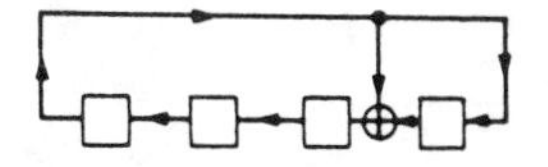

AS A POLYNOMIAL				AS A POWER OF ξ
0	0	0	1	1
0	0	1	0	ξ
0	1	0	0	ξ^2
1	0	0	0	ξ^3
0	0	1	1	ξ^4
0	1	1	0	ξ^5
1	1	0	0	ξ^6
1	0	1	1	ξ^7
0	1	0	1	ξ^8
1	0	1	0	ξ^9
0	1	1	1	ξ^{10}
1	1	1	0	ξ^{11}
1	1	1	1	ξ^{12}
1	1	0	1	ξ^{13}
1	0	0	1	ξ^{14}
0	0	0	1	$\xi^{15}=1$
...				...

Fig. 12. The elements of the field $GF(2^4)$.

Assuming that $h(x)$ is a primitive polynomial, the correspondence between the two forms can be obtained from a shift register. This shift register is the reverse of that in Fig. 2. E.g. if $h(x) = x^4 + x + 1$, the shift register is shown in Fig. 11. The states of this shift register provide a list of the elements of $GF(2^4)$, both as successive powers of ξ and as polynomials in x of degree at most three (see Fig. 12, where only the coefficients of the polynomials are given).

A finite field of order p^m exists for all primes p and positive integers m. Furthermore these are the only finite fields (see [44, ch. 4]).

If $\beta \in GF(p^m)$, the sum

$$T_m(\beta) = \beta + \beta^p + \beta^{p^2} + \cdots + \beta^{p^{m-1}}$$

is called the *trace* of β. The trace has the following properties:

i) $T_m(\beta) \in GF(p)$.
ii) $T_m(\beta + \gamma) = T_m(\beta) + T_m(\gamma), \quad \beta, \gamma \in GF(p^m)$.
iii) $T_m(\beta)$ takes on each value in $GF(p)$ equally often, i.e., p^{m-1} times, as β ranges over $GF(p^m)$.
iv) $T_m(\beta^p) = T_m(\beta)^p = T_m(\beta)$.
v) $T_m(1) \equiv m \pmod p$.

K. Pseudo-Random Sequences Form a Field

As before, let $\mathcal{S}_m$ be the set of pseudo-random sequences of length $n = 2^m - 1$, obtained from a primitive polynomial $h(x)$ of degree m. It is clear how to add two elements of $\mathcal{S}_m$. Multiplication is carried out using the polynomial representation given in Section II-I. That is, two polynomials are multiplied in the usual way (mod 2), and then x^n is replaced by 1, x^{n+1} by x, and so on. The result is another element of $\mathcal{S}_m$.

Property XIII – $\mathcal{S}_m$ is a Field: With this definition of addition and multiplication, $\mathcal{S}_m$ is isomorphic to the field $GF(2^m)$.

Let ξ be a primitive element of $GF(2^m)$ which is a zero of $h(x)$, as in the previous section. Then ξ^{-1} is a zero of $\tilde{h}(x)$, and $g(\xi^{-1}) \neq 0$. From Property XII, $b(\xi^{-1}) \neq 0$ for all nonzero $b(x) \in \mathcal{S}_m$.

The isomorphism between $\mathcal{S}_m$ and $GF(2^m)$ is given by

$$b(x) \in \mathcal{S}_m \xrightarrow{\varphi} b(\xi^{-1}) \in GF(2^m)$$

and

$$\gamma \in GF(2^m) \xrightarrow{\psi} (b_0 b_1 \cdots b_{n-1}) \in \mathcal{S}_m$$

where $b_i = T_m(\gamma\xi^i)$. For the proof of all this see [44, ch. 8].

The element $E(x) = \sum_{i=0}^{n-1} E_i x^i$ of $\mathcal{S}_m$ which maps onto $1 \in GF(2^m)$ must satisfy $E(x)^2 = E(x)$. This element is called the *idempotent* of $\mathcal{S}_m$ and has the property that $E_i = E_{2i}$ (subscripts modulo n) [27]. It is found as follows: Since n is odd, $x^n + 1$ has distinct factors over $GF(2)$. Therefore $g(x)$ and $\tilde{h}(x)$ are relatively prime and so there exist polynomials $u(x)$, $v(x)$ such that

$$u(x)g(x) + v(x)\tilde{h}(x) \equiv 1 \pmod 2 \tag{18}$$

and $\deg u(x) < m$ ([47], [70]). Setting $x^n = 1$ gives

$$g(x)\tilde{h}(x) = 0$$

and, multiplying (18) by $u(x)g(x)$,

$$(u(x)g(x))^2 = u(x)g(x).$$

Therefore $E(x) = u(x)g(x)$ is the idempotent of $\mathcal{S}_m$. Then

$$E(x) \xrightarrow{\varphi} 1$$

$$xE(x) \xrightarrow{\varphi} \xi^1.$$

For any $b(x) \in \mathcal{S}_m$ there is a $c(x) \in \mathcal{S}_m$ such that $b(x)c(x) = E(x)$. In fact if $b(x) = x^i E(x)$ (for every nonzero sequence in $\mathcal{S}_m$ is a cyclic shift of $E(x)$) then $c(x) = x^{n-i}E(x)$.

Example: With $h(x) = x^4 + x + 1$, $\tilde{h}(x) = x^4 + x^3 + 1$, and $g(x)$ given by (17), we find

$$x^3 g(x) + (1 + x^4 + x^6 + x^8 + x^{10})\tilde{h}(x) = 1$$

so that $E(x) = x^3 g(x)$, given by (14), is the idempotent.

III. Pseudo-Random Arrays

A. Two-Dimensional Arrays with Flat Autocorrelation Functions

These can be constructed by folding pseudo-random sequences. Take a number of the form $n = 2^{k_1 k_2} - 1$ such that $n_1 = 2^{k_1} - 1$ and[1] $n_2 = n/n_1$ are relatively prime and greater than 1. Then $n = n_1 n_2$. Examples are

$n = 15 = 2^4 - 1$ with $k_1 = k_2 = 2, \ n_1 = 3, \ n_2 = 5$

$n = 63 = 2^6 - 1$ with $k_1 = 3, \ k_2 = 2, \ n_1 = 7, \ n_2 = 9$

$n = 511 = 2^9 - 1$ with $k_1 = 3, \ k_2 = 3, \ n_1 = 7, \ n_2 = 73.$

The starting point is a pseudo-random sequence $\boldsymbol{a} =$

[1] $2^{k_1} - 1$ always divides $2^{k_1 k_2} - 1$.

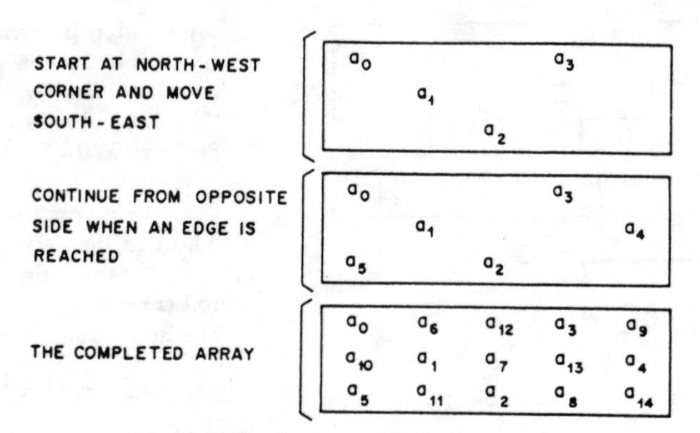

Fig. 13. Constructing a pseudo-random array.

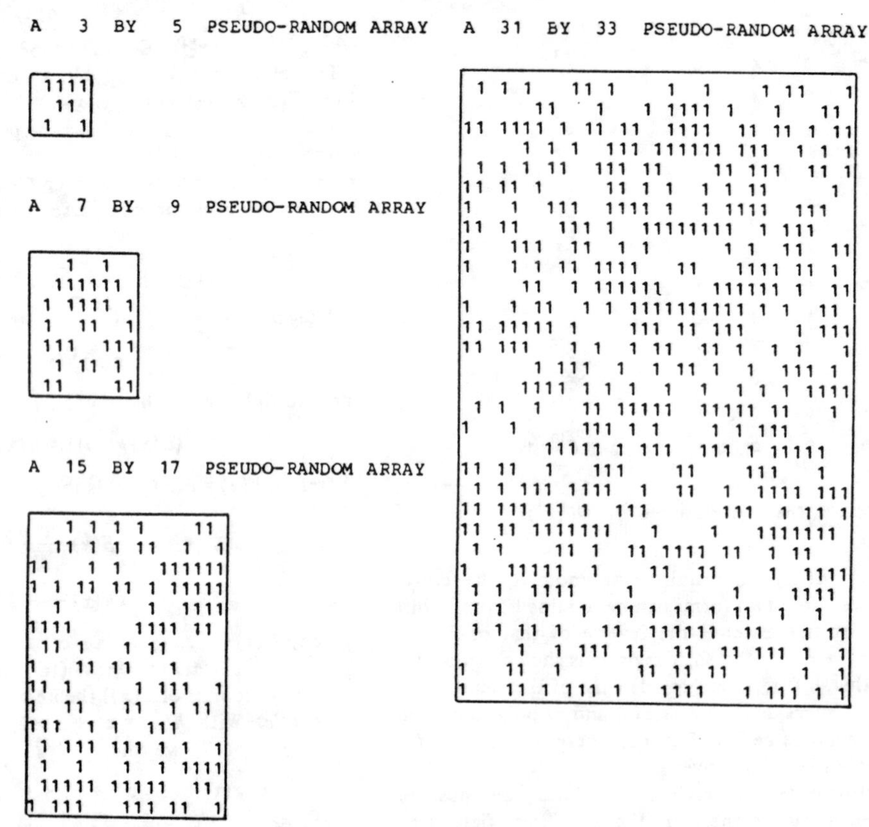

Fig. 14. Examples of pseudo-random arrays.

$a_0 a_1 \cdots a_{n-1}$ in $\mathcal{S}_{k_1 k_2}$, obtained from a primitive polynomial $h(x)$ of degree $m = k_1 k_2$. We use a to fill up an $n_1 \times n_2$ array, by writing a down the main diagonal and continuing from the opposite side whenever an edge is reached. Fig. 13 shows the construction when $n = 15$, $n_1 = 3$, $n_2 = 5$. Thus the pseudo-random sequence

$$000100110101111 \tag{19}$$

produces the array

$$\begin{bmatrix} 0 & 1 & 1 & 1 & 1 \\ 0 & 0 & 1 & 1 & 0 \\ 0 & 1 & 0 & 0 & 1 \end{bmatrix}. \tag{20}$$

This is the top array in Fig. 14.

An $n_1 \times n_2$ array formed in this way will be called a *pseudo-random array*. There are n different arrays, one from each pseudo-random sequence in $\mathcal{S}_{k_1 k_2}$. Fig. 14 shows examples of 3×5, 7×9, 15×17, and 31×33 pseudo-random arrays, the first three constructed from pseudo-random sequences in Fig. 7 and the fourth from the sequence defined by $h(x) = x^{10} + x^3 + 1$ (from Fig. 5). In Fig. 14 zeros have been replaced by blanks; in the first two arrays the left-hand column is completely blank.

The pseudo-random sequences in $\mathcal{S}_{k_1 k_2}$ at least look random to the eye, whereas the corresponding arrays have a conspicuous nonrandom feature, being symmetric about a column of zeros. Nevertheless we call them pseudo-random arrays because they share Properties III*, IV* and VII* below with an array formed by tossing a coin.

To state these properties some further notation is required. If the pseudo-random array is denoted by

$$b = \begin{bmatrix} b_{00} & b_{01} & \cdots & b_{0,n_2-1} \\ b_{10} & b_{11} & \cdots & b_{1,n_2-1} \\ \cdots & \cdots & & \cdots \\ b_{n_1-1,0} & & \cdots & b_{n_1-1,n_2-1} \end{bmatrix} \quad (21)$$

then

$$a_0 = b_{00}$$
$$a_1 = b_{11}$$
$$a_2 = b_{22}$$
$$\cdots$$

and

$$a_i = b_{i_1 i_2} \quad (22)$$

where

$$i \equiv i_1 \pmod{n_1}, \quad 0 \leqslant i_1 < n_1$$
$$i \equiv i_2 \pmod{n_2}, \quad 0 \leqslant i_2 < n_2. \quad (23)$$

Conversely, given i_1 and i_2 with $0 \leqslant i_1 < n_1$ and $0 \leqslant i_2 < n_2$, there is a unique value of i in the range $0 \leqslant i < n = n_1 n_2$ such that (23) holds, by the Chinese Remainder Theorem (since n_1 and n_2 are relatively prime)—see [47, p. 33] or [70, pp. 189–191].

These arrays are best described by polynomials. The array (21) is represented by

$$b(x, y) = \sum_{i=0}^{n_1-1} \sum_{j=0}^{n_2-1} b_{ij} x^i y^j. \quad (24)$$

E.g. (20) is

$$b(x, y) = (y + y^2 + y^3 + y^4) + x(y^2 + y^3) + x^2(y + y^4). \quad (25)$$

If we agree that $x^{n_1} = 1$ and $y^{n_2} = 1$, then multiplying $b(x, y)$ by x corresponds to a cyclic shift of the array downwards, and multiplying by y corresponds to a cyclic shift to the right.

B. Properties of Pseudo-Random Arrays

With n, n_1, n_2, k_1, k_2 as defined above, let $h(z)$ be a fixed primitive polynomial of degree $m = k_1 k_2$. Each of the pseudo-random sequences in $\mathcal{S}_m$ folds into a distinct $n_1 \times n_2$ pseudo-random array. Let $\mathcal{A}_m$ be the set of 2^m arrays so formed, together with the zero array. We briefly state the properties of these arrays.

Property XII–Polynomial Representation:* Suppose the pseudo-random sequence a folds up to produce the array b. The polynomial $b(x, y)$ describing b is obtained from $a(z) = \sum_{i=0}^{n-1} a_i z^i$ by replacing z by xy:

$$b(x, y) = a(xy). \quad (26)$$

(For this replaces each term $a_i z^i$ in $a(z)$ by $a_i x^{i_1} y^{i_2}$ where $i \equiv i_1 \pmod{n_1}$, $0 \leqslant i_1 < n_1$ and $i \equiv i_2 \pmod{n_2}$, $0 \leqslant i_2 < n_2$. Thus a_i goes into the square (i_1, i_2) of the array: $b_{i_1 i_2} = a_i$, in agreement with (22).)

E.g. (19) is described by

$$a(z) = z^3 + z^6 + z^7 + z^9 + z^{11} + z^{12} + z^{13} + z^{14}$$

and replacing z by xy with $x^3 = y^5 = 1$ we obtain

$$b(x, y) = a(xy) = y^3 + y + xy^2 + y^4 + x^2 y + y^2 + xy^3 + x^2 y^4$$

in agreement with (25).

Property I–The Shift Property:* If b (equation (21)) is a pseudo-random array in $\mathcal{A}_m$, so is any cyclic shift of b downwards or to the right,

$$\begin{bmatrix} b_{i0} & \cdots & b_{i,n_2-1} \\ \cdots & \cdots & \cdots \\ b_{i-1,0} & \cdots & b_{i-1,n_2-1} \end{bmatrix} \text{ or } \begin{bmatrix} b_{0j} & \cdots & b_{0,j-1} \\ \cdots & \cdots & \cdots \\ b_{n_1-1,j} & \cdots & b_{n_1-1,j-1} \end{bmatrix} \quad (27)$$

or any combination of these shifts.

(Since n_1 and n_2 are relatively prime there are integers μ and ν such that

$$\mu n_1 + \nu n_2 = 1. \quad (28)$$

E.g. if $n_1 = 3$, $n_2 = 5$, $2 \cdot 3 - 5 = 1$, and $\mu = 2$, $\nu = -1$. Then

$$z^{\nu n_2} = x^{\nu n_2} y^{\nu n_2} = x^{\nu n_2} = x^{\nu n_2} x^{\mu n_1} = x$$

and similarly $z^{\mu n_1} = y$. Therefore the arrays (27) are $x^i b(x, y) = z^{i \nu n_2} a(z)$ and $y^j b(x, y) = z^{j \mu n_1} a(z)$, both in $\mathcal{A}_m$.)

The pseudo-random arrays in $\mathcal{A}_m$ are represented by the polynomials $x^{i_1} y^{i_2} b(x, y)$, $0 \leqslant i_1 < n_1$, $0 \leqslant i_2 < n_2$, where $b(x, y)$ is any array in $\mathcal{A}_m$. Alternatively they are the polynomials $t(x, y)\gamma(x, y)$, where $\deg_x t(x, y) < n_1$, $\deg_y t(x, y) < n_2$, $\gamma(x, y) = g(xy)$, and $g(z)$ is the generator polynomial of $\mathcal{S}_m$. E.g. in $\mathcal{A}_4$ from (17) we obtain

$$\gamma(x, y) = (1 + y + y^3 + y^4) + x(1 + y^4) + x^2(y + y^3)$$

corresponding to

$$\begin{bmatrix} 1 & 1 & 0 & 1 & 1 \\ 1 & 0 & 0 & 0 & 1 \\ 0 & 1 & 0 & 1 & 0 \end{bmatrix}.$$

Property II–Recurrences:* Equation (9) implies that a pseudo-random array in $\mathcal{A}_m$ satisfies a recurrence along the diagonals. E.g. from (4), (20) satisfies

$$b_{i+4,i+4} = b_{i+1,i+1} + b_{i,i}.$$

It is also possible to find a *pair* of recurrences which generate the array, one for moving vertically and one horizontally. We illustrate this by two examples, but omit the general proof. For (20) the recurrences are

$$b_{i+2,j} = b_{i+1,j} + b_{i,j} \quad \text{(vertically)}$$
$$b_{i,j+2} = b_{i+2,j+1} + b_{i,j} \quad \text{(horizontally)}. \quad (29)$$

For the 15×17 array in Fig. 14 they are

$$b_{i+4,j} = b_{i+1,j} + b_{i,j} \quad \text{(vertically)}$$
$$b_{i,j+2} = b_{i+1,j+1} + b_{i,j+1} + b_{i,j} \quad \text{(horizontally)}. \quad (30)$$

From these recurrences the $k_1 \times k_2$ subarray in the North-West corner determines the entire array.

Property III–The Window Property:* If a $k_1 \times k_2$ window is slid over a pseudo-random array in $\mathcal{A}_m$, each of the $2^{k_1 k_2} - 1$ nonzero binary $k_1 \times k_2$ arrays is seen exactly once—see Fig. 15 for the case $n = 15$, $k_1 = k_2 = 2$. (The proof is given in the Appendix.)

Property IV–Half 0's and Half 1's:* Any array in $\mathcal{A}_m$ contains 2^{m-1} 1's and $2^{m-1} - 1$ 0's.

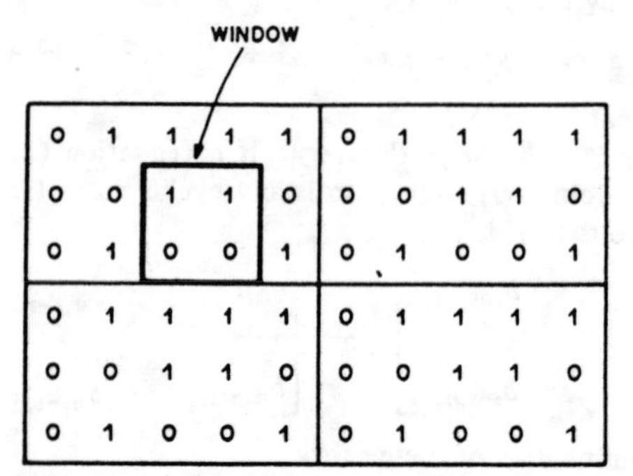

Fig. 15. The window property: Every nonzero 2 X 2 array is seen once.

i \ j	0	1	2	3	4	5	6	
0	1	$-\frac{1}{15}$	$-\frac{1}{15}$	$-\frac{1}{15}$	$-\frac{1}{15}$	1	$-\frac{1}{15}$	...
1	$-\frac{1}{15}$	$-\frac{1}{15}$	$-\frac{1}{15}$	$-\frac{1}{15}$	$-\frac{1}{15}$	$-\frac{1}{15}$	$-\frac{1}{15}$	...
2	$-\frac{1}{15}$	$-\frac{1}{15}$	$-\frac{1}{15}$	$-\frac{1}{15}$	$-\frac{1}{15}$	$-\frac{1}{15}$	$-\frac{1}{15}$	...
3	1	$-\frac{1}{15}$	$-\frac{1}{15}$	$-\frac{1}{15}$	$-\frac{1}{15}$	1	$-\frac{1}{15}$	...
	...				...			

Fig. 16. Autocorrelation function $\rho(i,j)$ of a pseudo-random array.

Property V–The Addition Property:* The sum of two arrays in $\mathcal{A}_m$ is again in $\mathcal{A}_m$.

Property VI–The Shift-and-Add Property:* The sum of $b(x,y) \in \mathcal{A}_m$ and any shift $x^i y^j b(x,y)$ is another array in $\mathcal{A}_m$. (These follow from Properties IV, V, VI.)

C. Autocorrelation Function

Let b (equation (21)) be any $n_1 \times n_2$ array of area $n = n_1 \times n_2$. The *(two-dimensional) autocorrelation function* of b, $\rho(i,j)$, is defined as follows. If A is the number of positions in which b and b shifted i places down and j to the right agree, and D is the number in which they disagree, then

$$\rho(i,j) = \frac{A-D}{n}, \qquad i,j = 0, \quad \pm 1, \pm 2, \cdots. \tag{31}$$

Again $A + D = n$. This is a doubly periodic function with $\rho(i,j) = \rho(i+n_1, j) = \rho(i, j+n_2)$, $\rho(0,0) = \rho(n_1, 0) = \cdots = 1$. For example, Fig. 16 shows the autocorrelation function of (20).

Property VII–Autocorrelation Function:* The two-dimensional autocorrelation function of a pseudo-random array in $\mathcal{A}_m$ is given by

$$\rho(0,0) = 1$$

$$\rho(i,j) = -\frac{1}{n}, \qquad 0 \leq i < n_1, \qquad 0 \leq j < n_2,$$
$$(i,j) \neq (0,0). \tag{32}$$

(From Properties VI*, IV*). Again this is best possible.

D. Other Constructions of Pseudo-Random Arrays

i) A pseudo-random sequence of length $n = 2^m - 1$ can be folded into an $n_1 \times n_2$ array as shown in Fig. 13 whenever $n = n_1 n_2$ and n_1, n_2 are relatively prime. The resulting array still has Properties XII*, I*, IV*, V*, and VI*.

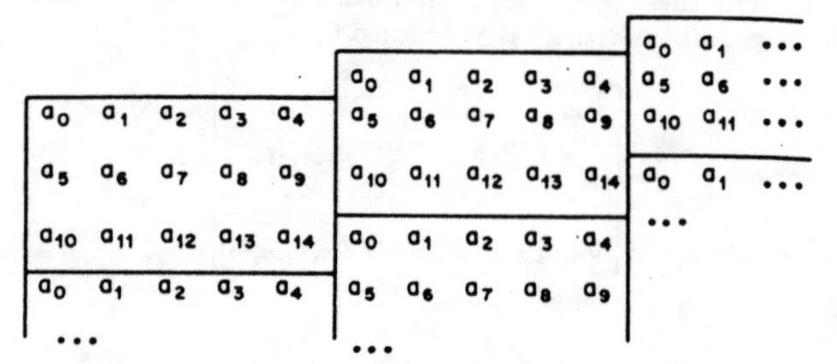

Fig. 17. Another construction of a pseudo-random array.

deg	$q = 3$	$q = 4$	$q = 8$
1	$x+1$	$x+\omega$	$x+\alpha$
2	x^2+x+2	$x^2+x+\omega$	$x^2+\alpha x+\alpha$
3	x^3+2x+1	$x^3+x^2+x+\omega$	$x^3+x+\alpha$
4	x^4+x+2	$x^4+x^2+\omega x+\omega^2$	$x^4+x+\alpha^3$
5	x^5+2x+1	$x^5+x+\omega$	$x^5+x^2+x+\alpha^3$
6	x^6+x+2	$x^6+x^2+x+\omega$	$x^6+x+\alpha$
7	$x^7+x^6+x^4+1$	$x^7+x^2+\omega x+\omega^2$	$x^7+x^2+\alpha x+\alpha^3$
8	x^8+x^5+2	$x^8+x^3+x+\omega$	
9	$x^9+x^7+x^5+1$	$x^9+x^2+x+\omega$	
10	$x^{10}+x^9+x^7+2$	$x^{10}+x^3+\omega(x^2+x+1)$	

Fig. 18. Primitive polynomials over $GF(q)$.

ii) Another way of making pseudo-random sequences into arrays has been suggested by Spann [63]. This applies when the whole plane is to be filled with copies of the array. Let $a_0 \cdots a_{n-1}$ be a pseudo-random sequence of length $n = 2^m - 1 = n_1 n_2$, where n_1 and n_2 may have a common factor. The infinite array $\{b_{ij}: i,j = 0, \pm 1, \cdots\}$ is formed by setting $b_{ij} = a_k$, where $k \equiv in_1 + j \pmod{n_2}$ and $0 \leq k \leq n-1$. Fig. 17 illustrates the construction in the case $n = 15$, $n_1 = 3$, $n_2 = 5$. The autocorrelation function of the $n_1 \times n_2$ rectangle $\{b_{ij}: 0 \leq i < n_1, 0 \leq j < n_2\}$ is not in general given by (32). However if we define the periodic autocorrelation function of the infinite array to be

$$\overline{\rho}(i,j) = \frac{1}{n} \sum_{r=0}^{n_1-1} \sum_{s=0}^{n_2-1} (-1)^{b_{rs} + b_{r+i,s+j}} \tag{33}$$

then $\overline{\rho}(i,j)$ *is* given by (32).

IV. Nonbinary Pseudo-Random Sequences and Displays

A. Arrays with Entries from an Alphabet of q Symbols

The same constructions can be used to obtain pseudo-random sequences and arrays with entries which, instead of being 0's and 1's, are taken from an alphabet of q symbols, for any q which is a prime or a power of a prime.

Let q be any prime power and let $GF(q)$ be the Galois field with q elements (see Section II-J). We need a polynomial $h(x)$ of degree m with coefficients from $GF(q)$ which i) is not the product of two such polynomials of lower degree, and ii) has as a zero a primitive element of $GF(q^m)$, say ξ. We call such a polynomial *primitive over* $GF(q)$. The primitive polynomials used in Section II are primitive over $GF(2)$. Fig. 18 gives a small table of primitive polynomials, using $GF(4) = \{0, 1, \omega, \omega^2\}$, with $\omega^2 + \omega + 1 = 0$, $\omega^3 = 1$, and

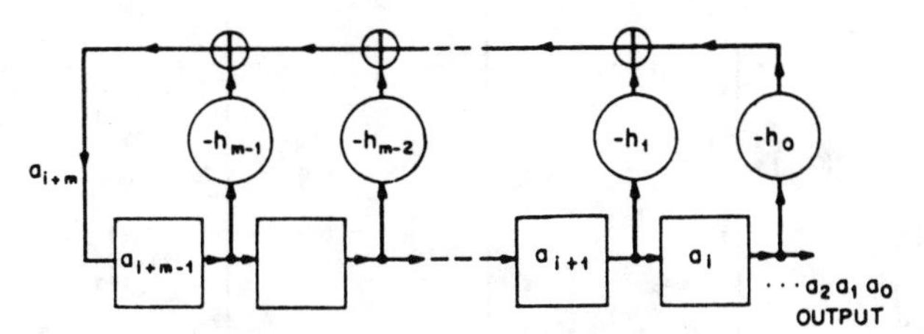

Fig. 19. Shift register specified by $h(x)$.

length 8 over GF(3)

01220211

length 26 over GF(3)

00101211201110020212210222

length 80 over GF(3)

00010021011120022010221101012122120122220002001202
221001102011220202121121021111

length 242 over GF(3)

00001000120011101002120221120201121101202111220010
10121211212012111120001100102012211102002210202221
20021102202201201111100002000210022202001210112210
10221220210122211002020212122121021222210002200201
021122201001120101112100122011011021022222

length 15 over GF(4)

011310221203323

length 63 over GF(4)

00110312223221020213100220123331332030321200330231
1121130101323

length 255 over GF(4)

00010122022311110013113123232223010320312020332213
30133112131030012101321003332011031000303110112333
30032332312121112030210231010221132203223313230200
31303213002221033023000202330331222200212212313133
31020130123030113321102112232120100232021320011130
22012

Fig. 20. Pseudo-random sequences over $GF(3)$ and $GF(4)$.

$GF(8) = \{0, 1, \alpha, \alpha^2, \cdots, \alpha^6\}$ with $\alpha^3 + \alpha + 1 = 0$, $\alpha^7 = 1$. For more extensive tables see [1], [7], [15], [32], [46], [52], [64].

Suppose

$$h(x) = x^m + h_{m-1}x^{m-1} + \cdots + h_1 x + h_0 \tag{34}$$

with $h_i \in GF(q)$, $h_0 \neq 0$. The shift register specified by $h(x)$ is shown in Fig. 19. In this figure the boxes contain elements of $GF(q)$, say $a_{i+m-1}, \cdots, a_i$. The feedback path then forms

$$a_{i+m} = -h_{m-1}a_{i+m-1} - h_{m-2}a_{i+m-2} - \cdots - h_1 a_{i+1} - h_0 a_i. \tag{35}$$

02132333113332312
00133210330123310
03002231331322003
01212032002302121
03320301111030233
03213111221113123
00211320110231120
01003312112133001
02323013003103232
01130102222010311
01321222332221231
00322130220312230
02001123223211002
03131021001201313
02210203333020122

Fig. 21. A 15×17 pseudo-random array with entries from the field of 4 elements.

Equation (35) is the recurrence describing the output sequence. This is an infinite sequence of period $q^m - 1$ (if the starting state is not zero), and each nonzero state appears once in a period. A segment of the output sequence of length $q^m - 1$ is called a *pseudo-random sequence over* $GF(q)$. Let $\mathcal{S}_m(q)$ be the set of such sequences, together with **0**. (See [3], [11], [14], [24], [25], [60], [76].)

For example, if $q = 4$, $m = 2$, and $h(x) = x^2 + x + \omega$ we obtain the pseudo-random sequence

$$0\ 1\ 1\ \omega^2\ 1\ 0\ \omega\ \omega\ 1\ \omega\ 0\ \omega^2\ \omega^2\ \omega\ \omega^2 \tag{36}$$

of length 15. Then $\mathcal{S}_2(4)$ consists of **0** and the 15 cyclic shifts of (36). Fig. 20 shows some pseudo-random sequences over $GF(3)$ and $GF(4)$, where in $GF(4)$ ω is replaced by 2 and ω^2 by 3.

Pseudo-random arrays can be obtained as in Section III. For example Fig. 21 shows a 15×17 array over $GF(4)$ obtained by folding the length 255 sequence in Fig. 20.

B. Some Properties.

Properties I, II, III, V, VI hold with obvious changes, while Property IV is replaced by:

*Property IV**:* In any pseudo-random sequence in $\mathcal{S}_m(q)$ 0 occurs $q^{m-1} - 1$ times and every nonzero element of $GF(q)$ occurs q^{m-1} times.

Instead of Property VII we have the following.

Property VIIa: A pseudo-random sequence in $\mathcal{S}_m(q)$ has the form

$$a = b, \gamma b, \gamma^2 b, \cdots, \gamma^{q-2} b, \tag{37}$$

where b is a sequence of length $(q^m - 1)/(q - 1)$ and γ is a primitive element of $GF(q)$. (This is because the states of the shift register can be made to correspond to a logarithm table of $GF(q)$. The details are omitted).

For example in (36), $b = 011\omega^2 1$ and $\gamma = \omega$. For some applications b is a more useful sequence than a.

Property VIIb: Let $a = (a_0 \cdots a_{n-1})$ be a pseudo-random sequence in $\mathcal{S}_m(q)$, and let $b = (a_s a_{s+1} \cdots a_{s-1}) = (b_0 \cdots b_{n-1})$ be a shift of a by s places. i) If s is not a multiple of $q - 1$, then among the $q^m - 1$ pairs (a_i, b_i), $(0, 0)$ occurs $q^{m-2} - 1$ times and every other pair of elements of $GF(q)$ occurs q^{m-2} times; ii) If $s = j(q - 1)$, then $(0, 0)$ occurs $q^{m-1} - 1$ times and the pairs $(\alpha, \gamma^j \alpha)$, for all nonzero α in $GF(q)$, occur q^{m-1} times.

In particular, if $s = 0$ (no shift) each (α, α), $\alpha \neq 0$, occurs q^{m-1} times. For example, if (36) is shifted once, thus

$$0\ 1\ 1\ \omega^2\ 1\ 0\ \omega\ \omega\ 1\ \omega\ 0\ \omega^2\ \omega^2\ \omega\ \omega^2$$
$$1\ 1\ \omega^2\ 1\ 0\ \omega\ \omega\ 1\ \omega\ 0\ \omega^2\ \omega^2\ \omega\ \omega^2\ 0$$

then (0, 0) does not occur and every other pair (0, 1), $(0, \omega)$, $\cdots$, (ω^2, ω^2) occurs once.

There are many ways of obtaining a real- or complex-valued sequence from a sequence in $\mathcal{S}_m(q)$. We shall just consider the case q = prime and the complex-valued sequence $\hat{a}$ obtained from $a \in \mathcal{S}_m(q)$ by replacing $r \in GF(q)$ by $e^{2\pi i r/q}$.

Property VIIc–Autocorrelation function: The autocorrelation function of $\hat{a}$ is given by

$$\rho(0) = 1, \qquad \rho(i) = -\frac{1}{q^m - 1}, \qquad 1 \leq i \leq q^m - 2.$$

(Immediate from Property VIIb.) For other autocorrelation functions, all of which can be calculated from Property VIIb, see [3], [11], [14], [25], [76].

V. Transmission Functions

The following question has arisen in certain optical applications (Knowlton [43]). For a given value q, say $q = 2, 3, 4, \cdots$, find a set of q $n_1 \times n_2$ arrays of 0's and 1's whose transmission function $\tau(r, s)$ is zero at the origin and approximately constant elsewhere. The *transmission function* τ is defined as follows. If the q arrays are (a_{ij}), (b_{ij}), (c_{ij}), $\cdots$ then

$$\tau(r, s) = \frac{1}{n_1 n_2} \sum_{i=0}^{n_1-1} \sum_{j=0}^{n_2-1} \cdot (1 - a_{ij})(1 - b_{i+r,j+s})(1 - c_{i+2r,j+2s}) \cdots \quad (38)$$

where the first subscript is taken modulo n_1, and the second modulo n_2, and the sum is evaluated as a real number. This is doubly periodic: $\tau(r, s) = \tau(r + n_1, s) = \tau(r, s + n_2)$.

Arrays with these properties can be obtained from pseudo-random sequences, as will be shown.

Case $q = 2$: Let $(a_{ij}) \in \mathcal{a}_m$ be an $n_1 \times n_2$ pseudo-random array as constructed in Section III, where $n = n_1 n_2 = 2^m - 1$, and let (b_{ij}) be its complement: $b_{ij} = 1 - a_{ij}$. Then $\tau(0, 0) = 0$ and

$$\tau(r, s) = \frac{1}{n} (\text{number of times } a_{ij} = 0 \text{ and } a_{i+r,j+s} = 1)$$
$$= \frac{2^{m-2}}{2^m - 1} (\text{from Properties I, IV, V}) \quad (39)$$

for all $0 \leq r \leq n_1 - 1$, $0 \leq s \leq n_2 - 1$, $(r, s) \neq (0, 0)$.

Case q = Prime Power: For ease of notation we take $q = 4$; the general case is similar. Let (A_{ij}) be an $n_1 \times n_2$ pseudo-random array over $GF(4)$ as constructed in Section IV, where $n = n_1 n_2 = 4^m - 1$, $m \geq 4$. The A_{ij}'s are 0, 1, ω or ω^2. The desired binary arrays are given by

$$a_{ij} = 1 \text{ if and only if } A_{ij} = 0$$
$$b_{ij} = 1 \text{ if and only if } A_{ij} = 1$$
$$c_{ij} = 1 \text{ if and only if } A_{ij} = \omega$$
$$d_{ij} = 1 \text{ if and only if } A_{ij} = \omega^2. \quad (40)$$

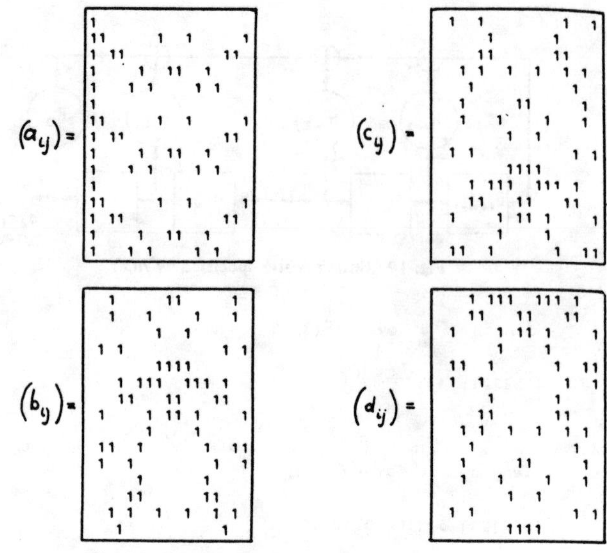

Fig. 22. Four 15 X 17 arrays obtained from Fig. 21.

Fig. 22 shows the four 15 X 17 arrays obtained in this way from Fig. 21.

The transmission function is given by

$$\tau(r, s) = \frac{1}{n_1 n_2} (\text{number of times } A_{ij} \neq 0, A_{i+r,j+s} \neq 1 \cdot A_{i+2r,j+2s} \neq \omega \text{ and } A_{i+3r,j+3s} \neq \omega^2). \quad (41)$$

Define t by $0 \leq t < n_1 n_2$, $t \equiv r (\text{mod } n_1)$ and $t \equiv s (\text{mod } n_2)$. Let $f(z)$ be the pseudo-random sequence of $\mathcal{S}_m(4)$ corresponding to (A_{ij})—see Section IV.

i) Suppose t is such that the sequences $f(z)$, $z^t f(z)$, $z^{2t} f(z)$, $z^{3t} f(z)$ are linearly independent. If $m = 4$, these are a set of generators for the code $\mathcal{S}_4(4)$. Therefore the columns of the array

$$\begin{matrix} f(z) \\ z^t f(z) \\ z^{2t} f(z) \\ z^{3t} f(z) \end{matrix} \quad (42)$$

contain each of the 255 nonzero 4-tuples from the set $\{0, 1, \omega, \omega^2\}$ exactly once. Then

$$\tau(r, s) = \frac{1}{255} (\text{number of times } f_i \text{ is } 1, \omega \text{ or } \omega^2; f_{i+t} \text{ is } 0, \omega \text{ or } \omega^2; f_{i+2t} \text{ is } 0, 1 \text{ or } \omega^2; \text{ and } f_{i+3t} \text{ is } 0, 1 \text{ or } \omega)$$
$$= \frac{3^4}{255} = \frac{81}{255}.$$

If $m > 4$ then

$$\tau(r, s) = \frac{81 \cdot 4^{m-4}}{4^m - 1}. \quad (43)$$

To test whether these sequences are linearly independent we

map them into $GF(4^m)$ using the mapping given in Section II-K. Now

$$f(\xi^{-1}),\quad \xi^{-t}f(\xi^{-1}),\quad \xi^{-2t}f(\xi^{-1}),\quad \xi^{-3t}f(\xi^{-1})$$

are linearly independent over $GF(4)$ if the equation

$$a_0 + a_1\delta + a_2\delta^2 + a_3\delta^3 = 0$$

where $\delta = \xi^{-t}$, $a_i \in GF(4)$, implies all $a_i = 0$. This means that the equation of least degree satisfied by δ over $GF(4)$ must have degree ≥ 4. Most elements of $GF(4^m)$ satisfy a minimal equation of degree m (which is the reason for the condition $m \geq 4$). However, the elements 1, $\xi^{(4^m-1)/3}$ and $\xi^{2(4^m-1)/3}$ satisfy an equation of degree 1. If m is divisible by 2 or 3 there are also elements which satisfy equations of degree 2 or 3.

ii) Suppose $t = (4^m - 1)/3$. Then $z^t f(z) = \omega^{\pm 1} f(z)$, $z^{2t} f(z) = \omega^{\pm 2} f(z)$, and $z^{3t} f(z) = f(z)$. The columns of (42) are of one of the types

$$\begin{matrix} 0 & 1 & \omega & \omega^2 \\ 0 & \omega & \omega^2 & 1 \\ 0 & \omega^2 & 1 & \omega \\ 0 & 1 & \omega & \omega^2 \end{matrix} \quad \text{or} \quad \begin{matrix} 0 & 1 & \omega & \omega^2 \\ 0 & \omega^2 & 1 & \omega \\ 0 & \omega & \omega^2 & 1 \\ 0 & 1 & \omega & \omega^2 \end{matrix}$$

and $\tau(r, s) = 4^{m-1}/(4^m - 1)$ or 0, respectively. If m is prime, i) and ii) exhaust all the values of t. (If m is divisible by 2 or 3 other values of $\tau(r, s)$ can occur and can be found in the same way.) When $m = 4$, the transmission function $\tau(r, s)$ of Fig. 21 is given by

$$\tau(0,0) = \tau(10,0) = 0, \quad \tau(5,0) = 0.502$$

$$\tau(r,0) = 0.314 \text{ for } 1 \leq r \leq 14, \quad r \neq 5, 10$$

$$\tau(r,s) = 0.318 \text{ for all other pairs } (r,s) \text{ in the range } 0 \leq r \leq 14, \quad 0 \leq s \leq 14. \tag{44}$$

VI. Summary

Pseudo-random sequences of 0's and 1's are constructed in Sections II-A and II-B; examples are shown in Fig. 7. They exist for all lengths of the form $n = 2^m - 1$, and their properties are discussed in Sections II-C through II-K. Properties I–IV, VI, and VII, are the most important.

Pseudo-random arrays of 0's and 1's of size $n_1 \times n_2$ exist whenever n_1 and n_2 are relatively prime, $n = n_1 n_2 = 2^{k_1 k_2} - 1$, and $n_1 = 2^{k_1} - 1$; examples are shown in Fig. 14. The construction is given in Section III-A and their properties in Sections III-B and III-C. Arrays with more general values of n_1 and n_2 are mentioned in Section III-D.

Pseudo-random sequences with entries from an alphabet of q symbols and length $q^m - 1$ are constructed in Seciton IV-A, examples are given in Fig. 20. These can also be formed into arrays, as illustrated in Fig. 21.

Section V discusses the transmission function (a kind of generalized autocorrelation function) of a set of binary arrays obtained from the pseudo-random sequences given in Section IV-A.

Appendix

Proof of the Window Property III:* Let α be a primitive element of $GF(2^{k_1 k_2})$. Let G be an $n \times k_1 k_2$ generator matrix for the simplex code $\mathcal{S}_{k_1 k_2}$ of length $n = 2^{k_1 k_2} - 1$, dimension $k_1 k_2$ and minimum distance $2^{k_1 k_2 - 1}$. The rows of G may be taken to be the binary $k_1 k_2$-tuples which are rectangular coordinates for the elements $1, \alpha, \alpha^2, \cdots, \alpha^{n-1}$ of $GF(2^{k_1 k_2})$. Of course $\alpha^n = 1$. Let $I(i,j)$ be the unique integer satisfying

$$I(i,j) \equiv i \pmod{n_1}$$

$$I(i,j) \equiv j \pmod{n_2}$$

and

$$0 \leq I(i,j) \leq n - 1.$$

The window property will follow from the following result.

Theorem: The rows of G which represent $\alpha^{I(i,j)}$ for $0 \leq i \leq k_1 - 1$ and $0 \leq j \leq k_2 - 1$ are linearly independent over $GF(2)$.

Proof: Suppose

$$\sum_{i=0}^{k_1-1} \sum_{j=0}^{k_2-1} c_{ij} \alpha^{I(i,j)} = 0, \qquad c_{ij} \in GF(2). \tag{45}$$

We must show that every $c_{ij} = 0$. Let Q denote the left-hand side of (45). With μ, ν as in (28) set $\beta = \alpha^{\nu n_2}$, $\gamma = \alpha^{\mu n_1}$. Then $\alpha = \beta\gamma$, $\alpha^{I(i,j)} = \beta^i \gamma^j$, and

$$Q = \sum_{j=0}^{k_2-1} \gamma^j \sum_{i=0}^{k_1-1} c_{ij}\beta^i.$$

Now n_1 and n_2 are the smallest positive integers such that $\beta^{n_1} = 1$, $\gamma^{n_2} = 1$. Since $n_1 = 2^{k_1} - 1$, $\beta \in GF(2^{k_1})$. Furthermore, β is a zero of an irreducible polynomial of degree k_1 over $GF(2)$, and of no polynomial of lower degree. The coefficient of γ^j in Q is an element of $GF(2^{k_1})$; hence it is necessary to find the lowest degree polynomial over $GF(2^{k_1})$ satisfied by γ. Let t be the least integer such that $\gamma^{2^t - 1} = 1$. Then $\alpha^{\mu n_1 (2^t - 1)} = 1$, so $\mu n_1 (2^t - 1)$ is divisible by $n_1 n_2$, and $2^t - 1$ is divisible by n_2. Now $n_2 = (2^{k_1 k_2} - 1)/(2^{k_1} - 1)$ has binary representation

$$1\overbrace{0\,0\cdots0}^{k_1-1}1\overbrace{0\,0\cdots0}^{k_1-1}1\cdots1\overbrace{0\,0\cdots0}^{k_1-1}1$$

with k_2 1's. The binary representation of $2^t - 1$ is $11\cdots1$ with t 1's. Comparing these we see that the least t for which n_2 is divisible by $2^t - 1$ is $t = k_1 k_2$. Thus $t = k_1 k_2$ is the least positive integer such that $\gamma^{2^t} = \gamma$, and the k_2 elements

$$\gamma, \gamma^{2^{k_1}}, \gamma^{2^{2k_1}}, \cdots, \gamma^{2^{(k_2-1)k_1}}$$

are distinct. The polynomial of least degree satisfied by γ over $GF(2^{k_1})$ is

$$\prod_{i=0}^{k_2-1} (x - \gamma^{2^{ik_1}})$$

of degree k_2. Now Q is a polynomial of degree $k_2 - 1$ in γ with coefficients from $GF(2^{k_1})$. Thus $Q = 0$ implies

$$\sum_{i=0}^{k_1-1} c_{ij}\beta^i = 0, \qquad 0 \leq j \leq k_2 - 1.$$

But this is a polynomial of degree $k_1 - 1$, and β satisfies no equation of degree less than k_1. Therefore, $c_{ij} = 0$ for all i, j. Q.E.D.

Proof of the Window Property III:* Suppose the codewords $a(z)$ of $\mathcal{S}_{k_1 k_2}$ are folded into $n_1 \times n_2$ arrays $b(x,y)$ as in

Section III-A. From the theorem, the $k_1 \times k_2$ array in the North-West corner of $b(x, y)$ can be chosen arbitrarily, and determines the rest of the codeword. If W is a $k_1 \times k_2$ window placed in the North-West corner of $b(x, y)$, as $a(z)$ runs through the nonzero codewords of $\mathcal{S}_{k_1 k_2}$ W sees each nonzero $k_1 \times k_2$ array once. If $b(x, y) \neq 0$, the arrays consist of all $x^i y^j b(x, y)$. Since the view of $x^i y^j b(x, y)$ seen by W is the same as the view of $b(x, y)$ seen by $x^{-i} y^{-j} W$, this proves the window property. Q.E.D.

Acknowledgment

We should like to thank Ken Knowlton for telling us about the physical problem [43] which stimulated this research, and Allen Gersho for helping to translate the physical problem into a nice mathematical one.

References

[1] J. D. Alanen and D. E. Knuth, "Tables of finite fields," *Sankhyā*, Series A, vol. 26, pp. 305-328, 1964.

[2] J. R. Ball, A. H. Spittle, and H. T. Liu, "High-speed m-sequence generation: a further note," *Electron. Lett.*, vol. 11, pp. 107-108, 1975.

[3] C. Balza, A. Fromageot, and M. Maniere, "Four-level pseudo-random sequences," *Electron. Lett.*, vol. 3, pp. 313-315, 1967.

[4] T. C. Bartee and D. I. Schneider, "Computation with finite fields," *Infom. Contr.*, vol. 6, pp. 79-98, 1963.

[5] L. D. Baumert, "Cyclic Hadamard matrices," *JPL Space Programs Summary*, vol. 37-43-IV, pp. 311-314 and 338, 1967.

[6] J. T. B. Beard, Jr., "Computing in GF(q)," *Math. Comput.*, vol. 28, pp. 1159-1166, 1974.

[7] J. T. B. Beard, Jr., and K. I. West, "Some primitive polynomials of the third kind," *Math. Comput.*, vol. 28, pp. 1166-1167, 1974.

[8] E. R. Berlekamp, *Algebraic Coding Theory*. New York: McGraw-Hill, 1968.

[9] —, "Some mathematical properties of a scheme for reducing the bandwidth of motion pictures by Hadamard smearing," *Bell Syst. Tech. J.*, vol. 49, pp. 969-986, 1970.

[10] E. R. Berlekamp and J. Justesen, "Some long cyclic linear binary codes are not so bad," *IEEE Trans. Inform. Theory*, vol. 20, pp. 351-356, 1974.

[11] P. A. N. Briggs and K. R. Godfrey, "Autocorrelation function of a 4-level m-sequence," *Electron. Lett.*, vol. 4, 232-233, 1963.

[11a] N. G. de Bruijn, "A combinatorial problem," *Nederl. Akad. Wetensch. Proc. Ser. A.*, vol. 49, pp. 758-764, 1946-*Indag. Math.*, vol. 8, pp. 461-467, 1946.

[12] D. Calabro and J. K. Wolf, "On the synthesis of two-dimensional arrays with desirable correlation properties," *Imform. Contr.*, vol. 11, pp. 537-560, 1968.

[13] D. E. Carter, "On the generation of pseudo-noise codes," *IEEE Trans. Aerosp. Electron. Syst.*, vol. 10, pp. 898-899, 1974.

[14] J. A. Chang, "Generation of 5-level maximal-length sequences," *Electron. Lett.*, vol. 2, p. 258, 1966.

[15] J. H. Conway, "A tabulation of some information concerning finite fields," *Computers in Mathematical Research*, R. F. Churchhouse and J. C. Herz, Eds. Amsterdam: North-Holland, 1968, pp. 37-50.

[16] I. G. Cumming, "Autocorrelation function and spectrum of a filtered, pseudo-random binary sequence," *Proc. Inst. Elec. Eng.* vol. 114, pp. 1360-1362, 1967.

[17] J. F. Dillon, "Elementary Hadamard difference sets," Ph.D. dissertation, Univ. Maryland, College Park. 1974.

[17a] J. V. Evans and T. Hagfors, Eds., *Radar Astronomy*. New York: McGraw-Hill, 1968.

[18] H. Feistel, W. A. Notz, and J. L. Smith, "Some cryptographic techniques for machine-to-machine data communications," *Proc. IEEE*, vol. 63, pp. 1545-1554, 1975.

[19] J. P. Fillmore and M. L. Marx, "Linear recursive sequences," *SIAM Review*, vol. 10, pp. 342-353, 1968.

[19a] H. Fredricksen, "The lexicographically least de Bruijn cycle," *J. Combinatorial Theory*, vol. 9, pp. 1-5, 1970.

[19b] H. Fredricksen, "A class of nonlinear de Bruijn cycles," *J. Combinatorial Theory*, vol. 19A, pp. 192-199, 1975.

[20] S. Fredricsson, "Pseudo-randomness properties of binary shift register sequences," *IEEE Trans. Inform. Theory*, vol. 21, pp. 115-120, 1975.

[21] P. R. Geffe, "An open letter to communication engineers," *Proc. IEEE*, vol. 55, p. 2173, 1967.

[22] —, "How to protect data with ciphers that are really hard to break," *Electronics*, vol. 46, pp. 99-101, Jan. 1973.

[23] E. N. Gilbert, unpublished.

[24] A. Gill, *Linear Sequential Circuits*. New York: McGraw-Hill, 1966.

[25] K. R. Godfrey, "Three-level m-sequences," *Electron. Lett.*, vol. 2, pp. 241-243, 1966.

[26] J. -M. Goethals, Factorization of cyclic codes, *IEEE Trans. Inform. Theory*, vol. 13, pp. 242-246, 1967.

[27] R. Gold, Characteristic linear sequences and their coset functions, *SIAM J. Appl. Math.*, vol. 14, pp. 980-985, 1966.

[28] S. W. Golomb, Ed., *Digital Communications with Space Applications*. Englewood Cliffs, N.J.,: Prentice-Hall, 1964.

[29] S. W. Golomb, *Shift Register Sequences*. San Francisco: Holden-Day, 1967.

[30] B. Gordon, "On the existence of perfect maps," *IEEE Trans. Inform. Theory*, vol. 12, pp. 486-487, 1966.

[31] D. Gorenstein and E. Weiss, "An acquirable code," *Inform. Contr.*, vol. 7, pp. 315-319, 1964.

[32] D. H. Green and I. S. Taylor, "Irreducible polynomials over composite Galois fields and their applications in coding techniques," *Proc. Inst. Elec. Eng.*, vol. 121, pp. 935-939, 1974.

[33] R. R. Green, "A serial orthogonal decoder," *JPL Space Programs Summary*, vol. 37-39-IV, pp. 247-253, 1966.

[34] —, "Analysis of a serial orthogonal decoder," *JPL Space Programs Summary*, vol. 37-53-III, pp. 185-187, 1968.

[35] M. Hall, Jr., *Combinatorial Theory*. Waltham, MA: Blaisdell, 1967.

[36] J. T. Harvey, "High-speed m-sequence generation," *Electron. Lett.*, vol. 10, pp. 480-481, 1974.

[37] M. Harwit, "Spectrometric imager," Applied Optics, vol. 10, pp. 1415-1421, 1971, and vol. 12, pp. 285-288, 1973.

[38] —, private communication.

[39] U. Henriksson, "On a scrambling property of feedback shift registers," *IEEE Trans. Commun.*, vol. 20, pp. 998-1001, 1972.

[40] T. Ikai and Y. Kojima," Two-dimensional cyclic codes," *Electron. Commun. in Japan*, vol. 57-A, pp. 27-35, 1974.

[41] H. Imai, "Two-dimensional Fire codes," *IEEE Trans. Inform. Theory*, Vol. 19, pp. 796-806, 1973.

[42] W. H. Kautz, Ed., *Linear Sequential Switching Circuits: Selected Papers*. San Francisco: Holden-Day, 1965.

[43] K. Knowlton, private communication.

[43a] D. E. Knuth, *The Art of Computer Programming*. Reading, MA.: Addison-Wesley, 1968, vol. 1, p. 379.

[43b] Y. I. Kotov, "Correlation function of composite sequences constructed from two M-sequences," *Radio Eng. Electron. Phys.*, vol. 19, pp. 128-130, 1974.

[43c] J. Lindner, "Binary sequences up to length 40 with best possible autocorrelation function," *Electron. Lett.*, vol. 11, p. 507, 1975.

[44]. F. J. MacWilliams and N. J. A. Sloane, *The Theory of Error-Correcting Codes*. Amsterdam: North-Holland Publishing, to appear.

[45] R. L. McFarland, "A family of difference sets in noncyclic groups," *J. Combinatorial Theory*, vol. 15A, pp. 1-10, 1973.

[46] K. Nakamura and Y. Iwadare, "Data Scramblers for multilevel pulse sequences," *NEC Research and Development*, No. 26, pp. 53-63, July 1972.

[47] I. Niven and H. S. Zuckerman, *An Introduction to the Theory of Numbers*. New York: Wiley, 1966, 2nd edition.

[48] T. Nomura and A. Fukuda, "Linear recurring planes and two-dimensional cyclic codes," *Electron. Commun. in Japan*, vol. 54A, pp. 23-30, 1971.

[49] T. Nomura, H. Miyakawa, H. Imai, and A. Fukuda, "A method of construction and some properties of planes having maximum area matrix," *Electron. Commun. in Japan*, vol. 54A, pp. 18-25, 1971.

[50] —, "Some properties of the $\gamma\beta$-plane and its extension to three-dimensional space," *Electron. Commun. in Japan*, vol. 54A, pp. 27-34, 1971.

[51] —, "A Theory of two-dimensional linear recurring arrays," *IEEE Trans. Inform. Theory*, vol. 18, pp. 775-785, 1972.

[52] W. W. Peterson and E. J. Weldon, Jr., *Error-Correcting Codes*. Cambridge, MA: M. I. T. Press, 1972, 2nd ed.

[52a] G. H. Pettengill, Radar Handbook, M. I. Skolnik, Ed. New York: McGraw-Hill, 1970, Ch. 33.

[53] E. C. Posner, "Combinatorial Structures" in *Planetary Reconnaissance, in Error Correcting Codes*, H. B. Mann, Ed. New York: Wiley, 1969, pp. 15-46.

[54] I. S. Reed and R. M. Stewart, "Note on the existence of perfect maps," *IEEE Trans. Inform. Theory*, vol. 8, pp. 10-12, 1962.

[55] P. D. Roberts and R. H. Davis, "Statistical properties of smoothed maximal-length linear binary sequences," *Proc. Inst. Elec. Eng.*, vol. 113, pp. 190-196, 1966.

[56] O. S. Rothaus, "On 'Bent' Functions," *J. Combinatorial Theory*, vol. 20A, pp. 300-305, 1976.

[57] J. E. Savage, "Some simple self-synchronizing digital data scramblers," *Bell Syst. Tech. J.*, vol. 46, pp. 449-487, 1967.

[58] P. H. R. Scholefield, "Shift registers generating maximum-length sequences," *Electron. Technol.*, vol. 37, pp. 389-394, 1960.

[59] M. R. Schroeder, "Sound diffusion by maximum-length se-

quences," *J. Acoust. Soc. Am.*, vol. 57, pp. 149-150, 1975.
[60] E. S. Selmer, "Linear recurrence relations over finite fields," Dept. of Math., Univ. of Bergen, Norway, 1966.
[61] N. J. A. Sloane, T. Fine, P. G. Phillips, and M. Harwit, "Codes for Multislit Spectrometry," *Appl. Opt.*, vol. 8, pp. 2103-2106, 1969.
[62] N. J. A. Sloane and M. Harwit, "Masks for Hadamard transform optics, and weighing designs," *Appl. Opt.*, vol. 15, pp. 107-114, 1976.
[63] R. Spann, "A Two-Dimensional correlation property of pseudo-random maximal-length sequences," *Proc. IEEE*, vol. 53, p. 2137, 1963.
[64] W. Stahnke, "Primitive binary polynomials," *Math. Comput.*, vol. 27, pp. 977-980, 1973.
[65] J. J. Stiffler, *Theory of Synchronous Communications.* Englewood Cliffs, NS: Prentice-Hall, 1971.
[66] M. H. Tai, M. Harwit, and N. J. A. Sloane, "Errors in Hadamard spectroscopy or imaging caused by imperfect masks," *Appl. Opt., vol. 14, pp.* 2678-2686, 1975.
[67] R. Theone and S. W. Golomb, "Search for cyclic Hadamard matrices," *JPL Space Programs Summary*, vol. 37-40-IV, pp. 207-208, 1966.
[67a] S. A. Tretter, "Properties of PN^2 sequences," *IEEE Trans. Inform. Theory*, vol. 20, pp. 295-297, 1974.
[68] S. H. Tsao, "Generation of delayed replicas of maximal-length linear binary sequences," *Proc. Inst. Elec. Eng.*, vol. 111, pp. 1803-1806, 1964.
[69] R. Turyn, "Sequences with small correlation," in *Error Correcting Codes*, H. B. Mann, Ed. New York: Wiley, 1969, pp. 195-228.
[70] J. V. Uspensky and M. A. Heaslet, *Elementary Number Theory.* New York: McGraw-Hill, 1939.
[71] W. D. Wallis, A. P. Street, and J. S. Wallis, "Combinatorics: Room squares, sum-free sets, Hadamard matrices," *Lecture Notes in Mathematics 292.* Berlin: Springer, 1972.
[72] G. D. Weathers, E. R. Graf, and G. R. Wallace, "The subsequence weight distribution of summed maximum length digital sequences," *IEEE Trans. Commun.*, vol. 22, pp., 997-1004, 1974.
[73] L. R. Welch, "Computation of finite Fourier series," *JPL Space Programs Summary*, vol. 37-37-IV, pp. 295-297, 1966.
[74] L. -J. Weng, "Decomposition of M-sequences and its applications," *IEEE Trans. Inform. Theory*, vol. 17, pp. 457-463, 1971.
[75] M. Willett, "The index of an m-sequence," *SIAM J. Appl. Math.*, vol. 25, pp. 24-27, 1973.
[76] N. Zierler, "Linear recurring sequences," *J. Soc. Ind. Appl. Math.*, vol. 7, pp. 31-48, 1959.
[77] ——, "Linear recurring sequences and error-correcting codes," *Error Correcting Codes*, H. B. Mann, Ed., New York: Wiley, 1969, pp. 47-59.
[78] N. Zierler and J. Brillhart, "On primitive trinomials (mod 2)," *Inform. Cont.*, vol. 13, pp. 541-554, 1968, and vol. 14, pp. 566-569, 1969.

Contributors

F. A. Benson (M'50-SM'62-F'75) was educated at the University of Liverpool, Liverpool, England, where he received the degrees of B. Eng. and M. Eng. He has also been awarded the degrees of Ph.D. and D. Eng. by the University of Sheffield, Sheffield, England.

After serving as a Member of the Research Staff at the Admiralty Signal Establishment, Witley, England, during the World War II, he became an Assistant Lecturer in Electrical Engineering at the University of Liverpool. Since 1949, he has been on the Staff at the University of Sheffield, first as a Lecturer, then as a Senior Lecturer, and later Reader in Electronics. He became Professor and Head of the Department of Electronic and Electrical Engineering in 1967. He has been a Pro-Vice-Chancellor in the University of Sheffield since 1972.

*

Chak Ming Chie (S'75) was born in Hong Kong, on August 29, 1951. He received the B.S.E.E. and B.A. (Math) degree from the University of Minnesota, MN, in 1972, and the M.S. degree in electrical engineering from the University of Southern California (USC), Los Angeles, in 1974.

From 1972 to 1974, he was a Teaching Assistant at USC. He is presently a Research Assistant working for a Ph.D. degree at USC with major interest in the synchronization aspect of digital communication systems.

Charles Elachi (M'71) was born in Rayak, Lebanon, on April 18, 1947. He received the "ingenieur" degree in radioelectricity with honors and the "Prix de la Houille Blanche" from the Polytechnic Institute of Grenoble, France, in 1968, and the M.S. and Ph.D. degrees in electrical engineering and business economics from the California Institute of Technology, Pasadena, in 1969 and 1971, respectively.

He has worked at the Physical Spectrometry Laboratory, University of Grenoble, France, on plasma in microwave cavities. He was a Teaching Assistant at Caltech in 1969. In 1970 he joined the Space Sciences Division, Jet Propulsion Laboratory, Pasadena, where he is presently Surpervisor of the Radar Science and Applications Group which is involved in investigating spacecraft borne scientific experiments for planetary and Earth studies using coherent radar techniques. He is a Principal Investigator on the Lunar Data Analysis and Synthesis Program. He is also involved in studying theoretical electromagnetic problems related to stratified media, space-time periodic media and DFB lasers. He is also affiliated with the University of California, Los Angeles. He has 60 papers, patents, reports, and conference presentations in the above fields.

In 1973, Dr. Elachi was the first recipient of the R. W. P. King award. He is a member of the Optical Society of America, AAAS and Sigma Xi.

*

Reuben Hackam received the B.Sc. degree from the Technion–Israel Institute of Technology, Haifa, Israel, in 1960, and the Ph.D. degree from the University of Liverpool, Liverpool, England in 1964.

Spread-Spectrum Technology for Commercial Applications

D. THOMAS MAGILL, MEMBER, IEEE, FRANCIS D. NATALI, SENIOR MEMBER, IEEE, AND GWYN P. EDWARDS, MEMBER, IEEE

Invited Paper

Only recently has our technology advanced to the point that commercial application of spread-spectrum signaling is economically feasible. This has motivated a number of companies and individuals to seek new ways to benefit from spread-spectrum techniques in commercial systems and products.

In this paper, we give a very brief overview of spread-spectrum signaling. We then consider applications to satellite mobile applications as well as indoor wireless applications. An overview is presented of several proposals to provide worldwide personal communications through satellite-based spread-spectrum systems, the tradeoffs to be considered, and the controversies involved. A description of a high-capacity wireless office telephone system serves to illustrate how spread-spectrum signaling may be useful in this environment. Finally, we describe a number of digital processing algorithms and devices that implement spread-spectrum signaling in a cost-effective manner.

I. Introduction

As our society has become increasingly interdependent, the transfer of information has become more and more important. Any product that can provide better, quicker, more mobile, or more convenient communications is destined for success. The general, personal desire to compute or communicate at any time, and from any place, has supported the growth of the cellular mobile telephone network, the emergence of the wireless PBX, and many other Personal Communications Service (PCS) experiments and proposals. The urge for untethered, mobile computing for professionals has led to the development of wireless Local-Area Networks (LAN's). Cordless telephones have become commonplace in the home as well as the work place.

The development of the spread-spectrum Global Positioning System (GPS) has spawned an increasing demand for mobile/portable position location products both for stand-alone use and in conjunction with a communications link. New products incorporating GPS are proliferating rapidly.

Manuscript received June 30, 1993; revised August 30, 1993.
The authors are with Stanford Telecom, Sunnyvale, CA 94089-1117.
IEEE Log Number 9215558.

As we attempt to make wireless technology available for a broader range of applications, we are faced with the challenges of spectrum overcrowding, privacy considerations, fading mobile channels, as well as the complexities of the propagation environment inside buildings, to name a few.

Many of these problems have been tackled for military applications where complex and costly solutions have often been considered acceptable, if the desired performance was achieved. Spread-spectrum (SS) communications has been one of the most intriguing and exciting technologies to emerge from these efforts. References [1] and [2] provide a brief introduction to SS signaling while [3] is an excellent three volume text.

SS may be defined as "a technique in which an auxiliary modulation waveform, independent of the information data, is employed to spread the signal energy over a bandwidth much greater than the signal information bandwidth." The signal is "despread" at the receiver using a synchronized replica of the auxiliary waveform.

SS signaling has properties which make it useful for

- signal hiding and noninterference with conventional systems
- anti-jam and interference rejection
- privacy
- accurate ranging
- multiple access
- multipath mitigation.

SS does not improve performance in the presence of Gaussian noise, and the signal does require increased bandwidth.

While some of the basic SS concepts were formulated earlier, World War II spurred researchers to find new ways of communicating that would be secure and work in the presence of jamming. Practical application of these techniques was limited by the technology available at the time (an excellent historical overview of the development

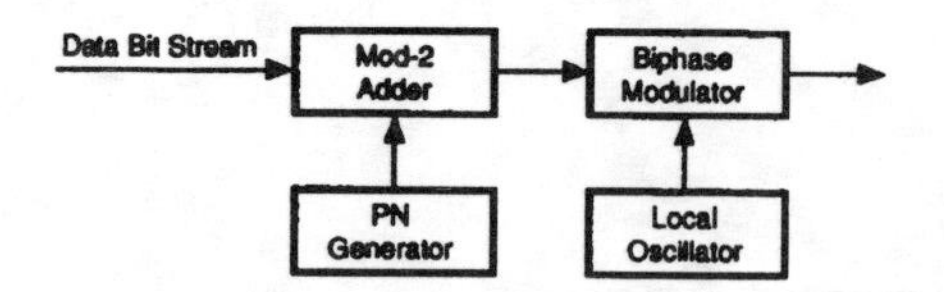

Fig. 1. Simple direct-sequence modulator.

of SS signaling is given in [3]). A general understanding of SS systems was slow to emerge after the war due to the strict classification that was imposed for many years, and to the inherent complexity of existing equipment, which served to limit its utility.

For many years SS was considered solely for military applications. However, rapid advances in LSI technology have made it possible to implement the complex functions required for SS within size and cost constraints that make it attractive for consumer products. In addition, the characteristics of SS signaling appear to be well-suited to mitigating the problems of mobile/portable personal communications.

This paper gives an overview of SS signaling, its advantages (and shortcomings), existing and proposed applications, and the technology that is making it all possible.

II. Direct-Sequence (DS) Spread Spectrum

There are two principal types of SS systems—frequency hopping (FH) and direct sequence (DS).

Direct-sequence signaling is accomplished by phase modulating the data signal with a pseudo-noise (PN), i.e., pseudo-random, sequence of zeros and ones, which are called chips. The chip modulation is most commonly binary phase-shift keying (BPSK) for simplicity and is achieved by mod-2 adding the PN chip sequence with the data as shown in Fig. 1. The number of PN chips per bit is a measure of the processing gain. Quaternary PSK (QPSK) chip modulation, while more complex, is sometimes used to prevent signal capture when a strong interferer drives the receiver into saturation.

Typically DS systems use BPSK data modulation since it is simple, and higher order PSK modulation formats offer no increase in processing gain. As we will see later QPSK data modulation can be useful for some specialized DS systems.

DS systems may be categorized as either long- or short-code systems. Long-code systems have a code length that is much longer than a data symbol so that a different chip pattern is associated with each symbol. The chip sequence is essentially random for a sufficiently long code. In code-division multiple access (CDMA) situations, the users will experience varying degrees of cross correlation which has an effect similar to random noise. In this case, it is generally necessary to implement accurate power control such that all user signals are received at nearly the same power level, otherwise, one signal can cause substantial interference to the others and reduce the capacity of the system. If the users have modest timing accuracy with respect to each other, it is possible for all users to share the same code with different time displacements for each user.

By contrast, short-code systems repetitively use the same sequence for each data symbol. Short-code systems normally use matched-filter detection while long-code systems use correlation detection. Thus short-code systems are capable of offering much more rapid acquisition and are well-suited for burst traffic, such as transaction systems.

Particular care must be taken in choosing short codes for the CDMA situation. Not only must the codes have good autocorrelation functions to permit reliable symbol synchronization, the codes must also have low cross correlation for all shifts. For short code lengths there are not many sequences which meet these properties. As a result, multiple access is often achieved by ALOHA or a carrier sense protocol in these systems.. The role of spread spectrum in these cases is usually to provide interference rejection and spectrum compatibility with other systems.

It is possible to achieve orthogonal CDMA operation with short codes if sufficiently precise timing is maintained. That is, the cross correlation between users is zero when the users are received in time synchronism. Typically this requires a star network configuration using orthogonal CDMA on the inbound (to the hub) links and orthogonal code-division multiplexing on the outbound links (the outbound signals are multiplexed on a single carrier in time synchronism). The orthogonal codes are frequently based on the Rademacher–Walsh functions but any orthogonal code set is acceptable. To maintain orthogonality on the inbound links it is necessary that all signals arrive at the hub timed to within a very small fraction of a code chip.

Forward error correction (FEC) encoding is usually employed in CDMA systems in order to increase the multiple-access capacity. Since the processing gain is not affected by the coding rate, performance is improved by the coding gain. Interleaving is also generally employed to combat burst errors due to interference or fading.

The CDMA system discriminates against multipath components that are delayed with respect to the line-of-sight path by more than one chip duration, thereby reducing fading due to multipath. Note that this capability is achieved without requiring the use of forward error correction coding. Obviously, a high chipping rate is required to discriminate against short delays.

One disadvantage associated with DS spread spectrum is its inability to deal with very-large interfering signals such as occur when an interferer is located very close to the receiver, i.e., the near/far problem. The interference is reduced by the available processing gain which is inadequate in many cases. One solution to this problem is to use notch filters, especially if the interference is narrowband. This can lead to excessive hardware complexity if there are a large number of interferers. Another approach is to employ a reduced chipping rate and multiple channels. Channel selection and filtering can help provide the desired rejection of the interference. If heavy, fixed interference is expected to be a problem, one may be better off with narrowband DS spread spectrum/FDMA than wideband DS spread spectrum. In this case, one may pay the price of increased susceptibility to multipath fading.

III. FH Spread Spectrum

FH spread spectrum divides the available bandwidth into N channels and hops between these channels according to a PN code known to both the modulator and demodulator. Since the hops generally result in a phase discontinuity (depending on the particular implementation), a noncoherent modulation format such as MFSK or differentially coherent PSK modulation is often employed. The increased energy efficiency associated with higher order MFSK, such as M equal to eight or sixteen, is gained at no increase in spread bandwidth. FH spread spectrum systems often employ forward error correction (FEC) coding to provide protection for those data bits transmitted at a hop frequency with interference or with a fade.

FH systems may be categorized as either slow- or fast-hopping (relative to the data symbol rate). With slow hopping there are multiple data symbols per hop and with fast hopping there are multiple hops per data symbol. Systems employing MFSK are generally fast hopping, while binary DPSK modulation is often used with slow hopping. In this case, the first symbol in a hop serves as a phase reference.

Fast hopping provides frequency diversity within a data symbol and, depending on the number of hops per symbol, coding may not be required. Note, however, that since MFSK detection is done on a per-hop basis, one encounters the loss associated with post-detection integration. One may wish to use FEC coding to compensate for this loss. By contrast, the coding system for slow hopping must provide frequency diversity through interleaving to deal with the error bursts.

IV. Comparison of DS- and FH-SS

The potential benefits of SS do not depend on how the spreading is achieved; however, there are many practical differences in both performance and implementation for different spreading techniques.

The bandwidth of the DS-SS signal is approximately twice the PN sequence clock rate. Wide spreading bandwidths require high clock rates which can present synchronization difficulties, as well as increased equipment cost and power consumption. The bandwidth of an FH system depends only on the tuning range and so can be hopped over a wide bandwidth with little difficulty.

Timing is generally much less critical in an FH system since hop rates usually range from a few hops per second to several thousand hops per second, compared to megahertz chipping rates in DS systems. In order to synchronize the transmitter/receiver pair, the receiver must search its initial time uncertainty until it is within a fraction of a hop or chip duration, at which time correlation will occur, and the signal can be detected. Since many fewer hops than chips occur in a given time interval, the FH code-acquisition times can be relatively short in comparison to some DS systems.

Another distinction between SS systems is that the spectrum of a DS signal looks relatively uniform (unless a very short code is employed) for observation bandwidths on the order of the symbol rate. The frequency hopper, on the other hand, is essentially a narrowband signal whose center frequency is changed frequently. This distinction may or may not be important, depending on the application.

Table 1 FCC Part 15, ISM Frequency Bands

Band	Bandwidth
902–928 MHz	26.0 MHz
2.4–2.4835 GHz	83.5 MHz
5.725–5.850 GHz	125.0 MHz

Both DS- and FH-SS can be useful for multipath mitigation when properly designed. In the case of DS spreading, multipath returns that are delayed by a chip period or longer relative to the desired return are essentially uncorrelated and do not contribute to multipath fading. In the FH case, one transmits over a wide enough bandwidth to ensure that the channel is frequency-selective, i.e., while some frequencies are "faded," others are not. As the signal hops around some hops are lost while others get through. The challenge is to design the signal with enough redundancy, and in some cases interleaving, to maintain an acceptable average error rate.

V. SS for Commercial Applications

The proliferation of SS products for commercial applications began with

- the emergence of the GPS System
- the successful demonstration of SS for cellular telephones.
- the allocation of the Instrumentation, Scientific, and Medical (ISM) bands in the USA by the FCC.

In 1989, the FCC mandated special use of the ISM bands so that an SS system is permitted to operate without license so long as the system does not interfere with an existing system in these frequency bands. A very concise summary of the frequency allocations and use restrictions for the ISM bands is given in Tables 1 and 2.

VI. Satellite Mobile Applications

Satellites have the capability to provide line-of-sight coverage of large geographical areas. As a result, they represent an attractive means of delivering mobile and portable services to large numbers of users who could not be economically served with terrestrial based stations.

The Global Positioning System (GPS) is a case in point. GPS receivers, which have the capability to calculate position to within about 100 ft anywhere in the world, now sell for under $1000. The burgeoning navigation market indicates that the technology involved to provide satellite-based mobile services has come of age. A direct sequence CDMA spread-spectrum signal was chosen for GPS because it provided a means of incorporating accurate ranging, data transmission, multipath mitigation, multiple access, interference rejection, and access security (for the

Table 2 FCC Part 15.247, Spread Spectrum Use of ISM Frequency Bands

	Direct Sequence	Frequency Hopped
Frequency Constraints	• At least 500 kHz must be occupied, up to a maximum of the allocated band.	• All subchannels must be occupied at some time. • Minimum number of channels is 50 in 902-MHz band and 75 in the other two bands. • Average dwell time on any band must not exceed. 4 s in any 20-s period for 902-MHz band and 0.4 s over a 30-s interval in the other bands. • Maximum bandwidth of a hopping channel is 500 MHz
Power	• 1 W maximum • Processing Gain (PG) of at least 10 dB (PG is the ratio of signal-to-noise ratio without spreading)	• 1 W maximum

high-accuracy P signal) in a convenient manner. This L-band system has proved to be both performance- and cost-effective.

The Qualcomm OmniTRACS system (Ku-band) provides a logical extension of this service with two-way data messaging as well as vehicle position reporting. The return link signal for this system employs a combination of direct-sequence CDMA and frequency hopping in order to ensure that users will not interfere with adjacent satellite systems.

The success of terrestrial mobile/portable cellular telephony has lead some industry experts to believe that a large market is waiting to be tapped by a similar satellite-based system that can provide service to areas that could not be economically served by conventional means. They expect personal communication and location services to grow to a multi-billion dollar industry within the decade. At the 1992 World Administrative Radio Conference (WARC-92), spectrum was allocated internationally for Mobile Satellite Service (MSS) in L-band 1610–1626.5 MHz (earth-to-space) and S-band 2483.5–2500 MHz (space-to-earth) on a primary basis. The band 1613.8–1626.5 MHz was also allocated for MSS downlinks on a secondary basis.

VII. Mobile Satellite Propagation

Traditionally, satellite systems have been used for point-to-point communications with directive antennas. This results in generally benign links with no significant fading (except for that due to rainfall attenuation). Until recently, the emphasis has been on power efficiency with bandwidth efficiency being of secondary importance. Initial shipboard mobile applications have also employed directional antennas. The omni-directional satellite mobile channel is radically different and presents some interesting design challenges.

L-band directional antennas are impractical at present for hand-held (and most mobile) units due to size and cost constraints. Omni-directional antennas do not reject scattered signals arriving from directions other than the direct path. It is these multipath signals which cause fading. Further, the user/satellite motion can result in the LOS path being obstructed by vegetation, terrain, or buildings which will attenuate the shadowed signal. Loo [4] has proposed that the channel be modeled for a rural environment as a log-normally-distributed direct path (due to foliage attenuation) and a Rayleigh-distributed multipath component.

Barts and Stutzman [5] suggest weighting the shadowed and unshadowed distributions by the appropriate fraction of shadowing and unshadowing to describe typical mixed path statistics, i.e.

$$G(R) = G_s(R)S + G_u(R)(1 - S)$$

where

$$G_s(R) = \int_R^\infty p_s(r)\, dr$$

is the cumulative fade distribution for the foliage-shadowed case and S is the fraction of shadowing. This model is generally agreed to be useful and provides a good fit to experimental data if the appropriate parameters are chosen. In fact, experimental propagation data are often described, for convenience, by giving the parameters for the above equations such that a best fit to the data is obtained [6], [7]. One of the advantages of the above modeling process is that it leads to convenient computer simulation algorithms [5], [8], [9].

Other researchers [10], [11] have chosen to model the shadowed signal as a Rayleigh distribution with a time-varying mean which is assumed to be log-normally distributed. Once again, this model is convenient for simulation purposes.

Goldhirsh and Vogel [12] have developed an Empirical Fade Distribution Equation (EFDE) for attenuation due to roadside trees in the mobile situation by curve fitting to various data [13]–[15]. The fade distribution curves generated using the EFDE are shown in Fig. 2 with elevation angle as a parameter. Similar data were presented by the US to a CCIR Study Group [13] for the heavily foliage-shadowed hand-held handset.

Terrestrial mobile systems in an urban environment often operate on multipath reflections that are stronger than the direct-path component. The satellite system is very different in this regard. Typically, the multipath delay spread is relatively small for appreciable received power. This is illustrated by data taken in downtown Chicago and the surrounding environment [16]. For example, it was found that only 1% of the time were multipaths with amplitude greater than −12 dB delayed by more than 100 ns relative to the direct path.

Table 3 Summary of Proposed System Parameters [17]

Company/System	Number of Satellites	Orbit Altitude (km)	Satellite Beams	Service Region
Constellation/Aries	48	1020	7	worldwide
Ellipsat /Ellipso	6, later 24	580 × 7800	8	worldwide
LQSS/Globalstar	48	1414	6	worldwide
Motorola/Iridium	66	780	48	worldwide
TRW/Odyssey	12	10 370	19	worldwide
AMSC	2	geostationary 62 W/139 W	4	N. America
Celsat/Celstar	2	geostationary 76 W/116 W	149	N. America

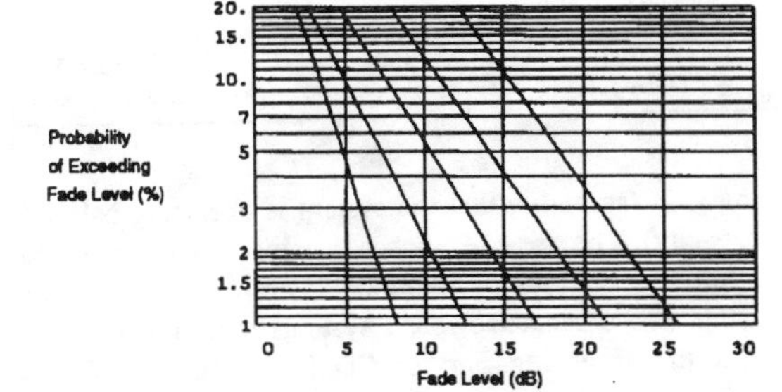

Fig. 2. Probability of exceeding a given fade level with elevation angle as a parameter based on the EFDE [12] for a heavily foliage-shadowed highway.

VIII. Personal Communication by Satellite

Several companies have filed applications with the FCC to provide satellite-based personal communications networks with worldwide coverage. The applicants have proposed to provide a variety of services including voice, data, paging, facsimile, and position determination to mobile or hand-held subscriber users. The design of these systems is driven by the perceived marketing requirement to support a hand-held subscriber unit which is similar in appearance and performance to a cellular telephone.

The proposed systems include, among others, low earth orbit (LEO) systems such as IRIDIUM (Motorola) and Globalstar (Loral/Qualcomm), the medium earth (MEO) orbit Odyssey (TRW) system, and geostationary earth orbit (GEO) as proposed by AMSC and Celsat for coverage of North America.

A summary of proposed system constellations is given in Table 3 [17]. The required number of LEO and MEO satellites for worldwide coverage ranges from 66 for Motorola's IRIDIUM to 12 for TRW's Odyssey. These systems represent a major escalation in the size of commercial systems. Present proposals estimate systems with worldwide coverage to cost anywhere from one to over three billion dollars, depending on the complexity and sophistication of the system, a very sizable sum of private investment dollars.

The currently envisioned frequency plans, modulation and channelization schemes are summarized in Table 4 [17] for some representative systems. Direct sequence spread-spectrum signaling is being proposed by nearly all of the potential system developers, the notable exception being Motorola's IRIDIUM. Potential advantages include multipath mitigation, increased system capacity, and band sharing between systems with a minimum of coordination. However, the actual gain to be realized by using spread spectrum in the proposed systems is a matter of some controversy and the issues are being hotly debated by the interested parties.

The ability of spread-spectrum signaling to combat multipath fading depends on the amplitude and delay spread of the multipath returns, as well as the spread signal chipping rate. Goldhirsh and Vogel [14] note that the severe fading experienced by the mobile satellite user at lower elevation angles in some environments is primarily due to shadowing as opposed to multipath. They noted that due to shadowing the probability of exceeding 20- and 10-dB fade levels was 1% and 10%, respectively. For the multipath fading channel at an elevation angle of 45°, the comparable fade levels were 6 and 3 dB, respectively. Rubow [18] made similar observations for an elevation angle of 15°. SS signaling has the potential for reducing fading due to multipath, but can do nothing, of course, to mitigate the attenuation due to shadowing. This leads to the conclusion that only modest gains in combating fading can be achieved by spreading.

Multipath mitigation is achieved only when the multipath delay is equal to (or greater than) a significant portion of a chip period. The multipath delay data of [16] suggest that a chipping rate of at least 10 MHz will be required to be effective for reducing fading. None of the systems summarized in Table 4 currently propose a spread bandwidth greater than 5.5 MHz, implying a chipping rate of about half of that value (TRW is investigating the advantages of full band spreading). It should be noted that the multipath delay data of Fig. 4, taken in downtown Chicago, are limited in scope and more data of this type are required before a full assessment can be made. In summary, it appears that multipath mitigation through spectrum spreading is not

Table 4 Summary of MSS System Parameters

Company/ System	Modulation	Multiple-Access Method (Forward Link)	Multiple-Access Method (Return Link)	Channelization (MHz)	Frequency Band (MHz)
Constellation	QPSK	Spread TDM	Channelized CDMA	16.5 forward 1 to 5 return	1610.1626.5 2483.5-2500
Ellipsat	OQPSK	Channelized CDMA	Channelized CDMA	1.1	1610.1626.5 2483.5-2500
LQSS	QPSK	Channelized CDMA	Channelized CDMA	1.25	1610.1626.5 2483.5-2500
Motorola	DE-QPSK	FDMA/TDMA	FDMA/TDMA	41.67 kHz	1616-1626.5 2483.5-2500
TRW	BPSK	Channelized CDMA	Channelized CDMA	5.5	1610.1626.5 2483.5-2500
AMSC	QPSK	CDMA (or FDMA/TDMA)	Channelized CDMA	5.5	1610.1626.5 2483.5-2500
Celsat	QPSK	Channelized CDMA	Channelized CDMA	1.25	1610.1626.5 2483.5-2500

of primary concern to any of the personal communication satellite systems presently proposed.

Another potential advantage of spectrum spreading is a possible increase in system capacity. Viterbi compared the spectral efficiency of CDMA with FDMA in [19] and found FDMA to be considerably superior to CDMA. He concluded that "When C/N_0 is at premium do not contribute further to the noise by having the users jam one another." Gilhousen *et al.* [9] claim that this result is reversed in the mobile satellite environment due to additional system considerations and the fact that, in general, the satellite system is more bandlimited than power-limited. Some of the factors that contribute to the CDMA capacity gain include: the voice activity factor, increased frequency reuse in conjunction with satellite multibeam antenna spatial discrimination, and the discrimination between multiple satellites with overlapping coverage. The actual system capacity is parameter-dependent and is very sensitive to received C/N_0 as well as the power control accuracy.

Some investigators, the authors among them, believe that the assumptions of [9] for voice activity factor, and gain due to polarization reuse are overly optimistic. Further, the importance of the link power limitation and percentage of users shadowed may have been underestimated. In reality, the capacity of the mobile satellite CDMA system is probably comparable to (or slightly less than) that of an FDMA or TDMA system. For example, satellite-user full duplex link capacities were estimated to be 2800 and 2300 for the CDMA Globalstar [20] and Odyssey [21] systems, respectively, as compared to 4070 links for the FDMA/TDMA IRIDIUM [22] system.

The "real" advantage of spread-spectrum signaling for mobile satellite systems may well be the ability to coexist with other users and systems with relatively little coordination. For example, CDMA users are relatively tolerant to unintentional interference by virtue of the inherent processing gain (assuming that the system is operating below full capacity). The use of "notches" against strong narrowband interferers can provide additional tolerance. Likewise, interference to other narrowband systems is minimized by the spectrum spreading. Further, CDMA systems can share the same band without coordination [17] (although individual system capacity is reduced). Once again, the effectiveness of system bandsharing is a matter of some controversy.

IX. Indoor Wireless Applications

A very large market is expected for indoor wireless communications. The economic advantages and convenience of untethered personal and computer communications are obvious and a sizable market already exists. However, the problems associated with the difficult propagation environment, interference, and the high user density in some buildings raise some formidable challenges for successful system implementation. Some of the channel considerations are discussed below and then an example spread-spectrum wireless telephone system is presented.

X. The Indoor Channel

A wide range of parameters exist for the indoor channels in which a wireless system must operate. The average path loss exponential with distance can vary from somewhat less than the nominal LOS value of two to greater than five, depending on the environment and on the building. The SS system must be designed to operate with a wide range of received signal levels (RSL). One can experience radical changes in the RSL with a motion of a few inches—something that a portable handset might encounter even in a "static" environment with relatively small displacements such as changing ears.

An excellent UHF indoor radio channel modeling tool called SIRCIM has been developed at the Virginia Polytechnic Institute by Seidel and Rappaport [23]. This model generates time-varying CW fading signal levels, fading distributions, path-loss values, and wide-band impulse responses for a variety of indoor environments.

Signals in an indoor environment encounter fading due to both attenuation and multipath. Saleh and Valenzuela [24] report experiments in a medium-size office building that show a delay spread range of 200 ns with an rms value of no more than 50 ns. This would indicate that an SS bandwidth on the order of 10 MHz or greater is required for significant multipath mitigation. It is often desirable to use bandwidths lower than this for reasons of interference to/from other systems, as well as timing and power dissipation considerations. In this case, antenna diversity is critical to achieving the desired performance.

We conducted some informal measurements of RSL's on the second floor of our headquarters office building at a frequency of approximately 1.6 GHz. For the power level and the ranges we used, we found that about 90% of the locations yielded acceptable performance. However, if at the same location we used the better of horizontal or vertical polarization, the number of acceptable locations was increased to approximately 99%. (Better coverage could have been obtained by increasing the transmitter power.) Our results indicate that the two polarizations fade independently and that the use of polarization (or spatial) diversity is a very effective means of improving performance. Fortunately, a handset is large enough to support two antennas. One configuration uses a whip near the speaker and an Alford loop near the microphone providing a combination of spatial and polarization diversity in a compact form.

Little information is present in the open literature on the interference environment indoors for the ISM bands. However, some measurements have been taken outdoors in several cities at a variety of locations [25]. There can be substantial narrowband interference at quite high power levels. Based on measurements at 850 MHz one can expect the outside interference to be reduced by about 14 dB [26] for the lowest ISM band. There are interior sources of interference sources such as microwave ovens and X-ray machines, which will not be reduced by the exterior wall attenuation. The interference level may be sufficiently high that there is not sufficient processing gain to negate the interference even if the entire available bandwidth is utilized. It is desirable to occupy only a limited portion of the ISM band such that filters can be used to avoid interference to/from other systems. Use of spread spectrum in the ISM bands is only permitted by the FCC under the condition of noninterference with existing services.

XI. Wireless PBX System

In this section we describe a wireless PBX system based on DS spread spectrum and digital signal processing technology. A wireless PBX consists of a base station which is connected to the PBX and a collection of handsets or subscriber terminals. The wireless PBX system has a star topology which is very desirable in that it permits centralized control of power, time, and frequency as well as centralized net management. All outbound (from the PBX or hub) signals can be code-division-multiplexed on a common carrier while the inbound signals use code-division multiple access. Maintaining relative power, time, and frequency on the outbound channels is easy since these channels are collocated. However, achieving these objectives for the inbound traffic is more difficult owing to their remote location.

The 902- to 928-MHz ISM band was selected and a frequency channelized design was adopted to minimize interference effects. For reasons of voice quality and cost, 32-kb/s ADPCM encoding was selected as the voice digitization technique. Omni-directional (in azimuth) antennas were assumed necessary to provide the desired coverage for mobile users.

A high priority is to permit as many simultaneous channels as possible in the available bandwidth. The system must operate under the propagation situations described in detail above, i.e., serious fading due to multipath and a large dynamic range. Fading is certainly possible since the user may be walking (typically less than 1 m/s) or moving in his chair. A few inches of motion can cause a 30-dB fade. The wireless PBX system must solve these serious propagation problems, support the desired number of users, and comply with the FCC requirements for the ISM bands (§ 15.247).

XII. Multiple-Access Technique and Channel Crosstalk

For spectral efficiency, an orthogonal CDMA system based on the orthogonal Rademacher–Walsh (RW) functions was selected. The approach assigned a RW function to each channel but also mod-2 added a common PN sequence to each channel on the basis of one PN chip per one RW chip. This step is necessary to meet the FCC spectral density requirements and has the beneficial effect of randomizing such that all channels have similar spectra and encounter similar degradations.

The RW functions provide, in theory, zero cross correlation between the various functions. In practice, finite-bandwidth effects and time-base errors will create finite correlation between the signals. Simulation runs were performed with several filter types and bandwidths to assess the sensitivity to filtering effects. Based on these results, a fourth-order, 0.1-dB-ripple, Chebychev filter was selected for transmitter and receiver. With the ratio of 3-dB bandwidth-to-chipping rate of 1, 0.8, and 0.6, the rms cross correlation between "orthogonal" channels was −34.5, −32.2, and −31.4 dB, respectively.

Simulations were performed for different RW set sizes, e.g., 16, 32, 64, 128. We found that for a given filter the average cross-correlation value was essentially independent of the set size. However, the number of potential interferors is directly proportional to the set size. Thus for the case of significant filtering, it is wise to use relatively small RW sets, e.g., 16 or 32, so as to minimize the impact of filtering.

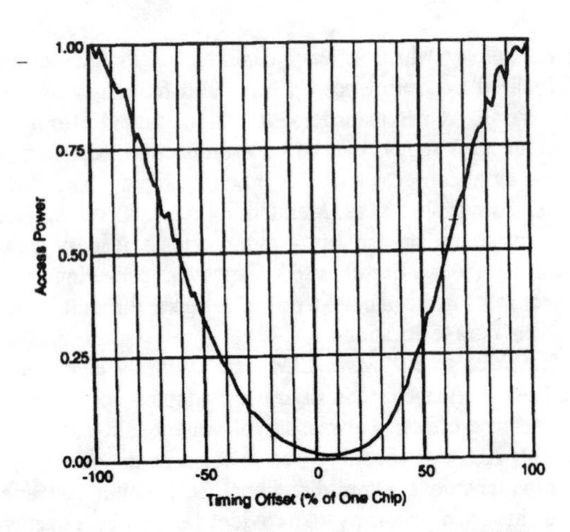

Fig. 3. Access noise power (normalized by processing gain) as a function of timing offset for the case of 16 simultaneous, equal power accesses.

The other major contributor to nonorthogonality is timing error. There are two types of timing error. First, there is a fixed timing error offset such as occurs on the inbound links to the station. The outbound links from the station are multiplexed and have no fixed time base error. Second, both inbound and outbound links encounter multipath spread. The delay spread sets a limit on the minimum crosstalk level which can be achieved no matter how accurate the network timing system is. We set up a simulation to evaluate the signal-to-interference ratio for the case of 300-ns delay spread with a uniform distribution and a mean offset of 2% of a chip duration. The simulation was performed for the following cases: 1) 16 chips/symbol at a 320-kchips/s rate and 15 other equal power users, 2) 64 chips/symbol at a 1.28-Mchips/s rate and 127 other users, and 3) 128 chips/symbol at a 2.56-Mchips/s rate and 63 other users. We found that the signal-to-interference ratios (in decibels) to be : 1) 23 dB, 2) 13 dB. and 3) 7 dB, respectively. From these results, which are based on reasonable assumptions, we conclude that the upper limit on the RW set size should be 32.

Figure 3 demonstrates the sensitivity to timing offset for the case of 16 chips/symbol when there are 15 other signals of equal power. For an offset of one chip there is no effective processing gain since the nominal value of 16 is reduced by a factor of 15 due to the other accesses. In order to provide adequate margin it is desirable to maintain a timing error better than one-tenth of a chip duration. The previous discussion on effects which contribute to channel crosstalk indicate that it is desirable to use the minimum chipping rate which will satisfy the FCC spectral density and bandwidth requirements.

XIII. Frame Structure

The previous section demonstrated that truly orthogonal CDMA cannot be achieved in a realistic environment. Further, the potential large variation in RSL's on the inbound links can make the system operation quite sensitive to channel crosstalk. Time-division duplexing (TDD) permits the same frequency to be used for both transmission and reception in both directions. The first half of the frame is used for outbound CDM transmissions and the second half carries the inbound CDMA transmissions. A TDD frame period of 10 ms was selected so that a subscriber handset can use the RSL from the base station as an accurate measure of the path loss and adjust the transmit power to achieve the desired signal level at the base station.

An advantage of the TDD format is that the antenna diversity system is simplified. The antenna diversity system uses a single omnidirectional at the base station and dual antenna diversity at the handset. Midway during the TDD frame the base station sends two identical pilot signals on the order wire (OW) channel. The handset sequentially receives the two signals—one-on-one antenna and the other on the other antenna. The handset measures the RSL and then selects the preferred antenna for one TDD frame.

XIV. Modulation Format

QPSK data modulation is employed since this doubles the number of frequency channels that could be obtained with BPSK data modulation. The QPSK signal is detected coherently using a block phase estimator [25] rather than the more familiar phase-lock loop techniques which perform poorly in fading environments [26].

For simplicity, BPSK chip modulation is used and spectral shaping achieved by post-modulation Chebychev filtering. Since there are 16 chips per data symbol the chipping rate is 640 kchips/s. With a normalized bandwidth of 0.78, the channel bandwidth is 1 MHz allowing 26 frequency channels in the lowest ISM band. This is more than enough to avoid interference problems and to support multiple adjacent cells with a frequency reuse plan. If a base station must support more than 15 simultaneous channels, then additional frequency channels must be employed.

XV. Network Control System

The base station serves as the point for centralized control of the handset frequency, transmit time base, and transmit power level. The base station monitors the power levels and the receive time bases of each of the RW channels and transmits time and power corrections on the outbound OW. The handset transmit frequencies are slaved to the base station carrier, there is no need for the base station to provide frequency corrections.

It is crucial for the base station to monitor the RSL of each access and to provide corrective information. While it is true that the handsets perform this function on an instantaneous feedforward basis, they rely on a constant receiver gain over a large-signal dynamic range. Temperature and aging effects will cause a slow variation in the gain and it is necessary for the base station to provide corrections if the RSL's are to be maintained to the desired accuracy of 1 or 2 dB.

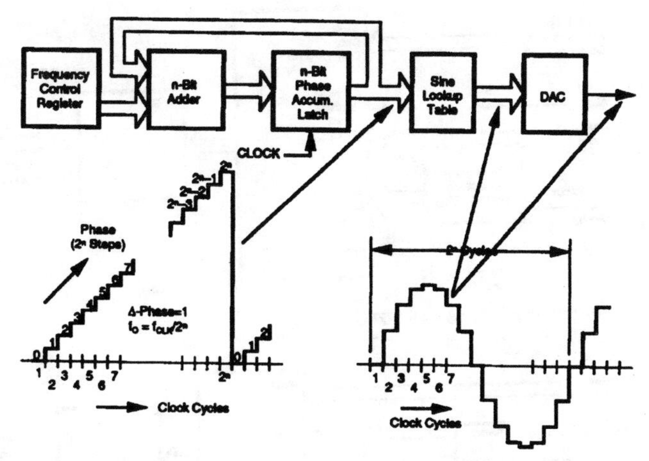

Fig. 4. Principle of Direct Digital Synthesis.

XVI. Technology for Spread-Spectrum Applications

A. Generating FH and DS Signals Using Direct Digital Synthesis

Direct digital synthesis (DDS) is an ideal technology for the generation of signals for both DS- and FH-SS systems due to the ease of modulating both the phase (PSK) and frequency (FSK and FH) of a DDS. A functional block diagram of a DDS is shown in Fig. 4.

A phase accumulator is used to address a sine look-up table (LUT), generating a digitized sine-wave output. This signal is usually fed into a digital-to-analog converter (DAC) to produce an analog output. The analog signal is then low-pass-filtered, resulting in a smooth continuous signal. The digital part of the DDS, i.e., the phase accumulator and LUT, is usually called a numerically controlled oscillator (NCO). The frequency of the output signal is determined by the number stored in the frequency control register, or phase step register. For example, in the n-bit system shown, if the phase step equals one the accumulator will count by ones, taking 2^n clock cycles to address the entire LUT and to generate one cycle of the output sine wave. This is the lowest frequency that the system can generate, and it is also the frequency resolution. Setting the phase step register equal to two results in the accumulator counting by twos, taking 2^n clock cycles to complete one cycle of the output sine wave. It can easily be shown that for any integer m, where $m < 2^{n-1}$, the number of clock cycles taken to generate one cycle of the output sine wave will be $2^n/m$, so that the output frequency will be

$$f_o = \frac{m \times f_{\text{clk}}}{2^n}.$$

A typical value for n is 32, although NCO's with 48-b resolution are also available, so that a DDS can have extremely high frequency resolution. For example, with $n = 48$ and $f_{\text{clk}} = 50$ MHz, the frequency resolution is 0.2 μHz! Note that when the value of the phase step register is changed, the output frequency changes instantaneously, i.e., from one clock cycle to the next, with no phase discontinuity in the output signal. In practice, all NCO's have a pipelined architecture resulting in a throughput delay such that the frequency change command does not occur immediately after the command. Frequency modulation for both hopping and FSK can easily be implemented by changing the phase step register value, and many NCO's incorporate features to simplify doing this, such as dual phase registers. These can be used to generate an FSK signal by loading the mark frequency into one register and the space frequency into the other; the FSK signal is then generated by selecting the appropriate register.

Phase modulation (PM) for PSK modulation is implemented by introducing an adder between the phase accumulator and the LUT. This modifies the absolute value of the phase without affecting the slope (which determines the frequency of the output signal). Modifying the MSB of the phase value changes the phase by 180°, modifying the next bit changes the phase by 90°, and so on, allowing simple BPSK or higher level PSK signals to be generated with the same ease.

Alternatively, by using a large number of bits for the PM function, linear PM can be implemented; 12 b is a typical value used. This capability can be used in conjunction with signal shaping for spectrum control. By filtering the data signal before modulation (using a raised cosine digital interpolating FIR filter, for example) the sidebands of the modulated carrier can be reduced to low levels without filtering at the output. This can be very advantageous in FH systems, where a narrowband filter cannot be used because of the need to hop the signal over a wide band. A similar technique can also be used when frequency modulating an NCO, since the modulation update rate capability of these devices is typically much greater than the data rate demands.

XVII. Using a DDS for Modulation and Carrier Generation in FH-SS.

As described in an earlier section, noncoherent FSK is usually used in FH-SS systems, and a DDS is ideal

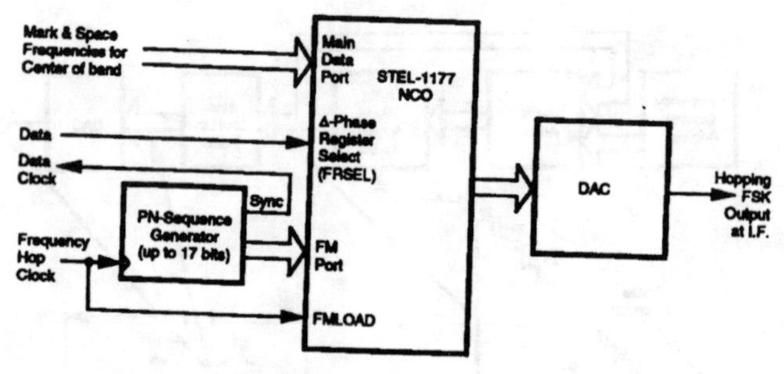

Fig. 5. Using an NCO to generate a frequency hopping FSK signal.

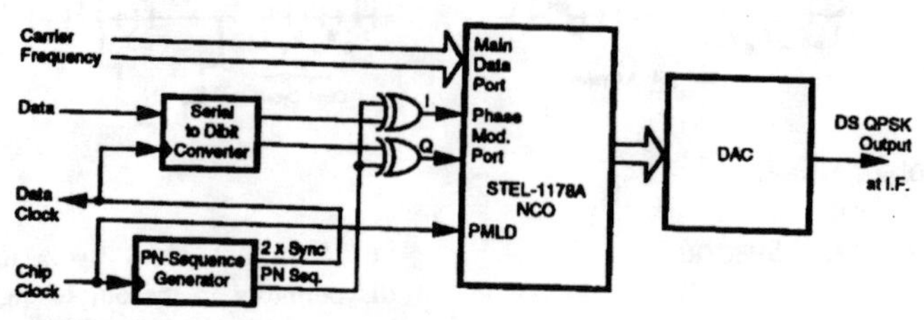

Fig. 6. Using a DDS to generate a QPSK direct sequence spread spectrum signal.

for implementing both the frequency hopping and the FSK functions. For example, at least one commercially available NCO incorporates dual-phase registers for easy FSK modulation as well as a separate FH port, as shown in Fig. 5.

With this NCO there are 2^{17} potential hop frequencies which can be changed at up to 15 MHz (the maximum hopping bandwidth is 7.5 MHz). Alternatively, the FH port, which has 16-b linearity, can be used in conjunction with an interpolating filter to shape the data themselves for spectrum control prior to modulation again taking advantage of the 15-MHz update rate capability. This device can also be phase-modulated with up to 12-b linearity, allowing it to be used for PSK FH-SS if required.

XVIII. Using a DDS for Modulation and Carrier Generation in DS-SS.

As described in an earlier section, PSK is commonly used in DS-SS systems. This is easily implemented with a DDS, as shown in Fig. 6 for BPSK chipping with either BPSK or QPSK data modulation.

XIX. Despreading and Demodulating Spread-Spectrum Signals.

Before a SS signal can be processed by a data demodulator it has to be despread, i.e., the spreading information has be removed. The way in which this is done depends on the spreading method, of course. Let us examine the methods for FH and DS SS in turn.

XX. Despreading FH SS Signals

The basic method of despreading an FH-SS signal involves generating a local-oscillator (LO) signal which hops in synchronism with the received signal carrier, resulting

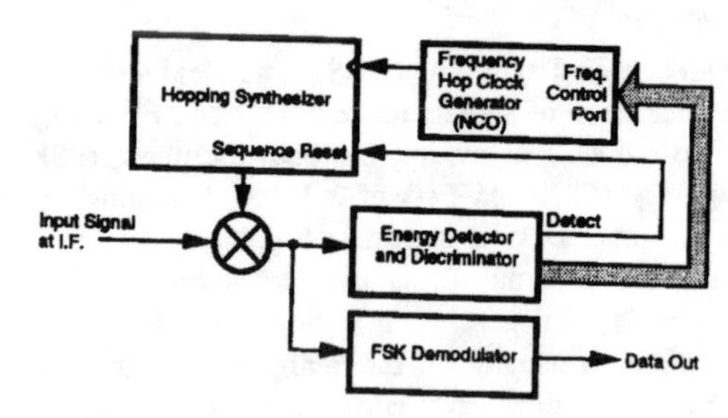

Fig. 7. Despreader for FH-SS.

in the signal being continuously converted to baseband as the frequency of the signal hops. Generating the hopping LO is readily accomplished using a DDS, in the same way as for generating the transmit signals described above. The difficult part, as with most SS signals, is synchronizing the hopping of the LO to the hopping IF signal.

There are two parts to the synchronization problem—initial acquisition and tracking. Details of the circuit implementation will depend on the FH and modulation formats but these functions are usually performed by the general approaches described below. Initial acquisition is generally achieved by a serial search procedure involving a single despreader as described above. However, rather than demodulating the data, the received energy over multiple symbols is compared with a threshold level. If the hopping patterns are aligned within $\pm 1/2$ hop duration, the detected power will exceed the threshold and synchronization acquisition will be declared and the FH receiver will enter the tracking mode.

In the tracking mode it is necessary for the FH receiver to develop a time discriminator characteristic. This entails generating early and late (e.g., by one-half hop duration) hopping patterns as well as the punctual hopping pattern.

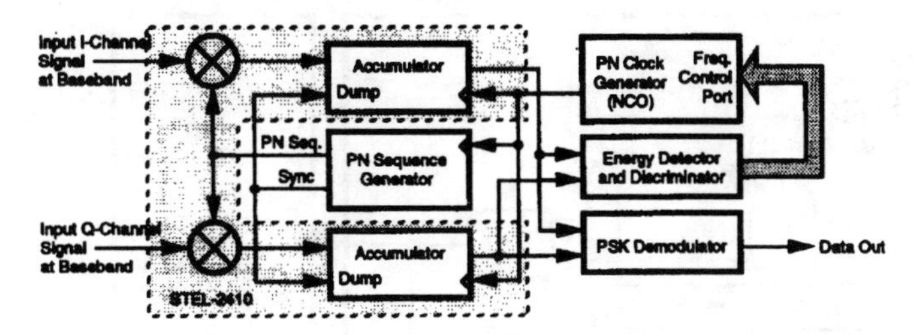

Fig. 8. Serial correlator despreader for direct sequence SS.

Time base tracking is obtained by taking the difference in detected energy levels in the early and late channels. When the received signal and receiver time bases are aligned the difference will be zero and NCO driving the PN generator(s) will maintain the correct frequency.

Before the advent of NCO's and DSP, the complexity of FH despreaders was sufficiently great that a single FH despreader was time-shared between the early and late positions. Since the FH despreader was never in the punctual position it was necessary to use quite small early/late time offsets to minimize the data detector loss. This meant that the time discriminator was quite narrow.

At present, through the use of DSP and ASIC technology, it is possible to implement parallel early, punctual, and late despreaders in a compact package. Thus it is possible to obtain optimal detection performance while maintaining a robust tracking loop. Figure 7 is a block diagram of a typical FH demodulator. Noncoherent or differentially coherent modulation techniques are generally used with FH.

XXI. Despreading DS-SS Signals

As with FH-SS signals, despreading DS-SS signals involves generating a replica of the spreading sequence at the receiver in synchronism with the sequence modulated on the incoming signal. The major difference is that with a digital implementation of DS-SS, the signals are generally despread at baseband, whereas FH signals are despread in the downconversion to baseband. Assuming that BPSK chipping is used, the spreader multiplies the transmitted signal by a given sequence of ±ones. When the despreader is correctly synchronized it will multiply the signal by the same sequence of ±ones, and since both $+1$ and -1 squared equal $+1$, the result is that the spreading sequence is removed from the signal. If the despreading sequence is not synchronized, or uses a different sequence, the signal will effectively be multiplied by a random sequence of ±oness so that the signal will not be despread.

There are two primary methods of implementing a despreader for DS-SS signals. The first uses a serial correlator–accumulator technique, and is shown in Fig. 8.

The complex correlator–accumulator is shown in the shaded region. The incoming complex baseband signal is multiplied by the despreading sequence and the samples are integrated over a symbol period. This generates the cross-correlation factor of the two signals over that period, so that when they are in phase the result will be a large sum, and when they are out of phase the sum will be much smaller, assuming that spreading sequences will good correlation properties have been selected. This technique has a lot in common with the method of despreading fast-hopping FH-SS signals described previously, since the major problem is synchronizing the locally generated reference sequence. As with the FH-SS system, the clock for the local sequence generator is first set to the nominal chipping rate of the incoming signal and the chipping sequences are made to precess with respect to each other until correlation occurs. When this occurs, the sum in the accumulator at the end of each symbol period will increase significantly. At this time the phase of the chipping sequence is adjusted to maximize the signal energy by adjusting the frequency of the clock for the local sequence generator under the control of the energy detector.

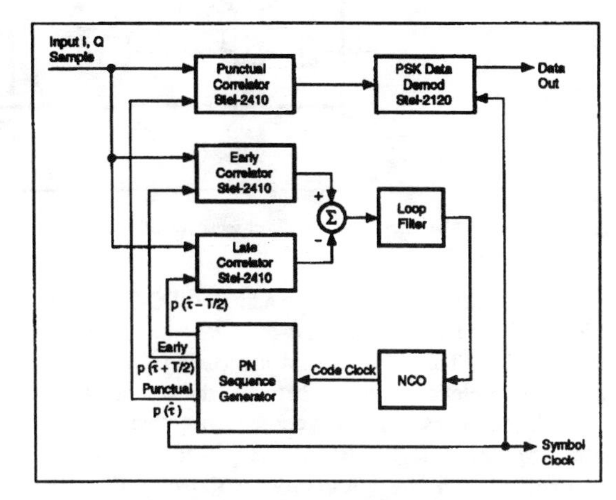

Fig. 9. Correlator despreader for direct-sequence SS.

One method of controlling the phase of the chipping sequence is the "delay-lock loop" technique [29]. Figure 9 illustrates how three commercially available correlator ASICs can be used to create a delay-lock discriminator and a correlation despreader.

One correlator is used as the "on-time" correlator, and the other two are used as "early" and "late" correlators, with the despreading sequences being fed into them half a chip early and late, respectively. The difference between the outputs of these two correlators forms a timing discriminator function going to zero when the timing is optimal. This method is used very frequently because of its excellent performance.

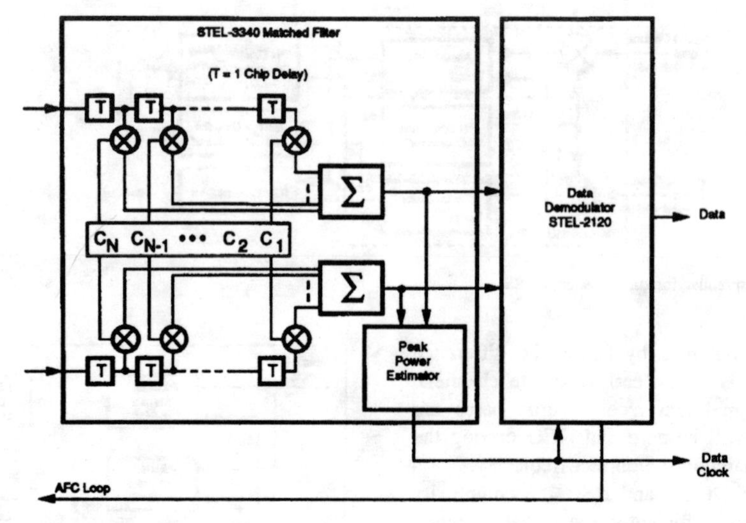

Fig. 10. Matched-filter despreader for direct-sequence SS.

The second method of despreading a DS-SS signal, which is most suitable for short-code systems, is with a matched filter. This is a parallel approach to computing the cross-correlation function between the reference and received sequences, with the consequence that it offers faster acquisition, at the expense of power consumption. The matched-filter based despreader is shown in Fig. 10.

In this system, the entire reference sequence is stored in a parallel register, instead of being generated serially. The incoming signal samples are also stored in another register of equal length, so that the $n-1$ previous samples of the signal are available, as well as the current sample, where n is the spreading code sequence length. The cross correlation between these two sequences (sum of products) is computed once per chip (as opposed to once per symbol in the correlator despreader). As the incoming signal moves down the signal register, chip by chip, its sequence will match the reference once per symbol, giving high cross correlation. At this time the symbol timing is automatically derived from the energy output of the correlation. Note that, depending on circumstances, no sequence acquisition process may be required; the signal is acquired during the first complete symbol received, i.e., when the threshold is crossed.

Although the code-matched filter technique is conceptually simpler than the correlator technique, its simplicity belies the complexity of its implementation, with two multipliers (typically 1 by 3 b) per chip of the code length. Thus this technique is limited in practice to relatively short spreading sequences, typically 64 chips or less, whereas the correlator despreader does not increase significantly in complexity as the length of the code increases, making it much more suitable for long-code system.

XXII. Conclusion

Spread-spectrum modulation offers several advantages such as precise timing, tolerance to interference, multipath amelioration, and the ability to share a common bandwidth with other signals. Until recently, the complexity of many spread-spectrum modems and navigation equipment had restricted their application to military use. However, due to the recent advances in the state of the art in digital signal processing and application-specific integrated circuits it is feasible now to use spread spectrum in low-cost commercial applications such as GPS receivers, satellite-based and terrestrial cellular telephone, and wireless PBX systems.

References

[1] Cook and Marsh, "An introduction to spread spectrum," *IEEE Commun. Mag.*, pp. 8–16, Mar. 1983.
[2] Pickholtz, Schilling, and Milstein, "Theory of spread-spectrum communications—A tutorial," *IEEE Trans. Commun.*, pp. 855–883, May 1982.
[3] Simon, *et al.*, *Spread Spectrum Communications.* Rockville, MD: Comput. Sci. Press, 1985.
[4] C. Loo, "A statistical model for a land mobile satellite link," *IEEE Trans. Vehic. Technol.*, pp. 122–127, Aug. 1985.
[5] Barts and Stutzman, "Modeling and simulation of mobile satellite propagation," *IEEE Trans. Antennas Propagat.*, pp. 375–382, Apr. 1992.
[6] C. Loo, "Measurements and models of a land mobile satellite channel and their applications to MSK signals," *IEEE Trans. Vehic. Technol.*, pp. 114–121, Aug. 1987.
[7] J. S. Butterworth, "Propagation measurements for land mobile satellite system at 1542 MHz," Commun. Res. Cent., Dept. Commun., CRC Tech. Note 724, Aug. 1984.
[8] Irvine and McLane, "Symbol-aided plus decision-directed reception for PSK/TCM modulation on shadowed mobile satellite fading channels," *IEEE J. Selected Areas Commun.*, pp. 1289–1299, Oct. 1992.
[9] Gilhousen *et al.*, "Increased capacity using CDMA for mobile satellite communication," *IEEE J. Selected Areas Commun.*, pp. 503–514, May 1990.
[10] Hansen and Meno, "Mobile fading—Rayleigh and lognormal superimposed," *IEEE Trans. Vehic. Technol.*, pp. 332–335, Nov. 1977.
[11] Lutz *et al.*, "The land mobile satellite communication channel—Recording, statistics, and channel model," *IEEE Trans. Vehic. Technol.*, pp. 375–386, May 1991.
[12] Goldhirsh and Vogel, "An overview of results derived from mobile-satellite propagation experiments," in *Proc. Int. Mobile Satellite Conf.*, (Ottawa, Ont., Canada, 1990), pp. 219–224.

[13] ——, "Roadside tree attenuation measurements at UHF for land-mobile satellite systems," *IEEE Trans. Antennas Propagat.*, vol. AP-35, pp. 589–596, 1987.
[14] ——, "Mobile satellite system fade statistics for shadowing and multipath from roadside trees at UHF and *L*-band," *IEEE Trans. Antennas Propagat.*, pp. 489–498, Apr. 1989.
[15] ——, "Mobile satellite system propagation measurements at *L*-band using MARECS-B2," *IEEE Trans. Antennas Propagat.*, pp. 259–264, Feb. 1990.
[16] USA delegates at the ICU, "Impact of propagation impairments on the design of LEO mobile satellite systems providing personal communication services," CCIR Study Groups, US WP-8D-14, Oct. 1992.
[17] "Final Report of the Majority of the Active Participants of Informal Working Group 1 to Above 1 GHz Negotiated Rulemaking Committee," MSSAC-41.6 (Final), IWG1-81 (Final), FCC, Apr. 1993.
[18] W. Rubow, "MOBILESTAR field test program," in *Proc. Mobile Satellite Conf.* (Pasadena, CA, May 1988), pp. 189–194.
[19] A. J. Viterbi, "When not to spread spectrum—A sequel," *IEEE Commun. Mag.*, vol. 23, pp. 12–17, Apr. 1985.
[20] "GLOBALSTAR System Application," presented by Loral Cellular Systems Corp. before the Federal Communications Commission, Washington, DC, June 1991.
[21] "Application of TRW Inc. For Authority to Construct a New Communications Satellite System Odyssey," presented by TRW Inc., before the Federal Communications Commission, Washington, DC, May 1991.
[22] "Application of Motorola Satellite Communications, Inc. for IRIDIUM," presented by Motorola Inc. before the Federal Communications Commission, Washington, DC, Dec. 1990.
[23] T. S. Rappaport, S. Y. Seidel, and K. Takamizawa, "Statistical channel impulse response models for factory and open plan building radio communication system design," *IEEE Trans. Commun.*, vol. 39, pp. 794–807, May 1991.
[24] A. A. M. Saleh and R. A. Valenzuela, "A statistical model for indoor multipath propagation," *IEEE J. Selected Areas Commun.*, vol. SAC-5, no. 2, pp. 128–137, Feb. 1987.
[25] Wepman *et al.*, "Spectrum usage measurements in potential PCS frequency bands," NTIA Rep. 91–279, Sept. 1991.
[26] E. H. Walker, "Penetration of radio signals into buildings in the cellular radio environment," *Bell Syst. Tech. J.*, vol. 62, no. 9, pp. 2719–2734, Nov. 1983.
[27] D. Richer, "A block estimator for offset QPSK signaling," in *Nat. Telecommunications Conf. Rec.*, vol. 2, pp. 30-6–30-11, 1975.
[28] F. M. Gardner, "Hang-up in phase-lock loops," *IEEE Trans. Commun.*, vol. COM-25, pp. 1210–1214, Oct. 1977.
[29] J. J. Jr. Spilker and D. T. Magill "The delay-lock discriminator–An optimum tracking device," *Proc. IRE*, pp. 1403–1416, Sept. 1961.

D. Thomas Magill (Member, IEEE) received the B.S.E. degree from Princeton University, Princeton, N J, in 1957, and the M.S. and Ph.D. degrees in 1960 and 1964, respectively, from Stanford University, Stanford, CA.

From 1958 to 1960 he worked in the field of ionospheric research as a Research Assistant at the Radioscience Laboratory, Stanford University. From 1960 to 1964 he worked in the Communications and Controls Research Department, Lockheed Aircraft Corporation, Palo Alto, CA, where he did research on and development of spread-spectrum systems including the implementation of the first delay-lock loop. From 1964 to 1970 he was employed by Philco-Ford Corporation, Palo Alto, CA, as a Staff Scientist and later as Department Manager of the Communication Sciences Department. At Philco-Ford he was involved in the development of spread-spectrum and time-division multiple access systems. From 1970 to 1983 he was employed by the Telecommunications Sciences Center, SRI International, Menlo Park, CA. As Associate Director of the center he worked in the areas of linear predictive vocoders and performance evaluation of frequency-division multiple-access systems as well as a wide variety of military and commercial projects. Since 1983 he has been with Stanford Telecommunications, Inc., Sunnyvale, CA, where he is presently a Technical Director in the Satellite Communications Operation. At Stanford Telecommunications he works in the development and analysis of time-division and spread-spectrum multiple access systems both for satellite and terrestrial wireless communication systems.

Francis D. Natali (Senior Member, IEEE) received the B.S.E.E. degree in 1960 from Rensslaer Polytechnic Institute, Troy, NY, and the M.S. and Ph.D. degrees in 1964 and 1967, respectively, from the State University of New York at Buffalo.

He started his career at the Sylvania Amherst Laboratories in Amherst, N Y, working on electronic countermeasures. In 1962 he joined the Systems Research Department of Cornell Aeronautical Laboratory, Ithaca, NY, where he worked on various problems associated with strategic weapons systems. In 1967 he became a member of the Communications Sciences Department at Philco-Ford Corporation in Palo Alto, CA. He has been employed by Stanford Telecommunications Inc. since 1973 where he is currently a Vice President and Chief Engineer. He has worked on a variety of problems associated with spread-spectrum signaling, satellite navigation (GPS), digital signal processing, and signal parameter estimation. His current interest is in applying his expertise to commercial systems.

Gwyn P. Edwards (Member, IEEE) was educated at the University of Wales, Bangor, UK, receiving the B.Sc. degree in electrical engineering, in 1965, and the Ph.D. degree in 1969.

He then spent 10 years at Pye TMC, a division of Philips Telecommunications where he lead a team of engineers working on the practical application of digital filters and demodulators. He joined American Microsystems, Inc., where he was a Senior Applications Engineer for Digital Signal Processing products and then spent 3 years at Racal-Vadic, where he was responsible for the introduction of new IC technologies for the companies data modem products. Since 1986 he has been with the ASIC and Custom products Division of Stanford Telecom where he is responsible for technical support for the division's products. He has published over 30 papers in the fields of ASIC design, DSP, and communications and has been involved with IEEE ASSP group technical subcommittees.

SECTION 1.2
MULTIPATH AND FADING

In this section, three papers have been selected to provide an overview and a general understanding of multipath and fading modelling and effects on spread-spectrum systems.

The first paper, "*Introduction to Spread-Spectrum Antimultipath Techniques and Their Application to Urban Digital Radio*" by Turin, discusses basic models for multipath propagation, presents receiver design for multipath channel, and analyzes interference due to multipath and error probability.

The second and third articles are Part I (Characterization) and Part II (Mitigation) of the paper "*Rayleigh Fading Channels in Mobile Digital Communication Systems*" by Sklar. These papers provide a tutorial on Rayleigh fading in the UHF band (300 MHz - 3GHz), which is the band used for cellular and personal communication systems. The first part addresses fundamental fading manifestations and degradation. The second part discusses various methods used to mitigate the degradation, which includes the RAKE receiver for direct-sequence spread-spectrum systems.

Introduction to Spread-Spectrum Antimultipath Techniques and Their Application to Urban Digital Radio

GEORGE L. TURIN, FELLOW, IEEE

Abstract—In a combination tutorial and research paper, spread-spectrum techniques for combating the effects of multipath on high-rate data transmissions via radio are explored. The tutorial aspect of the paper presents: 1) a heuristic outline of the theory of spread-spectrum antimultipath radio receivers and 2) a summary of a statistical model of urban/suburban multipath. The research section of the paper presents results of analyses and simulations of various candidate receivers indicated by the theory, as they perform through urban/suburban multipath. A major result shows that megabit-per-second rates through urban multipath (which typically lasts up to 5 μs) are quite feasible.

I. INTRODUCTION

SOME DIGITAL radio systems must operate through an extremely harsh multipath environment, in which the duration of the multipath may exceed the symbol length.[1] Two disciplines combine to shed light on receiver design for this environment: the modeling and simulation of multipath channels and the theory of multipath and other diversity receivers.

In this paper, we first present a tutorial review of pertinent aspects of both underlying disciplines, particularly in the context of spread-spectrum[2] systems. We then carry out rough analyses of the performances of two promising binary spread-spectrum antimultipath systems. Finally, since the analyses contain a number of oversimplifications that make them heuristic rather than definitive, we present results of computer simulations of the two proposed configurations and others, as they operate through simulated urban/suburban multipath. The simulation results highlight the importance of using realistic simulations of complex channels rather than simplified analyses, or they show that the analytic results, although based on standard assumptions, are unduly optimistic.

II. MODELING MULTIPATH PROPAGATION

Ultimately, a reliable multipath model must be based on empirical data rather than on mathematical axioms. Two types

Manuscript received February 8, 1979; revised October 9, 1979. This work was supported by the National Science Foundation under Grant ENG 21512 and SRI International under Advanced Research Projects Agency Contract MDA 903-78-C-0216.

The author is with the Department of Electrical Engineering and Computer Sciences and the Electronics Research Laboratory, University of California, Berkeley, CA 94720.

[1] An example is the ARPA Packet Radio network [38].

[2] In a spread-spectrum system, the bandwidth W of the transmitted signals is much larger than $1/T$, the reciprocal of the duration of the fundamental signalling interval, so $TW >> 1$. The transmitted spectrum is said to be "spread" since a signal lasting T seconds need not occupy more than the order of $W \cong 1/T$ Hz of bandwidth, in which case $TW \cong 1$. See [6] for references on the spread-spectrum concept.

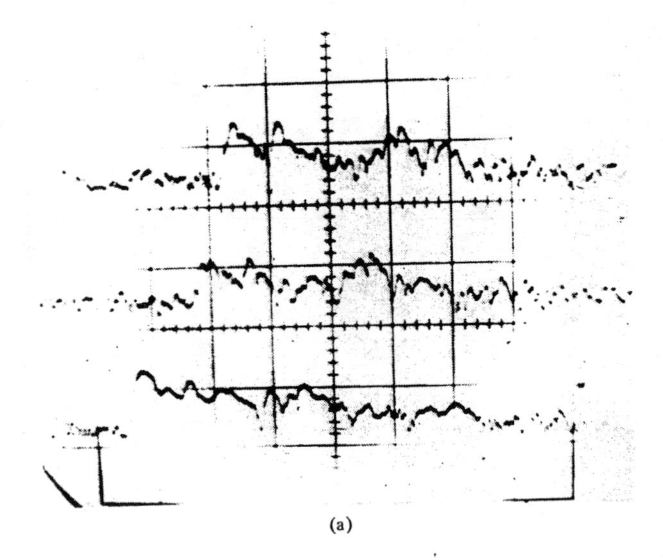

(a)

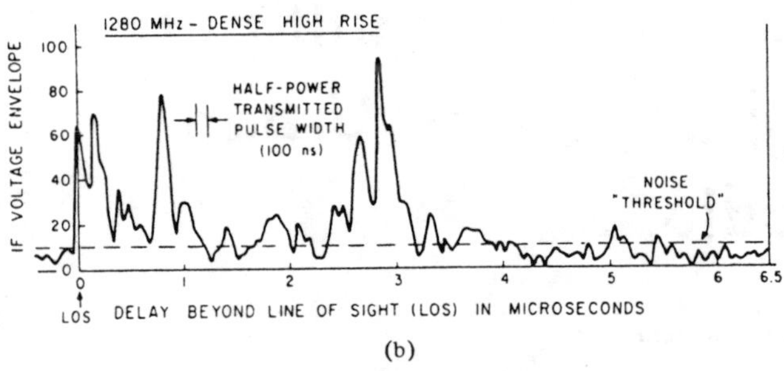

(b)

Fig. 1. Example of measured multipath profiles for a dense high-rise topography. (a) Top to bottom: 2920, 1280, 488 MHz. Vertical scale: 35 dB/cm. Horizontal scale: 1 μs/cm. Different apparent LOS delays are due to difference in equipment delays. (b) Middle trace of (a) on a linear scale.

of data are available. The more common type give the results of narrow-band or CW measurements, in which only a single fluctuating variable, a resultant signal strength, is measured [11]. Although the fluctuation of this strength variable depends on reception via multiple paths, these paths are not resolved by the measurements. We shall denote the results of such measurements as "fading" data rather than multipath data, because they determine a fading distribution of the single strength variable, e.g., Rayleigh, log-normal, Rice, etc.

Wide-band experimental data that characterize individual paths are less common [5], [10], [19], [33], [37]. In order to resolve two paths in such measurements, the sounding signal's bandwidth must be larger than the reciprocal of the difference between the paths' delays. Although bandwidths of 100 MHz or more have been used in exceptional circumstances to resolve path delay differences of less than 10 ns [10], the bulk of available data derives from 10-MHz bandwidths or less [5], [19], [33], [37]. In the latter measurements, paths separated by delays of more than 100 ns are resolved; multiple paths with smaller separations are seen as single paths.

The nature of the multipath measurements depends somewhat on the use envisioned for them. If understanding of the effect of the multipath channel on CW transmissions is required, measurements that show Doppler effects may be important [5], and these are reasonably related to a scattering-medium model of the channel [1], [12]. For high-rate packetized-data transmission, for vehicle-location sensing, and for other "bursty" transmissions, measurement of sequences of "impulse responses" of the propagation medium suffices.

The simulations of data reception that are presented in a later section are based on the "impulse response" approach, and it is to this type of model that we restrict ourselves. In order that the model and the simulations themselves be fully understood, we shall review here the experiments underlying the model. These were performed in urban/suburban areas [32], [33].

A. The Underlying Experiments

Pulse transmitters were placed at fixed, elevated sites in the San Francisco Bay Area. Once per second, these would simultaneously send out 100-ns pulses of carrier at 488, 1280, and 2920 MHz. The pulses were received in a mobile van that moved through typical urban/suburban areas, recording on a multitrace oscilloscope the logarithmically scaled output of the receiver's envelope detectors (see Fig. 1). Since the oscilloscope was triggered by a rubidium frequency standard that was synchronized with a similar unit at the transmitters prior to each experimental run, absolute propagation delays could be measured within experimental accuracies of better than 20 ns.

Four series of experiments were performed, in the following typical urban/suburban areas:

A) dense high-rise—San Francisco financial district,
B) sparse high-rise—downtown Oakland,
C) low rise—downtown Berkeley,
D) suburban—residential Berkeley.

In each area, regions of dimensions roughly 500–1000 ft (along the transmitter-receiver line of sight) by 2500–4000 ft (tangential to line of sight) were exhaustively canvassed, with care taken to include proper topographic cross sections: intersections, midblocks, points at which the transmitter site was visible or occluded, etc. About 1000 frames of data of the type shown in Fig. 1(a) were obtained in each area.

B. A Fundamental Model

The model upon which data reduction was based was one first posed in [27]. In this model, it is assumed that a transmission of the form

$$s(t) = \text{Re}\,[\sigma(t) \exp(j\omega_0 t)] \tag{1}$$

will be received as

$$r(t) = \text{Re}\,[\rho(t) \exp(j\omega_0 t)] + n(t) \tag{2}$$

where

$$\rho(t) = \sum_{k=0}^{K-1} a_k\, \sigma(t - t_k) \exp(j\theta_k). \tag{3}$$

In (1)–(3), $\sigma(t)$ is the complex envelope of the transmission, i.e., $|\sigma(t)|$ is its amplitude modulation and $\tan^{-1}[\text{Im}\,\sigma(t)/\text{Re}\,\sigma(t)]$ is its phase modulation. The transmission is received via K paths, where K is a random number that may vary from transmission to transmission. The kth path is characterized by three variables: its strength a_k, its modulation delay t_k, and its carrier phase shift θ_k. The waveform $n(t)$ is an additive noise component.

In the context of the spread-spectrum systems on which we shall concentrate, it is desirable to assume that all paths are

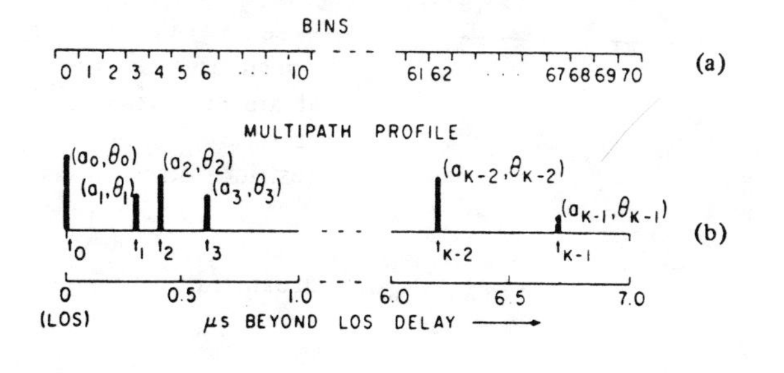

Fig. 2. Discrete-time model. (a) Division of excess delay axis into seventy one 100-ns bins. (b) A typical multipath profile. (c) Discrete-time path-delay indicator string, $\{\tau_l\}_{l=0}^{70}$.

resolvable, i.e., that

$$|t_k - t_l| > 1/W, \qquad \text{for all } k \neq l \tag{4}$$

holds, where W is the transmission bandwidth. Distinct paths in the physical medium that violate this resolvability condition are not counted separately, since they cannot be distinguished by a measurement using bandwidth W. Instead, any two paths—call them k_1 and k_2—for which $|t_{k_1} - t_{k_2}| < 1/W$ are considered as a single path in (3), with a common delay $t_k \cong t_{k_1} \cong t_{k_2}$ and a strength/phase combination given by

$$a_k \exp(j\theta_k) \triangleq a_{k_1} \exp(j\theta_{k_1}) + a_{k_2} \exp(j\theta_{k_2}).$$

It is the triplet $\{t_k, a_k, \theta_k\}$ that is to be determined for each "resolvable" path. To be sure, if a continuum of paths existed, it would be difficult uniquely to cluster the "subpaths" into paths. But many media, including the urban/suburban one, have a natural clustering, e.g., groups of facades on buildings, that make the model feasible.

C. A Discrete-Time Approximation to the Model

In addition to the additive random noise $n(t)$ in (2), the received signal $r(t)$ is therefore characterized by the random variables $\{t_k\}_0^{K-1}$, $\{a_k\}_0^{K-1}$, $\{\theta_k\}_0^{K-1}$ and K. The purpose of data reduction from the "multipath profiles" exemplified by Fig. 1 was to obtain statistics of these random variables upon which to base a simulation program. A number of generations of statistical models—based both on the data and on physical reasoning when the data were insufficient or undecisive—ensued [8], [9], [25], [26], [32]. The following final version emerged.

Each multipath profile starts with the line-of-sight (LOS) delay, which is chosen as the delay origin. Since the resolution of the original experiment is 100 ns, the delay axis is made discrete by dividing it into 100-ns bins, numbered from 0 to 70. Bin 0 is centered on LOS delay, subsequent bins being centered on multiples of 100 ns. The delay of any physical path lying in bin l is quantized to $100l$ ns, the delay of the bin's center. Fig. 2 shows the bin structure, a multipath profile, and the resulting discrete-time path-delay structure. Notice that only paths with delays less than 7.05 μs beyond LOS delay are encompassed in this model; experimental evidence shows that significant paths with larger delays are highly improbable.

The path-delay sequence $\{t_k\}_0^{K-1}$ is approximated by a string $\{\tau_l\}_0^{70}$ of 0's and 1's, as shown in Fig. 2(c). If a path

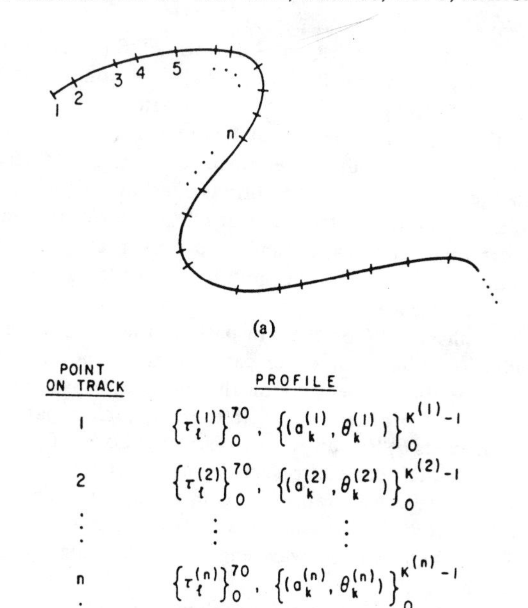

Fig. 3. Spatial variation of the multipath profile. (a) A vehicle's track, with spatial sample points. (b) The sequence of multipath profiles at the sample points on the track.

exists in bin l, $\tau_l = 1$; otherwise $\tau_l = 0$. In the sample string in Fig. 2(c), only $\tau_0, \tau_3, \tau_4, \tau_6, \cdots, \tau_{62}, \tau_{67}$ are nonzero, corresponding to the quantized path delays $\tilde{t}_0 = 0$, $\tilde{t}_1 = 300$ ns, $\tilde{t}_2 = 400$ ns, $\tilde{t}_3 = 600$ ns, $\cdots$, $\tilde{t}_{K-2} = 6200$ ns, $\tilde{t}_{K-1} = 6700$ ns. ($\tilde{t}_k$ is the value of t_k, as quantized to the nearest 100 ns.)

Associated with each nonzero τ_l is the corresponding (a_k, θ_k) pair. Thus the discrete-time model is completed by appending to the τ_l string a set of strength-phase pairs $\{(a_k, \theta_k)\}_{k=0}^{K-1}$, where the index k refers to the kth nonzero entry in the τ_l string. This is shown in Fig. 2(b).

The discrete-time model of Fig. 2 pictures the multipath profile at a single point in space. A *sequence* of such profiles is needed to depict the progression of multipath responses that would be encountered by a vehicle following a track such as shown in Fig. 3(a). One imagines points $1, 2, \cdots, n, \cdots$, arbitrarily placed on the track, at each of which a multipath response is seen. The discrete-time versions of these responses are arrayed in Fig. 3(b), where an additional spatial index n has been superscribed on all variables.

One begins to recognize the complexity of the model and of the required reduction of experimental data on realizing that in addition to the need for first-order statistics of the random variables $\tau_l^{(n)}$, $a_k^{(n)}$, $\theta_k^{(n)}$, and $K^{(n)}$ (where $0 \leq l \leq 70; 0 \leq k \leq K^{(n)} - 1; 1 \leq n < \infty$), there are two dimensions along which at least second-order statistics are necessary: temporal and spatial. For each profile (fixed n), there are temporal correlations of the delays, strengths and phases of the several paths; in addition, there are spatial correlations of these variables at neighboring geographical points.

The reduction of experimental data [9], [25], [33] and physical reasoning led to the following model, which was the basis for simulation.

1) The $\{\tau_l^{(n)}\}$ string of the nth profile is a modified Bernoulli sequence, in which the probability of a 1 in the lth place depends on: a) the value of l; b) whether a 1 or a 0 occurred in the $(l-1)$th place of the same profile; c) whether a 1 or a 0 occurred in the lth place of the $(n-1)$th profile.

2) The strength $a_k^{(n)}$ of the kth path of the nth profile is conditionally log-normally distributed,[3] the conditions being the values of strengths of the $(k - 1)$th path of the nth profile and of the path with the closest delay in the $(n - 1)$th profile; appropriate empirically determined correlation coefficients govern the influence these conditions exert on $a_k^{(n)}$. The mean and variance of the distribution of $a_k^{(n)}$ are also random variables, drawn from a spatial random process that reflects large-scale inhomogeneities in the multipath profile as the vehicle moves over large areas.

3) The phase $\theta_k^{(n)}$ of the kth path of the nth profile is independent of phases of other paths in the same profile, but has a distribution depending on the phase of a path with the same delay in profile $(n - 1)$, if there is such a path; if no such path exists $\theta_k^{(n)}$ is uniformly distributed over $[0, 2\pi)$.

4) The spatial correlation distances of the variables just described vary considerably, ranging from less than a wavelength for the θ_k's, through tens of wavelengths for the a_k's and τ_l's, to hundreds of wavelengths for the means and variances of the a_k's.

These statistics are more fully explained in [8], [9].

D. The Simulation Program

Hashemi's simulation program SURP, based on the statistics just outlined, generates sequences of multipath profiles, as depicted in Fig. 3. If one were to examine a sequence of such profiles, he would see paths appearing and disappearing at a rate depending on the spacing of points on the vehicle's track (Fig. 3(a)). Profiles at only slightly separated geographical points would look very similar, with high correlations of path delays and strengths (and, for *very* close points, phases). Profiles at greatly separated points would not only have grossly dissimilar $\{\tau_l\}$, $\{a_k, \theta_k\}$ strings, but the gross strength statistics of these strings (e.g., the average strength of the paths in a string) would be dissimilar, reflecting the spatial inhomogeneity incorporated into the model. The "motion picture" of simulated profiles just described is in fact very much like experimental data [37].

The simulation program can be run, using empirically determined parameters, for each of the three frequencies and four areas of the original experiment. Long sequences of strings were in fact generated for each of the twelve frequency/area combinations, assuming that the points on the vehicle track are uniformly spaced by distance d. An example of such sequences is given later in Fig. 25. For various values of d, the statistics of the simulated sequences were then compared with the original empirical statistics. Excellent agreement was obtained [8], [9]. (See Fig. 4 for examples.)

It should be noted that initial simulation experiments on urban/suburban radio ranging and location systems, using a rudimentary propagation simulation program preceding SURP, gave results which compared extraordinarily accurately with actual hardware experiments [34]. In particular, it was verified that although the data upon which the simulation program is based were taken in the San Francisco Bay area, one can expect simulation results that are not correspondingly restricted geographically. For example, use of the Area-A parameters in the program led to results that are as applicable to, say, downtown New York City or Chicago as to downtown San Francisco. This initial success encouraged the development of the more elaborate simulation capability just described.

Thus the sequence generated by SURP, described above, provide a data base with which to perform accurate experiments with urban/suburban radio systems, and the results of

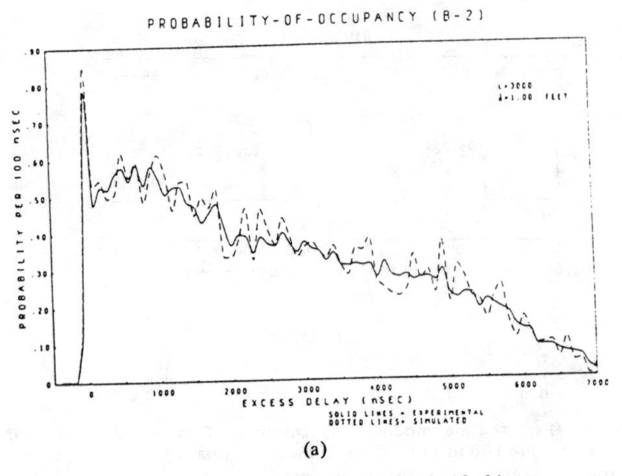

(a)

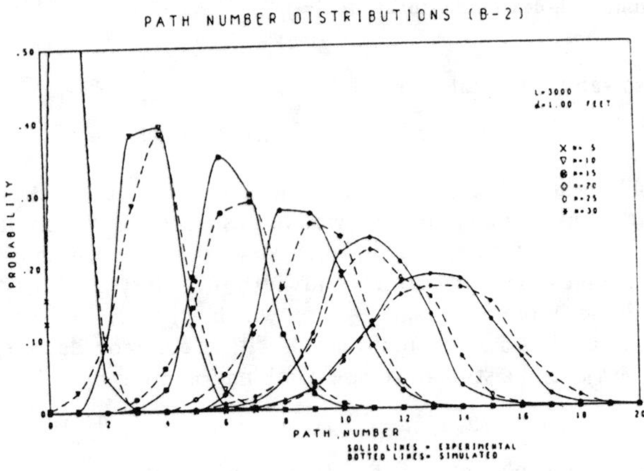

(b)

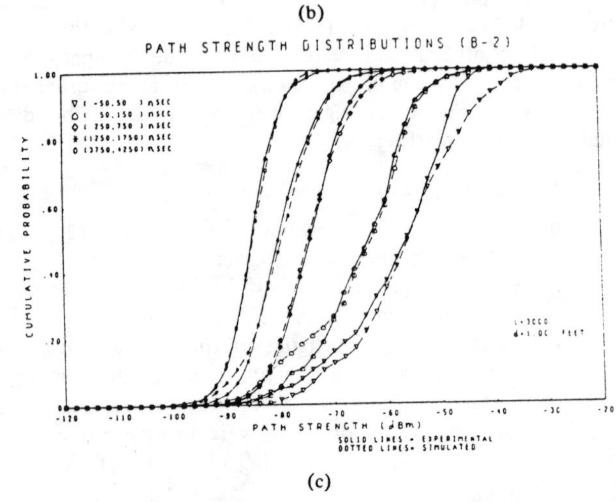

(c)

Fig. 4. Comparison of empirical statistics with statistics of simulation runs: sparse high-rise, 1280 MHz; 3000 simulation samples at 1-ft spacings. Solid curves: empirical; broken curves: simulation. (a) Ordinate is probability that a path occurs within ± 50 ns of abscissa value. (b) Ordinate is probability that there are the number of paths given by the abscissa within the first N bins. (c) Ordinate is the probability that the strength of a path in the indicated delay interval is less than the abscissa value.

[3] Actually, Suzuki [25] showed that paths with small delays (beyond LOS delay) were better modeled by Nakagami distributions, but Hashemi [9] was forced to approximate these log-normally because of the complexity of the simulation program.

these experiments can be expected to have wide applicability to typical urban/suburban topographies.

III. Design of Multipath Receivers

Multipath reception is one form of diversity reception, in which information flows from transmitter to receiver via the natural diversity of multiple paths rather than via the planned diversity of multiple frequency channels, multiple antennas, multiple time slots, etc. Thus instead of regarding the multipath phenomenon as a nuisance disturbance whose effects are to be suppressed, it should be regarded as an opportunity to improve system performance.

Two bodies of work in the literature are concerned with multipath receiver design. The older (see, e.g., [1], [4], [22], [23], [27]) concentrates on the explicit diversity structure of resolvable paths; its thrust is to take advantage of this structure by optimally combining the contributions of different paths. In its simplest form, this approach ignores the intersymbol interference that can be caused when the multipath medium delays a response from a transmitted symbol into intervals occupied by subsequent symbols, an approach that is justified only when the duration of the transmitted symbol is large compared with the duration of the multipath profile.

More recently [15]-[18], equalization techniques that were developed for data transmission over telephone lines [14] have been applied to the radio multipath problem. Here, receiver design concentrates on reduction of the effects of intersymbol interference, and the diversity-combining properties of the receiver are only implicit. This approach appears most suitable when the paths are not resolvable and when the symbol duration is much smaller than the multipath profile's "spread."

A melding of the two approaches is currently being worked on by L-F. Wei of ERL, UC Berkeley. Since we are concerned here with the case in which resolvability condition (4) is satisfied, we shall in this paper pursue only the former diversity-oriented approach, as modified to take into account the deleterious effects of intersymbol interference. Instead of indulging in general and complex derivations, however, we shall present results using a tutorial "building block" approach, employing intuitive arguments that are justified by references to more formal developments in the literature.

A. The Optimal Single-Path Receiver

We start with the simple case in which the channel comprises only one path: $K = 1$ in (3). We assume initially that the path strength a_0 and delay t_0 are known ($t_0 = 0$ for simplicity), but that the carrier phase θ_0 is unknown, being a random variable, uniformly distributed over $[0, 2\pi)$.[4]

Since the absence of multipath implies the absence of intersymbol interference (a point we discuss more fully later), we can concentrate on the reception of a single symbol, say over the interval $0 \leq t < T$. Knowledge of this interval of course implies some sort of synchronization procedure at the receiver, a question discussed below.

Suppose the received signal $r(t)$ is as in (2), with $0 \leq t < T$, and where, in (3), $K = 1$, $t_0 = 0$, a_0 is known, and θ_0 is random as described above. The transmitted signal $s(t)$ of (1) can be any of M possible waveforms

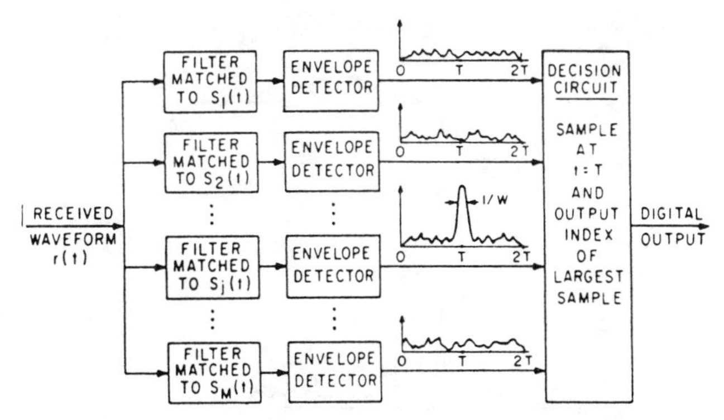

Fig. 5. Optimal noncoherent-phase receiver for single-path channel. (Equiprobable, equienergy signals, Gaussian noise.)

$$s_i(t) = \text{Re}\,[\sigma_i(t) \exp(j\omega_0 t)], \qquad i = 1, \cdots, M. \tag{5}$$

We assume that the transmitter has chosen among the s_i at random with equal probability and that the s_i have equal energy:

$$\int_0^T s_i^2(\tau)\,d\tau = \mathcal{E}, \qquad \text{for all } i. \tag{6}$$

The additive noise $n(t)$ of (2) is for simplicity assumed to be white and Gaussian, although non-Gaussian noise is also common in urban radio communication. (See [39] for a comprehensive survey of urban noise.)

It is well known that the optimal receiver—i.e., the receiver that decides which s_i was sent with minimum probability of error—has the form depicted in Fig. 5 (see, e.g., [36]). As shown there, $r(t)$ is passed through a bank of M filters, "matched" respectively to $s_i(t)$, $i = 1, \cdots, M$, i.e., having impulse responses $s_i(T - t)$, $0 \leq t < T$ [29]. The filter outputs are envelope detected and the envelopes sampled at $t = T$ and compared. The index $i = 1, \cdots, M$ of the largest sample is the receiver's output.

In Fig. 5, we have shown the outputs of the envelope detectors when $s_j(t)$ is the transmitted signal and when the received noise component $n(t)$ is negligible, assuming that the s_i have been "well chosen." This latter assumption means that if we define complex correlation function

$$\gamma_{ik}(t) \triangleq \int_0^T \sigma_i^*(\tau)\,\sigma_k(\tau - t)\,d\tau, \qquad i, k = 1, \cdots, M, \quad |t| \leq T \tag{7}$$

then [27]

$$|\gamma_{ik}(t)| \ll 2\mathcal{E}, \qquad \text{all } k \neq i, \qquad \text{all } i \tag{8a}$$

$$|\gamma_{ii}(t)| \ll 2\mathcal{E}, \qquad \text{for } |t| > \frac{1}{W}, \qquad \text{all } i \tag{8b}$$

where W is the bandwidth shared by all s_i, and, optimally but not necessarily,

$$\gamma_{ik}(0) = 0, \qquad \text{all } k \neq i, \qquad \text{all } i. \tag{8c}$$

In sketching the envelope detector outputs, we have also assumed that $TW \gg 1$, i.e., the signals are of the so-called

[4] Random path phases are assumed throughout this paper, since these generally change too rapidly in the mobile environment to make use of coherent-receiver techniques.

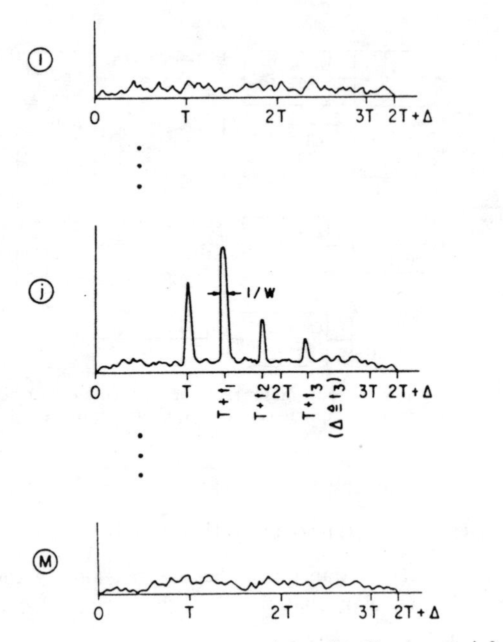

Fig. 6. Envelope detector output waveforms (small-noise case) for the receiver of Fig. 5; four-path channel.

spread-spectrum type [6]. None of the foregoing assumptions about the structure of the signal set $\{s_i(t)\}_{i=1}^{M}$ is necessary for the optimality of the single-path receiver of Fig. 5 to hold; but we shall invoke them when discussing multipath receivers later, as they become necessary or desirable.

The noisefree waveforms sketched in Fig. 5 are in fact given by [27]

$$e_l(t) \triangleq \tfrac{1}{2}\,|\gamma_{jl}(t-T)|, \qquad l=1,\cdots,M, \qquad 0 \leqslant t < 2T. \tag{9}$$

Conditions (8a, b) and $TW \gg 1$ assure that the jth output envelope $e_j(t)$ consists of a sharp "mainlobe" peak surrounded by low-level "sidelobes," while all other outputs have only low-level sidelobes. By careful signal selection, (8a, b) can be satisfied with the maximum sidelobe level in all these waveforms at a factor of about $2/\sqrt{TW}$ down from the mainlobe. Typically, for $TW = 100$, this means that the maximum sidelobe is about 17 dB or more down from the mainlobe.

If condition (8c) is also satisfied, the values of $e_l(T)$ for $l \neq j$ are zero at the sampling instant $t = T$, so that—in the absence of received noise—the receiver will not make an error. If the received noise is nonzero, the probability that the lth output exceeds the jth at $t = T$ for some $l \neq j$ is also nonzero, and it is this probability (of erroneous decision) that characterizes the receiver's performance.

A final feature of Fig. 5 is important. There, we have depicted the output waveforms when a single isolated symbol is sent during $0 \leqslant t < T$. If another symbol, say $s_l(t)$, were sent immediately afterward, in $T \leqslant t < 2T$, it is clear that the response to it would occur over the interval $T \leqslant t < 3T$. The mainlobe peak in the lth output would be centered exactly at $t = 2T$, precisely when all responses from the first symbol have died out. Thus on sampling the outputs at $t = 2T$, one would be able to make a decision based on the response to the second symbol alone, whence our previous statement that no intersymbol interference occurs in this single-path case.

B. The Optimal Multipath Receiver: Known Delays

If we should attempt to use the receiver of Fig. 5 when many paths are present ($K > 1$ in (3)), we would expect from the linearity of the medium and of the matched filters that the envelope detector output waveforms will look something like those in Fig. 6. Here, we have shown a four-path situation ($K = 4$).

The lth response in Fig. 6 is the envelope of the superposition of the several paths' contributions, and, when noise is absent, can be shown from (3) and (7) to be of the form [27]

$$e_l(t) \triangleq \frac{1}{2}\left|\sum_{k=0}^{K-1} a_k \exp(j\theta_k) \int_0^T \sigma_j^*(\tau)\,\sigma_l(\tau + T + t_k - t)\,d\tau\right|,$$
$$l = 1,\cdots,M, \qquad 0 \leqslant t < 2T + \Delta. \tag{10}$$

Under resolvability condition (4), the mainlobe peaks in the jth output $e_j(t)$ are distinct, and occur as shown at $t_0 = 0$, t_1, t_2, and $t_3 = \Delta$.[5] The heights of these peaks are proportional to the path strengths a_k. The sidelobes, both of $e_j(t)$ and of the other outputs (none of the latter having mainlobe peaks), are mixtures of sidelobes due to the several paths. We stress that Fig. 6 is drawn for the isolated transmission of a single waveform $s_j(t)$, $0 \leqslant t < T$.

The waveforms of Fig. 6 differ from those of Fig. 5 in several important respects.

1) Strong peaks are available in $e_j(t)$ at multiple times. If the decision circuit of Fig. 5 knows the values of the path delays $t_0, \cdots, t_{K-1}$, it can sample the contributions of all paths and combine them, affording the receiver the advantages of diversity reception, as discussed earlier. The ability to resolve the paths in Fig. 6 is the essence of the spread-spectrum approach. If we instead had $TW \cong 1$, the peaks in Fig. 6 would merge, and explicit diversity combination would no longer be available.

2) The sidelobe levels of all outputs is increased, since (10) shows the addition of multipath contributions.

3) The responses to the symbol sent during $0 \leqslant t < T$ now extends beyond $t = 2T$, thus overlapping with the responses to the next symbol, which is sent during $T \leqslant t < 2T$. That is, we now have intersymbol interference, caused by the multipath.

Effects 2) and 3) are deleterious, while 1) is favorable. As we shall see, however, the benefits of 1) usually far outweigh the deterioration caused by 2) and 3).

For the time being, we shall ignore the effects of intersymbol interference, and inquire into the structure of the optimum receiver for reception of a single symbol through multipath, assuming first that the path delays $\{t_k\}_0^{K-1}$ are known. However, we again assume random phases $\{\theta_k\}_0^{K-1}$, independently and uniformly distributed over $[0, 2\pi)$; we also assume that the path strengths $\{a_k\}_\theta^{K-1}$ are random, perhaps having different distributions.

Intuitively, one might expect under these conditions that the optimal receiver is still of the form of Fig. 5, but what

[5] The maximum excess delay anticipated in the channel—i.e., $\max t_{K-1} \cdot \min_{t_0}(t_{K-1} - t_0) \triangleq \Delta$—is called the multipath spread; it is by this amount that the waveforms of Fig. 6 can spread beyond those of Fig. 5. In the four-path example of Fig. 6, it is assumed that $t_3 - t_0$ achieves this maximum.

the decision circuit now samples each of its inputs at multiple times $T + t_k$, $k = 0, \cdots, K-1$, combines these samples for each input, and compares the resulting combined values; the decision would be the index of the largest combined value. Indeed this is the case, at least when (4) and (8b) hold so that the pulses in output j of Fig. 6 are distinct [27]. However, the optimal combining law is sometimes complicated, and depends on the statistics of the path strengths.

Suppose that the sample of the lth output envelope at time $T + t_k$ is x_{lk}. (In the absence of noise $x_{lk} = e_l(t_k)$ as given by (10).) Then, if all path strengths a_k are known, the optimal[6] combining of the samples is given by [27]

$$w_l = \sum_{k=0}^{K-1} \log_e I_0 \left(\frac{2a_k x_{lk}}{N_0} \right) \tag{11}$$

where I_0 is a Bessel function and where N_0 is the channel noise power density. If, on the other hand, the kth path strength is Rayleigh distributed with mean-square strength $\psi_k \triangleq E[a_k^2]$, and all path strengths are independent,[7] the optimal combining law is [27], [30]

$$w_l = \sum_{k=0}^{K-1} \frac{\psi_k x_{lk}^2}{N_0 + \psi_k \mathcal{E}} \tag{12}$$

where $\mathcal{E}$ is the common energy of the signals s_i, given by (6). More complicated combining laws for other strength distributions are given elsewhere [3], [27], [30]. In any case, a decision is made by comparing the w_l and favoring the largest.

Note that different combining laws accentuate the various samples in different ways. In (11), for example, the samples are approximately linearly combined, since $I_0(\cdot)$ increases approximately exponentially with its argument for large argument; but samples corresponding to paths with larger strengths are given more weight. In (12), the samples are square-law combined, thus accentuating the larger samples; but all samples from paths with large mean-square strength ($\psi_k \mathcal{E} \gg N_0$) are weighted equally while samples from weaker paths are suppressed. The essence of optimal combining laws is the relative accentuation of more credible data and the relative suppression of less credible data.

C. The Optimal Multipath Receiver: Unknown Delays

As indicated in Section II, the path delays $\{t_k\}_0^{K-1}$ and the number of paths K are often random variables, not known *a priori*. In order to determine the optimal receiver for this situation, we simplify somewhat from the path-delay model described in Section II. We now assume that the t_k's are independently chosen from a single common probability density distribution $p(t_k)$, $0 \leqslant t_k \leqslant \Delta$, and their indices subsequently reordered in order of increasing t_k. This model violates the assumptions in Section II in two respects. First, the resolvability condition (4) is not always met, since it is possible that two paths will be drawn from the distribution in such a way that $|t_k - t_l| < 1/W$. However, the probability that this will occur is small if there are no intervals of length $\leqslant 1/W$ in which substantial probability is concentrated; so $p(t_k)$ must be "diffuse," without high peaks and with $W\Delta \gg 1$.[8] Second, the method of generation of τ_l strings discussed in Section II (see Fig. 2) incorporates dependences among τ_l's for neighboring l's, which implies corresponding dependences in the associated delays t_k that are not incorporated in the simplified model just broached.

These variations from reality are not substantial for purposes of deriving a receiver that will be quasi-optimal for the real channel. The density distribution $p(t_k)$ of the simplified model can be determined from the empirical data described in Section II: it is just the path-occupancy curve exemplified by Fig. 4(a), as normalized to unit area by dividing it by the average number of paths $E(K)$. Were the simplified model used to generate the τ_l strings of Fig. 2, the path-number distributions would be Poisson distributions instead of the somewhat narrower distributions exemplified in Fig. 4(b).[9]

In deriving the optimal receiver for unknown delays, we assume that path strengths a_k and a_l are independent for all $k \neq l$, a deviation from the reality that paths whose delays are not greatly different generally have correlated strengths. As with our simplified delay model, we follow our intuition in assuming that the derived receiver, based on the simplified strength model, will be close to optimal in the real world.

With these mathematical simplifications, and assuming that (8b) holds,[10] the optimal receiver structure can be easily derived [4], [27]. This receiver computes the quantities

$$w_l \triangleq \int_T^{T+\Delta} p(t-T) F[x_l(t), t] \, dt, \qquad l = 1, \cdots, M \tag{13}$$

where $p(\cdot)$ is the path-delay density defined above, $x_l(\cdot)$ is the output envelope of the lth filter, and $F[\cdot,\cdot]$ is a time-varying nonlinear function. The w_l's are compared and a decision made favoring the index of the largest w_l.

The nonlinearities $F[\cdot,\cdot]$ depend on the path-strength statistics. If the path strengths are known, then (cf. (11))

$$F[x_l(t), t] = I_0 \left[\frac{2a(t-T)\, x_l(t)}{N_0} \right] \tag{14}$$

where $a(t)$ is strength of a path at delay t. If the path strengths are all Rayleigh distributed, with a path at delay t having mean-square strength $\psi(t)$, then (cf. (12))

$$F[x_l(t), t] = \exp \left[\frac{\psi(t-T)\, x_l^2(t)}{N_0 + \psi(t-T)\, \mathcal{E}} \right]. \tag{15}$$

Expressions for $F[\cdot,\cdot]$ for other path-strength statistics are given in [3], [4], [27].

In general, $F[x,t]$ is positive and monotone increasing in x, and can therefore be written as

$$F[x,t] = F[0,t] + \hat{F}[x,t] \tag{16}$$

[6] Strictly, the w_l's of (11), and of (12) below, are optimal only if (4) holds and if the $\gamma_{ii}(t)$ of (7) are identically zero for $|t| > 1/W$ rather than merely satisfying (8b); for practical purposes, satisfaction of (8b) suffices.

[7] The assumption of independent path strengths is a variation from the multipath model described in the previous section, and to that extent the resulting receiver is only quasi-optimal.

[8] It is tempting to apply subsequent results to a "channel sounding" receiver, in which $p(t_k)$ becomes an *a posteriori* distribution. But such a distribution would be highly peaked, and would therefore violate this "diffuseness" condition. We shall comment further on this point later, when discussing channel-sounding receivers.

[9] See [33], Fig. 5, for comparisons of these Poisson distributions with the actual empirical distributions.

[10] Again, as for known delays, we strictly should have $\gamma_{ii}(t) \equiv 0$ for $|t| > 1/W$, all i.

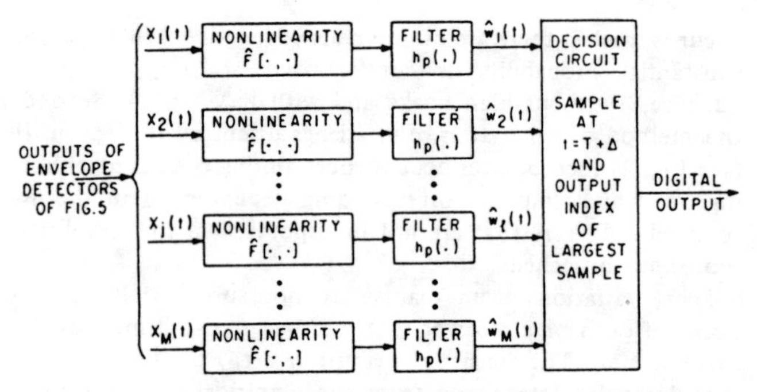

Fig. 7. Optimal noncoherent receiver for multipath channel with unknown delays. (Equiprobable, equi-energy signals; Gaussian noise; Poisson path delays, independent path strengths.)

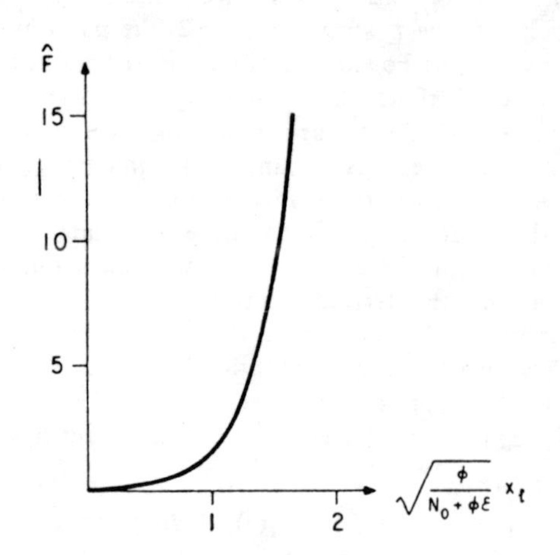

Fig. 8. The nonlinearities of Fig. 7 for Rayleigh-distributed path strengths with a common mean-square value ψ.

where $\hat{F}[x, t]$ is positive and monotone increasing in x. On substitution of (16) into (13), the $F[0, t]$ term leads to a component of w_l that is independent of l and can therefore be neglected in the comparison of the w_l's. That is, the receiver need only compare the quantities

$$\hat{w}_l \triangleq \int_T^{T+\Delta} p(t - T)\,\hat{F}[x_l(t), t]\,dt, \qquad l = 1, \cdots, M. \tag{17}$$

We note that these integrals can be realized by convolution, i.e., by generating the functions

$$\hat{w}_l(t) \triangleq \int_{t-\Delta}^{t} p[\Delta - (t - \tau)]\,\hat{F}[x_l(\tau), \tau]\,d\tau \tag{18}$$

and sampling them at $t = T + \Delta$. Thus $\hat{w}_l$ can be realized by passing $x_l(t)$ into a nonlinearity $\hat{F}$, passing the output of $\hat{F}$ into a filter with impulse response $h_p(t) = p(\Delta - t)$, and sampling the filter's output at $t = T + \Delta$. The decision circuit of Fig. 5 is therefore replaced by the circuitry of Fig. 7.

Notice that, in contradistinction to the combining laws of (11) and (12), in which the samples x_{lk}, $k = 0, \cdots, K - 1$, are combined linearly or quadratically, the combining law of (18) involves extreme nonlinearities of the exponential type. For example, if the path strengths are all Rayleigh distributed with a common mean-square value ψ, then the nonlinearity

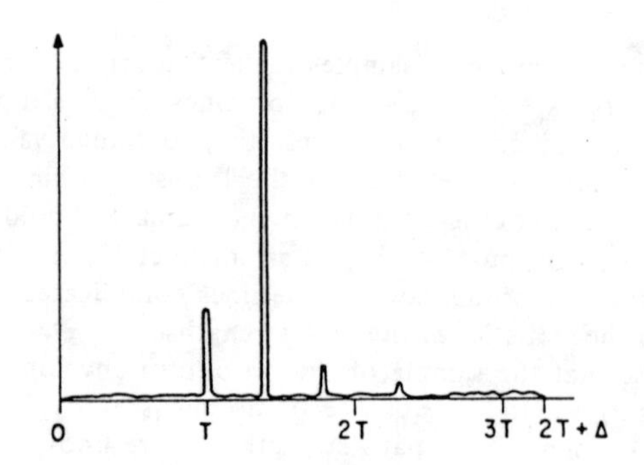

Fig. 9. Result of passage of jth waveform of Fig. 6 through the "Rayleigh" nonlinearity of Fig. 8.

in (18) is time-invariant, and of the form

$$\hat{F}[x_l, t] = \exp\left[\frac{\psi x_l^2}{N_0 + \psi \mathcal{E}}\right] - 1 \tag{19}$$

which is shown in Fig. 8. When passed through this nonlinearity, the jth envelope detector output of Fig. 6 is transformed into the waveform of Fig. 9. (On the same scale, the transformations of the other waveforms of Fig. 6 are negligibly small.)

The waveform of Fig. 9 illustrates a certain self-adaptivity implicit in (17). *A priori*, path delays are unknown. However, a large pulse in one of the $x_l(t)$'s at some instant t in $[T, T + \Delta]$ is convincing evidence that a path exists at delay $t - T$ (see Fig. 6); the larger the pulse, the more convincing is the evidence, and the more heavily the pulse is emphasized by the nonlinearity. On the other hand, whenever $x_l(t)$ is small, it is presumed to be caused by noise (or sidelobes) and it is strongly suppressed by the nonlinearity. The output of the nonlinearity thus presents data that are heavily adjusted by *a posteriori* evidence of the existence of paths. As shown by (17), these data are further weighted by the *a priori* knowledge of the probabilities of path occurrences implicit in the function $p(\cdot)$.

We stress that the illustrations in Figs. 6 and 9 are drawn for relatively large average SNR $\psi\mathcal{E}/N_0$, for which case the signal peaks are prominent and are highly emphasized with respect to the noise by the operation of $\hat{F}$. On the other hand, when $\psi\mathcal{E}/N_0$ is smaller, the noise-suppressing effect of the nonlinearities will not be great; that is, the various outputs $\hat{F}[x_l(t), t]$, $l = 1, \cdots, M$—even the jth—will be of the same scale. The signal peaks in the jth output will then contribute comparatively little to the jth integral in (17), and the receiver will be prone to error. One can see that known-path-delay receivers (such as these based on (11) and (12)), by only having to sample the envelope detector outputs at the positions of signal peaks and not the noise contributions at other instants, will perform better than unknown-path-delay receivers.

D. A DPSK Receiver

As previously mentioned, phase-coherent techniques have been avoided here because of the complexity of coherent receivers and because of the rapid time variations of path phases with vehicle motion. On the other hand, differentially coherent techniques clearly will show promise if the path phases do not vary appreciably over the interval during which two successive signals are sent; this is the usual case. Although

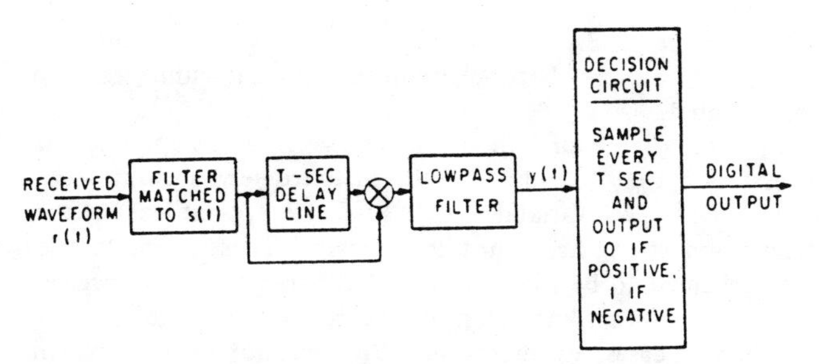

Fig. 10. Optimal DPSK receiver for the one-path case.

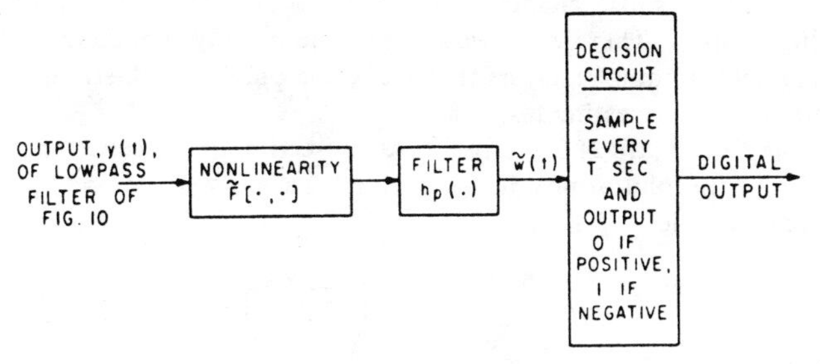

Fig. 11. Conjectured optimal DPSK receiver for multipath channel with unknown delays. (Equiprobable symbols; Gaussian noise; Poisson path delays; independent path strengths.)

the form of the optimal multipath receiver for differentially coherent signalling is not known, it is strongly suggested by that for the one-path case.

Let a signal of the form of $s(t)$ of (1) be differentially phase shift keyed: a binary 0 is sent by following a previous transmission—say $\pm s(t)$, $t \in [0, T)$—by a transmission of the same polarity in the succeeding interval—$\pm s(t - T)$, $t \in [T, 2T)$; a binary 1 is sent by changing polarity—$\pm s(t)$ followed by $\mp s(t - T)$. In the single-path case, the received waveform in $[0, T)$ is (cf. (1)-(3), with $K = 1$, $t_0 = 0$)

$$r_1(t) = \pm a_0 \operatorname{Re}[\sigma(t) \exp(j\omega_0 t) \exp(j\theta_0)] + n_1(t), \qquad 0 \leqslant t < T \quad (20)$$

where $n_1(t)$ is a noise waveform. If $d = 0$ or 1 is the data symbol sent, the received waveform in $[T, 2T)$ is

$$r_2(t) = \pm(-1)^d a_0 \operatorname{Re}[\sigma(t - T) \exp[j\omega_0(t - T)] \exp(j\theta_0)] + n_2(t), \qquad T \leqslant t < 2T \quad (21)$$

where $n_2(t)$ is a noise waveform and we have assumed that a_0 and θ_0 have not changed from their values during $[0, T)$. The optimal receiver in this case is well known [24] to have the form of Fig. 10. (The low-pass filter in Fig. 10 serves only to eliminate the double-frequency terms generated by the multiplier.)

By performing the operations shown in Fig. 10 on the signal components of (20) and (21), i.e., neglecting the noise components, one can easily show that the output of the low-pass filter has the form

$$y(t) = \tfrac{1}{2}(-1)^d a_0^2 |\gamma(t - 2T)|^2 \quad (22)$$

where

$$\gamma(t) \triangleq \int_0^T \sigma^*(\tau)\, \sigma(\tau - t)\, d\tau, \qquad |t| \leqslant T \quad (23)$$

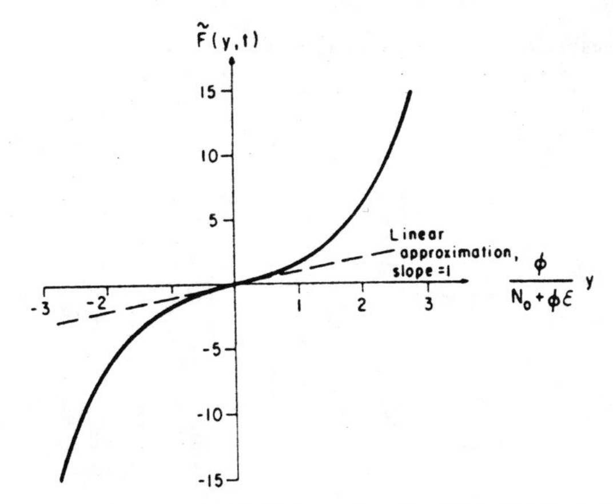

Fig. 12. The nonlinearity of Fig. 11 for Rayleigh-distributed path strengths with a common mean-square value ψ.

and $\gamma(t) \equiv 0$ elsewhere. On comparing (22), (23) with (7), (9), we see that $y(t)$ is twice the squared envelope of the response of the matched filter to the waveform $s(t)$, delayed by T seconds and keyed by $(-1)^d$. Thus in the absence of noise, the decision circuit will output exactly the input digit; when noise is present, errors will of course occur.

We now conjecture that for the case of multipath obeying the simplified model of Section III-C above, i.e., Poisson-distributed, resolvable path delays and independent path strengths, the optimal DPSK receiver bears the same relationship to Fig. 10 as the receiver of Fig. 7 bears to that of Fig. 5. More precisely, we conjecture that the optimal receiver has the form of Fig. 11. There, the nonlinearity

$$\tilde{F}(y, t) \triangleq \hat{F}(\sqrt{|y|}, t) \operatorname{sgn} y \quad (24)$$

is a bipolar version of $\hat{F}$, adjusted for the fact that the nonlinearity's input is related to the *square* of the matched filter output envelope, rather than, as in Fig. 7, the envelope itself. A graph of $\tilde{F}$ for the $\hat{F}$ of (19) is shown in Fig. 12; compare this to Fig. 8.

The samples taken every T sec by the decision circuit in Fig. 11 are timed to capture the extrema of the output of the path integrating filter $h_p(\cdot)$. Fig. 13 shows some appropriate waveforms illustrating this point. In the absence of noise, the output $y(t)$ of the product detector of Fig. 10, and therefore the input to the nonlinearity in Fig. 11, is approximately[11] of the form (cf. (22))

$$y(t) = \frac{1}{2}(-1)^d \sum_{k=0}^{K-1} a_k^2 |\gamma(t - 2T - t_k)|^2, \qquad 2T \leqslant t \leqslant 2T + \Delta \quad (25)$$

where Δ is, as before, the multipath spread. A sequence of such noiseless $y(t)$'s is shown in Fig. 13(c), for the input symbol sequence 10110 · · ·. The corresponding outputs of $\tilde{F}$ and $h_p(\cdot)$ are shown in Fig. 13(d) and (e), assuming

[11] In (25), we have assumed that the sidelobes of $\gamma(t)$ of (23) satisfy a condition of the form of (8a), so that "interpath interference" is negligible; i.e., the sidelobes due to path l are small at the peak due to path k, $l \neq k$, $l = 0, \cdots, K - 1$. We have also neglected intersymbol interference, i.e., assumed that the sidelobes due to all paths in one signaling interval are negligible insofar as they extend into adjacent intervals. We return to the questions of interpath and intersymbol interference below.

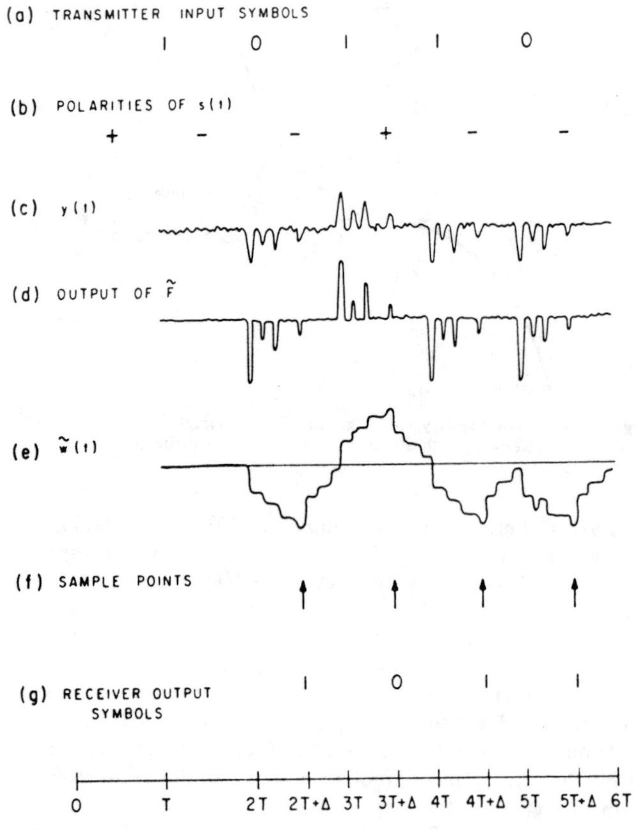

Fig. 13. Symbols and waveforms illustrating the operation of the DPSK multipath receiver of Fig. 11.

that

$$h_p(t) = \begin{cases} 1, & 0 \leqslant t \leqslant \Delta \\ 0, & \text{elsewhere.} \end{cases} \tag{26}$$

The output of $h_p(\cdot)$ is sampled at instants $nT + \Delta$, $n = 2, 3, \cdots$, and leads to the receiver outputs shown; these are delayed by $T + \Delta$ seconds from the corresponding input symbols. Of course, when noise is present, some of the output symbols will differ from the associated input symbols.

E. Channel-Sounding Receivers

Up until this point, our discussion of receiver design has been based upon *a priori* statistical knowledge of the channel. In many situations, measurements can be made of channel characteristics by use of sounding signals that enable the derivation of *a posteriori* statistics. Such sounding signals might be special signals used for sounding only, or might be the data signals themselves; in the latter case, sounding and data transmission occur simultaneously.

One's first impulse is simply to use the receiver structures discussed above for the case of unknown delays, but to base the characteristics of $\hat{F}$, $\tilde{F}$, and $h_p(\cdot)$ in Figs. 7 and 11 on *a posteriori* rather than *a priori* statistics. This would in fact be appropriate with regard to the path-strength distributions and the nonlinearities $\hat{F}$ and $\tilde{F}$ they determine. However, as indicated in footnote 8, as soon as the sounding signals enable very accurate estimation of the path delays, the diffuseness condition on the path-arrival distribution $p(t_k)$ no longer holds, and the derivation leading to (17) breaks down. Although expressions for optimal receivers using nondiffuse *a posteriori* delay distributions can be derived, they and the resulting receivers become inordinately and unnecessarily complicated.

An alternative approach is usually used. In this approach, it is assumed that sounding results in extremely accurate estimates of the path variables $\{t_k\}$, $\{a_k\}$, $\{\theta_k\}$, or at least of the delays and strengths, if not the phases. These estimates are then assumed to be exact, and used as parameters in a receiver that assumes exact knowledge of the associated variables.

In many cases, estimates of $\{\theta_k\}$ are not deemed worth making, either because of the complexity of the resulting receiver or—especially in the mobile receiver context—because these phase shifts change too rapidly to be tracked and used effectively. Therefore, we discuss here only noncoherent channel-sounding receivers that make use only of path-strength and path-delay estimates.

The "optimal" M'ary strength/delay-estimating receiver that follows the philosophy just described is clearly based on (11). The quantities

$$w_l = \sum_{k=0}^{K-1} \log_e I_0\left(\frac{2\hat{a}_k \hat{x}_{lk}}{N_0}\right) \tag{27}$$

must be calculated, where $\hat{a}_k$ is the estimate of the strength of the kth path and $\hat{x}_{lk}$ is a sample of the lth matched filter output envelope at $t = T + \hat{t}_k$, $\hat{t}_k$ being the estimate of the kth path's delay. The index $(l = 1, \cdots, M)$ for which w_l is maximum is the receiver's digital output. If the paths are strong enough with respect to the noise to be measured accurately, which is our assumption, then (27) can be approximated by

$$w_l \cong \frac{2}{N_0} \sum_{k=0}^{K-1} \hat{a}_k \hat{x}_{lk} \tag{28}$$

since $\log_e I_0(x) \cong x$ for large x. This approximation is the optimal linear diversity combiner of Brennan [2], and we shall use it henceforth.

The linear combiner of (28) can be realized through use of a transversal filter, as shown in Fig. 14. This filter incorporates a delay line Δ seconds long, which is tapped at least every $1/W$ seconds, for a minimum of $W\Delta$ taps. The input to the transversal filter is the output envelope of a matched filter, say $x_j(t)$ of Fig. 6. The output of the transversal filter is a weighted sum of certain tap outputs, the taps that are included in the sum depending on the path delay estimates $\hat{t}_k$.

The estimates $\hat{t}_k (k = 0, \cdots, K - 1)$ are used to turn on the amplifiers of those taps having the delays (measured from the *right-hand* end of the line) that most closely approximate the $\hat{t}_k$'s; i.e., K of the tap amplifiers are activated. The gains of the activated tap amplifiers are then set to be proportional to the associated strength estimates $\hat{a}_k$. Amplifier gains are shown in Fig. 14 for the four-path response assumed.

To explain how the transversal filter works, we have shown in Fig. 14 the voltage profile along the delay line that would occur if the delay line's input were the jth matched filter output envelope $x_j(t)$ of Fig. 6; this profile moves to the right with time and is shown at the instant $t = T + \Delta$. At this instant, signal peaks in the profile lie at delays t_0, t_1, $\cdots$, t_{K-1}, as measured from the line's right-hand end; these are shown in Fig. 14 for $t_0 = 0$ and $K = 4$, assuming that $t_3 = \Delta$, Δ being the maximum excess delay (beyond t_0) anticipated in the channel. If the delay estimates $\hat{t}_k$ are accurate, the activated taps will sample the profile close to these peaks, so at time $t = T + \Delta$ the transversal filter's out-

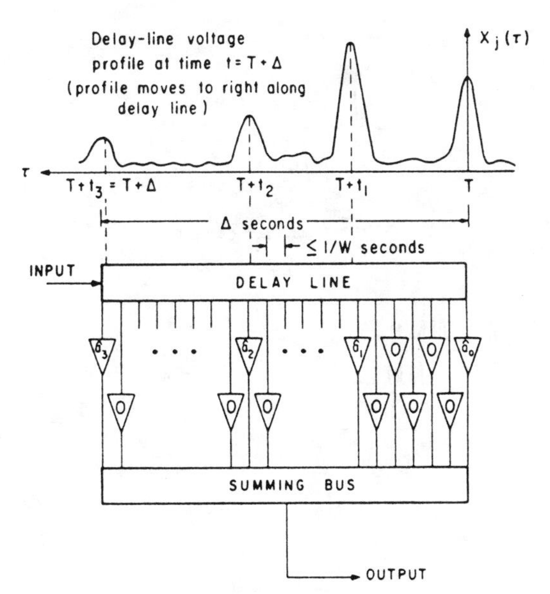

Fig. 14. A transversal filter used in realizing (28).

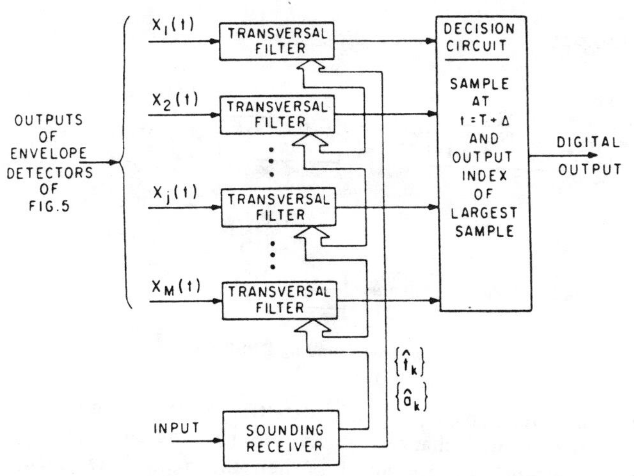

Fig. 16. A channel-sounding M'ary receiver (RAKE).

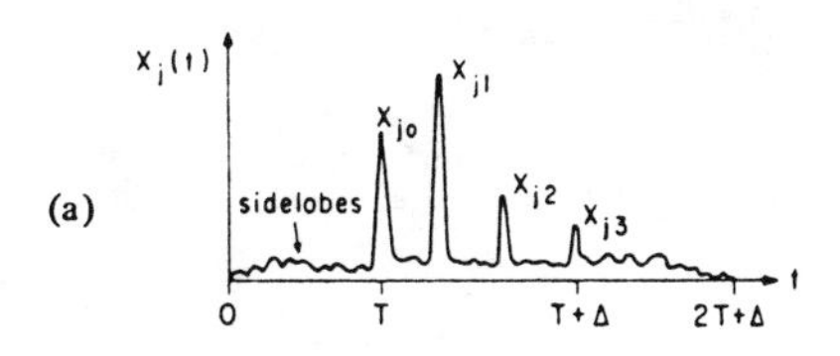

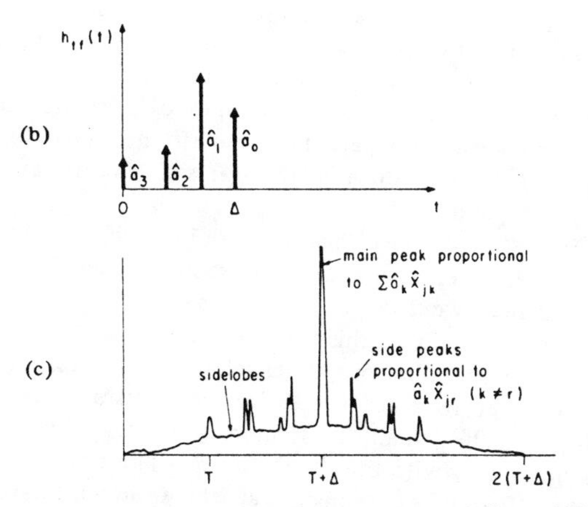

Fig. 15. Illustration of the operation of the transversal filter of Fig. 14: (a) Filter input; (b) Filter impulse response; (c) Filter output, obtained by convolving (a) and (b).

put will be approximately

$$\sum_{k=0}^{K-1} \hat{a}_k \hat{x}_j(T + \hat{t}_k) \cong \sum_{k=0}^{K-1} \hat{a}_k \hat{x}_{jk} \tag{29}$$

which is proportional to w_l, $l = j$, of (28).

Actually, the transversal filter is itself a matched filter of sorts. Note that the low-pass equivalent of the channel's impulse response, i.e., that relating $\sigma(t)$ and $\rho(t)$ of (1) and (3), is

$$h_m(t) = \sum_{k=0}^{K-1} a_k \exp(j\theta_k)\, \delta(t - t_k). \tag{30}$$

On the other hand, the impulse response of the transversal filter can clearly be written as

$$h_{tf}(t) = \sum_{k=0}^{K-1} \hat{a}_k \delta(t - \Delta + \hat{t}_k)$$

$$= \sum_{k=0}^{K-1} \hat{a}_k \delta(\Delta - t - \hat{t}_k). \tag{31}$$

The transversal filter is therefore matched to an estimate of the magnitude of the low-pass-equivalent channel impulse response.[12]

As $x_j(t)$ moves to the right in Fig. 14, the transversal filter convolves $x_j(t)$ with $h_{tf}(t)$, shown in Figs. 15(a) and 15(b), respectively, producing an output like that shown in Fig. 15(c). The main peak of this output occurs when $t = T + \Delta$, i.e., when $x_j(t)$ is aligned as in Fig. 14, with its peaks all located at activated taps. This main peak is proportional in height to (29). The output also has minor side peaks proportional to $\hat{a}_k \hat{x}_{jr} (k = 0, 1, 2, 3; r \neq k)$ and sidelobes accumulated from convolution of $h_{tf}(t)$ with the sidelobes of $x_j(t)$. It is clear that we need to sample only the main peak, to obtain (29).

A complete M'ary channel-estimating receiver can therefore be depicted as in Fig. 16. Each envelope detector output $x_l(t)$, $l = 1, \cdots, M$, enters a transversal filter of the form of Fig. 14, the parameters of which are driven by the estimates $\{\hat{t}_k\}$, $\{\hat{a}_k\}$ available from a sounding receiver. At time $t = T + \Delta$, the transversal filters' outputs are sampled and compared, and a decision made favoring the index of the largest. Such a receiver has been called a RAKE receiver [23] because of the tooth-like structure of the taps on the transversal filter's delay line.

The sounding receiver of Fig. 16 can itself be structured as a cascade of a matched filter, envelope detector and tapped delay line, where the matched filter is matched to a known waveform, perhaps containing more energy and lasting longer

[12] As discussed later in Section IV (see Fig. 20), if phase-coherent techniques were used, the transversal filter would become a bandpass filter matched to an estimate of the channel impulse response itself, i.e., the summand in (31) would contain the factor $\exp(-j\hat{\theta}_k)$. The cascade of the receiver's signal matched filter and the transversal matched filter would then form a receiver whose filter is matched to an estimate of the actual *received* signal, as modified by the channel.

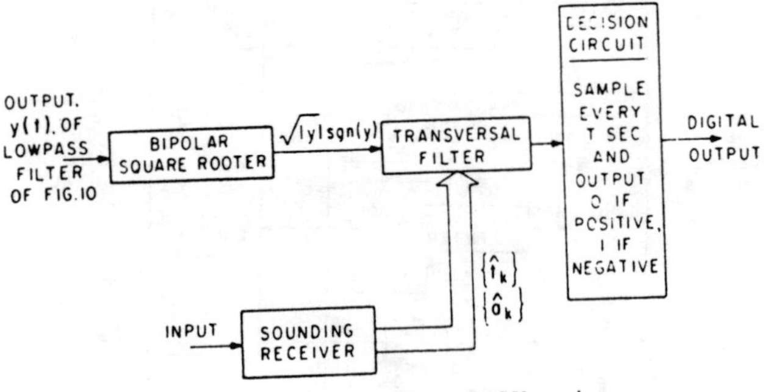

Fig. 17. A channel-sounding DPSK receiver.

than the data signals s_j. Prior to data transmission (and during it, if the channel changes rapidly enough), the sounding receiver "listens" for its signal. When the signal is received, the matched-filter output envelope will have multiple peaks, like the jth waveform of Fig. 6. This waveform will progress down a tapped delay line like that in Fig. 14. A threshold is triggered when the first peak reaches the line's right end, at which moment the voltages at each tap are sampled and held if they exceed another threshold, or set to zero if they do not. The frozen tap voltages, which are proportional to the path strengths at the tap's delays, are then used to set the gains of the associated taps in the data receiver's transversal filters.

A self-adaptive version of this receiver uses the data signals themselves for sounding. At the time of a decision favoring, say, signal j, the tap voltages of the jth transversal filter are frozen, just as in the sounding receiver discussed above; these voltages are then used to set the corresponding tap gains of all the transversal filters.[13] This self-adaptive version can be started by using a threshold triggering device of the type used in the separate sounding receiver.

We note that the sounding mechanism also performs a receiver synchronizing function, which, as previously mentioned, we have thus far ignored. Triggering of the threshold on the rightmost tap, in either the separate-sounding or self-adaptive realization, starts a clock that subsequently supplies properly timed sampling pulses to the decision circuit.

Finally, we note that a RAKE-like DPSK receiver can be structured along the same principles. This is shown in Fig. 17. Since the input in Fig. 17 is akin to the square of the inputs of Fig. 16, we have inserted a bipolar square-root operation before the transversal filter in the DPSK realization.

F. Intersymbol Interference

Heretofore, our discussion has ignored the effects of intersymbol interference by concentrating on isolated single transmissions (see, e.g., Fig. 6). Even in the DPSK case, where information is sent via the agreement or disagreement of the polarities of two successive transmissions, these transmissions were assumed to be sufficiently long or sufficiently spaced so that the time dispersion caused by the multipath channel causes little or no performance degradation (see Fig. 13).

Suppose now that each transmission of a choice from the signal set requires T seconds, and that successive transmissions occur at intervals of length T_s. As indicated by (10), the outputs of the matched filters in the receivers discussed above

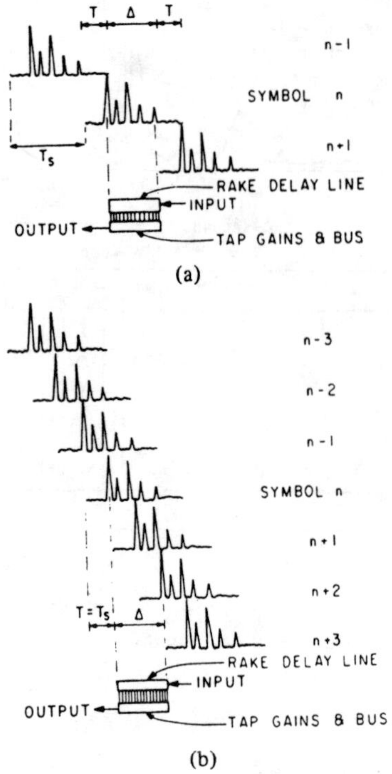

(a)

(b)

Fig. 18. Illustration of intersymbol interference (a) No intersymbol interference: $T_s \geqslant T + \Delta$. (b) Moderate intersymbol interference: $T_s = T$, $\Delta = 2T$.

will last $2T + \Delta$ seconds in response to each transmission of length T, raising the possibility of intersymbol interference. In order to help visualize intersymbol interference in the spread-spectrum ($TW \gg 1$) case of importance here, Fig. 18 shows sequences of output envelopes of a single matched filter in response to a periodic input of the signal to which it is matched. This diagram illustrates the interrelationships among T, T_s, and Δ.

In Fig. 18(a), we have shown a case in which there is no intersymbol interference, a situation requiring that $T_s \geqslant T + \Delta$. For reference, we have also shown a RAKE delay line and combiner, of length Δ; this is turned end for end compared to that of Fig. 14. The output envelopes of the response of a matched filter to three successive transmissions are shown, which should be visualized as entering the delay line from the right and moving with time to the left.[14] This output sequence is shown "frozen" at the instant at which the multipath pulses of the central (nth) transmission are perfectly aligned with the delay line and ready to be sampled. Notice that, at this instant, neither the $(n-1)$th nor the $(n+1)$th response is in the delay line to interfere with the nth.

In Fig. 18(b) a case of moderate intersymbol interference is shown. Here $T_s = T$ and $\Delta = 2T$, so $T_s = \frac{1}{3}(T + \Delta)$. A sequence of output envelopes in response to seven successive transmissions is shown, frozen at the instant when the multipath pulses of the central (nth) response are aligned with the delay line. However, in this case, parts of two predecessor and two successor responses ($n-2$, $n-1$, $n+1$, $n+2$) are also in the delay line, causing intersymbol interference. In general, a total of $2\lceil \Delta/T \rceil$ predecessor and successor symbols will inter-

[13] More elaborately, the nth tap gain can be set on the basis of a weighted average of the nth tap voltages held over a given number of past decisions.

[14] Of course, in practice these responses would have been superposed by the matched filter prior to envelope detection, and the actual output envelope would be a nonlinear combination of the three responses.

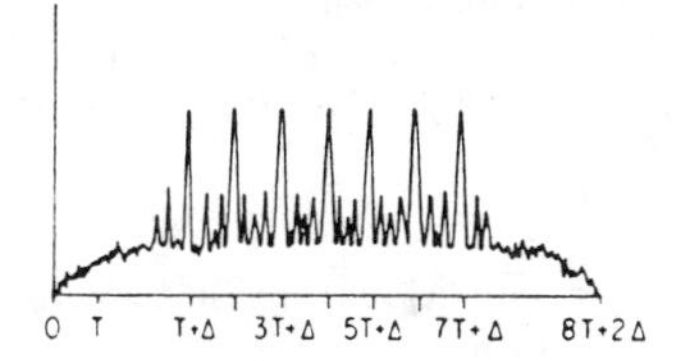

Fig. 19. The output of a RAKE transversal filter when the input consists of the seven pulse trains of Fig. 18(b).

fere with each symbol as it is sampled, where $[x]$ is the smallest integer greater than or equal to x.[15]

In order to avoid intersymbol interference completely, the condition $T_s \geqslant T + \Delta$ must be satisfied. Although we have some flexibility in decreasing T (subject to our requirement $TW >> 1$ and to limitations imposed on the bandwidth W), Δ is fixed by the channel. Thus, for a binary system, we cannot completely avoid intersymbol interference when the data rate is greater than roughly $1/\Delta$ b/s. For example, for binary transmission through urban multipath ($\Delta = 5$ μs, effectively), any transmission rate greater than about 200 kb/s will result in intersymbol interference, even in a spread-spectrum system.

Of course, it is well known that the effects of intersymbol interference can be ameliorated, so higher data rates can be achieved without undue deterioration of receiver performance. One approach, followed by Monsen [15]-[17], is based on classical equalization techniques developed for reduction of intersymbol interference on baseband landlines; it is particularly applicable when $TW \cong 1$ and $\Delta >> T$. In our case, when $TW >> 1$ and Δ/T is moderate, we follow a different approach, based on the RAKE receiver.

The basis of the RAKE approach is to recognize that, while multipath pulses from predecessor and successor symbols are on the RAKE delay line at the decision instant for the present symbol, as shown in Fig. 18(b), it is unlikely that they will appear at taps which are activated. Recall that the delay line has taps every $1/W$ seconds. Since $TW >> 1$, and we are now assuming that $\Delta/T > 1$, we will have $W\Delta >> 1$. There will thus be a large number, $W\Delta$, of taps. For example, if $TW = 100$ and $\Delta/T = 5$, then there are $W\Delta = 500$ taps on the delay line. On the other hand, only those taps at which pulses are expected are activated, so even if there are as many as 20 paths in the example above, only $20/500 = 2.5$ percent of the taps will be activated. These activated taps will be aligned to sample the multipath pulses of the central (nth) response in Fig. 18(b); but it is extremely unlikely that the pulses of adjacent responses on the line will also be aligned with the activated taps.

Yet another insight into the capability of the RAKE transversal filter to suppress intersymbol is given by looking at the filter's output. Although the total input to the delay line is really a nonlinear combination of the pulse trains shown in Fig. 18(b), for simplicity one can visualize the result of each pulse train sweeping to the left through the line and being convolved with the RAKE filter's impulse response, as depicted in Fig. 15. A sequence of seven such resulting convolutions, when properly combined, would look something like Fig. 19. Each of the major peaks there corresponds to the exact alignment of one of the pulse trains with the activated taps on the delay line, and it is these peaks that are sampled. The pedestal upon which the major peaks sit is composed of minor peaks, as in Fig. 15, and a general sidelobe "hash" level. We call this pedestal "multipath-induced interference."

The ability of the RAKE receiver to concentrate on the part of the matched-filter response due to the current symbol is peculiar to its tapped-delay-line structure. The integrating receivers of Figs. 7 and 11, by integrating over the interleaved responses to several symbols, cannot suppress intersymbol interference as RAKE does.

In summary, it appears that data rates much greater than $1/\Delta$ b/s can be achieved with RAKE receivers without undue deterioration of performance. We investigate this possibility both analytically and by simulation in the next several sections.

We close this subsection by noting one complication that arises in realization of a RAKE receiver when $\Delta/T > 1$. Recall that one alternative for estimating the path delays and strengths was self-adaptive, using the data signals themselves as sounding signals, and using the multipath profiles that are stretched out along the RAKE delay lines to aid in path-parameter estimation. As shown in Fig. 18(b), however, when the present (nth) signal's multipath profile is aligned with the delay line, adjacent signals can cause spurious pulses to appear on the line. These latter must be ignored in order for the RAKE receiver to avoid intersymbol interference, but—in the simple adaptive approach to path-parameter estimation outlined in the previous subsection—they also would tend to be identified as true paths and cause spurious activation of taps. Thus, when $\Delta/T > 1$, the self-adaptive approach to estimation is not straightforward. Work on self-adaptive systems with $\Delta/T > 1$, in which the data signals are used for channel estimation, is now ongoing at this Laboratory. Initial simulation results show that such systems can successfully adapt to and track a time-varying multipath profile.[16]

G. Comments on Optimality and Quasi-Optimality

From a theorist's point of view, we are in an unenviable position. Only in the case of an incoherent single-path channel disturbed solely by additive white Gaussian noise have we specified a receiver, i.e., Fig. 5, that is strictly optimal in anything approaching the real world. The multipath receivers we have discussed are not strictly optimal: either their optimality depends on invoking a large set of oversimplifying assumptions (e.g., the receivers specified by (11), (12), and (13)), or they have no sure claims to optimality at all (e.g., the receivers of Figs. 11, 16, and 17). The unfortunate fact is that we do not know the optimal receiver structure for the real-world multipath channels described in Section II, nor for nonideal signal sets, and such structures are unlikely to be analytically derivable. We are thus deprived of the theorist's benchmark: the knowledge of the optimal performance, beyond which no receiver can reach.

Nonetheless, we rely on a pragmatic faith that the multipath receivers we have specified, by virtue of the intuitive reasonableness of their derivation, cannot be too far off the mark, whence the appellation "quasi-optimal." That is, we believe their performance to be within a very few decibels of the unknown benchmark of optimality. We shall in a later section present evidence that supports this belief.

[15] In the DPSK case shown in Fig. 13, $T_s = T$ and $\Delta/T < 1$, so there is intersymbol interference by only one successor and one predecessor. The interference is due only to the sidelobes of the multipath pulses in neighboring transmission intervals, rather than to the pulses themselves as in Fig. 18(b), and the interference can be deemed negligible.

[16] Private communications from L-F. Wei.

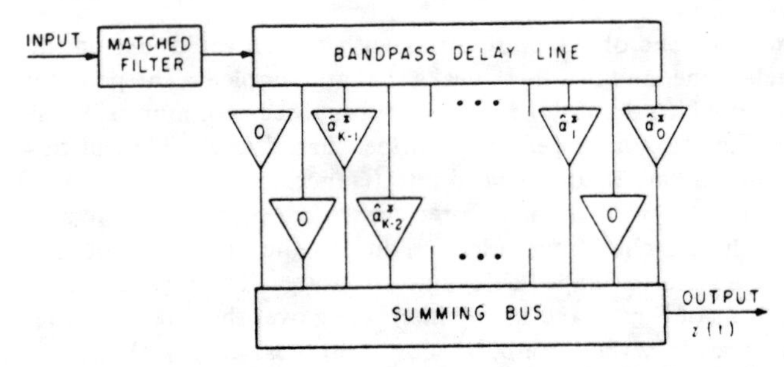

Fig. 20. Bandpass RAKE system used for "equivalent noise" analysis.

IV. Simplified Analysis of Multipath-Induced Interference

We have just concluded by a number of heuristic arguments that, even if $T_s << T + \Delta$, a RAKE receiver will work satisfactorily, despite multipath-induced interpath and intersymbol interference.[17] In the present section, we present an analysis that bolsters our confidence in this conclusion.

A. A Simplified Model

For simplicity, we consider a coherent receiver, since the noncoherent receivers discussed in Section III lead to complicated nonlinear analysis. More specifically, we analyze a binary, coherent PSK system whose receiver filters are shown in Fig. 20. Here, a bandpass matched filter is followed by a bandpass transversal filter, rather than (as previously) by a cascade of an envelope or DPSK detector and a low-pass transversal filter. As in the case of the low-pass transversal filter of Fig. 14, the tap gains in Fig. 20 are set by estimates of path strengths.[18] Now, however, these gains not only weight the tap outputs by estimates $\hat{a}_k$ of the path strengths, but also try to correct for the paths' phase shifts θ_k by phase shifting the tap outputs in the opposite directions, by amounts $-\hat{\theta}_k$. (See footnote 12.) We have written the nonzero gains in Fig. 20 in complex form

$$\hat{\alpha}_k^* = \hat{a}_k \exp(-j\hat{\theta}_k). \tag{32}$$

The lowpass-equivalent impulse response of the transversal filter is clearly of the form (cf. (31))

$$\hat{h}_{tf}(t) = \sum_{k=0}^{K-1} \hat{\alpha}_k^* \delta(\Delta - t - t_k). \tag{33}$$

We therefore see that $\hat{h}_{tf}(t)$ would be matched to the channel's impulse response, given by (30), if the estimates were perfect, i.e., if $\hat{a}_k = a_k$, $\hat{\theta}_k = \theta_k$, and $\hat{t}_k = t_k$, all k.

We continue in low-pass-equivalent form in the frequency domain. Let $\sigma(t)$, $0 \leq t \leq T$, be the low-pass equivalent of a signal transmitted through the channel (cf. (1)) and let the Fourier transform of $\sigma(t)$ be $S(f)$. Disregarding channel noise for the moment, the channel output then has spectrum $S(f)H_m(f)$, where $H_m(f)$ is the Fourier transform of $h_m(t)$ of (30), i.e.,

$$H_m(f) = \sum_{k=0}^{K-1} \alpha_k \exp(-j2\pi f t_k) \tag{34}$$

where $\alpha_k = a_k \exp(j\theta_k)$. The matched filter in Fig. 20 has low-pass-equivalent impulse response $\sigma(T - t)$ and transfer function $S^*(f) \exp(-j2\pi fT)$, so its output has spectrum $|S(f)|^2 H_m(f) \exp(-j2\pi fT)$. The transversal filter has transfer function

$$\hat{H}_{tf}(f) = \sum_{k=0}^{K-1} \hat{\alpha}_k^* \exp[-2\pi f(\Delta - \hat{t}_k)] \tag{35}$$

(the Fourier transform of (33)), whence its output has spectrum

$$\begin{aligned} W(f) &= |S(f)|^2 H_m(f) \hat{H}_{tf}(f) \\ &= |S(f)|^2 \sum_{k=0}^{K-1} \sum_{l=0}^{K-1} \alpha_k \alpha_l^* \\ &\quad \cdot \exp[-j2\pi f(t_k - \hat{t}_l)] \\ &\quad \cdot \exp[-j2\pi f(T + \Delta)]. \end{aligned} \tag{36}$$

We assume for simplicity that the channel-sounding procedure has led to very good estimates, so we set $\hat{a}_k = a_k$, $\hat{\theta}_k = \theta_k$ (hence, $\hat{\alpha}_k = \alpha_k$), and $\hat{t}_k = t_k$.[19] We also shift the time origin to the right by $T + \Delta$ seconds to eliminate the factor $\exp[-j2\pi f(T + \Delta)]$ in (36). Then (36) becomes

$$W(f) = |S(f)|^2 \sum_{k=0}^{K-1} \sum_{l=0}^{K-1} \alpha_k \alpha_l^* \exp[j2\pi f(t_k - t_l)]. \tag{37}$$

The corresponding time waveform is the inverse Fourier transform of $W(f)$, i.e.,

$$w(t) = \sum_{k=0}^{K-1} \sum_{l=0}^{K-1} \alpha_k \alpha_l^* \gamma(t - t_k + t_l), \qquad |t| \leq T + \Delta \tag{38}$$

where $\gamma(t)$ is the inverse Fourier transform of $|S(f)|^2$, and is given by (23).

For the case in which $\sigma(t)$ is sent only once through the channel, (38) is the low-pass-equivalent output of the transversal filter, shifted so that its peak lies at $t = 0$. (The true bandpass output is $\frac{1}{2} Re[w(t) \exp(j\omega_0 t)]$ and the output envelope, $\frac{1}{2}|w(t)|$, has a shape like that depicted in Fig. 15(c), where the central peak is now centered at $t = 0$.) If we now suppose that an infinitely long PSK sequence, $\sum_{n=-\infty}^{\infty} d_n \sigma(t - nT)$, $d_n = \pm 1$, is sent through the channel, the corresponding low-pass-equivalent output of the transversal filter will be

$$v(t) = \sum_{n=-\infty}^{\infty} d_n w(t - nT). \tag{39}$$

The output envelope, $\frac{1}{2}|v(t)|$, will now look like Fig. 19, where the result of only seven transmissions is shown. The central peak in Fig. 19 would correspond to the $n = 0$ term in (39), now centered on $t = 0$.

The analysis in this section is concerned with the degradation in performance caused by the multipath-induced interference. Ideally, this degradation should be measured in terms of error probability. However, the multipath statistics that govern the error probability are so complicated as to make this ideal goal unattainable. Instead, we shall perform an "equivalent noise" analysis.

[17] See footnote 11, for a definition of these terms.

[18] Unlike the case in Fig. 14, in Fig. 20 the largest estimated excess delay $\hat{t}_{K-1} - \hat{t}_0$ is less than Δ.

[19] As seems obvious, and is shown in Appendix C, if the channel-sounding signal's energy is much greater than the data signals' energy, this is a reasonable assumption.

We concentrate on the central peak in $v(t)$, i.e., on

$$v(0) = \sum_{n=-\infty}^{\infty} d_n\, w(-nT). \tag{41}$$

Noting from (38) that $w(t)$ is nonzero only in $|t| \leqslant T + \Delta$, we see that (41) only has $2N+1$ nonzero terms, where $N = \lceil \Delta/T \rceil$ is the smallest integer greater than or equal to T/Δ. Thus using (38), (41) becomes

$$v(0) = \sum_{n=-N}^{N} d_n \sum_{k=0}^{K-1} \sum_{l=0}^{K-1} \alpha_k \alpha_l^* \gamma(nT - t_k + t_l). \tag{42}$$

This is the signal component of the transversal filter output at $t = 0$. A component due to channel-noise, call it $n(0)$ adds to this, leading to a total output

$$z(0) = v(0) + n(0). \tag{43}$$

Our approach to evaluating the effect of multipath-induced interference is to determine how much its presence increases the variance of $z(0)$ above the value contributed by $n(0)$, i.e., how much "equivalent channel noise" is added by intersymbol interference.

In order to proceed, we make a number of simplifying assumptions about the multipath statistics that are akin to those made in Section III in deriving "optimal" receiver structures:

1) $\{a_k\}_0^{K-1}$, $\{\theta_k\}_0^{K-1}$ and $\{t_k\}_0^{K-1}$ are independent sets of random variables and all variables in each set are independent of each other;
2) the number of paths K is independent of $\{a_k\}_0^{K-1}$, $\{\theta_k\}_0^{K-1}$ and $\{t_k\}_0^{K-1}$;
3) all t_k's are equidistributed over $[0, \Delta]$ with probability density $p(t_k)$; see the discussion in Section III-C;
4) each θ_k is uniformly distributed over $(0, 2\pi]$.

As previously noted, these assumptions depart somewhat from reality, but will serve for the present heuristic analysis.

We also assume that the data sequence $\{d_n\}_{-N}^{N}$ consists of independent binary values, being +1 or −1 with equal probability, and that the channel noise is independent of the multipath variables.

B. The variance of z (0)

Attacking the signal component $v(0)$ in (43) first, suppose that the desired datum in $v(0)$, i.e., d_0, is +1. Then, using the assumptions discussed above, it is straightforward to show that

$$E[v(0)]\,\Big|_{d_0=+1} = \gamma(0) E_K\left[\sum_{k=0}^{K-1} \overline{|\alpha_k|^2}\right] = 2\mathcal{E} A^2. \tag{44}$$

Here, an overbar denotes "expectation," and E_K denotes expectation over the random variable K. We have used the fact that $\gamma(0) = 2\mathcal{E}$, $\mathcal{E}$ being the energy in the transmitted signal $s(t) = Re[\sigma(t) \exp(j\omega t)]$; we have also let

$$A^2 \triangleq E_K\left[\sum_{k=0}^{K-1} \overline{|\alpha_k|^2}\right]. \tag{45}$$

Not so straightforwardly, one can show (see Appendix A) that, when $W\Delta >> 1$,

$$\operatorname{var}[v(0)]\,\Big|_{d_0=+1} = (2\mathcal{E})^2 B^2 + \left[\int_{-\infty}^{\infty} |S(\nu)|^4 d\nu\right]\left[\sum_{n=-N}^{N} q(nT)\right] C^2 \tag{46}$$

where

$$B^2 = E_K\left[\sum_{k=0}^{K-1} \overline{|\alpha_k|^4} - \sum_{k=0}^{K-1} \overline{|\alpha_k|^2}^{\,2}\right] = E_K\left[\sum_{k=0}^{K-1} \operatorname{var}(|\alpha_k|^2)\right] \tag{47}$$

$$C^2 = E_K\left[\sum_{\substack{k=0\\k\neq l}}^{K-1} \sum_{l=0}^{K-1} \overline{|\alpha_k|^2}\;\overline{|\alpha_l|^2}\right] = E_K\left[\left(\sum_{k=0}^{K-1} \overline{|\alpha_k|^2}\right)^2 - \sum_{k=0}^{K-1} \overline{|\alpha_k|^2}^{\,2}\right] \tag{48}$$

and

$$q(t) = \int_0^{\Delta} p(\tau)\, p(\tau - t)\, d\tau. \tag{49}$$

The variance in (46) is, in fact, independent of the condition $d_0 = +1$. It is important to note that the two contributions to this variance have distinctly different interpretations. The first term is independent of N and would be present even if there were no multipath-induced interference; it reflects the variation of $v(0)$ around its mean due to the fluctuations in the strengths of the paths that have been combined by the RAKE receiver to decide on the datum d_0 in $v(0)$. The second term is due to multiple-induced interference; it reflects the fluctuations in $v(0)$ due to the presence of both interpath interference on the "present" transmission ($n = 0$) and intersymbol interference from neighboring transmissions ($n \neq 0$). The second term will vanish only if, with probability one, there is but one path ($K = 1$) in (48).

We next consider the term $n(0)$ in (43). It has zero mean, so its mean adds nothing to (44). Its variance is derived in Appendix B. Since $v(0)$ and $n(0)$ are assumed to be independent, the variance of $z(0)$ becomes

$$\operatorname{var}[z(0)] = \operatorname{var}[v(0)] + \operatorname{var}[n(0)] = (2\mathcal{E})^2 B^2 + \left[\int_{-\infty}^{\infty} |S(\nu)|^4 d\nu\right]\left[\sum_{n=-N}^{N} q(nT)\right] C^2 + 4\mathcal{E} N_0 A^2 \tag{50}$$

where N_0 is the (single-ended) power density of the channel noise, assumed to be white.

C. A Measure of Performance Degradation

As mentioned, only the second term in (50) is due to the multipath-induced interference of interest to us here. In the next section, for the purpose of estimating error probabilities, we shall model this interference as approximately Gaussian, equivalent to an additional amount of channel noise. In this framework, multipath-induced interference degrades system performance by the same amount as would an increase in channel noise power by a factor of

$$D = 1 + \frac{C^2}{4\mathcal{E}N_0A^2}\left[\int_{-\infty}^{\infty} |S(\nu)|^4 d\nu\right]\left[\sum_{n=-N}^{N} q(nT)\right]. \quad (51)$$

We now specialize (51) by some additional assumptions. First, suppose that $|S(f)|^2$ is flat over the band $|f| \leq W/2$ and zero elsewhere. Then, since the integral of $|S(f)|^2$ is equal to $2\mathcal{E}$,

$$|S(f)|^2 = \frac{2\mathcal{E}}{W}, \qquad |f| \leq \frac{W}{2} \quad (52)$$

and

$$\int_{-\infty}^{\infty} |S(f)|^4 df = \frac{4\mathcal{E}^2}{W}. \quad (53)$$

Second, suppose that $\Delta >> T$, so that we can use the following approximation:

$$\sum_{n=-N}^{N} q(nT) \cong \frac{1}{T}\int_{-\Delta}^{\Delta} q(t)dt$$

$$= \frac{1}{T}\left[\int_{0}^{\Delta} p(\tau)\, d\tau\right]^2 = \frac{1}{T}. \quad (54)$$

Finally, assume that all paths have identical strength statistics, so $\overline{|\alpha_k|^2} = \overline{|\alpha|^2}$, all k, whence

$$A^2 = \overline{K}\ \overline{|\alpha|^2} \quad (55)$$

$$C^2 = (\overline{K^2} - \overline{K})\ \overline{|\alpha|^2}^2 = \overline{K}^2\ \overline{|\alpha|^2}^2. \quad (56)$$

(The last equality in (56) follows from the assumption of independence and equidistribution of the t_k's over $[0, \Delta]$, implying that the t_k point process is Poisson, for which $\overline{K^2} - \overline{K} = \overline{K}^2$.)

With the foregoing assumptions, (51) becomes

$$D = 1 + \frac{\overline{K}}{TW} \cdot \frac{\overline{\mathcal{E}}}{N_0} \quad (57)$$

where $\overline{\mathcal{E}} \triangleq \overline{|\alpha|^2}\mathcal{E}$ is the mean energy per bit arriving per path. Thus the degradation due to multipath-induced interference increases with $\overline{K}\overline{\mathcal{E}}/N_0$ (as this interference from multiple paths increasing dominates the channel noise) and decreases with TW (as the sidelobes of $\gamma(t)$, hence the interference contribution by each interfering path, decrease).

As an example, suppose—using parameters typical of the simulation experiments in Section VI—that $\overline{K} \cong 25$, $TW = 127$ and $\overline{\mathcal{E}}/N_0 = 5$ dB. Then $D = 1.62$, or 2.1 dB, not much degradation to pay for the ability to increase the data rate by, say, a factor of 10 or more by making $T << \Delta$!

We close this section by noting that if the errors in the estimates $\{\hat{a}_k\}$, $\{\hat{\theta}_k\}$, $\{\hat{t}_k\}$ are taken into account, then, as shown in Appendix C, another term having value $\mathcal{E}\Delta/\mathcal{E}_x T$ is added to (57), where $\mathcal{E}_x$ is the sounding signal's energy. By ignoring estimation noise, we have implicitly assumed that $\mathcal{E}_x/\mathcal{E} >> \Delta/T$.

Although the analysis above was performed for a purely coherent PSK system, we shall assume in Section VI that the main results, (51) and (57), carry over approximately to a DPSK system.

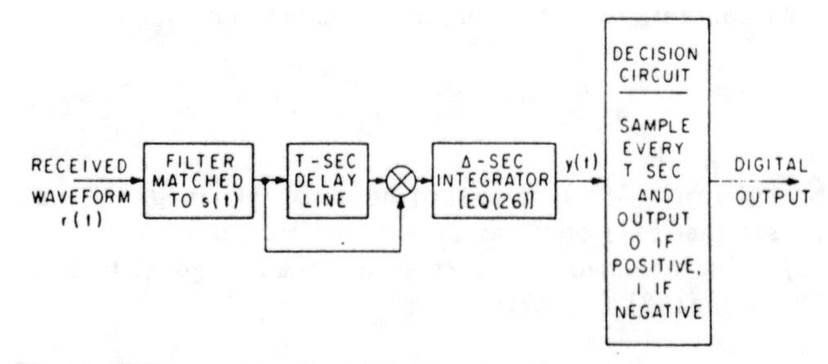

Fig. 21. PDI: A quasi-optimal DPSK receiver for unknown path delays.

V. Simplied Analysis of Error Probability

Exact evaluation of the error probability of a multipath receiver quickly becomes intractable as the assumed model for the multipath channel becomes more realistic. Most analyses assume independent, Rayleigh-distributed path strengths, although some also involve Rice [13] or Nakagami [4] strength distributions and/or correlated strengths [21], [31]. With some exceptions [4], [31], the path strengths are assumed to be equidistributed and the path delays known. None of the cited analyses takes into account the multipath-induced interference discussed in the previous section, although Monsen [15]-[17] and Morgan [18] consider such interference in a different context. Exact analytical evaluation of the error probability for the model in Section II—involving unknown, non-Poisson path delays; correlated Nakagami-distributed path strengths; spatial correlations and inhomogeneities; and multipath-induced interference—seems beyond tractability.

Another approach is to evaluate the error probability of various receivers by simulation techniques, using the channel simulation program described in Section II; the results of a few such simulation experiments are given in Section VI. In the present section, however, we shall present a much simplified analysis. This analysis will provide both intuitive insight into the performance of several receivers, and a basis for determining (in Section VI) the gap between reality and the results of rather standard analytical methods used. In our analysis, we assume lack of knowledge of path delays, but still use very much oversimplified statistics for the path delays and strengths. We temporarily suppress the effects of multipath-induced interference, taking them into account later in Section VI.

We limit ourselves here to a system using DPSK transmission and simplifications of the receivers of Fig. 11 and 17. The analysis depends heavily upon the work of Charash [3], [4].

A. Post-Detection Integrating Receiver

A simplification of the receiver of Fig. 11 is shown in Fig. 21, where the pertinent portion of Fig. 10 is also included. Notice that the (generally time-varying) nonlinearity $\widetilde{F}[\cdot, \cdot]$ has been eliminated, i.e., replaced by a time-invariant linear device. For the case of Rayleigh path strengths with equal mean-square values, the nature of the approximation is shown in Fig. 12 (see also (19) and (24)). By eliminating $\widetilde{F}$, we are giving up the strong-path-accentuation/noise-suppression features of the receiver that is shown by the waveforms of Fig. 13(c) and (d).

In Fig. 21, we also have taken the integrating filter's impulse response $h_p(t)$ to be of the form of (26). If this $h_p(t)$ were optimal, it would mean that paths occur with uniform density over the interval $[0,\Delta)$; see Section III-C, where we would have $p(t_k) = 1/\Delta, 0 \leq t_k \leq \Delta$. In actuality, paths do not occur uniformly, so this $h_p(t)$ is suboptimal. Since the lowpass filter of Fig. 10 serves only to eliminate the double-carrier-frequency terms in the multiplier output, its function is handled in Fig. 21 by $h_p(t)$, also a lowpass impulse response.

The decision circuit input in Fig. 21 will still look much like the waveform of Fig. 13(e), but it will now be an integration of $y(t)$ of Fig. 13(c), rather than of Fig. 13(d). It must still be sampled by the decision circuit at the instants shown in Fig. 13(f).

The simplified suboptimal receiver just discussed has also been considered by Fralick [7], who descriptively calls it a post-detection integrator (PDI), a name we shall use.

The error-probability performance of the PDI can be estimated from the work of Charash [3], [4]. He has given error-probability expressions and curves for a system having the following characteristics:

1) binary transmission using uniformly orthogonal, zero-sidelobe signals, i.e., ones for which $|\gamma_{12}(t)|$ of (7) is identically zero rather than merely satisfying (8a), and $|\gamma_{ii}(t)|$, $i = 1,2$, is identically zero for $|t| > 1/W$ rather than merely satisfying (8b);
2) a receiver of the form of Fig. 7 (with $M = 2$), but with the nonlinearity being a simple time-invariant square law device, $\hat{F}(x, t) = x^2$;
3) an integrating-filter impulse response, $h_p(t)$, of the form of (26).

Note that this system bears the same relationship to our DPSK/PDI system in the multipath case as the usual orthogonally keyed binary system bears to a standard DPSK system in the single-path case (cf. Figs. 5 and 10). We shall exploit this relationship to estimate the error probability for the DPSK/PDI system.

Charash also makes the following assumptions about the channel:

4) all paths are independently and identically Nakagami-distributed;
5) exactly K paths arrive at random over $[0,\Delta]$. Of course, to preserve resolvability condition 4), we must have $K << W\Delta$, where W is the transmission bandwidth;
6) $\Delta << T$, so there is no appreciable intersymbol interference;
7) the additive channel noise is white and Gaussian, with single-ended power density N_0.

Under these seven assumptions, Charash has derived an expression for the bit error probability $P_E(K)$ conditioned on the number of paths K. (The unconditional bit error probability would then be the average of $P_E(K)$ over the Poisson distribution of K that is implied by (5).)

As mentioned, our DPSK/DPI system is related to Charash's system in the same way as an ordinary single-path DPSK system is related to an ordinary single-path orthogonally keyed system. The performance of these latter, single-path systems are well known [24] to be exactly 3 dB separated, with the DPSK system having the advantage; i.e., the DPSK system requires 3 dB less SNR to achieve the same bit error probability as the orthogonally keyed system. We now conjecture that the same 3-dB separation holds (at least approximately) for the corresponding multipath systems.

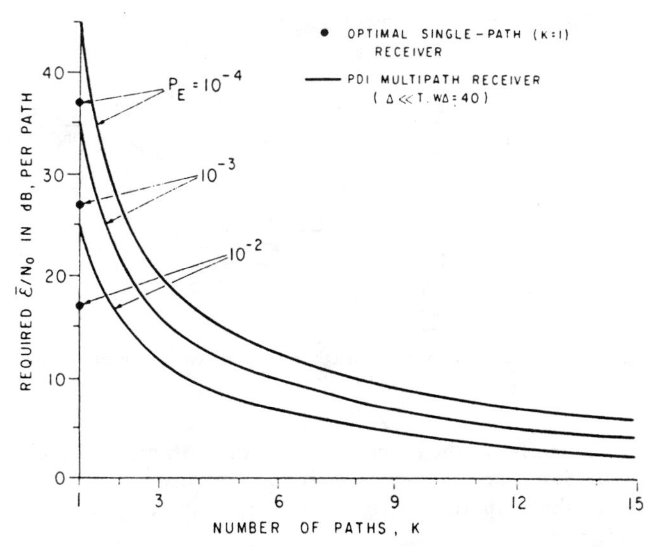

Fig. 22. Required $\bar{\mathcal{E}}/N_0$ per path for DSPK/PDI system: K independent, Rayleigh-fading paths.

Using this conjecture, we have used Charash's results to estimate the error-probability of the DPSK/PDI, of course still assuming that assumptions 3)-7) hold and that the DPSK signal has zero sidelobes. Fig. 22 shows some typical results, for Rayleigh fading (a special case of Nakagami fading) and for $W\Delta = 40$. The curves there show the SNR $\bar{\mathcal{E}}/N_0$ required per path (see (57) *et seq.*) in order to achieve a given value of $P_E(K)$, as a function of the number of paths K. Note that, since typical urban multipath spreads are $\Delta \cong$ 4-5 μs, $W\Delta = 40$ implies a system bandwidth of 8-10 MHz; and since we postulate that $\Delta << T$ for the PDI, we are assuming a spectrum spreading factor of $TW >> 40$.

On the $K = 1$ axis, we have shown for reference the performance of the optimal single-path DPSK receiver of Fig. 10.[20] One can see that if a multipath medium has a number of Rayleigh paths of more-or-less equal mean-square strengths, our DPSK/PDI system will do substantially (5-20 dB) better than a receiver that locks onto a single path, say the LOS path, and disregards the remaining paths. This comparison strikingly illustrates the diversity performance improvement that multipath channels can afford, as discussed at the beginning of Section III.

It must be stressed that the curves of Fig. 22 take into account neither intersymbol interference nor interpath interference (see footnote 11). The former is precluded by the assumption $\Delta << T$, the latter by the assumption of a zero-sidelobe signal.

B. A Digital RAKE Receiver

The channel-sounding receivers of Figs. 16 and 17 involve adjustment of the tap gains of the transversal filter(s) with estimates of path strengths. An arrangement that is much simpler to implement, although it will not perform as well, is as follows: connect each tap to the summing bus or not according as it is estimated that a path is present at that delay or not.

[20] This is derived by averaging the nonfading, single-path DPSK error probability [24] of $(1/2) \exp[-\mathcal{E}/N_0]$ over a Rayleigh distribution for $\sqrt{\mathcal{E}}$ having *rms* value $\sqrt{\psi}$, to obtain $P_E = [2(1 + \bar{\mathcal{E}}/N_0)]^{-1}$, where $\bar{\mathcal{E}} = \psi\mathcal{E}$.

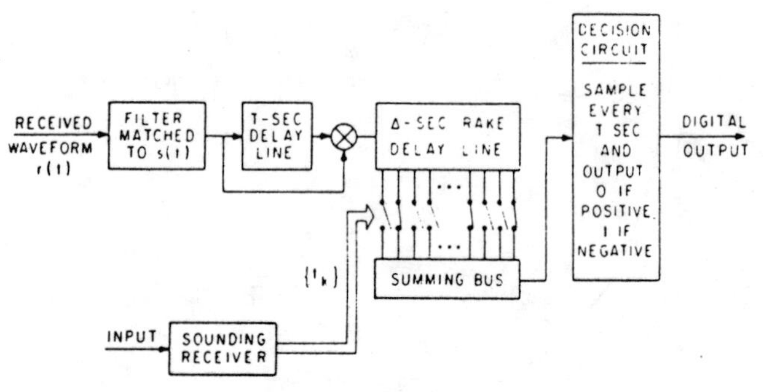

Fig. 23. DRAKE: A quasi-optimal DPSK receiver when path delays are estimated.

Only a transfer of the estimates $\{\hat{t}_k\}$ of the path delays is then needed from the sounding receiver to the transversal filter(s). We call this simplified configuration a digital RAKE (DRAKE) receiver.

Fig. 23 shows a DPSK/DRAKE receiver based on Fig. 17, in which, as a further simplification, the square rooter has been eliminated. The sounding receiver's estimates of path delays turn the associated tap switches on the transversal filter on, connecting the taps to the summing bus. Note that if all tap switches were turned an, Fig. 23 would revert approximately to the PDI of Fig. 21, with the Δ-second integration replaced by an approximating Δ-second discrete-time summation.

In analyzing the DPSK/DRAKE receiver, we assume that the sounding receiver's estimates $\hat{t}_k$ of the t_k are exact ($\hat{t}_k = t_k$), and that these estimates correspond exactly to available discrete tap delays in the transversal filter. The receiver then takes the form of an equal-weight path combiner for known path delays. Charash [3], [4] has given expressions and curves for such a receiver under assumptions 1), 4), 5), 6) and 7) of the previous subsection. We can convert these results to the DPSK format of interest to us by the same conjecture as used previously, i.e., that Charash's orthogonally keyed system performs 3 dB worse than our DPSK system. On this basis, curves of the required $\overline{\mathcal{E}}/N_0$ per path necessary to achieve a given P_E are shown in Fig. 24 as functions of the number of paths K. Only curves for the Rayleigh-fading case are shown here; these could also have been obtained from [31].

As in the case of Fig. 22, the curves of Fig. 24 take into account neither intersymbol interference nor the interpath interference in the current symbol. Both of these introduce an effective additional noise term. The performance degradation caused by this term has been estimated in Section IV for an ideal, coherent PSK/RAKE system. In the next section, we shall modify our error-probability analysis by assuming that this same degradation holds approximately for DPSK/DRAKE.

VI. Simulation Experiments

In order to evaluate the error probabilities of various receivers in an actual urban multipath environment and to assess the applicability of the simplied analyses of the previous two sections, a number of simulation experiments were performed. These experiments combined the urban propagation program SURP described in Section III-D and various binary modulator/demodulator programs described below. Two sets of experiments were performed: low rate at 78.7 kb/s and high rate at 787 kb/s. The experiments reported here were restricted to DPSK transmission and to the dense-high-rise environment

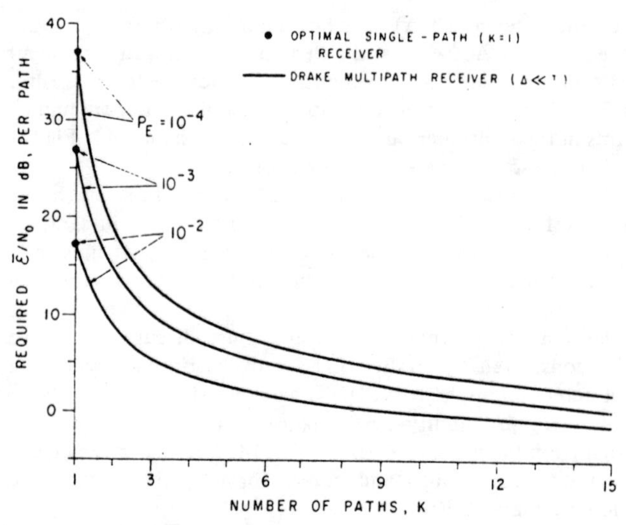

Fig. 24. Required $\overline{\mathcal{E}}/N_0$ per path for DPSK/DRAKE system: K independent, Rayleigh-fading paths.

(Area A); other experiments are in progress, and will be reported in detail elsewhere.[21]

A. Multipath Data Base

As a first step, a data base of urban/suburban multipath profiles were obtained by using SURP.[22] For each of the four areas A, B, C, D, defined in Section III-A, 150 000 profiles were generated along a track 3000-ft long (see Fig. 3), with uniform spacings of 0.02 ft, using L-band (1280 MHz) program parameters. In order to reduce the cost of the runs, new $\{\tau_l\}$ and $\{a_k\}$ strings were generated only at every fifth point, i.e., at 0.1-ft (0.12 wavelength) spacings, and used again for the subsequent four points. (This approximation is quite justified, since spatial correlation distances for $\{\tau_l\}$ and $\{a_k\}$ are tens of wavelengths.) However, since the correlation distance for the $\{\theta_k\}$ string is a fraction of a wavelength, a new $\{\theta_k\}$ string was generated at every point.

Some typical $\{\tau_l\}$ strings and a_k and θ_k values are shown in Fig. 25. One see the $\{\tau_l\}$ strings beginning to decorrelate after about one wavelength and showing substantial decorrelation after 10 wavelengths. Bin 13 happens to have a path in all $\{\tau_l\}$ strings shown, and the corresponding a_k and θ_k values generated for that bin are displayed. Again, for a_k, one sees partial decorrelation at one wavelength and substantial decorrelation after 10 wavelengths. On the other hand θ_k begins to decorrelate at 0.1 wavelength and shows substantial decorrelation at 1 wavelength.

The generated profiles were stored on tape. In addition, for future use in planned narrow-band experiments, the phasor sums $\Sigma_k\, a_k^{(n)} \exp[j\theta_k^{(n)}] = A_n \exp(j\psi_n)$ were computed and stored for each $n (n = 1, \cdots, 150\,000)$.

B. Low-Rate Experiments

Recall from Fig. 2 that the simulated multipath profiles are discrete-time impulse responses of a low-pass equivalent chan-

[21] The communication system simulation programs were written and run by M. Kamil. They are part of a Ph.D. thesis in progress, the aim of which is to explore urban mobile digital systems by simulation, in much more depth than the results given here.

[22] This data base was established by H. Hashemi.

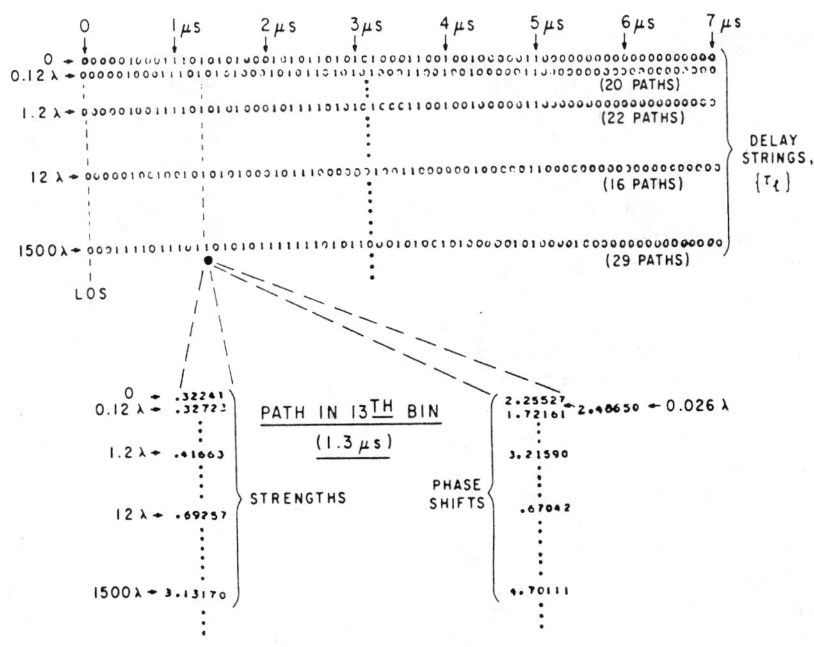

Fig. 25. Typical $\{\tau_l\}$ strings and a_k and θ_k values generated for L-band, using SURP.

nel filter, with time running in 100-ns increments. In designing the "low-rate" communication system simulation experiments, we postulated a DPSK transmitted signal whose basic low-pass-equivalent waveform $s(t)$ is a 127-chip maximal-length shift register (MLSR) sequence having a 100-ns chip duration.[23] Denoting the chip sequence by $\{s_j\}_0^{126}$, the low-pass-equivalent transmitted signal for an N-bit transmission is then of the form

$$w_i = \sum_{n=0}^{N} d_n s_{i-127n}, \qquad i = 0, \cdots, 127N - 1. \tag{58}$$

Here, s_j is defined to be zero for j outside $[0,126]$, and $d_n d_{n-1} = \pm 1$ according as bit $n (n = 1, \cdots, N)$ is 0 or 1, respectively (see Fig. 13(a), (b)).

The w_i sequence was convolved with a space-varying sequence of multipath profiles to form a channel output sequence that simulated the effect of vehicle motion. To this were added independent complex-valued Gaussian noise samples having variance σ_n^2, which was made a program parameter to allow adjustment of the SNR. The noisy received sequence was then convolved with the matched filter impulse response $\{s_{126-i}\}_{i=0}^{126}$.[24]

At this stage, one has a discrete-time representation of the matched-filter output in Fig. 21 or Fig. 23. This output was then DPSK demodulated by multiplying it by the complex-conjugate of a 127-chip-delayed version of the same output and taking the real part of the product. No nonlinear enhancement (e.g., as in Fig. 11) was attempted. The DPSK-demodulator output, which is a noisy discrete-time version of Fig. 13(c), was processed directly by various decision mechanisms, to be described.

The signal format we are using in this low-rate case implies the following parameters: signal bandwidth, $W = (100 \text{ ns})^{-1} = 10$ MHz; spectrum spread factor, $TW = 127$; bit rate, $1/T = 78.7$ kb/s. Further, since the multipath profiles are restricted to 71 samples spaced by 100 ns (see Fig. 2), we have $W\Delta = 70$. Thus $\Delta/T = 0.55$, so that intersymbol interference is negligible.

Notice that a vehicle moving even at 60 mph will move only about 0.001 ft (about 10^{-3} wavelength at L-band) per bit, so bit-to-bit changes in the phases $\{\theta_k\}$ negligibly affect the DPSK demodulation. We therefore simplified the experiments by "sampling" the transmissions periodically in space. That is, we imagined that the vehicle listened to 10 bits at each of 10 000 spatial positions spaced 0.3 ft (0.39 wavelength) apart, for a total of 100 000 bits per experiment. In this way, we obtained adequate data to determine bit error probabilities down to 10^{-4} - 10^{-5}, while guaranteeing a representative geographical cross section (3000 ft) of multipath responses and a representative ensemble of noise waveforms.

Seven decision mechanisms, operating on the simulated DPSK demodulator output, were evaluated. These are described below, and should be visualized in connection with the waveform $y(t)$ in Fig. 13(c).

1) First Path (FP): This is a classical mechanism, in which, in each bit interval, the first threshold crossing of $|y(t)|$ is examined, and a 0 or 1 is decided according as $y(t)$ is positive or negative. In our experiment, the threshold level on $|y(t)|$ was set so that it would be exceeded if the corresponding peak in the matched-filter output exceeded the *rms* value of the noise in the output. For simplicity in running the simulations, we idealized the mechanism by assuming that the first threshold crossing was always due to a signal peak. The position of the first signal-peak crossing was artificially estimated by having the program successively compare the noiseless path strengths

[23] The use of an MLSR sequence has a beneficial effect in our context: whenever $s(t)$ is sent twice with the same polarity, the right-hand sidelobe fluctuations of the matched filter's first response cancel the left-hand sidelobe fluctuations of the second response, thus considerably reducing interpath and intersymbol interference. This cancellation is a well-known property of MLSR sequences [29].

[24] In fact, for computational convenience, the order of the operations was quite different: the convolution of $\{s_j\}$ with $\{s_{126-j}\}$ was done first, then the modulation with the d_n sequence, then the convolution with the multipath profile, then the addition of appropriate noise.

$a_0, a_1, a_2, \cdots$ (which are known to the computer, but not to the receiver) with a threshold related to that set on $|y(t)|$. Because of this idealization, we expect our simulation result to be a lower bound on the error probability of the actual mechanism.

2) Largest Path (LP): Here, the largest peak in $|y(t)|$ in each bit interval is examined and the polarity of this peak determines the output. Again we idealized (not much of an idealization in this case) by assuming that the largest peak is a signal peak. Its position was identified as that of the largest a_k.

3) Post-Detection Integrator (PDI): This is the decision mechanism of Fig. 21. However, instead of integrating over a full $\Delta = 7$-μs interval, we integrated only over a 4-μs interval beyond the LOS path.

4) Adaptive Post-Detection Integrator (APDI): Here, the PDI integrator was made adaptive by letting the integration interval run only from the first crossing to the last crossing of $|y(t)|$ above a threshold. The threshold level and the method of determining the first and last crossing positions were identical to those used for the FP mechanism.

5) Weighted Post-Detector Integrator (WPDI): Recognizing that later paths in a multipath profile are likely to be weaker and less frequent than earlier paths, we here changed the impulse response of the integrator of Fig. 21 from the boxcar $h_p(t)$ of (26) to one with exponential weighting: $h_p(t) = \exp[-(\Delta - t)/\tau]$, $0 \leqslant t \leqslant \Delta$, where the time constant τ was set at 1.5 μs.

6) DRAKE: This is the receiver of Fig. 23. The switch positions were determined by assuming that the sounding receiver will identify as paths only those that are strong enough to cause $|y(t)|$ in the data receiver to exceed a threshold. The threshold level and the method for identifying above-threshold paths are as in the FP mechanism.

7) RAKE: Here, Fig. 23 is modified to restore the ideal tap amplifiers of Fig. 14 in place of the switches. The amplifier gains were determined by assuming that the sounding receiver's estimates of the a_k's are perfect. (See Appendix C.)

Mechanisms (1), (4), and (6) depend on there being at least one path above threshold. At very small SNR's, there were a small number of multipath profiles for which no path exceeded the threshold, and the demodulators deleted the corresponding sequence of 10 bits. Since these bits by definition are very noisy, ignoring them leads to an optimistic estimate of the error probability P_E. The effect was not serious, however, causing less than a 10 percent bias for $P_E \gtrsim 0.1$ and no noticeable bias for $P_E \lesssim 0.1$. But, in order to assure valid comparison of mechanisms (1), (4), and (6) with (2), (3), (5), and (7), which have no threshold but which would be adversely affected by inclusion of the noisy bits, we deleted these bits from the latter demodulator outputs also.

The empirical bit error probability curves determined from the simulation are shown in Fig. 26 as functions of both $\overline{\mathcal{E}}_{av}/N_0$ and $\overline{\mathcal{E}}_{LOS}/N_0$. Here, N_0 is the (single-ended) channel noise power density, as defined in Section IV. $\overline{\mathcal{E}}_{av}$ is the average energy received per bit per path, determined by calculating the mean-square path strength of all paths in all profiles used in the experiment. $\overline{\mathcal{E}}_{LOS}$ is the average energy received per bit on only LOS paths, determined by calculating the mean-square LOS path strength in all profiles having an LOS path. In actual system design, $\overline{\mathcal{E}}_{LOS}$ is a quantity derivable from the power budget, being determined from free-space and excess-attenuation calculations [33].

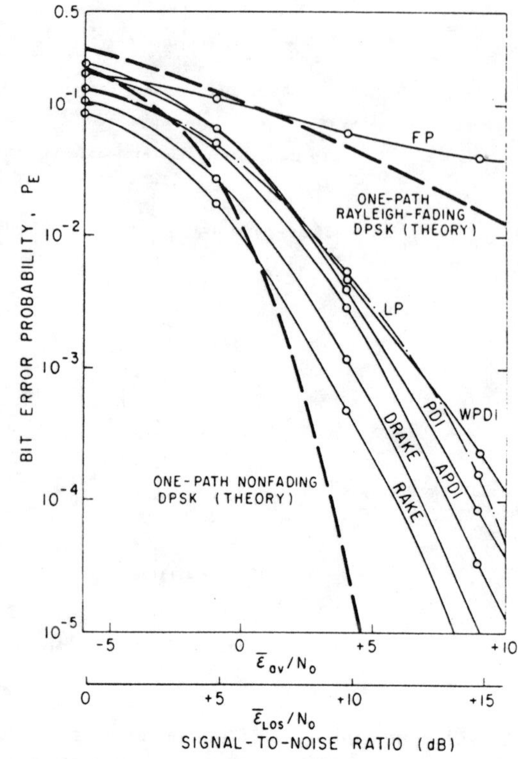

Fig. 26. Error probabilities for the low-rate case, from simulations; dense-high-rise topography ($TW = 127$, $\Delta/T = 0.55$, $R = 1/T = 78.7$ kb/s).

For comparison, also shown in Fig. 26 are classical error-probability curves for an optimal DPSK system operating through a single-path channel, either nonfading or Rayleigh fading (see footnote 20). For these curves, the abscissa is $\overline{\mathcal{E}}_{LOS}/N_0$, where $\overline{\mathcal{E}}_{LOS}$ is the average energy received per bit over the single path, the $\overline{\mathcal{E}}_{av}/N_0$ scale has no meaning for the comparison curves.

From Fig. 26, we can draw the following conclusions.

1) Comparison of the theoretical one-path Rayleigh-fading curve with the empirical first-path-above-threshold (FP) curve shows that path fading (at least of initial paths) is worse than Rayleigh, a conclusion already drawn by Suzuki [26] and Charash [4]. (The fact that the FP curve lies below the Rayleigh curve for small SNR merely shows the value of searching for an above-threshold path when the noise is substantial, rather than always utilizing the same path.) Comparing of the FP curve with Charash's single-path P_E curves [4], as appropriately modified for DPSK signalling, shows that the FP curve reflects Nakagami fading with parameter $m = 0.75$, rather than $m = 1$ (Rayleigh fading).

2) Comparison of the WPDI and PDI shows that a 4-μs boxcar integrator outperforms a 1.5-μs exponentially weighted integrator, at least for dense urban multipath. However, there is reason to believe that for other urban areas and for other time constants, this conclusion might be reversed.

3) An LP receiver, which "captures" the largest peak in $y(t)$, works overwhelmingly better than one that captures the first available peak. (This is the familiar "largest of" voting diversity effect.) In fact, the FP receiver works less than one dB worse than a PDI in the range $10^{-5} \leqslant P_E \leqslant 10^{-2}$, provided that the largest peak in $y(t)$ indeed corresponds to the largest path rather than to noise. However, we note that the PDI is

easier to implement than the FP receiver, among other things requiring less-accurate sampling and timing circuitry.

4) Progressing through successively more complex systems (PDI, APDI, DRAKE, and RAKE) gains of the order of 1 dB per step, provided that APDI, DRAKE, and RAKE have perfect estimates of the path information they need. It is this moderate gain in performance as one progresses through a sequence of substantially more and more sophisticated receivers that supports our faith in their quasi-optimality, as discussed in Section III-G. Changing from PDI (unknown delays) to DRAKE (known delays) results in only a 2-dB improvement, whence it seems certain that PDI is a very close to optimal for unknown delays. Again, changing from DRAKE (unknown strengths, known delays) to RAKE (known strengths, known delays) yields only 1 dB, whence it would seem that DRAKE is close to optimal for a delay-only channel-estimating receiver and RAKE is close to optimal for a delay-and-strength estimating receiver.

5) Because of their path-combining characteristics, by which they accumulate the energies of many almost-independent paths, the WPDI, PDI, APDI, DRAKE, and RAKE receivers operate at an enormous advantage over a receiver that utilizes only a fixed single fading path. At small SNR, they even moderately outperform a receiver that utilizes a single *nonfading* path;[25] while they lose this moderate advantage at large SNR, their performance still comes within a few decibels of the nonfading single-path case. (See Monsen [15] for a similar result obtained by equalization techniques.)

If equipment simplicity is taken into account as well as performance, it would seem desirable to choose the PDI receiver in this low-rate case, for at the expense of less than 2 or 3 dB in SNR the PDI can be substituted for the vastly more complicated DRAKE or RAKE. APDI, while relatively simple, outperforms PDI by less than 1 dB, scarcely worth the additional threshold, AGC and integrate-and-dump features of APDI.

As we have seen, however, the PDI structure is quite susceptible to intersymbol interference. Thus in the high-rate case ($\Delta >> T$), we must rely on RAKE and DRAKE, with all their complexities. We investigate the high-rate performance of RAKE and DRAKE in the following section.

C. High-Rate Experiments

In the high-rate experiments, we increased the data rate by a factor of 10, i.e., to $1/T = 787$ kb/s. Since Δ remains at 7 μs, we now have $\Delta/T = 5.5$, so considerable intersymbol interference exists: 6 predecessor and 6 successor symbols affect the present symbol.

We wish to maintain the same spectrum spreading factor, $TW = 127$, which means that W is now 100 MHz.[26] A problem is posed by this increased bandwidth, for it implies a resolution of $1/W = 10$ ns in the multipath model, while our fundamental data and channel simulation program are based on 100-ns resolution. Unfortunately, virtually no 10-ns-resolution data on the urban channel at the frequencies under consideration are available. (See [10], however.)

We have resolved this dilemma by an approximate extrapolation of our 100-ns data. Each 100-ns bin in Fig. 2(a) was divided into 10 subbins. If $\tau_l = 1$ for some $l \neq 0$, the corresponding path was placed at random in one of the 10 subbins; and if $\tau_0 = 1$, i.e., an LOS path exists, the path was always placed in the first subbin. Thus implicitly, we have structured the multipath to preclude an average of more than one path every 100 ns. While this restriction almost certainly varies from reality, it allows us to use SURP and the data base generated by it for the high-rate case. Hopefully, the resulting simulation results will not be too far off.[27]

In other respects, the high-rate experiment followed the low-rate experiment. DPSK signaling was postulated, using the same MLSR sequence, but now having a 10-ns chip duration and a 1.27-μs bit duration. The multipath impulse response with which the transmitted signal (58) is convolved is now 710 samples long rather than 71, but consists of more than 90 percent zeros. Only RAKE and DRAKE were evaluated, for, as seen in Sections III-F and IV, only they are capable of suppressing intersymbol interference. (WPDI, PDI, and APDI irretrievably scramble the responses to different symbols, while FP and LP present substantial difficulties in deciding which is the first or largest path of the current symbol amidst the interlaced responses to several symbols.)

The results of the high-rate experiment, again run for approximately 100 000 bits per point, are shown in Fig. 27, where the RAKE and DRAKE low-rate results are also shown for comparison.

We see that, as predicted in Section IV, the presence of substantial intersymbol interference does not drastically curtail system operation. In fact, the degradation formula (57), although derived for coherent RAKE and a much simplified multipath model, proves to be remarkably accurate. The average number of paths per profile encountered in the simulation run was 23.2. If we allow the simplified multipath model of Section IV to have $\overline{K} = 23.2$ paths and the $\overline{\mathcal{E}}/N_0$ parameter of that model to be identified with $\overline{\mathcal{E}}_{av}/N_0$ of Fig. 27, then, setting $TW = 127$, D of (57) is 1.3 dB at $\overline{\mathcal{E}}_{av}/N_0 = 3$ dB, 2.0 dB at $\overline{\mathcal{E}}_{av}/N_0 = 5$ dB and 2.8 dB at $\overline{\mathcal{E}}_{av}/N_0 = 7$ dB. These are almost exactly the gaps between the two RAKE curves shown in Fig. 27.

Again we see that changing from DRAKE to RAKE yields only a modest improvement (2-3 dB), and we therefore guess that the optimal receiver will not substantially outperform RAKE.

D. Comparison with the Simplified Theory

In Section V, we gave theoretical error-probability curves for the low-rate PDI and DRAKE demodulators; see Figs. 22 and 24. Since these curves were based on a number of very simple but often-used assumptions (uniformly distributed Poisson path arrivals; independent Rayleigh-distributed path strengths; no intersymbol interference; a zero-sidelobe DPSK

[25] Much of this small-SNR advantage over nonfading, single-path operation is real, reflecting the benefits of path combination when the noise is large. Some of the advantage, however, reflects the previously mentioned bias in the experiment at small SNR due to discarding certain "bad" profiles.

[26] This large bandwidth does not require exclusive assignment, however, as many systems with different codes could coexist in the same band; see [40] for a discussion of a spread-spectrum code-multiplexed digital-voice system proposed for urban use.

[27] Since we are dealing with an average of about 25 paths per profile, it is unlikely that use of a refined 10-ns model—having, say, 50–75 paths per profile but delivering the same total energy—would result in a substantially different error rate. This assertion follows from the well-known phenomenon that most diversity gain accrues from the addition of only a few diversity links, the point of diminishing returns being quickly reached.

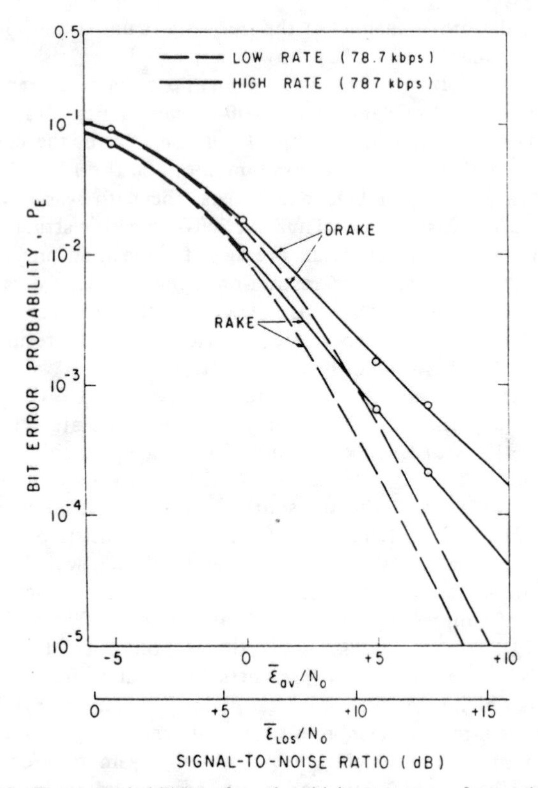

Fig. 27. Error probabilities for the high-rate case, from simulations; dense high-rise topography ($TW = 127$, $\Delta/T = 5.5$, $R = 1/T = 787$ kb/s).

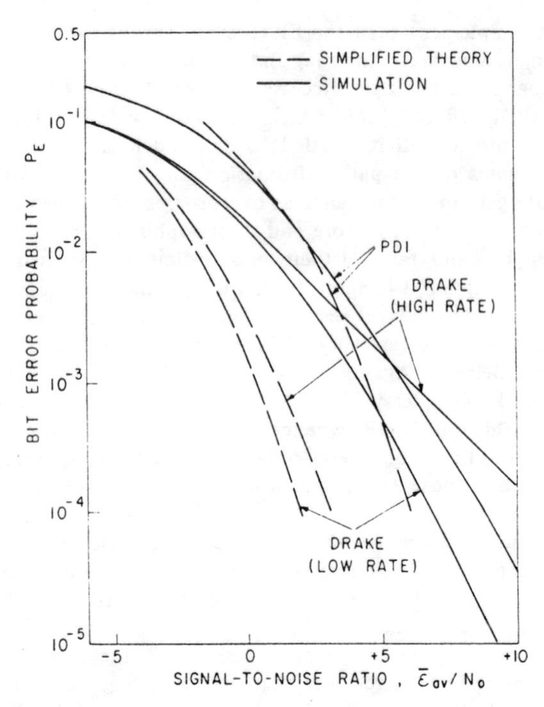

Fig. 28. Comparison of simplified theory with experiment.

signal, i.e., no interpath interference), it is of value to compare them with reality. The comparison is shown in Fig. 28. There, we have redrawn the simulation-derived low-rate PDI and DRAKE curves from Fig. 26 and the high-rate DRAKE curve from Fig. 27. Also shown are theoretical low-rate PDI and DRAKE curves taken from Figs. 22 and 24 (as extended from [3]), which were obtained by extrapolating Figs. 22 and 24 to a value of K equal to the *average* number of paths per profile (23.7) encountered in the simulation. The theoretical high-rate DRAKE curve was obtained from the corresponding low-rate curve by assuming that the degradation given by (57), derived for ideal coherent RAKE, applies approximately also to DRAKE; in (57), we used $\overline{K} = 23.7$.

We see that the simplified theory leads to results that are between 3 and 7 dB optimistic in neighborhood of $P_E = 10^{-4}$, and less so at larger error probabilities. This optimism is due to the combined effects of the following factors.

1) As we have noted in connection with the FP curves of Fig. 26, actual urban fading is worse than Rayleigh. Reference to the worse-than-Rayleigh curves of [3] show that the Rayleigh assumption leads to about 1-dB worth of optimism at $P_E = 10^{-4}$.

2) Actual urban paths are correlated, rather than independent; we estimate that assuming independence leads to another 1 dB of optimism.

3) Because of spatial inhomogeneities, actual path strengths have variances that are smaller in local areas than on a global basis. The simplified theory imagines that the global variances apply at each local site, leading to a larger-than-actual "path diversity" gain.

4) The actual DPSK signal used did not have zero sidelobes, so more interpath and intersymbol interference exists than in the simplified theory. We estimate from Section IV that ignoring this interference accounts for about 2.5 dB of optimism at $P_E = 10^{-4}$.

5) Finally, since the theoretical low-rate DRAKE curve is optimistic in estimating the required $\overline{\mathcal{E}}_{av}/N_0$ for a given P_E, use of (57) will lead to an optimistic value of D to be applied in obtaining the high-rate curve, thus compounding the error. The error is further compounded by the assumption that (57) applies to DRAKE, since Fig. 27 shows that the intersymbol interference is more damaging to DRAKE than to RAKE.

Thus the simplified theory, although based on common analytical assumptions, must be treated with caution, especially at $P_E < 10^{-2}$; that is, its optimism should be recognized in attempting to apply it to urban multipath.

Appendix A: Derivation of (46)

From (42) and using the assumptions listed just prior to that equation, we obtain

$$E[|v(0)|^2]|_{d_0=+1} = \sum_{n=-N}^{N} E_K \sum_{k,l,q,r=0}^{K-1} \overline{\alpha_k\alpha_l^*\alpha_q^*\alpha_r} \cdot \gamma(nT - t_k + t_l)\,\gamma^*(nT - t_q + t_r) \qquad \text{(A-1)}$$

where the overbars denote expectation and E_K denotes expectation over the random variable K. Note that, of all the combinations of the indices k, l, q and r in (A-1), in any instance where one is distinct from the others, the expectation $\overline{\alpha_k\alpha_l^*\alpha_q^*\alpha_r}$ will factor into the form $\overline{\alpha_k\alpha_l^*\alpha_q^*}\;\overline{\alpha_r}$ (where here r is the distinct index); this factorization follows by virtue of the independence of the α_k's. But, because of the assumed uniform distribution of the θ_k's and the independence of the a_k's and θ_k's, $\overline{\alpha}_k = E[a_k \exp(j\theta_k)] = 0$. Thus the only surviving terms in (A-1) are those in which indices are pairwise equal, and, of these, those where $k = r$ and $l = q$ vanish because $\overline{\alpha_k^2} = E[\alpha_k^2 \exp(j2\theta_k)] = 0$. Therefore,

$$E[|v(0)|^2] = \sum_{n=-N}^{N} E_K\left[\sum_{k=0}^{K-1} \overline{|\alpha_k|^4}\, |\gamma(nT)|^2 + \sum_{\substack{k=0\\k\neq l}}^{K-1}\sum_{l=0}^{K-1} \overline{|\alpha_k|^2}\,\overline{|\alpha_l|^2}\, |\gamma(nT)|^2 + \sum_{\substack{k=0\\k\neq l}}^{K-1}\sum_{l=0}^{K-1} \overline{|\alpha_k|^2}\,\overline{|\alpha_l|^2}\, \overline{|\gamma(nT - t_k + t_l)|^2}\right] \quad \text{(A-2)}$$

where we have noted that (A-1) is independent of d_0.

Since

$$\gamma(t) = \int_{-\infty}^{\infty} |S(f)|^2 \exp(j2\pi ft)\, df \quad \text{(A-3)}$$

we have

$$\overline{|\gamma(nT - t_k + t_l)|^2} = \int_{-\infty}^{\infty} |S(f)|^2\, |S(\nu)|^2 \exp[j2\pi(f-\nu)nT] \cdot \overline{\exp[j2\pi(f-\nu)t_k]} \cdot \overline{\exp[-j2\pi(f-\nu)t_l]}\, df\, d\nu. \quad \text{(A-4)}$$

If we define $p(t_k)$ to be the probability density distribution of the t_k's, then

$$\overline{\exp(j2\pi f t_k)} = \int_0^{\Delta} p(t_k) \exp(j2\pi f t_k)\, dt_k = P(f) \quad \text{(A-5)}$$

is the characteristic function of $p(t_k)$.

We note from (23) *et seq.*, that $|\gamma(nT)|^2 = 0$ for $n \neq 0$. Using this fact and (A-4) and (A-5), (A-2) becomes

$$E[|v(0)|^2] = (2\mathcal{E})^2\left[E_K \sum_{k=0}^{K-1} \overline{|\alpha_k|^4} + E_K \sum_{\substack{k=0\\k\neq l}}^{K-1}\sum_{l=0}^{K-1} \overline{|\alpha_k|^2}\,\overline{|\alpha_l|^2}\right] + \sum_{n=-N}^{N} E_K \sum_{k=0}^{K-1}\sum_{l=0}^{K-1} \overline{|\alpha_k|^2}\,\overline{|\alpha_l|^2} \cdot \iint_{-\infty}^{\infty} |S(f)|^2\, |S(\nu)|^2\, |P(f-\nu)|^2 \cdot \exp[j2\pi(f-\nu)nT]\, df\, d\nu \quad \text{(A-6)}$$

where we have used the fact that $\gamma(0) = 2\mathcal{E}$, $\mathcal{E}$ being the energy of the transmitted signal $s(t) = \text{Re}[\sigma(t)\exp(j\omega t)]$.

From (A-5) we see that for a reasonably diffuse $p(t_k)$, in the sense of Section III-C, the "width" of $P(f)$ must be of the order of $1/\Delta$. On the other hand, the width of $|S(f)|^2$ is of the order of W. But we are assuming $TW >> 1$ and $\Delta > T$, so $W >> 1/\Delta$. The situation is depicted in Fig. 29, where the factors $|S(f)|^2$ and $|P(f-\nu)|^2$ in the integrand of (A-6) are shown. If we assume that $|S(f)|^2$ is smoothly varying, the integral on f in (A-6) can be written approximately as

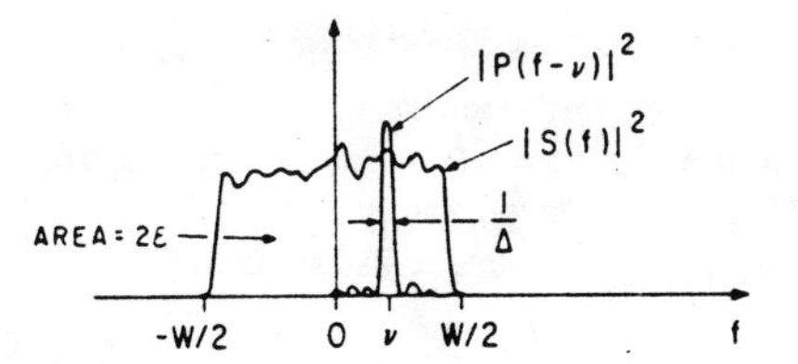

Fig. 29. Illustrating the integration in equation (A-6).

$$\int_{-\infty}^{\infty} |S(f)|^2\, |P(f-\nu)|^2 \exp(j2\pi fnT)\, df$$

$$\cong |S(\nu)|^2 \int_{-\infty}^{\infty} |P(f-\nu)|^2 \exp(j2\pi fnT)\, df$$

$$\cong \begin{cases} |S(\nu)|^2\, q(nT) \exp(j2\pi\nu nT), & |\nu| \leq \dfrac{W}{2} \\ 0, & |\nu| > \dfrac{W}{2} \end{cases} \quad \text{(A-7)}$$

where

$$q(t) = \int_{-\infty}^{\infty} |P(f)|^2 \exp(j2\pi ft)\, df = \int_0^{\Delta} p(\tau)\, p(\tau - t)\, d\tau. \quad \text{(A-8)}$$

The double integral in (A-6) then becomes

$$q(nT) \int_{-\infty}^{\infty} |S(\nu)|^4\, d\nu \quad \text{(A-9)}$$

so (A-6) becomes, after a little manipulation,

$$E[|v(0)|^2] = (2\mathcal{E})^2 E_K\left[\sum_{k=0}^{K-1} \left(\overline{|\alpha_k|^4} - \overline{|\alpha_k|^2}^{\,2}\right)\right] + (2\mathcal{E})^2 E_K\left[\left(\sum_{k=0}^{K-1} \overline{|\alpha_k|^2}\right)^2\right] + \left[\int_{-\infty}^{\infty} |S(\nu)|^4\, d\nu\right]\left[\sum_{n=-N}^{N} q(nT)\right] \cdot E_K \sum_{\substack{k=0\\k\neq l}}^{K-1}\sum_{l=0}^{K-1} \overline{|\alpha_k|^2}\,\overline{|\alpha_l|^2}. \quad \text{(A-10)}$$

The second line in (A-10) is recognizable as the square of $E[v(0)]$ of (44); subtracting it, we get var $[v(0)]$, as given in (46).

Appendix B: Derivation of the Last Term in (50)

Suppose that the channel noise, $n(t) = \text{Re}[\nu(t)\exp(j2\pi f_0 t)]$, is a bandpass white process having spectral density $N_0/2$ in a band around $\pm f_0$. The low-pass-equivalent noise $\nu(t)$ then has a spectral density $2N_0$ in a band of the same width around

$f = 0$ (see [30]). $\nu(t)$ passes through the receiver's cascade of the matched filter and transversal filter, which has overall transfer function $S^*(f)\,H_{tf}(f)$ (see (35)). Assuming that the transversal filter's taps are exactly set ($\hat{\alpha}_k = \alpha_k, \hat{t}_k = t_k$, all k), the noise output of this cascade has spectrum

$$U(f) = 2N_0|S(f)|^2 \left|\sum_{k=0}^{K-1} \alpha_k^* \exp(+j2\pi f t_k)\right|^2$$

$$= 2N_0|S(f)|^2 \sum_{k=0}^{K} \sum_{l=0}^{K-1} \alpha_k^* \alpha_l \exp[j2\pi f(t_k - t_l)]. \quad \text{(B-1)}$$

Using the simplified assumptions on multipath statistics listed in Section IV, we can easily show that the average of $U(f)$ is

$$\overline{U}(f) = 2N_0|S(f)|^2 A^2 \quad \text{(B-2)}$$

where A is as in (45). The variance of the noise at the output of the transversal filter is therefore

$$\int_{-\infty}^{\infty} \overline{U}(f)\, df = 4\mathcal{E}N_0A^2. \quad \text{(B-3)}$$

Appendix C: The Effect of Noisy Path Estimates

In the analysis of Section IV, we assumed that the channel-sounding system was capable of supplying perfect estimates of path variables and that the RAKE receiver was capable of using these exactly. In effect, we therefore assumed that the sounding receiver can supply $h_m(t)$ of (30) or $H_m(f)$ of (34) exactly and RAKE can then configure itself into a filter with transfer function $\hat{H}_{tf}(f) = H_m^*(f) \exp(-j2\pi f\Delta)$ (see (35)). We investigate the implications of these assumptions here.

Estimation of the impulse response of an unknown medium has been investigated in [28], where the optimum sounding signal and optimum estimating receiver are derived. Generally speaking, the optimal signal's energy is distributed in a way depending on the noise spectrum and average channel transmission function, while the optimal receiver uses a filter that is between being inverse to and matched to the optimum signal.

When the simplified channel model of Section IV is assumed, however, the optimal signal distributes its energy uniformly over the transmission band of the system and the optimal receiver is matched to this signal. In lowpass-equivalent form, the sounding receiver's matched-filter output is a noisy, band-limited version of (30), i.e.,

$$\hat{h}_m(t) = \sum_{k=0}^{K-1} a_k \exp(j\theta_k) \frac{\sin \pi W(t - t_k)}{\pi W(t - t_k)} + \xi(t) \quad \text{(C-1)}$$

where $\xi(t)$ is a complex low-pass output noise. In (C-1), the $(\sin x)/x$ pulses have been scaled to have the same areas as the corresponding impulses in (30). Hence, the Fourier transform of the first term of (C-1) agrees with that of (30) within the band $|f| \leqslant W/2$.

We approximate the RAKE transversal filter by assuming that its impulse response is

$$\hat{h}_{tf}(t) = \hat{h}_m^*(\Delta - t)$$

$$= \sum_{k=0}^{K-1} \alpha_k^* \frac{\sin \pi W(\Delta - t - t_k)}{\pi W(\Delta - t - t_k)} + \xi^*(\Delta - t) \quad \text{(C-2)}$$

instead of (33). The first term of (C-2) agrees with (33) in the band $|f| \leqslant W/2$ and therefore will act on RAKE input signals identically as (33); to implement it would require infinitesimally spaced taps on the RAKE delay line, however. The second term of (C-2) comprises the noise in the estimate supplied by the sounding receiver.

The second term in (C-2) is that responsible for the degradation in performance due to a noisy sounding estimate. As mentioned, the optimal sounding signal for our case has a flat spectrum,

$$|X(f)| = \sqrt{2\mathcal{E}_x/W}, \quad |f| \leqslant W/2 \quad \text{(C-3)}$$

where $\mathcal{E}_x$ is the sounding energy. In order to have the scaling of (C-2), the magnitude of the frequency characteristic of the sounding receiver's matched filter must be $\sqrt{W/2\mathcal{E}_x}$ over the band. The noise power density of $\xi(t)$ is thus

$$P_\xi(f) = \begin{cases} N_0W/\mathcal{E}_x, & |f| \leqslant W/2 \\ 0, & |f| > W/2 \end{cases} \quad \text{(C-4)}$$

since, from Appendix B, the low-pass-equivalent channel noise has power density $2N_0$.

Let

$$y(t) = \mathcal{F}^{-1}[H_m(f)|S(f)|^2 \exp(j2\pi fT)]$$

$$= \sum_{k=0}^{K-1} \alpha_k \gamma(t - t_k) \quad \text{(C-5)}$$

be the low-pass equivalent signal output of the data receiver's matched filter, in response to a single-signal transmission, as shifted so that its LOS peak occurs at $t = 0$. The RAKE transversal filter convolves this output with $\hat{h}_{tf}(t)$ to form $w(t)$, the signal component of which is specified by (36) and (38). $w(t)$ also has a component due to noisy estimation of channel parameters, obtained by convolving the $\xi^*(\Delta - t)$ of (C-2) with $y(t)$ of (C-5).[28] This component of $w(t)$ has the form

$$\int_0^\Delta \xi^*(\Delta - t)\, y(t - \tau)\, d\tau \quad \text{(C-6)}$$

which, when shifted to the left by Δ seconds, becomes

$$\int_0^\Delta \xi^*(\tau)\, y(t + \tau)\, d\tau. \quad \text{(C-7)}$$

Equation (C-7) gives the estimation-noise component of $w(t)$ when only a single signal is transmitted. When a sequence of signals is transmitted with PSK modulation, as discussed prior to (39), the estimation-noise component becomes

$$\epsilon(t) \triangleq \int_0^\Delta \xi^*(\tau) \sum_{n=-N}^{N} d_n y(t - nT + \tau)\, d\tau \quad \text{(C-8)}$$

where we have indicated that the responses to only $2N + 1$ symbols are in the RAKE delay line at one time.

At $t = 0$, when the signal component of the RAKE filter output peaks for the present symbol (see (42)), the estimation-noise component is

$$\epsilon(0) = \sum_{n=-N}^{N} d_n \int_0^\Delta \xi^*(\tau)\, y(\tau - nT)\, d\tau. \quad \text{(C-9)}$$

[28] There is also a term in $w(t)$ involving convolution of $\xi(t)$ with the noise in the data receiver's matched-filter output; we neglect this.

This has zero mean, since $\overline{\xi^*(t)} \equiv 0$, and variance

$$E[|\epsilon(0)|^2] = \sum_{n=-N}^{N} \sum_{m=-N}^{N} d_n d_m \iint_0^{\Delta} \overline{\xi^*(\tau)\, \xi(\sigma)}\, \overline{y(\tau - nT)\, y^*(\sigma - mT)}\, d\tau\, d\sigma. \quad \text{(C-10)}$$

Since $\overline{d_n d_m} = \delta_{mn}$ (independent data symbols), and (from (C-4)) $\overline{\xi^*(\tau)\, \xi(\sigma)} = (N_0 W/\mathcal{E}_x)\, \delta(\tau - \sigma)$, and (from (C-5)),

$$\overline{y(\tau - nT)\, y^*(\sigma - mT)} = E_K \sum_{k=0}^{K-1} \sum_{l=0}^{K-1} \overline{\alpha_k \alpha_l^*}\, \overline{\gamma(\tau - nT - t_k)\, \gamma^*(\sigma - mT - t_l)} \quad \text{(C-11)}$$

and using $\overline{\alpha_k \alpha_l^*} = \overline{|\alpha_k|^2}\, \delta_{kl}$ (from the simplified multipath model of Section IV), we have

$$E[|\epsilon(0)|^2] = \frac{N_0 W}{\mathcal{E}_x} \sum_{n=-N}^{N} \cdot E_K \sum_{k=0}^{K-1} \int_0^{\Delta} \overline{|\alpha_k|^2}\, \overline{|\gamma(\tau - nT - t_k)|^2}\, d\tau. \quad \text{(C-12)}$$

Again using the simplified multipath model of Section IV, in which all t_k's are independent and uniformly distributed over $[0, \Delta]$, we have

$$E[|\epsilon(0)|^2] = \frac{N_0 W A^2}{\mathcal{E}_x \Delta} \sum_{n=-N}^{N} \int_0^{\Delta} \int_0^{\Delta} |\gamma(\tau - nT - t_k)|^2\, d\tau\, dt_k \quad \text{(C-13)}$$

where A^2 is given by (45). Using the transformation $u = \tau - t_k$, $v = \tau + t_k$, whose Jacobian is $\frac{1}{2}$, equation (C-13) becomes

$$E[|\epsilon(0)|^2] = \frac{N_0 W A^2}{2\mathcal{E}_x \Delta} \sum_{n=-N}^{N} \int_{-\Delta}^{\Delta} du \int_{|u|}^{2\Delta - |u|} dv\, |\gamma(u - nT)|^2$$

$$= \frac{N_0 W A^2}{\mathcal{E}_x \Delta} \sum_{n=-N}^{N} \int_{-\Delta}^{\Delta} (\Delta - |u|)\, |\gamma(u - nT)|^2\, du. \quad \text{(C-14)}$$

Finally, we note that $|\gamma(t)|$ is a narrow pulse of width $1/W << \Delta$. Treating it as a delta function of area

$$\int_{-\infty}^{\infty} |\gamma(t)|^2\, dt = \int_{-\infty}^{\infty} |S(f)|^4\, df \quad \text{(C-15)}$$

(since the Fourier transform of $\gamma(t)$ is $|S(f)|^2$, the signal's low-pass-equivalent energy density spectrum), and using (53), we have

$$E[|\epsilon(0)|^2] \cong \frac{4 N_0 \mathcal{E}^2 A^2}{\mathcal{E}_x \Delta} \sum_{n=-N}^{N} (\Delta - |n|\, T). \quad \text{(C-16)}$$

When $T > \Delta$, so only the $n = 0$ term exists, equation (C-16) becomes

$$E[|\epsilon(0)|^2] \cong \frac{4 N_0 \mathcal{E}^2 A^2}{\mathcal{E}_x}. \quad \text{(C-17)}$$

More interestingly, when $\Delta >> T$, so we can approximate the sum in (C-16) with an integral of value Δ^2/T, we have

$$E[|\epsilon(0)|^2] \cong \frac{N_0 \mathcal{E}^2 A^2}{\mathcal{E}_x} \cdot \frac{\Delta}{T}. \quad \text{(C-18)}$$

$E[|\epsilon(0)|^2]$ is another term to be added to (50), adding to the degradation caused by multipath-induced interference and channel noise. Taking this term into account, the increase in the RAKE output noise variance, as given by (57), now becomes

$$D \cong 1 + \frac{\overline{K}}{TW} \frac{\overline{\mathcal{E}}}{N_0} + \frac{\mathcal{E}}{\mathcal{E}_x} \frac{\Delta}{T} \quad \text{(C-19)}$$

where we have assumed that $\Delta >> T$ and used (C-18).

We thus conclude that if the ratio of sounding-signal energy to data-signal energy satisfies

$$\frac{\mathcal{E}_x}{\mathcal{E}} >> \frac{\Delta}{T} >> 1 \quad \text{(C-20)}$$

then the estimation-noise contribution to the total RAKE output noise is negligible. If, as in the high-rate case of Section VI-C, $\Delta/T = 5.5$, then $\mathcal{E}_x$ should be 10 times or more greater than $\mathcal{E}$ for (C-20) to hold.

Acknowledgment

A heavy debt of gratitude is due the National Science Foundation for its support of underlying basic research on multipath channel characterization and simulation at the University of California (Berkeley). Analyses and simulations of particular systems were supported by the Advanced Research Projects Agency (ARPA) through SRI International. Finally, the author acknowledges with pleasure his students U. Charash, H. Hashemi, M. Kamil, and H. Suzuki, upon whose Ph.D. research this paper heavily relies.

References

[1] P. A. Bello, "Characterization of randomly time-variant linear channels," *IEEE Trans. Commun. Syst.*, vol. CS-11, pp. 360–393, Dec. 1963.
[2] D. G. Brennan, "On the maximum signal-to-noise ratio realizable from several noisy signals," *Proc. IRE*, vol. 43, p. 1530, Oct. 1955.
[3] U. Charash, "A study of multipath reception with unknown delays," Ph.D. Thesis, Dep. Elec. Eng. Comput. Sci., Univ. California, Berkeley, Jan. 1974.
[4] —, "Reception through Nakagami-fading multipath channels with random delays," *IEEE Trans. Commun.*, vol. COM-27, pp. 657–670, Apr. 1979.
[5] D. C. Cox, "A measured delay-doppler scattering function for multipath propagation at 910 MHz in an urban mobile environment," *Proc. IEEE*, vol. 61, pp. 479–480, Apr. 1973.
[6] R. C. Dixon, Ed., *Spread-Spectrum Techniques*. New York: IEEE Press, 1976.
[7] S. C. Fralick, "An improved packet radio demodulator for mobile operation," Temporary packet radio note 132, Stanford Res. Inst., Menlo Park, CA, Mar. 1975.
[8] H. Hashemi, "Simulation of the urban radio propagation channel," Ph.D. thesis, Dep. Elec. Eng. Comput. Sci., Univ. California, Berkeley, Aug. 1977.
[9] —, "Simulation of the urban radio propagation channel," *IEEE Trans. Veh. Technol.*, vol. VT-28, pp. 213–224, Aug. 1979.
[10] R. W. Hubbard *et al.*, "Measuring characteristics of microwave mobile channels," Nat. Telecommun. and Inform. Admin., Boulder, CO, Rep. 78-5, June 1978.
[11] W. C. Jakes, Jr., Ed., *Microwave Mobile Communications*. New York: Wiley, 1974.
[12] R. S. Kennedy, *Fading Dispersive Communication Channels*. New York: Wiley, 1969.
[13] W. C. Lindsey, "Error probability for incoherent diversity reception," *IEEE Trans. Inform. Theory*, vol. IT-11, pp. 491–499, Oct. 1965.
[14] R. W. Lucky *et al.*, *Principles of Data Communication*. New

York: McGraw-Hill, 1968.

[15] P. Monsen, "Feedback equalization for fading dispersive channels," *IEEE Trans. Inform. Theory*, vol. IT-17, pp. 56–64, Jan. 1971.

[16] —, "Digital transmission performance on fading dispersive diversity channels," *IEEE Trans. Commun.*, vol. COM-21, pp. 33–39, Jan. 1973.

[17] —, "Adaptive equalization of the slow fading channel," *IEEE Trans. Commun.*, vol. COM-22, pp. 1064–1075, Aug. 1974.

[18] D. R. Morgan, "Adaptive multipath cancellation for digital data communications," *IEEE Trans. Commun.*, vol. COM-26, pp. 1380–1390, Sept. 1978.

[19] D. L. Nielson, "Microwave propagation and noise measurements for mobile digital radio application," Packet radio note 4, Stanford Res. Inst., Menlo Park, CA, Jan. 1975.

[20] —, "Microwave propagation measurements for mobile digital radio application," *IEEE Trans. Veh. Technol.*, vol. VT-27, pp. 117–132, Aug. 1978.

[21] J. N. Pierce and S. Stein, "Multiple diversity with nonindependent fading," *Proc. IRE*, vol. 48, pp. 89–104, Jan. 1960.

[22] R. Price, "Optimum detection of random signals in noise with applications to scatter-multipath communication," *IRE Trans. Inform. Theory*, vol. IT-2, pp. 125–135, Dec. 1956.

[23] R. Price and P. E. Green, Jr., "A communication technique for multipath channels," *Proc. IRE*, vol. 46, pp. 555–570, Mar. 1958.

[24] M. Schwartz *et al.*, *Communication Systems and Techniques*. New York: McGraw-Hill, 1966.

[25] H. Suzuki, "A statistical model for urban radio propagation," Ph.D. thesis, Dep. Elec. Eng. Comput. Sci., Univ. of California, Berkeley, Apr. 1975.

[26] —, "A statistical model for urban radio propagation," *IEEE Trans. Commun.*, vol. COM-25, pp. 673–680, July 1977.

[27] G. L. Turin, "Communication through noisy, random-multipath channels," *IRE Nat. Conv. Rec.*, pt. 4, pp. 154–166, 1956.

[28] —, "On the estimation in the presence of noise of the impulse response of a random, linear filter," *IRE Trans. Inform. Theory*, vol. IT-3, pp. 5–10, Mar. 1957.

[29] —, "An introduction to matched filters," *IRE Trans. Inform. Theory*, vol. IT-6, pp. 311–329, June 1960.

[30] —, "On optimal diversity reception," *IRE Trans. Inform. Theory*, vol. IT-7, pp. 154–166, July 1961.

[31] —, "On optimal diversity reception, II," *IRE Trans. Commun. Syst.*, vol. CS-10, pp. 22–31, Mar. 1962.

[32] G. L. Turin *et al.*, "Urban vehicle monitoring: Technology, economics and public policy," vol. II: Technical Analysis and Appendices," report prepared under DHUD contract H-1030, Oct. 1970.

[33] G. L. Turin *et al.*, "A statistical model of urban multipath propagation," *IEEE Trans. Veh. Technol.*, vol. VT-21, pp. 1–9, Feb. 1972.

[34] G. L. Turin *et al.*, "Simulation of urban vehicle-monitoring systems," *IEEE Trans. Veh. Technol.*, vol. VT-21, pp. 9–16, Feb. 1972.

[35] G. L. Turin, "Simulation of urban radio propagation and of urban radio communication systems," *Proc. Int. Symp. Antennas and Propagat.*, Sendai, Japan, pp. 543–546, Aug. 1978.

[36] J. M. Wozencraft and I. M. Jacobs, *Principles of Communication Engineering*. New York: Wiley, 1965.

[37] W. R. Young, Jr. and L. Y. Lacy, "Echoes in transmission at 450 megacycles from land-to-car radio units," *Proc. IRE*, vol. 38, pp. 255–258, March 1950.

[38] R. E. Kahn *et al.*, "Advances in packet radio technology," *Proc. IEEE*, vol. 66, pp. 1468–1496, Nov. 1978.

[39] E. N. Skomal, *Man-Made Radio Noise*. New York: Van Nostrand Reinhold, 1978.

[40] G. R. Cooper and R. W. Nettleton, "A spread-spectrum technique for high-capacity mobile communications," *IEEE Trans. Veh. Technol.*, vol. VT-27, pp. 264–275, Nov. 1978.

ABSTRACT

When the mechanisms of fading channels were first modeled in the 1950s and 1960s, the ideas were primarily applied to over-the-horizon communications covering a wide range of frequency bands. The 3–30 MHz high-frequency (HF) band is used for ionospheric communications, and the 300 MHz–3 GHz ultra-high-frequency (UHF) and 3–30 GHz super-high-frequency (SHF) bands are used for tropospheric scatter. Although the fading effects in a mobile radio system are somewhat different than those in ionospheric and tropospheric channels, the early models are still quite useful to help characterize fading effects in mobile digital communication systems. This tutorial addresses Rayleigh fading, primarily in the UHF band, that affects mobile systems such as cellular and personal communication systems (PCS). Part I of the tutorial itemizes the fundamental fading manifestations and types of degradation. Part II will focus on methods to mitigate the degradation.

Rayleigh Fading Channels in Mobile Digital Communication Systems Part I: Characterization

Bernard Sklar, Communications Engineering Services

In the study of communication systems the classical (ideal) additive white Gaussian noise (AWGN) channel, with statistically independent Gaussian noise samples corrupting data samples free of intersymbol interference (ISI), is the usual starting point for understanding basic performance relationships. The primary source of performance degradation is thermal noise generated in the receiver. Often, external interference received by the antenna is more significant than thermal noise. This external interference can sometimes be characterized as having a broadband spectrum and is quantified by a parameter called antenna temperature [1]. The thermal noise usually has a flat power spectral density over the signal band and a zero-mean Gaussian voltage probability density function (pdf). When modeling practical systems, the next step is the introduction of bandlimiting filters. The filter in the transmitter usually serves to satisfy some regulatory requirement on spectral containment. The filter in the receiver often serves the purpose of a classical "matched filter" [2] to the signal bandwidth. Due to the bandlimiting and phase-distortion properties of filters, special signal design and equalization techniques may be required to mitigate the filter-induced ISI.

If a radio channel's propagating characteristics are not specified, one usually infers that the signal attenuation versus distance behaves as if propagation takes place over ideal free space. The model of free space treats the region between the transmit and receive antennas as being free of all objects that might absorb or reflect radio frequency (RF) energy. It also assumes that, within this region, the atmosphere behaves as a perfectly uniform and nonabsorbing medium. Furthermore, the earth is treated as being infinitely far away from the propagating signal (or, equivalently, as having a reflection coefficient that is negligible). Basically, in this idealized free-space model, the attenuation of RF energy between the transmitter and receiver behaves according to an inverse-square law. The received power expressed in terms of transmitted power is attenuated by a factor, $L_s(d)$, where this factor is called *path loss* or *free space loss*. When the receiving antenna is isotropic, this factor is expressed as [1]:

$$L_s(d) = \left(\frac{4\pi d}{\lambda}\right)^2 \qquad (1)$$

In Eq. 1, d is the distance between the transmitter and the receiver, and λ is the wavelength of the propagating signal. For this case of idealized propagation, received signal power is very predictable.

For most practical channels, where signal propagation takes place in the atmosphere and near the ground, the free-space propagation model is inadequate to describe the channel and predict system performance. In a wireless mobile communication system, a signal can travel from transmitter to receiver over multiple reflective paths; this phenomenon is referred to as *multipath* propagation. The effect can cause fluctuations in the received signal's amplitude, phase, and angle of arrival, giving rise to the terminology *multipath fading*. Another name, *scintillation*, which originated in radio astronomy, is used to describe the multipath fading caused by physical changes in the propagating medium, such as variations in the density of ions in the ionospheric layers that reflect high-frequency (HF) radio signals. Both names, fading and scintillation, refer to a signal's random fluctuations or fading due to multipath propagation. The main difference is that scintillation involves mechanisms (e.g., ions) that are much smaller than a wavelength. The end-to-end modeling and design of systems that mitigate the effects of fading are usually more challenging than those whose sole source of performance degradation is AWGN.

MOBILE RADIO PROPAGATION: LARGE-SCALE FADING AND SMALL-SCALE FADING

Fig. 1 represents an overview of fading channel manifestations. It starts with two types of fading effects that characterize mobile communications: large-scale and small-scale fading. Large-scale fading represents the average signal power attenuation or path loss due to motion over large areas. In

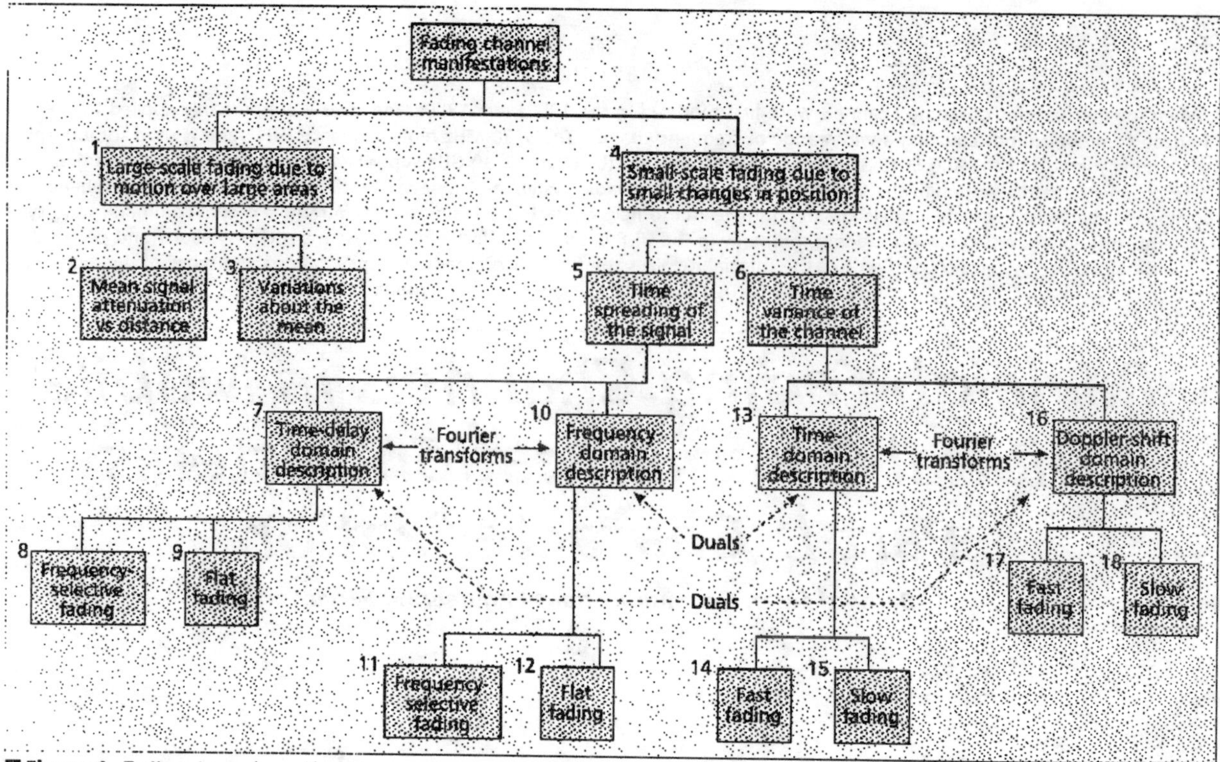

■ **Figure 1.** *Fading channel manifestations.*

Fig. 1, the large-scale fading manifestation is shown in blocks 1, 2, and 3. This phenomenon is affected by prominent terrain contours (hills, forests, billboards, clumps of buildings, etc.) between the transmitter and receiver. The receiver is often represented as being "shadowed" by such prominences. The statistics of large-scale fading provide a way of computing an estimate of path loss as a function of distance. This is described in terms of a mean-path loss (nth-power law) and a log-normally distributed variation about the mean. Small-scale fading refers to the dramatic changes in signal amplitude and phase that can be experienced as a result of small changes (as small as a half-wavelength) in the spatial separation between a receiver and transmitter. As indicated in Fig. 1, blocks 4, 5, and 6, small-scale fading manifests itself in two mechanisms, namely, time-spreading of the signal (or signal dispersion) and time-variant behavior of the channel. For mobile radio applications, the channel is time-variant because motion between the transmitter and receiver results in propagation path changes. The rate of change of these propagation conditions accounts for the fading rapidity (rate of change of the fading impairments). Small-scale fading is also called *Rayleigh fading* because if the multiple reflective paths are large in number and there is no line-of-sight signal component, the envelope of the received signal is statistically described by a Rayleigh pdf. When there is a dominant nonfading signal component present, such as a line-of-sight propagation path, the small-scale fading envelope is described by a Rician pdf [3]. A mobile radio roaming over a large area must process signals that experience both types of fading: small-scale fading superimposed on large-scale fading.

There are three basic mechanisms that impact signal propagation in a mobile communication system. They are reflection, diffraction, and scattering [3]:

- Reflection occurs when a propagating electromagnetic wave impinges on a smooth surface with very large dimensions compared to the RF signal wavelength (λ).
- Diffraction occurs when the radio path between the transmitter and receiver is obstructed by a dense body with large dimensions compared to λ, causing secondary waves to be formed behind the obstructing body. Diffraction is a phenomenon that accounts for RF energy traveling from transmitter to receiver without a line-of-sight path between the two. It is often termed *shadowing* because the diffracted field can reach the receiver even when shadowed by an impenetrable obstruction.
- Scattering occurs when a radio wave impinges on either a large rough surface or any surface whose dimensions are on the order of λ or less, causing the reflected energy to spread out (scatter) in all directions. In an urban environment, typical signal obstructions that yield scattering are lampposts, street signs, and foliage.

Figure 1 may serve as a table of contents for the sections that follow. We will examine the two manifestations of small-scale fading: signal time-spreading (signal dispersion) and the time-variant nature of the channel. These examinations will take place in two domains, time and frequency, as indicated in Fig. 1, blocks 7, 10, 13, and 16. For signal dispersion, we categorize the fading degradation types as frequency-selective or frequency-nonselective (flat), as listed in blocks 8, 9, 11, and 12. For the time-variant manifestation, we categorize the fading degradation types as fast- or slow-fading, as listed in blocks 14, 15, 17, and 18. The labels indicating Fourier transforms and duals will be explained later.

Figure 2 illustrates the various contributions that must be considered when estimating path loss for a link budget analysis in a cellular application [4]. These contributions are:

- Mean path loss as a function of distance due to large-scale fading
- Near-worst-case variations about the mean path loss (typically 6–10 dB), or large-scale fading margin
- Near-worst-case Rayleigh or small-scale fading margin (typically 20–30 dB)

In Fig. 2, the annotations "⊕ 1–2%" indicate a suggested area (probability) under the tail of each pdf as a design goal.

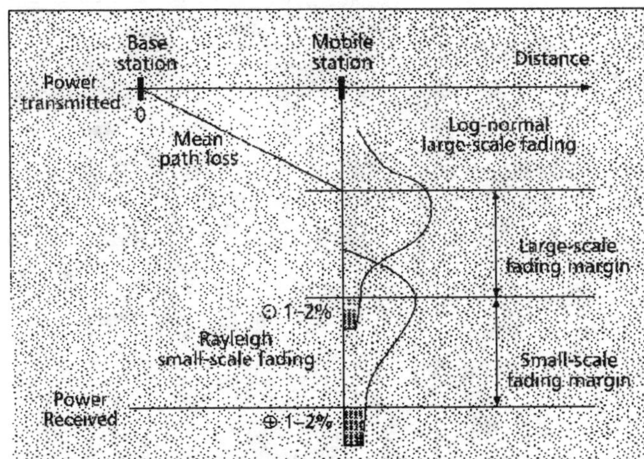

■ **Figure 2.** *Link-budget considerations for a fading channel.*

Hence, the amount of margin indicated is intended to provide adequate received signal power for approximately 98–99 percent of each type of fading variation (large- and small-scale).

A received signal, $r(t)$, is generally described in terms of a transmitted signal $s(t)$ convolved with the impulse response of the channel $h_c(t)$. Neglecting the degradation due to noise, we write

$$r(t) = s(t) * h_c(t), \tag{2}$$

where * denotes convolution. In the case of mobile radios, $r(t)$ can be partitioned in terms of two component random variables, as follows [5]:

$$r(t) = m(t) \times r_0(t), \tag{3}$$

where $m(t)$ is called the large-scale-fading component, and $r_0(t)$ is called the small-scale-fading component. $m(t)$ is sometimes referred to as the *local mean* or *log-normal fading* because the magnitude of $m(t)$ is described by a log-normal pdf (or, equivalently, the magnitude measured in decibels has a Gaussian pdf). $r_0(t)$ is sometimes referred to as *multipath* or *Rayleigh fading*. Figure 3 illustrates the relationship between large-scale and small-scale fading. In Fig. 3a, received signal power $r(t)$ versus antenna displacement (typically in units of wavelength) is plotted, for the case of a mobile radio. Small-scale fading superimposed on large-scale fading can be readily identified. The typical antenna displacement between the small-scale signal nulls is approximately a half wavelength. In Fig. 3b, the large scale fading or local mean, $m(t)$, has been removed in order to view the small-scale fading, $r_0(t)$, about some average constant power.

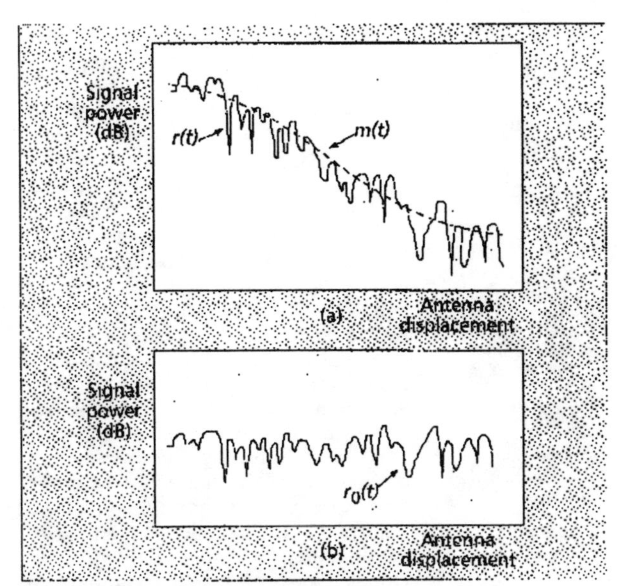

■ **Figure 3.** *Large-scale fading and small-scale fading.*

In the sections that follow, we enumerate some of the details regarding the statistics and mechanisms of large-scale and small-scale fading.

Large-Scale Fading: Path-Loss Mean and Standard Deviation

For the mobile radio application, Okumura [6] made some of the earlier comprehensive path-loss measurements for a wide range of antenna heights and coverage distances. Hata [7] transformed Okumura's data into parametric formulas. For the mobile radio application, the mean path loss, $\overline{L_p}(d)$, as a function of distance, d, between the transmitter and receiver is proportional to an nth power of d relative to a reference distance d_0 [3].

$$\overline{L_p}(d) \propto \left(\frac{d}{d_0}\right)^n \tag{4}$$

$\overline{L_p}(d)$ is often stated in decibels, as shown below.

$$\overline{L_p}(d)\ (\text{dB}) = L_s(d_0)\ (\text{dB}) + 10\, n \log\,(d/d_0) \tag{5}$$

The reference distance d_0 corresponds to a point located in the far field of the antenna. Typically, the value of d_0 is taken to be 1 km for large cells, 100 m for microcells, and 1 m for indoor channels. $\overline{L_p}(d)$ is the average path loss (over a multitude of different sites) for a given value of d. Linear regression for a minimum mean-squared estimate (MMSE) fit of $\overline{L_p}(d)$ versus d on a log-log scale (for distances greater than d_0) yields a straight line with a slope equal to $10n$ dB/decade. The value of the exponent n depends on the frequency, antenna heights, and propagation environment. In free space, $n = 2$, as seen in Eq. 1. In the presence of a very strong guided wave phenomenon (like urban streets), n can be lower than 2. When obstructions are present, n is larger. The path loss $L_s(d_0)$ to the reference point at a distance d_0 from the transmitter is typically found through field measurements or calculated using the free-space path loss given by Eq. 1. Figure 4 shows a scatter plot of path loss versus distance for measurements made in several German cities [8]. Here, the path loss has been measured relative to the free-space reference measurement at $d_0 = 100$ m. Also shown are straight-line fits to various exponent values.

The path loss versus distance expressed in Eq. 5 is an average, and therefore not adequate to describe any particular setting or signal path. It is necessary to provide for variations about the mean since the environment of different sites may be quite different for similar transmitter-receiver separations. Figure 4 illustrates that path-loss variations can be quite large. Measurements have shown that for any value of d, the path loss $L_p(d)$ is a random variable having a log-normal distribution about the mean distant-dependent value $\overline{L_p}(d)$ [9]. Thus, path loss $L_p(d)$ can be expressed in terms of $\overline{L_p}(d)$ plus a random variable X_σ, as follows [3]:

$$L_p(d)\ (\text{dB}) = L_s(d_0)\ (\text{dB}) + 10n\log_{10}(d/d_0) + X_\sigma\ (\text{dB}) \tag{6}$$

where X_σ denotes a zero-mean Gaussian random variable (in decibels) with standard deviation σ (also in decibels). X_σ is site- and distance-dependent. The choice of a value for X_σ is often based on measurements; it is not unusual for it to take on values as high as 6–10 dB or greater. Thus, the parameters needed to statistically describe path loss due to large-scale fading for an arbitrary location with a specific transmitter-receiver separation are:

- The reference distance d_0
- The path-loss exponent n
- The standard deviation σ of X_σ

There are several good references dealing with the measurement and estimation of propagation path loss for many different applications and configurations [3, 7–11].

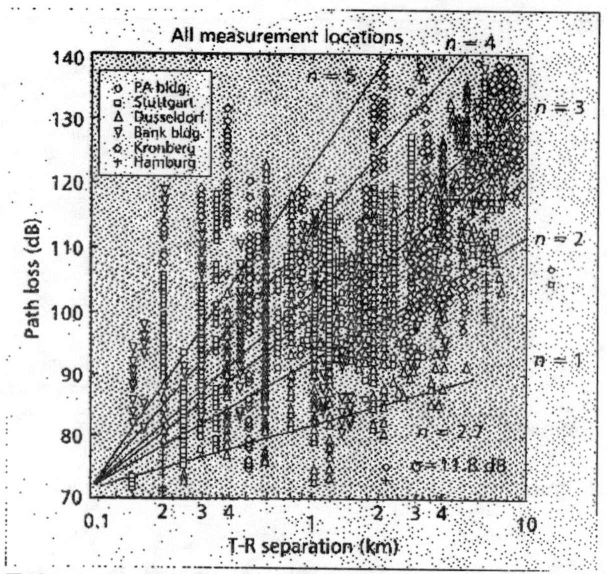

■ **Figure 4.** *Path loss versus distance measured in serveral German cities.*

SMALL-SCALE FADING: STATISTICS AND MECHANISMS

When the received signal is made up of multiple reflective rays plus a significant line-of-sight (nonfaded) component, the envelope amplitude due to small-scale fading has a Rician pdf, and is referred to as *Rician fading* [3]. The nonfaded component is called the *specular component*. As the amplitude of the specular component approaches zero, the Rician pdf approaches a Rayleigh pdf, expressed as

$$p(r) = \begin{cases} \frac{r}{\sigma^2}\exp\left[-\frac{r^2}{2\sigma^2}\right] & \text{for } r \geq 0 \\ 0 & \text{otherwise} \end{cases} \tag{7}$$

where r is the envelope amplitude of the received signal, and $2\sigma^2$ is the pre-detection mean power of the multipath signal. The Rayleigh faded component is sometimes called the *random* or *scatter* or *diffuse* component. The Rayleigh pdf results from having no specular component of the signal; thus, for a single link it represents the pdf associated with the worst case of fading per mean received signal power. For the remainder of this article, it will be assumed that loss of signal-to-noise ratio (SNR) due to fading follows the Rayleigh model described. It will also be assumed that the propagating signal is in the ultra-high-frequency (UHF) band, encompassing present-day cellular and personal communications services (PCS) frequency allocations — nominally 1 GHz and 2 GHz, respectively.

As indicated in Fig. 1, blocks 4, 5, and 6, small-scale fading manifests itself in two mechanisms:

- Time-spreading of the underlying digital pulses within the signal
- A time-variant behavior of the channel due to motion (e.g., a receive antenna on a moving platform)

Figure 5 illustrates the consequences of both manifestations by showing the response of a multipath channel to a narrow pulse versus delay, as a function of antenna position (or time, assuming a constant velocity of motion). In Fig. 5, we distinguish between two different time references — delay time τ and transmission or observation time t. Delay time refers to the time-spreading manifestation which results from the fading channel's nonoptimum impulse response. The transmission time, however, is related to the antenna's motion or spatial changes, accounting for propagation path changes that are perceived as the channel's time-variant behavior. Note that for constant velocity, as assumed in Fig. 5, either antenna position or transmission time can be used to illustrate this time-variant behavior. Figures 5a–5c show the sequence of received pulse-power profiles as the antenna moves through a succession of equally spaced positions. Here, the interval between antenna positions is 0.4 λ [12], where λ is the wavelength of the carrier frequency. For each of the three cases shown, the response-pattern differs significantly in the delay time of the largest signal component, the number of signal copies, their magnitudes, and the total received power (area) in the received power profile. Figure 6 summarizes these two small-scale fading mechanisms, the two domains (time or time-delay and frequency or Doppler shift) for viewing each mechanism, and the degradation categories each mechanism can exhibit. Note that any mechanism characterized in the time domain can be characterized equally well in the frequency domain. Hence, as outlined in Fig. 6, the time-spreading mechanism will be characterized in the time-delay domain as a multipath delay spread, and in the frequency domain as a channel coherence bandwidth. Similarly, the time-variant mechanism will be characterized in the time domain as a channel coherence time, and in the Doppler-shift (frequency) domain as a channel fading rate or Doppler spread. These mechanisms and

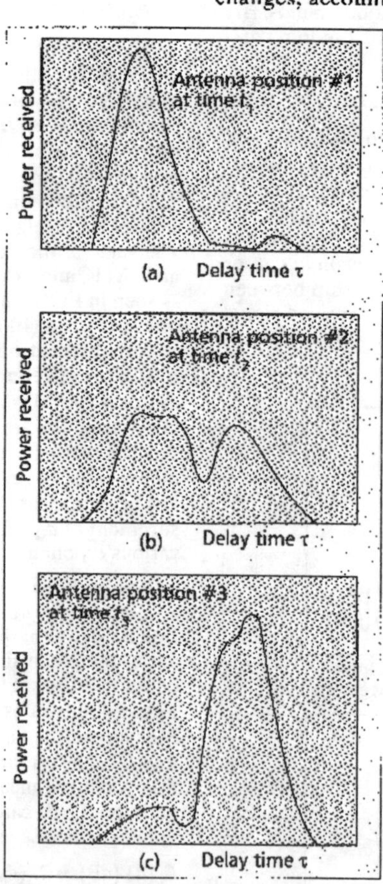

Figure 5. *Response of a multipath channel to a narrow pulse versus delay, as a function of antenna position.*

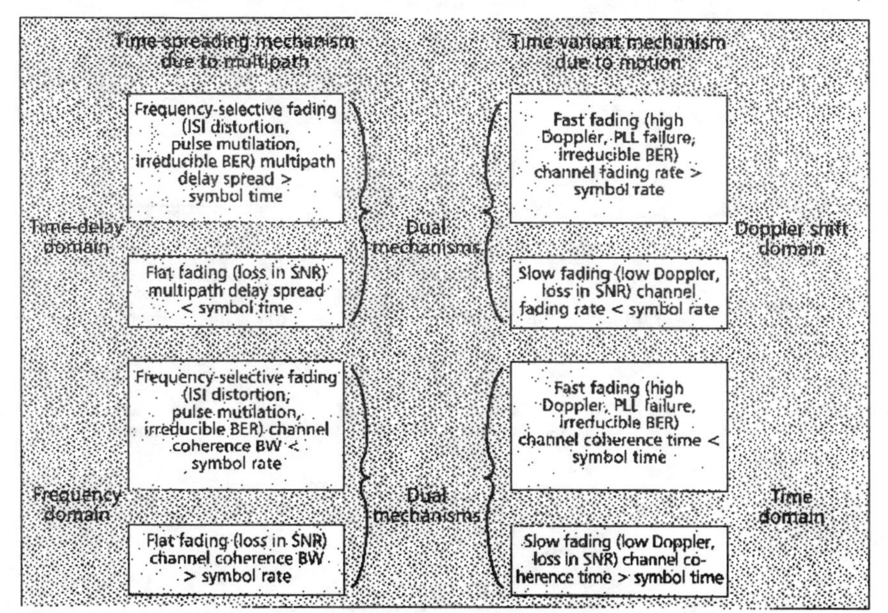

■ **Figure 6.** *Small-scale fading: mechanisms, degradation categories, and effects.*

their associated degradation categories will be examined in greater detail in the sections that follow.

Signal Time-Spreading Viewed in the Time-Delay Domain: Fig. 1, Block 7 — The Multipath Intensity Profile

A simple way to model the fading phenomenon was introduced by Bello [13] in 1963; he proposed the notion of wide-sense stationary uncorrelated scattering (WSSUS). The model treats signal variations arriving with different delays as uncorrelated. It can be shown [4, 13] that such a channel is effectively WSS in both the time and frequency domains. With such a model of a fading channel, Bello was able to define functions that apply for all time and all frequencies. For the mobile channel, Figure 7 contains four functions that make up this model [4, 13–16]. We will examine these functions, starting with Fig. 7a and proceeding counterclockwise toward Fig. 7d.

In Fig. 7a a *multipath-intensity profile*, $S(\tau)$ versus time delay τ, is plotted. Knowledge of $S(\tau)$ helps answer the question "For a transmitted impulse, how does the average received power vary as a function of time delay, τ?" The term "time delay" is used to refer to the excess delay. It represents the signal's propagation delay that exceeds the delay of the first signal arrival at the receiver. For a typical wireless radio channel, the received signal usually consists of several discrete multipath components, sometimes referred to as *fingers*. For some channels, such as the tropospheric scatter channel, received signals are often seen as a continuum of multipath components [14, 16]. For making measurements of the multipath intensity profile, wideband signals (impulses or spread spectrum) need to be used [16]. For a single transmitted impulse, the time, T_m, between the first and last received component represents the *maximum excess delay*, during which the multipath signal power falls to some threshold level below that of the strongest component. The threshold level might be chosen at 10 or 20 dB below the level of the strongest component. Note that for an ideal system (zero excess delay), the function $S(\tau)$ would consist of an ideal impulse with weight equal to the total average received signal power.

Degradation Categories due to Signal Time-Spreading Viewed in the Time-Delay Domain

In a fading channel, the relationship between maximum excess delay time, T_m, and symbol time, T_s, can be viewed in terms of two different degradation categories, *frequency-selective fading* and *frequency nonselective* or *flat fading*, as indicated in Fig. 1, blocks 8 and 9, and Fig. 6. A channel is said to exhibit frequency-selective fading if $T_m > T_s$. This condition occurs whenever the received multipath components of a symbol extend beyond the symbol's time duration. Such multipath dispersion of the signal yields the same kind of ISI distortion caused by an electronic filter. In fact, another name for this category of fading degradation is *channel-induced ISI*. In the case of frequency-selective fading, mitigating the distortion is possible because many of the multipath components are resolvable by the receiver. In Part II of this tutorial, several such mitigation techniques are described.

A channel is said to exhibit frequency nonselective or flat fading if $T_m < T_s$. In this case, all the received multipath components of a symbol arrive within the symbol time duration; hence, the components are not resolvable. Here, there is no channel-induced ISI distortion, since the signal time spreading does not result in significant overlap among neighboring received symbols. There is still performance degradation since the unresolvable phasor components can add up destructively to yield a substantial reduction in SNR. Also, signals that are classified as exhibiting flat fading can sometimes experience frequency-selective distortion. This will be explained later when viewing degradation in the frequency domain, where the phenomenon is more easily described. For loss in SNR due to flat fading, the mitigation technique called for is to improve the received SNR (or reduce the required SNR). For digital systems, introducing some form of signal diversity and using error-correction coding is the most efficient way to accomplish this.

Signal Time-Spreading Viewed in the Frequency Domain: Fig. 1, Block 10 — The Spaced-Frequency Correlation Function

A completely analogous characterization of signal dispersion can begin in the frequency domain. In Fig. 7b is seen the function $|R(\Delta f)|$, designated a spaced-frequency correlation function; it is the Fourier transform of $S(\tau)$. $R(\Delta f)$ represents the correlation between the channel's response to two signals as a function of the frequency difference between the two signals. It can be thought of as the channel's frequency transfer function. Therefore, the time-spreading manifestation can be viewed as if it were the result of a filtering process. Knowledge of $R(\Delta f)$ helps answer the question "What is the correlation between received signals that are spaced in frequency Δf

$= f_1 - f_2$?" $R(\Delta f)$ can be measured by transmitting a pair of sinusoids separated in frequency by Δf, cross-correlating the two separately received signals, and repeating the process many times with ever-larger separation Δf. Therefore, the measurement of $R(\Delta f)$ can be made with a sinusoid that is swept in frequency across the band of interest (a wideband signal). The *coherence bandwidth*, f_0, is a statistical measure of the range of frequencies over which the channel passes all spectral components with approximately equal gain and linear phase. Thus, the coherence bandwidth represents a frequency range over which frequency components have a strong potential for amplitude correlation. That is, a signal's spectral components in that range are affected by the channel in a similar manner as, for example, exhibiting fading or no fading. Note that f_0 and T_m are reciprocally related (within a multiplicative constant). As an approximation, it is possible to say that

$$f_0 \oplus 1/T_m \tag{8}$$

The maximum excess delay, T_m, is not necessarily the best indicator of how any given system will perform on a channel because different channels with the same value of T_m can exhibit very different profiles of signal intensity over the delay span. A more useful measurement of delay spread is most often characterized in terms of the root mean squared (rms) delay spread, σ_τ, where

$$\sigma_\tau = \sqrt{\overline{\tau^2} - (\bar{\tau})^2} \tag{9}$$

$\bar{\tau}$ is the mean excess delay, $(\bar{\tau})^2$ is the mean squared, $\overline{\tau^2}$ is the second moment, and σ_τ is the square root of the second central moment of $S(\tau)$ [3].

An exact relationship between coherence bandwidth and delay spread does not exist, and must be derived from signal analysis (usually using Fourier techniques) of actual signal dispersion measurements in particular channels. Several approximate relationships have been described. If coherence bandwidth is defined as the frequency interval over which the channel's complex frequency transfer function has a correlation of at least 0.9, the coherence bandwidth is approximately [17]

$$f_0 = \frac{1}{50\sigma_\tau} \tag{10}$$

For the case of a mobile radio, an array of radially uniformly spaced scatterers, all with equal-magnitude reflection coefficients but independent, randomly occurring reflection phase angles [18, 19] is generally accepted as a useful model for urban surroundings. This model is referred to as the *dense-scatterer* channel model. With the use of such a model, coherence bandwidth has similarly been defined [18], for a bandwidth interval over which the channel's complex frequency transfer function has a correlation of at least 0.5, to be

$$f_0 = \frac{0.276}{\sigma_\tau} \tag{11}$$

The ionospheric effects community employs the following definition [20]:

$$f_0 = \frac{1}{2\pi\sigma_\tau} \tag{12}$$

A more popular approximation of f_0 corresponding to a bandwidth interval having a correlation of at least 0.5 is [3]

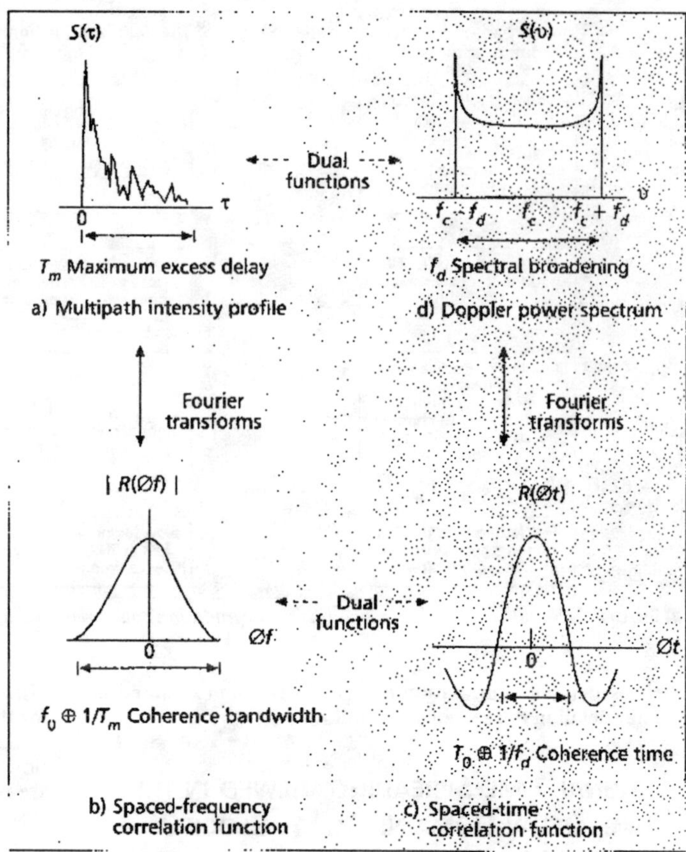

Figure 7. *Relationships among the channel correlation functions and power density functions.*

$$f_0 = \frac{1}{5\sigma_\tau} \tag{13}$$

Degradation Categories Due to Signal Time Spreading Viewed in the Frequency Domain

A channel is referred to as frequency-selective if $f_0 < 1/T_s \oplus W$, where the symbol rate, $1/T_s$ is nominally taken to be equal to the signal bandwidth W. In practice, W may differ from $1/T_s$ due to system filtering or data modulation type (quaternary phase shift keying, QPSK, minimum shift keying, MSK, etc.) [21]. Frequency-selective fading distortion occurs whenever a signal's spectral components are not all affected equally by the channel. Some of the signal's spectral components, falling outside the coherence bandwidth, will be affected differently (independently) compared to those components contained within the coherence bandwidth. This occurs whenever $f_0 < W$ and is illustrated in Fig. 8a.

Frequency-nonselective or flat fading degradation occurs whenever $f_0 > W$. Hence, all of the signal's spectral components will be affected by the channel in a similar manner (e.g., fading or no fading); this is illustrated in Fig. 8b. Flat-fading does not introduce channel-induced ISI distortion, but performance degradation can still be expected due to loss in SNR whenever the signal is fading. In order to avoid channel-induced ISI distortion, the channel is required to exhibit flat fading by ensuring that

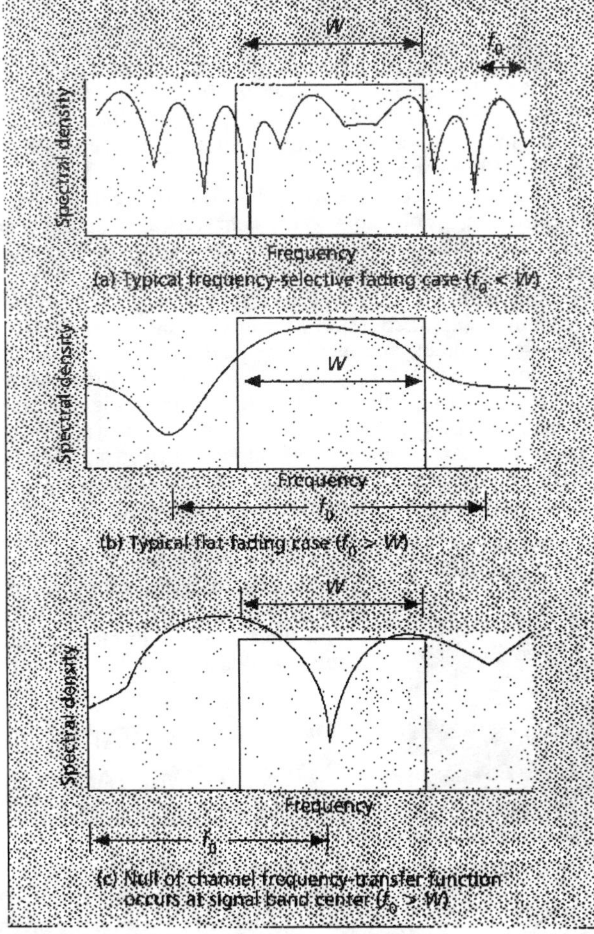

Figure 8. *Relationships between the channel frequency-transfer function and a signal with bandwidth* W.

$$f_0 > W \approx \frac{1}{T_s} \tag{14}$$

Hence, the channel coherence bandwidth f_0 sets an upper limit on the transmission rate that can be used without incorporating an equalizer in the receiver.

For the flat-fading case, where $f_0 > W$ (or $T_m < T_s$), Fig. 8b shows the usual flat-fading pictorial representation. However, as a mobile radio changes its position, there will be times when the received signal experiences frequency-selective distortion even though $f_0 > W$. This is seen in Fig. 8c, where the null of the channel's frequency transfer function occurs at the center of the signal band. Whenever this occurs, the baseband pulse will be especially mutilated by deprivation of its DC component. One consequence of the loss of DC (zero mean value) is the absence of a reliable pulse peak on which to establish the timing synchronization, or from which to sample the carrier phase carried by the pulse [18]. Thus, even though a channel is categorized as flat fading (based on rms relationships), it can still manifest frequency-selective fading on occasions. It is fair to say that a mobile radio channel, classified as having flat fading degradation, cannot exhibit flat fading all the time. As f_0 becomes much larger than W (or T_m much smaller than T_s), less time will be spent in conditions approximating Fig. 8c. By comparison, it should be clear that in Fig. 8a the fading is independent of the position of the signal band, and frequency-selective fading occurs all the time, not just occasionally.

Typical Examples of Flat Fading and Frequency-Selective Fading

Figure 9 shows some examples of flat fading and frequency-selective fading for a direct-sequence spread-spectrum (DS/SS) system [20, 22]. In Fig. 9, there are three plots of the output of a pseudonoise (PN) code correlator versus delay as a function of time (transmission or observation time). Each amplitude versus delay plot is akin to $S(\tau)$ versus τ in Fig. 7a. The key difference is that the amplitudes shown in Fig. 9 represent the output of a correlator; hence the waveshapes are a function not only of the impulse response of the channel, but also of the impulse response of the correlator. The delay time is expressed in units of chip durations (chips), where the chip is defined as the spread-spectrum minimal-duration keying element. For each plot, the observation time is shown on an axis perpendicular to the amplitude versus time-delay plane. Figure 9 is drawn from a satellite-to-ground communications link exhibiting scintillation because of atmospheric disturbances. However, Fig. 9 is still a useful illustration of three different channel conditions that might apply to a mobile radio situation. A mobile radio that moves along the observation-time axis is affected by changing multipath profiles along the route, as seen in the figure. The scale along the observation-time axis is also in units of chips. In Fig. 9a, the signal dispersion (one "finger" of return) is on the order of a chip time duration, T_{ch}. In a typical DS/SS system, the spread-spectrum signal bandwidth is approximately equal to $1/T_{ch}$; hence, the normalized coherence bandwidth f_0T_{ch} of approximately unity in Fig. 9a implies that the coherence bandwidth is about equal to the spread-spectrum bandwidth. This describes a channel that can be called frequency-nonselective or slightly frequency-selective. In Fig. 9b, where f_0T_{ch} = 0.25, the signal dispersion is more pronounced. There is definite interchip interference, and the coherence bandwidth is approximately equal to 25 percent of the spread spectrum bandwidth. In Fig. 9c, where f_0T_{ch} = 0.1, the signal dispersion is even more pronounced, with greater interchip interference effects, and the coherence bandwidth is approximately equal to 10 percent of the spread-spectrum bandwidth. The channels of Figs. 9b and 9c can be categorized as moderately and highly frequency-selective, respectively, with respect to the basic signaling element, the chip. In Part II of this article, we show that a DS/SS system operating over a frequency-selective channel at the chip level does not necessarily experience frequency-selective distortion at the symbol level.

Time Variance Viewed in the Time Domain: Fig. 1, Block 13 — The Spaced-Time Correlation Function

Until now, we have described signal dispersion and coherence bandwidth, parameters that describe the channel's time-spreading properties in a local area. However, they do not offer information about the time-varying nature of the channel caused by relative motion between a transmitter and receiver, or by movement of objects within the channel. For mobile radio applications, the channel is time-variant because motion between the transmitter and receiver results in propagation path changes. Thus, for a transmitted continuous wave (CW) signal, as a result of such motion, the radio receiver sees variations in the signal's amplitude and phase. Assuming that all scatterers making up the channel are stationary, whenever motion ceases, the amplitude and phase of the received signal remains constant; that is, the channel appears to be time-invariant. Whenever motion begins again, the channel once again appears time-variant. Since the channel characteristics are dependent on the positions of the transmitter and receiver, time variance in this case is equivalent to spatial variance.

Figure 7c shows the function $R(\Delta t)$, designated the *spaced-time* correlation function; it is the autocorrelation function of the channel's response to a sinusoid. This function specifies the extent to which there is correlation between the channel's response to a sinusoid sent at time t_1 and the response to a similar sinusoid sent at time t_2, where $\Delta t = t_2 - t_1$. The *coherence time*, T_0, is a measure of the expected time duration over which the channel's response is essentially invariant. Earlier, we made measurements of signal dispersion and coherence bandwidth by using wideband signals. Now, to measure the time-variant nature of the channel, we use a narrowband signal [16]. To measure $R(\Delta t)$ we can transmit a single sinusoid ($\Delta f = 0$) and determine the autocorrelation function of the received signal. The function $R(\Delta t)$ and the parameter T_0 provide us with knowledge about the fading rapidity of the channel. Note that for an ideal *time-invariant* channel (e.g., a mobile radio exhibiting no motion at all), the channel's response would be highly correlated for all values of Δt, and $R(\Delta t)$ would be a constant function. When using the dense-scatterer channel model described earlier, with constant velocity of motion, and an unmodulated CW signal, the normalized $R(\Delta f)$ is described as [19],

$$R(\Delta t) = J_0(kV\Delta t) \qquad (15)$$

where $J_0(\cdot)$ is the zero-order Bessel function of the first kind, V is velocity, $V\Delta t$ is distance traversed, and $k = 2\pi/\lambda$ is the free-space phase constant (transforming distance to radians of phase). Coherence time can be measured in terms of either time or distance traversed (assuming some fixed velocity of motion). Amoroso described such a measurement using a CW signal and a dense-scatterer channel model [18]. He measured the statistical correlation between the combination of received magnitude and phase sampled at a particular antenna location x_0, and the corresponding combination sampled at some displaced location $x_0 + \zeta$, with displacement measured in units of wavelength λ. For a displacement ζ of 0.38 λ between two antenna locations, the combined magnitudes and phases of the received CW are statistically uncorrelated. In other words, the state of the signal at x_0 says nothing about the state of the signal at $x_0 + \zeta$. For a given velocity of motion, this displacement is readily transformed into units of time (coherence time).

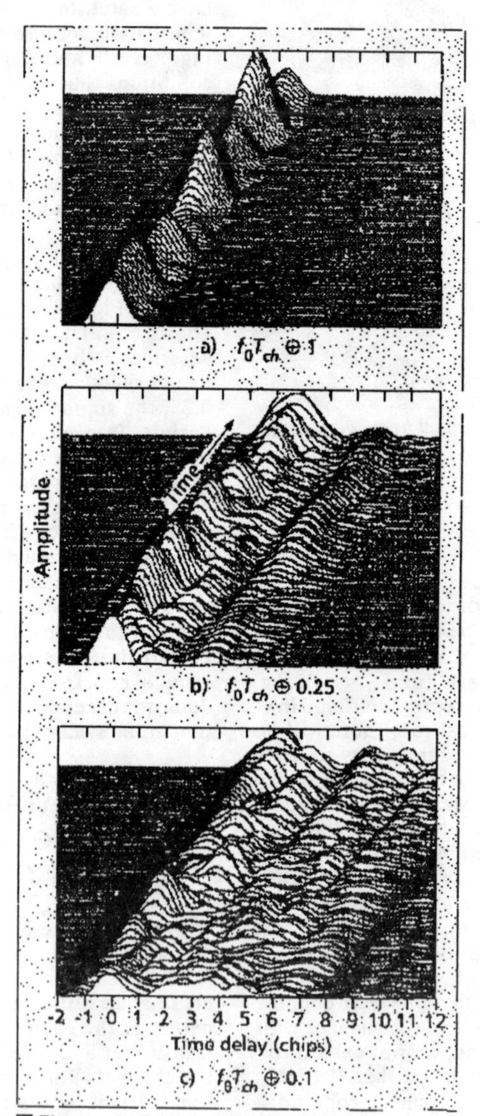

Figure 9. *DS/SS matched-filter output time-history examples for three levels of channel conditions, where* T_{ch} *is the time duration of a chip.*

The Concept of Duality

Two operators (functions, elements, or systems) are dual when the behavior of one with reference to a time-related domain (time or time-delay) is identical to the behavior of the other in reference to the corresponding frequency-related domain (frequency or Doppler shift).

In Fig. 7, we can identify functions that exhibit similar behavior across domains. These behaviors are not identical to one another in the strict mathematical sense, but for understanding the fading channel model it is still useful to refer to such functions as *duals*. For example, $R(\Delta f)$ in Fig. 7b, characterizing signal dispersion in the frequency domain, yields knowledge about the range of frequency over which two spectral components of a received signal have a strong potential for amplitude and phase correlation. $R(\Delta t)$ in Fig. 7c, characterizing fading rapidity in the time domain, yields knowledge about the span of time over which two received signals have a strong potential for amplitude and phase correlation. We have labeled these two correlation functions duals. This is also noted in Fig. 1 as the duality between blocks 10 and 13, and in Fig. 6 as the duality between the time-spreading mechanism in the frequency domain and the time-variant mechanism in the time domain.

Degradation Categories Due to Time Variance Viewed in the Time Domain

The time-variant nature of the channel or fading rapidity mechanism can be viewed in terms of two degradation categories listed in Fig. 6: *fast fading* and *slow fading*. The terminology "fast fading" is used to describe channels in which $T_0 < T_s$, where T_0 is the channel coherence time and T_s is the time duration of a transmission symbol. Fast fading describes a condition where the time duration in which the channel behaves in a correlated manner is short compared to the time duration of a symbol. Therefore, it can be expected that the fading character of the channel will change several times while a symbol is propagating, leading to distortion of the baseband pulse shape. Analogous to the distortion previously described as channel-induced ISI, here distortion takes place because the received signal's components are not all highly correlated throughout time. Hence, fast fading can cause the baseband pulse to be distorted, resulting in a loss of SNR that often yields an irreducible error rate. Such distorted pulses cause synchronization problems (failure of phase-locked-loop receivers), in addition to difficulties in adequately defining a matched filter.

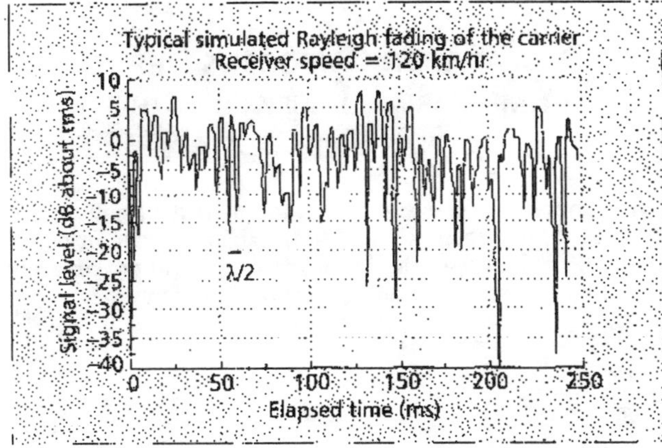

■ **Figure 10.** *A typical Rayleigh fading envelope at 900 MHz.*

A channel is generally referred to as introducing slow fading if $T_0 > T_s$. Here, the time duration that the channel behaves in a correlated manner is long compared to the time duration of a transmission symbol. Thus, one can expect the channel state to virtually remain unchanged during the time in which a symbol is transmitted. The propagating symbols will likely not suffer from the pulse distortion described above. The primary degradation in a slow-fading channel, as with flat fading, is loss in SNR.

TIME VARIANCE VIEWED IN THE DOPPLER-SHIFT DOMAIN: FIG. 1, BLOCK 16 — THE DOPPLER POWER SPECTRUM

A completely analogous characterization of the time-variant nature of the channel can begin in the Doppler shift (frequency) domain. Figure 7d shows a *Doppler power spectral density*, $S(v)$, plotted as a function of Doppler-frequency shift, v. For the case of the dense-scatterer model, a vertical receive antenna with constant azimuthal gain, a uniform distribution of signals arriving at all arrival angles throughout the range (0, 2π), and an unmodulated CW signal, the signal spectrum at the antenna terminals is [19]

$$S(v) = \frac{1}{\pi f_d \sqrt{1 - \left(\frac{v}{f_d}\right)^2}} \tag{16}$$

The equality holds for frequency shifts of v that are in the range $\pm f_d$ about the carrier frequency f_c, and would be zero outside that range. The shape of the RF Doppler spectrum described by Eq. 16 is classically bowl-shaped, as seen in Fig. 7d. Note that the spectral shape is a result of the dense-scatterer channel model. Equation 16 has been shown to match experimental data gathered for mobile radio channels [23]; however, different applications yield different spectral shapes. For example, the dense-scatterer model does not hold for the indoor radio channel; the channel model for an indoor area assumes $S(v)$ to be a flat spectrum [24].

In Fig. 7d, the sharpness and steepness of the boundaries of the Doppler spectrum are due to the sharp upper limit on the Doppler shift produced by a vehicular antenna traveling among the stationary scatterers of the dense scatterer model. The largest magnitude (infinite) of $S(v)$ occurs when the scatterer is directly ahead of the moving antenna platform or directly behind it. In that case the magnitude of the frequency shift is given by

$$f_d = \frac{V}{\lambda} \tag{17}$$

where V is relative velocity, and λ is the signal wavelength. f_d is positive when the transmitter and receiver move toward each other, and negative when moving away from each other. For scatterers directly broadside of the moving platform the magnitude of the frequency shift is zero. The fact that Doppler components arriving at exactly 0° and 180° have an infinite power spectral density is not a problem, since the angle of arrival is continuously distributed and the probability of components arriving at exactly these angles is zero [3, 19].

$S(v)$ is the Fourier transform of $R(\Delta t)$. We know that the Fourier transform of the autocorrelation function of a time series is the magnitude squared of the Fourier transform of the original time series. Therefore, measurements can be made by simply transmitting a sinusoid (narrowband signal) and using Fourier analysis to generate the power spectrum of the received amplitude [16]. This Doppler power spectrum of the channel yields knowledge about the spectral spreading of a transmitted sinusoid (impulse in frequency) in the Doppler shift domain. As indicated in Fig. 7, $S(v)$ can be regarded as the dual of the multipath intensity profile, $S(\tau)$, since the latter yields knowledge about the time spreading of a transmitted impulse in the time-delay domain. This is also noted in Fig. 1 as the duality between blocks 7 and 16, and in Fig. 6 as the duality between the time-spreading mechanism in the time-delay domain and the time-variant mechanism in the Doppler-shift domain.

Knowledge of $S(v)$ allows us to glean how much spectral broadening is imposed on the signal as a function of the rate of change in the channel state. The width of the Doppler power spectrum is referred to as the *spectral broadening* or *Doppler spread*, denoted by f_d, and sometimes called the *fading bandwidth* of the channel. Equation 16 describes the Doppler frequency shift. In a typical multipath environment, the received signal arrives from several reflected paths with different path distances and different angles of arrival, and the Doppler shift of each arriving path is generally different from that of another path. The effect on the received signal is seen as a Doppler spreading or spectral broadening of the transmitted signal frequency, rather than a shift. Note that the Doppler spread, f_d, and the coherence time, T_0, are reciprocally related (within a multiplicative constant). Therefore, we show the approximate relationship between the two parameters as

$$T_0 \approx \frac{1}{f_d} \tag{18}$$

Hence, the Doppler spread f_d or $1/T_0$ is regarded as the typical *fading rate* of the channel. Earlier, T_0 was described as the expected time duration over which the channel's response to a sinusoid is essentially invariant. When T_0 is defined more precisely as the time duration over which the channel's response to a sinusoid has a correlation greater than 0.5, the relationship between T_0 and f_d is approximately [4]

$$T_0 \approx \frac{9}{16\pi f_d} \tag{19}$$

A popular rule of thumb is to define T_0 as the geometric mean of Eqs. 18 and 19. This yields

$$T_0 = \sqrt{\frac{9}{16\pi f_d^2}} = \frac{0.423}{f_d} \tag{20}$$

For the case of a 900 MHz mobile radio, Fig. 10 illustrates the typical effect of Rayleigh fading on a signal's envelope amplitude versus time [3]. The figure shows that the distance traveled by the mobile in the time interval corresponding to two adjacent nulls (small-scale fades) is on the order of a half-wavelength ($\lambda/2$) [3]. Thus, from Fig. 10 and Eq. 17, the time (approximately the coherence time) required to traverse a distance when traveling at a constant velocity, V, is

$$T_0 \approx \frac{\lambda/2}{V} = \frac{0.5}{f_d} \tag{21}$$

Thus, when the interval between fades is taken to be $\lambda/2$, as in Fig. 10, the resulting expression for T_0 in Eq. 21 is quite close to the rule of thumb shown in Eq. 20. Using Eq. 21 with the parameters shown in Fig. 10 (velocity = 120 km/hr and carrier frequency = 900 MHz), it is straightforward to compute that the coherence time is approximately 5 ms and the Doppler spread (channel fading rate) is approximately 100 Hz. Therefore, if this example represents a voice-grade channel with a typical transmission rate of 10^4 symbols/s, the fading rate is considerably less than the symbol rate. Under such conditions, the channel would manifest slow-fading effects. Note that if the abscissa of Fig. 10 were labeled in units of wavelength instead of time, the figure would look the same for any radio frequency and any antenna speed.

Analogy between Spectral Broadening in Fading Channels and Spectral Broadening in Digital Signal Keying

Help is often needed in understanding why spectral broadening of the signal is a function of fading rate of the channel. Figure 11 uses the keying of a digital signal (e.g., amplitude shift keying or frequency shift keying) to illustrate an analogous case. Figure 11a shows that a single tone, cos $2\pi f_c t$ ($-° < t < °$), which exists for all time is characterized in the frequency domain in terms of impulses (at $\pm f_c$). This frequency domain representation is ideal (i.e., zero bandwidth), since the tone is pure and neverending. In practical applications, digital signaling involves switching (keying) signals on and off at a required rate. The keying operation can be viewed as multiplying the infinite-duration tone in Fig. 11a by an ideal rectangular (switching) function in Fig. 11b. The frequency domain description of the ideal rectangular function is of the form $(\sin f)/f$. In Fig. 11c, the result of the multiplication yields a tone, cos $2\pi f_c t$, that is time-duration-limited in the interval $-T/2$ " t " $T/2$. The resulting spectrum is obtained by convolving the spectral impulses in part a with the $(\sin f)/f$ function in part b, yielding the broadened spectrum in part c. It is further seen that, if the signaling occurs at a faster rate characterized by the rectangle of shorter duration in part d, the resulting spectrum of the signal in part e exhibits greater spectral broadening. The changing state of a fading channel is somewhat analogous to the keying on and off of digital signals. The channel behaves like a switch, turning the signal "on and off." The greater the rapidity of the change in channel state, the greater the spectral broadening of the received signals. The analogy is not exact because the on and off switching of signals may result in phase discontinuities, but the typical multipath-scatterer environment induces phase-continuous effects.

Degradation Categories Due to Time Variance Viewed in the Doppler Shift Domain

A channel is referred to as fast fading if the symbol rate, $1/T_s$ (approximately equal to the signaling rate or bandwidth W) is less than the fading rate, $1/T_0$ (approximately equal to f_d); that is, fast fading is characterized by

$$W < f_d \tag{22a}$$

or

$$T_s > T_0. \tag{22b}$$

Conversely, a channel is referred to as slow fading if the signaling rate is greater than the fading rate. Thus, in order to avoid signal distortion caused by fast fading, the channel must be made to exhibit slow fading by ensuring that the signaling rate must exceed the channel fading rate. That is,

$$W > f_d \tag{23a}$$

or

$$T_s < T_0. \tag{23b}$$

In Eq. 14, it was shown that due to signal dispersion, the coherence bandwidth, f_0, sets an *upper limit* on the signaling rate which can be used without suffering frequency-selective distortion. Similarly, Eq. 23 shows that due to Doppler spreading, the channel fading rate, f_d, sets a *lower limit* on the signaling rate that can be used without suffering fast fading distortion. For HF communication systems, when teletype or Morse code messages were transmitted at a low data rate, the channels were often fast fading. However, most present-day terrestrial mobile radio channels can generally be characterized as slow fading.

Equation 23 does not go far enough in describing what we desire of the channel. A better way to state the requirement for mitigating the effects of fast fading would be that we desire $W \gg f_d$ (or $T_s \ll T_0$). If this condition is not satisfied, the random frequency modulation (FM) due to varying Doppler shifts will limit the system performance significantly. The Doppler effect yields an irreducible error rate that cannot be overcome by simply increasing E_b/N_0 [25]. This irreducible error rate is most pronounced for any modulation that involves switching the carrier phase. A single specular Doppler path, without scatterers, registers an instantaneous frequency

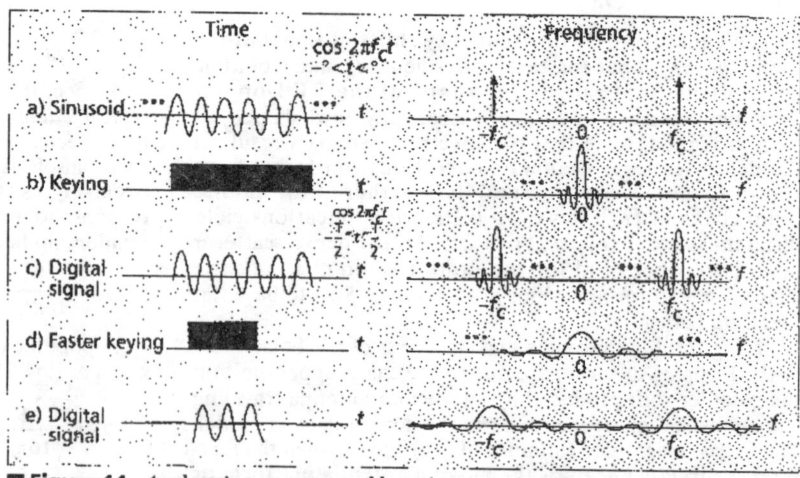

Figure 11. *Analogy between spectral broadening in fading and spectral broadening in keying a digital signal.*

shift, classically calculated as $f_d = V/\lambda$. However, a combination of specular and multipath components yields a rather complex time dependence of instantaneous frequency which can cause much larger frequency swings than $\pm V/\lambda$ when detected by an instantaneous frequency detector (a nonlinear device) [26]. Ideally, coherent demodulators that lock onto and track the information signal should suppress the effect of this FM noise and thus cancel the impact of Doppler shift. However, for large values of f_d, carrier recovery becomes a problem because very wideband (relative to the data rate) phase-locked loops need to be designed. For voice-grade applications with bit error rates of 10^{-3} to 10^{-4}, a large value of Doppler shift is considered to be on the order of 0.01 x W. Therefore, to avoid fast fading distortion and the Doppler-induced irreducible error rate, the signaling rate should exceed the fading rate by a factor of 100–200 [27]. The exact factor depends on the signal modulation, receiver design, and required error rate [3, 26–29]. Davarian [29] showed that a frequency-tracking loop can help to lower, but not completely remove, the irreducible error rate in a mobile system by using differential minimum-shift keyed (DMSK) modulation.

SUMMARY

In Part I of this article, the major elements that contribute to fading in a communication channel have been characterized. Figure 1 was presented as a guide for the characterization of fading phenomena. Two types of fading, large-scale and small-scale, were described. Two manifestations of small-scale fading (signal dispersion and fading rapidity) were examined, and the examination involved two views, time and frequency. Two degradation categories were defined for dispersion: frequency-selective fading and flat fading. Two degradation categories were defined for fading rapidity: fast and slow. The small-scale fading degradation categories were summarized in Fig. 6. A mathematical model using correlation and power density functions was presented in Fig. 7. This model yields a nice symmetry, a kind of "poetry" to help us view the Fourier transform and duality relationships that describe the fading phenomena. In Part II, mitigation techniques for ameliorating the effects of each degradation category will be treated, and methods that have been implemented in two mobile communication systems will be described.

REFERENCES

[1] B. Sklar, *Digital Communications: Fundamentals and Applications*, Ch. 4, Englewood Cliffs, NJ: Prentice Hall, 1988.

[2] H. L. Van Trees, *Detection, Estimation, and Modulation Theory, Part I*, Ch. 4, New York: Wiley, 1968.

[3] T. S. Rappaport, *Wireless Communications*, Chs. 3 and 4, Upper Saddle River, NJ: Prentice Hall, 1996.

[4] D. Greenwood and L. Hanzo, "Characterisation of Mobile Radio Channels," *Mobile Radio Communications*, by R. Steele, Ed., Ch. 2, London: Pentech Press, 1994.

[5] W. C. Y. Lee, "Elements of Cellular Mobile Radio Systems," *IEEE Trans. Vehic. Tech.*, vol. VT-35, no. 2, May 1986, pp. 48–56.

[6] Y. Okumura, E. Ohmori, and K. Fukuda, "Field Strength and Its Variability in VHF and UHF Land Mobile Radio Service," *Rev. Elec. Commun. Lab*, vol. 16, nos. 9 and 10, 1968, pp. 825–73.

[7] M. Hata, "Empirical Formulae for Propagation Loss in Land Mobile Radio Services," *IEEE Trans. Vehic. Tech.*, vol. VT-29, no. 3, 1980, pp. 317–25.

[8] S. Y. Seidel *et al.*, "Path Loss, Scattering and Multipath Delay Statistics in Four European Cities for Digital Cellular and Microcellular Radiotelephone," *IEEE Trans. Vehic. Tech.*, vol. 40, no. 4, Nov. 1991, pp.721–30.

[9] D. C. Cox, R. Murray, and A. Norris, "800 MHz Attenuation Measured in and around Suburban Houses," *AT&T Bell Lab. Tech. J.*, vol. 673, no. 6, July-Aug. 1984, pp. 921 54.

[10] D. L. Schilling *et. al.*, "Broadband CDMA for Personal Communications Systems," *IEEE Commun. Mag.*, vol. 29, no. 11, Nov. 1991, pp. 86–93.

[11] J. B. Andersen, T. S. Rappaport, S. Yoshida, "Propagation Measurements and Models for Wireless Communications Channels," *IEEE Commun. Mag.*, vol. 33, no. 1, Jan. 1995, pp. 42–49.

[12] F. Amoroso, "Investigation of Signal Variance, Bit Error Rates and Pulse Dispersion for DSPN Signalling in a Mobile Dense Scatterer Ray Tracing Model," *Int'l. J. Satellite Commun.*, vol. 12, 1994, pp. 579–88.

[13] P. A. Bello, "Characterization of Randomly Time-Variant Linear Channels," *IEEE Trans. Commun. Sys.*, vol. CS-11, no. 4, Dec. 1963, pp. 360–93.

[14] J. G. Proakis, *Digital Communications*, Chapter 7, New York: McGraw-Hill, 1983.

[15] P. E. Green, Jr., "Radar Astronomy Measurement Techniques," MIT Lincoln Lab., Lexington, MA, Tech. Rep. No. 282, Dec. 1962.

[16] K. Pahlavan and A. H. Levesque, *Wireless Information Networks*, Chs. 3 and 4, New York: Wiley, 1995.

[17] W. Y. C. Lee, *Mobile Cellular Communications*, New York: McGraw-Hill, 1989.

[18] F. Amoroso, "Use of DS/SS Signaling to Mitigate Rayleigh Fading in a Dense Scatterer Environment," *IEEE Pers. Commun.*, vol. 3, no. 2, Apr. 1996, pp. 52–61.

[19] R. H. Clarke, "A Statistical Theory of Mobile Radio Reception," *Bell Sys. Tech. J.*, vol. 47, no. 6, July-Aug. 1968, pp. 957 1000.

[20] R. L. Bogusch, "Digital Communications in Fading Channels: Modulation and Coding," Mission Research Corp., Santa Barbara, CA, Report no. MRC-R-1043, Mar. 11, 1987.

[21] F. Amoroso, "The Bandwidth of Digital Data Signals," *IEEE Commun. Mag.*, vol. 18, no. 6, Nov. 1980, pp. 13–24.

[22] R. L. Bogusch, *et. al.*, "Frequency Selective Propagation Effects on Spread-Spectrum Receiver Tracking," *Proc. IEEE*, vol. 69, no. 7, July 1981, pp. 787–96.

[23] W. C. Jakes, Ed., *Microwave Mobile Communications*, New York: Wiley, 1974.

[24] Joint TC of Committee T1 R1P1.4 and TIA TR46.3.3/TR45.4.4 on Wireless Access, "Draft Final Report on RF Channel Characterization," Paper no. JTC(AIR)/94.01.17-238R4, Jan. 17, 1994.

[25] P. A. Bello and B. D. Nelin, "The Influence of Fading Spectrum on the Binary Error Probabilities of Incoherent and Differentially Coherent Matched Filter Receivers," *IRE Trans. Commun. Sys.*, vol. CS-10, June 1962, pp. 160–68.

[26] F. Amoroso, "Instantaneous Frequency Effects in a Doppler Scattering Environment," *Proc. IEEE ICC '87*, Seattle, WA, June 7–10, 1987, pp. 1458 66.

[27] A. J. Bateman and J. P. McGeehan, "Data Transmission over UHF Fading Mobile Radio Channels," *IEE Proc.*, vol. 131, pt. F, no. 4, July 1984, pp. 364–74.

[28] K. Feher, *Wireless Digital Communications*, Upper Saddle River, NJ: Prentice Hall, 1995.

[29] F. Davarian, M. Simon, and J. Sumida, "DMSK: A Practical 2400-bps Receiver for the Mobile Satellite Service," Jet Propulsion Lab. Pub. 85-51 (MSAT-X Rep. No. 111), June 15, 1985.

BIOGRAPHY

BERNARD SKLAR [LSM] has over 40 years of experience in technical design and management positions at Republic Aviation Corp., Hughes Aircraft, Litton Industries, and The Aerospace Corporation. At Aerospace, he helped develop the MILSTAR satellite system, and was the principal architect for EHF Satellite Data Link Standards. Currently, he is head of advanced systems at Communications Engineering Services, a consulting company he founded in 1984. He has taught engineering courses at several universities, including the University of California, Los Angeles and the University of Southern California, and has presented numerous training programs throughout the world. He has published and presented scores of technical papers. He is the recipient of the 1984 Prize Paper Award from the IEEE Communications Society for his tutorial series on digital communications, and he is the author of the book *Digital Communications* (Prentice Hall). He is past Chair of the Los Angeles Council IEEE Education Committee. His academic credentials include a B.S. degree in math and science from the University of Michigan, an M.S. degree in electrical engineering from the Polytechnic Institute of Brooklyn, New York, and a Ph.D. degree in engineering from the University of California, Los Angeles.

ABSTRACT

In Part I of this tutorial, the major elements that contribute to fading and their effects in a communication channel were characterized. Here, in Part II, these phenomena are briefly summarized, and emphasis is then placed on methods to cope with these degradation effects. Two particular mitigation techniques are examined: the Viterbi equalizer implemented in the Global System for Mobile Communication (GSM), and the Rake receiver used in CDMA systems built to meet Interim Standard 95 (IS-95).

Rayleigh Fading Channels in Mobile Digital Communication Systems Part II: Mitigation

Bernard Sklar, Communications Engineering Services

We repeat Fig. 1, from Part I of the article, where it served as a table of contents for fading channel manifestations. In Part I, two types of fading, large-scale and small-scale, were described. Figure 1 emphasizes the small-scale fading phenomena and its two manifestations, time spreading of the signal or signal dispersion, and time variance of the channel or fading rapidity due to motion between the transmitter and receiver. These are listed in blocks 4, 5, and 6. Examining these manifestations involved two views, time and frequency, as indicated in blocks 7, 10, 13, and 16. Two degradation categories were defined for dispersion, frequency-selective fading and flat- fading, as listed in blocks 8, 9, 11, and 12. Two degradation categories were defined for fading rapidity, fast-fading and slow-fading, as listed in blocks 14, 15, 17, and 18.

In Part I, a model of the fading channel consisting of four functions was described. These functions are shown in Fig. 2 (which appeared in Part I as Fig. 7). A *multipath-intensity profile*, $S(\tau)$, is plotted in Fig. 2a versus time delay τ. Knowledge of $S(\tau)$ helps answer the question "For a transmitted impulse, how does the average received power vary as a function of time delay, τ?" For a single transmitted impulse, the time, T_m, between the first and last received components represents the *maximum excess delay* during which the multipath signal power falls to some threshold level below that of the strongest component. Figure 2b shows the function $|R(\Delta f)|$, designated a *spaced-frequency* correlation function; it is the Fourier transform of $S(\tau)$. $R(\Delta f)$ represents the correlation between the channel's response to two signals as a function of the frequency difference between the two signals. Knowledge of $R(\Delta f)$ helps answer the question "What is the correlation between received signals that are spaced in frequency $\Delta f = f_1 - f_2$?" The *coherence bandwidth*, f_0, is a statistical measure of the range of frequencies over which the channel passes all spectral components with approximately equal gain and linear phase. Thus, the coherence bandwidth represents a frequency range over which frequency components have a strong potential for amplitude correlation. Note that f_0 and T_m are reciprocally related (within a multiplicative constant). As an approximation, it is possible to say that

$$f_0 \oplus 1/Tm \qquad (1)$$

A more useful measurement of delay spread is most often characterized in terms of the root mean squared (rms) delay spread, σ_τ [1]. A popular approximation of f_0 corresponding to a bandwidth interval having a correlation of at least 0.5 is

$$f_0 \oplus 1/5\sigma_\tau \qquad (2)$$

Figure 2c shows the function $R(\Delta t)$, designated the *spaced-time* correlation function; it is the autocorrelation function of the channel's response to a sinusoid. This function specifies to what extent there is correlation between the channel's response to a sinusoid sent at time t_1 and the response to a similar sinusoid sent at time t_2, where $\Delta t = t_2 - t_1$. The *coherence time*, T_0, is a measure of the expected time duration over which the channel's response is essentially invariant.

Figure 2d shows a *Doppler power spectral density*, $S(\nu)$, plotted as a function of Doppler-frequency shift, ν; it is the Fourier transform of $R(\Delta t)$. The sharpness and steepness of the boundaries of the Doppler spectrum are due to the sharp upper limit on the Doppler shift produced by a vehicular antenna traveling among a dense population of stationary scatterers. The largest magnitude of $S(\nu)$ occurs when the scatterer is directly ahead of or directly behind the moving antenna platform. The width of the Doppler power spectrum is referred to as the *spectral broadening* or *Doppler spread*, denoted f_d and sometimes called the *fading bandwidth* of the channel. Note that the Doppler spread, f_d, and the coherence time, T_0, are reciprocally related (within a multiplicative constant). In Part I of this tutorial, it was shown that the time (approximately the coherence time) required to traverse a distance $\lambda/2$ when traveling at a constant velocity, V, is

$$T_0 = \frac{\lambda/2}{V} = \frac{0.5}{f_d} \qquad (3)$$

DEGRADATION CATEGORIES IN BRIEF

The degradation categories described in Part I are reviewed here in the context of Fig. 3, which summarizes small-scale fading mechanisms, degradation categories, and their effects. (This figure appeared in Part I as Fig. 6.) When viewed in the time-delay domain, a channel is said to exhibit *frequency-selective* fading if $T_m > T_s$ (the delay time is greater than the symbol time). This condition occurs whenever the received multipath components

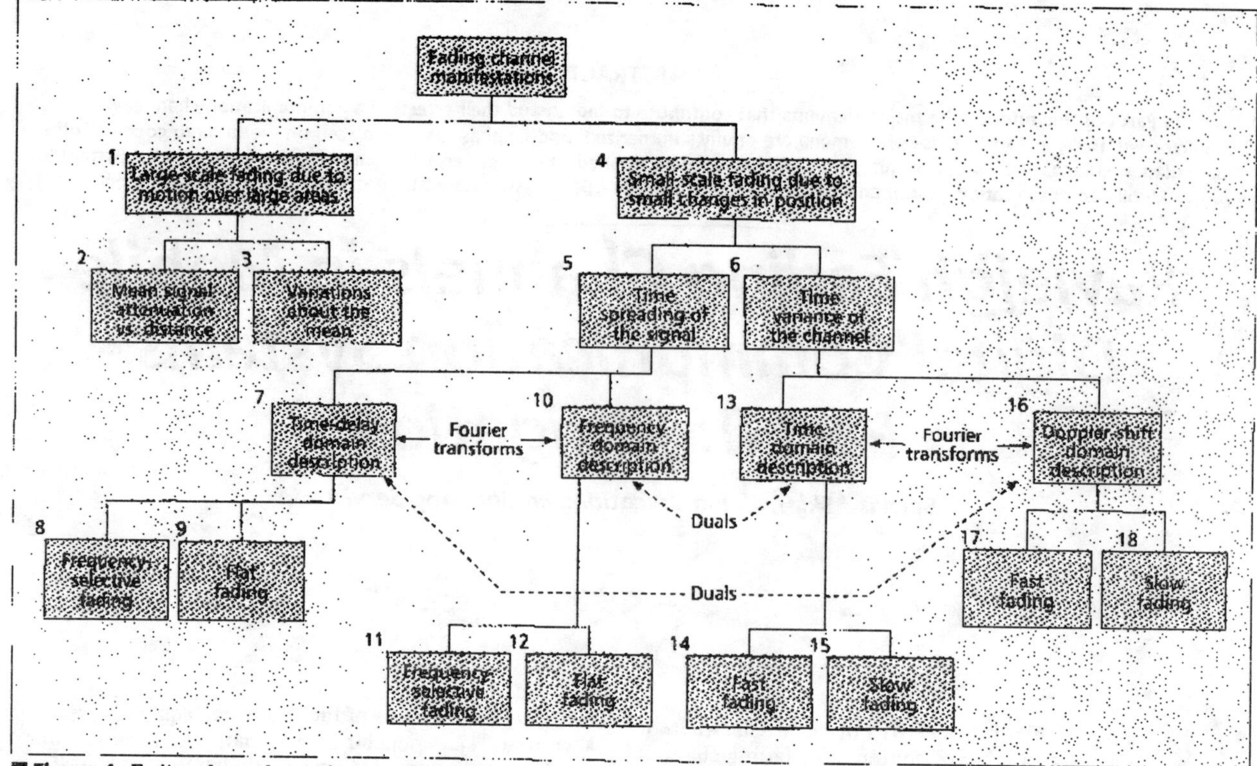

Figure 1. *Fading channel manifestations.*

of a symbol extend beyond the symbol's time duration, thus causing channel-induced intersymbol interference (ISI).

Viewed in the time-delay domain, a channel is said to exhibit *frequency-nonselective* or *flat fading* if $T_m < T_s$. In this case, all of the received multipath components of a symbol arrive within the symbol time duration; hence, the components are not resolvable. Here, there is no channel-induced ISI distortion, since the signal time spreading does not result in significant overlap among neighboring received symbols. There is still performance degradation since the unresolvable phasor components can add up destructively to yield a substantial reduction in signal-to-noise ratio (SNR).

When viewed in the frequency domain, a channel is referred to as frequency-selective if $f_0 < 1/T_s \oplus W$, where the symbol rate, $1/T_s$ is nominally taken to be equal to the signal bandwidth W. Flat fading degradation occurs whenever $f_0 > W$. Here, all of the signal's spectral components will be affected by the channel in a similar manner (e.g., fading or no fading). In order to avoid ISI distortion caused by frequency-selective fading, the channel must be made to exhibit flat fading by ensuring that the coherence bandwidth exceeds the signaling rate.

When viewed in the time domain, a channel is referred to as fast fading whenever $T_0 < T_s$, where T_0 is the channel coherence time and T_s is the symbol time. Fast fading describes a condition where the time duration for which the channel behaves in a correlated manner is short compared to the time duration of a symbol. Therefore, it can be expected that the fading character of the channel will change several times during the time a symbol is propagating. This leads to distortion of the baseband pulse shape, because the received signal's components are not all highly correlated throughout time. Hence, fast fading can cause the

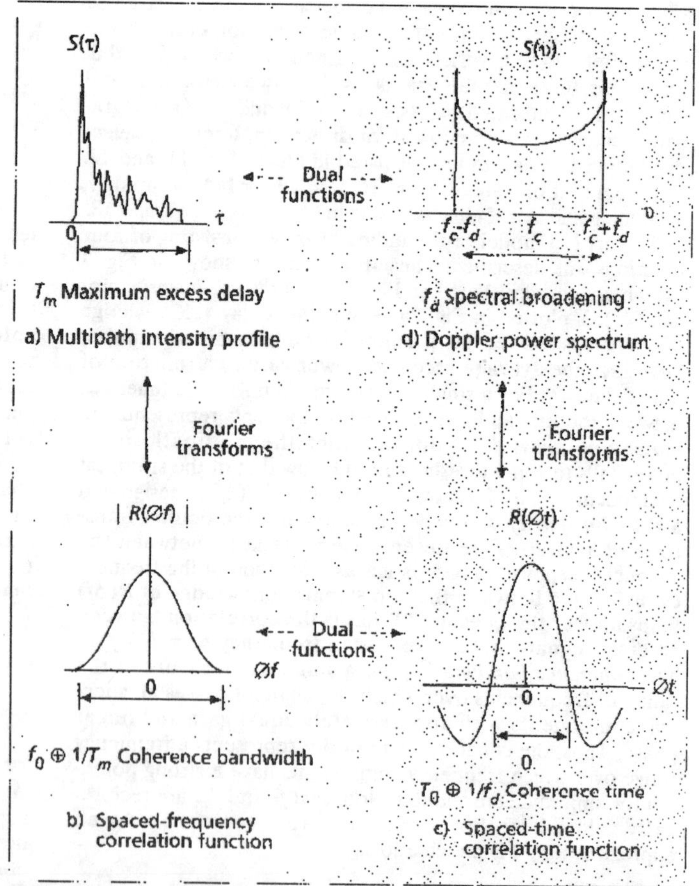

Figure 2. *Relationships among the channel correlation functions and power density functions.*

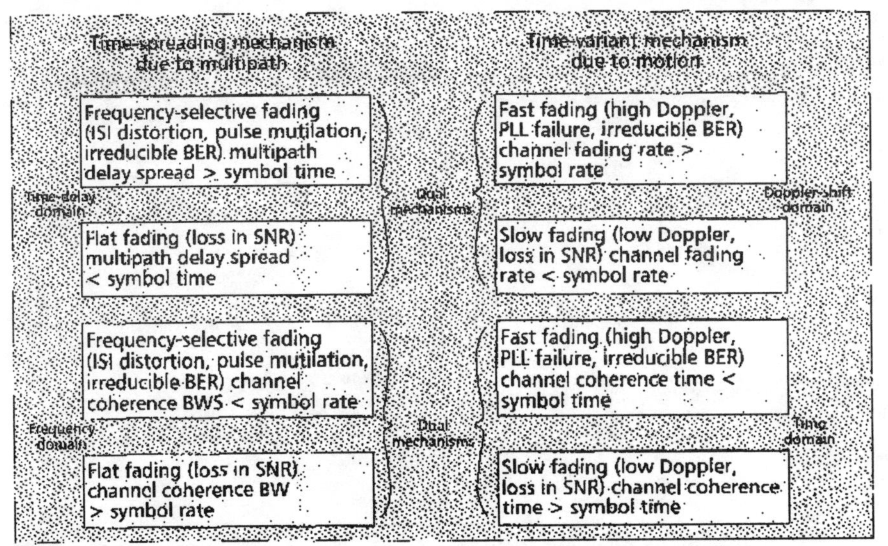

■ **Figure 3.** *Small-scale fading: mechanisms, degradation categories, and effects.*

baseband pulse to be distorted, resulting in a loss of SNR that often yields an irreducible error rate. Such distorted pulses typically cause synchronization problems, such as failure of phase-locked-loop (PLL) receivers.

Viewed in the time domain, a channel is generally referred to as introducing *slow fading* if $T_0 > T_s$. Here, the time duration for which the channel behaves in a correlated manner is long compared to the symbol time. Thus, one can expect the channel state to remain virtually unchanged during the time a symbol is transmitted.

When viewed in the Doppler shift domain, a channel is referred to as fast fading if the symbol rate, $1/T_s$, or the signal bandwidth, W, is less than the fading rate, $1/T_0$ or f_d. Conversely, a channel is referred to as slow fading if the signaling rate is greater than the fading rate. In order to avoid signal distortion caused by fast fading, the channel must be made to exhibit slow fading by ensuring that the signaling rate exceeds the channel fading rate.

Mitigation Methods

Figure 4, subtitled "the good, the bad, and the awful," highlights three major performance categories in terms of bit error probability, P_B, versus E_b/N_0. The leftmost exponentially shaped curve represents the performance that can be expected when using any nominal modulation type in additive white Gaussian noise (AWGN). Observe that with a reasonable amount of E_b/N_0, good performance results. The middle curve, referred to as the *Rayleigh limit*, shows the performance degradation resulting from a loss in SNR that is characteristic of flat fading or slow fading when there is no line-of-sight signal component present. The curve is a function of the reciprocal of E_b/N_0 (an inverse-linear function), so for reasonable values of SNR, performance will generally be "bad." In the case of Rayleigh fading, parameters with overbars are often introduced to indicate that a mean is being taken over the "ups" and "downs" of the fading experience. Therefore, one often sees such bit error probability plots with mean parameters denoted by $\overline{P_B}$ and $\overline{E_b}/N_0$. The curve that reaches an irreducible level, sometimes called an *error floor*, represents "awful" performance, where the bit error probability can approach the value of 0.5. This shows the severe distorting effects of frequency-selective fading or fast fading.

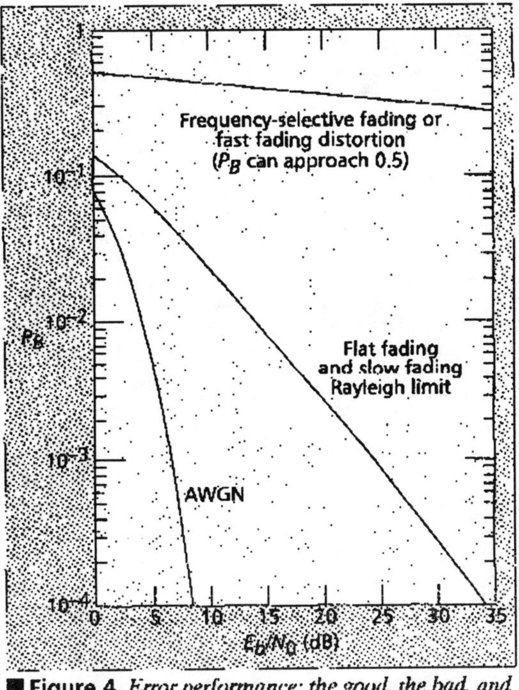

■ **Figure 4.** *Error performance: the good, the bad, and the awful.*

If the channel introduces signal distortion as a result of fading, the system performance can exhibit an irreducible error rate; when larger than the desired error rate, no amount of E_b/N_0 will help achieve the desired level of performance. In such cases, the general approach for improving performance is to use some form of mitigation to remove or reduce the distortion. The mitigation method depends on whether the distortion is caused by frequency-selective or fast fading. Once the distortion has been mitigated, the P_B versus E_b/N_0 performance should have transitioned from the "awful" bottoming out curve to the merely "bad" Rayleigh limit curve. Next, we can further ameliorate the effects of fading and strive to approach AWGN performance by using some form of diversity to provide the receiver with a collection of uncorrelated samples of the signal, and by using a powerful error correction code.

In Fig. 5, several mitigation techniques for combating the effects of both signal distortion and loss in SNR are listed. Just as Fig. 1 serves as a guide for characterizing fading phenomena and their effects, Fig. 5 can similarly serve to describe mitigation methods that can be used to ameliorate the effects of fading. The mitigation approach to be used should follow two basic steps: first, provide distortion mitigation; next, provide diversity.

Mitigation to Combat Frequency-Selective Distortion

- Equalization can compensate for the channel-induced ISI that is seen in frequency-selective fading. That is, it can help move the operating point from the error-performance curve that is "awful" in Fig. 4 to the one that is "bad." The process of equalizing the ISI involves some method of gathering the dispersed symbol energy back together into its original time interval. In effect, equalization involves insertion of a

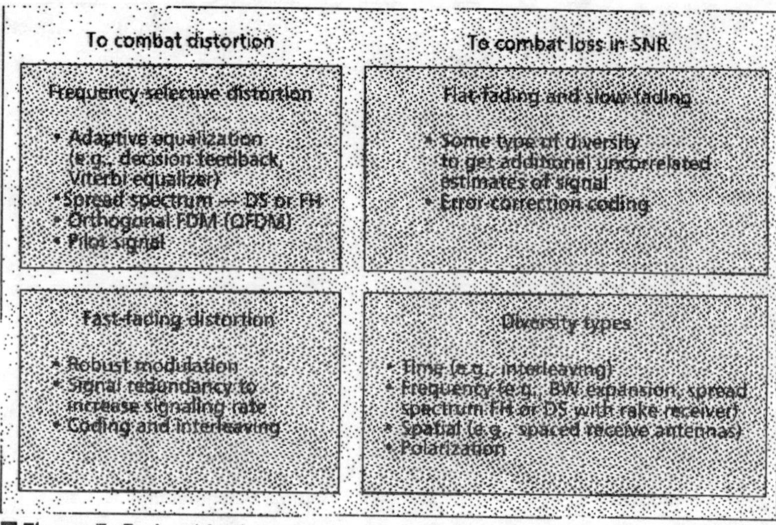

Figure 5. *Basic mitigation types.*

filter to make the combination of channel and filter yield a flat response with linear phase. The phase linearity is achieved by making the equalizer filter the complex conjugate of the time reverse of the dispersed pulse [1]. Because in a mobile system the channel response varies with time, the equalizer filter must also change or adapt to the time-varying channel. Such equalizer filters are therefore called *adaptive equalizers*. An equalizer accomplishes more than distortion mitigation; it also provides diversity. Since distortion mitigation is achieved by gathering the dispersed symbol's energy back into the symbol's original time interval so that it doesn't hamper the detection of other symbols, the equalizer is simultaneously providing each received symbol with energy that would otherwise be lost.

- The decision feedback equalizer (DFE) has a feedforward section that is a linear transversal filter [1] whose length and tap weights are selected to coherently combine virtually all of the current symbol's energy. The DFE also has a feedback section which removes energy that remains from previously detected symbols [1–4]. The basic idea behind the DFE is that once an information symbol has been detected, the ISI it induces on future symbols can be estimated and subtracted before the detection of subsequent symbols.
- The maximum likelihood sequence estimation (MLSE) equalizer tests all possible data sequences (rather than decoding each received symbol by itself) and chooses the data sequence that is the most probable of the candidates. The MLSE equalizer was first proposed by Forney [5] when he implemented the equalizer using the Viterbi decoding algorithm [6]. The MLSE is optimal in the sense that it minimizes the probability of a sequence error. Because the Viterbi decoding algorithm is the way in which the MLSE equalizer is typically implemented, the equalizer is often referred to as the *Viterbi equalizer*. Later in this article we illustrate the adaptive equalization performed in the Global System for Mobile Communications (GSM) using the Viterbi equalizer.
- Spread-spectrum techniques can be used to mitigate frequency-selective ISI distortion because the hallmark of any spread-spectrum system is its capability to reject interference, and ISI is a type of interference. Consider a direct-sequence spread-spectrum (DS/SS) binary phase shift keying (PSK) communication channel comprising one direct path and one reflected path. Assume that the propagation from transmitter to receiver results in a multipath wave that is delayed by τ_k compared to the direct wave. If the receiver is synchronized to the waveform arriving via the direct path, the received signal, $r(t)$, neglecting noise, can be expressed as

$$r(t) = Ax(t)g(t)\cos(2\pi f_c t) + \alpha Ax(t-\tau_k)g(t-\tau_k)\cos(2\pi f_c t + \theta), \tag{4}$$

where $x(t)$ is the data signal, $g(t)$ is the pseudonoise (PN) spreading code, and τ_k is the differential time delay between the two paths. The angle θ is a random phase, assumed to be uniformly distributed in the range $(0, 2\pi)$, and α is the attenuation of the multipath signal relative to the direct path signal. The receiver multiplies the incoming $r(t)$ by the code $g(t)$. If the receiver is synchronized to the direct path signal, multiplication by the code signal yields

$$Ax(t)g^2(t)\cos(2\pi f_c t) + \alpha Ax(t-\tau_k)g(t)g(t-\tau_k)\cos(2\pi f_c t + \theta) \tag{5}$$

where $g^2(t) = 1$, and if τ_k is greater than the chip duration,

$$\left|\int g^*(t)g(t-\tau_k)dt\right| \ll \int g^*(t)g(t)dt \tag{6}$$

over some appropriate interval of integration (correlation), where * indicates complex conjugate, and τ_k is equal to or larger than the PN chip duration. Thus, the spread-spectrum system effectively eliminates the multipath interference by virtue of its code-correlation receiver. Even though channel-induced ISI is typically transparent to DS/SS systems, such systems suffer from the loss in energy contained in all the multipath components not seen by the receiver. The need to gather up this lost energy belonging to the received chip was the motivation for developing the Rake receiver [7–9]. The Rake receiver dedicates a separate correlator to each multipath component (finger). It is able to coherently add the energy from each finger by selectively delaying them (the earliest component gets the longest delay) so that they can all be coherently combined.
- In Part I of this article, we described a channel that could be classified as flat fading, but occasionally exhibits frequency-selective distortion when the null of the channel's frequency transfer function occurs at the center of the signal band. The use of DS/SS is a good way to mitigate such distortion because the wideband SS signal would span many lobes of the selectively faded frequency response. Hence, a great deal of pulse energy would then be passed by the scatterer medium, in contrast to the nulling effect on a relatively narrowband signal (see Part I, Fig. 8c) [10].
- Frequency-hopping spread spectrum (FH/SS) can be used to mitigate the distortion due to frequency-selective fading, provided the hopping rate is at least equal to the symbol rate. Compared to DS/SS, mitigation takes place through a different mechanism. FH receivers avoid multipath losses by rapid changes in the transmitter frequency band, thus avoiding the interference by changing the receiver band position before the arrival of the multipath signal.
- Orthogonal frequency-division multiplexing (OFDM) can be used in frequency-selective fading channels to avoid the use of an equalizer by lengthening the symbol duration. The signal band is partitioned into multiple subbands, each exhibiting a lower symbol rate than the original band. The subbands are then transmitted on multiple

orthogonal carriers. The goal is to reduce the symbol rate (signaling rate), $W \oplus 1/T_s$, on each carrier to be less than the channel's coherence bandwidth f_0. OFDM was originally referred to as *Kineplex*. The technique has been implemented in the United States in mobile radio systems [11], and has been chosen by the European community under the name coded OFDM (COFDM), for high-definition television (HDTV) broadcasting [12].
- *Pilot signal* is the name given to a signal intended to facilitate the coherent detection of waveforms. Pilot signals can be implemented in the frequency domain as an in-band tone [13], or in the time domain as a pilot sequence which can also provide information about the channel state and thus improve performance in fading [14].

Mitigation to Combat Fast-Fading Distortion

- For fast fading distortion, use a robust modulation (noncoherent or differentially coherent) that does not require phase tracking, and reduce the detector integration time [15].
- Increase the symbol rate, $W \oplus 1/T_s$, to be greater than the fading rate, $f_d \oplus 1/T_0$, by adding signal redundancy.
- Error-correction coding and interleaving can provide mitigation, because instead of providing more signal energy, a code reduces the required E_b/N_0. For a given E_b/N_0, with coding present, the error floor will be lowered compared to the uncoded case.
- An interesting filtering technique can provide mitigation in the event of fast-fading distortion and frequency-selective distortion occurring simultaneously. The frequency-selective distortion can be mitigated by the use of an OFDM signal set. Fast fading, however, will typically degrade conventional OFDM because the Doppler spreading corrupts the orthogonality of the OFDM subcarriers. A polyphase filtering technique [16] is used to provide time-domain shaping and duration extension to reduce the spectral sidelobes of the signal set, and thus help preserve its orthogonality. The process introduces known ISI and adjacent channel interference (ACI), which are then removed by a post-processing equalizer and canceling filter [17].

Mitigation to Combat Loss in SNR

After implementing some form of mitigation to combat the possible distortion (frequency-selective or fast fading), the next step is to use some form of diversity to move the operating point from the error-performance curve that is "bad" in Fig. 4 to a curve that approaches AWGN performance. The term "diversity" is used to denote the various methods available for providing the receiver with uncorrelated renditions of the signal. Uncorrelated is the important feature here, since it would not help the receiver to have additional copies of the signal if the copies were all equally poor. Listed below are some of the ways in which diversity can be implemented:
- Time diversity — Transmit the signal on L different time slots with time separation of at least T_0. Interleaving, often used with error correction coding, is a form of time diversity.
- Frequency diversity — Transmit the signal on L different carriers with frequency separation of at least f_0. Bandwidth expansion is a form of frequency diversity. The signal bandwidth, W, is expanded to be greater than f_0, thus providing the receiver with several independently fading signal replicas. This achieves frequency diversity of the order $L = W/f_0$. Whenever W is made larger than f_0, there is the potential for frequency-selective distortion unless we further provide some mitigation such as equalization. Thus, an expanded bandwidth can improve system performance (via diversity) only if the frequency-selective distortion the diversity may have introduced is mitigated.
- Spread spectrum is a form of bandwidth expansion that excels at rejecting interfering signals. In the case of direct-sequence spread spectrum (DS/SS), it was shown earlier that multipath components are rejected if they are delayed by more than one chip duration. However, in order to approach AWGN performance, it is necessary to compensate for the loss in energy contained in those rejected components. The Rake receiver (described later) makes it possible to coherently combine the energy from each of the multipath components arriving along different paths. Thus, used with a Rake receiver, DS/SS modulation can be said to achieve path diversity. The Rake receiver is needed in phase-coherent reception, but in differentially coherent bit detection a simple delay line (one bit long) with complex conjugation will do the trick [18].
- FH/SS is sometimes used as a diversity mechanism. The GSM system uses slow FH (217 hops/s) to compensate for those cases where the mobile user is moving very slowly (or not at all) and happens to be in a spectral null.
- Spatial diversity is usually accomplished through the use of multiple receive antennas separated by a distance of at least 10 wavelengths for a base station (much less for a mobile station). Signal processing must be employed to choose the best antenna output or to coherently combine all the outputs. Systems have also been implemented with multiple spaced transmitters; an example is the Global Positioning System (GPS).
- Polarization diversity [19] is yet another way to achieve additional uncorrelated samples of the signal.
- Any diversity scheme may be viewed as a trivial form of repetition coding in space or time. However, there exist techniques for improving the loss in SNR in a fading channel that are more efficient and more powerful than repetition coding. Error correction coding represents a unique mitigation technique, because instead of providing more signal energy it reduces the required E_b/N_0 in order to accomplish the desired error performance. Error correction coding coupled with interleaving [15, 20–25] is probably the most prevalent of the mitigation schemes used to provide improved performance in a fading environment.

Summary of the Key Parameters Characterizing Fading Channels

We summarize the conditions that must be met so that the channel does not introduce frequency-selective and fast-fading distortion. Combining the inequalities of Eq. 14 and 23 from Part I of this article, we obtain

$$f_0 > W > f_d \qquad (7a)$$

or

$$T_m < T_s < T_0. \qquad (7b)$$

In other words, we want the channel coherence bandwidth to exceed our signaling rate, which in turn should exceed the fading rate of the channel. Recall from Part I that without distortion mitigation, f_0 sets an upper limit on the signaling rate, and f_d sets a lower limit on it.

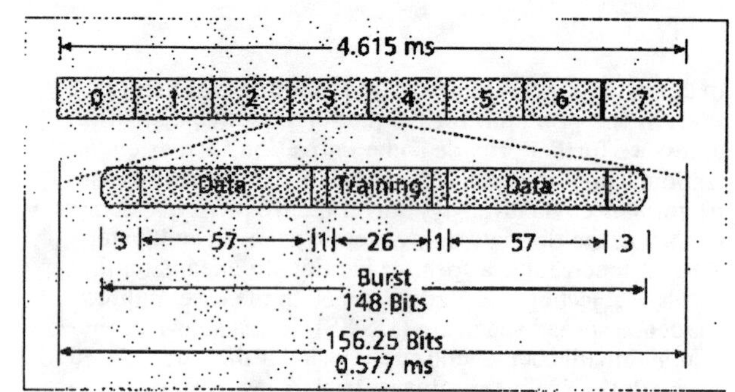

Figure 6. *The GSM TDMA frame and time slot containing a normal burst.*

FAST-FADING DISTORTION: EXAMPLE 1

If the inequalities of Eq. 7 are not met and distortion mitigation is not provided, distortion will result. Consider the fast-fading case where the signaling rate is less than the channel fading rate, that is,

$$f_0 > W < f_d. \qquad (8)$$

Mitigation consists of using one or more of the following methods (Fig. 5):

- Choose the modulation/demodulation technique that is most robust under fast-fading conditions. This means, for example, avoiding carrier recovery with PLLs since fast fading could keep a PLL from achieving lock conditions.
- Incorporate sufficient redundancy that the transmission symbol rate exceeds the channel fading rate. As long as the transmission symbol rate does not exceed the coherence bandwidth, the channel can be classified as flat fading. However, even flat-fading channels will experience frequency-selective distortion whenever a channel null appears at the band center. Since this happens only occasionally, mitigation might be accomplished by adequate error correction coding and interleaving.
- The above two mitigation approaches should result in the demodulator operating at the Rayleigh limit [15] (Fig. 4). However, there may be an irreducible floor in the error-performance versus E_b/N_0 curve due to the frequency modulated (FM) noise that results from the random Doppler spreading (see Part I). The use of an in-band pilot tone and a frequency-control loop can lower this irreducible performance level.
- To avoid this error floor caused by random Doppler spreading, increase the signaling rate above the fading rate still further (100–200 x fading rate) [26]. This is one architectural motive behind time-division multiple access (TDMA) mobile systems.
- Incorporate error correction coding and interleaving to lower the floor and approach AWGN performance.

FREQUENCY-SELECTIVE FADING DISTORTION: EXAMPLE 2

Consider the frequency-selective case where the coherence bandwidth is less than the symbol rate; that is,

$$f_0 < W > f_d. \qquad (9)$$

Mitigation consists of using one or more of the following methods (Fig. 5):

- Since the transmission symbol rate exceeds the channel fading rate, there is no fast-fading distortion. Mitigation of frequency-selective effects is necessary. One or more of the following techniques may be considered.
- Adaptive equalization, spread spectrum (DS or FH), OFDM, pilot signal. The European GSM system uses a midamble training sequence in each transmission time slot so that the receiver can learn the impulse response of the channel. It then uses a Viterbi equalizer (explained later) for mitigating the frequency-selective distortion.
- Once the distortion effects have been reduced, introduce some form of diversity and error correction coding and interleaving in order to approach AWGN performance. For direct-sequence spread-spectrum (DS/SS) signaling, a Rake receiver (explained later) may be used for providing diversity by coherently combining multipath components that would otherwise be lost.

FAST-FADING AND FREQUENCY-SELECTIVE FADING DISTORTION: EXAMPLE 3

Consider the case where the coherence bandwidth is less than the signaling rate, which in turn is less than the fading rate. The channel exhibits both fast-fading and frequency-selective fading, which is expressed as

$$f_0 < W < f_d \qquad (10a)$$

or

$$f_0 < f_d. \qquad (10b)$$

Recalling from Eq. 7 that f_0 sets an upper limit on the signaling rate and f_d sets a lower limit on it, this is a difficult design problem because, unless distortion mitigation is provided, the maximum allowable signaling rate is (in the strict terms of the above discussion) less than the minimum allowable signaling rate. Mitigation in this case is similar to the initial approach outlined in example 1.

- Choose the modulation/demodulation technique that is most robust under fast-fading conditions.
- Use transmission redundancy in order to increase the transmitted symbol rate.
- Provide some form of frequency-selective mitigation in a manner similar to that outlined in example 2.
- Once the distortion effects have been reduced, introduce some form of diversity and error correction coding and interleaving in order to approach AWGN performance.

THE VITERBI EQUALIZER AS APPLIED TO GSM

Figure 6 shows the GSM time-division multiple access (TDMA) frame, with a duration of 4.615 ms and comprising eight slots, one assigned to each active mobile user. A normal transmission burst, occupying one slot of time, contains 57 message bits on each side of a 26-bit midamble called a *training* or *sounding sequence*. The slot-time duration is 0.577 ms (or the slot rate is 1733 slots/s). The purpose of the midamble is to assist the receiver in estimating the impulse response of the channel in an adaptive way (during the time duration of each 0.577 ms slot). In order for the technique to be effective, the fading behavior of the channel should not change appreciably during the time interval of one slot. In other words, there should not be any fast-fading degradation during a slot time when the receiver is using knowledge from the midamble to compensate for the channel's fading behavior. Consider the example of a GSM receiver used aboard a high-speed train, traveling at a constant velocity of 200 km/hr (55.56 m/s). Assume the carrier frequency to be 900 MHz, (the wavelength is $\lambda = 0.33$ m). From Eq. 3, we can calculate that a half-wavelength is traversed in approximately the time (coherence time)

$$T_0 \approx \frac{\lambda/2}{V} \approx 3\text{ms} \qquad (11)$$

Therefore, the channel coherence time is over five times

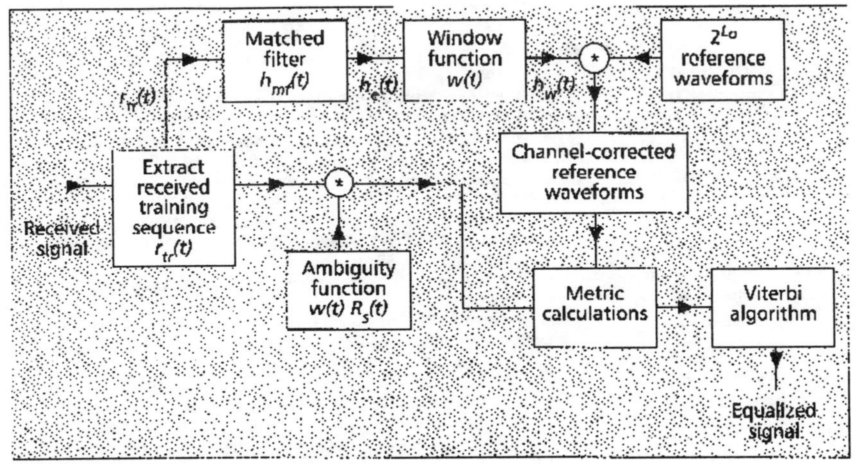

■ **Figure 7.** *The Viterbi Equalizer as applied to GSM.*

greater than the slot time of 0.577 ms. The time needed for a significant change in fading behavior is relatively long compared to the time duration of one slot. Note that the choices made in the design of the GSM TDMA slot time and midamble were undoubtedly influenced by the need to preclude fast fading with respect to a slot-time duration, as in this example.

The GSM symbol rate (or bit rate, since the modulation is binary) is 271 ksymbols/s and the bandwidth is W = 200 kHz. If we consider that the typical rms delay spread in an urban environment is in the order of $\sigma_\tau = 2\ \mu s$, then using Eq. 2 the resulting coherence bandwidth is $f_0 \oplus 100$ kHz. It should therefore be apparent that since $f_0 < W$, the GSM receiver must utilize some form of mitigation to combat frequency-selective distortion. To accomplish this goal, the Viterbi equalizer is typically implemented.

Figure 7 illustrates the basic functional blocks used in a GSM receiver for estimating the channel impulse response, which is then used to provide the detector with channel-corrected reference waveforms [27]. In the final step, the Viterbi algorithm is used to compute the MLSE of the message. A received signal can be described in terms of the transmitted signal convolved with the impulse response of the channel. We show this below, using the notation of a received training sequence, $r_{tr}(t)$, and the transmitted training sequence, $s_{tr}(t)$, as follows:

$$r_{tr}(t) = s_{tr}(t) * h_c(t) \quad (12)$$

where * denotes convolution. At the receiver, $r_{tr}(t)$ is extracted from the normal burst and sent to a filter having impulse response, $h_{mf}(t)$, that is matched to $s_{tr}(t)$. This matched filter yields at its output, an estimate of $h_c(t)$, denoted $h_e(t)$, developed from Eq. 12 as follows:

$$\begin{aligned} h_e(t) &= r_{tr}(t) * h_{mf}(t) \\ &= s_{tr}(t) * h_c(t) * h_{mf}(t) \quad (13) \\ &= R_s(t) * h_c(t) \end{aligned}$$

where $R_s(t)$ is the autocorrelation function of $s_{tr}(t)$. If $R_s(t)$ is a highly peaked (impulse-like) function, then $h_e(t) \oplus h_c(t)$.

Next, using a windowing function, $w(t)$, we truncate $h_e(t)$ to form a computationally affordable function, $h_w(t)$. The window length must be large enough to compensate for the effect of typical channel-induced ISI. The required observation interval L_o for the window, can be expressed as the sum of two contributions. The interval of length L_{CISI} is due to the controlled ISI caused by Gaussian filtering of the baseband pulses, which are then minimum shift keying (MSK) modulated. The interval of length L_C is due to the channel-induced ISI caused by multipath propagation; therefore, L_o can be written as

$$L_o + L_{CISI} + L_C. \quad (14)$$

The GSM system is required to provide mitigation for distortion due to signal dispersions of approximately 15–20 µs. The bit duration is 3.69 µs. Thus, the Viterbi equalizer used in GSM has a memory of 4–6 bit intervals. For each L_o-bit interval in the message, the function of the Viterbi equalizer is to find the most likely L_o-bit sequence out of the 2^{L_o} possible sequences that might have been transmitted. Determining the most likely L_o-bit sequence requires that 2^{L_o} meaningful reference waveforms be created by modifying (or disturbing) the 2^{L_o} ideal waveforms in the same way the channel has disturbed the transmitted message. Therefore, the 2^{L_o} reference waveforms are convolved with the windowed estimate of the channel impulse response, $h_w(t)$, in order to derive the disturbed or channel-corrected reference waveforms. Next, the channel-corrected reference waveforms are compared against the received data waveforms to yield metric calculations. However, before the comparison takes place, the received data waveforms are convolved with the known windowed autocorrelation function $w(t)R_s(t)$, transforming them in a manner comparable to that applied to the reference waveforms. This filtered message signal is compared to all possible 2^{L_o} channel-corrected reference signals, and metrics are computed as required by the Viterbi decoding algorithm (VDA). The VDA yields the maximum likelihood estimate of the transmitted sequence [6].

THE RAKE RECEIVER APPLIED TO DIRECT-SEQUENCE SPREAD-SPECTRUM SYSTEMS

Interim Specification 95 (IS-95) describes a DS/SS cellular system that uses a Rake receiver [7–9] to provide path diversity. In Fig. 8, five instances of chip transmissions correspond-

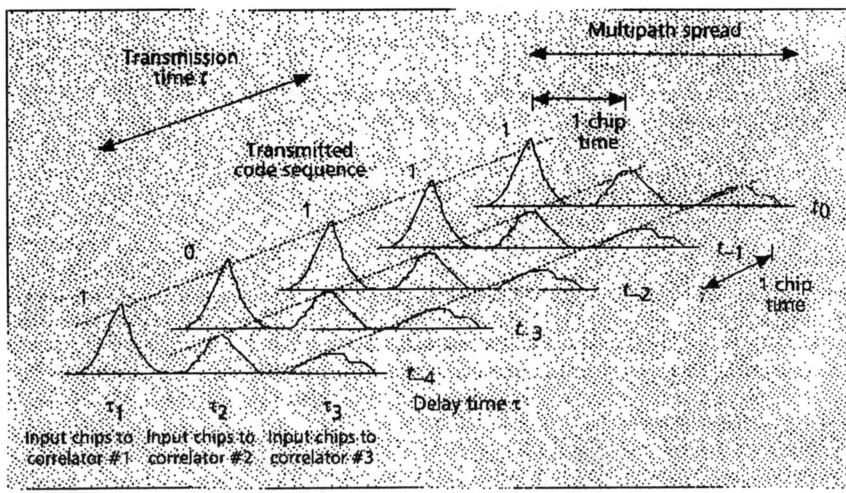

■ **Figure 8.** *An example of received chips seen by a three-finger Rake receiver.*

ing to the code sequence 1 0 1 1 1 are shown, with the transmission or observation times labeled t_{-4} for the earliest transmission and t_0 for the latest. Each abscissa shows three "fingers" of a signal that arrive at the receiver with delay times τ_1, τ_2, and τ_3. Assume that the intervals between the t_i transmission times and the intervals between the τ_i delay times are each one chip long. From this, one can conclude that the finger arriving at the receiver at time t_{-4}, with delay τ_3, is time-coincident with two other fingers, namely the fingers arriving at times t_{-3} and t_{-2} with delays τ_2 and τ_1, respectively. Since, in this example, the delayed components are separated by exactly one chip time, they are *just* resolvable. At the receiver, there must be a sounding device that is dedicated to estimating the τ_i delay times. Note that for a terrestrial mobile radio system, the fading rate is relatively slow (milliseconds) or the channel coherence time large compared to the chip time ($T_0 > T_{ch}$). Hence, the changes in τ_i occur slowly enough that the receiver can readily adapt to them.

Once the τ_i delays are estimated, a separate correlator is dedicated to processing each finger. In this example, there would be three such dedicated correlators, each processing a delayed version of the same chip sequence, 1 0 1 1 1. In Fig. 8, each correlator receives chips with power profiles represented by the sequence of fingers shown along a diagonal line. Each correlator attempts to match these arriving chips with the same PN code, similarly delayed in time. At the end of a symbol interval (typically there may be hundreds or thousands of chips per symbol), the outputs of the correlators are coherently combined, and a symbol detection is made. At the chip level, the Rake receiver resembles an equalizer, but its real function is to provide diversity.

The interference-suppression nature of DS/SS systems stems from the fact that a code sequence arriving at the receiver merely one chip time late, will be approximately orthogonal to the particular PN code with which the sequence is correlated. Therefore, any code chips that are delayed by one or more chip times will be suppressed by the correlator. The delayed chips only contribute to raising the noise floor (correlation sidelobes). The mitigation provided by the Rake receiver can be termed path diversity, since it allows the energy of a chip that arrives via multiple paths to be combined coherently. Without the Rake receiver, this energy would be transparent and therefore lost to the DS/SS system. In Fig. 8, looking vertically above point τ_3, it is clear that there is interchip interference due to different fingers arriving simultaneously. The spread-spectrum processing gain allows the system to endure such interference at the chip level. No other equalization is deemed necessary in IS-95.

Summary

In Part II of this article, the mathematical model and the major elements that contribute to fading in a communication channel have been briefly reviewed (from Part I of the tutorial). Next, mitigation techniques for ameliorating the effects of each degradation category were treated, and summarized in Fig. 5. Finally, mitigation methods that have been implemented in two system types, GSM and CDMA systems meeting IS-95, were described.

References

[1] T. S. Rappaport, *Wireless Communications*, Upper Saddle River, NJ: Prentice Hall, 1996.

[2] J. G. Proakis, *Digital Communications*, Ch. 7, New York: McGraw-Hill, 1983.

[3] R. L. Bogusch, F. W. Guigliano, and D. L. Knepp, "Frequency- Selective Scintillation Effects and Decision Feedback Equalization in High Data-Rate Satellite Links," *Proc. IEEE*, vol. 71, no. 6, June 1983, pp. 754–67.

[4] S. U. H. Qureshi, "Adaptive Equalization," *Proc. IEEE*, vol. 73, no. 9, Sept. 1985, pp. 1340–87.

[5] G. D. Forney, "The Viterbi Algorithm," *Proc. IEEE*, vol. 61, no. 3, Mar. 1978, pp. 268–78.

[6] B. Sklar, *Digital Communications: Fundamentals and Applications*, Ch. 6, Englewood Cliffs, NJ: Prentice Hall, 1988.

[7] R. Price and P. E. Green, Jr., "A Communication Technique for Multipath Channels," *Proc. IRE*, vol. 46, no. 3, Mar. 1958, pp. 555–70.

[8] G. L. Turin, "Introduction to Spread-Spectrum Antimultipath Techniques and Their Application to Urban Digital Radio," *Proc. IEEE*, vol. 68, no. 3, Mar. 1980, pp. 328–53.

[9] M. K. Simon *et al.*, *Spread Spectrum Communications Handbook*, McGraw Hill, 1994.

[10] F. Amoroso, "Use of DS/SS Signaling to Mitigate Rayleigh Fading in a Dense Scatterer Environment," *IEEE Pers. Commun.*, vol. 3, no. 2, Apr. 1996, pp. 52–61.

[11] M. A. Birchler and S. C. Jasper, "A 64 kbps Digital Land Mobile Radio System Employing M-16QAM," *Proc. 1992 IEEE Int'l. Conf. Sel. Topics in Wireless Commun.*, Vancouver, Canada, June 25–26, 1992, pp. 158–62.

[12] H. Sari, G. Karam, and I. Jeanclaude, "Transmission Techniques for Digital Terrestrial TV Broadcasting," *IEEE Commun. Mag.*, vol. 33, no. 2, Feb. 1995, pp. 100–109.

[13] J. K. Cavers, "The Performance of Phase Locked Transparent Tone-in-Band with Symmetric Phase Detection," *IEEE Trans. Commun.*, vol. 39, no. 9, Sept. 1991, pp. 1389–99.

[14] M. L. Moher and J. H. Lodge, "TCMP — A Modulation and Coding Strategy for Rician Fading Channel," *IEEE JSAC*, vol. 7, no. 9, Dec. 1989, pp. 1347–55.

[15] R. L. Bogusch, *Digital Communications in Fading Channels: Modulation and Coding*, Mission Research Corp., Santa Barbara, CA, Rep. no. MRC-R-1043, Mar. 11, 1987.

[16] F. Harris, "On the Relationship Between Multirate Polyphase FIR Filters and Windowed, Overlapped FFT Processing," *Proc. 23rd Annual Asilomar Conf. Signals, Sys., and Comp.*, Pacific Grove, CA, Oct. 30–Nov. 1, 1989, pp. 485–88.

[17] R. W. Lowdermilk and F. Harris, "Design and Performance of Fading Insensitive Orthogonal Frequency Division Multiplexing (OFDM) Using Polyphase Filtering Techniques," *Proc. 30th Annual Asilomar Conf. Signals, Sys., and Comp.*, Pacific Grove, CA, Nov. 3–6, 1996.

[18] M. Kavehrad and G. E. Bodeep, "Design and Experimental Results for a Direct-Sequence Spread-Spectrum Radio Using Differential Phase-Shift Keying Modulation for Indoor Wireless Communications," *IEEE JSAC*, vol. SAC-5, no. 5, June 1987, pp. 815–23.

[19] G. C. Hess, *Land-Mobile Radio System Engineering*, Boston: Artech House, 1993.

[20] J. Hagenauer and E. Lutz, "Forward Error Correction Coding for Fading Compensation in Mobile Satellite Channels," *IEEE JSAC*, vol. SAC-5, no. 2, Feb. 1987, pp. 215–25.

[21] P. I. McLane, *et. al.*, "PSK and DPSK Trellis Codes for Fast Fading, Shadowed Mobile Satellite Communication Channels," *IEEE Trans. Commun.*, vol. 36, no. 11, Nov. 1988, pp. 1242–46.

[22] C. Schlegel, and D. J. Costello, Jr., "Bandwidth Efficient Coding for Fading Channels: Code Construction and Performance Analysis," *IEEE JSAC*, vol. 7, no. 9, Dec. 1989, pp. 1356–68.

[23] F. Edbauer, "Performance of Interleaved Trellis-Coded Differential 8-PSK Modulation over Fading Channels," *IEEE JSAC*, vol. 7, no. 9, Dec. 1989, pp. 1340–46.

[24] S. Soliman and K Mokrani, "Performance of Coded Systems over Fading Dispersive Channels," *IEEE Trans. Commun.*, vol. 40, no. 1, Jan. 1992, pp. 51–59

[25] D. Divsalar and F. Pollara, "Turbo Codes for PCS Applications," *Proc. ICC '95*, Seattle, WA, June 18–22, 1995, pp. 54–59.

[26] A. J. Bateman and J. P. McGeehan, "Data Transmission over UHF Fading Mobile Radio Channels," *IEE Proc.*, vol. 131, pt. F, no. 4, July 1984, pp. 364–74.

[27] L. Hanzo and J. Stefanov, "The Pan-European Digital Cellular Mobile Radio System — Known as GSM," *Mobile Radio Commun.*, R. Steele, Ed., Ch. 8, London: Pentech Press, 1992.

Biography

Bernard Sklar [LSM] has over 40 years of experience in technical design and management positions at Republic Aviation Corp., Hughes Aircraft, Litton Industries, and The Aerospace Corporation. At Aerospace, he helped develop the MILSTAR satellite system, and was the principal architect for EHF Satellite Data Link Standards. Currently, he is head of advanced systems at Communications Engineering Services, a consulting company he founded in 1984. He has taught engineering courses at several universities, including the University of California, Los Angeles and the University of Southern California, and has presented numerous training programs throughout the world. He has published and presented scores of technical papers. He is a past Chair of the Los Angeles Council IEEE Education Committee. His academic credentials include a B.S. degree in math and science from the University of Michigan, an M.S. degree in electrical engineering from the Polytechnic Institute of Brooklyn, New York, and a Ph.D. degree in engineering from the University of California, Los Angeles.

SECTION 1.3
RECEIVERS AND THEIR PERFORMANCES

Spread-spectrum receiver performance and design have been an active area of research for quite some time. Initially, receivers were designed with the single-user concept (i.e. the receiver is designed by disregarding other users in the channel). Under additive white Gaussian noise, the receiver would be a matched filter type. There were various reports addressing the bit error probability calculation of single-user receiver in a multiple-access environment (i.e. performance when there are multiple-user interferences (MUI)). Various models have been used to represent the MUI and various methods have been proposed to calculate the performance. Today, research on spread-spectrum receivers has shifted to design and performance of multiple-user receivers (i.e. receivers which are designed by taking into account and make use of all the active users' signals).

The first paper, "*Direct-Sequence Spread-Spectrum Multiple-Access Communications Over Nonselective and Frequency-Selective Rician Fading Channels*" by Geraniotis, and the second paper, "*Error Probabilities for Binary Direct-Sequence Spread-Spectrum Communications with Random Signature Sequences*" by Lehnert and Pursley, provide a representative analysis of the error probability for single-user receiver in multiple-access channels. Here, the signature sequence (the pseudorandom sequence assigned to each user) is treated as a random signal. To more accurately characterize the MUI, there have been other reports in the literature that do not assume the signature sequence as a random signal in deriving the error probability.

The next three papers, "*Muli-User Detection for DS-CDMA Communications*" by Moshavi, "*Multiuser Detection for CDMA Systems*" by Duel-Hallen et al., and "*Minimum probability of error for asynchronous Gaussian Multiple-Access Channels*" by Verdu, deal with multi-user CDMA receivers. The paper by Verdu has become a classic paper on this subject. It derives the optimum receiver for a multiple-access channel with additive white Gaussian noise when the active users transmit signals asynchronously with respect to each other. The optimum receiver demodulates the data of all the users simultaneously, which needs the knowledge of the signature sequences of all the users. It is shown that the optimum receiver requires computation that grows exponentially as the number of users increases, which becomes prohibitive in practical situations. Consequently, several other suboptimum receivers have been proposed in the past decade. These suboptimum receivers require only a linear increase in computation. Discussions of the optimum and suboptimum receivers are given in the two tutorial papers by Moshavi and Duel-Hallen et al.

The requirement of the knowledge of all users' spreading sequences in the optimum and suboptimum multi-user receivers poses two problems: the receiver must be centralized and message security cannot be assured. To avoid this problem, the paper "*Single-User Detectors for Multiuser Channels*" by Poor and Verdu proposes single-user receivers which do not require other users' spreading sequence, but somehow take into account interferences from other users in designing the receiver. This is in contrast with conventional single-user receivers which completely ignore multiple-user interferences in designing the receiver. The resulting receivers are compromises between multi-user receivers and conventional single-user receivers.

Direct-Sequence Spread-Spectrum Multiple-Access Communications Over Nonselective and Frequency-Selective Rician Fading Channels

EVAGGELOS GERANIOTIS, MEMBER, IEEE

Abstract—An accurate approximation is obtained for the average probability of error in an asynchronous binary direct-sequence spread-spectrum multiple-access communications system operating over nonselective and frequency-selective Rician fading channels. The approximation is based on the integration of the characteristic function of the multiple-access interference which now consists of specular and scatter components. For nonselective fading, the amount of computation required to evaluate this approximation grows linearly with the product KN, where K is the number of simultaneous transmitters and N is the number of chips per bit. For frequency-selective fading, the computational effort grows linearly with the product KN^2. The resulting probability of error is also compared with an approximation based on the signal-to-noise ratio. Numerical results are presented for specific chip waveforms and signature sequences.

I. Introduction

THIS paper is concerned with the average probability of error for asynchronous binary direct-sequence spread-spectrum multiple-access (DS/SSMA) communications over nonselective and frequency-selective Rician fading channels. The results presented represent a generalization of the analysis of [7] which considered only additive white Gaussian noise (AWGN) channels.

The system model is described in detail in [6]. Throughout this paper we restrict attention to binary DS/SSMA systems with phase-shift-keyed (PSK) modulation and chip waveforms which are pulses of arbitrary shape with duration equal to the inverse of the chip rate. However, our results can be extended (using the results of [7]) to include other forms of direct-sequence modulation like quadriphase-shift-keying (QPSK), offset QPSK, and minimum-shift-keying (MSK).

The channel model considered in this paper is the Rician or specular-plus-Rayleigh fading channel model described in [1] for nonselective fading, and in [5] for selective fading. In the second case the fading is assumed to be frequency-selective [3]. Although our results can be extended to time-selective, as well as doubly-selective wide-sense-stationary uncorrelated-scattering (WSSUS) fading channels [2], [5], we restrict attention to frequency-selective channels, since these channels often arise in practice and, as shown in [5], single-user DS/spread-spectrum systems outperform biphase PSK systems over such channels.

In [5] the average signal-to-noise ratio at the output of the receiver of a binary DS/SSMA system operating over time and frequency-selective Rician fading channels was evaluated. In [4] the moment space bounding technique of [12] was applied to bound the probability of error of such systems. However, the results were limited because of unsurpassable difficulties in evaluating high-order moments of the interference. For the same reason, the approximation based on the Gauss quadrature rule, which was used in [10] and [11] for DS/SSMA systems over AWGN channels, is not computationally attractive in this case, since it requires calculation of moments which involve high-order channel autocovariance functions that may not be measurable or otherwise available. Similarly, the bounds of [13] were based on convexity properties of the complementary error function which are valid only in the AWGN case. Finally, the results for two users of [14] for nonselective fading channels and [15] for time-selective fading channels appear difficult to extend, so that they do not provide an efficient and accurate computational method for the multiuser case.

In this paper we obtain an approximation with a computational requirement which for nonselective fading is linear in the product KN and for frequency-selective fading is linear in the product KN^2, where K is the number of simultaneous transmitters and N is the number of chips per bit. This approximation is based on the integration of the characteristic function of the multiple-access interference component of the output of the correlation receiver. This component consists now of a fixed (or specular) part and a random (or scatter) part. This method, which we refer to as the characteristic-function method, was applied to DS/SSMA systems in [7] and DS/SS systems with specular multipath fading in [9], where it was shown to provide very accurate approximations to the average probability of error. Moreover, as shown in [8], any prespecified degree of accuracy can be achieved by using this approximation to obtain an expansion point for a Taylor series representation of the actual probability of error. In the selective fading case, however, it is computationally intractable to evaluate the moments of the interference required for the series expansion method; therefore, in the sequel we consider only the characteristic-function method.

We should also point out at this point that the DS/SSMA system under consideration employs conventional matched filter receivers to combat rather than utilize the interference (multiple-access faded and nonfaded components) by discriminating against it. This is in contrast to the work of [17] and [18], where diversity reception is used to combine the contributions of the faded (reflected in those cases) paths and use them in the binary decision.

This paper is organized as follows. Nonselective Rician channels are considered in Section II, and frequency-selective Rician channels are treated in Section III. In each of these sections the fading channel model is first presented. Next the

Paper approved by the Editor for Spread Spectrum of the IEEE Communications Society. Manuscript received October 9, 1984; revised January 29, 1986. This work was supported in part by the Office of Naval Research under Contracts N00014-84-K-0023 and N00014-86-K-0013, and by the Systems Research Center at the University of Maryland, College Park, through the National Science Foundation under Grant CDR-85-00108.

The author is with the Department of Electrical Engineering and the Systems Research Center, University of Maryland, College Park, MD 20742.

IEEE Log Number 8609559.

0090-6778/86/0800-0756$01.00 © 1986 IEEE

characteristic-function method is applied to each particular model and the computational requirements are considered. Finally, a simpler approximation which is based on the average signal-to-noise ratio is cited for the sake of comparison. In Section IV, numerical results are presented for specific chip waveforms and signature sequences.

II. Performance of DS/SSMA Communications Over Nonselective Rician Fading Channels

A. A System and Channel Model

The system model is described in detail in [6] and [9]. Here we repeat the basic elements of this model so that the concepts and notation which are necessary for the rest of the paper are introduced.

The kth *transmitted* signal for a binary DS/SSMA system with PSK modulation and arbitrary chip waveform can be expressed as

$$s_k(t)=\text{Re}\ \{x_k(t)\ \exp\ (j2\pi f_c t)\} \tag{1}$$

where Re stands for real part, $j = \sqrt{-1}$, f_c is the common carrier frequency, and the DS-spreaded signal $x_k(t)$ is given by

$$x_k(t)=\sqrt{2P}\ b_k(t)\Psi(t)a_k(t)\ \exp\ (j\theta_k). \tag{2}$$

In (2) P is the power in each of the K transmitted signals; the equal power assumption is made for convenience in presenting numerical results, and is not necessary for the methods considered in this paper. The phase angle θ_k introduced by the modulator is uniformly distributed in $[0, 2\pi]$. The shaping waveform $\Psi(t)$ that appears in (2) is periodic with period T_c (the duration of a chip), and is defined by $\Psi(t) = \psi(s)$ for $s = t(\text{mod } T_c)$ where $\psi(s)$ is a chip waveform of arbitrary shape (see [6] and [9]) which satisfies a time-limiting constraint and an average energy constraint. The data signal $b_k(t)$ consists of rectangular pulses of duration T which take values $+1$ and -1 with equal probability. The lth pulse has amplitude $b_l^{(k)}$. The information sequence $(b_l^{(k)})$ is modeled as an i.i.d. sequence. The code waveform $a_k(t)$ is a periodic sequence of rectangular pulses of amplitude $+1$ or -1 and duration T_c. The jth code pulse has amplitude $a_j^{(k)}$. We assume that there are N code pulses in each data pulse ($T = NT_c$) and that the period of the signature sequence $(a_j^{(k)})$ is N.

For a system with K asynchronous simultaneous transmitted signals, the received signal at a receiver which is interested in the ith signal is given by

$$r(t)=\sum_{k=1}^{K} y_k(t-\tau_k)+n(t). \tag{3}$$

In (3) τ_k is the time delay for the communication link between the kth transmitter and the ith receiver. The process $n(t)$ is a white Gaussian noise process with two-sided spectral density $N_0/2$. Finally, $y_k(t)$ is the fading channel output to input $s_k(t)$.

A *Rician nonselective fading channel* is described by the following input–output relationship:

$$y_k(t)=s_k(t)+\text{Re}\ \{\gamma_k A_l^{(k)}\ \exp\ (j\theta_l^{(k)})x_k(t)\ \exp\ (j2\pi f_c t)\} \tag{4}$$

for $lT \leqslant t < (l + 1)T$. In (4) γ_k is a nonnegative real number and the nonnegative random variable $A_l^{(k)}$ satisfies the normalization constraint

$$E\{[A_l^{(k)}]^2\}=1. \tag{5}$$

The attenuation of the signal strength due to the fading during the time interval $[lT, (l + 1)T)$ is thus represented by $\gamma_k A_l^{(k)}$, and the phase shift due to the fading is denoted by $\theta_l^{(k)}$. Therefore, for $\gamma_k > 0$ the output signal is the sum of a nonfaded version of the input signal (specular component) and a nondelayed faded version of the input signal (scatter component). For $A_l^{(k)}$ Rayleigh distributed and $\theta_l^{(k)}$ uniformly distributed in $[0, 2\pi]$, this model is discussed in [1]. Notice that from (4) and (3) the channel model reduces to the AWGN model if $\gamma_k = 0$. Similarly, for $\gamma_k \to \infty$ (very weak specular component in comparison to the faded component) the channel reduces to the Rayleigh nonselective channel.

Regarding the statistical dependence of the attenuation in different data bit intervals, we will consider the cases: i) $A_l^{(k)} = A_m^{(k)}$ for several values of adjacent l and m (the length of the burst is immaterial for our analysis as long as it is longer than three adjacent bits), and ii) $A_l^{(k)}$ and $A_m^{(k)}$ are independent if $l \neq m$. Case i) corresponds to a system with no interleaving (or partial interleaving) and a channel with fading statistics which remain invariant over the duration of several data bits. An example of case ii) arises in a system which is fully interleaved or a fast fading channel. Similar restrictions are imposed on the phase sequence $(\theta_l^{(k)})$.

The ith *receiver* is matched to the ith transmitter signal; it is assumed capable of acquiring time and phase synchronization (coherent demodulation) with the nonfaded component of the ith signal. Therefore, we may measure time delays and phase angles relative to τ_i and θ_i, respectively, and thus set $\tau_i = \theta_i = 0$. If there is not a nonfaded component, we assume that the phase of the faded component of the Rician nonselective fading channel [see (3)] changes very slowly so that the receiver can again acquire time and phase synchronization. In this case we should let $\theta_i + \theta_0^{(i)} = 0$.

Once time and phase acquisition is completed, the received signal is passed through a correlation receiver which outputs

$$Z_i=\int_0^T r(t)\Psi(t)a_i(t)\ \cos\ (2\pi f_c t)\ dt. \tag{6}$$

If Z_i is positive the receiver will decide that a 1 was sent; otherwise it will output a -1. In practical spread-spectrum systems, the common carrier frequency f_c and bit duration T are such that $f_c \gg T^{-1}$; consequently, the double-frequency terms in the integrand of (6) may be ignored.

Finally, concerning the interaction among the different communication links between the users, we may measure the delays τ_k with respect to τ_i, set $\tau_i = 0$, and assume that, for $k \neq i$, τ_k (mod T) is uniformly distributed in $[0, T]$. Then the delays τ_k, the phase angles θ_k, and the data streams $(b_l^{(k)})$ are assumed to be mutually independent for a given k, as well as for any two different transmitters. Finally, we assume that all communication *links fade independently*; therefore, all the random variables and/or random processes which characterize the fading are independent for any two different links.

B. Evaluation of Error Probability Via the Characteristic-Function Method

By ignoring the double-frequency terms in (6) we may write Z_i as

$$Z_i=(1/2E_bT)^{1/2}(b_0^{(i)}+\eta_i+\mathcal{I}_i). \tag{7}$$

In (7) $E_b \triangleq PT$ is the received energy per bit for the nonfaded component, the first term inside the parenthesis is the desired signal component (i.e., the ith information signal for $0 \leq t < T$), η_i represents the normalized contribution of the AWGN and of the faded component of the ith signal, and $\mathcal{I}_i$ stands for the normalized multiple-access interference (which now has nonfaded and faded components) due to all the other users besides the ith user. We actually have that

$$\eta_i=\eta+\gamma_i F_i \tag{8}$$

and

$$\mathcal{I}_i=\sum_{k\neq i}(I_{k,i}+\gamma_k F_{k,i}) \tag{9}$$

where η is a zero-mean Gaussian random variable with variance $N_0/2E_b$, F_i is the normalized interference due to the faded component of the ith signal, and $I_{k,i}$ and $F_{k,i}$ represent normalized interference due to the nonfaded and faded components of the kth ($k \neq i$) signal.

Due to the additive form of the multiple-access interference [see (9)], the characteristic-function method is recommended for the evaluation of the probability of error for the DS/SSMA systems under consideration. As in [7] we can write $\bar{P}_{e,i}$, the average probability of error for the ith receiver (the average to be considered with respect to all the random variables introduced by the transmitters and the channel) as

$$\bar{P}_{e,i}=1/2P\{Z_i\leq 0|b_0^{(i)}=+1\}+1/2P\{Z_i>0|b_0^{(i)}=-1\}$$

$$=1/2-1/2P\{-1<\eta_i+\mathcal{I}_i\leq 1\} \tag{10}$$

where η_i and $\mathcal{I}_i$ were defined above. Let ϕ_{η_i} and ϕ_i denote the characteristic functions of the random variables of η_i and $\mathcal{I}_i$, respectively. Since the distributions of η_i and $\mathcal{I}_i$ considered in this paper are symmetric, the characteristic functions ϕ_{η_i} and ϕ_i are real-valued even functions. Therefore, as in [7]–[9] we can write

$$\bar{P}_{e,i}=\bar{P}^0_{e,i}+\pi^{-1}\int_0^\infty u^{-1}(\sin u)\phi_{\eta_i}(u)[1-\phi_i(u)]\,du \tag{11}$$

where $\bar{P}^0_{e,i}$ is the average probability of error in the absence of multiple-access interference (i.e., when $\mathcal{I}_i = 0$ or $K = 1$). Because of (9) and the independence assumptions of Section II-A above, we can write for each real u

$$\phi_i(u)=\prod_{k\neq i}E\{\exp[ju(I_{k,i}+\gamma_k F_{k,i})]\}, \tag{12}$$

which facilitates the computation of the characteristic function of the multiple-access interference.

For the *Rician* nonselective fading channel of Section II-A we can easily derive that

$$F_i=b_0^{(i)}A_0^{(i)}\cos(\theta_0^{(i)}) \tag{13a}$$

$$I_{k,i}=T^{-1}[b_1^{(k)}R_{k,i}(\tau_k)+b_2^{(k)}\hat{R}_{k,i}(\tau_k)]\cos\theta_k' \tag{13b}$$

and

$$F_{k,i}=T^{-1}[b_1^{(k)}A_1^{(k)}R_{k,i}(\tau_k)\cos(\theta_k'+\theta_1^{(k)})+b_2^{(k)}A_2^{(k)}\hat{R}_{k,i}(\tau_k)\cos(\theta_k'+\theta_2^{(k)})] \tag{13c}$$

where

$$\theta_k'=\theta_k-2\pi f_c\tau_k.$$

Recall that $\tau_i = \theta_i = 0$ so that $\theta_i' = 0$. The arguments of the partial continuous cross-correlation functions $R_{k,i}$ and $\hat{R}_{k,i}$ (defined in [6]) should be considered modulo T. Let $b_k = (b_1^{(k)}, b_2^{(k)})$ denote the pair of the $(\lambda_k - 1)$th and λ_kth consecutive data bits (suppose that for $k \neq i$, $\lambda_k T \leq \tau_k < (\lambda_k + 1)T$). If, for simplicity, we assume that the $A_l^{(k)}$'s are Rayleigh distributed, then since $\theta_k' + \theta_l^{(k)} \pmod{2\pi}$ is also uniformly distributed in $[0, 2\pi]$, the random variables F_i and $F_{k,i}$ are conditionally zero-mean Gaussian when conditioned on $b_0^{(i)}$ and (b_k, τ_k), respectively.

Next we proceed to the evaluation of the characteristic functions of η_i and $\mathcal{I}_i$ which are necessary for the computation in (11). Let σ_i^2 denote the variance of η_i defined by (8); then

$$\sigma_i^2=(2E_b/N_0)^{-1}+1/2\gamma_i^2 \tag{14}$$

$$P^0_{e,i}=Q(\sigma_i^{-1}) \tag{15}$$

where Q is the complementary error function, and

$$\phi_{\eta_i}(u)=\exp(-1/2u^2\sigma_i^2). \tag{16}$$

To proceed further we need to evaluate the variance of the conditionally Gaussian random variable $F_{k,i}$ given by (13c). It turns out that for i) constant fading

$$\text{Var}\{F_{k,i}|b_k,\tau_k\}=1/2T^{-2}[b_1^{(k)}R_{k,i}(\tau_k)+b_2^{(k)}\hat{R}_{k,i}(\tau_k)]^2 \tag{17}$$

and for ii) independent fading

$$\text{Var}\{F_{k,i}|\tau_k\}=1/2T^{-2}\{R^2_{k,i}(\tau_k)+\hat{R}^2_{k,i}(\tau_k)\}. \tag{18}$$

For $lT_c \leq \tau < (l + 1)T_c$ and any arbitrary function h, define

$$\sigma^2(l,h,\tau)=1/2T^{-2}[h(l+1)R_\psi(\tau)+h(l)\hat{R}_\psi(\tau)]^2 \tag{19}$$

$$\sigma^2_{k,i}(l;\tau)=1/2T^{-2}[R^2_{k,i}(\tau)+\hat{R}^2_{k,i}(\tau)] \tag{20}$$

and

$$f(u;l,h;\tau)=\frac{2}{\pi}\int_0^{\pi/2}\cos\left\{\frac{u}{T}[h(l+1)R_\psi(\tau)+h(l)\hat{R}_\psi(\tau)]\cos\theta\right\}d\theta. \tag{21}$$

In (19) and (21) R_ψ and $\hat{R}_\psi$ are the continuous partial autocorrelation functions of the chip waveform. The functions $R_{k,i}$ and $\hat{R}_{k,i}$ depend on them and on the discrete cross-correlation function of the kth and ith signature sequences $C_{k,i}$. Then after some straightforward but cumbersome manipulations, (12) becomes

$$\phi_i(u)=\prod_{k\neq i}\left\{(2N)^{-1}\sum_{l=0}^{N-1}T_c^{-1}\int_0^{T_c}\{f(u;l,\theta_{k,i};\tau)\cdot\exp[-1/2u^2\gamma_k^2\sigma^2(l,\theta_{k,i};\tau)]+f(u;l,\hat{\theta}_{k,i};\tau)\cdot\exp[-1/2u^2\gamma_k^2\sigma^2(l,\hat{\theta}_{k,i};\tau)]\,d\tau\right\} \tag{22a}$$

for i) constant fading, and

$$\phi_i(u)=\prod_{k\neq i}\left\{(2N)^{-1}\sum_{l=0}^{N-1}T_c^{-1}\int_0^{T_c}[f(u;l,\theta_{k,i};\tau)+f(u;l,\hat{\theta}_{k,i};\tau)]\cdot\exp[-1/2u^2\gamma_k^2\sigma^2_{k,i}(l;\tau)]\,d\tau\right\} \tag{22b}$$

for ii) independent fading.

If the channel encountered by the ith transmitted signal is a *Rayleigh* nonselective fading channel (i.e., there is no nonfaded component of the ith signal), we need to modify the above results in the following way. We start with finite γ_i and define $\bar{E}_b \triangleq (1 + \gamma_i^2)PT$ to be the total received energy at the ith receiver in the absence of AWGN and multiple-access interference (i.e., $K = 1$), and $\bar{\gamma}_k^2 = (1 + \gamma_k^2)/(1+\gamma_i^2)$ to be the ratio of the total received energy from the kth signal to the total received energy from the ith signal. For large γ_i, we assume (see Section II-A) that the receiver acquires time and phase synchronization with the faded component of the ith signal, so that we may set $\theta_i + \theta_0^{(i)} = 0$ and, together with the definitions above, write (7)–(9) as

$$Z_i=(1/2\bar{E}_bT)^{1/2}\left[\eta+\frac{1}{\sqrt{1+\gamma_i^2}}(b_0^{(i)}\cos\theta_i+\gamma_i b_0^{(i)}A_0^{(i)})+\sum_{k\neq i}\frac{\bar{\gamma}_k}{\sqrt{1+\gamma_k^2}}(I_{k,i}+\gamma_k F_{k,i})\right]. \quad (23)$$

If we fix $\bar{E}_b$ and let $\gamma_i\to\infty$, (23) reduces to

$$Z_i=(1/2\bar{E}_bT)^{1/2}[b_0^{(i)}A_0^{(i)}+\eta_i+\mathcal{I}_i] \quad (24)$$

where $A_0^{(i)}$ is Rayleigh distributed (with second-order moment 1), $\eta_i=\eta$ is zero-mean Gaussian with variance $N_0/2\bar{E}_b$, and $\mathcal{I}_i$ is given by

$$\mathcal{I}_i=\sum_{k\neq i}\frac{\bar{\gamma}_k}{\sqrt{1+\gamma_k^2}}(I_{k,i}+\gamma_k F_{k,i}) \quad (25)$$

where $I_{k,i}$ and $F_{k,i}$ are as defined in (13b) and (13c).

A discussion of the computational requirements for formulas of the type of (11) can be found in [7] and [8], and since the situation here is very similar to that treated there, it will not be repeated. However, it is worth observing that since there is a summation of N terms and a product of K terms in (22a) and (22b), the computational effort for this method grows linearly with the product KN.

Starting from (24) and proceeding as for (10), we find that $P_{e,i}(A_0^{(i)})$, the conditional probability of error given $A_0^{(i)}$, can be expressed initially as

$$P_{e,i}(A_0^{(i)})=1/2-1/2P\{-A_0^{(i)}<n_i+\mathcal{I}_i\leq A_0^{(i)}\} \quad (26)$$

and finally [cf. (11)] as

$$P_{e,i}(A_0^{(i)})=P_{e,i}^0(A_0^{(i)})+\pi^{-1}\int_0^\infty u^{-1}\cdot\sin(A_0^{(i)}u)\phi_{\eta_i}(u)[1-\phi_i(u)]\,du. \quad (27)$$

Since η_i is now zero-mean Gaussian with variance

$$\sigma_i^2=(2\bar{E}_b/N_0)^{-1} \quad (28)$$

$\phi_{\eta_i}(u)$ is given by (16) with σ_i^2 defined by (28) and

$$P_{e,i}^0(A_0^{(i)})=Q(A_0^{(i)}/\sigma_i). \quad (29)$$

To obtain the average probability of error $\bar{P}_{e,i}$ we should average the second member of (27) with respect to the Rayleigh distributed random variable $A_0^{(i)}$. The result (see [9, Appendix A] for the proof) is

$$\bar{P}_{e,i}=\bar{P}_{e,i}^0+1/2\pi^{-1/2}\cdot\int_0^\infty\exp(-1/4u^2)\phi_{\eta_i}(u)[1-\phi_i(u)]\,du \quad (30)$$

where

$$\bar{P}_{e,i}^0=1/2\{1-[1+(\bar{E}_b/N_0)^{-1}]^{-1/2}\} \quad (31)$$

and $\phi_i(u)$ is given by (22a) and (23b) for case i) constant fading and ii) independent fading, respectively, evaluated [because of (25)] at $\bar{\gamma}_k u/\sqrt{1+\gamma_k^2}$ instead of u.

C. Approximation Based on Average Signal-to-Noise Ratio

For the sake of comparison we next cite an approximation based on SNR_i, the *average signal-to-noise ratio* at the output of the ith receiver. As suggested in [6]

$$\mathrm{SNR}_i\triangleq E\{Z_i\}/[\mathrm{Var}\{Z_i\}]^{1/2} \quad (32)$$

which in our case becomes

$$\mathrm{SNR}_i=[\mathrm{Var}\{\eta_i\}+\sum_{k\neq i}\mathrm{Var}(I_{k,i}+\gamma_{k,i}F_{k,i})]^{-1/2}. \quad (33)$$

For *Rician* nonselective fading (33) reduces to

$$\mathrm{SNR}_i=\left\{(2E_b/N_0)^{-1}+1/2\gamma_i^2+\sum_{k\neq i}(1+\gamma_k^2)T^{-3}[\mu_{k,i}(0)m_\psi+\mu_{k,i}(1)m'_\psi]\right\}^{-1/2} \quad (34)$$

and the resulting approximation to the average error probability is

$$\bar{P}_{e,i}^G=Q(\mathrm{SNR}_i). \quad (35)$$

In (34) the function $\mu_{k,i}(n)$ (see [6]) depends on the discrete cross-correlation function $C_{k,i}$ while m_ψ and m'_ψ (again see [6]) depend on R_ψ and $\hat{R}_\psi$. For *Rayleigh* nonselective fading (33) becomes

$$\mathrm{SNR}_i=\{(\bar{E}_b/N_0)^{-1}+\sum_{k\neq i}\bar{\gamma}_k^2T^{-3}[\mu_{k,i}(0)m_\psi+\mu_{k,i}(1)m'_\psi]\}^{-1/2} \quad (36)$$

and the resulting approximation is

$$\bar{P}_{e,i}^G=1/2\{1-[1+(\mathrm{SNR}_i)^{-2}]^{-1/2}\}. \quad (37)$$

Notice that, in contrast to the approximation based on the characteristic-function method, the approximations obtained via the signal-to-noise ratio do not distinguish between the constant and independent fading cases considered in Section II-A.

III. Performance of DS/SS Communications Over Frequency-Selective Rician Fading Channels

A. Channel Model

The input–output relationship for a wide-sense-stationary uncorrelated-scattering (WSSUS) frequency-selective fading channel can be expressed as

$$y_k(t)=s_k(t)+\mathrm{Re}\{u_k(t-\nu_kT)\exp[j2\pi f_c(t-\nu_kT)]\} \quad (38)$$

where

$$u_k(t)=\gamma_k\int_{-\infty}^{\infty}h_k(t,\zeta)x_k(t-\zeta)\,d\zeta. \quad (39)$$

The fading process $h_k(t,\zeta)$ (which can be thought of as the time-varying impulse response of a low-pass filter) is a zero-mean complex Gaussian random process which has autocovariance

$$E\{h_k(t,\zeta)h_k^*(s,\xi)\}=g_k(\zeta)\delta(\zeta-\xi) \quad (40a)$$

and satisfies the normalization constraint

$$\int_{-\infty}^{\infty}g_k(\zeta)\,d\zeta=1 \quad (40b)$$

and the necessary condition for stationary bandpass processes (see [2], and the discussion in [4, sect. 2.2])

$$E\{h_k(t,\zeta)h_k(s,\xi)\}=0. \quad (40c)$$

The first term of (38) is again termed the *specular* component, while the second term in (38) is now called the *diffuse* faded component. If γ_k is a positive real number, the channel is a Rician fading channel. If $\gamma_k \rightarrow \infty$ the faded component becomes dominant and the channel is a Rayleigh fading channel. This model is basically described in [2] and [3]; in particular the Rician fading channel is described in [4] and [5]. Here we consider a slightly more general model, since for the Rician case we allow [see (38)] an average delay $\nu_k T$ (ν_k is a nonnegative integer) between the nonfaded specular component and the diffused faded component.

The WSSUS fading model can be thought of as arising when the transmitted signal encounters a slowly moving scattering random medium which can be modeled as a layered scatterer consisting of a large number of layers of infinitesimal thickness. It is therefore a frequency-selective channel, i.e., it is dispersive only in frequency. The high-frequency (HF) ionospheric and the microwave tropospheric scatter channels fit the above description; [4, sect. 2.1] provides more examples of such channels.

We impose a limitation on the selectivity of the channel in order to facilitate subsequent analysis. In particular, as in [5] and [3] it is assumed that

$$g_k(\zeta)=0 \quad \text{for } |\zeta|>T \tag{41}$$

which is a constraint on the frequency selectivity of the channel that allows us to restrict attention to the intersymbol interference from the two adjacent data bits.

With regard to the average delay $\nu_k T$ between the nonfaded specular component and the faded scatter component of $y_k(t)$ in (38), we will consider the two extreme cases i) $\nu_k = 0$ (no delay) and ii) $\nu_k \geq 3$ (large delay). The choice of the integer 3 as the lower bound for ν_k in ii) is justified by the fact that, in view of (41), it uncorrelates the contributions of the nonfaded and faded components at the output of the receiver.

B. Evaluation of Error Probability Via the Characteristic-Function Method

The output of the matched filter Z_i is again given by (7) where η_i and I_i are given by (8) and (9), respectively. The random variable η in (8) is again zero-mean Gaussian with variance $N_0/2E_b$; the nonfaded component of the interference (due to the kth signal) $I_{k,i}$ is given by (13b), but the faded components F_i and $F_{k,i}$ of (8) and (9) are now found [combine (38) and (39) with (3) and (6)] to be

$$F_i=\text{Re}\{\tilde{F}_i\} \tag{42a}$$

$$\tilde{F}_i=T^{-1}\int_{-\infty}^{\infty} h_i(t-\nu_i T, \zeta)\Gamma_i(t, \zeta+\nu_i T)\exp(j\Phi_i)\, d\zeta dt \tag{42b}$$

$$F_{k,i}=\text{Re}\{\tilde{F}_{k,i}\} \tag{42c}$$

$$\tilde{F}_{k,i}=T^{-1}\int_0^T\int_{-\infty}^{\infty} h_k(t-\tau_k-\nu_k T, \zeta)\Gamma_{k,i} \cdot (t, \zeta+\tau_k+\nu_k T)\exp(j\Phi_k)\, d\zeta\, dt \tag{42d}$$

where

$$\Gamma_{k,i}(t, \zeta)=a_k(t-\zeta)a_i(t)b_k(t-\zeta)b_i(t)\Psi(t-\zeta)\Psi(t)$$

and

$$\Phi_k=\theta_k-2\pi f_c(\tau_k+\nu_k T).$$

The function $\Gamma_{i,i}$ is denoted by Γ_i. Since h_k is a complex Gaussian random field, the random variables F_i and $F_{k,i}$ are conditionally Gaussian random variables (conditioned on time delays, phase angles, and data streams). Recall that $\nu_k T$ ($1 \leq k \leq K$) accounts for the average delay between the nonfaded and the diffused components of the channel output (see Section III-A).

Next, we proceed to the evaluation of the average error probability $\bar{P}_{e,i}$ via the characteristic-function method. As in the nonselective fading case, we need to evaluate $\bar{P}_{e,i}$, ϕ_{η_i}, and ϕ_i for use in (11). We follow the same steps as before. The random variable F_i is zero-mean conditionally Gaussian with conditional variance

$$\text{Var}\{F_i|b_i'\}=1/2E\{\tilde{F}_i\tilde{F}_i^*\} = T^{-2}\int_0^T g_i(\zeta)[R_i^2(\zeta) + \hat{R}_i^2(\zeta)+(b_1^{(i)\prime}+b_2^{(i)\prime})R_i(\zeta)\hat{R}_i(\zeta)]\, d\zeta \tag{43}$$

where $b_i' = (b_1^{(i)}, b_2^{(i)})$ denotes the pair of the $(\nu_i - 1)$th and ν_ith consecutive data bits of the ith data stream. To derive (43) we used (42a), (42b), (40a)–(40c), and (41). The Gaussian random variable η_i has then conditional variance

$$\text{Var}\{\eta_i|b_i'\}=(2E_b/N_0)^{-1}+\gamma_i^2\ \text{Var}\{F_i|b_i'\}. \tag{44}$$

Then, if we define

$$[\sigma_i^{(1)}]^2=(2E_b/N_0)^{-1}+\gamma_i^2\int_0^T g_i(\zeta)[R_i^2(\zeta)+\hat{R}_i^2(\zeta)]\, d\zeta \tag{45a}$$

$$[\sigma_i^{(2)}]^2=(2E_b/N_0)^{-1}+\gamma_i^2 T^{-2}\int_0^T g_i(\zeta)[R_i(\zeta)+\hat{R}_i(\zeta)]^2\, d\zeta \tag{45b}$$

$$[\sigma_i^{(3)}]^2=(2E_b/N_0)^{-1}+\gamma_i^2 T^{-2}\int_0^T g_i(\zeta)[R_i(\zeta)-\hat{R}_i(\zeta)]^2\, d\zeta \tag{45c}$$

we can write

$$P_{e,i}^{\eta}=\sum_{j=1}^{3} c_j Q([\sigma_i^{(j)}]^{-1}) \tag{46}$$

and

$$\phi_{\eta_i}=\sum_{j=1}^{3} c_j \exp\{-1/2[u\sigma_i^{(j)}]^2\} \tag{47}$$

where $c_1 = 1/2$ and $c_2 = c_3 = 1/4$. Note that these two results are independent of the events $\nu_i = 0$ [no delay: case i) of Section III-A] or $\nu_i \geq 3$ [large delay: case ii) of Section III-A].

The next step is to evaluate ϕ_l through (12) with $I_{k,i}$ and $F_{k,i}$ given by (13b) and by (42c) and (42d), respectively. We first compute the conditional variance of $F_{k,i}$; this turns out to be

$$\text{Var}\{F_{k,i}|b_k', \tau\} =1/2T^{-2}\left\{\int_0^\tau g_k(\zeta-\tau+T)[b_1^{(k)\prime}R_{k,i}(\zeta)+b_2^{(k)\prime}\hat{R}_{k,i}(\zeta)]^2\, d\zeta + \int_0^T g_k(\zeta-\tau)[b_2^{(k)\prime}R_{k,i}(\zeta)+b_3^{(k)\prime}\hat{R}_{k,i}(\zeta)]^2\, d\zeta + \int_\tau^T g_k(\zeta-\tau-T)[b_3^{(k)\prime}R_{k,i}(\zeta)+b_4^{(k)\prime}\hat{R}_{k,i}(\zeta)]^2\, d\zeta\right\}. \tag{48}$$

In (48) $b_k' = (b_1^{(k)}, b_2^{(k)}, b_3^{(k)}, b_4^{(k)})$ $(k \neq i)$ is the quadruple which consists of the $(\mu_k - 2)$th, $(\mu_k - 1)$th, μ_kth, and $(\mu_k + 1)$th consecutive bits of the kth data stream, $\mu_k \triangleq \lambda_k + \nu_k$, and $\tau = \tau_k (\text{mod } T)$ where $lT_c \leq \tau < (l + 1)T_c$. The conditional variance takes on about eight different values which we denote by $[\sigma_{k,i}^{(j)}(l; \tau)]^2$ and $[\hat{\sigma}_{k,i}^{(j)}(l; \tau)]^2$ for $j = 1, 2, 3, 4$. Then the characteristic function ϕ_i is derived as

$$\phi_i(u) = \prod_{k \neq i} \Bigg\{ (8N)^{-1} \sum_{l=0}^{N-1} T_c^{-1}$$

$$\cdot \int_0^{T_c} \Bigg[\bar{f}(u; l, \theta_{k,i}; \tau) \sum_{j=1}^{4} \exp\{-1/2[u\gamma_k \sigma_{k,i}^{(j)}(l; \tau)]^2\}$$

$$+ \bar{f}(u; l, \hat{\theta}_{k,i}; \tau) \sum_{j=1}^{4} \exp\{-1/2[u\gamma_k \hat{\sigma}_{k,i}^{(j)}(l; \tau)]^2\} \Bigg] d\tau \Bigg\} \tag{49a}$$

for i) no delay ($\nu_k = 0$, all k), and

$$\phi_i(u) = \prod_{k \neq i} \Bigg\{ (8N)^{-1} \sum_{l=0}^{N-1} T_c^{-1}$$

$$\cdot \int_0^{T_c} [\bar{f}(u; l, \theta_{k,i}; \tau) + \bar{f}(u; l, \hat{\theta}_{k,i}; \tau)]$$

$$\cdot \sum_{j=1}^{4} [\exp\{-1/2[u\gamma_k \sigma_{k,i}^{(j)}(l; \tau)]^2\}$$

$$+ \exp\{-1/2[u\gamma_k \hat{\sigma}_{k,i}^{(j)}(l; \tau)]^2\}] \, d\tau \Bigg\} \tag{49b}$$

for ii) large delay ($\nu_k \geq 3$, all k). In case i) $I_{k,i}$ [given by (13b)] and $F_{k,i}$ above are correlated [$b_k' = (b_1^{(k)}, b_k^{(k)}, b_4^{(k)})$]. In case ii) $I_{k,i}$ and $F_{k,i}$ are uncorrelated (b_k and b_k' have no components in common). These facts are reflected in the form of ϕ_i in (49a) and (49b) (the function $\bar{f}$ and the sum with the four terms are the characteristic functions of $I_{k,i}$ and $F_{k,i}$ given τ, respectively). The function $\bar{f}$ is as defined in (21).

To see that the computational effort grows linearly with KN^2, notice that in (49a) and (49b), besides the product of $K - 1$ terms and the summation of N terms, the integrals split into summations with a total of N terms as well [e.g., $\int_0^\tau = \Sigma_{n=0}^{l-1} \int_{nT_c}^{(n+1)T_c} + \int_{lT_c}^{\tau}$, for $lT_c \leq \tau < (l + 1)T_c$].

C. Approximation Based on the Average Signal-to-Noise Ratio

The average signal-to-noise ratio in this case is given by [cf. (34)]

$$\text{SNR}_i = \{[\sigma_i^{(1)}]^2 + \sum_{k \neq i} (1 + \gamma_k^2)[\mu_{k,i}(0)m_\psi + \mu_{k,i}(1)m_\psi']\}^{-1/2} \tag{50}$$

where $\sigma_i^{(1)}$ is given by (45a) above. This is a slight generalization of the results of [5] for arbitrary chip waveforms. Notice that SNR_i and the resulting approximation (35) cannot distinguish between cases i) and ii) (no delay versus large delay) of the fading model of Section III-A.

IV. Numerical Results and Conclusions

All the signature sequences employed in this section are AO/LSE m-sequences of lengths $N = 31$ and 127. In particular, for a DS/SSMA system with K users the first K sequences of [15, Figs. A.1.(a) ($N = 31$) and A.1.(b) ($N = 127$)] are used.

TABLE I
PROBABILITY OF ERROR FOR DS/SSMA SYSTEMS WITH RICIAN NONSELECTIVE FADING ($N = 31$, $\gamma^2 = 0.1$)

	K=2				K=3				K=4			
E_b/N_0	$\bar{P}_{e,1}^G$	$\bar{P}_{e,1}^{(i)}$	$\bar{P}_{e,1}^{(ii)}$		$\bar{P}_{e,1}^G$	$\bar{P}_{e,1}^{(i)}$	$\bar{P}_{e,1}^{(ii)}$		$\bar{P}_{e,1}^G$	$\bar{P}_{e,1}^{(i)}$	$\bar{P}_{e,1}^{(ii)}$	
6	1.04	1.04	1.04	$(\times 10^{-2})$	1.25	1.26	1.26	$(\times 10^{-2})$	1.53	1.54	1.54	$(\times 10^{-2})$
8	3.83	3.86	3.86	$(\times 10^{-3})$	5.27	5.37	5.35	$(\times 10^{-3})$	0.73	0.75	0.74	$(\times 10^{-2})$
10	1.36	1.40	1.39	$(\times 10^{-3})$	2.23	2.34	2.31	$(\times 10^{-3})$	3.58	3.80	3.76	$(\times 10^{-3})$
12	5.18	5.43	5.37	$(\times 10^{-4})$	1.02	1.12	1.10	$(\times 10^{-3})$	1.93	2.16	2.11	$(\times 10^{-3})$
14	2.26	2.45	2.40	$(\times 10^{-4})$	5.37	6.25	6.05	$(\times 10^{-4})$	1.18	1.39	1.35	$(\times 10^{-3})$
16	1.17	1.32	1.28	$(\times 10^{-4})$	3.29	4.04	3.86	$(\times 10^{-4})$	0.82	1.02	0.97	$(\times 10^{-3})$

For $\bar{P}_{e,i}$, the approximation to the error probability which is based on the characteristic-function method, we use Simson's integration technique with the same parameters as in [7, Sect. III] ($L = 20$, $\epsilon < 10^{-14}$, $n_\theta = 10$, $n_\tau = 10$, and $n = 20$). We set $i = 1$ and consider $\bar{P}_{e,1}$ for the error probability of the receiver matched to the first signal.

First, we present numerical results for DS/SSMA systems with *nonselective* Rician and Rayleigh fading channels. In Table I the Gaussian approximation to the error probability $\bar{P}_{e,1}^G$ and the approximations $\bar{P}_{e,1}^{(i)}$ and $\bar{P}_{e,1}^{(ii)}$ obtained by the characteristic-function method are compared for a DS/SSMA system with sequences of length 31 and $K = 2$, 3, and 4 simultaneous users. The communications link between the kth transmitter and the receiver matched to the first signal is a Rician nonselective fading channel with relative power of the faded component $\gamma_k^2 = \gamma^2 = 0.1$, $1 \leq k \leq K$. The error probabilities $\bar{P}_{e,1}^{(i)}$ and $\bar{P}_{e,1}^{(ii)}$ correspond to the cases of i) slow fading and ii) fast fading discussed in Section II-A. We observe that the Gaussian approximation is somewhat optimistic when compared with the more accurate approximations based on the characteristic function, but still in good agreement with them. Notice that slow fading causes a slightly worse performance of the DS/SSMA system than fast fading. Recall that the Gaussian approximation results in the same value of the error probability for both slow and fast fading.

In Fig. 1 we plotted both $\bar{P}_{e,1}^{(i)}$ (for slow fading) and $\bar{P}_{e,1}^G$ versus E_b/N_0 for a DS/SSMA system with $K = 3$, $N = 31$, and Rician nonselective fading. The relative power of all faded components are $\gamma^2 = 0.01$, 0.05, 0.1, 0.2, and 0.4. Notice that as γ^2 increases the system degrades gracefully. For nonselective fading channels, DS/SS modulation enables the matched filter receiver to discriminate effectively against the nonfaded and faded components of the interfering signals but not against the faded component of the desired signal. This explains why the error probability is already larger than 10^{-3} for $\gamma^2 = 0.1$, 0.2, and 0.4. Finally, notice that the accuracy of the Gaussian approximation improves as γ^2 increases.

In Table II we compare the performance of DS/SSMA systems with rectangular and sine chip waveforms. The sine chip waveform used is $\psi(t) = \sqrt{2} \sin(\pi t/T_c)$, $0 \leq t \leq T_c$. The system parameters are $K = 3$ and $N = 31$. The Rician nonselective slow fading channel's parameter takes on the values $\gamma^2 = 0.01$ and 0.1. For moderately heavy fading ($\gamma^2 = 0.1$) the sine chip waveform results in a slightly better performance than the rectangular chip waveform, as both the Gaussian approximation $\bar{P}_{e,1}^G$ and the more accurate approximation $\bar{P}_{e,1}^{(i)}$ indicate. For light fading ($\gamma^2 = 0.01$) the Gaussian approximation still shows better performance for the sine chip waveform for all values of E_b/N_0; however, the more accurate approximation indicates that the sine chip waveform outper-

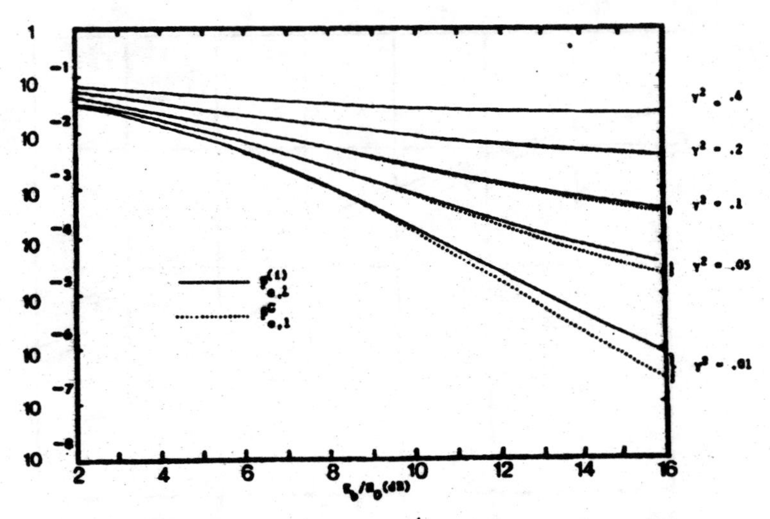

Fig. 1. Probability of error for a DS/SSMA system with Rician nonselective fading ($K = 3$, $N = 31$, and $\gamma^2 = 0.01, 0.05, 0.1, 0.2$, and 0.4).

TABLE II
PROBABILITY OF ERROR FOR DS/SSMA SYSTEMS WITH DIFFERENT CHIP WAVEFORMS AND RICIAN NONSELECTIVE FADING ($K=3$, $N=31$)

	$\bar{\gamma}^2=.01$					$\bar{\gamma}^2=.1$				
	rect		sine			rect		sine		
E_b/N_0(dB)	$\bar{P}^G_{e,1}$	$\bar{P}^{(i)}_{e,1}$	$\bar{P}^G_{e,1}$	$\bar{P}^{(i)}_{e,1}$		$\bar{P}^G_{e,1}$	$\bar{P}^{(i)}_{e,1}$	$\bar{P}^G_{e,1}$	$\bar{P}^{(i)}_{e,1}$	
6	5.19	5.25	4.92	4.99	$(\times 10^{-3})$	1.25	1.26	1.21	1.22	$(\times 10^{-2})$
8	1.06	1.11	0.95	1.02	$(\times 10^{-3})$	5.27	5.37	4.97	5.09	$(\times 10^{-3})$
10	1.52	1.78	1.24	1.61	$(\times 10^{-4})$	2.23	2.34	2.04	2.17	$(\times 10^{-3})$
12	1.70	2.56	1.20	2.41	$(\times 10^{-5})$	1.02	1.12	0.91	1.03	$(\times 10^{-3})$
14	1.80	4.07	1.05	4.28	$(\times 10^{-6})$	5.37	6.25	4.63	5.67	$(\times 10^{-4})$
16	1.23	0.83	1.03	1.00	$(\times 10^{-6})$	3.29	4.04	2.76	3.66	$(\times 10^{-4})$

TABLE III
PROBABILITY OF ERROR FOR A DS/SSMA SYSTEM WITH RAYLEIGH NONSELECTIVE FADING ($K = 3$, $N = 31$)

	$\bar{\gamma}^2=.01$				$\bar{\gamma}^2=.1$				$\bar{\gamma}^2=.2$			
$\bar{E}_b/N_0$	$\bar{P}^G_{e,1}$	$\bar{P}^{(i)}_{e,1}$	$\bar{P}^{(ii)}_{e,1}$		$\bar{P}^G_{e,1}$	$\bar{P}^{(i)}_{e,1}$	$\bar{P}^{(ii)}_{e,1}$		$\bar{P}^G_{e,1}$	$\bar{P}^{(i)}_{e,1}$	$\bar{P}^{(ii)}_{e,1}$	
10	6.42	6.43	6.43	$(\times 10^{-2})$	6.47	6.51	6.51	$(\times 10^{-2})$	6.53	6.60	6.60	$(\times 10^{-2})$
20	2.35	2.34	2.34	$(\times 10^{-2})$	2.55	2.43	2.43	$(\times 10^{-2})$	2.74	2.53	2.53	$(\times 10^{-2})$
30	0.85	0.78	0.78	$(\times 10^{-2})$	1.34	0.87	0.87	$(\times 10^{-2})$	1.72	0.98	0.98	$(\times 10^{-2})$
40	4.37	2.59	2.59	$(\times 10^{-3})$	1.14	0.35	0.35	$(\times 10^{-2})$	1.56	0.46	0.46	$(\times 10^{-2})$
50	3.70	0.89	0.89	$(\times 10^{-3})$	11.2	1.83	1.83	$(\times 10^{-3})$	15.6	2.87	2.87	$(\times 10^{-3})$
•	36.2	1.07	1.07	$(\times 10^{-4})$	11.2	1.05	1.05	$(\times 10^{-3})$	15.6	2.08	2.09	$(\times 10^{-3})$

forms the rectangular chip waveform only for low signal-to-noise ratios; the situation is reversed for higher signal-to-noise ratios.

In Table III we present results for a DS/SSMA system with $K = 3$, $N = 31$, and Rayleigh nonselective fading channels with relative power of faded components taking values $\bar{\gamma}^2 = 0.01, 0.1$, and 0.2. The system performance has now degraded considerably and higher signal-to-noise ratios were considered. Notice that the Gaussian approximation is conservative and, for large values of $\bar{\gamma}^2$, is off the actual value by one order of magnitude. On the other hand, the values of the accurate approximation for i) slow fading and ii) fast fading do not differ in the first four significant digits.

TABLE IV
PROBABILITY OF ERROR FOR A DS/SSMA SYSTEM WITH FREQUENCY-SELECTIVE FADING ($K = 3$, $\gamma^2 = 0.4$)

	N=31				N=127			
E_b/N_0(dB)	$\bar{P}^G_{e,1}$	$\bar{P}^{(i)}_{e,1}$	$\bar{P}^{(ii)}_{e,1}$		$\bar{P}^G_{e,1}$	$\bar{P}^{(i)}_{e,1}$	$\bar{P}^{(ii)}_{e,1}$	
6	6.34	6.40	6.40	$(\times 10^{-3})$	3.20	3.20	3.20	$(\times 10^{-3})$
8	1.57	1.62	1.62	$(\times 10^{-3})$	3.78	3.82	3.82	$(\times 10^{-4})$
10	3.10	3.42	3.41	$(\times 10^{-4})$	1.83	1.97	1.97	$(\times 10^{-5})$
12	5.54	6.97	6.95	$(\times 10^{-5})$	3.31	3.89	3.89	$(\times 10^{-7})$
14	1.05	1.61	1.59	$(\times 10^{-5})$	1.91	3.22	3.20	$(\times 10^{-9})$
16	2.47	4.75	4.73	$(\times 10^{-6})$	0.44	1.02	1.01	$(\times 10^{-10})$

Next, we consider DS/SSMA systems operating through Rician *frequency-selective* channels. In Table IV the Gaussian approximation is compared to the approximation obtained via the characteristic-function method for the cases i) zero average delay between nonfaded and faded components and ii) large delay (see Section III-A) and DS/SSMA systems with $K = 3$ and $N = 31$ and 127. The relative power of the diffuse component is $\gamma^2 = 0.4$ for all signals. Notice that the system performance for case ii) gives slightly lower error probability than case i) due to the more extensive randomization involved. The Gaussian approximation results in an optimistic estimate of the error probability for all cases, but it is satisfactorily close to the more accurate results. Recall that the Gaussian approximation takes the same value for the aforementioned cases i) and ii).

In Fig. 2, $\bar{P}^G_{e,1}$ and $\bar{P}^{(i)}_{e,1}$ are plotted versus E_b/N_0 for DS/SSMA systems with $N = 31$ and $K = 2$, 3, and 4. The autocovariance function of the frequency-selective WSSUS channel is triangular: $g_k(\tau) = (1/T)(1 - (|t|/T))$ for $|t| \leq T$ and 0 otherwise. The relative power of all the faded components is $\gamma^2 = 0.4$. Notice that for an error probability of 10^{-5} the Gaussian approximation is off by almost 1 dB for K

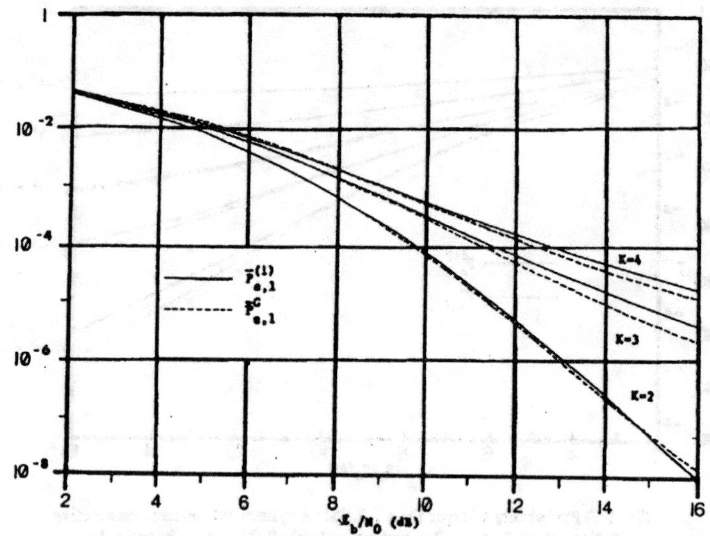

Fig. 2. Probability of error for a DS/SSMA system with frequency-selective fading (N = 31, γ^2 = 0.4, and K = 2, 3, and 4).

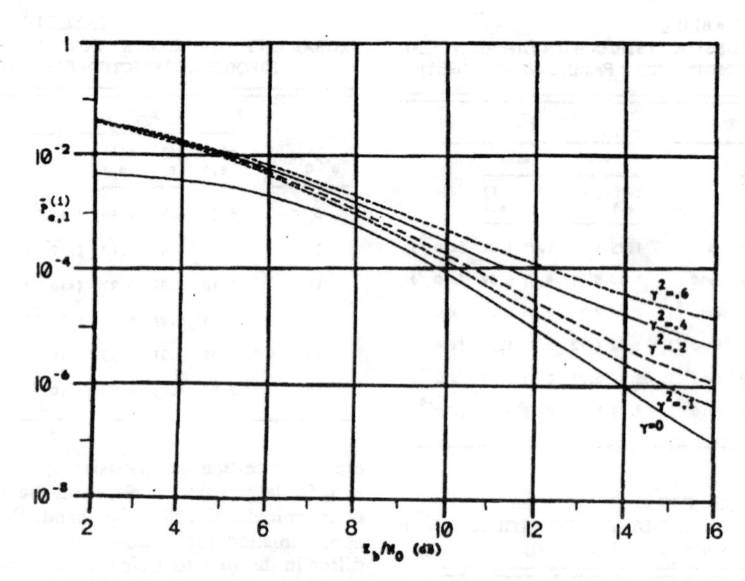

Fig. 3. Probability of error for a DS/SSMA system with frequency-selective fading (K = 3, N = 31, and γ^2 = 0, 0.1, 0.2, 0.4, and 0.6).

= 3. For fixed E_b/N_0 the Gaussian approximation is satisfactory for all cases, as also indicated by the results of Table IV.

Finally, in Fig. 3 we have plotted $\bar{P}^{(1)}_{e,1}$ versus E_b/N_0 for a DS/SSMA system with K = 3 and N = 31 and relative power in the diffuse component taking values γ^2 = 0, 0.1, 0.2, 0.3, and 0.4. Notice the graceful degradation of the system performance as γ^2 increases. The value γ^2 = 0 corresponds to DS/SSMA with an AWGN channel. A comparison of Figs. 1 and 3 verifies that DS/SS modulation is much more efficient against frequency-selective fading than against nonselective fading.

REFERENCES

[1] G. L. Turin, "Communication through noisy, random-multipath channels," in *IRE Nat. Conv. Rec.*, 1956, part 4, pp. 154–166.

[2] P. A. Bello, "Characterization of randomly time-invariant linear channels," *IEEE Trans. Commun. Syst.*, vol. CS-11, pp. 360–393, Dec. 1963.

[3] P. A. Bello and B. D. Nellin, "The influence of fading spectrum on the binary error probabilities of incoherent and differentially coherent matched filter receivers," *IRE Trans. Commun. Syst.*, vol. CS-11, pp. 160–168, June 1962.

[4] D. E. Borth, "Performance analysis of direct-sequence spread-spectrum multiple-access communication via fading channels," Ph.D. dissertation, Dep. Elec. Eng., Univ. Illinois, Urbana (also Coordinated Sci. Lab. Rep. R-880), Apr. 1980.

[5] D. E. Borth and M. B. Pursley, "Analysis of direct-sequence spread-spectrum multiple-access communication over Rician fading channels," *IEEE Trans. Commun.*, vol. COM-27, pp. 1566–1577, Oct. 1979.

[6] M. B. Pursley, "Spread-spectrum multiple-access communications," in *Multi-User Communication Systems*, G. Longo, Ed. New York: Springer-Verlag, 1981, pp. 139–199.

[7] E. A. Geraniotis and M. B. Pursley, "Error probability for direct-sequence spread-spectrum multiple-access communications—Part II: Approximations," Special Issue on Spread-Spectrum Communications, *IEEE Trans. Commun.*, vol. COM-30, pp. 985–995, May 1982.

[8] ——, "Error probability for binary PSK spread-spectrum multiple-access communications," in *Proc. Conf. Inform. Sci. Syst.*, Johns

Hopkins Univ., Baltimore, MD, Mar. 1981, pp. 238–244.

[9] ——, "Performance of coherent direct-sequence spread-spectrum communications over specular multipath fading channels," *IEEE Trans. Commun.*, vol. COM-33, pp. 502–508, June 1985.

[10] K.-T. Wu, "Average error probability for DS-SSMA communications: The Gram–Charlier expansion approach," in *Proc. 19th Annu. Allerton Conf. Commun., Contr., Comput.*, Sept. 1981, pp. 237–246; see also, "Direct sequence spread spectrum communications: Applications to multiple-access and jamming resistance," Ph.D. dissertation, Univ. Michigan, Ann Arbor, 1981.

[11] D. Laforgia *et al.*, "Bit error rate evaluation for multiple-access spread-spectrum systems," *IEEE Trans. Commun.*, vol. COM-32, pp. 660–669, June 1984.

[12] K. Yao, "Error probability of asynchronous spread spectrum multiple access communication systems," *IEEE Trans. Commun.*, vol. COM-25, pp. 803–809, Aug. 1977.

[13] M. B. Pursley, D. V. Sarwate, and W. E. Stark, "Error probability for direct-sequence spread-spectrum multiple-access communications—Part I: Upper and lower bounds," *IEEE Trans. Commun.*, vol. COM-30, pp. 975–984, May 1982.

[14] R. C. Hanlon and C. S. Gardner, "Error performance of direct sequence spread spectrum systems on non-selective fading channels," *IEEE Trans. Commun.*, vol. COM-27, pp. 1696–1700, Nov. 1979.

[15] C. S. Gardner and J. A. Orr, "Fading effects on the performance of a spread-spectrum multiple-access communication system," *IEEE Trans. Commun.*, vol. COM-27, pp. 143–149, Jan. 1979.

[16] M. B. Pursley and H. F. A. Roefs, "Numerical evaluation of correlation parameters for optimal phases of binary shift-register sequences," *IEEE Trans. Commun.*, vol. COM-27, pp. 1597–1604, Oct. 1979.

[17] G. L. Turin, "Introduction to spread-spectrum anti-multipath techniques and their application to urban digital radio," *Proc. IEEE*, vol. 68, pp. 328–353, Mar. 1980.

[18] J. S. Lehnert and M. B. Pursley, "Multipath diversity reception of coherent direct-sequence spread-spectrum communications," in *Proc. Conf. Inform. Sci. Syst.*, Johns Hopkins Univ., Baltimore, MD, Mar. 1983, pp. 770–775.

Evaggelos Geraniotis (S'76–M'82), for a photograph and biography, see p. 226 of the March 1986 issue of this TRANSACTIONS.

Error Probabilities for Binary Direct-Sequence Spread-Spectrum Communications with Random Signature Sequences

JAMES S. LEHNERT, MEMBER, IEEE, AND MICHAEL B. PURSLEY, FELLOW, IEEE

Abstract—**Binary direct-sequence spread-spectrum multiple-access communications, an additive white Gaussian noise channel, and a coherent correlation receiver are considered. An expression for the output of the receiver is obtained for the case of random signature sequences, and the corresponding characteristic function is determined. The expression is used to study the density function of the multiple-access interference and to determine arbitrarily tight upper and lower bounds on the average probability of error. The bounds, which are obtained without making a Gaussian approximation, are compared to results obtained using a Gaussian approximation. The effects of transmitter power, the length of the signature sequences, and the number of interfering transmitters are illustrated. Each transmitter is assumed to have the same power, although the general approach can accommodate the case of transmitters with unequal powers.**

I. INTRODUCTION

IN this paper we are concerned with binary direct-sequence spread-spectrum multiple-access (DS/SSMA) communications, an additive white Gaussian noise (AWGN) channel, and a coherent correlation receiver. We assume that all the signature sequences are randomly generated. Each transmitter is assumed to have the same power, although the general approach can accommodate the case of transmitters with unequal powers. An expression is given for the output of the correlation receiver in terms of a set of mutually independent random variables. An expression is also given for the density function for each of the random variables in the set. These expressions are then used to obtain arbitrarily tight upper and lower bounds on the average probability of error without making a Gaussian approximation. Previous results [1] deal only with moments of correlation parameters of random sequences. A complete approach for obtaining arbitrarily tight bounds on the error probability of DS/SSMA systems with random sequences is detailed in the present paper.

The mathematical models for the signals and the system are given in Section II. Expressions are developed for the multiple-access interference in Section III. Random variables characterizing various aspects of the multiple-access interference are expressed in terms of the random signature sequences, the data bits, the correlation functions of the chip waveform, and the propagation delays. The importance of conditioning on an appropriately selected set of random variables is discussed. In particular, conditioning on the correlation parameter $C_1(1)$, where C_1 denotes the aperiodic autocorrelation function for the first signature sequence, is critical in the analysis because it provides conditional independence for a key set of random variables. The results of Section III are applied to the determination of the characteristic function of the multiple-access interference in Section IV. In particular, an expression for the conditional characteristic function of the multiple-access interference given $C_1(1)$ is obtained. The main application of the results of Section III is developed in Section V. In this section, the upper and lower bounds on the average probability of error are developed. The multiple-access interference is modeled using two vectors that are employed to obtain upper and lower bounds for a system with two transmitters only. These results are then used to obtain the general results for an arbitrary number of transmitters. Numerical examples are given for illustration.

Paper approved by the Editor for Spread Spectrum of the IEEE Communications Society. Manuscript received August 6, 1985; revised August 4, 1986. This work was supported by the Army Research Office under Contract DAAG 29-84-K-0088 and by the NSF ERC for Intelligent Manufacturing Systems.

J. S. Lehnert is with the School of Electrical Engineering, Purdue University, West Lafayette, IN 47907.

M. B. Pursley is with the Coordinated Science Laboratory, University of Illinois, Urbana, IL 61801.

IEEE Log Number 8611691.

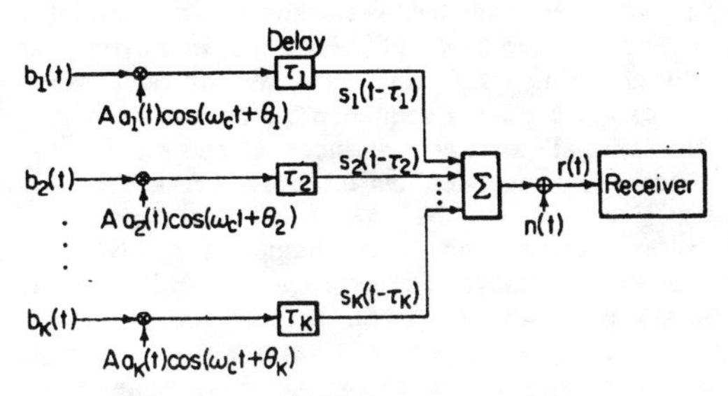

Fig. 1. DS/SSMA system model.

II. SYSTEM MODEL

The model under consideration is shown in Fig. 1. It is similar to the models defined in [2] and [3]. The received signal in this asynchronous binary DS/SSMA system is the sum of K spread-spectrum signals $s_k(t - \tau_k)$, $1 \le k \le K$, plus an AWGN process $n(t)$, which has two-sided spectral density $N_o/2$. The spread-spectrum signal $s_k(t - \tau_k)$ is given by

$$s_k(t-\tau_k)=\sqrt{2P}\, b_k(t-\tau_k)a_k(t-\tau_k) \cos (\omega_c t+\phi_k) \quad (1)$$

where $b_k(t)$ is the data signal, $a_k(t)$ is the spectral-spreading signal, τ_k is the time delay parameter that accounts for propagation delay and the lack of synchronism between the transmitters, ϕ_k is the phase angle of the kth carrier, and P is the power of the transmitted signal. Although equal power levels have been assumed for all transmitters, the results are easily modified to consider unequal power levels. Notice in Fig. 1 that $A = \sqrt{2P}$ and $\theta_k = \phi_k + \omega_c\tau_k$.

If a rectangular pulse is defined by $p_T(t) = 1$ for $0 \le t < T$ and $p_T(t) = 0$, otherwise, the kth data signal can be expressed as $b_k(t) = \Sigma_{j=-\infty}^{\infty} b_j^{(k)} p_T(t - jT)$, where the sequence $(b_j^{(k)})$ is the binary data sequence of the kth transmitter ($b_j^{(k)} \in \{-1, 1\}$ for each j). If the chip waveform $\psi(t)$ is defined to be time limited to the interval $[0, T_c)$ and normalized to have energy T_c, the spectral-spreading signal of the kth transmitter can be expressed as $a_k(t) = \sum_{j=-\infty}^{\infty} a_j^{(k)}\psi(t - jT_c)$, where $(a_j^{(k)})$ is a periodic binary sequence of elements from the set $\{-1, 1\}$. We assume that each bit is encoded with N chips, i.e., $T = NT_c$, and that the signature sequence $(a_j^{(k)})$ has period N. If the chip waveform $\psi(t)$ is the rectangular pulse function $p_{T_c}(t)$, the DS/SSMA system has the binary phase-shift-keyed signaling format.

The signature sequences are deterministic. Each transmitter–receiver pair has been designed to encode and decode data using a particular sequence. We assume in our model that the sequences have been randomly generated, however. Instead of carefully choosing a signature sequence for each transmitter–receiver pair, a signature sequence has been chosen at random from the set of all possible sequences. Each sequence in the set of all possible sequences has an equal probability of being chosen. Furthermore, the choice of a sequence for each transmitter–receiver pair is made independently of the choices made for all other pairs. The average probability of error that we obtain is an average with respect to all the possible combinations of signature sequences that might be selected.

It is useful to list some of the random variables that model the communication system and to describe their distributions. As in [2] and [3], there are K transmitter–receiver pairs indexed by integers k, $1 \le k \le K$. The kth data sequence is modeled as a sequence of independent and identically distributed random variables $(b_j^{(k)})$ such that $\Pr\{b_j^{(k)} = +1\} = \Pr\{b_j^{(k)} = -1\} = 1/2$. The kth signature sequence is periodic with period N. One period of the sequence is a random vector $[a_0^{(k)}, a_1^{(k)}, \cdots, a_{N-1}^{(k)}]$ of length N. The components of this vector $a_j^{(k)}$, $0 \le j \le N-1$, form a set of independent and identically distributed random variables such that $\Pr\{a_j^{(k)} = +1\} = \Pr\{a_j^{(k)} = -1\} = 1/2$. Because of the symmetry of the model, we need consider only the receiver that is listening to the first transmitter. Also, since only relative delays and phases are important, we set $\tau_1 = \phi_1 = 0$. The properties of an SSMA system and the stationarity of the noise $n(t)$ permit us to restrict attention to time delays modulo T and phase angles modulo 2π. Hence, for $k \neq 1$, we model the delay τ_k as a random variable that is uniformly distributed on $[0, T)$ and the phase ϕ_k as a random variable that is uniformly distributed on $[0, 2\pi)$. Finally, we assume that all of the random variables mentioned in this paragraph are mutually independent.

III. System Analysis

The decision statistic for a coherent correlation receiver is given by [3], [4]

$$Z_{\text{out}}^{(1)} = \eta + T\sqrt{P/2}\left[b_0^{(1)} + \sum_{k=2}^{K} I_{k,1}(\mathbf{b}_k, \tau_k, \phi_k)\right] \tag{2}$$

where

$$I_{k,1}(\mathbf{b}_k, \tau, \phi) = T^{-1}[B_{k,1}(\mathbf{b}_k, \tau)]\cos\phi \tag{3}$$

and

$$B_{k,1}(\mathbf{b}_k, \tau) = b_{-1}^{(k)}R_{k,1}(\tau) + b_0^{(k)}\hat{R}_{k,1}(\tau). \tag{4}$$

The random variable η is Gaussian with mean equal to zero and variance equal to $N_0T/4$, where $N_0/2$ is the two-sided density of white Gaussian noise. P denotes the power of each transmitter's signal, and T denotes the data bit duration. The vector $\mathbf{b}_k = (b_{-1}^{(k)}, b_0^{(k)})$ represents a pair of consecutive data bits of the kth signal. The functions $R_{k,m}(\tau)$ and $\hat{R}_{k,m}(\tau)$ are the continuous-time partial cross-correlation functions of the kth and the mth spectral-spreading waveforms defined in [2] and [3] by $R_{k,i}(\tau) = \int_0^\tau a_k(t-\tau)a_i(t)\,dt$ and $\hat{R}_{k,i}(\tau) = \int_\tau^T a_k(t-\tau)a_i(t)\,dt$ for $0 \le \tau \le T$.

The functions $R_{k,m}(\cdot)$ and $\hat{R}_{k,m}(\cdot)$ can be expressed in terms of the discrete aperiodic cross-correlation function $C_{k,m}(\cdot)$ and the continuous-time partial autocorrelation functions $R_\psi(\cdot)$ and $\hat{R}_\psi(\cdot)$ of the chip waveform. The discrete aperiodic cross-correlation function $C_{k,m}(\cdot)$ involves only the kth and mth signature sequences and is given by [2]

$$C_{k,m}(\lambda) = \begin{cases} \sum_{j=0}^{N-1-\lambda} a_j^{(k)}a_{j+\lambda}^{(m)}, & 0 \le \lambda \le N-1 \\ \sum_{j=0}^{N-1+\lambda} a_{j-\lambda}^{(k)}a_j^{(m)}, & 1-N \le \lambda < 0 \\ 0, & |\lambda| \ge N \end{cases} \tag{5}$$

where $a_i^{(k)}$ and $a_j^{(m)}$ are elements of the sequences $(a_i^{(k)})$ and $(a_j^{(m)})$. For convenience, we denote the discrete aperiodic autocorrelation function $C_{i,i}(\cdot)$ by $C_i(\cdot)$. The continuous-time partial autocorrelation functions of the chip waveform are defined as in [3] for $0 \le s \le T_c$ by $R_\psi(s) = \int_0^s \psi(t)\psi(t + T_c - s)\,dt$ and $\hat{R}_\psi(s) = \int_s^{T_c} \psi(t)\psi(t-s)\,dt$. For $s > T_c$ or $s < 0$, $R_\psi(s) = \hat{R}_\psi(s) = 0$. The functions $R_{k,m}(\cdot)$ and $\hat{R}_{k,m}(\cdot)$ depend on the kth and mth signature sequences only through the function $C_{k,m}(\cdot)$. The functions depend on the chip waveform only through the functions $R_\psi(\cdot)$ and $\hat{R}_\psi(\cdot)$. The dependence is governed [3] for $0 \le \tau \le T$ by the equations

$$R_{k,m}(\tau) = C_{k,m}(\gamma - N)\hat{R}_\psi(\tau - \gamma T_c) + C_{k,m}(\gamma + 1 - N)R_\psi(\tau - \gamma T_c) \tag{6}$$

and

$$\hat{R}_{k,m}(\tau) = C_{k,m}(\gamma)\hat{R}_\psi(\tau - \gamma T_c) + C_{k,m}(\gamma+1)R_\psi(\tau - \gamma T_c) \tag{7}$$

where $\gamma = \lfloor \tau/T_c \rfloor$.

We now return to the examination of (2). From (4), (6), and (7) we find

$$B_{k,1}(\mathbf{b}_k, \tau_k) = [b_{-1}^{(k)}C_{k,1}(\gamma_k - N) + b_0^{(k)}C_{k,1}(\gamma_k)]\hat{R}_\psi(S_k) + [b_{-1}^{(k)}C_{k,1}(\gamma_k + 1 - N) + b_0^{(k)}C_{k,1}(\gamma_k + 1)]R_\psi(S_k) \tag{8}$$

where $S_k = \tau_k - \gamma_k T_c$, and $\gamma_k = \lfloor \tau_k/T_c \rfloor$. Since τ_k is a random variable that is uniformly distributed on $[0, T)$, S_k is a random variable that is uniformly distributed on $[0, T_c)$, and γ_k is a random variable which is uniformly distributed on the set $\{0, \cdots, N-1\}$. At this point, in order to simplify notation we define $(y_i) = (a_i^{(1)})$, $(x_i) = (a_i^{(k)})$, $b_{-1} = b_{-1}^{(k)}$, $b_0 = b_0^{(k)}$, $\tau = \tau_k$, $\gamma = \gamma_k$, and $S = S_k$. When we need to consider data sequences, signature sequences, and delays from a number of users, we can restore the appropriate indexes. If we use (5) to expand (8), we obtain

$$B_{k,1}(\mathbf{b}_k, \tau) = \left[\sum_{j=0}^{\gamma-1} b_{-1}x_{j-\gamma+N}y_j + \sum_{j=\gamma}^{N-1} b_0x_{j-\gamma}y_j\right]\hat{R}_\psi(S) + \left[\sum_{j=0}^{\gamma} b_{-1}x_{j-\gamma-1+N}y_j + \sum_{j=\gamma+1}^{N-1} b_0x_{j-\gamma-1}y_j\right]R_\psi(S). \tag{9}$$

Equation (9) may be expanded further to obtain

$$B_{k,1}(b_k,\tau)=\left[\sum_{j=0}^{\gamma-1} b_{-1}x_{j-\gamma+N}y_j+\sum_{j=\gamma}^{N-2} b_0x_{j-\gamma}y_j + b_0x_{N-\gamma-1}y_{N-1}\right]\hat{R}_\psi(S) + \left[b_{-1}x_{N-\gamma-1}y_0+\sum_{j=0}^{\gamma-1} b_{-1}x_{j-\gamma+N}y_{j+1} + \sum_{j=\gamma}^{N-2} b_0x_{j-\gamma}y_{j+1}\right]R_\psi(S). \quad (10)$$

Finally, the terms of (10) can be rearranged to give

$$B_{k,1}(b_k,\tau)=b_{-1}\sum_{j=0}^{\gamma-1} x_{j-\gamma+N}(y_j\hat{R}_\psi(S)+y_{j+1}R_\psi(S)) + b_0\sum_{j=\gamma}^{N-2} x_{j-\gamma}(y_j\hat{R}_\psi(S)+y_{j+1}R_\psi(S)) + b_0x_{N-\gamma-1}y_{N-1}\hat{R}_\psi(S)+b_{-1}x_{N-\gamma-1}y_0R_\psi(S). \quad (11)$$

We are interested in an average performance over all of the possible combinations of K signature sequences of length N. Since there are 2^N possible sequences for each transmitter, there are 2^{KN} possible combinations to consider. This number can be too large to perform practical computations, even when the sequence length N is as small as 31. For this reason it is necessary to manipulate (11) to obtain an expression for the decision statistic that is useful for practical computations.

With the motivation of reducing complexity, we consider (11) conditioned on the signature sequence of the first receiver (y_j) and the random variable γ, which is uniformly distributed on the set $\{0,\cdots,N-1\}$. It is very important to condition the decision statistic on the first signature sequence (or, as we will later demonstrate, just the one parameter $C_1(1)$ of that sequence) before proceeding with the analysis. Without this conditioning, the random variables that model the multiple-access interference from the multiple transmitters are not a set of independent random variables, and our expression for the decision statistic loses its utility. We condition on $\gamma=\hat{\gamma}$ and $(y_j)=(\hat{y}_j)$. In order to simplify (11), we define a set of $N+1$ random variables Z_j, $0\le j\le N$, by

$$Z_j=\begin{cases} b_{-1}x_{j-\hat{\gamma}+N}\hat{y}_j, & j=0,\cdots,\hat{\gamma}-1\\ b_0x_{j-\hat{\gamma}}\hat{y}_j, & j=\hat{\gamma},\cdots,N-2\\ b_0x_{N-\hat{\gamma}-1}\hat{y}_{N-1}, & j=N-1\\ b_{-1}x_{N-\hat{\gamma}-1}\hat{y}_0, & j=N.\end{cases} \quad (12)$$

For any $\hat{\gamma}$ in the set $\{0,\cdots,N-1\}$, the random variables Z_j, $0\le j\le N$, are mutually independent and satisfy $\Pr\{Z_j=+1\}=\Pr\{Z_j=-1\}=1/2$. One might first toss a fair coin to determine $x_{N-\hat{\gamma}-1}$. A second toss determines b_{-1}, and therefore $Z_N=b_{-1}x_{N-\hat{\gamma}-1}\hat{y}_0$. A third toss determines b_0, and therefore $Z_{N-1}=b_0x_{N-\hat{\gamma}-1}\hat{y}_{N-1}$. The remaining $N-1$ tosses determine $x_{j-\hat{\gamma}+N}$ (and hence $Z_j=b_{-1}x_{j-\hat{\gamma}+N}\hat{y}_j$) for $j=0,\cdots,\hat{\gamma}-1$ and $x_{j-\hat{\gamma}}$ (and hence $Z_j=b_0x_{j-\hat{\gamma}}\hat{y}_j$) for $j=\hat{\gamma},\cdots,N-2$. Using the definition of the random variables Z_j, $0\le j\le N$, and the fact that $\hat{y}_j^2=1$ for each j, (11) can be simplified to

$$B_{k,1}(b_k,\tau)=\sum_{j=0}^{N-2} Z_j(\hat{R}_\psi(S)+\hat{y}_j\hat{y}_{j+1}R_\psi(S)) + Z_{N-1}\hat{R}_\psi(S)+Z_NR_\psi(S) \quad (13)$$

where the random variables Z_j, $0\le j\le N$, are mutually independent and satisfy $\Pr\{Z_j=+1\}=\Pr\{Z_j=-1\}=1/2$.

For notational convenience we define $f(s)=\hat{R}_\psi(s)+R_\psi(s)$ and $g(s)=\hat{R}_\psi(s)-R_\psi(s)$. We also define the set A to be the set of all nonnegative integers i less than $N-1$ such that $\hat{y}_i\hat{y}_{i+1}=1$ and the set B to be the set of all nonnegative integers i less than $N-1$ such that $\hat{y}_i\hat{y}_{i+1}=-1$ in order to strategically split the sum in (13). The structure of (13) can now be exploited by writing it as

$$B_{k,1}(b_k,\tau)=\sum_{j\in A} Z_jf(S)+\sum_{j\in B} Z_jg(S) + Z_{N-1}\hat{R}_\psi(S)+Z_NR_\psi(S). \quad (14)$$

If we restore the index k, which has been implicit in the preceding paragraphs, we obtain

$$B_{k,1}(b_k,\tau_k)=X_kf(S_k)+Y_kg(S_k)+P_k\hat{R}_\psi(S_k)+Q_kR_\psi(S_k) \quad (15)$$

where

$$X_k=\sum_{j\in A} Z_j \quad (16)$$

$$Y_k=\sum_{j\in B} Z_j \quad (17)$$

$P_k=Z_{N-1}$, and $Q_k=Z_N$.

The random variable X_k is a function of the elements of the set $\{Z_j:j\in A\}$, and the random variable Y_k is a function of the elements of the set $\{Z_j:j\in B\}$. Furthermore, the sets A, B, $\{N-1\}$, and $\{N\}$ are disjoint. Since the random variables Z_j for $0\le j\le N$ are mutually independent, the random variables X_k, Y_k, P_k, and Q_k are mutually independent.

By using (2), (3), and (15), we can express the decision statistic in the simplified form

$$Z_{\text{out}}^{(1)}=\eta+b_0^{(1)}T\sqrt{P/2}+\sqrt{P/2}\sum_{k=2}^{K} W_k \quad (18)$$

where

$$W_k=[P_k\hat{R}_\psi(S_k)+Q_kR_\psi(S_k) + X_kf(S_k)+Y_kg(S_k)]\cos\phi_k. \quad (19)$$

Furthermore, the random variables W_k, $2\le k\le K$, are mutually independent. This follows from the fact that these random variables are functions of elements in disjoint subsets of mutually independent random variables. The density functions of P_k and Q_k for $2\le k\le K$ are already known, and the density functions of X_k and Y_k can be determined by elementary combinatorial arguments. If we denote the cardinality of the set A by $|A|$ and the cardinality of the set B by $|B|$, the density function for the discrete random variable X_k is given by

$$p_{X_k}(j)=C\left(|A|,\frac{j+|A|}{2}\right)2^{-|A|}, \quad j=-|A|,-|A|+2,\cdots,|A|-2,|A| \quad (20)$$

and the density function for the discrete random variable Y_k is given by

$$p_{Y_k}(j)=C\left(|B|,\frac{j+|B|}{2}\right)2^{-|B|},$$

$$j=-|B|,\ -|B|+2,\ \cdots,\ |B|-2,\ |B| \quad (21)$$

where, in the above equations, the function $C(n, k)$ represents the binomial coefficient $\binom{n}{k}$. The densities $p_{X_k}(i)$ and $p_{Y_k}(j)$ are nonzero only for the discrete values specifically mentioned in (20) and (21).

It is helpful at this point to examine the progress in reducing the complexity of (2). The computational difficulty of considering all the 2^{KN} possible combinations of signature sequences has already been mentioned. If we examine (19), we see that, conditioned on the first signature sequence, the random variable W_k depends on the kth signature sequence only through the mutually independent random variables P_k, Q_k, X_k, and Y_k. There are two possible values for P_k and two possible values for Q_k. There are $|A| + 1$ possible values for X_k and $|B| + 1$ possible values for Y_k. This means, instead of considering the 2^N possibilities for the kth signature sequence, we need only consider $4(|A| + 1)(|B| + 1)$ possible combinations of the values of these four random variables. Since $|A| + |B| = N - 1$, the number $4(|A| + 1)(|B| + 1)$ is significantly smaller than 2^N for any practical number N of chips per data bit. In fact, this product is less than or equal to $(N + 1)^2$. Furthermore, the random variables W_k, $2 \le k \le K$, are mutually independent and identically distributed. If the $4(|A| + 1)(|B| + 1)$ possible combinations of the values of the random variables P_k, Q_k, X_k, and Y_k are considered in order to determine the density of W_2, the densities of the random variables W_k for $3 \le k \le K$ are determined as well. We can perform $K - 2$ convolutions to obtain the density of $W = \Sigma_{k=2}^{K} W_k$. In summary, we have shown for the case of two transmitters that, conditioned on knowing the first signature sequence, it is sufficient to consider at most $(N + 1)^2$ possible combinations of values of discrete random variables. For the case of K transmitters, $K-2$ convolutions can be performed to study the effects of the $K - 1$ interfering signature sequences. It is not necessary to consider the $2^{N(K-1)}$ possible combinations of interfering signature sequences.

It is important to recall that the expression for the decision statistic $Z_{\text{out}}^{(1)}$ of (18) is conditioned on the first signature sequence. However, (18) depends on the first sequence only through the parameter $C_1(1)$, the discrete aperiodic autocorrelation function of this sequence evaluated at argument 1. The densities of X_k and Y_k depend only on the parameters $|A|$ and $|B|$, and these parameters in turn depend on the first signature sequence only through the parameter $C_1(1)$. To see this, notice that $C_1(1) = |A| - |B|$. Since $|A| + |B| = N - 1$, $|A| = [N - 1 + C_1(1)]/2$ and $|B| = [N - 1 - C_1(1)]/2$.

At this point another key simplification can be stated. In order to obtain (18), it is not necessary to condition on the random vector which models one period of the first signature sequence. It is sufficient to condition on the single random variable $C_1(1)$. This is a significant simplification because, although there are 2^N possible signature sequences for the first receiver, there are only N possible values for $C_1(1)$. The 2^N signature sequences that the first receiver can use fall into N classes. The performance of the receiver is the same for all signature sequences in the same class. Combining this fact with the simplification mentioned earlier, we see that for the case of two transmitters, it suffices in our simplified model to consider no more than $N\cdot(N + 1)^2$ possible combinations of values of discrete random variables. The effects of additional transmitters can be studied by performing additional convolutions.

The problem of obtaining the distribution of the random variable $C_1(1)$ remains. However,

$$C_1(1)=\sum_{j=0}^{N-2} a_j^{(1)}a_{j+1}^{(1)}. \quad (22)$$

Equation (22) can be written as

$$C_1(1)=\sum_{j=0}^{N-2} c_j^{(1)} \quad (23)$$

where $c_j^{(1)} = a_j^{(1)}a_{j+1}^{(1)}$. Each $c_j^{(1)}$ is a random variable that indicates whether or not the next element of the first signature sequence is the same as the preceding element: $c_j^{(1)} = 1$ implies $a_j^{(1)} = a_{j+1}^{(1)}$ and $c_j^{(1)} = -1$ implies $a_j^{(1)} \ne a_{j+1}^{(1)}$. Since $\{a_j^{(1)} : 0 \le j \le N - 1\}$ is a set of independent and identically distributed random variables with $\Pr\{a_j^{(1)} = +1\} = \Pr\{a_j^{(1)} = -1\} = 1/2$, $\{c_j^{(1)} : 0 \le j \le N - 2\}$ is a set of independent and identically distributed random variables with $\Pr\{c_j^{(1)} = +1\} = \Pr\{c_j^{(1)} = -1\} = 1/2$. Routine combinatorial arguments show that the density function for the discrete random variable $C_1(1)$ is given by

$$p_{C_1(1)}(j)=C\left(N-1,\frac{j+N-1}{2}\right)2^{1-N},$$

$$j=1-N,\ 3-N,\ \cdots,\ N-3,\ N-1 \quad (24)$$

and $p_{C_1(1)}(j) = 0$, elsewhere.

We now summarize the results of this section. The decision statistic of the correlation receiver has been obtained in the form given by (18)–(21). This form is particularly useful for performing computations. Although there are 2^{KN} possible combinations of signature sequences that could be used in the system, we have shown that, to model the signature sequences for the case of two transmitters, it suffices to consider no more than $N\cdot(N + 1)^2$ possible combinations of values of discrete random variables. The effects of additional transmitters can be studied by performing additional convolutions. Furthermore, the random variables η, S_k, ϕ_k, $C_1(1)$, P_k, Q_k, X_k, and Y_k for k in the set $\{2, \cdots, K\}$ form a set of mutually independent random variables. This is also a significant computational aid.

IV. Characteristic Function of the Decision Statistic

It is sometimes preferable to work with the characteristic function of the decision statistic instead of the density functions of the random variables that define the decision statistic [5]. For this reason we evaluate the characteristic function in this section. The characteristic function of the decision statistic follows directly from the simplified expression given by (18)–(21). We begin by normalizing the decision statistic so that the magnitude of the desired signal component is 1. Next, the characteristic function of the multiple-access interference is evaluated. As was done in [5] for the case of deterministic sequences, this is then multiplied by the characteristic function of the thermal noise in order to obtain the overall characteristic function of the random component of the decision statistic.

We first normalize (18) so that the magnitude of the desired signal component is equal to 1; i.e., we divide both sides of (18) by $T\sqrt{P/2}$. The resulting normalized decision statistic is given by

$$Z_{\text{norm}}^{(1)}=\zeta+b_0^{(1)}+T^{-1}W \quad (25)$$

where $W = \Sigma_{k=2}^{K} W_k$ and ζ is a Gaussian random variable with variance $N_0/(2TP)$ (or $N_0/(2E_b)$, where E_b is the energy per bit). The characteristic function of ζ is given by

$$\Phi_\zeta(u)=\exp\left(\frac{-N_0u^2}{4E_b}\right). \quad (26)$$

Recall that η, and hence ζ, is a random variable that represents the effects of thermal noise at the receiver. The term $T^{-1}W$ represents the total multiple-access interference.

The first step in finding the characteristic function of the multiple-access interference is to find the characteristic function of the random variable W_2, which is defined in (19). Suppose $S_2 = \hat{S}_2$ and $\phi_2 = \hat{\phi}_2$. Since P_2, Q_2, X_2, Y_2, S_2, and ϕ_2 are mutually independent, the characteristic function of W_2, conditioned on S_2, ϕ_2, and $C_1(1)$, can be computed from (19)–(21) to be

$$\begin{aligned}\Phi_{W_2}(u;\ \hat{S}_2,\ \hat{\phi}_2,\ c)&=E\{e^{juW_2}|S_2=\hat{S}_2,\ \phi_2=\hat{\phi}_2,\ C_1(1)=c\}\\&=z(uT;\ \hat{S}_2,\ \hat{\phi}_2,\ |A|,\ |B|) \quad (27)\end{aligned}$$

where

$$\begin{aligned}z(u;\ s,\ \phi,\ i,\ j)=&\cos\,[uT^{-1}\hat{R}_\psi(s)\cos\phi]\\&\cdot\cos\,[uT^{-1}R_\psi(s)\cos\phi]\\&\cdot\{\cos\,[uT^{-1}f(s)\cos\phi]\}^i\\&\cdot\{\cos\,[uT^{-1}g(s)\cos\phi]\}^j. \quad (28)\end{aligned}$$

The characteristic function of W_2 conditioned on $C_1(1)$ is

$$\begin{aligned}\Phi_{W_2}(u;\ c)=&\frac{2}{\pi T_c}\int_0^{T_c}\int_0^{\pi/2}\cos\,[u\hat{R}_\psi(s)\cos\phi]\\&\cdot\cos\,[uR_\psi(s)\cos\phi]\ \{\cos\,[uf(s)\cos\phi]\}^{|A|}\\&\cdot\{\cos\,[ug(s)\cos\phi]\}^{|B|}\,d\phi\,ds. \quad (29)\end{aligned}$$

Conditioned on $C_1(1)$, the random variables W_k, $2 \le k \le K$, are mutually independent and identically distributed. Thus, the characteristic function of the random variable W conditioned on $C_1(1)$ is given by

$$\Phi_W(u;\ c)=\{\Phi_{W_2}(u;\ c)\}^{K-1}. \quad (30)$$

We denote the total multiple-access interference $T^{-1}W$ by Ξ. The characteristic function of Ξ conditioned on the random variable $C_1(1)$ is given by

$$\Phi_\Xi(u;\ c)=\left\{\Phi_{W_2}\left(\frac{u}{T};\ c\right)\right\}^{K-1} \quad (31)$$

or, using (29),

$$\Phi_\Xi(u;\ c)=\left\{\frac{2}{\pi T_c}\int_0^{T_c}\int_0^{\pi/2}z(u;\ s,\ \phi,\ |A|,\ |B|)\,d\phi\,ds\right\}^{K-1}. \quad (32)$$

The characteristic functions that have been defined so far have been conditioned on $C_1(1)$. We can express $|A|$ and $|B|$ in (32) in terms of $C_1(1)$ and average over the distribution specified by (24). The resulting characteristic function can be expressed as

$$\begin{aligned}\Phi_\Xi(u)=&2^{1-N}\sum_{i=0}^{N-1}C(N-1,\ i)\\&\cdot\left\{\frac{2}{\pi T_c}\int_0^{T_c}\int_0^{\pi/2}z(u;\ s,\ \phi,\ i,\ N-1-i)\,d\phi\,ds\right\}^{K-1}. \quad (33)\end{aligned}$$

An alternative expression for this characteristic function is given by

$$\Phi_\Xi(u)=2^{1-N}\sum_{i=0}^{N-1}C(N-1,\ i)[B(u,\ i)+B(u,\ i+1)]^{K-1} \quad (34)$$

where

$$\begin{aligned}B(u,\ i)=&\frac{1}{\pi T_c}\int_0^{T_c}\int_0^{\pi/2}\{\cos\,[uT^{-1}f(s)\cos\phi]\}^i\\&\cdot\{\cos\,[uT^{-1}g(s)\cos\phi]\}^{N-i}\,d\phi\,ds. \quad (35)\end{aligned}$$

This alternative expression can be obtained from (33) by applying the trigonometric identity

$$\begin{aligned}&2\cos\,[uT^{-1}\hat{R}_\psi(s)\cos\phi]\cos\,[uT^{-1}R_\psi(s)\cos\phi]\\&=\cos\,[uT^{-1}f(s)\cos\phi]+\cos\,[uT^{-1}g(s)\cos\phi] \quad (36)\end{aligned}$$

to (33).

The characteristic function of the total random component of the normalized decision statistic $Z^{(1)}_{\text{norm}}$ is given by the product of the characteristic functions $\Phi_\zeta(u)$ and $\Phi_\Xi(u)$. The magnitude of the signal component is 1. One could apply numerical integration to the characteristic function in order to obtain an approximation to the average probability of error; this is being pursued by Geraniotis.

V. Upper and Lower Bounds on the Average Probability of Error

In this section we illustrate the use of the simplified expression for the decision statistic of the correlation receiver by obtaining upper and lower bounds on the average probability of error. For an illustration, the case of two transmitters (i.e., $K = 2$), signature sequences of length 31, and a rectangular chip waveform [i.e., $\psi(t) = p_{T_c}(t)$] is chosen. The extensions of the methods to greater numbers of interfering transmitters and longer signature sequences, as well as to other detection schemes, are also discussed.

We proceed by first dividing the interval containing all possible values of Ξ, the total multiple-access interference, into a set of subintervals. For each subinterval, an upper bound is found on the conditional probability that the value of Ξ lies in that subinterval given $C_1(1)$. Next, the upper bound is averaged using the distribution of $C_1(1)$ in order to obtain an upper bound on the (unconditional) probability that the value of Ξ lies in that subinterval. The set of upper bounds corresponding to the set of subintervals forms a vector $\boldsymbol{u}$. This vector is used to obtain an upper bound on the average probability of error. In a similar way, a set of lower bounds is obtained that forms a vector $\boldsymbol{v}$. The vector $\boldsymbol{v}$ is used to obtain a lower bound on the average probability of error.

There are key differences between the approach of this section and other previously developed approaches [4]–[8] for obtaining the average probability of error of the correlation receiver. First of all, random sequences are considered here instead of deterministic sequences. Second, in the approach of this section, two vectors are evaluated. One vector is used to obtain an upper bound on the performance of the correlation receiver, and another vector is used to obtain a lower bound. The two bounds become tighter as the size of the vectors grows. This approach avoids the problem of evaluating the characteristic function of the random output of the receiver and the corresponding problem of evaluating integrals numerically. Furthermore, the upper and lower bounds together specify the accuracy of the resulting estimate of the average probability of error, and the accuracy can be improved by increasing the size of the vectors. The problem of ascertaining whether numerical integrations have been performed with enough accuracy to yield a good final result is eliminated.

Another benefit of the approach outlined in this section is that the computations are performed in a series of steps. The total computational requirement for a specific problem is just the sum of the requirements of the several steps. The computations of several of the steps need be performed only once, and they apply to several different problems. For example, by performing a certain amount of computation, we obtain the vectors u and v that model the mulitple-access interference at the output of a matched filter. This result is used to obtain upper and lower bounds on the average probability of error for coherent detection. The same result can be used to obtain upper and lower bounds on other forms of detection and signaling. Much of the required computation has already been completed in order to obtain upper and lower bounds on the average probability of error for coherent detection. Another example of this benefit is apparent when computing the average probability of error for various numbers of active transmitters. Once the computation has been performed for a given number of transmitters, some of the computation that is required to determine the performance for larger numbers of transmitters has already been completed.

Another feature of this approach is of great practical importance. Many of the computations that are required to obtain the vectors u and v involve standard operations on vectors. This is also true of the remaining steps involved in determining upper and lower bounds on the average probability of error from the vectors u and v. These standard vector operations can be performed very efficiently on an array processor. Standard routines from the library of an array processor have been used in order to obtain the numerical results of this paper.

A. The Vectors u and v

In this section we obtain the vectors u and v that model the multiple-access interference. Although the total multiple-access interference Ξ is a continuous random variable, two vectors are used to model this interference. This is possible because the density function for the random variable Ξ is nonzero only on an interval $[-k, k]$, where k is a constant that can be evaluated under the worst or the best interference conditions. The approach is to consider a normalized version of Ξ, $\hat{\Xi} = N\Xi$, and to partition the interval $[-Nk, Nk]$ into a number N_i of subintervals. If the number of subintervals per unit is N_u, the interval $[-Nk, Nk]$ is partitioned into $2NN_uk$ subintervals. Next, two vectors of length $2NN_uk$ are determined. Each component of the first vector is an upper bound on the probability that the value of the random variable $\hat{\Xi}$ lies in a corresponding subinterval. Each component of the second vector is a lower bound on the probability that the value of the random variable $\hat{\Xi}$ lies in a corresponding subinterval. There is a one-to-one correspondence between the components of each of the two vectors and the subintervals of the partition.

For an illustration, we consider a system with two transmitters and an odd number of chips per bit. The methods extend in an obvious way to systems with an even number of chips per bit. Although the extension to the case of more than two active transmitters is less obvious, it is described in the following. We normalize (18) by dividing each side by $T_c\sqrt{P/2}$. The normalized multiple-access interference $\hat{\Xi}$ is given by the expression $\hat{\Xi} = \Sigma_{k=2}^{K} B_k$, where $B_k = T_c^{-1} W_k$. For the situation under consideration, the normalized multiple-access interference is the random variable B_2. From (19) and the definitions of the random variables that appear in (19), it follows that the density function for B_2 is nonzero only on the interval $[-N, N]$ and symmetric about zero. We partition the interval $[-N, N]$ into $2NN_u$ subintervals and define two vectors u and v consisting of $2NN_u$ components. The vector u consists of the components u_i, $-NN_u \le i \le NN_u - 1$. The component u_i is an upper bound on the probability that the value of B_2 lies in the interval $[iN_u^{-1}, (i+1)N_u^{-1}]$. Similarly, the vector v consists of the components v_i, $-NN_u \le i \le NN_u - 1$. The component v_i is a lower bound on the probability that the value of B_2 lies in the interval $[iN_u^{-1}, (i+1)N_u^{-1}]$. Notice that, because of the symmetry of the density function of B_2, we can choose $u_i = u_{-i-1}$ and $v_i = v_{-i-1}$. The goal is to determine the components u_i, $0 \le i \le NN_u - 1$, and the components v_i, $0 \le i \le NN_u - 1$. Because of the symmetry of the density function of B_2, the complete vectors u and v have then been determined.

In the following we rely on the notion of conditional density functions. We express each random variable B_k, $2 \le k \le K$, as $B_k = A_k \cos \phi_k$, where $A_k = [P_k\hat{R}_\psi(S_k) + Q_kR_\psi(S_k) + X_kf(S_k) + Y_kg(S_k)]T_c^{-1}$. The density function $p_{A_2}(x)$ is shown in Fig. 2 for a rectangular chip waveform and $N = 31$. Since the density function $p_{A_2}(x)$ is symmetric about zero, it is shown for $x \ge 0$ only. Notice that the density consists of a component that is piecewise constant and a component consisting of impulses. In the Appendix we show that $p_{A_2}(x; c)$, the conditional density function for the random variable A_2 given $C_1(1) = c$, can be expressed in terms of conditional densities that have a certain form. The expression is

$$p_{A_2}(x;\ c) = E\{p_{A_2|\alpha}(x|\alpha;\ c)\} \tag{37}$$

where the expectation is a conditional expectation given $C_1(1) = c$. The random variable α can be viewed as an index of several conditional density functions. In the Appendix it is shown that the conditional density functions $p_{A_2|\alpha}(x|i;\ c)$ can be written as

$$p_{A_2|\alpha}(x|i;\ c) = \begin{cases} \delta(x-i), \\ \qquad i = -N,\ -N+2,\ \cdots,\ N-2,\ N \\ \dfrac{1}{2}p_2(x-i+1), \\ \qquad i = -N+1,\ -N+3,\ \cdots, \\ \qquad N-3,\ N-1 \end{cases} \tag{38}$$

where $\delta(x)$ is the unit impulse function, $p_2(x) = 1$ for $0 \le x \le 2$, and $p_2(x) = 0$ for $x < 0$ or $x > 2$. Each conditional density is either uniform on an interval or an impulse. Notice that the conditional densities $p_{A_2|\alpha}(x|i;\ c)$ do not depend on the parameter c. The density of the discrete random variable α, however, does depend on the parameter c. Notice that the index α takes on values in the set $\{-N, \cdots, N\}$. We denote the discrete density function for the random variable α by a vector $w(c)$ with components $w_i(c)$, $-N \le i \le N$. The component $w_i(c)$ is the conditional probability that the index α is equal to i given $C_1(1) = c$ (i.e., $w_i(c) = \Pr\{\alpha = i|C_1(1) = c\}$).

We now find the conditional density function for B_2 given α. It can be shown [9] that if the random variable A_2 is uniformly distributed on an interval $[a, b]$, where a and b are both positive, the density function for the random variable B_2 is given by

$$p_{B_2}(x;\ a,\ b) = \begin{cases} \dfrac{1}{\pi(b-a)} \ln \left[\dfrac{b}{a}\dfrac{(1+\sqrt{1-x^2/b^2})}{(1+\sqrt{1-x^2/a^2})}\right], & 0 \le |x| \le a \\ \dfrac{1}{\pi(b-a)} \ln \left[\dfrac{b}{|x|}(1+\sqrt{1-x^2/b^2})\right], & a \le |x| \le b \\ 0, & |x| > b. \end{cases} \tag{39}$$

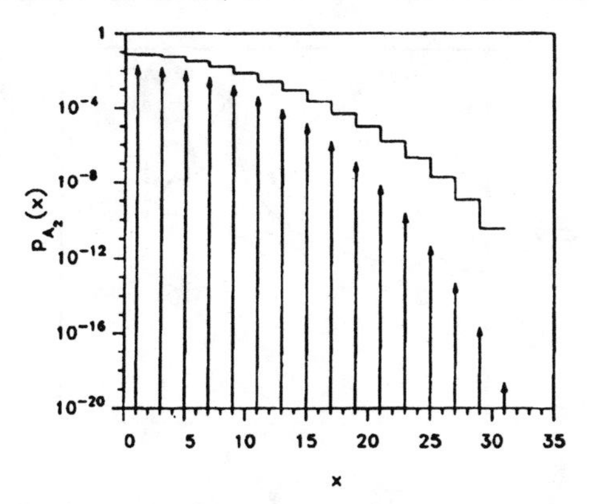

Fig. 2. Density function for the random variable, A_2 ($N = 31$, rectangular chip waveform).

It can also be shown [9] that if A_k is equal to a positive constant k, the density function for the random variable B_2 is given by

$$p_{B_2}(x;\ k)=\begin{cases}\dfrac{1}{k\pi\sqrt{1-x^2/k^2}}, & |x|<k\\ 0, & |x|\geq k.\end{cases} \tag{40}$$

(Notice that in the notation being used, the functions $p_{B_2}(x;\ a,\ b)$ and $p_{B_2}(x;\ k)$ are distinguished by the number of arguments that each function has.) Since the conditional densities of (38) are either impulses or else uniform on an interval, we can use the results of (38)–(40) to obtain the conditional density of B_2 given $\alpha = i$ and $C_1(1) = c$, i.e., to obtain $p_{B_2|\alpha}(x|i;\ c)$. The conditional density of B_2 given that $\alpha = i$ and $C_1(1) = c$ can be expressed as

$$p_{B_2|\alpha}(x|i;\ c)=\begin{cases}p_{B_2}(x;\ |i|), & i=-N,\ -N+2,\ \cdots,\ N-2,\ N\\ p_{B_2}(x;\ |i|-1,\ |i|+1), & i=-N+1,\ -N+3,\ \cdots,\ -4,\ -2,\ 2,\ 4,\ \cdots,\ N-3,\ N-1\\ \lim\limits_{\epsilon\to 0^+} p_{B_2}(x;\ \epsilon,\ 1), & i=0.\end{cases} \tag{41}$$

Notice that the conditional density $p_{B_2|\alpha}(x|i;\ c)$ does not depend on the parameter c.

We have found an exact expression for the conditional density of B_2 given α and $C_1(1)$. The exact conditional density function for B_2 given $C_1(1) = c$ can be obtained from the expression

$$p_{B_2}(x;\ c)=E\{p_{B_2|\alpha}(x|\alpha;\ c)\}. \tag{42}$$

We can also express (42) in terms of the components of the vector $\boldsymbol{w}(c)$ as

$$p_{B_2}(x;\ c)=\sum_{i=-N}^{N} w_i(c)p_{B_2|\alpha}(x|i;\ c). \tag{43}$$

Notice that for the case of multiple interfering transmitters, the conditional density function for the multiple-access interference Ξ given $C_1(1)$ can be found by performing $K-2$ convolutions since the random variables B_k, $2 \leq k \leq K$, form a set of conditionally independent and identically distributed random variables.

Instead of using the density of (43), we define two vectors $\boldsymbol{u}(i;\ c)$ and $\boldsymbol{v}(i;\ c)$, $-N \leq i \leq N$, as follows. The vector $\boldsymbol{u}(i;\ c)$ is the vector $\boldsymbol{u}$ that we have defined earlier given that $\alpha = i$ and $C_1(1) = c$. There are $2NN_u$ components of the vector $\boldsymbol{u}(i;\ c)$. The components are denoted by $u_j(i;\ c)$, $-NN_u \leq j \leq NN_u - 1$, and the component $u_j(i;\ c)$ is an upper bound on the probability that the multiple-access interference lies in the interval $[jN_u^{-1},\ (j+1)N_u^{-1}]$ given that $\alpha = i$ and $C_1(1) = c$. Similarly, the vector $\boldsymbol{v}(i;\ c)$ is the vector $\boldsymbol{v}$ that we have defined earlier given that $\alpha = i$ and $C_1(1) = c$. There are $2NN_u$ components of the vector $\boldsymbol{v}(i;\ c)$. The components are denoted by $v_j(i;\ c)$, $-NN_u \leq j \leq NN_u - 1$, and the component $v_j(i;\ c)$ is a lower bound on the probability that the multiple-access interference lies in the interval $[jN_u^{-1},\ (j+1)N_u^{-1}]$ given that $\alpha = i$ and $C_1(1) = c$.

In order to evaluate the vectors $\boldsymbol{u}(i;\ c)$ and $\boldsymbol{v}(i;\ c)$, we first consider the density function $p_{B_2|\alpha}(x|i;\ c)$ for i odd and positive. (An examination of (41) reveals that the problem is the same for negative i.) In this case, $p_{B_2|\alpha}(x|i;\ c) = p_{B_2}(x;\ i)$. The function $p_{B_2}(x;\ i)$ is strictly increasing and convex on the interval $[0,\ i)$. Since the function $p_{B_2}(x;\ i)$ is strictly increasing on the interval $[0,\ i)$, we choose the components $v_j(i;\ c)$ to be

$$v_j(i;\ c)=\begin{cases}N_u^{-1}p_{B_2}(jN_u^{-1};\ i), & 0\leq j\leq iN_u-1\\ 0, & iN_u\leq j\leq NN_u-1.\end{cases} \tag{44}$$

Since we can choose $v_j = v_{-j-1}$, $0 \leq j \leq NN_u - 1$, this gives a complete vector of lower bounds. Since the function is convex on the interval $[0,\ i)$, we choose the components $u_j(i;\ c)$ to be

$$u_j(i;\ c)=\begin{cases}N_u^{-1}[p(jN_u^{-1};\ i)+p(N_u^{-1}(j+1);\ i)]/2, & 0\leq j\leq iN_u-2\\ \dfrac{1}{2}-\sum\limits_{k=0}^{iN_u-2} v_k(i;\ c), & j=iN_u-1\\ 0, & iN_u\leq j\leq NN_u-1.\end{cases} \tag{45}$$

Since we can choose $u_j = u_{-j-1}$, $0 \leq j \leq NN_u - 1$, this gives a complete vector of upper bounds. Notice that the definition of the vector component $u_{iN_u-1}(i;\ c)$ must be considered as a special case.

In Fig. 3, the bounds that are given by (44) and (45) are illustrated for the case in which $N = 31$, $N_u = 2$, and $i = 5$. In this example, the interval $[0,\ 5]$ is divided into ten subintervals of width 1/2. The vector component $u_j(5;\ c)$, $0 \leq j \leq 9$, is the upper bound on the probability that the value of the random variable B_2 lies somewhere in the interval $[0.5j,\ 0.5(j+1)]$ given that $\alpha = 5$ and $C_1(1) = c$. It is given by the area under the top curve of Fig. 3 that corresponds to the interval $[0.5j,\ 0.5(j+1)]$ [except in the special case of the last component $u_9(5,\ c)$]. The vector component $v_j(i;\ c)$, $0 \leq j \leq 9$, is the lower bound on the probability that the value of the random variable B_2 lies in the interval $[0.5j,\ 0.5(j+1)]$ given that $\alpha = 5$ and $C_1(1) = c$. It is given by the area under the bottom curve of Fig. 3 that corresponds to the interval $[0.5j,\ 0.5(j+1)]$. Notice that, since the density is symmetric, only nonnegative arguments are shown.

We next consider the density function $p_{B_2|\alpha}(x|i;\ c)$ for i even and positive. In this case, $p_{B_2|\alpha}(x|i;\ c) = p_{B_2}(x;\ i-1,\ i+1)$ for $i > 0$, and $p_{B_2|\alpha}(x|i;\ c) =$

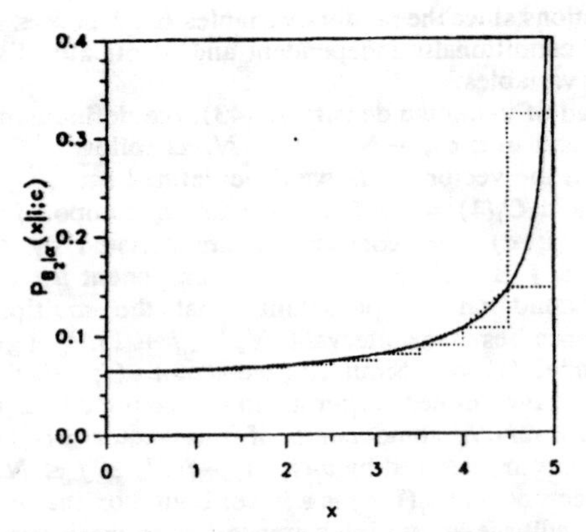

Fig. 3. Illustration of a bounding technique for the conditional density $p_{B_2|\alpha}(x \mid i; c)$ for $N = 31$, $N_u = 2$, and $i = 5$.

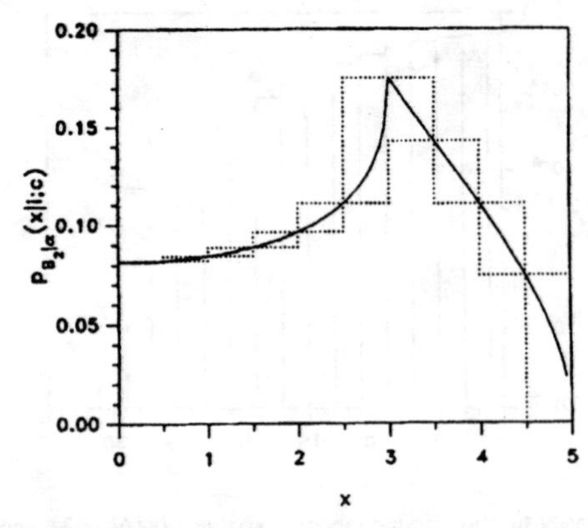

Fig. 4. Illustration of a bounding technique for the conditional density $p_{B_2|\alpha}(x \mid i; c)$ for $N = 31$, $N_u = 2$, and $i = 4$.

$\lim_{\epsilon \to 0^+} p_{B_2}(x; \epsilon, 1)$ for $i = 0$. The function $p_{B_2}(x; i-1, i+1)$ for $i > 0$ is strictly increasing on the interval $[0, i-1]$ and strictly decreasing on the interval $[i-1, i+1]$. Also, the function $p_{B_2}(x; \epsilon, 1)$ for $0 < \epsilon < 1$ is strictly decreasing on the interval $[\epsilon, 1]$. If the components $v_j(i; c)$ for $i > 0$ are given by

$$v_j(i;\ c)=\begin{cases} N_u^{-1} p_{B_2}(jN_u^{-1};\ i-1,\ i+1), & 0 \le j \le (i-1)N_u - 1 \\ N_u^{-1} p_{B_2}(N_u^{-1}(j+1);\ i-1,\ i+1), & (i-1)N_u \le j \le (i+1)N_u - 1 \\ 0, & (i+1)N_u \le j \le NN_u - 1 \end{cases} \tag{46}$$

and the components $v_j(0; c)$ are given by

$$v_j(0;\ c)=\begin{cases} N_u^{-1} \lim_{\epsilon \to 0^+} p_{B_2}(N_u^{-1}(j+1);\ \epsilon,\ 1), & 0 \le j \le N_u - 1 \\ 0, & N_u \le j \le NN_u - 1 \end{cases} \tag{47}$$

since we can choose $v_j = v_{-j-1}$, $0 \le j \le NN_u - 1$, a complete vector of lower bounds is obtained. If the components $u_j(i; c)$ for $i > 0$ are

$$u_j(i;\ c)=\begin{cases} N_u^{-1} p_{B_2}(N_u^{-1}(j+1);\ i-1,\ i+1), & 0 \le j \le (i-1)N_u - 1 \\ N_u^{-1} p_{B_2}(jN_u^{-1};\ i-1,\ i+1), & (i-1)N_u \le j \le (i+1)N_u - 1 \\ 0, & (i+1)N_u \le j \le NN_u - 1 \end{cases} \tag{48}$$

and the components $u_j(0; c)$ are

$$u_j(0;\ c)=\begin{cases} \frac{1}{2} - \sum_{k=1}^{N_u-1} v_k(0;\ c), & j=0 \\ N_u^{-1} \lim_{\epsilon \to 0^+} p_{B_2}(jN_u^{-1};\ \epsilon,\ 1), & 1 \le j \le N_u - 1 \\ 0, & N_u \le j \le NN_u - 1 \end{cases} \tag{49}$$

since we can choose $u_j = u_{-j-1}$, $0 \le j \le NN_u - 1$, then a complete vector of upper bounds is obtained.

In Fig. 4, the bounds that are given by (46) and (48) are illustrated for the case in which $N = 31$, $N_u = 2$, and $i = 4$. In this example, the interval $[0, 5]$ is again divided into ten subintervals of width 1/2. The vector component $u_j(4; c)$, $0 \le j \le 9$, is the upper bound on the probability that the value of the random variable B_2 lies somewhere in the interval $[0.5j, 0.5(j+1)]$ given that $\alpha = 4$ and $C_1(1) = c$. It is given by the area under the top curve of Fig. 4 that corresponds to the interval $[0.5j, 0.5(j+1)]$. The vector component $v_j(4; c)$, $0 \le j \le 9$, is the lower bound on the probability that the value of the random variable B_2 lies in the interval $[0.5j, 0.5(j+1)]$ given that $\alpha = 4$ and $C_1(1) = c$. It is given by the area under the bottom curve of Fig. 4 that corresponds to the interval $[0.5j, 0.5(j+1)]$. Notice again that, since the density is symmetric, only nonnegative arguments are shown.

Several alternatives to (44)–(49) could have been used to define the vectors $u(i; c)$ and $v(i; c)$. For instance, the densities given in (39) and (40) can be integrated to obtain the corresponding cumulative distribution functions in terms of the logarithmic function and inverse trigonometric functions. These distribution functions can be used to obtain tighter bounds than those depicted in Figs. 3 and 4—in fact, bounds for which $u(i; c) = v(i; c)$. Our choice of (44)–(49) illustrates that arbitrarily tight bounds on the average probability of error can be obtained without requiring that $u(i; c) = v(i; c)$.

In the preceding paragraphs two vectors of bounds conditioned on α and $C_1(1)$ have been obtained. We now average using the distribution of α. The vector of upper bounds given $C_1(1) = c$ can be expressed as

$$u(c) = E\{u(\alpha;\ c)\} \tag{50}$$

and the vector of lower bounds given $C_1(1) = c$ can be expressed as

$$v(c) = E\{v(\alpha;\ c)\}. \tag{51}$$

Equation (50) can be written in the alternative form

$$u(c) = \sum_{i=-N}^{N} w_i(c) u(i;\ c) \tag{52}$$

and (51) can be written in the alternative form

$$v(c) = \sum_{i=-N}^{N} w_i(c) v(i;\ c). \tag{53}$$

Since, for the case of two transmitters, the normalized multiple-access interference $\tilde{Z}$ conditioned on $C_1(1)$ is just the

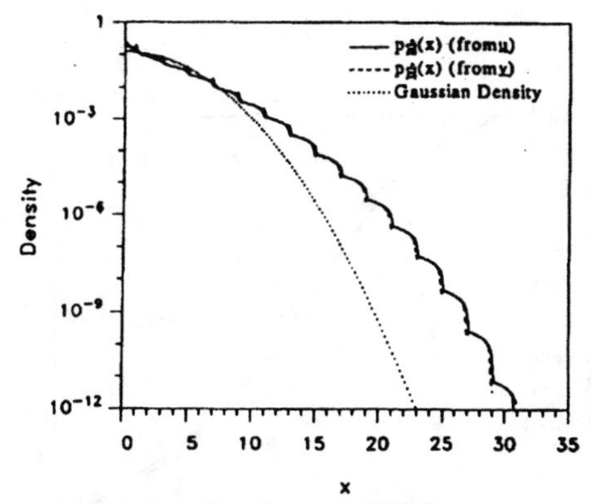

Fig. 5. Density functions for $\hat{Z}$ corresponding to the vectors u and v compared with the Gaussian density ($K = 2$, $N = 31$, $N_u = 10$, rectangular chip waveform).

random variable B_2, we have obtained the desired vectors of bounds given $C_1(1)$. Averaging these vectors using the distribution of $C_1(1)$ yields the overall upper and lower bounds on $\hat{Z}$, and hence Z. The final vector of upper bounds u is given by

$$u = E\{u(C_1(1))\} \tag{54}$$

and the final vector of lower bounds v is given by

$$v = E\{v(C_1(1))\} \tag{55}$$

where in (54) and (55) the expectation is performed using the distribution of $C_1(1)$ given by (24).

In Fig. 5, $N_u u_i$ and $N_u v_i$ are shown as a function of iN_u^{-1} (for i in the set $\{0, \cdots, NN_u - 1\}$, $K = 2$, $N = 31$, $N_u = 10$, and a rectangular chip waveform) in order to approximate the shape of the density function of the normalized multiple-access interference $\hat{Z}$. For comparison, the Gaussian density with the same variance as $\hat{Z}$ is plotted. Notice that, although the Gaussian density and the density of the normalized multiple-access interference $\hat{Z}$ agree fairly well for small arguments, the Gaussian density decays at a much faster rate for large arguments.

B. Upper and Lower Bounds on the Average Probability of Error

In this section we illustrate how the vectors u and v that have been described in Section V-A can be used to obtain bounds on the performance of various receivers. The correlation receiver is used for an example.

First, bounds are obtained on the performance of the correlation receiver when a constant interfering term is added to the output, but no multiple-access interference is present. In order to obtain these bounds, the decision statistic of the correlation receiver given in (18) is specialized to the case of one active transmitter. Equation (18) is normalized by dividing both sides by $T_c\sqrt{P/2}$ so that the signal component is equal to N. For the case under consideration, the only random component of the decision statistic is a Gaussian random variable with variance $\sigma^2 = N_0 N^2/(2E_b)$. Therefore, an upper bound on the average probability of error is given by

$$P_E^{(U)}(\beta) = Q_U((N+\beta)/\sigma) \tag{56}$$

where $Q_U(x)$ is an upper bound on the function $Q(x) = (2\pi)^{-1/2}\int_x^\infty e^{-u^2/2}\,du$ and β is the constant interfering term. Similarly, a lower bound on the average probability of error is given by

$$P_E^{(L)}(\beta) = Q_L((N+\beta)/\sigma) \tag{57}$$

where $Q_L(x)$ is a lower bound on the function $Q(x)$ and β is the constant interfering term.

Although the multiple-access interference is random, we have from Section V-A an upper bound on the probability that the value of the multiple-access interference lies in an interval $[iN_u^{-1}, (i+1)N_u^{-1}]$, $-NN_u \le i \le NN_u - 1$. This probability is given by the vector component u_i. To obtain bounds on the average probability of error of the receiver, we first assume that with probability u_i the multiple-access interference is equal to the value in the interval $[iN_u^{-1}, (i+1)N_u^{-1}]$ that yields the largest probability of error. Since the function $Q(\cdot)$ is a strictly decreasing function, for the correlation receiver this value is iN_u^{-1}. We next use (56) to obtain an upper bound on the average probability of error given by

$$P_E^{(U)} = \sum_{i=-NN_u}^{NN_u-1} u_i P_E^{(U)}(iN_u^{-1}). \tag{58}$$

Finally, we use (57) to obtain a lower bound on the average probability of error given by

$$P_E^{(L)} = \sum_{i=-NN_u}^{NN_u-1} v_i P_E^{(L)}((i+1)N_u^{-1}). \tag{59}$$

It is important to realize that this approach applies to other detection schemes. Suppose that the output of a bandpass filter that is matched to a signature signal is sampled in the detection process. If the performance can be determined when a constant interfering term is added to the output of the matched filter, the outlined approach gives the average performance when multiple-access interference is present.

In Fig. 6, the upper and lower bounds on the average probability of error of the correlation receiver are plotted as a function of E_b/N_0. The results are given for $K = 2$, $N = 31$, $N_u = 10$, and a rectangular chip waveform. For comparison, the result that is obtained if $\hat{Z}$ is modeled by a Gaussian random variable with the same variance as $\hat{Z}$ is also given. As expected from an observation of Fig. 5, the Gaussian approximation is fairly good for small values of E_b/N_0, but rather bad for large values of E_b/N_0.

C. Extension to Multiple Interfering Transmitters

In Section V-A, we obtained the vectors u and v that model the normalized multiple-access interference $\hat{Z}$. However, in Section V-A, only the case of one interfering transmitter ($K = 2$) is considered. In this section we describe the extension to $K > 2$.

In order to obtain the vectors corresponding to u and v for $K > 2$, we begin with the vectors $u(c)$ and $v(c)$ for a single interfering transmitter given that $C_1(1) = c$. This is necessary because the random variables B_k, $2 \le k \le K$, which model the multiple-access interference from the various transmitters, are not independent. However, they are conditionally independent given $C_1(1)$. Since the procedure for finding the vector corresponding to v for $K > 2$ is the same as it is for finding the vector corresponding to u for $K > 2$, we describe only the latter.

We begin by recalling that the vector $u(c)$ consists of the components $u_i(c)$, $-NN_u \le i \le NN_u - 1$. The component $u_i(c)$ is an upper bound on the probability that the value of the random variable B_2, modeling the interference from a single transmitter, lies in the interval $[iN_u^{-1}, (i+1)N_u^{-1}]$. Next, a set of discrete random variables D_k, $2 \le k \le K$, is defined such that $D_k = (i + 1/2)N_u^{-1}$ if and only if the random variable B_k lies somewhere in the interval $[iN_u^{-1}, (i+1)N_u^{-1}]$. Notice that the random variable B_k never differs from the random variable D_k by more than $N_u^{-1}/2$. Finally, we define the random variables $D = \sum_{k=2}^{K} D_k$ and

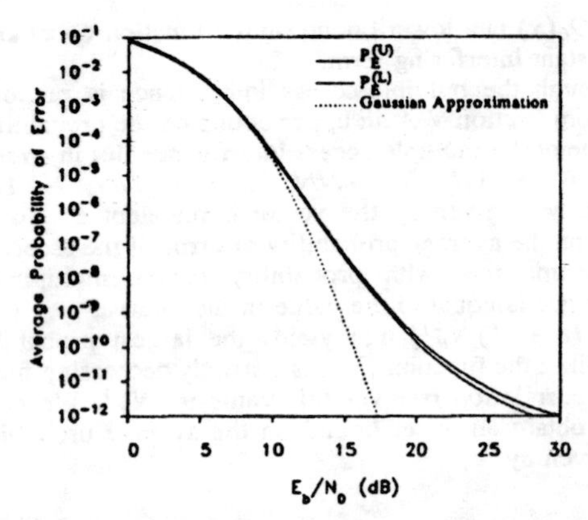

Fig. 6. Upper and lower bounds on the average probability of error of the correlation receiver versus E_b/N_0 compared with a Gaussian approximation ($K = 2$, $N = 31$, $N_u = 10$, rectangular chip waveform, random sequences).

$B = \Sigma_{k=2}^{K} B_k$, and notice a relationship between the random variables D and B. To do this, we first notice that the random variable D is the sum of the $K - 1$ terms D_k, $2 \le k \le K$, and that the random variable B is the sum of the $K - 1$ terms B_k, $2 \le k \le K$. Furthermore, each term D_k can differ from the corresponding term B_k by at most $N_u^{-1}/2$. Therefore, the sum D of terms can differ from the sum B of terms by at most $(K - 1)N_u^{-1}/2$.

We now proceed to obtain an upper bound on the discrete density for D. Since the random variable B is related to the random variable D, this also serves to model the density function for B. We note that, since the random variables B_k, $2 \le k \le K$, form a set of conditionally independent [given $C_1(1)$] and identically distributed random variables, the random variables D_k, $2 \le k \le K$, form a set of conditionally independent and identically distributed random variables. Therefore, we can obtain the discrete conditional density function for the random variable $D = \Sigma_{k=2}^{K} D_k$ given $C_1(1)$ by performing a series of discrete convolutions. The discrete conditional density function for D given that $C_1(1) = c$ is given by

$$p_D(k; c) = p_{D_2}(k; c) * p_{D_2}(k; c) * \cdots * p_{D_2}(k; c) \quad (60)$$

where in (60) there are $K - 2$ convolutions indicated and the asterisk denotes a convolution. Since the vector $u(c)$ is an upper bound on the discrete conditional density function for D_2 given that $C_1(1) = c$, we can replace the discrete conditional density function $p_{D_2}(k; c)$ in (60) with the appropriate components of $u(c)$. We can then obtain an upper bound on the discrete conditional density function for D given that $C_1(1) = c$. We next average using the distribution of $C_1(1)$ in order to obtain an overall bound on the discrete (unconditional) density function for the random variable D.

The upper bound on the discrete density function for D described in the previous paragraph yields a vector of upper bounds that is useful in modeling the density function for B. We can use this vector in place of u to obtain an upper bound on the average probability of error for the case of multiple interfering transmitters using the approach of Section V-B. The difference is that for the vector component corresponding to the subinterval $[iN_u^{-1}, (i + 1)N_u^{-1}]$, we choose as the argument of $P_E^{(U)}(\cdot)$ the value in the interval $[N_u^{-1}(2i + 2 - K)/2, N_u^{-1}(2i + K)/2]$ that yields the largest probability of error.

In Figs. 7–9, the upper and lower bounds on the average probability of error are plotted as a function of E_b/N_0 for the case of a rectangular chip waveform and $N_u = 10$. In Fig. 7, the results are plotted for the case $N = 31$ and $K = 2$ or 3. In

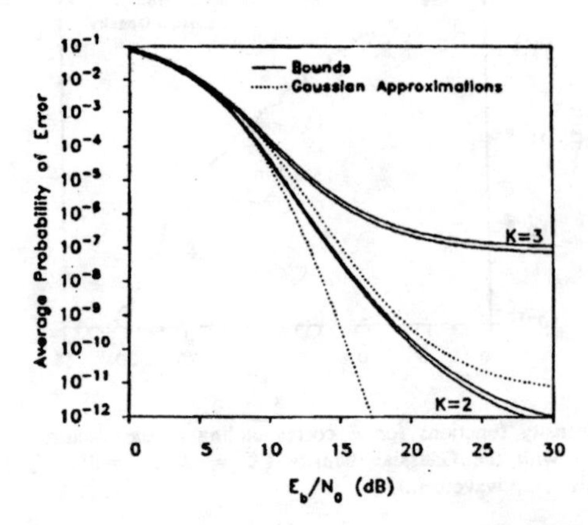

Fig. 7. Upper and lower bounds on the average probability of error of the correlation receiver versus E_b/N_0 compared with Gaussian approximations ($N = 31$; $K = 2$ or 3; $N_u = 10$; rectangular chip waveform; random sequences).

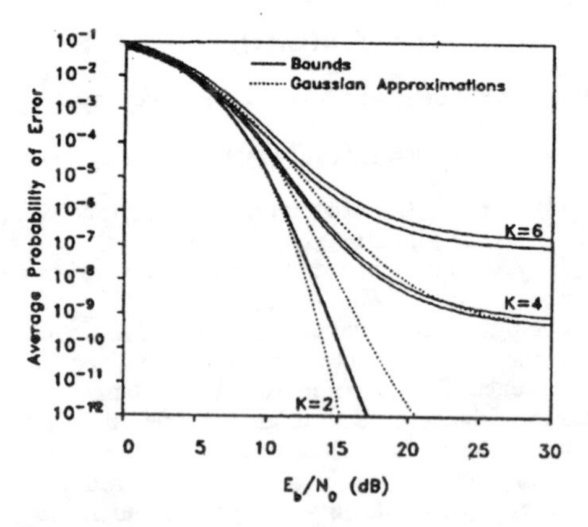

Fig. 8. Upper and lower bounds on the average probability of error of the correlation receiver versus E_b/N_0 compared with Gaussian approximations ($N = 63$; $K = 2$, 4, or 6; $N_u = 10$; rectangular chip waveform; random sequences).

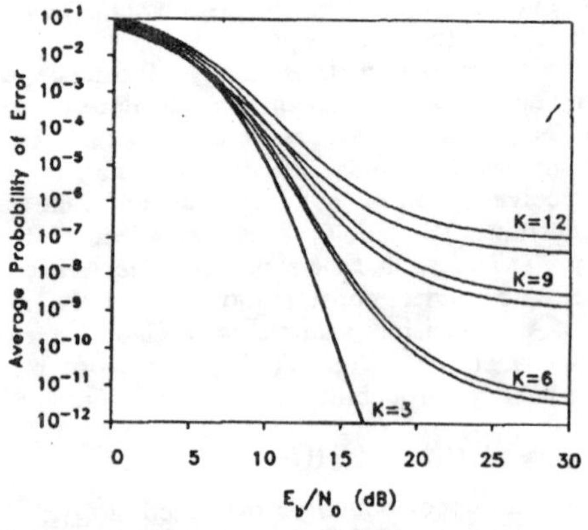

Fig. 9. Upper and lower bounds on the average probability of error of the correlation receiver versus E_b/N_0 ($N = 127$; $K = 3, 6, 9$, or 12; $N_u = 10$; rectangular chip waveform; random sequences).

Fig. 8, the results are plotted for the case $N = 63$ and $K = 2$, 4, or 6. In Fig. 9, the results are plotted for the case $N = 127$ and $K = 3, 6, 9$, or 12. In Figs. 7 and 8, results obtained when $\hat{\Xi}$ is modeled by a Gaussian random variable with the same variance as $\hat{\Xi}$ are also given for comparison.

VI. Conclusions

We have expressed in simplified form the decision statistic of a correlation receiver in a DS/SSMA communication system that employs randomly chosen signature sequences. For the case of two transmitters, a method of modeling the effects of the sequences has been obtained that involves considering no more than $N\cdot(N+1)^2$ possible combinations of values of a set of discrete random variables. The effects of additional transmitters can be studied by performing additional convolutions. We can analyze the performance of a receiver directly from the simplified expression for the decision statistic. Alternatively, the characteristic function of the decision statistic can be evaluated from the expression, and the receiver can be analyzed using the characteristic function.

We have also described and given an illustration of an approach for obtaining upper and lower bounds on the average probability of error of a DS/SSMA system. This approach involves conditioning the decision statistic on a set of discrete random variables and then analytically evaluating a set of conditional density functions. The final density function for the multiple-access interference is determined by combining the various conditional density functions with the appropriate weighting. Bounds are obtained that take the form of two vectors. One vector has components that are upper bounds on the probability that the value of the multiple-access interference lies in various subintervals. The other vector has components that are lower bounds on the probability that the value of the multiple-access interference lies in various subintervals. These vectors are used, together with the results on the performance of the receiver for a constant interfering term and a single active transmitter, to determine upper and lower bounds on the average probability of error.

The approach that we have outlined has several unique features and advantages over other methods of evaluating the average probability of error. Instead of focusing immediately on a single scalar, such as the average probability of error, in this approach two vectors are determined. Once obtained, these vectors can be used to examine the performance for greater numbers of interfering transmitters or for other signaling and demodulation techniques. An accurate approximation to the density function of the multiple-access interference can be obtained, and this information is much more general and useful than a single evaluation of the probability of error applied to one specialized system. A practical consideration of great importance is that the computations needed for the approach that we have outlined largely involve operations on vectors. These computations can be accomplished in an efficient manner by an array processor. Also, the computations can be separated into a series of steps so that the total computation required is just the sum of the requirements at the several steps. Furthermore, the approach yields bounds that can be made arbitrarily tight. This allows a careful study of the nature of possible approximations and of the range of system parameters for which the approximations are valid. This has been illustrated with a study of the Gaussian approximation.

Appendix
Form of Conditional Density of A_2

In Section V, we defined the random variable $A_2 = [P_2\hat{R}_\psi(S_2) + Q_2R_\psi(S_2) + X_2f(S_2) + Y_2g(S_2)]T_c^{-1}$. In this Appendix, we show that in (37) the conditional density functions $p_{A_2|\alpha}(x|i;\,c)$ can have a certain form. The conditional density functions $p_{A_2|\alpha}(x|i;\,c)$ of (37) for the case in which N is odd can be written as

$$p_{A_2|\alpha}(x|i;\,c)=\begin{cases}\delta(x-i), & i=-N,\ -N+2,\ \cdots,\ N-2,\ N\\ \frac{1}{2}p_2(x-i+1), & i=-N+1,\ -N+3,\ \cdots,\ N-3,\ N-1\end{cases} \tag{A.1}$$

where $\delta(x)$ is the unit impulse function, $p_2(x) = 1$ for $0 \le x \le 2$, and $p_2(x) = 0$, for $x < 0$ or $x > 2$.

In order to show this, we first establish that $|A_2| \le N$. First notice that $|T_c^{-1}f(S_2)| \le 1$ and $|T_c^{-1}g(S_2)| \le 1$ and recall from (20) and (21) that $|X_2| \le |A|$ and $|Y_2| \le |B| = N - 1 - |A|$. This means that $|[X_2f(S_2) + Y_2g(S_2)]T_c^{-1}| \le N - 1$. Also, since the term $[P_2\hat{R}_\psi(S_2) + Q_2R_\psi(S_2)]$ is equal to $f(S_2)$, $-f(S_2)$, $g(S_2)$, or $-g(S_2)$, $|[P_2\hat{R}_\psi(S_2) + Q_2R_\psi(S_2)]T_c^{-1}| \le 1$. Hence, $|A_2| \le N$. Next, notice that for any possible values of the discrete random variables P_2, Q_2, X_2, and Y_2, the expression for the random variable A_2 has the form

$$A_2=[af(S_2)+bg(S_2)]T_c^{-1} \tag{A.2}$$

where a and b are integers. For the rectangular chip waveform, $f(s) = T_c$ for s in the interval $[0, T_c]$. Hence, the expression of (A.2) reduces to

$$A_2=a+bg(S_2)T_c^{-1}. \tag{A.3}$$

Furthermore, since S_2 is uniformly distributed on the interval $[0,\ T_c]$, $g(S_2)T_c^{-1}$ is uniformly distributed on the interval $[-1, 1]$ for the rectangular chip waveform. Hence, if $b = 0$, the random variable A_2 is a constant. If $b \ne 0$, the random variable A_2 is uniform on an interval. By utilizing the densities of (20) and (21), it can be shown that when N is odd, the transition points of $p_{A_2}(x;\,c)$ from one constant level to another occur at the odd integers. When N is odd, it is also true that if A_2 is a constant, it is an odd integer. Therefore, the conditional density functions $p_{A_2|\alpha}(x|i;\,c)$ can have the form specified in (A.1).

References

[1] H. F. A. Roefs and M. B. Pursley, "Correlation parameters of random binary sequences," *Electron. Lett.*, vol. 13, pp. 488–489, Aug. 1977.

[2] M. B. Pursley, "Performance evaluation for phase-coded spread-spectrum multiple-access communication—Part I: System analysis," *IEEE Trans. Commun.*, vol. COM-25, pp. 795–799, Aug. 1977.

[3] M. B. Pursley, F. D. Garber, and J. S. Lehnert, "Analysis of generalized quadriphase spread-spectrum communications," in *Conf. Rec., IEEE Int. Conf. Commun.*, June 1980, vol. 1, pp. 15.3.1–15.3.6.

[4] M. B. Pursley, D. V. Sarwate, and W. E. Stark, "Error probability for direct-sequence spread-spectrum multiple-access communications—Part I: Upper and lower bounds," *IEEE Trans. Commun.*, vol. COM-30, pp. 975–984, May 1982.

[5] E. A. Geraniotis and M. B. Pursley, "Error probability for direct-sequence spread-spectrum multiple-access communications—Part II: Approximations," *IEEE Trans. Commun.*, vol. COM-30, pp 985–995, May 1982.

[6] D. Laforgia, A. Luvison, and V. Zingarelli, "Exact bit error probability with application to spread-spectrum multiple access communications," in *Conf. Rec., IEEE Int. Conf. Commun.*, June 1981, vol. 4, pp. 76.5.1–76.5.5.

[7] K. T. Wu and D. L. Neuhoff, "Average error probability for direct sequence spread-spectrum multiple access communication systems," in *Proc. 18th Annu. Allerton Conf. Commun. Contr. Comput.*, Oct. 1980, pp. 359–368.

[8] K. Yao, "Error probability of asynchronous spread spectrum multiple access communication systems," *IEEE Trans. Commun.*, vol. COM-25, pp. 803–809, Aug. 1977.
[9] P. G. Hoel, S. C. Port, and C. J. Stone, *Introduction to Probability Theory*. Boston, MA: Houghton Mifflin, 1971, pp. 119, 150–151.

★

James S. Lehnert (S'83–M'84) received the B.S. (highest honors), M.S., and Ph.D. degrees in electrical engineering, from the University of Illinois at Urbana-Champaign in 1978, 1981, and 1984, respectively.

From 1978 to 1984, he was a research assistant at the Coordinated Science Laboratory, University of Illinois, Urbana. He has held summer positions at Motorola Communications, Schaumburg, IL, in the Data Systems Research Laboratory, and Harris Corporation, Melbourne, FL, in the Advanced Technology Department. He is currently an Assistant Professor of Electrical Engineering at Purdue University, West Lafayette, IN. His current research work is in communications and information theory with emphasis on spread-spectrum communications.

Dr. Lehnert was a University of Illinois Fellow from 1978 to 1979 and an IBM Pre-Doctoral Fellow from 1982 to 1984.

★

Michael B. Pursley (S'68–M'68–SM'77–F'82), for a photograph and biography, see this issue, p. 12.

ABSTRACT

Direct-sequence code-division multiple access (DS-CDMA) is a popular wireless technology. in DS-CDMA communications, all of the users' signals overlap in time and frequency and cause mutual interference. The conventional DS-CDMA detector follows a single-user detection strategy in which each user is detected separately without regard for the other users. A better strategy is multi-user detection, where information about multiple users is used to improve detection of each individual user. This article describes a number of important multi-user DS-CDMA detectors that have been proposed.

Multi-User Detection for DS-CDMA Communications

Shimon Moshavi, Bellcore

The year is 2010, and the world has gone wireless. The wireless personal communicator is as common as the wireline telephone used to be, and it provides reliable and affordable communication, anywhere and anytime: in the car, restaurant, park, home, or office, or on the slopes of the Swiss Alps. Portable computers provide a vast array of integrated wireless services, such as voice, data, and video communications, movies and television programs on demand, and unlimited access to the treasures of cyberspace.

To bring this vision to fruition, major improvements in the current state of wireless technology are necessary. One type of wireless technology which has become very popular over the last few years is direct-sequence code-division multiple access (DS-CDMA). In this article we review multi-user detection, an area of research with the potential to significantly improve DS-CDMA communications.

Code-division multiple access (CDMA) is one of several methods of multiplexing wireless users. In CDMA, users are multiplexed by distinct codes rather than by orthogonal frequency bands, as in frequency-division multiple access (FDMA), or by orthogonal time slots, as in time-division multiple access (TDMA). In CDMA, all users can transmit at the same time. Also, each is allocated the entire available frequency spectrum for transmission; hence, CDMA is also known as spread-spectrum multiple access (SSMA), or simply spread-spectrum communications.

Direct-sequence CDMA is the most popular of CDMA techniques. The DS-CDMA transmitter multiplies each user's signal by a distinct code waveform. The detector receives a signal composed of the sum of all users' signals, which overlap in time and frequency. In a conventional DS-CDMA system, a particular user's signal is detected by correlating the entire received signal with that user's code waveform.

There has been substantial interest in DS-CDMA technology in recent years because of its many attractive properties for the wireless medium [1–4].[1] While DS- CDMA systems are only now beginning to be commercially deployed, these properties have led to expectations of large capacity increases over TDMA and FDMA systems. Air interface standards based on DS-CDMA, IS-95, and IS-665 [5] have been defined, and a strong commercial effort is currently underway to deploy cellular systems that use them. (See the article in this issue describing IS-665.)

Multiple access interference (MAI) is a factor which limits the capacity and performance of DS-CDMA systems. MAI refers to the interference between direct-sequence users. This interference is the result of the random time offsets between signals, which make it impossible to design the code waveforms to be completely orthogonal. While the MAI caused by any one user is generally small, as the number of interferers or their power increases, MAI becomes substantial.[2] The conventional detector does not take into account the existence of MAI. It follows a single-user detection strategy in which each user is detected separately without regard for other users.

Because of the interference among users, however, a better detection strategy is one of multi-user detection (also referred to as joint detection or interference cancelation). Here, information about multiple users is used jointly to better detect each individual user. The utilization of multi-user detection algorithms has the potential to provide significant additional benefits for DS-CDMA systems.

The next section contains a description of conventional DS-CDMA detection. In the third section we discuss multi-user detection, and we review the optimal multi-user sequence detector. We then review the two main classes of suboptimal detectors that have been proposed: linear multi-user detectors and subtractive interference cancellation multi-user detectors. This is followed by a summary and concluding remarks.

CONVENTIONAL DETECTION

In this section we take a more detailed look at the conventional detector and the effect of multiple access interference; but first we must define the mathematical system model.

RECEIVED SIGNAL MODEL

We begin with a mathematical description of a synchronous DS-CDMA channel. In a synchronous channel all bits of all users are aligned in time. In practical DS-CDMA applications, however, the channel is generally asynchronous (i.e., signals are randomly delayed — offset — from one another). The asynchronous channel is described in the next section.

To simplify the discussion, we make the assumption that all carrier phases are equal to zero. This enables us to use baseband notation while working only with real signals. To further simplify matters, we also assume that each transmitted signal arrives at the receiver over a single path (no multipath), and that the data modulation is binary phase-shift keying (BPSK [8]).

Assuming there are K direct-sequence users in a synchronous single-path BPSK real channel, the baseband received signal can be expressed as

$$r(t)=\sum_{k=1}^{K}A_k(t)g_k(t)d_k(t)+n(t) \quad (1)$$

where $A_k(t)$, $g_k(t)$, and $d_k(t)$ are the amplitude, signature code waveform, and modulation of the kth user, respectively, and $n(t)$ is additive white Gaussian noise (AWGN), with a two-sided power spectral density of $N_0/2$ W/Hz. The power of the kth signal is equal to the square of its amplitude, which is assumed to be constant over a bit interval. The modulation consists of rectangular pulses of duration T_b (bit interval), which take on $d_k = \pm 1$ values corresponding to the transmitted data. We assume a total of N transmitted bits. The code waveform consists of rectangular pulses of duration T_c ("chip" interval), which pseudorandomly take on ± 1 values, corresponding to some binary "pseudo-noise" (PN) code sequence [5, 8].

The rate of the code waveform, $f_c = 1/T_c$(chip rate), is much greater than the bit rate, $f_b = 1/T_b$. Thus, multiplying the BPSK signal at the transmitter by $g(t)$ has the effect of spreading it out in frequency by a factor of f_c/f_b, (hence, the codes are sometimes referred to as "the spreading codes.") The frequency spread factor of a direct-sequence system is referred to as the processing gain, PG. Hence, for the model of Eq. (1) there are PG chips per bit.

The Conventional Detector

The conventional detector for the received signal described in Eq. (1) is a bank of K correlators, as shown in Fig. 1. Here, each code waveform is regenerated and correlated with the received signal in a separate detector branch. The correlation detector can be equivalently implemented through what is known as matched filtering [8];[3] thus, the conventional detector is often referred to as the matched filter detector. The outputs of the correlators (or matched filters) are sampled at the bit times, which yields "soft" estimates of the transmitted data. The final ± 1 "hard" data decisions are made according to the signs of the soft estimates.

It is clear from Fig. 1 that the conventional detector follows a single-user detector strategy; each branch detects one user without regard to the existence of the other users. Thus, there is no sharing of multiuser information or joint signal processing (i.e., multi-user detection).

The success of this detector depends on the properties of the correlations between codes. We require the correlations between the same code waveforms (i.e., the autocorrelations) to be much larger than the correlations between different codes (i.e., the cross-correlations). The correlation value is defined as

$$\rho_{i,k}=\frac{1}{T_b}\int^{T_b} g_i(t)g_k(t)dt \quad (2)$$

Here, if $i = k$, $\rho_{k,k} = 1$, (i.e., the integrand must equal one since $g_i(t) = \pm 1$), and if $i \neq k$, $0 \le \rho_{i,k} < 1$. The output of the kth user's correlator for a particular bit interval is

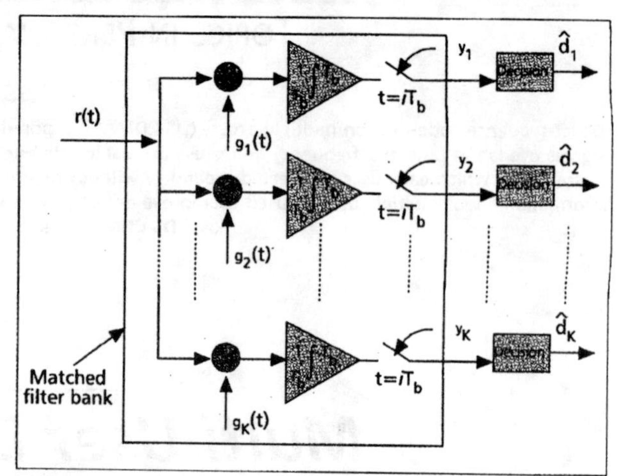

Figure 1. *The conventional DS-CDMA detector: a bank of correlators (matched filters).*

$$\begin{aligned} y_k &= \frac{1}{T_b}\int^{T_b} r(t)g_k(t)dt \\ &= A_k d_k + \sum_{\substack{i=1 \\ i\neq k}}^{K}\rho_{i,k}A_i d_i + \frac{1}{T_b}\int^{T_b} n(t)g_k(t)dt \\ &= A_k d_k + MAI_k + z_k \end{aligned} \quad (3)$$

In other words, correlation with the kth user itself gives rise to the recovered data term, correlation with all the other users gives rise to multiple access interference (MAI), and correlation with the thermal noise yields the noise term z_k. Since the codes are generally designed to have very low crosscorrelations relative to autocorrelations (i.e., $\rho_{i,k} \ll 1$), the interfering effect on user k of the other direct-sequence users is greatly reduced.[4,5]

Nevertheless, the existence of MAI has a significant impact on the capacity and performance of the conventional direct-sequence system. As the number of interfering users increases, the amount of MAI increases. In addition, the presence of strong (large-amplitude) users exacerbates the MAI of the weaker users, as can be seen by Eq. (3). Thus, the overall effect of MAI on system performance is even more pronounced if the users' signals arrive at the receiver at different powers: weaker users may be overwhelmed by stronger users. Such a situation arises when the transmitters have different geographical locations relative to the receiver, because the signals of the closer transmitting users undergo less amplitude attenuation than the signals of users that are further away. This is known as the near-far problem. (Note that this problem also arises due to fading.)

An analogy which helps to illustrate the effect of MAI is as follows. Consider that you are at a party where every conversation takes place in a different language. In general, your ear is reasonably good at picking out your own language and tuning out the other conversations. However, as the number of simultaneous conversations in the room increases, it becomes harder and harder to continue your own conversation. Similar difficulties arise if some of the other conversations get closer or louder, or if the person you are talking to moves further away or begins to whisper (the near-far effect).

Mitigating the Effect of MAI

Research efforts directed at mitigating the effect of MAI on the conventional detector have focused on several areas.

Code Waveform Design — This approach is aimed at the design of spreading codes with good cross-correlation properties. Ideally, if the codes were all orthogonal, then $\rho_{i,k} = 0$, and there would be no MAI term. However, since in practice most channels contain some degree of asynchronism, it is not possible to design codes that maintain orthogonality over all possible delays. So instead we look for codes that are nearly orthogonal, that is, have as low cross-correlation as possible (e.g., [11, 12]).

Power Control — The use of power control ensures that all users arrive at about the same power (amplitude), and therefore no user is unfairly disadvantaged relative to the others (e.g. [13]). In the IS-95 standard, the mobiles adjust their power through two methods. One method is for the mobiles to adjust their transmitted power to be inversely proportional to the power level it receives from the base station (open loop power control). The other method is for the base station to send power control instructions to the mobiles based on the power level it receives from the mobiles (closed loop power control) [5]. Power control is currently considered indispensable for a successful DS-CDMA system.

FEC Codes — The design of more powerful forward error correction (FEC) codes allows acceptable error rate performance at lower signal-to-interference ratio levels. This obviously has broad application, and provides benefits to more than just CDMA systems.

Sectored/Adaptive Antennas — Here, directed antennas are used that focus reception over a narrow desired angle range. Therefore, the desired signal and some fraction of the MAI are enhanced (through the antenna gain), while the interfering signals that arrive from the remaining angles are attenuated. The direction of the antenna can be fixed, as is the case for sectored antennas, or adjusted dynamically. In the latter case, adaptive signal processing is used to focus the antenna in the direction corresponding to a particular desired user(s). Applications for these techniques also extend well beyond CDMA. An overview of the work in this area can be found in [14].

MULTI-USER DETECTION

There has been great interest in improving DS-CDMA detection through the use of multi-user detectors. In multi-user detection, code and timing (and possibly amplitude and phase) information of multiple users are jointly used to better detect each individual user. The important assumption is that the codes of the multiple users are known to the receiver a priori.[6]

Verdú's seminal work [31], published in 1986, proposed and analyzed the optimal multiuser detector, or the maximum likelihood sequence detector (described later in this section). Unfortunately, this detector is much too complex for practical DS-CDMA systems. Therefore, over the last decade or so, most of the research has focused on finding suboptimal multiuser detector solutions which are more feasible to implement.

Most of the proposed detectors can be classified in one of two categories: linear multi-user detectors and subtractive interference cancellation detectors. In linear multi-user detection, a linear mapping (transformation) is applied to the soft outputs of the conventional detector to produce a new set of outputs, which hopefully provide better performance. In subtractive interference cancellation detection, estimates of the interference are generated and subtracted out. We discuss several important detectors in each category in the next two sections.

There are other proposed detectors, as well as variations of each detector, that are not covered here. There is also a large and growing literature dealing with extensions of the various multi-user algorithms to realistic environments.[7–9] The interested reader can find additional references and discussion in the survey articles [16–18].

It is interesting to note that there is a strong parallel between the problem of MAI and that of intersymbol interference (ISI). This point is made in [31], where the asynchronous K-user channel is identified with the single-user ISI channel with memory $K - 1$. The mathematical and conceptual similarity of the two problems is evident if one thinks of the $K - 1$ overlapping ISI symbols as separate users. Therefore, a number of multi-user detectors have equalizer counterparts, such as the maximum-likelihood, zero-forcing, minimum mean-squared error, and decision-feedback equalizers [8]. We will point out these similarities as we go along.

LIMITATIONS AND POTENTIAL BENEFITS

Before discussing the details of multi-user detection, it is important to examine some of the limitations that exist and potential benefits available. We focus on the cellular environment, although the ideas extend to other wireless applications.

In a cellular environment, there are two channels in a given coverage region: a central station, called a base station, transmits to mobiles (downlink), and the mobiles transmit to the base station (uplink). The coverage region associated with one base sation is referred to as a cell. Generally, the uplink and the downlink utilize different frequency bands. There are two main limitations on the benefits of multiuser detection for the cellular environment:

Existence of Other-Cell MAI — In cellular DS-CDMA systems, the same uplink/ downlink pair of frequency bands are reused for each cell. Thus, a signal transmitted in one cell may cause interference in neighboring cells. If this interference is not included in the multi-user detection algorithm, the potential gain is significantly reduced. (A similar effect occurs from uncaptured multipath signals [1].) An upper bound on the capacity increase is easily derived by comparing the total interference for systems with and without multi-user detection. If we neglect background noise, the total interference in a system without multi-user detection is $I = I_{MAI} + fI_{MAI}$, where I_{MAI} is MAI due to same-cell users, and f is the ratio of other-cell MAI to same-cell MAI (also referred to as the *spillover ratio*). For an ideal system where all same-cell MAI is eliminated, we are still left with interference $I = fI_{MAI}$. Since the number of users is roughly proportional to the interference [3], the maximum capacity gain factor would be $(1 + f)/f$ [1]. A typical value for f in cellular systems is 0.55 [1]; this translates to a maximum capacity gain factor of 2.8.

Difficulty in Implementing Multi-User Detection on the Downlink — Because issues of cost, size, and weight are much larger concerns for the mobiles than for the base station, it is not currently practical to include multi-user detection in mobiles. Instead, it has primarily been considered for use at the base station (for uplink reception of mobiles), where detection of multiple users is required in any case. However, improving the capacity of the uplink past that of the downlink does not improve the overall capacity of the system [17].

Despite these limitations, the use of multi-user detectors offers substantial potential benefits:

Significant Improvement in Capacity

- Although other-cell MAI causes the capacity improvement for the cellular environment to be bounded, the improvement is still significant.
- The bound can be improved by including signals from the

surrounding cells in the multi-user detection algorithm.[10]

- There are other applications, such as satellite communications, where the spillover ratio is much less than that of cellular communications.
- Although multi-user detection is not currently practical for the downlink, DS-CDMA systems are generally considered to be uplink limited [3]. In addition, with improvements in technology, techniques for improving downlink performance may become more practical (e.g., see Endnote 6).

More Efficient Uplink Spectrum Utilization — The improvement in the uplink allows mobiles to operate at a lower processing gain [15]. This leads to a smaller chunk of bandwidth required for the uplink; the extra bandwidth could then be used to improve the downlink capacity. Alternatively, for the same bandwidth the uplink could support higher data rates.

Reduced Precision Requirements for Power Control — Since the impact of MAI and the near-far effect is much reduced, the need for all users to arrive at the receiver at exactly the same power is reduced; thus, less precision is needed in controlling the transmitted power of the mobiles. Therefore, the additional complexity at the base station required for multi-user detection may allow reduced complexity at the mobiles [15].

More Efficient Power Utilization — The reduction of interference on the uplink may translate to some reduction in the required transmit power of the mobiles [15]. Alternatively, the same transmit power may be used to extend the size of the coverage region.

Matrix-Vector Notation

In discussing multi-user detection, it is convenient to introduce a matrix-vector system model to describe the output of the conventional detector. We begin with a simple example to help illustrate our discussion: a three user synchronous system. From Eq. (3), the output for each of the users for one bit is

$$\begin{aligned} y_1 &= A_1 d_1 + \rho_{2,1} A_2 d_2 + \rho_{3,1} A_3 d_3 + z_1 \\ y_2 &= \rho_{1,2} A_1 d_1 + A_2 d_2 + \rho_{3,2} A_3 d_3 + z_2 \\ y_3 &= \rho_{1,3} A_1 d_1 + \rho_{2,3} A_2 d_2 + A_3 d_3 + z_3 \end{aligned} \quad (4)$$

This can be written in the matrix-vector form

$$\begin{bmatrix} y_1 \\ y_2 \\ y_3 \end{bmatrix} = \begin{bmatrix} 1 & \rho_{2,1} & \rho_{3,1} \\ \rho_{1,2} & 1 & \rho_{3,2} \\ \rho_{1,3} & \rho_{2,3} & 1 \end{bmatrix} \begin{bmatrix} A_1 & 0 & 0 \\ 0 & A_2 & 0 \\ 0 & 0 & A_3 \end{bmatrix} \begin{bmatrix} d_1 \\ d_2 \\ d_3 \end{bmatrix} + \begin{bmatrix} z_1 \\ z_2 \\ z_3 \end{bmatrix} \quad (5)$$

or

$$\mathbf{y} = \mathbf{RAd} + \mathbf{z} \quad (6)$$

For a K user system, the vectors **d**, **z**, and **y**, are K-vectors that hold the data, noise, and matched filter outputs of all K users, respectively; the matrix **A** is a diagonal matrix containing the corresponding received amplitudes; the matrix **R** is a $K \times K$ correlation matrix, whose entries contain the values of the correlations between every pair of codes. Note that since $\rho_{i,k} = \rho_{k,i}$, the matrix **R** is clearly symmetric.

It is instructive to break up **R** into two matrices: one representing the autocorrelations, the other the crosscorrelations. Therefore, parallel to Eq. (3), the conventional matched filter detector output can be expressed as three terms:

$$\mathbf{y} = \mathbf{Ad} + \mathbf{QAd} + \mathbf{z} \quad (7)$$

where **Q** contains the off-diagonal elements (crosscorrelations) of **R**, that is, **R** = **I** + **Q** (**I** is the identity matrix). The first term, **Ad**, is simply the decoupled data weighted by the received amplitudes. The second term, **QAd**, represents the MAI interference.

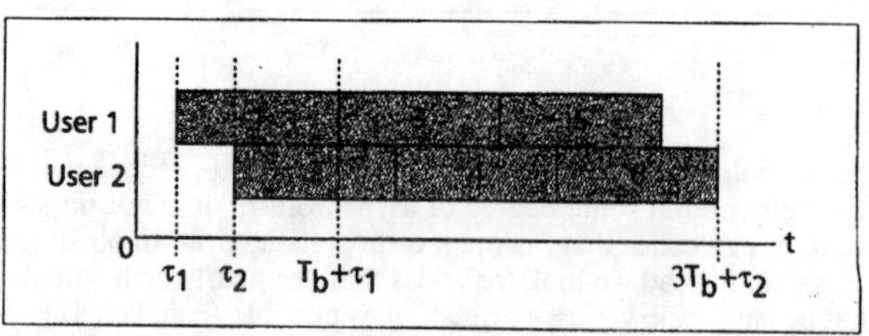

Figure 2. *Sample timing diagram for an asynchronous channel. There are 2 users and 3 bits per user.*

Asynchronous Channel

The detection problem in an asynchronous channel is more complicated than in a synchronous channel. In a synchronous channel, by definition, the bits of each user are aligned in time. Thus, detection can focus on one bit interval independent of the others (e.g., Eq. (3)); the detection of N bits of K users is equivalent to N separate "one-shot" detection problems. In most realistic applications, however, the channel is asynchronous and thus, there is overlap between bits of different intervals. Here, any decision made on a particular bit ideally needs to take into account the decisions on the 2 overlapping bits of each user; the decisions on these overlapping bits must then further take into account decisions on bits that overlap them and so on. Therefore, the detection problem must optimally be framed over the whole message [40].

The continuous-time model expressed in Eq. (1) can easily be modified for asynchronous channels by including the relative time delays (offsets) between signals. The received signal is now written as

$$r(t) = \sum_{k=1}^{K} A_k(t) g_k(t-\tau_k) d_k(t-\tau_k) + n(t) \quad (8)$$

where τ_k is the delay for user k.

The discrete-time matrix-vector model describing the asynchronous channel takes the same form as Eq. (6). However, now the equation must encompass the entire message; thus, assuming there are N bits per user, the size of the vectors and the order of the matrices are NK. The vectors **d**, **z**, and **y** hold the data, noise, and matched filter outputs of all K users for all N bit intervals, and the matrix **A** contains the corresponding received amplitudes. The matrix **R** now contains the partial correlations that exist between every pair of the NK code words and is of size $NK \times NK$. We use the term partial correlations because in an asynchronous channel, the codes for each bit only partially overlap each other.

An example helps to illustrate our discussion. Consider the timing diagram of Fig. 2, where there are a total of two users and 3 bits per user. The output of the conventional detector can be described using Eq. (6), where we treat the problem as if there were six users (each transmitting 1 bit over the interval $3T_b + \tau_2 - \tau_1$). The vectors **d**, **z**, and **y** hold the data, noise, and matched filter outputs associated with each of these 6 bits. The correlation matrix, **R**, is of dimension 6 x 6 and can be written as

$$R = \begin{bmatrix} 1 & \rho_{2,1} & 0 & 0 & 0 & 0 \\ \rho_{1,2} & 1 & \rho_{3,2} & 0 & 0 & 0 \\ 0 & \rho_{2,3} & 1 & \rho_{4,3} & 0 & 0 \\ 0 & 0 & \rho_{3,4} & 1 & \rho_{5,4} & 0 \\ 0 & 0 & 0 & \rho_{4,5} & 1 & \rho_{6,5} \\ 0 & 0 & 0 & 0 & \rho_{5,6} & 1 \end{bmatrix} \quad (9)$$

where $\rho_{i,K}$ is now the partial cross-correlation between the code associated with bit i and that associated with bit k; in other

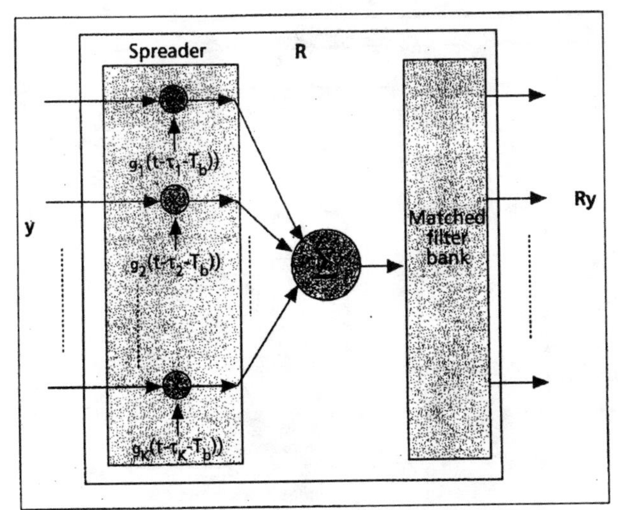

■ **Figure 3.** *One stage of the polynomial expansion detector. The input is the matched filter bank output vector* **y** *and the output is* **Ry**. *A diagram of the matched filter bank is pictured in Fig. 1.*

words, it denotes the cross-correlation between the overlapping part of code i and code k. Note that the 0 entrees correspond to the correlations between bits that do not overlap. For a typical message length N is much greater than K; hence, the correlation matrix is sparse because most of the NK bits do not overlap.

For the remainder of this article, an asynchronous channel is assumed unless otherwise stated. A more in-depth presentation of the mathematical details of the asynchronous channel can be found in [16, 18].

Maximum-Likelihood Sequence Detection

The detector which yields the most likely transmitted sequence, **d**, chooses **d** to maximize the probability that **d** was transmitted given that $r(t)$ was received, where $r(t)$ extends over the whole message. This probability is referred to as the *joint a posteriori probability*, $P(\mathbf{d} \,|\, (r(t), \text{for all } t))$ [8]. Under the assumption that all possible transmitted sequences are equally probable, this detector is known as the maximum-likelihood sequence (MLS) detector [8].[11–13]

The problem with the MLS approach is that here there are 2^{NK} possible **d** vectors; an exhaustive search is clearly impractical for typical message sizes and numbers of users. However, it turns out that MLS detection can be implemented for DS-CDMA by following the matched filter bank with a Viterbi algorithm [31].[14] This method parallels the use of the Viterbi algorithm to implement MLS detection in channels corrupted by intersymbol interference [8, 31]. Unfortunately, the required Viterbi algorithm has a complexity that is still exponential in the number of users, that is, on the order of 2^K.[15]

Another disadvantage of the MLS detector is that it requires knowledge of the received amplitudes and phases. These values, however, are not known a priori, and must be estimated (e.g. [34–37]).

Despite the huge performance and capacity gains over conventional detection, the MLS detector is not practical. A realistic direct-sequence system has a relatively large number of active users; thus, the exponential complexity in the number of users makes the cost of this detector too high. In the remainder of this article we look at various suboptimal multi-user detectors that are simpler to implement.

Linear Detectors

An important group of multi-user detectors are linear multi-user detectors. These detectors apply a linear mapping, **L**, to the soft output of the conventional detector to reduce the MAI seen by each user. In this section we briefly review the two most popular of these, the decorrelating and minimum mean-squared error detectors. We then examine the polynomial expansion detector, a linear detector recently proposed by the author that can efficiently implement both of the aforementioned detectors.

Decorrelating Detector

The decorrelating detector applies the inverse of the correlation matrix

$$\mathbf{L}_{dec} = \mathbf{R}^{-1} \tag{10}$$

to the conventional detector output in order to decouple the data. (Note that **R** can be assumed to be invertible for asynchronous systems [40].) From Eq. (6), the soft estimate of this detector is

$$\begin{aligned}\hat{\mathbf{d}}_{dec} &= \mathbf{R}^{-1}\mathbf{y} = \mathbf{A}\mathbf{d} + \mathbf{R}^{-1}\mathbf{z} \\ &= \mathbf{A}\mathbf{d} + \mathbf{z}_{dec}\end{aligned} \tag{11}$$

which is just the decoupled data plus a noise term. Thus, we see that the decorrelating detector completely eliminates the MAI. This detector is very similar to the zero-forcing equalizer [8] which is used to completely eliminate ISI.

The decorrelating detector was initially proposed in [38, 39]. It is extensively analyzed by Lupas and Verdu in [40, 41], and is shown to have many attractive properties. Foremost among these properties are [16, 40, 41]:

- **Provides substantial performance/capacity gains over the conventional detector under most conditions.**[16]
- **Does not need to estimate the received amplitudes.** In contrast, detectors that require amplitude estimation are often quite sensitive to estimation error. (Note that as in the case of most multi-user detectors, the need to estimate the received phases can also be avoided through the use of noncoherent detection .[9])
- **Has computational complexity significantly lower than that of the maximum likelihood sequence detector.** The per-bit complexity is linear in the number of users, excluding the costs of recomputation of the inverse mapping.

Other desirable features of the decorrelating detector are [16, 40, 41]:

- **Corresponds to the maximum likelihood sequence detector when the energies of all users are not known at the receiver.** In other words, it yields the joint maximum likelihood sequence estimation of the transmitted bits and their received amplitudes.
- **Has a probability of error independent of the signal energies.** This simplifies the probability of error analysis, and makes the decorrelating detector resistant to the near-far problem.
- **Yields the optimal value of the near-far resistance performance metric.**[17]
- **Can decorrelate one bit at a time.** For bit k, we only need apply the kth row of $\mathbf{R}^{-1}$ to the matched filter bank outputs.

Because of its many advantages, the decorrelating detector has probably received the most attention of any multi-user detector in the literature. Many additional references can be found in [16–18].[18]

A disadvantage of this detector is that it causes noise enhancement (similar to the zero-forcing equalizer [8]). The power associated with the noise term $\mathbf{R}^{-1}\mathbf{z}$ at the output of the decorrelating detector — Eq. (11) — is always greater than or equal to the power associated with the noise term at the output of the conventional detector — Eq. (6) — for each bit (proved in [44]). Despite this drawback, the decorrelating detector generally provides significant improvements over the conventional detector.[19]

A more significant disadvantage of the decorrelating detector is that the computations needed to invert the matrix **R** are difficult to perform in real time. For synchronous systems, the problem is somewhat simplified: we can decorrelate one bit at a time. In other words, we can apply the inverse of a $K \times K$ correlation matrix. For asynchronous systems, however, **R** is of order NK, which is quite large for a typical message length, N.

There have been numerous suboptimal approaches to implementing the decorrelating detector [16–18]. Many of them entail breaking up the detection problem into more manageable blocks [45–48, 79, 81] (possibly even to one transmission interval [16, 49]); the inverse matrix can then be exactly computed.[20] A K-input K-output linear filter implementation is also possible [40], where the filter coefficients are a function of the cross-correlations.[21,22]

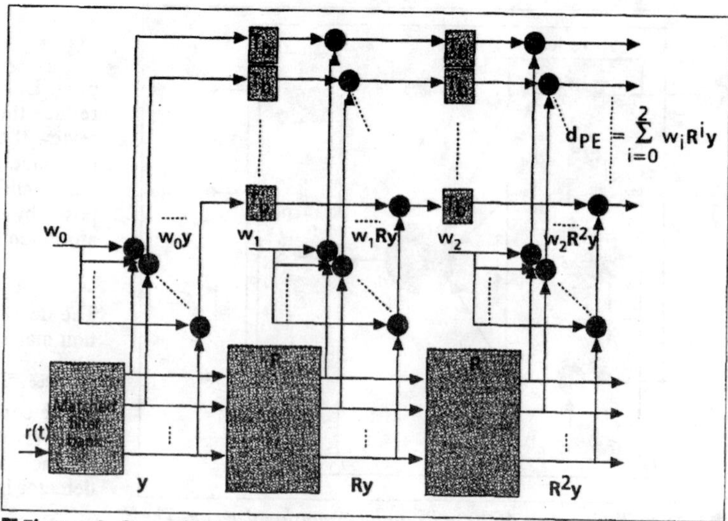

■ **Figure 4.** *General DS-CDMA polynomial expansion detector with 2 stages.*

Whichever suboptimal decorrelating detector technique is used, the computation required is substantial. Therefore, the use of codes that repeat each bit ("short" codes) is generally assumed so that the partial correlations between all signals are the same for each bit. This minimizes the need for recomputation of the matrix inverse or the filter coefficients from one bit interval to the next. Where recomputation cannot be avoided, (e.g., new user activation), research has been directed at trying to simplify the cost of recomputation (e.g. [52, 53]). The processing burden still appears to present implementation difficulties.

Minimum Mean-Squared Error (MMSE) Detector

The minimum mean-squared error (MMSE) detector [45] is a linear detector which takes into account the background noise and utilizes knowledge of the received signal powers. This detector implements the linear mapping which minimizes $E[|\mathbf{d} - \mathbf{Ly}|^2]$, the mean-squared error between the actual data and the soft output of the conventional detector. This results in [45, 84]

$$\mathbf{L}_{\mathrm{MMSE}} = [\mathbf{R} + (N_0/2)\mathbf{A}^{-2}]^{-1} \tag{12}$$

Thus, the soft estimate of the MMSE detector is simply

$$\hat{\mathbf{d}}_{\mathrm{MMSE}} = \mathbf{L}_{\mathrm{MMSE}}\,\mathbf{y} \tag{13}$$

As can be seen, the MMSE detector implements a partial or modified inverse of the correlation matrix. The amount of modification is directly proportional to the background noise; the higher the noise level, the less complete an inversion of **R** can be done without noise enhancement causing performance degradation. Thus, the MMSE detector balances the desire to decouple the users (and completely eliminate MAI) with the desire to not enhance the background noise. (Additional explanation can be found in [54].) This multi-user detector is exactly analogous to the MMSE linear equalizer used to combat ISI [8].

Because it takes the background noise into account, the MMSE detector generally provides better probability of error performance than the decorrelating detector. As the background noise goes to zero, the MMSE detector converges in performance to the decorrelating detector.[23]

An important disadvantage of this detector is that, unlike the decorrelating detector, it requires estimation of the received amplitudes. Another disadvantage is that its performance depends on the powers of the interfering users [45]. Therefore, there is some loss of resistance to the near-far problem as compared to the decorrelating detector.

Like the decorrelating detector, the MMSE detector faces the task of implementing matrix inversion. Thus, most of the suboptimal techniques for implementing the decorrelating detector are applicable to this detector as well.[24]

Polynomial Expansion (PE) Detector

The polynomial expansion (PE) detector [54, 55], applies a polynomial expansion in **R** to the matched filter bank output, **y**. Thus, the linear mapping for the PE detector is

$$\mathbf{L}_{\mathrm{PE}} = \sum_{i=0}^{N_s} w_i \mathbf{R}^i \tag{14}$$

and the soft estimates of **d** are given by

$$\hat{\mathbf{d}}_{\mathrm{PE}} = \mathbf{L}_{\mathrm{PE}}\,\mathbf{y} \tag{15}$$

For a given **R** and N_s, the weights (polynomial coefficients) w_i, $i = 0, 1, \ldots, N_s$ can be chosen to optimize some performance measure.

The structure which implements the matrix **R** is shown in Fig. 3, and the full detector (with two stages) is shown in Fig. 4. Each stage implements **R** by recreating the overall modulation (spreading), noiseless channel (summing), and demodulation (matched filtering) process. The fact that this implements **R** is clear from the expression for the noiseless conventional detector output, $\mathbf{y} = \mathbf{RAd}$ (Eq. (6)). Cascading these stages produces higher-order terms of the polynomial. A two-stage PE detector is shown in Fig. 4; the detector corresponding to Eq. (14) requires N_s stages.

It can be shown (by the Cayley-Hamilton Theorem) that the PE detector structure can exactly implement the decorrelating detector for finite message length, N [54]. However, for typical N this would require a prohibitive number of stages. As $N \rightarrow \infty$, infinite stages would be needed, with one bit delay required per stage.[25] Fortunately, good approximations can be obtained with a relatively small number of stages. Therefore, we can choose $\mathbf{w} = [w_0\ w_1 \ldots w_{N_s}]$ so that

$$p(\mathbf{R}) = \sum_{i=0}^{N_s} w_i \mathbf{R}^i \approx \mathbf{R}^{-i} \tag{16}$$

The resulting weights are used in the structure of Fig. 4 to yield a K-input K-output finite memory-length detector, which approximates the decorrelating detector.

The PE detector structure can also be used to approximate the MMSE detector, as described in [54].

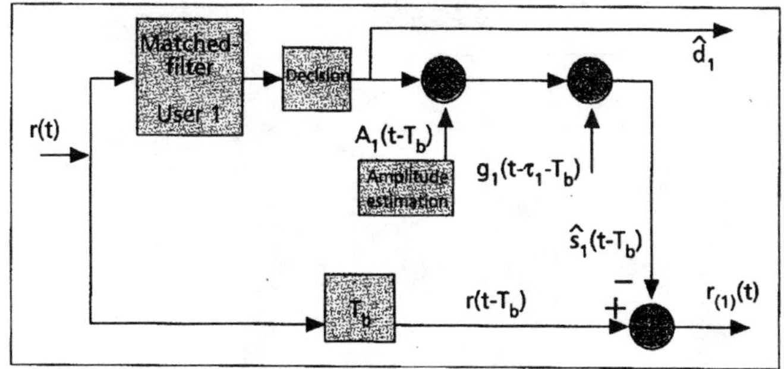

■ **Figure 5.** *SIC detection – first stage (hard decision).*

The polynomial expansion detector has a number of attractive features [54, 55]:

- **Can approximate the decorrelating and MMSE detectors.** As such, it can enjoy the desirable features of these two detectors, which were discussed earlier.
- **Has low computational complexity.** In approximating the decorrelating (or MMSE) detector, neither the matrix **R** nor its inverse must be explicitly calculated. Everything can be implemented on-line, using anything from analog hardware to DSP chips.
- **Does not need to estimate the received amplitudes (or phases).** This important feature, which is true for the decorrelating detector, is also true for the PE detector in approximating the decorrelating detector. (If the PE detector is approximating the MMSE detector, however, amplitude estimation will be necessary).
- **Can be implemented just as easily using long codes as short codes.** (See [51] which points out a problem with using short codes.)
- **Can use weights that work well over a large variation of system parameters.** As shown in [54], the use of additional stages in the PE detector (a higher order polynomial) allows more flexibility to use pre-computed weights that work well over a broad operating range. This minimizes or eliminates the need to adapt the weights to changes in the operating environment.
- **Has a relatively simple structure.** The types of system components used are the same as those of the conventional detector. The amount of system components increases linearly with the product of the number of users and the number of stages. As we will see in the next section, the structure is very similar to that of the parallel interference cancellation detector structure. In that structure, each stage contains a modulator (spreader) a *partial* summer, and a demodulator (matched-filter bank), which implements the matrix **Q** (**R** with its main diagonal removed).

Subtractive Interference Cancellation

Another important group of detectors can be classified as subtractive interference cancellation detectors. The basic principle underlying these detectors is the creation at the receiver of separate estimates of the MAI contributed by each user in order to subtract out some or all of the MAI seen by each user. Such detectors are often implemented with multiple stages, where the expectation is that the decisions will improve at the output of successive stages.

These detectors are similar to feedback equalizers [8] used to combat ISI. In feedback equalization, decisions on previously detected symbols are fed back in order to cancel part of the ISI. Thus, a number of these types of multi-user detectors are also referred to as decision-feedback detectors.

The bit decisions used to estimate the MAI can be hard or soft. The soft-decision approach uses soft data estimates for the joint estimation of the data and amplitudes, and is easier to implement.[26] The hard-decision approach feeds back a bit decision and is nonlinear; it requires reliable estimates of the received amplitudes in order to generate estimates of the MAI. If reliable amplitude estimation is possible, hard-decision subtractive interference cancellation detectors generally outperform their soft-decision counterparts. However, studies such as [56, 57] indicate that the need for amplitude estimation is a significant liability of the hard-decision techniques: imperfect amplitude estimation may significantly reduce or even reverse the performance gains available.

We briefly review several subtractive interference cancellation detectors below. Additional references can be found in two surveys which focus on these detectors [58, 59] and in the general surveys [16–18].

Successive Interference Cancellation (SIC)

The successive interference cancellation (SIC) detector [60, 68] takes a serial approach to canceling interference. Each stage of this detector decisions, regenerates, and cancels out one additional direct-sequence user from the received signal, so that the remaining users see less MAI in the next stage. (Note that the basic concept behind this approach can be found earlier in information theory [61–63].)

A simplified diagram of the first stage of this detector is shown in Fig. 5, where a hard-decision approach is assumed. The first stage is preceded by an operation which ranks the signals in descending order of received powers (not shown). The first stage implements the following steps:

1. Detect with the conventional detector the strongest signal, s_1.
2. Make a hard data decision on s_1.
3. Regenerate an estimate of the received signal for user one, $\hat{s}_1(t)$, using:
   - Data decision from step 2
   - Knowledge of its PN sequence
   - Estimates of its timing and amplitude (and phase)[9]
4. Cancel (subtract out) $\hat{s}_1(t)$ from the total received signal, $r(t)$, yielding a partially cleaned version of the received signal, $r_{(1)}(t)$.

Assuming that the estimation of $\hat{s}_1(t)$ in step 3 above was accurate, the outputs of the first stage are:

1. A data decision on the strongest user
2. A modified received signal without the MAI caused by the strongest user

This process can be repeated in a multistage structure: the kth stage takes as its input the "partially cleaned" received signal output by the previous stage, $r_{(k-1)}(t)$, and outputs one additional data decision (for signal s_k) and a "cleaner" received signal, $r_{(k)}(t)$.[27, 28]

The reasons for canceling the signals in descending order of signal strength are straightforward [17, 68]. First, it is easiest to achieve acquisition and demodulation on the strongest users (best chance for a correct data decision). Second, the removal of the strongest users gives the most benefit for the remaining users. The result of this algorithm is that the strongest user will not benefit from any MAI reduction; the weakest users, however, will potentially see a huge reduction in their MAI.[29]

The SIC detector requires only a minimal amount of additional hardware and has the potential to provide significant improvement over the conventional detector. It does, however, pose a couple of implementation difficulties. First, one additional bit delay is required per stage of cancellation. Thus, a trade-off must be made between the number of users that are canceled and the amount of delay that can be tolerated [64]. Second, there is a need to reorder the signals whenever the power profile changes [64]. Here, too, a trade-off must be made between the precision of the power ordering and the acceptable processing complexity.

A potential problem with the SIC detector occurs if the initial data estimates are not reliable. In this case, even if the timing, amplitude, and phase estimates are perfect, if the bit estimate is wrong, the interfering effect of that bit on the signal-to-noise ratio is quadrupled in power (the amplitude doubles, so the power quadruples). Thus, a certain minimum performance level of the conventional detector is required for the SIC detector to yield improvements; it is crucial that the data estimates of at least the strong users that are canceled first be reliable.

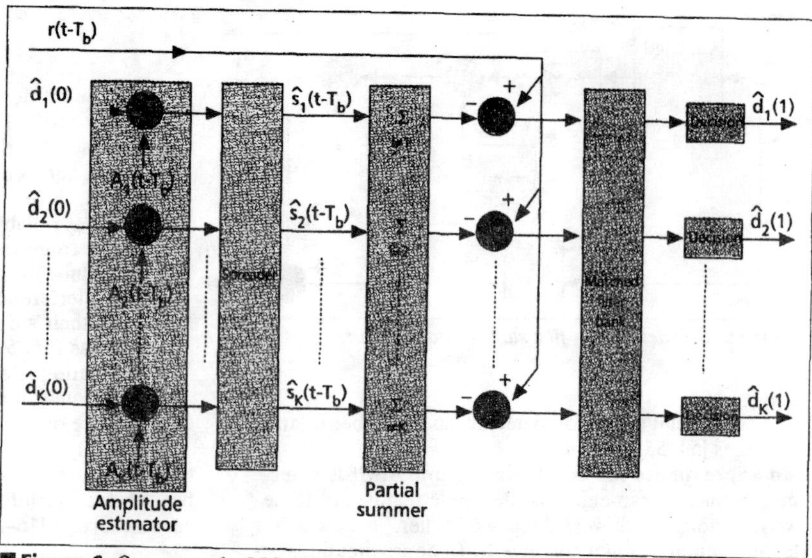

Figure 6. *One stage of a PIC detector (hard decision) for K users. The initial stage (conventional detector) is not shown; it introduces one bit delay, which is why the received signal and the amplitudes are delayed aby T_b. The spreader is defined in Fig. 3 and the matched filter bank is defined in Fig. 1.*

Parallel Interference Cancellation

In contrast to the SIC detector, the parallel interference cancellation (PIC) detector estimates and subtracts out all of the MAI for each user in parallel. The multistage PIC structure which we assume here was introduced in [67]. A basic one stage PIC structure is assumed in [68, 69] and several earlier references (see [18]).

The first stage of this detector is pictured in Fig. 6, where a hard-decision approach is assumed. The initial bit estimates, $\hat{d}_i(0)$, are derived from the matched filter detector (not shown), which we refer to as stage 0 of this detector. These bits are then scaled by the amplitude estimates and respread by the codes, which produces a delayed estimate of the received signal for each user, $\hat{s}_k(t - T_b)$. The partial summer sums up all but one input signal at each of the outputs, which creates the complete MAI estimate for each user.

Assuming perfect amplitude and delay estimation, the result after subtracting the MAI estimate for user k is

$$r(t-T_b)-\sum_{i\neq k}^{K}\hat{s}_i(t-T_b)=$$
$$d_k(t-\tau_k-T_b)A_k(t-\tau_k-T_b)g_k(t-\tau_k-T_b)+n(t-T_b)$$
$$+\sum_{i\neq k}^{K}\Big(d_i(t-\tau_i-T_b)-\hat{d}_i(t-\tau_i-T_b)\Big)A_i(t-\tau_i-T_b)g_i(t-\tau_i-T_b \qquad (17)$$

As shown in Fig. 6, the result of Eq. (17) (for $k = 1...K$) is passed on to a second bank of matched filters to produce a new, hopefully better, set of data estimates.

This process can be repeated for multiple stages. Each stage takes as its input the data estimates of the previous stage and produces a new set of estimates at its output. We can use a matrix-vector formulation to compactly express the soft output of stage $m + 1$ of the PIC detector for all N bits of all K users as [70]

$$\begin{aligned}\tilde{\mathbf{d}}(m+1) &= \mathbf{y}-\mathbf{QA\hat{d}}(m)\\ &= \mathbf{Ad}+\mathbf{QA}(\mathbf{d}-\mathbf{\hat{d}}(m))+\mathbf{z}\end{aligned} \qquad (18)$$

The term $\mathbf{QA\hat{d}}(m)$ represents an estimate of the MAI (7). (As usual, for BPSK, the hard data decisions, $\hat{\mathbf{d}}(m)$, are made according to the signs of the soft outputs, $\tilde{\mathbf{d}}(m)$.) Perfect data estimates, coupled with our assumption of perfect amplitude and delay estimation, result in the complete elimination of MAI.[30]

A number of studies have investigated PIC detection which utilizes soft decisions, such as [55, 72, 76, 86]. In [72] soft-decision PIC and SIC detectors are compared; since soft-decision SIC exploits power variation by canceling in order of signal strength, it is found to be superior in a non-power-controlled fading channel. On the other hand, soft-decision PIC is found to be superior in a well-power-controlled channel.

A number of variations on the PIC detector have been proposed for improved performance. These include the following.

Using the Decorrelating Detector as the First Stage [70] — The performance of the PIC detector depends heavily on the initial data estimates [67]. As we pointed out for the SIC detector, the subtraction of an interfering bit based on an incorrect bit estimate causes a quadrupling in the interfering power for that bit. Thus, too many incorrect initial data estimates may cause performance to degrade relative to the conventional detector (no cancelation may be better than poor cancelation). Therefore, using the decorrelating detector as the first stage significantly improves the performance of the PIC detector. (An additional benefit from this approach is that the perormance analysis is found to be much simplified.)[31, 32]

Using the Already Detected Bits at the Output of the Current Stage to Improve Detection of the Remaining Bits in the Same Stage [74] — Thus, the most up-to-date bit decisions available are always used. This contrasts with the standard PIC detector, which only uses the previous stage's decisions. This detector is referred to as a multistage decision feedback detector [74]. Proposals for the initial stage of this detector include a decision-feedback detector [74], the conventional detector [45], and the decorrelating detector [78].[33]

Linearly Combining the Soft-Decision Outputs of Different Stages of the PIC Detector [55] — This simple

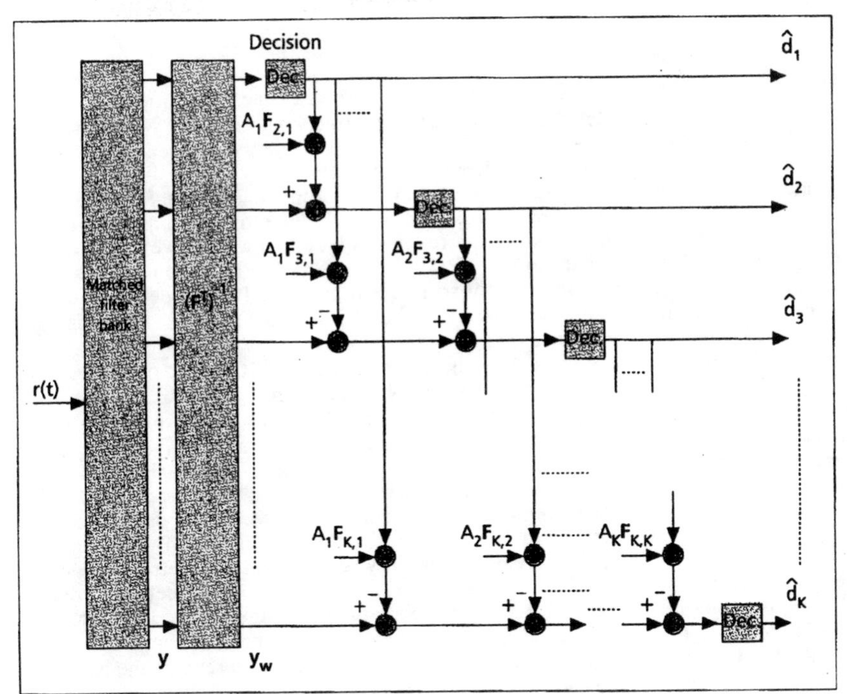

■ **Figure 7.** *The ZF-DF detector. A form of SIC is performed on the whitened matched filter output.*

modification yields very large gains in performance over the standard soft-decision PIC detector. The reason for this has to do with the extensive noise correlations that exist between outputs of different stages. The linear combination is made in such a way as to capitalize on the noise correlations and cause cancellation among noise terms.

Doing a Partial MAI Cancellation at Each Stage, with the Amount of Cancellation Increasing for Each Successive Stage [76] — Thus, the MAI estimate is first scaled by a fraction before cancellation; the value of the fraction increases for successive stages. This takes into account the fact that the tentative decisions of the earlier stages are less reliable than those of the later stages. Huge gains in performance and capacity are reported over the standard ("brute force") PIC detector. This recently proposed detector may be the most powerful of the subtractive interference cancellation detectors, and needs to be studied further.

Zero-Forcing Decision-Feedback (ZF-DF) Detector

The zero-forcing decision-feedback (ZF-DF) detector (also referred to as the decorrelating DF detector)[77–79, 81] performs two operations: linear preprocessing followed by a form of SIC detection. The linear operation *partially* decorrelates the users (without enhancing the noise), and the SIC operation decisions and subtracts out the interference from one additional user at a time, in descending order of signal strength. As we describe below, the initial partial decorrelation enables the SIC operation to be much more powerful.

The ZF-DF detector is based on a white noise channel model. A noise-whitening filter is obtained by factoring **R** by the Cholesky decomposition [83], $\mathbf{R} = \mathbf{F}^T\mathbf{F}$, where **F** is a lower triangular matrix. Applying $(\mathbf{F}^T)^{-1}$ to the matched filter bank outputs of Eq. (6) yields the white noise model [77]

$$\mathbf{y_w} = \mathbf{FAd} + \mathbf{z_w} \qquad (19)$$

where the covariance matrix of the noise term, $\mathbf{z_w}$, is $(N_0/2)\mathbf{I}$ (white noise). (This is similar to the white noise model that is derived for ISI chanels [8].)

In the white noise model of Eq. (19), the data bits are partially decorrelated. This can be shown to arise from the fact that the matrix **F** is lower triangular [77]. Thus, the output for bit one of the first user contains no MAI; the output for bit one of the second user contains MAI only from bit one of the first user, and is completely decorrelated from all other users; similarly, the output for user k at bit interval i is completely decorrelated from users $k + 1, k + 2, ..., K$, at time i, and from all bits at future time intervals.

The ZF-DF detector uses SIC detection to exploit the partial decorrelation of the bits in the white noise model. The soft output of bit one of the first user, which is completely free of MAI, is used to regenerate and cancel out the MAI it causes, thereby leaving the soft output of bit one of the second user also free of MAI (decorrelated). This process continues: for each iteration, the MAI contributed by one additional bit (the previously decorrelated bit) is regenerated and canceled, thereby yielding one additional decorrelated bit.

Prior to forming and applying $(\mathbf{F}^T)^{-1}$ to create the white noise model, the users are ordered according to their signal strength, thus insuring that interference cancellation takes place in descending order of signal strength. This maximizes the gains to be had from SIC detection, as discussed earlier.

A diagram of the ZF-DF detector is shown in Fig. 7, where we assume a synchronous channel for clarity.[34] In a synchronous channel we can deal with one bit interval at a time; hence, the size of the vectors and the order of **F** in Eq. (19) are reduced to K. Assuming perfect estimates of **F** and the received amplitudes, the soft output for the kth user is [77]

$$\hat{d}_k = y_{w,k} - \sum_{i=0}^{k-1} \mathbf{F}_{k,i} A_i \hat{d}_i \qquad (20)$$

where $\hat{d}_i$ = sign [$\hat{d}_i$] are the previously detected bits (of the stronger users), A_i is the received amplitude of this bit, and $\mathbf{F}_{k,i}$ is the (k, i)th element of **F**.

Under the assumption that all past decisions are correct, the ZF-DF detector eliminates all MAI and maximizes the signal-to-noise ratio [78].[35] It is analogous to the ZF-DF equalizer used to combat ISI.[36, 37]

An important difficulty with the ZF-DF detector is the need to compute the Cholesky decomposition[38] and the whitening filter $(\mathbf{F}^T)^{-1}$ (matrix inversion). Attempts to simplify its implementation are similar to those of the decorrelating detector.

The ZF-DF detector, like the other nonlinear detectors, has the disadvantage of needing to estimate the received signal amplitudes. If the soft outputs of the decorrelating detector are used to estimate the amplitudes, the ZF-DF detector is equivalent to the decorrelating detector [78]. If the amplitude estimates are more reliable than those produced by the decorrelating detector, the ZF-DF detector performs better than the decorrelating detector; if less reliable, however, the ZF-DF detector performs worse than the decorrelating detector.

SUMMARY AND CONCLUSION

Multiple access interference significantly limits the performance and capacity of conventional DS-CDMA systems. Much research has been directed at mitigating this problem through the design of multi-user detectors.

In multi-user detection, code and timing information of multiple users is jointly used to better detect each individual user. The optimum multi-user sequence detector is known, and provides huge gains in performance and capacity over the conventional detector; it also minimizes the need for power control. Unfortunately, it is too complex to implement for practical DS-CDMA systems.

Many simpler suboptimal multi-user detectors have been proposed in the last few years, all of which have the potential to provide substantial performance and capacity gains over the conventional detector. Most of the detectors fall into two categories: linear and subtractive interference cancellation.

LINEAR DETECTORS

Linear multi-user detectors, which include the decorrelating, minimum mean-squared error (MMSE), and polynomial expansion (PE) detectors, apply a linear transformation to the outputs of the matched filter bank to reduce the MAI seen by each user.

The decorrelating detector applies the inverse of the correlation matrix to the matched filter bank outputs, thereby decoupling the signals. It has many desirable features, including its ability to be implemented without knowledge of the received amplitudes.

The MMSE detector applies a modified inverse of the correlation matrix to the matched filter bank outputs. It yields a better error rate performance than the decorrelating detector, but it requires estimation of the received powers.

Both the decorrelating and MMSE detectors require nontrivial computations that are a function of the cross-correlations. This is particularly difficult for the case of long (time-varying) codes, where the cross-correlations change each bit. Many proposals for simplifying the necessary computations have been made, but difficulties remain.

The polynomial expansion detector applies a polynomial expansion in the correlation matrix to the outputs of the matched filter bank. This detector has the important advantage that it can efficiently approximate either the decorrelating or MMSE detectors; in doing so, neither the correlation matrix nor its inverse needs to be explicitly calculated. Like the decorrelating detector, it does not need to estimate the received amplitudes. Unlike the decorrelating detector, it can easily be implemented with long codes. Also, it appears that weights (polynomial coefficients) can be chosen that are fairly robust over a wide range of system parameters, thereby minimizing or eliminating the need for adaptation.

SUBTRACTIVE INTERFERENCE CANCELLATION DETECTORS

Subtractive interference cancellation detectors attempt to estimate and subtract off the MAI. These detectors include the successive interference cancellation (SIC), parallel interference cancellation (PIC), and zero-forcing decision-feedback (ZF- DF) detectors.

The bit decisions used to estimate the MAI may be either hard decisions or soft decisions. Soft decisions provide a joint estimate of data and amplitude and are easier to implement. If reliable channel estimates are available, however, hard-decision (nonlinear) schemes perform better than their soft-decision counterparts.

The SIC detector takes a serial approach to subtracting out the MAI: it decisions, regenerates, and cancels out one additional direct-sequence user at a time. In contrast, the PIC detector estimates and subtracts out all of the MAI for each user in parallel. Both of these detectors may be implemented with a variable number of stages.

From the work in [72], it appears that the SIC detector performs better than the PIC detector in a fading environment, while the reverse is true in a well-power-controlled environment, (although this work has been done specifically for the case of soft decisions). The PIC detector requires more hardware, but the SIC detector faces the problems of power reordering and large delays.

Various methods for improving PIC detection have been proposed. The recently proposed improved PIC detector of [76] may be the most powerful of the subtractive interference cancellation detectors, and needs to be studied further.

Several detectors combine linear preprocessing with subtractive interference cancellation. Examples are the ZF-DF detector and a PIC detector with a decorrelating detector as the first stage. A significant disadvantage of the ZF-DF detector is that it requires Cholesky factorization and matrix inversion.

A major disadvantage of nonlinear detectors is their dependence on reliable estimates of the received amplitudes. Studies such as [56, 57] indicate that imperfect amplitude estimation may significantly reduce or even reverse the gains to be had from using these detectors.

CONCLUSION

Multi-user detection holds much promise for improving DS-CDMA performance and capacity. Although multi-user detection is currently in the research stage, efforts to commercialize multi-user detectors are expected in the coming years as DS-CDMA systems are more widely deployed. The success of these efforts will depend on the outcome of careful performance and cost analyses for the realistic environment.

ACKNOWLEDGMENT

The author sincerely thanks Dr. Mahesh Varanasi, Dr. Ken Smolik, Dr. Zoran Siveski, Dr. Howard Sherry, Dr. Joseph Wilkes, and Dr. Larry Ozarow for their helpful comments. The author also expresses his warm appreciation to his wife, Adina, for her help in editing this article.

REFERENCES

[1] A. J. Viterbi, "The Orthogonal-Random Waveform Dichotomy for Digital Mobile Personal Communications," *IEEE Pers. Commun*, 1st qtr., 1994., pp. 18–24.
[2] W. C. Y. Lee, "Overview of Cellular CDMA," *IEEE Trans. Vehic. Tech.*, vol. 40, no. 2, May 1991, pp. 291–302.
[3] K. S. Gilhousen *et al.*, "On the Capacity of a Cellular CDMA System," *IEEE Trans. Vehic. Tech.*, vol. 40, no. 2, May 1991, pp. 303–12.
[4] R. L. Pickholtz, L. B. Milstein, and D. L. Schilling, "Spread Spectrum for Mobile Communications," *IEEE Trans. Vehic. Tech.*, vol. 40, no. 2, May 1991, pp. 313–22.
[5] V. K. Garg, K. Smolik, and J. E. Wilkes, *Applications of Code-Division Multiple Access (CDMA) in Wireless/Personal Communications*, Upper Saddle River, NJ: Prentice Hall, 1996.
[6] H. V. Poor and L. A. Rusch, "Narrowband Interference Suppression in Spread Spectrum CDMA," *IEEE Pers. Commun.*, 3rd qtr., 1994,pp. 14–27.
[7] L. A. Rusch and H. V. Poor, "Multi-User Detection Techniques for Narrowband Interference Suppression in Spread Spectrum Communications," *IEEE Trans. Commun.*, vol. 43, no. 2/3/4, Feb./Mar./Apr. 1995, pp. 1725–37.
[8] J. G. Proakis, *Digital Communications*, 2nd ed., New York: McGraw-Hill, 1989.
[9] M. B. Pursley, "Performance Evaluation for Phase-Coded Spread-Spectrum Multiple-Access Communication — Part I: System Analysis," *IEEE Trans. Commun.*, vol. COM-25, no. 8, Aug. 1977, pp. 795-99.
[10] K. Yao, "Error Probability of Asynchronous Spread Spectrum Multiple Access Communication Systems," *IEEE Trans. Commun.*, vol. COM-25, no. 8, Aug. 1977, pp. 803–9.
[11] D. V. Sarwate and M. B. Pursley, "Crosscorrelation Properties of Pseudorandom and Related Sequences," *Proc. IEEE*, vol. 68, no. 5, May

1980, pp. 593–619.
[12] R. Kohno, "Pseudo-Noise Sequences and Interference Cancellation Techniques for Spread Spectrum Systems," *IEICE Trans. Commun.*, vol. J74-B-I, no. 5, May 1991, pp. 1083–92. Reprinted in *Multiple Access Communications: Foundations for Emerging Technologies*, N. Abramson, ed., IEEE Press, 1993.
[13] J. M. Holtzman, "CDMA Power Control for Wireless Networks," *Third Generation Wireless Information Networks*, D. G. Goodman and S. Nanda, eds., Norwell, MA: Kluwer, 1992, pp. 299–311.
[14] J. C. Liberti, "Spatial Processing for High Tier Wireless Systems," Bellcore Pub. IM-558, Sept. 1996.
[15] S. Verdu, "Adaptive Multi-User Detection," *Code Division Multiple Access Communications*, S. G. Glisic and P. A. Leppanen, Eds., pp. 97-116, The Netherlands: Kluwer, 1995.
[16] S. Verdu, "Multi-User Detection," *Advances in Statistical Signal Processing*, vol. 2, JAI Press 1993, pp. 369–409.
[17] A. Duel-Hallen, J. Holtzman, and Z. Zvonar, "Multi-User Detection for CDMA Systems," *IEEE Pers. Commun.*, vol. 2, no. 2, Apr. 1995, pp. 46–58.
[18] S. Moshavi, "Survey of Multi-User Detection for DS-CDMA Systems," Bellcore pub., IM-555, Aug. 1996.
[19] Z. Zvonar and D. Brady, "Linear Multipath-Decorrelating Receivers for CDMA Frequency-Selective Fading Channels," *IEEE Trans. Commun.*, vol. 44, no. 6, June 1996, pp. 650–53.
[20] Z. Zvonar and D. Brady, "Suboptimal Multi-User Detector for Frequency-Selective Rayleigh Fading Synchronous CDMA Channels," *IEEE Trans. Commun.*, vol. 43, no. 2/3/4, Feb./Mar./Apr. 1995, pp. 154–57.
[21] Z. Zvonar, "Multi-User Detection and Diversity Combining for Wireless CDMA Systems," *Wireless and Mobile Communications*, J. Holtzman and D. Goodman, eds., Boston: Kluwer, 1994, pp. 51–65.
[22] Z. Zvonar, "Combined Multi-User Detection and Diversity Reception for Wireless CDMA Systems," *IEEE Trans. Vehic. Tech.*, vol. 45, no. 1, Feb. 1996, pp. 205–11.
[23] H. C. Huang and S. C. Schwartz, "A Comparative Analysis of Linear Multi-User Detectors for Fading Multipath Channels," *Proc. IEEE Globecom '94*, San Francisco, CA, Nov. 1994, pp. 11–15.
[24] Y. C. Yoon, R. Kohno, and H. Imai, "A Spread-Spectrum Multi-Access System with Cochannel Interference Cancellation for Multipath Fading Channels," *IEEE JSAC*, vol. 11, no. 7, Sept. 1993, pp. 1067–75.
[25] P. M. Grant, S. Mowbray, and R. D. Pringle, "Multipath and Cochannel CDMA Interference Cancellation," *Proc. IEEE 2nd Int'l. Symp. on Spread Spectrum Techniques and Apps. (ISSSTA '92)*, Yokohama, Japan, Dec. 1992, pp. 83–86.
[26] M. K. Varanasi and S. Vasudevan, "Multi-User Detectors for Synchronous CDMA Communications over Non-Selective Rician Fading Channels," *IEEE Trans. Commun.*, vol. 42, no. 2/3/4, Feb/Mar/Apr 1994, pp. 711–22.
[27] S. Vasudevan, and M. K. Varanasi, "Optimum Diversity Combiner Based Multi-User Detection for Time-Dispersive Rician Fading CDMA Channels," *IEEE JSAC*, vol. 12, no. 4, May 1994, pp. 580–92.
[28] M. K. Varanasi and B. Aazhang, "Optimally Near-Far Resistant Multi-User Detection in Differentially Coherent Synchronous Channels," *IEEE Trans. Info. Theory*, vol. 37, no. 4, July 1991, pp. 1006–18.
[29] M. K. Varanasi, "Noncoherent Detection in Asynchronous Multi-User Channels," *IEEE Trans. Info. Theory*, vol. 39, no. 1, Jan. 1993, pp. 157–76.
[30] Z. Zvonar and D. Brady, "Multi-User Detection in Single-Path Fading Channels," *IEEE Trans. Commun.*, vol. 42, no. 2/3/4, Feb./Mar./Apr. 1994, pp. 1729–39.
[31] S. Verdu, "Minimum Probability of Error for Asynchronous Gaussian Multiple-Access Channels," *IEEE Trans. Info. Theory*, vol. IT-32, no. 1, Jan. 1986, pp. 85–96.
[32] S. Verdu, "Minimum Probability of Error for Asynchronous Multiple Access Communication Systems," *Proc. IEEE MILCOM '83*, vol. 1, Nov. 1983, pp. 213–19.
[33] Z. Xie, C. K. Rushforth, and R. T. Short, "Multi-User Signal Detection Using Sequential Decoding," *IEEE Trans. Commun.*, vol. 38, no. 5, May 1990, pp. 578–83.
[34] H. V. Poor, "On Parameter Estimation in DS/SSMA Formats," *Lecture Notes in Control and Information Sciences - Advances in Communications and Signal Processing*, Springer Verlag, 1989, pp. 59–70.
[35] Z. Xie *et al.*, "Joint Signal Detection and Parameter Estimation in Multi-User Communications," *IEEE Trans. Commun.*, vol. 41, no. 7, Aug. 1993, pp. 1208–16.
[36] Y. Steinberg and H. V. Poor, "Sequential Amplitude Estimation in Multi-User Communications," *IEEE Trans. Info. Theory.*, vol. 40, no. 1, Jan. 1994, pp. 11–20.
[37] T. K. Moon *et al.*, "Parameter Estimation in a Multi-User Communication System," *IEEE Trans. Commun.*, vol. 42, no. 8, Aug. 1994, pp. 2553–60.
[38] K. S. Schneider, "Optimum Detection of Code Division Multiplexed Signals," *IEEE Trans. Aerospace Elect. Sys.*, vol. AES-15, no., Jan. 1979, pp. 181–85.
[39] R. Kohno, M. Hatori, and H. Imai, "Cancellation Techniques of Co-Channel Interference in Asynchronous Spread Spectrum Multiple Access Systems," *Elect. and Commun. in Japan*, vol. 66-A, no. 5, 1983, pp. 20–29.
[40] R. Lupas and S. Verdu, "Near-Far Resistance of Multi-User Detectors in Asynchronous Channels," *IEEE Trans. Commun.*, vol. 38, no. 4, Apr. 1990, pp. 496–508.
[41] R. Lupas and S. Verdu, "Linear Multi-User Detectors for Synchronous Code-Division Multiple-Access Channels," *IEEE Trans. Info. Theory*, vol. 35, no. 1, Jan. 1989, pp. 123–36.
[42] M. K. Varanasi, "Group Detection for Synchronous Gaussian Code-Division Multiple-Access Channels," *IEEE Trans. Info. Theory*, vol. 41, no. 4, July 1995, pp. 1083–96.
[43] M. K. Varanasi, "Parallel Group Detection for Synchronous CDMA Communication over Frequency Selective Rayleigh Fading Channels," *IEEE Trans. Info. Theory*, vol. 42, no. 1, Jan. 1996, pp. 116–28.
[44] R. Lupas-Golaszewski and S. Verdu, "Asymptotic Efficiency of Linear Multi-User Detectors," *Proc. 25th Conf. on Decision and Control*, Athens, Greece, Dec. 1986, pp. 2094–2100.
[45] Z. Xie, R. T. Short, and C. K. Rushforth, "A Family of Suboptimum Detectors for Coherent Multi-User Communications," *IEEE JSAC*, vol. 8, no. 4, May 1990, pp. 683–90.
[46] S. S. H. Wijayasuriya, G. H. Norton, and J. P. McGeehan, "A Near-Far Resistant Sliding Window Decorrelating Algorithm for Multi-User Detectors in DS-CDMA Systems," *Proc. IEEE Globecom '92*, Dec. 1992, pp. 1331–38.
[47] A. Kajiwara and M. Nakagawa, "Microcellular CDMA System with a Linear Multi-User Interference Canceller," *IEEE JSAC*, vol. 12, no. 4, May 1994, pp. 605–11.
[48] F. Zheng and S. K. Barton, "Near Far Resistant Detection of CDMA Signals via Isolation Bit Insertion," *IEEE Trans. Commun.*, vol. 43, no. 2/3/4, Feb./Mar./Apr. 1995, pp. 1313–17.
[49] A. Kajiwara and M. Nakagawa, "Crosscorrelation Cancellation in SS/DS Block Demodulator," *IEICE Trans.*, vol. E 74, no. 9, Sept. 1991, pp. 2596–2601.
[50] D. S. Chen and S. Roy, "An Adaptive Multi-User Receiver for CDMA Systems," *IEEE JSAC*, vol. 12, no. 5, June 1994, pp. 808–16.
[51] S. Vembu, and A. J. Viterbi, "Two Different Philosophies in CDMA — A Comparison," *Proc. IEEE Vehic. Tech. Conf. 1996 (VTC '96)*, Atlanta, GA, Apr. 28–May 1, 1996, pp. 869–73.
[52] S. S. H. Wijayasuriya, G. H. Norton and J. P. McGeehan, "A Novel Algorithm for Dynamic Updating of Decorrelator Coefficients in Mobile DS-CDMA," *Proc. 4th Int'l. Symp. on Pers., Indoor, and Mobile Radio Commun. (PIMRC '93)*, Yokohama, Japan, Oct. 1993, pp. 292–96.
[53] U. Mitra and H. V. Poor, "Analysis of an Adaptive Decorrelating Detector for Synchronous CDMA Channels," *IEEE Trans. Commun.*, vol. 44, no. 2, Feb. 1996, pp. 257–68.
[54] S. Moshavi, E. G. Kanterakis, and D. L. Schilling, "Multistage Linear Receivers for DS-CDMA Systems," *Int'l. J. Wireless Info. Networks*, vol. 3, no. 1, Jan. 1996. (patent pending).
[55] S. Moshavi, "Multistage Linear Detectors for DS-CDMA Communications," Ph.D. dissertation, Dept. Elec. Eng., City Univ. New York, NY, Jan. 1996. (patents pending)
[56] H. Y. Wu and A. Duel-Hallen, "Performance Comparison of Multi-User Detectors with Channel Estimation for Flat Rayleigh Fading CDMA Channels," *Wireless Pers. Commun.*, July/Aug. 1996.
[57] S. D. Gray, M. Kocic, and D. Brady, "Multi-User Detection in Mismatched Multiple-Access Channels," *IEEE Trans. Commun.*, vol. 43, no. 12, Dec. 1995, pp. 3080–89.
[58] R. Kohno, "Spatial and Temporal Filtering for Co-Channel Interference in CDMA," *Code Division Multiple Access Communications*, S. G. Glisic and P. A. Leppanen, eds., The Netherlands: Kluwer, 1995, pp. 117–46.
[59] J. M. Holtzman, "DS/CDMA Successive Interference Cancellation," *Proc. IEEE Int'l. Symp. on Spread Spectrum Techniques and Appl. 1994 (ISSSTA '94)*, Oulu, Finland, July 1994, pp. 69–78.
[60] A. J. Viterbi, "Very Low Rate Convolutional Codes for Maximum Theoretical Performance of Spread-Spectrum Multiple-Access Channels," *IEEE JSAC*, vol. 8, no. 4, May 1990, pp. 641–49.
[61] A. D. Wyner, "Recent Results in the Shannon Theory," *IEEE Trans. Info. Theory*, vol. IT-20, Jan. 1974, pp. 2–10.
[62] T. M. Cover, "Some Advances in Broadcast Channels," *Advances in Communication Systems*, A. J. Viterbi, ed., New York: Academic, 1975, pp. 229–60.
[63] A. B. Carleial, "A Case Where Interference Does Not Reduce Capacity," *IEEE Trans. Info. Theory*, vol. IT-21, Sept. 1975, pp. 569–70.
[64] I. Seskar, "Practical Implementation of Successive Interference Cancellation," presentation, WINLAB Semi-Annual Res. Rev., Rutgers Univ., NJ, Apr. 26, 1996.
[65] P. Patel and J. Holtzman, "Analysis of a Simple Successive Interference Cancellation Scheme in a DS/CDMA System," *IEEE JSAC*, vol. 12, no. 5, June 1994, pp. 796–807.
[66] M. Ewerbring, B. Gudmundson, G. Larsson, and P. Teder, "CDMA with Interference Cancellation: A Technique for High Capacity Wireless Systems," *Proc. ICC '93*, Geneva, Switzerland, 1993, pp. 1901–6.
[67] M. K. Varanasi and B. Aazhang, "Multistage Detection in Asynchronous Code-Division Multiple-Access Communications," *IEEE Trans. Commun.*, vol. 38, no. 4, Apr. 1990, pp. 509–19.
[68] R. Kohno *et al.*, "Combination of an Adaptive Array Antenna and a Canceller of Interference for Direct-Sequence Spread-Spectrum Multiple-Access System," *IEEE JSAC*, vol. 8, no. 4, May 1990, pp. 675–82.
[69] R. Kohno *et al.*, "An Adaptive Canceller of Cochannel Interference for Spread-Spectrum Multiple-Access Communication Networks in a Power

Line," *IEEE JSAC*, vol. 8, no. 4, May 1990, pp. 691–99.

[70] M. K. Varanasi and B. Aazhang, "Near-Optimum Detection in Synchronous Code-Division Multiple-Access Systems," *IEEE Trans. Commun.*, vol. 39, no. 5, May 1991, pp. 725–36.

[71] U. Fawer and B. Aazhang, "A Multi-User Receiver for Code Division Multiple Access Communications over Multipath Channels," *IEEE Trans. Commun.*, vol. 43, no. 2/3/4, Feb./Mar./Apr. 1995, pp. 1556–65.

[72] P. Patel and J. Holtzman, "Performance Comparison of a DS/CDMA System Using a Successive Interference Cancellation (IC) Scheme and a Parallel IC Scheme under Fading," *Proc. ICC '94*, New Orleans, LA, May 1994, pp. 510–14.

[73] B. Zhu, N. Ansari, and Z. Siveski, "Convergence and Stability Analysis of a Synchronous Adaptive CDMA Receiver," *IEEE Trans. Commun.*, vol. 43, no. 12, Dec. 1995, pp. 3073–79.

[74] T. R. Giallorenzi and S. G. Wilson, "Decision Feedback Multi-User Receivers for Asynchronous CDMA Systems," *Proc. IEEE Globecom '93*, Houston, TX, Dec. 1993, pp. 1677–82.

[75] B. S. Abrams, A. E. Zeger, and T. E. Jones, "Efficiently Structured CDMA Receiver with Near-Far Immunity," *IEEE Trans. Vehic. Tech.*, vol. 44, no. 1, Feb. 1995, pp. 1–13.

[76] D. Divsalar and M. Simon, "Improved CDMA Performance Using Parallel Interference Cancellation," JPL pub. 95-21, Oct. 1995 (patent pending).

[77] A. Duel-Hallen, "Decorrelating Decision-Feedback Multi-User Detector for Synchronous Code-Division Multiple Access Channel," *IEEE Trans. Commun.*, vol. 41, no. 2, Feb. 1993, pp. 285–90.

[78] A. Duel-Hallen, "A Family of Multi-User Decision-Feedback Detectors for Asynchronous Code-Division Multiple Access Channels," *IEEE Trans. Commun.*, vol. 43, no. 2/3/4, Feb./Mar./Apr. 1995, pp. 421–34.

[79] A. Klein, G. K. Kaleh, and P. W. Baier, "Zero Forcing and Minimum Mean-Square-Error Equalization for Multi-User Detection in Code-Division Multiple-Access Channels," *IEEE Trans. Vehic. Tech.*, vol. 45, no. 2, May 1996, pp. 276–87.

[80] G. K. Kaleh, "Channel Equalization for Block Transmission Systems," *IEEE JSAC*, vol. 13, no. 1, Jan. 1995, pp. 110–21.

[81] P. Jung and J. Blanz, "Joint Detection with Coherent Receiver Antenna Diversity in CDMA Mobile Radio Systems," *IEEE Trans. Vehic. Tech.*, vol. 44, no. 1, Feb. 1995, pp. 76–88.

[82] L. Wei and C. Schlegel, "Synchronous DS-SSMA System with Improved Decorrelating Decision Feedback Multi-User Detection," *IEEE Trans. Vehic. Tech.*, vol. 43, no. 3, Aug. 1994, pp. 767–72.

[83] G. W. Stewart, *Introduction to Matrix Computations*, New York: Academic, 1973.

[84] M. Honig, U. Madhow, and S. Verdu, "Blind Adaptive Multiuser Detection," *IEEE Trans. Info. Theory*, vol. 41, no. 4, July 1995, pp. 944–60.

[85] U. Madhow and M. L. Honig, "MMSE Interference Suppression for Direct Sequence Spread-Spectrum CDMA," *IEEE Trans. Commun.* vol. 42, no. 12, Dec. 1994, pp. 3178–88.

[86] R. M. Buehrer and B. D. Woerner, "Analysis of Adaptive Multistage Interference Cancellation for CDMA Using an Improved Gaussian Approximation," *Proc. IEEE MILCOM '95*, San Diego, CA, Nov. 1995, pp. 1195–99.

[87] A. J. Viterbi, *CDMA: Principles of Spread Spectrum Communciation*, Reading, MA: Addison-Wesley, 1995.

[88] H. Taub and D. L. Schilling, *Principles of Communication Systems*, 2nd ed., New York: McGraw-Hill, 1986.

ENDNOTES

1 These properties include: frequency reuse of one, resistance to multipath fading, multipath diversity combining (RAKE reception), soft capacity, soft handoff, natural usage of the voice activity cycle (VAC), ability to overlay on existing systems, ability to use forward error correction coding without overhead penalty, natural exploitation of sectored antennas and adaptive beamforming, ease of frequency management, low probability of detection and intercept (LPD and LPI), and jam resistance. See [1–4] for details.

2 Note that we focus here only on the effect of MAI, and not on the effect of narrowband (NB) interference. A good survey of work dealing with CDMA in the presence of NB interference can be found in[6]. See also [7] where multiuser detection is proposed for eliminating NB inteference.

3 The detector would consist of a band of K matched filters, where each filter is "matched" to a different code waveform. Matched filter detection and correlation detection are equivalent methods of implementing optimal detection where the only interference is from additive white Gaussian noise [8], that is, in a single-user channel.

4 The operation of the conventional detector can also be explained in the frequency domain. All signals arrive at the receiver spread in frequency by the processing gain factor, PG. This has the effect of reducing the power of each signal over any given narrow band of frequencies. After multiplying the received signal by the code of user k, the signal of user k is de-spread back to the original information bandwidth; the other signals, however, remain spread in frequency (i.e., $g_i(t)g_k(t)$ is equivalent to some new spreading code waveform). The integrator then acts as a low pass filter with cut-off at frequencies $\pm f_b$. Within this frequency range the de-spread signal is at full power, while the power of the interfering signals has been reduced by an amount proportional to the processing gain [88].

5 A popular approximation of the SNR at the output of the conventional detector is obtained by modeling the MAI as a Gaussian random variable [9]. Thus, for the conventional detector, the MAI can be lumped with the thermal noise for analysis, that is, it raises the noise floor. The resulting equation yields fairly accurate probability of error results for most reasonable system parameters (i.e., for K, PG, and probability of error not too small [10]). This equation is often used in analysis of DS-CDMA systems, e.g., [3, 4].

6 Another important area of research is the design of improved single-user detectors, where the code of only one (desired) user is known. Here detection is optimized in some way for the multi-user channel, where the general structure of the interference is known to be that of other direct-sequence users. As a substitute for the specific knowledge of the interfereing users' code waveforms, these detectors generally rely heavily on adaptive signal processing. They are also sometimes referred to as adaptive multi-user detectors. An overview of the work in this area can be found in [15].

7 Issues dealt with include multipath, fading, noncoherent detection, general modulation schemes, power variation and power control, coding, acquisition and tracking (code synchronization), channel estimation, multiple and adaptive antennas, complexity and cost, efficient suboptimal implementations, application to IS-95, and sensitivity and robustness (e.g., the effects of amplitude and phase estimation errors, delay tracking errors, and quantization errors).

8 Multipath is an important issue in multi-user detection. The bandwidth of a DS-CDMA signal is very wide (or equivalently, the chip duration is very small); hence more than one signal path can generally be resolved at the receiver [8]. This yields what is known as "multipath diversity." The conventional detector in this case takes the form of a bank of RAKE detectors [8], which allows it to take advantage of the availble diversity. The RAKE detector of each user has M "fingers," where each finger detects a different signal path through a matched filter. The RAKE receiver then combines the M outputs in some manner (e.g., maximal ratio or equal gain). The name "RAKE" comes from a similarity of this detector to an ordinary garden rake [8]. There has been much literature on multi-user detection in a multipath environment, e.g., [19–23] (for the decorrelating detector), [24, 25] (for the PIC detector), and [26, 27] (for the MLS detector and the decorrelating detector in a 2 path Rician fading channel). See [16–18] for additional references and discussion. One approach to multi-user detection in the presence of multipath is to maximal ratio combine the M corresponding signal paths for each user and then perform multi-user detection on the resulting K signals. A more common approach is to treat each path as a separate user with respect to the multi-user detection algorithm. Thus, first multi-user detection is performed on MK signals and then RAKE combining takes place on the corresponding M outputs for each user.

9 We are assuming BPSK modulation and thus coherent detection. However, in the IS-95 standard,a pilot signal is not availble on the uplink (it is available, however, in IS-665). Thus, a coherent reference is not avaialble for tracking the phase, and noncoherent detection is necessary [5]. Two basic works that consider noncoherent multi-user detection (for the decorrelating detector) are [28] for the synchronous channel and [29] for the asynchronous channel; other articles include [19, 21-23, 30] (for the decorrelating detector), and [65] (for the SIC detector). See [16–18] for additional references and discussion.

10 The ability to detect signals from multiple cells is already assumed in IS-95 for the implementation of soft-handoff [1]. Here base stations of neighboring cells may simultaneously transmit to, and receive from the same mobile user. Note that the value of 0.55 given for the spillover ratio in [1] actually already includes soft handoff users [87].

11 By definition, the maximum-likelihood sequence detector chooses **d** to maximize $P(r(t)|\mathbf{d})$; but if all **d** vectors are equally probable, this is equivalent to maximizing $P(\mathbf{d}|r(t))$ [8]. Thus, the MLS detector yields the most likely tranmsitted **d** vector as long as all possible **d** vectors are equally likely [8]. It can be shown that maximizing the probabilty $P(r(t)|\mathbf{d})$ is equivalent to maximizing the log likelihood function $\mathcal{L} = 2\mathbf{d}^T\mathbf{A}\mathbf{y} - \mathbf{d}^T\mathbf{ARAd}$ where $\mathbf{d} \in \{-1, 1\}^{NK}$ [41]. From this follows the well known result that **y** (the matched filter output over the whole message) is a sufficient statistic for optimum detection of the transmitted data [31].

12 MLS detection guarantees the most likely sequence (i.e. a global optimum). An alternate optimality crieteria is "minimum probability of error," which results from the maximation of the marginal a posteriori distributions, $P(d_{k,i}|r(t))$, $k = 1...K$, $i = 1...N$ (locally optimum) [31]. This is more difficult to implement. Fortunately, the bit error rate of the MLS detector turns out to be indistinguishable from the minimum probability of error for SNR regions of interest, that is, where the thermal noise is not dominant [16]; in the limit as the noise goes to zero, the MLS error rate is equivalent to that of the minimum error rate. A 2 user synchronous channel example which illustrates the difference between the MLS criteria and the minimum probability of error criteria is given in [16], and repeated here. Assume that the joint posterior

probablities $P(\{d_1, d_2\}|r(t))$, are given as $P(\{1, 1\}|r(t)) = 0.26$, $P(\{-1, 1\}|r(t)) = 0.26$, $P(\{1, -1\}r(t)) = 0.27$, and $P(\{-1, -1\}|r(t)) = 0.21$. The most likely sequence is {1, –1}; however, the most likely value of the second user's bit is 1.

13 Besides yielding the most likely transmitted sequence, this detector is also optimal in terms of the performance measures known as the asymptotic efficiency and the near-far resistance [31, 40, 41]. These metrics are covered in the surveys [16, 18].

14 In [31] a Viterbi implementation is proposed with path metrics that are a function of the user crosscorrelations, and that are similar to that of a single-user periodic time varying ISI channel with memory $K - 1$; the resulting Viterbi algorithm has 2^{K-1} states and a complexity per binary decision on the order of 2^K. Unfortunately, no algorithm is known to solve the maximization of the likelihood function $\mathcal{L}$ (see Endnote 11) in polynomial time in K (i.e., it is NP-hard) [31]. An illustrative example of MLS detection for an asynchronous 2 user DS-CDMA system is spelled out in [16]. Note that [58] cites two article by Kohno from 1982 and 1983 that also proposes a Viterbi altorithm implementation with a complexity per binary decision on the order of 2^K; both of these articles appear only in Japanese. Viterbi implemenations of higher complexity were also proposed in [32, 38].

15 A natural simplification of MLS detection is to replace the Viterbi algorithm with a sequential decoder, as is done for convolutional decoding [33]. Sequential decoding searches for the most likely path based on local metric values; in contrast, the Viterbi algorithm tracks and evaluates all possible paths. Although simpler, sequential decoding for DS-CDMA is still fairly difficult to implement.

16 As is discussed below, the decorrelating detector pays a noise enhancement penalty for eliminating the MAI. Thus, if the MAI is relatively low and the background noise power is relatively high, ignoring the MAI, as does the conventional detctor, may yield better performance [16, 40].

17 In brief, the near-far resistance [31, 40, 41] is a performance measure that indicates performance under worst-case conditions of interfering powers; it provides some quanitfication of the resistance of a detector's error performance to the power of the interfering users. A detector that is near-far resistant (i.e., the metric is not equal to zero), can achieve any given performance level in the multi-user enviroment, no matter how powerful the multi-user interference, provided that the desired user is supplied enough power. Both the maximum likelihood sequence detector and the decorrelating detector are guaranteed to be near-far resistant for linearly independent users (linearly dependent users, however, are not near-far resistant). Both detectors also yield the largest value of this metric for a given set of code waveforms. In contrast, the conventional detector is not near-far resistant, unless all waveforms are orthogonal. For more details on near-far resistance, see [16].

18 Two recent papers treat the decorrelating detector as a special case of what is termed "parallel group detectors" [42, 43]. These detectors bridge the gap in performance and complexity between the decorrelating detector, (which corresponds to the case of one user per group), and the MLS detector (which corresponds to the case of all users in one group).

19 As mentioned above, the decorrelating detector is the optimal sequence detector (linear or nonlinear) when the energies of the users are unknown. If they are known, however, there are linear detectors that provide better probability of error performance. This involves trading off some MAI reduction for less noise enhancement. An example of this is the MMSE detector discussed in the next subsection.

20 Degradation from the ideal decorrelating detector performance results because of the "edge effects" [45, 46]. Some proposals include a form of "edge correction" to mitigate this problem [45, 46]; other proposals involve physically separating the data sub-blocks, to entirely avoid the edge problem [47, 48, 79, 81]. The latter scheme, however, requires some time synchronization among users.

21 It is shown in [40] that for the case of short codes (codes that repeat each bit), and where the message length, N, approaches infinity, the decorrelating detector apporaches a K-input K-output linear time-invariant noncausal infinte memory-length filter. It is further shown in [40] that under mild conditions a stable unique realization of this filter exists. Since the filter has infinite memory-length and is non-causal, a practical implementation would require truncation to a finite length filter, and the insertion of sufficient delay. Since stability requires that the impulse response, $h(n)$, go to zero as $n \to \infty$, the more remote symbols will count less heavilty. Therefore, the approximation to the exact decorrelating filter will be good for a truncation window (filter memory) of sufficient length [40].

22 See also [50] where an adaptive decorrelating detector is proposed that avoids the need for computations with the correlation matrix.

23 On the other hand, as the noise gets very large, or the MAI amplitudes get very small, $\mathbf{L}_{MMSE} \approx (2/N_0)\mathbf{A}^2$. In this case, performance of the MMSE detector approaches that of the conventional detector [15, 45]. See Endnote 16.

24 For example, in [45] MMSE detection takes place on blocks of subsequences; in [85] "one-shot" MMSE detection is proposed, where detection is based only on observation over one transmission interval. MMSE detection has also received much attention lately because of its ability to be implemented adaptively, where the codes of the interfering users are not known, that is, improved single-user detection (e.g. [84, 85]). For more on this subject, see [15].

25 In this case, the PE detector structure can be thought of as being a K-input K-output linear infinite memory-length filter realization of the decorrelating detector.

26 Note that soft-decision subtractive interference cancellation detectors can usually also mathematically be classified as linear detectors.

27 The Wireless Information Network Laboratory (WINLAB) at Rutgers University, New Jersey, is currently implementing a protoype of the SIC detector which utilizes soft decisions [64]. A soft-decision SIC detector was initially investigated in [65].

28 A distinctly different SIC scheme that does cancellation in the Walsh-Hadamard spectral domain is discussed in [66]. Additional references on this approach can be found in [18].

29 Because of the cancellation order, this detector is most potent when there is significant power variation between each users' received signal. A specific geometric power distribution is derived in [60] that enables each user to see the same level of signal power to interference (+ noise) ratio, and produce the same probability of error. It is also shown in [60] that by using the SIC detector with this power profile, along with very low rate forward error correction (FEC) codes, it is possible for the composite bit rate of all users to approach the Shannon limit.

30 The multistage PIC algorithm is used in [71] as part of a joint parameter estimation and data detection scheme.

31 This detector can be considered to be a special case of the modified parallel group detectors introduced in [42] (corresponding to the case of one user per group).

32 An adaptive version of this detector that does not require explicit estimation of the received amplitudes is proposed in [73] for synchronous systems.

33 In [75] a PIC detector is proposed that is based entirely on feedback cancellation: the outputs of the correlators are continuously fed back during the correlation for cancellation.

34 Note that the cancelation takes place on the post correlation MAI terms. Although both the SIC and PIC detectors were described earlier with "pre-correlation" cancelation, they too can be equivalently implemented through "post-correlation" cancellation [24, 59, 65].

35 The ZF-DF detector can be considered to be a special case of the "sequential group detectors" introduced in [42] (corresponding to the case of one user per group). A general analysis is given there without the assumption that all past decisions are correct.

36 An MMSE-DF detector is proposed in [78, 79, 81] which is analogous to the MMSE-DF equalizer [8]. Here the feed-forward and feedback filters are chosen to minimize the mean square error under the assumption that all past decisions are correct. This detector is similar to the ZF-DF detector except that the feed-forward filter is obtained by Cholesky factoring the matrix $[\mathbf{ARA} + (N_0/2)\mathbf{I}]$. Like in equalization, the MMSE-DF detector outperforms the ZF-DF detector.

37 An improved ZF-DF detector is proposed for synchronous channels in [82] which feeds back more than one set of likely decision vectors along with their corresponding metrics. The approach of this detector is similar to that of sequential decoding.

38 The "Schur algorithm" with parallel processing is proposed for Cholesky factorization in [80]; it results in a complexity that is linear with the order of the matrix.

Biography

Shimon Moshavi [S '91] received the B.A. degree in physics from Yeshiva University in 1988, the M.S. degree in electrical engineering from City College of New York in 1994, and the Ph.D. degree in electrical engineering from City University of New York in January 1996. Since January 1996 he has been a research scientist at Bell Communications Research (Bellcore) in Red Bank, New Jersey, in the Wireless Systems Research Department (Tel: 908-758-5091, moshavi@bellcore.com). His current interests include communication theory, CDMA, multi-user detection, and wireless networks.

Modifying present systems could yield significant capacity increases

Multiuser Detection for CDMA Systems

ALEXANDRA DUEL-HALLEN, JACK HOLTZMAN, AND ZORAN ZVONAR

Spread spectrum has been very successfully used by the military for decades. Recently, spread-spectrum-based code division multiple access (CDMA), has taken on a significant role in cellular and personal communications. Multiple access allows multiple users to share limited resources such as frequency (bandwidth) and time. There are a number of multiple access schemes including more than one type of CDMA. We shall concentrate on one type, direct sequence CDMA (DS/CDMA). CDMA has been found to be attractive because of such characteristics as potential capacity increases over competing multiple access methods, anti-multipath capabilities, soft capacity, and soft handoff.

We shall not cover all the background of DS/CDMA, since that has been well explained in recent literature (e.g., [1]). In fact, this article may be viewed as a supplement to that literature with an update on some potential enhancements to the versions of DS/CDMA currently being developed [2-4]).[1] We will show that there is a natural modification of the present systems that is potentially capable of significant capacity increases. By "natural modification" we mean a modification that can be made conceptually clear, not that it is easy to implement. Indeed, the optimal multiuser detector is much too complex and most of the present research addresses the problem of simplifying multiuser detection for implementation. The objective of this article is to make the basic idea intuitive and then show how investigators are trying to reduce the idea to practice. We also indicate multiuser receiver structures with potentially acceptable levels of complexity and address potential obstacles for achieving theoretically predicted performance in practice. As a result of these investigations, an answer to the following question is expected: Is there a suboptimal multiuser detector that is cost effective to build with significant enough performance advantage over present day systems? A definitive answer is not yet available.

We will first review some salient features of CDMA systems needed for the discussion to follow.

Limitations of a Conventional CDMA System

A conventional DS/CDMA system treats each user separately as a signal, with the other users considered as either interference, e.g.,Multiple Access Interference (MAI), or noise. The detection of the desired signal is protected against the interference due to the other users by the inherent interference suppression capability of CDMA, measured by the processing gain. The interference suppression capability is, however, not unlimited and as the number of interfering users increases, the equivalent noise results in degradation of performance, i.e., increasing bit error rate (BER) or frame error rate. Even if the number of users is not too large, some users may be received at such high signal levels that a lower power user may be swamped out. This is the *near/far effect*: users near the receiver are received at higher powers than those far away, and those further away suffer a degradation in performance. Even if users are at the same distance, there can be an effective near/far effect because some users may be received during a deep fade. DS/CDMA systems are very sensitive to the near/far effect and the recent success of DS/CDMA has, in large part, been due to the successful implementation of relatively tight power control, with attendant added complexity. There are thus two key limits to present DS/CDMA systems:

- All users interfere with all other users and the interferences add to cause performance degradation.
- The near/far problem is serious and tight power control, with attendant complexity, is needed to combat it.

Multipath Propagation

One other aspect of CDMA that we need to review is the ability to combat multipath reception of signals [6]. Due to multiple reflections, the received signal contains delayed, distorted replicas of the original transmitted signal. First, consider what happens in a non-spread-spectrum system. When the multiple reflections, called multipath signals or simply multipaths, from one transmitted bit are received within the time duration of one bit, the received signal consists of the superposition of several signal replicas, each with its own amplitude and phase. It is important to recognize that this superposition is the addition of complex quantities. Due to the motion of the mobile (or, even of the base station in some systems), the relative phases of the received signals are continually changing. This results in successive reinforcement and interference of the superposed multipath signals, resulting in very large time variations in the received signal. Such variations are referred to as Rayleigh fading (or Rician fading, if there is a direct component in addition to the reflections). The variations due to Rayleigh fading are a serious

The work of the first two authors has been supported by an NSF TIE Project Award, No. EEC 9416209.

[1] *A comprehensive reference set on spread spectrum until 1985 is [5].*

 1070-9916/95/$04.00

cause of performance degradation and a communication system must be designed carefully, taking that into account.

We shall refer to systems where all of the multipath signals arrive within one bit interval as "narrowband." On the other hand, the bit rate may be so high that multipath signals from one bit arrive over a duration longer than that of one bit. Such systems will be called "wideband." The Rayleigh fading effect is less pronounced, because there are fewer multipath signals from one transmitted bit arriving during the bit duration.

CDMA systems are inherently wideband when the chip duration, as opposed to the longer bit duration, is compared to the time between multipath receptions. One can then combat multipath interference by multipath reception, whereby the different multipath arrivals are considered as independent receptions of the signal and are used to give a beneficial time diversity. This is usually done with a RAKE receiver, the name apparently taken from the action of a rake with a number of teeth pulling in a number of items simultaneously. So, instead of multipath being just a source of performance degradation, the multipaths are used to provide the benefit of diversity [7].

Interference Cancellation and Multiuser Detection

In a conventional CDMA system, all users interfere with each other. Potentially significant capacity increases and near/far resistance can *theoretically* be achieved if the negative effect that each user has on others can be canceled. A more fundamental view of this is multiuser detection, in which all users are considered as signals for each other. Then, instead of users interfering with each other, they are all being used for their mutual benefit by joint detection. The drawback of optimal multiuser detection is one of complexity so that suboptimal approaches are being sought. There is a wide range of possible performance/complexity combinations possible. Much of the present research is aimed at finding an appropriate tradeoff between complexity and performance.

Multiuser Detection in Cellular Systems

In a cellular system, a number of mobiles communicate with one base station (BS). Each mobile is concerned only with its own signal while the BS must detect all the signals. Thus, the mobile has information only about its own chip sequence while the base station has the knowledge of all the chip sequences. For this reason, as well as less complexity being tolerated at the mobile (where size and weight are critical), multiuser detection is currently being envisioned mainly for the BS, or in the reverse link (mobile to BS). It is important to realize, however, that the BS maintains information only on those mobiles in its own cell. This plays a role in the limitations on improvements to be expected in a multiuser detection system, to be discussed next.

Limitations to Improvements

Before we discuss multiuser improvements to the conventional DS/CDMA detector, it is important to define factors that limit such improvement [1, 8]. One factor is intercell interference in a system that cancels only the intracell interference[2] I. For intercell interference which is a fraction f of the intracell interference, the bound of capacity increase (all of the intracell interference is canceled) is $(1 + f)/f$. For $f = 0.55$, this factor is 2.8 [1]. Observe that with a sectorized antenna, it is conceivable to cancel users from another sector and thus improve the bound.

Another limiting factor is the fraction f_c of energy captured by a RAKE receiver. That is, a RAKE receiver with L branches or "fingers" will try to capture the power in the L strongest multipath rays, but there will be additional received power

We will show that there is a natural modification of the present systems that is potentially capable of significant capacity increases. As a result of these investigations, an answer to the following question is expected: Is there a suboptimal multiuser detector that is cost effective to build with significant enough performance advantage over present day systems?

in additional rays. For the conventional detector, this is self-interference. Reference [10] gives examples of the fraction of captured power. The fraction of captured power is a function of chip rate and delay spread as well as the number of RAKE branches. So, combining the two effects (measured by f and f_c), the total interference before cancellation is $(1 + f)I$ (neglecting the smaller self-interference due to uncaptured multipath power of the desired user). Cancellation removes at most $f_c I$ so the bound on improvement is $(1 + f)/(1 + f - f_c)$. For $f_c \approx 1$, the above bound of 2.8 on capacity improvement remains. For $f_c = 0.5$, the bound is reduced to 1.5.

It should be recognized that multiuser detection is used not only to increase capacity but also to alleviate the near/far problem, and the preceding bound does not account for that benefit. Relaxing the power control requirement actually translates into a capacity benefit which is, however, more difficult to quantify than by the above simple signal/interference argument. A multiuser detector could recapture part of this reduction by reducing variability (or relax the requirements on power control).

To put these constraints on improvements into further perspective, we are assuming here that multiuser detection is a candidate primarily for the reverse link for reasons given earlier. Since the reverse link is usually more limiting than the forward link,[3] increasing the reverse link capacity will improve the overall system capacity. But increasing it beyond the forward link capacity will not further increase the overall system capacity. Thus,

- The potential capacity improvements in cellular systems are not enormous (order of magnitude) but certainly nontrivial.
- Enormous capacity improvements only on the reverse link (the candidate for multiuser detection) would only be partly used anyway in determining overall system capacity.
- Hence, the cost of doing multiuser detection must be as low as possible so that there is a per-

[2] Intracell interference is from interferers in the same cell as the desired user while intercell interference is from interferers outside the cell. It has been proposed that intercell interference be canceled by explicitly communicating this information, or by adaptive or blind methods (see [9]). This research is at an earlier stage.

[3] Some cases in which the forward link appear to be limiting are given in [62].

formance/cost tradeoff advantage to multiuser detection.

The bottom line is that there are significant advantages to multiuser detection which are, however, bounded and a simple implementation is needed.

Historical Background

The idea of interference cancellation arises in many contexts, e.g., noise cancellation in speech [11] and adaptive interference canceling as in Chapter 12 of [12]. There are thus a number of non-CDMA references with ideas similar to those being currently studied for CDMA. We should distinguish between canceling noise which has no useful purpose (as in [11] and Chapter 12 of [12]) from canceling interference which is due to other signals which are themselves to be detected. A couple of non-CDMA examples in the latter category are [13-15]. The CDMA case considered here is of the second type, where the signals being canceled are of interest also. It should be remarked, however, that the first type of cancellation also is of importance in CDMA systems, e.g., in suppressing narrowband interference (this is not discussed in this article, but is discussed in Section 5 of [3]). Both types of interference cancellation have in common the goal of removing from a desired signal a noise-like interference. But in the second type (the type considered here), the fact that the signals being removed are themselves information carrying leads to a new viewpoint, that of *simultaneously* detecting all the information carrying signals.

In a conventional CDMA system, all users interfere with each other. In multiuser detection, all users are considered as signals for each other. The drawback of optimal multiuser detection is one of complexity so that suboptimal approaches are being sought.

The first CDMA interference cancellation references we are aware of are [16, 17]. Both of these papers delineate a number of ideas that are present in much of the ongoing research. Estimates based on mean square error and maximum likelihood are discussed in [16]. Reference [17] shows how cancellation is implemented by solving simultaneous equations, in essence, by inverting a key matrix. There were subsequently a number of papers with variants of the ideas of [16, 17]. Significant theoretical steps forward were taken in [18, 19] (with earlier references), in analyzing the structure and complexity of optimal receivers. This work triggered a new research effort on suboptimal algorithms. The strong connection between MAI and intersymbol interference (ISI) was also made in [18]. There are aspects of MAI, however, that are not shared by ISI:

- The near/far problem.
- MAI is affected by the relationship among user chip sequences (codes) as well as by the imperfections of the radio channel, while ISI is due only to the channel.

Thus, while equalization (used to combat ISI) will play a role in the multiuser detectors to be discussed, it should not be expected that it can be used without modification.

Recent survey papers include [8, 9] with many further references. The rest of the references here are cited as needed in the discussion.

Multiuser Detection: Concept and Techniques

The CDMA Channel Model and Approaches to Detection

A CDMA channel with K users sharing the same bandwidth is shown in Fig. 1. The signaling interval of each user is T seconds, and the input alphabet is antipodal binary: $\{+1, -1\}$. The objective is to detect those polarities, which contain the transmitted information. During the n-th signaling interval, the input vector is $x_n = (x_n^1, \ldots, x_n^K)^T$, where x_n^k is the input symbol of the k-th user. User k ($k = 1, \ldots, K$) is assigned a signature waveform (or code, or spreading chip sequence) $s_k(t)$ which is zero outside $[0, T]$ and is normalized

$$\int_0^T s_k(t)^2\, dt = 1$$

Pulse amplitude modulation is employed at the transmitter. The baseband signal of the k-th user is

$$u_k(t) = \sum_{i=0}^{\infty} x_i^k c_i^k s_k(t - iT - \tau_k), \quad (1)$$

where τ_k is the transmission delay, and c_i^k is the complex channel attenuation. According to (1), each user's signal travels along a single path, so this model does not illustrate multipath propagation. The effect of multipath is discussed in the section on noncoherent multiuser detection. For synchronous CDMA, the delay $\tau_k = 0$ for all users. For asynchronous CDMA, the delays can be different. The channel attenuation is a complex number

$$c_i^k = \sqrt{w_i^k}\, \exp(j\theta_i^k).$$

where w_i^k and θ_i^k are the received power and phase of the k-th user, respectively. The received signal (at baseband) is the noisy sum of all the users' signals:

$$y(t) = \sum_{k=1}^{K} u_k(t) + z(t), \quad (2)$$

where $z(t)$ is the complex additive white Gaussian noise (AWGN). The first step in the detection process is to pass the received signal $y(t)$ through a matched filter bank (or a set of correlators). It consists of K filters matched to individual signature waveforms followed by samplers at instances $nT + \tau_k$, $k = 1, \ldots, K$, $n = 1, 2, \ldots$. The outputs of the matched filter bank form a set of sufficient statistics about the input sequence x_n given $y(t)$ [18]. Thus, we will consider the equivalent discrete-time channel model which arises at the output of the matched filter bank.

For the rest of this section, we will concentrate on a very simplified DS/CDMA system. (There are a number of simplifications which will be exposed in the rest of the article. In fact, each relaxation of simplification will represent another factor to consider for the multiuser detection system.) The *simplifying assumptions* are as follows:

We Consider *Real* Channel Attenuations – The real model is convenient for analyzing coherent methods, and can be easily generalized to the complex case. In the following section, we extend our treatment to multiuser detectors for fading channels, where complex attenuations need to be considered.

Derivation of Multiuser Detectors is Presented for *Synchronous* CDMA System – The synchronous assumption considerably simplifies exposition and analysis and often permits the derivation of closed-form expressions for the desired performance measures. These are useful since similar trends are found in the analysis of the more complex asynchronous case. Furthermore, every asynchronous system can be viewed as an equivalent synchronous system with larger effective user population [20], which is often explored in burst CDMA communications. Moreover, synchronous systems are becoming more of practical interest since quasi-synchronous approach has been proposed for satellite [21] and microcell applications [22]. It should be recognized, however, that the transition from synchronous to asynchronous can considerably increase the complexity of multiuser detection. Throughout the paper, we address implementation issues and complexity increase for various detectors for the asynchronous model.

Certain Parameters are *Known* Exactly – Although multiuser detectors presented in this section take advantage of completely known amplitudes, and phases and delays do not appear in the treatment at all, the next section addresses noncoherent detectors which do not require the knowledge of amplitudes and phases. Sensitivity and robustness are discussed in a subsection titled "Issues in Practical Implementations."

For synchronous CDMA, the output signal $y(t)$ (2) for $nT \leq t < (n+1)T$ does not depend on the inputs of other users sent during past or future time intervals. Consequently, it is sufficient to consider a one-shot system with input vector $x_n = (x^1, \ldots, x^K)^T$, real positive channel attenuations (amplitudes) $c^1 = \sqrt{w^1}, \ldots, c^K = \sqrt{w^K}$ and real additive white Gaussian noise $z(t)$ with power spectral density N_0. The sampled output of the k-th matched filter (matched to the signature waveform of user k) is

$$y_k = \int_0^T y(t)s_k(t)dt = \int_0^T s_k(t)\left[\sum_{j=1}^K s_j(t)c^j x^j + z(t)\right]dt$$
$$= c^k x^k + \sum_{j \neq k}^K x^j c^j \int_0^T s_k(t)s_j(t)dt + \int_0^T s_k(t)z(t)dt \quad (3)$$

Note that y_k consists of three terms. The first is the desired information which gives the sign of the information bit x_k (which is exactly what is sought). The second term is the result of the multiple access interference (MAI), and the last is due to the noise. The second term typically dominates the noise so that one would like to remove its influence. Its influence is felt through the cross-

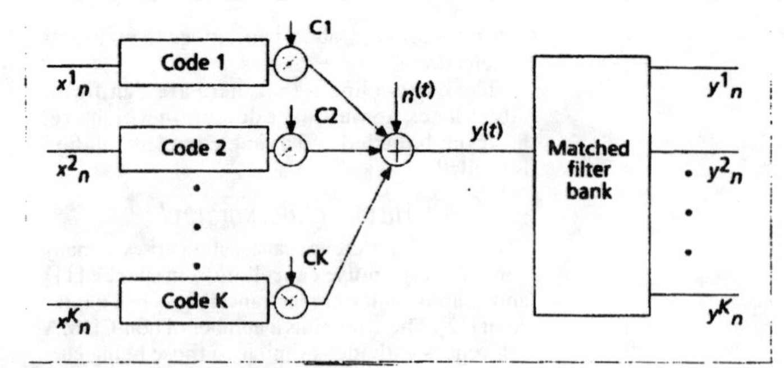

■ **Figure 1.** *The CDMA channel model.*

correlations between the chip sequences and the powers of users. If one knew the cross-correlations and the powers, then one could attempt to cancel the effect of one user upon another. This is, in fact, the intuitive motivation for interference cancellation schemes.

Suppose there are only two users in the system. Let r be the cross-correlation between the signature waveforms of the two users

$$r = \int_0^T s_1(t)s_2(t)dt.$$

In this case, the outputs of the matched filters are

$$y^1 = c^1x^1 + rc^2x^2 + z^1 \text{ and } y^2 = c^2x^2 + rc^1x^1 + z^2. \quad (4)$$

The MAI terms for users 1 and 2 are rc^2x^2 and rc^1x^1, respectively. If these terms were not present, the single user system would result. The bit error rate of the optimal detector for the single user system serves as a lower bound on the performance of any other detector. This single user bound is

$$P_k(E) = Q\left(\sqrt{\frac{w^k}{N_0}}\right), \quad (5)$$

where the Q-function

$$Q(x) = \frac{1}{\sqrt{2\pi}}\int_x^\infty \exp\left(\frac{-y^2}{2}\right)dy.$$

The conventional DS/CDMA uses the same approach as the optimal receiver for the single user system. It detects the bit from user k by correlating the received signal with the chip sequence of user k. Thus, the conventional detector makes its decision at the output of the matched filter bank:

$$\hat{x}^k = \mathrm{sgn}(y^k). \quad (6)$$

When MAI terms are significant, as shown in (3), the bit error rate of this detector is high. Note that MAI depends both on the cross-correlations and the powers of users. In the 2-user example above, if user 1 is much stronger than user 2 (the near/far problem), the MAI term rc^1x^1 present in the signal of the second user is very large, and can significantly degrade performance of the conventional detector for that user. A multiuser detector called a successive interference canceller (decision-directed) can remedy this problem as follows. First, a decision $\hat{x}^1$ is made for the stronger user 1

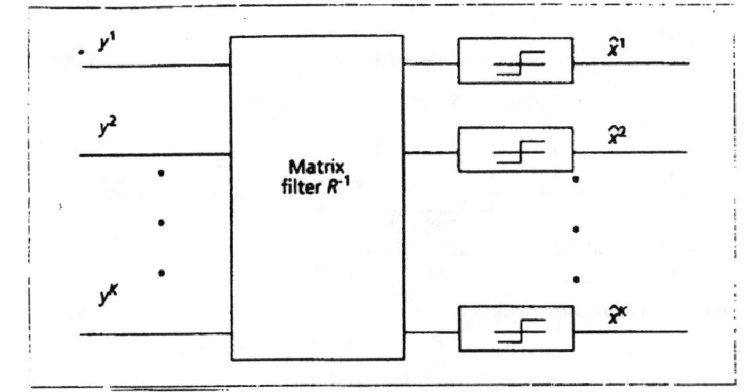

■ **Figure 2.** *The decorrelator for synchronous CDMA.*

using the conventional detector. Since user 2 is much weaker then user 1, this decision is reliable from the point of view of user 2. So, this decision can be used to subtract the estimate of MAI from the signal of the weaker user. The decision for user 2 is given by

$$\begin{aligned}\hat{x}^2 &= \mathrm{sgn}(y^2 - rc^1\hat{x}^1)\\ &= \mathrm{sgn}(c^2x^2 + rc^1(x^1 - \hat{x}^1) + z^2)\end{aligned} \quad (7)$$

Provided the decision of the first user is correct, all MAI can be subtracted from the signal of user 2.[4] If we fix the energy of the second user, and let the energy of the first user grow, the error rate of the successive interference canceller for the second user will approach the single-user bound. Thus, this detector is successful in combating the near/far problem. This simple example motivates the use of multiuser detectors for CDMA channels. Below, we will discuss several previously proposed multiuser detectors.

The Decorrelating Detector

As a step towards the most general formulation, consider the matrix version of the equivalent discrete time model (3). The output vector $y = [y^1, y^2, \ldots, y^K]^T$ can be expressed as

$$y = RWx + z, \quad (8)$$

where R and W are $K \times K$ matrices, and z is a colored Gaussian noise vector. The components of the matrix R are given by cross-correlations between signature waveforms

$$R_{k,j} = \int_0^T s_k(t)s_j(t)dt. \quad (9)$$

The second matrix W is diagonal with $W_{k,k}$ given by the channel attenuation c_k of the k-th user. For example, in a two-user system, the matrix

$$R = \begin{pmatrix} 1 & r \\ r & 1 \end{pmatrix},$$

where r is the cross-correlation between the signature waveforms of the users (9).

Inspection of (8) immediately suggests a method to solve for x, whose components x_k contain the bit information sought. If z was identically zero, we have a linear system of equations, $y = RWx$, the solution of which can be obtained by inverting R (it is invertible in most cases of interest [23]). With a non-zero noise vector z, inverting R is still an effective procedure and actually optimal in certain circumstances, to be discussed later. This results in

$$\tilde{y} = R^{-1}y = Wx + \tilde{z} \quad (10)$$

where it is seen that the information vector x is recovered but contaminated by a new noise term (Fig. 2). From (10), the signal of the k-th user is

$$\tilde{y}^k = c^k x^k + \tilde{z}^k. \quad (11)$$

The decision is $\hat{x}^k = \mathrm{sgn}(\tilde{y}^k)$.

Note that the decorrelating detector completely eliminates MAI. However, the power of the noise $\tilde{z}^k$ is $N_0 (R^{-1})_{k,k}$ which is greater than the noise power N_0 at the output of the matched filter (8). For example, for the two-user system with the cross-correlation r, the noise power at the output of the decorrelating filter is $N_0/(1-r^2)$. The error rate of the decorrelator is given by

$$P_k(E) = Q\left(\sqrt{\frac{w_k}{N_0 R^{-1}_{k,k}}}\right) \quad (12)$$

The performance of the decorrelating detector degrades as the cross-correlations between users increase. In the asynchronous case the decorrelating detector also reduces to matrix inversion in a burst type communications, or is given by linear, time-invariant K-input K-output filter for the infinite length transmitted data [20]. In both cases the complexity of the detector grows and several approaches have been proposed to reduce the complexity, as addressed later.

The decorrelator has several desirable features. It does not require the knowledge of the users' powers, and its performance is independent of the powers of the interfering users. This can be seen from (11). The only requirement is the knowledge of timing which is anyway necessary for the code despreading at the centralized receiver. Observe that neither signal nor noise terms depend on the powers of interferers. In addition, when users' energies are not known, and the objective is to optimize performance for the worst case MAI scenario, the decorrelator is the optimal approach [23]. In addition, the noncoherent version of the decorrelator has been developed (see the following section). These properties of the decorrelator make it very well suited for the near/far environment.

Multiuser detection is closely related to equalization for intersymbol interference (ISI) channels [7]. For example, the decorrelating detector is analogous to the zero-forcing equalizer. Similarly, the MMSE linear multiuser detector [24] (also given by a matrix inverse) is the multidimensional version of the MMSE linear equalizer for the single-user ISI channel. The linear structure of these detectors often limits their performance. In the following section, we will describe several non-linear approaches to multiuser detection.

The Optimal Detector

The objective of maximum-likelihood sequence estimation (MLSE) is to find the input sequence which maximizes the conditional probability, or likelihood of the given output sequence [7]. For the simplified synchronous CDMA problem discussed above, the maximum likelihood decision

[4] *Note, also, that r and c^1 are assumed to be known exactly for this example. We shall return to this issue.*

for the vector of bits x is given by

$$\hat{x} = \arg\left\{ \max_{x\in\{-1,+1\}^K} \left[2y^T Wx - b^T WRWb \right] \right\} \quad (13)$$

This equation dictates a search over the 2^K possible combinations of the components of the bit vector x. For asynchronous CDMA, the MLSE detector can be implemented using the Viterbi algorithm [18]. The path metrics of this algorithm were derived by identifying the asynchronous CDMA channel with a single-user channel with periodically time-varying ISI. The memory of this equivalent channel is K-1 (the number of interferers), and therefore the resulting Viterbi algorithm has 2^{K-1} states and requires K storage updates per transmission interval. Although the optimal detector has excellent performance, it is too complex for practical implementation, and we will not discuss it in greater detail. A suboptimal detector which uses a sequential decoder instead of the Viterbi algorithm was presented in [25].

Non-Linear Suboptimal Multiuser Detectors

In this section, we will consider several interference cancellation methods which utilize feedback to reduce MAI in the received signal. These algorithms can be broken into three classes:

- Multistage detectors, e.g., [24, 26-28, 33]
- Decision-feedback detectors [29-31].
- Successive interference cancellers (this idea is explicit or implicit in a number of papers).

Note: this classification is to facilitate exposition. The three categories are not actually disjoint and particular realizations of suboptimal detectors may use combinations of the three classes.

The first two classes of algorithms are decision-directed. They utilize previously made decisions of other users to cancel interference present in the signal of the desired user. These algorithms require estimation of channel parameters and coherent detection. The algorithms in the third class can use soft decisions (e.g., outputs of the correlation receivers as in [32]) rather than hard decisions to remove MAI components. They lend themselves to noncoherent implementation. The algorithms of the second and third classes employ successive interference cancellation (also proposed in [33]), which requires ordering of users according to their powers. The signals of stronger users are demodulated first and canceled from the signals of weaker users. This technique provides an efficient and practical solution to the near/far problem.

Several representatives from the three classes of non-linear detectors are described below.

Multistage Detectors – A multistage detector (Fig. 3) proposed in [26] uses (14) instead of (13):

$$\hat{x}_k(n) = \arg\left\{ \max_{\substack{x_k\in\{-1,+1\} \\ x_l = \hat{x}_l(n-1), l\neq k}} \left[2y^T Wx - b^T WRWb \right] \right\} \quad (14)$$

The n-th stage of this detector uses decisions of the $(n$-1)-st stage to cancel MAI present in the received signal. Thus, maximization is over one bit at a time, instead of over k bits, as in (13). Due to delay constraints, it is desirable to limit the number of stages to two. For example, consider a two-stage detector with the conventional first stage for the synchronous two-user system with the cross-correlation r(4). The decisions produced by the first stage (conventional) detector are $\hat{x}^1(1)$ and $\hat{x}^2(1)$ computed as in (6). The decisions of the second stage are $\hat{x}^1(2)= \text{sgn}[y^1 - rc^2\hat{x}^2(1)]$ and $\hat{x}^2(2) = \text{sgn}[y^2 - rc^1 \hat{x}^1(1)]$. The performance of this two-stage detector depends on the relative energies of the users. Clearly, if the first user is stronger than the second, the decisions of the second stage for user 2 agree with those of the decision-directed successive interference canceller, described in the last paragraph of the section on the CDMA channel model. Thus, for the weaker user, the second stage produces more reliable decisions than the first stage. However, for the stronger user, feedback might not be beneficial since the decision produced by the conventional detector for the weaker user is poor. More reliable two-stage detector results if the conventional detector in the first stage is replaced by the decorrelator [28]. This example illustrates the issues which play a role in the design of multistage detectors. In summary, the two important questions are:

- How to choose the initial stage.
- How to choose the subsequent stages of processing.

A discussion of different options for the initial and subsequent stages is given, along with further references, in [27].

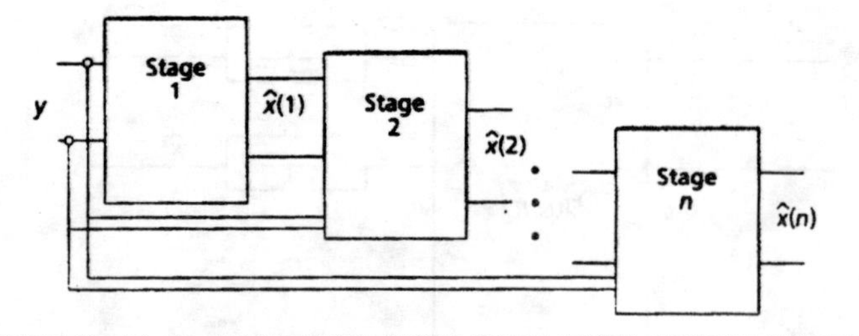

Figure 3. *The multistage detector.*

Decision-Feedback Detectors – The detectors proposed in [29-31] are multiuser decision-feedback equalizers, characterized by two matrix transformations: a forward filter and a feedback filter. These detectors are analogous to the decision-feedback equalizers employed in single user ISI channels [7]. However, in addition to equalization, the decision-feedback multiuser detectors employ successive cancellation. In each time frame, decisions are made in the order of decreasing user's strength, i.e., the stronger users make decisions first, allowing the weaker users to utilize these decisions. The sorting is performed by any multiuser detector with successive MAI cancellation. We will explain the rationale for using this particular order in the next section.

A diagram of the decorrelating (zero-forcing) decision-feedback detector for synchronous CDMA [30] is shown in Fig. 4. At the output of the sorter, users are ranked according to their

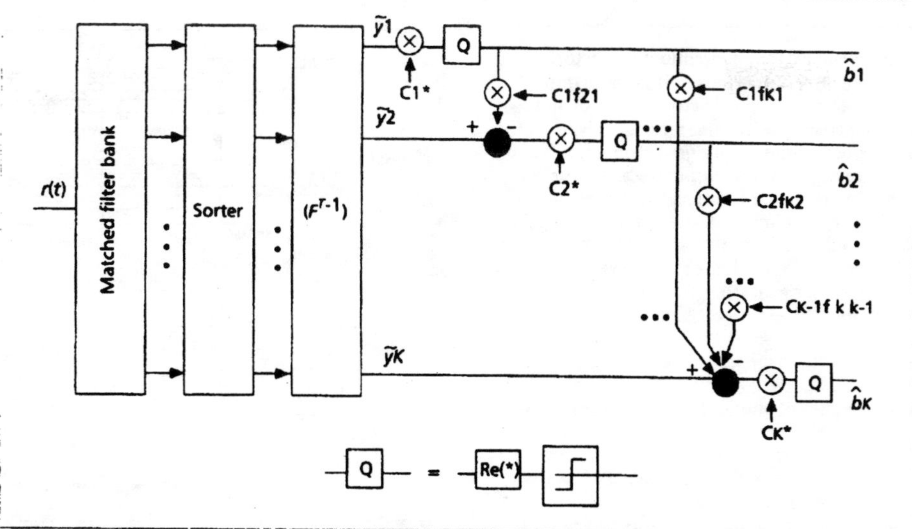

■ **Figure 4.** *The decorrelating decision-feedback detector for Synchronous CDMA.*

powers, so that the strongest user is ranked first, and the weakest is ranked last. Following the sorter, a noise whitening filter is applied. This filter is obtained by Cholesky factorization of the correlation matrix, which yields a resulting MAI matrix that is lower triangular. Consequently, at the output of the whitening filter, the signal of the k-th strongest user $\tilde{y}_k$ is given by:

$\tilde{y}_k$ = desired signal
+ MAI due to stronger users (1,..., k - 1) + noise.

In particular, the signal of the strongest user $\tilde{y}$ is not corrupted by MAI, and can be demodulated first. This decision is then used to subtract MAI from the signal of the second user, and so on. For the asynchronous CDMA, several decision-feedback detectors were derived in [29, 31].

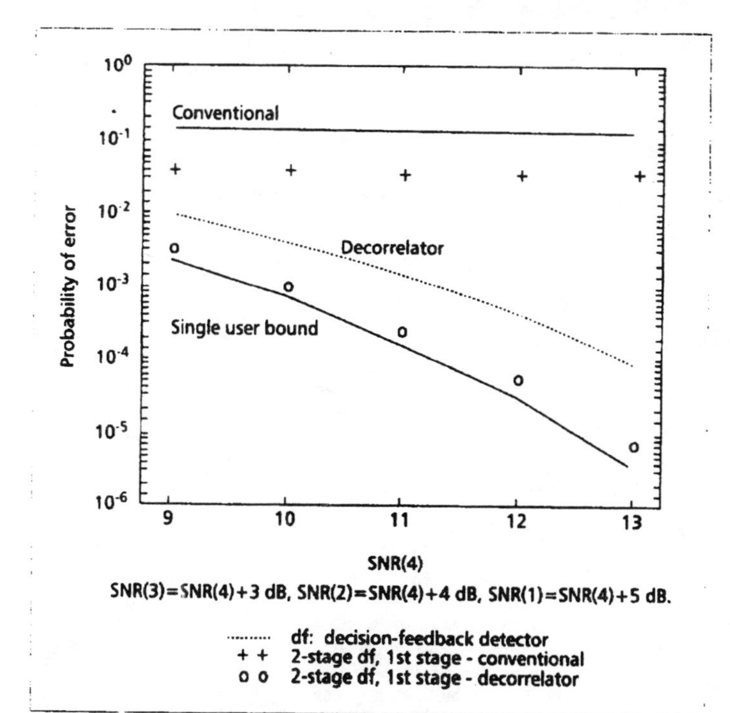

■ **Figure 5.** *Error rates for user 4 in the four-user system of [31].*

The performance of the decision-feedback detector is similar to that of the decorrelator for the strongest user, and gradually approaches the single user bound as the user's power decreases relative to powers of interferers. Thus, for the decision-feedback detector, performance advantages with respect to the conventional or the decorrelating detectors are greater for relatively weaker users. This is also the case for multistage detectors with the decorrelating first stage. Figure 5 depicts typical performance of several detectors for the weakest user in a bandwidth efficient system. The signature waveforms for this asynchronous four-user CDMA system were derived from Gold sequences of length 7 [31]. (see also [28, 30].) In Fig. 5, the conventional, decorrelating, decision-feedback and multistage detectors are compared for the weakest user (user 4). The powers of all users grow, but the differences between the powers remains the same. Note that the two-stage detector with the conventional first stage is interference-limited. Both the decision-feedback and the two-stage detector with the decorrelating first stage have excellent performance in this near/far scenario.

Successive Interference Cancellers – One approach to successive interference cancellation is to consider what would be the simplest augmentation to the conventional detector which would achieve some of the benefits of multiuser detection. This can be explained most simply by referring back to (3). In order to cancel the MAI, the factors $x^j c^j$ are needed, in addition to the cross-correlations. These can be obtained either with estimates of each of the factors x^j and c^j separately, i.e., separation of the bit estimate and power estimates. Alternately, one can estimate the product $x^j c^j$ directly by using the correlator output. We shall focus on the latter method because that requires the simplest augmentation to the conventional detector. It is found that using the correlator output to estimate $x^j c^j$ is sufficiently accurate to obtain improvement over the conventional detector.

As mentioned previously, it is important to cancel the strongest signal before detection of the other signals because it has the most negative effect. Also, the best estimate of signal strength is from the strongest signal for the same reason that the best bit decision is made on that signal: the strongest signal has the minimum MAI, since the strongest signal is excluded from its own MAI. This is the twofold rationale for doing successive cancellation in order of signal strength:

- Canceling the strongest signal has the most benefit.
- Canceling the strongest signal is the most reliable cancellation.

In a number of studies (see references in [8], it has been shown that this method of cancellation yields significant improvements over the conventional detector (but substantially less than the optimum multiuser detector).

Successive cancellation works by successively subtracting off the strongest remaining signal. An alternative (the parallel method) is to simultaneously subtract off all of the users' signals from all of the others. It is found [34] that when all of the users are received with equal strength, the parallel method outperforms the successive scheme (Fig. 6). When the received signals are of distinctly different strengths (the more important case), the successive method is superior in performance (Fig. 7). The important thing to note is that in both cases, both the successive and parallel interference cancellers outperform the conventional detector and the unequal power case is the more important case.

The successive cancellation must operate fast enough to keep up with the bit rate and not introduce intolerable delay. For this reason, it will presumably be necessary to limit the number of cancellations. The ability to limit the number of cancellations is consistent with the objective of controlling complexity by choosing an appropriate performance/complexity tradeoff. For more information and references on successive interference cancellation, see [8].

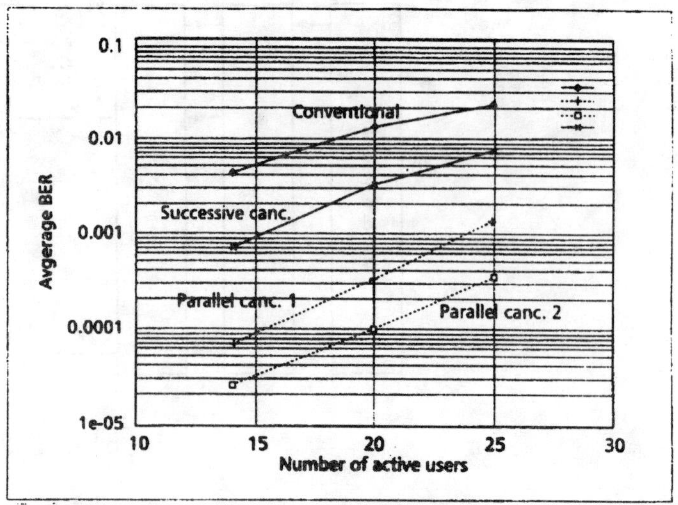

■ **Figure 6.** *BER vs. no. of active users under ideal power control (asynchronous).*

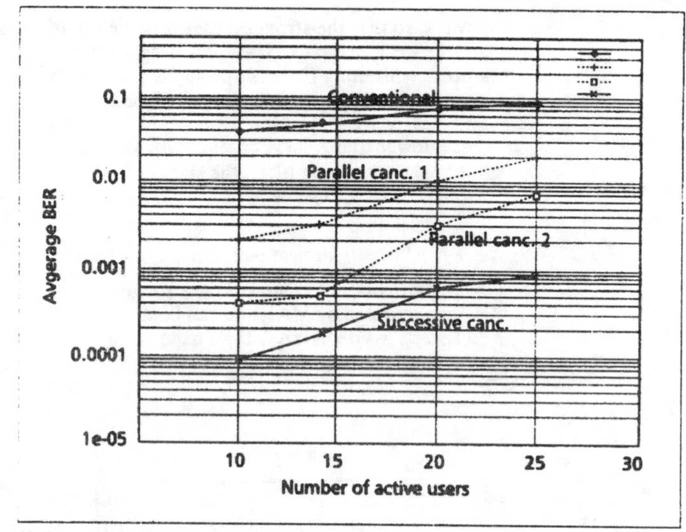

■ **Figure 7.** *BER vs. no. of active users under Rayleigh fading (asynchronous).*

Multiuser Detectors for Encoded Data

Error-control codes are essential for reliable performance of cellular systems. There has not been much work so far on performance of multiuser detectors for encoded signals. When convolutional codes are employed by all users, the MLSE detector is given by the Viterbi algorithm which is more complex than the optimal detector discussed previously for the uncoded case (due to the additional memory associated with each user) [35]. Since the MLSE detector is too complex to implement, several suboptimal methods were addressed in [33, 36-40]. In [33], successive cancellation technique was presented for a CDMA system with orthogonal convolutional coding. [37] discussed multistage detection for convolutionally encoded signals. The authors divided various approaches to multiuser detection for encoded signals into two classes. The first class contains the partitioned approaches, in which a multiuser detector precedes the decoder and does not utilize the decoded data. In the algorithms of the second class, the integrated approaches, the decoded symbols of the interferers are used for MAI cancellation in the signal of the desired user. Reduced complexity receivers, combined with decoders which incorporate reliability information, were presented in [36, 38]. Applicability of turbo-codes to CDMA mobile radio system using joint detection was demonstrated in [40]. Reference [39] addressed combined multistage detection and trellis-coded modulation.

One issue to be cognizant of is that error control allows operation at a lower SNR, which can reduce the improvements available with some multiuser detectors.

Noncoherent Multiuser Detection and Multipath Fading Channels

For the sake of simplicity of exposition, most of the discussion in the second section concentrated on coherent multiuser reception. The underlying assumption is that the multiuser receiver is able to estimate and track the phase of each active user in CDMA scenario. However, as stated

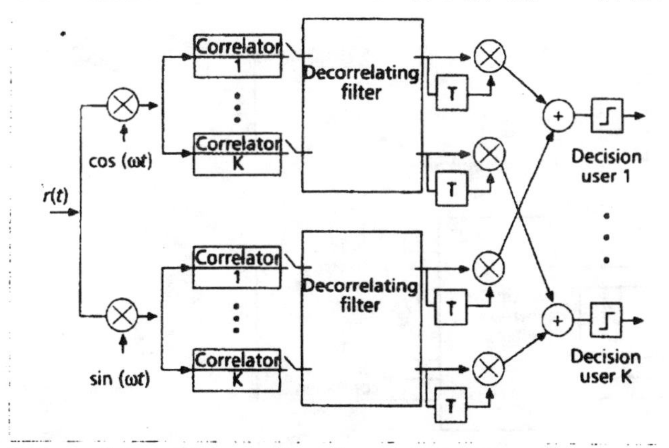

Figure 8. *Decorrelating multiuser detector for DPSK signals.*

earlier, the reverse link of a cellular CDMA system employs noncoherent reception since the pilot signal which provides a coherent reference is not available.[5] Therefore, the concept of noncoherent multiuser detection and multiuser receiver performance in multipath fading channels are of particular interest for practical CDMA systems.

As discussed in the section on multipath propagation, multipath fading presents a major limitation to the performance of wireless CDMA systems such as cellular mobile radio, indoor wireless communications, and personal communication services. In these systems MAI is enhanced by multipath propagation, and near/far effects are produced not only by the difference in the distance between the transmitter and receiver, but also by the fading on the propagation paths. While multipath propagation, usually encountered in an urban scenario, offers inherent diversity, certain scenarios in suburban areas may result in a single-path propagation [6, 41]. This depends on whether the individual multipaths can be resolved i.e., whether the chip duration is small enough relative to the separation between multipaths. In the case of single-path propagation, small but nonzero cross-correlations among chip sequences can cause a severe near/far problem in the presence of fading. When there is only a single fading path for each active user and interference is relatively strong, there are no means of diversity to overcome fading of the desired signal below the level of MAI, unless the explicit diversity using distributed antennas is introduced. Since multiuser receivers alleviate the near/far problem, they significantly improve the CDMA system performance in the single-path scenario [42-45, 61].

When the chip duration is small enough to resolve the different multipath receptions, multipath diversity is exploited to improve the performance in the presence of MAI. The conventional receiver in the case of multipath fading channel consists of a bank of RAKE receivers, one for each active user, at the base station and one for the desired user in a mobile. A summary of the research efforts on conventional reception techniques is given in [46]. Consequently, having in mind the multipath combining property of a RAKE receiver, multiuser techniques in multipath fading channels utilize some form of RAKE structure at the receiver front-end. With multipath resolution available, multiuser detection and multipath diversity reception are combined to provide the reliable receiver performance. An optimal MLSE receiver for multipath fading CDMA channel presented in [47] consists of the same front-end of coherent RAKE filters, followed by a dynamic programming algorithm of the Viterbi type. This optimal structure provides the same order of diversity and asymptotically has the same error probability as a RAKE receiver in the single-user case at the expense of high complexity which is again exponential in the number of active users.

All noncoherent multiuser techniques perform in-phase and quadrature (I & Q) demodulation before the detection. The noncoherent multiuser detection was first considered for differentially phase-shift keying (DPSK) systems [48]. The concept can be easily described as presented in Fig. 8. The decorrelating operation is performed both on in-phase and quadrature signal branches (after I & Q demodulation), so the phase information of the signal is preserved, although the phase is not being explicitly tracked. Since the decorrelating filter eliminates MAI, the same decision logic as in the single-user receiver can be applied for DPSK demodulation. Moreover, this type of detector was shown to be optimally near/far resistant.

The resulting expression for the error probability indicates that the performance loss compared to coherent reception is the same as in a single-user channel [48]. The realization and performance of this noncoherent decorrelating receiver do not depend on signal amplitudes and phases.

Since the RAKE receiver can be interpreted as a combiner of the correlators outputs and the combining method depends on the modulation type, the concept of noncoherent linear multiuser detection can be extended to the multipath fading scenario. To eliminate the effects of MAI prior to the combining process, linear multipath decorrelating receivers [49, 50] perform the decorrelating operation on KL correlator outputs, where L is the number of resolvable fading paths and K is the number of active users as depicted in Fig. 9. Consequently, the equal-gain diversity combining for DPSK signaling is performed on signals which suffer from less interference. The performance loss due to the noise enhancement in the presence of other active users is modest, for the typical mobile radio scenarios it is on the order of 3 dB compared to single-user RAKE performance over the whole range of SNRs [51]. Similar performance degradations are observed for the coherent linear multiuser receivers [52]. When a fraction of the multipath power is captured due to limited number of RAKE correlators, only a portion of MAI is eliminated by decorrelating operation and residual MAI may cause the performance degradation (recall the section on limitations to improvements). In that case additional antenna diversity was shown to be effective in reducing the effects of the residual MAI [51]. Another combination of interference cancellation and antenna diversity was analyzed in [53].

Interference cancellation techniques for multipath fading channels inherently employ the regeneration of the interfering signals. The major difference among numerous interference cancellation techniques is in the methods for the channel parame-

[5] *This is the case in [2]. Other approaches are currently being investigated.*

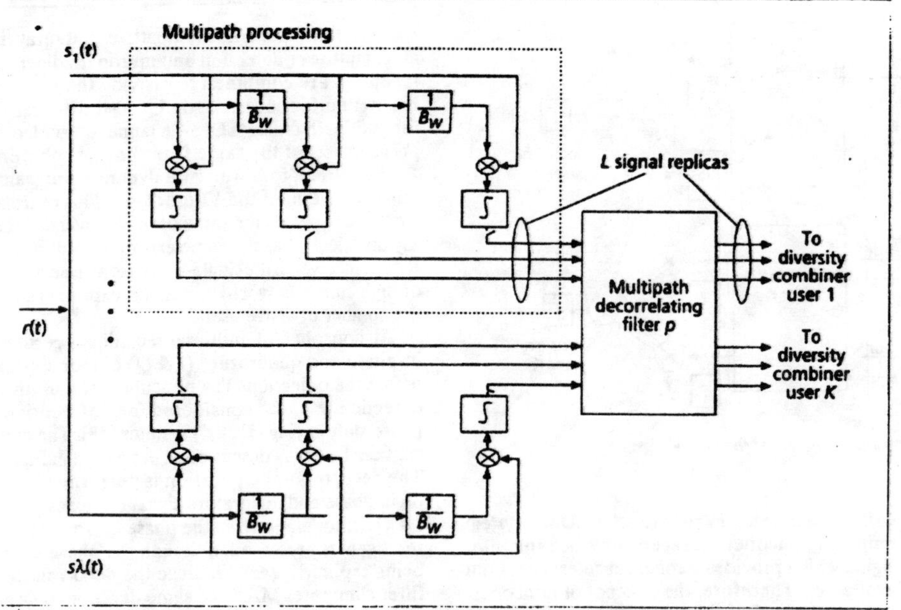

■ **Figure 9.** *Multipath decorrelating/multipath combining linear multiuser receiver for DPSK signals.*

ters estimation and interfering signal reconstruction. A successive interference cancellation receiver for noncoherent M-ary orthogonal modulation, which uses the outputs of the correlators to estimate the signal amplitudes and hence does not require any separate channel estimates, is given in [32]. A noncoherent version of successive interference cancellation employs a combination of M-ary orthogonal signaling and CDMA on the reverse link. After the *I* & *Q* demodulation at the receiver front-end, the received signal is correlated with respective *I* & *Q* chip sequences, the user's spreading code, and with all M-ary symbols obtained from Walsh functions. The interference cancellation algorithm starts by decoding the strongest user first. The amplitude of the decoded user is estimated from the correlator output and the strongest user's signal is regenerated using this estimate and the corresponding chip sequence, and canceled from the received signal. The cancellation is repeated until all users are decoded or until a limited number of cancellations are done. Since RAKE receivers are used in the front-end in CDMA multipath fading channels, each multipath arrival tracked by RAKE is canceled from the received signal using the appropriate correlator output as shown in Fig. 10.

Again, for any of the multipath multiuser detectors, the residual MAI problem could arise when the number of canceled paths is smaller than the total number of multipaths in the channel.

Issues in Practical Implementations

Complexity

Major obstacles to the application of the multiuser detectors in practical wireless systems are processing complexity and possible processing delay. The optimal multiuser MLSE receiver is clearly too complex for any application in a system with a large user population, and most of the present efforts are focused on implementation of suboptimal structures. However, even the suboptimal structures can lead to unacceptable levels of complexity. For example, the decorrelating detector in the asynchronous case results in a K-input K-output filter implementation which is stable but noncausal, so the appropriate delay has to be inserted [20]. In addition, changes in timing and addition/removal of new users results in time-varying coefficients of the detector.

The bottom line is that there are significant advantages to multiuser detection which are, however, bounded and a simple implementation is needed.

To overcome these difficulties, the sliding window decorrelating algorithm has been proposed as a practical alternative, both for infinite and finite data block lengths in asynchronous CDMA systems. Rather than having the total length of the received signal available for the construction of cross-correlation parameters, only a finite-length window of the signal is used for decorrelating operation with the correction of the edge effects. The resulting algorithm has to solve the linear system of equations described in [50]. The major cost in each iteration is due to recomputation of the linear system due to the change in relative delays among users, dynamic selection of multipath, and voice activity exploration. The computationally efficient algorithm for updating the coefficients of the decorrelating filter is proposed by exploiting the parallelism in the linear system solution [54]. The number of operations for the correction of the matrix filter coefficients is on the order of KN, where K is the number of active users in CDMA system, and N is the length of the sliding window. The readily available technology for high-speed zero-forcing equalizers makes the decorrelator

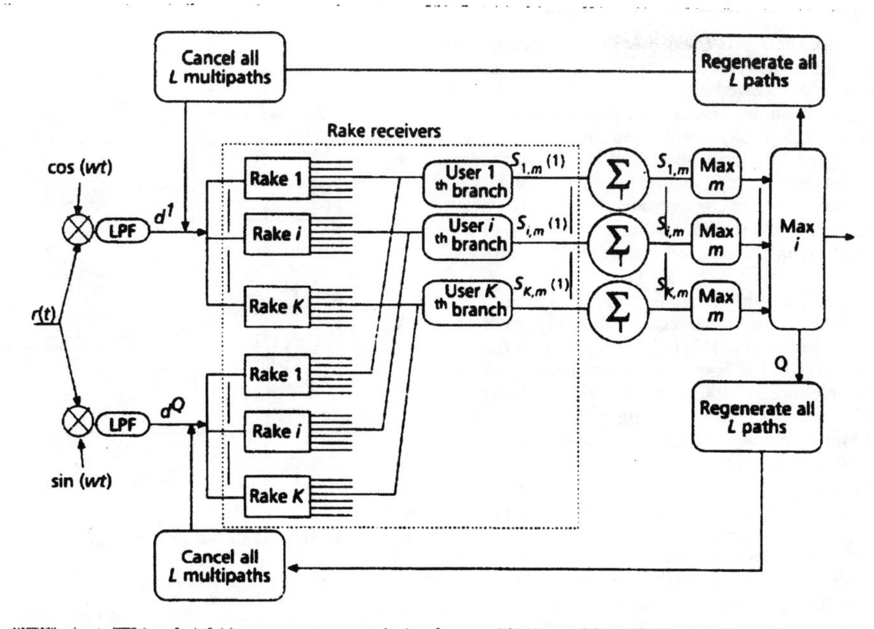

■ **Figure 10.** *Noncoherent receiver with interference cancellation and multipath combining.*

one of the simplest interference cancellation techniques to implement [54].

The successive interference cancellation scheme of the section on successive interference cancellers may be the simplest augmentation to the conventional detector. The major complexity in the multipath environment comes from tracking each multipath at the RAKE receiver, which involves, for each path M, coefficients corresponding to the possible M orthogonal symbols [32]. These coefficients are generated by a Walsh-Hadamard Transform (WHT) and the speed of performing the transform determines the processing delay of the canceller. To ensure real-time implementation, the cancellations, including the WHTs, must be done fast enough to keep up with the symbol rate in the CDMA system. The complexity of the successive interference cancellation scheme can be controlled by limiting the number of cancellations, i.e., one can achieve a compromise between performance and complexity. Observe that there is decreasing improvement with subsequent cancellations because cancellations are done in order of strength of the different users' signals.

Sensitivity and Robustness

Almost all of the discussions and analyses of multiuser detection have assumed a number of idealizations. While multiuser detection has been analyzed thoroughly for both AWGN and fading channels, less work has been reported on the underlying parameter estimation issues in CDMA channels, and consequently, on the impact of imperfect parameter estimates on the performance of multiuser receivers. Clearly, for any type of tracking error (frequency, amplitude, phase, or timing), the chip sequences being canceled will be offset and imperfect cancellation will be performed. For fading channel applications and noncoherent modulation on the reverse link, timing synchronization plays a vital role. It should also be emphasized that the chip tracking is essential for any type of DS/CDMA system and that tight code synchronization, within a fraction of a chip duration, is required for the reliable operation of the conventional detector. The pertinent question is whether the tracking error tolerable for the conventional detector is tolerable for the cancellation receiver, or how much tighter it must be for multiuser detection.

The problem of the impact of the synchronization errors on the multiuser receiver performance is inherently complex due to the nonlinear nature of the solution and can be analyzed only to a certain degree. Afterwards, one has to rely on simulations which brings another difficulty due to the coupling effect from various parameters of the CDMA system to the receiver performance. We will briefly summarize the current status in this area, which can be described more as the effort of establishing credential methodology, rather than trying to give a definite answer.

In the case of decorrelating detector, tracking errors result in the mismatch of the cross-correlation coefficients among different users. Analysis was performed in [55], assuming the Gaussian distribution of the tracking error and the performance of decorrelating detector was assessed by simulation using the standard deviation as the parameter. In this case, Gold codes of length 15 were employed, with three active CDMA users. The authors have shown that the performance degradation due to tracking errors is quite sensitive to the inequality of received powers.

The impact of tracking errors on the performance of successive interference cancellation receiver is analyzed in [56]. For numerical results presented here, the interference cancellation scheme was subject to pessimistic conditions:

- Did not use averaging of the correlator outputs for amplitude estimates which significantly improves cancellation performance.
- Assumed equal received powers (perfect power

control). The improvement over the conventional detector is much greater in the more realistic case of unequal received powers.

Figure 11 shows results for a processing gain (number of chips/bit) $N = 31$, and the total number of users is varied from five to 20. There are three curves each for the interference cancellation scheme and the conventional detector. Each curve represents different standard deviation e of tracking error, normalized with respect to chip duration ($e = 0$ is zero tracking error). The interference cancellation scheme retains superiority over the conventional detector. Similar types of results were found in [57] from simulation of the scheme of [58]. For Rayleigh fading on one path and mobile speed of 100 km/h it was reported that interference cancellation technique is less sensitive to imperfect power control and chip synchronization. For example, for error in power control with standard deviation of 1 dB with respect to nominal received power, an error of 5 percent in chip synchronization does not cause significant degradation in error probability.

While it is premature to draw any general conclusions about robustness of the interference cancellation receivers at this point, there are some promising results.

Several authors also addressed sensitivity of coherent multiuser detectors to channel mismatch (see references of [9] and [59-61]. While these investigations are still preliminary, they bring researchers closer to understanding performance advantages of multiuser detectors.

Concluding Remarks

The theoretical bases of optimal multiuser detection are well understood. Given the prohibitive complexity of optimum multiuser detectors, attention has been focused on suboptimal detectors, and the properties of these detectors are well understood by now. The next stages of investigation, involving implementation and robustness issues, are accelerating now and will lead to determination of the practical and economic feasibility of the multiuser detector. Initial studies of robustness show that robustness need not be a fatal flaw. Further investigations into practicality will include actual hardware implementations. This is the critical issue in answering the question posed at the end of the introduction to this article.

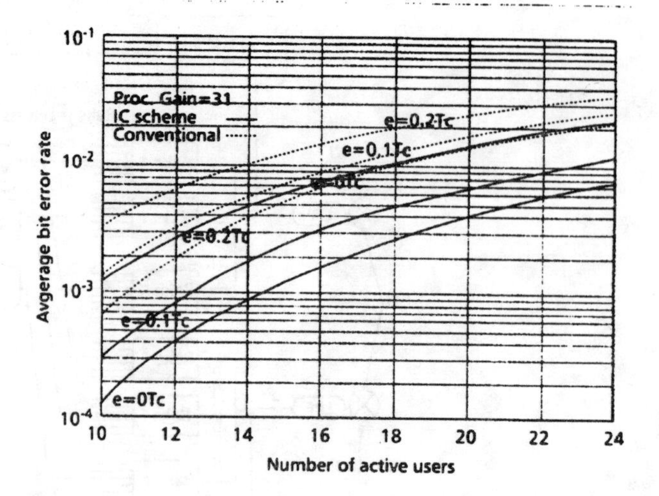

Figure 11. *BER performance degradation with tracking errors.*

References

[1] A. J. Viterbi, "The Orthogonal-Random Wave form Dichotomy for Digital Mobile Personal Communications," *IEEE Personal Commun.*. First Quarter 1994, pp. 18-24.

[2] K. S. Gilhousen *et al.*, "On the capacity of a Cellular CDMA System," *IEEE Trans. on Vehicular Tech.*, vol. VT-40, no. 2, May 1991, pp. 303-312.

[3] R. L. Pickholtz, L. B. Milstein and D. L.Schilling, "Spread Spectrum for Mobile Communications" *IEEE Trans. on Vehicular Tech.*, vol. VT-40, no. 2, May 1991, pp. 313-322.

[4] R. Kohno, R. Meidan, and L.B. Milstein, "Spread Spectrum Access Methods for Wireless Communications," *IEEE Commun. Magazine*, Jan. 1995, pp. 58-67.

[5] M. K. Simon *et al.*, "Spread Spectrum Communications, Vols. I-III, (Computer Science Press, 1985).

[6] G. Turin, "The Effects of Multipath and Fading on the Performance of Direct Sequence CDMA Systems," *IEEE JSAC*, vol. SAC-2, no. 4, July 1984, pp. 597-603.

[7] J. G. Proakis, "Digital Communications," 2nd Ed., (McGraw-Hill, 1989).

[8] J. M. Holtzman, "DS/CDMA Successive Interference Cancellation," Proc. of ISSSTA '94, Oulu, Finland, July 1994, pp. 69-78.

[9] S. Verdu, "Adaptive Multiuser Detection," Proc. of ISSSTA '94, Oulu, Finland, July 1994, pp. 43-50.

[10] L. F. Chang, "Dispersive Fading Effects in CDMA Radio Systems," Proc. of ICUPC '92 Dallas, TX, Sep. 1992, pp. 185-189.

[11] J. S. Lim, ed., Speech Enhancement, (Prentice-Hall, 1983).

[12] B. Widrow and S. D. Stearns, Adaptive Signal Processing, (Prentice-Hall, 1985).

[13] H. Nicolas, A. Giordano and J. Proakis, "MLD and MSE Algorithms for Adaptive Detection of Digital Signals in the Presence of Interchannel Interference," *IEEE Trans. on Info. Theory*, vol. IT-23, no. 5, Sep. 1977, pp. 563-575.

[14] J. Salz, "Digital Transmission Over Cross-Coupled Linear Channels," *AT&T Tech. J.*, vol. 64, no. 6, July-Aug. 1985, pp. 1147-1158.

[15] J. W. Carlin *et al.*, "An IF Cross-Pol Canceller for Microwave Radio Systems," *IEEE JSAC*, vol. SAC-5, no. 3, April 1987, pp. 502-514.

[16] K. S. Schneider, "Optimum Detection of Code Division Signals," *IEEE Trans. on Aerospace and Electronic Sys.*, vol. AES-15, no. 1, Jan. 1979, p. 181-185.

[17] R. Kohno, M. Hatori, and H. Imai, "Cancellation Techniques of Co-Channel Interference in Asynchronous Spread Spectrum Multiple Access Systems," *Electronics and Commun.*, vol. 66-A, no. 5, 1983, pp. 20-29.

[18] S. Verdu, "Minimum Probability of Error for Asynchronous Gaussian Multiple Access Channels," *IEEE Trans. on Info. Theory*, vol. IT-32, no. 1, Jan. 1986, pp. 85-96.

[19] S. Verdu, "Optimum Multiuser Asymptotic Efficiency," *IEEE Trans. on Commun.*, Vol. Com-34, No. 9, Sept. 1986, pp. 890-897.

[20] R. Lupas and S. Verdu, "Near-Far Resistance of Multiuser Detectors in Asynchronous Channel," *IEEE Trans. on Commun.*, vol. COM-38, no. 4, April 1990, pp. 496-508.

[21] R. De Gaudenzi, C. Elia, and R. Viola, "Bandlimited Quasi-Synchronous CDMA: A Novel Satellite Access Technique for Mobile and Personal Communication Systems," *IEEE JSAC*, vol. SAC-10, no. 2, Feb. 1992, pp. 328-343.

[22] A. Kajiwara and M. Nakagawa, "Microcellular CDMA System with a Linear Multiuser Interference Canceller," *IEEE JSAC*, vol. 12, no. 4, May 1994, pp. 605-611.

[23] R. Lupas and S. Verdu, "Linear Multiuser Detectors for Synchronous Code-Division Multiple-Access Channel," *IEEE Trans. on Info. Theory*, vol. IT-35, no. 1, Jan. 1989, pp. 123-136.

[24] Z. Xie, R. T. Short, and C. K. Rushforth, "A Family of Suboptimum Detectors for Coherent Multiuser Communications," *IEEE JSAC*, vol. SAC-8, no. 4, May 1990 pp. 683-690.

[25] Z. Xie, C. K. Rushforth, and R. T. Short, "Multiuser Signal Detection Using Sequential Decoding," *IEEE Trans. on Commun.*, vol. COM-38, no. 5, May 1990, pp. 578-583.

[26] M. K. Varanasi and B. Aazhang, "Multistage Detection in Asynchronous Code Division Multiple-Access Communications," *IEEE Trans. on Commun.*, vol. COM-38, no. 4, April 1990, pp. 509-519.

[27] T. R. Giallorenzi and S. G. Wilson, "Decision Feedback Multiuser Receivers for Asynchronous CDMA Systems," Proc. of GLOBECOM '93, Houston, TX, Nov.- Dec. 1993, pp. 1677-1681.

[28] M. K. Varanasi and B. Aazhang, "Near-Optimum Detection in Synchronous Code-Division Multiple Access Systems," *IEEE Trans. on Commun.*, vol. COM-39, May 1991, pp. 725-736.

[29] A. Duel-Hallen, "On Suboptimal Detection for Asynchronous Code-Division Multiple Access Channels," Proc. of the 26th Annual Conference on Information Sciences and Systems, Princeton University, Princeton, NJ, March 1992, pp. 838-843.

[30] A. Duel-Hallen, "Decorrelating Decision-Feedback Multiuser Detector for Synchronous Code-Division Multiple Access Channel," *IEEE Trans. on Commun.*, vol. COM-41, no. 2, Feb. 1993 pp. 285-290.

[31] A. Duel-Hallen, "A Family of Multiuser Decision-Feedback Detectors for Asynchronous Code-Division Multiple Access Channels," To appear in *IEEE Trans. on Commun.*, Feb. 1995.

[32] P. Patel and J. Holtzman, "Analysis of a Simple Successive Interference Cancellation Scheme in DS/CDMA System," *IEEE JSAC* - Special Issue on CDMA, vol. 12, no. 5, June 1994, pp. 796-807.

[33] A. Viterbi, "Very Low Rate Convolutional Codes for Maximum Theoretical Performance of Spread-Spectrum Multiple-Access Channels," *IEEE JSAC*, vol. SAC-8, no. 4, May 1990, pp. 641-649.
[34] P. Patel and J. Holtzman, "Performance Comparison of a DS/CDMA System using a Successive Interference Cancellation (IC) Scheme and a Parallel IC Scheme under Fading," Proc. of ICC'94, New Orleans, LA, May 1994, pp. 510-515.
[35] T. R. Giallorenzi and S. G. Wilson, "Trellis-Based Multiuser Receivers for Convolutionally Coded CDMA Systems," Proc. of the 31st Allerton Conference on Comm., Control and Computing, Oct. 1993.
[36] P. Hoeher, "On Channel Coding and Multiuser Detection for DS-CDMA,"Proc. of the 2nd International Conference on Universal

The properties of optimal and suboptimal multiuser detectors are well understood now. The next stages of investigation, involving implementation and robustness issues, are accelerating now and will lead to determination of the practical and economic feasibility of the multiuser detector. Initial studies of robustness show that robustness need not be a fatal flaw.

Personal Communications, Ottawa, Canada, Oct. 1993, pp. 641-646.
[37] T. R. Giallorenzi and S. G. Wilson, "Multistage Decision Feedback and Trellis-Based Multiuser Receivers for Convolutionally Coded CDMA Systems," Technical Report UVA/538341/EE93/102, Comm. Systems Lab., Dept of Electr. Eng., Univ. Of Virginia, May 1993.
[38] M. Nasiri-Kenari and C. K Rushforth, " An Efficient Soft-Decision Decoding Algorithm for Synchronous CDMA Communications with Error-Control Coding," Proceedings of the IEEE International Symposium on Information Theory, Trondheim, Norway, June 27-July 1, 1994, p. 227.
[39] U. Fawer and B. Aazhang, "Multiuser Reception for Trellis-Based Code Division Multiple Access Communications," Proceedings of MILCOM'94, Eatontown, NJ, Oct. 1994, pp. 977-981.
[40] P. Jung, M. Naβhan and J. Blanz, "Application of Turbo-Codes to CDMA Mobile Radio System Using Joint Detection and Antenna Diversity," Proceedings of VTC'94, Stockholm, Sweden, June 1994, pp. 770-774.
[41] W. Lee, "Overview of Cellular CDMA," *IEEE Trans. on Vehicular Technology*, vol. VT-40, no. 2, May 1991, pp. 291-302.
[42] Z. Zvonar and D. Brady, "Multiuser on in Single-Path Rayleigh Fading Channels," *IEEE Trans. on Commun.*, vol. COM-42, no.4, April 1994, pp.1729-1739.
[43] A. Duel-Hallen, "Performance of Multiuser Zero-Forcing and MMSE Decision Feedback Detectors for CDMA Channels," Conference Record of the Second Communication Theory Mini-Conference in conjunction with Globecom '93, Houston, TX, Dec. 1993, pp. 82-86.
[44] P. R. Patel and J. M. Holtzman, "Analysis of Successive Interference Cancellation in M-ary Orthogonal DS-CDMA System with Single Path Rayleigh Fading," Proc. of 1994 International Zurich Seminar on Digital Communications, Zurich, Switzerland, March 1994, pp. 150-161.
[45] H. Y. Wu, A. Duel-Hallen, "Performance of Multiuser Decision-Feedback Detectors for Flat Fading Synchronous CDMA Channels," Proc. of the 28-th Annual Conference on Information Sciences and Systems, Princeton University, Princeton, NJ, March 16-18, 1994, pp. 133-138.
[46] C. Kchao and G. Stuber, "Performance Analysis of a Single Cell Direct Sequence Mobile Radio System," *IEEE Trans. on Commun.*, vol. COM-41, no. 10, Oct. 1993, pp. 1507-1516.
[47] Z. Zvonar and D. Brady, "Optimum Detection in Asynchronous Multiple-Access Multipath Rayleigh Fading," Proc. of the 26th Annual Conference on Information Sciences and Systems, Princeton University, March 1992, pp. 826-831.
[48] M. Varanasi, "Noncoherent Detection in Asynchronous Multiuser Channels," *IEEE Trans. on Info. Theory*, vol. IT-39, no. 1, Jan. 1993, pp. 157-176.
[49] Z. Zvonar and D. Brady, "Coherent and Differentially Coherent Multiuser Detectors for Asynchronous CDMA Frequency-Selective Channels," Proc. MILCOM '92, San Diego, CA, Oct. 1992, pp. 442-446.
[50] S. Wijayasuriya, J. McGeehan and G. Norton, "RAKE Decorrelating Receiver for DS-CDMA Mobile Radio Networks," *Electronic Letters*, vol. 29, no 4, Feb. 1993, pp. 395-396.
[51] Z. Zvonar, "Multiuser Detection and Diversity Combining for Wireless CDMA Systems," Wireless and Mobile Communications, J. Holtzman and D. Goodman, eds., (Kluwer Academic Publishers, 1994), pp. 51-65.
[52] A. Klein and P. Baier, "Linear Unbiased Data Estimation in Mobile Radio System Applying CDMA," 1993. *IEEE JSAC*, vol. SAC-11, no. 7, Sep. 1993, pp. 1058-1066.
[53] R. Kohno *et al.*, "Combination of an Adaptive Array Antenna and a Canceller of Interference for Direct-Sequence Spread-Spectrum Multiple-Access System," *IEEE JSAC*, vol. SAC-8, no.4, May 1990, pp. 675-682.
[54] S. Wijayasuriya, G. Norton, and J. McGeehan, "A Novel Algorithm for Dynamic Updating of Decorrelator Coefficients in Mobile DS-CDMA," Proc. of the 4th International Symposium on Personal, Indoor and Mobile Radio Communications, Yokohama, Japan, Oct. 1993, pp. 292-296.
[55] E. Storm *et al.*, "Sensitivity Analysis of Near-Far Resistant DS-CDMA Receivers to Propagation Delay Estimation Errors," Proc. of VTC'94, Stockholm, Sweden, July 1994, pp. 757-761.
[56] F. C. Cheng and J. M. Holtzman, "Effect of Tracking Errors on DS/CDMA Successive Interference Cancellation," Proc. of Third Communication Theory Mini Conference in conjunction with Globecom '94, San Francisco, CA, Nov. 1994, pp. 166-170.
[57] L. Levi, F. Muratore and G. Romano, "Simulation Results for a CDMA Interference Cancellation Technique in a Rayleigh Fading Channel," Proc. of 1994 International Zurich Seminar on Digital Communications, Zurich, Switzerland, March 1994, pp 162-171.
[58] P. Dent, B. Gudmundson, and M. Ewerbring, "CDMA-IC: A Novel Code Division Multiple Access Scheme Based on Interference Cancellation," Proc. of PIMRC '92, Boston, MA, Oct. 1992, pp. 98-102.
[59] S. Gray, M. Kocic and D. Brady, "Multiuser Detection in Mismatched Multiple-Access Channels," To appear in *IEEE Trans. on Commun.*.
[60] Z. Zvonar, M. Stojanovic, "Performance of Multiuser Diversity Reception in Nonselective Rayleigh Fading CDMA Channels," Proc. of the Third Communication Theory Mini-Conference (CTMC '94), San Francisco, CA, Nov. 1994, pp. 171-175.
[61] H. Y. Wu, A. Duel-Hallen, "Channel Estimation and Multiuser Detection for Frequency-Nonselective Fading Synchronous CDMA Channels," Proceedings of the 32nd Annual Allerton Conference on Communication, Control and Computing, Monticello, IL, Sep. 1994.
[62] M. Wallace and R. Walton, "CDMA Radio Network Planning," Proc. of ICUPC '94, San Diego, CA, Sept. 27 - Oct. 4, 1994, pp. 62-67.

Biographies

ALEXANDRA DUEL-HALLEN received a B.S. in mathematics from Case Western Reserve University in 1982, an M.S. in computer, information, and control engineering from the University of Michigan in 1983, and a Ph.D. in electrical engineering from Cornell University in 1987. She worked for AT&T Bell Laboratories in Columbus, Ohio during the summer of 1982, and participated in the AT&T One Year on Campus Program at the University of Michigan during the 1982-1983 academic year. She received an AT&T Ph.D. Fellowship during 1985-1987. In 1987-1990 she was a visiting assistant professor at the School of Electrical Engineering, Cornell University. In 1990-1992, she was with the Mathematical Sciences Research Center, AT&T Bell Laboratories, Murray Hill, New Jersey. She joined the Department of Electrical and Computer Engineering at North Carolina State University, Raleigh, North Carolina, in January 1993 as an assistant professor. Her current research interests are in channel equalization and spread spectrum communications. She has served as an editor for Communication Theory for the *IEEE Transactions on Communications* since 1989. During 1994, she was a secretary of the Information Theory Society Board of Governors.

JACK M. HOLTZMAN [F '95] received a B.E.E. from City College of New York, an M.S. from U.C.L.A., and a Ph.D. from the Polytechnic Institute of Brooklyn. He worked for AT&T Bell Laboratories for 26 years on control theory, teletraffic theory, telecommunications, and performance analysis. At Bell Labs, he was supervisor of the Mathematical Analysis and Consulting Group and then Head of the Teletraffic Theory and System Performance Department. In 1990 he joined Rutgers University, where he is Professor of Electrical and Computer Engineering and Associate Director of the Wireless Information Network Laboratory (WINLAB), Piscataway, New Jersey. He is also the director of the Wireless Communications Certificate Program. His current areas of work are on spread spectrum, handoffs, resource management, propagation, and wireless system performance.

ZORAN ZVONAR received a Dipl. Ing. in 1986 and an M.S. in 1989, both from the Department of Electrical Engineering, University of Belgrade, Yugoslavia, and a Ph.D. in electrical engineering from Northeastern University, Boston, in 1993. From 1986 to 1989 he was with the Department of Electrical Engineering, University of Belgrade, Belgrade, Yugoslavia, where he conducted research in the area of telecommunications. From 1993 to 1994 he was a post-doctoral investigator at the Woods Hole Oceanographic Institution, where he worked on multiple-access communications for underwater acoustic local area networks. Since 1994 he has been with the Analog Devices, Communications Division, Wilmington, Massachusetts, where he is working on the design of wireless communications systems.

Minimum Probability of Error for Asynchronous Gaussian Multiple-Access Channels

SERGIO VERDÚ, MEMBER, IEEE

Abstract—**Consider a Gaussian multiple-access channel shared by K users who transmit asynchronously independent data streams by modulating a set of assigned signal waveforms. The uncoded probability of error achievable by optimum multiuser detectors is investigated. It is shown that the K-user maximum-likelihood sequence detector consists of a bank of single-user matched filters followed by a Viterbi algorithm whose complexity per binary decision is $O(2^K)$. The upper bound analysis of this detector follows an approach based on the decomposition of error sequences. The issues of convergence and tightness of the bounds are examined, and it is shown that the minimum multiuser error probability is equivalent in the low-noise region to that of a single-user system with reduced power. These results show that the proposed multiuser detectors afford important performance gains over conventional single-user systems, in which the signal constellation carries the entire burden of complexity required to achieve a given performance level.**

I. Introduction

CONSIDER a Gaussian multiple-access channel shared by K users who modulate simultaneously and independently a set of assigned signal waveforms without maintaining any type of synchronism among them. The coherent K-user receiver commonly employed in practice consists of a bank of optimum single-user detectors operating independently (Fig. 1). Since in general the input to every threshold has an additive component of multiple-access interference (because of the cross correlation with the signals of the other users), the conventional receiver is not optimum in terms of error probability. However, if the designer is allowed to choose a signal constellation with large bandwidth (e.g., in direct-sequence spread-spectrum systems), then the cross correlations between the signals can be kept to a low level for all relative delays, and acceptable performance can be achieved. Nevertheless, if data demodulation is restricted to single-user detection systems, then the cross-correlation properties of the signal constellation carry the entire burden of complexity required to achieve a given performance level, and when the power of some of the interfering users is dominant, performance degradation is too severe. For this reason and because of the availability of computing devices with increased capabilities, there is recent interest in investigating the degree of performance improvement achievable with more sophisticated receivers and, in particular, with the minimum error probability detector.

Manuscript received May 3, 1983; revised November 5, 1984. This work was supported by the U.S. Army Research Office under Contract DAAG-81-K-0062.

The author is with the Department of Electrical Engineering, Princeton, University, Princeton, NJ 08544.

IEEE Log Number 8406098.

Optimum multiuser detection of asynchronous signals is inherently a problem of sequence detection, that is, observation of the whole received waveform is required to produce a sufficient statistic for any symbol decision, and hence one-shot approaches (where the demodulation of each symbol takes into account the received signal only in the interval corresponding to that symbol) are suboptimal. The reason is that the observation of the complete intervals of the overlapping symbols of the other users gives additional information about the received signal in the bit interval in question, and since this reasoning can be repeated with the overlapping bits, no restriction of the whole observation interval is optimal for any bit decision. Furthermore, since the transmitted symbols are not independent conditioned on the received realization, decisions can be made according to two different optimality criteria, namely, selection of the sequence of symbols that maximizes the joint posterior distribution (maximum-likelihood sequence detection), or selection of the symbol sequence that maximizes the sequence of marginal posterior distributions (minimum-probability-of-error detection). Moreover, the simultaneous demodulation of all the active users in the multiple-access channel can be regarded as a problem of periodically time-varying intersymbol interference, because from the viewpoint of the coherent K-user detector, the observed process is equivalent to that of a single-user-to-single-user system where the sender transmits K symbols during each signal period by modulating one out of K waveforms in a round-robin fashion.

Earlier work on multiuser detection includes, in the case of synchronous users (which reduces to an m-ary hypothesis testing problem) the receivers of Horwood and Gagliardi [9] and Schneider [16], and, in the asynchronous case, the one-shot baseband detector obtained by Poor [12] in the two-user case, and the detectors proposed by Van Etten [18] and Schneider [16] for interference-channel models with vector observations. Results on the probability of error have been obtained only for the conventional single-user receiver ([8], [14], and the references therein).

In Section II we obtain a K-user maximum-likelihood sequence detector which consists of a bank of K single-user matched filters followed by a Viterbi forward dynamic

0018-9448/86/0100-0085$01.00 ©1986 IEEE

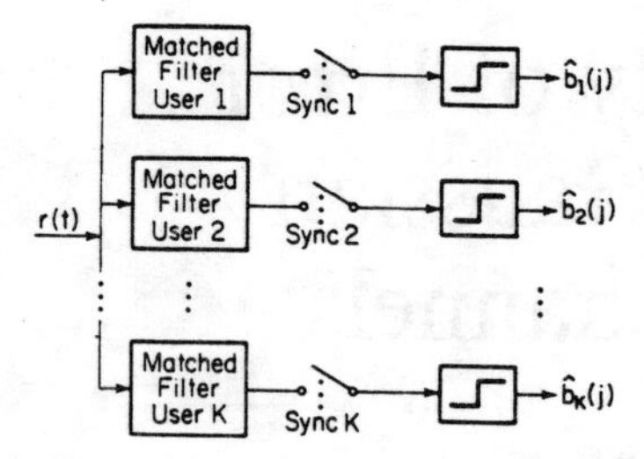

Fig. 1. Conventional multiuser detector.

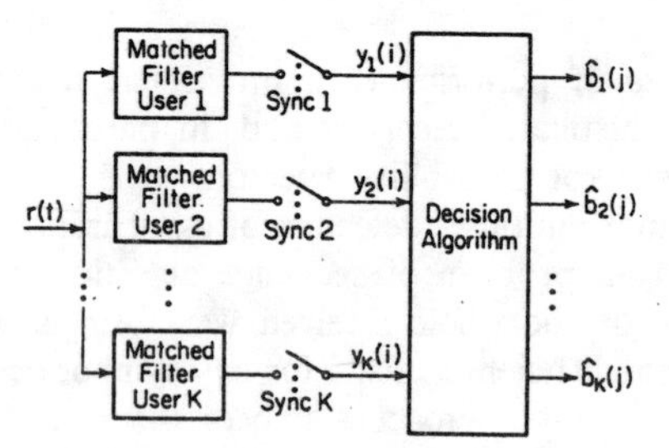

Fig. 2. Optimum K-user detector for asynchronous multiple-access Gaussian channel.

programming algorithm (Fig. 2) with 2^{K-1} states and $O(2^K)$ time complexity per bit (in the binary case). Section III is devoted to the analysis of the minimum uncoded bit error rate of multiuser detectors. This is achieved through various bounds that together provide tight approximations for all noise levels. A possible route to upper bound the error probability of the multiuser maximum-likelihood sequence detector is to generalize the approach taken by Forney in the intersymbol interference problem [3]. However, motivated by the more general structure of the multiuser problem, we introduce a different approach that, when applied to the intersymbol interference case, turns out to result in a bound that is tighter than the Forney bound. In Section IV several numerical examples illustrate the performance gains achieved by optimum multiuser detectors over conventional single-user systems.

II. Multiuser Maximum-Likelihood Sequence Detection

In this section we derive optimum decision rules for the following multiple-access model with additive linearly modulated signals in additive white Gaussian noise and scalar observations:

$$dr_t = S_t(\boldsymbol{b})\,dt + \sigma\,d\omega_t, \qquad t \in R \tag{1}$$

where

$$S_t(\boldsymbol{b}) = \sum_{i=-M}^{M} \sum_{k=1}^{K} b_k(i) s_k(t - iT - \tau_k), \tag{2}$$

the symbol interval duration is equal to T (assumed to be the same for all users), $\boldsymbol{b} = \{b(i) \in A_1 \times \cdots \times A_K,\ i = -M, \cdots, M\}$, and A_k, $s_k(t)$ ($= 0$ outside $[0, T]$), and $\tau_k \in [0, T)$ are the finite alphabet, the signal waveform, and the delay (modulo T with respect to an arbitrary reference), respectively, of the kth user, and ω_t is a standard Wiener process started at $t = -MT$. Without loss of generality, and for the sake of notational simplicity, we suppose that the users are numbered such that $0 \le \tau_1 \le \cdots \le \tau_K < T$. Note that even if all the transmitted symbols are assumed to be equiprobable and independent, there is not a unique optimality criterion due to the existence of several users. It is possible to select either the set of symbols that maximize the joint posterior distribution $P[\boldsymbol{b}|\{r_t,\ t \in R\}]$ (globally optimum or maximum-likelihood sequence detection) or those that maximize the marginal posteriori distributions $P[b_k(i)|\{r_t,\ t \in R\}]$, $i = -M, \cdots, M$, $k = 1, \cdots, K$ (locally optimum or minimum-error-probability detection). It is shown later that the maximum-likelihood sequence detector can be implemented by a signal processing front end that produces a sequence of scalar sufficient statistics, followed by a dynamic programming decision algorithm of the forward (Viterbi) type. It can be shown [22] that the multiuser detector that minimizes the probability of error has the same structure, but it uses a backward–forward dynamic programming algorithm instead [21]. The computational complexity of the various decision algorithms will be measured and compared by their *time complexity per binary decision* (TCB), that is, the limit as $M \to \infty$ of the time required by the decision algorithm to select the optimum sequence divided by the number of transmitted bits.

Since all transmitted sequences of symbols are assumed to be equiprobable, the maximum-likelihood sequence detector selects the sequence that maximizes

$$P[\{r_t, t \in R\}|\boldsymbol{b}] = C\exp(\Omega(\boldsymbol{b})/2\sigma^2) \tag{3}$$

where C is a positive scalar independent of $\boldsymbol{b}$ and

$$\Omega(\boldsymbol{b}) = 2\int_{-\infty}^{\infty} S_t(\boldsymbol{b})\,dr_t - \int_{-\infty}^{\infty} S_t^2(\boldsymbol{b})\,dt. \tag{4}$$

Therefore, the maximum-likelihood sequence detector selects among the possible noise realizations the one with minimum energy. Using the definition (2), we can express the first term in the right-hand side of (4) as

$$\int_{-\infty}^{\infty} S_t(\boldsymbol{b})\,dr_t = \sum_{i=-M}^{M} \boldsymbol{b}^T(i)\boldsymbol{y}(i), \tag{5}$$

where $y_k(i)$ denotes the output of a matched filter for the ith symbol of the kth user, that is,

$$y_k(i) = \int_{\tau_k+iT}^{\tau_k+iT+T} s_k(t - iT - \tau_k)\,dr_t. \tag{6}$$

Hence, even though $y_k(i)$ is not a sufficient statistic for the detection of $b_k(i)$, (4) and (5) imply that the whole sequence of outputs of the bank of K matched filters $\boldsymbol{y}$ is a sufficient statistic for the selection of the most likely sequence $\boldsymbol{b}$. This implies that the maximum-likelihood multiuser coherent detector consists of a front end of matched filters (one for each user) followed by a decision algorithm (Fig. 2), which selects the sequence $\boldsymbol{b}$ that maximizes (3) or, equivalently, (4). The efficient solution of this combinatorial optimization problem is the central issue in the derivation of the multiuser detector. The TCB of the exhaustive algorithm that computes (4) for all possible sequences has not only the inconvenient feature of being dependent on the block-size M, but it is so in an exponential way. Fortunately, $\|S(\boldsymbol{b})\|^2 = \int_{-\infty}^{\infty} S_t^2(\boldsymbol{b})\, dt$, the energy of the sequence $\boldsymbol{b}$, has the right structure to result in decision algorithms with significantly better TCB. The key to the efficient maximization of $\Omega(\boldsymbol{b})$ lies in its sequential dependence on the symbols $b_k(i)$, which allows us to put it as a sum of terms that depend only on a few variables at a time.

Suppose that we can find a discrete-time system $x_{i+1} = f_i(x_i, u_i)$, with initial condition x_{i_0}; a transition-payoff function $\lambda_i(x_i, u_i)$; and a bijection between the set of transmitted sequences and a subset of control sequences $\{u_i,\ i = i_0, \cdots, i_f\}$ such that $\Omega(\boldsymbol{b}) = \sum_{i=i_0}^{i_f} \lambda_i(x_i, u_i)$, subject to $x_{i+1} = f_i(x_i, u_i)$, x_{i_0}, and $\boldsymbol{b} \leftrightarrow \{u_i,\ i = i_0, \cdots, i_f\}$. Then the maximization of $\Omega(\boldsymbol{b})$ is equivalent to a discrete-time deterministic control problem with additive cost and finite input and state spaces, and therefore it can be solved by the dynamic programming algorithm either in backward or in forward fashion. Although the decision delay is unbounded because optimum decisions cannot be made until all states share a common shortest subpath, a well-known advantage in real-time applications of the forward dynamic programming algorithm (the Viterbi algorithm) is that little degradation of performance occurs when the algorithm uses an adequately chosen fixed finite decision lag. It turns out that there is not a unique additive decomposition of the log-likelihood function $\Omega(\boldsymbol{b})$, resulting in decision algorithms with very different computational complexities. It can be shown that the Viterbi algorithm suggested by Schneider [16] has 4^K states and $O(8^K/K)$ time complexity per bit, while the decision algorithm of the multiuser detector in [20] has 2^K states and TCB = $O(4^K/K)$. By fully exploiting the sequential dependence of the log-likelihood function on the transmitted symbols, it is possible to obtain an optimum decision algorithm that exhibits a lower time complexity per bit than the foregoing. This is achieved by the additive decomposition of the log likelihood function given by the following result.

Proposition 1: Define the following matrix of signal crosscorrelations:

$$G_{ij} = \begin{cases} \int_{-\infty}^{\infty} s_{i+j}(t - \tau_{i+j}) s_j(t - \tau_j - T)\, dt, & \text{if } i + j \le K \\ \int_{-\infty}^{\infty} s_{i+j-K}(t - \tau_{i+j-K}) s_j(t - \tau_j)\, dt, & \text{if } i + j > K \end{cases} \tag{7}$$

for $i = 1, \cdots, K - 1$ and $j = 1, \cdots, K$ (i.e., the entries of the column[1] G^j are the correlations of the signal of the jth user with the $K - 1$ preceding signals) and denote the received signal energies by $w_k = \int_0^T s_k^2(t)\, dt$, $k = 1, \cdots, K$. For any integer i, denote its modulo-K decomposition with remainder $\kappa(i) = 1, \cdots, K$, by $i = \eta(i)K + \kappa(i)$. Then we have

$$\Omega(\boldsymbol{b}) = \sum_{i=i_0}^{i_f} \lambda_i(x_i, u_i) \tag{8}$$

where $i_0 = 1 - MK$, $i_f = (M+1)K$, $u_i = b_{\kappa(i)}(\eta(i)) \in A_{\kappa(i)}$, and

$$\lambda_i(x, u) = u\left[2y_{\kappa(i)}(\eta(i)) - uw_{k(i)} - 2x^T G^{\kappa(i)}\right] \tag{9}$$

with

$$x_{i+1} = \left[x_i^2 x_i^3 \cdots x_i^{K-1} u_i\right]^T, \qquad x_{i_0} = \mathbf{0}. \tag{10}$$

Proof: Utilizing the foregoing modulo-K decomposition, it is easy to check that

$$S_t(\boldsymbol{b}) = \sum_{i=i_0}^{i_f} b_{\kappa(i)}(\eta(i)) s_{\kappa(i)}(t - \eta(i)T - \tau_{\kappa(i)}). \tag{11}$$

Hence we have

$$\begin{aligned} \|S(\boldsymbol{b})\|^2 = \sum_{i=i_0}^{i_f} \Bigg[& b_{\kappa(i)}^2(\eta(i)) \int_{-\infty}^{\infty} s_{\kappa(i)}^2(t - \eta(i)T - \tau_{\kappa(i)})\, dt \\ & + 2 \sum_{j=i_0}^{i-1} b_{\kappa(i)}(\eta(i)) b_{\kappa(j)}(\eta(j)) \cdot \int_{-\infty}^{\infty} s_{\kappa(i)}(t - \eta(i)T - \tau_{\kappa(i)}) \cdot s_{\kappa(j)}(t - \eta(j)T - \tau_{\kappa(j)})\, dt \Bigg] \\ = \sum_{i=i_0}^{i_f} \Bigg[& b_{\kappa(i)}^2(\eta(i)) w_{\kappa(i)} \\ & + 2 \sum_{l=1}^{K-1} b_{\kappa(i)}(\eta(i)) b_{\kappa(i-l)}(\eta(i-l)) \cdot \int_{-\infty}^{\infty} s_{\kappa(i)}(t - \eta(i)T - \tau_{\kappa(i)}) \cdot s_{\kappa(i-l)}(t - \eta(i-l)T - \tau_{\kappa(i-l)})\, dt \Bigg], \end{aligned} \tag{12}$$

where by agreement $b_{\kappa(i)}(\eta(i)) = 0$ for $i < i_0$. Now we show that the integral in the right-hand side of (12) is equal to $G_{K-l\,\kappa(i)}$. To that end we will examine separately the terms in which $\eta(i) = \eta(i - l)$ and $\eta(i) = \eta(i - l) + 1$. In the first case, $\kappa(i - l) = \kappa(i) - l$ and

$$\begin{aligned} & \int_{-\infty}^{\infty} s_{\kappa(i)}(t - \eta(i)T - \tau_{\kappa(i)}) \cdot s_{\kappa(i-l)}(t - \eta(i-l)T - \tau_{\kappa(i-l)})\, dt \\ & = \int_{-\infty}^{\infty} s_{\kappa(i)}(t - \tau_{\kappa(i)}) s_{\kappa(i)-l}(t - \tau_{\kappa(i)-l})\, dt \\ & = G_{K-l\,\kappa(i)}. \end{aligned} \tag{13}$$

[1] The columns of the matrix G and of the row vector x_i^T are denoted by superscripts.

In the second case, $\kappa(i-l) = \kappa(i) + K - l$ and

$$\int_{-\infty}^{\infty} s_{\kappa(i)}\big(t - \eta(i)T - \tau_{\kappa(i)}\big) \cdot s_{\kappa(i-l)}\big(t - \eta(i-l)T - \tau_{\kappa(i-l)}\big)\, dt = \int_{-\infty}^{\infty} s_{\kappa(i)}\big(t - T - \tau_{\kappa(i)}\big) s_{\kappa(i)+K-l}\big(t - \tau_{\kappa(i)+K-l}\big)\, dt = G_{K-l\,\kappa(i)}. \quad (14)$$

From (10) and $u_i = b_{\kappa(i)}(\eta(i))$ it is clear that $b_{\kappa(i-l)}(\eta(i-l)) = x_i^{K-l}$, and therefore

$$\|S(\boldsymbol{b})\|^2 = \sum_{i=i_0}^{i_f} b_{\kappa(i)}(\eta(i)) \cdot \left[b_{\kappa(i)}(\eta(i)) w_{\kappa(i)} + 2 \sum_{j=1}^{K-1} x_i^j G_{j\kappa(i)} \right]. \quad (15)$$

Using (15) and the fact that

$$\int_{-\infty}^{\infty} S_t(\boldsymbol{b})\, dr_t = \sum_{i=i_0}^{i_f} b_{\kappa(i)}(\eta(i)) y_{\kappa(i)}(\eta(i))$$

[see (5)], the sought-after decomposition (9) follows.

The algorithm resulting from the decomposition of Proposition 1 performs K times the number of stages of that derived in [20]; however, its asymptotic complexity is considerably better since it exploits fully the separability of the log-likelihood function. In this case the dimensionality of the state space is equal to 2^{K-1}, and each state is connected to two states in the previous stage (if all the users employ binary modulation), resulting in TCB = $O(2^K)$.[2] This decomposition was introduced in [19] and shows that the part of the transition metric that is independent of the matched filter outputs is periodically time-varying with a period equal to the number of users. The nature of this behavior can be best appreciated by particularizing it to the intersymbol interference problem, in which $s_i(t) = s_j(t)$ for all $t \in [0, T]$, $i \neq j$, and $\tau_{i+1} - \tau_i = T/K$. In this case, $w_i = w_j$ and $G^i = G^j$ for $i \neq j$, and (9) reduces to the Ungerboeck metric [17, eq. (27)] when each symbol suffers the interference of $K-1$ signals. So, (9) can be viewed as the generalization of the Ungerboeck metric to a problem of periodically time-varying intersymbol interference equivalent from the receiver viewpoint, to the asynchronous multiuser model (1)–(2). The assumption that the signals of all users have the same duration can be relaxed, resulting in a decomposition similar to (9); however, the periodicity of the transition metric is lost unless the ratio between every pair of signal periods is rational.

The Viterbi algorithm resulting from the additive decomposition of Proposition 1 requires knowledge of the partial cross correlations between the signals of every pair of active users, and, unless binary antipodal modulation is employed, they also require the received signal energies. However, the signal cross correlations depend on the received relative delays, carrier phases, and amplitudes and hence cannot be determined *a priori* by the receiver. Nevertheless, the basic assumption is that the signal waveform of each user is known and that the K-user coherent receiver locks to the signaling interval and phase of each active user. Then the required parameters G_{ij} can be generated internally by cross-correlating the normalized waveform replicas stored in the receiver with the adequate delays and phases supplied by the synchronization system and by multiplying the resulting normalized cross correlations by the received amplitudes of the corresponding users. Hence the only requirement (beyond synchronization) imposed by the need for the partial cross correlations is the availability (up to a common scale factor) of the K received signal amplitudes.

The decomposition in Proposition 1 can be generalized to the case where the modulation is not necessarily linear, that is, each symbol $u \in A_k$ is mapped into a different waveform $s_k(t; u)$. It can be shown, analogously to the proof of Proposition 1, that in this case the transition metric is

$$\lambda_i(x, u) = 2 y_{\kappa(i)}(\eta(i); u) - w_{\kappa(i)}[u] - 2 \sum_{l=1}^{K-1} G_{l\kappa(i)}[x^l, u], \quad (16)$$

where $y_k(i, u)$ is the output of the matched filter of the uth waveform of the kth user; $w_k[u] = \int_0^T s_k^2(t; u)\, dt$; and

$$G_{lk}[a, u] = \begin{cases} \int s_{l+k}(t - \tau_{l+k}; a) s_k(t - \tau_k - T; u)\, dt, & \text{if } l + k \leq K \\ \int s_{l+k-K}(t - \tau_{l+k-K}; a) s_k(t - \tau_k; u)\, dt, & \text{if } l + k > K. \end{cases} \quad (17)$$

III. Error Probability Analysis of Optimum Multiuser Detectors

A. Upper and Lower Bounds

This section is devoted to the analysis of the minimum uncoded bit error probability of multiuser detectors for antipodally modulated signals in additive white Gaussian noise, that is, attention is focused on the multiple-access model (1)–(2) in the case $A_1 = \cdots = A_K = \{-1, 1\}$. Denote the most likely transmitted ith symbol by the kth user given the observations by

$$\hat{b}_k^M(i) \in \arg \max_{b \in \{-1,1\}} P\big[b_k(i) = b \,|\, dr_t = S_t^M(\boldsymbol{b}) + \sigma\, d\omega_t,\ t \in R\big] \quad (18)$$

where $\boldsymbol{b} = \{b(i) \in \{-1,1\}^K,\ i = -M, \cdots, M\}$ is the transmitted sequence. The goal is to obtain the finite and

[2] If the alphabet sizes are arbitrary, then (see [22]) TCB = $O(K \Pi_{i=1}^K |A_i| / \Sigma_{i=1}^K \log |A_i|)$. If there are S clusters of synchronized users, then it is possible to reduce this complexity by a factor of S/K by taking into account that some of the cross correlations are equal to zero.

infinite horizon error probabilities,

$$P_k^M(i) = P[b_k(i) \neq \hat{b}_k^M(i)], \qquad i = -M,\cdots,M \quad (19)$$

$$P_k = \lim_{M\to\infty} P_k^M(i), \quad (20)$$

for arbitrary signal waveforms and relative delays. The existence of the limit for any given i in the right-hand side of (20) follows because $P_k^M(i)$ is equal to probability of error of an optimum detector for a multiuser signal with horizon equal to $M+1$ and complete information about $b(-M-1)$ and $b(M+1)$; hence $P_k^{M+1}(i) \geq P_k^M(i)$ for any positive integer M, $i \in \{-M,\cdots,M\}$ and $k \in \{1,\cdots,K\}$. The independence of the limit on i readily follows from the assumed stationarity of the noise and of the priors.

Our approach is to derive lower and upper bounds on the error rate for each user, which are tight in the low and high SNR regions and give a close approximation over the whole SNR range. The upper bounds are based on the analysis of two detectors that are suboptimum in terms of error probability, namely, the conventional single-user coherent detector and the K-user maximum-likelihood sequence detector. The lower bounds are the error probabilities of two optimum binary tests derived from the original problem by allowing certain side information.

The normalized difference between any pair of distinct transmitted sequences will be referred to as an error sequence, that is, the set of nonzero error sequences is[3]

$$E = \Big\{\epsilon = \big\{\epsilon(i) \in \{-1,0,1\}^K, i = -M,\cdots,M;\ \epsilon(j) \neq \mathbf{0} \text{ for some } j\big\}\Big\}.$$

The set of error sequences that are admissible conditioned on b being transmitted, that is, those that correspond to the difference between b and the sequence selected by the detector, is denoted by

$$A(b) = \{\epsilon \in E, 2\epsilon - b \in D\},$$

where

$$D = \Big\{b = \big\{b(i) \in \{-1,1\}^K, i = -M,\cdots,M\big\}\Big\}.$$

The admissible error sequences that affect the ith bit of the kth user of b are

$$A_k(b,i) = \{\epsilon \in A(b), \epsilon_k(i) = b_k(i)\}.$$

The number of nonzero components of an error sequence and the energy of a hypothetical multiuser signal modulated by the sequence ϵ are denoted, respectively, by

$$w(\epsilon) = \sum_{i=-M}^{M}\sum_{k=1}^{K}|\epsilon_k(i)|$$

and

$$\|S(\epsilon)\|^2 = \int_{-\infty}^{\infty}\left(\sum_{i=-M}^{M}\sum_{k=1}^{K}\epsilon_k(i)s_k(t-iT-\tau_k)\right)^2 dt.$$

[3]For the sake of notational simplicity, the explicit dependence of the sets of sequences on M is dropped when this causes no ambiguity.

Proposition 2: Define the following minimum distance parameters:

$$d_k^M(b,i) = \inf_{\epsilon \in A_k^M(b,i)} \|S(\epsilon)\|$$

and

$$d_{k,\min}^M(i) = \inf_{b \in D} d_k^M(b,i).$$

Then, the minimum error probability of the ith bit of the kth user is lower-bounded by

$$P_k \geq P_k^M(i) \geq P[d_k^M(b,i) = d_{k,\min}^M(i)]\,Q(d_{k,\min}^M(i)/\sigma) \quad (21)$$

and by

$$P_k \geq P_k^M(i) \geq Q(w_k/\sigma), \quad (22)$$

where $P[d_k(b,i) = d_{k,\min}(i)]$ is the *a priori* probability that the transmitted sequence is such that one of its congruent error sequences affects the ith bit of the kth user and has the minimum possible energy.

Proof: The basic technique for obtaining lower bounds on the minimum error probability is to analyze the performance of an optimum receiver that, in addition to observing $\{r_t, t \in R\}$, has certain side information.

To obtain the first lower bound (21), the following reasoning, analogous to that of Forney [2], can be employed. Suppose that if the transmitted sequence b is such that $d_k(b,i) = d_{k,\min}(i)$, the detector is told by a genie that the true sequence is either b or $b - 2\delta$, where δ is arbitrarily chosen by the genie (independently of the noise realization) from the set

$$\arg \min_{\epsilon \in A_k(b,i)} \|S(\epsilon)\|.$$

Under these conditions, the minimum error probability detector and the optimum sequence detector coincide and both reduce to a binary hypothesis test $\Omega(b) \gtrless \Omega(b-2\delta)$. The conditional probability of error given that b is transmitted is given by

$$P[\Omega(b-2\delta) > \Omega(b)|b \text{ transmitted}] = P\left[\sigma\int S_t(\delta)\,d\omega_t + \|S(\delta)\|^2 < 0\right] = Q(\|S(\delta)\|/\sigma) \quad (23)$$

where the foregoing equalities follow from (4) and the fact that $\int S_t(\delta)/\|S(\delta)\|\,d\omega_t$ is a zero-mean unit-variance Gaussian random variable. If the transmitted sequence b is such that $d_k(b,i) > d_{k,\min}(i)$, then the error probability of the receiver with side information is trivially bounded by zero, and the lower bound (21) follows. The lower bound (22) is the kth user minimum error probability if no other user was active or, equivalently, if the receiver knew the transmitted bits of the other users.

For any error sequence ϵ such that $\epsilon(j) = \epsilon(n) \neq \mathbf{0}$, for $j \neq n$, the sequence

$$\epsilon'(m) = \begin{cases}\epsilon(m), & m \leq j\\ \epsilon(m+n-j), & m > j\end{cases}$$

satisfies $\|S(\epsilon')\| \leq \|S(\epsilon)\|$. (Otherwise, one could construct a sequence with negative energy.) This implies that the infinite-horizon minimum distances, $d_k(\boldsymbol{b})$ and $d_{k\,\min}$ (i.e., one-half of the minimum rms of the difference between the signals of any pair of transmitted sequences that differ in any bit of the kth user), are achieved by finite-length error sequences, and the error rate of the kth user can be lower-bounded by

$$P_k \geq P\left[d_k(\boldsymbol{b}) = d_{k,\min}\right] Q(d_{k,\min}/\sigma). \quad (24)$$

Note that since $d_{k,\min} \leq d^M_{k,\min}(i)$, the bound (24) is at least as tight as (21) in the low-noise region.

The following upper bound, the error probability of the conventional single-user coherent receiver, is mainly useful in the low SNR region.

Proposition 3: Let $R_{ij} = G_{K-i\kappa(k+i)}$, for $i = 1,\cdots, K-1$ and $j = 1,\cdots,K$, that is, the entries of the column R^j are the correlations of the signal of the jth user with the $K-1$ posterior signals, and denote $I_k(\alpha,\beta) = \sum_{j=1}^{K-1}(\alpha_j G_{jk} + \beta_j R_{jk})/w_k$. The minimum error probability of the kth user is upper-bounded by

$$P^M_k(i) \leq P_k \leq E\left[Q\left(\sqrt{w_k}\left[1 + I_k(\alpha,\beta)\right]/\sigma\right)\right], \quad (25)$$

where the expectation is over the ensemble of independent uniformly distributed $\alpha \in \{-1,1\}^{K-1}, \beta \in \{-1,1\}^{K-1}$.

Proof: Suppose that $|i| < M$ and that the transmitted sequence is such that $b_k(i) = 1$. The kth matched filter output corresponding to the ith bit is a conditionally Gaussian random variable with variance equal to $w_k\sigma^2$ and mean given by

$$w_k + \sum_{l=k+1}^{K} b_l(i-1)G_{l-kk} + \sum_{l=1}^{k-1} b_l(i)G_{l-k+Kk}$$
$$+ \sum_{l=k+1}^{K} b_l(i)R_{l-kk} + \sum_{l=1}^{k-1} b_l(i+1)R_{l-k+Kk}.$$

Since the transmitted bits are assumed to be equiprobable and independent, we have

$$P^M_k(i) \leq E\left[Q\left(\sqrt{w_k}\left[1 + I_k(\alpha,\beta)\right]/\sigma\right)\right], \quad (26)$$

and because the right-hand side does not depend on M, (25) follows.

We turn to the derivation of an upper bound on the error probability of the K-user maximum-likelihood sequence detector. Our approach hinges on the following definition. An error sequence $\epsilon \in E$ is *decomposable* into $\epsilon' \in E$ and $\epsilon'' \in E$ if

1) $\epsilon = \epsilon' + \epsilon''$,
2)[4] $\epsilon' < \epsilon, \epsilon'' < \epsilon$,
3) $< S(\epsilon'), S(\epsilon'') > \geq 0$.

As an illustration consider the following simple two-user example: $s_1(t) = s_2(t) = 1$, $0 \leq t \leq T/2$; $s_1(t) = -s_2(t) = -1$, $T/2 < t \leq T$; and $\tau_2 - \tau_1 = T/2$. Then it is easy to check that the error sequence $\epsilon = \cdots, \mathbf{0}, [1\ 1]^T, [1\ 0]^T, \mathbf{0}, \cdots$ is decomposable into $\epsilon' = \cdots, \mathbf{0}, [1\ 0]^T, [1\ 0]^T, \mathbf{0}, \cdots$, and $\epsilon'' = \cdots, \mathbf{0}, [0\ 1]^T, [0\ 0]^T, \cdots$.

[4] We denote $\epsilon' < \epsilon$ if $|\epsilon'_j(i)| \leq |\epsilon_j(i)|$ for all $j = 1,\cdots,K$ and $i = -M,\cdots,M$.

Proposition 4: Denote the set of error sequences that affect the ith bit of the kth user by $Z_k(i) = \{\epsilon \in E, \epsilon_k(i) \neq 0\}$. Let $F_k(i)$ be the subset of *indecomposable* sequences in $Z_k(i)$. Then the minimum-error probability of the ith bit of the kth user is upper-bounded by

$$P^M_k(i) \leq \sum_{\epsilon \in F_k(i)} 2^{-w(\epsilon)} Q(\|S(\epsilon)\|/\sigma). \quad (27)$$

Proof: Formula (27) is an upper bound on the probability of error of the K-user maximum-likelihood sequence detector, that is, the receiver whose output is the sequence that maximizes $\Omega(\boldsymbol{b})$. This is the only property of the detector that we use for the proof of (27); it is not necessary to assume any specific decision algorithm, in particular that of Section II. Define the following sets of error sequences:

$$L = \left\{\epsilon \in E, \sigma \int S_t(\epsilon)\, dw_t \leq -\|S(\epsilon)\|^2\right\} \quad (28)$$

and

$$ML(\boldsymbol{b}) = \{\epsilon \in A(\boldsymbol{b}), \Omega(\boldsymbol{b} - 2\epsilon) \geq \Omega(\boldsymbol{d}), \text{ for all } \boldsymbol{d} \in D\}. \quad (29)$$

If $\boldsymbol{b}$ is the transmitted sequence and $\epsilon \in A(\boldsymbol{b})$, then it can be shown that $\Omega(\boldsymbol{b} - 2\epsilon) \geq \Omega(\boldsymbol{b})$ if and only if $\epsilon \in L$. Hence it follows that $ML(\boldsymbol{b}) \subset L$ and that if $A(\boldsymbol{b}) \cap L \neq \varnothing$, then the sequence detector outputs an erroneous sequence $\boldsymbol{b} - 2\epsilon$, where $\epsilon \in ML(\boldsymbol{b})$. Consider the following inclusions between events in the probability space on which the transmitted sequence $\boldsymbol{b}$, and the Wiener process $\{\omega_t, t \in R\}$ are defined:

$$\{b_k(i) \neq b^*_k(i)\} \subset \bigcup_{\epsilon \in E} \{\epsilon \in A_k(\boldsymbol{b},i) \cap ML(\boldsymbol{b})\}$$
$$\subset \bigcup_{\epsilon \in F_k(i)} \{\epsilon \in A_k(\boldsymbol{b},i) \cap L\}, \quad (30)$$

where $\boldsymbol{b}^*$ is the sequence selected by the detector. The first inclusion follows from the definitions of $A_k(\boldsymbol{b},i)$ and $ML(\boldsymbol{b})$ (the converse holds if there are no ties in the maximization of $\Omega(\cdot)$). The key to the tightness of the bound in (27) is the second inclusion in (30); to verify it, we show that for every $\epsilon \in Z_k(i)$ there exists $\epsilon' \in F_k(i)$ such that

1) $\{\epsilon \in A_k(\boldsymbol{b},i)\} \subset \{\epsilon' \in A_k(\boldsymbol{b},i)\}$
2) $\{\epsilon \in ML(\boldsymbol{b})\} \subset \{\epsilon' \in L\}$.

If $\epsilon \in F_k(i)$, then $\epsilon' = \epsilon$ satisfies 1) and 2). Otherwise, $\epsilon \in Z_k(i) - F_k(i)$. We now show by induction on $w(\epsilon)$, the weight of the sequence ϵ, that there exists $\epsilon^* \in F_k(i)$ such that ϵ is decomposable into $\epsilon^* + (\epsilon - \epsilon^*)$. If a sequence of weight two is decomposable, then it is so into its two components, both of which are indecomposable since they have unit weight. Now suppose that the claim is true for any sequence whose weight is strictly less than $w(\epsilon)$.

Find the (not necessarily unique) decomposition $\epsilon = \epsilon^1 + \epsilon^2$, $\epsilon^1 \in Z_k(i)$, with largest inner product, that is,

$$\langle S(\epsilon^1), S(\epsilon^2)\rangle \geq \langle S(\epsilon^a), S(\epsilon^b) > \geq 0$$

for any decomposition $\epsilon = \epsilon^a + \epsilon^b$. If $\epsilon^1 \in F_k(i)$, we have found the sought-after decomposition of ϵ. Otherwise, we can decompose ϵ^1 into $\epsilon^3 + \epsilon^4$ such that $\epsilon^3 \in F_k(i)$ because of the induction hypothesis. However, ϵ is indeed decomposable into ϵ^3 and $(\epsilon^2 + \epsilon^4)$, for otherwise the right-hand side of the equation

$$\langle S(\epsilon^2) + S(\epsilon^3), S(\epsilon^4)\rangle - \langle S(\epsilon^1), S(\epsilon^2)\rangle = 2\langle S(\epsilon^3), S(\epsilon^4)\rangle - \langle S(\epsilon^3), S(\epsilon^4) + S(\epsilon^2)\rangle \quad (31)$$

is strictly positive, contradicting the choice of $\epsilon^1 + \epsilon^2$ as the largest-inner-product decomposition of ϵ.

Since $\epsilon^* < \epsilon$, it follows that $\epsilon' = \epsilon^*$ satisfies property 1); to see that 2) is also fulfilled, let $\epsilon'' = \epsilon - \epsilon'$ and consider

$$\begin{aligned}\Omega(b - 2\epsilon) - \Omega(b - 2\epsilon'') &= 2\sigma\int[S_t(b - 2\epsilon) - S_t(b - 2\epsilon'')]\,dw_t \\ &\quad + \|S(b - 2\epsilon'') - S(b)\|^2 - \|S(b - 2\epsilon) - S(b)\|^2 \\ &= -4\left(\sigma\int S_t(\epsilon')\,dw_t + \|S(\epsilon')\|^2\right) - 8\langle S(\epsilon''), S(\epsilon')\rangle \end{aligned} \quad (32)$$

where (32) follows from the fact that b is the transmitted sequence. If $\epsilon \in ML(b)$, then it is necessary that $\Omega(b - 2\epsilon) \geq \Omega(b - 2\epsilon'')$; moreover, the decomposition of ϵ into $\epsilon' + \epsilon''$ implies that $\langle S(\epsilon'), S(\epsilon'')\rangle \geq 0$. Therefore, (32) indicates that $\epsilon' \in L$, and property 2) is satisfied (see Fig. 3).

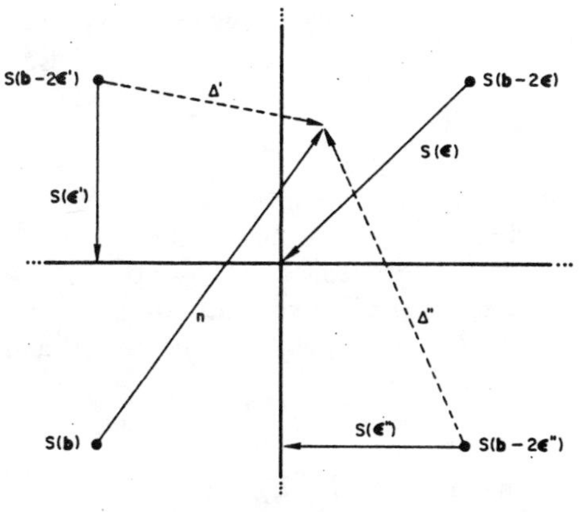

Fig. 3. If $b - 2\epsilon$ is most likely sequence *a posteriori* ($\epsilon \in ML(b)$) and $\langle S(\epsilon'), S(\epsilon'')\rangle \geq 0$, then $\|\Delta'\| \leq \|n\|$ and $\|\Delta''\| \leq \|n\|$, i.e., $\epsilon' \in L$ and $\epsilon'' \in L$. $S(b)$ is transmitted signal. n is projection of noise realization.

In order to take probabilities of the events in the right-hand side of (30), note that $\{\epsilon \in A_k(b, i)\}$ depends on the transmitted sequence but not on the noise realization; hence it is independent of the event $\{\epsilon \in L\}$. Finally,

$$P[\epsilon \in A_k(b, i)] = 2^{-w(\epsilon)} \quad (33)$$

and

$$P[\epsilon \in L] = Q(\|S(\epsilon)\|/\sigma) \quad (34)$$

are immediate from the respective definitions of $A_k(b, i)$ and L, and since the probability of the union of the events in the right-hand side of (3) is not greater than the sum of their probabilities, the upper bound of (27) follows.

One of the features of the foregoing proof is that it remains valid in the general case where the modulation is not necessarily linear, that is,

$$S_t(b) = \sum_{i=-M}^{M}\sum_{k=1}^{K} s_k(t - iT - \tau_k; b_k(i)), \qquad b \in D. \quad (35)$$

The only modification needed to state and prove Proposition 4 is to substitute $S_t(\epsilon)$ by

$$\hat{S}_t(\epsilon) = \frac{1}{2}\sum_{i=-M}^{M}\sum_{k=1}^{K} s_k(t - iT - \tau_k; \epsilon_k(i)) - s_k(t - iT - \tau_k; -\epsilon_k(i)). \quad (36)$$

It is interesting to particularize the foregoing result to the intersymbol interference problem. The Forney bound [3] corresponds to the sum of $2^{-w(\epsilon)}Q(\|S(\epsilon)\|/\sigma)$ over all *simple* sequences, that is, those containing no more than $L - 1$ consecutive zeros amid nonzero components (where L is the number of overlapping symbols). It turns out that a great proportion of simple sequences are, in fact, decomposable; however, all indecomposable sequences are simple. Hence the analysis of maximum-likelihood sequence detection of signals subject to intersymbol interference via decomposition of error sequences results in a bound that is tighter than Forney's result. This issue and the question of how to compute (27) up to any prespecified degree of accuracy are examined in [24].

In the case of bit-synchronous users ($\tau_1 = \cdots = \tau_K$), the derivation of optimum decision rules is, in contrast to Section II, a simple exercise. However, the analysis of the optimum synchronous receiver has basically the same complexity as the general case presented in this section. In particular, even though the one-shot model ($M = 0$) is sufficient, the approach of decomposition of error sequences (K-vectors in this case) is the most effective one.

B. Convergence of the Bounding Series

The limit of the right-hand side of (27) as $M \to \infty$ is an infinite series that bounds the multiuser infinite-horizon error probability. Since this upper bound is monotonic in the noise level, we can distinguish three situations depending on the actual energies, crosscorrelations, and relative delays of the signals—namely, divergence of the series for all noise levels, convergence for sufficiently low noise levels, and global convergence. Although most cases fall into the second category, we will show examples of the other two. We begin by proving a sufficient condition for local convergence of a series that overbounds (27).

Proposition 5: Partition the real line with the nontrivial semi-open intervals $R = \bigcup_{i \in Z} \Lambda_i$ defined by the points $\tau_k + iT$, $k = 1, \cdots, K$, $i \in Z$. Define $N(\epsilon)$ as the union of all intervals Λ_i for which $\int_{\Lambda_i} S_t^2(\epsilon)\, dt = 0$. Let G be a set of simple sequences such that for every pair of distinct finite sequences $\epsilon^1, \epsilon^2 \in G$ that satisfy $N(\epsilon^1) = N(\epsilon^2)$, there exist $j \in \{1, \cdots, K\}$ and $i \in Z$ such that $\epsilon_j^1(i) \neq \epsilon_j^2(i)$ and $(iT + \tau_j, (i+1)T + \tau_j) \not\subset N(\epsilon^1)$. Then there exists $\sigma_0 > 0$ such that

$$\sum_{\epsilon \in G} 2^{-w(\epsilon)} \exp\left(-\|S(\epsilon)\|^2 / 2\sigma^2\right) < \infty \tag{37}$$

for $0 < \sigma < \sigma_0$.

Proof: Define $G_{n,l} = \{\epsilon \in G$, for some $j \in Z$ the point of arrival of the first nonzero component of ϵ and the point of departure of the last nonzero component of ϵ define the interval $\bigcup_{i=j+1}^{j+n} \Lambda_i$ and exactly l of these intervals are such that $\int_{\Lambda_i} S_t^2(\epsilon)\, dt > 0\}$. For every $\epsilon \in G_{n,l}$, we have

$$(K-1) w(\epsilon) \geq n \tag{38}$$

because ϵ is simple and we cannot have more than $K-2$ zeros surrounded by nonzeros. Define

$$\alpha^2 = \min \left\{ r > 0 \mid \text{there exists } i \in Z \text{ and } \epsilon \in G \text{ with } r = \int_{\Lambda_i} S_t^2(\epsilon)\, dt \right\}. \tag{39}$$

Notice that the existence of α is guaranteed because the signal energies are nonzero and there is a finite number of distinct signal waveforms in each interval Λ_i. From the definition of $G_{n,l}$, (38), and (39), we have

$$\sum_{\epsilon \in G_{n,l}} 2^{-w(\epsilon)} \exp\left(-\|S(\epsilon)\|^2/2\sigma^2\right) \leq 2^{(n/1-K)} |G_{n,l}| \exp\left(-l\alpha^2/2\sigma^2\right). \tag{40}$$

Now we use the assumption in the theorem to overbound the cardinality of $G_{n,l}$ by

$$|G_{n,l}| \leq n \binom{n}{l} 3^{Kl}. \tag{41}$$

To show (41) let us see how many different sequences are congruent with every choice of j in the definition of $G_{n,l}$ (for which there are n possibilities, since $\epsilon_k(0) \neq 0$ for all $\epsilon \in G$) and with every distribution of the l nonzero-energy intervals (at most $\binom{n}{l}$ possibilities). The assumption of the result states that any pair of sequences whose nonzero-energy intervals coincide cannot differ only in symbols whose intervals have zero energy. Since $S_t(\epsilon)$, $t \in \Lambda_i$ depends on ϵ through K elements at most, the number of error sequences congruent with the aforementioned choice is bounded by 3^{Kl}, and (41) follows. Substituting (41) into (40), we obtain

$$\begin{aligned}
&\sum_{n=1}^{\infty} \sum_{l=0}^{n} \sum_{\epsilon \in G_{n,l}} 2^{-w(\epsilon)} \exp\left(-\|S(\epsilon)\|^2/2\sigma^2\right) \\
&\quad \leq \sum_{n=1}^{\infty} n 2^{n/1-K} \sum_{l=0}^{n} \binom{n}{l} \left(3^{Kl} \exp\left(-l\alpha^2/2\sigma^2\right)\right) \\
&\quad = \sum_{n=1}^{\infty} n 2^{n/1-K} \left(1 + 3^K \exp\left(-\alpha^2/2\sigma^2\right)\right)^n,
\end{aligned} \tag{42}$$

which converges for

$$0 < \sigma^2 < \frac{\alpha^2}{2} \left[K \ln 3 - \ln\left(2^{1/K-1} - 1\right)\right]^{-1}. \tag{43}$$

Local convergence of the Forney bound for any intersymbol interference problem was proved by Foschini [5]. Not every multiuser problem, however, results in a locally convergent bounding series. Admittedly, the implications of the sufficient condition of Proposition 5 on the waveforms and delays of the signal set are not readily apparent. As is shown by the next result, the sufficient condition of Proposition 5 is satisfied except in certain pathological cases where some degree of synchronism exists along with heavy correlation between the signals.

Proposition 6: Define the times of effective arrival and departure of the ith signal of the kth user as

$$\lambda_{iK+k}^a = \tau_k + iT + \sup\left\{\tau \in [0, T),\ \int_0^\tau s_k^2(t)\, dt = 0\right\}$$

and

$$\lambda_{iK+k}^d = \tau_k + iT + \inf\left\{\tau \in (0, T],\ \int_\tau^T s_k^2(t)\, dt = 0\right\},$$

respectively.

Suppose that a pair of distinct finite sequences $\epsilon^1, \epsilon^2 \in E$ exists such that $N(\epsilon^1) = N(\epsilon^2)$ and such that $\epsilon_k^1(i) \neq \epsilon_k^2(i)$ implies $(iT + \tau_k, (i+1)T + \tau_k) \subset N(\epsilon^1)$. Then the following two statements are true:

1) $S_t(\epsilon^1) = S_t(\epsilon^2)$, a.e.;
2) a pair of reals $\rho < \xi$ exists such that
   - $\rho = \lambda_i^a = \lambda_j^a$ for $i \neq j$,
   - $\xi = \lambda_i^d = \lambda_j^d$ for $i \neq j$, and
   - if $\epsilon_k^1(i) \neq \epsilon_k^2(i)$ then $\rho \leq \lambda_{iK+k}^a < \lambda_{iK+k}^d \leq \xi$.

Proof: 1) If $t \notin N(\epsilon^1) = N(\epsilon^2)$, then $S_t(\epsilon^1)$, $S_t(\epsilon^2)$ depend on their arguments only through those symbols that coincide; on the other hand,

$$\int_{N(\epsilon^1)} S_t^2(\epsilon^1)\, dt = 0 = \int_{N(\epsilon^2)} S_t^2(\epsilon^2)\, dt.$$

2) We show first that the effective arrival of the first symbol that differs, say $\epsilon_q(j)$, must be a point of effective multiarrival, that is, there exists $i > i_1 = jK + q$, $\lambda_i^a = \lambda_{i_1}^a$. If $\lambda_{i_1}^a$ is not a point of effective multiarrival, then we can select λ such that $\lambda_{i_1}^a < \lambda < \lambda_{iK+k}^a$, $\epsilon_k^1(i) \neq \epsilon_k^2(i)$ and

$$\int_{\lambda_{i_1}^a}^{\lambda} s_q^2(t - \tau_q - jT)\, dt > 0. \tag{44}$$

On the other hand, if $t \in (\lambda_{i_1}^a, \lambda)$, then

$$S_t(\epsilon^1) = \epsilon_q^1(j) s_q(t - \tau_q - jT) + \delta^1(t) \tag{45a}$$

and

$$S_t(\epsilon^2) = \epsilon_q^2(j) s_q(t - \tau_q - jT) + \delta^2(t), \tag{45b}$$

where $\delta^1(t) = \delta^2(t)$, a.e. in $(\lambda_{i_1}^a, \lambda)$ because the effective arrival of the rest of the unequal symbols is posterior to λ. Using 1) and (45), we obtain that $\epsilon_q^1(j) s_q(t - \tau_q - jT) = \epsilon_q^2(j) s_q(t - \tau_q - jT)$ a.e. in $(\lambda_{i_1}^a, \lambda)$, which contradicts (44) since $\epsilon_q^1(j) \neq \epsilon_q^2(j)$. Similarly, it can be shown that the effective departure of the last symbol that differs between

ϵ' and ϵ'' must be a point of effective multideparture, and 2) follows.

Proposition 6 implies that for asynchronous models, where the delays are independent and uniformly distributed, an interval of convergence exists for the bounding series with probability one. On the other hand, it is easy to see that for bit-synchronous models, the set $F_k(i)$ is finite so that (27) is finite for all noise levels in that case. Hence, the necessary ingredients for the everywhere divergence of (27) are the partial effective synchronism and the heavy cross correlation of the signal constellation. To illustrate this, consider the following example of divergence of the bounding series (27) for all noise levels.

Let $K = 6$, $s_k(t) = 1$, $t \in [0,1]$, $k = 1,\cdots,6$; $\tau_k - \tau_1 = 1/2$, $k = 2,\cdots,6$; $\tau_i - \tau_j = 0$, $i, j = 2,\cdots,6$. Consider the set of error sequences: $A_n = \{\epsilon \in E,\ \epsilon(i) = 0,\ i < 0$ and $i > n;\ \epsilon(i) \in \{[1\ -1\ 0\ 0\ 0\ 0]^T, [1\ 0\ -1\ 0\ 0\ 0]^T, [1\ 0\ 0\ -1\ 0\ 0]^T, [1\ 0\ 0\ 0\ -1\ 0]^T, [1\ 0\ 0\ 0\ 0\ -1]^T\},\ i = 0,\cdots,n-1;\ \epsilon(n) = [1\ 0\ 0\ 0\ 0\ 0]^T\}$. Notice that $\|S(\epsilon)\| = 1$, $w(\epsilon) = (2n+1)$, for all $\epsilon \in A_n$, and $|A_n| = 5^n$. It is straightforward to show that every sequence $\epsilon \in A_n$ is indecomposable; thus $A_n \in F_1$, for all $n > 1$, and the following inequality is true:

$$\sum_{\epsilon \in F_1} 2^{-w(\epsilon)} Q(\|S(\epsilon)\|/\sigma) > \sum_{n=1}^{\infty} \sum_{\epsilon \in A_n} 2^{-w(\epsilon)} Q(\|S(\epsilon)\|/\sigma) = Q(1/\sigma) \sum_{n=1}^{\infty} \sum_{\epsilon \in A_n} 2^{-w(\epsilon)} = \frac{1}{2} Q(1/\sigma) \sum_{n=1}^{\infty} (5/4)^n.$$

It follows that (27) diverges for any noise level.

C. Asymptotic Probability of Error

In this section we show that whenever the error probability upper bound (27) converges for sufficiently low noise, both bounds (24) and (27) are asymptotically tight as $\sigma \to 0$. In particular, we prove that for any $\delta > 0$ there exists $\sigma_0 > 0$ such that for all $\sigma < \sigma_0$,

$$C_k^L Q(d_{k,\min}/\sigma) \le P_k \le C_k^U (1+\delta) Q(d_{k,\min}/\sigma), \quad (46)$$

where

$$C_k^L = P[d_k(b) = d_{k,\min}] = P\left[\bigcup_{\substack{\epsilon \in F_k \text{ s.t.} \\ \|S(\epsilon)\| = d_{k,\min}}} \{\epsilon \in A(b)\} \right] \quad (47)$$

and

$$C_k^U = \sum_{\substack{\epsilon \in F_k \text{ s.t.} \\ \|S(\epsilon)\| = d_{k,\min}}} 2^{-w(\epsilon)} = \sum_{\substack{\epsilon \in F_k \text{ s.t.} \\ \|S(\epsilon)\| = d_{k,\min}}} P[\epsilon \in A(b)]. \quad (48)$$

The left-hand inequality of (46) was obtained in (24). Expression (47) follows because if $\epsilon \notin F_k$ has $\|S(\epsilon)\| = d_{k,\min}$, then it is decomposable into $\epsilon' + \epsilon''$ such that $\epsilon' \in F_k$, $\|S(\epsilon')\| = d_{k,\min}$ and $\{\epsilon \in A(b)\} \subset \{\epsilon' \in A(b)\}$. C_k^U defined in (48) is the sum of the coefficients in the bounding series (27) that correspond to the sequences with minimum energy.[5] The right-hand inequality of (46) follows from the following proposition.

Proposition 7: If $\sigma_0 > 0$ exists such that for all $0 < \sigma \le \sigma_0$

$$\sum_{\epsilon \in F_k} 2^{-w(\epsilon)} \exp\left(-\|S(\epsilon)\|^2/2\sigma^2\right) < \infty, \quad (49)$$

then

$$\lim_{\sigma \to 0} \sum_{\epsilon \in F_k} 2^{-w(\epsilon)} Q(\|S(\epsilon)\|/\sigma)/Q(d_{k,\min}/\sigma) = C_k^U. \quad (50)$$

Proof: First, we show that for any set $G \subset E$ and any constant $r \ge 0$ that satisfy

1) $\inf_{\epsilon \in G} \|S(\epsilon)\| > r$,
2) there exists σ_0 such that for all $0 < \sigma \le \sigma_0$, $\sum_{\epsilon \in G} 2^{-w(\epsilon)} \exp(-\|S(\epsilon)\|^2/2\sigma^2) < \infty$,

we have

$$\lim_{\sigma \to 0} \sum_{\epsilon \in G} 2^{-w(\epsilon)} Q(\|S(\epsilon)\|/\sigma)/Q(r/\sigma) = 0. \quad (51)$$

Consider the following inequalities:

$$\sum_{\epsilon \in G} 2^{-w(\epsilon)} Q(\|S(\epsilon)\|/\sigma)/Q(r/\sigma) \le \sum_{\epsilon \in G} 2^{-w(\epsilon)} \exp\left([r^2 - \|S(\epsilon)\|^2]/2\sigma^2\right)$$
$$\le \exp\left(\left[\inf_{\epsilon \in G} \|S(\epsilon)\|^2 - r^2\right]\left[1/2\sigma_0^2 - 1/2\sigma^2\right]\right) \cdot \sum_{\epsilon \in G} 2^{-w(\epsilon)} \exp\left([r^2 - \|S(\epsilon)\|^2]/2\sigma_0^2\right) \quad (52)$$

where the inequalities follow from $Q(\sqrt{x+y}) \le \exp(-x/2) Q(\sqrt{y})$ if $x \ge 0$ and $y \ge 0$, and 1) respectively. The left-hand side of (52) vanishes as $\sigma \to 0$ because as a result of 2) the series therein converges.

Particularizing this result to $r = \inf_{\epsilon \in F_k} \|S(\epsilon)\|$ and $G = \{\epsilon \in F_k$, such that $\|S(\epsilon)\| > r\}$, we see that (50) follows. Note that the infimum of $\|S(\epsilon)\|$ over both F_k and the subset $\{\epsilon \in F_k$, such that $\|S(\epsilon)\| > r\}$ are achieved because both sets have a finite subset whose elements have no greater energy than the rest of the elements in F_k and G. This implies that $r = d_{k,\min}$, and condition 1) is satisfied.

The high SNR-upper and lower bounds (46) to the kth user error probability differ by a multiplicative constant independent of the noise level. From (47) and (48) it can be seen that this constant is related to the degree of overlapping of the events $\{\{\epsilon \in A(b)\},\ \epsilon \in F_k$ and $\|S(\epsilon)\| = d_{k,\min}\}$. Typically, there exists only a pair of elements in F_k, $\{\epsilon, -\epsilon\}$, that achieve the minimum energy. Since $\{\epsilon \in A(b)\} \cap \{-\epsilon \in A(b)\} = \varnothing$ for all $\epsilon \in E$, it follows that in such case $C_k^L = C_k^U$.

An important performance measure for multiuser detectors in high SNR situations is the SNR degradation due to the existence of other active users in the channel, i.e., the

[5] It can be shown that the error probability of the multiuser maximum-likelihood sequence detector can be lower bounded by $C_k^U Q(d_{k,\min}/\sigma)$.

limit as $\sigma \to 0$ of the ratio between the effective SNR (that required by a single-user system to achieve the same asymptotic error probability) and the actual SNR. We denote this parameter as η_k, the kth user asymptotic efficiency[6] defined formally as

$$\eta_k = \sup\left\{0 \le r \le 1;\ P_k(1/\sigma) = o\left(Q\left(\sqrt{rw_k}/\sigma\right)\right)\right\}. \quad (53)$$

This parameter depends both on the signal constellation and on the multiuser detector employed. In the case of the minimum-error probability detector, (46) indicates that in the high SNR region the behavior of the kth user error probability coincides with that of an antipodal single-user system with bit-energy equal to $d^2_{k,\min}$. Therefore, the maximum achievable asymptotic efficiency is given by

$$\eta_k = d^2_{k,\min}/w_k. \quad (54)$$

Since the upper bound (46) is actually an upper bound on the error probability of the multiuser maximum-likelihood sequence detector, this detector achieves the maximum asymptotic efficiency, although it is not optimum in terms of error probability. The set of K-user asymptotic efficiencies emerge as the parameters that determine optimum performance for all practical purposes in the SNR region of usual interest. In a sequel to this paper, we derive analytical expressions, bounds, and numerical methods for the computation of these parameters which play a central role in the analysis and comparison of multiuser detectors.

IV. Numerical Examples

Two pairs of lower and upper bounds to the kth user minimum-error probability have been presented in Section III, and they have been shown to be tight asymptotically; nonetheless, it remains to ascertain the SNR level for which such asymptotic approximation is sufficiently accurate. In the sequel this question is illustrated by several examples of the computation of averages and extreme cases of the foregoing bounds with respect to the relative delays of asynchronous users. The first example is a baseband asynchronous system with two equal-energy users that employ a simple set of signal waveforms (Fig. 4). In this figure the upper bounds on the best and worst cases of the optimum detector are indistinguishable from each other, and for SNR higher than about 6 dB, from the single-user lower bound (which is also the minimum energy lower bound since $\eta_1 = 1$). Note also that the maximum interference coefficient, $\bar{I}_k = \max_{\alpha,\beta} I_k(\alpha,\beta)$ (recall Proposition 3) is one-third for all delays, and the performance of the conventional receiver varies only slightly with the relative delay.

In the next examples, we employ a set of spread-spectrum signals: three maximal-length signature sequences of length 31 generated to maximize a signal-to-multiple-access interference functional [7, table 5]. The average probability of error of the conventional receiver for equal-energy users

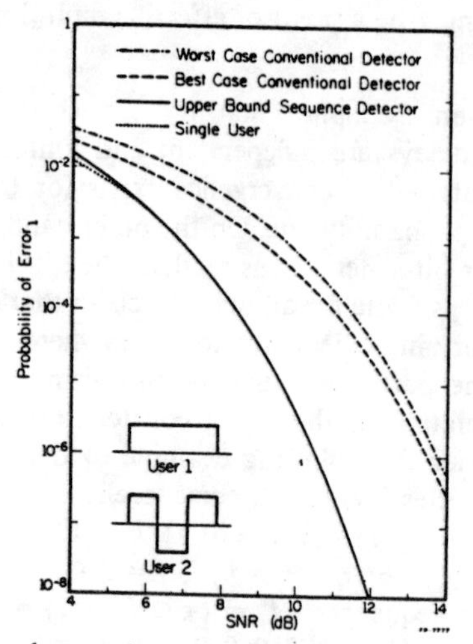

Fig. 4. Best and worst cases of error probability of user 1 achieved by conventional and optimum detectors.

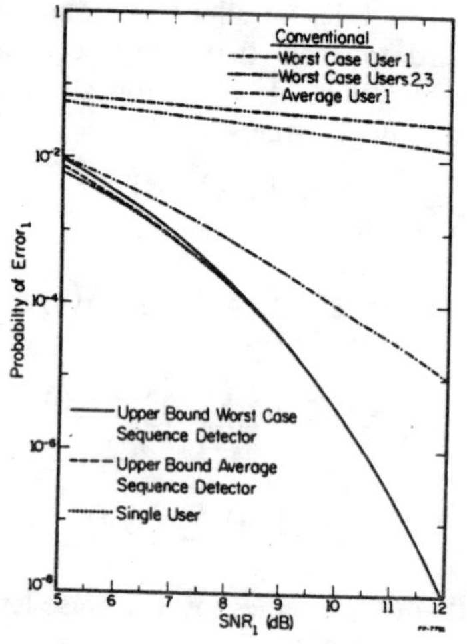

Fig. 5. Worst-case and average error probabilities achieved by conventional and optimum multiuser detectors with three active users employing m-sequences of length 31.

employing this signal set has been thoroughly studied previously [5], [14], and in Fig. 5 we reproduce (from [8, fig. 2]) the average error probability of user 1 achieved by the coherent conventional detector. Also shown in Fig. 5 are the worst cases of the conventional detector and upper bounds to the baseband worst-case and average minimum error probabilities for user 1. From the observation of Fig. 5, we can conclude that for error probabilities of 10^{-2} the average performance of the conventional detector is fairly close to the single-user lower bound, but the worst-case error probability is notably poor for the whole SNR range considered in the figure. Note, however, that since the signal set has good cross correlations for most of the relative delays, error probabilities close to the worst-case curve will occur with low probability. The worst-case and, especially, the average upper bounds on the optimum sequence detector performance are remarkably close to the single-user lower bound and show that the minimum-error

[6]This is not to be confused with Pitman's asymptotic measure of detection efficiency.

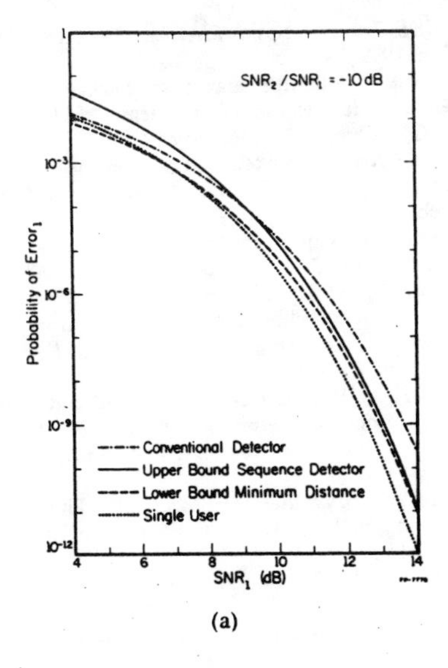

(a)

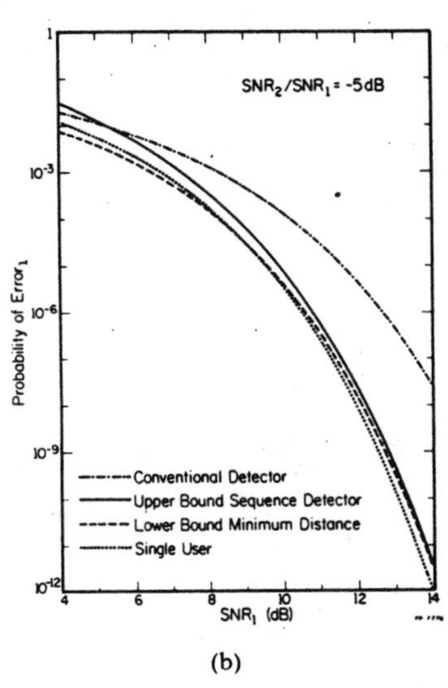

(b)

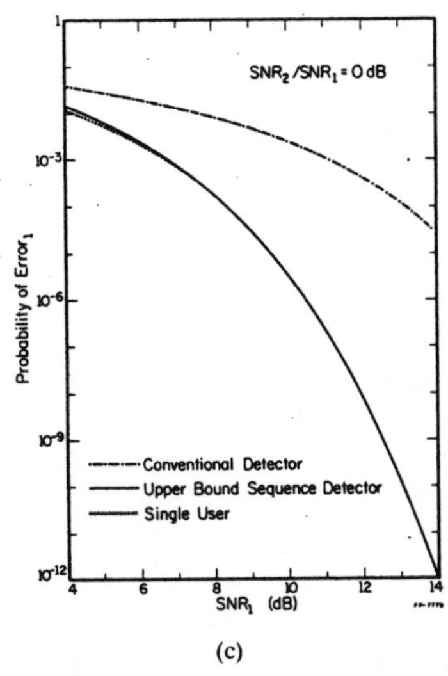

(c)

Fig. 6. Bounds on minimum error probability of user 1. Worst-case delays and two active users. (a) $w_2/w_1 = -10$ dB. (b) -5 dB. (c) 0 dB.

probability not only has a low average (around one order of magnitude better than that of the conventional detector, at 9 dB), but its dependence on the delays is negligible.

The next example investigates the *near-far* problem (i.e., the effects of unequal received energies) for two users that employ a subset of the previous set of maximal-length signature sequences. Bounds on the error probabilities corresponding to this example (worst-case relative delay between users 1 and 2) are calculated in Fig. 6 for three relative energies, namely, $\mathrm{SNR}_2/\mathrm{SNR}_1 = -10$ dB, -5 dB, and 0 dB. It is interesting to observe in the graphs corresponding to $\mathrm{SNR}_2/\mathrm{SNR}_1 = -10$ dB, -5 dB that all four bounds derived in this chapter play a role in some SNR interval; in particular, the error probability of the conventional detector is lower than the upper bound on the optimum sequence detector for small SNR. The opposite effect of an increase in the energy of the interfering users on the minimum and conventional probabilities of error is apparent: while the optimum sequence detector bounds become tighter and closer to the single-user lower bound, the conventional error probability grows rapidly until it becomes multiple-access limited (for $\mathrm{SNR}_2/\mathrm{SNR}_1 = 6.3$ dB).

The results presented here open the possibility of a trade-off between the complexities of the receiver and the signal constellation in order to achieve a fixed level of performance; the actual compromise being dictated by the relative power of each user at the receiver. In multipoint-to-multipoint problems, when some active users need not be demodulated at a particular location, such a trade-off is likely to favor a multiuser detector that takes into account only those unwanted users that are not comparatively weak. If the signal constellation has moderate cross-correlation properties and the energy of the kth user is not dominant, then its minimum distance is achieved by an error sequence with only one nonzero component, and the minimum error probability approaches asymptotically the single-user bit error rate. This implies that contrary to what is sometimes conjectured, the performance of the conventional receiver is not close to the minimum error even if signals with low cross correlations are employed.

References

[1] P. R. Chevillat, "*N*-User trellis coding for a class of multiple-access channels," *IEEE Trans. Inform. Theory*, vol. IT-27, no. 1, pp. 114–120.

[2] G. D. Forney, "Lower bounds on error probability in the presence of large intersymbol interference," *IEEE Trans. Commun.*, vol. COM-20, pp. 76–77, Feb. 1972.

[3] ——, "Maximum likelihood sequence estimation of digital sequences in the presence of intersymbol interference," *IEEE Trans. Inform. Theory*, vol. IT-18, pp. 363–378, May 1972.

[4] ——, "The Viterbi algorithm," *Proc. IEEE*, vol. 61, pp. 268–278, Mar. 1973.

[5] G. J. Foschini, "Performance bound for maximum-likelihood reception of digital data," *IEEE Trans. Inform. Theory*, vol. IT-21, pp. 47–50, Jan. 1975.

[6] R. Gagliardi, "*M*-link multiplexing over the quadrature communications channel," *IEEE Commun. Mag.*, vol. 22, pp. 22–30, Sep. 1984.

[7] F. D. Garber and M. B. Pursley, "Optimal phases of maximal

sequences for asynchronous spread-spectrum multiplexing," *Electron. Lett.*, vol. 16, pp. 756–757, Sep. 11, 1980.
[8] E. A. Geraniotis and M. B. Pursley, "Error probability for direct-sequence spread-spectrum multiple-access communications-Part II: Approximations," *IEEE Trans. Commun.*, vol. COM-30, pp. 985–995, May 1982.
[9] D. Horwood and R. Gagliardi, "Signal design for digital multiple access communications," *IEEE Trans. Commun.*, vol. COM-23, pp. 378–383, Mar. 1975.
[10] J. K. Omura, "Performance bounds for Viterbi algorithms," in *1981 IEEE Int. Commun. Conf. Rec.*, pp. 2.21–2.25.
[11] C. H. Papadimitrou and K. Steiglitz, *Combinatorial Optimization: Algorithms of Complexity*. Englewood Cliffs, NJ: Prentice-Hall, 1982.
[12] H. V. Poor, "Signal detection in multiple access channels," U.S. Army Research Office Proposal (Contract DAAG29-81-K-0062 to Coordinated Science Laboratory, Univ. of Illinois), 1980.
[13] J. G. Proakis, *Digital Communications*. New York: McGraw-Hill, 1983.
[14] M. B. Pursley, D. V. Sarwate, and W. E. Stark, "Error probability for direct-sequence spread-spectrum multiple-access communications—part I: Upper and lower bounds," *IEEE Trans. Commun.*, vol. COM-30, pp. 975–984, May 1982.
[15] J. E. Savage, "Signal detection in the presence of multiple-access noise," *IEEE Trans. Inform. Theory*, vol. IT-20, pp. 42–49, Jan. 1974.
[16] K. S. Schneider, "Optimum detection of code division multiplexed signals," *IEEE Trans. Aerosp. Electron. Syst.*, vol. AES-15, pp. 181–185, Jan. 1979.
[17] G. Ungerboeck, "Adaptive maximum likelihood receiver for carrier-modulated data transmission systems," *IEEE Trans. Commun.*, vol. COM-22, pp. 624–636, May 1974.
[18] W. Van Etten, "Maximum likelihood receiver for multiple channel transmission systems," *IEEE Trans. Commun.*, vol. COM-24, pp. 276–283, Feb. 1976.
[19] S. Verdú, "Optimum sequence detection of asynchronous multiple-access communications," in *Abstracts of Papers: IEEE 1983 Int. Symp. Inform. Theory*, Sep. 1983, p. 80.
[20] —, "Minimum probability of error for asynchronous multiple-access communication systems," in *Proc. 1983 IEEE Mil. Commun. Conf.*, vol. 1, Nov. 1983, p. 80, pp. 213–219.
[21] S. Verdú and H. V. Poor, "Backward, forward and backward-forward dynamic programming models under commutativity conditions," in *Proc. 23rd IEEE Conf. Decision Contr.*, Dec. 1984, pp. 1081–1086.
[22] S. Verdú, *Optimum Multi-user Signal Detection*, Ph.D. dissertation, Dep. Elec. Comput. Eng., Univ. of Illinois, Urbana-Champaign, Aug. 1984.
[23] A. J. Viterbi and J. K. Omura, *Principles of Digital Communication and Coding*. New York: McGraw-Hill, 1979.
[24] S. Verdú, "New bound on the error probability of maximum likelihood sequence detection of signals subject to intersymbol interference," in *Proc. 1985 Conf. Inform. Sci. Syst.*, Mar. 1985, pp. 413–418.

Single-User Detectors for Multiuser Channels

H. VINCENT POOR, FELLOW, IEEE, AND SERGIO VERDU, MEMBER, IEEE

Abstract—**Optimum decentralized demodulation for asynchronous Gaussian multiple-access channels is considered. It is assumed that the receiver is the destination of the information transmitted by only one active user, and single-user detectors that take into account the existence of the other active users in the channel are obtained. This approach is in contrast to both conventional demodulation, which is fully decentralized but neglects the presence of multiple-access interference, and globally optimum demodulation, which requires centralized sequence detection. The problem considered is one of signal detection in additive colored non-Gaussian noise, and attention is focused on one-shot structures where detection of each symbol is based only on the received process during its corresponding interval. Particular emphasis is placed on asymptotically optimum detectors for each of the following situations: 1) weak interferers, 2) CDMA signature waveforms with long spreading codes, and 3) low background Gaussian noise level.**

I. INTRODUCTION

THE conventional approach to the demodulation of code-division multiplexed multiuser digital communications is to demodulate each user as if it were the only user in the channel. The multiple-access capability of such systems is thus achieved by using complex signal constellations that exhibit favorable cross-correlation properties. (See, for example, [1] for a description of conventional multiple-access demodulation techniques.) However, recent work by Verdu [2] has shown that substantial performance gains can be achieved in coherent multiuser systems by using a receiver that takes advantage of the structure of the multiple-access interference. For example, this approach can be used to alleviate such limitations as the near/far problem in the direct-sequence spread-spectrum multiple-access (DS/SSMA) format. The performance gains realized by the receiver proposed in [2] are achieved by the use of simultaneous sequence detection of all users in the channel, a task that requires a centralized implementation and a high degree of software complexity (for example, the decision algorithm required is a dynamic program (DP) whose complexity is $O(2^K)$ where K is the number of users in the channel). Since the implementation costs of such fully centralized detection algorithms may be unacceptably high for many applications, and since network security restrictions may not permit the distribution of all user's signaling waveforms to all demodulating terminals, it is of interest to consider demodulators that lie between these two philosophies of conventional demodulation, in which other users' signaling waveforms to all demodulating terminals, it is optimum demodulation, in which all users in the channel are tracked and demodulated simultaneously. The performance results obtained in [2]–[4] indicate that an attractive compromise in practice is to use optimum multiuser demodulators for only a subset of the active users and simply neglect the presence of all other users. In order to take advantage of the superior performance achievable by multiuser detectors, the subset of active users to be taken into account at the receiver should contain all the users whose power is sufficient regardless of whether their messages are destined to that particular location.

In this paper, we consider the case of full decentralization where the receiver is constrained to track and lock to the signal of only one user, but unlike the conventional single-user detector, is optimized to take into account the structure of multiple-access interference in making decisions. We consider several design approaches that can be used to optimize these structures depending on the amount of information that one is able to assume to be known about the signature waveforms assigned to the interfering users.

This paper is organized as follows. In Section II, we will first discuss the general structure of optimum decentralized demodulators that simultaneously track and demodulate a group of D users from a total population of K users sharing a common communications channel where $D \leq K$. We then consider the structure of optimum *single-user* detectors ($D = 1$), and particularize to the case $K = 2$ to illustrate this structure. The results of Section II are for general antipodal signaling formats. In Section III, we turn to the development of specific results for single-user demodulation of DS/SSMA transmissions. In this modulation technique, which among coherent signaling formats is of particular practical interest, each signature waveform consists of a sequence of chip waveforms whose polarities are determined by a binary word assigned to each user. The specific structure of the direct-sequence format allows for the development of useful approximations to optimum single-user detection which are asymptotically exact as either the length of the spreading codes or the signal-to-background-noise ratio (SBNR) increases without bound. We also show that (with $K = 2$) even in the absence of any prior knowledge about either the spreading codes or the timing of the multiuser interference, the optimum single-user detector is *not* multiple-access noise-limited as the background thermal noise level vanishes. This is in contrast to the conventional detector, which can incur an irreducible error probability even in the absence of background noise. All of these results for DS/SSMA require only that the chip waveform (which is usually common to all users in a given network) be known. Thus, these techniques can be applied in secure networks where the distribution of one user's spreading code to other users is not desirable.

In Section IV, we return to general coherent signaling formats to consider the problem of optimum single-user detection in the presence of weak unlocked interfering users. We model this problem by assuming that the multiple-access interference is multiplied by a small amplitude factor ϵ. We then derive an expression for the likelihood ratio statistic for optimum symbol decisions on the locked user that is of the form of the conventional correlation statistic, modified by an ϵ^2 term involving signal cross-correlation functions, and then having higher order terms of order ϵ^4. The resulting locally optimum detector correlates the observation with a replica of the waveform of the user of interest, suitably smoothed to take into account the presence of multiple-access interference.

Paper approved by the Editor for Spread Spectrum of the IEEE Communications Society. Manuscript received February 17, 1987; revised July 27, 1987. This work was supported in part by the U.S. Army Research Office under Contract DAAL03-87-K-0062.

H. V. Poor is with the Department of Electrical and Computer Engineering and the Coordinated Science Laboratory, University of Illinois, Urbana, IL 61801.

S. Verdu is with the Department of Electrical Engineering, Princeton University, Princeton, NJ 08544.

IEEE Log Number 8717473.

0090-6778/88/0100-0050$01.00 © 1988 IEEE

II. Optimum Decentralized Detection for Multiuser Channels

Throughout this paper, we consider a received signal model of the form

$$r_t = S_t(b) + n_t, \qquad -\infty < t < \infty \tag{2.1}$$

where $\{n_t; -\infty < t < \infty\}$ represents white Gaussian noise with spectral height $N_0/2$, and where the received signal $S_t(b)$ is the superposition of transmissions received from K separate asynchronous users, i.e.,

$$S_t(b) = \sum_{k=1}^{K} \sum_{i=-M}^{M} b_k(i) s_k(t - iT - \tau_k) \tag{2.2}$$

where T is the symbol interval, $b_k(i)$ is the ith symbol of the kth user, τ_k is the relative delay (modulo T) with which the kth user's transmission is received, and $s_k(t)$ is the signature waveform assigned to the kth user. (It is assumed that $s_k(t)$ is zero for $t \notin [0, T]$.) Note that $(2M + 1)$ is the number of symbols per user in the given transmission, and b denotes the $K \times (2M + 1)$ matrix whose (k, i) entry is $b_k(i)$.

Suppose that we wish to demodulate some group of D users from the total population of K users where $D \le K$. For simplicity of notation, we assume that these D users of interest are labeled 1-D. Thus, we know $s_1, \cdots, s_D$ and $\tau_1, \cdots, \tau_D$, and the *maximum likelihood* demodulator chooses a symbol matrix $b_D = \{b_k(i); k = 1, \cdots, D\}_{i=-M}^{M}$ to maximize the log-likelihood function

$$\frac{2}{N_0}\int_{-\infty}^{\infty} r_t S_t^D(b_D)\, dt - \frac{1}{N_0}\int_{-\infty}^{\infty} [S_t^D(b_D)]^2\, dt + \log E\left\{\exp\left[\frac{2}{N_0}\int_{-\infty}^{\infty} [r_t - S_t^D(b_D)] S_t^{MA}\, dt - \frac{1}{N_0}\int_{-\infty}^{\infty} [S_t^{MA}]^2\, dt\right]\right\} \tag{2.3}$$

where

$$S_t^D(b_D) = \sum_{k=1}^{D} \sum_{k=-M}^{M} b_k(i) s_k(t - iT - \tau_k), \qquad -\infty < t < \infty \tag{2.4}$$

$$S_t^{MA} = S_t(b) - S_t^D(b_D) \tag{2.5}$$

and where the expectation is over the ensemble of all unknown quantities in S_t^{MA}, including delays, symbols, and (possibly) waveforms.

Note that, even if we ignore the complexity of computing the expectation in (2.3), the time-complexity-per-demodulated-bit (TCB) of brute-force maximization of the log-likelihood function is $O(|A|^{D(2M+1)}/DM|A|)$ where $|A|$ is the size of the symbol alphabet. Thus, unless some simpler algorithm can be found, simultaneous maximum likelihood sequence detection of D users is out of the question from a practical point of view. For the particular case of fully centralized detection ($K = D$), it turns out that a much simpler algorithm can indeed be found. In particular, for this case, the expectation term in (2.3) disappears (since $S_t^{MA} \equiv 0$), and the remaining terms can be decomposed in a way that allows maximization of (2.3) with a dynamic programming algorithm yielding a TCB of $O(|A|)^K$. (See [2], [3] for details of this analysis.)

Unfortunately, for $D < K$, the decomposition of (2.3) necessary for a dynamic programming solution is not possible because of the coupling among symbols in the expectation term. This means that maximum likelihood sequence detection is not generally computationally feasible (its TCB is exponential in the number of symbols per user) if all users' signaling waveforms are not known and locked. Thus, in considering decentralized demodulators in multiuser channels, we will restrict our attention to algorithms which demodulate only a single symbol at a time, i.e., we consider *one-shot detectors*. We also will restrict attention to the binary signaling case, in which $b_k(i) \in \{-1, +1\}$ for all i, k. Extensions to general alphabets are, in most cases, straightforward.

In the sequel, we will consider the case of full decentralization, i.e., single-user detection, which can be modeled by the binary hypothesis-testing problem

$$\begin{aligned} H_0&: r_t = s_1(t) + S_t^{MA} + n_t, \qquad 0 \le t \le T \\ H_1&: r_t = -s_1(t) + S_t^{MA} + n_t, \qquad 0 \le t \le T \end{aligned} \tag{2.6}$$

where $\{n_t; -\infty < t < \infty\}$ is the white Gaussian noise and where

$$S_t^{MA} = \sum_{k=2}^{K} [b_k^L s_k(t - \tau_k + T) + b_k^R s_k(t - \tau_k)], \qquad 0 \le t \le T \tag{2.7}$$

with b_k^L and b_k^R denoting the kth user's bits in the intervals $[-T + \tau_k, \tau_k)$ and $[\tau_k, T + \tau_k)$, respectively. We also assume that the receiver is coherent with user 1 so that $\{s_1(t); t \in [0, T]\}$ is a deterministic waveform, and that each user's signaling waveform is of the form

$$s_k(t) = (2w_k)^{1/2} a_k(t) \cos(\omega_c t + \theta_k) \tag{2.8}$$

where ω_c is known. We assume that $(\omega_c T/2\pi)$ is an integer large enough so that integrals of $2\omega_c$ components can be neglected.

Optimum (maximum likelihood/minimum error probability) decisions for (2.6) are based on comparing the likelihood ratio to a threshold. With this in mind, we give the following result.

Proposition 2.1: Suppose that the phase vector of the interfering users $\theta = (\theta_2, \cdots, \theta_K)$ is uniformly distributed on $[0, 2\pi)^{K-1}$ and is independent of $b_k = (b_k^L, b_k^R)$, $k = 2, \cdots, K$, $\tau = (\tau_2, \cdots, \tau_K)$ and $(a_k(t), t \in [0, T], k = 2, \cdots, K)$. If the dependence of $\|S^{MA}\|$ on θ can be neglected,[1] then the likelihood ratio for (2.6) can be written in the following form:

$$\exp\left[\frac{4w_1^{1/2}}{N_0}\int_0^T r_p(t) a_1(t)\, dt\right] \frac{E\left[\prod_{k=2}^{K} I_0((\rho_k^2(b_k, \tau_k, 1) + \psi_k^2(b_k, \tau_k))^{1/2}) \exp\left(-\sum_{j=2}^{k-1} \Gamma_{kj}(b_k, b_j, \tau_k, \tau_j)\right)\right]}{E\left[\prod_{k=2}^{K} I_0((\rho_k^2(b_k, \tau_k, -1) + \psi_k^2(b_k, \tau_k))^{1/2}) \exp\left(-\sum_{j=2}^{k-1} \Gamma_{kj}(b_k, b_j, \tau_k, \tau_j)\right)\right]} \tag{2.9}$$

where the expectation is over the ensemble of bits, delays, and possibly waveforms of the interfering users[2] and we use the notation

$$\rho_k(b_k, \tau_k, e) = \frac{2(w_k)^{1/2}}{N_0}\int_0^T \alpha_k(b_k^L, b_k^R, t - \tau_k) \cdot [r_p(t) - e w_1^{1/2} a_1(t)]\, dt \tag{2.10}$$

[1] For a waveform $x = \{x(t); 0 \le t \le T\}$, $\|x\|^2$ denotes $\int_0^T x^2(t)\, dt$.

[2] $I_0(\cdot)$ is the modified Bessel function of the first kind of order 0, i.e., $I_0(x) = 1/2\pi \int_0^{2\pi} \exp(x \cos \theta)\, d\theta$.

$$\psi_k(b_k, \tau_k)=\frac{2(w_k)^{1/2}}{N_0}\int_0^T \alpha_k(b_k^L, b_k^R, t-\tau_k)r_q(t)\,dt \quad (2.11)$$

$$\Gamma_{kj}(b_k, b_j, \tau_k, \tau_j)=\frac{2(w_k w_j)^{1/2}}{N_0}\int_0^T \alpha_k(b_k^L, b_k^R, t-\tau_k)\cdot\alpha_j(b_j^L, b_j^R, t-\tau_j)\,dt \quad (2.12)$$

$$r_p(t)=\sqrt{2}r_t\cos(\omega_c t+\theta_1) \quad (2.13)$$

$$r_q(t)=\sqrt{2}r_t\sin(\omega_c t+\theta_1) \quad (2.14)$$

$$\alpha_k(b, c, \lambda)=ba_k(\lambda+T)+ca_k(\lambda). \quad (2.15)$$

Proof: The likelihood ratio for (2.6) is equal to the ratio of expected values of conditionally Gaussian *a priori* densities where the expected value is taken with respect to all random quantities in S^{MA}; this is given straightforwardly by

$$LR=\frac{\exp\left[-\frac{1}{N_0}\|r-s_1\|^2\right]}{\exp\left[-\frac{1}{N_0}\|r+s_1\|^2\right]}\cdot\frac{E\left\{\exp\left[-\frac{1}{N_0}\|S^{MA}\|^2+\frac{2}{N_0}\langle r-s_1, S^{MA}\rangle\right]\right\}}{E\left\{\exp\left[-\frac{1}{N_0}\|S^{MA}\|^2+\frac{2}{N_0}\langle r+s_1, S^{MA}\rangle\right]\right\}} \quad (2.16)$$

where, for functions x and y on $[0, T]$, the notation $\langle x, y\rangle$ denotes $\int_0^T x(t)y(t)\,dt$. The first ratio in the above expression is readily shown to be equal to $\exp[(4(w_1)^{1/2}/N_0)^T\int_0^T r_p(t)a_1(t)\,dt]$. Now, neglecting the dependence of $\|S^{MA}\|$ on θ, we have for every b and τ

$$\frac{1}{N_0}\|S^{MA}\|^2=\left[\sum_{k=2}^{K}\left(\frac{w_k}{N_0}+\sum_{j=2}^{k-1}\Gamma_{kj}(b_k, b_j, \tau_k, \tau_j)\right)\right]. \quad (2.17)$$

So, it remains to show that for all $(\alpha_k, b_k, \tau_k, k=2, \cdots, K)$, we have

$$\int_0^{2\pi}\cdots\int_0^{2\pi}\exp\left[\frac{2}{N_0}\langle r-ex_1, S^{MA}\rangle\right]\frac{d\theta_2\cdots d\theta_K}{(2\pi)^{K-1}}=\prod_{k=2}^{K}I_0((\rho_k^2(b_k, \tau_k, e)+\psi_k^2(b_k, \tau_k))^{1/2}). \quad (2.18)$$

To this end, we note that the following sequence of equalities holds:

$$\frac{2}{N_0}\langle r-es_1, S^{MA}\rangle=\frac{2\sqrt{2}}{N_0}\sum_{k=2}^{K}\int_0^T(r(t)-es_1(t))w_k^{1/2}\alpha_k(b_k^L, b_k^R, t-\tau_k)\cdot\cos(\omega_c t-\omega_c\tau_k+\theta_k)\,dt$$

$$=\frac{2}{N_0}\sum_{k=2}^{K}w_k^{1/2}\int_0^T[(r_p(t)-2ew_1^{1/2}a_1(t)\cdot\cos^2(\omega_c t+\theta_1))\alpha_k(b_k^L, b_k^R, t-\tau_k)\cdot\cos(\theta_k-\omega_c\tau_k-\theta_1)-(r_q(t)-ew_1^{1/2}a_1(t)\cdot\sin(2\omega_c t+\theta_1))\alpha_k(b_k^L, b_k^R, t-\tau_k))\cdot\sin(\theta_k-\omega_c\tau_k-\theta_1)]\,dt$$

$$=\sum_{k=2}^{K}\rho_k(b_k, \tau_k, e)\cos(\theta_k-\omega_c\tau_k-\theta_1)-\psi_k(b_k, \tau_k)\sin(\theta_k-\omega_c\tau_k-\theta_1) \quad (2.19)$$

where the last equality follows by neglecting the integrals of the $2\omega_c$ terms. Equation (2.18) is immediate from (2.19) and the result follows. □

Note that the structure (2.9) consists of the single-user ($K=1$) correlation statistic

$$\frac{2w_1^{1/2}}{N_0}\int_0^T r_p(t)a_1(t)\,dt \quad (2.20)$$

used by conventional single-user receivers, modified by an additive correction term which accounts for the other users in the channel. Note that the received waveform enters this correction term through the sliding correlation statistics of (2.10) and (2.11).

The simplifying approximation in Proposition 2.1, which states that the energy of the multiple-access interference process is independent of the carrier phases, is certainly accurate when $\omega_c T$ is sufficiently large and the normalized (i.e, unit energy) cross correlations between the interfering users are low. We assume throughout this section and the following one that this independence is valid. If such an approximation is not assumed, then it can be shown straightforwardly that the multiplicative correction term in the likelihood ratio (2.16) is equal to

$$\frac{E\int_0^{2\pi}\cdots\int_0^{2\pi}\exp\left[\sum_{k=2}^{K}\rho_k(b_k, \tau_k, 1)\cos\alpha_k-\psi_k(b_k, \tau_k)\sin\alpha_k-\sum_{j=2}^{k-1}\Gamma_{kj}\cos(\alpha_k-\alpha_j)\right]d\alpha_2\cdots d\alpha_K}{E\int_0^{2\pi}\cdots\int_0^{2\pi}\exp\left[\sum_{k=2}^{K}\rho_k(b_k, \tau_k, -1)\cos\alpha_k-\psi_k(b_k, \tau_k)\sin\alpha_k-\sum_{j=2}^{k-1}\Gamma_{kj}\cos(\alpha_k-\alpha_j)\right]d\alpha_2\cdots d\alpha_K}. \quad (2.21)$$

Several variants of the general structure of (2.9) are of interest and will be considered here. One such variant is that in which the *modulation* waveforms of the interfering users $\{a_k(t);\ 2\le k\le K\}$ are known, and the remaining unknown quantities in $\{s_k(t);\ 2\le k\le K\}$ are all independent with the data bits and delays uniformly distributed in their ranges. In this case, the expectations in (2.9) reduce to

$$E\{(\cdot)\}=\frac{1}{(4T)^{K-1}}\sum_{b'\in 4^{K-1}}\cdot\int_{[0,T]^{K-1}}E\{(\cdot)|b', \tau\}\,d\tau_2\cdots d\tau_K \quad (2.22)$$

where $b'=(b_2, \cdots, b_K)$ and where the inner expectation is over the amplitudes. Thus, the computation of this likelihood ratio is of exponential complexity in K. Moreover, there will be a further substantial computational burden in computing the $(K-1)$-dimensional integral corresponding to averaging over the relative delays $\tau_2, \cdots, \tau_K$.

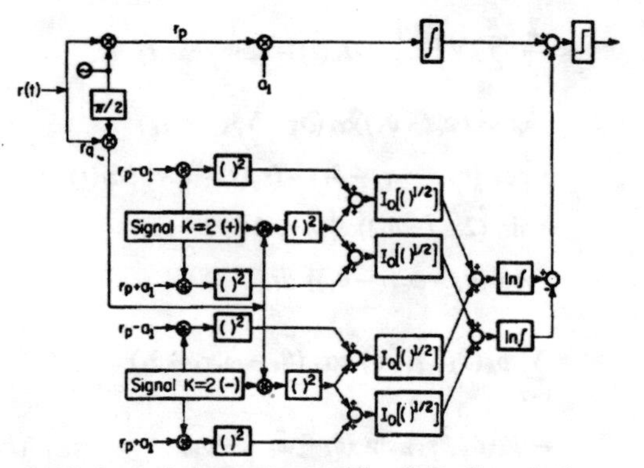

Fig. 1. Optimum single-user detector ($K = 2$). Replicas of $\alpha_2(+, +, -t)$ and $\alpha_2(+, -, -t)$ are generated by the blocks corresponding to the signal of the second user.

Fig. 1 illustrates the particularization of the demodulator derived in Proposition 2.1 to the two-user case. Note that the quadrature component of the input is used and that convolutions [required to generate (2.10) and (2.11)] and nonlinear memoryless operations are also needed.

For $K > 2$, the delay integrals do not appear to be obtainable in closed form. However, even if they could be, the exponential (in K) complexity of $\Sigma_{b'}$ shows that optimum one-shot single-user detection in a K-user channel is at least as computationally burdensome as centralized simultaneous sequence detection of fully locked users. However, one-shot single-user detection does not require tracking phases, delays, and amplitudes of all users, and thus may be preferred if these quantities are not stable for relatively long periods of time. Moreover, and perhaps more importantly, Proposition 2.1 also applies to situations in which the modulating *waveforms* of the interferers $\{a_k(t); k = 2, \cdots, K\}$ are not known. This situation is the norm for the noncentral nodes in many practical radio networks, and thus the centralized detection algorithm of [2] cannot be applied to such cases unless the receiver estimates the unknown signal cross correlations. Furthermore, as it is shown in Section III, an important reduction in the complexity of computing (2.9) results from the modeling of the modulation waveforms of the interfering users as being signature sequences.

III. Single-User Detectors for DS/SSMA Channels

In practice, one of the most important types of code-division multiple-access systems is direct-sequence spread spectrum. This corresponds to the particular case of the model (2.1), (2.2), and (2.8), in which the kth user's signature waveform is of the form

$$a_k(t)=\sum_{i=0}^{N-1} c_{ki}\psi(t-iT_c), \qquad 0\le t\le T \tag{3.1}$$

where $\{c_{ki}\}_{i=0}^{N-1}$ is a signature sequence of binary (± 1) digits, the chip waveform ψ is nonzero only on $[0, T_c]$, and the chip duration T_c is given by $T_c = T/N$. In many DS/SSMA multipoint-to-multipoint channels, it is frequently reasonable to assume that user 1 knows the chip waveforms of users 2–K, but not the specific signature sequences they employ. Since these sequences are usually chosen to be pseudonoise sequences, it is reasonable to model them (from the viewpoint of user 1) as independent sequences of independent, equiprobable binary digits. In this section, we apply this model for the interfering users in the likelihood ratio formula of Proposition 2.1. As we will see below, this affords a much more manageable form for the likelihood ratio in the limiting cases of practical interest, namely, when the number of chips is large and when the white Gaussian noise level is low.

A. Optimum Single-User Detection for Long Spreading Sequences

To study the large-N behavior of the likelihood ratio of Proposition 1, we first define the following functions:

$$\xi_e(\lambda)=\int_0^T [r_p(t)-ew_1^{1/2}a_1(t)]\psi(t-\lambda)\, dt \tag{3.2}$$

$$\phi(\lambda)=\int_0^T r_q(t)\psi(t-\lambda)\, dt \tag{3.3}$$

and

$$g_e(\lambda, \theta)=\xi_e(\lambda)\cos\theta-\phi(\lambda)\sin\theta \tag{3.4}$$

where the parameter e takes on the values $+1$ and -1 and g_e is abbreviated as g_+ and g_-, respectively, in the remainder of the section. Now fix $\tau \in (0, T)$ and suppose that $n \in \{1, \cdots, N\}$ is such that $(n-1)T_c < \tau \le nT_c$. Then define

$$d_{kj}=\begin{cases} b_k^R \quad c_{kj-n} & j-n\ge 0 \\ b_k^L \quad c_{kj-n+N} & j-n<0 \end{cases} \qquad j=0, \cdots, N \tag{3.5}$$

and notice that $g_e(\tau - T + iT_c, \theta) = 0$ for $i \le N - n - 1$ and $g_e(\tau + iT_c, \theta) = 0$ for $i \ge N - n + 1$ because $\psi(t) = 0$ for $t \notin [0, T_c]$. Then it follows that

$$\begin{aligned}
&\frac{N_0}{2w_k^{1/2}}[\rho_k(b_k, \tau, e)\cos\theta-\psi_k(b_k, \tau)\sin\theta] \\
&=\sum_{i=0}^{N-1} c_{ki}[b_k^L g_e(\tau-T+iT_c, \theta)+b_k^R g_e(\tau+iT_c, \theta)] \\
&=c_{kN-n}b_k^L g_e(\tau-nT_c, \theta)+c_{kN-n}b_k^R g_e(\tau-nT_c+T, \theta) \\
&\quad+\sum_{i=1}^{N-1} d_{ki}g_e(\tau+(i-n)T_c, \theta) \\
&=\sum_{i=0}^{N} d_{ki}g_e(\tau+(i-n)T_c, \theta)
\end{aligned} \tag{3.6}$$

and thus, the distribution of $I_0((\rho_k^2(b_k, \tau_k, e) + \psi_k^2(b_k, \tau_k))^{1/2})$ is the same modulo T_c when τ_k is uniformly distributed.

Let us now consider the particular case of a single interferer $K = 2$ which may also be used to approximate the situation in which we have a single *dominant* interferer. In this case, the correction term of the likelihood ratio is equal to

$$\frac{\displaystyle\int_0^{T_c}\int_0^{2\pi} E\exp\left(\frac{2w_2^{1/2}}{N_0}\sum_{i=0}^{N} d_{2i}g_+(iT_c-\lambda, \theta)\right) d\theta\, d\lambda}{\displaystyle\int_0^{T_c}\int_0^{2\pi} E\exp\left(\frac{2w_2^{1/2}}{N_0}\sum_{i=0}^{N} d_{2i}g_-(iT_c-\lambda, \theta)\right) d\theta\, d\lambda} \tag{3.7}$$

where the expectation is over the independent and equiprobable sequence $d_{2i} \in \{-1, 1\}$, $i = 0, \cdots, N$. The integrands in the numerator and denominator of (3.7) are products of hyperbolic cosines which do not lend themselves to further simplification. However, if N (the number of chips) is large, the distribution of the discrete random variable $\Sigma_{i=0}^{N} d_{2i}g_e(iT_c - \lambda, \theta)$ approximates the normal curve, and further simplification of (3.7) is possible. To justify this approximation, we show that for each θ, λ and each realization of

$$x(t)=[r_p(t)-ew_1^{1/2}a_1(t)]\cos\theta-r_q(t)\sin\theta$$

such that $\sup\{|x(t)| t \in (0, T)\} < \infty$, the following triangular array of random variables[3]

$$\zeta_{ni} = d_{2i} \int_{(i-1)T/n}^{iT/n} x(t-\lambda)\, dt \qquad i = 1, \cdots, n \quad (3.8)$$

satisfies the Lindeberg–Feller condition (e.g., [5])

$$\lim_{n\to\infty} \sum_{i=1}^{n} E[(\zeta_{ni}/e_n)^2 I\{|\zeta_{ni}| > \delta e_n\}] = 0 \qquad \text{for every } \delta > 0 \quad (3.9)$$

where $e_n^2 = E[\Sigma_{i=1}^n \zeta_{ni}^2]$. To check (3.9), first note that

$$\lim_{n\to\infty} \frac{n}{T} \sum_{i=1}^{n} \zeta_{ni}^2 = \lim_{n\to\infty} \frac{n}{T} \sum_{i=1}^{n} x^2(t_i^n) T^2/n^2$$

$$= \|x\|^2 \quad (3.10)$$

where $t_i^n \in ((i - 1/n)T, (iT/n))$ and the first equation in (3.10) uses the mean-value theorem on the integral of (3.8). Therefore, for every $\epsilon > 0$, there exists n_0 such that for all $1 \le i \le n$ and $n > n_0$,

$$I\{|\zeta_{ni}| > \delta e_n\} < I\left\{\frac{n}{T} \zeta_{ni}^2 > \delta^2(\|x\|^2 - \epsilon)\right\}. \quad (3.11)$$

But for each $\mu > 0$, we can find n_1 such that for $n > n_1$, we have

$$I\left\{\frac{n}{T} |\zeta_{ni}|^2 > \mu\right\} \le I\left\{\frac{T}{n} \sup_t{}^2 |x(t)| > \mu\right\} = 0. \quad (3.12)$$

Hence, only a finite number of terms on the left-hand side of (3.9) are nonzero, and since $\lim_{n\to\infty} \zeta_{ni}^2/e_n^2 = 0$, (3.9) follows.

If $\rho \cos\theta - \psi \sin\theta$ is a Gaussian random variable, then it is straightforward to check that

$$EI_0(\sqrt{\rho^2 + \psi^2}) = \exp\left(\frac{1}{4}(E[\rho^2] + E[\psi^2])\right)$$

$$I_0\left(\sqrt{\frac{1}{16}(E[\rho^2] - E[\psi^2])^2 + \frac{1}{4}E^2[\rho\psi]}\right). \quad (3.13)$$

Hence, using the Gaussian approximation[4] to the distribution of $\Sigma_{i=0}^N d_{2i} g_e(iT_c - \lambda, \theta)$, the correction term in (3.7) reduces to

$$\frac{\int_0^{T_c} \exp\left(\frac{w_2}{N_0^2}(\Xi_+(\lambda) + \Phi(\lambda))\right) I_0\left(\frac{w_2}{N_0^2}\sqrt{(\Xi_+(\lambda) - \Phi(\lambda))^2 + \Theta_+^2(\lambda)}\right) d\lambda}{\int_0^{T_c} \exp\left(\frac{w_2}{N_0^2}(\Xi_-(\lambda) + \Phi(\lambda))\right) I_0\left(\frac{w_2}{N_0^2}\sqrt{(\Xi_-(\lambda) - \Phi(\lambda))^2 + \Theta_-^2(\lambda)}\right) d\lambda} \quad (3.14)$$

where

$$\Xi_e(\lambda) = \sum_{i=0}^{N} \xi_e^2(iT_c - \lambda) \quad (3.15)$$

$$\Phi(\lambda) = \sum_{i=0}^{N} \phi^2(iT_c - \lambda) \quad (3.16)$$

and

$$\Theta_e(\lambda) = 2 \sum_{i=0}^{N} \xi_e(iT_c - \lambda)\phi(iT_c - \lambda). \quad (3.17)$$

This structure is illustrated in Fig. 2. Note that there is considerable simplification in this structure over that of Fig. 1. In particular, each of the "+" and "−" channels involves chip matched filtering of the in-phase and quadrature components followed by chip-rate sampling, quadratic accumulation, memoryless nonlinear transformation, and integration over the offset λ. This latter operation can be implemented in parallel form by decimating an M/T_c-rate sampler (rather than a $1/T_c$-rate sampler) where M is the number of points taken in the numerical computation of the integral.

Further simplification of the correction term (2.9) is also possible in the case $K > 2$ by using the Gaussian approximation. We can obtain an expression similar to (3.14) where the integration is now over the hypercube $[0, T_c]^{K-1}$. Analogously to the case $K = 2$, for each $\theta = (\theta_2, \cdots, \theta_K)$ and $\tau = (\tau_2, \cdots, \tau_K)$, the distribution of

$$\sum_{k=2}^{K} \rho_k(b_k, \tau_k, e) \cos\theta_k - \psi_k(b_k, \tau_k) \sin \cdot \theta_k - \sum_{j=2}^{k-1} \Gamma_{kj}(b_k, b_j, \tau_k, \tau_j)$$

is approximately Gaussian, and since Γ_{kj} is uncorrelated with ρ_k, ψ_k, ρ_j, and ψ_j, both the numerator and the denominator of the correction term in (2.9) are approximated by

$$\int_{[0,T_c]}^{K-1} \cdots \int d\tau_2 \cdots d\tau_k E \cdot \exp\left(-\sum_{k=2}^{K}\sum_{j=2}^{k-1} \Gamma_{kj}(b_k, b_j, \tau_k, \tau_j)\right) \cdot \sum_{k=2}^{K} EI_0(\sqrt{\rho_k^2(b_k, \tau_k, e) - \psi_k^2(b_k, \tau_k)}).$$

Now, $E\exp(-\Sigma_{k=2}^K \Sigma_{j=2}^{k-1} \Gamma_{kj}(b_k, b_j, \tau_k, \tau_j))$ depends only on τ and on the chip waveform, and since if $X \sim N(0, \sigma)$, then $E\exp X = \exp \sigma^2/2$, we have

$$E\exp\left(-\sum_{k=2}^{K}\sum_{j=2}^{k-1} \Gamma_{kj}(b_k, b_j, \tau_k, \tau_j)\right)$$

$$\approx \exp\left(\frac{1}{2}E\left[\left(\sum_{k=2}^{K}\sum_{j=2}^{k-1} \Gamma_{kj}(b_k, b_j, \tau_k, \tau_j)\right)^2\right]\right)$$

$$= \exp\left(\frac{1}{2}\sum_{k=2}^{K}\sum_{j=2}^{k-1} E\Gamma_{kj}^2(b_k, b_j, \tau_k, \tau_j)\right)$$

$$\approx \prod_{k=2}^{K}\prod_{j=2}^{k-1} \exp\left[\frac{w_k w_j}{N_0^2}\left(\frac{|\tau_k - \tau_j|^2 + (T_c - |\tau_k - \tau_j|)^2}{T_c}\right)\right] \quad (3.18)$$

[3] For the sake of notational simplicity, here we consider the case of a rectangular chip waveform. In this case, $\zeta_{Ni} = d_{2i} g_e(iT_c - \lambda, \theta)$.

[4] It should be noted that the use of a central-limit theorem here is quite different from the Gaussian approximations used in many previous analyses of conventional single-user receivers. Here, we do not claim that the multiple-access interference is asymptotically a white Gaussian process; however, we do show via the Lindeberg–Feller condition (3.9) that the decision statistics in (3.7) are conditionally Gaussian random variables as the number of chips per symbol goes to infinity.

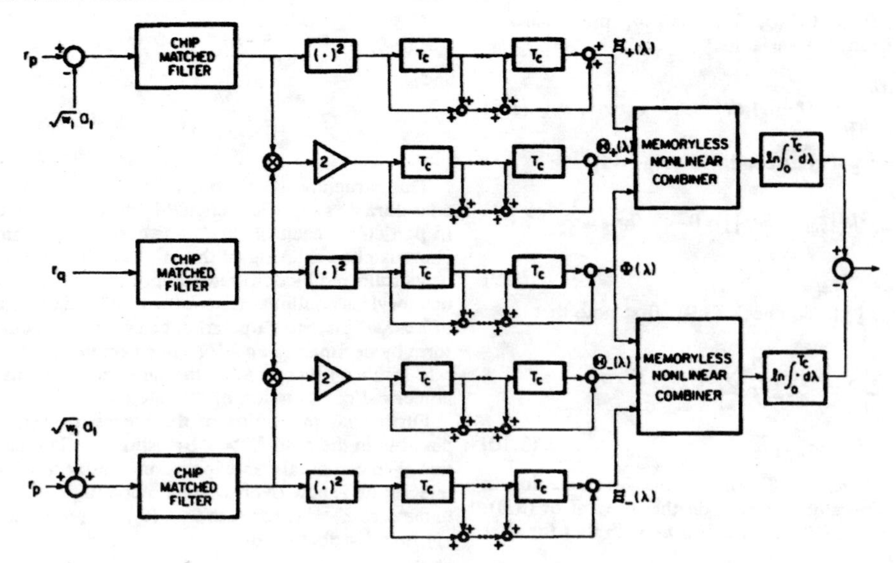

Fig. 2. Correction statistic for single-user detector with long spreading sequences.

where the last approximation follows by assuming that $\psi(t) = 1/T^{1/2}$ for $t \in [0, T_c]$ and by neglecting an $O(1/N^2)$ term in the exponent. Hence, the overall correction term is approximately equal to

$N_0 \to 0$.[5] Hence, rather than using (3.19), we must take the limit as $N_0 \to 0$ of the original likelihood ratio (2.9). As in the previous analysis, we will first focus attention on the case of a single interferer ($K = 2$).

$$\frac{\displaystyle\int_{[0,T_c]} \cdots \int_{[0,T_c]} \prod_{k=2}^{K} \exp\left(\frac{w_k}{N_0^2}(\Xi_+(\tau_k)+\Phi(\tau_k))\right) I_0\left(\frac{w_k}{N_0^2}\sqrt{(\Xi_+(\tau_k)-\Phi(\tau_k))^2+\Theta_+^2(\tau_k)}\right)}{\displaystyle\int_{[0,T_c]} \cdots \int_{[0,T_c]} \prod_{k=2}^{K} \exp\left(\frac{w_k}{N_0^2}(\Xi_-(\tau_k)+\Phi(\tau_k))\right) I_0\left(\frac{w_k}{N_0^2}\sqrt{(\Xi_-(\tau_k)-\Phi(\tau_k))^2+\Theta_-^2(\tau_k)}\right)}$$
$$\cdot\frac{\displaystyle\prod_{j=2}^{k-1} \exp\left[\frac{2w_k w_j}{NN_0^2}\left(\frac{|\tau_k-\tau_j|^2+(T_c-|\tau_k-\tau_j|)^2}{T_c}\right)\right] d\tau_2 \cdots d\tau_K}{\displaystyle\prod_{j=2}^{k-1} \exp\left[\frac{2w_k w_j}{NN_0^2}\left(\frac{|\tau_k-\tau_j|^2+(T_c-|\tau_k-\tau_j|)^2}{T_c}\right)\right] d\tau_2 \cdots d\tau_K}. \tag{3.19}$$

Notice that the term that couples the integrals in (3.19) is asymptotically independent of τ as $N \to \infty$. Hence, (3.19) approaches the product of $K - 1$ (3.14)-like terms (substituting w_2 by w_k). Thus, in this limiting case, implementation of the multiuser correction term in the likelihood ratio involves the implementation of only one chip-matched-filter/quadratic-accumulator section followed by multiple averaging channels, one for each different value of w_k. Fig. 2 shows an implementation of the correction statistic to be added to the output of the single-user matched filter in the case of a single interferer. The general structure is the same, except that the memoryless nonlinearities output a process for each interferer which is then passed through a separate logarithmic integrator.

B. Optimum Single-User Detection for High SBNR

We now turn to another limiting case of the single-user detector for which a simplified form of the likelihood ratio exists, namely, the case when the power spectral density of the additive Gaussian noise goes to zero. In the above case, we saw that when the rest of the parameters are fixed, we can use a Gaussian approximation as $N \to \infty$. However, for fixed N, the error between the expected values of the exponentials, according to the true and Gaussian distributions, diverges as

Since the spreading codes of the interfering users are modeled by the single-user receiver as equiprobable and independent binary sequences, the correction term of the likelihood ratio is given by (3.7) and the log-likelihood ratio is (except for a positive multiplicative constant) equal to

$$2\int_0^T r_p(t)a_1(t)\,dt + \frac{N_0}{2w_1^{1/2}}$$
$$\cdot \log \int_0^{T_c}\int_0^{2\pi} E\exp\left(\frac{2w_2^{1/2}}{N_0} d_{2i}g_+(iT_c-\lambda,\theta)\right) d\theta\, d\lambda$$
$$-\frac{N_0}{2w_1^{1/2}} \log \int_0^{T_c}\int_0^{2\pi}$$
$$\cdot E\exp\left(\frac{2w_2^{1/2}}{N_0}\sum_{i=0}^{N} d_{2i}g_-(iT_c-\lambda,\theta)\right) d\theta\, d\lambda \tag{3.20}$$

[5] This is due to the fact that as the variance goes to infinity, the error between the distributions accumulates on the tails (the true random variable is bounded) on which the expected value of the exponential largely depends.

where the expectation is over the independent and equiprobable sequence $d_{2i} \in \{-1, 1\}$, $i = 0, \cdots, N$. On taking the limit of the correction terms in (3.20), we obtain

$$\lim_{N_0\to 0} \frac{N_0}{2w_1^{1/2}} \log \int_0^{T_c}\int_0^{2\pi} \cdot E \exp\left(\frac{2w_2^{1/2}}{N_0}\sum_{i=0}^{N} d_{2i}g_e(iT_c-\lambda, \theta)\right) d\theta\, d\lambda$$

$$= \lim_{N_0\to 0} \frac{1}{w_1^{1/2}} \log \left[\int_0^{T_c}\int_0^{2\pi} \cdot \left[\exp\left(w_2^{1/2}\sum_{i=0}^{N}|g_e(iT_c-\lambda, \theta)|\right)\right]^{2/N_0} d\theta\, d\lambda\right]^{N_0/2}$$

$$- \lim_{N_0\to 0}\frac{N_0}{2w_1^{1/2}}(N+1)\log 2$$

$$= \frac{1}{w_1^{1/2}} \log \sup_{\lambda,\theta}\left\{\exp\left(w_2^{1/2}\sum_{i=0}^{N}|g_e(iT_c-\lambda, \theta)|\right)\right\}$$

$$= (w_2/w_1)^{1/2} \sup_{\substack{\lambda\in[0,T_c]\\ \theta\in[0,2\pi]}} \sum_{i=0}^{N}|g_e(iT_c-\lambda, \theta)|. \quad (3.21)$$

Therefore, in the limit as $N_0 \to 0$, the optimum single-user detector for $K = 2$ in the case of unknown interfering codes compares the test statistic

$$2\int_0^T r_p(t)a_1(t)\,dt + (w_2/w_1)^{1/2} \sup_{\substack{\lambda\in[0,T_c]\\ \theta\in[0,2\pi]}} \sum_{i=0}^{N}|g_+(iT_c-\lambda, \theta)|$$

$$-(w_2/w_1)^{1/2} \sup_{\substack{\lambda\in[0,T_c]\\ \theta\in[0,2\pi]}} \sum_{i=0}^{N}|g_-(iT_c-\lambda, \theta)| \quad (3.22)$$

to a zero threshold. Note that as might be expected, (3.22) is also the limiting form of the generalized likelihood ratio test or maximum likelihood detector (see Helstrom [6, p. 291], for example).

We now investigate the error probability of the test in (3.22) when $N_0 = 0$. It was shown in [4] that when the delays, phases, and waveforms of all users are known, the fully centralized optimum detector achieves perfect demodulation with probability 1 in the absence of background noise. This is a nontrivial result, as is illustrated by the behavior of the *conventional* single-user detector which becomes multiple-access limited, i.e., the limit of its error probability as $N_0 \to 0$ is nonzero for sufficiently powerful interfering users. However, as in the present case, the conventional detector does not have access to the delays, phases, or signature sequences of the interfering users. So, the question arises as to whether an optimum single-user detector can achieve error-free performance regardless of the energies of the interfering users without knowledge of those parameters. The answer, in the two-user case, is given in the affirmative by the following result which does not put any restrictions on the signature sequences.

Proposition 3.1: Suppose $K = 2$ and $w_1 > 0$. If $r(t) = bs_1(t) + S^{MA}(t)$, $b \in \{-1, 1\}$, then

$$\operatorname{sgn}\left[2\int_0^T r_p(t)a_1(t)\,dt + (w_2/w_1)^{1/2} \cdot \sup_{\substack{\lambda\in[0,T_c]\\ \theta\in[0,2\pi]}} \sum_{i=0}^{N}|g_+(iT_c-\lambda, \theta)| - (w_2/w_1)^{1/2} \cdot \sup_{\substack{\lambda\in[0,T_c]\\ \theta\in[0,2\pi]}} \sum_{i=0}^{N}|g_-(iT_c-\lambda, \theta)|\right] = b \quad (3.23)$$

with probability 1.

Proof: See Appendix.

In the general case of $K > 2$ users, the log-likelihood ratio is proportional to [cf. (2.16)]

$$2w_1^{1/2}\int_0^T r_p(t)a_1(t)\,dt + \frac{N_0}{2}\log\int_{[0,T_c]}\cdots\int_{K-1}\int_{[0,2\pi]}\cdots\int_{K-1} \cdot E\exp\left[-\frac{1}{N_0}\|S^{MA}\|^2 + \frac{2}{N_0}\langle r-s_1, S^{MA}\rangle\right] d\theta\, d\tau$$

$$-\frac{N_0}{2}\log\int_{[0,T_c]}^{K-1}\cdots\int\int_{[0,2\pi]}^{K-1}\cdots\int \cdot E\exp\left[-\frac{1}{N_0}\|S^{MA}\|^2 + \frac{2}{N_0}\langle r+s_1, S^{MA}\rangle\right] d\theta\, d\tau \quad (3.24)$$

where the expectation is over the independent sequences $d = \{d_{ki} \in \{-1, 1\};\ i = 0, \cdots, N, k = 2, \cdots, K\}$. As in (3.21), this expectation is dominated as $N_0 \to 0$ by the atom corresponding to the largest integrand, i.e.,

$$d^* \in \arg\max_d \Omega_e(d, \tau, \theta) \quad (3.25)$$

where

$$\Omega_e(d, \tau, \theta) = \langle r - es_1, S^{MA}(d)\rangle - \frac{1}{2}\|S^{MA}(d)\|^2 \quad (3.26)$$

and

$$S^{MA}(t, d) = \sum_{i=0}^{N}\sum_{k=2}^{K} d_{ki}(2w_k)^{1/2}\psi(t-(i-1)T_c-\tau_k) \cdot \cos(\omega_c t + \theta_k - \omega_c\tau_k). \quad (3.27)$$

Since there are $2^{K(N+1)}$ possible sequences, it is necessary to find an efficient way to carry out the maximization in (3.25). But (3.26) and (3.27) have the same structure as (2.3) and (2.4), respectively, so we can apply the results of [2] to carry out the maximization of (3.25) with linear complexity in N. On taking the limit of (3.24) as $N_0 \to 0$, we obtain the test statistic

$$2w_1^{1/2}\int_0^T r_p(t)a_1(t)\,dt + \sup_{\tau,\theta}\Omega_+^*(\tau, \theta) - \sup_{\tau,\theta}\Omega_-^*(\tau, \theta) \quad (3.28)$$

where $\Omega_e^*(\tau, \theta) = \Omega_e(d^*, \tau, \theta)$. Even if these quantities are obtained through efficient dynamic programming recursions as in [2], the main computational burden of (3.28) is the maximization over $[0, T_c]^{K-1}$ and $[0, 2\pi]^{K-1}$, which imposes severe limitations on its feasibility for even a moderate number of users. However, note that, in performing the maximization, the receiver is essentially acquiring the chip timing and carrier phases of the interfering users. Thus, in practice, it would normally be unnecessary to undergo a full search for the maximizing τ and θ in each symbol interval since these quantities will change little from symbol interval to symbol interval. For this reason, (3.28) might be reasonably efficient to implement in approximate form.

IV. Locally Optimum Single-User Detectors with Weak Interferers

We have seen in the preceding sections that in multiple-access environments with many users, the complexity of optimum detection is increased considerably (over centralized reception) when the unwanted users are unlocked. This is true even without sequence detection and regardless of whether the interfering waveforms are known. However, one of the main incentives for the study of optimum decentralized detectors is the situation in which all or some of the interfering users are comparatively weak, so that it may be impractical to provide reliable synchronization for them. The objective of this section is to derive locally optimum (up to a third-order approximation) decentralized detectors for reception in the presence of weak unlocked users. As we shall see, such detectors can be viewed as versions of the detector that would be optimum without the weak interferers, modified to be robust against small deviations from the nominal white Gaussian noise statistics caused by weak multiple-access interference. As in the preceding sections, we consider both the case in which the waveforms of the interfering users are known, and the case in which they are coded with binary signature sequences unknown to the receiver. We will see here that the *locally optimum* version takes care only of the nonwhiteness of the multiple-access noise.

The approach we follow to derive locally optimum decentralized demodulators is to obtain an asymptotic form of the log-likelihood ratio for signal detection in contaminated white Gaussian noise given by the following result.

Lemma 4.1: Consider the following pair of statistical hypotheses:

$$H_0:\ r_t = s_t^0 + \epsilon\tilde{n}_t + n_t \qquad t \in [t_p, t_f]$$

$$H_1:\ r_t = s_t^1 + \epsilon\tilde{n}_t + n_t \qquad t \in [t_p, t_f] \tag{4.1}$$

where s^1 and s^0 are deterministic finite-energy signals, $\{n_t\}$ is white Gaussian noise with spectral height σ^2, and $\{\tilde{n}_t, t \in [t_0, t_f]\}$ in a symmetric random process such that $\|\tilde{n}\| < B$ (a.s.) for some constant B, and whose correlation function is denoted by $C_{t,\lambda} = E[\tilde{n}_t\tilde{n}_\lambda]$, $(t, \lambda) \in [t_0, t_f]^2$. Then the log-likelihood ratio for (4.1) admits in the following expression:

$$\log LR(\epsilon) = \frac{1}{\sigma^2}\int_{t_0}^{t_f}\left[(s_t^1 - s_t^0) - \left(\frac{\epsilon}{\sigma}\right)^2 \cdot \int_{t_0}^{t_f} C_{t,\lambda}(s_\lambda^1 - s_\lambda^0)\,d\lambda\right]\left(r_t - \frac{1}{2}s_t^1 - \frac{1}{2}s_t^0\right)dt + O(\epsilon^4). \tag{4.2}$$

Proof: Using the Cameron-Martin likelihood ratio formula, we obtain

$$\log LR(\epsilon) = \log\frac{D_1(\epsilon)}{D_0(\epsilon)} \tag{4.3}$$

where

$$D_i(\epsilon) = E\left[\exp\left(-\frac{1}{2\sigma^2}\left(\|s^i + \epsilon\tilde{n}\|^2 - 2\int_{t_p}^{t_f}(s_t^i + \epsilon\tilde{n}_t)\,dr_t\right)\right)\right] \qquad i = 0, 1 \tag{4.4}$$

where the expectation is over the ensemble of sample functions of $\{\tilde{n}_t, t \in [t_p, t_f]\}$.

In order to derive (4.2), we take the Taylor series expansion of (4.3) around the origin. Since $\tilde{n}_t$ is a symmetric random variable, it follows that $D_i(-\epsilon) = D_i(\epsilon)$, and hence the odd terms in the Taylor expansion of $D_i(\epsilon)|_{\epsilon=0}$ and $\log D_i(\epsilon)|_{\epsilon=0}$ are equal to zero.

Using the fact that $\|\tilde{n}\| < B$ a.s. and the Schwarz inequality, it follows that the expectation of every coefficient in the series expansion of the exponential in (4.4) exists, and we can write

$$D_i(\epsilon) = D_i(0)\left[1 + \frac{\epsilon^2}{2}E\left[\left(\frac{1}{\sigma^2}\int_{t_p}^{t_f}\tilde{n}_t(r_t - s_t^i)\,dt\right)^2 - \frac{1}{\sigma^2}\|\tilde{n}\|^2\right] + O(\epsilon^4)\right]. \tag{4.5}$$

Now, since $\log(1 + x) = x + O(x^2)$, we obtain

$$\log\frac{D_1(\epsilon)}{D_0(\epsilon)} = \log\frac{D_1(0)}{D_0(0)} + \frac{\epsilon^2}{2}E\left[\left(\frac{1}{\sigma^2}\int_{t_p}^{t_f}\tilde{n}_t(r_t - s_t^1)\,dt\right)^2 - \left(\frac{1}{\sigma^2}\int_{t_p}^{t_f}\tilde{n}_t(r_t - s_t^0)\,dt\right)^2\right] + O(\epsilon^4) \tag{4.6}$$

and (4.2) follows straightforwardly. □

Notice that the stringent condition $\|\tilde{n}\| < B$ (a.s.) allows a straightforward proof of Lemma 4.1 and is satisfied in the case in which we are interested, namely,

$$\tilde{n}_t = \sum_{i=-M}^{M}\sum_{k=D+1}^{K} b_k(i)s_k(t - iT - \tau_k);\ b_k(i) \in \{-1, 1\}. \tag{4.7}$$

If the waveforms $\{a_k(t), k = D + 1, \cdots, K\}$ are known by the receiver, then the autocorrelation function of $\tilde{n}$ with support in $\mathbb{R}^2$ (for $M = \infty$) is equal to

$$C_{t,\lambda}^{MA} = \frac{1}{T}\cos(\omega_c(t - \lambda))\sum_{k=D+1}^{K} w_k R_k(t - \lambda) \tag{4.8}$$

where

$$R_k(t) = \int_0^T a_k(s - t)a_k(s)\,ds. \tag{4.9}$$

If the waveforms of the interfering users have the form in (3.1) and the code of each user is unknown by the receiver and assumed to be equiprobably distributed among all $\{-1, 1\}$ sequences of length N, then the autocorrelation is

$$C_{t,\lambda}^{MA} = \frac{1}{T}\cos(\omega_c(t - \lambda))\sum_{k=D+1}^{K} w_k\Psi(t - \lambda) \tag{4.10}$$

where the autocorrelation of the chip waveform is denoted by $\Psi(t) = \int_0^{T_c}\psi(s)\psi(s - t)\,dt$.

The one-shot single-user detector can be obtained readily

from the result of Lemma 4.1. Since the signal of user 1 is antipodally modulated, we have

$$s_t^1 - s_t^0 = 2\sqrt{2w_1}a_1(t)\cos(\omega_c t + \theta_1)$$

and (4.2) becomes

$$\log LR = \frac{4\sqrt{w_1}}{N_0}\int_0^T r_p(t) \cdot \left[a_1(t) - \frac{1}{N_0 T}\sum_{k=2}^{K} w_k \int_0^T a_1(\lambda)R_k(t-\lambda)\,d\lambda\right] dt + O(\max_{k>1} w_k^2). \tag{4.11}$$

Hence, the locally optimum one-shot single-user detector is a conventional correlation receiver in which $a_1(t)$ is replaced by $a_1(t) - (1/N_0T)\sum_{k=2}^K w_k \int_0^T a_1(\lambda)R_k(t-\lambda)\,d\lambda$, $t \in [0, T]$, i.e., the pseudosignal is the output in $[T, 2T]$ of a causal linear filter, driven by $a_1(t)$, and whose impulse response is equal to $\delta(t-T) - (1/N_0T)\sum_{k=2}^K w_k R_k(t-T)$. If the signature sequences are unknown, the impulse response is $\delta(t-T) - (1/N_0T)\sum_{k=2}^K w_k \Psi(t-T)$, which amounts to a mild smoothing of the signal replica of the user of interest.

The locally optimum detector that locks to D of K users is, in fact, a generalization of this conclusion. Using Lemma 4.1, it can be shown (see [3, ch. 5] for details) that the locally optimum D-user detector is a centralized detector whose correlators use replicas of the unmodified waveforms of the users of interest. However, the input is processed by a causal filter that whitens the interference due to unlocked users, and whose impulse response depends on the autocorrelation function and signal-to-noise ratio of each interfering signal. This requires a modification of the DP algorithm to account for the intersymbol interference introduced by the prefilter, and results in a complexity of $O(2^{2D})$ as opposed to $O(2^D)$ for the corresponding algorithm that neglects the additional $K - D$ interferers.

V. Summary

In this paper, we have obtained decentralized single-user detectors which take into account the presence of interfering users. The general decentralized demodulation problem is one of sequence detection in additive colored non-Gaussian noise, and results in nonlinear detectors whose decision algorithms do not admit recursive forms and hence are more complex than their centralized counterparts. Important reductions in complexity occur when attention is focused on one-shot single-user detectors.

The general form of the single-user likelihood ratio obtained in Proposition 2.1 is equal to the single-user likelihood ratio affected by a correction term which depends on both the in-phase and quadrature components of the input. Both the case where the baseband interfering waveforms are known and the case where they are coded by an unknown signature sequence have been studied.

Under the assumption that the assigned waveforms are signature sequences with N chips per bit, we have obtained limiting forms of the correction term for $N \gg 1$ and for $N_0 = 0$. In the first case, the correction term depends on the received waveform only through the functions $\Xi_\pm(\lambda)$, $\Phi(\lambda)$, and $\Theta_\pm(\lambda)$ which represent the l_2 norms and inner product, respectively, of the subintegrals of an N partition (with offset $\lambda \in [0, T_c]$) of the in-phase and quadrature components of the received noise process under both hypotheses. The correction term when $N_0 = 0$ is best illustrated in the single-interferer case where it is obtained through the maximization over the relative phase and delay of the l_1 norm of the above subintegrals. It has been shown that this detector (which assumes knowledge of only the chip waveform and energy of the interfering user) achieves perfect demodulation in the absence of Gaussian noise regardless of the energy of the interference, thus avoiding the multiple-access limitation that plagues the conventional receiver. Using dynamic programming, the single-user detector can be implemented in linear time in N; however, its main computational burden is the maximization over $[0, T_c]^{K-1}$ and $[0, 2\pi]^{K-1}$ needed in the correction term.

Using an asymptotic form of the log-likelihood ratio for signal detection in contaminated white Gaussian noise, we have derived locally optimum detectors up to a third-order approximation in the amplitude of the interfering users. The locally optimum one-shot detector has been shown to be a single-user correlation receiver which uses a smooth replica of the signal of interest. It has been shown in [3] that this approach can be generalized to the case of partial decentralization ($D > 1$), resulting in robustified versions of the centralized D-user receiver, which may offer substantial computational savings over the optimum K-user receiver.

Appendix

Proof of Proposition 3.1

We assume that the bit transmitted by user 1 is $b = 1$, the proof being identical in the antipodal case. For notational convenience and without lost of generality, we suppose that the relative delay of the interfering user is $0 < \tau_2 \le T_c$; then it follows that

$$\alpha_2(b_2^L, b_2^R, t-\tau_2) = \sum_{i=0}^{N} d_i \psi(t - iT_c + \lambda_2) \tag{A.1}$$

where $\lambda_2 = T_c - \tau_2$, $d_0 = c_{2N-1}b_2^L$, and $d_{i+1} = c_{2i}b_2^R$ for $i = 0, N-1$. Let $\beta = \theta_1 + \omega_c\tau_2 - \theta_2$; then it is easy to show that

$$\int_0^T r_p(t)a_1(t) = w_1^{1/2} + w_2^{1/2}\cos\beta \cdot \int_0^T a_1(t)\alpha_2(b_2^L, b_2^R, t-\tau_2)\,dt \tag{A.2}$$

$$g_+(iT_c - \lambda, \theta) = w_2^{1/2}\cos(\theta + \beta) \cdot \int_0^T \alpha_2(b_2^L, b_2^R, t-\tau_2)\psi(t - iT_c + \lambda)\,dt \tag{A.3}$$

and

$$g_-(iT_c - \lambda, \theta) = 2w_1^{1/2}\cos\theta \cdot \int_0^T a_1(t)\psi(t - iT_c + \lambda)\,dt + g_+(iT_c - \lambda, \theta). \tag{A.4}$$

We show now that

$$\sup_{\substack{\lambda \in [0, T_c] \\ \theta \in [0, 2\pi]}} \sum_{i=0}^{N} |g_+(iT_c - \lambda, \theta)| = w_2^{1/2}. \tag{A.5}$$

To that end, using (A.1) and (A.3), we obtain for every $\lambda \in [0, T_c]$

$$\sup_{\theta \in [0, 2\pi]} \sum_{i=0}^{N} |g_+(iT_c - \lambda, \theta)| = w_2^{1/2}\sum_{i=0}^{N}\left|\int_0^T \alpha_2(b_2^L, b_2^R, t-\tau_2)\psi(t - iT_c + \lambda)\,dt\right|$$

$$= w_2^{1/2} \sum_{i=0}^{N} \left| \int_0^T \sum_{j=0}^{N} d_j \psi(t - jT_c + \lambda_2) \psi(t - iT_c + \lambda) \, dt \right|$$

$$\leq w_2^{1/2} \int_0^T \sum_{i=0}^{N} \sum_{j=0}^{N} |\psi(t - jT_c + \lambda_2)| |\psi(t - iT_c + \lambda)| \, dt$$

$$\leq w_2^{1/2} \left(\int_0^T \sum_{j=0}^{N} \psi^2(t - jT_c + \lambda_2) \, dt \right)^{1/2}$$

$$\cdot \left(\int_0^T \sum_{i=0}^{N} \psi^2(t - iT_c + \lambda) \, dt \right)^{1/2} = w_2^{1/2} \quad \text{(A.6)}$$

where the last two equations follow from the Schwarz inequality and from the relationship $\int_0^{T_c} (\psi^2(t + s) + \psi^2(t - T_c + s)) \, dt = \int_0^{T_c} \psi^2(t) = 1/N, \ 0 \leq s \leq T_c$, respectively. But the right-hand side of (A.6) is achieved when $\lambda = \lambda_2$; hence, (A.5) follows. Consequently, in order to show that the sign of the log-likelihood ratio is positive, one has to prove that

$$2w_1 + w_2 + w_1^{1/2} w_2^{1/2} \int_0^T 2a_1(t) \alpha_2(b_2^L, b_2^R, t - \tau_2)$$
$$\cdot \cos \beta \, dt - w_2^{1/2} \sum_{i=0}^{N} |g_-(iT_c - \lambda, \theta)| > 0 \quad \text{(A.7)}$$

for all $\lambda \in [0, T_c]$ and $\theta \in [0, 2\pi]$. Using (A.3) and (A.4), we obtain

$$2w_1 + w_2 + w_1^{1/2} w_2^{1/2} \int_0^T 2a_1(t) \alpha_2(b_2^L, b_2^R, t - \tau_2) \cos \beta \, dt$$
$$- w_2^{1/2} \sum_{i=0}^{N} |g_-(iT_c - \lambda, \theta)|$$
$$= 2w_1 + w_2 + w_1^{1/2} w_2^{1/2}$$
$$\cdot \int_0^T 2a_1(t) \alpha_2(b_2^L, b_2^R, t - \tau_2) \cos \beta \, dt$$
$$- w_2^{1/2} \sum_{i=0}^{N} \left| \int_0^T (2w_1^{1/2} a_1(t) \cos \theta + w_2^{1/2} \right.$$
$$\left. \cdot \alpha_2(b_2^L, b_2^R, t - \tau_2) \cos (\theta + \beta)) \psi(t - iT_c + \lambda) \, dt \right|. \quad \text{(A.8)}$$

The last term on the right-hand side of the above equation can be bounded as follows:

$$\sum_{i=0}^{N} \left| \int_0^T (2w_1^{1/2} a_1(t) \cos \theta + w_2^{1/2} \alpha_2(b_2^L, b_2^R, t - \tau_2) \right.$$
$$\left. \cdot \cos (\theta + \beta)) \psi(t - iT_c + \lambda) \, dt \right|$$
$$\leq \sum_{i=0}^{N} \int_0^T |2w_1^{1/2} a_1(t) \cos \theta + w_2^{1/2} \alpha_2(b_2^L, b_2^R, t - \tau_2)$$
$$\cdot \cos (\theta + \beta)| \cdot |\psi(t - iT_c + \lambda)| \, dt$$
$$= \int_0^T |2w_1^{1/2} a_1(t) \cos \theta + w_2^{1/2} \alpha_2(b_2^L, b_2^R, t - \tau_2)$$
$$\cdot \cos (\theta + \beta)| \sum_{i=0}^{N} |\psi(t - iT_c + \lambda)| \, dt$$
$$\leq \left[\int_0^T (2w_1^{1/2} a_1(t) \cos \theta + w_2^{1/2} \alpha_2(b_2^L, b_2^R, t - \tau_2) \right.$$
$$\left. \cdot \cos (\theta + \beta))^2 \, dt \right]^{1/2} \left[\int_0^T \sum_{i=0}^{N} \psi^2(t - iT_c + \lambda) \, dt \right]^{1/2}$$
$$= \left[4w_1 \cos^2 \theta + w_2 \cos^2 (\theta + \beta) + 4w_1^{1/2} w_2^{1/2} \right.$$
$$\left. \cdot \cos \theta \cos (\theta + \beta) \int_0^T a_1(t) \alpha_2(b_2^L, b_2^R, t - \tau_2) \, dt \right]^{1/2}. \quad \text{(A.9)}$$

Since $|\int_0^T a_1(t) \alpha_2(b_2^L, b_2^R, t - \tau_2) \, dt| \leq 1$, we can denote $\int_0^T a_1(t) \alpha_2(b_2^L, b_2^R, t - \tau_2) \, dt = \cos \alpha$, and using (A.9), the right-hand side of (A.8) can be lower bounded by

$$2w_1 + w_2 + w_1^{1/2} w_2^{1/2} \int_0^T 2a_1(t) \alpha_2(b_2^L, b_2^R, t - \tau_2) \cos \beta \, dt$$
$$- w_2^{1/2} \sum_{i=0}^{N} \left| \int_0^T (2w_1^{1/2} a_1(t) \cos \theta + w_2^{1/2} \right.$$
$$\left. \cdot \alpha_2(b_2^L, b_2^R, t - \tau_2) \cos (\theta + \beta)) \psi(t - iT_c + \lambda) \, dt \right|$$
$$\geq 2w_1 + w_2 + 2w_1^{1/2} \cos \alpha \cos \beta$$
$$- w_2^{1/2} (4w_1 \cos^2 \theta + w_2 \cos^2 (\theta + \beta)$$
$$+ 4w_1^{1/2} w_2^{1/2} \cos \theta \cos (\theta + \beta) \cos \alpha)^{1/2}. \quad \text{(A.10)}$$

Now, since $2w_1 + w_2 + 2w_1^{1/2} w_2^{1/2} \cos \alpha \cos \beta > 0$, the sign of the right-hand side of (A.10) is equal to the sign of

$$(2w_1 + w_2 + 2w_1^{1/2} w_2^{1/2} \cos \alpha \cos \beta)^2$$
$$- [4w_1 w_2 \cos^2 \theta + w_2^2 \cos^2 (\theta + \beta)$$
$$+ 4w_2 w_1^{1/2} w_2^{1/2} \cos \theta \cos \alpha \cos (\theta + \beta)]$$
$$= (2w_1 + 2w_1^{1/2} w_2^{1/2} \cos \alpha \cos \beta)^2 + 4w_1 w_2 (1 - \cos^2 \theta)$$
$$+ w_2^2 (1 - \cos^2 (\theta + \beta)) + 4w_2 w_1^{1/2} w_2^{1/2}$$
$$\cdot \cos \alpha [\cos \beta - \cos \theta \cos (\theta + \beta)]$$
$$= (2w_1 + 2w_1^{1/2} w_2^{1/2} \cos \alpha \cos \beta)^2 + 4w_1 w_2 \sin^2 \theta$$
$$+ w_2^2 \sin^2 (\theta + \beta) + 4w_2 w_1^{1/2} w_2^{1/2} \sin \theta \sin (\theta + \beta) \cos \alpha$$
$$= (2w_1 + 2w_1^{1/2} w_2^{1/2} \cos \alpha \cos \beta)^2$$
$$+ (w_2 \sin (\theta + \beta) \cos \alpha + 2\sqrt{w_1 w_2} \sin \theta)^2$$
$$+ (w_2 \sin (\theta + \beta) \sin \alpha)^2. \quad \text{(A.11)}$$

Therefore, we have shown that (A.10) and, consequently, the left-hand side of (3.22), are nonnegative. Moreover, the right-hand side of (A.11) is equal to zero only if

$$2w_1 + 2w_1^{1/2} w_2^{1/2} \cos \alpha \cos \beta = 0, \quad \text{(A.12)}$$

but since $\beta = \theta_1 + \omega_c \tau_2 - \theta_2$ is uniformly distributed, (A.12) occurs with probability zero if $w_1 > 0$.

REFERENCES

[1] M. K. Simon, J. K. Omura, R. A. Scholtz, and B. K. Levitt, *Spread Spectrum Communications, Vol. 3.* Rockville, MD: Computer Science Press, 1985.

[2] S. Verdu, "Minimum probability of error for asynchronous Gaussian multiple-access channels," *IEEE Trans. Inform. Theory,* vol. IT-32, pp. 85–96, Jan. 1986.

[3] ——, "Optimum multi-user signal detection," Ph.D. dissertation, Dep. Elec. Comput. Eng., Univ. Illinois, Urbana–Champaign, Coordinated Sci. Lab., Urbana, IL, Rep. T-151, Aug. 1984.

[4] ——, "Optimum multiuser asymptotic efficiency," *IEEE Trans. Commun.,* vol. COM-34, pp. 890–897, Sept. 1986.

[5] K. L. Chung, *A Course in Probability Theory,* 2nd ed. New York: Academic, 1974.

[6] C. W. Helstrom, *Statistical Theory of Signal Detection,* 2nd ed. London: Pergamon, 1968.

H. Vincent Poor (S'72–M'77–SM'82–F'87) was born in Columbus, GA, on October 2, 1951. He received the B.E.E. (with highest honor) and M.S. degrees in electrical engineering from Auburn University, Auburn, AL, in 1972 and 1974, respectively, and the M.A. and Ph.D. degrees in electrical engineering and computer science from Princeton University, Princeton, NJ, in 1976 and 1977, respectively.

Since 1977 he has been with the University of Illinois, Urbana-Champaign, where, since 1984, he has been Professor of Electrical and Computer Engineering and Research Professor in the Coordinated Science Laboratory. He was an Academic Visitor in the Department of Electrical Engineering at London University's Imperial College of Science and Technology in 1985, and a Visiting Professor in the Department of Electrical Engineering, University of Newcastle, NSW, Australia in 1987. His research interests are primarily in the general area of statistical signal processing in communications and control, with emphasis on the particular problems of robustness in signal detection, estimation, and filtering; processing techniques for multiple-access communications; quantization of inferential data; and modeling of nonstandard noise environments. He is the author of numerous publications in these areas including the book *An Introduction to Signal Detection and Estimation* (Springer-Verlag, 1987), a coeditor (with I. F. Blake) of the volume *Communications and Networks: A Survey of Recent Advances* (Springer-Verlag, 1986), and the Editor of the JAI Press series *Advances in Statistical Signal Processing.* He is also a member of the Editorial Board of the new Springer-Verlag journal, *Mathematics of Control, Systems and Signals.* He serves regularly as a consultant in the areas of digital communications, signal processing, and radar systems for several industrial and governmental clients.

Dr. Poor is a member of Phi Kappa Phi, Tau Beta Pi, Eta Kappa Nu, and Sigma Xi. He has served in several official capacities in various local, regional, and international IEEE organizations. In 1981–1982 he was Associate Editor for Estimation of the IEEE TRANSACTIONS ON AUTOMATIC CONTROL, and in 1982–1985 he was Associate Editor for Detection and Estimation of the IEEE TRANSACTIONS ON INFORMATION THEORY. He is currently serving as a member of the Board of Governors of the IEEE Information Theory Group, of which he was Treasurer in 1983–1985. Last year, he served as Program Chairman for the 25th IEEE Conference on Decision and Control held in Athens, Greece, in December 1986, and he will serve as General Chairman of the 1989 American Control Conference.

Sergio Verdu (S'80–M'84) was born in Barcelona, Catalonia, Spain, on August 15, 1958. He received the Telecommunications Engineer degree (first in class) from the Polytechnic University of Barcelona in 1980 and the M.S. and Ph.D. degrees in electrical engineering from the University of Illinois, Urbana-Champaign, in 1982 and 1984, respectively.

From 1981 to 1984 he was a Research Assistant in the Coordinated Science Laboratory of the University of Illinois. In September 1984 he joined the faculty of Princeton University, Princeton, NJ, where he is an Assistant Professor of Electrical Engineering. His current research interests are in the areas of multiuser communication and information theory and detection and estimation.

Dr. Verdu is a recipient of the National University Prize of Spain, of an IBM Faculty Development Award, and of the 1987 Rheinstein Outstanding Junior Faculty Award from the School of Engineering and Applied Science of Princeton University. He is a member of Sigma Xi and other professional societies.

SECTION 1.4
ACQUISITION AND TRACKING OF PN SEQUENCE

The synchronization of the receiver-generated pseudonoise (PN) sequence to that of the incoming (received) one is one of the main task of the receiver. The process of synchronization is usually done in two steps: acquisition, which is a coarse synchronization, and tracking, which is a fine tuning. Acquisition attempts to align the locally generated PN sequence and the incoming one to within a specified range, normally within one chip time. After acquisition the tracking process is initiated to bring the phase difference to zero.

Traditionally acquisition involves correlation of a locally generated PN sequence with the incoming signal (which is noisy). When the phase difference is small, the correlation is high. Otherwise, the correlation is low. Basically, the acquisition subsystem searches for a local PN phase to identify one with a high correlation value, which corresponds to a phase difference within a specified range. To this end, the range of the phase is divided into a finite number of intervals, which are searched using one of several possible search strategies: serial search, parallel search, Z-search, or a hybrid scheme. Within each phase, the correlation may be performed with a pre-fixed integration duration (called fixed-dwell scheme), with multiple integrations (called multiple-dwell scheme), or with an integration duration which depends on the result of correlation (called sequential scheme).

The first four papers describe various acquisition schemes. The first of these, "*A unified approach to serial search spread-spectrum code acquisition-Part I: General Theory*" by Polydoros, discusses the use of flow graph technique to compute the mean and variance of the time to reach acquisition in serial search acquisition schemes. The second paper, "*Direct-Sequence Spread-Spectrum Parallel Acquisition in a Fading Mobile Channel*" by Sourour and Gupta, analyzes parallel acquisition when the channel is non-fading and Rayleigh fading. The third paper, "*Noncoherent Sequential Acquisition of PN Sequences for DS/SS Communications with/without Channel Fading*" by Tantaratana et al., describes a design method for sequential acquisition with serial search and showed that the acquisition time is reduced, compared to serial search with fixed duration of correlation. The fourth paper, "*A Closed-Loop Coherent Acquisition Scheme for PN Sequences Using an Auxiliary Sequence*" by Salih and Tantaratana, proposes the use of a different sequence (instead of the PN sequence) to correlate with the incoming signal in order to decide whether the phase difference is within the specified range. The sequence used is easily generated from the PN sequence. The use of the auxiliary sequence helps reduce the average acquisition time.

The last two papers in this subsection talks about tracking. The first of them, "*Decision-Directed Coherent Delay-Lock Tracking Loop for DS-Spread-Spectrum Signals*" by Gaudenzi and Luise, presents a decision-directed delay-lock tracking loop. The second paper, "*A New Tracking Loop for Direct Sequence Spread Spectrum Systems on Frequency-Selective Fading Channels*" by Sheen and Stuber, proposes a tracking loop which exploits the inherent multipath diversity of the channel.

A Unified Approach to Serial Search Spread-Spectrum Code Acquisition—Part I: General Theory

ANDREAS POLYDOROS, MEMBER, IEEE, AND CHARLES L. WEBER, SENIOR MEMBER, IEEE

Abstract—**The purpose of this two-part paper is threefold: 1) Part I discusses the code-acquistion problem in some depth and 2) also provides a general extension to the approach of analyzing serial-search acquisition techniques via transform-domain flow graphs; 3) Part II illustrates the applicability of the proposed theoretical framework by evaluating a matched-filter (fast-decision rate) noncoherent acquisition receiver as an example.**

The theory is formulated in a general manner which allows for significant freedom in the receiver modeling. The statistics of the acquisition time for the single-dwell [2], [3] and N-dwell [5] systems are shown to be special cases of this unified approach.

I. Introduction

ACCURATE synchronization plays a cardinal role in the efficient utilization of any spread-spectrum system. Typically, the process of synchronization between the spreading (incoming) pseudonoise code and the local despreading (receiver) code is performed in two steps: first, code acquisition, then tracking via one of the available code-tracking loops. The focus here is on the first part. In other words, *acquisition* is a process of successive decisions wherein the ultimate goal is to bring the two codes into coarse time alignment within one code-chip interval.

The purpose of the present two-part paper is threefold: 1) Part I discusses in some depth the overall acquisition concept and 2) provides a general extension to a previously well-known analytical approach (originally suggested in [2]), namely, the state diagram or flow graph technique; 3) Part II illustrates the applicability of the proposed theoretical framework by evaluating a matched-filter (fast-decision-rate) noncoherent acquisition receiver structure, whereupon a number of conclusions will be drawn.

The acquisition problem has attracted considerable research in the recent past (see [1]-[13], [18], [19] and the references therein). Since we are mostly interested in low signal-to-noise-ratio (SNR) environments (in the presence of strong interfering signals or deep noise, for example), we shall consider serial-search techniques [1]-[12] exclusively, as opposed to sequential-estimation techniques [18], [19] whose merits decrease with decreasing SNR, or any *maximum a posteriori probability* technique [16], [17] whose complexity is prohibitive.

The present theory, which is based strictly on transform-domain techniques, is formulated in a manner general enough to encompass past results [1]-[3] and more recent ones [4]-[7] as well as to allow for significant freedom when modeling the receiver structure. Thus, it accounts for arbitrary choices for the following: 1) cell logic (verification mode), 2) search logic (serial-search strategy), 3) prior information, and 4) the form of spectral spreading (direct-sequence [DS] or frequency-hopping [FH]). For exemplary purposes, Part I shows the results pertaining to a straight serial search only (see Section II) of a DS code-acquisition system with fixed dwell times. We note that Holmes and Woo [8], Weinberg [9], and Brown [10] have used combinatorial arguments to derive expressions for a more sophisticated serial-search strategy that can be regarded as a member of the class of "expanding-window" techniques. In a forthcoming paper, we shall indicate how the flow graph framework proposed herein can be generalized to provide results for any arbitrary serial-search strategy (see Fig. 1), such as Z-search, expanding-window, etc., in a unified manner.

Part I is organized as follows. Section II classifies the possible acquisition receiver structures and discusses various performance criteria. Section III provides an elaborate description of the straight serial-search acquisition procedure and defines pertinent notation. In Section IV, a generalized transform-domain flow graph approach to the acquisition problem is introduced and results are shown for the particular search considered; examples are also included which rederive previously published results as special cases. Finally, Section V offers a brief discussion on the possible extensions of the theory.

II. Receiver Structures and Performance Measures

A number of sources contribute to the randomness of the acquisition process, including 1) initial uncertainty about the code-phase offset, 2) channel distortion (e.g., fading channels) and additive interference (intentional or unintentional), 3) possible presence of random data, 4) unknown carrier phase (noncoherent receivers) and, possibly, carrier frequency offset (Doppler), and 5) front-end receiver additive white Gaussian noise (AWGN). Another, less evident source of randomness stems from the partial correlation between the received code and the local replica [21] in *direct-sequence spread-spectrum* (DS/SS) systems.

The set of design parameters for the acquisition procedure includes threshold settings, correlation times, number of tests per code chip, and system complexity as manifested by the choice of search strategy, verification logic, etc. Also implicit is the knowledge of important parameters such as the design SNR, code rate, code length, code uncertainty region, reset penalty time[1] and others. On the other hand, specification of satisfactory performance measures for the overall system is a more complicated task and very dependent on the particular application. However, since the dominant parameter of interest in most cases is the time which elapses prior to acquisition T_{acq}, one can distinguish two basic scenarios.

1) The case where, although the fastest possible acquisition is desired, no absolute time limit or termination time T_s exists on the acquisition time T_{acq}. This situation arises when data

Paper approved by the Editor for Communication Theory of the IEEE Communications Society for publication without oral presentation. Manuscript received June 1, 1982; revised September 22, 1983. This work was supported in part by the Office of Naval Research under Grant N00014-92-K-0328, by Axiomatix Corporation, and by the Army Research Office under Grant DAAG29-79-C-0054.

A. Polydoros is with the Department of Electrical Engineering, University of Southern California, Los Angeles, CA 90089, and with Axiomatix Corporation, Los Angeles, CA 90045.

C. L. Weber is with the Department of Electrical Engineering, University of Southern California, Los Angeles, CA 90089.

[1] By *reset penalty time* T_r, we mean the time required to realign the codes to a particular phase offset after an unsuccessful sweep of the entire uncertainty region.

0090-6778/84/0500-0542$01.00 © 1984 IEEE

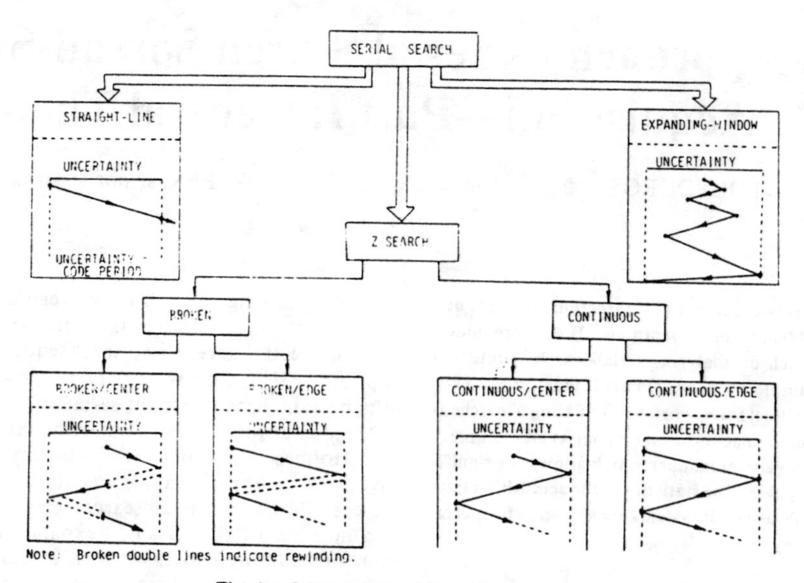

Fig. 1. Structure of serial-search strategies.

are always present in the received waveform as, for example, in the TDRSS-to-Space-Shuttle spread-spectrum link.

2) In certain systems where data transmission starts after a certain time interval T_s from the initial system turn-on, it is imperative that code acquisition be performed in that interval with very high probability; otherwise, communication is impossible. "Push-to-talk" spread-spectrum systems are examples where a fixed time is allotted to acquisition.

The above two possibilities constitute a first partition of the class of acquisition receivers. Of course, in both scenarios, the measure of performance is statistical in nature and is imbedded in the knowledge of the probability distribution function $F_{acq}(t)$ of the random time T_{acq}. Having $F_{acq}(t)$, one can then derive and optimize any meaningful performance parameter, such as the mean acquisition time $E\{T_{acq}\}$ (case 1) or the probability of prompt acquisition $\Pr\{T_{acq} \leq T_s\}$ (case 2).

Another important measure of system performance is the *overall probability of missing the code* $P_M{}^{ov}$ or its complement, *overall probability of detection* $P_D{}^{ov} = 1 - P_M{}^{ov}$. Clearly, if the acquisition process terminates only when the acquisition cell is identified, the detection probability $P_D{}^{ov}$ is one. A simple counterexample to that is case 2 above, when a stopping time exists. In the absence of other system-deadlock situations, $P_D{}^{ov}$ is exactly equal to $\Pr\{T_{acq} \leq T_s\}$. However, in addition to finite search time, another deadlock can sometimes be created due to a false-alarm situation (FA). By *false alarm*, we indicate the case where the acquisition mechanism erroneously decides that code synchronization has occurred and fine tracking via a code-tracking loop is initiated. The time required for the tracking loop to indicate false lock and the acquisition system to resume search is a random variable and can be modeled as such. In this case, we refer to the false-alarm state as a *returning state* associated with a *random penalty time* T_p. For computational ease, T_p can also be modeled as a fixed, known time [2], [4], [5]. In some other cases, however, the delay involved in initiating the tracking loop is catastrophic for the system operation. FA then corresponds to an *absorbing state*, whereupon reaching it results in complete loss of code acquisition (in other words, a final miss). In the presence of both *system-deadlock* situations, namely, finite acquisition stopping time T_s *and* absorbing false-alarm state, $P_M{}^{ov}$ is the sum of the two probabilities:

$$P_M{}^{ov} = \Pr\{\text{FA before } T_s\} + \Pr\{\text{neither FA nor ACQ before } T_s\}$$

where ACQ indicates the *correct acquisition-absorbing state*. Similarly, for case 1 (no stopping time), but with absorbing FA, it follows that $P_M{}^{ov} = \Pr\{\text{FA occurs some time}\}$.

The basic unit in any acquisition receiver is the decision-making device (detector). This can be classified as either *coherent* or *noncoherent*, as well as according to the specific statistical testing philosophy employed for each cell (offset) test, i.e., *Bayes, Neyman-Pearson,* or *other*. Furthermore, the dwell times[2] can be either *fixed* or *variable* (random), the latter case pertaining to the well-known sequential tests [20]. In addition, the first detection (testing) level may or may not be followed by a *verification mode* (logic) which is sometimes used to ensure the initial positive (synchronization) indication and prevent eventual false alarm. The penalty for that extra statistical security is, of course, additional average acquisition time. The presence of a verification mode is usually denoted by the nomenclature "multiple-dwell-time detectors" as opposed to "single-dwell-time detectors" for no verification. Single-dwell receivers can employ either full-period (FPC) or partial-period (PPC) correlation. In contrast, multiple-dwell detectors almost invariably employ short-time PPC at the first stages in order to expedite the acquisition process since the negative statistical effects of PPC can be removed at the higher stages of the verification mode. Such a verification structure can employ *immediate-rejection logic* or *nonimmediate-rejection logic*. The detectors considered in [5] and [3] can be viewed as examples of the former and latter categories, respectively. Finally, multiple-dwell systems can have a first-stage detector which employs either *passive* or *active* integration. A *hybrid* structure with a matched-filter (passive) first stage and active integration, majority-logic verification will be described and analyzed in the companion paper (Part II).

The remaining important aspect of the acquisition process is the *search strategy*, by which one denotes the particular procedure adopted by the receiver in its search through the uncertainty region.[3] Sweeping through the time uncertainty can be done either *continuously* or in *discrete steps*, the latter being the one we shall consider here. Thus, we shall assume that the time uncertainty region has been quantized into a finite number of elements (*cells*), through which the receiver is

[2] Dwell times are the time intervals allotted to each decision.

[3] Although this uncertainty is two-dimensional in nature (time/frequency), we shall be primarily concerned here with the time uncertainty; its frequency counterpart will be briefly discussed in Section V.

stepped. Which particular search strategy is selected by the receiver is very dependent on the nature of the uncertainty region, available prior information, statistical quality of the tests performed, availability of stepping and rewinding mechanisms, etc. A multitude of such search algorithms is shown in Fig. 1. If the uncertainty region covers the whole code period, a straight-line search would be appropriate. In this case, the cell following the last one in the uncertainty region is the exact cell from which the search was first started. On the other hand, prior information is sometimes available,[4] which suggests the use of more sophisticated strategies (Z-search or expanding-window search) for better results. Nonetheless, all the above search procedures share a common feature in that they are serial-search strategies; namely, all of them start from a specific cell and serially examine the remaining cells in some direction and in a prespecified order until the correct cell is found. Hence, serial-search techniques do not account for any additional information gathered during the past search time that could conceivably be used to alter the direction of search towards cells showing an increased *a posteriori* likelihood [16], [17]. The theoretical suboptimality of serial search, when compared to maximum *a posteriori* techniques, is compensated for by the relative simplicity of its implementation. Finally, let us note that all the above techniques–serial or not–base their individual decisions on likelihood ratios and, thus, are fundamentally different from the sequential-estimation techniques[5] [18] whose merit decreases with decreasing SNR.

The following theory provides a transform-domain framework for fixed dwell, straight-serial-search DS code acquisition with arbitrary verification logic schemes and arbitrary prior distributions. Extensions to random-dwell and/or frequency-hopping acquisition are briefly discussed in Section V, while a comprehensive study of arbitrary serial-search strategies (see Fig. 1) that generalizes the present concepts is the subject of a forthcoming paper.

III. Straight-Serial-Search Acquisition Procedure

We shall now describe and introduce notation pertaining to a generic straight-serial-search DS acquisition scheme. Let $c(t + \zeta T_c)$ be a ± 1-valued,[6] L-chip-long spreading code with chip time T_c seconds, delayed by ζT_c with respect to an arbitrary time reference. The signal $r(t)$ at the front end of the acquisition receiver contains the modulated code (possibly with data) plus noise, i.e.,

$$r(t) = \sqrt{2S}\, d(t) c(t + \zeta T_c) \cos(\omega_0 t + \omega_d t + \theta) + n(t). \tag{1}$$

In (1), ω_0 and θ are the nominal carrier radian frequency and random phase, respectively, S is the transmitter signal power, $n(t)$ represents AWGN with one-sided power spectral density (PSD) N_0 W/Hz, $d(t)$ is the data sequence, and ω_d is the Doppler radian frequency. In the ensuing discussion, we shall let $d(t) = 1$ and $\omega_d = 0$, namely, we shall set aside the effects of data present[7] and Doppler offset until considering them in Part II. Suffice it to say here that they account for implementation losses which should eventually be incorporated into the final system design.

Acquisition consists of aligning the (unknown) phase of the incoming code with the known phase ζT_c of a local identical PN code generated at the receiver to within a chip-time interval. Specifically, it is desired to position the local code such that the absolute phase offset normalized by T_c, $|p| \triangleq |\zeta - \tau|$, is less than 1. Having performed the above successfully, the coarse synchronization or acquisition mode terminates and fine synchronization or tracking begins.

In the following, H_1 denotes the hypothesis that $|p| < 1$, i.e., the codes are in synchronization (often designated as a "hit") and H_0 denotes the alternative hypothesis $|p| \geqslant 1$. Let Θ represent the total number of chips in the uncertainty region of T_u seconds, i.e., $\Theta \triangleq T_u/T_c$. The receiver, setting the local code at the beginning of the uncertainty region, serially examines the possible positions (phases) of the incoming code by properly mixing (correlating) the two codes. The exact number of positions to be examined depends on Θ and the quantization level of the uncertainty region which determines the size of the advancing step of the local code-updating mechanism. Letting q and Δ represent the number of positions (cells) and the advancing step size, respectively (the latter normalized to a fraction of the chip time, i.e., $\Delta = 1$ or $1/2$ or $1/4$, etc.), it follows that $q = \Theta/\Delta$. It also follows that, depending on the actual value of the offset p, up to either $2\Delta^{-1}$ or $2\Delta^{-1} - 1$ tests will be performed under hypothesis H_1.

When the receiver decision for a specific code offset is negative (H_0), the phases of the two codes (incoming and local) are automatically adjusted to the next offset position and the test is repeated. Which of the two code phases is physically relocated (advanced) with respect to the other is simply a matter of system design. For passive correlation (matched filtering), the incoming code phase moves continuously with respect to the fixed local code segment, while the opposite is also possible in active implementations. On the other hand, a tentative positive (H_1) synchronization indication from the detector does not usually result in immediate tracking, i.e., a verification mode almost always follows. For the specified interval of verification time, the two codes run in parallel at some fixed phase offset while a number of statistical tests are conducted. A final positive decision activates the tracking loop; otherwise, the code phases are readjusted and the cell-by-cell format is reinitiated.

As mentioned before, in the case where T_u equals the code period LT_c and the code has not been found in all q positions, the serial search can continue in a straight line and repeat the entire uncertainty region until synchronization is declared. A simple modification of the above, which includes a rewinding mechanism, is discussed in the companion paper (Part II). Fig. 2 summarizes the salient features of the acquisition procedure described above through a generic model. Note the conceptual difference between the cell logic (verification mode) and the search logic (search strategy).

IV. The Flow Graph Technique in Acquisition

The duality which exists between the state transition diagram of a discrete-time Markov process and the flow graphs of electrical systems was first pointed out and exploited in [22] and [23]. Holmes and Chen [2] suggested using the flow graph technique in the serial-dwell acquisition problem since the underlying process of the serial search with fixed dwell times is of the aforementioned Markovian nature. However, the flow graph developed in [2] and also used in [4], [5] is complicated and does not easily lend itself to generalizations. Here we consider the minimum number of states required to model the process, namely, $\nu + 2$. Of the total number of $\nu + 2$ states, $\nu - 1$ correspond to the offsets (cells) belonging to hypothesis H_0, while one state corresponds to the *collective state* H_1. Equivalently, any offset with $|p| < 1$ is included in the νth state H_1. We index these ν states in a circular arrangement, with the ith state ($i = 1, 2, \cdots, \nu - 1$) corresponding to the ith

[4] For example, through approximate transmitter/receiver timing coordination or the use of short preamble synchronizing codes [13].

[5] Note that sequential-estimation techniques [18] are not related to sequential detection [20].

[6] Generalizations to arbitrary chip pulse shapes are straightforward.

[7] In certain system operational designs, data are absent *a fortiori* during acquisition (see Section II).

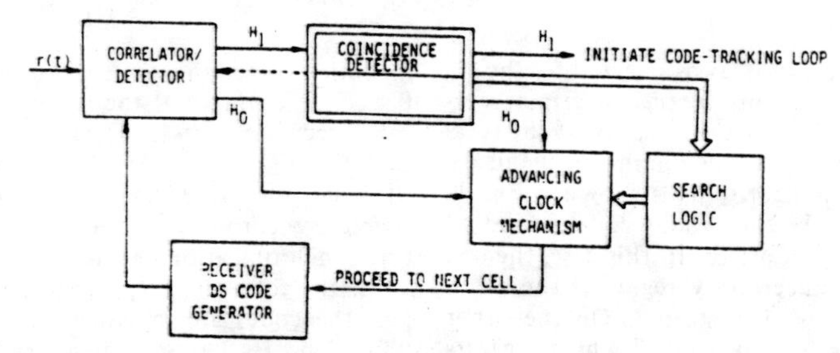

Fig. 2. Generic model for the serial-search code-acquisition receiver.

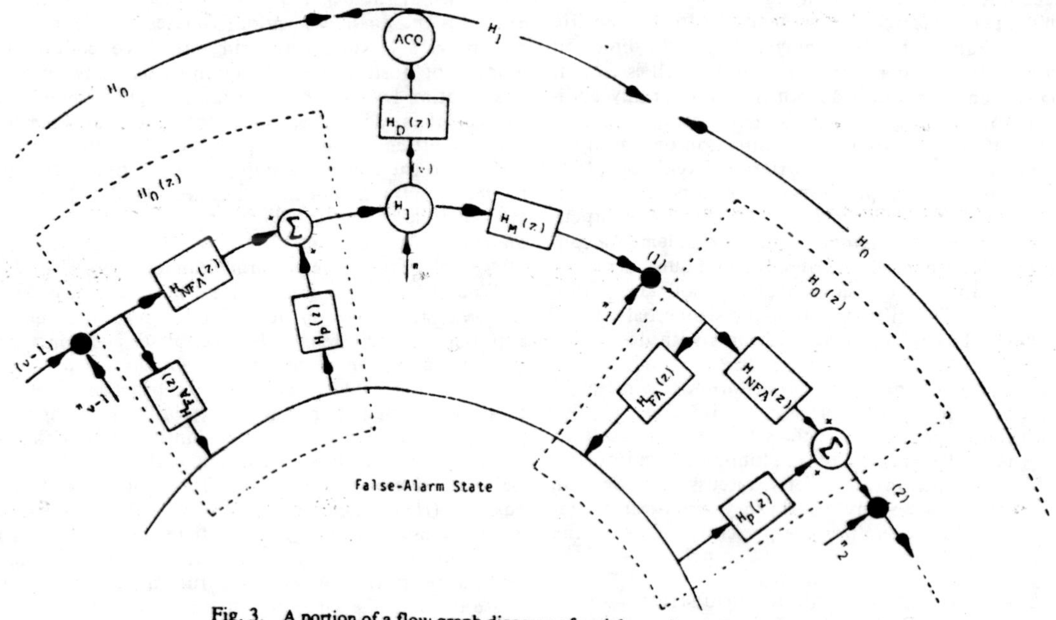

Fig. 3. A portion of a flow graph diagram of serial search acquisition with an arbitrary *a priori* distribution $\{\pi_j\}$ and various polynomial gains.

offset position to the right of H_1. A segment of that circle, including the states numbered $\nu - 1$, ν (or H_1), 1, and 2, is shown in Fig. 3. In terms of the notation in Section III, ν would equal either $1 + (q - 2\Delta^{-1})$ or $1 + (q - 2\Delta^{-1} + 1)$, depending on the actual offset p. The two remaining states are the *correct-acquisition* (ACQ) and *false-alarm* (FA) states, as shown in Fig. 3.

Entry into the search process can occur at any one of the ν states, according to some *a priori* distribution (π_j, $j = 1, 2, \cdots, \nu$) which reflects the designer's confidence about the initial relative position of the codes. Total uncertainty (uniform-prior distribution) ($\pi_j = 1/\nu$; $j = 1, 2, \cdots, \nu$) and worst-case location ($\pi_1 = 1$, $\pi_j = 0$; $j \neq 1$) considerations will then result as special cases.

Let $p_{ij}(n)$ indicate the probability that the Markov process will move from state i to state j in n steps and let z indicate the unit-delay operator. If the unit delay specifically corresponds to τ seconds, z is replaced by z^{τ} in the following. Sittler [22] has shown that the state transition diagram can be mapped to its equivalent flow graph if each transition branch from i or j in the Markovian diagram is assigned a gain equal to $p_{ij}z$, where $p_{ij} \triangleq p_{ij}(1)$ is the one-step transition probability and z represents the unit delay associated with that transition. Furthermore, if we define the generating function

$$\mathrm{P}_{ij}(z) \triangleq \sum_{n=0}^{\infty} p_{ij}(n) z^n$$

we note that $\mathrm{P}_{ij}(z)$ represents the transfer function from node i to node j of the flow graph. $\mathrm{P}_{ij}(z)$ is useful because it contains statistical information about the Markovian process. One can exploit systems concepts and derive $\mathrm{P}_{ij}(z)$ through flow graph reduction methods or Mason's formula.

In an effort to extend the previous notions, let us assign more general gains $H(z)$ to the different branches of our model in Fig. 3 as follows: $H_D(z)$ is the gain of the branch leading from node H_1 (νth node) to the node ACQ; $H_M(z)$ is the gain of the branch connecting H_1 with node 1 while $H_0(z)$ is the gain of the branch connecting any other two successive nodes $(i, i+1)$; $i = 1, \cdots, \nu - 1$. Furthermore, the process can move between any two successive nodes $(i, i+1)$ with $i \neq \nu$ either without false alarm [associated gain $H_{NFA}(z)$] or by first reaching the FA state [branch gain $H_{FA}(z)$], then pass from FA to node $i + 1$ [branch gain $H_P(z)$] so that

$$H_0(z) = H_{NFA}(z) + H_{FA}(z)H_P(z).$$

It should be emphasized that the gains just described include all possible paths by which the process can move along the "generalized" branch associated with that gain. So $H_{NFA}(z)$ models all paths between successive H_0 states (such as *partial* false alarm, for instance) which do not lead to the false tracking-loop initiation; the latter path is modeled by $H_{FA}(z)$. Similarly, both $H_D(z)$ and $H_M(z)$ include all paths leading to successful acquisition or miss, respectively. It is easily seen that those gains almost always contain more than one path each. As an example, consider the H_1 region flow graph of a full-chip advancing step-size receiver ($\Delta = 1$). The H_1 region will then contain two cells.[8] say, Q_1 and Q_2: cell Q_1 will correspond to a code offset $|p| < 1$ while Q_2 will correspond to the next offset $1 \leq |p| < 1$, still in H_1. If we let $H_D{}^{Q_i}(z)$, $H_M{}^{Q_i}(z)$; $i = 1, 2$ denote the individual detection and miss gains of each cell, we can then write the appropriate total gains as

$$H_D(z) = H_D{}^{Q_1}(z) + H_M{}^{Q_1}(z)H_D{}^{Q_2}(z) \tag{2a}$$

and

$$H_M(z) = H_M{}^{Q_1}(z)H_M{}^{Q_2}(z). \tag{2b}$$

The general functional form of these gains allows one to model a variety of verification modes (logic) of the detection unit. Examples of such different kinds of logic can be found in [2], [3], [5], [8], [12]. The present theory provides a framework for dealing with all those systems in a unified manner.

From the flow graph of Fig. 3, we want the generating function

$$\mathrm{P}_{ACQ}(z) \triangleq \sum_{n=0}^{\infty} p_{ACQ}(n)z^n \tag{3a}$$

associated with the probabilities $p_{ACQ}(n)$ that, starting from *some* initial state with a given prior probability, the process will reach ACQ in n steps. From the well-known transition property of Markov processes and the dual linearity property of the corresponding flow graph, it follows that

$$\mathrm{P}_{ACQ}(z) = \sum_{i=1}^{\nu} \pi_i \mathrm{P}_{i,ACQ}(z). \tag{3b}$$

Taking the structure of Fig. 3 into account,

$$\mathrm{P}_{ACQ}(z) = H_D(z) \sum_{i=1}^{\nu} \pi_i \mathrm{P}_{i\nu}(z). \tag{3c}$$

The transfer function $\mathrm{P}_{i\nu}(z)$ can be found from Fig. 3 by the loop-reduction method. Substituting into (3), it follows that

$$\mathrm{P}_{ACQ}(z) = \frac{H_D(z)}{1 - H_M(z)H_0{}^{\nu-1}(z)} \sum_{i=1}^{\nu} \pi_i H_0{}^{\nu-i}(z). \tag{4}$$

This is the basic result for the case of arbitrary $\{\pi_i\}$. The two special cases, namely, uniform and worst-case prior distribution, reduce to the following:

$$\mathrm{P}_{ACQ}(z) = \frac{H_D(z)H_0{}^{\nu-1}(z)}{1 - H_M(z)H_0{}^{\nu-1}(z)}, \quad \text{worst case} \tag{5}$$

and

$$\mathrm{P}_{ACQ}(z) = \frac{1}{\nu} \frac{H_D(z)(1 - H_0{}^{\nu}(z))}{(1 - H_M(z)H_0{}^{\nu-1}(z))(1 - H_0(z))}, \quad \text{uniform}. \tag{6}$$

If the complete statistical description of the acquisition process is desired, one should substitute the proper expressions for the different gains involved (see the examples at the end of this section), then expand the resulting expression into a power series in z by one of the methods described in [22]. Except for a few manageable cases (e.g., [4]), obtaining the probability distribution of the acquisition time in closed form is beyond the ability of these authors. A truncated distribution $p(n)$: $0 \leq n \leq N_{max}$ for a large N_{max} can be obtained numerically through a discrete Fourier transform (DFT) or a fast Fourier transform (FFT) algorithm since the generating function $\mathrm{P}(z)$, evaluated on the unit cycle $z = e^{j\omega}$, is nothing but the DFT of $\{p(n)\}$.

Alternatively, if the detection probability $P_D{}^{ov}$ and/or the first few moments of T_{acq} (e.g., the mean acquisition time $E\{T_{acq}\}$ and the variance var $\{T_{acq}\}$) constitute sufficient statistical information about the acquisition process, then (4)-(6) can be of direct value. Indeed, the probability of detection $P_D{}^{ov}$ for the no-stopping-time case is given by

$$P_D{}^{ov} = \Pr\{\text{reaching ACQ at any time}\} = \mathrm{P}_{ACQ}(z)|_{z=1}. \tag{7a}$$

From (3), the mean and variance of the acquisition time can be expressed as

$$E\{T_{acq}\} = \left.\frac{d\mathrm{P}_{ACQ}(z)}{dz}\right|_{z=1} \tag{7b}$$

and

$$\mathrm{var}\{T_{acq}\} = \left[\frac{d^2\mathrm{P}_{ACQ}(z)}{dz^2} + \frac{d\mathrm{P}_{ACQ}(z)}{dz} - \left(\frac{d\mathrm{P}_{ACQ}(z)}{dz}\right)^2\right]_{z=1}. \tag{7c}$$

Higher moments are similarly obtained. The mean $E\{T_{acq}\}$ for the two particular distributions of (5) and (6) are derived by substitution into (7b), from which we obtain

$$\underset{\text{(worst case)}}{E\{T_{acq}\}} = \begin{cases} \dfrac{P_D{}^{ov}}{H_D(1)}\Big\{H_D'(1) + H_M'(1)P_D{}^{ov} + (\nu - 1)\dfrac{H_0'(1)}{H_0(1)} \\ \quad \cdot (H_D(1) + H_M(1)P_D{}^{ov})\Big\} & \text{if } H_0(1) < 1 \\ \dfrac{1}{H_D(1)}[H_D'(1) + H_M'(1) + (\nu - 1)H_0'(1)] & \text{if } H_0(1) = 1 \end{cases}$$

and

$$\tag{8}$$

$$\underset{\text{(uniform)}}{E\{T_{acq}\}} = \begin{cases} \dfrac{P_D{}^{ov}}{H_D(1)}\Big\{H_D'(1) + \dfrac{\nu H_0'(1)H_0{}^{\nu-2}(1)}{1 - H_0{}^{\nu}(1)} \\ \quad \cdot \Big[-H_D(1)H_0(1) + P_D{}^{ov}\Big(\dfrac{1}{H_0{}^{\nu-2}(1)} \\ \quad + \nu H_M(1) + \dfrac{H_M'(1)}{H_0'(1)}(1 - H_0(1))H_0(1) \\ \quad - H_M(1)(1 + H_0(1))\Big)\Big]\Big\} & \text{if } H_0(1) < 1 \\ \dfrac{1}{H_D(1)}\Big[H_D'(1) + H_M'(1) + (\nu - 1)H_0'(1) \\ \quad \cdot \Big(1 - \dfrac{H_D(1)}{2}\Big)\Big] & \text{if } H_0(1) = 1. \end{cases} \tag{9}$$

[8] The only exception is when $p = 0$; however, this occurs with zero probability.

In (8) and (9), the primed quantities denote differentiation with respect to z, i.e., $H'(1) \triangleq dH(z)/dz$ at $z = 1$. Furthermore, in the case where $H_0(1) = 1$, note that (4) and (7a) combine to yield

$$P_D^{ov} = \frac{H_D(1)}{1 - H_M(1)} = 1$$

since, from Fig. 3, it is always true that $H_D(1) + H_M(1) = 1$. Hence, the case $H_0(1) = 1$ effectively signifies the existence of only one absorbing state, namely, ACQ.

We now address the limited-acquisition-time case. Let n_{max} denote the maximum number of transitions which the acquisition process is able to perform before the time limit of T_s seconds is reached. For example, if τ_D is the minimum dwell time, $n_{max} = \lfloor T_s/\tau_D \rfloor$, where $\lfloor \ \rfloor$ means "the integer part." It then follows that the probability of detection $P_D^{ov} \triangleq \Pr\{T_{acq} \leqslant T_s\}$ is equal to $F_{ACQ}(n_{max})$, where $F_{ACQ}(n)$ is the cumulative probability distribution function of the acquisition time. The generating function $\mathrm{F}_{ACQ}(z)$ of $F_{ACQ}(n)$ is easily shown to relate to $\mathrm{P}_{ACQ}(z)$ as $\mathrm{F}_{ACQ}(z) = (1 - z)^{-1}\mathrm{P}_{ACQ}(z)$. Therefore, in this case, a closed-form expression for P_D^{ov} is

$$P_D^{ov} = \Pr\{T_{acq} \leqslant T_s\} = F_{acq}(n_{max}) = \frac{1}{2\pi j} \oint_{\mathcal{D}} \frac{\mathrm{P}_{ACQ}(z)}{(1-z)z^{n_{max}+1}}\, dz. \tag{15}$$

In (15), $\mathrm{P}_{ACQ}(z)$ is given by (4), $j = \sqrt{-1}$ and $\mathcal{D}$ is a counterclockwise-closed contour in the region of convergence of $\mathrm{F}_{ACQ}(z)$ and encircles the origin of the z-plane. Let us note that the above equation (15) for P_D^{ov} is valid both when the FA state is absorbing and when it is not, since that information is imbedded in the expression for $\mathrm{P}_{ACQ}(z)$. It is well known that convergence of $\mathrm{P}_{ACQ}(z)$ occurs inside or on the unit circle $|z| = 1$; however, due to the factor $(1 - z)^{-1}$, the convergence region of $\mathrm{F}_{ACQ}(z)$ excludes the point $z = 1$. Hence, if we choose $\mathcal{D}$ to be a circle of radius $R < 1$ so that $z = Re^{j\theta}$, it follows from (15) that

$$P_D^{ov} = \frac{1}{\pi R^{n_{max}}} \int_0^{\pi} \mathrm{Re}\left\{\frac{\mathrm{P}_{ACQ}(Re^{j\theta})}{(1 - Re^{j\theta})e^{jn_{max}\theta}}\right\} d\theta \tag{16}$$

since the integral in $(0, \pi)$ is the complex conjugate of that in $(\pi, 2\pi)$: here $\mathrm{Re}\{\cdot\}$ denotes "the real part of."

In many applications, good approximations to P_D^{ov} can be devised that avoid calculating (16). To illustrate the idea, let $P_{FA,c}$ denote the total false-alarm probability per H_0 cell, and let $P_{d,r}$ denote the total detection probability *per single run* of the H_1 region. So, for the previous example of an H_1 region with two cells, Q_1 and Q_2, $P_{d,r}$ would be given by

$$P_{d,r} = P_d(Q_1) + (1 - P_d(Q_1))P_d(Q_2) \tag{17}$$

where the notation employed is obvious. Furthermore, let R_{max} denote the maximum number of runs through the H_1 region that, on the average, the system can sustain before it exceeds the time limit T_s. An approximate estimate for R_{max} is the ratio $T_s/E\{T^1\}$, rounded to the closest integer, where $E\{T^1\}$ is the mean time for an *unsuccessful* sweep of the entire uncertainty region of T_u seconds. Referring back to Fig. 3, it is seen that

$$E\{T^1\} = H_M'(1) + (\nu - 1)H_0'(1) \cong \nu H_0'(1) \tag{18}$$

where the last approximation emerges because, typically, $\nu \gg 1$. It is then a simple exercise to show that the probability of detection within R_{max} runs is given by

$$P_D^{ov} = P_{d,r}(1 - P_{FA,c})^{\nu-1} \cdot \left(\frac{1 - (1 - P_{d,r})^{R_{max}}(1 - P_{FA,c})^{(\nu-1)R_{max}}}{1 - (1 - P_{d,r})(1 - P_{FA,c})^{\nu-1}}\right). \tag{19}$$

We finally proceed to illustrate the applicability of the previous theory by means of two examples.

Example 1: First consider the single-dwell system described in [2]. The prior distribution is uniform ($\pi_j = 1/\nu$) and each *cell false alarm* (with probability $P_{FA,c}$) costs $K\tau_D$ seconds, where τ_D is the dwell time per H_0 cell without false alarm. The system acquires only at a single cell (the νth) with probability $P_{d,r}$. Note that $P_{d,r}$, being the probability of detection per run, is less than 1 as opposed to the overall detection probability P_D^{ov}, which equals 1 since ACQ is the single absorbing state. In terms of the general model, define

$$H_P(z) = z^{K\tau_D}, \quad H_M(z) = (1 - P_{d,r})z^{\tau_D},$$

$$H_D(z) = P_{d,r}z^{\tau_D}$$

and

$$H_{FA}(z) = P_{FA,c}z^{\tau_D}, \quad H_{NFA}(z) = (1 - P_{FA,c})z^{\tau_D}.$$

It then follows that

$$H_0(z) = (1 - P_{FA,c})z^{\tau_D} + P_{FA,c}z^{(K+1)\tau_D}$$

in which case $H_0(1) = 1$, as expected. Calculating the derivatives of the gains and substituting into (8) and (9) yields

$$E\{T_{acq}\} = \begin{cases} P_{d,r}^{-1}\left(1 + (1 + KP_{FA,c})\left(\dfrac{\nu - 1}{2}\right)(2 - P_{d,r})\right)\tau_D \\ \quad \text{(uniform distribution)} \\ P_{d,r}^{-1}(1 + (\nu - 1)(1 + KP_{FA,c}))\tau_D \\ \quad \text{(worst-case distribution).} \end{cases} \tag{20}$$

The uniform-distribution conclusion in (20) is the result given in [2].

Example 2: Here we indicate how the generating function in [5] for an immediate-rejection multiple-dwell system can be derived and extended to an arbitrary prior distribution by using the above generic model. To match the notation in [5], let τ_i; $i = 1, 2, \cdots, N$ be the sampling instants of an N-dwell system, at each of which instants the detector output is compared to the corresponding threshold η_i; $i = 1, 2, \cdots, N$. Failure to exceed any one threshold implies cell rejection and stepping to the next test position, while acquisition is declared (correctly or falsely), only if *all* thresholds are exceeded. Let z_1 correspond to delay τ_1 and z_i correspond to the additional delay $(\tau_i - \tau_{i-1})$; $i = 2, \cdots, N$ for each extended dwell. A false alarm costs $K(\tau_N - \tau_{N-1})$ in acquisition time. The general model is also applicable here if z is replaced by the vector $\mathbf{z} = (z_1, z_2,$

$\cdots, z_N)$ and the gains are defined as follows:

$$H_P(z) = z_N{}^K, \quad H_D(z) = \prod_{i=1}^{N} P_{d_{i|i-1}} z_i = P_{d,c} \prod_{i=1}^{N} z_i$$

$$H_M(z) = \sum_{j=2}^{N} \left(\prod_{i=1}^{j-1} P_{d_{i|i-1}} z_i \right) (1 - P_{d_{j|j-1}}) z_j + (1 - P_{d_{1|0}}) z_1$$

$$H_{FA}(z) = \prod_{i=1}^{N} P_{f_{i|i-1}} z_i = P_{FA,c} \prod_{i=1}^{N} z_i$$

$$H_{NFA}(z) = \sum_{j=2}^{N} \left(\prod_{i=1}^{j-1} P_{f_{i|i-1}} z_i \right) (1 - P_{f_{j|j-1}}) z_j + (1 - P_{f_{1|0}}) z_i.$$

In the previous expressions, $P_{d_{i|i-1}}$ and $P_{f_{i|i-1}}$ denote the detection and false-alarm probabilities of the ith dwell, conditioned on the event that all $i - 1$ previous dwells have been positive (i.e., the thresholds have been exceeded), and $P_{d,c}$ and $P_{FA,c}$ denote the detection and FA probabilities per cell, namely, the product of all N aforementioned conditional probabilities. Note that if the delays $\tau_1, (\tau_2 - \tau_1), \cdots, (\tau_N - \tau_{N-1})$ have a common integer divisor $\tau \neq 1$ such that $a_1 = \tau_1/\tau$, $a_i = (\tau_i - \tau_{i-1})/\tau$, $i = 2, \cdots, N$ are all integers, then only one variable z needs to be introduced by letting $z_1 = z^{a_1 \tau}$, $z_i = z^{a_i \tau}$, $i = 2, \cdots, N$. Such a substitution would facilitate further calculations. In any case, all the previous results hold, assuming that the scalar operations are replaced by their vector counterparts. So, for example, the mean acquisition time from (7b) is given by

$$E\{T_{acq}\} = \left. \frac{\partial^N P(z_1, z_2, \cdots, z_N)}{\partial z_1 \partial z_2 \cdots \partial z_N} \right|_{z_1 = 1, \cdots, z_N = 1} \tag{21}$$

and similarly for the other expressions.

V. Discussion

The theory developed in the previous section employed flow graph, transform-domain concepts in order to suggest a unified treatment of the code-acquisition problem. By so doing, the contributions of the various factors entering the overall picture of the system (for instance, *cell logic* versus *search logic*) were effectively decoupled. Let us recall that the basic features of the theory, as presented herein, are 1) fixed dwell times, 2) DS spreading, and 3) straight-serial-search techniques. The remaining aspects were left arbitrary. Now we will comment on the extension of the theory to different acquisition systems.

The alternative to fixed dwell times is random dwell times, wherein the cell-testing format is *sequential* [14], [15]. For a well-designed sequential test, the average time to reject an H_0 cell can be made smaller than the observation time in a fixed dwell system (for the same $P_{FA,c}$ and $P_{d,c}$), thus resulting in a smaller average T_{acq}. A key observation here is that the overall T_{acq} is the sum of independent random decision times T_c for successive cells. This in turn implies that the characteristic function (CF) $\Phi_{ACQ}(\omega)$ of T_{acq} [24], which is the dual of the generating function $P_{ACQ}(z)$ for fixed times, assumes expressions identical to those derived in Section IV, where the various gains $H(z)$ should now be replaced by the CF's of the random times pertaining to the corresponding paths. For instance, $H_0(z)$ should be replaced by $\Phi_0(\omega)$, the CF of the random time to reject an H_0 cell, with or without an absorbing FA state. Thus, the overall theoretical framework remains functional, provided that the probability density functions of the various random times involved can somehow be derived (by theory or simulation). Unfortunately, this is not always easy for sequential tests, particularly for noncoherent systems with finite truncation time [14].

In an FH environment [11]-[13], under a proper transformation of system parameters and detector structures, the concepts developed remain almost intact. Since the hopping rates in practical FH systems are much smaller than the code rates in a DS system, acquisition is more easily performed in the former case. Such assessments are appropriate only in benign environments and could be upset in the presence of strong interference (such as tone jamming). Work in that direction is certainly an interesting extension.

The assumption of serial search is perhaps the strongest. Although, due to space limitations, only the straight search was considered here, similar results can be derived for the whole family of search strategies (see Fig. 1) which share the *serial* feature (see Section II). Later it will be shown that such derivations can be based on the construction of *equivalent circular state diagrams* which assimilate the straight search.

Finally, let us note that, although we analyzed only the time-uncertainty case here, the basic procedures also apply to a two-dimensional time/frequency uncertainty region which is encountered whenever the accumulated oscillator frequency drift and/or Doppler offset are significant. A quantization of the compound time/frequency region will now produce two-dimensional cells which again have to be examined serially. Acquisition can be expedited by constructing a *bank* of filters (detectors), each tuned to a different cell frequency; this would then result in a parallel-processing serial search whereby two or more cells are examined at a time. The tradeoff between cost and effectiveness for such receiver structures, which would need to incorporate some conflict-resolution logic, is interesting and worth examining in the future.

References

[1] G. F. Sage, "Serial synchronization of pseudonoise systems," *IEEE Trans. Commun. Technol.*, vol. COM-12, pp. 123-127, Dec. 1964.

[2] J. K. Holmes and C. C. Chen, "Acquisition time performance of PN spread spectrum systems," *IEEE Trans. Commun.*, vol. COM-25, pp. 778-783, Aug. 1977.

[3] P. M. Hopkins, "A unified analysis of pseudonoise synchronization by envelope correlation," *IEEE Trans. Commun.*, vol. COM-25, pp. 770-778, Aug. 1977.

[4] D. M. DiCarlo and C. L. Weber, "Statistical performance of single-dwell serial-synchronization systems," *IEEE Trans. Commun.*, vol. COM-28, pp. 1382-1388, Aug. 1980.

[5] ——, "Multiple-dwell serial acquisition of direct-sequence code signals," *IEEE Trans. Commun.*, vol. COM-31, pp. 650-659, May 1983.

[6] A. Polydoros and C. L. Weber, "Worst-case considerations for coherent serial acquisition of PN sequences," in *Proc. Nat. Telecommun. Conf.*, Houston, TX, Dec. 1980, pp. 24.6.1-24.6.5.

[7] ——, "Rapid acquisition techniques for direct-sequence spread spectrum systems using an analog detector," in *Proc. Nat. Telecommun. Conf.*, New Orleans, LA, Dec. 1981, pp. A7.1.1-A7.1.5.

[8] J. K. Holmes and K. T. Woo, "An optimum PN code search technique for a given *a priori* signal location density," in *Proc. Nat. Telecommun. Conf.*, Birmingham, AL, Dec. 1978, pp. 18.6.1-18.6.5.

[9] A. Weinberg, "Search strategy effects on PN acquisition performance,"

in *Proc. Nat. Telecommun. Conf.*, New Orleans, LA, Dec. 1981, pp. F1.5.1–F1.5.5.

[10] W. R. Brown, "Performance analysis for the expanding-search PN acquisition algorithm," *IEEE Trans. Commun.*, vol. COM-30, pp. 424–435, Mar. 1982.

[11] J. Krebser, "Performance of FH synchronizers with constant search rate in the presence of partial-band noise," presented at the Int. Zurich Sem. Commun., Zurich, Switzerland, 1980.

[12] C. A. Putnam, S. S. Rappaport, and D. L. Schilling, "A comparison of schemes for coarse acquisition of frequency-hopped spread spectrum signals," in *Proc. Int. Conf. Commun.*, Denver, CO, June 1981, pp. 34.2.1–34.2.5.

[13] C. R. Cahn, "Spread spectrum applications and state-of-the-art equipment," in *AGARD-NATO Lect. Series 58 on Spread Spectrum Commun.*, May 28–June 6, 1973, paper 5.

[14] J. J. Bussgang and D. Middleton, "Optimum sequential detection of signals in noise," *IRE Trans. Inform. Theory*, vol. IT-1, pp. 5–18, Dec. 1955.

[15] I. Selin and F. Tuteur, "Synchronization of coherent detectors," *IRE Trans. Commun. Syst.*, vol. CS-11, pp. 100–109, Mar. 1963.

[16] C. Gumacos, "Analysis of an optimum synchronization search procedure," *IRE Trans. Commun. Syst.*, vol. CS-11, pp. 89–99, Mar. 1963.

[17] E. C. Posner, "Optimal search procedures," *IRE Trans. Inform. Theory*, vol. IT-11, pp. 157–160, July 1963.

[18] R. B. Ward and K. P. Yiu, "Acquisition of pseudonoise signals by recursion-aided sequential estimation," *IEEE Trans. Commun.*, vol. COM-25, pp. 784–794, Aug. 1977.

[19] C. C. Kilgus, "Pseudonoise code acquisition using majority-logic decoding," *IEEE Trans. Commun.*, vol. COM-21, pp. 772–773, June 1973.

[20] A. Wald, *Sequential Analysis*. New York: Wiley, 1947.

[21] L. V. Sarwate and M. B. Pursley, "Cross-correlation properties of pseudo-noise and related sequences," *Proc. IEEE*, vol. 68, pp. 598–619, May 1980.

[22] R. W. Sittler, "Systems analysis of discrete Markov processes," *IRE Trans. Circuit Theory*, vol. CT-3, pp. 257–266, Dec. 1956.

[23] W. H. Huggins, "Signal-flow graphs and random signals," *Proc. IRE*, vol. 45, January 1957.

[24] W. Feller, *An Introduction to Probability Theory and Its Applications*, 2nd ed., vol. I. New York: Wiley, 1950.

Andreas Polydoros (S'76–M'82) was born in Athens, Greece, in 1954. He received the Diploma in electrical engineering from the National Technical University of Athens, Athens, in 1977, the M.S.E.E. degree from the State University of New York at Buffalo in 1979, and the Ph.D. degree from the University of Southern California, Los Angeles, in 1982.

He has been an Assistant Professor at U.S.C. and a member of the Communication Sciences Institute since 1982. He has also been affiliated with Axiomatix Corporation, Los Angeles, since 1979, involved in a variety of digital telecommunication projects with current emphasis on spread-spectrum systems. He is currently participating in the development of a textbook, *Principles of Spread-Spectrum Systems*.

Charles L. Weber (S'57–M'65–SM'78) was born in Germantown, OH, on December 2, 1937. He received the B.S. degree from the University of Dayton, Dayton, OH, in 1958, the M.S. degree from the University of Southern California, Los Angeles, in 1960, and the Ph.D. degree from the University of California at Los Angeles in 1964.

Since 1965, he has been a member of the faculty of the University of Southern California, where he is Professor of Electrical Engineering and Director of the Communication Sciences Institute. His research activities have explored many aspects of communication systems theory, including the development of the theory of transmitter optimization for coherent and noncoherent digital communication systems, spread-spectrum and multiple-access communication systems, convolutional codes, and sonar signal processing. He is the author of the text, *Elements of Detection and Signal Design*, which presents the development of his theory of signal design for digital systems, and is currently co-developing a textbook, *Principles of Spread-Spectrum Systems*.

Dr. Weber is a member of Sigma Xi, Eta Kappa Nu, Pi Mu Epsilon, and Tau Beta Pi. Of the publications which have resulted from his research, one (with R. A. Scholtz) has been nominated by the IEEE AdCom for the annual Browder J. Thompson IEEE Paper of the Year Award.

Direct-Sequence Spread-Spectrum Parallel Acquisition in a Fading Mobile Channel

ESSAM A. SOUROUR AND SOMESHWAR C. GUPTA, SENIOR MEMBER, IEEE

Abstract—**Spread-spectrum systems offer high immunity to interference and hostile environments. However, this high immunity can be fully exploited only if precise receiver synchronization is performed. In the last few years, some parallel acquisition schemes have been suggested to reduce the mean acquisition time. Most of the work done, however, considered either the effect of additive white Gaussian noise (AWGN) only or AWGN and narrow-band jamming. In this paper, a parallel acquisition scheme for direct-sequence spread-spectrum systems is proposed, and its mean acquisition time performance is analyzed in both nonfading and Rayleigh fading environments. The parallel scheme is compared to the corresponding serial scheme, and a significant improvement of performance is shown. The derived results provide the tool to study the interaction of many of the design parameters in the system.**

I. Introduction

DIRECT-sequence spread-spectrum (DS-SS) systems are very useful in applications such as anti-interference, anti-eavesdrop, multiple access, and ranging. Recently, DS-SS has attracted attention for mobile communications applications [1]-[4]. To exploit the advantages of a DS-SS signal, the receiver must be able to synchronize the locally generated pseudonoise (PN) sequence with the incoming one. This is done in two steps, acquisition (coarse alignment) and tracking (fine alignment). In this paper, the focus is mainly on the acquisition system of DS-SS signals.

Most of the techniques that have been proposed for minimizing the mean acquisition time of DS-SS systems employ the serial search strategy, wherein the cells are tested one at a time. In the last few years, some compound (i.e., partially parallel) acquisition schemes were proposed in which a number of cells, less than the number of uncertainty region cells, is tested simultaneously. In [7], a scheme that employs a bank of surface acoustic wave (SAW) convolvers is discussed. In another work [8], an approach using a charge-coupled device (CCD) pseudonoise matched filter (PN-MF) is proposed, and in [6], a system that uses a bank of tapped analog delays, each connected to a sequence detection circuit, is described.

Very few published studies in this area have assumed any threat to the acquisition system other than the additive white Gaussian noise (AWGN). In [11], the effect of data modulation and a single tone jammer is analyzed for an I-Q baseband noncoherent MF. Another example is [5], which studies a parallel acquisition scheme for a frequency hopping spread-spectrum signal in a Gilbert model fading channel.

In this paper, a DS-SS acquisition scheme that utilizes a bank of N parallel I-Q noncoherent MF's is proposed. It can be looked at as an extension of the schemes described in [10], which uses one MF only, and [7] which uses a bank of SAW convolvers. An expression for the mean acquisition time is derived for the new system

Paper approved by the Editor for Fading, Dispersive, and Multipath Channels of the IEEE Communications Society. Manuscript received February 6, 1989. This work was supported by DARPA under Contract MDA 903-86-C-0182. This paper was presented at the 1989 IEEE Vehicular Technology Conference, San Francisco, CA, April 1989.

The authors are with the Department of Electrical Engineering, Southern Methodist University, Dallas, TX 75275.

IEEE Log Number 9036338.

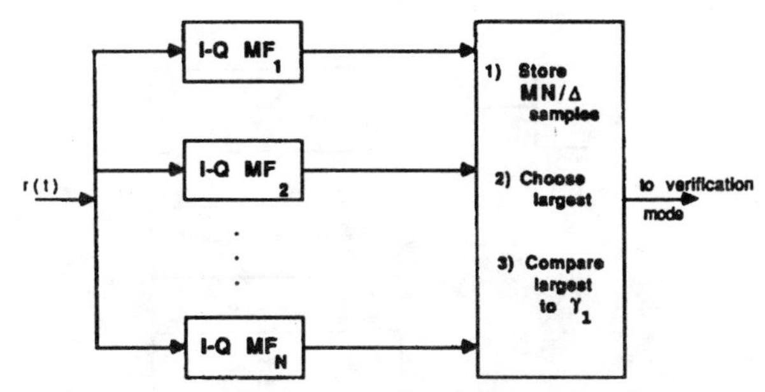

Fig. 1. Parallel matched filters for the search mode.

in terms of the probabilities of detection, missing, and false alarm. These parameters are first analyzed for a typical AWGN channel; then the Rayleigh fading channel encountered in a typical UHF or microwave land mobile radio channel is studied. The channel is assumed frequency nonselective, and the effect of data modulation and code Doppler is not considered in this paper. The performance of the parallel system is compared to the corresponding serial system, and it is shown that a significant improvement can be achieved.

The paper is organized as follows. The acquisition scheme is described in Section II, and the mean acquisition time expression is derived in Section III. Section IV considers the nonfading channel, while Section V considers the fading channel. Numerical results are presented in Section VI, and finally, the findings and conclusions are discussed in Section VII.

II. System Description

The system under consideration in this paper has two modes of operation: the search mode and the verification mode. The search mode is best described by referring to Fig. 1. It consists of a bank of N parallel I-Q passive noncoherent PN-MF's. One of the I-Q PN-MF's is shown in Fig. 2. The detailed hardware discussion of the I-Q PN-MF is given in [12] and [13]. The full period L of the PN code is divided into N subsequences, each of length $M = L/N$, and each I-Q PN-MF is loaded by (and hence matched to) one of the N subsequences. The total delay of each PN-MF is $T = MT_c$ where T_c is the chip duration and its total number of taps is M/Δ with ΔT_c delay between successive taps. A typical value for Δ is 1/2, and this value will be used throughout this paper.

In T seconds, MN/Δ samples are collected and stored from the N parallel MF's. Each sample corresponds to one possible phase in the uncertainty region where the uncertainty region is assumed to be the total PN sequence length L. if the largest of the MN/Δ samples exceeds a threshold γ_1, the corresponding phase is assumed, tentatively, to be the correct phase of the received signal, and the acquisition system moves to the verification mode. If γ_1 is not exceeded, new MN/Δ samples are collected, and so on. If γ_1 were set to zero, the search mode would have been similar to that in [7] where the largest sample is chosen as the one corresponding to the correct phase.

The function of the verification mode is to avoid a costly false alarm that can supply the tracking system with a wrong phase. The verification mode utilized here is similar to the one in [10]. Briefly, when a phase is selected by the search mode, one of the parallel I-Q

0090-6778/90/0700-0992$01.00

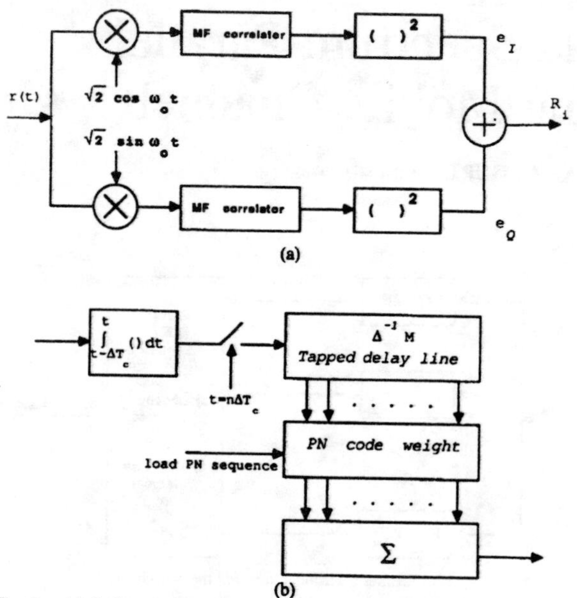

Fig. 2. (a) I-Q noncoherent matched filter. (b) Matched filter correlator of Fig. 2(a).

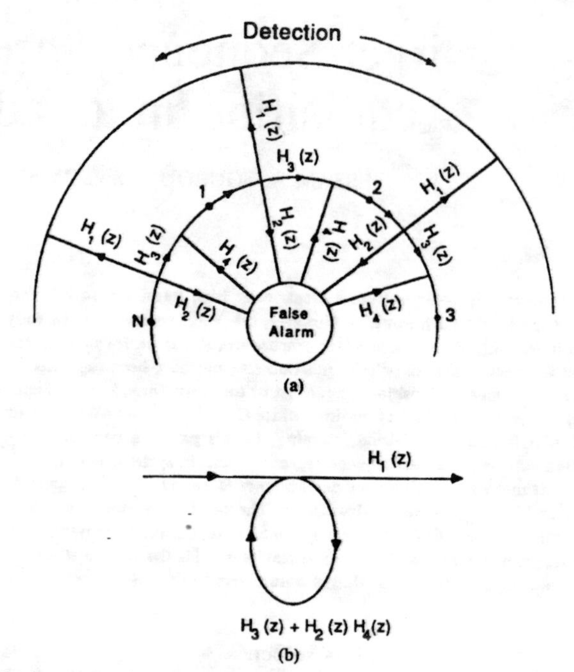

Fig. 3. (a) State transition diagram of the acquisition system. (b) Simplified state transition diagram.

MF's (or possibly a separate I-Q MF) is loaded with this phase. The receiver advances this local phase by the same rate as the incoming code so that both codes run in parallel, and a sample is taken each T seconds to ensure independence. If B out of A samples exceed a threshold γ_2, acquisition is declared and the tracking system is enabled; otherwise, the system goes back to the search mode. The two thresholds γ_1 and γ_2 are selected numerically to minimize the mean acquisition time.

It is worthy to note that in channels where the only deteriorating factor is the AWGN, it is advantageous to increase M, which means that less parallelism is better. However, practically M cannot be increased arbitrarily due to the possible existence of data modulation, frequency offset, code Doppler, or fading [8]. With the limitations on increasing M, the alternative to minimize the acquisition time is to increase the parallelism. Although it is known that a full parallel acquisition system yields the optimum performance, the condition $L = MN$ also puts a limitation on the total size of the system.

III. Mean Acquisition Time

Due to the Markovian nature of the acquisition process, the state transition diagram can be used in deriving the probability generating function of the acquisition time. In the first $T = MT_c$ seconds, the N parallel I-Q MF's are loaded with the incoming signal, and in the following T seconds, MN/Δ samples are tested as described in the previous section. There is a probability P_{M1} that none of the MN/Δ samples exceeds γ_1; in this case, new MN/Δ samples are collected in T more seconds. If one or more samples exceed γ_1, the largest sample is assumed, tentatively, to be corresponding to the incoming phase. This assumption can be correct with probability P_{D1} or false with probability P_{F1}. This phase is then loaded to the verification mode I-Q MF (called coincidence detector (CD) in [10]) which tests A samples (in AT seconds) against a threshold γ_2. If B out of the A tests exceed γ_2, this phase is declared to be the incoming phase and the system moves to the tracking mode. This happens with probability P_{D2} if it is actually the correct phase, and with probability P_{F2} if it is actually a wrong phase. If a correct phase is handed to the tracking system, the acquisition process is finished. But if a false phase is handled, it cannot be absorbed, and the system goes back to acquisition after JT penalty time.

The state transition diagram is shown in Fig. 3(a) where there is only one absorbing state, which is the detection state. Due to the similarity of all the I-Q MF's, Fig. 3(a) can be simplified to Fig. 3(b). It can be easily shown that

$$H_1(Z) = P_{D1}Z^T P_{D2}Z^{AT} \tag{1}$$

$$H_2(Z) = P_{F1}Z^T P_{F2}Z^{AT} \tag{2}$$

$$H_3(Z) = P_{M1}Z^T + P_{D2}Z^T(1 - P_{D1})Z^{AT} + P_{F1}Z^T(1 - P_{F2})Z^{AT} \tag{3}$$

$$H_4(Z) = Z^{JT} \tag{4}$$

and the generating function is

$$H(Z) = \frac{H_1(Z)}{1 - H_3(Z) - H_2(Z)H_4(Z)}. \tag{5}$$

Keeping in mind that $P_{M1} + P_{D1} + P_{F1} = 1$, we can see that the probability of acquisition $H(1) = 1$.

The mean acquisition time is given by

$$E[T_{acq}] = \frac{\delta}{\delta Z} \ln H(Z)|_{Z=1} \tag{6a}$$

which yields after some algebra

$$E[T_{acq}] = \frac{1 + A(1 - P_{M1}) + JP_F}{NP_D} LT_c \tag{6b}$$

where $P_D = P_{D1}P_{D2}$ and $P_F = P_{F1}P_{F2}$ denote the overall detection probability and false alarm probability, respectively.

In (6b), although we might see that increasing the number of parallelism N decreases the mean acquisition time, this is not always true. Increasing N decreases the MF's length M, which might affect P_D, P_F, and P_{M1} unfavorably. It is shown later that in a nonfading AWGN channel, it is advantageous, at low signal-to-noise ratio, to increase M as much as practically possible (reduce parallelism), while at high SNR, increasing the parallelism is advantageous. In a fading channel, increasing parallelism is shown to be advantageous in all cases.

IV. Nonfading Channel

The derivation of P_{D1} for a nonfading AWGN channel has been shown in [8]. In this section, we derive the other needed probability expressions in a similar way. The same assumptions of [8] which are also adopted in [7] are used, namely, 1) there is only one sample corresponding to the correct phase (one H_1 cell only), 2) all samples are independent, 3) $M \gg 1$ such that the correlation of the received sequence and local code yields zero when they are not in phase (H_0 cells), and 4) the uncertainty region is the full code length L.

The received signal can be written as

$$r(t) = \sqrt{2S}\, C(t - \beta T_c) \cos(\omega_0 t + \theta) + n(t) \tag{7}$$

where β = received code offset, $C(t)$ = PN sequence, θ = uniformly distributed random phase, ω_0 = carrier frequency in rad/s, S = received signal power, and $n(t)$ = narrow-band AWGN with zero mean and one-sided power spectral density of N_0. Referring to Fig. 2(a), it can easily be shown that the probability density function (pdf) of the H_1 sample is the noncentral χ^2 with two degrees of freedom:

$$p_R(y|H_1) = \frac{1}{2\sigma_n^2} \exp\left(-\frac{m^2 + y}{2\sigma_n^2}\right) I_0\left(\frac{m\sqrt{y}}{\sigma_n^2}\right) \tag{8}$$

where

$$\sigma_n^2 = N_0 M T_c / 2 \tag{9}$$

$$m^2 = M^2 T_c^2 S. \tag{10}$$

$I_0(\ )$ = modified Bessel function of the first kind and zero order. The pdf of the H_0 samples is the central χ^2 with two degrees of freedom:

$$p_R(x|H_0) = \frac{1}{2\sigma_n^2} \exp\left(-\frac{x}{2\sigma_n^2}\right). \tag{11}$$

The detection probability of the search mode P_{D1} is given by

$$P_{D1} = \int_{\gamma_1}^{\infty} p_R(y|H_1) \left[\int_0^y p_R(x|H_0)\, dx\right]^{\frac{L}{\Delta}-1} dy \tag{12}$$

and the missing probability of the search mode P_{M1} is given by

$$P_{M1} = \int_0^{\gamma_1} p_R(y|H_1)\, dy \left[\int_0^{\gamma_1} p_R(x|H_0)\, dx\right]^{\frac{L}{\Delta}-1}. \tag{13}$$

Substituting (8)–(11) and (12) and (13), one gets

$$P_{D1} = \frac{1}{2} e^{-Mv} \int_{\gamma_1'}^{\infty} e^{-x/2} I_0(\sqrt{2Mvx})[1 - e^{-x/2}]^{2L-1}\, dx \tag{14}$$

$$= \sum_{n=0}^{2L-1} \frac{1}{n+1} (-1)^n \binom{2L-1}{n} e^{-nMv/n+1} \cdot Q\left(\sqrt{\frac{2Mv}{n+1}}, \sqrt{(n+1)\gamma_1'}\right) \tag{15}$$

and

$$P_{M1} = (1 - e^{-\gamma_1'/2})^{2L-1} [1 - Q(\sqrt{2Mv}, \sqrt{\gamma_1'})] \tag{16}$$

where $Q(a, b)$ is the Marcum's Q function [14, p. 585], v is the chip energy-to-noise power spectral density (SNR/chip) defined as

$$v = \frac{ST_c}{N_0}, \tag{17}$$

and γ_1' is the normalized threshold of the search mode

$$\gamma_1' = \frac{\gamma_1}{\sigma_n^2}. \tag{18}$$

The false alarm probability of the search mode can be obtained from P_{D1} and P_{M1} as

$$P_{F1} = 1 - P_{D1} - P_{M1}. \tag{19}$$

The verification mode function was described earlier, and the analysis of its detection and false alarm probabilities is straightforward. It can be shown that

$$P_{D2} = \sum_{n=B}^{A} \binom{A}{n} P_1^n (1 - P_1)^{A-n} \tag{20}$$

and

$$P_{F2} = \sum_{n=B}^{A} \binom{A}{n} P_2^n (1 - P_2)^{A-n} \tag{21}$$

where P_1 and P_2 are the probabilities that H_1 and H_0 cells exceed γ_2, respectively. P_1 and P_2 are given by

$$P_1 = \int_{\gamma_2}^{\infty} p_R(y|H_1)\, dy \tag{22}$$

and

$$P_2 = \int_{\gamma_2}^{\infty} p_R(x|H_0)\, dx. \tag{23}$$

Substituting (8)–(11) in (22) and (23), one can get

$$P_1 = Q(\sqrt{2Mv}, \sqrt{\gamma_2'}) \tag{24}$$

and

$$P_2 = e^{-\frac{\gamma_2'}{2}} \tag{25}$$

where γ_2' is the normalized threshold of the verification mode and is given by

$$\gamma_2' = \frac{\gamma_2}{\sigma_n^2}. \tag{26}$$

To get the mean acquisition time for a certain SNR/chip for the case of a nonfading channel, the expressions derived above for P_{D1}, P_{D2}, P_{F1}, P_{F2}, and P_{M1} have to be substituted in (6b).

V. Rayleigh Fading Channel

It is common in the literature of bit error rate analysis in fading channels to consider the fading process approximately as a *constant* over the entire symbol interval, and independent from one symbol to another. This assumption is realistic when interleaving or interlacing is employed [14, p. 398]. In the fading channel considered here, the fading process is regarded as a constant over k successive chips, $k \ll M$, and these successive groups of k chips are correlated. Clearly, the smaller the value of k and the correlation coefficients among the chips, the faster the fade rate. Since the transmitted signal is a carrier biphase modulated by a PN code, the received signal in the fading channel described above can be written as

$$r(t) = \sum_{i=-\infty}^{\infty} \sqrt{2} x_{\lceil i/k \rceil}(t) \cos(\omega_0 t + d_i \pi) - \sqrt{2} y_{\lceil i/k \rceil}(t) \sin(\omega_0 t + d_i \pi) + n(t), \quad iT_c \le t \le (i+1)T_c \tag{29}$$

where $d_i = 0$ or 1 according to the PN code and $\lceil m/n \rceil$ = first integer $< m/n$ (i.e., if m/n is an integer, $\lceil m/n \rceil = (m/n) - 1$). $x_i(t)$ and $y_i(t)$ are independent zero-mean stationary Gaussian processes with variances

$$E[x_i^2(t)] = E[y_i^2(t)] = \sigma_s^2 \tag{30}$$

and autocorrelations

$$E[x_i(t)x_j(t)] = E[y_i(t)y_j(t)] = \rho_{|i-j|}\sigma_s^2, \qquad i \neq j \tag{31}$$

where $1 \geq \rho_1 \geq \rho_2 \geq \cdots \geq 0$ are the autocorrelation coefficients among the fading processes.

The outputs of the I and Q branches of the I-Q MF are, respectively,

$$e_I = xT_c + N_I \tag{32}$$

$$e_Q = yT_c + N_Q \tag{33}$$

where N_I, N_Q, x, and y are independent zero-mean Gaussian random variables (r.v.'s). Both N_I and N_Q have a variance σ_n^2, and since x and y are identically distributed, it suffices to define x as follows.

Under hypothesis H_1, and referring to Fig. 2(b), x is a summation of the $\lceil M/k \rceil + 2$ zero-mean Gaussian r.v.'s that constitute the vector

$$X' = (x_1 x_2 x_3 \cdots x_l x_{l+1} x_{l+2}) \tag{34}$$

where

$$l = \left\lceil \frac{M}{k} \right\rceil.$$

The two rv.'s x_1 and x_{l+2} are due to the incomplete groups of k chips with *constant* fading envelope at the two edges of the tapped delay line. Their variances are $q\sigma_s^2$ and $(k' - q)\sigma_s^2$, respectively. The r.v.'s $x_2, x_3, \cdots$ and x_{l+1} are due to the groups in the interior of the tapped delay line. They all have the same variance $k^2\sigma_s^2$.

The covariance matrix of the vector X' is given by

$$E(XX') = k^2\sigma_s^2 \cdot \begin{bmatrix} \left(\frac{q}{k}\right)^2 & \frac{q}{k}\rho_1 & \frac{q}{k}\rho_2 & \cdots & \frac{q}{k}\rho_l & \frac{q(k'-q)}{k^2}\rho_{l+1} \\ \frac{q}{k}\rho_1 & 1 & \rho_1 & \cdots & \rho_{l-1} & \frac{k'-q}{k}\rho_l \\ \frac{q}{k}\rho_2 & \rho_1 & 1 & \cdots & \rho_{l-2} & \frac{k'-q}{k}\rho_{l-1} \\ \vdots & \vdots & \vdots & \cdots & \vdots & \vdots \\ \frac{q}{k}\rho_l & \rho_{l-1} & \rho_{l-2} & \cdots & 1 & \frac{k'-q}{k}\rho_1 \\ \frac{q(k'-q)}{k^2}\rho_{l+1} & \frac{k'-q}{k}\rho_l & \frac{k'-q}{k}\rho_{l-1} & \cdots & \frac{k'-q}{k}\rho_1 & \left(\frac{k'-q}{k}\right)^2 \end{bmatrix} \tag{35}$$

where

$$k' = k\left(\frac{M}{k} - \left\lceil \frac{M}{k} \right\rceil\right)$$

and $q = 0, 1, 2, \cdots, k'$ is a uniformly distributed discrete r.v. that represents the number of chips belonging to an incomplete group of *constant* fading amplitude of the received PN sequence at one edge of the tapped delay line. From (35), one can get the conditional variance of x to be

$$E(x^2|H_1, q) = k^2\sigma_s^2\left[l + \left(\frac{k'}{k}\right)^2 + 2\sum_{j=1}^{l}\left(\frac{M}{k} - j\right)\rho_j\right] + 2(q^2 - k'q)(1 - \rho_{l+1})\sigma_s^2. \tag{36}$$

Averaging (36) over the uniform r.v. q, one can get

$$E(x^2|H_1) = W\sigma_s^2 \tag{37}$$

where

$$W = k^2\left[l + \frac{1}{3}\left(\frac{k'}{k}\right)^2(2 + \rho_{l+1}) + 2\sum_{j=1}^{l}\left(\frac{M}{k} - j\right)\rho_j + \frac{1}{3}\frac{k'}{k^2}(1 - \rho_{l+1})\right]. \tag{38}$$

Under hypothesis H_0, the MF coefficients (not matched to the incoming PN sequence) can be considered approximately independent, $\pm$ 1-valued random variables [10], [11]. Thus, x is a summation of M independent identically distributed (i.i.d.) Gaussian random variables with variances σ_s^2. The variance of x is given by

$$E(x^2|H_0) = M\sigma_s^2. \tag{39}$$

From (32)–(38), under hypothesis H_i, e_I and e_Q follow the Gaussian distribution

$$e_I \text{ and } e_Q|_{H_i} \sim G(0, \sigma_i^2), \qquad i = 0, 1 \tag{40}$$

where

$$\sigma_0^2 = MT_c^2\sigma_s^2 + \sigma_n^2 \tag{41}$$

$$\sigma_1^2 = WT_c^2\sigma_s^2 + \sigma_n^2 \tag{42}$$

and $G(\mu, \sigma^2)$ is a Gaussian distribution of mean μ and variance σ^2.

Referring to Fig. 2(a), under H_i, $i = 0, 1$, the sample R follows the χ^2 distribution with two degrees of freedom:

$$p_R(y|H_i) = \frac{1}{2\sigma_i^2}\exp\left(-\frac{y}{2\sigma_i^2}\right), \qquad i = 0, 1. \tag{43}$$

Substituting (43) in (12) and (13), we can easily get P_{D1}:

$$P_{D1} = \frac{1}{2\left(1 + \frac{W}{M}v\right)}\int_{\gamma_1'}^{\infty}\exp\left(-\frac{x}{2\left(1 + \frac{W}{M}v\right)}\right) \cdot \left[1 - \exp\left(-\frac{x}{2(1+v)}\right)\right]^{2L-1} dx \tag{44}$$

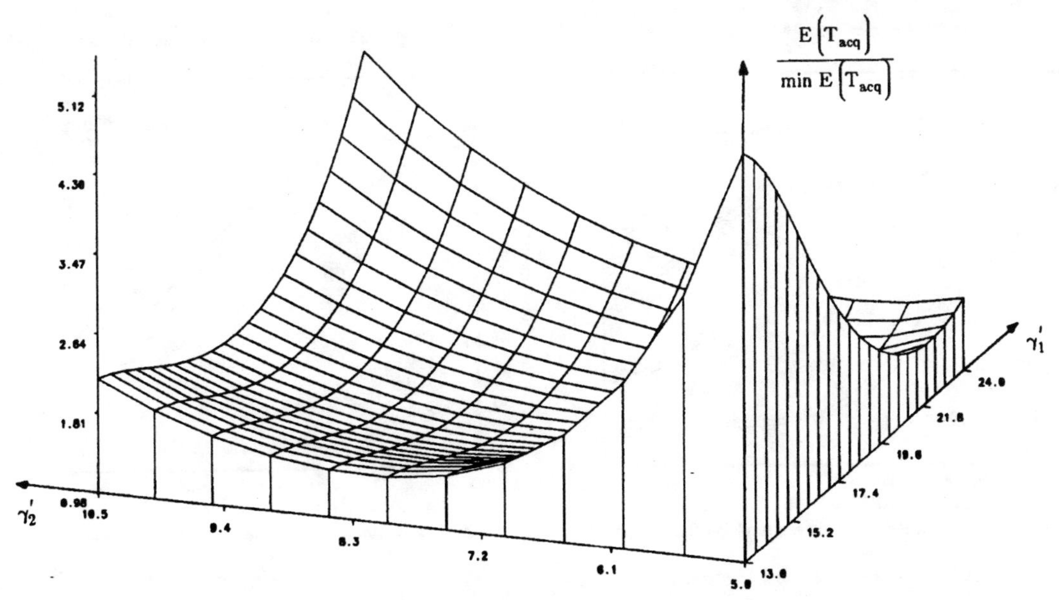

Fig. 4. Mean acquisition time/minimum mean acquisition time for a nonfading channel.

from which

$$P_{D1} = (1+v)\exp\left(-\frac{\gamma_1'}{2\left(1+\frac{W}{M}v\right)}\right)\sum_{n=0}^{2L-1}(-1)^n \cdot \binom{2L-1}{n}\frac{\exp\left(-\frac{n\gamma_1'}{2(1+v)}\right)}{1+n\left(1+\frac{W}{M}v\right)+v} \tag{45}$$

and P_{M1}:

$$P_{M1} = \left(1-\exp\left(-\frac{\gamma_1'}{2\left(1+\frac{W}{M}v\right)}\right)\right) \cdot \left(1-\exp\left(-\frac{\gamma_1'}{2(1+v)}\right)\right)^{2L-1} \tag{46}$$

where v is the SNR/chip defined as

$$v = \frac{2\sigma_s^2 T_c}{N_0}. \tag{47}$$

Finally, from (22), (23), and (43), we get P_1 and P_2 for the fading channel:

$$P_1 = \exp\left(-\frac{\gamma_2'}{2\left(1+\frac{W}{M}v\right)}\right) \tag{48}$$

$$P_2 = \exp\left(-\frac{\gamma_2'}{2(1+v)}\right). \tag{49}$$

From (20), (21), (48), and (49), we can get P_{D2} and P_{F2} for the fading channel.

In the above analysis, we assumed that the output samples of the I-Q MF of Fig. 2(a) are independent. This assumption is approximately true when $k \ll M$ and the correlation coefficients are small. The same assumptions also enable us to assume that the Markovian nature of the acquisition process still exists.

It is interesting to note that we put all ρ's $= 1$ or $q = 0$ and $k = M$ in (36), i.e., $W = M^2$; the mathematical expressions for P_{D1}, P_{M1}, and P_1 in this section are the same expressions that can result from integrating (15), (16), and (24) over a Rayleigh fading envelope, provided that in this case, $\sigma_0^2 = \sigma_n^2$. This could be the case if the fading process is constant over at least M successive chips. Also, it is noted that for a fade rate equivalent to $k = 1$ and all ρ's $= 0$, acquisition is impossible since $\sigma_0^2 = \sigma_1^2$ in this case.

VI. Numerical Results and Discussion

As an application for the proposed system, we consider a PN code with rate 1 Mchip/s and length $L = 1023$. Different values of the number of parallel I-Q MF are considered: $N = 16$, 11, and 8 corresponding to MF lengths $M = 64$, 93, and 128, respectively, where whenever L/N is not an integer, M is increased to the first integer $\geq L/N$. It has been mentioned earlier that M cannot be increased arbitrarily due to the possible existence of code Doppler or data modulation. Also, it is found that in a nonfading channel, it is advantageous to increase M. The values of M taken above are assumed to be within the maximum allowed practical limitations. The thresholds γ_1 and γ_2 are selected numerically to minimize the mean acquisition time for each value of SNR/chip. The parameters A and B of the verification mode are quoted from [10] to be 4 and 2, respectively, since it appeared in [10] that these values gave good performance for the verification mode, and it suffices for this paper to optimize γ_2. The penalty factor J is taken to be 1000 (i.e., penalty time $1000MT_c$ seconds). In calculating P_{D1}, numerical integration of (14) and (39) is used due to the computationally difficult expressions (15) and (40). This numerical problem is also noticed in [15, p. 494] for a similar mathematical expression. As an example for the Rayleigh fading channel, the correlation coefficients ρ_i are taken as ρ^i [16], [17]. Values of ρ of 0, 0.1, 0.2, 0.3, and 0.4 are considered in the results.

Fig. 4 shows the mean acquisition time relative to its optimum value at SNR/chip$=-15$ dB versus the normalized thresholds γ_1' and γ_2' for a nonfading channel. It is clear that an inappropriate choice of the thresholds can increase the mean acquisition time several times.

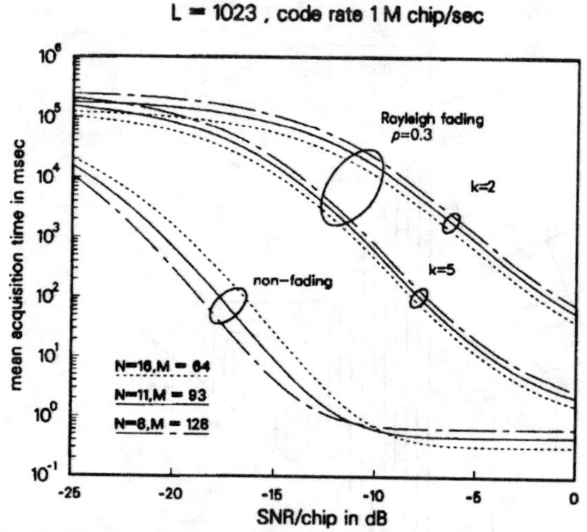

Fig. 5. Mean acquisition time for nonfading and Rayleigh fading channels.

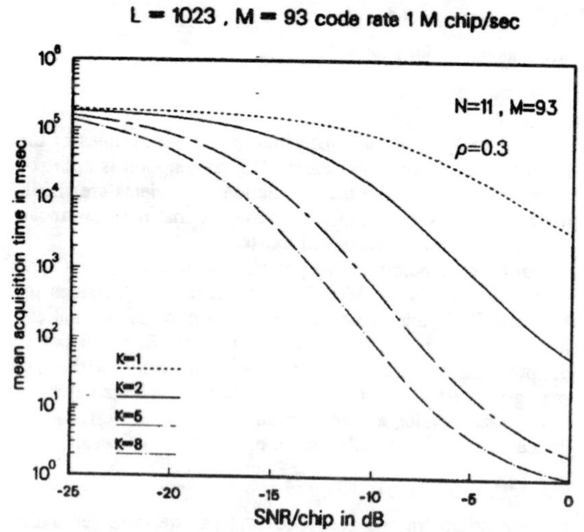

Fig. 6. Mean acquisition time for the Rayleigh fading channel, $\rho = 0.3$.

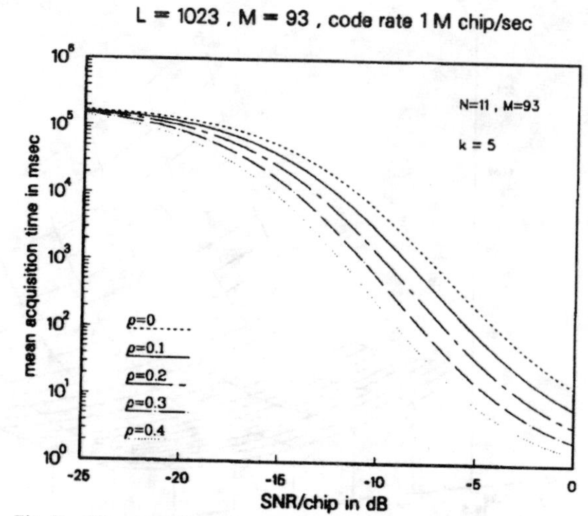

Fig. 7. Mean acquisition time for the Rayleigh fading channel, $k = 5$.

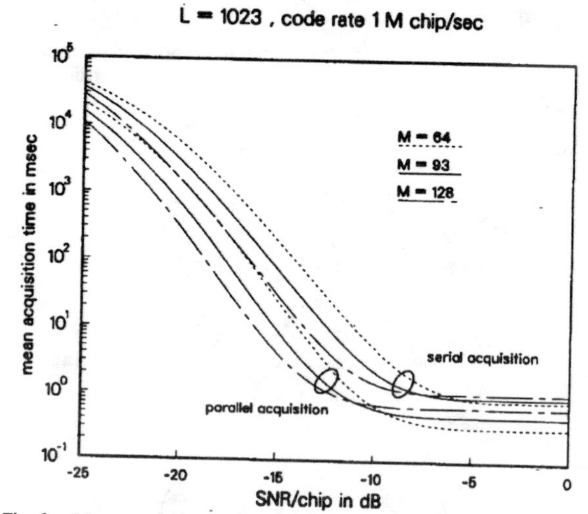

Fig. 8. Mean acquisition time for parallel acquisition and serial acquisition systems for a nonfading channel.

This sensitivity of the mean acquisition time to γ_1 and γ_2 is found to decrease when the SNR/chip increases (Fig. 4 flattens) Fig. 5 shows the mean acquisition time versus SNR/chip for both the nonfading and Rayleigh fading channels. A severe degradation in the mean acquisition time performance due to the fading channel can be observed, which is basically due to the effect of the fading environment on P_D and P_F. Also, it is worthwhile to note that in the nonfading channel, for SNR < -10 dB, it is advantageous to increase the MF length M, i.e., decrease the number of parallel MF's, while in the fading channel, it is seen that the opposite is true for the whole range of SNR shown. The effect of fade rapidity is shown in Figs. 6 and 7. As expected, for faster fade rates (lower k or ρ), the system is slower. It is also noted that changes in k affect the mean acquisition time more than similar changes in ρ.

To compare the proposed parallel acquisition system to the serial acquisition system, the serial acquisition system described in [9], [10], to which this paper is an extension, is reanalyzed using the parameters and assumptions in this paper (one H_1 cell, $\Delta = 1/2$, no code self-noise, penalty time $= JMT_c$, $L = 1023$, and code rate $= 1$ Mchip/s).

Fig. 8 presents the results of the serial and parallel acquisition systems in a nonfading channel, while Fig. 9 shows the comparison in the Rayleigh fading channel. For all values of M, there is about a

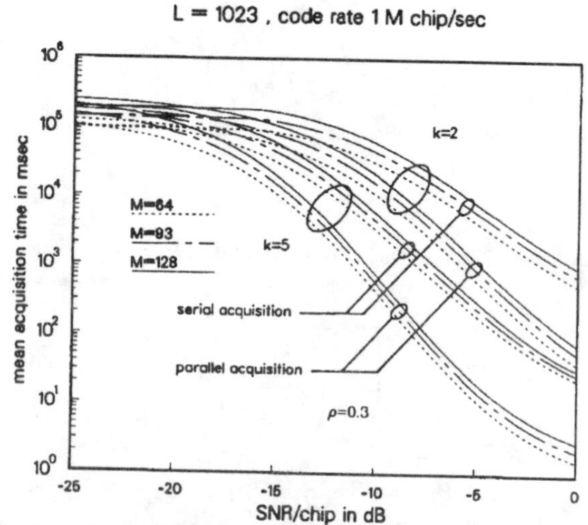

Fig. 9. Mean acquisition time for parallel and serial acquisition systems for a Rayleigh fading channel, $\rho = 0.3$.

2.5 dB savings in SNR if the parallel system is used instead of the serial one in a nonfading channel. In the fading channel ($k = 2$ and $k = 5$), this improvement can increase to about 4 dB, which makes the parallel acquisition system a good candidate for spread-spectrum systems operation in mobile fading channels.

VII. Conclusions

A parallel acquisition system which is an extension of the ones in [7], [10] is described, and an expression for the mean acquisition time in terms of detection, false alarm, and missing probabilities is derived. These probability expressions are derived in both nonfading and Rayleigh fading AWGN channels, and the mean acquisition time is shown for both cases. It is shown that in a nonfading channel, the designer should choose the MF length as big as practically possible, and then cover the rest of the uncertainty region with parallel MF's. In fading channels, the contrary is the right choice; for better performance, the designer needs to increase parallelism at the cost of the MF's length.

The degradation in mean acquisition times due to fading conditions seems to be great, which raises some doubts about the ability of DS-SS systems to work in such channels when frequency selectivity and code Doppler are taken into account in low SNR values and high fade rates.

Comparing the performance of the parallel acquisition system and the serial acquisition system, in a nonfading channel, the parallel system is shown to save about 2.5 dB, while in the Rayleigh fading channel, the improvement of the parallel system over the serial system reaches up to 4–5 dB.

References

[1] Special Issue on Mobile Spread-Spectrum Communication, *IEEE Trans. Veh. Technol.*, vol. VT-30, Feb. 1981.

[2] R. P. Eckert and P. M. Kelly, "Implementing spread spectrum technology in the land mobile radio services," *IEEE Trans. Commun.*, vol. COM-25, Aug. 1977.

[3] M. Mizuno, E. Moriyama, and Y. Kadokawa, "Spread spectrum communication systems for land mobile radio," Rev. Radio Res. Lab., Ministry of Posts and Telecommun., Japan, 1983.

[4] G. R. Cooper and R. W. Nettleton, "A spread spectrum technique for high-capacity mobile communications," *IEEE Trans. Veh. Technol.*, vol. VT-27, Nov. 1978.

[5] N. D. Wilson, S. S. Rappaport, and M. M. Vasudevan, "Rapid acquisition scheme for spread-spectrum radio in a fading environment," *Proc. IEE*, vol. 135, part F, Feb. 1988.

[6] A. K. El-Hakeem and S. A. M. Liebrecht, "A multiprocessing approach to spread spectrum code acquisition," in *Proc. IEEE MILCOM'85*, Boston MA, Oct. 1985.

[7] L. B. Milstein, J. Gevargiz, and P. K. Das, "Rapid acquisition for direct sequence spread spectrum communications using parallel SAW convolvers," *IEEE Trans. Commun.*, vol. COM-33, July 1985.

[8] Y. T. Su, "Rapid code acquisition algorithms employing PN matched filters," *IEEE Trans. Commun.*, vol. 36, June 1988.

[9] A. Polydoros and C. Weber, "A unified approach to serial search spread spectrum code acquisition—Part I: General theory," *IEEE Trans. Commun.*, vol. COM-32, May 1984.

[10] —, "A unified approach to serial search spread spectrum acquisition—Part II: A matched filter receiver," *IEEE Trans. Commun.*, vol. COM-32, May 1984.

[11] E. W. Siess and C. L. Weber, "Acquisition of direct sequence signals with modulation and jamming," *IEEE J. Select. Areas Commun.*, vol. SAC-4, Mar. 1986.

[12] S. S. Rappaport and D. M. Grieco, "Spread spectrum signal acquisition: Methods and Technology," *IEEE Commun. Mag.*, vol. 22, June 1984.

[13] M. K. Simon, J. K. Omura, R. A. Scholtz, and B. K. Levitt, *Spread Spectrum Communications, Vol. III*. Rockville, MD: Computer Science Press, 1985.

[14] M. Schwartz, W. R. Bennett, and S. Stein, *Communication Systems and Techniques*. New York: McGraw-Hill, 1966.

[15] J. G. Proakis, *Digital Communications*. New York: McGraw-Hill, 1983.

[16] X.-Y. Hou and N. Morinaga, "Detection performance of Rayleigh fluctuating targets in correlated Gaussian clutter plus noise," *Trans. IEICE*, vol. E71, Mar. 1988.

[17] I. Kanter, "Exact detection probability for partially correlated Rayleigh targets," *IEEE Trans. Aero. Electron. Syst.*, vol. AES-22, pp. 184–195, Mar. 1986.

Essam A. Sourour was born in Alexandria, Egypt, on March 16, 1959. He received the B.Sc. and M.Sc. degrees, both in electrical engineering, from Alexandria University, Alexandria, Egypt, in 1982 and 1985, respectively. He received the Ph.D. degree from the Department of Electrical Engineering, Southern Methodist University, Dallas, TX.

In 1982 he was a Teaching and Research Assistant in the Department of Electrical Engineering, Alexandria University. He has been a Teaching and Research Assistant in the Department of Electrical Engineering, Southern Methodist University from 1987–1990. He recently joined Bell-Northern Research, Inc., Richardson, TX. His research has been in the area of digital communications, with emphasis on spread spectrum and mobile communications.

Someshwar C. Gupta (S'61–M'63–SM'65) was born in Ludihana, India, on April 23, 1935. He received the B.A. (Hons.) and M.A. degrees in mathematics in 1951 and 1953, respectively, from Punjab University, India, the B.S. (Hons.) degree in electrical engineering from Glasgow University, Scotland, in 1957, and the M.S.E.E. and Ph.D. degrees in electrical engineering from the University of California, Berkeley, in 1962.

He has considerable industrial and teaching experience, and is currently Acting Dean of the School of Engineering and Applied Sciences, Southern Methodist University, Dallas, TX. He is author of the book, *Transform and State Variable Methods in Linear Systems* (Wiley, 1966) and coauthor of the books, *Fundamentals of Automatic Control* (Wiley, 1970) and *Circuit Analysis with Computer Applications to Problem Solving* (Matrix, 1972).

Dr. Gupta is a member of the Editorial Board of the *International Journal of System Sciences.*

Noncoherent Sequential Acquisition of PN Sequences for DS/SS Communications with/without Channel Fading

Sawasd Tantaratana, *Senior Member, IEEE*, Alex W. Lam, *Member, IEEE*, and Patrick J. Vincent

***Abstract*— In this paper, we study sequential acquisition schemes of m-sequences, based on the sequential probability ratio test (SPRT) and a truncated SPRT (TSPRT), with noncoherent demodulation. Most reported results on sequential acquisition schemes assume that independent samples are available for the decision process. The assumption of independent and identically distributed (i.i.d.) samples requires the integrator in the receiver to be reset periodically. This introduces loss in the effective signal-to-noise ratio, degrading the performance. In this paper, on the contrary, our two sequential schemes use continuous integration. It can be shown that the likelihood ratio is monotonic; consequently, the tests can be easily implemented in real time. Methods are proposed for designing the decision thresholds to achieve the desired false-alarm and miss probabilities. Performances of the proposed schemes are obtained and they suggest that the TSPRT is more desirable. The effect of slowly-varying channel fading is also investigated. Results show that fading does not affect the false alarm probabilities, but it can drastically reduce the probability of detecting the alignment of the two PN sequences, especially when the fading is severe.**

I. Introduction

In direct-sequence spread-spectrum (DS/SS) systems, the receiver must align the locally generated pseudonoise (PN) code with the incoming PN code. This code synchronization process is usually performed in two steps: acquisition and tracking. First, the acquisition process coarsely aligns the phases of the two PN sequences to within one chip or a fraction of a chip. The tracking process then takes over and perform fine adjustment until the phase difference becomes ideally zero.

Extensive research on PN code acquisitions has been carried out during the past two decades. See [1]-[16] for examples. Acquisition schemes can be classified into fixed-dwell, multiple-dwell, and sequential schemes [3][17]. Sequential schemes have the potential to achieve the best performance, but they are the least studied because of the difficulty in design and analysis. In most reports on sequential schemes, the decision process is based on independent and identically distributed (i.i.d.) random variables. To obtain i.i.d. samples, the integrator in each correlation arm is reset periodically, which could reduce the effective signal-to-noise ratio.

Since the acquisition process involves searching through the uncertainty phases of the PN sequence, acquisition schemes can also be classified into parallel, serial, and hybrid schemes [3][17]. A parallel scheme inspects all the uncertainty phases simultaneously and decides which is the most likely one. Each uncertainty phase is investigated by a path consisting of a correlator or a matched filter. If the period of the PN sequence and the timing uncertainty are large, parallel schemes would require excessive hardware and, hence, are impractical. On the other hand, a serial scheme inspects one uncertainty phase at a time and determines whether the phase of the local sequence and the phase of the incoming sequence are in alignment. If they are, the tracking circuit is initiated. Otherwise, the next uncertainty phase is examined. Hardware requirements of such schemes are minimal, but the complexity of the control process and the average time to reach acquisition increase. Various combinations of the serial search and the parallel search are possible.

In this paper, we study noncoherent sequential acquisition schemes using serial search. We use the sequential probability ratio test (SPRT) and a truncated SPRT (TSPRT) for the decision process, i.e., the process of deciding whether the local sequence is in alignment with the incoming sequence. Threshold design methods for these sequential schemes that would achieve the desired false-alarm and detection probabilities are presented. The average test lengths are obtained through simulations.

In Section II, the problem under consideration is introduced and the fixed-dwell, SPRT, and TSPRT noncoherent acquisition schemes are described. In Section III, threshold design methods are addressed. Section IV compares numerical results. Finally, a conclusion is presented in Section V.

II. Noncoherent Acquisition Schemes

In this section, we describe three noncoherent acquisition schemes. The acquisition system under consideration is depicted in Figure 1. We assume that the channel has slowly-varying fading and additive white Gaussian noise with two-sided power spectral density $N_0/2$. We also assume that there is no data modulation during the acquisition process. The input signal at the receiver is

$$r(t) = A_0 \psi a(t + i\Delta T_c)\cos(\omega_0 t + \theta) + n(t) \qquad (1)$$

Paper approved by James S. Lehnert, Editor for Modulation & Signal Design of the IEEE Communications Society. Manuscript received December 23, 1991; revised January 4, 1993, and May 2, 1994. This work was supported by the Army Research Office under Grant DAAL03-91-G-0001. This paper was presented in part at the Twenty-Ninth Annual Allerton Conference on Communication, Control, and Computing, University of Illinois, October 1991.

S. Tantaratana is with the Department of Electrical and Computer Engineering, University of Massachusetts at Amherst, Amherst, MA 01003-5110.

A. W. Lam and P.J. Vincent are with the Department of Electrical and Computer Engineering, Naval Postgraduate School, Monterey, CA 93943.

IEEE Log Number 9411640.

0090-6778/95$4.00

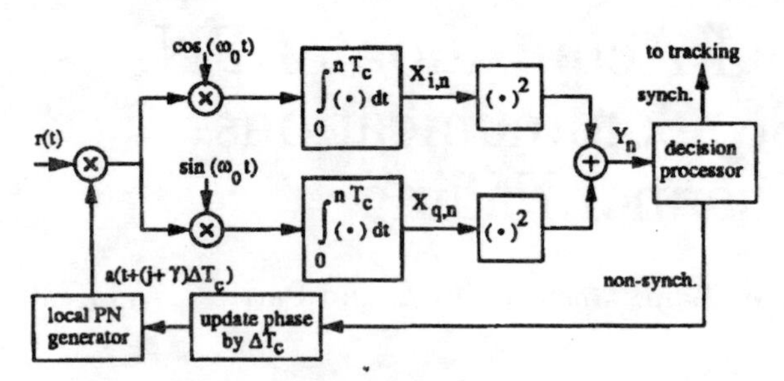

Fig. 1. Block diagram of noncoherent serial acquisition scheme.

where A_0 is the signal amplitude, $a(t)$ is the PN waveform, ψ is the fading random variable, i is an integral phase number, Δ is a value determining how much the timing of the local PN generator is updated during the search process, T_c is the chip duration, ω_0 and θ are the frequency and phase of the carrier, and $n(t)$ is the additive noise. The value of Δ is usually 1 or 1/2, meaning that the local timing is updated each time by T_c or $T_c/2$.

The fading is assumed to be Rician, i.e., ψ has a Rician probability density function (pdf) given by [18]

$$f_\psi(\psi) = 2\psi(1+r)e^{-r-\psi^2(1+r)} I_0\left(2\psi\sqrt{r(1+r)}\right), \quad \psi \geq 0, \tag{2}$$

where r is the ratio of the power in the direct component (s^2) and the power in the diffused component ($2\sigma^2$), i.e., $r = s^2/(2\sigma^2)$, with the constraint that $s^2 + 2\sigma^2 = 1$. Note that $s^2 = r/(1+r)$ and $\sigma^2 = 0.5/(1+r)$. When $r = \infty$ (or $\sigma^2 = 0$ and $s^2 = 1$), it corresponds to no fading. When $r = 0$ (or $s^2 = 0$ and $\sigma^2 = 0.5$), we have Rayleigh fading, for which case the pdf of ψ is given by $f_\psi(\psi) = 2\psi e^{-\psi^2}$, $\psi \geq 0$.

Let the locally generated PN waveform be $a(t + (j + \gamma)\Delta T_c)$, where j is an integer and $|\gamma| < 1/2$. The received signal $r(t)$ is first multiplied by the local PN waveform, followed by noncoherent demodulation (see Figure 1). The result Y_n is used by the decision processor to test if the local and the incoming PN waveforms are aligned to within $\Delta T_c/2$ seconds. When (coarse) alignment is achieved, we have $j = i$ and the tracking circuit is initiated to bring γ to zero. Otherwise, the phase of the local PN waveform is updated by ΔT_c seconds and the acquisition process continues.

The in-phase and the quadrature components, after integration over n chips, i.e., from $t = 0$ to nT_c, and ignoring the double-frequency terms, are

$$\begin{aligned} X_{i,n} &= \int_0^{nT_c} r(t)a(t+(j+\gamma)\Delta T_c)\cos(\omega_0 t)\, dt \\ &= \frac{A_0}{2}\psi T_c S_n \cos\theta \; + \; N_{i,n} \end{aligned} \tag{3}$$

$$\begin{aligned} X_{q,n} &= \int_0^{nT_c} r(t)a(t+(j+\gamma)\Delta T_c)\sin(\omega_0 t)\, dt \\ &= \frac{A_0}{2}\psi T_c S_n \sin\theta \; + \; N_{q,n} \end{aligned} \tag{4}$$

where

$$N_{i,n} = \int_0^{nT_c} n(t)a(t+(j+\gamma)\Delta T_c)\cos(\omega_0 t)dt \tag{5}$$

$$N_{q,n} = \int_0^{nT_c} n(t)a(t+(j+\gamma)\Delta T_c)\sin(\omega_0 t)dt \tag{6}$$

are independent Gaussian random variables with mean zero and variance $\sigma_n^2 = nT_cN_0/4$, and where

$$S_n = \frac{1}{T_c}\int_0^{nT_c} a(t+i\Delta T_c)a(t+(j+\gamma)\Delta T_c)dt \tag{7}$$

Note that T_cS_n is the partial autocorrelation over n chips of the PN waveform. Note also that $N_{i,n}$, $N_{q,n}$, and S_n depend on the length of integration. For a fixed n and given ψ, $X_{i,n}$ and $X_{q,n}$ are independent Gaussian random variables with variance σ_n^2 and means $(A_0/2)\psi T_cS_n\cos\theta$ and $(A_0/2)\psi T_cS_n\sin\theta$, respectively. However, $X_{i,n}$ and $X_{i,m}$ are statistically dependent, so are $X_{q,n}$ and $X_{q,m}$. The test statistic for deciding alignment or non-alignment is

$$Y_n = X_{i,n}^2 \; + \; X_{q,n}^2 \tag{8}$$

The test statistic Y_n is a chi-square random variable with pdf given by:

$$\begin{aligned} f_{Y_n}(y_n) &= \frac{(1+r)}{2\sigma_n^2[(1+r)+\lambda_n/2\sigma_n^2]} \\ &\quad \cdot \exp\left[-\frac{(1+r)(y_n/2\sigma_n^2)+r(\lambda_n/2\sigma_n^2)}{(1+r)+\lambda_n/2\sigma_n^2}\right] \\ &\quad \cdot I_0\left(\frac{2\sqrt{r(1+r)(\lambda_n/2\sigma_n^2)(y_n/2\sigma_n^2)}}{(1+r)+\lambda_n/2\sigma_n^2}\right) \end{aligned} \tag{9}$$

where $\lambda_n = (A_0^2/4)T_c^2S_n^2$ and $I_0(.)$ is the modified Bessel function of order zero. To show (9), we introduce a random variable $\tilde{\theta} \in (0, 2\pi)$ which is dependent on ψ through the joint pdf

$$f(\psi, \tilde{\theta}) = \frac{\psi}{2\pi\sigma^2} e^{-(\psi^2+s^2)/2\sigma^2} e^{\psi s\cos\tilde{\theta}/\sigma^2}, \quad \begin{array}{c}\psi \geq 0, \\ 0 \leq \tilde{\theta} < 2\pi\end{array}$$

where s and σ^2 are defined earlier. Note that integrating this joint pdf with respect to $\tilde{\theta}$ yields (2). Letting $\psi\cos\tilde{\theta} = s + U$ and $\psi\cos\tilde{\theta} = V$, we can show that U and V are independent Gaussian random variables, both with zero mean and variance σ^2. Now, rewrite Eqs.(3) and (4) as

$$\begin{aligned} X_{i,n} &= \sqrt{\lambda_n}\psi\cos\theta + N_{i,n} \\ &= \sqrt{\lambda_n}\psi\cos((\theta-\tilde{\theta})+\tilde{\theta}) + N_{i,n} \\ &= \sqrt{\lambda_n}[s\cos\theta' + U\cos\theta' - V\sin\theta'] + N_{i,n} \\ X_{q,n} &= \sqrt{\lambda_n}[s\sin\theta' + U\sin\theta' + V\cos\theta'] + N_{q,n} \end{aligned}$$

where $\theta' = \theta - \tilde{\theta}$ modulo 2π. Since θ is uniform in $(0, 2\pi)$, independent of ψ and $\tilde{\theta}$ (hence independent of U and V), it follows that θ' is also uniform in $(0, 2\pi)$ and independent of U and V. Given θ', $X_{i,n}$ and $X_{q,n}$ are uncorrelated, which implies that they are independent (conditioned on θ') since they are jointly Gaussian. Therefore, we can write the conditional joint pdf of $X_{i,n}$ and $X_{q,n}$ given θ'. Letting $Y_n = X_{i,n}^2 + X_{q,n}^2$ and removing the conditioning yields the chi-square pdf (9) for Y_n.

When there is no fading, Y_n is also a chi-square random variable, with pdf

$$f_{Y_n}(y_n) = \frac{1}{2\sigma_n^2} e^{-(y_n+\lambda_n)/2\sigma_n^2} I_0\left(\frac{\sqrt{\lambda_n y_n}}{\sigma_n^2}\right), \quad y_n \geq 0 \quad (10)$$

Comparing (9) and (10), we see that λ_n and σ_n^2 in the non-fading case are replaced with $r\lambda_n/(1+r)$ and $\sigma_n^2[(1+r)+\lambda_n/2\sigma_n^2]/(1+r)$, respectively, when there is fading. Since $r\lambda_n/(1+r) < \lambda_n$ and $\sigma_n^2[(1+r)+\lambda_n/2\sigma_n^2]/(1+r) > \sigma_n^2$, for $0 \leq r < \infty$, the fading reduces the effective signal-to-noise ratio by a factor of $r/[(1+r)+\lambda_n/2\sigma_n^2]$.

Based on Y_n, the decision processor must decide whether the local and the incoming PN waveforms are aligned to within $\Delta T_c/2$ of each other (i.e., $j = i$), or they differ by at least one chip (i.e., $|(j+\gamma) - i|\Delta T_c \geq T_c$), or neither in the case of a sequential test. Without loss of generality, we set $i = 0$ so the phase difference becomes $j + \gamma$. Equivalently, the decision processor has a task of testing the following hypotheses:

$$\begin{aligned} &H_0 \text{ (non-alignment)}: \quad |j+\gamma| \geq \tfrac{1}{\Delta} \text{ and } |\gamma| < \tfrac{1}{2} \\ &H_1 \text{ (alignment)}: \quad j = i = 0 \text{ and } |\gamma| < \tfrac{1}{2} \end{aligned} \quad (11)$$

Note that under H_1 we have $|j+\gamma| < 1/2$. It is possible to have $\frac{1}{2} < |j+\gamma| < \frac{1}{\Delta}$, which corresponds to a scenario between H_0 and H_1.

Since $|\gamma| < 1/2$, both H_0 and H_1 are composite hypotheses and the value of λ_n takes on different ranges under different hypothesis. Let $\lambda_{n,0}$ denote the largest value of λ_n under H_0 and $\lambda_{n,1}$ the smallest value of λ_n under H_1, where $\lambda_{n,1} > \lambda_{n,0}$. Using $\lambda_{n,1}$ and $\lambda_{n,0}$ we can write the likelihood ratio as:

$$\begin{aligned} \Lambda_n(y_n) &= \frac{f_{Y_n}(y_n|H_1)}{f_{Y_n}(y_n|H_0)} \\ &= \frac{(1+r)+(\lambda_{n,0}/2\sigma_n^2)}{(1+r)+(\lambda_{n,1}/2\sigma_n^2)} \exp\left[r\left(\frac{\lambda_{n,0}/2\sigma_n^2}{(1+r)+\lambda_{n,0}/2\sigma_n^2} - \frac{\lambda_{n,1}/2\sigma_n^2}{(1+r)+\lambda_{n,1}/2\sigma_n^2}\right)\right] \exp\left[(1+r)\frac{y_n}{2\sigma_n^2} \right. \\ &\quad \left. \cdot \left(\frac{1}{(1+r)+\lambda_{n,0}/2\sigma_n^2} - \frac{1}{(1+r)+\lambda_{n,1}/2\sigma_n^2}\right)\right] \\ &\quad \cdot \frac{I_0\left(\frac{2\sqrt{r(1+r)(\lambda_{n,1}/2\sigma_n^2)(y_n/2\sigma_n^2)}}{(1+r)+\lambda_{n,1}/2\sigma_n^2}\right)}{I_0\left(\frac{2\sqrt{r(1+r)(\lambda_{n,0}/2\sigma_n^2)(y_n/2\sigma_n^2)}}{(1+r)+\lambda_{n,0}/2\sigma_n^2}\right)} \end{aligned} \quad (12)$$

where y_n is a realization of Y_n. It can be shown that the likelihood ratio (12) is a monotonically increasing function of the variable y_n, provided that $\lambda_{n,1} > \lambda_{n,0}$, which is the case in our situation. When the fading is Rayleigh, (12) reduces to

$$\begin{aligned} \Lambda_n(y_n) &= \frac{1+(\lambda_{n,0}/2\sigma_n^2)}{1+(\lambda_{n,1}/2\sigma_n^2)} \exp\left\{\frac{y_n}{2\sigma_n^2}\right. \\ &\quad \left.\cdot \frac{(\lambda_{n,1}/2\sigma_n^2)-(\lambda_{n,0}/2\sigma_n^2)}{(1+(\lambda_{n,1}/2\sigma_n^2))(1+(\lambda_{n,0}/2\sigma_n^2))}\right\} \end{aligned} \quad (13)$$

When there is no fading, it becomes

$$\Lambda_n(y_n) = \exp\left(\frac{\lambda_{n,0}-\lambda_{n,1}}{2\sigma_n^2}\right) \frac{I_0(\sqrt{(\lambda_{n,1}/\sigma_n^2)(y_n/\sigma_n^2)})}{I_0(\sqrt{(\lambda_{n,0}/\sigma_n^2)(y_n/\sigma_n^2)})} \quad (14)$$

Next, we define the fixed-dwell, sequential, and truncated sequential decision schemes, based on the likelihood ratio (12).

In a fixed-dwell (single-dwell) or fixed-sample-size (FSS) decision scheme, the length of the integration is fixed *a priori* and a decision is based on the resulting test statistic. If the integration time is from 0 to MT_c, the FSS test compares $\Lambda_M(y_M)$ to a threshold τ and decide H_1 if it is $\geq \tau$ or H_0 otherwise. Since the likelihood ratio is monotonic, the FSS test can also be written as

$$\underline{\text{FSS:}} \qquad y_M \begin{cases} \geq \tau' = \Lambda_M^{-1}(\tau) & \text{say } H_1 \\ < \tau' & \text{say } H_0 \end{cases} \quad (15)$$

where $\Lambda_n^{-1}(.)$ is the inverse function of $\Lambda_n(.)$. The determination of M and τ will be discussed in the next section.

A sequential probability ratio test (SPRT) [19] compares the likelihood ratio $\Lambda_n(y_n)$ against two constant thresholds A and B, where $A > B > 0$, for $n = 1, 2, 3, \ldots$, until one of these thresholds is reached. If the upper threshold (A) is reached first, then H_1 is accepted. On the other hand, if the lower threshold (B) is reached first, H_0 is accepted instead. If, at time $t = nT_c$, neither of these thresholds is reached, then one more chip duration is integrated and the test continues for the next value of n. It is impractical for real-time implementation to compute the likelihood ratio at each n. To alleviate this problem, we rewrite the SPRT using the inverse likelihood ratio as

$$\underline{\text{SPRT:}} \qquad y_n \begin{cases} \geq A'(n) \equiv \Lambda_n^{-1}(A) & \text{say } H_1 \\ \leq B'(n) \equiv \Lambda_n^{-1}(B) & \text{say } H_0 \\ \text{otherwise,} & \text{continue} \end{cases} \quad (16)$$

Although the thresholds $A'(n)$ and $B'(n)$ are now functions of n, they can be pre-calculated and stored and the test (16) can be easily performed in real-time. Note that Y_n and Y_{n+1} are dependent random variables. However, Wald's inequalities [19], which relate Type I and Type II errors (i.e., false-alarm and miss probabilities, respectively) to A and B, still hold:

$$A \leq \frac{1-\beta}{\alpha} \quad \text{and} \quad B \geq \frac{\beta}{1-\alpha} \quad (17)$$

with α and β being the resulting false-alarm and miss probabilities, respectively. If the excess over the boundary is small when the test terminates, which is the case when the errors and the signal-to-noise ratio are small (which implies that the average test length is large), the inequalities (17) can be approximated by $A \approx \frac{1-\beta}{\alpha}$ and $B \approx \frac{\beta}{1-\alpha}$. Since $1-\beta \leq 1$ and $1-\alpha \leq 1$, (17) implies

$$\alpha \leq \frac{1}{A} \quad \text{and} \quad \beta \leq B \tag{18}$$

The inequalities in (18) are useful in choosing the values of the thresholds when strict upper bounds on the errors are desired. The bounds in (18) are tight when the errors and the signal-to-noise ratio are small.

The SPRT has been shown to be optimum, in the sense of minimizing the average test length under H_0 and H_1 among all tests for a given pair of Type I and Type II errors, if the samples are i.i.d. [20]. In the case of non-i.i.d. samples, as in our situation here, no optimality has been established. However, the performance is still good and much better than the FSS likelihood ratio test, which will be confirmed by numerical results in Section IV.

One drawback of the SPRT is that there is no upper bound on the test length. Therefore, it is possible that the decision processor can be stuck at a particular uncertainty phase for a long period of time, which is undesirable. Such events can occur (and did occur in our simulations), especially when the phase difference corresponds to neither H_0 nor H_1, namely, $1/2 < |j+\gamma| < 1/\Delta$. To avoid a lengthy test, we propose to put an upper bound on the test length, yielding a truncated SPRT, which can be described by

$$\underline{\text{TSPRT:}} \quad \text{if } n < \hat{n},\ y_n \begin{cases} \geq \hat{A}'(n) \equiv \Lambda_n^{-1}(\hat{A}) & \text{say } H_1 \\ \leq \hat{B}'(n) \equiv \Lambda_n^{-1}(\hat{B}) & \text{say } H_0 \\ \text{otherwise,} & \text{continue} \end{cases}$$

$$\text{if } n = \hat{n},\ y_{\hat{n}} \begin{cases} \geq \Lambda_{\hat{n}}^{-1}(\hat{\tau}) \equiv \hat{\tau}' & \text{say } H_1 \\ < \Lambda_{\hat{n}}^{-1}(\hat{\tau}) & \text{say } H_0 \end{cases} \tag{19}$$

Therefore, the test is truncated at $n = \hat{n}$ if it has not previously terminated. The design of the thresholds $\hat{A}$, $\hat{B}$, $\hat{\tau}$, and truncation point $\hat{n}$ will be addressed in the next section.

The average time to reach acquisition can be derived [21] in terms of the expected sample sizes, the detection and false-alarm probabilities, and the penalty time of a false alarm, using the flow-graph technique presented in [3]. If the detection and false-alarm probabilities and the penalty time are considered fixed, then the average acquisition time is a linear combination of the expected sample sizes at various phase offsets [21]. In comparing two acquisition schemes of the same search strategy (e.g., serial search having the same amount of phase update) but with different alignment tests, the test with the smaller expected sample sizes will have the smaller expected acquisition time, provided that the two schemes have the same false-alarm and miss probabilities and the same false-alarm penalty time. Therefore, we will compare acquisition schemes only in terms of the expected sample size.

Note that noncoherent single-dwell, two-dwell, and multiple-dwell schemes with continuous integration and reset integration were studied by Hall and Weber [9][10]. In previously reported sequential schemes [11]-[17], the integrators are reset periodically (integrate and dump), in order to obtain independent (or nearly independent) and identically distributed samples, so that design and analysis can be simplified. However, resetting the integrator induces loss in the effective signal-to-noise ratio at the decision processor, as pointed out by Hall and Weber [10]. This noncoherent combining loss may be intuitively explained as follows. Since the results of different integration intervals are squared and then summed, the signal component of the test statistic is enhanced (increased) under H_0 but it is essentially unchanged under H_1. Thus, the effective signal-to-noise ratio is reduced. On the other hand, if the integration is continuous, the signal component under H_0 would stay low because of the autocorrelation properties of PN sequences. The SPRT (16) and TSPRT (19) proposed here use continuous integration, which generates dependent and non-identically distributed samples. We would need to properly design the decision thresholds so that the decision processor can perform as desired. In the next section, methods are described for designing the thresholds of the SPRT (16) and the TSPRT (19).

III. Design of Decision Processors

Denote the chips of the PN sequence by c_k, $k = \ldots, 0, 1, 2, \ldots$, where $c_k = \pm 1$, and let the period be $N = 2^m - 1$, i.e., $c_{k+N} = c_k$ for all k. The PN waveform can be written as

$$a(t) = \sum_{k=-\infty}^{\infty} c_k p_{T_c}(t - kT_c) \tag{20}$$

where $p_{T_c}(t)$ is a rectangular pulse, equal 1 for $0 \leq t < T_c$ and zero elsewhere. Assuming that $i = 0$, S_n in (7) can be expressed as

$$S_n = \begin{cases} n(1-|\gamma|\Delta) + (|\gamma|\Delta)\sum_{k=0}^{n-1} c_k c_{k+\text{sgn}(\gamma)} \\ \qquad \equiv S_{n,1} \text{ under } H_1 \\ (1-\delta)\sum_{k=0}^{n-1} c_k c_{k+\ell} + \delta \sum_{k=0}^{n-1} c_k c_{k+1+\ell} \\ \qquad \equiv S_{n,0} \text{ under } H_0 \end{cases} \tag{21}$$

where $\text{sgn}(x)$ is 1 for $x \geq 0$ and -1 for $x < 0$, $\ell = \lfloor (j+\gamma)\Delta \rfloor$, and $\delta = (j+\gamma)\Delta - \lfloor (j+\gamma)\Delta \rfloor$. Here, $\lfloor x \rfloor$ denotes the integer part of x. Note that $0 \leq \delta < 1$ and that $\ell \neq 0$. Defining the (per-chip) signal-to-noise ratio as $\text{SNR} = A_0^2 T_c/(2N_0)$, we have

$$\frac{\lambda_n}{2\sigma_n^2} = \begin{cases} \frac{1}{n}(\text{SNR})S_{n,1}^2 & \text{under } H_1 \\ \frac{1}{n}(\text{SNR})S_{n,0}^2 & \text{under } H_0 \end{cases} \tag{22}$$

Since the receiver does not know the exact value of $S_{n,0}$ and $S_{n,1}$, some nominal values for the worst case must be used in designing the test. We use the following approximations for these values. For an m-sequence, if $i \neq 0$, $\sum_{k=0}^{n-1} c_k c_{k+i} = \sum_{k=0}^{n-1} c_{k+j}$ for some j. In addition, if the sequence length N is large, the chips are approximately independent and $\sum_{k=0}^{n-1} c_{k+j}$ is approximately Gaussian with mean 0 and standard deviation $\sqrt{n}$. Under H_1, with high probability $S_{n,1}$ is larger than its mean minus one standard deviation, which is $n(1-|\gamma|\Delta) - |\gamma|\Delta\sqrt{n} \approx n(1-|\gamma|\Delta)$ for large n. Therefore, we use $n(1-|\gamma|\Delta)$ as the nominal worst-case value for $S_{n,1}$. Under H_0, $S_{n,0} \approx \sum_{k=0}^{n-1} c_k c_{k+\ell} =$

$\sum_{k=0}^{n-1} c_{k+j}$ for some j. With high probability, $S_{n,0}$ is smaller than its mean plus its standard deviation, which is $0+\sqrt{n}$. Therefore, we use $\sqrt{n}$ as the nominal worst-case value for $S_{n,0}$. Numerical results show that these choices of the nominal values work well in satisfying the desired false-alarm and miss probabilities. With these nominal values, we have

$$\frac{\lambda_n}{2\sigma_n^2} \begin{cases} \approx n(\mathrm{SNR})(1-|\gamma|\Delta)^2 & \equiv \frac{\lambda_{n,1}}{2\sigma_n^2} \quad \text{under } H_1 \\ \lesssim \mathrm{SNR} & \equiv \frac{\lambda_{n,0}}{2\sigma_n^2} \quad \text{under } H_0 \end{cases} \tag{23}$$

We use these values for designing the tests.

The cumulative distribution function (cdf) of Y_n, corresponding to (9), is

$$F_{Y_n}(y_n) = 1 - Q\left(\sqrt{\frac{r(\lambda_n/\sigma_n^2)}{(1+r)+(\lambda_n/2\sigma_n^2)}}, \sqrt{\frac{(1+r)(y_n/\sigma_n^2)}{(1+r)+(\lambda_n/2\sigma_n^2)}}\right), \quad y_n \geq 0 \tag{24}$$

where $Q(\cdot,\cdot)$ is the Marcum's Q function, which is given by [22] $Q(\zeta,\xi) = \int_\xi^\infty xe^{-(x^2+\zeta^2)/2} I_0(\zeta x)\, dx$. There are iterative algorithms for calculating this function, see [23] for example. When there is no fading, the cdf becomes $F_{Y_n}(y_n) = 1 - Q(\sqrt{\lambda_n/\sigma_n^2}, \sqrt{y_n/\sigma_n^2})$, $y_n \geq 0$.

We now describe design methods for the thresholds of the FSS test, the SPRT, and the TSPRT, such that the decision probabilities are within the desired limits. We use α_d and β_d to denote the desired false-alarm and miss probabilities, respectively.

Consider the FSS test (15). Using the Q function, we can write the false-alarm and miss probabilities as

$$P_{fa} = Q(\sqrt{\lambda_{M,0}/\sigma_M^2}, \sqrt{\tau'/\sigma_M^2}) \tag{25}$$

$$P_{miss} = 1 - Q(\sqrt{\lambda_{M,1}/\sigma_M^2}, \sqrt{\tau'/\sigma_M^2}) \tag{26}$$

Given fixed P_{fa} and P_{miss}, this set of simultaneous nonlinear equations can be solved numerically for τ' and M.

For the SPRT (16), we can use the following to calculate the thresholds.

$$A = \frac{1}{\alpha_d} \quad \text{and} \quad B = \beta_d \tag{27}$$

From (18), we can see that the resulting false-alarm and miss probabilities are $P_{fa} \leq 1/A = \alpha_d$ and $P_{miss} \leq B = \beta_d$.

Next, consider the TSPRT (19). If α_{sprt} and β_{sprt} denote the decision probabilities of a SPRT with thresholds $\hat{A}$ and $\hat{B}$, and α_{fss} and β_{fss} denote the decision probabilities of a FSS likelihood ratio test with sample size $\hat{n}$ and threshold $\hat{\tau}$. It can be shown [24] that the decision probabilities of the TSPRT (19) are bounded by:

$$\alpha_{tsprt} \leq \alpha_{sprt} + \alpha_{fss} \tag{28}$$

$$\beta_{tsprt} \leq \beta_{sprt} + \beta_{fss} \tag{29}$$

To obtain $\alpha_{tsprt} \leq \alpha_d$ and $\beta_{tsprt} \leq \beta_d$, we can split each of the desired probabilities into a sum of two probabilities due, respectively, to the SPRT and the FSS test. In particular, we let $\alpha_d = p_0\alpha_d + (1-p_0)\alpha_d$ and $\beta_d = p_1\beta_d + (1-p_1)\beta_d$, where p_0 and p_1 are some constants in [0,1]. Letting

$$\alpha_{sprt} = p_0\alpha_d \quad \text{and} \quad \beta_{sprt} = p_1\beta_d \tag{30}$$

we can use (27) to get $\hat{A} = 1/[p_0\alpha_d]$ and $\hat{B} = p_1\beta_d$. Letting

$$\alpha_{fss} = (1-p_0)\alpha_d \quad \text{and} \quad \beta_{fss} = (1-p_1)\beta_d \tag{31}$$

we can design $\hat{\tau}$ and $\hat{n}$ using (25) and (26). The resulting TSPRT will have $\alpha_{tsprt} \leq \alpha_d$ and $\beta_{tsprt} \leq \beta_d$, according to (28)-(29). Note that if $p_0 = p_1 = 0$, the TSPRT becomes the FSS test since $\hat{A} = \infty$ and $\hat{B} = 0$. On the other hand, if $p_0 = p_1 = 1$, the TSPRT becomes the SPRT since $\hat{n} = \infty$. For values of p_0 and p_1 in between, the TSPRT can be viewed as a mixture of the SPRT and the FSS test [25]. Therefore, the choices of p_0 and p_1 control the degree of mixture between the SPRT and the FSS test. Suppose p_1 is held fixed. An increase in p_0 will decrease the upper threshold $\hat{A}$, while the lower threshold $\hat{B}$ is not affected. Since the TSPRT will be more likely to terminate at the upper threshold under H_1 than it will at the lower threshold under H_0, an increase in p_0 will reduce the ASN under H_1; the ASN under H_0 will be essentially unchanged. We can similarly argue that for a fixed p_0, an increase in p_1 will reduce the ASN under H_0, but its effect on the ASN under H_1 will be small.

We have addressed the decision thresholds design problems so that the resulting tests would have error probabilities within the desired values. Expressions for the average test lengths (or the expected sample size) of the SPRT and the TSPRT, however, are difficult to obtain. Therefore, simulations will be used.

IV. Performance Comparison

In this section, we compare numerical results on the expected sample size and the power function for the FSS test, the SPRT, and the TSPRT. The expected sample size, also called average sample number (ASN), is the average number of chips needed for the test to terminate. For the FSS test, the sample size M is constant. The power function is the probability of accepting H_1. Both the ASN and the power function are obtained as functions of the absolute phase offset $|j+\gamma|$. In all the results, the chip SNR is -10dB; the updating of the local PN sequence generator is half a chip, i.e., $\Delta = 1/2$; the value of γ used for the threshold design is 0.5, which is the worst case; and the m-sequence used for the simulations has a length of 1023, with primitive polynomial $1+x^2+x^5+x^6+x^{10}$. Figures 2 to 4 present results for the non-fading case, while Figures 5 to 7 are results for fading cases.

Figures 2(a) and 2(b) depict the ASN and power function of the FSS test, SPRT, and TSPRT with $\alpha_d = \beta_d = 0.005$. The TSPRT is designed with $p_0 = p_1 = 0.5$. When $|j+\gamma| = 0.5$, we have the hypothesis H_1 (since $j = 0$ and $\gamma = 0.5$). The hypothesis H_0 corresponds to $|j+\gamma| \geq 2.0$

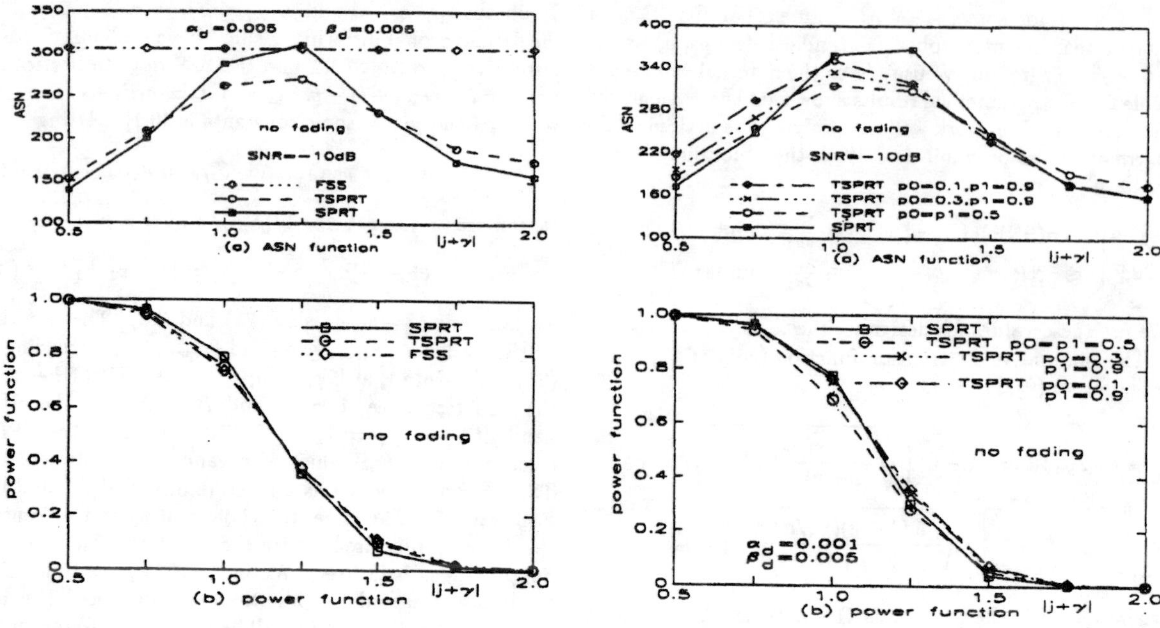

Fig. 2. ASN and power functions of the SPRT, TSPRT, and FSS test, without fading.

Fig. 4. ASN and power functions of the SPRT and TSPRT, no fading.

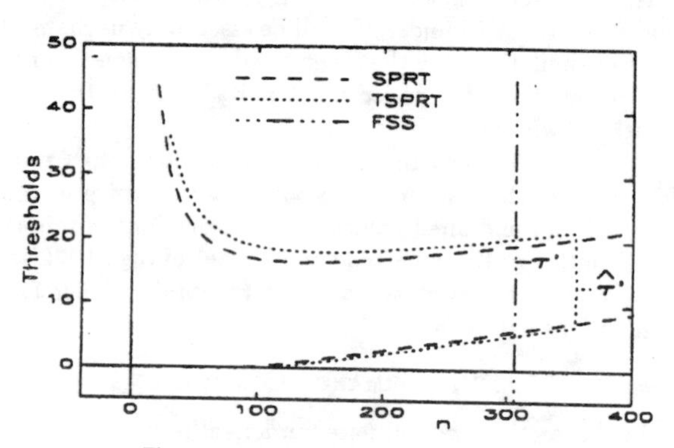

Fig. 3. Thresholds for the tests in Figure 2.

and it is represented by $|j+\gamma| = 2.0$ in the figures. When $0.5 < |j+\gamma| < 2$, the phase offset is small, but not small enough to be considered alignment. The results presented are the average over several initial phases i (see (1)). For all tests, the resulting P_{fa} and P_{miss} are no larger than the design values α_d and β_d, respectively, which suggests that the model of (23) works well. We can see from these figures that both the SPRT and the TSPRT are superior to the FSS test. The SPRT performs the best (has the smallest ASN) under H_0 and H_1. However, the TSPRT is the best when $|j+\gamma|$ is between 0.8 and 1.5, and it is only slightly worse than the SPRT under H_0 and H_1. Since the TSPRT has a bounded test length and the SPRT does not, the TSPRT is preferable to SPRT as a practical sequential test. The thresholds used for these three tests are plotted in Figure 3.

Among all the uncertainty phases to be tested in one period, one of them corresponds to H_1, one or a few of them (depending on the value of Δ) correspond to $0.5 < |j+\gamma| < 2.0$, while most of them correspond to H_0. Therefore, to reduce the expected acquisition time, we should minimize the ASN under H_0. This can be achieved by the TSPRT with large p_1 (close to 1). To demonstrate this, we plot the ASN and the power functions of the SPRT and the TSPRT designed for different p_0 and p_1 in Figure 4 with $\alpha_d = 0.001$ and $\beta_d = 0.005$. The TSPRT is designed for three cases: $p_0 = p_1 = 0.5$, $p_0 = 0.1$ and $p_1 = 0.9$, and $p_0 = 0.3$ and $p_1 = 0.9$. We can see that the ASN of the TSPRT under H_0 is essentially the same as that of the SPRT when $p_1 = 0.9$. Note that with a larger p_0 the ASN under H_1 is reduced, but the truncation point $\hat{n}$ increases, which becomes a tradeoff.

To see the effect of channel fading on the performance of a decision processor, we first design tests assuming no fading and $\alpha_d = \beta_d = 0.01$. Performances of these tests are then evaluated under channel fading. Results for the FSS test are plotted in Figure 5 for $r = 10$, $r = 1$, and $r = 0$, as well as $r = \infty$ (no fading). We observe that the false-alarm probability (i.e., at $|j+\gamma| = 2.0$) is essentially unchanged, but the detection probability (i.e., at $|j+\gamma| = 0.5$) is drastically reduced. For small r (severe fading), the reduction of the detection probability is significant. Results for the TSPRT with $p_0 = p_1 = 0.5$ are plotted in Figure 6. These results are obtained through simulations. In the simulations, the pdf of the fading random variable ψ in (2) is divided into 25 segments of equal areas. The mean value of each segment is used as the value of ψ in the simulations. Results from these 25 cases are averaged to obtained the data for plotting. Again, we ob-

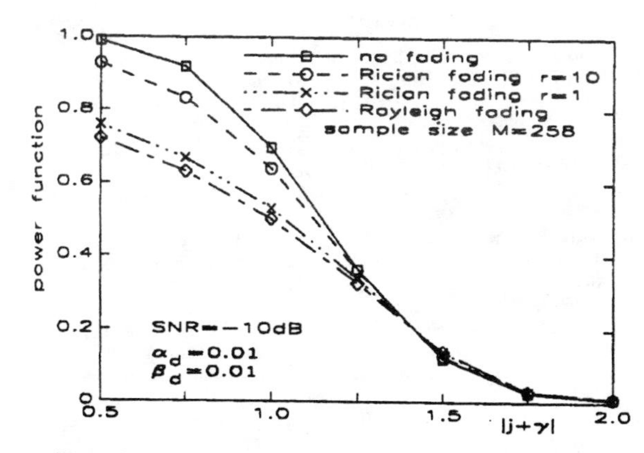

Fig. 5. Power functions of the FSS test with and without fading.

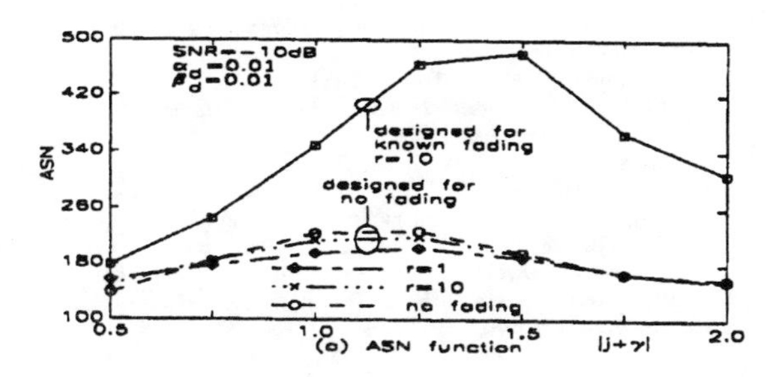

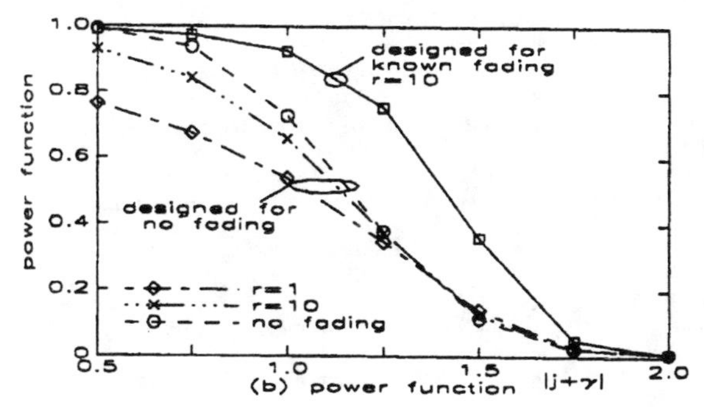

Fig. 6. ASN and power functions of the TSPRT with Rician fading.

serve that the false alarm probability and the ASN under H_0 are essentially unaffected by the fading, while the detection probability can be severely affected and the ASN under H_1 increases slightly. Results for the SPRT are not shown, but they behave similar to those of the TSPRT.

Next, we consider performances of tests designed with the knowledge of the fading, i.e., assume $f_\psi(\psi)$ is known, the sample size and the thresholds of the tests are adjusted accordingly to satisfy the desired false-alarm and detection probabilities. For the FSS test, the sample size increases to a very large number when the fading is severe. For example, the required sample size M increases from 258 (when there is no fading) to 509, 6617, and 8944 when $r = 10$, 1, and 0, respectively, to achieve $\alpha_d = 0.01$ and $\beta_d = 0.01$ at $\gamma = 0.5$. We show in Figure 6 the ASN and power function of the TSPRT designed for known fading, for $\alpha_d = \beta_d = 0.01$, $p_0 = p_1 = 0.5$, and $r = 10$. Compared to the case with no fading, we see again that the ASN has increased significantly, especially under H_0, to compensate for the fading. Since most of the uncertainty phases correspond to H_0, the expected acquisition time also increases significantly. This suggests that when there is fading, especially when it is severe, or when speed is critical, it does not pay to adjust the test to compensate the fading, because it takes a long time to reach a decision. Instead, it may be better to use a test designed under no fading and accept any reduction of the detection probability caused by the fading.

Finally, we present some results, for comparison, of a noncoherent sequential acquisition scheme using the SPRT with resetting integrators, which is the case in previous studies of sequential acquisition schemes. Suppose that the integrators (see Figure 1) are reset after every K chips, to obtain i.i.d. samples. We also assume, as in the analyses of previous investigations, that the partial correlation of the incoming and local sequences under H_0 is zero, i.e., $S_{K,0} = 0$, implying $\lambda_{K,0} = 0$. We used $\alpha_d = \beta_d = 0.01$, the same as in Figure 6, and performed simulations with no fading. When $K = 1$ (integrators reset after every chip), simulation shows that the detection probability is satisfied, but the false alarm probability is very large. This is because the out-of-phase autocorrelation of an m-sequence over only one chip can be +1 or −1 with almost equal probability. When the value is +1, it is the same as the in-phase autocorrelation over one chip; hence, it is indistinguishable from the case of phase alignment. To solve this problem, K is chosen to be larger than 1. We used $K = 20$ and $K = 30$. The thresholds were first calculated using Wald's formulas and then minor threshold adjustments (by trial and error) were made so that the desired false-alarm and detection probabilities are satisfied. For $K = 20$, the thresholds are −4.136 and 4.595. Under H_1, we obtain detection probability of 0.989 (at $\gamma = 0.5$) and ASN of 267.9. Under H_0 ($\delta = 0.5$), the false-alarm probability is 0.012, with ASN of 334.9. For $K = 30$, the thresholds are −4.136 and 4.136. The resulting detection probability is 0.990 and false-alarm probability is 0.011, with the ASN being 205.8 under H_1 and 268.8 under H_0. Comparing with the results in Figure 6 (with no fading), which has ASN of 137.8 under H_1 and 157.8 under H_0 (with detection probability of 0.992 and false-alarm probability of 0.010), we see that the SPRT with resetting integrators has much higher ASN to achieve the desired error probabilities.

V. Conclusion

We have studied noncoherent acquisition of PN sequences for DS/SS systems with and without channel fading. A SPRT and a TSPRT are proposed for testing whether or not the incoming and the local sequences are in (coarse) alignment. The test statistics are obtained from continuous integration and consequently they are dependent and non-identical from chip to chip. Design methods for the decision thresholds are proposed for the SPRT and the TSPRT. Both of these sequential schemes outperform the

FSS test. The SPRT has the smallest ASN under H_0 and H_1. However, the TSPRT has the advantages that its test length is bounded and that it can outperform the SPRT when the absolute phase offset is between 0.5 (H_1) and 2.0 (H_0). Furthermore, the TSPRT is almost as good as the SPRT under H_0 and H_1.

The effect of channel fading on the test performance was also investigated. It was found that the false-alarm probability and the ASN under H_0 are essentially unaffected by the fading. However, the detection probability can be severely reduced. If the tests are designed to compensate for the known fading, the sample size has to increase significantly to restore the detection probability.

ACKNOWLEDGEMENT

We would like to thank the anonymous reviewers for several useful comments, especially one of the reviewers who pointed out the proof for (9), which is more intuitively appealing than the authors' original proof.

REFERENCES

[1] R. Pickholtz, D. Schilling, and L. Milstein, "Theory of spread-spectrum communications-A tutorial," *IEEE Trans. Commun.*, Vol. COM-30, pp.855-884, May 1982.

[2] D.M. DiCarlo and C.L. Weber, "Multiple dwell serial search: performance and application to direct sequence code acquisition," *IEEE Trans. Commun.*, Vol. COM-31, pp. 650-659, May 1983.

[3] A. Polydoros and C.L. Weber, "A unified approach to serial search spread-spectrum code acquisition - Part I: general theory," *IEEE Trans. Commun.*, Vol. COM-32, pp. 542-549, May 1984.

[4] A. Polydoros and C.L. Weber, "A unified approach to serial search spread-spectrum code acquisition - Part II: a matched-filter receiver," *IEEE Trans. Commun.*, Vol. COM-32, pp. 550-560, May 1984.

[5] S.S. Rappaport and D.M. Grieco, "Spread-spectrum signal acquisition: methods and technology," *IEEE Commun. Magazine*, Vol. 22, pp. 6-21, June 1984.

[6] S. Davidovici, L.B. Milstein, and D.L. Schilling, "A new rapid acquisition technique for direct sequence spread-spectrum communications," *IEEE Trans. Commun.*, Vol. COM-32, pp. 1161-1168, Nov. 1984.

[7] A. Polydoros and M.K. Simon, "Generalized serial search code acquisition: the equivalent circular state diagram," *IEEE Trans. Commun.*, Vol. COM-32, pp. 1260-1268, Dec. 1984.

[8] E. Sourour, S.C. Gupta, and W. Refai, "Direct Sequence Spread Spectrum Serial Acquisition in a Nonselective and Frequency Selective Rician Fading Channel," *MILCOM Conf. Record*, pp. 5.7.1.-5.7.5, 1990.

[9] D. Hall and C.L. Weber, "Noncoherent sequential acquisition of DS waveforms," *MILCOM Conf. Record*, pp.13.3.1-5, 1986.

[10] D. Hall and C.L. Weber, "Two Methods for Noncoherent Acquisition of DS Waveforms," *MILCOM Conf. Record*, pp. 13.7.1-13.7.5, 1987.

[11] H. Meyr and G. Polzer, "A simple method for evaluating the probability density function of the sample number for the optimum sequential detector," *IEEE Trans. Commun.*, Vol. COM-35, pp. 99-103, Jan. 1987.

[12] G.M. Comparetto, "A generalized analysis for a dual threshold sequential detection PN acquisition receiver," *IEEE Trans. Commun.*, Vol. COM-35, pp. 956-960, Sept. 1987.

[13] Y.T. Su, "Rapid Code Acquisition Algorithms Employing PN Matched Filters," *IEEE Trans. on Communications*, Vol. 36, pp. 724-733, June 1988.

[14] Y.T. Su and C.L. Weber, "A Class of Sequential Tests and Its Applications," *IEEE Trans. on Communications*, Vol. 38, pp. 165-171, Feb. 1990.

[15] K.K. Chawla and D.V. Sarwate, "Coherent Acquisition of PN Sequences for DS/SS Systems Using a Random Sequence Model and the SPRT," *Proc. of the 1990 Conf. on Information Sciences and Systems*, Princeton Univ., New Jersey, pp.52-57, March 1990.

[16] Y.H. Lee and S. Tantaratana, "Acquisition of PN Sequences for DS/SS Systems Using a Truncated Sequential Probability Ratio Test," *Journal of the Franklin Institute*, Vol. 328, pp. 231-248, 1991.

[17] M.K. Simon, J. Omura, R. Scholtz, and K. Levitt, *Spread Spectrum Communication*, Volume III. Rockville, MD: Computer Science Press, 1985.

[18] D. Divsalar and M.K. Simon, "The design of trellis-coded MPSK for fading channel: performance criteria," *IEEE Trans. Commun.*, Vol. COM-36, pp. 1004-1012, Sept. 1988.

[19] A. Wald, *Sequential Analysis*, John Wiley and Sons, New York, 1947.

[20] E.L. Lehmann, *Testing Statistical Hypotheses*. New York: Wiley, 1959.

[21] A.W. Lam and S. Tantaratana, "Mean acquisition time for noncoherent PN sequence sequential acquisition schemes," *MILCOM Conf. Record*, pp. 784-788, 1993.

[22] C.W. Helstrom, *Statistical Theory of Signal Detection*. London: Pergamon Press, 1968.

[23] L.E. Brennan and I.S. Reed, "A Recursive Method of Computing the Q Function'" *IEEE Trans. on Inform. Theory*, Vol. IT-11, pp.312-313, April 1965.

[24] S. Tantaratana and A.W. Lam, "Noncoherent sequential acquisition for DS/SS systems," *Proc. 29th Annual Allerton Conf. on Comm., Control, and Computing*, Univ. of Illinois, Oct. 1991, pp. 370-379.

[25] S. Tantaratana and H.V. Poor, "Asymptotic efficiencies of truncated sequential tests," *IEEE Trans. Inform. Theory*, Vol. IT-28, pp. 911-923, 1982.

Sawasd Tantaratana (S'75-M'78-SM'94) received the B.E.E. degree with high distinction from the University of Minnesota, Minneapolis, in 1971, the M.S. degree in electrical engineering from Stanford University, Stanford, Calif., in 1972, and the M.A. and Ph.D. degrees in electrical engineering from Princeton University, Princeton, N.J., in 1976 and 1977, respectively.

Prior to joining the University of Massachusetts in 1986, he had been on the faculty of the King Mongkut's Institute of Technology, Thonburi, Thailand, and Auburn University, Alabama, and on the Technical Staff of AT&T Bell Laboratories. From 1980-1981 he was a Visiting Assistant Professor at the University of Illinois at Urbana-Champaign. Currently, he is an Associate Professor in the Department of Electrical and Computer Engineering, University of Massachusetts at Amherst.

Dr. Tantaratana is a member of Eta Kappa Nu and Sigma Xi.

Alex W. Lam (S'81-M'87) received the B.S., M.S., and Ph.D. degrees in electrical and computer engineering from the University of Illinois at Urbans-Champaign, IL, in 1982, 1984, and 1987, respectively.

From August 1987 to August 1990, he was an assistant professor of electrical and computer engineering at the University of Massachusetts, Amherst. Since September 1990, he has been with the Naval Postgraduate School, Monterey, CA, where he is currently an associate professor of electrical and computer engineering. His research interests are digital communications theory, spread spectrum, and time-frequency analysis of signals and systems.

Patrick J. Vincent was born in New York City on July 18, 1962. He received the degree of Bachelor of Science in Electrical Engineering from the Polytechnic Institute of New York in 1984 and the degree of Master of Science in Electrical Engineering from the Naval Postgraduate School in 1992.

Lieutenant Vincent has served in the United States Navy since 1983. He has completed qualification as engineer officer of a naval nuclear power plant. He served aboard the nuclear powered submarine USS Alabama in various capacities from 1986 until 1989 and is qualified in submarine warfare. His research interests include spread spectrum communications and code synchronization.

A Closed-Loop Coherent Acquisition Scheme for PN Sequences Using an Auxiliary Sequence

Murat Salih and Sawasd Tantaratana, *Senior Member, IEEE*

***Abstract*— We propose a closed-loop system for the acquisition of the pseudo-noise (PN) signal in direct-sequence spread-spectrum (DS/SS) systems. We introduce a novel idea of using an auxiliary signal, as opposed to the PN signal itself, for correlation with the incoming signal. The cross-correlation function of the auxiliary signal and the PN signal has a triangle shape that covers essentially the entire period of the PN signal. Consequently, their correlation provides the direction for the phase update of the local signal generator in the acquisition scheme. With coherent demodulation, the mean and variance of the acquisition time are derived under additive white Gaussian noise (AWGN). They are compared to those of the conventional serial-search acquisition system. Results suggest that the proposed system acquires the PN phase at least twice faster, with significantly smaller acquisition time variance, than the conventional system.**

I. Introduction

An important task of a direct-sequence spread-spectrum (DS/SS) receiver is to generate a local replica of the transmitted pseudo-noise (PN) sequence and to synchronize its phase to that of the incoming one. This synchronization process is typically accomplished in two stages: acquisition and tracking. In the first stage (acquisition), the phases of the two sequences are aligned to within a small range of one chip or less. Then, in the second stage the tracking circuitry makes the fine adjustment and maintains the two codes in synchronism by means of a closed loop operation. Many types of DS/SS acquisition and tracking systems have been reported and analyzed [1].

In most DS/SS acquisition techniques, the received signal and the local PN sequence are correlated to produce an output which is used for detecting whether the two codes are in (coarse) alignment or not. There are a variety of acquisition schemes, depending on the type of the detector and the search algorithm [1]. An acquisition scheme is either coherent or noncoherent, depending on whether or not the carrier phase is used during acquisition.

Without data modulation during acquisition, the baseband representation of a received DS/SS signal during acquisition is of the following form:

$$u(t) = \sqrt{2P}c(t-\tau) + z(t) \tag{1}$$

where P is the average signal power assumed to be known, τ is the unknown code phase to be estimated by an acquisition system, $z(t)$ is the zero mean AWGN with auto-correlation function $E\{z(t+\tau)z^*(t)\} = 2N_0\delta(\tau)$, and $c(t)$ is the PN code waveform expressed as

$$c(t) = \sum_{k=-\infty}^{\infty} c_k P_{T_c}(t-kT_c). \tag{2}$$

In this expression, c_k is the kth chip of a binary PN sequence which is periodic with period N, i.e., $c_k = c_{k+N}$ for all k, and $P_{T_c}(\cdot)$ is the unit-amplitude rectangular pulse in the interval $[0, T_c]$, where T_c is the chip duration. We assume that $\{c_k\}$ is an m-sequence.

A technique for PN synchronization is the maximum likelihood (ML) estimation of the phase. Practical implementations of the ML algorithm can be parallel, serial, or hybrid [2]. The receiver quantizes the phase uncertainty region into a number of discrete phases (cells). Locally generated PN waveforms having these phases are correlated with the received signal. In a parallel scheme, the correlations are performed in parallel and the phase estimate is the one corresponding to the largest correlator output. The number of correlation branches is equal to the number of cells, which is proportional to the length of the uncertainty region. Such a receiver is impractical for PN sequences with long periods. In a serial scheme, the local phases are checked for phase alignment one at a time. When there is no *a priori* information on the received PN code phase, the phases in the uncertainty region are usually searched in a uniform, unidirectional fashion from one end of the region to the other. When there is *a priori* information on the received PN code phase, the search starts in the region of highest code phase certainty and expands to regions of lesser certainty [3], [4]. This reduces the acquisition time compared to the uniform search, especially when the PN code is very long. A hybrid scheme is a mixture of the parallel and serial schemes.

The simplest of serial-search techniques [5]–[10] is the single-dwell system, where a single correlator with a fixed integration duration is used to examine each of the possible phases until a phase alignment is detected. When phase alignment is rejected, the local PN phase is updated to the next one. A coherent single-dwell acquisition system is shown in Fig. 1. The dwell time (integration duration) is nT_c seconds. It is assumed that each phase update is T_c.

Manuscript received April 27, 1995; revised November 22, 1995. This paper was presented in part at the 28th Asilomar Conference on Signals, Systems, and Computers, Pacific Grove, CA, October 31–November 2, 1994.

M. Salih was with the Department of Electrical and Computer Engineering, University of Massachusetts, Amherst, MA 01003-5110 USA. He is now with Karel Inc., Ankara, Turkey.

S. Tantaratana was with the Department of Electrical and Computer Engineering, University of Massachusetts, Amherst, MA 01003-5110 USA. He is now with the Telecommunication Laboratory, National Electronics and Computer Technology Center, Ministry of Science, Technology and Environment, Rama VI Road, Rajthevi, Bangkok, Thailand (email: tsawas@nwg.nectec.or.th).

Publisher Item Identifier S 0733-8716(96)05217-1.

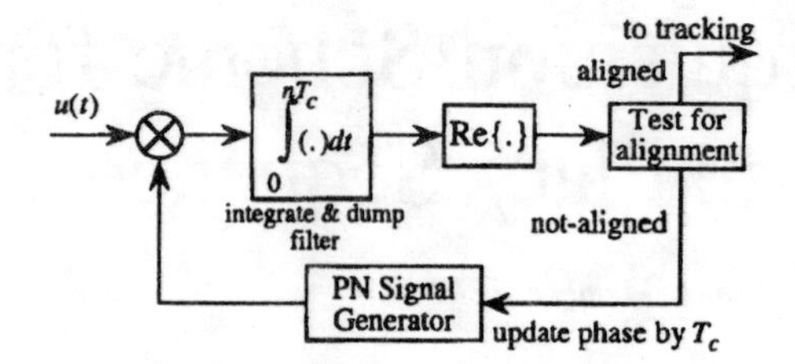

Fig. 1. Coherent fixed-dwell serial-search acquisition system.

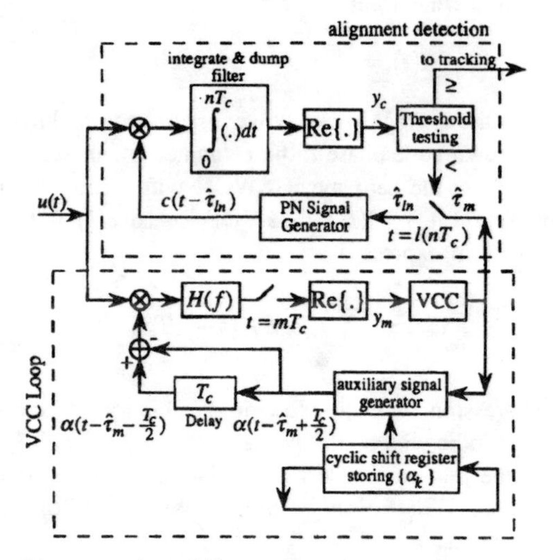

Fig. 2. The proposed acquisition system.

In this paper, we propose a closed-loop acquisition scheme which provides the direction for the local phase update. This is accomplished by using an auxiliary signal, as opposed to the PN signal itself, to correlate with the incoming signal [11].

II. The Proposed Acquisition System

We propose the acquisition system shown in Fig. 2. The system consists of two subsystems: the alignment detection subsystem and the voltage-controlled-clock (VCC) loop. In the following subsections, we describe in detail the operation of this proposed system.

A. Auxiliary Signal

The most important aspect of this system is that we use a new local waveform $\alpha(t)$. It is defined as

$$\alpha(t) = \sum_{m=-(N-3)/2}^{(N-3)/2} \left[\frac{N-1}{2} - |m|\right] c(t - mT_c) \quad (3)$$

$$= \sum_{k=-\infty}^{\infty} \alpha_k P_{T_c}(t - kT_c) \quad (4)$$

where

$$\alpha_k = \sum_{m=-(N-3)/2}^{(N-3)/2} \left[\frac{N-1}{2} - |m|\right] c_{k-m}. \quad (5)$$

The sequence $\{\alpha_k\}$ and the waveform $\alpha(t)$ are periodic with the respective periods N and NT_c, which are the same as the periods of $\{c_k\}$ and $c(t)$, respectively. As seen from (3), $\alpha(t)$ is a weighted sum of shifted versions of $c(t)$. The periodic cross-correlation function between $c(t)$ and $\alpha(t)$ is

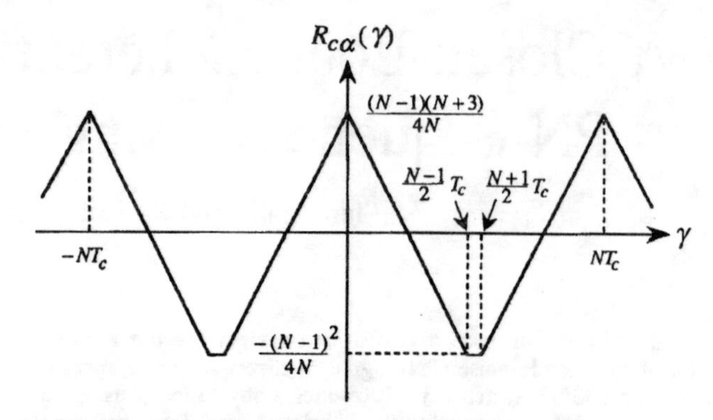

Fig. 3. The periodic cross-correlation function of $c(t)$ and $\alpha(t)$.

$$R_{c\alpha}(\gamma) = \frac{1}{NT_c}\int_0^{NT_c} c(t+\gamma)\alpha(t)dt$$

$$= \sum_{m=-(N-3)/2}^{(N-3)/2} \left[\frac{N-1}{2} - |m|\right] R_{cc}(\gamma + mT_c) \quad (6)$$

where $R_{cc}(\cdot)$ is the periodic autocorrelation function of $c(t)$ given by

$$R_{cc}(\gamma) = \begin{cases} 1 - \frac{N+1}{NT_c}|\gamma|, & |\gamma| \le T_c \\ -\frac{1}{N}, & T_c < |\gamma| \le \frac{NT_c}{2}. \end{cases} \quad (7)$$

It is straightforward to show that

$$R_{c\alpha}(\gamma) = \begin{cases} \frac{(N-1)(N+3)}{4N} - \frac{N+1}{NT_c}|\gamma|, & |\gamma| \le \frac{(N-1)T_c}{2} \\ -\frac{(N-1)^2}{4N}, & \frac{(N-1)T_c}{2} < \gamma < \frac{(N+1)T_c}{2} \end{cases} \quad (8)$$

which is plotted in Fig. 3. Note that $R_{cc}(\cdot)$ and $R_{c\alpha}(\cdot)$ have the same period of NT_c.

B. VCC Loop

The lower part of the receiver in Fig. 2 is the VCC loop. It is used for updating the phase of $\alpha(t)$ until it aligns with the phase of the received PN signal. It has a cyclic shift register for storing the auxiliary sequence $\{\alpha_k\}$. The shift register output drives the auxiliary signal generator to generate the auxiliary signal $\alpha(t + \frac{T_c}{2})$. The generator shifts the signal $\alpha(t + \frac{T_c}{2})$ in time by an amount equal to the code phase estimate formed by the VCC. All the signals shown in the VCC loop are signals at $t = mT_c$.

The received signal $u(t)$ is correlated with the difference of the early and late versions of the local code waveform $\alpha(t)$. The time separation between the early and late versions of the local code waveform $\alpha(t)$ is T_c seconds (i.e., one chip). The correlation filter has a frequency response $H(f)$ corresponding to an impulse response of

$$h(t) = \begin{cases} 1/M, & 0 \le t \le MT_c \\ 0, & \text{otherwise} \end{cases} \quad (9)$$

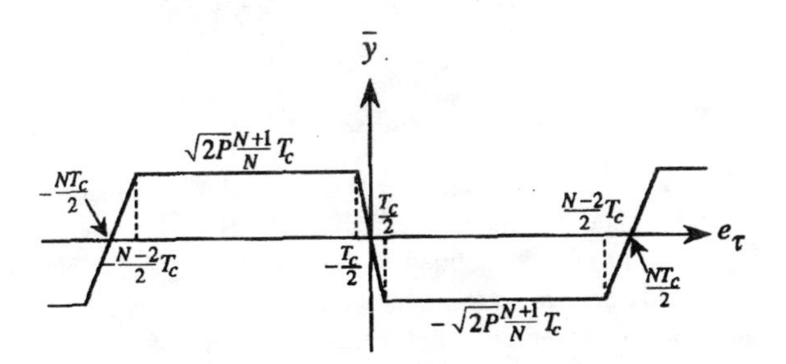

Fig. 4. The average VCC input signal for $M = 1$.

where M is some fixed integer. The filter output is sampled every T_c seconds, and the real-part of this signal is the error signal used to control the VCC. The VCC forms an estimate $\hat{\tau}_m$ at $t = mT_c$, which is updated every T_c seconds using the recursive formula

$$\hat{\tau}_{m+1} = \hat{\tau}_m + K_{\text{vcc}} y_{m+1}(\hat{\tau}_m, \tau), \quad m = 0, 1, 2, \cdots \tag{10}$$

where $\hat{\tau}_0$ is the initial code phase at $t = 0$, K_{vcc} is a constant, and $y_{k+1}(\hat{\tau}_k, \tau)$ is the VCC control signal at $t = (k+1)T_c$. The constant K_{vcc} determines the step size of the updating. The code phase estimate $\hat{\tau}_m$ formed by the VCC at $t = mT_c$ is used by the VCC loop until a new estimate $\hat{\tau}_{m+1}$ is formed at $t = (m+1)T_c$.

The input of the VCC is obtained as

$$y_{m+1}(\hat{\tau}_m, \tau) = \frac{1}{M} \sum_{k=m+1-M}^{m} [s_{k+1}(\hat{\tau}_k, \tau) + \eta_{k+1}] \tag{11}$$

where $s_{k+1}(\hat{\tau}_k, \tau)$ and η_{k+1} are defined as

$$s_{k+1}(\hat{\tau}_k, \tau) = \sqrt{2P} \int_{kT_c}^{(k+1)T_c} c(t-\tau) x(t - \hat{\tau}_k) dt \tag{12}$$

$$\eta_{k+1} = \text{Re}\left\{ \int_{kT_c}^{(k+1)T_c} z(t) x(t - \hat{\tau}_k) dt \right\} \tag{13}$$

where $x(t) = \alpha(t - \frac{T_c}{2}) - \alpha(t + \frac{T_c}{2})$. The integer M does not have to be an integer multiple of N. Therefore, the received signal is partially correlated with the difference of early and late versions of $\alpha(t)$. At $t = kT_c$, the auxiliary signal generator produces the signal $\alpha(t - \hat{\tau}_k + \frac{T_c}{2})$ using the sequence $\{\alpha_k\}$ and the phase estimate $\hat{\tau}_k$ from the VCC.

At $t = mT_c$, code phase estimation error $e_{\tau,m}$ is defined as

$$e_{\tau,m} = \hat{\tau}_m - \tau = (j_m + \gamma_m)T_c \tag{14}$$

for some integer j_m and $-0.5 < \gamma_m \le 0.5$. This error satisfies the recursion (10), i.e.,

$$e_{\tau,m+1} = e_{\tau,m} + K_{\text{vcc}} y_{m+1}(\hat{\tau}_m, \tau). \tag{15}$$

For convenience, we write the incoming phase as

$$\tau = (i + \beta)T_c \tag{16}$$

for some integer i and $0 \le \beta < 1$, and the phase estimate as

$$\hat{\tau}_m = (j'_m + \gamma'_m)T_c \tag{17}$$

where j'_m is an integer and $-0.5 < \gamma'_m \le 0.5$.

To obtain an expression for (12), we substitute (2) into (12) and use the definitions (14) and (16), yielding

$$\begin{aligned} s_{k+1}(\hat{\tau}_k, \tau) = \sqrt{2P}T_c \cdot \Bigg\{ & c_{k-i}\Bigg[\left(\frac{1.5 - \beta + \gamma_k}{2} - \frac{|\gamma_k + \beta - 0.5|}{2}\right) \cdot (\alpha_{k-i-1-j_k} - \alpha_{k-i-j_k}) \\ & + \left(\frac{0.5 - \beta - \gamma_k}{2} + \frac{|\gamma_k + \beta - 0.5|}{2}\right) \cdot (\alpha_{k-i-j_k} - \alpha_{k+1-i-j_k})\Bigg] + c_{k-i-1} \\ & \times \Bigg[\left(\frac{0.5 + \beta - \gamma_k}{2} - \frac{|\gamma_k + \beta - 0.5|}{2}\right) \cdot (\alpha_{k-i-1-j_k} - \alpha_{k-i-j_k}) \\ & + \left(\frac{\beta + \gamma_k - 0.5}{2} + \frac{|\gamma_k + \beta - 0.5|}{2}\right) \cdot (\alpha_{k-2-i-j_k} - \alpha_{k-1-i-j_k})\Bigg]\Bigg\}. \end{aligned} \tag{18}$$

The noise η_{k+1} in (13) is a zero-mean Gaussian random variable. Its variance can be calculated as

$$\sigma^2_{\eta_{k+1}} = N_0 T_c \left[(0.5 + \gamma'_k)(\alpha_{k-1-j'_k} - \alpha_{k-j'_k})^2 + (0.5 - \gamma'_k)(\alpha_{k-j'_k} - \alpha_{k+1-j'_k})^2\right]. \tag{19}$$

Over time, the VCC input signal follows its average value. For $M \ne 1$, derivation of this average signal is very difficult because $y_{m+1}(\hat{\tau}_m, \tau)$ in (11) depends on M previous code phase estimates of the VCC. Assuming that the noise is ergodic, the average VCC input signal $\bar{y}$ for $M = 1$ is

$$\bar{y} = \frac{1}{N} \sum_{m=0}^{N-1} E_{\eta_m}\{y_m(\hat{\tau}, \tau)\} = \frac{1}{N} \sum_{m=0}^{N-1} s_m(\hat{\tau}, \tau) \tag{20}$$

where $E_{\eta_m}\{.\}$ is the statistical expectation with respect to noise η_m and $\hat{\tau}$ is the code phase estimate. Equation (20) can be expressed in terms of the periodic cross-correlation function $R_{c\alpha}(\cdot)$ as

$$\bar{y} = \sqrt{2P}T_c \left[R_{c\alpha}\left(e_\tau + \frac{T_c}{2}\right) - R_{c\alpha}\left(e_\tau - \frac{T_c}{2}\right) \right] \tag{21}$$

where $e_\tau = \hat{\tau} - \tau$. It is plotted in Fig. 4. If $0 < e_{\tau,m} \le \frac{NT_c}{2}$, it is seen from Fig. 4 that the average VCC control signal for that code phase error is negative. Hence, with the recursion (15), we have $e_{\tau,m+1} < e_{\tau,m}$. For $-\frac{NT_c}{2} \le e_{\tau,m} < 0$, the average VCC control signal is positive, so that $|e_{\tau,m+1}| < |e_{\tau,m}|$. Therefore, the proposed receiver in Fig. 2 is able to extract the code phase estimate update direction from the incoming signal for the whole code phase uncertainty region $[-\frac{NT_c}{2}, \frac{NT_c}{2}]$. The pull-in range is NT_c and the estimation error goes to zero regardless of the initial error $e_{\tau,0}$ at $t = 0$.

The difference of early and late correlations in Fig. 2 causes the difference of the early and late versions of the periodic cross-correlation function in (21). Therefore, the wide pull-in range of the proposed receiver is due to the wide cross-correlation function $R_{c\alpha}(\cdot)$.

C. Alignment Detector

The upper part in Fig. 2 is the phase alignment detector. It has a PN signal generator. It correlates the incoming signal $u(t)$ with the local PN code waveform $c(t-\hat{\tau}_{ln})$ for a fixed dwell-time of nT_c seconds, where n is a constant. A new value of the code phase estimate $\hat{\tau}_m$ is fed to the detector every nT_c seconds, i.e., at $m = 0, n, 2n, \ldots$, and is tested for phase alignment. This continues until alignment is declared, which will initiate the tracking circuitry. Detection probability P_d is defined as the probability of declaring acquisition when the two phases are in fact aligned (H_1). Similarly, P_{fa} is defined as the probability of falsely declaring acquisition when the phases are not aligned (H_0). The hypotheses H_1 and H_0, for each integer l, are defined as

$$\begin{array}{lll} H_1: & |e_{\tau,ln}| \le \frac{T_c}{2} & \text{(alignment)} \\ H_0: & T_c \le |e_{\tau,ln}| \le \frac{NT_c}{2} & \text{(no alignment)} \end{array} \quad (22)$$

where $e_{\tau,ln}$ is given by (14) with $m = ln$. Substituting $e_{\tau,ln} = (j_{ln}+\gamma_{ln})T_c$ into the hypotheses definitions we obtain

$$\begin{array}{ll} H_1: & j_{ln} = 0 \text{ and } -0.5 < \gamma_{ln} \le 0.5 \\ H_0: & j_{ln} \in \{-\frac{N-1}{2}, \ldots, -2, 2, \ldots, \frac{N-1}{2}\} \\ & \text{and } -0.5 < \gamma_{ln} \le 0.5; \\ & \text{or } j_{ln} = -1 \text{ and } -0.5 < \gamma_{ln} \le 0; \\ & \text{or } j_{ln} = 1 \text{ and } 0 \le \gamma_{ln} \le 0.5. \end{array} \quad (23)$$

For simplicity, we assume $\tau = 0$ in the detector analysis without loss of generality. The correlator output $y_c(\hat{\tau}_{ln}, \tau)$ in Fig. 2 can be expressed as follows:

$$\begin{aligned} y_c(\hat{\tau}_{ln}, \tau) &= y_c(\hat{\tau}_{ln}, 0) = y_c(\hat{\tau}_{ln}) \\ &= \begin{cases} \sqrt{2P}T_c\lambda_1(n, \gamma_{ln}) + \eta \\ \sqrt{2P}T_c\lambda_0(n, \gamma_{ln}, j_{ln}) + \eta \end{cases} \end{aligned} \quad (24)$$

where

$$\lambda_1(n, \gamma_{ln}) = (1 - |\gamma_{ln}|)n + |\gamma_{ln}| \sum_{k=0}^{n-1} c_k c_{k-\mathrm{sgn}(\gamma_{ln})} \quad (25)$$

$$\begin{aligned} \lambda_0(n, \gamma_{ln}, j_{ln}) &= (1 - |\gamma_{ln}|) \sum_{k=0}^{n-1} c_k c_{k-j_{ln}} \\ &\quad + |\gamma_{ln}| \sum_{k=0}^{n-1} c_k c_{k-j_{ln}-\mathrm{sgn}(\gamma_{ln})} \end{aligned} \quad (26)$$

and η is a zero-mean Gaussian noise with variance $\sigma_\eta^2 = nN_0T_c$. Since both $\lambda_1(n, \gamma_{ln})$ and $\lambda_0(n, \gamma_{ln}, j_{ln})$ depend on the unknown γ_{ln}, in order to obtain acceptable performance for all γ_{ln} the test is designed under the worst case, which occurs when $\gamma_{ln} = 0.5$. For $\gamma_{ln} = 0.5$, $j_{ln} \in I$ under H_0, where $I = [-\frac{N-1}{2}, \cdots, -2, 1, 2, \cdots, \frac{N-1}{2}]$. Since $y_c(\hat{\tau}_{ln})$ is Gaussian with the same variance under both hypotheses, the optimum test is one which compares the observation $y_c(\hat{\tau}_{ln})$ against a threshold Γ, i.e.,

$$y_c(\hat{\tau}_{ln}) = \begin{cases} \ge \Gamma & \text{accept } H_1 \\ < \Gamma & \text{accept } H_0 \end{cases} \quad (27)$$

P_d and P_{fa} for this test are as follows:

$$P_d = \Phi\left(\frac{\lambda_1(n, 0.5) - \Gamma'}{\sqrt{n/\mathrm{SNR}}}\right) \quad (28)$$

$$P_{fa} = \frac{1}{N-2} \sum_{j \in I} \Phi\left(\frac{\lambda_0(n, 0.5, j) - \Gamma'}{\sqrt{n/\mathrm{SNR}}}\right) \quad (29)$$

where $\Gamma' = \Gamma/(\sqrt{2P}T_c)$, $\Phi(\cdot)$ is the distribution function of a standard Gaussian random variable, and SNR is the per-chip signal-to-noise ratio defined as $\mathrm{SNR} = 2PT_c/N_0$. Note that the P_{fa} is defined as the average probability over all $j_{ln} \in I$ since the receiver does not know the value of j_{ln}. Let the desired detection and false alarm probabilities for the receiver in Fig. 2 be P_d^* and P_{fa}^*, respectively, then (28) and (29) can be solved for n and Γ'.

For the phase alignment detector and the VCC loop to work well together in detecting phase alignment, the code phase estimate should not change more than one chip duration T_c in n iterations. This is required so that the alignment detector does not skip over the interval of length T_c that contains the incoming code phase. This imposes a limit on the updating step size K_{vcc} in (10). When a phase estimate is fed to the detector, the test for phase alignment lasts nT_c seconds. In the mean time, the VCC loop updates the code phase estimate n times. The limit on K_{vcc} can be calculated by observing from Fig. 4 (for $M = 1$) that $|\bar{y}|$ does not exceed $\sqrt{2P}T_c(N+1)/N$. For $M > 1$, it is found by computer simulations that $|\bar{y}|$ does not exceed $\sqrt{2P}T_c(N+1)/N$ either. Hence, we use

$$K_{vcc} \le \frac{N}{n(N+1)\sqrt{2P}}. \quad (30)$$

III. Performance Analysis

We can evaluate the average acquisition time of the proposed receiver using the flow graph technique [5]. A flow graph for the acquisition process is shown in Fig. 5, where the states are defined as

$$\begin{aligned} &\text{State 0: } |e_{\tau,m}| \le \frac{T_c}{2} \\ &\text{State } j, j = 1, 2, \ldots, \frac{N-1}{2}: \\ &|e_{\tau,m}| \le |e_{\tau,m-1}| \text{ and } (j-0.5)T_c < |e_{\tau,m}| \le (j+0.5)T_c \\ &\text{State } k', k = 1, 2, \ldots, \nu: \\ &|e_{\tau,m-1}| < |e_{\tau,m}| \text{ and } (k-0.5)T_c < |e_{\tau,m}| \le (k+0.5)T_c \\ &\text{State } S\text{: start (start acquisition)} \\ &\text{State } F\text{: finish (achieve acquisition).} \end{aligned} \quad (31)$$

The starting error $e_{\tau,0}$ is equally likely to be anywhere in $[-\frac{NT_c}{2}, \frac{NT_c}{2}]$. Therefore, the starting state can be any of the states $0, 1, \cdots, \frac{N-1}{2}$ with the respective probabilities $P_0, P_1, \cdots, P_{\frac{N-1}{2}}$, where $P_0 = \frac{1}{N}$, $P_1 = P_2 = \cdots = P_{\frac{N-1}{2}} = \frac{2}{N}$. The value of ν is a parameter. It represents the number of chips the phase has traveled past an undetected phase alignment before the phase updating changes to the correct direction. Note that ν is a random variable, but we model it as a parameter to ease the analysis.

In Fig. 5, z represents the unit delay operator. A unit delay for the proposed receiver is nT_c seconds. At any state, except state 0, there is either a false alarm or no false alarm. In the case of a false alarm, we assume that it takes one unit of time to

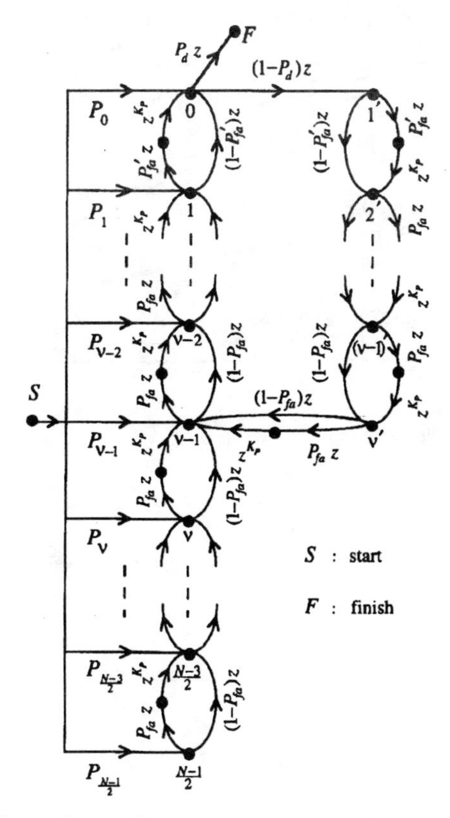

Fig. 5. System flow graph.

declare phase alignment, followed by K_p units of time ($K_p nT_c$ seconds) needed by the tracking receiver to realize that there is no phase alignment, i.e., $K_p nT_c$ is the penalty time of a false alarm. When there is no false alarm, it takes one unit of time to declare no phase alignment. When the system is at state j, $j = 1, 2, \cdots, \frac{N-1}{2}$, i.e., $(j - 0.5)T_c < |e_{\tau,m}| \leq (j + 0.5)T_c$, the code phase error magnitude is decreasing from one phase update to the next. State 0 occurs when the incoming code phase and the code phase estimate are within half a chip (i.e., the phases are in alignment). When the system is at state 0, it either detects or misses the alignment with probabilities P_d and $1 - P_d$, respectively. If alignment is detected, then the system goes into the finish state F, hence acquisition is achieved and tracking is initiated. If alignment is missed, then the magnitude of the error increases from one phase update to the next and the system goes through states $1', 2', \ldots, \nu'$. Due to the duration MT_c of the filter impulse response $h(t)$ in (9), the system reaches state ν' in roughly $\frac{M}{2}$ iterations after missing state 0. At state ν' the filter output changes polarity and the phase estimation error magnitude starts decreasing. Therefore, the next state after state ν' is the state $\nu - 1$. The phase estimation error magnitude decreases until it reaches state 0. This cycle goes on until acquisition is detected. ν can be found approximately by the following formula:

$$\nu \approx \left\lceil \frac{M}{2n} \right\rceil \tag{32}$$

where $\lceil \cdot \rceil$ is the integer part of a number. It is observed from the state transition diagram that the acquisition time gets longer as ν gets larger. Therefore, a small ν is preferred. However, a small ν means a small M. The filter with a small M cannot filter out the fluctuations on the VCC control signal caused by both noise and code chips, giving unsatisfactory operation. Therefore, M and ν should be sufficiently large. For a code period $N = 511$, $\nu \geq 10$ yields satisfactory performance.

Note that the false alarm probability for states 1 and $1'$, denoted by P'_{fa}, is different from P_{fa} for the other states. For states 1 and $1'$, we may have $\frac{T_c}{2} < |e_{\tau,m}| < T_c$, which is neither H_1 nor H_0. In this case $P'_{fa} \geq P_{fa}$. However, for large N, such as $N = 511$, assuming $P'_{fa} = P_{fa}$ has negligible effect on the results.

The generating function $U(z)$ is

$$U(z) = \sum_{m=0}^{\infty} P_{SF}(m) z^m \tag{33}$$

where $P_{SF}(m)$ is the probability of going from state S to state F in m steps. $U(z)$ is also the transformed signal at node F when a unit impulse is applied to node S. Applying flow graph reduction method yields

$$U(z) = \frac{\frac{2}{N} P_d z \left[\left(\sum_{l=0}^{\frac{N-1}{2}} H^l(z) \right) - 0.5 \right]}{1 - (1 - P_d) z H^{2\nu - 1}(z)} \tag{34}$$

where $H(z) = P_{fa} z^{K_p + 1} + (1 - P_{fa}) z$. Note that the probability of reaching acquisition (state F) once the acquisition process has started is one, which is equal to $U(1)$. $U(z)$ contains statistical information about the acquisition process. The average acquisition time $\bar{T}_{\text{acq}}$ can be derived from $U(z)$ as

$$\bar{T}_{\text{acq}} = \left[\sum_{m=0}^{\infty} m P_{SF}(m) \right] nT_c = \left. \frac{dU(z)}{dz} \right|_{z=1} nT_c. \tag{35}$$

Substituting (34) into (35), we obtain

$$\bar{T}_{\text{acq}} = nT_c \left[(P_{fa} K_p + 1) \frac{N^2 - 1}{4N} + \frac{1 + (1 - P_d)(2\nu - 1)(P_{fa} K_p + 1)}{P_d} \right]. \tag{36}$$

Note that the first term in (36) is the average time to reach state 0 from state S. The second term is the average time to reach state F from state 0. We compare this average acquisition time with that of the coherent serial-search single-dwell acquisition receiver shown in Fig. 1. Using the formulas in [5], the average acquisition time for Fig. 1 is

$$\bar{T}_{\text{acq,serial}} = \frac{2 + (2 - P_d)(N - 1)(P_{fa} K_p + 1)}{2P_d} nT_c. \tag{37}$$

In order to compare the two acquisition schemes, we define $R_1 = \bar{T}_{\text{acq,serial}} / \bar{T}_{\text{acq}}$. For $P_d \gg 4N/(N^2 - 1)$, R_1 can be approximated as

$$R_1 \approx \frac{(1 - 0.5P_d)N}{(1 - P_d)(2\nu - 1) + \frac{N^2 - 1}{4N} P_d} \tag{38}$$

which is plotted in Fig. 6 for $N = 511$ with different values of ν. We see that the proposed system acquires the incoming phase at least twice as fast as the serial-search system. For

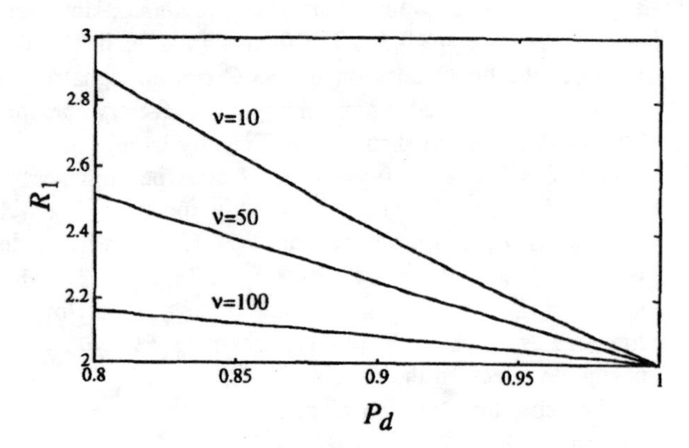

Fig. 6. The average acquisition time for $N = 511$ and $\nu = 10, 50, 100$.

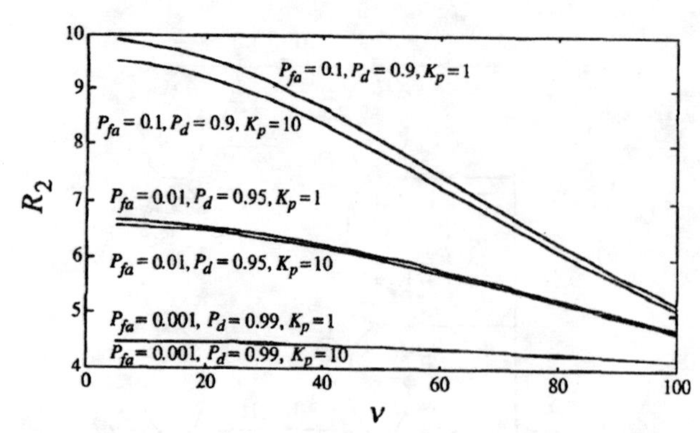

Fig. 7. Plots of the ratio R_2 versus ν for $N = 511$.

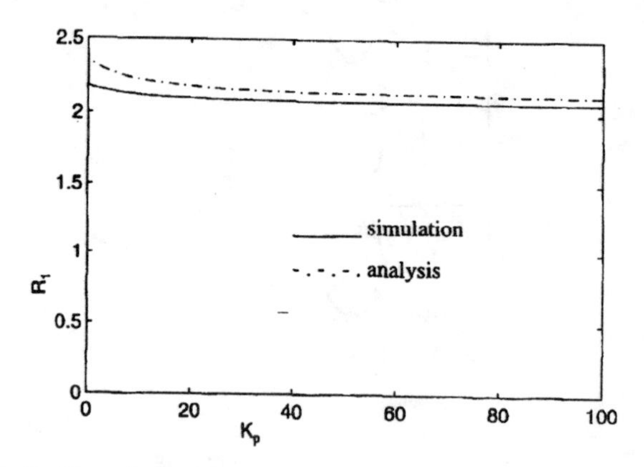

Fig. 8. The ratio R_1 obtained from analysis and simulation.

fixed P_d, as ν gets smaller, performance improvement of the proposed system over the single-dwell serial-search system increases.

The variance of the acquisition time can also be derived from $U(z)$

$$\sigma^2_{T_{\rm acq}} = \left[\frac{d^2U(z)}{dz^2} + \frac{dU(z)}{dz} - \left(\frac{dU(z)}{dz}\right)^2\right]_{z=1} (nT_c)^2. \tag{39}$$

Substituting (34) into (39) yields

$$\begin{aligned}\sigma^2_{T_{\rm acq}} = (nT_c)^2 \Bigg[& \frac{1-P_d}{P_d^2} + P_{fa}K_p(K_p+1) \\ & \times \left(\frac{N^2-1}{4N} + (2\nu-1)\frac{1-P_d}{P_d}\right) \\ & + (P_{fa}K_p+1) \\ & \times \left(\frac{N^2-1}{4N} + (2\nu-1)(1-P_d)\frac{2+P_d}{P_d^2}\right) \\ & + (P_{fa}K_p+1)^2 \\ & \times \Bigg(\frac{N^4-12N^3+2N^2+12N-3}{48N^2} \\ & + \frac{1-P_d}{P_d^2}(2\nu-1-P_d)(2\nu-1)\Bigg)\Bigg]. \end{aligned} \tag{40}$$

The variance of the acquisition time for the receiver in Fig. 1 can be found as

$$\begin{aligned}\sigma^2_{T_{\rm acq,serial}} = & (nT_c)^2 \\ & \times \Bigg[\frac{1-P_d}{P_d^2} + (N-1) \\ & \times \Bigg\{(P_{fa}K_p+1)\frac{4-2P_d-P_d^2}{2P_d^2} \\ & + P_{fa}K_p(K_p+1)\frac{2-P_d}{2P_d} + (P_{fa}K_p+1)^2 \\ & \times \left(\frac{(1-P_d)(N-1-P_d)}{P_d^2} + \frac{N-5}{12}\right)\Bigg\}\Bigg]. \end{aligned} \tag{41}$$

To compare the variances, we define $R_2 = \sigma^2_{T_{\rm acq,serial}}/\sigma^2_{T_{\rm acq}}$. This ratio is plotted versus ν in Fig. 7 for different values of P_{fa}, P_d, and K_p. It is observed that the acquisition time variance of the serial-search system is at least four times the acquisition time variance of the proposed system. For fixed ν and K_p, R_2 increases as P_{fa} gets larger and P_d gets smaller. For fixed P_{fa}, P_d, and K_p, we see that R_2 decreases as ν increases. When all other parameters are fixed, R_2 decreases as K_p increases.

The acquisition performances of the proposed receiver and the single-dwell serial-search receiver are also evaluated by computer simulations. A binary m-sequence with period $N = 511$ is used in the simulations. The numerical solution of (28) and (29) yields $n = 87$ for $P_d^* = 0.9$, $P_{fa}^* = 0.1$, and SNR $= -5$ dB. We use $M = 5220$ and calculate ν to be 30, using (32). Simulations were run to find $\bar{T}_{\rm acq}$, $\bar{T}_{\rm acq,serial}$, $\sigma^2_{T_{\rm acq}}$, and $\sigma^2_{T_{\rm acq,serial}}$ for various values of K_p. From these, R_1 and R_2 were computed. R_1 is plotted in Fig. 8. We also plot the analytical R_1, computed from (36) and (37), in the same figure. It is observed that the two curves for R_1 agree reasonably well. R_2 is plotted in Fig. 9 along with the analytical R_2, computed from (40) and (41). Except for small K_p, these two curves also agree reasonably well.

IV. Conclusion

A closed-loop coherent acquisition system for DS/SS systems is proposed. Unlike conventional acquisition schemes, this system provides the direction for the local phase update by using an auxiliary signal. The system is analyzed under additive white Gaussian noise channel. The mean and variance of the acquisition time are obtained using the flow graph

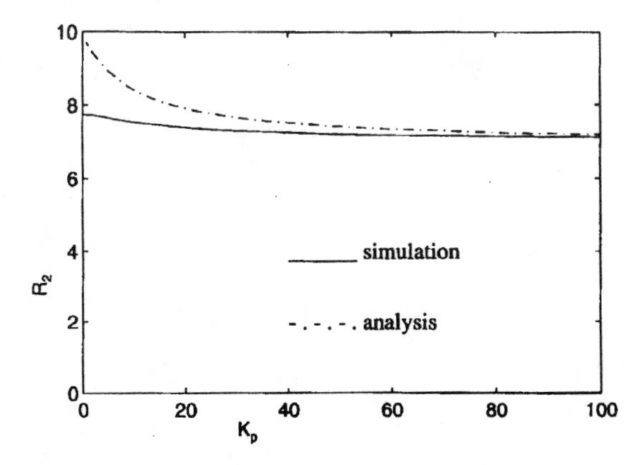

Fig. 9. The ratio R_2 obtained from analysis and simulation.

technique. Results are compared with those of the single-dwell serial-search system. On the average, the proposed system acquires the incoming phase at least twice faster. Moreover, the acquisition time variance of the proposed system is at most one fourth of that of the single-dwell system. Analytical results are verified by computer simulations.

REFERENCES

[1] K. Simon, J. Omura, R. Scholtz, and K. Levitt, *Spread Spectrum Communication*, vol. 3. Rockville, MD: Computer Science Press, 1985.

[2] R. Pickholtz, D. Schilling, and L. Milstein, "Theory of spread-spectrum communications—A tutorial," *IEEE Trans. Commun.*, vol. COM-30, no. 5, pp. 855–884, May 1982.

[3] A. Weinberg, "Generalized analysis for the evaluation of search strategy effects on PN acquisition performance," *IEEE Trans. Commun.*, vol. COM-31, no. 1, pp. 37–49, Jan. 1983.

[4] W. R. Braun, "Performance analysis for the expanding search PN acquisition algorithm," *IEEE Trans. Commun.*, vol. COM-30, no. 3, pp. 424–435, Mar. 1982.

[5] J. K. Holmes and C. C. Chen, "Acquisition time performance of PN spread-spectrum systems," *IEEE Trans. Commun.*, vol. 25, no. 8, pp. 778–783, Aug. 1977.

[6] D. M. DiCarlo and C. L. Weber, "Statistical performance of single-dwell serial synchronization systems," *IEEE Trans. Commun.*, vol. 28, no. 8, pp. 1382–1388, Aug. 1980.

[7] ——, "Multiple-dwell serial search: Performance and application to direct sequence code acquisition," *IEEE Trans. Commun.*, vol. COM-31, no. 5, pp. 650–659, May 1983.

[8] ——, "A unified approach to serial search spread-spectrum code acquisition—Part I: General theory," *IEEE Trans. Commun.*, vol. COM-32, no. 5, pp. 542–549, May 1984.

[9] A. Polydoros and C. L. Weber, "A unified approach to serial search spread-spectrum code acquisition—Part II: A matched-filter receiver," *IEEE Trans. Commun.*, vol. COM-32, no. 5, pp. 550–560, May 1984.

[10] S. Tantaratana, A. W. Lam, and P. J. Vincent, "Noncoherent sequential acquisition for DS/SS communication with/without fading," *IEEE Trans. Commun.*, vol. COM-43, pp. 1738–1745, Feb./Mar./Apr. 1995.

[11] M. Salih and S. Tantaratana, "A closed-loop coherent PN acquisition scheme for DS/SS systems using an auxiliary sequence," in *Proc. 28th Annu. Asilomar Conf. Signals, Syst. Computers*, Pacific Grove, CA, Nov. 1994.

Murat Salih was born in Ankara, Turkey, on January 20, 1965. He received the B.S. and the M.S. degrees in electrical engineering from the Middle East Technical University, in 1987 and 1989, respectively.

His interests include acquisition and tracking systems for spread-spectrum communications and channel equalization. He is now with Karel Inc., Ankara, Turkey.

Sawasd Tantaratana (S'75–M'78–SM'94) received the B.E.E. degree (with high distinction) from the University of Minnesota, Minneapolis, in 1971, the M.S. degree in electrical engineering from Stanford University, Stanford, CA, in 1972, and the M.A. and Ph.D. degrees in electrical engineering from Princeton University, Princeton, NJ, in 1976 and 1977, respectively.

Prior to joining the National Electronics and Computer Technology Center in Bangkok, Thailand, in January 1996, he had been on the faculties of the University of Massachusetts at Amherst, MA, Auburn University, AL, and the King Mongkut's Institute of Technology, Thonburi, Thailand, and on the Technical Staff of AT&T Bell Laboratories, Holmdel, NJ. From 1980–1981, he was a Visiting Assistant Professor at the University of Illinois at Urbana-Champaign.

Dr. Tantaratana is a member of Eta Kappa Nu and Sigma Xi.

Decision-Directed Coherent Delay-Lock Tracking Loop for DS-Spread-Spectrum Signals

Riccardo De Gaudenzi, *Member, IEEE*, and Marco Luise, *Member, IEEE*

Abstract— **In this paper we present a nonconventional joint data demodulation-pseudo-noise (PN) code tracking scheme for direct sequence (DS) spread-spectrum (SS) signals which solves problems of components imbalance and sensitivity with a remarkable hardware simplicity and no performance degradation.**

An integrate-and-dump Costas loop is employed for carrier recovery and data demodulation of the SS signal. Both data and carrier are then used to derive the baseband error signal of the code tracking loop. Moreover, a single passband correlator is used to perform the early–late correlation, leading to a hardware complexity equivalent to that of the tau–dither scheme, but without its loss in performance.

Results of a thorough theoretical analysis of the system in an AWGN environment are reported. We provide performance curves in terms of steady-state jitter and mean time to first lock loss. A superior jitter performance for low values of E_b/N_0 with respect to a traditional noncoherent DLL is shown, along with the potential gain of Manchester coding upon the more usual NRZ format.

I. Introduction and System Description

DELAY LOCK tracking of pseudo-noise (PN) codes for spread-spectrum (SS) systems is an already well-developed technique after the considerable research efforts provided in the past [1], [2]. Most preeminent and established schemes for this purpose are the noncoherent delay lock loop (DLL) (Fig. 1) [3], [4] and its time-shared modification, namely, the tau–dither loop (TDL) [5], [3]. This last structure can be regarded as the standard way of coping with the arm imbalance impairment the traditional DLL is subjected to, at the price of a slight performance degradation. A further modified noncoherent DLL tracking scheme with no balance problems and a simpler hardware configuration is presented in [6]. In the following, we introduce a coherent decision-directed DLL (CDD-DLL) which, enhancing the hardware simplification pursued in [6], retains the same absence of imbalance phenomena and provides a superior performance in terms of steady-state jitter with respect to the abovementioned schemes.

The block diagram of the proposed CDD-DLL is shown in Fig. 2. The key idea to attain a simple configuration is to combine as much as possible of the SS demodulator with the PN code tracking loop. Moreover, the early and late correlations intrinsic to the delay-lock technique are performed in a single step (mixer M2), thus avoiding the use of a double correlator [7, p. 481].

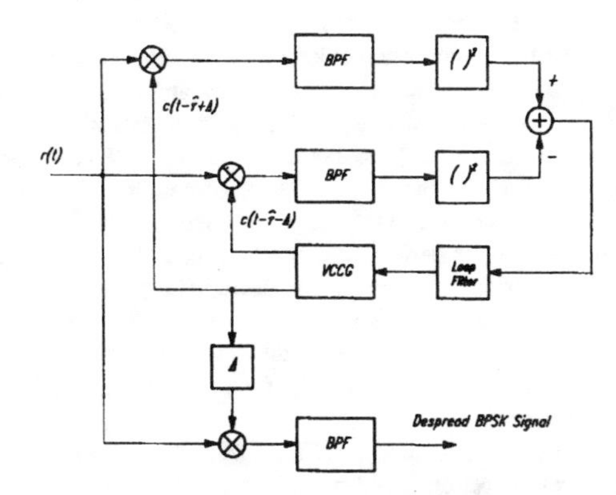

Fig. 1. Noncoherent traditional delay lock loop.

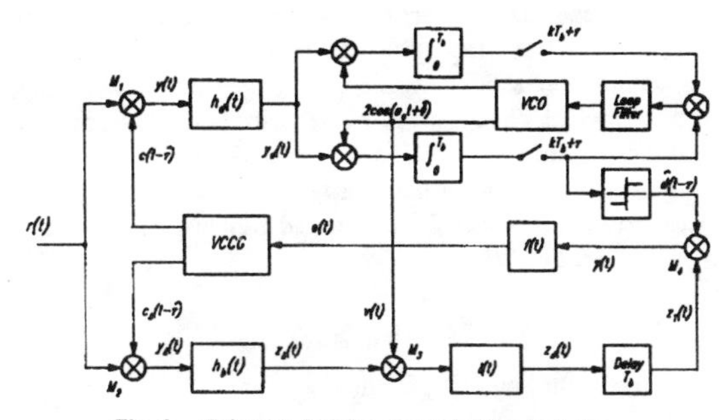

Fig. 2. Coherent decision-directed delay lock loop.

As is seen, the upper part of the scheme is a coherent BPSK demodulator for the DS/SS signal $r(t)$. After PN code despreading (mixer $M1$), the recovered BPSK signal $y_d(t)$ enters the carrier recovery and data demodulation subsystems, i.e., an integrate-and-dump (I & D) Costas loop [8]. In the lower arm, the received signal is correlated with a composite signal $c_\Delta(t-\hat{\tau})$ made up of the difference between the early and late replica of the spreading code. To derive a proper loop error signal, the correlation result $z_\Delta(t)$ is converted to baseband by means of the recovered carrier $v(t)$. Data modulation is then wiped out by means of the estimated binary data stream coming from the carrier/data unit (multiplier

Paper approved by the Editor for Spread Spectrum of the IEEE Communications Society. Manuscript received October 15, 1989; revised January 15, 1990.

R. De Gaudenzi is with the European Space Agency ESA, European Space Research and Technology Centre ESTEC, RF Systems Division, Postbus 299, 2200 AG Noordwijk, The Netherlands.

M. Luise is with Consiglio Nazionale delle Ricerche CNR Centro Studio Metodi Dispositivi Radiotrasmissioni CSMDR, Via Diotisalvi 2, 56126 Pisa, Italy.

IEEE Log Number 9144546.

$M4$). The resulting error signal $\gamma(t)$ enters the loop filter $f(t)$ whose output $e(t)$ finally drives the voltage controlled code generator (VCCG). The absence of squaring devices, as opposed to traditional noncoherent schemes, improves the tracking capabilities of the loop. In particular, no "squaring loss" in steady-state jitter is present and a broad linear error characteristic is obtained.

During acquisition phase, the code loop is kept open and VCCG is swept until an in-lock condition is detected by means of a conventional noncoherent lock detector [7]. VCCG is then frozen and thus carrier and clock loops are in a condition to acquire a valid reference from the despread signal [8]. Eventually, as soon as these synchronizers indicate an in-lock condition[1], the code loop is closed and coherent tracking starts.

The loop is well suited for a digital implementation, when an A/D conversion is performed just after mixer $M3$, so that the successive operations of delay and data modulation cancelling are of no concern. As a final remark, we underline again that the absence of separate chains for the error signals solves in an almost minimum-hardware configuration the problem of arm gain imbalance.

After this Introduction, the CDD–DLL system analysis is carried out in Section II where the loop equation is derived and examples of loop error characteristics (S-curves) are shown. In Section III we work out an expression for the steady-state tracking jitter variance under the usual assumptions of narrow-loop bandwidth and negligible self-noise contributions, while in Section IV we show how to compute the other fundamental parameter which characterizes the behavior of the loop, i.e., the mean time to lose lock (MTLL). Section V reports numerical results about normalized jitter variance and MTLL for a few cases of relevant practical interest, and conclusions are finally drawn in Section VI.

II. System Analysis

We model the input signal in Fig. 2 as

$$r(t) = \sqrt{2P}d(t-\tau)c(t-\tau)\cos(\omega_o t+\phi)+n(t) \quad (1)$$

where P is the average power of the signal, $d(t)$ is a binary data stream, $c(t)$ is the data-synchronous PN code sequence, ω_o is the carrier angular frequency, ϕ is the initial carrier phase and $n(t)$ is AWGN with double-sided power spectral density (PSD) $N_o/2$.

We further assume

$$d(t) = \sum_{i=-\infty}^{\infty} d_i q(t-iT_b) \quad (2)$$

$$c(t) = \sum_{k=-\infty}^{\infty} c_k p(t-kT_c) \quad (3)$$

where $\{d_i\}$ are i.i.d. symbols selected from the alphabet $\{-1,+1\}$, $q(t)$ is the NRZ pulse with bit duration T_b, c_k is the kth chip value ($c_k=\pm1$), $p(t)$ is the code pulse shape and T_c is the chip duration.

[1]Some digital clock recovery schemes are independent of carrier phase reference errors. They appear to be particularly useful in this case to shorten further the acquisition procedure.

The VCCG produces both a variable-delay replica of the spreading code, namely, $c(t-\hat{\tau})$, and the composite signal

$$c_\Delta(t-\hat{\tau}) \triangleq c(t-\hat{\tau}+\Delta) - c(t-\hat{\tau}-\Delta) \quad (4)$$

where Δ is a fixed time shift and $\hat{\tau}$ is the estimated code delay.

Let us now decompose the Gaussian noise $n(t)$ in (1) into its I–Q components

$$n(t) = n_c(t)\cos(\omega_o t+\phi) - n_s(t)\sin(\omega_o t+\phi) \quad (5)$$

where $n_c(t)$ and $n_s(t)$ are two mutually independent white Gaussian processes with PSD N_o. Let us also indicated with $\tilde{h}_b(t)$ the baseband equivalent of the IF filter $h_b(t)$,

$$h_b(t) \triangleq \mathrm{Re}\left\{\tilde{h}_b(t)\exp[j(\omega_o t+\phi)]\right\} \quad (6)$$

where we assume $\tilde{H}_b(0)=1$. At the output of the IF filter we have

$$\begin{aligned} z_\Delta(t) &= [r(t)c_\Delta(t-\hat{\tau})]\otimes h_b(t) = \\ &\left\{\left[\sqrt{2P}\,d(t-\tau)c(t-\tau)c_\Delta(t-\hat{\tau})\right]\otimes\tilde{h}_b(t)\right\}\cos(\omega_o t+\phi) \\ &+ w_{c\Delta}(t)\cos(\omega_o t+\phi) - w_{s\Delta}(t)\sin(\omega_o t+\phi) \end{aligned} \quad (7)$$

where

$$w_{c\Delta}(t) \triangleq [n_c(t)c_\Delta(t-\hat{\tau})]\otimes\tilde{h}_b(t) \quad (8)$$

$$w_{s\Delta}(t) \triangleq [n_s(t)c_\Delta(t-\hat{\tau})]\otimes\tilde{h}_b(t). \quad (9)$$

Let us now introduce the PN code average autocorrelation function

$$R_c(\theta) \triangleq \frac{1}{NT_c}\int_0^{NT_c} c(t)c(t+\theta)\,dt. \quad (10)$$

NT_c being the code length. As B_b is in most cases of the same order of magnitude as the data rate, it follows $B_b \ll 1/T_c$ and then (7) takes the form

$$\begin{aligned} z_\Delta(t) &\cong \sqrt{2P}d_F(t-\tau)[R_c(\Delta+\hat{\tau}-\tau) - R_c(-\Delta+\hat{\tau}-\tau)] \\ &\cdot \cos(\omega_o t+\phi) + w_{c\Delta}(t)\cos(\omega_o t+\phi) \\ &- w_{s\Delta}(t)\sin(\omega_o t+\phi) \end{aligned} \quad (11)$$

where the time-averaging in (10) is practically performed by the IF filter $H_b(f)$ and where $d_F(t) \triangleq d(t)\otimes\tilde{h}_b(t)$.

Assume now that the recovered carrier can be represented as

$$v(t) = 2\cos\left(\omega_o t+\hat{\phi}\right) \quad (12)$$

where $\hat{\phi}$ is the estimated carrier phase. The signal at the output of the ideal low-pass filter $l(t)$ is then $z_d(t) = \{z_\Delta(t)v(t)\}\otimes l(t)$ so that after the delay T_b we get

$$\begin{aligned} z_T(t) &= z_d(t-T_b) = \sqrt{2P}\,d_F(t-\tau-T_b)[R_c(\Delta+\hat{\tau}-\tau) \\ &- R_c(-\Delta+\hat{\tau}-\tau)]\cos\left(\phi-\hat{\phi}\right) + w_{c\Delta}(t-T_b)\cos\left(\phi-\hat{\phi}\right) \\ &- w_{s\Delta}(t-T_b)\sin\left(\phi-\hat{\phi}\right). \end{aligned} \quad (13)$$

Skipping all trivial details relative to data demodulation which is performed by successive despreading, filtering and I & D operations on the received signal, we find the following expression for the estimated data stream $\hat{d}(t-\tau)$

$$\hat{d}(t-\tau) = \sum_{i=-\infty}^{\infty} \hat{d}_{i-1} q(t-\tau-iT_b) \tag{14}$$

where we took into account the decision delay T_b introduced by the I & D device.

Using (14) we have

$$\begin{aligned}\gamma(t) &\triangleq z_T(t)\hat{d}(t-\tau)\\ &= \sqrt{2P}\, d_F(t-\tau-T_b)\hat{d}(t-\tau)[R_c(\Delta+\hat{\tau}-\tau)\\ &\quad - R_c(-\Delta+\hat{\tau}-\tau)] \cos\left(\phi-\hat{\phi}\right) + w_\Delta(t)\hat{d}(t-\tau)\end{aligned} \tag{15}$$

with $w_\Delta(t) \triangleq w_{c\Delta}(t-T_b)\cos(\phi-\hat{\phi}) - w_{s\Delta}(t-T_b)\cdot\sin(\phi-\hat{\phi})$.

We are now in a position to find the final expression of the open-loop error signal $e(t)$ at the input of the VCCG. Observe first that, due to the predominance of additive noise in the total loop noise for SS systems, we can retain only the dc component

$$M \triangleq \left\langle E\left\{d_F(t-T_b)\hat{d}(t)\right\}\right\rangle, \quad <\bullet> \triangleq \frac{1}{T_b}\int_0^{T_b}(\bullet)\,dt \tag{16}$$

of the useful error signal in (15), thus neglecting the so-called self-noise contribution $d_F(t-T_b)\hat{d}(t) - <E\{d_F(t-T_b)\hat{d}(t)\}>$.

Denoting with $P_e(\varepsilon)$ the bit error rate (BER) of the BPSK data demodulator when a fixed normalized code timing error ε

$$\varepsilon \triangleq \frac{\tau-\hat{\tau}}{T_c} \tag{17}$$

is present, it is found

$$\begin{aligned}M = M(\varepsilon) &= \left\langle E\left\{\left[d(t-T_b)\otimes\tilde{h}_b(t)\right]\hat{d}(t)\right\}\right\rangle\\ &= \int_{-\infty}^{\infty}\left\langle E\left\{d(t-\xi-T_b)\hat{d}(t)\right\}\right\rangle \tilde{h}_b(\xi)\,d\xi\\ &= (1-2P_e(\varepsilon))\tilde{H}_b(0)\\ &= 1-2Q\left[R_c(\varepsilon T_c)\sqrt{\frac{2E_b}{N_o}}\cos\left(\phi-\hat{\phi}\right)\right]\end{aligned} \tag{18}$$

where

$$Q(x) \triangleq \frac{1}{\sqrt{2\pi}}\int_x^{\infty}\exp\{-y^2/2\}\,dy.$$

The error signal $e(t)$ at the VCCG input takes eventually the form

$$e(t) = -\sqrt{2P}\,\eta M_o\cos\left(\phi-\hat{\phi}\right)$$

$$\cdot\left[S(\varepsilon) - \frac{N(t)}{\sqrt{2P}\,\eta M_o\cos\left(\phi-\hat{\phi}\right)}\right]\otimes f(t) \tag{19}$$

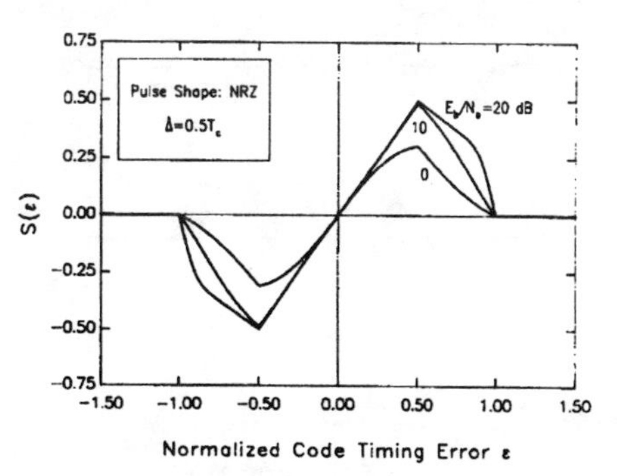

Fig. 3. Normalized loop error characteristic $S(\varepsilon)$—NRZ chip pulse.

where the following definitions have been made:

$$\eta \triangleq \frac{d}{d\varepsilon}[R_c(\Delta+\varepsilon T_c) - R_c(-\Delta+\varepsilon T_c)]\Big|_{\varepsilon=0} \tag{20}$$

$$S(\varepsilon) \triangleq -\frac{1}{\eta M_o}[R_c(\Delta+\varepsilon T_c) - R_c(-\Delta+\varepsilon T_c)]M(\varepsilon) \tag{21}$$

$$M_o \triangleq M(0) = 1-2Q\left[\sqrt{\frac{2E_b}{N_o}}\cos\left(\phi-\hat{\phi}\right)\right] \tag{22}$$

$$N(t) \triangleq w_\Delta(t)\,\hat{d}(t). \tag{23}$$

We can now write the stochastic differential equation which describes the dynamic behavior of the tracking loop presented in Fig. 2

$$\frac{d\varepsilon(t)}{dt} = Ke(t) \tag{24}$$

or, equivalently,

$$\frac{d\varepsilon}{dt} = -K\sqrt{2P}\,\eta M_o\cos\left(\phi-\hat{\phi}\right)$$

$$\cdot\left[S(\varepsilon) - \frac{N(t)}{\sqrt{2P}\,\eta M_o\cos\left(\phi-\hat{\phi}\right)}\right]\otimes f(t) \tag{25}$$

where K denotes the VCCG sensitivity.

Plots of the normalized error characteristic (NEC) of the loop $S(\varepsilon)$ are shown in Figs. 3 and 4 for the cases of NRZ and SPL (Manchester or bi-ϕ) chip pulse $p(t)$, respectively, for various values of the energy-per-bit to noise PSD ratio E_b/N_o. The term $R_c(\varepsilon T_c)$ in expression (18) of $M(\varepsilon)$ is responsible for the progressively rounded shape of $S(\varepsilon)$ for decreasing E_b/N_o's, i.e., for increasing BER. Moreover, for the same reason all loop NEC's are forced to zero at $\varepsilon = \pm 1$ and stay null beyond that value where BER is equal to 1/2.

In the following sections, we will make the additional assumption of a negligible recovered carrier phase error, i.e., $\phi - \hat{\phi} \cong 0$.

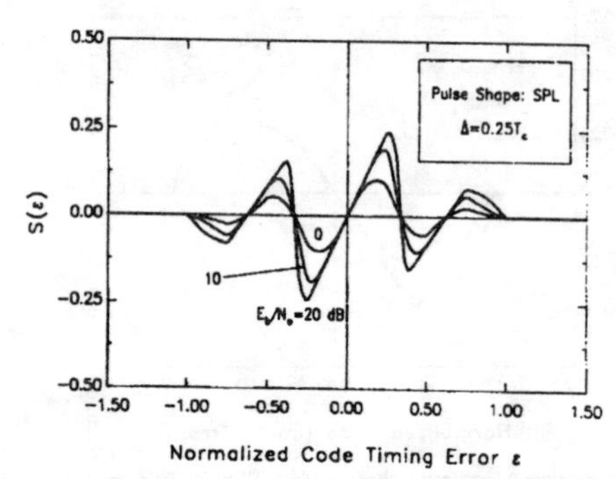

Fig. 4. Normalized loop error characteristic $S(\varepsilon)$—SPL chip pulse.

III. Computation of Steady-State Code Tracking Jitter

Equation (25) is the starting point for the evaluation of the mean square value of the normalized timing error ε (tracking jitter).

In the hypothesis of a high signal-to-noise ratio within the loop, the NEC of the loop $S(\varepsilon)$ can be conveniently replaced by its linear equivalent in the neighborhood of $\varepsilon = 0$ and (25) can be written as

$$\frac{d\varepsilon}{dt} = -K\sqrt{2P}\,\eta M_o\left[\varepsilon - \frac{N(t)}{\sqrt{2P}\,\eta M_o}\right] \otimes f(t). \quad (26)$$

Defining the loop transfer function $H(s)$ as [9]

$$H(s) \triangleq \frac{K\sqrt{2P}\,\eta M_o F(s)}{s + K\sqrt{2P}\,\eta M_o F(s)}, \quad (27)$$

the normalized tracking error variance σ_ε^2 can be evaluated from (26), yielding

$$\sigma_\varepsilon^2 = \frac{1}{2PM_o^2\eta^2}\int_{-\infty}^{\infty} S_N(f)|H(j2\pi f)|^2\, df \quad (28)$$

where $S_N(f)$ is the PSD of the noise $N(t)$.

Assuming a flat PSD of $N(t)$ within the loop bandwidth, we have

$$\sigma_\varepsilon^2 \cong \frac{1}{2PM_o^2\eta^2} S_N(0) 2B_L \quad (29)$$

with

$$2B_L \triangleq \int_{-\infty}^{\infty} |H(j2\pi f)|^2\, df \quad (30)$$

In Appendix I, it is found that

$$S_N(0) = 2N_o K_c K_d \quad (31)$$

where

$$K_c \triangleq R_c(0) - R_c(2\Delta) \quad (32)$$

$$K_d \triangleq \int_{-\infty}^{\infty} S_d(f)|\tilde{H}_d(f)|^2\, df \quad (33)$$

being $S_d(f)$ the PSD of the data signal $d(t)$ and $\tilde{H}_d(f)$ the baseband equivalent of the data filter $H_d(f)$. Insertion of (31) in (29) eventually yields

$$\sigma_\varepsilon^2 \cong \frac{2B_L N_o}{P}\frac{K_c K_d}{M_o^2\eta^2}. \quad (34)$$

We note from (27) and (30) that the actual loop bandwidth B_L depends on the BER of the demodulator via M_o, and hence it is a function of the ratio E_b/N_o. In Section V, when numerical results will be presented, we will therefore rely on the quantity

$$B_o \triangleq B_L|_{Eb/No=\infty} \quad (35)$$

as the BER-independent distinctive bandwidth parameter of the loop.

IV. Computation of the Mean Time To Lose Lock

We say that a lock loss occurs whenever the loop, starting from the stable equilibrium point $\varepsilon = 0$, reaches one of the nearest unstable equilibrium points in the NEC $S(\varepsilon)(\varepsilon = \pm 1$ for NRZ pulse, for instance).

Although the system described by (25) is not strictly sense Markovian, the first-passage problem related to the evaluation of the mean time to lose lock (MTLL) can be approximately solved as well by resorting to Fokker–Planck techniques [10]. In particular, the key hypothesis to the validity of the approximation is that the system noise bandwidth be much larger than the loop bandwidth [11], which is verified in all practical cases.

Let us define the function $q(\varepsilon; t)$ as the probability density function of all error signal processes $\varepsilon(t)$ which, after starting from value 0 at $t = 0$, have produced no lock loss at time t. In the case of a first order loop $q(\varepsilon; t)$ satisfies the following Fokker–Planck equation [2], [10], [7]:

$$\frac{\partial q(\varepsilon;t)}{\partial t} = \frac{\partial}{\partial\varepsilon}[K_1(\varepsilon)q(\varepsilon;t)] - \frac{1}{2}\frac{\partial^2}{\partial\varepsilon^2}[K_2(\varepsilon)q(\varepsilon;t)] \quad (36)$$

where

$$K_1(\varepsilon) \triangleq K\eta M_o\sqrt{2P}S(\varepsilon) \quad (37)$$

$$K_2(\varepsilon) \equiv K_2 \triangleq -K^2 2N_o K_c K_d. \quad (38)$$

As all trajectories $\varepsilon(t)$ of the system are supposed to start from the initial point $\varepsilon = 0$ for $t = 0$, we add to (36) the initial condition

$$q(\varepsilon; 0) = \delta(\varepsilon) \quad (39)$$

and as the probability of no sync loss vanishes for $t \to \infty$, we have

$$q(\varepsilon; \infty) = 0. \quad (40)$$

Furthermore, to solve the first-passage problem we set absorbing boundaries at $\varepsilon = \pm\varepsilon_c$:

$$q(\pm\varepsilon_c; t) = 0 \quad \forall\, t \quad (41)$$

where the boundary ε_c is application-dependent ($\varepsilon_c = 1$ for NRZ pulses and $\varepsilon_c = 1/3$ for SPL, see Figs. 3–4).

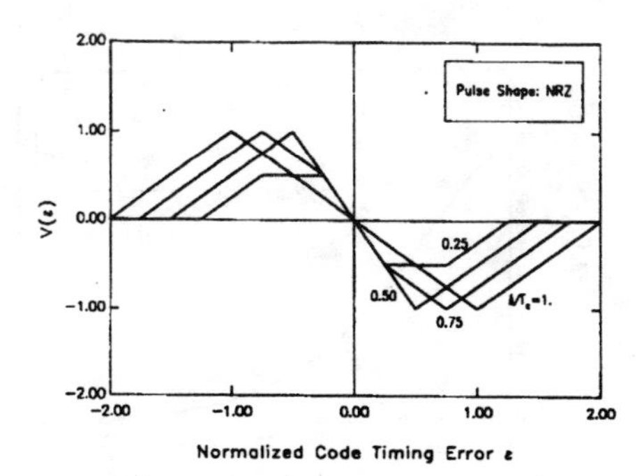

Fig. 5. Error function $V(\varepsilon)$—NRZ chip pulse.

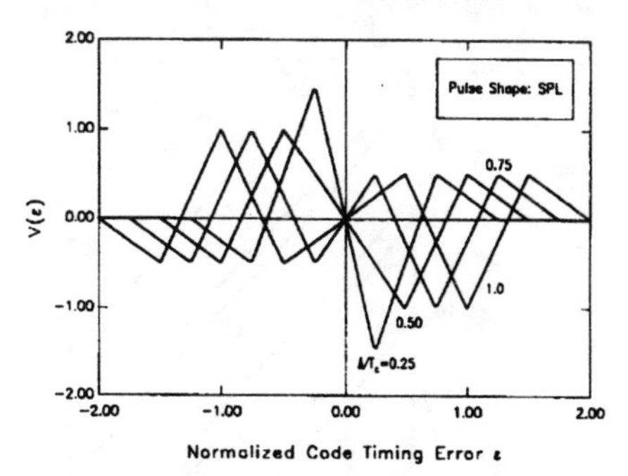

Fig. 6. Error function $V(\varepsilon)$—SPL chip pulse.

The MTLL can now be expressed as [7]

$$\overline{T} = \int_{t=0}^{\infty} \int_{\varepsilon=-\varepsilon_c}^{\varepsilon_c} q(\varepsilon; t)\, d\varepsilon\, dt. \tag{42}$$

We report in Appendix II the detailed solution of (36), which leads to the following final formula for $\overline{T}$:

$$B_o\overline{T} = \frac{\alpha}{8M_o} \int_{\varepsilon=-\varepsilon_c}^{0} \int_{\theta=-\varepsilon_c}^{\varepsilon} \exp[U(\theta) - U(\varepsilon)]\, d\theta\, d\varepsilon \tag{43}$$

where α is the loop signal-to-noise ratio (see Appendix II) and $U(\bullet)$ is a potential function defined as

$$U(x) \triangleq \frac{\alpha}{2} \int_{\theta=-\varepsilon_c}^{x} S(\theta)\, d\theta. \tag{44}$$

V. Numerical Results

Numerical computations of jitter variance and MTLL have been carried out in the two relevant cases of NRZ and SPL chip pulses $p(t)$. The particular choice of the value of Δ for each application, as already addressed in Figs. 3–4, is justified examining Figs. 5–6 where plots of the error function

$$V(\varepsilon) \triangleq [R(\Delta + \varepsilon T_c) - R(-\Delta + \varepsilon T_c)] \tag{45}$$

are shown for different values of Δ. It is apparent that the maximum slope in the neighborhood of $\varepsilon = 0$ is attained for $\Delta = 0.5T_c$ and $\Delta = 0.25T_c$ for NRZ and SPL pulses, respectively.

From Figs. 5–6 we can also derive the value of the parameter η in (20), namely, $\eta = 2$ for NRZ and $\eta = 6$ for SPL pulses. Moreover, it is easily found

$$K_c = \begin{cases} \frac{2\Delta}{T_c} & 0 \le \Delta \le T_c/2 \\ 1 & \Delta > T_c/2 \end{cases} \qquad \text{NRZ}$$

$$K_c = \begin{cases} \frac{6\Delta}{T_c} & 0 \le \Delta < T_c/4 \\ 2(1 - \Delta/T_c) & T_c/4 \le \Delta \le T_c/2 \\ 1 & \Delta > T_c/2 \end{cases} \qquad \text{SPL.} \tag{46}$$

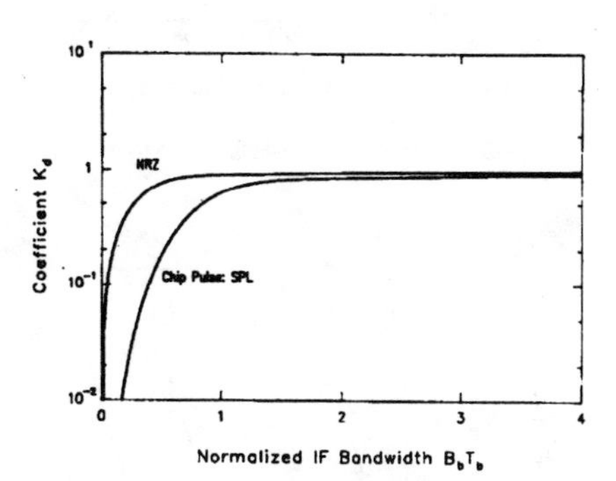

Fig. 7. Coefficient K_d.

Also, the coefficient K_d is plotted in Fig. 7 for a Butterworth IF filter $\tilde{H}_b(f)$ of the fourth order. Note that, for sufficiently wide bandwidths, it can safely be assumed $K_d \cong 1$.

Results from (34) are presented in Fig. 8 where the normalized tracking jitter variance σ_ε^2 is plotted as a function of E_b/N_o, for diverse loop bandwidths. Solid lines are relative to NRZ chip pulse whereas dashed ones refer to SPL. The gain of SPL with respect to NRZ is 8 dB in terms of *normalized* jitter. If the variance of the *unnormalized* timing error $\tau - \hat{\tau}$ is considered, the advantage of SPL is reduced to 2 dB only when this latter is constrained to occupy the same bandwidth as NRZ (i.e., half chip-rate). Note the almost straight shape of the jitter curve due to the absence of a loop "squaring loss." The slight bend for low values of E_b/N_o is caused by the "BER loss" term $1/M_o^2$ in (34).

MTLL results obtained via numerical integration of (43) are reported in Fig. 9. NRZ (solid line) pulse loses roughly 2 dB in MTLL performance with respect to SPL (dashed line). These results, derived as addressed in the previous Section for a first-order loop, are approximately valid for a second-order loop as well, provided a sufficiently large damping factor is assumed [12].

In both cases of jitter variance and MTLL curves, loop bandwidths have been normalized with respect to the bit duration T_b and can be converted to a spec in terms of the chip length T_c when the processing gain T_b/T_c is known.

Finally, we tried a direct comparison between jitter performances of the CDD–DLL and the traditional noncoherent

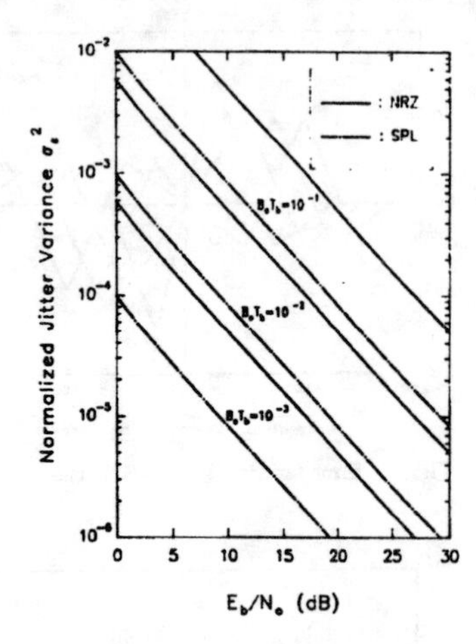

Fig. 8. Normalized tracking jitter variance σ_ϵ^2.

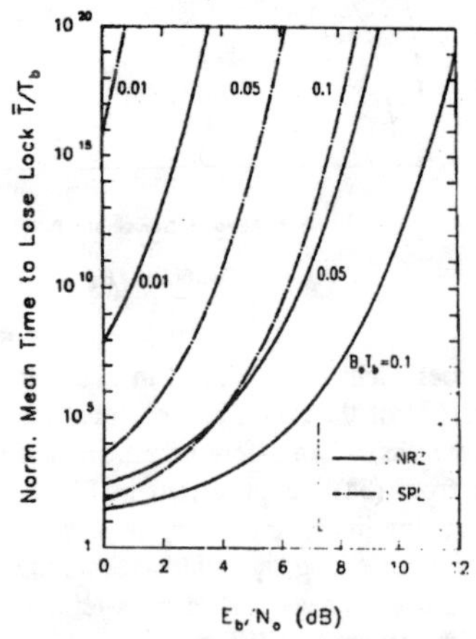

Fig. 9. Normalized mean time to lose lock $\bar{T}/T_b$—first-order loop.

DLL (TDLL). Fig. 10 shows the loss Γ in steady-state jitter variance experienced by TDLL with respect to CDD–DLL for NRZ chips and two-sided normalized IF bandwidth $B_{IF}T_b = 4$. We have used for the TDLL jitter the following simplified formula presented in [4]:

$$\sigma^2_{\epsilon TDLL} = \frac{2B_LN_o}{P}\frac{1}{4}\left[1 + \frac{B_{IF}T_b}{E_b/N_o}\right] \tag{47}$$

which agrees with the results in [3] if we neglect arm filtering effects. The jitter improvement of CDD–DLL due to its coherent structure is particularly relevant at low to medium E_b/N_o values and asymptotically vanishes as $E_b/N_o \to \infty$.

VI. Summary and Conclusions

An alternative structure of PN code coherent tracking loop for DS/SS signal has been proposed and analyzed as far as tracking jitter variance and mean time to lose lock are concerned. Also, it has been pointed out the potential hardware saving with respect to other more traditional noncoherent schemes.

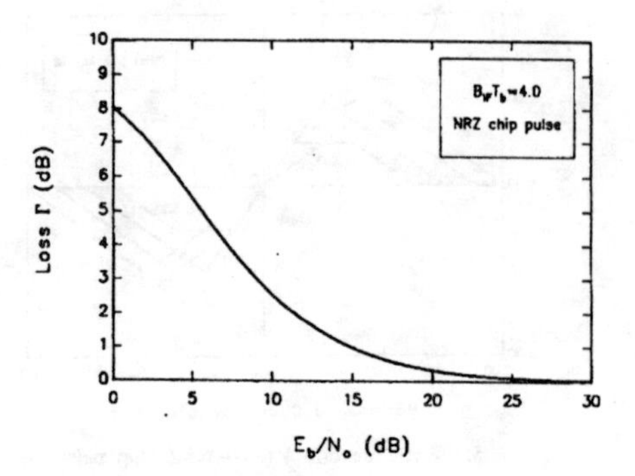

Fig. 10. Steady-state jitter variance loss Γ of traditional DLL with respect to CDD–DLL.

Results of the theoretical analysis confirm the expected absence of "squaring loss" in the expression of steady-state jitter variance, due to the lack of squaring devices in the loop. On the other hand, the "BER loss" introduced by the decision-directed nature of the proposed scheme has been evaluated and favorably compared to the abovementioned squaring loss.

The shaping induced upon the loop error characteristic by the phenomenon of incorrect data decisions has been clearly explained and taken into account in evaluating the mean time to lose lock performance of the loop by means of the Fokker–Planck (diffusion) equation.

Numerical computations have been carried out for various loop bandwidths and E_b/N_o ratios in the cases of NRZ and SPL (Manchester) chip pulses. The superior performance of the latter has been pointed out, together with possible impairments in terms of bandwidth requirements.

Appendix I

Our task is to evaluate the component at $f = 0$ of the PSD $S_N(f)$ of the loop noise $N(t)$.

Let us first compute the autocorrelation function R_N of $N(t)$

$$\begin{aligned} R_N(\theta) &\triangleq E\{N(t)N(t+\theta)\} \\ &= E\left\{\hat{d}(t)\hat{d}(t+\theta)\right\}E\{w_\Delta(t)w_\Delta(t+\theta)\} \\ &= R_{\hat{d}}(\theta)R_{w_\Delta}(\theta) \end{aligned} \tag{A1.1}$$

where

$$R_{\hat{d}}(\theta) = R_d(\theta) \triangleq E\{d(t)d(t+\theta)\}. \tag{A1.2}$$

The noise PSD $S_{w_\Delta}(f)$ [i.e., the Fourier transform (FT) of $R_{w_\Delta}(\theta)$] can be expressed as

$$S_{w_\Delta}(f) = |\tilde{H}_b(f)|^2 S_{n_\Delta}(f) \tag{A1.3}$$

with $S_{n_\Delta}(f)$ the FT of $R_{n_\Delta}(\theta)$,

$$\begin{aligned} R_{n_\Delta}(\theta) &\triangleq E\{n_c(t)c_\Delta(t-\hat{\tau})n_c(t+\theta)c_\Delta(t-\hat{\tau}+\theta)\} \\ &= E\{n_c(t)n_c(t+\theta)\}E\{c_\Delta(t-\hat{\tau})c_\Delta(t-\hat{\tau}+\theta)\} \end{aligned}$$

$$= N_o\delta(\theta)[2R_c(\theta) - R_c(\theta + 2\Delta) - R_c(\theta - 2\Delta)]$$
$$= 2N_oK_c\delta(\theta), \quad \text{(A1.4)}$$

$$K_c \triangleq R_c(0) - R_c(2\Delta). \quad \text{(A1.5)}$$

From (A1.3)–(A1.5) it is

$$S_{w_\Delta}(f) = 2N_0K_c|\tilde{H}_b(f)|^2 \quad \text{(A1.6)}$$

and, after the FT of both sides of (A1.1) is taken,

$$S_N(f) = S_d(f) \otimes S_{w_\Delta}(f)$$
$$= 2N_oK_c \int_{-\infty}^{\infty} S_d(f - v)|\tilde{H}_b(v)|^2\, dv \quad \text{(A1.7)}$$

so that

$$S_N(0) = 2N_oK_cK_d \quad \text{(A1.8)}$$

with

$$K_d \triangleq \int_{-\infty}^{\infty} S_d(f)|\tilde{H}_b(f)|^2\, df. \quad \text{(A1.9)}$$

Q.E.D.

Appendix II

In this appendix, we solve (36) in order to get an expression for the MTLL $\overline{T}$ of a first-order tracking loop.

To this aim, let us first introduce the auxiliary function

$$\Omega(\varepsilon) \triangleq \int_{t=0}^{\infty} q(\varepsilon; t)\, dt. \quad \text{(A2.1)}$$

Integrating both sides of (36) first with respect to time and then with respect to ε, we get

$$C - u(\varepsilon) = K_1(\varepsilon)\Omega(\varepsilon) - K_2\frac{d\Omega(\varepsilon)}{d\varepsilon} \quad \text{(A2.2)}$$

where $u(\bullet)$ is the unit step function and C is an arbitrary constant. Next we normalize (A2.2) by introducing the loop signal-to-noise ratio α

$$\alpha \triangleq \frac{\eta^2M_o^2P}{2B_LN_oK_cK_d} \quad \text{(A2.3)}$$

and by observing that from (30), in the case of a first order tracking loop [i.e., $f(t) = \delta(t)$], it is found

$$B_L = \frac{K\eta M_o\sqrt{2P}}{4} \quad \text{(A2.4)}$$

so that

$$C - u(\varepsilon) = 4B_LS(\varepsilon)\Omega(\varepsilon) + \frac{8B_L}{\alpha}\frac{d\Omega(\varepsilon)}{d\varepsilon}. \quad \text{(A2.5)}$$

Resolution of this first-order linear differential equation yields

$$\Omega(\varepsilon) = D\exp[-U(\varepsilon)] + \frac{\alpha}{8B_L}\exp[-U(\varepsilon)]$$
$$\cdot \int_{\theta=-\varepsilon_c}^{\varepsilon} [C - u(\theta)]\exp[U(\theta)]\, d\theta \quad \text{(A2.6)}$$

where D is a second arbitrary constant and $U(\bullet)$ is the potential function

$$U(x) \triangleq \frac{\alpha}{2}\int_{\theta=-\varepsilon_c}^{x} S(\theta)\, d\theta. \quad \text{(A2.7)}$$

Applying the boundary and initial conditions (39)–(41), we easily find $C = 1/2$ and $D = 0$, hence,

$$\Omega(\varepsilon) = \frac{\alpha}{8B_L}\exp[-U(\varepsilon)]\int_{\theta=-\varepsilon_c}^{\varepsilon}\left[\frac{1}{2} - u(\theta)\right]\exp[U(\theta)]d\theta. \quad \text{(A2.8)}$$

From (42) the MTLL has the form

$$\overline{T} = \int_{\varepsilon=-\varepsilon_c}^{\varepsilon_c} \Omega(\varepsilon)\, d\varepsilon$$
$$= \frac{\alpha}{8B_L}\int_{\varepsilon=-\varepsilon_c}^{\varepsilon_c} \exp[-U(\varepsilon)]\int_{\theta=-\varepsilon_c}^{\varepsilon}\left[\frac{1}{2} - u(\theta)\right]$$
$$\cdot \exp[U(\theta)]\, d\theta\, d\varepsilon. \quad \text{(A2.9)}$$

As in all applications $S(\bullet)$ is an odd function of its argument, the potential $U(\bullet)$ is an even function. The inner integral in (A2.9) is again an even function of ε as the integrand function is odd in θ. Then the overall function to be integrated in $d\varepsilon$ is even, so that

$$\overline{T} = 2\frac{\alpha}{8B_L}\int_{\varepsilon=-\varepsilon_c}^{0} \exp[-U(\varepsilon)]\int_{\theta=-\varepsilon_c}^{\varepsilon}\left[\frac{1}{2} - u(\theta)\right]$$
$$\cdot \exp[U(\theta)]\, d\theta\, d\varepsilon$$
$$= \frac{\alpha}{8B_oM_o}\int_{\varepsilon=-\varepsilon_c}^{0} \exp[-U(\varepsilon)]\int_{\theta=-\varepsilon_c}^{\varepsilon} \exp[U(\theta)]\, d\theta\, d\varepsilon \quad \text{(A2.10)}$$

which leads immediately to (45). The factor $2[\frac{1}{2} - u(\theta)]$ has been suppressed because constantly equal to 1 within the integration domain.

Finally, (A2.3), (A2.4), and (35) can be combined to yield the relation

$$\alpha = \frac{\eta^2\left[1 - 2Q\left(\sqrt{\frac{2E_b}{N_o}}\right)\right]\frac{E_b}{N_o}}{2B_oT_bK_cK_d} \quad \text{(A2.11)}$$

which allows to link a particular value of $\overline{T}$ computed via (A2.10) in correspondence of a particular loop signal-to-noise ratio α to the important parameter E_b/N_o.

Acknowledgment

The authors would like to thank Prof. U. Mengali of the University of Pisa for the helpful comments and suggestions about this work.

REFERENCES

[1] J. J. Spilker, "Delay lock tracking of binary signals," *IEEE Trans. Space. Electron. Telemet.*, vol. SET-9, pp. 1–8, Mar. 1963.

[2] H. Meyr, "Delay-lock tracking of stochastic signals," *IEEE Trans. Commun.*, vol. COM-24, Mar. 1976.

[3] M. K. Simon, "Noncoherent pseudonoise code tracking performance of spread spectrum receivers," *IEEE Trans. Commun.*, vol. COM-25, pp. 327–345, Mar. 1977.

[4] W. J. Gill, "A comparison of binary delay-lock loop implementations," *IEEE Trans. Aerosp. Electron. Syst.*, vol. AES-2, pp. 415–424, July 1966.

[5] H. P. Hartmann, "Analysis of a dithering loop for PN code tracking," *IEEE Trans. Aerosp. Electron. Syst.*, vol. AES-10, pp. 2–9, Jan. 1974.

[6] R. A. Yost and R. W. Boyd, "A modified PN code tracking loop: Its performance analysis and comparative evaluation," *IEEE Trans. Commun.*, vol. COM-30, pp. 1027–1036, May 1982.

[7] J. K. Holmes, *Coherent Spread Spectrum Systems.* New York: Wiley, 1982.

[8] M. K. Simon and W. C. Lindsey, "Optimum performance of suppressed carrier receivers with Costas loop tracking," *IEEE Trans. Commun.*, vol. COM-25, pp. 215–227, Feb. 1977.

[9] F. M. Gardner, *Phaselock Techniques.* New York: Wiley, 1979, 2nd ed.

[10] R. L. Stratonovitch, *Topics in the Theory of Random Noise, Vol. 1.* New York: Gordon and Breach, 1963.

[11] H. J. Kushner, "Diffusion approximation to output processes of nonlinear systems with wide-band inputs and applications," *IEEE Trans. Inform. Theory*, vol. IT-26, pp. 715–725, Nov. 1980.

[12] G. Ascheid and H. Meyr, "Cycle slips in phase-locked loops: A tutorial survey," *IEEE Trans. Commun.*, vol. COM-30, pp. 2228–2241, Oct. 1982.

Riccardo De Gaudenzi (M'88) was born in Rosignano M.mo, Italy, on November 8, 1960. He received the Doctor Engineer degree (cum laude) from the University of Pisa, Italy, in 1985.

From 1986 to 1988 he was with the European Space Agency (ESA), Stations and Communications Engineering Department, Darmstadt (RFA) where he was involved in satellite telecommunication ground systems design and testing. In particular he followed the development of two new ESA's satellite tracking systems. In 1988 he joined the ESA's Research and Technology Centre (ESTEC), Noordwijk, The Netherlands where he is holding a position as telecommunication engineer in the Electrical Systems Department. His present activity is mainly related with efficient digital modulation and access techniques, synchronization topics, and communication systems simulation software.

Marco Luise (M'90) was born in Livorno, Italy, in 1960. He received the Doctor Engineer (cum Laude) and Research Doctor degrees in electronic engineering from the University of Pisa, Italy, in 1984 and 1989, respectively.

In 1987, he spent one year at the European Space Research and Technology Centre (ESTEC), Noordwijk, The Netherlands, as a Research Fellow of the European Space Agency (ESA). He is now holding the position of Research Scientist of CNR, the Italian National Council for Research, at the Centro Studio Metodi Dispositivi Radiotrasmissioni (CSMDR), Pisa. His main research interests lie in the areas of communication theory, synchronization, and spread-spectrum systems.

A New Tracking Loop for Direct Sequence Spread Spectrum Systems on Frequency-Selective Fading Channels

Wern-Ho Sheen
National Chung Cheng University, Taiwan, ROC
Gordon L. Stüber
Georgia Institute of Technology, USA

Abstract: A new tracking loop is proposed for direct-sequence spread-spectrum signaling on a frequency-selective fading channel. By exploiting the inherent multipath diversity of the channel, the new tracking loop overwhelmingly outperforms the traditional noncoherent delay-locked loop (DLL) for all cases under consideration. Linear and nonlinear methods are employed to analyze the new tracking loop with perfect channel estimation. The performance degradation caused by imperfect channel estimation is determined by computer simulations. Over a range of signal-to-noise-ratios (SNRs) of practical interest, the simulation results show that the degradation is less than 2 dB.

I Introduction

Pseudo-Noise (PN) code synchronization is essential for spread spectrum systems [1]. PN code synchronization is usually accomplished in two steps: *code acquisition* and *code tracking*. Code acquisition is the initial search process that brings the phase of the locally generated code to within a chip duration of the incoming code. Code tracking is the process of achieving and maintaining fine alignment of the chip boundaries of the incoming and locally generated PN codes. Two different types of code tracking loops have been extensively employed in practical applications [1, 2, 3]. One is the delay-locked loop (DLL), and the other is its time-sharing version, the tau-dither loop (TDL). DLLs and TDLs have been mainly applied to spread spectrum signaling on additive white Gaussian noise (AWGN) channels [1, 2, 3]. However, the performance of these code tracking loops is severely degraded by the effects of multipath fading, multipath spread, and Doppler spread [4]. Hence, a good tracking loop for multipath fading channels has yet to be found. This paper proposes a new low complexity tracking loop for direct sequence (DS) spread spectrum signaling on frequency-selective multipath fading channels. By utilizing the inherent multipath diversity of the channel, the new tracking loop provides a much better loop performance than a conventional noncoherent DLL.

II Channel and Transmission Models

Fig. 1 depicts DS spread spectrum signaling on a multipath fading channel, where $m(t)$ is the binary data signal with a non-return-to-zero (NRZ) shaping function, c(t) is the spreading function, and w_c and P are the radian frequency and power of the carrier, respectively. The spreading function is an m-sequence with an NRZ shaping function. For a large period N, the autocorrelation of the spreading function $R_C(\xi) \triangleq \frac{1}{NT_C}\int_0^{NT_C} c(t)c(t+\xi T_C)dt$ is well approximated as

$$R_C(\xi) = \begin{cases} 1-|\xi| & \text{if } |\xi| \leq 1 \\ 0 & \text{otherwise} \end{cases} \tag{1}$$

where T_C is the chip duration.

The multipath fading channel is assumed to be a wide-sense stationary uncorrelated scattering (WSSUS) channel. Let $\tilde{g}(t,\zeta)$ denote the complex low-pass impulse response of the channel. Then the autocorrelation of $\tilde{g}(t,\zeta)$ is given by

$$\begin{aligned} R_{\tilde{g}}(\Delta t;\zeta_1,\zeta_2) &\triangleq \frac{1}{2}\,\mathrm{E}\left[\tilde{g}^*(t;\zeta_1)\tilde{g}(t+\Delta t;\zeta_2)\right] \\ &= \phi_{\tilde{g}}(\Delta t;\zeta_1)\,\delta(\zeta_1-\zeta_2) \end{aligned}$$

where $\phi_{\tilde{g}}(\Delta t;\zeta)|_{\Delta t=0}$ is the so called multipath intensity profile [5]. For Rayleigh fading channels, $\tilde{g}(t;\zeta_1)$, is a stationary zero-mean complex Gaussian random process in time t. The noise $n(t)$ in Fig. 1 is zero-mean AWGN with a two-sided power spectral density (PSD) equal to $N_0/2$ watts/Hz.

The noiseless bandpass received signal is

$$s(t) = \mathrm{Re}\{\sqrt{2P}\tilde{s}(t)e^{jw_ct}\}$$

where

$$\tilde{s}(t) = \int_{\tau(t)}^{\infty} m(t-\xi)c(t-\xi)\,\tilde{g}(t,\xi)\,d\xi \tag{2}$$

with $\tau(t)$ denoting the channel time-delay. In (2), we have assumed that $\tilde{g}(t,\xi)=0$ for $\xi \leq \tau(t)$. Since $c(kT_C+\zeta) = c(kT_C)\quad 0 \leq \zeta \leq T_C$ and $\tau(t)$ can be adjusted if necessary so that $t-\tau(t) = jT_C$ for some integer j, it can be shown that

$$\tilde{s}(t) = \sum_{k=0}^{L} m(t-\tau(t)-kT_C)\,c(t-\tau(t)-kT_C)\,g_k(t)\ , \tag{3}$$

where

$$g_k(t) = \int_{kT_C}^{(k+1)T_C} \tilde{g}(t,\tau(t)+\xi)\,d\xi$$

and $\phi_{\tilde{g}}(0,\zeta)=0$ for $\zeta > \tau(t)+(L+1)T_C$. From (3) the channel can be modeled by the T_C-spaced tapped delay line shown in Fig. 2 with complex tap gains $g_k(t) \triangleq g_{ck}(t)+jg_{sk}(t) = a_k(t)e^{j\theta_k(t)}$, where

$$a_k(t) = \sqrt{(g_{ck}(t))^2+(g_{sk}(t))^2}$$

0-7803-2486-2/95 $4.00 © 1995 IEEE

and

$$\theta_k(t) = \mathrm{Tan}^{-1}\frac{g_{sk}(t)}{g_{ck}(t)} .$$

The gains $a_k(t)$ is Rayleigh distributed and the phases $\theta_k(t)$ has uniform distribution in $(0, 2\pi]$, respectively. With no consideration of data modulation, then the received bandpass signal is

$$r(t) = \sqrt{2P}\sum_{k=0}^{L} a_k(t)c(t-\tau(t)-kT_C)\cos(w_c t+\theta_k(t))+n(t) .$$

III The New Code Tracking Loop

Loop Operation: The structure of the proposed tracking loop is shown in Fig. 3. The key feature of this new tracking loop is that it exploits the multipath diversity of the frequency-selective fading channel so as to obtain a better loop performance. The upper part of Fig. 3 is a channel-parameter estimation unit which consists of $L+1$ branches of gain-phase estimators. After despreading the received signal, the tap gains $a_k(t)$ and phases $\theta_k(t)$ are estimated by individual gain-phase estimators. The lower part of Fig. 3 is a coherent tracking loop that exploits the inherent multipath diversity of the channel. In each branch, say branch k, the received signal is first correlated with the *correlation function* $c_\Delta(t-\hat{\tau}(t)-kT_C)$ which is made up of the difference of the early and late replicas of the delayed spreading functions, i.e.,

$$c_\Delta(t-\hat{\tau}(t)) \triangleq c(t-\hat{\tau}(t)-\Delta T_C) - c(t-\hat{\tau}(t)+\Delta T_C) ,$$

where $\Delta \leq 1$ is called the early-late discriminator offset. Then, the correlated signal is converted to baseband through multiplication with the recovered carrier $\hat{a}_k(t)\cos(w_c t+\hat{\theta}_k(t))$. The desired code phase error-correcting signal $e(t)$ is obtained by summing the low-pass filtered signal $y_k(t)$ from the individual branches, and finally, the loop is closed by low-pass filtering the error signal which is used to drive the voltage control clock (VCC) for correcting the code phase error of the local PN code.

Stochastic Differential Equation: From Fig. 3, the received signal at the output of the kth band-pass filter (BPF) is

$$\begin{aligned} z_k(t) &\triangleq r(t)\, c_\Delta(t-\hat{\tau}(t)-kT_C)|_{BP} \\ &= \sqrt{2P}\sum_{m=0}^{L} a_m(t)h(\varepsilon, m+k-\Delta)\cos(w_c t+\theta_m(t)) \\ &\quad +\sqrt{2}\left\{\eta_k^c(t)\cos(w_c t) - \eta_k^s(t)\sin(w_c t)\right\} \end{aligned} \tag{4}$$

where

$$\begin{aligned} h(\varepsilon,\xi) &= R_C(\varepsilon-\xi) - R_C(\varepsilon+\xi) \\ \varepsilon &= \frac{\tau(t)-\hat{\tau}(t)}{T_C} \\ \eta_k^c(t) &= n_c(t)\, c_\Delta(t-\hat{\tau}(t)-kT_C) * \tilde{h}_l(t) \\ \eta_k^s(t) &= n_s(t)\, c_\Delta(t-\hat{\tau}(t)-kT_C) * \tilde{h}_l(t) \\ n(t) &= \sqrt{2}\left\{n_c(t)\cos(w_c t) - n_s(t)\sin(w_c t)\right\} \end{aligned}$$

with $\tilde{h}_l(t)$ denoting the equivalent lowpass impulse response of the BPF, and $*$ denoting the convolution operator. In (4), the PN code self-noise has been neglected, because the bandwidth of the BPF is assumed to be much smaller than the chip rate $1/T_C$, and the maximum Doppler shift of the channel has been assumed to be much smaller than the bandwidth of the BPF as in most practical applications. Therefore, the low-pass component $y_k(t)$ of the signal at the output of the multiplier is

$$\begin{aligned} y_k(t) &= \sqrt{\frac{P}{2}}\sum_{m=0}^{L} a_m(t)\hat{a}_k(t)h(\varepsilon, m+k-\Delta) \\ &\quad \cdot\cos(\theta_m(t)-\hat{\theta}_k(t)) + \frac{1}{\sqrt{2}}\hat{a}_k(t)\, w_k(t) \end{aligned}$$

where

$$w_k(t) = \eta_k^c(t)\cos(\hat{\theta}_k(t)) + \eta_k^s(t)\sin(\hat{\theta}_k(t)) .$$

Then the error signal $e(t)$ is given by

$$\begin{aligned} e(t) &\triangleq \sum_{k=0}^{L} y_k(t) \\ &= \sqrt{\frac{P}{2}}S(\varepsilon) + n_T(t) \end{aligned} \tag{5}$$

where

$$S(\varepsilon) = \sum_{k=0}^{L}\sum_{m=0}^{L} a_m(t)\hat{a}_k(t)h(\varepsilon, m+k-\Delta)\cos(\theta_m(t)-\hat{\theta}_k(t)) \tag{6}$$

and

$$n_T(t) = \frac{1}{\sqrt{2}}\sum_{k=0}^{L}\hat{a}_k(t)w_k(t) .$$

As shown in Fig. 3 the error signal at the output of the loop filter is used to drive a VCC for correcting the code phase error. The operation of the VCC is described by the equation

$$\frac{\hat{\tau}(t)}{T_C} = k_L\int_0^t f(t') * e(t')dt' \tag{7}$$

where k_L is gain of the VCC, and $f(t)$ is the impulse response of the loop filter. Combining (5) and (7) gives the following stochastic differential equation (SDE) which describes the dynamic behavior of the tracking loop:

$$\frac{d\varepsilon}{dt} = \beta_D - k_L\left\{\sqrt{\frac{P}{2}}S(\varepsilon) + n_T(t)\right\} * f(t) \tag{8}$$

where

$$\beta_D = \frac{1}{T_C}\frac{d\tau(t)}{dt}$$

is the code Doppler shift.

IV Loop Performance with Ideal Channel Estimation

In this section, the performance of the new code tracking loop is analyzed for a slowly time-varying channel ($\beta_D = 0$) with a perfect channel estimation (PCE). The performance degradation caused by imperfect channel estimation will be evaluated in the Section V.

Linear Analysis: Because of the assumptions of perfect channel estimation and very slowly time-varying channels, (6) becomes

$$\begin{aligned} S(\varepsilon) &= \sum_{k=0}^{L} a_k^2\, h(\varepsilon,\Delta) + \sum_{j=1}^{L}\sum_{m=0}^{L-j} a_m a_{m+j} \\ &\times \quad [h(\varepsilon, j+\Delta) - h(\varepsilon, j-\Delta)]\cos(\theta_m - \theta_{m+j}) \quad (9) \end{aligned}$$

From (1) and (9) it can be seen that the loop error characteristic $S(\varepsilon)$ is linear for $\varepsilon \le \Delta$ when $\Delta \le 1/2$, and linear for $\varepsilon \le 1-\Delta$ when $\Delta \ge 1/2$. Note that within this linear region $h(\varepsilon, 1+\Delta) = 0$, and $h(\varepsilon, j \pm \Delta) = 0$ for $j \ge 2$. Thus, for a tracking loop operating with a large enough SNR, the SDE of the loop becomes

$$\frac{d\varepsilon}{dt} \doteq \beta_D - k_L\sqrt{\frac{P}{2}}S'(0)\left\{\varepsilon + \frac{n_T(t)}{\sqrt{\frac{P}{2}}S'(0)}\right\} * f(t)$$

where

$$S'(0) = 2\left\{\sum_{k=0}^{L} a_k^2 - \sum_{k=0}^{L-1} a_k a_{k+1}\cos(\theta_k - \theta_{k+1})\right\} .$$

The closed-loop transfer function of the linearized model is defined as

$$\begin{aligned} H(s) &\triangleq \frac{\mathcal{L}\{\hat{\tau}(t)\}}{\mathcal{L}\{\tau(t)\}} \\ &= \frac{k_L\sqrt{\frac{P}{2}}S'(0)F(s)}{s + k_L\sqrt{\frac{P}{2}}S'(0)F(s)} , \end{aligned}$$

where $\mathcal{L}\{\cdot\}$ denotes the Laplace transform operator, and $F(s)$ is the transfer function of the loop filter. Since $\beta_D = 0$, the mean-square steady-state tracking error for a given channel vector $\mathbf{g} = (g_0, g_1, \ldots, g_L)$ is

$$\sigma^2_{\varepsilon|\mathbf{g}} = \frac{2}{PS'(0)^2}\int_{-\infty}^{\infty} S_{n_T}(f)\,|H(f)|^2\,df \qquad (10)$$

where $S_{n_T}(f)$ is the PSD of the total noise $n_T(t)$. In practice, $S_{n_T}(f)$ is approximately constant over the loop bandwidth and, therefore,

$$\sigma^2_{\varepsilon|\mathbf{g}} = \frac{2S_{n_T}(0)B_L}{PS'(0)^2} ,$$

where the loop bandwidth B_L is

$$B_L = \int_{-\infty}^{\infty} |H(f)|^2\,df .$$

For very slow fading channels the unconditional mean-square tracking error can be approximated by averaging (10) over all possible channel impulse responses, i.e.,

$$\sigma_\varepsilon^2 = \int_{\mathbf{g}} \sigma^2_{\varepsilon|\mathbf{g}}\, p(\mathbf{g})\, d\mathbf{g}$$

where $p(\mathbf{g})$ is the joint PDF of the channel impulse response. As to be seen in the following, the linear analysis is valid only when the received SNR is larger than a threshold value. If the SNR is below this threshold or if $\beta_D \ne 0$, then nonlinear theory must be employed for the performance analysis. To determine the threshold SNR and, hence, the applicable range for the linear analysis, we must compare results from the linear and nonlinear analysis.

Nonlinear Analysis: For simplicity, only the first order code tracking loop will be considered in the nonlinear analysis. Thus, the SDE in (8) becomes

$$\frac{d\varepsilon}{dt} = \beta_D - k_L\left\{\sqrt{\frac{P}{2}}S(\varepsilon) + n_T(t)\right\} .$$

Before we proceed, the following definitions are useful.

$\mathcal{R} \triangleq (\varepsilon_{\min}, \varepsilon_{\max})$: The *in-lock* region of the loop.

$P_{\mathrm{LD}}(\varepsilon) = 1 - P_{\mathrm{D}}(\varepsilon)$: The probability that the tracking unit declares an *out-of-lock* condition when the tracking error is still within the *in-lock* region $\mathcal{R}$.

$\pi(\varepsilon)$: The initial PDF of the code phase error when the synchronization process is switched from the acquisition unit to the tracking unit.

Let $p(\varepsilon|g)$ be the stationary PDF of the tracking error for a given channel g. Then it can be found that

$$p(\varepsilon|g) = \frac{Q(\varepsilon|g)}{\mathrm{E}[\tau^{tr}|g]} , \qquad \varepsilon_{\min} \le \varepsilon \le \varepsilon_{\max} \qquad (11)$$

where

$$Q(\varepsilon|g) = 2P_{\mathrm{D}}(\varepsilon)\frac{\exp[-u_0(\varepsilon)]}{k_2}\int_{\varepsilon_{\min}}^{\varepsilon} [d_0 - \Pi(x)]\exp[u_0(x)]\,dx$$

with

$$u_0(\varepsilon) = -2\int_{\varepsilon_{\min}}^{\varepsilon} \frac{k_1(x)}{k_2}\,dx \qquad (12)$$

$$k_2 = k_L^2\int_{-\infty}^{\infty} R_{n_T}(\xi)\,d\xi \qquad (13)$$

$$k_1(\varepsilon) = \beta_D - k_L\sqrt{\frac{P}{2}}S(\varepsilon) \qquad (14)$$

$$d_0 = \frac{\int_{\varepsilon_{\min}}^{\varepsilon_{\max}} \Pi(x)\exp[u_0(x)]\,dx}{\int_{\varepsilon_{\min}}^{\varepsilon_{\max}} \exp[u_0(x)]\,dx} \qquad (15)$$

$$\Pi(\varepsilon) = \int_{\varepsilon_{\min}}^{\varepsilon} \pi(x)\,dx \qquad (16)$$

and

$$\mathrm{E}[\tau^{tr}|g] = \int_{\varepsilon_{\min}}^{\varepsilon_{\max}} Q(x|g)\,dx .$$

In (13), $R_{n_T}(\xi) = \mathrm{E}[n_T(t)n_T(t+\xi)]$ is the autocorrelation function of the total noise $n_T(t)$. From (11) the conditional mean-square tracking error is obtained as:

$$\sigma^2_{\varepsilon|g} = \int_{\varepsilon_{\min}}^{\varepsilon_{\max}} \varepsilon^2 \, p(\varepsilon|g) \, d\varepsilon \ . \qquad (17)$$

By averaging (17) over the ensemble of channel impulse responses, the unconditional mean-square tracking error is:

$$\sigma^2_\varepsilon = \int_g \sigma^2_{\varepsilon|g} \, p(g) \, dg \ .$$

In general, there are no closed form expressions for σ^2_ε. Hence, numerical methods must be employed for evaluating this quantity.

Numerical Results: For simplicity only two-tap channels will be considered. Since the channels are assumed to be slowly time-varying, β_D is set to zero. The tracking loops are evaluated in terms of the averaged received SNR γ_b, the averaged nominal loop bandwidth $B_{LO} = \mathrm{E}\left\{B_L|_{\Delta=1/2}\right\}$, and the power ratio R of the first and second channel taps. Fig. 4 compares the mean-square tracking error obtained by using the linear and nonlinear analyses for the case of $\Delta = 1/2$. It is shown that the results obtained by the linear and nonlinear analyses agree very well if $\gamma_b \geq -12$ dB for $B_{LO}T_b = 0.01$ or $\gamma_b \geq -2$ dB for $B_{LO}T_b = 0.1$. However, for γ_b below these thresholds, the nonlinear theory must be used.

V Effects of Imperfect Channel Estimation

The new code tracking loop has been simulated on a computer to evaluate the effect of imperfect channel estimation. A traditional noncoherent DLL was also simulated for comparison. The simulated tracking loops had the following parameters.

- PN code: An m-sequence with generating polynomial of $g(x) = 1 + x^3 + x^7$
- Carrier frequency: 900 MHz
- Chip rate $R_C = \frac{1}{T_C}$: 1.27 M chips/s
- Bit rate $R_b = \frac{1}{T_b}$: 10 Kbps
- Sampling rate: 8 samples per chip period
- Total simulation time: 3000 T_b
- $\Delta = 0.5$
- The low-pass equivalent impulse response $\tilde{h}_l(t)$:

$$\tilde{h}_l(t) = \begin{cases} 1/T_b & \text{if } 0 \leq t \leq T_b \\ 0 & \text{otherwise} \end{cases}$$

- Fading channel: Jakes' two-dimensional isotropic scattering model (two independent paths with Rayleigh fading characteristics)
- Maximum Doppler frequency f_D: $f_D = 8.3$ Hz ($\beta_D =$ 0.0117598) and $f_D = 83$ Hz ($\beta_D = 0.117598$)

The channel estimator was the simplest one, where the equivalent low-pass channel impulse response was estimated by sampling the output of the receive filter in the Channel Parameter Estimation Unit of Fig. 3. Figs. 5 and 6 compare the analytical and simulation results for various f_D (β_D), R and B_{LO}. As evident, the new tracking loop overwhelmingly outperforms the traditional noncoherent DLL. Observe that for R as large as 10 dB, there still exists an irreducible performance floor for the traditional DLL (also, see [4]). For $\gamma_b \geq 5$ dB, the effect of imperfect channel estimation is less than 2 dB.

VI Conclusions

A new code tracking loop is proposed for direct sequence spread-spectrum signaling on frequency-selective fading channels. By exploiting the inherent multipath diversity of the channel, the new tracking loop significantly outperforms the traditional delayed locked loop. The new tracking loop is analyzed using linear and nonlinear methods, by assuming a slowly time-varying channel and perfect channel estimation. The performance degradation caused by imperfect channel estimation has been evaluated by computer simulations. For the simplest form of channel estimator, the degradation is less than 2 dB for the SNRs of practical interest. The effect of data modulation on the performance of the new tracking loops is not considered in this paper, but can be a subject for further study.

References

[1] M.K. Simon, J.K. Omura, R. A. Scholtz, and B.K. Levitt, **Spread Spectrum Communications** Vol. I-III. MD: Computer Science Press, 1985.

[2] W.J. Gill, "A comparison of binary delay-lock loop implementations," *IEEE Trans. Aerosp. Electr. Syst.*, Vol. AES-2, pp. 415-424, July 1966.

[3] H.P. Hartmann, "Analysis of a dithering loop for PN code tracking," *IEEE Trans. Aerosp. Electr. Syst.*, Vol. AES-10, pp. 2-9, January 1974.

[4] W-H. Sheen, and G.L. Stüber, "Effects of multipath fading on delay-locked loops for spread spectrum systems," *IEEE Trans. Commun.*, Vol. 42, pp. 1947-1956, February/March/April 1994.

[5] J.G. Proakis, **Digital Communication**. NY: MacGraw Hill, 1989.

[6] H. Meyr, "Delay-lock tracking of stochastic signals," *IEEE Trans. Commun.*, Vol. COM-24, pp. 331-339, March 1976.

[7] A. Polydoros and C.L. Weber, "Analysis and optimization of correlative code-tracking loops in spread-spectrum systems," *IEEE Trans. Commun.*, Vol. COM-33, pp. 30-43, January 1985.

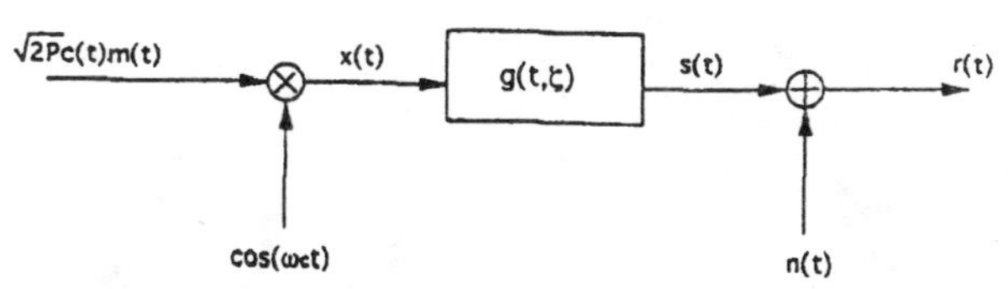

Fig. 1: Direct-sequence spread spectrum signaling on a multipath fading channel.

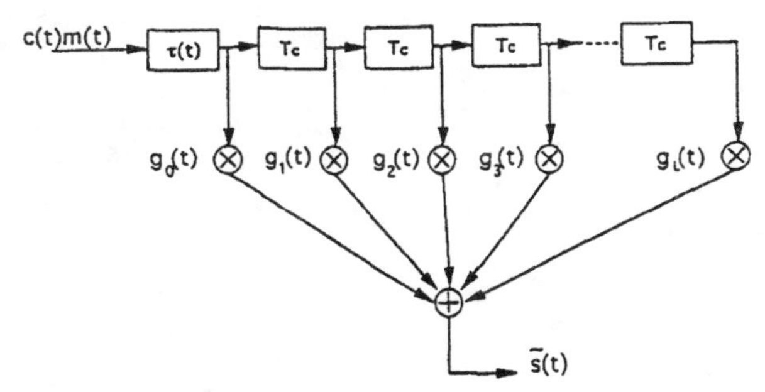

Fig. 2: Discrete-channel model for a time-varying WSSUS channel.

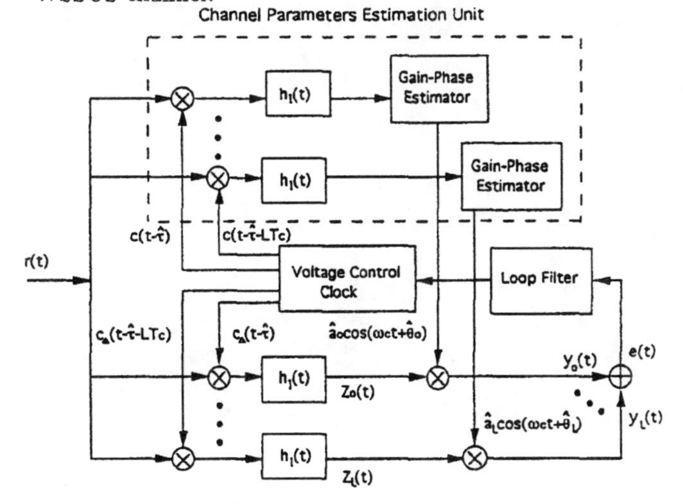

Fig. 3 The new tracking loop.

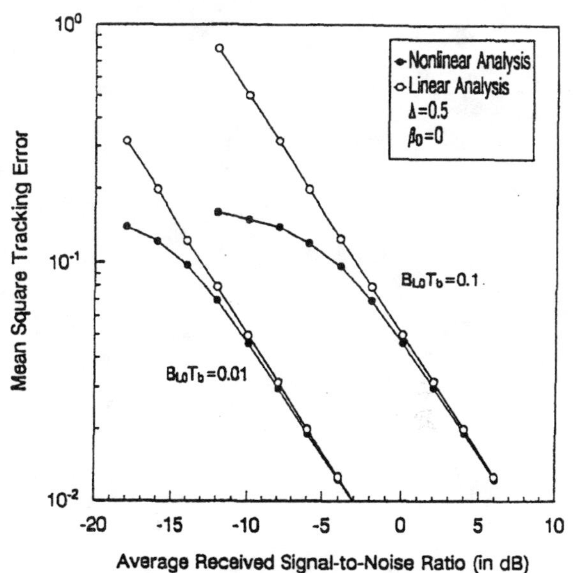

Fig. 4: Comparisons of linear and nonlinear analyses for the new tracking loop.

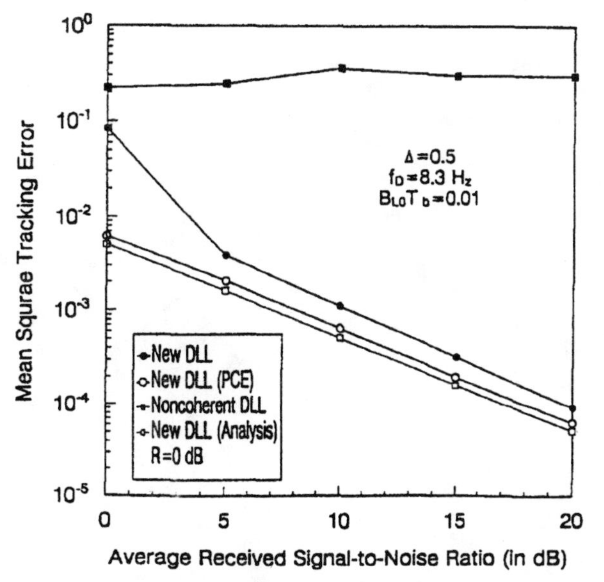

Fig. 5: The tracking-error performance of the new tracking loop and the conventional DLL.

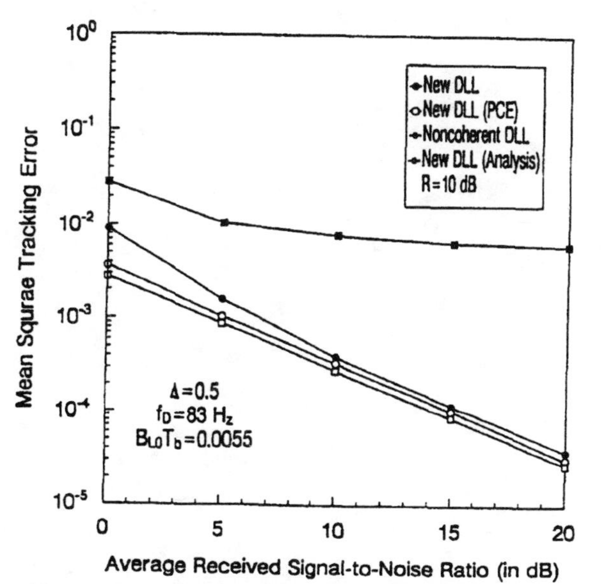

Fig. 6: The tracking-error performance of the new tracking loop and the conventional DLL.

SECTION 1.5
CODE DIVISION MULTIPLE ACCESS (CDMA) SYSTEM PERFORMANCE

In this section, we have selected six papers that provide a general overview of CDMA systems.

The first three papers in this subsection present some basic features of CDMA and consider its capacity. In FDMA and TDMA cellular systems, the capacity of each cell has a hard limit (i.e. when the capacity is reached), no more frequency channel or time slot is available for a new user. In contrast, signals from other users become interferences in a CDMA system. As the interferences increase, the performance of the system gradually degrades. The capacity of a CDMA cell may be defined as the number of active users which brings the performance to an "unacceptable" level. Due to the graceful degradation, CDMA capacity is soft-limited, rather than hard-limited. The first paper, "*Spread Spectrum for Mobile Communications*" by Pickholtz et al., obtains a simple formula for the capacity in terms of the energy/symbol to noise spectral density ratio. The second paper, "*Erlang Capacity of a Power Controlled CDMA System*" by Viterbi and Viterbi, derives the Erlang capacity. Fixing a blocking probability (the probability that a user requesting service is denied service), Erlang capacity is defined as the average number of active users which results in the blocking probability. The third paper, "*Reverse Link Performance of IS-95 Based Cellular Systems*" by Padovani, presents the reverse link transmission of the IS-95 CDMA system and computes the Erlang capacity of the reverse link.

The fourth paper, "*Spread Spectrum Access Methods for Wireless Communications*" by Kohno et al., gives an overview of CDMA characteristics and design considerations. The fifth paper, "*Design Study for a CDMA-Based Third-Generation Mobile Radio System*" by Baier et al., discusses some design study of the third generation of mobile ratio using CDMA, with emphasis on the flexibility of providing a wide range of services with variable bit rate. The sixth paper, "*Wireless Communications Going Into 21st Century*" by Schilling, paints a picture of the evolution of cellular and PCS systems into the 21st century.

Spread Spectrum for Mobile Communications

Raymond L. Pickholtz, *Fellow, IEEE*, Laurence B. Milstein, *Fellow, IEEE*, and Donald L. Schilling, *Fellow, IEEE*

Abstract—**The characteristics of spread spectrum that make it advantageous for mobile communications are described. The parameters that determine both the performance and the total capacity are introduced, and an analysis is presented which yields (approximately) the number of users that can simultaneously communicate while maintaining a specified level of performance. Spread-spectrum overlay, wherein a code division multiple access (CDMA) network shares a frequency band with narrow-band users, is analyzed, and it is seen that excision of the narrow-band signals from the CDMA receivers before despreading can improve both performance and capacity.**

I. Introduction

THE use of spread spectrum techniques for military communications has become quite commonplace over the past decade, and it was the potential use of spread spectrum in hostile environments that motivated most of the basic research in the subject for the past 20 years or more. In recent years, however, there has been an even greater interest in understanding both the capabilities and the limitations of spread-spectrum techniques for commercial applications. Entirely new opportunities such as wireless LAN's, personal communication networks, and digital cellular radios have created the need for research on how spread spectrum systems can be "reoptimized" for most efficient use in these nonhostile environments.

Indeed, in this new decade we will experience a revolution in telecommunication, which might be even greater than the PC revolution of the past decade. With this increased need to communicate, and with a "fixed" available spectrum, it becomes necessary to use this spectrum more effectively. Spread spectrum is a technique for efficiently using spectrum by allowing additional (spread spectrum) users to use the same band as other, existing users.

The new applications of spread-spectrum communications have characteristics which are quite different from those of interest in the past. That is, up until fairly recently, most interest in spread spectrum was dominated by the classical military scenario of intentional, smart jamming [1]. However, once the threat of having to design a system which is capable of combatting an intelligent adversary is removed, one can consider ways of improving the receiver design to make the system more efficient and more practical for commercial applications. Indeed, the new design philosophy will emphasize spectral efficiency, cost, reliability and complexity reduction.

Of the many potential uses for spread-spectrum communications in civilian applications, code division multiple access (CDMA) appears to be the most popular. This is especially true in what is arguably the hottest topic in communications today, mobile communications. In such situations, multipath is often a fundamental limitation to system performance, and spread spectrum is a well-known technique to combat multipath.

In addition to multiple accessing capability and multipath rejection, one of the major opportunities that arise when using spread spectrum communications is the possibility of overlaying low level direct sequence (DS) waveforms on top of existing narrow-band users, and hence increasing the overall spectrum capacity even more so than just through the use of the CDMA network. However, such a procedure must be done very carefully so as to not cause intolerable interference for either the existing narrow-band users or the spread spectrum users. One means of helping to ensure this is to use signal processing techniques to suppress those narrow-band users that occupy the spread bandwidth; this will clearly lessen the interference level that the DS spread-spectrum signals have to live with. However, it will also help the narrow-band users because the power levels of the CDMA users can now be correspondingly decreased precisely because they do not have to compete with the preexisting signals.

Finally, since spread sprectrum is obtained by the use of noise-like signals, where each user has a uniquely addressable code, privacy is inherent.

II. Review of Spread Spectrum Characteristics

The idea behind spread spectrum is to transform a signal with bandwidth B_s into a noise-like signal of much larger bandwidth B_{ss}. This is illustrated in Fig. 1, where the ordinate is power spectral density (watts/hertz) and the abscissa is the frequency axis (hertz). Assume the total power transmitted by the spread-spectrum signal is the same as that in the original signal. In this case, the power spectral density of the spread-spectrum signal is $P_s(B_s/B_{ss})$; the ratio B_{ss}/B_s, called the *processing gain*, is denoted by N, and is typically 10–30 dB. Hence the power of the radiated spread spectrum signal is spread over 10–1000 times the original bandwidth, while its power spectral density is correspondingly reduced by the same amount. It is this feature that gives

Manuscript received May 1, 1990; revised September 17, 1990. This paper was supported in part by the NSF I/UCRC for Ultra-High Speed Integrated Circuits and Systems, and by SCS Telecom, Inc.

R. L. Pickholtz is with the Department of Electrical Engineering and Computer Science, George Washington University, Washington, DC 20052.

L. B. Milstein is with the Department of Electrical and Computer Engineering, University of California at San Diego, La Jolla, CA 92093.

D. L. Schilling is with the Department of Electrical Engineering, City College of New York, New York, NY 10031.

IEEE Log Number 9144469.

0018-9545/91/0500-0313$01.00

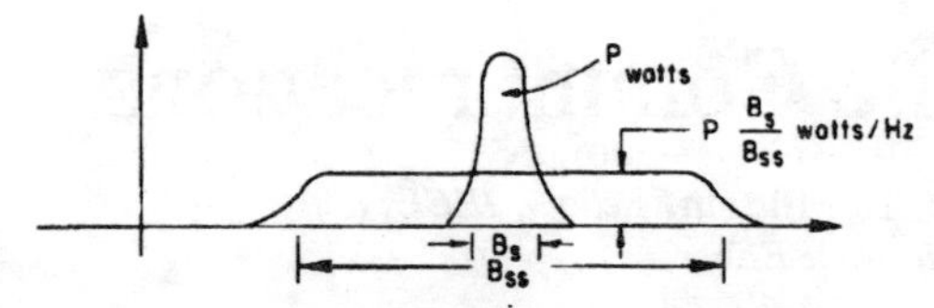

Fig. 1. Spectra of signal before and after spreading.

the spread spectrum signal the characteristic of causing little interference to a narrow-band user and is the basis of proposed *overlay* systems that operate spread-spectrum concurrently with existing narrow-band systems. The other requirement of the spread-spectrum signal is that it be "noise-like." That is, each spread-spectrum signal should behave as if it were "uncorrelated" with every other spread signal using the same band. In practice, the correlation used need not be zero (i.e., the signals are not completely orthogonal), and thus there can be many more such signals with this property than if they were all required to be truly orthogonal. The consequence of this will be seen in the following.

III. Basic Principles and Features

There are several ways to implement a spread spectrum system. Each requires:

- signal spreading by means of a code;
- synchronization between pairs of users;
- care to insure that some of the signals do not overwhelm the others (near-far problem);
- source and channel coding to optimize performance and total throughput.

The two most popular signal spreading schemes are:

- direct sequence spreading;
- frequency hopping.

Spread spectrum is a "second modulation technique." For example, in direct sequence, one starts with a standard digital modulation, such as binary phase shift keying (BPSK) and then applies the spreading signal. The despreading at the front end of the receiver then delivers the BPSK signal to the standard processor for such signals. In frequency hopping, the modulated signal is first generated. Then the spread spectrum technique consists of changing the center frequency of the transmitted signal every T_H seconds, so that the *hop rate* is $f_H = (1/T_H)$ hops/s. This could be done *slowly* (one hop per many symbols) or *fast* (many hops per symbol). The total frequency spread B_{ss} is equal to the total number N of distinct frequencies used for hopping times the frequency bin occupied for each such frequency. For slow hoppers, this is $N/T = B_{ss}$, or $N = B_{ss}/B_s$, where $B_s \sim 1/T$ and T is the duration of a data symbol. We observe that N is the processing gain defined above.

To make the FH signal "noise-like," the frequency hopping pattern is driven by a "pseudo-random" number generator having the property of delivering a uniform distribution for each frequency that is "independent" on each hop. So long as the intended receiver knows this pseudo-random

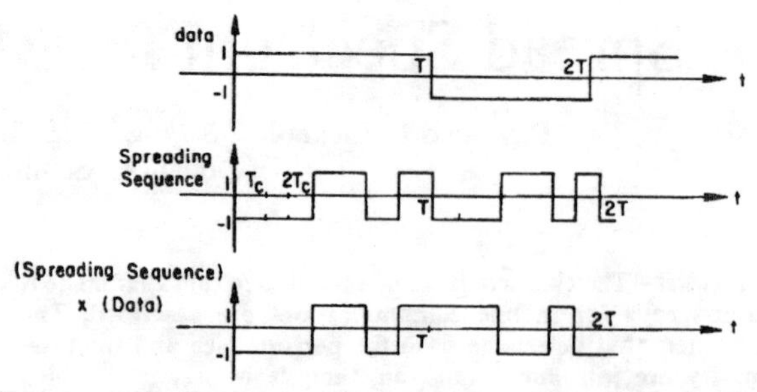

Fig. 2. Time waveforms involved in generating a direct sequence signal.

sequence, and can arrange to synchronize with the transmitter, the frequency hopped (FH) pattern may be dehopped (or despread) to receive the original signal, which is then processed normally. FH systems are very popular in military as well as commercial communications, and they enjoy certain advantages over DS (such as not requiring a contiguous band).

Returning to direct sequence, we see from Fig. 2 that the spreading is accomplished by multiplying the modulated information-bearing signal by a (usually) binary $\{\pm 1\}$ baseband code sequence waveform, $PN_i(t)$. The code sequence waveform may be thought of as being (pseudo) randomly generated so that each binary chip can change (with probability = 1/2) every T_c s. Thus the signal for the ith transmitter is

$$S_i(t) = PN_i(t)Ad(t)(\cos \omega_0 t + \phi), \quad (1)$$

where $d(t)$ is the data modulation (assumed to be ± 1 for BPSK signaling), A is the amplitude of the BPSK waveform, and ϕ is a random phase.

Since T_c is less than T (and frequently *much* less than T), the ratio of the spread bandwidth, B_{ss}, to the unspread bandwidth, B_s, is given by $B_{ss}/B_s = T/T_c = N$, the processing gain. It is clear that a receiver with access to $PN_i(t)$, and synchronized to the spread spectrum transmitter, can receive the data signal, $d(t)$, by a simple correlation. That is, in the interval $[0, T]$, if the data symbol is d_1, which can take on values ± 1, then

$$\frac{2}{T}\int_0^T PN_i(t)(\cos \omega_0 t + \phi)S_i(t)\,dt = Ad_1 = \pm A. \quad (2)$$

If the spreading sequence is properly designed, it will have many of the randomness properties of a fair coin toss experiment where "1" = heads and "−1" = tails. These properties include the following:

1) in a long sequence, about 1/2 the chips will be +1 and 1/2 will be −1;
2) a run of length r chips of the same sign will occur about $2^{-r}l$ times in a sequence of l chips;
3) the autocorrelation of the sequence $PN_i(t)$ and $PN_i(t + \tau)$ will be very small except in the vicinity of $\tau = 0$;
4) the cross correlation of any two sequences $PN_i(t)$ and $PN_j(t + \tau)$ will be small.

An important class of sequences called *maximal length linear feedback shift register sequences* are well known to exhibit properties 1), 2), and 3). In particular, the autocorrelation function

$$R_i(\tau) = \frac{1}{T_p}\int_0^{T_p} PN_i(t)PN_i(t+\tau)\,dt \tag{3}$$

is given by

$$R_i(\tau) = \begin{cases} 1 - \dfrac{\tau}{T_p}\left(1 + \dfrac{T_c}{T_p}\right), & 0 \le \tau \le T_c \\ -\dfrac{T_c}{T_p}, & T_c \le \tau \le (N-1)T_c \\ \tau - \dfrac{T_p - T_c}{T_c}\left(1 + \dfrac{T_p}{T_c}\right) - \dfrac{T_p}{T_c}, & \\ (N-1)T_c \le \tau \le NT_c, & \end{cases} \tag{4}$$

where T_p is the period of the sequence and $R_i(\tau)$ is also periodic with period T_p.

Each of these properties can be quite significant in a mobile communications system. For example, if $T = T_p$ and $N = T/T_c = 255$, then (1) tells us that a signal due to multipath, arriving τ s after the first signal, is attenuated by $R_i(\tau)$. In particular, if $T_c \le \tau \le T - T_c$, then the power of the multipath signal is reduced by $(T_c/T)^2 = (1/255)^2$ (or about 48 dB).

Note, however, that (3) refers to a full correlation (i.e., a correlation over the complete period of the spreading sequence). Since data are usually present on the signal, and since data transitions typically occur 50% of the time, (3) should really be replaced by

$$\pm\frac{1}{T_p}\int_0^{\tau} PN_i(t)PN_i(t+\tau)\,dt \pm \frac{1}{T_p}\int_{\tau}^{T_p} PN_i(t)PN_i(t+\tau)\,dt, \tag{5}$$

where the independent $\pm$ signs on the two terms of (5) correspond to the fact that they are due to different data symbols. When either both signs are plus or both signs are minus, (3) applies and, for the example presented above, the attenuation would indeed be 48 dB. However, when the two signs differ, (5) applies, indicating that we now have the sum of two partial correlations, rather than one total correlation. In particular, if $\tau = 63T_c$ (i.e., if, in our example, the multipath is delayed by about one quarter of the symbol duration), then, for one specific maximal length shift register sequence, the attenuation of the multipath can be shown to be reduced from 48 to 16 dB.

Finally, we focus on two issues which are crucial to deploying a successful spread spectrum system. The first is the requirement for acquisition of synchronization. The synchronization between code generators is essential either at initialization or if a dropout has occurred long enough for the clock to slip. No communication is possible until acquisition of synchronization has been achieved. This is usually done by means of a search for the chip epoch that results in a large correlation. Furthermore, a loop must keep the code generators in synchronization after acquisition has taken place. This is usually done with a *delay locked loop*, which takes advantage of the correlation property 3) above.

The second issue is *power control*. This is required because a close-in undesired transmitter can swamp a remote, desired transmission. While the powers are additive, the close-in transmitter has a $(d_d/d_u)^r$ advantage in power, where d_u is the distance to the undesired transmitter, d_d is the distance to the desired transmitter, and r is the propagation exponent. For example, for free space, if $r = 2$ and if $d_d/d_u = 20$, then the undesired signal has an advantage of $(20)^2$, or 26 dB. This intolerable condition may be mitigated by using power control to reduce the transmission power of a close-in user. For example, in a mobile cellular environment, one would attempt to separately control the forward power from the cell to each mobile and the reverse power from each mobile to the cell site.

IV. Code Division Multiple Access

Certainly the fundamental reason for the interest in spread spectrum communications for cellular radio is the fact that CDMA can allow many users to access the channel simultaneously, just as TDMA and FDMA can. The distinction, however, between CDMA and either TDMA or FDMA is that CDMA provides, in addition to the basic multiple accessing capability, the other attributes described above (such as privacy, multipath tolerance, etc.). These latter attributes are either not available with the use of the narrow-band waveforms which are employed with TDMA or FDMA, or are much more difficult to achieve. For example, one can typically implement a narrowband digital communication link that is tolerant of multipath interference by including in the receiver an adaptive equalizer, but this increases the complexity of the receiver, and may affect the ability to perform a smooth handover. Indeed, since the equalizer must continually adapt to an ever changing channel, it is a high-risk component of a TDMA system. Furthermore, CDMA degrades gracefully. No more than N users can simultaneously access a TDMA or FDMA system. However, if more than N users simultaneously access a CDMA system, the noise level, and hence the error rate, increases in proportion to the percent overload.

There are, of course, a number of disadvantages associated with CDMA, the two most obvious of which are the problem of "self-jamming," and the related problem of the "near-far" effect. The self-jamming arises from the fact that in an asynchronous CDMA network, the spreading sequences of the different users are not orthogonal, and hence in the despreading of a given user's waveform, nonzero contributions to that user's test statistic arise from the transmissions of the other users in the network. This is as distinct from either TDMA or FDMA, wherein for reasonable time or frequency guardbands, respectively, orthogonality of the received signals can be (approximately) preserved.

Given that such orthogonality cannot be preserved in CDMA, the obvious point of interest is to determine how much degradation in system performance results. From a quantitative point of view, there are many analyses and bounding techniques available to answer this question [2]–[6]. Qualitatively, we can easily see two major areas of concern for the specific application of digital cellular radio. The first is the propagation law; because these channels typically have multiple reflections associated with them, the net propagation law is not that which would be observed over a free-space channel. Measurements have indicated that the received power falls off roughly as the inverse of the distance between the transmitter and the receiver raised to a power that is somewhere between two and four, and because of the potentially large number of users causing the multiple access interference, there can be a noticeable difference in performance depending on which power law is used for performance computations.

Associated with this is the near-far problem, that is, signals closer to the receiver of interest are received with smaller attenuation than are signals located further away. This means that power control techniques must be used in the cell of interest. However, this still does not guarantee that interference from neighboring cells might not arrive with power levels higher than can be tolerated, especially if the waveforms in different cells are undergoing independent fading.

One final concern is the following: in CDMA, a smooth handover from one cell to the next requires acquisition by the mobile of the new cell before it relinquishes the old cell.

It is seen then that while the use of spread spectrum techniques offers some unique opportunities to system designers of digital cellular radios, there are issues to be concerned about as well, and some of these issues are addressed below. In particular, in what follows, we try to discuss some aspects of spread spectrum which are at least somewhat particular to portable communication networks.

A. Voice Activity Effects in CDMA and Capacity Estimates

CDMA systems tend to be self-interference limited. That is, in attempting to have many users communicate simultaneously, the mutual interference sets a limit on the number of simultaneously active users. To the extent that not every user in the network is always transmitting, the capacity of the system is increased. Note that this argument can also be applied to FDMA and TDMA systems, but there is a fundamental difference. In these latter systems, some type of central control is needed to implement a demand assignment protocol; in CDMA, no such control is necessary. Alternately, if such control is indeed used, one still achieves a performance advantage in CDMA because of this effect that is not achieved with either FDMA or TDMA. This is because with CDMA, if a given user stops talking, and no new user wants to access the channel at that particular instant of time, then all the remaining users on the channel experience less interference.

In light of the above, it is of interest to determine the effect of voice activity on the performance of a CDMA network ([7]). Toward that end, consider starting with the following approximate expression for the bit error rate (BER) of a CDMA system, which was originally derived in [8]:

$$P_e = \phi\left(-\left[\frac{\eta_0}{2E} + \frac{m}{3N}\right]^{-1/2}\right), \tag{6}$$

where E/η_0 is the ratio of energy-per-bit-to-noise power spectral density, $\phi(\cdot)$ is the cumulative normal function, N is the number of chips-per-bit, and m is the number of additional active users (i.e., the total number of active users is $m + 1$). Equation (6) is based upon the so-called "Gaussian approximation," i.e., the despread multiple access interference is approximated as a Gaussian random variable. This approximation has been studied in a variety of references (see, e.g., [9]–[11]), and has been found to hold reasonably well when long spreading sequences are used (i.e., when the period of the spreading sequence spans many data symbols) and when $N \gg 1$, $m \gg 1$, and P_e is not too small. This latter assumption is reasonable when forward error correction is used, since (6) corresponds to the channel BER, not the decoded BER. In particular, the Gaussian approximation is shown in [10] to hold, for fixed N, as K becomes large when thermal noise can be neglected, and in what follows, we will indeed be interested in the situation where the performance is limited by the multiple access interference, not by the thermal noise.

Note that m is a random variable, since it depends upon how many users are "on," and how many of the "on" users are talking. Therefore, (6) is more accurately designated as $P(e \mid m)$. To find the unconditional BER, which we will denote by $\bar{P}_e$, we observe the following: If $L + 1$ is the total number of users, if p is the probability that one of them is "on," and if α is the probability that one of the "on" users is talking, then

$$\bar{P}_e = \sum_{k=0}^{L} \sum_{m=0}^{k} \phi\left(-\left[\frac{\eta_0}{2E} + \frac{m}{3N}\right]^{-1/2}\right)\binom{L}{k} p^k (1-p)^{L-k} \cdot \left[\begin{matrix} k \\ m \end{matrix}\right] \alpha^m (1-\alpha)^{k-m}. \tag{7}$$

In order to avoid having to evaluate (7) (since L might be a large number), we can consider the following approximation: Define

$$M \triangleq \alpha L p, \tag{8}$$

i.e., M is the average number of "on" users who are talking. Then we approximate P_e by $\tilde{P}_e$, where

$$\tilde{P}_e \triangleq \phi\left(-\left[\frac{\eta_0}{2E} + \frac{M}{3N}\right]^{-1/2}\right). \tag{9}$$

Note that while we have arrived at (9) as an *approximation* to (7), in reality, (9) is, at times, a lower bound to (4). To see this, note that the second derivative of the $\phi(\cdot)$ function of (7) with respect to m is given by

$$\frac{d^2\phi\left(\left[\frac{\eta_0}{2E} + \frac{m}{3N}\right]^{-1/2}\right)}{dm^2} = \frac{\exp\left[-\frac{1}{2}\left[\frac{\eta_0}{2E} + \frac{m}{3N}\right]^{-1}\right]}{\sqrt{2}\pi(6N)^2\left[\frac{\eta_0}{2E} + \frac{m}{3N}\right]^{5/2}}$$

$$\cdot\left[\left[\frac{\eta_0}{2E}+\frac{m}{3N}\right]^{-1}-3\right]$$

$$\triangleq \beta\left[\left[\frac{\eta_0}{2E}+\frac{m}{3N}\right]^{-1}-3\right] \quad (10)$$

where β is a function of many parameters, but is always positive. Therefore, (10) will be greater than zero for all m if

$$\frac{L}{N} < 1 - \frac{3}{2}\left(\frac{\eta_0}{E}\right). \quad (11)$$

In turn, this implies that

$$\phi\left(-\left[\frac{\eta_0}{2E}+\frac{m}{3N}\right]^{-1/2}\right) \quad (12)$$

is convex U in m, and hence that (9) lower bounds (7) [12].

Another way to look at this problem is from an operational point of view. A user wanting to access a cell in order to place a call must first ask permission to do so. If the cell has a receiver available to accept the call, the mobile user and receiver, employing the same code, communicate. If no receiver is "free," a busy signal is returned to the user, who must keep trying until a receiver becomes available.

To obtain an initial (although certainly approximate) estimate of the number of receivers, say $\hat{m}$, allotted to a cell, it is necessary to make some modifications to (6), since (6) applies to an AWGN channel, whereas mobile channels experience fading. To account for the fading, we model the channel as a two-ray multipath channel with each ray undergoing independent Rayleigh fading and having equal average signal-to-noise ratio. We further assume that these two paths can be resolved and coherently summed to yield twofold maximal-ratio combining diversity. Then, assuming for simplicity a scenario where the interference does not fade, we can use standard techniques to show that the average probability of error is given by (see [13]) $P_e \approx 3/(16\Gamma^2)$, where Γ is the average signal-to-noise ratio on each ray and is assumed to be $\gg 1$. Note that this corresponds to somewhat of a worst-case situation, since we are assuming that the desired signal experiences fading but that the multiple access interference does not. Letting $\Gamma = 1/(\eta_0/E + 2\hat{m}/3N)$, for a given P_e, we can solve for $\hat{m}$. For example, suppose $P_e = 10^{-2}$. (Note that this is the channel error rate; through the use of forward error correction coding, the error rate can then be reduced to a satisfactory level.) Then Γ should be about 6 dB, and, assuming $E/\eta_0 \gg 1$, the number of users, $\hat{m}$, is approximately $\hat{m} \simeq 0.38N$, or about one-third of the processing gain.

To refine this estimate, as well as account for other sources of degradation, such as interference from adjacent cells, consider the following: In a digital communication system operating in Gaussian noise, the usual performance measure is E/η_0, the ratio of the energy/bit to the noise power spectral density. We may calculate this number by examining the output of a matched filter receiver (matched to one of the PN coded waveforms) whose input consists of the desired signal, thermal white noise with one side power density η_0 W/Hz, and interference consisting of m *other* PN coded waveforms with an activity factor $\alpha < 1$ and possibly other interference, such as users in adjacent cells, that increases the measured interference by $(1 + K)$. Applying the matched filter integration over one symbol time, $T = NT_c = E_s$ for a ± 1 signal, we get the *effective* energy/symbol to noise spectral density

$$\left(\frac{E_s}{\eta_0}\right)_{\text{eff}} = \frac{\frac{1}{2}(NT_c)^2}{NT_c\frac{\eta_0}{2} + \frac{NT_c^2}{3}\hat{m}(1+K)\alpha} = \frac{1}{\frac{\eta_0}{E_s} + \frac{2}{3N}\hat{m}(1+K)\alpha}, \quad (13)$$

where $\hat{m} \triangleq pL$.

Let $rE_b = E_s$ (we now use E_b as the energy per bit to distinguish it from E_s), where $r < 1$ is the dimensionless code-rate of an error correcting code (if used). Dividing both sides by r gives

$$\left(\frac{E_b}{\eta_0}\right)_{\text{eff}} = \frac{1}{\frac{\eta_0}{E_b} + \frac{2r}{3N}\hat{m}(1+K)\alpha}. \quad (14)$$

In order to guarantee a given performance, this quantity must be at least equal to the *required* value $(E_b/\eta_0)_{\text{req}}$. If we define the required $(E_b/\eta_0)_{\text{req}} \triangleq \lambda_{\text{req}}$, and $E_b/\eta_0 \triangleq \lambda_0$ (this is $(E_b/\eta_0)_{\text{eff}}$ without interference), and solve for $\hat{m}$, we obtain

$$\hat{m} \leq \left\lfloor \frac{3}{2\alpha(1+K)} \frac{N}{r\lambda_{req}}\left(1 - \frac{\lambda_{\text{req}}}{\lambda_0}\right)\right\rfloor. \quad (15)$$

A typical calculation might be based on the following parameters:

Signal-to-thermal noise ratio	$\lambda_0 = 30$ dB
Signal-to-noise ratio required before decoding	$\lambda_{\text{req}} = 6$dB
FEC code rate	$r = \frac{1}{2}$
Uncoded bit rate	$R_b = 4800$ b/ps
Transmitted bit rate after encoding	$R_s = 9600$ b/ps
Spread bandwidth	$B_{ss} = 12.5$ MHz[1]
Adjacent channel spill over factor	$K = 1/2$
Speaker activity factor	$\alpha = 1/2$
Processing gain	$N = \frac{B_{ss}}{2R_s} = 640$ ($\simeq 28$ dB)

For these numbers, we get from (15) $\hat{m} \leq 640$ active users

[1] 12.5 MHz is the bandwidth available for cellular transmission and for reception.

(with no frequency reuse). It should be pointed out, however, that this number may be off by a significant factor in any application, since the result is very sensitive both to the assumptions used in the modeling and to the values used for the illustrative parameters. Further, notice that although the number of *active* users is 640, there may be many more potential users who will (statistically) share the channel without the need to reassign codes or spectral space. For example, if the probability of a mobile being turned on is $p = 0.01$, there may be as many as 64 000 such potential users at any single cell.

V. Overlay Considerations and Narrow-Band Interference Rejection

As indicated above, one of the most attractive features of using spread spectrum in a mobile scenario is the ability to overlay the spread spectrum waveforms on top of existing narrow-band users. In a sense, this allows a different type of frequency reuse, or frequency sharing. Frequency sharing is a consequence of the fact that a DS waveform, spread over a sufficiently wide bandwidth, has a very low power spectral density, and hence the additional degradation its presence causes to a narrow-band user located somewhere in its bandwidth, over and above that caused by the thermal noise, can be made quite small.

This assumes, of course, that the power of the DS waveform is not too large. Since, in a "shared" environment, each DS user must be able to withstand the interference caused by the presence of the narrow-band users, it is of interest to make that interference as small as possible. That is, the larger the level of the narrow-band-to-spread spectrum interference, the larger must be the transmitted power of the spread spectrum signal, and hence the larger will be the spread spectrum-to-narrow-band interference.

As one means of potentially alleviating this seemingly vicious cycle, one can consider employing any one of a number of narrow-band interference suppression techniques [14]. These are techniques that are based upon signal processing schemes, and which attempt to make use of the key difference between a DS waveform and a narrow-band interfering signal, namely the great disparity in bandwidth. In the remainder of this section, we briefly describe two narrow-band interference rejection schemes and show some typical performance results for both the narrow-band-to-spread spectrum interference and the spread spectrum-to-narrow-band interference.

Let us assume that we are employing frequency sharing. Then each spread spectrum receiver receives wide-band spread spectrum waveforms in the presence of wide-band thermal noise and narrow-band interference (e.g., narrow-band microwave signals). Assume that, prior to despreading, the receiver attempts to accumulate, say, μ samples of the received waveform and, on the basis of those μ samples, predict the value of the $(\mu + 1)$st sample. The only component of the received signal that can be accurately predicted is the interference component. This is because both the DS signal and the thermal noise are wide-band processes, and hence their past histories are not of much use in predicting

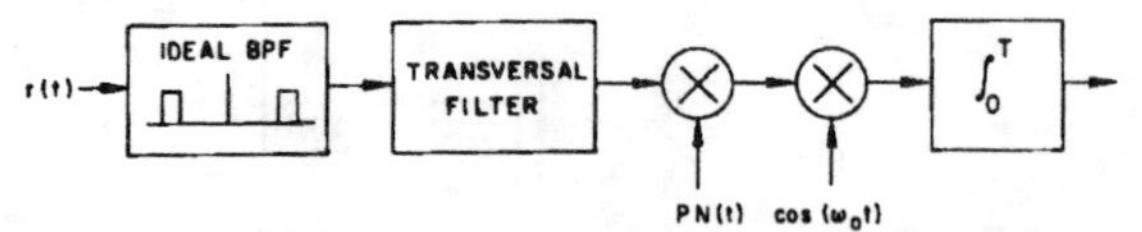

Fig. 3. Block diagram of direct sequence receiver which employs an interference suppression filter.

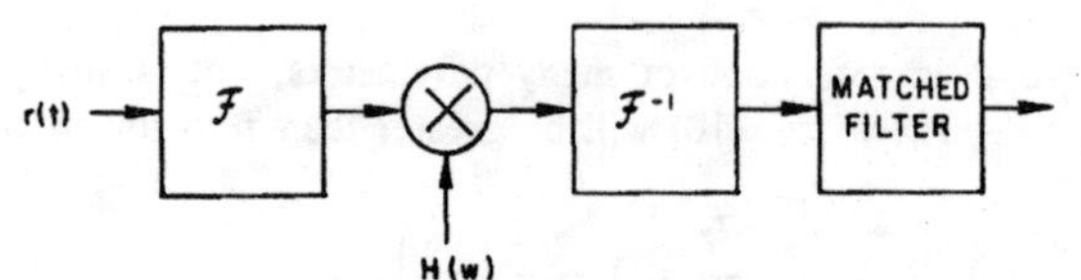

Fig. 4. Block diagram of transform domain processing receiver.

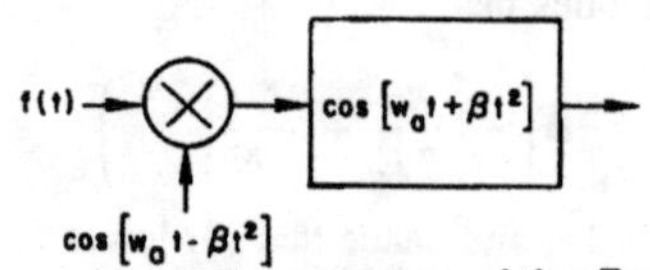

Fig. 5. Implementation used to generate a real-time Fourier transform.

future values. On the other hand, the interference, being a narrow-band process, *can* have its future values predicted from past values.

Therefore, if a prediction of the $(\mu + 1)$st value of the received waveform is made, and if this estimate is then subtracted from the actual received value of the $(\mu + 1)$st sample, the interference component will be significantly attenuated, whereas the DS signal and the thermal noise will have their values only slightly altered. A block diagram of a receiver employing this type of suppression filter is shown in Fig. 3 [15]. Similarly, if a two-sided transversal filter is used instead of a single-sided filter (sometimes referred to as a prediction-error filter), both past and future values of the received waveform can be used to predict the current value of the process. In reality, since the interfering narrow-band waveforms will not be at fixed locations (in frequency) as the mobile moves, some means of implementing an adaptive interference rejection filter is needed. Such a filter can be designed using any one of a number of adaptive algorithms, and analyses of system performance when the well-known LMS algorithm is used are presented in [16] and [17].

An alternate approach to narrow-band interference rejection is to use the technique of transform domain processing. This latter scheme is illustrated in Fig. 4. It consists of a Fourier transformer, a multiplier, an inverse Fourier transformer, and a matched filter. In essence, the filtering by the transfer function $H(\omega)$ is performed by multiplication followed by an inverse Fourier transformation, rather than by convolution. This multiplication, while ostensibly being performed in the "frequency domain," is, of course, accomplished by the Fourier transform device in real-time.

Clearly, the essence of the scheme is the ability to perform a real-time Fourier transform, and this operation, for example, can be accomplished using the block diagram of Fig. 5. The signal to be transformed, say $f(t)$, is multiplied by the waveform $\cos(\omega_a t - \beta t^2) P_T(t)$ (i.e., it is multiplied by a linear FM or chirp waveform), and the product is used as the input to a time-invariant linear filter whose impulse response

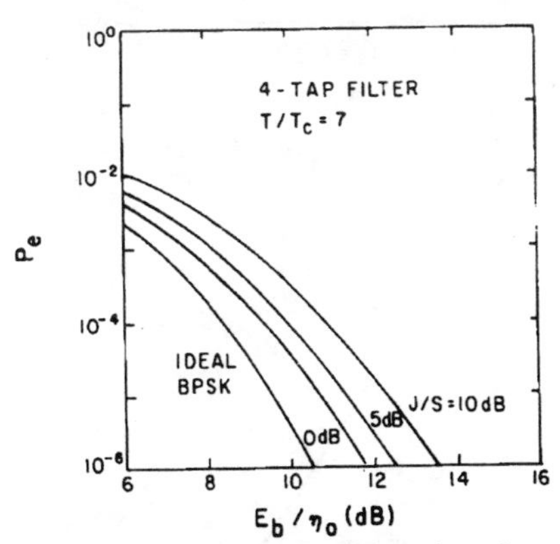

Fig. 6. Bit error rate results of direct sequence receiver when narrow-band interference occupying 10% of spread bandwidth is present.

is $\cos(\omega_a t + \beta t^2)P_{T_1}(t)$, where $P_a(x)$ is 1 for $0 \le x \le a$ and zero elsewhere (see, e.g., [18], [19] for details on this type of signal processing). That this type of processing does, indeed, result in the Fourier transform of a time-limited input signal $f(t)$ can be seen as follows. Denoting the output of the system shown in Fig. 5 by $f_0(t)$, we have, for $t \in [T, T_1]$,

$$\begin{aligned} f_0(t) = & \int_0^T f(\tau)\cos(\omega_a\tau - \beta\tau^2) \\ & \cdot \cos\left[w_a(t-\tau) + \beta(t-\tau)^2\right] d\tau \\ = & \frac{1}{2}\cos(w_a t + \beta t^2)\int_0^T f(\tau)\cos 2\beta t\tau \, d\tau \\ & + \frac{1}{2}\sin(w_a t + \beta t^2)\int_0^T f(\tau)\sin 2\beta t\tau \, d\tau \\ & + \frac{1}{2}\int_0^T f(\tau)\cos\left[2w_a\tau - 2\beta\tau^2\right. \\ & \left. + 2\beta t\tau - w_a t - \beta t^2\right] d\tau. \end{aligned} \quad (16)$$

Ignoring the third term of (16) (it is a double frequency term), we have

$$\begin{aligned} f_0(t) = & \frac{1}{2}F_R(2\beta t)\cos(w_a t + \beta t^2) \\ & + \frac{1}{2}F_I(2\beta t)\sin(w_a t + \beta t^2), \end{aligned} \quad (17)$$

where $F_R(w)$ and $F_I(w)$ are the real and imaginary part of the Fourier transform of $f(t)$, respectively.

To obtain a perspective on how these systems perform, consider Figs. 6 and 7, both taken from [20], and corresponding to the receiver of Fig. 3. These figures show curves of average probability of error versus the ratio of energy-per-bit-to-noise spectral density. The curves are parameterized by the ratio of interference power-to-signal power (denoted J/S), and the results shown in Fig. 6 correspond to the interference (modeled as a narrow-band Gaussian noise process) occupying 10% of the spread spectrum bandwidth, while those of Fig. 7 correspond to an interference bandwidth equal to 50% of the spread bandwidth. From these figures, it

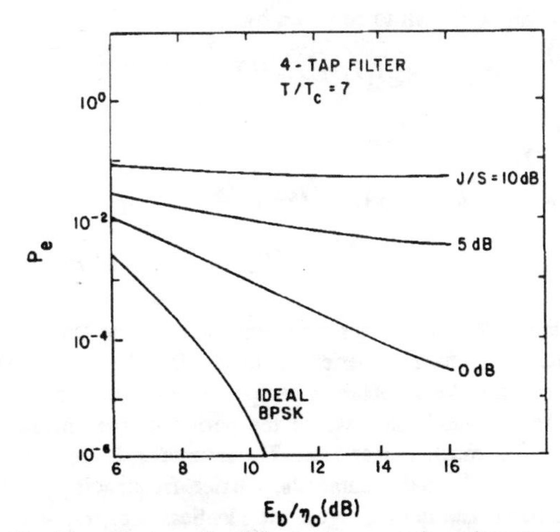

Fig. 7. Bit error rate results of direct sequence receiver when narrow-band interference occupying 50% of spread bandwidth is present.

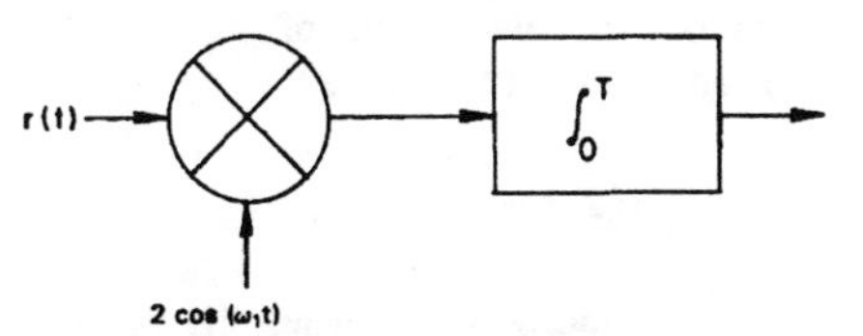

Fig. 8. Block diagram of BPSK receiver.

is evident that the rejection scheme works well if the narrow-band user does not occupy more than 10% of the bandwidth of the CDMA user, although the performance of the technique degrades very rapidly as the interference bandwidth increases beyond about 10%. However, we note that the results shown are uncoded error rates. Using forward error correction, the error rate can be significantly reduced.

Up to this point, we have concentrated on the interference that the narrow-band users induce onto the spread spectrum users. It is at least as important to quantify the opposite effect, the interference on the narrow-band waveforms caused by the spread spectrum overlay. Toward this end, consider the following simplified system shown in Fig. 8, which is used to model a BPSK receiver. The received waveform $r(t)$ is given by the sum of a conventional BPSK signal, a DS waveform, and AWGN. That is,

$$r(t) = A_1 d_1(t)\cos\omega_1 t + A_2 d_2(t-\tau)PN(t-\tau) \cdot \cos(\omega_2 t + \theta) + n_w(t), \quad (18)$$

where $d_1(t)$ and $d_2(t)$ are independent random binary sequences. The bit rate of $d_1(t)$ is $1/T$, and that of $d_2(t)$ is $1/T_2$. Each of the bit streams is assumed to be bipolar, taking values ± 1, $PN(t)$ is the spreading sequence for the DS waveform and consists of rectangular pulses of duration T_c s, A_1 and A_2 are constant amplitudes, τ is a random time delay, θ is a random phase, and $n_w(t)$ is AWGN of two-sided spectral density $\eta_0/2$.

If we assume the receiver of Fig. 8 is perfectly synchronized to the BPSK waveform, then the test statistic at the output can be shown to be given by

$$g_1(T_1) = A_1T_1 + A_2I(T_1) + N(T_1), \quad (19)$$

where

$$I(T_1) \triangleq \int_0^{T_1} d_2(t-\tau)PN(t-\tau) \cdot \cos\left[(\omega_2-\omega_1)t+\theta\right] dt. \quad (20)$$

Note that $I(T_1)$ determines the attenuation that the DS spread spectrum waveform experiences in passing through BPSK receiver. In order to obtain a perspective on how much that attenuation actually is, assume the period of the spreading sequence is much longer than T_1, assume τ is an integer multiple of T_c, and assume the carrier frequencies of the BPSK signal and the DS signal are identical (i.e., $\omega_1 = \omega_2$). The first assumption allows us to model the spreading sequence as a random binary sequence, and the second assumption in conjunction with the first allows us to ignore $d_2(t)$ (i.e., we can combine the effects of $d_2(t)$ and $PN(t)$). Hence $I(T)$ can be shown to reduce to

$$I(T_1) = \sum_{i=1}^{N_1} c_iT_c\cos\theta, \quad (21)$$

where the $\{c_i\}$ are independent, identically distributed random variables taking on values ± 1 with equal probability, and where N_1 is given by

$$N_1 \triangleq \frac{T_1}{T_c}. \quad (22)$$

Note that N_1 corresponds to the ratio of the duration of a data symbol of the narrow-band user to the duration of a chip of the spread spectrum user. The phase θ is a random variable uniformly distributed in $[-\pi, \pi]$.

If we assume that $N_1 \gg 1$, then $\sum_{i=1}^{N_1} c_i$ can be approximated as a zero-mean Gaussian random variable with variance equal to N_1. In turn, $I(T_1)$, conditioned upon θ, can be approximated as a conditional Gaussian random variable, having zero-mean and variance $N_1T_c^2\cos^2\theta$. Therefore, the conditional probability of error of the system can be approximated by

$$P(e|\theta) \simeq \phi\left(\frac{-A_1T_1}{\left[\eta_0T_1 + T_c^2A_2^2N_1\cos^2\theta\right]^{1/2}}\right)$$
$$= \phi\left[-\left[\frac{\eta_0}{2E_1} + \left(\frac{A_2}{A_1}\right)^2 N_1\left(\frac{T_c}{T_1}\right)^2\cos^2\theta\right]^{-1/2}\right]$$
$$= \phi\left(-\left[\frac{\eta_0}{2E_1} + \frac{S_2}{N_1S_1}\cos^2\theta\right]^{-1/2}\right), \quad (23)$$

and the average probability of error can be approximated by

$$P_e = \frac{1}{2\pi}\int_0^{2\pi} P(e|\theta)\,d\theta. \quad (24)$$

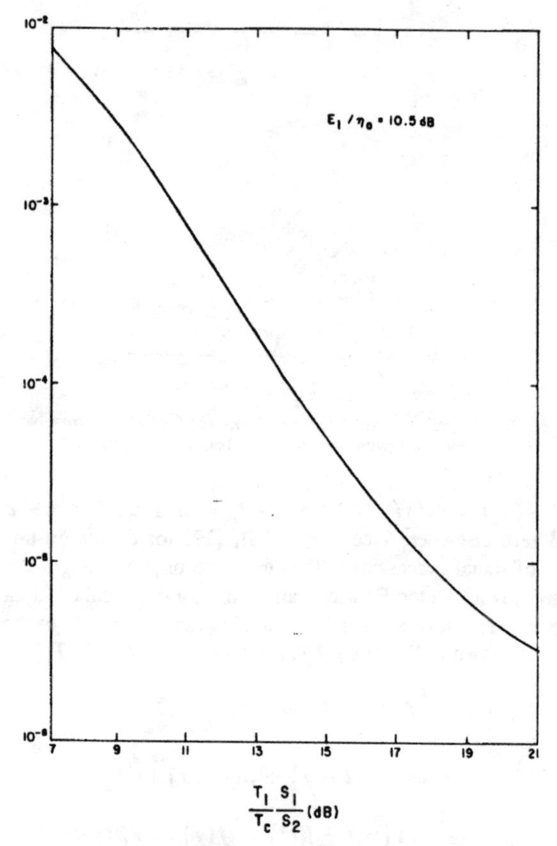

Fig. 9. Bit errror rate performance of BPSK receiver when operating in presence of direct sequence waveform.

In (23), $E_1 \triangleq A_1^2T_1/2$ is the energy-per-bit of the BPSK signal, $S_1 \triangleq A_1^2/2$ is the average power of the BPSK signal, and $S_2 \triangleq A_2^2/2$ is the average power of the PN signal. Interestingly, if these results are compared to those of [21], it will be seen that this problem is the same as that of detecting a DS waveform in the presence of sinusoidal interference, except in our problem it is the DS signal that represents the interference. However, the key point is that the mathematical expressions for average probability of error are the same for both problems.

To obtain some perspective as to how much interference the BPSK signal can withstand, consider Fig. 9, which shows the results of evaluating (24) for $E_1/\eta_0 = 10.5$ dB (i.e., for an average probability of error of the BPSK receiver 10^{-6} in the absence of interference). It can be seen that, if an error rate of 10^{-5} is desired in the presence of the spread spectrum overlay, then the value of N_1S_1/S_2 must be about 18 dB. For example, assume the narrow-band user occupies 1% of the spread bandwidth (i.e., assume $N_1 = 100$). Then the ratio of power in the BPSK waveform to power in the DS waveform can be as small as -2 dB, (i.e., the level of the BPSK signal can actually be less than that of the DS waveform and the performance of the BPSK receiver will still not degrade to a BER greater than 10^{-5}).

VI. Conclusion

As more people demand access to communication media, system planners will be required to inaugurate more efficient techniques of allocating spectrum than are currently used today. One such technique is spread spectrum. The natural attributes of spread spectrum waveform design, such as inherent security, multipath rejection, and multiple accessing capability, make spread spectrum very competitive for a variety of commercial applications, in particular, those involving mobile units.

In this paper, we have tried to emphasize that, in addition to the above properties, spread spectrum CDMA can also be used to share the spectrum with existing narrow-band users. Thus, rather than reallocating spectrum and disrupting existing users, spread spectrum "layers" can be allocated. As pointed out in the paper, there is a limit to how many layers can be allocated and that this limit can be extended by using interference rejection techniques. (Remember that these interference rejection techniques reduce the interference seen by each of the "layered" users, thereby allowing them to decrease their powers. This, in turn, further reduces their interference on the existing narrow-band users.)

Over the past five years, the use of spread spectrum has increased significantly, and it appears that this trend will increase even more dramatically during the coming decade.

References

[1] R. L. Pickholtz, D. L. Schilling, and L. B. Milstein, "Theory of spread-spectrum communications—A tutorial," *IEEE Trans. Commun.*, vol. COM-30, pp. 855-884, May 1982.

[2] D. E. Borth and M. B. Pursley, "Analysis of direct-sequence spread spectrum multiple-access communications over Rician fading channels," *IEEE Trans. Commun.*, vol. COM-27, pp. 1566-1577, Oct. 1979.

[3] M. B. Pursley, D. V. Sarwate, and W. Stark, "Error probability for direct sequence spread spectrum multiple-access communications—Part I: Upper and lower bounds," *IEEE Trans. Commun.*, vol. COM-30, pp. 975-984, May 1982.

[4] K. J. Wu and D. L. Neuhoff, "Average error probability for direct sequence spread spectrum multiple access communication systems," in *Proc. 18th Annu. Allerton Conf. Commun., Cont., and Comput.*, Oct. 1980, pp. 359-380.

[5] K. Yao, "Error probability of asynchronous spread spectrum multiple access communication systems," *IEEE Trans. Commun.*, vol. COM-25, pp. 803-809, Aug. 1977.

[6] R.-H. Dou and L. B. Milstein, "Error probability bounds and approximations for DS spread spectrum communication systems with multiple tone or multiple access interference," *IEEE Trans. Commun.*, vol. COM-32, pp. 493-502, May 1984.

[7] K. S. Gilhousen, I. M. Jacobs, R. Padovani, and L. A. Weaver, Jr., "Increased capacity using CDMA for mobile satellite communication," *IEEE J. Select. Areas Commun.*, vol. 8, pp. 503-514, May 1990.

[8] M. B. Pursley, "Performance evaluation for phase-coded spread spectrum multiple-access communication, Part I, System analysis," *IEEE Trans. Commun.*, vol. COM-25, pp. 795-799, Aug. 1977.

[9] J. S. Lehnert and M. B. Pursley, "Error probabilities for binary direct-sequence spread-spectrum communications with random signature sequences," *IEEE Trans. Commun.* vol. COM-35, pp. 87-98, Jan. 1987.

[10] R. K. Morrow, Jr. and J. S. Lehnert, "Bit-to-bit dependence in slotted DS/SSMA packet systems with random signature sequences," *IEEE Trans. Commun.*, vol. 37, pp. 1052-1061, Oct. 1989.

[11] M. B. Pursley, "The role of spread spectrum in packet radio networks," *Proc. IEEE*, vol. 75, pp. 115-13, Jan. 1987.

[12] R. G. Gallager, *Information Theory and Reliable Communication*. New York: Wiley, 1968, ch. 4.

[13] M. Schwartz, W. R. Bennett, and S. Stein, *Communication Systems & Techniques*. New York: McGraw-Hill, 1966, ch. 10.

[14] L. B. Milstein, "Interference rejection techniques in spread spectrum communications," *Proc. IEEE* vol. 76, pp. 657-671, June 1988.

[15] R. A. Iltis and L. B. Milstein, "Performance analysis of narrowband interference rejection techniques in DS spread spectrum systems," *IEEE Trans. Commun.*, vol. COM-32, pp. 1169-1177, Nov. 1984.

[16] —, "An approximate statistical analysis of the Widrow LMS algorithm with application to narrowband interference rejection," *IEEE Trans. Commun.*, vol. COM-33, pp. 121-130, Feb. 1985.

[17] N. J. Bershad, "Error probabilities for DS spread-spectrum systems using an ALE for narrowband interference rejection," *IEEE Trans. Commun.*, vol. 36, pp. 588-595, May 1988.

[18] L. B. Milstein and P. K. Das, "An analysis of a real-time transform domain filtering digital communication system, Part I: Narrowband interference rejection," *IEEE Trans. Commun.*, vol. COM-28, pp. 816-824, June 1980.

[19] —, "An analysis of a real-time transform domain filtering digital communication system—Part II: Wideband interference rejection," *IEEE Trans. Commun.*, vol. COM-31, pp. 21-27, Jan. 1983.

[20] E. Masry and L. B. Milstein, "Performance of DS spread-spectrum receivers employing interference-suppression filter under a worst-case jamming condition," *IEEE Trans. Commun.*, vol. 34, pp. 13-21, Jan. 1988.

[21] L. B. Milstein, S. Davidovici, and D. L. Schilling, "The effect of tone interfering signals on a direct sequence spread spectrum communication system," *IEEE Trans. Commun.*, vol. COM-30, pp. 436-336, Mar. 1982.

Raymond L. Pickholtz (S'54-A'55-M'60-SM'77-F'82) received the Ph.D. degree in electrical engineering from the Polytechnic Institute of Brooklyn (now New York) in 1966.

He is a Professor and former Chairman of the Department of Electrical Engineering and Computer Science at the George Washington University, Washington, DC. He is also President of Telecommunications Associates, a research and consulting firm specializing in communication system disciplines. He was a researcher at RCA Laboratories and at ITT Laboratories. He has been on the faculty of the Polytechnic Institute of Brooklyn and of Brooklyn College, and has been a Visiting Professor at the Université du Quebec and the University of California. He is the Editor of the Telecommunication Series for Computer Science Press. He was an Editor of the IEEE the TRANSACTIONS ON COMMUNICATIONS and Guest Editor for special issues on computer communication military communications, and spread spectrum. He has published scores of papers and holds six U.S. patents.

Dr. Pickholtz is a fellow of the American Association for Advancement of Science (AAAS). In 1990, he was elected President of the IEEE Communications Society.

Laurence B. Milstein (S'66-M'68-SM'77-F'85) received the B.E.E. degree from the City College of New York, New York, NY, in 1964, and the M.S. and Ph.D. degrees in electrical engineering from the Polytechnic Institute of Brooklyn, Brooklyn, NY, in 1966 and 1968, respectively.

From 1968 to 1974 he was employed by the Space and Communications Group of Hughes Aircraft Company, and from 1974 to 1976 he was a member of the Department of Electrical and Systems Engineering, Rensselaer Polytechnic Institute, Troy, NY. Since 1976 he has been with the Department of Electrical and Computer Engineering, University of California at San Diego, La Jolla, where he is a Professor and former Department Chairman, working in the area of digital communication theory with special emphasis on spread-spectrum communication systems. He has also been a consultant to both government and industry in the areas of radar and communications.

Dr. Milstein was an Associate Editor for Communication Theory for the IEEE TRANSACTIONS ON COMMUNICATIONS, and an Associate Technical Editor for the *IEEE Communications Magazine*. He is the Vice President for Technical Affairs of the IEEE Communications Society, is a member of the Board of Governors of the IEEE Information Theory Society, and is a member of Eta Kappa Nu and Tau Beta Pi.

Donald L. Schilling (S'56–M'58–SM'69–F'75) is the Herbert G. Keyser Distinguished Professor of Electrical Engineering at the City College of the City University of New York, New York, NY, where he has been a Professor since 1969. Prior to that, he was a Professor at the Polytechnic Institute of New York, Brooklyn, NY. He is also President of SCS Telecom, Inc. In this capacity, he directs programs dealing with research and development, and training, in the military and commercial aspects of telecommunications. He co-authored eight international bestselling texts, *Electronic Circuits: Discrete and Integrated* (1969); *Introduction to Systems, Circuits and Devices* (1973); *Principles of Communications Systems* (1971); *Digital Integrated Electronics* (1977); *Electronic Circuits: Discrete and Integrated* (2nd edition—1979); *Principles of Communications Systems* (2nd edition 1986); *Electronic Circuits: Discrete and Integrated*, (3rd edition); and *Dynamic Project Management: A Practical Guide for Managers and Engineers* (1989). He has published more than 140 papers in the telecommunications field. He is a well-known expert in the field of military communications systems and has made many notable contributions in spread spectrum communications systems, FM and phase locked systems and has directed research efforts in the performance of HF and Meteor Burst Communications. His algorithm for an adaptive DM is used on the space shuttle.

Dr. Schilling was President of the IEEE Communications Society from 1979–1981 and a member of the Board of Directors of the IEEE from 1981–1983. He was Editor of the IEEE TRANSACTIONS ON COMMUNICATIONS from 1968–1978. He was nominated and accepted as a member of the U.S. Army Science Board in November 1987. He is a member of Sigma Xi and has been an international representative for the IEEE in the Soviet Union where he was part of a Popov Society exchange, and in China where he led an IEEE delegation.

Erlang Capacity of a Power Controlled CDMA System

Audrey M. Viterbi and Andrew J. Viterbi

***Abstract*—This paper presents an approach to the evaluation of the reverse link capacity of a CDMA cellular voice system which employs power control and a variable rate vocoder based on voice activity. It is shown that the Erlang capacity of CDMA is many times that of conventional analog systems and several times that of other digital multiple access systems.**

I. Introduction

FOR any multiuser communication system, the measure of its economic usefulness is not the maximum number of users which can be serviced at one time, but rather the peak load that can be supported with a given quality and with availability of service as measured by the blocking probability (the probability that a new user will find all channels busy and hence be denied service, generally accompanied by a busy signal). Adequate service is usually associated with a blocking probability of 2 percent or less. The average traffic load in terms of average number of users requesting service resulting in this blocking probability is called the *Erlang capacity* of the system.

In virtually all existing multiuser circuit-switched systems, blocking occurs when all frequency slots or time slots have been assigned to a voice conversation or message. In code division multiple access (CDMA) systems, in contrast, users all share a common spectral frequency allocation over the time that they are active. Hence, new users can be accepted as long as there are receiver-processors to service them, independent of time and frequency allocations. We shall assume that a sufficient number of such processors is provided in the common base station such that the probability of a new arrival finding them all busy is negligible. Rather, blocking in CDMA systems will be defined to occur when the interference level, due primarily to other user activity, reaches a predetermined level above the background noise level of mainly thermal origin. While this interference-to-noise ratio could, in principle, be made arbitrarily large, when the ratio exceeds a given level (about 10 dB nominally), the interference increase per additional user grows very rapidly, yielding diminishing returns and potentially leading to instability. Consequently, we shall establish blocking in CDMA as the event that the total interference-to-background noise level exceeds $1/\eta$ (where $\eta = 0.1$ corresponding to 10 dB), and we determine the Erlang capacity which results in a 1 percent blocking probability. We emphasize, however, that this is a "soft blocking" condition, which can be relaxed as will be shown, as contrasted to the "hard blocking" condition wherein channels are all occupied.

Also, in conventional systems, a fraction of the time or frequency slots must be set aside for users to transmit requests for initiating service, and a protocol must be established for multiple requests when two or more users collide in simultaneously requesting service. In CDMA systems, even the users seeking to initiate access can share the common medium. Of course, they add to the total interference and hence lower the Erlang capacity to some degree. We shall demonstrate that this reduction is very small for initial access requests whose signaling time is on the order of a few percent of the average duration of a call or message.

Conclusions will be drawn regarding the relative increase in Erlang capacity of a direct sequence spread-spectrum CDMA system over existing FDMA and TDMA systems.

II. Conventional Blocking

In conventional multiple access systems, such as FDMA and TDMA, traffic channels are allocated to users as long as there are channels available, after which all incoming traffic is blocked until a channel becomes free at the end of a call. The blocking probability is obtained from the classical Erlang analysis of the $M/M/S/S$ queue, where the first M refers to a Poisson arrival rate of λ calls/s; the second M refers to exponential service time with mean $1/\mu$ s/call; the first S refers to the number of servers (channels); and the second S refers to the maximum number of users supported before blockage occurs.

The Erlang-B formula [1] gives the blocking probability under these conditions

$$P_{\text{blocking}} = \frac{(\lambda/\mu)^S/S!}{\sum_{k=0}^{S} (\lambda/\mu)^k/k!}$$

where λ/μ is the offered average traffic measured in Erlangs. Of course, the average active number of users equals $(\lambda/\mu)(1 - P_{\text{blocking}})$.

Thus, for the conventional *AMPS system* with 30 kHz channels $K = 7$ frequency reuse factor[1] and 3 sectors, the number

Manuscript received March 16, 1992; revised November 16, 1992. Part of this paper was presented at the IEEE International Symposium on Information Theory, San Antonio, TX, January 17–22, 1993.

The authors are with Qualcomm Inc., San Diego, CA 92121-1617.

IEEE Log Number 9208658.

[1] It should be noted that the reuse factor of 7 applies strictly only to North American analog systems. Digital TDMA systems have been proposed for other regions with lower reuse factors (4 or even 3). Field experience is not yet available, however, to support general use of such lowered reuse factors.

0733–8716/93$03.00

of channels (servers) in 12.5 MHz is

$$S_{\text{AMPS}} = \frac{12.5 \text{ MHz}}{(30 \text{ kHz})(7)(3)} = 19 \text{ channel/sector}.$$

Similarly, for a *3-slot TDMA system*, which otherwise uses the same channelization, sectorization, and reuse factor as AMPS, and is generally *called D-AMPS*,

$$S_{D-\text{AMPS}} = 57 \text{ channel/sector}.$$

The blocking probabilities $P(\text{blocking})$ as a function of the average number of calls offered at any instant λ/μ are obtained from the Erlang-B formula. This establishes that the offered traffic *Erlang capacity* per sector for a *blocking probability equal to 2 percent* is respectively,

$$(\lambda/\mu)_{\text{AMPS}} = 12.34 \text{ Erlangs},$$
$$(\lambda/\mu)_{D-\text{AMPS}} = 46.8 \text{ Erlangs}.$$

III. CDMA Reverse Link Erlang Capacity Bounds and Approximations

For the CDMA reverse link (or uplink), which is the limiting direction, blocking is defined to occur when the total collection of users both within the given sector (cell) and in other cells introduce an amount of interference density I_0 so great that it exceeds the background noise level N_0 by an amount $1/\eta$, taken to be 10 dB.

If there were always

1) a constant number of users N_u in every sector,
2) each (perfectly power controlled) user were transmitting continually, and
3) required the same E_b/I_0 (under all propagation conditions),

then, as established in [2] and [3], the number of users N_u would be determined by equating

$$N_u(\text{Signal Power/User}) + \text{Other Cell Interference} + \text{Thermal Noise} = \text{Total Interference}.$$

Taking

W = spread-spectrum bandwidth
R = data rate,
E_b = bit energy,
N_0 = thermal (or background) noise density,
I_0 = maximum total acceptable interference density (interference power normalized by W), and
f = ratio of other cell interference (at base station for given sector)-to-own sector interference[2] then the condition for nonblocking is

$$N_u E_b R(1+f) + N_0 W \leq I_0 W \quad (1)$$

whence it follows that

$$N_u \leq \frac{W/R}{E_b/I_0} \cdot \frac{1-\eta}{1+f} \quad (2)$$

[2]In [7], a tight upper bound on other-cell interference is derived, which gives $f = 0.55$ for the propagation parameters of interest.

where

$$\eta = N_0/I_0 = 0.1 \text{ (nominally)}. \quad (3)$$

In fact, however, none of the three assumptions above holds since

a) the number of active calls is a Poisson random variable with mean λ/μ (there is no hard limit on servers);
b) each user is gated on with probability ρ and off with $1-\rho$;
c) each user's required energy-to-interference E_b/I_0 ratio is varied according to propagation conditions to achieve the desired frame error rate ($\approx$1 percent).

For simplicity, we continue to assume that all cells are equally loaded (with the same number of users per cell and sector which are uniformly distributed over each sector).

With assumptions a), b), and c) replacing 1), 2), and 3), the condition for nonblocking, replacing (1) becomes

$$\sum_{i=1}^{k} \nu_i E_{bi} R + \sum_{j}^{\text{other cells}} \sum_{i=1}^{k} \nu_{i(j)} E_{b_{i(j)}} R + N_0 W \leq I_0 W \quad (4)$$

with k, the number of users/sector, being a Poisson random variable with mean λ/μ; and ν being the binary random variable taking values 0 and 1, which represents voice activity, with

$$P(\nu = 1) = \rho. \quad (5)$$

Dividing by $I_0 R$ and defining

$$\epsilon = E_b/I_0, \quad (6)$$

the nonblocking condition (4) becomes

$$Z \triangleq \sum_{i=1}^{k} \nu_i \epsilon_i + \sum_{j}^{\text{other cells}} \sum_{i=1}^{k} \nu_i^{(j)} \epsilon_i^{(j)} \leq (W/R)(1-\eta) \quad (7)$$

where η is given by (3). Hence, the blocking probability for CDMA becomes

$$P_{\text{blocking}} = \Pr[Z > (W/R)(1-\eta)]. \quad (8)$$

Setting this equal to a given value (nominally 1 percent) establishes the Erlang capacity of a CDMA cellular system. Again, we note that this is a "soft blocking" phenomenon which can be occasionally relaxed by allowing I_0/N_0, and consequently $1-\eta$, to increase.

Naturally, when condition (8) is exceeded, call quality will suffer. Thus, this probability is kept sufficiently low so that we can ensure high availability of good quality service. Conventional multiple access systems also are limited to providing good quality service only on the order of 90–99 percent of the time because of the variability of interference from just one or a few other users, in contrast with the CDMA case where quality depends on an average over the entire user population.

To evaluate this blocking probability, we must determine the distribution function of the random variable Z which in

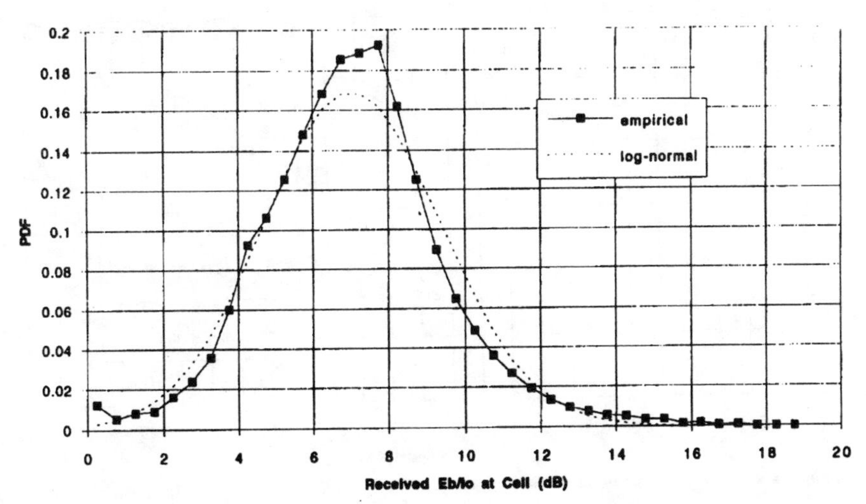

Fig. 1. Empirical E_b/I_0 probability density and log-normal approximation (m = 7.0 DB; sigma = 2.4 dB).

turn depends on the random variables ν, k, and ϵ, representing voice activity, number of users in a sector, and E_b/I_0 of any user, respectively. The distribution of ν is given by (5). Since k is Poisson, its distribution is given by

$$p_k = \Pr(k \text{ active users/sector}) = \frac{(\lambda/\mu)^k}{k!} e^{-\lambda/\mu} \tag{9}$$

where λ and μ are the arrival and service rates as defined in the previous section.

On the other hand, ϵ, the E_b/I_0 ratio of a single user, depends on the power control mechanism which attempts to equalize the performance of all users. It has been demonstrated that inaccuracy in power control loops are approximately log-normally distributed with a standard deviation between 1 and 2 dB [4]. However, since under some propagation conditions (e.g., with excessive multipath) higher than normal E_b/I_0 will be required to achieve the desired low error rates, the overall distribution, also log-normally distributed, will have a larger standard derivation. An example drawn from field trials with all cells fully loaded, in which E_b/I_0 is varied in order to maintain frame error rate below 1 percent is shown by the histogram of Fig. 1. As demonstrated by the dotted curve, the histogram is closely approximated by a log-normal probability density, with the mean and standard deviation of the normal exponent equal to 7 and 2.4 dB, respectively. Hence, we shall use the log-normal approximation[3]

$$\epsilon = 10^{x/10} \tag{10}$$

where x is a Gaussian variable with mean $m \approx 7$ dB and standard deviation $\sigma \approx 2.5$ dB. Note then that the first and second moments of ϵ are given by

$$E(\epsilon) = E(e^{\beta x}) = \exp\left[(\beta\sigma)^2/2\right]\exp(\beta m) \tag{11}$$

$$E(\epsilon^2) = E(e^{2\beta x}) = \exp\left[2(\beta\sigma)^2\right]\exp(2\beta m), \quad \beta = (\ln 10)/10. \tag{12}$$

Although all moments exist, the moment generating function of ϵ does not converge; hence the ordinary Chernoff bound for the blocking probability (8) cannot be obtained. In Appendix I, we derive a modified Chernoff upper bound obtained by treating the upper end of the distribution of ϵ separately. The result for a single sector (no interference from other cells) is

$$P_{\text{blocking}} < \min_{\substack{s>0\\ \tau>0}} \exp\left\{\rho(\lambda/\mu)\left[E\left(e^{s\epsilon'_{\tau}}\right) - 1\right] - sA\right\} + \rho(\lambda/\mu)Q(\tau/\sigma) \tag{13}$$

where[4]

$$A \triangleq \frac{(W/R)(1-\eta)}{\exp(\beta m)} = \frac{(W/R)(1-\eta)}{E_b/I_{0_{\text{median}}}} \tag{14}$$

$$E\left(e^{s\epsilon'_{\tau}}\right) = \int_{-\infty}^{\tau/\sigma} \exp\left[se^{\beta\sigma\zeta}\right]e^{-\zeta^2/2}\,d\zeta/\sqrt{2\pi} \tag{15}$$

$$Q(\tau/\sigma) = \int_{\tau/\sigma}^{\infty} e^{-\zeta^2/2}\,d\zeta/\sqrt{2\pi}. \tag{16}$$

This bound, obtained through numerical integration of (15), is plotted, for the single sector case, as the upper curve of Fig. 2.

[3] In the following, we could use the exact empirical distribution in numerically computing the Chernoff bound (13). Using the log-normal approximation gives us more flexibility in obtaining a general analytical result. Actually, the approximation is slightly pessimistic (an upper bound) since it appears from Fig. 1 that the empirical histogram is slightly skewed to the left of the log-normal approximation with the same mean and standard deviation. Fig. 1 is typical of data from a large number of field tests conducted in widely varying terrain in a large number of cities and in several countries.

[4] Note that m is the mean of the exponent of the log-normal variable. Since $e^{\beta m} = 10^{m/10}$, this is the value of E_b/I_0 corresponding to the mean (dB) value, but it is not the true mean. It can be shown by symmetry that it is the median, while the true mean is related to it by (11).

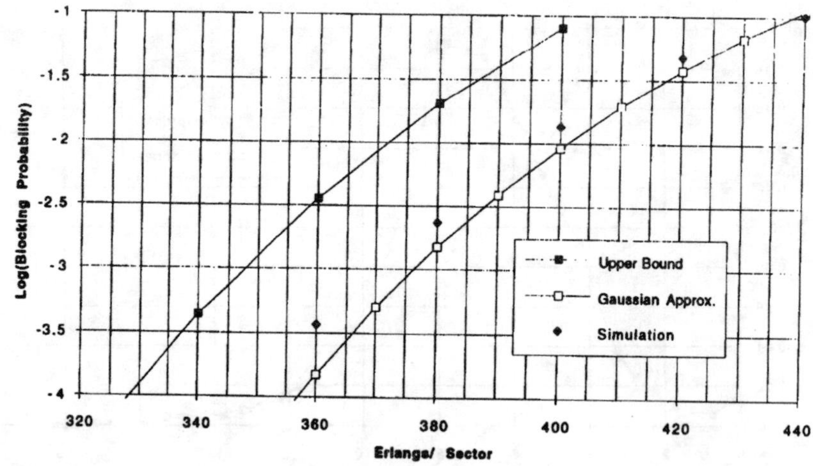

Fig. 2. Blocking probabilities for single cell interference (CDMA parameters: $W/R = 1280$; voice act. $= 0.4$; $I_0/N_0 = 10$ dB; median $E_b/I_0 = 7$ dB; sigma $= 2.5$ dB).

A much simpler approach is to assume a central limit theorem approximation for Z and to compute its mean and variance. Then

$$P_{\text{blocking}} \approx Q\left[\frac{A - E(Z')}{\sqrt{\text{Var } Z'}}\right] \tag{17}$$

where $Z' = Z/\exp(\beta m) = Z/(E_b/I_0)_{\text{median}}$.

Now since Z is the sum of k random variables, where k is itself a random variable, we have from [5], letting $\epsilon' = \epsilon/\exp(\beta m)$,

$$\begin{aligned} E(Z') &= E(k)E(\nu\epsilon') \\ &= (\lambda/\mu)\rho \exp\left[(\beta\sigma)^2/2\right] \end{aligned} \tag{18}$$

$$\text{Var}(Z') = E(k)\,\text{Var}(\nu\epsilon') + \text{Var}(k)[E(\nu\epsilon')]^2. \tag{19}$$

But since k is a Poisson variable, $E(k) = \text{Var}(k) = \lambda/\mu$, so that

$$\begin{aligned} \text{Var}(Z') &= \lambda/\mu\left[E(\nu\epsilon')^2\right] = \lambda/\mu E(\nu^2)E(\epsilon'^2) \\ &= (\lambda/\mu)\rho \exp\left[2(\beta\sigma)^2\right]. \end{aligned} \tag{20}$$

Using (18) and (20) in (17) yields the lower curve of Fig. 2.

It is commonly accepted that a Chernoff upper bound overestimates the probability by almost an order of magnitude. Moreover, a simulation involving on the order of one million frames was run to evaluate the tightness of the bound and the accuracy of the approximation. Results shown in Fig. 2 indicate excellent agreement with the approximation, with at most *1 percent discrepancy* in Erlang capacity for the nominal 1 percent blocking probability. Thus, we shall henceforth use only central limit approximations, given the much greater ease of computation, which will allow us to obtain more general and more easily employed results. We note that for any specific case, a strict upper bound can always be obtained by numerically computing the Chernoff bound based on the empirical distribution of E_b/I_0 as given in (A.3) of Appendix I. We note, however, that the approximation underestimates this upper bound by at most a few percent.

TABLE I
FACTOR f FOR SEVERAL VALUES OF SIGMA (FOURTH POWER LAW)

δ (dB)	f
0	0.44
2	0.43
4	0.45
6	0.49
8	0.55
10	0.66
12	0.91

IV. General Erlang Capacity Formula Including Other Cell Interference

Before deriving the general formula, we consider the effect of users in other cells which are power controlled by other base stations. It has been shown both analytically [3] and by simulation that the interference from surrounding cells increases the average level at the base station under consideration by a fraction between 1/2 and 2/3 of that of the desired cell's users when the propagation attenuation is proportional to the fourth power of the distance times a log-normally distributed component whose differential standard deviation is 8 dB. The results of an improved upper bound on the mean outer-cell interference fraction f as derived in [7] and shown in Table I.

We note also that each user which is controlled by an other-cell base station will also have an E_b/I_0 which is distributed according to the histogram of Fig. 1. Hence, this other-user interference can also be modeled by the same log-normal distribution as assumed for users of the desired cell. We can now modify (18) to accommodate other-cell users as well. The total number of other-cell users is generally much larger, but their average power is equivalent to that of kf users. We shall therefore model them as such, recognizing that this is somewhat pessimistic since, with a larger number of smaller received power users, the mean power will remain the same but the variance will be reduced. In any case, accepting this approach as an overbound, we find that the mean and variance (18) and (20) are simply increased by the factor $1 + f$. Thus, we may restate (17)–(20), when other-cell (power controlled) interfering users are included, as follows:

$$P_{\text{blocking}} \approx Q\left[\frac{A - E(Z')}{\sqrt{\text{Var } Z'}}\right] \tag{21}$$

$$E(Z') = (\lambda/\mu)\,\rho\,(1+f)\exp\left[(\beta\sigma)^2/2\right] \tag{22}$$

$$\text{Var}(Z') = (\lambda/\mu)\,\rho\,(1+f)\exp\left[2(\beta\sigma)^2\right]. \tag{23}$$

Inverting (21) yields the quadratic equation

$$x\alpha^4\left[Q^{-1}(P_{\text{blocking}})\right]^2 = [A - x\alpha]^2 \tag{24}$$

where $x = (\lambda/\mu)\,\rho\,(1+f)$, and $\alpha = \exp[(\beta\sigma)^2/2]$, while A is given by (14). Its solution is

$$x = \frac{A}{\alpha}\left[1 + \frac{\alpha^3 B}{2}\left(1 - \sqrt{1 + \frac{4}{\alpha^3 B}}\right)\right] \tag{25}$$

where

$$B = \frac{\left[Q^{-1}(P_{\text{blocking}})\right]}{A} = \frac{(E_b/I_0)_{\text{median}}\left[Q^{-1}(P_{\text{blocking}})\right]^2}{(W/R)(1-\eta)}. \tag{26}$$

Using (24) and (25), this can be expressed as a formula for Erlang capacity

$$\frac{\lambda}{\mu} = \frac{(1-\eta)(W/R)F(B,\sigma)}{\rho(1+f)(E_b/I_0)_{\text{median}}} \quad \text{Erlangs/Sector} \tag{27}$$

where

$$F(B,\sigma) = \exp\left[-(\beta\sigma)^2/2\right] \cdot \left\{1 + (B/2)\exp\left[3(\beta\sigma)^2/2\right] \cdot \left(1 - \sqrt{1 + 4\exp\left[-3(\beta\sigma)^2/2\right]/B}\right)\right\} \tag{28}$$

with B given by (26). Fig. 3 is a plot of $(Q^{-1})^2$—the factor of B which depends on P_{blocking}. Fig. 4 shows plots of $F(B,\sigma)$ as a function of B for several values of σ, the standard deviation in decibels of the power-controlled E_b/I_0. We note finally that the Erlang capacity, as given by (27), is the same as the capacity predicted under ideal assumptions (2) but increased by the inverse of the voice activity factor $1/\rho$ and decreased by the reduction factor $F(B,\sigma)$ given in Fig. 4. We note from Fig. 4 that a power control inaccuracy of $\sigma = 2.5$ dB causes only about as much capacity reduction as does the variability in arrival times and voice activity. Numerically, for $E_b/I_0 = 7$ dB, $\eta = 0.1$, $W/R = 31$ dB and $P_{\text{blocking}} = 1$ percent we obtain $B = 0.024$. Then, from Fig. 4, we note that $F(B,\sigma = 2.5) = 0.695$ while $F(B,\sigma = 0) = 0.86$. Thus, it follows that the incremental loss due to power control inaccuracy $F(B,\sigma = 2.5)/F(B,\sigma = 0) = 0.8$. This means that in this case a standard deviation of 2.5 dB in power control causes a reduction of only 1 dB in capacity. Fig. 3 and 4 also show that results are not very sensitive to the level of blocking probability. Doubling the latter only increases capacity by about 2 percent. Note finally that substituting $F(B,\sigma = 2.5) = 0.695$ in the general formula (27), we obtain, for the parameters noted above along with voice activity $\rho = 0.4$ and other-cell factor $f = 0.55$ (as obtained from Table I), an Erlang capacity $\lambda/\mu = 258$ Erlangs/sector. Thus, comparison to Section II gives the ratio

$$\frac{C_{\text{CDMA}}}{C_{\text{AMPS}}} \approx \frac{258}{12.34} \approx 20.9.$$

Based on the simulation results shown in Fig. 2, which suggest reduction of the approximate results by one to two percent, an estimate in excess of a twentyfold increase in Erlang capacity is thus justified.

V. Designing for Minimum Transmitted Power

In the preceding, the total received interference-to-noise I_0/N_0 was assumed fixed, and blocking probability was evaluated as a function of average user loading λ/μ. Now suppose the blocking probability is fixed at 1 percent, but for each user loading λ/μ, the minimum value of I_0/N_0 is determined to achieve this. This can be established by varying $\eta = N_0/I_0 < 1$ and determining from (27) the λ/μ value for which blocking probability equals 1 percent for each I_0/N_0 level.

More importantly, from I_0/N_0, we may determine the minimum value of received signal power-to-background noise for each user, and consequently the minimum transmitted power per user, given the link attenuation, the receiver sensitivity, and antenna gains. The minimum required received signal-to-background noise per user $S/(N_0W)$ is obtained by equating the average sum of the per-user ratios weighted by the average voice activity factor to the total other user interference-to-noise ratio

$$\rho\frac{\lambda}{\mu}\frac{S}{N_0W} = \left(\frac{I_0 - N_0}{N_0}\right).$$

Thus,

$$\frac{S}{N_0W} = \frac{1}{\rho(\lambda/\mu)}\left(\frac{I_0}{N_0} - 1\right) = \frac{(1/\eta) - 1}{\rho(\lambda/\mu)} \tag{29}$$

where $1/\eta = I_0/N_0$ is fixed, as noted above, and λ/μ is the resulting loading to achieve $P_{\text{blocking}} = 0.01$ for this value of η, as given by (26)–(28). Fig. 5 shows the per-user signal-to-background noise ratio as a function of the relative Erlang capacity for the same parameters used in the last section. The inverse of η, the interference-to-noise density I_0/N_0 ranges from 1 to 10 dB as the capacity varies over a factor of 8. The important point is that the per-user power can be reduced by about 8 dB for lightly loaded cells. It also helps to justify the choice of $I_0/N_0 = 10$ dB for blocking when all cells are heavily loaded, since per-user power requirements increase rapidly above this point.

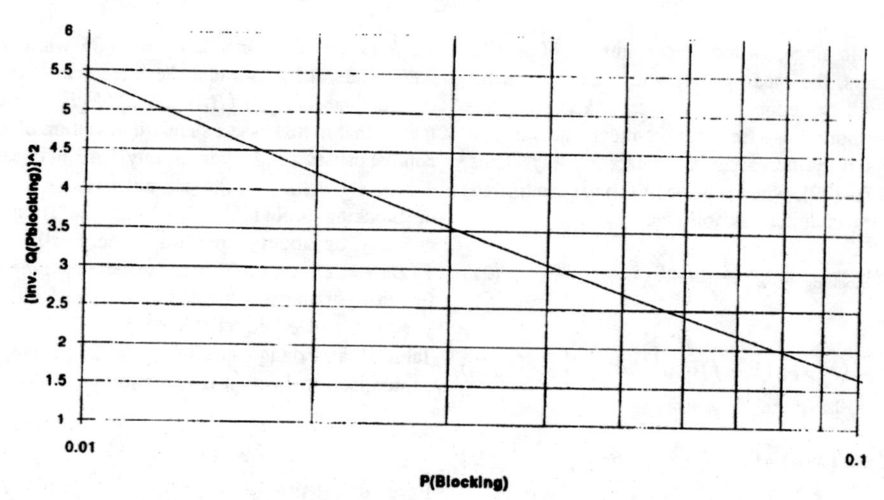

Fig. 3. Variable factor in B as a function of P (blocking).

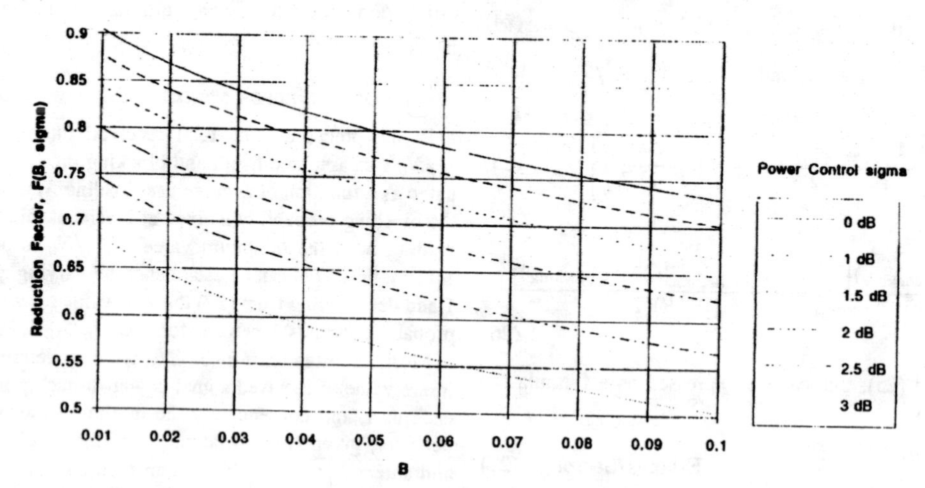

Fig. 4. Erlang capacity reduction factors.

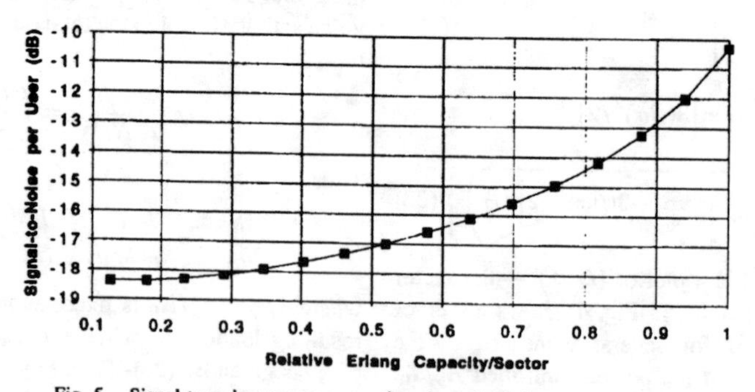

Fig. 5. Signal-to-noise per user as a function of relative sector capacity.

VI. Initial Access

Prior to initiating a call on the reverse link, a user must signal a request to the base station. In conventional systems, a time or frequency slot is allocated for the purpose of access request, and a protocol is provided for recovering from collisions which occur when two or more new users send access requests simultaneously. In CDMA, in which the allocated commodity is energy rather than time or frequency, access requests can share the common channel with ongoing users. The arrival rate of user requests is taken to be λ calls/s, which is the same as that for ongoing calls under the assumption that all requests are eventually served. Since newly arriving users are not power controlled until their requests are recognized, the initial power level will be taken to be a random variable uniformly distributed from 0 to a maximum value, corresponding to a (bit) energy level of E_M. Thus, the initial access energy level is γE_M where γ is a random variable with probability density function

$$p(\gamma) = \begin{cases} 1, & 0 < \gamma < 1 \\ 0, & \text{otherwise}. \end{cases} \qquad (30)$$

If this initial power level is not sufficient for detection,[5] and hence acknowledgment is not received, the user increases his power in constant decibel steps every frame until his request is acknowledged. Thus, the initial access user's power grows exponentially (i.e., linearly in decibels) with time, taken to be continuous since the frame time is only tens of milliseconds long. Hence, the energy as a function of time for initial access requests is

$$E(t) = \gamma E_M e^{\delta t} \qquad (31)$$

where δ is fixed. Letting τ denote the time required for an initial access user to be detected, it follows from (30) to (31) that

$$\begin{aligned}\Pr(\tau > T) &= \Pr(\gamma E_M e^{\delta T} < E_M) \\ &= \Pr(\gamma < e^{-\delta T}) = e^{-\delta T}, \qquad T > 0 \qquad (32)\end{aligned}$$

where the last equation follows from the fact that γ is uniformly distributed on the unit interval.

Thus, the "service time" for any initial access requests (time required for acceptance) for all users is exponentially distributed with mean $1/\delta$ s/message. The consequences of this observation are very significant. We have thus shown that with Poisson distributed arrivals and exponentially distributed service time, *the output distribution is also Poisson*. This then guarantees that the distribution for users initiating service is the same as for newly accessing users. Furthermore, at any given time, the power distribution for users which have not yet been accepted is also uniform. As for the service time distribution, according to (32), this is exponential with mean service time equal to $1/\delta$.

Thus, the total interference of (4) is augmented by the interference from initial accesses. When normalized as in (18), this introduces an additional term in the mean and variance of Z', (18) and (20), respectively. Calling this Z'', we have, ignoring for the moment surrounding cells,

$$E(Z'') = E(Z') + (\lambda/\delta)E(\gamma E_M/I_0)/e^{\beta m} \qquad (33)$$

$$\mathrm{Var}(Z'') = \mathrm{Var}(Z') + (\lambda/\delta)E(\gamma^2 E_M/I_0)/e^{2\beta m} \qquad (34)$$

where γ is uniformly distributed on the unit interval and is independent of E_M/I_0. Thus $E(\gamma) = 1/2$ and $E(\gamma^2) = 1/3$. We take $E_M/I_0 = \theta(E_b/I_0)$, and hence log-normally distributed with the same σ as E_b/I_0. Then $E(E_M/I_0) = \theta E(\epsilon), E[(E_M/I_0)^2] = \theta^2 E(\epsilon^2)$.

If surrounding cells support the same coverage load as the given cell, initial accesses will also produce the same relative interference. Introducing this term as well, we obtain, from (33) and (34),

$$\begin{aligned}E(Z'') &= \Big\{(\lambda/\mu)\,\rho\,\exp[(\beta\sigma)^2/2] \\ &\quad + (\lambda/\delta)(1/2)\,\theta\exp[(\beta\sigma)^2/2]\Big\}[1+f] \\ &= \rho(\lambda/\mu)\left(1+\frac{\theta\mu}{2\rho\delta}\right)\exp[(\beta\sigma)^2/2][1+f] \qquad (35)\end{aligned}$$

$$\begin{aligned}\mathrm{Var}(Z'') &= \Big\{(\lambda/\mu)\,\rho\,\exp[2(\beta\sigma)^2] \\ &\quad + (\lambda/\delta)(1/3)\,\theta^2\exp[2(\beta\sigma)^2]\Big\}[1+f] \\ &= \rho(\lambda/\mu)\left(1+\frac{\theta^2\mu}{3\rho\delta}\right)\exp[2(\beta\sigma)^2][1+f]. \qquad (36)\end{aligned}$$

If we take the arbitrary scale factor $\theta = 3/2$—a reasonable choice since this places the detection bit energy level at 50 percent higher than the operating bit energy level—we find that the effect is to increase the Erlang level (λ/μ) by the factor

$$F = 1 + \left(\frac{3}{4\rho}\right)\left(\frac{\mu}{\delta}\right) = 1 + 1.88(\mu/\delta) \quad \text{for } \rho = 0.4. \qquad (37)$$

Note that μ/δ is the ratio of mean detection time-to-mean message duration, which should be very small. If, for example, $\mu/\delta = 0.0055$, corresponding to a mean access time of 1 s (50 frame)[6] for a 3-min mean call duration, this means that the effect of initial accesses is to reduce the Erlang traffic by about 1 percent (a negligible cost considering the notable advantages).

VII. Conclusions

The foregoing analysis represents a conservative estimate of the Erlang capacity of direct sequence spread-spectrum CDMA. Classical blocking was replaced by the condition that the total interference exceeds the background noise by 10 dB, and Erlang capacity was defined as the traffic load corresponding to a 1 percent probability that this event occurs. The three random variables which contribute to this event were modeled as follows:

1) Poisson traffic arrival, exponentially distributed message length, and arbitrarily many servers ($M/M/\infty$ in the terminology of queuing theory [1]);

2) voice activity factor ρ equal to 40 percent, as established by extensive experimental measurements on ordinary speech conversations [6];

[5] We assume initially that the access request is detected (immediately) when the initial access user's energy reaches E_M, but it remains undetected until that point. Below, we shall take E_M to be itself a random variable, with distribution similar to E_b. It should also be noted that initial access detection is performed on an unmodulated signal, other than for the user's pseudorandom code which is known to the base station. It should thus be more robust than the demodulator performance at the same power level.

[6] Note that at a data rate of 9600 bits/s, each 20 ms voice frame contains 192 bits. For $E_b/I_0 \approx 7$ dB, the frame energy-to-interference level is about 32 dB—an ample value to virtually assure detection in one frame.

3) individual user's received bit energy-to-interference density E_b/I_0 whose distribution was determined empirically from field measurements and supported by analytical results [4].

The main conclusion is that the Erlang capacity of CDMA is about 20 times that of AMPS. This is based on establishing the blocking condition as the event that the total interference-to-noise-ratio exceeds 10 dB. This "soft" blocking condition can be relaxed to allow a maximum ratio of 13 dB. for example, for particularly heavily loaded sectors. When traffic is light, much lower interference-to-noise ratios can be imposed which translate into much lower mobile transmitted powers, a particularly valuable feature in prolonging battery life for portable subscriber units.

Finally, a significant byproduct of spectrum sharing is the very small Erlang capacity overhead required to accommodate initial access requests along with ongoing traffic, as demonstrated in the last section.

Appendix
Modified Chernoff Bound for a Single Sector

Starting with (8), without considering outer cells, we may bound (8) by introducing an arbitrary threshold on each term $\nu_i\epsilon'_i$, where $\epsilon' = \epsilon/e^{\beta m}$. Thus,

$$\begin{aligned} P_{\text{blocking}} &= \Pr\left(\sum_{i=1}^{k} \nu_i\epsilon_i > (W/R)(1-\eta)\right) \\ &= \Pr\left(\sum_{i=1}^{k} \nu_i\epsilon'_i > A\right) \\ &< \Pr\left(\sum_{i=1}^{k} \nu_i\epsilon'_i > A \mid \nu_i\epsilon'_i < T \quad \text{for all } i\right) \\ &\quad + \Pr(\nu_i\epsilon'_i > T \quad \text{for any } i) \end{aligned} \tag{A.1}$$

where A is given by (14).

Bounding the first term by its (conditional) Chernoff bound and the second by a union bound, we have, using (9),

$$\begin{aligned} P_{\text{blocking}} &< E_k E_{\epsilon'} E_\nu\left([\exp(s\nu\epsilon')]^k \mid \epsilon' < T\right)e^{-sA} \\ &\quad + E_k\left[\sum_{i=1}^{k} \Pr(\epsilon'_i > T)\Pr(\nu_i = 1)\right] \\ &= E_k\left(\left[\rho E_e^{s\epsilon'} + (1-\rho)\right]^k \,\middle|\, \epsilon' < T\right)e^{-sA} \\ &\quad + \rho(\lambda/\mu)\Pr(\epsilon' > T) \\ &= \exp\left[\rho(\lambda/\mu)E\left(e^{s\epsilon'}t\right) - 1\right]e^{-sA} \\ &\quad + \rho(\lambda/\mu)\Pr(\epsilon' > T) \end{aligned} \tag{A.2}$$

$$\begin{aligned} Ee^{s\epsilon'_T} &= \int_{-\infty}^{(\ln T)/\beta} \exp\left(se^{\beta\xi}\right)\frac{e^{-\xi^2/(2\sigma^2)}}{\sqrt{2\pi}\sigma}\,d\xi \\ &= \int_{-\infty}^{\tau/\delta} \exp\left(se^{\beta\sigma\zeta}\right)e^{-\zeta^2/2}\,d\zeta/\sqrt{2\pi} \end{aligned} \tag{A.3}$$

where $\tau = (\ln T)/\beta$, and

$$\begin{aligned} \Pr(\epsilon' > T) &= \Pr(\xi > (\ln T)/\beta) \\ &= \int_{(\ln T)/\beta}^{\infty} \frac{e^{-\xi^2/2\sigma^2}}{\sqrt{2\pi}\sigma} = Q(\tau/\sigma). \end{aligned} \tag{A.4}$$

Inserting (A.4) and (A.3) into (A.2), and minimizing $s > 0$ and $\tau > 0$, yields (13)–(16).

The total interference is then the sum of (B.3) and (B.7), where the latter is integrated over all regions not including the six central region pairs A and B. The result equals $N_u(1+f)$. By performing this numerical integration, the other-cell interference factor f was obtained for various values of μ and σ_0, as shown in Table I.

Acknowledgment

The authors are grateful to E. Zehavi for suggesting the approach to the modified Chernoff bound of the Appendix, and to the highly motivated QUALCOMM team whose tireless efforts proved the feasibility of CDMA technology, and particularly of the power control methods which led to the empirical results shown in Fig. 1.

References

[1] D. Bertsekas and R. Gallager, *Data Networks*. Englewood Cliffs, NJ: Prentice-Hall, 1987.
[2] G. R. Cooper and R. W. Nettleton, "A spread spectrum technique for high capacity mobile communications," *IEEE Trans. Veh. Technol.*, vol. VT-27, pp. 264–275, Nov. 1978.
[3] K. S. Gilhousen, I. M. Jacobs, R. Padovani, A. J. Viterbi, L. A. Weaver, Jr., and C. E. Wheatley, III, "On the capacity of a cellular CDMA system," *IEEE Trans. Veh. Technol.*, vol. 40, May 1991.
[4] A. J. Viterbi, A. M. Viterbi, and E. Zehavi, "Performance of power-controlled wideband terrestrial digital communications," to appear in *IEEE Trans. Commun.*, 1993.
[5] W. Feller, *An Introduction to Probability Theory and Its Applications, Vol. I, 2nd ed.* New York: Wiley, 1957.
[6] P. T. Brady, "A statistical analysis of on–off patterns in 16 conversations," *Bell Syst. Tech. J.*, vol. 47, pp. 73–91, Jan. 1968.
[7] A. J. Viterbi, A. M. Viterbi, and E. Zehavi, "Other-cell interference in cellular power-controlled CDMA," *IEEE Trans. Commun.*, to appear.

Audrey M. Viterbi received the B.S. degree in electrical engineering and computer science and the B.A. degree in mathematics from the University of California, San Diego, in 1979. She received the M.S. and Ph.D. degrees in electrical engineering from the University of California, Berkeley, in 1981 and 1985, respectively.

She was an Assistant Professor of Electrical and Computer Engineering at the University of California, Irvine, from 1985 to 1990. Since 1990 she has been a Staff Engineer at Qualcomm, Inc., San Diego, where she is currently involved in the development of the CDMA cellular telephone system. Her professional interests are in the area of modeling and performance evaluation of computer-communication networks and communications systems.

Andrew J. Viterbi received the S.B. and S.M. degrees from the Massachusetts Institute of Technology, Cambridge, in 1957, and the Ph.D. degree from the University of Southern California in 1962.

In his first employment after graduating from M.I.T., he was a member of the project team at C.I.T. Jet Propulsion Laboratory which designed and implemented the telemetry equipment on the first successful U.S. satellite, *Explorer I*. In the early 1960's at the same laboratory, he was one of the first communication engineers to recognize the potential and propose digital transmission techniques for space and satellite telecommunication systems. As a Professor in the UCLA School of Engineering and Applied Science from 1963 to 1973, he did fundamental work in digital communication theory and wrote two books on the subject, for which he received numerous professional society awards and international recognition. In 1968 he co-founded LINKABIT Corporation and served as its Executive Vice President from 1974 to 1982, and as President from 1982 to 1984. On July 1, 1985, he co-founded and became Vice Chairman and Chief Technical Officer of QUALCOMM, Inc., a company specializing in mobile satellite and terrestrial communication and signal processing technology. Since 1975 he has been associated with the University of California, San Diego, and since 1985, as Professor (quarter time) of Electrical and Computer Engineering.

Dr. Viterbi is a member of the U.S. National Academy of Engineering. He is past Chairman of the Visiting Committee for the Electrical Engineering Department of Technion, Israel Institute of Technology, past Distinguished Lecturer at the University of Illinois and the University of British Columbia, and he is presently a member of the M.I.T. Visiting Committee for Electrical Engineering and Computer Science. In 1986 he was recognized with the Annual Outstanding Engineering Graduate Award by the University of Southern California, and in 1990 he received an honorary Doctor of Engineering Degree from the University of Waterloo, Ont. Canada. He presented the Shannon Lecture at the 1991 International Symposium on Information Theory. He has received three paper awards, culminating in the 1968 IEEE Information Theory Group Outstanding Paper Award. He has also received four major international awards: the 1975 Christopher Columbus International Award (from the Italian National Research Council sponsored by the City of Genoa); the 1984 Alexander Graham Bell Medal (from IEEE, sponsored by AT&T) "for exceptional contributions to the advancement of telecommunications;" the 1990 Marconi International Fellowship Award; and was co-recipient of the 1992 NEC C&C Foundation Award.

The application of spread spectrum to PCS has become a reality

Reverse Link Performance of IS-95 Based Cellular Systems

ROBERTO PADOVANI

IS-95 "Mobile Station-Base Station Compatibility Standard for Dual-Mode Wideband Spread Spectrum Cellular System" is a digital cellular standard endorsed by the U.S. Telecommunications Industry Association/Electronic Industries Association (TIA/EIA) based on CDMA technology [1]. The first issue of *IEEE Personal Communications* presented a tutorial review of random-waveform access techniques such as those implemented in IS-95 based systems [2]. This article presents a tutorial review of the reverse link (subscriber to base station link) characteristics and its performance in terms of both coverage and capacity.

Reverse Link Waveform

The reverse link waveform generation is shown in Fig. 1. The speech coder[1] generates variable length packets according to the speech activity at a rate of one packet every 20 ms. The length of the packets in bits and the corresponding data rates are shown in Table 1.

The protocol supports a mix of services within the same high data rate packet. For example, the 171 information bits can be used to support simultaneous transmission of speech and data, speech and signaling, or data and signaling. A specific example is provided by the transmission of in-traffic signaling data, whereby in addition to the common technique of "blank-and-burst" where a speech frame is replaced by signaling information, a "dim-and-burst" approach can be used in which speech (or data) and signaling information share the same high data rate packet.

The packets formatted as described in Table 1 are then convolutionally encoded by a powerful rate 1/3 constraint length $K = 9$ code (thus the eight code tail bits of Table 1) with code

[1] *This discussion, which is limited to the physical layer aspects of the link, is also valid for data applications.*

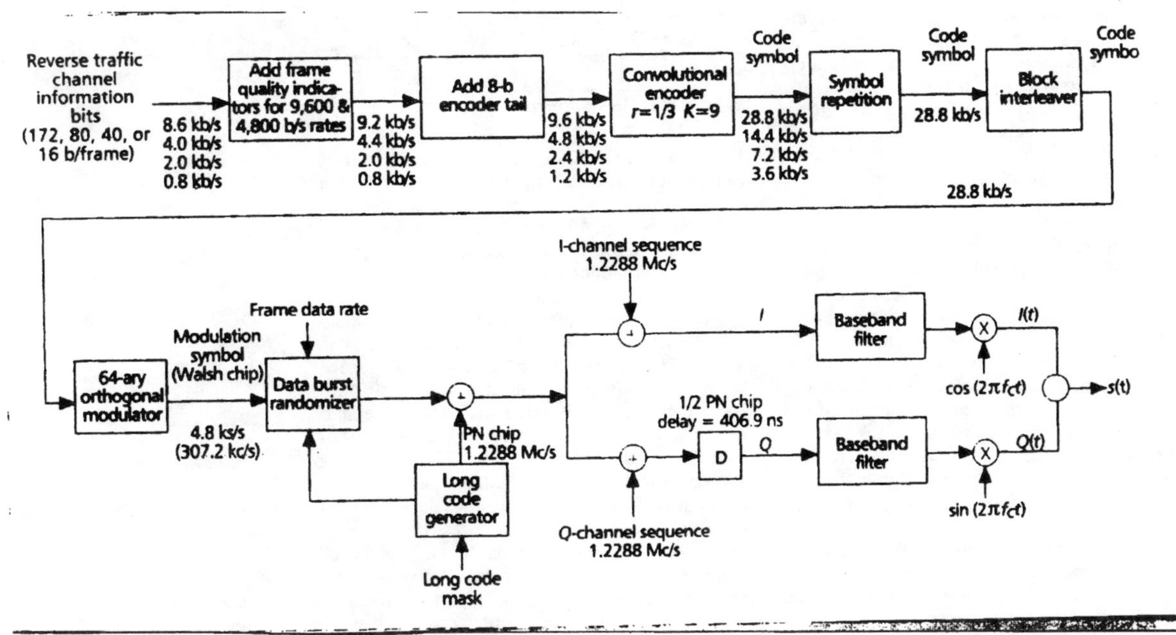

■ **Figure 1.** *Reverse link transmission.*

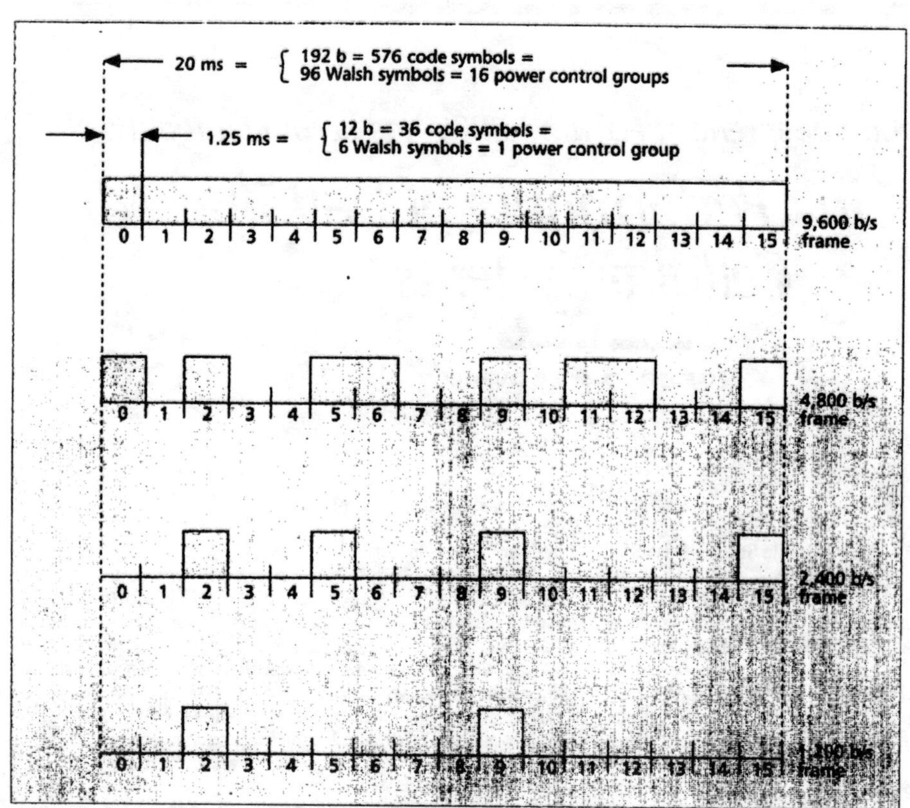

■ **Figure 2.** *Example of reverse link transmission for different data rates.*

generators 557, 663, and 711 (octal) and free distance d_{free} = 18. Thus all information bits are equally protected by the error correcting code. In addition, the two high rate packets are also protected by a 12-b and 8-b Cyclic Redundancy Code (CRC) for error detection, as shown in Table 1.

After convolutional encoding, the code symbols are interleaved by a block interleaver with span equal to one frame or 20 ms and modulated by a 64-ary orthogonal modulator. In other words, two information bits which after coding generate six code symbols, select after interleaving one of 64 orthogonal Walsh functions to be transmitted. Therefore, the orthogonal symbols have duration equal to T = 208.3 μs allowing the receiver to take advantage of the partial coherence of the channel.[2]

The final signal processing elements perform the direct-sequence spreading functions. First, the modulation symbols are spread by a subscriber unique PN sequence, i.e., the subscriber address (generated by a maximum length code of period 2^{42}-1) at a rate of 1,228,800 chips per second, i.e., 256 chips per orthogonal modulation symbol. Furthermore, the waveform is spread by a pair of PN codes (maximum length codes of period 2^{15}-1), common to all subscribers, in an OQPSK arrangement. The final waveform is then tightly filtered to generate a spectrum with 1.2288 MHz double-sided 3 dB bandwidth.

The actual transmission is then gated on-off pseudo-randomly[3] at 1.25 ms intervals, as shown in Fig. 2. This effectively reduces the transmit duty cycle from 100 percent for 9600 b/s transmission to 50 percent for 4800 b/s, 25 percent for 2400 b/s, and 12.5 percent for 1200 b/s. The overall result is the reduction of self-interference by a factor directly proportional to the average voice activity of the users, e.g., 40 percent voice activity corresponds to 4.0 dB interference reduction equivalent to a 2.5 time increase in the number of users.

Information [bits]	Signaling [bits]	CRC [bits]	Code tail [bits]	Total [bits]	Bit rate [b/s]
171	1	12	[illegible]	[illegible]	9,600
80	0	8	8	96	4,800
40	0	0	8	48	2,400
16	0	0	8	24	1,200

■ **Table 1.** *Packet structure.*

Two receiver structures are shown in Fig. 3. For early development and IS-95 field test validation the structure shown in Fig. 3a has been used. This structure implements a four-way RAKE receiver to demodulate the four strongest multipath components received on the two diversity antennas, as shown in Fig. 4. In this configuration a set of four ASIC's is employed each one implementing a complete reverse link demodulation path. The decision output from each of the active demodulators is then fed to an external microprocessor. The microprocessor combines the individual demodulator decisions, weighing each one by the relative strength of the respective multipath component, and generates a single stream of soft-decision inputs to the Viterbi decoder.

---

[2] *Coherence over the period T is guaranteed for vehicular applications in both cellular and PCS frequencies.*

[3] *This is the function performed by the data burst randomizer block in Fig 1.*

It is quite obvious that the above multipath diversity combining is sub-optimal since an independent decision on the transmitted orthogonal symbol is being made by each individual demodulator. The second demodulator architecture implemented in a new generation ASIC is shown in Fig. 3b. With this architecture, the multipath diversity receiver outputs are optimally combined by first combining the matched filter output energies. Note that this can be easily accomplished in this architecture since all four demodulators reside in the same device.

The second receiver architecture provides substantial performance improvements coupled with a much higher level of integration. Fig. 5 shows a comparison of the performance achieved in a Rayleigh channel with one, two, and four paths combining. Table 2 compares simulation results of the E_b/N_0 required for a 1 percent FER in a Rayleigh channel and two paths combining versus vehicle speed for the two architectures.

The final step in recovering the information

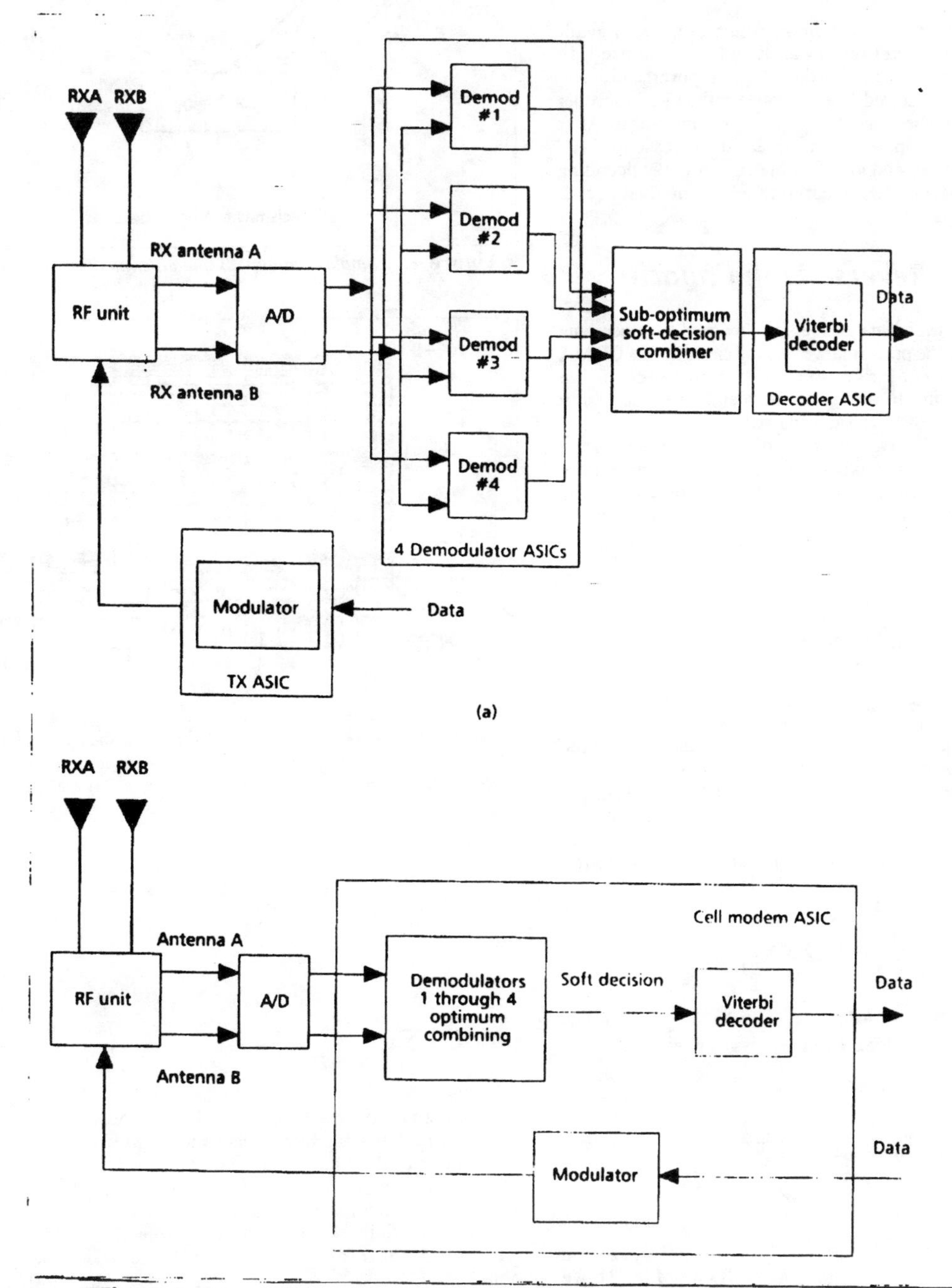

■ **Figure 3.** *a) Sub-optimal multipath diversity receiver architecture; b) optimal multipath diversity receiver architecture.*

	Eb/No [dB] architecture (a)	Eb/No [dB] architecture (b)
AWGN	4.0	2.6
8 [Km/h]	5.7	3.8
30 [Km/h]	7.9	5.5
100 [Km/h]	8.0	5.8

■ **Table 2.** *Comparison of Eb/No requirements in AWGN and Rayleigh fading for two-way combining and FER = 1 percent.*

consists of determining which of the four available packet types was actually transmitted. In order to accomplish this without any overhead penalty the received data is decoded by the Viterbi decoder four times once for each of the four hypothesis. After the multiple decoding, several metrics, such as CRC pass/fail and metrics obtained from the decoding process, are compared to select one final decoded packet.

Reverse Link Capacity

Several analyses have been carried out concerning the performance and the capacity of a CDMA cellular system [3-6]. In this section, we attempt to reconcile the results obtained from field tests with such analysis.

In the reverse link, one of the fundamental parameters to be analyzed and measured in determining the capacity is the total power received at the base station antennas. With M active users in one isolated sector the total received power C can be expressed as

$$C = N_o W + \sum_{i=1}^{M} \upsilon_i P_i , \quad (1)$$

where N_oW represents the background noise power in the bandwidth W, υ_i represents voice activity of the ith user, and P_i is the received power of the ith user. For IS-95, W=1.2288 MHz and $E\{\upsilon\}$ is taken to be equal to 0.4 (during the field tests the precise average voice activity of 40 percent was achieved by setting the mobile stations in test mode). By expressing the receive power relative to the background noise, we obtain

$$Z = \frac{C}{N_o W} = 1 + \sum_{i=1}^{M} \upsilon_i \frac{P_i}{N_o W} \quad (2)$$

Furthermore, the signal-to-noise plus interference ratio for a given user is given by,

$$\frac{E_{bi}}{N_o + I_o} = \frac{W}{R} \cdot \frac{\frac{P_i}{N_o W}}{1 + \frac{1}{WN_o} \sum_{j=1}^{M-1} \upsilon_j P_j} , \quad (3)$$

where R = 9,600 b/s. Combining Eqs. (2) and (3) and approximating M-1 with M in (3), we obtain

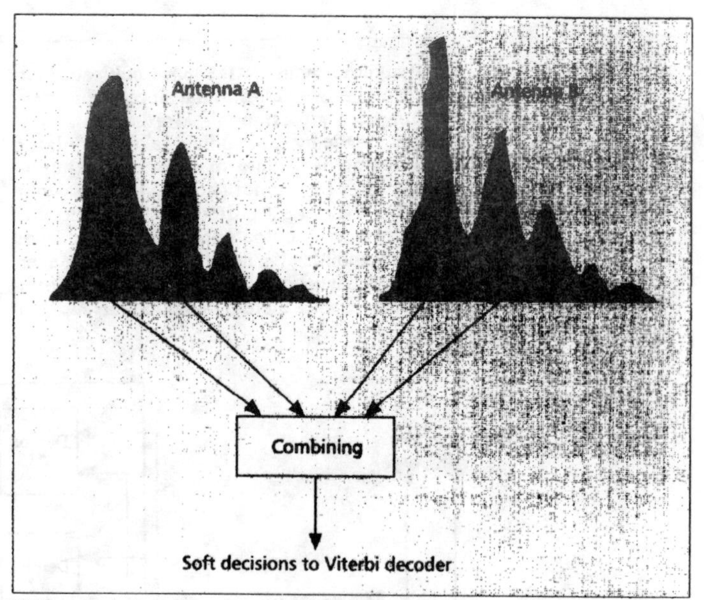

■ **Figure 4.** *Example of multipath and antenna diversity combining.*

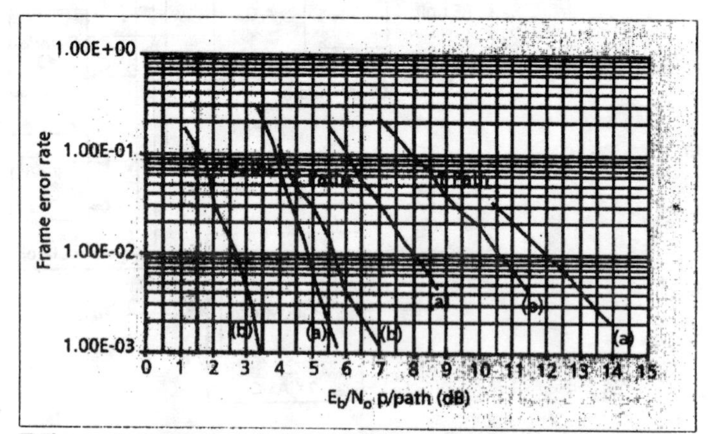

■ **Figure 5.** *Comparison of FER vs. E_b/N_o for the two architectures (a) and (b). Vehicle speed of 100 Km/h at a carrier frequency equal to 850 MHz. Results are shown for 1, 2, and 4 independent Rayleigh paths.*

$$Z \cong \frac{1}{1 - \frac{R}{W} \sum_{i=1}^{M} \upsilon_i \frac{E_{bi}}{N_o + I_o}} = \frac{1}{1 - X} \quad (4)$$

where

$$X = \frac{R}{W} \sum_{i=1}^{M} \upsilon_i \frac{E_{bi}}{N_o + I_o} \quad (5)$$

The signal-to-interference ratio is closely approximated by a lognormal distribution which has a mean εdB and a standard deviation σdB [2]. The voice activity υ is a quaternary random variable with mean $E\{\upsilon\}$. By central limit arguments, the variable X approaches a normal distribution.[4] Therefore, with $\beta = ln(10)/10$, we obtain

[4] *It should be clear that the approximation of Eq. (4) holds only if the probability of X > 1 is small.*

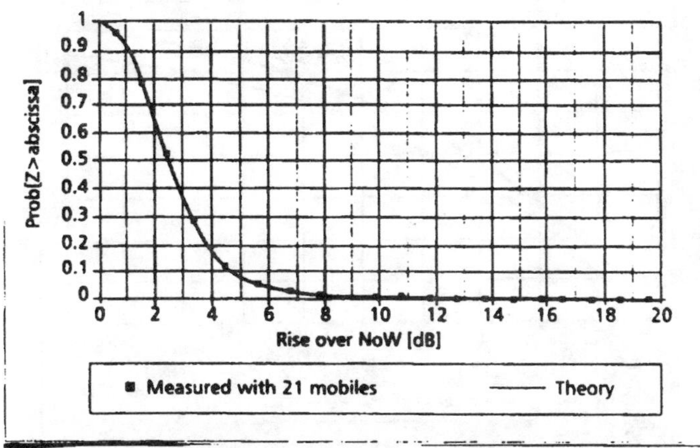

Figure 6. *Complementary cdf of the cell receiver power rise over background noise Z. Theoretical and measured results with 21 mobiles.*

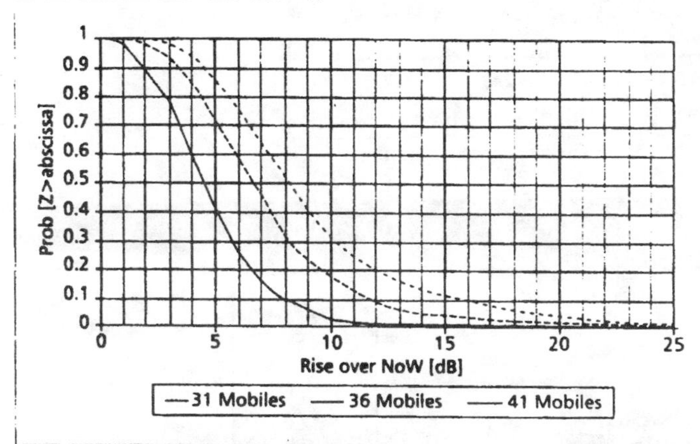

Figure 7. *Complementary cdf of the cell receiver power rise over background noise Z. Measure with 31, 36, and 46 mobiles.*

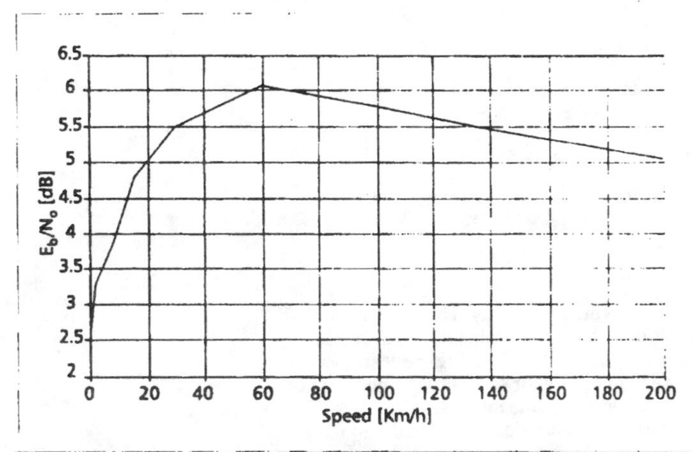

Figure 8. *Typical E_b/N_0 performance vs. vehicle speed for 850 MHz links to achieve a FER = 1 percent. Rayleigh channel with two independent path (speed = 0 corresponds to the AWGN channel with two paths.*

$$E\{X\} = \frac{R}{W} M\, E\{v\}\, e^{(\beta\sigma)^2/2 + \beta\varepsilon}, \tag{6}$$

$$\mathrm{Var}\{X\} = \left(\frac{R}{W}\right)^2 M e^{(\beta\sigma)^2 + 2\beta\varepsilon} \bullet \left[E\{v^2\}\, e^{(\beta\sigma)^2} - E\{v\}^2\right]. \tag{7}$$

Expressing Z in dB, i.e., $Z = -10 \log_{10}(1 - X)$, we can easily derive the distribution and density functions of the rise in dB over background noise, namely

$$P_z(z) = \frac{1}{\sqrt{2\pi}} \int_{-\infty}^{\frac{1-e^{-\beta z}-E(x)}{\sqrt{\mathrm{Var}(x)}}} e^{-y^2/2} dy \tag{8}$$

$$p_z(z) = \frac{1}{\sqrt{2\pi \mathrm{Var}(x)}} e^{-\frac{[1-e^{-\beta z}-E(x)]^2}{2\mathrm{Var}(x)}} \beta e^{-\beta z} \tag{9}$$

Returning now to the field test results, Fig. 6 compares the complementary cumulative distribution function of Z, i.e., $1 - P_Z(z)$, calculated from (8) and measured during a test involving $M = 21$ mobile units in an isolated sector.[5] The numerical values used for Eq. (8) are $M = 21$, $\varepsilon = 7.9$dB, and $\sigma = 2.4$dB. A large number of field tests performed in a variety of environments have shown similar performance to that of Fig. 6 with signal-to-noise requirements varying from $\varepsilon = 5$ dB to $\varepsilon = 8.5$ dB needed to maintain a frame error rate (FER) of 1 percent. Figure 7 shows another set of results obtained in an isolated sector. In this particular case the sector under test covers an eight-lane interstate freeway and all the mobiles involved in the test are placed on this freeway.

A distinct improvement in the signal-to-noise ratio requirement is obtained from low mobility users, e.g., pedestrian or in-building users, which are not experiencing the faster fading induced by vehicular motion. This is easily seen from Fig. 8 which shows the signal-to-noise required to achieve a 1 percent FER as a function of vehicle speed with one Rayleigh path per receive antenna. The shape of the curve shown in Fig. 8 is explained by the fact that at relatively low speeds power control is very effective in counteracting the slow fades whereas at higher speeds, where power control is not as effective in counteracting the fast fading, the effects of interleaving become increasingly beneficial.

Since each user is very accurately power controlled to the minimum signal-to-noise value necessary to achieve a given FER, low mobility users produce approximately one half the interference of high mobility users (typical signal-to-noise requirements for low mobility users is 4 to 5 dB). This has the obvious beneficial effect of increasing the capacity of the reverse link when the user population is a mix of high and low mobility users.

[5] *This particular test was conducted in a densely populated residential area in San Diego, California. The base station facilities and antennas subsystems were shared with those of the existing analog system.*

A complete analysis of the reverse link capacity must include the effects of other-cells interference and a model for the traffic load. These analysis have been carried out in detail in [4] and [6]. In the following we present a summary of the main result. The derivation uses the following parameters:

Median $E_b/(N_0+I_0)$: $\varepsilon = 7dB$ — This assumption combines the values measured in the field tests for high mobility users with the improvements achieved by the new receiver architecture described in the previous section.

$E_b/(N_0+I_0)$ standard deviation: $\sigma = 2.5dB$ — This value, induced by the closed loop power control, has been consistently measured in the field tests for high mobility users. Smaller values (σ = 1.5 dB) have been consistently measured in field tests for low mobility users.

Average voice activity: $E\{\upsilon\} = 0.4$, $Var\{\upsilon\} = 0.15$ — This is an estimate that should be refined as large commercial deployments are carried out and large population samples can be measured.

Other-cell interference fraction: $f = 0.55$ — This is the fraction of other-cell interference with respect to in-cell interference generated in an equally loaded network. This assumes a 4th power propagation law with 8 dB lognormal shadowing. Higher propagation exponents will reduce the factor f and lower exponents will increase it [6].

Traffic model — Poisson arrival rate of calls with parameter λ [calls/s] and exponential service time with parameter $1/\mu$ [s/calls], namely

$$Pr(k \text{ active users } / \text{ sector}) = \frac{(\lambda/\mu)^k}{k!}e^{-\lambda/\mu}$$

$$E\{k\} = \frac{\lambda}{\mu} \qquad \text{Var}\{k\} = \frac{\lambda}{\mu}$$

Given all the above assumptions we can now calculate the distribution of Z, i.e., the rise over background noise Z can now be expressed as

$$Z = \frac{C}{N_o W} = 1 + \sum_{i=1}^{k} \upsilon_i \frac{P_i}{N_o W} + \sum_{j}^{\text{other cells}} \sum_{i=1}^{k} \upsilon_i^{(j)} \frac{P_i^{(j)}}{N_o W} \qquad (10)$$

The distribution of Z is then given by Eq. (10) where now the mean and variance of X are given by

$$E\{X\} = \frac{R}{W} \bullet \frac{\lambda}{\mu} \bullet E\{\upsilon\} \bullet e^{(\beta\sigma)^2/2+\beta\varepsilon} \bullet (1+f) \qquad (11)$$

$$\text{Var}\{X\} = \left(\frac{R}{W}\right)^2 \bullet \frac{\lambda}{\mu} \bullet E\{\upsilon^2\} \bullet e^{2(\beta\sigma)^2+2\beta\varepsilon} \bullet (1+f) \qquad (12)$$

From Eqs. (10), (12), and (13) it is straightforward to calculate the offered load in Erlangs for a given blocking probability. The blocking probability is defined as the probability of Z exceeding a given value z in dB [4]. Figure 9 shows the offered load per sector versus z for 1 percent and 2 percent blocking probabilities.

Operationally, a value of z = 10 dB and 2 percent blocking probability are a good compromise between offered load and coverage. As seen in Fig. 9, this corresponds to 19 Erlangs/sector or approximately 27 voice channels per sector. Notice that this result applies to a single CDMA frequency assign-

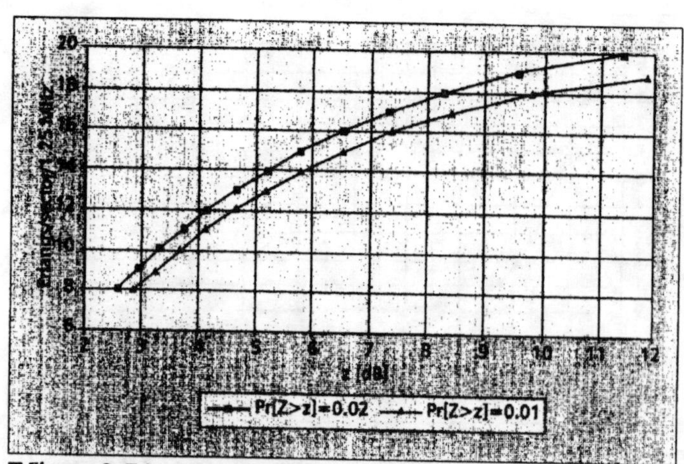

■ **Figure 9.** *Erlangs/sector/1.25 MHz with 1 percent and 2 percent blocking probabilities.*

System	Radio Capacity/ Sector	Erlang Capacity/ Sector
AMPS	19	12.3
IS-95	27•9 = 243	229
IS-95/AMPS	12.8 times	18.6 times

■ **Table 3.** *Reverse link capacity summary.*

Parameter	Value	Units
Power amplifier peak power	23.0	dBm
Subscriber antenna gain	0.0	dBi
Peak EIRP	23.0	dBm
Maximum isotropic path loss Single user — no shadowing	–155.2	dB
Base station antenna gain	12.0	dBi
Base station losses	–2.0	dB
Received signal strength	–122.2	dBm
Receiver noise figure	5.0	dB
Receiver noise density	–169.0	dBm/Hz
Data rate (9,600 b/s)	39.8	dB-Hz
Received Eb/No	7.0	dB
Probability of service at cell edge: Ps	90	%
LogNormal shadowing sigma	8.0	dB
Offered load	19	Erlangs/sector
Margin to achieve specified Ps with soft-handoff	7.7	dB
Maximum isotropic path loss	–147.5	dB

■ **Table 4.** *Link budget.*

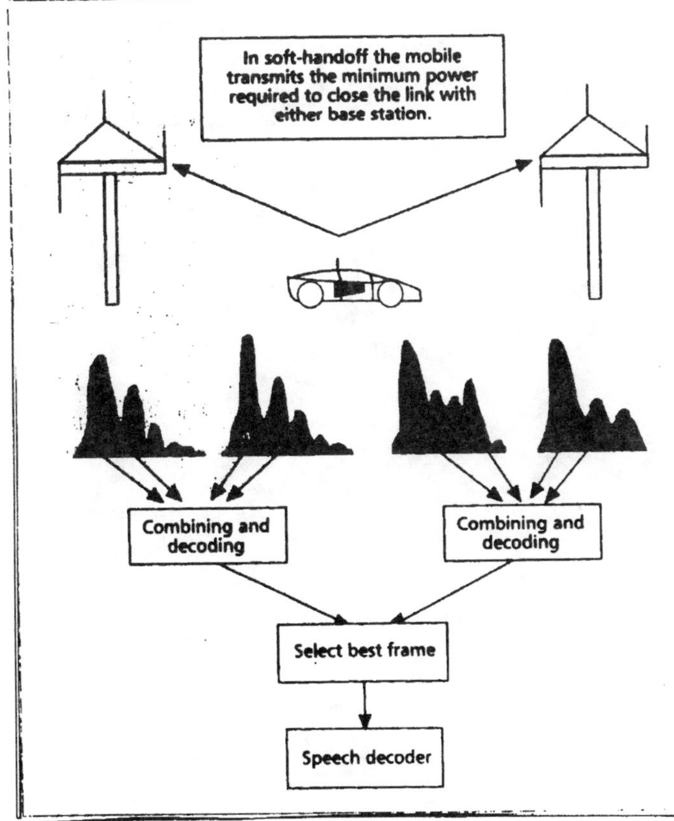

Figure 10. *Reverse link soft handoff.*

ment, i.e., one 1.25 MHz block. The US cellular spectrum allocation when completely converted to CDMA will utilize nine CDMA frequency blocks in a FDM arrangement.[6]

Finally Table 3 summarizes the capacity results and compares them to those of the analog AMPS system for the same 2 percent blocking probability.

[6] *The U.S. spectrum allocation for mobile station to base station transmission consists of 12.5 MHz for each of the two carriers servicing one market. The 12.5 MHz are not contiguous for either of the two carriers. The wireline carrier will deploy seven CDMA carriers in a block of 10 MHz and 2 more carriers in a block of 2.5 MHz. The non-wireline carrier will deploy eight CDMA carriers in a block of 11 MHz and one more carrier in a block of 1.5 MHz.*

Reverse Link Coverage

An additional advantage of the efficient waveform design implemented in the IS-95 standard is found in the increased coverage as compared to other digital cellular standards. In this section a simple link budget for an IS-95 based system is presented. The link budget is derived for Class III portable subscriber units with a minimum output power requirement of 200 mWatts [1].

Figure 10 shows the complete reverse link processing for a mobile station in soft handoff. In order to close the link, a mobile at the edge of coverage and in soft handoff with two (or multiple) base stations needs to transmit the minimum power required to achieve the desired SNR to either of the two (or multiple) base stations.

The margin required to achieve a given probability of service P_s at the cell border, taking into account the effects of lognormal shadowing and soft-handoff, is calculated in detail in [7]. Additionally, in [8] the effects of shadowing, soft-handoff, and offered traffic are combined to obtain the final margin shown in Table 4. The margin required by a mobile to overcome independent lognormal shadowing (with 8 dB sigma) between two base stations equally loaded at a level of 19 Erlangs is equal to 7.7 dB for a probability of service at the cell border equal to 90 percent. With the above assumptions, the maximum isotropic path loss that a portable unit can sustain equals 147.5 dB.

Conclusion

The application of spread spectrum to personal communication services has become a reality in just a few years. The approval of interim standard IS-95 and the imminent commercial deployment of IS-95 based networks around the world have sparked enormous interest in both academia and industry. Ongoing work in the areas of interference cancellation, antenna beam forming, advanced signal processing, and components integration will certainly improve on the performance reported in this tutorial presentation.

References

[1] TIA/EIA/IS-95 "Mobile Station-Base Station Compatibility Standard for Dual-Mode Wideband Spread Spectrum Cellular System," Telecommunication Industry Association, July 1993.
[2] A. J. Viterbi, "The Orthogonal-Random Waveform Dichotomy for Digital Mobile Personal Communications," *IEEE Personal Commun.*, vol. 1, no. 1, pp. 18-24, First Qrtr. 1994.
[3] K. S. Gilhousen *et al.*, "On the Capacity of a Cellular CDMA System," *IEEE Trans. Veh. Technol.*, vol. 40, pp. 303-311, May 1991.
[4] A. M. Viterbi and A. J. Viterbi, "Erlang Capacity of a Power Controlled CDMA System," *IEEE Jour. on Sel. Areas of Commun.*, vol. 11, pp. 892–890, Aug 1993.
[5] W. C. Y. Lee, "Overview of Cellular CDMA," *IEEE Trans. Veh. Technol.*, vol. 40, pp. 291-301, May 1992.
[6] A. J. Viterbi, A. M. Viterbi, and E. Zehavi, "Other-Cell Interference in Cellular Power-Controlled CDMA," *IEEE Trans. on Commun.*, vol. 42, no. 4, pp.1501-1504, April 1994.
[7] A. J. Viterbi *et al.*, "Soft Handoff Extends CDMA Cell Coverage and Increases Reverse Link Capacity," *IEEE Jour. on Sel. Areas of Commun.*, to be published.
[8] R. Vijayan, R. Padovani, and E. Zehavi, "The Effects of Lognormal Shadowing and Traffic Load on CDMA Cell Coverage," submitted for publications to *IEEE Trans. on Commun.*

Biography

Roberto Padovani received a Laurea degree from the University of Padova, Italy and M.S. and Ph.D. degrees from the University of Massachusetts, Amherst in 1978, 1983, and 1985, respectively, all in electrical and computer engineering. In 1984 he joined M/A-COM Linkabit, San Diego, where he was involved in the design and development of satellite communciation systems, secure video systems, and error-correcting coding equipment. In 1986, he joined QUALCOMM, Incorporated, San Diego, California, and is now vice president of system engineering in the Engineering Department. He has been involved in the design, development, and test of the CDMA cellular system which led to EIA/TIA IS-95 standard.

Spread Spectrum Access Methods for Wireless Communications

The authors present an overview of the characteristics of CDMA as it is currently being envisioned for use in wireless communications. There are many considerations in the design of such systems, and there are multiple designs being proposed.

Ryuji Kohno, Reuven Meidan, and Laurence B. Milstein

ver the past several years, code division multiple access (CDMA) has been shown to be a viable alternative to both frequency division multiple access (FDMA) and time division multiple access (TDMA), and the use of spread spectrum techniques (upon which CDMA is based) in wireless communications applications has become a very active area of research and development [1, 2]. While there does not appear to be a single multiple accessing technique that is superior to others in all situations, there are characteristics of spread spectrum waveforms that give CDMA certain distinct advantages. The two basic problems which the cellular mobile radio system designer is faced with are multipath fading of the radio link and interference from other users in the cellular reuse environment. Spread spectrum signals are effective in mitigating multipath because their wide bandwidth introduces frequency diversity. They are also useful in mitigating interference, again because of their wide bandwidth. The result of these effects is a higher capacity potential compared to that of non-spread access methods.

Consider the use of direct sequence (DS) spread spectrum. As is well-known, DS waveforms can be used to either reject multipath returns that fall outside of the correlation interval of the spreading waveform, or enhance overall performance by diversity combining multipath returns in a RAKE receiver [3]. The above will hold any time the spread bandwidth exceeds the coherence bandwidth of the channel, that is, when the channel appears frequency-selective to the spread spectrum signal. Alternately, in frequency hopped (FH) spread spectrum, frequency diversity is obtained through coding the data and interleaving it over multiple-hops.

Another consideration in using CDMA in cellular systems is the so-called reuse factor. For non-spread multiple accessing techniques (i.e., FDMA and TDMA), frequencies used in a given cell are typically not used in immediately adjacent cells. This is done so that a sufficient spatial isolation will exist to ensure cells using the same frequency will not cause excessive interference (i.e., co-channel interference) with one another. For example, in the analog AMPS system, a frequency reuse of one-in-seven is employed. However, with spread spectrum signaling, the possibility of a frequency reuse of one-in-one exists. Further, in a CDMA system, performance is typically limited by average (rather than worst-case) interference. For these reasons, in a multicell system, CDMA is anticipated to have a larger capacity than either FDMA or TDMA. Note, however, for an *isolated cell*, because of the nonorthogonality of the CDMA waveforms, the capacity of the cell is less than what would be the case with an orthogonal technique such as TDMA or FDMA.

CDMA also provides a natural way to exploit the bursty nature of a source for added capacity. In the case of a two-way telephone conversation, the voice activity of each participant is about 50 percent of the time. If transmission is discontinued during non-activity periods, in principle one can double the number of simultaneous conversations in the system.

The above attributes promise the potential of higher system capacity through spread spectrum techniques. Also, the one cell reuse pattern alleviates the problem of frequency planning required with the narrow band systems (although other planning issues may become necessary, like careful power planning and pilot timing for DS systems, or careful choice of hopping patterns for FH systems). As a result, CDMA has become a serious competitor in the cellular arena.

The ability of spread spectrum signals to combat interference has also been applied in a different arena, that of the so-called industrial, scientific, and medical (ISM) bands. The ISM bands are frequency bands which, by Part 18 of the U.S. FCC regulations, were originally designated for operation of equipment which "generate and use locally RF energy for industrial, scientific, and medical" applications, "excluding applications in the field of telecommunications." In view of the local nature of these RF radiations, it was later suggested to use these bands for telecommunications, also of a local nature, such as on-site communications. Spread spectrum techniques, both DS- and FH-based, were established, and a byproduct of the use of spread spectrum is the ability to allow unlicensed operation, per Part 15 of the FCC regulations.

RYUJI KOHNO is an associate professor in the Division of Electrical and Computer Engineering, Yokohama National University.

REUVEN MEIDAN is with Motorola Israel as chief scientist and director of advanced technology.

LAURENCE B. MILSTEIN is a professor with the Department of Electrical and Computer Engineering at the University of California at San Diego.

0163-6804/95/$04.00 1995 © IEEE

Direct Sequence Cellular CDMA

DS spread spectrum signals are generated by linear modulation with wideband PN sequences which are assigned to individual users as their signature codes. The wideband character is utilized to achieve enhanced performance in the presence of interference and multipath propagation. For a description of DS spread spectrum at a tutorial level, see [4]; for an in-depth treatment, see references such as [5] and [6].

In DS/CDMA, the time/frequency space is usually shared by the users in the following way. One frequency band is used for the base-to-mobile link (also called the forward or down link). A separate frequency band is used for mobile-to-base link (i.e., the reverse or up link). Except for this frequency division duplex (FDD) operation, no other frequency or time division between the mobiles takes place.

The two links, forward and reverse, differ in certain fundamental ways. On the forward link, a cell's common pilot can be used for channel estimation and time synchronization. Furthermore, the users can be orthogonalized. (However, the orthogonalization is not preserved between different paths of the multipath propagation, nor is it preserved between the forward links of different cells.) The reverse link, on the other hand, does not enjoy these features. For example, it typically cannot be orthogonalized, in view of the different locations as well as independent movements of the mobiles. It is the task of the designer to balance the two links.

Characteristics of DS/CDMA

Universal Frequency Reuse — FDMA and TDMA cellular systems must rely on spatial attenuation to control intercell interference. As a result, neighboring cells need to be assigned different frequencies to protect against excessive (co-channel) interference. In contrast, a DS/CDMA cellular system can apply a universal one-cell frequency reuse pattern. If the traffic requirement at a certain location increases, introduction of a new cell will be less restricted than in the case of either FDMA or TDMA. This ability to employ universal frequency reuse not only beneficially affects the capacity of the system, but also results in ease of frequency management (although other management issues may be required such as power management and pilot timing).

Power Control —

Reverse Link: The reverse link is typically designed to be asynchronous, and an asynchronous CDMA system is vulnerable to the "near-far" problem, that is, the problem of very strong undesired users' DS signals at a receiver swamping out the effects of a weaker, desired user's, DS signal. Such interference is due to the nonzero crosscorrelation of the PN sequences assigned to individual users in CDMA. A solution to the near-far problem is the use of power control, which attempts to ensure that all signals from the mobiles within a given cell arrive at the base of that cell with equal power. The primary use of power control is to maximize the total user capacity; an additional benefit is to minimize consumption of the transmitted power of a portable unit. The power control required must be accurate (typically within 1 dB), fast enough to compensate for Rayleigh fading of fast moving vehicles as well as changes in shadowing (closed loop control with an update rate on the order of 1000 b/s) and have a large dynamic range (80 dB).

Forward link: Since all the cell's signals can be received at the mobile with equal power, the forward link does not suffer from the near-far problem. However, power control can be applied by increasing the transmitted power to mobiles that suffer from excessive intercell interference. This usually will happen when a mobile reaches the cell's boundary. It should be noted that the forward link power control is of an entirely different nature relative to that of the reverse link; it requires only a limited dynamic range, and need not be fast.

Soft Handoff and Space Diversity — The universal frequency reuse presents a problem at the cell's boundaries, where the transmission from two or more cell sites are received at nearly equal levels. To resolve the problem on the forward link, the user's information is sent via two or more base stations, which is diversity-combined by the user's receiver. On the reverse link, the user's data is received by the corresponding base station receivers and selection diversity is performed through the fixed infrastructure, which carries the receptions from the spatially separated sites to a common place. Power control of the mobile is coordinated by that base station that receives the strongest signal; this ensures that excessive interference will not be generated. Note that the universal reuse pattern, together with a RAKE receiver, combine to achieve the desired result. The soft handoff mode is particularly helpful in a handoff transition from one cell to another, as it provides a "make before break" handoff transition.

Soft handoff can also be used between sectors of the same base site. The two sectors use the same frequencies and a soft handoff mode is used to cover the boundary region between adjacent sectors. This mode is often termed "softer handoff."

Further, soft handoff can be used for providing space diversity to mitigate multipath having a delay spread that is short compared to the signal's correlation time (e.g., for indoor application). Another case were such soft handoff is proposed is in the satellite-based case (see below) to provide satellite diversity to the mobiles.

Coding — Coding redundancy can be regarded as part of the spreading. In principle, one can envision using very low rate codes [12]. In the limit, a code of rate 1/G, where G is the processing gain, can be considered. In practice, code rates of 1/2 of 1/3 are typically used.

Source Burstiness (Voice Activity) — Multiple access interference (MAI) in CDMA is the most dominant factor in the limitation of capacity. A way to reduce the instantaneous MAI is to stop transmission when voice or data activity is absent. On two-way telephone conversations, numerous measurements have established that voice is active less than 50 percent of the time. Thus, if voice activity detection is employed, the capacity of a cellular CDMA system, in terms of numbers

While there does not appear to be a single multiple access technique that is superior to others in all sitruations, there are characteristics of spread spectrum waveforms that give CDMA distinctive advantages.

In a CDMA system, performance is typically limited by average (rather than worst-case) interference.

of users that are simultaneously served by the system, can be approximately doubled.

Antenna Gain — Fixed sectored antennas and phased arrays also accomplish reduction of MAI, and hence increase the user capacity. That is, the universal one cell reuse pattern can be applied to cells that have been subdivided into sectors. Typically, a three-sectored antenna pattern is employed.

Cellular User Capacity

For the forward link, a base station can employ synchronous transmission to all mobile users, as noted above. This is called synchronous CDMA, and orthogonal codes can be applied for the synchronized transmission to mitigate MAI from within a cell. We concentrate here on a coarse estimate of the capacity of the reverse link, which is typically an asynchronous link [2, 7-9]. We assume perfect power contol, and an identical set of users (e.g., all voice users requiring the same performance).

From Eq. (14) of [2], an effective energy-per-bit-to-noise spectral density ratio can be expressed as

$$\left(\frac{E_b}{\eta_0}\right)_{eff} = \frac{1}{\frac{\eta_0}{E_b} + \frac{2}{3G}(M-1)(1+K)\alpha}, \quad (1)$$

where we define the following terms:

$\left(\frac{E_b}{\eta_0}\right)_{eff} \triangleq$ energy-per-bit-to-total noise (i.e., thermal noise plus interference) spectral density ratio

$E_b \triangleq$ received energy-per-bit

$\eta_0 \triangleq$ single-sided noise spectral density

$G \triangleq$ processing gain, equal to the number of chips per information symbol (corresponding to a processing gain of rG per coded symbol if rate r forward error correction is employed)

$M \triangleq$ number of users per cell

$\alpha \triangleq$ voice activity factor

$K \triangleq$ adjacent cell spillover factor, given by the ratio of intercell interference to intracell interference.

and

$\frac{2}{3}$ is a coefficient that arises because rectangular pulses were assumed; it will change somewhat if the chip shape changes.

Eq. (1) relates the effective energy-per-bit-to-noise spectral density ratio to the number of users. It can be rearranged to yield the following expression for the number of users per cell:

$$M = 1 + \frac{3}{2}\frac{G}{(1+K)\alpha}\left\{\left[\left(\frac{E_b}{\eta_0}\right)_{eff}\right]^{-1} - \frac{\eta_0}{E_b}\right\} \quad (2)$$

Consider now some interpretations and comments on the above results:

• The concept of an effective energy-per-bit-to-noise spectral density ratio, presented in (1), "makes sense" in view of the fact that a DS/CDMA system operates in an environment where the multiple access noise is caused by a sum of a large number of users, which, by the central limit theorem, tends to be Gaussian. Hence, the DS/CDMA link can be characterized by a specific $\left(\frac{E_b}{\eta_0}\right)_{eff}$ required to achieve a required bit error rate (BER). The actual BER achieved, for a given $\left(\frac{E_b}{\eta_0}\right)_{eff}$, is a function of the characteristics of channels, such as the degree of multipath fading. From Eq. (1), we see that for other parameters held constant, when the value for $\left(\frac{E_b}{\eta_0}\right)_{eff}$ is decreased, the resulting number of users is increased.

• Equation (2) represents an average value for M. As the averaging is done with respect to several random variables, one should exercise care in the interpretation. For example, α represents the average voice activity duty cycle. However, the activity of a user is a binomial random variable. The average value for α alone does not indicate the outage probability of the system, and specifying outage probability, which may be a more meaningful way to establish the number of simultaneous users, might result in a lower capacity than does using average BER.

• K represents the reduction of capacity due to intercell interference. It is dependent on several factors, notably, intercell isolation. For terrestrial systems, this is a function of the propagation law exponent (falloff of received power with distance). The lower the propagation exponent, the higher K becomes (see [10]). For a propagation law exponent of 4, $K \simeq 0.5$. Alternately, for a satellite system, cell isolation is primarily obtained through the satellite multibeam antenna pattern. As an example, $K = 1$ is reported in [11]. Further, K is dependent on the users' spatial distribution. The numbers quoted above relate to a uniform distribution; for other distributions, K may change considerably. Finally, there is a difference between the K's of the forward and reverse links; for example, the use of soft handoff causes the forward link to suffer from a worse reuse factor, since mobiles that are in a soft handoff zone require links from two or more base stations.

• Thermal noise reduces capacity, as seen by the second term in Eq. (2). If the mobile power is high enough, we obtain the asymptotic capacity by neglecting the effect of thermal noise. Since $M \gg 1$,

$$M \simeq \frac{3}{2}\frac{G}{\left(\frac{E_b}{\eta_0}\right)_{eff}}\frac{1}{(1+K)\alpha} \quad (3)$$

• The minimum value required for $\left(\frac{E_b}{\eta_0}\right)_{eff}$, denoted $\left(\frac{E_b}{\eta_0}\right)_{min}$, is a key parameter, since it effects the capacity directly, as seen from Eqs. (2) or (3). It depends on the BER required, which, in turn, depends on many factors, e.g., the multipath characteristics of the link in relation to the signal's bandwidth. Multipath returns with time delays greater than the correlation time of the signal can be resolved by a RAKE receiver and constructively combined. While the precise multipath characteristics depend on the operating environment, as will be discussed below in the section on performance, the correlation time of the signal is roughly equal to the chip duration, and thus to ensure that the delay spread of the channel exceeds the correlation time of the signal, a high chip rate is desired, which implies a wide bandwidth. Also, $\left(\frac{E_b}{\eta_0}\right)_{min}$ depends upon the coding employed. Lastly, the minimum $\left(\frac{E_b}{\eta_0}\right)_{eff}$ depends on the modulation. A typical value for $\left(\frac{E_b}{\eta_0}\right)_{min}$ for outdoor cellular terrestial applications, a BER of 10^{-3}, convolutional coding with a rate of 1/3, constraint length $k = 9$, and soft decision decoding, is 7 dB.

• The ratio $M\alpha/G$ can be given the interpretation of efficiency, as it expresses the information bit rate communicated in a cell (one way) per one Hz of bandwidth. In the above example, the efficiency is 0.2 b/s/Hz/cell ($K = 0.5$ is assumed). This efficiency can, of course, be increased through the use of sectorization of the cell.

• The processing gain plays an important role in determining the efficiency of the system. Typically, the larger the processing gain, the better is the ability of the system to mitigate the multipath; for example, smaller delay spreads can be resolved by a RAKE receiver. Also, the "averaging ability" of the system is improved, since more users are sharing the frequency band. This will reduce the outage probability. On the other hand, larger processing gain can cause certain implementation complexities such as a larger power drain; further, there may be a limitation to the allocated spread bandwidth imposed by a regulatory agency. Typical values for G can range from 100 to 1,000.

• Since DS/CDMA systems are invariably interference limited, there exists the option of trading off capacity for coverage. In other words, as a cell's loading (i.e., M) decreases, its coverage can increase. This can be seen from Eq.(1); decreasing M in the denominator of (1) allows for a decrease in E_b without causing the overall denominator to increase. This is helpful in cellular system construction where the spatial distribution of the load is not uniform (which, in practice, is the case). In the core of a typical system, the cells are small, and are heavily loaded. At the fringes of the system, where the load is light, the cells can become large, and the coverage ability is increased. Alternatively, when the system is lightly loaded, less RF power is needed, which is advantageous, for example, in terms of battery life.

• Finally, we make the following observations. The estimate of Eq. (2) is very optimistic, in that it ignores several key effects, such as imperfect power control, and imperfect interleaving. A more realistic capacity estimate, accounting for some of these sources of degradation, is presented below in the section on performance. Further, the use of perfect rectangular chip shapes yields an optimistic result, since it does not account for bandlimiting. All of these latter effects have been treated in the literature (e.g., [12,-16]), and the severity of them depends upon the specific scenario. For example, if the spread bandwidth does not exceed the coherence bandwidth of the channel, multipath fading will cause noticeable degradation because the fading will be flat rather than frequency selective. Similarly, if propagation delay precludes the use of accurate power control, as in a satellite link, this too can cause a significant reduction in capacity.

Also, the simplified analysis presented above assumed a uniform distribution of users (i.e, M users in each cell). For a nonuniform distribution, orthogonal multiple accessing techniques such as TDMA and FDMA have an advantage, because they can take slot assignments from a lightly used cell and transfer them to a more heavily used adjacent cell. For a nonorthogonal CDMA system, this option also exists, but to a much more limited extent. In other words, as the total number of active users become concentrated in a single cell, an orthogonal system will outperform a nonorthogonal system.

Current DS Designs

To illustrate the actual designs currently being proposed, we will consider both the IS-95 standard [8], and the proposed broadband CDMA (BCDMA) system [17]. The former system employs BPSK data and QPSK spreading on the forward link, in conjunction with synchronous intracell transmissions, which means that the forward link is made orthogonal so that there is essentially no intracell multiple access interference. (There will, of course, be intracell cell interference because of multipath, since the delayed returns are no longer orthogonal to the dominant path.) The chip rate is about 1.228 Mchips/s, the spread bandwidth is about 1.25MHz, and forward error correction is employed with a rate - $\frac{1}{2}$ convolutional code. The use of a spread pilot tone allows coherent detection to take place.

On the reverse link, 64-ary binary orthogonal signaling is used in conjunction with a rate - $\frac{1}{3}$ convolutional code. The receiver employs non-coherent detection, and an adaptive power control scheme is employed to attempt to have all intracell users arrive at the base with the same received power level.

By contrast, the proposed BCDMA system spreads the data over a 10 MHz bandwidth, and, in one version [17], overlays the current analog AMPS system. Both forward and reverse links employ QPSK data with BPSK spreading, and coherent reception is used on both links. This latter characteristic is achieved by using spread pilot tones on both links. The forward link uses a common spread pilot tone, and the reverse link is designed so that each user transmits a spread pilot tone at a power level 6 dB below that of its information-bearing signal. Also, rate - $\frac{1}{2}$ convolutional coding is employed on both links. The chip rate is 8 Mchips/s, and, as with the IS-95 based design, adaptive power control is required on the reverse link.

While there are numerous detailed differences between the two systems (e.g., type of modulation, algorithm for implementing adaptive power control, etc.), the fundamental difference is the extent to which the spectrum is spread. As pointed out in [18], an advantage of a narrower spread is the flexibility to use a non-contiguous spectrum. That is, if a total spectral band of, say, B Hz is available, but not as a single contiguous band, a system employing a narrow spread bandwidth might be able to constructively use all the bandwidth by operating in an FDMA/CDMA mode. Alternately, advantages of a wider spread include the ability to more effectively use the multipath and operate with interference averaged over more mobiles for enhanced performance [18, 19].

The proposal to employ BCDMA is based upon spreading the spectrum of each user over the entire available RF bandwidth.

The capacity of DS/CDMA is dependent upon the accuracy in power control, voice activity gain, antenna gain, and other techniques of the physical layer.

There is another interesting difference between the two approaches, and that involves the manner in which one transitions from the current analog cellular system to one based upon CDMA. In the case of narrowband CDMA, the transition is basically one of replacement, i.e., to deploy a set of CDMA users in a given frequency band, that band must first be vacated by the narrowband signals. In the case of BCDMA, the spread spectrum waveforms overlay the narrowband signals, meaning that both sets of users simultaneously occupy the same frequency band. This is possible, because with AMPS, only $\frac{1}{7}$ ($\approx$ 14 percent) of the frequencies are used in any given location. Initially, when all AMPS users are present, the capacity of the BCDMA system is constrained; as time evolves, and more users convert to the digital technology, the number of AMPS users decreases, and the capacity of the BCDMA system increases.

The concept of an overlay has been proposed for both the 1.8-GHz PCS band [20] and the cellular band [17]. Clearly, the key consideration in such an overlay is whether the interference that one set of users imposes upon the other is within tolerable limits, and this is ensured by using notch filtering techniques in the CDMA transmitters and receivers. For example, if a narrowband notch is placed in the transmit spectrum of each CDMA waveform such that it coincides with the location of a narrowband signal, the interference from those CDMA waveforms to the overlaid narrowband waveforms will be minimized [17, 18]. Alternately, if narrowband notch filtering is employed at each CDMA receiver, the interference from the narrowband waveforms to the BCDMA signals will be reduced [22-24].

In Europe, DS/CDMA for the future Universal Mobile Telecommunications System (UMTS) has been studied through the Code Division Testbed (CODIT) [25] research project within the framework of the Research in Advanced Communications in Europe (RACE) program. CODIT is intended to be the basis of a third-generation system. The "generation count" considers the analog system as first generation, the current (GSM in Europe, IS-54 in the United States, and PDC in Japan) as well as near future (IS-95 in the United States) digital systems as second generation. Third generation extends beyond that to, say, the year 2000. As such, CODIT deviates in some key respects from second generation DS/CDMA systems. Its main requirement is variety of services with flexible coexistence.

Interference Cancellation

The capacity of DS/CDMA is dependent upon the accuracy in power control, voice activity gain, antenna gain, and other techniques of the physical layer. In order to increase capacity, two main obstacles have to be overcome. First, conventional spread spectrum matched filter receivers are suboptimal in the presence of multiple access interference. Second, they are highly sensitive to the near-far effect. To address these issues, many researchers have recently proposed and investigated adaptive spread spectrum receivers that can potentially combat these impediments.

The conventional receiver demodulates each signal with a single-user detector consisting of a matched filter followed by a threshold detector, and MAI caused by cross-correlation of the desired and interfering spreading sequences severely limits the error performance, in particular, because of the near-far problem. If the ultimate goal is to approach the information theoretic capacity, it is necessary to utilize a maximum likelihood multiuser receiver (MLMR) [26]. The MLMR consists of a bank of matched filters followed by a maximum likelihood sequence detector. Such a receiver displays resistance to the near-far problem, and thus results in error performance considerably superior to that of the conventional single user receiver. However, such schemes are prohibitive because of their extreme implementation complexity.

To achieve a reasonable level of complexity with some compromise in performance, several multiuser receivers based on interference suppression and cancellation techniques have been studied [27-30]. These receivers employ a multiuser detection strategy based on a set of appropriately chosen linear transformations on the outputs of a matched filter bank. The computational complexity increases linearly with the number of users, as opposed to exponentially in the MLMR. The receivers are often "near-far resistant," thus reducing the need for strict power control.

MAI in the literature on CDMA has generally been modeled as Gaussian noise. The proposed adaptive receivers that exploit the cyclostationary character of the interference can achieve considerably improved error rate. For example, consider the system of [31]. The receiver estimates replicas of the transmitted sequence for each user by multiplying the received signal with its despreading sequence. Then each estimated sequence is passed through a digital filter adaptively matched to the channel in order to compute the estimates of the MAI for each user, and the MAI estimates are subtracted from the original received signal. This MAI detection and cancellation is combined with an adaptive array antenna for temporal and spatial filtering.

Since all received signals from mobile terminals are demodulated at a base station, a multiuser receiver based on MAI cancellation is more feasible for the reverse link than it is for the forward link. An advanced DS/CDMA receiver, using such interference cancellations, might be considered as a possible candidate for a future generation CDMA cellular system. However, at present, more time is required to prove the feasibility of these interference cancellation techniques.

Frequency Hopping Cellular CDMA

We now consider the use of frequency hopping for wireless communications. Since, for practical reasons, virtually all FH systems proposed for this type of application are slow FH (SFH) systems (i.e., multiple bits are transmitted on each hop), we limit our discussion to SFH.

SFH was first proposed for cellular systems in the literature in [32, 33], and the use of such systems has been discussed more recently in [34]. The motivation behind FH is the same as DS, i.e., spread the spectrum, so that frequency diversity is obtained to help mitigate the multipath, and diversify the interferers as seen by any given user, so that the latter user's performance is determined by an average pattern of interference. This is achieved with the help of a sufficient degree of error correction coding, combined with bit interleaving and a proper choice of hopping sequences that cause fading, as well as the interferers, seen by any user, to be uncorrelated from hop to hop. Compared to DS, where it can be said the diversity effect is gained in parallel, with FH the diversity is achieved sequentially.

The main difference in the performance between DS and FH is linked to the different forms the intracell interference take in the two methods. While in DS, intracell interference is typically the dominant source of interference, for FH it can be approximately orthogonalized such that users within a cell do not interfere with one another. This can be accomplished by choosing hopping sequences that are orthogonal within the cells, combined with advancing/retarding the transmit time of the mobiles so that the time of arrival at the base receiver of the uplink bursts are time-synchronized. Clearly, FH/CDMA enjoys the same universal one cell reuse pattern as does DS/CDMA. The potential advantages of this type of FH system are as follows.

Less Total Interference — In [10], the ratio of intracell-to-intercell interference was estimated to be about two to one, under the assumption of a fourth power propagation law. It follows that the major interference is intracell, and cancelling it will mean a higher capacity potential.

Solving the Reverse Link, Near-Far Problem — Because the multipath delay spread is typically much smaller than the duration of a hop, the reverse link can be made approximately orthogonal in a given cell by appropriately aligning the transmission time of each burst, and allowing for some guard time. As the channels are orthogonal, the extensive power control arrangement, which is the basis for the DS system, is not needed. Power control for FH is employed only to reduce intercell interference, but for this purpose, the requirements on the power control system do not extend beyond the level currently carried out with existing systems, such as AMPS and GSM.

External Jamming — Most existing users in any given frequency band are narrowband. A certain level of out-of-band spurious emission is unavoidable and, in fact, is legally permitted. Such jamming from existing services to new mobile services, can, at times, be more benign to an FH system than it is to a DS system. That is to say, on the one hand, jamming to either system by an external interferer does not knock out specific channels; however, jamming reduces capacity. With an instantaneously narrowband system, such as FH, a saturation effect exists, i.e., a narrowband jammer cannot take out more capacity than determined by the ratio of its bandwidth to the system bandwidth. In DS, there is no such saturation, so that, in the limit, when the interferer gets very close to a base site, it can significantly degrade the capacity of the entire cell.

Frequency Agility — While a given DS/CDMA signal requires a wide and contiguous frequency band, FH is agile in the sense that the spectrum does not have to be contiguous. In particular, this allows the implementation of FH/CDMA for private land mobile operation (FCC, Part 90), where operational licenses are given on the basis of single and isolated narrowband channels. Furthermore, the instantaneous narrowband characteristic of FH/CDMA is beneficial in restricting out-of-band emissions. DS/CDMA, being a wideband signal, requires considerable guard bands at the fringes of its spectrum in order to control the level of emissions into adjacent bands.

For FH systems currently under consideration, typical values of the various parameters are summarized below:

- Coding is usually done with a rate one-half code, and a soft decoding metric is employed at the receiver.
- Bit interleaving is performed over 10 to 20 hops.
- If we allow a time delay of 40msec for the interleaving, a hopping rate of 250 hops/s (for an interleaving depth of 10 hops) to 500 hops/s (interleaving depth of 20 hops) is obtained.
- Power control is used to reduce intercell interference. As such, a dynamic range of, say, 30 dB, with a step of 3 dB, similar to what is employed by current analog systems, is sufficient. Also, voice activity detection can be employed to increase capacity.
- Hopping sequences are assigned for both intracell orthogonality and minimum correlation with respect to intercell interference [35]. The latter means that any two users in adjacent cells interfere only at one hop during the period of the hopping sequence.
- As the hop durations are relatively short compared to the coherence time of the Rayleigh fading, if we assume synchronized base stations, the individual hop's carrier-to-interference ratio (C/I) is fixed during the hop's duration. Hence, each hop is characterized by a C/I, and the link's performance is determined by a histogram of C/I values. The individual hop's C/I can be estimated and used as side information in a soft decision decoder. For example, one can use this procedure to erase hops which have been severely hit by adjacent cell users.

In principle, the hopping carrier can be either a single channel or multiple channels through a time division structure. The latter has an implementation advantage, as the mobile is engaged in transmission or reception only part of the time, and so the remaining idle time can be used to allow the synthesizer more acquisition time. For example, with GSM, the carrier is divided into eight timeslots. A mobile transmits on one timeslot, receives on another, and uses a third one for monitoring other carriers for mobile-assisted handoff. However, the ability of the system to average interference is reduced if timeslot hopping is not added. By timeslot hopping, we mean that the users multiplexed on a single time division carrier hop from one timeslot to another (from one TDMA frame to another) to ensure that there is sufficient randomization with respect to the interference seen by any user caused by the other users.

Cellular systems in the field which allow for the use of SFH are GSM and DCS 1800. The latter system is basically GSM shifted in frequency to the 1800 MHz range. In GSM, the carriers are separated by 200 kHz, and each multiplexes eight (voice coder) full-rate channels by time division. A TDMA frame is 4.615-ms long. Frequency hopping is

The main difference in the performance between DS and FH is linked to the different forms the intracell interference take in the two methods.

A fundamental difference between the satellite channel and the terrestrial channel is related to the overall time delay a signal experiences when transmitted over the channel.

specified as a system option, and is used on the basis of a TDMA frame, resulting in a hopping rate of 217 hops/s.

The air interface of GSM could benefit from some modifications [36] if the full advantages of its FH are to be achieved. Also, current GSM recommendations hop only the traffic carriers, leaving the common control carrier fixed. This makes the system easier to implement, particularly with regard to system acquisition by the subscriber unit. In [37], a technique to hop the GSM common control channel is described.

Performance on Terrestrial-Based and Satellite-Based Wireless Links

CDMA has been proposed for both terrestrial links and satellite links. However, there are key differences in the characteristics of the two types of links relative to the way they affect a CDMA system. We now show how the ideas presented above affect both terrestrial-and satellite-based systems.

Multipath

For either type of link, the two characteristics most relevant to CDMA performance are the coherence bandwidth, which is roughly the inverse of the multipath delay spread, and the fading amplitude statistics. The latter characteristics are typically described as being either Rician or Rayleigh, depending upon whether a dominant line-of-sight path does or does not exist between transmitter and receiver.

Terrestrial-Based Systems — For terrestial systems, we make a distinction between indoor and outdoor links; for outdoor links, we further distinguish between street level antennas (microcells) and high elevation antennas (macrocells), and also note that rural and urban macrocells display different delay spread characteristics [20, 38-40]. For a typical indoor environment, the coherence bandwidth is about 2 to 5 MHz for large offices and rooms (e.g., banks), and can be greater than 10 MHz for smaller rooms. Thus, a CDMA system designed to take advantage of the multipath enhancement characteristics of DS waveforms should be spreading over an RF bandwidth on the order of 10 MHz. Alternately, for outdoor terrestrial links, the multipath delay spread is typically much larger, so the coherence bandwidth is correspondingly smaller. This is especially true in rural areas, where the multipath delay spread is often on the order of several microseconds; thus, the spread bandwidth of the transmitted signal can be on the order of 1 MHz and multipath enhancement can be realized. At the other extreme, namely in the downtown area of most major cities, the multipath delay spread is once again small when antennas are placed on ground level, and thus a wider spread bandwidth (i.e., a higher chip rate) is required.

Satellite-Based Systems — Now consider a typical satellite link [14, 41]. There are many mobile satellite systems currently being proposed/designed, some of which employ CDMA. Of those that are envisioning using CDMA, most are low earth orbiting satellite (LEOS) systems, and for such systems, the multipath delay spread is on the order of 100 nanoseconds because of the inclined path of propagation, implying that the coherence bandwidth is about 10 MHz. Since the proposed spread bandwidth of most of these LEOS systems is less than 10 MHz, and can be as low as 1.25 MHz, they will be unable to take advantage of the multipath enhancement potential of the spread spectrum waveforms. Rather, they can attempt to implement a dual satellite diversity system, whereby each mobile unit will be in view of two satellites; the mobile will transmit to both satellites simultaneously, and the two satellites will then retransmit the waveforms to a fixed earth station, where they will be combined in a manner which yields a diversity gain.

Path Delay/Power Control

A second fundamental difference between the satellite channel and the terrestrial channel is related to the overall time delay a signal experiences when transmitted over the channel; this time delay affects the ability of the system to implement closed-loop power control. Recall that in order for the currently designed DS systems to function, there must be accurate power control so that all waveforms arrive at the receiver with roughly the same power. This power control invariably has both an open loop and a closed loop component; the open loop power control is useful for overcoming variations in power caused by both differences in distance and differences in levels of shadowing, since both such effects are roughly reciprocal (i.e., the loss in signal power from the mobile to the base is about the same as the loss from the base to the mobile). But the loss due to multipath fading is not reciprocal; because the frequency band for base-to-mobile transmission is separated from that for mobile-to-base transmission by an amount that exceeds the coherence bandwidth of the channel, the two bands experience independent fading. Therefore, closed loop power control must be used for this latter purpose. However, if the round trip time delay is greater than the duration of most of the deep fades, this latter technique will not be effective.

The proposed CDMA LEOS systems operate in a "bent pipe" mode, in the sense that all the satellite does is act as a transponder, which frequency translates the radio links. That is, a signal originating, say, at a mobile, goes from the mobile to the satellite, and then down to a gateway on the ground, and so the propagation delay far exceeds that of a terrestrial channel. For example, as pointed out in [42], the round trip time delay for LEOS systems is on the order of tens of milliseconds, yet for mobile velocities as low as 10-to20 miles-per-hour, fade rates in excess of 0.5 dB per millisecond have been measured, and individual fade events can vanish within the duration of a round trip delay. Therefore, while closed loop power control is effective in terrestrial systems, where the time delay is small, it is significantly less useful over LEOS links, and, of course, even less useful on a geostationary link.

Interleaving

From the previous paragraph, it is evident that for either type of system, the performance is velocity-related; slowly moving subscriber units are more accurately power-controlled than are rapidly moving subscriber units. However, accuracy of power control is not the only velocity-related issue. Because the error correction codes employed in these systems are designed to correct random errors, and because the effect of a fading channel is to cause correlated errors, interleaving must be employed at the transmitter, so that a deinterleaver at the receiver can be used to disperse the correlated symbols. How effective the interleaver/deinterleaver is in accomplishing this task is a function of its span; that is, if it is large relative to the number of consecutive fading symbols (determined by the duration of a fade), then it will be very effective, but if the number of correlated symbols exceeds the capacity of the interleaver, the deinterleaver will be relatively ineffective in enhancing performance.

Consider now the situation of a mobile unit. If the velocity of the unit is large, the multipath channel will be changing rapidly, and the interleaver/deinterleaver will function properly; however, at the other extreme, if the mobile unit is stationary, the interleaver will be relatively useless, because it cannot, as a practical matter, be made large enough to result in the dispersion of errors of an essentially constant channel. Thus, we see that there are counterbalancing effects on the performance of a CDMA system as a result of motion of the mobile unit. At slow speeds, the power control system functions well, but the interleaver is not effective; at high speeds, the interleaver is very effective, but the power control system does not perform well.

As an illustration, we will attempt to quantify the degradation due to imperfect power control by using the model and results of [14]. We assume a flat Rayleigh fading channel for shadowed users, and a flat Rician channel with a ratio of specular power-to-scatter power of 10 dB for non-shadowed users. Note that the fading on either the shadowed or the non-shadowed users can be interpreted as residual fading after the closed loop power control system has tracked out as much of the instantaneous fade as possible.

Assuming M instantaneous users per cell (i.e., ignoring voice activity detection) such that 30 percent of them are shadowed at any instant of time; a system employing a rate 1/3 convolutional code with soft decision decoding; perfect side information regarding the state of the channel; and DS spreading with a processing gain of $G = 150$, we show the effect of imperfect power control in Fig. 1. These curves are based upon the analysis presented in [14], and correspond to an interbeam interference factor (analogous to intercell interference in a terrestrial system) of $K = 1$ in Eq. (1). Because the power control can compensate for shadow loss (as opposed to being able to track out the more rapidly varying multipath fading), the average received signal power of the shadowed user is the same as that of the unshadowed user. However, the performance of a shadowed user is still much worse than that of an unshadowed one, because the former one experiences Rayleigh fading while the latter one experiences Rician fading. To equalize the performance of the two sets of users, one can overcompensate the shadowed user (i.e., boost its power by an amount greater than the power lost due to the shadowing). However, note that such overcompensation for a shadowed user results in a decrease of system capacity, since it results in additional interference to each non-shadowed user.

There are three curves shown in the figure, corresponding to values of M of 10, 20, and 40. The ordinate axis is the probability of error of a *shadowed* user, and the abscissa is the standard deviation of the power control error in decibels; all the results shown correspond to a ratio of energy-per-bit-to-thermal-noise power spectral density of 7 dB. Note that the effect of thermal noise is usually much more pronounced on a satellite link than it is on a terrestrial one, and thus we have chosen a nominal value of $\frac{E_b}{N_0} = 7dB$, as opposed to a much larger value. In particular, asymptotic capacities corresponding to no thermal noise are much less meaningful for satellite communications. To interpret the results, assume a decoded bit error rate (BER) of .001 is desired, corresponding to acceptable quality of voice transmission. From the figure, a standard deviation of 2 dB will limit the number of instantaneously active users per cell to about 10; if the standard deviation can be reduced to 1.5 dB, the capacity would double to about 20. Note that even with perfect power control (i.e., a standard deviation of zero), the system could not support 40 simultaneously active users. In particular, the value of capacity as predicted by Eq. (2) is now seen to be quite optimistic, as was indicated previously.

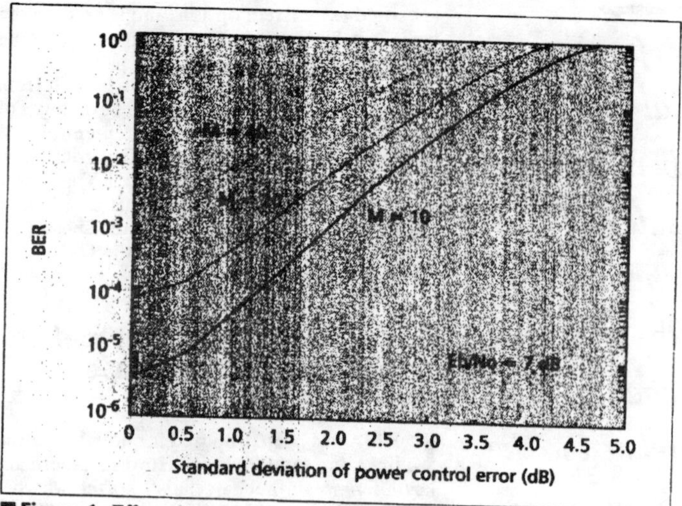

■ Figure 1. *Effect of imperfect power control on CDMA performance (from [14]).*

In order to achieve a larger capacity, one can employ diversity. If the overall system constraints allow for an increased spread bandwidth, this diversity can be achieved by spreading beyond the coherence bandwidth and then employing a RAKE receiver. If such additional spreading is not feasible, explicit (e.g., space) diversity can be used. An example of the use of such diversity is shown in Fig. 2, also taken from [14]. The conditions are the same as those which resulted in the curves of Fig. 1, except that dual satellite diversity with maximal-ratio combining is incorporated in the receiver. Note that now, if the standard deviation of the power control error is 2 dB, the system can support 40 simultaneously active users, and 10 users can be supported with a standard deviation as large as 3 dB.

Spread Spectrum and the ISM Bands

In Part 18 of the FCC regulations, three frequency bands are designated for equipment that generate, and use locally, RF energy for industrial, scientific

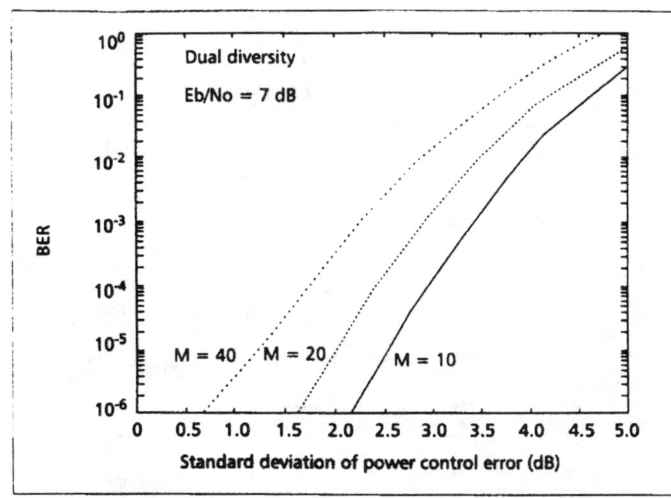

■ **Figure 2.** *Effect of imperfect power control when dual satellite diversity is employed (from [14]).*

and medical (ISM) applications, excluding telecommunications. The bands are 902-928 MHz, 2400-2483.5 MHz, and 5725-5850 MHz. Typical applications are industrial heating equipment, microwave ovens, medical diathermy equipment, and ultrasonic equipment (RF energy used for excitation). Since the RF radiation of these devices is localized to the immediate vicinity of the devices, it was determined that these same frequency bands were compatible with telecommunication applications for residences, offices, local area networks, etc. Further, an unlicensed operation per Part 15 of the FCC regulations was sought. Unlicensed operation has the advantage of rapid deployment, but also has the disadvantage of not ensuring interference-free operation. Spread spectrum modulation alleviates this latter vulnerability to interference. Regulation No. 15.247 set forth the parameters of the spread spectrum signals, both DS and FH, which opened these bands for telecommunication applications.

The first band to receive commercial attention was 902-928MHz. Products such as local area data networks, automatic vehicle location devices, and cordless telephones for residential use can be found on the market. A recent thrust of emerging systems is local area data networks in the band 2400-2483.5MHz. The main ISM equipment in the latter band is microwave ovens. This band is wider than the 900 MHz band and enjoys an international flavor in view of the worldwide distribution of microwave ovens. Also, currently, the band is virtually free of communication systems.

An IEEE 802.11 committee has been established to determine a standard for both the physical and the multiaccess communication layers for data LAN applications in the 2.4 GHz band. The basic data rate is 1 Mb/s, with an effort to double the rate to 2 Mb/s. Both DS and FH based physical layers are being pursued.

Discussion

It has become customary to express the capacity of the forthcoming digital systems in terms of the factor by which they increase the capacity over that of the current AMPS analog technology with a seven-cell reuse pattern. This probably was motivated by the CTIA specifying the target capacity of the future technologies in such terms. However, it appears that such a single number cannot be used to make a full comparison between the systems. For example, for CDMA systems, the capacity is a "soft" number, i.e., the capacity, which is set to a target performance of the users, can, in principle, be extended when necessary at the expense of some degradation in performance. This is not the case for the narrowband technologies, where a "hard" capacity holds. On the other hand, CDMA systems operate under the principle of equal performance to all users, and, in the case of full loading, all users are served at the minimal acceptable performance level. For the narrowband technologies, the performance of the users is not equalized; hence, when the capacity limit is reached, only a small fraction of the users will be subjected to the minimal performance level.

As previously mentioned, the users' spatial distribution is another factor that can change the comparison between DS/CDMA and a narrowband technology. Indeed, when the users' distribution is concentrated in "hot" spots, narrowband technologies have an advantage, since, being orthogonal, they enjoy higher capacity in isolated cell situations. On the other hand, for a uniform distribution, DS/CDMA is advantageous.

Other factors that can affect the comparison results are the multipath characteristics of the link and the radio link time delay. The nature of the user information also makes a difference. Bursty activity, like voice and certain data sources, lead to an advantage for a CDMA implementation. On the other hand, largely differing user requirements in terms of service (e.g., information rate, BER) tend to be better served by an orthogonal system. Further, the percentage of shadowed users in any given environment affects the capacity. In the example presented above, we assumed that 30 percent of the users were shadowed; had we assumed a different value, our capacity estimates would have also changed. Since introducing spread spectrum can add complexity, and thus cost, to a system, its use should be predicated on an anticipated net benefit to the system.

With the above ideas in mind, in this article we present an overview of the characteristics of CDMA as it is currently being envisioned for use in wireless communications. As should be evident from the previous sections, there are many considerations in the design of such systems, and, indeed, there are multiple designs being proposed.

For some DS systems, the designs are similar in the sense that they employ an orthogonal forward link within a given cell, but an asynchronous (and, thus, nonorthogonal) reverse link. On the other hand, the designs can also vary greatly. For example, in the cellular band used in the United States, the IS-95 standard uses a hybrid FDMA/CDMA approach of dividing the frequency band into multiple disjoint segments and then employing CDMA with a chip rate of about 1.228 Mchips/sec in each frequency band. However, the proposal to employ BCDMA is based upon spreading the spectrum of each user over the entire available RF bandwidth. With respect to FH, the European-based GSM system has the option of slow hopping, but that option has not been universally exploited as of yet.

Since CDMA has been proposed for use in satellite-based systems as well as terrestrial systems, we have attempted to point out the key differences in using CDMA over those two channels. In particular, it was pointed out that the increased propagation delay, coupled with the smaller multipath delay spread, make the use of CDMA over a LEOS channel inherently less attractive than it is over a terrestrial channel.

However, for a terrestrial link, CDMA appears to be especially well-suited. The combination of RAKE-enhanced performance over multipath fading channels, the simplicity with which voice activity detection can be employed, and the flexibility of complete frequency reuse within adjacent cells, suggests that CDMA will continue to be a strong competitor for wireless applications.

References

[1] D. L. Schilling, R. L. Pickholtz, and L. B. Milstein, "Spread spectrum goes commercial," *IEEE Spectrum*, pp. 40-45, Aug. 1990.

[2] R. L. Pickholtz, L. B. Milstein, and D. L. Schilling, "Spread spectrum for mobile communications," *IEEE Trans. Vehicular Tech.*, VT-40, pp. 313-322, May 1991.

[3] J. G. Proakis, Digital Communications, (McGraw-Hill, 1989).

[4] R. L. Pickholtz, D. L. Schilling, and L. B. Milstein, "Theory of spread spectrum communications - A tutorial," *IEEE Trans. Commun.*, pp. 855-884, May 1982.

[5] M. K. Simon *et al.*, "Spread Spectrum Communications, Volumes I-III, (Computer Science Press, 1985).

[6] R. F. Ziemer and R. L. Peterson, Digital Communications and Spread Spectrum Systems, (Macmillan, 1985).

[7] W. C. Y. Lee, "Overview of cellular CDMA," *IEEE Trans. Vehicular Tech.*, VT-40, no. 2, pp. 291-302, May 1991.

[8] K. S. Gilhousen *et al.*, "On the capacity of a cellular CDMA system," *IEEE Trans. Veh. Tech.*, VT-40, no. 5, pp. 303-312, May 1991.

[9] P. Jung, P. W. Baier and A. Steil, "Advantages of CDMA and spread spectrum techniques over FDMA and TDMA in cellular mobile radio applications," *IEEE Trans. Vehicular Tech.*, VT-42, no. 3, pp. 357-364, Aug. 1993.

[10] T. S. Rappaport and L. B. Milstein, "Effects of radio propagation path loss on CDMA cellular frequency reuse efficiency for the reverse channel," *IEEE Trans. Vehicular Tech.*, VT-41, pp. 231-242, Aug. 1992.

[11] P. Monsen, "Multiple access capacity in mobile user satellite system." To appear in *IEEE JSAC*.

[12] A. J. Viterbi, "Very low rate convolutional codes for maximum theoretical performance of spread spectrum multiple-access channels," *IEEE JSAC*, pp. 641-649, May 1990.

[13] M. E. Davis and L. B. Milstein, "Anti-jamming properties of a DS-CDMA multiple access noise rejecting receiver," 1993 IEEE Military Comm. Conf., pp. 1008-1012.

[14] B. R. Vojcic, R. L. Pickholtz and L. B. Milstein, "Performance of DS-CDMA with imperfect power control operating over a low earth orbiting satellite link," *IEEE JSAC*, pp. 560-567, May 1994.

[15] L. F. Chang, "Dispersive Fading Effect in CDMA Radio Systems," IEEE ICUPC '92, Dallas, Texas, Sept. 28 - Oct. 2, 1992 (see also *IEEE Electronics Lett.*, vol. 28, no. 19, pp. 1801-1802, Sept. 10, 1992).

[16] S. Ariyavisitakul, "Effects of Slow Fading on the Performance of a CDMA System," *IEEE Electronics Lett.*, Issue 17, Aug. 19, 1993.

[17] D. L. Schilling and E. Kanterakis, "Broadband-CDMA overlay of FM on TDMA in the cellular system," 1992 IEEE Global Telecomm. Conf., Mini-Conference Volume, pp. 61-65.

[18] T. Eng and L. B. Milstein, "Capacities of hybrid FDMA/CDMA systems in multipath fading," *IEEE JSAC*, pp. 938-951, June 1994.

[19] D. L. Noneaker and M. B. Pursley, "On the chip rate of CDMA systems with doubly selective fading and rake reception," *IEEE JSAC*, pp. 853-861, June 1994.

[20] L. B. Milstein *et al.*, "On the feasibility of a CDMA overlay for personal communications networks," *IEEE JSAC*, vol. 10, pp. 655-668, May 1992.

[21] M. Davis and L. B. Milstein, "Filtered spreading sequences for interference avoidance," International Conference on Universal Personal Communications, Oct., 1991.

[22] J. Wang and L. B. Milstein, "CDMA overlay situations for microcellular mobile communications," To appear in *IEEE Trans. Commun.*

[23] K. G. Filis and S. C. Gupta, "Coexistence of cellular CDMA and FSM: Interference suppression using filtered PN sequences," IEEE 1993 Global Telecomm. Conf., pp. 898-902.

[24] L. A. Rusch and H. V. Poor, "Narrowband interference suppression in CDMA spread spectrum communications," *IEEE Trans. Commun.*, April 1994, pp. 1969-1979.

[25] A. Baier *et al.*, "Design study for a CDMA-based thrid-generation mobile radio system," *IEEE JSAC*, May 1994, pp. 733-743.

[26] S. Verdu, "Minimum probability of error for asynchronous Gaussian multiple-access channels," *IEEE Trans. Info. Theory*, IT-31, no. 1, pp. 85-96, Jan. 1986.

[27] Z. Xie, R. T. Short and C. T. Rushforth, "A family of suboptimum detectors for coherent multiuser communications," *IEEE JSAC*, SAC-8, pp. 683-690, May 1990.

[28] R. Kohno *et al.*, "An adaptive canceller of co-channel interference for spread spectrum multiple access communication networks in a power line," *IEEE JSAC*, SAC-8, no. 4, pp. 691-699, May 1990.

[29] M. K. Varanasi and B. Aazhang, "Multistage detection in asynchronous code-division multiple-access communications," *IEEE Trans. Commun.*, COM-38, pp. 509-519, Apr. 1990.

[30] Y. C. Yoon, R. Kohno, and H. Imai, "A spread-spectrum multiaccess system with co-channel interference cancellation over multipath fading channels," *IEEE JSAC*, SAC-11, no. 7, pp. 1067-1075, Aug. 1993.

[31] R. Kohno *et al.*, "Combination of an adaptive array antenna and a canceller of interference for direct-sequence spread-spectrum multiple-access system," *IEEE JSAC*, SAC-8, no. 4, pp. 675-682, May 1990.

[32] G. R. Cooper and R. W. Nettleton, "A spread spectrum technique for high capacity mobile communications," *IEEE Trans. on Vehicular Tech.*, VT-27, pp. 264-275, Nov. 1978.

[33] D. Verhulst, M. Mouly and J. Szpirglas, "Slow frequency hopping multiple access for digital cellular radiotelephone," *IEEE JSAC*, SAC-2, no. 4, July 1984.

[34] N. Livneh *et al.*, "Frequency hopping CDMA for cellular radio," Proceedings International Commsphere Symposium, Herzilya, Israel, pp. 10.5.1-10.5.6, Dec. 1991.

[35] A. Lempel and H. Greenberger, "Families of sequences with optimal Hamming correlation properties," *IEEE Trans. on Info. Theory*, 1, Jan. 1974.

[36] R. Meidan, "Frequency hopped CDMA and the GSM system," Proceedings of the Fifth Nordic Seminar on Digital Mobile Radio Communications, DMR-V, Helsinki, Finland, Dec. 1992.

[37] R. Meidan, D. Rabe and M. Kotzin, "Hopping the common control channel of a GSM system," Proceedings of the Sixth Nordic Seminar on Digital Mobile Radio Communications, DMR-VI, Stockholm, Sweden, June 1994.

[38] D. L. Schilling *et al.*, "Field test experiments using broadband code division multiple access," *IEEE Commun. Mag.*, pp. 86-93, Nov. 1991.

[39] D. C. Cox, "Correlation Bandwidth and Delay Spread Multipath Propagation Statistics for 910 MHz Urban Mobiles Radio Channels," *IEEE Trans. Commun.*, vol. Com-23, pp. 1271-1280, Nov. 1975.

[40] D. C. Cox, "Delay Doppler Characteristics of Multipath Propagation at 910 MHz in a Suburban Mobile Radio Environment," *IEEE Trans. Ant. and Prop.*, vol. AP-20, pp. 625-635, Sept. 1972.

[41] K. S. Gilhousen *et al.*, "Increased capacity using CDMA for mobile satellite communication," *IEEE JSAC*, SAC-4 no. 4, pp. 503-514, May 1990.

[42] C. L. Devieux, "Systems implications of L-band fade data statistics for LEO mobile systems," International Mobile Satellite Communications Conference, June, 1993.

Biographies

RYUJI KOHNO received B.E. and M.E. degrees in computer engineering from Yokohama National University in 1979 and 1981, respectively, and a Ph.D. in electrical engineering from the University of Tokyo in 1984. Since 1988, he has been an associate professor in the Division of Electrical and Computer Engineering, Yokohama National University, Yokohama, Japan. Presently, he is a vice-chair of the Society of Spread-Spectrum Technology of the IEICE and a chair of the Program Committee of the 1992 IEEE International Symposium on Spread-Spectrum Techniques and Applications. His current research interests lie in the areas of adaptive signal processing, coding theory, spread spectrum systems, and various kinds of communication systems.

REUVEN MEIDAN [SM '92] received B.Sc. and M.Sc. degrees in electrical engineering from the Technion, Israel Institute of Technology, and a Ph.D. in applied mathematics from the State University of New York at Stony Brook. He has served on the academic staff of Tel-Aviv University and later was with the University of the Witwatersrand as a professor of computer sciences. Since 1981 he has been with Motorola Israel, Tel Aviv, Israel, first as a product division manager, and from 1990, as chief scientist and director of advanced technology. His current field of interest is wireless communications.

LAURENCE B. MILSTEIN [F '85] received B.E.E. from the City College of New York in 1964, and the M.S. and Ph.D. degrees in electrical engineering from the Polytechnic Institute of Brooklyn in 1966 and 1968, respectively. Since 1976, he has been with the Department of Electrical and Computer Engineering, University of California at San Diego, La Jolla, California, where he is a professor and former department chair, working in the area of digital communication theory with special emphasis on spread-spectrum communication systems. He was the vice-president for technical affairs in 1990 and 1991 of the IEEE Communications Society, and is currently a member of the Board of Governors of both the IEEE Communications Society and the IEEE Information Theory Society.

Increased propagation delay, coupled with the smaller multipath delay spread, make the use of CDMA over a LEOS channel inherently less attractive than over a terrestrial channel.

Design Study for a CDMA-Based Third-Generation Mobile Radio System

Alfred Baier, *Senior Member, IEEE*, Uwe-Carsten Fiebig, *Member, IEEE*, Wolfgang Granzow, Wolfgang Koch, *Member, IEEE*, Paul Teder, and Jörn Thielecke, *Member, IEEE*

***Abstract*— This paper focuses on a CDMA design study for future third-generation mobile and personal communication systems such as FPLMTS and UMTS. In the design study, a rigorous top down approach is adopted starting from the most essential objectives and requirements of universal third-generation mobile systems. Emphasis is laid on high flexibility with respect to the implementation of a wide range of services and service bit rates including variable rate and packet services. Flexibility in frequency and radio resource management, system and service deployment, and easy operation in mixed-cell and multioperator scenarios are further important design goals. The system concept under investigation is centered around an open and flexible radio interface architecture based on asynchronous direct-sequence CDMA with three different chip rates of approximately 1, 5, and 20 Mchip/s.**

The presented CDMA system concept forms the basis for an experimental test system (testbed) which is currently under development. This experimental system concept has been jointly established by the partners in the European RACE project R2020 (CODIT). The paper describes the radio transmission scheme and appropriate receiver principles and presents first performance results based on simulations.

I. Introduction

CODE Division Multiple Access (CDMA) is a promising technique for radio access in future cellular mobile and personal communication systems. CDMA in cellular systems offers some attractive features such as the potential for high spectrum efficiency, soft capacity, soft handover and macro diversity, low-frequency reuse cluster size, simplified frequency planning, and easy system deployment. This has been claimed and demonstrated in various system design studies, analyses and trials [1]–[3].

However, it is still an open issue in how far CDMA is the right choice for third-generation mobile and personal telecommunication systems such as the global FPLMTS (Future Public Land Mobile Telecommunication System) and the European UMTS (Universal Mobile Telecommunication System) being jointly standardized until the end of this century [4], [5]. A CDMA-based second-generation system standard (IS-95) [6] has now been adopted besides Digital AMPS (IS-54) [7] in the United States. However, objectives and system framework of third-generation systems go far beyond what is known from second-generation systems such as IS-54, GSM [8], or IS-95, especially with respect to:

- the wide range of services and service bit rates (up to 2 Mb/s) to be supported,
- the high quality of service requirements (e.g., toll quality speech, data services with BER less than 10^{-6}),
- operation in mixed-cell scenarios (macro, micro, pico, etc.),
- operation in different environments (indoor/outdoor, business/domestic, cellular/cordless, etc.),
- the required flexibility in frequency and radio resource management, system deployment, and service provision.

Present second-generation CDMA systems are primarily designed as low-rate voice and data microcellular systems and fall short of meeting important requirements of third-generation systems.

To explore the potential of CDMA for third-generation mobile systems and to come up with a system concept based on CDMA which is particularly tailored to the requirements of such systems, a multinational research project named CODIT (Code Division Testbed) has been set up within the European RACE Program [9]. In an extensive design study, the partners and subcontractors in the CODIT project, i.e.:

Philips Kommunikations Industrie AG (Germany)
Philips Research Laboratories (United Kingdom, France)
Deutsche Forschungsanstalt für Luft- und Raumfahrt e.V. (Germany)
Ericsson Radio Systems AB (Sweden)
Ericsson Business Mobile Networks (Netherlands)
Ascom Tech Ltd. (Switzerland)
British Telecommunications PLC (United Kingdom)
Centro Studi E Laboratori Telecomunicazioni, CSELT (Italy)
ITALTEL (Italy)
IBM (France, Switzerland, United Kingdom)
Matra Communication (France)
Telefonica de Espana (Spain)
Televerket/Telia Research (Sweden)

have jointly established a CDMA system concept which forms the basis for an experimental system (testbed) currently under development. It is the purpose of this paper to give an in-depth introduction to the system concept developed for the CODIT testbed, although it is not possible in such a contribution to cover all aspects in sufficient detail. Also, it should be noted that the system concept and design being presented is not final,

Manuscript received May 25, 1993; revised November 29, 1993. This work was performed within the RACE Project R2020 (Code Division Testbed—CODIT) with financial contribution from the Commission of the European Communities (CEC).

A. Baier, W. Granzow, W. Koch, and J. Thielecke are with Philips Kommunikations Industrie AG, D-90327 Nürnberg, Germany.

U.-C. Fiebig is with the German Aerospace Research Establishment, D-82234 Oberpfaffenhofen, Germany.

P. Teder is with Ericsson Radio Systems, S-16480 Stockholm, Sweden.

IEEE Log Number 9216730.

0733–8716/94$04.00 © 1994 IEEE

but may be revised and optimized during the course of the project.

Section II gives an overview of the system concept, and Section III a detailed description of the generic transmission scheme on the radio interface. Cellular aspects are considered in Section IV. An appropriate receiver concept for the uplink with an emphasis on the channel estimation problem is presented in Section V. First results on bit error rate performance will be reported in Section VI.

II. System Overview

A. Concept of an Open Multirate Radio Interface

The radio interface of third-generation mobile systems has to be capable of handling a wide selection of services with information bit rates ranging from a few kb/s to as much as 2 Mb/s [4], [5]. Considering the issue of frequency and radio resource management in the light of third-generation multiple operator scenarios, it is obvious that this can hardly be achieved with a single radio frequency (RF) channel bandwidth. This, in particular, holds for CDMA where even moderate spreading factors applied to high information bit rates result in an enormous RF channel bandwidth difficult to provide and handle in cellular radio systems.

Rather than a single-bandwidth system, a CDMA system with multiple RF channel bandwidths seems to be the appropriate way to implement an open and flexible radio interface as required for third-generation systems. In the CODIT project, an asynchronous direct-sequence (DS) CDMA technique with three different chip rates is investigated [10], [11], i.e., $R_{c1} = 1.023$ Mchip/s, $R_{c2} = 5.115$ Mchip/s, $R_{c3} = 20.46$ Mchip/s. The chip rates R_{c1} and R_{c2} are implemented in the testbed. The three chip rates correspond to three different RF channel bandwidths of approximately 1, 5, and 20 MHz, which are referred to as narrowband, mediumband, and wideband RF channels, respectively. A generic transmission scheme on the physical layer, using techniques such as channel coding, interleaving, and spreading, is capable of mapping each information bit rate R_b offered in the system onto at least one of the three chip rates. For every service, the parameters of the transmission scheme (e.g. coding rate, interleaving depth, chip rate, spreading factor, transmit power, etc.) may be adjusted such that the specific requirements of the service to be provided are met.

Obviously, the coding and spreading factor R_c/R_b and, hence, the coding and spreading gain achieved in the proposed transmission scheme varies with the information bit rate. Given a limited RF bandwidth or, equivalently, a fixed maximum chip rate, the system loses its CDMA characteristics more and more if the information bit rate is increased.

The multirate DS-CDMA radio interface is matched to multioperator scenarios, where numerous independent network operators will coexist in the same geographical area offering different bearer and teleservice to different user groups. This is illustrated in Fig. 1, which shows the spectrum allocation and utilization for four different example networks (uplink or downlink in a frequency duplex system). A network operator,

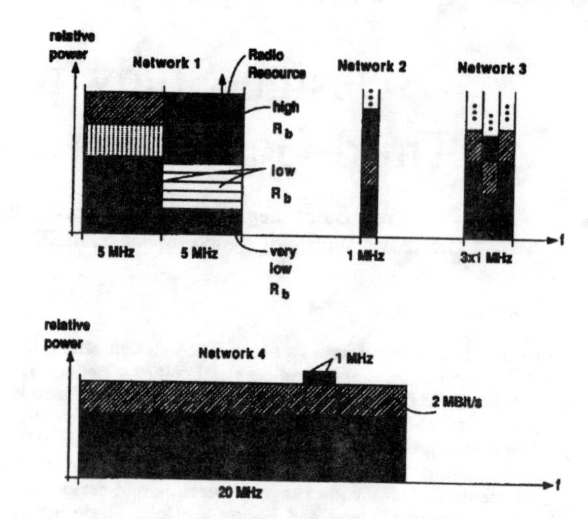

Fig. 1. Spectrum allocation and utilization for four example networks.

depending on his service profile and spectrum needs, can be assigned spectrum portions of 1, 5, or 20 MHz, multiples thereof and any combination of these. A mediumband or wideband RF channel (5 or 20 MHz) can be statically or dynamically split into several narrowband channels (e.g., 1 MHz) as indicated for network 1 in Fig. 1. To a certain extent, narrowband, mediumband, and wideband RF channels may even be overlaid in the same frequency band, cf. networks 1 and 4.

Low-rate services (e.g., speech), on the one hand, can optionally be implemented on narrowband or mediumband channels. Whereas 1 MHz could be the standard RF channel used for low-cost voice-only mobile phones, 5 MHz could be an option offering the user a higher grade of service through better exploitation of multipath and interferer diversity and allowing the network operator to fully exploit the inherent advantages of CDMA such as a single-cell cluster size [1], [2]. High-rate services (64 kb/s and above), on the other hand, will in any case require the use of RF channels of 5 or 20 MHz.

The radio resource to be managed by the network operator is defined by the allocated frequency bands and the interference budget (power spectral density) in these bands (rather than fixed frequency slots or time slots as in conventional FDMA or TDMA). The higher the bit rate and performance requirement for a specific service, the bigger the amount of radio resources (bandwidth times power spectral density) utilized for the respective connection, cf. Fig. 1. Since the interference budget in a given frequency band is not strictly limited, the system exhibits a soft capacity behavior typical for CDMA systems [1]. The soft capacity characteristic of a third-generation CDMA system is mainly determined by the service mix offered in the relevant frequency band.

B. Principle of Variable Rate Transmission

DS-CDMA has been shown to be well suited to support variable bit rate services, such as speech, in a spectrum efficient way [1]. A variable bit rate transmission generally

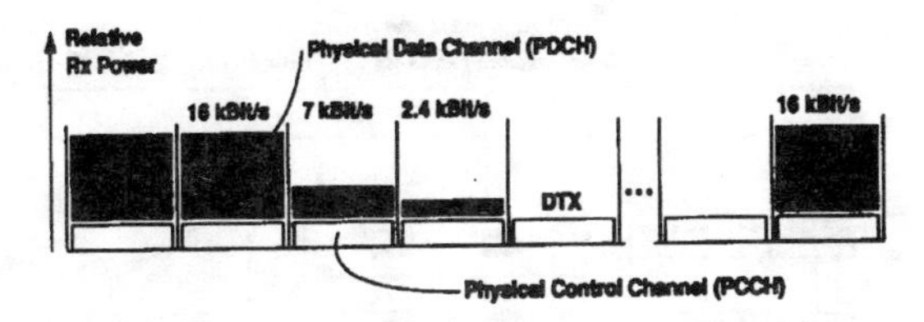

Fig. 2. Variable bit-rate transmission.

requires the provision of a control information specifying the instantaneous symbol rate.

In order to do this in regular time intervals, all physical channels are organized in frames of equal length, denoted as CDMA frames. Every frame carries an integer number of chips and an integer number of information bits. The amount of information bits per frame is denoted as a physical packet. For the experimental system, a frame length of 10 ms has been chosen which offers a sufficient degree of flexibility with respect to the data rate while the introduced transmission delay is kept to an amount which is considered to be tolerable for all expected services.

According to this frame structure, the bit rate control information is provided every CDMA frame by transmitting it on a separate physical channel. The physical channels carrying the data and the control information are denoted as Physical Data Channel (PDCH) and Physical Control Channel (PCCH), respectively. Spreading code and spreading factor of the PCCH are *a priori* known to the receiver.

Variable-rate transmission can be exploited to reduce the interference for other users. Since the chip rate is kept constant, a lower bit rate gives a higher spreading factor, thus allowing a lower transmit power. Fig. 2 illustrates this approach for a variable-rate data stream with 16 kb/s maximum bit rate.

This principle of variable-rate transmission also enables connection-oriented packet transmission. While the traffic channel carrying the user information is switched off between the packets, the link is maintained by the PCCH in order to keep track of channel variations and synchronization, to implement a closed loop power control scheme, and to monitor the link for making handover decisions.

C. Organization of Logical Channels

In a cellular mobile radio system, a number of logical channels are required which are mapped onto physical channels on the radio interface. In the CODIT testbed concept, we distinguish between Dedicated Channels, Common Control Channels and System Control Channels, cf. Fig. 3. These logical channels and their mapping onto physical channels will be briefly introduced in the sequel. A physical channel is characterized by a chip rate R_c, an RF channel (carrier frequency f_c), and a DS spreading code.

Dedicated Channels: Dedicated Channels are uniquely assigned to a specific mobile-to-base station link (uplink or downlink) when a connection is established. We distinguish between Traffic Channels (TCH) and Dedicated Control Channels (DCCH).

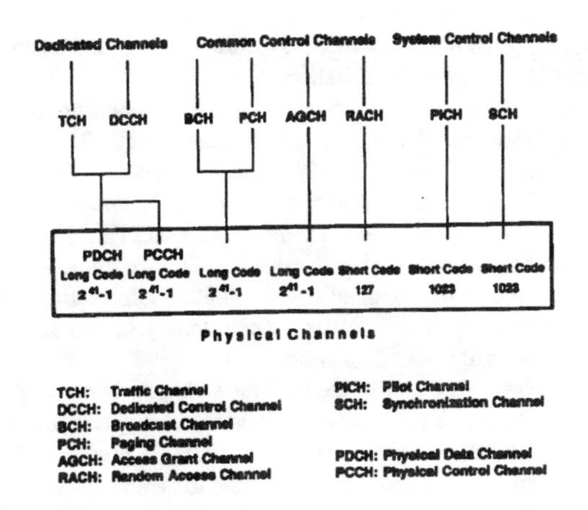

Fig. 3. Structure of logical channels.

Traffic Channels carry the user data to be transmitted on the radio interface, i.e., encoded speech, video, or data. The bit rate in the testbed is 0 to 144 kb/s, and may be variable on a frame-by-frame basis, cf. Section II-B.

Dedicated Control Channels carry all layer 2 and 3 control information to be exchanged between mobile and base station (connection control, mobility control, radio link control). The variable bit rate capabilities of the CDMA radio interface enable the implementation of a flexible DCCH with variable bit rates, e.g., in the range of 0–9.6 kb/s, which may replace the set of different fast and slow dedicated or associated control channels present in second-generation systems [7], [8].

Common Control Channels: The Broadcast Channel (BCH), the Paging Channel (PCH), and the Access Grant Channel (AGCH) are Common Control Channels used on the downlink only and available to all mobiles. They broadcast information specific for the radio cell and network rather than the link or connection, as well as paging and access grant messages. Using a variable multiframe structure, BCH, PCH, and AGCH may be multiplexed in a flexible way and can be transmitted on a joint physical downlink channel with a long PN spreading code unique to every base station.

In the CODIT testbed, a solution is preferred where a separate physical channel is devoted to the AGCH. This channel is to be decoded by the mobile station only during a random access attempt and, in connection with an associated PCCH, serves as a return channel to the mobile for closed-loop power control during random access.

The only Common Control Channel on the uplink is the Random Access Channel (RACH) used by a mobile for initial access to the system. For random access, the mobile sends a special RACH signal. The fast closed-loop power control becomes active during RACH transmission. The power control commands are transmitted over the AGCH. On the BCH, it is broadcast which short PN code (Gold sequence of length 127) should be used on the RACH. This short code enables easy and fast synchronization to the RACH signal in the base station.

System Control Channels: Two System Control Channels, i.e., the Pilot Channel (PICH) and the Synchronization Channel (SCH), are provided on the downlink to facilitate base station monitoring and identification, synchronization, and channel estimation in the mobile station.

The PICH is a separate physical channel broadcast on every RF channel and chip rate used in a radio cell. The PICH is characterized by a short PN spreading code (Gold code of length 1023 for R_{c1} and R_{c2}) which is unique to the radio cell (or base station) in the local area. This PN code is periodically sent (10 times within a CDMA frame at R_{c1} and 50 times at R_{c2}) without any modulating information data, thus simplifying pilot detection, synchronization, and channel estimation in the mobile station.

As the short PN code of the PICH is ambiguous with respect to the CDMA frame clock, an additional Synchronization Channel (SCH), synchronous to the PICH, is provided. The SCH marks the CDMA frame boundaries and gives time stamps relative to the long PN code used on the BCH. The SCH is transmitted on a separate physical channel using a short PN spreading code (length 1023) directly derived from the respective PICH code.

D. Physical Channels

TCH and DCCH are time multiplexed on layer 2 within every CDMA frame of 10 ms and then mapped onto a single physical channel denoted as the Physical Data Channel (PDCH).

Every PDCH is accompanied by a Physical Control Channel (PCCH) which carries physical layer control information. As already mentioned in Section II-B, the PCCH carries information about the actual spreading factor in the corresponding frame of the PDCH. It also conveys the information on how the PDCH frame is to be demultiplexed. On the downlink, the PCCH is also utilized for transmission of the power control information.

The PCCH has a fixed bit rate of 4 kb/s after encoding of the critical information and is transmitted frame synchronously with the PDCH (same chip rate and RF carrier as PDCH). PDCH and PCCH are distinguished by using different phases of a long PN spreading code.

III. Generic Transmission Scheme

The basis of the multirate CDMA concept considered in the CODIT project is a generic transmission scheme on the physical layer of the radio interface. Rather than optimizing and fixing the transmission scheme for a few selected services, a family of schemes with as much universality and commonality as possible is to be designed. This is achieved by starting from a common structure for the transmission scheme, with a few fixed parameters (e.g. a basic CDMA frame length of 10 ms), and introducing a set of free parameters to control and adjust individual parts of the scheme (e.g., the coding rate, interleaving depth, etc.). The free parameters have to be adjusted such that the requirements of the specific service are met.

Fig. 4 shows the block diagram of a generic transmitter (uplink or downlink) for one "user channel" allowing simultaneous transmission of traffic and control information. We denote this user channel as the dedicated information channel (DICH). Each signal processing block in Fig. 4 is characterized by a number of parameters which are adjusted by a configuration unit and determine the exact behavior of the respective block. When a connection is to be established, the radio resource manager of the system determines the desired chip rate R_c and RF carrier frequency f_c, taking into account the requirements of the requested service, the actual traffic situation in the radio network, as well as certain cell characteristics (environment, equipment constraints, etc.). Based on a service qualifier and the assigned chip rate R_c, the configuration unit determines the parameters for all signal processing blocks on the physical layer and, thus, configures the physical layer for the respective connection and service. The physical layer configuration can even be dynamically adapted to changing transmission and traffic conditions. The signal processing blocks of the transmission chain are described in the sequel.

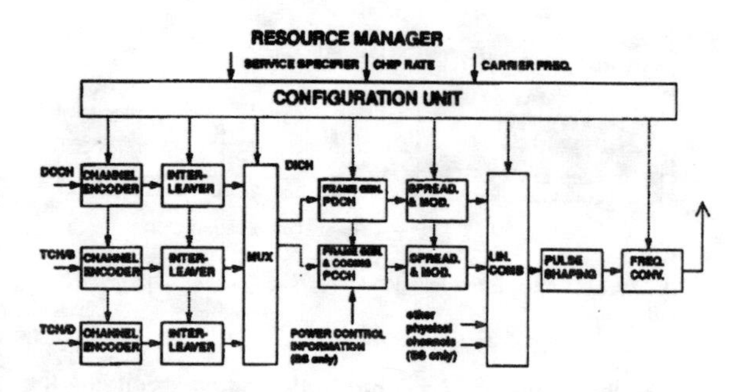

Fig. 4. Generic transmission scheme (traffic and control channels).

A. Coding and Interleaving for Speech

The coding and interleaving scheme strongly depends on service requirements and the possible channel characteristics. Different coding and interleaving schemes are used not only for traffic and control channels but also for different services supported by the traffic channels. Unequal error protection (UEP) and equal error protection (EEP) schemes for speech transmission and data transmission, respectively, are used.

In the case of speech transmission, the speech encoder unit provides three different classes of bits. Since the bits in each class—referred to as a protection class—require different protection levels, UEP techniques are suitable for speech transmission. In [12], it has been shown that the optimum code rate with respect to the required bit error rate does not depend strongly on E_b/N_0. Analysis and simulations taking into account among others synchronization aspects, interleaving depth, and the Doppler frequency show that the optimum code rate lies in the range of 1/2 and 1/8 for the system parameters in question. This result is due to the tradeoff between channel coding and spreading.

The UEP technique envisaged is based on convolutional codes having different rates. The choice of the rates of these codes depends on the significance of each protection class: the lowest rate code (1/3) is used for the most significant protection class, the highest rate code for the least significant protection class. For each speech data rate, a fixed set of codes is chosen by the system configuration unit. The various code rates are obtained by puncturing a "mother code" of rate 1/3.

Because error bursts are likely to occur due to the channel characteristics, interleaving techniques are employed. However, the interleaving depth is limited to one CDMA frame of 10 ms due to the tight delay constraints for speech transmission. In order to enable framewise decoding, tail bits are appended at the end of each CDMA frame. Decoding is accomplished using the soft-output Viterbi algorithm [13]. The soft outputs are passed to the speech decoder unit, where they can be utilized to improve error concealment techniques.

B. Coding and Interleaving for Data

In the case of data transmission, EEP techniques and a concatenated coding scheme based on inner convolutional codes of rate 1/2 and outer byte-oriented Reed-Solomon (RS) codes are employed providing bit error rates of 10^{-6} or less [14]. Inner and outer interleaver with a fixed total interleaving depth of 120 ms are used in order to mitigate the effects of fading. The inner convolutional code is decoded with soft decision techniques.

This channel coding approach can be used for both transparent and nontransparent data services. Transparent services rely solely on the forward error correction capabilities of the codes, whereas for nontransparent services radio link protocols are employed based on flow control, forward error correction, and automatic repeat request (ARQ) [14]. For nontransparent services, the channel coding approach can form the basis for a type-I hybrid ARQ scheme [15] exploiting the error detection capabilities of the RS code.

C. Time Division Multiplexing

Certain logical channels may be mapped together onto a single physical channel, either within a CDMA frame or in consecutive frames. The mapping is carried out with the multiplexer shown in Fig. 4.

In the CODIT testbed, multiplexing within a frame is permitted for the following logical channel combinations:

- TCH/Speech + DCCH
- TCH/Data + DCCH
- TCH/Speech + TCH/Data + DCCH

Data from the logical channels BCH and PCH are mapped onto one physical channel using a multiframe structure.

D. Frame Generation

The multiplexer generates two signals: the actual information bit stream and the control information, which is transmitted on the PDCH and PCCH, respectively. A frame generation unit in either signal path arranges the bits with the correct timing within the CDMA frames. The control information is encoded prior to the framing. In the case of downlink, the closed-loop power control information is inserted.

E. DS Spreading

A key element in the generic multirate CDMA transmission scheme is DS spreading. Spreading in a DS-CDMA system can be performed synchronously based on short periodic spreading codes, or asynchronously in connection with long-period pseudonoise (PN) codes. Although synchronous spreading allows the design of orthogonal or near-orthogonal signal sets if the data symbol length is fixed (or organized in powers of 2) [1], [16], this technique restricts the flexibility in applicable spreading factors and requires carefull code assignment, which becomes an essential problem during handover in a system with nonsynchronized base stations.

Asynchronous DS spreading with long PN codes, on the other hand, has some striking advantages in the context of third-generation mobile radio systems: the number of available codes is virtually unlimited, the flexibility with respect to multiple bit rates and variable spreading factors is extremely high, and the requirements with respect to interchannel and intercell synchronization are extremely low.

Two categories of spreading sequences are employed in the proposed multirate DS-CDMA system: Short PN sequences used on the PICH, SCH, and RACH, and one long PN sequence used with different phases for all other logical channels.

For the short PN sequences, good even autocorrelation and crosscorrelation properties are required in order to guarantee fast acquisition with a minimum false alarm probability. Since each base station is assigned its own PN sequence on both the PICH and SCH, a large code family size is required. A code family is a set of PN sequences with specific auto- and crosscorrelation properties. Another requirement for the PN sequences is the so-called balance property: a PN sequence is said to be balanced if the number of 1's within one period of the PN sequence differs from the number of zeros at most by 1. The balance property guarantees a favorable shape of the spectrum. Finally, a simple hardware implementation of the PN sequences in question is required since each mobile station should be able to generate all PN sequences within one code family.

Various types of PN sequences have been taken into account, including the well-known m sequences, Gold codes, and Kasami sequences. These PN sequences all have periods $2^n - 1$, where n is an integer. Since PN sequences having period 2^n better fit system clocks than PN sequences of period $2^n - 1$, extended m sequences and linear combinations of extended m sequences have been investigated. An extended m sequence is obtained, inserting an additional element within one period of the corresponding m sequence. In [17], the correlation behavior of extended m sequences is addressed, and it has been shown that extended m sequences provide autocorrelation properties similar to those of Gold codes. However, linear combinations of extended m sequences provide a rather undesirable behavior in that their peak magnitudes of both the even autocorrelation and crosscorrelation functions are up to more than two times larger than that of Gold codes.

Thus, on both the PICH and SCH, balanced Gold codes of period 1023 have been chosen. On the RACH, balanced Gold codes of period 127 are envisaged.

The long PN code used on PCCH, PDCH, BCH/PCH, and AGCH is an m sequence of period $2^{41} - 1$ as has been proposed in [1]. The generator is based on the polynomial

$f(x) = 1 + x^3 + x^{41}$, thus consisting of 41 cells. Different physical channels are distinguished by different phases of this code.

F. Modulation

The basic modulation scheme used in the system is QPSK. The data symbols are fed to both the in-phase (I) and quadrature (Q) branches and multiplied with different phases of the same long spreading sequence. If short spreading sequences are employed, two different PN sequences are used in the I and Q branches of the modulator.

On the downlink, QPSK with coherent detection is employed. On the uplink, offset QPSK (OQPSK) is employed with coherent detection for the PDCH and differential encoding and differentially coherent detection on bit levels for the PCCH [18, p. 216 ff.].

G. Power Control

On the uplink, a combination of an open-loop and closed-loop power control is applied in order to keep the received signal power from the mobile station at a desired level at the base station. The open-loop power control is mainly used to track the shadowing and distance attenuation. The closed-loop power control tracks fading at low Doppler frequencies.

For the closed-loop power control, the base station constantly observes the received signal strength. From the actual received power, it determines power control commands (transmitter power up or down requests) that are transmitted on a downlink PCCH to the mobile station. In the CODIT testbed, the power control information bit rate will be set to 2 kb/s.

H. Combining, Pulse Shaping, and Frequency Conversion

Multiple physical channels are combined linearly before pulse shaping is applied. In the testbed, pulse shaping with a root raised cosine Nyquist filter is carried out in the baseband. After frequency conversion and amplification, the signal is fed to the antenna. At the mobile station, usually only a PCCH/PDCH pair has to be combined and transmitted. At the base station, the entire set of channels to be broadcast in the same RF band can be combined at the baseband level before pulse shaping is applied.

IV. Cellular Aspects

A. Synchronization Aspects

The use of asynchronous DS spreading with long PN codes on the PDCH and PCCH results in very loose requirements with respect to interchannel and intercell synchronization.

Mobile stations synchronized to a base station (via PICH and SCH) may directly time align their Tx CDMA frames with the Rx CDMA frames at the location of the mobile. Since signals received at the base station from different mobiles need not be frame aligned (as in TDMA systems) or even symbol synchronous, no timing advance control loop has to be implemented.

Moreover, since asynchronous DS spreading is also used on the downlink, the individual links within a radio cell or between adjacent cells need not be synchronized with each other, neither on symbol level nor on frame level. This enables a very cost-efficient implementation of nonsynchronized base station subsystems and is crucial for system deployment in noncoordinated cordless telephone type environments. Besides the cost factor, it is important that the radio network be operated independently of external time base systems (such as GPS) not under the control of the mobile network operators.

B. Soft Handover and Macro Diversity

Soft handover and macro diversity are powerful techniques to combat shadowing effects in cellular mobile radio systems and improve the transmission performance at the cell boundaries. In the proposed DS-CDMA system, soft handover and macro diversity may be efficiently applied for adjacent radio cells using the same RF channels (assuming a one-cell frequency reuse cluster).

On the uplink, the signals sent out by a mobile station may be simultaneously received by two or more spatially separated base stations. Diversity combining takes place at a common node in the hierarchically organized base station subsystem and may be implemented as a frame-by-frame selection combining or, even more powerful, by maximum ratio combining using soft decoding information provided by the base station receivers.

On the downlink, two or more base stations may establish links and transmit the same information (on TCH and DCCH) from different locations to a mobile station in soft handover, cf. Fig. 5. Maximum ratio diversity combining takes place in the RAKE receiver of the mobile station, where the RAKE fingers are adjusted to the strongest rays in the delay power spectra identified for all active links. Determination of the strongest rays (channel estimation) is accomplished with the aid of the PICH's broadcast by every base station and continuously monitored by the mobile station. Also, the criterion to select an adjacent base station for macro diversity and to enter the soft handover mode is based on the monitored PICH's.

Although the base station subsystem may in general stay nonsynchronized, according to Section IV-A, frame synchronization at the mobile station during handover eases the implementation of the receiver. For interfrequency handover, frame synchronization is even a requirement. It can easily be achieved on a per-call basis [19]. For that purpose, the mobile station measures the time offset t_{12} between the SCH's of the first and second base station, cf. Fig. 5. This information is conveyed to the second (new) base station via the old link and network when a link to the new base station shall be established. On the old and new downlink PDCH/PCCH, the same RF channel, chip rate, and long PN codes are used. The frame timing on the new link is adjusted such that the time offset t_{12} measured by the mobile station for the SCH's is compensated for the PDCH's and PCCH's. Hence, the second

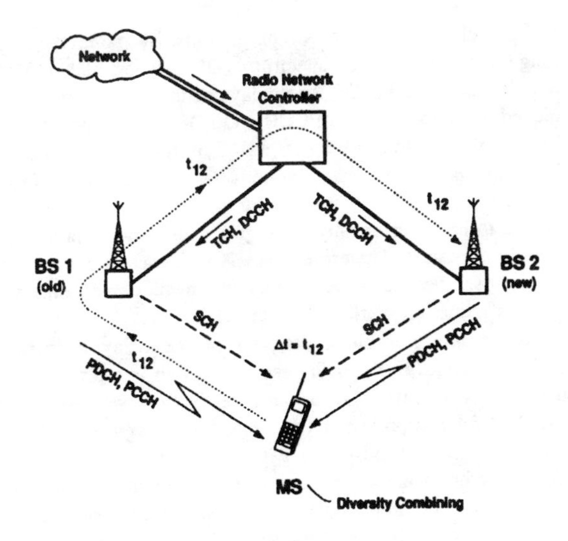

Fig. 5. Soft handover on the downlink.

base station is synchronized to the mobile station (just for this very link), rather than the mobile to the base station.

C. Interfrequency Handover

Hierarchically layered cell structures based on pico, micro, and macro cells, partly or in some cases even fully overlapping, will be vital for third-generation mobile systems. Although CDMA enables the efficient use of the same RF channels in adjacent cells of the same hierarchical layer (i.e., all micro cells), different RF channels have to be assigned to cells on different hierarchical layers in order to avoid power control problems and excessive interference. Soft handover and macro diversity, as outlined in Section IV-B, is no longer feasible between such cells. Instead, a preferably seamless handover between different RF channels is required (interfrequency handover).

Interfrequency handover presupposes that a mobile station is able to monitor pilot channels and to transmit and receive signals quasi-simultaneously on two different frequencies. This is easy in TDMA, but usually calls for a costly second radio transceiver in a CDMA mobile station. In the CODIT testbed, an alternative approach based on a time division technique denoted as "Compressed Mode" is explored.

During Compressed Mode, the PDCH and PCCH data of a 10 ms CDMA frame are squeezed into a signal burst of approximately 5 ms. The burst occupies just one half of the CDMA frame, cf. Fig. 6. Since the symbol length before DS spreading is halved but the chip rate R_c remains unchanged during Compressed Mode, the spreading factor is in effect reduced by a factor of two. In order not to deteriorate transmission performance, this is compensated for by doubling the instantaneous Tx power as indicated in Fig. 6.

The second half of a CDMA frame in Compressed Mode may then be used to switch the radio transceiver of the mobile station to another RF channel in order to monitor PICH's

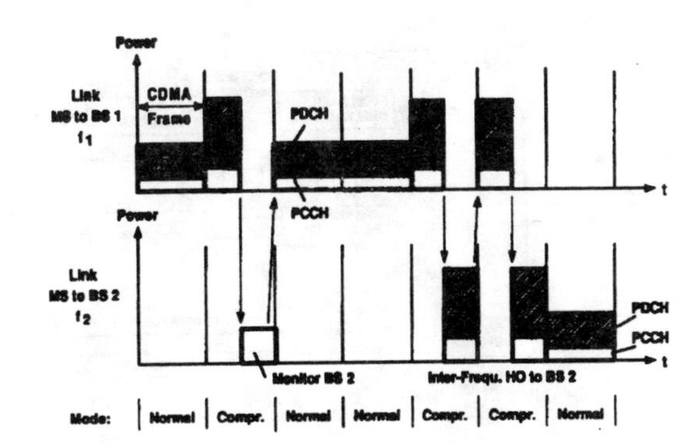

Fig. 6. Interfrequency handover using compressed mode.

of adjacent base stations operating on other RF channels or to establish a link to such a base station. A seamless interfrequency handover is implemented by establishing a link to the new base station in Compressed Mode and releasing the link to the old base station only after the new link is up and stable. Then, on the new RF channel, the Dedicated Channels return to normal mode utilizing the full CDMA frame length, cf. Fig. 6.

V. Receiver Concept

The system design reflects that the up- and downlink in a CDMA system differ significantly. Of course, this carries through to the receiver structures. On the downlink, a strong pilot channel can be utilized for channel sounding. This allows demodulation of both the PDCH and PCCH coherently. On the uplink, a strong pilot channel that is common to all users is not possible. This makes channel estimation on the uplink more difficult. For brevity, we will limit the following discussion to the base station receiver.

A. Receiver Structure

The structure of the base station receiver is sketched in Fig. 7. The received complex baseband signal is first filtered with the pulse shape matched filter MF and sampled at a rate of two samples per chip. Then, the signal is distributed to the RAKE demodulators and the channel estimation unit. As described in Section II-B, a PCCH frame contains relevant information about the structure of the concurrently transmitted PDCH. Particularly, the spreading factor used in the present frame on the PDCH is transmitted via the PCCH. Therefore, the PCCH needs to be decoded before the PDCH can be demodulated. This is the main reason for the frame buffer in front of the PDCH RAKE demodulator. Channel estimation is done on the PCCH,[1] which is transmitted continuously. Two decision feedback paths coming from the PCCH RAKE demodulator are provided to obtain the required input signals for the channel estimation unit, cf. Section V-B.

It is due to the concept of decoding the PCCH before demodulating the PDCH that there are two different detection

[1] The presence of the PDCH depends on its data rate, which might be zero due to DTX.

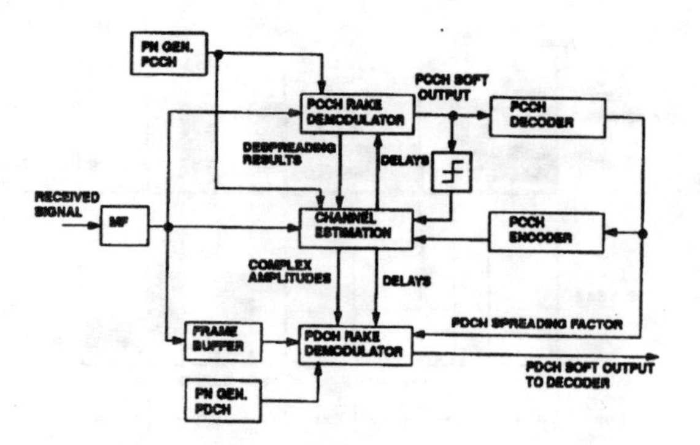

Fig. 7. Structure of a base station receiver.

schemes on the PDCH and PCCH: the PDCH is coherently demodulated, while on the PCCH the data are differentially encoded to allow for differentially coherent demodulation. To understand the reason, we have to consider the limitations inherent in the possible channel estimation schemes. For coherent detection, we need to know the delays *and* the complex amplitudes of the rays that are used in the demodulator. The delays can be assumed to change by less than half a chip interval, $T_c/2$ ($T_c = 200$ ns for the 5 Mchip/s case), within one frame of 10 ms. Therefore, delay estimation can be very precise (and it has to be). Obtaining precise amplitude estimates of the rays is more difficult, because the amplitudes change at a faster rate. The effective window length that is used for amplitude estimation should be shorter than the minimal coherence time of the channel, i.e., the inverse of (twice) the maximum Doppler frequency $f_{D,\max}$. Taking this into account with the period of the channel-encoded bits on the PCCH ($\approx 250\,\mu$s), a modulation scheme with differentially encoded data seems reasonable: in such a scheme, complex amplitude estimation is done implicitly on a single bit period [20, p. 300].

After differentially coherent demodulation of the PCCH and soft decision decoding, the PCCH can be looked upon as a pilot channel: decoding errors on the PCCH will unavoidably lead to a lost frame because the information transmitted on the PCCH, i.e., the correct spreading factor, is required for PDCH demodulation. Since the PCCH after decoding is a pilot channel, it can be used for complex amplitude estimation of the rays. The effective estimation window length is doubled compared to the situation before decoding the PCCH because now we can use not only the past for the estimation process, as in the differentially coherent demodulation scheme, but also the future signal parts to estimate the amplitude of a ray at a particular time t_0. This doubled window length is just enough to get fairly good estimates of the complex amplitudes and allow for coherent demodulation of the PDCH. The performance gain, compared to differentially coherent demodulation of the PCCH, will be addressed in Section VI.

B. Channel Estimation Unit

The channel estimation process separates into two steps: delay estimation and complex amplitude estimation.

The delays are estimated on a frame-by-frame basis: A long-term delay power spectrum (DPS) is estimated using one frame of the PCCH. The delays of the strongest rays are picked to demodulate the PCCH and PDCH in the next frame. Instead of estimating a long-term DPS, a short-term DPS could be estimated via a sliding window and the instantaneously strongest rays could be used for demodulation. However, this proves to be superior only for high signal-to-noise ratios that are of no interest. Therefore, frame-wise delay estimation has been chosen because it better fits the transmission scheme and permits reduction of the computational effort.

Fig. 8 shows a block diagram of the delay estimation scheme. Central to the delay estimation unit is a time-variant matched filter. Currently, for our investigations we applied the matched filter approach: all the received energy is exploited for estimating the power at a particular delay in the DPS. To reduce hardware complexity in a final implementation, the matched filter could be replaced by a set of correlators that use only parts of the received energy for this purpose. The filter coefficients are taken from the PCCH PN sequence, which is modulated like in the transmitter by differentially encoded bits. Since they are *a priori* unknown, they are fed back by the PCCH RAKE demodulator. In order to meet timing requirements in the hardware, the bits are taken from the PCCH demodulator before decoding, cf. first feedback path of Fig. 7. The received signal is delayed accordingly at the input of the channel estimation unit. At the output of the matched filter, we obtain per measurement interval an estimate $\hat{\underline{h}}$ of the channel impulse response vector $\underline{h}$. During a measurement interval, the filter coefficients of the time-variant matched filter are kept fixed. The length of the interval determines the length of the estimated impulse response vector. Therefore, the interval needs to be chosen according to the longest possible impulse response we have to take into account. The number of filter coefficients in the matched filter determines the correlation length or processing gain C_L. On the one hand, the gain C_L must be large enough to raise the impulse response above the noise floor; on the other hand, the correlation time (or length C_L) must be small compared to the coherence time of the channel, i.e., the inverse of (twice) the maximum Doppler frequency $f_{D,\max}$. A first short-term DPS estimate $\hat{\underline{\Phi}}_0$ is obtained from the impulse response vector $\hat{\underline{h}}$ by taking absolute values and squaring. The processing gain may have been insufficient for a good estimate. Therefore, several, i.e., N_Φ subsequent estimates $\hat{\underline{\Phi}}_0$, are averaged in the final estimate $\hat{\underline{\Phi}}$. The delay power spectrum $\hat{\underline{\Phi}}$ is searched for the strongest rays. The corresponding delays are sufficient to run the PCCH RAKE demodulator in the next frame.

For the coherently operating PDCH RAKE demodulator, we also need per finger, i.e., per demodulated ray, the continuously changing complex amplitude. It is obtained by processing the despreading results of the corresponding PCCH RAKE finger, cf. Fig. 7. The despreading results can be viewed as noisy samples of the time-variant complex amplitude modulated by the PCCH bits. The modulation can be removed perfectly because, after decoding the PCCH, the modulating bits can be obtained easily by re-encoding the information bits as in the transmitter. This makes the PCCH a pilot channel

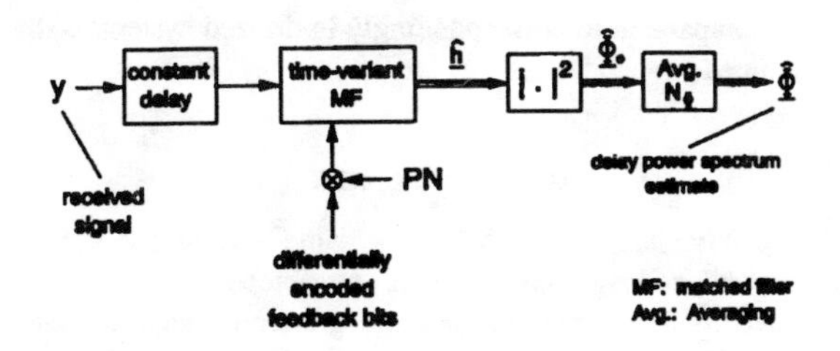

Fig. 8. Delay estimation unit.

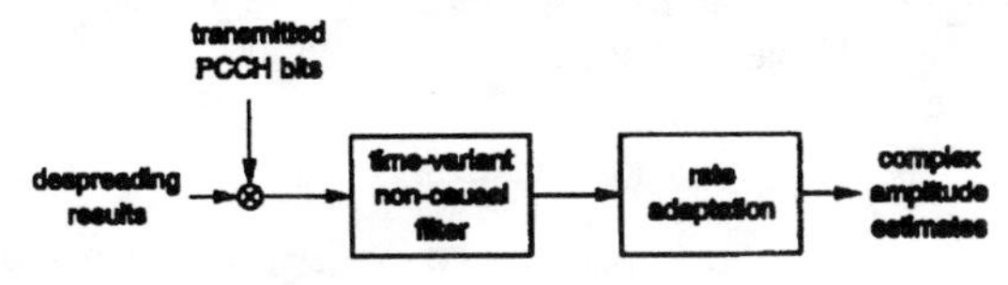

Fig. 9. Amplitude estimation for a single ray.

for PDCH demodulation. After removing the modulation, the despreading results are filtered to reduce the noise, cf. Fig. 9. A computationally efficient method is to run a recursive exponential window in forward and backward direction on the demodulated despreading results of a whole PCCH frame and combine the outcomes. The loss in performance is almost negligible compared to an optimal time-variant Wiener filter that exploits the unkown Doppler spectrum. Before the smoothed complex amplitude estimates can be used for PDCH demodulation, a rate adaptation is necessary because of the different bit rates on the PCCH and PDCH.

VI. Performance

To assess the performance of the CDMA receiver, simulations have been carried out and bit error rates (BER) versus the signal-to-noise ratio E_s/N_0 and E_b/N_0 have been recorded. Here, E_s and E_b denote the average received energy per *code* bit (symbol) and *information* bit, respectively. Other users in the CDMA system are assumed to cause noise-like interference. Therefore, instead of simulating other users, their influence can be subsumed in additive white Gaussian noise[2] of power spectral density N_0. For the simulations, a four-tap RAKE receiver has been used on a six-ray Rayleigh fading channel with classical Doppler spectra. The maximum Doppler frequency $f_{D,\max}$ was 244 Hz. This corresponds to a vehicle speed of 120 km/h at a carrier frequency of 2.2 GHz. All six rays have been equally strong. A chip rate of 1.023 Mchip/s has been assumed (narrowband case). On the PDCH, transmission with a coded data rate of 32 kb/s, i.e., an information bit rate of 16 kb/s, has been simulated. The corresponding spreading factor was $g = 31$ chip/bit $\approx$ 1.023 Mchip/32 kbit. On the PCCH, the spreading factor was $g = 248$. Per frame, a long-term power delay spectrum of length 50 μs has been estimated by using an impulse response

[2]Of course, this is an approximation. Due to pulse shaping, other users will cause an interference similar to colored noise. This has to be taken into account if a number of users equivalent to N_0 is to be calculated.

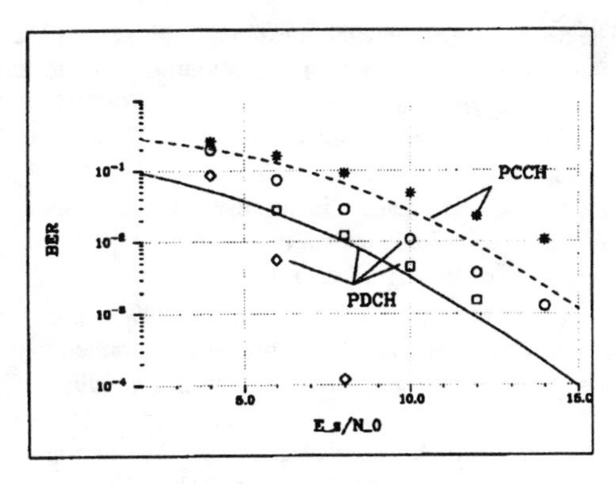

Fig. 10. Performance of the CDMA base station receiver.

correlation length of $C_L = 2 \cdot 248$ chips and averaging $N_\Phi = 20 (\approx 10\,\text{ms}/496\,\mu\text{s})$ short-term power delay spectra.

Fig. 10 shows simulation results without and including channel coding (markers). Theoretical bit error rates for four-path diversity in case of a six-ray Rayleigh fading channel [20] have been included for orientation (solid and dashed line). The curves have been calculated under simplified assumptions: the four fingers of the RAKE demodulators have been tuned to four of the six rays, there is no tap switching in the RAKE demodulators, in case of coherent demodulation the complex amplitudes of the rays are perfectly known, and interpath interference from the two not captured rays can be neglected. The latter is true only for low signal-to-noise ratios. The figure allows us to make several observations:

1) For bit error rates around 5–10 %, there is a potential gain of 5 dB in signal-to-noise ratio if we switch from differentially coherent demodulation (dashed line) to coherent demodulation (solid line).
2) There are slight deviations between the measured bit error rates on the PCCH (*) and the theoretical ones (dashed line). For low signal-to-noise ratios, this is due to wrong delay estimates; for high signal-to-noise ratios, this is due to interpath interference. In addition, the higher the signal-to-noise ratio, the more the time variance of the channel will be noticeable due to the long correlation length (large spreading factor) on the PCCH.
3) The circles (o) and boxes (□) show measured bit error rates for the coherently demodulated PDCH. In the case of boxes, ideal amplitude estimation has been assumed. The delays have been kept fixed. We notice that, for low signal-to-noise ratios, the measured bit error rates coincide well with the theoretically predicted ones. For high signal-to-noise ratios, we observe deviations due to interpath interference.
4) There is a loss of about 2–4 dB for bit error rates around 5-10 % if we switch from the "ideal" case (□) to the channel estimation scheme described in Section V (o). However, there is still a significant gain for these bit error rates compared to the differentially coherent case (*, PCCH).

5) The diamonds (◇) show the measured BER on the PDCH after decoding for a rate 1/2 convolutional code and interleaving within a single 10 ms frame. A BER of 10^{-3} is achieved at approximately $E_s/N_0 = 7$ dB, which corresponds to $E_b/N_0 = 10$ dB.
   Note: this performance is achieved without antenna diversity. Also, the four-path rake receiver captures only two thirds of the signal energy.

The receiver performance, in terms of the E_b/N_0 required to achieve a desired BER, is a very important parameter that significantly effects the spectral efficiency of a cellular radio system. For calculations of spectral efficiency and system capacity, a flat Rayleigh fading channel and ideal two-branch antenna diversity is commonly assumed [21], [22].

We have, therefore, carried out simulations with the respective assumptions. For a flat Rayleigh fading channel, the receiver employs a single RAKE arm in either antenna branch. The RAKE soft outputs are combined before decoding. A signal-to-noise ratio of approximately $E_b/N_0 = 6.2$ dB is required on the uplink PDCH to obtain after decoding the bit error rate BER $= 10^{-3}$ when a rate 1/2 convolutional code and interleaving within a single 10 ms frame is applied. This figure compares favorably with the 7 dB figure that is claimed for the system described in [21]. The achieved E_b/N_0 gain is due to the coherent detection. Note that this gain is set off by the energy devoted to the PCCH. For variable-rate speech channels, the PCCH requires on average approximately 15% of the PDCH symbol energy E_s. This PCCH overhead can be regarded as the cost for the variable-rate and multiplexing flexibility introduced into the system. Considering the gain due to coherent detection, the cost for the flexibility is extremely low.

VII. Conclusion

A CDMA system design study for third-generation mobile and personal communication systems has been presented and a number of novel techniques for cellular CDMA have been introduced:

- an open and flexible multirate CDMA radio interface architecture based on asynchronous direct-sequence code spreading with three different chip rates,
- a technique to implement variable bit rate and packet transmission in a very flexible way,
- the concept of a parallel code-multiplexed physical control channel for variable bit rate and fast closed-loop power control,
- a technique to implement soft handover in nonsynchronized base station subsystems,
- a time division technique (Compressed Mode) to enable handover between different radio frequency channels and different hierarchical cell layers (e.g., pico, micro, and macro cells).

These techniques are currently being explored within the CODIT project and will be validated in a testbed. Although the primary design goal was a highly flexible radio interface rather than optimizing spectral efficiency, first simulation studies on bit error rate performance on the uplink indicate that results comparable to correspondingly optimized systems will be obtained.

Acknowledgment

Since it is impossible to list by name and to honor adequately all colleagues who have contributed to the results presented in this paper, the authors gratefully acknowledge the fruitful cooperation within the project as a whole. In particular, however, the authors would like to thank G. Brismark (Ericsson), P. Chevillat (IBM), C. Günther (ASCOM), and P. Mège (Matra) who have helped to improve this paper by providing valuable comments.

References

[1] A. Salmesi and K. S. Gilhousen, "On the system design aspects of CDMA applied to digital cellular and personal communication networks," in *Proc. 41st IEEE Conf. Vehic. Technol.*, St. Louis, MO, May 1991.

[2] A. Baier and W. Koch, "Potential of CDMA for 3rd generation mobile radio systems," in *Proc. MRC Mobile Radio Conf.*, Nice, Italy, Nov. 1991.

[3] P. Dent, B. Gudmundsson, and M. Ewerbring, "CDMA-IC: A novel code division multiple access scheme based on interference cancellation," in *Proc. 3rd IEEE Int. Symp. Personal, Indoor and Mobile Radio Commun.*, Boston, MA, Oct. 1992.

[4] "Future public land mobile telecommunication systems," CCIR Recommend. 687.

[5] "Objectives and framework of the UMTS," Draft ETSI Tech. Rep. SMG-50101, Version 0.7.0, Jan. 1993.

[6] "Mobile station-base station compatability standard for dual-mode wideband spread spectrum cellular system," TIA/EIA Interim Standard IS-95, July 1993.

[7] "Cellular system dual-mode mobile station-base station compatability standard," TIA/EIA Interim Standard IS-54-B, Apr. 1992.

[8] "Physical layer on the radio path," ETSI/TC GSM/DCS-1800 Recommend. 05-Series.

[9] P. G. Andermo and G. Larsson, "Code division testbed, CODIT," in *Proc. 2nd Int. Conf. Universal Personal Commun.*, Ottawa, Ont. Canada, Oct. 1993.

[10] A. Baier, "Multi-rate DS-CDMA: A promising access technique for third-generation mobile radio systems," in *Proc. 4th Int. Symp. Personal, Indoor and Mobile Radio Commun.*, Yokohama, Japan, Sept. 1993.

[11] A. Baier, "Open multi-rate radio interface architecture based on CDMA," in *Proc. 2nd Int. Conf. Universal Personal Commun.*, Ottawa, Ont., Canada, Oct. 1993.

[12] P. Hoeher, "Tradeoff between channel coding and spreading in a mobile DS-CDMA system," submitted to *IEEE Trans. Vehic. Technol.*

[13] J. Hagenauer and P. Hoeher, "A Viterbi algorithm with soft-decision outputs and its application," in *Proc. GLOBECOM'89*, Dallas, TX, pp. 47.1.1–47.1.7.

[14] R. D. Cideciyan and E. Eleftheriou, "Performance of concatenated coding for data transmission in CDMA cellular systems," in *Proc. Int. Zurich Sem. Digital Commun.*, Zurich, Switzerland, Mar. 1994.

[15] S. Lin and J. Costello, *Error Control Coding—Fundamentals and Applications.* Englewood Cliffs, NJ: Prentice Hall, 1983.

[16] U. Grob, A. L. Welti, E. Zollinger, R. Küng, and H. Kaufmann, "Microcellular direct-sequence spread-spectrum radio system using *n*-path RAKE receiver," *IEEE J. Select. Areas Commun.*, vol. 8, pp. 772–780, June 1990.

[17] U.-C. Fiebig and M. Schnell, "Correlation properties of extended *M*-sequences," *Electron. Lett.*, vol. 29, no. 20, pp. 1753–1755, Sept. 1993.

[18] S. Benedetto, E. Biglieri, and V. Castellani, *Digital Transmission Theory.* Englewood Cliffs, NJ: Prentice Hall, 1987.

[19] H. Persson and P. Willars, "Techniques to provide seamless handover for a DS-CDMA system," in *Proc. RACE Mobile Telecommun. Workshop*, Metz, June 1993 (not publically available).

[20] J. G. Proakis, *Digital Communications.* New York: McGraw-Hill, 1989.

[21] K. S. Gilhousen, I. M. Jacobs, R. Padovani, A. J. Viterbi, L. A. Weaver, Jr., and C. E. Wheatley, "On the capacity of a cellular CDMA system," *IEEE Trans. Vehic. Technol.*, vol. 40, no. 2, pp. 303–312, 1991.

[22] W. Granzow and W. Koch, "Potential capacity of TDMA and CDMA cellular telephone systems," in *Proc. IEEE Second Int. Symp. Spread Spectrum Techniques and Applications (ISSSTA'92),* Yokohama, Japan, Nov. 1992, pp. 243–246.

Alfred Baier (M'84–SM'93) was born in Kaiserslautern, Germany, in 1956. He received the Dipl.-Ing. and Dr.-Ing. degrees in electrical engineering from the University of Kaiserslautern in 1982 and 1986

From 1982 to 1986, he was employed as a Teaching and Research Assistant at the University of Kaiserslautern. During this period, he was involved in the analysis and design of burst transmission spread-spectrum communication systems and digital matched filters. In 1986, he joined the Radio Communication Systems Division of Philips Kommunikations Industrie AG (PKI) in Nuremberg, Germany, where he was engaged in the development of the GSM system and in the analysis and design of future mobile radio communication systems. From 1992 to 1993, he headed the System Planning Department at PKI. Early in 1994, he joined Mannesmann Mobilfunk GmbH (German D2 Network) as Manager of the Radio Network Planning Department at their headquarters in Düsseldorf. In 1991, he was awarded the ITG Paper Prize.

Dr. Baier is a member of the Verband Deutscher Elektrotechniker (VDE) and the Informationstechnische Gesellschaft (ITG) in Germany.

Uwe Carsten Fiebig (M'90) received the Dipl.-Ing. degree in electrical engineering from the Technical University of Munich, Munich, Germany, in 1987 and the Dr.-Ing. degree from the University of Kaiserslautern, Kaiserslautern, Germany, in 1993.

In 1988, he joined the Institute of Telecommunications at DLR (German Aerospace Research), Oberpfaffenhofen, Germany. As a researcher, he is engaged in activities in the area of spread spectrum techniques considering both direct sequence and frequency hopping. Furthermore, he is concerned with propagation measurements of the 20, 30, and 40 GHz satellite channel.

Wolfgang Granzow was born in Steinhagen, Germany, in 1955. He received the Ing. (grad.) degree from Fachhochschule Bielefeld in 1976 and the Dipl.-Ing. and Dr.-Ing. degrees from the Technical University of Berlin, Berlin, Germany, in 1984 and 1990, respectively.

From 1976 to 1978, he was with Deutsche Telephonwerke AG, Berlin, Germany. From 1984 to 1989, he worked on problems related to speech processing and speech coding at the Technical University of Berlin. From 1989 to 1991, he visited the Speech Research Department of AT&T Bell Laboratories, Murray Hill, NJ. He is now with Philips Kommunikations Industrie AG in Nuremberg, Germany, where he is working in the Advanced Development Department, Radio Communication Systems, in the field of digital signal processing techniques for mobile radio transmitters and receivers.

Wolfgang Koch (M'83) was born in Hannover, Germany, in 1949. He received the Ing. (grad.) degree from the Fachhochschule Hannover in 1973, and the Dipl.-Ing. and Dr.-Ing. degrees from the Technical University of Hannover in 1977 and 1982, respectively.

From 1977 to 1982, he worked on source modeling of analytical speech signals at the Institut für Allgemeine Nachrichtentechnik of the TU Hannover. Since 1983, he has been with Philips Kommunikations Industrie AG in Nuremberg, Germany, where he has worked in the Group "Transmission" of the Advanced Development Department. Since 1988, he has lead this group. He was involved in the solution of transmission-related problems like channel coding, channel modeling, digital signal processing, and simulation techniques for mobile radio applications.

Dr. Koch is a member of the German VDE/ITG.

Paul Teder was born in Stockholm, Sweden, in 1966. He received the M.Sc. degree in electrical engineering from the Royal Institute of Technology, Stockholm, Sweden, in 1991.

Since 1990, he has been employed at the Research and Development Department of Ericsson Radio Systems, Stockholm, Sweden. He is presently involved in the European Research Project CODIT.

Jörn Thielecke (M'92) was born in Berlin, Germany, in 1957. He received the Dipl.-Ing. and Dr.-Ing. degrees in electrical engineering from the University of Erlangen-Nürnberg, Germany, in 1985 and 1991, respectively.

In 1982, he was awarded a one-year Fulbright scholarship to attend the Georgia Institute of Technology, Atlanta. From 1985 to 1991, he was a Research and Teaching Assistant at the University of Erlangen-Nürnberg. Since 1991, he has been employed in the Advanced Development Department for Radio Communication Systems at Philips Kommunikations Industrie AG in Nuremberg. His research interests lie in the field of digital signal processing for mobile radio communications and adaptive signal processing.

Wireless Communications Going Into the 21st Century

Donald L. Schilling, *Fellow, IEEE*

(Invited Paper)

Abstract—**The evolution of telecommunications, from the wired phone to personal communications services, is resulting in the availability of wireless products not previously considered practical.**

The user of a cellular or personal communication system wants to use *one phone* for all of his intended needs. Thus, a single portable phone should be able to operate in a residence, as a cordless phone; in a vehicle using a cellular system, in an office using a WPBX; and outside with wireless local access.

Broad-band code-division multiple access (B-CDMA) is a technique which allows PCS operation in the cellular frequency band in conjunction with existing cellular service, as well as in the PCS band (1850–1990 MHz in the U.S.). Using B-CDMA, high-quality voice with no dropped calls as well as data-rate-on-demand can be achieved, which will permit ISDN and multimedia communications at power levels which are much less than that required for other technologies.

This paper describes the present cellular and PCS environments, as well as the evolution of these environments into the 21st century, and explains how broad-band CDMA can provide the *one-phone* service required by business people as well as people at home.

Introduction

IT was about 20 years ago when Peter Goldmark introduced the concept of the wired city: the interconnection of computers, faxes, and telephones in the office and between offices. Today, even hotels provide data and voice mail service and electronic check-out. Indeed, this is the decade of wireless telecommunications. The last decade, the 1980's, was the decade of computers. During that decade, we all went out and purchased computers. We brought them home and then looked for an application. Most still use the computer primarily as a word processor. During this decade, we will purchase wireless telephones—and everyone *knows* what to do with a telephone.

Today, not only has Goldmark's dream become a reality, the wired city is taken for granted, and we are all hard at work redefining that dream, replacing wires by wireless radio transmission. Indeed, Fig. 1 is an excerpt of an 1865 *Boston Post* editorial which recognizes that wired communications is not a preferred approach. Our thinking about telecommunications has changed dramatically over the past five–ten years. In 1989, Shelby Bryan, CEO of Millicom, and this author went to see the then Chairman of the FCC, Mr. Alfred Sykes, requesting an experimental license. The license was used to prove that personal communications services (PCS) could be supplied in the frequency band 1850–1990 MHz in the U.S. and in similar frequency bands elsewhere in the world, using a technology new to commercial applications, broad-band code-division multiple access [1]. At about the same time, we saw the emergence of capitalism throughout the world. A requirement of capitalism is the need to have voice and data communications. In order to move quickly, as well as inexpensively, to provide such communications, countries began to employ a *wireless* loop solution using FDMA or TDMA. Fig. 2 shows how TDMA (or FDMA) is used and where B-CDMA can be used to advantage.

PCS means different things to different people. PCS means a *one-phone* solution to one's communication

Worth Noting

"Well-informed people know it is impossible to transmit the voice over wires and that, were it possible to do so, the thing would be of no practical value".

Excerpt from an 1865 *Boston Post* editorial.

Fig. 1. Wired communication should be replaced by wireless systems.

Manuscript received September 30, 1993.

The author is with InterDigital Communications Corporation, Great Neck, NY 11021.

IEEE Log Number 9403343.

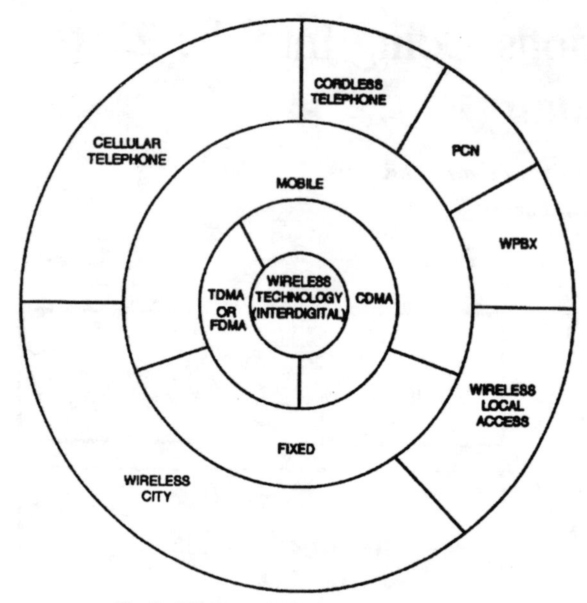

Fig. 2. Wireless technologies and applications.

needs. It should provide wired line voice quality and data rate on demand so that fax, modem, and video can be communicated. We call this multimedia communications, and it requires data rate capability to 144 kb/s. The *One Phone*, illustrated in Fig. 3, allows us to start making calls as soon as we awaken, as we exercise, and during breakfast. This is the cordless phone mode of operation. When we go outside, the call automatically shifts to a local base station at the curb—the wireless access mode. Next, we enter our car, and the call shifts to a cellular type of service. Finally, in the office, the *One Phone* shifts to a wireless PBX service.

The result is *One Phone*, which allows you to start talking early in the morning when you awaken and to continue talking until you go to sleep. A service provider's dream. What are the characteristics of the *One Phone*? Arther D. Little was commissioned by Millicom to determine what the consumer wanted. The results were loud and clear. As illustrated in Fig. 4, they are

• *Wired line quality and wireless convenience*—No dropped calls. Recent testing has demonstrated that current digital cellular quality is poorer than analog cellular.

• *Privacy*—When using a cellular phone, do not say anything you do not want the world to know. A thriving business is to tune in with your spectrum analyzer—a leading electronic distributor actually sells an inexpensive receiver which is readily adapted to receive cellular calls. You have all heard of the Mafia leader in Sicily who eluded the police for 22 years and was caught because he used a cellular phone.

• *Data rate on demand*—Business users require data up to the ISDN rate of 144 kb/s.

Within the next five years, cellular at 850 MHz and PCS at 1900 MHz will provide competitive services at competitive prices. Fig. 5 shows that demand is so great and the time to market so small that we vastly underestimate the market. In the U.S., the FCC probably will limit the cellular service providers' ability to provide PCS ser-

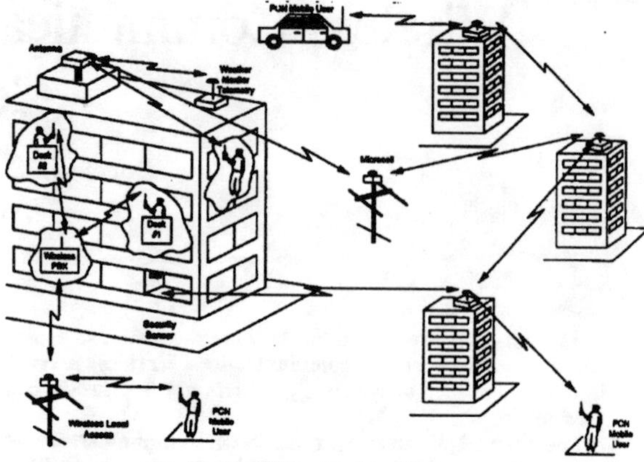

Fig. 3. PCS using *One Phone*.

QUALITY

- High quality voice using Adaptive DPCM
- Fax and modem interface
- Direct data service
- Low outage time
- Fast handoff in cellular applications
- Privacy

VALUE

- Largest number of users/square mile
- Lowest cost for handset and base

Fig. 4. Characteristics of B-CDMA.

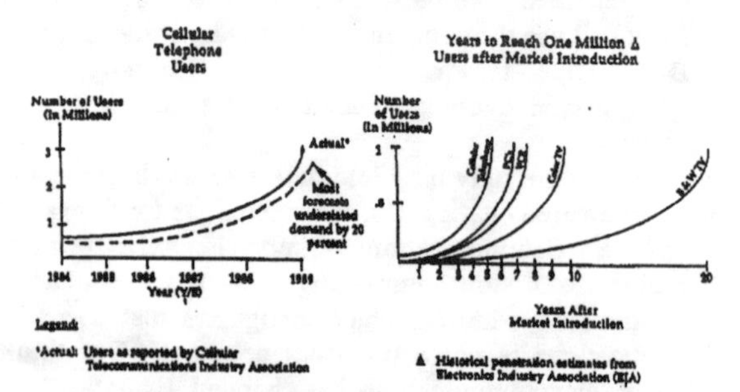

Fig. 5. User demand exceeds our imagination.

vice in the same geographical region in order to increase competition. In order to allow the cellular providers the ability to compete with the PCS providers, InterDigital Communications Corporation (IDC) has developed a cellular overlay system using broad-band CDMA technology, in which the B-CDMA, PCS system shares the same spectrum as the cellular users who employ FDMA or TDMA/FDMA technology [2]. Thus, by using the spectrum efficiently, a "first class" PCS system can be added to the present "economy class" cellular system.

The reason the cellular system is referred to as the "economy class" is not price, for as we all know, cellular service is expensive, but because there is a lack of privacy, poor quality voice (compared to wired line service), dropped calls due to fading, and minimal data handling capability. Testing of this cellular overlay system will begin in Des Moines, IA, with US Cellular as host.

In the United States, 140 MHz has been set aside, between 1850 and 1990 MHz for PCS use. Channels with three different bandwidths, 5, 10, and 15 MHz transmit and receive, will probably be auctioned in 1995 to service providers in different regions of the country. Since 1989, numerous field tests have been performed with service providers which verified the operation of PCS systems using B-CDMA. The demonstrated characteristics include the following.

- Privacy; almost no fading, outdoor or indoor; data on demand with multimedia data transmission capability; very high capacity—several hundred simultaneous users per antenna sector.
- Interoperability with microwave users.
- Indoor operation: In a major New York City hospital, X-ray data were transmitted from the X-ray room to the head radiologist's office, without coding, at 9.6 kb/s. Wireless operation was demonstrated in the offices of a major investment banking firm in New York by putting the base station in the entrance area and walking around and communicating throughout the office.
- Wireless local access was demonstrated with a major cable operator in San Diego, CA. Operation throughout the Wall Street area has also been demonstrated; this is the urban "jungle" where there are many people and narrow streets, and where it is critical that the system operates properly. All of these measurements were taken during the day, while people were at work, in the heart of traffic.

In a military environment, a broad-band CDMA system could result in *One Phone* being using by our troops. Indeed, in June 1993, demonstrations were performed at Fort Gordon showing that PCS, using B-CDMA, could be used to save troops' lives. To do this, video was transmitted at 64 kb/s from a simulated battlefield to Walter Reed Medical Hospital. As illustrated in Fig. 6, using two-way voice communication, the expert at Walter Reed would be able to give advice to the medical officer on the battlefield. This demonstration was repeated on the White House lawn to President Clinton and Vice President Gore, and again at Fort Benning.

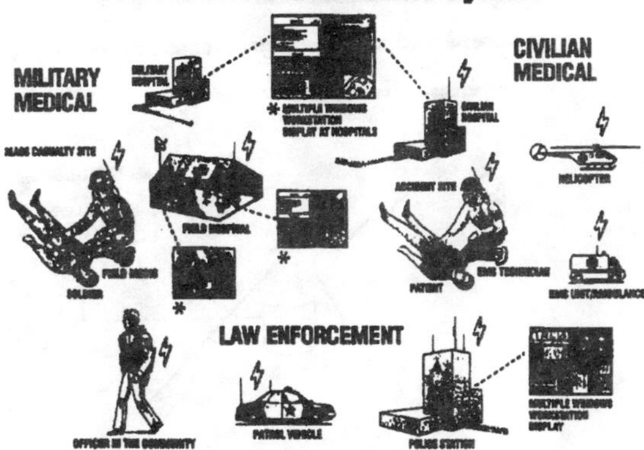

Fig. 6. PCS can be used for emergency medical service, law enforcement, and on the battlefield.

Satellite–Ground Communication: The LEO

Another example of a PCS system, for the 21st century, is the proposed service to be provided by the GSO, ICO, or LEO satellites. This service is intended for rural users, either mobile or fixed, or for urban users, to bypass the local telephone system. As an example of rural service, consider communication to a caravan or to a ship. Or consider communication to a new village in an area of Russia where oil exploration in starting. As an example of urban bypass communications, consider a system placed on top of a large office building, enabling all occupants to bypass the local telephone company when making a long distance call.

The differences between the different satellites systems focus on distance, number of satellites, capacity, and delay. All satellites use numerous spot beams to illuminate the ground. The GSO uses fewer satellites, but they are more distant from the ground, and therefore the user notices a significant delay. LEO satellites require more frequent handoff between satellites since, during a conversation, the mobile user may be illuminated in sequence by several satellites. Recently, the FCC allocated 16 MHz of bandwidth for LEO applications. About 5 MHz was allocated to Motorola's Iridium system, and the remaining 11 MHz was given to the CDMA systems. Ellipsat, which uses an elliptical orbit to minimize the number of satellites required, employs B-CDMA to maximize performance.

It is interesting to note that, using TDMA, the satellite systems cannot communicate with a user within a building. There just is not sufficient fade margin for such communication. On the other hand, a recent study showed that a CDMA paging service from a satellite could reach users within a building. Thus, after receiving the page, one must go outside to an unshaded spot or to the roof of a tall building in order to place a call using TDMA [3].

The Wireless Local Loop

Now, let us consider the wireless local loop application illustrated in Fig. 7. As mentioned earlier, healthy eco-

- USES EXISTING HOUSE WIRING FOR WIRED PHONES
- SUBSTITUTES FOR CORDLESS PHONE
- ROAM INSIDE, OUTSIDE, IN CAR, AROUND THE NEIGHBORHOOD
- COMPETES WITH LOCAL EXCHANGE CARRIER

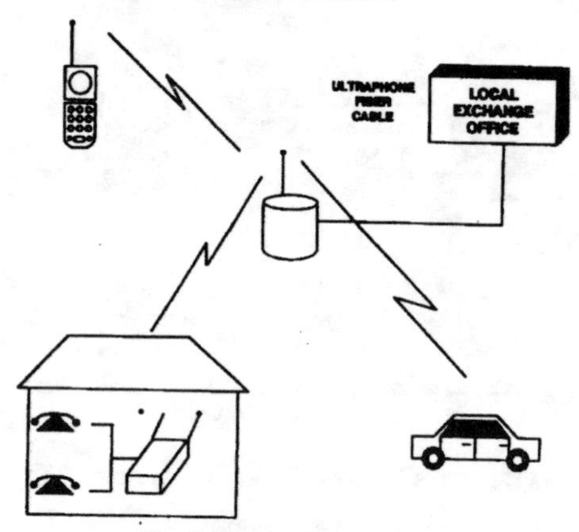

Fig. 7. Wireless local access.

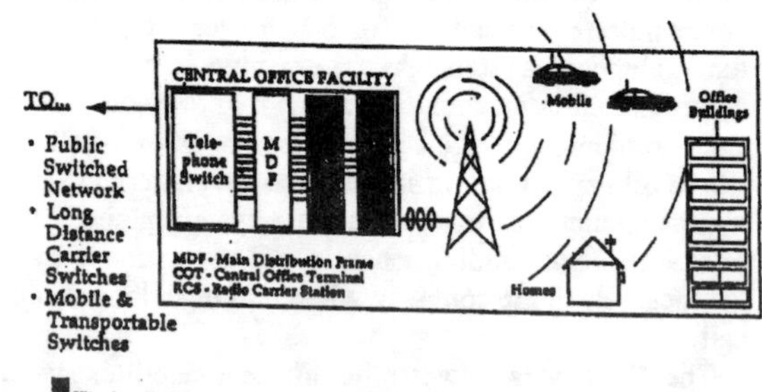

Fig. 8. The Ultraphone TDMA system provides wireless access.

nomic growth requires that regions of a country which previously had no telephone communications obtain good quality telephony immediately. For most of these countries, low cost, as well as immediate installation, are required. The solution to this problem is a wireless local loop. Such a system connects directly to the central office switch in a city, and then, using point-to-multipoint transmission, connects to remote subscriber units at designated factories, office buildings, and residences. Wireless coin phones are also possible connections. Technologies currently being used for this type of service are analog FDMA and TDMA. These systems are narrow band and typically operate in a large, single-cell environment, using the spectrum very inefficiently. For example, IDC offers a four-call/25 kHz channel, 16-PSK, TDMA system (Fig. 8) called the Ultraphone. This system encodes the voice using a 14.4 kb/s vocoder. The system provides almost wireline quality. Quitaque, TX, illustrated in Fig. 9, is the first wireless city in the U.S. GTE is the service provider, and they provide wireless access from all buildings to the switch using the Ultraphone. Voice, fax, and modem capability exists.

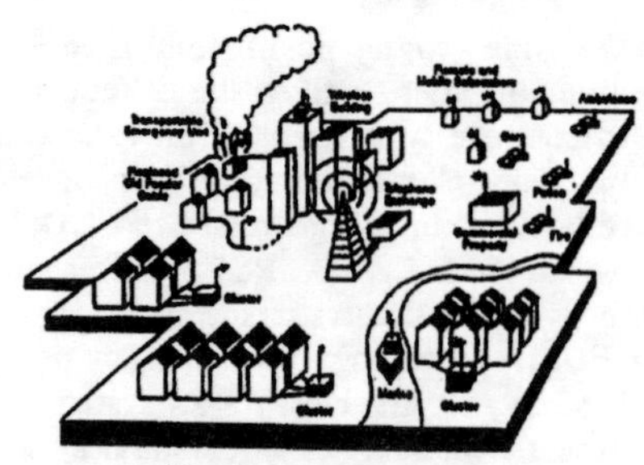

Fig. 9. The wireless city.

FDMA systems, with their lack of privacy, are not ideal solutions, nor are cellular systems which employ frequency reuse, which is needed in a multicell system, but which results in great spectral inefficiencies in a fixed, single-cell wireless loop unless the system is redesigned, using different filters, etc. If 5, 10, 20 MHz or more spectrum is available, a B-CDMA wireless local loop is ideal. Such a system is very desirable since it provides wired line quality voice, extreme privacy, and high data rates. It also offers mobility.

Multipath Fading—The Fundamental Limitation of a Wireless System

A major advantage of broad-band CDMA is its immunity to fading caused by multipath signals, and its ability to compensate for received signal variations due to fading and shadowing through the use of adaptive power control. The received signal generally consists of delayed versions of the transmitted signal which are the multipath signals. Those delayed versions of the signal that arrive within a chip duration and subtract cause the received signal to "fade." Those multipath signals arriving outside the chip duration result in an increase in the interference level since these components look like additional CDMA signals; however, they do not produce fading. Furthermore, the power contained in each of these greatly delayed signals is significantly reduced by the "processing gain" of the B-CDMA system. (The processing gain is the ratio of the bandwidth of the spread spectrum signal divided by the information bit rate.) The fade depth is inversely proportional to the bandwidth of the B-CDMA signal, i.e., the wider the signal bandwidth, the smaller the chip duration and the fewer the multipath components that fall within a chip duration. Hence, the probability of a frequency-flat fade decreases as the bandwidth increases [4], [5].

Fig. 10 presents the fade probability and fade depth as a function of signal bandwidth. Experimental results were obtained for bandwidths of 48, 30, 22, 11, 2, MHz and CW. These experiments were performed in an office, in the suburbs, and in downtown New York City. Note the increase in the fade depth as a function of signal bandwidth. Note particularly that of the fade depth varies only slightly for bandwidths exceeding 11 MHz, but increases sharply for bandwidths of 5 MHz or less. For example, a narrow-band, 30 kHz communication system would re-

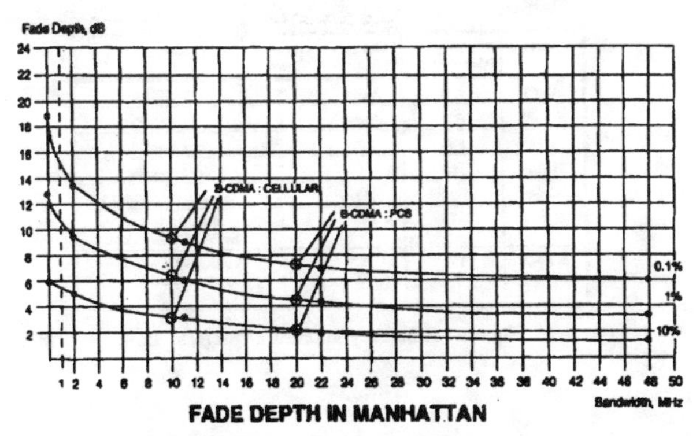

Fig. 10. Narrow-band systems must transmit more power than wide-band systems due to fading and shadowing.

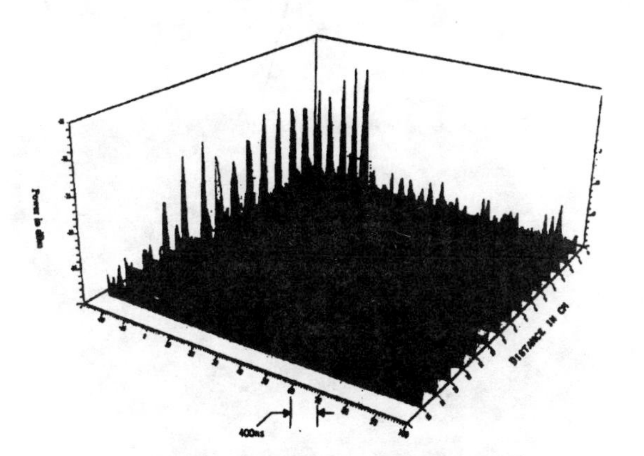

Fig. 11. Multipath signals are typically spaced 200 ns.

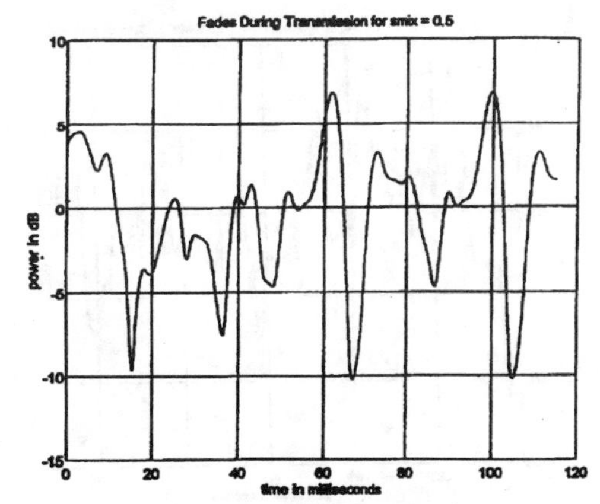

Fig. 12. A typical fading signal.

quire 12 dB more power than a broad-band, 20 MHz CDMA system communicating over the same path.

Diversity Reception

In a severe fading environment, diversity combining can often be used to improve performance. Time diversity and space diversity are often added to a direct sequence, spread spectrum CDMA which is a frequency diversity technique. Space diversity is usually achieved through the use of multiple antennas at the base station. However, future generation CDMA handsets will employ beam-steered antennas. Time diversity is readily achieved using a "RAKE" receiver. A "RAKE" receiver is one that receives each multipath signal, delays each by an appropriate amount, and then combines them following some algorithm. Experiments were performed in which a RAKE receiver was used to *display* each of the received multipath signals. Fig. 11 shows experimental results, obtained in Manhattan, using a RAKE which views the signal for 4 μs. In this experiment, a transmitter was placed on the top of a building located on W. 23rd St. and Ave of the Americas (6th Ave.) in New York City. The receiver was on the ground between 26th and 27th St. and 6th Ave., out of sight of the antenna. Note that the primary signal returns were 6–10 dB (or more) greater than any of the observed multipaths. Note that the multipath signals were spaced by more than 100 ns apart. Typical spacing in an urban area is found to be about 200 ns (the resolution of the system was 25 ns) [6].

In summary, we find that wider band B-CDMA systems suffer less fading since the chip duration is usually less than the multipath spread of the channel. In addition, far-out multipath components, such as those reflected from a distant "mountain," are most likely to be smaller than the primary signal component. The need for a time diversity system, such as RAKE, therefore diminishes as the bandwidth increases. Of course, a RAKE receiver could always be used to further enhance the performance of any CDMA system.

Adapative Power Control

Another technique used to compensate for fading or shadowing is adaptive power control. To illustrate this technique, consider that a spread spectrum base station receives all of the incoming signals simultaneously. Thus, if any siganl is received at a higher level than the others, that signal's receiver will have a higher signal-to-noise ratio, and therefore a lower bit error rate. The broad-band CDMA base station receivers ensure that each remote mobile transmits at the correct power level by telling each remote, every 500 μs, whether to increase or decrease its power. This technique is called adaptive power control (APC).

Fig. 12 shows a typical fading signal. The deep fades correspond to speeds up to 80 mi/h. Fig. 13 shows an algorithm by which the control voltage in the remote unit changes its power to follow command signals from the base to increase or to decrease this power. Note that the remote unit changes its power by a factor of 1.5 when going in the same direction, or by a factor of 0.5 when going in the opposite direction. In the implementation shown, the minimum step size was 0.25 dB and the maximum step size was 4 dB. The result was a bit error rate of 8×10^{-4}. The use of interleaving and FEC usually can correct these errors.

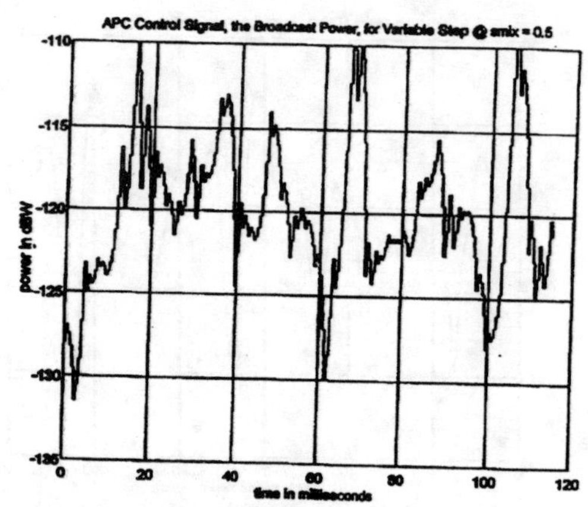

Fig. 13. The remote units transmitted power to counteract the fade of Fig. 12.

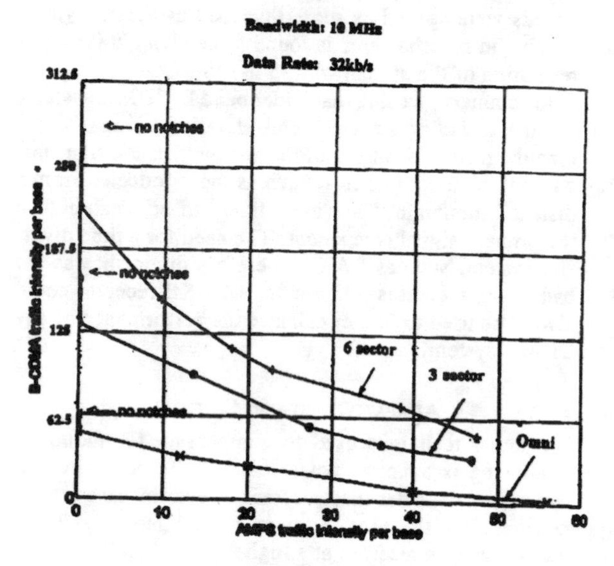

Fig. 14. B-CDMA capacity.

Broad-band CDMA wireless access permits users to receive communications within a building as well as while they are mobile, i.e., walking or driving. Adaptive power control (APC) with interleaving and forward error correction permits the B-CDMA system to maintain an acceptable, low bit error rate.

Conclusion

Fig. 14 illustrates the performance of a B-CDMA system operating at various data rates and bandwidths. These results were obtained by simulations which included shadowing, realistic antenna patterns, a multicellular environment, etc. [2].

Voice activity detection, which can result in doubling the number of voice users, is *not* employed. To illustrate the use of the table, consider that a bandwidth of 20 MHz is available for transmission and also for reception, and that a data rate of 64 kb/s is required. Then, using a six-sector antenna with a single cell, the top table shows that there can be $26.95 \times 20 = 539$ simultaneous transmissions taking place. With a 10:1 subscriber-user ratio, such a cell can provide high-quality service to 5390 subscribers.

Fig. 15 compares B-CDMA to other systems operating in a cellular environment. The efficiencies, in terms of simultaneous users/MHz, of FDMA, TDMA, and CDMA systems are compared. AMPS is considered as the baseline system, in which 56 simultaneous conversations can occur in a 25 MHz bandwidth. Hence, there are 2.24 users/MHz. Note that B-CDMA has the highest capacity since it minimizes the effect of multipath.

Technology	Efficiency (Users/MHz)	Voice Data Rate (kb/s)	Frequency Reuse	Comments
AMPS	2.24	Analog	7	BASELINE
NAMPS	6.72	Analog	7	x 3 AMPS
IS54	6.72	8	7	x 3 AMPS
ETDMA	18.0	4	7	x 8 AMPS; Includes VAD
GSM	5	13	4	Includes Fec, Rate 1/2
NCDMA	22.4	8	1	x 10 AMPS; Includes VAD and Variable Rate Coder
BCDMA	130 32	8 32	1 1	Includes VAD, and a 3 sector antenna

Fig. 15. Efficiency in terms of users/MHz.

Fig. 16. A "standard" wireless telephone.

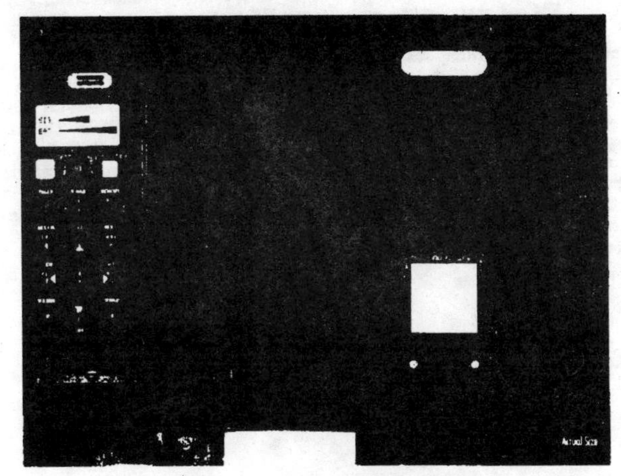

Fig. 17. A typical wireless handset.

Fig. 19. A high-quality video display and camera will be a part of everyone's handset.

Fig. 18. A wireless handset with PDA and computer capability.

Figs. 16–19 show the next generation of wireless telephones. Fig. 16 is a wireless phone that can be purchased at any department store. It plugs into the wall for power and has a battery backup. The handsets shown in Figs. 17–19 can be used to provide cellular telephone, cellular videophone, and full-featured touch-screen PDA.

By the 21st century, we will be a wireless society.

REFERENCES

[1] Spread Spectrum goes Commerical
[2] Broadband-CDMA Overlay
[3] Satellite paper—Bruno
[4] Schilling
[5] Vinko
[6] Schilling
[7] Broadband-CDMA Overlay

Donald L. Schilling (S'56-M'58-SM'69-F'75) is Vice Chairman of the Board of Directors of InterDigital Communications Corporation. He also serves as Executive Vice President and Chief Technical Officer of InterDigital Communications Corporation, and is Chairman, President, and Chief Executive Officer of InterDigital Telecom, Inc.

He was formerly Chairman of the Board, Chief Executive Officer, and President of SCS TELECOM/MOBILECOM, Inc., where he led the development of broad-band CDMA and began the field of PCS in the U.S. SCS also provided training as well as research and development in telecommunications for military clients. He is the coauthor of ten international best-selling textbooks and more than 250 papers in telecommunications and electronics. He is a past member of the U.S. Army Science Board. He retired in May 1992 as the Herbert G. Kayser, Distinguished Professor of Electrical Engineering at the City College of the City University of New York, where he had been a Professor since 1969. He is now President Emeritus. Prior to moving to CCNY, he was a Professor at the Polytechnic Institute of New York. He is widely recognized throughout the technical world, including the military and other government agencies, as an expert in telecommunications and electronic warfare. He is a frequent lecturer at national and international symposia, a consultant to government agencies, and the holder of over 20 patents in telecommunications technology. An internationally known expert in the field of communications systems, he has made many notable contributions in personal communications, meteor-burst communications systems, spread spectrum communications systems, FM and phase-locked systems, and HF systems. His design of an adaptive data modulator, used for voice coding, is used on the Space Shuttle.

Dr. Schilling was President of IEEE Communications Society from 1980 to 1981, and a member of the Board of Directors of the IEEE from 1982 to 1983. He was Editor-in-Chief of the IEEE TRANSACTIONS ON COMMUNICATIONS and Director of Publications from 1968 to 1978. During this period, he initiated IEEE COMMUNICATIONS MAGAZINE and IEEE JOURNAL ON SELECTED AREAS IN COMMUNICATIONS. During his term as President, he also initiated the MILCOM and INFOCOM Conferences. He is a member of Sigma Xi, and has been an international representative for the IEEE in Russia where he was part of the Popov Society exchange and in China where he led an IEEE delegation.

SECTION 1.6
FREQUENCY-HOPPING SPREAD-SPECTRUM SYSTEMS

This section contains three papers on frequency-hopping spread-spectrum systems. The first paper, "*Error Probability of Fast Frequency Hopping Spread Spectrum with BFSK Modulation in Selective Rayleigh and Selective Rician Fading Channels*" by Solaiman et al. evaluates the performance of a fast frequency hopping system with noncoherent binary FSK demodulation when the channel is selective Rayleigh and Rician fading. The second paper, "*Probability of Error in Frequency-Hop Spread-Spectrum Multiple-Access Communication Systems with Noncoherent Reception*" by Cheun and Stark, derives expressions for the error probability of frequency hopping spread spectrum system in a multiple-access environment. The hopping rate is at one hop per symbol and the modulation is binary FSK. The third paper, "*Optimum Detection of Slow Frequency-Hopped Signals*" by Levitt et al., derives the optimum receiver for slow frequency-hopping system with M-ary FSK modulation.

Error Probability of Fast Frequency Hopping Spread Spectrum with BFSK Modulation in Selective Rayleigh and Selective Rician Fading Channels

BASSEL SOLAIMAN, ALAIN GLAVIEUX, AND ALAIN HILLION

Abstract—**The performance of noncoherent reception in fast frequency hopped spread-spectrum (FFH-SS) communication systems operating through noisy, fading multipath channels is investigated. Systems operating with binary frequency-shift keying modulation (BFSK) and noncoherent demodulation are examined under the assumption of very slow fading. These analyses demonstrate the frequency hopping benefits in selective channels. Expressions are derived for the bit error rate in the context of selective Rayleigh and selective Rician fading channels, as a function of channel and system parameters.**

I. Introduction

Spread-spectrum communication systems have been widely studied and mainly used for military applications [1]. They provide a certain degree of protection against intentional or unintentional jamming. Recently, spread-spectrum systems have been found to be applicable in combatting multipath fading [2]. Both, direct sequence (DS) and frequency hopping (FH) schemes have been proposed for such applications [2]–[3]. Synchronous carrier recovery is a difficult task in a fading multipath environment, which leads to the consideration of noncoherent demodulation techniques. Fast frequency hopping spread-spectrum (FFH-SS) systems using frequency-shift keying (FSK) modulation of a binary bit stream appear attractive. So far, boundaries and approximations for the bit error probability have been proposed [3]. In this study, two important classes of fading models are considered: the class of Rayleigh selective fading channels which includes the additive white Gaussian noise, and the class of Rician selective fading channels. Furthermore, a distinction is made between selective channels and what are called highly selective channels. In selective channels, the multipath time delay spread $\tau_{\max}$ [4] (i.e., the maximum time delay that may occur along the communication link) is assumed to be less than the symbol duration T, whereas for highly selective channels $\tau_{\max}$ is greater than the symbol duration.

A brief outline of the paper is as follows. The model of the FFH-BFSK system is presented in Section II. Detailed description of the statistics of the envelope detector outputs and the evaluation of the bit error rate (BER) performance are given in Section III for both Rayleigh and Rician selective fading channels. In Section IV the simulation method and numerical results are presented. Conclusions are given in Section V.

II. System Model and Definitions

A. Transmitter Model

The data signal $a(t)$ at the input of the modulator is a sequence of nonoverlapping rectangular pulses of duration T. The amplitude of the lth pulse, $lT \le t < (l+1)T$, is a random variable a_l which may assume values from $\{-1, +1\}$ with the same probability. The signal, at the output of the FFH-BFSK modulator performing one hop per data bit, when $a(t)$ is the input, is given by

$$E_0(t) = \sqrt{\frac{2E_b}{T}} \sum_{l=-\infty}^{+\infty} P(t - lT) \cdot \cos\left[2\pi\left(f_c + f_l + a_l \frac{\beta}{2T}\right)t + \theta_l\right] \quad (1)$$

where $P(t)$ denotes a rectangular pulse of unit height and duration T, E_b is the energy per bit, f_c is the carrier frequency and $\beta/2T$ $\{\beta = 1, 2, \cdots\}$ is one-half the spacing between the two FSK tones. The phase angle θ_l is a random variable uniformly distributed between 0 and 2π. The frequency f_l (l Modulo K) assumes values from the set $F = \{F_0, F_0 + C/T, F_0 + 2C/T, \cdots, F_0 + (K-1)C/T\}$ where C is a positive integer and K denotes the number of frequencies used in frequency hopping. At this stage, it must be noted that K must be chosen in order to satisfy the inequality

$$(K-1)T \ge \tau_{\max} \quad (2)$$

where $\tau_{\max}$ denotes the multipath time delay spread. This way of determining K provides the guarantee that the data bit a_l, transmitted with a given frequency f_l, is not perturbed by an echo using the same frequency emanating from another data bit.

B. Channel Model and Received Signal

The channel model under consideration is similar to the model defined in [5]–[6] where the transmitted waveform is propagated through a noisy fading multipath channel. The multipath and the noise are assumed to be statistically independent. The multipath comprises M fading paths. Each path results from a different scattering channel (see Fig. 1). The mth path ($m = 0, 1, \cdots, M-1$) is associated with three random processes $r_{m,l}(t)$, τ_m and $\theta_{m,l}(t)$ that respectively describe the strength coefficient, the time delay and the carrier phase shift introduced by the path. The index l is due to the fact that the transmitted frequency changes every T seconds. Since the Doppler shift is a function of the frequency of the transmitted signal, the strength coefficient and the carrier phase shift of each path are also functions of the frequency f_l transmitted during $[lT, (l-1)T]$, [7]. It can be assumed, however, that the path characteristics do not vary significantly within a transmission period T so that these processes can be respectively described by the single random variables $r_{m,l}$, τ_m, and $\theta_{m,l}$. Hence, the received signal can be written as follows:

$$R(t) = \sqrt{\frac{2E_b}{T}} \sum_{m=0}^{M-1} \sum_{l=-\infty}^{+\infty} r_{m,l} P(t - lT - \tau_m) \cdot \cos\left[2\pi\left(f_c + f_l + a_l \frac{\beta}{2T}\right)t - \theta_{m,l}\right] + N(t). \quad (3)$$

Paper approved by the Editor for Radio Communications of the IEEE Communications Society. Manuscript received November 25, 1987; revised April 13, 1988, September 9, 1988, and November 14, 1988. This paper was presented in part at ICASSP'88, New York.

The authors are with Ecole Nationale Superieure des Telecommunications de Bretagne, BP 832, 29285 Brest Cedex, France.

IEEE Log Number 8933296.

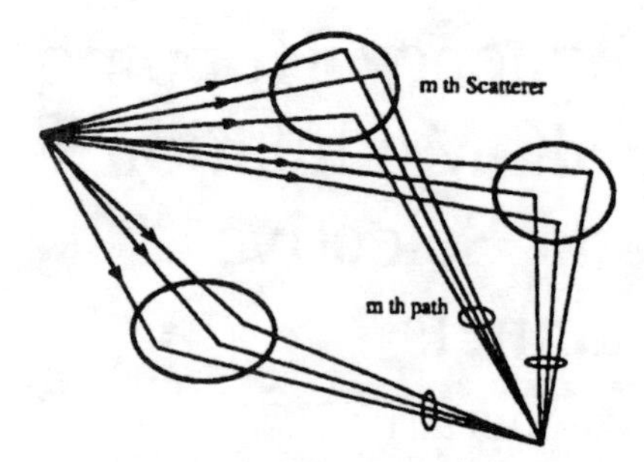

Fig. 1. Channel model.

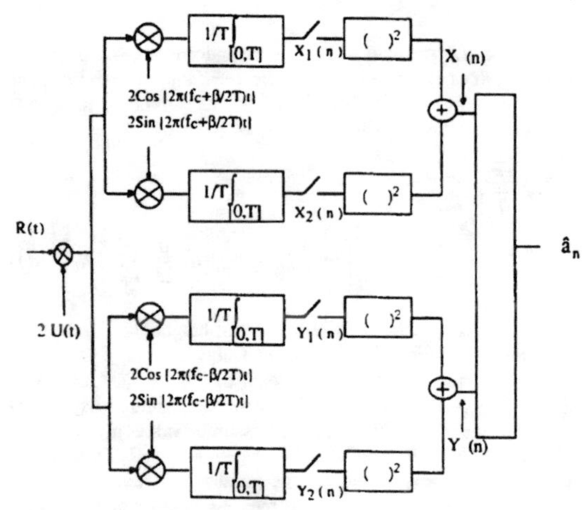

Fig. 2. Envelope detector structure.

Each $\theta_{m,l}$ is assumed to be uniformly distributed over $0, 2\pi$. All path strengths are independently and identically Rayleigh-distributed [7]–[8] according to

$$P_{r_{m,l}}(r) = \frac{2r}{b_m} \operatorname{Exp}\left\{-\frac{r^2}{b_m}\right\} \qquad r \geq 0, \text{ for all } l. \tag{4}$$

The noise term $N(t)$ is assumed to be additive, zero-mean, white Gaussian noise (AWGN) and statistically independent of the multipath.

C. Receiver Model

The FFH-BFSK receiver is depicted in Fig. 2. It consists of a frequency dehopper followed by an envelope detector. The dehopping signal $U(t)$ [given in (5)] introduces a phase signal φ_l which is constant during the time intervals between hops.

$$U(t) = \sum_{l=-\infty}^{+\infty} P(t - lT) \operatorname{Cos}(2\pi f_l t + \varphi_l) \qquad (l \text{ modulo } K). \tag{5}$$

The only assumption made is that the receiver is time synchronous with the main (first arriving) path (i.e., $\tau_0 = 0$).

The normalizing condition that the mean received energy per bit at the input of the envelope detector is equal to E_b is expressed as

$$E\left\{\frac{2E_b}{T} \int_{-\infty}^{+\infty} \left[\sum_{m=0}^{M-1} r_{m,l} P(t - lT - \tau_m) \cdot \operatorname{Cos}(2\pi(f_c + a_l\beta/2T)t + \theta_{m,l} - \varphi_l)\right]^2 dt\right\} = E_b$$

where $E\{.\}$ denotes the mathematical expectation.

Hence,

$$\sum_{m=0}^{M-1} b_m = 1. \tag{6}$$

The double-sided power spectral density of the noise at the input of the envelope detector is taken equal to $N_0/2$ W/Hz.

The square-law envelope detector has two branches each with in-phase and quadrature subbranches. The output of the branch corresponding to "+1" data bits is denoted by $X(n)$ and the output corresponding to "−1" data bits by $Y(n)$ where (see Fig. 2):

$$X(n) = [X_1(n)]^2 + [X_2(n)]^2$$
$$Y(n) = [Y_1(n)]^2 + [Y_2(n)]^2 \tag{7}$$

$X_1(n)$ and $Y_1(n)$ are the outputs of the in-phase subbranches, $X_2(n)$ and $Y_2(n)$ are the outputs of the quadrature subbranches:

$$X_1(n) = 4/T \int_{nT}^{(n+1)T} R(t)U(t) \operatorname{Cos}[2\pi(f_c + \beta/2T)t]\, dt \tag{8}$$

$$X_2(n) = 4/T \int_{nT}^{(n+1)T} R(t)U(t) \operatorname{Sin}[2\pi(f_c + \beta/2T)t]\, dt \tag{9}$$

$Y_1(n)$ [resp., $Y_2(n)$] has the same expression as $X_1(n)$ [resp., $X_2(n)$] after replacing $\beta/2T$ by $-\beta/2T$.

III. Bit Error Rate (BER) Performance

The bit error rate performance of the system under consideration is evaluated when a maximum likelihood detector is used. Both Rayleigh and Rician selective fading channels are considered. In the bit interval $[nT, (n+1)T]$ (synchronized to the main path), the received waveform $R(t)$ depends on the symbols $a_n, a_{n-1}, \cdots, a_{n-(K-1)}$. This, in fact, is due to the intersymbol interference (ISI) introduced by the multipath propagation. Therefore, in the following analyses we deal with the conditional probability of error P_e defined as

$$P_e = \frac{1}{2} \Pr\{X(n) > Y(n)/a_n = -1, \{a_{n-i}\}\} + \frac{1}{2} \Pr\{X(n) < Y(n)/a_n = 1, \{a_{n-i}\}\} \tag{10}$$

where $\{a_{n-i}\}$ represents the sequence $a_{n-1}, \cdots, a_{n-(K-1)}$.

Consequently, to obtain the probability of error $\bar{P}_e$, expressions of the conditional probability of error to be determined must be averaged over all possible patterns of $\{a_{n-i}\}$, i.e.,

$$\bar{P}_e = E_{\{a_{n-i}\}}\{p_e\}. \tag{11}$$

A. Statistics of the Envelope Detector Outputs

To evaluate the error probability of the system, a detailed description of the outputs of the envelope detector must be provided. First, substituting (3) and (5) in (8), $X_1(n)$ can be expressed as follows:

$$X_1(n) = D(n) + I(n) + N(n) \tag{12}$$

where

- $D(n)$ is the term due to the main path. It is called the desired signal component. Taking into account the definition of the function $P(t)$ and ruling out the high frequency terms, it turns out that

$$D(n) = \sqrt{2E_b/T}\, r_{0,n} \operatorname{Cos} \Psi_{0,n} \quad \text{if } a_n = 1$$
$$= 0 \quad \text{if } a_n = -1 \tag{13}$$

where

$$\Psi_{0,n} = \theta_{0,n} - \varphi_n.$$

- $I(n)$ is the term due to other paths. It is called the multipath interference signal component. Hence,

$$I(n) = 4/T\sqrt{2E_b/T}\int_{nT}^{(n+1)T}\sum_{m=1}^{M-1}\sum_{l=-\infty}^{\infty} r_{m,l}P(t - lT - \tau_m) \cdot \mathrm{Cos}\,[2\pi(f_c + f_l + a_l\beta/2T)t + \theta_{m,l}] \cdot \mathrm{Cos}\,(2\pi f_n t + \varphi_n)\,\mathrm{Cos}\,[2\pi(f_c + \beta/2T)t]\,dt. \quad (14)$$

To evaluate $I(n)$ we can write the time delay τ_m of the mth path as

$$\tau_m = JT + t_{m,J} \quad (15)$$

where $0 \le t_{m,J} \le T_0$ and $J \in \{0, 1, \cdots, K-2\}$. The parameter T_0 is in fact taken equal to T when dealing with highly selective fading channels ($\tau_{\max} > T$). When dealing with selective channels ($J = 0$, $\tau_{\max} < T$) T_0 could assume any value between 0 and T. The main reason for introducing the parameter T_0 is that it enables us to show the influence of the selectivity of the channel on the performance of the system. This may be obtained by using several values of T_0. However, the performance of the system in nonselective fading channels may be deduced from the performance of the system in selective fading channels by making T_0 tend to zero. In the following analyses, $I(n)$ is approximated by a Gaussian, zero-mean, random variable with variance α^2 (Appendix A)

$$\alpha^2 = 2E_b/T\sum_{J=0}^{K-2} S_J \left\{\frac{1}{A^2(n,J)} \cdot \left[1 - \frac{\mathrm{Sin}\,A(n,J) - \mathrm{Sin}\,[A(n,J)(1 - T_0/T)]}{A(n,J)T_0/T}\right] + \frac{1}{A^2(n,J+1)}\left[1 - \frac{\mathrm{Sin}\,[A(n,J+1)T_0/T]}{A(n,J+1)T_0/T}\right]\right\} \quad (16)$$

with

$$S_J = \sum_{m \in N_J} b_m \quad (17)$$

and

$$A(n,J) = 2\pi\left[(f_{n-J} - f_n)T + \frac{\beta}{2}(a_{n-J} - 1)\right] \quad (18)$$

where N_j denotes all paths arriving with time delays between $[JT, (J+1)T]$.

- $N(n)$ is the noise term given by

$$N(n) = \frac{4}{T}\int_{nT}^{(n+1)T} N(t)\,\mathrm{Cos}\,(2\pi f_n t + \varphi_n)\,\mathrm{Cos}\,[2\pi(f_c + \beta/2T)t]\,dt$$

is also a Gaussian, zero-mean, random variable with variance $\sigma^2 = N_0/T$. Finally, $X_1(n)$ can be written as follows:

$$X_1(n) = \left(\frac{1 + a_n}{2}\right)\sqrt{2E_b/T}r_{0,n}\,\mathrm{Cos}\,\Psi_{0,n} + Z_1(n) \quad (19)$$

where $Z_1(n) = I(n) + N(n)$ is a Gaussian, zero-mean, random variable with variance $\alpha^2 + \sigma^2$. The evaluation of $X_2(n)$, $Y_1(n)$, and $Y_2(n)$ can be made in the same way. Hence,

$$X_2(n) = -\left(\frac{1 + a_n}{2}\right)\sqrt{2E_b/T}r_{0,n}\,\mathrm{Sin}\,\Psi_{0,n} + Z_2(n) \quad (20)$$

$$Y_1(n) = \left(\frac{1 - a_n}{2}\right)\sqrt{2E_b/T}r_{0,n}\,\mathrm{Cos}\,\Psi_{0,n} + W_1(n) \quad (21)$$

$$Y_2(n) = -\left(\frac{1 - a_n}{2}\right)\sqrt{2E_b/T}r_{0,n}\,\mathrm{Sin}\,\Psi_{0,n} + W_2(n) \quad (22)$$

where $Z_2(n)$, $W_1(n)$, and $W_2(n)$ are Gaussian, zero-mean, random variables with variances, respectively, given by

$$E\{[Z_1(n)]^2\} = E\{[Z_2(n)]^2\} = \alpha^2 + \sigma^2$$
$$E\{[W_1(n)]^2\} = E\{[W_2(n)]^2\} = \xi^2 + \sigma^2. \quad (23)$$

ξ^2 has the same expression as α^2 given in (16) after replacing $(\beta/2)\,(a_{n-J} - 1)$ by $-(\beta/2)\,(a_{n-J} + 1)$ in (18) giving $A(n, J)$. Finally, to determine the bit error rate, two classes of fading channels are considered: the Rayleigh selective fading channel and the Rician selective fading channel.

B. Conditional BER (Pe1) in Rayleigh Selective Fading Channel

In a Rayleigh selective fading channel, all strength coefficients are Rayleigh distributed (4). In this case, and conditionally on a_n, $\{a_{n-i}\}$, $X(n)$ given in (7) is exponentially distributed with the parameter η^2 [9]. The probability density function (pdf) of $X(n)/a_n$, $\{a_{n-i}\}$ is expressed as

$$P_{X(n)/a_n,\,\{a_{n-i}\}}(x) = \frac{1}{2\eta^2}\,\mathrm{Exp}\left\{-\frac{x}{2\eta^2}\right\}, x \ge 0 \quad (24)$$

where

$$\eta^2 = \frac{1}{2}E\{X(n)/a_n, \{a_{n-i}\}\} = \left(\frac{1 + a_n}{2}\right)\frac{E_b}{T}b_0 + \alpha^2 + \sigma^2 \quad (25)$$

$Y(n)/a_n$, $\{a_{n-i}\}$ is also exponentially distributed with a parameter having the same expression as η^2 given in (25) after replacing $(1 + a_n)$ by $(1 - a_n)$ and α^2 by ξ^2. α^2 and ξ^2 depend on the data bit a_n and on the sequence $\{a_{n-i}\}$. Thus, conditionally on the sequence $\{a_{n-i}\}$ we denote

$$\alpha^2 = \alpha_+^2 \quad \xi^2 = \xi_+^2 \quad \text{if}\, a_n = +1$$
$$\alpha^2 = \alpha_-^2 \quad \xi^2 = \xi_-^2 \quad \text{if}\, a_n = -1. \quad (26)$$

Hence, after some mathematical manipulations, the conditional probability of error (10) of the FFH-BFSK modulation scheme with noncoherent detection in a Rayleigh selective fading channel may be expressed as

$$P_{e1} = \frac{(\alpha_-^2 + \xi_+^2)T/2N_0 + 1}{(E_b/N_0)b_0 + (\alpha_+^2 + \xi_+^2)T/N_0 + 2}. \quad (27)$$

The classical result relating the bit error rate of the BFSK modulation scheme to the signal-to-noise ratio in a nonselective Rayleigh fading channel is easily deduced by making T_0 tend to zero and taking $K = 2$. Therefore, α_-^2 and ξ_+^2 are reduced to zero and α_+^2, ξ_-^2 are reduced to $E_b S_0/T$. Hence, P_{e1} can be expressed as

$$P_{e1} = \frac{1}{2 + \frac{E_b}{N_0}}. \quad (28)$$

C. Conditional BER (Pe2) in a Rician Selective Fading Channel

In this case, the main path is scaled by a deterministic coefficient r_0 and all the other paths have Rayleigh-distributed strength coefficients. Distributions of $\sqrt{X(n)}$ and $\sqrt{Y(n)}$ are determined instead of those of $X(n)$ and $Y(n)$. This makes the evaluation of the bit error rate easier. Where $\sqrt{X(n)}$ is concerned, conditionally to $\{a_n = -1$ and $\{a_{n-i}\}\}$ it is Rayleigh-distributed (4) with parameter $(\alpha_-^2 + \sigma^2)$

and, conditionally to $\{a_n = +1$ and $\{a_{n-i}\}\}$, it is Rice-distributed [4] with parameter $(\alpha_+^2 + \sigma^2)$:

$$P_{\sqrt{X(n)}/a_n=1,\{a_{n-i}\}}(x) = \frac{x}{\gamma_+^2} \cdot \text{Exp}\left\{-\frac{1}{2\gamma_+^2}[x^2 + (2E_b/T)r_0^2]\right\} \cdot Io\left(x\frac{\sqrt{2E_b/T}}{\gamma_+^2}r_o\right), x \geq 0 \tag{29}$$

where

$$\gamma_+^2 = \alpha_+^2 + \sigma^2$$

and $Io(.)$ is the zeroth-order modified Bessel function defined as

$$Io(x) = \frac{1}{\pi}\int_0^\pi \text{Exp}\{-x \, \text{Cos}\,\varphi\} d\varphi. \tag{30}$$

The probability density function of $\sqrt{Y(n)}$ is determined in a similar way. Conditionally to $\{a_n = -1$ and $\{a_{n-i}\}\}$, it is Rice-distributed with parameter $(\xi_-^2 + \sigma^2)$ and, conditionally to $\{a_n = +1$ and $\{a_{n-i}\}\}$ it is Rayleigh-distributed with parameter $(\xi_+^2 + \sigma^2)$.

Therefore, the conditional probability of error P_{e2} (10) of the FFH-BFSK modulation scheme with noncoherent detection in a Rician selective fading channel can be expressed as follows:

$$P_{e2} = \frac{N_0/T + 1/2(\alpha_-^2 + \xi_+^2)}{2N_0/T + \alpha_-^2 + \xi_-^2} \text{Exp}\left\{-\frac{(E_b/2N_0)r_0^2}{1 + (\alpha_-^2 + \xi_-^2)T/2N_0}\right\}. \tag{31}$$

The bit error probability of the BFSK modulation scheme in a nonselective Rician fading channel can be easily deduced from (31) by making T_0 tend to zero and $K = 2$

$$P_{e2} = \frac{1}{2 + S_0E_b/N_0} \text{Exp}\left\{-\frac{b_0E_b/N_0}{2 + S_0E_b/N_0}\right\} \tag{32}$$

where b_0E_b is the energy of the nonfaded version of the transmitted signal, and S_0E_b represents the energy of the non-selective Rayleigh faded version of the transmitted signal. The classical result relating the bit error rate of the BFSK modulation scheme to the signal-to-noise ratio in a Gaussian channel is obtained by making S_0 equal to zero and $b_0 = 1$

$$P_{e2} = \frac{1}{2} \text{Exp}\left\{-\frac{E_b}{2N_0}\right\}. \tag{33}$$

IV. Numerical Results

In this section, the BER numerical evaluation of the FFH-BFSK modulation scheme in the Rayleigh selective fading channel and in the Rician selective fading channel are presented. The channel model has been specified previously (Section II-B). Furthermore, the path gain b_m is assumed to be the same for all the N_J paths, and S_J (17) decreases exponentially as a function of J [10]–[11]. However, this model is one of the simplest fading models to analyze. The performance of the system for this channel model serves as a reference with which the performance for more realistic but more complex models can be compared.

Two specific cases are dealt with. The first one is when all path delays are smaller than the symbol duration T. This is the case of Indoor Wireless Communications (IWC) [12] and [13] for example. The second case is when the multipath time delay spread is greater than the symbol duration T. This is the case of underwater acoustic channel [14], [15]. A channel following the first case is referenced

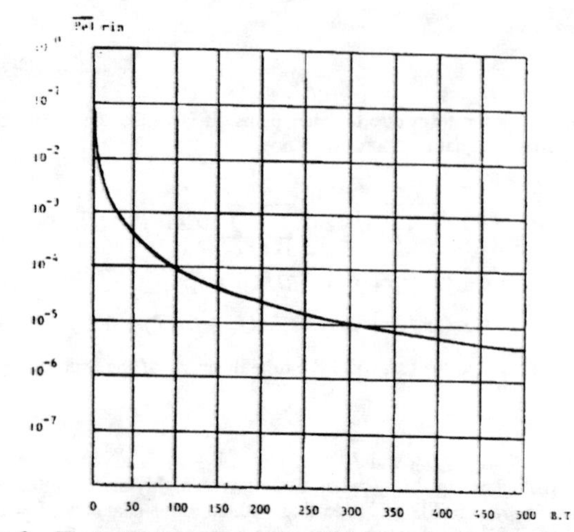

Fig. 3. The irreducible error probability $\overline{Pe1}_{min}$ in Rayleigh selective fading channel as a function of the spectral bandwidth $B.T$.

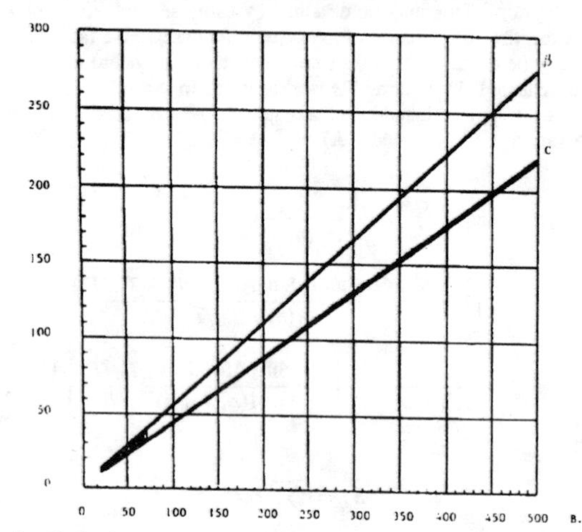

Fig. 4. Optimal values of C and β as a function of the spectral bandwidth $B.T$.

as "a selective fading channel," and one following the second case as "a highly selective fading channel."

A. Selective Fading Channels

Given that the number K of frequencies used in frequency hopping must satisfy (2), the set F consists of at least two frequencies ($K \geq 2$). It is enough for our purposes to have $K = 2$. In this case $F = \{F_0, F_0 + C/T\}$. In order to obtain numerical results, the values of the parameters C and β (where $\beta/2T$ is one-half the spacing between the two FSK tones) must be determined. The method used in determining C and β is described as follows. Since the key performance indicator in communications through a noisy fading multipath channel is the irreducible error probability (which is due to the intersymbol interference ISI introduced by the multipath propagation), this irreducible error probability is first determined from (27) [Rayleigh selective fading channel] by computing the limit of $\bar{P}_{e1}$ when the signal-to-noise ratio tends towards infinity:

$$\bar{P}_{e1}\infty = \text{Lim}\,\bar{P}_{e1} \quad \text{when} \quad \frac{E_b}{N_0} \to \infty. \tag{34}$$

The irreducible error probability is independent of the choice of b

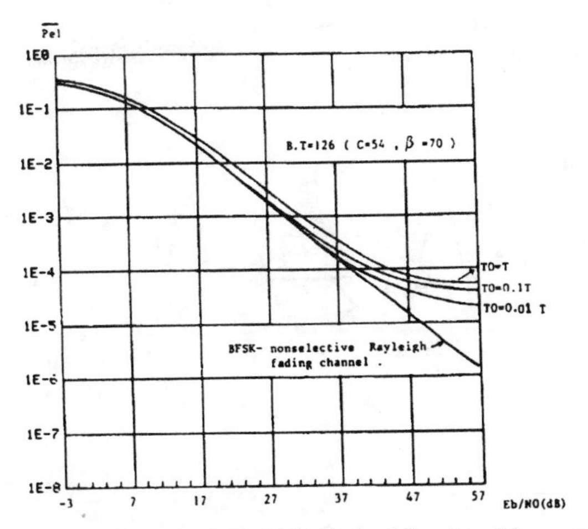

Fig. 5. Probability of error in Rayleigh selective fading channel for several values of T_0, compared to the probability of error of the BFSK modulation in nonselective Rayleigh fading channel. Optimal values of C and β were used to have $B.T = 126$. SIR = 0 dB.

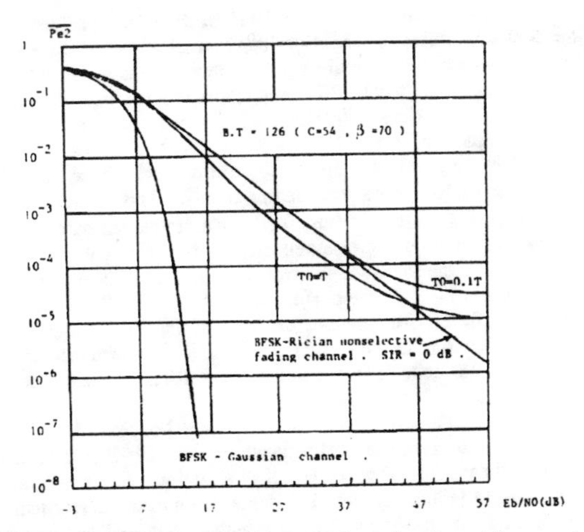

Fig. 7. Probability of error in Rician selective fading channel for two values of T_0, compared to the probability of error of the BFSK modulation in a Gaussian and in nonselective Rician fading channels. Optimal values of C and β were used to have $B.T = 126$. SIR = 0 dB.

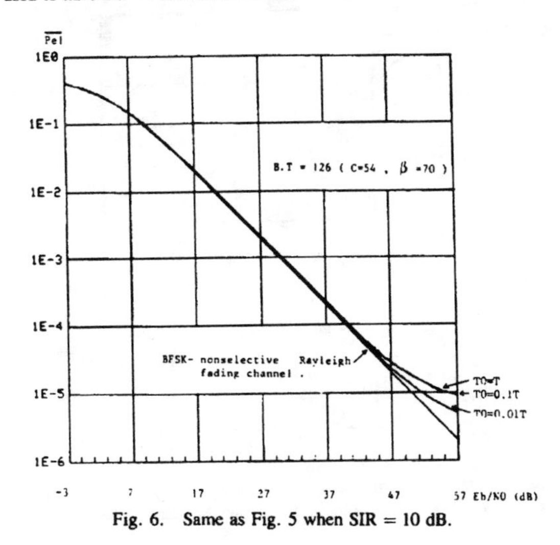

Fig. 6. Same as Fig. 5 when SIR = 10 dB.

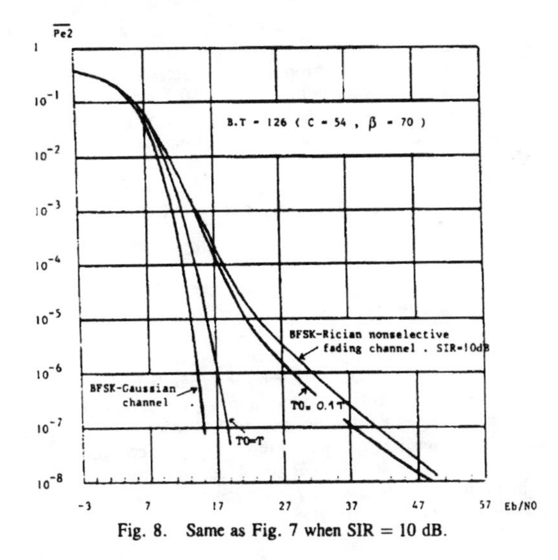

Fig. 8. Same as Fig. 7 when SIR = 10 dB.

(where $b = b_m$ for all m), since b affects the signal and the interference alike.

The spectral bandwidth, $B.T$ is approximately given by

$$B.T \approx (K-1)C + \beta + 2. \tag{35}$$

Therefore, for a given spectral bandwidth, the values of C and β are determined in order to minimize the irreducible error probability:

$$\bar{P}_{e1}\min = \operatorname{Min} C, \beta\{\bar{P}_{e1}\infty\}. \tag{36}$$

The minimum irreducible error probability $\bar{P}_{e1}$ min and the optimal values of C and β are plotted as functions of the spectral bandwidth $B.T$ in Fig. 3 and Fig. 4, respectively. It can be seen, for example, that if $10E-5$ is the irreducible error probability needed, a spectral bandwidth $B.T$ of at least 300 is required to obtain the value (Fig. 3). In this case $C = 131$ and $\beta = 167$ (Fig. 4).

Notice that for $B.T$ greater than 100, the minimum irreducible error probability does not decrease significantly. Clearly, the constant ratio C/β which is independent of the spectral bandwidth is due to the fact that the four tones used for the FFH-BFSK modulation scheme (with $K = 2$) must be spaced alike.

In the case of selective fading channels, the signal-to-interference ratio is defined as

$$\mathrm{SIR} = 10\,\mathrm{Log}\left(\frac{b_0}{S_0}\right). \tag{37}$$

Figs. 5 and 6 compare the performance of the FFH-BFSK modulation scheme in the case of a selective Rayleigh fading channel to the performance of the BFSK modulation scheme in a nonselective Rayleigh fading channel [4]. The probability of error $\bar{P}_{e1}$ has been plotted for three values of T_0 and this for two values of the signal-to-interference ratio (SIR) equal to 0 and 10 dB. This was done by assuming that C and β assume their optimum values to obtain a spectral bandwidth equal to 126.

For E_b/N_0 less than 40 dB and SIR equal to 0 dB, there is approximately a 2 dB difference between the performance of the FFH-BFSK in a selective Rayleigh fading channel and the performance of the BFSK modulation scheme in a nonselective Rayleigh fading channel. It can be seen from Fig. 6 where the signal-to-interference ratio is equal to 10 dB, that there is nearly no difference between the performance of the two systems.

Similar results for the FFH-BFSK modulation scheme are given in Figs. 7 and 8 in the case of selective Rician fading channels. Fig.

7 compares the performance of the FFH-BFSK to the performance of the BFSK modulation scheme in the case of nonselective Rician fading channel when the signal-to-interference ratio is equal to 0 dB. The performance of the BFSK modulation scheme in a Gaussian channel (33) [i.e., Rician nonselective fading channel when SIR = ∞] has been plotted, too.

Notice that the results are not much different from those obtained in the case of a selective Rayleigh fading channel. This means that for small signal-to-interference ratios, Rayleigh selective fading channels and Rician selective fading channels are nearly equivalent.

Fig. 8 where the signal-to-interference ratio is equal to 10 dB, shows that when $T_0 = T$, the performance of the FFH-BFSK system is not much different from the performance of the BFSK modulation scheme in a Gaussian channel and better than the performance of the BFSK modulation scheme in the nonselective Rician fading channel for the same SIR.

This result is due to the use of frequency hopping. However, when T_0 tends to zero the performance of the FFH-BFSK system approaches the performance the BFSK modulation scheme in a nonselective Rician fading channel. The degradation of the performance when T_0 tends to zero is due to the fact that the faded versions of the transmitted signal affect more and more the nonfaded (Specular) component.

B. Highly Selective Fading Channels

In this case, the multipath time delay spread $\tau_{\max}$ is greater than the symbol duration T. An example of such channels is the underwater acoustic channel where the propagation velocity is approximately 1500 m/s, which means that a time delay of 0.1 s corresponds to a 150 m distance. The choice of the frequency hopping pattern from the set F may be linear or pseudorandom. The linear frequency hopping pattern sequence is simply

$$F_0, F_0 + C/T, F_0 + 2C/T, \cdots, F_0 + (K-1)C/T, \quad F_0, F_0 + C/T, F_0 + 2C/T, \cdots.$$

Given the fact that the S_J's decrease exponentially as a function of J, it is preferable to space the successive frequencies by more than C/T. This can be performed by using a pseudorandom hopping pattern sequence. A method for determining a pseudorandom hopping pattern is presented in this paper and the system's performance for the linear and for the pseudorandom hopping patterns is given. The method used guarantees that the spacing between a given frequency and the X frequencies used just before frequency changes, will be at least equal to $P.C/T$ where P and X are positive integers. This may be achieved if the number K of frequencies used in frequency hopping conforms to the equation

$$(X+1)P + X = K \tag{38}$$

and must satisfy (2) at the same time. In this case, the frequency hopping pattern sequence is determined as follows.

If frequency f_n (n module K) assumes the value $F_0 + JC/T$ (J module K) from the set F then the frequency f_{n+1} ($n+1$ modulo K) is given by $F_0 + (J+P+1)C/T$ ($J+P+1$ modulo K). To evaluate the performance of the FFH-BFSK modulation scheme in highly selective fading channels, a realistic example used in underwater acoustic communications has been considered. In this example, the transmission of remote-controlled signal with low rate ($1/T = 300$ bits/s) is considered. The system under consideration is intended to work at low depths (about 100 m), in the neighborhood of offshore oil rigs. In this context, the multipath time delay spread of the channel ($\tau_{\max}$) is estimated at 0.1 s. At these depths, the acoustic frequencies to be used are of the order of a few hundred KHz. In our application, the acoustic transducer used has a spectral bandwidth ranging from ≈100 to 150 KHz. From (2), the number of frequencies to be used in frequency hopping is $K \geq 31$.

Given that the S_J's decrease exponentially, to ensure a good protection against paths with small delays, a pseudorandom hopping pattern is considered. In this example the values of K, X, and P have

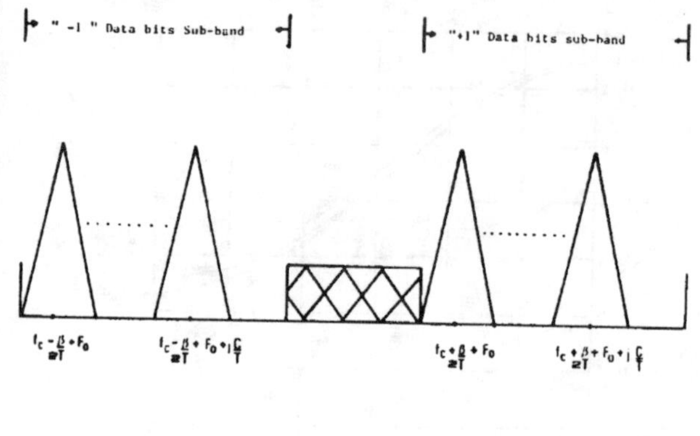

Fig. 9. Spectral band used in simulation.

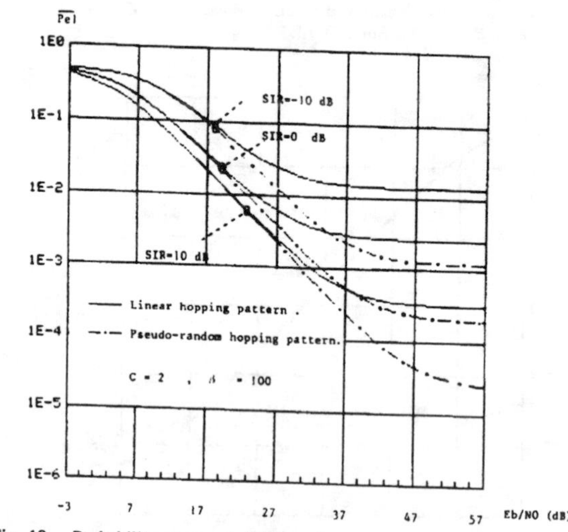

Fig. 10. Probability of error in highly selective Rayleigh fading channel for several values of SIR and for both linear and pseudorandom hopping patterns.

been chosen equal to 31, 3, and 7, respectively (that is, the spacing between a given frequency and the three frequencies used just before frequency changes is at least equal to $7C/T$). The method used for determining the optimal values of C and β is given in (Section IV-A) in the case of selective fading channels ($\tau_{\max} < T$). The same method can be used when dealing with highly selective fading channels. Simulation results show that when $\tau_{\max}$ becomes much greater than T, the FSK tones corresponding to "-1" data bits ($f_n + f_c - \beta/2T$) must be transmitted in the lower half band, and the FSK tones corresponding to "$+1$" data bits ($f_n + f_c + \beta/2T$) in the upper half band. The spectral band used in this example is given Fig. 9, where a central subband (of 11.4 KHz) has been reserved for synchronisation purposes. Therefore, C and β must take the values $C = 2$ and $\beta = 100$. Note that when dealing with highly selective fading channels, T_0 must be taken equal to T.

The signal-to-interference ratio for a highly selective channel is defined as

$$\mathrm{SIR} = 10\,\mathrm{Log}\left\{\frac{b_0}{\sum_{J=0}^{K-2} S_J}\right\}. \tag{39}$$

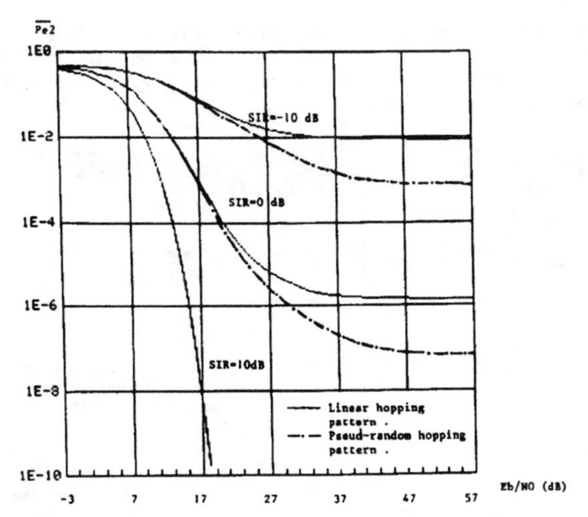

Fig. 11. Probability of error in highly selective Rayleigh fading channel for several values of SIR and for both linear and pseudorandom hopping patterns. $C = 2$, $\beta = 100$.

In Fig. 10, the BER of the FFH-BFSK modulation scheme in a highly selective Rayleigh fading channel has been plotted as a function of the signal-to-noise ratio. This was done for both linear and pseudorandom hopping patterns and for several values of SIR. Similar results are obtained in Fig. 11 in the case of a highly selective Rician fading channel. Note that for high signal-to-noise ratios, the performance of the system when using a pseudorandom hopping pattern is effectively better than the performance of the system when using a linear hopping pattern sequence. For small signal-to-noise ratios both hopping pattern sequences are equivalent.

V. Conclusion

In this study, the performance analysis of noncoherent reception in a FFH-BFSK spread-spectrum communication system in selective fading environment has been carried out. Expressions are derived for the BER in the context of selective Rayleigh and selective Rician fading channels. A distinction between selective channels and highly selective channels has been made as a function of the multipath time delay spread. When dealing with selective fading channels, the performance of the FFH-BFSK modulation scheme can be improved by increasing the spectral bandwidth used by the system. A pseudorandom hopping pattern has been proposed in order to improve the performance of the FFH-BFSK modulation scheme in highly selective fading channels. Since synchronous carrier recovery is a difficult task in a fading multipath environment, the use on a noncoherent demodulation technique (which performs less well than coherent demodulation techniques but makes the system easily implementable) has been considered. At present, a prototype of the system intended to be used in underwater acoustic channel is being produced.

Appendix A

The Variance of the Interference Signal Component

To evaluate $I(n)$ (14), the time delay τ_m of the mth path can be expressed as

$$\tau_m = JT + t_{m,J}$$

where $0 \leq t_{m,J} \leq T_0$ and $J \in \{0, 1, \cdots, (K-2)\}$. Ruling out the high frequency term and taking into account the definition of the function $P(t)$, $I(n)$ becomes

$$I(n) = 1/T\sqrt{2E_b/T}\sum_{J=0}^{K-2}\sum_{m\in N_J}\left\{\int_{nT+t_{m,J}}^{(n+1)T} r_{m,n-J}\,\mathrm{Cos}\,[2\pi(f_{n-J} - f_n + \beta/2T(a_{n-J}-1))t + \theta_{m,n-J} - \varphi_n]\,dt + \int_{nT}^{nT+t_{m,J}} r_{m,n-J-1}\,\mathrm{Cos}\,[2\pi(f_{n-J-1} - f_n + \beta/2T(a_{n-J-1}-1))t]\,dt\right\} \quad \text{(A-1)}$$

where N_J denotes all paths arriving with time delays between JT and $(J+1)T$. After some mathematical manipulations, it turns out that

$$I(n) = \sqrt{2E_b/T}\sum_{J=0}^{K-2}\sum_{m\in N_J} r_{m,n-J}\frac{\mathrm{Sin}\,[A(n,J)(n+1) + \Psi_{m,n-J}] - \mathrm{Sin}\,[A(n,J)(n + t_{m,J}/T) + \Psi_{m,n-J}]}{A(n,J)} + r_{m,n-J-1}\frac{\mathrm{Sin}\,[A(n,J+1)(n + t_{m,J}/T) + \Psi_{m,n-J-1}) - \mathrm{Sin}\,[A(n,J+1)n + \Psi_{m,n-J-1}]}{A(n,J+1)} \quad \text{(A-2)}$$

where

$$A(n,J) = 2\pi[(f_{n-J} - f_n)T + \beta/2(a_{n-J}-1)]$$

and

$$\Psi_{m,n-J} = \theta_{m,n-J} - \Psi_n.$$

$I(n)$ can be approximated by a Gaussian, zero-mean, random variable with variance α^2 (since it is the sum of independent and identically distributed random variables, using the central limit theorem). To determine α^2, the assumption is made that the variances of the strength coefficients $r_{m,J}$ are identical for all the N_J paths, and that the variables $t_{m,J}$ are uniformly distributed over $[0, T_0]$. Clearly, $\alpha^2 = E\{[I(n)]^2\}$ therefore, from the independence assumption, it turns out that

$$\alpha^2 = 2E_b/T \sum_{J=0}^{K-2} \sum_{m \in N_J} E \left\{ r_{m,n-J}^2 \left[\frac{\operatorname{Sin}[A(n,J)(n+1) + \Psi_{m,n-J}] - \operatorname{Sin}[A(n,J)(n + t_{m,J}/T) + \Psi_{m,n-J}]}{A(n,J)} \right]^2 \right.$$
$$\left. + r_{m,n-J-1}^2 \left[\frac{\operatorname{Sin}[A(n,J+1)(n + t_{m,J}/T) + \Psi_{m,n-J-1}] - \operatorname{Sin}[A(n,J+1)n + \Psi_{m,n-J-1}]}{A(n,J+1)} \right]^2 \right\}. \quad \text{(A-3)}$$

Taking into account that $E\{r_{m,n-J}^2\} = b_m$, it is straight forward to show that

$$\alpha^2 = 2E_b/T \sum_{J=0}^{K-2} S_J \left\{ \frac{1}{A^2(n,J)} \cdot \left[1 - \frac{\operatorname{Sin} A(n,J) - \operatorname{Sin}[A(n,J)(1 - T_0/T)]}{A(n,J)T_0/T} \right] + \frac{1}{A^2(n,J+1)} \left[1 - \frac{\operatorname{Sin}[A(n,J+1)T_0/T]}{A(n,J+1)T_0/T} \right] \right\} \quad \text{(A-4)}$$

with

$$S_J = \sum_{m \in N_J} b_m.$$

Acknowledgment

The authors wish to express their grateful acknowledge to P. Vandamme from the Centre National d'Etudes des Télécommunications (C.N.E.T) and P. Aubergier from the Ecole Nationale des Télécommunications de Bretagne (E.N.S.T BR) for their fruitful discussions and helpful suggestions during this work.

References

[1] R. C. Dixon, *Spread Spectrum Systems*. New York: Wiley, 1976.

[2] G. L. Turin, "Introduction to spread spectrum antimultipath techniques and their applications to urban digital radio," *Proc. IEEE*, vol. 68, pp. 328–353, 1980.

[3] E. A. Geraniotis and M. B. Pursley, "Error probabilities for slow frequency hopped spread spectrum multiple-access communications over fading channels," *IEEE Trans. Commun.*, vol. COM-30, pp. 996–1009, May 1982.

[4] J. G. Proakis, *Digital Communications*. New York, McGraw-Hill, 1983.

[5] G. L. Turin, "Communications through noisy, random multipath channels," *IRE Convention Rec.*, Part 4, pp. 154–166, 1956.

[6] H. W. Arnold and W. F. Bodtmann, "A hybrid multichannels hardware simulator for frequency selective mobile radio paths," *IEEE Trans. Commun.*, vol. COM-31, pp. 370–377, Mar. 1983.

[7] W. C. Jacks, Jr., Ed, *Microwave Mobile Communications*. New York: Wiley, 1974.

[8] J. M. Wozencraft and I. M. Jacobs, *Principles of Communication Engineering*. New York: Wiley, 1965.

[9] A. Papoulis, *Probability, Random Variables and Stochastic Processes*. New York: McGraw-Hill, 1965.

[10] D. C. Cox and R. P. Leck, "Distribution of multipath delay spread and average excess delay for 910 MHz urban mobile radio paths," *IEEE Trans. Antennas Propagat.*, vol. AP-23, pp. 206–213, Mar. 1975.

[11] D. C. Cox and R. P. Leck, "Correlation bandwidths and delay-spread multipath propagation statistics of 910 MHz urban mobile radio channels," *IEEE Trans. Commun.*, vol. COM-23, pp. 1271–1281, Nov. 1975.

[12] M. Kavehrad, "Performance of non-diversity receivers for spread spectrum in indoor wireless communications," *AT&T Tech. J.*, vol. 64, no. 6, pp. 1181–1210, July–Aug. 1985.

[13] —, "Direct sequence spread spectrum with DPSK modulation and diversity for indoor wireless communications," *IEEE Trans. Commun.*, vol. COM-35, pp. 224–236, Feb. 1987.

[14] G. Jourdain and G. Y. Jourdain, "Caractérisation du canal marin," Copenhague NATO 1980.

[15] R. Laval, "Characterisation déterministe et stochastique du canal de transmission acoustique sous-marin," Colloque GRETSI, Nice, 1975.

Bassel Solaiman was born in Damascus, Syria, in 1960, and received the telecommunication engineering degree from the Ecole Nationale Supérieure des Télécommunications de Bretagne (ENST BR), Brest, France, in 1984. He is currently a Ph.D. degree candidate in the fields of Digital Communications over fading channels and Underwater Acoustic Communications.

From 1984 to 1985 he has been a Research Assistant in the Communication Group at the Centre d'Etudes et de Recherches Scientifiques (CERS), Damascus, Syria. He joined the Digital Communication Laboratory at the ENST Br, Brest, France, in 1985.

Alain Glavieux was born in France in 1949, and received the engineering degree from the Ecole Nationale Supérieure des Télécommunications, Paris, France, in 1978.

He joined the Ecole Nationale Supérieure des Télécommunications de Bretagne, Brest, France, in 1979, where he is currently Head of Digital Communications Laboratory. His current interests are in the areas communications over fading channels and in underwater acoustic communications.

Alain Hillion was born in France in 1947, and received the Agregation de Mathématique degree from the Ecole Normale Supérieure in 1970, and the Doctorat d'Etat from Pierre et Marie Curie University, Paris, in 1980.

He is currently Professor and Head of Department of Mathématics and Communication Systems at the Ecole Nationale Supérieure des Télécommunications de Bretagne, Brest, France. His research interests include: mathematical statistics, pattern recognition, signal and image processing, and digital communications.

Probability of Error in Frequency-Hop Spread-Spectrum Multiple-Access Communication Systems with Noncoherent Reception

Kyungwhoon Cheun and Wayne E. Stark, *Member, IEEE*

***Abstract*—In this paper we develop expressions for the probability of error for asynchronous frequency-hop spread-spectrum multiple-access networks using Markov hopping patterns and binary frequency shift keying with one symbol transmitted per hop. The expressions are exact when there is one interfering user and orthogonal BFSK employed. They provide excellent approximations when there are more than one interfering user. The model employed takes into account the random phases, random data bits and accounts for the effects due to random delays of the interfering signals. The model also incorporates a finite number of different power levels in the network at the intended receiver. It is also shown that the error probability when Markov hopping patterns are employed is a good approximation to the error probability when memoryless hopping patterns are employed. Finally, by computing the channel capacity and the associated throughput, a simple hard decisions receiver is shown to perform much better than a receiver using perfect side-information to erase the symbols transmitted on hops that were hit when all the users have same power and one binary symbol is transmitted per hop.**

I. Introduction

IN THIS paper we consider an asynchronous frequency-hop spread-spectrum multiple-access (AFHSS–MA) network employing binary frequency shift keying (BFSK) to transmit one binary symbol per hop. Two approximations usually made in analyzing this type of systems are that the hits due to multiple-access interference are independent from hop to hop (the independence assumption) and that the error probability is upper bounded by $1/2$ whenever a hop is hit by multiple-access interference [1]–[3] (the 1/2-bound). Recently, it has been shown by Hegde and Stark [4] that though the hits exhibit an underlying Markov structure, the independence assumption is quite accurate. On the other hand, there has been no definitive solution to computing the probability of error when a hop is hit by multiple-access interference in an AFHSS–MA network.

In [5], a Gaussian approximation was used to approximate the probability of error. In [6], an exact expression for the probability of error was derived for orthogonal BFSK when a hop is hit by one interfering user, and Monte–Carlo simulations were performed to evaluate the error performance for the cases when there are a small number of interfering users (≤ 4). On the other hand in [7], the effect of nonorthogonality of the interfering signals due to the asynchronocity (i.e., random delays) of different users was neglected in the analysis by assuming that the frequency separation between the orthogonal BFSK signals are much larger than the minimum required to guarantee orthogonality (we will refer to this as the Geraniotis' assumption), and expressions for the error probability was derived based on this assumption. But in practice, it would be advantageous to use the minimum frequency separation required because for a fixed bandwidth, this would allow for a larger number of frequency slots and thus decrease the probability that a hop is hit by other interfering users. When minimum frequency separation is employed, an interfering signal arriving at the receiver with a nonzero delay will affect the outputs of both of the correlators even if orthogonal BFSK is employed and thus Geraniotis' assumption would no longer be valid. In [8], a Poisson model was used for the number of interfering users and error probability analysis was done under similar assumptions as in [7]. Very recently, Short and Rushforth [9], obtained an exact[1] expression for the error probability for AFHSS–MA using BFSK *conditioned* on the phases, the delays, the power levels, and the data bits of the interfering signals. This paper also incorporated the case when the FSK signals used are nonorthogonal. In most practical AFHSS–MA systems, it is natural to assume that the phases of the signals of different users as seen by a receiver are random variables statistically independent of each other and to all other random variables and are uniformly distributed on $[0, 2\pi)$. Also, when synchronization at the hop level is not achievable (asynchronous hopping), it is reasonable to assume that the delays of the interfering signals as seen by a receiver are also random variables that are independent of all other random variables and uniformly distributed on $(-T, T)$ where T is the duration of a hop. It is theoretically possible to average

Paper approved by the Editor for Multiple-Access Strategies of the IEEE Communications Society. Manuscript received February 16, 1989; revised June 26, 1991. This work was supported by The National Science Foundation under Grant ECS 8451266 and The Hughes Aircraft Company. This paper was presented in part at the 1989 Conference on Information Sciences and Systems at Johns Hopkins University and the 1990 International Symposium on Information Theory, San Diego, CA.

K. Cheun was with the Department of Electrical Engineering and Computer Science, University of Michigan, Ann Arbor, MI, 48109 and the Department of Electrical Engineering, University of Delaware, Newark, DE 19716. He is now with the Department of Electronics and Electrical Engineering, Pohang Institute of Technology, Pohang, Korea.

W. Stark is with the Department of Electrical Engineering and Computer Science, University of Michigan, Ann Arbor, MI, 48109.

IEEE Log Number 9144548.

[1] The only assumption made here is that the interference from other frequency-hop slots are negligible which is also the assumption made here.

the expression given in [9] over the phases, delays and the data bits of the interfering signals, but with K' interfering users, this would mean the evaluation of a $2K'$-fold integral and a K'-fold summation of the expression for the probability of error, which is not very attractive.

In this paper an AFHSS–MA system where each user employs Markov hopping patterns that are independent of each other is considered. Hence, given that a frequency slot is used for the nth hop of a user, the slot used for the $(n+1)$st hop of that user will be uniformly distributed among the available frequency slots other than the one used on the nth hop. It is assumed that the initial phases of the signals of the users are random, independent of each other and to any other random variable, and are uniformly distributed on $[0, 2\pi)$. Then assuming (for $K' \geq 2$) that the outputs of the two correlators are independent, we apply the results from [10], [11] to obtain an approximation for the probability of error (the results are exact for one interfering user). By assuming that the delays and the data bits of the users are independent of each other and are independent from user to user, and from all other random variables, a reasonably compact expression is obtained for the average probability of error given the power levels of the interfering users. The computational complexity of numerically evaluating this expression is dominated by computation of two expectations which are independent of K' (the number of interfering users in a given hop) and the signal-to-noise ratio. Thus once we have computed these two expectations, computing the error probability given a hop is hit by K' users for any signal-to-noise ratio is a simple task.

In the numerical results section, we compare our results to Monte-Carlo simulation results. It is shown that the approximation (assuming the outputs of the correlators are independent) provides an excellent fit to the actual error probability. It is shown that assuming Markov hopping patterns not only greatly simplifies the analysis and gives results that are easily computed, but also gives very good approximations to the probability of error when memoryless hopping patterns are employed. Channel capacity and the associated throughput based on our results for the probability of error are computed to reveal, quite surprisingly, that a simple hard decisions receiver performs better than a receiver using perfect side-information to erase the code symbols transmitted on a hop that was hit. Performance of synchronous FHSS–MA networks is also considered as a special case of asynchronous FHSS–MA networks where the delays of all the interfering users are zero. For this case the expressions for the probability of error given here are exact for all K' given that orthogonal BFSK is employed. It is shown that asynchronous systems have superior performance over synchronous systems. These observations are true for systems with one binary symbol transmitted per hop but may not be true for systems with multiple symbols transmitted per hop (a slow FH system).

In Section II, we give a brief description of our system model and present the analyses that leads to the expression for the probability of error. In Section III, it is shown that our results given excellent approximations to Monte–Carlo simulation results. Also, numerical results are given for the error probability, channel capacity and throughput and in Section IV, we draw conclusions.

II. System Description and Analysis

Before carrying out the analysis we define some notation that will be used throughout this paper. Let K be the number of active users in the network. Let N be the number of power level groups, that is, there are N different power levels as seen by a receiver, and let $\overline{K}_1, \cdots, \overline{K}_N$ be the number of interfering users in each power level group with $\sum_{i=1}^{N} \overline{K}_i = K - 1$. Let $\boldsymbol{K} = (\overline{k}_1, ..., \overline{k}_N)$ be the interference pattern vector with each component $\overline{k}_i$ representing the number of users from each power level group that hit the hop under consideration, and let $\boldsymbol{\alpha} = (\alpha_1, \cdots, \alpha_{K'+1})$ be the amplitude vector with the ith component being the received amplitude of the signal of the ith interfering user. Clearly, $\sum_{i=1}^{N} \overline{k}_i = K'$ is the total number of interfering users that hit the hop. Also let p_h denote the probability that a hop is hit when $K = 2$ and q be the number of frequency slots available to the network.

We assume Markov hopping patterns are employed so that the probability of a single interfering user overlapping a given hop for its entire duration is zero. The probability of error when Markov hopping patterns are employed will be shown to be a good approximation to the probability of error when memoryless hopping patterns are employed. This is because the probability of a memoryless hopping pattern generating hops at two consecutive frequency slots is very small for practical values of q, namely, $1/q^2$. For Markov hopping patterns, p_h is simply $2/q$ [12].

Now, consider a receiver perfectly synchronized with the first transmitter. Our goal is to compute the average probability that a binary symbol sent on a particular hop of duration T by the first transmitter will be received in error by this receiver given that BFSK modulation with noncoherent detection is used.

The channel model as seen by the first user is shown in Fig. 1. Also, the noncoherent demodulator which is optimal for AWGN channels is shown in Fig. 2 [13]. The complex low-pass equivalent of the received signal for the jth hop interval $[jT, (j+1)T)$ by the first receiver may be written as follows,

$$r(t) = \sum_{k=1}^{K'+1} \alpha_k p(t - jT - \tau_k) \exp(i2\pi b_k \Delta f(t - \tau_k) + i\varphi_k) + z(t) \qquad (1)$$

where $\alpha_k > 0, b_k \in \{-1, +1\}, \tau_k \in (-T, T)$, and $\varphi_k \in [0, 2\pi)$ are, respectively, the amplitude, data bit, the delay and the initial phase of the kth transmitter, $i = \sqrt{-1}$ and $p(t) = 1$ for $t \in [0, T]$ and zero, otherwise. The frequency separation between the two FSK signals is denoted by $2\Delta f$. The background noise is incorporated as $z(t)$ which is the equivalent low-pass Gaussian noise process with power spectral density N_0. Since the receiver is assumed to be perfectly synchronized with the first transmitter, $\tau_1 = 0$.

The outputs of the two correlators corresponding to $b_1 = 1$ and $b_1 = -1$ are denoted by U_1 and U_{-1}, and tedious algebra shows that given a $b_1 = +1$ was transmitted U_1, U_{-1}

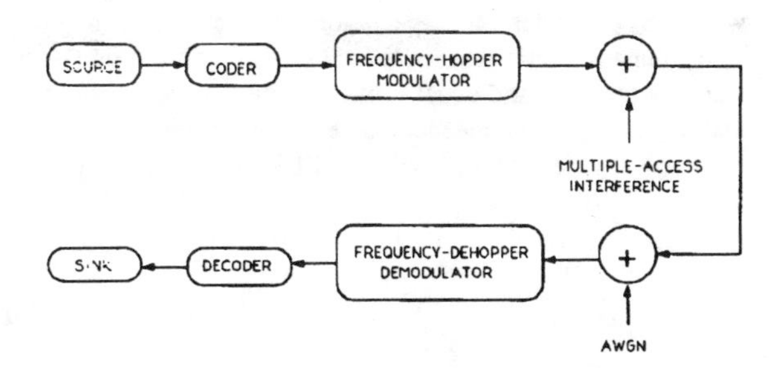

Fig. 1. The channel model as seen by the first user.

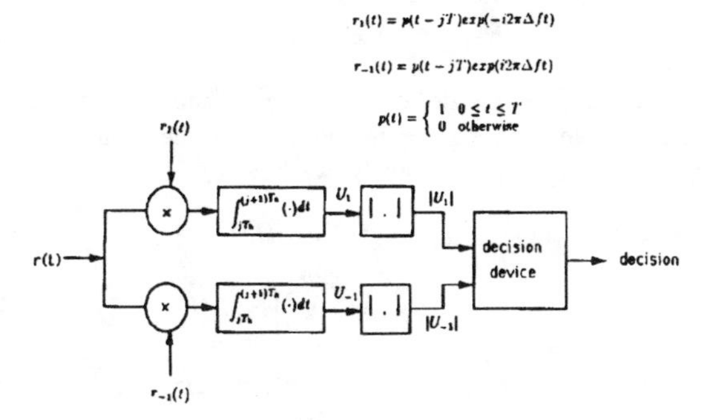

Fig. 2. Noncoherent demodulator.

can be written as

$$U_1 = z_1 + e^{i\varphi_1} + \sum_{k=2}^{K'+1} \left(\frac{\alpha_k}{\alpha_1}\right) \exp(i(\theta(1, p_k, b_k) + \varphi_k)) \cdot A(1, p_k, b_k) \quad (2)$$

$$U_{-1} = z_{-1} + \rho e^{i\varphi_1} + \sum_{k=2}^{K'+1} \left(\frac{\alpha_k}{\alpha_1}\right) \exp(i(\theta(-1, p_k, b_k) + \varphi_k)) \cdot A(-1, p_k, b_k) \quad (3)$$

where $\zeta = 2\Delta fT$ is the normalized separation between the two BFSK signals and $E_b = \alpha_1^2 T/2$. The term ρ is the complex correlation coefficient between the two BFSK signals defined as $\rho = \exp(-i\pi\zeta)\text{sinc}(\zeta)$ where $\text{sinc}(x) = \sin(\pi x)/(\pi x)$. Also, z_1 and z_{-1} are zero mean complex Gaussian random variables with $E\{z_1\bar{z}_1\} = E\{z_{-1}\bar{z}_{-1}\} = 1/(E_b/N_0), E\{z_1\bar{z}_{-1}\} = \rho$. The normalized delay of the kth signal is denoted by $p_k = \tau_k/T$. The magnitude and phase contributions of the kth interfering user, $A(l, p_k, b_k)$ and $\theta(l, p_k, b_k)$, $l \in \{-1, 1\}$, to the correlator outputs is given as follows:

$$A(l, p_k, b_k) = \begin{cases} q_k \,\text{sinc}(\zeta q_k) & b_k = l \\ q_k & b_k \neq l \end{cases} \quad (4)$$

$$\theta(l, p_k, b_k) = \begin{cases} -l\pi\zeta & b_k \neq l \\ -l\pi p_k \zeta & b_k = l \end{cases} \quad (5)$$

where $q_k = 1 - |p_k|$. We first compute the conditional probability of error conditioned on $p = \{p_2, p_3, \cdots, p_{K'+1}\}, b = \{b_2, b_3, \cdots, b_{K'+1}\}$ and K. When $K' = 1$ and $\zeta = 1$, the decision variables $|U_1|$ and $|U_{-1}|$ are independent (see Appendix). We assume that they are independent for $K' > 1$ and $\zeta < 1$. Under this assumption we can rewrite $|U_1|$ and $|U_{-1}|$ as

$$U_1 = z_1 + e^{i\varphi_1^{(1)}} + \sum_{k=2}^{K'+1} \left(\frac{\alpha_k}{\alpha_1}\right) \exp\left(i\varphi_k^{(1)}\right) A(1, p_k, b_k) \quad (6)$$

$$U_{-1} = z_{-1} + |\rho| e^{i\varphi_1^{(-1)}} + \sum_{k=2}^{K'+1} \left(\frac{\alpha_k}{\alpha_1}\right) \exp\left(i\varphi_k^{(-1)}\right) A(-1, p_k, b_k) \quad (7)$$

where $\{\varphi_k^{(l)}\}$ are independent and identically distributed (i.i.d.) with a uniform distribution on $[0, 2\pi)$. Now it is clear that U_1 and U_{-1} are sums of independent and spherically symmetric complex random variables and hence they themselves are spherically symmetric conditioned on p, b and K [10], [11].

Since the noncoherent detector chooses $\text{argmax}_{l=-1,1} \{|U_l|\}$ as its estimate, the probability of error conditioned on p, b and K can be written as follows:

$$\begin{aligned} P_e(p, b, K) = \; & \Pr\{|U_{-1}| > |U_1| \,|\, b_1 = +1, p, b, K\} \\ & \cdot \Pr\{b_1 = +1\} \\ & + \Pr\{|U_1| > |U_{-1}| \,|\, b_1 = -1, p, b, K\} \\ & \cdot \Pr\{b_1 = -1\}. \end{aligned} \quad (8)$$

Assuming that the data bits of the first user are equiprobable, $P_e(p, b, K)$ can be simplified to

$$P_e(p, b, K) = \Pr\{|U_{-1}| > |U_1| \,|\, b_1 = +1, p, b, K\}. \quad (9)$$

Hence, the probability of error can be approximated as folows (for $K' \geq 2$) by assuming that $|U_1|$ and $|U_{-1}|$ are independent [7], [10]

$$P_e(p, b, K) \simeq -\int_0^\infty \Phi_1(s) \frac{d\Phi_{-1}(s)}{ds} \, ds \quad (10)$$

where $\Phi_1(s)$ and $\Phi_{-1}(s)$ are the characteristic functions of U_1 and U_{-1}, respectively. Let us define $\overline{\Phi}_1(s)$ and $\overline{\Phi}_{-1}(s)$ to be

$$\overline{\Phi}_1(s) = \Pi_{k=2}^{K'} J_0(A(1, p_k, b_k)s) \quad (11)$$

$$\overline{\Phi}_{-1}(s) = \Pi_{k=2}^{K'} J_0(A(-1, p_k, b_k)s) \quad (12)$$

which are the characteristic functions of the contributions of the multiple-access interference on the decision variables, and $J_n(\cdot)$ denotes the Bessel function of the first kind of order n defined by [14]

$$J_n(z) = \frac{1}{\pi} \int_0^\pi \cos(z \sin(\theta) - n\theta) \, d\theta. \quad (13)$$

Then $\Phi_1(s)$ and $\Phi_{-1}(s)$ can be easily shown by

$$\Phi_1(s) = e^{-s^2/4\gamma} J_0(s) \overline{\Phi}_1(s) \quad (14)$$

$$\Phi_{-1}(s) = e^{-s^2/4\gamma} J_0(|\rho| s) \overline{\Phi}_{-1}(s) \quad (15)$$

where $\gamma = E_b/N_0$. Hence, $d\Phi_{-1}(s)/ds$ can be computed to be

$$\frac{d\Phi_{-1}(s)}{ds} = -e^{-s^2/4\gamma}\left[\left(\frac{s}{2\gamma}J_0(|\rho|s) + |\rho|J_1(|\rho|s)\right)\overline{\Phi}_{-1}(s) - J_0(|\rho|s)\frac{d\overline{\Phi}_{-1}(s)}{ds}\right]. \quad (16)$$

Now if we further define $\overline{\Phi}_{1,-1}(s)$ as

$$\overline{\Phi}_{1,-1}(s) = \overline{\Phi}_1(s)\left[\left(\frac{s}{2\gamma}J_0(|\rho|s) + |\rho|J_1(|\rho|s)\right)\overline{\Phi}_{-1}(s) - J_0(|\rho|s)\frac{\overline{\Phi}_{-1}(s)}{ds}\right], \quad (17)$$

then $P_e(p, b, K)$ can be written as

$$P_e(p, b, K) = \int_0^\infty e^{-s^2/2\gamma}J_0(s)\overline{\Phi}_{1,-1}(s)ds. \quad (18)$$

Straightforward computation yields

$$\frac{d\overline{\Phi}_{-1}(s)}{ds} = -\sum_{k=2}^{K'} A(-1, p_k, b_k)J_1(A(-1, p_k, b_k)s) \cdot \Pi_{l\neq k}J_0(A(-1, p_l, b_l)s). \quad (19)$$

Hence $\overline{\Phi}_{1,-1}(s)$ is given by

$$\overline{\Phi}_{1,-1}(s) = T_1(s)\cdot\Pi_{k=2}^{K'+1}T_2(p_k, b_k, s) + J_0(|\rho|s) \cdot \sum_{k=2}^{K'+1}[T_3(p_k, b_k, s)\cdot\Pi_{l\neq k}T_2(p_l, b_l, s)] \quad (20)$$

where

$$T_1(s) = \left[\frac{s}{2\gamma}J_0(|\rho|s) + |\rho|J_1(|\rho|s)\right] \quad (21)$$

$$T_2(p_k, b_k, s) = J_0(A(1, p_k, b_k)s)J_0(A(-1, p_k, b_k)s) \quad (22)$$

$$T_3(p_k, b_k, s) = A(-1, p_k, b_k)J_1(A(-1, p_k, b_k)s) \cdot J_0(A(1, p_k, b_k)s). \quad (23)$$

This completes the derivation of the probability of error conditioned on p, b, K (18).

Now, in order to average (18) over p and b, we need to evaluate $P_e(K) = E_{p,b}\{P_e(p, b, K)\}$ which can be written as

$$E_{p,b}\{P_e(p, b, K)\} = E_{p,b}\left\{\int_0^\infty e^{-s^2/2\gamma}J_0(s)\overline{\Phi}_{1,-1}(s)\right\}ds \quad (24)$$

$$= \int_0^\infty e^{-s^2/2\gamma}J_0(s)E_{p,b}\{\overline{\Phi}_{1,-1}(s)\}\,ds \quad (25)$$

where $E_{p,b}$ denotes the expectation with respect to p and b. Using our assumption that p_k's are i.i.d. and uniform on (−1,1) and b_k's are i.i.d. and equally likely to be −1 or +1, and also that p and b are independent of each other and to all other random variables, we can write $E_{p,b}\{\overline{\Phi}_{1,-1}(s)\}$ as follows:

$$E_{p,b}\{\overline{\Phi}_{1,-1}(s)\} = T_1(s)\cdot\Pi_{k=2}^{K'+1}E_2(k, s) + J_0(|\rho|s) \cdot \sum_{k=2}^{K'+1}[E_3(k, s)\cdot\Pi_{l\neq k}E_2(l, s)] \quad (26)$$

where

$$E_2(k, s) = E_{p,b}\{T_2(p, b, s)\} \quad (27)$$

$$E_3(k, s) = E_{p,b}\{T_3(p, b, s)\} \quad (28)$$

and we dropped the dependence of p_k's and b_k's on k in the notation. Now if we group the product and the summation into groups of equal power levels, we have

$$E_{p,b}\{\overline{\Phi}_{1,-1}(s)\} = T_1(s)\cdot\Pi_{l=1}^{N}E_2(l, s) + J_0(|\rho|s)\sum_{l=1}^{N}\left[\overline{k}_lE_3(l, s)\cdot E_2(l, s)^{\overline{k}_l-1} \cdot \Pi_{j\neq l}E_2(j, s)^{\overline{k}_j}\right] \quad (29)$$

where the parameters l and j in E_2 and E_3 now denote the power level group. Hence, for each point s we choose in our numerical integration of (25), we need to compute $E_2(l, s)$ and $E_3(l, s)$ for each power level of l. Note that these two terms are independent of K' and the signal-to-noise ratio.

It is worth while at this point to see how (29) would be simplified if all the users in the network had the same power. If we set $\alpha_k = \alpha_1$ for all $k = 1, 2, \cdots K$, then $E_{p,b}\{\overline{\Phi}_{1,-1}(s)\}$ simplifies to

$$E_{p,b}\{\overline{\Phi}_{1,-1}(s)\} = T_1(s)\cdot E_2(s)^{K'} + J_0(|\rho|s)\cdot K' \cdot E_3(s)\cdot E_2(s)^{K'-1} \quad (30)$$

where we further dropped the dependence of E_2, E_3 on the power levels. We may further simplify this expression for the case when orthogonal BFSK is employed, i.e., $\rho = 0$, as

$$E_{p,b}\{\overline{\Phi}_{1,-1}(s)\} = \left(\frac{s}{2\gamma}\right)E_2(s)^{K'} + K'\cdot E_3(s)\cdot E_2(s)^{K'-1}. \quad (31)$$

In this case we may compute $P_e(K) = P_e(K')$, the probability of error for a hop given that the hop is hit by K' users with the same power as the first user using (25) and (31) and thus compute the average probability of error P_e for the first user using

$$P_e = \sum_{K'=0}^{K-1}\binom{K-1}{K'}p_h^{K'}(1-p_h)^{(K-K')}P_e(K'). \quad (32)$$

Going back to the general case where the power levels are different, we need to further average (25) over K and thus the average probability of error can be written as[2] [7]

$$P_e = E_K\{P_e(K)\} \tag{33}$$

$$= \sum_{\bar{k}_1=0}^{\overline{K}_1} \cdots \sum_{\bar{k}_N=0}^{\overline{K}_N} P_e(K)\Pi_{l=1}^{N}\left\{\binom{\overline{K}_l}{\bar{k}_l} p_h^{\bar{k}_l} \cdot (1-p_h)^{\overline{K}_l-\bar{k}_l}\right\} \tag{34}$$

$$= \int_0^{\infty} e^{-s^2/2\gamma} J_0(s) \times \left[\sum_{\bar{k}_1=0}^{\overline{K}_1} \cdots \sum_{\bar{k}_N=0}^{\overline{K}_N} E_{p,b}\{\Phi_{1,-1}(s)\}\Pi_{l=1}^{N} \cdot \left\{\binom{\overline{K}_l}{\bar{k}_l} p_h^{\bar{k}_l}(1-p_h)^{\overline{K}_l-\bar{k}_l}\right\}\right] ds. \tag{35}$$

Now let us discuss the reason for considering nonorthogonal FSK where $|\rho| > 0$. The basic idea behind using nonorthogonal FSK in AFHSS–MA is to provide a tradeoff between the number of errors caused by background noise and multiple-access hits. If we assume that the error probability is 1/2 whenever a hop is hit, the fact that the system performance will be dominated by multiple-access interference rather than background noise for sufficiently large signal-to-noise ratios leads to the following idea. If we use nonorthogonal FSK instead of orthogonal FSK, we will be able to increase the number of slots available for a given fixed bandwidth. This in effect reduces p_h. This is done at the cost of higher probability of error since increasing $|\rho|$ increases the probability of error when the hop is not hit [13] (it also increases probability of error when the hop is hit). These competing factors will result in an optimum $|\rho|$ that minimizes the average error probability and since the multiple-access hits are far more detrimental than the quiescent errors under the 1/2-bound, the optimal $|\rho|$ is expected to be greater than zero under this assumption for a reasonably high signal-to-noise ratio. In general this optimum $|\rho|$ will depend on the number of users in the network and the signal-to-noise ratio. Since $\zeta = 2\Delta fT$, we may write

$$\zeta = \frac{\Delta f}{\Delta f_{\text{ortho}}} \tag{36}$$

where $\Delta f \le \Delta f_{\text{ortho}}$ is the frequency separation employed and $2f_{\text{ortho}} = 1/T$ is the minimum frequency separation needed for the signals to be orthogonal. Then the approximate number of slots available is given by

$$q_n = \left\lfloor \frac{q_{\text{ortho}}}{\frac{1}{2}(1+\zeta)} \right\rfloor \ge q_{\text{ortho}} \tag{37}$$

where q_{ortho} is the number of slots available when $\Delta f = \Delta f_{\text{ortho}}$ and $\lfloor x \rfloor$ denotes the largest integer not exceeding x. Thus, with Markov hopping patterns, p_h decreases from $2/q_{\text{ortho}}$ to $2/q$ as we decrease the frequency separation from Δf_{ortho} to Δf. Another way of making $|\rho| > 0$ is to fix $\Delta f = \Delta f_{\text{ortho}}$ and decrease the hop duration from T_{ortho} to T where T_{ortho} is the minimum hop duration required to guarantee that the two BFSK signals are orthogonal. This idea of using nonorthogonal signaling is interesting. Since it can be shown, using the 1/2-bound, that the gain achieved by making $|\rho| > 0$ is high when coding is employed and background noise is small. However when a more accurate approximation is used it is observed in Section III that we may actually degrade performance by using nonorthogonal BFSK. Significant gains may be obtainable for systems employing other modulation schemes such as slow frequency hopping where the 1/2-bound is thought to be more realistic.

[2]Loose upper and lower bound on P_e can be obtained by setting $P_e(K)$ equal to 1 or 0 for $K' \ge 2$.

A. Memoryless Hopping Patterns

Up to now, we have considered Markov hopping pattterns to simplify the analyses and to obtain results that are readily computed. We may follow similar steps as before and derive expressions for the probability of error when memoryless hopping patterns are employed [15]. But here, let us consider an upper bound and a lower bound on the error probability for the memoryless hopping pattern case. If we can show that the upper bound and the lower bound are tight and also that the error probability when Markov hopping patterns are employed falls in between these bounds, then we will have in effect shown that the probability of error when Markov hopping patterns are employed is a good approximation to the probability of error when memoryless hopping patterns are employed. For the case where orthogonal BFSK is employed and all the users in the network have the same power as seen by the first receiver, a simple bound can be derived where we assume that whenever a hop is overlapped by an interfering user for its entire duration, the error probability is 1 or 0 resulting in an upper bound and a lower bound, respectively. Then it is easy to see that the average error probability will be upper bounded as

$$P_e \le \sum_{K'=0}^{K-1} \binom{K-1}{K'} p_p^{K'}(1-p_p-p_f)^{K-1-K'} P_e(K') + \left[1-(1-p_f)^{K-1}\right] \tag{38}$$

and lower bounded as

$$P_e \ge \sum_{K'=0}^{K-1} \binom{K-1}{K'} p_p^{K'}(1-p_p-p_f)^{K-1-K'} P_e(K') \tag{39}$$

where p_p and p_f are the probability of partial and full hits given by [16]

$$p_p = \frac{2}{q}\left(1-\frac{1}{q}\right) \tag{40}$$

$$p_f = \frac{1}{q^2} \tag{41}$$

and $P_e(K')$ is the probability of error given that the hop is hit by K' partial hits which has already been computed. The numerical results given in the next section show that for all

K, the probability of error when Markov hopping patterns are assumed is indeed a good approximation to the probability of error when memoryless hopping patterns are employed.

B. Synchronous Hopping

An FHSS-MA network where the hopping instances of all the users are synchronized is a special case of the AFHSS-MA network considered above. The error performance of such a system may be obtained by setting all p_k's equal to zero in the expressions for the error probability derived for AFHSS-MA networks and setting $p_h = 1/q$. It is not hard to see that U_1 and U_{-1} are statistically independent in this case and the expressions for the probability of error derived here are exact for all K' when $\zeta = 1$.

C. Asymptotic Performance

It is also possible to consider the asymptotic average probability of error as $K, q \to \infty$ but with $K/q = \lambda$ fixed as in [2] for the case when orthogonal BFSK is used and all users have the same power level as seen by the first user. Let us first define $F(s)$ as follows:

$$F(s) = e^{-s^2/2\gamma} J_0(s)\left[\frac{s}{2\gamma} J_0(|\rho|s) + |\rho| J_1(|\rho|s)\right]. \quad (42)$$

Then we can show that the average error probability may be written as

$$P_e(K) = \sum_{K'=0}^{K-1} \binom{K-1}{K'} p_h^{K'} (1-p_h)^{(K-K')} \times \int_0^\infty [F(s)E_2(s)^{K'} + J_0(|\rho|s)\cdot K' \cdot E_3(s)E_1^{K'-1}(s)]\, ds. \quad (43)$$

Straightforward analysis shows that

$$\lim_{K,q\to\infty,\frac{K}{q}=\lambda} P_e(K) = \int_0^\infty e^{-2\lambda(1-E_2(s))} \cdot [F(s) + 2\lambda J_0(|\rho|s)E_3(s)]\, ds. \quad (44)$$

Unfortunately, evaluating this integral is rather difficult but computations using sample values for K, q, and λ show that the probability of error as a function of λ remains essentially unchanged for $q \geq 100$.

III. Numerical Results

First let us consider orthogonal BFSK, that is, the case when $\rho = 0$. We use (18) to compute the probability of error for a given hop as a function of the normalized delay p when there is one interfering user of the same power level with the data bit of the interfering user as the parameter [16]. For signal-to-noise ratio equal to 11 dB, this is shown in Fig. 3 for $p > 0$ along with simulation results where 10 000 errors were collected for each data point. The error probability is symmetric about $p = 0$ since it is a function of $|p|$. In this case, (18) provides exact results and the simulation is performed only to confirm the

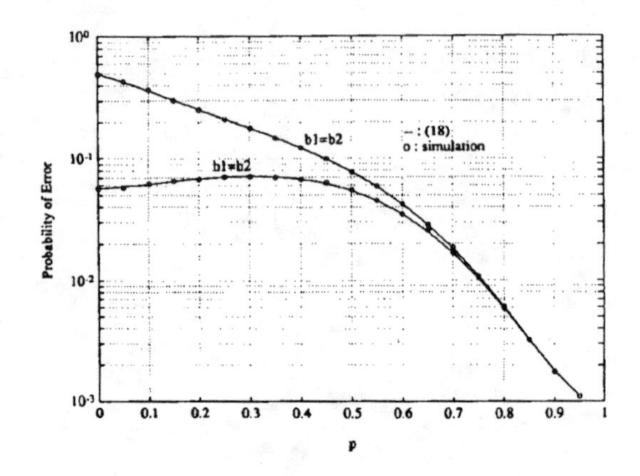

Fig. 3. Probability of error versus the normalized delay p with one interfering user. $E_b/N_0 = 11$ dB.

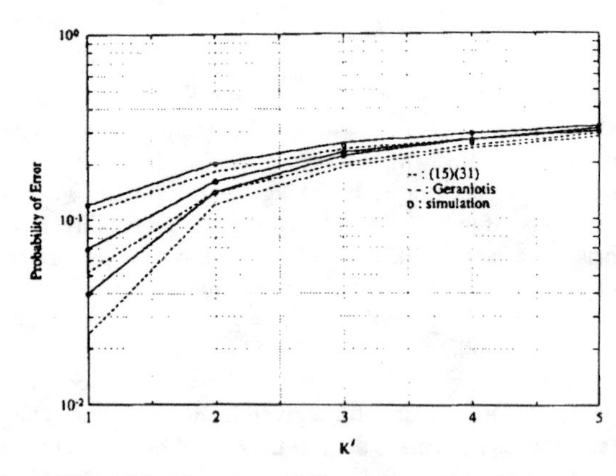

Fig. 4. Probability of error versus the number of interfering users for $E_b/N_0 = 8$, 13, and 19 dB. All users have same power and $\rho = 0$.

TABLE I
Comparison of Geraniotis' Approximation and (25), (31). $E_b/N_0 = 11$ dB

K'	Geraniotis'	(25), (31)	Simulation
1	6.88×10^{-2}	8.59×10^{-2}	8.50×10^{-2}
2	0.15	0.17	0.17
3	0.21	0.24	0.24
4	0.26	0.28	0.28
5	0.29	0.31	0.31

accuracy of the numerical computation. We note that the error probability is highly dependent on the data bit and the delay of the interfering user.

Fig. 4. shows the plots for the probability of error computed using (25) and (31) and Geraniotis' approximation which is obtained by setting

$$A(1, p_k, -1) = A(-1, p_k, 1) = 0 \quad (45)$$

for $K' = 1, \cdots, 5$ and $E_b/N_0 = 8$, 13, and 19 dB. Simulation results are also plotted where 20 000 errors were collected for each data point. Again orthogonal BFSK with $2\Delta f = 1/T$ and equal power levels are assumed in these plots.

TABLE II
AVERAGE PROBABILITY OF ERROR FOR TWO POWER LEVEL GROUPS. $E_b/N_0 = 11$ dB; $q = 100$

$\frac{\alpha_2}{\alpha_1}$	K	K_1, K_2	$P_e{}^*(35)$	*Assuming equal power	*1/2-bound
		(5, 5)	1.04×10^{-2}		
	11	(3, 7)	7.39×10^{-2}	1.79×10^{-2}	9.22×10^{-2}
		(7, 3)	1.34×10^{-2}		
0.5		(10, 10)	2.03×10^{-2}		
	21	(5, 15)	1.27×10^{-2}	3.49×10^{-2}	0.167
		(15, 5)	2.76×10^{-2}		
		(5, 5)	2.93×10^{-2}		
	11	(3, 7)	3.37×10^{-2}	1.79×10^{-2}	9.22×10^{-2}
		(7, 3)	2.47×10^{-2}		
1.5		(10 10)	5.57×10^{-2}		
	21	(5, 15)	6.58×10^{-2}	3.49×10^{-2}	0.167
		(15, 5)	4.54×10^{-2}		

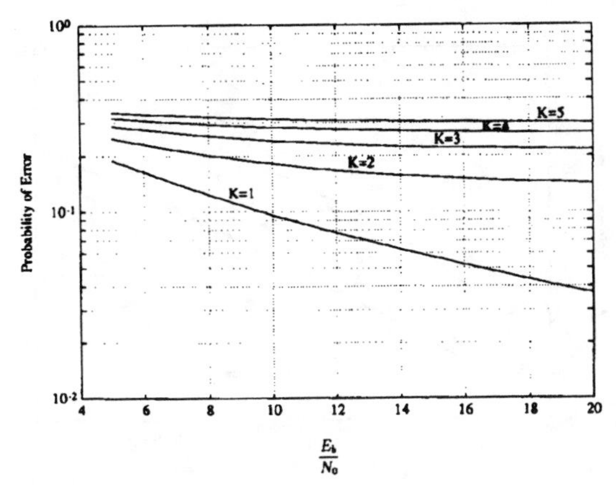

Fig. 5. Probability of error versus E_b/N_0 with K' as the parameter. All users have same power and $\rho = 0$.

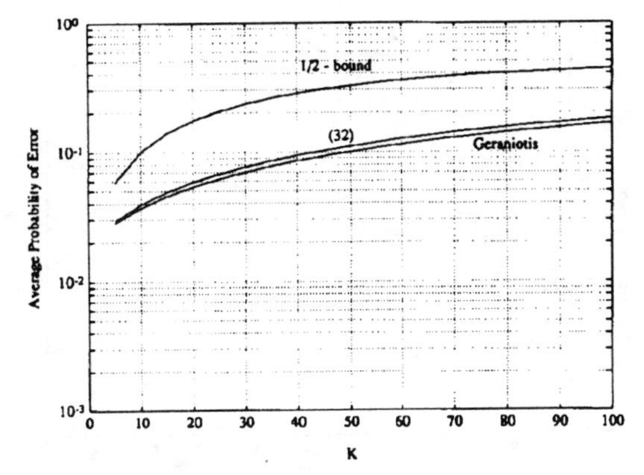

Fig. 6. Average probability of error given that there are K active users in the network with the same power level and $\rho = 0$. $E_b/N_0 = 8$ dB. $q = 100$.

We observe that the values obtained using the approximations (for $K' \geq 2$) derived in this paper fit nicely to the simulation results and Geraniotis' approximation gives optimistic results. These probabilities for $E_b/N_0 = 11$ dB are also tabulated in Table I. Note that the real error probabilities are much smaller than 1/2 for small K'. Since we have checked that the probability of error computed using the expressions developed in this paper fits the simulation results very closely for the signal-to-noise ratios considered, from here on, we will only consider our results using Geraniotis' assumption and compare them to the results obtained using the 1/2-bound. The probability of error as a function of E_b/N_0 is also plotted in Fig. 5.

Let $q = 100$ and consider the average probability of error. In Figs. 6–8 we show the average probability of error of a hop given that there are K users in the network for $E_b/N_0 = 8$, 13, and 19 dB computed using (32). We note that there could be up to an order of magnitude difference between the actual values and the ones obtained using the 1/2-bound for high signal-to-noise ratios. With error correction coding, the difference will be further amplified.

Next, we consider the case where there are two different power level groups P_1 and P_2. Let $\alpha^{(1)}$ and $\alpha^{(2)}$ be the magnitudes of the signals in each group as seen by the first receiver. We take P_1 to have the same power as the first user, i.e., $a^{(1)} = a_1$. We consider two cases where $\alpha^{(2)}/a_1 = 0.5$, $\alpha^{(2)}/a_1 = 1.5$. We tabulate the results for the average error probabilities for these cases for $K = 11$ and

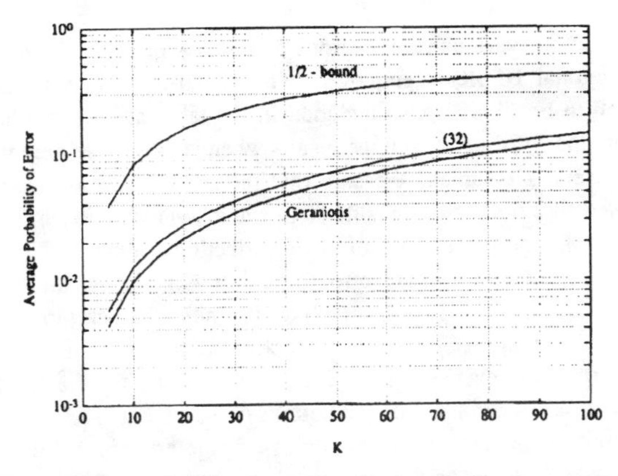

Fig. 7. Average probability of error given that there are K active users in the network with the same power level and $\rho = 0$. $E_b/N_0 = 13$ dB. $q = 100$.

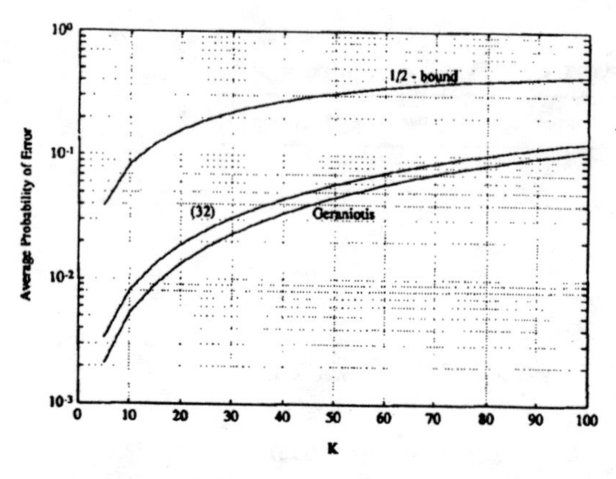

Fig. 8. Average probability of error given that there are K active users in the network with the same power level and $\rho = 0$. $E_b/N_0 = 19$ dB. $q = 100$.

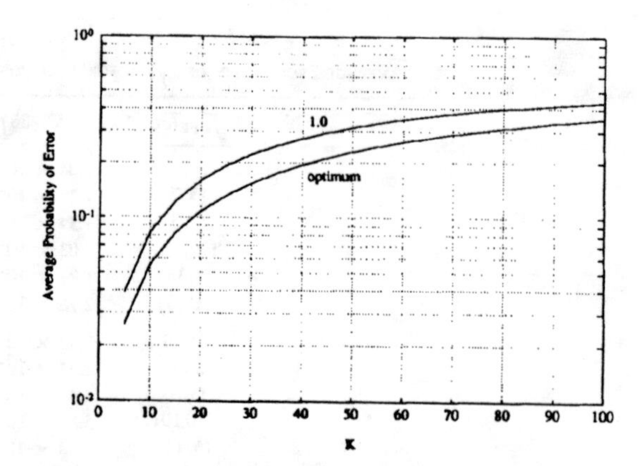

Fig. 10. Average probability of error using nonorthogonal BFSK using the 1/2-bound. All users have equal power. $E_b/N_0 = 19$ dB. $q = 100$.

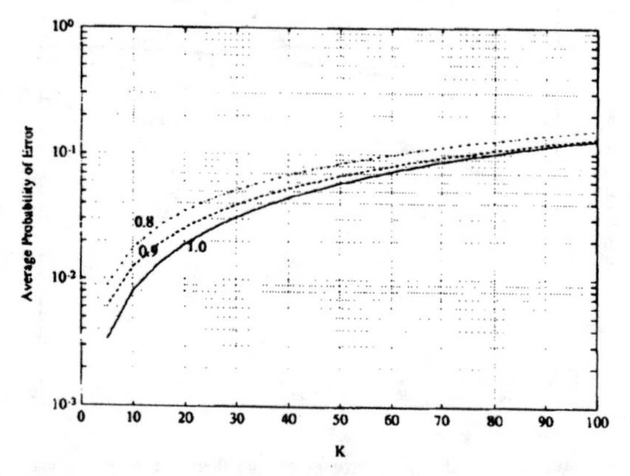

Fig. 9. Average probability of error using nonorthogonal BFSK. All users have equal power. $E_b/N_0 = 19$ dB. $q = 100$. $\mu = 1, 0.9, 0.8$.

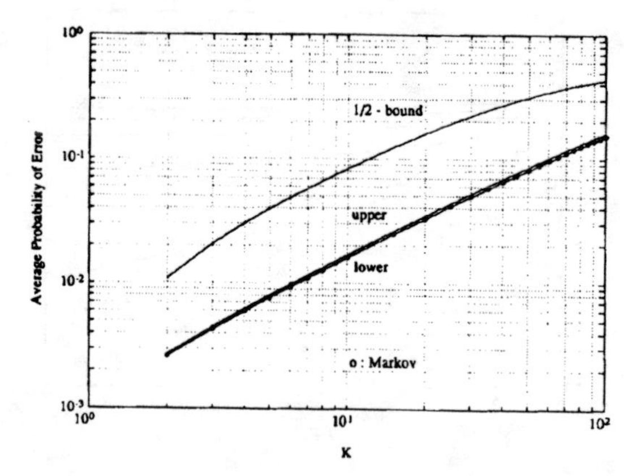

Fig. 11. Bounds on the average probability of error using independent hopping patterns. All users have same power and $\rho = 0$. $E_b/N_0 = 11$ dB. $q = 100$.

$K = 21$ for various combinations of $\overline{K}_1, \overline{K}_2$ in Table II when orthogonal BFSK is employed. The signal-to-noise ratio is taken to be 11 dB. Also included in the table are the results using the 1/2-bound, and the results when equal power among the users is assumed. We note that the errors resulting from these approximations are quite large and again they will further be amplified with the use of error correction coding.

In Fig. 9, the average error probability is shown for the cases when $\mu = 1, 0.9$, and 0.8. We note that employing nonorthogonal BFSK actually degrades performance. Fig. 10 shows similar curves when using the 1/2-bound for the error probability when hit. For the bottom curve of Fig. 10, μ was optimized for each K for minimum average error probability. This shows that conclusions we obtain using the 1/2-bound could be quite different from what we obtain by using our approximation to the error probability when $|\rho| > 0$. In this case, the difference arises from the fact that when the expressions for the probability of error developed in this paper are used, the decrease in the hit probability by using $\mu < 1$ does not sufficiently compensate for the increase in the error probability (both when the hop is hit and when the hop is not hit). More significant gains may be expected for systems where the symbol error probability is close to $(M-1)/M$ whenever a hop is hit such as a slow frequency-hopping system.

In Fig. 11, the bounds on the error probability when memoryless hopping patterns are employed given by (38), (39) are shown. These results verify that the probability of error when Markov hopping patterns are employed is indeed a very good approximation to the probability of error when memoryless hopping patterns are employed for sufficiently large q. The signal-to-noise ratio was taken to be 11 dB.

Most of the computing time needed to evaluate the average error probability is consumed in evaluating (27) and (28) which are independent of K' and the signal-to-noise ratio when Markov hopping patterns are employed. Hence, once these values are computed for each s and the power levels, it is a simple task to obtain the average error probability for different number of interfering users and signal-to-noise ratios. Hence the computing time increases approximately linearly with the number of different power levels and is almost irrelevant to the number of interfering users and the signal-to-noise ratio.

A. Channel Capacity and Associated Throughput

Now we consider the coded performance of AFHSS–MA

networks by considering the channel capacity and the normalized throughput as performance measures. When the detector simply makes hard decisions on each hop, then the resulting channel can be accurately modeled as a memoryless binary symmetric channel (BSC) shown in Fig. 12 (independence assumption). When the detector makes erasures by erasing those hops that were hit and makes hard decisions on those hops that were not hit, the resulting channel can be modeled as a memoryless binary symmetric errors and erasure channel (BSEEC) shown in Fig. 13. The channel capacity is the maximum code rate at which there exists a channel code (the best possible code) that achieves reliable communications over the channel. The channel capacity C_{BSC} and C_{BSEEC} for the BSC and the BSEEC are given by [17].

$$C_{\mathrm{BSC}} = 1 + (1-p)\log_2(1-p) + p\log_2 p, \tag{46}$$

$$C_{\mathrm{BSEEC}} = p_c \log_2\left(\frac{2p_c}{1-p_x}\right) + p_e \log_2\left(\frac{2p_e}{1-p_x}\right). \tag{47}$$

We define the normalized throughput of the channel associated with this best possible channel code as the average number of successfully transmitted data bits per channel symbol per frequency slot over the network using this code which is given by [3]

$$w_c = \frac{C \cdot K}{q} \tag{48}$$

where C is the capacity of the channel and K is the number of active users in the network.

In Figs. 14–15, we show the channel capacity and the associated normalized throughput for a system making hard decisions on each hop (system I) and a system using perfect side-information to erase those hops that were hit and making hard decisions on the symbols that were not hit (system II) for $E_b/N_0 = 8$ and 10 dB. It is assumed that all users have the same power as seen by the receiver and orthogonal BFSK is used. We note from these figures that, surprisingly, system I performs much better than system II which is more complicated to implement. An intuitive explanation for this is that the average probability of error when a hop is hit is not as high as 1/2 and system II will in effect make excessive erasures. Again, this observation may not hold for slow frequency-hop systems.

In Fig. 16, the normalized throughput associated with channel capacity for FHSS–MA networks with asynchronous and synchronous hopping are shown. Orthogonal BFSK and equal power levels among users are assumed. We note that even though p_h for the synchronous system is only half that of the asynchronous system, its performance is inferior to that of the asynchronous system. This is because even though the hops for the synchronous system are less likely to be hit than the asynchronous system, the probability of error once a hop is hit is much large for the synchronous system. Again, this observation may not hold for slow frequency-hop systems.

IV. Conclusion

In this paper, we derived accurate approximations for the error probability of an AFHSS–MA network employing Markov hopping patterns transmitting one BFSK modulated symbol per hop. We showed that the bound on the error probability of 1/2 whenever a hop is hit by multiple-access interference is quite pessimistic for a single bit per hop system. We conjecture that for a system employed M-ary FSK signaling and transmitting one symbol per hop, the usual bound on the error probability of $(M-1)/M$ is also very loose.

The performance of other detection schemes such as

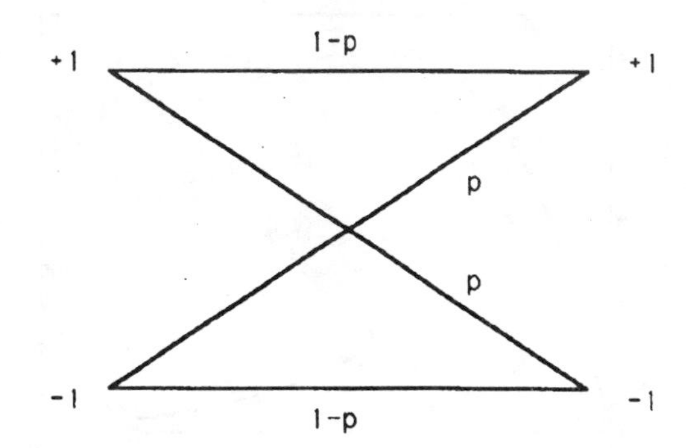

Fig. 12. BSC.

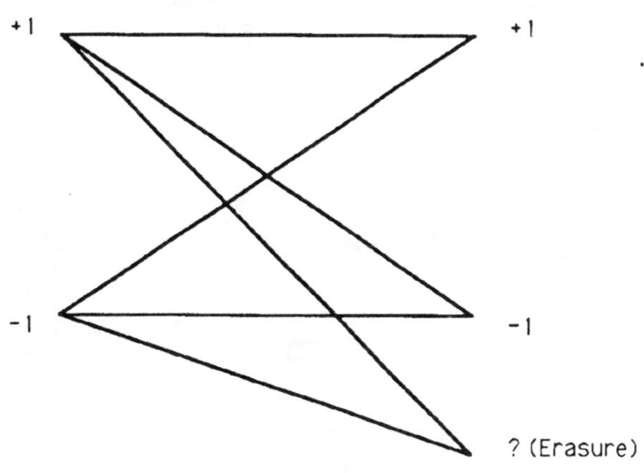

Fig. 13. BSEEC.

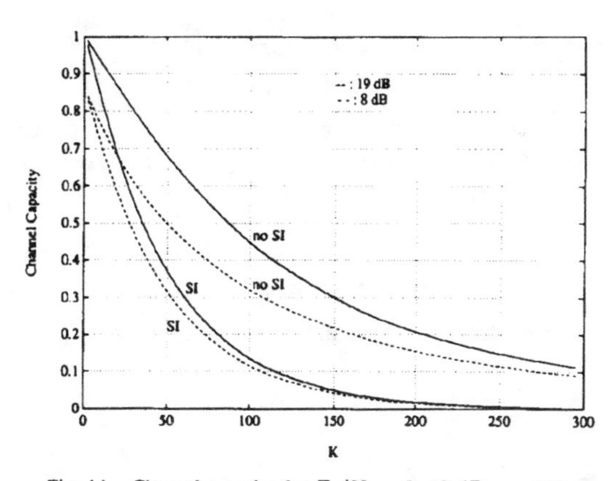

Fig. 14. Channel capacity for $E_b/N_0 = 8$, 19 dB, $q = 100$.

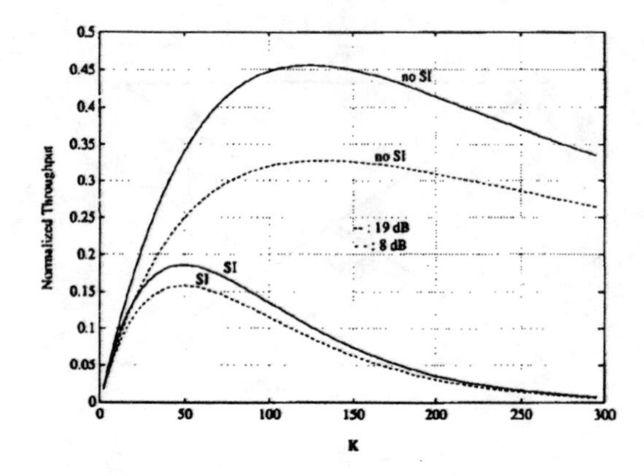

Fig. 15. Normalized throughput associated with channel capacity for $E_b/N_0 = 8, 19$ dB, $q = 100$

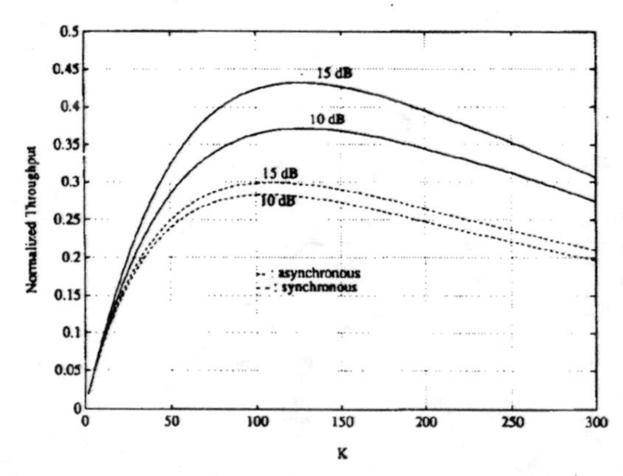

Fig. 16. Normalized throughput associated with channel capacity for synchronous and asynchronous systems for $E_b/N_0 = 10$ dB, $q = 100$.

the Viterbi ratio thresholding in obtaining imperfect side-information in AFHSS-MA networks can be analyzed using the results given in this paper [18].

APPENDIX

Here we show that $|U_1|$ and $|U_{-1}|$ given by (2), (3) are statistically independent given p_2, b_2 when $\rho = 0$ $K' = 1$. In this case U_1 and U_{-1} are given by

$$U_1 = z_1 + e^{i\varphi_1} + \left(\frac{\alpha_2}{\alpha_1}\right) \exp(i(\theta(1, p_2, b_2) + \varphi_2))A(1, p_2, b_2) \tag{49}$$

$$U_{-1} = z_{-1} + \left(\frac{\alpha_2}{\alpha_1}\right) \exp(i(\theta(-1, p_2, b_2) + \varphi_2))A(-1, p_2, b_2) \tag{50}$$

where $\theta(l, p_2, b_2), A(l, p_2, b_2), l \in \{-1, 1\}$ are constants. Let φ be a random variable independent of all the phase and noise terms in (49), (50) with a uniform distribution on $[0, 2\pi)$. Then $U_{-1}e^{i\varphi}$ is given by

$$U_{-1}e^{i\varphi} = z_{-1}e^{i\varphi} + \left(\frac{\alpha_2}{\alpha_1}\right) \exp(i(\theta(-1, p_2, b_2) + \varphi_2 + \varphi)) \cdot A(-1, p_2, b_2). \tag{51}$$

Since we consider all the phase terms $\mathrm{mod}(2\pi)$, $(\varphi_2 + \varphi)$ $\mathrm{mod}(2\pi)$ can be replaced with a random variable φ' which is uniformly distributed on $[0, 2\pi)$ and independent of both φ_2 and φ. This establishes that U_1 and $U_{-1}e^{i\varphi}$ are independent. Hence, since $|U_{-1}| = |U_{-1}e^{i\varphi}|$, $|U_1|$ and $|U_{-1}|$ are independent.

REFERENCES

[1] E. A. Geraniotis, "Coded FH/SS communications in the presence of combined partial-band noise jamming, Rician nonselective fading, and multiuser interference," *IEEE J. Select. Areas Commun.*, vol. SAC-5, pp. 194–214, Feb. 1987.

[2] S. Kim and W. Stark, "Optimum rate Reed-Solomon codes for frequency-hop spread-spectrum multiple-access communication system," *IEEE Trans. Commun.*, vol. 37, pp. 138–144, Feb. 1989.

[3] M. B. Pursley, "Frequency-hop transmission for satellite packet switching and terrestrial packet radio networks," *IEEE Trans. Inform. Theory*, vol. IT-32, pp. 652–667, Sept. 1986.

[4] M. Hegde and W. E. Stark, "Capacity of frequency-hop spread-spectrum multiple-access communication systems," *IEEE Trans. Commun.*, vol. 38, pp. 1050–1059, July 1990.

[5] E. Geraniotis, "Noncoherent hybrid DS/SFH spread-spectrum multiple-access communications," *IEEE Trans. Commun.*, vol. COM-34, pp. 862-872, Sept. 1986.

[6] C. M. Keller, "An exact analysis of hits in frequency-hopped spread-spectrum multiple-access communications," in *Proc. Conf. Inform. Sci. Syst.*, Mar. 1988, pp. 981–986.

[7] E. Geraniotis, "Multiple-access capability of frequency-hopped spread-spectrum revisited: an analysis of the effect of unequal power levels," *IEEE Trans. Commun.*, vol. 38, pp. 1066–1077, July 1990.

[8] R. Agusti, "On the performance analysis of asynchronous FH-SSMA communications," *IEEE Trans. Commun.*, vol. 37, pp. 488–499, May 1989.

[9] R. T. Short and C. K. Rushforth, "Probability of error for noncoherent frequency-hop spread-spectrum multiple-access communications," Unisys Tech. Rep., PX18829, May 1988.

[10] J. S. Bird, "Error performance of binary NCFSK in the presence of multiple tone interference and system noise," *IEEE Trans. Commun.*, vol. COM-33, pp. 203–209, Mar. 1985.

[11] R. D. Lord, "The use of Hankel transforms in statistics," *Biometrika*, vol. 41, pp. 44–45, 1954.

[12] E. A. Geraniotis and M. B. Pursley, "Error probabilities for slow-frequency-hop spread-spectrum multiple-access communications over fading channels," *IEEE Trans. Commun.*, vol. COM-30, pp. 996–1009, May 1982.

[13] J. G. Proakis, *Digital Communications.* New York: McGraw-Hill, 1983.

[14] M. Abramowitz and I. A. Stegun, *Handbook of Mathematical Functions.* New York: Dover, 1970 and 1972.

[15] K. Cheun, "Analysis of asynchronous frequency-hop spread-spectrum multiple-access networks," Ph.D. dissertation, EECS Dep. Univ. Michigan, Ann Arbor, MI 48109; also Tech. Rep. Commun. Signal Processing Lab., no. 268, EECS Dep. Univ. Michigan.

[16] E. A. Geraniotis, "Coherent hybrid DS-SFH spread-spectrum multiple-access communications," *IEEE J. Select. Areas Commun.*, vol. SAC-3, pp. 695–705, Sept. 1985.

[17] R. G. Gallager, *Information Theory and Reliable Communications.* New York: Wiley, 1968.

[18] K. Cheun and W. Stark,"Performance of convolutional codes with Viterbi ratio thresholding in asynchronous FHSS-MA," *Proc. 1990 Conf. Inform. Sci. Syst.*, Mar. 1990, pp. 86–91.

Kyungwhoon Cheun was born in Seoul, Korea on December 16, 1962. He received the B.A. degree in electronic engineering from Seoul National University in 1985 and the M.S. and Ph.D. degrees from the University of Michigan, Ann Arbor, MI, in 1987 and 1989, respectively, both in electrical engineering.

From 1987 to 1989, he was a Research Assistant at the Department of Electrical Engineering and Computer Science at the University of Michigan. From 1989 to July 1991 he was an Assistant Professor of Electrical Engineering at the University of Delaware. Since July 1991 he has been with the Department of Electronics and Electrical Engineering at Pohang Institute of Technology, Pohang Korea, where he is currently an Assistant Professor. His current research interests include coding theory and its applications, information theory and multiple-access networks with emphasis on spread-spectrum systems.

Wayne E. Stark (S'77–M'82) was born on February 26, 1956. He received the B.S. (with highest honors), M.S., and Ph.D. degrees in electrical engineering from the University of Illinois, Urbana-Champaign, in 1978, 1979 and 1982 respectively.

Since 1982 he has been a faculty member at the University of Michigan where he is currently Associate Professor of Electrical Engineering and Computer Science. His research interests include information theory, coding theory, communication theory especially for spread-spectrum communication systems.

Dr. Stark received the 1985 National Science Foundation's Presidential Young Investigator Award, was a member of the Board of Governors of the IEEE Information Theory Group and was an Editor for the IEEE TRANSACTIONS ON COMMUNICATIONS from 1984 to 1989. He participated in the organization of the 1986 IEEE International Symposium on Information Theory held in Ann Arbor, MI. For the 1989 calendar year he was on sabbatical at IBM Zurich Research Laboratory, Rushlikon, Switzerland.

IEEE TRANSACTIONS ON COMMUNICATIONS, VOL. 42, NO. 2/3/4, FEBRUARY/MARCH/APRIL 1994

Optimum Detection of Slow Frequency-Hopped Signals

Barry K. Levitt, *Member, IEEE*, Unjeng Cheng, *Member, IEEE*,
Andreas Polydoros, *Senior Member, IEEE* and Marvin K. Simon, *Fellow, IEEE*

ABSTRACT

Optimum detectors have previously been derived for fast frequency-hopped (FFH) signals with *M*-ary frequency-shift-keyed (MFSK) data modulation received in additive white Gaussian noise (FFH here implies that a single MFSK tone is transmitted per hop). This paper extends that work to the more analytically complex category of slow frequency-hopped (SFH) signals with multiple MFSK tones per hop. A special subset of the SFH/MFSK format that receives particular attention in this paper is the case of continuous-phase modulation (CPM) for which the carrier phase is assumed to be constant over the entire hop. A fundamental conclusion is that SFH/CPM modulation is advantageous not only to the communicator but also to a sophisticated noncoherent detector. By applying techniques developed in this paper to exploit the continuous-phase characteristic, an intercept receiver of reasonable complexity will perform appreciably better than traditional channelized detectors.

I. INTRODUCTION

Within both the commercial and military arenas, there is a trend toward the use of spread-spectrum (SS) communication systems in general, and frequency-hopped (FH) systems in particular, because of their inherent anti-jam or anti-interference and low-probability-of-intercept characteristics. There has been a concomitant interest from both sectors in the ability of unauthorized intercept receivers to detect FH emissions. Many *ad hoc* FH detection structures have been proposed which typically incorporate a fast-Fourier-transform (FFT) or other channelized front-end preprocessor. The performance of some of these signal processors have been tested in the field, simulated in computers, and occasionally analyzed, sometimes leading to questionable superiority claims. However, these exercises generally do not investigate fundamental performance limits unrestricted by practical implementation constraints and the notion of optimality in the classical average-likelihood ratio (ALR) detection theory sense [1], which is the subject of this paper.

The most general formulation of the classical signal detection problem seeks to determine whether a bandlimited signal $s(t)$ has been received in additive white Gaussian noise (AWGN) $n(t)$ with two-sided power spectral density $N_0/2$ based on the observation of the composite received signal $r(t)$ over the observation interval $(0, T)$:

$$r(t) = \begin{cases} s(t) + n(t); H_1 \\ \\ n(t); H_0 \end{cases} \tag{1}$$

where H_1 and H_0 are respectively the signal present and signal absent hypotheses. Absent any further *a priori* information about $s(t)$, the simple noncoherent (i.e., unknown carrier phase) energy detector or radiometer [2] is often used to distinguish between the two hypotheses [3, pp. 128-135]. For wideband signals, the radiometer performs much worse than more sophisticated detectors that exploit additional information about the characteristics of the received signal such as the modulation format. Nonetheless, the wideband radiometer is commonly used as a yardstick of minimum basic performance because of its simplicity and robustness to signal feature variations. For example, if s(t) is an FH signal, the radiometer does not require frequency channel and hop timing epoch synchronization.

Now suppose that $s(t)$ is, in fact, an FH signal with SS bandwidth W_{ss}, hop rate R_h, hop dwell time $T_h = 1/R_h$, received power S, and received energy per hop $E_h = ST_h$. For arbitrary baseband data modulation, the FH carrier naturally partitions W_{ss} into N_c contiguous, non-overlapping channels, each with bandwidth $W_m = W_{ss}/N_c$, such that the entire signal energy in any hop lies within a single channel. (As discussed below, these N_c FH channels need not correspond to the unmodulated FH carrier frequencies.) Nonetheless, for reasons of mutual orthogonality between the channels, the time-bandwidth product (TBP) T_hW_m is a positive integer. Also, for analytical simplicity, assume that the detector observation interval $(0, T)$ contains an integer number $N_h = T/T_h$ of complete hops. In the usual idealized signal detection problem formulation, the detector is assumed to have somehow achieved frequency and time synchronization with the received signal: i.e., it knows R_h, the location of the FH channels in the frequency domain, and hop epoch timing (in practice, it would obviously have to extract these characteristics from the received signal). Because of the channelized nature of the FH signal and the arbitrary data modulation, it is not unreasonable for an ad hoc detector to prefilter the received signal with a matching bank of N_c contiguous, non-overlapping energy detectors, each with bandwidth W_m and integration time T_h aligned with the received hops: this preprocessor is sometimes referred to as a channelized radiometer [3, pp. 135-140]. Conditioned on the observable $\boldsymbol{R} \equiv \{R_{ij}\}$, whose components denote the preprocessor outputs for the ith hop and jth FH channel, Woodring and Edell [4] showed that the ALR has the form

Paper approved by Sorin Davidovici, the Editor for Spread Spectrum of the IEEE Communications Society. Manuscript was received February 11, 1993; revised April 4, 1993. This work was performed at the Jet Propulsion Laboratory, California Institute of Technology under a contract with the National Aeronautics and Space Administration.

Unjeng Cheng, Marvin K. Simon, and Barry K. Levitt are with the Jet Propulsion Laboratory, California Institute of Technology, Pasadena, CA 91109. Andreas Polydoros is with the University of Southern California and is a consultant to the Jet Propulsion Laboratory.

IEEE Log Number 9401601.

$$\Lambda(R) \propto \prod_{i=1}^{N_h} \sum_{j=1}^{N_c} R_{ij}^{\frac{T_hW_m-1}{2}} I_{T_hW_m-1}\left(\frac{2}{N_0}\sqrt{E_hR_{ij}}\right)$$
$$= \prod_{i=1}^{N_h} \sum_{j=1}^{N_c} I_0\left(\frac{2}{N_0}\sqrt{E_hR_{ij}}\right);\ T_hW_m = 1 \tag{2}$$

Since R is not generally a sufficient observable for the original received signal $r(t)$, the Woodring-Edell (WE) FH detector is not strictly optimum in the classical noncoherent ALR sense (although it is essentially optimum for the special case $T_hW_m = 1$). The misperception that it is the *unconditionally* optimum FH detector for any TBP has been perpetuated in the literature (e.g., [5, Vol. III, pp. 290-295], [6]). The performance of the WE detector is computed in [4] using inaccurate central-limit theorem (CLT) approximations. However, the Bessel operation in (2) is highly nonlinear and CLT convergence is slow for Ricean or Rayleigh arguments (i.e., the square root of the noncentral or central chi-square random variables R_{ij}) because of the long tails in their probability density functions, which usually leads to optimistic performance results. In comparing the performance of the WE and other detectors, the results in this paper employ more accurate techniques described separately in [7] and [8].

A common suboptimal simplification of the WE (and other) detectors is the filter-bank combiner (FBC) wherein a hard decision is made as to the presence of a received signal in each FH channel on each observed hop (sometimes referred to as a "frequency-time cell" in FH detector terminology) based on a threshold comparison (e.g., [5, Vol. III, pp. 295-305], [9]). In the WE case, because the square root, scaling and Bessel function operations in each frequency-time cell are monotonic, each channelized preprocessor output random variable (RV) R_{ij} can equivalently be compared with a common threshold, thereby reducing postprocessing complexity. If this threshold is exceeded for any of the FH channels on a given hop, the presence of a signal on that hop is postulated and an OR gate generates a "1"; otherwise it outputs a "0". These intermediate hard decisions are then summed over all N_h observed hops and compared with a second integer threshold to decide between H_1 and H_0. The two thresholds are *jointly* optimized for best performance. The end result is that the FBC performs much better than the wideband radiometer because it is channelized to match the FH signal, and, in fact, does almost as well the WE detector but with considerably less computational complexity.

Beaulieu *et al.* [10] derived the optimum ALR detector for the special class of fast frequency-hopped (FFH) signals with M-ary frequency-shift-keyed (MFSK) data modulation.[1] In the noncoherent case, orthogonality constraints for FFH/MFSK signals dictate that the minimum TBP = 1: then, [10] shows that the ALR has the form of the WE unity TBP detection metric in (2), except that

$$R_{ij} \equiv \left(\int_{(i-1)T_h}^{iT_h} dt\ r(t)\sqrt{\frac{2}{T_h}}\cos(2\pi f_j t)\right)^2 + \left(\int_{(i-1)T_h}^{iT_h} dt\ r(t)\sqrt{\frac{2}{T_h}}\sin(2\pi f_j t)\right)^2 \tag{3}$$

i.e., the WE channelized radiometer preprocessor is replaced by a bank of inphase and quadrature (I&Q) noncoherent detectors which generate the R_{ij}s. However, the statistical difference between the outputs of the two preprocessors is so insignificant that the performance of the detectors is almost identical [1], [11]. Thus, the WE structure for unity TBPs is essentially the optimum noncoherent FFH/MFSK detector.

II. Derivation of Optimum SFH Detectors

A. SFH/MFSK Signals with Discontinuous Phase

The remainder of this paper is concerned with optimal coherent and noncoherent detection of slow frequency-hopped (SFH) signals, which has not been previously analyzed in the open literature. Particular emphasis is given below to the combination of SFH with continuous-phase modulation (CPM) formats such as continuous-phase MFSK (CPFSK). However, for comparison with previous work, SFH/MFSK signals in which the received carrier phase is discontinuous from symbol to symbol are considered first. The received signal term in (1) is now

$$s(t) = \sqrt{2S}\cos\left\{2\pi\left[f_j + \left(m - \frac{M-1}{2}\right)R_s\right]t - \theta_q\right\};$$
$$(i-1)T_h + (l-1)T_s \le t < (i-1)T_h + lT_s \tag{4}$$

for the lth data symbol within the ith hop, where T_s is the data symbol baud time, $R_s = 1/T_s$ is the symbol rate, $E_s = ST_s$ is the received signal energy per symbol, and there are $N_s = T_h/T_s$ symbols per hop. For noncoherent[2] detection, orthogonality requires that the MFSK tones have minimum separation R_s (i.e., the modulation index $h = 1$) corresponding to $N_c = W_{ss}/R_s$ FH channels. With the added non-overlapping M-ary band convention[3] illustrated in Fig. 1, the particular FH carrier frequency f_j on the ith hop is equally likely to be any of $G = N_c/M = W_{ss}/MR_s$ equally-spaced frequencies that are statistically independent (SI) from hop to hop. The lth data symbol on the ith hop is denoted by m, which is equally likely to be any integer in $\{0, 1, ..., M-1\}$ and is SI from symbol to symbol. Finally, the received carrier phase during the lth symbol in the ith hop, θ_q, is

[1] This paper adopts the widely accepted terminology that distinguishes between FFH and slow frequency-hopped (SFH) signals according to whether there is a single data symbol per hop (FFH) or multiple data symbols per hop (SFH)[5, Vol. II, pp. 62-64]. One of the reasons that the FFH/MFSK case is so much easier to analyze than the SFH/MFSK case is that it is statistically equivalent to an unmodulated FH carrier with arbitrary alphabet size M.

[2] Coherent detection would be an unrealistic idealization for SFH/MFSK signals where the phase is discontinuous over M-ary symbol transitions.

[3] In general, for FFH or SFH signals with MFSK data modulation, there is a difference between the N_c FH channels and the unmodulated carrier frequencies. For any particular carrier frequency, the M adjacent FH channels form an M-ary band (except for the special case of M independent FH carriers). If these bands are contiguous and non-overlapping as shown in Fig. 1, there are $G = N_c/M$ equally-spaced carrier frequencies or M-ary bands. In the alternative maximally-overlapping case (ignoring band edge effects), $G = N_c$ [5, Vol. II, Fig. 2.3]. This paper uses the non-overlapping format exclusively because it simplifies the SFH performance analysis.,

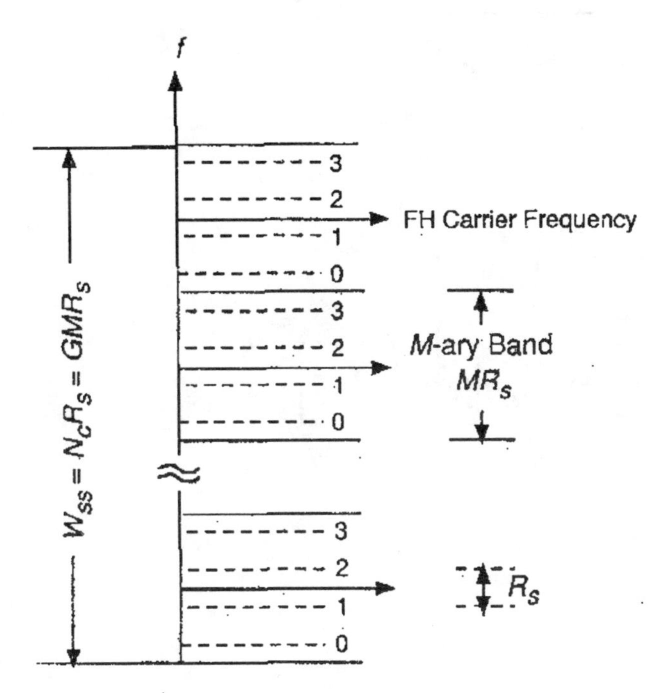

Fig. 1. Non-overlapping M-ary band convention shown for $M = 4$.

assumed to be constant, uniformly distributed over $(0, 2\pi)$, and SI from symbol to symbol.

Because all of the stochastic characteristics of the received signal are at least SI from hop to hop, and the FH carrier frequencies are equally likely on a given hop, the ALR can always be partitioned into the form

$$\Lambda[r(t)] \propto \prod_{i=1}^{N_h} \sum_{j=1}^{G} \Lambda_{ij}[r(t)|f_j] \tag{5}$$

where the condition in the argument of Λ_{ij} denotes that the FH carrier frequency over the ith hop is f_j. From [1, p. 253, (23)], it follows that

$$\Lambda_{ij}[r(t)|f_j] = E_{s(t)|f_j}\left\{\exp\left[\frac{2}{N_0}\int_{(i-1)T_h}^{iT_h} dt\; r(t)s(t) - \frac{E_h}{N_0}\right]\right\} \tag{6}$$

So the likelihood function only needs to be computed for a single representative frequency-time cell in which the SS aspects of the signal are suppressed, and this partition is valid for all SFH signals of interest.

Substituting the SFH/MFSK signal of (4) into (6) and averaging over the random phase yields the representative frequency-time cell likelihood function

$$\Lambda_{ij}[r(t)|f_j] \propto \prod_{l=1}^{N_s} \sum_{m=0}^{M-1} I_0\left(\frac{2}{N_0}\sqrt{E_s R_{ijlm}}\right) \tag{7}$$

where

$$R_{ijlm} = \left(\int_{(i-1)T_h+(l-1)T_s}^{(i-1)T_h+lT_s} dt\; r(t)\sqrt{\frac{2}{T_s}}\cos\left\{2\pi\left[f_j+\left(m-\frac{M-1}{2}\right)R_s\right]t\right\}\right)^2 + \left(\int_{(i-1)T_h+(l-1)T_s}^{(i-1)T_h+lT_s} dt\; r(t)\sqrt{\frac{2}{T_s}}\sin\left\{2\pi\left[f_j+\left(m-\frac{M-1}{2}\right)R_s\right]t\right\}\right)^2 \tag{8}$$

In conjunction with (5), (7) and (8) define the noncoherent ALR detector for SFH/MFSK signals with discontinuous phase, which is illustrated in Fig. 2.

A comparison of (2) and (3) with (5), (7) and (8) illustrates the differences between the optimum noncoherent detectors for FFH/MFSK and SFH/MFSK signals. Both detectors form I & Q components at the symbol level, which undergo identical operations culminating with the zeroth-order Bessel function: this is done for each of the N_c FH channels, however the postprocessing is different. In the FFH case, since there is one symbol per hop, all N_c Bessel outputs are summed for each hop. In the SFH case, the N_c channels are partitioned into $G = N_c/M$ M-ary bands corresponding to the G FH carrier frequencies; the Bessel outputs are summed within each M-ary-band, and, since there are N_s symbols per hop and the carrier phase is assumed to be SI from symbol to symbol, the M-ary outputs are multiplied over all of the symbols in each hop and the resultant is summed over the G carrier frequencies. From a complexity perspective, the only difference is the product over the N_s symbols in each M-ary band on each hop in the SFH case: the single larger summation over the N_c channels for FFH signals is equivalent to the summations over the M symbols associated with each carrier frequency followed by the summation over the G carrier frequencies for SFH signals.

B. SFH/CPM Signals

Consider SFH signals with arbitrary data phase modulation that is continuous over each hop, which is representative of a broad class of signals of interest. The general SFH/CPM signal can be written as

$$s(t) = \sqrt{2S}\cos\left[2\pi f_j t + \phi(d_n, t) - \theta_q\right]; \quad (i-1)T_h \le t < iT_h \tag{9}$$

where $\phi(d_n, t)$ is the CPM component which depends on the particular M-ary data sequence d_n in the ith hop. The number of distinct data sequences that can occur on a given hop is $N_d = M^{N_s}$, and, in the absence of specific channel coding information to the contrary, all sequences are assumed to be equally likely. For noncoherent detection, the received carrier phase θ_q is uniformly distributed over $(0, 2\pi)$ and is assumed to be constant over the entire hop but SI from hop to hop; in the coherent case, we can set $\theta_q = 0$. The separation of the G FH carrier frequencies $\{f_j\}$ is assumed to be uniform and consistent with orthogonality requirements and the contiguous, non-overlapping M-ary band convention of Fig. 1.

The general SFH/CPM signal of (9) can be inserted into (6) to determine the ijth likelihood function term. In particular, for coherent detection with $\theta_q = 0$,

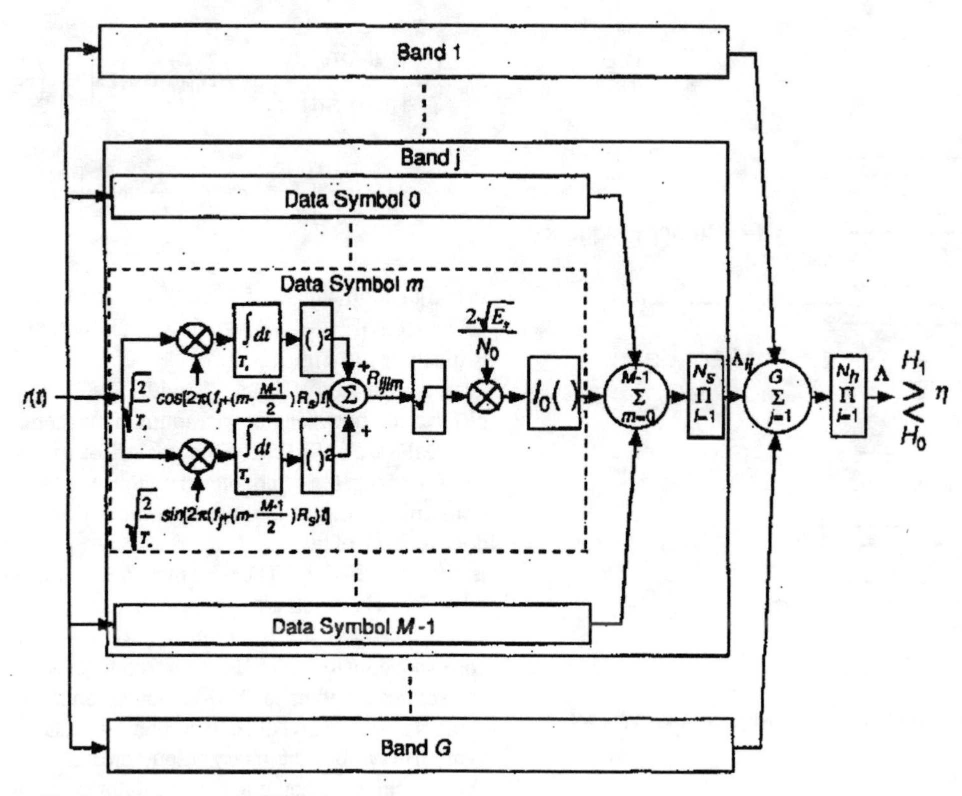

Fig. 2. Noncoherent ALR SFH/MFSK detector when received carrier phase is constant over each symbol but discontinuous (i.e., SI) from symbol to symbol.

$$\Lambda_{ij}\left[r(t)\middle|f_j\right] \propto E_{d_n}\left[\exp\left(\frac{2}{N_0}\sqrt{E_h}\int_{(i-1)T_h}^{iT_h} dt\ r(t)\sqrt{\frac{2}{T_h}}\cos\left[2\pi f_j t+\phi(d_n,t)\right]\right)\right]$$

$$\propto \sum_{n=1}^{N_d}\exp\left(\frac{2}{N_0}\sqrt{E_h}\int_{(i-1)T_h}^{iT_h} dt\ r(t)\sqrt{\frac{2}{T_h}}\cos\left[2\pi f_j t+\phi(d_n,t)\right]\right) \tag{10}$$

which, in conjunction with the universally applicable FH partition expression of (5), defines the optimum coherent detector for any SFH/CPM signal.

In the noncoherent case, the likelihood function must also be averaged over the random phase so that

$$\Lambda_{ij}\left[r(t)\middle|f_j\right] \propto \sum_{n=1}^{N_d} I_0\left(\frac{2}{N_0}\sqrt{E_h R_{ijn}}\right) \tag{11}$$

where

$$R_{ijn} \equiv \left(\int_{(i-1)T_h}^{iT_h} dt\ r(t)\sqrt{\frac{2}{T_h}}\cos\left[2\pi f_j t+\phi(d_n t)\right]\right)^2 + \left(\int_{(i-1)T_h}^{iT_h} dt\ r(t)\sqrt{\frac{2}{T_h}}\sin\left[2\pi f_j t+\phi(d_n t)\right]\right)^2 \tag{12}$$

Combining this result with (5) yields the noncoherent ALR detector, again valid for any SFH/CPM signal: the ijth representative frequency-time cell is illustrated in Fig. 3. While theoretically correct, (11)-(12) is not the most practical noncoherent detector implementation. The Bessel operation in (11) results from the expectation over the unknown phase θ_q in each hop. As observed in [8], since the expectations over this phase and the data sequence in each frequency-time cell are linear, the phase expectation can be performed last. In particular, θ_q can be inserted into (10) and the expectation of this expression can be computed over this random phase:

$$\Lambda_{ij}\left[r(t)\middle|f_j\right] \propto E_{\theta_q}\left[\sum_{n=1}^{N_d}\exp\left(\frac{2}{N_0}\sqrt{E_h}\int_{(i-1)T_h}^{iT_h} dt\ r(t)\sqrt{\frac{2}{T_h}}\cos\left[2\pi f_j t+\phi(d_n,t)-\theta_q\right]\right)\right]$$

$$\equiv \frac{1}{Q}\sum_{q=1}^{Q}\sum_{n=1}^{N_d}\exp\left(\frac{2}{N_0}\sqrt{E_h}\int_{(i-1)T_h}^{iT_h} dt\ r(t)\sqrt{\frac{2}{T_h}}\cos\left[2\pi f_j t+\phi(d_n,t)-\theta_q\right]\right) \tag{13}$$

where the continuous phase expectation has been approximated by an average over Q discrete phases uniformly spaced over the sufficient range $[0, \pi)$.

From a complexity perspective, a tremendous penalty must be paid to optimally detect an SFH signal with continuous phase over each hop if the coherent or noncoherent ALR is implemented as in (10) or (13) respectively. Since the number of data patterns N_d that must be examined within each M-ary band for each FH carrier frequency on each hop grows exponentially with the number of symbols per hop N_s, the complexity can be impractical even for binary ($M=2$) data modulation unless N_s is sufficiently small. It is shown later in the SFH/CPFSK case that

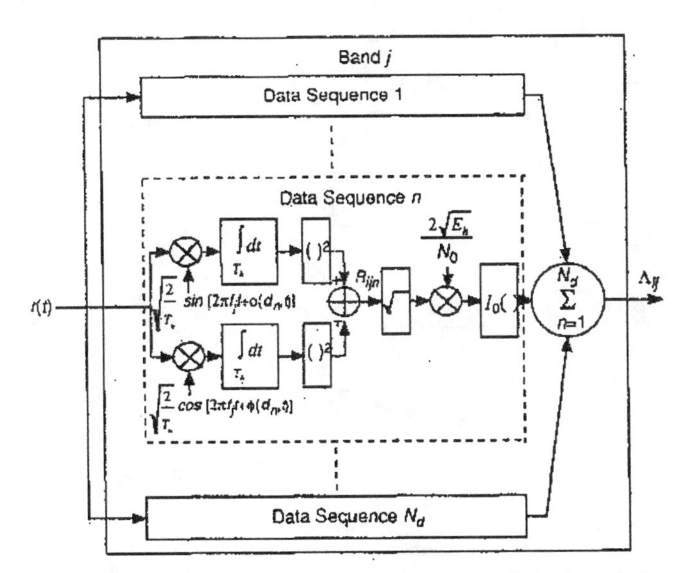

Fig. 3. Representative frequency-time cell for ith hop and jth M-ary band of noncoherent ALR detector for arbitrary SFH/CPM signals.

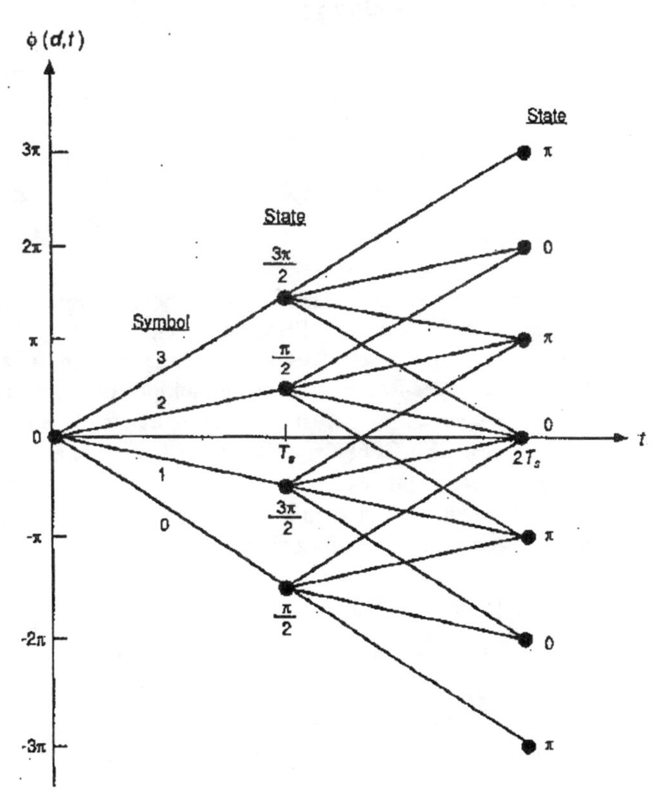

Fig. 4. CPFSK phase tree for the special case $M = 4$, $h = 1/2$.

this increased complexity is counterbalanced by a dramatic improvement in noncoherent ALR detector performance relative to the discontinuous-phase SFH/MFSK case of Fig. 2. It will also be seen that for certain modulation indices h, the coherent and noncoherent ALR SFH/CPFSK detectors can be implemented in such a way that the complexity grows linearly rather than exponentially with N_s [8].

C. SFH/CPFSK Signals

For arbitrary modulation index h, CPFSK signals are defined by (9) with

$$\phi(d_n,t) = \phi\left[d_n,(i-1)T_h\right] + 2\pi h R_s \int_{(i-1)T_h}^{t} d\tau \;\; p(d_n,\tau); \qquad (14)$$

$$(i-1)T_h \le t < iT_h$$

where the M-ary data stream is specified by

$$p(d_n,t) = m - \frac{M-1}{2}; \qquad (15)$$

$$(i-1)T_h + (l-1)T_s \le t < (i-1)T_h + lT_s$$

and $m \in \{0, 1, ..., M-1\}$ is the lth data symbol in the ith hop, $l = 1, 2, ..., N_s$ for the nth data sequence d_n. Equations (14) and (15) define a phase tree for all possible data sequences on the ith hop with arbitrary modulation index and data alphabet size; for example, Fig. 4 illustrates the phase tree for the special case $M = 4$, $h = 1/2$. The slope of a particular phase tree branch determines the frequency offset for the corresponding symbol relative to the FH carrier frequency. The frequency transmitted during a given symbol duration is $f_j + [m - (M-1)/2]hR_s$ so that the spacing between adjacent M-ary tones is hR_s. The branches of the phase tree are labelled with the data symbol m and the symbol transitions with the phase state; since $\phi(d_n, t)$ appears in the argument of a sinusoid, these phase states can be reduced to the range $[0, 2\pi)$ by expressing the actual phase modulo 2π as shown in Fig. 4. For the special case of integer sub-multiple modulation indices, i.e., $h = 1/K$ where K is integer, the phase tree alternates between two disjoint sets of K possible states on each successive symbol transition. This reduces the phase tree with M^{N_s} branches for each hop to a much simpler trellis with KMN_s branches as shown in Fig. 5 for $M = 4$, $h = 1/2$: that is, the data-dependent phase-tree complexity has been changed from an exponential to a linear dependence on N_s.

For coherent detection, the minimum CPFSK tone separation for orthogonality is $h = 1/2$, referred to as minimum shift-keying (MSK). However, for simplicity, first consider the special case $h = 1$, corresponding to the minimum separation for noncoherent detection. For this particular modulation index, the phase state ψ_k at each symbol transition is completely deterministic independent of the data sequence, alternating between 0 and π: i.e., at the beginning of the lth symbol baud, $\psi_k = (l-1)\pi \bmod 2\pi$. Then the general SFH/CPM signal of (9) simplifies to the mathematical form of (4) with the addition of ψ_k for the initial phase state of the lth symbol (an important difference here is that the received carrier phase θ_q is constant over the entire hop for CPFSK modulation whereas it was only constant over each symbol in the discontinuous-phase MFSK case). In summary, for SFH/CPFSK modulation with $h = 1$, the received signal is independent and identically distributed (i.i.d.) from symbol to symbol conditioned on the FH carrier frequency f_j and the received hop carrier phase θ_q. As is shown later, this is a sufficient condition for drastically simplifying the ALR SFH/CPM detector structure in both the coherent and noncoherent cases.

For coherent detection with $h = 1$ and $\theta_q = 0$, this conditional i.i.d. characteristic allows (10) to simplify as follows:

$$\Lambda_{ij}[r(t)|f_j] \propto E_{d_n}\left[\exp\left(\sum_{l=1}^{N_s}\frac{2}{N_0}\sqrt{E_s}\int_{(i-1)T_h+(l-1)T_s}^{(i-1)T_h+lT_s} dt\; r(t)\sqrt{\frac{2}{T_s}}\cos\left\{2\pi\left[f_j+\left(m-\frac{M-1}{2}\right)R_s\right]t+(l-1)\pi\right\}\right)\right]$$

$$\propto \prod_{l=1}^{N_s}\sum_{m=0}^{M-1}\exp\left(\frac{2}{N_0}\sqrt{E_s}\int_{(i-1)T_h+(l-1)T_s}^{(i-1)T_h+lT_s} dt\; r(t)\sqrt{\frac{2}{T_s}}\cos\left\{2\pi\left[f_j+\left(m-\frac{M-1}{2}\right)R_s\right]t+(l-1)\pi\right\}\right) \quad (16)$$

This is a very dramatic reduction in operational complexity relative to the general coherent SFH/CPM detector of (10): M^{N_s} exponential functions must still be computed, but instead of M^{N_s} additions, only N_s M-fold sums must be multiplied (i.e., N_sM mathematical operations). So, using the configuration of (16) to coherently detect SFH/CPFSK signals with $h = 1$, the complexity now grows linearly with N_s.

The same reduction in complexity can be realized in the noncoherent detection case. As a special case of (13),

$$\Lambda_{ij}\left[r(t)\middle|f_j\right] \propto \sum_{q=1}^{Q}\prod_{l=1}^{N_s}\sum_{m=0}^{M-1}\exp\left(\frac{2}{N_0}\sqrt{E_s}\int_{(i-1)T_h+(l-1)T_s}^{(i-1)T_h+lT_s} dt \right.$$

$$\left. \times r(t)\sqrt{\frac{2}{T_s}}\cos\left\{2\pi\left[f_j+\left(m-\frac{M-1}{2}\right)R_s\right]t+(l-1)\pi-\theta_q\right\}\right) \quad (17)$$

It was found in [8] that sufficient accuracy could be achieved with $Q \sim 16$. This structure for the ijth frequency-time cell of the optimum noncoherent SFH/CPFSK detector is illustrated in Fig. 6. As noted in [8] and confirmed by comparing Figs. 2 and 6, the computational complexity of this essentially optimum detector for *continuous-phase* MFSK modulation is on the order of that of the optimum detector for the *discontinuous-phase* case.

Earlier, it was implied that the conditional i.i.d. feature of SFH/CPFSK modulation with $h = 1$ is sufficient but not necessary to reduce the complexity of the coherent or noncoherent ALR detectors from an exponential to a linear dependence on N_s; in fact, similar results can be realized for *all* integer sub-multiple modulation indices. For example, in the $M = 4$, $h = 1/2$ (MSK) case, Fig. 4 shows that the phase state ψ_k alternates between $\{\pi/2, 3\pi/2\}$ and $\{0, \pi\}$ for the beginning of the even and odd symbol bauds respectively. However, instead of a fully-connected trellis with alternating pairs of states, it may be more convenient to consider a 4-state trellis which is semi-connected as shown in Fig. 5.

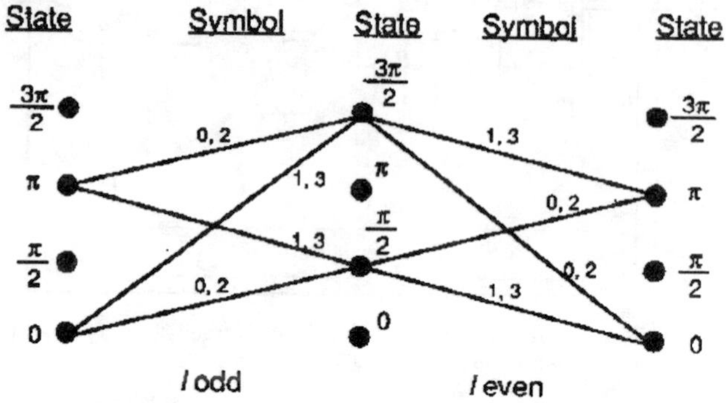

Fig. 5. CPFSK phase trellis for lth symbol bauds when $M = 4$, $h = 1/2$.

Consider the general case of coherent detection of CPFSK signals with $h = 1/K$ and arbitrary M. Analogous to Fig. 5, the CPFSK phase trellis is semi-connected with $2K$ states at each symbol transition. Let d_l denote a particular data sequence over the first l symbols of the ith hop, and let $D_{l,k}$ denote the set of all such data sequences that end at the kth phase state, ψ_k. As a variation on (10), define the conditional partial metric

$$\Delta_{l,k} \equiv \sum_{d_l \in D_{l,k}} \exp\left(\frac{2}{N_0}\sqrt{E_s}\int_{(i-1)T_h}^{(i-1)T_h+lT_s} dt \right.$$

$$\left. \times r(t)\sqrt{\frac{2}{T_s}}\cos\left[2\pi f_j t+\phi(d_l,t)\right]\right) \quad (18)$$

where the average in (18) is only over those l-symbol sequences d_l that lie in $D_{l,k}$.[4]

Note that

$$\Lambda_{ij}\left[r(t)\middle|f_j, d_{N_s} \in D_{N_s,k}\right] \propto \Delta_{N_s,k} \quad (19)$$

so that

$$\Lambda_{ij}\left[r(t)\middle|f_j\right] \propto \sum_{k=1}^{2K}\Delta_{N_s,k} \quad (20)$$

where $\Delta_{N_s,k} = 0$ for every other k. The conditional partial metric $\Delta_{l,k}$ should not be interpreted as a conditional partial ALR, nor should (18) be used to compute it since the number of data sequences in $D_{l,k}$ grows exponentially with the sequence length l. In fact, the phase trellis can be used to establish a recursive expression for the lth conditional partial metric $\Delta_{l,k}$ in terms of the $(l-1)$th metrics. Let $M_{k',k}$ be the set of all data symbols m in the trellis diagram that connect the k'th state at the beginning of the lth symbol baud to the kth state at the end. Then (18) yields the recursive relationship

[4] There is a similarity here to the classical Viterbi algorithm. The difference is that instead of making a maximum-likelihood estimate of the data sequence by discarding all but one of the branches that end at a given state, the ALR signal detector averages over all of these branches. An MLR receiver (see Section III. B) that jointly detects the CPFSK signal and estimates the imbedded data sequence would, in fact, employ the original Viterbi algorithm.

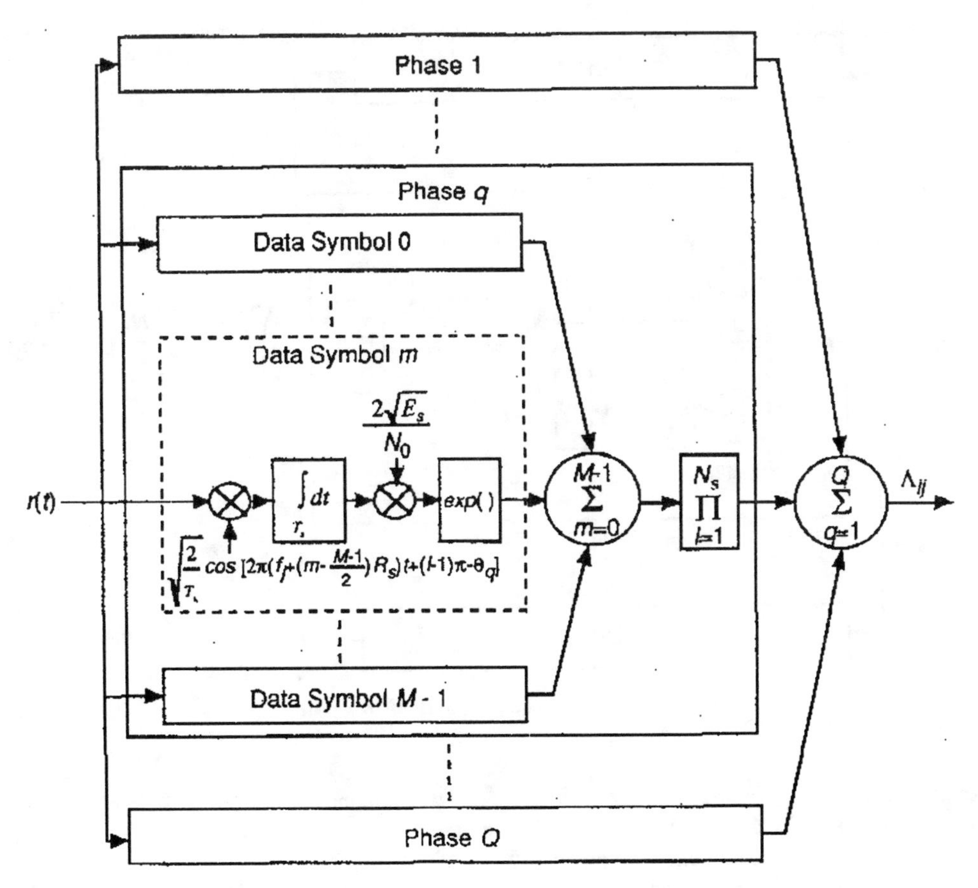

Fig. 6. Representative frequency-time cell for efficient implementation of noncoherent ALR detector for SFH/CPFSK signals with $h = 1$.

$$\Delta_{l,k} = \sum_{k'=1}^{2K} \Delta_{l-1,k'} \sum_{m \in M_{k',k}} \exp\left(\frac{2}{N_0}\sqrt{E_s} \int_{(i-1)T_h+(l-1)T_s}^{(i-1)T_h+lT_s} dt \times r(t)\sqrt{\frac{2}{T_s}} \cos\left\{ 2\pi\left[f_j + \left(m - \frac{M-1}{2} \right) R_s \right] t + \psi_{k'} \right\} \right) \tag{21}$$

and, for the initial symbol baud $l = 1$,

$$\Delta_{1,k} \propto \sum_{m \in M_{k',k}} \exp\left(\frac{2}{N_0}\sqrt{E_s} \int_{(i-1)T_h}^{(i-1)T_h+T_s} dt \times r(t)\sqrt{\frac{2}{T_s}} \cos\left\{ 2\pi\left[f_j + \left(m - \frac{M-1}{2} \right) R_s \right] t \right\} \right) \tag{22}$$

where k' is the particular state for which the phase $\psi_{k'} = 0$. Equations (21) and (22) can now be applied to compute $\Delta_{l,k}$ for $l = 2, 3, \ldots, N_s$, and then (20) can be used to calculate the ijth ALR function.

III. Suboptimum SFH Detectors

A. FBC and WE

For comparison with the optimum noncoherent detectors derived above, the traditional noncoherent FFH FBC and WE detector are extended in this paper to the SFH/MFSK case with minimum tone separation R_s for orthogonality ($h = 1$ again) and non-overlapping M-ary bands.

As shown in Fig. 7 for the FBC, for each data symbol on each hop within each M-ary band, the received signal is processed by a simple noncoherent, unmodulated-carrier, I & Q demodulator, and the outputs are then squared and summed. Within each frequency-time cell, these SI chi-square RVs with 2 degrees of freedom are summed over the N_s data symbols in each hop creating a chi-square RV R_{ij} with $2N_s$ degrees of freedom; these are compared with a threshold η which is the same for all cells. If this threshold is exceeded for any of the G bands on a given hop, the OR gate generates a "1"; otherwise, it outputs a "0". These numbers are then summed over all N_h observed hops and compared with a second integer threshold L. The two thresholds are jointly optimized to minimize the probability of a miss (P_M), for the selected probability of false alarm (P_{FA}) and received signal-to-noise ratio (SNR). Although suboptimum, it shall be shown that this structure performs almost as well as the ALR noncoherent detector for SFH/MFSK signals with discontinuous phase.

In the WE case, the metric of (2) is used with $W_m = MR_s$, N_c replaced by G, and the radiometer-derived R_{ij}s replaced by the I & Q generated R_{ij}s of Fig. 7 calculated for each of the $G = N_c/M$ bands. Since the WE detector is optimum conditioned on this channelized preprocessor, it can be expected to outperform the FBC, and this is confirmed below.

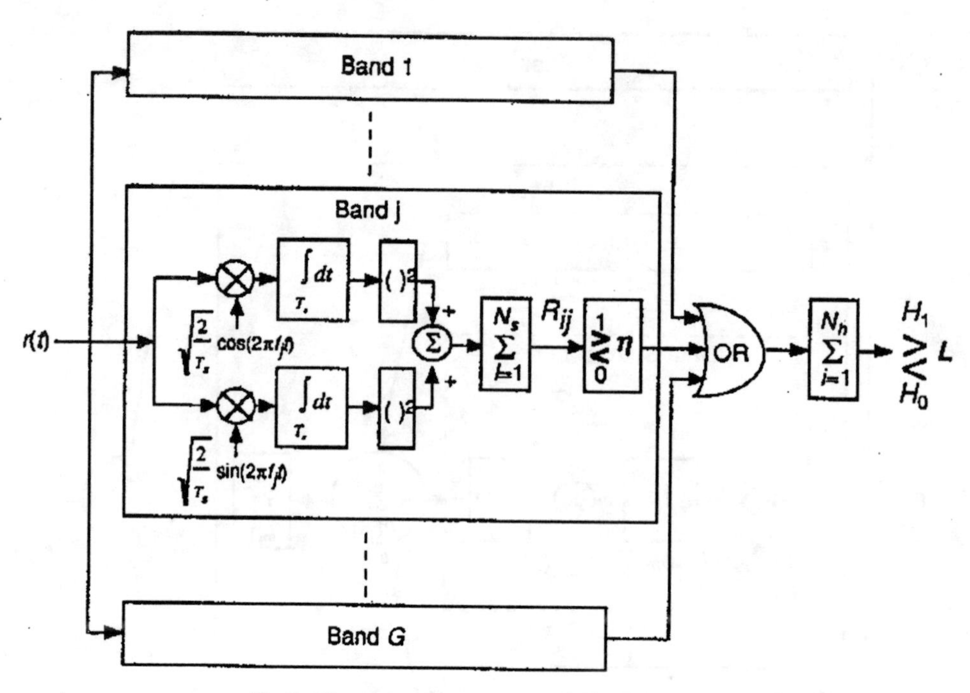

Fig. 7. Filter-bank combiner for SFH/MFSK signals.

B. Maximum-Likelihood Ratio FH Detectors

In deriving the ALR detectors for the various SFH signals considered, it was assumed that the data sequences within each hop and the FH pattern were unknown and equally likely. Furthermore, in the noncoherent case, the received carrier phase was assumed to be unknown and uniformly distributed over $[0, 2\pi)$. Consequently, the optimum detectors had to average over these signal characteristics. An alternative approach is to jointly detect the presence of the signal while simultaneously estimating one or more of these unknown signal attributes (sometimes referred to as "feature extraction") under H_1 using a maximum-likelihood ratio (MLR)[5] signal processor. By definition, an ALR receiver must perform better from a pure detection perspective than any MLR structure; however, in many cases the detection performance is almost the same, the MLR format may be easier to analyze or less complex to implement, and the derived signal feature is of interest to the intercept receiver.

For example, consider the previously derived ALR coherent and noncoherent detectors for SFH/CPFSK signals with $h = 1$. The summation (average) over the Q discrete phases in the noncoherent detector of Fig. 6 could be replaced by a maximization function to jointly detect the signal and determine the (discretized) received carrier phase. Mathematically, this MLR detector would replace (17) by the ijth likelihood function

$$\Lambda_{ij}\left[r(t)\middle|f_j\right] \propto \max_{q=1}^{Q}\left[\prod_{l=1}^{N_s}\sum_{m=0}^{M-1}\exp\left(\frac{2}{N_0}\sqrt{E_s}\int_{(i-1)T_h+(l-1)T_s}^{(i-1)T_h+lT_s} dt\right.\right.$$
$$\left.\left.\times r(t)\sqrt{\frac{2}{T_s}}\cos\left\{2\pi\left[f_j+\left(m-\frac{M-1}{2}\right)R_s\right]t+(l-1)\pi-\theta_q\right\}\right)\right] \tag{23}$$

To simultaneously detect the signal and determine the FH pattern, (5) is replaced by

$$\Lambda\left[r(t)\right] \propto \prod_{l=1}^{N_s}\max_{j=1}^{G}\left\{\Lambda_{ij}\left[r(t)\middle|f_j\right]\right\} \tag{24}$$

where Λ_{ij} is given by (16) or (17) for the coherent or noncoherent cases respectively. Finally, for joint signal and data detection (sometimes referred to as a "copy" function), the summation over the M data symbols in (16) or (17) is replaced by a maximization. In particular, for coherently detecting the signal while estimating the imbedded data sequence, use the ijth likelihood function

$$\Lambda_{ij}\left[r(t)\middle|f_j\right] \propto \prod_{l=1}^{N_s}\max_{m=0}^{M-1}\left[\exp\left(\frac{2}{N_0}\sqrt{E_s}\int_{(i-1)T_h+(l-1)T_s}^{(i-1)T_h+lT_s} dt\right.\right.$$
$$\left.\left.\times r(t)\sqrt{\frac{2}{T_s}}\cos\left\{2\pi\left[f_j+\left(m-\frac{M-1}{2}\right)R_s\right]t+(l-1)\pi\right\}\right)\right] \tag{25}$$

IV. Performance Example for Various SFH/BFSK Detectors

To illustrate the relative performance of the derived optimum and suboptimum SFH/MFSK detectors, consider the following example. As was done throughout this paper, assume an idealized situation in which the detector has acquired perfect side information about all the usual received signal parameters including hop timing epoch. Of course, it is not assumed that the detector has *a priori* knowledge about the particular FH pattern or data sequence. For the ALR and MLR receivers, unless stated otherwise, CPFSK data modulation is assumed to demonstrate the extent to which the continuous-phase characteristic can be exploited by a properly designed detector.

[5] Van Trees calls these metrics generalized likelihood ratios [1, p. 92].

For ease of computational complexity, consider only the case of binary (BFSK) data modulation. Using the previously defined notation, the received SFH/BFSK signal is assumed to have the parameters

$$W_{ss} = 10 \text{ MHz} \qquad R_h = 100 \text{ hops/sec}$$
$$T = 0.2 \text{ sec} \qquad R_s = 10 \text{ Ksymbols/sec}$$

which implies that for coherent detection ($h = 1/2$),

$$G = W_{ss}/R_s = 1000 \text{ binary bands}$$

and for noncoherent detection ($h = 1$),

$$G = W_{ss}/2R_s = 500 \text{ binary bands}$$

Furthermore,

$$N_h = R_h T = 20 \text{ consecutive hops}$$
$$N_s = R_s/R_h = 100 \text{ symbols/hop}$$

The performance of various SFH/BFSK detectors discussed in this paper was plotted for these signal characteristics in Fig. 8 as P_M versus the received SNR $\gamma = S/N_0W_{ss}$ for fixed $P_{FA} = 10^{-3}$, based on a combination of analytical and computer simulation techniques. These detectors include the wideband radiometer, WE, FBC, noncoherent ALR discontinuous- and continuous-phase, and coherent ALR and MLR discontinuous-phase structures, where, in this example, the MLR receiver jointly detects the signal and estimates the data.

There is understandably a strong motivation to simplify the performance computation task by liberally applying CLT arguments to approximate sums of large numbers of i.i.d. RVs by a single Gaussian RV (GRV). However, as was noted earlier in conjunction with the WE detector, it is well known that when the PDFs of the i.i.d. RVs have long tails, the convergence of their sum to a GRV is very slow, rendering the CLT approximation inaccurate and often leading to overly optimistic performance results. For this reason, the performance of most of the detectors in Fig. 8 was based on novel computer simulation techniques described in depth in [7] and [8] and which will not be repeated here.

Looking at the curves in Fig. 8, the radiometer performance was simply based on the closed-form approximation of [5, Vol. III, (4.4) and (4.5)] for which the CLT approach is accurate. As a performance benchmark, it achieves $P_M = 10^{-2}$ (for $P_{FA} = 10^{-3}$) at $\gamma = -24.2$ dB.

The FBC curve was determined for the structure of Fig. 7 using a purely analytical approach. In the ijth frequency-time cell, R_{ij} is a central (for the $G-1$ noise-only bands under H_1 or all G bands under H_0) or non-central (for the remaining signal-plus-noise band under H_1) chi-square RV with $2N_s = 200$ degrees of freedom. The probability that each of these SI RVs exceeds the common threshold η can be readily calculated; then the probability that the OR gate generates a "0" or a "1" on a given hop can be computed. The final summation over the $N_h = 20$ observed hops is then a well-defined binomial RV, and the probability that it exceeds the second integer threshold L (jointly optimized with η) can be calculated as in [5, Vol. III, pp. 295-300]. Actually, for the parameters in this example, $L = 5$ was optimum over most of the range in Fig. 8. In particular, the FBC required $\gamma = -32.4$ dB at the benchmark value of P_M which is almost 8 dB better than the radiometer performance. As noted earlier, although it is suboptimum, the FBC performs as well as it does primarily because it is channelized to match the hypothesized received FH signal.

The WE performance in Fig. 8 was based on the approach discussed in Section III.A. using computer simulation tech-

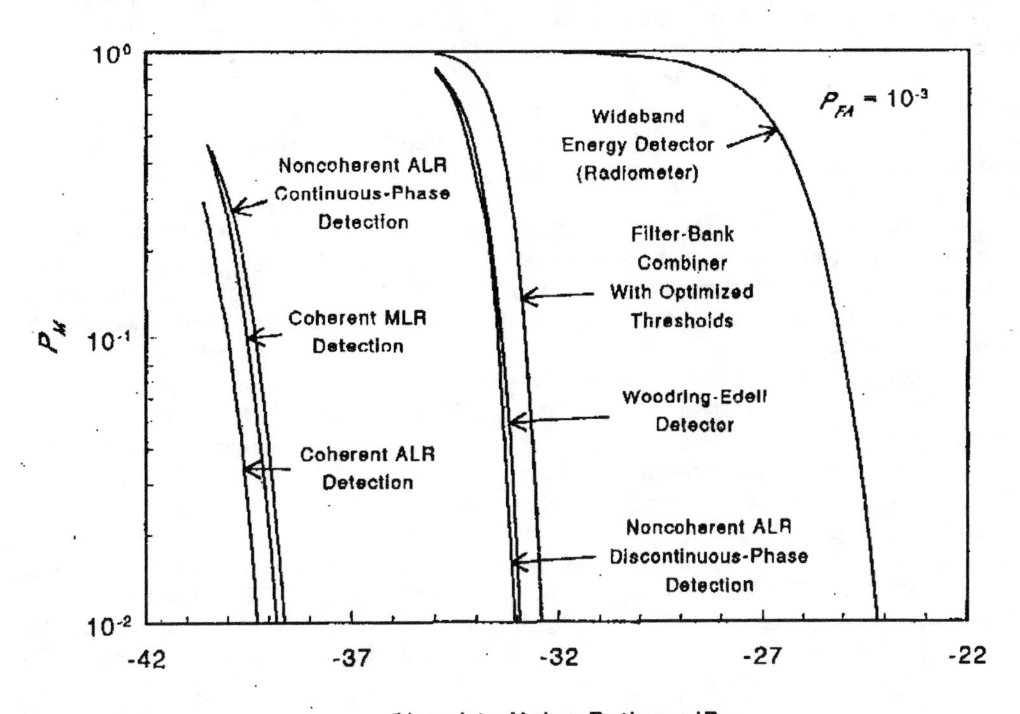

Fig. 8. Performance of SFH/BFSK detectors.

niques described in [8]. Surprisingly, its performance is only negligibly inferior to the noncoherent ALR discontinuous-phase detector over the entire range in Fig. 8.

The noncoherent ALR SFH/BFSK performance curve was based on the detector of Fig. 2 using computer simulation techniques described in [8]. It is moderately better than the FBC performance: in particular, at $P_M = 10^{-2}$, it requires $\gamma = -33.0$ dB, which is only 0.7 dB better than the FBC.

Just as all of the channelized detectors perform significantly better than the wideband energy detector, Fig. 8 shows that the continuous-phase characteristic results in another major performance improvement for ALR noncoherent detectors. The computer-simulated performance curve for the latter case is based on the structure of Fig. 6 with $Q = 16$ discrete phases uniformly spaced over $[0, \pi)$. It requires $\gamma = -38.7$ dB at $P_M = 10^{-2}$, which is remarkably almost 6 dB better than the discontinuous-phase BFSK case. This is an indication of the degree to which optimum noncoherent detectors can exploit CPM signaling.

Finally, there are the coherent ALR and MLR detectors for SFH/BFSK signals with $h = 1/2$. The tones are now spaced $R_s/2$ apart and there are $G = 1000$ M-ary bands. For analytical convenience, the performance was simulated for signals with discontinuous but known phase. That is, the received signal is given by (4) with $M = 2$, R_s replaced by $R_s/2$, and $\theta_q = 0$, and the ijth likelihood functions for the ALR and MLR[6] detectors are given by (16) and (25) respectively, again with $M = 2$ and R_s replaced by $R_s/2$. It must be stressed that this assumption of symbol-by-symbol phase coherence in the receiver is not meant to be a realizable condition: rather it is intended only as a vehicle for comparison with the ALR noncoherent detector performance. From this perspective, it is evident that both coherent detectors are only marginally better than the noncoherent ALR CPFSK detector. In particular, at $P_M = 10^{-2}$, the coherent MLR and ALR detectors perform only 0.2 dB and 0.6 dB better respectively than the ALR noncoherent detector.

Note that the performance curves in Fig. 8 are grouped into three quality levels:

(1) The worst is the radiometer which is noncoherent and unchannelized.

(2) The intermediate detectors, which include the FBC, WE and noncoherent ALR for discontinuous phase and perform about 9 dB better than the radiometer, are all channelized to match the FH frequencies but are noncoherent and not designed to exploit continuous-phase signals.

(3) Finally, the noncoherent ALR continuous-phase and coherent ALR and MLR detectors perform about 15 dB better than the radiometer.

It is worthwhile mentioning that all of the computer simulated curves in Fig. 8 were CPU-intensive. Each typically is based on a spline curvefit through 5-6 performance simulation points, with each point representing about 40,000 trials requiring approximately 2 days to generate on a SPARCstation using the techniques described in [7] and [8].

[6] Although, as discussed in Section III. B, there are several possible MLR structures, this example specifically refers to joint coherent detection of the SFH/BFSK signal and the imbedded data.

V. Conclusions

This paper began with a brief overview of previously published optimum and suboptimum intercept receivers for FFH signals intercepted in AWGN. It then extended this work to ALR and MLR detection of SFH signals, with particular attention devoted to CPM baseband data modulation formats. Practical trellis implementation structures of optimum detection algorithms were derived for the special case of SFH/CPFSK signals with integer sub-multiple modulation indices. Novel techniques for accurate computer simulation of the performance of these detectors were used to compare their relative capabilities. A fundamental observation is that CPM signals can be exploited by sophisticated albeit practical noncoherent detectors, with performance comparable to the best coherent detection schemes.

References

[1] H. L. Van Trees, *Detection, Estimation, and Modulation Theory*, Part I, John Wiley, New York, NY, 1968.

[2] H. Urkowitz, "Energy detection of unknown deterministic signals," *Proceedings of the IEEE*, Vol. 55, No. 4, pp. 523-531, April 1967.

[3] D. J. Torrieri, *Principles of Military Communication Systems*, Artech House, Dedham, MA, 1981.

[4] D. G. Woodring and J. D. Edell, "Detectability calculation techniques," U. S. Naval Research Laboratory, Report No. 1977-1, September 1977.

[5] M. K. Simon, J. K. Omura, R. A. Scholtz and B. K. Levitt, *Spread Spectrum Communications*, Computer Science Press, Rockville, MD, 1985.

[6] W. E. Snelling, "New methods for the detection and interception of frequency-hopped waveforms," Johns Hopkins University Applied Physics Laboratory, Technical Memorandum JHU/APL TG 1378, p. 15, November 1990.

[7] U. Cheng, M. K. Simon, A. Polydoros and B. K. Levitt, "Statistical models for evaluating the performance of coherent slow frequency-hopped M-FSK intercept receivers," Jet Propulsion Laboratory Report D-10534, Part 1, July 30, 1992; *IEEE Transactions on Communications*, Vol. 42, No. 2, February 1994.

[8] U. Cheng, M. K. Simon, A. Polydoros and B. K. Levitt, "Statistical models for evaluating the performance of noncoherent slow frequency-hopped M-FSK intercept receivers," Jet Propulsion Laboratory Report D-10534, Part 2, January 15, 1993; *IEEE Transactions on Communications*, Vol. 42, No. 4, April 1994.

[9] R. A. Dillard, "Detectability of spread-spectrum signals," *IEEE Transactions on Aerospace and Electronic Systems*, Vol. AES-15, No. 4, pp. 526-537, July 1979.

[10] N. C. Beaulieu, W. L. Hopkins and P. J. McLane, "Interception of frequency-hopped spread-spectrum signals," *IEEE Journal on Selected Areas of Communications*, Vol. 8, No. 5, pp. 854-855, June 1990.

[11] J. V. DiFranco and W. L. Rubin, *Radar Detection*, Prentice-Hall, Englewood Cliffs, NJ, 1968.

Barry K. Levitt (S'68-M'72) was born in Montreal, Canada on July 26, 1943. He received the B.Eng. degree in Engineering Physics from McGill University, Montreal, in 1965, and the S.M., E.E. and Ph.D. degrees in Electrical Engineering from the Massachusetts Institute of Technology, Cambridge, MA, in 1967, 1967 and 1971 respectively. His doctoral dissertation was on variable-rate optical communication through the turbulent atmosphere.

Upon graduation from MIT, he joined the Jet Propulsion Laboratory, where he is presently a Technical Group Leader for military communications programs in the Communications Research Section. His contributions at JPL have been in many diverse areas, including frame synchronization for deep-space missions, system management and analysis of the Radio Frequency Interference Surveillance System (RFISS), analytical support for the Search for Extra-Terrestrial Intelligence (SETI), design of low-probability-of-intercept communication radios for unmanned aerial vehicles, analysis of spread-spectrum communication systems, rain compensation techniques for mobile satellite communication links, and development of optimal signal processing algorithms for the detection and characterization of wideband emissions. He has published many journal and conference papers as a result of his research, and is a co-author with Dr. Marvin Simon, Dr. Jim Omura and Dr. Robert Scholtz of the three-volume reference text, *Spread Spectrum Communications*.

Dr. Levitt is an active member of the IEEE Communication Theory Committee, and served as General Chairman of the Communication Theory Workshop in 1982 and Co-Chairman in 1986.

Unjeng Cheng (S'78-M'82) received the B.S. degree in Electrical Engineering from the National Taiwan University, Taipei, Taiwan in 1976, the M.S. degree in Material Science from the University of Southern California in 1978, and the Ph.D. degree from the University of Southern California in 1981.

From 1981 to 1986, he was with Axiomatix Corporation in Los Angeles, CA, where he was working in the area of error-correcting codes and spread-spectrum communication. From 1986 to 1989, he as with the Jet Propulsion Laboratory in Pasadena, CA, where he was working in the area of communication networks and spread-spectrum code acquisition. From 1989 to 1991, he was with Space Computer Corporation in Santa Monica, CA, where he was working in the area of image processing and computer vision. Since 1991, has been with the Jet Propulsion Laboratory. His research interests are error correcting codes, spread-spectrum communication, communication networks, image processing, computer vision, and software engineering.

Dr. Cheng is the author of "Mathematical Functions Library", a software library including many frequently-used mathematical functions, published by Wiley.

Andreas Polydoros (M'78, SM'92) was born in Athens, Greece, in 1954. He was educated at the National Technical University of Athens, Greece (Diploma in EE, 1977), State University of New York at Buffalo (MSEE, 1979) and the University of Southern California (Ph.D., EE, 1982). He has been a faculty member in the Electrical Engineering/Systems Department and the Communications Sciences Institute (CSI) at the University of Southern California since 1982. He is currently a Professor and the Co-Director of CSI.

His general area of scientific interest is statistical communication theory with applications to spread-spectrum systems, signal detection and classification, and multi-user radio networks. He has over a dozen years of teaching, research and extensive consulting experience on these topics, both for the government and industry.

Professor Polydoros is the recipient of a 1986 NSF Presidential Young Investigator Award. He was the Associate Editor for Communications of the *IEEE Trans. Inform. Theory* (1987-88), the Guest Editor of the July 1993 Special Issue on "Digital Signal Processing in Communications" for *Digital Signal Processing: A Review Journal*. He is a co-Guest Editor for an upcoming Special Issue on "CDMA" of the *IEEE Journal on Selected Areas in Communications*.

Marvin K. Simon is currently a Senior Research Engineer at the Jet Propulsion Laboratory, Pasadena, California and Lecturer in Communications at the California Institute of Technology, Pasadena California. Dr. Simon has worked extensively for the last 25 years in the area of modulation, coding, and synchronization for space, satellite, and radio communications. The fruits of his research have been successfully applied to the design of many of NASA's deep space and near-earth missions for which he holds 8 patents and over 15 NASA Tech Briefs. He is a Fellow of the IEEE, Fellow of the IAE, and winner of a NASA Exceptional Service Medal both in recognition of outstanding contributions in analysis and design of space communications systems. In addition, he is listed in Marquis Who's Who in America.

He has published over 110 papers on the above subjects and is co-author of several textbooks including, *Telecommunication Systems Engineering* (Prentice-Hall, 1973 and reprinted by Dover Press, 1991), *Phase-Locked Loops and Their Application* (IEEE Press, 1978) *Spread Spectrum Communications, Vols. I, II, and III* (Computer Science Press, 1984), and *Trellis Coded Modulation with Applications* (Macmillan, 1991). His work has also appeared in the textbook *Deep Space Telecommunication Systems Engineering* (Plenum Press, 1984). He is the co-recipient of the 1986 Prize Paper Award in Communications for the *IEEE Transactions on Vehicular Technology*. He is currently completing a new two-volume set of books entitled Digital Communication Techniques.

UNIT 2
SPREAD SPECTRUM APPLICATIONS IN CELLULAR MOBILE

This unit is devoted to articles dealing with application of spread spectrum in cellular mobile systems. The spread spectrum technique in the form of Code Division Multiple Access (CDMA) opened a new horizon in the rapid development of cellular mobile communications. Researches throughout the world producing hundreds of papers in CDMA techniques gave a tremendous boost to the second generation cellular mobile and personal communication systems. In this section, we have selected some leading papers mainly concerning performance evaluation, power control issues, overlay capability, microcellular technique, and some traffic aspects in cellular CDMA.

SECTION 2.1
CDMA OVERVIEW AND PERFORMANCE

The first article, "*Overview of Cellular CDMA*" by Lee, gives a complete overview on cellular CDMA, touching on the issues of narrowband and wideband propagation in mobile radio environment, key elements of cellular design, capacity calculation and other important issues. The next paper, "*On the Capacity of a Cellular CDMA System*" by Gilhousen et al., puts emphasis on the capacity-improvement capability of CDMA in cellular telephony due to its interference suppression feature. The overall performance of a cellular DS-CDMA system under a flat Rayleigh fading channel is investigated in the third article, "*Performance Evaluation for Cellular CDMA*" by Milstein et al. The next two papers, "*Effects of Radio Propagation Path Loss on DS-CDMA Cellular Frequency Reuse Efficiency for the Reverse Channel*" by Rappaport and Milstein, and "*Multipath Propagation Effects on a CDMA Cellular System*" by Chan, deal with the evaluation of range of variation of frequency reuse efficiency due to the effects of radio propagation path loss and evaluation of reverse link performance on the basis of SNR measurement, respectively. A complete analysis of a single-cell DS-CDMA and a multiple-cell DS-CDMA is presented in the next two articles: "*Analysis of a Direct-Sequence Spread-Spectrum Cellular Radio System*" by Kchao and Stuber and "*Analysis of a Multiple-Cell Direct-Sequence CDMA Cellular Mobile Radio System*" by Stuber and Kchao. Approximate expressions for the area-averaged bit error probability and the area-averaged outage probability are obtained, keeping in view the effects of path loss, log-normal shadowing, multipath fading, multiple access interference and background noise.

Overview of Cellular CDMA

William C. Y. Lee, *Fellow, IEEE*

***Abstract*—This paper is a general description of code division multiple access (CDMA). The analysis of power control schemes in CDMA is an original work. The wide-band wave propagation in the cellular environment presents an interesting result (the short-term fading reduction over the wide band-signal in cellular). Also less fading in urban areas than in suburban areas. The advantages of using CDMA listed in this paper have excited the cellular industry. Radio capacity is the key issue in selecting CDMA and is carefully described in this paper.**

I. Introduction

The development of the code division multiple access (CDMA) scheme is mainly for capacity reasons. Ever since the analog cellular system started to face its capacity limitation in 1987, the promotion of developing digital cellular systems for increasing capacity has been carried out. In digital systems, there are three basic multiple access schemes, frequency division multiple access (FDMA), time division multiple access (TDMA), and code division multiple access (CDMA). In theory, it does not matter whether the spectrum is divided into frequencies, time slots, or codes, the capacity provided from these three multiple access schemes is the same. However, in the cellular system, we might find that one may be better than the another. Especially in the North American Cellular System, no additional spectrum will be allocated for digital cellular. Therefore, the analog and digital systems will co-exist in the same spectrum. Also, the problem of transition from analog to digital is another consideration. Although the CDMA has been used in satellite communications, the same CDMA system cannot be directly applied to the mobile cellular system. In order to design a cellular CDMA system, we first need to understand the mobile radio environment; then study whether the characteristics of CDMA are suitable for the mobile radio environment or not; and finally describe the natural beauty of applying CDMA in cellular systems.

II. Mobile Radio Environment

The propagation of a narrow-band carrier signal is a conventional means of communication. However, in a CDMA system, the propagation of a wide-band carrier signal is used. Therefore, we first describe the propagation of the narrow-band wave, then of the wide-band wave.

A. Narrow-Band (NB) Wave Propagation

A signal transmitted from the cell-site and received by either a mobile unit or a portable unit would propagate over a particular terrain configuration between two ends. Therefore, the effect of the terrain configuration generates a different long-term fading characteristic which follows a log-normal variation appearing on the envelope of the received signal, as shown in Fig. 1. Since the antenna height of a mobile or portable unit is close to the ground, three effects are observed [1]. First, the signal received is not only from the direct path but also from the strong reflected path due to the fact that the antennas of the mobile units are close to the ground. These two paths create an excessive path loss which is 40 dB/dec (fourth power law applied), i.e., doubling the path loss in decibels of the free-space path loss. Second, under the low antenna height condition at the mobile units, the human-made structures surrounding them would generate the multipath fading on the received signal called Rayleigh fading, as shown in Fig. 1. The multipath fading causes the burst error in digital transmission. The average duration of fades $\bar{t}$ as well as the level crossing rates $\bar{n}$ at 10 dB below the average power of a signal is a function of vehicle speed V and wavelength λ.

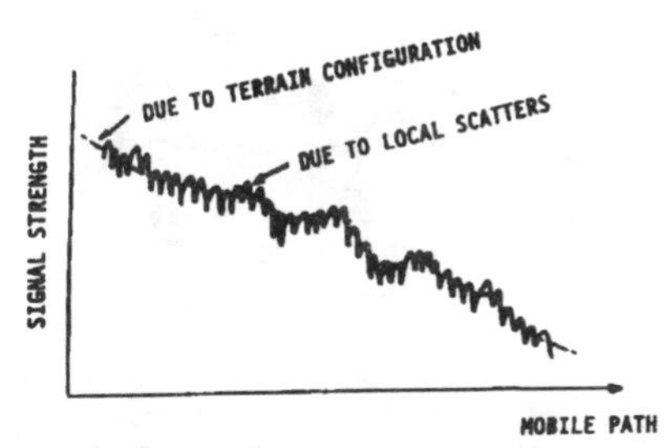

Fig. 1. Mobile radio environment.

$$\bar{t} = 0.132\left(\frac{\lambda}{V}\right) \text{ s} \tag{1}$$

$$\bar{n} = 0.75\left(\frac{V}{\lambda}\right) \text{ crossings/s.} \tag{2}$$

For a frequency of 850 MHz and a speed of 15 m/h then $\bar{t}$ = 6 ms and $\bar{n}$ = 16 crossings/s. Third, a time delay spread phenomenon exists due to the time dispersive medium. In a mobile radio environment a single symbol, transmitted from one end and received at the other end, receives not only its own symbol but also many echoes of its symbol. The time delay spread intervals are measured from the first symbol to the last detectable echo, which are different in human-made environments. The average time delay spread due to the local scatterers in suburban areas is 0.5 μs and in urban areas is 3 μs. These local scatterers are in the near-end region as illustrated in Fig. 2, and the time delay spread corresponding to this region is illustrated in Fig. 3. There are other types of

Manuscript received August 1, 1990; revised October 1, 1990. This paper was presented at the 1990 IEEE GLOBECOM Conference, San Diego, CA.
The author is with PacTel Cellular, Irvine, CA 92714.
IEEE Log Number 9144471.

0018-9545/91/0500-0291$01.00

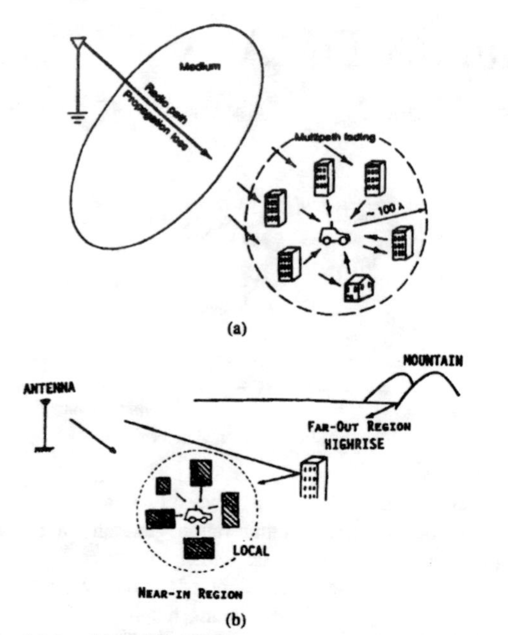

Fig. 2. (a) A mobile radio environment—two parts: propagation loss and multiple fading. (b) Time-delay spread scenario.

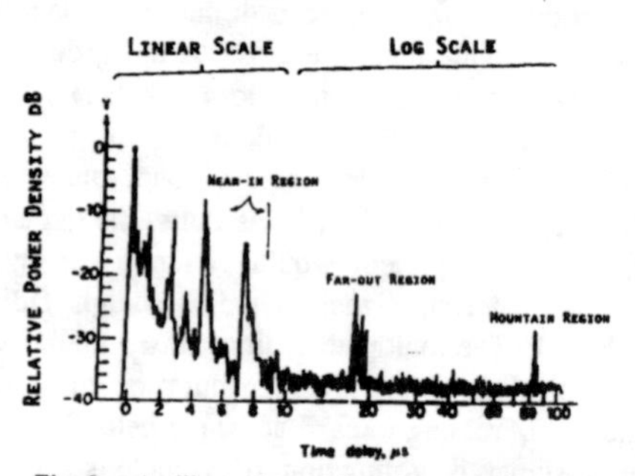

Fig. 3. An illustration on time-delay spread.

time delay spreads as illustrated in Fig. 2. One kind of delayed wave is due to the reflection of the high-rise buildings (far-out region), and one kind of delayed wave is due to the reflection from the mountains. Their corresponding time delays are illustrated in Fig. 3. In certain mountain areas, the time delay spread can be up to 100 μs. These time delay spreads would cause intersymbol interference (ISI) for data transmission [2]. In order to avoid the ISI, the transmission rate R_b should not exceed the inverse value of the delay spread Δ if the mobile unit is at a standstill (nonfading case),

$$R_b < 1/\Delta \tag{3}$$

or R_b should not exceed the inverse value of $2\pi\Delta$ if the mobile unit is in motion (fading case)

$$R_b < 1/(2\pi\Delta). \tag{4}$$

If the transmission rate R_b is higher than (3) or (4), both FDMA and TDMA need equalizers which are capable of reducing the ISI to a certain degree depending on the hardship of the time delay spread length and the wave arrival distribution [3]–[5]. An FDMA system always requires less transmission rate than a TDMA system if both systems offer the same radio capacity. Usually an FDMA system can get away from using an equalizer as long as its transmission rate does not exceed too much above 10 kilosamples per second. The CDMA system does not need an equalizer but a simpler device called a correlator will be used. It will be described later.

B. Wide-Band Wave Propagation [6]

1) *Path Loss:* Suppose that a transmitted power P_t in watts is used to send a wide-band signal with a bandwidth B in hertz along a mobile radio path r. The power spectrum over the bandwidth B is $S_t(f)$, then the P_t can be expressed as

$$P_t = G_t \int_{f_0 - \frac{B}{2}}^{f_0 + \frac{B}{2}} S_t(f)\, df. \tag{5}$$

The received power

$$P_r = \frac{P_t}{4\pi r^2} \times C(r.f) \times A_e(f) \tag{6}$$

where

$$C(r,f) = \text{medium characteristic} = k/(r^2 f) \tag{7}$$

$$A_e(f) = \text{effective aperture of the receiving antenna} = \frac{c^2 G_r}{4\pi f^2} \tag{8}$$

k is a constant factor, c is the speed of light, G_t and G_r are the gains of the transmitting and receiving antennas, respectively. Substituting (5), (7), and (8) into (6), we obtain

$$P_r = \frac{kc^2 G_R G_t}{(4\pi r^2)^2} \int_{f_0 - \frac{B}{2}}^{f_0 + \frac{B}{2}} S_t(f) \frac{1}{f^3}\, df. \tag{9}$$

For simplicity but without losing much generality, let

$$S_t(f) = \text{constant}, \quad \text{for } f_0 - B/2 \le f \le f_0 + B/2. \tag{10}$$

Then (9) becomes

$$P_r = \frac{kc^2 G_t G_R}{(4\pi r^2)^2} \cdot \frac{1}{f_0^3 \left[1 - \left(\frac{B}{2f_0}\right)^2\right]^2}. \tag{11}$$

Equation (11) is a general formula. For a narrow-band signal, $B \ll f_0$, then (11) becomes

$$P_r = \frac{kc^2 G_t G_R}{(4\pi r^2)^2 f_0^3} \quad \text{(narrow-band)}. \tag{12}$$

From (11), we may find the B/f_0 ratio for the case of 1-dB difference in path loss between narrow-band and wide-band.

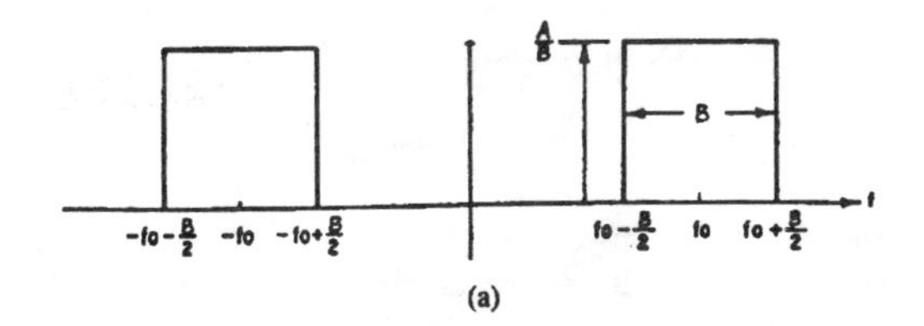

(a)

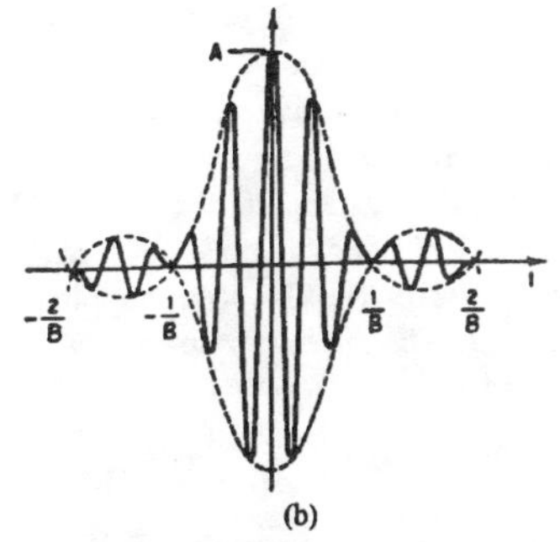

(b)

Fig. 4. Band-limited impulse. (a) Spectrum. (b) Waveshape.

That means by solving the denominator of (11) as follows:

$$10\log\left[1-\left(\frac{B}{2f_0}\right)^2\right]^2 = -1\ \text{dB}$$

we obtain

$$B = 0.66f_0.$$

In most wide-band applications, B will not be wider than $f_0/2$. Therefore, the narrow-band propagation path loss should be applied to the wide-band propagation path loss.

2) Multipath Fading Characteristic on Wide-Band: The wide-band pulse signaling $S_0(t)$ can be expressed as [7]

$$S_0(t) = A\frac{\sin(\pi Bt)}{\pi t} \tag{13}$$

where A is the pulse amplitude shown in Fig. 4.

The received signal can be represented as

$$S(t) = (A/B)\sum_{m=-\infty}^{\infty} b_m(t)\frac{\sin \pi B\left(t-\dfrac{m}{B}\right)}{\pi\left(t-\dfrac{m}{B}\right)}. \tag{14}$$

The pulsewidth of 1/B is the time interval of the pulse occupied. Count all b_m that are not vanishing over a range of a finite number of m which is corresponding to a time delay spread Δ. Then the effective number of diversity branches, M, can be approximated by

$$M = \frac{\Delta + \dfrac{1}{B}}{\dfrac{1}{B}} = B\cdot\Delta + 1. \tag{15}$$

The effective number of diversity varies according to the human-made structures. The M is larger in the urban area than in the suburban area. Letting $\Delta = 0.5\ \mu s$ for suburban and $\Delta = 3\ \mu s$ for urban, and $B = 30$ kHz for narrow-band and 1.25 MHz for wide-band, we find the effective number of diversity M in the following table.

Human-made environment	M diversity branches	
	$B = 30$ kHz	$B = 1.25$ MHz
$\Delta = 0.5\ \mu s$ Suburban	1.015	1.625
$\Delta = 3\ \mu s$ Urban	1.09	4.75

The wider the bandwidth, the less the fading. For $B = 1.25$ MHz, the fading of its received signal is reduced as if the diversity-branch receiver which equals $M = 1.625$ (between a single branch and two branches) is applied in suburban areas, and $M = 4.75$ (between four and five branches) is applied in urban areas. The wide-band signal would provide more diversity gain in urban areas than in suburban areas. For $B = 30$ kHz, no effective diversity gain is noticeable on its narrow-band received signal.

III. Key Elements in Designing Cellular

The frequency reuse concept guides the cellular system design.

A. Cochannel Interference Reduction Factor (CIRF)

The minimum separation between two cochannel cells, D_s, is based on a cochannel interference reduction factor q which is expressed as

$$q = D_s/R \tag{16}$$

where R is the cell radius. The value of q is different for each system. For analog cellular systems, $q = 4.6$ is based on the channel bandwidth $B_c = 30$ kHz and the carrier-to-interference ratio (C/I) equals 18 dB.

B. Handoffs

The handoff is a unique feature in cellular. It switches the call to a new frequency channel in a new cell site without either interrupting the call or alerting the user. Reducing unnecessary handoffs and making necessary handoffs successfully are very important tasks for the cellular system operators in analog systems or in future FDMA or TDMA digital systems.

C. Frequency Management and Frequency Assignment

Based on the minimum distance D_s, the number of cells k, in a cell reuse pattern may be obtained,

$$K = (D_s/R)^2/3 = q^2/3. \tag{17}$$

The total allocated channels will be divided by K. There are K sets of frequencies; each cell operates its own set of frequencies managed by the system operator. This is the frequency management task. During a call process different frequencies are assigned to different calls. This is the frequency assignment task. Both tasks are critically impacted by interference and capacity.

D. Reverse-Link Power Control

The reverse-link power control is for reducing near-end to far-end interference. The interference occurs when a mobile unit close to the cell site can mask the received signal at the cell site so that the signal from a far-end mobile unit is unable to be received by the cell site at the same time. It is a unique type of interference occurring in the mobile radio environment.

E. Forward-Link Power Control

The forward-link power control is used to reduce the necessary interference outside its own cell boundary.

F. Capacity Enhancement

The capacity of cellular systems can be increased by handling q in two conditions.

1) Within standard cellular equipment—the value of q shown in (16) remains a constant. Reduce the cell radius R, thus D_s reduces. For a smaller D_s the same frequency can be used more often in the same geographical area: that is why we are trying to use small cells (sometimes called microcells or picocells) to increase capacity.
2) Chosen from different cellular systems—many different types of radio equipment can be chosen. Search for those cellular systems which can provide smaller values of q. When q shown in (16) is smaller, D_s can be less, even if the cell radius remains unchanged. We believe that q is smaller in properly designed digital cellular systems than q in analog systems. Choosing a smaller new q of a new system, we can increase the same amount of capacity without reducing the size of the cell based on the old q of an old system. That is why we are choosing a new digital system to replace the old analog system.

Reducing the size of cells in a system requires more cells. It is always costly. Therefore, the development of digital cellular systems properly is the right choice.

IV. Spreading Techniques in Modulation

Spreading techniques in modulation are generally used in military systems for antijamming purposes. In general, there are two techniques: 1) spectrum spreading (spread spectrum) and 2) time spreading (time hopping) stated as follows:

A. Spread Spectrum (SS) Techniques

There are two general spread spectrum techniques, direct sequence (DS) and frequency hopping (FH).

1) Direct Sequence: In direct sequence, each information bit is symbolized by a large number of coded bits called chips. For example, if an information bit rate $R = 10$ kb/s is used and it needs an information bandwidth $B = 10$ kHz, and if each bit of 10 kb/s is coded by 100 chips, then the chip rate is 1 Mb/s which needs a DS bandwidth, $B_{ss} = 1$ MHz. The bandwidth is thus spreading from 10 kHz to 1 MHz. The spectrum spreading in DS is measured by the processing gain (PG) in decibels

$$PG = 10 \log \frac{B_{ss}}{B} \text{ (in dB)}. \tag{18a}$$

Then the PG of the above example is 20 dB. Or we say that this SS system has 20 dB processing gain. The first DS experiment was carried out in 1949 by DeRosa and Rogoff who established a link between New Jersey and California.

2) Frequency Hopping: An FH receiver would equip N frequency channels for an active call to hop over those N frequencies with a determined hopping pattern. If the information channel width is 10 kHz and there are 100 channels to hop, $N = 100$, the FH bandwidth $B_{ss} = 1$ MHz. The spectrum is spreading from 10 kHz (no hopping) to 1 MHz (frequency hopping). The spectrum spreading in FH is measured by the PG as

$$PG = 10 \log N \quad \text{(in dB)}. \tag{18b}$$

Then the PG of the above example is 20 dB. The total hopping frequency channels are called chips. There are two basic hopping patterns; one called fast hopping which makes two or more hops for each symbol. The other called slow hopping which makes two or more symbols for each hop. In general, the transmission data rate is the symbol rate. The symbol rate is equal to the bit rate at a binary transmission. Due to the limitation of today's technology, the FH is using a slow hopping pattern.

B. Time Hopping

A message transmitted with a data rate of R requiring a transmit time interval T is now allocated at a longer transmission time interval T_s. In time T_s the data are sent in bursts dictated by a hopping pattern. The time interval between bursts t_n also can be varied. The time spreading data rate R_s is always less than the information bit rate R. Assume that N bursts occurred in time T, then

$$R_s = \left(\frac{T_s}{T}\right) R = \left(1 - \frac{\sum_1^N t_n}{T}\right) R. \tag{19}$$

V. Description of DS Modulation

The spread spectra (DS and FH) are used for reducing intentional interference (enemy jamming), and now we are using it for increasing capacity instead of reducing the intentional interference. Immediately we realize that the FH with a slow hopping does not serve the purpose of increasing capacity. The slow hopping is to let good channels downgrade and bad channels upgrade. In order to have a system design for capacity, all the channels have to be deployed only marginally well. If bad channels do occur in this high capacity SS system, the system does not provide normal channels with excessive signal levels which can average with the poor signal levels of those bad channels to within an acceptable quality level. It just pulls down all the channels to an unacceptable level. The proper way should be either drop the

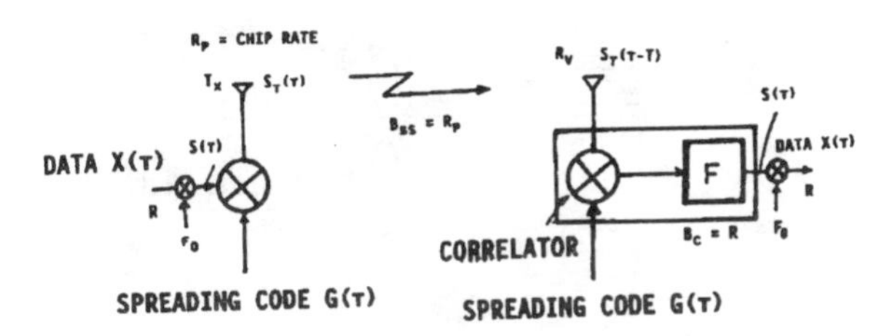

Fig. 5. Basic spread—spectrum technique.

bad channels or correct the bad channels by other means. The fast hopping does help increase the capacity because of its advantage of applying diversity but the technology to have fast hopping at 800 MHz is not available.

1) Basic DS Technique: The basic DS technique is illustrated in Fig. 5. The data $x(t)$ transmitted with a data rate R is modulated by a carrier f_0 first, then by a spreading code $G(t)$ to form a DS signal $S_t(t)$ with a chip rate R_p which takes a DS bandwidth B_{ss}. The DS signal $S_t(t - T)$ after a propagation delay T is received and goes through a correlator using the same spreading code $G(t)$ prestored in it to despread the DS signal. Then the despread signal $S(t - T)$ is obtained. After demodulating it by f_0, $x(t)$ is recovered. Take a constant-envelop signal modulated on a carrier f_0 at transmitting end shown in Fig. 5. Let $x(t)$ be a data stream modulated by a binary phase shift keying (BPSK) that

$$x(t) = \pm 1 \tag{20}$$

modulated by a binary shift keying

$$S(t) = x(t)\cos(2\pi f_0 t). \tag{21}$$

At the transmitting end, the spreading sequence $G(t)$ modulation also uses BPSK

$$G(t) = \pm 1 \tag{22}$$

then

$$S_t(t) = x(t)G(t)\cos(2\pi f_0 t). \tag{23}$$

At the receiving end, the $S_t\ (t - T)$ is received after T seconds propagation delay. The despreading processing then takes place. The signal $S(t - T)$ coming out from the correlator is

$$S(t-T) = x(t-T) \cdot G(t-T)G(t-\hat{T})\cos(2\pi f_0(t-T)) \tag{24}$$

where $\hat{T}$ is the estimated propagation delay generated in the receiver. Since $G(t) = \pm 1$,

$$G(t-T)G(t-\hat{T}) = 1 \tag{25}$$

from a good correlator $T = \hat{T}$. Then

$$S(t-T) = x(t-T)\cos(2\pi f_0(t-T)). \tag{26}$$

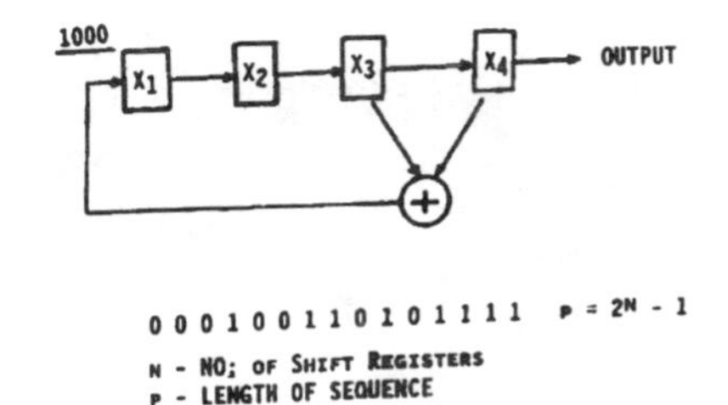

Fig. 6. PN code (linear maximal length sequence) generator.

After it is demodulated by the carrier frequency f_0, the data $x(t - T)$ then are recovered as shown in Fig. 5.

2) Pseudonoise (PN) Code Generator: Pseudonoise code coming from a PN sequence is a deterministic signal [8]. For example, the sequence 00010011010111 is a PN sequence. It contains three properties.

a) *Balance property:* 7 zeros and 8 ones. The numbers of zeros and ones of a PN code are different only by one.

b) *Run property:* There are four "zero" runs (or "one" runs): runs = 4.
1/2 of runs (i.e., 2) of length 1; i.e., two single "zeros (or ones)."
1/4 of runs (i.e., 1) of length 2; i.e., one "2 consecutive zeros (or ones)."
1/8 of runs (i.e., 0.5) of length 3; i.e., one "3 consecutive zeros (or ones)." In the above example, 1/8 of runs cannot be counted for too short a code.

c) *Correlation property:* Let D denote the "difference," and S denote the "same" by comparing two PN codes as follows:

0 0 0 1 0 0 1 1 0 1 0 1 1 1 1
1 0 0 0 1 0 0 1 1 0 1 0 1 1 1.
D S S D D S D S D D D D S S S

The value of the correlation of two N-bit sequences can be obtained by counting the number N_d of D's and the number N_s of S's and inserting them into the following equation:

$$P = \frac{1}{N}(N_s - N_d) = \frac{1}{15}(7-8) = -\frac{1}{15}. \tag{27}$$

Then the correlation of a 15-b PN code is $-1/15$. The PN code generator of a four-shift register is shown in Fig. 6. The modulo 2 adder is summing the shift register X_3 and the shift register X_4. The summing signal then feeds back to the shift register X_1. Suppose that a 4-b sequence 1000 is fed into the shift register X_1. The output PN sequence from this PN code generator is 00010011010111. The code length L of any PN code generator is dependent upon the number of shift registers N:

$$L = 2^N - 1. \tag{28}$$

The PN sequence generated in Fig. 6 is also called the linear maximal length sequence. For $N = 4$, L is 15.

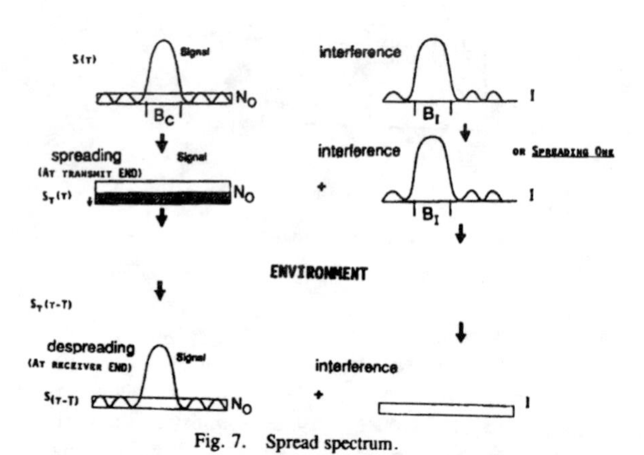

Fig. 7. Spread spectrum.

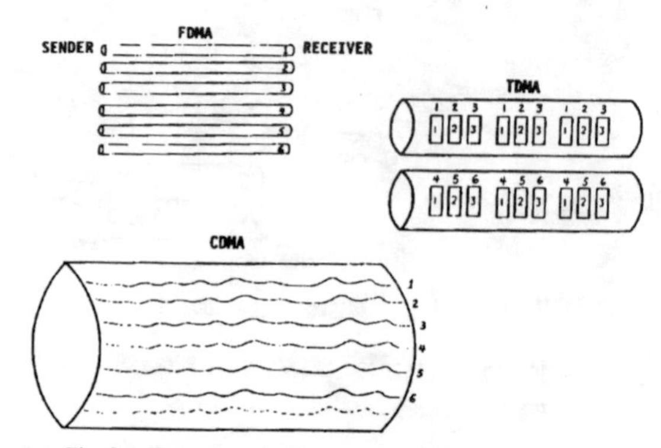

Fig. 8. Illustration of different multiple access systems.

3) Reduction of Interference by a DS signal: The signal $S(t)$ of Fig. 5, before the spreading processing can be illustrated in both the frequency and time domains, is shown in Fig. 7. After spreading $S(t)$ with a given $G(t)$, the output $S_t(t)$ is transmitted out while the interference in the air could be a narrow-band signal or a DS signal with a different $G_I(t)$. When $S_t(t - T)$ is received after a propagation delay T, it is despreading with the same $G(t)$ and obtaining $S(t - T)$. The interference signal would spread to an SS signal by the $G(t)$ if it was a narrow-band signal, or stay as an SS signal because $G(t)$ and $G_I(t)$ do not agree with each other. Thus as a result, a low level of interference within the desired signal bandwidth B_c can be achieved.

VI. Multiple Access Schemes

The multiple access schemes are used to provide resources for establishing calls. There are five multiple access schemes. *FDMA* serves the calls with different frequency channels. *TDMA* serves the calls with different time slots. *CDMA* serves the calls with different code sequences. *PDMA* (polarization division multiple access) serves the calls with different polarization. PDMA is not applied to mobile radio [6]. *SDMA* (space division multiple access) serves the calls by spot beam antennas. The calls in different areas covered by the spot beams can be served by the same frequency—a frequency reuse concept. In the cellular system, the first three multiple access schemes can be applied. The illustration of the differences among three multiple access schemes are shown in Fig. 8. Assume that a set of six channels is assigned to a cell. In FDMA, six frequency channels serve six calls. In TDMA, the channel bandwidth is three times wider than that of FDMA channel bandwidth. Thus two TDMA channel bandwidths equal six FDMA channel bandwidths. Each TDMA channel provides three time slots. The total of six time slots serve six calls. In CDMA, one big channel has a bandwidth equal to six FDMA channels. The CDMA radio channel can provide six code sequences and serve six calls. Also, CDMA can squeeze additional code sequences in the same radio channel, but the other two multiple access schemes cannot. Adding additional code sequences, of course, degrades the voice quality.

A. Carrier-to-Interference Ratio (C/I)

In analog systems, only FDMA can be applied. The C/I received at the RF is closely related to the S/N at the baseband which is related to the voice quality. In digital systems, all three, FDMA, TDMA, and CDMA can be applied. The C/I received at the RF is closely related to the E_b/I_0 at the baseband.

$$\begin{aligned} \text{C/I} &= (E_b/I_0)(R_b/B_c) \\ &= (E_b/I_0)/(B_c/R_b) \end{aligned} \tag{29}$$

where E_b is the energy per bit; I_0 is the interference power per hertz, R_b is the bit per second, and B_c is the radio channel bandwidth in hertz. In digital FDMA or TDMA there are designated channels or time slots for calls. Thus R_b equals B_c and E_b/I_0 at the baseband is always greater than one, then C/I is also greater than one, i.e., a positive value in decibels. In CDMA, all the coded sequences say N, share one radio channel; thus B_c is much greater than R_b. The notation B_c is often replaced by B_{ss} which is the spread-spectrum channel. Within the radio channel, any one code sequence is interfered with N-1 of other code sequences. Therefore, the interference level is always higher than the signal level. C/I is less than one, i.e., a negative value in decibels.

B. Capacity of Cellular FDMA and TDMA [9]

In FDMA or TDMA, each frequency channel or each time slot is assigned to one call. During the call period, no other calls can share the same channel or slot. In this case, the cochannel interference would come from a distance of $D_s = q R$. Assume that the worst case of having six cochannel interferers (see Fig. 9) and the fourth power law pathloss are applied. The capacity of the cellular FDMA and cellular TDMA can be found by the radio capacity m expressed as

$$m = \frac{B_t/B_c}{K} = \frac{M}{\sqrt{\frac{2}{3}\left(\frac{C}{I}\right)_s}} \text{ number of channels/cell} \tag{30}$$

where

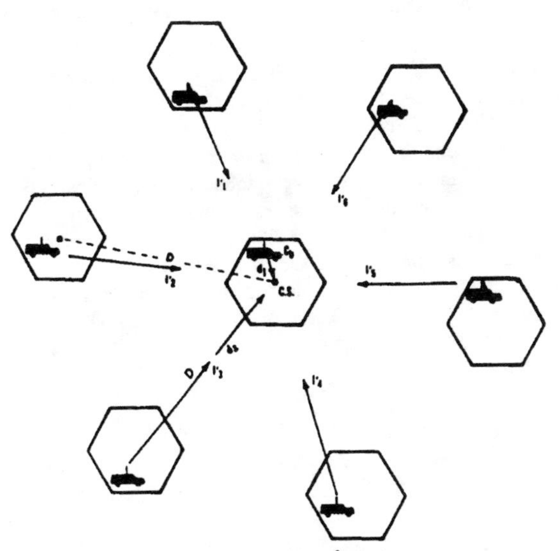

Fig. 9. Cochannel interference.

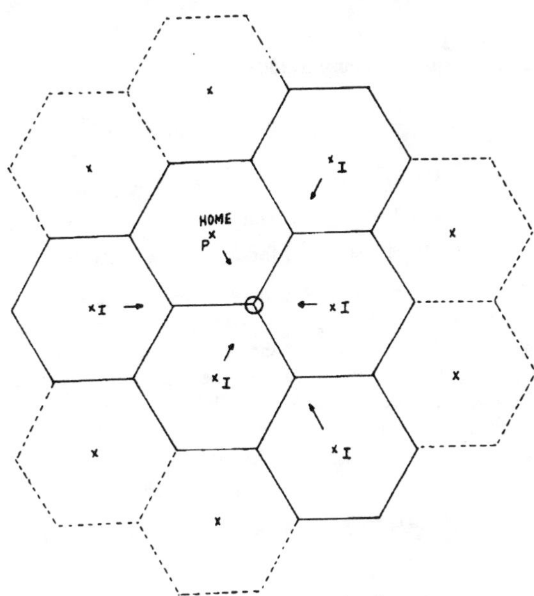

Fig. 10. CDMA system and its interference.

B_t total bandwidth (transmitted or received),

B_c channel bandwidth (transmitted or received) or equivalent channel bandwidth,

$M = B_t / B_c$ total number of channels or equivalent channels,

$(C/I)_s$ minimum required carrier-to-interference ratio per channel or per time slot.

Equation (30) can be directly applied to both analog FDMA and digital FDMA systems. In TDMA systems, B_c is an equivalent channel bandwidth. For example, a TDMA radio channel bandwidth of 30 kHz with three time slots can have an equivalent channel bandwidth of 10 kHz (B_c = 10 kHz). Therefore the minimum required $(C/I)_s$ of each time slot turns out to be the same as $(C/I)_s$ of the TDMA equivalent channel. The radio capacity is based on two parameters, B_c and $(C/I)_s$ as shown in (30). It has the same two parameters as appear in Shannon's channel capacity formula. The difference between (30) and Shannon's is that the two parameters are related in the former one and independent in the latter one. The $(C/I)_s$ of radio capacity can be found based on a standard voice quality as soon as the channel bandwidth B_c is given.

C. Radio Capacity of Cellular CDMA

Cellular CDMA is uniquely designed to work in cellular systems. The primary purpose of using this CDMA is for high capacity. In cellular CDMA, there are two CIRF values. One CIRF is called adjacent CIRF, $q_a = D_s/R = 2$. It means that the same radio channel can be reused in all neighboring cells. The other CIRF is called self-CIRF, $q_s = 1$. It means that different code sequences use the same radio channel to carry different traffic channels. The two CIRF's are shown in Fig. 10. With the smallest value of CIRF, the CDMA system is proven to be the most efficient frequency-reuse system we can find.

1) Required $(C/I)_s$ in Cellular CDMA: $(C/I)_s$ can be found from (29) depending on the value of E_b/I_0 which is measured at the baseband determined by the voice quality. For example, the vocoder rate is R_b = 8 kb/s and the total wide-band channel bandwidth B_t = 1.25 MHz, then if E_b/I_0 is determined as follows:

$$E_b/I_0 = 7 \text{ dB, then } (C/I)_s = 0.032$$

$$E_b/I_0 = 4.5 \text{ dB, then } (C/I)_s = 0.01792.$$

The radio capacity of this system can be derived as follows. It can be calculated based on the forward link, and can also be further improved by the power control schemes.

2) Without Power Control Scheme: The radio capacity is calculated from the forward link C/I ratio. The $(C/I)_s$ received by a mobile unit at the boundary of a CDMA cell shown in Fig. 10 can be obtained based on nine interfering cells as follows:

$$(C/I)_s = \frac{\alpha \cdot R^{-4}}{\underbrace{\alpha(M-1)\cdot R^{-4}}_{\text{within the cell}} + \underbrace{\alpha \cdot 2M \cdot R^{-4}}_{\text{two closest adjacent cells}} + \underbrace{\alpha \cdot 3M \cdot (2R)^{-4}}_{\text{three intermediate-range cells}} + \underbrace{\alpha \cdot 6M (2.633R)^{-4}}_{\text{six distant cells}}}$$

$$= \frac{1}{3.3123M - 1} \tag{31}$$

where α is a constant factor, M is the number of traffic channels. $(C/I)_s$ can be determined based on E_b/I_0 and R_b/B_s as shown in (29). Then M can be found from (31)

$$(C/I)_s = 0.032 \quad M = 9.736$$

$$(C/I)_s = 0.01792 \quad M = 17.15.$$

The radio capacity defined in (30)

$$m = \frac{M}{K} \text{ number of traffic channels/cell.} \quad (32)$$

In this case $K = q_s^2/3 = 4/3 = 1.33$. Therefore,

$$m = M/1.33 = 7.32 \text{ traffic channels/cell for } E_b/I_0 = 7 \text{ dB}$$
$$= 12.9 \text{ traffic channels/cell for } E_b/I_0 = 4.5 \text{ dB}.$$

3) With Power Control Scheme: We can increase the radio capacity by using a proper power control scheme. The power control scheme used at the forward link of each cell can reduce the interference to the other adjacent cells. The less the interference generated in a cell, the more the value of M increases. In (31), we notice that if we can neglect all the interference, then, as shown in Fig. 10

$$(\text{C/I})_s = \frac{R^{-4}}{(M-1)R^{-4}} = \frac{1}{M-1} \quad (33)$$

for

$$(\text{C/I})_s = 0.032 \quad M = 30.25$$
$$(\text{C/I})_s = 0.01792 \quad M = 54.8.$$

Comparing (31) with (33), the total number of traffic channels M is drastically reduced due to the existence of interference. However, since interference is always existing in the adjacent cells, we can only reduce it by using a power control scheme. By using a power control scheme the total power after combining all traffic channels should be considered in two cases. a) The necessary power delivery to the close-in mobile unit and b) The total power reduced at the boundary.

a) The necessary power delivery to a close-in mobile unit. The transmitted power at the cell site for the jth mobile unit is P_j, which is proportional to r_j^n.

$$P_j \propto r_j^n \quad (34)$$

where r_j is the distance between the cell site and the jth mobile unit. n is a number. In examining the number n, we find that the power control scheme of using $n = 2$ in (34) can provide the optimum capacity and also meet the requirement that the forward link signal can still reach the near-end mobile unit at distance r_j from the cell site with a reduced power

$$P_j = P_R\left(\frac{r_j}{R}\right)^2 \quad (35)$$

where P_R is the power required to reach those mobile units at the cell boundary R. The M mobile units served by M traffic channels are assumed uniformly distributed in a cell. Then

$$p(M_l) = kr_l, \quad 0 \le r_l \le R \quad (36)$$

where $M = \sum_{l=1}^{L} M_l$. There are L groups of mobile units. Each one of L is equally circled around the cell site. Where M_e is the number of mobile units in the lth group depending on its location. k is a constant. Equation (36) indicates that fewer mobile units are closely circling around the cell site, more mobile units are at the outside ring of the cell site. Assume that the distance r_0 is from the cell site to a desired mobile unit, also assume that r_0 is a near-in distance between the mobile unit and the cell site. With the help of (34) and (35) the power transmitted from the cell site, P_t, is equal to

$$P_t = \sum^{M_1} P_1 + \sum^{M_2} P_2 + \sum^{M_3} P_3 + \cdots + \sum^{M_L} P_L$$
$$= P_R\left[\sum^{kr_1}\left(\frac{r_1}{R}\right)^2 + \sum^{kr_2}\left(\frac{r_2}{R}\right)^2 + \cdots + \sum^{kr_L}\left(\frac{r_L}{R}\right)^2\right]$$
$$= P_R\left[kr_1\left(\frac{r_1}{R}\right)^2 + kr_2\left(\frac{r_2}{R}\right)^2 + \cdots + kr_L\left(\frac{r_L}{R}\right)^2\right]. \quad (37)$$

Since r_L is the distance from the cell site to the cell boundary, $r_L = R$ then (37) becomes

$$P_t = P_R k \int_0^R \frac{r^3}{R^2}\,dr = P_R k \frac{R^2}{4}. \quad (38)$$

The total number of mobile units M can be obtained as

$$M = \sum_{l=1}^{L} M_l = k(r_1 + r_2 + \cdots + R)$$
$$= k\int_0^R r\,dr = k\frac{R^2}{2}. \quad (39)$$

Substituting (39) into (38):

$$P_t = P_R k\left[\frac{M}{2k}\right] = P_R\frac{M}{2}. \quad (40)$$

If the full power P_R is applied to every channel, then

$$P_t = MP_R. \quad (41)$$

Comparing (40) and (41), the total transmitted power reduces to one-half by using the power control scheme of (35). The $(\text{C/I})_s$ of a mobile unit at a distance of r_0 which is close to the cell site is

$$(\text{C/I})_{s1} = \frac{P_R(r_0/R)^2 \cdot r_0^{-4}}{P_R(M/2) \cdot r_0^{-4}} = \frac{(r_0/R)^2}{(M/2)}. \quad (42)$$

The interference from the adjacent cells can be neglected in (42) in this case.

b) The total power is reduced at the cell boundary. The $(\text{C/I})_s$ of a mobile unit at a distance R which is at the cell boundary can be obtained similarly to (31).

$$(\text{C/I})_{s2} = \frac{P_R}{P_R\left[\frac{M-1}{2} + 2\frac{M}{2} + 3\left(\frac{M}{2}\right)\cdot(2)^{-4} + 6\left(\frac{M}{2}\right)(2.633)^{-4}\right]} = \frac{1}{1.656M}. \quad (43)$$

The values of M and m can be found from (43) for the case of applying the power control scheme.

$$M = 18.87,\ m = 14.19,\ (\mathrm{C/I}) = 0.032\ (-15\ \mathrm{dB})$$
$$M = 23.7,\ m = 28.33,\ (\mathrm{C/I}) = 0.01792\ (-17\ \mathrm{dB}). \tag{44}$$

At this time, $(\mathrm{C/I})_s$ received by the mobile unit at the distance r_0 from (42) should be checked with (43) to see whether it is valid or not.

$$(\mathrm{C/I})_{s1} = \frac{(r_0/R)^2}{M/2} = \frac{3.3(r_0/R)^2}{3.3(M/2)} \geq \frac{1}{1.656M}. \tag{45}$$

In (44), the power reduction ratio $(r/R)^2$ has to be not less than 0.302 for those mobile units located less than the distance r_0 which is $0.55R$. If we set the lowest power to be $0.302 P_R$ then the total power has to be changed.

$$P_t = P_R k\left[\frac{r_0^2}{R^2} r_1 + \frac{r_2^3}{R^2} + \frac{r_3^3}{R^2} + \cdots\right]$$
$$= P_R k\left[\left(\frac{r_0}{R}\right)^2 \int_0^{r_0} r\,dr + \int_{r_0}^{R} \frac{r^3}{R^2}\,dr\right]$$
$$= P_R k \frac{R^2}{4}\left[1 + \left(\frac{r_0}{R}\right)^4\right]. \tag{46}$$

For $r_0/R = 0.55$, then $(r_0/R)^4 = 0.0913$.
The transmitted power P_t in (46) has to be adjusted as

$$P_t = P_R k(R^2/4) \times 1.0913 = P_R(M/2) \times 1.0913. \tag{47}$$

Equation (47) indicates that by setting the condition of the lowest power per traffic channel to be $0.302 P_R$ at the cell site to serve the mobile units within and equal to the distance r_0, $r_0 = 0.55\ R$, the total power at the cell site is slightly increased by 1.0913 times as compared with (38). Under the adjusted transmitted power P_t as shown in (47), the actual values of M and m are reduced.

$$M = 18.87/1.0913 = 17.3,\quad m = 13 \quad \text{for } (\mathrm{C/I})_s = 0.032$$
$$M = 33.7/1.0913 = 30.9,\quad m = 25.96 \quad \text{for } (\mathrm{C/I})_s = 0.1792. \tag{48}$$

Comparing (48) with (44), we find no significant change of M and m when the adjusted transmitted power is applied.

D. Comparison of Different Cases in CDMA

Table I lists the performance of five different cases:

Case 1: No adjacent cell interference is considered (this is not a real case).

Case 2: No power control, adjacent cell interference is considered.

Case 3: Power control with $n = 1$, adjacent cell interference is considered.

Case 4: Power control with $n = 2$, adjacent cell interference is considered.

Case 5: Power control with $n = 3$, adjacent cell interference is considered.

In Table I, Case 1 is not a real case. In Case 2, without power control, the performance is poor. The power control schemes are used in Cases 3–5. In these cases, in order to provide the minimum transmitted power at the cell site for serving those mobile units within or equal to the distance of r_0, the total transmitted power at the cell site increases as indicated under the heading "after adjusting the transmitted power." Comparing the number of channels per cell m among Cases 3–5, we found that Case 4 has two channels more than Case 3 but one channel less than Case 5. However, Case 5 is harder to implement than Case 4. One channel gained in Case 5 over Case 4 can be washed out in the practical situation. When the power control schemes of $n > 3$ are used, no further improvement in radio capacity is found. Therefore, we conclude that $n = 2$ in Case 4 is a better choice.

VII. Reduction of Near-Far Ratio Interference in CDMA

In CDMA, all traffic channels are sharing one radio channel. Therefore a strong signal received from a near-in mobile unit will mask the weak signal from a far-end mobile unit at the cell site. To reduce this near-far ratio interference, a power control scheme should be applied on the reverse link. As a result, the signals received at the cell site from all the mobile units within a cell remain at the same level. The scheme is described as follows. The power transmitted from each mobile unit has to be adjusted based on its distance from the cell site, as

$$P_j = P_R\left(\frac{r_j}{R}\right)^4 \tag{49}$$

where P_R, r, and R are mentioned previously, and a fourth power rule is applied in (49). Neglecting the interfering signals from adjacent cell, the C/I received from a mobile unit J, at the cell site can be obtained as

$$\mathrm{C/I} = \frac{P_R\left(\frac{r_J}{R}\right)^4 (r_J)^{-4}}{\sum_{1}^{M-1} P_R\left(\frac{r_j}{R}\right)^4 (r_j)^{-4}} = \frac{1}{M-1}. \tag{50}$$

The C/I of (48) has to be greater than or equal to the required $(\mathrm{C/I})_s$,

$$\mathrm{C/I} \geq (\mathrm{C/I})_s. \tag{51}$$

Applying (51) in (50), we obtain

$$M = 30.25,\ m = 22.74,\quad \text{for } (\mathrm{C/I})_s = 0.032\ (-15\ \mathrm{dB})$$
$$M = 54.5,\quad m = 41.2,\quad \text{for } (\mathrm{C/I})_s = 0.01792\ (-17\ \mathrm{dB}).$$

The number of channels M obtained from the reverse link is much higher than that from the forward channel as shown in Table I. It indicates that the effort is to increase the number of channels on the forward link for more radio capacity.

TABLE I

	Adjacent Cell Interfering				No Adjacent Cell Interference is Considered
	No Power Control	Power Control Schemes			
Performance in Different Cases	Case 2 $N = 0$	Case 3 $N = 1$	Case 4 $N = 2$	Case 5 $N = 3$	Case 1
Power Control due to the distance from the cell site	P_R	$P_R(r_j/R)$	$P_R(r_j/R)^2$	$P_R(r_j/R)^3$	P_R
R_0	N/A	$0.303R$	$0.55R$	$0.7R$	N/A
Before adjusting the TX power					
Total Transmitted Power at the Cell Site	MP_R	$P_R(2M/3)$	$P_R(M/2)$	$P_R(2M/5)$	MP_R
The $(C/I)_s$ Received at R_0	$\frac{1}{M-1}$	$(r_0/R)/(2M/3)$	$(r_0/R)^2/(M/2)$	$(r_0/R)3/(2M/5)$	$\frac{1}{M-1}$
At R (Cell Boundary)	$\frac{1}{3.3123M-1}$	$\frac{1}{2.2M}$	$\frac{1}{1.656M}$	$\frac{1}{1.32M}$	$\frac{1}{M-1}$
M at $(C/I)_s = 0.032$	9.736	14.2	18.87	23.67	30.25
$(C/I)_s = 0.0179$	17.15	25.36	33.7	42.27	54.8
M at $(C/I)_s = 0.032$	7.32	10.67	14.19	17.8	22.74
$= 0.0179$	12.9	19	28.33	31.78	41.2
After adjusting the TX power					
Total Transmitted Power at the Cell Site		$P_R(2M/3) \times 1.0139$	$P_R(M/2) \times 1.09$	$P_R(2M/5) \times 1.25$	
The $(C/I)_s$ Received at $R \leq R_0$		$(r_0/R)/[(2M/3) \times 1.0139]$	$(r_0/R)^2/[(M/2) \times 1.09]$	$(r_0/R)^3/[2M/5 \times 1.25]$	
at $R > R_0$		$(r/R)/[(2M/3) \times 1.0139]$	$(r/R)^2/[(M/2) \times 1.09]$	$(r/R)^3/[(2M/5) \times 1.25]$	
at R		$\frac{1}{2.23M}$	$\frac{1}{1.8M}$	$\frac{1}{1.65M}$	
M at $(C/I)_s = 0.032$ (−15 dB)		14.2	17.3	19	
at $(C/I)_s = 0.010792$ (−17.4 dB)		25.36	31	33.8	
m at $(C/I)_s = -15$ dB		10.67	13	14	
at $(C/I)_s = -17.4$ dB		19	23.3	25.4	

VIII. Natural Attributes of CDMA [10]

There are many attributes of CDMA which are of great benefit to the cellular system.

1) Voice activity cycles: The real advantage of CDMA is the nature of human conversation. The human voice activity cycle is 35%. The rest of the time we are listening. In CDMA all the users are sharing one radio channel. When users assigned to the channel are not talking, all others on the channel benefit with less interference in a single CDMA radio channel. Thus the voice activity cycle reduces mutual interference by 65%, increasing the true channel capacity by three times. CDMA is the only technology that takes advantage of this phenomenon. Therefore, the radio capacity shown in (48) can be three times higher due to the voice activity cycle. It means that the radio capacity is about 40 channels per cell for C/I = −15 dB or $E_b/I_0 = 7$ dB.
2) No equalizer needed: When the transmission rate is much higher than 10 kb/s in both FDMA and TDMA, an equalizer is needed for reducing the intersymbol interference caused by time delay spread. However, in CDMA, only a correlator is needed instead of an equalizer at the receiver to despread the SS signal. The correlator is simpler than the equalizer.
3) One radio per site: Only one radio is needed at each site or at each sector. It saves equipment space and is easy to install.
4) No hard handoff: Since every cell uses the same CDMA radio, the only difference is the code sequences. Therefore, no handoff from one frequency to another frequency while moving from one cell to another cell. It is called a soft handoff.
5) No guard time in CDMA: The guard time is required in TDMA between time slots. The guard time does occupy the time period for certain bits. Those waste bits could be used to improve quality performance in TDMA. In CDMA, the guard time does not exist.
6) Sectorization for capacity: In FDMA and TDMA, the utilization of sectorization in each cell is for reducing the interference. The trunking efficiency of dividing channels in each sector also decreases. In CDMA, the sectorization is used to increase capacity by introducing three radios in three sectors and therefore, three times the capacity is obtained as compared with one radio in a cell in theory.
7) Less fading: Less fading is observed in the wide-band

signal while propagating in a mobile radio environment. More advantage of using a wide-band signal in urban areas than in suburban areas for fading reduction as described in Section II-B.

8) Easy transition: In a situation where two systems, analog and CDMA, have to share the same allocated spectrum, 10% of the bandwidth (1.25 MHz) will increase two times (= 0.1×20) of the full bandwidth of FM radio capacity as shown below. Since only 5% (heavy users) of the total users take more than 30% of the total traffic, the system providers can let the heavy users exchange their analog units for the dual mode (analog/CDMA) units and convert 30% of capacity to CDMA on the first day of CDMA operations.

9) Capacity advantage: Given that

   $B_t = 1.25$ MHz, the total bandwidth

   $B_{ss} = 1.25$ MHz the CDMA radio channel

   $B_c = 30$ kHz for FM

   $B_c = 30$ kHz and three time slots for TDMA.

   Capacity of FM $1.25/30 = 41.6$

   total numbers of channels $= \dfrac{1.25 \times 10^6}{30 \times 10^3} = 41.67$ channels

   the cell reuse pattern $K = 7$

   the radio capacity $m_{FM} = \dfrac{41.67}{7} = 6$ channels/cell.

   Capacity of TDMA

   total number of channels $\dfrac{1.25 \times 10^6}{10 \times 10^3} = 125$ channels

   the cell reuse pattern $K = 4$ (assumed)

   the radio capacity $m_{TDMA} = \dfrac{125}{4} = 31.25$ channels/cell.

   Capacity of CDMA

   total number of channels / cell, $m = 13$

   the cell reuse pattern $K = 1.33$

   the radio capacity, take (48) at $E_b/I = 7$ dB add voice activity cycle and sectorization,

   $m_{CDMA} = 13 \times 3 \times 3 \approx 120$ channels/cell.

   Therefore,

$$m_{CDMA} = 20 \times m_{FM}$$
$$= 4 \times m_{TDMA}.$$

10) No frequency management or assignment needed: In FDMA and TDMA, the frequency management is always a critical task to carry out. Since there is only one common radio channel in CDMA, no frequency management is needed. Also, the dynamic frequency would implement in TDMA and FDMA to reduce real-time interference, but needs a linear broad-band power amplifier which is hard to develop. CDMA does not need the dynamic frequency assignment.

11) Soft capacity: In CDMA, all the traffic channels share one CDMA radio channel. Therefore, we can add one additional user so the voice quality is just slightly degraded as compared to that of the normal 40-channel cell. The difference in decibels is only $10 \log \frac{41}{40}$ which is 0.24 dB down in C/I ratio.

12) Coexistence: Both systems, analog and CDMA can operate in two different spectras, and CDMA only needs 10 % of bandwidth to general 200 % of capacity. No interference would be considered between two systems.

13) For microcell and in-building systems: CDMA is a natural waveform suitable for microcell and in-building because of being susceptible to the noise and the interference.

IX. Conclusion

The overview of CDMA highlights the potential of increasing capacity in future cellular communications. This paper describes the mobile radio environment and its impact on narrow-band and wide-band propagation. The advantage of having CDMA in cellular systems is depicted. The concept of radio capacity in cellular is also introduced. The power control schemes in CDMA have been carefully analyzed. The natural attributes of CDMA provide the reader with the reasons that cellular is considering using it. This paper leads the reader to understand two CDMA papers [11], [12], which are analyzed in more depth in this issue, and to build interest in CDMA by reading other references [13]-[19].

References

[1] W. C. Y. Lee, *Mobile Cellular Telecommunication System.* New York: McGraw-Hill, 1989, ch. 4.

[2] —, *Mobile Communications Engineering.* New York: McGraw-Hill, 1982, pp. 340-399.

[3] J. G. Proakis, "Adaptive equalization for a TDMA digital mobile radio," *IEEE Trans. Veh. Technol.*, pp. 333-341, this issue.

[4] S. N. Crozier, D. D. Falconer, and S. Mahmond, "Short-block equalization techniques employing channel estimation for fading time-dispersive channels," in *Proc. IEEE Veh. Technol. Conf.*, San Francisco, CA, 1989, pp. 142-146.

[5] P. Monsen, "Theoretical and measured performance of a DEF modem on a fading multipath channel," *IEEE Trans. Commun.*, vol. COM-25, pp. 1144-1153, Oct. 1977.

[6] W. C. Y. Lee, *Mobile Communications Design Fundamentals*, New York: Howard W. Sams, 1986, p. 274.

[7] M. Schwartz, W. R. Bennett, and S. Stein, *Communications Systems and Techniques*, New York: McGraw-Hill, 1966, p. 561.

[8] B. Sklar, *Digital Communications, Fundamentals and Applications.* Englewood Cliffs, NJ: Prentice-Hall, 1988, p. 546.

[9] W. C. Y. Lee, "Spectrum efficiency in cellular," *IEEE Trans. Veh. Technol.*, vol. 38, pp. 69-75, May 1989.

[10] PacTel Cellular and Qualcomm, "CDMA cellular—The next generation," a pamphlet distributed at CDMA demonstration, Qualcomm, San Diego, CA, Oct. 20-Nov. 7, 1989.

[11] K. S. Gilhousen, I. M. Jacobs, R. Padovani, A. J. Viterbi, L. A. Weaver, and C. E. Wheatley, "On the capacity of a cellular CDMA system," *IEEE Trans. Veh. Technol.*, pp. 303-312, this issue.

[12] R. L. Pickholtz, L. B. Milstein, and D. L. Schilling, "Spread spectrum for mobile communications," *IEEE Trans. Veh. Technol.*, pp. 313-322, this issue.

[13] A. J. Viterbi, "When not to spread spectrum—A sequel," *IEEE Communications Mag.*, vol. 23, pp. 12-17, Apr. 1985.

[14] L. B. Milstein, R. L. Pickholtz, and D. L. Schilling, "Optimization of the processing gain of an FSK-FH system," *IEEE Trans. Commun.*, vol. COM-28, pp. 1062-1079, July 1980.

[15] G. K. Huth, "Optimization of coded spread spectrum system performance," *IEEE Trans. Commun.*, vol. COM-25, pp. 763-770, Aug. 1977.

[16] M. K. Simon, J. K. Omura, R. A. Scholtz, and B. K. Levitt, *Spread Spectrum Communications*, vol. 2. Rockville, MD: Computer Science Press, 1985.

[17] R. L. Pickholtz, D. L. Schilling, and L. B. Milstein, "Theory of spread-spectrum communications—A tutorial," *IEEE Trans. Commun.*, vol. COM-30, pp. 855–884, May 1982.

[18] R. A. Scholtz, "The origins of spread spectrum communications," *IEEE Trans. Commun.*, vol. COM-30, pp. 882–854, May 1982.

[19] A. J. Viterbi, "Spread spectrum communications—Myths and realities," *IEEE Communications Mag.*, pp. 11–18, May 1979.

William C. Y. Lee (M'64–SM'80–F'82) received the B.Sc. degree from the Chinese Naval Academy, Taiwan, and the M.S. and Ph.D. degrees from The Ohio State University, Columbus, in 1954, 1960, and 1963, respectively.

From 1959 to 1963 he was a Research Assistant at the Electroscience Laboratory, The Ohio State University. He was with AT&T Bell Laboratories from 1964 to 1979 where he was concerned with the study of wave propagation and systems, millimeter and optical waves propagation, switching systems, and satellite communications. He developed a UHF propagation model for use in planning the Bell System's new Advanced Mobile Phone Service and was a pioneer in mobile radio communication studies. He applied the field component diversity scheme over mobile radio communication links. While working in satellite communications, he discovered a method of calculating the rain rate statistics which would affect the signal attenuation at 10 GHz and above. He successfully designed a 4 × 4 element printed circuit antenna for tryout use. He studied and set a 3-mm wave link between the Empire State Building and Pan American Building in New York City, experimentally using the newly developed IMPATT diode. He also studied the scanning spot beam concept for satellite communication using the adaptive array scheme. From April 1979 until April 1985 he worked for ITT Defense Communications Division and was involved with advanced programs for wiring military communications system. He developed several simulation programs for the multipath fading medium and applied them to ground mobile communication systems. In 1982 he was Manager of the Advanced Development Department, responsible for the pursuit of new technologies for future communication systems. He developed an artificial intelligence application in the networking area at ITT and a patent based on his work was issued in March 1991. He joined PacTel Mobile Companies in 1985, where he is engaged in the improvement of system performance and capacity. He is currently the Vice President of Research and Technology at PacTel Cellular, Irvine, CA. He has written more than 100 technical papers and three textbooks, all in the mobile radio communications area. He was the founder and the first co-chairman of the Cellular Telecommunications Industry Association Subcommittee for Advanced Radio Techniques involving digital cellular standards. He has received the "Distinguished Alumni Award" from The Ohio State University, and the "Avant Garde" award from the IEEE Vehicular Technology Society in 1990.

On the Capacity of a Cellular CDMA System

Klein S. Gilhousen, *Senior Member, IEEE,* Irwin M. Jacobs, *Fellow, IEEE,* Roberto Padovani, *Senior Member, IEEE,* Andrew J. Viterbi, *Fellow, IEEE,* Lindsay A. Weaver, Jr., and Charles E. Wheatley III, *Senior Member, IEEE*

Abstract—**The use of spread spectrum or code division techniques for multiple access (CDMA) has long been debated. Certain advantages, such as multipath mitigation and interference suppression are generally accepted, but past comparisons of capacity with other multiple access techniques were not as favorable. This paper shows that, particularly for terrestrial cellular telephony, the interference suppression feature of CDMA can result in a many-fold increase in capacity over analog and even over competing digital techniques.**

I. Introduction

Spread-spectrum techniques, long established for antijam and multipath rejection applications as well as for accurate ranging and tracking, have also been proposed for code division multiple access (CDMA) to support simultaneous digital communication among a large community of relatively uncoordinated users. Yet, as recently as 1985 a straightforward comparison [1] of the capacity of CDMA to that of conventional time division multiple access (TDMA) and frequency division multiple access (FDMA) for satellite applications suggested a reasonable edge in capacity for the latter two more conventional techniques. This edge was shown to be illusory shortly thereafter [2] when it was recognized that since CDMA capacity is only interference limited (unlike FDMA and TDMA capacities which are primarily bandwidth limited), any reduction in interference converts directly and linearly into an increase in capacity. Thus, since voice signals are intermittent with a duty factor of approximately 3/8 [3], capacity can be increased by an amount inversely proportional to this factor by suppressing (or squelching) transmission during the quiet periods of each speaker. Similarly, any spatial isolation through use of multibeamed or multisectored antennas, which reduces interference, also provides a proportional increase in capacity. These two factors, voice activity and spatial isolation, were shown to be sufficient to render CDMA capacity at least double that of FDMA and TDMA under similar assumptions for a mobile satellite application [2].

While previous comparisons primarily applied to satellite systems, CDMA exhibits its greatest advantage over TDMA and FDMA in terrestrial digital cellular systems, for here isolation among cells is provided by path loss, which in terrestrial UHF propagation typically increases with the fourth power of the distance. Consequently, while conventional techniques must provide for different frequency allocation for contiguous cells (only reusing the same channel in one of every 7 cells in present systems), CDMA can reuse the same (entire) spectrum for all cells, thereby increasing capacity by a large percentage of the normal frequency reuse factor. The net improvement in capacity, due to all the above features, of CDMA over digital TDMA or FDMA is on the order of 4 to 6 and over current analog FM/FDMA it is nearly a factor of 20.

The next section deals with a single cell system, such as a hubbed satellite network, and develops the basic expression for capacity. The subsequent two sections derive the corresponding expressions for a multiple cell system and determine the distribution on the number of users supportable per cell. The last section presents conclusions and system comparisons.[1]

II. Single Cell CDMA Capacity

The network to be considered throughout consists of numerous mobile (or personal) subscribers communicating with one or multiple cell sites (or base stations) which are interconnected with a mobile telephony switching office (MTSO), which also serves as a gateway to the public switched telephone network. We begin by considering a single cell system, which can also serve as a model for a satellite system whose "cell site" is a single hub.

Each user of a CDMA system occupies the entire allocated spectrum, employing a direct sequence spread spectrum waveform. Without elaborating on the modulation and spreading waveform, we assume generic CDMA modems at both subscriber units and the cell site with digital baseband processing units as shown in Fig. 1 for the transmitter sides of each. These consist of (digital) forward-error correction (FEC), modulation and (direct sequence) spreading functions, preceding the (analog) amplification and transmission functions. Each of the digital functions can be performed using binary sequences in the subscriber modulator.

At the cell-site transmitter, the spread signals directed to the individual subscribers are added linearly and phase randomness is assured by modulating each signal with independent pseudorandom sequences on each of the two quadrature phases. The weighting factors $\emptyset_1, \emptyset_2, \cdots, \emptyset_N$ can be taken to be equal for the time being, but for the multiple cell case they will provide power control based on considerations to be

Manuscript received May 1, 1990; revised September 14, 1990. This paper was presented at the 1990 IEEE GLOBECOM Conference, San Diego, CA.

The authors are with QUALCOMM, Inc., 10555 Sorrento Valley Road, San Diego, CA 92121-1617.

IEEE Log Number 9144470.

[1] It should be noted that our purpose is not to evaluate or optimize modem performance for the channels under consideration. Rather, assuming an efficient modulation and FEC code for the given channels, we shall establish conditions under which the modems will achieve an acceptable level of performance, particularly in terms of the maximum number of users supportable per cell.

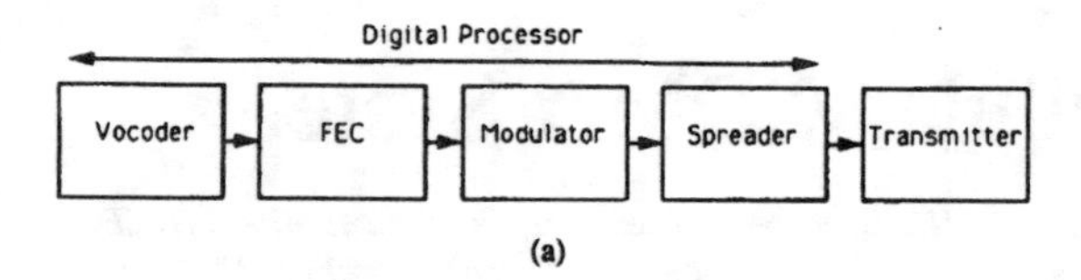

(a)

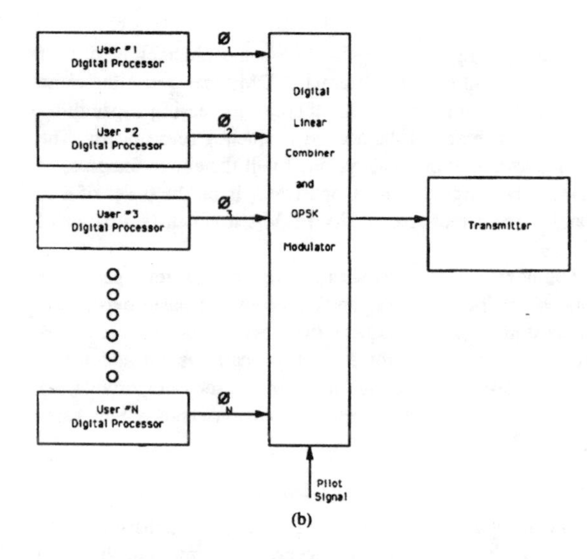

(b)

Fig. 1. Cellular system simplified block diagram. (a) Reverse link subscriber processor/transmitter. (b) Forward link cell-site processor/transmitter.

described later. The receiver processors in both subscriber and cell-site receivers provide the inverse baseband functions, which are of course considerably more complex than the transmitter baseband functions.

One other key feature of the cell-site transmitter is the inclusion of a pilot signal in the forward (cell-site-to-subscriber) direction. This provides for acquisition by the mobile terminals, including initial power control by the mobile, which adjusts its output power inversely to the total signal power it receives. Power control is a basic requirement in CDMA and will be expanded on in a later section.

We note also that the pilot signal is used by the subscriber demodulator to provide a coherent reference which is effective even in a fading environment since the desired signal and the pilot fade together. In the subscriber-to-cell-site (reverse) direction, no pilot is used for power efficiency considerations, since unlike the forward case, an independent pilot would be needed for each signal. A modulation consistent with, and relatively efficient for, noncoherent reception is, therefore, used for the reverse direction.

Without elaborating further on the system implementation details, we note that for a single cell site with power control, all reverse link signals (subscribers-to-cell site) are received at the same power level. For N users, each cell-site demodulator processes a composite received waveform containing the desired signal having power S and $(N-1)$ interfering signals each also of power S. Thus the signal-to-noise (interference) power is

$$\mathrm{SNR} = \frac{S}{(N-1)S} = \frac{1}{N-1}.$$

Of greater importance for reliable system operation is the bit energy-to-noise density ratio, whose numerator is obtained by dividing the desired signal power by the information bit rate, R, and dividing the noise (or interference) by the total bandwidth, W. This results in

$$E_b/N_0 = \frac{S/R}{(N-1)S/W} = \frac{W/R}{N-1}. \quad (1)$$

This paper does not explicitly address modulation techniques and their performance. Rather, an E_b/N_0 level is assumed which ensures operation at the level of bit error performance required for digital voice transmission. Among the factors to be considered in establishing the modulation and the resulting required E_b/N_0 level are phase coherence, amplitude fading characteristics and power control techniques and their effectiveness, particularly for the reverse link. One of the lesser considerations, albeit one of the most cited, is the probability distribution of the interfering signals. While Gaussian noise is often assumed, this is not strictly necessary to establish the E_b/N_0 requirements. Nonetheless, the assumption is quite reasonable when powerful forward error-correcting codes are employed, particularly at low code rates, because in such cases decisions are based on long code sequence lengths over which the interfering signal sequence contributions are effectively the sums of a large number of binomial variables, which closely approximate Gaussian random variables.

Equation (1) ignores background noise, η, due to spurious interference as well as thermal noise contained in the total spread bandwidth, W. Including this additive term in the denominator of (1) results in a required

$$E_b/N_0 = \frac{W/R}{(N-1) + (\eta/S)}. \quad (2)$$

This implies that the capacity in terms of number of users supported is

$$N = 1 + \frac{W/R}{E_b/N_0} - \frac{\eta}{S} \quad (3)$$

where W/R is generally referred to as the "processing gain" and E_b/N_0 is the value required for adequate performance of the modem and decoder, which for digital voice transmission implies a BER of 10^{-3} or better. In words, the number of users is reduced by the inverse of the per user signal-to-noise ratio (SNR) in the total system spread bandwidth, W. In a terrestrial system, the per user SNR is limited only by the transmitter's power level. As will be justified below, we shall assume SNR just below unity corresponding to a reduction in capacity equivalent to removing one user. The background noise, therefore establishes the required received signal power at the cell site, which in turn fixes the subscriber's power or the cell radius for a given maximum transmitter power.

For the reverse (subscriber-to-cell-site) direction, noncoherent reception and independent fading of all users is assumed. With dual antenna diversity, the required $E_b/N_0 = 7$ dB for a relatively powerful (constraint length 9, rate 1/3) convolutional code. Since the forward link employs coherent demodulation by the pilot carrier which is being tracked, and since its multiple transmitted signals are synchronously combined, its performance in a single cell system will be much superior to that of the reverse link. For a multiple cell system however, other cell interference will tend to equalize performance in the two directions, as will be described below.

All this leaves us at the point of our previous conclusions [1], only worse because of the Rayleigh fading encountered in terrestrial mobile applications. In the next section we begin to remedy the situation.

III. Augmented Performance Through Sectorization and Voice-Activity Monitoring

Short of reducing E_b/N_0 through improved coding or possibly modulation, which rapidly reaches the point of diminishing returns for increasing complexity (and ultimately the unsurmountable Shannon limit), we can only increase capacity by reducing other user interference and hence the denominator of (1) or (2). This can be achieved in two ways.

The first is the common technique of sectorization, which refers to using directional antennas at the cell site both for receiving and transmitting. For example, with three antennas per cell site, each having 120° effective beamwidths, the interference sources seen by any antenna are approximately one-third of those seen by an omnidirectional antenna. This reduces the $(N - 1)$ term in the denominator of (2) by a factor of 3 and consequently, in (3) N is increased by nearly this factor. Henceforth, we shall take N_s to be the number of users per sector and the interference to be that received by one sector's antenna. Using three sectors, the number of users per cell $N = 3N_s$.

Secondly, voice activity can be monitored, a function which virtually already exists in most digital vocoders, and transmission can be suppressed for that user when no voice is present. Extensive studies show that either speaker is active only 35 % to 40 % of the time [3]. We shall assume for this the "voice activity factor," $\alpha = 3/8$ throughout. On the average, this reduces the interference term (in the denominator of (2)) from $(N - 1)$ to $(N - 1)\alpha$. Below, we will find through a more careful analysis that the net improvement in capacity due to voice activity is reduced from 8/3 to about 2 due to the fact that with a limited number of calls per sector, there is a nonnegligible probability that an above average number of users are talking at once. We ignore this in this preliminary discussion but include it in the results described below. Thus with sectorization and voice activity monitoring, the average $\overline{E_b}/N_0$, is increased relative to (2) to become[2]

$$\frac{\overline{E_b}}{N_0} = \frac{W/R}{(N_s - 1)\alpha + (\eta/S)}. \tag{4}$$

[2] These arguments leading to (4) were first advanced by Cooper and Nettleton [4].

This suggests that the average number of users per cell is increased by almost a factor of 8. In fact, because of variability in E_b/N_0 this increase will need to be backed off to a factor of 5 or 6. We shall return to this variability issue and other more precise results after we consider multiple cell interference in the next section. For now, note from (3) and (4) that this is enough to bring the number of users/cell up to the processing gain, $N \approx W/R$ users/cell which makes CDMA at least competitive with other multiple access techniques (FDMA or TDMA) on a single cell basis. As we will presently show, in multiple-cell systems additional advantages accrue through frequency reuse of the same spectrum in all cells. To assess this advantage, we must first consider the power control techniques and their effect on multicell interference.

IV. Reverse Link Power Control in Multiple-Cell Systems

As should be clear by now, power control is the single most important system requirement for CDMA, since only by control of the power of each user accessing a cell can resources be shared equitably among users and capacity maximized. In a single cell system, the principle is straightforward, though the implementation may not be. Prior to any transmission, each of the subscribers monitors the total received signal power from the cell site. According to the power level it detects, it transmits at an initial level which is as much below (above) a nominal level in decibels as the received pilot power level is above (below) its nominal level. Experience has shown that this may require a dynamic range of control on the order of 80 dB. Further refinements in power level in each subscriber can be commanded by the cell site depending on the power level it receives from the subscriber.

The relatively fast variations associated with Rayleigh fading may at times be too rapid to be tracked by the closed-loop power control but variations in relative path losses and shadowing effects, which are modelled as an attenuation with log-normal distribution, will generally be slow enough to be controlled. Also, while Rayleigh fading may not be the same for forward and reverse links, log-normal shadowing normally will exhibit reciprocity. For the forward link, no power control is required in a single cell system, since for each subscriber any interference caused by other subscriber signals remains at the same level relative to the desired signal; inasmuch as all signals are transmitted together and hence vary together, there are no resulting degradations due to fading assuming the background noise may be neglected.

In multiple-cell CDMA systems, the situation becomes more complicated in both directions. First, for the reverse link, subscribers are power controlled by the base station of their own cell. Even the question of cell membership is not simple. For it is not minimum distance which determines which base station (cell site) the subscriber joins, but rather the maximum pilot power among the cell sites the subscriber receives. In any case the interference level from subscribers in the other cells varies not only according to the attenuation in the path to the subscriber's cell site, but also inversely to the attenuation from the interfering user to his own cell site, which through power control by that cell site may increase, or decrease, the interference to the desired cell site. These issues will be treated in the next section.

As for the forward link for a multiple cell system, interference from neighboring cell sites fade independently of the given cell site and thereby degrade performance for any level of interference. This becomes a particularly serious problem in the region where two or even three cell transmissions are received at nearly equal strengths. Techniques for mitigating this condition are treated in Section VI.

V. Reverse Link Capacity for Multiple Cell CDMA

Recalling that power control to a given mobile is exercised by the cell whose pilot signal power is maximum to that mobile, it follows that if the path loss were only a function of distance from the cell site, then the mobile would be power controlled by the nearest cell site, which is situated at the center of the hexagon in which it lies, as shown in Fig. 2(a) for an idealized placement of cell sites. In fact, the loss is proportional to other effects as well, the most significant being shadowing. The generally accepted model is an attenuation which is the product of the fourth power of the distance and a log-normal random variable whose standard deviation is 8 dB. That is, the path loss between the subscriber and the cell site is proportional to $10^{(\xi/10)}r^{-4}$ where r is distance from subscriber to cell site and ξ is a Gaussian random variable with standard deviation $\sigma = 8$ and zero mean. Fast fading (due largely to multipath) is assumed not to affect the (average) power level.

We note that other propagation exponents can be found in different environments. In fact, within a single cell the propagation may vary from inverse square law very close to the cell antenna to as great as the inverse 5.5 power far from the cell in a very dense urban environment such as Manhattan. The present analysis is primarily concerned with interference from neighboring and distant cells so the assumption of inverse fourth law propagation is a reasonable one.

The interference from transmitter within the given subscriber's cell is treated as before; that is, since each user is power controlled by the same cell site, it arrives with the same power S, when active. Thus given N subscribers per cell, the total interference is never greater than $(N-1)S$, but on the average it is reduced by the voice activity factor, α. Subscribers in other cells, however, are power controlled by other cell sites (Fig. 2(a)). Consequently, if the interfering subscriber is in another cell and at a distance r_m from its cell site and r_0 from the cell site of the desired user, the other user, when active, produces an interference in the desired user's cell site equal to

$$\frac{I(r_0, r_m)}{S} = \left(\frac{10^{(\xi_0/10)}}{r_0^4}\right)\left(\frac{r_m^4}{10^{(\xi_m/10)}}\right) = \left(\frac{r_m}{r_0}\right)^4 10^{(\xi_0-\xi_m)/10} \le 1 \quad (5)$$

where the first term is due to the attenuation caused by distance and blockage to the given cell site, while the second term is the effect of power control to compensate for the corresponding attenuation to the cell site of the out-of-cell interferer.[3] Of course ξ_0 and ξ_m are independent so that the difference has zero mean and variance $2\sigma^2$. For all values of the above parameters, the expression is less than unity, for otherwise the subscriber would switch to the cell site which makes it less than unity (i.e., for which the attenuation is minimized).

Then, assuming a uniform density of subscribers, and normalizing the hexagonal cell radius to unity, and since the average number of subscribers/cell is $N = 3N_s$, the density of users is

$$\rho = \frac{2N}{3\sqrt{3}} = \frac{2N_s}{\sqrt{3}} \text{ per unit area.} \quad (6)$$

Consequently, the total other-cell user interference-to-signal ratio is

$$I/S = \iint \psi\left(\frac{r_m}{r_0}\right)^4 \{10^{(\xi_0-\xi_m)/10}\} \cdot \varnothing(\xi_0 - \xi_m, r_0/r_m)\rho\, dA \quad (7)$$

where m is the cell-site index for which

$$r_m^4 10^{-\xi_m} = \min_{k \neq 0} r_k^4 10^{-\xi_k} \quad (8)$$

and

$$\varnothing(\xi_0 - \xi_m, r_0/r_m) = \begin{cases} 1, & \text{if } (r_m/r_0)^4 10^{(\xi_0-\xi_m)/10} \le 1 \\ & \text{or } \xi_0 - \xi_m \le 40\log_{10}(r_0/r_m) \\ 0, & \text{otherwise} \end{cases} \quad (9)$$

and ψ is the voice activity variable, which equals 1 with probability α and 0 with probability $(1-\alpha)$. To determine the moment statistics of the random variable I, the calculation is much simplified and the results only slightly increased if for m we use the smallest distance rather than the smallest attenuation. Thus (7), with (9), holds as an upper bound if in place of (8) we use that value of m for which

$$r_m = \min_{k \neq 0} r_k. \quad (8')$$

In Appendix I, it is shown that the mean or first moment, of the random variable I/S is upper bounded (using (8′) rather than (8) for m) by the expression

$$E(I/S) = \alpha \iint \frac{r_m^4}{r_0} f\left(\frac{r_m}{r_0}\right)\rho\, dA$$

where

$$f\left(\frac{r_m}{r_0}\right) = \exp\left[(\sigma \ln 10/10)^2\right]\left\{1 - Q\left[\frac{40}{\sqrt{2\sigma^2}} \cdot \log_{10}\left(\frac{r_0}{r_m}\right) - \sqrt{2\sigma^2}\frac{\ln 10}{10}\right]\right\} \quad (10)$$

[3] Cooper and Nettleton [4] employed similar geometric arguments to compute interference, but did not consider log-normal statistical variations due to blockage.

and

$$Q(x) = \int_x^{\infty} e^{-y^2/2}\, dy / \sqrt{2\Pi}.$$

This integral is over the two-dimensional area comprising the totality of all sites in the sector (Fig. 2(a)). The integration, which needs to be evaluated numerically, involves finding for each point in the space the value of r_0, the distance to the desired cell site and r_m, which according to (8′), is the distance to the closest cell site, prior to evaluating at the given point the function (10). The result for $\sigma = 8$ dB is

$$E(I/S) \leq 0.247N_s.$$

Calculation of the second moment, $\text{var}(I/S)$ of the random variable requires an additional assumption on the second-order statistics of ξ_0 and ξ_m. While it is clear that the relative attenuations are independent of each other, and that both are identically distributed (i.e., have constant first-order distributions) over the areas, their second-order statistics (spatial correlation functions) are also needed to compute $\text{var}(I)$. Based on experimental evidence that blockage statistics vary quite rapidly with spatial displacement in any direction, we shall take the spatial autocorrelation functions of ξ_0 and ξ_m to be extremely narrow in all directions, the two-dimensional spatial equivalent of white noise. With this assumption, we obtain in Appendix I that

$$\text{var}(I/S) \leq \iint \left(\frac{r_m}{r_0}\right)^8 \left[\alpha\, g\left(\frac{r_m}{r_0}\right) - \alpha^2 f\left(\frac{r_m}{r_0}\right)\right] \rho\, dA$$

where

$$g\left(\frac{r_m}{r_0}\right) = \exp\left[\left(\frac{\sigma \ln 10}{5}\right)^2\right]\left\{1 - Q\left[\frac{40}{\sqrt{2\sigma^2}} \cdot \log_{10}\left(\frac{r_0}{r_m}\right) - \sqrt{2\sigma^2}\left(\frac{\ln 10}{5}\right)\right]\right\}. \quad (11)$$

This integral is also evaluated numerically over the area of Fig. 2(a), with r_m defined at any given point by condition (8′). The result for $\sigma = 8$ dB is $\text{var}(I/S) \leq 0.078N_s$. The above argument also suggests that I, as defined by (7), being a linear functional on a two-dimensional white random process, is well modelled as a Gaussian random variable.[4]

We may now proceed to obtain a distribution on the total interference, both from other users in the given cell, and from other-cell users on the desired user's reverse link transmission. With sectorization, variable voice activity and the other-cell interference statistics just determined, the received E_b/N_0 on the reverse link of any desired user becomes the random variable

$$E_b/N_0 = \frac{W/R}{\sum_{i=1}^{N_s-1} \chi_i + (I/S) + (\eta/S)} \quad (12)$$

where N_s is the users/sector and I is the total interference from users outside the desired user's cell. This follows easily

[4]Of course, it can never be negative, but since the ratio of mean-to-standard deviation is approximately $\sqrt{N_s}$, with typical values of $N_s > 30$ the approximating Gaussian distribution is nearly zero for negative values.

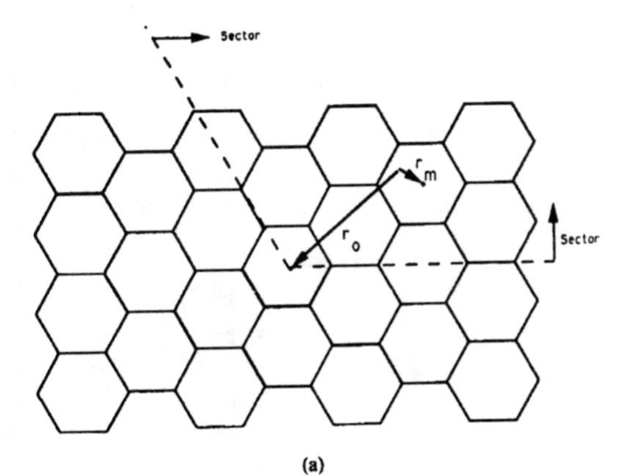

(a)

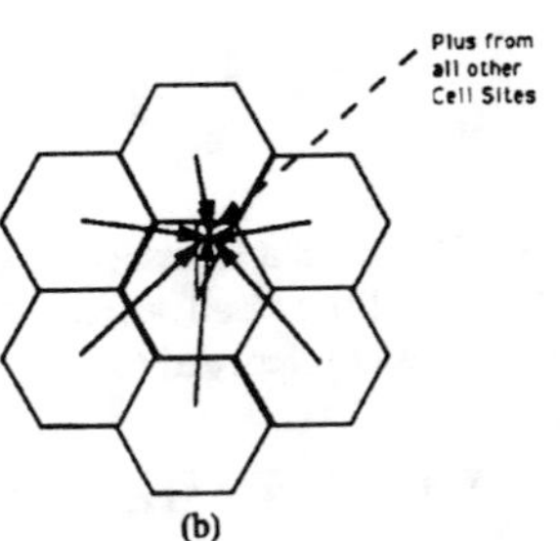

(b)

Fig. 2. Capacity calculation geometrics. (a) Reverse link geometry. (b) Forward link allocation geometry.

from (2) with the recognition that the $N_s - 1$ same sector normalized power users, instead of being unity all the time, now are random variables χ_i with distribution

$$\chi_i = \begin{cases} 1, & \text{with probability } \alpha \\ 0, & \text{with probability } 1 - \alpha. \end{cases} \quad (13)$$

The additional term I represents the other (multiple) cell user interference for which we have evaluated mean and variance,

$$E(I/S) \leq 0.247N_s \quad \text{and} \quad \text{var}(I/S) \leq 0.078N_s \quad (14)$$

and have justified taking it to be a Gaussian random variable. The remaining terms in (12), W/R and S/η, are constants.

As previously stated, with an efficient modem and a powerful convolutional code and two-antenna diversity, adequate performance (BER $< 10^{-3}$) is achievable on the reverse link with $E_b/N_0 \geq 5$ (7 dB). Consequently, the required performance is achieved with probability $P = \Pr(\text{BER} < 10^{-3}) = \Pr(E_b/N_0 \geq 5)$. We may *lower bound* the probability of achieving this level of performance for any desired fraction of users at any given time (e.g., $P = 0.99$) by obtaining an upper bound on its complement, which according to (12), depends on the distribution of χ_i and I, as follows

$$1 - P = \Pr(\text{BER} > 10^{-3}) = \Pr\left(\sum_{i=1}^{N_s} \chi_i + I/S > \delta\right) \quad (15)$$

where

$$\delta = \frac{W/R}{E_b/N_0} - \frac{\eta}{S}, \quad E_b/N_0 = 5.$$

Since the random variable χ_i has the binomial distribution

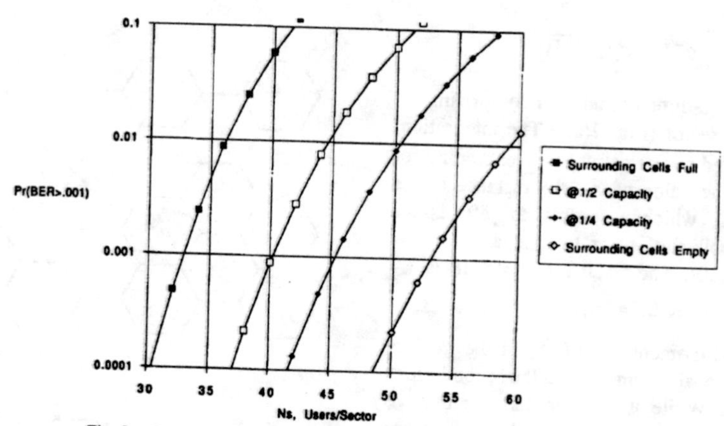

Fig. 3. Reverse link capacity/sector. (W = 1.25 MHz, R = 8 kb/s, voice activity = 3/8).

given by (13) and I/S is a Gaussian variable with mean and variance given by (14) and all variables are mutually independent, (15) is easily calculated to be

$$1 - P = \sum_{k=0}^{N_s-1} \Pr\left(I/S > \delta - k \middle| \sum x_i = k\right) \Pr\left(\sum x_i = k\right)$$
$$= \sum_{k=0}^{N_s-1} \binom{N_s - 1}{k} \alpha^k (1-\alpha)^{N_s-1-k} \cdot Q\left(\frac{\delta - k - 0.247N_s}{\sqrt{0.078N_s}}\right). \quad (16)$$

This expression is plotted for $\delta = 30$ (a value chosen as discussed in the conclusion) and $\alpha = 3/8$, as the leftmost curve of Fig. 3. The rightmost curve applies to a single cell without other cell interference ($I = 0$), while the other intermediate curves assume that all cells other than the desired user's cells are on the average loaded less heavily (with averages of 1/2 and 1/4 of the desired user's cell).

We shall discuss these results further in the concluding section, and now concern ourselves with forward link performance.

VI. Multiple-Cell Forward Link Capacity with Power Allocation

As noted earlier, although with a single cell no power control is required, with multiple cells it becomes important, because near the boundaries of cells considerable interference can be received from other cell-site transmitters fading independently.

For the forward link, power control takes the form of power allocation at the cell-site transmitter according to the needs of individual subscribers in the given cell. This requires measurement by the mobile of its relative SNR, defined as the ratio of the power from its own cell site transmitter to the total power received. Practically, this is done by acquiring (correlating to) the highest power pilot and measuring its energy, and also measuring the total energy received by the mobile's omnidirectional antenna from all cell site transmitters. Both measurements can be transmitted to the selected (largest power) cell site when the mobile starts to transmit. Suppose then that based on these two measurements, the cell site has reasonably accurate estimates of S_{T_1} and $\sum_{i=1}^{K} S_{Ti}$, where

$$S_{T_1} > S_{T_2} > \cdots > S_{T_K} > 0 \quad (17)$$

are the powers received by the given mobile from the cell site sector facing it, assuming all but K (total) received powers are negligible. (We shall assume hereafter that all sites beyond the second ring around a cell contribute negligible received power, so that $K \leq 19$.) Note that the ranking indicated in (17) is not required of the mobile—just the determination of which cell site is largest and hence which is to be designated T_1.

The ith subscriber served by a particular cell site will receive a fraction of S_{T_1} the total power transmitted by its cell site, which by choice and definition (17) is the greatest of all the cell site powers it receives, and all the remainder of S_{T_1} as well as the other cell site powers are received as noise. Thus its received E_b/N_0 can be lower bounded by

$$\left(\frac{E_b}{N_0}\right)_i \geq \frac{\beta \varnothing_i S_{T_1}/R}{\left[\left(\sum_{j=1}^{K} S_{T_j}\right)_i + \eta\right]/W} \quad (18)$$

where S_{T_j} is defined in (17), β is the fraction of the total cell site power devoted to subscribers ($1 - \beta$ is devoted to the pilot) and $\varnothing_i$ is the fraction of this devoted to subscriber i. Because of the importance of the pilot in acquisition and tracking, we shall take $\beta = 0.8$. It is clear that the greater the sum of other cell-site powers relative to S_{T_1}, the larger the fraction $\varnothing_i$ which must be allocated to the ith subscriber to achieve its required E_b/N_0. In fact, from (18) we obtain

$$\varnothing_i \leq \frac{(E_b/N_0)_i}{\beta W/R} \left[1 + \left(\frac{\sum_{j=2}^{K} S_{T_j}}{S_{T_1}}\right)_i + \frac{\eta}{(S_{T_1})_i}\right] \quad (19)$$

where

$$\sum_{i=1}^{N_s} \varnothing_i \le 1 \tag{20}$$

since βS_{T_1} is the maximum total power allocated to the sector containing the given subscriber and N_s is the total number of subscribers in the sector. If we define the relative received cell-site power measurements as

$$f_i \triangleq \left(1 + \sum_{j=2}^{K} S_{T_j}/S_{T_1}\right)_i, \qquad i = 1,\cdots, N_s \tag{21}$$

then from (19) and (20) it follows that their sum over all subscribers of the given cell site sector is constrained by

$$\sum_{i=1}^{N_s} f_i \le \frac{\beta W/R}{E_b/N_0} - \sum_{i=1}^{N_s} \frac{\eta}{S_{T_{1_i}}} \triangleq \delta'. \tag{22}$$

Generally, the background noise is well below the total largest received cell site signal power, so the second sum is almost negligible. Note the similarity to δ in (15) for the reverse link. We shall take $\beta = 0.8$ as noted above to provide 20 % of the transmitted power in the sector to the pilot signal, and the required $E_b/N_0 = 5$ dB to ensure BER $\le 10^{-3}$. This reduction of 2-dB relative to the reverse link is justified by the coherent reception using the pilot as reference, as compared to the noncoherent modem in the reverse link. Note that this is partly offset by the 1-dB loss of power due to the pilot.

Since the desired performance (BER $\le 10^{-3}$) can be achieved with N_s subscribers per sector provided (22) is satisfied with $E_b/N_0 = 5$ dB, capacity is again a random variable whose distribution is obtained from the distribution of variable f_i. That is, the BER can not be achieved for all N_s users/sector if the N_s subscribers combined exceed the total allocation constraint of (22). Then following (15),

$$1 - P = \Pr(\text{BER} > 10^{-3}) = \Pr\left(\sum_{i=1}^{N_s} f_i > \delta'\right). \tag{23}$$

But unlike the reverse link, the distribution of the f_i, which depends on the sum of ratios of ranked log-normal random variables, does not lend itself to analysis. Thus we resorted to Monte Carlo simulation, as follows.

For each of a set of points equally spaced on the triangle shown in Fig. 2(b), the attenuation relative to its own cell center and the 18 other cell centers comprising the first three neighboring rings was simulated. This consisted of the product of the fourth power of the distance and the log-normally distributed attenuation

$$10^{(\xi_k/10)} r_k^{-4}, \qquad k = 0, 1, 2 \cdots, 18.$$

Note that by symmetry, the relative position of users and cell sites is the same throughout as for the triangle of Fig. 2(b). For each sample, the 19 values were ranked to determine the maximum (S_{T_1}), after which the ratio of the sum of all other 18 values to the maximum was computed to obtain $f_i - 1$. This was repeated 10 000 times per point for each of 65 equally spaced points on the triangle of Fig. 2(b). From this, the histogram of $f_i - 1$ was constructed, as shown in Fig. 4.

From this histogram the Chernoff upper bound on (23) is obtained as

$$\begin{aligned} 1 - P &=< \min_{s>0} E \exp\left[s \sum_{i=1}^{N_s} f_i - s\delta'\right] \\ &= \min_{s>0}\left[(1 - \propto) + \propto \sum_k P_k \exp(sf_k)\right]^{N_s} e^{-s\delta'} \end{aligned} \tag{24}$$

where P_k is the probability (histogram value) that f_i falls in the kth interval. The result of the minimization over s based on the histogram of Fig. 4, is shown in Fig. 5.

VII. Conclusions and Comparisons

Figs. 3 and 5 summarize performance of reverse and forward links. Both are theoretically pessimistic (upper bounds on probability). Practically, both models assume only moderately accurate power control.

The parameters for both links were chosen for the following reasons. The allocated total spread bandwidth $W = 1.25$ MHz represents ten percent of the total spectral allocation, 12.5 MHz, for cellular telephone service of each service provider. which as will be discussed below, is a reasonable fraction of the band to devote initially to CDMA and also for a gradual incremental transition from analog FM/FDMA to digital CDMA. The bit rate $R = 8$ kb/s is that of an acceptable nearly toll quality vocoder. The voice activity factor, 3/8, and the standard sectorization factor of 3 are used. For the reverse channel, the received SNR per user $S/\eta = -1$ dB reflects a reasonable subscriber transmitter power level. In the forward link, 20 % of each site's power is devoted to the pilot signal for a reduction of 1 dB ($\beta = 0.8$) in the effective processing gain. This ensures each pilot signal (per sector) is at least 5 dB above the maximum subscriber signal power. The role of the pilot, as noted above, is critical to acquisition, power control in both directions and phase tracking as well as for power allocation in the forward link. Hence, the investment of 20 % of total cell site power is well justified. These choices of parameters imply the choices $\delta = 30$ and $\delta' \approx 38$ in (16) and (24) for reverse and forward links, respectively.

With these parameters, according to Fig. 3, the reverse link can support over 36 users/sector or 108 users/cell, with 10^{-3} bit error rates better than 99 % of the time. This number becomes 44 users/sector or 132 users/cell if the neighboring cells are kept to half of this loading. The forward link according to Fig. 5, can do the same or better for 38 users/sector or 114 users/cell.

Clearly, if the entire cellular allocation is devoted to CDMA, these numbers are increased tenfold. Similarly, if a lower bit rate vocoder algorithm is developed, or if narrower sectors are employed, the number of users may be increased further.

Remaining with the parameters assumed, interesting comparisons can be drawn to existing analog FM/FDMA cellular systems as well as other proposed digital systems. First, the former employs 30-kHz channel allocation, and assuming 3

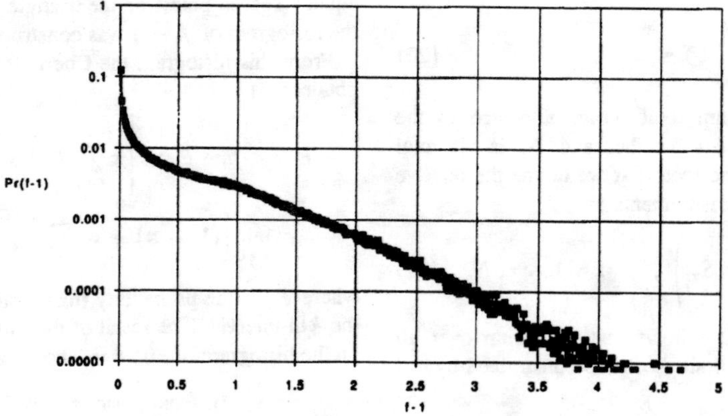

Fig. 4. Histogram of forward power allocation.

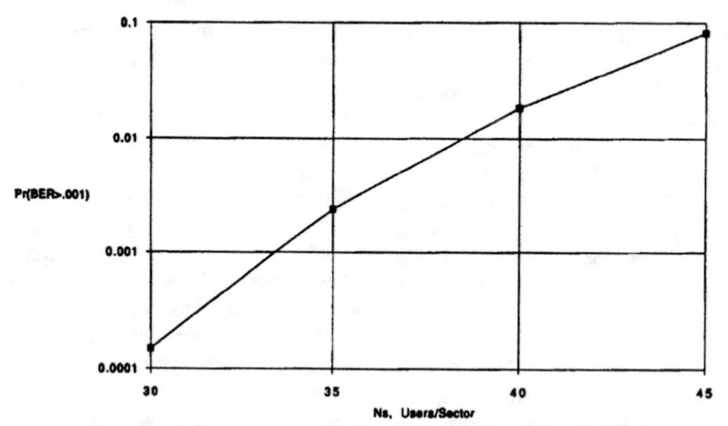

Fig. 5. Forward link capacity/sector. ($W = 1.25$ MHz, $R = 8$ kb/s, voice activity = 3/8, pilot power = 20 %).

sectors/cell, requires each of the six contiguous cells in the first ring about a given cell to use a different frequency band. This results in a "frequency reuse factor" of 1/7. Hence, given the above parameters, the number of channels in a 1.25-MHz band is slightly less than 42, and with a frequency reuse factor of 1/7, this results in slightly less than 6 users/cell for a 1.25-MHz band. Thus, CDMA offers at least an eighteenfold increase in capacity. Note further that use of CDMA over just ten percent of the band supports over 108 users/cell whereas analog FM/FDMA supports only 60 users/cell using the entire 12.5 MHz band. Thus by converting *only 10 % of the band* from analog FDMA to digital CDMA, overall capacity is increased *almost threefold*.

Comparisons of CDMA with other digital systems are more speculative. However, straightforward approaches such as narrower frequency channelization with FDMA or multiple time slotting with TDMA can be readily compared to the analog system. The proposed TDMA standard for the U.S. is based on the current 30-kHz channelization but with sharing of channels by three users each of whom is provided one of three TDMA slots. Obviously, this triples the analog capacity but falls over a factor of 6 short of CDMA capacity.

In summary, properly augmented and power-controlled multiple-cell CDMA promises a quantum increase in current cellular capacity. No other proposed scheme appears to even approach this performance. Other advantages of CDMA not treated here include inherent privacy, flexibility in supporting multiple services and multiple voice and data rates, lower average transmit power requirements and soft limit on capacity, since if the bit error rate requirement is relaxed more users can be supported. With all these inherent advantages, CDMA appears to be the logical choice henceforth for all cellular telephone applications.

Appendix I
Reverse Link Outer-Cell Interference

Outer-cell normalized interference, I/S, is a random variable defined by (7), (8), and (9), and upper bounded by replacing (8) by (8'). Then the upper bound on its first moment, taking into account also the voice activity factor of the outer-cell subscribers, α, becomes

$$E(I/S) \le \iint_{\text{sector}} \left(\frac{r_m}{r_0}\right)^4 E(\psi \cdot 10^{\chi/10} \varnothing(\chi, r_0/r_m)) \rho \, dA$$

where r_m is defined by (8′) for every point in the sector, $\psi = 1$ with probability α and 0 with probability $(1 - \alpha)$, and $\chi = \xi_0 - \xi_m$ is a Gaussian random variable of zero mean and variance $2\sigma^2$ with $\varnothing(\chi, r_0/r_m)$ defined by (9),

$$\varnothing(\chi, r_0/r_m) = \begin{cases} 1, & \text{if } \chi \leq 40\log(r_0/r_m) \\ 0, & \text{otherwise.} \end{cases}$$

The expectation is readily evaluated as

$$\alpha f\left(\frac{r_m}{r_0}\right) \triangleq E(\psi)E(e^{\chi \ln 10/10}\varnothing(\chi, r_0/r_m))$$

$$= \alpha \int_{-\infty}^{40\log r_0/r_m} e^{x \ln 10/10} \frac{e^{-x^2/4\sigma^2}}{\sqrt{4\Pi\sigma^2}} dx$$

$$= \alpha\, e^{(\sigma \ln 10/10)^2} \int_{-\infty}^{40\log(r_0/r_m)} \cdot \frac{\exp\left[-\frac{1}{2}(x/\sqrt{2\sigma^2} - \sqrt{2\sigma^2}\ln 10/10)^2\right]}{\sqrt{2\Pi(2\sigma^2)}} dx$$

$$= \alpha\, e^{(\sigma \ln 10/10)^2}\left\{1 - Q\left[\frac{40\log(r_0/r_m)}{\sqrt{2\sigma^2}} - \sqrt{2\sigma^2}\frac{\ln 10}{10}\right]\right\}$$

which yields (10).

To evaluate $\mathrm{var}(I/S)$, assuming the "spatial whiteness" of the blockage variable, we have

$$\mathrm{var}(I/S) \leq \iint_{\text{sector}} \left(\frac{r_m}{r_0}\right)^8 \cdot \mathrm{var}(\psi \cdot 10^{\chi/10}\varnothing(\chi, r_0/r_m))\rho\, dA.$$

Rewriting the variance in the integral as

$$E(\psi^2)E[10^{2\chi/10}\varnothing^2(\chi, r_0/r_m)] - \{E(\psi)E[10^{\chi/10}\varnothing(\chi, r_0/r_m)]\}^2 = \alpha g\left(\frac{r_m}{r_0}\right) - \alpha^2 f^2\left(\frac{r_m}{r_0}\right)$$

where $f(r_m/r_0)$ was derived above and

$$\alpha g\left(\frac{r_m}{r_0}\right) = E(\psi^2)E[e^{\chi \ln 10/5}\varnothing^2(\chi, r_0/r_m)]$$

$$= \alpha\, e^{(\sigma \ln 10/5)^2}\left\{1 - Q\left[\frac{40\log(r_0/r_m)}{\sqrt{2\sigma^2}} - \frac{\sqrt{2\sigma^2}\ln 10}{5}\right]\right\}$$

which yields (11).

ACKNOWLEDGMENT

The authors gratefully acknowledge the contribution of Dr. Audrey Viterbi and Dr. Jack Wolf.

REFERENCES

[1] A. J. Viterbi, "When not to spread spectrum—A sequel," *IEEE Communications Mag.* vol. 23, pp. 12-17, Apr. 1985.

[2] K. S. Gilhousen, I. M. Jacobs, R. Padovani, and L. A. Weaver, "Increased capacity using CDMA for mobile satellite communications," *IEEE Trans. Select. Areas Commun.*, vol. 8, pp. 503-514, May 1990.

[3] P. T. Brady, "A statistical analysis of on-off patterns in 16 conversations," *Bell Syst. Tech. J.*, vol. 47, pp. 73-91, Jan. 1968.

[4] G. R. Cooper and R. W. Nettleton, "A spread spectrum technique for high capacity mobile communications," *IEEE Trans. Veh. Technol.*, vol. VT-27, pp. 264-275, Nov. 1978.

Klein S. Gilhousen (M'86-SM'91) was born in Coshocton, OH, in 1942. He received the B.S. degree in electrical engineering from the University of California, Los Angeles, in 1969.

In 1985, he became a cofounder and Vice President for Systems Engineering for QUALCOMM, Inc., San Diego, CA. his professional interests include satellite communications, cellular telephone systems, spread spectrum systems, communications privacy, communications networks, video transmission systems, error correcting codes and modem design. He holds six patents in these areas with five more applied for. Prior to joining QUALCOMM, he was Vice President for Advanced Technology at M/A-COM LINKABIT San Diego, CA, from 1970 to 1985 and Senior Engineer at Magnavox Advanced Products Division, Torrance, CA from 1966 to 1970.

Irwin M. Jacobs (S'55-M'60-F'74) received the B.E.E. degree in 1956 from Cornell University, Ithaca, NY, and the M.S. and Sc.D. degrees in electrical engineering from the Massachusetts Institute of Technology, Cambridge, in 1957 and 1959, respectively.

On July 1, 1985, he became a founder and the Chairman and President of QUALCOMM, Inc. From 1959 to 1966, he was an Assistant/Associate Professor of Electrical Engineering at M.I.T. and a staff member of the Research Laboratory of Electronics. During the academic year 1964-1965, he was a NASA Resident Research Fellow at the Jet Propulsion Laboratory. In 1966, he joined the newly formed Department of Applied Electrophysics, now the Department of Electrical Engineering and Computer Science, at the University of California, San Diego (UCSD). IN 1972, he resigned as Professor of Information and Computer Science to devote full time to LINKABIT Corporation. While at M.I.T., he coauthored a basic textbook in digital communications, *Principles of Communication Engineering*, published first in 1965, and still in active use. He retains academic ties through memberships on the Cornell University Engineering council, the visiting committees of the M.I.T. Laboratory for Information and Decision Systems, as Academic/Scientific member of the Technion International Board of Governors, and as a Board Member of the UCSD Green Foundation for Earth Sciences. He is a past Chairman of the Scientific Advisory Group for the Defense Communications Agency, and of the Engineering Advisory Council for the University of California. He has served on the governing boards of the IEEE Communications Society, the IEEE Group on Information Theory, and as General Chairman of NTC'74. In 1980, he and Dr. A. Viterbi were jointly honored by the American Institute of Aeronautics and Astronautics (AIAA) with their biannual award "for an outstanding contribution to aerospace communications." In 1984, he received the Distinguished Community Service Award for the Anti-Defamation League of B'nai B'rith. The local American Electronics Association's First Annual ExcEL Award was presented to Dr. Jacobs in 1989 for excellence in electronics and his "dedication and innovation, which have set the highest standards in the local electronics industry."

Dr. Jacobs was elected a member of the National Academy of Engineering for "Contributions to communication theory and practice, and leadership in high-technology product development." He is a member of Sigma Xi, Phi Kappa Phi, Eta Kappa Nu, and the Association for Computing Machinery (ACM).

Roberto Padovani (S'83-M'84-SM'91) received the Laurea degree from the University of Padova, Italy, and the M.S. and Ph.D. degrees from the University of Massachusetts, Amherst, in 1978, 1983, and 1985, respectively, all in electrical and computer engineering.

In 1986 he joined QUALCOMM, Inc. and he is now Director of System Engineering in the Engineering Department. As a member of the engineering department of QUALCOMM, Inc., he has been involved in the design and development of CDMA modems for the mobile satellite channel, various satllite modems, and VLSI Viterbi decoders. He is currently involved in the development of the CDMA digital cellular telephone system. In 1984 he joined M/A-COM Linkabit, San Diego where he was involved in the design and development of satellite communication systems, secure video systems, and error-correcting coding equipment.

Andrew J. Viterbi (S'54-M'58-SM'63-F'73) received the S.B. and S.M. degrees in electrical engineering from the Massachusetts Institute of Technology, Cambridge, in 1957, and the Ph.D. degree in electrical engineering from the University of Southern California, Los Angeles, in 1962.

He has devoted approximately equal segments of his career to academic research, industrial development, and entrepreneurial activities. In 1985, he became a founder and Vice Chairman and Chief Technical Officer of QUALCOMM, Inc., a company concentrating on mobile satellite communications for both commercial and military applications. In 1968, he cofounded LINKABIT Corporation. He was Executive Vice President of LINKABIT from 1974 to 1982. In 1982, he took over as President of M/A-COM LINKABIT, Inc. From 1984 to 1985, he was appointed Chief Scientist and Senior Vice President of M/A-COM, Inc. After graduating from M.I.T., he was a member of the project team at C.I.T. Jet Propulsion Laboratory which designed and implemented the telemetry equipment on the first successful U.S. satellite, Explorer I. From 1963-1973 he was a Professor with the UCLA School of Engineering and Applied Science. He did fundamental work in digital communication theory and wrote books on the subject, for which he received numerous professional society awards and international recognition. These include three paper awards, culminating in the 1968 IEEE Information Theory Group Outstanding Paper Award. He has also received three major society awards: the 1975 Christopher Columbus International Award (from the Italian National Research Council sponsored by the City of Genoa); the 1980 Aerospace Communications Award jointly with Dr. I. Jacobs (from AIAA); and the 1984 Alexander Graham Bell Medal (from IEEE sponsored by AT&T) "for exceptional contributions to the advancement of telecommunications." He has a part-time appointment as Professor of Electrical and Computer Engineering at the University of California, San Diego.

Dr. Viterbi is a member of the National Academy of Engineering.

Lindsay A. Weaver, Jr., received the S.B. and S.M. degrees from the Massachusetts Institute of Technology, Cambridge, in 1976 and 1977, respectively.

He is Vice President of Engineering at QUALCOMM, Inc. He was a key member of the design teams at QUALCOMM for the Mobile Satellite CDMA voice system, the OmniTRACS mobile satellite messaging system (hybrid frequency hopping and direct sequence), and the CDMA cellular telephone system. He has also lead projects developing FDMA modems, Viterbi decoders, highspeed packet switches, and satellite video scrambling.

Charles E. Wheatley III (SM'91) received the B.S. degree in physics from the California Institute of Technology, Pasadena, in 1956, the M.S. degree in electrical engineering from the University of Southern California, Los Angeles, in 1958 and the Ph.D. degree in electrical engineering from the University of California, Los Angeles in 1972.

He joined QUALCOMM, Inc., in 1987 as Principal Engineer, and has worked on both government and commercial programs, concentrating on system performance issues. The last two years have been spent working on RF hardware and system design for CDMA cellular phone applications. He has over 30 years of experience in RF satellite-based communications systems. His areas of expertise include time/frequency, anti-jam and LPI, all of which he has applied to a wide variety of systems. Prior to joining QUALCOMM, he held the position of Technical Assistant Vice President at M/A COM LINKABIT, Inc. in San Diego, CA.

Performance Evaluation for Cellular CDMA

Laurence B. Milstein, *Fellow, IEEE*, Theodore S. Rappaport, *Senior Member, IEEE*, and Rashad Barghouti

***Abstract*—In this paper, we consider the performance of a cellular radio direct-sequence code-division multiple access system. The base-to-mobile link is modeled as a flat Rayleigh fading channel, with all signals transmitted from a given base station fading in unison. For the mobile-to-base link, we use a similar model, except that the waveforms from all users are assumed to experience independent fading. Finally, we show the effects of imperfect power control.**

I. Introduction

Because of its well-known ability to both combat multipath and allow multiple users to simultaneously communicate over a channel, spread-spectrum techniques are being considered for use in cellular communication networks [1]-[3]. In this paper, we look at the performance of a code-division multiple-access (CDMA) cellular system when all users employ coherent BPSK modulation and direct-sequence (DS) spreading. Consistent with current CDMA designs, we assume that there are separate frequency bands for the mobile-to-base link and the base-to-mobile link, but otherwise there is no frequency separation. That is, we assume that all base stations in the system reuse one frequency band, and all subscribers reuse a separate band. We also assume that all transmitters, whether in bases or in mobiles, employ omnidirectional antennas. These two assumptions imply that any mobile in the system experiences interference from all base stations, but does not experience interference from the transmissions of other mobiles. Similarly, a given base station experiences interference from all mobiles in the system, but not from other base stations.

In the next section, we consider the base-to-mobile link. Under the assumption of perfect synchronization, we derive the average probability of error of a user which is straddling the boundary between adjacent cells. This particular point is chosen because it represents worst-cast conditions in the following sense: A mobile on the boundary of two or more cells receives the same nominal (i.e., average unfaded) power from its own base station as it does from its immediate neighbor base stations, since the distances to the various base stations are the same. Hence, the level of multiple-access interference from those cells is maximum. We assume the propagation path loss is a function of distance, and is the same in all cells.

The channel itself is modeled as undergoing flat Rayleigh fading. That is, we assume the coherent bandwidth of the channel exceeds the spread bandwidth of the signal. This assumption, while typically incorrect for a large degree of spreading (i.e., a channel will appear frequency selective to such a DS waveform and, thus, the power over the spread bandwidth will undergo only slight variation), is made for two reasons: one is analytic simplicity, and the other is the expectation that it will lead to worst-case results, since a frequency-selective channel will not undergo the broad deep fades that the flat channel will.

In Section III, we analyze the mobile-to-base link. The same propagation model as used for the base-to-mobile link is considered, with the following exception: In the base-to-mobile link, since all waveforms from a given base traverse the same path, they all fade in unison. However, for the mobile-to-base links, each mobile signal traverses a radio path different from that of the other mobiles' signals, and thus they all are assumed to fade independently of one another.

In the system model used to generate the results summarized above, perfect power control within each base station was assumed. In Section IV, this condition is relaxed so that the sensitivity of the system to imperfect power control at the mobiles can be assessed. We make no attempt to analyze any specific scheme; rather, we allow signals to arrive at the receiver-of-interest with random power levels, and then determine the resulting degradation in performance.

Finally, Section V presents our conclusions and suggestions for additional research that is needed to allow for a full understanding of this problem.

Manuscript received September 1, 1991; revised November 19, 1991 and December 30, 1991. This work was supported in part by the Office of Naval Research under Grant N00014-91-J-1234 and the NSF I/UCRC for Ultra-High Speed Integrated Circuits and Systems at UCSD, and by the Mobile and Portable Radio Research Group Industrial Affiliates Program and DARPA ESTO at Virginia Tech.

L. B. Milstein and R. Barghouti are with the Department of Electrical and Computer Engineering, University of California at San Diego, La Jolla, CA 92093.

T. S. Rappaport is with the Department of Electrical Engineering, Virginia Polytechnic Institute and State University, Blacksburg, VA 24061.

IEEE Log Number 917007.

II. Base-to-Mobile Link

It is desired to determine the performance of mobile users in a cellular CDMA system which are straddling the boundary between two adjacent cells. The significance of mobiles being at the edge of a cell is very clear; in general, due to propagation path loss, a mobile in the interior of a given cell experiences a power advantage in the reception of the signal transmitted from its own base station relative to signals received from the base stations in

0733-8716/92$03.00

neighboring cells (which are further away from the mobile). However, when the mobile is at the boundary between two cells, this advantage disappears.

To analyze this problem, consider the following model. The two cells shown in Fig. 1 are used, and a mobile with an omnidirectional antenna is assumed to be located on the boundary of the cells and to be receiving energy from both base stations. The signal from either base is composed of K DS waveforms, all of which are asynchronous with one another. However, the composite signal from each base station is assumed to independently undergo flat fading with either a Rayleigh or log-normal distribution. That is, because we are considering the base-to-mobile link and all signals that arrive from the base at a given mobile propagate over the same path, we assume they all fade in unison.

With the above model, the received waveform at the mobile is given by

$$r(t) = \sum_{i=1}^{K} A\alpha_1\, d_i(t - \tau_i)PN_i(t - \tau_i) \cos(w_0 t + \theta_i) + \sum_{i=K+1}^{2uc} A\alpha_2\, d_i(t - \tau_i)PN_i(t - \tau_i) \cdot \cos(w_0 t + \theta_i) + n_w(t) \quad (1)$$

where A represents the unfaded amplitude of any of the received signals, α_1 and α_2 are independent random variables representing the fading, $d_i(t)$ is the binary data of the ith signal, $PN_i(t)$ is the spreading sequence of the ith signal, and τ_i and θ_i represent the time delay and rf phase, respectively, of the ith signal. The delays are uniformly distributed in $[0, T]$, where T is the bit duration, the rf phases are uniformly distributed in $[0, 2\pi]$, and all delays and phases are assumed independent of one another and independent of the data. Finally, the noise $n_w(t)$ is additive white Gaussian noise (AWGN) having two-sided power spectral density $\eta_0/2$.

If we assume that each user employs a long spreading sequence (i.e., one that spans many bits), that the processing gain L is such that

$$L \triangleq \frac{T}{T_c} >> 1 \quad (2)$$

where T_c is the chip duration, and that $K >> 1$, then, using the Gaussian approximation developed in [4] and discussed in many other references (see, e.g., [3], [5]–[7]), we can approximate the conditional probability of error of the system, conditioned upon α_1 and α_2, as

$$P(e|\alpha_1, \alpha_2) = \phi\left(-\alpha_1\left[\frac{\eta_0}{2E} + \frac{K}{3L}(\alpha_1^2 + \alpha_2^2)\right]^{-1/2}\right) \quad (3)$$

where

$$\phi(x) \triangleq \frac{1}{\sqrt{2\pi}} \int_{-\infty}^{x} e^{-y^2/2}\, dy \quad (4)$$

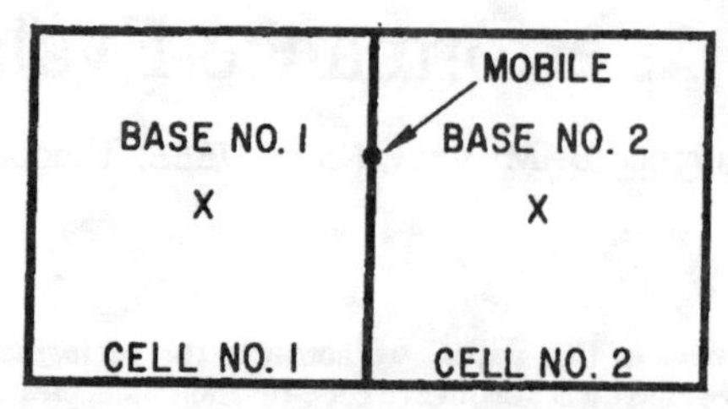

Fig. 1. Mobile on boundary of two cells.

and where $E = A^2T/2$ is the energy per bit. If we further assume that $E/\eta_0 >> 1$, we can approximate (3) as

$$P(e|\alpha_1, \alpha_2) \approx \phi\left(-\left[\frac{K}{3L}\left(1 + \frac{\alpha_2^2}{\alpha_1^2}\right)\right]^{-1/2}\right). \quad (5)$$

Define $u \triangleq \alpha_2/\alpha_1$ and $\nu \triangleq u^2$. If α_1 and α_2 are independent and identical Rayleigh random variables, then it can be shown that the probability density of ν is given by

$$f_\nu(\nu) = \begin{cases} \dfrac{1}{(\nu + 1)^2} & \nu \ge 0 \\ 0 & \text{elsewhere.} \end{cases} \quad (6)$$

Hence, the average probability of error at a subscriber receiver is given by

$$P_e = \int_0^\infty \frac{1}{(\nu + 1)^2}\, \phi\left(-\sqrt{\frac{3L}{K(1 + \nu^2)}}\right) d\nu. \quad (7)$$

On the other hand, if α_i, $i = 1, 2$, has the log-normal density, then

$$f_{\alpha_i}(\alpha_i) = \frac{\beta}{\sqrt{2\pi}\sigma\alpha_i} \exp\left[-\frac{1}{2\sigma^2}(\beta \ln(u) - d)^2\right] \quad (8)$$

where $\beta \triangleq 20/\ln 10$ and where d and σ^2 are the standard log-normal parameters, and it can be shown that

$$f_u(u) = \begin{cases} \dfrac{\beta}{2\sqrt{\pi}\sigma u} \exp\left[-\dfrac{\beta^2}{4\sigma^2}(\ln(u))^2\right] & u \ge 0 \\ 0 & \text{elsewhere} \end{cases} \quad (9)$$

and, hence, that

$$P_e = \int_0^\infty \phi\left(-\sqrt{\frac{3K}{1 + u^2}}\right) f_u(u)\, du. \quad (10)$$

Finally, if we assume an (n, k) block code capable of correcting all combinations of e and fewer errors, in conjunction with ideal interleaving, are used at the output of the demodulator, then the final decoded average bit error rate (BER) can be approximated by

$$P_b \approx \frac{1}{n} \sum_{i=e+1}^{n} i \binom{n}{i} P_e^i (1 - P_e)^{n-i}. \quad (11)$$

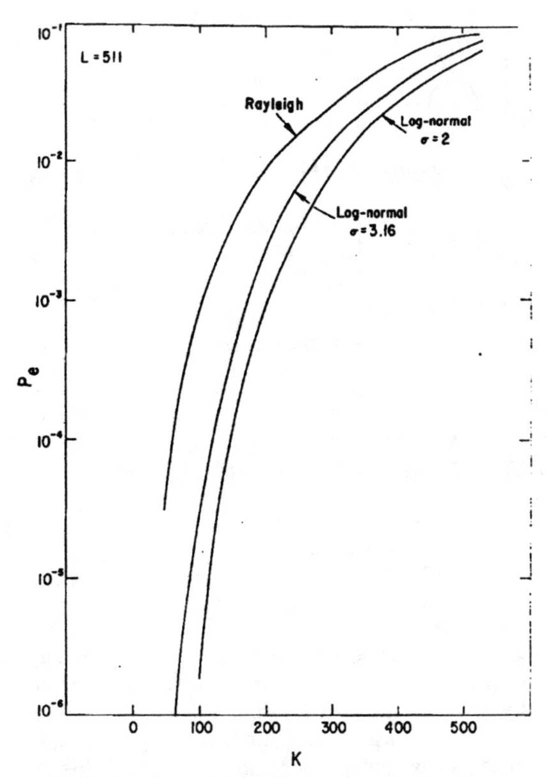

Fig. 2. Average probability of error versus number of users.

Equations (7) and (10) were evaluated numerically to obtain values for P_e, as a function of K, and those values were used in (11) to obtain the average decoded BER at a mobile receiver. Typical performance results are shown in Fig. 2, wherein BER is plotted against the number of active users in each of the two cells, K. There are three curves shown in Fig. 2, one corresponding to Rayleigh fading and the other two corresponding to log-normal fading. In all cases, the processing gain was $L = 511$, and the forward error correction was provided by the (23, 12) Golay code, for which $e = 3$.

If we assume that, for satisfactory voice communications, a decoded BER of 10^{-3} or better is required, we see that for Rayleigh fading, K can be about 100; for log-normal fading with $\sigma = 3.16$ (i.e., 10 dB variance) K can be about 170, while if $\sigma = 2$, K can increase to about 200.

Note that these results do not account for speaker activity factors. If, for example, the active users are only transmitting, in some average sense, 50% of the time, the above values of K will double. Also note that these results do not account for diversity reception.

Suppose we now extend the above model to include additional cells. From Fig. 3, it is clear the worst-case location for a mobile is at the point of intersection of all four cells; for this case, the received average power from the base stations of each of the three interfering cells equals that from the mobile's own base station.

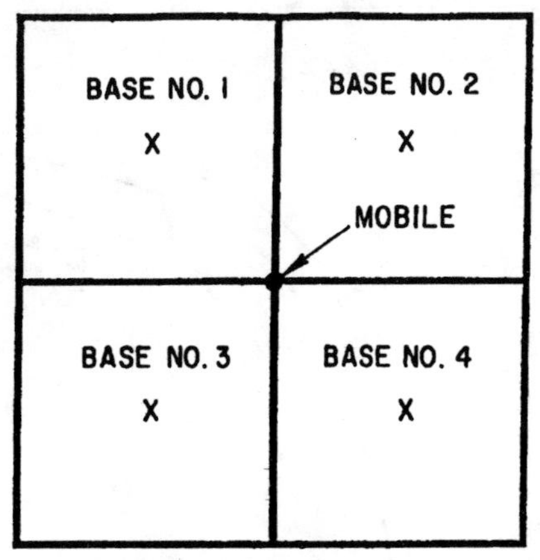

Fig. 3. Mobile at intersection of four cells.

Following the same procedure as before, we now have a conditional probability of error of

$$P(e \mid \alpha_1, \alpha_2, \alpha_3, \alpha_4) \simeq \phi\left(-\frac{\alpha_1 AT}{\left[\eta_0 T + \frac{A^2 K T_c^2}{3L}(\alpha_1^2 + \alpha_2^2 + \alpha_3^2 + \alpha_4^2)\right]^{1/2}}\right) = \phi\left(-\frac{\alpha_1}{\left[\frac{\eta_0}{2E} + \frac{K}{3L}(\alpha_1^2 + \alpha_2^2 + \alpha_3^2 + \alpha_4^2)\right]^{1/2}}\right). \tag{12}$$

Again, if we assume $E/\eta_0 >> 1$, we have

$$P(e \mid \alpha_1, \alpha_2, \alpha_3, \alpha_4) \simeq \phi\left(-\left[\frac{K}{3L}\left(1 + \frac{\alpha_2^2 + \alpha_3^2 + \alpha_4^2}{\alpha_1^2}\right)\right]^{-1/2}\right). \tag{13}$$

Letting

$$x \triangleq \alpha_2^2 + \alpha_3^2 + \alpha_4^2 \tag{14}$$

it is clear that x is a chi-squared random variable with six degrees-of-freedom, and hence the density function of x is given by

$$f_x(x) = \begin{cases} \frac{x^2}{16\sigma^6} \exp(-x/2\sigma^2) & x \geq 0 \\ 0 & \text{elsewhere.} \end{cases} \tag{15}$$

Consequently, with $z \triangleq x/\alpha_1^2$, it is straightforward to show that

$$f_z(z) = \frac{3z^2}{(z+1)^4} \tag{16}$$

and, hence, the average probability of error for three interfering cells is given by

$$P_e = \int_0^\infty \phi\left(-\left[\frac{K}{3L}(1+z)\right]^{-1/2}\right) \frac{3z^2}{(z+1)^4}\, dz. \tag{17}$$

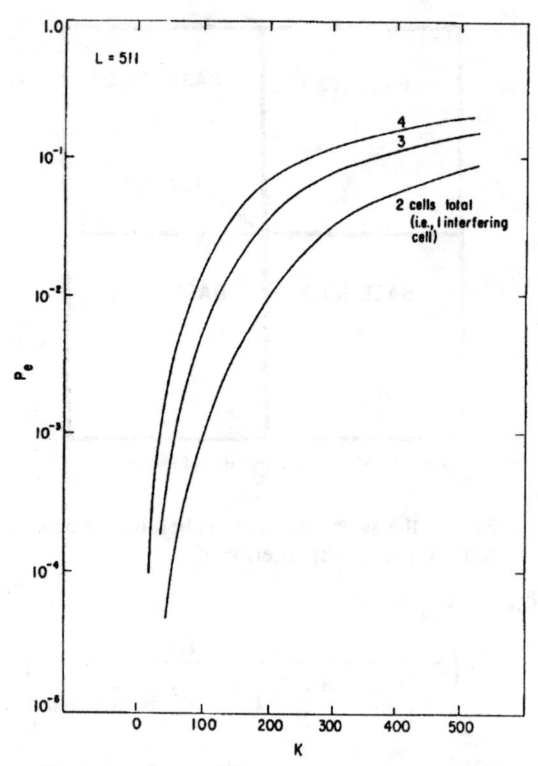

Fig. 4. Average probability of error versus number of users.

The decoded BER for this case is shown in Fig. 4 for a Rayleigh fading channel. Also shown in Fig. 4 is the curve of Fig. 2 for the Rayleigh channel when only a single interfering cell is considered, as well as results which correspond to two interfering cells. From Fig. 4, it can be seen that, with one interfering cell, for an average BER of 10^{-3}, 105 users are simultaneously supported. However, K drops to 61 users for two interfering cells and to 50 users for three interfering cells. It is clear that the increased degradation is quite significant. However, this latter situation of three adjacent interfering cells is probably too pessimistic since, in actual cell layouts, the cell geometry is closer to a hexagon than it is to a square. Hence, only three cells can intersect at a point, not the four cells considered above.

Consider now the effect of diversity on the flat fading channel we are considering. Assume, for simplicity, the two-cell model of Fig. 1, and assume second-order space diversity. The system block diagram is shown in Fig. 5, and the received waveforms $r_1(t)$ and $r_2(t)$ are given by

$$r_1(t) = \sum_{i=1}^{K} A\alpha_{11}\, d_i(t - \tau_{i1})PN_i(t - \tau_{i1}) \cos(\omega_0 t + \theta_{i1}) + \sum_{i=K+1}^{2K} A\alpha_{21}\, d_i(t - \tau_{i1})PN_i(t - \tau_{i1}) \cdot \cos(\omega_0 t + \theta_{i1}) + n_{w_1}(t) \tag{18}$$

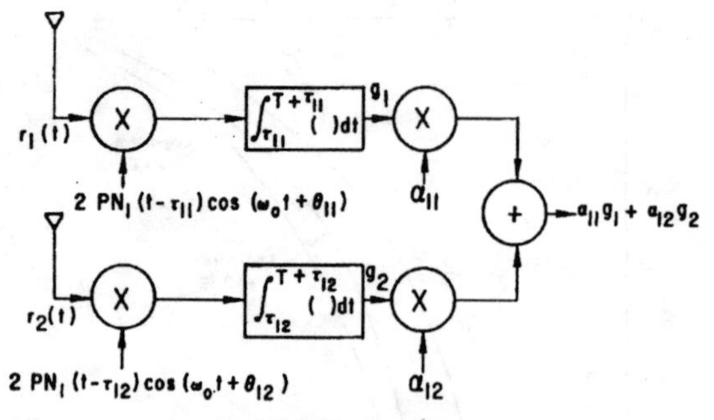

Fig. 5. Diversity receiver.

and

$$r_2(t) = \sum_{i=1}^{K} A\alpha_{12}\, d_i(t - \tau_{i2})PN_i(t - \tau_{i2}) \cos(\omega_0 t + \theta_{i2}) + \sum_{i=K+1}^{2K} A\alpha_{22}\, d_i(t - \tau_{i2})PN_i(t - \tau_{12}) \cdot \cos(\omega_0 t + \theta_{i2}) + n_{w_2}(t) \tag{19}$$

respectively. The two AWGN processes, $n_{w_1}(t)$ and $n_{w_2}(t)$, have spectral density $\eta_0/2$ and are assumed to be statistically independent, and the four Rayleigh random variables which represent the fading, α_{11}, α_{12}, α_{21} and α_{22}, are taken to be independent and identically distributed with parameter σ^2.

If we define

$$I_{ji} \triangleq \int_{\tau_{1j}}^{T+\tau_{1j}} PN_i(t - \tau_{ij})\, d_i(t - \tau_{ij})PN_1(t - \tau_{1j})\, dt, \quad j = 1, 2 \tag{20a}$$

and

$$\phi_{ij} \triangleq \theta_{1_j} - \theta_{ij}, \quad j = 1, 2 \tag{20b}$$

then the test statistics out of the two integrators in Fig. 5 are given by

$$g_1 = AT\alpha_{11} + \alpha_{11} \sum_{i=2}^{K} AI_{1i} \cos\phi_{i1} + A\alpha_{21} \sum_{i=K+1}^{2K} I_{1i} \cos\phi_{i1} + n_1 \tag{21}$$

and

$$g_2 = AT\alpha_{12} + \alpha_{12} \sum_{i=2}^{K} AI_{2i} \cos\phi_{i2} + A\alpha_{22} \sum_{i=K+1}^{2K} I_{2i} \cos\phi_{i2} + n_2. \tag{22}$$

Assuming perfect maximal-ratio combining (i.e., assuming g_1 and g_2 are appropriately aligned in time and weighted by α_{11} and α_{12}, respectively), the final test statistic is given by

$$g = \alpha_{11}g_1 + \alpha_{12}g_2 = AT(\alpha_{11}^2 + \alpha_{12}^2) + N \tag{23}$$

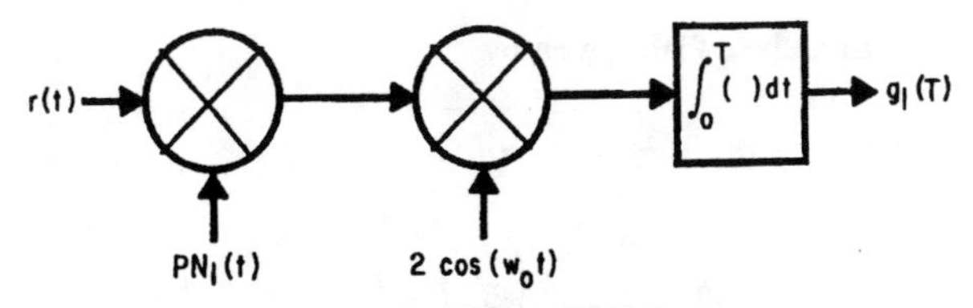

Fig. 6. Block diagram of receiver.

where

$$N \triangleq A\alpha_{11}^2 G_{11} + A\alpha_{11}\alpha_{21}G_{21} + A\alpha_{12}^2 G_{12} + A\alpha_{12}\alpha_{22}G_{22} + \alpha_{11}n_1 + \alpha_{12}n_2. \quad (24)$$

In (24),

$$G_{11} \triangleq \sum_{i=2}^{K} I_{1i}\cos\phi_{i1} \quad (25a)$$

$$G_{21} \triangleq \sum_{i=K+1}^{2K} I_{1i}\cos\phi_{i1} \quad (25b)$$

$$G_{12} \triangleq \sum_{i=2}^{K} I_{2i}\cos\phi_{i2} \quad (25c)$$

and

$$G_{22} \triangleq \sum_{i=K+1}^{2K} I_{2i}\cos\phi_{i2}. \quad (25d)$$

For simplicity, assume $\cos\phi_{i_1}$ is uncorrelated with $\cos\phi_{i_2}$, due, in part, to independent noise sources in the respective carrier recovery loops, as well as independent channel phases in the two diversity links which are not completely tracked out by the carrier recovery loops. Using this approximation and assuming further that $K \gg 1$, then, conditioned on α_{11}, α_{12}, α_{21}, and α_{22}, N is approximately Gaussian with zero-mean and conditional variance

$$\text{var}(N|\alpha_{11}, \alpha_{12}, \alpha_{21}, \alpha_{22}) \simeq \frac{A^2LT_c^2K}{3}[\alpha_{11}^4 + \alpha_{11}^2\alpha_{21}^2 + \alpha_{12}^4 + \alpha_{12}^2\alpha_{22}^2] + (\alpha_{11}^2 + \alpha_{12}^2)\eta_0 T. \quad (26)$$

The conditional probability of error is then

$$P(e|\alpha_{11}, \alpha_{12}, \alpha_{21}, \alpha_{22}) = \phi\left(-\frac{\alpha_{11}^2 + \alpha_{12}^2}{\left[\frac{K}{3L}(\alpha_{11}^4 + \alpha_{12}^4 + \alpha_{11}^2\alpha_{21}^2 + \alpha_{12}^2\alpha_{22}^2) + \frac{\eta_0}{2E}(\alpha_{11}^2 + \alpha_{12}^2)\right]^{1/2}}\right). \quad (27)$$

If we now define

$$Z \triangleq \alpha_{11}^2\alpha_{21}^2 + \alpha_{12}^2\alpha_{22}^2 \quad (28)$$

it is straightforward to show that

$$f_{Z|\alpha_{11},\alpha_{12}}(Z|\alpha_{11}, \alpha_{12}) = \frac{1}{2\sigma^2(\alpha_{11}^2 - \alpha_{12}^2)}\left[\exp\left(-\frac{Z}{2\sigma^2\alpha_{11}^2}\right) - \exp\left(-\frac{Z}{2\sigma^2\alpha_{12}^2}\right)\right]. \quad (29)$$

Hence, the average probability of error, P_e, is given by

$$P_e = \int_0^\infty\int_0^\infty\int_0^\infty \phi\left(-\frac{X_1^2 + X_2^2}{\sqrt{\gamma_1}\left[X_1^4 + X_2^4 + X_3 + \frac{\gamma_3}{\gamma_1\sigma^2}(X_1^2 + X_2^2)\right]^{1/2}}\right) \cdot \frac{\exp\left(-\frac{X_3}{2X_1^2}\right) - \exp\left(-\frac{X_3}{2X_2^2}\right)}{2(X_1^2 - X_2^2)} X_1\exp\left(-\frac{X_1^2}{2}\right) X_2\exp\left(\frac{-X_2^2}{2}\right) dX_1\, dX_2\, dX_3 \quad (30)$$

where

$$\gamma_1 \triangleq \frac{K}{3L} \quad (31)$$

and

$$\gamma_3 \triangleq \frac{\eta_0}{2E}. \quad (32)$$

As a cross-check on (30), note that if we let $\gamma_1 = 0$ (i.e., we remove the multiple access interference), (30) reduces to

$$P_e = \int_0^\infty\int_0^\infty \phi\left(-\left[\frac{2E\sigma^2}{\eta_0}(X_1^2 + X_2^2)\right]^{1/2}\right) X_1 \cdot \exp\left(-\frac{X_1^2}{2}\right) X_2\exp\left(-\frac{X_2^2}{2}\right) dX_1, dX_2. \quad (33)$$

If we now change to polar coordinates, it can be shown in a straightforward manner that

$$P_e = \frac{1}{2} - \frac{1}{2}\frac{d}{\sqrt{1 + d^2}} - \frac{1}{4}\frac{d}{(1 + d^2)^{3/2}} \quad (34)$$

where

$$d \triangleq \frac{2E\sigma^2}{\eta_0}. \quad (35)$$

Equation (34) is the well-known result for second-order diversity (see, e.g., [8]).

TABLE I

K	BER Without Diversity	BER with Second-Order Diversity	
		$d = 20$	$d = 200$
100	9.1×10^{-4}	3.9×10^{-6}	9×10^{-7}
200	9.5×10^{-3}	1.9×10^{-4}	1.3×10^{-4}
300	2.6×10^{-2}	2×10^{-3}	1.4×10^{-3}
400	5.7×10^{-2}	7.3×10^{-3}	5.7×10^{-3}
500	8.6×10^{-2}	1.6×10^{-2}	1.4×10^{-2}

Returning to (30), because there does not appear to be a way to significantly simplify it, it was evaluated numerically by making repeated use of a one-dimensional Gaussian quadrature formula. For the special case of $\gamma_1 = 0$, the numerical result was checked against (34) and was found to be in excellent agreement.

To see what is the effect of dual diversity, consider Table I. The entries in this table were obtained by assuming the (23, 12) Golay code was used in conjunction with the dual-order diversity (again, assuming ideal interleaving). It is seen that now one can increase the number of active users by about a factor of three (i.e., from about 100 with no diversity to about 300 with diversity and with $d = 200$).

III. Mobile-to-Base Link

Consider now the mobile-to-base link and, for simplicity, consider initially just two adjacent cells as shown in Fig. 1. Assuming K users per cell, we model the received waveform at base station #1 as

$$r(t) = \sum_{i=1}^{K} A\alpha_i\, d_i(t - \tau_i)PN_i(t - \tau_i) \cos(\omega_0 t + \theta_i) + \sum_{i=K+1}^{2K} A\left(\frac{d_{2_i}}{d_{1_i}}\right)^{r/2} \alpha_i\, d_i(t - \tau_i)PN_i(t - \tau_i) \cdot \cos(\omega_0 t + \theta_i) + n_w(t) \tag{36}$$

$$P(e|\alpha_1) = \phi\left(-\frac{A\alpha_1 T}{\left[\eta_0 T + \frac{A^2}{3}\sigma^2 T_c^2 L\left(K - 1 + \sum_{i=K+1}^{2K} \beta_i^2\right)\right]^{1/2}}\right) = \phi\left(-\frac{\alpha_1}{\left[\frac{\eta_0}{2E} + \frac{\sigma^3}{3L}\left(K - 1 + \sum_{i=K+1}^{2K} \beta_i^2\right)\right]^{1/2}}\right). \tag{42}$$

where the notation is the same as in the previous section, except now each transmitted mobile signal is assumed to experience independent fading. Specifically, we model the $\{\alpha_i\}$ as independent Rayleigh random variables, each with parameter σ^2. Also, d_{2_i} is the distance from the ith mobile in cell #2 to its own base station, while d_{1_i} is the distance from the same mobile to the base station in cell #1; thus, $d_{2_i} \le d_{1_i}$. Finally, r is the path loss exponent which describes how the received power falls off with distance [9].

Using the receiver of Fig. 6, we see that the test statistic of receiver #1 is given by

$$g_1(T) = A\alpha_1 T + A \sum_{i=2}^{K} \alpha_i I_i(T) \cos\theta_i + A \sum_{i=K+1}^{2K} \alpha_i \left(\frac{d_{2_i}}{d_{1_i}}\right)^{r/2} I_i(T) \cos\theta + N(T) \tag{37}$$

where, as before, $N(T)$ is a zero-mean Gaussian random variable with variance $\eta_0 T$, and

$$I_i(T) \triangleq \int_0^T d_i(t - \tau_i)PN_i(t - \tau_i)PN_1(t)\, dt \tag{38}$$

and where θ_1 and τ_1 have been set equal to zero. If we now define

$$\beta_i \triangleq \begin{cases} 1 & 2 \le i \le K \\ \left(\frac{d_{2_i}}{d_{1_i}}\right)^{r/2} & K + 1 \le i \le 2K \end{cases}$$

we can express (19) as

$$g_I^{uc}(T) = A\alpha_1 T + A \sum_{i=2}^{2K} \alpha_i\beta_i I_i(T) \cos\theta_i + N(T). \tag{39}$$

Consider the term

$$g_I(T) \triangleq \sum_{i=2}^{2K} \alpha_i\beta_i I_i(T) \cos\theta_i. \tag{40}$$

It is shown in the Appendix that, as K becomes arbitrarily large, $g_I(T)$ can be taken to be a Gaussian random variable with zero-mean and variance given by

$$\sigma_I^2 = \frac{\sigma^2}{3} LT_c^2\left[K - 1 + \sum_{i=K+1}^{2K} \beta_i^2\right]. \tag{41}$$

Therefore, the conditional probability of error, conditioned upon α_1, is approximated by (42).

Upon averaging (24) over the density of α_1, we obtain the average probability of error as

$$P_e = \frac{1}{2}\left[1 - \left[1 + \frac{\eta_0}{2E\sigma^2} + \frac{1}{3L}\left((K - 1) + \sum_{i=K+1}^{2K} \beta_i^2\right)\right]^{-1/2}\right]. \tag{43}$$

Finally, the extension of (25) to any number of cells is straightforward. For example, consider the expanded cell

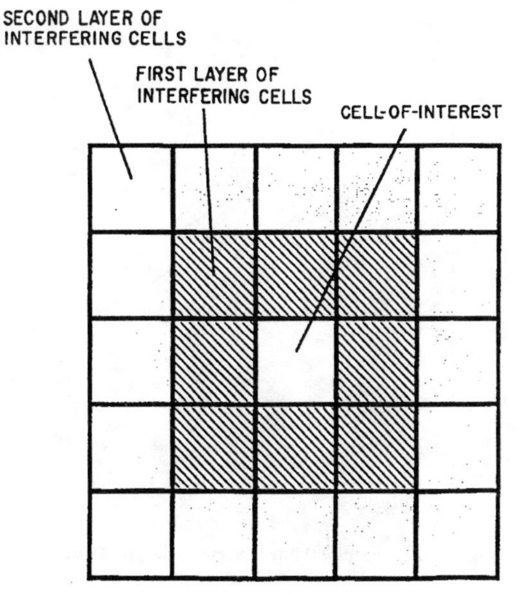

Fig. 7. Multiple layers of interfering cells.

diagram of Fig. 7. If we want to include the interference effects of, say, M_1 cells in addition to the cell-of-interest, the term

$$\sum_{i=K+1}^{2K} \beta_i^2$$

in the denominator of the $\phi(\cdot)$ function is replaced with the term

$$\sum_{i=K+1}^{(M_1+1)K} \beta_i^2$$

where the K β_i's that correspond to, say, the jth cell, $2 \leq j \leq M_1 + 1$, are defined by

$$\beta_i = \frac{d_{j_i}}{d_{1_i}}. \tag{44}$$

As before, d_{1_i} is the distance from the ith mobile, in this case assumed located in the jth cell, to the base station in cell #1, and d_{j_i} is the distance from that same mobile to the base station in the jth cell.

If we now assume these results correspond to the channel error rate of a receiver employing, as before, the (23, 12) Golay code, we can use (11) to obtain an approximation to the decoded BER. Figs. 8 and 9 show curves of decoded BER versus K, the number of active users, for $E/\eta_0 = 30\ dB$ and $L = 511$. The curves are parameterized by r, and Fig. 8 corresponds to just a single layer of cells around the cell-of-interest, while Fig. 9 corresponds to two layers of interfering cells. It is clear from the figures that for those scenarios where $r = 2$ applies, accounting for only a single layer of cells yields results which are much too optimistic; alternately, if $r = 4$ is applicable, the difference in performance between ac-

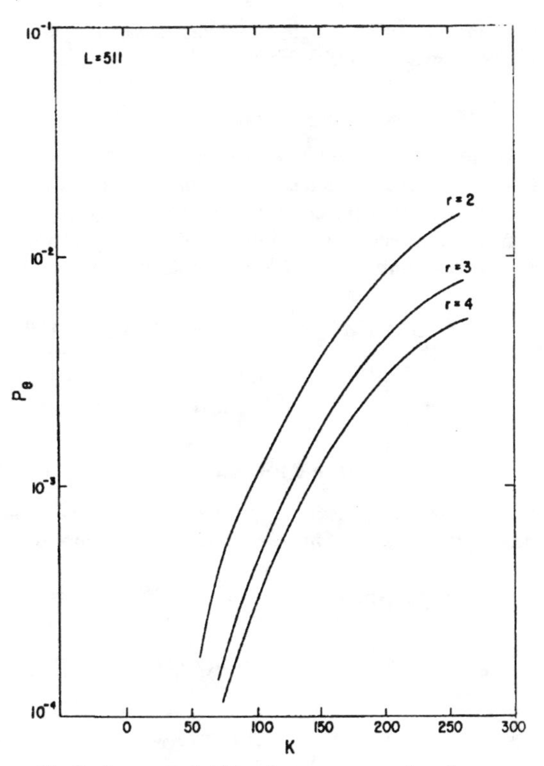

Fig. 8. Average probability of error versus number of users.

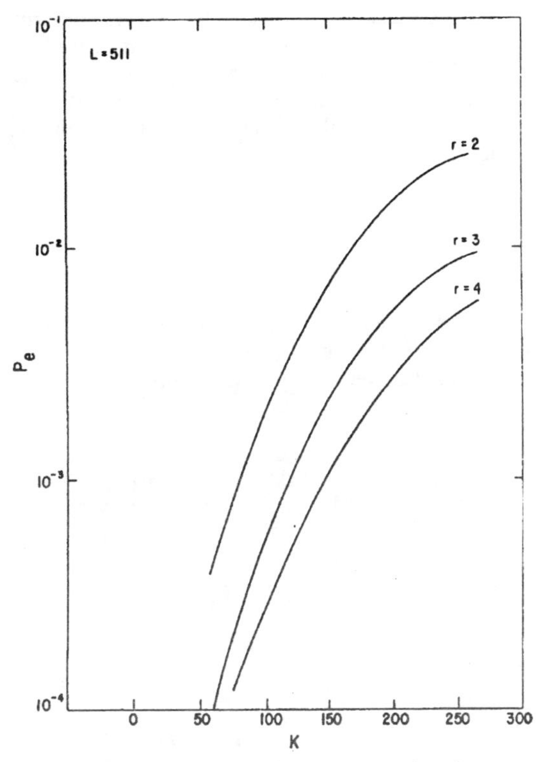

Fig. 9. Average probability of error versus number of users.

counting for only a single layer and accounting for both layers is insignificant.

IV. Effects of Imperfect Power Control

Up to this point, we have assumed that the power control [2, 3, 10] at each mobile within each cell is perfect. To obtain some perspective on the sensitivity of the system to imperfect power control, consider again the mobile-to-base link, and consider rewriting (36) as

$$r(t) = \sum_{i=1}^{K} A_i \alpha_i \, d(t - \tau_i) PN_i(t - \tau_i) \cos (w_0 t + \theta_i)$$
$$+ \sum_{i=K+1}^{2K} A_i \left(\frac{d_{2i}}{d_{1i}}\right)^{r/2} \alpha_i \, d_i(t - \tau_i) PN_i(t - \tau_i)$$
$$\cdot \cos (w_0 t + \theta_i) + n_w(t) \tag{45}$$

where now A_i, representing the received unfaded amplitude of the ith user at his own base station, is no longer a constant.

The test statistic now becomes

$$g_1(T) = A_1 \alpha_1 T + \sum_{i=2}^{K} A_i \alpha_i I_i(T) \cos \theta_i$$
$$+ \sum_{i=K+1}^{(M_1+1)K} A_i \alpha_i \beta_i I_i(T) \cos \theta_i + N(T) \tag{46}$$

where M_1, as before, is the number of cells contributing nonnegligible interference, and β_i is defined by (44).

To proceed further requires a more specific description of the $\{A_i\}$. For our purposes, assume that $A_1 = A$ and $A_i = \lambda_i A$, $i > 1$, where the $\{\lambda_i\}$ are i.i.d random variables. In particular, we assume they are uniformly distributed around the desired value A. That is, for some value V, let the density of λ_i be given by

$$f_{\lambda_i}(\lambda) = \begin{cases} \dfrac{1}{2V} & A - V \leq \lambda \leq A + V \\ 0 & \text{elsewhere.} \end{cases} \tag{47}$$

With these assumptions, it is straightforward to show that the central-limit theorem used in the Appendix still applies, and that the average probability of error is now given by

$$P_e = \frac{1}{2}\left\{1 - \left[1 + \frac{\eta_0}{2E\sigma^2} + \frac{1 + \frac{\Gamma}{3}}{3L}\left[(K-1) + \sum_{i=K+1}^{(M_1+1)K} \beta_i^2\right]\right]^{-1/2}\right\} \tag{48}$$

where

$$\Gamma \triangleq \frac{V}{A}. \tag{49}$$

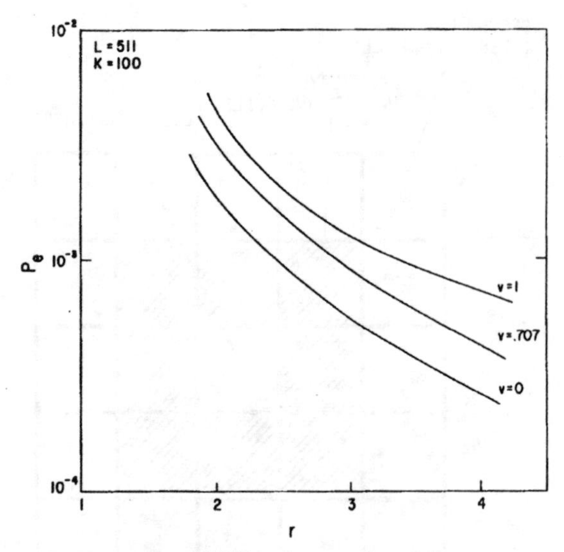

Fig. 10. Average probability of error versus path loss exponent.

In Fig. 10, the performance of the system versus the propagation law exponent is shown for the case of 100 users, each operating with a processing gain of 511 and a signal-to-noise ratio of 30 *dB*. The three curves correspond to $V = 0$ (i.e., perfect power control), $V = 0.707$, and $V = 1$. Note that $V = 0.707$ corresponds to the average power of any user varying by $\pm 50\%$ about its nominal value, whereas $V = 1$ corresponds to a variation of $\pm 100\%$.

V. Conclusions

In this paper, we have evaluated the performance of both the base-to-mobile link and the mobile-to-base link of a cellular CDMA system. The channel was modeled as a flat fading Rayleigh channel, and the maximum number of users per cell was determined such that the average probability of error did not exceed some predetermined level. For example, if a decoded error rate of 10^{-3} is required and the propagation law exponent is taken to be 3, then with perfect power control, a processing gain of 511 allows about 120 users to simultaneously access the channel (again, as pointed out above, this does not include effects of voice activity). On the other hand, if imperfect power control effects are considered, then, assuming the received power can vary by 50% about the nominal value, the number of simultaneous users drops to about 105.

To extend these results further, there are several obvious topics from which to choose. The channel should be modeled as exhibiting frequency selectivity, there should be a more accurate model to account for imperfect power control, and better path loss models are required. Finally, other coding schemes should be considered to provide an overall optimization of performance.

APPENDIX

It is desired to show that the random variable

$$g_I(T) = \sum_{i=2}^{2K} \alpha_i \beta_i I_i(T) \cos \theta_i \quad \text{(A1)}$$

of (40) is asymptotically normal. Recall the following:

a) The $\{\alpha_i\}$ are independent Rayleigh random variables with parameter σ^2.

b) The $\{\theta_i\}$ are i.i.d random variables uniformly distributed in $[0, 2\pi]$.

c) The term $I_i(T)$ is given by (20). The data $d_i(t)$ is assumed to be an independent random binary sequence for each i. If the period of each spreading sequence is sufficiently large, it can also be accurately modeled as a random binary sequence. Hence, if T/T_c is an integer, we can ignore the data (i.e., it can be combined with the chips of the spreading sequence) and we can represent $I_i(t)$ as

$$I_i(T) = \sum_{j=0}^{L} [b_{2j-1}^{(i)} \tau + b_{2j}^{(i)} (T_c - \tau)] \quad \text{(A2)}$$

where the $\{b_j\}$ are i.i.d binary random variables taking values ± 1 (i.e., if, in a given T_c-second interval, for a given user, the data symbol is d and the chip symbol is c, then $b = dc$) and τ is uniformly distributed in $[0, T_c]$.

To show that $g_I(T)$ is asymptotically normal, we invoke the Liapounoff version of the Central Limit Theorem [11], which states that

$$z \triangleq \sum_{i=1}^{J} x_i \quad \text{(A3)}$$

is asymptotically normal if the following condition holds:

$$\gamma \triangleq \frac{\left[\sum_i^J E(|x_i - \mu_i|^3)\right]^{1/3}}{\left[\sum_i^J E(x_i - \mu_i)^2\right]^{1/2}} \xrightarrow[J \to \infty]{} 0 \quad \text{(A4)}$$

where $\mu_i \triangleq E(x_i)$.

Note that α_i, $I_i(T)$, and $\cos \theta_i$ are all statistically independent, and their statistics (e.g., their moments) are independent of i. Consequently, since both the variance and third-central moment of all three are finite, we can express γ of (A4) as

$$\gamma = C \frac{\left(\sum_i \beta_i^3\right)^{1/3}}{\left(\sum_i \beta_i^2\right)^{1/2}} \quad \text{(A5)}$$

where C is a constant which depends upon the statistics of α_i, $I_i(T)$, and $\cos \theta_i$, but is independent of i. Finally, we define $d_{2\,\min}$ as the smallest value of d_{2i}, and further assume that $d_{2\,\min} > 0$ (i.e., we assume the shortest distance in a cell between any mobile and its own base station is greater than zero), and note that $\beta_{\max} = 1$. Then, with $\beta_{\min}$ associated with $d_{2\,\min}$, we have:

$$\gamma \le \frac{\left[\sum_{i=2}^{2K} \beta_{\max}^3\right]^{1/3}}{\left[\sum_{i=2}^{2K} \beta_{\min}^2\right]^{1/2}} = \frac{(2K-1)^{1/3}}{(2K-1)^{1/2}\beta_{\min}} = \frac{1}{(2K-1)^{1/6}\beta_{\min}} \xrightarrow[K \to \infty]{} 0. \quad \text{(A6)}$$

REFERENCES

[1] W. C. Y. Lee, "Overview of cellular CDMA," *IEEE Trans. Vehic. Technol.*, pp. 291-302, May 1991.

[2] K. S. Gilhousen, I. M. Jacobs, R. Padovani, A. J. Viterbi, L. A. Weaver, Jr., and C. E. Wheatley, III, "On the capacity of a cellular CDMA system," *IEEE Trans. Vehic. Technol.*, pp. 303-312, May 1991.

[3] R. L. Pickholtz, L. B. Milstein, and D. L. Schilling, "Spread spectrum for mobile communications," *IEEE Trans. Vehic. Technol.*, pp. 313-322, May 1991.

[4] M. B. Pursley, "Performance evaluation for phase-coded spread spectrum multiple-access communication, Part I: System analysis," *IEEE Trans. Commun.*, vol. COM-25, pp. 795-799, Aug. 1977.

[5] J. S. Lehnert and M. B. Pursley, "Error probabilities for binary direct-sequence spread-spectrum communications with random signature sequences," *IEEE Trans. Commun.*, vol. 35, pp. 87-98, Jan. 1987.

[6] R. K. Morrow, Jr. and J. S. Lehnert, "Bit-to-bit dependence in slotted DS/SSMA packet systems with random signature sequences," *IEEE Trans. Commun.*, vol. 37, pp. 1052-1061, Oct. 1989.

[7] M. B. Pursley, "The role of spread spectrum in packet radio networks," *Proc. IEEE*, vol. 75, pp. 116-134, Jan. 1987.

[8] J. G. Proakis, *Digital Communications*, 2nd Edition. New York: McGraw-Hill, 1989, p. 753.

[9] T. S. Rappaport and L. B. Milstein, "Effects of path loss and fringe user distribution on CDMA cellular frequency reuse efficiency," in *Proc. IEEE Global Telecom. Conf.*, Dec. 1990, pp. 404-6.1-404-6.7.

[10] K. S. Gilhousen, I. M. Jacobs, R. Padovani, and L. A. Weaver, Jr., "Increased capacity using CDMA for mobile satellite communication," *IEEE J. Select. Areas Commun.*, pp. 503-514, May 1990.

[11] D. P. Fraser, *Nonparametric Methods in Statistics*. New York: Wiley, 1957, pp. 213-214.

Laurence B. Milstein (S'66-M'68-SM'77-F'85), for a photograph and biography, see this issue, p. 667.

Theodore S. Rappaport (S'83-M'87-SM'91) was born in Brooklyn, NY, on November 26, 1960. He received the B.S.E.E, M.S.E.E, and Ph.D. degrees from Purdue University, West Lafayette, IN, in 1982, 1984, and 1987, respectively.

In 1983 and 1986, he was employed with Harris Corporation, Melbourne, FL. From 1984 to 1988, he was with the NSF Engineering Research Center for Intelligent Manufacturing Systems at Purdue University. In 1988, he joined the Electrical Engineering faculty of Virginia Tech where he is an Assistant Professor and Director of the Mobile and Portable Radio Research Group. He conducts research on mobile radio communication system design and RF propagation prediction through measurements and modeling, and consults often in these areas. He guides a number of graduate and undergraduate students in mobile radio communications, and has authored or coauthored more than 40 technical papers in the areas of mobile radio communications and propagation, vehicular navigation, ionospheric propagation, and wideband communications. He holds a U.S. patent for a wideband antenna, and is co-inventor of SIRCIM, an indoor radio channel simulator that has been adopted by over 40 companies and universities. He is an active member of the IEEE Communications and Vehicular Technology Societies, and serves as Senior Editor of the IEEE JOURNAL ON SELECTED AREAS IN COMMUNICATIONS.

Dr. Rappaport is a member of the IEEE COMSOC Radio Committee, and is a Fellow of the Radio Club of America. He is also a member of ASEE, Eta Kappa Nu, Tau Beta Pi, Sigma Xi, and is a life member of the ARRL. In 1990, he received the Marconi Young Scientist Award for his contributions to indoor radio communications.

Rashad Barghouti received the B.S. degree in electrical engineering from the University of California, San Diego, in 1989.

Since graduation, he has been involved in graduate studies and research projects with faculty members at UCSD. Since 1990, he has been employed by AUDRE, Inc., in San Diego as a Systems R&D Engineer working on the development of image understanding and enhancement systems. His research interests lie in the areas of digital communications and two-dimensional digital signal processing.

Effects of Radio Propagation Path Loss on DS-CDMA Cellular Frequency Reuse Efficiency for the Reverse Channel

Theodore S. Rappaport, *Senior Member, IEEE,* and Laurence B. Milstein, *Fellow, IEEE*

***Abstract*—Analysis techniques that quantitatively describe the impact of propagation path loss and user distributions on wireless direct-sequence code-division multiple-access (DS-CDMA) spread spectrum systems are presented. First, conventional terrestrial propagation models which assume a d^4 path loss law are shown to poorly describe modern cellular and personal communication system channels. Then, using both a two-ray propagation model and path loss models derived from field measurements, we analyze the impact of path loss on the frequency reuse efficiency of DS-CDMA cellular radio systems. Analysis is carried out for the reverse (subscriber-to-base) channel using a simple geometric modeling technique for the spatial location of cells, and inherent to the geometry is the ability to easily incorporate nonuniform spatial distributions of users and multiple layers of surrounding cells throughout the system. Our analysis shows the frequency reuse efficiency (F) of the reverse channel with a single ring of adjacent cells can vary between a maximum of 71% in d^4 channels with a favorable distribution of users, to a minimum of 33% in d^2 channels with a worst case user distribution. For three rings of adjacent users, F drops to 58% for the best d^4 case, and 16% for the worst d^2 case. Using the two-ray model, we show that F can vary over a wide range of values due to the fine structure of propagation path loss. The analysis techniques presented here can be extended to incorporate site-specific propagation data.**

I. Introduction

WIDESPREAD use of CDMA for personal wireless communications is likely to occur over the next several years. In 1990, more than a dozen experimental license applications were tendered by the FCC for U.S. wireless communication systems using direct-sequence spread spectrum code division multiple access (DS-CDMA). U.S. CDMA field trials are currently underway by major cellular and personal communications companies using existing cellular spectrum in the 800–900 MHz and 1.8 GHz bands. The virtues of spread spectrum in a frequency reuse mobile radio environment have been discussed by only a handful of researchers over the past two decades (see [1]-[3] for example), and in light of recent claims that CDMA could afford several times more capacity than other access techniques, analyses and simulation methods that accurately predict the capacity of CDMA systems in real world propagation environments are warranted.

Manuscript received April 1, 1991; revised October 15, 1991. This work was supported in part by DARPA/ESTO and the Mobile and Portable Radio Research Group (MPRG) Industrial Affiliates Program at Virginia Tech, and the NSF I/UCRC for Ultra-High Speed Integrated Circuits and Systems at the University of California at San Diego. Portions of this paper were presented at GLOBECOM '90, San Diego, CA.

T. S. Rappaport is with the Mobile and Portable Radio Research Group, Bradley Department of Electrical Engineering, Virginia Polytechnic Institute and State University, Blacksburg, VA 24061–0111.

L. B. Milstein is with the Department of Electrical and Computer Engineering, University of California, La Jolla, CA 92093-0407.

IEEE Log Number 9200218.

This paper studies one quantitative cellular performance measure: frequency reuse efficiency on the reverse (subscriber-to-base) channel of a CDMA cellular system that employs total frequency reuse. That is, we analyze the performance of a cellular system that allows all users in all cells to share the same reverse channel carrier frequency. In our work, we model a cellular radio system by using a concentric circle geometry that easily accommodates various distributions on the number or location of users within cells. Within each cell, ideal power control is assumed on the reverse channel, so that within a cell of interest, the same power level is received at the base from all subscribers within that cell. Subscribers in other cells are power-controlled within their own cell, and it is their radiated power that propagates as interference to the base station receiver in the cell of interest. This interference raises the noise level, and thereby reduces the number of users that can be supported at a specified average performance level.

Holtzman [9] has shown how imperfect power control in the subscriber unit degrades the frequency-reuse efficiency of a CDMA cellular system. However, except for [11], we are not aware of analyses that quantitatively describe how propagation path loss or user distributions affect the frequency reuse efficiency of such a system. As is shown subsequently, both propagation path loss and distribution of users can greatly impact frequency reuse efficiency of CDMA, even with perfect power control.

This paper is organized into five sections. Section II presents a brief overview of propagation models and demonstrates with both the classic two-ray model and measured data how modern cellular systems undergo path loss that increases with distance to the second or third power, as opposed to the commonly used d^4 model. For emerging microcellular systems, we propose a path loss law that incorporates a close-in free space reference power level to which all further users are referenced. Section III presents a frequency reuse analysis technique that uses concentric circles to model the location of adjacent cells, as well as the distribution of users within each cell. We also show how the concentric circle geometry can be equated to the conventional hexagonal geometry used in [1], [3]. In Section IV, we use this analysis technique in conjunction with path loss models developed in Section II to arrive at reverse channel frequency reuse factors for a variety of propagation channels and user distributions. Section V summarizes the results of this work.

II. Propagation

Accurate propagation modeling is vital for accurately predicting the coverage and capacity of cellular radio systems. This is particularly true in evolving CDMA systems, since capacity is interference limited instead of power limited. Unlike conventional cellular radio system design which strives for excellent base station coverage in each cell, and then relies on judicious frequency management within a service area to mitigate cochannel interference, CDMA requires no frequency planning.[1]

While more recent work has considered small scale fading and imperfections in power control [15], in this work we use a simple small scale shadow fading model, neglect frequency-selective fading effects, and consider average path loss as a function of distance between base and mobile. Thus, we consider very slow flat fading channels with no resolvable multipath components. Since we are interested in analyzing the average frequency reuse efficiency of DS-CDMA systems in different path loss environments, the use of an average distance-dependent path loss model that neglects small scale shadow fading of individual users is adopted. That is, we simplify the analysis by assuming the total received power from N independent transmitters located at a radial distance r from the base receiver is, on average, N times the power from a single subscriber at r as predicted by the path loss model. This is a valid approach for the case of any symmetric shadowing distribution about the mean path loss, and a sufficiently large number of users. However, field measurements often show that shadow fading is log-normally distributed about the mean, which is not a symmetric distribution in terms of absolute power levels. Thus, while our analysis offers insight into how different cell sizes and different path loss models impact frequency reuse, our analysis does not address second-order shadowing effects. The simulation methodology in Section IV, however, can be adopted to include shadowing effects and is described subsequently.

First, we consider the classic two-ray model (direct path and single ground reflection) to see how received signal power varies with distance between transmitter and receiver due to a simple site-specific channel model. Then, we show measured data which agrees with a path loss power law of distance to the nth power, where n is a constant that ranges between 2 and 4.

Fig. 1 illustrates the classic two-ray propagation geometry that is used to model long-haul terrestrial microwave systems. There is a π radian phase shift induced by the ground reflection [10]. This geometry does not consider other multipath or shadowing effects other than the ground reflected path. The individual LOS and ground reflected components are considered to have amplitudes which do not vary with time, and which individually undergo free space propagation as the separation distance (d) between transmitter and receiver increases. The path loss is due to the envelope produced as the RF phases of each signal component vectorially combine in different ways as the mobile travels over space. We shall show shortly that in real channels which have many multipath components that fade and are shadowed [4], [5], measured large scale path loss is comparable to those predicted from the simple geometry in Fig. 1, so long as a free space reference distance close to the transmitter is used [11], [14].

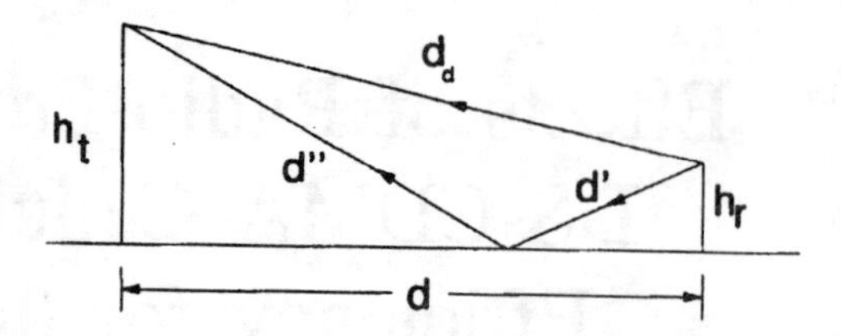

Fig. 1. Geometry for the two-ray (ground reflection) path loss model.

Historically, the geometry in Fig. 1 has been used to describe the large scale site-specific propagation phenomenon observed in terrestrial and analog cellular systems, systems which have deliberately aimed to maximize coverage distance. Traditional use of the two-ray model predicts the received signal several kilometers away, based on a received signal level at 1 km separation. It is only recently that the two-ray model of Fig. 1 has been used to model path loss in microcellular channels where small coverage areas are desired to increase frequency reuse, and thus capacity [6], [14]. Measurements in cellular and microcellular channels that provide log-distance path loss models using reference distances closer than 1 km have also been reported recently [7].

At a distance r from a transmitting antenna, if an electric field intensity given by $E = (A/r)cos(2\pi ft)$ V/m (without ground reflection) is measured in free space, then for the two-ray model of Fig. 1, the received electric field envelope at an antenna located at a distance r from the transmitting antenna is given by

$$|E(r)| = \frac{2A}{r}\sin\left(\frac{\theta}{2}\right) \quad (1)$$

where

$$\theta = \frac{2\pi}{\lambda}\left(\sqrt{(h_t+h_r)^2+r^2} - \sqrt{(h_t-h_r)^2+r^2}\right) \approx \frac{4\pi h_t h_r}{\lambda r}, \quad \text{for } r \gg h_t h_r \quad (2)$$

where h_t is the height of the transmitter antenna in meters, h_r is the height of the receiver antenna in meters, and λ is the carrier wavelength in meters. Equation (1) assumes unity gain antennas at both transmitter and receiver.[2] From (1), the

[1] CDMA can exploit resolvable multipath components using path diversity and combining (RAKE reception); however, energy in multipath components in frequency selective channels can add to the interference level of the system. Urban wide-band measurements have shown that channels can induce many delayed multipath components that have strengths within 10 dB of the strongest signal [4],[5]. We do not consider the effect of frequency-selective fading in this analysis.

[2] This analysis assumes that both the transmitter and receiver antennas offer equal gains to the direct and ground-reflected components. For very small values of r, say less than a few tens of meters, (1)–(4) are not valid and the true antenna gain patterns must be considered [14]. It can be shown that at very close range, the path loss is dominated by the direct path. The authors thank Dr. Joseph Shapira for pointing this out.

TABLE I
EXPERIMENTAL RESULTS OF WIDE-BAND PROPAGATION MEASUREMENTS IN SIX CELLULAR AND MICROCELLULAR CHANNELS; BEST-FIT EXPONENT VALUES WERE COMPUTED ASSUMING A 100 m FREE SPACE REFERENCE DISTANCE

	Antenna Height (m)	n	σ (dB)	Maximum T-R Separation (km)	Maximum rms Delay Spread (μs)	Maximum Excess Delay Spread 10 dB (μs)
Hamburgh	40	2.5	8.3	8.5	2.7	7.0
Stuttgart	23	2.8	9.6	6.5	5.4	5.8
Dusseldorf	88	2.1	10.8	8.5	4.0	15.9
Frankfurt (PA Building)	20	3.8	7.1	1.3	2.9	12.0
Frankfurt (Bank Building)	93	2.4	13.1	6.5	8.3	18.4
Kronbreg	50	2.4	8.5	10.0	19.6	51.3
All (100 m)		2.7	11.8	10.0	19.6	51.3
All (1 km)		3.0	8.9	10.0	19.6	51.3

received power is proportional to $|E(r)|^2$, and is given by

$$P(r) \propto |E(r)|^2 = \frac{4A^2}{r^2}\left(\frac{1}{2} - \frac{1}{2}\cos\theta\right). \quad (3a)$$

When the value of A is known exactly, (3a) can be used to determine the absolute electric field (and thus absolute received power) as a function of the transmitter-receiver (T-R) separation distance r. The value of A in (3a) represents the measured electric field intensity at a convenient close-in distance, which is in the far field of the transmitting antenna, i.e., the measured electric field at a distance of 1 m, 100 m, 1 km, or some other convenient distance from the transmitting antenna.

The absolute received power level as a function of r is given exactly by

$$P(r) = |E(r)|^2 * A_e/377 \quad (3b)$$

where A_e is the effective area of the receiver antenna, 377 Ω is the intrinsic impedance of free space, and A_e is proportional to the receiver antenna gain [12].

When the T-R separation, r, is much greater than $h_t h_r$, (3a) simplifies to

$$P(r) \propto \frac{A^2}{r^4}\left(\frac{4\pi h_t h_r}{\lambda}\right)^2. \quad (4)$$

The denominator of (4) suggests that power increases as distance to the fourth power, and this is sometimes quoted in the literature. This is also the propagation law assumed in [3]. It must be noted, though, that (4) is an asymptotic relationship, which holds when r is typically much larger than 1 km. As shown subsequently, (4) does not relate the actual power of distant users to close-in users, although such a model is required for meaningful CDMA system analysis and design. Rather, (4) relates the received power of very distant users to other distant users. Thus, we propose the use of a close-in free-space reference distance, α, to which the received signal strengths at all farther distances may be compared.

A. Path Loss Models

The geometry in Fig. 1 and the propagation equations (1)–(3) can be used to determine the distance-dependence of received power, and consequently the distance-dependence of an appropriate power law exponent for a two-ray flat-fading channel. If we let α denote the distance between the closest subscriber and the base station, and let α be the distance at which A in (3a) is measured, then the power of all subscribers with T-R separations greater than α can be related to the known power at α.

Consider DS-CDMA where all in-cell transmitting subscribers are power controlled to have identical signal levels P_n arriving at the base receiver. If we assume all subscribers are located at least α m away from the base station, and further assume that subscribers at α have transmitted power $P(\alpha)$, the transmitted power $P(r)$ of a subscriber at distance$r (r > \alpha)$ can be described by a log-distance model [5]

$$P(r)_{\text{dB}} = P(\alpha)_{\text{dB}} + \log_{10}\left(\frac{r}{\alpha}\right)^n. \quad (5a)$$

For this work, we assume that $P(\alpha)$ is controlled based on a free space propagation channel (direct path only) between the base and all subscribers located α m from the transmitter. The free space channel assumption for the base to the closest subscriber is not strict, and in real channels will not necessarily be valid since it depends on the antenna heights and patterns, and multipath. However, for analysis of CDMA frequency reuse, it is necessary to relate all of the power levels to some known power level $P(\alpha)$, which we consider to be caused by close-in free space propagation. For an interference limited system, one in fact may use any reasonable constant for $P(\alpha)$.[3]

The propagation path loss model in (5a) implies that if $P(\alpha)$is specified and $P(r)$ is known, either from a model (i.e., (3)) or measurement, then the path loss exponent n depends on both α and r. This can be seen be rewriting (5a)

$$n(\alpha, r) = \frac{P(r)_{\text{dB}} - P(\alpha)_{\text{dB}}}{+10\log_{10}(r/\alpha)}. \quad (5b)$$

For modern systems, α is on the order of several tens of meters, particularly in microcellular systems within heavily populated areas. Once α and $P(\alpha)$are specified, the system path loss PL in decibels, referenced to the closest user from the base

[3]The actual value of $P(\alpha)$ does have system design ramifications, since the absolute power levels will determine coverage areas. In this paper, we assume that base stations are suitably spaced so that the cellular system is not limited by link budget, but by interference.

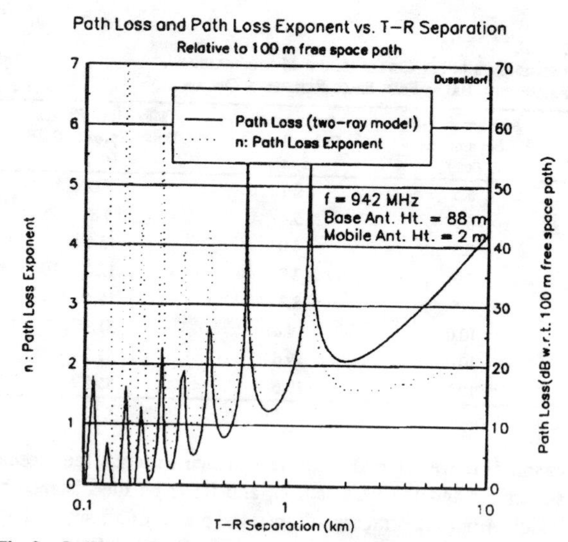

Fig. 2. Path loss and path loss exponent values for a cellular radio system as a function of T-R separation (r)and close-in reference distance α. Path loss is computed using a two-ray propagation model from (1)–(3), and path loss exponent is computed from (5b) using $f = 942$ MHz, $h_t = 88$ m, $h_r = 2$ m.

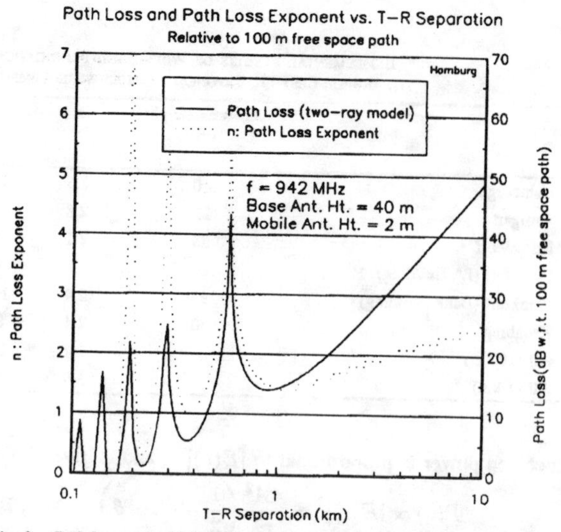

Fig. 3. Path loss and path loss exponent values for a cellular radio system as a function of T-R separation (r)and close-in reference distance α. Path loss is computed using a two-ray propagation model from (1)–(3), and path loss exponent is computed from (5b) using $f = 942$ MHz, $h_t = 40$ m, $h_r = 2$ m.

station, can be expressed as

$$PL(\alpha, r)_{dB} = P(r)_{dB} - P(\alpha)_{dB} = 10 \log_{10}\left(\frac{r}{\alpha}\right)^{n(\alpha,r)}. \quad (5c)$$

Figs. 2–4 give examples of $PL(\alpha, r)$ and $n(\alpha, r)$ for three different channels that use different antenna heights and frequencies in the 900 MHz band. Additional examples can be found in [11]. Figs. 2–4 were derived by using the two-ray model (3) to compute $P(r)$, (5b) to compute $n(\alpha, r)$, and (5c) to compute path loss, where the value of $P(\alpha)$ was assumed to be due to free space propagation ($n = 2$) over a 100 m path to the base. An important observation is that for a particular antenna configuration, the path loss exponent value, n, is certainly not constant at close-in distances, although it converges to a constant several kilometers away from the base station. It can be seen that at typical microcell fringe distances of between $d = 0.5$ km and $d = 2$ km, path loss exponents approach values between about 2.0 and 3.0 when referred to a 100 m reference. Large signal increases and deep fades caused by phase combining of the LOS and ground reflected paths occur and are predicted by the two-ray model. It should be clear that the d^4 path loss power law is a poor model at distances less than several kilometers from a terrestrial base station. This can be readily seen by observing Fig. 2 and noticing the path loss at 10 km (42 dB) is only 27 dB down from the path loss at 1 km (15 dB). In a fourth power law, this difference would be 40 dB. Between 0.1 and 1 km, the error in the fourth power model is even more striking. Even with very low antennas, Fig. 4 shows that between 0.1 and 1 km, the two-ray model predicts only 23 dB path loss isolation.

Another model assumes the power law exponent, n, in (5) is not a function of r, but is a simple constant. The proper value for n is found by minimizing the mean square error of the best fit line on a scatter plot such as shown in Fig. 5 [5]. Recent experiments have shown that the appropriate

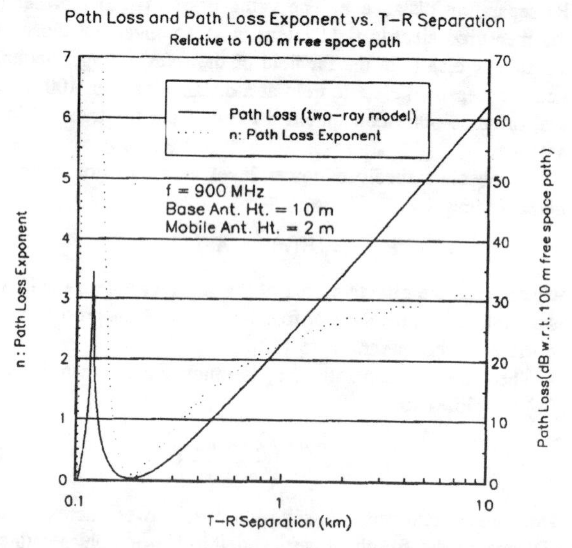

Fig. 4. Path loss and path loss exponent values for a cellular radio system as a function of T-R separation (r)and close-in reference distance α. Path loss is computed using a two-ray propagation model from (1)–(3), and path loss exponent is computed from (5b) using $f = 900$ MHz, $h_t = 10$ m, $h_r = 2$ m.

value of n is a function of the close-in reference distance used (see [5]–[8], [14]), and that d^4 predicts too much attenuation. In [5], a 500 ns probe was used to obtain an experimental database of about 6000 power delay profile measurements in six cell sites. A simple propagation model that assumes path loss is log-normally distributed about a mean path loss that falls off as a function of distance to a fixed exponent [12] was used to model the data. The measured averaged path loss provided the best (i.e., minimum mean square error) fit with a $d^{3.0}$ law when the free space reference distance was 1.0 km, but when a 0.1 km reference distance was used, the best fit was $d^{2.7}$. Fig. 5 shows the measured data and

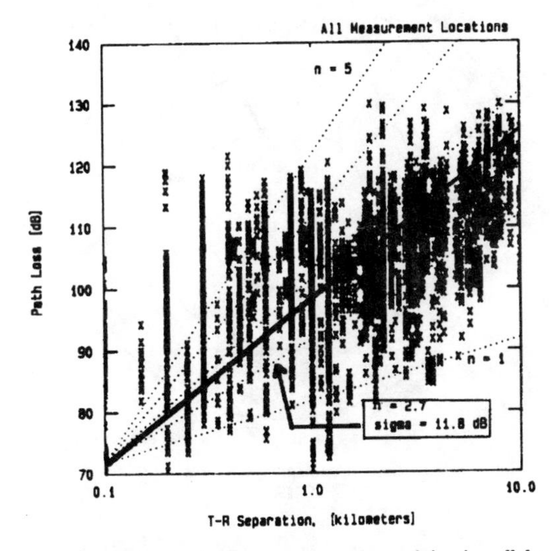

Fig. 5. Scatter plot of absolute path loss measured in six cellular and microcellular systems. The different asymptotic values for n have been computed from a 100 m reference distance using (5) in the paper. The path loss law of $d^{2.7}$ is the best linear regression fit to a log distance law of the form d with a free space leverage point at α = 100 m. Note that at distances less than 1 km, there are several locations which undergo deep fades, just as in Figs. 2–4 [5].

the minimum mean square error $d^{2.7}$ path loss law for a 100 m free space reference distance. Measurements in [5], [14] verify that measured average path loss obeys a distance power law of between d^2 to d^4 when referenced to a 100 m free space reference, and spot measurements are log-normally distributed in decibels about the average due to shadowing, with standard deviations ranging between 7 and 13 dB. Table I gives channel parameters[4] from measurements in six cellular and microcellular systems, and includes the minimum-mean-squared-error fit of n for a d^n mean path loss law [5].

When scatter plots of measured path loss data are compared with predicted path loss from a two-ray model using the correct antenna heights for base and mobile, there is reasonable agreement between the predicted location of deep fades within a kilometer from the base station [5],[6]. For example, Fig. 3 shows the two-ray prediction using a base antenna height that is approximately equal to the average base antenna height used for the measurements described in Table I and Fig. 5. By comparing path loss generated by the analytic model (i.e., Fig. 3) with measured data shown in Fig. 5, one can see both figures show large path loss values (deep fades) occurring within the first few kilometers around a base station. For distances further from the base station, the analytical model and the best fit line for the average measured path loss have comparable path loss exponents. In Section IV, both the analytical two-ray model and a simple d^n path loss model are used to compute frequency reuse efficiencies for CDMA, but the techniques shown subsequently can be used to predict performance in any distant-dependent propagation model, and may include shadow fading, as well.

[4] In Table I, maximum rms delay spread denotes the largest measured square root of the second central moment of spatially-averaged multipath power delay profiles chosen from all measurement runs in a geographic region. Maximum excess delay spread denotes the largest measured propagation delay of a multipath component, relative to the first arriving signal, having strength within 10 dB of the first arriving signal in a spatially averaged multipath power delay profile chosen from all measurement runs in a geographic region [5].

III. Propagation Effects On Frequency Reuse

We now find the cochannel interference on the reverse channel of a cellular DS-CDMA system by finding loose upper and tight lower bounds on the amount of out-of-cell (adjacent cell subscribers) interference and the in-cell interference (caused by other subscribers in the cell of interest). The adjacent cell subscribers are power-controlled by their own cell base station, and all in-cell subscribers are power controlled to provide the same signal level at the base receiver as any other in-cell subscriber.

The ratio of in-cell noise to total received noise is a figure of merit, called the frequency-reuse factor, which has values between 0 and 1. For DS-CDMA cellular, the frequency reuse factor is denoted by f, and is given by

$$f = \frac{N_0}{(N_0 + M_1 N_{a1} + M_2 N_{a2} + M_3 N_{a3} + \cdots)} \tag{6a}$$

and F is the frequency reuse efficiency, given by

$$F = f * 100\%. \tag{6b}$$

For conventional channelized cellular radio systems (including the U.S. Analog Mobile Phone standard), f is given by the number of channels assigned to a single cell divided by the total number of channels allocated to the system. This is because no cochannel interference comes from within the cell, but only from adjacent cochannel cells. The reuse factor is then dictated by the distance between cochannel cells and the path loss of the channel. For a d^4 path loss model, it can be shown that the minimum separation for a specific cochannel interference level (i.e., 18 dB) yields a seven-cell reuse geometry, or $f = 1/7$ [13]. For n values less than 4, f is less than 1/7 for channelized cellular systems.

In contrast, DS-CDMA deliberately allows for in-cell interference but does not require frequency coordination between any cell. In (6a), N_0 denotes the received noise power from all but one of the subscriber transmissions in the cell of interest (we assume one subscriber transmission is signal, not noise), and N_{ai} denotes the noise power received from all of the subscribers in one of the adjacent M_i cells located in the ith surrounding ring of the cell of interest. In general, f will be a random variable, since the noise powers from all surrounding cells will be a function of the random locations of adjacent users as well as random shadowing and voice activity, despite the fact that power control is employed in all cells. In this work, we neglect voice activity (i.e., assume all subscribers are transmitting simultaneously), assume perfect power control, and restrict the analysis to solving for *average* values of N_{ai} since we assume the shadowing of each user in each cell is symmetrically distributed about the average path loss. Note from (6a) that if only the single cell of interest is operating with a particular number of users, then a fixed signal-to-noise ratio will exist at the base station receiver for each received subscriber signal. When adjacent cells are then considered, the total noise level at the desired base station increases by a factor $1/f$. If independent users and uncorrelated spreading codes are assumed, then the noise powers add linearly after despreading at the receiver, and the increased noise will have the same effect as decreasing the number of users by a factor f to maintain the same signal-to-noise ratio as for the single cell case.

A. Assumptions

We make the following reasonable assumptions for analysis. 1) The spatial distribution of subscribers within each adjacent cell is identical, although our technique can be used to individually specify distributions in each adjacent cell (our approach accommodates various spatial distributions of users by using simple weighting factors); 2) all cells have equal area A, and there are K users per unit area. Thus, each cell has $U = KA$ users; 3) the distance of the closest subscriber to its own base is ($\alpha < d$); 4) α is the same for all cells; 5) the distance of the farthest in-cell user from the base station of interest is d; 6) the cell of interest has radius d; 7) the power received from subscribers > $7d$ (outside of the third ring) from the desired cell base is ignored; 8) ideal power control is assumed within each cell; 9) omnidirectional antennas are used at base and mobiles; 10) the received powers from all users add at the base receiver.

Fig. 6 shows the geometry used for analysis, where only the first ring of adjacent cells is shown. Our analysis technique can consider an unlimited number of surrounding cells, although here we present frequency-reuse factors for up to three surrounding rings. The center cell is the cell of interest, which contains the desired subscriber transmitting to the base in the center of the figure. Each subscriber will also interfere with adjacent base receivers, and adjacent subscribers interfere with the desired base receiver. We use wedges from an annullus made by concentric circles to represent adjacent cells in surrounding cell rings. For equal loading, the adjacent cells in each ring must each contain an area and number of users equal to that of the center cell. The geometry of the first surrounding layer is used to define the angle θ_1 over which the cell wedges of the first ring span. Because of the concentric circular geometry, all cells beyond the first ring may be related to the area and span of the first ring cells.

The geometry of Fig. 6 was chosen for analytic simplicity and differs from the traditional hex geometry in two ways. First, since the concentric circular geometry uses wedges as cells, a uniform distribution of subscribers over space means that more users will be located at the outer part of the adjacent cells (farther away from desired base), and thus will yield optimistic results for f (as compared with a uniform user distribution in a hexagonal cellular layout) if the number of users within the cell are not scaled in some way. As shown shortly, this is easily accounted for by scaling the number of users in the inner and outer parts of each adjacent cell wedge. In fact, a variety of scaling factors can be used to determine various user distributions and corresponding f values. Second, to exploit a simple geometric relationship using the law of cosines while maintaining tractability, we assume there are small concentric forbidden zones of width 2α in all of the surrounding rings where users cannot be located. Then, we locate two points on the inner and outer edge of the forbidden zone which fall on a line drawn radially from the central base station. The distance of an adjacent cell subscriber to its own base is then closely approximated by the distance from the subscriber to the point on the closest edge of the forbidden zone. Fig. 7 shows the geometry for computing distances for subscribers in both the inner and outer portion of an adjacent cell. This assumption can be made with virtually no impact on results when $\alpha < d$. While the geometry of Fig. 7 does not describe the physical shape of cells in urban streets with sharp losses around street corners, it offers significant analytical simplicity that is useful for general system design.

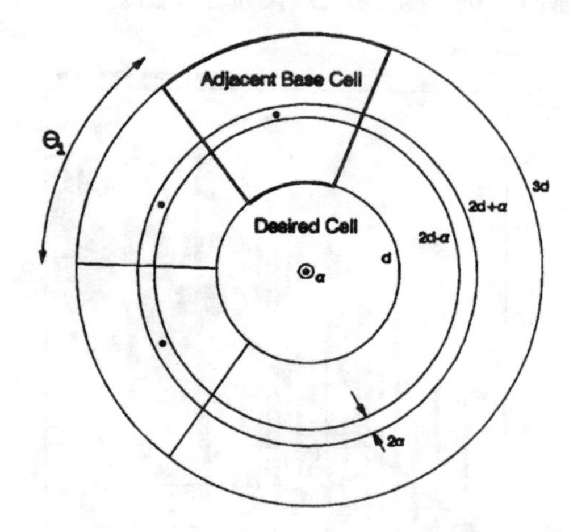

Fig. 6. Simple cell geometry used to analyze frequency reuse efficiency for CDMA.

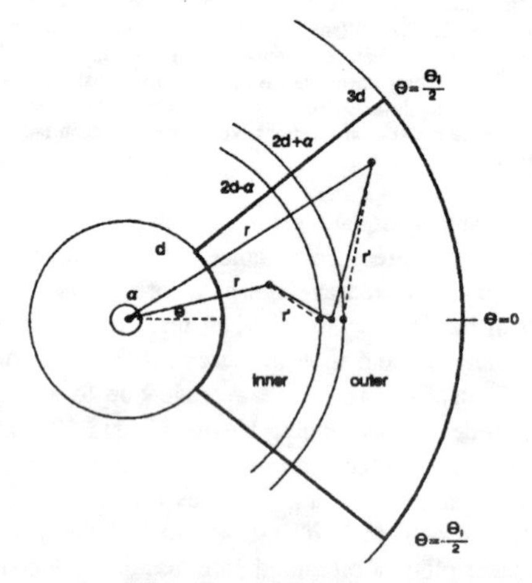

Fig. 7. By breaking the adjacent cells into two sectors (inner and outer sector), the impact of the distribution of users close to, and far from, the desired cell can be varied by using weighting coefficients. This figure shows the geometry used to relate the distances of adjacent cell subscribers to their own base cell with the distances to the base station within the center cell of interest.

B. Cell Geometries and User Distribution Weighting Factors

The area of the central cell of interest is equal to the area of all surrounding cells, and is given by

$$A = \pi d^2 - \pi \alpha^2 \approx \pi d^2. \quad (7a)$$

Let A_i denote the total area of the ith cell ring, and θ_i denote the angular span of each wedge-shaped cell in the ith ring. Neglecting α, the first surrounding ring has a total area A_1 occupied by the M_1 closest surrounding cells, and each cell wedge spans θ_1 given by (7b)

$$A_1 \approx \pi(3d)^2 - \pi(d)^2 = M_1 A$$
$$\theta_1 = 2\pi/M_1. \quad (7b)$$

From (7a) and (7b), $M_1 = 8$, thus the angular span of each closest cell is $\theta_1 = 2\pi/8 = \pi/4$ radians. For the second and all subsequent surrounding layers, it can be shown that

$$A_i = M_i A = iM_1 A = i8A, \qquad i \geq 1$$
$$\theta_i = \theta_1/i = \pi/4i. \quad (7c)$$

To account for various user distributions within the cells, it is possible to weight the number of subscribers in the inner sector and outer sector of each cell, so that the combination of users in both cell sectors sums to the desired number of users. We denote $W_{i\text{in}}$ as the weighting factor for the inner sector of a cell in the ith surrounding ring, and $W_{i\text{out}}$ as the weighting factor for the outer sector of a cell in the ith surrounding ring. The desired weighting factors are easy to obtain by first breaking the ith surrounding ring into inner and outer rings with areas that sum to A_i . Then, $W_{i\text{in}}$ and $W_{i\text{out}}$ are solved by adjusting the number of subscribers within inner and outer cell sectors to represent various spatial distributions on the users while still maintaining the desired number of users and equal area within each cell.

IV. Example Of Use Of Weighting Factors

In the first surrounding ring, the areas in the inner and outer sectors of each cell are:

$$A_{1\text{in}}/M_1 \approx \left[\pi(2d)^2 - \pi(d)^2\right] / 8 = 3A/8 \quad (7d)$$

$$A_{1\text{out}}/M_1 \approx \left[\pi(3d)^2 - \pi(2d)^2\right] / 8 = 5A/8. \quad (7e)$$

For each cell to possess equal area and $U = KA$ users, the weighting factors must satisfy

$$U = KA = KW_{1\text{in}}A_{1\text{in}}/M_1 + KW_{1\text{out}}A_{1\text{out}}/M_1$$
$$= KA[3/8W_{1\text{in}} + 5/8W_{1\text{out}}]. \quad (7f)$$

In a hexagonal (honeycomb) cellular geometry with a uniform distribution of users, half of the adjacent cell users will be closer than $r = 2d$ and half will be farther than $r = 2d$ from the central base station. To obtain a user distribution from the cellular geometry in Figs. 6 and 7, which has half of the users in an adjacent cell closer than $2d$ and half of the users farther than $2d$ from the center base station, (7d) and (7e) are solved and yield $W_{1\text{in}} = 4/3$ and $W_{1\text{out}} = 4/5$. Similarly, using (7a)–(7c) and extending (7f) to subsequent rings, one finds that $W_{2\text{in}}$ is 8/7, $W_{2\text{out}}$ is 8/9, $W_{3\text{in}}$ is 12/11, and $W_{3\text{out}}$ is 12/13. By multiplying the interference powers supplied by various cell sectors with the appropriate weighting factors, our method provides an equivalent method for solving a hexagonal cellular system with uniformly distributed users [11].

To achieve higher values of F, it is desirable to place most users far away from the center base. If no weighting is performed (i.e., if $W_{i\text{in}} = W_{i\text{out}} = 1$), this occurs naturally from the concentric circle geometry. For lower values of F, all users in every adjacent cell will be in the inner sectors, and no users will be in the outer sectors. This situation is given by $W_{i\text{out}} = 0$ for all i, and $W_{1\text{in}} = 8/3$, $W_{2\text{in}} = 16/7$, and $W_{3\text{in}} = 24/11$.

C. Comparison to Traditional Hexagonal Cell Geometry

To compare analysis results from the concentric circle geometry with traditional hexagonal cell structures considered in [1],[3],[12], one needs to relate the number of surrounding hexagonal cells to the number of surrounding wedge-shaped cells within a specified area surrounding the center cell. A hexagonal cell with major radius d occupies an area of $A_{\text{hex}} = 3\sqrt{3}\, d^2/2 = 2.598d^2$, whereas cells in Fig. 6 possess area $A = \pi d^2$. Thus, the first ring of eight wedge-shaped cells shown in Figs. 6 and 7 occupies the same area as would $8A/A_{\text{hex}}$, or 9.666, hexagonal cells. From (7c), the second ring of 16 wedge-shaped cells occupies the same area as $16A/A_{\text{hex}}$ or 19.333 hexagonal cells. The third ring of 24 wedge-shaped cells occupies the same area as 29 hexagonal cells.

It can be shown that (7g) relates the number of rings surrounding the circular center cell (Fig. 6) to the equivalent number of surrounding hexagonal cells that would surround the same center cell.

$$N_{\text{hex}} = \sum_{i=1}^{I} 8i\pi/2.598. \quad (7g)$$

In (7g), I is the total number of surrounding rings and N_{hex} is the total number of equivalent surrounding hexagonal cells. It is readily seen that the first surrounding ring of cells in the geometry proposed here will possess a greater area (i.e., 9.666 surrounding hexagonal cells) and larger number of users, and consequently will predict lower F values than the conventional "first layer" of six surrounding cells in the classic hexagonal (honeycomb) geometry. For two and three surrounding rings, the geometry of Fig. 6 can be equated to the case of 29 and 58 surrounding hexagonal cells, respectively.

V. Analysis

We presume all users obey a propagation path loss law that relates path loss to distance to the nth power relative to the reference free space level $P(\alpha)$, as given in (5a). However, we consider two distinct types of propagation models, one which uses the result of (5b) and the two-ray model to yield an exponent that is a function of distance, and another that assumes large scale path loss having the form d^n where the path loss exponent is assumed to be constant over distance.

All subscribers are assumed to be under power control within their own cells such that all desired mobile signals arrive with the same power P_n as the closest user within their own cell. Thus, all subscriber transmitter powers are referenced to the power transmitted by the *closest* subscriber within the cell, which is denoted as P_α .

Under perfect power control within the center cell, the received power of each of the in-cell subscribers at the base receiver is P_n , and the interfering power N_0 due to all but the single desired in-cell subscriber is given by

$$N_0 = P_n(U-1) \approx P_n U = P_n K A \tag{8a}$$

regardless of the path loss law within the cell (assuming there is sufficient dynamic range on the power transmitted by the subscriber). Without loss of generality in (8a), we replace P_n by P_α , the power received at the base due to the closest subscribers. Since we assume the same close-in propagation occurs within α m of every base station in the system, all signal levels may be referenced to the received power at a radial distance α pfrom any base station. Then, for the center cell of interest,

$$N_0 = P_\alpha K A. \tag{8b}$$

For simplicity, P_α may be assumed to be unity so that N_0 is equal to the number of users in each cell. Furthermore, K can be arbitrarily set to $1/A$, so that the noise due to in-cell users is unity, regardless of the number of users, and all subsequent cells add to the unit noise level produced by the center cell of interest. Each subscriber in the center cell has a transmitted power equal to $P_\alpha(r/\alpha)^n$, where r is the distance from mobile to base, and n is the propagation path loss exponent in (5). Note that n may be a function of r and α, as in (5b) and Figs. 2–4, or n may be assumed constant over distance, as shown in Fig. 5 and Table I.

The adjacent cell subscribers are under power control within their own cell and are a distance r' from their own base station. As shown in Fig. 7 for the first ring (i=1), the relationship between r' and r is given by (9) using the law of cosines. The true transmitter power for each subscriber in the adjacent cell can be related by P_α and r', where r' is related to r and θ by

$$\begin{aligned} r' &= \left(r^2 \sin^2\theta + (2*d*i-\alpha-r*\cos\theta)^2\right)^{1/2}, \\ &\qquad \text{for } d*i \le r \le 2*d*i-\alpha \\ r' &= \left(r^2 \sin^2\theta + (r*\cos\theta-2*d*i-\alpha)^2\right)^{1/2}, \\ &\qquad \text{for } 2*d*i+\alpha \le r \le d(2*i+1). \end{aligned} \tag{9}$$

Let $P_{ai}(r,\theta,\alpha)$ denote the power received at the center base station from a subscriber at a particular location in an adjacent ith layer cell. Then, the received power from each adjacent cell user is given by

$$P_{ai}(r,\theta,\alpha) = P_\alpha (r'/\alpha)^{n(r',\alpha)} (\alpha/r)^{n(r,\alpha)} \tag{10}$$

where r' is a function of θ as given by (9). The total interference power, N_{ai}, contributed by all subscribers in a cell in the ith surrounding ring is found by summing up the received powers from each user in the adjacent cell

$$N_{ai} = \sum_{u \epsilon A_i/M_i} P_{ai}(r,\theta,\alpha). \tag{11a}$$

It is clear from (10) that P_{ai} values in (11a) are functions of the specific locations of adjacent cell users as given by (9) and (10), and that the summation in (11a) represents a summation over a geographic cell region. To compute (11a), one can use any spatial distribution of users and any distance-dependent path loss. In real-world situations, individual values of P_{ai} will exhibit shadow fading (e.g. log-normal fading) and this can be modeled statistically in (11a) in a Monte Carlo fashion. In our work, though, we have assumed symmetric shadow fading of P_{ai} at specific T-R separations about the distance-dependent mean.

Equation (11a) is sufficiently general to solve for the power contributed from adjacent cells obeying any distance-dependent propagation path loss law. One simple way of manipulating a specified distribution of users is to employ weighting factors described in Section III-B. With weighting factors, the total noise powers contributed by users within the inner and outer sectors of adjacent cells can be easily scaled without the need to recompute the interference powers contributed by individual users. Thus, bounds on average adjacent cell noise can be obtained easily. To use the weighting factors, (11a) can be written as two separate summations as given in (11b):

$$\begin{aligned} N_{ai} = \mathbf{W}_{1\text{in}} &\sum_{u \epsilon A_{i\text{in}}/M_i} P_{ai}(r,\theta,\alpha) \\ + \mathbf{W}_{1\text{out}} &\sum_{u \epsilon A_{i\text{out}}/M_i} P_{ai}(r,\theta,\alpha). \end{aligned} \tag{11b}$$

Note that if $W_{1\text{in}}$ and $W_{1\text{out}}$ are unity, then (11b) is identical to (11a). As discussed in Section III-B, if $W_{1\text{in}} = 4/3$ and $W_{1\text{out}} = 4/5$, then a uniform spatial distribution of users in (11a) can be used in (11b) to provide the same N_{ai} as a hexagonal cellular system.

A. Results

Equation (11b) was solved numerically using a uniform spatial distribution of users for specific values of close-in reference distance and cell radius d. Then, the frequency reuse factor was computed from (6), where three different user distribution weightings in (11b) gave rise to various values for f using identical channels.

All results in this work were found using 20 000 uniformly spaced users in each adjacent cell for $d = 2$ km, and 400 000 equally spaced users within $d = 10$ km cells. Equation (6a) and symmetry were exploited to limit computations to a single cell within each ring.

Tables II and III give values of frequency reuse factor (f) as a function of path loss exponent (assumed constant over distance) for $\alpha = 0.1$ km and $\alpha = 0.05$ km, respectively, for the cases of cell boundaries at $d = 2$ km and $d = 10$ km. Only one ring of adjacent cells was considered for results given in the tables. Note that f values for $n = 4$ are comparable to those

TABLE II
FREQUENCY REUSE FACTOR FOR CDMA SYSTEMS AS A FUNCTION OF PROPAGATION EXPONENT. ONE RING OF ADJACENT CELLS CONSIDERED FOR $D = 2$ km AND $D = 10$ km, $\alpha = 0.1$ km

		Frequency Reuse Factor		
		Lower Bound	Hex Case	Upper Bound
d (km)	n	$W_1 = 3.0$, $W_2 = 0.0$	$W_1 = 1.38$, $W_2 = 0.78$	$W_1 = 1.0$, $W_2 = 1.0$
2.0	2	0.326	0.434	0.471
2.0	3	0.423	0.571	0.625
2.0	4	0.499	0.666	0.721
10.0	2	0.310	0.422	0.457
10.0	3	0.399	0.553	0.606
10.0	4	0.466	0.638	0.698

TABLE III
FREQUENCY REUSE FACTOR FOR CDMA SYSTEMS AS A FUNCTION OF PROPAGATION EXPONENT; ONE RING OF ADJACENT CELLS CONSIDERED FOR $D = 2$ km AND $D = 10$ km, $\alpha = 0.05$ km

		Frequency Reuse Factor		
		Lower Bound	Hex Case	Upper Bound
d (km)	n	$W_1 = 3.0$, $W_2 = 0.0$	$W_1 = 1.38$, $W_2 = 0.78$	$W_1 = 1.0$, $W_2 = 1.0$
2.0	2	0.316	0.425	0.462
2.0	3	0.408	0.558	0.613
2.0	4	0.479	0.646	0.707
10.0	2	0.308	0.419	0.455
10.0	3	0.396	0.550	0.603
10.0	4	0.462	0.634	0.695

given in [1],[3], where a hexagonal cellular system geometry with a comparable number of adjacent users was considered. For weighting factors that offer uniform user distributions comparable to the hexagonal geometry, we see for $d = 10$ km, f is 64% for $n = 4$, 55% for $n = 3$, and 42% for $n = 2$. A loose upper bound and tight lower bound on f are also given in Tables II and III. Weighting coefficients listed in Tables II and III differ slightly from those given in Section III-B since the term $\pi\alpha^2$ was considered in calculations which led to tabular entries, but was neglected for the general results given in (7). For a propagation model that has a constant n over distance, it is evident that f is not sensitive to α or d, but is highly sensitive to the value of the path loss exponent. Note that in free space channels ($n = 2$) with only one surrounding ring of adjacent cells, f is more than 3 dB down from the $n = 4$ result.

Although not shown in the tables, it is worth noting that when the first surrounding ring and just the inner portion of the second surrounding ring were considered in our analysis, we found $f = 0.606$ for the d uniform (hexagonal) user distribution case, which is comparable to results obtained in [3] using 18 surrounding hex cells. Using (7g) it can be shown that 18 hexagonal cells occupy the same area as the first ring

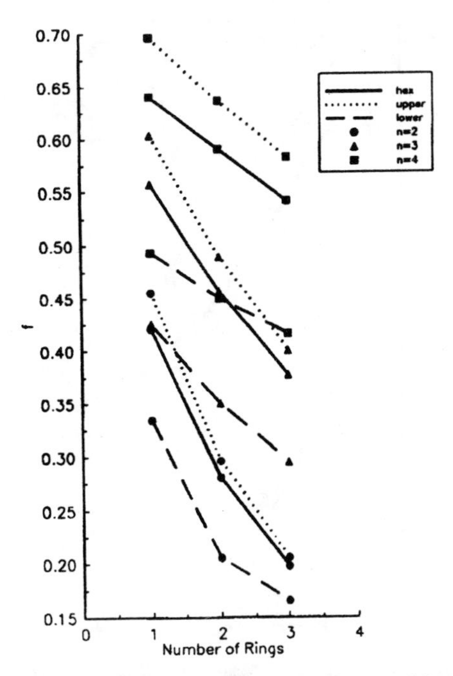

Fig. 8. Plot of the frequency reuse factor, f, as a function of the path loss exponent, n, and the number of rings surrounding the cell of interest.

and the inner portion of the second ring. Thus, our analysis yields identical results to average results in [3] which uses hexagonally shaped cells.

Fig. 8 extends the results shown in Tables II and III to include the interference effects of the second and third rings. One can see how additional rings effect the frequency reuse efficiency of CDMA. Fig. 8 shows f as a function of the number of adjacent cell rings for the case where path loss exponents remain constant with distance, with n values ranging from 2 to 4. The three weightings given in Section IV (hex, upper bound, lower bound) were used to give representative ranges on f. For the data shown in Fig. 8, $= 0.1$ km and $d = 10$ km were assumed. Fig. 8 shows that f drops to below 0.2 when $n = 2$ and three surrounding rings are considered. However, for $n = 4$, the hexagonal case (uniform distribution) has a frequency reuse efficiency of 54% with three layers. Note that for three surrounding rings, the $n = 3$ channel affords 3 dB more reuse than does the $n = 2$ channel, and the $n = 2$ reuse degrades by 3 dB when the surrounding layers are increased from one to three. These data indicate how the propagation environment can impact the estimate of frequency reuse.

In wireless channels, deep fading between an adjacent base and one of its subscribers will dramatically impact that subscriber's power control, and thus the power received from that unit at the cell of interest [15]. This is seen from (10), since if an adjacent user is in a deep fade while transmitting to its own base, the required transmitter power will be large and could strongly interfere with the base of interest.

We consider three different two-ray propagation channels that have a distance dependent path loss exponent. These channels were derived using (1)–(3), and possess the path loss

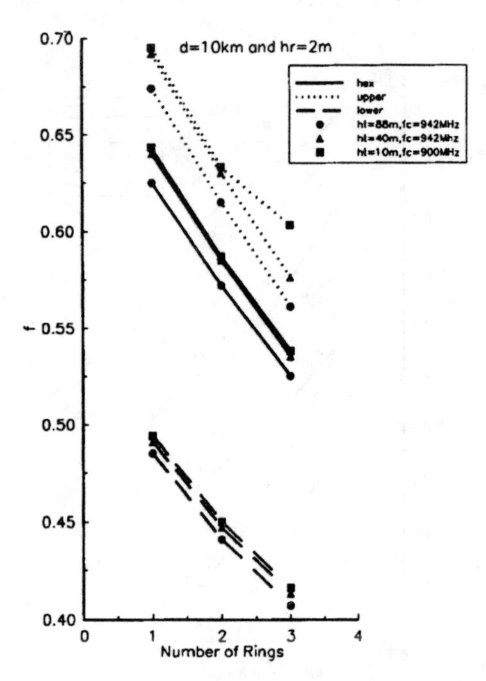

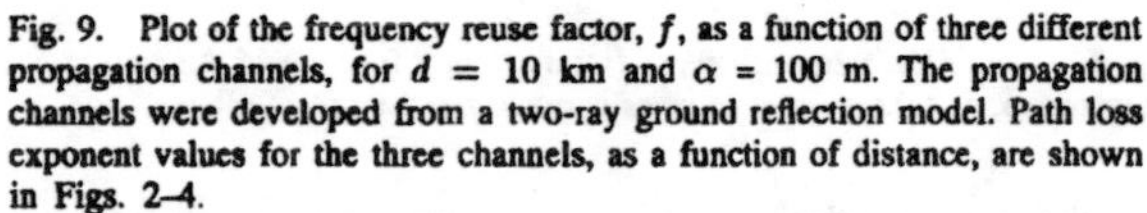
Fig. 9. Plot of the frequency reuse factor, f, as a function of three different propagation channels, for $d = 10$ km and $\alpha = 100$ m. The propagation channels were developed from a two-ray ground reflection model. Path loss exponent values for the three channels, as a function of distance, are shown in Figs. 2–4.

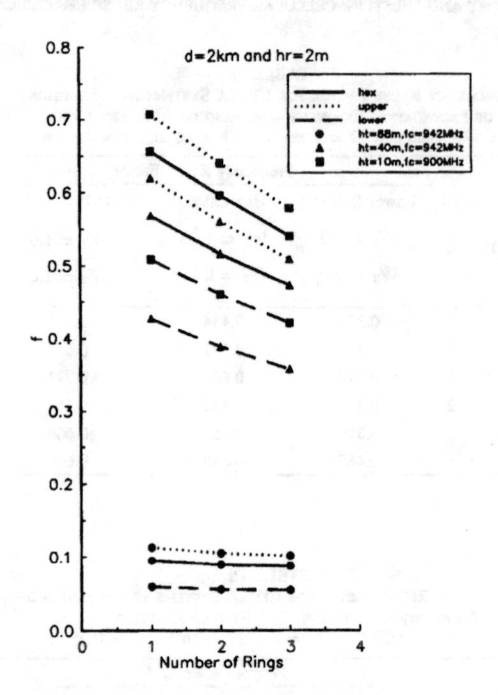

Fig. 10. Plot of the frequency reuse factor, f, as a function of three different propagation channels, for $d = 2$km and $\alpha = 100$ m. The propagation channels were developed from a two-ray ground reflection model. Path loss exponent values for the three channels, as a function of distance, are shown in Figs. 2–4.

responses given in Figs. 2–4. We assume that all base stations in the cellular system have identical channel (i.e., path loss) characteristics over distance. Further, to reduce the impact that deep fades have on results, we assume that within a 2 km radius of each base station, each base station's subscribers can have path loss fades no more than 40 dB below the reference power $P(\alpha)$. This fade limit ensures our analysis yields realistic results, since subscribers in an actual system cannot transmit excessive power levels to overcome path loss.

Figs. 9 and 10 show how cell size and a distance-dependent propagation path loss exponent can impact frequency reuse. Fig. 9 shows values of f for the case of large cells ($d = 10$ km, $\alpha = 100$ m). It is clear from Figs. 2–4 that when $d = 10$ km, all adjacent cells are located in the asymptotic region of d^4 path loss from the cell of interest. Fig. 9 indicates f ranges between 0.4 and 0.7, regardless of the particular two-ray channel or user distribution. By comparing Fig. 8 results with those in Fig. 9, one can see that for all user distributions, the three different distance-dependent analytical channels yield frequency reuse factors close to those offered by channels using a d^4 path loss law. This is due to the fact that the cell size is large, and natural isolation of adjacent noise power due to path loss occurs. The frequency reuse factors obtained for each of the three two-ray channels are, at most, only 5% lower than results obtained for the d^4 propagation law.

Fig. 10 shows how the same three two-ray propagation channels used to produce results in Fig. 9 can dramatically alter the frequency reuse when small cells are considered. For $d = 2$ km and α=100 m, path loss values shown in Figs. 2–4 were used to determine values of f. For 2 km cells, it is clear that the two-ray channels described by Figs. 2 and 3 have adjacent subscribers that provide stronger signal levels than those predicted by a d^4 path loss law. In addition, the proportion of deep faded regions within the adjacent 2 km cells is much greater than in 10 km cells. This means that a greater percentage of adjacent users in the smaller cells will need to provide greater transmitter power to overcome their own in-cell fades. While the channel represented in Fig. 4 offers frequency-reuse values virtually unchanged from those found for the same channel in a 10 km cellular system, it is readily seen that the channels portrayed by Figures 2 and 3 substantially degrade f for a 2 km system.

Fig. 10 shows the impact that close-in propagation has on frequency reuse. As seen in Fig. 2, there are many deep nulls within a 2 km radius from the base, and the first ring of subscribers will offer signal levels only 21 to 35 dB down from $P(\alpha)$. Fig. 10 shows how both of these factors lead to values of f (with three rings and uniform distribution of users in each cell) near 0.1 for the channel portrayed in Fig. 2. This is more than six times (8 dB) lower than the frequency reuse obtained using the exact same channel except with $d = 10$ km. Also, the channel portrayed in Fig. 2 provides f values that are 8 dB lower than results obtained for $d = 2$ km using the channel portrayed in Fig. 4. Fig. 10 shows the channel portrayed by Fig. 3 does not degrade f as severely, but the resulting frequency reuse factor (with three rings and uniform distribution of users in each cell) of 0.47 is 0.6 dB less than that provided by the same channel in a 10 km cell system. These results explicitly show that if CDMA is implemented in a propagation environment that has deep fades or shadowing within cells, but little isolation between adjacent cells, the frequency reuse efficiency can be severely degraded. It is worth emphasizing that the analytical channels used here are based on a particular two-ray model that uses antenna height as a parameter (see (3)). However, it should be clear that if some other physical mechanism, such as shadowing, provides the same path loss behavior as shown in Figs. 2–4, then the results in Fig. 10 will still hold, even if the physical propagation mechanisms and antenna heights are different than those which lead to (3).

From this work it becomes clear that accurate site-specific propagation models which could accurately predict the values of P_{ai} in (11a) are needed. Based on this analysis, it seems one of the major challenges that lies ahead for implementors of DS-CDMA is to strive to predict suitable base station locations that can exploit the propagation environment, so as to offer rapid path loss decay to adjacent cells while maintaining low path loss within a desired coverage area. As site-dependent propagation prediction tools are developed, it will become possible to incorporate an accurate path loss exponent that is a function of location. In addition, it should become possible to predict, *a priori*, base station placements that will exploit known shadowing properties of building structures to increase the isolation between cells. At such a time, analysis techniques such as the ones presented here will be useful for analysis, simulation, and automated installation of high-capacity CDMA systems.

VI. Conclusion

This paper has studied the frequency reuse efficiency of the reverse channel in a CDMA cellular radio system employing total frequency reuse. We considered two propagation models, both of which related path loss to the log of the T-R separation distance. One model assumed a path loss law of d^n , where n varied between 2 and 4, but was not a function of T-R separation. Experiments confirmed this range of n occurs in real channels. The other model was analytically based on the two-ray ground reflection model, and provided a path loss exponent that varied with distance between the transmitter and receiver. We showed that, even if the two ray model fails to explain the physical causes of propagation in typical urban mobile radio systems, it provides reasonable agreement to measured data, particularly within the range of 100 m to 2 km from the base antenna.

We developed an analysis technique for frequency reuse based upon the geometry of concentric circles. This geometry provides a very straightforward approach to solving for frequency reuse, and allows the use of scaler weighting factors to redistribute the location of users with any cell. In this paper, three different sets of weighting factors were used to provide average frequency reuse values that were bounded above and bounded below the result obtained for hexagonal cell sites that have a uniform distribution of users. We showed how the concentric circle geometry is related to the traditional mosaic of hexagonal cells.

General expressions for computing the frequency reuse factor for channels that possess distance-dependent path loss exponents were given and were used to solve for three specific channels. Expressions for computing frequency reuse in channels which have fixed path loss exponents were also given and used to solve for channels with exponents ranging between 2 and 4. For exponents that are not a function of location, the results showed that frequency reuse is highly dependent upon the propagation path loss exponent, particularly between $n = 2$ and $n = 3$. When three rings of adjacent users were considered (equivalent to 58 adjacent hexagonal cells), a uniform distribution of users yielded values of f that ranged from 0.197 when $n = 2$ to 0.541 when $n = 4$.

For distance-dependent path loss exponents, we found that, depending upon the isolation between cells, and the nature of fading or shadowing within the cell of interest, the frequency reuse factor can range between about 0.1 and 0.7. Such a wide range on frequency reuse suggests that accurate propagation prediction techniques are warranted for analysis, simulation, and expert system design of future CDMA personal communication systems.

Acknowledgment

The authors thank Morton Stern of Motorola, Inc. for discussions about this work.

References

[1] G. R. Cooper and R. W. Nettleton, "A spread-spectrum technique for high-capacity mobile communications," *IEEE Trans. Veh. Technol.*, vol. VT-27, pp. 264–275, Nov. 1978.

[2] K. S. Gilhousen, I. M. Jacobs, R. Padovan, and L. A. Weaver, Jr., "Increased capacity using CDMA for mobile satellite communication," *IEEE J. Select. Areas Commun.*, vol. 8, pp. 503–512, May 1990.

[3] K. S. Gilhousen *et al.* "On the capacity of a cellular CDMA system," *IEEE Trans. Veh. Technol.*, , vol. 40, pp. 303–312, May 1990.

[4] T S Rappaport, S. Y. Seidel, and R. Singh, "900 MHz multipath propagation measurements for U. S. digital cellular radiotelephone," *IEEE Trans. Veh. Technol.*, vol. 39, pp. 132–139, May 1990.

[5] S. Y. Seidel, T. S. Rappaport, S. Jain, M.L. Lord, and R. Singh, "Path loss, scattering, and multipath delay statistics in four European cities for digital cellular and microcellular radiotelephone," *IEEE Trans. Veh. Technol.*, vol. 40, pp. 721–730, Nov. 1991.

[6] A. J. Rustako, N. Amitay, G. J. Owens, R.S Roman, "Radio propagation measurements at microwave frequencies for microcellular mobile and personal communications," in *Proc. 1989 IEEE Int. Commun. Conf.*, Boston, MA, pp. 15.5.1–15.5.5.

[7] P. Harley, "Short distance attenuation measurements at 900 MHz and 1.8 GHz using low antenna heights for microcells," *IEEE J. Select. Areas Commun.*, vol. 7, pp. 5–11, Jan. 1989.

[8] S. Mockford, A. M. D. Turkmani, and J. D. Parsons, "Local mean signal variability in rural areas at 900 MHz," in *Proc. 1990 IEEE Veh. Technol. Conf.*, Orlando, FL, pp. 610–615.

[9] J. Holtzman, "Power control and its effects on CDMA wireless systems," 1990 WINLAB Workshop," Rutgers Univ., Piscataway, NJ, Oct. 1990.

[10] K. Bullington, "Radio propagation for vehicular communications," *IEEE Trans. Veh. Technol.*, vol. VT–26, pp. 295–308, Nov. 1977.

[11] T. S. Rappaport and L. B. Milstein, "Effects of path loss and fringe user distribution on CDMA cellular frequency reuse efficiency," in *Proc. IEEE GLOBECOM '90*, San Diego, CA, pp. 500–506.

[12] W. C. Y. Lee, *Mobile Communications Engineering* New York: McGraw-Hill, 1982.

[13] ——"Spectrum efficiency in cellular," *IEEE Trans. Veh. Technol.*, vol. 38, pp. 69–75, May 1989.

[14] K. L. Blackard, M-J. Feuerstein, T. S. Rappaport, S. Y. Seidel, and H. H. Xia, "Path loss and delay spread models as functions of antenna height for microcell system design," presented at 42nd IEEE Veh. Technol. Conf., Denver, CO, May 1992.

[15] L. B. Milstein, T. S. Rappaport, and R. Barghouti, "Performance evaluation for cellular CDMA," *IEEE J. Select. Areas Commun.*, vol. 10, pp. 680–689, May 1992.

Theodore S. Rappaport (S'83–M'84–S'85–M'87–SM'91) was born in Brooklyn, NY, on November 26, 1960. He received the B.S.E.E., M.S.E.E., and Ph.D. degrees from Purdue University, West Lafayette, IN, in 1982, 1984, and 1987, respectively.

In 1988 he joined the electrical engineering faculty of Virginia Polytechnic Institute and State University, Blacksburg, where is he is an Associate Professor and Director of the Mobile and Portable Radio Research Group. He conducts research in mobile radio communication system design, simulation, and RF propagation through measurements and modeling. He has authored or coauthored technical papers in the areas of mobile radio communications and propagation, vehicular navigation, ionospheric propagation, and wide-band communications.

Dr. Rappaport holds a U.S. patent for a wide-band antenna, and is co-inventor of SIRCIM, an indoor radio channel simulator that has been adopted by more than 50 companies and universities. In 1990 he received the Marconi Young Scientist Award for his contributions in indoor radio communications and was named an NSF Presidential Faculty Fellow in 1992. He is an active member of the IEEE Communications and Vehicular Technology Societies and serves as a Senior Editor of the IEEE JOURNAL ON SELECTED AREAS IN COMMUNICATIONS. He is a Registered Professional Engineer in the State of Virginia, and is a Fellow of the Radio Club of America. He is also President of TSR Technologies, a cellular radio and paging test equipment manufacturer.

Laurence B. Milstein (S'66–M'68–SM'77–F'85) received the B.E.E. degree from the City College of New York, New York, NY, in 1964, and the M.S. and Ph.D. degrees in electrical engineering from the Polytechnic Institute of Brooklyn, Brooklyn, NY, in 1966 and 1968, respectively.

From 1968 to 1974 he was employed by the Space and Communications Group of Hughes Aircraft Company, and from 1974 to 1976 he was a member of the Department of Electrical and Systems Engineering, Rensselaer Polytechnic Institute, Troy, NY. Since 1976 he has been with the Department of Electrical and Computer Engineering, University of San Diego, La Jolla, CA, where he is a Professor and former Department Chairman, working in the area of digital communication theory with special emphasis on spread-spectrum communications systems. He has also been a consultant to both government and industry in the areas of radar and communication.

Dr. Milstein was an Associate Editor for Communication Theory for the IEEE TRANSACTIONS ON COMMMUNICATIONS, an Associate Editor for Book Reviews for the IEEE TRANSACTIONS ON INFORMATION THEORY, and an Associate Technical Editor for the *IEEE Communications Magazine*. He has been on the Board of Governors of the IEEE Communications Society, and was the Vice President of Technical Affairs in 1990 and 1991. He is currently a member of the Board of Governors of the IEEE Information Theory Society, and is a member of Eta Kappa Nu and Tau Beta Pi.

Multipath Propagation Effects on a CDMA Cellular System

Norbert L. B. Chan, *Member, IEEE*

***Abstract*— The performance of the reverse link of a code division multiple access cellular system is evaluated. At the base station, the signal from each user is demodulated by a coherent BPSK RAKE receiver. Parameters for the model of the impulse response of the channel were taken from measurements of the digital cellular channel in Toronto. The signal-to-noise ratio (SNR) for a received signal is used to measure the performance of the reverse link. The variation in SNR of received signals at the base station should be as small as possible to reduce interference in the network. A power control scheme to lower the variation in SNR of the received signals is analyzed. The effects of lowering the bandwidth of the transmitted signal were also investigated.**

I. Introduction

SPREAD SPECTRUM signalling, which was primarily used in military applications, is now finding applications in mobile communications [1]. In a multipath environment, spread spectrum signalling allows one to make use of the multiple paths so that the energy in each path can be combined to provide a performance gain over a standard one path receiver. The RAKE receiver proposed by Price and Green [2] is a spread spectrum receiver that can demodulate multiple paths. In addition to dealing effectively with multipath, spread spectrum signalling allows for multiple access. This is done through CDMA, in which each user is assigned a code.

This paper investigates the effects of multipath propagation on a CDMA cellular system. Previous work by Qualcomm [3] includes an analysis of the capacity of a cellular CDMA system, taking into account path loss and shadowing.

In this paper, we focus on the performance of the mobile-to-base link. All mobiles and base stations use omnidirectional antennae for transmission. As in current CDMA designs, we assume all base stations transmit on the same frequency band and all mobiles transmit on another frequency band. These two frequency bands are assumed to be sufficiently spaced apart so that in the mobile-to-base link, a base station will only receive interference from other mobiles and not other base stations. We assume a BPSK direct sequence modulation scheme with coherent detection. At the base station, the signal from each user is demodulated by a coherent BPSK RAKE receiver. Parameters for our model of the impulse response of the channel are taken from impulse response measurements of the cellular channel in Toronto [4].

Manuscript received August 18, 1993; revised December 1, 1993. This research was supported in part by a contract from Bell Mobility Cellular and in part by a grant from the Information Technology Research Center of Ontario, ITRC.

The author is with Computing Devices Canada Limited, NE Calgary, Alberta, Canada T2E 8P2.

IEEE Log Number 9403808.

In a cellular system with many users, large sets of codes must be found that have the following two properties: 1) low cross-correlation to reduce multiple access interference, and 2) small minor autocorrelation peaks that attenuate rapidly as the delay from the main correlation peak increases, so that there is negligible interference from multiple received paths; this type of interference is termed interpath interference. We assume that the spreading codes are long and model them as random. The interference analysis of Sousa [5] for a BPSK signalling scheme in a noncellular context is extended to a multipath environment.

The RAKE receiver that we are considering is different from that of Turin [6] and Proakis [7]. Proakis uses a wide-sense stationary uncorrelated scattering (WSSUS) model of the channel, which may not be accurate in a mobile environment because the channel is not wide-sense stationary. Turin discusses a number of RAKE receivers which use a discrete time model of the channel. We consider a BPSK RAKE receiver that is provided with time delay estimates of the arriving paths, and it is only at these delays where the received signal is despread and combined in the maximal ratio sense. Such a RAKE receiver will only require a limited number of taps. We compare the performance of an optimum RAKE receiver that demodulates all received paths with a RAKE receiver that only demodulates the first few strongest paths.

The signal-to-noise ratio (SNR) is used as a measure of performance of the reverse link. The difference between the highest and lowest received SNR is called the variation in SNR and it is desired to have the variation in SNR as low as possible. The variation in SNR can be lowered through power control.

A power control scheme is essential for any CDMA system to operate efficiently. In a power control scheme, the transmitted power from all mobiles are adjusted such that all signals are received with equal power at the base station. Holtzman [8] considers the reduction in capacity of CDMA wireless networks if the power control scheme is not perfect and the received signal powers from the mobiles are not equal. In this paper, a perfect power control scheme is assumed and an attempt is made to verify whether power adjustment commands sent from the base station to the mobile are sent at a high enough rate to compensate for the effects of fading.

Finally, we consider the effects of reducing the bandwidth of the transmitted signal. Two paths that are separated by a small delay may become indistinguishable at the receiver and are demodulated as a single path because of the reduced bandwidth. The minimum time delay between received paths

0018-9545/94$04.00

required at the receiver before the paths can be distinguished is called the time resolution. We compare the variation in SNR of RAKE receivers at different bandwidths.

The paper is organized as follows. Section II briefly describes the procedures used for the impulse response measurements taken in the Toronto area. Section III considers the performance of a BPSK RAKE receiver. Section IV considers the effect of reducing the bandwidth of the transmitted signal. A peak detection algorithm is used to find the new paths at a lower bandwidth. Section V investigates the variation of the received signal over small distances and provides a rough calculation to see if present power control schemes can react to fast fading. Finally, Section VI provides the major conclusions.

II. Impulse Response Measurements

We use data from impulse response measurements of the cellular channel in Toronto. The measuring procedures are described in greater detail in [4]. The transmitter was placed close to existing base stations to simulate real propagation conditions. The receiver was placed in a van, which followed a planned route, or "run." In each run, a specific number of measurements were taken. At each measurement location, the impulse response was measured while the van was stationary. The measurements were performed during the day, with normal traffic conditions.

The measurements were first processed to find the number of received paths using a technique called thresholding. Multipath echoes above a set threshold were declared to be valid echoes. For the results presented in this paper, we choose the same threshold level that is used in [11].

A sliding correlator at a 910 MHz carrier frequency was used for the measurements. The PN code chip rate was 10 MHz and the code length was 511 chips, allowing for a time resolution to resolve multipath components of 0.1 μs. Two types of measurements will be analyzed in this paper. One type is called survey measurement; measurements were taken at a nonuniform spacing of 30–300 feet in a run. Analysis of the survey measurements will provide insight into the long-term variation of the received signal. The other type of measurement is the track measurement; the van was placed on a track 20 feet long and measurements were taken every two inches. Analysis of the track measurements will provide insight into the short-term variation of the signal.

For each measurement, the ideal impulse response will be of the form

$$h(t) = \sum_{i=0}^{n} a_i \exp(\jmath\theta_i)\delta(t - \tau_i) \qquad (1)$$

where $n + 1$ is the number of received paths, and a_i, τ_i, and θ_i are the amplitude, time delay and phase of the i-th path, respectively. The parameters a_i, θ_i and τ_i of (1) could be modeled by statistical distributions. However, the more accurate the modeling, the more complicated the resulting analysis. In this paper, the parameters, a_i, θ_i and τ_i are taken from the processed measurements. As well, since the measurements were taken of the forward link, we assume that the propagation conditions are the same on the reverse link, although there is a 45 MHz separation between the two frequency bands. An example of a survey measurement is the run along Bay Street which is typical of an urban environment, shown in the map of Toronto in Fig. 1.

The impulse response given by (1) may change depending on the time of day and the existing traffic conditions. Another type of measurement which was performed and not analyzed in this paper is the temporal measurement; three separate impulse response measurements at the same location were taken over a duration of two minutes. The temporal measurements were taken to see if the impulse response would change over time. In most cases, the impulse response was found not to change drastically [4], and we assume the impulse response of (1) is an accurate representation of the channel.

III. Received Signal and RAKE Receiver Model

With multipath propagation, the received BPSK signal for a single user is given by

$$r(t) = \sum_{i=0}^{n} a_i d(t - \tau_i) c(t - \tau_i) \cos(\omega_c t + \theta_i) + n(t) \qquad (2)$$

where $n + 1$ is the number of received components and a_i, τ_i and θ_i are the signal amplitude, time delay and phase, respectively, of the i-th component, $c(t)$ is the spreading code, and $d(t)$ is the binary data. The data, $d(t)$ is given by

$$d(t) = A \sum_{k=-\infty}^{\infty} d_k h(t - kT) \qquad (3)$$

where T is the data symbol duration, $\{d_k\}$ is a data sequence with equal probable values equal to ± 1, $h(t)$ is a rectangular pulse with unit amplitude, and A is the amplitude of the data signal. The data symbols $\{d_k\}$ are assumed to be random to model a general binary data stream from the transmitter. The spreading code, $c(t)$ is given by

$$c(t) = \sum_{k=-\infty}^{\infty} c_k h_c(t - k\tau) \qquad (4)$$

where τ is the chip period, $\{c_k\}$ is the spreading code sequence with equal probable values of ± 1, and $h_c(t)$ is the chip pulse shape. The carrier frequency is ω_c; the phase difference of the i-th path, θ_i, from the carrier is due to propagation delay and Doppler shift. Finally, $n(t)$ is additive white Gaussian noise (AWGN) with power spectral density $\frac{N_0}{2}$.

At the base station, the coherent BPSK RAKE receiver of Fig. 2 is used to demodulate the received signal of (2). The weights, w_i ($i = 0$ to n) are set depending on what type of combining is used. The synchronization circuits, time delay, and phase estimation circuits are not shown. We assume perfect time delay and phase estimates. However, in a fast fading environment, there may be significant error in the phase estimates. A DPSK RAKE receiver used for indoor multipath channels is discussed in [9].

The signal-to-noise ratio (SNR) at the output of the RAKE receiver is given by

$$SNR = \frac{(\sum_{i=0}^{n} w_i s_i)^2}{\sum_{i=0}^{n} w_i^2 E(n_i^2)}, \qquad (5)$$

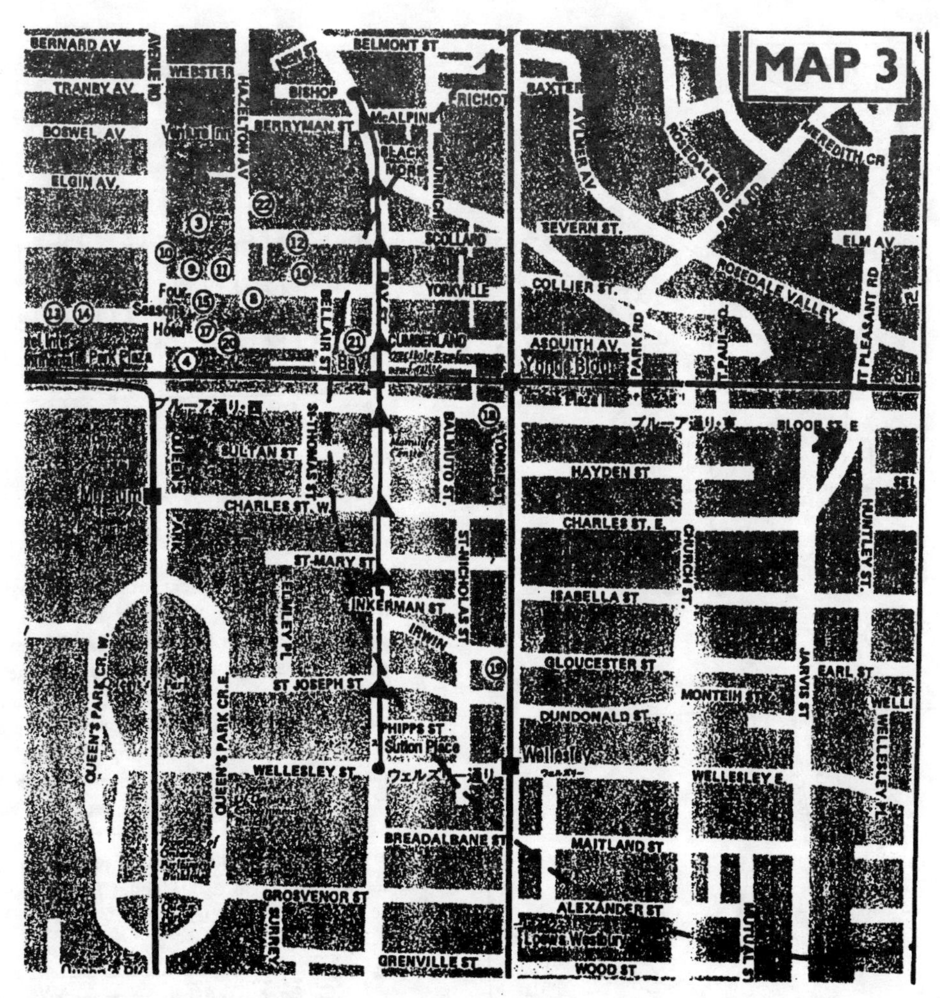

Fig. 1. The route taken by the van for the run along Bay Street. The dotted line displays the theoretical coverage of the cell site which has a radius of 1 km, the transmitter is denoted by the "X" and the solid line displays the route taken by the van.

where s_i and n_i are the signal and noise component, respectively, of the output of the integrator of the i-th branch of the RAKE receiver, and $E(\cdot)$ denotes expectation. Using maximal ratio combining, the optimum weights are derived in Schwarz *et al.* [13] and are given by

$$w_k = C\frac{s_k}{E(n_k^2)}, \tag{6}$$

where C is an arbitrary constant. In maximal ratio combining, accurate estimates of s_k and $E(n_k^2)$ for $k = 0$ to n are required. It is assumed the environment is a slowly fading one, so that the terms s_k and $E(n_k^2)$ do not greatly change over a series of received bits. The estimates of s_k and $E(n_k^2)$ can therefore be found by averaging over the series of past received bits. Letting $C = 1$ and substituting (6) into (5) results in an SNR of

$$SNR = \sum_{i=0}^{n} \frac{s_i^2}{E(n_i^2)}, \tag{7}$$

which is the sum of the individual SNR's of each branch of the RAKE receiver. For the k-th branch, the signal power is given by

$$s_k^2 = a_k^2 T^2 \tag{8}$$

and the noise power is given by

$$E(n_k^2) = \frac{T^2}{3N}\sum_{\substack{i=0\\ i\neq k}}^{n} a_i^2 + N_0 T + \frac{T^2}{3N}(MAI) \tag{9}$$

using the interference analysis of Sousa [5]. Here, N is the processing gain and MAI is the noise power from other users, called multiple access interference.

In this paper, we set $N_0 = 0$ and assume, as in a loaded cell, that the MAI is much greater than the interpath interference, or

$$\frac{T^2}{3N}(MAI) \gg \frac{T^2}{3N}\sum_{\substack{i=0\\ i\neq k}}^{n} a_i^2. \tag{10}$$

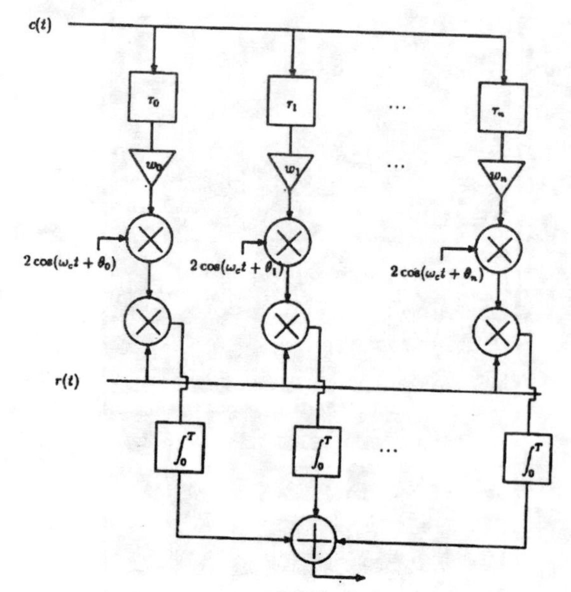

Fig. 2. A BPSK RAKE receiver.

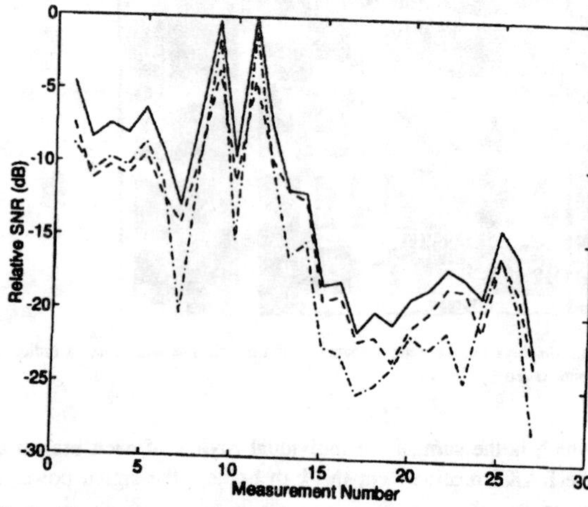

Fig. 3. The solid line shows the SNR of a RAKE receiver with maximal ratio combining using (7). The dotted line shows the SNR of a RAKE receiver with equal gain combining using (5) with $w_i = 1$ for all i. The dash–dot line shows the SNR of a standard receiver using (5), with $w_p = 1$, where the p-th path is the strongest and all other weights are set to zero.

A comparison of maximal ratio combining (MRC), equal gain combining, and a standard receiver using the Bay Street run is shown in Fig. 3. The SNR values are normalized to the highest received SNR in the run. This plot can be interpreted as the received SNR at the base station with no power control. In Fig. 3, maximal ratio combining gives the best performance. The standard receiver will perform better than equal gain combining if most of the signal power is contained in one path and the rest of the paths contain little signal power, because in equal gain combining, the paths with small signal power contribute more noise, lowering the SNR.

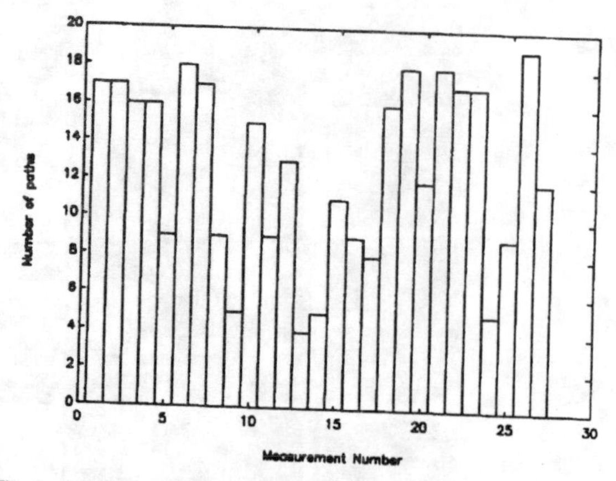

Fig. 4. Distribution of the number of received paths on the Bay Street run at a bandwidth of 10 MHz. The threshold for the paths is the same threshold used in [4].

A bar chart of the number of received paths for each measurement of the Bay Street run is shown in Fig. 4. In some instances, there can be as many as 19 received paths. In practice, it is not always worthwhile to combine all received paths, because many correlators would be required and some paths may contain very little signal power. In the next section, we consider the effects of lowering the bandwidth of the transmitted signal, and hence the number of paths. We will also consider the performance of RAKE receivers with different number of taps. We refer to a RAKE receiver that combines L of the total received paths as an L paths RAKE receiver.

For multiple users in a one cell CDMA system, we can write the SNR of the kth user as

$$SNR_k = \sum_{i=0}^{n_k} \frac{(a_{ki}T)^2}{\frac{T^2}{3N}\sum_{\substack{b=0\\b\neq i}}^{n_k} a_{kb}^2 + \frac{No}{T} + \frac{T^2}{3N}\sum_{\substack{c=1\\c\neq k}}^{K}\sum_{l=0}^{n_c} a_{cl}^2}, \tag{11}$$

where $n_i + 1$ is the number of received paths for the ith user, and a_{ki} is the signal strength of the kth user on the ith path. To study the effects of multiple users, extra users can be modeled by the different measurements along a run. For example, measurement 1 of the Bay Street run can be taken to be the user of interest, while measurements 2–27 can be taken to be interfering users that are transmitting simultaneously to the same base station. A histogram of the relative SNR of any one measurement in the Bay Street run versus all the remaining measurements is shown in Fig. 5. The variation in SNR is approximately 25 dB. The variation in SNR signifies the need for power control, in which case the variation in SNR can be reduced.

For the impulse response of (1), the received power P_R is proportional to the sum of the individual powers on each path, or

$$P_R \propto \sum_{i=0}^{n} a_i^2. \tag{12}$$

To have a particular user received at a desired power level P_D, the base station transmits power adjustment signals to the user

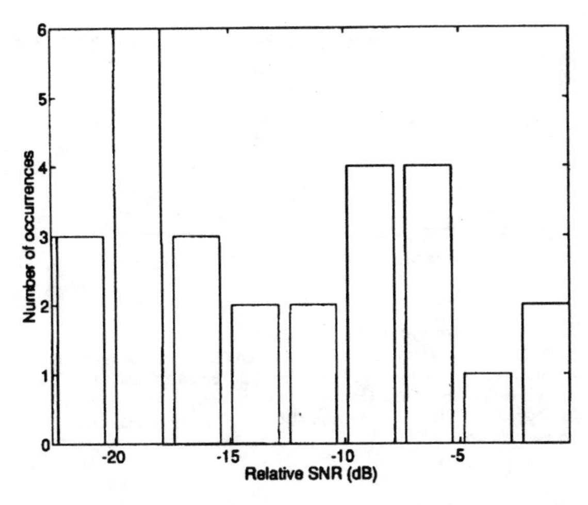

Fig. 5. Histogram of the SNR's of the received signals, where the SNR of the k-th measurement is given by (11).

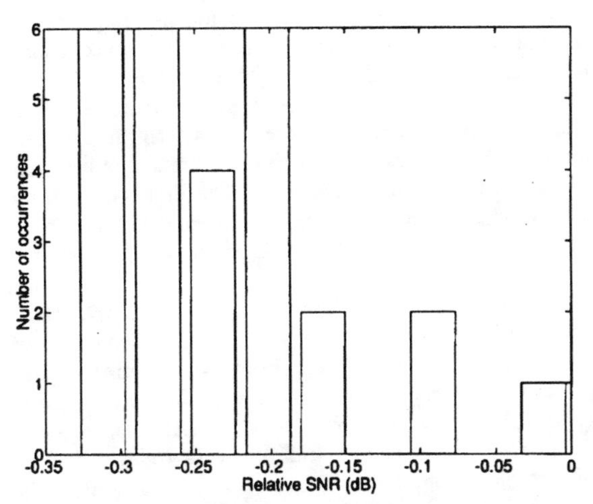

Fig. 6. Histogram of the SNR's of the received signals with power control, where the SNR of each measurement is given by (11).

if the received power level deviates from P_D. The transmitted power of the user will be adjusted so that the impulse response of (1) becomes

$$h_n(t) = \sum_{i=0}^{n} a_i \sqrt{\frac{P_D}{P_R}} \delta(t - \tau_i) \exp(j\theta_i). \tag{13}$$

Using this power control scheme, a histogram of the relative SNR of any one user versus all other users acting as interference is shown in Fig. 6. The variation in SNR is approximately 0.35 dB, and is nonzero due to the interpath interference.

With an L paths RAKE receiver, the power control scheme will not result in all users being received at a power level of P_D because only L paths can be tracked at a time. The power in the other paths not being tracked will also change, which could cause interference to the network. We make the assumption that power control is based on the total power received in the L paths that are being tracked.

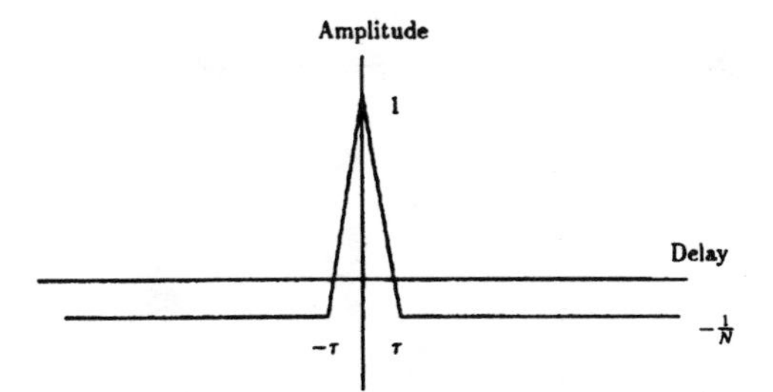

Fig. 7. The impulse response $w(t)$ of the band limiting filter used to model the reduction of the bandwidth of the transmitted signal.

IV. Reduction of Transmitted Signal Bandwidth

We modeled each of the 10 MHz impulse responses from the measurements as if they were performed at an infinite bandwidth, and represented them by (1). However, when the bandwidth of the transmitted signal is decreased, the time resolution is reduced and paths will interfere with each another. If the bandwidth of the transmitted signal is lowered, the impulse response of the channel can be modeled by

$$h(t) * w(t) = \sum_{i=0}^{n} a_i \exp(j\theta_i) w(t - \tau_i) \tag{14}$$

where $w(t)$ is the impulse response of an appropriate band limiting filter. We choose $w(t)$ to be the autocorrelation function of a maximal length linear shift register sequence (m-sequence), as shown in Fig. 7, which corresponds to using the sliding correlator at a lower chip rate to make the measurement at a lower bandwidth. Since we are using random codes, we assume that the side peaks of the autocorrelation of a random code have small amplitudes which can be neglected. We also assume that correlating over a partial period of the sequence does not lead to significant side peaks.

A typical impulse response at a bandwidth of 10 MHz given by (1) is shown in Fig. 8. If the bandwidth of the transmitted signal is reduced to 5 MHz, the new time resolution, τ_{new}, is 0.2 μs, and using (14), the reduced bandwidth impulse response of Fig. 8 is shown in Fig. 9. The new time resolution is related to the bandwidth B by

$$\tau_{new} = \frac{1}{B}. \tag{15}$$

At a reduced bandwidth, the processing gain N will also be reduced.

The problem that remains is how to find the paths in Fig. 9. The paths are most likely to be the "peaks" in the impulse response. A peak detection algorithm is used to find the new paths. However, samples of the impulse response in Fig. 9 must be separated by at least τ_{new} so that the noise is not correlated from sample to sample, and there would be no gain in SNR due to the combining. The peak detection algorithm finds peaks from largest to smallest with the constraint that samples must be separated by at least τ_{new}. The algorithm is set to stop when the number of detected peaks is equal to the number of paths in the original 10 MHz impulse response. Generally, there will be less observable peaks at a lower bandwidth, but the extra peaks found at the lower bandwidth

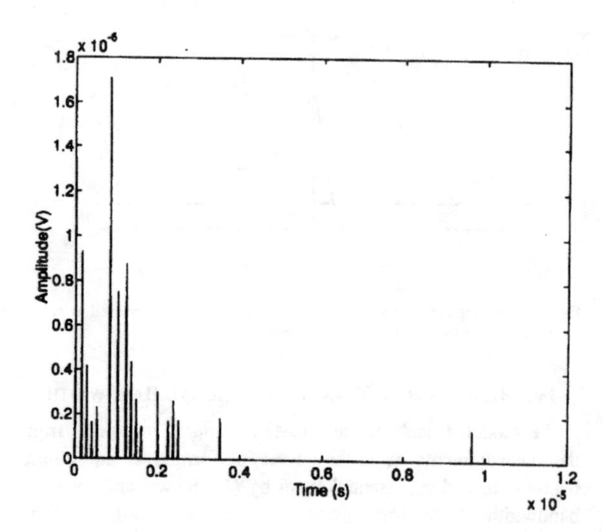

Fig. 8. Magnitude of typical impulse response at a bandwidth of 10 MHz.

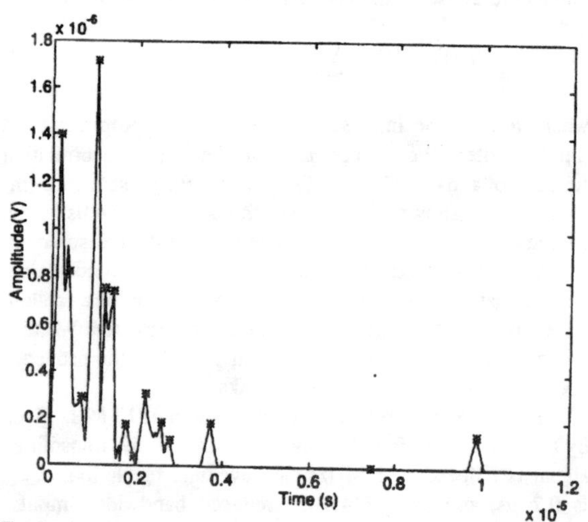

Fig. 9. 5 MHz impulse responses with detected peaks. The *'s represent the amplitudes of the detected peaks.

will have very small amplitudes and when maximal ratio combined, contribute very little signal power. In Fig. 9, the *'s represent the amplitudes of the detected peaks which are input to the RAKE combiner. At a lower bandwidth, there will be a smaller number of detected paths. If there are $k+1$ detected paths, the SNR for a RAKE receiver using maximal ratio combining is given by

$$SNR = \frac{3N}{MAI}\sum_{i=0}^{k} a_i^2, \quad (16)$$

which is proportional to the received power.

For the survey measurements, to obtain a statistical distribution of impulse responses, each impulse response measurement in a particular run was used to generate a family of impulse responses (generally 3–5 new impulse responses), because the number of measurement points in a run were small. Each

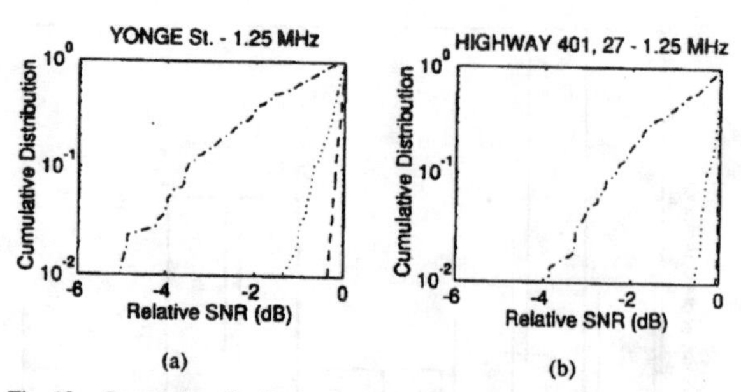

Fig. 10. Cumulative distribution of SNR's on Yonge Street survey measurement in Bond Place cell site using a RAKE receiver that combines the five strongest received paths (dashed line), three paths (dotted line) and one path (dash-dot line), using (16) with power control scheme. The cumulative distribution of SNR's on Highways 401 and 27 in the Widdicombe cell site are also shown.

family of impulse responses would have the same signal amplitudes, a_k, and time delays, τ_k of the original impulse response, but phases would be randomly generated uniformly from 0 to 2π for each path in the impulse response. The random phases will show the degree of phase cancellation of the received signal as the bandwidth is reduced.

With the power control scheme discussed in Section III, the cumulative distribution of the SNR's for different RAKE receivers for a typical dense urban environment at bandwidth of 1.25 MHz is shown in Fig. 10(a). This plot is taken from a survey run in Yonge Street. The normalized SNR's are ordered from smallest to largest on the x-axis, and the fraction of the total number of impulse responses, or cumulative distribution, is plotted on the y-axis. The graph in Fig. 10(b) depicts a typical suburban environment, which is taken from a survey run on Highways 401 and 27. Because we assume a perfect power control scheme, the cumulative distribution of the all-paths RAKE receiver would be a vertical line, which we normalize to 0 dB, and coincides with the right vertical axis of each plot. From these plots, a rough estimate of the fading margin, or increase in SNR required to achieve a desired SNR level, can be determined. For example, on Yonge Street, the three-paths RAKE receiver performs approximately 1.2 dB worse than the all-paths RAKE receiver at a bandwidth of 1.25 MHz. It was observed in [10] that the fading margin for a three-paths RAKE receiver is generally less than 1.8 dB at a bandwidth of 1.25 MHz. The fading margin is generally smaller for suburban and rural areas.

V. Short-Term Variation of Received Signal

The track measurements, which give similar information as the survey measurements in terms of the cumulative distribution of SNR's, also give insight into how the received signal varies with small changes in distance. For the track measurements, the phases, θ_k were taken directly from the data. Rather than show all the results of all the measurements, we choose two measurements to illustrate some ideas. Using the King Street track measurement as a typical dense urban environment, the SNR of a RAKE receiver at a bandwidth of 10 MHz using (16) is shown in Fig. 11. Fig. 11(b) shows the SNR's at bandwidth of 1.25 MHz.

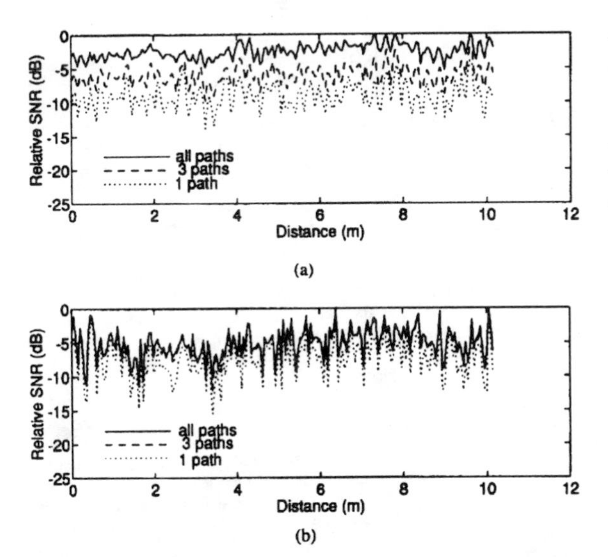

Fig. 11. Relative SNR's of King Street track measurement using a RAKE receiver that combines all received paths, three paths and one path, using (16) at a bandwidth of 10 MHz (a) and 1.25 MHz (b), with no power control.

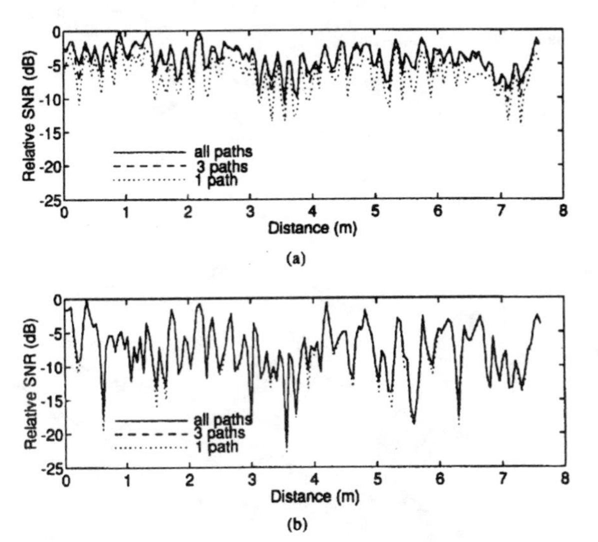

Fig. 12. Relative SNR's of Hawsberry Street track measurement using a RAKE receiver that combines all received paths, three paths and one path, using (16) at a bandwidth of 10 MHz (a) and 1.25 MHz (b), with no power control.

The track measurement on Hawsberry Street is used to show some features of a typical suburban environment. The SNR of a RAKE receiver at a bandwidth of 10 MHz is shown in Fig. 12. Fig. 12(b) shows the SNR's at a bandwidth of 1.25 MHz.

In both Figs. 11 and 12, as the bandwidth is reduced, with no power control scheme, the variation in SNR increases. Another important observation to make is that at a bandwidth of 1.25 MHz, a three-paths RAKE receiver performs almost as well as an all-paths RAKE receiver. Generally, at a lower bandwidth, there will be a smaller number of resolvable paths and greater degree of phase cancellation, resulting in a greater variation in the received signal power. The variation in received power is also smaller if the RAKE receiver tracks more paths.

For their power control scheme, Qualcomm proposes that power adjustment signals are sent from the base station to the mobile every 1.25 ms [14]. As well, the rate of increase of transmitted power level is limited to 1 dB every 1.25 ms. (The power level can generally be decreased much quicker through an open loop power control scheme.) To determine if this rate is adequate enough in the Toronto environment, a rough calculation can be performed as follows. We note that measurements are spaced apart by two inches or 0.0508 m. If we assume a maximum vehicle speed of 50 km/h or 13.89 m/s, then in 1.25 ms, the mobile would have moved 0.01736 m. Within a span of 0.0508 m, the distance between measurements, $0.0508 / 0.01736 = 2.93 \approx$ three power adjustment signals can be made, allowing for a 3 dB increase in power. However, in Fig. 12, the signal power can drop by as much as 12 dB over a distance of two inches. A fading margin may be required for the power control scheme to react fast enough to signal changes due to multipath fading. Turin [12] has made the observation that it would be difficult for power control schemes to react fast enough to multipath fading.

VI. Conclusions

In this paper, we have considered the performance of the reverse link of a CDMA cellular system. The reduction of the transmitted signal bandwidth is investigated and a rough calculation procedure is proposed to evaluate the effectiveness of power control schemes.

With a reduction in the bandwidth of the transmitted signal, with no power control, there will be a wider variation in the SNR of the received signals at the base station because there will be less resolvable paths and a greater degree of phase cancellation. With a power control scheme at a bandwidth of 1.25 MHz, a three-paths RAKE receiver performs, at most, 1.8 dB worse than an all-paths RAKE receiver in the environments observed. This also means that most of the power is contained in the strongest three paths at a bandwidth of 1.25 MHz. The importance of this is that CDMA systems with complexity constraints will not be operating at a significant disadvantage, if there is an adequate power control scheme. However, lower bandwidth CDMA systems will suffer a performance loss compared to higher bandwidth CDMA systems.

It would be difficult at this time to estimate the effectiveness of the Qualcomm power control scheme without detailed information about its operation. However, a rough calculation procedure has been proposed in this paper which takes into account the speed of the mobile when evaluating present power control schemes.

Acknowledgment

The author would like to thank Professor E. S. Sousa of the University of Toronto for suggesting the research topic and many helpful discussions. The many discussions with Dr. V. M. Jovanović regarding the experimental data is also gratefully acknowledged. In addition, the author would like to thank the anonymous reviewers for their suggestions and comments.

REFERENCES

[1] R. L. Pickholtz, L. B. Milstein and D. L. Schilling, "Spread spectrum for mobile communications," *IEEE Trans. Veh. Technol.*, vol. 40, pp. 313–322, May 1991.

[2] R. Price and P. E. Green, Jr., "A communication technique for multipath channels," *Proc. IRE*, vol. 46, pp. 555–570, Mar. 1958.

[3] K. S. Gilhousen, I. M. Jacobs, R. Padovani, A. J. Viterbi, L. A. Weaver, Jr. and C. E. Wheatley, III, "On the capacity of a cellular CDMA system," *IEEE Trans. Veh. Technol.*, vol. 40, pp. 303–312, May 1991.

[4] E. S. Sousa, V. M. Jovanović and C. Daigneault, "Delay spread measurements for the digital cellular channel in Toronto," *IEEE Trans. Veh. Technol.*, vol. 43, no. 4, pp. 837–847, Nov. 1994.

[5] E. S. Sousa, "Interference modeling in a direct sequence spread spectrum packet radio network," *IEEE Trans. Commun.*, vol. 38, no. 9, pp. 1475–1482, Sept. 1990.

[6] G. L Turin, "Introduction to antimultipath techniques and their application to urban digital radio," *Proc. IEEE*, vol. 68, pp. 328–353, Mar. 1980.

[7] J. G. Proakis, *Digital Communications*, 2nd ed. New York: McGraw-Hill, 1989, p. 270.

[8] J. M. Holtzman, "CDMA power control for wireless networks," presented at the 1991 Proc. 2nd WINLAB Workshop, Rutgers Univ., Piscataway, NJ.

[9] U. Grob, A. L. Welti, E. Zollinger, R. Küng and H. Kaufmann, "Microcellular direct-sequence spread spectrum radio system using N-path RAKE receiver," *IEEE J. Select. Areas Commun.*, vol. 8, pp. 772–780, June 1990.

[10] N. Chan, "The effects of multipath propagation on a code division multiple access cellular system," M.A.Sc. thesis, Univ. of Toronto, Apr. 1993.

[11] E. S. Sousa and V. M. Jovanović, "Delay spread measurements for the digital cellular channel in Toronto," presented at the 3rd IEEE Symp. on Personal, Indoor and Mobile Radio Communications, Boston, MA, Oct. 1992.

[12] G. L. Turin, "The effects of multipath and fading on the performance of direct sequence spread CDMA systems," *IEEE Trans. Veh. Technol.*, vol. VT-33, pp. 213–219, Aug. 1984.

[13] M. Schwarz, W. R. Bennet and S. Stein, *Communication Systems and Techniques*. New York: McGraw-Hill, 1966, ch. 10.

[14] "An overview of the application of code division multiple access to digital cellular systems and personal cellular networks," Qualcomm, p. 36, May 1992.

Norbert L. B. Chan (S'90–M'94) was born in Fredericton, New Brunswick, Canada, on August 3, 1969. He received the B.Sc. degree from Queen's University in 1991 and the M.A.Sc. degree from the University of Toronto in 1993, both in electrical engineering.

He is currently working in the Communications Systems Division at Computing Devices Canada Limited, where he is working on reducing the interference between frequency hopping radios mounted on a vehicle. His research interests include CDMA cellular systems and adaptive interference cancellers.

Analysis of a Direct-Sequence Spread-Spectrum Cellular Radio System

Chamrœun Kchao, *Student Member, IEEE,* and Gordon L. Stüber, *Member, IEEE*

Abstract—**The uplink and downlink performance of a digital cellular radio system that uses direct sequence code division multiple access is evaluated. Approximate expressions are derived for the area averaged bit error probability while accounting for the effects of path loss, log-normal shadowing, multipath-fading, multiple-access interference, and background noise. Three differentially coherent receivers are considered; a multipath rejection receiver, a RAKE receiver with predetection selective diversity combining, and a RAKE receiver with postdetection equal gain combining. The RAKE receivers are shown to improve the performance significantly, except when the channel consists of a single faded path. Error correction coding was also shown to substantially improve the performance, except for slowly fading channels.**

I. Introduction

RECENTLY, there has been increased interest in the commercial applications of spread-spectrum signaling. Direct sequence (DS) code division multiple access (CDMA) has been proposed for enhancing the capacity of the North American cellular mobile telephone system [1]. This paper evaluates the uplink and downlink performance of a cellular radio system that employs DS CDMA. The analysis accounts for the effects of path loss, log-normal shadowing, multipath-fading, multiple-access interference, and background noise. It is intended to illustrate some of the typical characteristics of a DS CDMA cellular radio system, and demonstrate how the performance depends on certain aspects of the propagation environment. The analysis is restricted to a single isolated cell. Additional results for a multiple cell mobile radio environment are available in [32].

A very large body of literature exists on the analysis of CDMA cellular systems. Much of this literature, such as the initial work by Cooper and Nettleton [2], considers frequency-hopped CDMA. Here we discuss some of the existing literature on the analysis of DS CDMA cellular systems.

Most analyses of DS CDMA have resulted in Gaussian approximations, and various upper and lower bounds on the bit error probability for additive white Gaussian noise (AWGN) channels [3]–[7]. For specular multipath-fading channels, Geraniotis and Pursley have approximated the bit error probability of single user direct-sequence systems that use coherent [8] and noncoherent [9] detection. This analysis has been extended to DS CDMA for a multipath rejection receiver [10]. A few attempts have been made to evaluate the performance of a multipath combining receiver. Lehnert and Pursley [11], [12] have concluded that a multipath combining receiver can overcome the increased effect of multiple-access interference due to multipath, provided that the receiver has knowledge of the amplitudes, delays, and phases of the multiple received signal replicas.

Lam and Steele [13] have evaluated the performance of multipath combining receivers by simulation. For this purpose, they used the frequency-selective, slowly fading, multipath channel model developed by Hashemi [14]. A channel sounding technique was studied that allowed the receiver to lock onto and combine the strongest M signal paths. Despite the use of channel sounding, multipath diversity, and antenna diversity, it was reported that a processing gain of 31 could only support three simultaneous transmissions at a bit error rate less than 10^{-2}. In contrast, the same system could support twelve simultaneous transmissions on an AWGN channel.

Turin [15] has also estimated the maximum number of allowable users for a DS CDMA system that uses differentially coherent multipath combining receivers. For an AWGN channel, it was reported that the maximum number of allowable users is about 10–20% percent of the processing gain for a bit error rate of 10^{-5}–10^{-3}. However, for an urban multipath-fading channel without power control, this figure drops to 1–5% even with ideal multipath combining receivers. Xiang [16] also observed the same effect. If the channels consist of a single faded link then DS CDMA may be unusable, a conclusion also reached by Gardner and Orr [17]. Power control will only partially recover the reduction in capacity due to multipath-fading, because power control cannot compensate for rapid signal fluctuations caused by multipath-fading [15]. Finally, Rappaport and Milstein [18], [19] have examined the effect of path loss and user distribution on the performance of a DS CDMA cellular system. They showed that reuse efficiency can range from a high of 0.68 for a fourth law path loss exponent to a low of 0.42 for a square law path loss exponent.

This paper differs from existing literature in several respects. It differs from Rappaport and Milstein [18], [19] by concentrating on the effects of multipath-fading rather than path loss. It differs from Turin [15] by using a different method of analysis, a more extensive treatment of power control, and a discussion of both the uplink and downlink performance. Finally, it differs from Xiang [16] by using a different channel model. Xiang

Paper approved by L. J. Cimini, Jr., the Editor for Mobile Communications of the IEEE Communications Society. Manuscript received October 31, 1990; revised July 19, 1991. This work was supported by Bell South Enterprises, Inc. This paper was presented in part at the 1991 IEEE Military Communications Conference, McLean, VA, November 4–7, 1991.

The authors are with the School of Electrical Engineering, Georgia Institute of Technology, Atlanta, GA 30332.

IEEE Log Number 9211342.

models the channel as a random number of discrete Nakagami faded multipaths. We model the channel as consisting of a continuum of multipaths that results in a fixed number of resolvable Rayleigh faded paths.

The remainder of this paper is organized as follows. Section II discusses the system and channel model. The uplink performance is evaluated in Section III for three differentially coherent receivers; a multipath rejection receiver, a RAKE receiver with predetection selective diversity combining, and a RAKE receiver with postdetection equal gain combining. Power control and error correction coding are also considered. Section IV discusses the downlink performance for a multipath rejection receiver and a RAKE receiver with predetection selective diversity combining.

II. System and Channel Model

The system consists of a single isolated circular cell of radius R with a centrally located base-station. There are K mobile transceivers that are using direct-sequence CDMA to establish a full-duplex channel with the base-station. The mobile transceivers are assumed to be uniformly distributed throughout the cell area.

Propagation at UHF/VHF frequencies used in cellular mobile radio systems is largely influenced by path loss, shadowing, and multipath-fading. Each of these phenomenon is caused by a different underlying physical principle, and each should be accounted for when evaluating the performance of a cellular system.

A. Multipath-Fading

Mobile radio channels are effectively modeled as a continuum of multipath components. The equivalent low-pass channel has the time-variant impulse response [20]

$$c(\tau;t) = \alpha(\tau;t)\exp\{-j2\pi f_c\tau\} \tag{1}$$

where f_c is the carrier frequency. A widely used model for multipath-fading channels is the wide sense stationary uncorrelated scattering (WSSUS) model where the low-pass impulse response $c(\tau;t)$ is a complex Gaussian random process having the autocorrelation

$$\frac{1}{2}E[c^*(\tau_1;t)c(\tau_2;t+\Delta t)] = Q(\tau_1,\Delta t)\delta(\tau_1-\tau_2). \tag{2}$$

The function $Q(\tau) \equiv Q(\tau,0)$ is called the multipath intensity profile (MIP), and gives the average power output of the channel as a function of the time delay τ. The MIP can assume various forms, but quite often a mobile radio channel is well characterized by the exponential MIP [21]

$$Q(\tau) = \psi\exp\{-\tau/T_m\}, \quad 0 \le \tau \le T_{\max}. \tag{3}$$

This same model was the outcome of the European COST207 multipath propagation study and was subsequently included in the specifications for the GSM system [22]. In our analysis it is assumed that all channels have the same MIP, but in reality every channel will have a distinct MIP.

The low-pass impulse response of the channel is commonly modeled as a tapped delay line with a tap spacing equal to T_c [23]–[26], i.e.,

$$c(\tau;t) = \sum_{n=-\infty}^{\infty} c_n(t)\delta(\tau - nT_c) \tag{4}$$

where the $\{c_n(t)\}$ are the tap gains. For a continuum of multipath components, the $c_n(t)$ are complex Gaussian random processes. For a WSSUS channel the $c_n(t)$ are uncorrelated and, hence, independent. If the $c_n(t)$ have zero-mean, then the magnitudes $|c_n(t)|$ are Rayleigh distributed at any time t. Since the total multipath spread is $T_{\max}$, the tapped delay line can be truncated at $L = \lfloor T_{\max}/T_c \rfloor + 1$ taps, where $\lfloor x \rfloor$ is the largest integer contained in x. For a slowly varying channel $c_n(t) = c_n$ for the duration of several data bits.

Assuming the above tapped delay line model, the instantaneous received bit energy-to-background noise ratio is

$$\lambda = \sum_{n=0}^{L_i-1} \lambda_n \tag{5}$$

where $\lambda_n = |c_n|^2\lambda$ and $\lambda \triangleq E_b/N_o$ is the received bit energy-to-background noise ratio in the absence of multipath-fading. If the $|c_n|$ are Rayleigh distributed, then the λ_n have the exponential probability distribution function (pdf)

$$f_{\lambda_n}(\lambda_n) = \frac{1}{\Lambda_n}\exp\left\{-\frac{\lambda_n}{\Lambda_n}\right\} \tag{6}$$

where

$$\Lambda_n = E\left[|c_n|^2\right]\lambda. \tag{7}$$

The Λ_n are related to the MIP. From (3), the average power output of the channel at delay nT_c is

$$Q(nT_c) = \psi\exp\{-n/\varepsilon\} \tag{8}$$

where $\varepsilon = T_m/T_c$ is the delay spread relative to a chip duration. Following the development in [26], suppose that $Q(nT_c + \tau) \approx Q(nT_c)$ for $0 \le \tau \le T_c$, meaning that the channel impulse response is stationary for delay intervals of length T_c. Then the average received bit-energy-to-background noise ratio associated with the nth path is

$$\Lambda_n = T_cQ(nT_c)\lambda. \tag{9}$$

Taking the expectation of both sides of (5), and using (8) and (9) gives

$$\Lambda = \sum_{n=0}^{L-1}\Lambda_n = \lambda T_c\psi\frac{1-\exp\{-L/\varepsilon\}}{1-\exp\{-1/\varepsilon\}} \tag{10}$$

where Λ is interpreted as the total average received bit energy-to-background noise ratio. It follows that

$$\Lambda_n = \frac{(1-\exp\{-1/\varepsilon\})\exp\{-n/\varepsilon\}}{1-\exp\{-L/\varepsilon\}}\Lambda. \quad (11)$$

When obtaining numerical results it will be assumed, somewhat arbitrarily, that $\varepsilon = L - 1/2$.

B. Multiple-Access Interference

For direct-sequence CDMA, the self-interference and multiple-access interference at the front end of the receiver matched to the desired signal can be modeled as additional broadband Gaussian noise. For a nonfaded channel, a rigorous comparison of this Gaussian approximation with exact error probabilities has been undertaken for deterministic sequences [4], [6], and random sequences [5], [7], under the assumption of coherent detection. Similar results have also been obtained for differentially coherent detection [27]. These results show that the Gaussian approximation becomes very optimistic with decreasing K, and increasing Λ. For multipath-fading channels, a rigorous comparison of a Gaussian approximation with more accurate approximations has also been made [8]–[10]. These results show that a Gaussian approximation is satisfactory, even for small K and large Λ, giving only slightly optimistic results.

C. Shadowing

Shadowing is caused by terrain features such as buildings and hills. Hilly terrain causes diffraction loss, while buildings cause scattering losses. The effect is a very slow change in the *local* mean value Λ. Shadowing is often modeled as being log-normal, meaning that $\Lambda^d = 10\log_{10}\Lambda$ is normally distributed [28], [21], [29], [30], [31].[1] Defining $\overline{\Lambda}^d \triangleq E[\Lambda^d]$, the conditional pdf of Λ^d is

$$f_{\Lambda^d|\overline{\Lambda}^d}\left(\Lambda^d|\overline{\Lambda}^d\right) = \frac{1}{\sqrt{2\pi}\,\sigma}\exp\left\{-\frac{1}{2\sigma^2}\left(\Lambda^d - \overline{\Lambda}^d\right)^2\right\}. \quad (12)$$

Typically, σ ranges from 6 to 12 dB. Four our purpose $\sigma = 8$ dB will be used. If shadowing is neglected, then $f_{\Lambda^d|\overline{\Lambda}^d}(\Lambda^d|\overline{\Lambda}^d) = \delta(\Lambda^d - \overline{\Lambda}^d)$.

D. Path Loss

A multitude of path loss prediction models exist for UHF/VHF land mobile radio in flat, urban, suburban, open, and hilly terrains [21], [29], [28]. Most of these models are empirical with a few exceptions. The simplest theoretical model, and the one used in this paper, assumed that the value of $\overline{\Lambda}$ in (12) is

$$\overline{\Lambda} = \begin{cases} \left(\frac{k}{\beta}\right)^\alpha, & r < \beta \\ \left(\frac{k}{r}\right)^\alpha, & r \geq \beta \end{cases} \quad (13)$$

where r is the distance between the transmitter and receiver, and k is a constant of proportionality. The path loss exponent α ranges from 2 in free space to 4 in a dense urban area. The parameter β is included because $\lim_{r\to 0} r^{-\alpha} = \infty$.

For a uniformly located mobile in a circular cell of radius R the pdf of the mobile location, in cylindrical coordinates, is $p(r,\theta) = r/(\pi R^2)$, $0 \leq r \leq R$, and $0 \leq \theta \leq 2\pi$. By using a univariate transformation, the pdf of $\overline{\Lambda}$ in (13) is

$$f_{\overline{\Lambda}}(\overline{\Lambda}) = \frac{\beta^2}{R^2}\delta\left(\overline{\Lambda} - \left(\frac{k}{\beta}\right)^\alpha\right) + \frac{2k^2}{\alpha R^2}\overline{\Lambda}^{-\frac{\alpha+2}{\alpha}},$$
$$\left(\frac{k}{R}\right)^\alpha \leq \overline{\Lambda} \leq \left(\frac{k}{\beta}\right)^\alpha. \quad (14)$$

It is convenient to make the following definitions:

- $\Lambda_f \triangleq \left(\frac{k}{R}\right)^\alpha$, the total average received bit energy-to-background noise ratio at distance R, which occurs when a mobile is on the cell fringe.
- $\zeta \triangleq \beta/R$.

Using these definitions the pdf of $\overline{\Lambda}$ becomes

$$f_{\overline{\Lambda}}(\overline{\Lambda}) = \zeta^2\delta\left(\overline{\Lambda} - \frac{\Lambda_f}{\zeta^\alpha}\right) + \frac{2}{\alpha}\Lambda_f^{\frac{2}{\alpha}}\overline{\Lambda}^{-\frac{\alpha+2}{\alpha}}, \quad \Lambda_f \leq \overline{\Lambda} \leq \frac{\Lambda_f}{\zeta^\alpha}. \quad (15)$$

When obtaining numerical results, it will be assumed that $\zeta = 0.1$ and $\alpha = 4$.

III. Uplink Performance Analysis

A. Multipath Rejection Receiver

Consider a multipath rejection receiver that processes the signal received over the zero path, and rejects the signal received over all other paths.[2] In the mobile radio environment, the probability of multipaths with significant power will decrease with increasing delay relative to the dominant path. Hence, the self-interference due to multipath can be minimized by using spreading codes that have small autocorrelation sidelobes in the time intervals during which delayed signals with significant power are expected. For large delays, the stringent requirements on the autocorrelation function can be relaxed. The spreading codes still must have small cross-correlation sidelobes over all delays, because the uplink transmissions are asynchronous. It is easy to find reasonably large sets of sequences that satisfy these properties. For example, a set of $2^m + 1$ Gold sequences can be generated of length $2^m - 1$. These sequences have autocorrelation values from the set $\{2^m - 1, -1, t_m - 2, -t_m\}$, and cross-correlation values form the set $\{-1, t_m - 2, -t_m\}$ where

$$t_m = \begin{cases} 2^{(m+1)/2} + 1, & m \text{ odd} \\ 2^{(m+2)/2} + 1, & m \text{ even} \end{cases}. \quad (16)$$

Of these $2^m + 1$ sequences, $2^{m-n+1} + 1$ will have their first autocorrelation sidelobe ($t_m - 2$ or t_m) at least n chip durations from the main lobe. Consequently, these $2^{m-n+1} + 1$ sequences will introduce negligible self-interference if they are used on a channel having n or fewer paths.

[1] The superscript d implies units of decibels.

[2] Here we assume the exponential MIP in (3) so that the zero path is the dominant path.

In the subsequent analysis, the self-interference due to multipath will be neglected, under the assumption that appropriate codes are used. This is a very important assumption, because a significant performance degradation will result from a poor choice of spreading codes, e.g., completely random sequences. Then by treating the multiple-access interference as additional Gaussian noise, the instantaneous received bit energy-to-*total* noise ratio associated with transmitter $\jmath$ is

$$R_{\jmath} = \frac{\left|c_0^{\jmath}\right|^2 E_b}{N_o + \eta^{-1}\sum_{(i,n)\in U}|c_n^i|^2 E_b} = \frac{\lambda_0^{\jmath}}{1+\eta^{-1}\sum_{(i,n)\in U}\lambda_n^i} \quad (17)$$

where $U = \{(i,n) : 1 \le i \le K, 0 \le n \le L-1, \text{ and } i \ne \jmath\}$, η is the processing gain, and $\lambda_n^j \triangleq \left|c_n^j\right|^2 E_b/N_o$. For a (nonfaded) AWGN channel,

$$R_{\jmath} = \frac{\Lambda^{\jmath}}{1+\eta^{-1}\sum_{\substack{i=1\\ i\ne j}}\Lambda^i}. \quad (18)$$

To compute the bit error probability, first define the mean vector

$$\overline{\mathbf{\Lambda}} = \left(\overline{\Lambda}_0^1,\cdots,\overline{\Lambda}_0^K;\overline{\Lambda}_1^1,\cdots,\overline{\Lambda}_1^K;\cdots;\overline{\Lambda}_L^1,\cdots,\overline{\Lambda}_L^K\right), \quad (19)$$

and the *local* mean vector

$$\mathbf{\Lambda} = \left(\Lambda_0^1,\cdots,\Lambda_0^K;\Lambda_1^1,\cdots,\Lambda_1^K;\cdots;\Lambda_L^1,\cdots,\Lambda_L^K\right). \quad (20)$$

The joint pdf of $\overline{\mathbf{\Lambda}}$, denoted by $p_1(\overline{\mathbf{\Lambda}})$, and the joint conditional pdf of $\mathbf{\Lambda}$, denoted by $p_2(\mathbf{\Lambda}|\overline{\mathbf{\Lambda}})$, can be obtained by using (11), (12), and (15). The bit error probability associated with transmitter $\jmath$, as a function of $\mathbf{\Lambda}$, is

$$P_{\jmath}(\mathbf{\Lambda}) = \int_0^{\infty} G(r_{\jmath}) f_{R_{\jmath}|\mathbf{\Lambda}}(r_{\jmath}|\mathbf{\Lambda})\, dr_{\jmath} \quad (21)$$

where $G(r_{\jmath})$ is the bit error probability as a function of the instantaneous bit energy-to-total-noise ratio, and $f_{R_{\jmath}|\mathbf{\Lambda}}(r_{\jmath}|\mathbf{\Lambda})$ is the conditional pdf of $R_{\jmath}$. For illustration purposes, DPSK signaling is considered where

$$G(r_{\jmath}) = \frac{1}{2}\exp\{-r_{\jmath}\}. \quad (22)$$

Other types of modulation and detection can be analyzed by defining the appropriate function $G(r_{\jmath})$ at this point. Averaging (21) over the joint pdf of $\overline{\mathbf{\Lambda}}$ and the joint conditional pdf $p_2(\mathbf{\Lambda}|\overline{\mathbf{\Lambda}})$ will account for the random mobile locations and shadowing, and gives the area averaged bit error probability[3]

$$P = \int_{\Lambda_f}^{\Lambda_f/\varsigma^{\alpha}} \int_0^{\infty} P_{\jmath}(\mathbf{\Lambda}) p_2(\mathbf{\Lambda}|\overline{\mathbf{\Lambda}}) p_1(\overline{\mathbf{\Lambda}})\, d\mathbf{\Lambda}\, d\overline{\mathbf{\Lambda}}. \quad (23)$$

[3] Note that the area averaged bit error probability is the same for every transmitter and, therefore, the dependency on $\jmath$ is removed.

In order to evaluate (21), $f_{R_{\jmath}|\mathbf{\Lambda}}(r_{\jmath}|\mathbf{\Lambda})$ is required. For a (nonfaded) AWGN channel this is simply

$$f_{R_{\jmath}|\mathbf{\Lambda}}(r_{\jmath}|\mathbf{\Lambda}) = \delta\left(r_{\jmath} - \frac{\Lambda^{\jmath}}{1+\eta^{-1}\sum_{\substack{i=1\\ i\ne j}}\Lambda^i}\right). \quad (24)$$

For the multipath-fading channel model defined in Section II-A

$$f_{R_{\jmath}|\mathbf{\Lambda}}(r_{\jmath}|\mathbf{\Lambda}) = \frac{1}{\Lambda_0^{\jmath}}\exp\left\{-\frac{r_{\jmath}}{\Lambda_0^{\jmath}}\right\} \cdot \sum_{(i,n)\in U} \mathcal{N}_{i,n}\left(\frac{\left(\frac{r_{\jmath}}{\Lambda_0^{\jmath}}\frac{\Lambda_n^i}{\eta}+1+\frac{\Lambda_n^i}{\eta}\right)}{\left(\frac{r_{\jmath}}{\Lambda_0^{\jmath}}\frac{\Lambda_n^i}{\eta}+1\right)^2}\right) \quad (25)$$

where

$$\mathcal{N}_{i,n} = \prod_{(j,k)\in U\backslash(i,n)} \frac{\Lambda_n^i}{\Lambda_n^i - \Lambda_k^j}, \quad (26)$$

and $U\backslash(i,n) = \{(j,k)\in U : (j,k) \ne (i,n)\}$.

Using (21), (22), and (25) gives

$$P_{\jmath}(\mathbf{\Lambda}) = \frac{1}{2}\sum_{(i,n)\in U}\mathcal{N}_{i,n} \cdot \left\{1 - \frac{\eta\Lambda_0^{\jmath}}{\Lambda_n^i}\exp\left\{\frac{\eta\Lambda_0^{\jmath}}{\Lambda_n^i}+\frac{\eta}{\Lambda_n^i}\right\}E_1\left(\frac{\eta\Lambda_0^{\jmath}}{\Lambda_n^i}+\frac{\eta}{\Lambda_n^i}\right)\right\} \quad (27)$$

where $E_n(\cdot)$ is the exponential integral of order n.

1) Power Control: Power control is a technique that is essential for the uplink of cellular systems employing DS CDMA. Here, we assume that the mobiles can adjust their transmitted power to compensate for the effects of path loss and shadowing only. Power control is not used to compensate for the rapid signal fluctuations introduced by multipath-fading.

It is assumed that the total transmitted power from all K mobiles in a cell is the same, regardless of whether or not power control is used. With power control, the total transmitted power is divided among the mobiles so that same signal power is received at the base-station for every mobile. In this case Λ and Λ_f are related by

$$\Lambda = \frac{KR^{\alpha}}{\sum_{i=1}^K d(r_i)}\Lambda_f \quad (28)$$

where

$$d(r_i) = \begin{cases} 1, & r_i < \beta \\ (r_i)^{\alpha}, & r_i \ge \beta \end{cases}. \quad (29)$$

Suppose that ideal power control is used so that $\Lambda_n^j = \Lambda_n$, $1 \le j \le K$. Since all channels have the same MIP, the λ_n^j, $1 \le j \le K$ are independent and identically distributed, i.e.,

$$f_{\lambda_n^j}(\lambda_n) = \frac{1}{\Lambda_n}\exp\left\{-\frac{\lambda_n}{\Lambda_n}\right\}, \quad 1 \le j \le K. \quad (30)$$

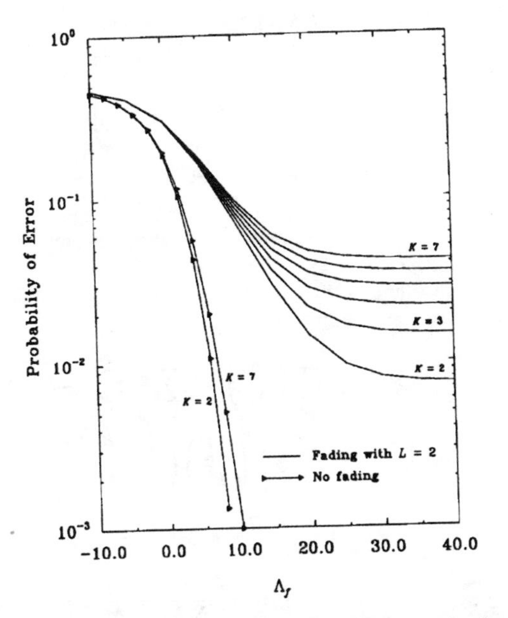

Fig. 1. Uplink bit error probability against Λ_f with power control and a multipath rejection receiver, for varying numbers of simultaneous transmissions; $L = 2$, $\eta = 100$.

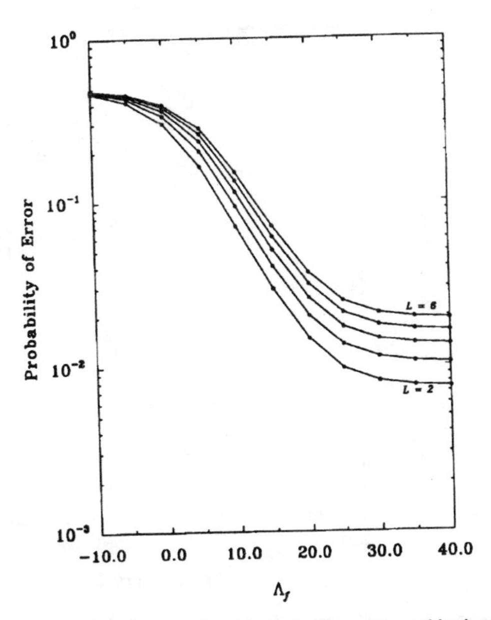

Fig. 2. Degradation in uplink bit error probability with a multipath rejection receiver due to increased multiple-access interference from multipath, for varying numbers of channel paths; $K = 2$, $\eta = 100$.

It follows that the pdf of $R_{\bar{j}}$ in (17) is

$$f_{R_{\bar{j}}}(r_{\bar{j}}) = \left\{\prod_{n=0}^{L-1}\left(\frac{\eta}{\Lambda_n}\right)^{K-1}\right\}\sum_{n=0}^{L-1}\sum_{k=0}^{K-2}\frac{1}{k!}\mathcal{D}_n^k\frac{1}{\Lambda_0} \cdot \exp\left\{-\frac{r_{\bar{j}}}{\Lambda_n}\right\}\left\{\frac{\frac{r_{\bar{j}}}{\Lambda_0}+\frac{\eta}{\Lambda_n}+K-1-k}{\left(\frac{r_{\bar{j}}}{\Lambda_0}+\frac{\eta}{\Lambda_n}\right)^{K-k}}\right\} \quad (31)$$

where

$$\mathcal{D}_n^k = \frac{d^k}{ds^k}\left\{\prod_{\substack{i=0\\ i\neq n}}^{L-1}\left(\frac{1}{s+\frac{\eta}{\Lambda_i}}\right)^{K-1}\right\}\Bigg|_{s=-\frac{\eta}{\Lambda_n}} \quad (32)$$

Since Λ is not random when power control is being used, substituting (31) and (22) into (21) gives the area averaged bit error probability

$$P = \frac{1}{2}\left\{\prod_{n=0}^{L-1}\left(\frac{\eta}{\Lambda_n}\right)^{K-1}\right\}\sum_{n=0}^{L-1}\sum_{k=0}^{K-2}\frac{1}{k!}\mathcal{D}_n^k\left(\frac{\Lambda_n}{\eta}\right)^{K-1-k} \cdot\left\{1-\frac{\eta\Lambda_0}{\Lambda_n}\exp\left\{\frac{\eta\Lambda_0}{\Lambda_n}+\frac{\eta}{\Lambda_n}\right\}E_{K-1-k}\left(\frac{\eta\Lambda_0}{\Lambda_n}+\frac{\eta}{\Lambda_n}\right)\right\}. \quad (33)$$

Fig. 1 plots the bit error probability with power control, as given by (33) for a multipath-fading channel, and by (21)–(23), and (24) for a nonfaded channel. The performance degradation due to multipath-fading is significant. Fig. 2 further illustrates the deleterious effect of multipath. In fact, an increase in the number of channel paths has an effect similar to an increase in the number of simultaneous transmissions.

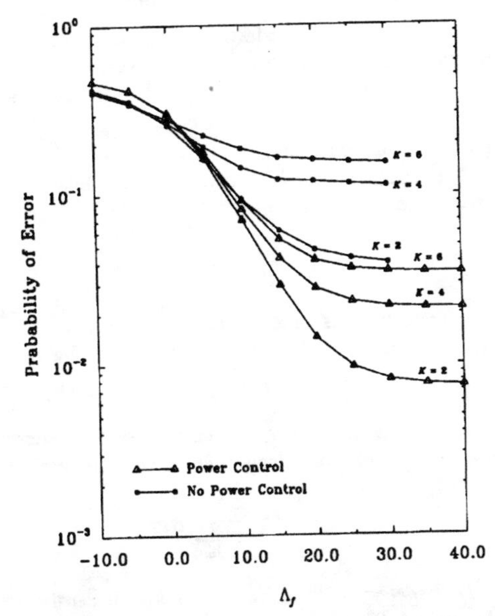

Fig. 3. Uplink bit error probability with a multipath rejection receiver, and with and without power control, for varying numbers of simultaneous transmissions; $L = 2$, $\eta = 100$.

Fig. 3 shows the bit error probability without power control, obtained from (23) and (27), and demonstrates the necessity of using power control to maintain an acceptable performance. The failure to use power control will result in a large reduction in capacity. Fig. 4 further illustrates this point by showing the bit error probability with and without power control for an interference limited environment (no background noise).

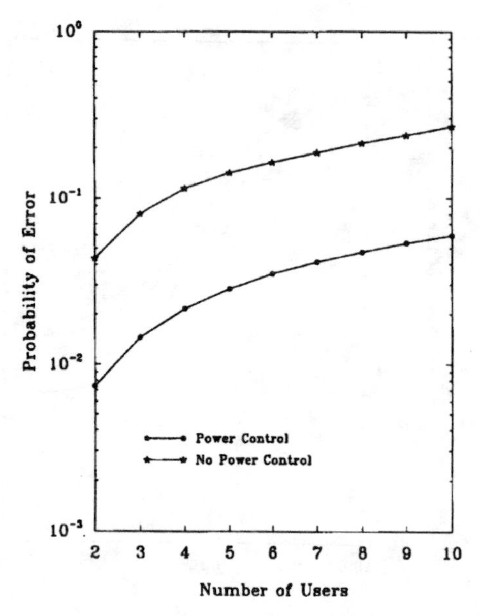

Fig. 4. Uplink bit error probability with a multipath rejection receiver, and with and without uplink power control, for an interference limited environment (negligible background noise); $\Lambda_f = \infty$, $L = 2$, $\eta = 100$.

B. RAKE Receivers with Power Control

The bit error rate performance can be improved by using multipath diversity techniques. In this section, we evaluate the performance of two different $\mathcal{L}$-tap RAKE receivers; one uses predetection selective diversity combining, and the other uses postdetection equal gain diversity combining.

The instantaneous bit energy-to-*total* noise ratio that is received over path t from transmitter $\jmath$ is

$$R_{\jmath,t} = \frac{\lambda_t^{\jmath}}{1+\eta^{-1}\sum_{(i,n)\in U}\lambda_n^i}. \tag{34}$$

The instantaneous processed bit energy-to-*total* noise ratio with predetection selective diversity combining is

$$R_{\jmath}^s = \max\{R_{\jmath,t} : 0 \le t \le \mathcal{L}-1\}. \tag{35}$$

Likewise, the total instantaneous processed bit energy-to-*total* noise ratio with postdetection equal gain combining is

$$R_{\jmath}^c = \sum_{t=0}^{\mathcal{L}-1} R_{\jmath,t}. \tag{36}$$

1) Predetection Selective Combining: It can be shown that the pdf of $R_{\jmath}^s$ is

$$f_{R_{\jmath}^s}(r_{\jmath}^s) = \prod_{n=0}^{L-1}\left(\frac{\eta}{\Lambda_n}\right)^{K-1}\sum_{t=1}^{\mathcal{L}}\sum_{j_1,\cdots,j_t}(-1)^{t+1}\sum_{n=0}^{L-1}\sum_{k=0}^{K-2} \cdot\left(\frac{\hat{\mathcal{D}}_n^k}{k!}\right)\times \mathcal{B}_t\exp[-r_{\jmath}^s\mathcal{B}_t] \cdot\left(\frac{r_{\jmath}^s\mathcal{B}_t+\frac{\eta}{\Lambda_n}+K-1-k}{\left(r_{\jmath}^s\mathcal{B}_t+\frac{\eta}{\Lambda_n}\right)^{K-k}}\right) \tag{37}$$

where

$$\hat{\mathcal{D}}_n^k = \frac{d^k}{ds^k}\left\{\prod_{\substack{i=0\\i\neq n}}^{L-1}\left(\frac{1}{s+\frac{\eta}{\Lambda_i}}\right)^{K-1}\right\}\Bigg|_{s=-\frac{\eta}{\Lambda_n}}, \tag{38}$$

$$\mathcal{B}_t = \frac{1}{\Lambda_{j_1}^{\jmath}}+\cdots+\frac{1}{\Lambda_{j_t}^{\jmath}}, \tag{39}$$

and $\sum_{j_1,\cdots,j_t}$ is summed over the t-element subsets of $\{0,1,2,\cdots,\mathcal{L}-1\}$.

Substituting (38) and (22) into (21) gives the area averaged bit error probability

$$P = \frac{1}{2}\prod_{n=0}^{L-1}\left(\frac{\eta}{\Lambda_n}\right)^{K-1}\sum_{t=1}^{\mathcal{L}}\sum_{j_1,\cdots,j_t}(-1)^{t+1} \cdot\sum_{n=0}^{L-1}\sum_{k=0}^{K-2}\left(\frac{\hat{\mathcal{D}}_n^k}{k!}\right)\left(\frac{\Lambda_n}{\eta}\right)^{K-1-k} \cdot\left\{1-\frac{\eta}{\mathcal{B}_t\Lambda_n}\exp\left\{\frac{\eta}{\mathcal{B}_t\Lambda_n}+\frac{\eta}{\Lambda_n}\right\} \cdot E_{K-1-k}\left(\frac{\eta}{\mathcal{B}_t\Lambda_n}+\frac{\eta}{\Lambda_n}\right)\right\}. \tag{40}$$

2) Postdetection Equal Gain Combining: It can be shown that the pdf of $R_{\jmath}^c$ is

$$f_{R_{\jmath}^c}(r_{\jmath}^c) = \prod_{n=0}^{L-1}\left(\frac{\eta}{\Lambda_n}\right)^{K-1}\sum_{t=0}^{\mathcal{L}-1}\sum_{n=0}^{L-1}\sum_{k=0}^{K-2}\left(\frac{\hat{\mathcal{D}}_n^k}{k!}\right)\left(\frac{\mathcal{C}_t}{\Lambda_t^{\jmath}}\right) \cdot\exp\left\{-\frac{r_{\jmath}^c}{\Lambda_t^{\jmath}}\right\}\left(\frac{\frac{r_{\jmath}^c}{\Lambda_t^{\jmath}}+\frac{\eta}{\Lambda_n}+K-1-k}{\left(\frac{r_{\jmath}^c}{\Lambda_t^{\jmath}}+\frac{\eta}{\Lambda_n}\right)^{K-k}}\right) \tag{41}$$

where

$$\mathcal{C}_t = \prod_{\substack{i=0\\i\neq t}}^{\mathcal{L}-1}\frac{\Lambda_t^{\jmath}}{\Lambda_t^{\jmath}-\Lambda_i^{\jmath}}. \tag{42}$$

For differential detection and postdetection equal gain combining, the bit error probability as a function of the total instantaneous processed bit energy-to-*total* noise ratio is [20]

$$G(r_{\jmath}^c) = \frac{1}{2^{2\mathcal{L}-1}}\exp\{-r_{\jmath}^c\}\sum_{t=0}^{\mathcal{L}-1}b_t(r_{\jmath}^c)^t \tag{43}$$

where

$$b_k = \frac{1}{k!}\sum_{\ell=0}^{\mathcal{L}-1-k}\binom{2\mathcal{L}-1}{\ell}. \tag{44}$$

Substituting (42) and (43) into (21), gives the area averaged bit error probability

$$P = \frac{1}{2^{2\mathcal{L}-1}}\prod_{n=0}^{L-1}\left(\frac{\eta}{\Lambda_n}\right)^{K-1}\sum_{t=0}^{\mathcal{L}-1} \cdot\frac{\mathcal{C}_t}{\Lambda_t}\sum_{n=0}^{L-1}\sum_{k=0}^{K-2}\left(\frac{\hat{\mathcal{D}}_n^k}{k!}\right)\left(\frac{\Lambda_n}{\eta}\right)^{K-1-k}$$

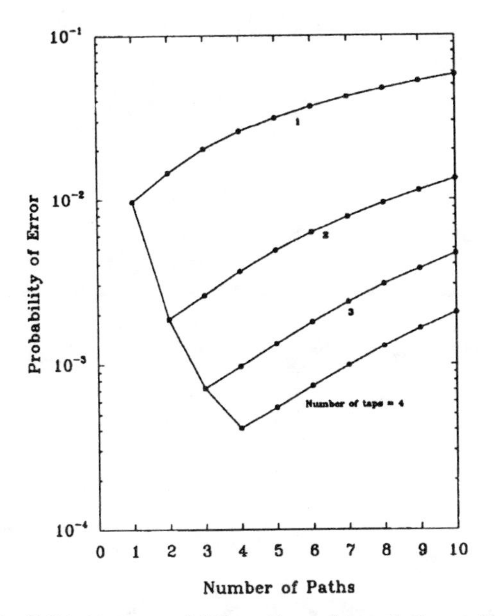

Fig. 5. Uplink bit error probability with an $\mathcal{L}$-tap selective combining RAKE receiver and power control, for varying numbers of RAKE taps in an interference limited environment; $\Lambda_f = \infty$, $K = 3$, $\eta = 100$.

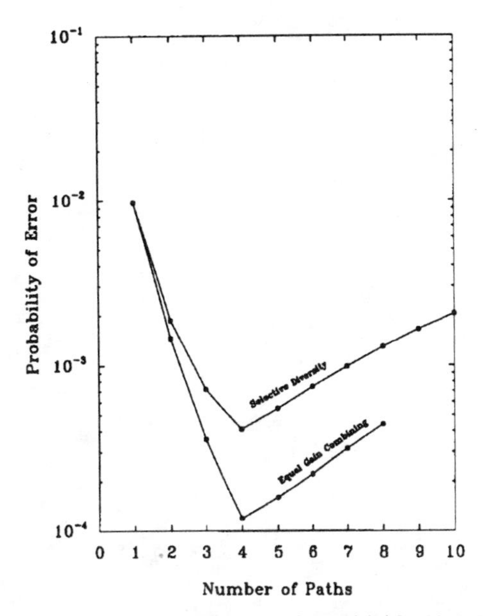

Fig. 6. Comparison of selective and equal gain combining for a 4-tap RAKE receiver with power control, in an interference limited environment; $\Lambda_f = \infty$, $K = 3$, $\eta = 100$.

$$
\cdot \sum_{t'=0}^{\mathcal{L}-1} b_{t'} \left(\frac{\eta\Lambda_t}{\Lambda_n}\right)^{t'+1} \sum_{\ell=0}^{t'} \binom{t'}{\ell} (-1)^{t'-\ell}
\cdot \left\{ \frac{1}{\left(\frac{\eta}{\Lambda_n} + \frac{\eta\Lambda_t}{\Lambda_n}\right)} + \left(\frac{\ell+1-K+k}{\left(\frac{\eta}{\Lambda_n} + \frac{\eta\Lambda_t}{\Lambda_n}\right)} + \frac{\Lambda_n(K-1+k)}{\eta} \right) \right.
\left. \cdot \left(\frac{\exp\left[\frac{\eta}{\Lambda_n} + \frac{\eta\Lambda_t}{\Lambda_n}\right]}{\left(\frac{\eta}{\Lambda_n} + \frac{\eta\Lambda^t}{\Lambda_n}\right)^{\ell+1-K+k}} \right) \Gamma\left(\ell+1-K+k, \frac{\eta}{\Lambda_n} + \frac{\eta\Lambda_t}{\Lambda_n}\right) \right\} \tag{45}
$$

where $\Gamma(\cdot,\cdot)$ is the incomplete gamma function defined by

$$
\Gamma(\alpha, x) = \int_x^\infty e^{-t} t^{\alpha-1}\, dt. \tag{46}
$$

Fig. 5 plots the bit error probability for an $\mathcal{L}$-tap RAKE receiver with predetection selective diversity combining. Fig. 6 compares the performance of predetection selective diversity combining and postdetection equal gain combining. As expected, with differential detection, postdetection equal gain combining performs somewhat better. These results show that a RAKE receiver can provide a significant improvement in performance provided that the channel has the right amount of dispersion. However, if the channel either consists of a single faded path or is very dispersive, then the performance can be quite poor.

C. Error Correction Coding

It is well known that the performance of a digital communication system can usually be improved by using error correction coding. However, for mobile communications, fading induced channel memory can degrade the coded performance. Traditionally, this is remedied by using enough interleaving to destroy the channel memory. Unfortunately, a large interleaving delay is unacceptable for voice communication. Even a moderate interleaving delay is undesirable, because it makes echo cancellation necessary.

In a radio mobile environment, the fading rate depends on vehicle speed. At high vehicle speeds, the fading rate is fast enough so that a small amount of interleaving will effectively destroy the channel memory. However, at low vehicle speeds, the delay requirements imposed by voice transmission will preclude the use of interleaving, so the channel will have memory. The effect of vehicle speed on the performance of error-correction codes is illustrated here by considering t error correcting BCH (n,k) codes. Two extreme cases are considered; slow fading, where the bit energy-to-*total* noise ratio is random but constant over the duration of an entire codeword, and fast fading, where the channel is memoryless.

1) Fast Fading: With fast fading, the probability of decoded bit error is approximately

$$
P \approx \frac{1}{n} \sum_{i=e+1}^{n} i \binom{n}{i} P_u^i (1-P_u)^{n-i} \tag{47}
$$

where P_u is the code symbol error probability. To obtain P_u, we simply use the previous expressions for the uncoded bit error probability and replace Λ with $r\Lambda$ where $r = k/n$ is the code rate.

2) Slow Fading: With slow fading, the decoded bit error probability is approximately

$$
P \approx \int_0^\infty \left\{ \frac{1}{n} \sum_{i=e+1}^{n} i \binom{n}{i} G(r_j)^i (1-G(r_j))^{n-i} \right\}
\cdot f_{R_j^{s,c}}\left(r_j^{s,c} | \mathbf{\Lambda}\right) dr_j^{s,c}. \tag{48}
$$

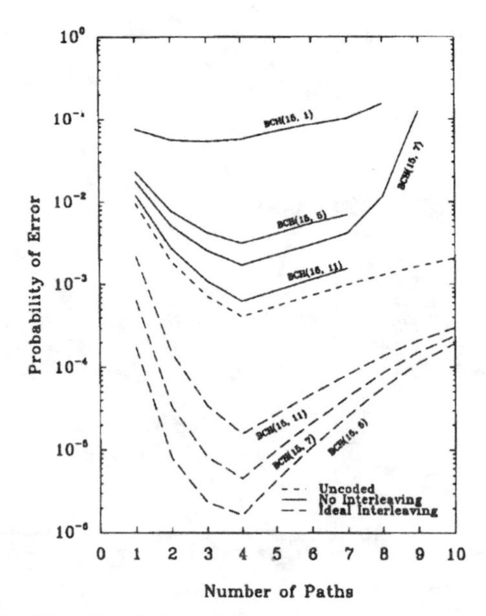

Fig. 7. Bit error probability for a 4-tap RAKE receiver with selective combining and BCH (15, k) error correcting codes, in an interference limited environment; $K = 3$, $\eta = 100$.

For selective diversity this results in

$$P \approx \frac{1}{2n}\left\{\prod_{\ell=0}^{L-1}\left(\frac{\eta}{\Lambda_\ell}\right)^{K-1}\right\}\sum_{e=\tau+1}^{n} e\binom{n}{e}\sum_{e'=0}^{n-e}\binom{n-e}{e'}(-1)^{e'} \cdot\sum_{t=1}^{\mathcal{L}}\sum_{j_1,\cdots,j_t}(-1)^{t+1}\sum_{\ell=0}^{L-1}\sum_{k=0}^{K-2}\left(\frac{\hat{\mathcal{D}}_\ell^k}{k!}\right)\left(\frac{\Lambda_\ell}{\eta}\right)^{K-1-k} \cdot\left\{1-\frac{\eta(e+e')}{\mathcal{B}_t\Lambda_\ell}\exp\left\{\frac{\eta(e+e')}{\mathcal{B}_t\Lambda_\ell}+\frac{\eta}{r\Lambda_\ell}\right\} \cdot E_{K-1-k}\left(\frac{\eta(e+e')}{\mathcal{B}_t\Lambda_\ell}+\frac{\eta}{r\Lambda_\ell}\right)\right\}. \tag{49}$$

Since both coding and processing gain result in bandwidth expansion, comparisons must be made on the basis of equal bandwidth. When coding is used, the processing gain is $\eta = r\eta^*$ where r is the code rate, and η^* is the processing gain for an uncoded system.

Fig. 7 shows the performance of BCH $(15, k)$ codes for a 4-tap RAKE receiver with predetection selective diversity combining. Results are shown for both fast fading and slow fading. Observe that when the channel is memoryless (fast fading), the performance is improved significantly as the code rate decreases. Hence, coding is more effective than processing gain. However, when the channel has memory (slow fading), coding actually results in a higher bit error probability. In this case, processing gain is more effective than coding.

IV. Downlink Performance Analysis

For the downlink of a cellular radio system, all signals experience identical fading when transmitted to a particular receiver. That is, $\lambda_n^j = \lambda_n$, $1 \le j \le K$. As a result, (34) reduces to

$$R_{\hat{j}} = \frac{\lambda_t^{\hat{j}}}{1+\eta^{-1}(K-1)\sum_{n=0}^{L-1}\lambda_n}. \tag{50}$$

Define the local mean vector

$$\mathbf{\Lambda} \triangleq (\Lambda_0, \cdots, \Lambda_L). \tag{51}$$

Suppose that $\Lambda_p \neq \Lambda_k$ for $p \neq k$, as is the case for an exponential MIP. Then the conditional pdf of $R_{\hat{j}}^s$ for a RAKE receiver with predetection selective diversity combining is

$$f_{R_{\hat{j}}^s|\mathbf{\Lambda}}(r_{\hat{j}}^s) = \sum_{t=1}^{\mathcal{L}}\sum_{j_1,\cdots,j_t}(-1)^{t+1}\sum_{n=0}^{L-1}\mathcal{N}_n\mathcal{B}_t\exp\{-r_{\hat{j}}^s\mathcal{B}_t\} \cdot\left(\frac{\left(r_{\hat{j}}^s\mathcal{B}_t(K-1)\frac{\Lambda_n}{\eta}+1+(K-1)\frac{\Lambda_n}{\eta}\right)}{\left(r_{\hat{j}}^s\mathcal{B}_t(K-1)\frac{\Lambda_n}{\eta}+1\right)^2}\right). \tag{52}$$

where

$$\mathcal{N}_n = \prod_{\substack{k=0\\k\neq n}}^{L-1}\frac{\Lambda_n}{\Lambda_n-\Lambda_k} \tag{53}$$

and $\mathcal{B}_t$ is defined in (39). Using (52), (21), and (22) gives

$$P_{\hat{j}}^s(\mathbf{\Lambda}) = \frac{1}{2}\sum_{t=1}^{\mathcal{L}}\sum_{j_1,\cdots,j_t}(-1)^{t+1}\sum_{n=0}^{L-1}\mathcal{N}_n \cdot\left\{1-\frac{\eta}{\mathcal{B}_t(K-1)\Lambda_n} \cdot\exp\left\{\frac{\eta}{\mathcal{B}_t(K-1)\Lambda_n}+\frac{\eta}{(K-1)\Lambda_n}\right\} \cdot E_1\left(\frac{\eta}{\mathcal{B}_t(K-1)\Lambda_n}+\frac{\eta}{(K-1)\Lambda_n}\right)\right\}. \tag{54}$$

The performance of a multipath rejection receiver can be easily obtained by setting $\mathcal{L} = 1$ in (54), leading to

$$P_{\hat{j}}(\mathbf{\Lambda}) = \frac{1}{2}\sum_{n=0}^{L-1}\mathcal{N}_n\left\{1-\frac{\eta\Lambda_0}{(K-1)\Lambda_n} \cdot\exp\left\{\frac{\eta\Lambda_0}{(K-1)\Lambda_n}+\frac{\eta}{(K-1)\Lambda_n}\right\} \cdot E_1\left(\frac{\eta\Lambda_0}{(K-1)\Lambda_n}+\frac{\eta}{(K-1)\Lambda_n}\right)\right\}. \tag{55}$$

Corresponding expressions can be obtained for the bit error probability with equal gain combining, but they are omitted here for brevity. In all cases, the area averaged bit error probability is given by (23).

Fig. 8 plots the bit error probability as given by (55) for a multipath-fading channel, and by (21)–(23), and (24) for a nonfaded channel. Similar to the uplink, the performance degradation due to multipath-fading is significant. Fig. 9 plots the bit error probability for a multipath rejection receiver and an $\mathcal{L}$-tap RAKE receiver with predetection selective diversity combining. This figure shows that multipath diversity can

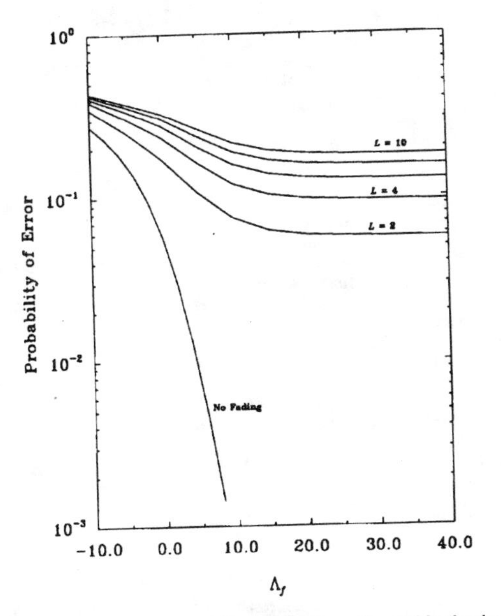

Fig. 8. Downlink bit error probability against Λ_f with a multipath rejection receiver, for varying number of channel paths; $K = 10$, $\eta = 100$.

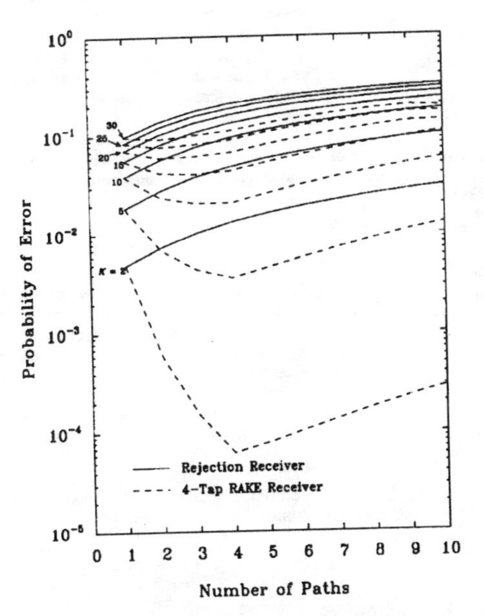

Fig. 9. Downlink bit error probability with a multipath rejection receiver and a 4-tap RAKE receiver with selective combining, for varying numbers of simultaneous transmissions in an interference limited environment; $\Lambda_f = \infty$, $\eta = 100$.

offer a significant improvement in the performance, but the degree of improvement diminishes rapidly with the number of simultaneous transmissions. Once again, if the channel either consists of a single faded path or is very dispersive, then the performance can be quite poor.

Finally, Fig. 10 illustrates the effect of log-normal shadowing on the downlink performance. It is seen that in the

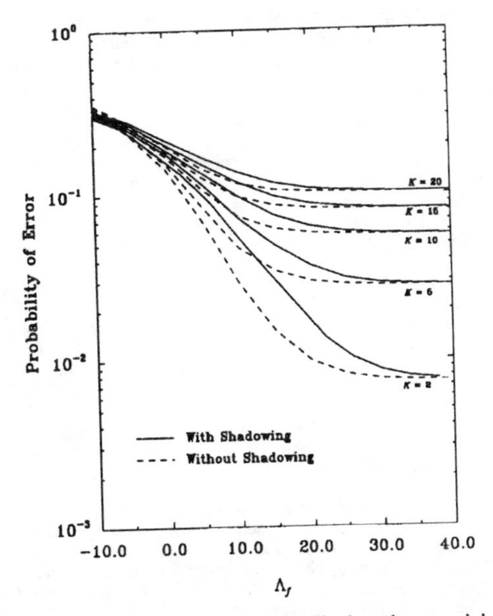

Fig. 10. Increase in downlink bit error probability due to log-normal shadowing with a multipath rejection receiver, for varying numbers of simultaneous transmissions; $L = 2$, $\eta = 100$, $\sigma = 8$ dB.

region of moderate Λ_f, shadowing will cause an increase in the area averaged bit error probability. However, for an interference limited environment (large Λ_f), shadowing does not have much effect.

V. Concluding Remarks

This paper has analyzed the performance of three differentially coherent spread-spectrum receivers in a cellular radio environment; a multipath rejection receiver, a RAKE receiver with predetection selective diversity combining, and a RAKE receiver with postdetection equal gain combining. It was shown that multipath-fading can have a severe effect on the performance of a multipath-rejection receiver, especially when the channel is highly dispersive. A RAKE receiver was shown to improve the performance significantly for a moderately dispersive channel. However, if the channel consists of a single faded path, a RAKE receiver cannot provide a performance improvement. In this case, antenna diversity may be required to guarantee an acceptable level of performance. It was also shown that error correction coding can provide a significant improvement in performance, provided that the channel fading rate is fast enough. However, for slowly varying channels it may not be desirable to use coding.

References

[1] D. L. Schilling, R. Pickholtz, and L. Milstein, "Spread spectrum goes commercial," *IEEE Spectrum*, pp. 40–45, Aug. 1990.

[2] G. R. Cooper and R. Nettleton, "A spread-spectrum technique for high-capacity mobile communications," *IEEE Trans. Vehic. Technol.*, vol. VT-27, pp. 264–275, Nov. 1978.

[3] M. B. Pursley, D. Sarwate, and W. Stark, "Error probability for direct-sequence spread-spectrum multiple-access communications—Part I: Upper and lower bounds," *IEEE Trans. Commun.*, vol. COM-30, pp. 975–984, 1982.

[4] E. A. Geraniotis and M. Pursley, "Error probability for direct-sequence spread-spectrum multiple-access communications—Part II: Approximations," *IEEE Trans Commun.*, vol. COM-30, pp. 985–995, May 1982.

[5] J. S. Lehnert and M. B. Pursley, "Error probabilities for binary direct-sequence spread-spectrum communications with random signature sequences," *IEEE Trans Commun.*, vol. COM-35, pp. 87–98, Jan. 1987.

[6] J. S. Lehnert, "An efficient technique for evaluating direct-sequence spread-spectrum multiple-access communications," *IEEE Trans Commun.*, vol. 37, pp. 851–858, Aug. 1989.

[7] R. K. Morrow and J. S. Lehnert, "Bit-to-bit error dependence in slotted DS/SSMA packet systems with random signature sequences," *IEEE Trans. Commun.*, vol. 37, pp. 1052–1061, Oct. 1989.

[8] E. Geraniotis and M. Pursley, "Performance of coherent direct-sequence spread-spectrum communications over specular multipath fading channels," *IEEE Trans Commun.*, vol. COM-33, pp. 502–508, June 1985.

[9] E. Geraniotis and M. B. Pursley, "Performance of noncoherent direct-sequence spread-spectrum communications over specular multipath fading channels," *IEEE Trans. Commun.*, vol. COM-34, pp. 219–226, Mar. 1986.

[10] E. Geraniotis, "Direct-sequence spread-spectrum multiple-access communications over nonselective and frequency-selective rician fading channels," *IEEE Trans. Commun.*, vol. COM-34, pp. 756–764, Aug. 1986.

[11] J. Lehnert and M. Pursley, "Multipath diversity reception of coherent direct-sequence spread-spectrum communications," in *Proc. Conf. Inform. Sci. Syst.*, The John Hopkins Univ., Baltimore, MD, Mar. 1983.

[12] ——, "Multipath diversity reception of spread spectrum multiple-access communications," *IEEE Trans. Commun.*, vol. COM-35, pp. 1189–1198, Nov. 1987.

[13] W. H. Lam and R. Steele, "Spread-spectrum communications using diversity in an urban mobile radio environment," IEEE Colloquium on methods of combating multipath effect in wide band digital cellular mobile systems, London, England, Oct. 1987.

[14] H. Hashemi, "Simulation of the urban mobile radio propagation channel," Ph.D. dissertation Dep. Elec. Eng. Comp. Sci., Univ. California, Berkley, Aug. 1977.

[15] G. L. Turin, "The effects of multipath and fading on the performance of direct-sequence CDMA systems," *IEEE Trans. Vehic. Technol.*, vol. VT-33, pp. 213–219, Aug. 1984.

[16] H. Xiang, "Binary code-division multiple-access systems operating in multipath fading noisy channels," *IEEE Trans. Commun.*, vol. COM-33, pp. 775–784, Aug. 1985.

[17] C. S. Gardner and J. A. Orr, "Fading effects on the performance of a spread-spectrum multiple-access communication system," *IEEE Trans. Commun.*, vol. COM-27, pp. 143–149, Jan. 1979.

[18] T. Rappaport and L. Milstein, "Effects of path loss and fringe user distribution on CDMA cellular frequency reuse efficiency," in *GLOBECOM'90*, San Diego, CA, 1990.

[19] ——, "Effects of radio propagation path loss on CDMA cellular frequency reuse efficiency for the reverse channel," *IEEE Trans. Vehic. Technol.*, vol. 41, pp. 231–242, Aug. 1992.

[20] J. G. Proakis, *Digital Communications*. New York: McGraw-Hill, 1983.

[21] W. C. Y. Lee, *Mobile Communications Design Fundamentals*. New York: Howard W. Sams, 1986.

[22] COST 207 TD(86)51-REV 3 (WG1), "Proposal on channel transfer functions for be used in GSM tests late 1986," September 1986. Paris.

[23] A. A. Giordano and F. M. Hsu, *Least Square Estimation with Applications to Digital Signal Processing*. New York: Wiley, 1985.

[24] F. Ling and J. G. Proakis, "Adaptive lattice decision-feedback equalizers–Their performance and application to time-variant multipath channels," *IEEE Trans. Commun.*, vol. COM-33, pp. 348–356, Apr. 1985.

[25] E. Eleftheriou and D. D. Falconer, "Adaptive equalization techniques for HF channels," *IEEE J. Select. Areas Commun.*, vol. SAC-5, pp. 238–247, Feb. 1987.

[26] H. Ochsner, "Direct-sequence spread-spectrum receiver for communication on frequency-selective fading channels," *IEEE J. Select. Areas Commun.*, vol. SAC-5, pp. 188–193, Feb. 1987.

[27] E. Geraniotis, "Performance of noncoherent direct sequence spread-spectrum multiple-access communications," *IEEE J. Select. Areas Commun.*, vol. SAC-3, pp. 687–694, Sept. 1985.

[28] K. Feher, *Advanced Digital Communications*. Englewood Cliffs, NJ: Prentice-Hall, 1987.

[29] W. C. Y. Lee, *Mobile Communications Engineering*. New York: McGraw-Hill, 1982.

[30] R. Muammar and S. C. Gupta, "Cochannel interference in high-capacity mobile radio systems," *IEEE Trans. Commun.*, vol. COM-30, pp. 1973–1978, Aug. 1982.

[31] R. C. French, "The effect of fading and shadowing on channel reuse in mobile radio," *IEEE Trans. Vehic. Technol.*, vol. VT-28, pp. 171–181, Aug. 1979.

[32] G. L. Stüber and C. Kchao, "Analysis of a multiple-cell direct-sequence CDMA cellular mobile radio system," *IEEE J. Select. Areas Commun.*, vol. 10, pp. 669–679, May 1992.

Chamrœun Kchao (S'87) was born in Phnom Penh, Cambodia, on April 23, 1966. He received the B.S. degree in electrical engineering from Georgia Institute of Technology in 1987. He is currently working towards the Ph.D. degree.

He has co-oped with IBM, Atlanta, GA, for several quarters between 1985 and 1987. From 1987 to 1988, he was a teaching assistant at the Georgia Institute of Technology. He spent the 1989 calendar year at Centre d'Etude et de Recherche de Lorraine, Metz, France. Currently, he is a research assistant at the Georgia Institute of Technology. His research area is communication theory, with emphasis on spread-spectrum communications and cellular mobile radio.

Gordon L. Stüber (S'81–M'86) was born in Atikokan, Ont., Canada, on June 7, 1958. He received the B.A.Sc. and Ph.D. degrees in electrical engineering from the University of Waterloo, Waterloo, Ont., Canada, in 1982 and 1986, respectively.

In 1986 he joined the School of Electrical Engineering at the Georgia Institute of Technology, Atlanta, GA, where he is currently an Associate Professor. His research interests include spread spectrum communications, cellular mobile radio systems, adaptive equalization, and error control coding.

IEEE JOURNAL ON SELECTED AREAS IN COMMUNICATIONS, VOL. 10, NO. 4, MAY 1992

Analysis of a Multiple-Cell Direct-Sequence CDMA Cellular Mobile Radio System

Gordon L. Stüber, *Member, IEEE*, and Chamroeun Kchao, *Student Member, IEEE*

***Abstract*—The performance of a multiple-cell direct-sequence code division multiple-access cellular radio system is evaluated. Approximate expressions are obtained for the area-averaged bit error probability and the area-averaged outage probability for both the uplink and downlink channels. The analysis accounts for the effects of path loss, multipath fading, multiple-access interference, and background noise. Two types of differentially coherent receivers are considered: a multipath rejection receiver and a RAKE receiver with predetection selective combining. Macroscopic base station diversity techniques, and uplink and downlink power control are also topics of discussion.**

I. Introduction

An earlier paper [1] considered the performance of a single-cell direct-sequence code division multiple access (DS CDMA) mobile radio system. This paper complements this previous work by considering the performance of a DS CDMA system in a multiple-cell mobile ratio environment. It is not intended to provide a rigorous analysis of the specific DS CDMA cellular system proposed by Qualcomm [2]. Rather, it is intended to illustrate some of the generic properties and characteristics of a multiple-cell DS CDMA mobile radio system.

A number of papers have appeared in recent literature that evaluate multiple-cell DS CDMA land-mobile ratio systems, and a few are mentioned here. Lee [3] has provided an overview of cellular DS CDMA. Gilhousen *et al.* [2] have provided a capacity analysis of a DS CDMA system that accounts for the effects of path loss and log-normal shadowing, but does not explicitly consider the effects of multipath fading. Pickholtz *et al.* [4] have considered the use of DS CDMA for sharing a frequency band with narrowband users. Rappaport and Milstein [5], [6] have examined the effect of path loss and user distribution on the performance of a DS CDMA cellular system. Vojčić *et al.* [7] have compared the capacity of TDMA and DS CDMA for microcellular radio channels characterized by multipath fading. Finally, Simpson and Holtzman [8] and Ariyavisitakul and Chang [9] have studied the effect of adaptive power control and interleaving on the performance of a coded DS CDMA cellular system.

Parts of this paper consider some of the same issues addressed in the above literature, but with a different approach and emphasis. In this paper, the basic approach is to evaluate the area-averaged error probability, and to evaluate the error probability as a function of the distance from the base station. A Gaussian approximation is used for the multiple-access interference that accounts for the variation in multiple-access interference due to multipath fading. By using this approach, we consider some of the more common issues such as the use of RAKE receivers and uplink (mobile-to-base) power control, but also treat some peculiar features of a multiple-cell DS CDMA system. One such feature is macroscopic base-station diversity, where the uplink transmissions are received simultaneously by several base stations [10]. Usually, the base stations that are closest to a mobile are the most effective for this purpose. Sometimes, the more distant base stations are more effective, depending on the severity of shadowing and the locations of the interfering mobiles. In this paper, two macroscopic diversity combining techniques are considered: majority logic combining and selective diversity combining.

It is well-known that power control is essential for the uplink of a DS CDMA cellular system. For a multiple-cell DS CDMA system, it is also possible that power control can improve the downlink (base-to-mobile) performance. The reason is the corner effect, where a mobile in a cell corner is equidistant from three base stations and will experience an increase in multiple-access interference. Nettleton and Alavi [11], [12] and Lee [3] have shown that downlink power control can be quite effective for channels without multipath fading. This paper will determine if a similar claim can be made for multipath-fading channels.

In DS CDMA cellular systems, every cell uses the same carrier frequency or set of carrier frequencies. This limits the reduction in multiple-access interference that can be gained from propagation path loss. Cell sectoring is effective for reducing multiple-access interference. It is common practice to use 120° cell sectors. For either the uplink or downlink, 120° cell sectoring reduces the multiple-access interference by roughly a factor of three. On the downlink, only one-third of the mobiles are in each cell sector (on average). On the uplink, the number of interfering mobiles is reduced by a factor of three (on

Manuscript received September 12, 1991; revised November 20, 1991 and December 19, 1991. This work was supported by BellSouth Enterprises, Inc.

G. L. Stüber is with the School of Electrical Engineering, Georgia Institute of Technology, Atlanta, GA 30332.

C. Kchao is with TRW Electronic Systems Group, Redondo Beach, CA 90278.

IEEE Log Number 9107009.

0733-8716/92$03.00 © 1992 IEEE

average). Multiple-access interference can also be reduced by using voice activity detection, where a source does not transmit during the inactive voice periods [13]. This paper assumes the use of 120° cell sectoring and provides some results on voice activity detection.

The remainder of the paper is organized as follows. Section II presents the system and channel model. General expressions are derived for the area-averaged bit error probability and area-averaged outage probability in Section III. Section IV discusses the uplink performance and addresses the issues of power control and macroscopic base-station diversity. Section V considers the downlink performance and discusses the use of downlink power control. Finally, Section VI provides some concluding remarks.

II. System and Channel Model

The cellular layout is described by a uniform planar grid of hexagonal cells of radius R.[1] Each cell contains a centrally located base station. The cells are divided into 120° sectors, where each sector employs the same carrier frequency. The mobiles are uniformly distributed throughout the system area with a density of K mobiles per cell. Since cell sectoring is used, the number of mobiles in sector c, denoted by K_c, is a random variable.[2] The effects of voice activity detection are modeled by assuming that each transmitter is independently active with probability p, so that the number of active transmitters in each sector has a (K_c, p) binomial distribution.

The channel model accounts for the effects of path loss, multipath fading, multiple-access interference, and background noise. Details of the channel model are available in [1], and only a summary is presented here.

A. Multipath Fading

Mobile radio channels are effectively modeled as a continuum of multipath components, and the lowpass impulse response of the channel $c(\tau; t)$ is commonly modeled as a tapped delay line with a tap spacing equal to the pseudonoise chip rate T_c [14]–[17], i.e.,

$$c(\tau; t) = \sum_{n=-\infty}^{\infty} c_n(t)\, \delta(\tau - nT_c) \tag{1}$$

where the tap gains $\{c_n(t)\}$ are complex Gaussian random processes. For a wide-sense stationary channel with uncorrelated scattering, the $c_n(t)$ are uncorrelated and, because they are Gaussian, independent. For a total multipath spread $T_{\max}$, the tapped delay line can be truncated at $L = \lfloor T_{\max}/T_c \rfloor + 1$ taps, where $\lfloor x \rfloor$ is the largest integer contained in x. For a slowly varying channel, $c_n(t) = c_n$ for the duration of several tens of data bits.

Assuming the above tapped delay line model, the total *instantaneous* received bit energy-to-background noise ratio is:

$$\lambda = \sum_{n=0}^{L-1} \lambda_n \tag{2}$$

where $\lambda_n = |c_n|^2\theta$ and $\theta \triangleq E_b/N_0$ is the received bit energy-to-background noise ratio in the absence of multipath fading. If the $|c_n|$ are Rayleigh distributed, then the λ_n are exponentially distributed with mean value $\Lambda_n = E[|c_n|^2]\theta$. The Λ_n are related to the multipath intensity profile (MIP) of the channel, $Q(\tau)$. Following the development in [17], suppose that $Q(nT_c + \tau) \approx Q(nT_c)$ for $0 \le \tau \le T_c$, meaning that the channel impulse response is stationary for delay intervals of length T_c. Then:

$$\Lambda_n = T_c Q(nT_c)\theta, \qquad 0 \le n \le L-1. \tag{3}$$

The channel MIP can assume various forms, but a mobile radio channel is well characterized by the exponential MIP:

$$Q(nT_c) = \psi \exp\{-n/\epsilon\}, \qquad 0 \le n \le L-1 \tag{4}$$

where ψ is a constant that specifies the total average received signal power, and ϵ is a measure of the delay spread of the channel relative to a chip duration. This model has been included in the specification for the GSM-system [18]. Our analysis will assume that all channels have the same MIP, but in reality all channels will have a unique MIP. Taking the expectation of both sides of (2) and using (3) and (4) gives:

$$\Lambda = \sum_{n=0}^{L-1} \Lambda_n = \theta T_c \psi \frac{1 - \exp\{-L/\epsilon\}}{1 - \exp\{-1/\epsilon\}} \tag{5}$$

where Λ is interpreted as the total *average* received bit energy-to-background noise ratio. It follows from (3)–(5) that:

$$\Lambda_n = \frac{(1 - \exp\{-1/\epsilon\}) \exp\{-n/\epsilon\}}{1 - \exp\{-L/\epsilon\}} \Lambda, \qquad 0 \le n \le L-1. \tag{6}$$

When obtaining numerical results it will be assumed, somewhat arbitrarily, that $\epsilon = L - \frac{1}{2}$.

B. Path Loss

A multitude of path loss prediction models exist for UHF/VHF land-mobile radio [19]–[21]. A simple theoretical model assumes that:

$$\Lambda = \begin{cases} \zeta^{-\alpha}\Lambda_f, & 0 \le \rho \le \zeta \\ \rho^{-\alpha}\Lambda_f, & \zeta \le \rho \le 1 \end{cases} \tag{7}$$

where

$\rho \triangleq$ normalized distance between the receiver and transmitter ($\rho = 1$ when the actual distance is equal to R)

$\Lambda_f \triangleq$ the value of Λ when $\rho = 1$

$\alpha \triangleq$ propagation path loss exponent.

The parameter ζ is included, because $\lim_{\rho \to 0} \rho^{-\alpha}\Lambda_f = \infty$. The path loss exponent ranges from $\alpha = 2$ in free

[1] R is the distance from the center to a corner of a cell.

[2] Since the same carrier frequency is used for every sector of every cell, references will be made to sectors rather than to cells.

space to $\alpha = 4$ in a dense urban area. When obtaining numerical results, it will be assumed that $\zeta = 0.1$ and $\alpha = 4$.[3]

C. Self-Interference and Multiple-Access Interference

As mentioned in [1], it is desirable to find reasonably large sets of spreading sequences that have small autocorrelation sidelobes in the time intervals during which delayed signals with significant power are expected. This will minimize the self-interference due to multipath. The spreading sequences must have small cross-correlation sidelobes over all delays, because the uplink transmissions are asynchronous. A subset of the Gold sequences can be constructed having these desirable correlation properties. In particular, there are $2^m + 1$ Gold sequences of length $2^m - 1$ that have autocorrelation values from the set $\{2^m - 1, -1, t_m - 2, -t_m\}$, and crosscorrelation values from the set $\{-1, t_m, -2, -t_m\}$, where

$$t_m = \begin{cases} 2^{(m+1)/2} + 1, & m \text{ odd} \\ 2^{(m+2)/2} + 1, & m \text{ even.} \end{cases} \tag{8}$$

Of these $2^m + 1$ sequences, $2^{m-n+1} + 1$ of them will have their first autocorrelation sidelobe ($t_m - 2$ or $-t_m$) at least n chip durations from the main lobe. Consequently, this subset of sequences will introduce negligible self-interference if used on a channel having n or fewer significant paths.[4]

The spreading sequences are usually deterministic and periodic, as in the above example. Nevertheless, the analysis of DS CDMA is often simplified by assuming that the spreading sequences are completely random. With this approach, the multiple-access interference at the front end of the receiver matched to the desired signal is modeled as additional broadband Gaussian noise [22]–[26]. The presence or absence of chip and phase alignment of the co-users can be accounted for by scaling the processing gain [26]. The effect of using deterministic rather than random spreading sequences can also be accounted for by scaling the processing gain, but in a more complicated fashion [22].

Two Gaussian approximations are considered in this paper. We do not compare the Gaussian approximations with more accurate approximations, but the results in [23]–[25] tend to suggest that a Gaussian approximation is quite accurate.

III. General Performance Analysis

Let $\lambda_n^{i(c)}$ be the instantaneous received bit energy-to-background noise ratio that is associated with path n and transmitter i located in sector c. If the self-interference due to multipath is neglected, and the multiple-access interference is treated as additional broadband Gaussian noise, then the instantaneous received bit energy-to-*total* noise ratio that is associated with transmitter $\hat{j}$ and path t is [22][5]:

$$R_{\hat{j},t} = \frac{\lambda_t^{\hat{j}(0)}}{1 + \eta^{-1} \sum_{c,i,n \in U} \lambda_n^{i(c)}} \tag{9}$$

where η is the processing gain. In the denominator of (9), the "1" accounts for the background noise, while the summation accounts for the multiple-access interference from all other co-users sharing the same bandwidth. The set U in (9) is defined as:

$$U \triangleq \bigcup_{c \in A} U_c \tag{10}$$

where

$$U_c = \{(c, i, n)\}: k_{ci} = 1, 1 \le i \le K_c, 0 \le n \le L - 1, \text{ and } (c, i) \ne (0, \hat{j})\} \tag{11}$$

and $k_{ci} = 1$ if transmitter i in sector c is active, and $k_{ci} = 0$ otherwise. The set A in (10) differs for the uplink and downlink. For the uplink, multiple-access interference is caused by mobiles that are located in the shaded sectors of Fig. 1, where only the adjacent cells are shown. For the downlink, multiple-access interference is caused by base stations transmitting to mobiles that are located in the shaded regions of Fig. 2, where only the adjacent cells are shown. In either case, A is the appropriate set of shaded sectors.

The instantaneous *processed* bit energy-to-total noise ratio depends on the type of receiver that is used. For a multipath rejection receiver, this is[6]:

$$R_{\hat{j}}^r = R_{\hat{j},0}. \tag{12}$$

Likewise, if an $\mathcal{L}$-tap RAKE receiver is used with predetection selective diversity combining, then:

$$R_{\hat{j}}^s = \max\{R_{\hat{j},t}: 0 \le t \le \mathcal{L} - 1\}. \tag{13}$$

In the sequel, the bit error probability of a multipath rejection receiver can be obtained by simply setting $\mathcal{L} = 1$ in the various expressions that will be derived. Similar to the development in [1], it is also possible to use a RAKE receiver with postdetection equal gain combining. Although this type of diversity combining provides slightly better performance, it will not be considered here for the sake of brevity.

To compute the bit error probability and outage probability, define the mean vector:

$$\Lambda \triangleq (\Lambda_0^{1(0)}, \cdots, \Lambda_{L-1}^{1(0)}; \cdots; \Lambda_0^{K_0(0)}, \cdots, \Lambda_{L-1}^{K_0(0)}; \cdots; \Lambda_0^{i(c)}, \cdots, \Lambda_{L-1}^{i(c)}; \cdots) \tag{14}$$

where $\Lambda_n^{i(c)} = E[\lambda_n^{i(c)}]$. The vector Λ is random because the mobile locations are random, and the joint probability

[3] The numerical results will not be applicable to a microcellular environment, where $2 \le \alpha \le 3$ is more realistic [5].

[4] References to a channel path refer to a channel tap in the tapped delay line channel model.

[5] Unless otherwise indicated, it is always assumed that the reference mobile is located in sector 0.

[6] Here we assume the exponential MIP in (4) so that the zero path is the dominant path.

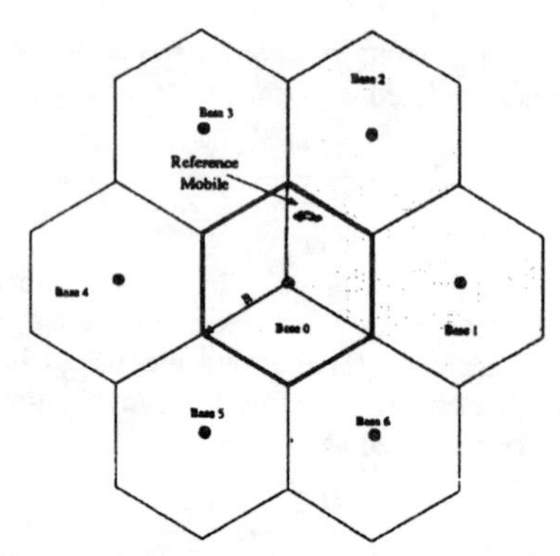

Fig. 1. Uplink transmissions from mobiles located in the shaded area will cause multiple-access interference with the uplink transmission from the reference mobile.

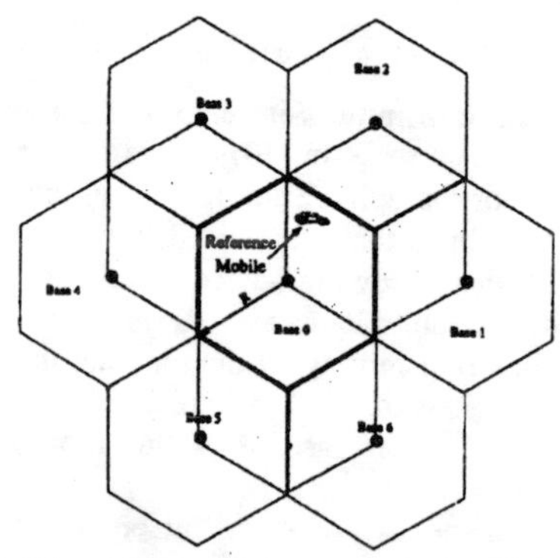

Fig. 2. Downlink transmissions to mobiles located in the shaded areas will cause multiple-access interference with the downlink transmission to the reference mobile.

density function (PDF) of Λ is denoted by $f(\Lambda)$. The elements of Λ are related by (6). In order to account for the random user voice activity, let the binary vector:

$$K_l = (k_{01}, k_{02}, \cdots, k_{0K_0}, k_{11}, k_{12}, \cdots, k_{1K_1}, \cdots, k_{c1}, k_{c2}, \cdots, k_{cK_c}, \cdots) \quad (15)$$

represent the lth activity pattern. Again, $k_{ci} = 1$ if transmitter i in sector c is active, and $k_{ci} = 0$ otherwise. It is always assumed that the reference transmitter is active, i.e., $k_{0\hat{j}} = 1$. Assuming that all mobiles are independently active with probability p, the probability of the lth activity vector is:

$$p(K_l) = \prod_{c \in A} \prod_{\substack{j=1 \\ (c,j) \neq (0,\hat{j})}}^{K_c} p k_{cj}(1-p)^{1-k_{cj}}. \quad (16)$$

The bit error probability associated with transmitter $\hat{j}$, as a function of Λ and K_l, is:

$$P_{\hat{j}}(\Lambda, K_l) = \int_0^\infty G(r_{\hat{j}}^s) f_{R_{\hat{j}}^s|\Lambda}(r_{\hat{j}}^s | \Lambda)\, dr_{\hat{j}}^s \quad (17)$$

where $G(r_{\hat{j}}^s)$ is the bit error probability as a function of the instantaneous processed bit energy-to-total noise ratio, and $f_{R_{\hat{j}}^s|\Lambda}(r_{\hat{j}}^s | \Lambda)$ is the conditional PDF of $R_{\hat{j}}^s$. The conditional PDF $f_{R_{\hat{j}}^s|\Lambda}(r_{\hat{j}}^s | \Lambda)$ assumes various forms that will be derived throughout the remainder of the paper. The function $G(r_{\hat{j}}^s)$ depends on the type of detection and diversity combining being used. If either a differentially coherent multipath rejection receiver or a differentially coherent RAKE receiver with predetection selective combining is used, then [27]:

$$G(r_{\hat{j}}^s) = \tfrac{1}{2} \exp\{-r_{\hat{j}}^s\}. \quad (18)$$

Note that (18) implies that differential detection is performed on the signal after despreading, rather than on the chip sequence. Other types of receivers, e.g., coherent receivers, can be analyzed by defining the appropriate function $G(r_{\hat{j}})$ at this point.

The area-averaged bit error probability is obtained by averaging over the density of Λ to account for the random mobile locations, and the distribution of K_l to account for the random user voice activity. This leads to:

$$P = \int_0^\infty \left\{ \sum_l p(K_l) P_{\hat{j}}(\Lambda, K_l) \right\} f(\Lambda)\, d\Lambda. \quad (19)$$

When obtaining numerical results, the averaging over Λ can be accomplished by generating a large number of uniformly distributed mobile locations. For each set of mobile locations, a large number of activity vectors are generated, and $P_{\hat{j}}(\Lambda, K_l)$ is calculated for each activity vector. Finally, an empirical average of the $P_{\hat{j}}(\Lambda, K_l)$'s is formed.

Sometimes it is desirable to compute the area-averaged outage probability, which is defined as the area-averaged probability that $P_{\hat{j}}(\Lambda, K_l)$ will exceed a constant P^*. This is given by:

$$O(P^*) = \int_0^\infty \left\{ \sum_l p(K_l) u(P_{\hat{j}}(\Lambda, K_l) - P^* \right\} f(\Lambda)\, d\Lambda \quad (20)$$

where $u(\cdot)$ is the unit step function.

Error correction coding can also be used to improve the performance, as discussed in [8], [1], [9]. Error correction coding is not considered in this paper, again for brevity, but it is easy to include in our formulation using the approach in [1]. We simply note that the effectiveness of error correction coding depends upon how well the combination of adaptive power control and delay constrained interleaving can create a memoryless coding channel, under conditions of varying Doppler spread. This issue has been partially addressed in [8], [9], but it is not completely resolved.

IV. Uplink Performance Analysis

A. Uplink Performance Without Power Control

For the uplink channel, the channel tap gains $c_n^{i(c)}$ are uncorrelated for all c, n, and i. Furthermore, if power control is not used, then the $\Lambda_n^{i(c)}$ will be distinct $\forall$ c, i, and n.[7] After some algebraic details [28], the conditional density of R_j^s in (13) is:

$$f_{R_j^s|\Lambda}(r_j^s|\Lambda) = \sum_{t=1}^{\mathcal{L}} \sum_{\vartheta_1,\cdots,\vartheta_t} (-1)^{t+1} \sum_{(c,i,n)\in U} N_{c,i,n} B_t \cdot \exp\{-r_j^s B_t\} \cdot \left(\frac{\left(r_j^s B_t \frac{\Lambda_n^{i(c)}}{\eta} + 1 + \frac{\Lambda_n^{i(c)}}{\eta}\right)}{\left(r_j^s B_t \frac{\Lambda_n^{i(c)}}{\eta} + 1\right)^2} \right) \tag{21}$$

where

$$N_{c,i,n} = \prod_{(m,k,l)\in U\setminus\{(c,i,n)\}} \frac{\Lambda_n^{i(c)}}{\Lambda_n^{i(c)} - \Lambda_l^{k(m)}} \tag{22}$$

$$B_t = \frac{1}{\Lambda_{\vartheta_1}^{j(0)}} + \cdots + \frac{1}{\Lambda_{\vartheta_t}^{j(0)}} \tag{23}$$

and $\Sigma_{\vartheta_1,\cdots,\vartheta_t}$ is the summation over the distinct t-element subsets of $\{0, 1, 2, \cdots, \mathcal{L} - 1\}$.

Using (21) and (18) in (17) gives:

$$P_j^s(\Lambda, K_l) = \frac{1}{2} \sum_{t=1}^{\mathcal{L}} \sum_{\vartheta_1,\cdots,\vartheta_t} (-1)^{t+1} \sum_{(c,i,n)\in U} N_{c,i,n} \times \left\{1 - \frac{\eta}{B_t \Lambda_n^{i(c)}} \exp\left\{\frac{\eta}{B_t \Lambda_n^{i(c)}} + \frac{\eta}{\Lambda_n^{i(c)}}\right\} \cdot E_1\left(\frac{\eta}{B_t \Lambda_n^{i(c)}} + \frac{\eta}{\Lambda_n^{i(c)}}\right)\right\} \tag{24}$$

where $E_n(x) = \int_1^\infty e^{-xt} t^{-n}\, dt$ is the exponential integral of order n.

Sometimes the above expressions exhibit numerical instability when $|U|$, the cardinality of U, is very large (there are many cells, mobiles, and/or channel paths). Numerical instability was observed to occur when $\Sigma_c K_c L \gtrsim 400$. Under this condition, the conditional PDF in (21) will not numerically integrate to unity. Fortunately, when $|U|$ is large, the central limit theorem can be invoked and a Gaussian approximation can be used. The mean and variance of the random variable $\chi = \eta^{-1} \Sigma_{c,i,n\in U} \lambda_n^{i(c)}$ in (9) are:

$$\mu = \eta^{-1} \sum_{c,i,n\in U} \Lambda_n^{i(c)} \tag{25}$$

and

$$\sigma^2 = \eta^{-2} \sum_{c,i,n\in U} [\Lambda_n^{i(c)}]^2 \tag{26}$$

respectively. The random variable χ is then approximated as being Gaussian. Notice from (9) and (13) that the accuracy of the Gaussian approximation depends on $|U|$, but does not depend on the diversity order $\mathcal{L}$. Because the multiple-access interference χ cannot assume a negative value, a further approximation is used. In particular, only the positive portion of the Gaussian PDF is used and is normalized by the factor $Q(-\mu/\sigma)$ to form a valid PDF. This further approximation is reasonably accurate when $Q(-\mu/\sigma)$ is close to unity. The condition under which this occurs can be roughly stated as follows. Suppose that the path loss is neglected so that the $\Lambda_n^{i(c)}$ are equal. In this case, $\mu/\sigma = \sqrt{|U|}$. Even when the total number of interfering mobiles is quite small, $Q(-\mu/\sigma) \approx 1$, e.g., when $|U| = 6$, $\mu/\sigma = 2.45$, and $Q(-\mu/\sigma) = 0.993$. From this argument, it follows that the conditional PDF in (21) has the approximate form [28]:

$$f_{R_j^s|\Lambda}(r_j^s|\Lambda) \approx \sum_{t=1}^{\mathcal{L}} \sum_{\vartheta_1,\cdots,\vartheta_t} (-1)^{t+1} \frac{B_t \sigma}{\sqrt{2\pi} Q(-\mu/\sigma)} \cdot \left[\exp\left\{-\frac{\mu^2}{2\sigma^2} - B_t r_j^s\right\} + \sqrt{2\pi}\left(\frac{1 + \mu - B_t \sigma^2 r_j^s}{\sigma}\right) \cdot \exp\left\{-(1 - \mu) B_t r_j^s + \frac{B_t^2 \sigma^2}{2} (r_j^s)^2\right\} \times Q\left(\frac{B_t \sigma^2 r_j^s - \mu}{\sigma}\right)\right]. \tag{27}$$

Using (27) and (18) in (17) gives:

$$P_j^s(\Lambda, K_l) = \frac{1}{2} \sum_{t=1}^{\mathcal{L}} \sum_{\vartheta_1,\cdots,\vartheta_t} (-1)^{t+1} \frac{B_t}{Q(-\mu/\sigma)} \cdot \left[\frac{\exp\left\{-\frac{\mu^2}{2\sigma^2}\right\}}{\sqrt{2\pi}(1 + B_t)} \sigma + \int_0^\infty (1 + \mu - B_t \sigma^2 r_j^s) \cdot \exp\left\{-((1 + \mu) B_t + 1) r_j^s + \frac{B_t^2 \sigma^2}{2} (r_j^s)^2\right\} \times Q\left(\frac{B_t \sigma^2 r_j^s - \mu}{\sigma}\right) dr_j^s\right]. \tag{28}$$

B. Uplink Performance with Average Power Control

Suppose that the emitted power levels from all mobiles are adjusted so that the base stations receive the same total average signal power from each mobile in a cell sector. Since the average received interference from mobiles in other cells is a constant, the ratio of the *average received total bit energy* to the *average received total noise ratio* will also be the same for every mobile in a cell sector. This power control technique is similar in some ways to the intracell power balancing scheme proposed by Nettleton and Alavi [11]. Let $\Lambda^{(c)}$ be the total average received bit energy-to-background noise ratio at base station c from any mobile in sector c. The total average *transmitted* bit

[7] This is certainly true for all c and i, but it also assumes distinct values of Λ_n in (3). This latter assumption is made throughout this paper, but does not result in any loss of generality.

energy-to-background noise ratio from all mobiles in sector c is:

$$\Psi_c = \Lambda^{(c)} \sum_{i=1}^{K_c} d(\hat{r}_{(c)}^{i(c)}) \tag{29}$$

where

$$d(\hat{r}) = \begin{cases} 1, & 0 \le \hat{r} \le \zeta \\ (\hat{r}R)^{\alpha}, & \zeta \le \hat{r} \le 1 \end{cases} \tag{30}$$

and $\hat{r}_{(c)}^{i(c)}$ is the normalized distance between the sector c base station, and mobile i in sector c. In order to make a fair comparison to the case when power control is not used, Ψ_c is related to Λ_f by $\Psi_c = K_c R^{\alpha} \Lambda_f$. Then, by using (29):

$$\Lambda^{(c)} = \frac{K_c R^{\alpha}}{\sum_{i=1}^{K_c} d(\hat{r}_{(c)}^{i(c)})} \Lambda_f. \tag{31}$$

Finally, the total average received bit energy-to-background noise ratio, at the sector 0 base station, from mobile i in sector c is:

$$\Lambda^{i(c)} = \Lambda^{(c)} \frac{d(\hat{r}_{(c)}^{i(c)})}{d(\hat{r}_{(0)}^{i(c)})} \tag{32}$$

and $\hat{r}_{(0)}^{i(c)}$ is the normalized distance between the sector 0 base station and mobile i in cell c.

1) Error Probability: Once again, the channel tap gains $c_n^{i(c)}$ are uncorrelated for all i, c, and n. With perfect average power control, $\Lambda_n^{i(0)}$ is the same for all mobiles in sector 0, and is simply denoted by $\Lambda_n^{(0)}$. However, the $\Lambda_n^{i(c)}$'s with $c \neq 0$ are all distinct. As in Section IV-A, it is possible to derive the exact conditional density $f_{R_j^s|\Lambda}(r_j^s|\Lambda)$ and the corresponding expression for $P_j^s(\Lambda, K_l)$. We refer the interested reader to [28] for these expressions. Unfortunately, the exact expression for $P_j^s(\Lambda, K_l)$ exhibits numerical instability even when $|U|$ is relatively small; in this case, when $|U| \ge 36$. As before, a Gaussian approximation can be used to remedy this problem, provided that $|U|$ is large enough. In this case, the mean and variance of the Gaussian random variable $\chi = \eta^{-1} \sum_{c,i,n \in U} \lambda_n^{i(c)}$ in (9) are:

$$\mu = \eta^{-1} \sum_{i=1}^{K_0} k_{0i} \Lambda_n^{(0)} + \eta^{-1} \sum_{c,i,n \in V} \Lambda_n^{i(c)} \tag{33}$$

and

$$\sigma^2 = \eta^{-2} \sum_{i=1}^{K_0} k_{0i} [\Lambda_n^{(0)}]^2 + \eta^{-2} \sum_{c,i,n \in V} [\Lambda_n^{i(c)}]^2 \tag{34}$$

respectively. The set V in (33) and (34) is defined as:

$$V \triangleq \bigcup_{\substack{c \in A \\ c \neq 0}} V_c \tag{35}$$

where

$$V_c = \{(c, i, n): k_{ci} = 1, 1 \le i \le K_c, 0 \le n \le L - 1\} \tag{36}$$

and A is the set of shaded sectors in Fig. 1.

The values of $\Lambda^{i(c)}$ obtained from (32) are used with (6) to obtain μ and σ, which are in turn used in (29) to approximate the error probability $P_j^s(\Lambda, K_l)$. Finally, the area-averaged bit error probability and area-averaged outage probability are obtained from (19) and (20), respectively.

C. Macroscopic Base-Station Diversity

Suppose that the signal from each mobile is received and processed by the three closest base stations, thereby providing macroscopic diversity.[8] The system area is effectively partitioned into a uniform triangular grid as shown in Fig. 3, such that the base stations on the vertices of each triangle are used to process the signals from all mobiles located within each triangle. For example, in Fig. 3 the signal from the reference mobile is processed by base stations 0, 2, and 3.

Let $P_j(\Lambda^d, K_l)$, $d = 0, 1, 2$ be the bit error probabilities corresponding to the base stations that are processing the signal from mobile $\hat{j}$, and let Λ^d be the mean vector Λ corresponding to base station d.

1) Majority Logic Combining: Since each of the base stations provides an independent estimate of an information bit, a majority logic decision can be made. In this case, the value of $P_j(\Lambda, K_l)$ used in (19) and (20) is:

$$\begin{aligned} P_j(\Lambda, K_l) = {} & P_j(\Lambda^0, K_l)P_j(\Lambda^1, K_l)P_j(\Lambda^2, K_l) \\ & + [1 - P_j(\Lambda^0, K_l)]P_j(\Lambda^1, K_l)P_j(\Lambda^2, K_l) \\ & + P_j(\Lambda^0, K_l)[1 - P_j(\Lambda^1, K_l)]P_j(\Lambda^2, K_l) \\ & + P_j(\Lambda^0, K_l)P_j(\Lambda^1, K_l)[1 - P_j(\Lambda^2, K_l)]. \end{aligned} \tag{37}$$

2) Selective Combining: An alternative to majority logic combining is selective combining, where a selection is made in favor of the base station that provides the smallest bit error probability $P_j(\Lambda^d, K_l)$. The resulting bit error probability used in (19) and (20) is:

$$P_j(\Lambda, K_l) = \min \{P_j(\Lambda^0, K_l), P_j(\Lambda^1, K_l), P_j(\Lambda^2, K_l)\}. \tag{38}$$

D. Numerical Results

All numerical results assume a processing gain of $\eta = 1000$. Only the multiple-access interference from the reference cell and the first tier of surrounding cells is included in the calculations. The multiple-access interference from the higher tiers of cells is neglected, since it has relatively little impact on the performance [29].

[8] If shadowing is neglected, then the three closest base stations will always be the most effective.

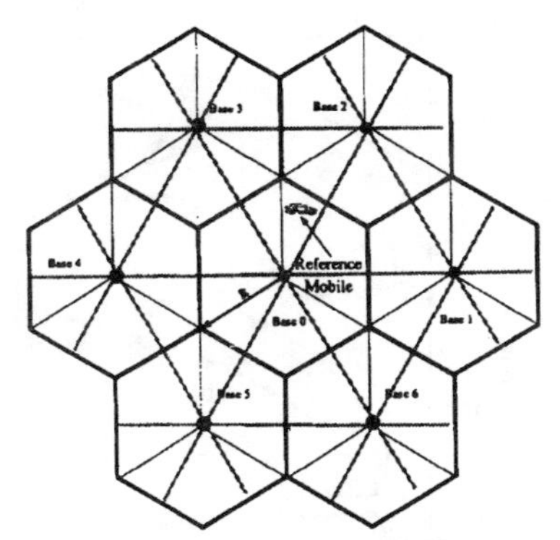

Fig. 3. Triangular coverage areas will result when the three closest base stations are used to provide macroscopic diversity for the uplink transmissions.

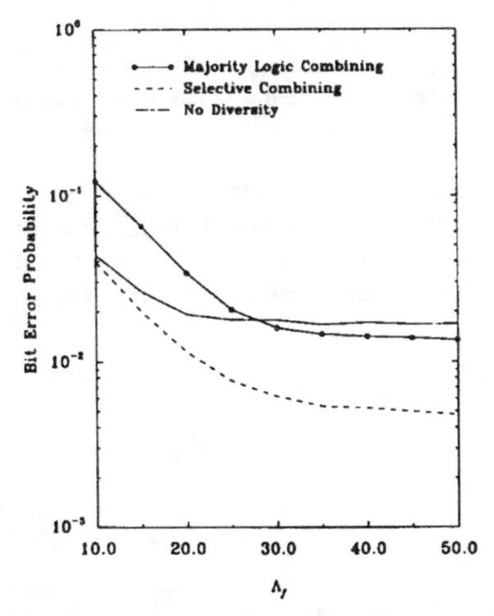

Fig. 4. Uplink area-averaged bit area probability against Λ_f. Macroscopic base-station diversity is used with a multipath rejection receiver. Uplink power control is not used; $K = 5$, $L = 2$, and $p = 1$.

Fig. 4 plots the area-averaged bit error probability with base station diversity, but without uplink power control. The exact conditional PDF in (21) and the corresponding error probability in (24) were used. A significant improvement is achieved by using selective diversity. This improvement will probably diminish for large K, but still be restored if shadowing is present. Majority logic combining is seen to improve the performance slightly at large Λ_f, but it is actually worse at small Λ_f. The reason is that the bit error probabilities on the diversity branches are not equal.

Fig. 5 plots the area-averaged bit error probability against the number of mobiles per cell for $\Lambda_f \rightarrow \infty$.[9] Results are shown with and without base station diversity, and with and without average power control. As shown in Fig. 5, (24) and (28) are the same for large K but differ significantly for small K.

Fig. 6 plots the area-averaged bit error probability

[9]The condition $\Lambda_f \rightarrow \infty$ means that the background noise is negligible when compared to the multiple-access interference.

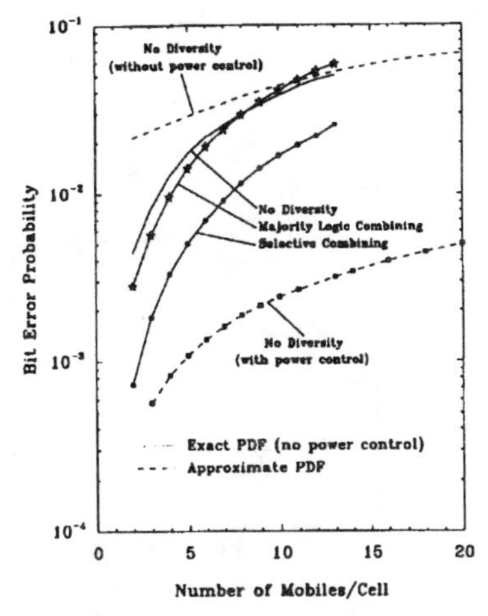

Fig. 5. Uplink area-averaged bit error probability against the number of mobiles per cell. Macroscopic base station diversity is used with a multipath rejection receiver, and a comparison is made between error probabilities given by (24) and (28); $L = 2$, $p = 1$.

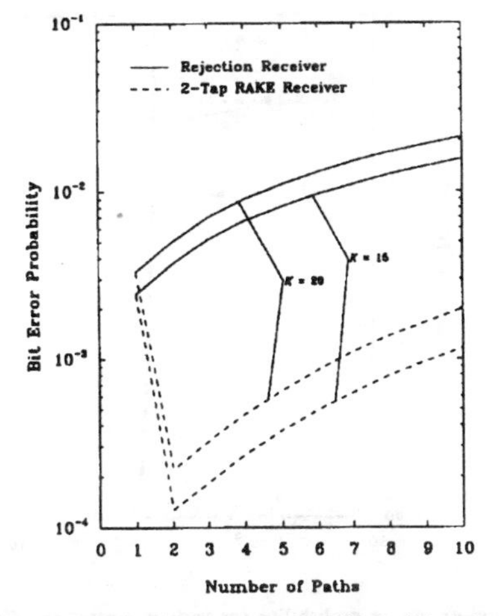

Fig. 6. Uplink area-averaged bit error probability against the number of channel paths. Results are shown for a multipath rejection receiver and a two-tap RAKE receiver with selective combining. Average uplink power control is used; $p = 1$.

against the number of channel paths, using (28). Even a small amount of multipath diversity improves the performance significantly, provided that the channel has the right amount of dispersion. However, if the channel either consists of a single faded path or is very dispersive, then the performance can be severely degraded because of the lack of diversity or the increased multiple-access interference, respectively.

Figs. 7 and 8 plot the bit error probability and outage probability as a function of the distance from the base station, using (28). It is seen that uplink average power control keeps the error probability and outage probability almost constant regardless of the distance from the base station.

Finally, Fig. 9 shows the effect of using voice activity

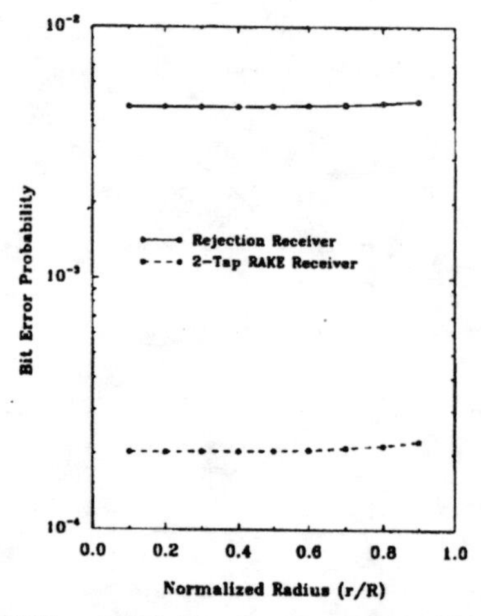

Fig. 7. Uplink bit error probability as a function of the normalized distance from the base station. Results are shown for a multipath rejection receiver and a two-tap RAKE receiver with selective combining. Average uplink power control is used; $K = 20, p = 1$.

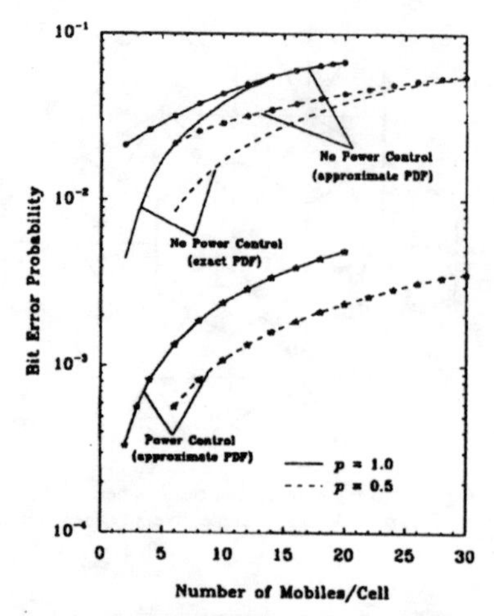

Fig. 9. Uplink area-averaged bit error probability against the number of mobiles per cell, with voice activity detection. Results are shown for a multipath rejection receiver, and with and without average uplink power control; $L = 2$.

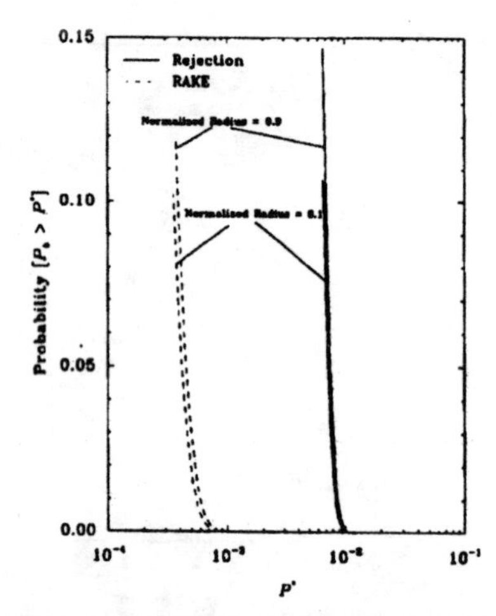

Fig. 8. Uplink outage probability for various normalized distances from the base station. Results are shown for a multipath rejection receiver and a two-tap RAKE receiver with selective combining. Average uplink power control is used; $K = 20, p = 1$.

detection. Results are shown with the exact conditional PDF $f_{R_j^{\ddagger}|\Lambda}(r_j^{\ddagger}|\Lambda)$ and its Gaussian approximation. The instantaneous bit error probability depends, in a rather complicated way, on the number and location of active mobiles. Despite this fact, it is verified that a voice activity factor of p simply increases the capacity by a factor of $1/p$ even for small K.

V. Downlink Performance Analysis

A. Downlink Performance Without Power Control

For the downlink channel, $c_n^{i(c)}$ is uncorrelated with $c_{\hat{n}}^{i(\hat{c})}$ for $(c, n) \neq (\hat{c}, \hat{n})$. However, all signals received from the same base station will have propagated over the same channel. Therefore, for fixed (c, n), the $c_n^{i(c)}$ are all equal. Without power control, $\Lambda_n^{i(c)}$ is the same for all mobiles in sector c and is denoted by $\Lambda_n^{(c)}$. But the $\Lambda_n^{(c)}$ are all distinct with respect to c and n. After some algebraic details [28], the conditional PDF of $R_j^{\ddagger}$ in (13) is:

$$f_{R_j^{\ddagger}|\Lambda}(r_j^{\ddagger}|\Lambda) = \sum_{t=1}^{\mathcal{L}} \sum_{\vartheta_1, \cdots, \vartheta_t} (-1)^{t+1} \sum_{c\in A} \sum_{n=0}^{L-1} \hat{N}_{c,n} B_t \cdot \exp\{-r_j^{\ddagger} B_t\} \cdot \left(\frac{\left(r_j^{\ddagger} B_t \frac{\hat{\Lambda}_n^{(c)}}{\eta} + 1 + \frac{\hat{\Lambda}_n^{(c)}}{\eta} \right)}{\left(r_j^{\ddagger} B_t \frac{\hat{\Lambda}_n^{(c)}}{\eta} + 1 \right)^2} \right) \tag{39}$$

where

$$\hat{N}_{c,n} = \prod_{i\in A} \prod_{\substack{k=0 \\ (i,k)\neq(c,n)}}^{L-1} \frac{\hat{\Lambda}_n^{(c)}}{\hat{\Lambda}_n^{(c)} - \hat{\Lambda}_k^{(i)}} \tag{40}$$

$$\hat{\Lambda}_n^{(c)} = \begin{cases} (K_0 - 1)\Lambda_n^{(c)}, & c = 0 \\ K_c \Lambda_n^{(c)}, & c \neq 0 \end{cases}. \tag{41}$$

By using (39) and (18), (17) becomes:

$$P_j^{\ddagger}(\Lambda, K_l) = \frac{1}{2} \sum_{t=1}^{\mathcal{L}} \sum_{\vartheta_1, \cdots, \vartheta_t} (-1)^{t+1} \sum_{c\in A} \sum_{n=0}^{L-1} \hat{N}_{c,n} \times \left\{ 1 - \frac{\eta}{B_t \hat{\Lambda}_n^{(c)}} \exp\left\{ \frac{\eta}{B_t \hat{\Lambda}_n^{(c)}} + \frac{\eta}{\hat{\Lambda}_n^{(c)}} \right\} \cdot E_1\left(\frac{\eta}{B_t \hat{\Lambda}_n^{(c)}} + \frac{\eta}{\hat{\Lambda}_n^{(c)}} \right) \right\}. \tag{42}$$

Finally, (19) and (20) give the area-average bit error probability and the area-averaged outage probability, respectively.

B. Downlink Performance with Average Power Control

One approach for implementing downlink power control is to maintain the same ratio of the *average received bit energy* to the *average received total noise* for every mobile in a cell sector. This is similar in some ways to the intracell power balancing scheme proposed by Alavi and Nettleton [12]. Here, we call it intrasector power balancing.

1) Intrasector Power Balancing: Let $\Psi(i, c)$ be the total average *transmitted* bit energy-to-background noise ratio, corresponding to a total average *received* bit energy-to-background noise ratio of $\Lambda(i, c)$ at mobile i in sector c. The sum of the total transmitted average bit energy-to-background noise ratios for sector c is:

$$\Psi_c = \sum_{i=1}^{K_c} \Psi(i, c). \tag{43}$$

To make a fair comparison to the case when power control is not used, Ψ_c and Λ_f are related by $\Psi_c = K_c R^{\alpha} \Lambda_f$. Let $\hat{r}_{(b)}^{i(c)}$ be the normalized distance between the sector b base station and mobile i in sector c. The sum of the total average received bit energy-to-background noise ratios from all base stations, at mobile i in sector c, is:

$$\sum_{b \in A} \frac{\Psi_b}{d(\hat{r}_{(b)}^{i(c)})} \tag{44}$$

where $d(\cdot)$ has been defined in (30). Assume that the signal from transmitter i of the sector c base station is intended for mobile i in sector c. Then the average received bit energy-to-average received total noise ratio, at mobile i in sector c, is:

$$S^{i(c)} = \frac{\dfrac{\Psi(i, c)}{d(\hat{r}_{(c)}^{i(c)})}}{\eta^{-1}\left(\sum_{b \in A} \dfrac{\Psi_b}{d(\hat{r}_{(b)}^{i(c)})} - \dfrac{\Psi(i, c)}{d(\hat{r}_{(c)}^{i(c)})}\right) + 1}. \tag{45}$$

Rearranging this expression gives:

$$\Psi(i, c) = \frac{S^{i(c)}}{1 + \eta^{-1} S^{i(c)}} d(\hat{r}_{(c)}^{i(c)}) \left(\eta^{-1} \sum_{b \in A} \frac{\Psi_b}{d(\hat{r}_{(b)}^{i(c)})} + 1\right). \tag{46}$$

However, from (43),

$$\Psi_c = \sum_{i=1}^{K_c} \frac{S^{i(c)}}{1 + \eta^{-1} S^{i(c)}} d(\hat{r}_{(c)}^{i(c)}) \left(\eta^{-1} \sum_{b \in A} \frac{\Psi_b}{d(\hat{r}_{(b)}^{i(c)})} + 1\right). \tag{47}$$

With intrasector power balancing, the power transmitted to each mobile in a sector is adjusted so that $S^{i(c)} = S^{(c)}$, $i = 1, \cdots, K_c$. Therefore, (47) gives:

$$S^{(c)} = \frac{\Psi_c}{\sum_{i=1}^{K_c} d(\hat{r}_{(c)}^{i(c)}) \left(\eta^{-1} \sum_{b \in A} \dfrac{\Psi_b}{d(\hat{r}_{(b)}^{i(c)})} + 1\right) - \eta^{-1} \Psi_c}. \tag{48}$$

By using (48) with $\Psi_l = K_l R^{\alpha} \Lambda_f$, the $S^{(c)}$ for all c can be obtained. After doing so, the $\Psi(i, c)$ for all c and $1 \leq i \leq K_c$ can be generated from (46). Finally, if the bit error probability of mobile $\hat{j}$ in sector 0 is of interest, then:

$$\Lambda^{i(c)} = \frac{\Psi(i, c)}{d(\hat{r}_{(c)}^{\hat{j}(0)})}. \tag{49}$$

2) Error Probability: Once again, for the downlink with power control, $c_n^{i(c)}$ is uncorrelated with $c_{\hat{n}}^{\hat{i}(\hat{c})}$ for $(c, n) \neq (\hat{c}, \hat{n})$, and $c_n^{i(c)} = c_n^{\hat{i}(c)}$, $\forall i, \hat{i}$. However, with power control $\Lambda_n^{i(c)} \neq \Lambda_{\hat{n}}^{\hat{i}(\hat{c})}$ whenever $(c, i, n) \neq (\hat{c}, \hat{i}, \hat{n})$. It follows that the conditional PDF of $R_j^{\$}$ in (13) is [28]:

$$f_{R_j^{\$}|\Lambda}(r_j^{\$}|\Lambda) = \sum_{l=1}^{\mathcal{L}} \sum_{\vartheta_1, \cdots, \vartheta_l} (-1)^{l+1} \sum_{c \in A} \sum_{n=0}^{L-1} \tilde{N}_{c,n} B_l \cdot \exp\{-r_j^{\$} B_l\} \cdot \left(\frac{\left(r_j^{\$} B_l \dfrac{\tilde{\Lambda}_n^{(c)}}{\eta} + 1 + \dfrac{\tilde{\Lambda}_n^{(c)}}{\eta}\right)}{\left(r_j^{\$} B_l \dfrac{\tilde{\Lambda}_n^{(c)}}{\eta} + 1\right)^2}\right) \tag{50}$$

where

$$\tilde{N}_{c,n} = \prod_{\substack{i \in A \\ (i,k) \neq (c,n)}} \prod_{k=0}^{L-1} \frac{\tilde{\Lambda}_n^{(c)}}{\tilde{\Lambda}_n^{(c)} - \tilde{\Lambda}_k^{(i)}} \tag{51}$$

$$\tilde{\Lambda}_n^{(c)} = \begin{cases} \sum\limits_{\substack{i=1 \\ i \neq j}}^{K_0} \Lambda_0^{i(c)}, & c = 0 \\ \sum\limits_{i=1}^{K_c} \Lambda_n^{i(c)}, & c \neq 0. \end{cases} \tag{52}$$

By using (50) and (18), (17) becomes

$$P_j^{\$}(\Lambda, K_l) = \frac{1}{2} \sum_{l=1}^{\mathcal{L}} \sum_{\vartheta_1, \cdots, \vartheta_l} (-1)^{l+1} \sum_{c \in A} \sum_{n=0}^{L-1} \tilde{N}_{c,n} \times \left\{1 - \frac{\eta}{B_l \tilde{\Lambda}_n^{(c)}} \exp\left\{\frac{\eta}{B_l \tilde{\Lambda}_n^{(c)}} + \frac{\eta}{\tilde{\Lambda}_n^{(c)}}\right\} \cdot E_1\left(\frac{\eta}{B_l \tilde{\Lambda}_n^{(c)}} + \frac{\eta}{\tilde{\Lambda}_n^{(c)}}\right)\right\}. \tag{53}$$

The values of $\Lambda_n^{i(c)}$ in (53) are obtained from (49) and (6). Finally, (19) and (20) give the area-averaged bit error probability and the area-averaged outage probability, respectively.

C. Numerical Results

Again, all numerical results assume a processing gain of $\eta = 1000$, and only the multiple-access interference from the reference base station and the first tier of interfering base stations is included in the calculations. All numerical results were obtained by using the exact conditional PDF's for $f_{R_j^{\$}|\Lambda}(r_j^{\$}|\Lambda)$ and the associated error probabilities in (42) and (53).

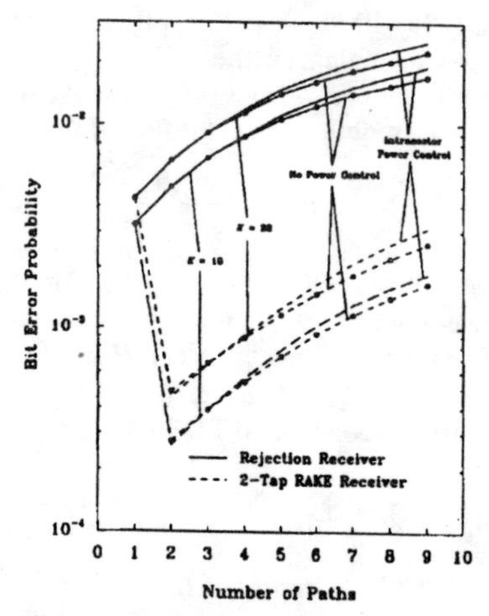

Fig. 10. Downlink area-averaged bit error probability against the number of channel paths. Results are shown for a multipath rejection receiver and a two-tap RAKE receiver with selective combining. The effect of using intrasector power balancing is also shown; $p = 1$.

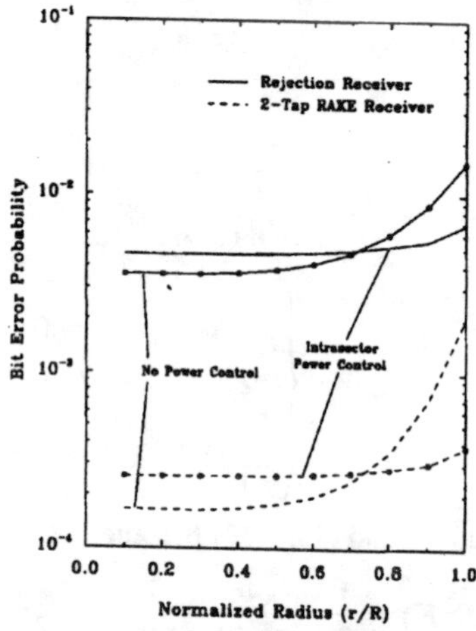

Fig. 11. Downlink bit error probability as a function of the normalized distance from the base station. Results are shown for a multipath rejection receiver and a two-tap RAKE receiver with selective combining. The effect of using intrasector power balancing is also shown; $K = 15, p = 1$.

Fig. 10 plots the bit error probability against the number of channel paths. Results are shown without power control and with intrasector power balancing. Again, a small amount of multipath diversity improves the performance significantly. However, downlink power control offers very little improvement in the area-averaged bit error probability.

Fig. 11 plots the bit error probability as a function of the normalized distance from the base station. From this diagram, it is apparent that downlink power control is quite effective for combating the corner effect, and its usefulness tends to be more apparent when diversity is used.

VI. Concluding Remarks

This paper has illustrated some of the typical properties and characteristics of a DS CDMA mobile radio system. For the uplink channel, a significant improvement in performance can be obtained by using macroscopic base-station diversity. Our results have also shown that average uplink power control will provide a nearly uniform bit error probability and outage probability throughout the system area. A significant improvement can also be obtained from various microscopic diversity techniques and voice activity detection. For the downlink channel, downlink power control does not change the area-averaged bit error probability. However, it is quite effective for combating the corner effect so as to provide a more uniform bit error probability and outage probability throughout the system area. The usefulness of downlink power control becomes more apparent when microscopic diversity is used.

References

[1] C. Kchao and G. L. Stüber, "Analysis of a direct-sequence spread-spectrum cellular radio system," *IEEE Trans. Commun.*, to appear.

[2] K. S. Gilhousen, I. M. Jacobs, R. Padovani, L. A. Weaver, Jr., and C. E. Wheatley III, "On the capacity of a cellular CDMA system," *IEEE Trans. Vehic. Technol.*, vol. 40, pp. 303-312, May 1991.

[3] W. C. Y. Lee, "Overview of cellular CDMA," *IEEE Trans. Vehic. Technol.*, vol. 40, pp. 291-302, May 1991.

[4] R. L. Pickholtz, L. B. Milstein, and D. L. Schilling, "Spread spectrum for mobile communications," *IEEE Trans. Vehic. Technol.*, vol. 40, pp. 313-322, May 1991.

[5] T. S. Rappaport and L. Milstein, "Effects of path loss and fringe user distribution on CDMA cellular frequency reuse efficiency," in *Proc. GLOBECOM'90*, San Diego, CA, 1990, pp. 404.6.1-404.6.7.

[6] T. Rappaport and L. Milstein, "Effects of radio propagation path loss on DS-CDMA cellular frequency reuse efficiency for the reverse channel," *IEEE Trans. Vehic. Technol.*, to appear.

[7] B. R. Vojčić, R. L. Pickholtz, and I. S. Stojanović, "Comparison of TDMA and CDMA in microcellular radio systems," in *Proc. ICC'91*, Denver, CO, 1991, pp. 28.1.1-28.1.5.

[8] F. Simpson and J. Holtzman, "CDMA power control, interleaving, and coding," in *Proc. 41st IEEE Vehic. Technol. Conf.*, Saint Louis, MO, 1991, pp. 362-367.

[9] S. Ariyavisitakul and L. F. Chang, "Signal and interference statistics of a simulated CDMA system with fast adaptive power control," submitted to *IEEE Trans. Commun.*

[10] L. F. Chang and J. C.-I. Chuang, "Diversity selection using coding in a portable radio communications channel with frequency-selective fading," *IEEE J. Select. Areas Commun.*, vol. 7, pp. 89-98, Jan. 1989.

[11] R. Nettleton and H. Alavi, "Power control for a spread spectrum cellular mobile radio system," in *Proc. 33rd IEEE Vehic. Tech. Conf.*, Toronto, Canada, May 1983, pp. 242-246.

[12] H. Alavi and R. Nettleton, "Downstream power control for a spread spectrum cellular mobile radio system," in *Proc. GLOBECOM'82*, Miami, FL, 1982, pp. A3.5.1-3.5.5.

[13] K. S. Gilhousen, I. M. Jacobs, R. Padovani, and L. A. Weaver, Jr., "Increased capacity using CDMA for mobile satellite communications," *IEEE J. Select. Areas Commun.*, vol. 8, pp. 503-514, May 1990.

[14] A. A. Giordano and F. M. Hsu, *Least Square Estimation with Applications to Digital Signal Processing*. New York: Wiley, 1985.

[15] F. Ling and J. G. Proakis, "Adaptive lattice decision-feedback equalizers—Their performance and application to time-variant multipath channels," *IEEE Trans. Commun.*, vol. COM-33, pp. 348-356, Apr. 1985.

[16] E. Eleftheriou and D. D. Falconer, "Adaptive equalization techniques for HF channels," *IEEE J. Select. Areas Commun.*, vol. 5, pp. 238-247, Feb. 1987.

[17] H. Ochsner, "Direct-sequence spread-spectrum receiver for communication on frequency-selective fading channels," *IEEE J. Select. Areas Commun.*, vol. 5, pp. 188-193, Feb. 1987.

[18] COST 207 TD(86)51-REV 3 (WG1), "Proposal on channel transfer functions to be used in GSM tests late 1986," Sept. 1986, Paris.

[19] W. C. Y. Lee, *Mobile Communications Design Fundamentals.* Indianapolis, IN: Sams, 1986.
[20] W. C. Y. Lee, *Mobile Communications Engineering.* New York: McGraw-Hill, 1982.
[21] K. Feher, *Advanced Digital Communications.* Englewood Cliffs, NJ: Prentice-Hall, 1987.
[22] C. L. Weber, G. K. Huth, and B. H. Batson, "Performance considerations of code division multiple-access systems," *IEEE Trans. Vehic. Technol.*, vol. VT-30, pp. 3-9, Feb. 1981.
[23] E. A. Geraniotis and M. B. Pursley, "Performance of coherent direct-sequence spread-spectrum communications over specular multipath fading channels," *IEEE Trans. Commun.*, vol. COM-33, pp. 502-508, June 1985.
[24] E. Geraniotis and M. B. Pursley, "Performance of noncoherent direct-sequence spread-spectrum communications over specular multipath fading channels," *IEEE Trans. Commun.*, vol. COM-34, pp. 219-226, Mar. 1986.
[25] E. Geraniotis, "Direct-sequence spread-spectrum multiple-access communications over nonselective and frequency selective Rician fading channels," *IEEE Trans. Commun.*, vol. COM-34, pp. 756-764, Aug. 1986.
[26] R. K. Morrow and J. S. Lehnert, "Bit-to-bit error dependence in slotted DS/SSMA packet systems with random signature sequences," *IEEE Trans. Commun.*, vol. 37, pp. 1052-1061, Oct. 1989.
[27] J. G. Proakis, *Digital Communications.* New York: McGraw-Hill, Inc., 1989.
[28] C. Kchao, "Direct sequence spread spectrum cellular radio," Ph.D. Thesis, Georgia Institute of Technology, 1991.
[29] G. L. Stüber and C. Kchao, "Spread spectrum cellular radio," Georgia Institute of Technology, OCA Project E21-662 for Bell South Enterprises, Dec. 1990.

Gordon L. Stüber (S'81-M'86) was born in Atikokan, Ont., Canada, on June 7, 1958. He received the B.A.Sc. and Ph.D. degrees in electrical engineering from the University of Waterloo, Waterloo, Ont., Canada, in 1982 and 1986, respectively.

In 1986, he joined the School of Electrical Engineering at the Georgia Institute of Technology, Atlanta, where he is currently an Associate Professor. His research interests include personal and mobile communications, spread-spectrum communications, adaptive equalization, and error control coding.

Chamroeun Kchao (S'87) was born in Phnom Penh, Cambodia, on April 23, 1966. He received the B.S. and Ph.D. degrees in electrical engineering from the Georgia Institute of Technology, Atlanta, in 1987 and 1991, respectively.

He co-oped with IBM, Atlanta, for several quarters between 1985 and 1987. From 1987 to 1988, he was a Teaching Assistant at the Georgia Institute of Technology. He spent the 1989 calendar year at Centre d'Etude et de Recherche de Lorraine, Metz, France. In 1991, he joined TRW, Redondo Beach, CA, where he is currently a Senior Member of the Technical Staff. His research interests are in communication theory, with an emphasis on spread-spectrum communications and cellular radio.

SECTION 2.2
CDMA POWER CONTROL ISSUES

The papers in this subsection deal with power control issues of cellular CDMA. The paper "*Cellular CDMA Networks Impaired by Rayleigh Fading: System Performance with Power Control*" [Tonguz and Wang] contains detail analysis and numerical results on BER performance for forward as well as reverse link for both with and without power control in a cellular CDMA network. The next paper: "*Spectral Efficiency of a Power-Controlled CDMA Mobile Personal Communication System*" by Gass Jr. et al. investigates the spectral efficiency of a CDMA system employing rapid closed-loop power control on a channel with doubly selective fading. The next three papers "*Capacity, Throughput, and Delay Analysis of a Cellular DS-CDMA System with Imperfect Power Control and Imperfect Sectorization*" by Jansen and Prasad, "*Performance Analysis of CDMA with Imperfect Power Control*" by Cameron and Woerner, and "*Effects of Imperfect Power Control and User Mobility on a CDMA Cellular Network*" by Priscoli and Sestini deal with the performance study of cellular CDMA under imperfect power control. The last paper in this subsection, "*The Capacity of a Spread Spectrum CDMA System for Cellular Mobile Radio with Consideration of System Imperfections*" by Newson and Heath makes a critical analysis of the claim that CDMA can offer capacity improvement. In doing this, the paper assesses the sensitivity of the CDMA system to typical propagation conditions, power control errors, and realistic antenna patterns and shows that the capacity of a CDMA system may be significantly reduced under non-ideal conditions.

Cellular CDMA Networks Impaired by Rayleigh Fading: System Performance with Power Control

Ozan K. Tonguz, *Member, IEEE,* and Melanie M. Wang, *Member, IEEE*

***Abstract*— The effect of power control on the performance of cellular code division multiple access (CDMA) networks is investigated. The bit error rate (BER) of a mobile unit at any given distance from the base station is obtained for both the forward link and the reverse link, taking into account Rayleigh fading and UHF propagation attenuation. To this end, a simple multiple-cell-system model is developed. Our work shows the dependence of performance gain on the different user spatial distributions under power control. Also, it is shown that power control in the forward link of CDMA networks may lead to a dismal performance degradation for close-in users. To alleviate this problem, the adjusted power control scheme is revisited. Two different performance criteria are proposed for this purpose, and the optimal threshold values corresponding to these criteria are computed. Results indicate that, although both threshold criteria are acceptable, the second criterion is more recommendable from a BER performance point of view, because it circumvents the so-called service "hole" problem which may occur when one uses the first criterion. The analysis for both the forward and reverse links leads to closed-form signal-to-interference ratio (SIR) results, which not only facilitate physical insight into the impact of key system parameters on the system performance of CDMA networks with power control, but may also eliminate the need for cumbersome Monte-Carlo-type simulations. It is shown that fourth-order power control ($n = 4$) leads to at least a threefold enhancement in capacity for $BER = 10^{-3}$.**

I. Introduction

In RECENT years, there has been a growing interest in the code division multiple access (CDMA) scheme for cellular communication networks. The well-known ability of CDMA to both combat multipath fading and allow multiple users to access a channel simultaneously, and also its potential for meeting the ever-increasing capacity demand [1]–[12], are the major reasons for this increasing interest. However, as pointed out in recent studies [1]–[5], because of the near–far problem and adjacent cell interference, to fully exploit the potential advantages of CDMA, power control must be used [1], [2], [13]–[15].

In the CDMA scheme, all traffic channels are sharing one radio channel; some mobile units are close to the base station and some are not. Therefore, because of the radio channel path losses, a strong signal received from a near-in mobile unit will mask the weak signal from a far-end mobile unit at the base station—this is known as the near–far problem [1]–[5], [16]–[20]. This intolerable situation may be mitigated by using power control on the reverse link to reduce the transmission power of a close-in user so that the resource can be shared equally, or else CDMA will be unusable [1], [2], [5], [6]. Power control to a given mobile is exercised by the cell whose pilot power is maximum to that mobile. It follows that if the path loss is only a function of distance from the base station, then the mobile is power controlled by the nearest cell site, which is situated at the center of the hexagon in which it lies (see Fig. 1). Fast fading (due largely to multipath) is assumed not to affect the (average) power level. As for the forward link, while no power control is required in a *single cell system* [3], since all signals are propagating together and hence fade together, there is no resultant performance degradation; this, however, is no longer true for a *multiple-cell* CDMA. In cellular CDMA, the same radio channel can be reused in all the neighboring cells. The interference caused by those neighboring cells fade independently of the given base station, and thereby degrade the performance. This can be a serious problem for those mobile units traveling close to the cell boundary. Therefore, power control on the forward link should also be used to reduce the interference from the neighboring cells [1], [7], [10]. In that case, power control takes the form of power allocation at the base station transmitter according to the needs of individual subscribers in the given cell. This requires measurement by the mobile of its relative SNR defined as the ratio of the power from its own base station transmitter to the total power received. Practically, this is done by acquiring the highest power pilot and measuring its energy, and also measuring the total energy received by the antenna of the mobile unit from all cell site transmitters. Both measurements can be transmitted to the selected (largest power) base station when the mobile starts to transmit [2].

While power control is recognized as the most important system requirement for cellular CDMA, relatively few papers have addressed this issue. The main objectives of this study are: 1) to examine and contrast the use of power control in the forward and reverse links of CDMA networks; 2) to provide a simple and accurate closed-form result for the system performance of CDMA networks with power control, taking into account different system parameters (such as spatial traffic distribution, Rayleigh fading effect, threshold adjustment value, number of subscribers in the cell, etc.), which may eliminate the need for cumbersome Monte-Carlo-type simulations, as well as provide valuable physical insight into

Manuscript received June 9, 1993; revised August 30, 1993.

O. K. Tonguz is with the Personal Communications Research Group, Department of Electrical and Computer Engineering, State University of New York at Buffalo, Buffalo, NY 14260 USA.

M. M. Wang was with the Personal Communications Research Group, Department of Electrical and Computer Engineering, State University of New York at Buffalo, Buffalo, NY. She is now with GTE Personal Communications Services, Mobile Data Group, Atlanta, GA 30319 USA.

IEEE Log Number 9403020.

0018–9545/94$04.00 © 1994 IEEE

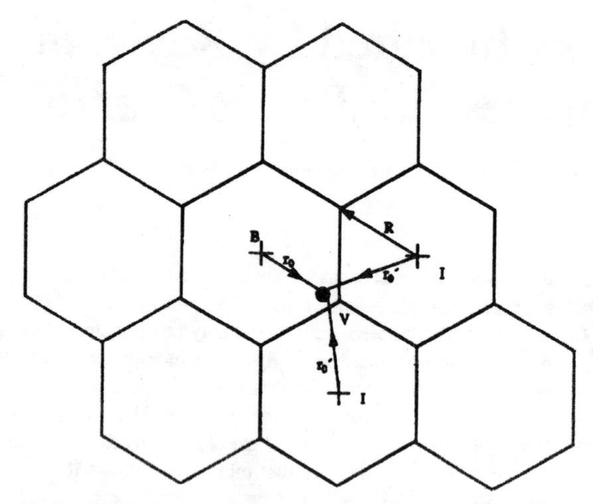

Fig. 1. Forward-link CDMA system geometry, showing "home" cell B and neighboring cells I.

the impact of key system parameters on system performance; and 3) to examine and quantify the increase in capacity due to the use of power control in CDMA networks. Hence, this paper complements other studies on cellular CDMA systems by offering a detailed analysis, based on the bit error probability of a mobile unit *as a function of the distance from the base station* in the presence of Rayleigh fading, both with and without the power control scheme. The closed-form results presented in the paper facilitate a clear physical insight into the effect of power control and key system parameters on the performance of CDMA networks.

The rest of this paper is organized as follows. In Section II, first the forward link (base station to mobile) is considered. Contrary to the approaches taken by the recent studies on power control in CDMA networks, where it is assumed that either no specific knowledge about the "test" unit is available or that the "test" unit is at the edge of a cell (which corresponds to the worst case), our analysis method takes into account the dynamic position of the "test" unit. Bit error probabilities, conditioned on the position of the "test" unit, are developed for two different user distributions to investigate the effect of power control. Using the model developed, the adjusted power control method proposed in [1], [3], and [7] is revisited, and the threshold values for two different performance criteria are obtained. In Section III, the effect of power control on the reverse link is studied essentially using the same models with necessary adjustments. Again, bit error probability results are given conditioned on the position of the "test" unit. A physical interpretation of the main results is presented in Section IV. Finally, Section V contains the conclusions of the paper, while the Appendix contains the necessary auxiliary material.

II. Power Control on the Forward Link

This section is organized as follows. First, the CDMA system and channel model is developed. Then, a performance analysis without power control is presented. In Subsection C, forward link bit error rate (BER) expressions are derived assuming a uniform and truncated-bell spatial traffic densities when power control is used. Adjusted power control is investigated in Subsection D using two different performance criteria. Finally, in Subsection E, numerical results and a discussion are presented.

A) The CDMA System and Channel Model

In this paper, the cellular CDMA channel is modeled as undergoing Rayleigh frequency nonselective fading with a fourth-order propagation path loss. The system employs coherent BPSK modulation and direct-sequence (DS) spread-spectrum signaling; also, it is completely synchronized (slotted). Both base stations and mobiles are transmitter and receiver antennas whose patterns are omnidirectional in the horizontal plane. First, let us assume the following scenario which is shown in Fig. 1: there are K users in each cell, that is, $K-1$ interfering users plus one "test" unit in the "home" cell, and K interfering users from each neighboring cell. The "test" mobile is moving along route BV, from the vertex of the hexagon toward the base station. This route is taken because it represents the worst-case situation [20]. When the mobile unit is still close to the vertex of the "test" cell, the signal it receives consists of the "desired" signal plus $K-1$ interference signals from its own cell denoted as B, together with another K interference signals from each of the two closest interfering cells denoted as I in Fig. 1. The interference from the other cells is taken to be negligible. Only a single path is assumed for each base station to the "test" mobile link, and all signals fade in unison within each path. Hence, the received signal at the mobile receiver r_o distance away from the base station transmitter, with or without power control, can be written similar to those given in [21]–[24] as

$$\begin{aligned} s(t|r_0) = n(t) &+ \sum_{i=0}^{K-1} \sqrt{2P_R\xi(r_i)g(r_0)}\chi_1 a_i(t-\tau_i) \\ &\cdot b_i(t-\tau_i)\cos(\omega_c t+\phi_i) \\ &+ \sum_{j=1}^{K} \sqrt{2P_R\xi(r_j)g(r_o')}\chi_2 a_j(t-\tau_j) \\ &\cdot b_j(t-\tau_j)\cos(\omega_c t+\phi_j) \\ &+ \sum_{k=1}^{K} \sqrt{2P_R\xi(r_k)g(r_0')}\chi_3 a_k(t-\tau_k) \\ &\cdot b_k(t-\tau_k)\cos(\omega_c t+\phi_k). \end{aligned} \quad (1)$$

The first term in (1) represents the channel noise, which is an additive white Gaussian process (AWGN) with two-sided power spectral density $N_0/2$. The remaining terms represent the signal received from the "test" base station and two adjacent interfering cells, respectively, where the desired signal is denoted as the 0th signal component. P_R is the transmitted power at the base station designed to reach those mobile units at the cell vertices $r = R$, and $\xi(r)$ is the power control function which contains the information on whether a power control is used or what type of power control method is used. $g(r_0)$ and $g(r_0')$ are functions of the propagation path loss in the "home" cell and the two adjacent cells and are

BER(r_i)

given as

$$g(r_0) = \frac{1}{r_0^4} \tag{2a}$$

and

$$g(r_0') = \frac{1}{(r_0')^4} \tag{2b}$$

where r_0' is the distance from the two interfering base station transmitters to that mobile unit which can be calculated from Fig. 1 as

$$r_0' = \sqrt{3R^2 - 3Rr_0 + r_0^2}. \tag{3}$$

χ_m (for $m = 1, 2, 3$) are i.i.d. Rayleigh random variables with variance 1, representing the multiple access fading effect in the channel. The $a_m(t)$ is the spectral spreading sequence with length N, where N is usually also the processing gain. The $b_m(t)$ is the data signal, τ_m is the time delay parameter which accounts for propagation delay and the lack of synchronism between the signals within a certain time slot, and ϕ_m is the phase angle for the mth signal ($m = i, j, k$). When $E_b/N_0 \gg 1$, where E_b is the bit energy, the value of the channel noise term $n(t)$ is very small compared to the other terms in expression (1). Hence, to simplify the analysis, it is neglected in all our formulations.

The Gaussian approximation method is used in this paper. Although there are more accurate approximations and analysis techniques that have been reported, the results in [6]–[9] and [12] tend to suggest that the Gaussian approximation technique is quite accurate when the number of users K is sufficiently large. Besides, Gaussian approximation simplifies the analysis as well as the end results.

B) Performance Analysis without Power Control

With no power control, the function $\xi(r)$ in (1) is simply $\xi(r) = 1$. The output of a coherent correlation receiver, matched to the desired signal during the lth bit interval $[0, T]$, can be approximated as

$$Z(r_0) = \sqrt{\frac{P_R g(r_0)}{2}}T\left[\chi_1 b_0^{(l)} + \sum_{i=1}^{K-1}\chi_1 I_{i,0}\right] + \sqrt{\frac{P_R g(r_0')}{2}}T\left[\sum_{j=1}^{K}\chi_2 I_{j,0} + \sum_{k=1}^{K}\chi_3 I_{k,0}\right] \tag{4}$$

where $I_{m,0}$, for ($m = i, j, k$), represents the multiple-access interference term due to the mth signal and is given as

$$I_{m,0} = \frac{1}{T}[B_{m,0}(b_m, \tau)]\cos\phi \tag{5}$$

where

$$B_{m,0}(b_m, \tau) = b_m^{(-1)}R_{m,0}(\tau) + b_m^{(0)}\hat{R}_{m,0}(\tau) \tag{6}$$

and the vector $b_m = (b_m^{-1}, b_m^0)$ represents a pair of consecutive data bits of the mth signal. The functions $R_{m,0}(\tau)$ and $\hat{R}_{m,0}(\tau)$ are the continuous-time partial cross-correlation functions of the mth and the 0th spectral-spreading waveforms defined in [17]–[20]. The term $I_{m,0}$ has been widely studied [8], [17]–[20], [25] and is proved to be approximately Gaussian assuming $K \gg 1$ and processing gain (or chip rate) $N \gg 1$. Its variance has been shown to be $var[I_{m,0}] = 1/3N$ for rectangular chip pulses. Hence, the bit error probability, with respect to a given distance r_0 conditioned on χ_1, χ_2, χ_3, is

$$P_e(r_0|\chi_1, \chi_2, \chi_3) = Q\left(\frac{1}{\sqrt{\frac{1}{3N}\left[K - 1 + K\frac{g(r_0')}{g(r_0)}\left(\frac{\chi_2^2 + \chi_3^2}{\chi_1^2}\right)\right]}}\right). \tag{7}$$

Here

$$Q(x) \triangleq \frac{1}{\sqrt{2\pi}}\int_x^\infty e^{-y^2/2}dy. \tag{8}$$

Letting

$$\alpha = \frac{\chi_2^2 + \chi_3^2}{\chi_1^2} \tag{9}$$

it is obvious that $\chi_1^2, \chi_2^2, \chi_3^2$ are chi-squared random variables, each with two degrees-of-freedom. So, the numerator of (9) is actually a chi-squared random variable with four degrees-of-freedom. Hence, the probability density function (pdf) of the random variable α can be found as (see the Appendix)

$$f_\alpha(\alpha) = \frac{2\alpha}{(\alpha + 1)^3}. \tag{10}$$

Using expression (10), one can get the average probability of error as

$$P_e(r_0) = \int_0^\infty Q\left(\frac{1}{\sqrt{\frac{1}{3N}\left(K - 1 + K\frac{g(r_0')}{g(r_0)}\alpha\right)}}\right) \cdot f_\alpha(\alpha)d\alpha. \tag{11}$$

The extreme case of $r_0 = R$ agrees with the formulations in [12]. Notice that this result is obtained assuming the "test" unit is still close to the cell vertex. When the mobile unit gets farther away from the vertex, the majority of interference comes not only from the two nearest interfering cells I, but the other four neighboring cells as well. However, it is well known that the effect of the adjacent cell interference is very small in that case anyway; hence, one can still use (11) to predict the probability of error throughout the whole range of distance without being too optimistic.

C) Forward Link BER with Power Control

Consider now the effect of using nth-order-of-distance power control on the forward link. In order to reduce the neighboring cell interference, the transmitted power at the base station for the mth mobile is now proportional to r_m^n, and the power control function in (1) now becomes

$$\xi(r_m) = \left(\frac{r_m}{R}\right)^n \tag{12}$$

where r_m is the distance between the base station and the mth mobile unit, and n is a number and is usually chosen between 1 and 4, depending on the cost of implementation and the need

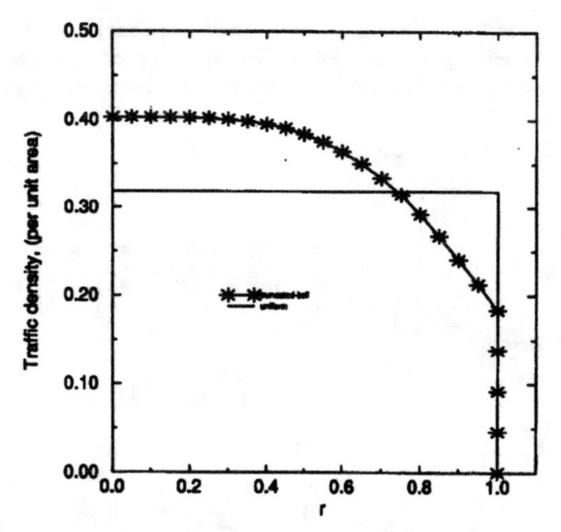

Fig. 2. Traffic density per unit area versus distance for uniform and truncated-bell spatial pdf.

for capacity gain. The decision statistic at the output of base station receiver can be rewritten as

$$\hat{Z}_P(r_0) = \sqrt{r_0^n}b_0^{(l)} + \sum_{i=1}^{K-1}\sqrt{r_i^n}I_{i,0} + \sqrt{\frac{g(r_0')}{g(r_0)}} \cdot \left[\sum_{j=1}^{K}\sqrt{r_j^n\frac{\chi_2}{\chi_1}}I_{j,0} + \sum_{k=1}^{K}\sqrt{r_k^n\frac{\chi_3}{\chi_1}}I_{k,0}\right]. \quad (13)$$

It is straightforward to show that the random variable (rv) $\Sigma_m\sqrt{r_m^n}I_{m,0}(m = i, j, k)$ is also zero-mean Gaussian whose variance depends on the spatial distribution of the mobile units around the base station. The locations of the mobiles are assumed to be independent rv's. In this paper, we consider two models for the user spatial distribution (see Fig. 2) defined as follows.

1) Uniform Spatial Traffic Density: The pdf of the distance r_m between a mobile and the base station is

$$f_r(r_m) = \begin{cases} \frac{2r_m}{R^2}, & 0 \le r_m \le R \\ 0, & \text{otherwise.} \end{cases} \quad (14)$$

Let $\gamma = \sqrt{r_m^n}I_{m,0}$. The variance of γ can be derived as

$$\sigma^2_{\gamma,\text{uniform}} = \int_0^R \sigma^2_{\gamma|r_m} f_r(r_m)dr_m = \text{var}[I_{m,0}]\int_0^R (\sqrt{r_m^n})^2\frac{2r_m}{R^2}dr_m = \frac{2R^n}{n+2}\cdot\frac{1}{3N}. \quad (15)$$

If we define

$$\tilde{\sigma}^2_{\gamma,\text{uniform}} = \sigma^2_{\gamma,\text{uniform}} \cdot 3N \quad (16)$$

then the average probability of error with respect to a certain distance r_0 after applying power control scheme becomes

$$P_e(r_0) = \int_0^\infty Q\left(\sqrt{\frac{r_0^n}{\tilde{\sigma}^2_{\gamma,\text{uniform}}\frac{1}{3N}\left(K-1+K\frac{g(r_0')}{g(r_0)}\alpha\right)}}\right) \cdot f_\alpha(\alpha)d\alpha. \quad (17)$$

2) Truncated Bell Spatial Traffic Density Centered at the Base Station: This is a more realistic assumption for traffic distribution compared to the commonly used ideal uniform distribution[1](see Fig. 2). In this case, the corresponding pdf of the distance r_m between a mobile and the base station is

$$f_r(r_m) = \frac{2r_m}{R^2}ce^{-\pi(r_m/R)^4/4}, \quad 0 \le r_m \le R \quad (18)$$

where c is a constant such that

$$\int_0^R \frac{2r_m}{R^2}ce^{-\pi(r_m/R)^4/4}dr_m = 1. \quad (19)$$

Then, (17) becomes

$$P_e(r_0) = \int_0^\infty Q\left(\sqrt{\frac{r_0^n}{\tilde{\sigma}^2_{\gamma,\text{bell}}\frac{1}{3N}\left(K-1+K\frac{g(r_0')}{g(r_0)}\alpha\right)}}\right) \cdot f_\alpha(\alpha)d\alpha \quad (20)$$

and $\tilde{\sigma}^2_{\gamma,\text{bell}}$ is calculated for different values of n as

$$\tilde{\sigma}^2_{\gamma,\text{bell}} = \begin{cases} c\frac{2R^2}{\pi}(1-e^{-(\pi/4)}), & n=2 \\ c\frac{2R^3}{\pi}\left(\frac{\sqrt{2}}{4}\pi^{-(1/4)}\Gamma\left(\frac{1}{4},0,\frac{\pi}{4}\right)-e^{-(\pi/4)}\right), & n=3 \\ c\frac{2R^4}{\pi}\left[erf\left(\frac{\sqrt{\pi}}{2}\right)-e^{-(\pi/4)}\right], & n=4. \end{cases} \quad (21)$$

D) Use of Adjusted Power Control

Since forward link power control takes the role of reducing adjacent-cell interference to improve the performance in the home cell, it is obvious that power control is not only useless but also "harmful" to users at a close-in distance, in which case the reduction in the desired signal far exceeds that in the interference caused by users from neighboring cells. To avoid this problem, an adjusted power control method is usually used and has been discussed in [1] and [7], in which the power control function becomes

$$\xi(r) = \begin{cases} (r_t/R)^n, & 0 \le r < r_t \\ (r/R)^n, & r_t \le r \le R \end{cases} \quad (22)$$

where r_t is the threshold point where, for those users whose distance is less than this value, a constant transmitting power is used by the base station. To evaluate the system BER under the adjusted power control method, the decision statistic can

[1] The truncated-bell distribution may not be realistic for the cells of certain urban business areas. Such a nonuniform distribution may be more realistic throughout the service area (in a metropolitan area) and not inside the cells (i.e., a densely populated cell in the center, with the density tapering off in more distant cells).

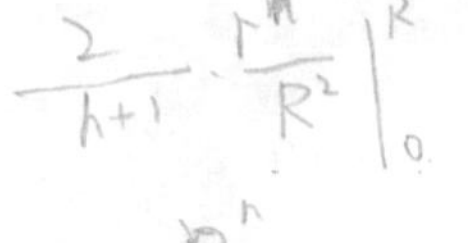

be rewritten as

$$\hat{Z}_P(r_0) = \sqrt{\xi(r_0)}\,b_0^{(l)} + \sum_{i=1}^{K-1}\sqrt{\xi(r_i)}\,I_{i,0} + \sqrt{\frac{g(r_0')}{g(r_0)}}$$

$$\left[\sum_{j=1}^{K}\sqrt{\xi(r_j)}\frac{\chi_2}{\chi_1}I_{j,0} + \sum_{k=1}^{K}\sqrt{\xi(r_k)}\frac{\chi_3}{\chi_1}I_{k,0}\right]. \quad (23)$$

Again, if we let $\gamma = \sqrt{\xi(r_m)}I_{m,0}$ and $\tilde{\sigma}^2 = 3N\cdot\sigma^2$,

$$\begin{aligned}\tilde{\sigma}^2_{\gamma,\text{uniform}} &= \int_0^R \tilde{\sigma}^2_{\gamma|r_m} f_r(r_m)\,dr_m \\ &= \int_0^{r_t}\left(\frac{r_t}{R}\right)^n \frac{2r}{R^2}dr + \int_{r_t}^{R}\left(\frac{r}{R}\right)^n\frac{2r}{R^2}dr \\ &= \left(\frac{r_t}{R}\right)^{n+2} + \frac{2}{n+2}\left(1-\left(\frac{r_t}{R}\right)^{n+2}\right) \\ &= \frac{n}{n+2}\left(\frac{r_t}{R}\right)^{n+2} + \frac{2}{n+2}. \end{aligned} \quad (24)$$

Similarly, for truncated-bell spatial traffic distribution,

$$\begin{aligned}\tilde{\sigma}^2_{\gamma,\text{bell}} &= \int_0^R \tilde{\sigma}^2_{\gamma|r_m} f_r(r_m)\,dr_m \\ &= \int_0^{r_t}\left(\frac{r_t}{R}\right)^n\frac{2r}{R^2}ce^{-\pi(r/R)^4/4}dr \\ &\quad + \int_{r_t}^{R}\left(\frac{r}{R}\right)^n\frac{2r}{R^2}ce^{-\pi(r/R)^4/4}dr \\ &= \left(\frac{r_t}{R}\right)^n erf\left(\frac{\sqrt{\pi r_t^2}}{2}\right) + D(r_t,n) \end{aligned} \quad (25)$$

where $D(\bullet)$ denotes the integration $\int_{r_t}^{R}(r/R)^n 2r/R^2c \cdot e^{-\pi(r/R)^4/4}dr$, which is a function of n dependent incomplete Gamma function. Therefore, $P_e(r)$ can be written in the same form as before, i.e.,

$$P_e(r_0) = \int_0^\infty Q\left(\sqrt{\frac{\xi(r_0)}{\tilde{\sigma}^2_{\gamma,\substack{\text{uniform}\\ \text{bell}}}\frac{1}{3N}\left(K-1+K\frac{g(r_0')}{g(r_0)}\alpha\right)}}\right) \cdot f_\alpha(\alpha)d\alpha. \quad (26)$$

E) Numerical Results and Discussion

Numerical results of the previous analysis are provided here. In all cases, we normalize the cell radius R to unity and assume a processing gain of $N = 1000$.[2]Note that r_0 is the distance of the "test" unit from the base station of the "home" cell along the particular route BV (see Fig. 1). The analysis presented in this paper, however, is not specific to route BV. Therefore, to denote the generic distance of the "test" unit from the base station in its "home" cell, we will use r in the remainder of this paper. Also, results are computed without considering any error correcting ability in the system. Figs. 3 and 4 show the bit error probability with respect to the distance of a mobile user from its base station with power control [no threshold adjustment, see (12)] and without power control, respectively. A different power control exponent n and number of users K are used to show how $P_e(r)$ varies. In Fig. 3, a uniform user spatial distribution is used; whereas in Fig.

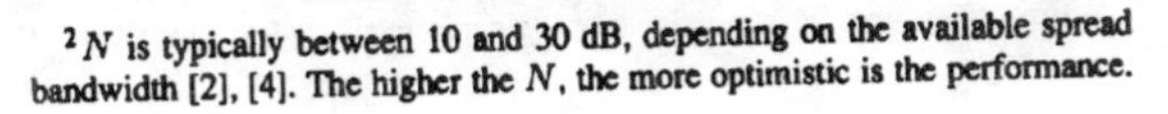

[2] N is typically between 10 and 30 dB, depending on the available spread bandwidth [2], [4]. The higher the N, the more optimistic is the performance.

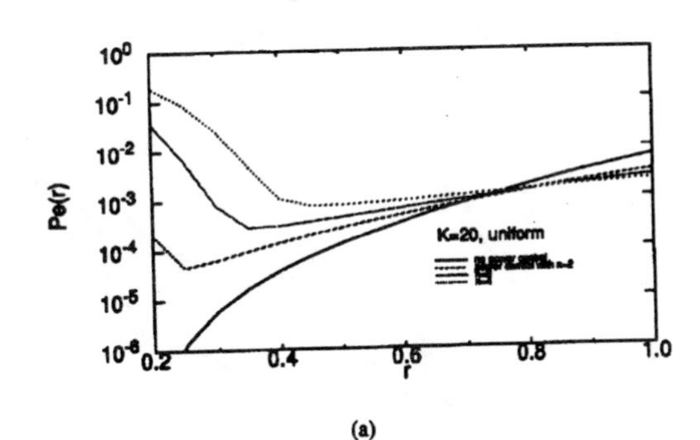

(a)

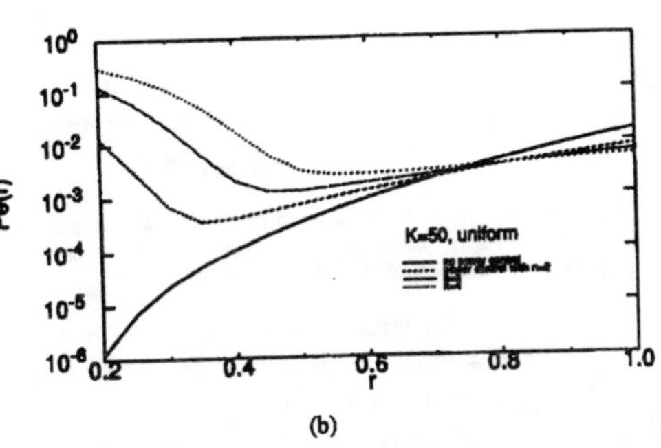

(b)

Fig. 3. Bit erorr probability $P_e(r)$ for forward link with power control ($n = 2, 3, 4$) compared to $P_e(r)$ without power control. Users are assumed to be uniformly distributed. The number of users is (a) $K = 20$ and (b) $K = 50$.

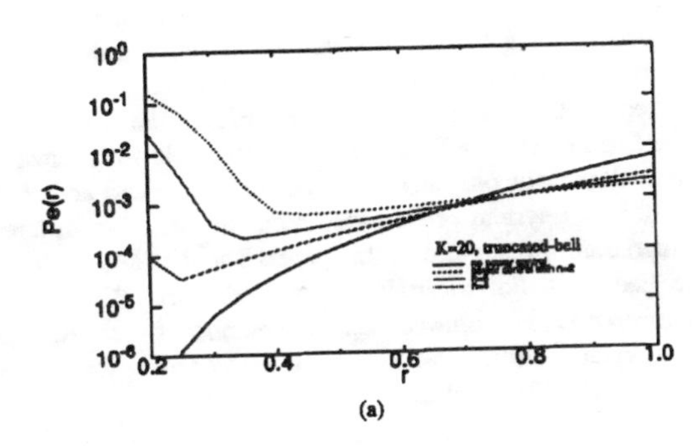

(a)

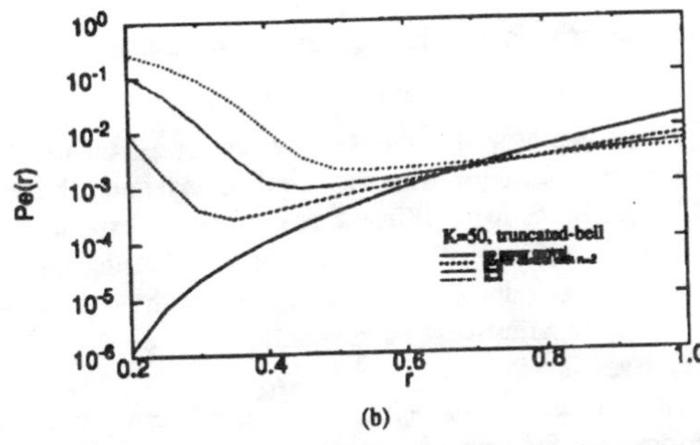

(b)

Fig. 4. Bit error probability $P_e(r)$ for forward link with power control ($n = 2, 3, 4$) compared to $P_e(r)$ without power control. Users are assumed to have a truncated-bell spatial distribution. The number of users is (a) $K = 20$ and (b) $K = 50$.

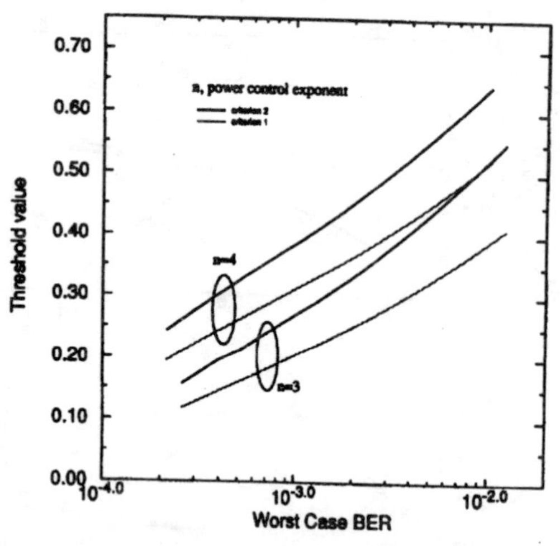

Fig. 5. Threshold value r_t for adjusted power control, using the two criteria discussed in the paper, shown as a function of required worst-case BER for edge users, given the power control exponent n.

4, a truncated-bell-shaped distribution is assumed. Observe that forward link power control can offer the near-edge users more than half an order of magnitude improvement on the bit error rate (BER) performance. However, this improvement in BER for near-edge users comes about at the expense of a significant performance degradation for close-in users. This again proves that, at a close-in distance, the adjacent cell interference no longer plays a major role on the bit error probability of a forward channel; hence, an adjusted power control method [see (21)] must be employed in order not to reduce the desired signal far more than the interference itself, resulting in a BER that is unacceptable for a close-in user. Obviously, one can choose this threshold value according to different system performance requirements. Two different criteria are investigated in this paper. The first one is similar to Lee's [1]. For a certain required edge-user BER, which corresponds to an allowed maximum number of users K in the cell, value r_{t1} (threshold 1) is obtained by setting the $P_e(r_{t1})$ in (17) or (20) equal to

$$P_e(r_{t1}) = P_e(r = 1). \tag{27}$$

The other criteria used is to choose r_{t2} (threshold 2) such that

$$P_e(r_{t2}) = \min P_e(r). \tag{28}$$

Fig. 5 shows how this threshold value, based on the above two criteria, changes with the required maximum edge-user BER. In Fig. 6, for a different power control exponent n (in this paper $0 \leq n \leq 4$), we show the $P_e(r)$ using the above two threshold values, respectively, along with the case where no threshold-adjustment power control is used ($r_t = 0$). The edge BER is set to be 10^{-3} in Fig. 6(a), (b), and (c). The results indicate that there is no obvious difference in the BER performance for near-edge users when one uses either (27) or (28). But for close-in users, (28) gives a much better result. Moreover, using (28), bit error rate increases monotonically with the distance r and, therefore, no service "hole," such as the one observed in [7] where the first criterion defined by (27) was used, will occur. Clearly, although both criteria are

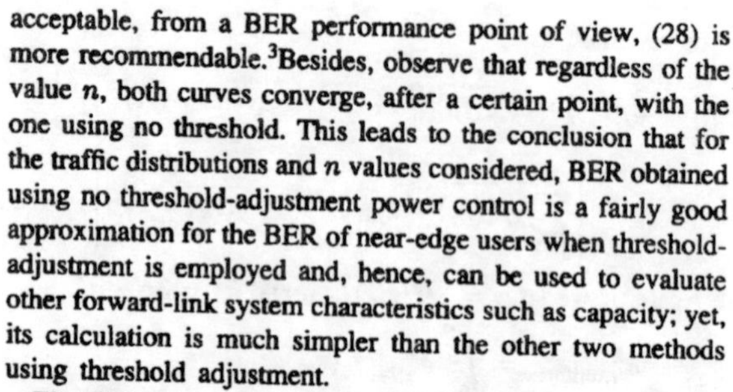
acceptable, from a BER performance point of view, (28) is more recommendable.[3] Besides, observe that regardless of the value n, both curves converge, after a certain point, with the one using no threshold. This leads to the conclusion that for the traffic distributions and n values considered, BER obtained using no threshold-adjustment power control is a fairly good approximation for the BER of near-edge users when threshold-adjustment is employed and, hence, can be used to evaluate other forward-link system characteristics such as capacity; yet, its calculation is much simpler than the other two methods using threshold adjustment.

Fig. 7 shows how the edge (or worst-case) bit error probability degrades when the number of users in the cell increases for uniform and truncated-bell spatial traffic distributions, respectively. Observe that power control can enhance the worst-case (cell-boundary) BER performance of the forward link significantly. Specifically, both figures clearly show that for a fixed number of users per cell (e.g., 30 users/cell), the BER performance improvement using $n = 4$ is roughly half an order of magnitude compared to no power control case, which is a significant improvement.

III. Power Control on Reverse Link

This section is organized as follows. First, a performance analysis of reverse link without power control is presented. To this end, a simple system model associated with the reverse link scenario under consideration is defined (see Fig. 9). In Subsection B, the impact of power control on the performance is analyzed. Finally, numerical results and a discussion are provided.

A) Performance Analysis without Power Control

The CDMA system model considered in this subsection is basically the same as the one used for the forward link analysis; except for reverse channel, the desired signal and the interfering signal arrive at base station through different paths and, therefore, are subject to independent Rayleigh fading. Hence, the output of the base station receiver, matched to the desired signal, is

$$Z(r_0) = \sqrt{\frac{P_R g(r_0)}{2}} T \chi_0 b_0^{(i)} + \sum_{i=1}^{K-1} \sqrt{\frac{P_R g(r_i)}{2}} T \chi_i I_{i,0} + m \cdot \sum_{j=1}^{k} \sqrt{\frac{P_R g(r_{j'})}{2}} T \chi_j I_{j,0}. \tag{29}$$

The first term in (29) represents the desired signal from the "test" mobile which is located at distance r_0 away from the receiver of the base station. The second term represents the interfering signal from "home" cell users, and r_i is their distance from the base station. The third term denotes the interfering signal from the adjacent interfering cells, where $r_{j'}$ is the distance of those users to the "test" base station (see Fig. 8), and m is the number of adjacent interfering

[3] Note, however, that when one uses (28), all the users do not see the same quality of service. For $n = 4$, for example, users in the range of $0 < r < 0.4$ see a much better quality of service than that of $0.4 < r < 1.0$; whereas using (27) provides approximately the same quality of service (5×10^{-4} BER).

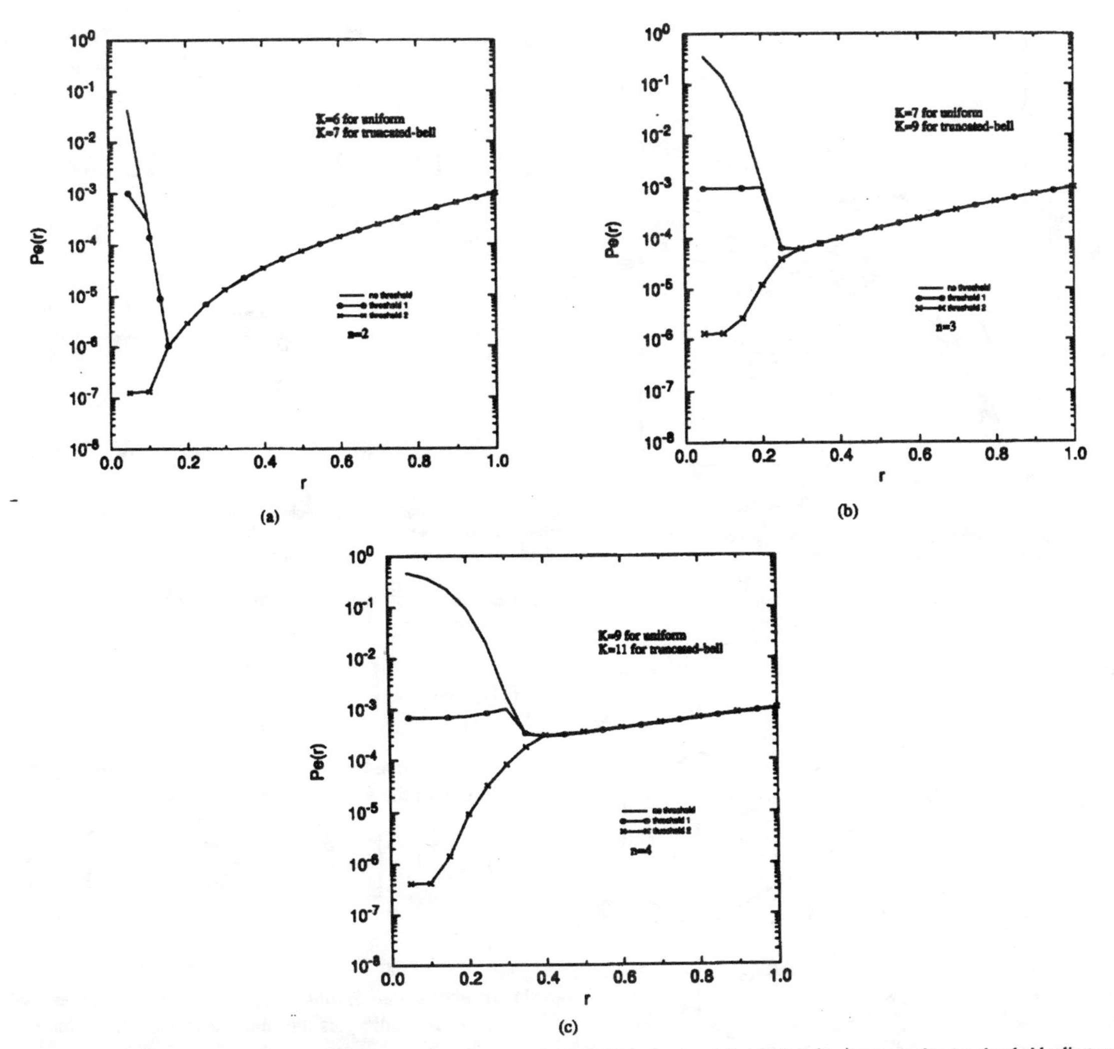

Fig. 6. Bit error probability $P_e(r)$ for forward-link power control using threshold 1 (r_{t1}) and threshold 2 (r_{t2}) compared to no threshold adjustment $(r_t = 0)$ given a required edge-user BER of 10^{-3} and the power control exponent (a) $n = 2$, (b) $n = 3$, (c) $n = 4$.

cells being considered. Using the Central Limit Theorem, it is easy to prove that the second and third term of (29) are both approximately Gaussian with zero mean when K is sufficiently large. The next step in the analysis is to compute their variance. Let $\gamma = \Sigma_{i=1}^{K-1}\sqrt{g(r_i)}\chi_i I_{i,0}$ and $\delta = \Sigma_{j=1}^{K}\sqrt{g(r_{j'})}\chi_j I_{j,0}$. Again, we proceed to examine the performance using two different user spatial distributions.

1) *Uniform Spatial Traffic Density*: In this case, the variance of γ will be

$$\sigma^2_{\gamma,\text{uniform}} = \sum_{i=1}^{K-1}\sigma^2_{I_{i,0}}\int_0^\infty \chi_i^2 f(\chi_i)d\chi_i\int_0^R g(r_i)f(r_i)dr_i$$
$$= \sum_{i=1}^{K-1}\frac{2}{3N}\left(\int_0^\tau \tau^{-4}\frac{2r_i}{R^2}dr_i + \int_\tau^R r_i^{-4}\frac{2r_i}{R^2}dr_i\right)$$
$$= \frac{2(K-1)}{3N}\frac{1}{R^2}\left(\frac{2}{\tau^2}-\frac{1}{R^2}\right). \tag{30}$$

τ is a small fraction of Rn usually used to prevent $g(r_i)$ from being infinite when r_i is very close to 0. To derive the variance of δ, we first consider $g(r_{j'})$. From the geometry shown in Fig. 9, it is easy to see that

$$r_{j'}^2 = 3R^2 + r_j^2 - 2\sqrt{3}Rr_j\cos\theta \tag{31}$$

where r_j is the distance of the adjacent interfering user to its own base station, and θ is uniformly distributed between $(0, 2\pi)$. Thus, the expected value of $g(r_{j'})$ is

$$E[g(r_{j'})] = E[g(r_j,\theta)]$$
$$= \int_0^R \frac{2r_j}{R^2}dr_j\int_0^{2\pi}\frac{\left(\sqrt{3R^2+r_j^2-2\sqrt{3}Rr_j\cos\theta}\right)^{-4}}{2\pi}d\theta$$

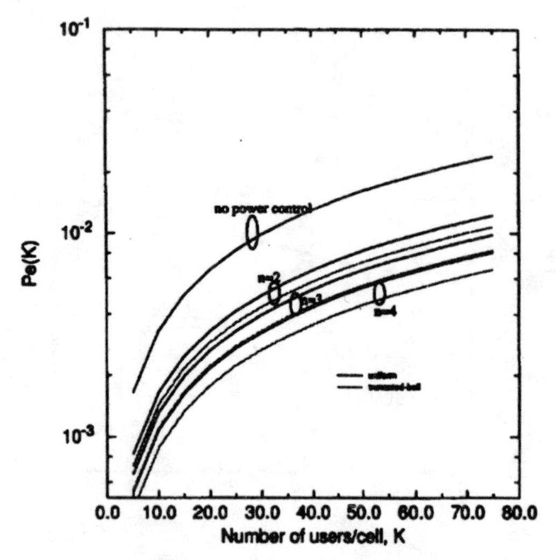

Fig. 7. Bit error probability as a function of users per cell, K, for a uniform and a truncated-bell traffic distribution.

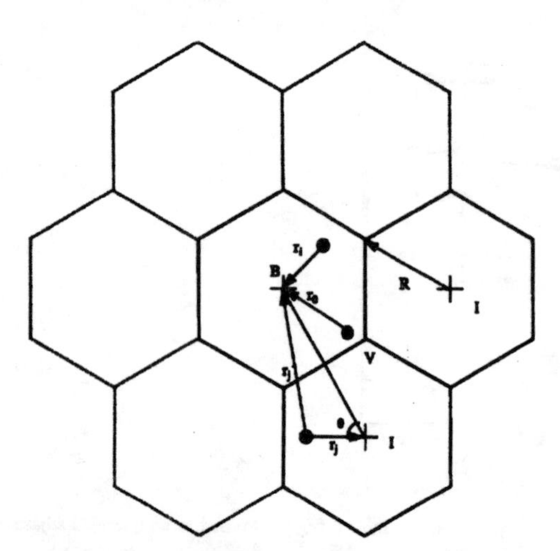

Fig. 8. Reverse-link CDMA system geometry where "test" user is r_0 distance away from the base station.

$$= \int_0^R \frac{2r_j}{R^2} \frac{3R^2 + r_j^2}{(3R^2 - r_j^2)^3} dr_j = \frac{1}{4R^4}. \tag{32}$$

Therefore, it follows that

$$\sigma_{\delta,\text{uniform}}^2 = \sum_{j=1}^{K} \frac{2}{3N} \int_0^R \left(\sqrt{g(r_{j'})}\right)^2 f(r_{j'})\, dr_{j'} = \frac{2K}{3N} E[g(r_{j'})] = \frac{2K}{3N} \frac{1}{4R^4}. \tag{33}$$

Then, the bit error probability for a user at a distance r_0 from the base station in a reverse channel (mobile to base link), conditioned on χ_0, is

$$P_e(r_0|\chi_0) = Q\left(\frac{\sqrt{g(r_0)}\chi_0}{\sqrt{\sigma_{\gamma,\text{uniform}}^2 + m \cdot \sigma_{\delta,\text{uniform}}^2}}\right). \tag{34}$$

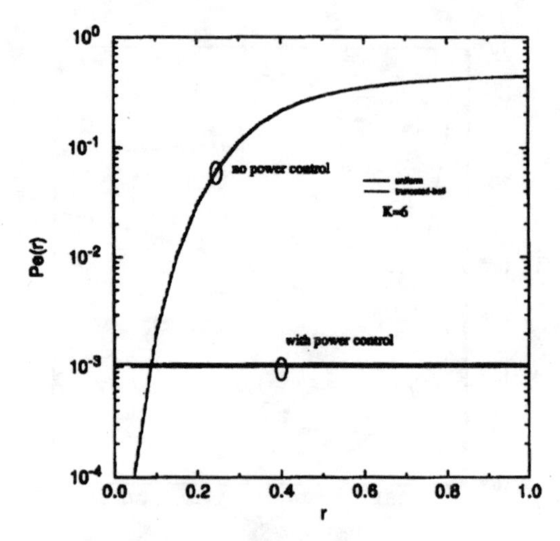

Fig. 9. Reverse-link bit error probability $P_e(r)$ with and without using power control. Number of users per cell K is assumed to be 6.

To get rid of the condition, we again average (34) over the pdf of χ_0 to obtain the average bit error probability

$$P_e(r_0) = \frac{1}{2}\left\{1 - \left[1 + \frac{\sigma_{\gamma,\text{uniform}}^2 + m \cdot \sigma_{\delta,\text{uniform}}^2}{g(r_0)}\right]^{-1/2}\right\}. \tag{35}$$

1) *Truncated-Bell Spatial Traffic Density*: The average bit error probability has the same form as in (35), that is,

$$P_e(r_0) = \frac{1}{2}\left\{1 - \left[1 + \frac{\sigma_{\gamma,\text{bell}}^2 + m \cdot \sigma_{\delta,\text{bell}}^2}{g(r_0)}\right]^{-1/2}\right\} \tag{36}$$

with $\sigma_{\gamma,\text{bell}}^2$ and $\sigma_{\delta,\text{bell}}^2$ obtained by replacing the uniform pdf in (32) and (33) with bell spatial pdf instead. The result turns out to be

$$\sigma_{\gamma,\text{bell}}^2 = \frac{2(K-1)}{3N} \cdot c\left[\left(\frac{1}{r^4} + \frac{\pi}{2}\right)\text{erf}\left(\frac{\sqrt{\pi}}{2}r^2\right) + \frac{1}{r^2}e^{-\frac{\pi r^4}{4}} - \frac{\pi}{2}\text{erf}\left(\frac{\sqrt{\pi}}{2}\right) - e^{-\pi/4}\right] \tag{37}$$

and

$$\sigma_{\delta,\text{bell}}^2 = \sum_{j=1}^{K} \text{var}\left[\sqrt{\frac{1}{r_{j'}^4}}\chi_j I_{j,0}\right] = \frac{2K}{3N}\int_0^R \frac{2r_j}{R^2} c e^{-\frac{\pi(r_j/R)^4}{4}} dr_j$$

$$\cdot \int_0^{2\pi} \frac{(\sqrt{3R^2 + r_j^2 - 2\sqrt{3}Rr_j \cos\theta})^{-4}}{2\pi} d\theta$$
$$= \frac{2K}{3N}\int_0^R \frac{3R^2 + r_j^2}{(3R^2 - r_j^2)^3}\frac{2r_j}{R^2} ce^{-\frac{\pi(r_j/R)^4}{4}} dr_j \quad (38)$$

which can be evaluated only numerically [27].

The results of (35) and (36) are shown in Fig. 9, where R has been set to unity and $\tau = 0.01R$. For simplicity, and also for the results to be compatible with the forward link case, we assume $m = 2$ in our calculations. Unlike the forward link case, where only the neighboring cell interference affects the signal-to-noise ratio as the "test" mobile moves away from(toward) the base station, in a reverse link or mobile-to-base link, the desired signal fades(increases) as the "test" mobile moves away from(toward) the base station; whereas the interference from the home cell as well as the neighboring cells stays statistically the same and, therefore, causes the so-called near–far problem. It is clear from the figure that CDMA is virtually useless for a reverse link unless some sort of power control is applied [1]–[6], [9], [11], [12], [26]. This is addressed in the following subsection.

B) Performance Analysis with Power Control

For the reverse link, since a much more severe near–far problem exists as opposed to the forward link, we are going to use the fourth-order-of-distance power control, which is the most commonly used form for the reverse link power control [1]–[6]. Hence, we define the power control function as

$$\xi(r) = \left(\frac{r}{R}\right)^4 \quad (39)$$

and, accordingly, the output of the receiver thus becomes

$$Z(r_0) = \sqrt{\frac{P_R g(r_0)\left(\frac{r_0}{R}\right)^4}{2}} T\chi_0 b_0^{(l)} + \sum_{i=1}^{K-1}\sqrt{\frac{P_R g(r_i)\left(\frac{r_i}{R}\right)^4}{2}} T\chi_i I_{i,0} + m\cdot\sum_{j=1}^{K}\sqrt{\frac{P_R g(r_{j'})\left(\frac{r_j}{R}\right)^4}{2}} T\chi_j I_{j,0}. \quad (40)$$

Equivalently, the decision statistic is

$$\tilde{Z}(r_0) = \chi_0 + \sum_{i=1}^{K-1}\chi_i I_{i,o} + m\cdot\sum_{j=1}^{K}\sqrt{\left(\frac{r_j}{r_{j'}}\right)^4}\chi_j I_{j,0}. \quad (41)$$

And in the case of using power control, we find

$$\sigma_\gamma^2 = \sum_{i=1}^{K-1}\text{var}[\chi_i I_{i,0}] = \frac{2(K-1)}{3N} \quad (42)$$

and

$$\sigma_\delta^2 = \sum_{j=1}^{K}\text{var}[\chi_j I_{j,0}]\int\int\left(\frac{r_j}{r_{j'}(r_j,\theta)}\right)^4 f(r_j,\theta)dr_j d\theta$$

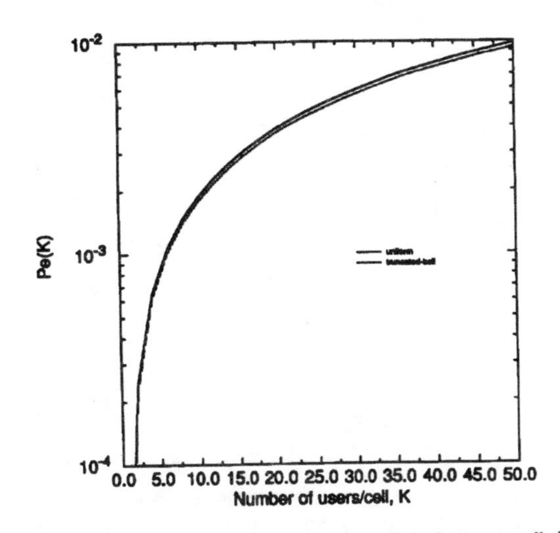

Fig. 10. Bit error probability as a function of number of users per cell, K, for both uniform and truncated-bell traffic distribution.

$$= \begin{cases} \frac{2K}{3N}\left(12\log_e\left(\frac{3}{2}\right) - \frac{19}{4}\right), & \text{for uniform traffic density} \\ \frac{2K}{3N}\int_0^R r_j^4 \frac{3R^2 + r_j^2}{(3R^2 - r_j^2)^3}\frac{2r_j}{R^2} ce^{-\frac{\pi(r_j/R)^4}{4}} dr_j, & \text{for truncated-bell traffic density.}\end{cases} \quad (43)$$

From Fig. 9, it can be seen that when power control is applied on the reverse channel, the $P_e(r)$ curve is almost flat. Figs. 10 and 11 show the bit error probability when using reverse-link power control for different number of users per cell K and a various number of interfering cells m in the first tier, respectively. Again, in all cases, we compare the performance of a uniform traffic distribution to that of a truncated-bell one. The results in Figs. 10 and 11 tend to suggest that although a truncated-bell user distribution offers a better performance, nevertheless, the improvement is much less than that of a forward-link channel. This actually agrees with the difference in the forward and reverse channel behavior as discussed earlier—where, for a reverse link, the adjacent cell interference is no longer the dominant factor on the SIR.

IV. Physical Interpretation of Results

In this paper, the impact of power control on the system performance of CDMA systems in the forward and reverse links is examined and quantified. To that end, a simple expression is provided for computing the BER of mobile units (subscribers) as a function of their distance from the base station (see (26), for instance, for the forward link case). The analysis presented takes into account the spatial distribution of users in that cell, number of users, power control function, and interference from the "home" and adjacent cells.

Since the physical insight provided by the closed-form results for the forward and reverse links are essentially the same, for briefness, we just discuss the results for the forward-link case as an example. It is straightforward, however, to

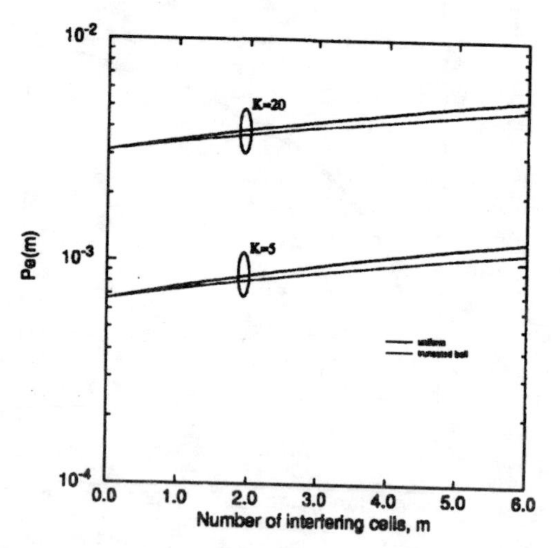

Fig. 11. Bit error probability as a function of number of interfering cells in the first tier, for a given K, for both uniform and truncated-bell traffic distribution.

extend these arguments to the results of the reverse link of CDMA networks.

The bit error probability in the forward link can be computed using (26). To appreciate the physical insight provided by (26), let us study it in more detail. After normalizing the cell radius $R = 1$, expression (26) can be written, after transformations, as (44), at the bottom of the page, where

$$\int_0^1 \xi(r)f(r)dr = \int_0^{r_t} r_t^n f(r)dr + \int_{r_t}^1 r^n f(r)dr = r_t^n \int_0^{r_t} f(r)dr + \int_{r_t}^1 r^n f(r)dr. \quad (45)$$

It is clear from (26) and (44) that the BER is determined by the $Q(\cdot)$ function, and the outer integral is just for averaging the Rayleigh fading effect. Furthermore, the argument of the $Q(\cdot)$ function represents the signal-to-interference ratio (SIR) of the CDMA network under investigation. Hence, the numerator represents the signal power, while the denominator represents the interference power at the decision gate of the receiver. To see the impact of certain system parameters on system performance, let us further assume that the spatial traffic distribution at hand is a uniform one. Using (24), it can be shown that the SIR in the forward link is (46), at the bottom of the page.

A careful inspection of (46) reveals that, for a fixed processing gain N, threshold distance r_t, "test" unit distance from the base station r_0, and power control exponent n, as K (the number of interferers) increases, the interference power increases and, therefore, SIR decreases, which implies that the BER performance of the CDMA network under investigation deteriorates. Similarly, if N, K, n, and r_t (with threshold-2 criterion, for example) are fixed, then as r_0 decreases, the interference power decreases much faster than the signal power and, hence, SIR increases, which implies that the

$$P_e(r_0) = \begin{cases} \displaystyle\int_0^\infty Q\left(\sqrt{\frac{r_t^n}{\frac{1}{3N}\int_0^1 \xi(r)f(r)dr\left(K-1+K\left(\frac{1}{\frac{3}{r_0^2}-\frac{3}{r_0}+1}\right)^2\alpha\right)}}\right) f_\alpha(\alpha)d\alpha, & \text{for } r_0 \le r_t \\ \displaystyle\int_0^\infty Q\left(\sqrt{\frac{r_0^n}{\frac{1}{3N}\int_0^1 \xi(r)f(r)dr\left(K-1+K\left(\frac{1}{\frac{3}{r_0^2}-\frac{3}{r_0}+1}\right)^2\alpha\right)}}\right) f_\alpha(\alpha)d\alpha, & \text{for } r_t < r_0 \le 1 \end{cases} \quad (44)$$

$$SIR = \frac{r_0^{n/2}}{\left[\frac{1}{3N}\left(\frac{n}{n+2}r_t^{n+2}+\frac{2}{n+2}\right)\cdot\left\{K-1+K\left(\frac{1}{\frac{3}{r_0^2}-\frac{3}{r_0}+1}\right)^2\alpha\right\}\right]^{0.5}} \quad (46)$$

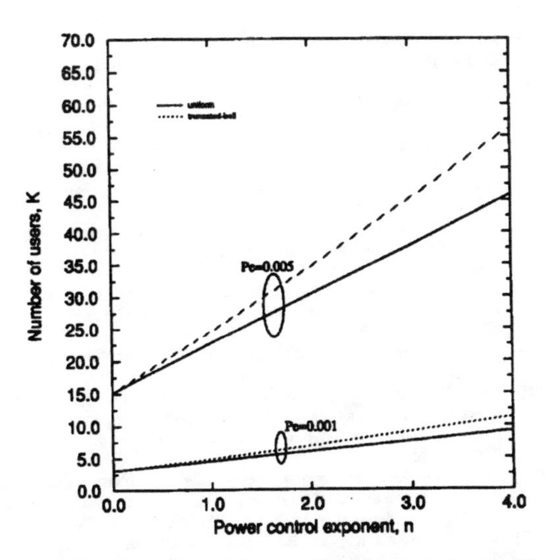

Fig. 12. Cell capacity K versus power control exponent n for two different values of edge performance, using uniform and truncated-bell traffic pdf's.

BER performance improves [see, e.g., Fig. 6(b) and (c)]. Specifically, the $K-1$ term in $\{\cdot\}$ of the denominator of (46) represents the interference of the "home" interferers, while the remaining term in the $\{\cdot\}$ denotes the interference from the adjacent cells. Observe that home interference is r_0 independent, whereas adjacent cell interference is inversely proportional to the distance of the "test" unit from the base station. Thus, expression (46) verifies the intuitive result that as r_0 decreases, the contribution of adjacent cell interference decreases, while the contribution of "home" interferers remains statistically the same. Furthermore, it should be noted that for a fixed BER level (say, BER=10^{-3}), SIR is fixed ($SIR = 3.09$). Again, for a fixed value of threshold, N, and r, expression (46) indicates that K (number of subscribers in the "home" cell and adjacent cells) can be increased by increasing power control factor n from $n = 0$ to $n = 4$. Theoretically speaking, K can be further increased by choosing $n > 4$, at least up to a critical value of n. Note, however, that $n > 4$ needs a much larger dynamic range of the transmission power and also a faster tracking speed which, in turn, means a more complicated and expensive system whose cost cannot be compensated by the capacity gain; hence, usually $n \leq 4$ is used [1]. This is illustrated in Fig. 12 for uniform and truncated-bell-shaped distributions. As expected, the figure illustrates a threefold or larger increase in capacity which highlights the significance of power control in CDMA.

It should be noted that, compared to the previously published results, the capacity enhancement predicted by this work is more conservative. There are two main reasons for this discrepancy. One reason is the fact that some of the recent studies on this subject neglect the impact of Rayleigh fading on the system performance of CDMA networks. As shown in this paper, this may be a critical source of degradation in the system performance evaluations. The other main reason is the use of error correction codes to improve system performance, which has been studied by several researchers. In this paper,

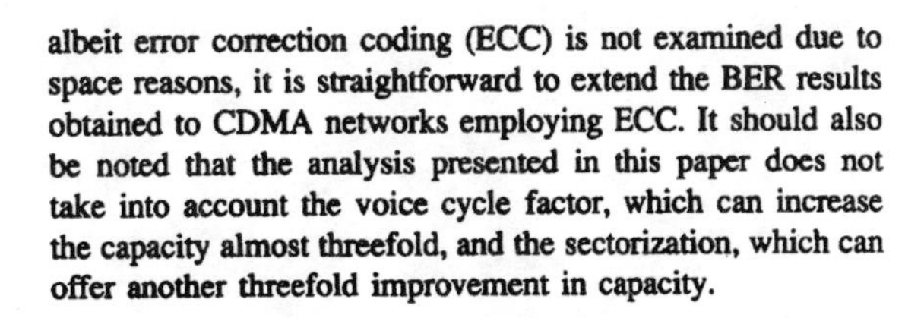
albeit error correction coding (ECC) is not examined due to space reasons, it is straightforward to extend the BER results obtained to CDMA networks employing ECC. It should also be noted that the analysis presented in this paper does not take into account the voice cycle factor, which can increase the capacity almost threefold, and the sectorization, which can offer another threefold improvement in capacity.

V. Conclusions

The impact of power control on the system performance of CDMA networks is examined and quantified for both the forward and reverse links. The analysis considers two different spatial user distributions for the derivations, and takes into account the position of the "test" mobile unit, number of users per cell, power control function, Rayleigh fading, and interference from the "home" and adjacent cells.

The analysis presented leads to a simple closed-form SIR result which facilitates valuable physical insight into the impact of key system parameters on the performance of the CDMA system under consideration. Also, for both the forward and the reverse links, the closed-form SIR result enables one to see the interrelationship of key system parameters without the need for often tedious simulations. One of the main strengths of the approach presented is that it is easy to extend it to other spatial user distributions for different geographical areas.

Results indicate that, for the forward link case, power control can be used to suppress the adjacent cell interference (which has a major adverse effect on the received SIR) and, hence, to improve the system performance near the cell boundary substantially. Moreover, the adjusted-power control is revisited using two different performance criteria. The second threshold criterion proposed is found to perform much better than the first one, which is similar to Lee's proposal, for close-in users. Besides, it also alleviates the service "hole" problem associated with the first criterion observed by [7]. For the reverse link case, theoretically speaking, power control mitigates the near–far problem since the BER is flat throughout the whole range of cell distance, and the performance difference caused by different traffic distributions is also less significant than the forward link case.

The capacity enhancement predictions of this work are less optimistic (roughly three times at the cell boundary) than previously reported results, mainly due to the fact that Rayleigh fading is taken into account in the analysis. Capacity, however, can somewhat be further increased via the use of error-correction coding.

Appendix

Proof of Expression (10): The Rayleigh model describes a continuous r.v. (χ) produced from two Gaussian random variables by the following transformation. Let a and b be independent Gaussian r.v.'s with zero mean and variance $\sigma^2 = 1$

$$\chi = +\sqrt{a^2 + b^2}. \quad \text{(A1)}$$

Then, $\chi_1^2, \chi_2^2, \chi_3^2$ are chi-squared random variables, each with two degrees-of-freedom. So, the numerator $\chi_2^2 + \chi_3^2$ is a chi-squared random variable with four degrees-of-freedom. Hence, let $x = \chi_2^2 + \chi_3^2$ and $y = \chi_1^2$; then,

$$\alpha = \frac{\chi_2^2 + \chi_3^2}{\chi_1^2} = \frac{x}{y} \tag{A2}$$

and

$$\begin{aligned} f_\alpha(\alpha) &= \int_{-\infty}^{+\infty} y f(\alpha y, y) dy \\ &= \int_0^{+\infty} y \frac{1}{\Gamma(1)2} e^{-\alpha y/2} \frac{1}{\Gamma(2)2^2} y e^{-y/2} dy \\ &= \frac{2\alpha}{(\alpha+1)^3}. \end{aligned} \tag{A3}$$

This result can be extended to a more general case where

$$\alpha = \frac{\sum_{i=1}^{n} \chi_i^2}{\chi_0^2} = \frac{n\alpha}{(\alpha+1)^{n+1}}. \tag{A4}$$

ACKNOWLEDGMENT

The authors wish to thank Dr. F. Simpson of Motorola for his valuable comments on the manuscript. The constructive criticisms and helpful suggestions of the two reviewers are also acknowledged.

REFERENCES

[1] W. C. Y. Lee, "Overview of cellular CDMA," *IEEE Trans. Veh. Technol.*, vol. 40, pp. 291–302, May 1991.
[2] K. S. Gilhousen, I. M. Jacobs, R. Padovani, A. J. Viterbi, L. A. Weaver, and C. E. Wheatley, "On the capacity of a cellular CDMA system," *IEEE Trans. Veh. Technol.*, vol. 40, pp. 303–312, May 1991.
[3] W. C. Y. Lee, "Power control in CDMA," in *Proc. IEEE 42nd Veh. Technol. Conf.*, 1991, pp. 77–79.
[4] R. L. Pickholtz, L. B. Milstein, and D. L. Schilling, "Spread spectrum for mobile communications," *IEEE Trans. Veh. Technol.*, vol. 40, pp. 313–321, May 1991.
[5] G. L. Turin, "The effects of multipath and fading on the performance of direct-sequence CDMA system," *IEEE Trans. Veh. Technol.*, vol. 33, pp. 213–219, Aug. 1984.
[6] J. M. Holtzman, "CDMA power control for wireless networks," in *Proc. IEEE 42nd Veh. Technol. Conf.*, 1992, pp. 299–311.
[7] R. R. Gejji, "Forward-link-power control in CDMA cellular systems," *IEEE Trans. Veh. Technol.*, vol. 41, pp. 532–536, Nov. 1992.
[8] J. M. Holtzman, "A simple, accurate method to calculate spread-spectrum multiple access error probabilities," *IEEE Trans. Commun.*, vol. 40, pp. 461–464, Mar. 1992.
[9] M. A. Mokhtar and S. C. Gupta, "Power control considerations for DS/CDMA personal communication systems," *IEEE Trans. Veh. Technol.*, vol. 41, pp. 479–487, Nov. 1992.
[10] H. Alavi and R. W. Nettleton, "Downstream power control for a spread spectrum cellular mobile radio system," in *Proc. Globecom '82*, 1982, pp. 84–88.
[11] R. W. Nettleton and H. Alavi, "Power control for a spread spectrum cellular mobile radio system," in *Proc. 33rd IEEE Veh. Technol. Conf.*, 1983, pp. 242–246.
[12] L. B. Milstein, T. S. Rappaport, and R. Barghouti, "Performance evaluation for cellular CDMA," *IEEE J. Select. Areas Commun.*, vol. 10, pp. 680–689, May 1992.
[13] J. Zander, "Performance of optimum transmitter power control in cellular radio systems," *IEEE Trans. Veh. Technol.*, vol. 41, pp. 57–62, Feb. 1992.
[14] ——, "Distributed cochannel interference control in cellular radio systems," *IEEE Trans. Veh. Technol.*, vol. 41, pp. 305–311, Aug. 1992.
[15] M. M. Wang and O. K. Tonguz, "Forward link power control for cellular CDMA networks," *IEE Electron. Lett.*, vol. 29, pp. 1195–1197, June 1993.
[16] D. J. Goodman and A. A. M. Saleh, "The near/far effect in local aloha radio communications," *IEEE Trans. Veh. Technol.*, vol. 36, pp. 19–27, Feb. 1987.
[17] D. Raychaudhuri, "Performance analysis of random access packet-switched code division multiple access systems," *IEEE Trans. Commun.*, vol. 29, pp. 895–901, June 1981.
[18] J. M. G. Linnartz, R. Hekmat, and R. Venema, "Near-far effects in land mobile random access networks with narrow-band rayleigh fading channels," *IEEE Trans. Veh. Technol.*, vol. 41, pp. 77–90, Feb. 1992.
[19] J. C. Arnbak and W. van Blitterswijk, "Capacity of slotted ALOHA in a Rayleigh fading channel," *IEEE J. Select. Areas Commun.*, vol. 5, pp. 261–269, Feb. 1987.
[20] V. H. Macdonald, "The cellular concept," *Bell Syst. Tech. J.*, vol. 58, pp. 17–41, Jan. 1979.
[21] M. B. Pursley, D. V. Sarwate, and W. E. Stark, "Error probability for direct-sequence spread-spectrum multiple-access communications—Part I: Upper and lower bounds," *IEEE Trans. Commun.*, vol. 30, pp. 975–984, May 1982.
[22] M. B. Pursley, "Performance evaluation for phase-coded spread-spectrum multiple-access communication—Part I: System analysis," *IEEE Trans. Commun.*, vol. 25, pp. 795–799, Aug. 1977.
[23] E. A. Geraniotis and M. B. Pursley, "Performance of noncoherent direct-sequence spread-spectrum communications over specular multipath fading channels," *IEEE Trans. Commun.*, vol. 34, pp. 219–226, Mar. 1986.
[24] J. S. Lehnert and M. B. Pursley, "Error probability for binary direct-sequence spread-spectrum communications with random signature sequences," *IEEE Trans. Commun.*, vol. 35, pp. 87–98, Feb. 1987.
[25] R. K. Marrow, Jr., "Bit-to-bit error dependence in slotted DS/SSMA packet systems with random signature sequences," *IEEE Trans. Commun.*, vol. 37, pp. 1052–1061, Oct. 1989.
[26] G. L. Stüber and C. Kchao, "Analysis of a multiple-cell DS CDMA cellular mobile radio system," *IEEE J. Select. Areas Commun.*, vol. 10, pp. 669–679, May 1992.
[27] M. Abramowitz and I. A. Stegun, *Handbook of Mathematical Functions*. New York: Dover, 1972.

Ozan K. Tonguz (S'86–M'90) was born in Nicosia, Cyprus, in May 1960. He received the B.Sc. degree in electronic engineering from the University of Essex, Essex, England, in 1980; and the M.Sc. and Ph.D. degrees in electrical engineering from Rutgers University, New Brunswick, NJ, in 1986 and 1990, respectively.

In 1981 he returned to Cyprus. After two years of mandatory military service (1981–1983), he was an Assistant Lecturer at the Eastern Mediterranean University of Northern Cyprus, during the academic year of 1983–1984. In September 1984 he joined the Department of Electrical and Computer Engineering, Rutgers University. Between January 1988 and May 1990 he was a visiting doctoral student at the Advanced Lightwave Systems Division of Bell Communications Research, Red Bank, NJ, where he conducted research for his Ph.D. thesis. From May 1990 to August 1990 he was a member of Technical Staff at Bellcore, and did work on coherent lightwave technology and optical amplifiers. He joined the Department of Electrical and Computer Engineering, State University of New York at Buffalo, in September 1990, as an Assistant Professor. At SUNY/Buffalo, he leads substantial research activity in the broad area of telecommunications. His current research interests are in mobile and personal communication systems, broadband optical networks, coherent lightwave systems, and fiber in the local loop applications. He has published in the areas of optical communications and mobile cellular radio communications, and is the author of more than 40 technical papers in leading technical journals and conferences in these areas.

Dr. Tonguz has acted as a Reviewer for various IEEE and IEE Transactions and Journals. He is a member of the Optical Society of America and Eta Kappa Nu.

Melanie M. Wang (S'92–M'93) was born in Shanghai, China. She transfered from Shanghai Jiao-Tong University and received the B.S. degree (summa cum laude) in May 1991, and the M.S. degree in May 1993, both in electrical and computer engineering, from the State University of New York at Buffalo.

From 1990 to 1991 she was a Student Assistant in the Superconductivity Lab at the State University of New York at Buffalo. From 1992 to 1993 she conducted research work in the Personal Communications Research Lab in the Electrical and Computer Engineering Department, where she was also a Teaching Assistant. Since July 1993 she has been with GTE Service Corporation. Currently she is involved in designing an air-to-gorund digital packet data network for GTE Airfone, Inc. Her research interests are in wireless mobile communications, CDMA for wireless networks, and the application of fiber optic communications to wireless networks.

Spectral Efficiency of a Power-Controlled CDMA Mobile Personal Communication System

John H. Gass Jr., *Student Member, IEEE*, Daniel L. Noneaker, *Member, IEEE*, and Michael B. Pursley, *Fellow, IEEE*

***Abstract*— The spectral efficiency is determined for a direct-sequence code-division multiple-access (CDMA) communications system that employs rapid closed-loop power control on a channel with doubly selective fading. The sensitivity of the spectral efficiency to the chip rate of the direct-sequence waveform and the number of taps in the rake receiver is considered. The effectiveness of the power-control method is also examined for different power-control delays, delay spectra, and Doppler spreads. The implications of the results for the design of personal communication systems are discussed.**

I. Introduction

SEVERAL years ago, direct-sequence (DS) spread spectrum modulation was proposed for use in mobile cellular code-division multiple-access (CDMA) systems [1]. That proposal led to the development of an air interface standard [2] for DS CDMA in cellular frequency bands. DS CDMA has also been proposed for personal communication systems (PCS) that will operate in recently allocated PCS frequency bands [3], [4]. Several air interfaces employing CDMA or a hybrid of CDMA and time-division multiple-access are under consideration as standards for PCS [5].

A multiple-access system that employs DS spread spectrum modulation is susceptible to *near–far interference* [6]. Near–far interference occurs when the receiver input includes one or more other CDMA signals that are stronger than the desired signal. The near–far effect can be reduced by adapting the power of each transmitter to changes in the channel response or the interference environment. This adaptation is referred to as *power control* [7]. The reverse (mobile to base station) link of a cellular CDMA system is particularly susceptible to near–far interference, although intercell interference can cause near–far interference on the forward link as well.

Important characteristics of a CDMA system include the chip rate [1], [3], the receiver structure and complexity, and the method of implementing power control. The manner in which each of these features of the system design affects performance is dependent upon the speed of the mobile units and the characteristics of the channels that are likely to be encountered in the intended application. The purpose of this paper is to examine the effects of the power-control technique on the performance of systems with different chip rates and receiver structures. In particular, we consider power control used to reduce intracell near–far interference on the reverse link. The susceptibility of each system to rapid fading and power-control feedback delay is considered in detail.

Manuscript received February 1995; revised August 1995. This work was supported in part by the Holcombe Endowment at Clemson University and by the Army Research Office under grant DAAH04-94-G-0154. J. H. Gass is the recipient of a National Science Foundation Graduate Research Fellowship. Part of this research was presented at the 32nd Annual Allerton Conference on Communications, Control, and Computing, Monticello, IL, September 1994 and at the IEEE 45th Vehicular Technology Conference, Chicago, IL, July 1995.

The authors are with Clemson University, Clemson, SC 29634-0915 USA.

Publisher Item Identifier S 0733-8716(96)01901-4.

In this paper, we consider power control implemented by varying the transmitted power of the mobile units so that an adequate signal-to-interference ratio (SIR) is maintained at the receiver for each transmission. The SIR is estimated and the power of each mobile transmitter is changed so that the SIR remains near a specified value. We focus on *closed-loop power control*, in which the estimates are formed at the base station receiver and commands to adjust the transmitted power are sent from the base station to the mobile unit. Power control does not eliminate the near–far effect, because the control is based on imperfect estimates of the SIR, and there may be some delay before the transmitter power can be adjusted. The amount of residual variation in the SIR depends on the technique used for power control, the rate of variation of the channel response, the chip rate of the system, and the type of receiver that is employed.

For a multipath fading channel, DS spread spectrum produces a signal waveform that permits an appropriately designed receiver to resolve some or all of the multipath components of the received signal. DS spread spectrum with a sufficiently high chip rate permits discrimination against unwanted multipath components in a standard correlation receiver or recombination of these components in a rake receiver [8]. The probability of error is determined in [9] for the demodulator output in a CDMA system that employs binary differential phase-shift keying (DPSK), noncoherent demodulation, and a rake receiver with equal-gain square-law combining. The performance of this system is shown to depend on both the characteristics of the channel and the chip rate of the system. The system analyzed in [9] uses an *average-power control* that compensates only for long-term variations in the net path loss, such as shadowing. In contrast, closed-loop power control is intended to compensate for short-term variations in the strength of the received signal. The effect of closed-loop power control on system performance is considered in [1], [7], and [10] for receivers that employ rake reception. In [1], [7], and [10], however, the focus is on a CDMA system with a specified chip rate and the difference in multipath resolution capability obtained with different chip rates is not addressed.

In this paper, we extend the investigation in [9] to consider a system that uses closed-loop power control. We examine the effectiveness of closed-loop power control for systems of different chip rates and with different numbers of taps in the rake receiver. Since our intent is to focus on channels that are subjected to short-term fading, we assume that average-power control compensates fully for long-term variations in the net path loss within the constraint imposed by the limitation on maximum transmission power. We consider a single cell in isolation, although the analysis can be extended to incorporate the effects of interference from nearby cells. As in [9], a CDMA system with a normalized chip rate of 50 chips per symbol is taken as a representative low chip rate system and a chip rate of 400 is used as an example of a high-chip-rate system.

The base station in a cellular system must include one receiver for each supported mobile transmission. In practice, the circuit complexity required to implement a rake receiver limits the number of taps for most applications. A receiver must make decisions regarding power control based on the strength of only those components of the received signal that are captured by the few taps of the receiver. For some channels, the limited number of taps may be insufficient to collect the signal energy that is received from every propagation path. Though the uncollected signal energy is not used in the generation of the power-control command for the corresponding mobile transmitter, it does contribute to the interference at the receiver for each of the other mobile transmitters in the cell. The result is that two transmissions which produce equal signal powers at the outputs of the corresponding rake receivers may yield dramatically different levels of multiple-access interference. Under some circumstances, the uncollected signal energy at the receivers can have a significant effect on the performance of the system. Unlike previous investigations of power control in mobile CDMA systems, we account for this effect in our analysis.

It is not the intent of this paper to model or analyze any one specific CDMA system. Rather, the goal is to consider a basic CDMA system that has the features that are important for the performance trade-offs described. Any CDMA system proposed for implementation would likely differ in the details of the power control from the model considered here. The system would also certainly have error-control coding and interleaving, neither of which is included in or necessary for the investigation reported in this paper. The performance measures employed in this paper are based on the average probability of error at the demodulator output, and the averaging accounts for the short-term variations in the fading channel. Any improvement that can be made in this error probability leads to a reduction in the error probability at the output of the decoder. In this paper, the absolute performance is far less important than relative performance. The trade-offs discussed in this paper are also applicable to CDMA systems with coding and interleaving.

II. Doubly Selective Fading Channels

Many mobile communications channels have a few strong propagation paths (perhaps only one) and a number of weaker paths [3]. For a DS system with a low chip rate, the arrival times for multiple multipath components commonly fall within one chip interval. The composite signal formed by such components may be modeled as a single diffuse fading signal that has a Rician fading amplitude with a strong Rayleigh component and a relatively weak specular component. If the DS signal has a chip interval that is much smaller than the smallest of the differential delays for the various propagation paths, however, a correlation receiver can lock onto a single multipath component. Thus, the channel may appear to the correlation receiver to produce a single component that exhibits Rician fading with a strong specular component and a relatively weak Rayleigh component.

We model the channel in a manner that reflects the phenomenon described above and permits tractable analysis of the performance of the system. The channel model consists of several clusters of paths, each cluster having a single specular path and a continuum of Rayleigh-fading paths with arrival times centered around the arrival time of the specular path.

This channel is a special case of the Gaussian wide-sense-stationary uncorrelated-scattering (WSSUS) channel [11]. The channel process for the kth transmission is represented as a complex Gaussian random process $h_k(t, \xi)$ characterized by its specular part

$$E[h_k(t, \xi)] = \sum_{i=1}^{M_k} \rho_{i,k}\delta(\xi - \xi_{i,k})$$

and the autocovariance function of its diffuse part

$$\text{Cov}\,[h_k(t, \xi), h_k(x, \alpha)] = \sum_{i=1}^{M_k} 2\sigma_{i,k}^2 d_k(t-x) g_{i,k}(\xi)\delta(\alpha - \xi)$$

where t and x are time variables and ξ and α are path delay variables. The number of path delay clusters in the channel is M_k. The parameter $\rho_{i,k}$ is the amplitude of the ith specular component, d_k is the *time-correlation function* of the kth channel, and $g_{i,k}$ is the *delay spectrum* of the ith cluster of paths of the kth channel. The parameter $\xi_{i,k}$ is the path delay for the ith specular component of the kth channel. For convenience we assume that $\rho_{i,1}$ is real for all i.

Several other parameters are used to describe each channel. In our examples, we consider rectangular delay spectra, so that the delay spectra are defined by

$$g_{i,k}(\xi) = \begin{cases} \dfrac{1}{\mu_{i,k}T}, & |\xi - \xi_{i,k}| < \mu_{i,k}T/2 \\ 0, & \text{otherwise} \end{cases}$$

where T is the data pulse interval for each mobile transmitter and $\mu_{i,k}$ is the normalized *total delay spread* of the ith cluster of channel k. The total delay spread of the ith cluster is the support of the function $g_{i,k}$. It is assumed that the support of $g_{i,k}$ and the support of $g_{j,k}$ are disjoint for distinct i and j. The time-correlation function d_k of the kth channel is characterized by the half-power bandwidth B_{d_k} of the Fourier transform of d_k, and the normalized half-power bandwidth

$D_T^{(k)} = B_{d_k}T$ is referred to as the *Doppler spread* of the kth channel [12]. In our examples, we consider exponential time-correlation functions. Another parameter of interest is the *power ratio*, defined by $\zeta_{i,k} = \rho_{i,k}^2/(2\sigma_{i,k}^2)$, which is the ratio of the specular power to the average diffuse power in the components of the received signal associated with the ith cluster of paths of channel k.

III. System Description

We consider a cell in which K mobile units transmit simultaneously and all transmitters employ binary DPSK modulation and the same chip rate. Each signal is represented in its equivalent complex-valued baseband form, and the transmitted signal from the kth mobile unit is

$$s_k(t) = \sum_{n=-\infty}^{\infty} \sqrt{2P_{n,k}}b_n^{(k)}p_t(t - nT - \mathrm{T}_k)c_k(t - \tau_k)e^{j\phi_k}$$

where $P_{n,k}$ is the signal power during the nth data pulse interval, $b_n^{(k)}$ is the sequence of differentially-encoded binary data, c_k is the spreading waveform [13], τ_k is the delay at the transmitter of the signal, and ϕ_k is the carrier-phase offset at the transmitter. The function p_t is equal to one for arguments between O and T, and it is equal to zero otherwise. Without loss of generality, the delay and carrier phase of transmitter one are assumed to be zero. The duration of a data pulse is denoted by T and the duration of a chip is denoted by T_c, so that the normalized chip rate [9] is

$$R = \frac{T}{T_c}.$$

The energy in the nth data pulse is $\mathcal{E}_{n,k} = P_{n,k}T$. The chip waveform [13] for the DS signal is rectangular. The baseband representation of the signal received at the base station is

$$r(t) = \sum_{k=1}^{K}\left[\int_{-\infty}^{\infty} h_k(t,\xi)s_k(t-\xi)\,d\xi\right] + n(t) \qquad (1)$$

where $n(t)$ is a white Gaussian process that results from thermal noise in the receiver.

In the base station, the rake receivers for all mobile transmitters are of the same form and are illustrated by considering the receiver for transmitter one. In the remainder of this paper, we omit the relevant subscript k, or superscript k, from the parameters and functions when referring to transmitter one. The rake receiver uses post-detection equal-gain square-law combining [9] and the ith tap is synchronized to the specular component of the received signal due to the ith cluster of the channel. Differentially coherent demodulation is employed, so the receiver makes one bit decision based on two consecutive pulses. The sampled matched-filter output at the ith tap during the nth pulse interval is denoted $U_{i,n}$, and it is given by

$$U_{i,n} = (\sqrt{2T})^{-1}\int_{nT}^{(n+1)T} r(t+\xi_i)c_1^*(t)\,dt.$$

The receiver makes a binary decision during the $(m+1)$th pulse interval based on the sign of the real-valued decision statistic

$$Z = \sum_{i=1}^{L}[U_{i,(m+1)}U_{i,m}^* + U_{i,(m+1)}^*U_{i,m}] \qquad (2)$$

where L is the number of taps in the rake receiver.

IV. Channel Estimation and Power Control

In closed-loop power control, the base station obtains an estimate of some characteristic of the signal received from the mobile unit. Possible estimated quantities include the signal energy and the SIR at the output of a correlator or rake receiver [10], [14]. The estimate is compared to a target value and the receiver determines the change in transmitter power needed to meet the target. A command to increase or decrease the transmitter power by the appropriate amount is sent to the mobile unit via a control channel. Any closed-loop power-control system has an unavoidable delay before the power is adjusted. This delay is caused by processing and transmission. Thus, rapid channel variation may make the estimate obsolete. In addition, multiple-access interference and thermal noise at the receiver may produce inaccurate estimates.

The power-control system considered in this paper uses estimates of the SIR at the output of the rake receiver in the base station. An estimate of the SIR, denoted SIR_e, is formed at the receiver, which then compares the estimate to the target value SIR_t. A command is sent to the mobile unit to change its level of transmitted power by a multiplicative factor Δ, given by

$$\Delta = \frac{SIR_t}{SIR_e}. \qquad (3)$$

Consider the estimator for the signal from mobile transmitter one. If the nth data pulse has polarity b_n, the signal component of the sampled output of the ith tap during that interval is $b_n a_{i,n}\sqrt{\mathcal{E}_n}$, where $\mathcal{E}_n$ is the energy transmitted during the nth data pulse. Here, $a_{i,n}$ denotes the signal component of the ith tap output if the pulse is transmitted with unit energy. Without loss of generality, we consider an estimate of the SIR for $n = 0$. The total energy from the signal component at the output of the receiver is $\|\underline{a}\|^2\mathcal{E}_0$, where the vector

$$\underline{a} = (a_{1,0}, \cdots, a_{L,0}).$$

The composite interference is modeled as additive white Gaussian noise (AWGN) with single-sided power-spectral density I. In practice, the received signal passes through an automatic gain-control element in the IF subsystem of the receiver and the effect of the gain control is to normalize the energy at each tap output with respect to I. The estimator considered in this paper is assumed to provide a perfect estimate of the SIR, so that $SIR_e = \|\underline{a}\|^2\mathcal{E}_0/I$. The power-control command is received at the mobile unit with some fixed delay, however, and the power level remains fixed for the duration of several symbols. Thus, SIR_e differs from the current value of the SIR because of variations in the channel. In our analysis, we consider a signal received during the mth and $(m+1)$th pulse intervals that is transmitted at a power level based on an estimate made during the zeroth interval.

Ideally, the mobile unit responds to the power-control command by changing the transmitted power by the factor Δ. In reality, there is an upper limit on the transmitter power. If $P_{\max}$ is the maximum power the transmitter can produce, then the transmitted power is

$$P(\underline{a}) = \min\{P_0\Delta, P_{\max}\}. \tag{4}$$

The corresponding energy per data pulse is $\mathcal{E}(\underline{a}) = P(\underline{a})T$. The quantities $P(\underline{a})$ and $\mathcal{E}(\underline{a})$ depend on $\underline{a}$ through the dependence of Δ on $\underline{a}$, and the notation that we use is chosen to emphasize that dependence.

The same power-control technique is used for each mobile transmitter. If the nth data pulse from the kth transmitter has unit energy, then the amplitude of the signal component at the output of the ith tap of the receiver matched to transmitter k is $a_{i,n}^k$. The vector $\underline{a}^k$ is defined as

$$\underline{a}^k = (a_{1,0}^k, \cdots, a_{L,0}^k).$$

The corresponding power-control command results in a transmitter power denoted $P(\underline{a}^k)$ and an energy per data pulse denoted $\mathcal{E}(\underline{a}^k)$. The value of each of the signal amplitudes varies with channel conditions. Thus, it is a random variable with a distribution determined by the fading process. In the following sections, boldface notation (e.g., $\mathbf{a}_{i,0}^k$) is used to represent a random variable and normal type (e.g., $a_{i,0}^k$) is used to denote a value taken on by the random variable.

V. System Performance

In a cellular CDMA system, an important measure of the quality of a voice connection is the probability of error on the reverse link. A connection is considered useful only if the probability of error is below a specified maximum value. The performance of the system is measured by its ability to maintain adequate link quality for various channel conditions under the greatest possible traffic load.

We consider two measures of system performance. One performance measure is expressed in terms of the strength of the *specular component* of the received signal due to the first cluster [9]. This measure, denoted by SIR_r, is the value of SIR_ρ required for a mobile transmission to achieve a specified average probability of error at the output of the demodulator where

$$SIR_\rho = E[\rho_1^2 \mathcal{E}(\underline{\mathbf{a}})/I]. \tag{5}$$

SIR_r is the performance measure used in [9]. The other performance measure is S, the *spectral efficiency* of the cell, which is the bandwidth-normalized aggregate bit rate attainable without any link exceeding the specified average probability of error. The bit rate of each transmitter is $1/T$ b/s and the bandwidth of the system is $2/T_c$ Hz. If $K_{\max}$ denotes the maximum number of mobile transmitters that can exist in the cell under given circumstances, the spectral efficiency is

$$\begin{aligned} S &= (K_{\max}/T)/(2/T_c) \text{ b/s/Hz} \\ &= K_{\max}/2R \text{ b/s/Hz.} \end{aligned} \tag{6}$$

The performance of the system is assumed to be limited by multiple-access interference. The analysis is simplified by approximating the aperiodic autocorrelation function of each spreading sequence by an ideal autocorrelation function [9]. The effects of the nonideal characteristics of the actual autocorrelation function are approximated as being equivalent to the presence of an additional interfering transmitter. For a system with interference-limited performance, the net effect of the two approximations on the probability of error is negligible.

The derivation of the probability of error is presented here for transmission one and the probability of error for each of the other transmissions is obtained in the same manner. Recall that the subscript k, or superscript k, is omitted from the parameters for transmission one. Consider the detection of the bit that is encoded in the transition from the mth to the $(m+1)$th data pulse. The transmitted power during these pulses is based on a signal-strength estimate generated by the base station during the zeroth data pulse interval, and this power level is given by (4). The L taps of the rake receiver produce statistically independent outputs, and $\{\mathbf{a}_{i,0}\}$, $\{\mathbf{a}_{i,m}\}$, and $\{\mathbf{a}_{i,m+1}\}$ are jointly Gaussian complex-valued random variables for $i = 1, 2, \cdots, L$. The variance of $\mathbf{a}_{i,0}$ is denoted by $\hat{\sigma}_i^2$ and the covariance of $\mathbf{a}_{i,0}$ and $\mathbf{a}_{i,n}$ is denoted by $\hat{c}_{i,n}^2$ for $n = m$ and $n = (m+1)$.

A simple expression is obtained for each of these second moments by making the approximation that the channel is constant over the duration of a pulse. (The validity of this approximation is discussed in [15].) For closed-loop power control, the variance of $\mathbf{a}_{i,0}$ is

$$\hat{\sigma}_i^2 \approx 2\alpha_i^2\sigma_i^2$$

and the covariance of $\mathbf{a}_{i,0}$ and $\mathbf{a}_{i,n}$ is given by

$$\hat{c}_{i,n}^2 \approx 2\alpha_i^2\sigma_i^2\, d(mT)$$

where

$$\alpha_i = \sqrt{\min\{\mu_i - (R\mu_i^2/2) + (R^2\mu_i^3/12), 2/3R\}/\mu_i}$$

indicates the fraction of diffuse signal energy collected from the ith cluster.

Conditioned on $\underline{\mathbf{a}}$ and on $\mathbf{b}_m$ and $\mathbf{b}_{(m+1)}$, the polarity of the mth and $(m+1)$th data pulses, $U_{i,n}$ has a complex Gaussian distribution with a mean value of

$$b_n\overline{\rho}_i = b_n\mathcal{E}^{1/2}(\underline{a})\,[\rho_i + \hat{c}_{i,n}^2(a_{i,0} - \rho_i)/\hat{\sigma}_i^2]$$

and a variance of

$$2\overline{\sigma}_i^2 + I = \mathcal{E}(\underline{a})\,[2\alpha_i^2\sigma_i^2 - \hat{c}_{i,n}^4/\hat{\sigma}_i^2] + I$$

for $n = m$ and $n = m+1$. The expression for I is derived in the Appendix. The covariance of the sampled outputs of the ith tap in the mth and $(m+1)$th pulse intervals is

$$\text{Cov}\,[U_{i,m}, U_{i,(m+1)}|\,\underline{\mathbf{a}}] \approx 2b_m b_{m+1} d(T)\overline{\sigma}_i^2.$$

The average probability of error is

$$\overline{P}_b = E[P_b(\underline{\mathbf{a}})] \tag{7}$$

where $P_b(\underline{\mathbf{a}})$ is the probability of error conditioned on $\underline{\mathbf{a}}$. Conditioned on $\underline{\mathbf{a}}$, the decision statistic Z in (2) is a Hermitian quadratic form in jointly Gaussian, complex-valued random variables [16], so $P_b(\underline{a})$ may be determined using the methods described in [17]. If the channel is doubly selective and the $\overline{\sigma}_i$ are distinct for $1 \leq i \leq L$, the average probability of error conditioned on $\underline{a}$ is [17]

$$P_b(\underline{a}) = \sum_{j=L+1}^{2L} A_j \prod_{i=1}^{L} \frac{\tilde{\sigma}_j^2}{\tilde{\sigma}_j^2 + \tilde{\sigma}_i^2} \exp\left[-2\overline{\rho}_i^2/(\tilde{\sigma}_j^2 + \tilde{\sigma}_i^2)\right] \tag{8}$$

where

$$A_j = \prod_{i \geq L+1, i \neq j} \frac{\tilde{\sigma}_j^2}{\tilde{\sigma}_j^2 - \tilde{\sigma}_i^2}$$

and

$$\tilde{\sigma}_i^2 = \begin{cases} 2[1 + d(T)]\overline{\sigma}_i^2 + I, & 1 \leq i \leq L \\ 2[1 - d(T)]\overline{\sigma}_{i-L}^2 + I, & L < i \leq 2L. \end{cases}$$

If the $\overline{\sigma}_i$ are not all distinct or if the channel is purely frequency selective, the probability of error is given by an expression that is obtained in the same manner as (8) and given in [9].

To determine the spectral efficiency of the cell in a given set of circumstances, the smallest value of SIR_t that produces an acceptable bit error probability for each transmission is determined. The corresponding value of SIR_r for each transmission is then obtained from (5). The values of SIR_r and the interference as given in (A1) determine the maximum number of transmitters $K_{\max}$. The spectral efficiency is obtained from $K_{\max}$ using (6).

VI. Numerical Results

In the numerical results that are presented, we consider parameter values that are appropriate for two CDMA systems of practical interest. One is a digital cellular system with a carrier frequency of 900 MHz and a channel bit rate of 25 kb/s. The other is a PCS system operating at 2 GHz that also has a bit rate of 25 kb/s. The feedback delays considered for the power-control commands span the range of delays that are of practical interest, and they are shown in Table I. In the examples, two channels are considered that are typical of the circumstances encountered in urban mobile communications. The delay profiles of the channels are given in Fig. 1. Channel A consists of three identically distributed clusters and each cluster has a specular component of amplitude ρ and a power ratio of ζ. Neither the second nor the third cluster of channel B has a specular component ($\rho_2 = \rho_3 = 0$), and the second and third clusters each have a diffuse component with an average power that is 6 dB less than that of the diffuse component of the first cluster ($\sigma_2 = \sigma_3 = \sigma/2$). The Doppler spreads considered in the examples are shown in Table II along with the corresponding speed of the mobile unit for either system. The discussion of the numerical results focuses on chip rates of 50 and 400 as representative low and high chip rates, respectively. A value of 10^4 is used for $P_{\max}$, unless otherwise stated, and this value is normalized with respect to the interference power spectral density. In each example that

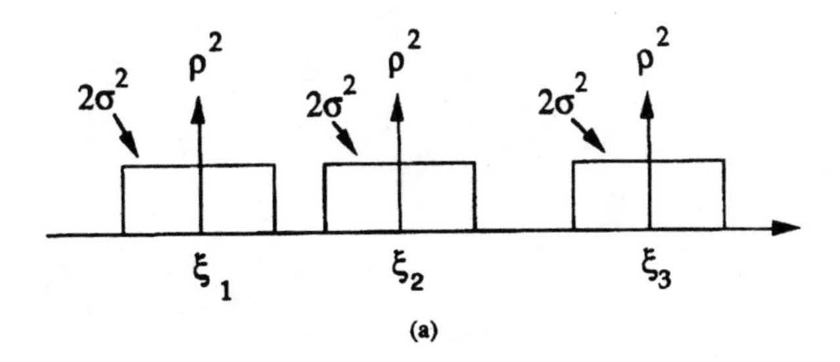

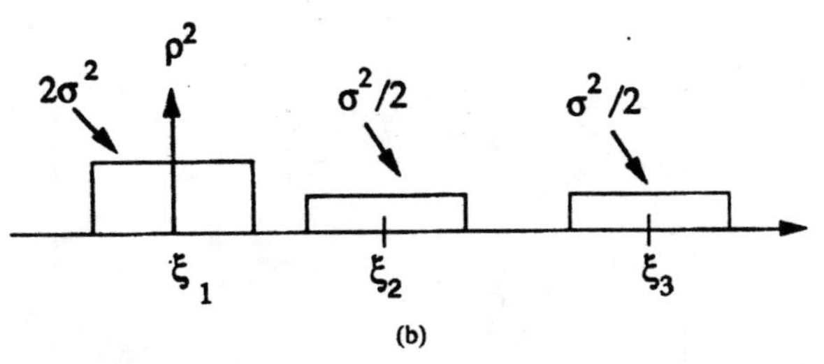

Fig. 1. Delay spectra of channels A and B ($\mu_1 = \mu_2 = \mu_3 = 0.05$, $\zeta_1 = -3$ dB); (a) channel A and (b) channel B.

TABLE I
VALUES OF m AND THE CORRESPONDING DELAYS FOR A 25 kb/s SYSTEM

m	Delay (ms)
31.25	1.25
62.5	2.5
125	5.0
250	10.0

TABLE II
VALUES OF D_T AND THE CORRESPONDING SPEEDS FOR A 25 kb/s SYSTEM

D_T	Speed (km/h)	
	900 MHz	2 GHz
0.00025	7.5	3.375
0.0005	15	6.75
0.0011	33	15
0.002	60	27
0.004	120	54

we consider, all reverse-link channels have the same delay spectrum and the same Doppler spread.

Figure 2 shows the value of SIR_ρ required to achieve a bit error probability of 10^{-2} for channel B and a correlation receiver as a function of the chip rate for different power-control delays. The Doppler spread is 0.002. A probability of error of 10^{-2} at the output of the demodulator, if used together with forward error-correction coding, is sufficient to obtain adequate performance in most voice applications. The spectral efficiency of the same system for the same bit error probability, power-control delays, and Doppler spread is shown in Fig. 3. The relative performance of the system with different delays depends on which measure is used. In particular, the calculation of the spectral efficiency accounts

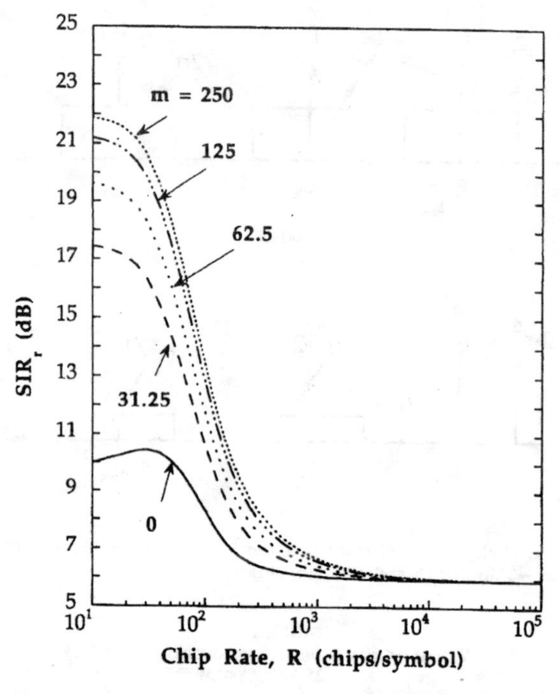

Fig. 2. Required SIR_ρ for channel B, $L = 1$ ($D_T = 0.002$, $\overline{P}_b = 10^{-2}$).

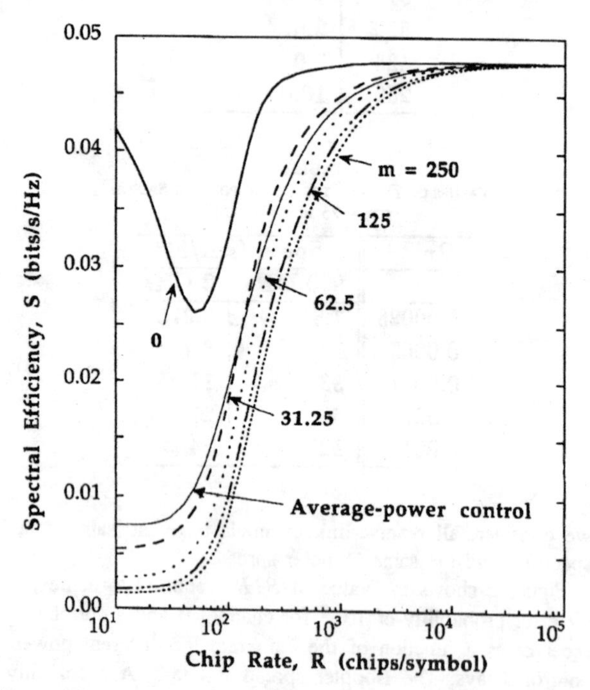

Fig. 3. Spectral efficiency for channel B, $L = 1$ ($D_T = 0.002$, $\overline{P}_b = 10^{-2}$).

for the energy received from the second and third clusters, while the calculation of SIR_r does not. Spectral efficiency is used as the performance measure in the remaining figures.

Increasing the chip rate of the system improves the multipath-rejection capability of each correlator and reduces

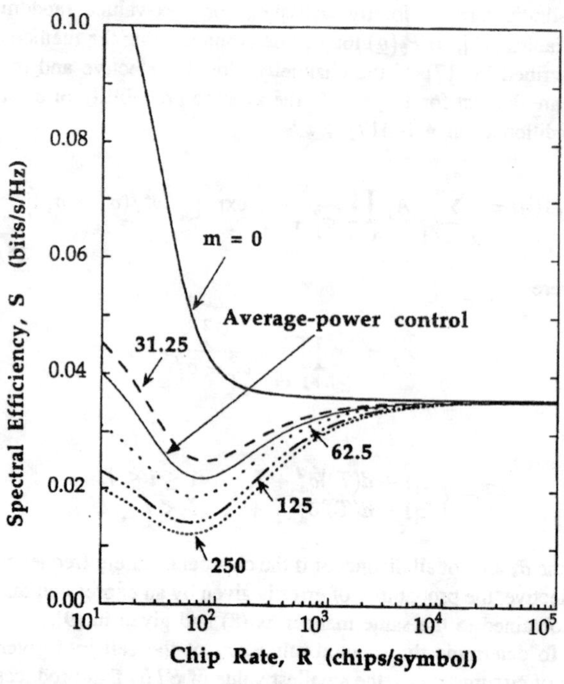

Fig. 4. Spectral efficiency for channel B, $L = 3$ ($D_T = 0.002$, $\overline{P}_b = 10^{-2}$).

the diffuse signal component that appears at the output of the rake receiver. The average amount of signal energy available for a decision is decreased, while the variance in the energy is also decreased. Thus, an increase in the chip rate produces two counteracting effects on the distribution of the decision statistic and, consequently, on the probability of error and the spectral efficiency of the system. The relative importance of the two opposing effects depends on the energy ratio of each cluster and the effectiveness of power control and rake combining in reducing the variance of the received signal energy for low chip rates. If one effect is dominant over one range of chip rates and the other over the complementary range, the performance of the system is a nonmonotonic function of the chip rate, as is seen in Fig. 3 for zero feedback delay. In contrast, if one effect is dominant over the entire range of chip rates, the performance is a monotonic function of the chip rate, as is seen for $m = 31.25$ in Fig. 3. Thus, no single chip rate provides the best performance under all circumstances.

In Figs. 3–6, the effect of the power-control feedback delay on the spectral efficiency is shown as a function of the chip rate for channels A and B and both a correlation receiver and a three-tap rake receiver. For comparison, each figure also includes the spectral efficiency of the average-power-control system examined in [9]. For the low chip rate system, there is a significant degradation in performance as the delay increases. The performance is very sensitive to delay with a low chip rate and a correlation receiver, as shown for channel B in Fig. 3. The multipath diversity provided with a three-tap rake receiver reduces the sensitivity to delay, as seen by comparing Fig. 3 to Fig. 4. Similar results are observed for

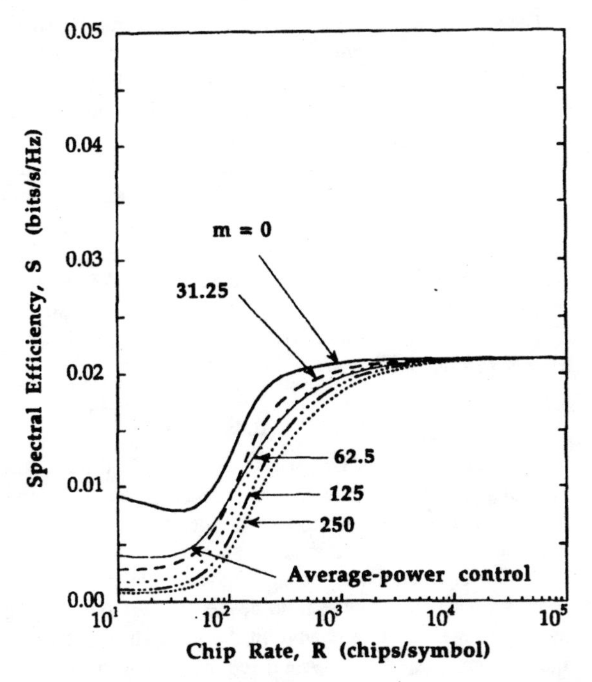

Fig. 5. Spectral efficiency for channel A, $L = 1$ ($D_T = 0.0011$, $\overline{P}_b = 10^{-2}$).

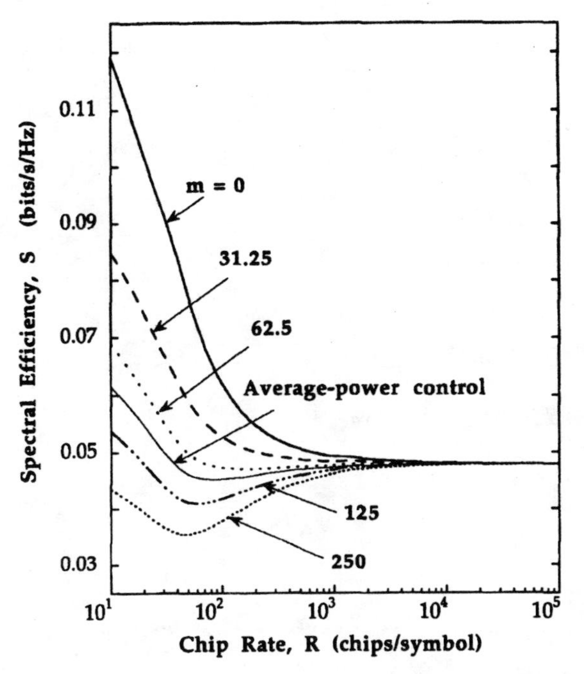

Fig. 6. Spectral efficiency for channel A, $L = 3$ ($D_T = 0.0011$, $\overline{P}_b = 10^{-2}$).

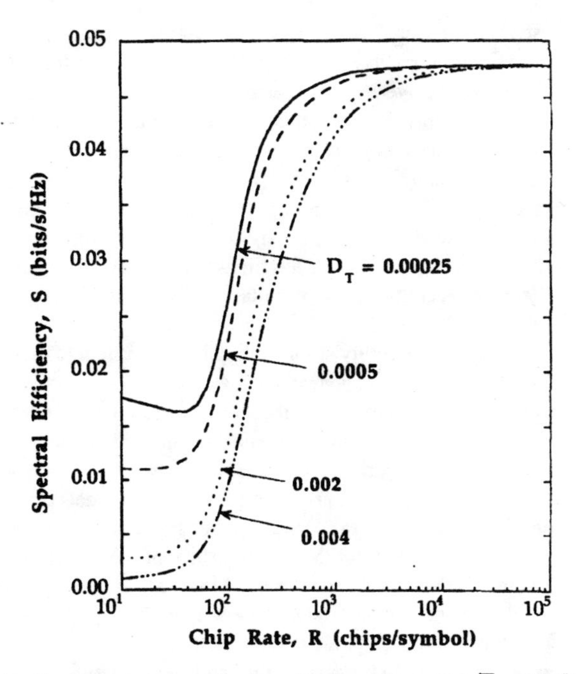

Fig. 7. Spectral efficiency for channel B, $L = 1$ ($m = 62.5$, $\overline{P}_b = 10^{-2}$).

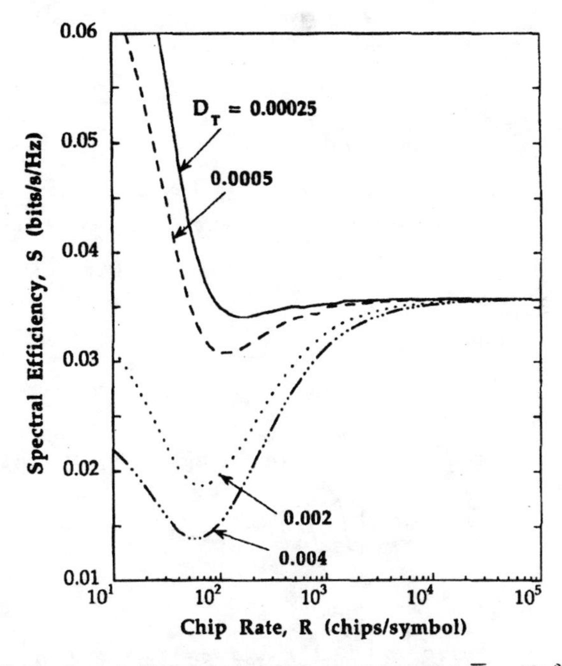

Fig. 8. Spectral efficiency for channel B, $L = 3$ ($m = 62.5$, $\overline{P}_b = 10^{-2}$).

the low chip rate and channel A, as shown in Figs. 5 and 6. In addition, the weak signal components at the output of the second and third taps for channel B provide only modest diversity, so the degradation in the performance of the rake receiver with large delays is greater for channel B than for channel A. These results demonstrate that the effectiveness of rapid power control for a low chip rate system in the presence of fading is highly dependent on the timeliness of the estimate.

In contrast, Figs. 3–6 illustrate that the performance of the high-chip-rate system shows little sensitivity to the delay with either a correlation receiver or a rake receiver for either channel A or B. Furthermore, the rake receiver provides better performance than the correlation receiver for the high-chip-rate system and channel A, but it yields poorer performance than the correlation receiver for channel B. The poor performance of the rake receiver for channel B is due to the small amount of signal energy at the output of the second and third taps and the use of suboptimal equal-gain square-law combining. Even with optimal combining, however, the additional taps are of minimal benefit with this channel for the high-chip-rate system because there is little signal energy to exploit at the output of those taps. Thus, the high-chip-rate system benefits from rake reception only if each tap resolves a strong specular component.

Because of the ability of the high-chip-rate system to resolve a stable specular component with a correlation receiver, the high chip rate is superior to the low chip rate for correlation reception and a wide range of channel parameters and feedback delays. This is illustrated in Figs. 3 and 5. In contrast, the low chip rate provides better performance with a rake receiver than the high chip rate if the Doppler spread and feedback delay are sufficiently small for the power control of the low chip rate system to adapt to changes in the SIR. This is seen in Figs. 4 and 6. If the channel changes more quickly than the power control of the low chip rate system can respond to the changes, however, the high-chip-rate system with a correlation receiver provides superior performance to the low chip rate system with a three-tap rake receiver.

The spectral efficiency is shown in Figs. 7 and 8 for channel B and both a correlation receiver and a rake receiver for different values of the Doppler spread. The effect of varying the Doppler spread is similar to the effect of varying the power-control delay, as seen by comparing Fig. 7 to Fig. 3 and Fig. 8 to Fig. 4. A wide range of applications is envisioned for PCS, including the support of both low-speed pedestrian traffic and high-speed vehicles. The results shown here indicate that for mobiles operating at high speed, a low chip rate system will provide adequate performance only if a rake receiver is employed and the power-control feedback delay is small. In contrast, the high-chip-rate system is far less sensitive to the effects of high mobile speed.

VII. Conclusions

Our analysis of a CDMA system with closed-loop power control demonstrates that the performance does not depend in a simple manner on the chip rate of the system. The chip rate that yields the greatest spectral efficiency depends upon the characteristics of the channel, the complexity of the rake receiver, and the time needed for the power-control system to respond to a change in the SIR. Thus, determining the best chip rate for a system requires some knowledge of the range of channel conditions that may be encountered in the application of interest, the method by which closed-loop

$$\begin{aligned}\tilde{i}_{l,k} &= \frac{1}{\sqrt{2T}} \int_{mT}^{(m+1)T} r_{l,k}(t+\xi_j)c_1^*(t)\,dt \\ &= T^{-1} \int_{mT}^{(m+1)T} \int_{\Xi} h_k(t+\xi_j, \xi)\mathcal{E}^{1/2}(\underline{\mathrm{a}}^k)\left[\sum_{i=-\infty}^{\infty} b_i^{(k)} c_1^*(t)\, c_k(t-\xi+\xi_j-\tau_k)e^{j\phi_k}\right] p_t(t-iT-\mathrm{T}_k)\,d\xi\,dt \\ &\approx T^{-1}\mathcal{E}^{1/2}(\underline{\mathrm{a}}^k) \int_{\Xi} h_k(mT, \xi)I_{l,k}[\tau_k, \phi_k, b^{(k)}, c_k, \xi_j, \xi]\,d\xi. \end{aligned} \tag{A2}$$

$$\begin{aligned} &E[|i_{l,k}|^2 | \mathbf{a}_{l,0}^k = a_{l,0}^k] \\ &= T^{-2}E\left\{\int_{\Xi}\int_{\Xi} h_k(mT, \xi)I_{l,k}[\tau_k, \phi_k, b^{(k)}, c_k, \xi_j, \xi]h_k^*(mT, \eta)I_{l,k}^*[\tau_k, \phi_k, b^{(k)}, c_k, \xi_j, \eta]\,d\eta\,d\xi | \mathbf{a}_{l,0}^k = a_{l,0}^k\right\} \\ &= T^{-2}E\left(\int_{\Xi}\int_{\Xi} h_k(mT, \xi)h_k^*(mT, \eta)E\{|I_{l,k}[\tau_k, \phi_k, b^{(k)}, c_k, \xi_j, \xi]|^2\}\,d\eta\,d\xi \,|\, \mathbf{a}_{l,0}^k = a_{l,0}^k\right) \\ &= (2/3R)E\left[\int_{\Xi}\int_{\Xi} h_k(mT, \xi)h_k^*(mT, \eta)\,d\eta\,d\xi | \mathbf{a}_{l,0}^k = a_{l,0}^k\right]. \end{aligned} \tag{A3}$$

$$\begin{aligned} &E[|i_{l,k}|^2 | \mathbf{a}_{l,0}^k = a_{l,0}^k] \\ &= 2/3R\left\{E\left[\int_{\Upsilon}\int_{\Upsilon} h_k(mT, \xi)h_k^*(mT, \eta)\,d\eta\,d\xi\right] + \int_{\xi_{l,k}-A}^{\xi_{l,k}+A}\int_{\xi_{l,k}-A}^{\xi_{l,k}+A} h_k(mT, \xi)h_k^*(mT, \eta)\,d\eta\,d\xi | \mathbf{a}_{l,0}^k = a_{l,0}^k]\right\} \\ &= 2/3R\left\{\int_{\Upsilon} 2\sigma_{l,k}^2/\mu_{l,k}T\,d\xi + E\left[\left|\int_{\xi_{l,k}-A}^{\xi_{l,k}+A} h_k(mT, \xi)\,d\xi\right|^2 |\mathbf{a}_{l,0}^k = a_{l,0}^k\right]\right\}. \end{aligned} \tag{A5}$$

$$\begin{aligned} E[|i_{l,k}|^2 | \mathbf{a}_{l,0}^k = a_{l,0}^k] &= (2/3R)(\mu_{l,k}T - 2A)\,2\sigma_{l,k}^2/\mu_{l,k}T + (2/3R) \\ &\cdot E\left[\left|\int_{\xi_{l,k}-A}^{\xi_{l,k}+A} h_k(mT, \xi)Ct_{T_c}(\xi)\,d\xi + \int_{\xi_{l,k}-A}^{\xi_{l,k}+A} h_k(mT, \xi)y(\xi)\,d\xi\right|^2 |\mathbf{a}_{l,0}^k = a_{l,0}^k\right]. \end{aligned} \tag{A6}$$

power control is implemented, and the economics of hardware implementation.

We have examined the effect of the chip rate on the system performance for a range of channel parameters appropriate for modeling the land-mobile communications channel. Rake receivers with as many as three taps are considered. The comparison of systems with normalized chip rates of 50 and 400 shows that the performance of the higher-chip-rate system is far less sensitive to the Doppler spread and the power-control delay than the lower-chip-rate system. As a result, the higher chip rate achieves superior performance if the velocities of the mobile transmitters are high or the power-control feedback delay is large. If the signal is diffuse but can be resolved into several components of nearly equal energy, the lower-chip-rate system can employ a rake receiver to obtain better performance than the higher chip rate for low mobile velocities and a small power-control delay.

The lower-chip-rate system benefits from the use of multiple taps in the rake receiver for a wider range of channel conditions than the higher-chip-rate system. With a chip rate of 400, a rake receiver is of value only if there are two or more strong specular components that are of nearly equal strength. In contrast, for the channels considered in this study, the system with a chip rate of 50 not only benefits from the use of rake reception but requires the diversity protection that it provides for adequate system performance. This diversity protection is especially critical to the performance of the low chip rate system for high mobile velocities or a large power-control feedback delay.

Appendix
Derivation of Interference Power-Spectral Density

The effect of multiple-access interference on the decision statistic is modeled as equivalent AWGN, so that the interference is completely characterized by the equivalent power-spectral density I. From (1), the received signal can be represented as

$$r(t) = \sum_{k=1}^{K} \sum_{l=1}^{M_k} r_{l,k}(t) + n(t)$$

where

$$r_{l,k}(t) = \int_{\xi_{l,k}-\mu_{l,k}T/2}^{\xi_{l,k}+\mu_{l,k}T/2} h_k(t,\xi) s_k(t-\xi)\, d\xi$$

denotes the contribution to $r(t)$ from the lth cluster of channel k. Since interference-limited performance is considered, the thermal noise component $n(t)$ is assumed to be zero.

The variance of the interference term in the sampled output of the jth tap (at tap delay ξ_j) is also equal to I and is given by

$$I = \operatorname{Var}\left[\sum_{k=2}^{K} \sum_{l=1}^{M_k} \tilde{i}_{l,k}\right] \tag{A1}$$

$$\begin{aligned}
&E\left[\left|\int_{\xi_{l,k}-A}^{\xi_{l,k}+A} h_k(mT,\xi) C t_{T_c}(\xi)\, d\xi + \int_{\xi_{l,k}-A}^{\xi_{l,k}+A} h_k(mT,\xi) y(\xi)\, d\xi\right|^2 \Big| \mathbf{a}_{l,0}^k = a_{l,0}^k\right] \\
&= E\left[\left|\int_{\xi_{l,k}-A}^{\xi_{l,k}+A} \tilde{h}_k(mT,\xi) C t_{T_c}(\xi)\, d\xi + C\rho_{l,k} + \int_{\xi_{l,k}-A}^{\xi_{l,k}+A} h_k(mT,\xi) y(\xi)\, d\xi\right|^2 \Big| \mathbf{a}_{l,0}^k = a_{l,0}^k\right] \\
&= E\Bigg[\Bigg|\int_{\xi_{l,k}-A}^{\xi_{l,k}+A} [d_k(mT)\tilde{h}_k(0,\xi) + \sqrt{1-d_k(mT)^2} X(\xi)] C t_{T_c}(\xi)\, d\xi + d_k(mT) C\rho_{l,k} \\
&\quad + [1-d_k(mT)] C\rho_{l,k} + \int_{\xi_{l,k}-A}^{\xi_{l,k}+A} h_k(mT,\xi) y(\xi)\, d\xi\Bigg|^2 \Big| \mathbf{a}_{l,0}^k = a_{l,0}^k\Bigg] \\
&= E\left\{\left| C d_k(mT) a_{l,0}^k + C\sqrt{1-d_k(mT)^2} \int_{\xi_{l,k}-A}^{\xi_{l,k}+A} X(\xi) t_{T_c}(\xi)\, d\xi + [1-d_k(mT)] C\rho_{l,k} + \int_{\xi_{l,k}-A}^{\xi_{l,k}+A} h_k(mT,\xi) y(\xi)\, d\xi\right|^2\right\} \\
&= [C d_k(mT) |a_{l,0}^k|]^2 + 2 d_k(mT) \rho_{l,k} C\{[1-d_k(mT)] C + y(\xi_{l,k})]\} \operatorname{Re}\{a_{l,0}^k\} \\
&\quad + \{[1-d_k(mT)] C + y(\xi_{l,k})\}^2 \rho_{l,k}^2 + [1-d_k(mT)^2] C^2 2\alpha_{l,k}^2 \sigma_{l,k}^2 + 2Y\sigma_{l,k}^2 \\
&= |C d_k(mT) a_{l,0}^k + [1-d_k(mT)C] \rho_{l,k}|^2 + \{[1-d_k(mT)^2] C^2 \alpha_{l,k}^2 + Y\} 2\sigma_{l,k}^2.
\end{aligned} \tag{A7}$$

$$\begin{aligned}
\operatorname{Var}(\tilde{i}_{l,k} | \underline{\mathbf{a}}^k = \underline{a}^k) &= (2/3R)\, \mathcal{E}(\underline{a}^k) \{(\mu_{l,k}T - 2A)\, 2\sigma_{l,k}^2/\mu_{l,k}T + [C d_k(mT) |a_{l,0}^k|]^2 + 2 d_k(mT) \rho_{l,k} C [1-d_k(mT)C] \\
&\quad \cdot \operatorname{Re}\{a_{l,0}^k\} + [1-d_k(mT)C]^2 \rho_{l,k}^2 + [1-d_k(mT)^2] C^2 2\alpha_{l,k}^2 \sigma_{l,k}^2 + 2Y\sigma_{l,k}^2\}.
\end{aligned} \tag{A8}$$

$$\begin{aligned}
\operatorname{Var}(\tilde{\mathrm{i}}_{l,k}) &= (2/3R)\, E[\mathcal{E}(\underline{\mathbf{a}}^k)] \{(\mu_{l,k}T - 2A)\, 2\sigma_{l,k}^2/\mu_{l,k}T + [1-d_k(mT)C]^2 \rho_{l,k}^2 + [1-d_k(mT)^2] C^2 2\alpha_{l,k}^2 \sigma_{l,k}^2 + 2Y\sigma_{l,k}^2\} \\
&\quad + (2/3R)\, E[\mathcal{E}(\underline{\mathbf{a}}^k) |\mathbf{a}_{l,0}^k|^2] [C d_k(mT)]^2 + (2/3R)\, E[\mathcal{E}(\underline{\mathbf{a}}^k) \operatorname{Re}\{\mathbf{a}_{l,0}^k\}] [2 d_k(mT) \rho_{l,k} C] [1-d_k(mT)C].
\end{aligned} \tag{A9}$$

where $\tilde{i}_{l,k}$ is the contribution of $r_{l,k}(t)$ to the multiple-access interference in the sample at the jth tap. For the power-control feedback delays considered in this paper, the effect of variation of the interferer's channel within the span of one detection interval is negligible. Thus, for a decision interval that occurs m pulse intervals after the power-control estimate for transmitter k is generated, $\tilde{i}_{l,k}$ is given by (A2), shown at the bottom of page 566, where Ξ denotes the interval $[\xi_{l,k} - \mu_{l,k}T/2, \xi_{l,k} + \mu_{l,k}T/2]$ and

$$I_{l,k}[\tau_k, \phi_k, b^{(k)}, c_k, \xi_j, \xi] = \int_{mT}^{(m+1)T} \left[\sum_{i=-\infty}^{\infty} b_i^{(k)} c_k(t - \xi + \xi_j - \tau_k) e^{j\phi_k}\right] c_1^*(t)\, dt.$$

For convenience, we define

$$i_{l,k} = \tilde{i}_{l,k}/\mathcal{E}^{1/2}(\underline{a}^k)$$

so that

$$i_{l,k} \approx T^{-1} \int_{\Xi} h_k(mT, \xi) I_{l,k}[\tau_k, \phi_k, b^{(k)}, c_k, \xi_j, \xi]\, d\xi.$$

Since ϕ_k is uniformly distributed over $[0, 2\pi]$, $\tilde{i}_{l,k}$ and $i_{l,k}$ have zero mean. Thus, the variance of the $\tilde{i}_{l,k}$ conditioned on $\underline{\mathbf{a}}^k$ is

$$\mathrm{Var}\,(\tilde{i}_{l,k} | \underline{\mathbf{a}}^k = \underline{a}^k) = \mathcal{E}(\underline{a}^k) E[|i_{l,k}|^2 |\, \mathbf{a}_{l,0}^k = a_{l,0}^k].$$

The random variable $I_{l,k}[\tau_k, \phi_k, b^{(k)}, c_k, \xi_j, \xi]$ is independent of the fading process, ξ, and $\mathbf{a}_{l,0}^k$, and from [18]

$$E\{|I_{l,k}[\tau_k, \phi_k, b^{(k)}, c_k, \xi_j, \xi]|^2\} = \frac{2T_cT}{3}.$$

From this, (A3) is obtained, shown at the bottom of page 566.

The estimate $a_{l,0}^k$ can be expressed as

$$a_{l,0}^k = \int_{\xi_{l,k}-A}^{\xi_{l,k}+A} h_k(0, \xi) t_{T_c}(\xi)\, d\xi \qquad \text{(A4)}$$

where $A = \min\{\mu_{l,k}T/2, T_c\}$ and

$$t_{T_c}(\xi) = \begin{cases} 1 - \dfrac{|\xi - \xi_{l,k}|}{T_c}, & \text{if } |\xi - \xi_{l,k}| \le T_c \\ 0, & \text{otherwise.} \end{cases}$$

Equation (A5) follows, shown at the bottom of page 566, where Υ denotes the portion of Ξ outside of $[\xi_{l,k} - A, \xi_{l,k} + A]$. Then

Let the function y be defined as

$$y(\xi) = 1 - Ct_{T_c}(\xi)$$

where

$$C = \begin{cases} 3/2, & \text{if } A = T_c \\ (1 - \mu_{l,k}T/4T_c)/\alpha_{l,k}^2, & \text{if } A = \mu_{l,k}T/2. \end{cases}$$

Then y is orthogonal to t_{T_c} on the interval $[\xi_{l,k} - A, \xi_{l,k} + A]$ and

$$y(\xi) + Ct_{T_c}(\xi) = 1.$$

Therefore, (A5) can be rewritten as (A6), shown at the bottom of page 566.

Let $\tilde{h}_k(mT, \xi)$ denote the diffuse (zero-mean) component of $h_k(mT, \xi)$. We can express $\tilde{h}_k(mT, \xi)$ as

$$\tilde{h}_k(mT, \xi) = d(mT)\tilde{h}_k(0, \xi) + \sqrt{1 - d(mT)^2} X(\xi)$$

where $X(\xi)$ and $\tilde{h}_k(0, \xi)$ are independent and identically distributed random processes. Substitution of this expression into (A6) yields (A7), shown at the bottom of page 567, where

$$Y = \begin{cases} T_c/2\mu_{l,k}T, & \text{if } A = T_c \\ 1 - (1 - \mu_{l,k}T/4T_c)^2/\alpha_{l,k}^2, & \text{if } A = \mu_{l,k}T/2. \end{cases}$$

Therefore, (A8) and (A9) follow, shown at the bottom of page 567.

Substitution of (A9) into (A1) yields the desired expression.

REFERENCES

[1] K. S. Gilhousen, I. M. Jacobs, R. Padovani, A. J. Viterbi, L. A. Weaver Jr., and C. E. Wheatley III, "On the capacity of a cellular CDMA system," *IEEE Trans. Veh. Technol.*, vol. 40, pp. 303–312, May 1991.

[2] IS-95, "Mobile station—Base station compatibility standard for dual-mode wideband spread spectrum cellular system," Telecommunications Industry Association/Electronics Industry Association Interim Standard, July 1993.

[3] D. L. Schilling and L. B. Milstein, "Broadband CDMA for indoor and outdoor personal communication," in *Proc. Fourth Int. Symp. Personal, Indoor, Mobile Commun.*, Sept. 1993.

[4] R. L. Pickholtz, L. B. Milstein, and D. L. Schilling, "Spread spectrum for mobile communications," *IEEE Trans. Veh. Technol.*, vol. 40, pp. 313–322, May 1991.

[5] C. Cook, "Development of air interface standards for PCS," *IEEE Personal Commun.*, pp. 30–34, vol. 1, no. 4, 1994.

[6] M. B. Pursley, "The role of spread spectrum in packet radio networks," *Proc. IEEE*, vol. 75, pp. 116–134, Jan. 1987.

[7] A. J. Viterbi, A. M. Viterbi, and E. Zehavi, "Performance of power-controlled wideband terrestrial digital communication," *IEEE Trans. Commun.*, vol. 41, pp. 559–569, Apr. 1993.

[8] R. Price and P. E. Green, "A communication technique for multipath channels," in *Proc. IRE*, vol. 46, Mar. 1958, pp. 555–570.

[9] D. L. Noneaker and M. B. Pursley, "On the chip rate of CDMA systems with doubly selective fading and rake reception," *IEEE J. Select. Areas Commun.*, vol. 12, no. 5, pp. 853–861, June 1994.

[10] S. Ariyavisitakul and L. F. Chang, "Signal and interference statistics of a CDMA system with feedback power control," *IEEE Trans. Commun.*, vol. 41, pp. 1626–1634, Nov. 1993.

[11] P. A. Bello, "Characterization of randomly time variant linear channels," *IEEE Trans. Commun. Syst.*, vol. CS-11, pp. 360–393, Dec. 1963.

[12] P. A. Bello and B. D. Nelin, "The influence of fading spectrum on the binary error probabilities of incoherent and differentially coherent matched filter receivers," *IRE Trans. Commun. Syst.*, vol. CS-10, pp. 160–168, June 1962.

[13] D. V. Sarwate and M. B. Pursley, "Crosscorrelation properties of pseudorandom and related sequences," *Proc. IEEE*, vol. 68, pp. 593–619, May 1980.

[14] A. J. Viterbi, *Principles of Spread Spectrum Communications*. Reading, MA: Addision-Wesley, 1995.

[15] D. L. Noneaker and M. B. Pursley, "The effects of spreading sequence selection on DS spread spectrum with selective fading and two forms of rake reception," in *Proc. Global Telecommun. Conf.*, vol. 4, Dec. 1992, pp. 66–70.

[16] G. L. Turin, "The characteristic function of Hermitian quadratic forms in complex normal variables," *Biometrika*, vol. 47, pp. 199–201, June 1960.

[17] D. L. Noneaker, "The performance of direct-sequence spread spectrum communications with selective fading channels and rake reception," Ph.D. dissertation, Dept. Elec. and Comp. Eng., Univ. of Illinois, Urbana-Champaign, Aug. 1993.

[18] J. S. Lehnert and M. B. Pursley, "Multipath diversity reception of spread spectrum multiple-access communications," *IEEE Trans. Commun.*, vol. COM-35, pp. 1189–1198, Nov. 1987.

John H. Gass Jr. (S'92) was born in New Orleans, LA, on October 24, 1969. He received the B.S. degree (with honors) from the California Institute of Technology, Pasadena, CA, in 1991, and the M.S. degree from the University of Illinois, Urbana-Champaign, in 1993, both in electrical engineering.

He held a research assistantship with the Coordinated Science Laboratory, Urbana, IL, from 1992 to 1993. He was also a Research Assistant in the Department of Electrical and Computer Engineering at Clemson University, Clemson, SC, from 1993 to 1995 where he is currently working toward the Ph.D. He has been a part-time engineer with the ITT Aerospace/Communications Division since May 1995. His research interests are in the areas of spread spectrum and multiple-access communications and communication over fading channels.

Mr. Gass was a National Science Foundation Graduate Research Fellow from 1992 to 1995. He is a member of Eta Kappa Nu and Tau Beta Pi.

Daniel L. Noneaker (S'90–M'93) was born in Montgomery, AL, on December 10, 1957. He received the B.S. degree (with high honors) from Auburn University, Alabama, in 1977, and the M.S. degree from Emory University, Atlanta, GA, in 1979, both in mathematics. He received the M.S. degree in electrical engineering from the Georgia Institute of Technology, Atlanta, in 1984, and he was awarded the Ph.D. degree in electrical engineering from the University of Illinois, Urbana-Champaign, in 1993.

He has industrial experience in both hardware and software design for communication systems. He was with Sperry-Univac, Salt Lake City, UT, from 1979 to 1982, and the Motorola Government Electronics Group, Scottsdale, AZ, from 1984 to 1988. He was a Research Assistant in the Coordinated Science Laboratory, University of Illinois, Urbana, IL, from 1988 to 1993. Since August 1993, he has held the position of Research Associate/Assistant Professor in the Department of Electrical and Computer Engineering at Clemson University, Clemson, SC. He is currently engaged in research on wireless communications for both military and commercial applications with emphases on spread spectrum communications, narrowband modulation, and error-control coding for fading channels. He has published several papers concerning the design and analysis of spread spectrum multiple-access systems for voice and data transmission.

Michael B. Pursley (S'68–M'68–SM'77–F'82) was born in Winchester, IN, on August 10, 1945. He received the B.S. degree (with highest distinction) and the M.S. degree from Purdue University, Lafayette, IN, in 1967 and 1968, respectively, and the Ph.D. degree in electrical engineering from the University of Southern California, Los Angeles, in 1974.

He has several years of industrial experience with the Nortronics Division of Northrop Corporation and the Space and Communications Group of the Hughes Aircraft Company. He was Hughes Doctoral Fellow and a Research Assistant in the Department of Electrical Engineering at the University of Southern California. From January through June of 1974, he was an Acting Assistant Professor in the System Science Department of the University of California, Los Angeles. From June 1974 to July 1993, he was with the Department of Electrical and Computer Engineering and the Coordinated Science Laboratory at the University of Illinois, Urbana, where he has held the rank of Professor since 1980. His research is in the general area of communications and information theory with emphasis on spread spectrum communications, correlation properties of sequences, mobile radio networks, and multiple-access communication theory.

Dr. Pursley has served as Holcombe Endowed Professor at Clemson University, Clemson, SC, since 1993. He is a member of Phi Eta Sigma, Tau Beta Pi, and the Institute of Mathematical Statistics. He was a member of the Board of Governors of the IEEE Information Theory Group during 1977–1984 and 1989–1991, and in 1982 he served as the President of the Group. He served as Program Chairman for the 1979 IEEE International Symposium on Information Theory which was held in Grignano, Italy, and he was a member of the Editorial Board of the PROCEEDINGS OF THE IEEE from 1984 to 1991. He was awarded an IEEE Centennial Medal in 1984. He was Co-Chairman for the 1995 IEEE International Symposium on Information Theory.

Capacity, Throughput, and Delay Analysis of a Cellular DS CDMA System With Imperfect Power Control and Imperfect Sectorization

Michel G. Jansen, *Student Member, IEEE*, and Ramjee Prasad, *Senior Member, IEEE*

***Abstract*— An analytical model is developed to evaluate the performance of a cellular slotted DS CDMA system in terms of user capacity, throughput, and delay for the reverse link, i.e., from mobile to base station, considering interference from both home cell and adjacent cells. The user capacity is studied for voice communications and the throughput and delay are investigated for data communications. The effect of both imperfect power control and imperfect sectorization on the performance is investigated. It is shown that the system is rather sensitive to small power control errors and that voice activity monitoring and sectorization are good methods to improve the performance of cellular DS CDMA systems.**

I. Introduction

Among several types of Code Division Multiple Access (CDMA), Direct Sequence CDMA (DS CDMA) is a strong candidate for future personal communications because this technique offers inherent multipath diversity and robustness to varying channel conditions [1]–[4]. In order to obtain a reasonable performance using DS CDMA systems, additional considerations are required. Three techniques proposed in the literature to increase the performance of a DS CDMA system are power control in order to combat the near-far effect; sectorization; and voice activity monitoring in order to decrease the interference power detected at the receiving antenna. In practice power control errors occur and due to antenna imperfections, sectorization is not always perfect.

In this paper a general analytical model is presented that can be used to evaluate the influence of both imperfect power control and imperfect sectorization on the performance of two types of CDMA systems, namely voice-oriented and data-oriented systems. For the first type, the maximum user capacity is an important performance measure while for the second one, throughput and delay characteristics are appropriate performance parameters. An analytical model is developed assuming that the data-oriented system is slotted and simultaneous transmission of two or more packets is permitted. If the actual number of transmitting packets exceeds the value of the maximum capacity, then all packets are destroyed. In both systems only the reverse link, i.e., from mobile to base station is considered.

Manuscript received January 11, 1994; revised March 9, 1994; accepted April 22, 1994.

The authors are with the Telecommunications and Traffic Control Systems Group, Delft University of Technology, 2600 GA Delft, The Netherlands.

IEEE Log Number 9407341.

In Section II the principles of CDMA are discussed briefly and the performance enhancing techniques mentioned above are described. In Sections III and IV, respectively, the models for the capacity analysis and the throughput and delay analysis are explained. Section V shows the computational results and, finally, in Section VI conclusions and recommendations are presented.

II. Cellular DS CDMA

In a direct sequence spread spectrum system the data signal at the transmitter is multiplied by a pseudo-random user-specific code sequence waveform. At the receiver the original data can be recovered by correlating the received spread spectrum signal with the correct code sequence. If code sequences with good correlation properties are used, after the correlation operation the unwanted signals appear as noise-like signals with a very low power density spectrum. In order to establish the spreading effect, the duration T_c of a code symbol (chip) must be chosen much smaller than the duration T_b of a data symbol. Therefore the bandwidth B_{ss} of the transmitted spread spectrum signal is much larger than the bandwidth B_s of the original data signal. The ratio between transmitted and original bandwidth is defined as the processing gain:

$$\mathrm{PG} \triangleq \frac{B_{ss}}{B_s}. \tag{1}$$

In practice, this number depends on the specific modulation method. We will however use the following approximation:

$$\mathrm{PG} \triangleq \frac{B_{ss}}{B_s} \approx \frac{T_b}{T_c} \tag{2}$$

where T_b is the bit duration and T_c is the chip duration.

Since two code sequences with a relative delay of more than two chip times usually have a low correlation value compared to the fully synchronized situation, DS CDMA offers the possibility to distinguish between paths with a relative delay of more than two chip times. This is called the inherent diversity of DS CDMA, implying that it is possible to resolve a number of paths separately using only one receiver. This property makes DS CDMA suitable for applications in mobile radio environments that are usually corrupted with severe multipath effects.

It is well known that one of the most serious problems faced by a DS CDMA system is the multi-user interference. Because all users are transmitting in the same frequency band and

0018-9545/95$04.00 © 1995 IEEE

the crosscorrelations of the codes are rarely zero, the signal-to-interference ratio and hence the performance decreases as the number of users increases, which shows that DS CDMA is an interference-limited rather than a noise-limited system. The near-far effect plays an especially important role when considering multi-user interference. The near-far effect can be explained by considering the reverse link. Due to the path-loss law (which implies that the received power decreases as the transmitter-receiver distance increases), a close user will dominate over a user located at the cell boundary. The path-loss law is usually described by assuming that the received power P_r is proportional to the distance d to the power β:

$$P_r \sim d^{-\beta} \tag{3}$$

where β is the path-loss law exponent. In the case of free space propagation, the value of β is 2. In a practical mobile radio environment the value of β is in the range of 3 to 5, which is caused by the fact that the radio waves are reflected and partially absorbed by objects between receiver and transmitter and by the surface of the earth. In order to combat the near-far effect, power control can be used. There are two other important techniques to reduce the multi-user interference in a cellular DS CDMA system. The first one is sectorization, which is established by using sectorized antennas at the base station, implying that only interference signals are received from within a limited angle. The second important technique is voice activity monitoring. A system using voice activity monitoring prevents a mobile from radiating power during speech pauses. A complication in a cellular system is the fact that a base station not only receives interference from users within its own cell (intra-cell interference) but also from users in other cells (inter-cell interference).

A. Power Control

Power control can be established by letting the base station continuously transmit a (wideband) pilot signal that is monitored by all mobile terminals. According to the power level detected by the mobile, the mobile adjusts its transmission power. Hence mobiles near the cell boundary transmit at a lower power level than mobiles located close to the base station. This is open loop power control. It is also possible to use closed loop power control as is suggested in [5] and [6]. In this case the base station measures the energy received from a mobile and controls the transmit power of the mobile by sending a command over a (low data rate) command channel. In a practical power control system, power control errors occur, implying that the average received power at the base station may not be the same for each user signal.

The performance of a power control system depends on the power control algorithm, speed of the adaptive power control system, dynamical range of the transmitter, spatial distribution of users, and propagation statistics (such as fading and shadowing). All these factors influence the probability density function (pdf) of the received power. Since the objective here is to investigate the influence of power control imperfections on the system performance, the explicit influence of the factors mentioned above on the pdf of the average received power is not investigated and it can be a subject for future study.

Investigations on DS CDMA have shown that the pdf of the received power P due to the combined influence can be assumed to be log normal [1], [2].

$$f(P) = \frac{1}{\sqrt{2\pi}\sigma P}\exp\left[-\frac{\ln^2(P)}{2\sigma^2}\right] \tag{4}$$

where the imperfection in the power control system is determined by the logarithmic standard deviation σ of the log-normal power distribution of the received signal. In the case of perfect power control the logarithmic standard deviation is 0 dB.

There are a number of motivations for choosing a log-normal distribution for the individual received power. First, the received power at the base station depends on a lot of independent factors. It can be assumed that each factor gives a contribution in dB to the received power. Using the central limit theorem it is clear that the logarithm of the received power is Gaussian-distributed, implying that the received power is lognormally distributed. The second motivation is that this model provides a means to investigate the influence of imperfect power control analytically. A third is data provided in [5], [6], where it is shown that the received signal to noise ratio is log-normally distributed with standard deviation between 1 and 2 dB. Assuming constant noise power, this implies that the received power can be modeled as a log-normal variable.

If a number of k mobiles is transmitting and if the power of each mobile is controlled individually, then the total received interference power P_I is the summation of k independent identically log-normally distributed random variables denoted by P_i:

$$P_I = \sum_{i=1}^{k} P_i. \tag{5}$$

Fenton [10] showed that the pdf of P_I for k users is approximately log-normal with the following logarithmic mean $m_I(k)$ and logarithmic variance $\sigma_I^2(k)$:

$$\sigma_I^2(k) = \ln\left(\frac{1}{k}e^{\sigma^2} + \frac{k-1}{k}\right) \tag{6}$$

$$m_I(k) = \ln(k) + m + \frac{\sigma^2}{2} - \frac{1}{2}\ln\left(\frac{k-1}{k} + \frac{1}{k}e^{\sigma^2}\right). \tag{7}$$

This method is valid for a logarithmic standard deviation σ less than 4 dB.

B. Sectorization

The multi-user interference received at the base station can be reduced by dividing a cell into a number of sectors by means of directional antennas. In the case of perfect directional antennas there is a sharp separation between the sectors. Due to overlap and sidelobe anomalies of practical antennas the base station still receives some interference from users in other sectors. This effect can be studied by modeling

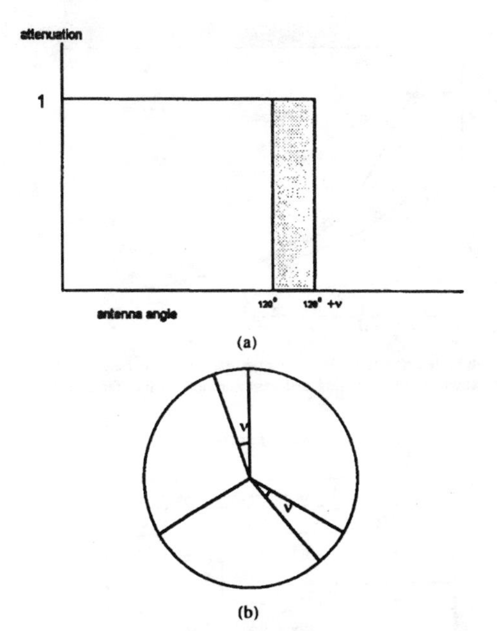

Fig. 1. (a) Radiation pattern model for imperfect directional antenna with opening angle 120°. (b) Sector coverage with an imperfect directional antenna with opening angle 120°.

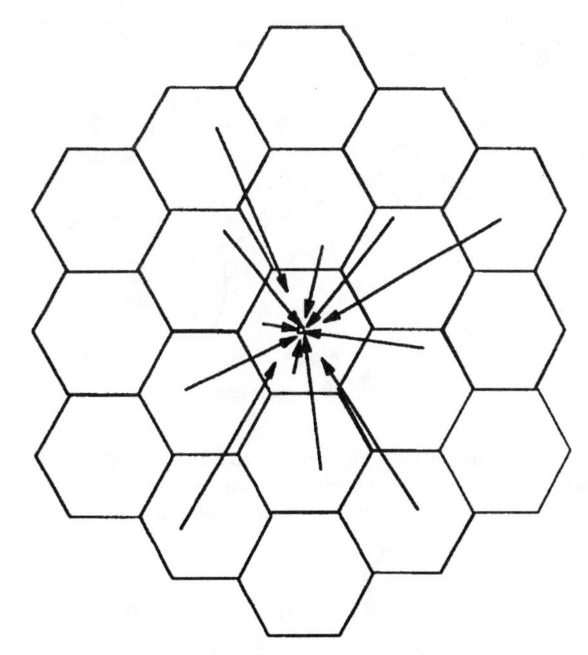

Fig. 2. Inter-cell interference in a cellular system.

the real antenna gain pattern for instance by a parabolic function. For simplicity, however, we assume the simplified antenna radiation pattern shown in Fig. 1(a) and (b) where the antenna imperfections are modeled by the overlap angle v. It is suggested to transform the real antenna gain patterns to this uniform model, analogously as the method for describing noise by the noise equivalent bandwidth. If D is the number of sectors and v is the overlap angle, then a relation can be derived between the ratio of the total interference power received in a sectorized system and the total interference power received in a non-sectorized system denoted by F_s.

$$F_s \triangleq \frac{\tilde{P}_{\text{sectorized}}}{\tilde{P}_{\text{non-sectorized}}} = \left(\frac{1}{D} + \frac{2v}{360}\right) \tag{8}$$

where D is an integer. From (8) it is clear that $v = 0$ corresponds with perfect sectorization and the combination $D = 1$ and $v = 0$ corresponds to the situation without sectorization.

C. Voice Activity Monitoring

Voice activity monitoring implies that the transmitter is not active during silent periods in human speech. It is possible to detect a silent period in the speech signal and let the transmitter stop transmitting during this period. Voice activity factors between 35% and 40% are reported in [3] and [4]. Studies done in Europe suggest that the total activity due to voice and background noise is higher in a mobile environment than in a wireline system and can have values between 50 and 60% [7]. When using voice activity monitoring, the probability that k out of n interferers are active can be described by a binomial distribution:

$$P(n,k) = \binom{n}{k} a^k (1-a)^{n-k} \tag{9}$$

where a is the voice activity factor. The effect of the voice activity factor on the CDMA capacity is investigated in this paper.

D. Inter-Cell Interference

In a cellular system, each base station not only receives interference from mobiles in the home cell (intra-cell interference) but also from terminals located in adjacent cells (Fig. 2). This kind of interference is called inter-cell interference. The principle of the CDMA protocol allows each cell to use the same frequency band, removing the need for a mobile to change its frequency when moving into another cell (soft handover). In order to model the interference received from terminals in other cells, we investigate here the interference power received by the base station in the home cell (H) from mobile terminals in an other cell (O) with base station separation distance d (Fig. 3).

We have adopted a generalized version of the approach as described in [9]. For analytical convenience the hexagonal cell structure is approximated by circles with radius R. The power of terminals in cell H and cell O is controlled by the base station in H and O, respectively. In the case of perfect power control, the power received by a base station from each mobile terminal in its service area is constant. This power is denoted by S_p. We assume that users are uniformly distributed in each cell with user density

$$\eta = \frac{N}{\pi R^2} \tag{10}$$

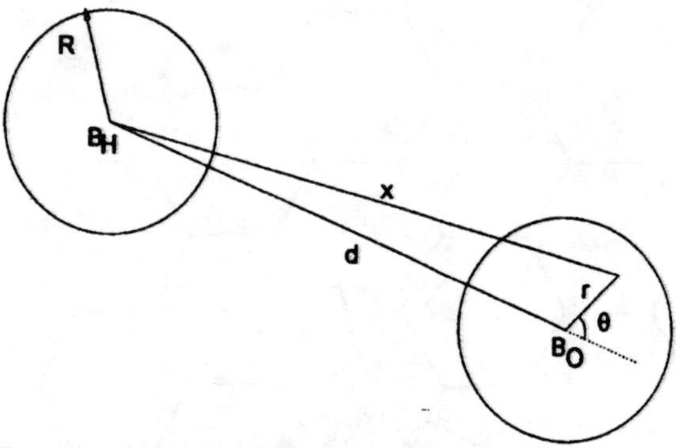

Fig. 3. Interference from terminals in a distant cell.

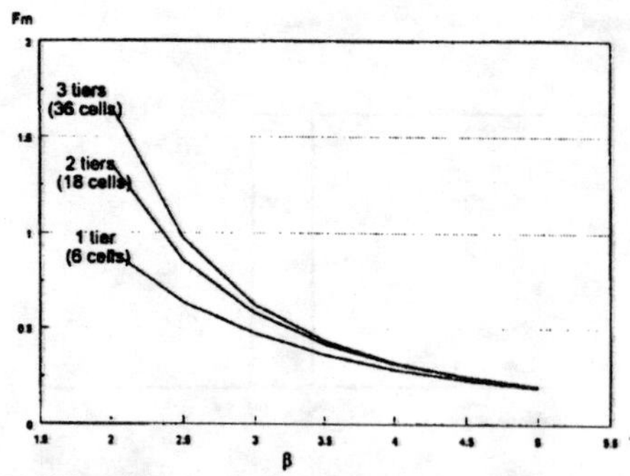

Fig. 4. Multiple cell interference reduction factor F_m as function of the path loss law exponent β considering different numbers of interfering cells.

where N is the number of users per cell and R is the cell radius. The power control strategy is such that a terminal at a distance r from the base station transmits with power

$$P_T(r) = S_p \cdot r^\beta \tag{11}$$

implying that the power received by the base station is

$$P_R(r) = S_p \cdot r^\beta \cdot r^{-\beta} = S_p \tag{12}$$

where β is the path loss law exponent that is two in case of free space loss and lies between three and five for mobile systems. In [9] the special case of $\beta = 4$ was considered while we derive an expression which is valid for all values of β. From Fig. 3 it can be seen that the power received by the base station denoted by $P_{\mathrm{RB}}(d)$ from a mobile users in a cell at distance d is obtained by integrating over the cell area (A):

$$P_{\mathrm{RB}}(d) = \int S_p\left(\frac{r}{x}\right)^\beta \eta dA \tag{13}$$

where

$$x = \sqrt{d^2 + r^2 + 2dr\cos\theta} \tag{14}$$

Using Fig. 3 and substituting (10) and (14), (13) can be written as

$$P_{\mathrm{RB}}(d) = \frac{2N \cdot S_p}{\pi R^2}\int_0^R dr \cdot r^{\beta+1}\int_0^\pi \frac{d\theta}{(d^2 + r^2 + 2dr\cos\theta)^{\beta/2}}. \tag{15}$$

In [6] a closed expression was found for the case $\beta = 4$, which we obtain from (15):

$$P_{RB}(d) = 2 \cdot N \cdot S_p \cdot \left[2\left(\frac{d}{R}\right)^2 \ln\left(\frac{\left(\frac{d}{R}\right)^2}{\left(\frac{d}{R}\right)^2 - 1}\right) - \frac{4\left(\frac{d}{R}\right)^4 - 6\left(\frac{d}{R}\right)^2 + 1}{2\left(\left(\frac{d}{R}\right)^2 - 1\right)^2}\right]. \tag{16}$$

Considering now a hexagonal cellular structure (Fig. 2), it is possible to compute the total received interference from all the cells in the system. The interference correction factor due to multiple cell interference F_m is defined as the ratio of the interference power received from the outer cells (I_m) and the

TABLE I

Number of tiers	Interference correction factor F_m β=2	β=3	β=4	β=5
1	0.904	0.417	0.284	0.191
2	1.365	0.579	0.312	0.199
3	1.669	0.625	0.319	0.200
5	2.079	0.667	0.323	0.201
10	2.665	0.701	0.326	0.201
15	3.018	0.714	0.326	0.201

interference power generated by users in the home cell (I_h). Using (12), F_m can be written as

$$F_m \triangleq \frac{I_m}{I_h} = \frac{I_m}{(N-1)S_p} \tag{17}$$

where N is the number of users per cell and S_p is the power received from a user in the case of perfect power control.

The value of I_m depends on the value of β and on the number of cell tiers considered. From (15) it follows that for a large number of users ($N \gg 1$), the ratio F_m is a constant because I_m is proportional to N. Fig. 4 and Table I show the interference correction factor F_m as function of the path loss law exponent β considering several number of cell tiers around the home cell. It is clear that for larger values of β, the influence of the outer cell tiers is decreasing with increasing tier number. For $\beta = 4$ no significant difference was found between considering 10 or 15 tiers. In both cases, we found $F_m = 0.326$. This value is less than the value of 0.66 reported in [4] that was obtained by simulation assuming log-normal shadowing. The value 0.326 was found analytically for the situation without shadowing and is confirmed by results presented in [9]. The value 0.326 is used in all the calculations in this paper.

III. Capacity Analysis

In a speech-oriented DS CDMA system, after setting-up a call each user is allowed to transmit data continuously. A very important design issue for such a system is the user capacity. A service provider will be interested in the maximum number of users that can be served simultaneously by the system with a predefined performance in a certain geographic area.

In a digital communication system such as DS CDMA, the bit error rate (BER) is an appropriate performance measure. BER calculations offer the possibility for detailed system considerations but often require complex models that can sometimes disturb the basic understanding of the system. In order to obtain a general idea about the influence of power control, sectorization, and voice activity on the capacity of a cellular DS CDMA system, a simplified model is proposed in this paper. Note that for a more detailed system investigation it is better to rely on BER models since the simplified model does not consider the effects of multipath in general.

The performance model used in this paper is based on a minimum required signal-to-interference ratio at the receiver in order to obtain a prespecified bit error rate performance. In [2] it was mentioned that irrespective of modulation method, the minimal signal to interference ratio after correlation for a bit error rate better than 10^{-3} is $E_b/N_0 = 5$ (7 dB). The threshold value γ for the signal-to-interference ratio before correlation is

$$\gamma = \frac{1}{\mathrm{PG}} \cdot \left(\frac{E_b}{N_0}\right)_{\min} \tag{18}$$

where PG is the processing gain, which is equivalent to the number of chips per bit in this case. In a system with multiple cells and (imperfect) sectorization, the signal-to-interference ratio at the receiver is

$$\frac{C}{I} = \frac{P_i}{P_I(1+F_m)\cdot F_s} \tag{19}$$

with

P_i: Received power from desired terminal in home cell.
P_I: Intra-cell interference power from users in home cell (single isolated cell without sectorization).
F_s: Interference correction factor due to imperfect sectorization (8).
F_m: Interference correction factor due to inter-cell interference (17).

The failure probability P_f is given by

$$P_f = \Pr\left(\frac{P_i}{P_I} < \gamma(1+F_m)\cdot F_s\right). \tag{20}$$

In the case of imperfect power control, the received power P_i from each mobile is modeled as a log-normal variable with zero logarithmic mean and standard deviation σ. The pdf of the total interference signal produced by $(k-1)$ other active terminals is approximated by a lognormal variable with logarithmic mean $m_I(k-1)$ and standard deviation $\sigma_I(k-1)$ ((6) and (7), respectively) computed according to the method of Fenton [10]. After some mathematical manipulations the failure probability for k active users can be written as

$$P_f(k) = \frac{1}{2} + \frac{1}{2}\mathrm{erf}\left(\frac{\ln[\gamma(1+F_m)\cdot F_s] - m_I(k-1)}{\sqrt{2(\sigma^2 + \sigma_I^2(k-1))}}\right) \tag{21}$$

where erf $(\cdot)$ is the error function, γ is the threshold ratio for the signal-to-interference ratio, F_m is the interference correction factor due to multiple cell interference, F_s is the interference correction factor due to sectorization and $m_I(k-1)$ and $\sigma_I(k-1)$ are the logarithmic mean and the logarithmic standard deviation respectively of the total power received from $(k-1)$ users in case of imperfect power control with individual power control error σ. Assuming that there are n users per cell ($n \geq k$), each with voice activity a, then the failure probability is

$$P_f^a(n) = \sum_{k=1}^{n} P_f(k) \cdot \binom{n}{k} a^k (1-a)^{n-k}. \tag{22}$$

The user capacity is now defined as the maximum number of users per cell with voice activity a such that the failure probability is less than 0.01.

IV. Throughput and Delay Analysis

For transmission of computer data, a packet communications schedule can be more efficient than using a circuit switched protocol. In this section we consider a slotted system, implying that that each user is allowed to transmit packets only at fixed time instants. In a packet network, throughput and delay are appropriate parameters, rather than maximum user capacity. The throughput determines the average number of successfully received packets per time slot, given a certain amount of traffic. For a certain amount of throughput it is important to know what will be the average delay of a packet. The throughput is defined as the average number of successfully received packets per time slot:

$$S = \sum_{k=1}^{N} k P_t(k) P_s(k) \tag{23}$$

where N is the number of users per cell, $P_s(k)$ is the packet success probability and $P_t(k)$ is the probability of a packet transmitted with $k-1$ other packets.

It is assumed that a total of n_f independent failures can occur within one time slot. This number in general depends on the length of a time slot T_{slot} and the average time between two failures T_{av} and is defined as:

$$n_f \triangleq \frac{T_{\mathrm{slot}}}{T_{\mathrm{av}}}. \tag{24}$$

Now the packet success probability $P_s(k)$ defined as the probability of a test packet not being destroyed by $(k-1)$ interfering packets is

$$P_s(k) = (1 - P_f(k))^{n_f} \tag{25}$$

with $P_f(k)$ being the failure probability given by (21). Although it is difficult to implement power control to packet transmission due to short packet duration, the present analysis is based on power control for simplicity.

Because the number of simultaneous terminals is limited, the offered traffic is assumed to have a binomial distribution. The probability of a packet transmitted with $k-1$ other packets P_k is then given by

$$P_t(k) = \binom{N}{k}\left[\frac{G}{N}\right]^k \left[1 - \frac{G}{N}\right]^{N-k} \tag{26}$$

where G is the average number of offered packets per time slot. Here it is assumed that the offered traffic consists of new packets and retransmission packets.

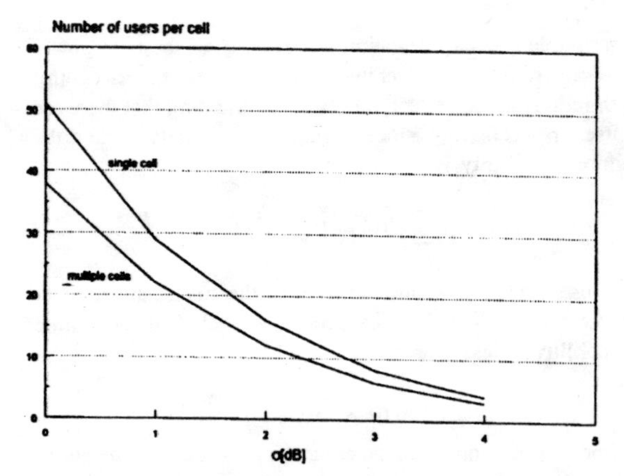

Fig. 5. Capacity as function of power control imperfection for a single cell and multiple cell system. Processing gain PG = 255, no voice activity monitoring ($a = 1$), no sectorization ($D = 1$).

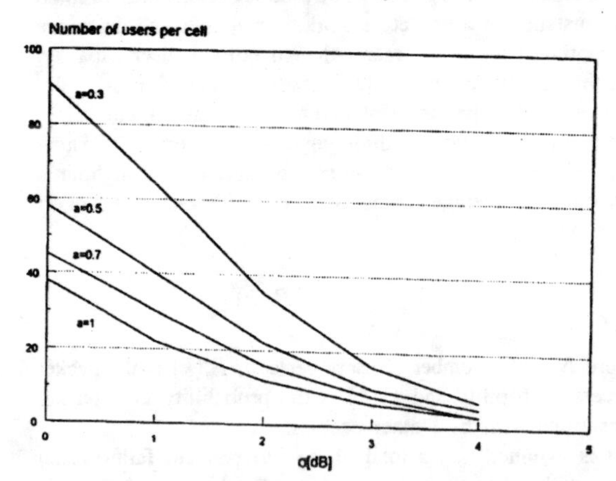

Fig. 6. Capacity as a function of power control imperfection for several values of the voice activity factor (a) Processing gain PG = 255, no sectorization ($D = 1$), multiple cells.

Another important performance parameter for a slotted packet network is the delay, which is defined as the average time between the generation and successful reception of a packet. Assuming a positive acknowledgment scheme and no transmission errors in the acknowledgment packets, the expression for the delay in a slotted DS CDMA system is [11]–[13]:

$$D = 1.5 + \left[\frac{G}{S} - 1\right] \cdot (\lfloor \delta + 1 \rfloor + 1) \quad (27)$$

where δ is the mean of the retransmission delay, which is uniformly distributed over the range from which the retransmission delay is selected, S is the throughput and G is the offered traffic. Note that the delay in (27) is normalized on the slot time.

V. RESULTS

All figures presented is this section are valid for the reverse link. First the results are given for the capacity, i.e., the maximum number of users per cell. For all results, arbitrary

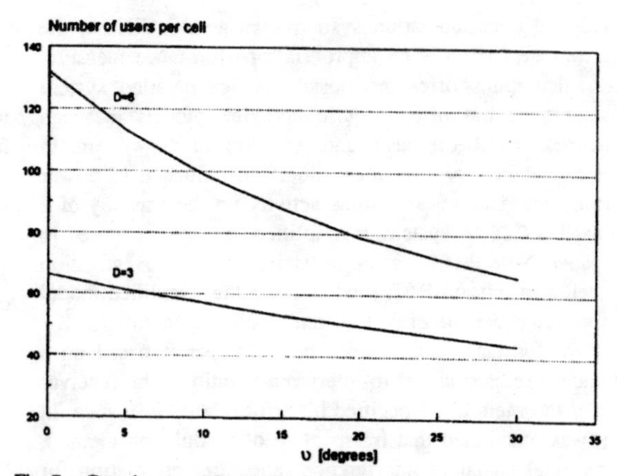

Fig. 7. Capacity as a function of overlap angle for sectorization with sectorization degree $D = 3$ and $D = 6$. Processing gain PG = 255, power control imperfection $\sigma = 1$ dB, no voice activity monitoring ($a = 1$), multiple cells.

PN sequences with 255 chips per bit are assumed, implying a processing gain PG = 255. Fig. 5 shows the influence of power control imperfections on the capacity for a DS CDMA system with only one cell and for a system with multiple cells. In the multiple cell case the inter-cell interference was modeled as a fraction of the intra-cell interference using (17) where we assumed a path loss law exponent $\beta = 4$. In case of perfect power control, the maximum capacity shows a decrease of 25% when considering multiple cells instead of a single cell. For lower values of β the decrease in capacity due to multiple cell interference will be higher. Fig. 5 also shows that the system is very sensitive to power control errors; a power control error of only 1 dB gives a capacity reduction of about 60%. A power control error of 1 dB is interpreted as the standard deviation of 1 dB around the desired power level. In [5] it was concluded that a standard deviation of 2.5 dB causes a reduction in capacity of only 20%. In [5], however, it was assumed that an outage occurs if the interference power density exceeds the background noise level by 10 dB whereas in our paper an outage occurs if the signal-to-noise-ratio after correlation exceeds a predefined value ($E_b/N_0 = 5$ (7 dB)). This explains the difference in capacity reduction. Fig. 6 shows that the capacity decreases as the voice activity factor increases. If the processing gain is increased, i.e., if the number of chips per bit is increased, then the maximum number of users increases at the cost of more required bandwidth (since the chip duration must decrease in order to keep the data rate fixed). As expected, the capacity increases with the increase in the degree of sectorization, which is shown in Fig. 7. Furthermore it can be seen from Fig. 7 that the capacity decreases as the overlap angle due to imperfect sectorized antennas increases. The model with the uniform antenna gain pattern and the overlap angle is a drastic simplification of real antenna gain patterns but it simplifies the calculations tremendously. It suggests to transform real antenna gain patterns into uniform patterns analogously to modeling thermal filtered noise by the noise equivalent bandwidth.

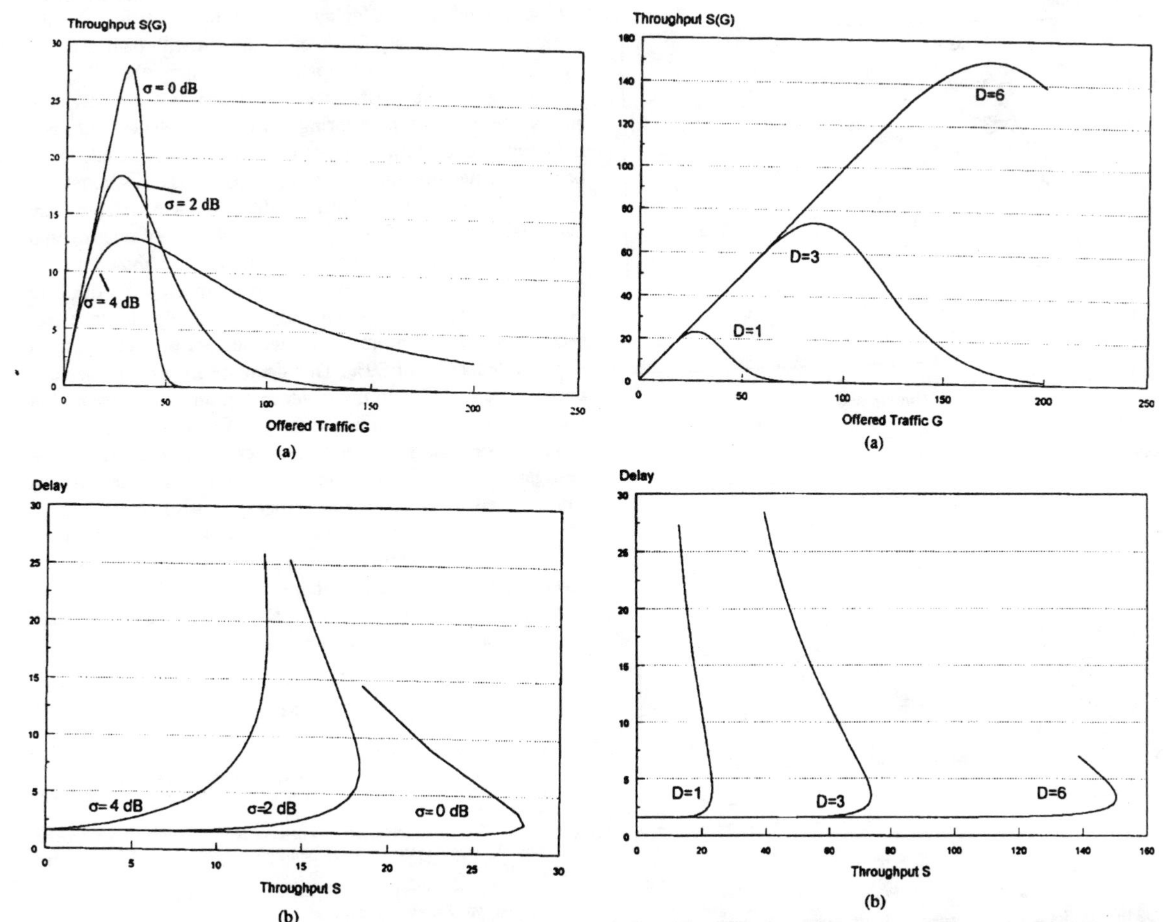

Fig. 8. (a) Influence of power control error on throughput. Processing gain PG = 255, 200 terminals per cell, and no sectorization. (b) Influence of power control error on the delay. Processing gain PG = 255, 200 terminals per cell, and no sectorization.

Fig. 9. (a) Influence of sectorization on throughput. Power control error $\sigma = 1$ dB, processing gain PG = 255, 200 terminals per cell, and perfect sectorization. (b) Influence of sectorization on delay. Power control error $\sigma = 1$ dB, processing gain PG = 255, 200 terminals per cell, and perfect sectorization.

In a packet-oriented DS CDMA system the throughput and delay characteristics are more appropriate performance parameters than the maximum user capacity. For all the following figures it is assumed that on the average one fade per packet occurs ($n_f = 1$) and that the processing gain PG again is 255. The throughput is given per cell and the delay is measured in time slot units. For the delay calculations we assumed a mean retransmission delay $\delta = 10$ slot times. Fig. 8(a) and (b) shows the influence of power control imperfections on the throughput and delay, respectively. It is clear that again the performance decreases as the power control error increases. A decrease in performance implies here that the maximum throughput is lower and that the delay corresponding to a throughput value is higher. From Fig. 8(a) it is clear that for high traffic loads the throughput increases for higher power control errors. This can be explained by considering that in the case of perfect power control the failure probability always equals 1 if the number of users exceeds the maximum capacity. This is not the case if we have imperfect power control ($\sigma \geq 0$ dB). Because of the tail of the log-normal distribution, there is always a probability that a few users are successful although the total number of users exceeds the maximum capacity. Then the failure probability decreases as the variance in the received power due to power control errors increases. Note that this effect occurs in the overload region.

The influence of sectorization (perfect) on the throughput and delay is depicted in Fig. 9(a) and (b). As expected, the performance increases with increasing number of sectors. From Fig. 9(a) it can be seen that the maximum throughput is better than proportional, i.e., the maximum throughput for $D = 3$ and $D = 6$ is higher than 3.S at $D = 1$ and 6.S at $D = 1$ respectively. In order to explain this it is first of all important to observe that the average number of offered packets at the maximum throughput is below the maximum capacity (maximum number of simultaneously transmitting users). This is caused by the fact that already for values of average offered traffic less than the maximum capacity there is a probability that the actual number of offered packets exceeds the maximum capacity. The maximum capacity of the cell is of course exactly proportional to the

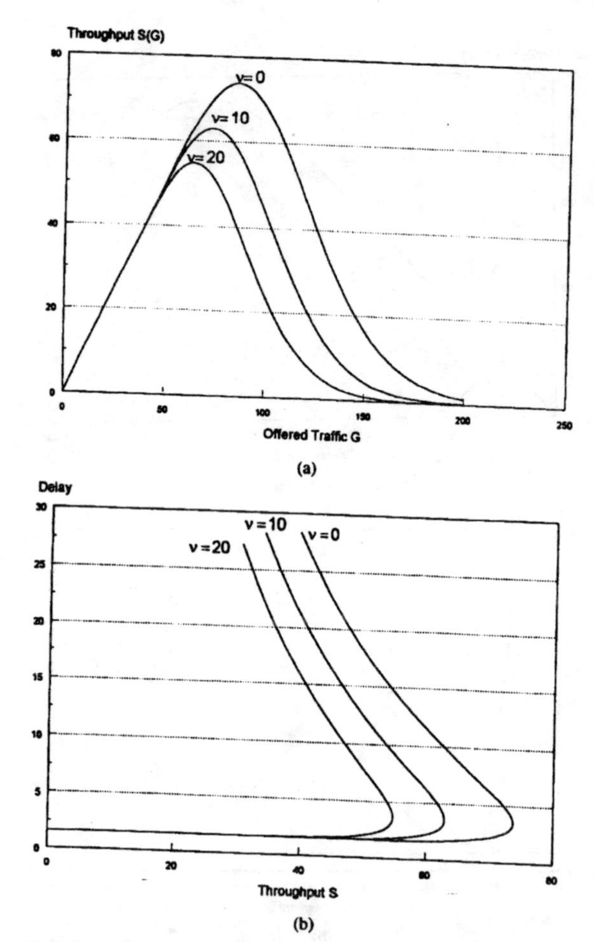

Fig. 10. (a) Influence of overlap angle (in degrees) on throughput. Power control error $\sigma = 1$ dB, processing gain PG = 255, 200 terminals per cell, and $D = 3$ sectors. (b) Influence of overlap angle (in degrees) on delay. Power control error $\sigma = 1$ dB, processing gain PG = 255, 200 terminals per cell, and $D = 3$ sectors.

number of sectors in the cell. Since the total user population for each cell is the same for the sectorization schemes we considered ($N = 200$), the expression for the probability of having k simultaneous packets in one entire cell ($P_t(k)$) is the same. The probability of increasing a maximum user capacity of 50 users is then higher than the probability of increasing a maximum user capacity of 150 users. This implies that the difference between the maximum offered traffic and the maximum capacity decreases with an increasing degree of sectorization and explains that the maximum throughput increases more than proportionally in our model.

Finally, Fig. 10(a) and (b) shows the influence of antenna imperfections on the throughput and delay of a slotted DS CDMA system. The performance decreases as the overlap angle due to imperfect directional antennas increases.

VI. Conclusion and Recommendations

A general model has been presented in this paper that is suitable to evaluate the reverse link performance of a cellular DS CDMA system in terms of maximum user capacity, throughput, and delay. The multiple cell interference is modeled analytically for an environment with path loss law exponent β. In order to keep the model comprehensible and to investigate global effects of power control, sectorization and voice activity monitoring, effects of shadowing, and multipath propagation have not been considered. Results have been presented in this paper for a path loss law exponent $\beta = 4$. Voice-oriented as well as data-oriented systems were evaluated and the effects of imperfect power control and imperfect sectorization were studied. It was shown that a power control error of 1 dB leads to a capacity loss of 50 to 60%. Furthermore, an overlap degree of 15° due to imperfect sectorization in a system with three sectors per cell causes a capacity reduction of 20%. This decrease in performance can not only be seen due to the lower maximum user capacity but also due to lower maximum throughput and higher delay.

The model described in this paper is appropriate for investigating general system properties in a very clear way and can be used to obtain a first general idea of the effects of voice activity monitoring, power control, and sectorization on the performance. In order to obtain more detailed information, however, it is recommended that shadowing effects, multipath statistics, and consideration of BER performance be included. This will be the subject of future study.

References

[1] R. Prasad, M. G. Jansen, and A. Kegel, "Capacity analysis of a cellular direct sequence code division multiple access system with imperfect power control," *IEEE Trans. Commun.*, vol. E76-B, no. 8, pp. 894–905, Aug. 1993.

[2] ——, "Effect of imperfect power control on a cellular code division multiple access system," *Elect. Lett.*, vol. 28, no. 9, pp. 848–849, Apr. 23, 1992.

[3] W. C. Y. Lee, "Overview of cellular CDMA," *IEEE Trans. Commun.*, vol. 40, pp. 291–302, May 1991.

[4] K. S. Gilhousen, J. M. Jacobs, R. Padovani, A. J. Viterbi, L. A. Weaver, and C. E. Wheatly, III, "On the capacity of a cellular CDMA system," *IEEE Trans. Commun.*, vol. 40, pp. 302–312, May 1991.

[5] A. M. Viterbi and A. J. Viterbi, "Erlang capacity of a power controlled CDMA system," *IEEE J. Select. Areas Commun.*, vol. 11, pp. 892–899, Aug. 1993.

[6] A. J. Viterbi, A. M. Viterbi, and E. Zehavi, "Performance of power controlled wideband terrestrial digital communication," *IEEE Trans. Commun.*, vol. 41, pp. 559–569, Apr. 1993.

[7] H. J. Braun, G. Cosier, D. Freeman, A. Gilloire, D. Sereno, C. B. Southcott, and A. Van der Krogt, "Voice control of the Pan-European digital mobile radio system," *CSELT Tech. Rep.*, vol. XVIII, no. 3, pp.183–187, June 1990.

[8] M. G. Jansen and R. Prasad, "Throughput analysis of a slotted CDMA system with imperfect power control," in *Inst. Elect. Eng. Colloquium on Spread Spectrum Techniques for Radio Commun. Syst.*, London, U.K., Apr. 27, 1993, pp. 8.1–8.4.

[9] K. I. Kim, "CDMA cellular engineering issues," *IEEE Trans. Vehicular Technol.*, vol. 42, pp. 345–349, Aug. 1993.

[10] L. F. Fenton, "The sum of a log-normal probability distribution in scattered transmission systems," *IRE Trans.*, vol. C5-8, pp. 57–67, Mar. 1960.

[11] L. Kleinrock and F. A. Tobagi, "Packet Switching in radio channels: Part I—Carrier sense multiple access modes and their throughput-delay characteristics," *IEEE Trans. Commun.*, vol. COM-23, Dec. 1975.

[12] R. Prasad, C. A. F. J. Wijffels, and K. L. A. Sastry, "Performance analysis of slotted CDMA with DPSK modulation diversity and BCH-Coding in indoor radio channels," *AEÜ*, vol. 46, no. 6, pp. 375–382, Nov. 1992.

[13] C. A. F. J. Wijffels, H. S. Misser, and R. Prasad, "A micro-cellular CDMA system over slow and fast Rician fading channels with forward error correcting coding and diversity," *IEEE Trans. Vehicular Technol.*, vol. 42, pp. 570–580, Nov. 1993.

Michel G. Jansen received the Ir (M.Sc. E.E.) degree in electrical engineering from Delft University of Technology, The Netherlands in 1992. His dissertation was on the effects of imperfect power control on the capacity of DS CDMA networks.

After his graduation he joined the Telecommunications and Traffic-Control Systems Group of Delft University of Technology as a post-graduate student, where he is participating in the research program on CDMA radio networks.

Ramjee Prasad (M'88–SM'90) received the B.Sc. Eng. degree from Bihar Institute of Technology, Sindri, India, and the M.Sc. Eng. and Ph.D. degrees from Birla Institute of Technology (BIT), Ranchi, India, in 1968, 1970, and 1979, respectively.

He joined BIT as a Senior Research Fellow in 1970 and became an Associate Professor in 1980. During 1983–1988 he was with the University of Dar es Salaam (UDSM), Tanzania, where he became a Professor in Telecommunications at the Department of Electrical Engineering in 1986. Since February 1988, he has been with the Telecommunications and Traffic Control Systems Group, Delft University of Technology, The Netherlands, where he is actively involved in the area of mobile, indoor, and personal radio communications. While he was with BIT, he supervised many research projects in the area of microwave and plasma engineering. At UDSM, he was responsible for the collaborative project "Satellite Communications for Rural Zones" with Eindhoven University of Technology, The Netherlands. He has published more than 200 technical papers. His current research interest lies in packet communications, adaptive equalizers, spread-spectrum CDMA systems, and multimedia communications. He has also presented tutorials on mobile and indoor radio communications at various universities, technical institutions, and IEEE conferences. He is also a member of a working group of European cooperation in the field of scientific and technical research (COST-231) project "Evolution of Land Mobile Radio (including personal) Communications" as an expert for The Netherlands.

Dr. Prasad is listed in *Who's Who in the World*. He was Organizer and Interim Chairman of the IEEE Vehicular Technology/Communications Society Joint Chapter, Benelux Section. Now he is the elected Chairman of the same chapter. He is also founder of the IEEE Symposium on Communications and Vehicular Technology (SCVT) in the Benelux and was Symposium Chairman of SCVT'93. He has served as a member of advisory and program committees of several IEEE international conferences. He is one of the Editors-in-Chief of a new journal on *Wireless Personal Communications* and is also a member of the editorial boards of other international journals including IEEE COMMUNICATIONS MAGAZINE. He was the Technical Program Chairman of the PIMRC'94 International Symposium held in The Hague, The Netherlands, September 19–23, 1994 and also of the Third Communication Theory Mini-Conference in conjunction with GLOBECOM'94, San Francisco, CA, November 27–30, 1994. He is a Fellow of the Institute of Electrical Engineers, a Fellow of the Institution of Electronics and Telecommunication Engineers, and a Member of the New York Academy of Sciences and The Netherlands Electronics and Radio Society.

Performance Analysis of CDMA with Imperfect Power Control

Rick Cameron and Brian Woerner, *Member, IEEE*

***Abstract*—The standard correlation receiver for code-division multiple access (CDMA) systems is susceptible to the near–far problem. Power control techniques attempt to overcome near–far effects by varying transmitted power levels to ensure that all signals are received with equal power levels. Since these algorithms cannot perfectly compensate for power fluctuations in a mobile communications channel, the capacity of the system is reduced for a given bit-error rate (BER). This paper examines the performance of a CDMA system using imperfect power control by extending analytical techniques that account for multiple access interference. Single cell capacity is compared with systems employing perfect power control.**

I. INTRODUCTION

THERE has been considerable recent interest in using code-division multiple access (CDMA) technology to improve the capacity of cellular telephone systems over that of the current analog system [1]. However, CDMA systems that employ correlation receivers suffer from the near–far problem, necessitating the use of power control techniques. The purpose of power control is to ensure that the received signal strength of all users are equal. However, this cannot be accomplished perfectly with practical systems, resulting in capacity loss due to the nonoptimal performance of the correlation receiver under these circumstances. We present an analytical model for evaluating the effect of imperfect power control. We compare the single cell capacity of these systems to systems with perfect power control, where the single cell capacity is the number of users that can be supported for a given bit-error rate (BER).

Many researchers have studied the performance of direct-sequence spread-spectrum (DS/SS) systems in additive white Gaussian noise (AWGN) channels [2]–[8]. However, this analysis does not account for power control, which is necessary in the mobile environment. There is an additional body of work that considers the effects of power control in spread-spectrum systems [9]–[11]. However, these papers typically use the Gaussian approximation to model the multiple access interference. The Gaussian approximation can be significantly optimistic for low BER's [5] or in the case of the near–far problem [12]. In this paper, we show how the analytic techniques of [5] can be extended to accommodate the effects of imperfect power control without introducing a Gaussian approximation. As a result, we are able to accurately model both multiple access interference and the effects of power control. We present performance results for a CDMA system employing a correlation receiver.

Power control is used to adjust for fading of the received signal. The transmitted signal must have enough power to overcome fading and keep the received power constant over time. In practice, there are physical limits as to how fast power control algorithms can respond, so ideal compensation is not possible. On the reverse channel (mobile to base station), the near–far problem exists when one transmitting mobile is much farther from the receiving base station than another mobile. The base station will receive a much stronger signal from the nearby mobile, causing a large amount of interference to the signal of the distant mobile. CDMA cellular systems use open and closed loop power control techniques on the reverse link to compensate for this problem [13]. The power control mechanisms for CDMA cellular systems have been tested during field trials, which have shown that received signal power has a log-normal distribution with a standard deviation in the neighborhood of 1 dB above the desired power setpoint. This verifies simulation results predicted by [14]. Since the power at any instant is the result of many small incremental adjustments, the central limit theorem implies that there may be a theoretical foundation for this empirical result [15].

Although power control is one technique for alleviating the near–far problem for the widely used correlation receiver, alternative CDMA receiver designs are capable of overcoming the near–far problem. Near–far resistance may be formally defined as the ratio between the signal-to-noise ratio (SNR) of a single user system to the SNR of a multiuser receiver at the same BER. Verdu has demonstrated an optimum BER CDMA receiver for the AWGN channel which has a near–far resistance of one [16]. Although this structure is too complex for most practical implementations, other researchers are investigating suboptimal approaches which also exhibit near–far resistance. These approaches include decorrelating receivers [17], [18], and parallel [19] and successive cancellation [20] of multiple access interference. In the conclusion of this paper, it is suggested that even tight power control may be insufficient to overcome the near–far problem by itself.

The remainder of this paper is organized as follows. Section II presents the system model. Section III describes the analytic technique and Section IV presents numerical results. Section V concludes this paper.

II. SYSTEM MODEL

We consider a DS/SS multiple access system with binary phase shift keyed (BPSK) signaling and a correlation receiver based on [21]. There are K users in the system. User k's

Paper approved by B. Aazhang, the Editor for Spread Spectrum Networks. Manuscript received October 3, 1994; revised January 16, 1995 and October 6, 1995. This work was supported by NSF Contract NCR-9 211 344, and by the MPRG Industrial Affiliate Foundation. This paper was presented at the 1992 IEEE Vehicular Technology Conference, Denver, CO, May 1992.

The authors are with the Mobile and Portable Radio Research Group, Bradley Department of Electrical Engineering, Virginia Polytechnic Institute and State University, Blacksburg, VA 24061-0350 USA.

Publisher Item Identifier S 0090-6778(96)05498-0.

received signal is represented by

$$s_k(t) = \sqrt{2P_k} b_k(t-\tau_k) a_k(t-\tau_k) \cos(\omega_c(t-\tau_k)+\theta_k) \quad (1)$$

where k is the user, P_k is the power of the signal, $b_k(t-\tau_k)$ represents the data signal, $a_k(t-\tau_k)$ represents the spreading signal, and $\cos(\omega_c(t-\tau_k)+\theta_k)$ represents the modulating waveform. The received power is log-normally distributed according to [22]

$$f_{P_k}(P_k) = \frac{20\, log(e)}{\sqrt{2\pi}\sigma_P P_k} \exp\left(-\frac{(10\log P_k)^2}{2\sigma_P^2}\right) \quad (2)$$

where σ_P is the standard deviation of the received power expressed in decibels. The random phase θ_k is uniformly distributed over $[0, 2\pi)$, while the random delay τ_k is uniformly distributed over $[0, T)$. We assume that P_k is independent for each user and independent of θ_k and τ_k. The data signal is given by $b_k(t) = \Sigma_{i=-\infty}^{\infty} b_{k,i} p_T(t-iT)$, where $b_{k,i} \in \{+1, -1\}$ is an infinite sequence of data bits and $p_T(t)$ is a rectangular pulse with unity amplitude and duration T. The spreading signal $a_k(t)$ is given by $a_k(t) = \Sigma_{j=-\infty}^{\infty} a_{k,j} p_{T_c}(t-iT_c)$, where $a_{k,j} \in \{+1, -1\}$. The duration of each encoded data bit is T, while the duration of each chip in the signature sequence is T_c. The total number of chips per bit, N, is given by $N = T/T_c$.

Without loss of generality, assume that we are trying to receive the signal from user 1 and that $\theta_1 = 0$ and $\tau_1 = 0$. When a single propagation path exists, the received signal for K transmitters is given by

$$r(t) = n(t) + s_1(t) + \sum_{k=2}^{K} s_k(t-\tau_k) \quad (3)$$

where $n(t)$ is additive white Gaussian noise (AWGN) with two-sided power spectral density (PSD) $N_o/2, s_1(t)$ is the desired signal, and $s_k(t-\tau_k)$ is the received signal from user k. A block diagram of this receiver is shown in Fig. 1. A synchronous receiver recovers the transmitted data bit by correlating the received signal with the signature sequence of user 1 to form a decision statistic Z_1, given by

$$Z_1 = \int_0^T r(t) a_1(t) \cos(\omega_c t)\, dt. \quad (4)$$

The estimate of the original data bit from user 1, $\hat{b}_0^{(1)}$, is determined from Z_1 by

$$\hat{b}_0^{(1)} = \begin{cases} 1, & Z_1 \geq 0 \\ -1, & Z_1 < 0. \end{cases} \quad (5)$$

III. Analysis of Multiple Access Interference

We now focus attention on the modeling of multiple access interferers with varying power levels. Our goal is to find upper- and lower-bounds on the average probability of error. This method is based upon the approach for determining multiple access interference for users that was first presented in [5], and refined in [6], and we show how this analysis may be modified to account for imperfect power control [23]. This method generates discrete approximations to the probability distributions of all interfering and desired components. We generate best and worst case distributions, so that arbitrarily tight upper- and lower-bounds on the probability of error can be found. This approach was initially presented in [12] and is presented in detail here.

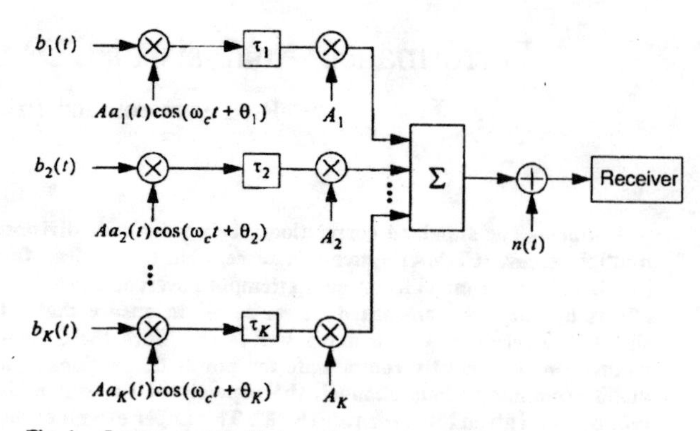

Fig. 1. System model with imperfect power control..

To modify the analysis in [6] to account for imperfect power control, there are several key differences that must be accounted for. First, the ranges over which the interference is defined are not bounded, and must be set to a level that achieves the desired numerical accuracy. Second, all interference levels must be mapped according to a log-normal distribution. Finally, even the desired signal is now a variable.

A. Analytical Techniques for Imperfect Power Control

We begin with the expression for the normalized multiple access interference distribution, B_k, that is defined as

$$B_k = \frac{\cos\phi_k}{T_c}\left[b_{-1}^{(k)} R_k(\tau) + b_0^{(k)} \hat{R}_k(\tau)\right] \quad (6)$$

where $\phi_k = \theta_k - \omega_c\tau_k$ and $R_k(\tau)$ and $\hat{R}_k(\tau)$ are the continuous-time partial cross-correlation functions of the kth and desired spectral-spreading waveforms defined in [5]. This assumes ideal power control. We know that with imperfect power control, received power levels may be modeled by a log-normal distribution [13]. Therefore, by multiplying each distribution by the appropriate random variable (rv), we can determine the new interference distribution that accounts for the effects of power control. We can perform a best and worst case analysis by finding the best and worst case values of B_k over a given interval. The worst case will assume the largest possible value of interference for that interval, whereas the best case analysis will assume the smallest possible value.

B. Worst Case Analysis

We wish to model the effects of imperfect power control on the probability distributions of the normalized interference B_k for users $k, 2 \leq k \leq K$. Initially, we know that the probability that user k's interference lies within a given interval is given by $d_i^{(k)}$ for $2 \leq k \leq K$ [6]

$$d_i^{(k)} = \Pr\left(B_k \in \left[\frac{i-1/2}{N_u}, \frac{i+1/2}{N_u}\right]\right) \quad (7)$$

for $-NN_u \leq i \leq NN_u$, where N_u is the number of subintervals per chip. The larger the value of N_u, the tighter the bounds on the probability of error. We then model the effects of power control by computing $J_k = A_k B_k$ for a given value of B_k, where $A_k = \sqrt{P_k/P}$ and P is the desired power setpoint of the power control algorithm. We now compute the probability that J_k falls in a given interval using

$$y_k(i) = \Pr\left(J_k \in \left[\frac{i-1}{N_u}, \frac{i}{N_u}\right]\right) \tag{8}$$

for $-NN_u \leq i \leq NN_u$. The lower- and upper-bounds on i are not fixed, as the normal distribution only asymptotically approaches zero as its argument approaches $\pm\infty$, and, thus, the bounds must be chosen so that the probability that the value of J_k lies outside these bounds is sufficiently small. For this case, limits of NN_u have been found to be sufficient.

The next step is to compute the probability that the power P_k is at a given level based on its log-normal distribution, using worst case values for B_k of $i - 1/2$, as

$$P_{i,j} = \Pr\left[\frac{j-1}{i-1/2} \leq A_k \leq \frac{j}{i-1/2}\right] \tag{9}$$

for $-NN_u \leq i \leq NN_u$ and $-NN_u \leq j \leq NN_u$, where $P_{i,j}$ is the probability that a point which lay in the ith bin before adjustment for power control lies within the jth bin after this adjustment. Since the imperfect power control is being modeled with a log-normal distribution, we can solve for $P_{i,j}$ using

$$P_{i,j} = Q\left(\frac{j}{i-1/2}\right) - Q\left(\frac{j-1}{i-1/2}\right) \tag{10}$$

where $Q(\cdot)$ is the standard Q function given by

$$Q(x) = \frac{1}{\sqrt{2\pi}} \int_x^\infty e^{-u^2/2}\, du. \tag{11}$$

We now wish to compute the probability distribution function for the new interference distributions $J_k, 2 \leq k \leq K$. We define a discrete rv Y_k to approximate the continuous rv J_k. We set Y_k to be the worst case values of J_k for a given interval. We then compute the discrete interference distribution of $Y_k, \boldsymbol{y}_k$, where $\boldsymbol{y}_k = [y_k(-NN_u), \cdots, y_k(NN_u)]$ and

$$y_k(j) = \sum_{i=-NN_u}^{NN_u} d_k(i) P_{i,j} \tag{12}$$

for $-NN_u \leq j \leq NN_u$.

We must also model the effects of imperfect power control on user 1's desired component. User 1 will not have a known correlation value of N, but a probability distribution similar to that of the other users, given by J_1. There is only one nonzero value of B_1, given by $B_1 = N$. Therefore, we wish to calculate

$$y_1(i) = \Pr\left(J_1 \in \left[\frac{i}{N_u}, \frac{i+1}{N_u}\right]\right) \tag{13}$$

for $0 \leq i \leq 4NN_u$. While the lower value of i will always be fixed at zero, the upper-bound is not a definite value and needs to be large enough to ensure that the probability that J_1 lies outside this upper-bound is very small. A value of $4NN_u$ proved to be suitable for the normal distributions studied here.

Since $d_1(i) = 1$ for $i = NN_u$ and is zero otherwise, we find $y_1(i)$ using

$$y_1(i) = \Pr\left[\frac{i}{NN_u} \leq A_1 \leq \frac{i+1}{NN_u}\right] \tag{14}$$

for $0 \leq i \leq 4NN_u$. In solving for the total interference distribution, we must convolve the interference distributions of each user. After convolution, we have worst case interference values of

$$Y(i) = \frac{i}{N_u} \tag{15}$$

for $-(K-1)(NN_u+1) \leq i \leq (K-1)(NN_u-1) + 4NN_u$.

C. Average Probability of Error

We now wish to compute upper- and lower-bounds on the average probability of error. Since we know the probability that the interference is at a given level, the only random component is the Gaussian noise. The constant interfering term will be $Y(i)$. The expected value of this term is zero, since user one's received power does not have a determinist ic value. The probability of error is given by

$$P_E(\beta) = Q\left(\frac{\beta}{\sigma_N}\right) \tag{16}$$

where σ_N^2 is the scaled SNR [6] and β is the constant interfering term. The upper-bound on the probability of error is given by

$$P_E^{(U)} = \sum_{i=-(K-1)(NN_u+1)}^{(K-1)(NN_u-1)+4NN_u} y(i) P_E\left(\frac{i}{N_u}\right). \tag{17}$$

For the best case, a parallel analysis shows that the lower-bound on the probability of error is given by

$$P_E^{(L)} = \sum_{i=1-(K-1)(NN_u-1)}^{(K-1)(NN_u+1)+4NN_u+1} y(i) P_E\left(\frac{i}{N_u}\right). \tag{18}$$

IV. Numerical Results

Results were generated using $N = 128, N_u = 5$, and $1 \leq K \leq 20$. It was assumed that the chip period $T_c = 1$ μs. The signature sequences were generated randomly. The standard deviations of the received power σ_P were allowed to be 1, 1.4, and 2 dB. Simulation and field trial results have suggested the standard deviation of the received power level is between 1 and 2 dB [13]–[14]. The BER versus E_b/N_o, as parameterized by K, for σ_P of 1, 1.4, and 2 dB are plotted in Figs. 2, 3, and 4, respectively, where $E_b = PT$. The average probability of error versus the number of users K has been plotted for the worst case in Fig. 5 for a fixed E_b/N_o of 12 dB for all possible values of σ_P, including the ideal case of $\sigma_P = 0$ dB. These results show that when using a correlation receiver, even slight variances in the received power levels seriously reduce capacity.

Capacity has been sharply reduced due to imperfect power control. Table I summarizes, for the worst and best cases, the capacity loss from the case of perfect power control for

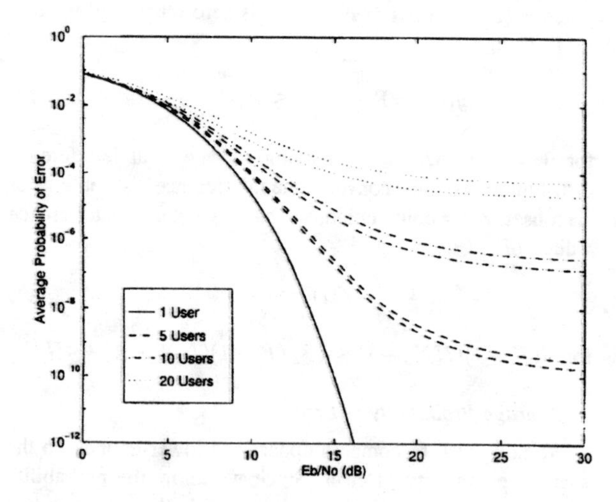

Fig. 2. Log-normal power control, $\sigma_P = 1$ dB.

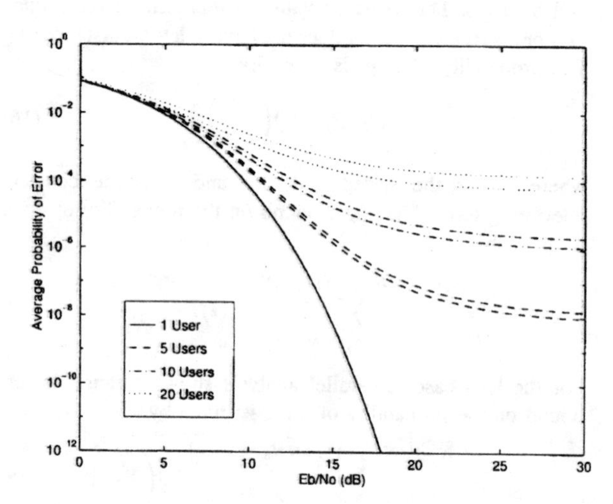

Fig. 3. Log-normal power control, $\sigma_P = 1.4$ dB.

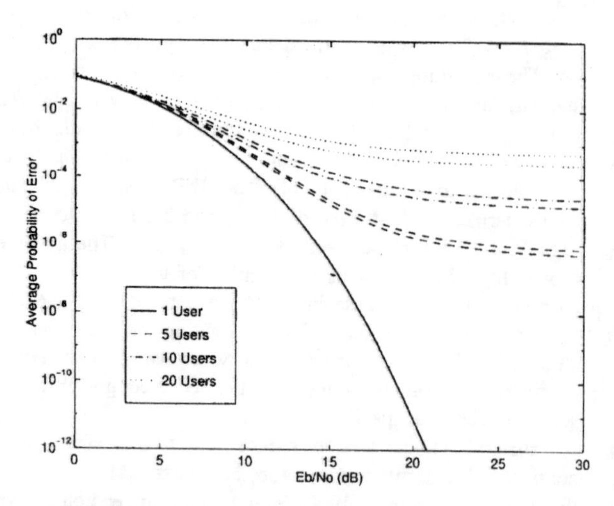

Fig. 4. Log-normal power control, $\sigma_P = 2$ dB.

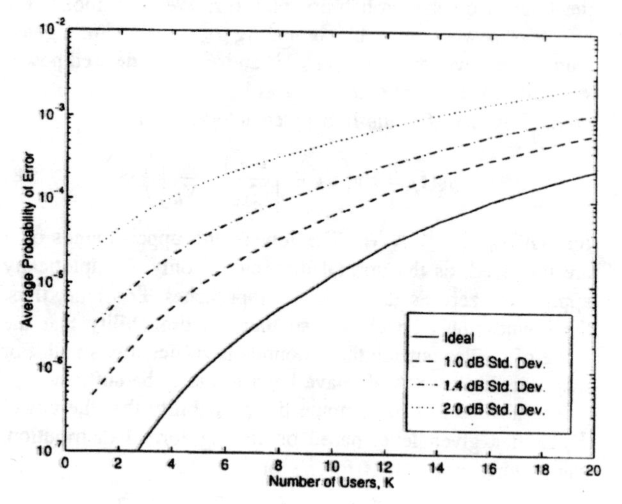

Fig. 5. BER versus number of users K.

TABLE I
CAPACITY LOSS FOR $E_b/N_o = 12$ dB AND BER $= 10^{-3}$

σ_P (dB)	Worst Case		Best Case	
	K_{max}	Capacity Loss (%)	K_{max}	Capacity Loss (%)
0	25	-	30	-
1	21	16	27	10
1.4	17	32	22	27
2	11	56	13	57

$E_b/N_o = 12$ dB and BER $= 10^{-3}$. The capacity loss is referenced to the ideal best case for the best case of all received power level variances, and similarly for the worst case. The maximum number of users that are allowed for this BER is given by $K_{\max}$. For $\sigma_P = 1$ dB, capacity is reduced in the neighborhood of 15%. For $\sigma_P = 1.4$ dB, capacity is reduced by approximately 30%. For $\sigma_P = 2$ dB, capacity is reduced nearly 60%. Thus, tight power control is a necessity in a CDMA system employing correlation receivers, and even standard deviations on the order of 1.4 dB can result in almost a 1/3 drop in capacity. Similar large drops in capacity have been shown if the received power levels are modeled as having uniform distributions about the mean [24].

V. DISCUSSION AND CONCLUSION

We have modeled imperfect power control in a CDMA system. Results show that even accurate power control mechanisms have a significant loss in capacity from the case of perfect power control. Proposed power control systems have a log-normal distribution with standard deviations less than 2 dB. Even with this elaborate system, the capacity reduction will be between 10% and 30%. If less stringent power control algorithms are used and the standard deviation increases to 2 dB, then capacity drops by nearly 60%. Clearly, stringent power control measures are needed to maintain high capacity levels in a multiple-access DS/SS system. This also suggests the need for receiver designs that can combat the near–far problem more effectively than the correlation receiver. Techniques such as decorrelators [17], [18], parallel interference cancellation [19], or successive interference cancellation receivers [20] could result in a significant capacity increase.

REFERENCES

[1] K. S. Gilhousen, I. M. Jacobs, R. Padovani, A. J. Viterbi, L. A. Weaver, and C. E. Wheatley, "On the capacity of a cellular (CDMA) system," *IEEE Trans. Veh. Technol.*, vol. 40, no. 2, pp. 303–312, May 1991.

[2] M. B. Pursley, "Performance evaluation for phase-coded spread-spectrum multiple-access communication—part I: System analysis," *IEEE Trans. Commun.*, vol. COM-25, no. 8, pp. 795–799, Aug. 1977.

[3] M. B. Pursley, D. V. Sarwate, and W. E. Stark, "Error probability for direct-sequence spread-spectrum multiple access communications—part I: Upper- and lower-bounds," *IEEE Trans. Commun.*, vol. COM-30, pp. 975–984, May 1982.

[4] E. A. Geraniotis and M. B. Pursley, "Error probability for direct-sequence spread-spectrum multiple access communications—part II: Approximations," *IEEE Trans. Commun.*, vol. COM-30, pp. 985–995, May 1982.

[5] J. S. Lehnert and M. B. Pursley, "Error probabilities for binary direct-sequence spread-spectrum communications with random signature sequences," *IEEE Trans. Commun.*, vol. COM-35, pp. 87–97, Jan. 1987.

[6] J. S. Lehnert, "An efficient technique for evaluating direct-sequence spread-spectrum multiple access communications," *IEEE Trans. Commun.*, vol. 37, no. 8, pp. 851–858, Aug. 1989.

[7] R. K. Morrow and J. S. Lehnert, "Bit-to-bit error dependence in slotted DS/SSMA packet systems with random signature sequences," *IEEE Trans. Commun.*, vol. 37, pp. 1052–1061, Oct. 1989.

[8] J. M. Holtzman, "A simple, accurate method to calculate spread spectrum multiple access error probabilities," *IEEE Trans. Commun.*, vol. 40, no. 3, pp. 1633–1636, Mar. 1992.

[9] C. L. Weber, G. K. Huth, and B. H. Batson, "Performance considerations of code-division multiple access systems," *IEEE Trans. Veh. Technol.*, vol. VT-30, no. 1, pp. 3–10, Feb. 1981.

[10] F. Simpson and J. M. Holtzman, "Direct sequence (CDMA) power control, interleaving, and coding," *IEEE J. Select. Areas Commun.*, vol. 11, no. 7, pp. 1085–1095, Sept. 1993.

[11] C. Kchao and G. L. Stüber, "Analysis of a direct-sequence spread-spectrum cellular radio system," *IEEE Trans. Commun.*, vol. 41, no. 10, pp. 1507–1516, Oct. 1993.

[12] R. A. Cameron and B. D. Woerner, "An analysis of CDMA with imperfect power control," in *Proc. 1992 Veh. Technol. Conf.*, Denver, CO, May 1992, pp. 977–980.

[13] A. J. Viterbi and R. Padovani, "Implications of mobile cellular CDMA," *IEEE Commun. Mag.*, vol. 30, no. 12, pp. 38–41, Dec. 1992.

[14] A. J. Viterbi, A. M. Viterbi, and E. Zehavi, "Performance of power-controlled wideband terrestrial digital communication," *IEEE Trans. Commun.*, vol. 41, no. 4, pp. 559–569, Apr. 1993.

[15] K. S. Shanmugan and A. M. Breipohl, *Random Signals: Detection, Estimation, and Data Analysis*. New York: Wiley, 1988.

[16] S. Verdu, "Minimum probability of error for asynchronous Gaussian multiple access channels," *IEEE Trans. Inform. Theory*, vol. IT-32, no. 1, pp. 85–96, Jan. 1986.

[17] R. Lupas and S. Verdu, "Linear multiuser detectors for synchronous code-division multiple access channels," *IEEE Trans. Inform. Theory*, vol. 35, no. 1, pp. 123–136, Jan. 1989.

[18] ——, "Near–far resistance of multiuser detectors in asynchronous channels," *IEEE Trans. Commun.*, vol. 38, no. 4, pp. 496–508, Apr. 1990.

[19] M. K. Varanasi and B. Aazhang, "Multistage detection in asynchronous code-division multiple access communications," *IEEE Trans. Commun.*, vol. 38, no. 4, pp. 509–519, Apr. 1990.

[20] P. Patel and J. Holtzman, "Analysis of a simple successive interference cancellation scheme in a DS/CDMA system," *IEEE J. Select. Areas Commun.*, vol. 12, no. 5, pp. 796–807, June 1994.

[21] M. B. Pursley, "Spread spectrum multiple access communications," in *Multi-User Communication Systems*. New York: Springer-Verlag, 1981, pp. 139–199.

[22] W. C. Y. Lee, *Mobile Communications Design Fundamentals*. New York: Wiley, 1993.

[23] R. A. Cameron, "Performance analysis of CDMA systems in multipath channels," Master's thesis, Virginia Polytechnic Institute and State University, Blacksburg, VA, May 1993.

[24] L. B. Milstein, T. S. Rappaport, and R. Barghouti, "Performance evaluation for cellular CDMA," *IEEE J. Select. Areas Commun.*, vol. 10, no. 4, pp. 680–689, May 1992.

Effects of Imperfect Power Control and User Mobility on a CDMA Cellular Network

Francesco Delli Priscoli and Fabrizio Sestini, *Member, IEEE*

***Abstract*—Code-division multiple-access (CDMA) is one of the major candidate access techniques for third generation systems. In recent years, a great deal of effort has been devoted to the study of the capacity it can support. This paper presents analytical derivations which allow the determination of the link availability in the presence of user mobility and power control imperfections in a CDMA network; moreover, it provides the guidelines which permit the implementation of a simple and flexible simulation tool which is independent of the specific CDMA implementations. As a matter of fact, the reported concepts can be applied to any asynchronous CDMA system, i.e., they hold both for the American Standard IS-95 and for the European Community Standard developed in the framework of the RACE CODIT Project.**

I. INTRODUCTION

CODE-DIVISION multiple-access (CDMA) is one of the major candidate access techniques for the third generation of mobile systems [2]–[6] and, hence, a great deal of effort has been devoted to the study of the effects which impact on the capacity of a CDMA cellular network (e.g., see [7]–[14]).

This paper deals with the effects of user mobility [12], [13] and power control imperfections [8]–[11] on the capacity of CDMA cellular networks. The most meaningful parameter for evaluating these effects is the link availability. Clearly, once a target link availability is set, these effects reflect on the maximum network capacity.

The paper aims at providing general concepts and derivations which leave out of considerations the specific CDMA implementation. So, the analysis only considers the basic CDMA-related concepts (i.e., shadowing, call dynamics, voice activation, sectorization), while more specific implementation-dependent issues (e.g., traffic mix, soft handover, CDMA Tx/Rx chain, etc.) are not taken into account. Clearly, the reported concepts can be particularized so that the specific issues are considered.

In particular, the main results of this paper are as follows below.

1) Analytical derivations which allow the determination of the link availability in the presence of user mobility and power control imperfections are presented. The guidelines for exploiting such results for implementing a simulator are exposed as well (Section II). In particular, Viterbi guidelines are followed [1], [9]. Nevertheless, the analysis is deepened in order to make possible the introduction of user mobility and power control imperfections; moreover, differently from [1] and [9], the number of users, the users position, the number of calls, the received power and the shadowing conditions vary as time varies.
2) A simple traffic model (suitable to be simulated) is presented which takes into account the presence of large and sudden user aggregations by considering the presence of "hot cells" where some congestion has occurred (Section III). Based on this traffic model, on the analytical derivations reported in Section II and on the parameters provided by Viterbi in [1], some interesting results are disclosed, which highlight how user mobility and power control imperfections impact on link availability (Section IV).

Manuscript received May 1, 1995; revised November 22, 1995. This paper has been presented in part at the IEEE International Conference on Communications, Seattle, WA, June 1995, and the Fifth IEEE Symposium on Personal, Indoor, and Mobile Communications, The Hague, The Netherlands, September 1994.

F. Delli Priscoli is with the Dip. di Informatica e Sistemistica, University of Rome "La Sapienza," Italy (email: dellipri@riscdis.ing.uniroma1.it).

F. Sestini is near the Dip. INFOCOM, University of Rome "La Sapienza," Italy (email:f.sestini@ieee.org).

Publisher Item Identifier S 0733-8716(96)05230-4.

II. ANALYTICAL DERIVATIONS

A. General Assumptions

The considered propagation model assumes that by transmitting from a point P_1 at time t, the received power in a point P_2 is proportional to $10^{(\xi(t)/10)}r^{-4}$ where r is the distance from P_1 to P_2 and $\xi(t)$ is a random Gaussian variable with standard deviation $\sigma_\xi = 8$ dB and zero mean. As a matter of fact, we assume that the propagation exponent is equal to four and that the factor $10^{(\xi(t)/10)}$ takes into account the shadowing, at time t, in the link from P_1 to P_2 [1].

Both in the uplink and in the downlink, a voice activation mechanism is foreseen. Following [1], we assume that the voice activity factor α is equal to 3/8. Lastly, we assume that all calls are vocal ones. An in-call mobile station (MS) in the ON state of the voice activation process is referred to as an *active* MS.

In the considered cellular network, the base stations (BS's) are regularly disposed at the centers of regularly disposed hexagonal cells (see Fig. 1).

The MS's are provided with omnidirectional antennas. The BS's are assumed to use three directional antennas whose transmitting/receiving gains $G_i(\theta)$, $i = 1, 2, 3$ are

$$G_i(\theta) = \begin{cases} G, & \text{if } 120(i-1)^\circ < \theta < 120i^\circ \\ 0, & \text{if } 120i^\circ < \theta < 120(i+2)^\circ \end{cases} \quad i = 1, 2, 3$$

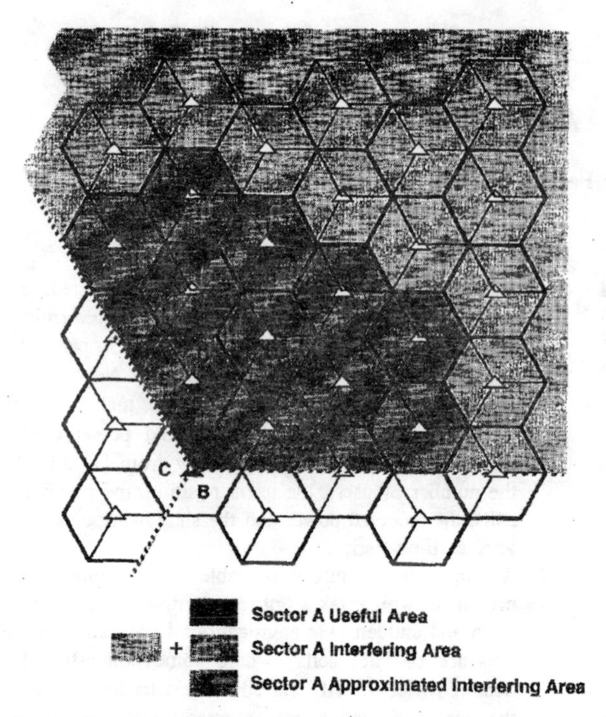

Fig. 1. Reference cellular network layout.

Therefore, each BS produces a partition of the cellular network in three areas, hereinafter referred to as *sector areas*, isolated one another with respect to the signals transmitted/received by the BS. As a result, the number of different sector areas is three times the number of BS's.

Each BS transmits a different *reference carrier* toward each sector area produced by it. Therefore, a one-to-one association exists between sector areas and reference carriers: the ith reference carrier is transmitted toward the ith sector area. All reference carriers are transmitted by the BS's with the same power level.

The ith *sector useful area* at time t is the locus where at time t the ith reference carrier is received with the highest power with respect to the other reference carriers. Hereinafter, for the sake of brevity and uniformity with the literature, the ith sector useful area is simply referred to as ith *sector*. The BS transmitting the ith reference carrier is referred to as the ith *sector BS*. An MS j which, at time t, is in the point P of the cellular network is *served* by the sector i whose associated reference carrier has the highest power in P at time t. It should be noted that due to the presence of the shadowing phenomena, at time t, the serving sector BS of an MS j is not necessarily the BS *closest* to the MS j.

The ith *sector interfering area* at time t is the locus which is the difference between the ith sector area and the ith sector.

For instance, in Fig. 1, the three sector areas produced by the black BS are separated by dotted lines. This figure also shows the partition in sectors of the cellular network *in the absence of shadowing*. In particular, the three sectors which have the same sector BS are delimited by hexagonal solid lines, e.g., sectors A, B, C have the same sector BS, i.e., the black BS. Lastly, the sector A useful area and the sector A interfering area are highlighted.

Let $r_j(t)$ and $r_{ij}(t)$ be the distances, at time t, of the jth MS from its serving sector BS and from the ith sector BS, respectively. Let $\xi_j(t)$ and $\xi_{ij}(t)$ be Gaussian random variables with zero mean and standard deviation $\sigma_\xi = 8$ dB relevant to the shadowing, at time t, in the link from the jth MS to the serving sector BS and to the ith sector BS, respectively. Then, for any couple (i, j), the following relations can be written:

$$\frac{10^{\xi_j(t)/10}}{r_j(t)^4} \geq \frac{10^{\xi_{ij}(t)/10}}{r_{ij}(t)^4} \tag{1}$$

where each of the previous relations holds with the sign of equality only if the serving sector BS and the ith sector BS coincide.

Note that relations (1) can be rewritten in the form

$$\chi_{ij}(t) \leq 40 \log_{10} \frac{r_{ij}(t)}{r_j(t)} \tag{2}$$

where $\chi_{ij}(t) = \xi_{ij}(t) - \xi_j(t)$ is a Gaussian random variable with zero mean and variance $\sigma_\chi^2 = 2\sigma_\xi^2$.

Both the uplink and the downlink employ a direct-sequence spread-spectrum (DS/SS) waveform. Two separate frequency bands (*spread bands*) are present for the uplink and the downlink. Each uplink signal occupies the entire uplink spread band W_u; likewise, each downlink signal occupies the entire downlink spread band W_d.

Due to the *near–far effect*, a very critical issue in CDMA is power control; as a matter of fact, several studies [8]–[11] show that in the presence of an imperfect power control, system performance can be remarkably affected. This is the reason why, in the considered analysis, an *imperfect* power control mechanism is considered. The level of imperfection of the mechanism depends on several causes (e.g., the power control algorithms, the speed of the equipment involved in the adaptive power control mechanism, the dynamic range of the transmitter, the MS distribution, the propagation statistics, etc.). In this paper, following [8]–[10], the imperfections of the power control mechanisms are taken into account by assuming that the energy per information bit, hereinafter referred to as $E_{bj}(t)$, received at time t by a BS in consequence of the transmission of the jth MS served by this BS, varies around the desired value E_{bd}, according to a log normal distribution. This means that

$$E_{bj}(t) = E_{bd} 10^{\varepsilon_j(t)/10} \tag{3}$$

where $\varepsilon_j(t)$ is a Gaussian random variable with zero mean and standard deviation σ_ε. Clearly, the higher values of σ_ε corresponds to more imperfect power control mechanisms; in particular, $\sigma_\varepsilon = 0$ dB means ideal power control.

Following [1], [8], and [9] we assume that the uplink is more critical than the downlink. The basic consideration which validates this assumption is that in the downlink a coherent detection is possible thanks to the presence of the reference carrier, which acts as a pilot carrier. Therefore, in the following, only the uplink is considered.

B. Uplink Budget Analysis

This section presents a simplified analysis of the uplink budget based on the assumption that the noise affecting the link from the useful MS to its serving BS is only caused by the thermal noise and by the self-noise. Therefore, at a certain time t, the link is available in the ith sector only if the following relation holds:

$$\left[\frac{E_b}{N_0}\bigg|_{\text{required}}\right]^{-1} > \left[\frac{E_b}{N_0}\bigg|_{\text{thermal}}\right]^{-1} + \left[\frac{E_{bu}(t)}{N_{0SNi}(t)}\right]^{-1} \quad (4)$$

where the term on the left side is the E_b/N_0 ratio required for achieving the performance target in terms of bit-error rate (BER), while the first and the second term on the right side represent the E_b/N_0 ratio relevant to the thermal noise and to the self noise, respectively. In particular, $E_{bu}(t)$ is the energy per bit received, at time t, by the ith sector BS in consequence of the transmission of the useful MS (MS u), while $N_{0SNi}(t)$ is the self-noise spectral density, at time t, at the ith sector BS caused by the interfering MS's.

As for the computation of $E_{bu}(t)$, the discussion at the end of the previous section applies; in the present case, in (3) the generic jth MS must be replaced by the useful MS u, i.e.,

$$E_{bu}(t) = E_{bd}10^{\varepsilon_u(t)/10}$$

where $\varepsilon_u(t)$ is a Gaussian random variable with zero mean and standard deviation σ_ε.

As for the computation of $N_{0SNi}(t)$, following [1], it is assumed that the power received by the ith sector BS in consequence of the transmission of the jth interfering MS appears at the BS receiver as white noise spread over the whole uplink bandwidth. Based on this assumption, the self-noise spectral density in the ith sector is

$$N_{0SNi}(t) = \sum_{j=1}^{M_i(t)-1} E_{bj}(i,t)R_b/W + \sum_{j=1}^{M_{INi}(t)} E_{bij}(t)R_b/W \quad (5)$$

where the two terms on the right side are the contributions to the self-noise spectral density of the interfering MS's in the ith sector and in the ith sector interfering area, respectively, $M_i(t)$ is the number of active MS's at time t served by the ith sector BS, $M_{INi}(t)$ is the number of active MS at time t in the ith sector interfering area, $E_{bj}(i,t)$ is the energy per bit received, at time t, by the BS serving the ith sector, in consequence of the transmission of the jth interfering active MS served by the ith sector, and $E_{bij}(t)$ is the energy per bit received, at time t, by the BS serving the ith sector, in consequence of the transmission of the jth interfering active MS roaming in the ith sector interfering area.

Note that $E_{bj}(i,t)$ can be computed according (3) since it refers to the energy per bit received by the BS serving the jth MS, namely the ith sector BS.

On the contrary, $E_{bij}(t)$ cannot be computed according (3), as, by definition, the jth MS roaming in the ith sector interfering area is not served by the ith sector BS. Hence, in this case, the energy per bit provided by (3) is the one received from the BS serving the jth MS and not the one received from the ith sector BS. Then

$$E_{bj}(t) = K(t)\frac{10^{\xi_j(t)/10}}{r_j(t)^4}$$

where $K(t)$ depends on the link parameters (e.g., MS transmitting power, antenna gains, receiver noise temperature, etc). Since

$$E_{bij}(t) = K(t)\frac{10^{\xi_{ij}(t)/10}}{r_{ij}(t)^4} \quad (6)$$

then, by eliminating $K(t)$ from the previous two relations

$$E_{bij}(t) = E_{bj}(t)\left(\frac{r_j(t)}{r_{ij}(t)}\right)^4 10^{\chi_{ij}(t)/10}. \quad (7)$$

Set

$$M_{EQSi}(t) = \frac{\sum_{j=1}^{M_i(t)-1} 10^{\varepsilon_j(t)/10}}{10^{\varepsilon_u(t)/10}} = \sum_{j=1}^{M_i(t)-1} 10^{\eta_j(t)/10} \quad (8)$$

$$M_{EQINi}(t) = \frac{\sum_{j=1}^{M_{INi}(t)} 10^{\varepsilon_j(t)/10}\left(\frac{r_j(t)}{r_{ij}(t)}\right)^4 10^{\chi_{ij}(t)/10}}{10^{\varepsilon_u(t)/10}} = \sum_{j=1}^{M_{INi}(t)} 10^{\eta_j(t)/10}\left(\frac{r_j(t)}{r_{ij}(t)}\right)^4 10^{\chi_{ij}(t)/10} \quad (9)$$

$$\delta = \frac{W_b}{R_b}\left(\left[\frac{E_b}{N_0}\bigg|_{\text{required}}\right]^{-1} - \left[\frac{E_b}{N_0}\bigg|_{\text{thermal}}\right]^{-1}\right) \quad (10)$$

where $\eta_j(t) = \varepsilon_j(t) - \varepsilon_u(t)$ is a random Gaussian variable with zero mean and variance $\sigma_\eta^2 = 2\sigma_\varepsilon^2$.

Then, taking into account (3), (5), (7), (8), (9), and (10), relation (4) can be rewritten

$$M_{EQSi}(t) + M_{EQINi}(t) < \delta. \quad (11)$$

For simulation purposes, following [1], we have assumed $\delta = 30$.

Note the following.

1) $M_{EQSi}(t)$ represents the equivalent number of interfering MS's roaming in the ith sector; as a matter of fact, if the power control mechanism was ideal, this term would yield $M_i(t) - 1$.
2) $M_{EQINi}(t)$ represents the number of active MS's in the ith sector which would produce a self-noise power density equivalent to that produced by the MS's in the ith sector interfering area.

At time t, $M_{EQSi}(t)$ is a random variable, hereinafter referred to as $M_{EQSi}(t) \mid t$, which, as is evident from (8), is the summation of $M_i(t) - 1$ log-normally distributed random variables; then, its computation by means of simulation does not yield any particular problem.

On the contrary, the computation of the term $M_{EQINi}(t)$ cannot be easily performed from the (9) due to *two problems*. On the one hand, the actual value of this term depends on a large number of statistically independent random processes (the voice activation process, the power control process, and the shadowing process of each call-in-progress MS roaming in the ith sector interfering area) and, on the other hand, it depends on the distances, at time t, of the active MS's in the ith sector interfering area from their serving sector BS and from the ith sector BS. These two problems can be solved as indicated in the next section.

C. Computation of $M_{\text{EQIN}i}(t)$

As for the *first problem* addressed at the end of Section II-B, it is assumed that at a certain time t, the value assumed by $M_{\text{EQIN}i}(t)$ is a random Gaussian variable (as argued in [1], since $M_{\text{EQIN}i}(t)$ is the sum of a large number of variables from random access processes), hereinafter referred to as $M_{\text{EQIN}i}(t) \mid t$, whose expected value and variance can be calculated as follows.

Relation (9) can be rewritten in the form

$$M_{\text{EQIN}i}(t) \mid t = \sum_{j=1}^{N_{\text{IN}i}(t)} \phi_j(t) 10^{\eta_j(t)/10} \left(\frac{r_j(t)}{r_{ij}(t)}\right)^4 10^{\chi_{ij}(t)/10} \tag{12}$$

where $\phi_j(t)$ is the voice activity random variable, at time t, relevant to the jth MS, which equals one with probability α and zero with probability $1-\alpha$, regardless of t and j and $N_{\text{IN}i}(t)$ is the number of call-in-progress MS's in the ith sector interfering area. Note that (12) is subject to the constraint (2).

Let us observe that given two random variables, x and $y = \log_{10} x$, the expected value m_x is related to the logarithmic expected value (i.e., m_y), and the second moment $m_x^{(2)}$ is related to the logarithmic variance (i.e., σ_y^2), by means of the following relations

$$m_x = \exp(\beta^2\sigma_y^2/2)\exp(\beta m_y) \tag{13}$$

$$m_x^{(2)} = \exp(2\beta^b\sigma_y^2)\exp(2\beta m_y) \tag{14}$$

where $\beta = (\ln 10)/10$.

By considering (2) and (12), simple computations, which take into account (13), yield the expected value $M_{\text{EQIN}i}(t) \mid t$

$$E[M_{\text{EQIN}i}(t) \mid t] = \alpha \exp(\beta^2\sigma_\eta^2/2) \sum_{j=1}^{N_{\text{IN}i}(t)} \left(\frac{r_j(t)}{r_{ij}(t)}\right)^4 f\left(\frac{r_j(t)}{r_{ij}(t)}\right) \tag{15a}$$

where $\beta = (\ln 10)/10$ and

$$f\left(\frac{r_j(t)}{r_{ij}(t)}\right) = \int_{-\infty}^{40\log_{10}\frac{r_{ij}(t)}{r_j(t)}} 10^{x/10} \frac{\exp\left(-\frac{x^2}{2\sigma_\chi^2}\right)}{\sqrt{2\pi\sigma_\chi^2}} dx. \tag{15b}$$

We assume the following: 1) The random variables $\chi_{ij}(t) (j = 1, 2, \ldots)$ are statistically independent—which means that shadowing undergone by different mobiles is independent. As a matter of fact, shadowing characteristics are assumed to vary very rapidly with spatial displacement in any direction [1]. 2) The random variables $\eta_j(t)$ are statistically independent of the random variables $\chi_{ij}(t)$. As a matter of fact, as stated in Section II-A, the power control imperfection process, different for each mobile, depends on several causes which are either independent of the shadowing process, or their actual dependence is so complex and subject to so many random variables that the assumption of statistical independence of the two processes appears a reasonable approximation. Thus, according to simple computations which take into account (14), the variance of the random variable $M_{\text{EQIN}i}(t) \mid t$ is

$$\begin{aligned} \text{Var}[M_{\text{EQIN}i}(t) \mid t] = &\sum_{j=1}^{N_{\text{IN}i}(t)} \left(\frac{r_j(t)}{r_{ij}(t)}\right)^8 \\ &\times \left[\alpha \exp(2\beta^2\sigma_\eta^2) g\left(\frac{r_j(t)}{r_{ij}(t)}\right) \right. \\ &\left. -\alpha^2 \exp((\beta\sigma_\eta)^2) f^2\left(\frac{r_j(t)}{r_{ij}(t)}\right)\right] \end{aligned} \tag{16a}$$

where $f(x)$ is given by (15b) and $g(x)$ is

$$g\left(\frac{r_j(t)}{r_{ij}(t)}\right) = \int_{-\infty}^{40\log_{10}\frac{r_{ij}(t)}{r_j(t)}} 10^{x/5} \frac{\exp\left(-\frac{x^2}{2\sigma_\chi^2}\right)}{\sqrt{2\pi\sigma_\chi^2}} dx. \tag{16b}$$

As for the *second problem* addressed at the end of Section II-B, in order to make the computation of (15) and (16) easier, instead of considering the whole ith sector interfering area, only the cells of the ith sector interfering area at distance $D \leq 5R$ (where R is the hexagon radius) from the sector i BS are taken into account. In the following, such an area will be referred to as the ith sector approximated interfering area and will be denoted as IAi. In the case shown in Fig. 1 (i.e., absence of shadowing) and referring to sector A, this assumption corresponds to only considering the sectors in the area depicted in medium gray, instead of the whole sector A interfering area. As a matter of fact, as can be confirmed by simple geometrical computations, it is reasonable assuming that, as a first approximation, the contribution of the in-call MS's in the remaining part of the ith sector interfering area can be neglected due to the high distances of these MS's from the ith sector BS.

Now assume that in the ith sector approximated interfering area IAi the shadowing effects can be neglected as for BS control. This is the same as saying that all sectors in the area IAi are shaped as the regular hexagons whose centers are the relevant BS's (this is just the situation shown in Fig. 1); in reality, as already noted, the shape of each sector varies in dependence on the shadowing phenomena. It is worth noting that both (15) and (16) have been determined without performing this approximation as, for their computation, relation (2) (which accounts for the shadowing phenomena) has been taken into account; however, for the present calculation, it is no longer possible to take (2) into account in an easy way. This approximation results in tight (as demonstrated in [1]) upper bounding of the random variable $M_{\text{EQIN}i}(t) \mid t$.

Lastly, assume that in the area IAi the call-in-progress MS's are uniformly distributed, i.e., the call-in-progress MS density in the ith sector approximated interfering area IAi is $2N_{\text{IA}i}(t)/(3\sqrt{3}R^2)$, where $N_{\text{IA}i}(t)$ is the number of call-in-progress MS's roaming, at time t, in a cell of the ith sector approximated interfering area IAi.

TABLE I
CONSIDERED VALUES OF σ_ε AND CORRESPONDING VALUES OF K_1 AND K_2

σ_ε	K_1	K_2
0.00	0.0412	0.0129
1 dB	0.0434	0.0161
2 dB	0.0509	0.0311
3 dB	0.0664	0.0920

In order to simplify the simulations whose results are reported in Section IV, we accounted for the value of $N_{\mathrm{IA}i}(t)$ only averaging the actual number of call-in-progress MS's in the six sectors of the two cells nearest to sector i (labeled as $D, E \cdots I$ in Fig. 1), that are the ones which mainly contribute to the interference, being at the shortest distance from sector i.

Thanks to these last assumptions, (15) and (16) can be rewritten in the form

$$E[M_{\mathrm{EQIN}i}(t) \mid t] = K_1 N_{\mathrm{IA}i}(t) \tag{17}$$

$$\mathrm{Var}[M_{\mathrm{EQIN}i}(t) \mid t] = K_2 N_{\mathrm{IA}i}(t) \tag{18}$$

with

$$K_1 = \frac{2}{3\sqrt{3}R^2}\alpha \exp((\beta\sigma_\eta)^2/2) \times \int_{\mathrm{IA}i}\left(\frac{r_j(P)}{r_{ij}(P)}\right)^4 f\left(\frac{r_j(P)}{r_{ij}(P)}\right)dP,$$

$$K_2 = \frac{2}{3\sqrt{3}R^2}\int_{\mathrm{IA}i}\left(\frac{r_j(P)}{r_{ij}(P)}\right)^8 \times \left[\alpha \exp(2(\beta\sigma_\eta)^2) g\left(\frac{r_j(P)}{r_{ij}(P)}\right) - \alpha^2 \exp((\beta\sigma_\eta)^2) f^2\left(\frac{r_j(P)}{r_{ij}(P)}\right)\right]dP$$

where the integrals are surface integrals extended to the approximated interfering area IAi. Note that the surface integrals in the previous relations are proportional to R^2 and, hence, K_1 and K_2 are independent of R.

By numerically computing the coefficients, K_1 and K_2 for the selected value of α, we get the results reported in Table I as a function of σ_ε.

In conclusion, with the considered reasonable assumptions and approximations, at a certain time t, $M_{\mathrm{EQIN}i}(t)$ can be considered as a Gaussian random variable whose expected value and variance are given by (17) and (18), respectively, where the coefficients K_1 and K_2 are given by Table I as a function of σ_ε, i.e., of the imperfections of the power control mechanism.

D. Computation of the Link Availability L_A

The link availability L_A is defined as the probability that (4) holds averaged on all times and on all cellular network sectors. By assuming the ergodicity of the random processes, the link availability represents the fraction of time during which the performance target in terms of BER is respected.

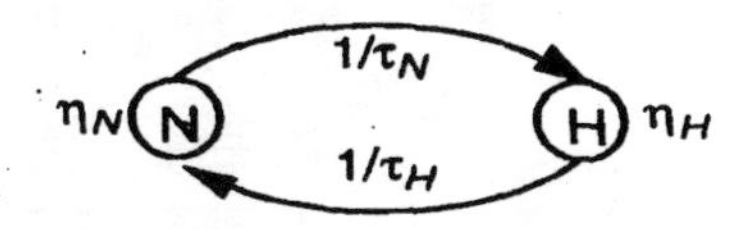

Fig. 2. Considered mobility model.

Since the various sectors have the same statistical characteristics, in order to compute the link availability it is sufficient to determine the probability that (4) holds in a given sector, e.g., the ith sector.

In Sections II-B and II-C, we have shown that (4) holds if and only if (11) holds where, at time t, $M_{\mathrm{EQS}i}(t)$ can be regarded as a random variable $M_{\mathrm{EQS}i}(t) \mid t$ equal to the sum of $M_i(t) - 1$ log-normally distributed random variables with zero mean and variance $2\sigma_\varepsilon^2$; $M_{\mathrm{EQIN}i}(t)$ can be regarded as a Gaussian random variable $M_{\mathrm{EQIN}i}(t) \mid t$ whose expected value and variance are given by (17) and (18), respectively.

Let us consider a function $\Omega(t,i)$ which, at time t, equals to one or zero according to whether at that time (11) holds or does not hold. Then, taking into account the assumed process ergodicity, the link availability has the following expression:

$$L_A = \mathrm{Probability}\{M_{\mathrm{EQS}i}(t) \mid t + M_{\mathrm{EQIN}i}(t) \mid t < \delta\} = \lim_{\Delta t \to \infty} \frac{1}{\Delta t}\int_{-\Delta t/2}^{\Delta t/2} \Omega(t,i)dt. \tag{19}$$

In the simulations, the link availability is computed according to the right hand side of (19), apart from the fact that due to obvious simulation requirements, the temporal integration cannot be done on an infinite temporal interval.

III. MOBILITY AND TRAFFIC MODELS

A. Mobility Model

User mobility is assumed to take place on a regular rhomboidal grid of sectors that tessellates the plane, as shown in Fig. 1. Each time a MS leaves a sector, it moves to any adjacent one with the same probability.

Hereinafter, the hexagonal area covered by the three sectors served by a same BS is referred to as a cell. In the simulations a regular 9×9 grid of cells is considered; opposite boundaries are assumed to be adjacent, so that the simulated area has a thoroidal structure.

The negative exponential probability density is widely used as a simple way to describe user cell crossing times, but it fails to account for highly variable user densities. As a matter of fact, sudden aggregations of users can be encountered within a cell area, especially in a microcellular environment, where cell sizes are such that we can not rely on a significant spatial average. Then, the cell crossing time statistics have to account for that variability, while keeping analytically tractable.

A good solution is a two-state Markov modulated Poisson process (MMPP), where one state ("N" state) corresponds to a normal cell, characterized by relatively high average values of the user speed and hence low cell crossing times; on the contrary, the other state ("H" state) corresponds to a hot

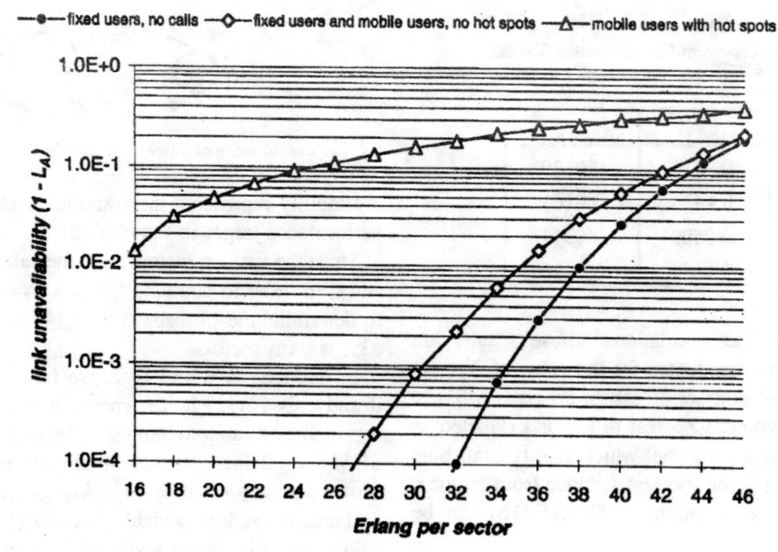

Fig. 3. Link unavailability versus mean call offered traffic per sector for several traffic and mobility models.

cell, i.e., a cell where some user congestion has occurred, so that average user speed is low and cell crossing takes a relatively long time. Each cell can dwell in either state for an exponentially distributed time, independently of one another.

In summary, the considered mobility model is based on the following assumptions.

1) Whenever a MS leaves a cell, it moves to any adjacent one with the same probability.
2) Each cell, independently of one another, can dwell either in a normal state (N state) or in an hot state (H state) for exponentially distributed times with means equal to τ_N and τ_H, respectively.
3) Normal cells are characterized by relatively high average values of the MS speed and hence by a low cell crossing time—conversely, hot cells are characterized by relatively low average values of the user speed and hence by an high cell crossing time: they are cells where some MS congestion has occurred. Both crossing times are assumed to be exponentially distributed with means equal to η_N and η_H ($\eta_N < \eta_H$), respectively.

Therefore, the considered mobility model is characterized by four parameters, i.e., τ_N, τ_H, η_N, η_H (see Fig. 2). In the following, this set of parameters will be indicated as $\Theta = (\tau_N, \tau_H, \eta_N, \eta_H)$.

In the simulations, the following cases are considered:

1) $\Theta_F = (\infty, \infty, -, -)$, which corresponds to the case of *fixed* mobiles, i.e., absence of user mobility;
2) $\Theta_S = (60, 60, -, -)$, which corresponds to a case of *smooth* user mobility, i.e., pure negative exponential cell crossing times—thus, in this case, hot cells are not considered;
3) $\Theta(\tau_H) = (60, 600, 9\tau_H, \tau_H)$, which corresponds to the case of *Markov modulated* user mobility in which, on the average, 10% cells are hot (*hot-spot* user mobility)—it should be noted that, in our simulation this corresponds to an average of 8.1 hot cells out of the 81 considered cells.

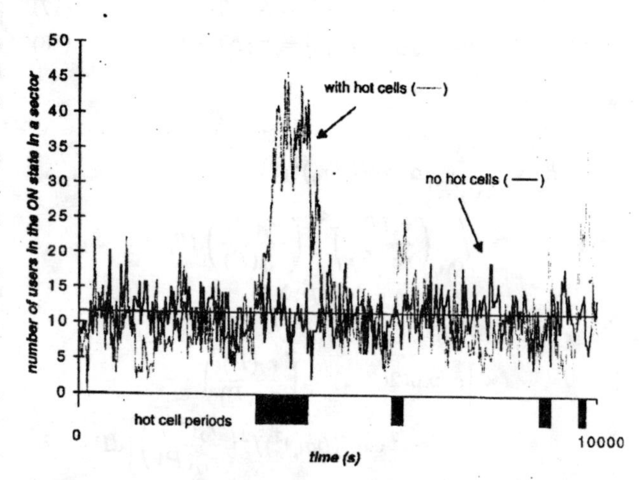

Fig. 4. Comparison of the trends of the number of users in the ON state in a sector with and without hot cells. The horizontal line represents the mean value.

B. Traffic Model

As for the offered traffic, we assume that each user behaves as a Markovian source of call attempts. Therefore, as long as a user is not engaged in a call, time intervals between successive call attempts are independent of one another and of the system state and exponentially distributed with mean $1/\lambda$. We assume $1/\lambda = 4800$ s.

Once a user is engaged, it holds the call for an exponentially distributed time with mean value $1/\mu$, unless the call has to be torn-down. We assume $1/\mu = 120$ s. Hence, each user offers a traffic of 25 mErl.

Moreover, we assume that the mean durations of the OFF state and of the ON state in the voice activation process are 1.666 s and 1 s, respectively.

Finally, the mobility process, the traffic process, and the voice activation process are assumed to be independent of each other.

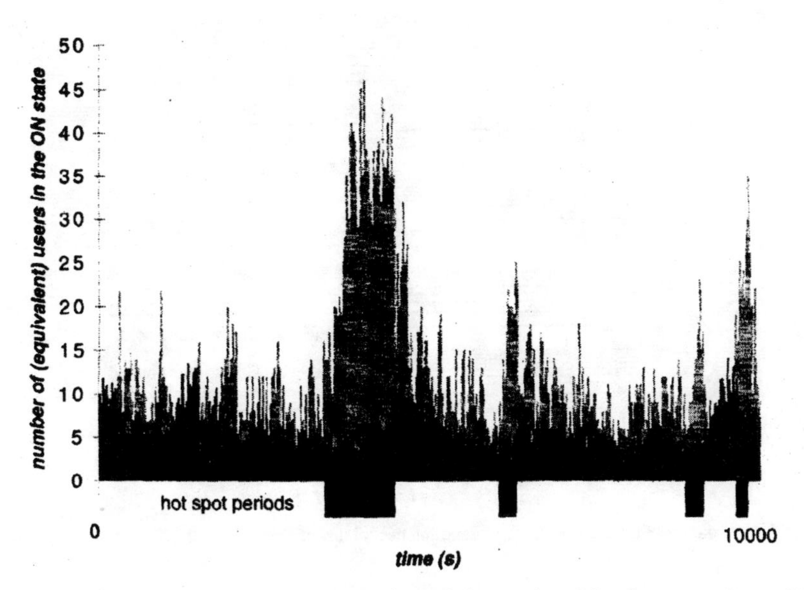

Fig. 5. Number of users in the ON state in a sector (clear gray) and the equivalent number of interfering users in the ON state from the cells of the interfering area (dark gray).

IV. Numerical Results

This section presents some results achieved via simulation by applying the analytical derivations deduced in Section II and the mobility and traffic models considered in Section III.

Fig. 3 plots the link unavailability $1-L_A$ as a function of the mean offered traffic per sector A for the case $\Theta = \Theta_F$ (fixed users), $\Theta = \Theta_S$ (mobile users, no hot cells) and $\Theta = \Theta(180)$ (mobile users with hot cells and mean duration of the "H" state equal to 180 s) and in the hypothesis of perfect power control ($\sigma_\varepsilon = 0$). Clearly, the values reported on the horizontal axis in Fig. 3 represent the average number of in-call MS's per sector; moreover, taking into account the assumption that each user offers a traffic of 25 mErl, by multiplying these values by 40 the average number of MS's (either in-call or in stand-by) roaming in a sector is obtained; lastly, taking into account that we have assumed a voice activity factor α equal to 3/8, by multiplying the values on the horizontal axis in Fig. 3 by 3/8, the average number of active MS's per sector is obtained. The black dotted curve refers to the case studied in [1], i.e., the case in which neither the mobility dynamics, nor the call dynamics, nor power control imperfections are considered.

The curves relevant to the cases $\Theta = \Theta_F$ and $\Theta = \Theta_S$ in Fig. 3 are indistinguishable each other and are represented by the same line. This means that a smooth user mobility does not cause any significant capacity impairment with respect to the fixed user case. On the contrary, a strong impairment is caused by the presence of hot cells; for instance, Fig. 3 shows that, for a mean offered traffic of 30 Erl per sector, the link availability decreases from about 0.9992 in the case $\Theta = \Theta_S$ (no hot cells), to about 0.86 in case $\Theta = \Theta(180)$ (with hot cells).

The rationale behind this impairment can be explained from an intuitive point of view by considering the plots reported in Fig. 4. Such plots represent the fluctuations of $M_i(t)$, i.e., the number of MS's in the ON state in the ith sector during a time interval of 10 000 s in the cases $\Theta = \Theta_S$ (no hot cells) and $\Theta = \Theta(180)$ (with hot cells) for an average number of in-call MS's per sector equal to 30, i.e., for an average number of active MS's equal to $30\alpha = 11.25$ ($\alpha = 3/8$); the horizontal line just reports this last average value.

From the figure it is evident that in the case $\Theta = \Theta(180)$, during the periods in which the cell is hot, the MS's tend to accumulate in such sector, so that much higher peaks are present for $M_i(t)$ than in the case $\Theta = \Theta_S$.

Assume, for the moment, a perfect power control (i.e., assume $\eta_j(t) = 0\ \forall j$) and neglect the contribution to self-noise of the MS's in the interfering area (i.e., assume $M_{\text{EQIN}i}(t) = 0$); then (11) becomes $M_i(t) < \delta + 1 = 31$. Since the link availability is the percentage of time in which (11) holds, the link is available for the time periods in which $M_i(t)$ is lower than the threshold $\delta + 1 = 31$. It is evident from Fig. 4 that the link can be unavailable just during the hot cell periods and that the probability to reach the aforementioned threshold during a certain hot cell period is directly related to the hot cell period duration. This explains the significant worsening of link availability when hot spots are introduced.

The situation is only marginally different if the contribution of the MS's in the ith interfering area is considered. As a matter of fact, Fig. 5 plots in clear gray the fluctuations of $M_i(t)$ and in dark gray the fluctuations of the equivalent number of in-call MS's from the ith sector interfering area (such a number is equal to $M_{\text{EQIN}i}(t)$ if a perfect power control is considered) during a time interval of 10 000 s in the case $\Theta = \Theta(180)$ (with hot cells), for an average number of in-call MS's per sector equal to 30. The considered time interval is the same as the one shown in Fig. 4. Thus, the plots relevant to $M_i(t)$ (in the hot cell case) in Figs. 4 and 5 are

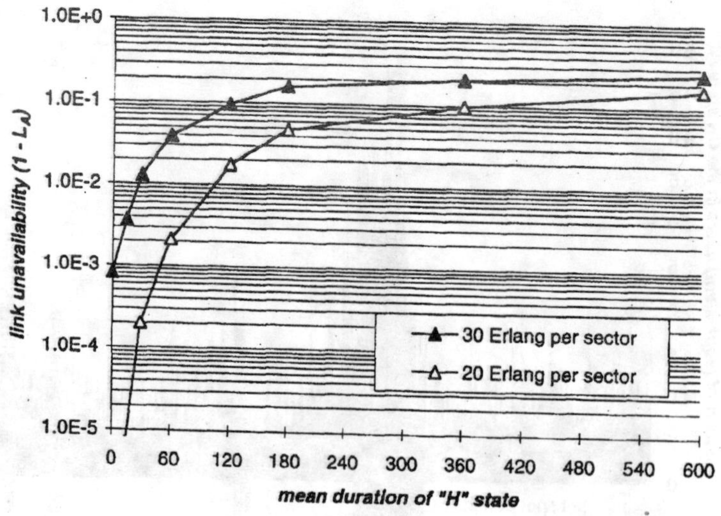

Fig. 6. Link unavailability versus the mean duration of the "H" state for two values of the mean call offered traffic per sector.

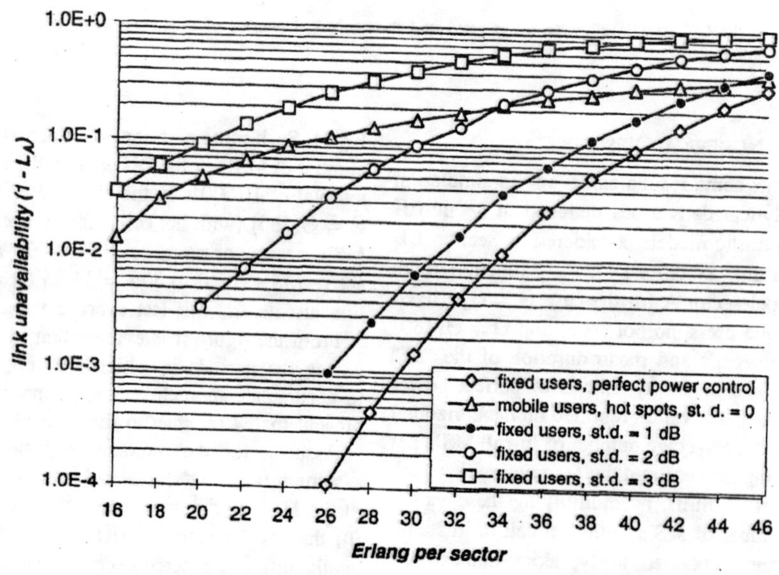

Fig. 7. Effects of power control imperfections (marked in dB of standard deviation σ_ε) on the link unavailability versus mean call offered traffic per sector.

identical. It is evident from Fig. 5 that the main contribution to self-noise in the ith sector is given by the MS's served by such sector; then, the situation is similar to the one shown in Fig. 4, i.e., the link can be unavailable only during hot cell periods.

Finally, the introduction of power control imperfections does not meaningfully alter the above phenomenon since it is independent of user mobility.

Fig. 6 plots the link unavailability $1 - L_A$ as a function of the mean duration of the "H" state τ_H for the cases $A = 20$ Erl per sector and $A = 30$ Erl per sector (i.e., for an average of 20 and 30 in-call MS's per sector) and $\Theta = \Theta(\tau_H)$. Clearly, the case $\tau_H = 0$ corresponds to the absence of hot cells (smooth user mobility).

The exponential flattening of the curves shown in Fig. 6 can be explained as follows. By increasing τ_H, more and more MS's tend to accumulate in the hot cells and, hence, L_A decreases. However, by increasing τ_H, the corresponding L_A decreases are lower and lower since, on the one hand, the cells adjacent to the hot ones tend to be emptied, thus decreasing the number of incoming handovers and, on the other hand, the high number of MS's in the hot cell increases the number of outcoming handovers. As τ_H tends to infinity, a statistical equilibrium between incoming and outcoming handovers is asymptotically achieved. As one could expect, the aforementioned statistical equilibrium is practically achieved when τ_H reaches η_H.

Fig. 7 plots the link unavailability $1 - L_A$ as a function of the mean offered traffic per sector A for the case $\Theta = \Theta_F$ (fixed users) and $\Theta = \Theta_S$ (mobile users, no hot cells) for values of the standard deviation σ_ε which can be found in the literature [8]–[10] and which have been reported in Table I. It can be noted that imperfect power control remarkably impairs performance. In fact, from the figure one can see that considering a link availability performance target equal to 0.01, a standard deviation $\sigma_\varepsilon = 1$ dB produces a capacity decrease of about 10%, while a standard deviation $\sigma_\varepsilon = 2$ dB produces a capacity decrease of more than 30%. In particular, the effect of a standard deviation equal to 3 dB is worse than the effect of a bursty user mobility, when hot cells are present ($\Theta = \Theta(180)$).

V. Conclusion

This paper deduces some basic analytical derivations by which an asynchronous CDMA cellular network is based on. Such derivations, together with appropriate mobility and traffic models, permit one to compute (by simulation) the effects of user mobility and imperfect power control on the network capacity. The approach is general enough so as to permit one to particularize the various derivations to the considered CDMA implementation.

Some simulation results are also shown; even if the obtained results depend on the selected particular CDMA implementation and on the relevant selected parameters, some general conclusions can be reached. As a matter of fact, the paper shows that the performance results, in terms of link availability, achieved by Viterbi in [1] for fixed users (i.e., in absence of user mobility for permanently in-call users—i.e., in absence of call dynamic and perfect power control), are remarkably worsened when a bursty user mobility is introduced and/or imperfect power control is considered. The paper also gives an intuitive insight of the phenomena which produces the aforementioned link availability decrease.

Acknowledgment

The authors wish to thank A. Baiocchi and V. Carducci for their significant suggestions and contributions to this work.

References

[1] K. S. Gilhousen and A. J. Viterbi, "On the capacity of a cellular CDMA system," *IEEE Trans. Veh. Technol.*, vol. 40, no. 2, pp. 303–312, May 1991.

[2] A. J. Viterbi and R. Padovani, "Implications of mobile cellular CDMA," *IEEE Commun. Mag.*, vol. 30, no. 12, pp. 38–41, Dec. 1992.

[3] P. G. Andermo and G. Larsson, "Code division testbed, CODIT," in *Int. Conf. Univ. Personal Commun.*, Ottawa, Canada, Oct. 1993, pp. 397–401.

[4] A. Baier and W. Koch, "Potential of CDMA for third generation mobile systems," in *MCR Proc.*, Nice, France, 1991.

[5] A. Baier, U. C. Fiebig, W. Granzow, W. Koch, P. Teder, and J. Thelecke, "Design study for a CDMA-based third-generation mobile radio system," *IEEE J. Select. Areas Commun.*, vol. 12, no. 4, May 1994.

[6] L. B. Milstein *et al.*, "On the feasibility of a CDMA overlay for personal communications networks," *IEEE J. Select. Areas Commun.*, vol. 10, no. 4, pp. 655–668, May 1992.

[7] K. S. Gilhousen and Alii, "Increased capacity using CDMA for mobile satellite communication," *IEEE J. Select. Areas Commun.*, vol. 8, no. 4, pp. 503–514, May 1990.

[8] G. Falciasecca, E. Gaiani, M. Missiroli, F. Muratore, V. Palestini, and G. Riva, "Influence of propagation parameters and imperfect power control on cellular CDMA capacity," *CSELT Tech. Rep.*, vol 20, no. 6, pp. 545–551, Dec. 1992.

[9] A. J. Viterbi and A. J. Viterbi, "Erlang capacity of a power controlled CDMA system," *IEEE J. Select. Areas Commun.*, vol. 11, no. 6, pp. 892–900, Aug. 1993.

[10] M. J. Jansen and R. Prasad, "Capacity, throughput, and delay analysis of a cellular DS CDMA system with imperfect power control and imperfect sectorization," *IEEE Trans. Veh. Technol.*, vol. 44, no. 1, pp. 67–75, Feb. 1995.

[11] A. J. Viterbi, A. J. Viterbi, and E. Zehavi, "Performance of power-controlled wideband terrestrial digital communication," *IEEE Trans. Commun.*, vol. 41, no. 4, pp. 559–569, Apr. 1993.

[12] F. Delli Priscoli and F. Sestini, "Fixed and adaptive blocking thresholds in CDMA cellular networks," in *Proc. IEEE ICC'95*, Seattle, WA, June 1995.

[13] A. Baiocchi, F. Delli Priscoli, and F. Sestini, "Effects of user mobility on the capacity of a CDMA cellular network," *Eur. Trans. Telecommun.*, vol. 7, no. 4, July/Aug. 1996.

[14] F. Delli Priscoli, "Study on an asynchronous CDMA satellite system for personal communications," in *Proc. ICC'94*, New Orleans, May 1994.

Francesco Delli Priscoli was born in Rome, Italy, in 1962. He received the Dr. Eng. degree in electronic engineering (summa cum laude) and the Ph.D. degree in system engineering from the University of Rome "La Sapienza," Italy, in 1986 and 1991, respectively.

From 1986 to 1991, he worked in the "Studies and Experimentation" Department of Telespazio, Rome, Italy. He was responsible for many tasks relevant to studies sponsored by the European Space Agency (ESA) concerning the design of advanced satellite systems based on FDMA, TDMA, ATM, CDMA. Since 1987, he has been cooperating with the University of Rome "La Sapienza," where he has been researching the nonlinear control system field and he has been teaching in the course "Automatic Controls." Since 1991 he is a Researcher at the University of Rome "La Sapienza." He has been researching in the field of advanced access techniques for mobile cellular systems (CDMA, PRMA, OFDM). In particular, he has been coping with the possible evolutions of the GSM cellular system, i.e., the dynamic channel allocation strategies, the effects of user mobility on the capacity of cellular systems, control decentralization, the evolution toward UMTS, and the application to satellite systems. He is presently deeply involved in several projects relevant to the mobile area of the ACTS European Community programme.

Fabrizio Sestini (M'91) was born in Pisa, Italy, in 1964. He received the Dr. Eng. degree (summa cum laude) in electronics engineering and the Ph.D. degree in information and communication engineering from the University of Rome "La Sapienza," Italy, in 1989 and 1993, respectively.

Since 1989, he has been involved in several activities related to the Telecommunication Project of the Italian National Research Council (CNR), including the design and development of an experimental ATM switching node test-bed. He has been with RAI in the development of several initiatives concerning databroadcasting systems and services on terrestrial and satellite television networks. His current research interests, in the field of cellular radio mobile networks for personal communication systems, include the design and performance evaluation of dynamic channel allocation strategies suitable for TDMA and PRMA based cellular networks, the study of different access techniques (CDMA, PRMA) and of the impact of user mobility on the performance of cellular systems.

The Capacity of a Spread Spectrum CDMA System for Cellular Mobile Radio with Consideration of System Imperfections

Paul Newson and Mark R. Heath

Abstract—**Recently, there has been much interest in the use of spread spectrum code division multiple access (CDMA) techniques for cellular mobile radio. To date, spread spectrum has been used mainly for military applications, in which the inherent transmission security and immunity to deliberate jamming are important. Spread spectrum systems, however, possess various other features such as multiple access and multipath rejection capability, which make their use attractive within the mobile radio environment. However, the current interest has been principally motivated by recent work [1] in which it is claimed that the CDMA option may offer capacity improvement over more conventional frequency and time division multiple access (FDMA) (TDMA) techniques. Within this paper, the relative capacities of a basic FDMA and CDMA system are examined. It is shown that, in the absence of capacity-enhancing features such as voice activity detection and cell sectorization, the capacity of each system is comparable. The paper then assesses the sensitivity of the CDMA system to typical propagation conditions, power control errors, and realistic antenna patterns and shows that the capacity of a CDMA system may be significantly reduced under nonideal conditions.**

I. Introduction

THE principle upon which the CDMA system is based is that all users have access to the entire available system bandwidth simultaneously. Of course, the interference caused to individual users when using conventional modulation techniques would be unacceptable; hence, spread spectrum modulation must be used [2]–[4]. Within the CDMA system, a relatively low bit rate data signal is forced to occupy the complete available channel by multiplying it by a much higher rate spreading sequence. It is this wideband signal which is transmitted. Multiple access capability is achieved by assigning to each user a unique spreading sequence which, ideally, is designed to have low cross correlation with all other allocated sequences. Since correlation between individual code sets is low, the desired message sequence can be reconstructed at the receiver by correlating the received signal with a locally generated version of the allocated code sequence. Correlation of the received signal results in a "processing gain" which is equal to the system spreading ratio W/b, where W is the total system bandwidth and b is the baseband bit rate.

In order to maximize the number of users which may operate within the system, it is necessary to overcome the "near/far" problem, in which the signal from a wanted distant transmitter may be swamped by a local interfering signal. This can be achieved by ensuring that the average received power level of each user is exactly equal. Consequently, accurate transmission power control is an important aspect of the system design. Within many capacity analyses of the CDMA system [1], [5], [6], power control is assumed to be ideal. In practice, however, the power control process is likely to exhibit some degree of error. This will inevitably degrade the apparent performance of the system to individual users and, therefore, reduce the number of users which may operate in the system with acceptable performance.

Within this paper, after this brief introduction, the capacities of a single-cell CDMA and FDMA systems are compared. This analysis is then extended to the multicell environment and the sensitivity of the CDMA system capacity to propagation conditions is investigated. The effect of system errors, such as power control error, on the capacity of the CDMA system uplink is then considered. Finally, the overall capacity as determined by consideration of the system downlink is analyzed. Throughout this work, the theoretical arguments are supported, where appropriate, by computer simulation results.

II. Basic Capacity Comparisons of CDMA with FDMA/TDMA Systems

Throughout this paper, the scenario assumed is that of a cellular radio system in which the geographical location is divided into various cells, each of which is served by a fixed base station. Each of the users within the cell communicates with the fixed network via a radio link to the base station.

A. FDMA Single-Cell Capacity

Assuming that the FDMA system is not thermal noise limited, then the number of channels available in a single-cell FDMA system, n_ν, is given by:

$$n_\nu = \frac{W}{b} \times m_f, \tag{1}$$

where m_f is the modulation efficiency, ie., the ratio of the baseband bit rate to the channel spacing adopted.

B. CDMA Single-Cell Capacity

In order to derive the maximum allowable number of users of the CDMA system, several simplifying assumptions are

Manuscript received May 28, 1993; revised December 2, 1993.
The authors are with BT Laboratories, Martlesham Heath, Ipswich, Suffolk, UK.
IEEE Log Number 9216731.

0733–8716/94$04.00 © 1994 IEEE

made. First, the radio channel is assumed to be time invariant and to contribute no multipath interference. Second, it is assumed that the receiver comprises a perfect matched filter which is both time and phase locked to the desired signal. Finally, it is assumed that the cross correlation between any two spreading codes is uniformly small compared with their common energy.

Assuming that system power control is ideal and that the number of system users is relatively high, it can be shown [1] that the number of users, n_u, which may be supported on the uplink of a single-cell CDMA system may be approximated, in the absence of capacity-enhancing features such as voice activity detection (VAD) and cell sectorization, by:

$$n_u \approx \frac{W}{b} \times m_c \times \frac{1}{(E_b/N_0)_{\text{req}}}, \quad (2)$$

where m_c is the CDMA system modulation efficiency, and $(E_b/N_0)_{\text{req}}$ is the signal-to-noise ratio (SNR) or, more specifically, the ratio of the signal bit energy to the single-sided interference power spectral density (PSD) required in order for an individual user to achieve an acceptable level of error rate performance. This figure represents the apparent SNR after matched filter detection [3]. Assuming that both the FDMA and CDMA systems employ modulation techniques of similar efficiency, then the relative system capacity is given by:

$$\frac{n_u}{n_\nu} \approx \frac{1}{(E_b/N_0)_{\text{req}}}. \quad (3)$$

It can be seen that, within the single-cell system, the capacity of the CDMA system is smaller than that of a corresponding FDMA system by a factor of $(E_b/N_0)_{\text{req}}$. Within the CDMA system, it is therefore of importance to minimize the system E_b/N_0 requirement in order to reduce the disadvantage suffered by the CDMA system within a single cell.

III. Multi-Cell Capacity Comparisons of CDMA with FDMA/TDMA

A. FDMA Multicell Capacity

In an FDMA system, in order to maximize the system capacity it is necessary to reuse the same frequency channel many times within the coverage area as a whole. Isolation between users is provided by the path loss between cells in which the same frequency is used. However, relatively large cochannel protection margins only permit the same frequencies to be used in one cell per cluster of cells. Thus, the number of channels available per cell for an FDMA/TDMA system N_ν is given by:

$$N_\nu = \frac{W}{b} \times m_f \times F_\nu, \quad (4)$$

where F_ν is the frequency reuse efficiency. If it is assumed that, for a typical FDMA system, frequencies are reused every seven cells, then a reuse efficiency of 0.14 may be obtained. It should be noted that a reuse efficiency of approximately 0.25 to 0.33 has been quoted for the GSM system [7]. However, since the modulation efficiency adopted within the GSM system is approximately 1/2 (this results from the fact that GSM can support eight 13 kb/s channels on a 200 kHz carrier), any increase in reuse efficiency is effectively cancelled. Consequently, the assumption $m_f \times F_\nu = 0.14$ remains reasonable.

B. CDMA Uplink Multicell Capacity

In a CDMA system, the number of users within each cell is critically dependent upon the interference received at the base station. In a multicell CDMA system, received interference will consist of interference from mobiles within the same cell and, in addition, interference due to mobiles within surrounding cells. Since the total interference is increased, the number of users which may be supported within any particular cell is reduced from the single-cell case and is given by:

$$N_u \approx \frac{W}{b} \times m_c \times \frac{1}{(E_b/N_0)_{\text{req}}} \times F_u, \quad (5)$$

where F_u is the uplink frequency reuse efficiency of the CDMA system and is given by the ratio of users within a multicell system to the number within a single cell. Neglecting additive noise, in the presence of interference from other cells the E_b/N_0 of the system is given by:

$$E_b/N_0 \approx \frac{W/b\, m_c}{(N_u + N_u I_u/P_u)}, \quad (6)$$

where I_u is the total interference generated from surrounding cells, and P_u is the total received signal power from all mobiles within an individual cell. In order to maintain the same level of E_b/N_0 as obtained within the single-cell system $(E_b/N_0)_{\text{req}}$, it is clear that the number of system users must be reduced such that:

$$n_u = N_u + N_u I_u/P_u. \quad (7)$$

Hence, the system reuse efficiency is given by:

$$F_u = N_u/n_u = \frac{1}{(1 + I_u/P_u)}. \quad (8)$$

In order to determine the interference levels received from adjacent cells, consider the uplink of the cell structure depicted in Fig. 1. Within the following analysis, it is assumed that ideal power control is employed, the number of system users is large, shadow fading is absent, and the hexagonal cells may be approximated by circular cells of equal area. Following the analysis of [8], power control at each base station ensures that the total power received from all mobiles within that cell is limited to P_u. Hence, any user within the cell is power controlled such that the transmitted power, $W_{i,j}(r)$, is given by:

$$W_{i,j}(r) = (P_u/N_u) r_{i,j}{}^{\alpha}, \quad (9)$$

where α is the decay law index and $r_{i,j}$ is the distance between the jth mobile and the base station to which it is operating, denoted[1] X_i. At base station X_0, the received interference

[1] The index i represents the cell index.

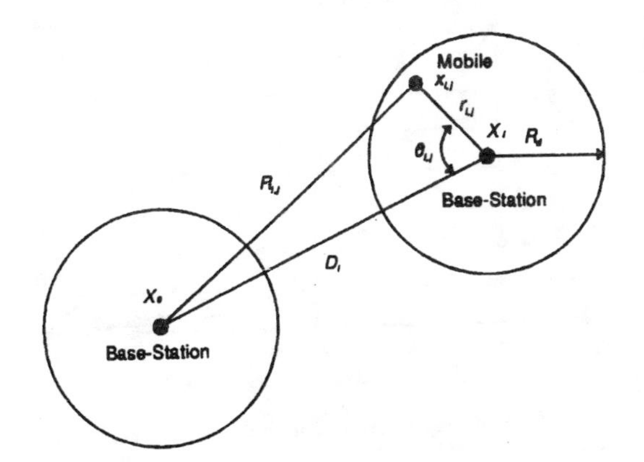

Fig. 1. Uplink interference calculation.

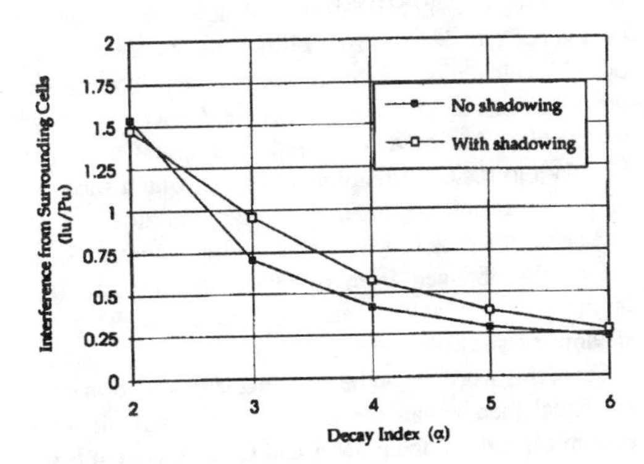

Fig. 2. Relationship between decay index and interference received from surrounding cells within a multiple-cell CDMA system.

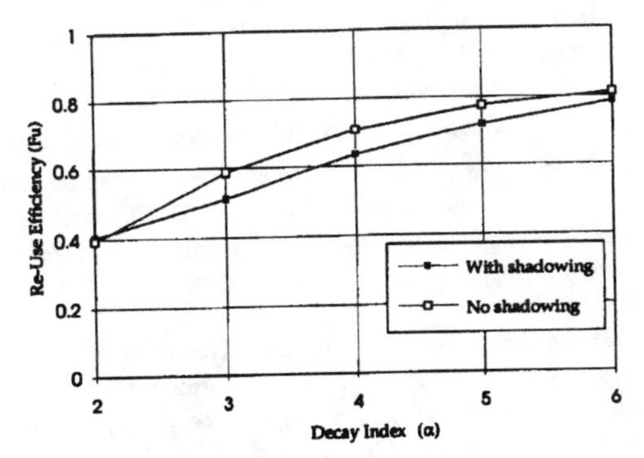

Fig. 3. Frequency reuse efficiency for a multiple-cell CDMA system.

from user $x_{i,j}$ is, therefore, given by:

$$I(X_0, x_{i,j}) = \frac{(P_u/N_u) r_{i,j}{}^{\alpha}}{R_{i,j}{}^{\alpha}}, \quad (10)$$

where $R_{i,j}$ is the distance between the mobile, $x_{i,j}$, and base station X_0. If the users are assumed to be uniformly distributed within the cells, the interference power per unit area received at base station X_0 from cell X_i is given by:

$$dI(X_0, X_i) = \frac{P_u r^{\alpha} dA}{\pi R_d{}^2 R^{\alpha}} = \frac{P_u r^{\alpha} dA}{\pi R_d{}^2 (D_i{}^2 + r^2 - 2D_i r \cos\theta)^{\alpha/2}}, \quad (11)$$

where r is the distance from any element of area to the base station to which a mobile positioned within that element would operate, R is the distance from the same element of area to the base station X_0, R_d is the radius of the cell, and $dA = r dr d\theta$. The total interference received from all adjacent cells is, therefore, given by:

$$I_u = \frac{P_u}{\pi R_d{}^2} \sum_{i=1}^{N_c - 1} \int_0^{2\pi} \int_0^{R_d} \frac{r^{\alpha+1} dr d\theta}{(D_i{}^2 + r^2 - 2D_i r \cos\theta)^{\alpha/2}}, \quad (12)$$

where N_c is the number of interfering cells within the coverage area. Performing this integral numerically for an arbitrary value of R_d and two tiers of surrounding cells (i.e., $N_c = 18$) for a range of values of decay index results in the plots of Fig. 2, in which the interference from surrounding cells, expressed as a proportion of the interference generated within the center cell, i.e., I_u/P_u, is plotted against decay index α. Substituting the results into (8) yields the plots for frequency reuse efficiency versus decay index shown in Fig. 3.

If the mobiles are subject to shadow fading [9], then the path loss between a mobile user and the base station to which it is operating is given by $PL_{i,j} = (r_{i,j}{}^{\alpha} 10^{-\gamma_{i,j}/10})$, where $\gamma_{i,j}$ is a zero mean Gaussian random variable with standard deviation of σ_s, which is assumed here to be 8 dB, and represents the shadow component between base station i and user j. If shadowing is included within this analysis, then the received interference at base station X_0 due to system user $x_{i,j}$ is given by:

$$I(X_i, x_{i,j}) = \frac{(P_u/N_u) r_{i,j}{}^{\alpha} 10^{(\delta_{i,j} - \gamma_{i,j})/10}}{R_{i,j}{}^{\alpha}}, \quad (13)$$

where $\delta_{i,j}$ is the shadowing component between mobile $x_{i,j}$ and base station X_0. The presence of shadowing means that if a mobile operates to the base station for which the path loss is least, i.e.,

$$r_{i,j}{}^{\alpha} 10^{-\gamma_{i,j}/10} = \min_{k \neq 0} r_{k,j}{}^{\alpha} 10^{-\gamma_{k,j}/10}, \quad (14)$$

then the mobile may not necessarily operate to its nearest base station. Moreover, if

$$\left(\frac{R_{i,j}{}^{\alpha}}{r_{i,j}{}^{\alpha}}\right) 10^{(\gamma_{i,j} - \delta_{i,j})/10} > 1, \quad (15)$$

then the particular user $x_{i,j}$ will operate to base station X_0 and should not, therefore, be included within the computation of the interference from surrounding cells I_u. In this instance, the total interference from users outside the coverage area of base station X_0 is given by:

$$I_u = \frac{P_u}{\pi R_d{}^2} \sum_{i=1}^{N_c - 1} \int_0^{2\pi} \int_0^{R_d} \frac{r^{\alpha+1} \left(10^{(\delta_i - \gamma_i)/10} . f(\delta_i - \gamma_i, r/R)\right) dr d\theta}{(D_i{}^2 + r^2 - 2D_i r \cos\theta)^{\alpha/2}}. \quad (16)$$

where $f(\delta_i - \gamma_i, r/R))$ is unity if the condition described by (15) is not satisfied and zero elsewhere. As previously pointed out [1], calculation of the statistics of the interference under the conditions described is not simple. Hence, within this work the problem has been addressed by Monte-Carlo simulation. Simulations result in the plots of Fig. 2, which show the mean of I_u/N_u computed under the same conditions as described earlier for the case in which shadowing is absent. Substituting the results obtained from these simulations into (8) results in the plots for system reuse efficiency in the presence of shadowing shown in Fig. 3.

From this work, it can be seen that if modulation efficiencies are equal then within the multicell environment the relative system capacity of the FDMA and CDMA system is given by:

$$\frac{N_u}{N_\nu} = \frac{F_u}{F_\nu (E_b/N_0)_{\text{req}}}. \qquad (17)$$

Assuming a fourth-order propagation law with 8 dB shadow component and an FDMA reuse efficiency of $F_\nu = 1/7$, then, from the previous results, it can be deduced that $\frac{N_u}{N_\nu} \approx \frac{4.2}{(E_b/N_0)_{\text{req}}}$ which, for a second-order decay index, is reduced to $\frac{N_u}{N_\nu} \approx \frac{2.7}{(E_b/N_0)_{\text{req}}}$. From these examples, it can be seen that in order to provide equal FDMA and CDMA capacities for a fourth-order power law, the E_b/N_0 requirement of the CDMA system is approximately 6.2 dB and for a second order law is 4.3 dB.

System Error Rate Performance: The level of E_b/N_0 effectively defines the error rate performance of the system and is, therefore, of critical importance. Many factors must be taken into account when establishing this. Foremost of these are the multipath and fading characteristics of the channel, the modulation technique adopted, the transmission filter characteristics, and the system coding gain. An examination of each of these factors is beyond the scope of this paper; however, recent work [1] has reported that by using a powerful convolutional coding scheme for forward error correction (FEC), under nonideal conditions an E_b/N_0 of 7 dB may be achieved. Assuming that this figure is achievable, substitution into (17) and using the results for reuse efficiency derived earlier reveals that $N_u/N_\nu \approx 0.84$ for a fourth-order decay index and $N_u/N_\nu \approx 0.54$ for a second-order law.

Modulation Efficiency: The modulation efficiency assumed throughout this work is unity for both the FDMA and CDMA systems. However, this figure may be increased by use of efficient modulation schemes [10]. While it is true that these schemes may be applied to both FDMA and CDMA techniques, it must be noted that the out-of-band radiation due to a CDMA system may be significantly greater than that due to a comparable FDMA system. This is due to the relative bandwidths available to each user which, within the CDMA system, is the entire allocated spectrum. For this reason, careful consideration must be given to the design of transmission filters and the modulation scheme employed within a CDMA system.

C. CDMA Features and Capacity Enhancements

Unequal Cell Loading: One attribute often claimed for the CDMA system is that, with unequal loading of cells, the capacity of more heavily loaded cells may be increased because less interference is received from lightly loaded neighboring cells. Assuming that surrounding cells are all equally loaded and that the proportion of users within each of the surrounding cells to the number of users within the center cell is given by ψ, then (8) may be rewritten:

$$N_u/n_u = \frac{1}{1 + \psi I_u/P_u}. \qquad (18)$$

Substituting the values of received interference from the surrounding cells given in Fig. 2 into (18) results in the figures for reuse efficiency against the percentage loading of surrounding cells in Fig. 4. The results given assume that shadowing is absent. Curves are shown for three values of propagation decay index ($\alpha = 2$, $\alpha = 4$, and $\alpha = 6$). It can clearly be seen that, as the loading in adjacent cells is decreased, the capacity of the center cell can exceed normal capacity[2]. For example, assuming an inverse fourth-power propagation law, the capacity of the center cell is able to operate at 130% of its original capacity provided that adjacent cells are operating at 26% of their normal capacity. From these results, CDMA does appear to take advantage of nonequal cell loadings. Gains are, however, relatively modest since CDMA exhibits very good frequency reuse efficiency. In comparison, FDMA systems which incorporate dynamic channel assignment (DCA) techniques [11], [12] might be expected to yield higher capacity gains through nonequal cell loadings.

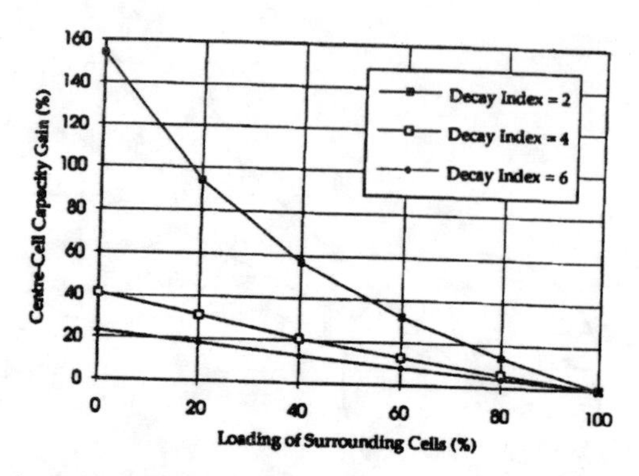

Fig. 4. Effect of nonequal cell loading on the capacity of the center cell in a multiple-cell CDMA system.

The capacity of the CDMA system may be significantly enhanced by generally reducing the level of interference apparent within the system. Two techniques have been proposed which enable this to be achieved. These are now discussed.

Multisector Base Station Antennas: The effective interference experienced at each base station may be reduced by the use of sectorization [1]. It can be seen that, by adopting 120° sectored antennas at each base station, the system capacity may be increased assuming ideal antenna patterns by a nominal factor of three. However, the capacity gains may be significantly reduced from this figure if more realistic antenna patterns are used.

[2] Here, the term "normal capacity" is used to describe the conditions in which adjacent cells operate at maximum loading.

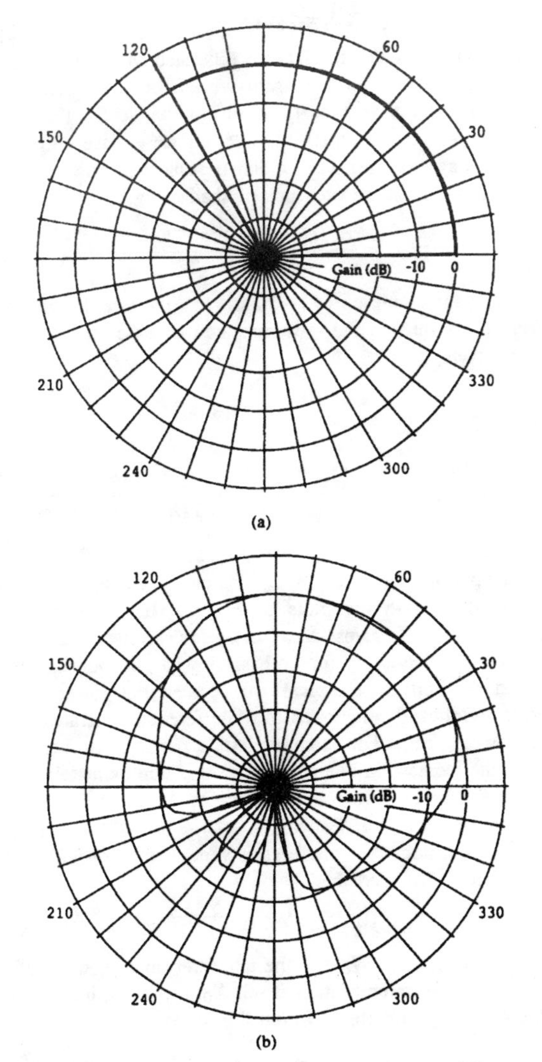

Fig. 5. Antenna patterns assumed for 120 sectors. (a) Ideal 120° antenna patterns. (b) Realistic 120° antenna patterns.

In order to investigate the effects of realistic antenna patterns on system capacity, computer models of the system have been developed. Within these models, three sectored base stations are used and each mobile is assumed to operate to its best sectored transceiver. Two antenna characteristics have been incorporated into the model, both of which are depicted in Fig. 5. The first pattern under consideration is an ideal antenna pattern in which the antenna gain is equal throughout its 120° range; outside this range, a gain of zero is assumed. The second pattern is a more "typical" pattern in which gain is less uniform and, in addition, power is radiated outside the nominal beam width of 120°.

Fig. 6 shows the number of mobiles which can operate to a single sector as a percentage of the single-cell capacity for

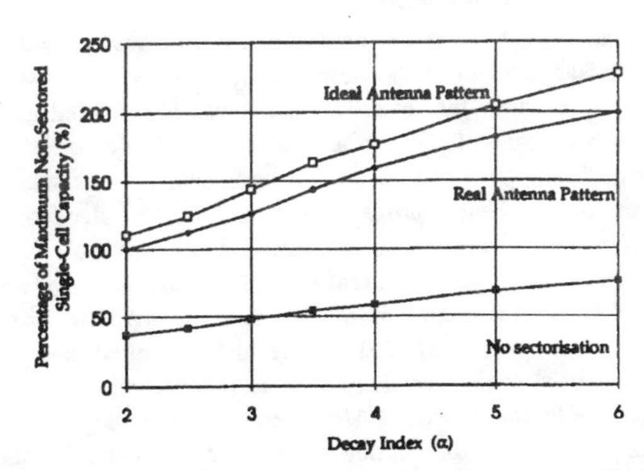

Fig. 6. Effect of sectorization on the capacity of a multiple-cell CDMA system.

a 90% availability[3]. Log-normal shadowing is assumed with standard deviation of $\sigma_s = 8$dB, and surrounding cells are assumed to operate at 100% loading. Three curves are shown. These represent the no sectored case, the ideal antenna pattern, and the more realistic antenna. It is clear that the use of non-ideal antenna characteristics results in a reduction in capacity with respect to the ideal antenna case of approximately 12%, which is largely independent of decay index. This gives a gain in capacity due to sectorization with the real antenna patterns of 2.6. It is to be expected that sectorization gain will be critically dependent upon the antennas used and, therefore, care must be taken in the selection of these.

Voice Activity Detection: It has been claimed that the duty cycle of speech is only 37.5%. Therefore, by restricting transmission to periods in which the voice is active, general levels of interference may be reduced and, therefore, capacity increased. Within a practical system, capacity gains of approximately 2 may be reasonably expected from this source [1].

Modified Capacity Equation: Assuming that both voice activity detection (VAD) and cell sectorization are used within the CDMA system, then the number of users per cell in a multicell CDMA system may be approximated:

$$N_u \approx \frac{W}{b} \times m_c \times \frac{1}{(E_b/N_0)_{\text{req}}} \times F_u \times D \times G. \qquad (19)$$

where D is the capacity gain due to VAD, and G is the gain due to the use of sectored antennas. Substitution of the figures quoted earlier for these factors into (19) (i.e., $D = 2$ and $G = 3$) reveals that if an E_b/N_0 requirement of 7 dB is assumed then for a fourth-order decay law the CDMA system has a capacity advantage of approximately 5; this being reduced to roughly 3.2 for a second-order law.

IV. CDMA Uplink Capacity Reduction Due to Power Control Errors

The capacity equations presented so far for the CDMA system have assumed that the power control process is ideal.

[3] Within this work, system availability is defined to be the expected number of system links for which the E_b/N_0 figure obtained is greater than the minimum acceptable level expressed as a percentage of the total number of system users.

In a practical system, power control nonidealities will result in some misadjustment of the power received by system users. Within this work, the misadjustment error is modeled as a random variable with log-normal distribution. This assumption is valid if the power control process compensates for shadowing and distance-based propagation loss since Doppler fading will result in significant instantaneous power control error. To reduce this within DS-CDMA systems, fast acting power control (FAPC) is often employed; indeed, as can be inferred from earlier work [13]–[15] and results presented later, if it is not used system capacity will be reduced significantly. Under conditions in which FAPC is employed, the log-normal assumption for the distribution of power control error is not, in general, accurate. However, analysis using this assumption remains of value since it is possible to derive a simple relationship between the magnitude of the power control error and the system capacity. This allows approximate assessment of the impact of nonidealities within the FAPC process such as finite step size, relative signal strength measurement error, synchronization error within the receiver, and residual fast fading not compensated for by the FAPC process to be made.

Power control errors have two effects. Since all users are subject to power control error, the statistical characteristics of the interference will alter from that found for ideal power control. Second, and more importantly, the power level of the required signal will be subject to variation. Since both the signal and interference components may be represented as random variables, it is clear that for any individual user there is a finite probability that the E_b/N_0 of the radio link on which it is operating may fall below the minimum required for acceptable performance. Within this work, any links for which this occurs are said to be in outage. Therefore, system availability may be expressed:

$$P_a = \Pr\{(E_b/N_0) > (E_b/N_0)_{\text{req}}\}. \tag{20}$$

If a user is subject to power control error, then the power of the user $x_{i,j}$ is given by:

$$W_{i,j}(r) = (P_u/N_u)r_{i,j}{}^{\alpha}10^{(-\gamma_{i,j}-\zeta_{i,j})/10}, \tag{21}$$

where $\zeta_{i,j}$ represents the power control error and is assumed to be a zero mean Gaussian deviate of standard deviation σ_p. The probability that a particular radio link within a cell served by base station X_0 is in outage may be expressed as:

$$P_o = 1 - P_a = \Pr\{(E_b/N_0)_{0,0} < (E_b/N_0)_{\text{req}}\}, \tag{22}$$

or, alternatively,

$$P_o = \Pr\left\{\left(P_u/N_u 10^{(-\zeta_{0,0})/10}\right) < \beta\left(\left(\sum_{j=1}^{N_u-1} P_u/N_u 10^{(-\zeta_{0,j})/10}\right) + I_u\right)\right\}, \tag{23}$$

where

$$\beta = \frac{10^{(E_b/N_0)_{\text{req}}/10}}{(W/bm_c)}.$$

The right-hand side (RHS) of the inequality of (23) comprises two terms: the interference generated within surrounding cells and the interference from within the same cell. First, consider the interference from surrounding cells. Following the analysis of Section III, the interference at base station X_0 due to a system user $x_{i,j}$, which is subject to power control error, is given in the presence of shadowing by:

$$I_u(X_0, x_{i,j}) = \frac{P_u/N_u r_{i,j}{}^{\alpha} 10^{(\delta_{i,j}-\gamma_{i,j}-\zeta_{i,j})/10}}{R_{i,j}{}^{\alpha}}. \tag{24}$$

Using the condition of (14) and (15), the total interference from users outside the coverage area of base station X_0 is then given by:

$$I_u = \frac{P_u}{\pi R_d{}^2}\sum_{i=1}^{N_c-1} \int_0^{2\pi}\int_0^{R_d} \frac{r^{\alpha+1}\left(10^{(\delta_i-\gamma_i-\zeta_i)/10} f_0(\delta_i-\gamma_i, R/r)\right)dr\,d\theta}{(D_i{}^2+r^2-2D_i r\cos\theta)^{\alpha/2}}. \tag{25}$$

Computing the mean of I_u/P_u using Monte-Carlo simulation results in the plots of Fig. 7. Here, the mean of I_u/P_u is plotted against decay index for various values of the standard deviation σ_p for both shadowed and nonshadowed instances. From these plots, it can be seen that power control error significantly increases the interference generated from surrounding cells. For example, with a power control error of 4 dB the increase in interference level is approximately 50%; however, for a power control error of 2 dB the increase is reduced to around 15%.

Next, consider the interference generated from within the center cell. Provided that

$$R_{i,j}{}^{\alpha}10^{-\delta_{i,j}/10} < r_{k,j}{}^{\alpha}10^{-\gamma_{k,j}/10}, \tag{26}$$

where $k = 1, 2, \ldots N_c - 1$, the interferer under consideration lies within the coverage area of cell X_0; hence, the interference due to a single interferer within this cell is given by:

$$I(X_0, x_{0,j}) = P_u/N_u 10^{-\zeta_{0,j}/10}. \tag{27}$$

The total interference generated at cell X_0 due to users within that cell may, therefore, be expressed as:

$$I_c = \frac{P_u}{\pi R_d{}^2}\int_0^{2\pi}\int_0^{R_d} r\cdot 10^{-\zeta_0/10} f_1(\delta_0-\gamma_i, R/r)\,dr\,d\theta. \tag{28}$$

where $f_1(\delta_i - \gamma_i, R/r)$ is unity if the condition expressed by (26) is satisfied and elsewhere is zero. Performing this integral for various values of power control error standard deviation results in the plots of Fig. 8. From this plot, it can be seen that the mean interference generated within the same cell increases as a function of power control error by a factor similar to that found for interference generated out with the center cell. This is not surprising, since each interference process is subject to the same dependencies. For this reason, the increase in total interference may be expressed as a common factor Γ, and may be written:

$$I_t = \Gamma\beta[I_c + I_u]. \tag{29}$$

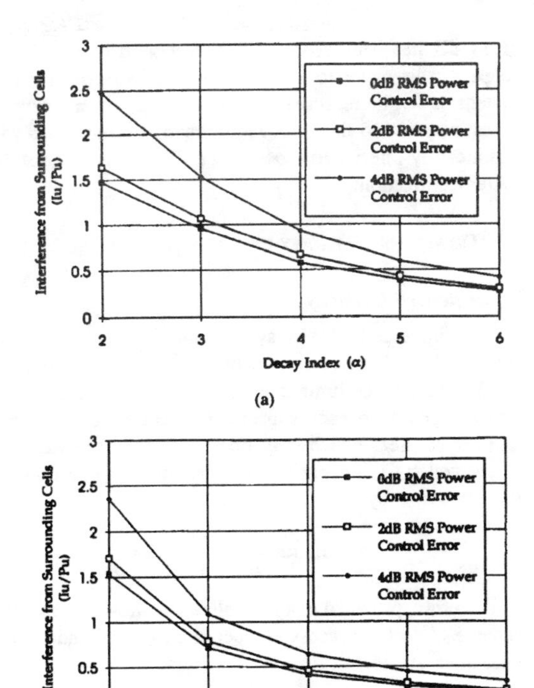

Fig. 7. Effect of power control error on interference from surrounding cells within a multiple-cell CDMA system. (a) Without shadowing. (b) Shadowing of $\sigma_s = 8$ dB.

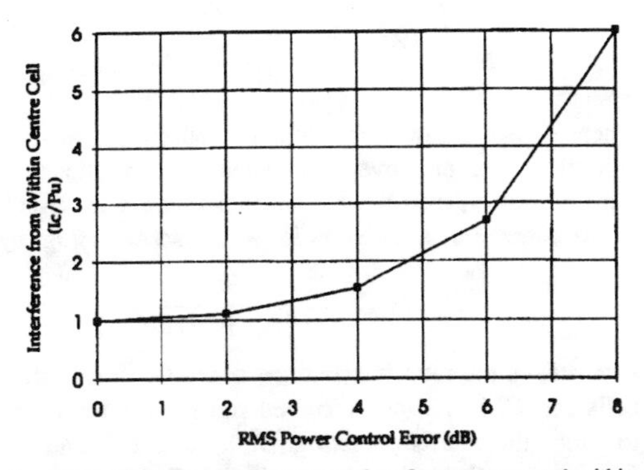

Fig. 8. Effect of power control error on interference generated within the center cellof a multiple-cell CDMA system. Without shadowing and decay index, $\alpha = 4$.

The total interference is, therefore, given by the addition of these independent random variables. Hence, the mean and variance of the total interference become $E[I_t] = \Gamma\beta(E[I_u] + E[I_c])$ and $Var[I_t] = E[I_t]^2 - (\Gamma\beta)^2(E[{I_u}^2] + E[{I_c}^2])$. Moreover, by applying Central Limit Theorem arguments it can be seen that the total interference I_t can be assumed to be normally distributed. The probability that a particular link will obtain acceptable performance, i.e., the system availability is given by:

$$\Pr(S_p < I_t) = \int_{-\infty}^{\infty} \int_{-\infty}^{i_t} f_{S_p,I_t}(s_p, i_t) ds_p di_t, \qquad (30)$$

where $f_{S_p,I_t}(s_p, i_t)$ is the joint density distribution of the total interference, I_t, and the desired signal, S_p, which is given by $S_p = (P_u/N_u 10^{-\zeta_{0,0}/10})$. Calculation of the error probabilities may, however, be significantly simplified by recognizing that since $Var(\beta I_t) = \beta^2 Var(I_t)$ and that normally $\beta \ll 1$, the variance of the interference is much smaller than the variance of the required signal, i.e., $Var[\beta I_t] \ll Var[S_p]$. Consequently, since the user power is a log-normally distributed variable, the probability that an individual link is in outage may be approximated as:

$$P_o \approx \Pr\{S_p < E[I_t]\}$$
$$\approx \frac{P_u/N_u}{(2\pi)^{1/2}\sigma_p} \int_{-\infty}^{E[I_t]} \exp\left(\frac{(\log_{10} s_p - m)^2}{2\sigma_p^2}\right) ds_p. \qquad (31)$$

Since there is no analytical solution to this integral, the probabilities must be computed numerically. For the purpose of computing error probabilities, it is, however, more convenient to express this equation in log form. Hence, S_p must be replaced by $S_{db} + \zeta_0$ where $S_{db} = 10\log_{10}(P_u/N_m)$ and becomes a normally distributed variable of mean S_{db} and standard deviation σ_p. The interference variable, I_t, is then given by $I_{db} = 10\log_{10} I_t$. Unfortunately, since I_{dB} is the log of a random variable its statistical characteristics alter from the linear case. It is, therefore, of importance to recognize the difference in the log and linear averages [16]. The variance of I_{dB} will also alter. However, in most practical cases it remains small compared with σ_p and can, therefore, be neglected. Using this approximation, the outage probability can be approximated as:

$$P_o \approx \frac{1}{(2\pi)^{1/2}} \int_x^{\infty} e^{-t/2} dt = \operatorname{erfc}[x], \qquad (32)$$

where $z = \frac{S_{dB} - E[I_{dB}]}{\sigma_p}$. From (32), it is possible to compute the increase in system E_b/N_0 required in order to obtain a given level of availability for varying levels of RMS power control error. This is plotted for availabilities of 98% and 90% in Fig. 9, which shows the decrease in E_b/N_0 for both the single-cell and multicell cases and serves to further illustrate the point that the increase in apparent interference due to power control error occurs to a similar extent within the center cell and surrounding cells.

A decrease in the system E_b/N_0 requirement may be obtained in several ways. However, assuming that little further gain is achievable using additional processing,[4] the only way in which this may be achieved is by reducing the total system interference and, consequently, the number of system users N_u. Therefore, under these conditions the capacity of a multicell CDMA system may be expressed:

$$N_u \approx \frac{W}{b} \times m_c \times \frac{1}{E_b/N_0} \times F_u \times D \times G \times \frac{1}{P}. \qquad (33)$$

[4] e.g., by use of error correcting codes or more efficient modulation schemes.

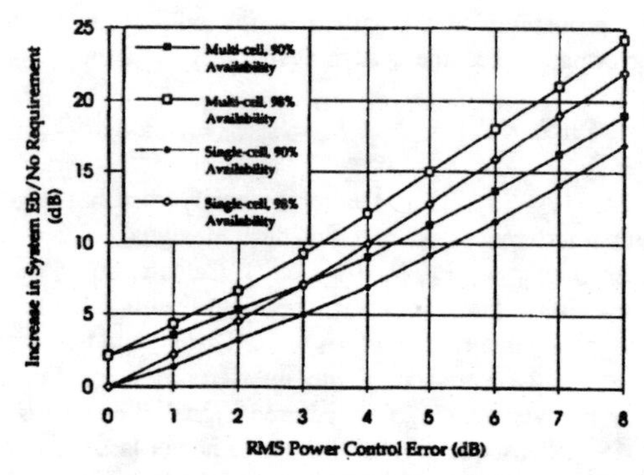

Fig. 9. Increase in system E_b/N_0 due to power control error. Decay index $\alpha = 4$.

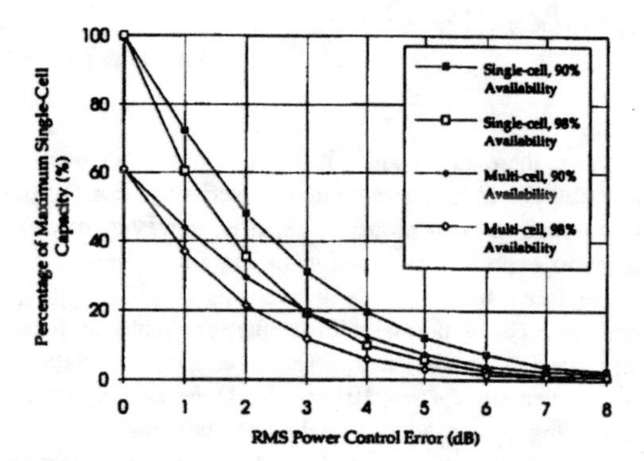

Fig. 10. Relationship between power control error and capacity in a CDMA system. No shadowing, decay index $\alpha = 4$.

where P is the capacity reduction factor due to errors within the power control process and is given by:

$$P = \frac{(E_b/N_0)_{\text{inc}}}{(E_b/N_0)_{\text{req}}}. \tag{34}$$

where $(E_b/N_0)_{\text{inc}}$ is the increase in system E_b/N_0 requirement due to power control errors. Substituting the values for $(E_b/N_0)_{\text{inc}}$ obtained from Fig. 9 into (33) results in the plots of Fig. 10, which shows the capacity reduction of a CDMA system due to the power control error for both single-cell and multicell system. These results indicate that a CDMA system is extremely sensitive to power control errors and that for a decay index of 4 and RMS power control error of 2 dB, the ratio of users of CDMA and FDMA systems N_u/N_ν is reduced by a factor of greater than two.

The capacity comparisons have largely been made without consideration of the relative system availabilities offered by each technique. However, this aspect of system performance has a significant impact on any conclusions being drawn. Unfortunately, it is not simple to compare generic CDMA and FDMA systems on the basis of availability since several system-specific assumptions must be made. The most important of these is the minimum level of SNR required to enable acceptable performance to be obtained within the FDMA system. This is, of course, dependent upon the system under consideration. However, for the GSM system a figure of 11 dB has been reported [7]. Although it is beyond the scope of this paper to assess the outage performance of this system, it can be reasonably expected that, given a seven cell repeat pattern and this SNR requirement, outage levels will be significantly inferior to those of the CDMA system for 100% system utilization.

V. Downlink Capacity of a Practical CDMA System

A. Single-Cell Capacity

In a single-cell CDMA system, the downlink differs from the uplink in that power control is not essential (assuming a totally interference-limited system). Hence, if the base station transmit power to each subscriber is exactly equal, then the number of users which may operate to the base station of a single cell n_d is approximated, assuming a large number of system users, by:

$$n_d \approx \frac{W/b\,m_c}{(E_b/N_0)_{\text{req}}}. \tag{35}$$

Therefore, provided that the uplink and downlink require the same E_b/N_0 for successful operation, uplink and downlink capacities are identical.

B. Multicell Capacity

In a multicell CDMA system, a mobile in a given cell will experience interference from its own base station but will also experience interference from surrounding base stations. This additional interference will degrade capacity with respect to the single-cell case. The number of users which can operate to a given base station in a multicell system, N_d, is approximated by:

$$N_d = \frac{W}{b} \times m_c \times \frac{1}{(E_b/N_0)_{\text{req}}} \times F_d, \tag{36}$$

where F_d is the downlink frequency reuse efficiency. If power control is not employed and each link is allocated equal transmit power, then the E_b/N_0 of a single user $x_{i,j}$ is given, in the absence of additive noise and shadow fading, by:

$$\{E_b/N_0\}_{x_{i,j}} \approx \frac{W/b\,m_c}{(N_d + N_d I_d/P_d)}, \tag{37}$$

where I_d is the total interference generated from surrounding cells and P_d is the total received signal from the base station to which the user is operating. Both the interference received from surrounding cells, I_d, and the interference received from the base station to which the mobile is operating, I_b, are of course dependent upon the position of the mobile within the cell. Both sources of interference are, therefore, functions of the distance, r, between the mobile and the base station to which it is operating and may be expressed, respectively, as:

$$\begin{aligned} I_d(r) &= P_t \sum_{i=1}^{N_c-1} R_i^{-\alpha} \\ &= P_t \sum_{i=1}^{N_c-1} (D_i^2 - 2D_i r\cos\theta_i + r^2)^{-\alpha/2}, \end{aligned} \tag{38}$$

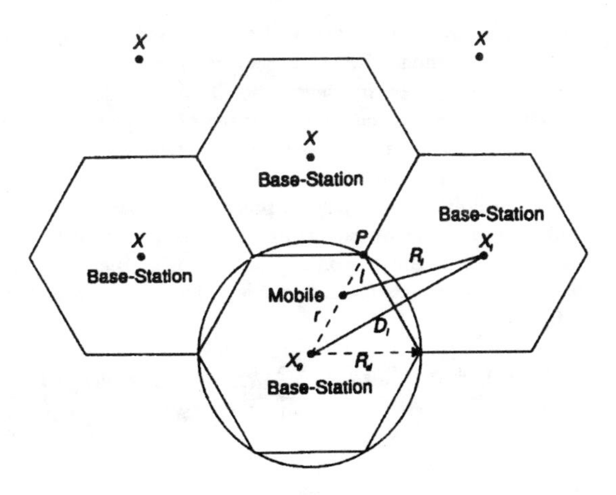

Fig. 11. Downlink interference calculation.

$$I_b(r) = \frac{P_t(N_d - 1)r^{-\alpha}}{N_d}, \tag{39}$$

where P_t is the total power emitted from each base station, and each of the other terms are as defined within Fig. 11. The received signal power at the output of the matched filter, $S_{i,j}$, is also dependent upon the radius r and may be expressed as:

$$S_{i,j}(r) = \frac{W/b m_c P_t r^{-\alpha}}{N_d}. \tag{40}$$

It can be seen that the E_b/N_0 received at a mobile, say $x_{0,j}$, is a function of r and is given by:

$$(E_b/N_0)_{x_{0,j}}(r) \approx \frac{W/b m_c}{(N_d-1)r^{-\alpha} + N_d \sum_{i=1}^{N_c-1} R_i^{-\alpha}}, \tag{41}$$

and may be approximated, for a large number of system users, by:

$$(E_b/N_0)_{x_{0,j}}(r) \approx \frac{W/b m_c}{N_d}\left(1 + r^{\alpha} \sum_{i=1}^{N_c-1} R_i^{-\alpha}\right)^{-1}. \tag{42}$$

As can be inferred from (42), worst-case E_b/N_0 occurs at the cell boundaries and, more specifically, at the point denoted within Fig. 11 as point p. Computing the received E_b/N_0 as a function of r for different decay indices as a mobile travels linearly from the base station to point p along the path denoted l results in the plots of E_b/N_0 against distance of Fig. 12. Within these computations, two tiers of surrounding cells are considered and the power transmitted from individual base stations is assumed to be equal. From these plots, it can be seen that at point p the received E_b/N_0 is between 0.18 and 0.32 times that of the maximum obtainable level for decay indices of 2 and 6, respectively. In order to obtain 100% availability, that is a level of at least $(E_b/N_0)_{\text{req}}$ at all positions within the cell, it can be seen that the reduced number of system users is given by:

$$N_d = \frac{W/b M_c}{(E_b/N_0)_{\text{req}}}\Delta, \tag{43}$$

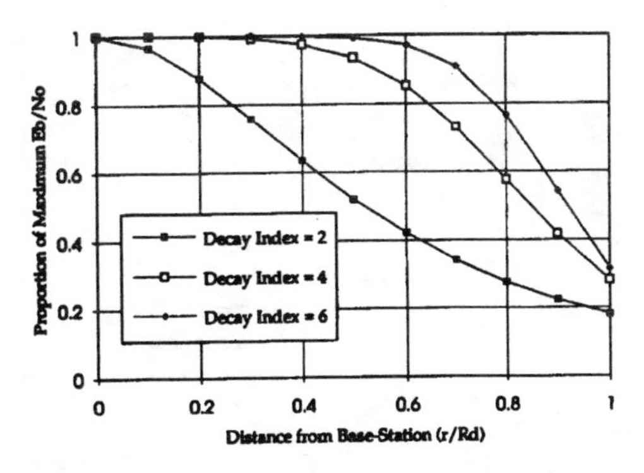

Fig. 12. Variationin downlink E_b/N_0 as a function of distance from the base station.

where Δ is given by:

$$\Delta = \left(1 + R_d^{\alpha} \sum_{i=1}^{N_c-1} R_{m_i}^{-\alpha}\right)^{-1}, \tag{44}$$

and R_{m_i} is the distance between each base station and a mobile at point p. Substituting the values derived from Fig. 12 into (43) yields the plots of reuse efficiency against the decay index shown in Fig. 13. From these, it can be seen that without downlink power control the number of system users is significantly reduced from that obtainable on the uplink in which power control is necessarily employed.

C. Downlink Power Control

Power control may be employed to maintain a given level of E_b/N_0 at each mobile user. In the absence of shadowing, the power transmitted to an individual system user is dependent upon its distance from the base station to which it is operating and may be derived by considering the E_b/N_0 figure of a mobile user as given by (37). Then, using (38) and (39) the power control factor $\varepsilon(r)$, which is defined here to be the user power at a given radius $P_s(r)$, over the nominal user power adopted when the user is very close to the base station may be given by:

$$\varepsilon(r) = \frac{P_s(r)}{P_{\text{nom}}},$$
$$\varepsilon(r) = \lambda\left(1 + r^{\alpha} \sum_{i=1}^{N_c-1} R_i^{-\alpha}\right), \tag{45}$$

where λ is given by:

$$\lambda = \frac{(E_b/N_0)_{\text{req}} N_d}{W/b M_c}. \tag{46}$$

As can be inferred from (45) and (46) within this work it is assumed that power control is a function of C/I and not of received signal strength alone. In the absence of interference from other base stations, the power control factor $\varepsilon(r)$ is unity irrespective of the position of the mobile within the cell. It can be seen, therefore, that if the system is subject to interference from other cells then $\varepsilon(r) \geq 1$. Consequently, the average total power transmitted from each base station, P_{av}, will increase with respect to the no interference case by a factor equal to the average level of $\varepsilon(r)$. Since the total interference within the

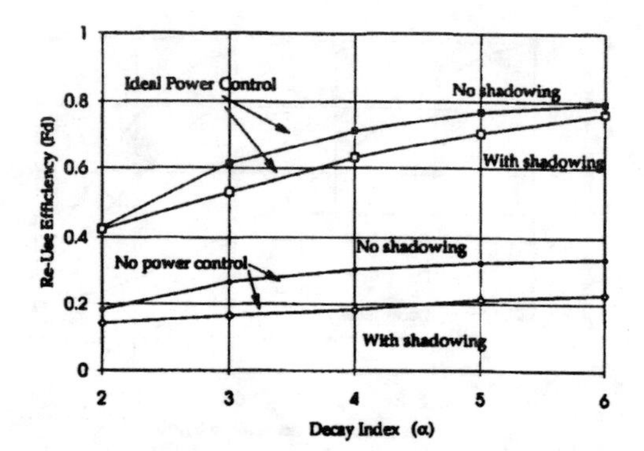

Fig. 13. Downlink frequency reuse efficiency in a multiple-cell CDMA system. Availability = 100%.

system is increased, in order to obtain similar levels of E_b/N_0 as were assumed within the single-cell system, the number of users, N_d, must be reduced from the single-cell case by a factor of:

$$\frac{N_d}{n_d} = \frac{1}{E[\varepsilon(r)]}. \tag{47}$$

The expectation $E[\varepsilon(r)]$ may be obtained by averaging the power control factor level over the complete cell area and may, therefore, be approximated by the following integration:

$$E[\varepsilon(r)] \approx \frac{1}{\pi R_d^2} \int_0^{2\pi} \int_0^{R_d} \left(r + r^{\alpha+1} \sum_{i=1}^{N_c-1} R_i^{-\alpha} \right) dr d\theta, \tag{48}$$

Computing this figure for a range of decay indices and substituting the results into (47) yields the figures for frequency reuse efficiency of Fig. 13. Comparing the reuse efficiency of the system in which no power control is employed with that of the system incorporating power control, it can be seen that in the absence of shadowing, power control results in a gain in reuse efficiency of greater than two. In addition, it can be inferred from the results of this analysis that the downlink power control dynamic range does not have to be large; for the decay indices considered, a maximum range of less than 7 dB is required. In contrast, the dynamic range of the uplink power control process has to be considerably greater.

If users are subject to shadowing, then both the signal and interference from within the same and adjacent cells will be altered. In this instance, these signals may be expressed respectively as:

$$S(r) = \frac{W/b\, m_c\, P_t r^{-\alpha} \cdot 10^{\delta_{(r,\theta)}/10}}{N_d}, \tag{49}$$

$$I_b(r) = \frac{P_t (N_d - 1) r^{-\alpha} \cdot 10^{\delta_{(r,\theta)}/10}}{N_d}, \tag{50}$$

and

$$I_d(r) = P_t \sum_{i=1}^{N_c-1} R_i^{-\alpha} \cdot 10^{\gamma_{(R_i,\Theta_i)}/10}. \tag{51}$$

where $\delta_{r,\theta}$ represents the shadowing component between base station X_0 and a mobile at radius r and angle θ from the base station, and γ_{R_i,Θ_i} represents the shadow component between base station X_i and a mobile at radius R_i and angle Θ_i from that base station. The addition of the shadowing component has two implications. Firstly, general interference levels will increase; hence, capacity degradation from the ideal case is inevitable. Secondly, the dynamic range of the power control process must be increased. It should be noted, however, that the dynamic range will remain relatively small compared with that of the uplink power control process. In order to perform similar reuse efficiency calculations to those described for the case in which shadowing is absent, it is again necessary to compute mean values for Δ and $E[\varepsilon_i]$. This can be achieved as follows:

$$\Delta = \left(1 + R_d^{\alpha} \cdot 10^{-\delta_{(r,\theta)}/10} \cdot \sum_{i=1}^{N_c-1} R_{m_i}^{-\alpha} \cdot 10^{\gamma_{(R_i,\Theta_i)}/10} \right)^{-1}, \tag{52}$$

$$E[\varepsilon_i] = \frac{1}{\pi R_d^2} \cdot \int_0^{2\pi} \int_0^{R_d} \left(r + r^{\alpha+1} \cdot 10^{-\delta_{(r,\theta)}/10} \sum_{i=1}^{N_c-1} R_i^{-\alpha} \cdot 10^{\gamma_{(R_i,\Theta_i)}/10} \right) \cdot f(\delta_{r,\theta} - \gamma_{R_i,\Theta_i}, r/R_i) dr d\theta, \tag{53}$$

where $f(\delta_{(r,\theta)} - \gamma_{(R_i,\Theta_i)}, r/R_i)$ is unity if

$$\left(\frac{r}{R_i} \right)^{\alpha} \cdot 10^{(\delta_{(r,\theta)} - \gamma_{(R_i,\Theta_i)})} > 1.0. \tag{54}$$

and zero elsewhere. Computing these values numerically for various values of decay index results in the plots for reuse efficiency of Fig. 13. As can be seen, the degradation in reuse efficiency due to the effects of shadow fading is similar irrespective of whether or not power control is incorporated into the system. Comparing the reuse efficiencies of the uplink and downlink, it can be seen that if both links employ power control then the reuse efficiency of each link is comparable. Consequently, provided that each link requires a similar E_b/N_0 figure for acceptable performance the capacities of each link is similar.

C. Downlink Pilot Signal

In a CDMA system, it is desirable for each base station to transmit a pilot signal. The pilot is critical to system performance in three main areas: it effectively defines the cell boundaries, it may be used for initial system acquisition, and it may be used to aid demodulation. Purely in terms of ensuring successful system operation, it would be beneficial to have as large a pilot power as possible. Unfortunately, the pilot is a spread spectrum signal and, as such, causes interference on the downlinks in every cell (thereby reducing capacity). Obviously, the larger the pilot signal power the greater the interference caused to the mobiles in each cell and the larger the capacity reduction. It is important to note that the pilot power has to be significantly higher than a single subscriber's power to ensure that a strong pilot signal is available even when the system is operating at full capacity. Being a spread spectrum signal, the pilot experiences an identical processing gain to subscriber signals at the receiver. However, it should also be noted that subscriber signals received will have the advantage of coding gain (e.g., the use of FEC coding effectively improves the SNR) which is not applicable with the pilot signal.

In a single-cell CDMA system, the base station will transmit at the same power level $P_{\max}$ to each subscriber. The base station also transmits a pilot spread spectrum signal at a power level P_{pilot}. If the pilot power is set to a level, say F_p times that of the maximum user power, then the received E_b/N_0 of any individual user may be expressed:

$$E_b/N_0 \approx \frac{(W/b)P_{\max}}{(n_d - 1)P_{\max} + F_p P_{\max}}, \quad (55)$$

where n_d represents the number of downlink system users in the single cell. Hence, the total number of users that may be supported within a cell may be approximated by:

$$n_d \approx \frac{W/b}{E_b/N_0} - F_p. \quad (56)$$

It can be seen that the pilot decreases capacity by F_p users, although this may be offset by a decrease in downlink E_b/N_0 requirement if the pilot is used to facilitate coherent demodulation. It should be noted that the choice of the factor F_p is not dependent upon the processing gain W/b, its value being simply determined by the required SNR for the pilot. In other words, the pilot is a fixed overhead. In a low processing gain system, where the number of users per cell is small, the pilot transmission constitutes a significant part of the total power radiated from the base station. However, given a very large processing gain system, the pilot will constitute a much less significant part of the base station power budget.

In a multicell CDMA system, power control has been found to be essential in order to gain satisfactory capacity. As has been shown, downlink power control approximately doubles capacity with respect to a nonpower-controlled system. However, with downlink power control, base stations do not transmit all their signals at maximum power. Consequently, since the pilot is the equivalent of F_p full-power users, it is to be expected that the capacity reduction due to the pilot is larger than in the nonpower-controlled case. Computer modeling results have shown that, when power control is implemented, capacity per cell is reduced by approximately $2F_p$ users. Hence, the relative effect of the pilot remains the same.

Within the previous discussion, the capacity reduction due to the pilot has been considered for a system which does not incorporate capacity-enhancing features such as VAD and cell sectorization. However, since the pilot must be transmitted continuously within each of the sectors, the relative effect of the pilot on system capacity is unaltered by the implementation of these features.

VI. Conclusion

The capacity of a CDMA system has been compared with that of a basic FDMA system. It has been found that, within a single-cell environment, the capacity of the CDMA system is significantly smaller than that of a corresponding FDMA system. In a multicell system, however, the relative frequency reuse efficiency of each technique is such that the multicell capacity of each system is very similar. The capacity of the CDMA system may be increased using techniques such as VAD and cell sectorization. Under ideal conditions, these features have been shown to result in a five-fold increase in capacity. The capacity of the CDMA system has, however, been shown to be sensitive to both propagation conditions and errors within the power control process. Consequently, within the CDMA system it is essential that accurate FAPC is employed.

In summary, a CDMA system which incorporates the capacity-enhancing features described may offer capacity advantage over a basic FDMA system. However, it may justifiably be argued that each of the capacity-enhancing features may also be applied to the FDMA system perhaps resulting in analogous capacity gain. Therefore, in order to obtain a reasonable analysis of the relative capacities of each system it is necessary to compare the CDMA system with a more advanced FDMA or TDMA system [17], [18].

References

[1] K. S. Gilhousen *et al.*, "On the capacity of a cellular CDMA system," *IEEE Trans. Vehic. Technol.*, vol. VT-40, no. 2, pp. 303–312, May 1991.

[2] R. C. Dixon, *Spread Spectrum Systems.* New York: Wiley, 1984.

[3] J. G. Proakis, *Digital Communications.* New York: McGraw-Hill, 1989.

[4] R. L. Pickholtz *et al.*, "Theory of spread spectrum communications—A tutorial," *IEEE Trans. Commun.*, vol. COM-30, no. 5, pp. 855–884, May 1982.

[5] J. S. Lehnert, "An efficient technique for evaluating direct-sequence spread-spectrum multiple access communications," *IEEE Trans. Commun.*, vol. COM-37, no. 8, pp. 851–858, Aug. 1989.

[6] J. S. Lehnert and M. B. Pursley, "Error probabilities for binary direct-sequence spread spectrum communication with random signature sequences," *IEEE Trans. Commun.*, vol. COM-35, no. 1, pp. 87–98, Jan. 1987.

[7] K. Raith and J. Uddenfeldt, "Capacity of digital cellular TDMA systems," *IEEE Trans. Vehic. Technol.*, vol. 40, no. 2, pp. 323–332, May 1991.

[8] G. R. Cooper and Nettleton, "A spread spectrum technique for high capacity mobile communications," *IEEE Trans. Vehic. Technol.*, vol. VT-27, pp. 264–275, Nov. 1978.

[9] W. C. Y. Lee, *Mobile Communications Engineering.* New York, NY: McGraw-Hill, 1982.

[10] S. Haykin, *Digital Communications.* New York: Wiley, 1988, ch. 7.

[11] W. Nettleton and G. R. Sclhloemer, "A high capacity assignment technique for cellular mobile telephone systems," in *Proc. IEEE Vehic. Technol. Conf.*, San Fransisco, CA, May 1–3 1989.

[12] I. R. Brodie, "Performance of dynamic channel assignment techniques in a cellular environment," in*Proc. IEEE Int.Conf. Select. Topics in Wireless Commun.*, Vancouver, Canada, June 1992.

[13] M. R. Heath and P. Newson, "On the capacity of spread-spectrum CDMA for mobile radio," in*Proc. IEEE Vehic. Technol. Conf.*, Denver, CO, May 10–13, 1992.

[14] A. Baier and W. Koch, "Potential and limitations of CDMA for 3rd generation mobile radio systems," in*Proc. MRC Mobile Radio Conf.*, Nice, France, Nov. 1991.

[15] E. Kudoh and T. Matsumoto, "Effect of transmitter power control imperfections on capacity in DS/CDMA cellular mobile radois," in *Proc. Int. Conf. on Commun.*, Chicago, IL, June 1992, pp. 237–242.

[16] S. C. Schwartz and Y. S. Yeh, "On the distribution function and moments of power sums with log-normal components," *Bell Syst. Tech. J.*, vol. 61, pp. 1441–1462, Sept. 1982.

[17] B. Gudmundson, J. Skold, and J. K. Ugland, "A comparison of CDMA and TDMA systems," in *Proc. IEEE Vehic. Technol. Conf.*, Denver, CO, pp. 732–735, May 10–13, 1992.

[18] W. Granzow and W. Koch, "Potential capacity of TDMA and CDMA cellular telephone systems," in *Proc. IEEE Second Int. Symp. on Spread Spectrum Techn. and Applicat. (ISSSTA '92)*, Yokohama, Japan, Nov. 1992, pp. 243–246.

Paul Newson received the M.Sc. and Ph.D. degrees in electrical engineering from the University of Edinburgh, Scotland, in 1988 and 1992, respectively.

Since joining BT Laboratories, Martlesham Heath, England, in 1991 as a research engineer, he has been involved in the mathematical modeling of both TDMA and CDMA cellular radio systems. His current research interests include digital signal processing techniques for cellular radio systems and adaptive algorithms for cellular network optimization. He is currently an Honorary Editor for the *IEEE Proceedings on Communications*.

Mark R. Heath received the B.Sc. and Ph.D. degrees in electrical and electronic engineering from the University of Leeds, England, in 1986 and 1989, respectively.

Since joining BT Laboratories in 1989, he has been involved in many aspects of cordless and cellular communications, including modeling and analysis of present and future mobile systems. In particular, he has been actively involved in the development and implementation of new quality improvement techniques for existing TACS and GSM cellular networks. He is currently developing advanced cellular planning and optimization tools.

SECTION 2.3
CDMA OVERLAY

The papers in this section describe cellular overlay issues including microcellular configurations. The first paper, "*CDMA Cellular Engineering Issues*" by Kim, investigates the possible interference scenario during the period of transition from the analog system to digital CDMA cellular system. The next paper, "*Overlay of Cellular CDMA on FSM*" by Filis and Gupta investigates the spectrum sharing capability of a cellular CDMA overlaid on the fixed service microwave band. The paper "*Microcellular Engineering in CDMA Cellular Networks*" by Shapira discusses the parameters involved in the engineering of a heterogeneous CDMA network including macro and micro cells. Factors that determine the size of the cell, the soft handoff zone, and the capacity of the cell clusters are analyzed, and engineering techniques for overlay-underlay cell clustering are outlined. The last two papers, "*A Micro-Cellular CDMA System Over Slow and Fast Rician Fading Radio Channels with Forward Error Correcting Coding and Diversity*" by Wijffels et al. and "*CDMA Overlay Situations for Microcellular Mobile Communications*" by Wang and Milstein, consider microcell CDMA performance.

CDMA Cellular Engineering Issues

Kyoung Il Kim, *Senior Member, IEEE*

Abstract—**In this paper, the frequency reuse efficiency for the proposed code division multiple access (CDMA) cellular system is analytically derived, and it is shown that the cell capacity of a fully loaded multiple cell system is about 75% of what would be available for a single cell system. In addition, an engineering issue in transitioning from analog advanced mobile phone system (AMPS) to CDMA system is discussed. Calculating excessive interference power due to a CDMA channel, it is shown that when a CDMA channel is introduced in a cell, there should be at least one ring of buffer cells, in which the analog channels falling into the CDMA band are not reused.**

I. Introduction

As a larger capacity is needed due to the high growth rate of cellular mobile phone subscribers, a transition from analog to digital cellular system is expected. As an alternative to the digital system based on time division multiple access (TDMA) method [1], a digital cellular system using code division multiple access (CDMA) technique was recently proposed [2], and it has drawn much attention [3]–[6]. Among various advantages of the CDMA system over the analog and/or TDMA system, the ease of frequency planning is a major advantage [5]. Traditional frequency planning associated with the analog system is a difficult process, and by some it is even considered an art.

Frequency planning is a result of the frequency reuse, which is needed for a frequency division multiple access (FDMA) system using only a limited number of radio channels to handle a large volume of traffic over a wide area. Frequency planning involves; partitioning the service area into smaller areas called cells, forming a cluster of K adjacent cells, and allocating a $1/K$ of the total available radio frequencies to each cell in the cluster. Then the K-cell cluster pattern is repeated so that the same frequencies are used again over other geographical areas. The number of cells in one cluster, K, is called the *frequency reuse factor*, and it determines the amount of the cochannel interference as well as the number of radio frequencies available per cell. As such, K is chosen such that it is small enough to assign more channels to a cell (i.e., large cell capacity) and at the same time it is large enough to have as little cochannel interference as possible, i.e., K is chosen as the minimum possible value that satisfies the desired carrier to cochannel interference power ratio (C/I). As an example, in the case of the analog AMPS, $K = 7$ is most frequently used for the desired $C/I = 18$ dB [7].

In terms of the radio frequency usage, the CDMA system can be seen as a one-cell system in that the same frequency is reused over all cells, and so K appears to be 1. However the system will still use the traditional cellular configuration in order to cover a large service area. When a service area is consisting of many cells, the channel capacity per CDMA cell decreases as a result of the excessive interference from other cells, resulting in the number of traffic channels per cell smaller than what would be available for a single isolated cell case. As discussed above, the capacity reduction in the multicell configuration is directly analogous to K in the case of an FDMA system since it has one traffic channel per one radio channel, but we need to directly look at the interference in order to examine the capacity reduction in a multicell CDMA system since a radio channel contains many CDMA traffic channels.

In general, the channel capacity per cell in the multicell configuration is limited by the interference. The efficiency of the cellular approach applied to different systems can be compared more meaningfully by comparing the capacity reduction with respect to a one-cell case, instead of merely comparing the traditional frequency reuse factor, K. Thus we define the multicell *frequency reuse efficiency* by

$$K' = \frac{\text{Total traffic channels for single isolated cell system}}{\text{Total traffic channels per cell for multiple cell system}}. \tag{1}$$

Note that, for an FDMA system, $K = K'$.

On the other hand, the proposed CDMA system [2] can provide a smooth transition to a digital cellular system in that a portion of the cellular spectrum, in increment of 1.25 MHz, can be converted to CDMA system. The spectral band of 1.25 MHz is 10% of the entire AMPS cellular band. Therefore, we expect to see both the analog and CDMA systems operating together during a transition period. Although a CDMA channel with 1.25 MHz replaces 6 analog channels in an analog-CDMA mixed cell, included also in the same band are the analog channels that should be allocated in the surrounding cells. It thus follows that once a 1.25-MHz band is converted to a CDMA channel, it becomes an interferer to the analog channels in the surrounding cells that fall into the same band. The question then arises; where can we use, in the surrounding analog cells, the channels in the CDMA band? We examine this question by analyzing the interferences a CDMA channel generates, i.e., the impact of a CDMA channel on the analog channels in terms of C/I ratio.

It is the objective of this paper to examine the frequency reuse efficiency of the CDMA system so as to compare it with that of the currently used analog system, and to investigate the amount of excessive CDMA interference received by analog channels to determine the size of guard zone that may be required during the period of transition from the analog to digital CDMA cellular system. We assume that the received

Manuscript received September 1, 1992; revised November 6, 1992.
The author is with AT&T Bell Laboratories, Whippany, NJ 07981.
IEEE Log Number 9208609.

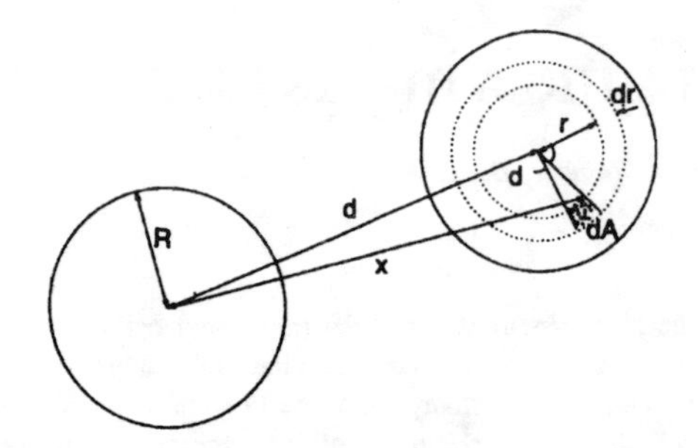

Fig. 1. Plot for the computation of the CDMA interference.

C/I ratios for both the forward and reverse links are the same, and the analyses will be carried out for only one direction; the reverse link.

II. CDMA Interference Power

In this section we will derive the CDMA mobile power received at a cell other than that to which the mobile unit belongs. We assume that N mobile units are uniformly distributed in a circular cell of radius R so that the density of mobile units is

$$\rho = \frac{N}{\pi R^2}. \tag{2}$$

Assuming a perfect power control, the received power at the base station would be the same for each mobile station. Let S_c be the power of a CDMA mobile unit received at the base station of its own cell. Then, referring to Fig. 1 and assuming the wave propagation with path loss proportional to the 4th power of distance, we find the total mobile power from a cell having N mobile stations uniformly distributed in it to a base station at distance $d = kR$ as follows:

$$P(d) = \int \alpha S_c \frac{r^4}{x^4} \rho \, dA \tag{3}$$

where the integration is over the one cell area, α is the voice activity factor, and

$$x = \sqrt{d^2 + r^2 + 2dr\cos\theta}. \tag{4}$$

From Fig. 1, and by using (2) and (4), (3) becomes

$$\begin{aligned} P(d) &= 2\int_0^{\pi}\int_0^{R} \alpha S_c \left(\frac{r}{x}\right)^4 \frac{N}{\pi R^2}\, r\, dr\, d\theta \\ &= \frac{2\alpha N S_c}{\pi R^2}\int_0^R dr\, r^5 \int_0^{\pi} \frac{d\theta}{(d^2 + r^2 + 2dr\cos\theta)^2}. \end{aligned} \tag{5}$$

The integration can be evaluated analytically (see, e.g., Appendix I), and is given by

$$P(d = kR) = 2\alpha N S_c \left[2k^2 \ln\left(\frac{k^2}{k^2 - 1}\right) - \frac{4k^4 - 6k^2 + 1}{2(k^2 - 1)^2} \right]. \tag{6}$$

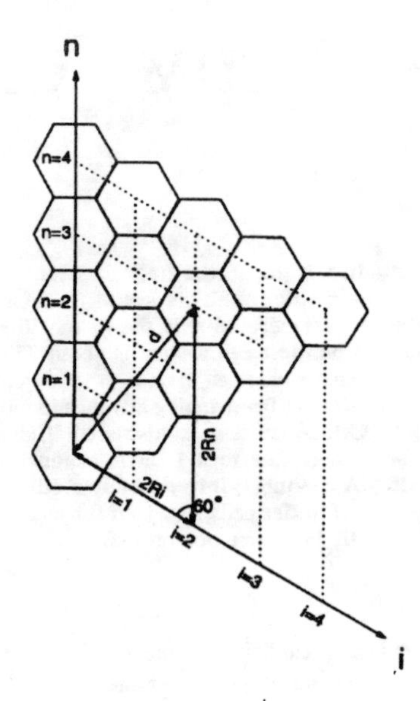

Fig. 2. Coordinates for inter-base station distance calculation.

III. Frequency Reuse Efficiency

Let N_c be the capacity of an isolated single cell CDMA system; that is, a CDMA cell can serve N_c traffic channels simultaneously at the minimally required bit error rate level. In this case, the relationship between the N_c and the required C/I should satisfy [3]

$$\left(\frac{C}{I}\right)_{\text{req}} = \frac{\alpha S_c}{(N_c - 1)\alpha S_c} = \frac{1}{N_c - 1}. \tag{7}$$

For the same performance in terms of the required C/I, the capacity of a multiple cell CDMA system, N, will now have to satisfy

$$\left(\frac{C}{I}\right)_{\text{req}} = \frac{\alpha S_c}{(N-1)\alpha S_c + I_{oc}} = \frac{1}{(N-1) + I_{oc}/\alpha S_c} \tag{8}$$

where I_{oc} the total interference power from other cells. From (1), (7) and (8), we have

$$K' = \frac{N_c}{N} = 1 + \frac{I_{oc}}{\alpha N S_c}. \tag{9}$$

Given a cell, we will refer to the cells immediately contiguous to it as ring-1 cells, those contiguous to the ring-1 cells as ring-2 cells, etc. When all CDMA cells are loaded with N active mobile units, the total interference power received at a base station from all mobiles in other cells will consist of the sum of the power from 6 ring-1 cells, 12 ring-2 cells and so on. From Fig. 2, it can be shown that the distance between the base station of a ring-n cell and that of the cell under consideration is $d_{R,n,i} = 2R\sqrt{n^2 + i^2 - ni}$. The nth ring consists of 6 cells at distance $d_{R,n,i}$ for each i, $i = 1, 2, \cdots, n$; thus the total number of cells in ring-n is $6n$. Thus using (6), and defining

$$L = 4(n^2 + i^2 - ni) \tag{10}$$

we find I_{oc} as follows:

$$I_{oc} = \sum_{n=1}^{\infty}\sum_{i=1}^{n} 6P(2R\sqrt{n^2+i^2-ni})$$
$$= 12\alpha N S_c \sum_{n=1}^{\infty}\sum_{i=1}^{n}\left[2L\ln\left(\frac{L}{L-1}\right) - \frac{4L^2-6L+1}{2(L-1)^2}\right]. \quad (11)$$

We also define the summand in (11) as follows, as it appears several times throughout this paper:

$$G(L) \equiv 2L\ln\left(\frac{L}{L-1}\right) - \frac{4L^2-6L+1}{2(L-1)^2}. \quad (12)$$

Then from (9) and (11), we obtain

$$K' = 1 + 12\sum_{n=1}^{\infty}\sum_{i=1}^{n} G(L). \quad (13)$$

By numerically calculating the second term in the above expression, it can be shown that $K' \approx 1.33$ with 100 terms (i.e., $1 \le n \le 100$). Comparing it with $K' = 7(= K)$ of the analog FDMA system, we can see that the cellular approach is much more efficient with the CDMA system in that the cell capacity reduction with respect to a single cell system is only 25% (or 0.33/1.33) for the CDMA system whereas it is 86% (or 6/7) for the FDMA system.

IV. Impact on the Frequency Reuse of Analog Channels

In this section, we will consider the excessive interference generated by a CDMA channel when a portion of cellular spectrum is converted so as to employ a CDMA system, and discuss its impact on the performance of the analog system. For numerical calculations, we will use the following assumptions.

1) Omni cells for both CDMA and analog;
2) Hexagonal cellular geometry;
3) $K = 7$ for the analog system;
4) CDMA utilizes the voice activity factor, $\alpha = 0.5$;
5) 30 traffic channels per 1.25 MHz CDMA radio channel[1];

A. The Case for One CDMA Cell

In this section, we will consider the case where only one cell is converted to analog-CDMA mixed cell, and examine the effect of the CDMA radio channel to analog AMPS channels in the neighboring cells. Referring to Fig. 3, suppose that a CDMA channel is allocated to the cell 1. We will assume that the power spectra of the CDMA signals are flat over the AMPS frequency bands that they have displaced. Using (6), we find the interference from the CDMA mobiles to ring-n cell as

$$I_{\text{CDMA}}(n, i) = P(d_{R,n,i}) \times \frac{W_f}{W_c} = 2\alpha N S_c G(L)\frac{W_f}{W_c} \quad (14)$$

where L and $G(L)$ are respectively defined by (10) and (12), N is the total number of CDMA mobile units in the cell, and

[1]This corresponds to a five times capacity improvement over the analog AMPS with omni cells, and approximately a 15 times capacity improvement for 3-sectored cells.

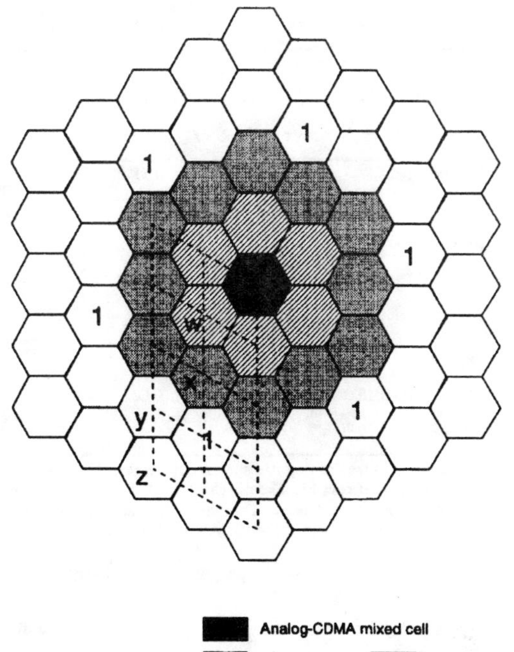

Fig. 3. The case for one CDMA cell.

W_f and W_c are spectral bandwidths of an analog and CDMA channels, respectively. Then for the worst case of an analog mobile at the edge of its cell, the C/I ratio for the ring-n-cell analog channels in the CDMA band is given by

$$\frac{C}{I} = \frac{P_f/R^4}{6(P_f/D^4) + I_{\text{CDMA}}(n, i)}$$
$$= \frac{(C/I)_A}{1 + \frac{(D/R)^4}{6}\frac{S_c}{P_f/R^4}\frac{W_f}{W_c}2\alpha N G(L)} \quad (15)$$

where P_f is the analog mobile transmit power, D is the distance between the two closest analog cochannel cells, and $(C/I)_A$ is the carrier to cochannel interference power ratio for the analog traffic channel without any CDMA channels. As mentioned before, $(C/I)_A$ of the AMPS is typically set to 18 dB, resulting in $K = 7$. In the above expression, we have assumed that the analog cochannel interferences from the cells other than the closest cochannel cells are negligible. We will refer to the closest cochannel cells as tier-1 cells. Using the relationship between R, D and the frequency reuse factor, K, that is given by [8]

$$D/R = \sqrt{3K} \quad (16)$$

we find that $D/R = \sqrt{21}$ when $K = 7$. In this case, substituting numerical values, i.e., $\alpha = 0.5$, $N = 30$, $W_f = 30$ KHz and $W_c = 1.25$ MHz, into the above expression, we get

$$10\log(C/I)\ [\text{dB}] = 10\log(C/I)_A - \Delta\ [\text{dB}] \quad (17)$$

where Δ, the term representing performance degradation due

TABLE I
PERFORMANCE DEGRADATION DUE TO A CDMA CELL

	Δ [dB]			
(n, i)	$S_c = S_f$	$S_c = 0.5 \times S_f$	$S_c = 0.1 \times S_f$	Remark
(1, 1)	3.53	2.11	0.51	ring-1 cell (w)
(2, 1)	0.33	0.17	0.03	ring-2 cell (x)
(3, 2)	0.054*	0.027	0.005	ring-3 cell (y)
(4, 2)	0.018	0.009	0.002	ring-4 cell (z)

* The cell 1 in the 3rd ring is a special case in that it has only five tier-1 analog cochannel cells because the CDMA cell 1 does not contribute analog cochannel interference. Thus the Δ for the analog channels of the ring-3 cell 1 that are within the CDMA band is given by

$$\Delta = 10 \log \left(\frac{5}{6} + 52.92 \frac{S_c}{S_f} G(L) \right) \text{[dB]}$$

which, with $S_c = S_f$ and $(n, i) = (3, 1)$, reduces to -0.73 dB.

to the introduction of a CDMA channel, is

$$\Delta = 10 \log \left(1 + 52.92 \frac{S_c}{S_f} G(L) \right) \text{[dB]} \tag{18}$$

where $S_f = P_f/R^4$ is the analog mobile power received by the base station.

We expect that the CDMA mobile power would be much less than that for an analog mobile unit,[2] i.e., $S_c \ll S_f$. Thus the worst case (i.e., the largest CDMA-to-analog cochannel interference) would occur when $S_c = S_f$. For the cells in up to ring-4, and for the three cases of $S_c = S_f$, $S_c = 0.5 \times S_f$ and $S_c = 0.1 \times S_f$, we show the numerical values of Δ in Table I. Also indicated in the last column of the table are the cells used for the numerical computations with reference to Fig. 3.

The performance degradation of the analog channels in ring-1 cell is more than 3 dB for the worst case ($S_c = S_f$), which suggests that those channels may not be allocated. Assuming that less than 1 dB degradation is acceptable, we find that one ring of buffer cells would be appropriate, i.e., the regular frequency reuse pattern can be applied to all analog cells except for the cells closest to the CDMA cell. In this case, it is interesting to note that there may be no capacity improvement during a transitioning period, if a CDMA channel is introduced in only one omni cell. This is because 42 analog channels in the 1.25-MHz CDMA band are removed from the CDMA cell and a ring of guard zone whereas a CDMA radio channel may not provide that many traffic channels.

[2] This is a safe assumption based on 1991 CDMA field tests in San Diego. The test results show that the mean transmit power of a CDMA mobile unit is about 10 mW [9], which is 1/60 of the peak power of a class III analog AMPS mobile unit [7]. Without dynamic power control, an analog mobile unit will transmit at its peak power.

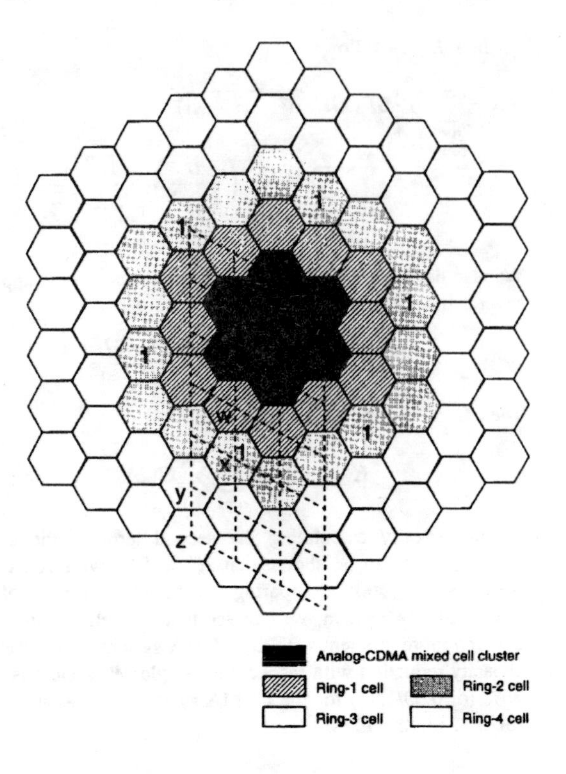

Fig. 4. The case for a cluster of CDMA cells.

B. The Case for a Cluster of CDMA Cells

Suppose now that a 1.25-MHz CDMA channel is assigned to all of the cells in a 7-cell cluster as shown in Fig. 4. The interferences to the analog channels in the cells surrounding the CDMA cell cluster are discussed below. Again, we will use the following values for numerical computations; $\alpha = 0.5$, $N = 30$, $W_f = 30$ KHz, and $W_c = 1.25$ MHz.

As an example, consider the cell w in Fig. 4, which is one of the closest cells to the CDMA cell cluster. This is a ring-1 cell with respect to the cluster. But, this cell can also be considered as a ring-1 cell with respect to the cells 5 and 6 in the CDMA cluster, a ring-2 cell with respect to the cells 1, 4 and 7, and a ring-3 cell with respect to the cells 2 and 3. In general, we see that the CDMA channel interferences to ring-n cells will consist of the interferences from two ring-n cells, three ring-$(n+1)$ cells and two ring-$(n+2)$ cells. Thus the total CDMA interference power to a ring-n cell from the CDMA cell cluster is

$$\begin{aligned} I'_{\text{CDMA}} = {} & I_{\text{CDMA}}(n, i_1) + I_{\text{CDMA}}(n, i_2) + I_{\text{CDMA}}(n+1, j_1) \\ & + I_{\text{CDMA}}(n+1, j_2) + I_{\text{CDMA}}(n+1, j_3) \\ & + I_{\text{CDMA}}(n+2, k_1) + I_{\text{CDMA}}(n+2, k_2) \end{aligned} \tag{19}$$

where $I_{\text{CDMA}}(n, i)$ is defined by (14). Note that the cells in the CDMA cluster would not contribute analog cochannel interference. Thus the number of tier-1 analog cochannel cells is five for the cells in up to ring-3, and six for the cells in

TABLE II
COORDINATE VALUES FOR (19)

n	(n, i_1)	(n, i_2)	$(n+1, j_1)$	$(n+1, j_2)$	$(n+1, j_3)$	$(n+2, k_1)$	$(n+2, k_2)$
1 (w)	(1, 0)	(1, 1)	(2, 0)	(2, 1)	(2, 2)	(3, 1)	(3, 2)
2 (x)	(2, 0)	(2, 1)	(3, 0)	(3, 1)	(3, 2)	(4, 1)	(4, 2)
3 (y)	(3, 1)	(3, 2)	(4, 1)	(4, 2)	(4, 3)	(5, 2)	(5, 3)
4 (z)	(4, 1)	(4, 2)	(5, 1)	(5, 2)	(5, 3)	(6, 2)	(6, 3)

TABLE III
PERFORMANCE DEGRADATION DUE TO A CDMA CELL CLUSTER*

	Δ [dB]			
n	$S_c = S_f$	$S_c = 0.5 \times S_f$	$S_c = 0.1 \times S_f$	Remark
1	5.47	3.39	0.42	ring-1 cell (w)
2	−0.025	−0.39	−0.71	ring-2 cell (x)
3	−0.59	−0.69	−0.77	ring-3 cell (y)
4	0.06	0.03	0.006	ring-4 cell (z)

* The negative values of Δ for $n = 2, 3$, which imply a performance improvement, are due to that the cells in ring-2 and -3 have 5 tier-1 cochannel cells instead of 6, since the CDMA cluster does not contribute analog cochannel interferences.

ring-4 and beyond. Thus

$$\frac{C}{I} = \begin{cases} \dfrac{(C/I)_A}{\frac{5}{6} + \frac{(D/R)^4}{6} \frac{1}{S_f} I'_{\text{CDMA}}} & n = 1, 2, 3 \\ \dfrac{(C/I)_A}{1 + \frac{(D/R)^4}{6} \frac{1}{S_f} I'_{\text{CDMA}}} & n \geq 4. \end{cases} \tag{20}$$

In this case, the performance degradation due to the CDMA cluster becomes

$$\Delta = \begin{cases} 10 \log \left(\frac{5}{6} + \frac{(D/R)^4}{6} \frac{1}{S_f} I'_{\text{CDMA}} \right) & n = 1, 2, 3 \\ 10 \log \left(1 + \frac{(D/R)^4}{6} \frac{1}{S_f} I'_{\text{CDMA}} \right) & n \geq 4. \end{cases} \tag{21}$$

The coordinate values of the cells used for the evaluation of (19) are shown in Table II, and the numerical values for the performance degradation due to the CDMA cluster, Δ, are shown in Table III.

The worst case performance degradation of 5.47 dB in ring-1 cells would not be acceptable, thus the analog channels within the CDMA band should not be used in ring-1 cells. Again, if we assume that less than 1 dB of degradation would be acceptable, we see that one ring of guard zone around the CDMA cell cluster would be required.

V. CONCLUSION

In this paper we have discussed some engineering issues that have to be taken into consideration when designing and introducing a digital CDMA cellular system. Specifically, the focus has been on the frequency reuse efficiency, and the impact of a CDMA channel on the performances of the analog channels. Comparing the frequency reuse efficiency of the CDMA and the traditional FDMA systems, it has been shown that the cellular approach is significantly more efficient with the CDMA system. Analyses, in terms of C/I ratio, have been carried out to discuss the performance degradation that can result from the presence of a CDMA channel. The analysis results show that the interference to the analog channels in cells closest to the CDMA cell would be too large to guarantee the desired performance of $C/I = 18$ dB. This suggests that when a cell (or a cluster of cells) is converted so as to employ a CDMA channel, there should be at least one ring of buffer around the cell/cluster, in which the analog channels falling into the CDMA band are not reused.

VI. APPENDIX I

We evaluate the following integral that is in (5):

$$A = \int_0^R dr r^5 B(r) \tag{22}$$

where

$$B(r) = \int_0^\pi \frac{d\theta}{(d^2 + r^2 + 2dr\cos\theta)^2} = \pi \frac{d^2 + r^2}{(d^2 - r^2)^3}. \tag{23}$$

Thus

$$\begin{aligned} A &= \pi \int_0^R dr \frac{d^2 r^5 + r^7}{(d^2 - r^2)^3} \\ &= \pi \left[\frac{-(d^6 + r^6) + 5d^2 r^4 - 2d^4 r^2}{2(d^2 - r^2)^2} - 2d^2 \ln(d^2 - r^2) \right] \Bigg|_{r=0}^{r=R} \\ &= \pi \left[2d^2 \ln\left(\frac{d^2}{d^2 - R^2} \right) - \frac{R^2(4d^4 - 6R^2 d^2 + R^4)}{2(d^2 - R^2)^2} \right]. \end{aligned} \tag{24}$$

When $d = kR$, this becomes

$$A = \pi R^2 \left[2k^2 \ln\left(\frac{k^2}{k^2 - 1} \right) - \frac{4k^4 - 6k^2 + 1}{2(k^2 - 1)^2} \right]. \tag{25}$$

ACKNOWLEDGMENT

The author is grateful to Wesley L. Shanks for his helpful comments.

REFERENCES

[1] Telecommunications Industry Association, "Dual-Mode Mobile Station—Base Station Compatibility Standard," Project Number 2398, Oct. 5, 1990.
[2] Qualcomm, Inc., *CDMA Digital CAI Standard, Draft Revision 1.1*, 1991.
[3] K. S. Gilhousen, I. M. Jacobs, R. Padovani, A. J. Viterbi, L. A. Weaver, and C. E. Wheatley, "On the capacity of a cellular CDMA system," *IEEE Trans. Veh., Tech.*, vol. 40, pp. 303–312, May, 1991.
[4] K. K. Ho, "Architectural design of a code division multiple access cellular system," in *Proc. IEEE 42nd VTS Conf.*, vol. 1, Denver, CO, May 10–13, 1992, pp. 47–50.
[5] C. K. Kwabi, M. P. McDonald, L. N. Roberts, W. L. Shanks, N. P. Uhrig, and C. J. Wu, "Operational advantages of the AT&T CDMA cellular system," in *Proc. IEEE 42nd VTS Conf.*, vol. 1, Denver, CO, May 10–13, 1992, pp. 233–235.
[6] K. I. Kim, "On the error probability of a DS/SSMA system with a noncoherent M-ary orthogonal modulation," in *Proc. IEEE 42nd VTS Conf.*, vol. 1, pp. 482–485, Denver, CO, May 10–13, 1992.
[7] W. C. Y. Lee, *Mobile Cellular Telecommunications Systems*. New York: McGraw-Hill, 1989.
[8] V. H. MacDonald, "The cellular concept," *The Bell System Tech. J.*, vol. 58, pp. 15–41, Jan. 1979.
[9] Qualcomm, Inc., *Notes for CDMA Technology Forum*, San Diego, CA, Jan. 16–17, 1992.

Kyoung Il Kim (S'84–M'86–SM'90) received the B.S.E. and M.S.E. degrees in electronic engineering from the Seoul National University, Seoul, Korea, in 1977 and 1982, respectively. He received the Ph.D. degree in electrical engineering from The University of Texas at Austin in 1986.

From 1977 to 1982 he was a Member of Research Staff at the Agency for Defense Development, Daejeon, Korea, where he worked on airborne telemetry systems. From 1986 to 1990 he was an Assistant Professor of Electrical Engineering at the University of Nevada, Las Vegas. In 1990 he joined AT&T Bell Laboratories, Whippany, NJ, where he is currently a Member of Technical Staff in wireless base station systems engineering department. His technical interests are concerned with the modeling and analysis of communication systems, and the development of digital signal processing algorithms and systems.

Overlay of Cellular CDMA on FSM

Konstandinos G. Filis and Someshwar C. Gupta

***Abstract*—The spectrum sharing capability of a CDMA PCN (48 MHz bandwidth), overlayed on the fixed service microwave (FSM) band (1850–1990 MHz), is investigated. This overlay may cause two types of interference: 1) interference from the overlayed PCN on the FSM systems, 2) interference from the microwave system to the PCN users. In the first case, both analog and digital systems are considered, because they are both in operation in the above frequency band. Furthermore, for the digital microwave system performance degradation, most of the modulation methods employed for digital transmission are considered. In the second case, the performance degradation of both the forward and the reverse PCN links are investigated, and, for the forward link, two options are considered: 1) incorporating power control, and 2) a no-power-control scheme. Next, real Fourier transformers and notch filters in the PCN receiver and transmitter are employed, in order to minimize both types of interference, and the above analysis is repeated.**

I. Introduction

About six years ago, in 1987, a new communication system was proposed by Dr. D. Cox [1], called Universal Personal Communications. Since then, a lot of new ideas and alternatives have been proposed concerning the new system, but the main objectives remain the same, including 1) tetherless, private, sequre, and good quality portable communications everywhere via the use of small and lightweight handsets, 2) efficient use of radio frequencies, 3) reliable and economical services. The universal services that may be provided by this system are usually called personal communications services (PCS), and the network providing these services is called the personal communications network (PCN). It consists of many base stations, arranged in a cellular configuration, each serving a cell of variable size. The cell size may vary from very small to very large according to the traffic requirements in each area. In this study we consider a small-cell PCN (183-m radius [1]) in an urban environment with large buildings.

Multiple access can be provided by means TDMA, FDMA, or CDMA. In this paper we use direct sequence broadband code division multiple access (DS BCDMA). This technique encompasses, theoretically, all the spread spectrum (SS) advantages including antijaming, antiinterference, low probability of intercept, no hard limit on the number of active users, privacy, system access anytime with no blocking calls, and spectrum sharing with conventional systems [2]–[5]. In this paper we investigate the spectrum sharing capability of a CDMA PCN (48-MHz bandwidth), overlayed on the fixed service microwave (FSM) band (1850–1990 MHz). This band is divided into a number of line of sight (LOS) channels employing analog and digital systems of a 10-MHz, or less, bandwidth.

This overlay may cause two types of interference: 1) interference from the overlayed PCN on the FSM systems, 2) interference from the microwave system to the PCN users. In the first case, we consider both analog and digital systems, because they are both in operation in the above frequency band. Furthermore, for the digital microwave system performance degradation we consider most of the modulation methods employed for digital transmission.

In the second case, we investigate the performance degradation of a CDMA PCN due to the interference produced by the FSM signal. We also consider the cochannel (CDMA) interference, within the SS system. We examine both the forward and the reverse PCN links, and, for the forward link, we consider two options: one incorporating power control, and a no-power-control scheme.

Next, a CDMA PCN transmitter is designed, employing filtered pseudonoise (PN) sequences that reduces the interference to the FSM systems. Real Fourier transformers and notch filters are used to filter only the PN sequences.

Finally, a CDMA PCN receiver is designed, employing filtered PN sequences, which reduces the interference from the FSM systems. The performance of the overlayed filtered PCN under FSM interference is evaluated in terms of the CDMA capacity reduction. Both the forward and the reverse links of the filtered system are studied.

II. Propagation in the Mobile Radio Environment

The two-ray-path model for free-space transmission is depicted in Fig. 1. The received power at the mobile receiver is given by [4]:

$$P_r = P\left(\frac{\lambda}{4\pi d}\right)^2 |1 + R(\phi)e^{j\Delta}|^2 = P\left(\frac{\lambda}{4\pi d}\right)^2 \left\{[R(\phi)+1]^2 - 4R(\phi)\sin^2\frac{\Delta}{2}\right\} \quad (1)$$

where

$$R(\phi) = \frac{\varepsilon \sin\phi - \sqrt{\varepsilon - \cos^2\phi}}{\varepsilon \sin\phi - \sqrt{\varepsilon - \cos^2\phi}} \quad (2)$$

and Δ is the phase difference between the two waves given by

$$\Delta = \frac{2\pi}{\lambda}\Delta d = \frac{2\pi}{\lambda}[\sqrt{(h_b+h_m)^2+d^2} - \sqrt{(h_b-h_m)^2+d^2}] \quad (3)$$

$$\phi = \tan^{-1}\frac{h_b+h_m}{d}. \quad (4)$$

Manuscript received February 12, 1993; revised April 21, 1993.

The authors are with the Department of Electrical Engineering, Southern Methodist University, Dallas, TX 75275.

IEEE Log Number 9212487.

0018-9545/94$04.00 © 1994 IEEE

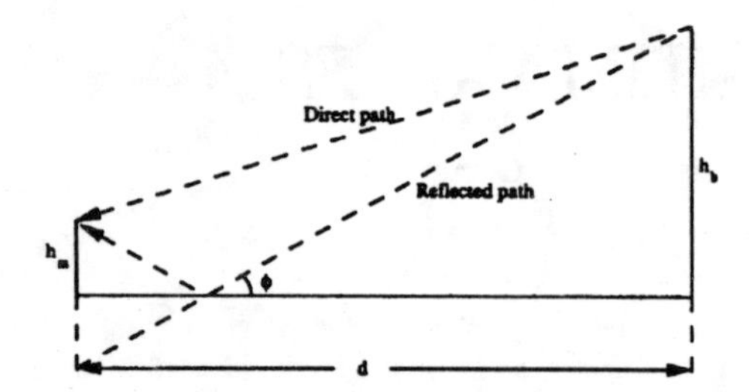

Fig. 1. Two-ray-path model for free-space transmission.

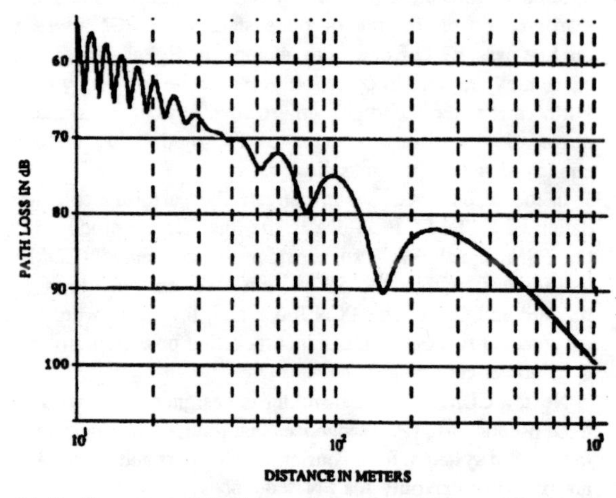

Fig. 2. Free space path loss between 7.6-m base and 1.5-m mobile. $\varepsilon = 15$.

Δd is the path difference between the direct and reflected paths, h_b and h_m are the base and the mobile antenna heights, r is the distance between the two antennas, ϕ is the angle of incidence, P is the transmitted power from the base, λ is the carrier signal wavelength, and ε is the dielectric constant.

For the derivation of (1), no approximations have been made; therefore, it can be used for all distances. Fig. 2 is the computer evaluation of (1), assuming 1.5 and 7.6 m mobile and base antenna heights. The path loss model of Fig. 2 is used in the estimation of FSM interference on the PCN forward link with power control (see Section VI-A(2).

III. CDMA PCN and FSM—Description of the Systems

A. Description of CDMA

The CDMA signal is given by [2]

$$C(t) = \sum_{k=1}^{M} \sqrt{2P_k}\, d_k(t-\tau_k) p_k(t-\tau_k) \cos(\omega_0 t + \theta_k) \quad (5)$$

where M is the number of simultaneous PCN users, P_k is the received power of each signal, τ_k is the time delay of the kth signal uniformity distributed into $[0, T]$, θ_k is the phase angle of each signal uniformly distributed on $[0, 2\pi]$, d_k is a sequence of nonoverlapping rectangular pulses each of amplitude +1 or -1, and T seconds duration, and ω_0 is the carrier frequency of the signal. p_k is the spreading code of the kth signal given by

$$p_k(t) = \sum_{i=-\infty}^{\infty} p_{ik}\xi(t - iT_s) \quad (6)$$

where p_{ik} is one chip of the random binary sequence $\{p_{ik}\}$, independent of $\{p_i\}$ for every k, which consists of statistically independent symbols, each taking the value +1 or -1 with equal probability, T_s is the spreading code chip duration, and ξ is the chip waveform, assumed rectangular in this paper.

The PSD of the spreading sequence is the following line spectrum [2]:

$$S_p(f) = \sum_{\substack{k=-\infty \\ k\neq 0}}^{\infty} \delta\left(f - k\frac{f_s}{L}\right)\frac{L+1}{L^2} \cdot \left(\frac{\sin \pi f/f_s}{\pi f/f_s}\right)^2 + \frac{1}{L^2}\delta(f) \quad (7)$$

where f_s is the spreading code rate, and $L = 2^r - 1$, where r is the number of shift registers used to generate the sequence. For all practical purposes, the PSD of the total CDMA signal, for the case $1/T \leq f_s/2L$, can be approximated as in [8], [13] by

$$S_C(f) = \begin{cases} \sum_{i=1}^{M} \frac{P_i T}{2}\frac{L+1}{L^2} \cdot \left\{\left[\frac{\sin \pi(f-f_0)/f_s}{\pi(f-f_0)/f_s}\right]^2 + \left[\frac{\sin \pi(f+f_0)/f_s}{\pi(f+f_0)/f_s}\right]^2\right\}, & |f_0| + \frac{f_s}{2L} < f < |f_0| - \frac{f_s}{2L} \\ \sum_{i=1}^{M} \frac{P_i T}{2L^2} \cdot \left\{\left[\frac{\sin \pi(f-f_0)T}{\pi(f-f_0)T}\right]^2 + \left[\frac{\sin \pi(f+f_0)T}{\pi(f+f_0)T}\right]^2\right\}, & f \in \left[|f_0| - \frac{f_s}{2L}, |f_0| + \frac{f_s}{2L}\right]. \end{cases} \quad (8)$$

The PCN considered in this paper has a cellular configuration of radius 183 m and employs mobile units with a transmitted power of 100 μW. The mobile antenna height range is 1.5 to 2.2 m, and the base antenna one is 7.6 to 9.15 m.

IV. Description of FSM

The FSM signal $i(t)$ is caused by a microwave LOS radio link, which can be either analog or digital, and, if digital, can employ any kind of modulation technique (most of the times M-PSK or M-QAM). The only restrictions we impose on $i(t)$ are that it is independent of the CDMA signal and that its time averaged power is bounded:

$$\frac{1}{T_i}\int_0^{T_i} i^2(t)\, d(t) \leq P_i \quad (9)$$

where P_i indicates a finite number, and T_i is FSM data bit duration.

TABLE I
CDMA INTERFERENCE RECEIVED BY FSM ANTENNA ON BORESIGHT

d (m)	PL (dB)	G (dB)	NUMBER OF CELLS	CDMA INTERFENCE - 10 log M (dBm)
183	102	4	1	- 162.3
549	116.5	10	6	- 163
915	123	19	12	- 157.5
1281	127.5	21.5	18	- 157.8
1647	131	28	24	- 153.5
2013	133	29	30	- 153.6
2379	136	30	36	- 154.8
2745	137	30	42	- 155.1
3111	139	30.5	48	- 156

The normalized PSD of the FSM signal will be given by

$$S_i(f) = \begin{cases} 1, & |f \pm f_i| \leq \frac{(1-\alpha)}{2T_i} \\ \frac{1}{2}\left\langle 1 + \cos\left\{\frac{\pi[2(f \pm f_i)T_i - 1 + \alpha]}{2\alpha}\right\}\right\rangle, & \frac{(1-\alpha)}{2T_i} \leq |f \pm f_i| \leq \frac{(1+\alpha)}{2T_i} \\ 0, & \text{elsewhere} \end{cases} \quad (10)$$

where α is the cosine rolloff factor which takes values between 0, and 1 [7], and f_i is the carrier frequency of the microwave transmitter. The FSM antenna height is considered to be 127 m, and its transmitted power 27 dBm. For the path loss between the antennas of the two systems we use the data of [4] for an urban area with large buildings.

V. INTERFERENCE ON FSM CAUSED BY THE CDMA PCN OVERLAY

We consider M simultaneous users per cell, uniformly distributed on circles around the FSM antenna. In order to find the number of PCN interfering users, we consider an angle of $\pm 0.5°$ along the boresight, and radii from 183 m to 3111 m in steps of 366 m according to Table I. The CDMA interference in Table I corresponds to 100 μW transmitting-power-mobile-units, d is the distance between the microwave antenna and the mobile PCN units, PL is the path loss between the microwave and the PCN mobile antennas, G is the microwave antenna gain on the boresight for each particular distance [4], and we have already subtracted the 28.8 dB PG of the CDMA system ($L = 750$).

A. CDMA Interference on Analog Microwave Systems

In order to evaluate the performance degradation of analog FSM systems due to the CDMA overlay, we use a simple FM receiver, consisting of a predestortion filter, a discriminator, and a postdetection filter. The receiver is corrupted by CDMA interference and AWGN. The SNR at the output of this receiver is given by (11) below [8], where N is the number of interfering PCN users, $n(t)$ is the AWGN with one-sided PSD of N_0, f_D is the frequency deviation constant, K_D is the discriminator constant, $m(t)$ is the FM message, P_F is the received power of the FM signal, W is the bandwidth of the low pass filter, and f_1 is the carrier frequency separation of the two systems.

For simulation, we consider a single sideband supressed carrier (SSSC-FM) microwave line-of-site radio link where each channel is limited to a 4-kHz bandwidth, and 600 channels are multiplexed and transmitted on a single RF carrier. The peak RF carrier frequency deviation is 4-MHz and a 10-MHz IF bandwidth is used to accomodate this deviation. The frequency deviation for each channel is given by [9]:

$$f_D = Kf_j = KjW, \qquad j = 1, \cdots, v \quad (12)$$

where v is the number of channels,

$$K = \left[\frac{2f_{SSSC}^2}{3W^2 v(v+1)(2v+1)}\right]^{1/2} \quad (13)$$

and f_{SSSSC} is the peak frequency deviation of the SSSC-FM system RF carrier. Each of the messages is assumed scaled, so that $\overline{m^2(t)} = 1$. The PCN, under consideration, employs 100 μW, transmitted power handsets, 48-MHz system bandwidth, and 366-m cell diameter.

$$\mathrm{SNR}_o = \frac{f_D^2 \overline{m^2(t)} P_F}{\dfrac{\sum_{i=1}^{N} P_i T(L+1) K_D^2}{2L^2} \displaystyle\int_{-W}^{W} f^2 \left[\frac{\sin \pi(f - f_1)/f_s}{\pi(f - f_1)/f_s}\right]^2 df + \frac{N_0 W^3}{3}} \quad (11)$$

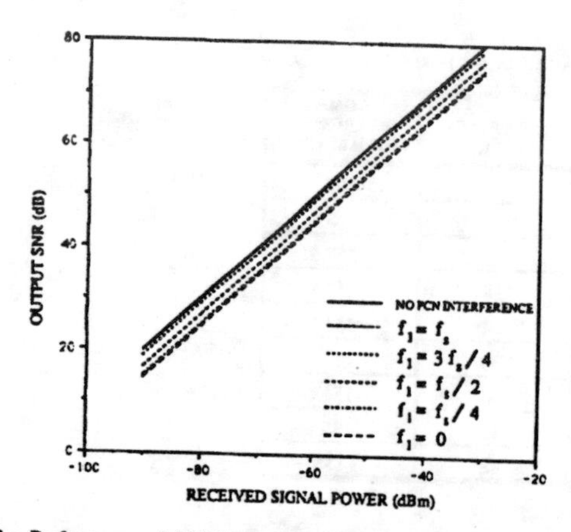

Fig. 3. Performance of SSSC-FM FDM with overlayed PCN serving 150 100 μW simultaneous users. f_1 is the carrier frequency separation between the two systems. $f_s = 24\,10^6$ Hz, $L = 750, 1/T = 32\,000$ bps.

For this system, the output SNR of the jth channel will be

$$\text{SNR}_{oj} = \frac{K^2W^2P_F}{\dfrac{\sum_{i=1}^{N} P_iT(L+1)}{2L^2j^2}\Lambda + N_0W^3}, \qquad j = 1, \cdots, v \tag{14}$$

where

$$\Lambda = \int_{W(j-1/2)}^{W(j+1/2)} f^2 \left[\frac{\sin \pi(f - f_1)/f_s}{\pi(f - f_1)/f_s}\right]^2 df + \int_{-W(j+1/2)}^{-W(j-1/2)} f^2 \left[\frac{\sin \pi(f - f_1)/f_s}{\pi(f - f_1)/f_s}\right]^2 df \tag{15}$$

The results are depicted in Fig. 3 for 150 simultaneous PCN users and different values of frequency separation.

B. CDMA Interference on Digital Microwave Systems

For evaluating the performance degradation of digital microwave systems, the CDMA interference is again taken from Table I. In order for us to be able to use the well-established formulas for the probability of error in AWGN, for the systems under consideration, we need to treat the CDMA interference as white noise. Of course, this is a questionable approximation, but the reasonably flat spectrum of the CDMA PSD in the middle, together with the fact that its bandwidth is five times larger than the bandwidth of the microwave systems, gives us a certain degree of confidence that we can get some measure of the digital microwave performance degradation because of the availability of the probability of error formulas in AWGN.

1) Performance Degradation of m-PSK FSM due to CDMA Overlay: For $m = 2$, the bit error probability (BEP) for the 2-level phase shift keying (PSK) system is computed by (16) on the following page [10]. For $m = 4$, the symbol error probability (SEP) is shown in (17). For $m > 4$, the SEP is computed by (18) on the next page where B_i is the bandwidth of the microwave signal, and E_b is its energy per bit. The results are depicted in Fig. 4 for $m = 8$, for 150 simultaneous PCN users per cell, and for several values of frequency separation.

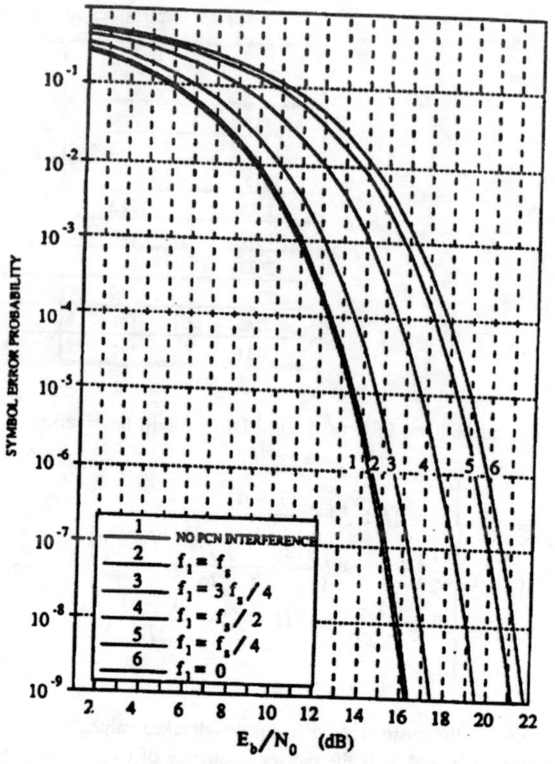

Fig. 4. Performance of 8-PSK with overlayed PCN serving 150 100 μW simultaneous users. f_1 is the carrier frequency separation between the two systems. $f_s = 24\,10^6$ Hz, $L = 750, 1/T = 32\,000$ bps.

2) Performance Degradation of m-QAM FSM due to CDMA Overlay: The SEP for the m-level quadrature amplitude modulation (QAM) is given by (19) on the folowing page. The results are depicted in Fig. 5 for $m = 64$.

VI. Interference on CDMA PCN Caused by FSM

The total signal at the input of the CDMA PCN receiver will be

$$x(t) = s(t) + c(t) + i(t) + n(t) \tag{20}$$

where $s(t)$ is the desired SS signal, $c(t)$ is the CDMA interference, $i(t)$ is the FSM interference, and $n(t)$ is the white noise. The coherent receiver multiplies this signal by $2p(t)\cos(\omega_0 t + \theta)$ and then filters the result in a band $1/2T$ (double sided). $p(t)$ is the spreading code of the desired signal, given by (6). After the multiplication we will have

$$\begin{aligned} y(t) &= 2[s(t)p(t) + c(t)p(t) + i(t)p(t) + n(t)p(t)] \\ &\quad \cdot \cos(\omega_0 t + \theta) \\ &= y_1(t) + y_2(t) + y_3(t) + y_4(t). \end{aligned} \tag{21}$$

Since the terms in the above summation are independent, the PSD of $y(t)$, $S_y(f)$, will be the sum of the PSD's of each individual term. Also, the autocorrelation function of each term will be the product of the autocorrelation functions of each input signal and the one of the receiver spreading sequence. Therefore, the PSD of each term will be the convolution of the PSD's of each input signal and the one of the receiver spreading sequence. The interfering PSD's, at the correlator input, will be given by

$$S_2(f) = S_c(f) * S_p(f) = \int_{-\infty}^{\infty} S_c(v)S_p(f-v)\,dv \quad (22)$$

which is the PSD of CDMA interference,

$$S_3(f) = S_i(f) * S_p(f) = \int_{-\infty}^{\infty} S_i(v)S_p(f-v)\,dv \quad (23)$$

which is the PSD of FSM interference,

$$S_4(f) = S_n(f) * S_p(f) = \int_{-\infty}^{\infty} S_n(v)S_p(f-v)\,dv \quad (24)$$

which is the PSD of AWGN interference. Substitution of (7)–(8) into (22) yields the PSD of the interference caused by the CDMA signal at the input of the correlator of the SS receiver, $S_2(f)$:

$$S_2(f) = \sum_{\substack{k=-\infty \\ k\neq 0}}^{\infty} \sum_{i=1}^{M-1} P_i T \frac{(L+1)^2}{2L^4} \left[\frac{\sin \pi \left(f - k\frac{f_s}{L}\right)\Big/ f_s}{\pi\left(f - k\frac{f_s}{L}\right)\Big/ f_s} \right]^2 \cdot \left[\frac{\sin \pi \left(k\frac{f_s}{L}\right)\Big/ f_s}{\pi\left(k\frac{f_s}{L}\right)\Big/ f_s} \right]^2 + \sum_{i=1}^{M-1} \frac{P_i T}{2L^4}\left(\frac{\sin \pi f/f_s}{\pi f/f_s}\right)^2. \quad (25)$$

The PSD of the interference (normalized) caused by the FSM signal at the input of the correlator $S_3(f)$ can be found by substituting (7) and (10) into (23):

$$S_3(f) = S_{3,1}(f) + S_{3,2}(f) \quad (26)$$

where

$$S_{3,1}(f) = \begin{cases} \dfrac{A}{L^2}, & |f| \le \dfrac{(1-\alpha)}{2T_i} \\ \dfrac{A}{2L^2}\left\langle 1 + \cos\left\{\dfrac{\pi[2(f)T_i - 1 + \alpha]}{2\alpha}\right\}\right\rangle, & \dfrac{(1-\alpha)}{2T_i} \le |f| \le \dfrac{(1+\alpha)}{2T_i} \\ 0, & \text{elsewhere} \end{cases} \quad (27)$$

and

$S_{3,2}(f) =$

$$P_e = \frac{1}{2}\operatorname{erfc}\left(\frac{E_b B_i L^2}{\sum_{i=1}^{N} P_i T(L+1) \int_{f_1 - B_i/2}^{f_1 + B_i/2} \left(\frac{\sin \pi f/f_s}{\pi f/f_s}\right)^2 df + B_i L^2 N_0} \right)^{1/2}. \quad (16)$$

$$P_e = \left[1 - \frac{1}{2}\operatorname{erfc}\left(\frac{E_b B_i L^2}{\sum_{i=1}^{N} P_i T(L+1) \int_{f_1 - B_i/2}^{f_1 + B_i/2} \left(\frac{\sin \pi f/f_s}{\pi f/f_s}\right)^2 df + B_i L^2 N_0} \right)^{1/2} \right]^2. \quad (17)$$

$$P_e = \operatorname{erfc}\left[\left(\frac{(\log_2 m) E_b B_i L^2}{\sum_{i=1}^{N} P_i T(L+1) \int_{f_1 - B_i/2}^{f_1 + B_i/2} \left(\frac{\sin \pi f/f_s}{\pi f/f_s}\right)^2 df + B_i L^2 N_0} \right)^{1/2} \sin\frac{\pi}{m} \right] \quad (18)$$

$$P_e = 1 - \left[1 - \left(1 - \frac{1}{\sqrt{m}}\right)\operatorname{erfc}\left(\frac{1.5(m-1)^{-1}(\log_2 m) E_b B_i L^2}{\sum_{i=1}^{N} P_i T(L+1) \int_{f_1 - B_i/2}^{f_1 + B_i/2} \left(\frac{\sin \pi f/f_s}{\pi f/f_s}\right)^2 df + B_i L^2 N_0} \right)^{1/2} \right]^2. \quad (19)$$

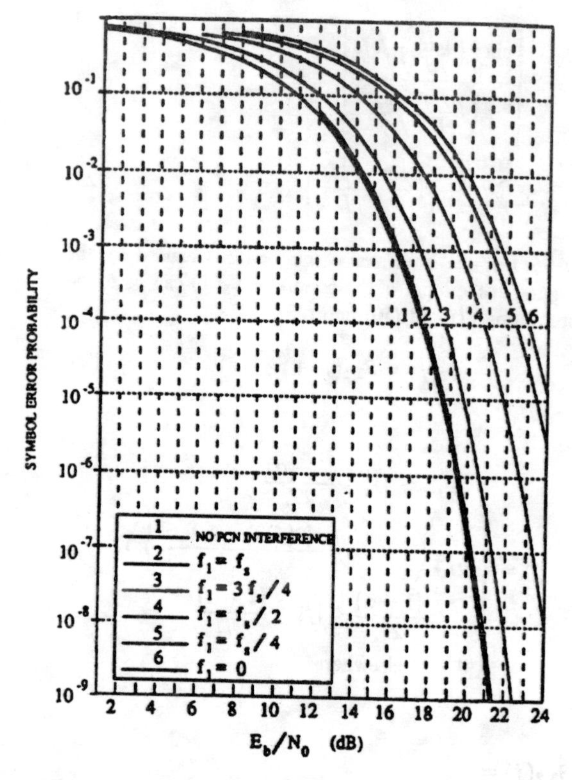

Fig. 5. Performance of 64-QAM with overlayed PCN serving 150 100 μW simultaneous users. f_1 is the carrier frequency separation between the two systems. $f_s = 24\,10^6$ Hz, $L = 750, 1/T = 32\,000$ bps.

$$\begin{cases} \sum_{\substack{k=-\infty \\ k\neq 0}}^{\infty} A\frac{L+1}{L^2}\left[\frac{\sin\pi\left(k\frac{f_s}{L}\right)/f_s}{\pi\left(k\frac{f_s}{L}\right)/f_s}\right]^2, & \left|f \pm k\frac{f_s}{2L}\right| \le \frac{(1-\alpha)}{2T_i} \\ \sum_{\substack{k=-\infty \\ k\neq 0}}^{\infty} \frac{A}{2}\frac{L+1}{L^2}\left[\frac{\sin\pi\left(k\frac{f_s}{L}\right)/f_s}{\pi\left(k\frac{f_s}{L}\right)/f_s}\right]^2 \cdot\left\langle 1+\cos\left\{\frac{\pi\left[2\left(f\pm k\frac{f_s}{L}\right)T_i - 1 + \lambda\right]}{2\alpha}\right\}\right\rangle, & \frac{(1-\alpha)}{2T_i} \le \left|f\pm k\frac{f_s}{L}\right| \le \frac{(1+\alpha)}{2T_i} \\ 0, & \text{elsewhere.} \end{cases} \tag{28}$$

Finally, the PSD of the interference caused by the AWGN at the correlator input will be

$$S_4(f) = \sum_{\substack{k=-\infty \\ k\neq 0}}^{\infty} \frac{N_0(L+1)}{2L^2}\left[\frac{\sin\pi\left(k\frac{f_s}{L}\right)/f_s}{\pi\left(k\frac{f_s}{L}\right)/f_s}\right]^2 + \frac{N_0}{2L^2}. \tag{29}$$

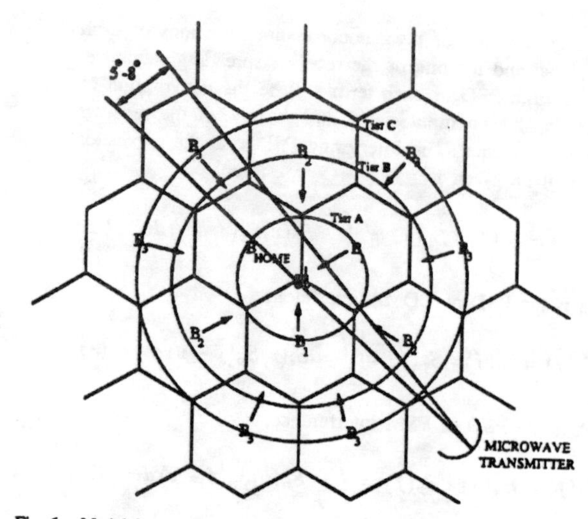

Fig. 6. Model for the performance evaluation of the PCN forward link.

The total interfering power at the output of the CDMA receiver will be

$$P_{to} = \underbrace{\int_{-1/2T}^{1/2T} S_2(f)\,df}_{\text{CDMA interference}} + \underbrace{\int_{-1/2T}^{1/2T} S_3(f)\,df}_{\text{FSM interference}} + \underbrace{\int_{-1/2T}^{1/2T} S_4(f)\,df.}_{\text{white noise}} \tag{30}$$

A. Performance Degradation of the PCN Forward Link Due to FSM Interference

1) Forward Link Without Power Control: For the forward link (interference to mobile) we consider the model of Fig. 6. The mobile receiver is located at the border point of three adjacent cells. This is considered the worst position for the mobile, because the signal coming from its home base is as weak as possible, while the signals coming from the two closest bases are as strong as they can get.

The desired power from the home base is -96 dBm, assuming PCN transmitting power of -10 dBm, and path loss -86 dBm, [4, fig. 3], between the base and mobile antenna at 183 m, in an area with large buildings.

The CDMA interference from the tier A bases will be $10\log(3M-1) - 96$ (dBm), where M is the number of simultaneous users per cell. The CDMA interference from tier B bases will be $10\log(3M) - 100$ (dBm) at a distance off 366 m, and the interference from tier C bases will be $10\log(6M) - 102$ (dBm).

The received interference from the FSM will be $27 - PL + G$ (dBm), where we assume microwave transmitting power of 27 dBm, and where PL, and G can be found from Table I, when the PCN receiver is located on the FSM antenna boresight. Thermal noise will have minimal effect on the total received interference.

In order to find the output interference, we substitute the above values in (25)–(28), and then we evaluate (30). The output SNR, in this case, assuming spreading codes of length

TABLE II
OUTPUT INTERFERING POWER FROM FSM AND PCN CAPACITY REDUCTION

d (m)	PL (dB)	G (dB)	I_o (dBm)	P/I_o (dB)	CAPACITY REDUCTION FOR 3dB SNR
183	102	4	- 99.7	3.7	85.2%
549	116.5	10	- 108.2	12	11.7%
915	123	19	- 105.7	9	21.42%
1281	127.5	21.5	- 107.7	11	13.26%
1647	131	28	- 104.7	8	27%
2013	133	29	- 105.7	9	21.43%
2379	136	30	- 107.7	11	13.26%
2745	137	30	- 108.7	12	10.7%
3111	139	30.5	- 110.2	14	7.65%

750, and PCN message rate of 32 000 bps, was calculated as

$$\mathrm{SNR}_o = \frac{2.512\,10^{-13}}{6.4\,10^{-16}M + I_o} \tag{31}$$

where I_o is the output interference power of the FSM signal given in (30) and tabulated as shown in Table II.

For the capacity reduction consider that maximum capacity for 3 dB SNR is 196 users. P is the desired signal received power. Next, using again the Gaussian approximation, we compute the forward link probability of error for a BPSK PCN receiver.

2) Forward Link With Power Control: We assume that the base distributes equal power to all mobile users. The users are located on circles $r_1, r_2, \cdots, R$ (R is the cell radius) around the base and they are uniformly distributed so that $M_j = kr^j$. From Fig. 2, under the worst conditions, the power reaching a mobile at the cell border will be

$$P_{rR} = P_{tbR}\left(\frac{0.15}{4\pi}\right)^2 R^{-2.5}. \tag{32}$$

The power received at a mobile on any other circle j is

$$P_{rj} = P_{tbj}\left(\frac{0.15}{4\pi}\right)^2 r_j^{-2.5} \tag{33}$$

where P_{tbR} and P_{tbj} are the transmitted powers from the base to reach R, and j located mobiles, respectively. Equations (32) and (33) must be equal. Therefore,

$$P_{tbj} = P_{tbR}\left(\frac{r_j}{R}\right)^{2.5}. \tag{34}$$

The total power transmitted from the base will be

$$P_{tb} = P_{tbR}k\int_0^R \frac{r^{3.5}}{R^{2.5}}dr = P_{tbR}\frac{M}{2.25}. \tag{35}$$

Therefore, the power control scheme of (34) increases the PCN capacity by a factor of 2.25. The probability of error is computed for a BPSK PCN receiver. The results are depicted in Fig. 7.

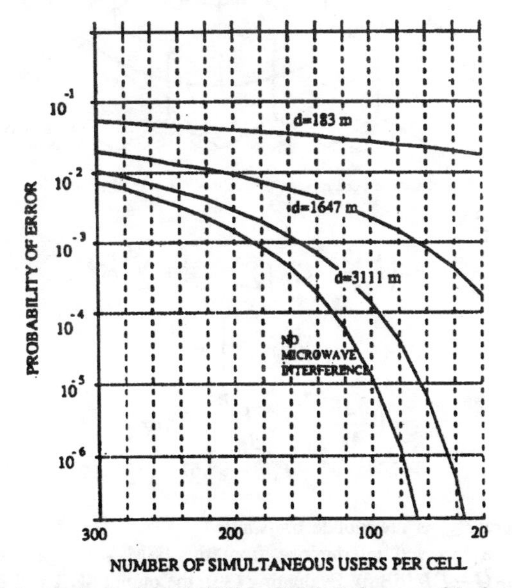

Fig. 7. Interference from the FSM on CDMA: Performance of the forward link, with power control, of a PCN, serving mobile units of 100 μW transmitting power each ($L = 750, f_s = 24\,10^6$ Hz, $1/T = 32\,000$ bps). d is the distance from the microwave antenna.

B. Performance Degradation of the PCN Reverse Link Due to FSM Interference

For the reverse link (interference to the base) we consider the model of Fig. 8. The desired power will be again −96 dBm. The home CDMA interference (tier A) will be 10 log$(M-1) - 96$ (dBm). From tiers B and C there are 12 interfering channels (6 from B, and 6 from C). There are, therefore, $12M$ simultaneous users that are assumed uniformly distributed so that $M_j = kr^j$.

The total CDMA interference at the base (from tiers B and C) is

$$P_{ib} = \sum^{M_R} P_R + \sum^{M_2} P_2 + \sum^{M_3} P_3 + \cdots + \sum^{M_{3R}} P_{3R}$$

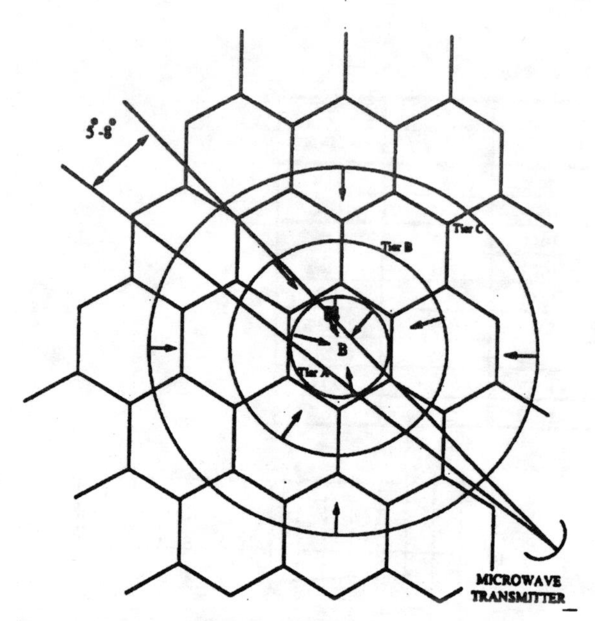

Fig. 8. Model for the performance evaluation of the PCN reverse link.

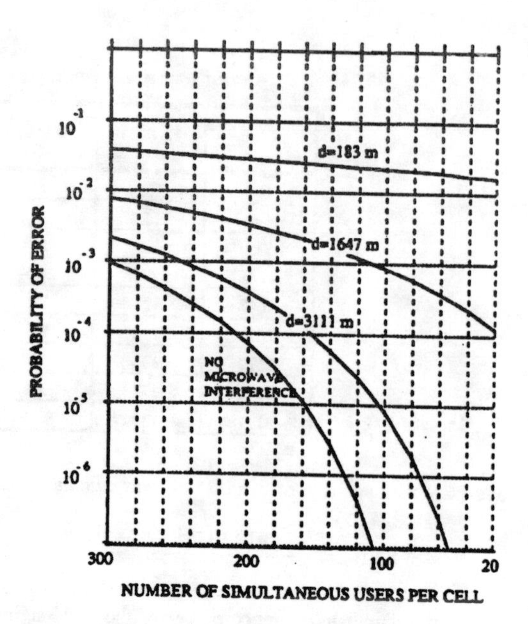

Fig. 9. Interference from the FSM on CDMA: Performance of the reverse link, with power control, of a PCN, serving mobile units of 100 μW transmitting power each ($L = 750, f_s = 24\,10^6$ Hz, $1/T = 32\,000$ bps). d is the distance from the microwave antenna.

$$= \left(\frac{0.15}{4\pi}\right)^2 P_{\rm tm}\left[\sum^{kr_R} r_R^{-3.3} + \sum^{kr_2} r_2^{-3.3} + \cdots + \sum^{kr_{3R}} r_{3R}^{-3.3}\right]$$

$$= k\left(\frac{0.15}{4\pi}\right)^2 P_{\rm tm}\int_R^{3R} r^{-2.3}\,dr$$

$$= 10\log M + P_{\rm tm} - 98.61 \quad ({\rm dBm}) \tag{36}$$

where $P_{\rm tm}$ is the mobile transmitted

The received interference from the FSM will again be $27 + G - PL$ (dBm). Evaluating (30), the output SNR will be

$$\text{SNR}_o = \frac{2.512\,10^{-13}}{1.737\,10^{-16}M + I_o} \tag{37}$$

where I_o is given by Table II. The performance of the reverse link is depicted in Fig. 9.

VII. Interference on FSM Caused by Filtered PCN

In this section, a surface acoustic wave (SAW) CDMA PCN transmitter is designed, employing notch filters and the real Fourier transformers of [11]–[12], which reduces the interference to the FSM systems. The model of the transmitter is depicted in Fig. 10. The PSD's of the unfiltered and filtered PN sequences are depicted at points 1 and 2, respectively. The filtered code is then multiplied by the data, and fed into BPSK transmitter. The resulting signal is a SS one, having a PSD resembling that at point 3 (with center frequency f_o). It can be seen that part of this PSD is almost zero. A proper selection of the notch center and width of the filter will produce a PSD that may reduce the interference to the FSM.

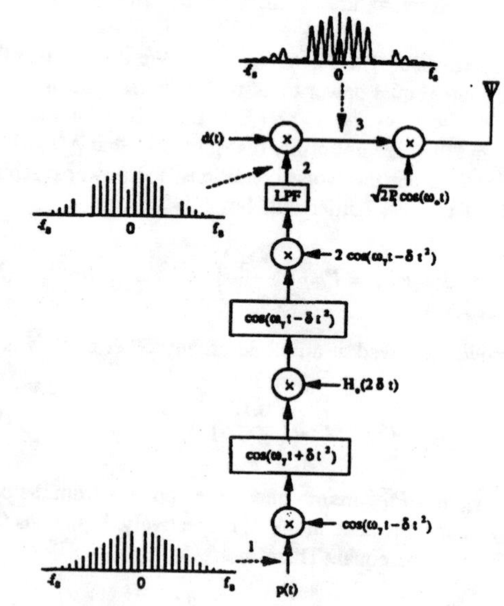

Fig. 10. Model of SS transmitter employing filtered spreading sequences and the associated PSD's.

The signal at point 2 will be [13]

$$p_F(t) = \int_{2\delta T}^{2\delta T_F} H_R(\omega)\{P_R(\omega)\cos[\omega t'] - P_I(\omega)\sin[\omega t']\}\,d\omega$$
$$= F^{-1}\{H_R(\omega)P(\omega)\} = h_R(t) * p(t) \tag{38}$$

where $t' = t - T_F, T_F$ is the length of the tapped delay line used to implement the transformer, [11]–[12], $\delta = \omega_s/2T_F =$

$\pi f_s/T_F, H(\omega)$, and $P(\omega)$ are the Fourier transforms of the notch filter and the PN sequence, respectively. Subscripts R and I denote real and imaginary parts. $H(\omega)$ is related to $H_c(2\delta\tau)$ by

$$H_c(2\delta t) = 4H_R(2\delta t)\cos(2\delta t T_F) + 4H_I(2\delta t)\sin(2\delta t T_F) \quad (39)$$

ω_γ is an arbitrary carrier frequency selected from the range $[2\delta T, 2\delta T_F]$. The notch filter is of the form

$$H_R(f) = \prod_{f_s}(f) - \prod_{\Delta B/2}(f - f_1) - \prod_{\Delta B/2}(f + f_1) \quad (40)$$

where

$$\prod_k(f) = \begin{cases} 1, & |f| \le k \\ 0, & \text{elsewhere} \end{cases} \quad (41)$$

and ΔB is the width of the notch.

A. Filtered CDMA Interference on Analog Microwave Systems

The SNR at the output of the jth channel of an SSSC-FM FSM is given by [13]

$$\text{SNR}_{Foj} = \frac{W^2 P_1 K^2}{\sum_{i=1}^{N} \frac{P_i T}{2} \frac{L+1}{j^2 L^2} \Lambda_F + N_0 W^3}, \quad j = 1, 2, \cdots, v \quad (42)$$

where

$$\Lambda_F = \int_{W(j-1/2)}^{W(j+1/2)} f^2 \left[\frac{\sin \pi(f - f_1)/f_s}{\pi(f - f_1)/f_s}\right]^2 H(f - f_1)\, df + \int_{-W(j+1/2)}^{-W(j-1/2)} f^2 \left[\frac{\sin \pi(f - f_1)/f_s}{\pi(f - f_1)/f_s}\right]^2 H(f + f_1)\, df \quad (43)$$

and the notch filter is given by (40) with $\Delta B = 8$ MHz. The results are depicted in Fig. 11 for 150 simultaneous filtered PCN users.

B. Filtered CDMA Interference on Digital Microwave Systems

1) Performance Degradation of m-PSK FSM: The probability of error for $m = 2$ is given by [13]

$$P_e = \frac{1}{2}\text{erfc}\left[\sum_{i=1}^{N} \frac{P_i T}{B_1} \frac{L+1}{E_b L^2} \left\{ \int_{f_1 - B_1/2}^{f_1 + B_1/2} \left[\frac{\sin \pi f/f_s}{\pi f/f_s}\right]^2 df - \int_{f_1 - \Delta B/2}^{f_1 + \Delta B/2} \left[\frac{\sin \pi f/f_s}{\pi f/f_s}\right]^2 df + \int_{f_1 - \Delta B/2}^{f_1 + \Delta B/2} [S_p(f) H_R(f)] * S_d(f) \right\} + \frac{N_0}{E_b}\right]^{-1/2}. \quad (44)$$

For $m = 4$, the symbol error probability is

$$P_e = \left[1 - \frac{1}{2}\text{erfc}\left(\sum_{i=1}^{N} \frac{P_i T}{B_1} \frac{L+1}{E_b L^2} \left\{ \int_{f_1 - B_1/2}^{f_1 + B_1/2} \left[\frac{\sin \pi f/f_s}{\pi f/f_s}\right]^2 df - \int_{f_1 - \Delta B/2}^{f_1 + \Delta B/2} \left[\frac{\sin \pi f/f_s}{\pi f/f_s}\right]^2 df + \int_{f_1 - \Delta B/2}^{f_1 + \Delta B/2} [S_p(f) H_R(f)] * S_d(f) \right\} + \frac{N_0}{E_b}\right)^{-1/2}\right]^2. \quad (45)$$

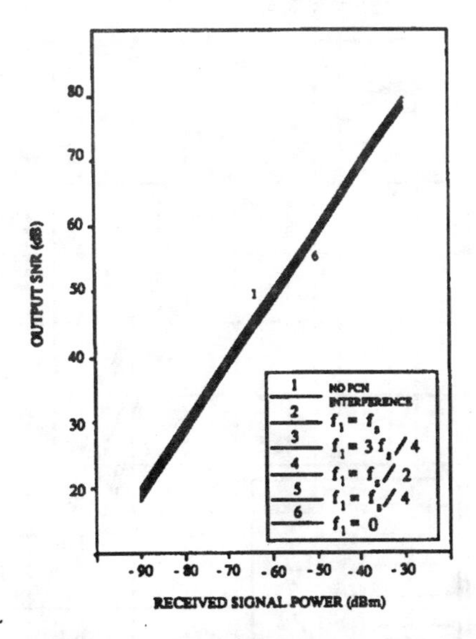

Fig. 11. Performance of SSSC-FM FDM with overlayed filtered PCN serving 150 100 μW simultaneous users. f_1 is the carrier frequency separation between the two systems. $f_s = 24\,10^6$ Hz, $L = 750, 1/T = 32\,000$ bps, notch center = f_1, notch width = 8 MHz.

For $m = 8$ the symbol error probability becomes

$$P_e = \text{erfc}\left(\sum_{i=1}^{N} \frac{P_i T}{0.44 B_1} \frac{L+1}{E_b L^2} \left\{ \int_{f_1 - B_1/2}^{f_1 + B_1/2} \left[\frac{\sin \pi f/f_s}{\pi f/f_s}\right]^2 df - \int_{f_1 - \Delta B/2}^{f_1 + \Delta B/2} \left[\frac{\sin \pi f/f_s}{\pi f/f_s}\right]^2 df + \int_{f_1 - \Delta B/2}^{f_1 + \Delta B/2} [S_p(f) H_R(f)] * S_d(f) \right\} + \frac{N_0}{0.44 E_b}\right)^{1/2}. \quad (46)$$

In the above equations, $S_p(f)$ is given by (7), $H_R(f)$ is given by (40), and $S_d(f)$ is the PSD of the PCN data usually given by $S_d(f) = T[\sin(\pi f T)/\pi f T]^2$. The results are depicted in Fig. 12 for $m = 8$, 150 simultaneous PCN users, employing notch filter of 8-MHz width and center frequency f_1.

2) Performance Degradation of m-QAM FSM: The SEP for m-QAM is given by [13]

$$P_e = 1 - \left[1 - 0.875\,\text{erfc}\left(\sum_{i=1}^{N} \frac{P_i T}{0.143 B_1} \frac{L+1}{E_b L^2} \cdot \left\{ \int_{f_1 - B_1/2}^{f_1 + B_1/2} \left[\frac{\sin \pi f/f_s}{\pi f/f_s}\right]^2 df - \int_{f_1 - \Delta B/2}^{f_1 + \Delta B/2} \left[\frac{\sin \pi f/f_s}{\pi f/f_s}\right]^2 df + \int_{f_1 - \Delta B/2}^{f_1 + \Delta B/2} [S_p(f) H_R(f)] * S_d(f) \right\} + \frac{N_0}{0.143 E_b}\right)^{1/2}\right]^2 \quad (47)$$

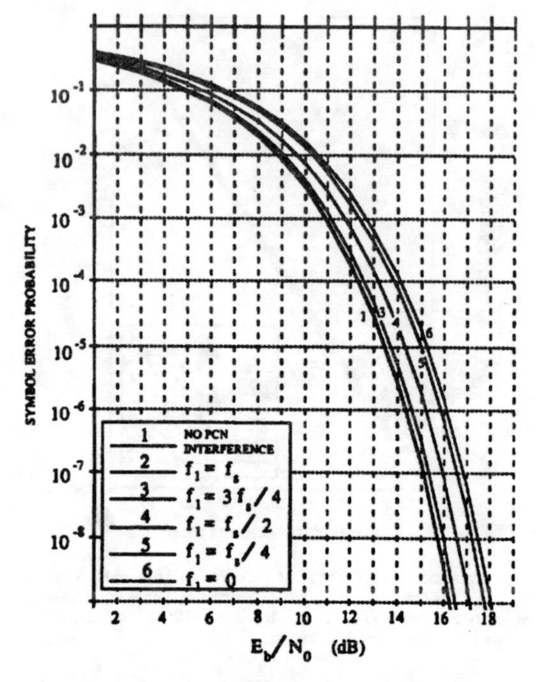

Fig. 12. Performance of 8-PSK with overlayed filtered PCN serving 150 100 μW simultaneous users. f_1 is the carrier frequency separation between the two systems. $f_s = 24\,10^6$ Hz, $L = 750, 1/T = 32\,000$ bps, notch center $= f_1$, notch width = 8 MHz.

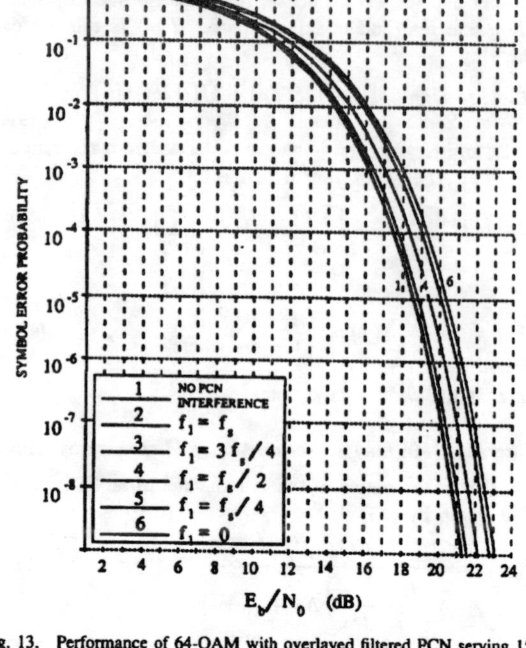

Fig. 13. Performance of 64-QAM with overlayed filtered PCN serving 150 100 μW simultaneous users. f_1 is the carrier frequency separation between the two systems. $f_s = 24\,10^6$ Hz, $L = 750, 1/T = 32\,000$ bps, notch center $= f_1$, notch width = 8 MHz.

The probability of error is depicted in Fig. 13, for $m = 64$.

VIII. Interference on Filtered CDMA PCN Caused by FSM

The model of Fig. 14 is considered. The total signal at the input of the CDMA PCN receiver, after frequency translation, will be

$$x(t) = a(t) + b(t) + i(t) + n(t) \tag{48}$$

where $a(t)$ and $b(t)$ are the filtered CDMA desired signal and interference. The interfering signal at the receiver output will be [13]

$$\begin{aligned} r(t) &= \int_0^T [b(t) + i(t) + n(t)][h_R(t) * p(t)]\, dt \\ &= \int_0^T p(k)[b(k) * h_R(k) + i(k) * h_R(k) \\ &\quad + n(k) * h_R(k)]\, dk. \end{aligned} \tag{49}$$

Therefore, the receiver of Fig. 14 multiplies the interfering PSD's with the transfer function of the notch filter $H_R(\omega)$ and then convolves the product with the PSD of the PN sequences $S_p(f)$. The multiplications of the PSD's of the desired signal and the CDMA interference, with the transfer function of the notch filter, will have no effect, provided the transmitter and receiver employ the same filters. Therefore, the desired signal and the CDMA interference remain unchanged. However, the multiplication of the FSM and AWGN PSD's with the transfer function of the notch filter, reduces the interference caused by the FSM and AWGN.

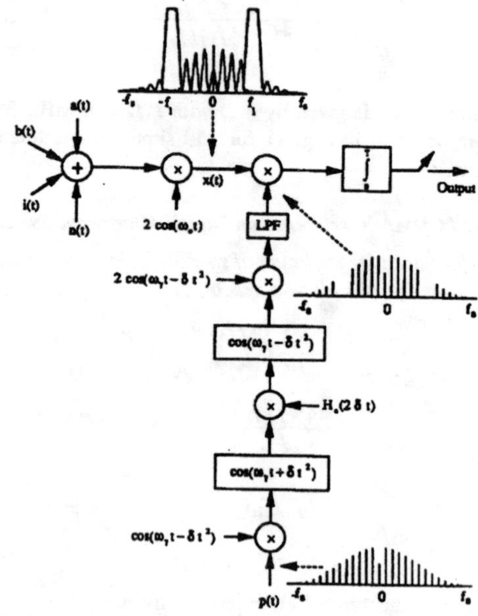

Fig. 14. Model of SS receiver employing filtered spreading sequences and the associated PSD's.

TABLE III
FSM INTERFERENCE AND FILTERED PCN CAPACITY REDUCTION

d (m)	PL (dB)	G (dB)	I_{oF} (dBm)	P/I_{oF} (dB)	CAPACITY REDUCTION FOR 3dB SNR
183	102	4	-110.11	14.11	7.65%
549	116.5	10	-118.61	22.41	1.00%
915	123	19	-116.11	19.41	1.82%
1281	127.5	21.5	-118.11	21.41	1.10%
1647	131	28	-115.11	18.41	2.30%
2013	133	29	-116.11	19.41	1.82%
2379	136	30	-118.11	21.41	1.10%
2745	137	30	-119.11	22.41	0.86%
3111	139	30.5	-120.61	24.41	0.57%

In this case, the interfering PSD's at the correlator input will be

$$S_2'(f) = [S_c(f)H_R^2(f)] * S_p(f) = \int_{-\infty}^{\infty} [S_b(v)H_R^2(v)]S_p(f-v)\,dv \tag{50}$$

$$S_3'(f) = [S_i(f)H_R(f)] * S_p(F) = \int_{-\infty}^{\infty} [S_i(v)H_R(v)]S_p(f-v)\,dv \tag{51}$$

$$S_4'(f) = [S_n(f)H_R(f)] * S_p(f) = \int_{-\infty}^{\infty} [S_n(v)H_R(v)]S_p(f-v)\,dv \tag{52}$$

where $S_c(f)$ is given by (8), $S_i(f)$ by (10), $S_p(f)$ by (7), $H_R(f)$ by (40), and $S_n(f)$ is the PSD of AWGN with a one-sided amplitude of N_0. Substitution yields

$$S_2'(f) = \sum_{\substack{k=-\infty\\k\neq 0}}^{\infty} \sum_{i=1}^{M-1} P_i T \frac{(L+1)^2}{L^4} \left[\frac{\sin\pi\left(f-k\frac{f_s}{L}\right)/f_s}{\pi\left(f-k\frac{f_s}{L}\right)/f_s}\right]^2 \cdot H_R(f)\left[\frac{\sin\pi\left(k\frac{f_s}{L}\right)/f_s}{\pi\left(k\frac{f_s}{L}\right)/f_s}\right]^2 + \sum_{i=1}^{M-1}\frac{P_i T}{L^4}\left(\frac{\sin \pi f/f_s}{\pi f/f_s}\right)^2 H_R(f) \tag{53}$$

$$S_3'(f) = [S_{3,1}(f) + S_{3,2}(f)]H_R(f) \tag{54}$$

where $S_{3,1}(f)$ and $S_{3,2}(f)$ are given by (27) and (28), and

$$S_4'(f) = \sum_{\substack{k=-\infty\\k\neq 0}}^{\infty} \frac{N_0(L+1)}{L^2}\left[\frac{\sin\pi\left(k\frac{f_s}{L}\right)/f_s}{\pi\left(k\frac{f_s}{L}\right)/f_s}\right]^2 \cdot H_R(f) + \frac{N_0}{L^2}H_R(f). \tag{55}$$

The total interfering power at the PCN receiver output will be

$$P_{to}' = \int_{-1/2T}^{1/2T} S_2'(f)\,df + \int_{-1/2T}^{1/2T} S_3'(f)\,df + \int_{-1/2T}^{1/2T} S_4'(f)\,df. \tag{56}$$

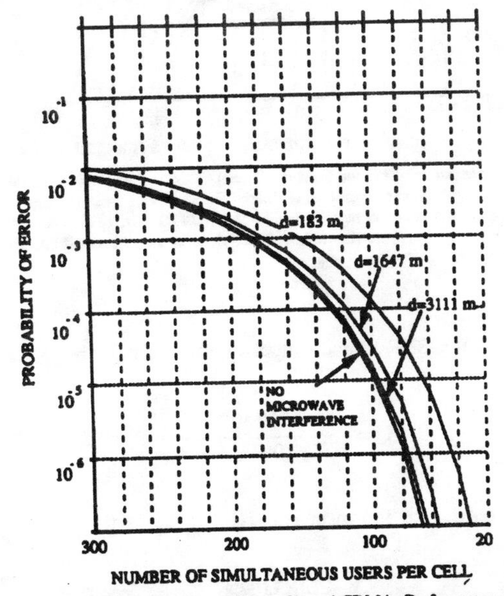

Fig. 15. Interference from the FSM on filtered CDMA: Performance of the forward link, with power control, of a filtered PCN, serving mobile units of 100 μW transmitting power each ($L = 750$, $f_s = 24\,10^6$ Hz, $1/T = 32\,000$ bps, notch center = f_1, notch width = 8 MHz). d is the distance from the microwave antenna.

The output filtered FSM interference will be

$$I_{oF} = \int_{-1/2T}^{1/2T} S_3'(f)\,df. \tag{57}$$

Evaluation of the above equation yields Table III. Repeating the analysis of Section V, the SNR at the output of the filtered PCN mobile (with power control) becomes [13]

$$\mathrm{SNR} = \frac{2.512\,10^{-13}}{2.84\,10^{-16}M + I_{oF}}. \tag{58}$$

The probability of error is again computed for a BPSK PCN receiver. The results are shown in Fig. 15.

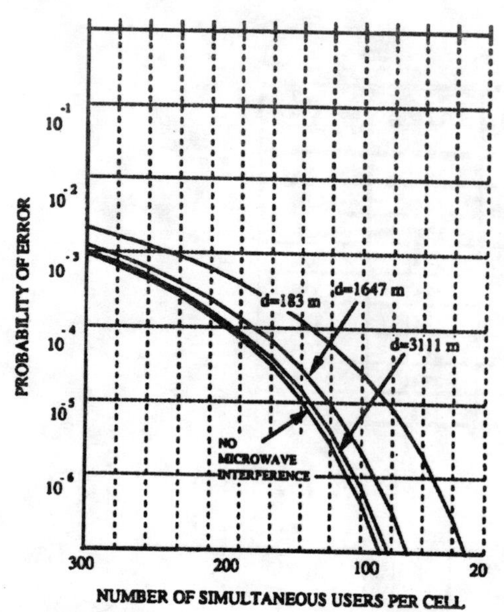

Fig. 16. Interference from the FSM on filtered CDMA: Performance of the reverse link, with power control, of a filtered PCN, serving mobile units of $100\,\mu$W transmitting power each ($L = 750$, $f_s = 24\,10^6$ Hz, $1/T = 32\,000$ bps, notch center = f_1, notch width = 8 MHz). d is the distance from the microwave antenna.

Finally, the SNR at the output of the base (with power control) is given by

$$\mathrm{SNR} = \frac{2.512\,10^{-13}}{1.737\,10^{-16}M + I_{0F}} \tag{59}$$

and the results are shown in Fig. 16.

IX. Conclusion

This paper considered the problem of overlaying a broadband CDMA PCN on the FSM systems of the 1850 to 1990-MHz band. We investigated both cases of interference: from the CDMA to the microwave systems, and from the microwave systems to the PCN. In the first case, we had to consider both analog and digital systems, because they are both in operation in the above frequency band. Furthermore, in the digital microwave system performance degradation we had to consider most of the modulation methods employed for digital transmission. In the second case, we investigated the performance degradation of a CDMA PCN due to the interference produced by the FSM signal. We also considered the cochannel (CDMA) interference within the SS system. We examined both the forward and the reverse PCN links, and, for the forward link, we considered two options: one incorporating power control, and a no-power-control scheme. Next, we repeated the above analysis, using real Fourier transformers and notch filters, to filter the PN sequences of the PCN transmitter and receiver.

For the unfiltered case, the probability of error graphs show, unquestionably, a considerable performance degradation for both systems. In order for the systems to coexist in the same frequency band, their carrier frequencies must be separated by at least $3f_s/4$, and their antennas by at least 3 km. Although this performance degradation may be interpreted by the SS system as a reduction in capacity and/or a degradation in quality, we may not assume the same for the mirowave systems.

However, for the filtered case, the probability of error graphs show that almost any frequency separation is allowable, and the PCN users can get as close as 183 m to the microwave antenna (with a PCN bit error probability of 10^3).

Finally, the results in this paper are for the on-the-boresight case for the microwave antenna, in both receiving and transmitting modes, which represents the worst case of receiving and delivering interference.

References

[1] D. C. Cox, "Universal digital portable radio communications," *Proc. IEEE*, vol. 75, no. 4, Apr. 1987.

[2] R. L. Pickholtz, D. L. Schilling, and L. B. Milstein, "Theory of spread-spectrum communications," *IEEE Trans. Commun.*, vol. 30, no. 5, May 1982.

[3] G. R. Cooper *et al.*, "Cellular land mobile radio: Why spread spectrum," *IEEE Commun. Mag.*, vol. 17, no. 2, 1979.

[4] L. B. Milstein *et al.*, "On the feasibility of a CDMA overlay for personal communications networks," *IEEE J. Select Areas Commun.*, pp. 655–667, May 1992.

[5] G. R. Cooper *et al.*, "Cellular land mobile radio: Why spread spectrum," *IEEE Commun. Mag.*, vol. 17, no. 2, 1979.

[6] A. Papoulis, *Probability, Random Variables, and Stochastic Processes*, 2nd ed. New York: McGraw-Hill, 1984.

[7] A. A. R. Townsend, *Digital Line-of-Sight Radio Links—A Handbook.* Englewood Cliffs, NJ: Prentice-Hall, 1988.

[8] K. G. Filis and S. C. Gupta, "Coexistence of DS CDMA PCN and analog FM: Performance degradation of the SSSC-FM channels due to spread spectrum overlay," in *3rd Int. Symp. Personal, Indoor, and Mobile Radio Communications*, Boston, Mass., Oct. 1992.

[9] J. J. Downing, *Modulation Systems and Noise.* Englewood Cliffs: Prentice-Hall, 1964.

[10] J. G. Proakis, *Digital Communications.* New York: McGraw-Hill, 1989.

[11] L. B. Milstein and P. K. Das, "An analysis of a real-time transform domain filtering digital communication system–Part I: Narrow-band interference rejection," *IEEE Trans. Commun.*, vol. COM-28, no. 6, pp. 816–824, June 1980.

[12] L. B. Milstein and P. K. Das, "Spread spectrum receiver using surface acoustic wave technology," *IEEE Trans. Commun.*, vol. COM-25, no. 8, pp. 841–847, Aug. 1977.

[13] K. G. Filis, "Overlay of cellular DS CDMA on FSM," Southern Methodist University, Dallas, TX, Ph.D. dissertation, May 1993.

Konstandinos Filis was born in Kozani, Greece, in 1963. He received the B.S.E.E. degree from Ohio University, Athens, OH, in 1985, the M.S.E.E. degree from Purdue University, W. Lafayette, IN, in 1987, and the Ph.D. degree in electrical engineering from Southern Methodist University, Dallas, TX, in 1993.

From 1987 to 1989 he served in the Greek army as a Communications Technician. From 1989 to 1990 he was with Antenna TV, Athens, Greece. From 1990 to 1993 he was a Research and Teaching Assistant at Southern Methodist University. His areas of interest include personal communications, spread spectrum, microwave radio links, and antennas.

Dr. Filis is a member of HKN, TBΠ, and the Technical Chamber of Greece.

Someshwar C. Gupta received the B.A. (Hons.) and M.A. degrees in mathematics from Punjab University, India, in 1951 and 1953, respectively, the B.S. (Hons.) degree in electrical engineering from Glasgow University, Scotland, in 1957, and the M.S.E.E. and Ph.D. degrees in electrical engineering from the University of California, Berkeley, in 1962.

He has considerable industrial, administrative, and teaching experience, and is presently Cecil H. Green Professor of Electrical Engineering at Southern Methodist University, Dallas, TX. He is author of the book *Transform and State Variable Methods in Linear Systems* (Wiley, 1966) and coauthor of the books *Fundamental of Automatic Control* (Wiley, 1970) and *Circuit Analysis with Computer Applications to Problem Solving* (Matrix, 1972). He is also a member of the Editorial Board of the *International Journal of Systems Sciences.*

Microcell Engineering in CDMA Cellular Networks

Joseph Shapira, *Fellow, IEEE*

***Abstract*—The demand for cellular radio service is growing rapidly, and in heavily populated areas the need arises to shrink the cell sizes and "scale" the clustering pattern. The extension of the service into the PCN domain, mostly in-buildings and in pedestrian areas, further enhances this trend. The vision of the "third generation" cellular systems incorporates micro- and picocells for pedestrian use, with macrocells for roaming mobiles. Connectivity between all these cells, while maximizing total system capacity, is the main challenge facing the "third generation architects." The CDMA cellular system, which shares the same frequency channel across the system (reuse pattern of one) and applies soft handoff between the cells, has already shown, both by analysis and by tests, to have full connectivity between the microcells and the overlaying macrocells without capacity degradation. The parameters involved in the engineering of a heterogeneous CDMA network are discussed in this paper. Factors that determine the size of the cell, the soft handoff zone, and the capacity of the cell clusters are analyzed, and engineering techniques for overlay–underlay cell clustering are outlined.**

I. INTRODUCTION

A. Why Microcells?

CELLULAR radio communication was designed in order to increase the teletraffic capacity in the service area. The demand for the service is growing rapidly, and in the heavily populated areas, the service is reaching its limit, what is known as "the brick wall." This trend is further enhanced by the increasing penetration of handheld portable units into service, which serve a pedestrian population with a much higher density.

The obvious solution for the systems of that vintage is to shrink the cell size and "scale" the cell clustering pattern. This trend has been ongoing in the heavily populated areas. Further shrinking the cells surfaces new challenges:

- The smaller cell coverage requires lower antenna heights, getting in the urban areas below the roof levels and changing the propagation regime. The heavy cabling connecting the cell site to the antenna has to be reduced to enable the positioning of the antennas at lamp-post heights in the streets.
- The small cells require a much denser wire lines backhaul infrastructure to the cells. The cost of the infrastructure has to decrease substantially to make the microcell network economical.
- The real estate for the cell sites are overly expensive in the urban areas and the need arises for a more compact, unattended cell site.

Manuscript received March 23, 1993; revised May 10, 1993.
The author is with Qualcomm Israel Limited, Haifa 34980, Israel.
IEEE Log Number 9213562.

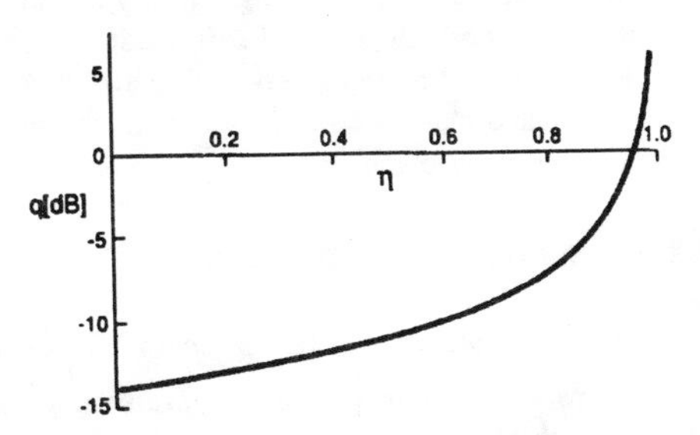

Fig. 1. The required SNR at the cell site as a function of the Load, with $E_b/N_o = 7$ dB.

Microcells are thus identified with the technical solutions proposed to cope with the aforementioned issues. Architectures proposed introduce the concept of remoting the antennas from the cell site, thus allowing for many microcells to be served by a single attended center. One scheme advocates a linear transformation of the RF to optical frequencies and relaying the signals via fibers to a center [1], [2]. Others propose a down conversion to IF (70 MHz), which is then relayed by microwaves or by fibers [3]. These architectures are frequently identified with "the microcell technology." They all are in their development or pilot deployment.

B. Types of Microcells

Microcells can be classified as the following:

1) Hot Spots: These are service areas with a higher teletraffic density or areas that are poorly covered. A hot spot is typically isolated and embedded in a cluster of larger cells.

2) Downtown Clustered Microcells: These occur in a dense, contiguous area that serves pedestrians and mobiles. They are typically found in an "urban maze of street canyons," with antennas located far below building height.

3) In-Building, 3-D Cells: These serve office buildings and pedestrians. This environment is highly clutter dominated, with an extremely high density and relatively slow user motion, and a strong concern for the power consumption of the portable units.

C. Microcells and the CDMA System

The CDMA system does not have the capacity shortage, which is the main driving force leading to the microcell solution. Microcells might, however, be the architecture of choice in certain cases, as, for example, when only a limited band is available in the transition period, or in-buildings, where coverage may become a leading factor. The CDMA system is then found to be very suitable to this architecture:

0018-9545/94$04.00 © 1994 IEEE

- Low transmission power and simple power amplifier and up/down conversion enable simple remote units. The distributed antenna, consisting of a series of distributed radiators with propagation delays inserted between them, enables the CDMA cell to uniformly illuminate a heterogeneous environment while using the diversity of the "artificial multipath" to maintain a high quality signal.
- The use of a common frequency and soft handoff alleviates a major problem of roaming between microcells and between micro- and macrocells and eliminates the need for complex frequency planning.

II. The Power Control and the Cell Boundary

A. Reverse Link Power Control—The Power Control Equation

The CDMA capacity equation [4] for the reverse (mobile to cell site) link may be recast into

$$\frac{q\dfrac{W}{R_b}}{1+\left(\dfrac{n}{F}-1\right)dq}=\frac{E_b}{N_o} \tag{1}$$

where W is the CDMA bandwidth, R_b is the data rate, q is the signal received at the cell site from a single user, referred to the system noise N_oW (SNR threshold per user at the cell site), n is the number of active users (number of calls), d is the voice activity factor, defined as the average data rate divided by 9600 (the nominal data rate), i is the interference introduced by surrounding cells/sectors, referred to the self interference in the cell/sector, F is the frequency reuse factor $1+i=1/F$, E_b is the energy per bit, and N_o is the noise spectral density.

The power control of all the transmitters in the cell causes the interference to the cell site to change proportionally to the change of q, the signal-to-noise ratio. In an environment where all the cells are equally loaded and controlled to the same SNR q, the external interference is also proportional to q. When (1) is solved for q [5]

$$q=\frac{1}{\dfrac{W}{R_b}\cdot\dfrac{N_o}{E_b}-\left(\dfrac{n}{F}-1\right)d}=\frac{1}{\dfrac{W}{R_b}\cdot\dfrac{N_o}{E_b}(1-\eta)} \tag{2}$$

$$\eta=\frac{\left(\dfrac{n}{F}-1\right)d}{\dfrac{W}{R_b}\cdot\dfrac{N_o}{E_b}}. \tag{3}$$

The capacity bound in each cell appears as a pole in the power-set equation in Fig. 1. Assuming a spread spectrum processing gain W/R_b of 128 ($\Rightarrow$ 21 dB) and a required E_b/N_o of 7 dB, the threshold SNR is then -14 dB for a single user. Additional users add interference that diminish this advantage, thus forming the load curve.

Here, η is the fractional load of the cell in Fig. 1, while

$$n_{\text{pole}}[\eta=1]=\left(1+\frac{W}{R_b}\cdot\frac{N_o}{E_b}\cdot\frac{1}{d}\right)F \tag{4}$$

is the upper bound on the number of calls in a cell/sector for a set E_b. A convenient nominal capacity is set at $\eta=0.9$, in which case each user is received at the cell site with SNR $= -4.2$ dB. This nominal setting is a tradeoff of capacity versus the sensitivity of the power control to changes in the load. The effect of the system parameters on the cell capacity is exemplified in Table I. F, the reuse factor, depends on the interference from the surrounding cells, and for a uniformly loaded system may vary mostly between .6 and .67. This depends on the placement of the surrounding cells, the propagation conditions between the cells, and on the distribution of the users in the surrounding cells. The required E_b/N_o depends on the multipath conditions, and may vary from 5 dB to more than 8 dB.

TABLE I
THE EFFECT OF SYSTEM PARAMETERS ON THE CAPACITY

	Uniformly loaded Omni cells	Isolated Omni cell
F	.667	1
E_b/N_o	7 dB	6 dB
W/R_b	128	128
$n\ [\eta=.9]$	39	72

B. Soft Handoff and the Definition of the Cell Boundary

The setting of adjacent cells is interconnected via the handoff condition which determines the cell boundary. A CDMA mobile in the handoff zone is controlled by both adjacent cells. Its power level is always set to match the lower power requirement between these cells by responding to the pilot received with the highest power. The boundary between the cells is thus defined by equal power requirements from both cells. Obviously, this balance depends on the power setting in each cell, which, in its turn, depends on the number of active mobiles in the cell and on the additional interference from the neighboring cells.

The boundary between the cells is defined as the equilibrium point where the transmission power required from the user by both cells is the same P_m.

At the boundary

$$\frac{q_1}{T_1}=\frac{q_2}{T_2}=\frac{P_m}{N_oW}\equiv p_m \tag{5}$$

where T_1, T_2 are the transmission loss from the user to each cell site. Normalized power $p\equiv P/N_oW$ will be used throughout the rest of the paper. For the sake of clarity of the following discussion, we shall assume a simple relation $T_i = A_iR_i^{\alpha_i}$ where A, α are constants of the propagation model, including the gains of the respective antennas, and R is the cell radius. When these constants are equal in both cells, the boundary condition becomes

$$q_1R_1^{\alpha}=q_2R_2^{\alpha} \tag{5a}$$

which is depicted in Fig. 2. Note that the y-axis is logarithmic (linear in dB), while the x-axis is linear. A change in the threshold setting of either cell causes the curve to shift (from shaded to dotted in Fig. 2) and the boundary (curve crossing)

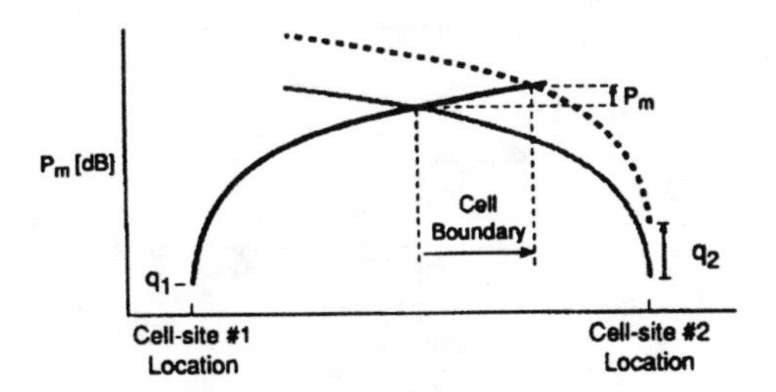

Fig. 2. The cell boundary and its dynamics.

to shift so as to shrink the coverage of the cell with the higher q (cell #2 in Fig. 2). Load is thus automatically shared throughout the system, and cells that are heavily loaded shrink and hand users off to the neighboring cells. This process is instantaneous, as both cells are engaged to the user while within the soft handoff window.

C. The Pilot Control and Handoff Condition

The soft handoff of the CDMA system requires link balancing, to ensure smooth handoffs and avoid "orphan" cases, where the user is in one cell in one direction and in another one in the other direction. This is done by the user's detection and comparison of the pilot signals from the different (two or more) cells around. The forward link cell boundary is determined at the equal power point and the soft handoff zone—within a predetermined ratio of received pilots' power.

The pilot is received without code protection, in order to avoid delay in the control. The pilot link equation is therefore [6]

$$\frac{E_c}{I_o} = \frac{\zeta P_c T(r)}{N_o W + I_{oc}(r)W + P_c T(r)(1-\zeta)} = \frac{q_m}{1 + \frac{I_{oc}(r)}{N_o} + q_m\left(\frac{1-\zeta}{\zeta}\right)} \tag{6}$$

where E_c is the required energy per chip for a given service level, I_o is the total power spectral density, ζ is the portion of the cell site power allocated to the pilot, P_c is the cell site transmission power, $T(r)$ is the transmission loss, $I_{oc}(r)$ is the power spectral density received from other cells, and q_m is the required receive SNR threshold at the mobile

$$q_m = \zeta \frac{P_c T(r)}{N_o W} = \zeta p_c T(r) \tag{6a}$$

where p_c is the normalized total transmission power from the cell site.

Now,

$$q_m = \frac{\frac{E_c}{I_o}\left(1 + \frac{I_{oc}(r)}{N_o}\right)}{\left(1 - \frac{1-\zeta}{\zeta}\frac{E_c}{I_o}\right)}. \tag{7}$$

Equating the transmission loss on both links ((6a) and (5), and (6) and (1))

$$T(r) = \frac{q_m}{\zeta p_c} = \frac{q}{p_m} = \frac{\frac{E_c}{I_0}\left(1 + \frac{I_{0c}(r)}{r_0}\right)}{\zeta p_c\left[1 - \frac{1-\zeta}{\zeta}\frac{E_c}{I_0}\right]} = \frac{\frac{E_b}{N_0}}{p_m \frac{W}{R_b}(1-\eta)} \tag{8}$$

and

$$\zeta = \frac{\frac{E_c}{I_0}}{1 + \frac{E_c}{I_0}}\left[1 + \frac{p_m}{p_c}\frac{\frac{W}{R_b}}{\frac{E_b}{N_0}}\left(1 + \frac{I_{0c}(r)}{N_0}\right)(1-\eta)\right]. \tag{9}$$

At the edge of the coverage, the transmission loss is the highest the system can support, and the mobile transmits its maximum power

$$P_m \Rightarrow P_{m\ \max}.$$

ζ is then shown to depend linearly on the cell remaining capacity $(1-\eta)$, in order to maintain the required balance between the links. The pilot power is therefore controlled as a linear decreasing function of the cell load.

III. Heterogeneous Cell Clustering

A. Parameters Governing the Cell Size

The cell boundary is defined at the intersection of the transmission curves (Fig. 2). The slopes of these curves vary with the distance from the cell-site, and depend both on the propagation conditions and on the cell-site antenna placement, height, and beam shape. In a generic flat-Earth model, the average transmission loss is maintained constant to a distance $R = 1.4GH$, then follows a R^{-2} curve to $8H/\lambda$, beyond which it follows an R^{-4} trend. Here, G represents the product of the gains of the cell site and the mobile antennas in the vertical plane, λ is the operating wavelength, H is the height of the cell site antenna, and the mobile antenna is assumed at 1.5 m.

The position of the curve is determined by the q at the cell site (reverse link) and by the corresponding cell site pilot transmission power (ζp_c, forward link).

A smaller cell has to have a steeper transmission curve slope or a higher q (and a corresponding lower ζp_c), or both. A microcell can be accommodated inside a macrocell if their transmission curves intersect on both sides of the microcell (see Fig. 3). Note that the outward boundary of the microcell stretches further out as the difference in the slopes of the transmission curves of the micro- and macrocell is smaller. This causes the soft handoff zone to stretch further out in that direction too.

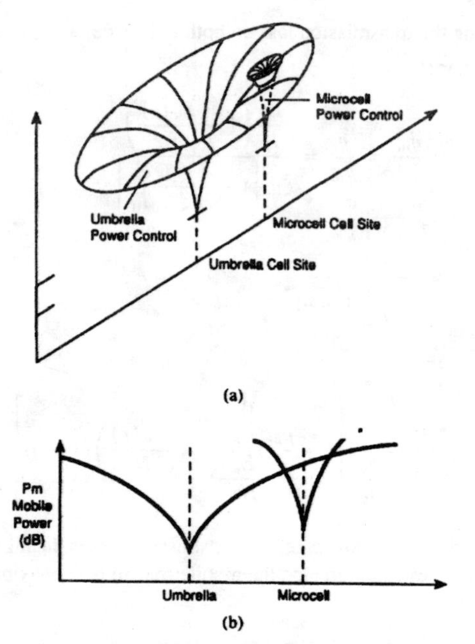

Fig. 3. A microcell in an umbrella cell—mobile transmission under power control. (a) Isometric view. (b) Vertical cut through (a).

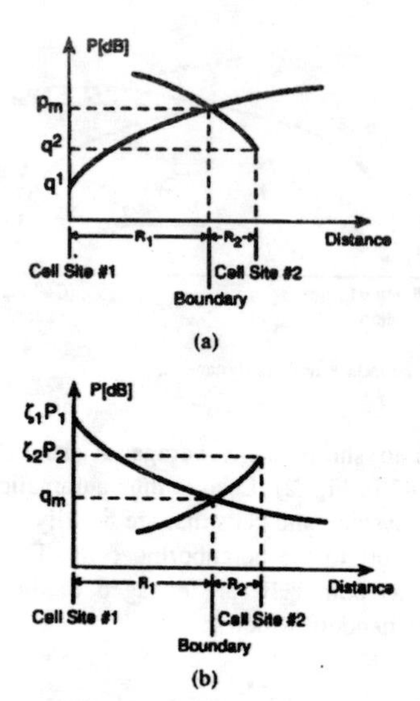

Fig. 4. Cell boundary condition and balancing. (a) Reverse link. (b) Forward link—Pilot power.

A properly designed cell should cover beyond the breakpoint. There the transmission loss depends on

$$T \propto \frac{H^2}{R^4} \tag{10}$$

and, from (5)

$$\frac{T_1}{T_2} = \left(\frac{H_1}{H_2}\right)^2 \left(\frac{R_2}{R_1}\right)^4 \tag{11}$$

$$\frac{R_1}{R_2} = \left[\left(\frac{H_1}{H_2}\right)^2 \frac{q_2}{q_1}\right]^{1/4}. \tag{12}$$

The reverse link boundary condition is shown in Fig. 4(a).

Similarly, pilot transmission powers from both cells have to match at the mobile

$$q_m = \zeta_1 p_1 T_1 = \zeta_2 p_2 T_2 \tag{13}$$

$$\frac{\zeta_1 p_1}{\zeta_2 p_2} = \frac{T_2}{T_1} = \left(\frac{H_2}{H_1}\right)^2 \left(\frac{R_1}{R_2}\right)^4 \tag{14}$$

$$\frac{R_1}{R_2} = \left[\frac{\zeta_1 p_1}{\zeta_2 p_2}\left(\frac{H_1}{H_2}\right)^2\right]^{1/4} \tag{15}$$

and the link balance equation (equating (12) and (15)) is

$$\frac{q_1}{q_2} = \frac{\zeta_2 p_2}{\zeta_1 p_1} \tag{16}$$

which relates the ratio of cell transmission powers to their desensitization. This is depicted in Fig. 4(b). Here, $p_{1,2}$ are the total powers transmitted by the cell sites, $\zeta_{1,2}$ are the portion of the total power allocated to the pilot in each cell, and q_m is the signal received at the mobile.

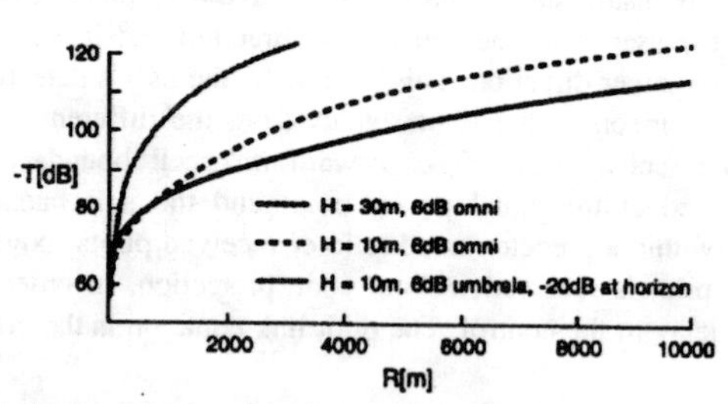

Fig. 5. Transmission loss from the cell site for different antenna heights and tilts.

B. Antenna Height and Beam Tilt

The minimum transmission loss obtainable for a mobile close to the cell-site antenna is higher for the higher antenna. The lower antenna generates a steeper curve (dotted), as the breakpoint to the R^{-4} regime is closer to the antenna in that case, and the curves therefore cross. Tilting the antenna beam down brings the R^{-4} zone even closer as shown shaded in Fig. 5.

C. Cell Directivity

Cells may be made directional by directive cell-site antennas. A sector is an example of a directive cell. The cell site in such a cell is not in the center of the coverage area, but rather near the perimeter.

A steeper crossing of the microcell and the umbrella cell transmission-loss curves are obtained by directing the microcell beam toward the umbrella cell site (as shown in Fig. 6), which helps in controlling the microcell coverage and the extent of the soft hand-off zone. The effect of the directivity is equivalent to the reduction of the transmission power, and desensitizing the receiver by that amount in the direction of the back lobe. This is shown in Fig. 6. A directional antenna

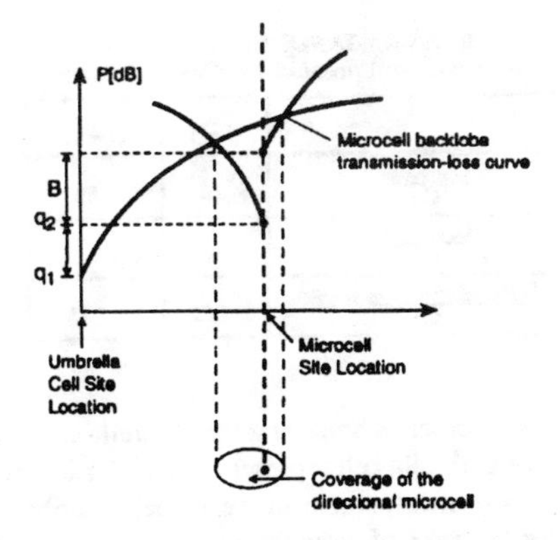

Fig. 6. Power control and cell boundary for a directional microcell under an umbrella cell.

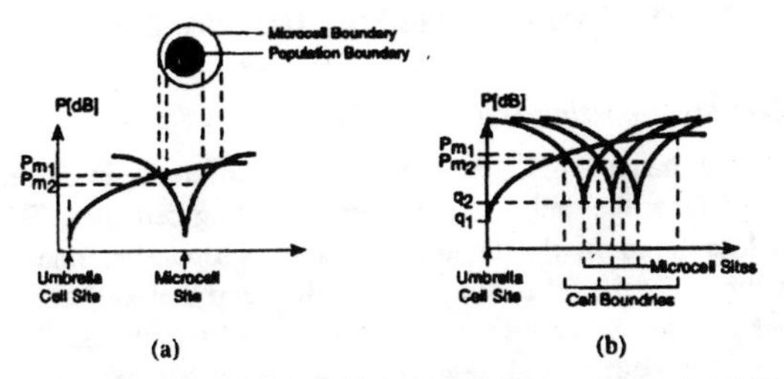

Fig. 7. Population distribution in limited portable power microcell. (a) A hot spot. (b) Macrocell bounding a cluster of microcells.

radiates some residual radiation into its back lobe. For many antennas and installations, this residual radiation may be 10 to 20 dB below the main lobe. This front-to-back ratio is termed B in Fig. 6.

D. Desensitization

The maximum coupling between the cell site and the user's unit depends on the cell-site antenna height and its gain. This minimum loss is at least 70 dB for cells with antenna heights of 30 m (see Fig. 5). The antenna height for microcells is as low as 6 m, which yields a minimum loss of 50–55 dB. In buildings, where the height is only 3 to 4 m, it becomes 35–40 dB.

Such a low transmission loss makes the cell-site receiver susceptible to interference from various sources and to saturation by nearby units. It is therefore necessary to desensitize the microcell. However, the amount of desensitization required is less than the excess coupling due to the lower antenna placement in the microcell.

Desensitization also affects the range of the microcell for a given portable transmit power, but capacity rather than range is the primary issue considered in the design of a microcell. The coverage for a portable with limited maximum power may not reach the boundary with the neighbor macrocell in certain configurations, and the portable may not be able to roam from a microcell to a macrocell while mobiles from the microcell can roam throughout the microcell as shown in Fig. 7.

Since the CDMA system uses the minimum power required to make its link, the ample peak power to cross the boundary to the master cell may be reserved in the unit to extend its range whenever needed without continuously draining power from the unit.

The reduced minimal transmission loss reduces the transmission power from the portable to microwatts (which is negligible in terms of conservation of battery life). Desensitization therefore, does not affect the average power consumption in the microcell.

IV. Overlay–Underlay Cells Clustering

A. Overlay–Underlay Hierarchy

Hierarchical cell clustering (overlay–underlay) is generally used in three situations: for smoothing coverage in areas where topography or obstructions create shadowed regions, for accommodating regions with higher user concentrations, or for local private networks within the coverage of the public network.

A smooth handoff between the cells and between the macro- and microcells is important for roaming purposes. This is a difficult task to achieve for systems which are using hard handoff. Even when mobile assisted handoff (MAHO) schemes are used, smooth handoff is often unattainable because the boundary between the cells is not as clearly defined between microcells where it is obstruction dominated, as it is between large cells. A high power margin is required therefore in these systems to overcome "ping-pong" handoff attempts which inadvertently reduce capacity.

The soft handoff in CDMA assures smooth handoffs due to its diversity combining between the cells over the transition zone. Power is thus reduced, instead of being boosted as in other systems. The extended zone of soft handoff does require additional channel cards in the cell sites though. Techniques which reduce this effect will be discussed in Section V.

B. Capacity of a "Hot Spot" Microcell and of the Umbrella Cell

Coverage of a "hot spot" area is sometimes required within the coverage area of a larger cell. Such an arrangement disturbs the frequency planning for systems with a frequency reuse pattern higher than one. They have to clear the frequencies used by the inner cells in a radius comparable to the reuse pattern of the large outer cells (apply a guard zone around the inner cell against the large and more powerful outer cells). This results in a loss of capacity to the overlay system.

The situation is different for the CDMA system, where the underlay cells and the overlay cells share the same frequency and their reuse efficiency is determined by the soft handoff between them. The total capacity in this scenario is more than twice that of a loaded cell in a uniform environment with other cells (C_o), as discussed in the Appendix.

The microcell capacity is higher than that of a loaded cell in such an environment. That is because the umbrella cell around it has a much sparser population and contributes very little interference. The umbrella cell loses some of its capacity due to the additional interference from the microcell. That loss diminishes, however, when the microcell's users'

TABLE II
CAPACITY OF AN UMBRELLA CELL WITH A HOT SPOT UNDERLAY

	D = 1	D = 10
$\frac{C_{umbrella}}{C_0}$	.822	.98
$\frac{C_{microcell}}{C_0}$	1.467	1.46
$\frac{C_{umbrella} + C_{microcell}}{C_0}$	2.28	2.44

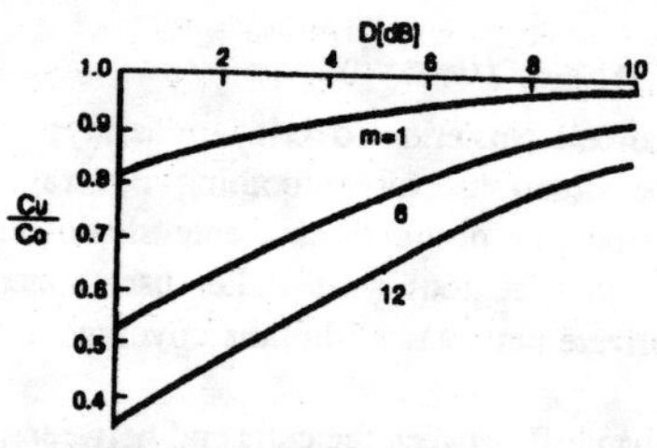

Fig. 8. The umbrella cell capacity loss as a function of D—the power ratio, and of m—the number of microcells in the underlay.

population does not reach the cell boundary, but is enclosed within a smaller radius, as shown in Fig. 7(a). This is indeed the typical case for a "hot spot microcell," where the high local concentration motivates the microcell emplacement. The transmission power of the portable units in such a case does not exceed p_p, which is lower than the power required at the boundary p_m by a factor $p_m/p_p = D$. The total capacity is more than twice that of a loaded cell in a uniform environment of other cells (C_0).

An example will clarify the trend. Let $U = 5$ be the ratio of the umbrella cell radius to that of the microcell, $i_{i0} = 0.5$ be the normalized interference to the umbrella cell from the surrounding cells. Then, (see (A5), (A6)). The capacity of the umbrella cell degrades to .822 C_0 for $D = 1$, but is almost not affected for $D = 10$. On the other hand, the capacity of the embedded microcell is higher than C_0 and not affected by the choice of D. The compound capacity of the umbrella and the embedded microcell exceeds $2C_0$ by 10 to 20%. Theses are summarized in Table II.

C. Capacity—Underlaid Microcell Cluster

Urban microcells may be clustered under the overlay of an umbrella cell. In this scenario, the microcells are in a uniform cluster and their capacity is C_0. The umbrella cell loses some of its capacity due to the microcells' loading, but this loss becomes negligible when the power of the portables in the microcells is bounded below the transmission level at the cell boundary. This trend is shown in Fig. 8.

An example with $i_{i0} = 0.5, U = 5, D = \{1, 10\}, m = 12$ is summarized in Table III (see (A7), (A8)). The capacity degradation of the umbrella cell due to the additional interference from the microcell cluster—is minimized by proper coverage design of the microcell cluster (choice of D). For $D = 10$ the loss is only 15%. The compound capacity quoted in the table is pessimistic, and should actually exceed 13. This is because the

TABLE III
CAPACITY OF AN UNDERLAY MICROCELL CLUSTER

	D = 1	D = 10
$\frac{C_{umbrella}}{C_0}$	.36	.85
$\frac{C_{microcell}}{C_0}$	1.	1.
$\frac{C_{umbrella} + m\,C_{microcell}}{C_0}$	12.36	12.85

capacity of the microcells bordering the umbrella cell is higher than C_0, as the umbrella cell represents a smaller interference. This has not been incorporated in the model described in the Appendix for the sake of simplicity.

V. THE EXTENT OF THE SOFT HANDOFF ZONE

A. Overlay–Underlay Soft Handoff Layout

The soft handoff in CDMA relies on the difference in power received from the mobile by the corresponding cell sites. The path loss trend in the contiguous cells is opposite: one is growing while the other one is decreasing, and their difference grows fast, as shown on the left-hand side of the microcell in Fig. 3. The path loss difference grows much slower on the outer boundary of the embedded microcell, which results in a larger soft handoff zone.

The contour of the boundary between cells may be defined by (12) (reverse link) or (15) (forward link), as shown in Fig. 9. Here, $R_1 = \sqrt{x^2 + y^2}$, $R_2 = \sqrt{(x + R)^2 + y^2}$. These contours are circles, the diameter of which may be calculated from their intersection with the x-axis. We now normalize $R = 1$ such that

$$\left(\frac{x}{x+1}\right)^2 = \left(\frac{H_1}{H_2}\right)^2 \sqrt{\frac{q_2}{q_1}}. \tag{17}$$

The soft handoff window is defined by the assigned ratio of pilot power received by the mobile from the two cells (forward link) or the ratio of mobile transmit powers requested by both cells. This ratio κ is typically 3 dB. The respective contours of the soft handoff zone become

$$\left(\frac{R_1}{R_2}\right)^2 = \left(\frac{H_1}{H_2}\right) \sqrt{\frac{q_2}{q_1}} \left\{\kappa; \frac{1}{\kappa}\right\} \tag{18}$$

and their crossing with the x-axis are at

$$\left(\frac{x}{x+1}\right)^2 = \left(\frac{H_1}{H_2}\right) \sqrt{\frac{q_2}{q_1}} \left\{\kappa; \frac{1}{\kappa}\right\}. \tag{19}$$

The ratio of the handoff zone to the microcell coverage is calculated by comparing the areas of the respective circles. This is drawn in Fig. 10 as a function of

$$\left(\frac{H_1}{H_2}\right) \sqrt{\frac{q_2}{q_1}}.$$

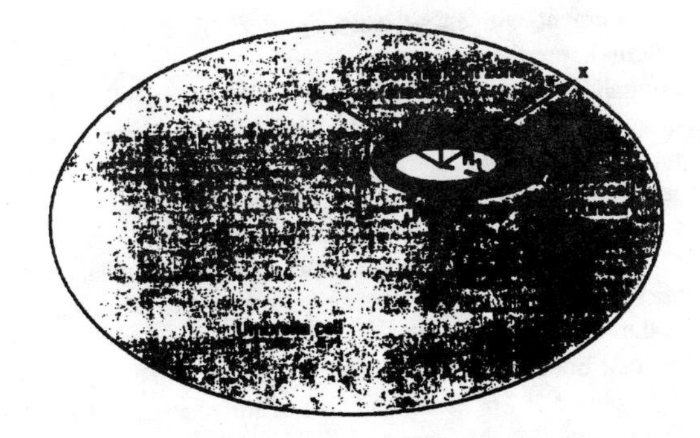

Fig. 9. The umbrella-microcell boundary and handoff zone.

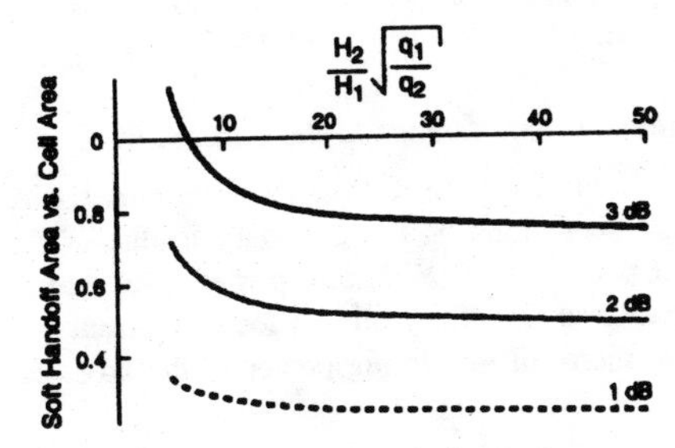

Fig. 10. Soft handoff zone as a function of window size.

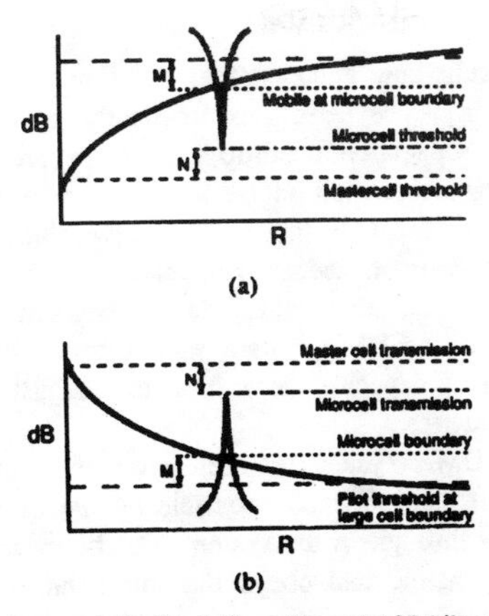

Fig. 11. Overlay—underlay handoff conditions. (a) Mobile power—Reverse link. (b) Pilot power—Forward link.

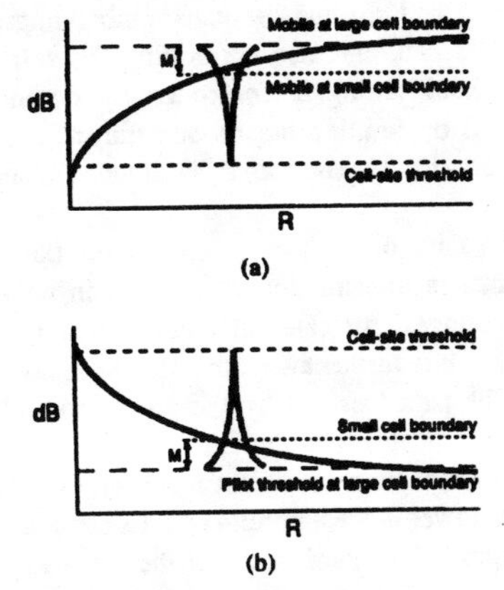

Fig. 12. Power scaling in the microcell. (a) Mobile power—Reverse link. (b) Pilot power—Forward link.

B. Soft Handoff Zone Reduction

The soft handoff zone may be further reduced by the following processes:

- Lowering the antenna appears to significantly effect propagation up to a height ratio of 20. Further improvement is obtained when the antenna is lowered below the building level and the path-loss exponent for the lower antenna is higher than 4.
- An equivalent effect can be obtained by raising the threshold of the microcell (desensitization) and lowering its cell transmission power accordingly.
- Tilting the antenna has the same affect as lowering the antenna. It brings the breakpoint to the r^{-4} slope closer to the antenna.
- Changing the soft handoff window has a major effect on the extent of its zone, as shown in Fig. 10.
- By building a directional cell with its lobe directed toward the umbrella cell site, the isolation in the outward direction is enhanced, and the outward soft handoff zone is reduced appreciably (see Fig. 6).

VI. Handoff Balancing

The handoff condition determines the coverage of each cell in a cell cluster. Clustering cells of different sizes require a respective higher path-loss gradient in the smaller cells to match this condition, or a desensitization of the small cell.

Overlay–underlay clustering poses a special situation. The large umbrella cell has two boundaries: one that is internal, with the microcell, and one that is external, with other large cells. Moreover, the cell's pilot power has to meet the handoff window condition at the external boundary, and thereby, the pilot power is much higher at the internal boundary (see Fig. 11). The forward and the reverse link formations (Fig. 11(a) and (b)) have to be identical in order to maintain the same cell boundaries in both directions.

These two conditions lead to the following:

- The overlay–underlay soft handoff window has a much higher SNR than usual (margin M above the noise), and has to be specified on a comparative basis.
- The reduced transmission power by the microcell site has to be balanced by comparable desensitization of its receiver. Otherwise, it will have to transmit the same power as the large cell.
- When the microcell transmits the same power, its mobiles (portables) may be desensitized according to the margin M (see Fig. 11).

The microcell power may be scaled N dB lower than the umbrella cell, as shown in Fig. 12. The microcell receiver has to be desensitized by the same ratio N.

VII. Summary and Conclusions

Microcells are a conceptual solution to high teletraffic area density. The density is frequently nonuniform and so are the propagation conditions in these areas. These lead to heterogeneous cell clustering and to interweaving cell boundaries. Whereas these are unfavorable conditions for systems relying on frequency allocation and hard handoff, and detract from their nominal efficiency, they suit well the unique features of the CDMA system: nonuniform density may assist in increasing capacity; the soft handoff adjusts to nonregular cell boundaries.

The CDMA system features power control in each cell and a soft handoff, made possible by the usage of a single frequency throughout the system. The boundary between the cells is dynamic and obeys the soft handoff condition. Its location depends on the momentary threshold setting in each cell and on the transmission loss curves. Proper control of these parameters allows for the clustering of heterogeneous cells, sizing and shaping the cells, embedding a microcell in an umbrella cell, and also controlling the soft handoff zone. The threshold setting is controlled by desensitization, and is balanced by limiting the mobile transmission power. The transmission loss is controlled by antenna placing and beam shaping.

The capacity of each cell depends on the external interference due to transmission of mobiles in neighboring cells. That is strongest for external mobiles that are near the cell boundary, while further away, their transmission decreases fast both by the path loss and by the power control in their own cells. A microcell embedded in an umbrella cell encounters less interference than in a uniform cluster, as the density of the mobiles is lower in the umbrella cell. Its capacity is thus higher than nominal. The interference of the embedded microcell to the umbrella cell may also be made small by proper design, whereby its boundary with the umbrella cell extends beyond the high teletraffic density area.

Appendix
Capacity Gain in Overlay–Underlay Microcells

A. Introduction

User concentration might dictate the installation of one or more small cells ("hot spot microcells") within the coverage of a large cell. Such an arrangement requires new frequency planning in systems with frequency reuse pattern greater than one, and a subsequent loss of capacity of the large cell cluster.

An overlay–underlay cell arrangement is possible in the same radio channel with the CDMA system. The capacity of the microcell and of the umbrella cell are addressed here.

B. The Model

A simple model will be used here and is illustrated in Fig. 3 for the sake of a simple and clear analysis.

Here, R is the radius of the large cell; r is the radius of the microcell; $U = R/r$; N is the nominal capacity of an isolated cell; $C = N/(1+i)$ is the capacity of a cell incurring external interference i; C_0 is the capacity of a cell in a uniform environment with other cells; $i_l = P_l/N$ is the normalized external interference to the large cell; $i_m = P_m/N$ is the normalized external interference to the microcell; i_{lm} is the normalized interference to the large cell from the microcell; i_{lo} is the normalized interference to the large cell from the surrounding large cells; i_{mo} is the normalized interference to a microcell in a cluster of other microcells; D is the ratio of the mobile transmission power at the boundary of the microcell/large cell and the maximum power of the portable in the microcell; δ is the ratio of the user population radius to the cell boundary radius (see Fig. 10). In a path-loss regime $r^{-\alpha}, D = \delta^{-\alpha}$, m is the number of microcells in the cluster. The capacity of the large cell is $N/(1+i_{lo}+i_{lm})$, while the capacity of the microcell is $N/(1+i_m)$.

As $U \gg 1$, one may assume that the microcell is embedded in a uniform distribution of equal power transmitters in the large cell.

C. Microcell Interference to the Large Cell

The handoff condition equalizes the transmission power of the microcell units on the boundary to that of the large cell at that point. As $U \gg 1$, that power level may be assumed constant over the microcell rim for the current analysis.

The microcell interfering power to the large cell is, therefore,

$$P_{lm} = \frac{mN}{D(1+im)\cdot \pi r^2}\cdot 2\pi \int_o^{\delta r} (x/r)^4 x\, dx = \frac{mN}{3D(1+i_m)}$$

$$i_{lm} = \frac{m}{3D(1+i_m)} = i_l - i_{lo} \tag{A1}$$

where r is the user density within the microcell.

D. The Large Cell Interference to the Microcell

As the large cell is much larger than the microcell $U \gg 1$, the microcell may be seen as embedded in a uniform distribution of equal power transmitters. The interference power is then

$$P_{ml} = \frac{N}{(1+i_l)\cdot\pi(R^2-r^2)}\cdot 2\pi\int_r^\infty (r/x)^4 x\, dx$$

$$= \frac{1}{(U^2-1)}\cdot\frac{N}{(1+i_l)}$$

$$i_m = \frac{1}{(U^2-1)(1+i_l)} \quad (\text{for } m=1)$$

$$i_m \cong \frac{1}{(U^2-1)(1+i_l)} + im_0 \quad (\text{for } m \gg 1). \tag{A2}$$

E. The Capacity of the Microcell and the Large Cell

Now let us define $x = 1+i_l, x_0 = 1+i_{l0}, y = 1+i_m, y_0 = 1+i_{m0}, A = U^2-1$.

From (A1), (A2)

$$x - x_0 = \frac{m}{3Dy} \tag{A3}$$

$$y = 1 + \frac{1}{Ax} \quad [m=1]; \quad y = y_0 \quad [m \gg 1] \tag{A4}$$

which is solved for x

$$x = \frac{1}{2}\left(x_0 + \frac{1}{3D} - \frac{1}{A}\right) 1 + \left\{ \sqrt{1 + \frac{4x_0}{A\left(x_0 + \frac{1}{3D} - \frac{1}{A}\right)^2}} \right\} \quad [m = 1] \tag{A5}$$

$$y = 1 + \frac{1}{Ax} \tag{A6}$$

$$x = x_0 + \frac{m}{3Dy_0} \quad [m \gg 1] \tag{A7}$$

$$y = y_0 \tag{A8}$$

$$\frac{C_{\text{umbrella}}}{C_0} = \frac{1 + i_{l0}}{x} \tag{A9}$$

$$\frac{C_{\text{microcell}}}{C_0} = \frac{1 + i_{m0}}{y}. \tag{A10}$$

REFERENCES

[1] T. S. Chu and M. J. Gans, "Fiber optic microcellular radio," in *Proc. 41st IEEE VTS Conf.*, St. Louis, MI, May 1991, pp. 339–344.

[2] L. J. Greenstein, N. Amitai, T. S. Chu, L. J. Climini, Jr., "Microcells in personal communications systems," *IEEE Commun. Mag.*, vol. 30, no. 12, pp. 76–88, Dec. 1992.

[3] W. C. Y. Lee, "Efficiency of a new microcell system," in *Proc. 42nd IEEE VTS Conf.*, Denver, CO, May 1992, pp. 37–42.

[4] K. S. Gilhousen, I. M. Jacobs, R. Padovani, A. J. Viterbi, L. A. Weaver, Jr., and C. E. Wheatley, "On the capacity of a cellular CDMA system," *IEEE Trans. Veh. Technol.*, vol. 40, pp. 303–312, May 1991.

[5] J. Shapira and R. Padovani, "Spatial topology and dynamics in CDMA cellular radio," in *Proc. 42nd IEEE VTS Conf.*, Denver, CO, May 1992, pp. 213–216.

[6] S. Soliman, C. Wheatley, and R. Padovani, "CDMA reverse link open-loop power control," in *Proc. GLOBECOM'92*.

Joseph Shapira (S'71–M'73–SM'83–F'91) received the B.Sc. and M.Sc. degrees in electrical engineering from the Technion–Israel Institute of Technology, Haifa, Israel, in 1961 and 1967, and the Ph.D. degree in electrophysics from the Polytechnic University (then Polytechnic Institute of Brooklyn), Brooklyn, NY, in 1974.

He joined RAFAEL, Israel Armament Development Authority, in 1963. There, he undertook a broad spectrum of research, management, and staff positions. His last position was Senior Fellow of RAFAEL. Prior to that, he was the Director of Advanced Strategic Programs Directorate, Deputy Director of the Missiles and Guidance Division, and Head and Founder of the Electromagnetics Department. He also was an Advisor to the Director of RAFAEL and a Research Fellow. During several leaves-of-absence from RAFAEL, he was Special Assistant to the Chief of Research and Development of Israel Ministry of Defense from 1979 to 1980, Manager of the Optronics Systems Operation in ELOP—the Electrooptics Industries of Israel from 1980 to 1981, and Visiting Scientist in Bell Telephone Laboratories, from 1982 to 1983, and in Scientific Atlanta in 1978. He joined QUALCOMM Incorporated in 1989. He studied and developed propagation models for mobile cellular communications, and radio aspects and network engineering for the CDMA cellular system that was invented in QUALCOMM. In 1991, he became the Director of Israel Office of QUALCOMM, and then the President of QUALCOMM Israel, Limited, when the company was established in October 1992. He is also affiliated with the Technion as an Adjunct Senior Teaching Fellow in the Electrical Engineering Faculty. He is the president of the Israel National Committee for Radio Science (INC/URSI) since 1981.

Dr. Shapira is a Member in the Scientific Committee for Telecommunication of the International Union for Radio Science (URSI). He won the IEEE Antennas and Propagation Society Best-Paper Award in 1974 for "Ray analysis of conformal antenna arrays," coauthored with L. B. Felsen and A. Hessel. In 1980, he was awarded the A. D. Bergman Prize (presented by the President of Israel) for his contributions in electromagnetics technology in Israel.

A Micro-Cellular CDMA System Over Slow and Fast Rician Fading Radio Channels with Forward Error Correcting Coding and Diversity

Caspar A. F. J. Wijffels, Howard Sewberath Misser, *Member, IEEE,* and Ramjee Prasad, *Senior Member, IEEE*

***Abstract*—The microcellular radio environment is characterized by a Rician fading channel. The use of a slotted code division multiple access (CDMA) scheme is considered in single- and multi-microcell systems. The throughput and delay performance of a slotted CDMA network are analyzed for slow and fast Rician fading radio channels using differential phase shift keying (DPSK) modulation. The application of selection diversity (SD) and maximal ratio combining (MRC) improve the performance for both slow and fast fading. It is also shown that the use of forward error correcting (FEC) codes enhances the system performance. Computational results are presented for maximum rms delay spread in the order of 2 μs and data rates of 32 and 64 kbit/s. A comparative analysis of macro-, micro- and pico-cellular CDMA systems is also presented.**

I. Introduction

RECENTLY, the code division multiple access (CDMA) scheme has drawn the attention of many researchers for its application in cellular radio systems [1]–[10]. Combining code division multiple access with micro-cells may provide high capacity radio access to cellular systems. In contrast to conventional cells (macro-cells) which are of large size (2–20 km diameter) with the antenna radiating rather large ower (0.6–10 W) from the top of tall buildings, micro-cells are relatively much smaller in size (0.4–2 km diameter) with their antennas at street lamp elevations operating at relatively low power (less than 20 mW). Measured results [11]–[14] indicate that the received signal envelope distributions for micro-cellular channels are always Rician. The spectrum efficiency of a micro-cellular radio system is evaluated in [15], [16] using analog FM communication.

This paper presents the performance analysis of a slotted CDMA network in a micro-cellular mobile radio system using DPSK modulation. The influence of forward error correcting coding, namely the (15,7) BCH coding and diversity schemes, *viz.*, selection diversity and maximal ratio combining on the performance parameters are investigated. The performance of the micro-cellular system is also compared with the performance of macro-cellular and pico-cellular systems. In the case of macro-cellular systems, the channel is Rayleigh fading [17] and the rms delay spread is higher than that for micro-cellular systems (e.g., 8 μs) [18]. Pico-cells are the smallest size (20–400 m diameter). The pico-cells are modeled by a Rician fading channel, and the rms delay spread is also minimal in the range of 50 ns–250 ns [19], [20]. Pico-cells are suitable for indoor wireless communications, with antennas placed on top of a bookshelf and radiating power in the order of a few mW.

This paper is organized as follows. Briefly the cellular system is described in Section II. Section III presents the throughput analysis of a slotted CDMA network. The packet success probability and average delay are evaluated in Section IV for a fast Rician fading channel using selection diversity/maximal ratio combining and BCH forward error correcting codes. The performance of micro-cellular radio systems for slotted CDMA networks is computed for slow Rician fading channels in Section V. The performance of macro-, micro-, and pico-cellular radio systems is compared in Section VI. Finally, the conclusions are given in Section VII.

II. Cellular System

In a direct sequence CDMA cellular system, the cluster size is one because every cell uses the same carrier frequency or set of carrier frequencies. Therefore, in a cellular CDMA system each base station receives not only interference from the users located within its own cell but also from users located in the surrounding cells, as is shown in Fig. 1. To simplify the performance evaluation, an approximation is made by assuming that every cell causes equal average interference power. If the average interference power of the home cell is p_i, the total multi-user interference power p_t can be approximated by

$$P_t \approx P_i(1 + 6R_u^{-n}) \tag{1}$$

where R_u is the (normalized) reuse distance and defined as the ratio of the distance between the centers of two adjacent cells D and the cell radius R, and n is the path loss exponent. Here n is assumed to be four.

It should be noted that the interference from only adjacent cells from the first tier is considered in this paper. The interference due to the second and higher order tiers are neglected. The influence of multi-user interference on the performance of a multi-micro-cellular CDMA system can be investigated using (1). Although (1) is obtained assuming an equal average

Manuscript received September 19, 1991; revised December 17, 1991, and February 9, 1993.

C. A. F. J. Wijffels is with Alcatel Telecom Systems B.V., The Netherlands.
H. S. Misser is with the Dr. Neher Laboratories of the Royal Dutch PTT, Leidschendam, The Netherlands.
R. Prasad is with the Telecommunications and Traffic Control Systems Group, Delft University of Technology, 2600 GA Delft, The Netherlands.

IEEE Log Number 9210676.

0018–9545/93$03.00 © 1993 IEEE

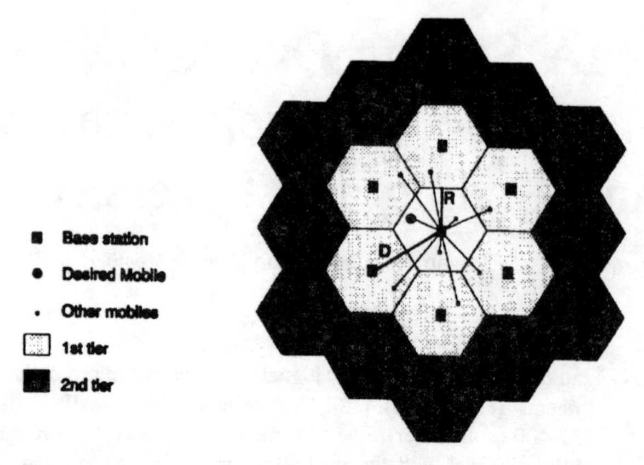

Fig. 1. Reverse link (mobile-to-base) interference in a cellular CDMA system.

interference power from all cells and considering (normalized) reuse distance to simplify the performance analysis, an instructive result can be obtained by this approximation without introducing a serious error.

III. Throughput Analysis

The chief appealing feature of a CDMA system is its soft capacity; i.e., there is no hard limit on the system capacity. However, the CDMA technique faces plenty of challenges, e.g., near-far problem, cross-correlation multiple access interference problem, etc. Therefore, in a CDMA system, collisions are possible and occur when more users access the system than the allowable system capacity. For any CDMA system, there is a pronounced threshold where the system performance degradation as a function of the number of users ceases to be gradual and performance degrades rapidly [24], [25]. The point of destructive interference or contention is called the CDMA threshold. The CDMA threshold is, therefore, defined as the maximum number of users that can transmit simultaneously within a shared direct-sequence frequency band without destructive interference occurring. In this paper, the system degradation is assumed to be a step function located at the system capacity of C users.

Let us assume that the number of users transmitting data packets simultaneously within a specific time slot in slotted CDMA is k. In a CDMA system, as soon as the number of active users rises above the users' threshold capacity C, fatal collisions occur, and all the packets are destroyed. The packets are lost due to excessive bit errors when the number of simultaneous users exceeds C. The lost packets are rescheduled and retransmitted after a sufficient time delay. Each user transmits a specific code to identify the user at the base station of a single cell system in a micro-cellular radio network. The micro-cellular network consists of numerous mobile terminals, transmitting packets to base stations situated in the home cell.

The (normalized) throughput is defined as the average number of successfully received packets per time slot, normalized to the users' capacity C, and is given by:

$$S \triangleq \frac{1}{C}\sum_{k=1}^{C} kP_k P_{rk}(k). \tag{2}$$

where P_k is the probability of a packet being transmitted with $k-1$ other packets and $P_{rk}(k)$ is the packet success probability.

Assuming a binomial arrival distribution of the offered traffic G, P_k is written as:

$$P_k = \binom{C}{k}\left[\frac{G}{C}\right]^k\left[1-\frac{G}{C}\right]^{C-k}. \tag{3}$$

The offered traffic G is defined as the average number of transmissions (new generated packets plus retransmitted packets) per time slot by k users. A binomial arrival distribution of the offered traffic (3) seems to be a quite appropriate assumption for a CDMA system because there is a finite number of users sharing a direct sequence frequency band, and the performance of the system for a certain chip length depends on the number of users (PN codes). With the binomial arrival model it is further assumed that the probability that a new packet is generated is the same as the probability of retransmission of a packet.

The (complex) lowpass equivalent impulse response of the passband channel for the link between the kth user and the base station is written as:

$$h_k(\tau) = \sum_{l=1}^{L}\beta_{lk}\delta(\tau-\tau_{lk})\exp(j\gamma_{lk}) \tag{4}$$

where β, τ and γ are the path gain, time delay, and phase of each path, respectively. The index lk refers to the lth path of the kth user and $j=\sqrt{-1}$. L is the maximum number of resolvable paths given by [19], [20]:

$$L = \left\lfloor\frac{T_m}{T_c}\right\rfloor + 1. \tag{5}$$

Here, $\lfloor x\rfloor$ is the largest integer smaller than or equal to x, T_m is the rms delay spread, and T_c is the chip duration. The number of paths L may be assumed either fixed or randomly changing, depending on a fixed T_m or not. Here fixed values for L are used (T_m is assumed to be fixed). We assume that the path phase of the received signal, $(\omega_c\tau_{lk}+\gamma_{lk})$, is an independent random variable uniformly distributed over $[0, 2\pi]$. The path delay also is an independent random variable and is assumed uniform over $[0, T_b]$ where T_b is the data bit duration. N, T_b and T_c are related by:

$$N = \frac{T_b}{T_c}. \tag{6}$$

Equation (6) enables us to fit one code sequence length in one data bit interval. The base station is assumed to provide average power control within its particular cell, to overcome the near-far problem. Further, it is assumed that $T_m < T_b$ to avoid intersymbol interference.

According to measurement results reported in [11] for the micro-cellular radio environment, β_{lk} is assumed to be an independent Rician random variable with probability density function (pdf) given by:

$$p_{\beta lk}(t) = \frac{r}{\sigma^2}\exp\left(-\frac{r^2+S_p^2}{2\sigma^2}\right)I_0\left(\frac{S_p r}{\sigma^2}\right)$$
$$0 \le r < \infty,\ S_p \ge 0,\ \text{and } R = \frac{S_p^2}{2\sigma^2}. \quad (7)$$

Here, $I_n()$ is the modified Bessel function of the first kind and nth order, S_p is the peak value of the specular radio signal, and σ^2 is the average of the fading signal. For convenience, we make use of normalized path gains of (7) by letting $v = r/\sigma$ which relates the transformed pdf $p_{\beta n,lk}(v)$ with $p_{\beta\, lk}(r)$ as follows:

$$P_{\beta n,lk}(v) = p_{\beta\, lk}(r)\left(\frac{dv}{dr}\right) = \frac{1}{\sigma}P_{\beta\, lk}(r). \quad (8a)$$

Using (7) and (8a), the transformed pdf $p_{\beta n,lk}(v)$ can be written as

$$p_{\beta n,lk}(v) = v\exp(-(v^2+s^2)/2)I_o(sv)$$
$$0 \le v < \infty,\ s \ge 0,\ \text{and } s = \frac{S_p}{\sigma} = \sqrt{2R}. \quad (8)$$

R is the characteristic parameter for the fading channel and known as the Rice parameter, which depends on the micro-cellular environment. Rice parameter R is defined as the ratio of dominant power (e.g., line-of-sight power) and the power received over specular paths. Typical values for R in a micro-cellular network are 7 dB and 12 dB [11]. In the transmitter a direct sequence code waveform is used to amplitude modulate the data waveform. The resulting waveform modulates a RF carrier. The transmitted signal of the kth user is then:

$$S_k(t) = Aa_k(t)b_k(t)\cos(\omega_c t + \theta_k) \quad (9)$$

where A is the amplitude of the transmitted signal, $a_k(t)$ is the chip waveform of the kth user ($a_k^i \in \{-1, 1\}$), $b_k(t)$ is the data waveform of the kth user ($b_k^i \in \{-1, 1\}$), ω_c is the angular carrier frequency, and θ_k is the carrier phase for the kth user.

The receiver consists of a matched filter, a DPSK demodulator and diversity processing components. The receiver input signal at the antenna for a certain user can be written as:

$$r(t) = \sum_{k=1}^{K}\sum_{l=1}^{L} A\beta_{lk}a_k(t-\tau_{lk})b_k(t-\tau_{lk}) \cdot\cos(\omega_c t + \theta_{lk}) + n(t) \quad (10)$$

where $\theta_{lk} = (\omega_c\tau_{lk} + \gamma_{lk})$ and $n(t)$ is white Gaussian noise with two-sided power spectral density $N_o/2$. Now, considering the above described channel, transmitter and receiver model, the throughput is evaluated for slow and fast fading channels using both types of diversity schemes and BCH (15,7) FEC coding.

IV. Fast Rician Fading Channel

Consider k users, simultaneously transmitting data packets in a fixed time slot to a central base station. Each user employs an unique PN (pseudo-random) code of N rectangular pulses to spread each data bit. In the analysis, Gold codes have been used as direct-sequence spread-spectrum codes.

In the case of fast fading, the channel variations are fast relative to the signalling interval. Thus each signalling symbol undergoes fading independently from other symbols. Therefore, when we send data in packets in a fast fading channel, all data bits undergo fading independently. Thus the packet success probability using *selection diversityis* given by:

$$P_{rk}(k) = [1 - P_{er\ s}(k)]^{N_d} \quad (11)$$

where $P_{er\ s}(k)$ is the bit error probability using selection diversity scheme, and N_d is the number of data bits per packet. $P_{er\ s}(k)$ can be obtained as follows [19]–[21]:

$$P_{er\ s}(k) = \int_{-\infty}^{\infty}\int_{-\infty}^{\infty}\int_{0}^{\infty} P_{sd}(\beta_{\max}, \mu, \mu_0) \cdot p_{\beta_{\max}}(\beta_{\max})p_\mu(\mu)p_{\mu_0}(\mu_0)\,d\beta_{\max}\,d\mu\,d\mu_0 \quad (12)$$

where

$$P_{sd}(\beta_{\max}, \mu, \mu_0) = \frac{1}{2}\left[1 - \frac{\mu}{\mu_0}\right]\exp\left[-\frac{A^2\beta_{\max}^2 T_b^2}{\mu_0}\right] \quad (13)$$

$\mu_0 = \text{var}(z_0 \mid \{\tau_{lk}\}, L)$, $\mu = E[(z_0 - m)(z_{-1} - m)^* \mid \{\tau_{lk}\}, L]$ and $m = E[z_0 \mid \beta_{\max}, \beta_1^0] = E[z_{-1} \mid \beta_{\max}, b_1^{-1}]$.

z_0 and z_{-1} are the envelopes of the signal at the current sampling instant and the previous sampling instant, respectively. μ_0 and μ are approximated by Gaussian random variables, and p_μ and $p_{\mu 0}$ are Gaussian pdf's. μ can be easily removed analytically as follows:

$$\int_{-\infty}^{\infty}\left[1 - \frac{\mu}{\mu_0}\right]p_\mu\,d\mu = 1 - \frac{E_\tau(\mu)}{\mu_0} \quad (14)$$

leaving a double integral to evaluate. μ_0 consists of the influence of the desired signal, multi-user interference (1) and white Gaussian noise. A derivation of the bit error probability for selection diversity (12) is given in Appendix A.

$p_{\beta_{\max}}(v)$ is the pdf of selecting the strongest path, given by

$$p_{\beta_{\max}}(v) = M[1 - Q(s, v)]^{M-1}v\exp\left[-\frac{s^2+v^2}{2}\right]I_0(sv) \quad (15)$$

where M is the order of diversity and $Q(s, v)$ is the Marcum Q-function [21].

In order to establish the accuracy of (12), a Monte Carlo simulation was conducted to assess the performance, in terms of bit error probability, of a direct sequence spread spectrum multiple access radio system in a cellular Rician fading channel using differential phase shift keying modulation without approximations. Fig. 2 compares the simulated results with the results obtained using (12) for $R = 6.8$ dB, $M = 4$, and L as a parameter in terms of bit error probability for a single cell system. It can be seen from Fig. 2 that the simulation results fit the analytical results very well. Thus the results obtained using

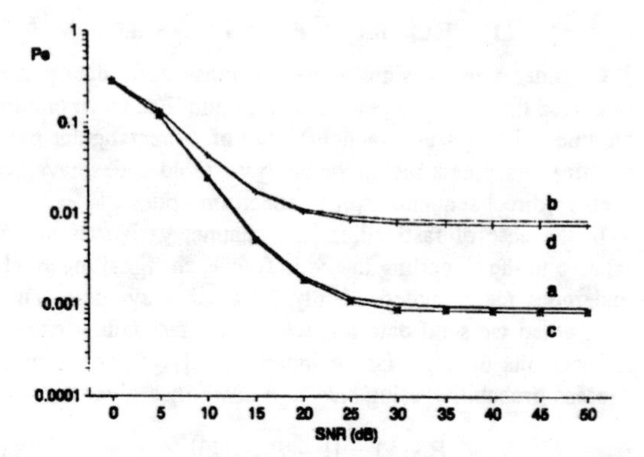

Fig. 2. A comparison between simulation and analytical bit error probability P_c for a cellular system with a single cell, $R = 6.8$ dB, $M = 4$. (a) $L = 1$, analytical (b) $L = 4$, analytical (c) $L = 1$, simulation (d) $L = 4$, simulation.

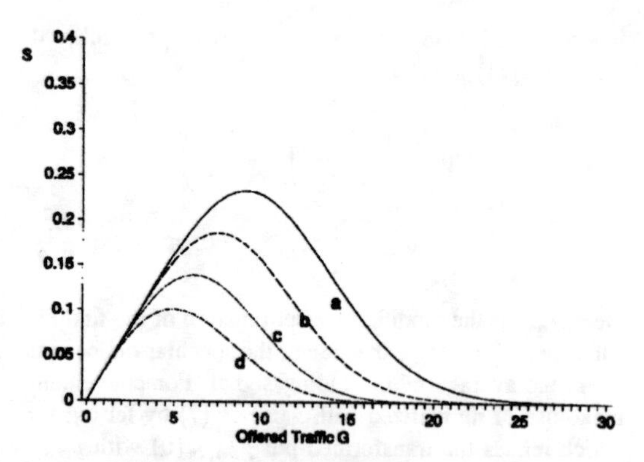

Fig. 3. A comparison between a single-cell system and a multi-cell system for fast Rice fading. Micro-cell, Selection diversity, $N = 225$, $R = 12$ dB, $C = 30$, $N_d = 42$, $M = 2$. (a) $L = 16$, single-cell, (b) $L = 16$, multi-cell, (c) $L = 32$, single-cell, (d) $L = 32$, multi-cell.

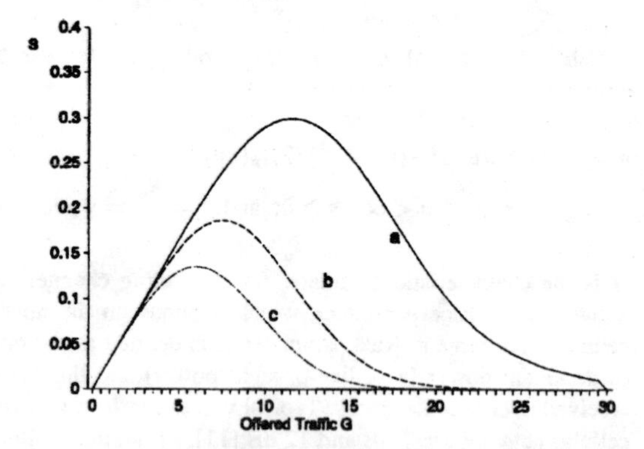

Fig. 4. The influence of the number of resolvable paths L on the throughput for fast Rice fading. Multi-microcell, Selection diversity, $N = 255$, $R = 12$ dB, $C = 30$, $N_d = 42$, $M = 2$. (a) $L = 8$, (b) $L = 16$, (c) $L = 24$.

(12) represent the true performance of a DS/SSMA system in a micro-cellular environment.

All numerical results were obtained using numerical integration methods from the Turbo Pascal Numerical Toolbox (adaptive Gaussian quadrature). The Bessel function was approximated, using the formulas and tables given in [23]. For the calculation of the variance of μ and μ_0 a specific set of Gold codes was used. The partial cross-correlation functions were calculated using the specific Gold codes and averaging over the (uniform) path delay distribution. For calculating the moments of μ and μ_0, these calculations had to be done for all resolvable paths of all users. After calculation of the statistical moments, the bit error probability, packet success probability, throughput, and delay were calculated.

In Fig. 3, a comparison is given between single- and multi-cell systems. The performance of the multi-user systems can be evaluated by substituting μ_0 by $(1+6R_u^{-n})\mu_0$ in (12)–(14) resulting from (1). As expected, multi-cell systems have an inferior throughput performance compared with single-cell systems, due to the interference introduced by the surrounding cells. Also it is seen that in Fig. 3 the best performance is for the single-cell system with 16 resolvable paths and the worst is for the multi-cell system with 32 resolvable paths.

Fig. 4 depicts the normalized throughput versus offered traffic for a multi-cell system with a code sequence length N of 255 chips, using selection diversity. From Table I and (5) we see that the number of resolvable paths L, depends on the rms delay spread T_m and the data rate R_b. Thus it is seen that as the number of resolvable paths increases, due to a change in either delay spread (higher T_m) or data rate (higher R_b), the performance of the system, in terms of throughput, decreases. This means that we can interpret curves a and b from Fig. 4 in two ways. The first is that both curves have the same data rate R_b of 32 kbit/s and different delay spreads T_m, say 1 and 2 μs, and the second is that the delay spread is constant (1 μs) and data rates R_b are 32 and 64 kbit/s. This means that in buildings with higher delay spreads, systems that use the same data rate will have a lower throughput than in buildings with lower delay spreads if the change in delay spread causes a change in the number of resolvable paths.

The influence of the Rice factor R on the performance is depicted in Fig. 5. It is seen that Rayleigh fading channels yield a far worse performance than Rician fading channels, and the performance increases with a higher value of R. Since the value of R depends on the environment of the pico-cell (i.e., the construction materials and layout of the building), this means that part of the (maximum) system performance is determined by the morphology of building.Fig. 6 shows the influence of the order of diversity M on the throughput using selection diversity. It is seen that a higher order of diversity increases the performance. The order of diversity M in case of selection diversity is usually chosen to be such that $M \leq LN_a$. Here N_a is the number of antennas used. In the case of *maximal ratio combining* with DPSK, the receiver which is set for a reference user combines all the spread spectrum correlation peaks of the demodulated signals noncoherently and forms the decision variable. Maximal ratio combining with DPSK is also known as predetection combining (PDC). The packet success probability using maximal ratio combining is obtained using equations (7.4.13) and (1.1.115) of [21]:

$$P_{rk}(k) = [1 - P_{erm}(k)]^{N_d} \tag{16}$$

TABLE I

IN THIS TABLE, THE HIGHER LIMIT VALUE OF THE RANGE OF THE RMS DELAY SPREAD T_m IS GIVEN (IN ns), FOR L RESOLVABLE PATHS. THE LOWER LIMIT VALUE FOR L PATHS IS FOUND IN THE TABLE FOR $L-1$ PATHS. E.G., WHEN 738 ns $< T_m <$ 984 ns, $N = 127$, AND $R_b = 32$ kbit/s, THERE ARE FOUR RESOLVABLE PATHS ($L = 4$)

	$R_b = 32$ kbit/s			$R_b = 64$ kbit/s		
$>L$	$N = 127$	$N = 255$	$N = 511$	$N = 127$	$N = 255$	$N = 511$
	Tm(ns)	Tm(ns)	Tm(ns)	Tm(ns)	Tm(ns)	Tm(ns)
1	246	123	61	123	61	31
2	492	245	122	246	123	61
3	738	368	183	369	184	92
4	984	490	244	492	245	122
5	1230	613	305	615	306	153
6	1476	735	366	738	368	183
7	1722	858	427	861	429	214
8	1969	980	488	984	490	244
9	2215	1103	549	1107	551	275
10	2461	1225	610	1230	613	305
11	2707	1348	671	1353	674	336
12	2953	1471	732	1476	735	366
13	3199	1593	793	1599	797	397
14	3445	1716	854	1722	858	427
15	3691	1838	916	1845	919	458
16	3937	1961	977	1969	980	488
17	4183	2083	1038	2092	1042	519
18	4429	2206	1099	2215	1103	549
19	4675	2328	1160	2338	1164	580
20	4921	2451	1221	2461	1225	610
21	5167	2574	1282	2584	1287	641
22	5413	2696	1343	2707	1348	671
23	5659	2819	1404	2830	1409	702
24	5906	2941	1465	2953	1471	732
25	6152	3064	1526	3076	1532	763
26	6398	3186	1587	3199	1593	793
27	6644	3309	1648	3322	1654	824
28	6890	3431	1709	3445	1716	854
29	7136	3554	1770	3568	1777	885
30	7382	3676	1831	3691	1838	916
31	7628	3799	1892	3814	1900	946
32	7874	3922	1953	3937	1961	977
33	8120	4044	2014	4060	2022	1007
34	8366	4167	2075	4183	2083	1038
35	8612	4289	2136	4306	2145	1068
36	8858	4412	2197	4429	2206	1099
37	9104	4534	2268	4552	2267	1129
38	9350	4657	2319	4675	2328	1160
39	9596	4779	2380	4798	2390	1190
40	9843	4902	2441	4921	2451	1221

where
$$P_{er\ m}(k) = \int_0^\infty P_2(\gamma_b)p(\gamma_b)\,d\gamma_b \tag{17}$$

$$P_2(\gamma_b) = \frac{1}{2^{2M-1}}\exp(-\gamma_b)\sum_{k=0}^{M-1} p_k\gamma_b^k \tag{18}$$

$$p_k = \frac{1}{k!}\sum_{n=0}^{M-1-k}\binom{2M-1}{n} \quad \text{and} \quad \gamma_b = \frac{E_b}{N}\sum_{k=1}^{M}\beta_k^2 \tag{19}$$

$$p(\gamma_b) = \frac{1}{2E_b/N}\left[\frac{\gamma_b}{(E_b/N)s_M^2}\right]^{(M-1)/2} \cdot\exp\left(-\frac{s_M^2+\gamma_b N/E_b}{2}\right) I_{M-1}(\sqrt{\gamma_b N/E_b}\,s_M) \tag{20}$$

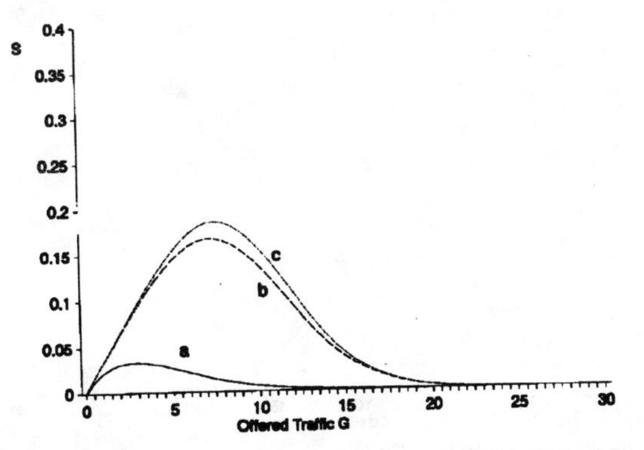

Fig. 5. The effect of the Rice factor R on the throughput for fast Rice fading. Multi-micro-cell, Selection diversity, $N = 255$, $C = 30$, $N_d = 42$, $M = 2$, $L = 16$. (a) Rayleigh, $R = 0$, (b) Rice, $R = 7$ dB, (c) Rice, $R = 12$ dB.

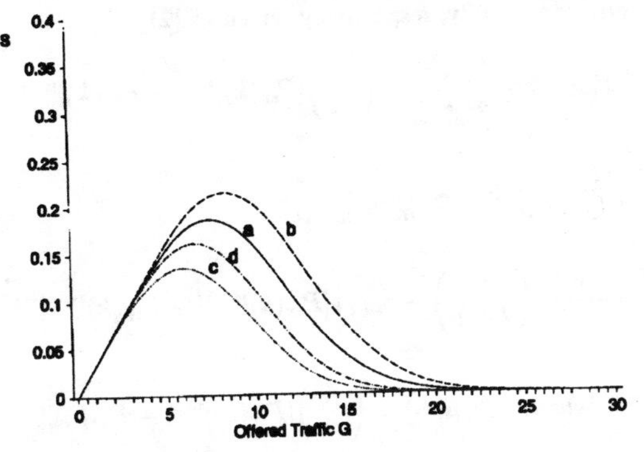

Fig. 6. The influence of the order of diversity M (and number of resolvable paths) on the throughput for fast Rice fading. Multi-micro-cell, Selection diversity, $N = 255$, $R = 12$ dB, $C = 30$, $N_d = 42$. (a) $M = 2$, $L = 16$, (b) $M = 4$, $L = 16$, (c) $M = 2$, $L = 24$, (d) $M = 4$, $L = 24$.

Here M is the order of diversity, β_k is the path gain of the kth combined path, N is the sum of the Gaussian noise and assumed multi-user interference, $s_M^2 = Ms^2$, $s = \sqrt{(2R)}$, and γ_b is the sum of the signal-to-noise ratios of the M combined paths. In case of maximal ratio combining, the order of diversity is chosen such that $M \leq L$.

Fig. 7 shows a comparison of the performance of systems using both selection diversity and maximal ratio combining. From Fig. 7 we clearly see that MRC is superior to SD, and the former achieves a higher maximum throughput (channel capacity).

The influence of *forward error correcting (FEC) coding* on the performance of CDMA systems is studied using BCH coding (n, k) block-encoding which encodes k data bits into n encoded bits (one block), and can correct up to t errors per block. With BCH (15,7) FEC coding, the number of errors per block, t, that can be corrected is 2, and $P_{ec}(k)$ is the bit error probability after decoding. We use two bounds, namely, $P_{ec1}(k)$ and $P_{ec2}(k)$ as upper and lower bounds for the bit

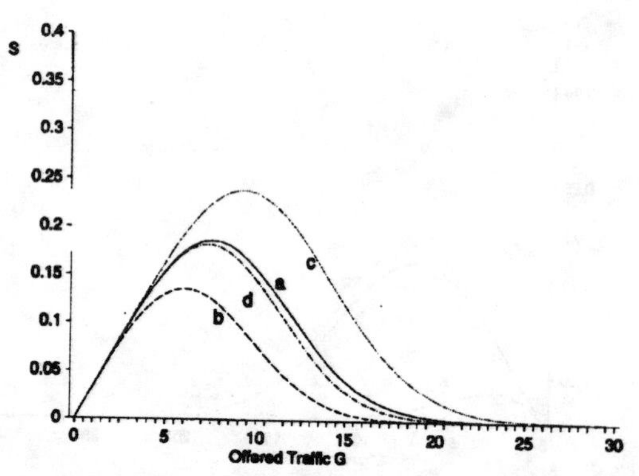

Fig. 7. Throughput curves for different diversity methods (Maximal Ratio Combining and Selection Diversity) and fast Rice fading. Multi-micro-cell, $N = 255$, $R = 12$ dB, $C = 30$, $N_d = 42$, $M = 2$. (a) SD, $L = 16$, (b) SD, $L = 24$, (c) MRC, $L = 16$, (d) MRC, $L = 24$.

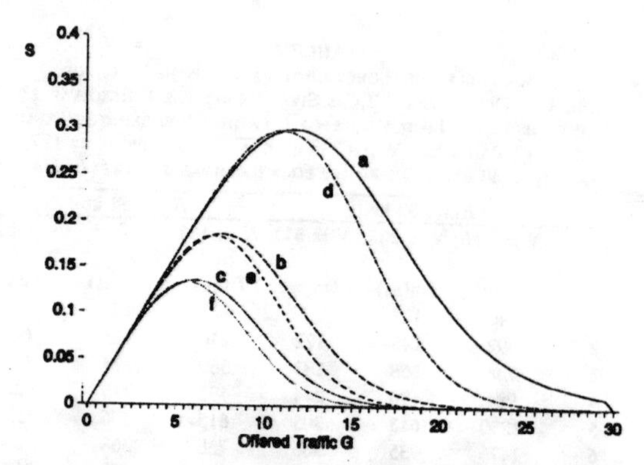

Fig. 8. Influence of FEC coding on the throughput for different numbers of resolvable paths and fast Rice fading. Multi-micro-cell, Selection diversity, $N = 255$, $R = 12$ dB, $C = 30$, $N_d = 42$, $M = 2$, No coding, (a) $L = 8$, (b) $L = 16$, (c) $L = 24$, and BCH-(15,7) FEC coding (d) $L = 16$, (e) $L = 32$, (f) $L = 48$.

error probability, respectively, given in [21]:

$$P_{ec1}(k) = \frac{1}{n} \sum_{m=t+1}^{n} m \binom{n}{m} [P_{er}(k)]^m [1 - P_{er}(k)]^{n-m} \quad (21)$$

$$P_{ec2}(k) = \frac{1}{n} \sum_{m=t+2}^{n} m P(m, n) + \frac{t+1}{n} \left[\binom{n}{t+1} - \beta_{t+1} \right] [P_{er}(k)]^{t+1} [1 - P_{er}(k)]^{n-t-1} \quad (22)$$

where $P(m, n) = \binom{n}{m} [P_{er}(k)]^m [1 - P_{er}(k)]^{n-m}$

and $\beta_{t+1} = 2^{n-k} - \sum_{i=0}^{t} \binom{n}{i}$.

In the computational results, the upper bound on the bit error probability, $P_{ec1}(k)$, is used while comparing systems with and without forward error correcting coding.

Using (11), (12), and (21), the packet success probability for fast fading with selection diversity and BCH (n, k) FEC coding is given by:

$$P_{rk}(k) = [1 - P'_{ec1}(k)]^{N_d} \quad (23)$$

where

$$P'_{ec1}(k) = \frac{1}{n} \sum_{m=t+1}^{n} m \binom{n}{m} [P_{er\ s}(k)]^m [1 - P_{er\ s}(k)]^{n-m} \quad (24)$$

The packet success probability for fast fading with maximal ratio combining and BCH (n, k) FEC coding is obtained using (16)–(21):

$$P_{rk}(k) = [1 - P''_{ec1}(k)]^{N_d} \quad (25)$$

where

$$P''_{ec1}(k) = \frac{1}{n} \sum_{m=t+1}^{n} m \binom{n}{m} [P_{er\ m}(k)]^m [1 - P_{er\ m}(k)]^{n-m} \quad (26)$$

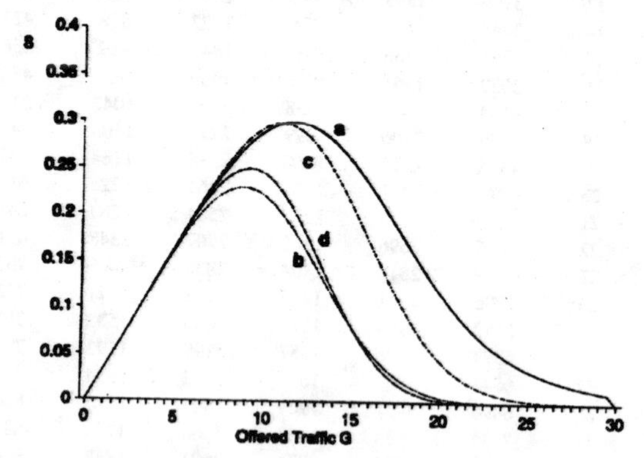

Fig. 9. Influence of FEC coding on the throughput, with varying packet lengths and fast Rice fading. Multi-micro-cell, Selection diversity, $N = 255$, $R = 12$ dB, $C = 30$, $M = 2$. No coding, $L = 8$, (a) $N_d = 42$, (b) $N_d = 168$, and BCH-(15,7) FEC coding, $L = 16$, (c) $N_d = 42$, (d) $N_d = 168$.

In Figs. 8 and 9 we can see the influence on the performance due to the introduction of FEC coding. Comparing curves b and d in Fig. 8 shows that there is significant improvement in the performance of the system due to the FEC coding. Further, it is seen from Fig. 8 and Table I that the maximum throughput is equal for the system with and without FEC coding, although the system with FEC coding shows superior performance beyond the point of maximum throughput (comparing curves d, e, and f with a, b, and c, respectively).

The effect of the packet length on the system performance with and without FEC coding is shown in Fig. 9. As expected, an increase in the packet length decreases the performance in both cases. Further, it is seen that a high value of packet length improves the performance of the system with FEC coding much more than the system without coding (curves d and b).

The average delay of the system is defined as the number of slot times it takes for a packet to be successfully received. The average delay (in slots), assuming immediate

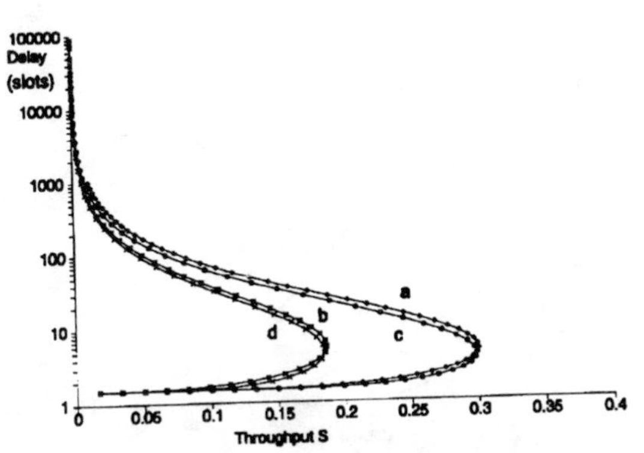

Fig. 10. Throughput-delay curves for fast Rice fading. The delay is given in slot-times. Multi-micro-cell, Selection diversity, $N = 255$, $R = 12$ dB, $C = 30$, $N_d = 42$, $M = 2$. No coding, (a) $L = 8$, (b) $L = 16$, and BCH-(15,7) FEC coding (c) $L = 16$, (d) $L = 32$.

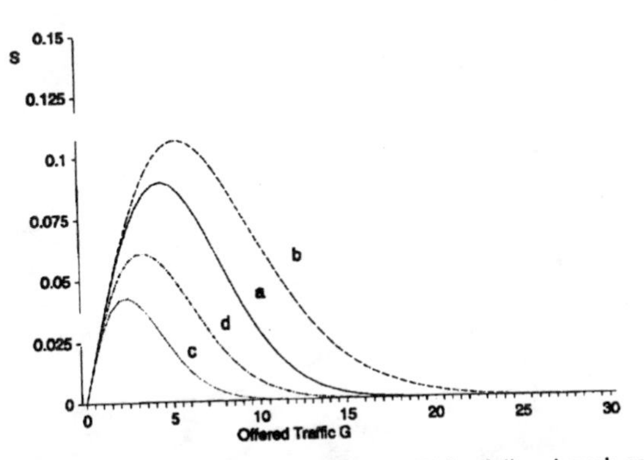

Fig. 11. A comparison between fast and slow Rician fading channels and the effect on the throughput. Multi-micro-cell, Maximal ratio combining, $N = 127$, $R = 12$ dB, $C = 30$, $M = 2$, $L = 6$. No coding, (a) fast, $N_d = 128$, (b) slow, $N_d = 128$, (c) fast, $N_d = 1024$, (d) slow, $N_d = 1024$.

acknowledgment, can be found using [22].

$$D = (1.5 + d) + \left[\frac{G}{CS} - 1\right](\lfloor\delta + 1\rfloor + 1 + 2d) \qquad (27)$$

where $(G/CS - 1)$ is the average number of retransmissions needed for a packet to be successfully received, δ is the mean of the retransmission delay, and $2d$ is the round trip propagation delay, all normalized to the duration of one slot time. It is seen that the minimum delay is $1.5+d$ slot times, for new packets that are immediately successfully received. The round trip propagation delay $2d$ can be neglected for pico-cells, but it can not be neglected for macro-cells. In the case of micro-cells, the round tirp propagation time is in the order of 10 μs. Therefore, the value of $2d$ is very much dependent on the slot time, which equals N_dT_b. For $R_b = 32$ or 64 kbit/s and the value of N_d between 42 and 168, $2d$ is then in the order of 0.01 (slot-times).

The throughput-delay characteristics for various system configurations are shown in Fig. 10. The throughput increases and the delay decreases with the decrease in L for the system with and without coding. Coding enhances the performance, i.e., high throughput and low delay (curves b and c).

V. Slow Rician Fading Channel

Challen characteristics are constant Channel for a (very) long period compared to a signaling interval for the slow fading transmission case. When we send data in packets, this definition is interpreted as that in a slow fading channel; all bits of a packet are received with the same average power.

Assuming a slow Rician fading channel with DPSK modulation and using (13)–(15), $P_{rk}(k)$ for systems using selection diversity can be found.

$$P_{rk}(k) = \int_{-\infty}^{\infty}\int_{-\infty}^{\infty}\int_{0}^{\infty} [1 - P_{sd}(\beta_{\max}, \mu, \mu_0)]^{N_d} \cdot p_{\beta_{\max}}(\beta_{\max}) p_\mu(\mu) p_{\mu_0}(\mu_0)\, d\beta_{\max}\, d\mu\, d\mu_0 \qquad (28)$$

Using (18)–(20), the packet success probability for slow fading channels with maximal ratio combining is given by

$$P_{rk}(k) = \int_0^\infty [1 - P_2(\gamma_b)]^{N_d} p(\gamma_b)\, d\gamma_b \qquad (29)$$

Fig. 11 compares the performance due to the fast and slow fading channels. It is seen that slow fading channels yield better performance than the performance due to fast fading channels. The effect of the packet length on the throughput is also shown in Fig. 11. As expected, a larger packet length results in a decreased throughput.

For slow fading channels, with selection diversity and BCH(n, k) coding, the packet success probability is obtained using (13)–(15) and (28)

$$P_{rk}(k) = \int_{-\infty}^{\infty}\int_{-\infty}^{\infty}\int_{0}^{\infty} [1 - P_{ec\ s}(k)]^{N_d} \cdot p_{\beta_{\max}}(\beta_{\max}) p_\mu(\mu) p_{\mu_0}(\mu_0)\, d\beta_{\max}\, d\mu\, d\mu_0 \qquad (30)$$

where

$$P_{ec\ s}(k) = \frac{1}{n}\sum_{m=t+1}^{n} m\binom{n}{m} \cdot [P_{sd}(\beta_{\max}, \mu, \mu_0)]^m [1 - P_{sd}(\beta_{\max}\mu, \mu_0)]^{n-m} \qquad (31)$$

Similarly, for slow fading channels, with maximal ratio combining and BCH (n, k) coding, the packet success probability is derived using (18)–(20) and (29)

$$P_{rk}(k) = \int_0^\infty [1 - P_{ec\ m}(k)]^{N_d} p(\gamma_b)\, d\gamma_b \qquad (32)$$

where

$$P_{ec\ m}(k) = \frac{1}{n}\sum_{m=t+1}^{n} m\binom{n}{m}[P_2(\gamma_b)]^m[1 - P_2(\gamma_b)]^{n-m} \qquad (33)$$

The improvement due to FEC coding is shown in Fig. 12. It is seen that the BCH (15,7) code improves the performance significantly. From comparing Fig. 11 with Fig. 12, we also see that a higher Rice factor $R = 12$ dB (curve d, Fig. 11), yields a higher throughput than $R = 7$ dB (curve b, Fig. 12).

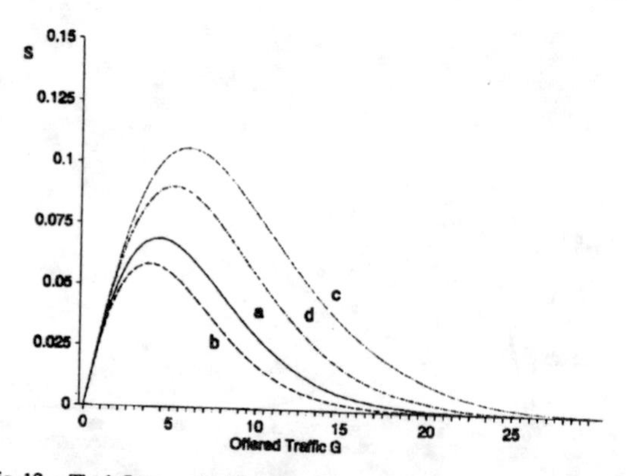

Fig. 12. The influence of FEC coding on the throughput for slow Rice fading. Multi-micro-cell, Maximal ratio combining, $N = 127$, $R = 7$ dB, $C = 30$, $M = 2$. No coding, $L = 6$, (a) $N_d = 512$, (b) $N_d = 1024$, and BCH-(15,7) FEC coding, $L = 12$, (c) $N_d = 512$, (d) $N_d = 1024$.

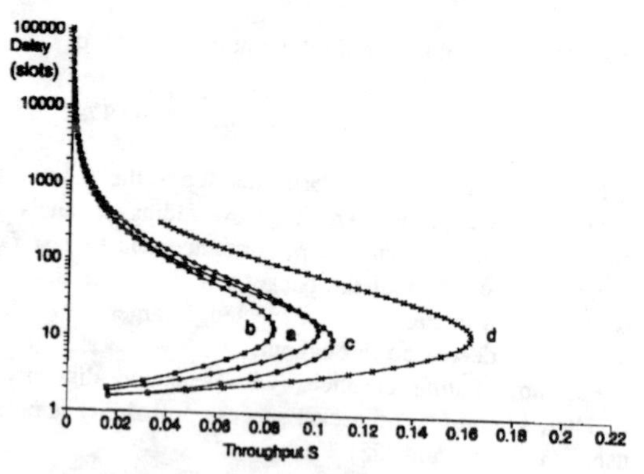

Fig. 13. Throughput-delay curve for slow Rice fading. The delay is given in slot-times. Mulit-micro-cell, Maximal ratio combining, $N = 127$, $C = 30$, $M = 2$. (a) $L = 6$, $R = 7$ dB, $N_d = 128$, (b) $L = 6$, $R = 7$ dB, $N_d = 256$, (c) $L = 6$, $R = 12$ dB, $N_d = 128$, (d) FEC coding, $L = 12$, $R = 7$ dB, $N_d = 128$.

The average delay in slow fading channels can be calculated using (27). Fig. 13 shows some throughput delay results for various system configurations. The effects of the Rice factor and packet length on the performance are observed as expected (curves *a*, *b*, and *c*), i.e., a higher R and lower N_d causes higher throughput and lower delay. FEC coding yields much enhanced performance (curve *d*).

VI. Comparison Between Macro-, Micro- and Pico-Cellular Radio Systems

In a macro-cellular system, the channel is Rayleigh fading because of the relatively large size of the cell. This implies that $S = 0$, i.e., $R = 0$, and (7) reduces to

$$P_{\beta_{1k}} = \frac{r}{\sigma^2} \exp\left[-\frac{r^2}{2\sigma^2}\right]. \tag{34}$$

Furthermore, the rms delay spread is in the order of 8 μs [18]. From Table I, we can see that, with $R_b = 32$ kbit/s and $N = 127$, the number of resolvable paths $L = 33$ if $T_m = 8$ μs.

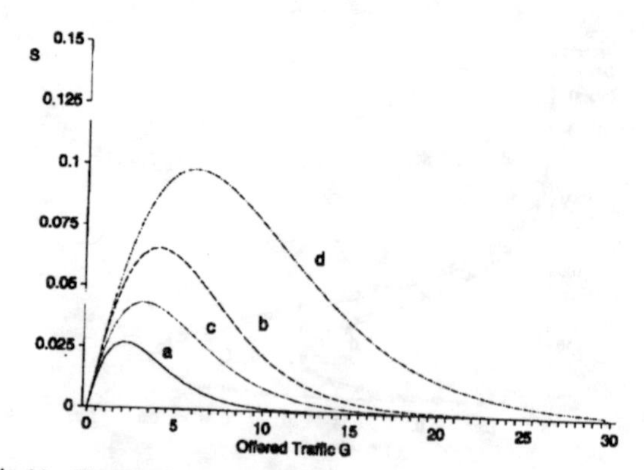

Fig. 14. Throughput curves for fast and slow Rayleigh fading (Multi-macro-cell-environment). Maximal ratio combining, $N = 127$, $R = 0$, $C = 30$, $N_d = 32$, $M = 4$. (a) fast, $L = 32$, (b) fast, $L = 16$, (c) slow, $L = 32$, (d) slow, $L = 16$.

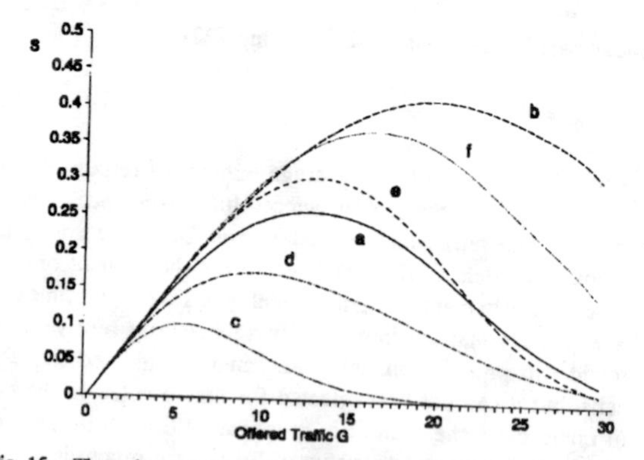

Fig. 15. Throughput curves for fast and slow Rician fading (Multi-pico-cell-environment). Maximal ratio combining, $N = 127$, $R = 6.8$ dB, $C = 30$, $N_d = 128$. $M = 2$, $L = 2$, (a) fast, (b) slow. $M = 2$, $L = 4$, (c) fast, (d) slow. $M = 4$, $L = 4$, (e) fast, (f) slow.

In Fig. 14, the throughput results are shown for fast and slow Rayleigh fading channels, in a macro-cellular environment. As in micro-cells, macro-cellular slow fading channels also yield a higher throughput than the throughput due to fast fading channels.

In the indoor environment, pico-cells are used which are characterized by Rician fading channels and having a rms delay spread of 50 ns–200 ns. In these pico-cells, the Rice factor R is found to be 6.8 dB and 11 dB, depending on the type and construction material of the building [19].

In Fig. 15, throughput results for fast and slow Rician fading channel are depicted for an indoor, pico-cellular surrounding. As seen in Fig. 15, the same conclusions apply for slow and fast fading channels in pico-cells as for macro-, and micro-cells. Note, however, that in pico-cells the system performance (channel capacity) is far better than the macro- and micro-cellular performance. This can also be seen from the following figure, where comparisons among macro-cellular, micro-cellular, and pico-cellular CDMA systems are shown (Fig. 16). In Fig. 16, the system capacity is $C = 30$. As expected, pico-cells yield the best performance, due to the (much) smaller rms delay spread.

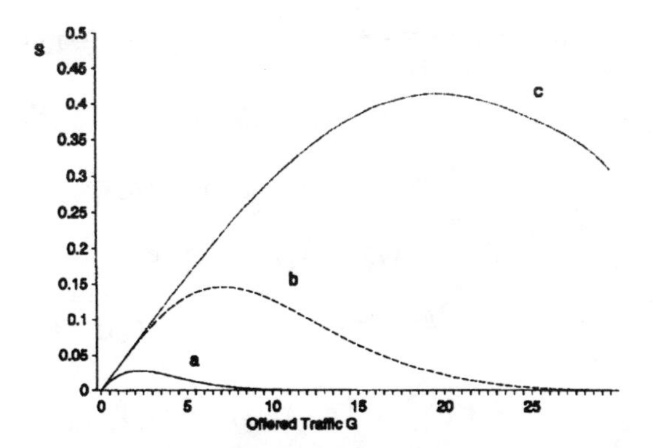

Fig. 16. Throughput curves for slow fading in Mulit-macro-, Multi-micro- and Multi-pico-cells. Maximal ratio combining, $N = 127$, $C = 30$, $R_b = 32$ kbit/s, $N_d = 128$. (a) Macro-cell, $R = 0$, $M = 4$, $L = 32$ (7.63 μs $< T_m <$ 7.87 μs), (b) Micro-cell, $R = 7$ dB, $M = 4$, $L = 8$ (1.72 μs $< T_m <$ 1.97 μs), (c) Pico-cell, $R = 6.8$ dB, $M = 2$, $L = 2$ (246 ns $< T_m <$ 492 ns).

VII. Conclusions

The throughput and average delay are evaluated for a micro-cellular CDMA system using DPSK modulation with two types of diversity schemes and BCH (15,7) forward error correcting codes. The computational results are obtained for both slow and fast Rician fading micro-cellular channels. The numerical results show that in fast fading channels, the throughput is lower than the throughput in slow fading channels (Figs. 11, 14, and 15). Furthermore, it was observed that a higher users' capacity C results in a lower channel capacity (maximum throughput). Larger packet lengths (Figs. 9, 11, and 12) and lower signal-to-noise ratios also result in a worse performance.

The Rician fading channel offers a better performance than the Rayleigh fading channel. A higher Rice factor results in a better performance (Figs. 5 and 13). Diversity substantially improves the performance. A higher order of diversity M yields higher throughput. As expected from theory, the performance due to maximal ratio combining is found to be superior to selection diversity in the case of micro-cellular CDMA system also (Figs. 6 and 7).

Performance degrades when an increase in data rate and/or delay spread results in a higher number of resolvable paths L (Fig. 4). A larger code sequence length N improves the performance (although the number of resolvable paths increases). Forward error correcting coding improves the performance, and performance is significantly enhanced in the case of increased packet length (Figs. 8, 9, 10, 12, and 13).

After comparing the performance of micro-cells with macro- and pico-cells, it is observed that due to the difference in rms delay spread T_m, micro-cells yield better performance than macro-cells, but worse than pico-cells (Figs. 14, 15, and 16).

Appendix A
Derivation of Bit Error Probability for Selection Diversity: Equation (12)

As in [26], (10) can be divided into two parts: in-phase and a quadrature component. Selecting user 1 as the reference user, the output (in-phase and quadrature component) of the matched filter of user 1 at the sampling point ($t = T_b$) is given as:

$$g_x(T_b) = \sum_{k=1}^{K}\sum_{l=1}^{L} A\beta_{lk}\cos(\phi_{lk}) \cdot [b_k^{-1}R_{1k}(\tau_{lk}) + b_k^0\hat{R}_{1k}(\tau_{lk})] + \eta$$

$$g_y(T_b) = \sum_{k=1}^{K}\sum_{l=1}^{L} A\beta_{lk}\sin(\phi_{lk}) \cdot [b_k^{-1}R_{1k}(\tau_{lk}) + b_k^0\hat{R}_{1k}(\tau_{lk})] + \nu \quad \text{(A1)}$$

where g_x and g_y are the in-phase and the quadrature component, respectively, b_k^{-1} and b_k^0 are the previous and current data bit, respectively, and:

$$R_{1k}(\tau) = \int_0^{\tau} a_k(t-\tau)a_1(t)\,dt$$

$$\hat{R}_{lk}(\tau) = \int_{\tau}^{T_b} a_k(t-\tau)a_1(t)\,dt \quad \text{(A2)}$$

The noise samples η and ν are independent, zero-mean Gaussian random variables with identical variance $\sigma_n^2 = N_oT_b/\sigma^2$. Let us assume without loss of generality that the receiver synchronizes to the jth path of user 1, so that $\tau_{j1} = 0$ and $\phi_{j1} = 0$ [26]. The complex envelope of the signal at the current sampling instant then is

$$z_0 = A\beta_{jl}T_bb_1^0 + \sum_{k=1}^{K} A(b_k^{-1}X_k + b_k^0\hat{X}_k) + j\sum_{k=1}^{K} A(b_k^{-1}Y_k + b_k^0\hat{Y}_k) + (\eta_1 + j\nu_1) \quad \text{(A3)}$$

where

$$X_1 = \sum_{\substack{l=1\\l\neq j}}^{L} R_{11}(\tau_{ll})\beta_{ll}\cos(\phi_{ll});$$

$$\hat{X}_1 = \sum_{\substack{l=1\\l\neq j}}^{L} \hat{R}_{11}(\tau_{ll})\beta_{ll}\cos(\phi_{ll})$$

$$Y_1 = \sum_{\substack{l=1\\l\neq j}}^{L} R_{11}(\tau_{ll})\beta_{ll}\sin(\phi_{ll});$$

$$\hat{Y}_1 = \sum_{\substack{l=1\\l\neq j}}^{L} \hat{R}_{11}(\tau_{ll})\beta_{ll}\sin(\phi_{ll}) \quad \text{(A4)}$$

and

$$X_{\substack{k\\k\neq 1}} = \sum_{l=1}^{L} R_{1k}(\tau_{lk})\beta_{lk}\cos(\phi_{lk});$$

$$\hat{X}_{\substack{k\\k\neq 1}} = \sum_{l=1}^{L} \hat{R}_{1l}(\tau_{lk})\beta_{lk}\cos(\phi_{lk})$$

$$Y_{\substack{k\\k\neq 1}} = \sum_{l=1}^{L} R_{1k}(\tau_{lk})\beta_{lk}\sin(\phi_{lk});$$

$$\hat{Y}_{\substack{k \\ k\neq 1}} = \sum_{l=1}^{L} \hat{R}_{1k}(\tau_{lk})\beta_{lk} \sin(\phi_{lk}). \tag{A5}$$

Involving only b_k^{-1} and b_k^0 means that it is assumed that $\tau_{lk} \geq 0$. This assumption can be made without loss of generality for two reasons: 1) all bits, except b_1^{-1} and b_1^0, are 1 or -1 with equal probability; 2) it is not important for the cross-correlation of two spread spectrum codes what the sign of the phase difference between the codes is. DPSK demodulation is now achieved by taking the real part of $z_o z_{-1}^*$, where z^* denotes complex conjugate of z.

The selection diversity scheme is based on selecting the strongest of L resolvable paths. By using multiple antennas, the highest possible order of diversity, i.e. number of paths to choose from, can be increased to $M_{\max} = kL$ where $M_{\max}$ is the maximum order of diversity and k is the number of antennas. To derive the PDF of the strongest path ($\beta_{\max}$), it is essential to note that the cumulative density function (CDF) of $\beta_{\max}$ is just the CDF of the PDF in (8) raised to the power of the order of diversity, M, hence:

$$p_{\beta_{\max}}(\nu) = M[1 - Q(s,\nu)]^{M-1}\nu \exp\left(-\frac{\nu^2+s^2}{2}\right) I_o(s\nu) \tag{A6}$$

where Q is the Marcum Q-function. Designate the decision variable for DPSK modulation $\xi = \mathrm{Re}[z_0 z_{-1}^*]$, where $\mathrm{Re}[a]$ denotes the real part of a. The decision variable obtained from demodulation of the strongest path is written as $\xi_{\max}$. Now the bit error probability in the case of selection diversity is defined as

$$P_e \Delta P(\xi_{\max} < 0 \mid b_1^0 b_1^{-1} = 1) \Delta P(\xi_{\max} > 0 \mid b_1^0 b_1^{-1} = -1). \tag{A7}$$

In the analysis it is assumed that $b_1^0 b_1^{-1} = 1$; i.e., $b_1^0 = b_1^{-1} = 1$.

If we assume that all path delays are given and β_{max} is correctly selected, the bit error probability can be obtained using equation (7A.26) of [21] as:

$$\begin{aligned} P_e \mid \beta_{\max}, \{\tau_{lk}\} &= Q(a,b) \\ &- \frac{1}{2}\left[1 + \frac{\mu}{\sqrt{\mu_0\mu_{-1}}}\right] I_o(ab) \exp\left(-\frac{a^2+b^2}{2}\right). \end{aligned} \tag{A8}$$

Here a, b, μ, μ_0, and μ_{-1} are defined as ($b_1^0 = b_1^{-1} = 1$):

$$\begin{aligned} &a \Delta \frac{m}{\sqrt{2}}\left(\frac{1}{\sqrt{\mu_0}} - \frac{1}{\sqrt{\mu_{-1}}}\right), \quad b \Delta \frac{m}{\sqrt{2}}\left(\frac{1}{\sqrt{\mu_0}} + \frac{1}{\sqrt{\mu_{-1}}}\right), \\ &m \Delta E[z_0 \mid \beta_{\max}, b_1^0] = E[z_{-1} \mid \beta_{\max}, b_1^{-1}] \\ &\mu_0 \Delta \mathrm{var}(z_0 \mid \{\tau_{lk}\}), \quad \mu_{-1} \Delta \mathrm{var}(z_{-1} \mid \{\tau_{lk}\}), \\ &\mu \Delta E[(z_0 - m)(z_{-1} - m)^* \mid \{\tau_{lk}\}] \end{aligned} \tag{A9}$$

with $E[\ ]$ denoting statistical average and $\mathrm{var}(\ )$ denoting variance.

Using (A3) and the assumption that $b_1^0 = b_1^{-1} = 1$, M, μ_0, μ_{-1} and μ are obtained as

$$\begin{aligned} m &= A\beta_{\max}T_b b_1^0 = A\beta_{\max}T_b b_1^{-1} \\ \mu_0 &= A^2 \sum_{k=1}^{K} E[X_k^2 + \hat{X}_k^2 + Y_k^2 + \hat{Y}_k^2 \mid \{\tau_{lk}\}] \\ &\quad + 2A^2 E[X_1\hat{X}_1 + Y_1\hat{Y}_1 \mid \{\tau_{lk}\}] + 2\sigma_n^2 \\ \mu_{-1} &= A^2 \sum_{k=1}^{K} E[X_k^2 + \hat{X}_k^2 + Y_k^2 + \hat{Y}_k^2 \mid \{\tau_{lk}\}] + 2\sigma_n^2 \\ \mu &= A^2 E\left[\sum_{k=1}^{K}(X_k\hat{X}_k + Y_k\hat{Y}_k) + \hat{X}_1^2 + \hat{Y}_1^2 \mid \{\tau_{lk}\}\right]. \end{aligned} \tag{A10}$$

Note that all path gains involved in μ_0, μ_{-1}, and μ have a Rician distribution as given in (8). The distribution of $\beta_{\max}$ is given in (A6).

As can be seen from (A10), the only difference between μ_0 and μ_{-1} is constituted by the second term of μ_0. If the number of simultaneously transmitting users, K, is large, the contribution of this term to μ_0 becomes relatively small [26]. If we drop this term from μ_0, a in (A9) becomes zero. Equation (A8) can then be simplified to

$$\begin{aligned} &P_e \mid \beta_{\max}, \{\tau_{lk}\} \\ &= Q(0,b) - \frac{1}{2}\left[1 + \frac{\mu}{\sqrt{\mu_0\mu_{-1}}}\right] I_0(0) \exp\left(-\frac{b^2}{2}\right) \\ &= \frac{1}{2}\left[1 - \frac{\mu}{\mu_0}\right] \exp\left(-\frac{m^2}{\mu_0}\right) \end{aligned} \tag{A11}$$

since $I_0(0) = 1$ and $Q(0,b) = \exp(-b^2/2)$. Now to remove the conditioning on τ_{lk}, we approximate μ and μ_0 by Gaussian variables and integrate (A11) over μ and μ_0. Likewise, the conditioning on $\beta_{\max}$ is removed. Thus (A11) reduces to

$$\begin{aligned} P_e = &\int_{-\infty}^{\infty}\int_{-\infty}^{\infty}\int_{0}^{\infty} \frac{1}{2}\left[1 - \frac{\mu}{\mu_0}\right] \exp\left(-\frac{A^2\beta_{\max}^2 T_b^2}{\mu_0}\right) \\ &\cdot p_{\beta_{\max}}(\beta_{\max}) p_{\mu_0}(\mu_0) P\mu(\mu)\, d\beta_{\max}\, d\mu_0\, d\mu. \end{aligned} \tag{A12}$$

For the calculation of the variance of μ and μ_0, a specific set of Gold codes is used. The partial cross-correlation functions are calculated using the specific Gold codes and averaging over the (uniform) path delay distributions. For calculating the moments of μ and μ_0, these calculations have to be done for all resolvable paths of all users.

ACKNOWLEDGMENT

The author are thankful to the anonymous reviewers for their valuable suggestions.

REFERENCES

[1] W. C. Y Lee, "Power control in CDMA," in *Proc. 41st IEEE Veh. Technol. Conf.* May 1991, pp. 77–80.
[2] W. C. Y. Lee, "Theory of wideband radio propagation," in *Proc. 41st IEEE Veh. Technol. Conf.*, May 1991, pp. 285–288.
[3] W. C. Y. Lee "Overview of cellular CDMA," in *Proc. 41st IEEE Veh. Technol. Conf.*, May 1991, pp. 291–302.

[4] K. S. Gilhousen, I. M. Jacobs, R. Padovani, A. J. Viterbi, L. A. Weaver, and C. E. Wheatley, "On the capacity of a cellular CDMA system," *IEEE Trans. Veh. Technol.*, vol. 40, pp. 303–312, May 1991.
[5] A. Salmasi and K. S. Gilhousen, "On the system design aspects of code division multiple access (CDMA), applied to digital cellular and personal communications networks," in *Proc. 41st IEEE Veh. Technol. Conf.*, May 1991, pp. 57–62.
[6] R. L. Pickholtz, L. B. Milstein, and D. L. Schilling, "Spread spectrum for mobile communications," *IEEE Trans. Veh. Technol.*, vol. 40, pp. 313–322, May 1991.
[7] D. L. Schilling, L. B. Milstein, R. L. Pickholtz, M. Kullback, and F. Miller, "Spread spectrum for commercial communications," *IEEE Commun. Mag.*, pp. 66–79, Apr. 1991.
[8] M. A. Beach, A. Hammer, S. A. Allpress, J. P. McGeehan, and A. Bateman, "An evaluation of direct sequence CDMA for future mobile communication networks," in *Proc. 41st IEEE Veh. Technol. Conf.*, pp. 63–70, May 1991.
[9] H. Yamamjura, R. Kohno, and H. Imai, "Interference cancellation and capacity of a cellular CDMA system based on multihop fast frequency hopping," in *Proc. 41st IEEE Veh. Technol. Conf.*, pp. 71–76, May 1991.
[10] H. S. Misser and R. Prasad, "Spectrum efficiency of a mobile cellular radio system using direct sequence spread spectrum," in *Proc. IEEE TENCON '91*, New Delhi, India, August 28–30, 1991.
[11] R. J. C. Bultitude and G. K. Bedal, "Propagation characteristics in microcellular urban mobile radio channels at 910 MHz," *IEEE J. Select. Areas Commun.*, vol. 7, pp. 31–39, Jan. 1989.
[12] E. Green "Radio link design for micro-cellular systems," *British Telecom. Techn.*, vol. 8, pp. 85–96, Jan. 1990.
[13] H. Harley, "Short distance attenuation measurements at 900 MHz and 1.8 GHz using low antenna heights for micro-cells," *IEEE J. Select. Areas Commun.*, vol. 7, pp. 5–10, Jan. 1989.
[14] S. T. S. Chia, R. Steel, E. Green, and A. Baran, "Propagation and bit error ratio measurements for a micro-cellular system," *J. Inst. Electric Radio Eng.*, vol. 57, no. 6 (Supplement), pp. s255–266, Nov./Dec. 1987.
[15] R. Prasad, A. Kegel, and J. Olsthoorn, "Spectrum efficiency for microcellular mobile radio systems," *Electron. Lett.*, vol. 27, pp. 423–425, Feb. 1991.
[16] R. Prasad and A. Kegel, "Spectrum efficiency of micro-cellular systems," *Proc. 41st IEEE Veh. Technol. Conf.*, May 1991, pp. 357–361.
[17] R. Prasad and A. Kegel, "Improved assessment of interference limits in cellular radio performance," *IEEE Trans. Veh. Technol.*, vol. 40, pp. 412–419, May 1991.
[18] S. Y. Seidel and T. S. Rappaport, "path loss and multipath delay statistics in four European cities for 900 MHz cellular and micro-cellular communications," *Electron. Lett.*, vol. 26, pp. 1713–1714, Sept. 1990.
[19] R. Prasad, H. S. Misser, and A. Kegel, "Performance analysis of direct sequence spread spectrum multiple access communication in an indoor Rician fading channel with DPSK modulation," *Electron. Lett.*, vol. 26, pp. 1366–1367, Sept. 1990.
[20] H. S. Misser, C. A. F. J. Wijffels, and R. Prasad, "Throughput analysis of CDMA with DPSK modulation and diversity in indoor Rician fading radio channels," *Electron. Lett.*, vol. 27, pp. 601–602, March 1991.
[21] J. G. Proakis, *Digital Communications*, 2nd ed. New York: McGraw-Hill, 1989.
[22] L. Kleinrock and F. A. Tobagi, "Packet switching in radio channels—Part I—Carrier Sense Multiple-Access modes and their throughput-delay characteristics," *IEEE Trans. Commun.*, vol. 23, pp. 1400–1416, 1975.
[23] M. Abramowitz and I. A. Segun, Eds., *Handbook of Mathematical Functions with Formulas, Graphs, and Mathematical Tables.* New York: Dover, 1968.
[24] J. M. Musser and J. N. Daigle, "Throughput analysis of an asynchronous code division multiple access (CDMA) system,' in *Proc. Int. Conf. Commun. (ICC)*, 1982, pp. 2F.2.1.–2F.2.7.
[25] C. L. Weber, G. K. Huth, and B. H. Batson, "Performance considerations of code division multiple-access systems," *IEEE Trans. Veh. Technol.*, vol. 30, pp. 3–9, Feb. 1981.
[26] M. Kavehrad and B. Ramamurthi, "Direct-sequence spread spectrum with DPSK modulation and diversity for indoor wireless communication," *IEEE Trans. Commun.*, vol. Com-35, pp. 224–236, Feb. 1987.

Caspar A. F. J. Wijffels was born in Roermond, The Netherlands, on March 29, 1966. He received the M.Sc. E.E. degree from Delft University of Technology, The Netherlands, in 1991. His dissertation was on throughput, delay and stability analysis of a slotted code division multiple access (CDMA) system for indoor wireless communication.

After his graduation, he joined the Telecommunications and Traffic-Control Systems Group of Delft University of Technology to do further investigations on CDMA. He is currently with Alcatel Telecom Systems B.V. in The Netherlands as a Product Manager for radio communications and defense systems.

Howard S. Misser (S'89–M'91) was born in Paramaribo, Surinam, on May 11, 1968. He received the M.Sc. degree in electrical engineering from Delft University, Delft, The Netherlands, in 1990.

He worked at the Telecommunications and Traffic-Control Systems Group of the same university as a Research Fellow. He is currently with the Dr. Neher Laboratories of the Royal Dutch PTT in Leidschendam, The Netherlands. His research interests are in the fields of broadband network technologies and architecture, radio and mobile communications, and spread-spectrum communication.

Ramjee Prasad (M'88–Sm'90) for a photograph and biography, please see page 281 of the August 1993 issue of this TRANSACTIONS.

CDMA Overlay Situations for Microcellular Mobile Communications

Jiangzhou Wang, *Senior Member, IEEE*, and Laurence B. Milstein, *Fellow, IEEE*

***Abstract*—Direct sequence code division multiple access communications is a promising approach to cellular mobile communications, which operates in an environment characterized by multipath Rician fading. In this paper, the CDMA network is assumed to share common spectrum with a narrowband microwave user. Because of the presence of the narrowband waveform, an interference suppression filter at each CDMA receiver is employed to reject the narrowband interference. The problem of interference from adjacent cells is also considered. Average power control is assumed to combat the near/far problem, and multipath diversity, in conjunction with simple interleaved channel coding, is considered for improving the performance of the CDMA system.**

I. INTRODUCTION

Code division multiple access (CDMA) is becoming a very attractive technique for personal communications networks (PCN) and microcellular mobile communications [1-7]. One reason is that the CDMA code sequences inherently combat multipath which is a major concern in mobile radio communications. Also, since the CDMA signals spread their power over a large bandwidth, the effect of such a transmission on a narrowband receiver is often just an imperceptible rise in its noise level. Hence, it may be possible to overlay a CDMA system on an existing set of microwave narrowband users without adversely affecting either. Moreover, because of the broadband noise-like character of direct sequence (DS) CDMA signals, it is possible to use newly developed signal processing techniques to perform narrowband interference suppression [14]. This will clearly reduce the narrowband interference for CDMA users, but it will also help the narrowband users, since it is then possible to operate the CDMA transmissions at a reduced power level.

As is well known, microcellular mobile radio can increase the capacity of future cellular systems. The chief advantages of the microcells are the enhanced spectrum efficiency and low power consumption of the handsets. Microcells, as the name indicates, are relatively much smaller in size (0.2-1 km radius) than the conventional cells (1 - 10 km radius). Also, microcells operate at lower power and have their antennas at streetlamp elevations. In [3], it is shown that in multipath models the propagation path (per path) loss exponent for close-in cells can be between two and three in microcellular radio, instead of four in conventional cellular and terrestrial radio. Also, the propagation measurements in [4] characterize the microcellular environments as a multipath Rician (rather than Rayleigh) fading channel.

Because of the presence of the narrowband waveform, a suppression filter at each CDMA receiver is employed to reject the interference. The problem of interference from adjacent cells is considered. Also, diversity is used in conjunction with block interleaved coding for improving the BER performance. The outline of the paper is as follows. The system and channel models are described in Section II. Section III derives an approximation to the bit error rate (BER) performance of the DS-CDMA overlay system, and this is followed by both representative numerical results and a discussion of these results in Section IV. Finally, conclusions are presented in Section V.

This paper was approved by James S. Lehnert, the Editor for Modulation & Signal Design of the IEEE Communications Society. Manuscript received: December 9, 1991; revised: January 11, 1993.

This work was supported by the Office of Naval Research under Grant N00014-91-J-1234 and the National Science Foundation Industry/University Cooperative Research Center on Ultra-High Speed Integrated Circuits and Systems at the University of California, San Diego.

Jiangzhou Wang is with Digital Communications Division, Rockwell Corporation, Newport Beach, CA 92658-8902

Laurence B. Milstein is with the Department of Electrical & Computer Engineering, University of California, San Diego, La Jolla, CA 92093

IEEE Log Number 9410093.

II. SYSTEM AND CHANNEL MODELS

A. *Transmitter Model*

The transmitter model of a DS system with interleaved coding and PSK modulation consists of a channel encoder, an interleaver and a DS spreader (or modulator). The transmitted signal of the kth user in the CDMA system takes the form

$$S_k(t)=\mathrm{Re}\left\{\sqrt{2P_k}\, b_k(t)\, a_k(t) \exp\left[j(2\pi f_0 t+\theta_k)\right]\right\}, \quad (1)$$

where $\mathrm{Re}\{\cdot\}$ stands for real part, f_0 denotes the carrier frequency, which is common to all users, and P_k and θ_k are the transmitted power and phase of the kth user, respectively. The symbol $b_k(t)$ is the kth interleaved and coded information bit; the bit rate $R_b = 1/T_b$, where T_b denotes the duration of a coded symbol. The information bit rate R_i (or the rate of source bits) and information bit duration T_i are given by $R_i = 1/T_i = (m/n)R_b$ and $T_i = T_b/(m/n)$, respectively, where m/n is the rate of the code. The λth square pulse of $b_k(t)$ has amplitude $b_k^{(\lambda)}$, taking values from $\{+1,-1\}$. The random signature sequence waveform $a_k(t)$ has a chip rate $1/T_c$, and the jth chip of $a_k(t)$ is denoted by $a_j^{(k)}$. We assume that there are N chips (processing gain) of the random signature sequence in each coded bit ($T_b = NT_c$). Therefore, the spread spectrum bandwidth (B_s) is approximately given by

$$B_s = 2T_c^{-1} = 2N(T_i m/n)^{-1}. \qquad (2)$$

B. Channel Model

The mobile radio channel can be modeled as a discrete multipath fading channel. "Discrete multipath" means that the received signal is the sum of a finite number of delayed versions of the transmitted signal, whereas "fading" refers to the fact that, for a given transmitted power, the received power fluctuates in a random fashion. Because of the multipath nature of the channel, the fading is frequency-selective.

The propagation measurements in [4] characterize the microcellular environments as a multipath Rician fading channel, where the received signal consists of several specular components plus several Rayleigh-fading components. The multipath Rician fading channel between the kth user and the receiver of interest (namely the receiver in the base station of what we refer to as the first cell) is modeled by the complex lowpass equivalent impulse response

$$h_k(t) = \frac{1}{(d_{1,k})^{\gamma/2}} \sum_{l=1}^{L} \left[A_{kl}\exp(j\eta_{kl}) + \beta_{kl}\exp(j\mu_{kl}) \right] \delta(t-\tau_{kl}), \qquad (3)$$

where $d_{1,k}$ ($d_{1,k} \neq 0$) is the distance between the kth user and the first cell base station, and γ is the propagation path loss exponent. In (3), A_{kl} ($0 \leq A_{kl} \leq 1$) and η_{kl} are gain and phase of the specular component of the lth path from the kth user, respectively; they are assumed to be deterministic and nearly constant over the duration of at least one bit. The random gain β_{kl} and random phase μ_{kl} of the fading component of the lth path of the kth user have a Rayleigh distribution with $E[\beta_{kl}^2] = 2\rho_{kl}$, and a uniform distribution in $[0, 2\pi]$, respectively. The path delay, τ_{kl}, is uniformly distributed in $[0, T_b]$. We assume that there are L paths associated with each user. The gains, delays and phases of different paths and/or of different users are all statistically independent. We define the quantity H_{kl}, as the ratio of the specular component power to the Rayleigh fading power. That is,

$$H_{kl} \equiv A_{kl}^2 / E[\beta_{kl}^2] = A_{kl}^2 / (2\rho_{kl}). \qquad (4)$$

The microcellular (outdoor) channel measurements from [4] show that Rician distributions occur with H_{kl} values ranging from 7 dB to 12 dB, instead of 2dB to 7dB for an indoor radio channel [7].

The interference in the channel is assumed to be a non-fading narrowband BPSK signal, and is given by

$$J(t) = \mathrm{Re}\left\{ \sqrt{2J}\, d(t) \exp\left[j\,[2\pi(f_0 + \Delta)t + \Theta] \right] \right\}, \qquad (5)$$

where Δ stands for the offset of the interference carrier frequency from the carrier frequency (f_0) of the CDMA signals. The parameters J and Θ denote the received interference power and phase, respectively. The information sequence $d(t)$ has the bit rate T_j^{-1}, where T_j denotes the duration of one bit. Therefore, the interference bandwidth is approximately $B_j = 2T_j^{-1}$ (we assume $B_j < B_s$). The ratio (p) of the interference bandwidth to the spread spectrum bandwidth and the ratio (q) of the offset of the interference carrier frequency to half of the spread spectrum bandwidth are defined as, respectively,

$$p = B_j / B_s = T_c / T_j \qquad (6)$$

and

$$q = \Delta / (B_s/2) = \Delta T_c. \qquad (7)$$

Therefore, the received signal $r(t)$ can be represented as

$$r(t) = \mathrm{Re}\left\{ \sqrt{2P} \sum_{k=1}^{CK} \sqrt{\varepsilon(\gamma, c_k, k)} \sum_{l=1}^{L} [A_{kl} exp(j\phi_{kl}) + \beta_{kl} exp(j\psi_{kl})] \cdot b_k(t-\tau_{kl}) a_k(t-\tau_{kl}) \exp(j2\pi f_0 t) \right\} + J(t) + n(t), \qquad (8)$$

where $n(t)$ is an AWGN process with two-sided power spectral density $N_0/2$, $\phi_{kl} \equiv \theta_k + \eta_{kl} - 2\pi f_0 \tau_{kl}$, $\psi_{kl} \equiv \theta_k + \mu_{kl} - 2\pi f_0 \tau_{kl}$. C stands for the number of cells, each one containing K users, and c_k denotes the c_kth cell (the integer portion of $1+(k-1)/K$, $c_k = 1, \cdots, C$). The first cell ($c_k = 1$) is defined as the cell of interest, and P and $\varepsilon(\gamma, c_k, k)$ are defined as

$$P = P_k / (d_{c_k,k})^{\gamma} = \text{constant} \qquad (9)$$

and

$$\varepsilon(\gamma, c_k, k) = (d_{c_k,k} / d_{1,k})^{\gamma}, \qquad (10)$$

respectively, where $d_{c_k,k}$ is the distance of the kth mobile user to its own base station (the c_kth cell), and $d_{c_k,k} \neq 0$. The constant P means each base station provides adaptive power control to all K users of its own cell so that the received signals from its cell have approximately equal power to avoid the near/far problem.

C. Receiver Model

As shown in Fig. 1(a), the receiver contains a suppression filter, a bank of bandpass (BP) matched filters, a bank of U parallel PSK demodulators and diversity circuits a hard decision device, a deinterleaver and a hard decision decoder. A detailed model of the PSK demodulators and diversity circuit is shown in Fig. 1(b).

Notice that, for simplicity, we have ignored any bandpass filtering at the front-end of the suppression filter. We take the suppression filter to be a double-sided Wiener filter with M taps on each side. Its impulse response is $\sum_{m=-M}^{M} \alpha_m \delta(t - mT_c)$, where $\alpha_0 = 1$ and $\alpha_m = \alpha_{-m}$. The resulting suppression filter output is given by

$$r_f(t) = \sum_{m=-M}^{M} \alpha_m r(t - mT_c), \qquad (11)$$

where $f_0 T_c$ is taken to be an integer.

The signal $r_f(t)$ enters a bank of BP matched filters, whose impulse responses are matched to consecutive T_b-second segments of $2a_i(t)\cos(2\pi f_0 t)\, P_{T_b}(t - vT_b)$, where $P_{T_b}(t) = 1$ for $0 \leq t < T_b$ and is 0 otherwise, $v = 1, \cdots, V$,

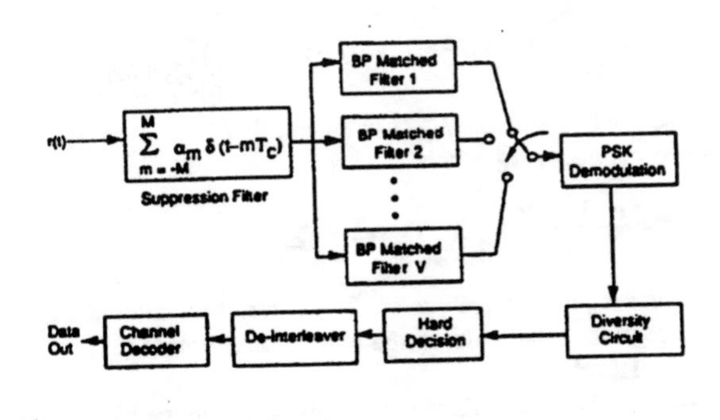

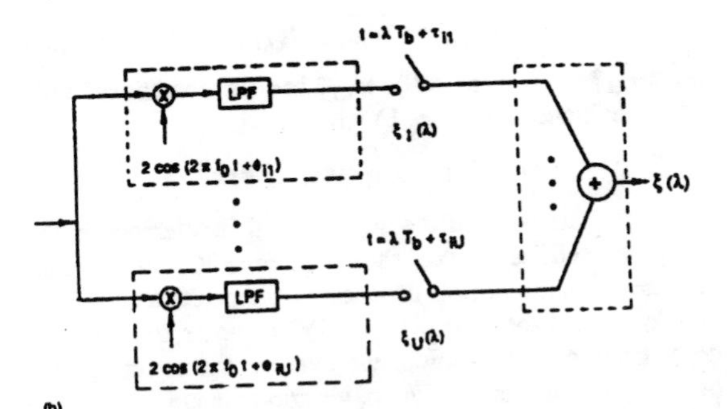

Fig. 1. (a) Receiver model of a CDMA overlay system. (b) A detailed model of the PSK demodulators and the diversity circuit.

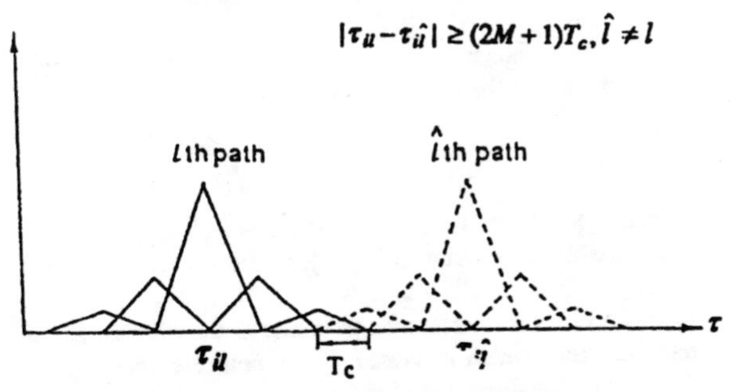

Fig. 2. The matched filter output envelope in CDMA overlay situation with suppression filters (2M+1=5) over a multipath channel.

and V is the number of matched filters. Also it is assumed that the ith user of the first cell is the reference user. During each bit duration, the BP signal at the output of the correct matched filter contains $2M+1$ narrow pulses, each of width $2T_c$, centered at instants $\lambda T_b+\tau_{il}+mT_c$ for $m=-M, \cdots, 0, \cdots, M$, which are caused by the $2M+1$ taps of the suppression filter and the lth ($l=1, 2, \cdots, L$) path of the reference user. Of the $2M+1$ peaks, the middle path of the reference user. Of the $2M+1$ peaks, the middle peak (m=0) is the largest (or main) peak, due to the zero-th tap of the suppression filter, whereas sidelobe peaks (or earlier and later peaks) are due to the taps excluding the zero-th tap. We assume that $|\tau_{il}-\tau_{i\hat{l}}|\geq(2M+1)T_c$ for $\hat{l}=1, \cdots, L$, $\hat{l}\neq l$ (see Fig. 2), since if $|\tau_{il}-\tau_{i\hat{l}}|<(2M+1)T_c$, the overlay of the outputs due to different paths might preclude the paths from being resolvable. We further assume that the receiver only uses the combination of the main peaks of the resolvable paths of the reference user to form the diversity. Paths from non-reference users do not give rise to regular peaks, because their spreading sequences are different from $a_i(t)$.

In any given symbol interval, the output of the appropriate BP matched filter enters U conventional PSK demodulators, $1\leq U\leq L$, (Fig. 1(b)), corresponding to U different resolvable paths of the reference user. The coherent receiver makes use of carrier reference signals $\{2\cos(2\pi f_0 t+\phi_{il})\}$, which are in phase with the specular components of the different signals from the reference user. These carrier reference signals can be derived from the BP matched filter outputs by means of carrier synchronization circuits (one circuit per considered path). The output of the lth, $1\leq l\leq U$, demodulator is a lowpass signal, which has $2M+1$ peaks at the instants $\lambda T_b+\tau_{il}+mT_c$, $m=-M, \cdots, 0, \cdots, M$. The desired component of the lowpass output signal (i.e., the main peak, m=0) of the lth demodulator is sampled at the instant $t=\lambda T_b+\tau_{il}$, and this gives rise to the random variable $\xi_l(\lambda)$. Note that the time difference between any two samples is at least equal to $(2M+1)T_c$. The U random variables, $\xi_l(\lambda)$, $l=1,\cdots,U$, are directly summed to form the decision variable $\xi(\lambda)$.

III. SYSTEM PERFORMANCE

The lth random variable, $\xi_l(\lambda)$, for detection can be written as

$$\xi_l(\lambda)=\int_{\lambda T_b+\tau_{il}}^{(\lambda+1)T_b+\tau_{il}} r_f(t)a_i(t-\tau_{il}) \quad (12)$$

where $2\cos(2\pi f_0 t+\phi_{il})$ is the recovered carrier of the specular component of the lth resolvable path of the reference user and $a_i(t-\tau_{il})$ is the random spreading sequence of the lth path of the reference user. Since the high frequency terms are removed by the lowpass filter following the mixer of the demodulator, the above expression reduces to

$$\xi_l(\lambda)=\sqrt{2P}\,A_{il}T_b b_i^{(\lambda)}+D_l(\lambda)+F_l(\lambda)+N_l(\lambda)+J_l(\lambda)+\sum_{\substack{\hat{l}=1\\ \hat{l}\neq l}}^{L} I_{i,l,\hat{l}}$$

$$+\sum_{\substack{k=1\\ k\neq i}}^{K}\sum_{\hat{l}=1}^{L} I_{k,l,\hat{l}}+\sum_{k=K+1}^{CK}\sum_{\hat{l}=1}^{L} W_{k,l,\hat{l}}\,, \quad (13)$$

where the terms on the right hand side of (13) are described in detail below.

--- The first term is the desired signal, corresponding to the specular component of the lth path of the reference user and the zero-th tap of the suppression filter. The power of this term equals

$$S_c=2PA^2T_b^2\,, \quad (14)$$

where it is assumed that $A_{il}=A$ for all i and l. For most of the remaining terms, we are interested in their conditional variances, conditioned on $b_i^{(\lambda)}$ and $a_i(t)$. Let us consider them one at a time.

--- $D_l(\lambda)$ is an interference term due to the fading component of the lth path of the reference user and the zero-th tap of the suppression filter, and is given by

$$D_l(\lambda)=\sqrt{2P}\,\beta_{il}T_b b_i^{(\lambda)}\cos(\mu_{il}-\eta_{il})\,. \tag{15}$$

Note that this term is a conditional zero-mean Gaussian random variable with variance $2P\rho T_b^2$, where it is assumed that $\rho_{il}=E(\beta_{il}^2)/2=\rho$ for all i and l.

--- $F_l(\lambda)$ is a self interference term due to the lth path of the reference user which is caused by the non-reference taps of the suppression filter; it is given by

$$F_l(\lambda)=\sqrt{2P}\,[A_{il}+\beta_{il}\cos(\mu_{il}-\eta_{il})]\sum_{\substack{m=-M\\m\neq 0}}^{M}\alpha_m \cdot\int_{\lambda T_b+\tau_a}^{(\lambda+1)T_b+\tau_a} b_i(t-mT_c-\tau_{il})a_i(t-mT_c-\tau_{il})a_i(t-\tau_{il})dt\,. \tag{16}$$

For relatively small narrowband interference bandwidths (relative to the spread bandwidth of the CDMA signals), this term does not have a significant effect, and thus it will be ignored in what follows.

--- $N(\lambda)$ is due to the thermal noise; its conditional variance equals $N_0T_b\sum_{m=-M}^{M}\alpha_m^2$.

--- $J_l(\lambda)$ is due to the BPSK narrowband interference, and is given by

$$J_l(\lambda)=\sum_{m=-M}^{M}\alpha_m\int_{\lambda T_b+\tau_a}^{(\lambda+1)T_b+\tau_a}\sqrt{2J}\,d(t-mT_c) \cdot\cos[2\pi\Delta(t-mT_c)+\hat{\Theta}]a_i(t-\tau_{il})dt\,, \tag{17}$$

where $\hat{\Theta}=\Theta+\phi_{il}$ and f_0T_c is assumed to be an integer. This term is very significant, and will be treated separately.

--- $I_{i,l,\hat{l}}\ (\hat{l}\neq l)$ is a multipath interference term due to the $\hat{l}$th path of the reference user and is given by

$$I_{i,l,\hat{l}}=\sqrt{2P}\,[A_{i\hat{l}}\cos(\phi_{i\hat{l}}-\phi_{il})+\beta_{i\hat{l}}\cos(\psi_{i\hat{l}}-\phi_{il})]\sum_{m=-M}^{M}\alpha_m \cdot\int_{\lambda T_b+\tau_a}^{(\lambda+1)T_b+\tau_a} b_i(t-\tau_{i\hat{l}}-mT_c)a_i(t-\tau_{i\hat{l}}-mT_c)a_i(t-\tau_{il})dt \tag{18}$$

For a large number of CDMA users (i.e., for $K\gg 1$), the effect of the multipath of the desired user is very small, since it roughly acts as one additional user. However, its inclusion in the analysis greatly complicates that analysis, and so, as was the case with $F_l(\lambda)$, it will be ignored.

--- $I_{k,l,\hat{l}}\ (k\neq i)$ is a multi-access interference term due to the $\hat{l}$th path of the kth user in the first cell (i.e., c=1 and k=1, 2, ···, K; $k\neq i$) and is given by

$$I_{k,l,\hat{l}}=\sqrt{2P}\,[A_{k\hat{l}}\cos(\phi_{k\hat{l}}-\phi_{il})+\beta_{k\hat{l}}\cos(\psi_{k\hat{l}}-\phi_{il})]\sum_{m=-M}^{M}\alpha_m \cdot\int_{\lambda T_b+\tau_a}^{(\lambda+1)T_b+\tau_a} b_k(t-\tau_{k\hat{l}}-mT_c)a_k(t-\tau_{k\hat{l}}-mT_c)a_i(t-\tau_{il})dt\,. \tag{19}$$

The random variable $I_{k,l,\hat{l}}$ has a conditional variance equal to approximately σ_0^2, where

$$\sigma_0^2=\frac{P(A^2+2\rho)}{3N}\left[2\sum_{m=-M}^{M}\alpha_m^2+\sum_{m=-M}^{M}\alpha_m\alpha_{m+1}\right]T_b^2\,, \tag{20}$$

where $\alpha_m=0$ for $|m|>M$. Therefore, the conditional variance of the total multi-access interference term in (13) equals $(K-1)L\sigma_0^2$.

--- $W_{k,l,\hat{l}}$ is the interference term due to the $\hat{l}$th path of the kth user in the c_kth adjacent cell (k=K+1, ···, CK and $c_k\neq 1$), given by

$$W_{k,l,\hat{l}}=\sqrt{\varepsilon(\gamma,c_k,k)}\,I_{k,l,\hat{l}}\,, \tag{21}$$

where $I_{k,l,\hat{l}}$ is given by (19). $W_{k,l,\hat{l}}$ has the conditional variance, conditioned on $\varepsilon(\gamma,c_k,k)$, given by

$$E[W_{k,l,\hat{l}}^2 \mid \varepsilon(\gamma,c_k,k)]=\varepsilon(\gamma,c_k,k)\sigma_0^2\,. \tag{22}$$

An exact description of the adjacent-cell interference in the hexagonal cellular model, shown in Fig. 3(a), is difficult, because the value of (22) depends on the position of the kth mobile user in the c_kth cell ($c_k\neq 1$). In order to avoid the dependence on the position, we average the result of (22) over the area of c_kth cell. Also, we approximate the hexagonal cell with a circular cell of equal area as in [5] (a hexagonal cell inscribed in a circular cell of radius $1.035r$ is equal in area to a circular cell of radius r), as shown in Fig. 3(b). (Note that there appears to be an error in the calculation of the radius of the circular cell in [5].) The factor $\varepsilon(\gamma,c_k,k)$, for $c_k=2,\cdots,C$, is given by

$$\varepsilon(\gamma,c_k,k)=\left[\frac{x^2}{(x\cos\theta)^2+(y+x\sin\theta)^2}\right]^{\gamma/2}, \tag{23}$$

where y is the distance between the c_kth cell base station and the first cell base station. As shown in Fig. 3(a), the distance between any cell base station in the first layer of cells and the first cell base station is identical ($y=\sqrt{3}*1.035r=1.793r$), where r is the radius of the circular cell, and the first layer contains six cells. The second layer of cells contains twelve cells, of which six have the distance $y=2\sqrt{3}*1.035r=3.586r$ from the first cell base station and the other six have the distance $y=3*1.035r=3.105r$ from the first cell base station. Finally, the third layer of cells contains eighteen cells, of which six have the distance $y=3\sqrt{3}*1.035r=5.379r$ and the other twelve have the distance $y=\sqrt{4.5^2+0.75^2}*1.035r=4.74r$ from the first cell base station. Referring to Fig. 3(b), the average of (23) over the c_kth cell area is given by

$$\varepsilon(\gamma,c_k)=E[\varepsilon(\gamma,c_k,k)]$$

$$=\frac{1}{\pi r^2}\int_0^r\int_0^{2\pi}\left[\frac{x^2}{(x\cos\theta)^2+(y+x\sin\theta)^2}\right]^{\gamma/2}x\,dx\,d\theta$$

$$=\frac{1}{\pi}\int_0^1\int_0^{2\pi}\left[\frac{\hat{x}^2}{(\hat{x}\cos\theta)^2+(y/r+\hat{x}\sin\theta)^2}\right]^{\gamma/2}\hat{x}\,d\hat{x}\,d\theta\,. \tag{24}$$

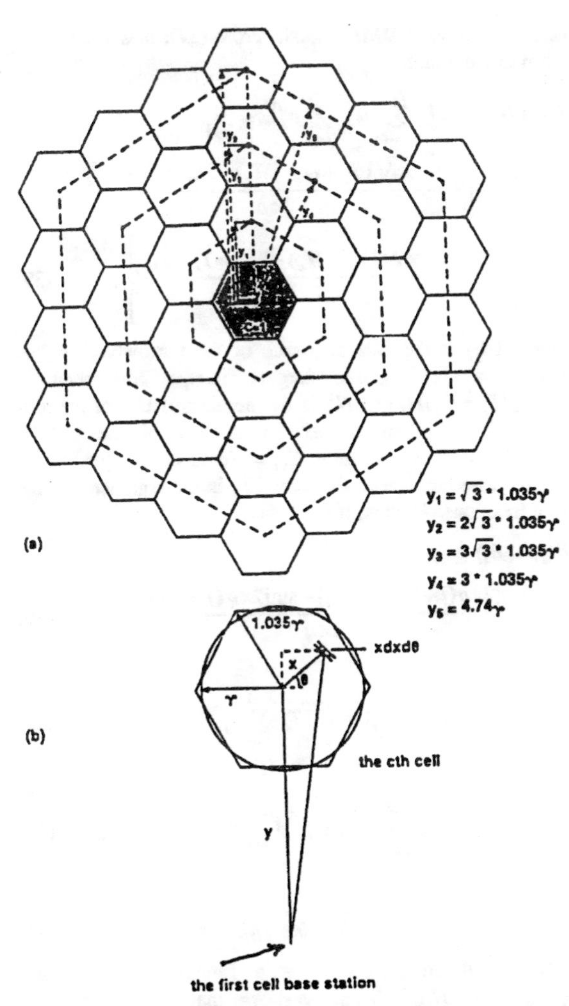

Fig. 3. (a) The microcellular model. (b) The hexagonal cell approximated by a circular cell of equal area.

Table I shows the result of (24) for γ=2,3,4 and for the various values of y. Therefore, the conditional variance of the total adjacent-cell interference involved with G layers of cells is approximated by

$$\sigma^2=\zeta(\gamma)KL\sigma_0^2 . \tag{25}$$

In (25), the factor $\zeta(\gamma)$ is due to all adjacent cells and is given by

$$\zeta(\gamma)=\sum_{j=1}^{G}\zeta_j(\gamma) , \tag{26}$$

where $\zeta_j(\gamma)$ is defined as

$$\zeta_j(\gamma)=\sum_{c_k\in s_j}\varepsilon(\gamma,c_k) , \tag{27}$$

and where s_j is the set of all cells in the jth layer.

Note that the factors $\varepsilon(\gamma,c_k)$, given by (24), are independent of cell size because we normalize distance to the cell radius. Thus, the total factor $\zeta(\gamma)$, given by (26), is due only to the number of layers of cells (or, equivalently, the number of cells).

A. *Average Probability of Error.*

To obtain an approximation to the average probability of error, we will use the general approach of [8] - [11], although the presence of the narrowband interference complicates the derivation. Before proceeding, let us first simplify the notation. The final test statistic is given by

$$\psi(\lambda)=\sum_{l=1}^{U}\xi_l(\lambda) , \tag{28}$$

where $\xi_l(\lambda)$ is given in (13). Consider rewriting (13) (after neglecting $F_l(\lambda)$ and $\sum_{\substack{l=1\\l\neq 1}}^{L} I_{i,l,\hat{l}}$) as

γ	First Layer of cells		Second Layer of cells			Third Layer of cells			$\zeta(\gamma)$		
	$\varepsilon(\gamma,C)$	$\zeta_1(\gamma)$	$\varepsilon(\gamma,C)$		$\zeta_2(\gamma)$	$\varepsilon(\gamma,C)$		$\zeta_3(\gamma)$	$\zeta_1(\gamma)$	$\sum_{j=1}^{2}\zeta_j(\gamma)$	$\sum_{j=1}^{3}\zeta_j(\gamma)$
	y=1.793r		y=3.105r	y=3.586r		y=4.74r	y=5.379r				
2	0.1986	1.192	0.0559	0.0411	0.582	0.023	0.0177	0.382	1.192	1.774	2.16
3	0.125	0.75	0.016	0.0099	0.155	0.0041	0.0027	0.065	0.75	0.905	0.97
4	0.0935	0.561	0.005	0.0026	0.046	0.0008	0.0005	0.013	0.561	0.607	0.62

Table I. The result of Evaluating Eq. (24).

$$\xi_l(\lambda) = s_l + n_l + j_l , \tag{29}$$

where

$$s_l \triangleq \sqrt{2P}\, A\, T_b\, b_l^{(\lambda)} , \tag{30}$$

$$n_l \triangleq D_l(\lambda) + N_l(\lambda) + \sum_{\substack{k=1 \\ k \neq i}}^{K} \sum_{l=1}^{L} I_{k,l,\hat{i}} + \sum_{k=K+1}^{CK} \sum_{l=1}^{L} W_{k,l,\hat{i}} , \tag{31}$$

and

$$j_l \triangleq J_l(\lambda) . \tag{32}$$

Also, define

$$S_T \triangleq \sum_{l=1}^{U} s_l , \tag{33}$$

$$N_T \triangleq \sum_{l=1}^{U} n_l \tag{34}$$

and

$$J_T \triangleq \sum_{l=1}^{U} j_l . \tag{35}$$

Note that, following the same arguments as in [7], conditioned on $b_l^{(\lambda)}$ and $a_i(t)$, N_T is asymptotically normal as $K \to \infty$, with zero conditional mean and conditional variance approximately given by

$$\mathrm{var}(N_T) \triangleq \sigma_T^2 = U\Bigg\{ N_0 T_b \sum_{m=-M}^{M} \alpha_m^2 + P\rho T_b^2 + P(A^2+2\rho)T_b^2 \frac{[1+\zeta(\gamma)]KL-L}{3N} \sum_{m=-M}^{M} (2\alpha_m^2 + \alpha_m \alpha_{m+1}) \Bigg\} \tag{36}$$

Therefore, we have for the conditional bit error rate, assuming $b_l^{(\lambda)} = +1$,

$$P(e \mid a_i(t), J_T) = \phi\left[-\frac{S_T + J_T}{\sigma_T} \right] , \tag{37}$$

where

$$\phi(x) \triangleq \frac{1}{\sqrt{2\pi}} \int_{-\infty}^{x} \exp[-y^2/2]\, dy . \tag{38}$$

Now let us consider J_T. For future use, let us tie the processing gain, N, and number of CDMA users, K, together, in the sense of letting $N = N(K)$ in such a manner that N increases monotonically with K. This can be justified physically by noting that any CDMA system must increase its processing gain as the number of users increases in order to maintain a satisfactory performance. However, the consequence of increasing processing gain is that either RF bandwidth increases or information rate decreases (or both). For the purposes of this paper, we assume T_b is constant and that T_c decreases as N increases (i.e., we assume constant information rate but we allow the RF bandwidth to increase). For simplicity, assume that there are an integer number of bits of narrowband waveform in T_b seconds, and ignore any timing offset between the bits of the BPSK signal and the bits of the reference CDMA signal. From (17), it is straightforward to show that

$$j_l = J_l(\lambda) = \sqrt{2J} \sum_{m=-M}^{M} \alpha_m \sum_{j=0}^{N-1} a_{j+m}^{(i)}\, d_{[pj]} \cdot \left[\frac{\sin[2\pi\Delta(\lambda T_b + (j+1)T_c) - 2\pi\Delta m T_c + \hat{\theta}_l]}{2\pi\Delta} - \frac{\sin[2\pi\Delta(\lambda T_b + jT_c) - 2\pi\Delta m T_c + \hat{\theta}_l]}{2\pi\Delta} \right] , \tag{39}$$

where $\{d_j\}$ is the data sequence of the narrowband BPSK waveform, $[x]$ is the largest integer less than x, $p \triangleq T_c/T_j = N_1/N$, and N_1 is the number of bits of the narrowband BPSK signal per bit of any of the CDMA waveforms. If we assume $\Delta T_b = Nq$ is an integer, where $q \triangleq \Delta T_c$, we can ignore the term λT_b in the arguments of the sine functions. Further, if we define

$$\beta(j,m,\hat{\theta}_l) \triangleq \frac{\sin[2\pi q(j+1-m) + \hat{\theta}_l] - \sin[2\pi q(j-m) + \hat{\theta}_l]}{2\pi q} , \tag{40}$$

then

$$j_l = \sqrt{2J}\, T_c \sum_{j=0}^{N-1} x_j , \tag{41}$$

where

$$x_j \triangleq \sum_{m=-M}^{M} \alpha_m a_{j+m}^{(i)}\, d_{[pj]}\, \beta(j, m, \hat{\theta}_l) . \tag{42}$$

Recall that

$$\hat{\theta}_l = \theta + \phi_{il} . \tag{43}$$

Conditioned on $\theta + \phi_{il}$, x_j is a function of $d_{[pj]}$ and $\{a_{j-M}^{(i)}, \cdots, a_{j+M}^{(i)}\}$. It can be shown that the sequence $\{x_j\}$ is a 2M - dependent sequence ([12]). For example, consider the simple case of $\Delta = 0$, $M = 1$ and $1/p = \frac{N}{N_1} = 5$. Then, assuming $a_j^{(i)} = d_j = 0$ for $j < 0$, we have the following sequence:

$$x_0 = \alpha_0\, d_0\, a_0 + \alpha_1\, d_0\, a_1$$

$$x_1 = \alpha_{-1}\, d_0\, a_0 + \alpha_0\, d_0\, a_1 + \alpha_1\, d_0\, a_2$$

$$x_2 = \alpha_{-1}\, d_0\, a_1 + \alpha_0\, d_0\, a_2 + \alpha_1\, d_0\, a_3$$

$$x_3 = \alpha_{-1}\, d_0\, a_2 + \alpha_0$$

$$x_4 = \alpha_{-1}\, d_0\, a_3 + \alpha_0$$

$$x_5 = \alpha_{-1}\, d_1\, a_4 + \alpha_0$$

$$x_6 = \alpha_{-1}\, d_1\, a_5 + \alpha_0$$

$$x_7 = \alpha_{-1}\, d_1\, a_6 + \alpha_0$$

$$\cdots$$

Note that in the sequence $\{x_0, x_1, x_2, \cdots\}$, if any $2M = 2$ consecutive x_i's are removed, the remaining two subsequences (the one to the left of the removed x_i's and the one to the right) are statistically independent.

Further, if $2\pi q = \frac{k_1 \pi}{k_2}$, where k_1 and k_2 are integers, then as $N \to \infty$, the sequence $\{x_j\}$ satisfies the conditions of the Hoeffding-Robbins version of the central limit theorem for dependent random variables ([12]). To illustrate this, consider again the above example. From [12], for a sum of m-dependent random variables to be asymptotically normal, it is necessary that

$$\lim_{P \to \infty} \frac{1}{P} \sum_{h=1}^{P} A_{i+h} = A \tag{44}$$

exist uniformly for all i, where

$$A_i \triangleq 2 \sum_{j=0}^{m-1} cov\{x_{i+j}, x_{i+m}\} + \mathrm{var}\{x_{i+m}\}. \tag{45}$$

In our example, $m = 2M = 2$ and

$$\mathrm{var}(x_i) = \sum_{j=-M}^{M} \alpha_j^2 \tag{46}$$

for all $i \geq 1$ (i.e., the variance term is independent of i as long as $i \geq 1$). The covariance term can be shown to take on one of three possible values as long as $i > 2$.

$$\sum_{j=0}^{1} \mathrm{cov}\{x_{i+j}, x_{i+2}\} = \begin{cases} 0, & i = 3+5k \\ \alpha_{-1}\alpha_0 + \alpha_0\alpha_1, & i = 4+5k \\ \alpha_{-1}\alpha_1 + \alpha_{-1}\alpha_0 + \alpha_0\alpha_1, & \text{otherwise} \end{cases} \tag{47}$$

where $k = 1, 2, 3, \cdots$. Since the variance term is independent of i for $i \geq 1$, and the sum of covariances is periodic in i for $i \geq 2$, it can be seen that $\frac{1}{P}\sum_{h=1}^{P} A_{i+h}$ equals the same value, independent of i, as $P \to \infty$.

If we call the conditional variance of j_l, conditioned on $\hat{\theta}_l$, $\sigma_j^2(\hat{\theta}_l)$, then

$$\sigma_j^2(\hat{\theta}_l) = (2J T_b^2/N^2) \sum_{m_1=-M}^{M} \sum_{m_2=-M}^{M} \alpha_{m_1}\alpha_{m_2} \cdot \sum_{j_1=0}^{N-1}\sum_{j_2=0}^{N-1} \beta(j_1, m_1, \hat{\theta}_l)\beta(j_2, m_2, \hat{\theta}_l) \cdot \delta(j_1+m_1, j_2+m_2)\,\delta([pj_1], [pj_2]), \tag{48}$$

where $\delta(i,j)$ is the Kronecker delta function. Further, since j_{l_1} is uncorrelated with j_{l_2} for $l_1 \neq l_2$, the variance of J_T is $\sigma_J^2(\bar{\phi}_i) \triangleq \sum_{l=1}^{U} \sigma_j^2(\hat{\theta}_l)$, where $\bar{\phi}_i \triangleq \{\hat{\theta}_1, \hat{\theta}_2, \cdots, \hat{\theta}_U\}$.

Finally, to obtain our approximation to the average BER, consider the following: From (37), we must average the conditional error rate over $a_i(t)$ and J_T. However, from (36), σ_T^2 is not a function of the $\{a_j^{(i)}\}$, and from (48), $\sigma_j^2(\hat{\theta}_l)$ is only a function of the random phase $\hat{\theta}_l$. Therefore,

$$P(e \mid a_i(t), J_T) = P(e \mid \bar{\phi}_i) = \frac{1}{\sqrt{2\pi}\,\sigma_J(\bar{\phi}_i)} \int_{-\infty}^{\infty} \phi\left[-\frac{S_T + J_T}{\sigma_T}\right] \cdot \exp\{-J_T^2/2\sigma_J^2(\bar{\phi}_i)\}\, dJ_T = \phi\left[-\frac{S_T}{\sqrt{\sigma_J^2(\bar{\phi}_i) + \sigma_T^2}}\right], \tag{49}$$

and

$$P_e = E_{\bar{\phi}_i}\{P(e \mid \bar{\phi}_i)\}, \tag{50}$$

where $E_{\bar{\phi}_i}\{\cdot\}$ means expectation over the random vector $\bar{\phi}_i$. The total signal-to-noise ratio, conditioned on $\bar{\phi}_i$, is approximately given by

$$SNR(\bar{\phi}_i) \triangleq \frac{1}{2}\frac{S_T^2}{\sigma_T^2 + \sigma_J^2} = U\Bigg\{\left[H\frac{E_b}{N_0}\right]^{-1} \sum_{m=-M}^{M} \alpha_m^2 + \frac{1}{H} + \left(1+\frac{1}{H}\right)\frac{[1+\zeta(\gamma)]KL - L}{3N} \sum_{m=-M}^{M} (2\alpha_m^2 + \alpha_m\alpha_{m+1}) + \frac{J/S}{N^2}\cdot\frac{2}{U}\sum_{l=1}^{U}\Bigg[\sum_{m_1=-M}^{M}\sum_{m_2=-M}^{M} \alpha_{m_1}\alpha_{m_2} \sum_{j_1=0}^{N-1}\sum_{j_2=0}^{N-1} \cdot \beta(j_1, m_1, \hat{\theta}_l)\beta(j_2, m_2, \hat{\theta}_l) \cdot \delta(j_1+m_1, j_2+m_2)\,\delta([pj_1],[pj_2])\Bigg]\Bigg\}^{-1}, \tag{51}$$

where $\bar{E}_b = 2P\rho T_b$ is the average energy per coded bit, $J/S = J/(PA^2)$ denotes the interference power-to -useful signal power ratio, and $H = A^2/(2\rho)$.

B. Channel Coding

We are interested in the performance of two simple block codes used in conjunction with the diversity. These are the BCH (15,7) code and the Golay (23,12) code. The former corrects two channel errors, while the latter corrects the three channel errors in each codeword. Assuming perfect interleaving, (i.e., ignoring the correlation introduced by both the suppression filter and the fading channel), the resulting bit error rate (BER) after hard decision decoding is approximately given by [13]

$$P_b \approx \frac{1}{n} \sum_{i=t+1}^{n} i \binom{n}{i} (P_e)^i (1-P_e)^{n-i}, \tag{52}$$

where n is the number of bits in a codeword and t denotes the number of errors that the code can correct.

C. Determination of the Suppression (Wiener) Filter Coefficients

The coefficients of the suppression filter can be determined by solving the following Wiener-Hopf equation:

$$\sum_{\substack{m=-M \\ m\neq 0}}^{M} \alpha_m \hat{\rho}[(n-m)T_c] + \hat{\rho}(nT_c) = 0, \tag{53}$$

where $n=-M, \cdots, M, n \neq 0$, and $\hat{\rho}(\cdot)$ is the low-pass version (normalized to the received useful power PA^2 of the reference user) of the autocorrelation function of the input signal to the suppression filter and is given by

$$\hat{\rho}(lT_c) = \begin{cases} (1+1/H)[1+\zeta(\gamma)]KL+2N[\bar{E}_b/N_0]^{-1}+J/S, & l=0 \\ (J/S)(1-|l|p)\cos(2\pi lq), & |l|\leq[1/p] \\ 0, & |l|\geq[1/p] \end{cases} \quad (54)$$

where p and q are given by (6) and (7), respectively.

IV. NUMERICAL RESULTS

In this section, we present numerical results for the BER of a CDMA system in the presence of the narrowband interference. Unless noted otherwise, it is assumed that the ratio of the interference bandwidth to the spread spectrum bandwidth is 10% (p=0.1), the ratio of the offset of the interference carrier frequency to half of the spread spectrum bandwidth is 20% (q=0.2), the ratio of the specular component power to the fading component power, H, is 7 dB, and the processing gain, N, is 255. Note that where the BER is plotted as a function of the ratio $\bar{E}_i/N_0$, this corresponds to the energy-per -bit to noise spectral density of an uncoded system.

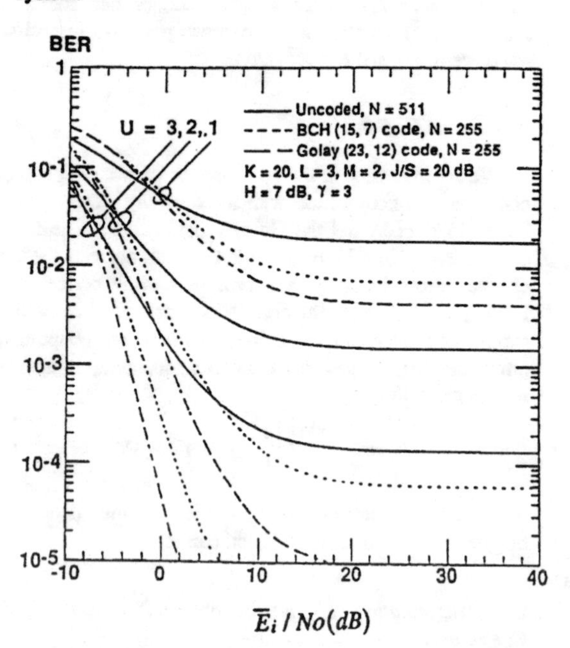

Fig. 4. The BER performance of DS-CDMA overlay system with diversity plus coding.

In Fig. 4, the decoded BER for H=7 dB is shown. Also, the uncoded BER is plotted in order to allow for comparison. Note that the uncoded and coded systems are compared for the same spread spectrum bandwidth $B=2N/(T_i m/n)$. In our example, N=255 for the coded systems, whereas N=511 for the uncoded system. As expected, both the uncoded and coded BER decreases as the order, U, of diversity increases. The performance advantage obtained by using diversity becomes larger as the error correcting capability of the code increases for the same spread bandwidth.

In order to investigate the relative effect of the narrowband interference on the performance of the CDMA system with respect to the multiple-access and adjacent-cell interference, Fig. 5 illustrates the asymptotic ($\bar{E}_i/N_0 \to \infty$)

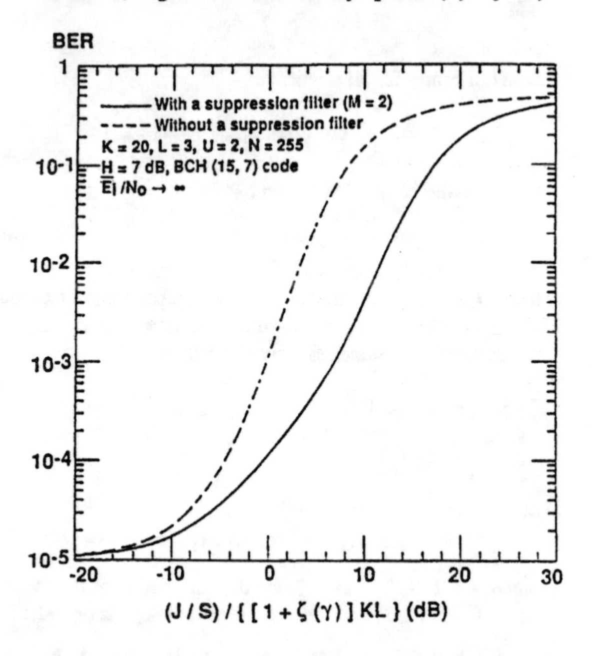

Fig. 5. The asymptotic ($\bar{E}_i/N_0 \to \infty$) decoded BER of DS-CDMA overlay system as a function of $(J/S)/[(1+\zeta(\gamma))KL]$.

BER of the CDMA system both with or without the suppression filter as a function of $(J/S)/[(1+\zeta(\gamma))KL]$. Note that the ratio of the narrowband interference to the multiple-access and adjacent-cell interference can be roughly represented by $(J/S)/[(1+\zeta(\gamma))KL]$. As expected, the BERs with or without suppression filters increase as J/S increases. When J/S is small, the multiple-access and adjacent-cell interference dominates; in this case, the BERs of the systems both with and without a suppression filter are very close, so that the suppression filter is unnecessary. When J/S is large, the narrowband interference is on the order of the multiple-access and adjacent-cell interference; in this case, the system without a suppression filter degrades significantly, whereas the system is much more tolerant of the interference when the suppression filter is present.

In Fig. 6, the asymptotic BER of the system with and without the suppression filter is shown as a function of the ratio (p) of the interference bandwidth to the spread spectrum bandwidth. It is seen from the figure that the CDMA system with a double-sided Wiener filter can suppress the narrowband interference very effectively (assuming, of course, that p does not become too large).

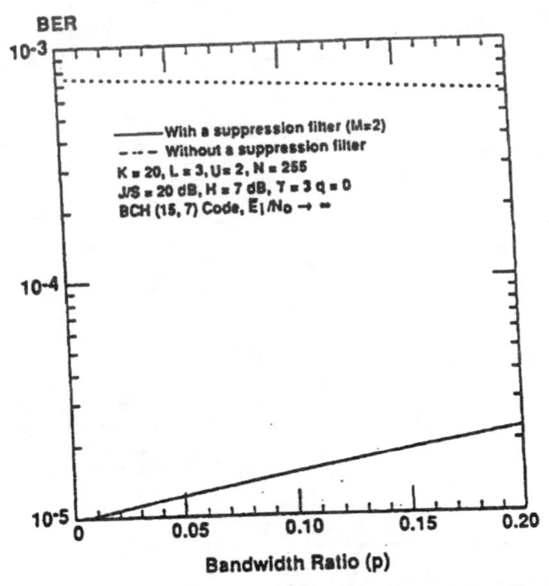

Fig. 6. The asymptotic ($\overline{E}_i/N_0 \leftarrow \infty$) BER of DS-CDMA system with and without suppression filters as a function of the bandwidth ratio (p) of the interference BW to the spread spectrum BW.

Fig. 7 illustrates the asymptotic BER for the system employing the suppression filter as a function of the number of taps on each side, M. It is seen from the figure that the BER decreases first when $M = 1$ and then keeps a steady value with increasing M. For $M \geq 1$, the narrow-band interference is reduced by the suppression filter, and increasing M beyond 1 does not noticeably reduce the narrowband interfer-

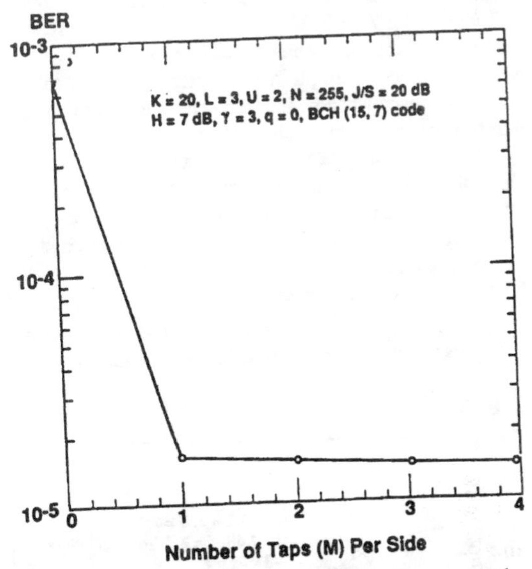

Fig. 7. The asymptotic ($\overline{E}_i/N_0 \rightarrow \infty$) BER of DS-CDMA systems with a suppression filter as a function of number of taps on each side.

Table II (K=20, L=3 U=2, N=255, M=2, q=0, H=7 dB, γ=3, J/S=20 dB, BCH (15, 7) code)

Number of taps per side (M)	Coefficients (α_m)									Q	W
	α_{-4}	α_{-3}	α_{-2}	α_{-1}	α_0	α_1	α_2	α_3	α_4		
0	0	0	0	0	1	0	0	0	0	0.379	0.36
1	0	0	0	-0.302	1	-0.302	0	0	0	0.079	0.33
2	0	0	-0.165	-0.213	1	-0.213	-0.165	0	0	0.056	0.35
3	0	-0.1	-0.131	-0.185	1	-0.185	-0.131	-0.1	0	0.053	0.35
4	-0.063	-0.083	-0.119	-0.174	1	-0.174	-0.119	-0.083	-0.063	0.054	0.35

$$Q = \frac{2J/S}{N^2} \sum_{m_1=-M}^{M} \sum_{m_2=-M}^{M} \alpha_{m_1} \alpha_{m_2} \sum_{j_1=0}^{N-1} \sum_{j_2=0}^{N-1} E_{\Phi_l} \left[\beta(j_1, m_1, \hat{\theta}_l)\, \beta(j_2, m_2, \hat{\theta}_l)\right] \delta(j_1+m_1, j_2+m_2)\, \delta\left[[pj_1], [pj_2]\right]$$

$$W = (1+H^{-1}) \frac{[1+\zeta(\gamma)]\,KL-L}{3N} \sum_{m=-M}^{M} \alpha_m (2\alpha_m + \alpha_{m+1})$$

Table II. Tap Coefficients and Their Effects on the Interference

ence, because $\hat{\rho}(T_c)$ and $\hat{\rho}(-T_c)$ are almost identical to $\hat{\rho}(0)$ in (54). Thus, the suppression filter with M=1 is sufficient to reject the interference in this case (small p and q=0), and more taps than one on each side are unnecessary. Table II shows the tap coefficients of the suppression filters for different numbers of taps each side ($M = 0, 1, 2, 3$). As shown in Table II, for $M = 0$, $Q \approx W$, where Q roughly represents the effect of the narrowband interference and W represents the effect of the multiple access interference; in this case, both the narrowband interference and multiple-access interference have comparable effects. However, for $M \geq 1$, $Q \ll W$ and W does not vary very much. Therefore, we conclude that when the carrier frequencies of the narrowband interference and CDMA signals are the same, the suppression filter with three total taps is sufficient to reduce the narrowband interference, and increasing the number of total taps beyond three is not necessary.

Fig. 8 illustrates the BER performance of the systems with the suppression filter as a function of the ratio (q) of the

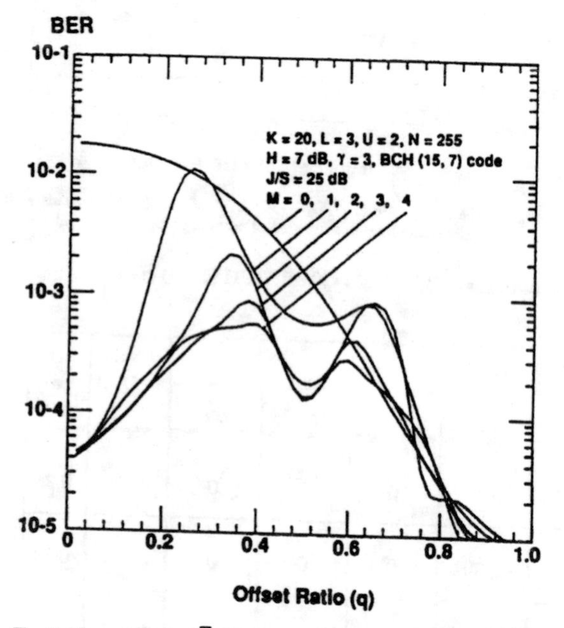

Fig. 8. The asymptotic ($\bar{E}_i/N_0 \rightarrow \infty$) BER of the system with double-sided filters as a function of the offset ratio (q) of the interference carrier frequency to the half spread spectrum BW.

offset of the interference carrier frequency to the half spread spectrum bandwidth for different number of the taps on each side. It is seen that when q is very small, the BERs of the system with double-sided filters for $M \geq 1$ are almost identical, as was the case in Fig. 7. However, the BER performance improves as M increases when q increases, and the BER for large M is also more robust to a change in q. Also, for $q \approx 0.25$, the BERs of the systems with the double-sided suppression filter and without the suppression filter are almost identical, because the low-pass version of the autocorrelation of the interference is zero, (i.e., $\rho_j(lT_c) = 0$ for $l \neq 0$), so that all off-center tap coefficients are zero.

In Fig. 9, the asymptotic BER of the DS-CDMA system with a suppression filter is plotted as a function of the number

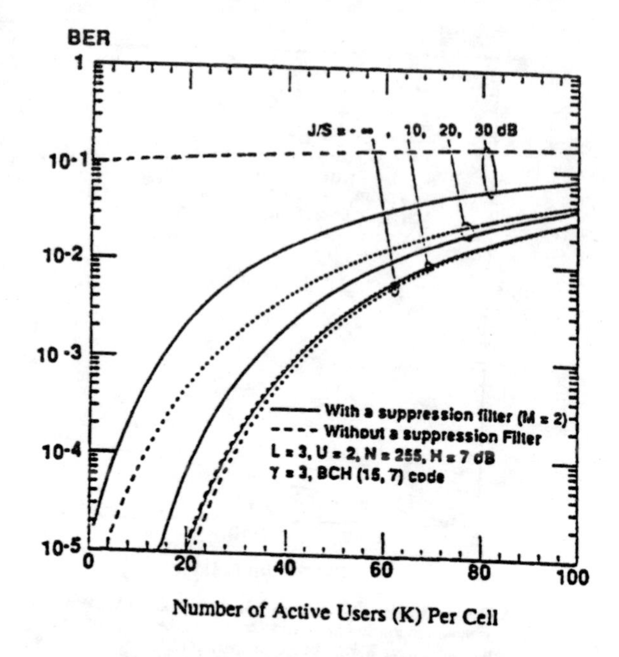

Fig. 9. The asymptotic decoded BER of DS-CDMA overlay systems as a function of the number of active users per cell.

of active users, K, for various values of J/S. Also, the BER of the system without the suppression filter is shown for comparison. It is seen that at large J/S, for a given BER, the system with the suppression filter can support many more users

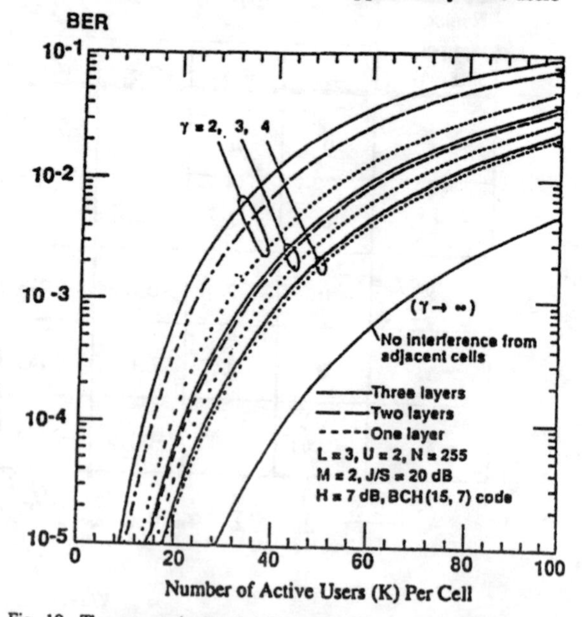

Fig. 10. The asymptotic decoded BER of DS-CDMA overlay system as a function of active users (K) per cell for various values of the propagation loss exponent (γ).

than can the system without the suppression filter. When the interference is such that $J/S \leq 10dB$, the multiple-access and adjacent-cell interference dominates so that the suppression filter is not needed, as was seen in Fig. 5.

Fig. 10 illustrates the asymptotic BER of the CDMA system versus K, the number of active users per cell, for various values of the propagation loss exponent, γ, (γ=2,3,4). Also, the BER in the absence of adjacent cell interference (*i.e.*, $\gamma \rightarrow \infty$) is shown for comparison. In order to show the effect of the adjacent-cell interference on the performance of the system for different values of γ, a single layer of cells, two layers of cells and three layers of cells are considered. It is clear from this figure that when γ =2, the BER computed by accounting for only a single layer of cells is too optimistic; alternately, if $\gamma \geq 3$, the difference in BER performance between accounting for only a single layer and accounting for more layers is insignificant. Further, for a given γ, the gap in the BER between two and three layers is much less than that between one and two layers. Indeed, these results can be seen from the $\zeta_j(\gamma)$ entries of Table I, and are consistent with the results presented in [6].

V. CONCLUSIONS

An approximate technique has been described for determining the performance of a DS-CDMA system operating over a multipath Rician fading channel and sharing common spectrum with various narrowband waveforms. It was shown that

(1) When the narrowband interference power-to-signal power ratio (J/S) is small with respect to the multiple-access and adjacent-cell interference, the suppression filter is unnecessary, whereas when J/S is on the order of the multiple-access and adjacent-cell interference, the suppression filter provides significant enhancement in performance;

(2) When the ratio (q) of the offset of the interference carrier frequency to half of the spread spectrum bandwidth is very small, that is, when the interference carrier frequency is very close to the spread spectrum carrier frequency, spread spectrum systems with double-sided suppression filters only need three total taps. However, the double-sided filter with a large number of taps (i.e., $M>1$) is preferable when the ratio is large.

(3) For large narrowband interference, the system with a suppression filter can support many more users than can the system without a suppression filter;

(4) As is well known, the number of users supported by the CDMA overlay system increases as the propagation loss exponent (γ) increases. For typical exponents, three layers of cells are sufficient for considering the adjacent cell interference.

REFERENCES

[1] R. L. Pickholtz, L. B. Milstein and D. L. Schilling, "Spread spectrum for mobile communications," *IEEE Trans. Vehicular Tech.*, vol. VT-40, pp. 313-322, May 1991.

[2] D. L. Schilling, L. B. Milstein, R. L. Pickholtz and F. Miller, "CDMA for personal communications networks," *Proc. of MILCOM'90*, pp. 28.2.1-28.2.4, October 1990.

[3] T. S. Rappaport and L. B. Milstein, "Effects of path loss and fringe user distribution on CDMA cellular frequency reuse efficiency," *Proc. of GLOBECOM'90*, pp. 404.6.1-404.6.7, December 1990.

[4] R. J. Bultitude and G. K. Bedal, "Propagation characteristics on microcellular urban mobile radio channels at 910 MHz," *IEEE J. Select. Areas Commun.*, vol. SAC-7, pp.31-39, Jan. 1989.

[5] G. R. Cooper and R. W. Nettleton, "A spread-spectrum technique for high-capacity mobile communications," *IEEE Trans. Vehicular Tech.*, vol. VT-27, pp.264-275, Nov. 1978.

[6] L. B. Milstein, T. S. Rappaport and R. Barghouti, "Performance evaluation for cellular CDMA," *IEEE Journal on Select. Areas in Commun.*, vol. SAC-10, pp. 680-689, May 1992.

[7] J. Wang, M. Moeneclaey and L. B. Milstein, "DS-CDMA with predetection diversity for indoor radio communications," *IEEE Trans. Comm.*, vol. COM-42, pp. 1929-1938, April 1994.

[8] M. B. Pursley, "Performance evaluation for phase-coded spread spectrum multiple access communications, Part I: system analysis," *IEEE Trans. on Commun.*, vol. Com-25, pp. 295-299, Aug. 1977.

[9] J. S. Lehnert and M. B. Pursley, "Error probabilities for binary direct sequence spread spectrum multiple access communications with random signature sequences," *IEEE Trans. Commun.*, vol. COM-35, pp. 87-98, Jan. 1987.

[10] R. K. Morrow, Jr. and J. S. Lehnert, " Bit-to-bit error dependence in slotted DS/SSMA packet systems with random signature sequences," *IEEE Trans. Commun.*, vol. COM-37, pp. 1052-1061, Oct. 1989.

[11] M. B. Pursley, "The role of spread spectrum in packet radio networks," *Proceedings of IEEE*, vol. 75, pp. 116-134, Jan. 1987.

[12] D. A. S. Fraser, *Nonparametric Methods in Statistics*, Wiley 1957, pp. 215-218.

[13] E. R. Berlekamp, "The technology of error-correcting codes," *Proceedings of IEEE*, vol. 68, pp. 564-593, May 1980.

[14] L. B. Milstein, "Interference rejection techniques in spread spectrum communications," *Proceedings of IEEE*, vol. 76, pp. 657-671, June 1988.

Jiangzhou Wang (M'91-SM'94) was born in Hubei, China, on November 15, 1961. He received the B.S. and M.S. degrees from Xidian University, Xian, China, in 1983 and 1985, respectively, and the Ph.D. degree (with Greatest Distinction) from the University of Ghent, Belgium, in 1990, all in electrical engineering.

From December 1990 to June 1992, he was a Postdoctoral Fellow in the Department of Electrical and Computer Engineering, University of California at San Diego. Since July 1992, he has been a Senior System Engineer in the Digital Communications Division, Rockwell Corporation, Newport Beach, California, where he is currently working on the development and system design of wireless communications. He has contributed one pending US patent, entitled "Variable multithreshold detection for 0.3-GMSK," which can be applied to wireless systems, such as GSM and Mobitex.

Laurence B. Milstein (S'66-M'68-SM'75-F'85) received the B.E.E. degree from the City College of New York, New York, NY, in 1964, and the M.S. and Ph.D. degrees in electrical engineering from the Polytechnic Institute of Brooklyn, Brooklyn, NY, in 1966 and 1968, respectively.

From 1968 to 1974, he was employed by the Space and Communications Group of Hughes Aircraft Company, and from 1974 to 1976 he was a member of the Department of Electrical and Systems Engineering, Rensselaer Polytechnic Institute, Troy, NY. Since 1976, he has been with the Department of Electrical and Computer Engineering, University of California at San Diego, La Jolla, CA, where he is a Professor and former Department Chairman, working in the area of digital communication theory with special emphasis on spread-spectrum communication systems. He has also been a consultant to both government and industry in the areas of radar and communications. He was an Associate Editor for Communication Theory for the *IEEE Transactions on Communications*, an Associate Editor for Book Reviews for the *IEEE Transactions on Information Theory*, and an Associate Technical Editor for the *IEEE Communications Magazine*, and is currently a Senior Editor for the *IEEE Journal on Selected Areas in Communications*. He was the Vice President for Technical Affairs in 1990 and 1991 of the IEEE Communications Society, and is currently a member of the Board of Governors of both the IEEE Communications Society and the IEEE Information Theory Society.

SECTION 2.4
CDMA TRAFFIC ISSUES AND CDMA ADVANTAGES

The first three papers in this section, "*SIR-Based Call Admission Control for DS-CDMA Cellular Systems*" by Liu and Zarki, "*Congestion Relief on Power-Controlled CDMA Networks*" by Jacobsmeyer, and "*Performance Analysis of Soft Handoff in CDMA Cellular Networks*" by Su et al., consider some traffic issues, such as studying SIR based call admission control algorithms in DS-CDMA by taking both the radio propagation and traffic loading variations into consideration, investigating the congestion problem in a power-controlled CDMA system, and investigating the soft handoff performance.

Finally, this section concludes with the paper "*Advantages of CDMA and Spread Spectrum Techniques over FDMA and TDMA in Cellular Mobile Radio Applications*" by Jung et al. which discusses several advantages of CDMA over comparable other multiple access techniques.

SIR-Based Call Admission Control for DS-CDMA Cellular Systems

Zhao Liu, *Student Member, IEEE*, and Magda El Zarki, *Member, IEEE*

***Abstract*—Signal-to-interference ratio (SIR)-based call admission control (CAC) algorithms are proposed and studied in a DS-CDMA cellular system. Residual capacity is introduced as the additional number of initial calls a base station can accept such that system-wide outage probability will be guaranteed to remain below a certain level. The residual capacity at each cell is updated dynamically according to the reverse-link SIR measurements at the base station. A 2^k factorial experimental design and analysis via computer simulations is used to study the impact of the parameters used in the algorithms. The influence of these parameters on system performance, namely blocking probability and outage probability, is then examined via simulation. The performance of the algorithms is compared together with that of a fixed call admission control scheme (fixed CAC) under both homogeneous and hot spot traffic loadings. The results show that SIR-based CAC always outperforms fixed CAC even under overload situations, which is not the case in FDMA/TDMA cellular systems. The primary benefit of SIR-based CAC in DS-CDMA cellular systems, however, lies in improving the system performance under hot spot traffics.**

I. Introduction

A LARGE number of papers in recent literature has been dedicated to the evaluation of physical level DS-CDMA cellular system capacity with respect to the effects of radio propagation and power control [1]-[7]. In all of these papers, however, a constant number of calls per cell was assumed and the effect of call arrival process was not taken into account. Huang [8] evaluated the signal-to-interference ratio (SIR) distribution of a DS-CDMA cellular system by using computer simulations; however, the impact of variations in the traffic loading is not explicitly considered.

Dynamic channel allocation (DCA), on the other hand, has been extensively studied for FDMA/TDMA cellular systems as a means of increasing capacity and adapting to traffic loading variations [9]-[13]. For DS-CDMA cellular system, however, the problem of channel allocation can be viewed as call admission control. In this paper, we propose and study SIR-based call admission control (CAC) algorithms in DS-CDMA cellular systems.

The performance of the system is evaluated by taking both the radio propagation and the traffic loading variations into consideration. Since the SIR measurement that is required for these algorithms already formed part of the operational mechanism of the proposed DS-DCMA cellular systems [14], their implementation will add very little additional overhead to the system.

The remainder of the paper is organized as follows: In Section II, the DS-CDMA cellular system model, the radio propagation model, and the traffic model used in this study are presented. In Section III, the reverse-link SIR is derived for the DS-CDMA cellular system model and then SIR-based CAC algorithms are introduced. In Section IV, a 2^k experimental design [15] with simulation is used to study the influence of the parameters on system performance. After selecting the appropriate design parameters for the algorithms, their performance is evaluated and compared with that of fixed CAC. Finally, a discussion of the results and conclusion are given in Section V.

Manuscript received June 30, 1993; revised November 30, 1993. This paper was presented at the Fourth International Symposium on Personal, Indoor, and Mobile Radio Communications, Yokohama, Japan, September 8-11, 1993.

The authors are with the Department of Electrical Engineering, University of Pennsylvania, Philadelphia, PA 19104 USA.

IEEE Log Number 9215436.

II. System Description and Models

A. DS-CDMA Cellular System Model

In DS-CDMA cellular systems, just as in FDMA/TDMA cellular systems, the whole service area is divided into cells and each cell is served by a base station. However, in DS-CDMA cellular systems, the same radio channel is reused in every cell. The separation of different users' signals is achieved by means of signature sequences used to spread the spectrum. The DS-CDMA cellular system model and the associated assumptions used in this paper are summarized below:

1) Service area is divided into K hexagonal cells of equal size, a base station with an omni antenna is located at the center of each cell.
2) Separate frequency bands are used for the reverse link and the forward link, so that the mobiles only experience interference from the base stations and the base stations only experience interference from the mobiles.
3) Compared with the intrasystem interference, the background Gaussian noise can be ignored, i.e., the signal-to-noise ratio is basically given by the SIR.
4) Each mobile is power controlled with respect to the base station of its home cell.
5) A mobile arriving to the system will choose its home cell such that the radio propagation attenuation between the mobile and the base station of its home cell is minimized.
6) Voice activity detection is not modeled; each mobile admitted to the system will transmit continuously until the completion of the call.

7) Mobility of the mobiles is not modeled, they are assumed to be uniformly distributed over the service area within each cell.

B. Radio Propagation Model

For cellular systems operating at UHF/VHF frequency bands, radio propagation is largely influenced by three nearly independent factors: path loss with distance, log-normal shadowing, and multipath fading [16]. The generally accepted radio propagation model for DS–CDMA cellular systems is a log-normal distribution of shadowing with its mean the path loss of the αth power of the distance [2], [16]. Supposing a mobile is at distance r from a base station, the average received field strength $\Gamma(r)$ in real value can be expressed as:

$$\Gamma(r) = 10^{\xi/10} r^{-\alpha} \tag{1}$$

where ξ in decibels has a normal distribution with zero mean and standard deviation of σ, which is independent of the distance and ranges $5 \sim 12$ dB with a typical value of 8 dB. Typical values of α in a cellular environment are $2.7 \sim 4.0$ [5].

C. Communication Traffic Model

The initial call arrival process to a cell k is modeled as an independent Poisson process with mean arrival rate λ_k. The calls arriving to each cell are assumed to be uniformly distributed over the service area in that cell. Handover calls are not explicitly modeled. The justification for this is that, with reverse power control, the need and acceptance of handover calls should always result in reducing the system-wide interference. The call duration is modeled as an exponentially distributed random variable. We would like to point out that, because of the shadowing, when a mobile initiates a call in a cell, it does not necessarily mean that the call will be served by that cell. The home cell of a call will be chosen as the one with the smallest radio propagation attenuation to the mobile.

III. SIR AND SIR-BASED CAC ALGORITHMS

A. Reverse Link Interference and SIR

After assuming a DS–CDMA cellular system with K cells $\kappa = \{1, \cdots, K\}$ and $n_k\ (k \in \kappa)$ calls in progress in cell k, the total power received by the base station in cell k is the summation of the power from all the mobiles in the system:

$$\begin{aligned} I(k) &= \sum_{h=1}^{K} \sum_{i=1}^{n_h} I_i(h,k) \\ &= S n_k + S \sum_{h \neq k} \sum_{i=1}^{n_h} \left(\frac{r_{ih}}{r_{ik}}\right)^{\alpha} 10^{(\xi_{ik} - \xi_{ih})/10} \end{aligned} \tag{2}$$

where $I_i(h,k)$ is the power received by the base station in cell k from a mobile i transmitting to the base station of its home cell h, r_{ih} is the distance of the mobile i to the base station of its home cell h, and r_{ik} is the distance of the mobile i to the base station of cell k; S is the power level received by a mobile's home cell base station.

The signal-to-interference ratio measured at the base station of cell k, SIR_k, is the ratio of the desired mobile's signal power to the summation of the power from all other mobiles:

$$\begin{aligned} \mathrm{SIR}_k &= \frac{S}{I(k) - S} \\ &= \frac{1}{n_k - 1 + \left(\sum_{h \neq k} \sum_{i=1}^{n_h} \left(\frac{r_{ih}}{r_{ik}}\right)^{\alpha} 10^{(\xi_{ik} - \xi_{ih})/10} \right)}. \end{aligned} \tag{3}$$

It can be seen that SIR_k is a random variable. The randomness comes from three stochastic processes, namely radio propagation, traffic variation, and mobile distribution. It can also be seen that the SIR measured locally contains aggregated information about congestion at the local cell and other cells in the system.

B. SIR-Based CAC Algorithms

We begin with a definition of the term residual capacity used in this work. In a DS–CDMA cellular system, the residual capacity in a cell is defined as the additional number of initial calls the base station can accept such that the system-wide outage probability, defined as the probability that an acceptable transmission quality cannot be maintained, will be guaranteed to be below a certain level. Since the acceptance of handover calls in a DS–CDMA cellular system will result in reducing the system-wide interference, from the interference point of view, it can be viewed that handover calls to a cell will not request residual capacity from the cell.

Two distributed SIR-based CAC algorithms for the DS–CDMA cellular system are proposed and studied. The first algorithm, denoted as Algorithm I, is a totally localized algorithm, and the CAC decision is solely based on the SIR measurement at the local cell base station. The second algorithm, denoted as Algorithm II, utilizes, the local SIR measurement and the SIR measurements of the adjacent cells (immediate neighbor cells).

- *Algorithm I*

1) The base station in each cell $k \in \kappa$ makes periodic measurement of its reverse link SIR_k;
2) Residual capacity $\mathcal{R}_k$ is then estimated and updated according to the formula:

$$\mathcal{R}_k = \begin{cases} \left\lfloor \frac{1}{\mathrm{SIR_{TH}}} - \frac{1}{\mathrm{SIR}_k} \right\rfloor & \text{if } \left\lfloor \frac{1}{\mathrm{SIR_{TH}}} - \frac{1}{\mathrm{SIR}_k} \right\rfloor > 0 \\ 0 & \text{otherwise} \end{cases} \tag{4}$$

where $\mathrm{SIR_{TH}}$ is the SIR threshold at the base station receiver input. It is a design parameter in the algorithm; $\lfloor X \rfloor$ denotes the greatest integer less than or equal to X.
3) For each initial call request in cell $k \in \kappa$, the base station checks the value of the residual capacity $\mathcal{R}_k$: if $\mathcal{R}_k > 0$, the new call is accepted and the residual capacity is reduced by one; otherwise, the call is rejected.

Note we should always have $\mathrm{SIR_{TH}} \geq \mathrm{SIR_0}$, where $\mathrm{SIR_0}$ is the minimum SIR for proper operation.

- *Algorithm II*

1) The base station in each site $k \in \kappa$ makes periodic measurement of its reverse link SIR_k and requests the reverse link SIR measurement from each of its adjacent cells;
2) The residual capacity $\mathcal{R}_k$ at cell $k \in \kappa$ is then estimated and updated according to the formula:

$$\mathcal{R}_k = \begin{cases} \min\{\mathcal{R}_k^{(j)} | j \in \kappa(k)\} & \text{if } \min\{\mathcal{R}_k^{(j)} | j \in \kappa(k)\} > 0 \\ 0 & \text{otherwise} \end{cases} \tag{5}$$

where

$$\mathcal{R}_k^{(j)} = \begin{cases} \left\lfloor \dfrac{1}{\mathrm{SIR_{TH}}} - \dfrac{1}{\mathrm{SIR}_k} \right\rfloor & \text{if } j = k \\ \left\lfloor \dfrac{1}{\beta}\left(\dfrac{1}{\mathrm{SIR_{TH}}} - \dfrac{1}{\mathrm{SIR}_j}\right) \right\rfloor & \text{if } j \in \kappa_{(k)}(k) \end{cases} \tag{6}$$

where
$\min\{X\}$ denotes the minimum value of X;
$\kappa(k)$ denotes a subset of cells including cell k and its adjacent cells;
$\kappa_{(k)}(k)$ denotes a subset of cells including only cell k's adjacent cells;
β denotes the estimate of the interference coupling between adjacent cells. It is a design parameter. We call β the adjacent cell interference coupling coefficient.

3) For each initial call request in cell $k \in \kappa$, the base station checks the value of residual capacity $\mathcal{R}_k$: if $\mathcal{R}_k > 0$, the new call is accepted and the residual capacity is reduced by one; otherwise, the call is rejected;

The β represents the estimation of normalized interference to a base station caused by a mobile transmitting to the base station of one of its adjacent cells. Note that a value of β may be obtained from the mean value of the adjacent cell mobile interference, i.e., $\beta = E[I_i(h,k)]$ where $h \in \kappa_{(k)}(k)$. By a simulation study, we obtained this mean value $\beta = 0.074917$. It was also noted from that study that there was a relatively large standard deviation $STD[I_i(h,k)] = 0.162705$ associated with this mean value. It was, therefore, decided to make it a design parameter for our algorithms.

The reason that the reverse-link SIR is used for the calculation of the residual capacity is that it has been shown, with power control and synchronized transmission on the forward link, the capacity of the DS–CDMA cellular system is limited by its reverse link [2].

For Algorithm I, only the locally measured reversed-link SIR is available to estimate the residual capacity. From (3), given the the measured SIR_k at the base station of cell k and the required SIR threshold $\mathrm{SIR_{TH}}$, the additional number of calls $\mathcal{R}_k$ cell k can accept such that the local SIR will still remain above the threshold $\mathrm{SIR_{TH}}$ is bounded by:

$$\frac{1}{n_k + \mathcal{R}_k - 1 + \left(\sum_{h\neq k}\sum_{i=1}^{n_h}\left(\dfrac{r_{ih}}{r_{ik}}\right)^{\alpha} 10^{(\xi_{ik}-\xi_{ih})/10}\right)} \geq \mathrm{SIR_{TH}} \tag{7}$$

From (3) and (7), the constraint on $\mathcal{R}_k$ when only the local SIR is considered is given by:

$$\mathcal{R}_k \leq \frac{1}{\mathrm{SIR_{TH}}} - \frac{1}{\mathrm{SIR}_k}. \tag{8}$$

For Algorithm II, the locally measured reverse-link SIR is used to calculate the potential residual capacity just as (8) for Algorithm I. Then, the impact of the interference (resulting from acceptance of new calls) on the SIR's of its adjacent cells is estimated by means of the adjacent cell interference coupling coefficient β. Since β estimates the normalized interference from a mobile to the adjacent cells' base stations, the constraints on $\mathcal{R}_k$ from the adjacent cells' SIR levels are given by:

$$\mathcal{R}_k \leq \frac{1}{\beta}\left(\frac{1}{\mathrm{SIR_{TH}}} - \frac{1}{\mathrm{SIR}_j}\right), \quad j \in \kappa_{(k)}(k). \tag{9}$$

The residual capacity in a given cell k is then calculated by considering all the constraints from the local cell SIR and its adjacent cells' SIRs.

Although the SIR measured at the local cell represents the exact effect of system-wide loading on the capacity at the local cell, it might underestimate the effect of the local cell's loading on other cells in the system. So, for Algorithm I, in order to prevent possible excessive interference, the SIR threshold $\mathrm{SIR_{TH}}$ must be chosen to be greater than that of the SIR required for digital voice transmission ($\mathrm{SIR_0}$). Selecting $\mathrm{SIR_{TH}}$ such that $\mathrm{SIR_{TH}} > \mathrm{SIR_0}$ will lead to a less greedy admission scheme, one that protects the overall behavior of the system. For Algorithm II, however, the use of parameter β and the adjacent cells' SIR measurement in the algorithm estimate the effect of the cell's loading on it adjacent cells, which shows a cell's "politeness" to its neighbor cells when estimating the residual capacity. Therefore, the SIR threshold $\mathrm{SIR_{TH}}$ can be chosen close or equal to that of $\mathrm{SIR_0}$.

C. Performance Measures

The performance is measured by the blocking probability and the outage probability. Blocking probability is defined in our model as:

$$P_{\mathrm{BLK}}(k) = \Pr\{\mathcal{R}_k = 0\}. \tag{10}$$

Outage probability is defined as:

$$P_{\mathrm{OTG}}(k) = \Pr\left\{\frac{E_b}{N_0} < \mathrm{EIR_0}\right\}. \tag{11}$$

This is because the residual capacity is estimated by distributed algorithms rather than a centralized global algorithm. Another performance measure is the Erlang capacity, and it is defined as the offered load in Erlangs for a given cell k when the blocking probability reaches $P_{\mathrm{BLK}}(k) = 0.02$.

IV. Simulation

The performance of both algorithms was studied by simulations. The DS–CDMA cellular system model presented in Section II was used in the simulation with $K = 49$ cells as shown in Fig. 1. The outer ring of 24 cells was

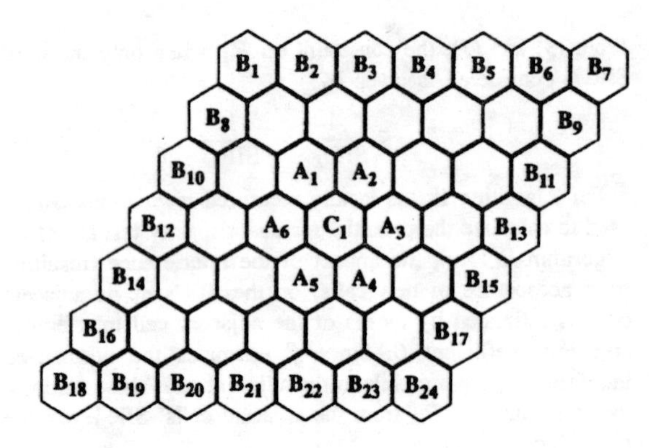

Fig. 1. The cellular layout for the simulation.

used as the boundary for the system. The call admission scheme used in those cells was fixed CAC. Different CAC schemes were then applied to each of the remaining 25 inner cells to evaluate performance. Samples were taken from the central cell (C1) and its adjacent cells (A1 ~ A6). The voice encoder bit rate is assumed to be $R_S = 32$ kb/s and spread spectrum bandwidth to be $B_{SS} = 12.5$ MHz. This gives a spread spectrum processing gain of $PG = 390.625$ (26 dB). The radio propagation attenuation model with path loss exponent $\alpha = 4$ and standard deviation $\sigma = 8$ dB was used. The independent Poisson call arrival process to each cell k is uniformly distributed over the service area with mean arrival rate λ_k as the parameter for load variation. Call duration is assumed to be exponentially distributed with mean 3 minutes. A receiver output bit-energy-to-interference density ratio $\text{EIR}_0 = 7$ dB for a bit error rate $\text{BER} = 10^{-3}$ is assumed.

For simplicity, it was assumed in this study that the SIR values are available instantaneously for each incoming call request. In real life, an update interval will be used. It should be pointed out that an 1.25 ms interval for reverse-link signal strength measurement has been suggested [17], while the minimum average interarrival time between calls in our study is 38 ms. Our assumption, therefore, is not unreasonable.

A. Effects of Design Parameters

2^3 and 2^4 factorial experimental designs [15] were used with simulation to quantify the main effects of the design parameters and loading conditions on the performance of Algorithms I and II, respectively.

2^k factorial experimental design is a method to determine the effect of k factors, each of which have two alternatives or levels, on the performance metrics selected as response. The result of a 2^k factorial design study provides the mean value, and the associated percentage of variation, of the performance metrics with respect to the two-level variations of the factors. The higher the percentage of variation explained by the factor(s), the larger the impact the corresponding factor(s) has(ve) on the performance of the system.

In our study, two performance metrics [blocking probability (P_B) and outage probability (P_{OTG})] are measured at both a central cell and its adjacent cells. The factors chosen for this experimental design study are the design parameters SIR_{TH} and β, the loading conditions of offered load at central cell, and load variations. The following observations can be drawn from this study.

- For both algorithms, the outage probability is much more sensitive to the variation in SIR_{TH} than the blocking probability. For Algorithm I, a variation in SIR_{TH} of 0.1 dB results in a 42.59% decrease in outage probability but only a 1.26% increase in blocking probability at the central cell. For Algorithm II, the same variation in SIR_{TH} results in a 31.46% decrease in outage probability but only a 0.97% increase in blocking probability at the central cell. The sensitivity of the outage probability might be due to the assumption of perfect reverse-link power control and the fact that the variation is for the threshold of SIR rather than the individually measured SIR level.
- For Algorithm I, a central cell loading of 30% higher than that of adjacent cells results in a 45.79% increase in blocking probability and a 14.01% increase in outage probability at the central cell, and only a 1.35% increase in blocking probability and a 0.77% increase in outage probability at its adjacent cells. For Algorithm II, the same load variation in the central cell results in only a 26.92% increase in blocking probability and 9.72% increase in outage probability at the central cell, and a 4.86% increase in blocking probability and a 0.12% decrease in outage probability at its adjacent cells.
- For Algorithm II, a variation in the β from 0.074917 to 0.74917 results in a 0.27% increase in blocking probability and a 12.13% decrease in outage probability at the central cell, and a 13.04% increase in blocking probability and a 17.12% decrease in outage probability at the adjacent cells.

The first observation we can make is that the tradeoff between blocking probability and outage probability is not balanced. SIR_{TH} can be increased to improve the outage performance without compromising too much of the blocking performance. The second observation is that, by changing the β, tradeoff in blocking performance can be made between the hot spot cell and its adjacent cells. In other words, the system behaves in a more fair manner by equalizing the blocking probability.

Based upon these observations, the performance of both algorithms with respect to the various values of the design parameters are evaluated by simulation. First, under homogeneous loadings, the performance of both algorithms is evaluated when SIR_{TH} is varied. The simulations are run under both a normal loading of 45 Erlangs per cell and an overloading of 80 Erlangs per cell. The β for Algorithm II is taken as 0.074917. The results are shown in Figs. 2 and 3 for blocking and outage, respectively. It can be seen that, for both algorithms, when SIR_{TH} is increased, the outage probabilities are greatly improved while the blocking probabilities remain almost unchanged. The performance of Algorithm II is then evaluated with respect to β under both homogeneous and hot spot loadings. Figs. 4 and 5 show the results of these

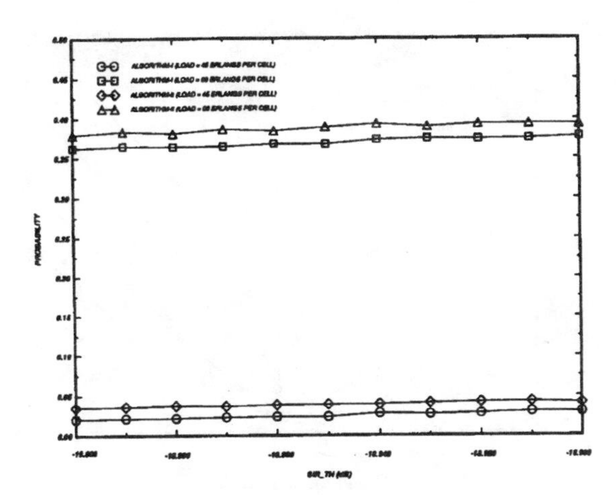

Fig. 2. Blocking probability versus signal-to-interference ratio threshold SIR_{TH} with homogeneous load.

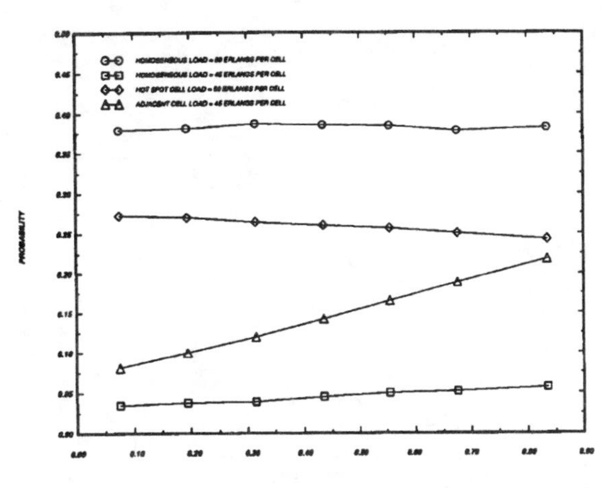

Fig. 4. Blocking probability versus adjacent cell interference coupling coefficient β for Algorithm II.

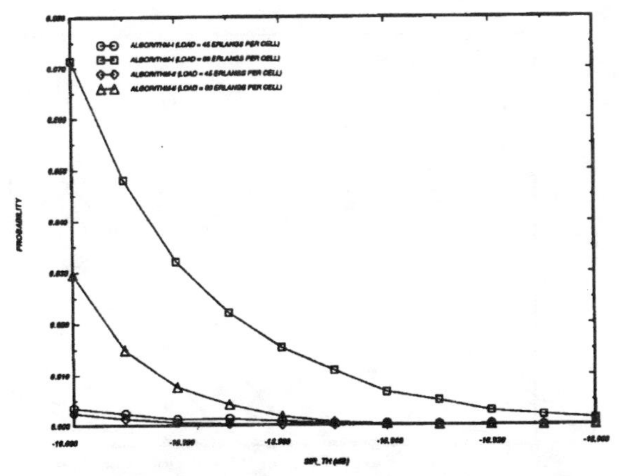

Fig. 3. Outage probability versus signal-to-interference ratio threshold SIR_{TH} with homogeneous load.

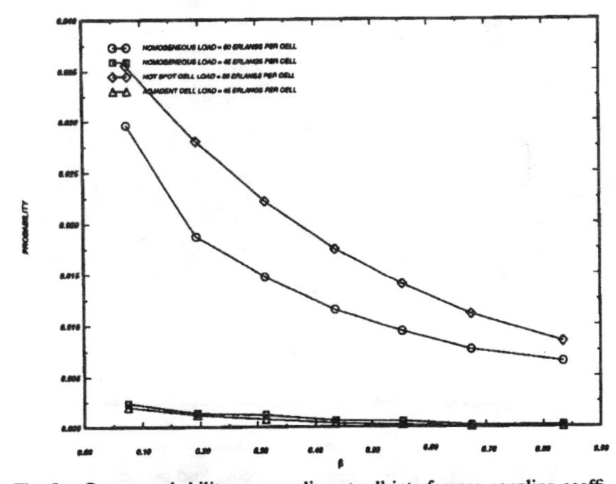

Fig. 5. Outage probability versus adjacent cell interference coupling coefficient β for Algorithm II.

evaluations. It can be seen that, under homogeneous loadings, the outage performance can be improved when β is increased while the blocking probability remains almost the same. Under hot spot loadings, however, it can be seen that when β is increased the blocking performance of the hot spot cell is improved at the expense of the adjacent cells' performance. The blocking performance of the hot spot cell and its adjacent cells tends to merge. As was said earlier, this trend makes the system behave more fairly.

B. Selection of Design Parameters

The selection of design parameters can be viewed as a process of compromising the tradeoff between blocking and outage performance. If a higher outage probability can be tolerated, a lower level of SIR_{TH} can be chosen, more calls will be admitted, and the blocking probability will be decreased. Otherwise, fewer calls will be admitted and outage probability will be decreased at the expense of a higher blocking probability. For Algorithm II, the choice of β results in a similar tradeoff between blocking and outage performance under homogeneous loading but has the additional effect of equalizing the blocking performance of the hot spot cell and its adjacent cells. The criterion we used to select the design parameters was to guarantee that the outage probability must be less than 2% under all loading situations. With this criterion, a $SIR_{TH} = -18.95$ dB is chosen for Algorithm I and a $SIR_{TH} = -19$ dB and $\beta = 0.5$ for Algorithm II.

C. Performance Comparison

The performance of SIR-based CAC algorithms, with parameters as chosen in Section IV-B, are compared via simulation together with a fixed CAC scheme under both homogeneous and hot spot loadings. For an outage probability less than 2% under all loading conditions, the number of calls a fixed CAC can admit per cell is found by simulation to be 49 calls per cell. For the purpose of performance comparison, the offered load for the hot spot cell is taken to be 50% higher than that of any one of its adjacent cells. The length of the

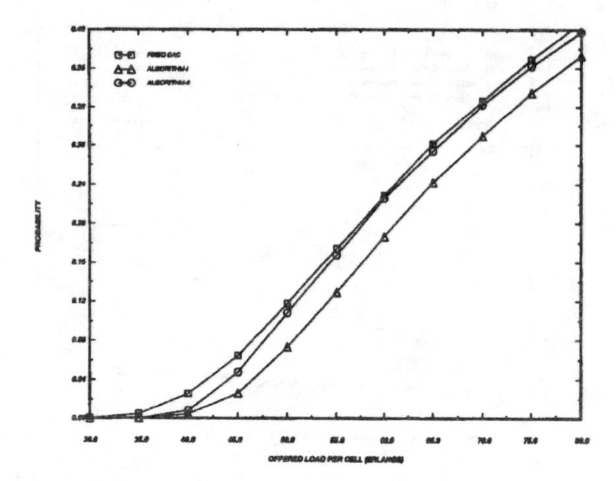

Fig. 6. Blocking probability versus homogeneous load per cell.

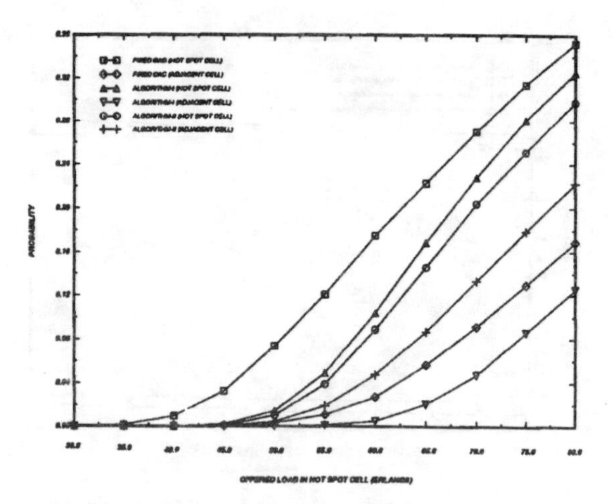

Fig. 8. Blocking probability versus load in a hot spot cell, where the load in the hot spot cell is 50% higher than that of any of its adjacent cells.

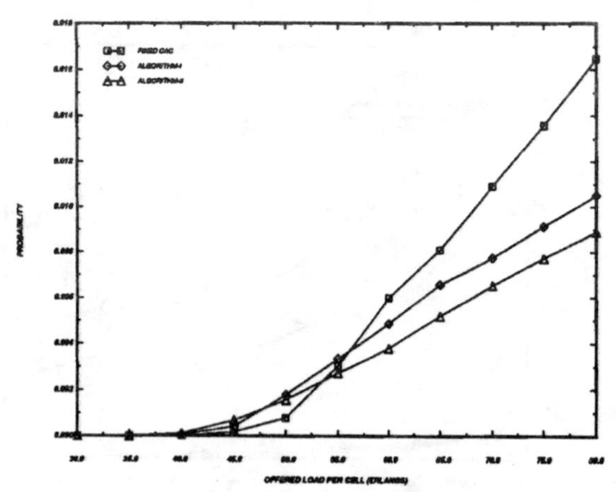

Fig. 7. Outage probability versus homogeneous load per cell.

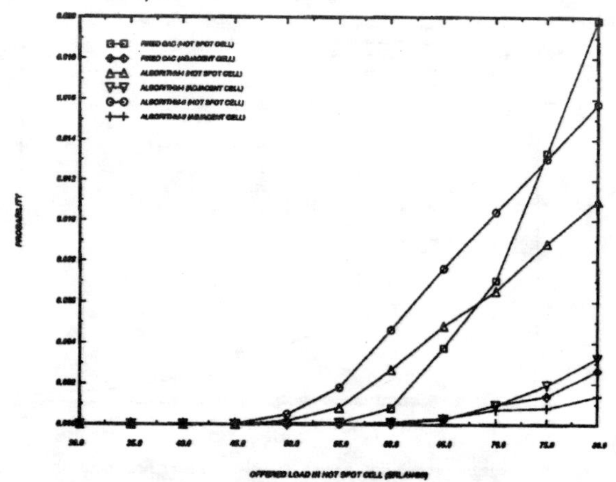

Fig. 9. Outage probability versus load in a hot spot cell, where the load in the hot spot cell is 50% higher than that of any of its adjacent cells.

simulations were chosen so that the 95% confidence intervals are within $\pm 1.0 \times 10^{-4}$ and $\pm 1.0 \times 10^{-5}$ for blocking and outage probability estimations, respectively.

Figs. 6–9 show the results of the performance comparison. The following observations can be made from these results.

1) Under homogeneous loading, Algorithm I outperforms Algorithm II and both algorithms outperform the fixed CAC scheme.
2) Under hot spot loading, Algorithm II outperforms Algorithm I and both algorithms outperform the fixed CAC scheme.
3) The performance improvement of both algorithms over the fixed CAC scheme is over the entire range of the offered loading, from normal loading to overloading. In the overloading range, both the blocking and outage performances of SIR-based CAC's are better than those of the fixed CAC scheme.
4) Under hot spot loading, Algorithm II tends to balance the blocking probabilities between the hot spot cell and its adjacent cells, which leads to improved fairness of performance between mobiles in the hot spot cell and mobiles in the adjacent lightly loaded cells;

By the definition in Section III-C, the Erlang capacity can be obtained under homogeneous loading and hot spot loading situations. Under homogeneous loadings, the Erlang capacity for fixed CAC, Algorithm I, and Algorithm II are 39, 44, and 42 Erlangs, respectively. Under hot spot loadings, the Erlang capacity for the hot spot cell for fixed CAC, Algorithm I, and Algorithm II are 43, 52 and 53 Erlangs, respectively.

Another observation that can be made here is that, in a DS–CDMA cellular system, fixed CAC designed for homogeneous loading situations gives a higher Erlang capacity under hot spot loadings than under homogeneous loadings. This is due to the fact that arriving processes are uniformly distributed over the service area within each cell, however, an arriving mobile will choose its home cell according to the radio attenuation and not to the geographical location. Because of shadowing, the average number of hot spot cell mobiles choosing an adjacent cell as their home cell will be greater than that of a lightly loaded adjacent cell mobile choosing a hot spot cell as their home cell.

V. Conclusion

The following conclusion can be drawn from this study. Very simple SIR-based CAC algorithms can be used in the DS–CDMA cellular system to improve system performance to traffic variations (hot spots). In the proposed algorithms, the signal-to-interference ratio threshold SIR_{TH} can be selected so as to satisfy a desired requirement for outage performance without sacrificing too much in blocking performance. The adjacent cell interference coupling coefficient β can be chosen to improve fairness under hot spot loading. Although CAC algorithms improve the system capacity under homogeneous loadings, the improvement is not as significant as that obtained when DCA is used for FDMA/TDMA cellular systems. This is due to the fact that no frequency reuse constraint is imposed in a DS–CDMA cellular system, thereby making the trunking efficiency of fixed CAC in DS–CDMA systems much higher than that of FCA in FDMA/TDMA systems. Therefore, the improvement in trunking efficiency obtained by using dynamic CAC as opposed to fixed CAC in a DS–CDMA cellular system is not significant. However, compared to the fixed CAC scheme, even for homogeneous loading, the SIR-based CAC algorithms do not deteriorate the system performance under overloading conditions; as a matter of fact, they give better performance. This is not the case for FCA in FDMA/TDMA cellular systems. However, the main objective in using SIR-based CAC in a DS–CDMA cellular system lies in improving the system performance under hot spot loadings as it can deal with varying traffic conditions very effectively.

Although a fairly extensive study was conducted in this paper, several parameters still remain to be studied. These include the impact of the update period, the value of α, the duration of calls, and dealing with handover calls.

References

[1] W. C. Y. Lee, "Overview of cellular CDMA," *IEEE Trans. Vehic. Technol.*, vol. 40, pp. 291–302, May 1991.

[2] K. S. Gilhousen, I. M. Jacobs, R. Padovani, A. J. Viterbi, L. A. Weaver, Jr., and C. E. Wheatley, III, "On the capacity of a cellular CDMA system," *IEEE Trans. Vehic. Technol.*, vol. 40, pp. 303–312, May 1991.

[3] G. L. Stuber and C. Kchao, "Analysis of a multiple-cell direct-sequence CDMA cellular mobile radio system," *IEEE J. Select. Areas Commun.*, vol. 10, pp. 669–679, May 1992.

[4] L. B. Milstein, T. S. Rappaport, and R. Barghoun, "Performance evaluation for cellular CDMA," *IEEE J. Select. Areas Commun.*, vol. 10, pp. 680–688, May 1992.

[5] T. S. Rappaport and L. B. Milstein, "Effect of radio propagation path loss on DS–CDMA cellular frequency reuse efficiency for the reverse channel," *IEEE Trans. Vehic. Technol.*, vol. 41, pp. 231–242, Aug. 1992.

[6] R. Prasad, M. Jansen, and A. Kegel, "Performance analysis of a direct sequence code division multiple access system considering multiple cells," in *Proc. IEEE ISSSTA'92*, Yokohama, Japan, pp. 15–18.

[7] R. R. Gejji, "Forward-link-power control in CDMA cellular systems," *IEEE Trans. Vehic. Technol.*, vol. 41, pp. 479–487, Nov. 1992.

[8] C.-C. Huang, "Computer simulation of a direct sequence spread spectrum cellular radio architecture," *IEEE Trans. Vehic. Technol.*, vol. 41, pp. 479–487, Nov. 1992.

[9] D. E. Everitt and N. W. Macfadyen, "Analysis of multicellular mobile radio telephone systems with loss," *British Telecom. Technol. J.*, vol. 1, pp. 37–45, 1983.

[10] D. Everitt and D. Manfield, "Performance analysis of cellular mobile communication systems with dynamic channel assignment," *IEEE J. Select. Areas Commun.*, vol. 7, pp. 1172–1180, Oct. 1989.

[11] R. Beck and H. Panzer, "Strategies for handover and dynamic channel allocation in micro-cellular mobile radio systems," in *Proc. IEEE Vehic. Tech. Conf.*, May 1989, pp. 178–185.

[12] R. W. Nettleton, "A high capacity assignment method for cellular mobile telephone systems," in *Proc. IEEE Vehic. Tech. Conf.*, May 1989, pp. 359–367.

[13] L. J. Cimini, Jr., G. J. Foschini, and C.-L. I, "Call blocking performance of distributed algorithms for dynamic channel allocation in microcells," in *Proc. IEEE ICC'92*, 1992, pp. 1327–1332.

[14] A. Salmasi, "An overview of code division multiple access (CDMA) applied to the design of personal communications networks," in S. Nanda and D. Goodman, Eds., *Third Generation Wireless Information Networks.* Boston, MA: Kluwer, 1992, pp. 277–298.

[15] R. Jain, *The Art of Computer Systems Performance Analysis.* New York: Wiley, 1991.

[16] G. L. Stuber, L-B. Yiin, and E. M. Long, "Outage control in digital cellular system," *IEEE Trans. Vehic. Technol.*, vol. 40, pp. 177–187, Feb. 1991.

[17] IS-95 "Mobile station-base station compatibility standard for dual-mode wideband spread spectrum cellular system," 1993.

Zhao Liu (S'93) received the B.E.E. degree from the Nanjing Institute of Posts and Telecommunications, Nanjing, P.R.C. in 1982, and the M.S.E. degree in electrical engineering from the University of Pennsylvania, Philadelphia, in 1993.

He is currently working towards the Ph.D. degree in electrical engineering at the University of Pennsylvania. His interests are in wireless personal communication networks.

Magda El Zarki (M'89) received the B.E.E. degree from Cairo University, Cairo, Egypt, in 1979, and the M.S. and Ph.D. degrees in electrical engineering from Columbia University, New York, NY, in 1981 and 1987, respectively.

She worked from 1981 to 1983 as a Communication Network Planner in the Department of International Telecommunications, Citibank. She joined Columbia University in 1983 as a Research Assistant in the Computer Communications Research Laboratory where she was involved in the design and development of an integrated local area network testbed called MAGNET. Currently, she holds the position of Assistant Professor in the Department of Electrical Engineering at the University of Pennsylvania, where she is involved in doing research in telecommunication networks. She also holds a secondary appointment in the Department of Computer and Information Sciences. In January 1993, she was appointed as a part-time Professor of Telecommunication Networks in the Faculty of Electrical Engineering at Delft University of Technology, Delft, The Netherlands.

Ms. El Zarki is a member of the Association for Computing Machinery and Sigma Xi. She is actively involved in many of their sponsored conferences.

Congestion Relief on Power-Controlled CDMA Networks

Jay M. Jacobsmeyer, *Senior Member, IEEE*

***Abstract*—A digital cellular radio code-division multiple-access (CDMA) system can only support a finite number of users before the interference plus noise power density, I_0, received at the cellular base station causes an unacceptable frame-error rate. Once the maximum interference level is reached, new arrivals should be blocked. In a power-controlled CDMA system, the base station can direct mobiles to reduce their power and data rate to reduce interference and allow more users on the system. This approach is employed in TIA IS-95 with respect to the time-varying voice activity on cellular voice channels. In this paper, we investigate an alternative technique where we adjust the power and data rate of mobile data users to the time-varying interference level to allow more users on a congested system. This scheme was simulated for various proportions of voice and data users and offered traffic levels. Blocking probabilities are reduced in some cases by two orders of magnitude. Message wait time, now a random variable, may exceed the wait time for a constant rate system at high traffic levels. If the cellular carrier has a maximum blocking requirement, an adaptive rate/power system can increase capacity. For example, a base station that normally supports 26.4 Erlangs offered traffic with 2% blocking can support 33.5 Erlangs with the same blocking probability if adaptive rates and power control are used. Thus, the adaptive rate system can increase capacity by 27%.**

I. Introduction

CELLULAR radio code-division multiple-access (CDMA) systems are capacity limited by the maximum tolerable level of interference plus noise power density, I_0, received at the cellular base station. Although there is no hard limit on the number of mobile users served, there is a practical level of I_0 that should not be exceeded if we wish to maintain low error rate communication. Once this maximum level is reached, new arrivals should be blocked. In a power-controlled CDMA system, the base station can direct mobiles to reduce their power and bit rate to reduce interference and allow more users on the system. This approach is used today in systems that detect voice activity and adjust the vocoder rate and mobile transmit power downward during pauses in speech. Unfortunately, this method is at the mercy of the statistics of the voice users and can only increase capacity in an average sense. Also, the advantage of voice activity detection diminishes as the proportion of cellular data users grows. However, unlike voice users, data users rarely require real-time two-way communications. Thus, we can reduce the power and bit rate of data users (maintaining a constant energy per bit, E_b) to temporarily relieve congestion and allow more users on the system. This approach can be applied to power-controlled CDMA systems like the North American standard, TIA IS-95 [2].

Manuscript received April 28, 1995; revised November 22, 1995. This work was supported by the National Science Foundation under Award 9361690.

The author is with the Pericle Communications Company, Colorado Springs, CO 80949 USA (email: 73251.1500@compuserve.com).

Publisher Item Identifier S 0733-8716(96)05478-9.

The remainder of this paper is organized as follows: Section II introduces the multiple-access model. In Section III, we address the arrival and server process models. An analytical expression for blocking probability in a simplified four-server model is derived in Section IV. Section V presents simulation results for blocking probability and message wait time for a larger, 35-server system. Section VI concludes the paper.

II. CDMA Model

The wideband system under investigation employs CDMA in a cellular radio environment. Mobile users are perfectly power-controlled to ensure equal received power at the base station regardless of the position of each mobile or the mean propagation conditions. The system can tolerate a maximum interference power level before service quality drops below an acceptable level. This maximum interference power threshold is a constant.

There are two types of mobile users: voice and data. Voice users require a full-rate channel at all times. Data users operate at a time-varying rate designed to maximize throughput and minimize blocking. The data rate and transmitted power level are adjusted in the same direction to maintain a constant energy per bit, E_b. It is not practical to allow the data rate to be lowered to an infinitely small value, so the minimum data rate is set to $r_{\min}$, a constant. The number of busy channels at time t is given by $k = k_v + k_d$, where k_v and k_d are the number of active mobile voice users and data users, respectively. The maximum interference power threshold corresponds to a maximum number of full rate (and full power) users, m, the base station can support. If an arrival occurs when all full-rate channels or their equivalents are busy and the data rate is at its minimum, the arrival is blocked and cleared from the system. No queuing is allowed at the base station.

In [1], Viterbi and Viterbi develop a model for outage probability as a function of the multiple-access interference (MAI) power. Viterbi's model is a $M/M/\infty$ queue with voice activity factor, ρ ($\rho \cong .4$). They did not consider the effects of data users. Because the capacity of a CDMA system is soft, Viterbi and Viterbi prefer *outage* probability to blocking probability. The outage probability is defined as the probability that the interference plus noise power density, I_0, exceeds the noise power density, N_0 by a factor, $1/\eta$, where η takes on typical values between 0.25 and 0.1 [5, p. 204]. The resulting expression for outage probability is simply the tail of the Poisson distribution [5, p. 205]

$$P_{\text{out}} < e^{-\rho\lambda/\mu} \sum_{k=K_0'}^{\infty} (\rho\lambda/\mu)^k/k! \tag{1}$$

where K_0' satisfies the outage condition

$$\sum_{j=2}^{m} v_j < \frac{(W/R)(1-\eta)}{E_b/I_0} = K_0' \tag{2}$$

and v_j is the binary random variable indicating whether the jth voice user is active at any instant. For example, for a process gain of 128, $\eta = 0.1$, and $E_b/I_0 = 5$, $K_0' = 23$. If voice activity is 100%, the maximum number of users supported is $m = K_0' + 1 = 24$.

In contrast, our system is an $M/M/m$ Erlang loss system. The blocking probability for voice users is given by the Erlang B formula [3], rather than the Poisson distribution, but for $m > 20$ and blocking probabilities of a few percent or less, the Poisson distribution and Erlang B are practically identical. From (1), we see that the net effect of the voice activity detection is to reduce the offered traffic by a factor ρ. Thus, if we are comparing like systems, we can ignore voice activity detection by letting $\rho = 1$ and $v_j = 1$ for all j. Because our power control is perfect, the system performance is a function of the total interference power, not the total number of users. Thus, we can increase the number of users beyond the limit of Viterbi [5] as long as we reduce the power and bit rate so the maximum interference power level is not exceeded.

III. Arrival and Server Process

We shall compare our results to a conventional multiserver (Erlang) loss system where all users operate at a constant rate. Users arrive at the base station according to a Poisson random process with rate λ. There are m servers (channels), each with independent, identically distributed service times. Arrivals who find all m servers busy are turned away and lost to the system. (They get a busy signal.) No queuing of arrivals is allowed. Service times (call durations) are exponentially distributed with mean $1/\mu$. For an m server system, the probability that k servers are busy is given by the Erlang B formula [3]

$$P_k = \frac{\left(\frac{\lambda}{\mu}\right)^k}{k!} \left(\sum_{j=0}^{m} \frac{\left(\frac{\lambda}{\mu}\right)^j}{j!} \right)^{-1} \tag{3}$$

and the mean number of busy servers is

$$E[K] = \frac{\lambda}{\mu}(1 - P_m) \tag{4}$$

where P_m is the probability that m servers are busy. The probability P_m is also known as the *blocking probability*, P_b.

In the adaptive rate/power system under investigation, voice users arrive according to a Poisson process with rate λ_v, and remain on the system for an exponentially distributed call duration with mean $(1/\mu_v)$, independent of the arrival process. Data users also arrive according to a Poisson process with rate λ_d, and have exponentially distributed message sizes of mean N bits. The data user arrival process is independent of the voice user arrival process. The call duration for data users is a function of the normalized data rate, R, a random variable. The mean call duration at full rate is $(1/\mu_d) = N/r_f$ seconds where r_f is the full rate in bits per second. The total offered traffic is $\lambda/\mu = \lambda_v/\mu_v + \lambda_d/\mu_d$.

Data users may operate at fractional normalized rates, R, where $R \geq r_{\min}$. All data users operate at the same rate at time t. The normalized (to r_f) operating rate of data users is given by the following expression

$$R = \begin{cases} \frac{m-k_v}{k_d}, & \frac{m-k_v}{k_d} \geq r_{\min} \\ r_{\min}, & \frac{m-k_v}{k_d} < r_{\min} \end{cases} \tag{5}$$

where k_v is the number of voice users and k_d is the number of data users. The variables k_d and k_v are nonnegative integers. The number of active equivalent full-rate channels is given by

$$k_f = k_v + R\,k_d. \tag{6}$$

If there are any data users in the system, all of the capacity is used, and $k_f = m$. When a user (voice or data) attempts to access the system when the equivalent number of full-rate channels is m, all data users will lower their power levels and data rates proportionally. This process creates more full-rate channels and avoids blocking. If all data users are already at the lowest allowable data rate, $r_{\min}$, the call is blocked. We shall show in the next section that the probability of blocking will be much lower than for the conventional constant rate case.

IV. Analytical Results

To illustrate the server process, consider the a simplified example with $m = 4$ and $r_{\min} = 1/2$. The state diagram for this system is shown in Fig. 1. The state probabilities for the system are found by solving the system of linear equations formed by the balance equations (e.g., see [4, p. 351]) plus the expression

$$\sum_{j=0}^{24} P_j = 1. \tag{7}$$

Note that there are nine blocking states: 7, 8, 14, 15, 19, 20, 22, 23, and 24. States 8, 16, 20, 23, and 24 are blocking states to both voice and data arrivals. States 7, 14, 19, and 22 are blocking states to voice arrivals only. The blocking probability is given by

$$P_b = p_v P_7 + P_8 + p_v P_{14} + P_{15} + p_v P_{19} + P_{20} + p_v P_{22} + P_{23} + P_{24} \tag{8}$$

where p_v is the probability that a voice arrival occurs before a data arrival. Because the voice and data interarrival times are exponentially distributed with rates λ_v and λ_d, respectively, the probability p_v is given by [4, p. 195]

$$p_v = \frac{\lambda}{\lambda_v + \lambda_d}. \tag{9}$$

Equation (8) can be used to compute the system blocking probability for various arrival rates and service times. For

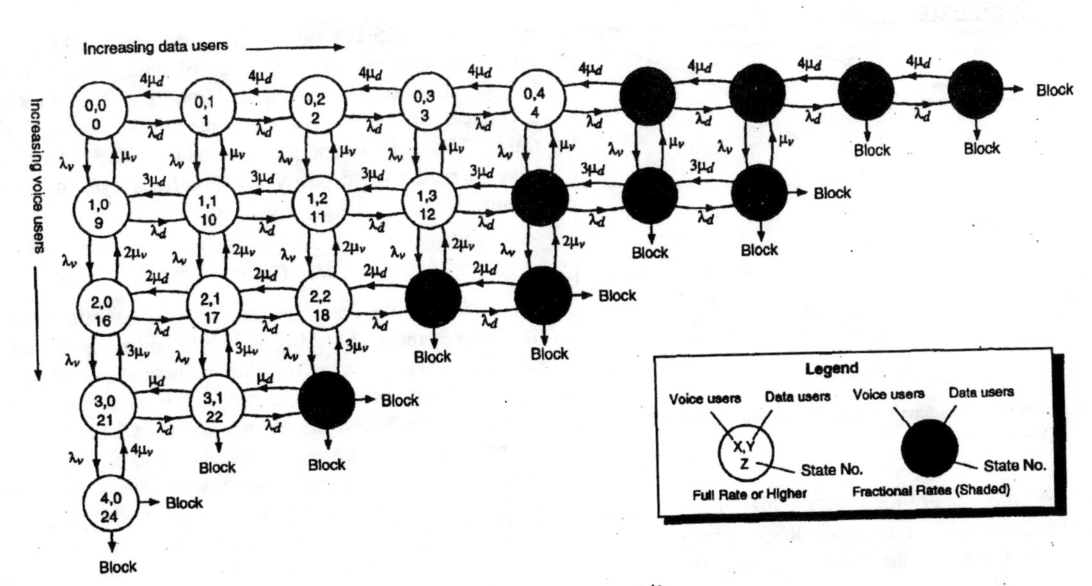

Fig. 1. State diagram for $m = 4$ system with $R = (m - k_v)/k_d$ and $r_{\min} = 1/2$.

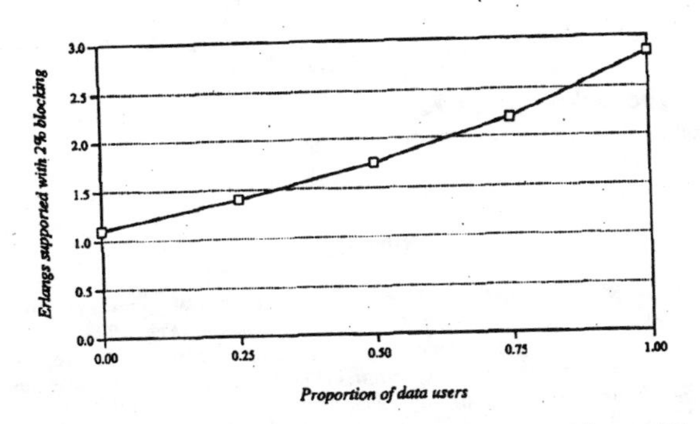

Fig. 2. Offered traffic supported at 2% blocking ($m = 4$, $1/\mu_v = 100$ s, $1/\mu_d = 10$ s).

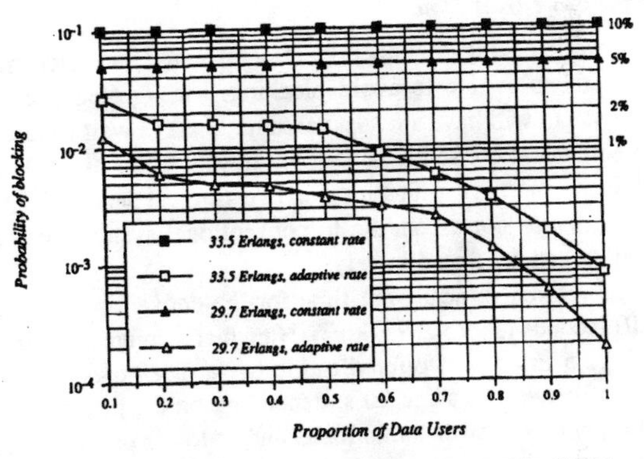

Fig. 3. Probability of blocking with adaptive rate modem in CDMA system ($m = 35$, $1/\mu_v = 200$ s, $1/\mu_d = 10$ s).

example, consider a total offered traffic of $\lambda/\mu = 2$ Erlangs with $\lambda_v = 0.01$ arrivals/s, $1/\mu_v = 100$ s, $\lambda_d = 0.1$ arrivals/s, and $1/\mu_d = 10$ s (50% voice traffic, 50% data traffic). The blocking probability for these parameters is 3%. The blocking probability for a conventional Erlang B system with this same traffic load is 9.5% [from (3)]. Thus, we have lowered the blocking probability by more than a factor of three by adapting the data rate to the traffic load.

If the cellular service provider has a blocking probability objective, a variable rate system can improve the system capacity. For example, we have plotted the offered traffic supported at a blocking rate of 2% for our $m = 4$ variable rate system in Fig. 2. Note that when half of the traffic is data, the capacity of the system is increased from 1.1 Erlangs to 1.8 Erlangs, an improvement of 60%.

V. SIMULATION RESULTS

Analytical results are more difficult to achieve for larger systems. To gain results for a more realistic server size, we simulated an $m = 35$ channel system with a minimum rate of $r_{\min} = 1/3$. The simulations used a mean call duration for voice of $1/\mu_v = 200$ s, a full rate of 9.6 kb/s, and a mean packet size for data of 96 kb ($1/\mu_d = 10$ s). Two performance measures were evaluated: blocking probability and message wait time.

A. Blocking Probability

In Fig. 3, we have plotted the blocking probability for two offered traffic loads before and after incorporating adaptive data rate communications. These two traffic loads are 33.5 and 29.7, corresponding to conventional system blocking probabilities of 10% and 5%, respectively. The abscissa of Fig. 3 is the proportion of data users in the system. For example, if the proportion of data users is 0.3, the fraction of offered traffic (λ/μ) comprising data users is 30% and the fraction comprising voice users is 70%.

Note that the blocking probability decreases as the proportion of data users increases. This result satisfies our intuition, since a higher proportion of data users means we have more flexibility and "headroom." The reduction in blocking probability can be quite dramatic, especially when the offered traffic

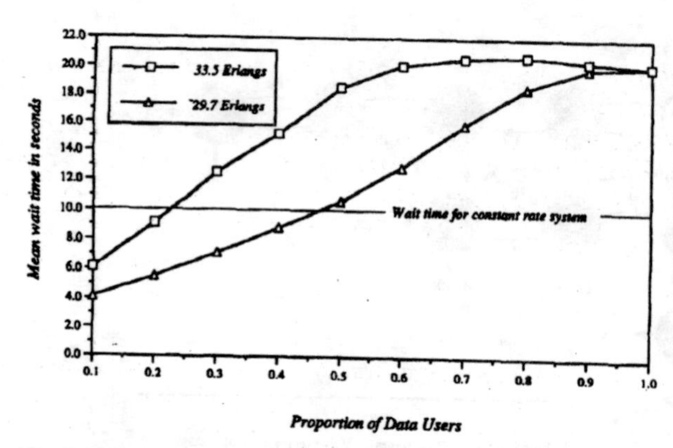

Fig. 4. Mean message wait time for adaptive rate mobile in CDMA system ($m = 35$, $1/\mu_v = 200$ s, $1/\mu_d = 10$ s).

is low or the proportion of data users is high. For example, at an offered traffic of 29.7 Erlangs (5% blocking before adaptive rate) and 80% data users, the blocking probability is reduced by a factor of 50.[1]

B. Message Wait Time

Lower blocking does not come free. Because data users must lower their bit rate to avoid blocking, a congested adaptive rate system will have a net increase in message wait time over the equivalent conventional system. In Fig. 4, we have plotted the mean message wait time for two traffic levels, 33.5 and 29.7 Erlangs, corresponding to conventional system blocking of 10% and 5%, respectively.

The reference mean wait time for the conventional system is 10 seconds (96 kb at 9.6 kb/s). Note that a congested system with 33.5 Erlangs of offered traffic has a mean message wait time of between 6 s and 20 s, depending on the proportion of data users. For lower traffic loads and/or lower proportions of data users, the mean wait time can actually be less than the reference mean wait time.

Thus, there is an inherent tradeoff between wait time and blocking. When the proportion of data users is low (< 25%), even a congested system with 33.5 Erlangs will have a net reduction in message wait time. But the blocking probability will be highest. As the proportion of data users increases, the message wait time increases while the blocking probability decreases.

C. Net Capacity Improvements

From Fig. 3, we see that an adaptive data rate system can reduce the blocking probability of a power-controlled CDMA system with 33.5 Erlangs offered traffic from 10% to 2% or less, depending on the proportion of data users. Consulting (3), or an Erlang B table, we find that 2% blocking on a conventional system corresponds to 26.4 Erlangs offered traffic. Thus, if the operator's standard is 2% blocking, we have increased the capacity of the cellular system from 26.4 Erlangs to 33.5 Erlangs, a gain of 27%.

[1] The reader may notice that our model allows a maximum of 35×3 or 105 channels but the IS-95 system is limited to 61 Walsh codes for traffic (plus one each for pilot, sysnc, and paging). The probability of having more than 61 active full rate channels is small, about 6×10^{-6}, for data traffic loads as high as 16.75 Erlangs and total offered traffic of 33.5 Erlangs. This problem does note exist on system that use a larger, nonorthigonal, code set like the Gold codes.

VI. Conclusion

By adapting the data rate and power of mobile data terminals, we can reduce blocking probability on a power-controlled CDMA system. The achievable reductions depend on the offered traffic level and the proportion of mean data users on the system. When the offered traffic is 33.5 Erlangs and the proportion of data users is 20%, blocking probability is reduced by a factor of six. Message wait time is now a random variable and congested systems may cause delays of up to a factor of two relative to conventional CDMA systems. If the cellular operator has a maximum blocking probability standard of 2%, this approach can increase system capacity by 27% on a typical 1.25 MHz CDMA system.

Acknowledgment

The author thanks K. Budka of AT&T Bell Laboratories, who provided several helpful suggestions on solving the analytical problem posed in Section IV.

References

[1] A. M. Viterbi and A. J. Viterbi, "Erlang capacity of a power controlled CDMA system," *IEEE J. Select. Areas Commun.*, Aug. 1993.
[2] TIA IS-95, "Mobile station-base station compatibility standard for dual-mode wideband spread spectrum cellular system," Telecommunications Industry Association, July 1993.
[3] D. R. Cox and W. L. Smith, *Queues*. London, England: Methuen & Co., 1961.
[4] S. M. Ross, *Introduction to Probability Models* New York: Academic, 1985.
[5] A. J. Viterbi, *CDMA Principles of Spread Spectrum Communication.* Reading, MA: Addison-Wesley, 1995.

Jay M. Jacobsmeyer (S'85–M'87–SM'91) was born in Okaloosa County, FL, in 1959. He received the B.S. degree in electrical engineering (magna cum laude) from Virginia Tech, Blacksburg, VA, in 1981, and the M.S. degree in electrical engineering from Cornell University, Ithaca, NY, in 1987.

From 1981 to 1990, he served on active duty with the United States Air Force as a Communications Officer. He is currently a member of the Air Force Reserve. From 1990 to 1993, he was a Senior Staff Engineer with ENSCO, Inc., heading the Colorado Springs Office. He co-founded Pericle Communications Company in 1992 and in June, 1993, he assumed his current position of Chief Technical Officer. His current research interests are communication theory and information theory and their applications to fading communications channels.

Mr. Jacobsmeyer is a member of Eta Kappa Nu and the Armed Forces Communications-Electronics Association (AFCEA).

Performance Analysis of Soft Handoff in CDMA Cellular Networks

Szu-Lin Su, *Member, IEEE*, Jen-Yeu Chen, and Jane-Hwa Huang

***Abstract*—The code-division multiple-access (CDMA) scheme has been considered as one possible choice of the future standards for cellular networks because of its various advantages. Since there can be only one carrier frequency being used in CDMA systems, a handoff scheme with diversity, a so-called "soft handoff," was proposed for higher communication quality and capacity. In this paper, a mathematical model is developed to analyze the soft handoff process. Markov conception is applied to describe the system's statistic behavior in steady state. System performances such as blocking probability, handoff refused probability, and channel efficiency are also determined. It is concluded that the larger the area of soft handoff region is, the better users in the cellular network will feel.**

I. Introduction

RECENTLY, code-division multiple-access (CDMA) has become a most promising technology for the future cellular networks due to its various advantages. Because there can be only one frequency being used in CDMA systems, a handoff scheme with diversity, a so-called "soft handoff," is proposed for higher communication quality [1], [11]. In brief, a handoff process in which the mobile unit can commence communication with a target base station without interrupting the communication with the current serving base station is called soft handoff, i.e., "make before break." The traditional handoff scheme which requires the mobile unit to break the communication with the current base station before establishing a new communication with the target base station is called hard handoff (break before make). The handoff region of the hard handoff scheme is, in general, very narrow. This causes frequent hard handoffs (frequent "communication breaks"), a so-called "ping-pong effect," when a mobile unit drives in and out of the cell's boundary. The fading effects of the channel will also make the "ping-pong effect" be more serious. To reduce this problem, a hysteresis scheme was suggested in hard handoff processes [2]. However, the hysteresis scheme requires larger transmitting power to ensure a broader handoff region and satisfactory communication quality. This also introduces larger interference to other users in a CDMA system and will reduce system's capacity. In soft handoff, since the procedure is "make before break," no matter how frequently a mobile unit drives in and out of the handoff region, communication "breaks" seldomly occur. Moreover, the adoption of diversity reception in soft handoff also leads to a better communication quality without as large a transmitting power as that required in hard handoff to maintain the broader handoff region. In other words, by using the soft handoff scheme and a proper power control strategy as proposed in IS-95, the required transmitted power as well as interference can be reduced to improve the communication quality and system's capacity [12], [13]. Furthermore, larger handoff regions of soft handoff also introduces a longer mean queueing time to get a new channel from the target base station.

The investigations of hard-handoff processes have been extensively proposed in [2]–[10]. In these papers, analytical models were developed and used to derive the characteristics of system behavior. The birth–death process conception was applied in the analysis and the iterative method was adopted to find implicit handoff parameters in statistical equilibrium systems [3]–[10]. Prioritized handoff procedures with queue and without queue were presented in [3] and [6]. The introduction of the queueing concept in the handoff process can reduce the probability of forced termination [3], [6], [7], [9], [10]. In [4], [7], and [10], the system performances regarding simultaneous multiple handoff requirements from a moving vehicle (bus, train, etc.) were analyzed. The performance of the cellular system with mixed platform types (i.e., different in the vehicle mobility and number of communication ports on the platform) were shown in [5], [8], [9], and [10].

However, there are still very few discussions on system performance of the soft handoff process. In the following sections we will extend the mathematical model as stated in [3], [6], [7], [9], and [10] to analyze the performance of soft handoff scheme.

II. System Model

In the analytic model, we assume that one mobile unit carries one call only. It means that there is no bulk handoff arrival. The soft handoff process is as defined in the US IS-95 standard, but we only consider that there are at most two different sources in diversity reception. Each cell will reserve C_h channels out of a total of C available channels exclusively for handoff calls, a so-called cut-off priority, because a suddenly forced termination during a call session will be more upsetting than a failure to connect. Every handoff requirement is assumed to be perfectly detected in our model and the assignment of the channel is instantaneous

Manuscript received May 17, 1995; revised October 15, 1995. This work was supported by the National Science Council of Taiwan, R.O.C., under Grant NSC 85-2221-E-006-015. This paper was presented in part at the 1995 IEEE International Symposium on Personal, Indoor and Mobile Radio Communications (PIMRC'95), Toronto, Canada, September 1995.

S.-L. Su and J.-H. Huang are with the Department of Electrical Engineering, National Cheng Kung University, Tainan, Taiwan, R.O.C. (email: ssl@eembox.ncku.edu.tw).

J.-Y. Chen is with the Network Planning Lab., TL, Ministry of Transportation and Communication, Taiwan, R.O.C.

Publisher Item Identifier S 0733-8716(96)05242-0.

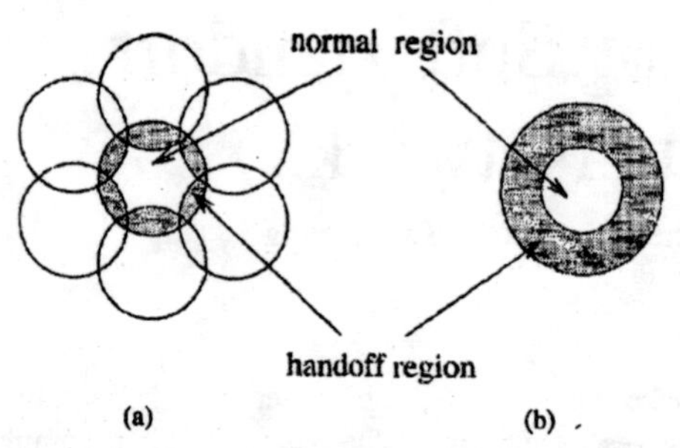

Fig. 1. Handoff region and normal region of: (a) cellular network and (b) simplified cell.

if the channel is available. From the intrinsic property of soft handoff process, a handoff requirement must not be denied immediately when there is no available channel. Instead, the handoff requirement shall be put into a queueing list, as proposed in [3]. In this paper, it is assumed that the allowable maximum queue length is equal to Q.

The *call duration time* T_c is assumed to be exponentially distributed with mean $\bar{T}_c = \mu_c^{-1}$. We may divide a cell into two regions: the normal and handoff regions. The normal region is surrounded by the handoff region (as shown in Fig. 1) and we assume their shapes are a circle and a ring, respectively, to simplify the analysis. A handoff process shall be carried out for any call which enters the handoff region from outside the cell, and the diversity reception from different base stations is possible only in this area. Both new call arrivals and handoff arrivals are assumed to be Poisson distributed with rates Λ_n and Λ_h. By the assumption that the location of a new generated call is uniformly distributed all over a cell, the new call arrival rates in two different regions are

$$\Lambda_{n1} \triangleq \text{the new call arrival rate in handoff region} = a \cdot \Lambda_n$$

and

$$\Lambda_{n2} \triangleq \text{the new call arrival rate in normal region} = (1-a) \cdot \Lambda_n \quad (1)$$

where

$$a = \frac{\text{area of handoff region}}{\text{area of a cell}}.$$

We assume that when a new call is generated in the handoff region, it can ask for channel assignment from both the cell of interest and a neighboring cell with which it can communicate. If the cell of interest has a channel available for such a generated call, the new call is considered as a successful new call in the analysis. If not, the new generated call is said to be blocked from the point of view of the cell of interest. However, for the latter case, the new call may successfully enter the cellular system by getting a channel assignment from the neighboring cell and becoming a handoff arrival for the cell of interest. Of course, if the new call cannot get channel assignment from either of the two cells, the call is said to be blocked by this cellular system, and the mobile unit shall try its attempt later.

Both the *dwell times* of a call in two distinct regions are assumed to be exponentially distributed. The mean dwell time in handoff region is $\bar{T}_{d1} = \mu_d^{-1}$ and that in normal region is $\bar{T}_{d2} = \mu_D^{-1}$. Since the relations between the mean dwell time in the whole cell $\bar{T}_{dc}$ and the mean dwell time in each region shall be a function of the covered area of each region, we can find the related functions in Appendix A, if all the mobile units are assumed to be "random walking" in a cell which is very suitable for the urban area. According to these related functions, we can get the values of $\bar{T}_{d1}$ and $\bar{T}_{d2}$ for given values of $\bar{T}_{dc}$ and "a."

III. System Analysis

From the description of the system model, it is clear that we can analyze the system performance by the birth–death process as in [3]. The state in this process is defined as

$$s = (v_1, v_2, q) \quad v_1, v_2 \geq 0 \quad \text{and} \quad 0 \leq q \leq Q$$

where v_1 is the number of active calls in the handoff region, v_2 is the number of active calls in the normal region, and q is the number of mobile units in queue. $N(s)$ is the total number of active channels at state s of a cell, i.e., $N(s) = v_1 + v_2$ and $N(s) \leq C$. We note that new calls will be blocked when $N(s) \geq C - C_h$ and the handoff arrival will be put in queue while all the channels are in use, i.e., $N(s) \geq C$. Since the queue is not used when $v_1 + v_2 \leq C$ and it is used only when $v_1 + v_2 = C$, so the two-dimensional (2-D) Markov chain is enough to describe the state transitions. An example with $C = 5$, $C_h = 2$, and $Q = 3$ is shown in Figs. 2 and 3.

A. Driving Processes and State Transitions

There are five conditions which can cause a state transition. They are: New call arrival, call completion, handoff arrival, handoff departure, and region transitions (from the normal region to the handoff region and vice versa).

It is noted that a region transition in which a mobile unit drives into the handoff region from the normal region of the current communicating cell and asks for a channel from the neighboring cell is also a handoff arrival on the aspect of the neighboring cell. Besides, if a mobile unit drives from the handoff region to the normal region in a cell, it is a handoff departure of the neighboring cell. If a mobile unit drives out of the handoff region of a cell, according to its moving direction, it can be a true handoff departure or only a region transition of the cell of interest. We assume the rate of the former is $\mu_{do} = (1 - bk) \cdot \mu_d$ and that of the latter is $\mu_{dk} = bk \cdot \mu_d$, where the parameter "$bk$" is the average probability of moving back to the normal region of the cell of interest for a mobile unit in the handoff region. The value of bk is a function of the value of "a" and the related function between "bk" and "a" is described in Appendix B. An example of detailed state transitions is shown in Figs. 2 and 3.

B. Flow Equilibrium Equations and Steady-State Probabilities

Let $p(i, j, q)$ be the steady-state probability of the state $s = (i, j, q)$. There is a flow equilibrium equation for each

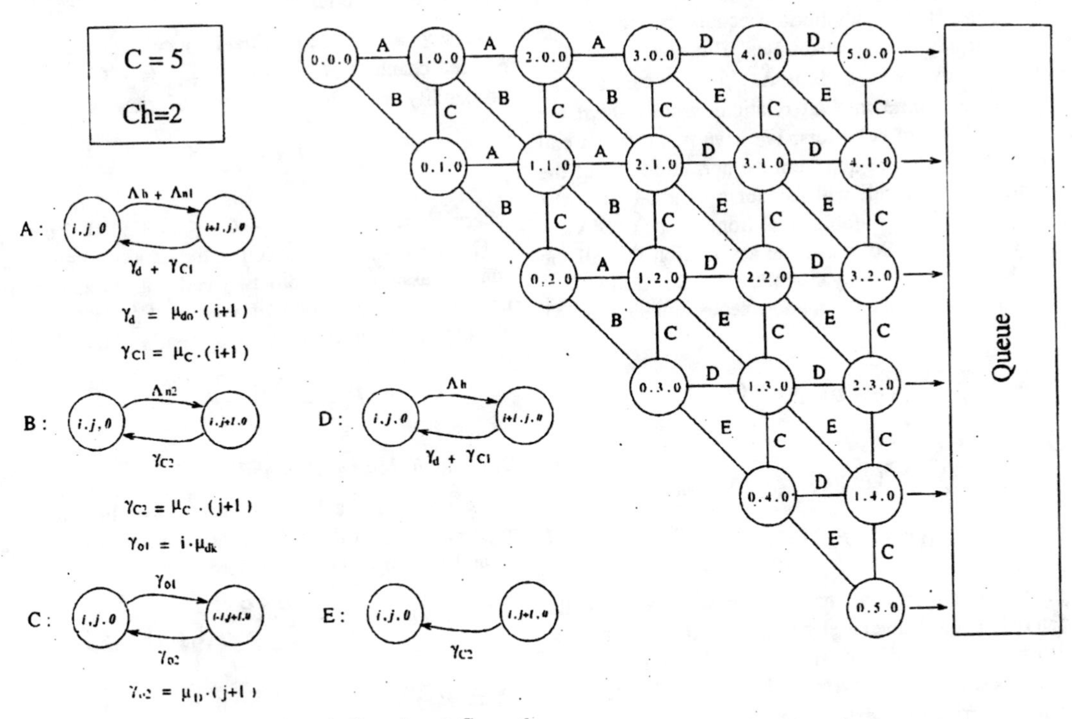

Fig. 2. The birth–death processes of a cell with $C = 5$ and $C_h = 2$.

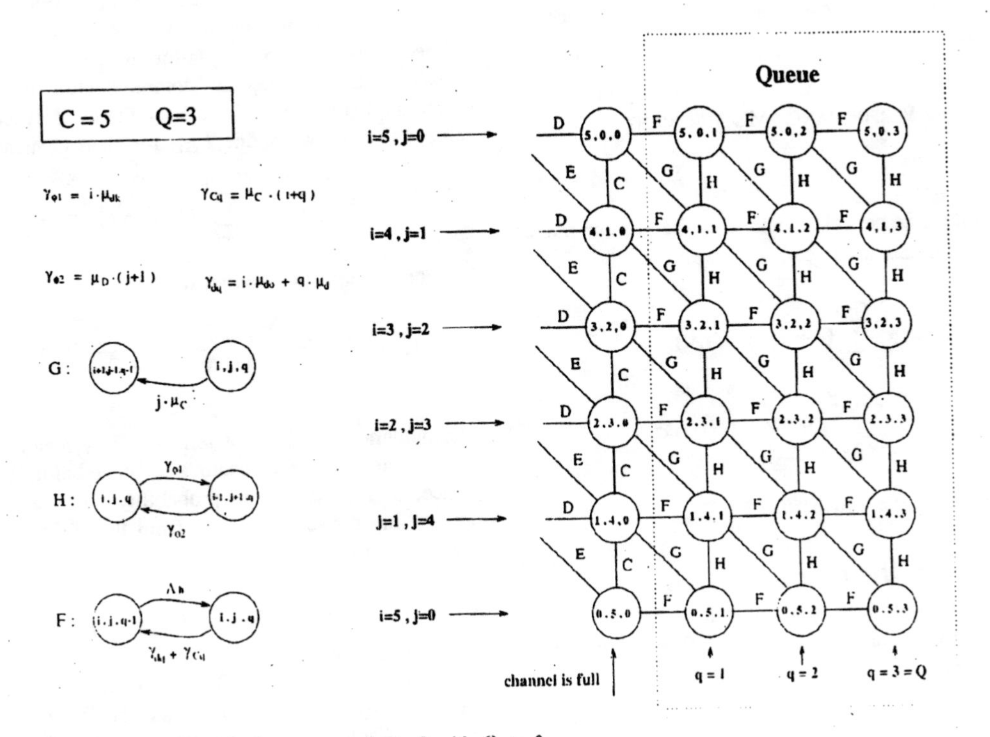

Fig. 3. The queueing part of the birth–death processes of Fig. 2 with $Q = 3$.

state, i.e., the total rate of flowing into a state will be equal to the total rate of flowing out from it. If the total number of states is Ω, there are $\Omega - 1$ linearly independent flow equilibrium equations. Besides, the sum of the steady-state probabilities is

$$\sum_{i=0}^{C}\sum_{j=0}^{C-i}\sum_{q=0}^{Q} p(i, j, q) = 1. \tag{2}$$

By solving this 2-D birth–death process from the chosen linearly independent flow equilibrium equations and (2), we can find the steady-state probabilities. But the parameters used in the Markov chain, such as Λ_h, $\mu_D, \cdots$ etc., are not arbitrarily chosen. Some implicit relations among them are restricted by the system behaviors. We have noted that a call which enters the handoff region from the normal region of the current communicating cell will ask for a channel from the neighboring cell. It is a handoff arrival from the point of view of the neighboring cell. So, under the assumption that all the cells have the same probability characteristics in steady state, the parameters shall satisfy the homogeneous equation as [4]

$$\Lambda_h = \sum_{i=0}^{C}\sum_{j=0}^{C-i}(j\cdot\mu_D)\cdot p(i,j,0) + \sum_{i=0}^{C}\sum_{q=1}^{Q}(j\cdot\mu_D)\cdot p(i,j,q)\,|_{j=C-i} + \Lambda_{n1}\cdot(1-P_B)\cdot P_B \tag{3}$$

where the last term represents the rate of new generated calls which can only get a channel assignment from the neighboring cell and become a handoff arrival for the cell of interest, and P_B is the blocking probability from the cell's point of view which will be derived in next section.

Utilizing a computer program, an iterative approach like that described in [4] and [7] is executed to find the stationary state probabilities.

IV. Performance Measurement

A. Carried Traffic

The carried traffic per cell is defined as the average number of occupied channels in the cell of interest, i.e.,

$$A_c = \sum_{i=0}^{C}\sum_{j=0}^{C-i}(i+j)\cdot p(i,j,0) + \sum_{i=0}^{C}\sum_{q=1}^{Q} C\cdot p(i,j,q)\,|_{j=C-i} \tag{4}$$

where Q is the queue length.

B. Carried Handoff Traffic

The carried handoff traffic per cell is defined as the average number of occupied channels in the handoff region of the cell of interest, i.e.,

$$A_{\text{ch}} = \sum_{i=0}^{C}\sum_{j=0}^{C-i} i\cdot p(i,j,0) + \sum_{i=0}^{C}\sum_{q=1}^{Q} i\cdot p(i,j,q)\,|_{j=C-i}. \tag{5}$$

C. Blocking Probability

A new call will be blocked by a cell when the total number of used channels $N(s) = (i+j) \geq C - C_h$, so the blocking probability from the cell's point of view is

$$P_B = \sum_{B_0} p(i,j,q) \tag{6}$$

where $B_0 = \{(i,j,q) \mid C - C_h \leq (i+j), 0 \leq q \leq Q\}$.

However, since a new cell in the handoff region can ask for channel assignment from two cells, its blocking probability from the system's point of view is P_B^2. Hence, the average blocking probability from the system's point of view is

$$P_{BS} = \frac{\Lambda_{n1}\cdot P_B^2 + \Lambda_{n2}\cdot P_B}{\Lambda_n}. \tag{7}$$

D. Handoff Refused Probability

We consider the unsuccessful probability of the handoff requirement from three different points of view as classes A, B, and C. For class A, we define

$$P_{HA} = \sum_{i=0}^{C}\sum_{q=0}^{Q} p(i,j,q)\,|_{j=C-i} \tag{8}$$

where P_{HA} is the probability that channels are all occupied when an active mobile unit drives into a cell. In this case, the handoff requirement is unsuccessful at temporal moment and it may be put into queue for latter trial or, if the queue is full, it is refused forever. (Of course, failure of the handoff requirement does not mean "immediate communication drop.") For the condition that the queue is full and the handoff requirement is refused forever, we define P_{HC} for class C such that

$$P_{HC} = \sum_{j=0}^{C-C_h} p(i,j,Q)\,|_{i=C-j}. \tag{9}$$

In another sense, we define, for class B

$$P_{HB} = \frac{\sum_{j=0}^{C-C_h}\sum_{q=1}^{Q} q\cdot(\mu_c+\mu_d)\cdot p(i,j,q)\,|_{i=C-j}}{\Lambda_h\cdot(1-P_{HC})} + P_{HC}. \tag{10}$$

The numerator of the first term of P_{HB} means the average leaving rate of the mobiles units in queue before being served. Hence, the P_{HB} is the total probability that a handoff call does not get any service ultimately from the cell of interest. It is also a sense of refused handoff.

E. Channel Efficiency (Trunk Resource Efficiency)

For assessing the channel efficiency, we define the "efficiency" of each channel by its "necessity." It means that the efficiency of a channel equals one if the corresponding active call holds only one channel and the efficiency reduces to one half if the active call holds two channels. Since we only consider the case that there are at most two different sources in diversity reception in this analysis, each active call in the handoff region will hold at most two channels from two neighboring cells, one from each. We can distinguish the active calls in the handoff region into two classes by the number of its carried channels, as follows.

1) There are three different conditions in which an active call in the handoff region will carry two channels. The following are their generated rates, respectively:

$$F_{2n} \triangleq \text{the generated rate of new calls with two channels} = \Lambda_{n1} \cdot (1 - P_B)^2 \quad (11)$$

$$F_{2h} \triangleq \text{the arrival rate of handoff calls with two channels} = \Lambda_h \cdot (1 - P_{HB}) \quad (12)$$

$$F_{2t} \triangleq \text{the transition rate of active calls from the normal region which will hold two channels} = \Gamma \cdot (1 - P_{HB}) \quad (13)$$

where Γ is the total rate of calls transiting from the normal region to the handoff region

$$\Gamma = \sum_{i=0}^{C}\sum_{j=0}^{C-i} (j \cdot \mu_D) \cdot p(i, j.0) + \sum_{i=0}^{C}\sum_{q=1}^{Q} (j \cdot \mu_D) \cdot p(i, j, q)\,|_{j=C-i}. \quad (14)$$

2) Any other active call which has been acquired by the cell of interest will hold exactly one channel. The total generated rate of this kind of call in the handoff region is

$$F_1 = \Lambda_{n1} \cdot (1 - P_B) \cdot P_B + \Gamma \cdot P_{HB}. \quad (15)$$

Thus, the average number of active calls in the handoff region which hold two channels is

$$\kappa_2 = A_{\text{ch}} \cdot \frac{F_{2n} + F_{2h} + F_{2t}}{F_{2n} + F_{2h} + F_{2t} + F_1}. \quad (16)$$

Hence, the total channel efficiency in a cell can be shown as

$$E = 1 - \frac{1}{2} \cdot \left(\frac{\kappa_2}{A_c}\right). \quad (17)$$

V. Numerical Examples

Some numerical results have been generated. The system parameters used in the numerical examples are $C = 12$, $C_h = 2$, $\mu_c = 0.01$, $\mu_{\text{dc}} = (T_{\text{dc}})^{-1} = 0.03$, and $Q = 4$. Figs. 4–10 show the relations between new call arrival rate and performance measurements for different values of ratio "a," the fraction of occupied area of the handoff region in a cell. Besides, as proposed in [12] and [13], the channel capacity in a cell of a CDMA system can be increased by the diversity reception in the handoff region. Hence, we consider two conditions in our numerical examples.

1) *Condition 1:* There is a fixed channel capacity $C = 12$ in a cell regardless of the value of "a." In this case, the voice/data quality will be better for larger "a" because of the higher signal-to-interference ratio (SIR) from the diversity reception in the larger area.

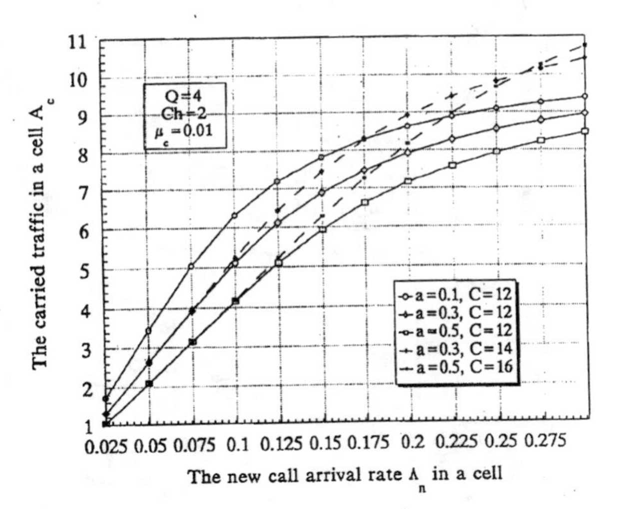

Fig. 4. The carried traffic in a cell versus the new call arrival rate Λ_n.

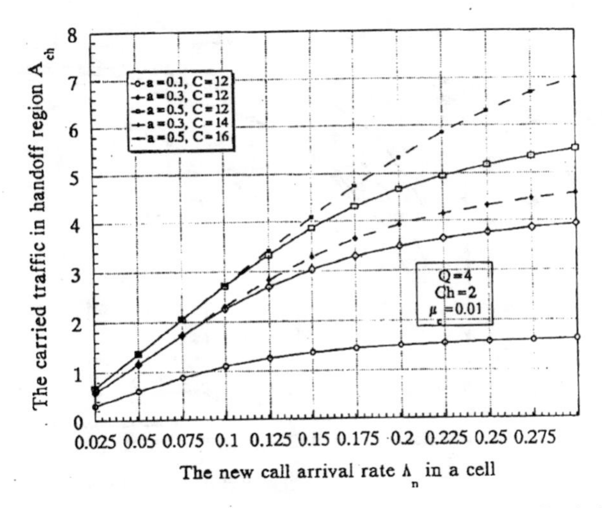

Fig. 5. The carried traffic in the handoff region versus the new call arrival rate Λ_n.

2) *Condition 2:* The channel capacity in each cell is increased under the requirement of the same voice/data quality when the value of "a" becomes larger. Here, we adopt $C = 14, 16$ when $a = 0.3, 0.5$, respectively, as suggested in [12].

In Condition 1, the higher value of "a" introduces lower blocking probability and handoff refused probabilities, i.e., the better system performances, due to less traffic in the whole cell. Besides, the higher value of "a" also results in a lower handoff refused probability for class B (P_{HB}) because, when "a" is larger, the dwell time T_d for the same mobility of the active users will be increased accordingly. In other words, a handoff arrival call in queue can stay with a longer dwell time in the handoff region to wait for the release of a channel. Since an active call in the handoff region may hold two channels, a higher value of "a" will introduce a larger number of carried handoff calls and a lower trunk efficiency as shown in Figs. 5 and 6, respectively.

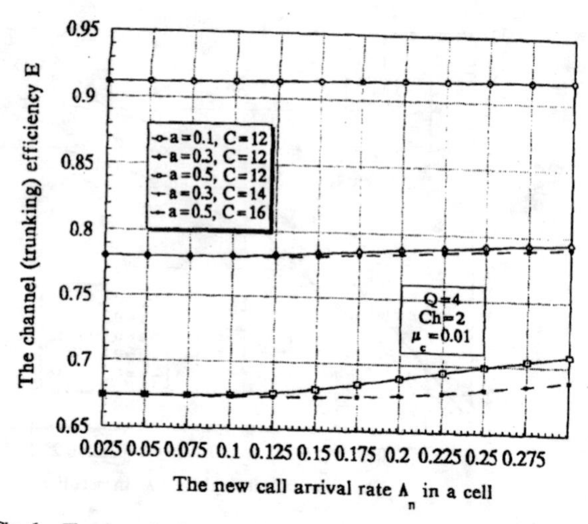

Fig. 6. The channel efficiency versus the new call arrival rate Λ_n.

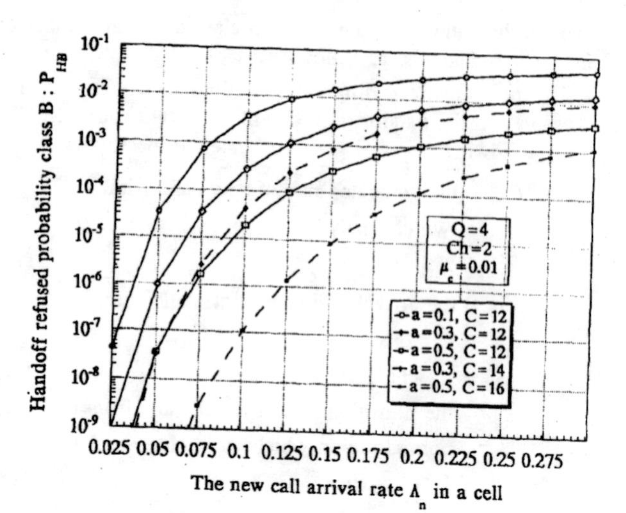

Fig. 9. The handoff refused probability class B versus the new call arrival rate Λ_n.

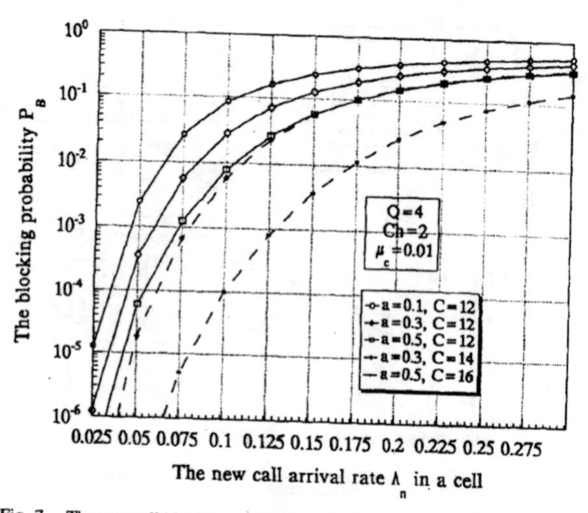

Fig. 7. The new call blocking probability versus the new call arrival rate Λ_n.

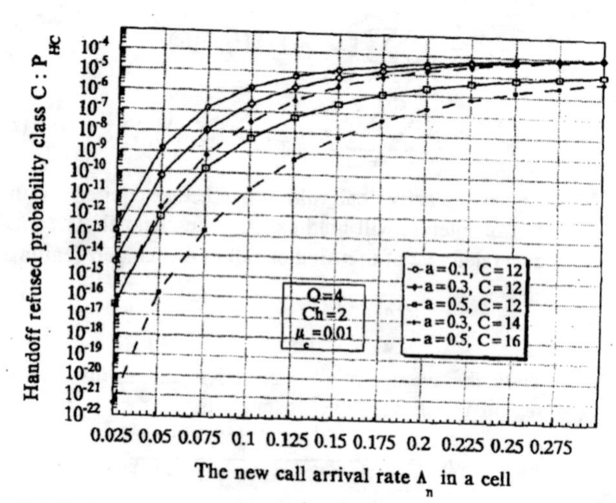

Fig. 10. The handoff refused probability class C versus the new call arrival rate Λ_n.

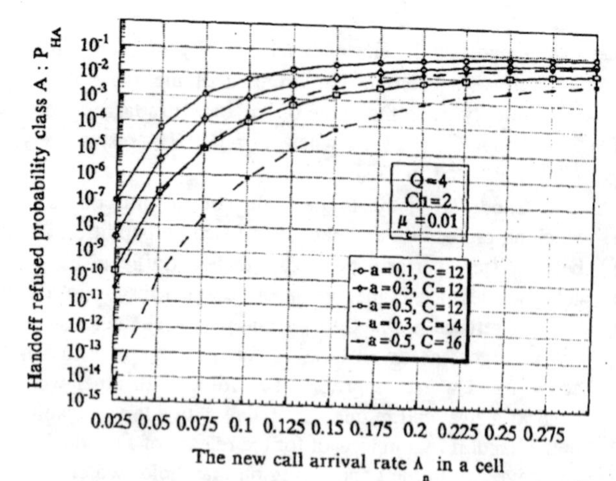

Fig. 8. The handoff refused probability class A versus the new call arrival rate Λ_n.

Although the carried traffic in a cell for Condition 2 is larger than that for Condition 1, there is also a larger cell channel capacity to achieve lower blocking or handoff refused probabilities at the same value of "a." The trunk efficiency of Condition 2 is a little bit lower than that of Condition 1 for the same value of "a," as shown in Fig. 6.

In conclusion, Condition 2, which keeps the same communication quality of voice/data when the ratio "a" increases, can achieve a better system performance than Condition 1. On the other hand, Condition 1 can have a higher communication quality of voice/data than Condition 2.

VI. Conclusion

The performance of soft handoff in CDMA cellular networks is analyzed by using the Markov chain in this paper. When soft handoff leads to a larger handoff region and better system performance than hard handoff, it may also cause a lower channel (trunk) efficiency due to the multiple-channel

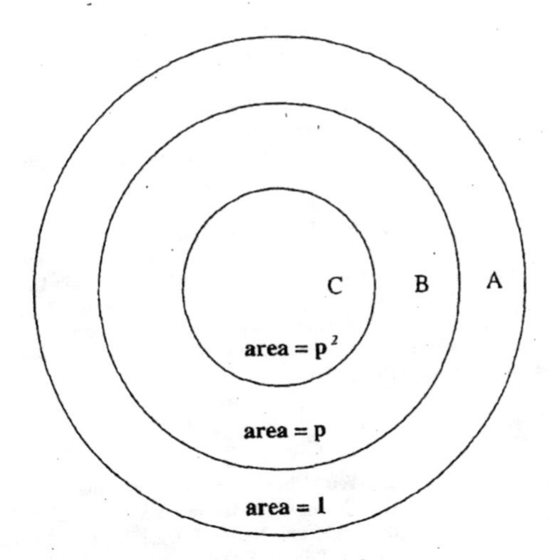

Fig. 11. Three cocentered circles A, B, and C.

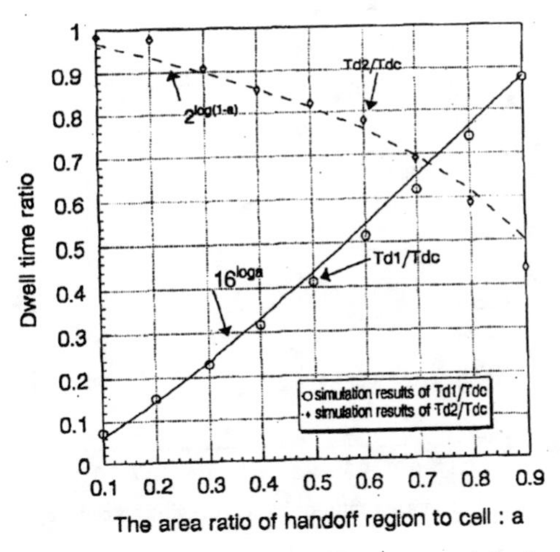

Fig. 12. Simulation results and approximated relations of the dwell time ratio.

assignment to a mobile unit. With soft handoff, we can keep the same channel capacity in each cell to get higher voice/data quality or keep the same voice/data quality to get higher channel capacity in each cell when "a" becomes larger. No matter what the case is, the system performance measurements (blocking and handoff refused probabilities) except channel efficiency will be better while "a" is larger. Consequently, in this paper, we have shown that the larger the area of the soft handoff region, the better the users in the cellular network feel.

Appendix A
Related Functions of Dwell Times

Consider three cocentered circles whose areas are $1, p$, and p^2, respectively, as shown in Fig. 11, where $0 < p < 1$. Since the circles are in the same regular shape and the mobile units are assumed to be randomly walking in the whole area, we can conclude that the ratio of the average dwell times in two distinct circles shall be a function of the ratio of the area only, i.e.,

$$\begin{aligned} T_C &= f(p^2) \cdot T_A \\ &= f(p) \cdot T_B \\ &= f(p) \cdot f(p) \cdot T_A \end{aligned} \tag{18}$$

where T_A, T_B, and T_C are the average dwell times of three distinct circles A, B, and C, respectively, and $f(p)$ is the related function. Thus $f(p^2) = f^2(p)$. Similarly, we can get $f(p^n) = f^n(p)$ for any positive integer n. From this particular formula, we assume $f(p) = x^{\log_{10} p}$, where x is a proper chosen constant.

For the cell of interest as shown in Fig. 1(b), if the fraction of the occupied area of the handoff region is "a," then we can conclude the relation between the dwell time of the whole cell ($\bar{T}_{\rm dc}$) and the dwell time of the normal region ($\bar{T}_{d2}$) is

$$\begin{aligned} \bar{T}_{d2} &= f(1-a) \cdot \bar{T}_{\rm dc} \\ &= x^{\log_{10}(1-a)} \cdot \bar{T}_{\rm dc}. \end{aligned} \tag{19}$$

We have done the computer simulation to obtain the values of $\bar{T}_{d2}/\bar{T}_{\rm dc}$ and $\bar{T}_{d1}/\bar{T}_{\rm dc}$ which are shown in Fig. 12. From the simulation results, we find that $x = 2$ is a proper value in function $f(p)$. We also find that another related function $g(a) = 16^{\log a}$ is a good approximation of the values of $\bar{T}_{d1}/\bar{T}_{\rm dc}$.

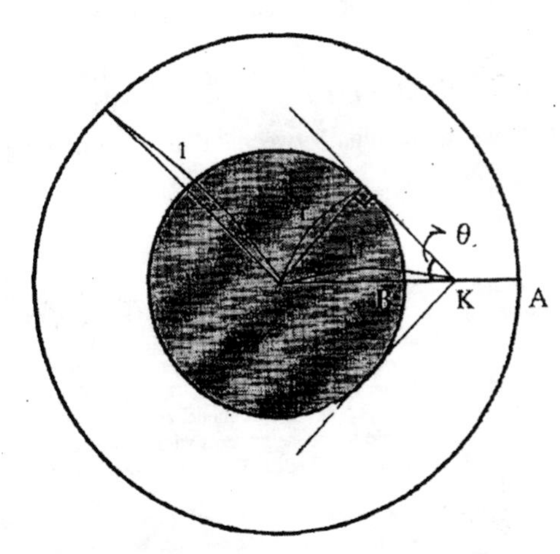

Fig. 13. The mobile unit is at point K which has a distance r_1 from the center. The radius of the normal region (inner circle) is r and the radius of the whole cell (outer circle) is one.

Appendix B
Derivation of the Probability of Moving Back From the Handoff Region to the Normal Region b_k

If a mobile unit is at the point K in the handoff region, as shown in Fig. 13, the probability that it will drive toward the handoff region is

$$b_k(r_1) = \frac{\theta}{\pi} = \frac{\sin^{-1}\left(\frac{r}{r_1}\right)}{\pi}.$$

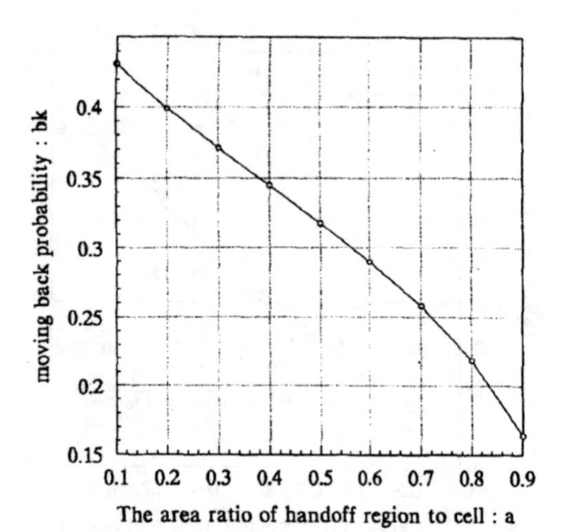

Fig. 14. The probability *bk* versus the ratio *a*.

Thus, the average probability of moving back to normal region for all the mobile units in the handoff region is

$$b_k = \frac{2}{\pi \cdot (1 - r^2)} \int_r^1 r_1 \cdot \sin^{-1}\left(\frac{r}{r_1}\right) dr_1. \tag{20}$$

Fig. 14 shows the relation between "a" and the moving back probability "b_k."

REFERENCES

[1] *Mobile Station-Base Station Compatibility Standard for Dual-Mode Wideband Spread Spectrum Cellular System*, EIA/TIA/IS-95 Interim Standard, Telecommunication Industry Association, July 1993.

[2] A. Murase, I. C. Symington, and E. Green, "Handover criterion for macro and microcellular systems," in *Proc. VTC'91*, 1991, pp. 524–530.

[3] D. Hong and S. S. Rappaport, "Traffic model and performance analysis for cellular mobile radio telephone systems with prioritized and nonprioritized hand-off procedures," *IEEE Trans. Veh. Technol.*, vol. VT-35, pp. 77–92, 1986.

[4] S. S. Rappaport, "The multiple-call hand-off problem in high-capacity cellular communications systems," *IEEE Trans. Veh. Technol.*, vol. 40, pp. 546–557, 1991.

[5] ——, "Modeling the hand-off problem in personal communications networks," in *Proc. VTC'91*, 1991, pp. 517–523.

[6] S. Tekinay and B. Jabbari, "A measurement-based prioritization scheme for handover and channel in mobile cellular networks," *IEEE J. Select. Areas Commun.*, vol. 10, no. 8, pp. 1343–1350, 1992.

[7] S. L. Su and L. S. Chen, "Exploration of handoff procedure for mobile cellular communication system," in *ISCOM'93*, Taipei, Taiwan, 1993, pp. 32–39.

[8] S. S. Rappaport, "Blocking, hand-off and traffic performance for cellular communication systems with mixed platforms," *IEE Proc.-I*, vol. 140, no. 5, pp. 389–401, Oct. 1993.

[9] C. Purzynski and S. S. Rappaport, "Traffic performance analysis for cellular communication systems with mixed platform types and queued hand-offs," in *Proc. VTC'93*, 1993, pp. 172–175.

[10] ——, "Multiple-call hand-off problem with queued hand-offs and mixed platform types," *IEE Proc.-Comm.*, vol. 142, no. 1, pp. 31–39, Feb. 1995.

[11] C. M. Simmonds and M. A. Beach, "Network planning aspects of DS-CDMA with particular emphasis on soft handoff," in *Proc. VTC'93*, 1993, pp. 846–849.

[12] S. C. Swales *et al.*, "Handoff requirements for a third generation DS-CDMA air interface," in *IEE Colloquium, Mobility Support Personal Commun.*, 1993.

[13] A. J. Viterbi, A. M. Viterbi, K. S. Gilhousen, and E. Zehavi, "Soft handoff extends CDMA cell coverage and increases reverse link capacity," *IEEE J. Select. Areas Commun.*, vol. 12, no. 8, pp. 1281–1288, Oct. 1994.

Szu-Lin Su (M'90) was born in Tainan, Taiwan, R.O.C., in 1954. He received the B.S. and M.S. degrees from National Taiwan University, Taiwan, R.O.C., in 1977 and 1979, respectively, and the Ph.D. degree from the University of Southern California, in 1985, all in electrical engineering.

From 1979 to 1989, he was a member of the Chung Shan Institute of Science and Technology, Taiwan, R.O.C., working on the design of digital communication and network systems. Since 1989, he has been with National Cheng Kung University, Tainan, Taiwan, R.O.C., where he is currently an Associate Professor of Electrical Engineering. His research interests are in the areas of digital communication, mobile communication networks, satellite communications, and coded modulation techniques.

Jen-Yeu Chen was born in Hualian, Taiwan, R.O.C., in 1970. He received the B.S. and M.S. degrees from National Cheng Kung University, Tainan, Taiwan, R.O.C., both in electrical engineering, in 1993 and 1995, respectively.

In 1995, he joined the Network Planning Lab., TL, Ministry of Transportation and Communication, Taiwan, R.O.C., as an Assistant Researcher. His research interests include mobile radio communications, traffic and performance analysis, voice/data integrated networks, and network management.

Jane-Hwa Huang was born in Tainan, Taiwan, R.O.C., in 1971. He received the B.S. degree in electrical engineering from National Cheng Kung University, Tainan, Taiwan, R.O.C., in 1994, where he is currently working toward the M.S. degree.

His research interests include traffic and performance analysis and mobile radio communications.

Advantages of CDMA and Spread Spectrum Techniques over FDMA and TDMA in Cellular Mobile Radio Applications

Peter Jung, *Member, IEEE,* Paul Walter Baier, *Senior Member, IEEE,* and Andreas Steil

***Abstract*—In this paper, a unified theoretical method for the calculation of the radio capacity of multiple-access schemes such as FDMA (frequency-division multiple access), TDMA (time-division multiple access), CDMA (code-division multiple access) and SSMA (spread-spectrum multiple access) in noncellular and cellular mobile radio systems shall be presented for AWGN (additive white Gaussian noise) channels. The theoretical equivalence of all the considered multiple-access schemes is found.**

However, in a fading multipath environment, which is typical for mobile radio applications, there are significant differences between these multiple-access schemes. These differences are discussed in an illustrative manner revealing several advantages of CDMA and SSMA over FDMA and TDMA. Furthermore, novel transmission and reception schemes called coherent multiple transmission (CMT) and coherent multiple reception (CMR) are briefly presented.

I. Introduction

In cellular mobile radio systems, the problem of multiple access can be solved by the basic multiple-access schemes FDMA (frequency-division multiple access), TDMA (time-division multiple access), CDMA (code-division multiple access) and SSMA (spread-spectrum multiple access) or by combinations thereof [1], [2]. When selecting a multiple-access scheme, perhaps the most important question is the number of admissible users per cell for a given available total bandwidth, for given radio propagation conditions, and for a required transmission quality. This number is termed the cellular radio capacity. In several papers [3]–[7], the cellular radio capacity of cellular radio systems using special multiple-access schemes has been studied. However, a unified theory of cellular radio capacity which is applicable independently of the used multiple-access scheme has not yet been presented.

Independently of the used multiple-access scheme, all reasonably well-designed multiple access schemes are theoretically equivalent if AWGN (additive white Gaussian noise) channels are considered. This shall be demonstrated in Sections II and III, both for noncellular and for cellular systems by the calculation of the radio capacity. When pursuing such a principle aim, effects such as transmitter power control, receiver synchronization, and channel estimation, which are important in practical system designs, cannot be considered because such effects are beyond the scope of a paper dealing with basic considerations. Nevertheless, the authors believe that this paper is helpful even for the practically oriented engineer when comparing the cellular radio capacity of competing multiple-access schemes.

If the channel exhibits time variance and frequency selectivity, which are typical for a mobile radio environment, the situation is different. In order to give an impression of this issue, Section IV shall present a brief introduction to fading multipath channels in mobile radio. In Section V, the diversity potential of the different multiple-access schemes shall be presented. In the case of such channels, multiple-access schemes having at lest a CDMA or SSMA component are superior to other multiple-access schemes because by CDMA and SSMA, the frequency selectivity of the radio channel, which severely impairs the system performance, can be averaged out.

In Section VI, further advantages of CDMA and SSMA over FDMA, and TDMA shall be discussed. Furthermore, novel transmission and reception schemes called coherent multiple transmission (CMT) and coherent multiple reception (CMR) are briefly presented.

II. Basic Multiple-Access Problem

Firstly, the basic multiple-access problem is considered for a noncellular system. This problem consists in dividing up the available frequency-time space among z users in such a way that there is no interference between the users. If a transmission interval of duration T is considered, and if the available total bandwidth of the noncellular system is B, a function $\Phi_\mu(t)$, $\mu = 1, 2, \cdots, z$, from a finite set of z orthonormal bandpass functions of ensemble duration T and ensemble bandwidth B can be exclusively assigned to each transmitter. The simultaneous limitation of both the duration and the bandwidth of these functions $\Phi_\mu(t)$ can be understood on the basis of the uncertainty principle, see e.g., [8]. Such an assignment is assumed in what follows. In this case, the number of admissible transmitters is

$$z \leq 2BT. \tag{1}$$

In the following, it is assumed that the time–bandwidth product BT is an integer, and therefore the equality in (1) holds:

$$z = 2BT. \tag{2}$$

Manuscript received February 18, 1991; revised September 26, 1991.

The authors are with the Research Group for RF Communications, University of Kaiserslautern, D-6750, Kaiserslautern, Germany.

IEEE Log Number 9207169.

0018–9545/93$03.00 © 1993 IEEE

Theoretically, an infinite number of sets of z orthonormal functions $\Phi_\mu(t)$ with ensemble duration T and ensemble bandwidth B exist. The most usual kinds of orthonormality are frequency orthonormality, time orthonormality and code orthonormality, which correspond to FDMA, TDMA, CDMA/SSMA, respectively. Also, combinations of FDMA, TDMA, CDMA, and SSMA are possible. Due to the close relationship between CDMA and SSMA, SSMA shall not be treated separately, although in contrast to CDMA in the case of SSMA, usually

$$z << 2BT \tag{3}$$

holds.

It is assumed that each transmitter uses its function $\Phi_\mu(t)$ as its individual carrier signal and that transmitter μ transmits information using the signal

$$s_\mu(t) = \sum_{\nu=-\infty}^{\infty} x_{\mu\nu}\Phi_\mu(t-\nu T), \qquad \mu = 1, 2, \cdots, z \tag{4}$$

where the factors $x_{\mu\nu}$ are samples of a Gaussian process $\{x_{\mu\nu}\}$ with zero mean. The signals $x_{\mu\nu}\Phi_\mu(t-\nu T)$ transmitted by transmitter μ generate the signals

$$a \cdot x_{\mu\nu}\Phi_\mu(t-\nu T - t_0), \qquad 0 < a < 1; \qquad a, t_0 \in \mathbf{R} \tag{5}$$

which represent the attenuated and time-delayed versions of the transmitted signals at the corresponding receiver μ . The average energy E_μ per received signal $a \cdot x_{\mu\nu}\Phi_\mu(t-\nu T-t_0)$ is assumed to be the same for all μ, i.e., with $\mathrm{E}\{\cdot\}$ denoting the expectation

$$E = E_\mu = \mathrm{E}\Big\{(a \cdot x_{\mu\nu})^2\Big\}, \qquad \mu = 1, 2, \cdots, z. \tag{6}$$

In the case of FDMA, each function $\Phi_\mu(t)$ uses the total ensemble duration T; therefore, the duration T_u of the individual function $\Phi_\mu(t)$ is given by

$$T_u = T. \tag{7}$$

In this case, the bandwidth B_u of the individual function $\Phi_\mu(t)$ assumes the value

$$B_u = \frac{B}{z} = \frac{1}{2T}. \tag{8}$$

In the case of TDMA, each function $\Phi_\mu(t)$ uses the total ensemble bandwidth B, i.e., the bandwidth B_u of the individual function $\Phi_\mu(t)$ is

$$B_u = B. \tag{9}$$

In this case, the duration T_u of the individual function $\Phi_\mu(t)$ is only the zth part of the ensemble duration T:

$$T_u = \frac{T}{z} = \frac{1}{2B}. \tag{10}$$

With CDMA and SSMA, each function $\Phi_\mu(t)$ uses the total ensemble duration T as well as the total ensemble bandwidth B; therefore,

$$T_u = T \tag{11}$$

and

$$B_u = B \tag{12}$$

are valid.

On account of their orthogonality, the signals from the z transmitters can be perfectly separated in the corresponding receivers by correlation or matched filtering, if all transmitters and receivers are synchronized, which shall be the case as already mentioned above. After the separation, with the average energy E of the received signals and with the one-sided spectral power density N_0 of the thermal noise, the signal-to-noise ratio γ is given by

$$\gamma = \frac{E}{N_0/2}. \tag{13}$$

The average information transmitted per signal $x_{\mu\nu}\Phi_\mu(t-\nu T)$ is $0.5 \cdot \log_2(1+\gamma)$ if the thermal noise is assumed to be Gaussian [9]. Because every T seconds a signal $x_{\mu\nu}\Phi_\mu(t-\nu T)$ is transmitted, the channel capacity per user assumes the value

$$C = \frac{1}{2T} \cdot \log_2(1+\gamma). \tag{14}$$

Substituting $2T$ in (14) by z/B (see (2)), yields

$$z \cdot \frac{C}{B} = \log_2(1+\gamma). \tag{15}$$

For a given available total bandwidth B, a required channel capacity C per user and a given signal-to-noise ratio γ, the number z of admissible users can be calculated from (15).

If the signals from the z transmitters are not perfectly separated at the corresponding receivers, the signal-to-noise ratio is given by

$$\gamma = \frac{E}{f \cdot E + N_0/2} \tag{16}$$

where the term $f \cdot E$ represents the interference from the $(z-1)$ other users. Nonperfect signal separation is the case if e.g., the signals $\Phi_\mu(t)$ are not perfectly orthogonal and at the same time the signal separation is performed by conventional matched filtering instead of applying optimum unbiased estimation. In such cases, f may assume rather large values, which leads to a considerable decrease of γ.

III. Cellular Systems in the Case of Ideal Radio Propagation

In the case of cellular systems, the set of $2BT$ orthornormal functions $\Phi_\mu(t)$ has to be subdivided into subsets of size

$$z = \frac{2BT}{r} \tag{17}$$

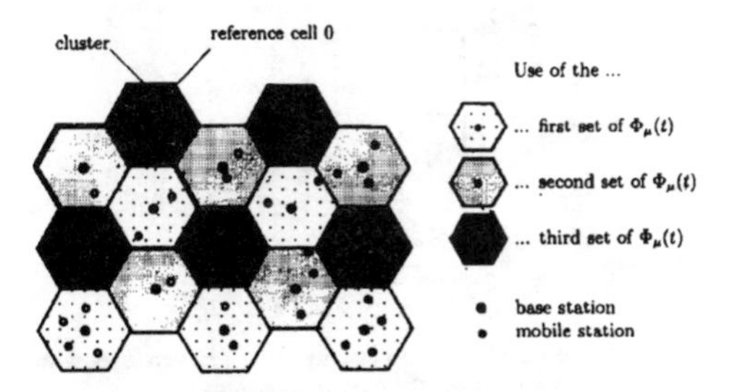

Fig. 1. Part of a cellular system with hexagonal cells for $r = 3$.

with r being the reuse factor and z being the number of users per cell. A fraction $1/r$ of the cells use the same subset of orthonormal functions $\Phi_\mu(t)$. Whereas it is possible in each cell to separate the transmitted signals $x_{\mu\nu}\Phi_\mu(t-\nu T)$ originating from this cell from one another, it is impossible to separate signals coming form the other cells using the same orthonormal functions $\Phi_\mu(t)$. Rather, these signals have to be treated as interference which has a similar effect as thermal noise [1].

Regular cellular systems are considered in which each cell contains a base station in the center of the cell and z mobile stations communicating with the base station of the cell. Conventionally, the shape of a cell in such a cellular scheme is assumed to be a regular hexagon [1], [10]. In Fig. 1, a part of a cellular system with such hexagonal cells is schematically shown for the case $r = 3$. The base stations and the mobile stations of those cells displayed with the same texture use the same set of orthonormal functions $\Phi_\mu(t)$. Each group of three neighboring cells using disjointed sets of $\Phi_\mu(t)$ is combined to form a cluster.

It is assumed that, by a suitable power control, all mobile stations belonging to a certain cell receive equal powers from their base station, and that all mobile stations generate equal powers in the receiver of their base station. Cell 0 with its base station BS_0 is taken as the reference cell. A worst-case situation is considered in which the interference power arriving at the receivers in cell 0 from extra-cell transmitters, i.e., from transmitters in cells other than cell 0, is maximum. In order to obtain this worst-case situation, both in the case of the mobile stations calling the base stations (uplink) and in the case of the base stations calling the mobile stations (downlink), all extra-cell transmitters must transmit maximum power, which means that the extra-cell mobile stations must have maximum distance R from their base stations (Condition I). In addition to this condition, the intra-cell and extra-cell mobile stations have to be arranged in such a way that the interference power received by the receivers in cell 0 is maximized (Condition II).

Following the discussion of the preceding section, the total interference power I in a cellular interference scenario as referred to here is given by

$$I = \frac{E}{T} \cdot f(r, \alpha) \tag{18}$$

with α being the attenuation coefficient [1]. In (18), the function $f(r,\alpha)$ is introduced, for which the term cellular interference function is proposed. In order to determine the interference power I, $f(r,\alpha)$ must be calculated.

In what follows, only the uplinks in a hexagonal cellular system comparable to the one shown in Fig. 1 are considered. The cellular interference function $f(r,\alpha)$ is dependent on the distances between the extra-cell transmitters and the base station of the reference cell 0. Due to the regular structure of such a hexagonal cellular system, setting out from cell 0, the cellular system can be divided into six 60° segments. In the above-mentioned worst-case situation, these 60° segments are equivalent. Therefore, only one 60° segment must be evaluated. Now it is advantageous to introduce an affine coordinate system with the basis vectors

$$\vec{n}_0 = D \cdot \begin{pmatrix} \cos(0 \cdot 60°) \\ \sin(0 \cdot 60°) \end{pmatrix} = D \cdot \begin{pmatrix} 1 \\ 0 \end{pmatrix} \tag{19}$$

and

$$\vec{n}_2 = D \cdot \begin{pmatrix} \cos(2 \cdot 60°) \\ \sin(2 \cdot 60°) \end{pmatrix} = \frac{D}{2} \cdot \begin{pmatrix} -1 \\ \sqrt{3} \end{pmatrix} \tag{20}$$

where D denotes the distance between the two adjacent cells with base stations and mobile stations using the same set of $\Phi_\mu(t)$ [1]. According to [1], with the cell radius R_{cell}, D is given by

$$D = R_{\text{cell}} \cdot \sqrt{3r}\,. \tag{21}$$

These cells are called co-channel cells [1]. Fig. 2 gives a graphical representation of the considered situation. Each base station is associated with a unique vector $\vec{R}_{uv}$ with

$$\vec{R}_{uv} = u \cdot \vec{n}_0 + v \cdot \vec{n}_2, \quad u \in \mathbf{N}, \quad v \in \mathbf{N}_0, \quad v < u \tag{22}$$

of length

$$R_{uv} = R_{\text{cell}} \cdot \sqrt{3r(u^2+v^2-uv)}, \quad u \in \mathbf{N}, \quad v \in \mathbf{N}_0, \quad v < u. \tag{23}$$

In order to simplify the mathematical solution, the hexagons are approximated by circular regions of radius R which have the same area as the hexagons. The cell radius R_{cell} can then be expressed in the following way:

$$R_{\text{cell}} = \frac{\sqrt{2\pi\sqrt{3}}}{3} \cdot R\,. \tag{24}$$

Substituting (24) into (23) yields

$$R_{uv} = R \cdot \sqrt{\frac{2\pi\sqrt{3}}{3} r(u^2+v^2-uv)}, \quad u \in \mathbf{N}, \quad v \in \mathbf{N}_0, \quad v < u. \tag{25}$$

The distances $d_{uv}^{(r)}$ between the extra-cell transmitters in the co-channel cells and the base stations of the reference cell 0 are now given by

$$d_{uv}^{(r)} = R_{uv} - R = R \cdot \left(\sqrt{\frac{2\pi\sqrt{3}}{3} r(u^2+v^2-uv)} - 1 \right), \quad u \in \mathbf{N}, \quad v \in \mathbf{N}_0, \quad v < u. \tag{26}$$

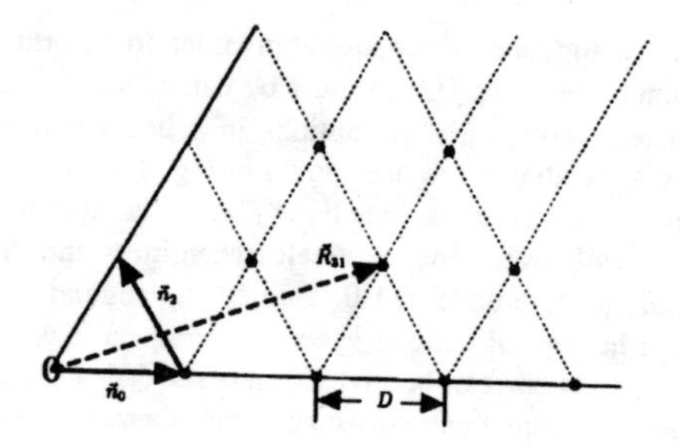

Fig. 2. Affine coordinate system in a 60° segment of a cellular system similar to the one in Fig. 1, used for the determination of $\vec{R}_{uv}$, $d_{uv}^{(r)}$, and $f(r,\alpha)$.

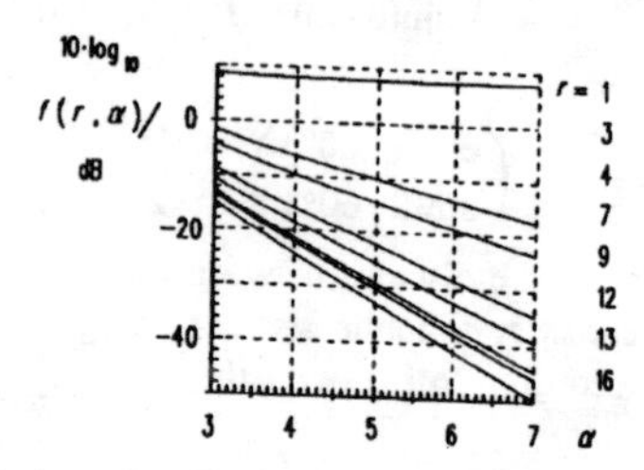

Fig. 3. Cellular interference function $f(r,\alpha)$ versus α with r as a parameter.

Using (26), the cellular interference function $f(r,\alpha)$ of the whole cellular system yields

$$\begin{aligned} f(r,\alpha) &= 6 \cdot \sum_{u=1}^{\infty}\sum_{v=0}^{u-1}\left(\frac{R}{d_{uv}^{(r)}}\right) \\ &= 6 \cdot \sum_{u=1}^{\infty}\sum_{v=0}^{u-1}\left(\frac{R}{d_{uv}^{(r)}}\right)\left(\sqrt{\frac{2\pi\sqrt{3}}{3}r(u^2+v^2-uv)}-1\right)^{-\alpha}, \\ & \qquad r>1 \end{aligned} \tag{27}$$

in the case of the uplink. In order to reduce the error arising form the circular approximation of the hexagons in the case of $r = 1$, the interference power equal to $6E/T$ resulting from the six neighboring cells of the reference cell 0 is considered separately. In this case, $f(1,\alpha)$ is given by

$$f(1,\alpha) = 6 \cdot \left\{1 + \sum_{u=2}^{\infty}\sum_{v=0}^{u-1}\left(\sqrt{\frac{2\pi\sqrt{3}}{3}(u^2+v^2-uv)}-1\right)^{-\alpha}\right\}. \tag{28}$$

For the considered cellular system, Fig. 3 shows $f(r,\alpha)$ according to (27) and (28) versus α with r as a parameter. As expected, $f(r,\alpha)$ decreases with increasing r and α.

If the extra-cell interference is modeled as white noise over the available total bandwidth B, the one-sided spectral interference power density assumes

$$\begin{aligned} I_0 &= \frac{z \cdot I}{B/r} = \frac{f(r,\alpha)\cdot r \cdot z \cdot E/T}{B} \\ &= 2E \cdot f(r,\alpha) \end{aligned} \tag{29}$$

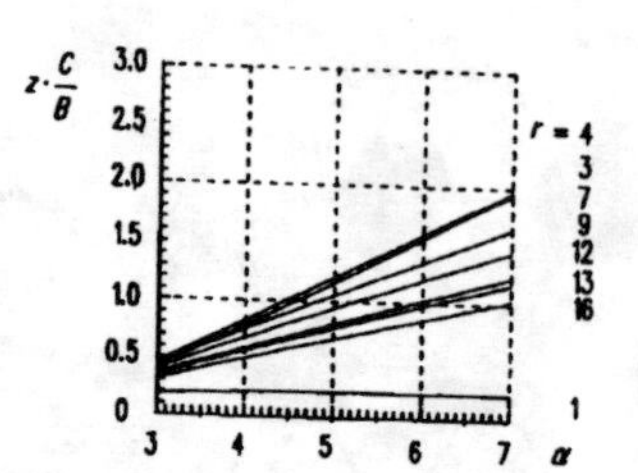

Fig. 4. Normalized cellular radio capacity $z \cdot C/B$ versus α with r as a parameter for the case of vanishing N_0.

by using (17) and (18). Now, instead of (13) the expression:

$$\gamma = \frac{E}{I_0/2 + N_0/2} \tag{30}$$

is obtained for the signal-to-noise ratio after the signal separation. Substituting (29) into (30) yields

$$\gamma = \frac{E}{E \cdot f(r,\alpha) + N_0/2}. \tag{31}$$

With this expression for γ, and considering the fact that the available time–bandwidth product per user is BT/r, instead of (15), the expression:

$$z \cdot \frac{C}{B} = \frac{1}{r} \cdot \log_2\left(1 + \frac{E}{E \cdot f(r,\alpha) + N_0/2}\right) \tag{32}$$

is obtained. For each quadruple r, α, N_0 and E, the expression $z \cdot C/B$ attains a certain value. It is recommended to term this quantity the normalized cellular radio capacity. If no thermal noise has to be considered, (32) reduces to

$$z \cdot \frac{C}{B} = \frac{1}{r} \cdot \log_2\left(1 + \frac{1}{f(r,\alpha)}\right). \tag{33}$$

By substituting $f(r,\alpha)$ according to (27) and (28) into (33), the normalized cellular radio capacity $z \cdot C/B$ for the cellular system shown in Fig. 1 is obtained. In Fig. 4, $z \cdot C/B$ is depicted versus α with r as a parameter for vanishing N_0. As expected, $z \cdot C/B$ increases with increasing α. With respect to the dependence on r, it can be stated by inspection of Fig. 4 that the maximum normalized cellular radio capacity $(z \cdot C/B)|_{max}$ is obtained for r equal to four for the considered example.

The theoretical results for AWGN channels presented in this section are independent of the chosen multiple-access scheme. Nevertheless, the situation changes for fading multipath channels typical for mobile radio applications. This shall be discussed in the following two sections.

IV. Fading Multipath Radio Channels

The mobile radio channel can be characterized by its time-variant impulse response $\underline{h}(\tau,t)$ [11], [12]. A short impulse sent into the channel results in a finely structured response of duration T_M. Typical experimental results are given in [13] and the references therein. The parameter T_M is called delay window. The spreading of the transmitted impulse is caused by the fact that the transmitted signal reaches the receiver via a number of different paths on account of reflections, diffractions and scattering [11], [12]. The order of the delay window T_M is

$$T_M \approx \begin{cases} 0.3\ \mu s, & \text{for indoor channels,} \\ 10\ \mu s, & \text{for outdoor channels} \end{cases} \tag{34}$$

which corresponds to maximum path differences of 100 m and 3 km, respectively.

In order to characterize the mean energy spread caused by the mobile radio channel, the delay spread S_D is introduced as the standard deviation of the delay time parameter τ (see below):

$$S_D = \sqrt{\frac{1}{P_m}\int_0^{T_M} \tau^2 |\underline{h}(\tau,t)|^2 \, d\tau - \left(\frac{1}{P_m}\int_0^{T_M} \tau |\underline{h}(\tau,t)|^2 \, d\tau\right)}$$

$$P_M = \int_0^{T_M} |\underline{h}(\tau,t)|^2 \, d\tau . \tag{35}$$

The order of the delay spread S_D is

$$S_D \approx \begin{cases} 10 \cdots 50\ \text{ns}, & \text{for indoor channels} \\ 0.1 \cdots 5.0\ \mu s, & \text{for outdoor channels.} \end{cases} \tag{36}$$

Due to the motion of the mobile transceiver and the scatterers in the surrounding environment, the dependence of $\underline{h}(\tau,t)$ on the time t results. The strength of the time dependence is closely related to the velocity v of the mobile stations, which also causes a Doppler shift of the transmitted frequency spectrum. Nevertheless, the impulse responses $\underline{h}(\tau, t+\Delta)$ and $\underline{h}(\tau,t)$ are similar for small increments Δ because $\underline{h}(\tau,t)$ is varying continuously versus time. However, the impulse response $\underline{h}(\tau,t)$ may be completely different after a certain minimum time T_{coh} has elapsed, which is typical for the channel. The parameter T_{coh} is called coherence time of the channel [14]. The time dependence of the impulse response results from the motion of the transmitter and/or the receiver [11], [12]. With the velocity v of the mobile station and the wavelength λ_0 at center frequency, T_{coh} can be approximated by

$$T_{coh} \approx \frac{\lambda_0}{2 \cdot v} . \tag{37}$$

The approximation (37) is based on the fact that the impulse response $\underline{h}(\tau,t)$ may look entirely different when the mobile station has changed its position by half the wavelength λ_0. With λ_0 equal to 0.3 m and v equal to 5 km/h for the indoor channel and v equal to 50 km/h for the outdoor channel, one obtains from (37)

$$T_{coh} \approx \begin{cases} 110\ \text{ms}, & \text{for the indoor channel} \\ 11\ \text{ms}, & \text{for the outdoor channel.} \end{cases} \tag{38}$$

The typical symbol durations T in mobile speech communication are smaller than T_{coh} according to (38) [1]. Therefore, for the duration T of one symbol the channel can be considered as time invariant.

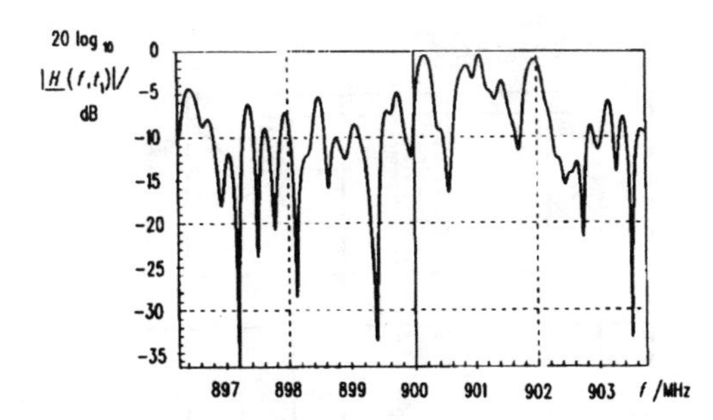

Fig. 5. Part of the absolute value of a sample $\underline{H}(f, t_1)$ of the transfer function $\underline{H}(f,t)$ for a typical urban channel.

By Fourier transform of the time-variant impulse response $\underline{h}(\tau,t)$, the time-variant transfer function

$$\underline{H}(f,t) = \int_{-\infty}^{\infty} \underline{h}(\tau,t) \cdot \exp[-j2\pi f \tau] \, d\tau \tag{39}$$

of the channel is obtained. In Fig. 5, part of a sample of the transfer function $\underline{H}(f,t)$ around a center frequency of 900 MHz for a typical urban channel at a time instant t_1 is displayed. The fine structure of $\underline{H}(f,t)$ along the frequency axis is determined by the delay spread S_D and the delay window T_M and has the approximate structure:

$$B_{coh} \le 1/T_M, \qquad B_{coh} \approx 1/[8S_D] \tag{40}$$

[12], [14]. For a typical urban channel the parameters S_D and T_M assume the following values:

$$T_M \approx 3 \cdots 5\ \mu s, \qquad S_D \approx 1\ \mu s \tag{41}$$

thus resulting in

$$125\ \text{kHz} \approx B_{coh} \le 333\ \text{kHz}. \tag{42}$$

The width B_{coh} is termed coherence bandwidth [14]. With T_M and S_D according to (34) and (36), respectively, one obtains typical values for the coherence bandwidth as follows:

$$B_{coh} \approx \begin{cases} 3\ \text{MHz}, & \text{for indoor channels} \\ 0.1\ \text{MHz}, & \text{for outdoor channels.} \end{cases} \tag{43}$$

The transfer function $\underline{H}(f,t)$ can be considered as a sample function of a two-dimensional stationary and ergodic complex process. The width of the main autocorrelation peak of this process can be approximated by B_{coh} in the f-direction and by T_{coh} in the t-direction. Consequently, if a grid of widths B_{coh} and T_{coh}, respectively, is imposed on the frequency–time plane (see Fig. 6) the values of $\underline{H}(f,t)$ in different time–frequency elements of this grid are mutually uncorrelated.

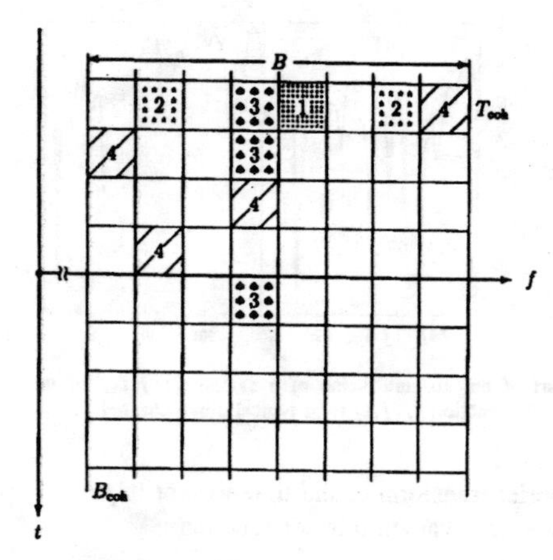

Diversity principles:
1 none
2 spectral
3 temporal
4 spectral and temporal

Fig. 6. Uncorrelated time-frequency elements.

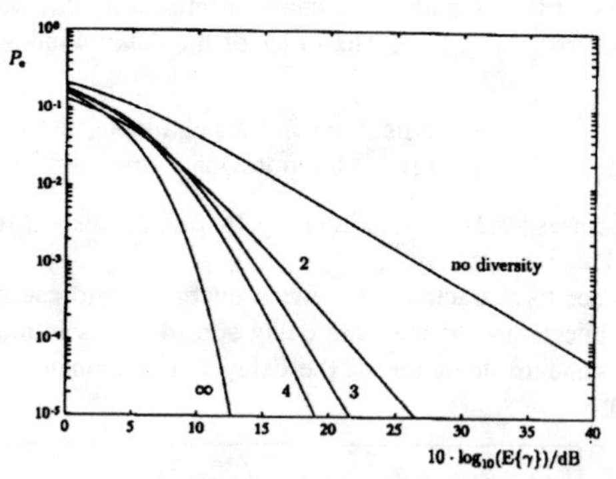

Fig. 7. Typical plot of P_e versus $E\{\gamma\}$ with L as a parameter.

V. Combatting Degradation Caused by Fading Multipath Radio Channels

The time and frequency dependences of the channel tend to degrade the system performance. This will be explained by the use of Fig. 6. In a fading multipath environment, with the average value of the received energy E per signal $x_{\mu\nu}\Phi_\mu(t-\nu T)$, the signal-to-noise ratio with the constant average value

$$E\{\gamma\} = \frac{E}{I_0/2} \tag{44}$$

and with variance $\mathrm{var}\{\gamma\}$ prevails at the receiver for an interference-limited cellular system (cf. Section III).

If the energy E is entirely concentrated within a single time–frequency element, (see case 1 in Fig. 6), for a constant $E\{\gamma\}$ the actual γ assumes quite different values, depending on the transfer function $\underline{H}(f,t)$ in the considered time–frequency element. Therefore, $\mathrm{var}\{\gamma\}$ will be maximum. If E is distributed over two, three, or four time-frequency elements, the diversity parameter L being equal to two, three, and four, respectively (see cases 2–4 in Fig. 6) for still-constant average $E\{\gamma\}$, the variance $\mathrm{var}\{\gamma\}$ will be reduced. The distribution of E over several time-frequency elements is termed diversity [15]. Diversity can be achieved by distributing the transmitted symbol energy along the frequency axis (see case 2 in Fig. 6) along the time axis (see case 3 in Fig. 6) or along both axes (see case 4 in Fig. 6).

In order to give a quantitative discussion, spectral diversity in the case of a multipath channel consisting of L paths that can be resolved in the receiver, with L depending on B_u, is considered. It can be shown that

$$\mathrm{var}\{\gamma\} = \left(\frac{E}{I_0/2}\right)^2 \cdot \frac{1}{L}. \tag{45}$$

[16]. Obviously, diversity reduces the dependence of the actual γ on the actual channel state.

The error probability P_e in the receiver is a strongly nonlinear convex function of γ. Consequently, for constant $E\{\gamma\}$, P_e decreases rapidly with decreasing $\mathrm{var}\{\gamma\}$ which is equivalent to increasing diversity. Fig. 7 shows a typical plot of P_e as a function of $E\{\gamma\}$ with the diversity as a parameter for the case of coherently-detected binary orthogonal frequency shift-keying (FSK) [16], [17]. For infinite diversity, the minimum error probability P_e is obtained. As a conclusion it can be stated that the variance $\mathrm{var}\{\gamma\}$ is reduced by diversity, which results in a decreasing error probability P_e.

In contrast to pure FDMA, the schemes of pure TDMA and of pure CDMA and SSMA are approaches to keep down the variance $\mathrm{var}\{\gamma\}$ of γ at receivers operating in time-variant, frequency-selective channels, as long as B_u is considerably larger than the coherence bandwidth B_{coh} of the channel. However, CDMA and SSMA have a number of additional advantages which are not encountered in TDMA. Those advantages are:

- The Euclidean distances between symbols are virtually invariant to time displacements of the symbols, i.e., the distances do not decrease rapidly when time-displaced versions of the symbols are faced. Therefore, problems of intersymbol interference (ISI) and co-channel interference are less severe in CDMA and SSMA than in TDMA [2], [17].
- In order to maintain the required temporal order among the symbols, a complicated system organization is necessary in TDMA, but not in CDMA and SSMA [2], [17].
- CDMA and SSMA permit a CW-like operation of the transmitter power stages which leads to favorable circuitry [2], [17].

VI. Further Advantages of CDMA and SSMA

The invariance of the Euclidean distances between symbols to time displacements entails a number of further advantages of CDMA and SSMA. One main advantage is that coherent multiple transmission and reception can be realized (cf. [17]).

In Fig. 8, three base stations $BS_{1,2,3}$ and one mobile station MS are depicted. In conventional systems, the mobile station

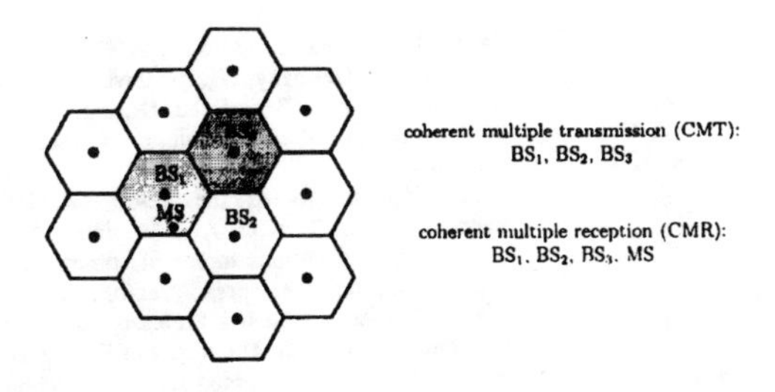

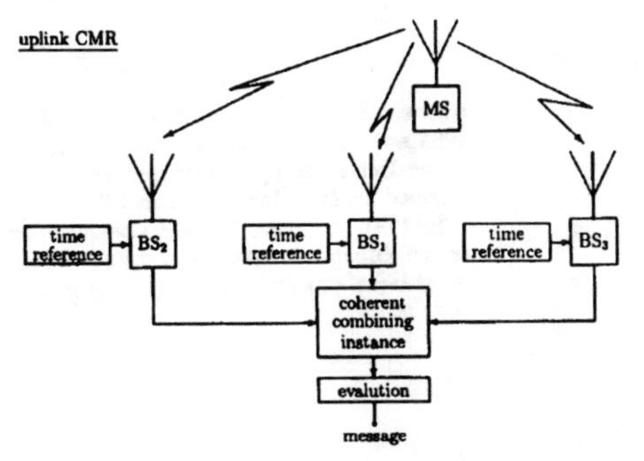

Fig. 8. Coherent multiple transmission (CMT) and reception (CMR).

MS communicates with one of the base stations $BS_{1,2,3}$ and is handed over to another base station, if, by doing so, the communication quality can be improved.

In coherent multiple transmission (CMT), the base stations $BS_{1,2,3}$ surrounding the mobile station MS simultaneously transmit to the mobile station MS. All signals arriving at the mobile station MS are coherently combined by coherent multiple reception (CMR). CMR in the uplink is obtained if the signal transmitted by the mobile station MS is simultaneously received by several base stations and if the received signals are coherently combined to obtain the message (see Fig. 8). As a presupposition for CMT and CMR, reliable and fast digital communication between base stations is required, e.g., via optical-fiber links. However, it should be emphasized that this digital communication between base stations does not have to fulfill exact analog timing conditions if exact time standards, e.g., Rubidium clocks, are available at the base stations as shown in Fig. 8. In this case, the signals transmitted to the coherent combining instance can be supplied with the information of their absolute time of arrival at the base stations. This information can be used to perform the coherent combination digitally.

CMT and CMR are used to improve the exploitation of the transmitted power and therefore reduce the necessary transmitted power and the electromagnetic load of the air. The achievable gains by CMT and CMR are shown in Table I. With CMT there is also a reduction of the carrier-to-interference ratio C/I by approximately 3 dB due to the diminution of the transmission power in the base stations $BS_{1,2,3}$ by factor 3 [17]. With CMT, only three base stations in the first tier [17] contribute to the interference, whereas without CMT there are six interferes. The latter situation is considered in [5].

TABLE I
MINIMUM GAINS IN dB BY CMT AND CMR

	Antenna at $BS_{1,2,3}$	
	Omnidirectional	Directional (60°)
Downlink	0	7.8
Uplink	4.8	12.6

Especially when directional antennas having an angular beamwidth of 60° are used by the base stations $BS_{1,2,3}$, which is feasible in the configuration shown in Fig.8, the gains are considerable. Additional favorable features of CMT and CMR are reduced shadowing and the possibility to locate the mobile station MS.

VII. CONCLUSIONS

In the present paper, a unified theoretical approach to the calculation of the normalized cellular radio capacity for multiple-access schemes in cellular mobile radio applications has been introduced in the case of AWGN channels. The considered multiple-access schemes FDMA, TDMA, CDMA, and SSMA are theoretically equivalent for AWGN channels. However, there are significant differences among these multiple-access schemes for fading multipath radio channels. These differences have been discussed in an illustrative way, revealing several advantages of CDMA and SSMA over FDMA and TDMA. In addition to the already presented advantages of CDMA and SSMA, there are further benefits:

- graceful degradation;
- less timing organization than TDMA;
- reduction of ISI and self-interference;
- additional gain by CMT and CMR;
- possibility of position location of MS;
- less bandwidth expansion due to Doppler spread than FDMA;
- less bandwidth expansion due to forward error correction than FDMA;
- independence of actual channel state; and
- potential exploitation of military research results.

ACKNOWLEDGMENT

The authors wish to thank Dipl.-Ing. Markus Naßhan and Karl-Heinz Eckfelder for their support during the preparation of this manuscript.

REFERENCES

[1] S. B. Rhee, (ED.): Special Issue on Digital Cellular Technologies, *IEEE Trans. Veh. Technol.*, vol. 40, 1991.
[2] M. K. Simon, J. K. Omura, R. A. Scholtz, and B. K. Levitt, *Spread Spectrum Communications, Volume III.* Rockville, MD: Computer Science, 1985, Ch. 5.
[3] D. N. Hatfield, "Measures of spectral efficiency in land mobile radio," *IEEE Trans. Electromagn. Compat.*, vol. EMC-19, pp. 266–268, 1977.
[4] G. R. Cooper and R. W. Nettleton, "A spread-spectrum technique for high-capacity mobile communications," *IEEE Trans. Veh. Technol.*, vol. VT-27, pp. 264–275, 1978.
[5] W. C. Y. Lee, "Spectrum efficiency in cellular," *IEEE Trans. Veh. Technol.*, vol. 38, pp. 69–75, 1989.
[6] W. C. Y. Lee, "Estimate of channel capacity in Rayleigh fading environment," *IEEE Trans. Veh. Technol.*, vol. 39, pp. 187–189, 1990.

[7] K. S. Gilhousen, I. M. Jacobs, R. Padovani, A. J. Viterbi, L. A. Weaver, and C. E. Wheatley, "On the capacity of a cellular CDMA system," *IEEE Trans. Veh. Technol.*, vol. 40 pp. 303–312, 1991.

[8] H. Baher, *Analog and Digital Signal Processing*. New York: Wiley, 1990.

[9] R. G. Gallager, *Information Theory and Reliable Communication*. New York: Wiley, 1968.

[10] W. C. Y. Lee, *Mobile Cellular Telecommunication Systems*. New York: McGraw-Hill, 1989.

[11] J. G. Proakis, *Digital Communications*. 2nd Ed. New York: McGraw-Hill, 1989.

[12] S. Stein, J. J. Jones, *Modern Communication Principles*. New York: McGraw-Hill, 1967.

[13] T. S. Rappaport, S. Y. Seidel, R. Singh, "900-MHz multipath propagation meausrements for U.S. digital celluar radiotelephone," *IEEE Trans. Veh. Technol.*, vol. 39, pp. 132–139, 1990.

[14] W. C. Y. Lee, *Mobile Communications Engineering*. New York: McGraw-Hill, 1982.

[15] R. S. Kennedy, *Fading Dispersive Communication Channels*. New York: Wiley-Interscience, 1969.

[16] P. W. Baier and W. Kleinhempel, "Wide band systems," *AGARD EPP Lecture Series*, no. 172, on "Propagation limitations for systems using band-spreading," Boston, Paris, Rome, June 1990.

[17] P. Jung and P. W. Baier, "CDMA and spread spectrum techniques versus FDMA and TDMA in cellular mobile radio applications," in *Conf. Proc. 21st EuMC'91*, Stuttgart, Germany, Sept. 9–12, pp. 404–409, 1991.

Peter Jung (S'91–M'92) was born in 1964 in Germany. From 1983 until 1993, he studied physics and electrical engineering at the University of Kaiserslautern, Germany. He received the Dipl.-Phys. and Ph.D. (Dr.-Ing.) degrees in 1990 and 1993, respectively.

From 1990 until 1992, he was with the Microelectronics Centre (ZMK) of the University of Kaiserslautern, where he was engaged in the design and implementation of Viterbi equalizers for mobile radio applications. In 1992 he joined the RF Communications Research Group. His present research interests are signal processing, such as adaptive interference cancellation, and multiple-access techniques for mobile radio systems.

Dr. Jung is a student member of VDE-GME, VDE-OTG and AES.

Paul Walter Baier (M'82–SM'87) was born in 1938 in Germany. He received the Dipl.-Ing. degree in 1963 and the Dr.-Ing. degree in 1965 from the Munich Institute of Technology, Germany.

In 1965, he joined the Telecommunications Laboratories of Siemens AG, Munich, where he was engaged in various topics of communications engineering, including spread-spectrum techniques. Since 1973, he has been a Professor for Electrical Communications at the University of Kaiserslautern, Germany. His present research interests are spread-spectrum techniques, mobile radio systems, and digital radar-signal processing.

Dr. Baier is a member of VDE-ITG and of the German U.R.S.I. section.

Andreas Steil was born in 1964 in Germany. From 1985 until 1991, he studied electrical engineering at the University of Kaiserslautern, Germany. He received his Dipl.-Ing. degree in 1991.

In 1991, he joined the RF Comunications Research Group. His present research interests are novel signal-processing techniques for mobile radio systems.

UNIT 3
SPREAD SPECTRUM APPLICATIONS IN MOBILE SATELLITE

The development of mobile satellite communications concept over the recent years has realistically established the idea of "any where, any time" for personal communications systems. Spread-spectrum technology, by its right, is claiming a strong presence in this form of mobile communication systems through the Code Division Multiple Access technique. The principle of application of CDMA in mobile satellite communications is similar to that in terrestrial cellular mobile, apart from fundamental difference in channel conditions and strong Doppler effects.

The first paper in this unit, "*Increased Capacity Using CDMA for Mobile Satellite Communication*" by Gilhousen et al., demonstrates an economically superior solution to satellite mobile communications by increasing the system capacity by the use of CDMA. The next paper, "*Direct-Sequence Spread Spectrum in a Shadowed Rician Fading Land-Mobile Satellite Channel*" by van Nee et al., presents a performance analysis of land-mobile satellite communications using direct-sequence spread spectrum with BPSK modulation, in terms of bit error, outage, and message success probability. Numerical results are obtained to compare systems with and without spectrum spreading under light, average, and heavy shadowing. The following paper, "*Open-Loop Power Control Error in a Land Mobile Satellite System*" by Monk and Milstein, addresses the power control issue in land-mobile satellite system. They propose an open-loop adaptive power control scheme for combating large scale shadowing and distance losses. The paper, "*Performance of DS-CDMA with Imperfect Power Control Operating Over a Low Earth Orbiting Satellite Link*" by Vojcic et al., presents an analytical derivation to obtain the average probability of error of a single user under imperfect power control and dual-order diversity. The next paper, "*A Performance Comparison of Orthogonal Code Division Multiple-Access Techniques for Mobile Satellite Communications*" by Gaudenzi et al., makes a performance comparison between the Walsh-Hadamard functions-based Qualcomm proposed CDMA system and a Gold functions-based quasi-orthogonal CDMA system in mobile satellite communications. The last paper in this section, "*Design Study for a CDMA-Based LEO Satellite Network: Downlink System Level Parameters*" by Glisic et al., presents and discusses the performance of a CDMA based LEO satellite network for mobile satellite communications. The main elements for the analysis touched upon in this paper are: the channel model, the pilot carrier frequency estimation for Doppler compensation, and multipath and multisatellite diversity combining.

Increased Capacity Using CDMA for Mobile Satellite Communication

KLEIN S. GILHOUSEN, MEMBER, IEEE, IRWIN M. JACOBS, FELLOW, IEEE, ROBERTO PADOVANI, MEMBER, IEEE, AND LINDSAY A. WEAVER, JR.

Abstract—**In this paper the performance of a spread-spectrum CDMA system in a mobile satellite environment is analyzed. Comparisons to single-channel-per-carrier FDMA systems are also carried out which show that the CDMA approach provides greater capacity. Results from computer simulations, laboratory tests, and field tests of a prototype modem are also presented. The tests results show excellent performance of the modem in the mobile environment and also the feasibility of the spread-spectrum approach to satellite mobile communications.**

INTRODUCTION

THIS paper will demonstrate that spread-spectrum CDMA systems provide an economically superior solution to satellite mobile communications by increasing the system capacity compared to single-channel-per-carrier FDMA systems. Following the comparative analysis of CDMA and FDMA systems, the paper describes the design of a modem that was developed to test the feasibility of the approach and the performance of a spread-spectrum system in a mobile environment. Results of extensive computer simulations as well as laboratory and field tests results are presented.

The bit error rate (BER) of the rate 1/3, constraint length 9, convolutionally encoded BPSK modem proved to be within 0.3 dB from theory in a AWGN channel. A BER of 10^{-3} which is adequate for voice was achieved at $Eb/No = 2.5$ dB. The powerful convolutional code combined with interleaving and a robust modulation also provide excellent performance in Rician fading and lognormal shadowing.

The paper is organized as follows. In Section I, a review of previous comparisons of the maximum throughput achievable by CDMA and FDMA systems is given. In Section II, the comparisons between CDMA and FDMA are rederived for mobile satellite systems. In Section III, a description of the spread-spectrum modem and the performance results are given.

I. SPECTRAL EFFICIENCY OF CDMA AND FDMA

In [1], Viterbi compared the spectral efficiency of CDMA and FDMA as a function of total carrier-to-noise power ratio. The comparison was carried out for uncoded BPSK (QPSK) as well as convolutionally coded systems with various code rates. We review here the main results of [1].

In a multiple access satellite system uplink from a number M of similar mobile user terminals, each assumed to provide equal incident power at the satellite, the total received carrier power at the satellite C is given by

$$C = ME_bR_b \tag{1}$$

where

E_b = Energy per bit of information.
R_b = Each User's information rate.

If we divide C by N_oW_s where N_o is the single-sided thermal noise spectral density and W_s is the total occupied system bandwidth we obtain

$$\frac{C}{N_oW_s} = M\frac{E_b}{N_o}\frac{R_b}{W_s}. \tag{2}$$

Link performance is measured by the spectral efficiency η of the transponder or link as a function of C/N_oW_s, and is defined as

$$\eta \equiv M\frac{R_b}{W_s} = \frac{\frac{C}{N_oW_s}}{\frac{E_b}{N_o}}. \tag{3}$$

In a spread-spectrum system the total noise is determined by the sum of the thermal noise N_o and the mutual interference noise spectral density I_o. The desired bit error rate performance of the link is determined by the $E_b/(N_o + I_o)$ or the ratio of the bit energy to the single-sided total noise power spectral density. The total noise $N_o + I_o$ is given by the following:

$$N_o + I_o = N_o + (M - 1)E_c \tag{4}$$

where E_c = energy/chip, defined as $E_c = E_bR_b/W_s$, resulting in

$$N_o + I_o = N_o + (M - 1)\frac{E_bR_b}{W_s}. \tag{5}$$

Dividing E_b by $N_o + I_o$, we obtain the following expression:

$$\frac{E_b}{N_o + I_o} = \frac{E_b/N_o}{1 + \frac{M-1}{M}\frac{C}{NoWs}}. \tag{6}$$

Manuscript received February 15, 1989; revised October 2, 1989. This work was supported in part by Hughes Aircraft Company. This paper was presented in part at the Mobile Satellite Conference, Pasadena, CA, May 1988.

The authors are with QUALCOMM, Inc., San Diego, CA 92121.

IEEE Log Number 9034779.

0733-8716/90/0500-0503$01.00 © 1990 IEEE

Solving (6) for E_b/N_o and substituting into (3), we obtain the following expression for the CDMA spectral efficiency:

$$\eta_{CDMA} = \frac{\dfrac{C}{N_o W_s}}{\dfrac{E_b}{N_o + I_o}\left(1 + \dfrac{C}{N_o W_s}\left(\dfrac{M-1}{M}\right)\right)}. \quad (7)$$

For a system with a large number of users M we obtain:

$$\eta_{CDMA} \approx \frac{\dfrac{C}{N_o W_s}}{\dfrac{E_b}{N_o + I_o}\left(1 + \dfrac{C}{N_o W_s}\right)} \text{ bits/sec/Hz.} \quad (8)$$

For a single-channel-per-carrier FDMA system, there is only one user per bandwidth segment. Therefore, $E_b/(N_o + I_o) = E_b/N_o$ and (3) becomes

$$\eta = \frac{\dfrac{C}{N_o W_s}}{\dfrac{E_b}{N_o}} \quad \text{if} \quad \frac{MR_b}{W_s} < r \log_2 (m) G_{FDMA} \quad (9)$$

for the power limited case, and the following expression for the bandwidth limited case:

$$\eta = r \log_2 (m) G_{FDMA} \quad \text{if} \quad \frac{MR_b}{W_s} \geq r \log_2 (m) G_{FDMA} \quad (10)$$

where

r = code rate,
m = signal constellation dimension ($m = 2$ for BPSK, $m = 4$ for QPSK, etc.),
R_b = each user's information rate, and
G_{FDMA} = FDMA guardband factor.

The FDMA guardband factor allows margin for adjacent channel interference. For the purpose of comparison, the 5 KHz channelization of the L-band spectrum assigned to mobile services, as proposed by the American Mobile Satellite Consortium (AMSC) in [4] is assumed. Thus, in the previous example $G_{FDMA} \approx 0.5$, i.e., an octal symbol rate of 2400 sps in the 5 KHz channel.[1]

For links that are neither power nor interference limited, the FDMA signal constellation dimension m can be allowed to increase, improving the spectral efficiency. Conversely, for links that are extremely power or interference limited, the proper design choice for FDMA would employ PSK and low rate coding. In this case, the efficiency of use of the satellite resource by CDMA and FDMA are nearly equal.

The spectral efficiencies as given by (8), (9), and (10) are shown in Fig. 1 for two representative systems[2] transmitting digitized voice at 4800 bps and achieving a BER of 10^{-3} on a AWGN channel. The first is the experimental spread-spectrum CDMA system described in this paper with a rate 1/3, $K = 9$ convolutional code and $E_b/(N_o + I_o) = 2.5$ dB. Note that 1.2 dB capacity margin has been added to the required $E_b/(N_o + I_o)$ of 2.5 dB. See the Appendix for a discussion of capacity margin. The FDMA system chosen for comparison uses rate 2/3 trellis coded 8-DPSK, as proposed in [3], with $E_b/N_o = 8.4$ dB for BER of 10^{-3}.

In Fig. 1, attention should be focused on the range of $C/N_o W_s$ around 10 dB, which corresponds to a satellite downlink power of 60 dBW, a 7 MHz system bandwidth and mobile terminal G/T of -22 dB, obtained with a 4 dBi omni-directional antenna and a 400°K receiver. These are typical baseline conditions for a large mobile satellite system for the 1.5/1.6 GHz band.

It is seen in Fig. 1 that the FDMA system has about twice the capacity of the CDMA system at $C/N_o W_s = 10$ dB, in agreement with the conclusions of [1]; "When C/N_o is at premium do not contribute further to the noise by having the users jam one another."

In (1), above it was assumed that each of the M user uplink signals arrived at the satellite with the same incident power. The system must provide an active power control means to cause this to happen. In the experimental system, the hub-to-mobile link includes a pilot carrier modulated by a separate, short PN code for rapid initial acquisition and for carrier and time tracking. The mobile units measure the received pilot carrier power level to determine an initial estimate of the mobile's path loss for setting the return link transmitter power. After a two-way link is established, the hub measures the signal-to-noise ratio of the mobile signal and provides a power adjustment command to "fine tune" the mobile transmitter power.

II. CDMA Versus FDMA: The Mobile Satellite Channel

The conclusion of the comparison developed in [1] and in Section I will be *reversed* when the two systems are compared in a mobile satellite environment [2] that provides additional system features and, in general, is more bandwidth limited than power limited. The four major factors that alter the result of the comparison are

1) voice activity,
2) spatial discrimination provided by satellite multibeam antennas,
3) cross-polarization frequency reuse, and
4) discrimination between multiple satellites providing co-coverage.

Voice services will likely occupy the largest percentage of the mobile communication channels. The voice activity

[1]A smaller guardband factor could be achieved with sharper filtering. In this example, we chose to compare the system proposed in [2] which uses a 100% roll-off factor. If a smaller roll-off is used, then the FDMA spectral efficiency increases accordingly.

[2]This paper focuses on the design parameters that are suitable for a large multisatellite system such as has been described in the *Proceedings of the Mobile Satellite Conference*, Pasadena, CA, May 3-5, 1988.

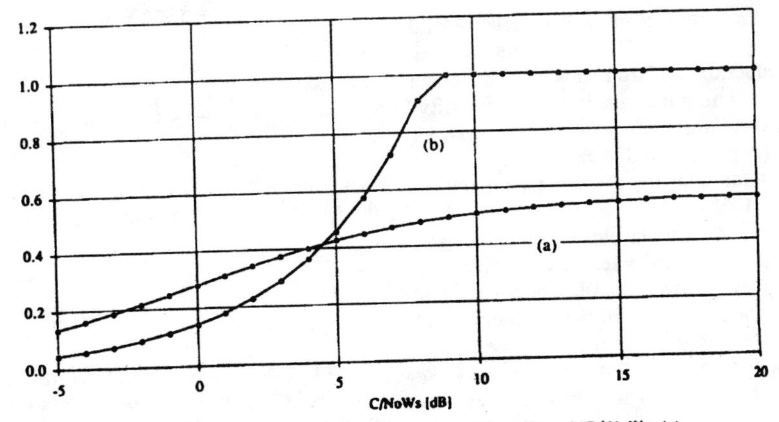

Fig. 1. Spectral efficiency in bits/s/Hz as a function of C/N_oW_s: (a) CDMA with rate 1/3, $K = 9$ code and $E_b/N_o = 2.5$ dB. (b) FDMA with trellis code, rate 2/3 $K = 5$ 8-DPSK and $E_b/N_o = 8.4$ dB.

factor will greatly reduce the self-noise of the spread-spectrum system and utilize the satellite downlink power more efficiently. CDMA voice services will use voice activated carrier transmission, so that when a user is listening or pausing during a conversation the carrier is turned off and thus does not contribute to the system self-noise. Conventional telephone practice [5] for satellite circuits indicates that a given user will be talking approximately 35% of the time. This has the effect of moving the CDMA curve of Fig. 1 upward by a factor of 2.85, or 4.6 dB.

In FDMA, the voice activity factor does not increase the capacity when the system is bandwidth limited but only reduces the satellite downlink power when operating in a power limited mode. This has the effect of shifting the FDMA efficiency curve of Fig. 1 to the left by the voice activity factor, i.e., by 4.6 dB for a 35% activity factor, for the hub-to-mobile link only. FDMA systems are unable to exploit voice activity factor to improve the capacity of bandwidth limited mobile-to-hub links because of the delays inherent with synchronous orbit satellites. Note that use of voice activity gating together with very low E_b/N_o CDMA modems may alter the conclusions of [1] even for conventional satellites without frequency reuse capabilities.

The capacity of the mobile satellite system is further improved by multiple-beam satellite antennas which allow a degree of frequency reuse. For example, the coverage shown in Fig. 2 may be used by first generation satellites.

An FDMA system will gain in capacity from the use of such an antenna through frequency reuse. For the baseline design described in [4], antenna coverage can be designed to provide frequency reuse every 3° or every fourth beam. Fig. 2 indicates that the full spectrum can be used twice, doubling the capacity of FDMA through frequency reuse. A much larger antenna [6], considered for satellites of later generations, could provide a four-fold frequency reuse for FDMA. Ultimately, a frequency reuse pattern with a minimum of three frequency sets could be employed, resulting in a reuse factor equal to the one-third of the number of beams.

Fig. 2. Satellite antenna coverage of the continental U.S.

A CDMA system can provide substantially greater frequency reuse capability than FDMA. On the mobile-to-satellite link, the value of I_o is equal to the total signal power received at the satellite from all active terminals weighted by the antenna beam pattern of each beam.

Equivalent to the concept of "noise bandwidth" in linear filters we can define an "equivalent noise beamwidth" B of an antenna where

$$B = \frac{\int_{-\pi/2}^{\pi/2} G(\theta)\, d\theta}{G(\theta)_{\max}} \qquad (11)$$

and $G(\theta)$ is the antenna gain at offset angle θ, and $G(\theta)_{\max}$ is the maximum gain of the antenna. The frequency reuse capability of a given antenna design in conjunction with CDMA is greater than for FDMA because the beams can be spaced on intervals of B degrees and the entire frequency band reused every B degrees. For example, with a 7 × 2.5 m antenna as described in [4], $B = 1.4°$ and assuming a uniform distribution of users within the continental U.S., the worst case beam will receive a total interference power equal to only 20% of the total uplink interference. The reduction in I_o from areas not in a given beam allows a corresponding increase in system capacity. Ultimately, with a very large number of beams, CDMA can reuse the entire frequency band in each antenna beam. This is a factor of three superior to the best obtainable frequency reuse with FDMA.

The CDMA system can reuse the entire frequency band again by utilizing the two opposite senses of circular polarization. The frequency reuse is possible because the I_o affecting a given channel is the sum of the I_o generated by the users with the same polarization plus the I_o generated by the users of opposite polarization *attenuated by the cross-polarization isolation*. This will increase capacity by 60% even with cross-polarization attenuation of only 6 dB.

On the other hand, polarization isolation usually cannot be exploited by a FDMA system employing small mobile antennas because such antennas provide only limited cross-polarization isolation, usually less than the necessary cochannel C/I required by FDMA.

Considering the above modifications, we can now recalculate the spectral efficiencies as a function of average carrier-to-noise power for the two system previously considered. The value of the total noise, thermal plus other users, affecting a CDMA user, taking into account voice activity, antenna discrimination, and cross-polarization attenuation, modifying (4) above is as follows:

$$N_o + I_o = N_o + a\rho V(M-1)\frac{E_b R_b}{W_s} \tag{12}$$

where

a = 1/number of antenna beams
ρ = polarization isolation factor = 1 + χpol attenuation/2
V = voice activity factor.

The average carrier-to-noise power received at the satellite is given by

$$\frac{C}{N_o W_s} = VM\frac{R_b}{W_s}\frac{E_b}{N_o} \tag{13}$$

resulting in spectral efficiency given by

$$\eta_{\text{CDMA}} = \frac{MR_b}{W_s} = \frac{\dfrac{C}{N_o W_s}}{V\dfrac{E_b}{N_o}}$$

$$= \frac{\dfrac{C}{N_o W_s}}{V\dfrac{E_b}{N_o + I_o}\left(1 + a\rho\dfrac{C}{N_o W_s}\left(\dfrac{M-1}{M}\right)\right)} \tag{14}$$

$$\eta_{\text{CDMA}} \approx \frac{\dfrac{C}{N_o W_s}}{V\dfrac{E_b}{N_o + I_o}\left(1 + a\rho\dfrac{C}{N_o W_s}\right)}. \tag{15}$$

Asymptotically, for $C/N_o W_s \rightarrow \infty$, corresponding to the bandwidth limited situation, (15) becomes

$$\text{Max } \eta_{\text{CDMA}} = \frac{1}{Va\rho\dfrac{E_b}{N_o + I_o}}. \tag{16}$$

It is emphasized that the above result for CDMA spectral efficiency is critically dependent on the $E_b/(N_o + I_o)$ required by the modem in the actual fading environment. This is discussed in detail in the next section.

The spectral efficiency of the FDMA system in the power limited region becomes

$$\eta_{\text{FDMA}} = 2\frac{\dfrac{C}{N_o W_s}}{V\dfrac{E_b}{N_o}} \tag{17}$$

where the factor of two accounts for the frequency reuse. Asymptotically, in the bandwidth limited region (17) becomes

$$\text{Max } \eta_{\text{FDMA}} = 2r\log_2(m)G_{\text{FDMA}} \tag{18}$$

The spectral efficiencies calculated in (16), (17), and (18) are shown in Fig. 3. Again, the CDMA system uses an $E_b/(N_o + I_o)$ = 2.5 dB with added capacity margin as for Fig. 1. The FDMA system assumes $m = 8$, $r = 2/3$, providing E_b/N_o = 8.4 dB. As before, attention should be focused on the region around $C/N_o W_s$ = 10 dB, of the satellite-to-mobile link which is generally the most power limited link.

Table I shows a link budget for both systems. The link budgets are consistent with Fig. 3 in showing almost threefold greater capacity for the CDMA system with respect to the FDMA system for a typical value of $C/N_o W_s$. The details and assumptions that led to the link budget of Table I are given in Appendix I.

Multiple satellites provide another way of improving the CDMA capacity. First, note that in Fig. 4(a), that if $C/N_o W$ were increased by 3 dB by the addition of the second satellite, that the capacity would increase by about 33% with no additional processing required. A possibility for additional capacity increase is the coherent combining of signals transmitted between a terminal and all satellites in view. The coherent combining will result in a capacity gain approaching the increased number of satellites.

This technique is not available for FDMA because the nulls in the resulting interference patterns cannot, in general, be made to correspond to locations of cochannel interference signals. Furthermore, because the FDMA system is operating in the bandwidth limited regime, the

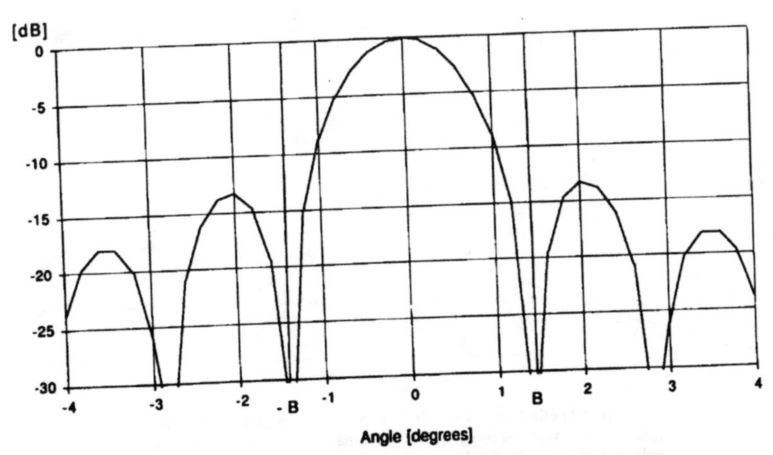

Fig. 3. Normalized gain of a 7 × 2.5 m antenna and "equivalent noise beamwidth" B.

TABLE I
CDMA AND FDMA LINK BUDGETS

Hub-to-Mobile	CDMA	FDMA
Frequency	1549.5 MHz	1549.5 MHz
RF Power	28.1 dBW	28.1 dBW
Power Loss	-1.8 dB	-1.8 dB
Spacecraft Antenna Gain	33.8 dB	33.8 dB
EIRP	60.1 dBW	60.1 dBW
Total Capacity	5045 Erlangs	1400 Erlangs
Voice Duty Cycle	35%	35%
-10•log(N•Duty Cycle)	-32.5 dB	-26.9 dB
Pilot Power	-0.2 dB	N/A
EIRP/channel	27.4 dBW	33.2 dBW
Path Loss	-188.3 dB	-188.3 dB
Polarization Loss	-0.5 dB	-0.5 dB
Mobile Antenna Gain	4.0 dB	4.0 dB
Data Rate	4,800 bits/sec	4,800 bits/sec
-10 log(Data Rate)	-36.8 dB/Hz	-36.8 dB/Hz
Eb	-194.2 dBW/Hz	-188.4 dBW/Hz
LNA Temperature	290 °K.	290 °K.
Antenna Noise	100 °K.	100 °K.
Total Thermal Noise	390 °K.	390 °K.
Thermal Noise Density, No	-202.7 dBW/Hz	-202.7 dBW/Hz
EIRP + Mobile Gain - Losses	-124.7 dBW	N/A
% of Satellite Power in Beam	20%	N/A
10•log(% of Sat. Power)	-7.0 dB	N/A
Cross-Polarization Isolation	6.0 dB	N/A
10•log((1+1/Cross-Pol)/2)	-2.0 dB	N/A
Spreading Bandwidth	7 MHz	N/A
-10•log(Spreading BW)	-68.5 dB/Hz	N/A
Pseudo-Noise Density, Io	-202.2 dBW/Hz	N/A
Thermal, Eb/No	8.5 dB	14.3 dB
Pseudo-Noise, Eb/Io	8.0 dB	N/A
Combined, Eb/(No+Io)	5.2 dB	14.3 dB
Capacity Margin/Fading Margin	-1.3 dB	-2.0 dB
Modem Implementation Loss	-0.3 dB	-0.4 dB
Eb/(No+Io) Minimum	2.2 dB	8.0 dB
Excess Link Margin	3.9 dB	3.9 dB

Mobile-to-Hub	CDMA	FDMA
Frequency	1651.0 MHz	1651.0 MHz
HPA Power	3.0 dBW	3.0 dBW
E/S Antenna Gain	4.0 dB	4.0 dB
Power Loss	-1.0 dB	-1.0 dB
Path Loss	-188.8 dB	-188.8 dB
Spacecraft Antenna Gain	33.8 dB	33.8 dB
Polarization Loss	-0.5 dB	-0.5 dB
Data Rate	4,800 bits/sec	4,800 bits/sec
-10•log(Data Rate)	-36.8 dB/Hz	-36.8 dB/Hz
Eb	-186.4 dBW/Hz	-186.4 dBW/Hz
Spreading Bandwidth	7 MHz	N/A
10•log(Data Rate/Spr. BW)	-31.6 dB	N/A
Cross-Polarization Isolation	0.63	N/A
% of Users in Beam	20%	N/A
Voice Duty Cycle	35%	35%
Total Capacity	5,045 Erlangs	N/A
Effective Interference	223 Erlangs	N/A
10•log(above)	23.5 dB	N/A
Pseudo-Noise Density, Io	-194.5 dBW/Hz	N/A
Total Sat. Noise Temperature	1190.0 °K.	1190.0 °K.
Thermal Noise Density, No	-197.8 dBW/Hz	-197.8 dBW/Hz
Thermal, Eb/No	11.5 dB	16.1 dB
Pseudo Noise, Eb/Io	8.2 dB	N/A
Combined, Eb/(No+Io)	6.5 dB	16.1 dB
Fading Margin	-2.5 dB	-2.0 dB
Modem Implementation Loss	-0.3 dB	-0.4 dB
Eb/(No+Io) Minimum	2.2 dB	8.0 dB
Excess Link Margin	6.1 dB	5.7 dB

increased available downlink power does not serve to increase capacity. Thus, an FDMA system operating in the bandwidth limited mode will not benefit from additional satellites as a way of increasing capacity unless every mobile terminal is equipped with a costly directive antenna capable of providing sidelobe rejection to result in adequate C/I performance in the adjacent satellite. The capacity increase for multiple satellites has not been included in the above equations, Figs. 1 and 3 or the link budgets.

III. Performance of a Direct-Sequence Spread-Spectrum Modem in the Mobile Environment

Recently, several studies [7]–[10] have been conducted to characterize the propagation effects for mobile satellite communications. A generally agreed upon conclusion is that the channel can be approximated by a Rician distribution, with a ratio of specular to diffuse component $K =$ 10 dB, and a lognormal distributed shadowing process affecting the direct path. In [7], a best fit to measured data

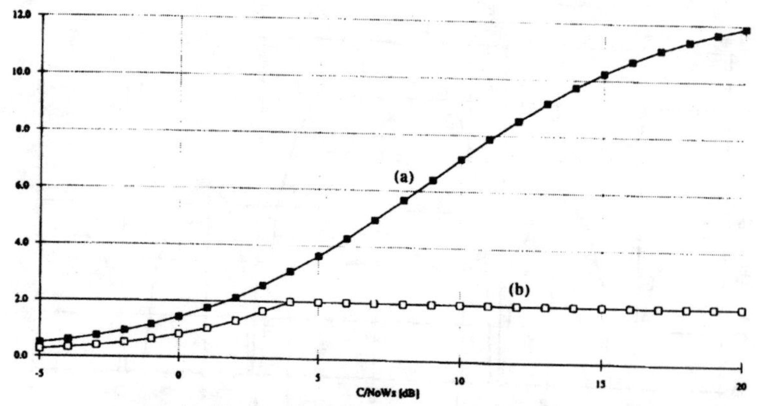

Fig. 4. Spectral efficiency in bits/s/Hz as a function of $C/N_o W_s$ taking into account: voice activity factor, antenna discrimination factor, and polarization reuse: (a) CDMA with rate 1/3, $K = 9$ code and $E_b/N_o = 2.5$ dB. (b) FDMA with trellis code rate 2/3, $K = 5$, 8-DPSK, and $E_b/N_o = 8.4$ dB.

affected by shadowing conditions was found with a log-normal distribution with mean value = −7.5 dB and a standard deviation of 3 dB. This model is shown in Fig. 5. With the above assumption, the received signal can be expressed as follows:

$$r(t) = \text{Re}\left\{x(t)e^{j(2\pi f_c t + \theta)}\right\} \tag{19}$$

where the complex envelope $x(t)$ is given by

$$x(t) = z(t)\,d(t) + w(t). \tag{20}$$

Here the quantity $d(t)$ can be expressed as

$$d(t) = \sqrt{E_s} \sum_{i=-\infty}^{\infty} a_i p(t - iT) \tag{21}$$

where the a_i's represent the binary ±1 data sequence, $p(t)$ is the channel pulse, and E_s represents the signal energy per coded symbol, i.e., $E_s = rE_b$ with r the code rate. In (12), $w(t)$ represents a stationary zero-mean complex Gaussian process with in-phase and quadrature components with single-sided spectral density N_o. The process $z(t)$ combines the effects of fading and shadowing. The process $z(t)$ can be expressed in terms of its real and imaginary part as follows:

$$z(t) = [s(t) + i(t)] + jq(t) \tag{22}$$

where $i(t)$ and $q(t)$, representing the Rayleigh fading, are wide sense stationary zero-mean Gaussian random processes with spectral density $R(f)$ and variance $1/2K$, and $s(t)$ represents the shadowing process. The random process $s(t)$ can be expressed more conveniently as

$$s(t) = 10^{(\gamma(t) + m_s)/20} \tag{22}$$

where $\gamma(t)$ is a zero-mean stationary Gaussian process with spectral density $R_\gamma(f)$ and standard deviation σ = 3 dB and m_s = −7.5 dB. Throughout the rest of the paper, we will assume that $R(f)$ and $R_\gamma(f)$ are 6th-order Butterworth spectra with cutoff frequencies F_c = 10 cycles/m and F_c = 5 cycles/m, respectively, i.e., corresponding to bandwidths of 150 Hz and 75 Hz at a vehicle speed of 15 m/s. With the above assumptions, the in-phase and quadrature components of the output of the matched filter are

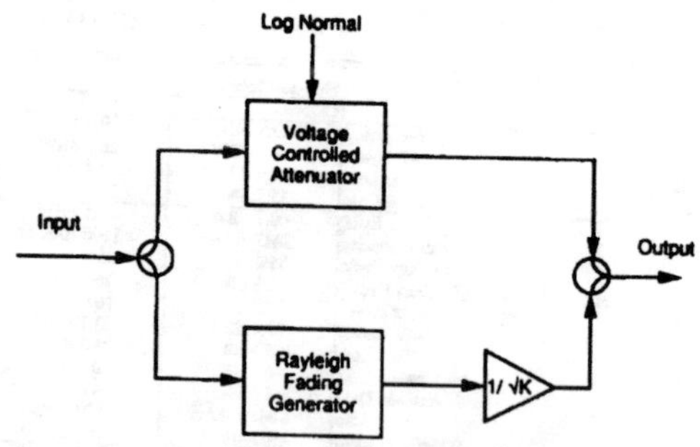

Fig. 5. Fading channel model of [7].

$$r_k^{(I)} = \sqrt{\left(\frac{2E_s}{N_o}\right)}\, a_k[(s_k + i_k)\cos(\theta) + q_k \sin(\theta)] + w_k^{(I)} \tag{23}$$

$$r_k^{(Q)} = \sqrt{\frac{2E_s}{N_o}}\, a_k[q_k \cos(\theta) - (s_k + i_k)\sin(\theta)] + w_k^{(Q)}. \tag{24}$$

Notice that in (24) the noise variables have been normalized to unit variance and, in the absence of shadowing, i.e., $s_k = 1$, (24) represents a Rician channel.

The feasibility of the CDMA approach to mobile communications was tested through the development of a direct-sequence spread-spectrum modem capable of supporting four data rates: 2400, 4800, 9600, and 16000 bps. A block diagram of the demodulator is shown in Fig. 6.

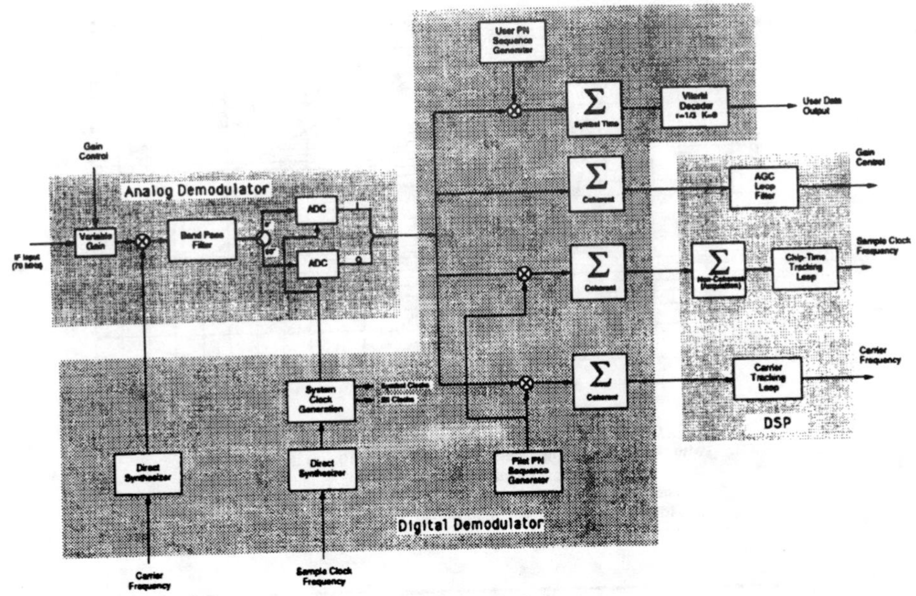

Fig. 6. Demodulator block diagram.

The coded BPSK modem was designed to occupy a 9 MHz bandwidth with a chip rate of 8 Mcps, corresponding to spread-spectrum processing gains of 35.3 dB, 32.3 dB, 29.2 dB, and 27 dB for the respective four data rates. A rate 1/3 constraint length $K = 9$ convolutional code with Viterbi decoding was used which provides a coding gain of 4.5 dB at a BER = 10^{-3}.

Interleaving following the convolutional encoder and deinterleaving prior to the Viterbi decoder is a necessary operation in a fading environment. The interleaver (in this case a block interleaver was chosen) has the effect of spreading adjacent coded symbols affected by fading, thus allowing the Viterbi decoder to correct most of the errors generated by the fades. Since voice transmission is the primary application of the system, long interleaver depths cannot be used, since additional delays are introduced.

A simulation was performed to evaluate the effects of different interleaving depths. The fading model used was the one shown in Fig. 5. The results are summarized in Fig. 7 which shows the cumulative fade depth probability for two interleaver depths, namely, 0.26 and 0.50 m, corresponding to 25 and 50 ms at a vehicle speed of about 35 km/h, and for the case where no interleaver is used. Since a 50 ms interleaver generates an additional 100 ms delay which was considered excessive for voice communication, the 0.26 m interleaver was selected. At a vehicle speed of 15 km/h, a full 25 ms vocoder frame was interleaved.

In order to ease the acquisition process of the mobile units, a pilot signal is transmitted by the Hub terminal. In contrast with the voice modulated signal, which is spread by means of a very long PN sequence, the pilot signal consists of an unmodulated, short PN sequence of length 4095. The carrier phase and PN timing relationship between the pilot signal and the voice carrier is held constant so that the pilot can be used as the reference for demodulation of the voice signal. The pilot carrier power is set about 6 dB above the power in a single voice carrier, providing very robust demodulation performance in the mobiles. Note that a single pilot carrier serves for all the voice modulated carriers in a particular antenna beam.

The initial pilot PN code acquisition of the mobile unit is accomplished with the search/lock strategy described in [11]. In both the search and lock mode a number N_C of pseudonoise chips were coherently combined. The resultant values were then incoherently combined N_I times in the search mode and N_I' times in the lock mode. The derivation of the false alarm and detection probabilities of such a scheme are given in [12]. For convenience, we report the results here

$$P_{FA} = Q\left(\frac{T - N_I}{\sqrt{N_I}}\right), \tag{25}$$

$$P_D = Q\left(\frac{Q^{-1}_{(P_{FA})} - \sqrt{N_I}\,\dfrac{N_C E_c}{N_o + I_o}}{\sqrt{1 + 2\,\dfrac{N_C E_c}{N_o + I_o}}}\right). \tag{26}$$

In (26), $Q(\bullet)$ is the complementary cumulative Gaussian distribution and T is a threshold to be set to achieve specific values of P_{FA} and P_D. The values of N_C and N_I were then selected to optimize the mean acquisition time and P_D. Figs. 8 and 9 show the probability of detection and the mean acquisition time as a function of N_C, for a

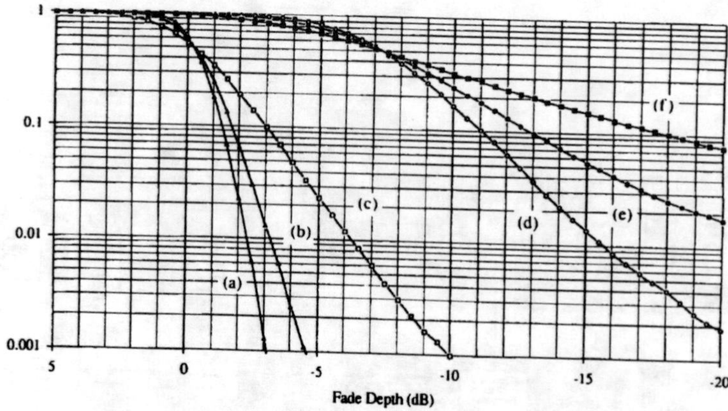

Fig. 7. Cumulative probability of fade depth simulation results obtained with the model of Fig. 5. Rician channel $K = 10$: (a) interleaver depth = 0.5 m; (b) interleaver depth = 0.26 m; (c) without interleaver. Fading channel $K = 10$ with lognormal shadowing $m_s = -7.5$ dB and $\sigma = 3$ dB: (d) interleaver depth = 0.50 m; (e) interleaver depth = 0.26 m; (f) without interleaver.

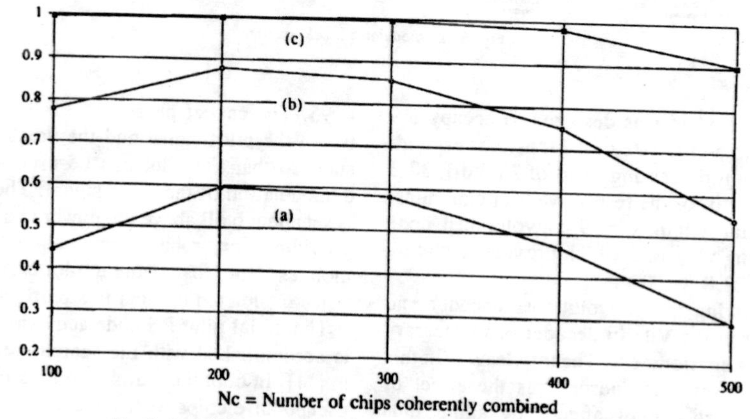

Fig. 8. Probability of detection as a function of the number of chips coherently combined. $P_{FA} = 0.01$ and total number of chips combined $N_C \cdot N_I = 5000$. (a) Pilot $E_b/(N_o + I_o) = -25$ dB, (b) Pilot $E_b/(N_o + I_o) = -23$ dB, (c) Pilot $E_b/(N_o + I_o) = -20$ dB.

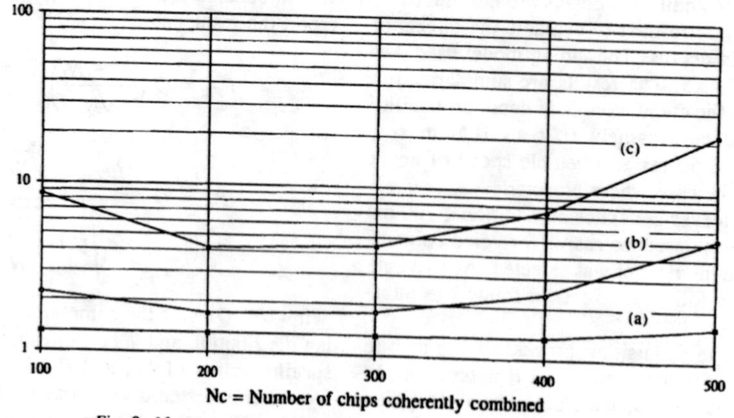

Fig. 9. Mean acquisition time in seconds for a chip rate of 8 MHz. (a) Pilot $E_b/(N_o + I_o) = -20$ dB. (b) Pilot $E_b/(N_o + I_o) = -23$ dB. (c) Pilot $E_b/(N_o + I_o) = -25$ dB.

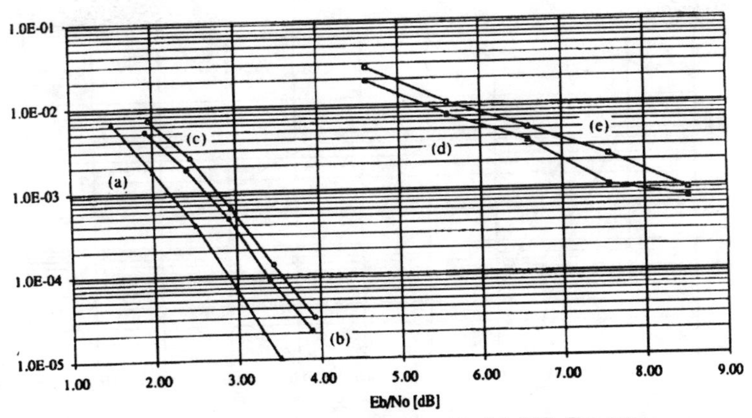

Fig. 10. Laboratory tests results: hub-to-mobile link BER. Data rate = 9600 bps, pilot $E_c/(N_o + I_o) = -20$ dB, $I_o/N_o = 10$ dB. (a) Gaussian channel theory. (b) Gaussian channel modem BER. (c) Rician fading channel $K = 10$ vehicle speed = 48 km/h. (d) Rician fading channel $K = 10$ with lognormal shadowing $m_s = -7.5$ dB $\sigma = 3$ dB, vehicle speed = 96 km/h. (e) Rician fading channel $K = 10$ with lognormal shadowing $m_s = -7.5$ dB $\sigma = 3$ dB vehicle speed = 48 km/h.

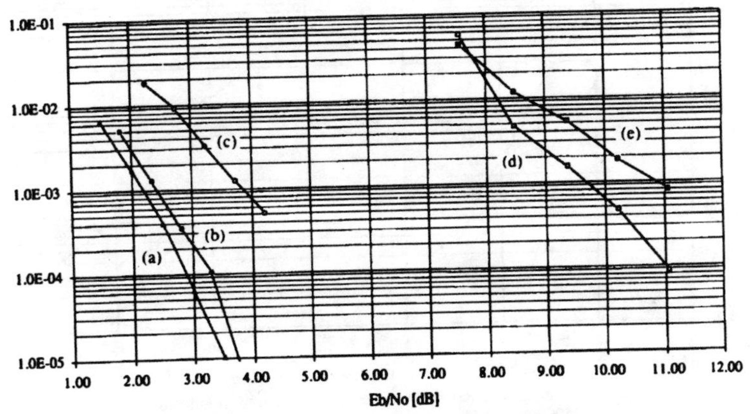

Fig. 11. Laboratory tests results: mobile-to-hub link BER, data rate = 9600 bps. (a) Gaussian channel theory. (b) Gaussian channel modem BER. (c) Rician fading channel $K = 10$ vehicle speed = 48 km/h. (d) Rician fading channel $K = 10$ with lognormal shadowing, $m_s = -7.5$ dB, $\sigma = 3$ dB, vehicle speed = 96 km/h, DPSK. (e) Rician fading channel $K = 10$ with lognormal shadowing $m_s = -7.5$ dB, $\sigma = 3$ dB, vehicle speed = 48 km/h, DPSK.

given value of P_{FA} and the maximum frequency offset a mobile unit would experience, for various values of the signal-to-noise ratio of the pilot signal. A value of $N_C = 200$ was selected with $N_I = 25$ and $N_I' = 50$.

In the mobile-to-hub link, the pilot signal scheme is not feasible. It is replaced by short preambles preceding every vocoder frame. The preambles are used to acquire frequency, phase, timing as well as signal levels of the mobile units. This link was also provided with a DPSK fallback modulation mode to be used when fading conditions do not allow coherent demodulation.

Laboratory measurements were conducted to test the BER of the modem. A channel fading simulator, as shown in Fig. 5, was used to simulate the fading and shadowing processes. Among other parameters the channel simulator allows the operator to simulate a given vehicle speed by selecting the appropriate bandwidths of the fading and shadowing processes.

The BER performance of the modem is shown in Figs. 10 and 11 for both directions of the link. From Figs. 10 and 11 it is clear that the hub-to-mobile link is more robust. The reason is twofold. First, the presence of the pilot signal allows the mobile to track the faded carrier with little degradation. Second, in the hub-to-mobile link, when the signal is shadowed so is the interference from the other users, whereas in the mobile-to-hub link when

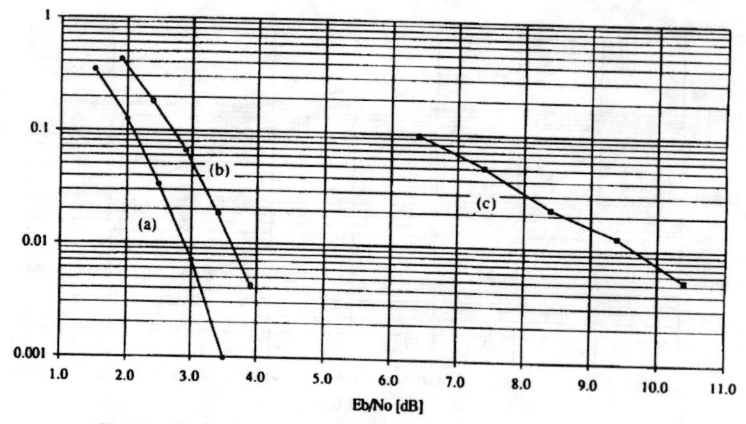

Fig. 12. Computer simulation frame error rate: hub-to-mobile link. $I_o/N_o = 10$ dB, data rate = 9600 bps. pilot $E_c/(N_o + I_o) = -20$ dB. (a) Gaussian channel. (b) Rician fading channel $K = 10$ vehicle speed = 54 km/h. (c) Rician fading channel $K = 10$ with shadowing $m_s = -7.5$ dB $\sigma = 3$ dB vehicle speed = 54 km/h.

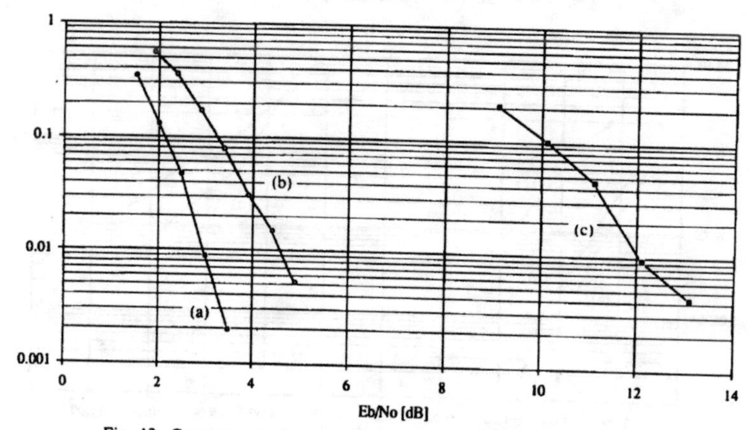

Fig. 13. Computer simulation frame error rate: mobile-to-hub link, data rate = 9600 bps. (a) Gaussian channel. (b) Rician fading channel $K = 10$ vehicle speed = 54 km/h. (c) Rician fading channel $K = 10$ with shadowing $m_s = -7.5$ dB, $\sigma = 3$ dB, vehicle speed = 54 km/h, DPSK.

the wanted signal is shadowed the interference from the other users may not be shadowed. A ratio of $I_o/N_o = 10$ dB was always assumed in the tests.

A degradation of 6.3 dB is observed in the mobile unit in faded and shadowed conditions, $K = 10$, $m_s = -7.5$ dB, $\sigma = 3$ dB, and a vehicle speed of 48 km/h. A degradation of 8.6 dB was observed in the mobile-to-hub link for the same channel conditions in the DPSK mode.

Computer simulations were also carried out to measure the frame error rate. A frame is defined here as a 25 ms vocoder frame and a frame error was counted when the frame contained at least one bit error. The results are shown in Figs. 12 and 13.

The modem was implemented with a single TMS32020 digital signal processor in six 5.2 × 1.6 cm wirewrap boards clearly indicating the feasibility and cost effectiveness of the approach.

Extensive field tests were also performed and the results are reported in [13]. The major conclusions of the field tests were that in all conditions the mobile-to-hub link never needed to fall back to the DPSK mode, when the BPSK mode was provided with a link margin of 2 dB, and that such a link margin proved to be sufficient to provide good quality voice in both directions. Concurrent tests using FDMA analog (ACSSB) and digital modems proved to require 6–9 dB additional C/N_o to provide the same level of quality in the fading environment.

Conclusions

This paper has shown that CDMA systems can provide greater capacity than FDMA for mobile satellite communications. Often CDMA is dismissed as a viable approach for mobile satellite communications and the results of [1] are used to argue in favor of FDMA systems.

We have shown that although the conclusions of [1] still hold, several factors that apply to mobile communication services can be exploited by a CDMA system thus shifting the results of the comparison to FDMA in favor of a CDMA approach. This paper shows that the capacity of a CDMA system is asymptotically about seven times greater than an FDMA system currently proposed and about three times greater when compared using current estimates for available satellite EIRP, antenna gains, etc.

The feasibility of a direct-sequence spread-spectrum approach to mobile communications has been tested through the development of a modem capable of supporting digitized voice at four different rates. Performance results are reported here and results of a comparative test to an ACSSB modem are reported in [13] as well as the results of a test of the spread-spectrum modem conducted at C-Band using an existing satellite.

Appendix

Table I shows a representative link budget for both CDMA and FDMA communicating 4800 bps to and from mobile terminals equipped with omnidirectional antennas. The assumptions on satellite position, available RF power, and antenna gain are those of [4].

The two major parameters computed through the link budget are the system capacity, i.e., the total number of simultaneous users the system can support, and the excess link margin. The calculation of the excess link margin is different for CDMA and FDMA. In the FDMA case, the excess link margin (ELM) is given by

$$\text{ELM} = \text{available } E_b/N_o - \text{fading margin} - \text{modem loss} - \text{minimum } E_b/N_o \quad \text{(A1)}$$

where minimum E_b/N_o is the theoretical signal-to-noise ratio required to support a BER $= 10^{-3}$, modem loss is the implementation loss, the fading margin is the degradation obtained in a Rician fading channel with $K = 10$, and available E_b/N_o is bit energy-to-noise density ratio provided by the channel.

In the CDMA case, we first calculate a capacity margin (CM) and then the ELM. The CM is defined as the amount by which the system capacity must be reduced in order to provide the marginal user additional $E_b/(N_o + I_o)$ and, at the same time, keep the $E_b/(N_o + I_o)$ of the nominal user a constant. To calculate the CM we have assumed the following distribution of users with the margin defined as the additional $E_b/(N_o + I_o)$ required for a BER $= 10^{-3}$ compared to an unfaded user.

	Required Margins [dB]	
	Mobile-to-Hub	Hub-to-Mobile
94% $K = 10$	1.6	0.7
6% $K = 10$, $m_s = -7.5$ dB, $\sigma = 3$ dB	8.6	6.3

The CM is the percent of users weighted by the above margins. Straightforward calculations show that the CM for the hub-to-mobile link is 1.3 dB and 2.5 dB for the mobile-to-hub link.

Finally, the ELM is the solution to the following equation:

$$10^{-(Eb/No - \text{Modem Loss})/10} = 10^{-(Eb/Io + CM)/10} + 10^{-(Eb/No - ELM)/10} \quad \text{(A2)}$$

where we have assumed the ELM directly effects E_b/N_o but does not have any effect on E_b/I_o.

Acknowledgment

The authors would like to thank M. Hurst and R. Blakeney who designed the modem software and B. Johnston, and K. Moallemi who designed, respectively, the RF and digital portions of the modem.

References

[1] A. J. Viterbi, "When not to spread spectrum—A sequel," *IEEE Commun. Mag.*, vol. 23, pp. 12-17, Apr. 1985.
[2] I. M. Jacobs *et al.*, "Comparison of CDMA and FDMA for the MobileStar(sm) system," in *Proc. Mobile Satellite Conf.*, May 3-5 1988, Pasadena, CA, pp. 283-290.
[3] D. Divsalar and M. K. Simon, "Trellis coded MPSK modulation techniques for MSAT-X," in *Proc. Mobile Satellite Conf.*, May 3-5 1988, Pasadena, CA, pp. 283-290.
[4] C. E. Agnew *et al.*, "The AMSC mobile satellite system," in *Proc. Mobile Satellite Conf.*, May 3-5 1988, Pasadena, CA, pp. 3-10.
[5] J. M. Fraser, "Engineering aspects of TASI," *Bell Syst. Tech. J.*, vol. 38, pp. 353-365, Mar. 1959.
[6] W. Rafferty, K. Dessouky, and M. Sue, "NASA's mobile satellite development program," in *Proc. Mobile Satellite Conf.*, May 3-5 1988, Pasadena, CA, pp. 11-22.
[7] J. S. Butterworth, "Propagation measurements for land mobile satellite systems at 1542 MHz," Commun. Res. Cent., Tech. Note no. 723, Dep. Commun., Canada, Aug. 1984.
[8] W. L. Stutzman *et al.* "Mobile satellite propagation measurements and modeling: A review of results for systems engineers," in *Proc. Mobile Satellite Conf.*, May 3-5 1988, Pasadena, CA, pp. 107-117.
[9] J. Godhirsh and W. J. Vogel, "Propagation effects by roadside trees measured at UHF and L-band for mobile satellite systems," in *Proc. Mobile Satellite Conf.*, May 3-5 1988, Pasadena, CA, pp. 87-94.
[10] D. C. Nicholas, "Land mobile satellite propagation results," in *Proc. Mobile Satellite Conf.*, May 3-5 1988, Pasadena, CA, pp. 125-131.
[11] P. M. Hopkins, "A unified analysis of pseudonoise synchronization by envelope correlation," *IEEE Trans. Commun.*, vol. COM-25, pp. 770-778, Aug. 1977.
[12] M. K. Simon *et al.*, *Spread Spectrum Communications, Vol. III*. Rockville, MD: Computer Science, 1985, pp. 31-37.
[13] W. Rubow, "MobileStar field test program," in *Proc. Mobile Satellite Conf.*, May 3-5 1988, Pasadena, CA, pp. 189-194.

Klein S. Gilhousen (M'86) was born in Coshocton, OH, in 1942. He received the B.S. degree in electrical engineering from the University of California, Los Angeles, in 1969.

In 1985, he became a cofounder and Vice President for Systems Engineering for QUALCOMM, Inc., San Diego, CA. His professional interests include satellite communications, cellular telephone systems, spread-spectrum systems, communications privacy, communications networks, video transmission systems, error correcting codes and modem design. He holds four patents in these areas with three more applied for. Prior to joining QUALCOMM, Mr. Gilhousen was Vice President for Advanced Technology at M/A-COM LINKABIT, San Diego, CA, from 1970 to 1985, and Senior Engineer at Magnavox Advanced Products Division, Torrance, CA, from 1966 to 1970.

Irwin M. Jacobs (S'55-M'60-F'74) received the B.E.E. degree in 1956 from Cornell University and the M.S. and Sc.D. degrees in electrical engineering from the Massachusetts Institute of Technology in 1957 and 1959, respectively. On July 1, 1985, he became a founder and the Chairman and President of QUALCOMM, Inc.

From 1959 to 1966, he was an Assistant/Associate Professor of Electrical Engineering at M.I.T. and a staff member of the Research Laboratory of Electronics. During the academic year 1964-1965, he was a NASA Resident Research Fellow at the Jet Propulsion Laboratory. In 1966, he joined the newly formed Department of Applied Electrophysics, now the Department of Electrical Engineering and Computer Science, at the University of California, San Diego (UCSD). In 1972, he resigned as Professor of Information and Computer Science to devote full time to LINKABIT Corporation. While a M.I.T., he coauthored a basic textbook in digital communications, *Principles of Communication Engineering*, published first in 1965 and still in active use. He retains his academic ties through memberships on the Cornell University Engineering Council, the visiting committees of the M.I.T. Laboratory for Information and Decision Systems, as Academic/Scientific member of the Technion International Board of Governors, and as a Board Member of the UCSD Green Foundation for Earth Sciences. He is a past Chairman of the Scientific Advisory Group for the Defense Communications Agency and of the Engineering Advisory Council for the University of California. He has served on the Governing Boards of the IEEE Communications Society, the IEEE Group on Information Theory, and as General Chairman of NTC'74.

In 1982, Dr. Jacobs was elected to membership in the National Academy of Engineering for "contributions to communication theory and practice, and leadership in high-technology product development." He is a member of Sigma Xi, Phi Kappa Phi, Eta Kappa Nu, Tau Beta Pi, and of the Association for Computing Machinery (ACM). In 1980, he and Dr. A. Viterbi were jointly honored by the American Institute of Aeronautics and Astronautics (AIAA) with their biannual award "for an outstanding contribution to aerospace communications." In 1984, he received the Distinguished Community Service Award for the Anti-Defamation League of B'nai B'rith. The local American Electronics Association's First Annual ExcEL Award was presented to Dr. Jacobs in 1989 for excellence in electronics and his "dedication and innovation, which have set the highest standards in the local electronics industry."

Roberto Padovani (S'83-M'84) received the Laurea degree from the University of Padova, Italy, and the M.S. and Ph.D. degrees from the University of Massachusetts, Amherst, in 1978, 1983, and 1985, respectively.

In 1984 he joined M/A-COM LINKABIT, San Diego, CA, where he was involved in the design and development of satellite communication systems, secure video systems, and error-correcting coding equipment. In 1986, he joined QUALCOMM, Inc., and he is now a Director in the Engineering Department. His current research interests include satellite communications, digital cellular telephone systems, and error-correcting codes. He is also a part-time instructor of communication courses with the University of California, San Diego.

Lindsay A. Weaver, Jr., received the S.B. and S.M. degrees from M.I.T. in 1976 and 1977, respectively.

He is Vice President of Engineering at QUALCOMM, Inc. He was a key member of the design teams at QUALCOMM for the Mobile Satellite CDMA voice system, the OmniTRACS® mobile satellite messaging system (hybrid frequency hopping and direct sequence), and the CDMA cellular telephone system. He has also lead projects developing FDMA modems, Viterbi decoders, high-speed packet switches, and satellite video scrambling.

Direct-Sequence Spread Spectrum in a Shadowed Rician Fading Land-Mobile Satellite Channel

Richard D. J. van Nee, Howard S. Misser, *Member, IEEE*, and Ramjee Prasad, *Senior Member, IEEE*

***Abstract*—The performance of a direct sequence spread-spectrum land-mobile satellite transmission system, using binary phase shift keying (BPSK) modulation, is analyzed. The satellite channel is modeled as having shadowed Rician fading characteristics. The bit error probability is evaluated, considering both the envelope and the phase variation. Assuming a Gaussian approximation for the interference, numerical results are obtained for both spread-spectrum and narrowband land-mobile satellite communication systems with BPSK modulation. A comparison of the two systems is made for light, average, and heavy shadowing.**

I. Introduction

The application of spread-spectrum modulation in the field of land-mobile satellite communications offers code division multiple access (CDMA), resistance to multipath fading, and low peak-to-average power ratio. In addition, the properties of low probability of interception (LPI), antijam resistance, and message privacy and security are attractive in some applications. Furthermore, as shown in [1] and [2], spread-spectrum CDMA systems can provide greater capacity than FDMA for mobile satellite communications. In general, land-mobile satellite systems allow a wide range of services, including voice, data, position-finding, and paging services, interconnection to the public switched telephone network, and the possibility of private networks [3]. Accordingly, numerous research papers (e.g., [1]–[10]) have been published recently to study the land–mobile satellite communication channel and its effects on such systems.

This paper presents a performance analysis of land–mobile satellite communications using direct sequence spread-spectrum with BPSK modulation, in terms of bit error, outage, and message success probability. The channel model adopted in the analysis is characterized by the combined effect of Rician fading and lognormal shadowing [4]–[6]. Section II formulates and extends the channel model to the use of spread-spectrum modulation. The receiver model and performance analysis follow in Sections III and IV, respectively. Numerical results are presented in Section V. Finally, our conclusions are given in Section VI.

Manuscript received March 28, 1991; revised October 4, 1991.

R. D. J. van Nee and R. Prasad are with Telecommunications and Traffic Control Systems Group, Delft University of Technology, 2600 GA Delft, The Netherlands.

H. S. Misser is currently with PTT Research Dr. Neher Laboratories, Leidschendam, The Netherlands.

IEEE Log Number 9105068.

II. Channel Model

A statistical propagation model for a narrowband channel in rural and suburban environments was developed in [4]–[6], assuming that the line-of-sight signal strength is lognormally distributed and the composite multipath signal is Rayleigh distributed. The resulting probability distribution of the received signal envelope r is given by:

$$p_B(r) = \frac{r}{b_o\sqrt{2\pi d_o}} \int_0^\infty \exp\left(-\frac{(\ln(z) - \mu_o)^2}{2d_o} - \frac{(r^2 + z^2)}{2b_o}\right) \frac{I_o(rz/b_o)}{z}\, dz \qquad (1)$$

where $I_n(\cdot)$ is the modified Bessel function of the first kind and nth order, b_o is the average scattered power due to multipath, μ_o is the mean value due to shadowing, and d_o is the variance due to shadowing.

The probability density function of the received signal phase ϕ was found to be approximately Gaussian [6]:

$$p_\phi(\phi) = \frac{1}{\sqrt{2\pi\sigma_\phi^2}} \exp\left[-\frac{(\phi - \mu_\phi)^2}{2\sigma_\phi^2}\right] \qquad (2)$$

where μ_ϕ and σ_ϕ^2 are the mean and variance of the received signal phase, respectively.

The above model is valid for a narrowband system. If spread-spectrum modulation is used with a chip duration less than the delay spread of the channel, the multipath power is partially reduced by the correlation operation in the receiver. The envelope and the phase distribution functions remain the same, but the values for b_o and σ_ϕ are reduced.

The impulse response of the channel can be written as:

$$h(t) = \sum_{m=1}^{M} \beta_m\, \delta(t - \tau_m) e^{j\theta_m} \qquad (3)$$

where β, τ, and θ are the gain, time delay, and phase of the mth path, respectively. The first path is the line-of-sight and therefore its propagation statistics are described by (1) and (2). The other paths have a Rayleigh path gain distribution and uniformly distributed phase, since the direct line-of-sight is suppressed by the correlation operation. The parameters of the various path distributions can be found if the power-delay profile is known. For the present study, the power-delay profile is

0733-8716/92$03.00 © 1992 IEEE

considered as:

$$P(\tau) = cb_o \exp(-c\tau) \tag{4}$$

where $1/c$ is the delay spread. For a rural environment, a typical value of $1/c$ is 0.65 μs [7].

Due to the correlation operation, the multipath power ($= b_o$) is reduced. For path m, it can be approximated as:

$$b_{mo} = b_o[1 - \exp(-cT_c)] \exp[-c(m-1)T_c] \tag{5}$$

where T_c is the chip duration.

Also, the phase variance of the first path will be decreased, since it is determined by the amount of multipath power and the statistics of the line-of-sight propagation. Using (4.5-19) of [11], the phase distribution function $p(\phi)$ of a lognormally shadowed Rician signal can be derived and given as:

$$p(\phi) = \frac{1}{\sqrt{8\pi^3 d_o}} \int_0^\infty \exp\left(\frac{-z^2}{2b_o} - \frac{(\ln(z) - \mu_o)^2}{2d_o}\right) \cdot \frac{(1 + G\sqrt{\pi}\exp(G^2)[1 + \text{erf}(G)])}{z} dz \tag{6}$$

where

$$G \triangleq \frac{z\cos(\phi)}{\sqrt{2b_o}}$$

and erf $(\cdot)$ is the error function [12]: erf $(G) \triangleq 2/\sqrt{\pi}\int_0^G \exp(-t^2)\,dt$. Now, the phase variance in the case of spread-spectrum modulation can be determined by:

$$\sigma_\phi^2 = \int_{-\pi}^{\pi} \phi^2 p(\phi)\, d\phi. \tag{7}$$

Note that the mean phase μ_ϕ is zero because $p(\phi) = p(-\phi)$.

Equations (6) and (7) are used only to determine the effect of spread-spectrum modulation on the phase variance. The combined effect of phase variation and envelope fading [6] is evaluated by using (2) as an approximation for (6).

III. Receiver Model

The spread-spectrum receiver model is shown in Fig. 1. The total received signal is:

$$r(t) = \sum_{k=1}^{K}\sum_{m=1}^{M} A\beta_{mk} a_k(t - \tau_{mk})\, b_k(t - \tau_{mk}) \cdot \cos((\omega_c + \omega_{mk})t + \phi_{mk}) + n(t) \tag{8}$$

where m and k denote the path and user number, respectively, and A is the transmitted signal amplitude, which is assumed to be constant and identical for all users. For user k, $\{a_k\}$ is the spread-spectrum code, $\{b_k\}$ is the data sequence, $\omega_c + \omega_{mk}$ is the carrier plus doppler angular frequency, ϕ_{mk} is the carrier phase, and $n(t)$ is white Gaussian noise with two-sided power spectral density $N_o/2$.

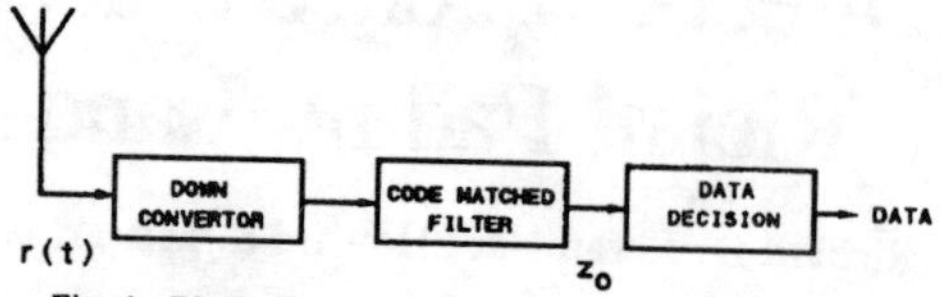

Fig. 1. Block diagram of the spread-spectrum receiver.

This signal is converted to baseband and correlated with a particular user code. If the receiver locks on the first path of user one, a signal sample of the correlation output can be written as:

$$z_o = A\beta_{11}\cos\phi_{11} T_b b_1^0 + \sum_{k=1}^{K} A(b_k^{-1}X_k + b_k^0\hat{X}_k) + \eta_1 \tag{9}$$

where b_k^{-1} and b_k^0 are the previous and current data bit, respectively, and

$$X_1 = \sum_{m=2}^{M} R_{11}(\tau_{m1})\beta_{m1}\cos\phi_{m1}$$

$$\hat{X}_1 = \sum_{m=2}^{M} \hat{R}_{11}(\tau_{m1})\beta_{m1}\cos\phi_{m1}$$

$$X_k = \sum_{m=1}^{M} R_{1k}(\tau_{mk})\beta_{mk}\cos\phi_{mk}$$

$$\hat{X}_k = \sum_{m=1}^{M} \hat{R}_{1k}(\tau_{mk})\beta_{mk}\cos\phi_{mk}$$

$$R_{1k}(\tau) = \int_0^\tau a_k(t-\tau)a_1(t)\,dt$$

$$\hat{R}_{1k}(\tau) = \int_\tau^{T_b} a_k(t-\tau)a_1(t)\,dt.$$

In (9), the carrier phase ϕ_{mk} is Gaussian distributed for $m = k = 1$, but uniformly distributed for the other users and paths, because the transmitters are assumed to have arbitrary phases. β_{mk} has a shadowed Rician distribution for $m = 1$, and a Rayleigh distribution otherwise. η_1 is a zero-mean Gaussian variable with variance $N_o T_b$.

IV. Performance Analysis

The bit error probability and outage probability are considered to be the two basic performance measures of digital systems. The message success probability is also an important consideration. All three performance measures are discussed here.

A. Bit Error Probability

Assuming that the data bits -1 and 1 are equiprobable, the bit error probability p_e can be expressed as:

$$p_e = P(z_o < 0 | b_1^0 = 1). \tag{10}$$

The interference can be approximated by Gaussian noise if KM is large, so p_e can be written as:

$$p_e = \int_{-\infty}^{\infty} p_e(x) p_u(x)\, dx. \quad (11)$$

Here,

$$p_e(x) = P\,(\text{Gaussian noise} < -x) = \frac{1}{2}\,\text{erfc}\left(\frac{x}{\sigma\sqrt{2}}\right). \quad (12)$$

erfc $(\cdot)$ is the complementary error function [12], p_u is the pdf of the first term of z_o, and σ^2 is defined as the total variance of the Gaussian noise:

$$\sigma^2 \triangleq N_o T_b + \sigma_i^2 \quad (13)$$

with σ_i^2 as the interference power.

1) Gaussian Approximation: Using the Gaussian approximation, only the mean and variance of the interference term $\Sigma_{k=1}^{K} A(b_k^{-1} X_k + b_k^0 \hat{X}_k)$ need to be evaluated. Since all terms of the summation are independent and symmetrically distributed, the mean reduces to zero. For the same reason, all the cross terms in the calculation of the second moment become zero, so the variance is:

$$\sigma_i^2 = \sum_{m=2}^{M} A^2 E([b_1^{-1} R_{11}(\tau_{m1}) + b_1^0 \hat{R}_{11}(\tau_{m1})]^2) E([\beta_{m1} \cos \phi_{m1}]^2) + \sum_{k=2}^{K} \sum_{m=1}^{M} A^2 E([b_k^{-1} R_{1k}(\tau_{mk}) + b_k^0 \hat{R}_{1k}(\tau_{mk})]^2) E([\beta_{mk} \cos \phi_{mk}]^2). \quad (14)$$

Note that the first and second term of the right-hand side of (14) represent the variance for $k = 1$ and $k > 1$, respectively.

The second moment of $\beta \cos \phi$ can be calculated as follows: If $m = 1$, then β has a shadowed Rician distribution, which can be viewed as a Rician distribution with a variable Rician parameter $s^2/2b_o$. The product of this Rician variable with the cosine of a uniformly distributed variable gives a Gaussian variable with a mean of $s/\sqrt{2}$ and a variance of b_o. Thus $E((\beta_{mk} \cos \phi_{mk})^2) = b_o + s^2/2$, on the condition that s is constant. This condition can be removed by integrating over the independent lognormal distribution of s:

$$E((\beta_{1k} \cos \phi_{1k})^2) = \int_0^{\infty} (b_{1o} + s^2/2) p(s)\, ds. \quad (15)$$

After integration, (15) reduces to:

$$E((\beta_{1k} \cos \phi_{1k})^2) = b_{1o} + \tfrac{1}{2} \exp(2d_o + 2\mu_o). \quad (16)$$

If $m > 1$, then β has a Rayleigh distribution. In that case, $s = 0$ and:

$$E((\beta_{mk} \cos \phi_{mk})^2) = b_{mo}. \quad (17)$$

Referring to [14], one obtains the second moment of the cross correlation for any user k and for any spread-spectrum code:

$$E[R_{1k}^{2q}(\tau_{1k}) \hat{R}_{1k}^{2(p-q)}(\tau_{1k})] = \frac{T_c^{2p+1}}{T_b} \sum_{n=0}^{N-1} \sum_{i=1}^{2q} \frac{(-1)_i}{i+1} \frac{\binom{2q}{i}}{\binom{2(p-q)+i+1}{i+1}} \frac{B_{n1k}^i}{\hat{B}_{n1k}^{i+1}} [(A_{n1k} + B_{n1k})^{2q-i} (\hat{A}_{n1k} + \hat{B}_{n1k})^{2(p-q)+i+1} - A_{n1k}^{2q-i} \hat{A}_{n1k}^{2(p-q)+i+1}] \quad (18)$$

where

$$A_{n1k} = C_{1k}(n-N) \qquad \hat{A}_{n1k} = C_{1k}(n)$$

$$B_{n1k} = C_{1k}(n+1-N) - C_{1k}(n-N)$$

$$\hat{B}_{n1k} = C_{1k}(n+1) - C_{1k}(n)$$

$$C_{1k}(n) = \begin{cases} \sum_{j=0}^{N-1+n} a_k^j a_1^{j+n} & 0 \le n \le N-1 \\ \sum_{j=0}^{N-1+n} a_k^{j-n} a_1^j & 1-N \le n < 0 \end{cases}$$

and p and q are dummy variables.

For Gold codes, a simplified technique for evaluating the variance of the cross correlations can be used using [13], [16]. Now, one gets:

$$E([b_k^{-1} R_{1k}(\tau_{mk}) + b_k^0 \hat{R}_{1k}(\tau_{mk})]^2) = 2T_b^2/3N \quad (19)$$

where N is the Gold code length, which is assumed here to be equal to T_b/T_c. Thus, (19) simplifies the computation of (14) for Gold codes.

2) Distribution of the Wanted Signal: The first term of z_o in (9) consists of β with a shadowed Rician distribution, multiplied by the cosine of a Gaussian distributed phase. Using [11], the total distribution function of β cos (ϕ) can be written as:

$$p_u(x) = 2 \int_{\text{abs}(x)}^{\infty} p_\beta(r) p_\phi\,(\arccos\,(x/r))\, \frac{dr}{\sqrt{r^2 - x^2}}. \quad (20)$$

Substituting (1) and (2) in (20) and using (11) and (12), p_e is expressed as:

$$p_e = 0.5 \int_0^{\infty} \text{erfc}\left(\frac{xT_b}{\sigma\sqrt{2}}\right) p_u(x)\, dx \quad (21)$$

with

$$p_u(x) = \int_{\text{abs}(x)}^{\infty} \int_0^{\infty} \frac{r}{b_o \sqrt{2\pi^2 d_o \sigma_\phi^2}} \exp\left(-\frac{(\ln(z) - \mu_o)^2}{2d_o} - \frac{(r^2+z^2)}{2b_o} - \frac{\arccos^2(x/r)}{2\sigma_\phi^2}\right) \cdot \frac{I_o(rz/b_o)}{z\sqrt{r^2 - x^2}}\, dz\, dr. \quad (22)$$

Equation (22) is valid in the case that the receiver locks on the line-of-sight signal. This requires the bandwidth of the carrier tracking loop to be much smaller than the fading bandwidth of the received signal. If the bandwidth of the tracking loop is larger than the fading bandwidth, ϕ_{11} is approximately zero because the receiver will lock on the phase of the total signal. In that case, the distribution function p_u is equal to the shadowed Rician distribution p_β. However, with a larger tracking bandwidth the loop noise increases, so there is always a certain phase error. Therefore, in the bit error probability calculations one gets an upper bound by using (22), and a lower bound by using $p_u = p_\beta$.

B. Outage Probability

The outage probability P_{out} is the probability that the instantaneous bit error probability exceeds a certain threshold, denoted by p_o, and can be written as:

$$P_{out} = P(p_e \geq p_o) = P(x \leq x_o) = \int_{-\infty}^{x0} xp_u(x)\, dx. \tag{23}$$

Here x_o is the value of the amplitude x at which the instantaneous bit error probability is equal to p_o.

C. Message Success Probability

The message success probability p_s is defined as the probability that, in a received message of L bits, all possible errors can be corrected.

$$p_s = \sum_{j=0}^{n_e} p_e^j (1 - p_e)^{L-j} \binom{L}{j}. \tag{24}$$

In (24), p_e is the bit error probability and n_e is the number of errors that can be corrected by the error correcting code.

V. Numerical Results

A. Bit Error Probability

By calculating (22) using the Gaussian integration technique, the bit error probability (21) is evaluated using the Newton–Cotes technique [12]. Fig. 2 shows the bit error probability for light, average, and heavy shadowing, according to the measured values found in [8] and reproduced in Table I. The mean phase μ_ϕ is zero in all cases.

The plots in Fig. 2 are for the narrowband case, i.e., without the use of spread-spectrum modulation. The major difference between these plots and those in [8] is that, in [8], the effects of phase and envelope fading were calculated separately, while in this paper their combined effect is shown. Further, in [8] an upperbound for the envelope fading was calculated by an approximation of the complementary error function, while we present an exact analysis.

If we compare Fig. 2 with the Fig. 2 in [8], where only envelope fading is considered, it appears that for low signal-to-noise ratio Fig. 2 gives higher values for the bit

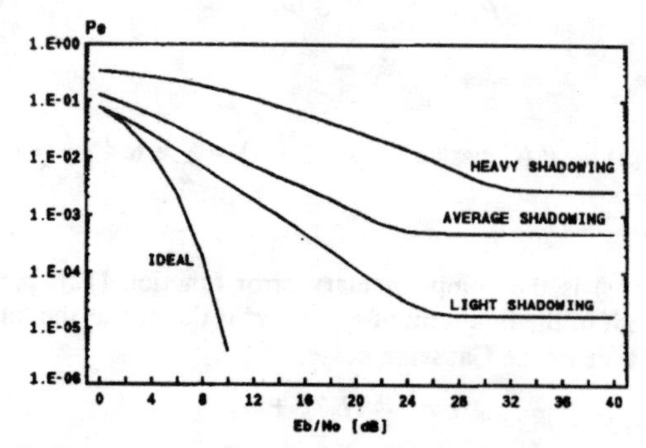

Fig. 2. Bit error probability for ideal BPSK and for narrowband BPSK with light, average, and heavy shadowing.

TABLE I
Channel Model Parameters

	Light	Average	Heavy
b_o	0.158	0.126	0.0631
μ	0.155	−0.115	−3.91
$\sqrt{d_o}$	0.115	0.161	0.806
σ	0.36	0.45	0.52

error probability. However, this is due to a mathematical error in the derivation of the upperbound of the bit error probability in [8]. Equation (25) is the correct upperbound of the bit error probability, which should replace (17) in [8].

$$p_e \leq \frac{1}{\sqrt{8\pi d_o}} \frac{\sigma^2}{b_o + \sigma^2} \int_0^\infty \frac{1}{z} \cdot \exp\left(\frac{-(\ln(z) - \mu)^2}{2d_o} - \frac{z^2}{2(b_o + \sigma^2)}\right) dz. \tag{25}$$

In addition, with the data in Table I, the power of the received signal in the case of light shadowing is greater than one, which results in a smaller bit error probability than theoretically possible at low signal-to-noise ratios. To repair this, we scaled the noise power by multiplying it with 1.585, which is the signal power for light shadowing. Without this scaling, all plots for light shadowing would shift 2 dB to the left.

In all three cases of shadowing, the bit error probability converges to the irreducible error probability p_{irr}, caused by the phase variation. A rather complicated expression for this error is given in (27) in [8]. By considering p_{irr} as the chance that the received bit in the absence of noise is negative while the transmitted bit is positive, p_{irr} can be simply given as:

$$p_{irr} = \int_{-\infty}^{0} p_u(x)\, dx, \quad b_1^0 = 1. \tag{26}$$

The irreducible error probabilities in Fig. 2 do correspond with those found in [8].

Next, the effect of spread-spectrum modulation is com-

TABLE II
CALCULATED CHANNEL MODEL PARAMETERS

	σ_ϕ Narrowband	σ_ϕ Spread-Spectrum	b_o Spread-Spectrum
Light	0.40	0.14	0.023
Average	0.47	0.16	0.018
Heavy	1.55	1.42	0.009

puted. First, the bit error probability is shown for $T_c = 0.1\ \mu s$, $1/T_b = 2400$ b/s, $N = T_b/T_c = 4095$ and $K = 1$. The fading parameters are adjusted according to (5) and (7). In Table II, the modified values for b_o and σ_ϕ are shown. Also, the calculated values for σ_ϕ for narrowband operation are given, which show a considerable difference with the measured values [6], [8] in the case of heavy shadowing. This difference may be due to the filtering of the received signal. It can be expected that σ_ϕ should be almost $\pi/2$ for heavy shadowing, because the distribution function approaches a Rayleigh pdf. This implies an almost uniformly distributed phase. However, in order to compare with [8], we have used the measured value for heavy shadowing ($\sigma_\phi = 0.52$) in all calculations. Assuming that, for heavy shadowing, σ_ϕ decreases by the same amount as for light and average shadowing, it is given by (0.52/2.9) in the presence of spread-spectrum modulation. Comparing Figs. 2 and 3, it is seen that spread-spectrum modulation yields better performance than narrowband modulation for light and average shadowing. For heavy shadowing, the performance is worse at signal-to-noise ratios below 36 dB. The reason for this is that for light and average shadowing most of the signal power is received via the line-of-sight, so if the multipath power is reduced by the use of spread-spectrum modulation, a less perturbed signal will be obtained. In the case of heavy shadowing, however, the direct line-of-sight power is much smaller than the multipath power, resulting in an approximately Rayleigh faded signal. The use of spread-spectrum modulation now decreases the total signal power considerably, with the result that the resulting bit error probability increases. In this case, diversity techniques can be used to improve the bit error probability [14], [15], which is not considered here.

If the receiver is able to lock onto the total phase of the first path, then only the envelope fading has to be considered, which is investigated in Figs. 4–6, where the bit error probability is compared to the narrowband and spread-spectrum modulation. It can be seen from Figs. 2–6 that the bit error probability for narrowband modulation changes considerably by removing the phase variation, while in the case of spread-spectrum modulation there is much less change. The phase variance, which is reduced by approximately a factor 9 by the use of spread-spectrum modulation (7), exerts a nonnegligible influence only near the irreducible error probability level ($\leq 10^{-9}$ for spread-spectrum modulation).

It is seen from Fig. 7 that the light and average shadowed Rician distributions can be represented by the normal Rician distribution with a Rice factor of 5.3 and 4.1 dB, respectively, with an error of less than 5% for bit error probability values between 10^{-1} and 10^{-5}. Outside this interval, the error tends to increase. It can be expected that matching will be possible only in a certain range, since the two distributions functions are different in nature because of the shadowing. For heavy shadow-

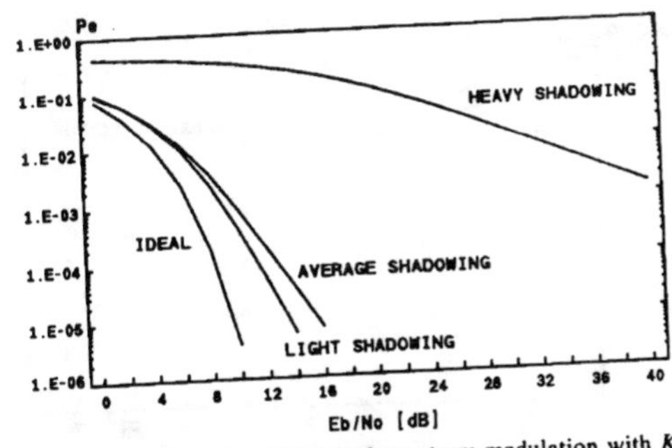

Fig. 3. Bit error probability for spread-spectrum modulation with $K = 1$ user, chip length $T_c = 0.1\ \mu s$, Gold code length $N = 4095$, and bit rate $1/T_b = 2400$ b/s.

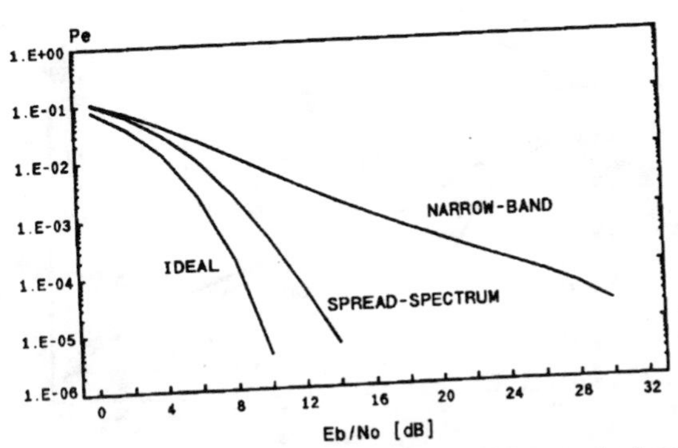

Fig. 4. Comparison of the bit error probability with narrowband and spread-spectrum modulation with light shadowing and envelope fading only for $K = 1$ user, chip length $T_c = 0.1\ \mu s$, Gold code length $N = 4095$, and bit rate $1/T_b = 2400$ b/s.

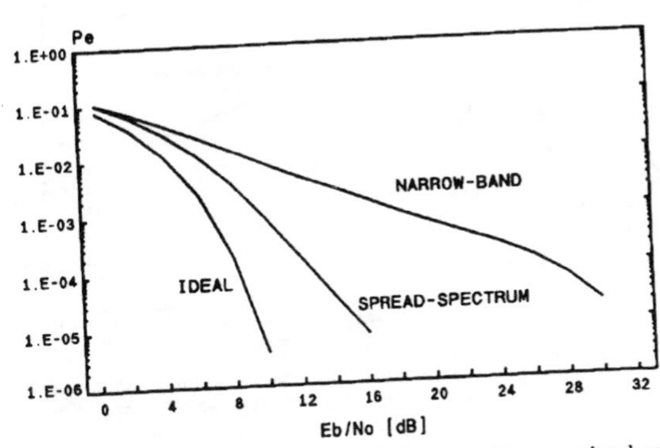

Fig. 5. Comparison of the bit error probability with narrowband and spread-spectrum modulation with average shadowing and envelope fading only for $K = 1$ user, chip length $T_c = 0.1\ \mu s$, Gold code length $N = 4095$, and bit rate $1/T_b = 2400$ b/s.

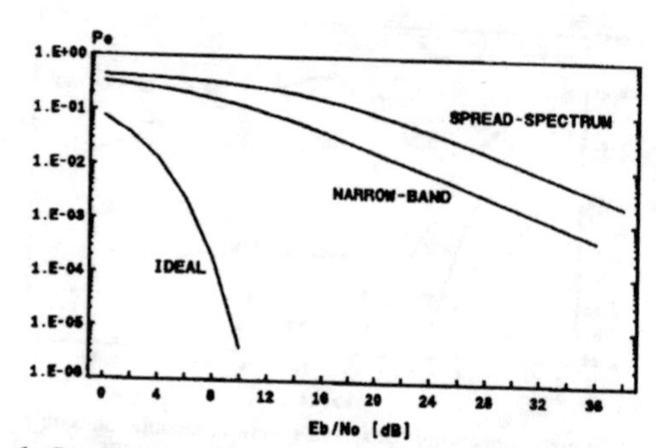

Fig. 6. Comparison of the bit error probability with narrowband and spread-spectrum modulation with heavy shadowing and envelope fading only for $K = 1$ user, chip length $T_c = 0.1$ μs, Gold code length $N = 4095$, and bit rate $1/T_b = 2400$ b/s.

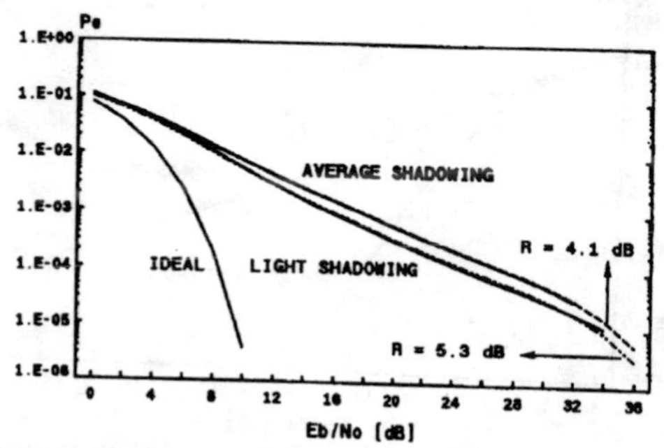

Fig. 7. Comparison of the shadowed Rician distribution with the normal Rician distribution for light and average shadowing with narrowband transmission.

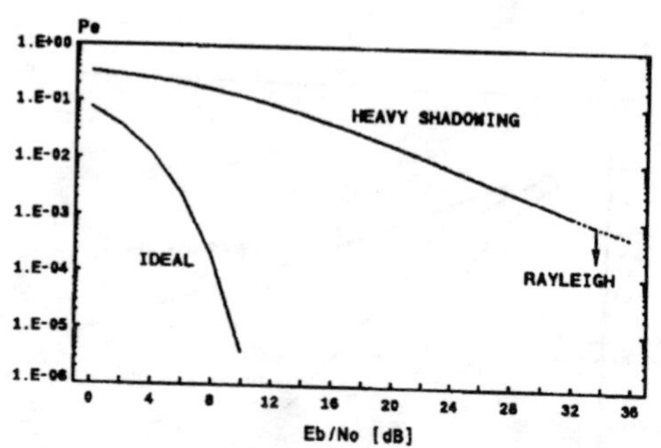

Fig. 8. Comparison of the shadowed Rician distribution with the Rayleigh distribution for heavy shadowing with narrowband transmission.

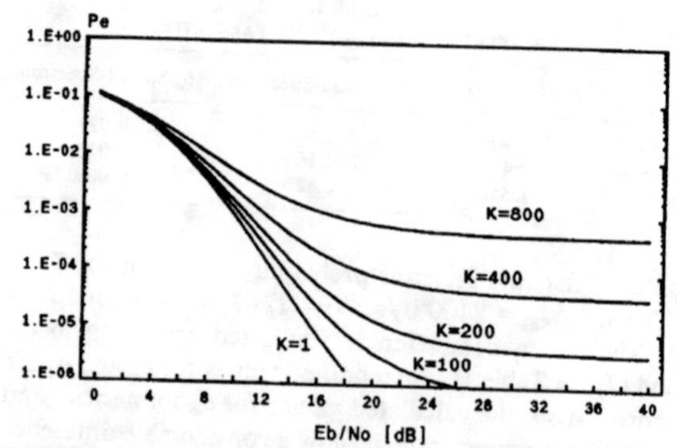

Fig. 9. Bit error probability for spread-spectrum modulation with average shadowing for chip length $T_c = 0.1$ μs, Gold code length $N = 4095$, bit rate $1/T_b = 2400$ b/s, and K as a parameter.

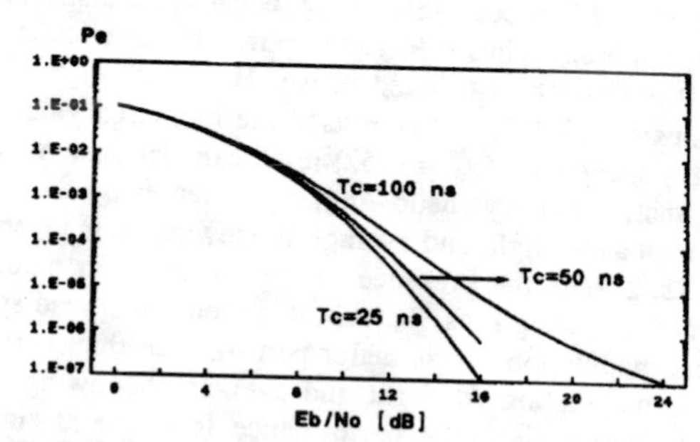

Fig. 10. Bit error probability for spread-spectrum modulation with average shadowing for $K = 1$, $N = 4095$, $T_b = T_c N$, and T_c as a parameter.

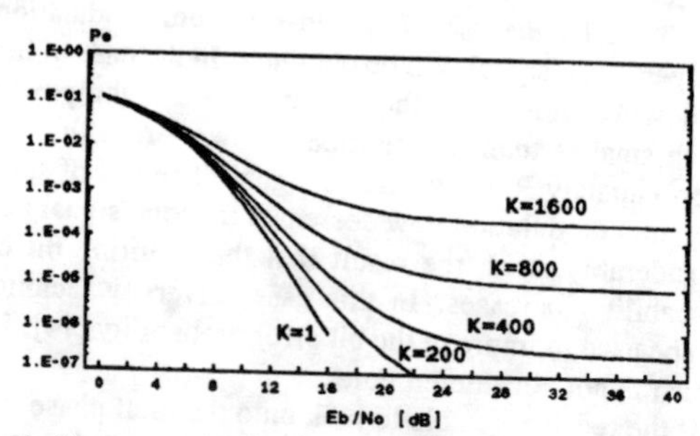

Fig. 11. Bit error probability for spread-spectrum modulation with average shadowing for $T_b = 1/2400$ s, $N = 8191$, $T_c = T_b/N$, and K as a parameter.

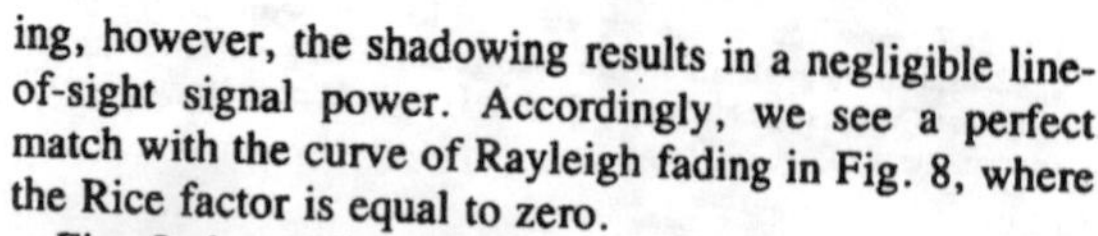

ing, however, the shadowing results in a negligible line-of-sight signal power. Accordingly, we see a perfect match with the curve of Rayleigh fading in Fig. 8, where the Rice factor is equal to zero.

Fig. 9 shows the results for spread-spectrum modulation with average shadowing and the number of users as a parameter. To maintain a bit error probability of 10^{-3}, the signal-to-noise ratio has to be increased by about 0.5 dB for $K = 100$ users, 1 dB for $K = 200$, and 2 dB for $K = 400$ as compared to the signal-to-noise ratio for a single user ($K = 1$). Note that E_b/N_o is the average signal-to-noise ratio, which does not include interference power. So increasing E_b/N_o decreases the bit error probability, until the irreducible bit error probability caused by the interference power is reached.

Fig. 10 shows that as T_c decreases, the bit error probability decreases because of the decreased multipath power.

Fig. 11 shows the bit error probability for $N = 8191$, $T_b = 1/2400$ s, $T_c = T_b/N$ and K as a parameter. The plot for $K = 1$ in Fig. 11 is the same as the one for $T_b = 1/4800$ s in Fig. 10, because they have the same values

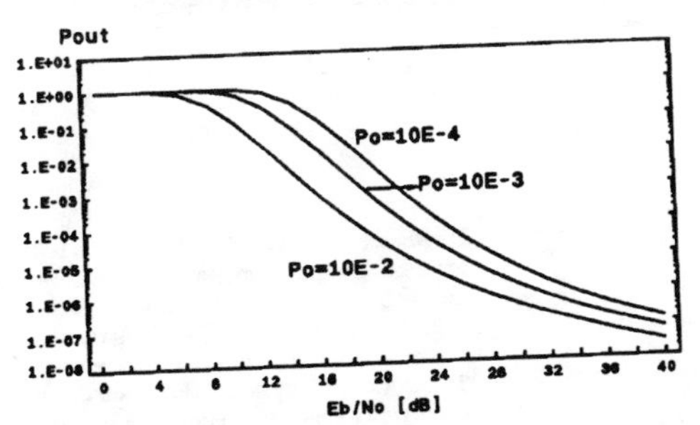

Fig. 12. Outage probability for spread-spectrum modulation with average shadowing for $K = 1$, $T_b = 1/2400$ s, and $T_c = T_b/4095$.

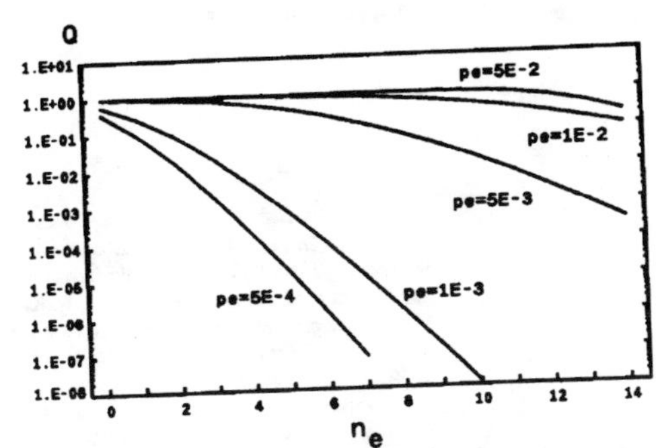

Fig. 13. Message error probability as a function of n_e for $K = 1$, $T_c = 0.1$ μs, $N = 4095$, $T_b = 1/2400$ s, and $L = 1024$.

for T_c. Comparing Figs. 9 and 11, it is seen that for particular values of p_e and E_b/N_o the number of users increases more than twice by doubling N, due to the reduced cross-correlation power and the reduced multipath power.

B. Outage Probability

The outage probability (23) is shown in Fig. 12 for $T_c = 0.1$ μs, $T_b = 1/2400$ s, $K = 1$ and average shadowing. To achieve an outage probability of 10^{-2}, for instance, the signal-to-noise ratio has to be 13, 16.5, or 19 dB for a threshold p_o of 10^{-2}, 10^{-3}, or 10^{-4}, respectively.

C. Message Success Probability

Instead of the message success probability, we have presented the computational results for the message error probability Q which is defined as $Q \triangleq 1 - p_s$.

Fig. 13 depicts the message error probability as a function of n_e with parameter p_e for $L = 1024$, $T_c = 0.1$ μs, $T_b = 1/2400$ s, $N = 4095$, $K = 1$ and for average shadowing.

In Fig. 14, Q is plotted as a function of the number of users K at $E_b/N_o = 10$ dB for $K = 1$. The other parameters are the same as in Fig. 13. If a maximum of 512 users, for instance, must be supported with a message success probability of at least 0.99, then the error correcting code must be able to correct 8 errors per 1024 bits.

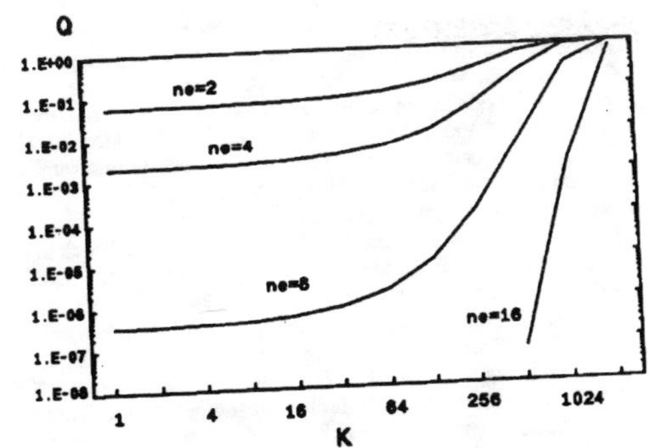

Fig. 14. Message error probability as a function of K for $E_b/N_o = 10$ dB, $T = 0.1$ μs, $N = 4095$, $T_b = 1/2400$ s, and $L = 1024$.

VI. Conclusions

The shadowed Rician channel model given in [4]–[6] has been expanded to a wideband model. The performance of a land-mobile satellite channel has been evaluated in terms of the bit error, outage and message success probability for light, average, and heavy shadowing, using a direct-sequence spread-spectrum transmission system with BPSK modulation. The main conclusion that can be drawn from this analysis is that the spread-spectrum system yields better performance than the narrowband transmission if the line-of-sight path is dominant.

It is shown that for bit error probability calculations, the shadowed Rician distribution can be approximated in a limited but useful range of the bit error probability by a normal Rician distribution with a Rice factor of 5.3, 4.1, and $-\infty$ dB for light, average, and heavy shadowing, respectively. When spread-spectrum modulation is used, these Rice factors will increase because of the multipath rejection capability of the correlation operation in the receiver. As a result, spread-spectrum modulation yields better performance than narrowband transmission, except for heavy shadowing, i.e., where the line-of-sight signal power is smaller than the multipath power. In the latter case, diversity techniques can be added to improve the performance. Further, it is recommended to study the influence of very low rate convolutional codes [17] on the performance of spread-spectrum multiple access land-mobile satellite channels.

Acknowledgment

The authors are grateful to Prof. J. C. Arnbak for his fruitful comments.

References

[1] K. G. Johanssen, "Code division multiple access versus frequency division multiple access channel capacity in mobile satellite communication," *IEEE Trans. Vehic. Technol.*, vol. 39, pp. 17–26, Feb. 1990.

[2] K. S. Gilhousen, I. M. Jacobs, R. Padovani, L. A. Weaver, Jr., "Increased capacity using CDMA for mobile satellite communications," *IEEE J. Select. Areas Commun.*, vol. 8, pp. 503–514, May 1990.

[3] J. D. Kiesling, "Land mobile satellite systems," *Proc. IEEE*, vol. 78, pp. 1107–1115, July 1990.
[4] C. Loo, "A statistical model for a land–mobile satellite link," *IEEE Trans. Vehic. Technol.*, vol. VT-34, pp. 122–127, Aug. 1985.
[5] C. Loo, E. E. Matt, J. S. Butterworth, and M. Dufour, "Measurements and modeling of land–mobile satellite signal statistics," presented at 1986 Vehic. Technol. Conf., Dallas, TX, May 20–22, 1986.
[6] C. Loo, "Measurements and models of a land–mobile satellite channel and their applications to MSK signals," *IEEE Trans. Vehic. Technol.*, vol. VT-36, pp. 114–121, Aug. 1987.
[7] J. van Rees, "Measurements of the wide-band radio channel characteristics for rural, residential, and suburban areas," *IEEE Trans. Vehic. Technol.*, vol. 36, pp. 2–6, Feb. 1987.
[8] C. Loo, "Digital transmission through a land–mobile satellite channel," *IEEE Trans. Commun.*, vol. 38, no. 5, May 1990.
[9] P. J. Mclane, P. H. Wittke, P. K. M. Ho, and C. Loo, "PSK and DPSK trellis codes for fast fading, shadowed mobile satellite communication channels," *IEEE Trans. Commun.*, vol. 36, pp. 1242–1246, Nov. 1988.
[10] A. C. M. Lee and P. J. Mclane, "Convolutionally interleaved PSK and DPSK trellis codes for shadowed, fast fading mobile satellite communication channels," *IEEE Trans. Vehic. Technol.*, vol. 39, pp. 37–47, Feb. 1990.
[11] P. Beckmann, *Probability in Communication Engineering*. New York: Harcourt, Brace & World, 1967.
[12] M. Abramowitz and I. A. Stegun, Eds., *Handbook of Mathematical Functions*. New York: Dover, 1965.
[13] M. B. Pursley, "Performance evaluation for phase-coded spread-spectrum multiple-access communication—Part I: System analysis," *IEEE Trans. Commun.*, vol. COM-25, no. 8, pp. 795–799, Aug. 1977.
[14] M. Kavehrad and B. Ramamurthi, "Direct-sequence spread spectrum with DPSK modulation and diversity for indoor wireless communications," *IEEE Trans. Commun.*, vol. 35, no. 2, pp. 224–236, Feb. 1987.
[15] R. Prasad, H. S. Misser, and A. Kegel, "Performance analysis of direct sequence spread-spectrum multiple access communication in an indoor Rician-fading channel with DPSK modulation," *Electron. Lett.*, vol. 26, pp. 1366–1367, Aug. 1990.
[16] M. Kavehrad and P. J. Mclane, "Performance of low-complexity channel coding and diversity for spread-spectrum in indoor, wireless communication," *AT&T Tech. J.*, vol. 64, pp. 1927–1965, Oct. 1985.
[17] A. J. Viterbi, "Very low rate convolutional codes for maximum theoretical performance of spread-spectrum multiple-access channels," *IEEE J. Select. Areas Commun.*, vol. 8, pp. 641–649, May 1990.

Richard D. J. van Nee was born in Schoonoord, the Netherlands, on January 17, 1967. He received the M.Sc. degree in electrical engineering (with distinction) from the University of Twente, Twente, The Netherlands, in 1990.

He is currently working towards the Ph.D. degree at Delft University, The Netherlands, where he studies the influence of channel fading on satellite communication and navigation.

Mr. van Nee received the Ashtech Award for the best student paper at the Institute of Navigation GPS-91 conference, held in Albuquerque, New Mexico, in September 1991, for his paper on the effects of multipath propagation on spread-spectrum code tracking errors.

Howard S. Misser (S'89–M'91) was born in Paramaribo, Surinam, on May 11, 1968. He received the M.Sc. degree in electrical engineering from Delft University, Delft, The Netherlands, in 1990.

He worked at the Telecommunications and Traffic-Control Systems Group of the same university as a Research Fellow. He is currently with Dr. Neher Laboratories of the Royal Dutch PTT in Leidschendam, The Netherlands. His research interests are in the fields of broadband network technologies and architectures, radio and mobile communication, and spread-spectrum communication.

Ramjee Prasad (M'89–SM'90) was born in Babhnaur (Gaya), Bihar, India, on July 1, 1946. He received the B.Sc. (Eng.) degree from the Bihar Institute of Technology, Sindri, India, and the M.Sc. (Eng.) and Ph.D. degrees from the Birla Institute of Technology (BIT), Ranchi, India, in 1968, 1970, and 1979, respectively.

He joined BIT as a Senior Research Fellow in 1970 and became Associate Professor in 1980. From 1983 to 1988, he was with the University of Dar es Salaam (UDSM), Tanzania, where in 1986 he became Professor in Telecommunications at the Department of Electrical Engineering. Since February 1988, he has been with the Telecommunications and Traffic Control Systems Group, Delft University of Technology, The Netherlands, where he is actively involved in the area of mobile and indoor radio communications. While he was with BIT, he supervised many research projects in the area of microwave and plasma engineering. At UDSM, he was responsible for the collaborative project "Satellite Communications for Rural Zones" with the Eindhoven University of Technology. He has published over 80 technical papers. His current research interests are in packet communications, adaptive equalizers, spread-spectrum systems, and telematics. He has served as a member of advisory and program committees of several IEEE international conferences. He has also presented tutorials on mobile and indoor radio communications at various universities, technical institutions, and IEEE conferences. He is also a member of a Working Group of the European Co-Operation in the field of scientific and technical research for Project (COST-231) as an expert for the Netherlands.

Prof. Prasad is a Fellow of the Institution of Electronics and Telecommunication Engineers.

Open-Loop Power Control Error in a Land Mobile Satellite System

Anton M. Monk, *Student Member,* and Laurence B. Milstein, *Fellow, IEEE*

***Abstract*— In order to combat large scale shadowing and distance losses in a land mobile satellite system, an adaptive power control (APC) scheme is essential. Such a scheme, implemented on the uplink, ensures that all users' signals arrive at the base station with equal average power as they move within the satellite spot beam—an important requirement in a CDMA system. Because of the lengthy round-trip delay on a satellite link, closed-loop power control systems are only of marginal benefit. Therefore, an *open-loop* APC scheme is proposed to counteract the effects of shadowing and distance loss. A fairly general channel model, consisting of log-normal shadowing and Rician fading, is assumed. This can be applied to a specific two-state land mobile satellite channel model, involving shadowed intervals with Rayleigh fading and unshadowed intervals with Rician fading. It is found that the power control error can be approximated by a log-normally distributed random variable. To quantify the performance of the APC, the standard deviation of the power control error in decibels is analyzed as a function of the specular power-to-scatter power ratio, the measurement time and the vehicle velocity. To illustrate the usefulness of the results, we analyze the effect of the power control error on the system capacity of a CDMA mobile satellite link.**

I. Introduction

Land mobile satellite systems are usually required to use adaptive power control (APC) between the mobiles and the satellite. In the cases of time or frequency orthogonal sytems, such as time-division multiple-access (TDMA) or frequency-division multiple-access (FDMA), the main purpose of power control is to maintain a constant average performance for all users, and to minimize the required transmit power. In a direct-sequence (DS) code-division multiple-access (CDMA) system, accurate APC is critical for a good aggregate system performance [1], since the multiple-access users act as interferers to each other. In this paper, we concentrate on developing an analytical model for the power control error (PCE). We analyze the APC on the uplink, or reverse link—traditionally, the weaker of the two links.

On a land mobile satellite link, the channel fluctuates rapidly, due to multipath fading. (The fading on the uplink is usually uncorrelated with that on the downlink, since the uplink and downlink frequency bands are separated by more than the coherence bandwidth of the channel.) In addition, the signal may suffer large scale, more slowly varying shadowing losses, due to blockage by buildings, hills, trees, etc., which are usually assumed identical on uplink and downlink. Other uncorrelated losses, such as knife-edge diffraction, exist. These effects, which can degrade the performance of a power control system, are not included here. One possible form of power control is a closed-loop APC scheme. In such a system, information regarding the state of the channel on the uplink is relayed from the base-station back to the mobile. The mobile then makes use of this information by modifying its transmitted signal strength to compensate for the channel on the uplink. If the round trip delay between the mobile and base-station is smaller than the correlation time of the channel, then such a scheme can compensate for the rapid multipath fading. In a terrestrial environment, a closed-loop APC scheme may be feasible under most conditions; however, even with low earth orbit satellites (LEOS), the round trip delay is around 10 ms for a 400 nautical mile (to the subsatellite point) satellite, and up to 60 ms for 800 nautical mile satellites at lower elevation angles [2]. Closed-loop APC is, thus, typically not feasible in a land mobile satellite system.

An alternate form of APC is an open-loop scheme, which relies on channel state information obtained on the opposite link. For example, the mobile would estimate the state of the channel on the downlink and use this as an estimate of the uplink channel state. Such a scheme would only be able to compensate for large scale variations in the channel, such as shadowing, which are the same on both links. An open-loop APC algorithm should therefore provide an estimate of the shadowing component, and attempt to minimize the effect of the fading component by averaging it out. It should be noted that, even with a fast open-loop APC scheme, some form of closed-loop power control is necessary to compensate for slowly moving mobile units, which may, for example, be caught in a deep fade on the uplink. In such a case, open-loop APC might be unable to detect the presence of the uplink fade. Also, closed-loop APC is needed to control long-term component drift effects.

The land mobile satellite channel has been characterized in [3]. A simple, yet elegant, model has been shown to give good agreement with the observed channel. The model is described by the time-share of shadowing, A, which is the fraction of time that a user is shadowed. During a shadowed interval, the channel is modeled as log-normal with frequency-

Manuscript received January 15, 1994; revised July 14, 1994. This work was supported in part by TRW Military Electronics & Avionics Division under Grant NB8541VK2S and by the MICRO Program of the State of California. This paper was presented in part at Globecom '94 and Milcom '94.

A. M. Monk is with ComStream Corporation, San Diego, CA 92121 USA.

L. B. Milstein is with the Department of Electrical and Computer Engineering, University of California-San Diego, La Jolla, CA 92093-0407 USA.

IEEE Log Number 9407503.

0733-8716/95$04.00

nonselective Rayleigh fading. During the fraction, $1-A$, of unshadowed time, the channel is modeled as frequency-nonselective Rician, with specular power to scatter power ratio K. In the unshadowed case, multipath fading occurs because, in addition to the direct path from the satellite, there also exists many scattered paths which are reflected from the surrounding terrain. We analyze the general case of log-normal shadowing with Rician fading. This can then be applied to either of the two channel states described above, by varying the standard deviation of the shadowing, σ_ζ, and the value of the Rice-factor, K ($\sigma_\zeta = 0$ dB corresponds to no shadowing and $K=0$ corresponds to Rayleigh fading).

To quantify the performance of the APC, the standard deviation of the power control error in decibels is derived. This is shown to be a critical function of both the measurement time used in the algorithm, and the vehicle velocity. The statistics of the power control error are analyzed and found to be closely approximated by a log-normal random variable.

The paper is organized as follows.The system model and various approximations are described in Section II. In Section III, we analyze the APC performance on a frequency-nonselective Rician channel. In Section IV, we illustrate the use of these results in overall system performance analysis by considering a convolutionally coded CDMA land mobile satellite link and computing the effect of the power control error on capacity. Finally, Section V presents conclusions.

II. System Model and Approximations

We model the channel as frequency-nonselective with Rician statistics, and consider shadowing, modeled with the well-known log-normal distribution. Consistent with some current commercial CDMA systems, the base station is assumed to transmit a pilot tone consisting of a pure spreading sequence (i.e., no data). The pilot tone is used as a coherent reference by the mobiles, and is typically transmitted at a power level which is greater than that of the data. An estimate of the received pilot power, which should be more accurate than a direct estimate of the received data power, can thus easily be converted to an estimate of the data power by a conversion factor. We make use of a standard coherent correlation receiver, and estimate the received shadowing *power* at the mobile by summing consecutive squared outputs of the correlator matched to the pilot sequence.

Also, we consider a single spot beam with the forward link using orthogonal spreading sequences. We assume that multiple-access interference can be ignored due to the orthogonality of the users' sequences. As an example, consider a spread bandwidth of 2 MHz, corresponding to a 1000 ns chip time (T_c). Typical delay spreads over land-mobile satellite channels have been measured around 100 ns, i.e., 0.1 T_c. We can therefore say that the orthogonality of the users is maintained over such a flat fading channel.

Let the received signal amplitude, after downconversion, be given by

$$r(t) = \alpha(t)S(t)c(t) + n(t) \tag{1}$$

where $\alpha(t)$ is a Rician distributed random process representing fading, $S(t)$ represents the amplitude of the shadowing component and is assumed to be lognormally distributed, i.e., $S(t) = e^{\zeta(t)}$, where $\zeta(t)$ is Gaussian distributed with mean $\bar{\zeta}$ (dependent on base-to-mobile distance) and standard deviation σ_ζ[1], and $c(t)$ is the spreading sequence of the user of interest, composed of binary square chips with a rectangular waveform, with chip time equal to T_c. Thermal noise after the downconversion is represented by the quantity $n(t)$, which is zero-mean Gaussian process, with two-sided power spectral density N_0. The amplitude of the correlator output at time kT, where T is the bit time, is given by

$$\begin{aligned} X(k) &= \frac{1}{T}\int_{(k-1)T}^{kT} r(t)c(t)dt \\ &= \frac{1}{T}\int_{(k-1)T}^{kT} [\alpha(t)e^{\zeta(t)} + n(t)c(t)]dt. \end{aligned} \tag{2}$$

In order to obtain tractable analytical solutions, it will be necessary to make some approximations. We will assume that $\alpha(t)$ (equivalently, $\alpha^2(t)$) is relatively constant over one bit time (verified in Section II-A. below). Since $\zeta(t)$ is much more slowly varying than $\alpha(t)$ (verified below), it too is assumed constant over a bit time. Both these variables will be approximated by their value at the midpoint of the bit time. Thus we have, from (2), that

$$X(k) \approx \alpha_k e^{\zeta_k} + N_k \tag{3}$$

where $\alpha_k = \alpha(kT - T/2)$, $\zeta_k = \zeta(kT - T/2)$, and N_k is Gaussian with mean zero and variance $\sigma_N^2 = N_0/T$.

A. Signal Statistics

We need to know the second-order statistics of $\alpha(t)$ and $\zeta(t)$. The Rician probability density function is given by

$$f_\alpha(\alpha) = \frac{\alpha}{\sigma^2}\exp\left(-\frac{\alpha^2 + A_s^2}{2\sigma^2}\right)I_0\left(\frac{\alpha A_s}{\sigma^2}\right) \tag{4}$$

where A_s is the amplitude of the specular path, and $2\sigma^2$ is the average power of the diffuse component. The total received multipath power, P, is given by the mean of the Rician envelope squared, which is

$$P = \mathbf{E}\{\alpha^2(\tau)\} = 2\sigma^2(1+K) \tag{5}$$

where $K = A_s^2/2\sigma^2$ is the ratio of the peak power in the specular component to the power in the diffuse component (the Rice-factor). K has been measured between 4 dB and 18 dB in [3]. We will consider the $K = 10$ dB case.

It is known [4] that with $\alpha(t)$ Rician distributed, the correlation function of $\alpha^2(t)$ is given by

$$\begin{aligned} R_{\alpha^2}(\tau) &= \mathbf{E}\{\alpha^2(t)\alpha^2(t+\tau)\} \\ &= 4\sigma^4\rho(\tau)[\rho(\tau) + 2K\cos 2\pi f_0\tau] + 4\sigma^4(1+K)^2 \\ &= (P/(1+K))^2\rho(\tau)[\rho(\tau) + 2K\cos 2\pi f_0\tau] + P^2. \end{aligned} \tag{6}$$

[1] This mean and standard deviation are in units of nepers. To convert to the more commonly used unit of decibels, multiply by 20/ ln 10.

The quantity f_0 is the Doppler shift of the specular path. Since we assume a coherent downlink, we will take f_0 to be equal to zero. The quantity $\rho(\tau)$ is the normalized autocovariance function of the complex Gaussian random process whose envelope is the diffuse Rayleigh process. In deriving (6), it is assumed that the horizontal components of the received electromagnetic waves have uniform angles of arrival.

If the received waves are travelling only horizontally, i.e., there is no vertical component, then, from [5]

$$\rho(\tau) = \mathrm{J}_0(2\pi f_d|\tau|) \tag{7}$$

where $f_d = v/\lambda$ is the one-sided Doppler bandwidth, or the maximum Doppler shift, v is the velocity of the mobile relative to the base station, and λ is the wavelength (we assume that the system bandwidth is much smaller than the absolute value of the carrier frequency). $\mathrm{J}_0(x)$ is a Bessel function of the first kind of zero order.

In [6], the more general case of angles of arrival in the vertical plane is examined for realistic distributions. It is found that for vertical angles of arrival less than 45°, $\rho(\tau)$ is quite close to (7).

Consider the approximation of $\alpha(t)$, made in (3), i.e., the assumption that $\alpha(t)$ (equivalently $\alpha^2(t)$) is relatively constant over a bit time. The accuracy of this approximation can be verified by computing the normalized autocovariance of $\alpha^2(t)$, obtained from (5) and (6). It is found that, for values of $f_d t < 0.05$, the normalized autocovariance of $\alpha^2(t)$ is greater than 95% (for $f_d t = 0.1$, the autocovariance drops to around 80%); thus, if we maintain the requirement that $f_d T < 0.05$, the fading changes very slowly over the period of a single bit.

Much less well-known are the time-varying statistics of the Gaussian shadowing component $\zeta(t)$. However, analysis of empirical results, reported in [7], shows the following covariance function for $\zeta(t)$ on a land mobile channel

$$\begin{aligned} C_\zeta(\tau) &= \mathbf{E}\{\zeta(t)\zeta(t+\tau)\} - \mathbf{E}^2\{\zeta(t)\} \\ &= \sigma_\zeta^2 e^{-v|\tau|/X_c} \end{aligned} \tag{8}$$

where X_c is the correlation distance. In [7], X_c has been measured as hundreds of feet for conventional terrestrial cells, and tens of feet for terrestrial microcells. In either case, a comparison of the autocovariances of the fading and shadowing shows that the shadowing process is much more slowly varying than the Rician fading process, for velocities in the range of interest. The standard deviation of the shadowing in decibels, σ_ζ(dB), has been measured between 2 dB and 6 dB, and the mean of the shadowing has been measured between −7 dB and −16 dB, for the land mobile satellite channel [3]. We will consider two shadowed cases, corresponding to *light* shadowing (σ_ζ(dB) = 3 dB, or $\sigma_\zeta = 5 \cdot \ln 10/20$ nepers, and $\overline{\zeta}$ (dB) = −7 dB) and *heavy* shadowing (σ_ζ (dB) = 5 dB and $\overline{\zeta}$ (dB) = −11 dB). As mentioned in [8], the above covariance function is an idealization. However, it serves as a method to quantify the effect of vehicle movement on the shadowing parameters. Note that the moments of a log-normal random variable are easily calculated as a function of the mean and variance of the underlying Gaussian distribution. Specifically, if x is Gaussian, with mean $\overline{x}$ and variance σ_x^2, then

$$\mathbf{E}\{e^x\} = \exp\left(\overline{x} + \sigma_x^2/2\right). \tag{9}$$

III. Power Control Error

In this section, we examine the proposed power control algorithm, and define the power control error mathematically. The algorithm produces an estimate of the received power at the mobile by averaging squared outputs of the correlator. In order to estimate the received power, we sum $X^2(k)$, weighted by a constant, c_1, to be defined later. Thus, the estimate at time mT is given by (3) as

$$\begin{aligned} \widehat{S^2}(mT) &= \frac{c_1}{m}\sum_{k=1}^{m} X^2(k) \\ &= \frac{c_1}{m}\sum_{k=1}^{m} \alpha_k^2 e^{2\zeta_k} + 2\frac{c_1}{m}\sum_{k=1}^{m} \alpha_k e^{\zeta_k} N_k \\ &\quad + \frac{c_1}{m}\sum_{k=1}^{m} N_k^2. \end{aligned} \tag{10}$$

The mobile transmitter uses this estimate to modify its transmitted amplitude. Specifically, its transmit amplitude is inversely proportional to $(\widehat{S^2}(mT))^{1/2}$. As explained in the Introduction, the round trip delay on a satellite link is on the order of 10ms or more. Assuming, for example, a 10 kb/s data rate, the delay on the reverse link is greater than $\Delta = 100$ symbol intervals. The received signal at the base station will contain a term similar to (1), except weighted by the inverse of $(\widehat{S^2}(mT))^{1/2}$, and including the effects of the asynchronous multiple-access interference (MAI), the signal amplitude at time $m'T = (m + \Delta)T$ (ignoring spreading sequence and carrier) is

$$\begin{aligned} r'(m'T) &= \frac{\alpha'(m'T)S(m'T)}{(\widehat{S^2}(mT))^{1/2}} + \mathrm{MAI}(m'T) + n'(m'T) \\ &= \gamma\alpha'(m'T) + \mathrm{MAI}(m'T) + n'(m'T) \end{aligned}$$

where γ is defined as the power control error, and the superscript primes indicate reverse link. Notice that the shadowing terms on both links are the same, as explained above, and the fading and noise terms on the two links are different. We can write $\gamma = 1/\sqrt{\epsilon}$, where

$$\begin{aligned} \epsilon = \frac{c_1}{m}\sum_{k=1}^{m} \alpha_k^2 e^{2\zeta_k - 2\zeta_{m+\Delta}} + 2\frac{c_1}{m}\sum_{k=1}^{m} \alpha_k e^{\zeta_k - 2\zeta_{m+\Delta}} N_k \\ + \frac{c_1}{m}\sum_{k=1}^{m} e^{-2\zeta_{m+\Delta}} N_k^2. \end{aligned} \tag{11}$$

We wish to find the distribution of either γ or ϵ. However, due to the presence of the lognormal terms in (11), this does not appear to be feasible. It has been suggested that the distribution of the power control error can be approximated as log-normal [8]. To justify the use of the log-normal approximation we have resorted to simulation techniques. Fig. 1

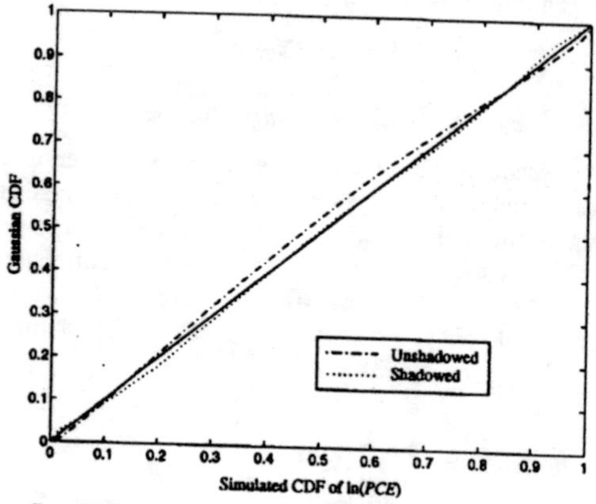

Fig. 1. Comparison of the C.D.F. of the logarithm of the simulated power control error with a Gaussian C.D.F. $m = 25, f_dT = 0.02, E/N_0 = 10\text{dB}$.

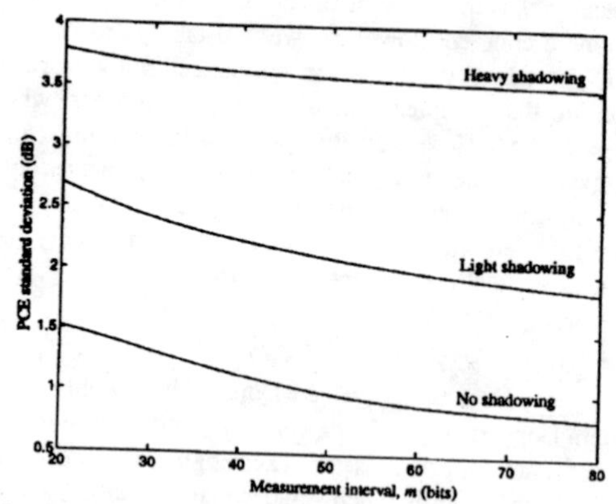

Fig. 2. Variation of PCE standard deviation with measurement interval. $X_c = 300, f_dT = 0.02, E/N_0 = 10\text{dB}$.

compares the cumulative distributions of the simulated power control error in nepers with a Gaussian cumulative distribution function. The two c.d.f.'s are plotted versus each other, so that a perfect Gaussian c.d.f. appears as a straight line. The plots show the unshadowed ($K = 10$ dB) and shadowed ($K = -\infty$ dB, $\sigma_\zeta = 5$ dB, $\overline{\zeta} = -11$ dB, $X_c/\lambda = 300$) cases, and it can be seen that, for both cases, the simulated PCE lies close to the straight line and thus justifies our use of the log-normal distribution as an *approximation* for analytical purposes.

We will therefore follow the approach of [8] end assume that the power control error, γ, is lognormally distributed, i.e., $\gamma = e^\delta$, where δ is Gaussian. It remains, then, to find the statistics of δ, which, as will be shown below, rely only on the first and second moments of either γ or $\epsilon = 1/\gamma^2$.

A. Evaluation of the Log-Normal Parameters

In order to find the parameters of the lognormal distribution which approximates the distribution of ϵ, we must find the first two moments of ϵ. Let $\mathbf{E}\{\epsilon\} = c_1M_1$, and $\mathbf{E}\{\epsilon^2\} = c_1^2V_1$. Then, from (5), (8), (9) and (11)

$$M_1 = P\frac{1}{m}\sum_{k=1}^{m} e^{4\sigma_\zeta^2 - 4C_\zeta((m+\Delta-k)T)} + \sigma_N^2 e^{-2\overline{\zeta}+2\sigma_\zeta^2}. \quad (12)$$

V_1 consists of four terms. The first three correspond to the individually squared terms of (11). The fourth term corresponds to the crossterm from the first and third terms of (11). All other crossterms from (11) result in zero expected values, since they contain odd-order Gaussian expectations. From (6), (8), (9), and (11)

$$\begin{aligned} V_1 = &\frac{1}{m^2}\sum_{k=1}^{m}\sum_{i=1}^{m} R_{\alpha^2}((k-i)T)\exp(\sigma_{\zeta_a}^2/2) \\ &+ \frac{4}{m^2}P\sigma_N^2\sum_{k=1}^{m}\exp(-2\overline{\zeta}+\sigma_{\zeta_b}^2/2) \\ &+ \left(\frac{3}{m}+\frac{m^2-m}{m^2}\right)\sigma_N^4\exp(-4\overline{\zeta}+8\sigma_\zeta^2) \\ &+ \frac{2}{m}P\sigma_N^2\sum_{k=1}^{m}\exp(-2\overline{\zeta}+\sigma_{\zeta_c}^2/2) \end{aligned} \quad (13)$$

where

$$\begin{aligned} \sigma_{\zeta_a}^2 &= \mathbf{Var}\{2\zeta_k - 4\zeta_{m+\Delta} + 2\zeta_i\} \\ &= 24\sigma_\zeta^2 - 16C_\zeta((m+\Delta-k)T) \\ &\quad - 16C_\zeta((m+\Delta-i)T) + 8C_\zeta((k-i)T) \\ \sigma_{\zeta_b}^2 &= \mathbf{Var}\{2\zeta_k - 4\zeta_{m+\Delta}\} \\ &= 20\sigma_\zeta^2 - 16C_\zeta((m+\Delta-k)T) \end{aligned} \quad (14)$$

and $\sigma_{\zeta_c}^2 = \sigma_{\zeta_b}^2$.

As mentioned above, we take $\gamma = e^\delta$, or, in terms of ϵ, $\epsilon = e^{-2\delta}$. The first two moments of ϵ, which were derived in (12) and (13), are related to those of δ by

$$\mathbf{E}\{\epsilon\} = c_1M_1 = e^{-2\overline{\delta}+2\sigma_\delta^2} \quad (15)$$

and

$$\mathbf{E}\{\epsilon^2\} = c_1^2V_1 = e^{-4\overline{\delta}+8\sigma_\delta^2}. \quad (16)$$

From (15) and (16), we find that

$$\overline{\delta} = \frac{1}{2}\ln\left(\sqrt{V_1}/c_1M_1^2\right) \quad (17)$$

$$\sigma_\delta = \frac{1}{2}\sqrt{\ln(V_1/M_1^2)}. \quad (18)$$

The purpose of the filter constant c_1 is now apparent. We would like the power control error to have a mean of 0 dB, corresponding to an unbiased estimate. By choosing c_1 appropriately, as in [8] (i.e., $c_1 = \sqrt{V_1}/M_1^2$), this condition is met, and then the log-normal distribution is specified only by the power control error standard deviation, $\sigma_\delta(20/\ln 10)$ dB.

B. Discussion

Fig. 2 shows the effect of the measurement interval on the power control error standard deviation, σ_δ, for various shadowed and unshadowed scenarios. The top curve is the heavy shadowing case mentioned in Section II. The middle curve corresponds to the lightly shadowed case, and the bottom curve is for unshadowed users. From the plot, one can see that even the unshadowed users experience a nontrivial degree of power control error, i.e., it would be a mistake to assume

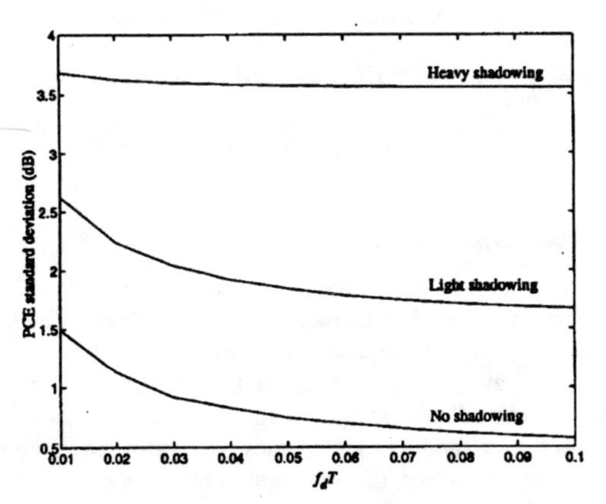

Fig. 3. Variation of PCE standard deviation with f_dT. $m = 40$, $E/N_0 = 10\text{dB}, X_c/\lambda = 300$.

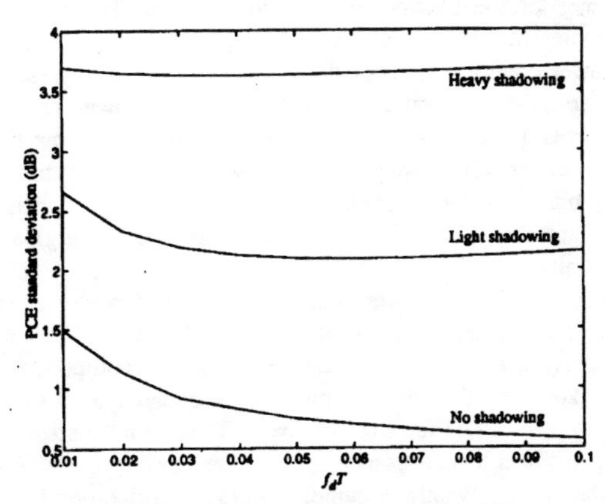

Fig. 4. Variation of PCE standard deviation with f_dT. $m = 40$, $E/N_0 = 10\text{dB}, X_c/\lambda = 60$.

that the unshadowed users have negligible PCE. It can also be seen that the heavily shadowed users have particularly bad PCE performance.

In Fig. 3, the normalized one-sided Doppler bandwidth, f_dT, is varied, holding the measurement interval constant ($m = 40$). The relative positions of the curves are the same as that in Fig. 2, and we have set the normalized correlation distance for the shadowed cases equal to 300 (i.e., very slowly varying). As f_dT increases, corresponding to an increase in the vehicle speed, σ_δ decreases. The reason for this is that an increase in Doppler bandwidth results in an increase in the fading rate of the multipath. Heuristically, this means that we are trying to average out a signal which is fluctuating more rapidly (recall that we are not trying to track the rapidly fluctuating multipath, only the more slowly varying shadowing), and thus the variance of our estimate will decrease. The same explanation is valid when the measurement interval increases in Fig. 2, i.e., over the span of the measurement interval, we observe more fluctuations of the multipath, which results in a reduced variance of the estimate.

In Fig. 4, we examine the effect of decreasing the normalized shadowing correlation distance to $X_c/\lambda = 60$. The plot shows the same initial behavior as the previous plot, for low values of f_dT. However, we see that as the vehicle speed (or f_dT) increases beyond a certain point, σ_δ starts to increase. (There is naturally no change in the unshadowed users from the previous plot.) This can be explained as follows. If the vehicle is moving too fast, when the correlation distance is small, then the local mean of the received signal (modeled by the lognormal shadowing random process) varies more rapidly. Thus, the estimate generated over m symbols at time mT is not a good representation of the channel at time $(m + \Delta)T$. The same effect would be seen if the measurement interval is made too large, when the correlation distance is small.

IV. Effect of PCE on System Capacity

The purpose of this section is to show how the results in the previous sections can be applied to the calculation of system capacity in a land mobile satellite link. We derive the performance of a shadowed user on the uplink, assuming each user employs convolutional coding with perfect interleaving and soft decision Viterbi decoding. The system model is presented below.

A. System Model

We assume a single spot beam with N_u multiple-access users. The channel model was described in the Introduction. To reiterate, the received signal is modeled as Rayleigh and log-normal for a fraction, A, of the time that it is shadowed. For the fraction $1 - A$ that the signal is not shadowed, the channel is modeled as Rician. For the unshadowed case, the Rice-factor, K, is taken to be 10 dB.

We will examine the performance of a shadowed user, and assume, for simplicity, that the shadowing is changing sufficiently slowly so that it can be considered constant over the interval of observation, i.e., we take the shadowing correlation distance to be infinite.

The received signal at the base station, after downconversion, is given by

$$r(t) = \sum_{k=1}^{K} e^{x_k} R_k d_k(t - \tau_k) c_k(t - \tau_k) \cos(\omega_c t + \theta_k) + n_w(t). \tag{19}$$

The thermal noise term, $n_w(t)$, is additive white Gaussian with two-sided power spectral density $N_0/2$. $c_k(t)$ is the spreading sequence of the ith user, generated at rate $1/T_c$, and $d_k(t)$ represents the coded binary data of the ith user, generated at rate $1/T$. The processing gain is defined as $L = T/T_c$, and the spreading sequences for all users are assumed to be random binary sequences. The parameters τ_k and θ_k are random variables representing delays and phases that are independent and uniformly distributed in $[0, T_c]$ and $[0, 2\pi]$, respectively, and ω_c is the carrier frequency in radians per second. The term x_k corresponds to the power control error in nepers of the kth user and is Gaussian distributed with zero mean and standard deviation in nepers of σ_δ, for the fraction of shadowed time, and σ_{δ_u} for the fraction of unshadowed time. The quantity R_k represents the received signal amplitude in the absence of power control error. For the shadowed case, it represents a Rayleigh random variable, and for the unshadowed case, it represents a Rician random variable. Finally, we assume that each user employs a long spreading sequence (i.e., spanning many bits), that the processing gain, L, is much greater than unity, and that the number of users, N_u, is also much greater than unity.

In the absence of power control, the received power of the unshadowed Rician component is $P_u = 2\sigma^2(1+K)$, while the received power in the shadowed Rayleigh component is $P_s = 2\sigma^2$, i.e., the received power is attenuated by a factor $c = 1/(1+K)$. The effect of the APC is to normalize the average power level to compensate for the loss in power due to the shadowing. We also assume, as in [9], that the mobile has the capability to overcompensate its transmit power, to attempt to overcome the additional degradation caused by the Rayleigh fading. Thus, we take the average received power of shadowed users to be $2\sigma^2(1+K)p$, where p is the overcompensation parameter. If $p = 1$, then there is no overcompensation, and the received powers of unshadowed and shadowed users are the same.

Based on the above assumptions, we can write the joint density of the received signal amplitude and power control error (in nepers) as the mixture density (see [2])

$$\begin{aligned} f_{x,R}(x_k, R_k) = {} & A\frac{\exp(-x_k^2/2\sigma_{\delta_s}^2)}{\sqrt{2\pi\sigma_{\delta_s}^2}} \frac{R_k}{\sigma^2(1+K)p} \\ & \times e^{-R_k^2/2\sigma^2(1+K)p} U(R_k) \\ & + (1-A)\frac{\exp(-x_k^2/2\sigma_{\delta_u}^2)}{\sqrt{2\pi\sigma_{\delta_u}^2}} \frac{R_k}{\sigma^2} e^{-R_k^2/2\sigma^2} \\ & \times e^{-K} I_0\left(\sqrt{\frac{2R_k^2 K}{\sigma^2}}\right) U(R_k) \end{aligned} \quad (20)$$

where $U(x)$ is the unit-step function.

A standard ideal correlator receiver is assumed. After demodulation, the test statistic for user 1, at time T, is given by

$$g(T) = e^{x_1} R_1 d_0 + \sum_{k=2}^{N_u} e^{x_k} R_k I_k(T) \cos\theta_k + N'(T) \quad (21)$$

where d_0 is the data bit of the first user, $N'(T)$ is a zero-mean Gaussian random variable with variance N_0/T, and

$$I_k(T) = \frac{1}{T}\int_{\tau_1}^{\tau_1+T} d_k(t-\tau_k) c_k(t-\tau_k) c_1(t-\tau_1)\,dt.$$

It can be shown that, as N_u becomes arbitrarily large, $g(T)$ can be taken to be a Gaussian random variable [10]. For square pulses, the variance of $I_k(T)$ is given by $\frac{2}{3}LT_c^2$, and thus the variance of $g(T)$ is

$$\sigma_g^2 = \frac{N_0}{T} + g_2\frac{N_u - 1}{3L} \quad (22)$$

where g_2 is the second moment of $e^{x_k} R_k$, given by

$$\begin{aligned} g_2 &= \int_0^{\infty}\int_{-\infty}^{\infty} e^{2x_k} R_k^2 f_{x,R}(x_k, R_k)\, dx_k\, dR_k \\ &= A e^{2\sigma_{\delta_s}^2} 2\sigma^2(1+K)p + (1-A) e^{2\sigma_{\delta_u}^2} 2\sigma^2(1+K). \end{aligned} \quad (23)$$

B. Coded Performance

We assume that the channel can be well-estimated, i.e., the base station has perfect knowledge of the uplink channel—a common assumption for analytical purposes. We also assume ideal interleaving with respect to the fading. This results in independent Rayleigh fading for each code symbol of a shadowed user and clearly precludes the case where a mobile is stationary; however, this assumption is also a common one for analytical purposes. In contrast, the power control error is assumed to be much more slowly changing, and thus the interleaving is assumed ineffective with respect to the PCE. In other words, over the range of code symbols observed for purposes of calculating the bit error probability, P_b, the PCE is assumed constant.

The approach taken is similar to that of [2]. The path metric after n nodes is given by $\sum_{k=i}^{n} R_{1i} g_i(T) c_{1i}^{(m)}$, where the subscript i corresponds to the ith coded symbol time, and $c_{1i}^{(m)}$ is the ith symbol of the mth path in the decoding trellis. In order to evaluate the well known transfer function bound on the bit error probability, we are required to compute the pairwise error probability, $P(d)$, i.e., the probability that a path metric with d symbol errors exceeds the correct path metric. An upper bound on the bit error probability is

$$P_b \leq \sum_{d=d_f}^{\infty} \beta_d P(d) \quad (24)$$

where d_f is the free distance of the convolutional code, and $\{\beta_d\}$ are the code weights obtained from the code transfer function [11]. We make use of the first 18 code weights [12]. The result is, therefore, an approximation to (24); however, since the sum converges fairly rapidly for bit error rates less than 10^{-3}, the approximation is quite good below that value.

The conditional pairwise error probability, conditioned on x_1 and R_{1i}, is given by

$$P(d \mid x_1, R_{\{1i\}}) = Q\left(\sqrt{2z}\right) \quad (25)$$

where

$$z = \frac{\exp(2x_1)}{2\sigma_g^2}\sum_{i}^{d} R_{1i}^2$$

and the summation is over only those symbols which differ in d out of n positions with the correct path. The quantity z is a conditional chi-square random variable, conditioned on x_1, with $2d$ degrees of freedom. Removing the conditioning of $\{R_{1i}\}$ on (25) is obtained by integrating over the above chi-square distribution. The result is [13, p. 722]

$$P(d \mid x_1) = \left(\frac{1-\nu}{2}\right)^d \sum_{k=0}^{d-1}\binom{d-1+k}{k}\left(\frac{1+\nu}{2}\right)^k \quad (26)$$

where

$$\nu = \sqrt{\frac{\exp(2x_1)\overline{\gamma}_c}{1+\exp(2x_1)\overline{\gamma}_c}}$$

and

$$\overline{\gamma}_c = \frac{2\sigma^2(1+K)p}{2\sigma_g^2} = \left[\frac{N_0}{2E} + 2\frac{N_u - 1}{3L}\left(Ae^{2\sigma_{\delta_s}^2} + (1-A)e^{2\sigma_{\delta_u}^2}/p\right)\right]^{-1}. \tag{27}$$

We define the energy per information bit of the unshadowed users as $E_b = 2\sigma^2(1+K)T/2r$, where r is the code rate. Thus the energy per information bit of the shadowed users is given by $E = pE_b$, i.e., if E_b/N_0 in the unshadowed case is nominally 7 dB, and the overcompensation parameter is $p = 2$, then the shadowed users have $E/N_0 = 10$ dB. Note that the processing gain, $L = T/T_c$, is defined relative to the coded symbols.

The unconditional pairwise probability of error cannot be found in closed form, due to the presence of the log-normal component, and must be calculated by numerically integrating (26) over the Gaussian density function of the power control error in nepers, i.e.

$$P(d) = \int_{-\infty}^{\infty} P(d \mid x_1)\frac{1}{\sqrt{2\pi\sigma_\delta^2}}\exp(-x_1^2/2\sigma_\delta^2)dx_1. \tag{28}$$

The performance for unshadowed users can be found in a similar way, except that the computation of the conditional pairwise error probability, conditioned on the shadowing, cannot be found in closed form, as in (26). However, asymptotically tight bounds have been found (see, for example, [14]). These results are not presented here.

C. Results

In all cases, we have assumed the use of the optimal constraint length 8, rate 1/3 convolutional code [12]. We also chose the shadowing percentage parameter to be $A = 0.3$, as in [2].

Figs. 5 and 6 show the performance of shadowed users versus the standard deviation of the power control error in decibels for $E/N_0 = 10$ dB ($p = 2$) and $E/N_0 = 13$ dB ($p = 4$), respectively, (see (27)). The plots are parameterized by the number of users, normalized by the processing gain, where we have assumed that $N_u - 1 \approx N_u$ (true for a large number of users). The curves should be examined in conjunction with the results derived above for the APC performance. For example, given a vehicle speed of 60 mph, a 1.8 GHz link, and a 10 kb/s bit rate, we obtain $f_dT \approx 0.02$. From Fig. 2, we see that for a measurement time of about 50 T, the standard deviation of the power control error is around 1 dB for unshadowed users, and between about 2 dB and 4 dB for shadowed users. From Fig. 5, it can be seen that the performance of the shadowed users is very poor, even with a 3 dB boost in received power over unshadowed users. For example, assuming shadowed users have a 3 dB PCE standard deviation, then no shadowed users

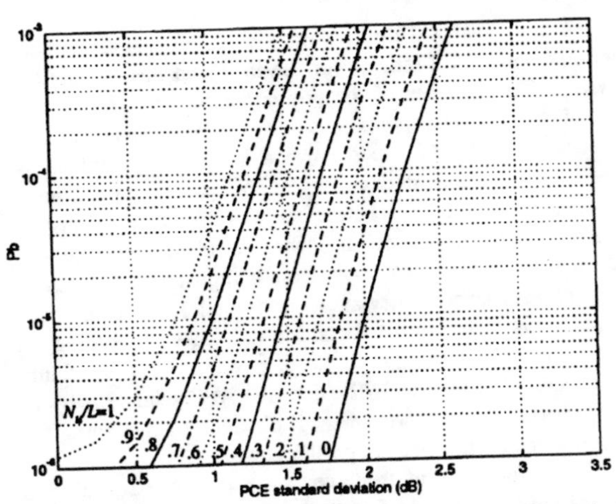

Fig. 5. Upper bound on the bit error probability versus PCE standard deviation. E/N_0 = 10dB (overcompensation factor, p = 2).

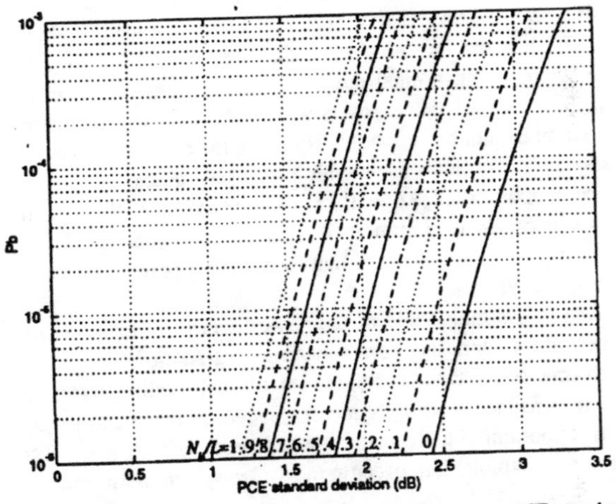

Fig. 6. Upper bound on the bit error probability versus PCE standard deviation. E/N_0 = 13dB (overcompensation factor, p = 4).

can be supported at a bit error rate of even 10^{-2}. Alternatively, if we look at a PCE standard deviation of 2 dB, i.e., lightly shadowed users, it can be seen that $N_u/L \approx 0.45$ can be supported (recall that L is the processing gain relative to coded symbols). In addition, although overcompensation of shadowed users' received power helps the shadowed user performance, it hurts the performance of unshadowed users, and the result is a decrease in the overall system capacity, as described in [9]. From Fig. 6, the performance has been improved so that shadowed users with a power control error standard deviation of 3 dB can be supported at a BER of 10^{-3} when $N_u/L \approx 0.15$. However, if the system can support a maximum PCE standard deviation of 2 dB, then the number of users can be increased by a factor of approximately 7.

Although the bound on the coded bit error rate is not accurate above 10^{-3}, it can be seen that, if the trend of the curves continues above 10^{-3}, relaxing the BER requirement for shadowed users will increase the system capacity.

V. Conclusions

We have analyzed the performance of an adaptive open-loop power control scheme for use in a land mobile satellite system. The statistics of the power control error have been found to be well approximated by a log-normal random process, and the APC performance has been quantified by deriving the standard deviation of the power control error in terms of various parameters, such as Rice-factor, Doppler spread, measurement time, shadowing correlation distance, and shadowing standard deviation. Results indicate typical power control error standard deviations of around 1 dB for unshadowed users, and 2–4 dB for shadowed users, depending on the degree of shadowing, choice of power control algorithm, and received pilot-tone SNR. Since the APC scheme is open-loop, performance improvement is obtained by averaging out the fast fading fluctuations on the downlink.

The performance of shadowed users is highly sensitive to power control error. In fact, in order to support a sufficient number of shadowed users, it may be necessary to boost their received power to 3–6 dB above that of the unshadowed users. For example, with shadowed user power overcompensated by 3 dB, and if shadowed users (30%) are assumed to have a PCE standard deviation of 2 dB, then at 10^{-3} BER, the number of users, normalized by the processing gain, can be $N_u/L \approx 0.45$. However, users with even 3 dB PCE standard deviation cannot be supported. Boosting the overcompensation factor to 6 dB allows users with 3 dB PCE standard deviation to be supported ($N_u/L = 0.15$). Alternatively, if the system can support a maximum PCE standard deviation of 2 dB, then the number of users that can be supported increases dramatically.

Due to the magnitude of the power control error for shadowed users, and the corresponding degradation in system performance for those users, the use of dual satellite diversity, where possible, is strongly recommended, as discussed in [9]. Any opportunity to have an unshadowed path to a satellite should be exploited, in order to maximize system performance.

Acknowledgment

The authors wish to thank Dr. A. J. Goldsmith and Prof. J. Holtzman for their ideas and suggestions, and the reviewers for their insightful comments and constructive criticism.

References

[1] A. M. Monk and L. B. Milstein, "A CDMA cellular system in a mobile base station environment," in *Proc. IEEE GLOBECOM*, vol. 4, Nov. 1993, pp. 65–69.

[2] B. R. Vojcic, R. L. Pickholtz, and L. B. Milstein, "Performance of DS-CDMA with imperfect power control operating over a low earth orbiting satellite link," *J. Select. Areas Commun.*, vol. 12, pp. 560–567, May 1994.

[3] E. Lutz, D. Cygan, M. Dippold, F. Dolainsky, and W. Papke, "The land mobile satellite communication channel—recording, statistics and channel model," *IEEE Trans. Veh. Technol.*, vol. 40, pp. 375–386, May 1991.

[4] T. Aulin, "A modified model for the fading signal at a mobile radio channel," *IEEE Trans. Veh. Technol.*, vol. VT-28, pp. 182–203, Aug. 1979.

[5] R. H. Clarke, "A statistical theory of mobile radio reception," *Bell Syst. Tech. J.*, vol. 47, pp. 957–1000, July 1968.

[6] J. D. Parsons and A. M. D. Turkmani, "Characterisation of mobile radio signals: model description," in *IEE Proc.—I*, vol. 138, Dec. 1991, pp. 549–556.

[7] M. Gudmundson, "Correlation model for shadow fading in mobile radio systems," *Electr. Letts.*, vol. 27, pp. 2145–2146, Nov. 1991.

[8] A. J. Goldsmith, L. J. Greenstein, and G. J. Foschini, "Error statistics of real-time power measurements in cellular channels with multipath and shadowing," *IEEE Trans. Veh. Technol.*, vol. 43, pp. 439–446, Aug. 1994.

[9] B. R. Vojcic, L. B. Milstein, and R. L. Pickholtz, "Power control versus capacity of a CDMA system operating over a low earth orbiting satellite link," in *Proc. IEEE GLOBECOM*, vol. 4, Nov. 1993, pp. 40–41.

[10] A. M. Monk, *Increased Performance and Flexibility in a CDMA System*, Ph.D. dissertation, Univ. of CA, San Diego, 1994.

[11] A. J. Viterbi, "Convolutional codes and their performance in communication systems," *IEEE Trans. Commun. Technol.*, vol. COM-19, pp. 751–771, Oct. 1971.

[12] J. Conan, "The weight spectra of some short low-rate convolutional codes," *IEEE Trans. Commun.*, vol. COM-32, pp. 1050–1053, Sept. 1984.

[13] J. G. Proakis, *Digital Communications*, 2nd ed. New York: McGraw-Hill, 1989.

[14] Y.-L. Chen and C.-H. Wei, "On the performance of rate 1/2 convolutional codes with QPSK on Rician fading channels," *IEEE Trans. Veh. Technol.*, vol. 39, pp. 161–170, May 1990.

Anton M. Monk (S'89) received the B.S. degree from the University of California at San Diego, La Jolla, CA, in 1989, the M.S. degree from the California Institute of Technology, Pasadena, in 1990, and the Ph.D. degree from the University of California at San Diego, in 1994, all in electrical engineering.

Laurence B. Milstein (S'66–M'68–SM'77–F'85) received the B.E.E. degree from the City College of New York, New York, NY, in 1964, and the M.S. and Ph.D. degrees in electrical engineering from the Polytechnic Institute of Brooklyn, Brooklyn, NY, in 1966 and 1968, respectively.

From 1968 to 1974, he was with the Space and Communications Group of Hughes Aircraft Company, and from 1974 to 1976, he was a member of the Department of Electrical and Systems Engineering, Rensselaer Polytechnic Institute, Troy, NY. Since 1976, he has been with the Department of Electrical and Computer Engineering, University of California at San Diego, La Jolla, CA, where he is a Professor and former Department Chairman, working in the area of digital communication theory with special emphasis on spread-spectrum communication systems. He has also been a consultant to both government and industry in the areas of radar and communications.

Dr. Milstein has been an Associate Editor for Communication Theory for the IEEE Transactions on Communications, an Associate Editor for Book Reviews for the IEEE Transactions on Information Theory, and is currently Senior Editor for the IEEE Journal on Selected Areas in Communications. He was the Vice President for Technical Affairs in 1990 and 1991 of the IEEE Communications Society and is currently a member of the Board of Governors of both the IEEE Communications Society and the IEEE Information Theory Society. He is a member of Eta Kappa Nu and Tau Beta Pi.

IEEE JOURNAL ON SELECTED AREAS IN COMMUNICATIONS, VOL. 12, NO. 4, MAY 1994

Performance of DS-CDMA with Imperfect Power Control Operating Over a Low Earth Orbiting Satellite Link

Branimir R. Vojcic, *Member, IEEE*, Raymond L. Pickholtz, *Fellow, IEEE*, and Laurence B. Milstein, *Fellow, IEEE*

***Abstract*—The analysis of both performance and capacity of direct sequence CDMA in terrestrial cellular systems has been addressed in the technical literature. It has been suggested that CDMA be used as a multiple access method for satellite systems as well, in particular for multispot beam Low Earth Orbit Satellites (LEOS). One is tempted to argue that since CDMA works well on terrestrial links, it will nominally work as well on satellite links. However, because there are fundamental differences in the characteristics of the two channels, such as larger time delays from the mobile to the base station and smaller multipath delay spreads on the satellite channels, the performance of CDMA on satellite links cannot always be accurately predicted from its performance on terrestrial channels. In this paper, we analytically derive the performance of a CDMA system which operates over a low earth orbiting satellite channel. We incorporate such effects as imperfect power control and dual-order diversity to obtain the average probability of error of a single user.**

I. Introduction

It is desired to determine the effect of imperfect power control on a satellite-based mobile communication system employing CDMA with coding and interleaving. It is assumed that the satellite is in a low earth orbit and that it segments the ground being illuminated by employing multiple spot beams. Direct sequence (DS) CDMA is used, and perfectly coherent reception is assumed. The two characteristics of the satellite channel most relevant to system performance are small multipath delay spread and the large propagation delay.

Consider the delay due to the large propagation time of the satellite channel. This is of fundamental importance to cellular communications based on a DS-CDMA signal structure because of the need for accurate power control. Power control systems can be open loop or closed loop. In the former case, the mobile adjusts its own transmit power based on the level of its receive power. In the latter case, the base station commands the mobile to either increase or decrease its power; the comand from the base station is based upon the level of the signal it receives from the mobile. Since the round-trip delay in a frequency translating LEOS systems is on the order of 10 to 20 ms, closed-loop power control is much less effective than it is on a terrestrial channel. In turn, this implies that a satellite-based system will have to rely much more extensively on open loop control, and the accuracy with which such open loop control can be made to work is yet to be demonstrated. Further, the large time delay inherent with satellite communications can lead to a secondary effect, namely imperfect interleaving. This effect arises because satisfactory voice communications can tolerate just a certain amount of delay; since the satellite link starts out with a nontrivial delay, additional delay due to the interleaving might have to be less than what it would be over a terrestrial link, and therefore the decoder is more susceptible to correlated fades in the channel symbols.

With respect to multipath delay spread, recall that for a terrestrial suburban or rural channel, the delay might be on the order of several microseconds. In low-altitude satellite channels, the delay spread is more typically less than 100 ns; thus, the coherence bandwidth of the satellite multipath channel is at least 10 MHz. This means that any CDMA system design that seriously intended to make constructive use of the spreading to combat multipath would have to be spreading by an amount greater than 10 MHz. The paper is organized in the following manner. The channel model is described in Section II, and the average probability of error of an uncoded system is derived in Section III. In Section IV, the results of using dual diversity are presented. The analyses of both Sections III and IV are extended in Section V to incorporate soft decision convolutional coding; both perfect and imperfect interleaving are considered. Numerical results are presented in Section VI, and our conclusions are given in Section VII.

II. Channel Model

The channel model is taken from [1] to be Rayleigh a fraction B of the time, corresponding to when the signal is shadowed, and Rician the fraction $1-B$ of the time. That is, the probability density function of the received amplitude is given by the mixture density

$$f_R(R) = B\frac{R}{\sigma^2}e^{-\frac{R^2}{2\sigma^2}} + (1-B)\frac{R}{\sigma^2}e^{-\frac{R^2+A_s^2}{2\sigma^2}} I_0\left(\frac{RA_s}{\sigma^2}\right), \quad (1)$$

where A_s is the amplitude of the specular component of the Rician part of the density, $2\sigma^2$ is the average power in the scatter component of the fade, and $I_0(\cdot)$ is the modified Bessel function of the first kind and zeroth order.

Manuscript received June 28, 1993; revised December 30, 1993. This paper was presented at MILCOM '93, Boston, MA, October 11-15, 1993.

B.R. Vojcic and R.L. Pickholtz are with the Department of Electrical Engineering and Computer Science, The George Washington University, Washington, D.C. 20052 USA.

L.B. Milstein is with the Electrical and Computer Engineering Department, University of California at San Diego, La Jolla, CA 92093 USA.

IEEE Log Number 9215432.

0733-8716/94$04.00 © 1994 IEEE

In order to conform to the notation introduced in [1], we define:

$$c = \frac{A_s^2}{2\sigma^2}, \quad (2)$$

and also set $A_s = 1$, so that $c = 1/2\sigma^2$. Then:

$$f_R(R) = 2BRce^{-cR^2} + 2(1-B)Rce^{-c(R^2+1)}I_0(2Rc). \quad (3)$$

For simplicity in our analysis, we assume that power control is normalized to a user which is not shadowed. That is, assume in the absence of either fading or shadowing of any type, a signal arrives at the receiver with a nominal power denoted S_{nom}. Then, in the absence of shadowing, but in the presence of the Rician fade, the average received power is

$$S_{av_{ns}} = S_{\text{nom}}(A_s^2 + 2\sigma^2) = S_{\text{nom}}\left(1 + \frac{1}{c}\right). \quad (4)$$

If the user is shadowed, the received power is given by

$$S_{av_s} = S_{\text{nom}}(2\sigma^2) = S_{\text{nom}}\left(\frac{1}{c}\right) \quad (5)$$

Assuming that the power control algorithm is not fast enough to track power variations due to the pure multipath fading (i.e., fading not caused by shadowing) but is fast enough to track the variations due to shadowing, then during that fraction of the time when the signal is shadowed, its transmitted power is multiplied by

$$p = \frac{S_{av_{ns}}}{S_{av_s}} = \frac{S_{\text{nom}}(1+\frac{1}{c})}{S_{\text{nom}}(\frac{1}{c})} = 1 + c \quad (6)$$

III. Analysis of Average Probability of Error

Consider the model of the mobile-to-base link in [2], but altered to account for the fact that we are interested in a satellite link and [2] was concerned with a terrestrial link. Assume each spot beam has K simultaneously active users, and there are $J+1$ spot beams which are simultaneously operating. Then, the received signal is given by

$$\begin{aligned} r(t) = &\sum_{i=1}^{K} A_i R_i d_i(t-\tau_i) PN_i(t-\tau_i)\cos(\omega_0 t + \Theta_i) \\ &+ \sum_{i=K+1}^{(J+1)K} A_i R_i \beta_i d_i(t-\tau_i) PN_i(t-\tau_i) \\ &\times \cos(\omega_0 t + \Theta_i) + n_w(t), \end{aligned} \quad (7)$$

where $PN_i(t)$ is the spreading sequence of the ith user, $d_i(t)$ is the binary data of the ith user, the sets $\{\tau_i\}$ and $\{\Theta_i\}$ represent independent time delays and *rf* phases, respectively, and the $\{R_i\}$ correspond to independent flat fading on each user. The parameters $\{\beta_i\}$ represent the discrimination due to spot beam antenna patterns. Note that, for simplicity, it is assumed that all users in the spot beam of interest are illuminated with a gain of unity[1]. The $\{A_i\}$ represent the effect of imperfect power control; if power control was perfect, we would have $A_i = A$, all i. Finally, $n_w(t)$ is additive white Gaussian noise of two-sided power spectral density $N_0/2$.

[1]This is valid for isoflux antennas. However, it should not be inferred that all users in the footprint (other spots) are thereby equally illuminated.

Assuming the ideal correlation receiver, the test statistic for user #1 is given by

$$\begin{aligned} g_1(T) = &A_1 R_1 T d_0 + \sum_{i=2}^{K} A_i R_i I_i(T)\cos\Theta_i \\ &+ \sum_{i=K+1}^{(J+1)K} A_i R_i \beta_i I_i(T)\cos\Theta_i + N(T), \end{aligned} \quad (8)$$

where T is the data bit duration, $N(T)$ is a zero-mean Gaussian random variable with variance $N_0 T$, and

$$I_i(T) = \int_0^T d_i(t-\tau_i) PN_i(t-\tau_i) PN_1(t) dt \quad (9)$$

For large K and large L, where L is the processing gain defined as $L = T/T_c$, T_c being the chip duration, we can model the total multiple access interference as a zero-mean Gaussian random variable. In particular, if we define the second moment of $A_i R_i$ to be $m_2 = E[A_i^2 R_i^2]$, then the variance of the multiple access interference is given by

$$\sigma_I^2 = m_2 \frac{LT_c^2}{3}\left(K - 1 + \sum_{i=K+1}^{(J+1)K} \beta_i^2\right), \quad (10)$$

where rectangular data symbols and rectangular chips of the spreading sequence have been assumed in (10).

To evaluate the second moment needed in (10), we modify the probability density of (1) in the following manner. Since the inaccuracy of the power control is a function of whether or not the user is shadowed (i.e., the power control error is, in general, different for a shadowed user than for a nonshadowed user), we have a joint mixture density of received signal amplitude and power control error of the form

$$f_{R_i,A_i}(R_i, A_i) = Bf_2(A_i)f_3(R_i) + (1-B)f_1(A_i)f_4(R_i), \quad (11)$$

where

$$f_1(A_i) = \begin{cases} \frac{1}{2V_1}, & A - V_1 \le A_i \le A + V_1 \\ 0, & \text{elsewhere} \end{cases}, \quad (12)$$

$$f_2(A_i) = \begin{cases} \frac{1}{2V_2}, & A - V_2 \le A_i \le A + V_2 \\ 0, & \text{elsewhere} \end{cases}, \quad (13)$$

$$f_3(R) = \frac{R}{\sigma^2(1+c)} e^{-\frac{R^2}{2(1+c)\sigma^2}} = \frac{2Rc}{1+c} e^{-\frac{cR^2}{1+c}}, \quad (14)$$

and

$$f_4(R) = 2Rce^{-c(R^2+1)} I_0(2Rc). \quad (15)$$

Note that $f_3(R)$ is the Rayleigh component of (1), but adjusted to account for the power control of (6). Further, note that the reason the probability densities expressed by (12) and (13) are taken to be uniform is to have them correspond to a maximal degree of uncertainty for a given maximum power control error. That is, since the results on the accuracy with which open-loop power control for a low earth orbiting satellite system can be implemented do not appear to be available in the literature, we chose the uniform distribution with a maximum absolute power control error.

If the expected value of $A_i^2 R_i^2$ is now computed from (11), one obtains:

$$m_2 = E[A_i^2 R_i^2] = A^2\left(1+\frac{1}{c}\right) \times \left[(1-B)\left(1+\frac{\gamma_1^2}{3}\right) + B\left(1+\frac{\gamma_2^2}{3}\right)\right], \tag{16}$$

where

$$\gamma_1 = \frac{V_1}{A} \tag{17}$$

and

$$\gamma_2 = \frac{V_2}{A} \tag{18}$$

If we now define the normalized random variable $\lambda_1 = A_1/A$, then conditioned on λ_1 and R_1, the probability of error is given by

$$P(e \mid R_1, A_1) = Q\left(\frac{\lambda_1 R_1}{\sigma_{\text{tot}}}\right), \tag{19}$$

where

$$\sigma_{\text{tot}}^2 = \frac{N_0}{2E} + \frac{1+\frac{1}{c}}{3L}\left[1+(1-B)\frac{\gamma_1^2}{3} + B\frac{\gamma_2^2}{3}\right] \times \left(K-1+\sum_{i=K+1}^{(J+1)K}\beta_i^2\right), \tag{20}$$

$$E = \frac{A^2 T}{2} \tag{21}$$

and

$$Q(x) = \frac{1}{\sqrt{2\pi}}\int_x^\infty e^{-\frac{y^2}{2}}dy \tag{22}$$

The average probability of error is now given by:

$$P_e = \int\int Q\left(\frac{\lambda_1 R_1}{\sigma_{\text{tot}}}\right) f_{R_1,A_1}(R_1, A_1)\,dR_1\,dA_1 \tag{23}$$

Using (11) in (23), after a considerable amount of algebra, the average probability of error can be shown to be given by

$$P_e = (1-B)P_1 + BP_2, \tag{24}$$

where

$$\begin{aligned} P_1 = \frac{1}{2\gamma_1}\Bigg\{&(1+\gamma_1)Q_m(u_1,w_1) - (1-\gamma_1)Q_m(u_2,w_2) \\ &- e^{-\frac{c(c+2a_1)}{2(c+a_1)}} I_0\left(\frac{c^2}{2(c+a_1)}\right) \times \left[\frac{1+\gamma_1}{2}\left(1+\sqrt{\frac{a_1}{a_1+c}}\right) + \sqrt{\frac{c^2\sigma_{\text{tot}}^2}{2(c+a_1)}}\right] \\ &+ e^{-\frac{c(c+2a_2)}{2(c+a_2)}} I_0\left(\frac{c^2}{2(c+a_2)}\right) \times \left[\frac{1-\gamma_1}{2}\left(1+\sqrt{\frac{a_2}{a_2+c}}\right) + \sqrt{\frac{c^2\sigma_{\text{tot}}^2}{2(c+a_2)}}\right]\Bigg\}, \end{aligned} \tag{25}$$

and

$$P_2 = \frac{1}{2}\left[1 - \frac{1}{2\gamma_2}\left(\sqrt{2c\sigma_{\text{tot}}^2 + (1+\gamma_2)^2} - \sqrt{2c\sigma_{\text{tot}}^2 + (1-\gamma_2)^2}\right)\right]. \tag{26}$$

In (25), $Q_m(a,b)$ represents the Marcum Q function defined as:

$$Q_m(a,b) = \int_b^\infty e^{-\frac{a^2+x^2}{2}} I_0(ax)\,x\,dx. \tag{27}$$

Also,

$$a_1 = \frac{(1+\gamma_1)^2}{2\sigma_{\text{tot}}^2}, \tag{28}$$

$$a_2 = \frac{(1-\gamma_1)^2}{2\sigma_{\text{tot}}^2}, \tag{29}$$

$$u_1 = \left[\frac{c\left(c+2a_1-2\sqrt{a_1(c+a_1)}\right)}{2(c+a_1)}\right]^{1/2}, \tag{30}$$

$$w_1 = \left[\frac{c\left(c+2a_1+2\sqrt{a_1(c+a_1)}\right)}{2(c+a_1)}\right]^{1/2}, \tag{31}$$

and u_2 and w_2 are obtained by replacing a_1 with a_2 in (30) and (31), respectively.

IV. Effect of Second-Order Diversity

Considering further the mobile-to-base link, assume the base station employs dual diversity. The diversity is accomplished by assuming that each mobile is always illuminated by two satellites. The transmission from the mobile is received by each of the satellites and retransmitted to a gateway on the ground; it is at the gateway that the diversity reception is actually performed. Using the same notation as in (7), it can be shown that the final test statistic, with maximal-ratio combining, is given by

$$\begin{aligned} g(T) = {}& AR_{11}^2 T d_0 + AR_{11}\sum_{i=2}^{(J+1)K} R_{i1}\beta_i I_{i1}\cos\theta_{i1} + R_{11}N_1 \\ &+ AR_{12}^2 T d_0 + AR_{12}\sum_{i=2}^{(J+1)K} R_{i2}\beta_i I_{i2}\cos\theta_{i2} + R_{12}N_2, \end{aligned} \tag{32}$$

where R_{ij} represents the effect of fading on the ith user as seen by the jth antenna of user #1, where $i = 2,3,\ldots,(J+1)K$, and $j = 1,2$. The remaining symbols in (32) are obvious extensions of the corresponding symbols in (7). For example,

$$I_{ij} = \int_{\tau_{ij}}^{T+\tau_{ij}} PN_i(t-\tau_{ij})d_i(t-\tau_{ij})PN_1(t-\tau_{1j})dt, \tag{33}$$

for $i = 2,3,\ldots,(J+1)K$ and $j = 1,2$.

Following the derivation for the base-to-mobile link in [2], it can be shown that for our case (i.e., the mobile-to-base link), the conditional probability of error, conditioned upon R_{11} and

R_{12}, is approximately given by

$$P(e \mid R_{11}, R_{12}) = Q\left(\sqrt{\frac{R_{11}^2 + R_{12}^2}{\frac{N_0}{2E} + \frac{1+1/c}{3L}\sum_{i=2}^{(J+1)K}\beta_i^2}}\right), \quad (34)$$

where we have initially assumed perfect power control. Since $R_{11}^2 + R_{12}^2$ is a chi-squared random variable with four degrees of freedom, we have:

$$P_e = \int_0^\infty c^2 x e^{-cx} Q\left(\sqrt{\frac{x}{\sigma_T^2}}\right) dx, \quad (35)$$

where

$$\sigma_T^2 = \frac{N_0}{2E(1+c)} + \frac{1/c}{3L}\left(K - 1 + \sum_{i=K+1}^{(J+1)K} \beta_i^2\right) \quad (36)$$

Performing the integration indicated in (35) yields:

$$P_e = \frac{1}{2} - \frac{3}{4(1+2c\sigma_T^2)^{1/2}} + \frac{1}{4(1+2c\sigma_T^2)^{3/2}} \quad (37)$$

V. Coded Performance

In this section, we derive the performance of the mobile-to-base link when convolutional encoding and interleaving is employed at the transmitter, and soft decision decoding is used at the receiver. We evaluate cases of perfect as well as imperfect interleaving. Consider first the case when the interleaving is perfect. The model is the same as in (7), except that now the data $d_i(t)$ corresponds to an encoded sequence.

To evaluate the system performance, we use the standard transfer function bound approach [3]. As is well known, this requires us to compute the pairwise error probability of the channel. Therefore, assume a given sequence $\{c_i\}$ of n-coded symbols is transmitted, and it is desired to find the probability that an alternate sequence, say $\{\overline{c_i}\}$, having Hamming distance d from $\{c_i\}$, is decoded. Denoting this probability as $P_{c\bar{c}}(d)$, we have the conditional probability of a pairwise error, conditioned upon the sequence of faded amplitudes $\{R_{1i}\}$ and the power control error A_1 (assumed the same for the entire codeword), given by

$$P_{c\bar{c}}(A_1, \{R_{1i}\}; d) = P\left(\sum_{i=1}^{d} R_{1i} g_{1i} c_{1i} \le \sum_{i=1}^{d} R_{1i} g_{1i} \overline{c_{1i}} \mid A_1, \{R_{1i}\}\right). \quad (38)$$

In (38), R_{1i} is the faded amplitude of the ith symbol of the desired signal (assumed to be waveform #1) having code sequence $\{c_{1i}\}$ and such that the ith symbol of $\{c_{1i}\}$ differs from the ith symbol of $\{\overline{c_{1i}}\}$; there are d such symbols on a given path of length n through the trellis, and, for notational simplicity, we label them from $i = 1$ to $i = d$, assuming a memoryless channel. Similarly, g_{1i} is the correlator output voltage of the entire received waveform at the ith relevant sampling instant (i.e., g_{1i} is $g_1(T)$ of (8) at the appropriate sampling time) and A_1 incorporates the power control error.

From the analysis presented in Section III, it is straightforward to show that

$$P_{c\bar{c}}(A_1, \{R_{1i}\}; d) = Q\left(\sqrt{2z}\right), \quad (39)$$

where

$$z = \frac{\lambda_1^2}{2\sigma_{tot}^2}\sum_{i=1}^{d} R_{1i}^2. \quad (40)$$

To obtain $P_{c\bar{c}}(d)$, we must average (39) over λ_1 and $\{R_{1i}\}$. To do this, we assume the system must be designed to handle the most disadvantaged user, namely one who is always shadowed. In that case, applying average power control, the signal sees Rayleigh fading, although the multiple access interference still sees the combination Rayleigh/Rician fading.

With this assumption, we note that z is conditionally a chi-squared random variable, conditioned on λ_1. If (39) is averaged over λ_1 where, analogous to (12) to (13), we take λ_1 to be uniform over $[1-\gamma, 1+\gamma]$, we obtain, after some algebra

$$\begin{aligned} P_{c\bar{c}}(\{R_{1i}\}; d) = \frac{1}{2\gamma}\Big[&(1+\gamma)Q\left(\sqrt{2z'(1+\gamma)^2}\right) \\ &- (1-\gamma)Q\left(\sqrt{2z'(1-\gamma)^2}\right)\Big] \\ &- \frac{1}{\sqrt{2\pi}}\left[\frac{(1+\gamma)}{\sqrt{2z'(1+\gamma)^2}} e^{-(1+\gamma)^2 z'} \right. \\ &\left. - \frac{(1-\gamma)}{\sqrt{2z'(1-\gamma)^2}} e^{-(1-\gamma)^2 z'}\right], \end{aligned} \quad (41)$$

where:

$$z' = \frac{1}{2\sigma_{tot}^2}\sum_{i=1}^{d} R_i^2 \quad (42)$$

and $\gamma = V/A$.

The unconditional pairwise error probability, $P_{c\bar{c}}(d)$, is given by

$$\begin{aligned} P_{c\bar{c}}(d) = \frac{1}{2\gamma}[&(1+\gamma)(G_1(d) - H_1(d)) \\ &- (1-\gamma)(G_2(d) - H_2(d))], \end{aligned} \quad (43)$$

where

$$G_1(d) = \left(\frac{1-\mu_1}{2}\right)^d \sum_{k=0}^{d-1} \binom{d-1+k}{k} \left(\frac{1+\mu_1}{2}\right)^k \quad (44)$$

$$\mu_1 = \sqrt{\frac{h_1}{1+h_1}} \quad (45)$$

$$h_1 = (1+\gamma)^2 \overline{z_c} \quad (46)$$

$$H_1(d) = \frac{(2d-3)!!}{(d-1)! 2^d (1+h_1)^d \mu_1} \quad (47)$$

and

$$\overline{z_c} = \frac{1+1/c}{2\sigma_{tot}^2} \quad (48)$$

To obtain $G_2(d)$ and $H_2(d)$, we replace h_1 with $h_2 = (1-\gamma)^2 \overline{z_c}$ in (44) and (47), respectively.

Finally, for a given convolutional code, the union bound on the probability of decoding error is given as [3]

$$P_e \le \sum_{d=d_{free}}^{\infty} a(d) P_{c\bar{c}}(d), \quad (49)$$

where d_{free} is the minimum distance of the code and coefficients $\{a(d)\}$ depend upon the specific code structure. These parameters have been tabulated in [4], or they may be found (with some effort) for any convolutional code using the state diagram transfer function [3].

Consider now the effect of imperfect interleaving. Depending upon the duration of a typical fade, this can be a serious problem, because the deinterleaving delay is limited by what is perceived to be acceptable for voice communications; indeed, this problem is exacerbated in satellite communications relative to what it is in terrestrial communications because of the additional delay due to the satellite channel.

We assume an interleaver which has n rows and m columns, and in which data is read in vertically and read out horizontally. Thus, for perfect interleaving, the fade can be correlated over no more than m coded symbols. However, to illustrate the severe degradation that could result from imperfect interleaving, consider the simplified case whereby the fade is perfectly correlated over 2 m symbols and is uncorrelated for any greater separation of symbols. This means that, after deinterleaving, adjacent pairs of channel symbols would have identical fade multipliers R_{1i} but would be uncorrelated from other pairs of coded symbols. In addition, we upper bound the pairwise error probability by assuming that of d positions that differ from the correct path, those symbols can be broken up into $d/2$ pairs of correlated fades. In other words, we ignore the possibility that the d symbols can be sufficiently dispersed so that there are more than $d/2$ independent fades affecting them. Intuitively, this corresponds to a worst-case scenario.

With this formulation, it is straightforward to show that (38), corresponding to perfect interleaving, reduces to (for d even):

$$P_{c\bar{c}}(A_1, \{R_{1i}\}; d) < Q\left(\frac{\sqrt{2}A_1 T}{\sigma_{\text{tot}}}\sqrt{\sum_{i=1}^{d/2} R_{1,2i-1}^2}\right). \tag{50}$$

One can now repeat the derivation leading to (43) and then use (49) to upper bound the performance.

Finally, let us consider the effect of dual diversity in conjunction with forward error correction coding and imperfect power control. For the ith coded symbol, we have a correlator output sample with the form of (32). Therefore, analogous to the development of (39), we use as the decoding metric $[R_{11}^{(i)}g_{11}^{(i)} + R_{12}^{(i)}g_{12}^{(i)}]c_{1i}$, where $R_{jk}^{(i)}$ is the fade amplitude of the ith coded symbol of the jth user at the kth antenna, $g_{1k}^{(i)}$ is the test statistic at the kth diversity branch corresponding to the ith coded symbol of user #1, and c_{1i} is the actual ith code symbol of user #1.

The conditional pairwise error probability, analogous to (39), is given by

$$\begin{aligned} P_{c\bar{c}}(A_1, \{R_{11}^{(i)}, R_{12}^{(i)}\}; d) &= P\Big(\sum_{i=1}^{d}\left(R_{11}^{(i)}g_{11}^{(i)} + R_{12}^{(i)}g_{12}^{(i)}\right)c_{1i} \\ &\le \sum_{i=1}^{d}\left(R_{11}^{(i)}g_{11}^{(i)} + R_{12}^{(i)}g_{12}^{(i)}\right)\overline{c_{1i}} \mid A_1, \{R_{11}^{(i)}, R_{12}^{(i)}\}\Big) \\ &= Q\left(\sqrt{2z}\right), \end{aligned} \tag{51}$$

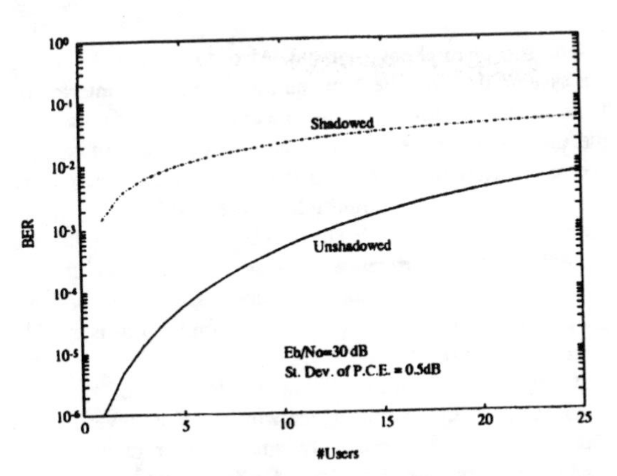

Fig. 1. Number of users per spot beam; uncoded CDMA.

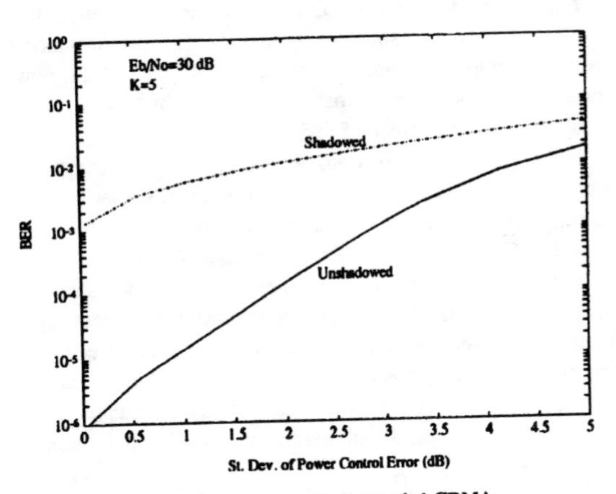

Fig. 2. Effect of imperfect power control; uncoded CDMA.

where now:

$$z = \frac{1}{2\sigma_{\text{tot}}^2}\sum_{i=1}^{d}\left(R_{11}^{(i)^2} + R_{12}^{(i)^2}\right). \tag{52}$$

Therefore, we can once again use (43) and (49) to provide an upper bound to the average probability of error.

VI. Numerical Results

In this section, we will present numerical results on the performance of a DS-CDMA system with imperfect power control, operating over a LEOS channel. We illustrate the system performance when the interleaving is both perfect and imperfect, as described in the previous section. We also demonstrate how important it is for dual diversity to be employed.

In Figs. 1 and 2, the performance of an uncoded DS-CDMA system is considered. The processing gain is chosen to be $L = 150$, and $E_b = E(1 + 1/c)$ represents average received energy per bit. We assume $c = 10$ as a typical value in a LEOS channel [1] and we choose the parameter $B = 0.3$; this value is chosen as a compromise between a large value, representing an urban area, and a smaller value, representing a suburban or rural environment. Also, to get specific results, we assume that the effect of the multiple access interference from all spot beams other than the one designed to illuminate the user of interest contributes an equal amount of multiple access interference as due to the users in the desired spot beam [5]. That is, if the multiple access interference from the users in the desired spot beam is denoted by KI, then the total multiple access interference is taken to be $2KI$. Of the two curves in Figs. 1 and 2, one corresponds to a user which is always shadowed and the other corresponds to a user which is never shadowed.

It can be seen from Fig. 1 that unshadowed and shadowed users exibit unequal performance with average power control. That is, even for a relatively small power control error, shadowed users cannot achieve $BER = 10^{-3}$. Hence, the system is underdesigned for the assumed fading channel. In Fig. 2, the power control error is varied, given a fixed number of users per spot beam $K = 5$. The performance of unshadowed users is more sensitive to variations of the power control error, because the performance of shadowed users is dominated by the effect of fading.

In the rest of this section, we concentrate on the performance of coded DS-CDMA in a worst-case scenario, which corresponds to a case in a shadowed state for an extended period of time (e.g., remaining stationary under a tree or driving in an urban canyon area). For this case, we ignore the percentage of time that the user of interest experiences Rician fading (i.e., when it is not shadowed), although we average the effect of multiple access interference over both shadowed and nonshadowed conditions. A convolutional code of rate 1/3 and constraint length 8 with maximum likelihood decoding in the receiver is assumed.

Consider Figs. 3–6, all of which correspond to $E_b/N_0 = 7$ dB and a processing gain $L = 50$, defined now as the number of chips per coded symbol. Figs. 3 and 4 correspond to perfect interleaving, while Figs. 5 and 6 show analogous results when the interleaving is imperfect. In all cases, BER is plotted against the maximum power control error, V/A.

When just a single signal is available (i.e., when we do not employ diversity), the system performance is shown in Fig. 3. If we assume an achievable open-loop power control error is V/A in the vicinity of 0.4 (which corresponds to a standard deviation of approximately 2 dB) and we choose a BER of 10^{-3} as our goal, then the system can accommodate about $K = 10$ simultaneous users per spot beam. In order to increase the capacity, suppose now that we have a system whereby we employ second-order diversity. Performance results for such a scenario are shown in Fig. 4; it is now seen more than $K = 40$ users can be accommodated with the same power control error as before.

As one would expect, the situation is much more pessimistic if the interleaving is imperfect. Figs. 5 and 6 correspond to the same systems as in Figs. 3 and 4, respectively, except now imperfect interleaving is incorporated in the model used to generate the results. It is seen, for example, from Fig. 6 that $K = 10$ is just barely achievable even with dual-order diversity if the standard deviation of power control error is 2 dB.

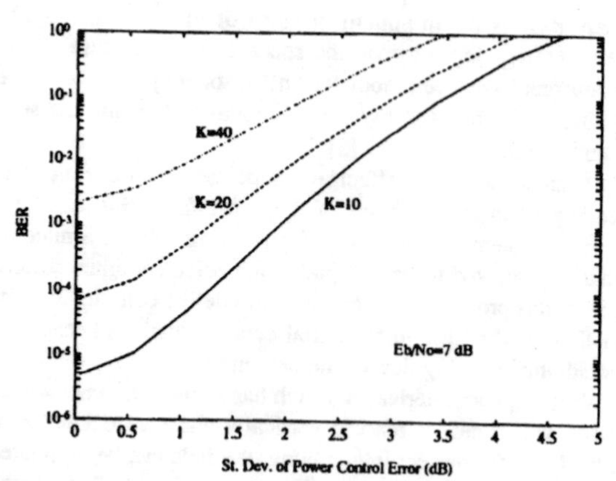

Fig. 3. Effect of imperfect power control; coded CDMA.

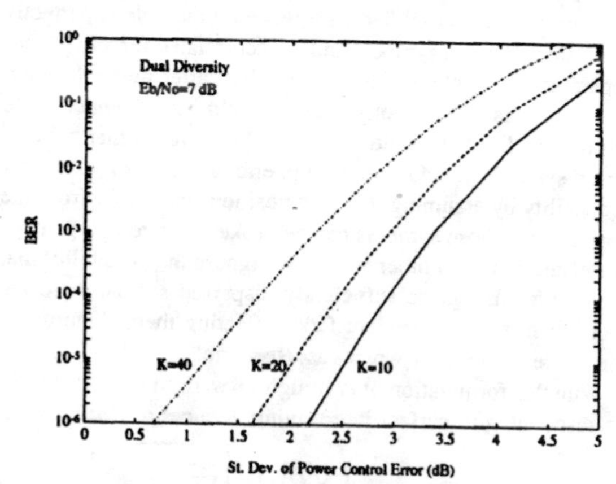

Fig. 4. Effect of imperfect power control; coded CDMA with diversity.

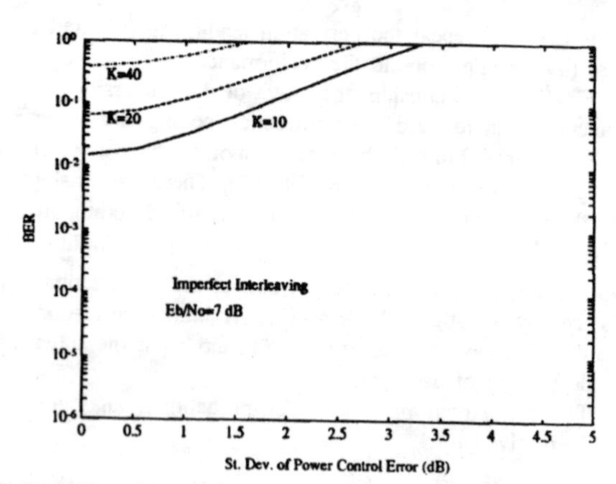

Fig. 5. Effect of imperfect power control; coded CDMA with imperfect interleaving.

To see the effect of of varying E_b/N_0, consider Figs. 7–10, the first two of which were derived assuming perfect interleaving and the last two showing what happens if the interleaving is imperfect. In all figures, BER is plotted versus the number of simultaneously active users per spot beam, and

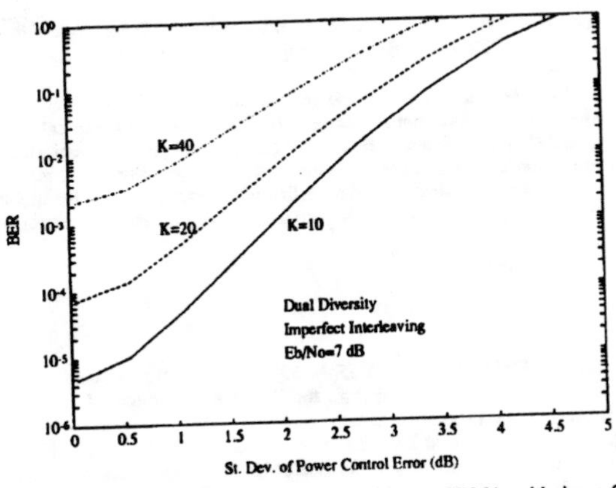

Fig. 6. Effect of imperfect power control; coded CDMA with imperfect interleaving and diversity.

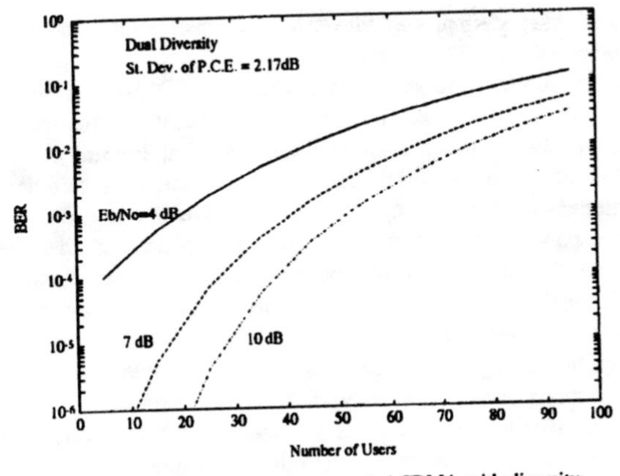

Fig. 8. Number of users per spot beam; coded CDMA with diversity.

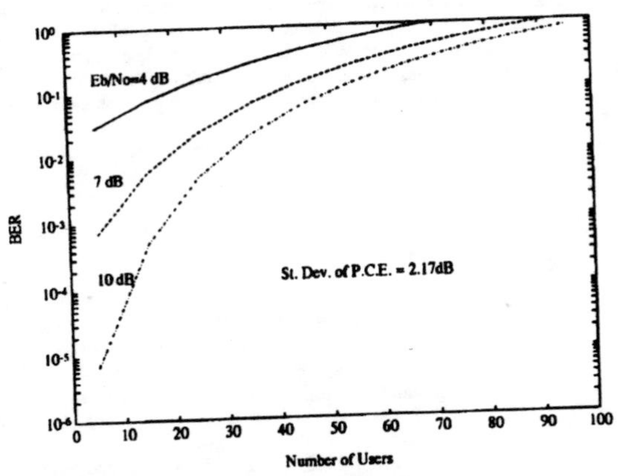

Fig. 7. Number of users per spot beam; coded CDMA.

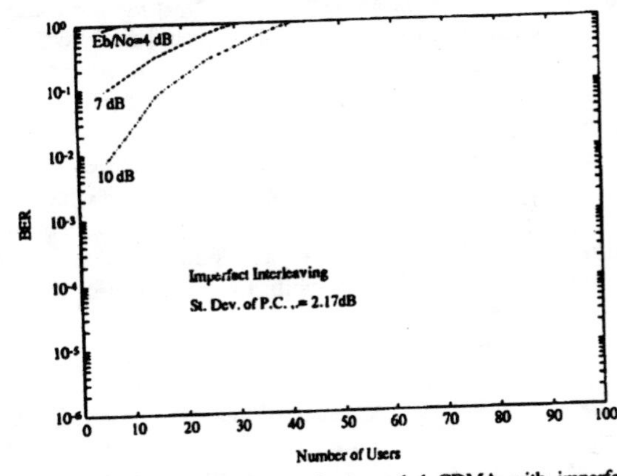

Fig. 9. Number of users per spot beam; coded CDMA with imperfect interleaving.

the system parameters are the same as those of Figs. 3–6, except that the power control error is kept constant at $V/A = 0.414$ (which corresponds to a standard deviation of 2.17 dB). Finally, each figure shows three curves, corresponding to $E_b/N_0 = 4$, 7, and 10 dB.

Suppose we concentrate on the 10 dB curves. Without diversity, Fig. 7 shows that the system can support about $K = 18$ users at a BER of 10^{-3}; the use of dual diversity increases K to about 51, as seen from Fig. 8.

Of course, once again the effect of imperfect power control can be severe. Comparing the corresponding curves of, say, Fig. 8 with those of Fig. 10, it is seen that the capacity of $K = 51$ with perfect interleaving is reduced to about $K = 18$ because of imperfect interleaving.

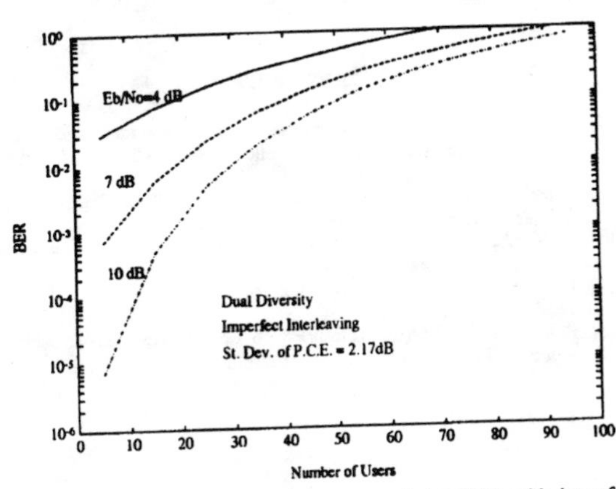

Fig. 10. Number of users per spot beam; coded CDMA with imperfect interleaving and diversity.

VII. CONCLUSIONS

In this paper, we have presented an analysis of the performance of a DS-CDMA cellular system when used over a LEOS channel. In particular, we have pointed out certain dissimilarities between a satellite channel and a terrestrial channel which result in DS-CDMA performing worse on the former channel than it does on the latter one. Of these differences in channel characteristics, the two we have concentrated on are the larger propagation time for a signal going from a mobile to its base station, and the relatively small multipath delay spread of the LEOS channel.

Our results have shown that while DS-CDMA is a viable multiple accessing technique for LEOS operations, the nature of the LEOS channel makes DS-CDMA a less-inviting choice than does the terrestrial channel. In particular, if sufficient interleaving can be employed, if dual diversity is used, and if a power control system can be implemented so that the standard deviation of the power control error is about 2 dB or less, DS-CDMA will result in good performance. If any of these design conditions are noticeably violated, significant degradation can result. For example, under the conditions listed previously, the system can accommodate about 40 simultaneous users per spot beam. However, with other parameters the same but with the maximum normalized power control error increased from 0.4 to 0.52 (corresponding to a standard deviation increase from 2 dB to 2.8 dB), the number of users per beam decreases to 20; if the normalized power control error is 0.59 (or 3.2 dB standard deviation), the number of users halves again to $K = 10$. As a perspective, it should be noted that, typically, the worst-case scenario for power control error does not correspond to the worst-case situation for imperfect interleaving. The power control error is invariably greater for a rapidly moving mobile, but such a mobile will experience very effective interleaving. Alternately, while the interleaving for a slowly moving mobile will typically be relatively ineffective, the power control system will operate very accurately.

Finally, there remains more research to be done to complete the study. A realistic power control scheme should be analyzed to determine the actual statistics of the power control error; also, a more accurate model should be used to quantify the degradation due to imperfect interleaving. In addition, while we have emphasized BER in our analysis, an alternate criterion is outage probability, and this should be determined and compared to the BER results presented in this paper.

REFERENCES

[1] E. Lutz *et al.*, "The land mobile satellite communication channel—Recording, statistics, and channel model," *IEEE Trans. Vehic. Technol.*, vol. 40, no. 2, May 1991.

[2] L. Milstein *et al.*, "Performance evaluation for cellular CDMA," *IEEE J. Select. Areas Commun.*, vol. 10, no. 4, May 1992.

[3] A. Viterbi and J. Omura, *Principles of Digital Communications and Coding*. New York: McGraw-Hill, 1979.

[4] J. Odenwalder, "Optimal decoding of convolutional codes," Ph.D. Thesis, U.C.L.A., 1970.

[5] P. Monsen, "CDMA performance with slow adaptive power control," *Report M.2 to Motorola Satel. Commun.*, Jan. 1993.

Branimir R. Vojcic (M'87) received the B.S., M.S., and D.Sc. degrees from the University of Belgrade, Belgrade, Yugoslavia, in 1980, 1986, and 1989, respectively.

Since 1991, he has been at the George Washington University, Washington, DC, where he is an Assistant Professor in the Department of Electrical Engineering and Computer Science. He teaches courses in digital communications, communication theory, and computer networks. His current research interests are the performance evaluation and modeling of terrestrial and satellite mobile communications, spread spectrum, equalization, and wireless data networks.

Raymond L. Pickholtz (S'54–A'55–M'60–SM'77–F'82) received the Ph.D. degree in electrical engineering from the Polytechnic Institute of Brooklyn, Brooklyn, NY, in 1966.

He was a Researcher at RCA Laboratories and ITT Laboratories. He was on the faculty of the Polytechnic Institute of Brooklyn and Brooklyn College. He was a Visiting Professor at the Université du Quebec and the University of California. He is a Fellow of the American Association for the Advancement of Science. He was an Editor of the IEEE TRANSACTIONS ON COMMUNICATIONS, and Guest Editor for special issues on computer communications, military communications, and spread spectrum systems. He is Editor of the *Telecommunications Series* for Computer Science Press. He has published scores of papers and holds six U.S. patents. He is President of Telecommunications Associates, a research and consulting firm specializing in communication system disciplines. He was elected a member of the Cosmos Club and a Fellow of the Washington Academy of Sciences in 1986. In 1984, he received the IEEE Centennial Medal. In 1987, he was elected Vice President, and in 1990 and 1991 as President, of the IEEE Communications Society.

Laurence B. Milstein (S'66– M'68–SM'77–F'85) received the B.E.E. degree from the City College of New York, New York, NY, in 1964, and the M.S. and Ph.D. degrees in electrical engineering from the Polytechnic Institute of Brooklyn, Brooklyn, NY, in 1966 and 1968, respectively.

From 1968 to 1974, he was employed by the Space and Communications Group of Hughes Aircraft Company, and from 1974 to 1976 he was a member of the Department of Electrical and Systems Engineering, Rensselaer Polytechnic Institute, Troy, NY. Since 1976, he has been with the Department of Eelctrical and Computer Engineering, University of California at San Diego, La Jolla, CA, where he is a Professor and former Department Chairman, working in the area of digital communication theory with special emphasis on spread-sprectrum communication systems. He has also been a consultant to both government and industry in the areas of radar and communications. He was an Associate Editor for Communication Theory for the *IEEE Transactions on Communications*, an Associate Editor for Book Reviews for the *IEEE Transactions on Information Theory*, and an Associate Technical Editor for the *IEEE Communications Magazine*, and is currently a Senior Editor for the *IEEE Journal on Selected Areas in Communications*. He was the Vice President for Technical Affairs in 1990 and 1991 of the IEEE Communications Society, and is currently a member of the Board of Governors of both the IEEE Communications Society and the IEEE Information Theory Society. He is also a member of Eta Kappa Nu and Tau Beta Pi.

A Performance Comparison of Orthogonal Code Division Multiple-Access Techniques for Mobile Satellite Communications

Riccardo De Gaudenzi, *Member, IEEE*, Tobias Garde, Filippo Giannetti, *Member, IEEE*, and Marco Luise, *Member, IEEE*

***Abstract*— In recent years, code division multiple-access (CDMA) techniques have received a great deal of attention for mobile terrestrial/satellite communication systems. Primarily considered for the noteworthy features of low power flux density emission and robustness to interference and multipath, CDMA is known to bear reduced bandwidth and power efficiency when compared to traditional TDMA and FDMA due to the intrinsic cochannel self-noise. Early attempts to increase the capacity of CDMA-based systems for commercial applications relied on voice activation and frequency reuse. More recently, practical solutions to implement (synchronous) orthogonal CDMA signaling are being developed independently in Europe and in the USA. This paper is focused on the comparative performance analysis of those two orthogonal CDMA schemes in the operating conditions of a mobile satellite communications system. In particular, the two CDMA systems are compared in the presence of that and frequency-selective multipath fading and a typical satellite transponder nonlinearity. Most numerical results are derived through a time-domain system simulation that confirms and integrates the theoretical findings.**

I. INTRODUCTION

RECENT adoption of code division multiple-access (CDMA) for commercial communication systems has generated a considerable stimulus for the study and the development of high-efficiency low-cost CDMA satellite networks. Key features of CDMA are the well-known robustness to narrow-band interference and to frequency-selective multipath fading, the low power flux density emission and the possibility to exploit diversity reception [1]. Earlier CDMA systems were mainly restricted to low-capacity military and professional communications where cost and capacity were not the main issues. On the contrary, commercial satellite communications call for low-cost user equipments on one side, and techniques for enhancing the achievable power and spectral efficiency of the link on the other. Different practical techniques to overcome the problems related to the reduced spectrum/power efficiency of CDMA have been recently proposed [2]–[4]. Such multiplexing schemes yield spectral and power efficiency similar to conventional FDMA. Both solutions aim at minimizing the CDMA self-noise effect by employing symbol- and chip-synchronous (quasi-)orthogonal spreading sequences. In particular, the access technique described in [2] has been adopted for the US terrestrial digital cellular system and has been proposed for the GLOBALSTAR [5] satellite-based personal communication services (PCS), while the technique in [3], [4] will be employed in the European Mobile Satellite Business Network (MSBN) service to be operated in the L-band via the EMS payload of the satellite ITALSAT II [6].

The goal of this work is to perform a comparison of the two systems in a realistic mobile satellite scenario. The results of such an analysis may prove helpful in the design of future PCS networks. In the following, we investigate the bit error rate (BER) of a traditional single-user matched-filter receiver in the presence of frequency-selective and flat fading and for a typical nonlinear satellite channel. Consideration of a more elaborate receiver structure, such as the RAKE combiner [7] and/or multiuser detection technique [8] is outside the scope of this work and will not be pursued further.

After this introduction, Section II describes the signal formats of the two (quasi-)orthogonal CDMA schemes to be analyzed, while Section III deals with the analysis and simulation of the performance degradation experienced on a typical mobile-satellite channel, encompassing frequency selective multipath fading and a typical transponder nonlinearity. A summary and a few concluding remarks are finally reported in Section IV.

II. OUTLINE OF THE TWO CDMA SYSTEMS

A. Modulator Description

The baseband equivalent model of the generalized BPSK/QPSK direct-sequence spread-spectrum (DS/SS) CDMA modulator is depicted in Fig. 1. As is apparent, the incoming binary data stream $\{d_k^l\}$ of the lth user is split in the in-phase/quadrature parallel streams $\{d_{p,k}^l\}$ and $\{d_{q,k}^l\}$, each one being spread with the respective chip sequence $c_{p,|i|_L}^l$ and $c_{q,|i|_L}^l$. Both chip-rate spread sequences are then shaped by a Nyquist-square-root raised-cosine filter with roll-off factor ρ[1]. The two baseband components are then upconverted to frequency f_0 and added to give the lth bandpass transmitted signal.

[1] Throughout the paper we assume $\rho = 0.4$.

Manuscript received January 15, 1994; revised September 20, 1994.

R. De Gaudenzi and T. Garde are with the European Space Agency, European Space Research and Technology Centre, RF System Division, 2200 AG Noordwijk, The Netherlands.

F. Giannetti and M. Luise are with the University of Pisa, Dipartimento di Ingegneria della Informazione, 56126 Pisa, Italy.

IEEE Log Number 9407516.

0733-8716/95$04.00

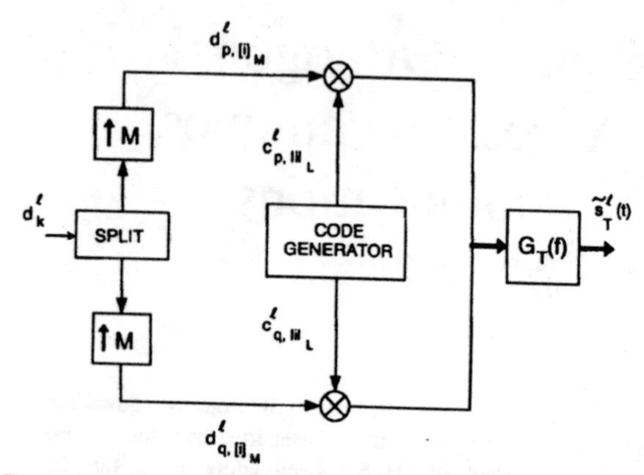

Fig. 1. Generalized BPSK/QPSK DS/SS-CDMA modulator.

By defining the operators $|i|_L \triangleq i \bmod L$ and $\lfloor i \rfloor_M \triangleq \mathrm{int}\{i/M\}$, we can express the baseband equivalent of such a signal as

$$\tilde{s}_T^l(t) = \sqrt{P_S^l} \sum_{i=-\infty}^{+\infty} \left(c_{p,|i|_L}^l d_{p,\lfloor i \rfloor_M}^l + j\, c_{q,|i|_L}^l d_{q,\lfloor i \rfloor_M}^l \right) \cdot g_T(t - iT_c) \tag{1}$$

where P_S^l is the average transmitted power of the lth user, $c_{p,i}^l$ and $c_{q,i}^l$ are the ith chip for the spreading sequences on the in-phase (I) and quadrature (Q) branch respectively, $d_{p,k}^l$ and $d_{q,k}^l$ are the kth data symbols on the I and Q branches, T_c is the chip interval, L is the spreading code period, M is the so-called spreading factor (the ratio between the symbol and chip periods, T_s and T_c, respectively) and finally $g_T(t)$ is the impulse response of the transmission filter. The chip interval T_c is related to the bit interval T_b through the so-called processing gain $G_p \triangleq T_b/T_c$. Two independent spreading sequences are used for the I and Q components in order to cope best with nonideal conditions (i.e. noisy carrier reference, imperfect carrier synchronization, nonlinear distortions etc.). Equation (1) fits the transmitted signal of different CDMA systems currently in use or in advanced testing status, whose modulators closely resemble the one depicted in Fig. 1, with minor modifications only. The following analysis is relevant to a satellite outbound link with all user signals perfectly synchronized in time, phase and frequency. This assumption, which is easily verified when the CDM multiplex is uplinked by the same hub station or by uplink stations synchronized in time and frequency [9], relies on the fact that the outbound link is often more demanding in terms of satellite power resources exploitation. The case of an inbound link with non ideal synchronization (which has been analyzed in [9]) will not be pursued in the following. Such a multipoint-to-point link might also take advantage of (suboptimum) multiuser detection techniques—which are becoming of affordable complexity for a central hub station [10]—to minimize the effect of asynchronous cochannel self-noise.

1) Walsh-Hadamard Functions-Based Orthogonal DS/SS-CDMA System: This particular system was introduced and patented a few years ago by Qualcomm Inc. [2]. We will call it for brevity "QC system." The user signal is obtained from (1) letting

$$d_{p,\lfloor i \rfloor_M}^l = d_{q,\lfloor i \rfloor_M}^l = d_{\lfloor i \rfloor_M}^l \tag{2}$$

$$c_{p,|i|_L}^l = c_{p,|i|_L} w_{|i|_M}^l, \quad c_{q,|i|_L}^l = c_{q,|i|_L} w_{|i|_M}^l \tag{3}$$

where w_i^l are binary symbols from the Walsh-Hadamard (WH) sequence assigned to the lth carrier [11] with period $M = 2^{N_d}$, N_d an integer, while $c_{p,i}$ and $c_{q,i}$ are symbols from two different binary augmented pseudo-noise (PN) sequences common to all users, whose period[2] is L. The WH sequence duration is thus exactly equal to one symbol so that the WH code properties are not affected by data transitions. To ensure perfect users' orthogonality, the start epochs of the WH spreading sequences are synchronized in the master station. Also, the following relationship between the two chip sequence periods holds true: $L/M = n$ with n an integer. For the system described in [2], $M = 64$, $L = 32768$ and, as is apparent from (2), the I and Q data symbols are identical. The (long) overlay PN sequence protects the WH-spread signal from possible asynchronous interference due to multipath distortion or adjacent cells/beams signals, and aids in the code acquisition process[3]. It also allows for the discrimination of distinct sources of the same multiplex signal (e.g. two separate satellites or two terrestrial cells) to perform handover and/or diversity reception.

2) Gold Functions-Based Quasi-Orthogonal DS/SS-CDMA System: An independently developed quasi-orthogonal CDMA system introduced and patented by the European Space Agency (ESA) [9] shows conceptual similarities. It also tends to minimize the self-noise by proper synchronization of the start epochs of the users spreading sequences, and forcing the spreading sequence period to be equal to the symbol period. The main difference from the previous system is that a single spreading sequence is used for code acquisition and users' orthogonalization. Also, truly baseband QPSK symbol signaling with independent I and Q spreading by means of two different Gold sequences is employed. The main advantage of this signaling technique lies in the reduced bandwidth occupancy for a given bit rate and spreading factor. The small residual self-noise due to the partial orthogonality of the Gold sequences [9] results to be negligible as far as both the BER performance and the overall system efficiency are concerned.

The transmitted signal in such a system is obtained from (1) by letting $L = M = 2^{N_d} - 1$, $N_d = 5, \cdots, 9$ and

$$c_{p,|i|_L}^l = g_{p,|i|_L}^l, \quad c_{q,|i|_L}^l = g_{q,|i|_L}^l \tag{4}$$

where $g_{p,i}^l$ and $g_{q,i}^l$ are binary symbols from preferentially-phased Gold sequences with period L [9].

[2] The PN duration is "artificially" extended to $L = 2^{N_{pn}}$ as described in [2].

[3] As is known, the cross-correlation function of the WH sequences exhibits high secondary peaks, thus making code acquisition ambiguous.

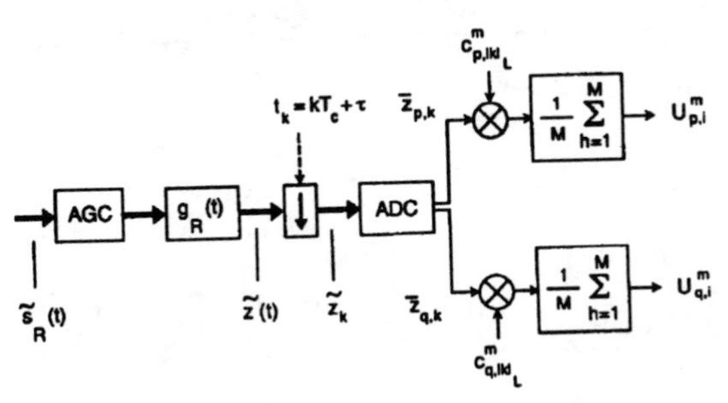

Fig. 2. BPSK/QPSK DS/SS-CDMA demodulator.

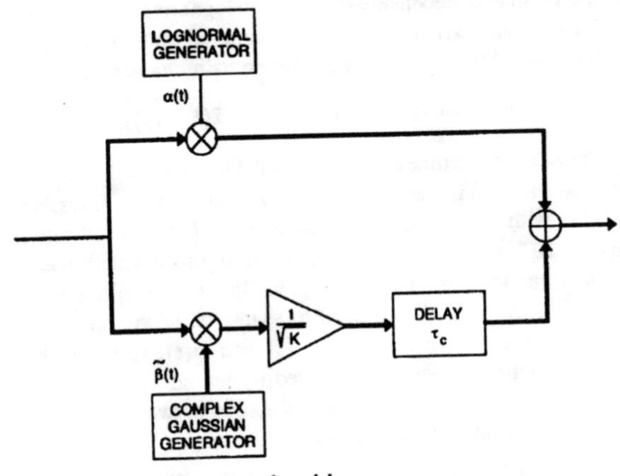

Fig. 3. Satellite fading channel model.

B. Ideal System Performance

The block diagram of the CDMA demodulator is depicted in Fig. 2. The received signal is first downconverted to IF where an automatic gain control (AGC) adjusts the signal amplitude so as to keep the total received signal power constantly equal to a reference value. This guarantees that the signal amplitude stays within the dynamic range of the subsequent analog-to-digital converters. After IF (analog) matched filtering, the received signal is then baseband-converted and its I and Q components are sampled at one sample per chip. The resulting I/Q samples are passed to the analog-to-digital converters (ADC) and the digitized samples are then despread with the locally generated code replica and then accumulated over a symbol period. Finally, PSK demodulation takes place.

When the systems above operate over an additive white Gaussian noise (AWGN) channel with ideal chip timing and carrier phase recovery, neglecting the effect of A/D conversion (whose influence can be taken into account following the approach in [12]), the BER of the demodulator for the mth channel is easily evaluated as [9]

$$p_b^m = Q\left(\sqrt{\frac{2E_b^m/N_0}{1+\mu_m \frac{(N-1)}{2G_p^2}\frac{E_b^m}{N_0}}}\right) \tag{5}$$

where $E_b^m \triangleq P_S^m T_b$ and $Q(x) \triangleq 1/\sqrt{2\pi}\int_x^\infty \exp(-y^2/2)\,dy$. In (5), μ_m amounts to zero for the QC system, while it represents the so-called cross-correlation factor of the preferentially-phased Gold codes [9] for the ESA system

$$\mu_m \triangleq \frac{1}{(N-1)}\sum_{i=1,\,i\neq m}^{N}\frac{P_S^i}{P_S^m}\left[\sum_{k=0}^{M-1} g_{p,k}^m\, g_{p,k}^i\right]^2$$
$$= \frac{1}{(N-1)}\sum_{i=1,\,i\neq m}^{N}\frac{P_S^i}{P_S^m}. \tag{6}$$

While (5) holds approximately for the ESA system, in the case of the QC system the expression is exact and, since $\mu_m = 0$, it reduces to the well-known BER formula of the BPSK matched-filter receiver. By straightforward manipulations of (5) the system bandwidth efficiency (defined as the ratio between the total transmitted bit rate and the total bandwidth occupancy) can be derived as a function of the total signal-to-noise ratio CNR. Numerical results show that the two orthogonal systems attain essentially to the same efficiency, which can be as greater as a factor five than that of asynchronous CDMA for medium-to-high CNR's. The peculiar I-Q spreading employed by both systems provides the system robustness to carrier tracking errors as a BPSK signal although QPSK modulation is adopted.

III. Performance Over The Mobile-Satellite Channel

The behavior of the two (quasi-)orthogonal CDMA systems over the nonideal mobile satellite channel is investigated in the sequel following, whenever possible, a simplified analytical approach based on straightforward extension of (5). A complete time domain system simulator based on the TOPSIM-IV tool [13] has also been developed to verify the theoretical findings, and to analyze the system in the nonlinear regime (due to the satellite transponder nonlinearity). The CDMA interference is simulated by generation of N "real" independent DS-SS signals. To reduce the computational effort, semianalytical techniques have been widely applied [14]. Furthermore, the semianalytical bit error rate results are validated by less efficient error counting. Systematic errors, possibly affecting the semianalytical approach, have been duly considered in the estimation process.

A. Mobile Fading Channel

The two CDMA systems described above exploit sequence (quasi-)orthogonality to achieve higher efficiency when compared to conventional CDMA systems. Unfortunately, it is known that this advantage may vanish in the presence of multipath fading channel [15]. In particular, the multipath fading experienced by mobile satellite systems is related to satellite elevation, propagation conditions and aerial directional characteristics, causing the problem to be extremely involved. Some sort of simplification has to be adopted to obtain first-approximation results about the behavior of the two systems in such an environment. In the following we will restrict therefore our consideration to a suited simplified two-ray model.

The channel model we assume is depicted in Fig. 3. It is a simple extension of the Loo narrowband model [16] which has been devised out of extensive satellite test campaigns with reduced elevation angles. The received signal is made of a line-of-sight component affected by lognormal shadowing and a delayed Rayleigh-distributed scattered component. The channel has thus the following time-domain characterization

$$\tilde{y}(t) = \alpha(t)\,\tilde{x}(t) + \frac{1}{\sqrt{K}}\tilde{\beta}(t)\,\tilde{x}(t-\tau_c) \tag{7}$$

where τ_c is a fixed time delay, $\alpha(t)$ is the lognormal shadowing process and $\tilde{\beta}(t)$ is a zero-mean, unit-power complex Gaussian process with independent I-Q components. More specifically, $\alpha(t) = 10^{\gamma(t)/20}$ where $\gamma(t)$ is a Gaussian process with mean μ_γ and variance σ_γ^2. Two three-pole Butterworth filters with bandwidth B_γ and B_β are used to shape the power spectrum of the two Gaussian processes $\gamma(t)$ and $\tilde{\beta}(t)$, respectively. The channel is further characterized by the so-called carrier-to-multipath ratio C/M defined as the ratio between the direct and reflected powers, respectively

$$\left[\frac{C}{M}\right]_{dB} = 10\,\log_{10} K + \mu_\gamma + \frac{\sigma_\gamma^2}{20\log_{10} e}. \tag{8}$$

The delayed path is responsible for the frequency-selectivity characteristic of the fading channel, and must be taken into account as long as $T_c > \tau_c$ (typically, $R_c \triangleq 1/T_c > 1$ Mchip/s). For low data rates and/or spreading factors, it may happen that $\tau_c < T_c$, so that frequency-flat fading essentially occurs. Model (7) is also representative of more complex situations with several propagation paths, provided that an equivalent C/M is considered.

B. Evaluation of the BER on the Fading Channel

We assume a framework of slow fading variation with respect to the bit period T_b, i.e. B_γ, $B_\beta \ll 1/T_b$. In this context, we further distinguish between frequency-flat fading ($\tau_c = 0$) and frequency-selective fading ($\tau_c > T_c$). In the former case, the BER of the CDMA demodulator is

$$p_b^m = \int_0^\infty Q\left(z\sqrt{\frac{2E_b^m/N_0}{1+\frac{(N-1)}{2G_p^2}\frac{E_b^m}{N_0}\mu_m}}\right) \cdot\left\{\int_0^\infty f_{Z|A}(z)\cdot f_A(\alpha)d\alpha\right\}dz \tag{9}$$

while in the latter case the BER amounts to

$$p_b^m(\tau_c) \simeq \int_0^\infty\int_0^\infty f_A(\alpha)f_\Lambda(\lambda) \cdot Q\left(\alpha\sqrt{\frac{2E_b^m/N_0}{1+\frac{(N-1)}{2G_p^2}\frac{E_b^m}{N_0}\left[\mu_m + \lambda^2\mu_m(\tau_c)\right]}}\right)d\alpha\,d\lambda \tag{10}$$

where

$$f_A(\alpha) = \frac{20\log e}{\alpha\,\sigma_\gamma\sqrt{2\pi}}\exp\left\{-\frac{1}{2\,\sigma_\gamma^2}(20\log(\alpha)-\mu_\gamma)^2\right\}, \quad \alpha \geq 0$$

$$f_{Z|A}(z) = Kz\,\exp\left\{-\frac{z^2+\alpha^2}{2/K}\right\}I_0(Kz\alpha), \quad z \geq 0$$

$$f_\Lambda(\lambda) = 2K\,\lambda\,\exp\{-K\,\lambda^2\}, \quad \lambda \geq 0.$$

$f_A(\alpha)$ and $f_\Lambda(\lambda)$ are the probability density functions (PDF's) of the lognormal process $\alpha(t)$ and of the amplitude $\lambda(t) = |\tilde{\beta}(t)|/\sqrt{K}$ of the reflected ray, respectively, while $f_{Z|A}(z)$ is the PDF of the received signal envelope Z, conditioned on the amplitude A of the direct (lognormal) ray. When comparing (10) to (5) derived for the AWGN channel, we notice that, apart from the average on the fading PDF's, an additional term appears in the denominator of the Q-function argument, due to the delayed multipath component. The coefficient $\mu_m(\tau_c)$ represents an extension of (6) for the case of asynchronous interference and is trivially dependent on the relative delay τ_c. The expression of $\mu_m(\tau_c)$ is quite tedious for the case of Nyquist-shaped chips and will not be reported here [15].

Let us first analyze the bit error rate variation as a function of the multipath delay τ_c. Fig. 4 shows some results of a Monte

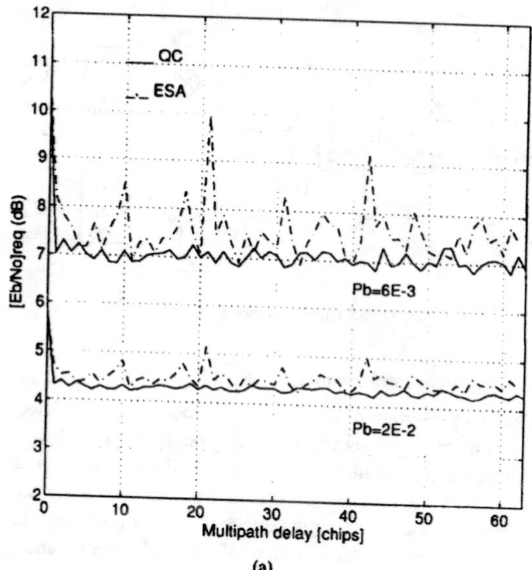

(a)

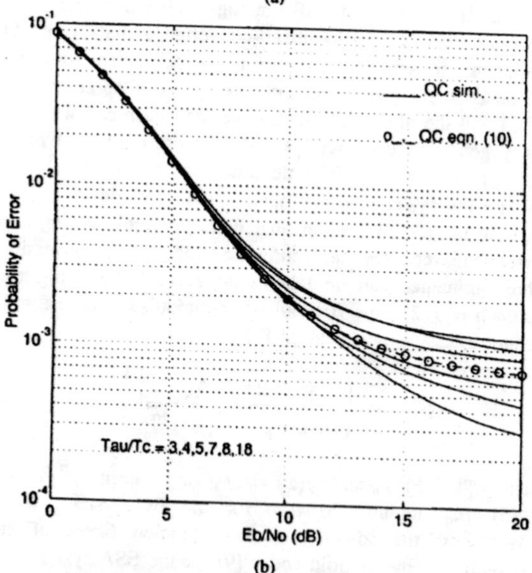

(b)

Fig. 4. QC CDMA simulation results. (a) BER versus E_b/N_0 for different multipath delays compared to (10). (b) E_b/N_0 required to achieve a given BER versus the multipath delay τ_c.

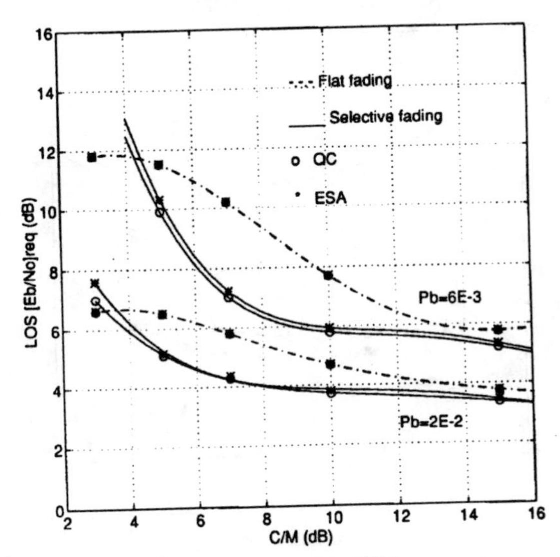

Fig. 5. Required E_b/N_0 as a function of the C/M.

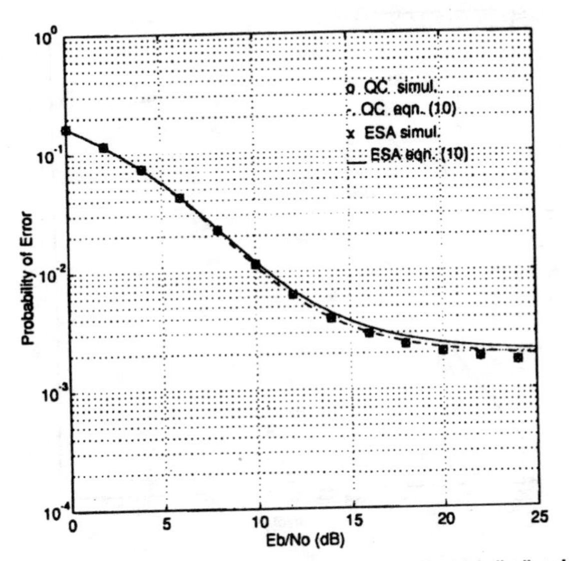

Fig. 6. CDMA system BER with Lognormal LOS plus Rayleigh distributed multipath.

Carlo simulation of the QC system with a clear line of sight (LOS) signal (i.e. (7) with μ_γ=0, σ_γ=0) and $C/M = 7$ dB. In all of the following simulations we have assumed 50% system loading ($N = M/2$), $N_d = 6$, $B_\beta T_s = 0.1$. An uncoded target BER of $2 \cdot 10^{-2}$ and $6 \cdot 10^{-3}$ has been assumed considering voice and data applications, respectively. As predicted by (10), the BER over the multipath fading channel is dependent on the actual delay τ_c (Fig. 4(a)). However, as Fig. 4(b) shows, this dependence is fairly weak when the operative BER is above the floor, apart from the case $\tau_c = 0$ that causes a sharp change (from frequency-selective to frequency-flat) in the characteristic of the channel. Observing Fig. 4(a) we notice that (10) turns out to be a good estimate of the average BER curve if we neglect the dependence of $\mu(\tau_c)$ on τ_c, assuming $\mu(\tau_c) = 2G_p$ as analytically derived in [15]. The results for the ESA system depicted in Fig. 4(b) are close to the QC system but shows a larger dependence on the multipath delay τ_c. This is due to the absence of the overlaying PN sequence which helps in randomizing the multipath delay effect.

The E_b/N_0 value required to achieve a given BER is plotted in Fig. 5 as a function of C/M for the same fading parameters as Fig. 4, for $\tau_c = 34 \cdot T_c$ and $\tau_c = 0$. In the case of flat fading, the performance degradation in terms of required line-of-sight E_b/N_0 to achieve the target BER slowly increases with the inverse of C/M. In the case of selective fading, the required E_b/N_0 for the two target BER's shows a reduced sensitivity to multipath up to $C/M = 7$ dB. The simulation results show that beyond $C/M = 7$ dB the effect of the (asynchronous) multipath interference becomes dominant causing a BER floor and thus reducing the advantage of orthogonal multiplexing. Under heavy multipath conditions (i.e. $C/M \leq 5$ dB), time-diversity reception provided by a RAKE receiver [7], possibly combined with powerful FEC techniques, might anyway give the required performance boost to make the link available. In Fig. 6 the case of combined Rayleigh multipath and lognormal LOS shadowing is considered. The BER simulations were performed with $C/M = 7$ dB, $B_\gamma T_s = 0.1$, $\mu_\gamma = -3$, $\sigma_\gamma = 2$, $B_\beta T_s = 0.1$, $\tau_c/T_c = 34$. Simulation results for the two systems are compared to (10), simplified as above. A remarkable agreement is shown.

C. Transponder Nonlinearity

When dealing with a satellite-based transmission system we have to assess also the impact of the satellite transponder nonlinearity on the overall performance. This aspect reveals particularly critical in a multiuser system. The goal of an efficient system design is in this respect the computation of the input (or equivalently output) back-off (IBO or OBO) of the amplifier operating point which gives optimum on-board power utilization for a fixed target BER.

The nonlinear on-board high-power amplifier (HPA) is modeled as a typical L-band travelling wave tube (TWT) whose AM/AM and AM/PM characteristics are shown in Fig. 7. Both characteristics are normalized to the saturation point, whose input level is thus set to 0 dB. Uplink AWGN and on-board filtering will be neglected in the following.

Assume that our transponder is operating with a given IBO (thus a given OBO), and that our specifications call for the particular value of the link BER P_b. This BER is attained with a corresponding specific value of the E_b/N_0 ratio, that is in general greater than the theoretical value on the AWGN channel due to nonlinear distortions. We have thus two different reasons leading to a waste of the available satellite power: the E_b/N_0 degradation due to residual nonlinear distortions, and the OBO we are forced to impose not to let distortions grow beyond acceptable values. The two degradation mechanisms can be put together by introducing the parameter $[E_b/N_0]_{sat}$, defined as the sum of the E_b/N_0 value required to obtain a BER equal to P_b at a given OBO, and the value of the OBO itself: $[E_b/N_0]_{sat}$ (dB) $= E_b/N_0$ (dB) $+ OBO$ (dB). $[E_b/N_0]_{sat}$ can be regarded as the E_b/N_0 value of a fictitious

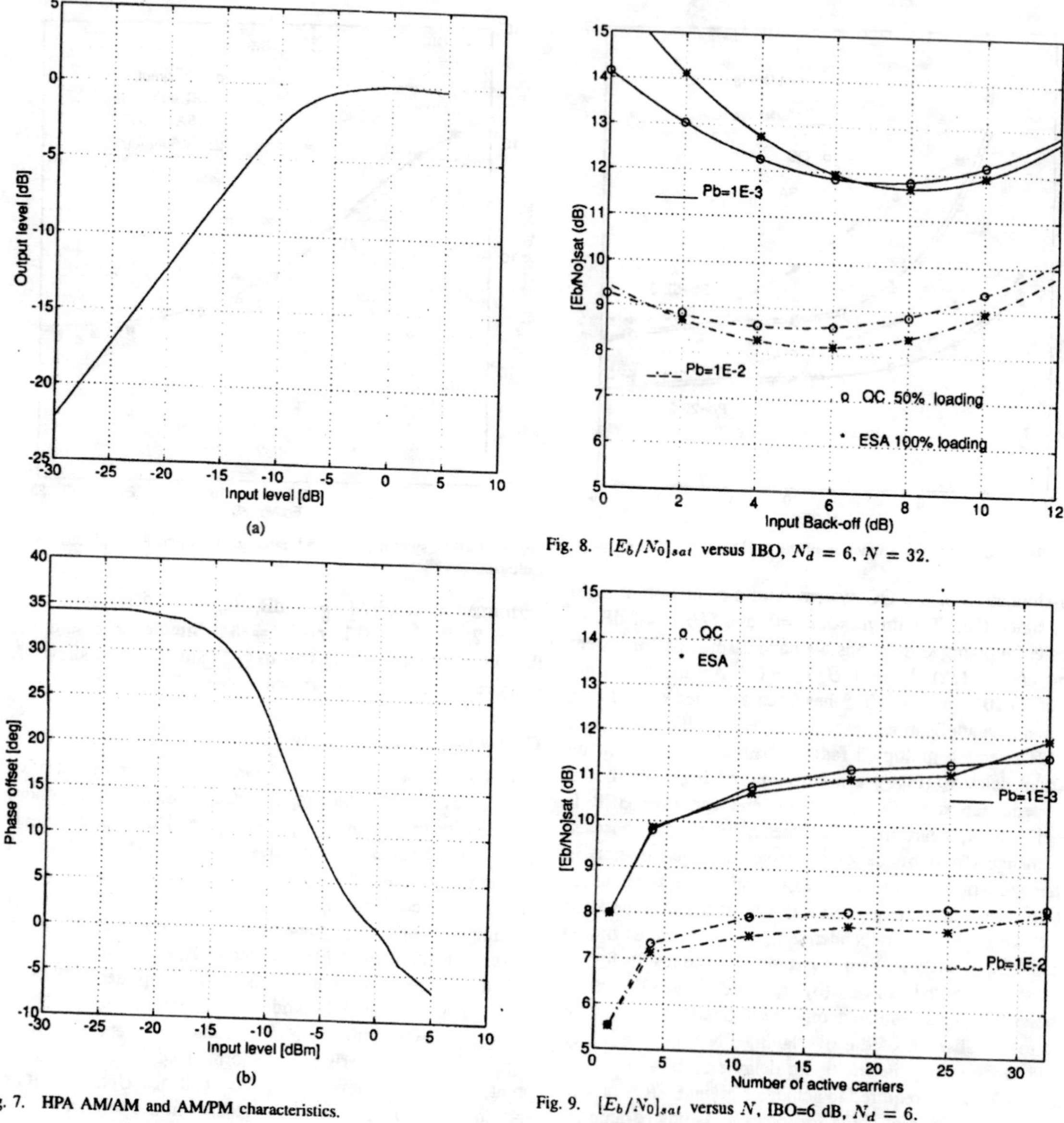

Fig. 7. HPA AM/AM and AM/PM characteristics.

Fig. 8. $[E_b/N_0]_{sat}$ versus IBO, $N_d = 6$, $N = 32$.

Fig. 9. $[E_b/N_0]_{sat}$ versus N, IBO=6 dB, $N_d = 6$.

transmission at amplifier saturation (0 dB OBO) with the same noise level as the real transmission with BER P_b and the real OBO. Clearly, an increase of the OBO causes a decrease of the E_b/N_0 degradation due to nonlinear distortions, but leads to more inefficient on-board power utilization. The optimum HPA operating point can be found as a trade-off condition between the two contrasting effects; it is in fact given by that value of IBO that minimizes $[E_b/N_0]_{sat}$ for a given P_b, i.e., that value of the IBO that leads to optimum on-board power utilization. The optimization can be performed resorting to the above mentioned semianalytical simulation techniques. In particular, our simulation is carried out with no AWGN source, but with the transponder model only. In this condition, we can measure the different values of the signal samples at the input of the decision device, corrupted by nonlinear distortion only $U_i^m = U_{p,i}^m + jU_{q,i}^m$. Since the AWGN is injected in the down-link, i.e., at the nonlinearity output, we can easily evaluate through standard analysis the BER of the link *conditioned* on the particular observed values of $U_{p,i}^m$ and $U_{q,i}^m$ by a simple formula. The goal of the simulation is thus to perform the *average* of the conditional BER expressions, to yield the final overall BER curve in the presence of AWGN and nonlinearity [17]. Following this method, we have derived the required $[E_b/N_0]_{sat}$ to achieve a given BER as a function of the HPA IBO. The phase shift caused by the satellite HPA is recovered in the demodulator by means of a narrowband feedforward decision-directed maximum likelihood phase estimator at the despreader output. This solution revealed optimum for minimizing the

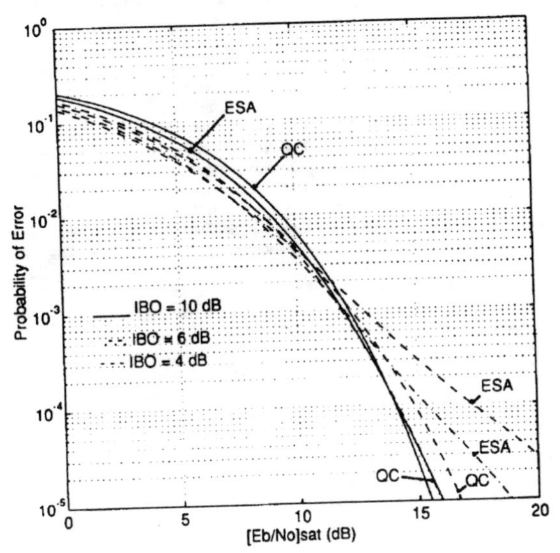

Fig. 10. Dependence on IBO of BER versus $[E_b/N_0]_{sat}$, $N_d = 6$, $N = 32$.

effect of CDMA interference on the carrier phase estimator. Fig. 8 shows the simulation results for the two systems with similar spreading factor, 50% and 100% loading for the QC and the ESA system respectively. It turns out that the ESA system performs slightly better although in a condition of double loading compared to the QC system. The higher loading is also the reason why an higher IBO is required by the ESA system to achieve the best performance. Fig. 9 shows the dependence of the required $[E_b/N_0]_{sat}$ on the number of active multiplexed users N. We observe that the degradation is almost constant for a loading greater than 20%. Finally, to further enlighten the effect of the nonlinearity on system performance, Fig. 10 shows the BER curves of the two systems, as a function of E_b/N_0 for different fixed values of the IBO.

IV. Conclusions

The body of the results presented in the previous sections does not allow to state clearly the superiority of any of the two schemes in comparison to the other. A minimum advantage of the WH-functions-based system on the frequency-selective fading channel is counterbalanced by a somewhat decreased sensitivity to implementation losses and nonlinear distortions for the Gold-functions-based scheme. No substantial difference is observed as far as overall bandwidth efficiency and carrier phase offset sensitivity are concerned. Further system level issues should be taken into consideration for an in-depth comparison. For instance, initial code acquisition is presumably slower in the WH-functions-based system due to the considerable length of the superimposed PN sequence. On the contrary, other high-level functions such as soft-handover or diversity reception may reveal easier in the WH-functions-based system due to its intrinsic carrier discrimination capability via the same (signature) long PN sequence. However, if required, a similar approach might be adopted for the Gold-based system, further reducing the slight performance difference observed.

References

[1] W. C. Y. Lee, "Overview of cellular CDMA," *IEEE Trans. Veh. Technol.*, vol. 40, no. 2, pp. 291–302, May 1991.

[2] "System and method for generating signal waveforms in a CDMA cellular telephone system," Qualcomm Inc., Patent PCT/US91/04400-WO 92/00639, Jan. 9, 1992.

[3] "Code distribution multiple access communication system with user voice activated carrier and code synchronization," European Space Agency, Patent PCT/EP90/01276-WO 91/02415, Feb. 21, 1991.

[4] "Method and device for multiplexing data signals in a satellite code division multiple access system based on PSK," European Space Agency, Patent PCT/EP92/02001 Aug. 28, 1992.

[5] R. A. Wiedemann and A. J. Viterbi, "The GLOBALSTAR mobile satellite system for worldwide personal communications," in *Proc. 3rd Int. Mobile Satellite Conf. IMSC '93* (Pasadena, CA), June 16–18, 1993, pp. 291–296.

[6] R. Rogard, "LMSS: From low data rate to voice services," in *Proc. 14th AIAA Int. Commun. Satellite Syst. Conf.* (Washington, DC), Mar. 1992, pp. 284–393.

[7] G. L. Turin, "Introduction to spread-spectrum anti-multipath techniques and their application to urban digital radio," *IEEE Proc.*, vol. 68, pp. 328–353, Mar. 1980.

[8] S. Verdu, "Multiuser demodulation," in *Proc. 3rd Int. ESA Workshop on DSP Appl. Space Commun.* (Noordwijk, The Netherlands), Sept. 1992.

[9] R. De Gaudenzi, C. Elia, and R. Viola, "Bandlimited quasi-synchronous CDMA: A novel satellite access technique for mobile and personal communication systems," *IEEE J. Select. Areas Commun.*, vol. 10, no. 2, pp. 328–343, Feb. 1992.

[10] Z. Xie, R. T. Short, and C. K. Rushfort, "A family of suboptimum detectors for coherent multiuser communications," *IEEE J. Select. Areas Commun.*, vol. 8, no. 4, pp. 683–690, May 1990.

[11] N. Ahmed and K. R. Rao, *Orthogonal Transforms for Digital Signal Processing*. New York: Springer-Verlag, 1975.

[12] R. De Gaudenzi, F. Giannetti, and M. Luise, "The effect of signal quantization on the performance of DS/SS-CDMA demodulators," in *Proc. IEEE GLOBECOM '94* (San Francisco CA), Nov./Dec. 1994, pp. 9.9.4–9.9.8.

[13] V. Castellani *et al.*, "TOPSIM-IV: An advanced versatile user-oriented software package for computer-aided analysis and design of communication systems," *ESA J.*, vol. 16, no. 2, pp. 135–158, 1992.

[14] M. C. Jeruchim, "Techniques for estimating the bit error rate in the simulation of digital communication systems," *IEEE J. Select. Areas Commun.*, vol. 2, no. 1, Jan. 1984.

[15] E. Colzi, "Multipath effects on band-limited quasi-synchronous CDMA systems: Analysis and countermeasures," European Space Agency ESTEC Working Paper EWP-1650, Feb. 1992.

[16] C. Loo, "A statistical model for a land-mobile satellite link," *IEEE Trans. Veh. Technol.*, vol. VT-34, pp. 1222–1227, Aug. 1985.

[17] M. Luise, R. Reggiannini, and G.M. Vitetta, "Efficient modulation techniques for satellite telemetry—Receiver design and performance evaluation," *Int. J. Satellite Commun.*, pp. 325–342, Nov./Dec. 1992.

Riccardo De Gaudenzi (M'89) was born in Rosignano M.mo, Italy in 1960. He received the Doctor Engineer degree (cum laude) in electronic engineering from the University of Pisa, Pisa, Italy, in 1985.

From 1986 to 1988 he was with the European Space Agency (ESA), Stations and Communications Engineering Department, Darmstadt, Germany, where he was involved in satellite telecommunication ground systems design and testing. In particular, he followed the development of two new ESA's satellite tracking systems. In 1988, he joined the ESA's Research and Technology Centre (ESTEC), Noordwijk, The Netherlands, where he is Senior Telecommunication Engineer in the Electrical Systems Department. He is presently responsible for the definition and development of advanced satellite communication systems. His actual interest is mainly related to efficient digital modulation and access and mobile satellite services, synchronization topics, and communication systems simulation techniques.

Tobias Friboe Garde was born in Copenhagen, Denmark, in 1966. He received the M.Sc. degree in electrical engineering from the Technical University of Denmark, Lyngby, in 1992.

Since 1993, he has been with the European Space Agency's Research and Technology Centre (ESA/ESTEC), Noordwijk, The Netherlands, where he is with the Electrical Systems Department, working on analysis and simulation of digital synchronization algorithms and SS-CDMA techniques for mobile communications.

Mr. Garde is a member of the Danish Society of Chemical, Civil (construction), Electrical and Mechanical Engineers.

Filippo Giannetti (M'93), was born in Pontedera, Italy, on September 16, 1964. He received the Doctor Engineer degree (cum laude) from the University of Pisa, Pisa, Italy, and the Research Doctor degree from the University of Padova, Padova, Italy, in 1989 and 1993, respectively, both in electronic engineering.

In 1992, he spent a research period at the European Space Agency Research and Technology Centre (ESA/ESTEC), Noordwijk, The Netherlands, where he was engaged in several activities in the field of digital satellite communications. He is currently a Research Scientist in the Department of Information Engineering, University of Pisa. His main research interests are in mobile and satellite communications, synchronization, and spread-spectrum systems.

Marco Luise (M'88) was born in Livorno, Italy, in 1960. He received the Doctor Engineer (cum laude) and Research Doctor degrees in electronic engineering from the University of Pisa, Pisa, Italy, in 1984 and 1989, respectively.

In 1987, he spent one year at the Eurpoean Space Research and Technology Centre (ESTEC), Noordwijk, The Netherlands, as a Research Fellow of the European Space Agency (ESA). Since 1988 to 1991, he was a Research Scientist for CNR, the Italian National Research Council, at the Centro Studio Metodi Dispositivi Radiotrasmissioni (CSMDR), Pisa. He chaired the fifth and sixth editions of the Tirrenia International Workshop on Digital Communications, in 1991 and 1993, respectively. He is now holding the position of Associate Professor in the Department of Information Engineering at the University of Pisa. His main interests lie in the broad area of communication theory, with particular emphasis on mobile and satellite communication systems and spread-spectrum communications.

Design Study for a CDMA-Based LEO Satellite Network: Downlink System Level Parameters

Savo G. Glisic, *Senior Member, IEEE*, Jaakko J. Talvitie, *Member, IEEE*, Timo Kumpumäki, Matti Latva-aho, *Student Member, IEEE*, Jari H. Iinatti, *Member, IEEE*, and Torsti J. Poutanen

***Abstract*— The performance analysis of a new concept of a code-division multiple-access (CDMA) based low earth orbit (LEO) satellite network for mobile satellite communications is presented and discussed. The starting point was to analyze the feasibility of implementing multisatellite and multipath diversity reception in a CDMA network for LEO satellites. The results will be used to specify the design parameters for a system experimental test bed. Due to the extremely high Doppler, which is characteristic of LEO satellites, code acquisition is significantly simplified by using a continuous wave (CW) pilot carrier for Doppler estimation and compensation. The basic elements for the analysis presented in this paper are: the channel model, the pilot carrier frequency estimation for Doppler compensation, and multipath and multisatellite diversity combining.**

I. Introduction

THE modern trend in the analysis of mobile satellite communications is very much oriented toward the implementation of the code-division multiple-access (CDMA) concept. Project examples are summarized in Table I. One can see that most of the projects are considering low earth orbit (LEO) constellations due to a limited link power budget. Unfortunately, this type of satellite constellation generates another problem due to severe Doppler in carrier and chip frequency. In addition to this, due to low minimum elevation angles (see Table I), multipath and shadowing are also a problem that has to be carefully addressed.

The existence of carrier and code Doppler result in a prolonged acquisition process or increased hardware complexity due to the need for a two-dimensional (delay and frequency) search of code synchronization. A number of papers have been published addressing this issue [1]–[5]. In our system, we suggest Doppler compensation prior to code synchronization, and analyze conditions under which this solution is feasible. The approach is based on the utilization of a continuous wave (CW) pilot carrier, and the analysis, based on known frequency estimation methods [8]–[16], provides values for the minimum power of the pilot carrier needed to provide reliable Doppler estimation. This is equivalent to the information of how much system capacity is to be sacrificed for the pilot carrier.

The multipath and shadowing problem can be solved by using multipath and multisatellite diversity reception. In this analysis, we start with the model of the channel and then based on the channel delay profile we analyze how much improvements we can expect from using a multipath RAKE receiver. To combat the shadowing problem, we consider feasibility of using multisatellite diversity reception. The overall analysis demonstrates that CDMA for LEO satellites is a feasible approach, and detailed system parameters are presented along with discussion of the possible system improvements.

In this initial analysis, we have not considered all aspects of receiver performance. Code acquisition, code tracking, and carrier frequency tracking and phase estimation are issues that we have not addressed in this paper. Also, the overall bit-error rate (BER) performance of the receiver with multisatellite combining is not discussed. However, all these issues are currently being analyzed in detail.

The paper is organized as follows: A general system model is presented in Section II, and the channel model is presented in Section III. Pilot carrier frequency estimation is discussed in Section IV. Multipath diversity (RAKE) and multisatellite diversity reception are discussed in Section V. The results are presented and discussed in Section VI, and the conclusions are presented in Section VII.

II. Model of Multisatellite CDMA Diversity Reception

In this paper, we analyze the downlink and the mobile unit receiver. We assume a synchronous channel and the spectrum of the transmitted signal as presented in Fig. 1(a). Analytically the transmitted downlink signal for the mth satellite can be expressed in time domain as

$$\begin{aligned} s_m(t) &= s'_{tm} + s_c + s_{sm} \\ &= \sum_{k=1}^{K} \mathrm{Re}\{C(k,m)\exp[(\omega_0 t + \varphi_m)]\} \\ &\quad + \mathrm{Re}\{\exp[j(\omega_0+\omega_c)t+\varphi_m]\} + P_m(t) \qquad (1) \end{aligned}$$

where the first term represents K traffic channels, the second term the pilot carrier for Doppler estimation, and the third term a pilot channel signal for synchronization purposes.

Manuscript received May 1, 1995; revised October 4, 1995. This work was supported by the European Space Agency under Contract 10885/94/NL/NB and it was carried out in cooperation with Elektrobit Ltd., Oulu, Finland.

S. G. Glisic, J. J. Talvitie, T. Kumpumäki, M. Latva-aho, and J. H. Iinatti are with the Telecommunication Laboratory, University of Oulu, Finland (email: savo.glisic@ee.oulu.fi).

T. J. Poutanen is with Elektrobit Ltd., Oulu, Finland.

Publisher Item Identifier S 0733-8716(96)05668-5.

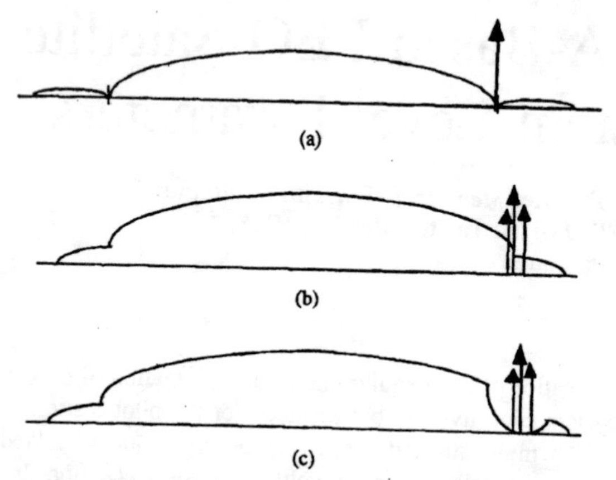

Fig. 1. Signal spectra. (a) Transmitted signal for synchronous channel (pilot carrier located to the spectral null), (b) received signal for synchronous channel at the input of the demodulator, and (c) received signal for synchronous channel at the input of the demodulator with prenotching.

Parameter $C(k,m)$ is a complex signal that can be expressed as

$$C(k,m) = C_r(k,m) + jC_i(k,m) = S_m(C_{kmr}d_{kmr} + jC_{kmi}d_{kmi}) \quad (2)$$

where S_m (real) stands for the "satellite code" which is a long code used to separate signals coming from different satellites, C_{km} (complex) stands for the kth user "channelization code" used to separate different users within the same satellite, and d_{km} (complex) is the convolutionally encoded information of the kth user of the mth satellite. C_{km} may be either the Walsh function or the Gold sequence. In general, different satellites may have different sets of users in the traffic channel, but all of them have the specific user we are interested in.

Let's say user $k = 1$ is common for all satellites, i.e., $d_{1m} = d_1, \forall m$. Because all other signals will appear as interference from the point of view of user 1, it is irrelevant whether or not $d_{km} = d_k, \forall m$ and $\forall k$. For notation simplicity, we will assume that this condition holds. The channelization codes C_k are the same for each satellite, i.e., $C_{km} = C_k, \forall m$. The structure of the satellite (long) code is the same for each satellite but initial states can be different, $S_m = S(t - \Delta_m)$ where Δ_m is a predetermined time shift. This is not necessary, and due to different delays of the signals coming from different satellites with the same initial state, $S_m = S, \forall m$ can be used. With these simplifications, (1) can be used with $C = C_r + jC_i = S(C_{kr}d_{kr} + jC_{ki}d_{ki})$.

A CW pilot is placed in between two CDMA signals (in the spectral null, see Fig. 1) so that zero or rather a low level of interference between the traffic channels and a pilot tone should be expected. In order to be able to operate with a low level of the pilot, the CDMA signal should be prenotched so that the estimation of the pilot signal frequency will not be affected by the variation of the number of users in the network.

This will be represented as $s_{mt} = s'_{mt} * f_{tn}$, where f_{tn} is the impulse response of the prenotch filter and $*$ stands for convolution. A multiple satellite signal spectrum, received by the mobile unit for a synchronous channel is presented in Fig. 1(b). This signal can be presented as

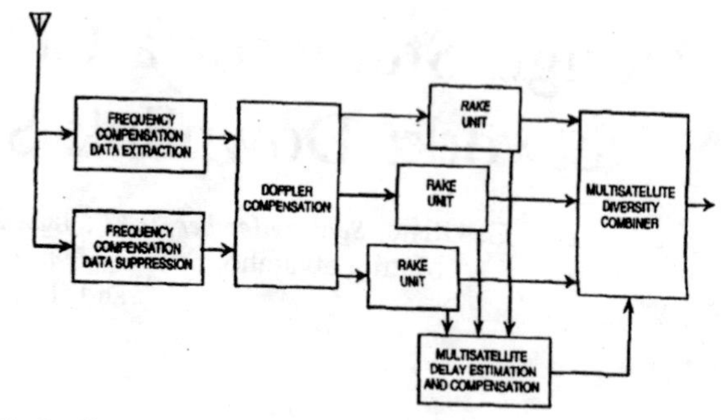

Fig. 2. General block diagram of the CDMA satellite receiver.

$$s'_r = \sum_m s'_{rm} = \sum_m s_m * f_{cm} \quad (3)$$

where f_{cm} is the channel impulse response for the transmission between satellite m and the receiver.

The overall system performance would be further improved if for LEO satellites the position of the different transmitters in the orbit could be such that the differential Doppler, Δf_D (difference in carrier Doppler) is larger than the bit rate R. For any other satellite constellation, where Δf_D for each pair of satellites is not larger than R, an additional frequency offset could be introduced in the pilot carrier of each satellite. An additional characteristic of the downlink signal structure is that all users are chip, symbol, and frame synchronous.

Instead of using a pilot tone, information about the Doppler can be distributed in the network from the central node by using an frequency shift keying (FSK) signal with a level higher than the level of the CDMA signal. This approach requires complex coordination within the network. Under these conditions, the receiver structure for three satellites is shown in Fig. 2. The operation of the receiver can be described as follows.

The first step is to detect the Doppler shift for each satellite. Independent of whether this information is sent by the network in the form of an FSK or M-ary FSK signal, or whether a pilot carrier is used, the detector may be based on fast Fourier transform (FFT). This block is designated in Fig. 2 as the "frequency compensation data extraction" block. Three separate frequency downconversions are performed by using local carriers $f_0 = \hat{f}_{Dm}$, where $\hat{f}_{Dm}$ is the estimated Doppler for the mth satellite (in our study $m = 1, 2, 3$).

Prior to frequency downconversion, frequency compensation data (FSK signal or CW pilot) can be suppressed by using an adaptive narrowband interference canceller based on the least mean square (LMS) algorithm [6], [7]. If the Doppler range is narrow compared with the signal bandwidth, a simple passive notch filter can be used for these purposes. In general, this can be analytically represented as $s_r = s'_r * f_{rn}$, where f_{rn} is the impulse response of the receiver notch filter. The third option is not to suppress this signal but to use only the inherent processing gain of the system. This will slightly reduce the system capacity.

Each frequency downconversion will produce a sum of three signals, one of which will be with essentially no Doppler, and the other two which will contain a residual Doppler.

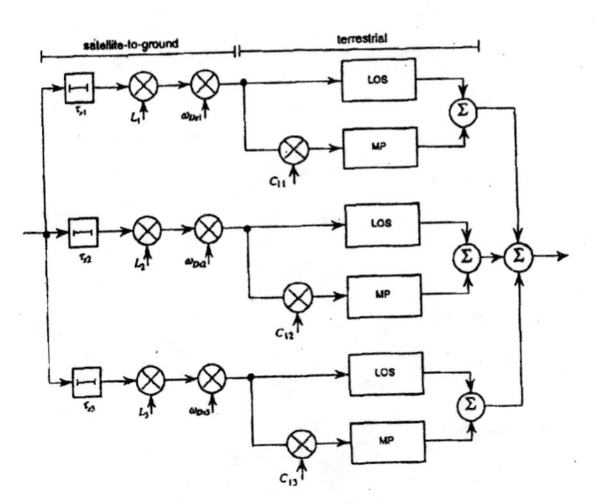

Fig. 3. Overall structure of the channel model.

If the residual Doppler is larger than the bit rate, these two signals will be additionally suppressed in the correlation process so that, in further processing, only one signal from a corresponding satellite will be dominant and signals from the other satellites will be suppressed. The additional separation between two signals from different satellites is based on different delays so that the long code will be able to separate such signals. In general, further processing is based on a RAKE receiver. Once a sample of the decision variable is formed at the output of a RAKE unit, this sample is combined in the multisatellite combiner with the outputs of the other RAKE units to receive the final decision about the bit being received.

Prior to combining, the differential delays between the satellites have to be estimated so that the samples of the same bit are combined for the final decision. This function is formally represented as a separate block. In practice, this means proper synchronization of the long code in each RAKE unit.

III. Channel Model

In this section, we propose a wideband land mobile satellite channel model that is based on preliminary results from a wideband measurement campaign. The model is a statistical (synthetic) model, with part of the parameters being deterministic functions of the satellite orbit geometry. The model is a general-purpose model in the sense that it is adaptable to different system configurations, i.e., different satellite constellations and user terminal types. Multisatellite reception with up to three visible satellites is included in the model as three separate satellite links. Each satellite link model consists of a satellite-to-ground part (deterministic) and a terrestrial part (statistical). The proposed model is restricted to the down-link only and includes the propagation channel and antenna effects.

The model parameters are functions of the satellite constellation (LEO), the user terminal type (hand-held, portable and vehicle-mounted), and the elevation angle. Furthermore, two environment types, rural and urban, are considered. An overall block diagram of the channel model is shown in Fig. 3.

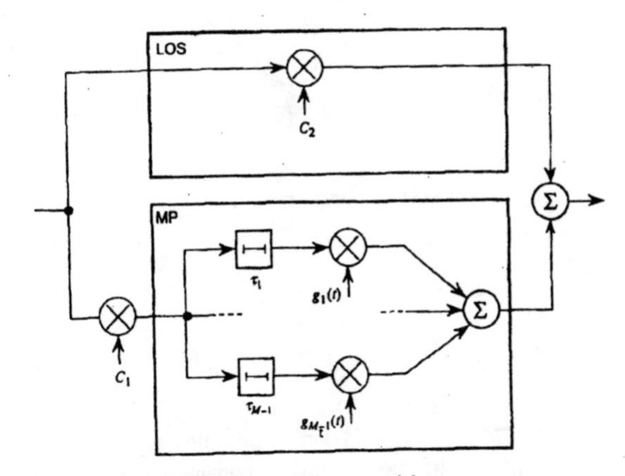

Fig. 4. Detailed block diagram of the terrestrial part.

The satellite-to-ground part is modeled by a delay τ_{sm}, an overall attenuation L_m (including both free-space loss, and antenna pattern effects), and a Doppler shift ω_{Dsm}. The modeling of the terrestrial part of propagation is based on a combination of currently available narrowband models [17], [18] and a wideband tapped delay line model. The approach is to extend the narrowband models to wideband models by replacing the Rayleigh component in the narrowband models with a tapped delay line model.

The terrestrial part structure includes M_t taps, divided to the line-of-sight (LOS) signal and an overall multipath (MP) component. A general block diagram of the LOS and MP blocks of the terrestrial part is shown in Fig. 4, where the coefficients $g_i(t)$ are uncorrelated complex Gaussian processes. Note that the first tap $(i = 0)$ is reserved for the LOS path, and the $M_t - 1$ remaining taps for the multipath part (MP). The general structure shown in Fig. 4 is used in slightly different configurations for three separate terrestrial environment cases: rural, urban unshadowed (urban good state), and urban shadowed (urban bad state). The real-valued coefficients C_1 and C_2 have different meanings and values in these configurations.

A. Rural Channel

In the rural narrowband case, the model of Loo [17] is adopted. An additive combination of Rayleigh and log-normal fading is therefore assumed in this case. The probability density function (pdf) $p(r)$ of the received signal envelope is given [17] by

$$p(r) = \frac{r}{\sigma_R^2 \sigma_{LN} \sqrt{2\pi}} \int_0^\infty \frac{1}{z} \times \exp\left[-\frac{(\ln(z) - \mu_{LN})^2}{2\sigma_{LN}^2} - \frac{r^2 + z^2}{2\sigma_R^2}\right] I_0\left(\frac{rz}{\sigma_R^2}\right) dz \quad (4)$$

where I_0 is the zeroth order modified Bessel function of the first kind. The parameters for the model are the mean μ_{LN} and the standard deviation σ_{LN} of the log-normal fading, and the average power of the Rayleigh fading $2\sigma_R^2$. The model corresponding to (4) is presented in Fig. 5, with a real-valued

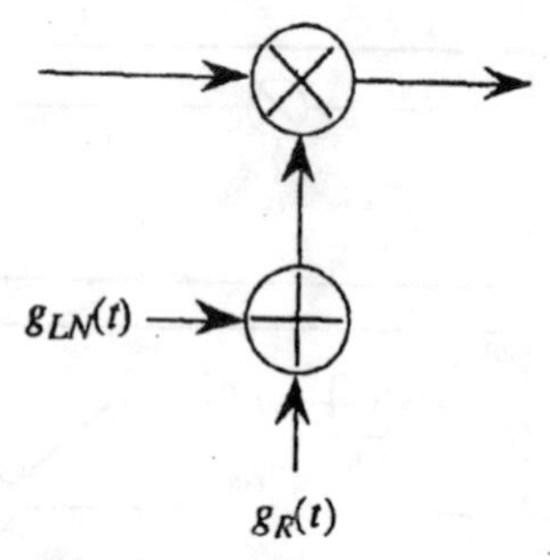

Fig. 5. Terrestrial channel model by Loo [17].

log-normal process $g_{LN}(t)$ and a complex-valued Rayleigh process $g_R(t)$. The model can be considered to consist of Rician fading, where the Rice-factor (LOS to multipath power ratio) is modulated by a log-normal process.

Expanding the narrowband model of Fig. 5 to the wideband structure shown in Fig. 4, we replace C_2 by $g_{LN}(t)$ and C_1 by G_{mp}, where the parameter G_{mp} controls the multipath power. Requiring that the overall average power of the MP part is equal to unity, we have that $G_{mp} = \sqrt{2\sigma_R^2}$.

B. Urban Channel

The terrestrial part in the urban case is modeled after Lutz *et al.* [18], as shown in Fig. 6. The model consists of a combination of Rician fading (left branch in Fig. 6) and log-normally shadowed Rayleigh fading (right branch in Fig. 6). In this case, note that the log-normal and Rayleigh processes are multiplicative, not additive—which is different from the Loo model.

The Lutz model is characterized by four parameters; the mean μ_{LN} and standard deviation σ_{LN} of the log-normal distribution, the Rice-factor c, and the time-share of shadowing A (giving the average share of time spent in the shadowed state). The parameter A is defined using the average durations of the good and the bad states (D_g and D_b, respectively, as shown in Fig. 6) as [18]

$$A = \frac{D_b}{D_g + D_b}. \tag{5}$$

The overall pdf of the received signal power is defined in [18] by combining the Rician and the Rayleigh/log-normal densities, weighted by $1 - A$ and A, respectively.

Expanding the narrowband model of Fig. 6 to the wideband structure shown in Fig. 4, we replace in the good state C_2 by unity and C_1 by $c^{-1/2}$. In the bad state, we replace C_2 by zero and C_1 by $g_{LN}(t)$. For simplicity, it is again required that the overall average power of the MP part is equal to unity.

C. The Multipath Part

From the above, it is clear that the MP part is the same in all model configurations. Each resolvable multipath signal, i.e., each tap in the model, is by itself a sum of several unresolved, reflected and scattered signal components, and is therefore subject to severe fading. This unresolved multipath is included in the time-variant path coefficients $g_i(t)$. The coefficients are

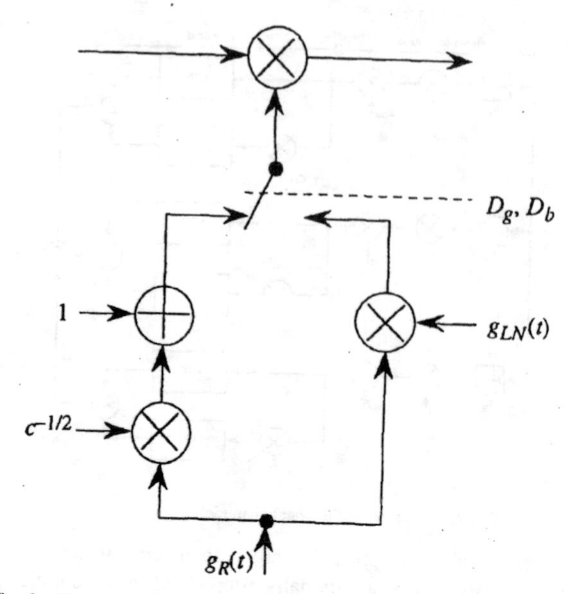

Fig. 6. Terrestrial channel model by Lutz *et al.* [18].

assumed to be complex zero-mean Gaussian signals, resulting in a Rayleigh-distributed amplitude. The fading rate of the path coefficients is determined by the bandwidth (total Doppler spread) of the signals $g_i(t)$.

The power spectra of the path coefficients depend on the type of environment considered. For an urban area, the spectra are all of the classical (Jakes) type [18]. For a received unmodulated carrier this doppler spectrum is defined as [19]

$$S(f) = \begin{cases} \dfrac{2\sigma_R^2}{2\pi f_{Dm}\sqrt{1-\left(\frac{f}{f_{Dm}}\right)^2}}, & \text{for } -f_{Dm} < f < f_{Dm} \\ 0, & \text{otherwise} \end{cases} \tag{6}$$

where f_{Dm} is the maximum terrestrial Doppler shift.

For rural environments, a spectrum with a truncated first-order Butterworth low-pass characteristic is adopted [17], with f_{Dm} as the 3 dB bandwidth. For a received unmodulated carrier this doppler spectrum is defined as

$$S(f) = \begin{cases} \dfrac{1}{1+\left(\frac{f}{f_{Dm}}\right)^2}, & \text{for } -f_{Dm} \le f \le f_{Dm} \\ 0, & \text{otherwise.} \end{cases} \tag{7}$$

Since the amplitude response of the filter is at f_{Dm} only 3 dB below the zero frequency value, the filter specified by (7) is effectively a close approximation of an ideal low-pass filter with a rectangular amplitude response.

IV. Pilot Carrier Frequency Estimation

The proposed system initialization is based on pilot carriers which are placed in the spectral null frequency of the wideband satellite transmitted signal. Different satellites are seen at different carrier frequency offsets at the mobile station. Hence, the Doppler shifts from different satellites can be estimated from the pilot carriers. Modern spectral analysis provides several advanced techniques for spectral line estimation [10], which are shown to be powerful—especially for short data records. In practice, the classical spectral estimation methods are satisfactory in most applications. The maximum likelihood estimate for a single frequency is the maximum value of the periodogram [10]. The periodogram-based spectral estimators lead to FFT algorithms which are quite practical and hardware components are commercially available.

Regardless of the implementation of intermediate frequency (IF) stages and sampling technique used in the receiver frontend, the pilot carriers have to be filtered before spectral analysis. The filter bandwidth has to be at least 100 kHz, which is approximately twice the maximum Doppler shift. The filter should also include some kind of windowing in order to avoid spectral leaking due to a finite length data record [8]. The window selection for this kind of application is discussed in [9]. From the code acquisition point of view, the frequency should be estimated at the accuracy of a few hundred Hz.

A 1024 point FFT will be used in the analysis, which gives the frequency estimate with ± 100 Hz accuracy at a certain probability. For a 200 kHz sampling frequency and 1024 point FFT, new frequency estimates are produced once every 5 ms when processing the samples on a block-by-block basis. The maximum frequency change during the processing due to Doppler rate is 1 Hz. Since the frequency change is very small compared with the accuracy requirements, some extra post-processing can be used to further decrease the probability of error in frequency detection. It should be noted that the best performance is obtained by using a nonlinear post-processor, which can be based, for example on multiple decision and majority logic selection. The post-processor also has to select the three strongest frequencies for receiving signals from different satellites. The post-processing problem is not elaborated in this paper.

Let us start the with probability of pilot carrier detection from one satellite. Each signal component at the output of the FFT block contains either pure noise or signal plus noise. If we are interested only in pilot carrier detection from one satellite, we will start with an assumption that there will be only one spectral component including the signal and all the other components are pure noise. Later on, we will modify the analysis to include the fact that, due to terrestrial Doppler spreading, even when one tone is transmitted the received signal consists of a number of spectral lines. So, the probability of correct detection is

$$p_c = \Pr[|s+n_i| > |n_j|], \quad \forall j \neq i \tag{8}$$

where i, j is the index of the spectral line. The probability of detecting one correct spectral line out of M lines is

$$\begin{aligned} p_c &= \{1 - \Pr[|s+n_i| < |n_j|]\}^{M-1}, \quad \forall j \neq i \\ &= \{1 - p_f\}^{M-1} \end{aligned} \tag{9}$$

where p_f is the probability of false detection. The envelope pdf for the noise bins is a Rayleigh distribution, and for the signal plus noise bins a Rician distribution.

By limiting our study to the Rician channel, the overall estimation error probability has to be evaluated by further averaging p_f over the Rician distribution due to fading

$$\begin{aligned} p_\varepsilon = 1 - \Bigg\{1 - \int_0^\infty \Bigg[\frac{1}{2}\exp\left(\frac{-x^2}{2\sigma^2}\right) \cdot \frac{x}{\sigma_R^2} \\ \cdot \exp\left(\frac{-(x^2+\gamma^2)}{2\sigma_R^2}\right) \cdot I_0\left(\frac{x\gamma}{\sigma_R^2}\right)\Bigg] dx\Bigg\}^{M-1} \end{aligned} \tag{10}$$

where γ^2 is the power of the LOS signal component $2\sigma_R^2$ and σ^2 is the noise power.

In (10), the first term under the integral is probability p_f which is then averaged over the Rician distribution of the signal envelope. The total received signal power is given by $2\sigma_R^2 + \gamma^2$, and $\gamma^2/2\sigma_R^2$ is the Rice factor. Two special cases can be seen from (10). In pure Rayleigh fading ($\gamma^2 = 0$), (10) will become

$$P_\varepsilon = 1 - \left\{1 - \frac{1}{2 + \frac{\sigma_R^2}{\sigma^2}}\right\}^{M-1} \tag{11}$$

where the quotient $1/(2 + \sigma_R^2/\sigma^2)$ is the error probability p_f in a Rayleigh-fading channel. If there is no multipath propagation, (10) will become

$$P_\varepsilon = 1 - \left\{1 - \frac{1}{2}\exp\left(-\frac{\gamma^2}{2\sigma^2}\right)\right\}^{M-1} \tag{12}$$

where the expression $0.5\exp(-\gamma^2/2\sigma^2)$ is p_f in an additive white Gaussian noise (AWGN) channel.

If the received signal consists of M_1 spectral components due to terrestrial Doppler spreading, then M_1 bins will consist of signal plus noise $s_i + n_i$, $i = 1, 2, \cdots, M_1$, and $M - M_1$ bins will consist of noise only. The estimation process will now be looking for the M_1 largest signals out of the M bins. If the M_1 largest outputs are grouped in M_1 consecutive bins, then all signal components are correctly detected and we have an estimate with a precision $\Delta f = 2f_D/m$, where f_D is the maximum expected Doppler shift. If M_1 largest outputs are not grouped in M_1 consecutive bins, we choose the largest output and have an estimate with precision $M_1\Delta f$.

If for simplicity we assume that all signal spectral components are of the same amplitude then (9) for Δf precision becomes

$$p_c(\Delta f) = \left[(1-p_f)^{M-M_1}\right]^{M_1} \tag{13}$$

and for $M_1\Delta f$ precision

$$p_c(M_1\Delta f) = 1 - \left[1 - (1-p_f)^{M-M_1}\right]^{M_1} \tag{14}$$

where $p_f = \Pr\{|s_b + n_i| < |n_j|\}$, and s_b is the signal component per bin. Further averaging in accordance with (10)–(12) depends on the signal distribution, which in turn depends on the satellite elevation angle. Extension of this analysis to the case when the amplitudes of the signal spectral components are not the same is straightforward.

Equations (10)–(12) were used in the numerical analysis to find the performance of the FFT based frequency estimator in the good urban LEO satellite channel for different elevations angles. Let us now derive the equation for pilot carrier-to-user signals ratio. The FFT input signal-to-noise ratio (SNR) is

$$\mathrm{SNR} = \frac{P}{I' \cdot (a_1 + a_2 + a_3)KS + N_0 W_d} \tag{15}$$

where P is the power of pilot carrier, a_m is the signal power ratio of the mth satellite compared to the satellite whose frequency is being estimated, S is the power of one user signal, N_0 is the thermal noise density, W_d is the filter bandwidth preceding FFT (twice the maximum Doppler range), and

$$I' = \int_{R_c - W_d}^{R_c} \left(\frac{\sin(\pi R_c x)}{\pi R_c x} \right)^2 dx \tag{16}$$

which is the relative power of the wideband user signal near the spectral null (for $R_c = 1.8$ MHz, $I' = 6.087 \cdot 10^{-4}$, and for $R_c = 9.8$ MHz, $I' = 5.408 \cdot 10^{-7}$). The pilot carrier to user signals ratio is

$$\frac{P}{KS} = \mathrm{SNR} \cdot \left(1 + \frac{G}{K} \frac{W_d}{R_c} \frac{N_0}{E_b} \right) \tag{17}$$

where $I = I' \cdot (a_1 + a_2 + a_3)$, R_c is the wideband signal bandwidth and G is the processing gain. At $E_b/N_0 = 0$ dB (value specified for the test bed design) the (17) becomes

$$\frac{P}{KS} = \mathrm{SNR} \cdot \left(1 + \frac{G}{K} \frac{W_d}{R_c} \right). \tag{18}$$

V. Multipath and Multisatellite Combining

In this section, the benefit of RAKE combining of terrestrial multipath signals is discussed. The analysis is based on the calculation of the average bit-error probabilities (BEP) with and without RAKE with path strength coefficients having distributions discussed in Section III. A channel model with M_p propagation paths is used. In the analysis, the urban channel has been studied, neglecting the log-normal fading. Thus, the LOS path has been assumed to be fixed during the good channel state ($1 - A$ part of the time) and absent during the bad channel state (A part of the time). All other paths are assumed to be Rayleigh fading. Maximum ratio combining is assumed in the RAKE receiver and the number of RAKE arms is equal to the number of propagation paths. So, the upper limit on system performance is obtained.

The amplitudes of M_p propagation paths are denoted as α_i, $i = 0, 1, \cdots, M_p - 1$. The first path ($i = 0$) is the direct path (LOS), the one the receiver tries to lock on in the no RAKE case.

In the no RAKE case the instantaneous SNR is given by

$$\mathrm{SNR}_{nr} = \frac{Y \alpha_0^2}{1 + \frac{Y}{G} \sum_{i=1}^{M_p - 1} \alpha_i^2} \tag{19}$$

where $Y = E_b/N_0$, $N_0 = N_{th} + N_{mu}$ (thermal noise power density + multi-user interference power density), and G = processing gain. The channel thermal noise density is assumed to be such that equal noise power is received in each RAKE arm.

With maximum ratio combining, the signal amplitude in each RAKE arm is multiplied by a weighting factor w_i equal to the corresponding path strength α_i. Hence, the instantaneous SNR becomes

$$\mathrm{SNR}_r = \frac{Y \left(\sum_{j=1}^{M_p - 1} w_j \alpha_j \right)^2}{\sum_{j=0}^{M_p - 1} w_j^2 + \frac{Y}{G} \sum_{j=0}^{M_p - 1} \sum_{\substack{i=0 \\ i \neq j}}^{M_p - 1} (w_j \alpha_j)^2}. \tag{20}$$

Equations (19) and (20) were used to define the instantaneous BEP for a given SNR. For coherent reception (PSK) we have $P_e = 0.5 \cdot \mathrm{erfc}\sqrt{0.5 \cdot \mathrm{SNR}}$.

When the path strengths are not fixed but have a certain distribution, the average BEP can be calculated by averaging the instantaneous BEP over the pdf's of these path strength coefficients. In the urban case, neglecting for this analysis the slow log-normal fading, all reflected paths are Rayleigh fading, so that the pdf of α can be defined as $p(\alpha) = (\alpha/\sigma_\alpha^2)\exp(-\alpha^2/2\sigma_\alpha^2)$, where $2\sigma_\alpha^2$ corresponds to the average power of each path. To get the average BEP, the following integral must be evaluated

$$P_{av} = \int_0^\infty \int_0^\infty \cdots \int_0^\infty P_e(\alpha) \cdot p(\alpha_0) \cdot p(\alpha_1) \cdots p(\alpha_{M-1}) d\alpha_0 d\alpha_1 \cdots d\alpha_{M-1} \tag{21}$$

where $\alpha = (\alpha_0, \alpha_1, \cdots, \alpha_{M_p - 1})$.

For simplicity only the two or three strongest paths ($M_p = 2$ or 3) were taken into account in our analysis. To receive the final average BEP in the urban case, the bad and good channel states were also taken into account. Thus, the final form used in the calculations for three paths becomes

$$P_a = (1 - A) \cdot \int_0^\infty \int_0^\infty P_e(\alpha_1, \alpha_2) \cdot p(\alpha_1) \cdot p(\alpha_2) d\alpha_1 d\alpha_2 + A \cdot \int_0^\infty \int_0^\infty \int_0^\infty P_e(\alpha_0, \alpha_1, \alpha_2) \cdot p(\alpha_0) \cdot p(\alpha_1) \cdot p(\alpha_2) d\alpha_0 d\alpha_1 d\alpha_2 \tag{22}$$

and for the two path case

$$P_a = (1 - A) \cdot \int_0^\infty P_e(\alpha_1) \cdot p(\alpha_1) d\alpha_1 + A \cdot \int_0^\infty \int_0^\infty P_e(\alpha_0, \alpha_1) \cdot p(\alpha_0) \cdot p(\alpha_1) d\alpha_0 d\alpha_1. \tag{23}$$

The multipath diversity combining gain is then defined as

$$G_{mc} = \frac{\mathrm{SNR}_0}{\mathrm{SNR}_{mc}} \tag{24}$$

where SNR_0 and SNR_{mc} are the SNR's required to achieve the same BEP without and with multipath combining, respectively.

The gain obtained by multisatellite diversity combining can be evaluated similar to the multipath combining gain analyzed above. In the case of other combining strategies than the maximum ratio combining (for example, selection combining), results available in the literature can be applied in (20).

An initial estimate of the gain obtained by utilizing multisatellite diversity can be evaluated as follows. Let us define the probability of a single link to be unavailable to be equal to the parameter A (time-share of shadowing) [18], which was

TABLE I
SATELLITE COMMUNICATION CONCEPTS

SYSTEM	Number of Satellites	Number of Orbits	Altitude [km]	Orbit inclination [dgr]	Minim. elev. angle [dgr]	Number of beams/ satell.	Transmitter power [W]	Estimated start of service	Band-width [MHz]
GLOBAL STAR	48	8	1389	47	5 - 10	6	875	1999	
ODYSSEY	12	3	10 335	55	> 22	19 - 37	1800	1999	16.5
CONSTEL-LATION	48	4	1020	90	7.5	7	250	1996	16.5
ELLIPSO	24	1 (HEO)	429 / 2903	63.4	> 5	4	900	1996	16.5
ARCHI-MEDES	6	1 (8 or 16 hours orbit)	1000 / 26 784	63.4	> 40	4 - 7	2700		

shown to be a function of the elevation angle $A = A(\theta)$ and thus also a function of time t, $A = A(t)$. Then the probability that all links are not available is given by

$$A_{M_S} = \prod_{m=1}^{M_S} A_m(\theta_m) \tag{25}$$

where $A_m(\theta_m)$ is the probability of link m at elevation θ_m to be unavailable, and M_s is the number of satellites whose signals are combined. For a given satellite constellation, the mutual relation between the elevation angles θ_m is well defined, and if one of these parameters is known at each time instant, then all of them are known. Therefore, the parameter A_{M_s} should be understood as a function of time, although not shown explicitly by the notation used.

By defining the availability of the overall link to be equal to the probability of having the LOS component available in at least one of the satellite links, the result is a system availability given as

$$av_{M_S} = 1 - A_{M_S}. \tag{26}$$

The multisatellite availability gain is then defined as

$$G_{ms} = \frac{av_{M_s}}{av_1}. \tag{27}$$

VI. RESULTS

A. Channel Model

A realistic LEO satellite constellation has been considered as detailed in the below sections.

1) Constellation: 48 satellites in eight circular polar orbits (inclination 52°), orbit altitude 1389 km, and the maximum number of visible satellites is three.

2) Model: The parameters of each branch depend on the elevation angle of each satellite.

The user terminal type and environment type (rural or urban) are taken into account in generating the parameter values for each model configuration. To narrow down the wide variety of different satellite geometries and to simplify the computations involved, the following assumptions have been made.

1) The rotation of the earth is not taken into account.
2) The satellite-to-ground parameter values are derived from realistic satellite scenarios.
3) The velocities of the user terminals are:

$$\begin{aligned} \text{hand-held terminal (HH)}: &\quad 5 \text{ km/h } (1.4 \text{ m/s}),\\ \text{portable terminal (PT)}: &\quad 5 \text{ km/h } (1.4\text{m/s})^1,\\ \text{vehicular terminal (VH)}: &\quad 100 \text{ km/h}(27.8 \text{ m/s}); \end{aligned}$$

4) The carrier frequency equals 2.2 GHz (S band).
5) In the satellite-to-ground part, time-variance is due to satellite movement only. In the terrestrial part, time-variance is only due to the movement of the terminal. The environment is assumed to be fixed.
6) The antenna orientation is assumed to be fixed in all cases, such that the 0° direction of the gain pattern (which usually is the direction of maximum gain) points toward the zenith.

Values for the satellite-to-ground part parameters were derived from realistic satellite orbit data. Global ranges of values (over complete orbit periods) for the parameters and their rates are listed in Table II. The minimum elevation angle was chosen to be 10°.

From the point of view of receiver performance, it is also of interest to know the maximum parameter value differences

[1] This means that the time-variance of the terrestrial environment is modeled by a moving PT terminal in a fixed environment. From the point of view of the channel model this is equivalent to a fixed PT terminal in a changing environment.

TABLE II
GLOBAL RANGES OF PARAMETER VALUES FOR SATELLITE-TO-GROUND PARAMETERS S BAND, OMNIDIRECTIONAL ANTENNA

Parameter	range
Delay	4.6...10.7 ms
Delay rate	-18.2...17.9 µs/s
Attenuation	162.1...169.4 dB
Attenuation rate	≤ ±0.015 dB/s
Doppler shift	-40.1...39.3 kHz
Doppler rate	-198...-13 Hz/s

TABLE III
MAXIMUM SATELLITE-TO-GROUND PART PARAMETER DIFFERENCES BETWEEN TWO SATELLITES AT ANY TIME INSTANT

Parameter	difference
Delay	6.1 ms
Attenuation	7.3 dB
Doppler shift	75.5 kHz
Doppler rate	168 Hz/s

between two visible satellites at any time instant. These are listed in Table III. Note that in both Tables II and III, the values do not include the effects of the antennas.

To get the parameter values for the terrestrial part as functions of the elevation angle, the model parameters fitted against preliminary wideband measurement data were used as a starting point. Least-squares polynomial fits relating the parameters to the elevation angle were then computed. In making the fits, the differences in terminal antenna patterns were taken into account. The HH antenna was assumed to have a vertically more omnidirectional-type pattern, whereas the PT and VH antennas were assumed to be similar, with a vertically more directive pattern. The resulting fits are as follows.

The rural model results are

$$\begin{aligned} G_{mp\mathrm{HH}}(\theta) &= -2.5319 \cdot 10^{-3}\theta + 0.3485 \\ G_{mp\mathrm{VH/PT}}(\theta) &= -3.2300 \cdot 10^{-3}\theta + 0.2981 \\ \mu_{LN}(\theta) &= 6.0763 \cdot 10^{-6}\theta^3 - 1.3809 \cdot 10^{-3}\theta^2 \\ &\quad + 0.1039\theta - 2.6468 \\ \sigma_{LN}(\theta) &= -5.6742 \cdot 10^{-3}\theta + 0.6419. \end{aligned} \tag{28}$$

The urban model results are

$$\begin{aligned} A(\theta) &= -0.0177\theta + 1.0095 \\ A &= 0 \quad \text{for} \quad \theta \geq 57.0339^\circ \\ \sigma_{LN}(\theta) &= -0.0979\theta + 8.2036 \text{ dB} \\ \sigma_{LN} &= 0 \text{ dB for } \theta \geq 83.8321^\circ \\ c_{\mathrm{HH}}(\theta) &= 0.2393\theta - 4.2679 \text{ dB} \\ c_{\mathrm{VH/PT}}(\theta) &= 0.3282\theta - 3.9554 \text{ dB}. \end{aligned} \tag{29}$$

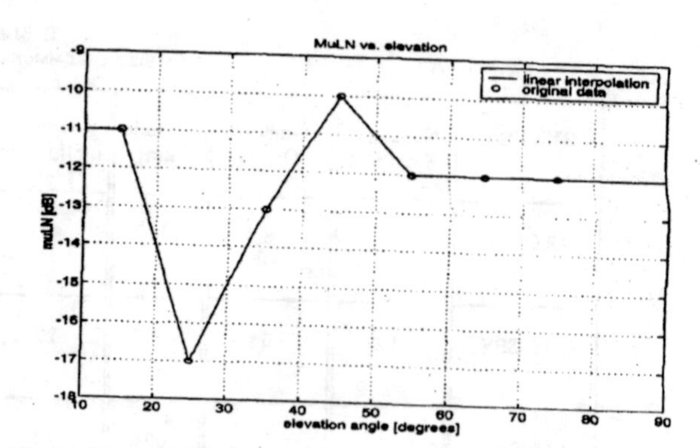

Fig. 7. Dependence of bad urban case parameter μ_{LN} on elevation angle.

The bad urban case parameter μ_{LN} could not be fitted to any reasonable polynomial. Therefore, the values of μ_{LN} are linearly interpolated between original data points, as shown in Fig. 7.

Computing the resulting parameter values for elevations 10°–90°, we get the parameter ranges shown in Table IV for the rural case, and in Table V for the urban case.

For the MP part parameters, the same preliminary measurement data was used to define delay profiles for the rural and urban cases. Using a tap spacing of 0.1 μs, the number of taps M_t was defined to be four in the rural and six in urban case. Remembering the requirement of unity for the overall average power of the MP part, the resulting MP delay profiles for taps $1, \cdots, M_t - 1$ are shown in Tables VI and VII for the rural and urban cases, respectively. In Tables VI and VII, P_i represents the average power of tap i.

B. Pilot Carrier Frequency Estimation

Using the results from Section IV, the required SNR for a frequency estimation error rate (FEER) of 10^{-3} with different Rice factors is presented in Table VIII.

The Rice factor 3.7 dB corresponds to 33° and 23° elevation angles, 5 dB corresponds to 39° and 27°, and 10 dB corresponds to 60° and 42° elevation angles in the HH and VH/PT terminal cases, respectively. The pilot carrier to all user signals ratios from one satellite, required to get a certain FEER for different chip rates, are presented in Tables IX–XI. The relative powers were assumed to be: $a_1 = 0$ dB, $a_2 = 10$ dB, and $a_3 = 10$ dB. The processing gains ($G = 64$ and $G = 512$) correspond to 28.8 kHz and 2.6 kHz symbol rates for $R_c = 1.8$ MHz; $G = 512$ and $G = 8192$ correspond to 19.2 kHz and 1.2 kHz symbol rates for $R_c = 9.8$ MHz. Those are the best and worst cases from the signal-to-noise ratio (E_b/N_0) point of view.

The final decision about the frequency value can be obtained by taking D successive primary decisions and using majority logic for the final decision. Let us suppose that during the time interval needed for D primary decisions the value of the frequency is in the same frequency slot. Thus, the best decision would be if we take the slot with the largest number of primary positive decisions. A modification of this approach is to put a threshold d and accept for the final decision a

TABLE IV
RANGES OF PARAMETER VALUES FOR RURAL TERRESTRIAL MODEL PARAMETERS

Parameter	range
G_{mpHH}	0.3232...0.1206
$G_{mpVH/PT}$	0.2658...0.0074
μ_{LN}	-1.7400...-0.0533
σ_{LN}	0.5851...0.1312

TABLE V
RANGES OF PARAMETER VALUES FOR URBAN TERRESTRIAL MODEL PARAMETERS

Parameter	range
A	0.83...0
c_{HH}	-1.8...17.3 dB
$c_{VH/PT}$	-0.7...25.6 dB
σ_{LN}	7.2...0 dB

TABLE VI
RURAL MP PART PARAMETERS

i	τ_i [μs]	P_i
1	0.1	0.704
2	0.2	0.225
3	0.3	0.071

TABLE VII
URBAN MP PART PARAMETERS

i	τ_i [μs]	P_i
1	0.1	0.588
2	0.2	0.247
3	0.3	0.106
4	0.4	0.041
5	0.5	0.018

frequency slot with the number of primary decisions larger than the threshold $d = D/2 + 1$.

If the probability of error of a primary decision is p_f, the final decision will be correct with the probability of

$$p_c = \sum_{k=\frac{D}{2}+1}^{D} \left(\frac{D!}{K!(D-k)!}\right)(1-p_f)^k \, p_f^{D-k}. \quad (30)$$

Hence, the probability of an incorrect final decision can be expressed as

$$p_e = 1 - p_c. \quad (31)$$

In Table XII, the probabilities of incorrect final decisions p_e are presented as a function of parameter D and probability of error of primary decision (p_f). It can be seen that even for a rather high probability of incorrect primary decision we can have a reliable final estimation of the frequency.

We should be aware that due to Doppler within D observation intervals, the frequency can move from one slot to the next (this is equivalent to having a few spectral components in the signal). For that reason we will be looking not only into one slot but into a number of adjacent slots (cluster) at the same time.

C. Multipath Combining

In the analysis, six delay profiles for an urban environment were chosen, corresponding to elevation angles 10°, 15°, 25°, 35°, 45°, and 55°. The value of A for each elevation angle was taken from Section III. The resulting SNR gains G_{mc} as a function of elevation angle are presented in Fig. 8.

In the rural case, the improvement obtained by using RAKE is insignificant even at low elevation angles. Thus, it would be beneficial to use a maximum ratio RAKE receiver for multipath combining only in urban environment, at elevation angles lower than or equal to 45°.

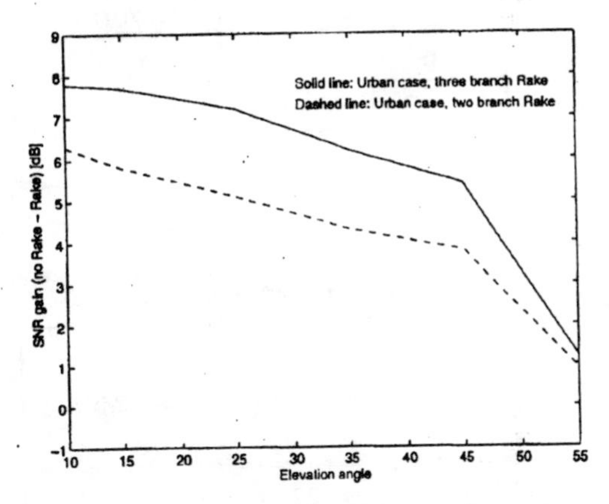

Fig. 8. SNR gain to achieve average BER of 10^{-2} (RAKE versus no RAKE).

It should be noted that the calculations made in this section are based on channel delay profiles in which the delay differences between taps are 0.1 μs. Basically, this means that if we want to resolve multipath components, the chip rate should be at least 10 MHz. On the other hand, the maximum excess delay defined in the channel model equals 0.5 μs. This means that it does not make sense to use RAKE if the chip rate is of the order of 2 MHz or below.

As a final conclusion, it is beneficial to use multipath RAKE only when the following conditions apply simultaneously: the elevation is low (≤45°), the mobile terminal is situated in an urban area, and the chip rate is high (≥10 MHz).

D. Multisatellite Combining

The evaluation of the multisatellite availability gain G_{ms} is complicated by the large amount of different situations

TABLE VIII
REQUIRED SNR'S TO ACHIEVE 10^{-3} FEER FOR DIFFERENT RICE FACTORS, FOR $M = 1024$

Rice factor [dB]	SNR [dB] (after FFT)	SNR–10log(*M*) [dB] (before FFT)
+∞	14	–16
–∞	60	30
15.4	16	–14
10	28	–2
5	52	22
3.7	54	24
0	56	26

TABLE IX
REQUIRED PILOT CARRIER TO ALL USER SIGNAL RATIOS TO ACHIEVE A CERTAIN FEER FOR ELEVATION ANGLES 23° (VH/PT) AND 33° (HH), $M = 1024$ AND $K = 100$ FOR $R_c = 1.8$ MHz AND $K = 1000$ FOR $R_c = 9.8$ MHz

FEER	*SNR* [dB] (before FFT)	*P/KS* [dB] R_c=1.8MHz K=100, G=64	*P/KS* [dB] R_c=1.8MHz K=100, G=512	*P/KS* [dB] R_c=9.8MHz K=1000, G=512	*P/KS* [dB] R_c=9.8MHz K=1000, G=8192
10^{-3}	24	10.8	18.6	1.2	13.2
10^{-2}	15	1.8	9.6	-7.8	4.2
10^{-1}	5	-8.2	-0.4	-17.8	-5.8

TABLE X
REQUIRED PILOT CARRIER TO ALL USER SIGNAL RATIOS TO ACHIEVE A CERTAIN FEER FOR ELEVATION ANGLES 27° (VH/PT) AND 39° (HH), $M = 1024$ AND $K = 100$ FOR $R_c = 1.8$ MHz AND $K = 1000$ FOR $R_c = 9.8$ MHz

FEER	*SNR* [dB] (before FFT)	*P/KS* [dB] R_c=1.8MHz K=100, G=64	*P/KS* [dB] R_c=1.8MHz K=100, G=512	*P/KS* [dB] R_c=9.8MHz K=1000, G=512	*P/KS* [dB] R_c=9.8MHz K=1000, G=8192
10^{-3}	22	7.5	16.5	-0.8	11.2
10^{-2}	12	-2.5	6.5	-10.8	1.2
10^{-1}	2	-12.5	-3.5	-20.8	-8.8

(satellite geometries) offered by any satellite system. It should be remembered that, for a given constellation, the elevations of the visible satellites as functions of time are governed by the location (latitude and longitude) of the mobile terminal on the globe. The satellite data used in this study is applicable to one point on the globe and thus is only an example.

Given a constellation and a point on earth, the value obtained for the multisatellite combining gain G_{ms} is thus a function of two parameters: time (giving the elevations of the visible satellites), and the choice of the single satellite with availability av_1 see (27), against which the multisatellite availability is compared. The gain could also be considered as a function of the elevation of the single satellite.

It is intuitively clear that the extremes of the gain G_{ms} are obtained in the following situations. If we choose a range of time and a satellite to be compared against such, that the elevations will be low, av_1 will have small values, and the gain will be large. If we choose a range of time and a satellite such that the elevations will be high, av_1 will have large values, and the gain will be small.

An example of the multisatellite availability gains in a realistic LEO situation is shown in Fig. 9. The upper plot represents the availability gains (G_{ms}) versus time, the solid line corresponding to the maximum gain and the dashed line to the minimum gain. The maximum gain is given by choosing for the single satellite availability av_1 at each time instant the lowest visible satellite. The minimum gain is given by choosing for the single satellite availability av_1 at each time instant the highest visible satellite. The lower plot shows the gain range, i.e., difference of the maximum and minimum gains, versus time.

TABLE XI
REQUIRED PILOT CARRIER TO ALL USER SIGNAL RATIOS TO ACHIEVE A CERTAIN FEER FOR ELEVATION ANGLES 42° (VH/PT) AND 60° (HH), $M = 1024$ AND $K = 100$ FOR $R_c = 1.8$ MHz AND $K = 1000$ FOR $R_c = 9.8$ MHz

FEER	SNR [dB] (before FFT)	P/KS [dB] R_c=1.8MHz K=100, G=64	P/KS [dB] R_c=1.8MHz K=100, G=512	P/KS [dB] R_c=9.8MHz K=1000, G=512	P/KS [dB] R_c=9.8MHz K=1000, G=8192
10^{-3}	-2	-16.5	-7.6	-24.8	-12.8
10^{-2}	-8	-22.5	-13.6	-30.8	-18.8
10^{-1}	-13	-27.5	-18.6	-35.8	-23.8

TABLE XII
PROBABILITIES p_e AS A FUNCTION OF D AND p_f

D	p_f	p_e
10	0.1	$1.6 \cdot 10^{-3}$
	0.2	$3.2 \cdot 10^{-2}$
	0.3	$1.5 \cdot 10^{-1}$
20	0.1	$7.1 \cdot 10^{-6}$
	0.2	$2.6 \cdot 10^{-3}$
	0.3	$4.8 \cdot 10^{-2}$
30	0.1	$3.6 \cdot 10^{-8}$
	0.2	$2.3 \cdot 10^{-4}$
	0.3	$1.7 \cdot 10^{-2}$
40	0.1	$1.9 \cdot 10^{-10}$
	0.2	$2.2 \cdot 10^{-5}$
	0.3	$6.3 \cdot 10^{-3}$

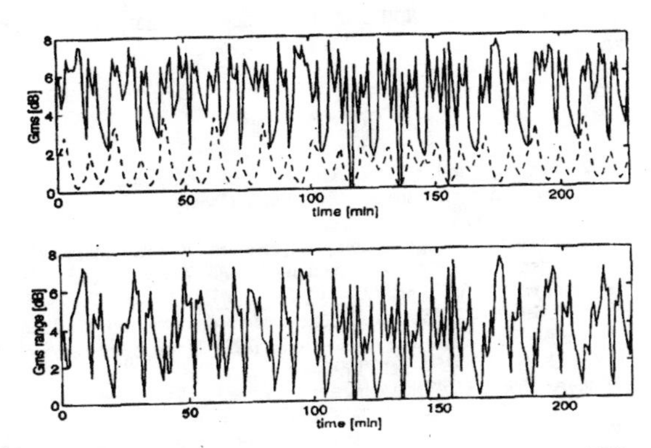

Fig. 9. Example of multisatellite availability gains in a realistic LEO constellation.

At the time instants where the gain range goes to unity (0 dB), the maximum and minimum gains coincide, and there is only one satellite visible to the mobile terminal. It is seen from Fig. 9 that comparing always to the best (highest) satellite, the multisatellite availability gain as defined by (27) never exceeds 4 dB in the LEO example case. On the other hand, comparing to the worst (lowest) satellite, the gain can reach almost 8 dB. Since the example includes data from over one complete orbit period, the gain values shown in Fig. 9 can be considered to be representative.

VII. CONCLUSION

Within this paper, a study of the design parameters for a CDMA-based LEO satellite network has been presented and discussed. The analysis investigates possible gains when using multipath and multisatellite diversity reception. In order to simplify the acquisition process, Doppler compensation based on a CW pilot carrier is used. The minimum pilot carrier power needed for a certain quality of Doppler estimation is also studied.

As a first step in the analysis, a channel model was presented, and its parameter values were determined by using preliminary measurement results. Using these results, it was found that a multipath RAKE would be beneficial only in an urban environment, at low elevation angles, and at chip rates of the order of 10 MHz or above. The expected combining gain is in the range 4, $\cdots$, 8 dB for elevation angles 45°, $\cdots$, 10°. For rural environments, the gains are insignificant.

The evaluation of possible gains obtained by combining multisatellite signals is based on an equivalent availability of LOS signal component av_{M_s}, which is by definition the probability that at least one out of M_s signals coming from the visible satellites has the LOS component. The multisatellite availability gain is still difficult to evaluate as a general result due to the large number of different satellite geometries within a given constellation. For that reason we rather talk about a possible range of the multisatellite availability gain. In LEO case, the link availability is increased one to seven times by using multisatellite reception, corresponding to multisatellite availability gains of 0 to 8 dB.

It was also shown that Doppler estimation based on a CW pilot carrier located at the first spectral null of the CDMA signal is possible with an accuracy at ±100 Hz with a probability better that 0.999 for all elevation angles down to 23°. This was achieved with a pilot carrier to all user signals power ratio (worst case) of −0.4 dB for a chip rate of 1.8 MHz, and −5.8 dB for a chip rate of 9.8 MHz. For elevation angles above 30°, this ratio becomes negligible.

As a final conclusion, the system concept proposed in this paper for a CDMA receiver for a LEO satellite network is feasible. Based on the results presented, it is possible to specify initial system parameter values for a receiver test bed design.

ACKNOWLEDGMENT

The authors would like to express their gratefulness to M. Luz de Mateo and S. Buonomo of the Technology Centre of the European Space Agency, and to H. Hakalahti of Elektrobit Ltd. for their valuable contributions to the work. The authors would also like to thank the European Space Agency for providing realistic satellite orbit data and wideband channel measurement data, and for assistance in channel parameter extraction.

REFERENCES

[1] A. Fuxjaeger and R. Iltis, "Acquisition of timing and doppler-shift in a direct-sequence systems," *IEEE Trans. Commun.*, vol. 42, no. 10, pp. 2870–2880, Oct. 1994.

[2] W. Hurd *et al.*, "High dynamic GPS receiver using maximum likelihood estimations and frequency tracking," *IEEE Trans. Aerosp. Electron. Syst.*, vol. AES-23, pp. 425–437, July 1987.

[3] M. Cheng *et al.*, "Spread spectrum code acquisition in the precence of doppler shifts and data modulation," *IEEE Trans. Commun.*, vol. 38, no. 2, pp. 241–250, Feb. 1990.

[4] A. Aftelak *et al.*, "Data-relay system spread spectrum receiver and modem," in *Proc. 2nd. Eur. Conf. Satellite Commun.*, 1991, pp. 393–397..

[5] M. Thompson *et al.*, "Non-coherent PN code acquisition in direct sequence spread spectrum systems using a neural network," in *MILCOM'93 Conf. Rec.*, 1993, pp. 30–39.

[6] S. G. Glisic *et al.*, "Rejection of frequency sweeping signal in DS spread spectrum systems using complex adaptive filters," *IEEE Trans. Commun.*, vol. 43, no. 1, pp. 136–146, Jan. 1995.

[7] S. G. Glisic *et al.*, "Rejection of FH signal in DS spread spectrum systems using complex adaptive filters," *IEEE Trans. Commun.*, vol. 43, no. 5, pp. 1982–1992, May 1995.

[8] F. J. Harris, "On the use of windows for harmonic analysis with the discrete fourier transform," *Proc. IEEE*, vol. 66, no. 1, pp. 51–83, Jan. 1978.

[9] S. Holm, "Optimum FFT-based frequency acquisition with application to COSPAS-SARSAT," *IEEE Trans. Aerosp. Electron. Syst.*, vol. 29, no. 2, pp. 464–475, Apr. 1993.

[10] S. M. Kay and S. L. Marple, "Spectrum analysis—A modern perspective," *Proc. IEEE*, vol. 69, no. 11, pp. 1380–1419, Nov. 1981.

[11] B. Kedem and S. Yakowitz, "Practical aspects of a fast algorithm for frequency detection," *IEEE Trans. Commun.*, vol. 42, no. 9, pp. 2760–2767, Sept. 1994.

[12] R. J. Kenefic and A. H. Nuttall, "Maximum likelihood estimation of the parameters of a tone using real discrete data," *IEEE J. Oceanic Eng.*, vol. OE-12, no. 1, pp. 279–280, Jan. 1987.

[13] L. C. Palmer, "Coarse frequency estimation using the discrete fourier transform," *IEEE Trans. Inform. Theory*, vol. IT-20, pp. 104–109, Jan. 1974.

[14] D. C. Rife and R. R. Boorstyn, "Single-tone parameter estimation from discrete-time observations," *IEEE Trans. Inform. Theory*, vol. IT-20, no. 5, pp. 591–598, Sept. 1974.

[15] ——, "Multiple tone parameter estimation from discrete-time observations," *The Bell Syst. Technical J.*, vol. 55, no. 9, pp. 1389–1410, Nov. 1976.

[16] V. A. Vilnrotter *et al.*, "Frequency estimation techniques for high dynamic trajectories," *IEEE Trans. Aerosp. Electron. Syst.*, vol. 25, no. 4, pp. 559–577, July 1989.

[17] C. Loo, "A statistical model for a land mobile satellite link," *IEEE Trans. Veh. Technol.*, vol. VT-34, no. 3, pp. 122–127, 1985.

[18] E. Lutz *et al.*, "The land mobile satellite communication channel, recordings, statistics, and channel model," *IEEE Trans. Veh. Technol.*, vol. 40, no. 2, pp. 375–386, 1991.

[19] Commision of the EC, COST 207, Digital Land Mobile Radio Communications, Final Report, EUR 12160 EN, Office for Official Publication of the EC, Luxembourg, p. 388, 1989.

Savo G. Glisic (M'90–SM'94) is a Professor of Electrical Engineering at the University of Oulu, Finland, and Vice President of the Globalcomm Institute of Technology. From 1976 to 1977, he was a Visiting Scientist at the Cranfield Institute of Technology, Cranfield, England, and from 1986 to 1987, a Visiting Scientist at the University of California, San Diego. He has been active in the field of spread spectrum for 20 years and has published a number of papers and four books. Currently, he is doing consulting in this field in Europe, the USA, and Australia. He served as Technical Program Chair of the Third IEEE International Symposium on Spread Spectrum Techniques and Applications ISSSTA'94 and is the Technical Program Chairman of the Eighth IEEE Symposium on Personal, Indoor and Mobile Radio Communications PIMRC'97.

Jaakko J. Talvitie (S'88–M'92) was born in April 1961. He received the M.S. and Lic. Tech. degrees from the University of Oulu, Finland, both in electrical engineering, in 1987 and 1993, respectively.

From 1987 to 1994, he was with the Telecommunication Laboratory, University of Oulu, Finland, working as a Research Engineer and Project Manager in spread-spectrum and channel modeling research projects, and as a Research Manager. During this period, he was also involved in teaching, working as a Teaching Assistant, an acting Associate Professor, and an Acting Professor. During the second half of 1994, he was the Head of the Laboratory. In 1995, he joined Elektrobit Ltd., Oulu, Finland, working as a Project Manager and Research Engineer in a research project dealing with CDMA multisatellite diversity receiver design for mobile satellite systems, on a contract with the European Space Agency. In September 1995, he took up his present post as the Director of the Centre for Wireless Communications at the University of Oulu, Finland. His main interests are currently in wireless and personal communication systems and techniques, especially spread spectrum and code-division multiple-access, and in radio channel measurement and modeling.

Timo Kumpumäki received the M.Sc. (E.E.) degree from the University of Oulu, Finland, in 1993.

Since 1993, he has been a Research Scientist in the Telecommunication Laboratory, the University of Oulu, Finland, where he is currently pursuing the Ph.D. degree in electrical engineering. His main research interests are in spread spectrum, synchronization and satellite communication.

Matti Latva-aho (S'96) was born on January 25, 1968, in Kuivaniemi, Finland. He received the M.Sc. (E.E.) and Lic.Tech. degrees from the University of Oulu, Finland, in 1992 and 1996, respectively.

From 1992 to 1993, he was a Research Engineer at Nokia Mobile Phones Ltd., Oulu, Finland. Since October 1993, he has been a Research Scientist in the Telecommunication Laboratory, the University of Oulu, Finland, where he is currently pursuing the Ph.D. degree in electrical engineering. His main research interests are in spread spectrum communications, synchronization, statistical signal processing and multiuser communications.

Jari H. Iinatti (M'95) was born in Oulu, Finland, in 1964. He received the Master of Science degree in electrical engineering and the Lic. Tech. degree from the University of Oulu, Finland, in 1989 and 1993, respectively.

Since 1989, he has been a Research Scientist at the Telecommunication Laboratory, the University of Oulu, Finland, where he belongs to its spread-spectrum research group. He is currently working toward the Doctor of Technology degree. His interests are in spread-spectrum communications, especially in code acquisition problems.

Torsti J. Poutanen was born in July 1959. He received the M.S. degree in technical physics from the Helsinki University of Technology, Finland, in 1983.

From 1983 to 1990. he was with Nokia Mobile Phones Ltd., Finland, working a s a Research Engineer and Group Manager in RF system and circuit design projects for cellular user terminals. In 1990, he joined Elektrobit Ltd., Oulu, Finland, working as a RF Design Engineer and Project Manager in various design projects dealing with wireless communication equipment involving CDMA. His main interests are currently in wireless communication systems and techniques, especially spread-spectrum techniques and algorithms for all-digital communication receivers. He is also actively involved with radio channel modeling and simulation.

UNIT 4
SPREAD SPECTRUM APPLICATIONS IN INDOOR WIRELESS

With the ever-growing trend in cellular mobile communications, indoor wireless communication has drawn extensive attention of many researchers. The in-building propagation characteristics have proved to be a complex problem requiring separate treatment. Due to its inherent multipath-combating property, spread spectrum became a leading contender as an access technique in indoor wireless. Consequently, CDMA for indoor environments has been attracting much attention.

The first three papers in this unit address the performance issue of CDMA in indoor wireless. The paper "*Performance of DS-CDMA over Measured Indoor Radio Channels Using Random Orthogonal Codes*" by Chase and Pahlavan analyzes the performance of CDMA using random orthogonal codes over multipath indoor radio channels with channel measurements from five different buildings. The average probability of error as a function of SNR is used as the performance criterion. The next two papers, "*Performance Evaluation of Direct-Sequence Spread Spectrum Multiple Access for Indoor Wireless Communication in a Rician Fading Channel*" by Prasad et al. and "*Performance of BPSK and TCM Using the Exponential Multipath Profile Model for Spread-Spectrum Indoor Radio Channels*" by Bargallo and Roberts, evaluate the performance considering Rician fading channel and Rayleigh fading with an exponential multipath profile, respectively.

The paper "*Decision Feedback Equalization for CDMA in Indoor Wireless Communications*" by Abdulrahman et al. gives an analysis on the optimum performance of the DFE receiver, showing the advantages of the system over others in terms of capacity improvements. The next paper "*Effects of Diversity, Power Control, and Bandwidth on the Capacity of Microcellular CDMA Systems*" by Jalali and Mermelstein deals with the capacity evaluation in in-building and microcellular systems. The next two papers "*Hybrid DS/SFH Spread-Spectrum Multiple Access with Predetection Diversity and Coding for Indoor Radio*" by Wang and Moeneclaey and "*Performance Analysis of a Hybrid DS/SFH CDMA System Using Analytical and Measured Pico Cellular Channels*" by Cakmak et al. discuss the use of hybrid direct sequence and slow frequency hopping systems in indoor environments. "*A CDMA-Distributed Antenna System for In-Building Personal Communications Services*" by Xia et al. evaluates a CDMA PCS distributed antenna system in 1.8 GHz band. Measurement and modelling results are presented on coverage, voice quality, reduction of transmit power and path diversity. The last paper in this section "*Traffic Handling Capability of a Broadband Indoor Wireless Network Using CDMA Multiple Access*" by Zhang and Falconer presents a study of a broadband indoor wireless network supporting high-speed traffic using CDMA. The impacts of the base station density, traffic load, average holding time, and variable traffic sources on the system performance are examined.

Performance of DS-CDMA Over Measured Indoor Radio Channels Using Random Orthogonal Codes

Mitchell Chase, *Member, IEEE,* and Kaveh Pahlavan, *Senior Member, IEEE,*

***Abstract*—Direct sequence spread spectrum, with its inherent resistance to multipath interference, has become a commercial reality for indoor wireless communications and has been proposed for personal communication networks. To allow multiple users within the limited bandwidths allocated by the FCC, code-division multiple-access (CDMA) is needed. This paper analyzes the performance of CDMA systems using random orthogonal codes over fading multipath indoor radio channels using channel measurements from five different buildings. The effect of RAKE receiver structure is studied, as is the effect of average power control. The average probability of error as a function of signal-to-noise ratio is used as the performance criteria. Results are compared with models for Rayleigh fading channels.**

I. Introduction

The increasing popularity of portable personal communications has created a need for solutions to the interference and poor reception quality caused by the limited frequencies available and the close proximity of multiple users. One radio frequency (RF) technique already in commercial use for wireless indoor communications [1] is being investigated for portable personal communications. Direct sequence code-division multiple-access (DS-CDMA) communications offer an alternative to conventional RF indoor communications [2]. Spread spectrum provides resistance to the multipath caused by walls, ceilings, and other objects between the transmitter and receiver. Spread spectrum can overlay existing systems because of the low spectral density level. In the U.S. the FCC has assigned three bands for nongovernmental applications of spread spectrum which enhance this alternative for indoor channels [3]. The only reservation concerning DS-CDMA communications is the efficiency of the bandwidth utilization in fading multipath channels [4]. Coding and diversity combining can improve the bandwidth efficiency of DS-CDMA communications [5]–[7]. Diversity can be external (i.e., multiple antennas) or implicit (internal). Implicit diversity combining makes use of the inherent diversity from multipath reception and can be achieved with a RAKE receiver. Combinations of implicit and external diversities can be used to improve performance.

The performance of a system that transmits alternating sequences to minimize intersymbol interference with a multipath combining receiver over statistically modeled channels was considered in [8]. Another method for improving the bandwidth efficiency is the use of M-ary signaling. The bandwidth efficiency of M-ary orthogonal codes over nonfading channels is discussed in [9], and over fading channels in [10]. This paper analyzes such a system using measured multipath fading channel data. The particular examples use the measured multipath characteristics from five manufacturing areas [11] and a bandwidth permitted by the FCC [3]. A RAKE receiver structure is investigated and compared with theoretical results for Rayleigh fading channels.

Manuscript received November 30, 1992; revised March 26, 1993.
M. Chase is with Comdisco Systems, Inc., Natick, MA 01760.
K. Pahlavan is with Worcester Polytechnic Institute, Worcester, MA.
IEEE Log Number 9211017.

II. System Model

A. Transmitter:

The system (Fig. 1) [9] consists of K users, each assigned a set of sequences $V^{(k)}$ consisting of M-ary orthogonal sequences, each of length N;

$$V^{(k)} = \{V_1^{(k)}, V_2^{(k)}, \cdots, V_M^{(k)}\} \quad (1)$$

where

$$V_\lambda^{(k)} = \{V_{\lambda,0}^{(k)}, V_{\lambda,1}^{(k)}, \cdots, V_{\lambda,N-1}^{(k)}\} \quad (2)$$

and $V_{\lambda,n}^{(k)} = \exp(j\Theta_{\lambda,n}^{(k)})$ is a complex rth root of unity (r-phase modulation). M-ary equally likely data symbols are transmitted at a rate of one every T seconds. The complex envelope of the signal transmitted by the kth user to send the λ th symbol during the period $[0, T)$ is

$$X_k(V_\lambda^{(k)}, t) = A\tilde{X}_k(V_\lambda^{(k)}, t) \quad (3)$$

where

$$\tilde{X}_k(V_\lambda^{(k)}, t) = \sum_{n=0}^{N-1} V_{\lambda,n}^{(k)} P_{T_c}(t - nT_c - \Delta_k)\exp(j\theta_k) \quad (4)$$

and A is the amplitude of the transmitted signal. The chip duration T_c is $1/W$, W being the available bandwidth. P_{T_c} is the chip waveform, θ_k the carrier phase, and Δ_k represents the asynchronous transmission delay time between transmitters. Δ_k for $k \neq 0$ is uniformly distributed in $[0, T_c)$ with $\Delta_0 = 0$, and θ_k is uniformly distributed in $[0, 2\pi)$. The chip waveform is defined for $0 \leq t < T_c$, is zero outside the range, and is normalized so that the energy per chip is equal to T_c.

The overall transmitted signal for the kth user is

$$U_k(t) = \sum_{s=-\infty}^{\infty} X_k(V_m^{(k)}, t - sT); \qquad m = 1, 2, \cdots, M. \quad (5)$$

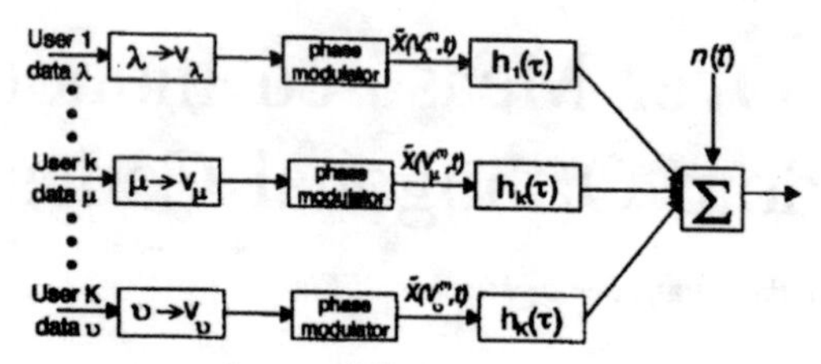
Fig. 1. System block diagram.

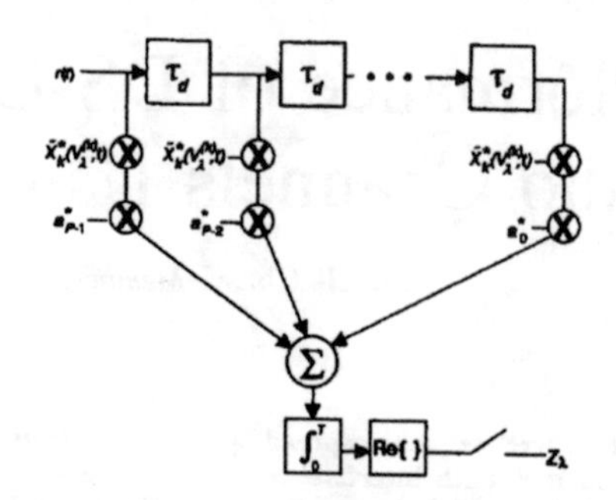
Fig. 2. RAKE receiver with maximal ration combining for the kth user and the λth symbol.

B. Channel:

The fading multipath indoor channels are assumed to be discrete and time-invariant with channel impulse response for each user given by [11]:

$$h_k(\tau) = \sum_{l=0}^{L_k-1} \alpha_l^{(k)} \delta(\tau - \tau_l^{(k)}) \exp(j\phi_l^{(k)}) \tag{6}$$

where $\alpha_l^{(k)}$, $\tau_l^{(k)}$, and $\phi_l^{(k)}$ are the path gain, delay and phase, respectively, $\delta(t)$ is the unit impulse function, and L_k is the number of multipaths for profile k. The path gain and delay for each channel are determined from actual measurements [11]. The path phase $\phi_l^{(k)} = 2\pi f_c \tau_l^{(k)}$ is assumed to be a uniformly distributed random variable in $[0, 2\pi)$ incorporating the carrier phase θ_k. The received signal is

$$r(t) = \sum_{k=0}^{K-1} \sum_{l=0}^{L_k-1} \alpha_l^{(k)} \exp(j\phi_l^{(k)}) U_k(t - \tau_l^{(k)}) + n(t) \tag{7}$$

where

$$n(t) = n_r(t) + jn_i(t); \qquad j = \sqrt{-1} \tag{8}$$

$n(t)$ is AWGN with power spectral density N_0.

C. Receiver:

To combat the multipath distortion and achieve better performance, diversity combining is needed [7]. A matched filter receiver structure is investigated that takes advantage of the implicit diversity of the received signal.

III. Receiver Structure

A. RAKE Receiver:

The optimum receiver for wideband fading multipath signals is a RAKE matched filter [12]. This receiver takes advantage of the implicit diversity of the multipaths in the received signal. Consider the coherent RAKE matched filter with maximal-ratio combining structure as shown in Fig. 2. The decision variable, Z_λ, for the first user and the λ th symbol is (Appendix A):

$$Z_\lambda = \mathrm{Re}\left\{ A \sum_{p=0}^{P-1} a_p \exp(-j\varphi_p) \sum_{k=0}^{K-1} \sum_{l=0}^{L_k-1} \alpha_l^{(k)} \exp(j\phi_l^{(k)}) \cdot \sum_{s=-\infty}^{\infty} R_{\nu,\lambda}^{(k,1)}\left(\tau_l^{(k)} - \frac{p}{W} + \Delta_k + sT\right) \right\} + \eta_\lambda \tag{9}$$

The receiver for M-ary signaling is shown in Fig. 3. The distribution of the correlation function $R_{\nu,\lambda}^{(k,1)}(\cdot)$ is approximately Gaussian by the central limit theorem. Similarly, the statistics of the difference of decision variables are assumed to follow a Gaussian distribution. Thus the probability of error can be expressed as

$$\Pr(\epsilon) \approx \frac{M-1}{2} \operatorname{erfc}\left(\sqrt{\frac{\gamma}{2}}\right). \tag{10}$$

. The energy per bit is

$$E_b = \frac{A^2}{2} \frac{NT_C}{\log_2 M}. \tag{11}$$

Hence, γ can be written as

$$\gamma = \gamma_{\mathrm{RAKE}} = \frac{\left[\sum_{p=0}^{P-1} \sum_{l=0}^{L_1-1} a_p \alpha_l^{(1)} \cos(\phi_l^{(1)} - \varphi_p) R_{\lambda,\lambda}^{(1,1)}\left(\tau_l^{(1)} - \frac{p}{W}\right)\right]^2}{\Omega_{\mathrm{RAKE}}} \tag{12}$$

for the RAKE receiver and

$$\Omega_{\mathrm{RAKE}} = \sum_{p=0}^{P-1} \sum_{l=0}^{L_1-1} [a_p \alpha_l^{(1)} \cos(\phi_l^{(1)} - \varphi_p)]^2 Y_a\left(\tau_l^{(1)} - \frac{p}{W}\right) + \sum_{k=1}^{K-1} \sum_{p=0}^{P-1} \sum_{l=0}^{L_k-1} [a_p \alpha_l^{(k)} \cos(\phi_l^{(k)} - \varphi_p)]^2 Y_b\left(\tau_l^{(k)} - \frac{p}{W}\right) + \frac{N_0}{E_b} \frac{(NT_c)^2}{\log_2 M} \sum_{p=0}^{P-1} |a_p|^2 \tag{13}$$

where $Y_a(\cdot)$ and $Y_b(\cdot)$ are related to the auto- and cross-correlation functions of the transmitted sequences (Appendix B). The first term represents the self-interference, the second term interference from other users, and the last term is noise.

B. Predicted Results:

The results of the simulations are compared with the theoretical performance for coherent detection of BFSK in fading channels [12]

$$\Pr(\epsilon) = \frac{1}{2}\left[1 - \sqrt{\frac{\bar{\gamma}}{2 + \bar{\gamma}}}\right] \tag{14}$$

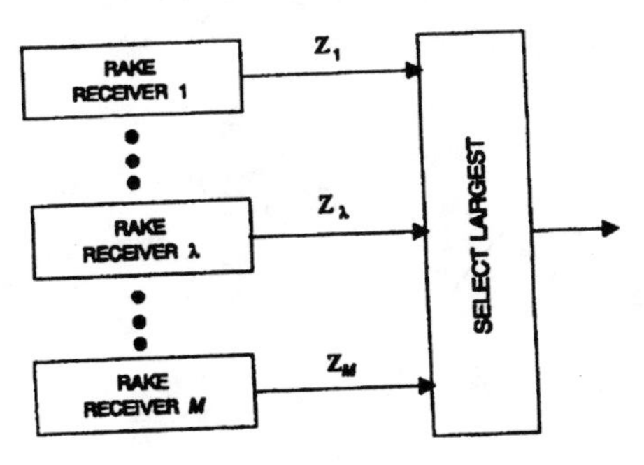

Fig. 3. M-ary RAKE receiver structure.

where the average signal-to-noise ratio is defined as

$$\overline{\gamma} = \frac{E_b}{N_0} E(a^2). \tag{15}$$

When diversity combining techniques are employed, the predicted performance for coherent detection of BFSK is [12]

$$\Pr(\epsilon) = \left(\frac{1-\mu}{2}\right)^L \sum_{k=0}^{L-1} \binom{L-1+k}{k} \left(\frac{1+\mu}{2}\right)^k \tag{16}$$

where

$$\mu = \sqrt{\frac{\overline{\gamma}}{2+\overline{\gamma}}} \tag{17}$$

and L is the order of diversity.

IV. Channel Measurements

A. Simulated

To confirm the validity of the system model and simulation program, two sets of computer generated channel profiles were used. The first set simulated a Rayleigh fading channel with one received path at $t = 0$. In the second set, each channel was assigned two path gains, one each at delays of $t = 0$ and $t = 1/W$. The path delay and tap spacing are ideal, thus allowing comparison with the expression of performance for fading channels (14). The amplitudes followed a Rayleigh distribution [6] with the parameter $\rho = 0.637$ and $E\{f(x)\} = 1.0$:

$$f(x) = \begin{cases} \frac{x}{\rho}\exp\left(-\frac{x^2}{2\rho}\right); & x \geq 0 \\ 0; & x < 0 \end{cases}. \tag{18}$$

B. Measured Channel Data:

Channel measurements were made at five different locations [10]. An RF oscillator on a mobile transmitter was used to generate a 910-MHz signal modulated by a train of 3-ns pulses with a 500-ns repetition period. The stationary base equipment comprised the receiver coupled to a digital storage scope and a personal computer. Both the transmitter and the receiver used vertically polarized quarter-wave dipole antennas placed about 1.5 m above floor level. The transmitter was moved to various locations in the site, and the received multipath profiles were stored in the computer. Each profile was an average of 64 channel snapshots collected in a 15 to 20-second period. The distance between the transmitter and the receiver varied between 1 and 50 meters. A total of about 300 profiles were collected from measurements made in five areas on three different manufacturing floors. The channel data represents a snapshot in time, providing an estimate of the performance, albeit at a particular time and location.

TABLE I
Multipath Statistics of the Five Measured Buildings

Area	RMS Multipath Delay Spread (ns)		
	Mean	Median	Maximum
A	16.6	15.25	40
B	29.0	31.62	60
C	73.1	52.57	150
D	33.1	19.37	146
E	52.4	48.90	152

Area A[1] with 54 profiles was a typical electronics shop floor having circuit board design equipment, soldering and chip mounting stations. Area B[2] with 48 profiles included test equipment and storage areas for common electronic equipment, partitioned by metallic screens. Area C[3] with 45 profiles was an automobile assembly line "jungle" comprising all kinds of welding and body shop equipment. Area D[4] with 66 profiles was a vast open area used for final inspection of the new cars coming off of the assembly line. Area E[5] with 75 profiles had grinding machines, huge ovens, transformers, and other heavy machinery. The multipath statistics of the five areas are shown in Table I.

V. Results and Discussion

The receiver analysis for one user with $M = 2$ and $N = 256$ over one- and two-path simulated Rayleigh fading channels is shown in Fig. 4. A one-tap coherent RAKE receiver was analyzed for the single path profiles, and a two-tap model for the two-path profiles. Simulations from 50000 computer generated channel profiles for each case were averaged and compared with the predicted results. The signal energies were adjusted for the profiles and the predicted results such that $E\{\alpha^2\} = 1.28$. The results for both one- and two-path models closely agree with predicted results until $\Pr(\epsilon) \approx 10^{-5}$, where errors due to machine underflow start to appear.

The effect of a RAKE receiver synchronized to the first received path is shown in Fig. 5. The performance over one of the measured areas (Area A) with various combinations of tap spacing and tap length is shown for the case of one active transmitter ($K = 1$, $M = 2$, $N = 256$, and 25 MHz bandwidth). Increasing receiver complexity by reducing tap spacing while adding more taps yields diminishing returns after the implicit diversity from the channel is realized by the receiver.

The performance of one active transmitter over all of the measured areas is shown in Fig. 6 ($K = 1$, $M = 2$, $N = 256$, and $W = 25$ MHz). A RAKE receiver synchronized to the

[1] Infinet Inc., N. Andover, MA
[2] Infinet Inc., N. Andover, MA
[3] General Motors, Framingham, MA
[4] General Motors, Framingham, MA
[5] Norton Company, Worcester, MA

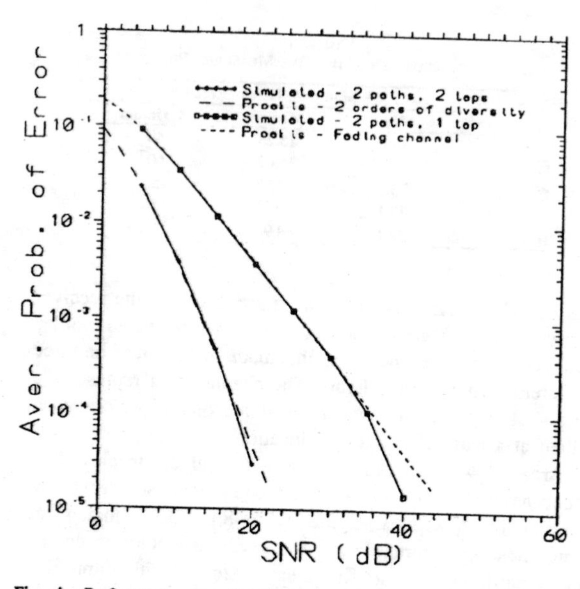

Fig. 4. Performance over Rayleigh simulated channel compared with predicted results.

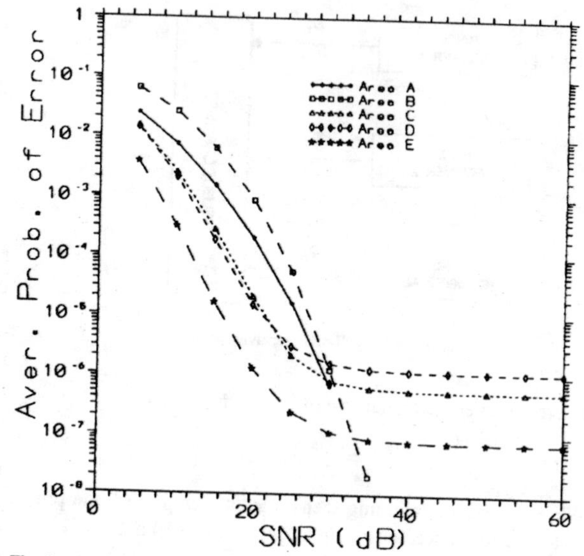

Fig. 6. Performance over measured area with one active transmitter using an 8-tap RAKE receiver with 10 ns tap spacing ($K = 1$, $M = 2$, $N = 256$, $W = 25$ MHz, no power control).

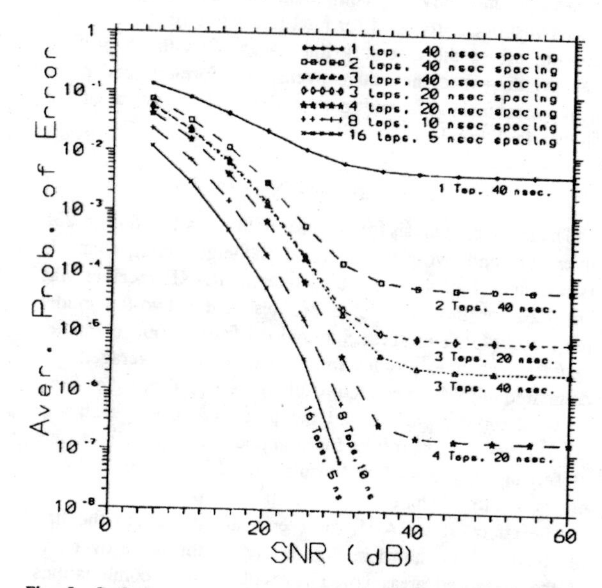

Fig. 5. Performance over Area A with one active transmitter using RAKE receivers of various combinations of tap spacing and tap length ($K = 1$, $M = 2$, $N = 256$, $W = 25$ MHz, no power control).

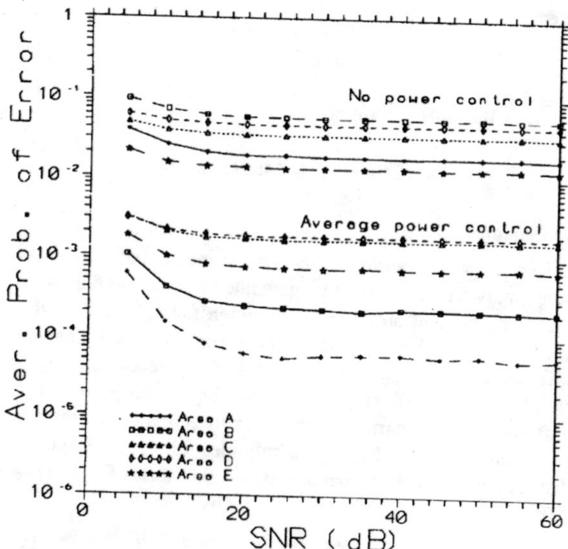

Fig. 7. Performance over measured channels with five active transmitters with and without power control using an 8-tap RAKE receiver with 10 ns tap spacing ($K = 5$, $M = 2$, $N = 256$, $W = 25$ MHz).

first received path with eight taps spaced 10 ns apart was utilized. Areas B, C, and D initially show better performance with this receiver until approximately 30 dB. The receiver structure chosen does not realize the implicit diversity inherent in those areas with higher multipath delay spread. Additional taps would be necessary to achieve the performance possible over these channels. The performance over Areas A and B is initially limited by the nature of the channels—metallic partitions and other structures limiting performance at lower signal energies.

The degradation of performance in multi-user systems is shown in Fig. 7. The same receiver as used in Fig. 5 is employed. The results without power control are uniformly poor. When the average energy of each *profile* from each area of the channel impulse response is normalized to unity, the performance improves by up to a factor of approximately 100. The effect is more pronounced in the areas with metallic partitions (Areas A and B) than in the other areas. The performance is related to the RAKE receiver structure and the multipath spread.

VI. Conclusions

The performance over indoor channels is highly dependent on the channel characterization as well as the receiver structure. Generalizations about the channel response will result in approximate results that may not be suitable for performance determinations without further study.

CDMA is an attractive solution to the problems encountered in portable personal communications. In addition to the attractions of spread spectrum, CDMA allows multiple users to share limited bandwidth. The traditional assumption of Rayleigh distributed fading multipath channels is not necessarily appropriate for the indoor radio channel. It is also apparent that each channel profile results in different received power levels. Uniform power assumptions for transmitter-receiver paths cannot be used. Power control is necessary for a multi-user system.

Appendix A
RAKE Receiver with Maximal Ratio Combining

The multiple-user M-ary received signal is

$$r(t) = \sum_{k=0}^{K-1}\sum_{l=0}^{L_k-1} \alpha_l^{(k)} \exp(j\phi_l^{(k)}) U_k(t-\tau_l^{(k)}) + n(t) \quad \text{(A1)}$$

where K is the number of simultaneous active transmitters, L_k is the number of multipaths received from the kth user and $\alpha_l^{(k)}$, $\tau_l^{(k)}$, and $\phi_l^{(k)}$ are the path gain, delay, and phase, respectively.

The decision variable for a RAKE receiver with maximal ratio combining for the first user and the λth symbol is (A2) and (A3), .

$$Z_\lambda = \text{Re}\left\{ \int_0^T \sum_{p=0}^{P-1} a_p \exp(-j\varphi_p) \tilde{X}_1^*\left(V_\lambda^{(1)}, t - \frac{p}{W}\right) r(t)\, dt \right\} \quad \text{(A2)}$$

$$\begin{aligned} Z_\lambda = \text{Re}&\left\{ \sum_{p=0}^{P-1} a_p \exp(-j\varphi_p) \sum_{k=0}^{K-1}\sum_{l=0}^{L_k-1} \alpha_l^{(k)} \cdot \exp(j\phi_l^{(k)}) \int_0^T \tilde{X}_1^*\left(V_\lambda^{(1)}, t-\frac{p}{W}\right) U_k\left(t-\tau_l^{(k)}\right) dt \right\} \\ &+ \text{Re}\left\{ \sum_{p=0}^{P-1} \alpha_p \exp(-j\varphi_p) \int_0^T \tilde{X}_1^*\left(V_\lambda^{(1)}, t-\frac{p}{W}\right) n(t)\, dt \right\} \end{aligned} \quad \text{(A3)}$$

Let

$$\eta_\lambda = \text{Re}\left\{ \sum_{p=0}^{P-1} a_p \exp(-j\varphi_p) \cdot \int_0^T \tilde{X}_1^*\left(V_\lambda^{(1)}, t-\frac{p}{W}\right) n(t)\, dt \right\}. \quad \text{(A4)}$$

Then, expanding $U_k(t-\tau_l^{(k)})$ the decision variable becomes

$$\begin{aligned} Z_\lambda = \text{Re}&\left\{ \sum_{p=0}^{P-1} \alpha_p \exp(-j\varphi_p) \sum_{k=0}^{K-1}\sum_{l=0}^{L_k-1} a_l^{(k)} \exp(j\phi_l^{(k)}) \cdot \int_0^T \tilde{X}_1^*\left(V_\lambda^{(1)}, t-\frac{p}{W}\right) \right. \\ &\left. \cdot \sum_{s=-\infty}^{\infty} X_k(V_\nu^{(k)}, t-\tau_l^{(k)} - sT)\, dt \right\} + \eta_\lambda. \end{aligned} \quad \text{(A5)}$$

Looking at the integral in (A5),

$$I = \int_0^T \tilde{X}_1^*\left(V_\lambda^{(1)}, t-\frac{p}{W}\right) \sum_{s=-\infty}^{\infty} X_k(V_\nu^{(k)}, t-\tau_l^{(k)} - sT)\, dt. \quad \text{(A6)}$$

Expanding,

$$I = A\int_0^T \tilde{X}_1^*\left(V_\lambda^{(1)}, t-\frac{p}{W}\right) \sum_{s=-\infty}^{\infty} \tilde{X}_k(V_\nu^{(k)}, t-\tau_l^{(k)} - sT)\, dt \quad \text{(A7)}$$

$$I = A\int_0^T \sum_{n=0}^{N-1} V_{\lambda,n}^{(1)^*} P_{T_c}\left(t-nT_c-\frac{p}{W}\right) \cdot \sum_{s=-\infty}^{\infty}\sum_{m=0}^{N-1} V_{\nu,m}^{(k)} P_{T_c}(t-mT_c-\tau_l^{(k)} - \Delta_k - sT)\, dt \quad \text{(A8)}$$

let $u = n - m$ and

$$\begin{aligned} y &= t - mT_c - \Delta_k - sT - \tau_l^{(k)} \\ &= t - (m+sN)T_c - \Delta_k - \tau_l^{(k)} \end{aligned} \quad \text{(A9)}$$

then, for $P_{T_c}(t)$ a rectangular pulse

$$I = A\sum_{m=0}^{N-1-u} V_{\nu,m}^{(k)} V_{\lambda,m+u}^{(1)^*} \int_0^T \sum_{s=-\infty}^{\infty} P_{T_c} \cdot \left(y - uT_c + \tau_l^{(k)} - \frac{p}{W} + \Delta_k + sT\right) P_{T_c}(y)\, dy. \quad \text{(A10)}$$

For $P_{T_c}(t)$ a rectangular pulse,

$$P_{T_c}(t) = \begin{cases} 1; & 0 \le t < T_c \\ 0; & \text{elsewhere} \end{cases} \quad \text{(A11)}$$

then

$$\int_0^{T_c} P_{T_c}(t) P_{T_c}(t-x)\, dt \begin{cases} = 0; & |x| \ge T_c \\ \ne 0; & -T_c < x < T_c \end{cases}. \quad \text{(A12)}$$

Therefore, for $I \ne 0$,

$$-T_c < \tau_l^{(k)} - \frac{p}{W} + \Delta_k + sT - uT_c < T_c \quad \text{(A13)}$$

Recall $T = N \cdot T_c$ so that

$$(u - sN - 1)T_c < \tau_l^{(k)} - \frac{p}{W} + \Delta_k < (u - sN + 1)T_c \quad \text{(A14)}$$

let

$$x = \tau_l^{(k)} - \frac{p}{W} + \Delta_k. \quad (A15)$$

If $x \geq 0$, then the following condition must be met:

$$0 \leq x < T_c \quad (A16)$$

or

$$(u - sN)T_c < \tau_l^{(k)} - \frac{p}{W} + \Delta_k < (u - sN + 1)T_c. \quad (A17)$$

Let $d = u - sN$

$$dT_c < \tau_l^{(k)} - \frac{p}{W} + \Delta_k < (d+1)T_c. \quad (A18)$$

For $x < 0$,

$$(d-1)T_c < \tau_l^{(k)} - \frac{p}{W} + \Delta_k < dT_c. \quad (A19)$$

Therefore, I can be expressed as

$$I = AR_{\nu,\lambda}^{(k,1)}\left(\tau_l^{(k)} - \frac{p}{W} + \Delta_k\right) \quad (A20)$$

where $R_{\nu,\lambda}^{(k,1)}$ is defined in Appendix B. The decision variable can now be written as

$$Z_\lambda = \text{Re}\left\{A\sum_{p=0}^{P-1} a_p \exp(-j\varphi_p)\sum_{k=0}^{K-1}\sum_{l=0}^{L_k-1}\alpha_l^{(k)} \cdot \exp(j\phi_l^{(k)})R_{\nu,\lambda}^{(k,1)}\left(\tau_l^{(k)} - \frac{p}{W} + \Delta_k\right)\right\} + \eta_\lambda. \quad (A21)$$

The tap gains are computed as follows:

$$\begin{aligned} a_p \exp(+j\varphi_p) &= E\left\{\sum_{l=0}^{L_1-1}\alpha_l^{(1)}\exp(j\phi_l^{(1)})\int_0^T \tilde{X}_1^* \cdot \left(V_\lambda^{(1)}, t - \frac{p}{W}\right)X\left(t - \tau_l^{(1)}\right)dt\right\} \\ &= E\left\{\sum_{l=0}^{L_1-1}\alpha_l^{(1)}\exp(j\phi_l^{(1)})R_{\lambda,\lambda}^{(1,1)}\left(\tau_l^{(1)} - \frac{p}{W}\right)\right\} \\ &= \sum_{l=0}^{L_1-1}\alpha_l^{(1)}\exp(j\phi_l^{(1)})E\left\{R_{\lambda,\lambda}^{(1,1)}\left(\tau_l^{(1)} - \frac{p}{W}\right)\right\}. \end{aligned} \quad (A22)$$

The difference between decision variables for the first user is

$$\begin{aligned} Z_\zeta - Z_\lambda = \text{Re}\Bigg\{&A\sum_{p=0}^{P-1} a_p\exp(-j\varphi_p)\sum_{k=0}^{K-1}\sum_{l=0}^{L_k-1}\alpha_l^{(k)} \cdot \exp(j\phi_l^{(k)})\Big[R_{\nu,\zeta}^{(k,1)}\left(\tau_l^{(k)} - \frac{p}{W} + \Delta_k\right) \\ &- R_{\nu,\lambda}^{(k,1)}\left(\tau_l^{(k)} - \frac{p}{W} + \Delta_k\right)\Big]\Bigg\} + \eta_\zeta - \eta_\lambda. \end{aligned} \quad (A23)$$

Assuming the statistics of the difference between decision variables follow a Gaussian distribution, the codes are random and orthogonal [9], and the λ th symbol is transmitted,

$$\begin{aligned} E\{Z_\zeta - Z_\lambda\} &= \text{Re}\left\{-A\sum_{p=0}^{P-1}a_p\exp(-j\varphi_p)\sum_{l=0}^{L_1-1}\alpha_l^{(1)} \cdot \exp(j\phi_l^{(1)})R_{\lambda,\lambda}^{(1,1)}\left(\tau_l^{(1)} - \frac{p}{W}\right)\right\} \\ &= -A\sum_{p=0}^{P-1}\sum_{l=0}^{L_1-1}a_p\alpha_l^{(1)} \cdot \cos(\phi_l^{(1)} - \varphi_p)R_{\lambda,\lambda}^{(1,1)}\left(\tau_l^{(1)} - \frac{p}{W}\right) \end{aligned} \quad (A24)$$

$$\begin{aligned} \text{Var}\{Z_\zeta - Z_\lambda\} &= A^2\sum_{p=0}^{P-1}\sum_{l=0}^{L_1-1}[a_p\alpha_l^{(1)}\cos(\phi_l^{(1)} - \varphi_p)]^2 \cdot Y_a\left(\tau_l^{(1)} - \frac{p}{W}\right) \\ &+ A^2\sum_{p=0}^{P-1}\sum_{k=1}^{K-1}\sum_{l=0}^{L-1}[a_p\alpha_l^{(k)}\cos(\phi_l^{(k)} - \varphi_p)]^2 \cdot Y_b\left(\tau_l^{(k)} - \frac{p}{W}\right) + N_0 N T_c\sum_{p=0}^{P-1}|a_p|^2. \end{aligned} \quad (A25)$$

Therefore, the signal-to-noise ratio (SNR) for the RAKE receiver is

$$\gamma_{\text{RAKE}} = \frac{\left[\sum_{p=0}^{P-1}\sum_{l=0}^{L_1-1}a_p\alpha_l^{(1)}\cos(\phi_l^{(1)} - \varphi_p)\,R_{\lambda,\lambda}^{(1,1)}\left(\tau_l^{(1)} - \frac{p}{W}\right)\right]^2}{N_{SI} + N_{MI} + N_I} \quad (A26)$$

where

$$N_{SI} = \sum_{p=0}^{P-1}\sum_{l=0}^{L_1-1}[a_p\alpha_l^{(1)}\cos(\phi_l^{(1)} - \varphi_p)]^2 Y_a\left(\tau_l^{(1)} - \frac{p}{W}\right) \quad (A27)$$

$$N_{MI} = \sum_{p=0}^{P-1}\sum_{k=1}^{K-1}\sum_{l=0}^{L_k-1}[a_p\alpha_l^{(k)}\cos(\phi_l^{(k)} - \varphi_p)]^2 Y_b\left(\tau_l^{(k)} - \frac{p}{W}\right) \quad (A28)$$

$$N_I = \frac{N_0}{E_b}\frac{(NT_c)^2}{2\log_2 M}\sum_{p=0}^{P-1}|a_p|^2 \quad (A29)$$

where $Y_a(\cdot)$ and $Y_b(\cdot)$ are defined in Appendix B. N_{SI} is the self-interference of the first user, N_{MI} is the mutual interference from the $K-1$ other users, and N_I is the noise.

APPENDIX B

The aperiodic auto- and cross-correlation function of the pseudonoise sequence vectors are defined as:

$$C(V_\nu^{(k)}, V_\lambda^{(1)})(l) = \begin{cases} \sum_{n=0}^{N-1-l} V_{\nu,n}^{(k)} V_{\lambda,n+l}^{(1)*}; & 0 \le l \le N-1 \\ \sum_{n=0}^{N-1+l} V_{\nu,n-l}^{(k)} V_{\lambda,n}^{(l)*}; & 1-N \le l < 0 \\ 0 & |l| \ge N \end{cases} \quad \text{(B1)}$$

where l is an integer. The phase of $V_{\lambda,n}^{(k)}$ is approximated by a uniform continuous distribution on $[0, 2\pi)$. For random sets of orthogonal codes, the first and second moments of the aperiodic auto- and cross-correlation functions are:

$$E\{C(V_\nu^{(k)}, V_\lambda^{(1)})(l)\} = \begin{cases} N; & k=1, \nu=\lambda, l=0 \\ 0; & \text{otherwise} \end{cases} \quad \text{(B2)}$$

$$E\{|C(V_\nu^{(k)}, V_\lambda^{(1)})(l)|^2\} = \begin{cases} N^2; & k=1, \nu=\lambda, l=0 \\ N-|l|; & k=1, 0<|l|\le N-1 \\ N-|l|; & k\neq 1, 0\le|l|\le N-1 \\ 0; & \text{otherwise} \end{cases}. \quad \text{(B3)}$$

For the chip waveform:

$$H_p(s) = \int_0^s P_{T_c}(t) P_{T_c}(t+T_c-s)\,dt$$
$$\hat{H}_p(s) = \int_s^{T_c} P_{T_c}(t) P_{T_c}(t-s)\,dt \quad \text{(B4)}$$

For P_{T_c}, a rectangular pulse and $0 \le s < T_c$:

$$H_p(s) = s; \quad \hat{H}_p(s) = T_c - s. \quad \text{(B5)}$$

Then, if $dT_c \le \tau_l + \Delta_k < (d+1)T_c$, Δ_k for $k \neq 0$ uniformly distributed in $[0, T_c)$ and $\Delta_0 = 0$:

$$E\{\Delta_k\} = \frac{T_c}{2}; \quad k \neq 0$$
$$E\{(\Delta_k)^2\} = \frac{T_c^2}{3}; \quad k \neq 0 \quad \text{(B6)}$$

Then

$$\begin{aligned} R_{\lambda,\lambda}^{(1,1)}(\tau_l^{(1)}) &= H_p(\tau_l^{(1)} - dT_c) \\ &\cdot [C(V_\nu^{(1)}, V_\lambda^{(1)})(d-N+1) + C(V_\lambda^{(1)}, V_\lambda^{(1)})(d+1)] \\ &+ \hat{H}_p(\tau_l^{(1)} - dT_c) \\ &\cdot [C(V_\nu^{(1)}, V_\lambda^{(1)})(d-N) + C(V_\lambda^{(1)}, V_\lambda^{(1)})(d)] \quad \text{(B7)} \end{aligned}$$

$$\begin{aligned} R_{\nu,\lambda}^{(k,1)}(\tau_l^{(k)}) &= H_p(\tau_l^{(k)} - dT_c) \\ &\cdot [C(V_\xi^{(k)}, V_\lambda^{(1)})(d-N+1) + C(V_\nu^{(k)}, V_\lambda^{(1)})(d+1)] \\ &+ \hat{H}_p(\tau_l^{(k)} - dT_c) \\ &\cdot [C(V_\xi^{(k)}, V_\lambda^{(1)})(d-N) + C(V_\nu^{(k)}, V_\lambda^{(1)})(d)] \quad \text{(B8)} \end{aligned}$$

$$\begin{aligned} R_{\lambda,\nu}^{(1,1)}(\tau_l^{(1)}) &= H_p(\tau_l^{(1)} - dT_c) \\ &\cdot [C(V_\xi^{(1)}, V_\nu^{(1)})(d-N+1) + C(V_\lambda^{(1)}, V_\nu^{(1)})(d+1)] \\ &+ \hat{H}_p(\tau_l^{(1)} - dT_c) \\ &\cdot [C(V_\xi^{(1)}, V_\nu^{(1)})(d-N) + C(V_\lambda^{(1)}, V_\nu^{(1)})(d)] \quad \text{(B9)} \end{aligned}$$

$$\begin{aligned} R_{\xi,\nu}^{(k,1)}(\tau_l^{(k)}) &= H_p(\tau_l^{(k)} - dT_c) \\ &\cdot [C(V_\mu^{(k)}, V_\nu^{(1)})(d-N+1) + C(V_\xi^{(k)}, V_\nu^{(1)})(d+1)] \\ &+ \hat{H}_p(\tau_l^{(k)} - dT_c) \\ &\cdot [C(V_\mu^{(k)}, V_\nu^{(1)})(d-N) + C(V_\xi^{(k)}, V_\nu^{(1)})(d)] \quad \text{(B10)} \end{aligned}$$

Evaluating the statistics of the difference of the correlation terms;

$$\begin{aligned} E\{R_{\lambda,\nu}^{(1,1)}(\tau_l^{(1)}) - R_{\lambda,\lambda}^{(1,1)}(\tau_l^{(1)})\} &= -R_{\lambda,\lambda}^{(1,1)}(\tau_l^{(1)}) \\ &= \begin{cases} -N(T_c - \tau_l^{(1)}); & d=0 \\ -N\tau_l^{(1)}; & d=-1 \\ 0; & \text{elsewhere} \end{cases} \end{aligned} \quad \text{(B11)}$$

$$\begin{aligned} Y_a(\tau_l^{(1)}) &= \mathrm{Var}\{R_{\lambda,\nu}^{(1,1)}(\tau_l^{(1)}) - R_{\lambda,\lambda}^{(1,1)}(\tau_l^{(1)})\} \\ &= H_p^2(\tau_l^{(1)} - dT_c) \\ &\cdot (2\mathcal{E}_c(d-N+1) + \mathcal{E}_c(d+1) + \mathcal{E}_a(d+1)) \\ &+ \hat{H}_p^2(\tau_l^{(1)} - dT_c) \\ &\cdot (\mathcal{E}_c(d) + 2\mathcal{E}_c(d-N) + \mathcal{E}_a(d)) \end{aligned} \quad \text{(B12)}$$

and, for Δ_k $(k \neq 0)$ uniformly distributed in $[0, T_c)$ see (B13)–(B17) that follows.

$$E\{R_{\beta,\nu}^{(k,1)}(\tau_l^{(k)} + \Delta_k) - R_{\nu,\lambda}^{(k,1)}(\tau_l^{(k)} + \Delta_k)\} = 0 \quad \text{(B13)}$$

$$\begin{aligned} Y_b(\tau_l^{(k)}) &= \mathrm{Var}\{R_{\beta,\nu}^{(k,1)}(\tau_l^{(k)} + \Delta_k) - R_{\nu,\lambda}^{(k,1)}(\tau_l^{(k)} + \Delta_k)\} \\ &= 2[\mathcal{E}_b(d+1) + \mathcal{E}_b(d-N+1)] \\ &\cdot \left[(\tau_l^{(k)} - dT_c)^2 + T_c\left(\frac{T_c}{3} + \tau_l^{(k)} - dT_c\right)\right] \\ &+ 2[\mathcal{E}_b(d) + \mathcal{E}_b(d-N)] \\ &\cdot \left[((1+d)T_c - \tau_l^{(k)})^2 + T_c\left(\frac{T_c}{3} + \tau_l^{(k)} - (1+d)T_c\right)\right] \end{aligned} \quad \text{(B14)}$$

where

$$\mathcal{E}_a(l) = \mathrm{Var}\{C(V_\lambda^{(1)}, V_\lambda^{(1)})(l)\} = \begin{cases} N-|l|; & 0<|l|\le N-1 \\ 0; & \text{otherwise} \end{cases} \quad \text{(B15)}$$

$$\mathcal{E}_b(l) = \mathrm{Var}\{C(V_\nu^{(k)}, V_\lambda^{(1)})(l)\} = \begin{cases} N-|l|; & 0\le|l|\le N-1 \\ 0; & \text{otherwise} \end{cases} \quad \text{(B16)}$$

$$\mathcal{E}_c(l) = \mathrm{Var}\{C(V_\nu^{(1)}, V_\lambda^{(1)})(l)\} = \begin{cases} N-|l|; & 0<|l|\le N-1 \\ 0; & \text{otherwise} \end{cases} \quad \text{(B17)}$$

References

[1] K. Pahlavan, "Wireless communications for office information networks," *IEEE Commun. Mag.*, vol. 23, pp. 19–27, June 1985.

[2] P. Ferert, "Application of spread-spectrum radio to wireless terminal communications," in *Proc. IEEE Nat. Telecommun. Conf.*, Dec. 1980, pp. 244–248.

[3] M. J. Marcus, "Recent U.S. regulatory decisions on civil use of spread spectrum," in *Proc. IEEE GLOBECOM*, Dec. 1985, pp. 16.6.1–16.6.3.

[4] K. Pahlavan, "Wireless intra-office networks," *ACM Trans. Office Inform. Networks*, July 1988.

[5] M. Kavehrad and P. J. McLane, "Performance of low-complexity channel coding and diversity for spread spectrum in indoor wireless communications," *AT&T Tech. J.*, vol. 64, no. 8, pp. 1927–1965, Oct. 1985.

[6] M. Kavehrad and B. Ramamurthi, "Direct sequence spread spectrum with DPSK modulation and diversity for indoor wireless communications," *IEEE Trans. Commun.*, vol. COM-35, pp. 224–236, Feb. 1987.

[7] M. Chase and K. Pahlavan, "Spread spectrum multiple-access performance of orthogonal codes in fading multipath channels," *IEEE MILCOM*, Oct. 1988, pp. 5.4.1–5.4.5.

[8] J. S. Lehnert and M. B. Pursley, "Multipath diversity reception of spread-spectrum multiple-access communications," *IEEE Trans. Commun.*, vol. COM-35, pp. 1189–1198, Nov. 1987.

[9] P. K. Enge and D. V. Sarwate, "Spread spectrum multiple-access performance of orthogonal codes: linear receivers," *IEEE Trans. Commun.*, vol. COM-35, pp. 1309–1319, Dec. 1987.

[10] K. Pahlavan and M. Chase, "Spread-spectrum multiple-access performance of orthogonal codes for indoor radio communications," *IEEE Trans. Commun.*, vol. 38, pp. 574–577, May 1990.

[11] R. Ganesh and K. Pahlavan, "On the modeling of the fading multipath indoor radio channel," *IEEE GLOBECOM*, 1989.

[12] J. G. Proakis, *Digital Communications*. New York: McGraw-Hill, 1989.

Mitchell Chase (S'73–M'78) was born in Brooklyn, New York on October 10, 1951. He received the B.E. degree in electrical engineering from the City College of New York, New York City, in 1974, the M.S. degree in biomedical engineering from Iowa State University, Ames, IA, in 1976, the M.B.A. degree from Northeastern University, Boston, MA, in 1982, and the Ph.D. degree in electrical engineering from Worcester Polytechnic Institute, Worcester, MA, in 1993.

Since 1978 he has held a variety of positions at several companies developing products ranging from agricultural electronics to flight simulators and automated systems for the identification of blood cells. He is currently an Applications Engineer with Comdisco Systems, Inc. His current research interests are in the areas of spread spectrum and wireless communications. He has authored several publications in this area as well as in the areas of digital and image signal processing, graphics, and biomedical engineering.

Kaveh Pahlavan is a Professor of Electrical and Computer Engineering and the Director of the Center for Wireless Information Network Studies at the Worcester Polytechnic Institute, Worcester, MA. His basic research in the past few yers has been focused on indoor radio propagation modeling and analysis of the multiple access and transmission methods for wireless local networks. His previous research background is on modulation, coding, and adaptive signal processing for digital communication over voice-band and fading multipath radio channels. He has contributed to numerous technical papers, has presented many tutorials and short courses in various countries and has been a consultant to many industries including CNR Inc., GTE Laboratories, Steinbrecher Corporation, Simplex Company, and WINDATA Inc. Before joining WPI, he was the Director of Advanced Development at Infinite Inc., Andover, MA, working on voice band data communications. He started his career as an Assistant Professor at Northeastern University, Boston, MA. He is the Editor-in-Chief of the *International Journal on Wireless Information Networks*. He was the program chairman and organizer of the IEEE Wireless LAN Workshop, Worcester, MA, May 9, 10, 1991 and the organizer and the technical program chairman of the 3rd IEEE International Symposium on Personal, Indoor, and Mobile Radio Communications, Boston, MA, October 19–21, 1992. He is a Member of Eta Kappa Nu and a Senior Member of the IEEE Communication Society.

Performance Evaluation of Direct-Sequence Spread Spectrum Multiple-Access for Indoor Wireless Communication in a Rician Fading Channel

RAMJEE PRASAD, *Senior Member, IEEE*, HOWARD SEWBERATH MISSER, *Member, IEEE*, and ADRIAAN KEGEL, *Member, IEEE*.

***Abstract*—The bit error probability in a Rician fading channel is evaluated for indoor wireless communications considering Direct-Sequence Spread Spectrum Multiple Access (DS/SSMA) with Differential Phase Shift Keying (DPSK) modulation and two types of diversity: selection diversity and maximal ratio combining. The performance of the indoor radio system is also obtained in terms of outage probability and bandwidth efficiency. The analysis is done for a star-connected multiple access radio network. Furthermore the influence of three types of Forward Error Correcting (FEC) codes namely, the (15,7) BCH code, the (7,4) Hamming code and the (23,12) Golay code, on the performance is studied. Computational results are presented for suitable values of Rician parameters in an indoor environment and using Gold codes as spread spectrum codes.**

I. Introduction

Indoor Wireless communication has recently drawn the attention of many researchers [1–14,16] due to its significant advantages over the conventional cabling: modiltity of users, elimination of wiring and rewiring, drastical reduction of wiring in new buildings, flexibility of changing or creating new communication services, time and cost saving, and reduction of the down time of services. Much attention is being paid to the use of Direct-Sequence Spread-Spectrum (DS/SS) modulation for indoor wireless multiple access communication, over multipath fading channels [1–8,16]. DS/SS modulation provides both multiple access capability and resistance to multipath fading.

In this paper, we obtain the performance of Direct Sequence Spread Spectrum Multiple Access (DS/SSMA) with DPSK, in an indoor Rician fading radio channel. The performance of a DS/SSMA system in indoor Rayleigh fading channels using DPSK or CPSK was considered in [1–4]. However, recent multipath measurements of the indoor radio channel at 800/900 MHz and 1.75 GHz characterize the indoor environment as a frequency selective, Rician fading channel [9,10]. Measurements in factory environments have also indicated that Rician distribution fits the experimental data [11]. A recent paper by Wang and Moeneclaey [16] has appeared on a similar topic. That paper [16] addressed the performance of Hybrid DS/SFH-SSMA systems in indoor Rician-channels using maximal ratio combining and coding. Thus, there is a small overlap between [16] and this paper as far as the performance using maximal ratio combining is concerned.

To avoid the need for synchronous carrier recovery at the receiver, which is a difficult task in a multipath fading environment, DPSK is used as modulation scheme. Also, selection diversity and maximal ratio combining are used to combat the multipath fading.

Radiowave propagation measurements showed that the maximum rms delay spread at 850 MHz, 1.7 GHz and 4.0 GHz did not exceed 270 nsec. in the larger buildings and 100 nsec. in the smaller buildings [12]. In this paper the performance is evaluated for the rms delay spread in the range of 50 nsec. to 250 nsec.

In the performance analysis average power control is assumed to make sure that all signals arrive at the base station with the same average power. As explained in [3] this can be accomplished as follows. The base station transmits a signal common to all users. The users monitor the average level of this signal. This information can now be used to adjust the transmitted power at the user location.

The paper is organized as follows. In section II the system model is introduced. In section III the performance is derived in terms of bit error probability, outage probability and bandwidth efficiency. Numerical results and discussions are presented in section IV. In section V the conclusions can be found.

Paper approved by James S. Lehnert, the Editor for Modulation & Signal Design of the IEEE Communications Society. Manuscript received: December 20, 1990; revised: September 25, 1992; January 7, 1993; November 15, 1993. This paper was presented in part at the First IEEE International Symposium on spread spectrum Techniques and Applications, King's College London, September 1990.

R. Prasad and A. Kegel are with the Telecommunications and Traffic-Control Systems Group, Delft University of Technology, P.O. Box 5031, 2600 GA Delft, The Netherlands. H. S. Misser is with Dr. Neher Laboratories, PTT Research, P.O. Box 421, 2260 AK Leidschendam, The Netherlands.

IEEE Log Number 9410091.

II. System Model

In this section we present the transmitter, the channel and the receiver model. The transmitter and receiver models are similar to those in [3].

A. Transmitter Model

K active users may simultaneously transmit to a base station using DS/SSMA with DPSK modulation. The (differentially encoded) data waveform of user k is denoted as

$$b_k(t) = \sum_{j=-\infty}^{\infty} b_k^j P_{T_b}(t-jT_b)\ ,\ b_k^j \in (1,-1) \tag{1}$$

where P_T is a rectangular pulse of unit height and duration T, and T_b is the duration of one data bit. Each user has a unique spread spectrum code of N chips that fits into one data bit, i.e., $T_b = NT_c$ where T_c is the chip duration. The spread spectrum code of user k is

$$a_k(t) = \sum_i a_k^i P_{T_c}(t-iT_c)\ ,\ a_k^i \in (1,-1) \tag{2}$$

where $i = ..,-1, 0, 1, ...$ and $a_k^i = a_k^{i+N}$.

Now the signal at the output of the kth transmitter can be written as

$$S_k(t) = Aa_k(t)b_k(t)\cos(\omega_c t + \theta_k) \tag{3}$$

where A is the amplitude of the carrier, ω_c is the common angular carrier frequency, and θ_k is the carrier phase for the kth user.

B. Channel model

We assume that the signal bandwidth is much larger than the coherence bandwidth of the radio channel which assures us of the existence of multiple resolvable paths. The (complex) lowpass equivalent impulse response of the bandpass channel for the link between the kth user and the base station is written as

$$h_k(t) = \sum_{l=1}^{L} \beta_{lk}\delta(t-\tau_{lk})\exp(j\gamma_{lk}) \tag{4}$$

Here β is the path gain, τ is the path delay, γ is the path phase, and L is the number of resolvable paths. The index lk refers to the lth path of the kth user, and $j = \sqrt{-1}$. The number of paths may be either fixed or randomly changing. Here fixed values for L are assumed. The number of paths is upper bounded by [4]

$$L = \left\lfloor \frac{T_m}{T_c} \right\rfloor + 1 \tag{5}$$

where T_m is the rms delay spread and T_c is the duration of a code chip.

We assume that the path phases on arrival at the receiver, $(\omega_c\tau_{lk} + \gamma_{lk})$, are independently uniformly distributed over $[0, 2\pi]$. We also assume that the path delays are independently uniformly distributed over $[0, T_b]$. Unlike [1–4], where β_{lk} was assumed to be Rayleigh distributed, we shall assume that the path gains are independent Rician distributed random variables. This is in accordance with recent measurements done in office [9,10] and factory [11] buildings.

The Rician probability density function (PDF) is given as

$$p_\beta(r) = \frac{r}{\sigma^2}\exp\left(-\frac{r^2+S^2}{2\sigma^2}\right)I_0\left(\frac{Sr}{\sigma^2}\right)$$

$$0 \le r \le \infty,\ S \ge 0 \tag{6}$$

$$R = \frac{S^2}{2\sigma^2}$$

where $I_0()$ is the modified Bessel function of the first kind and zero order, S is the peak value of the diffuse radio signal due to the superposition of the dominant (line-of-sight) signal and the time invariant scattered signals reflected from walls, ceiling and stationary inventory, σ^2 is the average signal power that is received over specular paths.

From [10] we know that typical values for R are 6.8 dB and 11 dB. R = 6.8 dB corresponds to a 30-year-old brick building with reinforced concrete and plaster and R = 11 dB corresponds to a building having the same construction, but with an open-office interior floor plan, and non-metallic ceiling tiles throughout.

C. Receiver model

Using equation (3) and (4) the received signal can be written as

$$r(t) = \sum_{k=1}^{K}\sum_{l=1}^{L} A\beta_{lk}a_k(t-\tau_{lk})b_k(t-\tau_{lk}) \cdot \cos(\omega_{c}t + \phi_{lk}) + n(t) \tag{7}$$

$$\phi_{lk} = \omega_c\tau_{lk} + \gamma_{lk}$$

where $n(t)$ is the white Gaussian noise with two-sided power spectral density $N_o/2$. The receiver model consists of a matched filter, a DPSK demodulator and diversity processing components, as shown in figure 1.

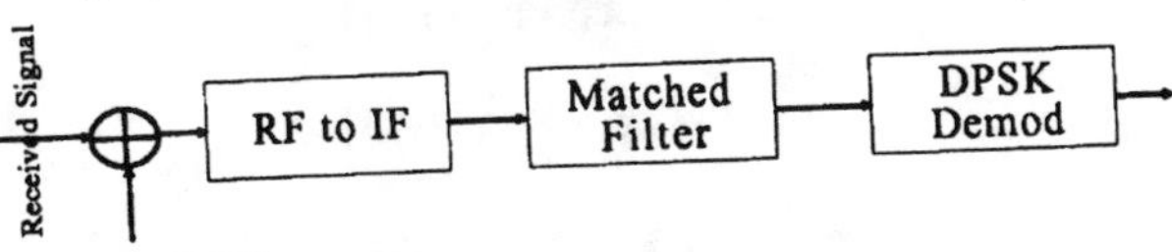

Fig. 1. Block diagram of the spread spectrum receiver using DPSK modulation.

As in [3] equation (7) can be divided into two parts: in-phase and a quadrature component. Selecting user 1 as the reference user, the output (in-phase and quadrature component) of the matched filter of user 1 at the sampling point ($t = T_b$) is given as

$$g_x(T_b) = \sum_{k=1}^{K}\sum_{l=1}^{L} A\beta_{lk}\cos(\phi_{lk})[b_k^{-1}R_{1k}(\tau_{lk}) + b_k^0\hat{R}_{1k}(\tau_{lk})] + \eta$$
$$g_y(T_b) = \sum_{k=1}^{K}\sum_{l=1}^{L} A\beta_{lk}\sin(\phi_{lk})[b_k^{-1}R_{1k}(\tau_{lk}) + b_k^0\hat{R}_{1k}(\tau_{lk})] + \nu \tag{8}$$

where g_x and g_y are the in-phase and the quadrature component, respectively, b_k^{-1} and b_k^0 are the previous and current data bit, respectively, and

$$R_{1k}(\tau) = \int_0^{\tau} a_k(t-\tau)a_1(t)\,dt$$
$$\hat{R}_{lk}(\tau) = \int_{\tau}^{T_b} a_k(t-\tau)a_1(t)\,dt \tag{9}$$

The noise samples η and ν are independent, zero-mean Gaussian random variables with identical variance $\sigma_n^2 = N_oT_b$.

Let us assume without loss of generality that the receiver synchronizes to the jth path of user 1, so that $\tau_{j1} = 0$ and $\phi_{j1} = 0$ [3]. The complex envelope of the signal at the current sampling instant then is

$$z_0 = A\beta_{j1}T_bb_1^0 + \sum_{k=1}^{K} A(b_k^{-1}X_k + b_k^0\hat{X}_k) + j\sum_{k=1}^{K} A(b_k^{-1}Y_k + b_k^0\hat{Y}_k) + (\eta_1 + j\nu_1) \tag{10}$$

where

$$X_1 = \sum_{\substack{l=1\\l\neq j}}^{L} R_{11}(\tau_{l1})\beta_{l1}\cos(\phi_{l1})\ ;\quad \hat{X}_1 = \sum_{\substack{l=1\\l\neq j}}^{L} \hat{R}_{11}(\tau_{l1})\beta_{l1}\cos(\phi_{l1})$$
$$Y_1 = \sum_{\substack{l=1\\l\neq j}}^{L} R_{11}(\tau_{l1})\beta_{l1}\sin(\phi_{l1})\ ;\quad \hat{Y}_1 = \sum_{\substack{l=1\\l\neq j}}^{L} \hat{R}_{11}(\tau_{l1})\beta_{l1}\sin(\phi_{l1}) \tag{11}$$

and

$$\underset{k\neq 1}{X_k} = \sum_{l=1}^{L} R_{1k}(\tau_{lk})\beta_{lk}\cos(\phi_{lk})\ ;\quad \underset{k\neq 1}{\hat{X}_k} = \sum_{l=1}^{L} \hat{R}_{1l}(\tau_{lk})\beta_{lk}\cos(\phi_{lk})$$
$$\underset{k\neq 1}{Y_k} = \sum_{l=1}^{L} R_{1k}(\tau_{lk})\beta_{lk}\sin(\phi_{lk})\ ;\quad \underset{k\neq 1}{\hat{Y}_k} = \sum_{l=1}^{L} \hat{R}_{1k}(\tau_{lk})\beta_{lk}\sin(\phi_{lk}) \tag{12}$$

Involving only b_k^{-1} and b_k^0 means that it is assumed that $\tau_{lk} \geq 0$. This assumption can be made without loss of generality for two reasons: 1) all bits, except b_1^{-1} and b_1^0, are 1 or −1 with equal probability; 2) it is not important for the cross-correlation of two spread spectrum codes what the sign of the phase difference between the codes is.

DPSK demodulation is now achieved by taking the real part of $z_oz^*_{-1}$, where z^* denotes complex conjugate of z.

III. Performance Analysis

In this section the bit error probability and the outage probability for both selection diversity and maximal ratio combining are derived. The bit error probability with FEC coding and the bandwidth efficiency are also obtained.

A. Selection diversity

1) Bit error probability: The selection diversity scheme is based on selecting the strongest of L resolvable paths. By using multiple antennas the highest possible order of diversity, i.e. number of paths to choose from, can be increased to M_{max} = kL where M_{max} is the maximum order of diversity and k is the number of antennas. To derive the PDF of the strongest path (β_{max}), it is essential to note that the cumulative density function (CDF) of β_{max} is just the CDF of the PDF in equation (6) raised to the power of the order of diversity, M, hence

$$
\begin{aligned}
P_{\beta_{max}}(r) = M[1 - Q(\frac{s}{\sigma}, \frac{r}{\sigma})]^{M-1} \frac{r}{\sigma^2} \\
\cdot \exp(-\frac{r^2+s^2}{2\sigma^2}) I_o(\frac{sr}{\sigma^2})
\end{aligned}
\tag{13}
$$

where Q(a,b) is the Marcum Q-function. Designate the decision variable for DPSK modulation $\xi = \mathrm{Re}[z_0 z^*_{-1}]$, where Re[a] denotes the real part of a. The decision variable obtained from demodulation of the strongest path is written as ξ_{max}. Now the bit error probability in the case of selection diversity is defined as

$$
\begin{aligned}
P_e \triangleq P(\xi_{max} < 0 \mid b_1^0 b_1^{-1} = 1) \\
\triangleq P(\xi_{max} > 0 \mid b_1^0 b_1^{-1} = -1)
\end{aligned}
\tag{14}
$$

In the analysis it is assumed that $b_1^0 b_1^{-1} = 1$, i.e. $b_1^0 = b_1^{-1} = 1$.

If we assume that all path delays are given and β_{max} is correctly selected, the bit error probability can be obtained using equation (7A.26) of [15]

$$
\begin{aligned}
P_e \mid \beta_{max}, \{\tau_{lk}\} = Q(a,b) - \frac{1}{2}\left[1 + \frac{\mu}{\sqrt{\mu_0 \mu_{-1}}}\right] \\
\cdot I_o(ab) \exp\left(-\frac{a^2+b^2}{2}\right)
\end{aligned}
\tag{15}
$$

Here a, b, μ, μ_0 and μ_{-1} are defined as ($b_1^0 = b_1^{-1} = 1$)

$$
\begin{aligned}
a \triangleq \frac{m}{\sqrt{2}}\left(\frac{1}{\sqrt{\mu_0}} - \frac{1}{\sqrt{\mu_{-1}}}\right), \; b \triangleq \frac{m}{\sqrt{2}}\left(\frac{1}{\sqrt{\mu_0}} + \frac{1}{\sqrt{\mu_{-1}}}\right), \\
m \triangleq E[z_0 \mid \beta_{max}, b_1^0] = E[z_{-1} \mid \beta_{max}, b_1^{-1}] \\
\mu_0 \triangleq var(z_0 \mid \{\tau_{lk}\}), \; \mu_{-1} \triangleq var(z_{-1} \mid \{\tau_{lk}\}), \\
\mu \triangleq E[(z_0 - m)(z_{-1} - m)^* \mid \{\tau_{lk}\}]
\end{aligned}
\tag{16}
$$

with E[] denoting statistical average and var() denoting variance.

Using equation (10) and the assumption that $b_1^0 = b_1^{-1} = 1$, M, μ_0, μ_{-1} and μ are obtained as

$$
\begin{aligned}
m = A\beta_{max} T_b b_1^0 = A\beta_{max} T_b b_1^{-1} \\
\mu_0 = A^2 \sum_{k=1}^{K} E\left[X_k^2 + \hat{X}_k^2 + Y_k^2 + \hat{Y}_k^2 \mid \{\tau_{lk}\}\right] \\
+ 2A^2 E\left[X_1 \hat{X}_1 + Y_1 \hat{Y}_1 \mid \{\tau_{lk}\}\right] + 2\sigma_n^2 \\
\mu_{-1} = A^2 \sum_{k=1}^{K} E\left[X_k^2 + \hat{X}_k^2 + Y_k^2 + \hat{Y}_k^2 \mid \{\tau_{lk}\}\right] + 2\sigma_n^2 \\
\mu = A^2 E\left[\sum_{k=1}^{K} \left(X_k \hat{X}_k + Y_k \hat{Y}_k\right) + \hat{X}_1^2 + \hat{Y}_1^2 \mid \{\tau_{lk}\}\right]
\end{aligned}
\tag{17}
$$

Note that all path gains involved in μ_0, μ_{-1} and μ have a Rician distribution as given in Equation (6) except for the path gains associated to the paths of user 1. The path gains of user 1 (except for the strongest path β_{max}) have a conditional Rician PDF where the conditioning is on β_{max}. The PDF becomes zero for path gains larger than β_{max}. The distribution of β_{max} is given in Equation (13).

As can be seen from equation (17), the only difference between μ_0 and μ_{-1} is constituted by the second term of μ_0. If the number of simultaneously transmitting users, K, is large the contribution of this term to μ_0 becomes relatively small [2,3]. If we drop this term from μ_0, a in equation (16) becomes zero. Equation (15) can then be simplified to

$$
\begin{aligned}
P_e \mid \beta_{max}, \{\tau_{lk}\} = Q(0,b) - \frac{1}{2}\left[1 + \frac{\mu}{\sqrt{\mu_0 \mu_{-1}}}\right] I_0(0) \\
\exp\left(-\frac{b^2}{2}\right) = \frac{1}{2}\left[1 - \frac{\mu}{\mu_0}\right] \exp\left(-\frac{m^2}{\mu_0}\right)
\end{aligned}
\tag{18}
$$

since $I_0(0) = 1$ and $Q(0,b) = \exp(-b^2/2)$. Now to remove the conditioning on τ_{lk}, we approximate μ and μ_0 by Gaussian variables and integrate equation (18) over μ and μ_0. Likewise the conditioning on β_{max} is removed. Thus equation (18) reduces to

$$
\begin{aligned}
P_e = \int_{-\infty}^{\infty} \int_{0}^{\infty} \frac{1}{2}\left[1 - \frac{\mu}{\mu_0}\right] \exp\left(-\frac{A^2 \beta_{max}^2 T_b^2}{\mu_0}\right) \\
\cdot P_{\beta_{max}} P_{\mu_0} \, d\beta_{max} \, d\mu_0
\end{aligned}
\tag{19}
$$

For the calculation of the variance of μ and μ_0 a specific set of Gold codes are used. The partial cross-correlation functions are calculated using the specific Gold codes and averaging over the (uniform) path delay distributions. For calculating the moments of μ and μ_0, these calculations have to be done for all resolvable paths of all users.

2) Outage probability: Outage probability is defined as the probability that the instantaneous bit error probability exceeds a preset threshold. We denote the threshold value as ber_o. The instantaneous value of the bit error probability can be obtained using equation (19). The averaging over β_{max} should then be removed and a fixed value for β_{max}, β, should be substituted. Equation (19) then transforms to

$$P_e(\beta) = \int_{-\infty}^{\infty} \frac{1}{2}\left[1 - \frac{\mu}{\mu_0}\right] \exp\left(-\frac{A^2\beta^2T_b^2}{\mu_o}\right) p_{\mu_0}(\mu_0)\, d\mu_0 \tag{20}$$

The outage probality in the case of selection diversity can then be calculated as follows

$$P_{out} = P(0 \le \beta_{max} \le \beta_0) = P(ber(\beta) \ge ber_o)$$

$$= \int_0^{\beta_0} M\left[1 - Q\left(\frac{s}{\sigma}, \frac{v}{\sigma}\right)\right]^{M-1} \frac{v}{\sigma^2} \exp\left(-\frac{v^2+s^2}{2\sigma^2}\right) I_0\left(\frac{sv}{\sigma^2}\right) dv \tag{21}$$

Here β_o is the value of β at which the instantaneous bit error probability is equal to ber_o. The integrated function is just the PDF of β_{max} where M is the order of diversity. For a given value of β_o, the outage probability decreases as the order of diversity increases. Also a higher signal-to-noise ratio results in a lower outage probability because for a given ber_o, the value of β_o obtained from (20) decreases as the signal to noise ratio increases.

B. Maximal ratio combining

1) Bit error probability: With Maximal-Ratio Combining (MRC) of order M the decision variable is the weighted sum of the demodulation results of M copies of the signal. The weights are taken equal to the corresponding complex-valued (conjugate) channel gain $\beta_i \exp(-j\gamma_i)$. The effect of this multiplication is to compensate for the phase shift in the channel and to weight the signal by a factor that is proportional to the signal strength. A reason for using DPSK, as in our case, is that the time variations in the channel parameters are sufficiently fast to preclude the implementation of synchronous carrier recovery schemes. However, the channel variations must on the other hand be sufficiently slow so that the channel phase shifts (γ_i) do not change appreciably over two consecutive signalling intervals (DPSK detection). We assume that the channel parameters $\{\beta_i \exp(-j\gamma_i)\}$ remain constant over two succesive signalling intervals. Under that condition we do not need estimates of the channel parameters because the signals are automaticaly weighted. The decision variable in the case of MRC/DPSK is

$$\xi = Re\left[\sum_{i=1}^{M} (AT_b\beta_i b_1^{\,0} + N_{1i})(AT_b\beta_i b_1^{\,-1} + N_{2i})^*\right] \tag{22}$$

Here N_{1i} and N_{2i} are Gaussian random variables. If we assume, just as in the case of selection diversity, the multi-user interference to be Gaussian, N_{1i} and N_{2i} are the sum of the Gaussian noise power and the multi-user interference power. The multi-user interference is calculated in the same way as for selection diversity. Equation (22) is just the sum of the demodulated signals of M paths. This method is therefore sometimes called predetection combining or post-demodulation combining. We will now make two assumptions. First, it can be assumed that N_{1i} and N_{2i} are independent. We have learned from the case of selection diversity that μ, which is defined in Equation (16), is nearly zero. This implies that the cov[N_{1i}, N_{2i}^*] is negligible compared with var[N_{1i}] and var[N_{2i}], and therefore the above assumption is reasonable [3]. Secondly, we assume that the pairs (N_{1i},N_{2i}) and (N_{1j},N_{2j}) are independent for i ≠ j. Mathematically this is not correct because the delays $\{\tau_{lk}\}_i$ are not independent of $\{\tau_{lk}\}_j$. However, since each set of delays is taken with reference to a different time origin (corresponding to the arrival time of the signal on the corresponding combined path), and also considering that any two resolved paths are separated by at least a chip time period, the assumption is physically reasonable [3]. With these assumptions, the bit error probability can be obtained using equations (7.4.13) and (1.1.115) of [15]

$$P_e = \int_0^{\infty} P_2(\gamma_b) p(\gamma_b)\, d\gamma_b \tag{23}$$

with

$$P_2(\gamma_b) = \frac{1}{2^{2M-1}} \exp(-\gamma_b) \sum_{n=0}^{M-1} p_k \gamma_b^k\,;$$

$$p_k = \frac{1}{k!} \sum_{n=0}^{M-1-k} \binom{2M-1}{n}\,; \quad \gamma_b = \frac{E_b}{N} \sum_{k=1}^{M} \beta_k^2$$

$$p(\gamma_b) = \frac{1}{2\sigma^2 \frac{E_b}{N}} \left[\frac{\gamma_b}{\frac{E_b}{N} s_M^2}\right]^{\frac{M-1}{2}} \cdot \exp\left(-\frac{s_M^2 + \gamma_b \frac{N}{E_b}}{2\sigma^2}\right) I_{M-1}\left(\frac{\sqrt{\gamma_b \frac{N}{E_b}}\, s_M}{\sigma^2}\right) \tag{24}$$

where $E_b = A^2T_b/2$, β_k is the path gain of the *kth* combined path, *N* is the sum of the Gaussian noise and the multi-user interference which is assumed to be Gaussian, $I_\alpha(x)$ is the αth-order modified Bessel function of the first kind and $s_M^2 = Ms^2$, γ_b is the sum of the signal-to-noise ratios of the M combined paths, and P2 is the bit error probability in the case of multichannel reception over time invariant paths, i.e., the β_k's are equal constants, with MRC/DPSK. Therefore, $P_2(\gamma_b)$ can be considered a conditional error probability. $p(\gamma_b)$ is the PDF of γ_b in the case of a Rician fading channel.

2) Outage probability: The outage probability is then obtained as

$$P_{out} = \int_0^{\gamma_{b_o}} p(\gamma_b)\, d\gamma_b \quad ; P_2(\gamma_{b_o}) = ber_o \tag{25}$$

where ber_o is the preset bit error threshold.

C. Bit error probability with FEC coding

The specific codes considered are the (15,7) BCH code, the (7,4) Hamming code and the (23,12) Golay code. (n,k) means that k bits are transformed to a block of n bits by coding. From coding theory we know that a code with hamming distance d_{min} can correct at least $t = (d_{min} - 1)/2$ errors [15]. For the BCH code $d_{min} = 5$, for the Hamming code $d_{min} = 3$ and for the Golay code $d_{min} = 7$, which means that they can correct two, one and three errors respectively. The probability of having m errors in a block of n bits is

$$P(m,n) = \binom{n}{m} P_e^{\,m} (1-P_e)^{n-m} \tag{26}$$

Now the probability of having more than t errors in a code block of n bits is

$$P_{ec} = \sum_{m=t+1}^{n} \binom{n}{m} P_e^{\,m} (1-P_e)^{n-m} \tag{27}$$

An approximation for the bit error probability after decoding is given in [2] as

$$P_{ec1} = \frac{1}{n} \sum_{m=t+1}^{n} m \binom{n}{m} P_e^{\,m} (1-P_e)^{n-m} \tag{28}$$

Since the block codes can correct at least t errors, equations (27) and (28) are actually upperbounds on the block error and bit error probability, respectively. Suppose we place a sphere of radius t around each of the possible transmitted code words in the code space. Codes where all these spheres are disjoint and where every received code word falls in one of these spheres, are called perfect codes. They can correct $t = (d_{min} - 1)/2$ errors. For these codes equation (28) is not an upperbound but the exact block error probability. The (7,4) Hamming code and the (23,12) Golay code are examples of these codes [15]. Codes where all the spheres of radius t are disjoint and where every received code word is at most at distance t+1 from one of the possible transmitted code words, are called quasi perfect codes. They can sometimes correct t+1 errors. The bit error probability can be derived, using the block error probability that is given in [15], as

$$P_{ec2} = \frac{1}{n} \sum_{m=t+2}^{n} m P(m,n) + \frac{t+1}{n} \left[\binom{n}{t+1} - \beta_{t+1} \right] P_e^{\,t+1} (1-P_e)^{n-t-1} \tag{29}$$

$$\beta_{t+1} = 2^{n-k} - \sum_{i=0}^{t} \binom{n}{i}$$

Like perfect codes, quasi perfect codes are optimum on the binary symmetric channel in the sense that they result in a minimum error probability among all codes having the same block length and the same number of information bits. Therefore P_{ec2} is a lowerbound for all non perfect linear block codes. Since the (15,7) BCH is neither a perfect code nor a quasi perfect code, P_{ec1} and P_{ec2} are respectively, an upperbound and an lowerbound for the bit error probability with the (15,7) BCH coding.

In equations (26)–(29) the channel bit error probability, P_e, is used. If the data rate, the rms delay spread and the spread spectrum code length are given, then the number of resolvable paths, L, can be calculated using equation (5). Note however that if the data rate is fixed, the signalling rate should be increased if FEC codes are used. This decreases the chip duration, T_c, which causes an increase in L. Since a higher value of L means more multi-user interference, the *channel bit error probability* is higher if FEC codes are used. The FEC code can therefore only be useful if it is able to decrease this increased channel error probability to a value that is smaller than the value of the channel error probability without FEC coding. If the channel is very noisy, due to thermal noise or multi-user interference, FEC codes might worsen the performance. This happens when the channel error probability becomes so large, that the probability of having more errors in a code block than can be corrected becomes too large.

In order to make a fair comparison between the performance for a system with and without FEC coding it is necessary to have equal transmitted power in both cases. Since for FEC coding more bits should be transmitted, the channel SNR will be lower in that case. To be more precise, the SNR in the case of FEC coding is k/n times the SNR in the case of no FEC coding.

D. Bandwidth efficiency

If the number of simultaneous users is fixed, the performance of a DS/SSMA system can be improved by using diversity, FEC codes or longer spread spectrum codes with lower crosscorrelations. In most of the cases the improvement in performance is paid for by an increase in the bandwidth. Considering this aspect, another measure of performance, bandwidth efficiency, is introduced. The bandwidth efficiency is defined as

$$BE \triangleq \frac{K_{max} r_b}{W} = \frac{K_{max} r_c}{N} \ (bits/Hz) \tag{30}$$

where K_{max} is the maximum number of simultaneous users for which the bit error probability is less than a preset value ber_{be}, r_b is the data rate, N is the spread spectrum code length, r_c is the code rate of the FEC code used and $W = Nr_b/r_c$ is the total bandwidth. The code rate of a (n,k) FEC code is defined as $r_c = k/n$. If no FEC code is used, then $r_c = 1$.

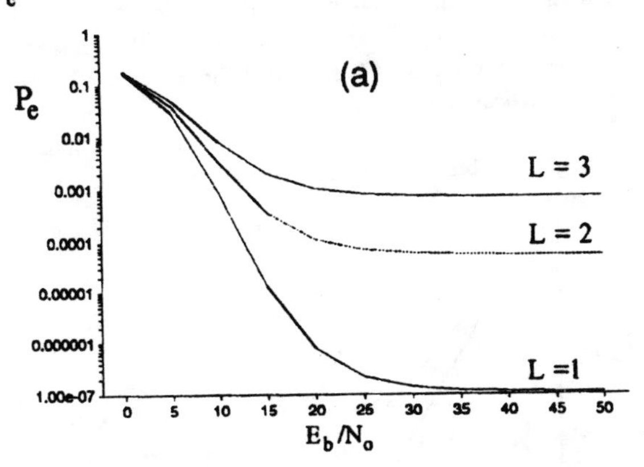

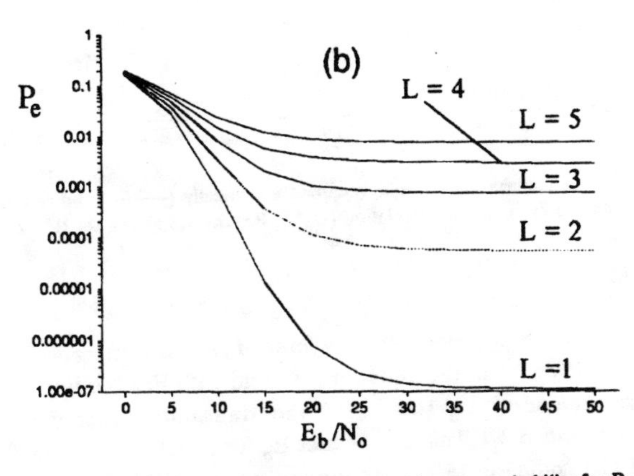

Fig. 2. The effect of the delay spread on the bit error probability for R = 6.8 dB, M = 8 (selection diversity), N = 127, K = 15 and (a) r_b = 64 kbit/s and (b) r_b = 144 kbit/s.

IV. Computational Results

In this section we present results obtained from numerical evaluation of bit error probability, outage probability and bandwidth efficiency, for both types of diversity. All results were obtained using Gold codes.

1) Bit error probability with diversity: The bit error probability is evaluated as a function of the signal-to-noise ratio E_b/N_o. We assume that $\omega_c T_b = 2\pi l$, where l is an integer. This means that the transmitted power per bit is $E_b = A^2T_b/2$. We consider code lenghts of N = 127 chips and of N = 255 chips. If not specified otherwise the number of simultaneous transmitting users is K = 15. First we consider the effect of the delay spread on the performance.

In Figure 2 the effect of the delay spread on the performance is shown for selection diversity with N = 127 and for three different bit rates (r_b). An important conclusion drawn from Figure 2 is that degradation in the performance only occurs if an increased value of the data rate or delay spread causes the number of paths L to increase. This happens because an increase in L increases the multi-user interference.

Figure 3 depicts the effect of the order of diversity in the case of selection diversity. It is confirmed that diversity can significantly improve the performance.

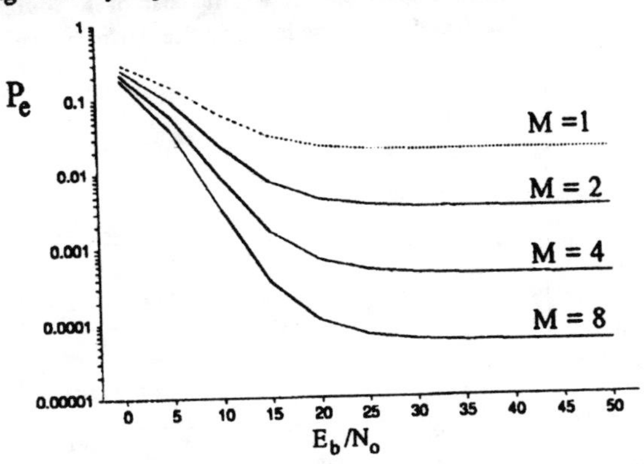

Fig. 3. The effect of the order of diversity on the bit error probability for selection diversity with R = 6.8 dB, L = 2 and N = 127.

Figure 4 shows the effect of R on the performance for selection diversity with M = 4 and L = 2. An increase in the Rician parameter R results in: 1) a better wanted signal, and 2) a worse (stronger) interference signal, i.e., more interference noise.

It is seen that for sufficiently high SNR (in this case for an SNR higher than 16 dB) this results in a better performance for R = 11 as compared to R = 6.8. For lower SNR the increased interference noise in combination with the thermal noise plays a dominant role, which in this case causes the performance to deteriorate.

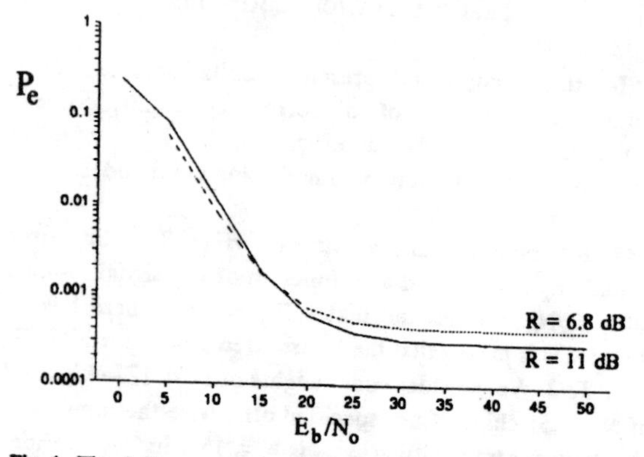

Fig. 4. The influence of the Rician parameter R on the bit error probability for selection diversity with M = 4, L = 2 and N = 127.

We now compare the performance for N = 127 and N = 255. Let T_m = 185 ns and r_b = 64 kbit/s. If N = 127 then L = 2 and if N = 255 then L = 4. As opposed to Figure 2, performance is enhanced by an increase in the value of L. This happens because the Gold codes of length 255 have better (lower) cross-correlations. Thus, the decrease in interference power per interference signal is such that even though the number of interfering signals (KL − 1) increases, the total amount of interference power decreases. Therefore, it is seen from Figure 5 that at the cost of a doubled bandwidth, codes of 255 chips improve the performance.

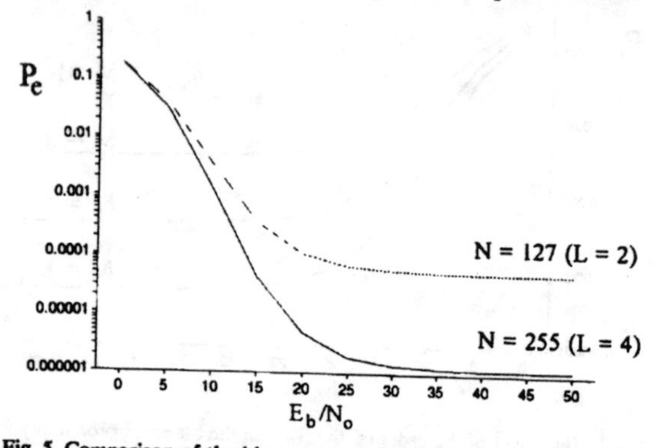

Fig. 5. Comparison of the bit error probability of N = 127 and N = 255 for selection diversity with M = 8, r_b = 64 kbit/s, T_m = 185 ns and R = 6.8 dB.

Because in the case of Rician fading there is a dominant (line-of-sight) signal component, we expect the performance to be better than in the case of Rayleigh fading. Comparing the results of [3] for the Rayleigh fading case with our results, confirms this.

In Figure 6 the performance of selection diversity is compared to the performance of maximal ratio combining. As expected, maximal ratio combining yields significantly better performance than selection diversity except for very low SNR. This especially holds for higher orders of diversity.

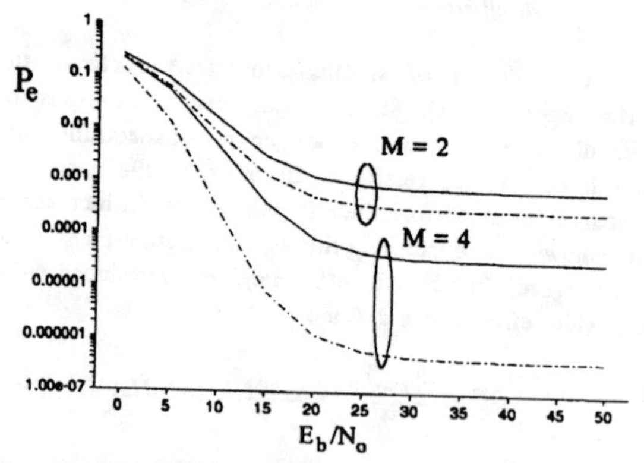

Fig. 6. Comparison of the bit error probability with selection diversity (———) and maximal ratio combining (---) for R = 6.8 dB, M = 2,4 and N = 255.

The accuracy of the calculation method is illustrated in Figure 7. The plot compares results obtained analytically with results obtained by computer simulation. The computer simulation was conducted wihtout the Gaussian assumption (see after equation (18)) made for the analytical performance analysis. It is seen that the difference between the analytical and simulation results is insignificant. Therefore, the analytical method used is a valid and fast technique to obtain the performance.

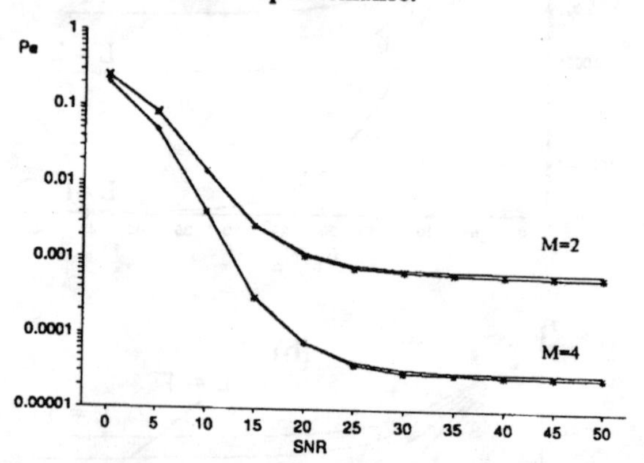

Fig. 7. Comparison of results obtained analytically (———) and results obtained by computer simulation (-*-*-). Results are shown for R = 6.8 dB, L = 4, N = 255, and M = 2.4.

2) Bit error probability with forward error correcting coding and diversity: In the plots for a system with FEC coding the signal energy E_b stands for the transmitted energy per *information bit*. This means that E_b for a system with (15,7) BCH coding is 15/7 times the energy transmitted per code symbol.

We have calculated the coding results with M = 1 for the following cases: a) R = 6.8 dB, L = 1 and N = 255, b) R = 6.8 dB, L = 5 and N = 255. Besides the curves for coding without diversity, Figure 8 also contains the curves for selection diversity with M = 2, 3 and 4.

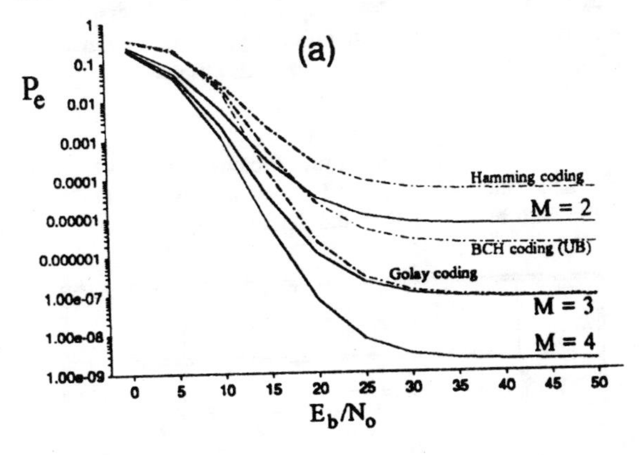

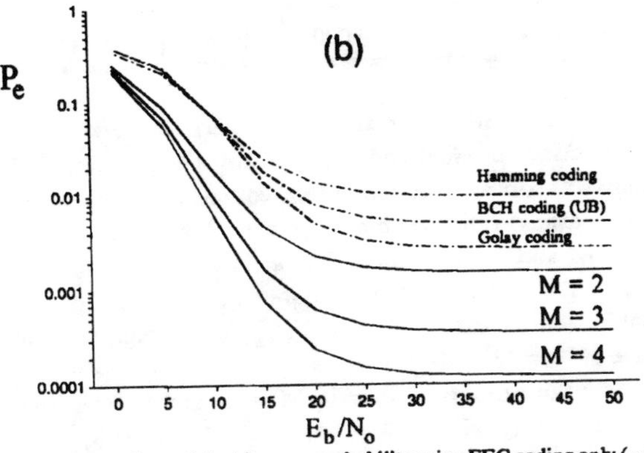

Fig. 8. Comparison of the bit error probability using FEC coding only (---) and selection diversity only (———) for R = 6.8 dB, N = 255 and (a) L = 1 and (b) L = 5.

It is seen that for low values of L (such as L = 1) the use of FEC coding without diversity can lead to acceptable bit error probabilities. However, for larger values of L (L = 5) the use of FEC coding only is not sufficient. In that case diversity is necessary either in combination with FEC coding or not. Combining selection diversity with FEC coding is attractive when the bit error probability in the case of M = L (maximum order of diversity with one antenna) is not sufficiently low. Using FEC coding then avoids the need for installing additional antennas.

We have already seen that both longer spread spectrum codes and FEC codes improve the performance at the expense of an increased bandwidth. It is then interesting to compare the performance for N = 127 + FEC coding with the performance for N = 255 without coding (Figure 9). Since all three FEC codes approximately double the required signal bandwidth, the two cases mentioned above require approximately the same amount of bandwidth. It is seen that in the case of Golay coding the bandwidth efficiency of N = 127 + FEC coding is higher since the performance is better than in the case of N = 255 without coding for equal bandwidth. Calculations have also shown that the performance is more sensitive to the (interference) noise level if FEC codes are used. This implies that if the number of users becomes too large, the use of spread spectrum codes of 255 chips will offer better performance than codes of 127 chips in combination with FEC codes.

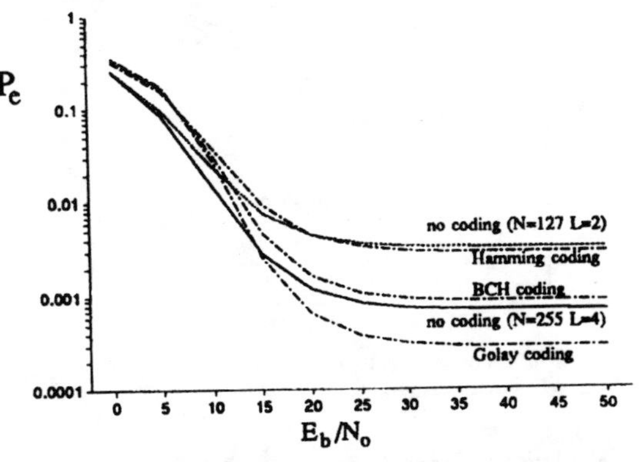

Fig. 9. Bit error probabilities with r_b = 64 kbit/s, T_m = 150 ns, R = 6.8 dB and selection diversity with M = 2 for 1) N = 127 + FEC coding and L = 2, and 2) N = 255 without FEC coding and L = 4.

3) Outage probability: We now consider the second measure of performance, the outage probability which is defined as the probability that the instantaneous bit error probability exceeds a certain threshold. Unless specified otherwise, the number of simultaneous transmitting users is K = 15. Using equation (21) we computed the outage probability for selection diversity with R = 6.8 dB, L = 5 and N = 255 for three values of the bit error threshold

TABLE I

The outage probability as a function of ber_o and E_b/N_o for selection diversity with M = 1,4 and R = 6.8 dB, L = 5 and N = 255

E_b/N_o / ber_o	10 dB	20 dB	30 dB	
1E–1	1.7E–1	4.4E–2	3.6E–2	M=1
	7.7E–4	3.8E–6	1.6E–6	M=4
1E–2	6.4E–1	2.3E–1	1.9E–1	M=1
	1.74E–1	2.6E–3	1.2E–3	M=4
1E–3	9.1E–1	4.7E–1	4.0E–1	M=1
	6.8E–1	4.9E–2	2.6E–2	M=4

(ber_o) and three different values of the signal to noise ratio. This was done for M = 1 (no diversity) and M = 4. The results are shown in Table I.

In Figure 10 the effect of the delay spread is depicted for N = 255, M = 4, R = 6.8 dB, L = 1,5,10 and $ber_o = 10^{-2}$. As expected, the outage probability increases as the number of resolvable paths increases due to increase in delay spread for N=255.

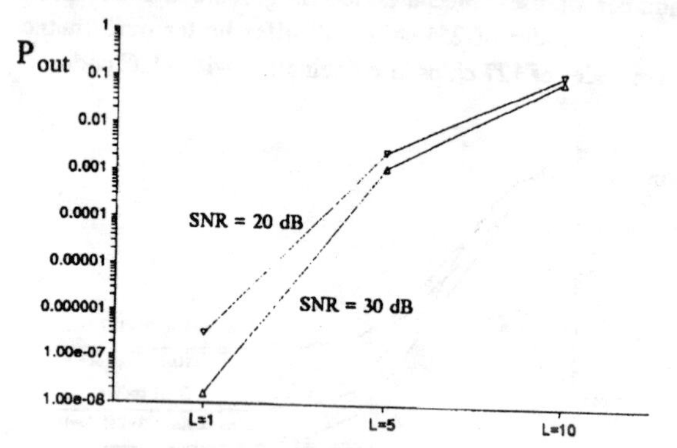

Fig. 10. The influence of the delay spread on the outage probability for R = 6.8 dB, selection diversity with M = 4 and $ber_o = 10^{-2}$.

Calculations confirm that maximal ratio combining yields better performance than selection diversity in terms of outage probability.

4) Bandwidth efficiency: Using equation (30) the bandwidth efficiency is evaluated for both selection diversity and maximal ratio combining. Unless specified otherwhise $ber_{be} = 10^{-4}$ and the signal-to-noise ratio E_b/N_o = 20 dB.

In Table II the bandwidth efficiency is given for selection diversity with four different values of the number of resolvable paths L = 1,2,4 and 8, and three orders of diversity M = 1,2 and 4. Gold codes of N = 255 chips are used. There are no results shown for those cases where the error probability is always greater than $ber_{be} = 10^{-4}$. We see from Table II that an increase in the order of diversity gives an increase in the bandwidth efficiency, and that an increase in L gives a decrease in the bandwidth efficiency.

TABLE II
The bandwidth efficiency as a function of M and L for selection diversity with N = 255, R = 6.8 dB and $ber_{be} = 10^{-4}$

L / M	L = 1	L = 2	L = 4	L = 8
1	——	——	——	——
2	0.098	0.055	0.031	0.012
4	0.214	0.090	0.051	0.031

In Table III we make a comparison between the performance of Gold codes of 127 chips and codes of 255 chips with forward error correcting coding, for both selection and maximal ratio combining. We presume that the data rate, r_b, and the rms delay, T_m, are the same in both the cases. This means that the number of resolvable paths is doubled in the case of N = 255 (L_{255}) as compared to the case of N = 127 (L_{127}). Results are shown for $L_{255} = 2L_{127} = 4$ for three orders of diversity M = 1,2 and 4. From Table III we see that N = 127 + FEC codes + diversity, offers

TABLE III
Bandwidth efficiency for selection diversity and maximal ratio combining with FEC coding for N = 127,255 and M = 1,2,4

	N = 127			N = 255		
	Hamming	BCH	Golay	Hamming	BCH	Golay
M = 1	0.004	0.015	0.021	0.011	0.016	0.025
M = 2	0.027	0.040	0.066	0.025	0.027	0.035
M = 4	0.040	0.055	0.082	0.031	0.033	0.041
↑ SELECTION DIVERSITY ↑						
↓ MAXIMAL RATIO COMBINING ↓						
M = 1	0.004	0.015	0.021	0.11	0.016	0.025
M = 2	0.041	0.048	0.069	0.034	0.037	0.041
M = 4	0.112	0.107	0.136	0.067	0.071	0.080

superior performance in terms of bandwidth efficiency, as compared to N = 255 + FEC codes + diversity.

We now compare the case N = 255, L = 4 with selection diversity (Table II) with the case N= 127, L = 2 with FEC codes and diversity (Table III). These two cases require approximately the same amount of bandwidth. It is seen that N = 127 + FEC codes is superior to N = 255 in terms of bandwidth efficiency.

V. Conclusions

We have evaluated the performance of a star-connected DS/SS system for indoor wireless applications. The indoor radio channel was assumed to be of the Rician fading type, and DPSK modulation was used. Two types of diversity, selection diversity and maximal ratio combining were considered. The performance was assessed in terms of bit error probability, outage probability and bandwidth efficiency.

In general it can be stated that diversity improves the performance (bit error probability, outage probability, bandwidth efficiency) significantly, and that maximal ratio combining yields superior performance as compared to selection diversity.

It was seen that the performance is quite sensitive to the value of the rms delay spread and the bit rate at which data is transmitted. In general the performance is affected (worsens) if an increased value of the rms delay spread or bit rate causes the number of resolvable paths, L, to increase.

Methods to improve the performance for a given bit rate and rms delay spread are investigated using diversity (selection diversity and maximal ratio combining), FEC coding (Hamming codes, BCH codes, and Golay codes), and longer spread spectrum codes. These methods can also be combined. The drawback of diversity is that if it is necessary to have an order of diversity, M, larger than the number of resolvable paths, then the use of multiple antennas is required. The drawback of FEC codes and longer spread spectrum codes is that more bandwidth is required. In all cases additional hardware and logic are required for implementation of the system. From the results the following can be concluded.

1) Instead of installing multiple (two or more) antennas per receiver FEC coding can be used either independent or in combination with diversity to further improve the performance at the cost of an increased system bandwidth. The two options should be compared in terms of implementation complexity and costs.

2) FEC coding can not completely replace for diversity. The performance with FEC coding deteriorates faster than the performance for diversity if the number of paths or simultaneously active users increases.

3) Using spread spectrum codes of 255 chips improve the bit error probability at the cost of a higher system bandwidth as compared to spread spectrum codes of 127 chips. However, the improvement of the bit error probability is not such that the bandwidth efficiency is improved.

4) If the interference power is below a certain limit, a system with FEC coding and spread spectrum codes of 127 chips can compete with a system with spread spectrum codes of 255 chips (note that these two system require the same amount of bandwidth). This holds for both bit error probability and bandwidth efficiency. If the number of paths is too large or the number of active users crosses a certain limit, the system with spread spectrum codes of 255 chips but without FEC codes, yields better performance.

References

[1] M. Kavehrad, "Performance of Nondiversity Receivers for Spread Spectrum in Indoor Wireless Communication," *AT & T Technical Journal*, vol. 64, no. 6, pp. 1181–1210, July–August 1985.

[2] M. Kavehrad and P. J. McLane, "Performance of Low-Complexity Channel Coding and Diversity for Spread Spectrum in Indoor, Wireless Communication," *AT & T Technical Journal*, vol. 64, no. 8, pp. 1927–1965, October 1985.

[3] M. Kavehrad and B. Ramamurthi, "Direct-Sequence Spread Spectrum with DPSK Modulation and Diversity for Indoor Wireless Communication," *IEEE Transactions on Communications*, vol. Com-35, no. 2, pp. 224–236, February 1987.

[4] M. Kavehrad and P. J. McLane, "Spread Spectrum for Indoor Digital Radio," *IEEE Communications Magazine*, vol. 25, no. 6, pp. 32–40, June 1987.

[5] M. Kavehrad and G.E. Bodeep, "Design and Experimental Results for a Direct-Sequence Spread-Spectrum Radio Using Differential Phase-Shift Keying Modulation for Indoor, Wireless Communications," *IEEE Journal on Selected Areas in Communications*, vol. SAC-5, no. 5, pp. 815–823, June 1987.

[6] K. Pahlavan, "Wireless Communication for Office Information Networks," *IEEE Communications Magazine*, vol. 23, no. 6, pp. 19–26, June 1985.

[7] K. Pahlavan and M. Chase, "Spread Spectrum Multiple-Access Performance of Orthogonal Codes for Indoor Radio Communications," *IEEE Transactions on Communications*, vol. 38, no. 5, pp. 574–577, May 1990.

[8] R. Prasad, H.S. Misser, A. Kegel, "Performance Ananlysis of Direct-Sequence Spread-Spectrum Multiple Access communication in an Indoor Rician-Fading channel with DPSK modulation," *Electronics Letters*, vol. 26. no. 17, pp. 1366–1367, August 1990.

[9] R.J.C. Bultitude, "Measurement, Characterization and Modeling of Indoor 800/900 MHz Radio Channels for Digital Communications," *IEEE Communications Magazine*, vol. 25, no. 6, pp. 5–12, June 1987.

[10] R.J.C. Bultitude, S.A. Mahmoud and W.A. Sullivan, "A comparison of indoor radio propagation characteristics at 910 MHz and 1.75 GHz," *IEEE journal on selected areas in communications*, vol. 7, no. 1, January 1989.

[11] T.S. Rappaport and C.D. McGillem, "UHF Fading in Factories," *IEEE Journal on Selected Areas in Communications*, vol. 7, no. 1, pp. 40–48, January 1989.

[12] D.M.J. Devasirvatham, C. Banarjee, M.J. Krain and Rappaport, "Multi-Frequency Radiowave propagation Measurements in the Portable Radio Environment," *Proceedings IEEE International Conference on Communications (ICC'90), Atlanta*, vol. 4 of 4, pp. 335.1.1–7, April 1990.

[13] R. Prasad, "Throughput analysis of slotted code division multiple access for indoor radio channels," *Proc. IEEE Symposium on spread spectrum techniques and applications*, King's College, London, pp. 12–17, September 1990.

[14] A. Zigic and R. Prasad, "Computer simulation of indoor data channels with a linear adaptive equalizer," *Electronic Letters*, vol. 26, no. 19, pp. 1596–1597, September 1990.

[15] J.G. Proakis, "Digital Communications (second edition)," *McGraw-Hill Book Company*, New York, 1989.

[16] J. Wang and M. Moeneclaey, "Hybrid DS/SFH-SSMA with predetection diversity and coding over indoor radio multipath Rician-fading channels," *IEEE Transactions on Communications*, vol. COM–40, October, 1992.

Ramjee Prasad (M'88-SM'90) was born in Babhnaur (Gaya), Bihar, India on July 1, 1946. He received the B.Sc. (Eng.) from Bihar Institute of Technology, Sindri, India, and the M.Sc. (Eng.) and Ph.D. degrees from Birla Institute of Technology (BIT), Ranchi, India in 1968, 1970 and 1979, respectively.

He joined BIT as Senior Research Fellow in 1970 and became Associate Professor in 1980. During 1983-1988 he was with the University of Dar es Salaam (UDSM), Tanzania, where he became Professor in Telecommunications at the Department of Electrical Engineering 1986. Since February 1988, he has been with the Telecommunications and Traffic Control Systems Group, Delft University of Technology, The Netherlands, where he is actively involved in the area of mobile, indoor and personal radio communications. While he was with BIT, he supervised many research projects in the area of Microwave and Plasma Engineering. At UDSM he was responsible for the collaborative project "Satellite Communications for Rural Zones" with Eindhoven University of Technology, The Netherlands. He has published over 200 technical papers. His current research interest lies in packet communications, adaptive equalizers, spread-spectrum CDMA systems and multi-media communications.

He has served as a member of advisory and programme committees of several IEEE international conferences. He has also presented tutorials on Mobile and Indoor Radio Communications at various universities, technical institutions and IEEE conferences. He is also a member of a working group of European co-operation in the field of scientific and technical research (COST-231) project dealing with "Evolution of Land Mobile Radio (including personal) Communications" as an expert for the Netherlands.

Prof. Prasad is listed in the US Who's Who in the World. He was Organizer and Interim Chairman of IEEE Vehicular Technology/Communications Society Joint Chapter, Benelux Section. Now he is the elected chairman of the joint chapter. He is also founder of the IEEE Symposium on Communications and Vehicular Technology (SCVT) in the Benelux and he was the Symposium Chairman of SCVT'93. He is the Co-ordinating Editors and one of the Editors-in-Chief of a Kluwer international journal on "Wireless Personal Communications" and also a member of the editorial board of other international journals including *IEEE Communications Magazine*. He was the Technical Program Chairman of PIMRC'94 International Symposium held in The Hague, The Netherlands during September 19-23, 1994 and also of the Third Communication Theory Mini-Conference in conjunction with GLOBECOM'94 held in San Francisco, California during November 27-30, 1994. He is a Fellow of IEE, a Fellow of the Institution of Electronics & Telecommunication Engineers, a Senior Member of IEEE and a Member of the New York Academy of Sciences and of NERG (The Netherlands Electronics and Radio Society).

Howard S. Misser (S'89-M'91) was born in Paramaribo, Surinam, on May 11, 1968. He received the M.Sc. degree in Electrical Engineering from Delft University, Delft, The Netherlands, in 1990.

He worked at the Telecommunications and Traffic-Control Systems Group of the same university as a Research Fellow. He is currently with the Dr. Neher Laboratories of the Royal Dutch PTT in Leidschendam, The Netherlands. His research interests are in the fields of broadband network technologies and architecture, radio and mobile communications, and spread-spectrum communication.

Adriaan Kegel was born in The Netherlands on November 4, 1932. He received the M.Sc. degree from Delft University of Technology, The Netherlands, in 1972. He joined Philips in 1954 where he worked on the development of microwave link equipment as a Radio Engineer. In 1964 he joined Delft University of Technology where he became an Associate Professor in 1985. In 1970 he became Chairman of the "Working Group Indonesia", which developed a low cost educational broadcasting system. For this development the working group received an award from the Scientific Radio Foundation Veder. After completion of the Indonesian Project in 1974, he continued his research work on the coding and transmission of graphical information. This work led to the development of several systems (e.g. Vidibord) and a number of publications (e.g. on Differential Chain Coding). He took retirement from his active service in the field of telecommunications in December 1993. His current interest is in photography and nature.

Mr. Kegel is a member of NERG (the Netherlands Electronics and Radio Society), a member of the FITCE (Federation of Telecommunication Engineers of the European Community) and a member of IEEE.

Performance of BPSK and TCM Using the Exponential Multipath Profile Model for Spread-Spectrum Indoor Radio Channels

Juan M. Bargallo, *Member, IEEE*, and James A. Roberts, *Senior Member, IEEE*

***Abstract* — A common approach to analyzing the performance of a spread-spectrum communication system in fading is to assume that the multipath profile is a constant function of delay. However, different multipath profile models can lead to significant differences in predicted system performance, so caution should be exercised when choosing a particular model for analysis. Several previous studies show that a good fit to the experimentally measured multipath profile of indoor wireless channels is an exponential function. Assuming Rayleigh fading and an exponential multipath profile, in this paper we derive a closed form expression for the bit error rate of biphase-shift-keying and an upper bound for trellis-code-modulation both with combined spread-spectrum and antenna diversity. Comparison of these results with simulated results based on actual indoor channel measurements shows that the exponential profile model differs from the simulation model by only 1-2 dB whereas the constant profile model differs by as much as 5 dB.**

I. INTRODUCTION

The indoor wireless environment is characterized by impairments such as signal fading produced by multipath propagation, dispersion due to delay spread, and shadowing caused by attenuation in walls and other objects. Spread-spectrum (SS) transmission is among the mitigation techniques employed. One advantage of SS systems is diversity gain: a signal with spread bandwidth W much larger than the coherence bandwidth of the channel, will resolve the multipath components and provide the receiver with several independently fading replicas of the transmitted signal. Spread-spectrum is also an effective multiple access scheme.

The optimum receiver for the wideband fading channel and a spread-spectrum signal is the RAKE correlator [1] which can be implemented as a tapped delay line with taps dependent on the channel impulse response and the spreading codes. The tapped delay line receiver collects the signal energy from all the received signal paths that fall within the span of the receiver delay (assumed to be equal to the maximum delay spread of the channel). Demodulation and combination of the resolvable multipath components with the RAKE correlator provides the diversity improvement.

SS diversity has been extensively used in mobile radio and indoor wireless systems [2]-[6]. Previous analytical results assumed that all resolved multipath components had the same average power [3]-[7]. However, extensive measurements of indoor wireless channel characteristics have shown that the average power of the multipath components decreases with increasing delay [8]-[10]. Whenever researchers used the more realistic situation of unequal average power for resolved paths, performance results had to be obtained numerically [11]-[13].

A good technique for achieving greater improvement in performance is the combination of conventional diversity combining, e.g. multiple antennas, and SS diversity. Again, analytical performance results previously presented for this situation assumed equal average power for all multipath components and computed the average bit error rate (BER) performance as that for an equivalent system utilizing only conventional diversity combining with the effective diversity order equal to the antenna diversity order multiplied by the number of resolved multipath components [3]-[4]. A few studies numerically evaluated the performance of systems with combined antenna and SS diversity with unequal path strengths [4].

Ungerboeck [14], demonstrated that trellis coded modulation (TCM) can achieve coding gains of 3-6 dB in the additive white Gaussian noise (AWGN) channel while avoiding the bandwidth expansion of traditional error correcting coding methods. Recent research has shown that TCM can be an effective way to combat signal fading when transmitting over indoor wireless channels [15]-[17]. It was shown in [17] that the combination of TCM and diversity yields a significant reduction in required transmit power compared to equivalent uncoded systems assuming a slowly varying, Rayleigh fading, discrete multipath channel. Equal average power was assumed for all paths. It was also shown that the use of diversity reduces the need for interleaving of symbols in the coded system, since it combines approximately independent samples of the channel. For the slowly fading channel, interleaving can result in large delays which may preclude voice applications.

In this paper we consider the transmission of DS-SS with either biphase-shift-keying (BPSK) or TCM M-PSK modulation over the indoor wireless channel, again modeled

This paper was approved by Roger Peterson, the Editor for Spread-Spectrum Systems of the IEEE Communications Society. Manuscript received: October 12, 1993; revised April 25, 1994. This work was supported by a grant from Nellcor Inc., Lenexa, KS. This paper was presented in part at Globecom '93, Houston, TX, December 1993.

J.M. Bargallo is with Mobile Systems International, Arlington, VA.

J.A. Roberts is with the Telecommunications and Information Sciences Laboratory, Department of Electrical Engineering and Computer Science, The University of Kansas, Lawrence, KS.

IEEE Log Number 9410094. 0090-6778/95$4.00

as a discrete multipath slowly fading Rayleigh channel. However, in this case the resolved multipath components are assumed to have unequal average power. We present closed form solutions and upper bounds to the average BER performance of BPSK and TCM M-PSK respectively, for both SS diversity alone and combined antenna and SS diversity. Only AWGN is assumed, but the effects of multiple-access interference can be included if the number of simultaneous users is large enough such that the combined interfering signal is approximately AWGN as well [18]. An example of this approach is in [17].

The analytical results are evaluated and compared for three different types of multipath power profiles: constant average power profile, exponentially decaying power, and profiles derived from the indoor channel model developed by Rappaport et al [19].

It is shown that there are significant differences in the performance between the constant and the exponential profile and that the performance for the exponential profile is reasonably close to that predicted employing the model in [19], which is based on experimental measurements.

II. SPREAD SPECTRUM DIVERSITY ALONE

Consider the transmission over the indoor radio channel of a SS signal with spread bandwidth much larger than the coherence bandwidth of the channel. No antenna diversity is used.

The SS signal resolves the multipath components with time resolution $1/W$ where W is the spread bandwidth. We assume the use of a RAKE correlator and that perfect knowledge of the channel tap gains is available. The availability of different resolved signal paths at the receiver allows us to use the discrete multipath fading channel. Since the signal envelope fading in the indoor environment has been found to be Rayleigh when the transmitter or receiver or both are mobile, we will use this worst case model. The low-pass equivalent impulse response of the channel is given by

$$h(t)=\sum_{k=1}^{L}\beta_k\delta(t-\tau_k)e^{j\phi_k} \tag{1}$$

where β_k, τ_k, and ϕ_k are the path gain, time delay and phase of the kth path. The path gains are independent identically distributed (i.i.d.) Rayleigh random variables with mean square values $E[\beta_k^2]$. The path delays and phases are assumed independent and uniformly distributed, the former over $[0, T]$ and the latter over $[0, 2\pi]$ [3]. The number of resolvable multipath components, L, is given by

$$L=\lfloor T_m/T_c\rfloor+1 \tag{2}$$

where T_c is the chip period, T_m is the delay spread in the channel, and $\lfloor\cdot\rfloor$ denotes the integer part [1]. We assume that the spread bandwidth is $W=1/T_c$. For the TCM case, the RAKE receiver is followed by demodulation and Viterbi decoding, after which an estimate of the original data sequence is produced. It has been shown in [1] that the RAKE receiver with perfect estimates of the channel tap weights is equivalent to a maximal-ratio combiner in a system with Lth order diversity.

A. General results

The average BER performance of different modulation schemes for the case with SS diversity alone and unequal values of signal-to-noise ratio (SNR) in all paths has been derived by Proakis [1]. The result for BPSK is:

$$\overline{P}_b=\frac{1}{2}\sum_{k=1}^{L}\pi_k\left[1-\sqrt{\frac{\overline{\gamma}_k}{1+\overline{\gamma}_k}}\right] \tag{3}$$

where

$$\pi_k=\prod_{i=1,\, i\neq k}^{L}\frac{\overline{\gamma}_k}{\overline{\gamma}_k-\overline{\gamma}_i}. \tag{4}$$

$\overline{\gamma}_k$ is the average bit energy-to-noise density ratio for the kth path defined by

$$\overline{\gamma}_k=E[\beta_k^2]\frac{E_b}{N_0} \tag{5}$$

and E_b/N_0 is the bit energy-to-noise density ratio in the absence of fading. Equation (3) is obtained by averaging the expression for BER performance of BPSK for a fixed value of the fading amplitude (AWGN channel performance) over the probability density function (PDF) corresponding to the total E_b/N_0, γ_b, for maximal ratio combining of L i.i.d. Rayleigh random variables with different mean square values (average power). This PDF is [1]

$$p(\gamma_b)=\sum_{k=1}^{L}\frac{\pi_k}{\overline{\gamma}_k}e^{\gamma_b/\overline{\gamma}_k},\ \gamma_b\geq 0 \tag{6}$$

For a fixed value of the fading amplitude, the BER performance for M-PSK TCM is upper bounded by [17], [20]

$$P_b\leq\frac{1}{2m}\sum_{j=1}^{\infty}b_j\,erfc\left(\sqrt{\frac{md_j^2}{4}\gamma_b}\right) \tag{7}$$

where b_j is the number of error events having Euclidean distance d_j from the all zeros path. For large γ_b, the performance is dominated by the error event with the smallest Euclidean distance, and (7) is closely approximated by

$$P_b\cong\frac{b_{d_{free}}}{2m}erfc\left(\sqrt{\frac{d_{free}^2 m}{4}\gamma_b}\right),\ \gamma_b>>1 \tag{8}$$

where $b_{d_{free}}$ is the number of paths with Euclidean distance d_{free} from the reference path. Assuming that γ_b is constant over a decoding span, the approximate average BER performance for M-PSK TCM with SS diversity alone and unequal path powers can be obtained by averaging (8) over the PDF in (6) and we obtain

$$\overline{P}_b \cong \frac{b_{d_{free}}}{2m}\sum_{k=1}^{L}\pi_k\left(1-\sqrt{\frac{\overline{\gamma}_k}{1+\overline{\gamma}_k}}\right) \quad (9)$$

where π_k is again given by (4), $\overline{\gamma}_k$ by (5), and m is the number of information bits per encoded symbol.

The average E_b/N_0 and the average E_b/N_0 per channel are related by

$$\overline{\gamma}_b = \sum_{k=1}^{L}\overline{\gamma}_k \quad (10)$$

for BPSK and

$$\frac{md_{free}^2}{4}\overline{\gamma}_b = \sum_{k=1}^{L}\overline{\gamma}_k \quad (11)$$

for M-PSK TCM. In reality, the average signal-to-noise ratios of the resolved paths are random variables, functions of the random path delays of the indoor channel.

B. Exponential Multipath Profile

In the exponential profile model, the signal-to-noise ratios of the different resolved signal paths decrease exponentially with increasing path delays. Examples of the use of exponential profiles for indoor wireless channel modeling are given in [11], [21], and [22].

We develop the exponential profile as follows:

$$E[\beta_k^2] = E[\beta_1^2]e^{-\tau_k/\varepsilon}, \; 1 \le k \le L \quad (12)$$

where β_1 is the amplitude of the first arriving path which we assume without loss of generality to have a delay $\tau_1 = 0$ ns. The maximum delay spread is defined as the range of signal path delays within which most of the multipath profile energy is contained. Therefore, proceeding in a similar manner to [11] we make the following assumption:

$$\tau_L = 5\varepsilon = T_m \quad (13)$$

that is, the last component has a delay equal to the maximum delay spread which we assume to be five times the time constant of the exponential function which for all practical purposes includes most of the signal energy. Intermediate multipath components have delays equally spaced between the first and last components. Taking the above into account we may replace τ_k in (12) by

$$\tau_k = \frac{(k-1)T_m}{L-1} \quad (14)$$

where k is a bin number indicating the multipath component number. Then from (5) and (14) we obtain

$$\overline{\gamma}_k = \overline{\gamma}_1 e^{-(k-1)/\eta} \quad (15)$$

where $\eta = (L-1)/5$. Substitution of (15) into (10) and (11) gives

$$\overline{\gamma}_1 = \frac{\overline{\gamma}_b}{\sum_{k=1}^{L} e^{-(k-1)/\eta}} \quad (16)$$

for BPSK and

$$\overline{\gamma}_1 = \frac{md_{free}^2\overline{\gamma}_b}{4\sum_{k=1}^{L} e^{-(k-1)/\eta}} \quad (17)$$

for M-PSK TCM, respectively.

Finally, substituting (15) and (16) into (3) and (4) and after some algebra, we obtain an expression for the average BER for BPSK, when the multipath profile has an exponential shape:

$$\overline{P}_b = \frac{1}{2}\sum_{k=1}^{L}\pi_k\left[1-\sqrt{\frac{\overline{\gamma}_b}{\delta_k+\overline{\gamma}_b}}\right] \quad (18)$$

with

$$\pi_k = \prod_{i=1,\, i\ne k}^{L}\frac{1}{1-e^{(k-i)/\eta}} \quad (19)$$

and

$$\delta_k = \frac{e^{(k-1)/\eta}\left(1-e^{-L/\eta}\right)}{\left(1-e^{-1/\eta}\right)}. \quad (20)$$

Similarly, substituting (15) and (17) into (9) and (4) we obtain the corresponding expression for M-PSK TCM (valid for large average E_b/N_0)

$$\overline{P}_b \cong \frac{b_{d_{free}}}{2m}\sum_{k=1}^{L}\pi_k\left(1-\sqrt{\frac{md_{free}^2\overline{\gamma}_b}{4\delta_k+md_{free}^2\overline{\gamma}_b}}\right) \quad (21)$$

where π_k and δ_k are again given by (19) and (20), respectively.

III. COMBINED ANTENNA AND SPREAD SPECTRUM DIVERSITY

We now derive the BER performance for DS-SS systems with BPSK and TCM M-PSK modulation for the case when both antenna and SS diversity are used. We denote the number of antennas (antenna diversity order) by L_a. Following the notation in the previous section, L is the SS diversity order. We call β_{ij} the amplitude of the jth resolved path at antenna i. We assume that the β_{ij}s are statistically independent random variables (a reasonable assumption if the antennas are spaced far enough apart and the spread bandwidth is much larger than the coherence bandwidth of the channel).

A. General results for BPSK

Consider first the performance of BPSK. Following the analysis for the ideal RAKE receiver in [1, pp. 729], it can be shown that the probability of error conditioned on a fixed set of path gains is

$$P_b = \frac{1}{2} erfc(\sqrt{\gamma_b}) \tag{22}$$

where γ_b is the instantaneous E_b/N_0 given by

$$\gamma_b = \sum_{i=1}^{L_a} \sum_{j=1}^{L} \gamma_{ij} \tag{23}$$

γ_{ij} is the instantaneous E_b/N_0 for the ith antenna and jth multipath component. For Rayleigh distributed path gains, γ_{ij} is exponentially distributed with mean $\bar{\gamma}_{ij}$ which is the average E_b/N_0 for the ith antenna and jth path.

If we assume that the diversity antennas are located on the same base station or radio port, then all antennas are affected in similar manner by shadowing [7]. On the other hand, we are assuming that the antennas are spaced far enough apart so that the signals detected in the different antennas are independent. Thus we conclude: 1) For the same multipath component number (same path delay), the instantaneous values of E_b/N_0, γ_{ij}, may be different in different antennas, 2) However, for the same path delay, the average E_b/N_0, $\bar{\gamma}_{ij}$ is the same for all antennas. Taking the above into account we can define

$$\bar{\gamma}_{1j} = \bar{\gamma}_{2j} = \cdots = \bar{\gamma}_{L_a j} = \bar{\gamma}_{cj}, \quad 1 \le j \le L \tag{24}$$

where $\bar{\gamma}_{cj}$ is the average E_b/N_0 for the jth multipath component in all antennas.

The unconditional error probability is determined by first computing the PDF of γ_b followed by averaging of (22) over $p(\gamma_b)$. γ_b is the sum of $L_a L$ statistically independent random variables whose means meet the condition (24). The usual approach is to compute the characteristic function of γ_b as the product of the characteristic function of the statistically independent random variables followed by an inverse Fourier transformation to obtain the PDF [1]. Since in this case the PDFs of the γ_{ij}s are defined only for $\gamma_{ij} \ge 0$, we can follow a similar procedure employing the Laplace transform which results in an easier evaluation of the PDF of γ_b. The Laplace transform of the exponential PDF is given by

$$\phi_{\gamma_{ij}}(s) = \frac{1/\bar{\gamma}_{ij}}{(s + 1/\bar{\gamma}_{ij})} \tag{25}$$

The Laplace transform of the PDF of γ_b is then the product of the Laplace transforms of the γ_{ij}s:

$$\phi_{\gamma_b}(s) = \prod_{i=1}^{L} \prod_{j=1}^{L_a} \frac{1/\bar{\gamma}_{ij}}{s + 1/\bar{\gamma}_{ij}} \tag{26}$$

or if we substitute (24) into (26) we obtain

$$\phi_b(s) = \rho \prod_{j=1}^{L} \frac{1}{(s + 1/\bar{\gamma}_{cj})^{L_a}} \tag{27}$$

where

$$\rho \equiv \prod_{j=1}^{L} 1/\bar{\gamma}_{cj}^{L_a} \tag{28}$$

A closed form solution for the inverse Laplace transform of (27) can be obtained with conventional partial fraction expansion and the method of residues. It is

$$p(\gamma_b) = \rho \sum_{k=1}^{L} \sum_{l=1}^{L_a} \frac{\phi_{kl}(-\bar{\gamma}_{ck}^{-1})}{(L_a - 1)!(l-1)!} \gamma_b^{L_a - l} e^{-\gamma_b/\bar{\gamma}_{ck}}, \quad \gamma_b \ge 0 \tag{29}$$

with

$$\phi_{kl}(s) = \frac{d^{l-1}}{ds^{l-1}} \left[1 \Big/ \prod_{j=1, j \ne k}^{L} \left(s + \frac{1}{\bar{\gamma}_{cj}} \right)^{L_a} \right] \tag{30}$$

For $L=1$ and $L_a = 1$, this result reduces to (7.4.13) and (7.5.26) in [1] respectively.

The closed form solution for the average BER for BPSK, obtained by averaging (22) over the PDF in (29), is given by

$$\bar{P}_b = \rho \sum_{k=1}^{L} \sum_{l=1}^{L_a} \sum_{j=1}^{L_a - l} \binom{L_a - l + j}{j} \frac{\bar{\gamma}_{ck}^{L_a - l + 1} \phi_{kl}(-\bar{\gamma}_{ck}^{-1})}{(l-1)!} \cdot \left(\frac{1 - \mu_k}{2} \right)^{L_a - l + 1} \left(\frac{1 + \mu_k}{2} \right)^{j} \tag{31}$$

where

$$\mu_k = \sqrt{\frac{\bar{\gamma}_{ck}}{1 + \bar{\gamma}_{ck}}} \tag{32}$$

and $\bar{\gamma}_{ck}$ is related to the average signal-to-noise ratio per bit by

$$\bar{\gamma}_b = L_a \sum_{k=1}^{L} \bar{\gamma}_{ck} \tag{33}$$

Observe that for $L=1$ (antenna diversity alone), (31) reduces to (7.4.15) in [1] with

$$\mu_1 = \mu = \sqrt{\frac{\bar{\gamma}_c}{1 + \bar{\gamma}_c}} \tag{34}$$

For $L_a = 1$ (SS diversity alone) (31) reduces instead to (7.5.28) in [1].

B. General results for TCM

Similarly, we can obtain the performance of M-PSK TCM by averaging (8) over the PDF in (29). The corresponding expression is

$$\overline{P}_b \cong \frac{\rho b_{d_{free}}}{m} \sum_{k=1}^{L} \sum_{l=1}^{L_a} \sum_{j=1}^{L-1} \binom{L_a - l + j}{j} \frac{\overline{\gamma}_{ck}^{L_a - l + 1} \phi_{kl}(-\overline{\gamma}_{ck}^{-1})}{(l-1)!} \cdot \left(\frac{1-\mu_k}{2}\right)^{L_a - l + 1} \left(\frac{1+\mu_k}{2}\right)^j \tag{35}$$

where μ_k is again given by (32) and

$$\overline{\gamma}_b = \frac{4L_a}{md_{free}^2} \sum_{k=1}^{L} \overline{\gamma}_{ck} \tag{36}$$

For $L=1$, (36) reduces to (27) in [17]. For $L_a=1$, (36) reduces to (9).

C. Results for the Exponential Multipath Profile

When the multipath profile is assumed to be exponential, we have

$$\overline{\gamma}_{ck} = \overline{\gamma}_{c1} e^{-(k-1)/\eta} \tag{37}$$

Substitution of (37) into (33) and (36) yields

$$\overline{\gamma}_{c1} = \frac{\overline{\gamma}_b}{L_a \sum_{k=1}^{L} e^{-(k-1)/\eta}} \tag{38}$$

for BPSK and

$$\overline{\gamma}_{c1} = \frac{md_{free}^2 \overline{\gamma}_b}{4L_a \sum_{k=1}^{L} e^{-(k-1)/\eta}} \tag{39}$$

for M-PSK TCM, respectively.

If we now substitute (37) and (38) into (31) and (32), we find that the average BER for BPSK with combined antenna/SS diversity and exponential multipath profile is given by

$$\overline{P}_b = \rho \sum_{k=1}^{L} \sum_{l=1}^{L_a} \sum_{j=1}^{L-1} \binom{L_a - l + j}{j} \frac{\overline{\gamma}_b^{L_a - l + 1} \phi_{kl}(-L_a \delta_k / \overline{\gamma}_b)}{(L_a \delta_k)^{L_a - l + 1} (l-1)!} \cdot \left(\frac{1-\mu_k}{2}\right)^{L_a - l + 1} \left(\frac{1+\mu_k}{2}\right)^j \tag{40}$$

where

$$\phi_{kl}(s) = \frac{d^{l-1}}{ds^{l-1}} \left[1 \Big/ \prod_{j=1, j \neq k}^{L} \left(s + \frac{L_a \delta_j}{\overline{\gamma}_{cj}} \right)^{L_a} \right] \tag{41}$$

$$\mu_k = \sqrt{\frac{\overline{\gamma}_b}{\overline{\gamma}_b + L_a \delta_k}} \tag{42}$$

For M-PSK TCM the corresponding expressions are (valid for large average E_b/N_0)

$$\overline{P}_b \cong \frac{\rho b_{d_{free}}}{m} \sum_{k=1}^{L} \sum_{l=1}^{L_a} \sum_{j=1}^{L-1} \binom{L_a - l + j}{j} \cdot \frac{\left(md_{free}^2 \overline{\gamma}_b\right)^{L_a - l + 1} \phi_{kl}\left(-L_a \delta_k / md_{free}^2 \overline{\gamma}_b\right)}{(L_a \delta_k)^{L_a - l + 1} (l-1)!} \cdot \left(\frac{1-\mu_k}{2}\right)^{L_a - l + 1} \left(\frac{1+\mu_k}{2}\right)^j \tag{43}$$

where now

$$\rho \equiv \left(\frac{4L_a L}{md_{free}^2}\right)^{L_a L} \frac{1}{\overline{\gamma}_b^{L_a L}} \prod_{j=1}^{L} \delta_j^{L_a} \tag{44}$$

$$\phi_{kl}(s) = \frac{d^{l-1}}{ds^{l-1}} \left[1 \Big/ \prod_{j=1, j \neq k}^{L} \left(s + \frac{4L_a \delta_j}{md_{free}^2 \overline{\gamma}} \right)^{L_a} \right] \tag{45}$$

and

$$\mu_k = \sqrt{\frac{md_{free}^2 \gamma_b}{md_{free}^2 \gamma_b + 4L_a \delta_k}} \tag{46}$$

δ_k is given in both cases by (20).

IV. SEMI-ANALYTIC BER COMPUTATION

The expressions (31) and (35) are valid for any multipath profile, where all the paths have different average power. Therefore it is possible to obtain more accurate estimates of the performance by computing (31) and (35) using random multipath profiles which are generated with one of the several models available in the literature for indoor wireless channels. This will give us the BER performance conditioned on a particular random profile. We can then compute the unconditioned average BER by repeating the above procedure for a large number of simulated profiles and by averaging the results.

In this paper we consider the statistical channel model developed by Rappaport *et al.* [19]. This model is based on a large set of measurements taken in a wide variety of buildings. The channel parameters that are modeled include the number, arrival times and amplitudes of individual multipath components.

The number of multipath components at a given location is modeled with a Gaussian distribution. The probabilities of multipath arrivals are modeled as piecewise functions of RMS delay spread [19]. The mean amplitude of each multipath component is generated from a log-normal distribution based on a $d^{n(t)}$ power law where d is the transmitter-receiver distance in meters and n(t) is the power law exponent which is a function of delay spread. Inside buildings n typically ranges between 2 and 4.

The statistical models have been incorporated into a propagation simulation program called SIRCIM which accurately recreates the statistics of measured wide-band impulse responses for a wide range of buildings and topographies [23]. SIRCIM models the case of a stationary transmitter and a mobile receiver. It neglects temporal fading effects caused by other moving objects in the channel, which results in less severe fading (modeled by a Rician PDF) as opposed to the spatial fading caused by receiver movement (modeled by a Rayleigh PDF).

When using normalized random multipath profiles generated by SIRCIM, the average BER versus E_b/N_0 conditioned on a given profile α, for DS-SS systems with BPSK modulation and combined antenna and SS diversity, $\overline{P}_b(\overline{\gamma}_b/\alpha)$, is given by (40)-(42) where δ_i is replaced by

$$\theta_i = \frac{\sum_{j=1}^{L} \alpha_j}{\alpha_i} \tag{47}$$

where α_i is the power of the ith group component in the simulated multipath profile. In Rappaport's model the multipath profiles are generated over a delay spread range of 500 ns with a time resolution of 7.8 ns (corresponding to the pulse duration of the signal that was used during the measurements of channel characteristics). The number of multipath components is equal to the number of 7.8 ns bins in the 500 ns range, that is 64. However the maximum number of multipath components that the RAKE receiver can resolve from the received signal with bandwidth W is given by (2). Therefore, to compute the average BER with (40) it is necessary to group the multipath components in the profiles generated by SIRCIM to obtain the number of paths that can actually be resolved by the receiver. The procedure employed to do the above is similar to the one in [4]. For M-PSK TCM we compute (34) - (32) with δ_i again replaced by (47).

Once we have computed (40) or (43) for a reasonably large number of multipath profiles (fifty profiles were employed in the examples shown later), the unconditioned average BER can be computed as follows:

$$\overline{P}_b = \frac{1}{N}\sum_{p=1}^{N} P_b(\overline{\gamma}_b/\alpha_p) \tag{48}$$

where N is the number of multipath profiles used in the averaging procedure and we have assumed that all profiles are equally likely.

V. NUMERICAL RESULTS

In this section we present an example using these analytical results. We select three representative measured values of T_m: 50, 150, and 250 ns [3]-[4]. These values are for a 2 GHz indoor channel. We assume W=20 MHz. Table I shows the relation between the maximum delay spread values, the number of resolved multipath components, the time constant of the exponential decaying profile (e) and the parameter η.

TABLE I

Delay Spread T_m (ns)	Number of Resolved Paths L	ε (ns)	η
50	2	10	0.2
150	4	30	0.6
250	6	50	1.0

For purposes of illustration we select BPSK and a simple 4-state QPSK trellis code [14]. In AWGN, this code provides an asymptotic coding gain of 4 dB over the corresponding uncoded system using BPSK while maintaining the same throughput and bandwidth. The code parameters are $b_{d_{free}} = 1$ and $d_{free}^2 = 10$. We also assume that dual antenna diversity is used ($L_a = 2$).

Fig. 1 shows the BER performance for uncoded BPSK and both the constant and exponential profiles for different values of L. Fig. 2 shows similar results for TCM-QPSK (4 states). Note that in general the exponential profile model results in worse error performance than the constant profile assumption. The reason for this is that the diversity improvement provided by subsequent multipath components depends on the power level of the components which is lower for the later paths in the exponential profile. The difference is large for L=2 (8.5 dB for BPSK at $\overline{P}_b = 10^{-6}$) but decreases for larger values of L (4 dB for L=6 and BPSK modulation at $\overline{P}_b = 10^{-6}$).

In Fig. 3 we illustrate the BER performance for BPSK with combined antenna and SS diversity (constant and exponential profiles) for different values of L. The antenna diversity order is $L_a = 2$. Fig. 4 shows similar results for TCM-QPSK (4 states). Again in this case the exponential profile gives worse error performance. The difference in performance between the two profiles is smaller for the case with both SS and antenna diversity than for the case of SS diversity alone: for L=2 and BPSK the difference is about 6.5 dB at $\overline{P}_b = 10^{-6}$. As in the SS diversity case the difference decreases for larger values of L (3 dB for L=6 and BPSK modulation at $\overline{P}_b = 10^{-6}$).

Fig. 5 illustrates the BER performance of DS-SS systems with BPSK and combined antenna and SS diversity for the constant, exponential and SIRCIM generated profiles (hard partitioned buildings, obstructed topography [22]) and two values of L. The antenna diversity order is two. Observe that the performance results obtained with the realistic SIRCIM channel model based on actual measurements are fairly close to the results obtained with the simpler exponential profile based performance equation and differ considerably from the ones corresponding to the constant profile model.

Fig. 6 illustrates similar results (and conclusions) for a DS-SS systems system with TCM-QPSK (4 states) and combined antenna and SS diversity. The importance of these results is that they show that it is possible to predict BER performance in the indoor wireless environment with a good level of accuracy with the simple closed form analytical solutions presented in this study.

VI. CONCLUSIONS

In this paper we have analyzed the performance of DS-SS systems with both SS diversity alone and combined antenna/SS diversity on the indoor wireless Rayleigh fading channel. A closed form solution for BPSK and upper bounds for TCM M-PSK were derived assuming that the multipath profile of the channel is exponential.

A more involved semi-analytic computation of the average BER employing SIRCIM simulated random multipath profiles based on actual measurements was also presented. It was shown that the results corresponding to the exponential profile are considerably closer to the more accurate results obtained with the simulated profiles (for the specific type of building and topography under study) than those obtained assuming a constant profile. This indicates that the analytical expressions presented in this paper combined with the exponential assumption are useful tools in the estimation of the average BER of SS systems for transmission over the indoor wireless Rayleigh fading channel.

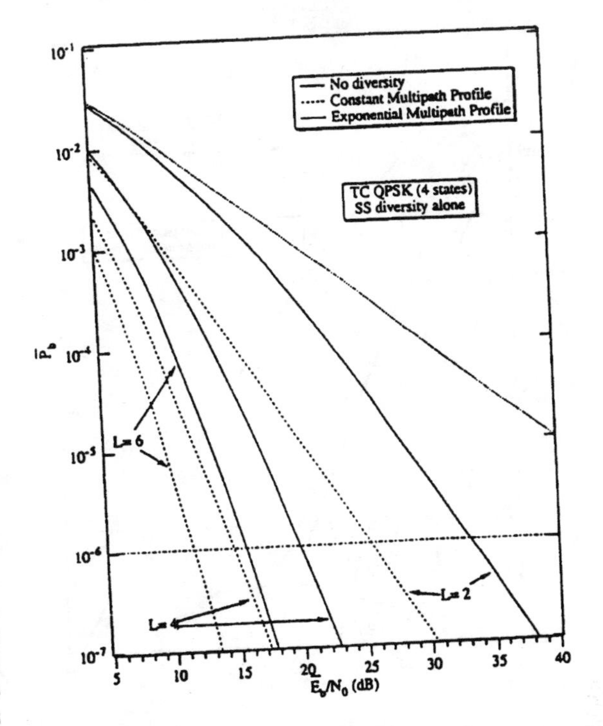

Fig. 2: Performance of TCM-QPSK (4 states) with spread spectrum diversity for a channel modeled with both constant and exponential multipath profiles.

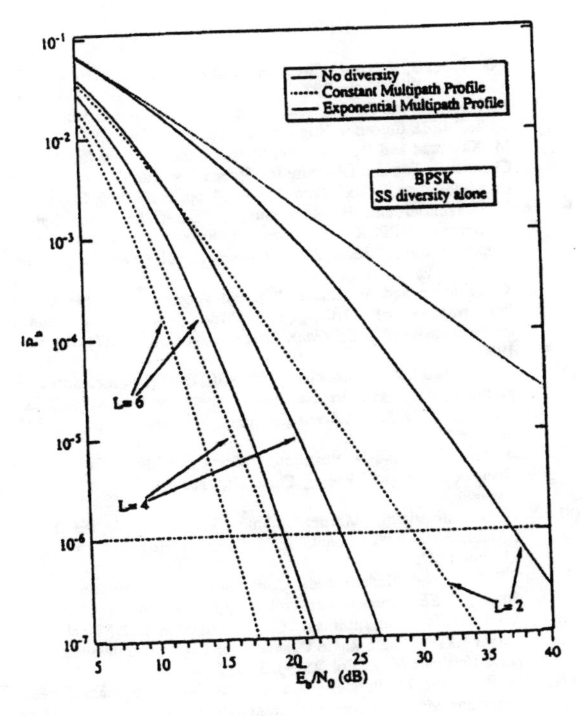

Fig. 1: Performance of BPSK with spread spectrum diversity for a channel modeled with both constant and exponential multipath profiles.

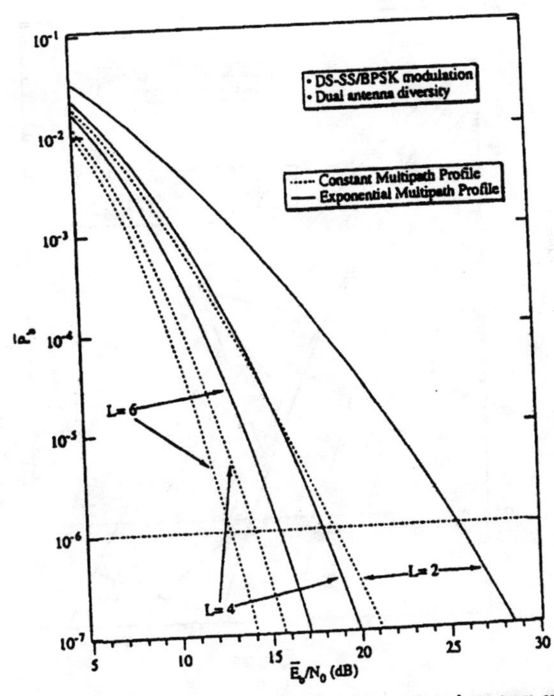

Fig. 3: Performance of DS-SS/BPSK and combined spread spectrum and antenna diversity for a channel modeled with both constant and exponential multipath profiles.

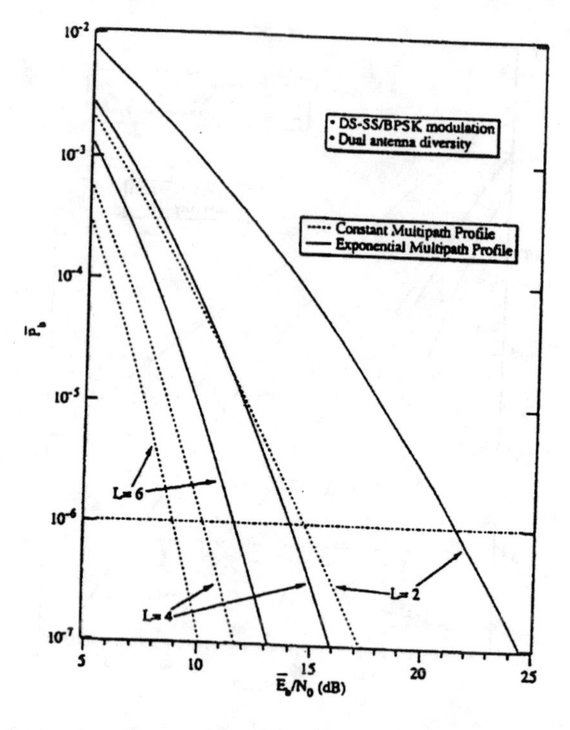

Fig. 4: Performance of DS-SS/TCM-QPSK (4 states) and combined spread spectrum and antenna diversity for a channel modeled with both constant and exponential multipath profiles.

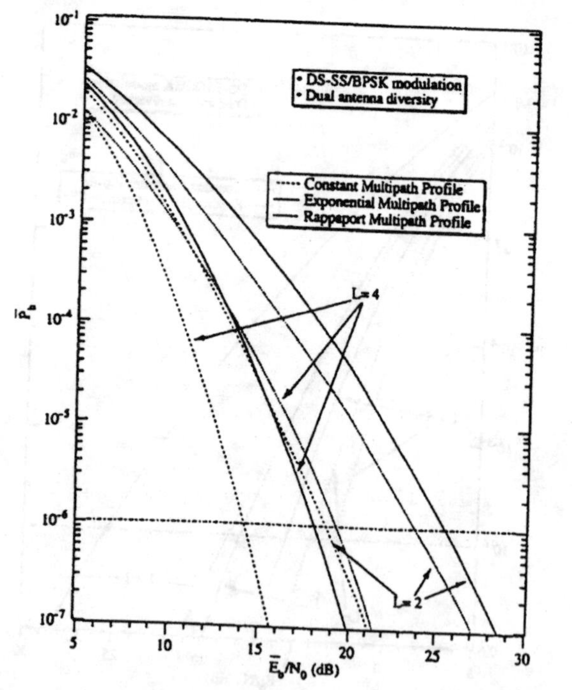

Fig. 5: Performance of DS-SS/BPSK with combined spread spectrum and antenna diversity for a channel modeled with constant, exponential and simulated random multipath profiles.

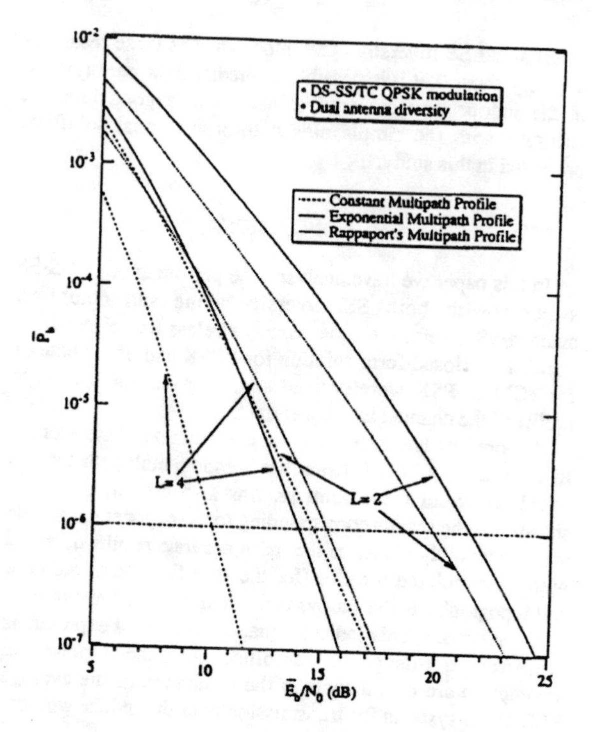

Fig. 6: Performance of DS-SS/TCM-QPSK (4 states) with combined spread spectrum and antenna diversity for a channel modeled with constant, exponential and simulated random multipath profiles.

REFERENCES

[1] J.G. Proakis, *Digital Communications*, (2nd Ed.). New York: McGraw-Hill, 1989, pp. 792-795.

[2] M. Kavehrad and P.J. McLane, "Spread Spectrum for Indoor Digital Radio," *IEEE Commun. Mag.*, vol. 25, pp. 32-40, June 1987.

[3] M. Kavehrad and P.J. McLane, "Performance of Low Complexity Channel Coding and Diversity for Spread Spectrum Indoor Wireless Communication," *AT&T Tech. J.*, vol. 64, pp. 1927-1965, Oct. 1985.

[4] M. Kavehrad and B. Ramamurthi, "Direct-Sequence Spread Spectrum with DPSK Modulation and Diversity for Indoor Wireless Communications," *IEEE Trans. Commun.*, vol. COM-35, pp. 224-236, Feb. 1987.

[5] K. Pahlavan and M. Chase, "Spread-Spectrum Multiple-Access Performance of Orthogonal Codes for Indoor Radio Communications," *IEEE Trans. Commun.*, vol. 38, pp. 574-577, May 1990.

[6] J. Wang and Marc Moeneclaey, "Hybrid DS/SFH Spread-Spectrum Multiple Access with Predetection Diversity and Coding for Indoor Radio," *IEEE J. Select. Areas Commun.*, Vol. SAC-10, pp. 705-713, May 1992.

[7] A. Abu-Dayya and N. Beaulieu, "Micro- and Macro Diversity on Shadowed Nakagami Fading Channels," submitted to *IEEE Trans. Commun.*

[8] D. Devasirvatham, "Multipath Time Delay Spread in the Digital Portable Radio Environment," *IEEE Commun. Mag.*, vol. 25, pp. 13-21, June 1987.

[9] T.S. Rappaport, "Indoor Radio Communications for Factories of the Future," *IEEE Commun. Mag.*, vol. 27, pp. 15-24, May 1989.

[10] R. Bultitude, "Measurement, Characterization and Modeling of Indoor 800/900 MHz Radio Channels for Digital Communications," *IEEE Commun. Mag.*, Vol. 25, pp. 5-12, June 1987.

[11] D. Borth and M. Pursley, "Analysis of Direct-Sequence Spread-Spectrum Multiple-Access Communication over Rician Fading Channels," *IEEE Trans. Commun.*, vol. COM-27, pp. 1566-1577, Oct. 1979.

[12] E. Geraniotis and M. Pursley, "Performance of Coherent Direct-Sequence Spread Spectrum Communications over Specular Multipath Fading Channels," *IEEE Trans. Commun.*, vol. COM-33, pp. 502-508, June 1985.

[13] W. Yung, "Direct-Sequence Spread-Spectrum Code Division Multiple Access Cellular Systems in Rayleigh Fading and Log-Normal Shadowing Channel," *in Proc. ICC'91*, June 1991, pp. 871-876.

[14] G. Ungerboeck, "Channel Coding with Multilevel/Phase signals," *IEEE Trans. on Inform. Theory*, vol. IT-28, pp. 55-67, Jan. 1982.

[15] R.V. Paiement and J.Y. Chouinard, "Simulated Application of Trellis Coded Modulation to the Indoor Wireless Channel," *in Proc. 1991 Veh. Technol. Conf.*, St. Louis, MO, May 1991, pp. 216-221.

[16] S.A. Hanna, M. El-Tanany, and S.A. Mahmoud, "Simulated Performance of Coded Digital Transmission in a Multi-User Indoor Radio Communication Environment," *in Proc. 1991 Veh. Technol. Conf.*, St. Louis, MO, May 1991, pp. 234-239.

[17] J.M. Bargallo and J.A. Roberts, "Trellis Coding with Diversity for DS-SS CDMA Communication over Indoor Radio Channels," *in Proc. Globecom '92*, Orlando, FL, Dec. 1992, pp. 1813-1818.

[18] J.M. Bargallo, G.J. Minden, and J.A. Roberts, *Advanced Indoor Radio System (AIRS)*, TR-02228-1, Telecommunications and Information Sciences Laboratory, The University of Kansas Center for Research, Inc., April 1993.

[19] T.S. Rappaport, S.Y. Seidel and K. Takamizawa, "Statistical Channel Impulse Response Models for Factory and Open Plan Building Radio Communication System Design," *IEEE Trans. Commun.*, vol. COM-39, May 1991, pp. 794-807.

[20] S.G. Wilson and Y.S. Leung, "Trellis-Coded Phase Modulation on Rayleigh Channels," *in Proc. ICC'87*, Seattle, WA, June 1987, pp. 739-743.

[21] A.M. Saleh and R.A. Valenzuela, "A Statistical Model for Indoor Multipath Propagation," *IEEE J. Select. Areas Commun.*, vol. SAC-5, pp. 128-137, Feb. 1987.

[22] T.A. Sexton and K. Pahlavan, "Channel Modeling and Adaptive Equalization of Indoor Radio Channels," *IEEE J. Select. Areas Commun.*, vol. 7, pp. 114-121, Jan. 1989.

[23] S. Seidel and T.S. Rappaport, A Users Manual *for SIRCIM* (Version 1.0), Virginia Polytechnic Institute and State University, Blacksburg, VA, October 19, 1990.

Juan M. Bargallo (S'90-M'93) was born in Buenos Aires, Argentina, in 1964. He received the Bachelor of Engineering and Electronics Engineer degrees from the Technological Institute of Buenos Aires in 1988 and 1989, respectively, and the M.S. degree in electrical engineering from the University of Kansas, Lawrence, in 1992.

From 1990 to 1992 he worked as a Graduate Research Assistant at the Telecommunications and Information Sciences Laboratory, performing sponsored research in the areas of trellis coded modulation, indoor wireless communications, and spread spectrum techniques.

From 1993 to 1994 he was with LCC, Arlington, VA, where he worked on the design of AMPS, IS-54, and GSM cellular mobile networks. Since July 1994, he has been with Mobile Systems International, where he is currently involved in the design of CDMA-based Personal Communication Systems (PCS) and the analysis of interference from PCS into Private Operational Fixed Services. His current research interests include performance analysis of multiple access techniques, wireless channel propagation modeling, and combined modulation and coding techniques.

Mr. Bargallo is a member of the IEEE, the Communications Society, the Information Theory Society, and the Vehicular Technology Society.

James A. Roberts (S'65-M'66-SM'78) was born in Vandalia, Illinois on March 11, 1944. He received the B.S. degree with highest distinction from the University of Kansas, Lawrence, KS, in 1966, the S.M. degree from the Massachusetts Institute of Technology, Cambridge, MA, in 1968, and the Ph.D. degree from Santa Clara University, Santa Clara, CA, in 1979, all in electrical engineering.

He began his career as an electronic design engineer with RCA, Burlington, MA, working on the NASA Apollo project. In 1969, he joined ESL, Inc., in Sunnyvale, CA, and worked as a research engineer, systems engineer, and project manager. His areas of interest were satellite communications, fading channel communications performance, system simulation, spread-spectrum communications, error correction coding, and signal processing. He was promoted to section head, department manager, and laboratory manager. While at ESL, he taught graduate applied mathematics and electrical engineering courses at Santa Clara University. TRW acquired ESL in 1978, and in 1983 he transferred to TRW in Denver and managed a large software project. With the continuing growth and diversification of the Denver activity, the company formed TRW Denver Operations with Dr. Roberts as the manager.

In 1990, he became Professor and Chairman of Electrical and Computer Engineering (now Electrical Engineering and Computer Science) at the University of Kansas. His research interests are communications theory, wireless communication systems, and information theory and coding. He is a consultant and director for several companies that develop radio communications products.

Decision Feedback Equalization for CDMA in Indoor Wireless Communications

Majeed Abdulrahman, *Student Member, IEEE,* Asrar U. H. Sheikh, *Senior Member, IEEE,* and David D. Falconer, *Fellow, IEEE*

***Abstract*—Commercial interest in Code Division Multiple Access (CDMA) systems has risen dramatically in the last few years. It yields a potential increase in capacity over other access schemes, because it provides protection against interference, multipath, fading, and jamming. Recently, several interference cancellation schemes for CDMA have been proposed but they require information about all interfering active users or some channel parameters. In this paper, we present an adaptive fractionally spaced decison feedback equalizer (DFE) for a CDMA system in an indoor wireless Rayleigh fading environment. This system only uses information about the desired user's spreading code and a training sequence. An analysis on the optimum performance of the DFE receiver shows the advantages of this system over others in terms of capacity improvements. A simulation of this system is also presented to study the convergence properties and implementation considerations of the DFE receiver. Effects on the performance because of sudden birth and death of users in the CDMA system and bit error rate performance of the DFE receiver is also presented.**

I. Introduction

CODE Division Multiple Access (CDMA) system is a spread spectrum system in which many users could share the same bandwidth. These users are assigned different spreading codes to spread their signals over a bandwidth which is much wider than their transmitted data bandwidth. Direct sequence spread spectrum (DS/SS) is a CDMA technique in which the transmitted data stream is multiplied by a higher rate spreading code to spread the information's energy over a wider bandwidth. In the past, spread spectrum techniques have been implemented in military communication systems since they provide protection against interference, multipath, fading, and jamming. Commercial interest in mobile and portable communciations has increased demands to allocate more bandwidth for radio communications, or use more efficient transmission schemes to provide services for wireless communications.

One CDMA system that has received much attention recently is that designed and implemented by Qualcomm Incorporated [1]. This system shows considerable improvement in capacity over other FDMA and TDMA systems, since it exploits multipath resolution, voice activity, and antenna sectorization as well as the spreading gain. The main gain in capacity comes from the fact that the same bandwidth could be used in neighboring cells of the cellular system, which is difficult in FDMA or TDMA systems. Using 1.25 MHz bandwidth, a data rate of about 9600 bits per second and E_b/N_0' (where the noise spectral density N_0' includes interference) of about 6 dB, the Qualcomm system could provide 120 voice channels in a given cell that uses 120° sectored antennas and voice activity detection. This capacity, however, drops to 33 channels if an omnidirectional antenna is used and voice activities of users are not considered [2].

CDMA systems using linear and nonlinear interference canceling techniques have shown even better performance if the receiver has knowledge of the spreading codes of all users, received powers of some of the interferers, or some channel parameters [3]–[5]. All spreading codes pertaining to a given cell in a cellular radio system are available at the receiver, but it is difficult to provide codes from other cells to the base station; it is also difficult to provide any interfering codes to individual portable receivers. Security is also of a concern when a receiver possesses knowledge of the spreading codes of other users in a given cell. Unpredictable activities of some users, such as an asynchronous mode of data transmission, make it more difficult to estimate their received powers over a given time interval. Estimation of channel parameters is also difficult, especially for the fast varying channel of mobile communications.

Considering the drawbacks of the assumptions of these proposed systems for CDMA, we propose the use of a fractionally spaced decision feedback equalizer (DFE) for CDMA systems. The choice of a fractionally spaced DFE receiver for CDMA was supported by previous research showing that a DFE receiver improves the performance in a near-end crosstalk (NEXT) environment of digital subscriber loops (DSL) [6]. All users in the DSL had the same transmission rate, and the cyclostationarity of the interference was exploited by the DFE in improving the overall performance. In a CDMA system, all users are transmitting at the same chip rate of the spreading code. This suggests the use of fractionally spaced DFE to suppress interference and increase overall capacity.

The DFE combines the functions of RAKE reception [7] to exploit diversity resulting from multipath, and performs interference and intersymbol interference (ISI) cancellation. The DFE minimizes the effects of interference as well as ISI by trying to force zeros in the impulse responses of the interferers at the decision instants. The number of interferers

Manuscript received June 13, 1993; revised November 29, 1993.

M. Abdulrahman is with Bell-Northern Research, Ottawa, Ont., Canada K1Y 4H7.

A. U.-H. Sheikh and D.D. Falconer are with the PCS Research Laboratory, Department of Systems and Computer Engineering, Carleton University, Ottawa, Canada, K1S 5B6.

IEEE Log Number 9215437.

that theoretically can be canceled increases with the signal's excess bandwidth. For a baseband bandwidth that is S times the data symbol rate, up to $2S-1$ interferers can be canceled [8], [9]. In practice, the maximum number of interferers that can be suppressed with satisfactory performance may be lower than this as a result of factors such as the finite number of equalizer taps, enhancement of background noise by the equalizer, and sensitivities to tap-weight inaccuracy. The use of a DFE instead of a linear equalizer may reduce the noise enhancement effect, and also affords the forward linear filter greater flexibility in handling ISI as well as cochannel interference. In this paper, we investigate the capabilities of practical DFE-based receivers for suppressing interference in CDMA systems with small processing gain, and which have only crude power control and must contend with multipath.

Two configurations of the DFE receiver have been suggested for CDMA [10], [11]. One uses information about the desired user's spreading code, while the other assumes no knowledge of any spreading code. In both configurations, we do not assume any information about channel parameters or user's activities but a known training sequence of data symbols is assumed in most cases. This reduces the complexity of the CDMA system, and provides security for users which do not want other systems to have access to their spreading code.

Linear CDMA receivers using transversal filters have also been investigated in [12] and [13] for the case of large spreading gains. The receiver has knowledge of only the spreading sequence of the desired user. Two receiver configurations were considered to reduce the complexity of the receiver. One configuration uses a bank of filters where each filter is a shifted version of the desired user's matched filter, while the other configuration samples the output of the matched filter multiple times in a symbol period. The output of either configuration is used to minimize the Mean Squared Error (MSE) of the receiver.

In Section II, we describe the proposed DFE receiver for a CDMA system. This provides detailed information about both transmitter and receiver parts of the system. This section also details the fading channels used and different parameters assumed for this particular system. Section III presents the optimum Minimum Mean Squared Error (MMSE) performance of the DFE receiver, and provides insight into the potential increases in capacity for this type of receiver. Section IV shows the results of simulation of the DFE receiver, which gives information about convergence and implementation aspects of the proposed CDMA system. Finally, Section V summarizes the conclusions, and the Appendix presents the MMSE analysis.

II. System Description

Fig. 1 shows the transmitter part of the CDMA system used in this paper. The transmitted signal of user k is a spread spectrum BPSK signal, represented as:

$$X_k(t) = \sqrt{2}\,\mathrm{Re}\left(\sum_{\lambda} b_{\lambda}^{k} a_k(t-\lambda T_b - \Omega_k) e^{j2\pi f_0 t}\right) \quad (1)$$

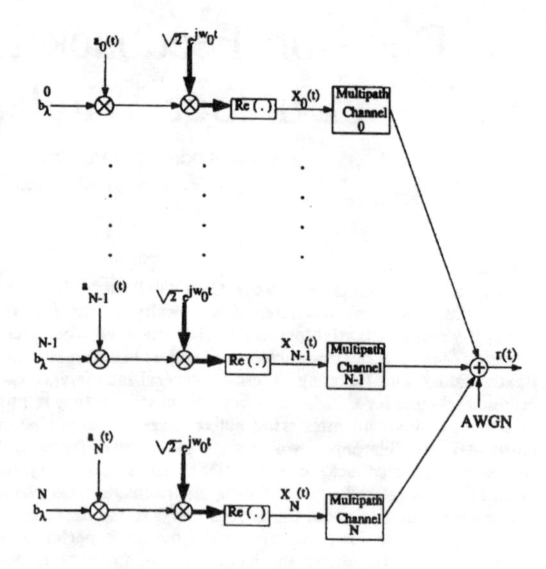

Fig. 1. Block diagram of $N+1$ CDMA transmitters.

where b_{λ}^{k} is real-valued transmitted data symbol ± 1, $a_k(t)$ is the spreading sequence of user k, T_b is the symbol period, Ω_k is the delay of user k with respect to the desired user, and f_0 is the carrier frequency. The spreading sequence of user k is defined as:

$$a_k(t) = \sum_{\lambda=0}^{p-1} a_{\lambda}^{k}\, \Psi(t-\lambda T_c) \quad (2)$$

where a_{λ}^{k} is the spreading code, T_c is the chip period, p is the length of the spreading code in terms of chip periods, and $\Psi(t)$ is the transmitted pulse shape. Each user's transmitted signal is assumed to pass through a frequency-selective Rayleigh fading channel. Additive White Gaussian Noise (AWGN), resulting from receiver thermal noise, is also considered in this system. The received chip-energy-to-noise ratio, E_c/N_0, which is also the receiver's input signal-to-thermal noise ratio, is assumed to be 6 dB. With a spreading gain of 8, the corresponding E_b/N_0 ratio is 15 dB. Note that N_0 does not include interference from other users.

Fig. 2 shows the block diagram of receiver "0." The received signal, from $N+1$ users, after demodulation is represented as:

$$d(t) = \sum_{k=0}^{N} d_k(t) + n(t) \quad (3)$$

where

$$d_k(t) = \sum_{\lambda} b_{\lambda}^{k} \sum_{m=0}^{M_k} C_{mk} a_k(t-\lambda T_b - \Omega_k - t_{mk}). \quad (4)$$

M_k is the number of paths of user k, the complex quantity C_{mk} is the amplitude and phase variation of the mth path of user k, and t_{mk} is the reception time of the mth path of user k. Two receiver configurations were considered. In one configuration, the demodulated signal passes through a spreading sequence matched filter (SSMF) to the desired user

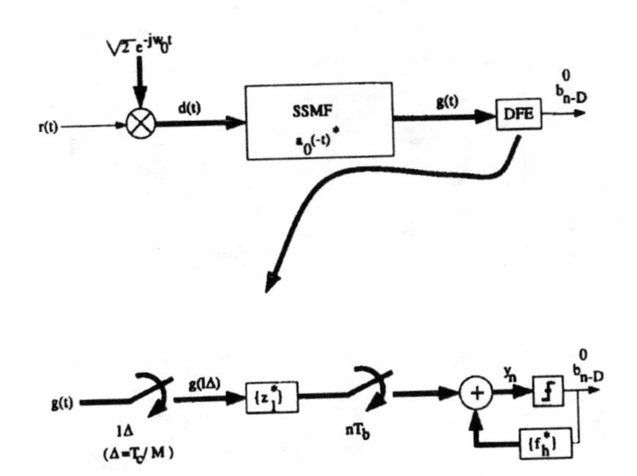

Fig. 2. Block diagram of a CDMA receiver for user "0".

before being received by the DFE. In the other receiver configuration, the SSMF is replaced by a lowpass filter (LPF) which has a bandwidth equal to the spreading code bandwidth since the DFE is sampling at multiples of the chip rate of the spreading code.

The transmitted signal of each user is assumed to be received at the input of the DFE receiver through three different paths ($M_k = 3; k = 0, 1, \cdots$). To determine the strength of each path, two types of channels were assumed. In one case, which we will call the *constant variance* model, all paths are assumed to have equal mean and variance, which represents a pessimistic type of channel. The other channel model has the variance of a given path decreasing by $e^{-((L-1/3))}$, where L is the time, measured in the chip periods of the spreading sequence, with respect to the arrival time of the first path. We will call this type of channel the *exponential variance* model. The location of the paths and delays of the various users are determined randomly in the first six chips of a bit period, so the delay spread is assumed to be equal to $6T_c$ or 3/4 of a bit period (T_b). In an indoor wireless system, where delay spreads on the order of 100 ns may be encountered this would be representative of a 10 Mb/s system; it could also be typical of lower bit rate systems in outdoor cellular environments. Since the total received powers of all users are equal only in an average sense in both of these models, they correspond to the case of very slow acting (or long-term average) power control. We have also assumed steady-state channels, which means that the channel coefficients do not change during the data transmission period of a user.

The input to decision device is equal to:

$$y_n = \sum_{l=0}^{N_z} Z_l^* g(nT_b - l\Delta) + \sum_{h=1}^{N_f} f_h^* b_{n-h-D}^0 \tag{5}$$

where N_z+1 and N_f are the numbers of forward and feedback taps, respectively, Z_l is the setting of the lth tap of the forward filter, f_h the setting of the hth tap of the feedback filter, $\Delta = (T_c/M)$ where M is an integer, and D is the delay assumed in the system. In order to determine the optimum performance of the DFE receiver, the forward and feedback taps of the DFE are optimized to minimize the MSE, where the error is:

$$e_n = y_n - b_{n-D}^0. \tag{6}$$

Details of derivation of the MMSE are given in the Appendix.

The processing gain (PG), or ratio of spread bandwidth to information bandwidth, of a CDMA system is usually chosen in the order of 100. Since we are using a DFE receiver for equalizing the effects of fading and interference and the forward filter of the DFE performs its sampling at multiples of the chip rate of the spreading code, then the PG should be moderate because higher sampling rate increases the complexity of the overall system. In this work, we have decided to use a processing gain equal to 8. Note that if we define N_0' to be the equivalent power spectal density of interference plus thermal noise, the signal-to-noise ratio per bit, E_b/N_0', of a conventional CDMA receiver is [1]:

$$\frac{E_b}{N_0'} = \frac{PG}{N + N_0/E_c} \tag{7}$$

where PG is the processing gain, and $N+1$ is the total number of users. With $PG = 8$ and $E_c/N_0 = 6$ dB, $(E_b/N_0') = (32/4N+1) = 2.75$, 3.9, and 5.5 dB for 4, 3, and 2 users, respectively.

We have chosen Walsh or Hadamard codes [14] of length 8 to be used as the spreading sequences of the CDMA users. These codes have zero crosscorrelation for users with one received path and equal delay. We do not claim that Walsh codes are the optimum spreading codes for our system since we are dealing with multipath fading channels. More research is needed to find the best spreading codes of the DFE receiver, which is ignored in this paper. The structure of the forward filter of the DFE receiver is similar to that of a RAKE receiver [7]. The RAKE receiver works on combining the energies of the desired user received through the multipath channel without considering the presence of interferers. Using MMSE criteria, the forward filter of the DFE receiver works on maximizing the desired user's energy and minimizing the effects of existing interference in the channel, AWGN, and ISI.

The performance of the proposed system is evaluated by studying the MMSE of the DFE receiver. Since $MSE_{dB} \cong -SNR_{dB}$ at the output of the receiver, $SNR_{dB} = (E_b/N_0')_{dB}$ where $(E_b/N_0')_{dB}$ is the receiver's bit-energy-to-interference plus noise ratio, which is about 3 to 6 dB for good performance of a coded coherent system [14], then an MSE_{dB} on the order of -6 dB is required for good performance of the proposed DFE receiver. The MSE presented in the optimum and convergence results of the DFE receiver in the following sections is the average of the MSE values obtained for 200 different independent constant variance (or exponential variance) model channels.

III. Optimum Performance of DFE Receivers

Using the derivations given in the Appendix, the MMSE was calculated in terms of the number of forward filter taps. The number of feedback taps was fixed at 2, since the results showed that MMSE does not improve for a number of feedback taps greater than 2. The DFE receiver was optimized

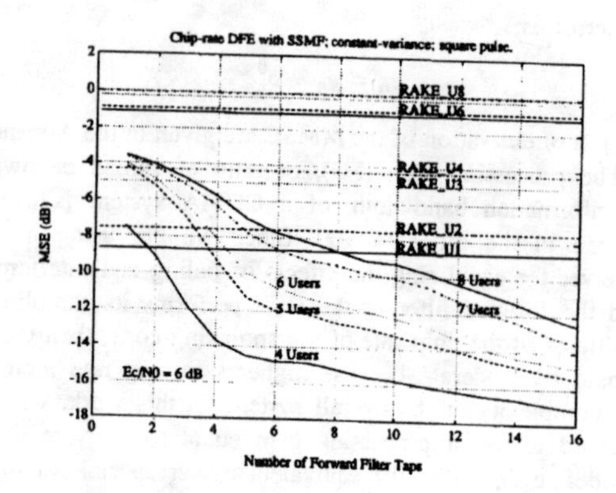

Fig. 3. Chip-rate DFE using SSMF, constant variance model, and square transmitted pulses.

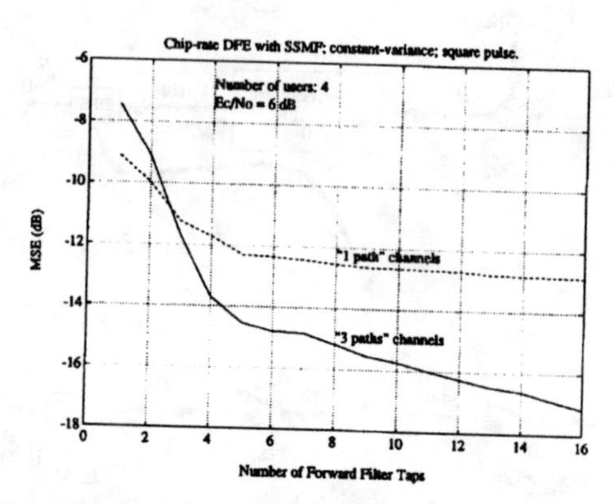

Fig. 4. Multipath channel effects on the performance of the DFE receiver. The "1 path" channels are flat fading channels with the same total average received power as that of one path in the "3 paths" channels.

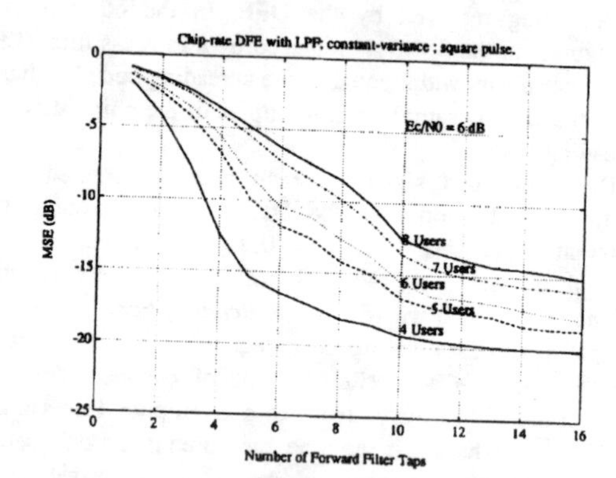

Fig. 5. Chip-rate DFE using LPF, constant variance model, and square transmitted pulses.

to the best sampling phase and decision delay. Preliminary MMSE results of the DFE receiver were shown in [10]. Fig. 3 give the MMSE performance of a chip-rate DFE receiver that uses an SSMF, square transmitted pulses, and constant variance model.

Performance of the conventional RAKE receiver with three taps, which coherently weights the desired signal's multipath components, is also shown for comparison. The RAKE receiver results for a different number of users are also averages of the MSE's of the 200 different channels that were used. The results show that the DFE receiver achieves an MSE of -6 dB for the case of 8 users with only 6 forward taps. A margin of error should be considered, however, when implementing this system, so it is necessary to have a higher number of forward taps in order to maintain good MSE performance. Assuming a margin of error of around 5 dB, good performance is achieved when 14 taps are used in the forward filter of the DFE receiver.

It was also noticed that the drop in MMSE from 5 to 6 users is smaller than that from 4 to 5 users. This was also noticed for 7 and 8 users and 6 and 7 users cases. The reason was attributed to the spreading codes of these users. When we calculated the MMSE after reassigning the spreading sequences to different users, the results showed a change in overall performance from those seen in Fig. 3. This shows the importance of using good codes in improving overall system performance. Fig. 3 also shows the advantages of our proposed DFE receiver over the RAKE receiver. The forward filter of the DFE simultaneously performs the functions of RAKE, and minimizes the effects of interference, fading, multipath, and AWGN.

The claim that the DFE performs the job of a RAKE is further supported by the results given in Fig. 4, where the DFE receiver gives a better performance for channels with three independently fading paths than for a flat fading channel (with only one fading path). The gains achieved when using multipath channels are due to two reasons. The received average power due to multipath channels is three times that received due to flat fading channels, since the average power of a received path is the same in either case. Another reason is the ability of the DFE to perform implicit diversity by combining the received powers of all paths to improve the performance. Good MMSE results were also achieved when sampling the forward filter of the DFE at twice the chip rate of the spreading code. However, the DFE with twice chip-rate sampling requires more taps than that needed for chip-rate DFE. The choice of either sampling rate becomes dependent on design limitations and performance criteria.

Fig. 5 shows the MMSE performance for a chip rate DFE receiver with square pulses and constant variance model, but replacing the SSMF with a LPF. The results show somewhat better performance than that achieved in Fig. 3 for the case of SSMF. A SSMF at the input of the receiver would maximize the desired user energy for the case of one received path and equal delays between all users, especially when using Walsh codes for spreading the data signals. However, the orthogonality of Walsh codes is lost when using multipath channels and the matched filter may then merely act to increase the effective impulse response durations of the transmission channels, which may explain the improvement in the overall performance of the LPF receiver over that of the SSMF receiver.

To exhibit the effects of pulse shaping on performance, Fig. 6 shows the MMSE performance for a chip-rate DFE receiver with SSMF, constant variance model, and square-root raised-

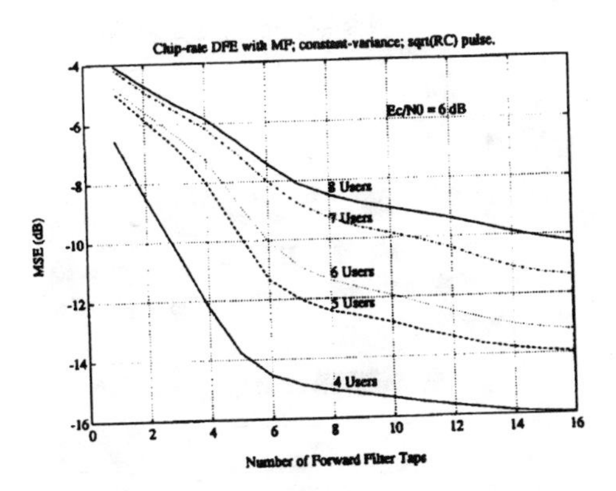

Fig. 6. Chip-rate DFE using SSMF, constant variance model, and square-root raised-cosine pulses.

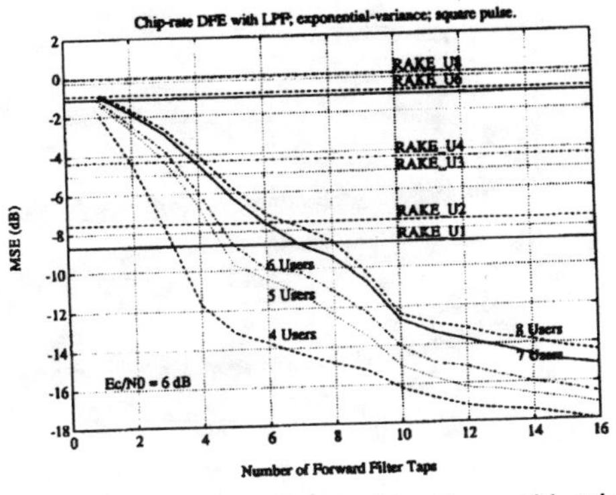

Fig. 7. Chip-rate DFE using LPF, exponential variance model, and square pulses.

cosine transmitted pulses. The results show that at least 14 taps are needed for acceptable performance. When the number of users is increased, the drop in performance in Fig. 6 is similar to that drop achieved for chip-rate DFE with SSMF and square transmitted pulses in Fig. 3. Comparing the results in these two figures, we see that the performance of the receiver with square transmitted pulses is better than that obtained for square-root raised-cosine pulses for any number of users, presumably as a result of the raised cosine pulse bandlimiting. This means that more forward taps are needed when using square-root raised-cosine pulses to cancel the ISI caused by this pulse shaping.

Using the exponential variance model, Fig. 7 shows the MMSE of the chip rate DFE receiver when using square pulses and LPF. Comparing these results to those in Fig. 5, the exponential variance model gives an MMSE which is almost 2 dB below that calculated for the constant variance model. The overall MMSE variation, however, between different numbers of users is less than that seen for the constant variance model.

This result is related to the fact that the average power of the first path in the exponential case is equal to that of any path in the constant variance model. Each of the second and third paths in the exponential variance model has an average power that is decaying exponentially. This means that the exponential variance model has an overall smaller average power, somewhat less potential diversity gain, and less ISI than the constant variance model, which explains the lower MMSE results and less variation in performance for a higher number of users.

IV. Simulation Results

In this section, we present the results of simulating the proposed DFE receiver. These results will be compared with their counterparts in the previous section to show the convergence time needed to achieve an acceptable system performance. Looking at the MMSE results in the previous section, we have decided that 14 forward filter taps would be needed for chip rate sampling of the DFE. Preliminary results of the simulation were also presented in [11]. To simplify the simulation of the DFE receiver, we have used fixed sampling phase and equalizer delay, even though they may be suboptimal for some channels. Five trials were performed to measure the MSE of each of the 200 channels, with a different data sequence for each user in a trial. A training sequence of 500 data bits was also used. It was found that most channels perform well with a shorter training sequence, but we decided to fix the length of the training sequence at 500 bits to simplify the simulation of the proposed system.

The Normalized Least-Mean-Squared (NLMS) algorithm was used to update the coefficients of the forward and feedback taps of the DFE [15]. The NLMS equations are:

$$\mathbf{f}_f[n+1] = \mathbf{f}_f[n] + \frac{\mu}{||\mathbf{m}_f[n]||^2} e[n]\mathbf{m}_f[n]$$
$$\mathbf{f}_b[n+1] = \mathbf{f}_b[n] - \frac{\mu}{||\mathbf{m}_f[n]||^2} e[n]\mathbf{m}_b[n] \quad (8)$$

where $\mathbf{f}_f[n]$ and $\mathbf{f}_b[n]$ are the forward and feedback coefficients vectors at time $n, \mathbf{m}_f[n]$ is the vector of received signals at time $n, \mathbf{m}_b[n]$ is the vector of decision bits used in the feedback filter, $e[n]$ is the error, μ is the step size which ranges between zero and 2, and $|| \cdot ||$ is the norm of a given vector[1].

Fig. 8 shows the convergence of the simulated DFE receiver (averaged over 200 channels) with chip rate sampling, SSMF, square transmitted pulses, and the constant variance model. Note that "FF taps" and "FB taps" in the figure refer to the number of forward and feedback taps of the DFE, respectively. Comparing these results with those in Fig. 3, we find that the "4 Users" case is close to its calculated MMSE result. However, the performance is worse by almost 3 dB than its MMSE value for five and more users. The sudden improvement in performance at 300 bits is related to a change in the step size parameter, μ, of the NLMS algorithm. We started the adaptation with $\mu = 1$, and then we changed it to $\mu = 0.1$ after transmitting 300 data bits. This "gear-shifting" NLMS proved to provide better performance, or faster convergence, than fixing the step size. As expected, the convergence of the DFE receiver is worse than optimum performance since the parameters of the simulation were fixed for all channels, while optimum sampling and equalizer delays were chosen

[1] Preliminary results reported in [11] showed that blind start-up adaptation was viable. Further results on this will be reported later.

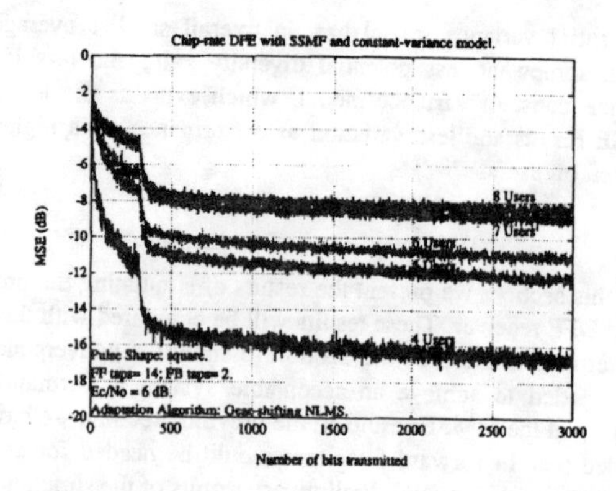

Fig. 8. Convergence time of chip-rate DFE using SSMF, square pulses, and constant variance model.

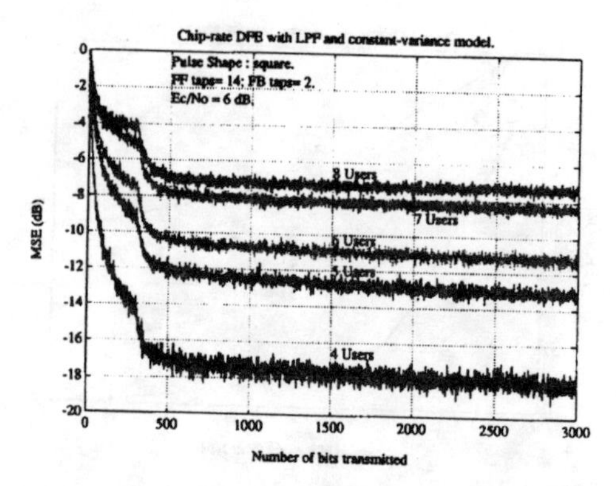

Fig. 9. Convergence time of chip-rate DFE using LPF, square pulses, and constant variance model.

for all channels when calculating the optimum MMSE of the receiver. When sampling the DFE at twice the chip rate of the spreading code but using 14 forward taps as in the chip-rate DFE, the convergence was slower than that of chip-rate DFE. Adding two extra forward taps improved the performance by almost 1.5 dB, which shows the importance of choosing the proper forward filter length to achieve good performance.

Fig. 9 shows the convergence time of the chip-rate DFE with LPF, square pulses, and constant variance model. There is a 2 dB difference between the simulated and calculated MMSE for the case of four users but the convergence is worse for a higher number of users, where the simulation gives an MSE which is worse by at least 5 dB than that calculated in Fig. 5. Similar results were achieved for DFE with twice chip rate sampling, but the performance improves as more forward taps are added to the DFE. This is expected since the forward filter is trying to perform the job of the SSMF, and cancel the effects of interference and AWGN. However, it requires more forward taps to achieve its purpose, so it becomes an implementation issue in deciding on the number of forward taps to use to achieve the acceptable MSE performance of the DFE receiver. The simulation results also show that the SSMF receiver with a number of users greater than 6 gives better performance than that achieved by the LPF system. This is in contrast to the results obtained in the optimum performance of the system. The reason is again related to the parameter optimization that was ignored in the simulation, which causes lower performance in the LPF receiver that uses the same number of forward taps as the SSMF receiver and has no information about the spreading sequence of the desired user. This shows that a DFE receiver with a LPF requires a longer time to converge, since it is trying to perform the job of the matched filter and equalization at the same time.

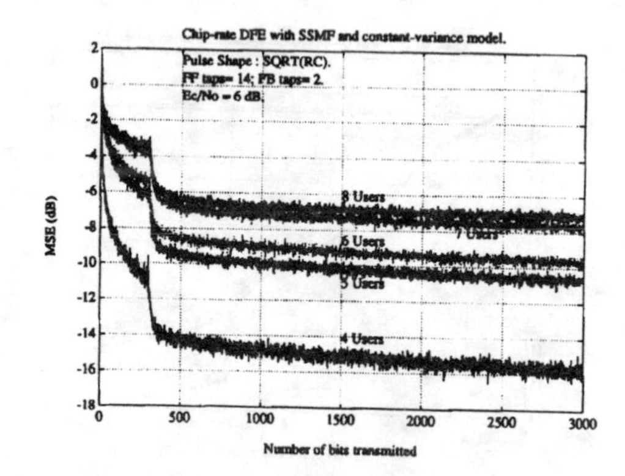

Fig. 10. Convergence time of chip-rate DFE using SSMF, square-root raised-cosine pulses, and constant variance model.

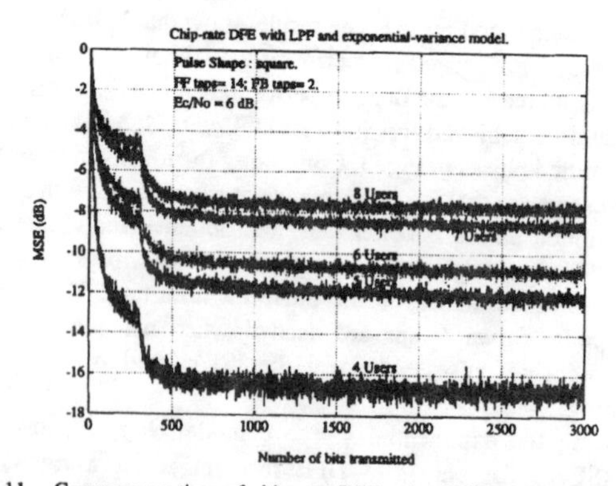

Fig. 11. Convergence time of chip-rate DFE using LPF, square pulses, and exponential variance model.

The effect of pulse shaping on the convergence time of the DFE can be seen in Fig. 10, where square-root raised-cosine pulses were used for the chip-rate DFE with SSMF and constant variance model. Comparing these results with the calculated MMSE in Fig. 6, the "4 Users" case reaches the expected MMSE while the other cases are worse by almost 3 dB than their calculated MMSE. Using the exponential variance model, Fig. 11 shows the convergence time of chip-rate DFE with LPF and square pulses. Similar to Fig. 9, the convergence time for 4 users gives an MSE close to the calculated MMSE after a short time of data bit transmission. However, the performance is worse by at least 4 dB for a number of users higher than 4.

In presenting the MMSE and convergence results of the DFE receiver, we assumed that all users start transmitting within one

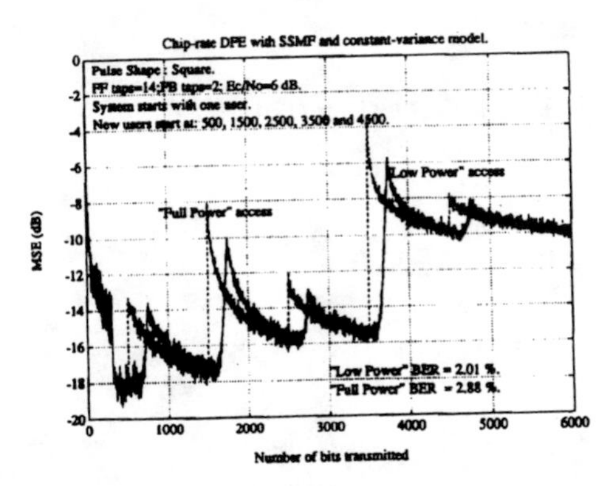

Fig. 12. Birth of interferers performance for chip-rate DFE with SSMF, constant variance model, and square pulses.

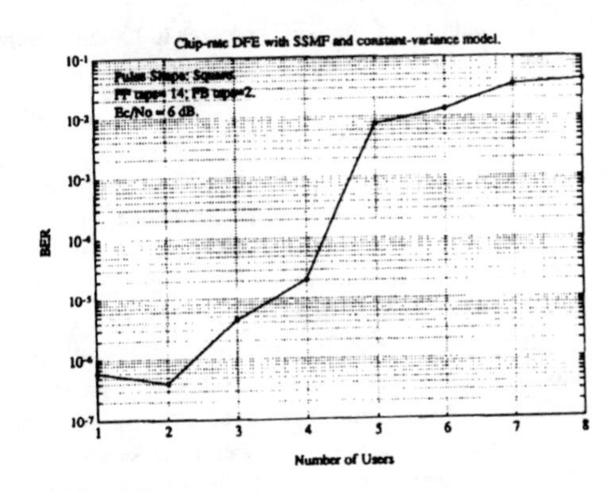

Fig. 14. Bit error rate of chip-rate DFE with SSMF, constant variance model, and square pulses.

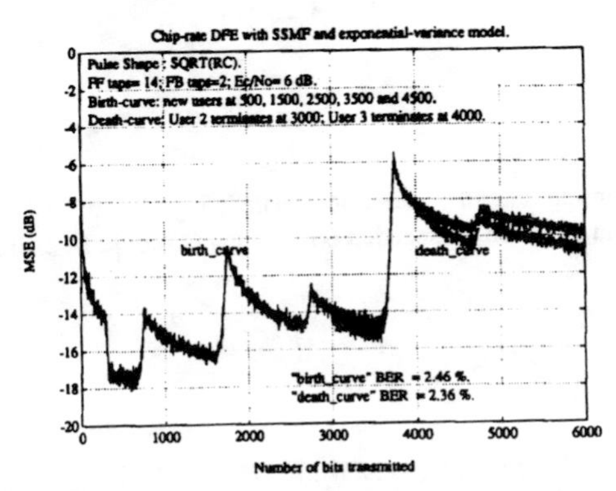

Fig. 13. Death of interferers performance for chip-rate DFE with SSMF, exponential variance model, and square-root raised-cosine pulses.

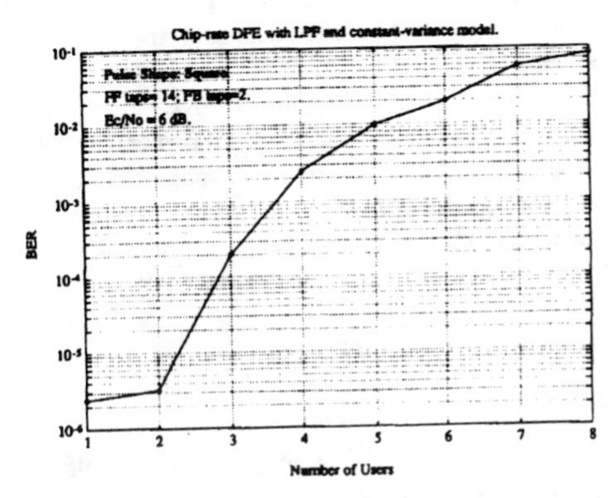

Fig. 15. Bit error rate of chip-rate DFE with LPF, constant variance model, and square pulses.

bit period. In practice, users transmit and terminate their access to the system at different times. Fig. 12 shows the case of sudden access or birth of users in the DFE receiver that uses a SSMF, constant variance model, and square pulses. The system starts with one user, and new users are given access starting at 500 bits with another new user every 1000 bits to a maximum of six users in the system. A user usually transmits at full power at the moment of establishing access in the system. This case is presented in the "Full Power" curve of Fig. 12. This situation causes a sudden drop in system performance, which might lead to a MSE lower than that required for good performance. This happened when the fifth user established access to the system, where the MSE went below the required -6 dB. This lead us to suggest the use of the "Low Power" access scheme, where a new user is asked to increase his power linearly over a short period of time.

Fig. 12 shows the case where a new user increases his power linearly over 250 bits. The time to establish full power could be used to transmit control information before actual data transmission could start. Comparing both access schemes for birth of new users, we find that the Low Power access scheme gives an improvement of almost 2 dB for the case of the third and fifth users access to the system. Fig. 13 shows the effects of sudden termination or death of users on the performance of the DFE receiver. The "birth curve" represents the case of users accessing the receiver that uses a SSMF, exponential variance model, and square-root raised-cosine pulses. The "death curve" shows the performance of the DFE receiver when users 2 and 3 terminate their access at 3000 and 4000 bits, respectively. As expected, the termination of access improves the MMSE performance of the DFE receiver.

Looking at the bit error rate (BER) of the proposed DFE receiver, Fig. 14 shows that four users provide a good BER when using a SSMF at the input of the DFE. In calculating the BER, one trial was performed for each channel and each user transmits 50,000 data bits. Fig. 15 shows the BER when the SSMF is replaced by a LPF. The results are worse in this case because the LPF receiver uses the same number of forward taps as the SSMF receiver and has no knowledge of the spreading code of the desired user but still provides good BER for four users. BER in the order of 10^{-2} is acceptable in the system for voice transmissions. The BER of the LPF case and the corresponding overall MSE could be improved by adding more forward taps. It was also noticed that the results of MSE and BER are similar in providing the same conclusion in terms of the capability of the DFE receiver to handle a number of users equal to half or more of the processing gain.

V. Conclusions

We have proposed the use of fractionally spaced DFE for CDMA systems. The optimum performance, convergence time, and BER results show the capabilities of this receiver in minimizing the effects of interference, multipath, fading, and AWGN in a slow acting (and, therefore, imprecise) power control environment. Two receiver configurations were considered: one that has knowledge of desired user spreading code (SSMF receiver), and the other has no information about all spreading codes (LPF receiver). A study of the optimum performance showed that the LPF receiver gives better MMSE performance than that obtained by the SSMF receiver. However, the opposite is true when a simulation of both receivers was performed. One reason was that parameter optimization was ignored in the simulation in order to simplify the implementation. Using a PG equal to 8, the receiver could easily service up to four users, which is equal to 50% of the processing gain. As shown by the 'RAKE' curves of Figs. 3 and 7, a conventional coherent RAKE receiver with this processing gain and equally imprecise power control would handle at most one or two users. Of course, both the DFE and conventional RAKE systems can benefit from exploiting voice activity and sectored antennas, as suggested in [1].

Appendix

MMSE Analysis for Fractionally Spaced DFE Receiver in CDMA Systems

Equation (4) represents the signal received at the input of the DFE receiver. This equation could be rewritten as:

$$d_k(t) = \sum_{\lambda} b_{\lambda}^{k} h_k(t - \lambda T_b) \tag{9}$$

where

$$h_k(t) = \sum_{m=0}^{M_k} C_{mk} a_k(t - \Omega_k - t_{mk}). \tag{10}$$

This received signal is passed through an SSMF (or LPF) before being processed by the equalizer part of the receiver. The received signal at the input of the DFE is presented as:

$$g(t) = \sum_{k=0}^{N} g_k(t) + \vartheta(t) \tag{11}$$

where

$$\begin{aligned} g_k(t) &= \int d_k(\tau) a_0(\tau - t)^* \, d\tau \\ &= \sum_{\lambda} b_{\lambda}^{k} V_k(t - \lambda T_b) \end{aligned} \tag{12}$$

$$\begin{aligned} V_k(t) &= \int h_k(\tau) a_0(\tau - t)^* \, d\tau \\ &= \sum_{m=0}^{M_k} C_{mk} \int a_k(\tau - \Omega_k - t_{mk}) a_0(\tau - t)^* \, d\tau \\ &= \sum_{m=0}^{M_k} C_{mk} R_{0K}(t - \Omega_k - t_{mk}) \end{aligned} \tag{13}$$

$$R_{0k}(t) = \int a_k(\tau) a_0(\tau - t)^* \, d\tau \tag{14}$$

and

$$\vartheta(t) = \int n(\tau) a_0(\tau - t)^* \, d\tau. \tag{15}$$

Assuming that all b_{λ}^{k} are uncorrelated; $E(b_{\lambda}^{k}) = 0$; $E([b_{\lambda}^{k}]^2) = 1$; and there are no decision errors, then (5) could be expressed as:

$$y_n = \mathbf{U^H X_n} \tag{16}$$

where

$$\mathbf{U^H} = [Z_0^*, Z_1^*, \cdots, Z_{N_z}^*, f_1^*, f_2^*, \cdots, f_{N_f}^*] \tag{17}$$

$$\begin{aligned} \mathbf{X_n^H} &= [g(nT_b)^*, g(nT_b - \Delta)^*, \cdots, \\ &\quad g(nT_b - N_z\Delta)^*, b_{n-1-D}^0, \cdots, b_{n-N_f-D}^0] \\ &= [\mathbf{G_n^H} \quad \mathbf{b_{n-1-D}^T}] \end{aligned} \tag{18}$$

and $\mathbf{H}$ is the Hermitian transpose of a matrix.

Similar to the analysis given in [6], the MMSE is found to be equal to:

$$\text{MMSE} = 1 - \mathbf{V^H A^{-1} V} = 1 - \mathbf{V^H U_{opt}} \tag{19}$$

where $\mathbf{U_{opt}}$ gives the optimum settings of the forward and feedback coefficients of the DFE receiver, $\mathbf{A} = E(\mathbf{X_n X_n^H})$ and $\mathbf{V} = E(\mathbf{X_n} b_{n-D}^0)$.

To calculate the MMSE, matrix $\mathbf{A}$ and vector $\mathbf{V}$ need to be calculated. Vector $\mathbf{V}$ could be written as:

$$\mathbf{A_2} = \begin{bmatrix} V_0((D+1)T_b) & \cdots & V_0((D+N_f)T_b) \\ V_0((D+1)T_b - \Delta) & \cdots & V_0((D+N_f)T_b - \Delta) \\ \vdots & & \vdots \\ V_0((D+1)T_b - N_z\Delta) & \cdots & V_0((D+N_f)T_b - N_Z\Delta) \end{bmatrix} \tag{22}$$

$$\mathbf{A_1} = \begin{bmatrix} E(g(nT_b)g(nT_b)^*) & \cdots & E(g(nT_b)g(nT_b - N_z\Delta)^*) \\ E(g(nT_b - \Delta)g(nT_b)^*) & \cdots & E(g(nT_b - \Delta)g(nT_b - N_z\Delta)^*) \\ \vdots & & \vdots \\ E(g(nT_b - N_z\Delta)g(nT_b)^*) & \cdots & E(g(nT_b - N_z\Delta)g(nT_b - N_z\Delta)^*) \end{bmatrix} \tag{23}$$

$$\mathbf{V} = E(\mathbf{X_n}\, b^0_{n-D}) = \begin{bmatrix} E(g(nT_b)b^0_{n-D}) \\ E(g(nT_b-\Delta)b^0_{n-D}) \\ \vdots \\ E(g(nT_b-N_z\Delta)b^0_{n-D}) \\ E(b^0_{n-1-D}b^0_{n-D}) \\ \vdots \\ E(b^0_{n-N_f-D}b^0_{n-D}) \end{bmatrix} = \begin{bmatrix} V_0(DT_b) \\ V_0(DT_b-\Delta) \\ \vdots \\ V_0(DT_b-N_z\Delta) \\ 0 \\ \vdots \\ 0 \end{bmatrix} \quad (20)$$

and the $\mathbf{A}$ matrix could be calculated as:

$$\mathbf{A} = E(\mathbf{X_n X_n^H}) = \begin{bmatrix} E(\mathbf{G_n G_n^H}) & E(\mathbf{G_n b^T_{n-1-D}}) \\ E(\mathbf{b_{n-1-D} G_n^H}) & \mathbf{I} \end{bmatrix} = \begin{bmatrix} \mathbf{A_1} & \mathbf{A_2} \\ \mathbf{A_2^H} & \mathbf{I} \end{bmatrix} \quad (21)$$

where (see top of previous page)

$$E(g(nT_b - i\,\Delta)g(nT_b - j\,\Delta)^*) = \sum_{k=0}^{N}\sum_{\lambda} V_k(\lambda T_b - i\,\Delta)V_k(\lambda T_b - j\,\Delta)^* + 2\,\sigma_n^2 R_{00}((j-i)\Delta) \quad (24)$$

σ_n^2 is the variance of AWGN, and $\mathbf{I}$ is a N_f by N_f identity matrix.

REFERENCES

[1] K. S. Gilhousen, I. M. Jacobs, R. Padovani, A. J. Viterbi, L. A. Weaver, Jr., and C. E. Wheatley III, "On the capacity of a cellular CDMA system," *IEEE Trans. Vehic. Technol.*, vol. 40, no. 2, pp. 303–312, May 1991.

[2] A. Salmasi and K. S. Gilhousen, "On the system design aspects of code division mulitple access (CDMA) applied to digital cellular and personal communications networks," in *IEEE 41st Vehic. Technol. Conf.*, 1991, pp. 57–62.

[3] R. Lupas and S. Verdu, "Near-far resistance of multiuser detectors in asynchronous channels," *IEEE Trans. Commun.*, vol. 38, no. 4, pp. 496–508, Apr. 1990.

[4] Z. Xie, R. T. Short, and C. K. Rushforth, "A family of suboptimum detectors for coherent multiuser communciations," *IEEE J. Select. Areas Commun.*, vol. 8, no. 4, pp. 683–690, May 1990.

[5] R. Kohno, H. Imai, M. Hatori, and S. Pasupathy, "Combination of an adaptive array antenna and a canceller of interference for direct-sequence spread-spectrum multiple-access system," *IEEE J. Select. Areas Commun.*, vol. 8, no. 4, pp. 675–682, May 1990.

[6] M. Abdulrahman and D. D. Falconer, "Cyclostationary crosstalk suppression by decision feedback equalization on digital subscriber loops," *IEEE J. Select. Areas Commun.*, vol. 10, no. 3, pp. 640–649, Apr. 1992.

[7] G. L. Turin, "Introduction to spread-spectrum antimultipath techniques and their application to urban digital radio," *Proc. IEEE.*, vol. 68, no. 3, pp. 328–353, Mar. 1980.

[8] B. R. Petersen and D. D. Falconer, "Minimum mean square equalization in cyclostationary and stationary interference—Analysis and subscriber line calculations," *IEEE J. Select. Areas Commun.*, vol. 9, no. 6, pp. 931–940, Aug. 1991.

[9] D. D. Falconer, M. Abdulrahman, N. W. K. Lo, B. R. Petersen, and A. U. H. Sheikh, "Advances in equalization and diversity for portable wireless systems," *Digit. Signal Process.*, vol. 3, no. 3, pp. 148–162, Mar. 1993.

[10] M. Abdulrahman, D. D. Falconer, and A. U. H. Sheikh, "Equalization for interference cancellation in spread spectrum multiple access systems," in *IEEE 42nd Vehic. Technol. Conf.*, 1992, pp. 71–74.

[11] M. Abdulrahman, A. U. H. Sheikh, and D. D. Falconer, "DFE convergence for interference cancellation is spread spectrum multiple access systems," in *IEEE 43rd Vehic. Technol. Conf.*, 1993, pp. 807–810.

[12] U. Madhow and M. L. Honig, "Minimum mean squared error interference suppression for direct sequence spread-spectrum code division multiple access," in *Proc. ICUPC*, 1992.

[13] U. Madhow and M. L. Honig, "Error probability and near-far resistance of minimum mean squared error interference suppression schemes for CDMA," in *Proc IEEE GLOBECOM*, 1992, pp. 1339–1343.

[14] J. G. Proakis, *Digital Communications.* New York: McGraw-Hill, 1989.

[15] G. C. Goodwin and K. S. Sin, *Adaptive Filtering Prediction and Control.* Englewood Cliffs, NJ: Prentice-Hall, 1984.

Majeed Abdulrahman (S'89) received the B.A.Sc. degree in electrical engineering from the University of Ottawa, Ottawa, Ontario, Canada in 1988, and the M.Eng. and Ph.D. degrees in electrical engineering from Carleton University, Ottawa, in 1989 and 1994, respectively.

During his undergraduate studies at the University of Ottawa, he worked with Bell-Northern Research, Ottawa. He was the recipient of the NSERC federal scholarship for the maximum period of four years during his graduate studies at Carleton University. He is currently working with Bell-Northern Research in Ottawa. His research interests are in spread spectrum systems and its applications for indoor wireless communications, cellular system design, adaptive equalization, digital subscriber lines, and ISDN.

Asrar U. H. Sheikh (M'70–M'82–SM '84) received the B.S. degree from the University of Engineering and Technology, Lahore, Pakistan, and received the M.Sc. and Ph.D. degrees from the University of Birmingham, Birmingham, England, in 1966 and 1969, respectively.

He held lectureships at universities in Pakistan, Iran, and Libya between 1969 and 1975 before returning to Birmingham as a Research Fellow. In 1981, he joined Carleton University, Ottawa, Ontario, Canada as an Associate Professor. At present, he is a Professor in the Department of Systems and Computer Engineering. From 1988 to 1991, he was Associate Chairman for Graduate Studies. Presently, he is the Director of the PCS Laboratory at Carleton University. Currently, he is leading a major research project on Universal, Secure, and Efficient Telecommunications (USETS). He has published extensively in the areas of mobile and personal communications, and impulsive noise. His current interests are in signal processing in communications, digital cellular systems, DS and FH spread spectrum systems, and personal communication systems. He has been consultant to several private and government agencies. He is a co-recipient of the Paul Adorian Premium from the IERE for his work on impulsive noise characterization. He is also a Professional Engineer licensed in the Province of Ontario.

David D. Falconer (M'68–SM'83–F'86), for photograph and biography, see this issue, p. 653.

IEEE JOURNAL ON SELECTED AREAS IN COMMUNICATIONS, VOL. 12, NO. 5, JUNE 1994

Effects of Diversity, Power Control, and Bandwidth on the Capacity of Microcellular CDMA Systems

Ahmad Jalali and Paul Mermelstein

***Abstract*—We evaluate the capacity and bandwidth efficiency of microcellular CDMA systems. Power control, multipath diversity, system bandwidth, and path loss exponent are seen to have major impact on the capacity. The CDMA system considered uses convolutional codes, orthogonal signaling, multipath/antenna diversity with noncoherent combining, and fast closed-loop power control on the uplink (portable-to-base) direction. On the downlink (base-to-portable), convolutional codes, BPSK modulation with pilot-signal-assisted coherent reception, and multipath diversity are employed. Both fast and slow power control are considered for the downlink. The capacity of the CDMA system is evaluated in a multicell environment taking into account shadow fading, path loss, fast fading, and closed-loop power control. Fast power control on the downlink increases the capacity significantly. Capacity is also significantly impacted by the path loss exponent. Narrowband CDMA (system bandwidth of 1.25 MHz) requires artificial multipath generation on the downlink to achieve adequate capacity. For smaller path loss exponents, which are more likely in microcellular environments, artificial multipath diversity of an order of as high as 4 may be needed. Wideband CDMA systems (10 MHz bandwidth) achieve greater efficiencies in terms of capacity per MHz.**

I. INTRODUCTION

CODE division multiple access (CDMA) is one of the techniques under consideration for personal communications systems [1], [2]. Power control and multipath diversity are among the major issues in system design which significantly affect the capacity of a CDMA system. Other factors which have a significant impact on the capacity are the system bandwidth and the path loss exponent. To achieve high-bandwidth efficiency and low system complexity, the impact of these factors on the system capacity needs to be assessed. The CDMA system under consideration uses convolutional codes, orthogonal signaling, multipath and antenna diversity with noncoherent combining, and closed-loop power control on the uplink (portable-to-base). The downlink (base-to-portable) employs convolutional codes and BPSK modulation with pilot-assisted coherent reception. The performance of the uplink transceiver has been studied in [3]–[5], but the performance of the downlink has received less careful attention. We focus on detailed capacity analysis of the uplink and downlink and assess the impact on the capacity of these factors. Capacity is defined as the number of simultaneous variable rate voice calls that can be supported without exceeding a given probability of outage. In this paper, system bandwidths of 1.25 and 10 MHz are considered for inbuilding and microcellular applications.

Manuscript received June 28, 1993; revised November 21, 1993. This work was supported in part by the Canadian Institute for Telecommunications Research under the Canadian government's Networks of Centers of Excellence program.

A. Jalali is with Bell Northern Research, Verdun, Quebec H3E 1H6 Canada.
P. Mermelstein is with INRS-Telecommunication, Verdun, Quebec H3E 1H6 Canada.

IEEE Log Number 9215584.

The objective of this paper is restricted to evaluating the capacity of various system design alternatives. Numerous other considerations such as complexity, power dissipation, and service requirements enter into decisions on selecting a multiple access structure for system development. Such decisions are beyond the scope of this contribution.

Closed-loop power control based on received signal strength, where power is adjusted at the portable every 1.25 ms based on commands from the base, has been suggested for the uplink direction. This power control scheme is intended to overcome the uplink near/far problem. It is, however, fast enough to partially mitigate multipath fading at low Doppler frequencies as well [6]. Power control based on signal-to-interference power ratios (SIR) may result in improved performance in the uplink direction [7]. In the downlink, power control may also be used to increase the system capacity [8]–[10]. Without power control on the downlink, the base station transmits the same amount of power to all portables. However, those portables which are near the base station receive a stronger signal from their own base and less interference from other bases. By reducing the power transmitted to these portables yet satisfying the SIR requirements, the overall capacity is increased [8]–[10]. The power control in the downlink may be designed to follow the slow variations in received SIR due to shadow fading and path loss (slow power control), or to adapt to both slow variations as well as fast variations due to fast multipath fading (fast power control). In this paper, both slow and fast power control are considered on the downlink.

Diversity is a major contributor to increasing the capacity in wireless access systems. At the base station, antenna diversity may be employed to achieve the desired capacity. At the portable, multipath diversity is more feasible. For inbuilding and microcellular systems, which we are considering, the rms delay spread is small enough that the channel of the 1.25 MHz systems may be modeled as narrowband (a single fading path) [11], [12]. Narrowband CDMA systems require artificial multipath generation on the downlink in order to achieve reasonable capacity. Artificial multipath is generated by using multiple transmit antennas at the base adequately separated, and with delays of at least one chip duration inserted between them. Capacity of the narrowband system is evaluated as a function of order of multipath diversity as well as the path loss exponent. In microcellular systems, the path loss exponent may be in the range of 2 to 3 [13]. It is found that capacity decreases significantly as the path loss exponent decreases.

The choice of system bandwidth has a wide impact on the bandwidth efficiency of a CDMA system, defined as the number of simultaneous calls supported per MHz of assigned spectrum in each cell. Wideband CDMA systems have been proposed previously [14]. However, the large differences in the designs associated with the wideband and narrowband proposals have prevented a direct capacity comparison. Wideband systems are generally preferable due to the greater amount of statistical multiplexing they can support resulting from the reduced variance of the momentary transmission requirements about the average requirements. In addition, due to the higher number of users, the variations about average of the multiuser interference power will be smaller which in turn provides higher capacity. However, sharing spectrum and power on systems provided by different service providers becomes quite difficult. Thus, practical limits may be placed on the available system bandwidth not only by the cost of the implementation technology but also by the regulatory aspects of subdividing the available spectrum so that several service providers may coexist in a limited geographic area. The results of this study quantify the impact on bandwidth efficiency of spectrum division as opposed to spectrum sharing.

Qualitative results are presented in [15] on the performance of a CDMA cellular system. A narrowband CDMA cellular system with different coding and modulation than considered here has been studied in [16]. In assessing the impact of these factors on the capacity, we consider details such as shadow fading, path loss exponent, fast fading, channel model, system bandwidth, and the effects of the power control algorithms on the received signal and the multiuser interference. A combination of analysis and simulation is used to arrive at representative results. We first determine the signal-to-noise (SNR) ratio needed to achieve the required grade of service such as the maximum tolerable bit error rates. Transmissions from randomly distributed portables are then simulated to determine the multiuser interference statistics. These statistics are then used to compute the maximum number of users per cell that can be supported for a given probability of outage.

A brief review of spread spectrum needed for our discussion is given in Section II. Section III describes the transceiver structure and the simulation model used for performance evaluation. Results on the performance of the transceivers are presented in Section IV. Section V describes the computation of the capacity for the uplink and downlink directions. Capacity estimates are presented in Section VI. The paper is concluded in Section VII.

II. Spread Spectrum and CDMA Signals

We consider quaternary direct-sequence systems where the same data is modulated onto the in-phase and quadrature channels using different PN codes. In this section, we quote the results from [17] which are relevant to our study. We consider a channel with $L \geq 1$ resolvable paths. We let 0 denote the index of the desired user under consideration. The received quaternary direct-sequence signal corresponding to the pth path of the zeroth user may be represented by:

$$\begin{aligned} d_0p(t) = {} & \sqrt{S_{0p}}m_0(t)c_0(t)\cos(\omega t + \theta_{0p}) \\ & + \sqrt{S_{0p}}m_0(t)c_0^q(t)\sin(\omega t + \theta_{0p}) + I_{0p}(t) \quad (1) \end{aligned}$$

where S_{ij} is the received signal power from the jth path of the ith user, $m_i(t)$ is the ith user's data modulation, $c_i(t)$ and $c_i^q(t)$ are the ith user's spreading waveforms for the in-phase and quadrature channels, θ_{ij} is the carrier phase offset assumed to be uniformly distributed in $[0, 2\pi]$, and $I_{0j}(t)$ is the multiuser interference on the jth path of the zeroth (desired) user. The spreading waveform of the ith user is given by:

$$c_i(t) = \sum_{k=-\infty}^{\infty} c_i(k)\psi(t - kT_c) \quad (2)$$

where $\psi(t)$ is the chip waveform which we take to be a square pulse of chip duration T_c, and $\{c_i(k)\}$ is the in-phase pseudonoise (PN) binary code sequence.

Effect of using short PN codes in conjunction with orthogonal waveforms on the performance of a CDMA system has been studied in [18]. Here, we consider the PN codes with long periods. Because of the large period of PN codes considered, the PN sequence $\{c_i(k)\}$ is assumed to be a random sequence. Moreover, we assume that the data symbols $m_i(t)$ form a random sequence and can, therefore, be incorporated into the PN sequence. Then, assuming that relative delays of the different resolvable paths are at least one chip duration, the interference signal corresponding to the jth path of the ith user (interferer) $i \geq 1$ is given by:

$$\begin{aligned} Y_{ij}(t) = {} & \sqrt{S_{ij}}r_{ij}(t - \tau_{ij})\cos(\omega t + \theta_{ij}) \\ & + \sqrt{S_{ij}}r_{ij}^q(t - \tau_{ij})\sin(\omega t + \theta_{ij}) \end{aligned}$$

where $r_{ij}(t)$ is defined the same as (2) but with a random sequence $\{r_{ij}(k)\}$, and τ_{ij} is the chip delay of the jth path of the ith user with respect to the desired user, assumed to be uniformly distributed in $[0, T_c]$. The multiuser interference signal seen on the pth path of the zeroth (desired) user is then given by:

$$I_{0p}(t) = \sum_{i=1}^{N}\sum_{j=1}^{L} Y_{ij}(t) + \sum_{\substack{j=1 \\ j\neq p}}^{L} Y_{0j}(t) \quad (3)$$

where N is the total number of interferers. The second term in (3) corresponds to the self interference caused by the multipath components of the desired user. The received signal power, S_{ij}, is given by:

$$S_{ij} = x_i^2\delta_i\alpha_{ij}^2\Lambda_i^2 \quad (4)$$

where $x_i^2, \delta_i, \alpha_{ij}^2$, and Λ_i^2 are the power variations due to fast power control, slow power control, fast fading on the jth path, and shadow (slow) fading and path loss, respectively. Note that we have assumed that the shadow fading and path loss are the same for all multipath components of the same siganl. Then, the relative powers of the different multipath components are given by ratios of $\overline{\alpha_{ij}^2}$, where $\overline{\alpha_{ij}^2}$ is the average of α_{ij}^2. In fact, we take the sum of $\overline{\alpha_{ij}^2}$ over all j to be 1.

A. Noncoherent Reception

In the uplink of the CDMA system studied in this paper, a correlation receiver with noncoherent square law combining is employed. It is also possible to provide symbol-assisted coherent reception on the uplink. In Section IV, where discuss the transceiver performance, we will discuss the gains in using coherent reception on the uplink of the particular CDMA system under consideration. The components of the signal energy corresponding to the in-phase channel are recovered by correlating the received signal with $2c_0(t)\cos\omega t$ and $2c_0^q(t)\cos\omega t$, and those of the quadrature channel similarly by correlating with $2c_0(t)\sin\omega t$ and $2c_0^q(t)\sin\omega t$. For M-ary signaling, there are M correlator outputs corresponding to each diversity branch. The output of the pth diversity branch of the 0th user corresponding to the transmitted signal is:

$$\nu_{0p} = (2\sqrt{S_{0p}}T_s\cos\theta_{0p} + X_{0p}^I)^2 + (2\sqrt{S_{0p}}T_s\sin\theta_{0p} + X_{0p}^Q)^2 \quad (5)$$

where X_{0p}^I and X_{0p}^Q are the in-phase and quadrature components of the multiuser interference and T_s is the symbol duration. The variances of X_{0p}^I and X_{0p}^Q are given by $2\sigma^2_{X_{0p}^I}$ and $2\sigma^2_{X_{0p}^Q}$, respectively. It follows from (1) and (3) and the results in [17] that, conditioned on the received signal power and on the relative chip delays and phases of the different interferers,

$$\sigma^2_{X_{0p}^I} = \sum_{i=1}^{N}\sum_{j=1}^{L} GS_{ij}[R_\psi^2(\tau_{ij}) + R_\psi^2(T_c - \tau_{ij})] + \sum_{\substack{j=1\\ j\neq p}}^{L} GS_{0j}[R_\psi^2(\tau_{0j}) + R_\psi^2(T_c - \tau_{0j})] \quad (6)$$

where G is the processing gain. $\sigma^2_{X_{0p}^Q}$ is the same as (6) but with sine replaced by cosine. $R_\psi(s)$, the partial autocorrelation of the chip waveform, is given by:

$$R_\psi(s) = \int_0^s \psi(t)\psi(t + T_c - s)dt, \quad 0 \le s < T_c. \quad (7)$$

Note that $R_\psi(s) = s$ for square chip waveforms.

The second term in (6) represents the self interference caused by multipath components of the desired user's signal. The third term is the self interference caused by the quadrature component of the signal that is being demodulated. We assume that the interference term at the output of the correlator is Gaussian. This Guassian assumption is valid for large processing gains [17]. Although we are assuming that the interference term at the output of the correlator is Gaussian, its variance given by (6) is taken to be time varying. In fact, (6) is a function of variations in the interference power due to the fast fading, shadow fading, path loss, relative chip delays, and phases of the different users. This is made more clear in Section V, where we discuss capacity computation.

B. Coherent Reception

In the downlink of the CDMA system, pilot-signal-assisted coherent reception is considered. Orthogonal PN codes are used. Since signals are received synchronously at the portable, the orthogonality of the PN codes implies that signals within one diversity branch do not create any interference to one another. The total transmitted power by the base is normalized to 1. A fraction of the power is allocated to the pilot signal. A fraction of the remainder of the power is allocated to each portable communicating with that base. In the downlink, S_{ij} denotes the power received from the jth path of the signal transmitted by the ith base station. Since the signals received on the jth path from the ith base station are synchronous, S_{ij} may be written as a sum of powers corresponding to the pilot signal and to each portable. We let S_{ij}^π and S_{ij}^k denote the components of S_{ij} which belong to the pilot signal and the kth portable, respectively. Moreover, we let 0 be the index of the portable under consideration and the base station with which the portable is communicating.

The received signal on the pth diversity branch is correlated with $2c_0(t)\cos(\omega t+\widehat{\theta_{op}})$ to recover the in-phase signal energy, where $\widehat{\theta_{0p}}$ is the phase estimate generated from the pilot signal. The quadrature component of the signal energy is recovered similarly by correlating the received signal with $2c_0^q(t)\sin(\omega t + \widehat{\theta_{0p}})$. The output of the pth diversity branch corresponding to the transmitted symbol is then given by

$$\nu_{0p} = \pm 2\sqrt{S_{0p}^0}T_s\cos\widetilde{\theta_{0p}} + Z_{0p}^I + Z_{0p}^Q \quad (8)$$

where $\widetilde{\theta_{op}}$ is the error in the phase recovery. Note that in (8) the outputs of the inphase and quadrature channels have been combined. Z_{0p}^I and Z_{0p}^Q are the interference terms from the inphase and quadrature channels at the output of the pth diversity branch. Their variances can be shown, using the results in [17], to be given by

$$\sigma^2_{Z_{0p}^I} = \sigma^2_{Z_{0p}^Q} = \sum_{i=1}^{N_b}\sum_{j=1}^{L} GS_{ij}[R_\psi^2(\tau_{ij}) + R_\psi^2(T_c - \tau_{ij})] + \sum_{\substack{j=1\\ j\neq p}}^{L} GS_{0j}[R_\psi^2(\tau_{0j}) + R_\psi^2(T_c - \tau_{0j})] \quad (9)$$

where N_b is the number of interfering base stations.

III. Transceiver Description

In this section, we describe the channel coding, modulation, and power control algorithms for the CDMA system under consideration. We also present simulation models of the power control algorithms as well as the phase recovery algorithm of the downlink. We assume variable rate voice sources with (uncoded) data rates of 9600 b/s when speech activity is noted and 1200 b/s otherwise. Wireline speech quality at these rates is expected to be achieved in the near future. Frame duration of 20 ms is considered. The processing gains are 128 and 1024 for the 1.25 and 10 MHz systems, respectively.

A. Uplink

The transceiver structure used in the uplink is described in detail in [3], [4]. We employ a combination of convolutional coding and orthogonal signaling. The user information is encoded using a rate 1/3 convolutional code. The code used has constraint length 9. Groups of six coded bits are mapped onto 64-ary orthogonal waveforms. The resulting symbols are spread onto the in-phase and quadrature channel using different PN codes. Two types of block interleaving are possible: bit *interleaving*, where the encoded bits are first interleaved and then mapped onto 64 orthogonal symbols, and *symbol interleaving*, where interleaving follows orthogonal mapping. In this paper, we compare the capacity performance of bit and symbol interleaving techniques. The interleaver table sizes are 16×6 for symbol interleaving and 32×18 for bit interleaving techniques, respectively.

The receiver at the base station employs antenna and multipath diversity with square-law combining. The combiner is followed by deinterleaving and soft decision Viterbi decoding. Closed-loop power control is considered for the uplink. A power control bit is transmitted by the base station to the portable every T_p ms requesting an incremental change in the transmitted power. To generate the power control command bit, the received signal power is estimated from the decision values at the output of the combiner and compared against a threshold. The signal power estimator is similar to that described in [3]. The signal energy per symbol is estimated by:

$$E_s = \frac{1}{n_s}\sum_{j=1}^{n_s} \max_k\{d_{j,k}\} - \frac{\overline{n^2}}{M-1}. \quad (10)$$

$$\overline{n^2} = \frac{1}{n_s}\sum_{j=1}^{n_s} \sum_{k=1, k\neq k_{\max}}^{M} d_{j,k}. \quad (11)$$

$k_{\max}$ is the index of the largest decision value for the given symbol interval, n_s is the number of symbols in the power control measurement interval of T_p ms, and $d_{j,k}$ is the kth decision value at the output of the diversity combiner during the jth symbol interval. The quantity in (10) is compared against a threshold, and a power control bit is generated accordingly to change the transmitted signal power by a fixed increment. Here, the threshold is chosen so that the signal power per antenna is normalized to 1. Possible channel errors are modeled in the simulations by introducing the random errors into the power control command bits.

B. Downlink

In the downlink, the user data is encoded using a rate 1/2 convolutional code followed by a bit interleaver. The interleaved encoded bits are spread onto the in-phase and quadrature channels using different PN codes and transmitted using BPSK. Soft decision Viterbi decoding is employed at the receiver. The PN codes used within the same cell for the different users are chosen to be orthogonal.

As mentioned in Section III-B, coherent reception using a pilot signal is considered in the downlink. The real and imaginary components of the channel impulse response are estimated by averaging the correlator outputs over the sampling interval of T_p ms (24 coded symbols when $T_p = 1.25$ ms), i.e., by:

$$\frac{1}{2n_s}\sum_{j=1}^{n_s}(2\sqrt{S_{0p}^{\pi}}T_s\cos\theta_{0p} + X_{0p}^{I}(j)) \quad (12)$$

$$\frac{1}{2n_s}\sum_{j=1}^{n_s}(2\sqrt{S_{0p}^{\pi}}T_S\sin\theta_{0p} + X_{0p}^{Q}(j)) \quad (13)$$

where S_{0l}^{π} was defined in Section II-B. The pilot signal components in (12) and (13) are recovered as described in Section II-A; they are obtained noncoherently because phase estimates are not yet available. The variance of X_{0p}^{I} is given by $2\sigma^2_{X_{0p}^I}$ where $\sigma^2_{X_{0p}^I}$ is given by (6), that of X_{0p}^{Q} by $2\sigma^2_{X_{0p}^Q}$ where $\sigma^2_{X_{0p}^Q}$ is given by (6) but with sine replaced by cosine.

Power control in the downlink is based on the SIR seen at the portable. We need to estimate the interference powers on each of the diversity branches given by (9). This is done by averaging the square difference of the in-phase and quadrature correlator outputs corresponding to the pth diversity branch. That is, the interference variance is estimated by:

$$\widehat{\sigma^2_{Z_{0p}^I}} = \widehat{\sigma^2_{Z_{0p}^Q}} = \frac{1}{2n_s}\sum_{j=1}^{n_s}(Z_{0p}^{I}(j) - Z_{0p}^{Q}(j))^2 \quad (14)$$

where $Z_{0p}^{I}(j)$ and $Z_{0p}^{Q}(j)$ are the interference observations at the jth sampling instance.

The received signal energy from one diversity path is simulated by averaging the squares of the correlator outputs given by (8) and substracting from it the estimate of the variance of the interference at the output of correlator given by (14). The total received signal energy is obtained by summing all diverse branches. The SIR is estimated using the estimates of the signal and interference powers. The resulting SIR estimate is then compared against a threshold, and a power control bit is generated. The threshold is chosen so that the signal power summed over all diversity branches is normalized to 1. The frequency with which the power control bits are generated depends on whether fast or slow power control is being employed. For fast power control, intended to control fast fading, power control bits are generated every 1.25 ms. In simulating the transceiver, maximal ratio combining is used [22]. Estimates of the channel gain are made using (12) and (13) and multiplied by the correlator output before adding the outputs of the different diversity branches.

IV. Transceiver Performance

This section presents simulation results on the performance of the transceiver simulation models described in Section III. The figure of merit is the required SNR to acheive a bit error ratio (BER) of less than 10^{-3}. The required SNR values reported in this section will be used in Section VI to evaluate the system capacity. Note that the required SNR values reported here do not take the effect of fast power control gain into account. In other words, the signal power used in computing the SNR is taken before the fast power control gain. The fast power control gain is taken into account in Section V when the capacity is computed.

The performance of the uplink has been studied extensively in [3]–[5]. In particular, it has been shown that symbol interleaving results in at least 1 dB improvement in the required SNR [4], [5]. Our simulations also confirm this result [20]. Since we are mainly interested in inbuilding and microcellular applications at a carrier frequency of about 2 GHz, we only present the transceiver performances for a slowly fading channel with Doppler frequency of 2 Hz. This Doppler frequency corresponds to movements of about 1 km/hr at the carrier of 2 GHz. The transceiver performance was studied for power control step sizes of 0.5 and 1 dB. The 0.5 dB power control step size results in smaller required SNR for the slowly fading channel considered here [20].

A. Channel Models

In this section, we define the channel models that are used in the remainder of the paper. We consider system bandwidths of 1.25 and 10 MHz. For the microcellular and indoor applications considered, we model the channel of the 1.25 MHz system by a single fading path and that of the 10 MHz system by three resolvable fading paths. We refer to the 1.25 MHz system as *narrowband* and the 10 MHz as *wideband*. For the wideband system, we need to consider a power distribution for the three resolvable paths. We approximate the possible power distributions on the three paths by the four power distributions of $(0, -3, -6)$ dB, $(0, -3, -10)$ dB, $(0, -6, -10)$ dB, and $(0, -10, -10)$ dB, where the powers of the three paths are normalized by that of the first path. In the absence of statistical information obtained from actual channel measurements, we assume that these four power distributions are equally probable.

B. Simulation Results

The following assumptions apply to the simulation results that are presented in the remainder of this paper.

- Power control command bits are generated every $T_p = 1.25$ ms when fast power control is employed.
- The power control processing and transmission delays are $2T_p$.
- The power control step size is set at 0.5 dB.
- The channel error rate suffered by the power control bits is assumed to be 10%.
- On the downlink, channel impulse response estimates are made every T_p as described in Section III-B.
- Each resolvable path is assumed to undergo independent Rayleigh fading. Jakes' model [19] is used to generate a sequence of time-correlated Rayleigh random numbers to simulate the channel. Doppler frequency of 2 Hz is simulated.

Artificial multipath diversity is generated on the downlink as described in Section I. For the narrowband system n transmitting antennas result in n equal power independently fading signals at the portable. Table I provides the SNR required to achieve a BER of less than 10^{-3} for the downlink

TABLE I
DOWNLINK REQUIRED SNR, DOPPLER FREQUENCY OF 2 Hz

Channel Model	No. Transmit Antennas at the Base	SNR (dB), Slow Power Control	SNR (dB), Fast Power Control
1.25 MHz System		(Pilot-SNR = 9 dB)	(Pilot-SNR = 6 dB)
Single Path	2	15.7	5.5
	3	11.6	4.7
	4	9.1	4.7
10 MHz System		(Pilot-SNR = 12 dB)	(Pilot-SNR = 12 dB)
Three Paths			
(0, -3, -6) dB	2	7.2	5.2
(0, -3, -10) dB	2	7.7	5.5
(0, -6, -10) dB	2	8.1	5.7
(0, -10, -10) dB	2	8.8	6.1

TABLE II
UPLINK REQUIRED SNR PER ANTENNA, TWO RECEIVE ANTENNAS, DOPPLER FREQUENCY OF 2 Hz

Channel Model	SNR (dB), Symbol Interleaving	SNR (dB), Bit Interleaving
1.25 MHz System Single Path	2.85	3.9
10 MHz System Three Paths		
(0, -3, -6) dB	4.2	5.3
(0, -10, -10) dB	4.2	5.3

transceiver as a function of the number of transmit antennas used at the base station, power control strategy, and the channel model assumed. For the wideband (10 MH) system, two transmit antennas are assumed at the base. This results in repeating each resolvable path in the channel twice, all assumed to be independently fading. Therefore, for the 10 MHz system, the portable's Rake receiver sees six independently fading paths. Pilot-SNR in Table I is the value of SNR seen when demodulating the pilot signal. The larger the value of pilot-SNR, the smaller the required SNR due to better coherent reception (which increases capacity). However, large values of pilot-SNR require that large fractions of transmitted power be allocated to the pilot signal (which reduces capacity). The values of pilot-SNR is Table I correspond to those that result in approximately the best capacity. Note that the fraction of power allocated to the pilot signal must be computed such as to ensure that the given pilot-SNR is observed in at least 95% of the cell. The procedure is described in more detail in [21]. In the case of the wideband system, a higher value of pilot-SNR is possible which results (Table I) in a further improvement in the required SNR compared to the narrowband (1.25 MHz) model.

Table II presents required SNR to achieve BER of less than 10^{-3} for the uplink transceiver with bit and symbol interleaving. Table II shows that, as the order of multipath diversity is increased, the required SNR is also increased. Moreover, the required SNR values are practically the same for the two power distributions of $(0, -3, -6)$ dB and $(0, -10, -10)$ dB for the wideband (10 MHz) system. The reason is that the combination of fast power and two-antenna diversity appear quite effective to mitigate fading on the uplink, and the extra multipath results in little diversity gain. The increase in the required SNR with increased diversity is then due to the well-known noncoherent combining loss of the receiver [22]. As mentioned earlier, symbol-assisted coherent reception may be employed on the uplink. It has been reported [23] that, with symbol-assisted coherent reception, about 2 dB reduction in required SNR is achieved when bit interleaving is employed; the coherent reception gain is about 1 dB when symbol interleaving is used. If coherent reception is used on the uplink, then the uplink capacity would increase by 25% and 60% in the cases of symbol interleaving and bit interleaving, respectively. As will be discussed in Section VI, the capacity of the CDMA system under consideration is downlink limited even when using noncoherent combining at the base.

V. Capacity Estimation

This section presents the procedures used to estimate the uplink and downlink capacities.

A. Uplink

To compute the uplink capacity, we consider a grid of 11 × 11 square cells. Our simulation assumes that N portables are uniformly distributed within the given cell layout. Probability of outage is defined as the probability that the BER exceeds 10^{-3}. We compute the number of simultaneous users N that can be accommodated within the grid, subject to the constraint that the probability of outage be less than 0.01. Capacity is defined as the number of simultaneous users per cell, which is N divided by the number of cells in the cell layout. To this end, we compute the interference seen at the receiver of the base station in the cell in the center of the layout from the N users in the grid of 11 × 11 cells. We are assuming that the interference contributed by users more than five cells away from the center cell can be ignored. Note that the variance of the interference given by (6) depends on the diversity branch index. In order to simplify the computations, we change N to $N+1$ and ignore the second term. This approximation may result in overestimation or underestimation of the capacity by at most one user. Then, the variance of the interference term at the output of the correlator is the same for all diversity branches and is given by:

$$\sigma^2 = \sum_{i=1}^{N+1}\sum_{j=1}^{L} GS_{ij}[R_\psi^2(\tau_{ij}) + R_\psi^2(T_C - \tau_{ij})]. \tag{15}$$

Since the received signal power per antenna is normalized to 1 in the transceiver simulations, the SIR per antenna is given by:

$$\mathrm{SIR} = \frac{1}{G}\,\frac{T_s^2}{\sigma^2}. \tag{16}$$

Note that the T_s^2 in the numerator of (16) is due to the fact that σ^2 is the variance of the interference at the output of the correlator and must be normalized by T_s^2 to give the interference power. In order not to exceed the probability of outage, we must have:

$$\mathrm{Prob}\left(\mathrm{SIR} > \frac{(\mathrm{SNR})_R}{G}\right) > 0.99. \tag{17}$$

where $(\mathrm{SNR})_R$ is the required SNR per antenna to achieve the target BER. Using (15) and (16), the computation of the probability in (17) is reduced to that of:

$$\mathrm{Prob}\left(\sum_{i=1}^{N+1}\sum_{j=1}^{L} S_{ij}g(\tau_{ij}) < \frac{G}{(\mathrm{SNR})_R}\right) > 0.99 \tag{18}$$

where, for square chip waveforms, we have:

$$g(\tau_{ij}) = \frac{1}{T_c^2}(R_\psi^2(\tau_{ij}) + R_\psi^2(T_c - \tau_{ij})) = \left(1 + \frac{2\tau_{ij}^2}{T_c^2} - \frac{2\tau_{ij}}{T_c}\right).$$

To completely specify the received signal power S_{ij} which was defined by (4) in Section II-A, we need to specify the distribution on the shadow fading as well as the slow power control gain δ_i. The shadow fading is assumed to be log-normally distributed. Then, the path loss and shadow fading between the ith portable and k base station is proportional to:

$$\Lambda_{ik}^2 = \frac{10^{Y_{ik}/10}}{r_{ik}^e}$$

where Y_{ik} is a Gaussian random variable, r_{ik} is the distance between the portable and the base station, and e is the path loss exponent. To determine the slow power control gain, we need to specify the slow power control algorthim as well as the algorithm that chooses the base station with which the portable communicates. For each portable, we compute the signal strength of the pilot signal received from the nearest nine base stations and assign the portable to the base station with the strongest signal.

Let k_0 be the index of the base station to which the portable is assigned. We assume that the power control algorithm equalizes the shadow fading and path loss perfectly. This assumption is reasonable because the shadow fading is expected to vary much more slowly than multipath fading and the variations in signal power due to power control are taken into account by including the effects of the fast power control algorithm into our simulations. Since the interferer is power controlled by its own base station, i.e., the base station indexed by k_0, the slow power control gain is the inverse of shadow fading and path loss in the direction from the interferer to base k_0. S_{ij} is then given by:

$$S_{ij} = x_i^2\alpha_{ij}^2\chi_i\Lambda_{io}^2/\Lambda_{ik_0}^2 \tag{19}$$

where index 0 denotes the cell in the center of the cell layout where we are computing the interference, and χ_i is a two-valued random variable which represents the speech activity factor and equals 1 with probability p_{saf} and 1/8 otherwise.

To compute the probability in (18), we need to compute the probability that the given portable is an out-of-cell interferer (an interferer which is communicating with a base station other than the one in the center cell), the cumulative probability distribution function (CDF) of the random variable $\Lambda_{io}^2/\Lambda_{ik_0}^2$ for the out-of-cell interferers, and the CDF of

$$x_i^2 \quad \text{and} \quad \sum_{j=1}^{L}\alpha_{ij}^2 g(\tau_{ij}). \tag{20}$$

For an in-cell interferer, $\Lambda_{io}^2/\Lambda_{ik_0}^2 = 1$ because of the assumption of perfect average power control. For the out-of-cell interferers, the two random variables in (20) are independent because an out-of-cell interferer is being power controlled by another cell whose fading is independent of the second term in (20). However, for an in-cell interferer, i.e., an interferer which is communicating with the center cell where we are computing the interference, the two terms in (20) are dependent and we need to compute the CDF of the product of the two terms. The CDF of the two terms in (20) for these two cases are obtained via simulating the closed-loop power control algorithm.

The CDF of $\Lambda_{io}^2/\Lambda_{ik_0}^2$ and the probability that a given portable is an out-of-cell interferer are obtained separately via simulations. We place a portable randomly in the given cell layout, assign the portable to a base station, and gather statistics on $\Lambda_{io}^2/\Lambda_{ik_0}^2$. Once the CDF of the random variables and the probability that a given portable is an out-of-cell interfer are obtained, then the probability in (18) can be computed accurately via Monte-Carlo simulation in a reasonable computing time.

B. Downlink

In the downlink direction, the base stations allocate a fraction of the total transmitted power to each portable with which they are communicating based on the SIR seen by that portable. Without loss of generality, we normalize the maximum power transmitted by a base station to 1. In computing the capacity, we consider a cell layout of $K \times K$ square cells and focus on portables that are communicating with the cell in the center of the layout. The algorithm that assigns a base station to a given portable was described in Section V-A.

We first compute the capacity with slow power control. In order to compute the downlink capacity, we must determine the fraction of power that is allocated to each portable. Let δ_i be the fraction of total power allocated to the ith user by the base when there is speech activity (the user transmits at the highest rate). The SIR seen by the ith portable, averaged over fast fading, is given, using (9), by:

$$\mathrm{SIR} = \delta_i \sum_{m=1}^{L} \cdot \frac{\overline{\alpha_{0m}^2}\Lambda_{i0}^2 P_o}{\sum_{\substack{j=1\\ j\neq m}}^{L} \overline{\alpha_{0j}^2}\Lambda_{i0}^2 g(\tau_{0j}) P_o + \sum_{k=1}^{N_b}\sum_{j=1}^{L} \overline{\alpha_{kj}^2}\Lambda_{ik}^2 g(\tau_{kj}) P_k}. \quad (21)$$

where N_b is the number of interfering cells, 0 is the index of the base station with which the portable is communicating, and P_k is the total power that is transmitted by the kth base station. Note that P_k is the sum of the fractions of power allocated to the pilot signal and to each portable communicating with the kth base station. Note that, as mentioned previously, the maximum value of P_k is taken to be 1. In this paper, a cell layout of 5×5 cells is considered. Then, $N_b = 5^2 - 1$. Also, note that if artificial multipath is generated, then the relative chip delays in the denominator of (21) will be the same for the different copies of the same natural multipath component. The approximation that the interference in the denominator of (21) is averaged over fast fading overestimates the capacity. This approximation, however, becomes accurate for large number of users.

To ensure that BER is less than 10^{-3}, SIR must be greater than the required $(\mathrm{SNR})_R/G$. That is, we must have:

$$\frac{f_i(\mathrm{SNR})_R}{G} \le \delta_i$$

where f_i is the reciprocal of the summation in (21). As mentioned, δ_i is the fraction of power allocated to a portable during the highest data rate. The actual fraction of power allocated to a portable is $\delta_i\chi_i$, where χ_i represents speech activity as defined in Section V-A. Note that since the total power transmitted by the base station was normalized to 1, the sum of $\delta_i\chi_i$ over all portables must be less than β, where $1-\beta$ is the fraction of power allocated to the pilot signal. The total number of portables that can be simultaneously supported while achieving a BER of less than 10^{-3} at least 99% of the time must then satisfy:

$$\mathrm{Prob}\left(\sum_{i=1}^{N}\chi_i f_i \le \frac{\beta G}{(\mathrm{SNR})_R}\right) > 0.99. \quad (22)$$

So far, we have considered slow or average power control. Let x_i^2 be the gain of the fast power control algorithm for the ith user. Then, we must have:

$$\sum_{i=1}^{N}\chi_i x_i^2 \delta_i \le \beta$$

which implies that

$$\mathrm{Prob}\left(\sum_{i=1}^{N}\chi_i x_i^2 f_i \le \frac{\beta G}{(\mathrm{SNR})_R}\right) > 0.99. \quad (23)$$

must hold in order to achieve the required grade of service.

In order to compute (22) for the case of slow power control and (23) for fast power control, we need CDF of $\chi_i f_i$ and $\chi_i x_i^2 f_i$, respectively. The algorithm, when using fast power control, is given later. With slow power control, the algorithm is the same except that (22) is used instead of (23) and the fast power control gain x_i^2 is set to 1. The CDF of x_i^2 is obtained from simulation of the transceiver. The difficulty in computing the CDF of $\chi_i x_i^2 f_i$ is that it depends on the CDF of P_k and vice versa. Given the CDF of P_k, one can compute a corresponding CDF for the $\chi_i x_i^2 f_i$ and vice versa. We let ν_i denote the CDF of $\chi_i x_i^2 f_i$. We implement an iterative algorithm to compute the capacity. At each step of the iteration, we take $P_{k_0} = 1$ when computing the CDF of ν_i, where k_0 is the index of the base station (cell) with which the portable is communicating.

1) Let $P_k = 1$ for all k.
2) Compute the number of portables per cell, N, such that (23) is achieved. Based on the computed N, obtain the CDF of ν_i using observations of $\chi_i x_i^2 f_i$.
3) Increment N by 1.
4) Update the CDF of P_k using the value of N from 3 and CDF of ν_i from 2.

5) Compute the probability in (23) and update the CDF of ν_i. Test the distribution of ν_i to ensure that it has converged for the value of N from step 3. If so, go to step 7, else go to step 6.
6) Update the CDF of P_k using the value of N from step 3 and the CDF of ν_i computed in step 5. Go to step 5.
7) If the probability is less than 0.99, then decrement N by 1 and exit, else go to 3.

In step 3 of the algorithm, N may be incremented by more than 1 in order to reduce the number of iterations. However, the stopping rule of the algorithm in step 7 must then be modified. In fact, in order to speed up the algorithm, we incremented N in step 3 by more than 1 at each iteration. When incrementing the value of N by more than 1 at each iteration, the following modification was made. If the probability in step 7 is less than 0.99, then we enter another loop which consists of steps 3 through 7 (with the difference that, in step 3, N is decremented by 1 and in step 7 the algorithm stops when the probability exceeds 0.99).

VI. CAPACITY RESULTS

To estimate the system capacity, we assume that the standard deviation of the log-normally distributed shadow fading is 8 dB. Speech activity is $p_{saf} = 0.45$, i.e., 45% of the time 9.6 kb/s transmission is employed and 55% of the time 1.2 kb/s transmission is used. The fraction of power allocated to the pilot signal, i.e., $1 - \beta$, is 35%, 20%, and 10% for the narrowband system with slow downlink power control, narrowband system with fast downlink power control, and the wideband system, respectively. These values correspond to the pilot–SNR values in Table I. In each case, the fraction of power allocated to the pilot signal is computed such that the corresponding pilot–SNR is achieved at least 95% of the time. The procedure is discussed in more detail in [21].

A. Multipath Diversity on the Uplink

In order to evaluate the impact of multipath diversity on the uplink capacity, the capacity of the 1.25 MHz system was evaluated with symbol interleaving and two receive antennas assuming the single path and the three models described in Section IV-A. The results for the single-path model are shown in column 3 of Table III. The capacity values computed using the three-path model are 8, 27, and 54 for path loss exponents of 2, 3, and 4, respectively. Note that, although the required SNR is higher for the three-path model, the overall capacity of the three-path model is slightly higher because the variations of the fast power control (thereby the multiuser interference) are smaller for the three-path model. Therefore, on the uplink, multipath diversity does not result in significant gain in capacity. The reason, as described in Section IV-B, is that the combination of two receive antennas and fast power control are quite effective to mitigate fading for the low Doppler frequency (2 Hz) considered here and the additional multipath diversity does not result in a significant gain in capacity.

TABLE III
UPLINK CAPACITY (USERS/CELL) FOR THE 1.25 AND 10 MHz SYSTEMS, TWO RECEIVE ANTENNAS

Path Loss Exponent	Users/Cell, 1.25 MHz, Bit Interleaving	Users/Cell, 1.25 MHz, Symbol Interleaving	Users/Cell, 10 MHz, Bit Interleaving	Users/Cell, 10 MHz, Symbol Interleaving
2	5	7	95	130
3	16	25	250	340
4	36	51	410	540

TABLE IV
DOWNLINK CAPACITY OF THE 1.25 MHz SYSTEM WITH SLOW POWER CONTROL

Path Loss Exponent	Users/Cell 2 Transmit Antennas	Users/Cell 3 Transmit Antennas	Users/Cell 4 Transmit Antennas
2	1	1	3
3	1	3	6
4	2	5	10

B. Path Loss and Interleaving Techniques

Table III evaluates the capacity of the uplink as a function of bandwidth, interleaving schemes, and path loss exponent. The path loss exponent significantly affects the uplink capacity. As the path loss exponent is increased from 2 to 3, the capacity is also increased by as much as 100%. In Tables IV and V, we observe a similar sensitivity to the path loss exponent for the downlink. We have considered a path loss model with a single exponent. In [24], it has baeen suggested that a two-slope path loss propagation model may be more appropriate. Since the parameters of the model are environment dependent and differ from location to location, we have used the single exponent model to study the sensitivity to path loss exponent. Since it is not reasonable to choose a set of parameters for the propagation model which accurately approximate different environments, we use the range of capacities for the single exponent model with exponents in the range of 3–4 to provide an indication of the expected capacity.

C. Fast Power Control on Downlink

The capacity of the slow and fast power control cases can be compared from Tables IV and V for different orders of artificial multipath diversity for the 1.25 MHz system. The fast power control algorithm for the downlink was described in Section III-B. Fast power control results in a significant capacity increase compared to that of slow power control. Even with fast power control on the downlink, capacity of the narrowband system is downlink limited.

D. Narrowband and Wideband Systems

In Section IV, we assumed that the channel of the 10 MHz system consists of three resolvable independently Rayleigh fading paths. For the downlink of the wideband CDMA system, we use artificial multipath of the order of 2 (two transmit antennas at the base). Without artificial multipath, the downlink capacity would be quite low. As Table II illustrates, the performance of the uplink transceiver is not sensitive to the power distribution of the three paths. The performance of the downlink transceiver, however, is found to be sensitive to this power distribution. Since we do not have enough actual channel measurement data to determine the probability

TABLE V
DOWNLINK CAPACITY OF THE 1.25 MHz SYSTEM WITH FAST POWER CONTROL

Path Loss Exponent	Users/Cell 2 Transmit Antennas	Users/Cell 3 Transmit Antennas	Users/Cell 4 Transmit Antennas
2	4	12	14
3	8	19	24
4	10	27	33

distribution of the relative powers of the three paths, we use the SNR required by the power distribution (0, −10, −10) dB for our downlink capacity estimates. This power distribution is expected to be conservative and would, therefore, result in an underestimate of the downlink capacity. In computing the multiuser interference statistics as outlined in Sections V-A and V-B, however, the three-path channel model of Section IV-A is used both for the uplink and downlink.

As can be inferred from (9), the weaker multipath components see larger interference variances. The optimal linear (maximal ratio) combiner for the white noise environment normalizes the output of each diversity branch by an estimate of the interference power at that branch [19]. This further reduces the effective gain of the weaker multipath components relative to the main (strongest) path that the Rake receiver at the portable sees. In simulating the performance of the downlink (portable) Rake receiver, we must use this effective gain. A simulation was carried out to assess the effect of this normalization on the effective gain of the Rake receiver branches. It was assumed that the portables receive three resolvable paths from each base station with a power distribution of (0, −3, −6) dB. The received signal power of the second and third paths were normalized by the corresponding interference power seen on each path. Statistics were collected on the resulting normalized power distribution. About 25% of the portables observed effective power distribution of less than −4 and −7 dB for the second and third paths; 5% of the time the second and third paths were lower by −5 and −8 dB. Note that, in the transceiver simulations, it is assumed that the interference powers are the same on all diversity branches. Therefore, the normalized power distributions must be used in the transceiver simulations. Considering the fact that the original power distribution of (0, −3, −6) dB is rather optimistic, the (0, −10, −10) dB power distribution used may not be too conservative. Therefore, in computing the downlink capacity we use 8.8 dB required SNR with slow power control and 6.1 dB with fast power control.

The bandwidth efficiencies of the narrowband and wideband systems, defined as the number of users per cell per MHz, for the uplink can be inferred from Table III by normalizing the capacity of the 1.25 MHz system by 1.25 and those of the 10 MHz system by 10. The bandwidth efficiency of the downlink are given in Table VI. On the downlink, the bandwidth efficiency of the narrowband system with artificial multipath diversity of an order of 4 is compared to that of the wideband system with artificial diversity of an order of 2. Then, the complexity of the Rake receiver is roughly the same in the two cases. Note that the wideband system would need a Rake with six branches, but the performance with four branches would be adequate. The wideband system is seen to be more efficient for both the uplink and downlink. Since more users are multiplexed onto the same bandwidth in the case of the wideband system, the variations of the multiuser interference power are smaller. This results in increased bandwidth efficiency in the case of the wideband system. Even in the wideband system, artificial multipath diversity of an order of 2 is needed on the downlink in order to achieve practically useful capacities. Fast power control on the downlink further increases the downlink capacity.

E. Comparison with Previous Results

Note that the downlink capacity results reported in [21] are smaller than those in this paper. The reason is that, in [21], the downlink capacity was computed assuming $P_k = 1$ for all k (Section V-B). This overestimated the interference from other cells, thereby underestimating the capacity.

VII. CONCLUSIONS

The capacity achievable in a CDMA microcellular wireless access system was assessed as a function of path loss exponent, order of diversity, power control, and system bandwidth. In the simulations, a Doppler frequency of 2 Hz was assumed for the carrier frequency of 2 GHz. Capacity is found to be significantly affected by the path loss exponent. For path loss exponents near 2, a diversity of an order as high as 4 may be needed for the narrowband (1.25 MHz) system to achieve practical capacities. For the same order of diversity, the downlink capacity is generally lower than that of the uplink even if fast power control is employed on the downlink. Even in the wideband system, artificial multipath diversity of an order of 2 is required on the downlink in order to achieve adequate capacity. Fast power control on the downlink would further increase the downlink capacity. Since a three-path channel model was assumed for the 10 MHz system, a Rake with six branches would be needed in the portable. Therefore, to achieve adequate capacity on the downlink a Rake with at least four branches would be needed.

The paper roughly quantifies the increased bandwidth efficiencies available with the wideband system. As more users are multiplexed onto the same bandwidth, the variations of the multiuser interference power is reduced. As a result, the wideband system is more bandwidth efficient.

The capacity estimates are based on rough assumptions concerning the microcellular delay spread and channel power profiles. They will have to be refined by taking into account the range of propagation conditions one can expect to encounter both indoors and out. Also, losses in capacity due to soft handoff have not been considered. Nonetheless, the results serve to provide rough guidelines to the design of high-capacity CDMA systems.

The 10 MHz bandwidth system achieves significant improvement in capacity. As seen in Table VI, the uplink capacity with two receive antennas accommodates about 340–540 voice calls with symbol interleaving and 250–410 calls with bit interleaving. From Table VI, we see that this exceeds the downlink capacity unless fast power control is used. With fast

TABLE VI
DOWNLINK BANDWIDTH EFFICIENCY OF 1.25 AND 10 MHz SYSTEMS

Path loss exponent	1.25 MHz, Fast Downlink Power Control, 4 Transmit Antennas, Users/Cell/MHz	10 MHz, Slow Downlink Power Control, 2 Transmit Antennas, Users/Cell/MHz	10 MHz, Fast Downlink Power Control, 2 Transmit Antennas, Users/Cell/MHz
2	11.2	9.5	13.5
3	19.2	16.5	23.5
4	26.4	24.0	35.0

power control and an artificial multipath of an order of 2, downlink capacity in the range of 235–350 calls is possible. The downlink capacity is still lower than that of uplink for path loss exponents of 3 and 4 (Tables III and VI).

The total information throughput of the cell in the wideband system can be estimated from the average voice transmission requirements of 5 kb/s to be 1.2 to 1.5 Mb/s. These throughput values appear sufficiently large so as to support integrated voice and data services on a shared spectrum basis [25]. A typical scenario may call for 50 data calls at 4.8 kb/s and five video/multimedia calls at 64 kb/s, requiring 0.56 Mb/s. Sufficient capacity would remain to accommodate more than 100 voice calls. If significant demand exists for even higher rate data calls within a microcell, systems of even wider bandwidth appear necessary. Alternatively, further capacity enhancements may be achievable through techniques such as sectorization.

The spectrum made available for wireless services is a scarce resource. It is important that the available bandwidth be utilized efficiently, without unduly increasing the cost and complexity of personal communications systems. These results indicate possible design directions.

ACKNOWLEDGMENT

The authors would like to thank Prof. W. Krzymien for his careful review of the paper. His many comments helped us to improve the paper significantly.

REFERENCES

[1] A. Salmasi and K. S. Gilhousen, "On the system design aspects of code division multiple access (CDMA) applied to digital cellular and personal communications networks," in *Proc. VTC'91*, pp. 57–62.
[2] K. S. Gilhousen *et al.*, "On the capacity of a cellular CDMA system," *IEEE Trans. Vehic. Technol.*, vol. 40, no. 5, pp. 303–312, 1991.
[3] L. F. Chang and S. Ariyavisitakul, "Performance of a CDMA radio communications system with feedback power control and multipath dispersion," in *Proc. GLOBECOM'91*, pp. 1017–1021.
[4] F. Ling and D. D. Falconer, "Combined orthogonal/convolutional coding for a digital cellular CDMA system," in *Proc. VTC'92*, pp. 63–66.
[5] L. F. Chang and N. R. Sollenberger, "Comparison of two interleaving techniques for CDMA radio communications systems," in *Proc. VTC'92*, pp. 275–278.
[6] S. Ariyavisitakul and L. F. Chang, "Signal and interference statistics of a CDMA system with feedback power control," in *Proc. GLOBECOM'91*, pp. 1490–1495.
[7] S. Ariyavisitakul, "SIR-based power control in a CDMA system," in *Proc. GLOBECOM'92*, pp. 868–873.
[8] W. C. Y. Lee, "Power control in CDMA," in *Proc. VTC*, 1991, pp. 77–80.
[9] R. R. Gejji, "Forward-link-power control in CDMA cellular systems," in *Proc. VTC*, 1992, pp. 981–984.
[10] H. Alavi and R. Nettleton, "Downstream power control for a spread spectrum cellular mobile radio system," in *Proc. GLOBECOM'82*, pp. 3.5.1–3.5.5.
[11] A. A. M. Saleh and R. A. Valenzuela, "A statistical model for indoor multipath propagation," *IEEE J. Select. Areas Commun.*, vol. 5, no. 2, pp. 128–137, 1987.
[12] R. J. C. Bultitude and G. K. Bedal, "Propagation characteristics on microcellular urban mobile radio channels at 910 MHz," *IEEE J. Select. Areas Commun.*, vol. SAC-7, 1989.
[13] T. S. Rappaport and M. L. Milstein, "Effects of radio propagation path loss on DS–CDMA cellular frequency reuse efficiency for the reverse channel," *IEEE Trans. Vehic. Technol.*, vol. 41, no. 3, pp. 231–242, 1992.
[14] D. L. Schilling, "Broadband spread spectrum multiple access for personal and cellular communications," in *Proc. VTC*, 1993, pp. 819–821.
[15] G. L. Stuber and C. Kchao, "Analysis of a multiple-cell direct-sequence CDMA cellular mobile radio system," *IEEE J. Select. Areas Commun.*, vol. 10, no. 4, pp. 669–679, 1992.
[16] L. Milstein, T. S. Rappaport, and R. Barghouti, "Performance evaluation of cellular CDMA," *IEEE J. Select. Areas Commun.*, vol. 10, no. 4, pp. 680–689, 1992.
[17] D. J. Torrieri, "Performance of direct-sequence systems with long pseudonoise sequences," *IEEE J. Select. Areas Commun.*, vol. 10, no. 4, pp. 770–781, 1992.
[18] A. K. Elhakeem *et al.*, "Modified SUGAR/DS: A new CDMA system," *IEEE J. Select. Areas Commun.*, vol. 10, no. 4, pp. 690–704, 1992.
[19] W. C. Jakes, Jr., Ed., *Microwave Mobile Communications*. New York: Wiley, 1974.
[20] A. Jalali and P. Mermelstein, "Effects of multipath and antenna diversity on the uplink capacity of a CDMA wireless system," in *Proc. GLOBECOM'93*.
[21] A. Jalali and P. Mermelstein, "Power control and diversity for the downlink of CDMA systems," in *Proc. 2nd Int. Conf. Univ. Personal Commun.*, pp. 980–984.
[22] J. G. Proakis, *Digital Communications*. New York: McGraw-Hill, 1983.
[23] F. Ling, "Coherent detection with reference-symbol based channel estimation for direct sequence CDMA uplink communications," in *Proc. VTC'93*, pp. 400–403.
[24] D. L. Schilling *et al.*, "Broadband CDMA for personal communications systems," *IEEE Commun. Mag.*, vol. 29, no. 11, pp. 86–93, Nov. 1991.
[25] P. Mermelstein, A. Jalali, and H. Leib, "Integrated services on wireless access networks," in *Proc. ICC*, 1993, pp. 863–867.

Ahmad Jalali received the Ph.D. degree in electrical engineering from McGill University in 1990.

During 1990, he worked on congestion control issues for ATM networks at INRS-Telecommunications. Since 1990, he has been working at Bell-Northern Research on different aspects of wireless access techniques.

Paul Mermelstein was born in Czechoslovakia in 1939. He received the B.Eng. degree in engineering physics from McGill University, Montreal, Quebec, Canada, in 1959, and the S.M., E.E., and D.Sc. degrees from the Massachusetts Institute of Technology, Cambridge, MA, in 1960, 1963, and 1964, respectively.

From 1964 to 1973, he was a member of the Technical Staff in the Speech and Communications Research Department of Bell Laboratories, Murray Hill, NJ. From 1973 to 1977, he was a member of the Research Staff of Haskins Laboratories, conducting research in speech analysis, perception, and recognition. From 1977 to 1986, he was Manager of Speech Communication Systems at Bell-Northern Research in Montreal. He currently serves as Manager of Personal Communications at BNR. He is a past Associate Editor for Speech Processing of the *Journal of the Acoustical Society of America* and Editor for Speech Communication of the IEEE TRANSACTIONS ON SPEECH COMMUNICATIONS. Since 1977, he has held appointments as Visiting Professor at INRS-Télécommunications (Université du Québec) and Auxiliary Professor of Electrical Engineering at McGill University. He is currently the Major Program Leader in personal and mobile communication of the Canadian Institute of Telecommunications Research.

Hybrid DS/SFH Spread-Spectrum Multiple Access with Predetection Diversity and Coding for Indoor Radio

Jiangzhou Wang and Marc Moeneclaey

Abstract—**This paper derives close upper and lower bounds on the average bit error probability for hybrid direct-sequence/slow-frequency-hopped spread-spectrum multiple-access (DS/SFH-SSMA) systems with noncoherent DPSK demodulation, employing predetection diversity [selection combining (SC) and equal gain combining (EGC)] in conjunction with interleaved channel coding [Hamming (7, 4) code and BCH (15, 7) code] operating through indoor radio channels. A multipath Rayleigh fading model is assumed for the indoor radio channel. Our results show that the DS portion of the modulation combats the multipath interference, whereas the FH portion is a protection against large multiaccess interference. It is shown that, for the considered types of channel coding, the use of predetection diversity is still essential for obtaining a satisfactory error performance.**

I. Introduction

In addition to its multiple-access capabilities, spread-spectrum multiple access (SSMA) simultaneously provides considerable effectiveness in combating various types of hostile or nonhostile interference. Recently, there has been an increased interest in direct-sequence SSMA (DS-SSMA) for applications that involve multipath fading [1]–[7]. This is partly due to the time diversity inherent in spread-spectrum signals, which is easily exploited with present technology. For example, a matched-filter-based receiver using a surface acoustic wave (SAW) device was analyzed in [1], [2] and made an experiment on in [7]. As DS-SSMA performance deteriorates when the different users have considerably different power levels (the "near/far" problem), the overriding shortcoming for DS-SSMA is the need for average power control. Although in star LAN's average power can be controlled [8], such control may not be possible in wider-area fully connected networks and random-access networks.

For a low spectral sidelobe level of the transmitted signal, a frequency-hopped (FH) system can solve the near/far problem by reducing the probability that two or more users have the same instantaneous carrier frequency. Unfortunately, FH-SSMA is more vulnerable to multipath interference than DS-SSMA.

Manuscript received June 15, 1989; revised September 15, 1989 and April 15, 1990. This work was supported in part by the Belgian National Fund for Scientific Research.

The authors are with the Communication Engineering Laboratory, University of Ghent, B-9000 Ghent, Belgium.

IEEE Log Number 9107005.

A hybrid form of DS and slow frequency hopping (SFH) is attractive because it can combine the antimultipath effectiveness of DS systems with the good antipartial-band-jamming and the good anti-near/far problem features of FH systems. A disadvantage of hybrid systems is the increased complexity of their transmitters and receivers.

In this paper, we are concerned with the bit error rate of noncoherent differential-phase-shift-keying (DPSK) hybrid DS/SFH-SSMA systems in predetection diversity plus interleaved coding, operating through indoor radio multipath Rayleigh fading channels. Predetection diversity means that the receiver, in some way, combines the signals that have traveled via different paths before making a decision about the transmitted data bits. Interleaving/deinterleaving of order two is used in order to split any error burst of length two at the output of DPSK demodulator over two codewords.

The paper includes the following sections. In Section II, the transmitter, channel, and receiver models are described. The upper and lower bounds on the average error probability of hybrid systems employing interleaved coding and predetection diversity are derived in Section III. This is followed by some numerical results in Section IV. Finally, conclusions and remarks are given in Section V.

Our paper can be viewed as an extension of [2] and [9], which considered DS-SSMA over the multipath Rayleigh fading channel and DS/SFH SSMA over the additive white Gaussian noise channel, respectively.

II. System Model

The transmitter for the kth user ($1 \leq k \leq K$, where K is the number of active users) consists of five parts: the channel encoder (the use of a block code is assumed), the interleaver, the differential encoder, the DS modulator (or spreader), and the frequency hopper.

For noncoherent DPSK modulation, the output of the kth DS modulator $C_k(t)$ is given by:

$$C_k(t) = \mathrm{Re}\,\{\sqrt{2P}\, b_k(t) a_k(t) \exp[j(2\pi f_c t + \theta_k)]\} \tag{1}$$

where Re $\{\cdot\}$ stands for real part and $j = \sqrt{-1} \cdot f_c$ and P are the carrier frequency and the transmitted power, which are common to all users, and θ_k is the phase angle introduced by the kth DPSK modulator. The kth data signal $b_k(t)$ is a differentially encoded version of the kth interleaved and channel-coded information sequence, with a coded bit rate $1/T$; $a_k(t)$ is a PN code waveform with a rate $1/T_c$. We assume that there are N code pulses during each coded bit ($T = NT_c$) and the period of the PN sequence is N. The λth NRZ square pulse of $b_k(t)$ has amplitude $b_\lambda^{(k)}$, randomly taking values from $\{+1, -1\}$ with equal probability; the γth NRZ square pulse of $a_k(t)$ has amplitude $a_\gamma^{(k)}$, also randomly taking values from $\{+1, -1\}$.

The DS signal $C_k(t)$ is frequency hopped according to the kth hopping pattern $f_k(t)$, which is derived from a sequence $\{f_J^{(k)}\}$ of frequencies from a set $S = \{v_1, v_2, \cdots, v_q\}$ of q not necessarily equally spaced frequencies with minimum spacing W. It is assumed that $\{f_J^{(k)}\}$ is a Markov sequence with:

$$\text{Prob}\,[f_{J+1}^{(k)} = v_i | f_J^{(k)} = v_j] = \begin{cases} 1/(q-1) & i \neq j \\ 0 & i = j \end{cases}. \quad (2)$$

Let T_h denote the duration of a single hopping interval (dwell time) and $f_J^{(k)}$ denote the frequency used by the kth user during the Jth dwell time. We assume that $W \geq 2T_c^{-1}$ so that there is negligible overlapping of the DS signals when hopped to adjacent frequencies. The number of data bits transmitted per hop $N_b = T_h/T$ is a positive integer. In the case of slow frequency hopping, we have $N_b >> 1$.

The transmitted signal of the kth user takes the form:

$$S_k(t) = \sqrt{2P}\, b_k(t) a_k(t) \cos\{2\pi[f_c + f_k(t)]t + \theta_k + \alpha_k(t)\} \quad (3)$$

where $\alpha_k(t)$ represents the phase waveform introduced by the kth frequency hopper; it takes on a constant value $\alpha_J^{(k)}$ during the Jth dwell time.

Indoor radio communications is made particularly difficult because of the severe multipath nature of the channel, coupled with large propagation losses. Multipath measurements of the indoor radio channel at about 1 GHz in and around office buildings [11] showed that it can be characterized by very slowly varying, multipath Rayleigh fading. Unlike the urban mobile radio channel, a deep fade at a given frequency in the indoor channel can last for up to several seconds, or even minutes. That is, the fading rate in an indoor radio environment is slow computed to the bit rate, so the random parameters associated with the channel do not vary significantly over two consecutive bit intervals; therefore, DPSK is a convenient modulation. We assume that there are L paths associated to each user. The ℓth path of the kth user is characterized by three random variables: the gain $\beta_{k\ell}$, delay $\tau_{k\ell}$, and phase $r_{k\ell}$. The gains, delays, and phases of different paths and/or of different users are all statistically independent. We assume that $\beta_{k\ell}$ has a Rayleigh distribution with $E[\beta_{k\ell}^2] = 2\rho_{k\ell}$, and that the path delay $\tau_{k\ell}$ is uniformly distributed in $[0, T]$. That is, we restrict our attention to channels for which the multipath time delay spread is less than the coded bit duration T. Further, we assume that the path phase $r_{k\ell}$ is uniformly distributed in $[0, 2\pi]$. Finally, the channel introduces additive white Gaussian noise $n(t)$, with two-sided power spectral density $N_0/2$. Hence, the received signal $r(t)$ can be represented as:

$$r(t) = \sqrt{2P} \sum_{k=1}^{K} \sum_{\ell=1}^{L} \beta_{k\ell} b_k(t - \tau_{k\ell}) a_k(t - \tau_{k\ell}) \cdot \cos\{2\pi[f_c + f_k(t - \tau_{k\ell})]t + \phi_{k\ell}(t)\} + n(t) \quad (4)$$

where

$$\phi_{k\ell}(t) = \theta_k + \alpha_k(t - \tau_{k\ell}) - 2\pi[f_c + f_k(t - \tau_{k\ell})]\tau_{k\ell} - r_{k\ell}.$$

Let us consider the receiver which must detect the information transmitted by the ith user (which will be denoted as the reference user). This receiver knows the hopping sequence $f_i(t)$ and the PN sequence $a_i(t)$. The receiver consists of a dehopper, a bandpass matched filter, a DPSK demodulator, a diversity circuit, a hard decision device, a deinterleaver, and a hard-decision decoder. A detailed model of the frequency dehopper, the bandpass matched filter, and the DPSK demodulator is shown in Fig. 1.

The received signal $r(t)$ enters a bandpass filter which removes out-of-band noise. The mixer of the dehopper performs the appropriate frequency translation. We assume that the hopping pattern of the receiver is synchronized with the hopping pattern of the signal associated with the jth path of user i (which will be denoted as the reference path). The dehopper introduces a phase $d_i(t)$, which is constant over a hopping interval; $d_J^{(i)}$ stands for the constant phase during the Jth hop. The bandpass filter which follows the mixer removes high frequency terms. The resulting dehopper output signal is given by:

$$r_d(t) = \sqrt{P/2} \sum_{k=1}^{K} \sum_{\ell=1}^{L} \beta_{k\ell}\, \delta[f_k(t - \tau_{k\ell}), f_i(t - \tau_{ij})] \cdot b_k(t - \tau_{k\ell}) a_k(t - \tau_{k\ell}) \cdot \cos[2\pi f_c t + \Phi_{k\ell}(t)] + \hat{n}(t) \quad (5)$$

where $\hat{n}(t)$ can be considered as white Gaussian noise with two-sided power spectral density $N_0/8$. The phase waveform $\Phi_{k\ell}(t)$ is defined as $\Phi_{k\ell}(t) = \phi_{k\ell}(t) + d_i(t)$, while $\delta(u, v) = 0$ for $u \neq v$ and $\delta(u, v) = 1$ for $u = v$ (both u and v are real). Note that the dehopper suppresses, at any instant t, all path signals whose frequency at instant t differs from $f_i(t - \tau_{ij})$. Evidently, the reference path signal is not suppressed; the other path signals from the reference user are suppressed only during a part of the first or last bit of a hop, depending on the relative delay of the considered path with respect to the reference path. Path signals from users different from the reference user contribute to the dehopper output only during those time intervals for which their frequency accidentally equals the frequency of the reference path signal.

The dehopper output signal $r_d(t)$ enters a bandpass filter, whose impulse response is matched to a T-seconds

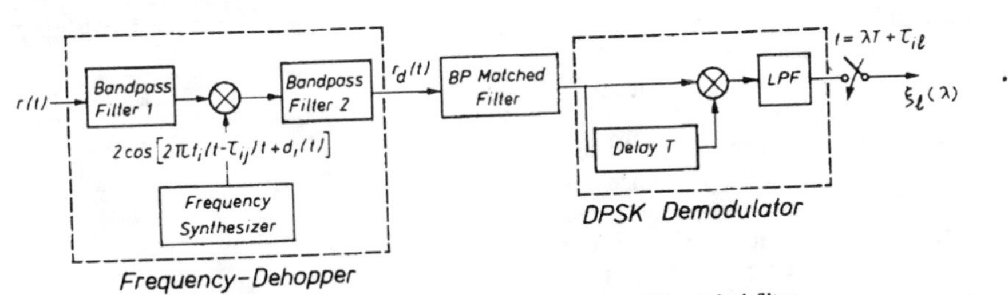

Fig. 1. A detailed model of the frequency dehopper, BP matched filter, and DPSK demodulator.

segment of $a_i(t) \cos(2\pi f_c t)$. Such a filter is commonly implemented by means of a surface acoustic wave (SAW) device. The bandpass signal at the matched filter output contains narrow peaks of width $2T_c$, at instants $\lambda T + \tau_{ij}(\lambda = 0, \pm 1, \pm 2, \cdots)$ which are caused by the reference path signal. The other $L - 1$ path signals from the reference user give rise to peaks at the instants $\lambda T + \tau_{i\ell}(\ell \neq j)$. However, there is no peak during the first or last bit of a hop, because each of these $L - 1$ path signals is partly suppressed by the dehopper. Path signals from nonreference users do not give rise to peaks, because their PN sequence is different from $a_i(t)$.

The matched filter output signal enters a conventional DPSK demodulator. The lowpass filter following the mixer removes the frequency terms at $2f_c$. The DPSK demodulator output is a lowpass signal, which has peaks at the same instants $\lambda T + \tau_{i\ell}$ as the peaks in the matched filter output signal; a positive (negative) peak indicates that the corresponding channel encoder output bit is likely to be a logical one (logical zero). We assume that the delay differences between path signals from the same user are larger than $2T_c$, so these peaks do not overlap; this assumption is justified when the PN sequence period N is large. The DPSK demodulator output signal is sampled at the instants $\lambda T + \tau_{i\ell}$, and this gives rise to the decision variables $\xi_\ell(\lambda)$.

In order to make a hard decision about the λth interleaved encoded bit, one could examine the sign of the decision variable $\xi_j(\lambda)$ corresponding to the reference path signal. However, the decision variables $\xi_\ell(\lambda)$, $\ell = 1, \cdots, L$ are independent, identically distributed random variables so they can be combined into a single decision variable, in order to improve the reliability of the decision; this is called predetection diversity. In the case of selection combining (SC) of order M (with $1 \leq M \leq L$), the decision variable $\xi_{SC}(\lambda)$ is the maximum of M variables $\xi_\ell(\lambda)$, whereas equal gain combining (EGC) of order M uses a decision variable $\xi_{EGC}(\lambda)$ which is the arithmetical average of M variables $\xi_\ell(\lambda)$. Then, the sign of the decision variable $\xi_{EGC}(\lambda)$ or $\xi_{SC}(\lambda)$ is used as a hard decision of the λth interleaved encoded bit. In a practical implementation, the decision variable $\xi_{EGC}(\lambda)$ is approximated by integrating the DPSK demodulator output over an interval depending on the delay spread, thereby replacing a sum of samples by a continuous-time integration [2]. This approximation avoids the need for time synchronization to each of the different path signals from the reference user. Note that predetection diversity cannot be used for the first and last bit of a hop, because the corresponding nonreference path signals from the reference user are partially suppressed by the dehopper. Therefore, the first and last bit of each hop are dummy bits, whereas the other bits are a differentially encoded version of the interleaved channel encoded bits. The second bit of each hop serves as a phase reference bit needed for differential demodulation. Hence, only $N_b - 3$ channel encoded bits are involved in a hop of N_b bits.

Because of DPSK demodulation, the errors at the output of the hard decision device tend to occur in bursts of length two. In order to avoid such a burst affecting two bits of a single codeword, interleaving (at the channel encoder output) and deinterleaving (at the output of hard decision device) of order two is applied. The interleaving/deinterleaving makes the errors within a single codeword statistically independent, because these errors belong to different error events.

The bits at the output of the deinterleaver are fed to a hard decision decoder. The hard decision decoder looks for the valid codeword which has minimum Hamming distance with respect to the codeword at the deinterleaver output.

III. System Performance

The decision variables $\xi_\ell(\lambda)$ at the DPSK demodulator output can be written as:

$$\xi_\ell(\lambda) = \text{Re}\,[Z_\ell(\lambda) Z_\ell^*(\lambda - 1)] \tag{6}$$

where

$$Z_\ell(\lambda) = \int_{\lambda T + \tau_{i\ell}}^{(\lambda+1)T + \tau_{i\ell}} r_d(t) \cdot \exp(-j2\pi f_c t) a_i(t - \lambda T - \tau_{i\ell})\, dt \tag{7}$$

and * denotes complex conjugate. The random variable $Z_\ell(\lambda)$ consists of the following four terms.

- A useful term $D_\ell(\lambda)$, which is due to the ℓth path signal from the references user. This useful term is given by:

$$D_\ell(\lambda) = \sqrt{P/8T}\,\beta_{i\ell} \cdot b_\lambda^{(i)} \cdot \exp\{j[\theta_i + \alpha_j^{(i)} - d_j^{(i)} - r_{i\ell}]\}. \tag{8}$$

• A complex Gaussian noise term $N_\ell(\lambda)$, which is due to the noise term $\hat{n}(t)$ at the dehopper output. The real and imaginary parts of $N_\ell(\lambda)$ are statistically independent, and have the same variance $N_0T/16$.

• A multipath interference term, which is due to the $L - 1$ other path signals from the reference user. The contributions from different paths are uncorrelated.

• A multiaccess interference term, which is due to the path signals from the $K - 1$ nonreference users. The contributions from different paths are uncorrelated.

When N, the period of the PN sequence, is large, the multipath interference term and the multiaccess interference term consist of a large number of statistically independent contributions with the same distribution, so that each interference term can be well approximated by a Gaussian random variable. Combining both interference terms and the noise term into a single Gaussian random variable, we will be able to derive analytical results for the bit error probability of the receiver. But first, we will carefully examine the multipath and multiaccess interference terms.

1) The Multipath Interference Term: Fig. 2(a) schematically represents the ℓth, the earliest, and the latest path signals from the reference user during the Jth hop. It is clear that during each of the nondummy bits of the ℓth path signal, the $L - 1$ other path signals from the reference user are at the same frequency as the ℓth path signal: each nondummy bit of the ℓth path signal is fully hit $L - 1$ times. The contribution from the ℓth path signal, assuming that $\tau_{i\ell'} > \tau_{i\ell}$, is given by:

$$I_{i\ell'} = \sqrt{P/8}\, \beta_{i\ell'} \exp\{j[\theta_i + \alpha_J^{(i)} - d_J^{(i)} - r_{i\ell'}]\} \cdot \left[b_{\lambda-1}^{(i)} \int_{\tau_{i\ell}}^{\tau_{i\ell'}} a_i(t - \tau_{i\ell}) a_i(t - \tau_{i\ell'})\, dt + b_\lambda^{(i)} \int_{\tau_{i\ell'}}^{\tau_{i\ell}+T} a_i(t - \tau_{i\ell}) a_i(t - \tau_{i\ell'})\, dt \right]. \quad (9)$$

When the period N of the PN sequence is large, it can be assumed that the path delay difference $\tau_{i\ell} - \tau_{i\ell'}$ is larger than T_c. In this case, both integrals in (9) are zero-mean random variables, so that $I_{i\ell'}$ is an interference term. The real and imaginary parts of $I_{i\ell'}$ are uncorrelated, and have a variance equal to $P\rho_{i\ell'}T^2/(12N)$, where $\rho_{i\ell'} = E[\beta_{i\ell'}^2]/2$. This result is an extension of [10], which considers hybrid DS/FSH SSMA over the additive white Gaussian noise channel. The same result is obtained when $\tau_{i\ell'} > \tau_{i\ell}$. Hence, the multipath intereference term, due to the $L - 1$ other paths from the reference user, is a complex-valued random variable with uncorrelated real and imaginary parts, each having a variance equal to $(L - 1)P\rho_0T^2/(12N)$, where we have assumed that all path signals of the reference user have the same path power ρ_0.

2) The Multiaccess Interference: Fig. 2(b) schematically represents an arbitrary bit from the reference path signal, and the earliest and latest path signals during the Jth hop of a nonreference user $k (k \neq i)$. An exact statistical description of the multiaccess interference is most difficult, because the number of interfering path signals and their amount of interference depend on the relative delays between the considered reference path bit and all path signals from all nonreference users. In order to avoid the dependence on relative delays, we will determine simple upper and lower bounds for the multiple-access interference, which do not involve the relative delays. These bounds have been introduced in [12].

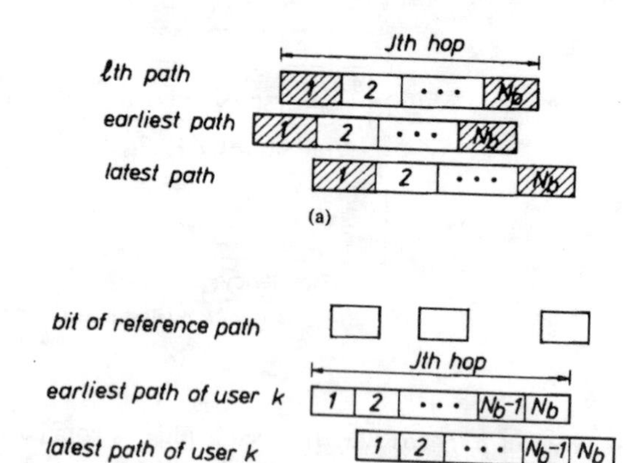

Fig. 2. (a) The illustration of the multipath interference (hatched: dummy bit). (b) The illustration of the multiaccess interference.

The upper (lower) bound is obtained by overestimating (underestimating) the interference from the nonreference user k, in the case where the considered reference bit starts during the first or last bit of the Jth hop of the earliest path signal of user k. The upper bound assumes that such a reference path bit is fully hit by the L path signals from user k, provided that the Jth hopping frequency or the $(J \pm 1)$th hopping frequency ($J - 1$ when the reference path bit starts during the first bit of the Jth hop; $J + 1$ when the reference path bit starts during the last bit of the Jth hop) equals the frequency of the reference path bit; otherwise, there is no interference from user k. The lower bound assumes that there is no interference, in the case where the considered reference path bit starts during the first or last bit of the Jth hop of the earliest path signal of the user. For the $N_b - 2$ other positions of the considered reference path bit, both the upper and lower bounds take into account the actual L full hits from user k.

Hence, both the upper and lower bounds assume that the considered reference path bit is fully hit L times by the nonreference user k, with probability P_h, and is not hit by that user with probability $1 - P_h$, where:

$$P_h = \begin{cases} (1 - 2/N_b)/q & \text{lower bound} \\ (1 - 2/N_b)/q + (2/N_b)(2/q) = (1 + 2/N_b)/q & \text{upper bound} \end{cases} \quad (10)$$

and q denotes the number of different hopping frequencies. When $N_b >> 1$ (this is the case of slow frequency hopping), we expect the upper and lower bounds to be very tight. Taking into account all $K - 1$ nonreference users, the considered reference path bit is fully hit $k_h L$ times, where k_h is a random variable indicating the number of interfering nonreference users; k_h takes the values $0, 1, \cdots, K - 1$, with:

$$P(m) = \text{Prob}\,[k_h = m] = \binom{K-1}{m} P_h^m \cdot (1 - P_h)^{K-1-m}. \quad (11)$$

Conditioned on the random variable k_h, the real and imaginary parts of the multiaccess interference term are uncorrelated, and each has a variance equal to $P\rho_1 T^2 k_h L/(12N)$ (we assume that all $K - 1$ nonreference users have the same average path power ρ_1, so that the multiaccess interference power depends only on the number of hits).

The above upper and lower bounds correspond to the multiaccess interference to a single bit from the reference path signal, and will be used for evaluating the bit error probability at the output of the hard decision device (or, equivalently, the bit error probability in the absence of channel coding). Using these bounds for evaluating the bit error probability after hard decision decoding is very complicated, because the various bits from the reference path signal which constitutes a single codeword are in general not hit the same number of times, so that the various bits within a single codeword have different error probabilities. This problem can be circumvented by deriving separate upper and lower bounds for the multiaccess interference to a block of two interleaved codewords. Using a similar reasoning as above, these bounds assume that all bits of a block of two interleaved codewords of the considered reference path signal are fully hit L times by a nonreference user k, with probability $\hat{P}_h$, and none of these bits is hit by that user with probability $1 - \hat{P}_h$, where:

$$\hat{P}_h = \begin{cases} [1 - (2n + 1)/N_b]/q & \text{lower bound} \\ [1 + (2n + 1)/N_b]/q & \text{upper bound} \end{cases} \quad (12)$$

and n denotes the number of bits in a codeword. These bounds are expected to be very tight when $N_b >> 2n$ (which is fulfilled in the case of slow frequency hopping). Taking into account all $K - 1$ nonreference users, the considered block is fully hit $k_h L$ times. The distribution of k_h is denoted by $\hat{P}(m)$, and is obtained by replacing P_h in (11) by $\hat{P}_h$:

$$\hat{P}_m = \text{prob}\,[k_h = m] = \binom{K-1}{m} \hat{P}_h^m (1 - \hat{P}_h)^{K-1-m}. \quad (13)$$

3) Bit Error Rate Performance: Collecting the above results, the complex-valued random variable $Z_\ell(\lambda)$, given by (7), consists of a useful term $D_\ell(\lambda)$, given by (8), and a total interference term (Gaussian noise + multipath interference + multiaccess interference). When the period of the PN sequence is large, the total interference is well approximated by a Gaussian random variable; its real and imaginary parts are statistically independent, and have the same variance σ^2, given by:

$$\sigma^2 = N_0 T/16 + PT^2[k_h L\rho_1 + (L - 1)\rho_0]/(12N) \quad (14)$$

where k_h denotes the number of interfering nonreference users; the distribution of k_h is given by (11). For purely DS ($q = 1$, $P_h = 1$, and $k_h = K - 1$), we have:

$$\sigma^2 = N_0 T/16 + PT^2[(K - 1)L\rho_1 + (L - 1)\rho_0]/(12N). \quad (15)$$

It can be shown that the total interference term in $Z_\ell(\lambda - 1)$ is statistically independent from and has the same distribution as the total interference term in $Z_\ell(\lambda)$. Also, it can be verified that the random variables $\xi_\ell(\lambda)$, $\ell = 1, \cdots, L$ are independent and, because of identical reference signal path powers, identically distributed.

The bit error rate P_e in the absence of coding (or, equivalently, at the hard decision output) is given by:

$$P_e = \sum_{k_h=0}^{K-1} P(k_h) P_e(k_h) \quad (16)$$

where $P(k_h)$, the distribution of the number of interfering nonreference users, is given by (11), and $P_e(k_h)$ is the bit error probability in the absence of coding, conditioned on the number of interfering nonreference users.

In the presence of coding, the bit error rate after hard decision decoding is given by:

$$P_b = \sum_{k_h=0}^{K-1} \hat{P}(k_h) P_b(k_h) \quad (17)$$

where $\hat{P}(k_h)$ is given by (13), and $P_b(k_h)$ is the bit error probability after decoding, conditioned on the number of interfering nonreference users. This conditional bit error probability after decoding is approximately given by [14]:

$$P_b(k_h) = \frac{1}{n} \sum_{i=t+1}^{n} i \binom{n}{i} [P_e(k_h)]^i [1 - P_e(k_h)]^{n-i} \quad (18)$$

where $P_e(k_h)$ is the conditional bit error probability at the hard decision output, and t denotes the number of transmission errors that the block code can correct.

In the following, we derive analytical expressions for $P_e(k_h)$ in the cases of EGC diversity and SC diversity.

A. EGC Diversity

In the case of EGC diversity of order M, the decision variable $\xi_{\text{EGC}}(\lambda)$ is the average of M random variables $\xi_\ell(\lambda)$. Assuming that the total interference in $Z_\ell(\lambda)$ is nearly Gaussian, the conditional error probability $P_e(k_h)$ follows from [13]:

$$P_e(k_h) = \frac{1}{2^{2M-1}(M - 1)!(1 + r_c)^M} \cdot \sum_{m=0}^{M-1} C_m (M - 1 + m)! \left(\frac{r_c}{r_c + 1}\right)^m \quad (19)$$

where

$$C_m = \frac{1}{m!} \sum_{n=0}^{M-m-1} \binom{2M-1}{n}.$$

The quantity r_c is the average signal-to-noise ratio per combined path, with "noise" taken as the total interference:

$$r_c = \left\{ \left(\frac{\overline{E}_b}{N_0} \right)^{-1} + 2[k_h L\gamma + (L-1)]/(3N) \right\}^{-1} \quad (20)$$

where $\overline{E}_b = 2\rho_0 PT$ represents the average received energy per bit per combined path, and $\gamma = \rho_1/\rho_0$ is the ratio between the average path powers of a nonreference user and the reference user. For purely DS ($q = 1$, $P_h = 1$, and $k_h = K - 1$), we have $P_e = P_e(k_h = K - 1)$ (16) and

$$r_c = \left\{ \left(\frac{\overline{E}_b}{N_0} \right)^{-1} + 2[(K-1)L\gamma + (L-1)]/(3N) \right\}^{-1}. \quad (21)$$

When diversity is not exploited (i.e., $M = 1$), we obtain from (19):

$$P_e(k_h)|_{M=1} = \frac{1}{2(1 + r_c)} \quad (22)$$

which is the familiar expression for the bit error probability over the fading channel [13].

B. SC Diversity

In the case of SC diversity of order M, the decision variable $\xi_{SC}(\lambda)$ is the maximum of M random variables $\xi_\ell(\lambda)$. When the total disturbance in $Z_\ell(\lambda)$ is relatively small, the largest of the M random variables $\xi_\ell(\lambda)$ is with high probability the one with the largest $\beta_{i\ell}$. The distribution of $\beta^2_{\max}$, the square of the largest of M independent identically distributed Rayleigh random variables $\beta_{i\ell}$, has been derived in [1]:

$$P_{\beta^2_{\max}}(y) = M \sum_{m=0}^{M-1} \binom{M-1}{m} \frac{(-1)^m}{2\rho_0} \cdot \exp\left[\frac{y}{2\rho_0}(m+1) \right], \quad y > 0. \quad (23)$$

As the above distribution is a weighed sum of exponential distributions, $P_e(k_h)$ is given by:

$$P_e(k_h) = M \sum_{m=0}^{M-1} \binom{M-1}{m} \frac{(-1)^m}{(m+1)} \cdot \frac{1}{2[1 + r_c/(m+1)]}. \quad (24)$$

Again, for $M = 1$, we obtain (22).

IV. Numerical Results

In this section, we provide some representative numerical results about the bit error rate (BER) performance of DS/SFH SSMA systems. Unless noted otherwise, we assume $\gamma = 1$. We have verified that the upper and lower bounds resulting from (10) are very close for $N_b > 100$. In the following, we use the upper bound as an approximation of the true BER.

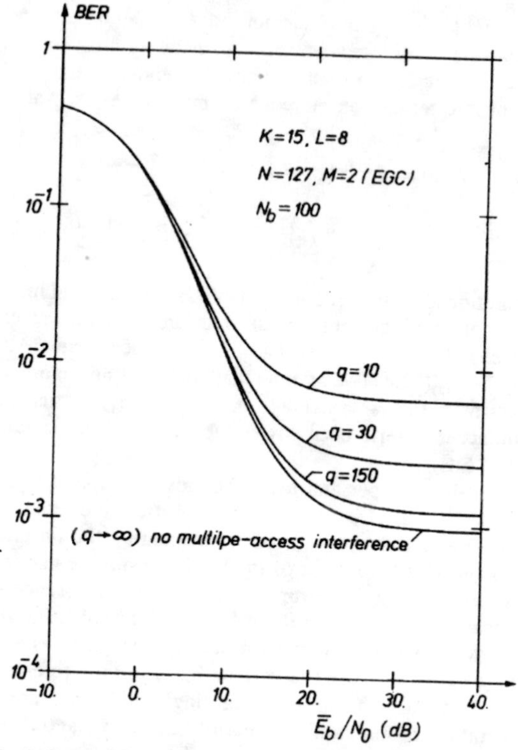

Fig. 3. Uncoded BER of hybrid systems with EGC for a varying number (q) of available hopping frequencies, and a fixed number ($K = 15$) of users.

The uncoded BER in the case of second-order EGC is shown in Fig. 3, for different values of q, the number of available hopping frequencies. Increasing q reduces the effect of multiaccess interference, because of the decreasing probability of a hit. When q is much larger than the number of users (K), the BER at high $\overline{E}_b/N_0$ is dominated by the multipath interference while the multiaccess interference is negligible. At low $\overline{E}_b/N_0$, the BER is essentially independent of q because additive noise is the dominating disturbance.

Fig. 4 shows the uncoded BER in the cases of EGC and SC, for various orders (M) of diversity. It is clearly seen that increasing the order of diversity considerably reduces the BER. For a given $M (M > 1)$, EGC performs better than SC, and the advantage of EGC over SC increases with increasing M.

The BER after hard decision decoding is shown in Figs. 5 and 6, for various diversity orders (M), for both EGC and SC. These results correspond to the use of a Hamming (7, 4) code (Fig. 8), which is able to correct single errors ($t = 1$), and BCH (15, 7) code (Fig. 9), which is

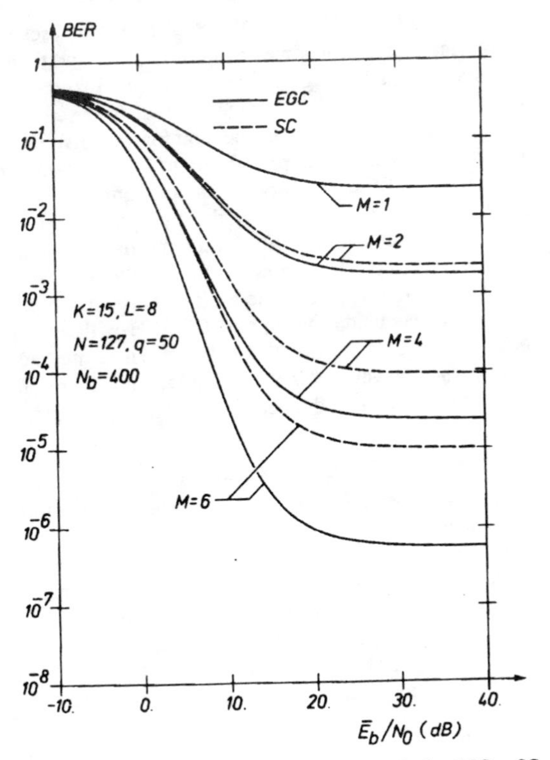

Fig. 4. Performance of uncoded hybrid systems employing EGC or SC.

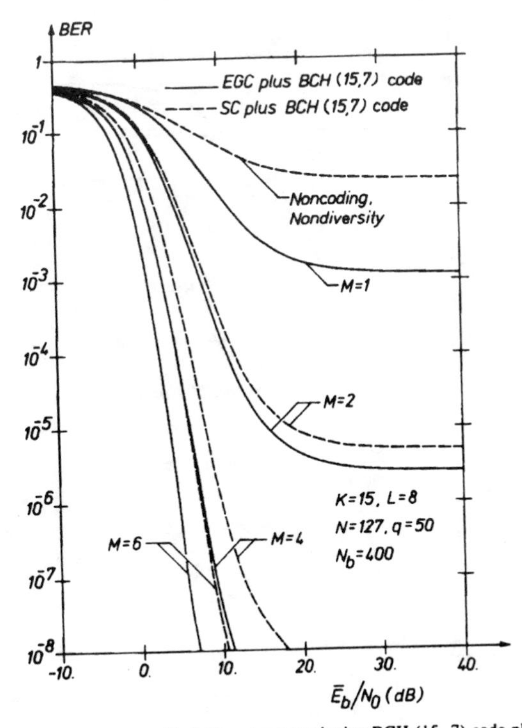

Fig. 6. Performance of hybrid systems employing BCH (15, 7) code plus EGC or SC for indoor radio.

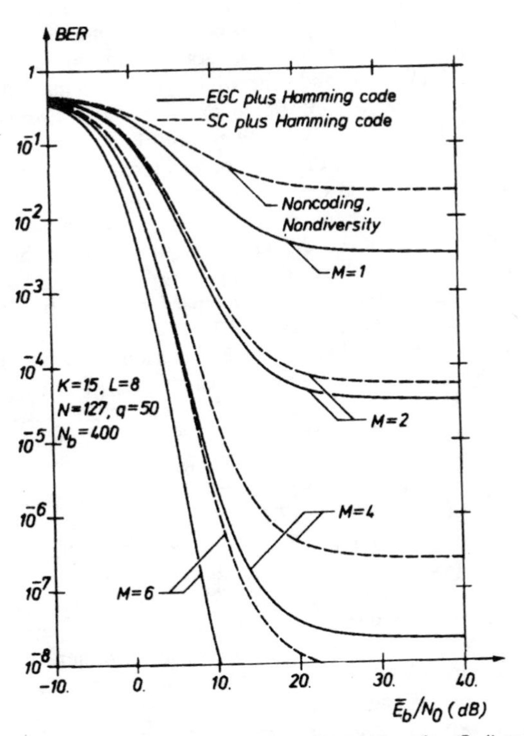

Fig. 5. Performance of hybrid systems employing Hamming (7, 4) code plus EGC or SC for indoor radio.

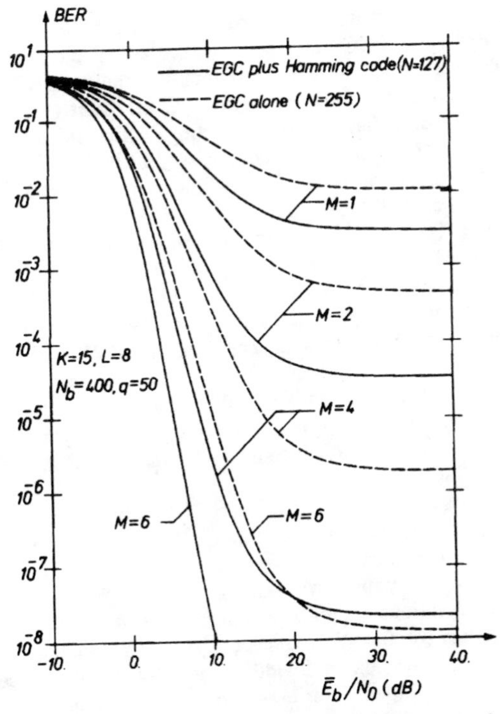

Fig. 7. Comparison of bit error rate (BER) of hybrid systems employing EGC plus Hamming (7, 4) code to EGC alone for a given transmission bandwidth.

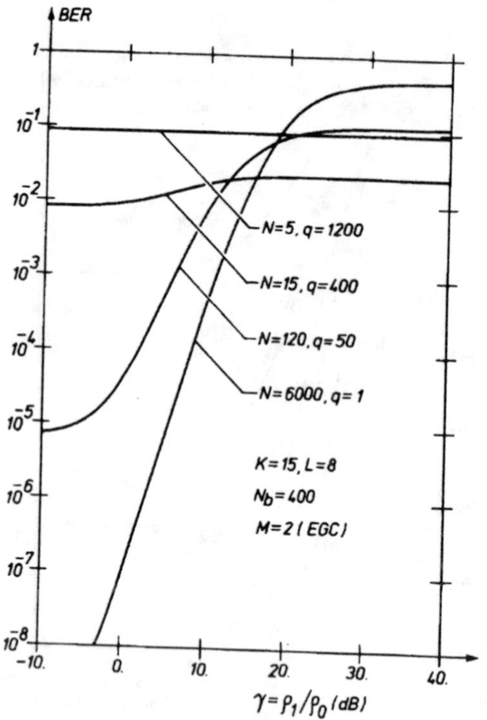

Fig. 8. The asymptotic ($\overline{E}_b/N_0 \to \infty$) BER for second-order EGC plus Hamming (7, 4) code.

able to correct up to two errors ($t = 2$). As for the uncoded case, EGC has a performance advantage over SC when $M > 1$. For a given diversity technique (EGC or SC) and a given diversity order M, the BCH code performs better than Hamming coding, which in turn performs better than the uncoded case. For both considered codes, the use of a diversity technique is essential for obtaining a small BER (say, less than 10^{-5}).

Fig. 7 compares the BER for various orders of EGC, with and without coding, under the assumption of a given transmission bandwidth and a given information bit rate. More specifically, we consider the cases of no coding combined with a PN sequence with period $N = 255$ and Hamming (7, 4) coding with $N = 127$, yielding (nearly) the same transmission bandwidth. We conclude that, for a given transmission bandwidth and a given information bit rate, coding plus diversity performs better than diversity alone.

The asymptotic ($\overline{E}_b/N_0 \to \infty$) BER for second-order EGC combined with Hamming (7, 4) coding is shown in Fig. 8, as a function of the nonreference user to reference user average path power ratio $\gamma = \rho_1/\rho_0$. We have considered various combinations of numbers of hopping frequencies (q) and PN sequence periods (N), yielding the same transmission bandwidth (constant value of qN). For $\gamma < 1$, multipath interference from the reference user is the dominating disturbance; the BER performance improves with increasing N, so that pure DS yields the smallest BER. For $\gamma >> 1$, the BER first decreases with increasing q, reaches some minimum value, and then increases with increasing q. Indeed, for small q, the probability of a hit ($\approx 1/q$) is large, so that for $\gamma >> 1$ the multiaccess interference is the dominating disturbance; increasing q reduces the probability of a hit and therefore decreases the BER. For large q, the probability of a hit is so small, that even for $\gamma >> 1$ the multipath interference causes more bit errors than does the multiaccess interference. For a given transmission bandwidth, increasing q means reducing N, so that the BER performance becomes worse. We conclude that hybrid DS/SFH with an appropriate value of q is less vulnerable to large multiaccess interference power levels than pure DS and, therefore, is better suited to solve the near/far problem.

V. Conclusions and Remarks

From the above numerical results, we draw the following conclusions.

1) The upper and lower bounds on the BER are valid for multipath fading channels in which the maximum delay spread does not exceed a coded bit duration; the extension to a larger delay spread is straightforward. We have verified that the bounds are very tight for a large number of coded bits per hop (i.e., slow frequency hopping).

2) When the number of frequencies (q) is much larger than the number of users, the effect of the multipath disturbance from the reference user is much more important than the multiple-access disturbance from other users (assuming that all users have the same average path power).

3) Nondiversity and noncoding hybrid systems will not be able to operate well for indoor radio. The performance of hybrid systems with predetection diversity improves significantly as the diversity order increases. Also, performance of hybrid systems employing EGC is better than that of hybrid systems employing SC for the same diversity order ($M \neq 1$); the advantage of EGC over SC increases with increasing diversity order.

4) Interleaved channel coding used with various diversity techniques for hybrid SSMA for indoor radio performs better than diversity techniques alone for the same spread-spectrum transmission bandwidth, because the combination can be optimized. Using more powerful codes would give even greater gains in performance, at the expense of additional complexity.

5) For a given system bandwidth, pure DS performs better than hybrid DS/SFH when the multipath interference from the reference user is the dominating disturbance; when the multiaccess interference from the other users is very strong, hybrid DS/SFH with a suitable number of hopping frequencies performs better than pure DS.

In this paper, the indoor radio channel is assumed to be multipath Rayleigh fading, which corresponds to measurements in and around office buildings [12]. Recently, propagation measurements in factories have been reported in [15] where the fading model is Rician instead of Rayleigh. The performance analysis for SSMA over a multipath Rician fading channel will be the subject of a future paper.

ACKNOWLEDGMENT

The authors are grateful to the anonymous reviewers for their helpful comments and valuable suggestions which considerably improved this paper.

REFERENCES

[1] M. Kavehrad and P. J. Mclane, "Performance of low-complexity channel coding and diversity for spread spectrum in indoor wireless communications," *AT&T Tech. J.*, vol. 64, no. 10, pp. 1927–1965, Oct. 1985.

[2] M. Kavehrad and B. Ramamurthi, "Direct-sequence spread spectrum with DPSK modulation and diversity for indoor wireless communications," *IEEE Trans. Commun.*, vol. 35, pp. 224–236, Feb. 1987.

[3] G. L. Turin, "Introduction to spread spectrum antimultipath techniques and their applications to urban digital radio," *Proc. IEEE*, vol. 68, pp. 328–353, Mar. 1980.

[4] E. A. Geraniotis and M. B. Pursley, "Performance of coherent direct sequence spread spectrum communication over specular multipath fading channels," *IEEE Trans. Commun.*, vol. 30, pp. 502–508, June 1985.

[5] J. S. Lehnert and M. B. Pursley, "Multipath diversity reception of spread spectrum multiple access communication systems," *IEEE Trans. Commun.*, vol. 35, pp. 1189–1198, Nov. 1987.

[6] H. Xian, "Binary code division multiple access operating in multipath fading noise channels," *IEEE Trans. Commun.*, vol. COM-33, pp. 775–784, Aug. 1985.

[7] M. Kavehrad and G. Bodeep, "Design and experimental results for a direct-sequence spread-spectrum radio using differential phase keying modulation for indoor wireless communications," *IEEE J. Select. Areas Commun.*, vol. 5, pp. 815–823, June 1987.

[8] M. Kavehrad and P. L. Mclane, "Spread spectrum for indoor digital radio," *IEEE Commun. Mag.*, vol. 25, pp. 32–40, June 1987.

[9] E. A. Geraniotis, "Noncoherent hybrid DS-SFH spread-spectrum multiple-access communications," *IEEE Trans. Commun.*, vol. COM-34, pp. 862–872, Sept. 1986.

[10] —, "Coherent hybrid DS-SFH spread spectrum multiple-access communications," *IEEE J. Select. Areas Commun.*, vol. SAC-3, pp. 695–705, Sept. 1985.

[11] A. A. M. Saleh and R. A. Valenzuela, "A statistical model for indoor multipath propagation," *IEEE J. Select. Areas Commun.*, vol. 5, pp. 138–146, Feb. 1987.

[12] M. Moenclaey and J. Wang, "Performance of hybrid DS/SFH spread spectrum multiple access with predetection diversity for indoor radio," *AEÜ*, vol. 45, pp. 11–16, Jan. 1991.

[13] J. G. Proakis, *Digital Communications*. New York: McGraw-Hill, 1983.

[14] E. R. Berlekamp, "The technology of error-correcting codes," *Proc. IEEE*, vol. 68, pp. 564–593, May 1980.

[15] T. S. Rappaport and C. D. McGillen, "UFH fading in factories," *IEEE J. Select. Areas Commun.*, vol. 7, pp. 40–48, Jan. 1989.

Jiangzhou Wang was born in Hubei, China, on November 15, 1961. He received the B.S. and M.S. degrees, both in electrical engineering, from Xidian University, Xian, China, in 1983 and 1985, respectively. Currently, he is working towards the Ph.D. degree at the University of Ghent, Ghent, Belgium.

From 1986 to 1988, he was a Ph.D. candidate at Southeast University (former Nanjing Institute of Technology), Nanjing, China. Since March 1989, he has been a Research Assistant at the Communication Engineering Laboratory of the University of Ghent. His research interests are in spread-spectrum communications, mobile radio, multiuser communications, and synchronization problems.

Marc Moeneclaey was born in Ghent, Belgium, on October 15, 1955. He received the diploma of electrical engineering and the Ph.D. degree from the University of Ghent, Ghent, Belgium, in 1978 and 1983, respectively.

Since October 1978, he has been working at the Electronics Laboratory of the University of Ghent, first as a Research Assistant for the Belgian National Fund for Scientific Research (NFWO) then as a Senior Research Assistant for the NFWO. Since August 1985, he has been working at the Communication Engineering Laboratory, University of Ghent, as a Research Associate for the NFWO. His main research interests are statistical communication theory, synchronization problems, digital networks, and satellite communication.

Performance Analysis of a Hybrid DS/SFH CDMA System Using Analytical and Measured Pico Cellular Channels

Fevzi Çakmak, *Student Member, IEEE*, René G. A. Rooimans, *Member, IEEE*, and Ramjee Prasad, *Senior Member, IEEE*

***Abstract*—This paper presents the throughput and delay analysis of a packet-switched code division multiple access (CDMA) network based on the hybrid direct sequence (DS)/slow frequency hopping (SFH) spread-spectrum multiple access (SS MA) technique with Q-, B-, and D-PSK modulation using analytical and measured pico cellular channels. The performance of the hybrid DS/SFH, DS, and SFH multiple access techniques have been compared in a pico cellular personal communications network (PCN) environment. Multipath and multiple access interference are considered. The performance is evaluated for a given delay spread and a fixed bandwidth. The effects of forward error correction (FEC) coding and diversity techniques, such as selection diversity and maximal ratio combining on the performance, are also investigated.**

I. Introduction

CODE division multiple access (CDMA) is a strong future candidate as a multiple access scheme for a pico cellular personal communications network (PCN). CDMA is not only attractive from the point of multiple access capabilities, which is needed when a large number of users want to transmit simultaneously, but also for combating multipath fading, which occurs in pico cellular radio channels. In the family of CDMA, direct sequence (DS) and slow frequency hopping (SFH) techniques have their individual features. The DS CDMA technique is a good answer to the problem of multipath fading [1]–[6]; it suffers, however, from the near–far effect. On the other hand, the SFH CDMA technique is more sensitive to multipath interference, but suffers less from the near–far effect. As a consequence the combination of direct sequence and slow frequency hopping, hybrid DS/SFH CDMA, is an attractive solution for combating the multipath effect, and the multiple access interference, and for reducing the near–far problem [7]–[9].

The present paper has the following three main technical contributions. First, the paper evaluates the throughput and the delay characteristics of a packet-switched CDMA network based on hybrid DS/SFH, DS, and SFH. Secondly, the performance of of hybrid DS/SFH CDMA is evaluated and compared using Q-, B-, and D-PSK modulation. These two evaluations are performed using an analytical model in which the radio channel has been modeled as a slow Rician-fading channel. Thirdly, the throughput and delay of hybrid DS/SFH CDMA are obtained using the measured delay profiles in a pico cellular environment and the performance is compared with the results obtained with the analytical model.

An analytical framework of random access spread-spectrum networks has been presented in [10], in which link-level and topological issues are discussed. We use this analytical framework in this paper and we additionally consider the physical level of communication in the CDMA network. This means that the effects of fading, the use of various modulation types such as Q- , B-, and D-PSK, the use of forward error correction (FEC) codes and the effect of diversity on the throughput and delay performance of the CDMA network are evaluated. The results of this paper can be applied to develop pico cellular wireless personal communications systems and wireless local area networks (LAN's).

The paper is structured as follows. In Section II, analytical and measured channel models are described. In Section III, the expression for the packet success probabilities for the different modulation types are given. In Section IV, the performance results are given and discussed. Finally conclusions are given in Section V.

II. Channel Models

We consider a centralized network with C users and one coordinating base station, consisting of C spread-spectrum transceivers (transmitters/receivers). Each transmitter/receiver pair is identified by its own spread-spectrum code and hopping pattern. So, from a topological viewpoint we are dealing with paired-off users, which means there is no topological "competition" between active transmitters. The multi-user interference is the only disruptive interaction between pairs [10]. The network has the following characteristics: i) Data is transmitted in packets of N_p bits; ii) the system is slotted; iii) a time slot is equal to the packet duration; iv) the system has a positive acknowledgment scheme. In order to be able to give some closed expressions for the throughput, we have to make the following five assumptions: i) All users are identical from a statistical point of view and the same holds for the transceivers at the base station; ii) the averaged received power at the base station is equal for each user; iii) the acknowledgment are almost costless from a capacity point of view, fully reliable and

Manuscript revised December 8, 1995. This paper was presented in parts at PIMRC'93, PIRMC'94, GLOBECOM'94, and GLOBECOM'95.

The authors are with the Telecommunications and Traffic Control Systems Group, Delft University of Technology, 2600 GA Delft, The Netherlands.

Publisher Item Identifier S 0733-8716(96)02520-6.

instantaneous; iv) the channel is memoryless; v) the system is in a stable state.

The link between the kth user and the base station is characterized by its lowpass equivalent impulse response, which is given as

$$h_k(t) = \sum_{l=1}^{L} \beta_{kl}\delta(t - \tau_{kl}) \exp(j\gamma_{kl}) \tag{1}$$

where kl refers to the lth path of user k.

A. Analytical Channel Model Assuming Identically Distributed Path Gains

In case of the analytical independent identically distributed (i.i.d.) channel the following assumptions are made concerning the channel parameters: i) the path gains β_{kl} are Rician distributed; ii) the delays τ_{kl} are uniformly distributed random variables over $[0, T]$; iii) the phases γ_{kl} are uniformly distributed over $[0, 2\pi]$. The channel is assumed to be of the slow fading type, which means that the random variables do not change considerably for the duration of one packet.

The probability density function (pdf) of the Rician distributed path gains β_{kl} is given by

$$p_\beta(r) = \frac{r}{\sigma_r^2} \exp\left(-\frac{r^2 + A^2}{2\sigma_r^2}\right) I_0\left(\frac{rA}{\sigma_r^2}\right) \tag{2}$$

with $r \geq 0$ and $A \geq 0$. Here $I_0()$ represents the modified Bessel function of the first kind and zero order. The Rice parameter R is defined as the ratio of the power associated with the specular signal component $A^2/2$ and the power associated with the scattered components σ_r^2. This parameter incorporates the radio characteristics of the environment.

When the signal bandwidth is much larger than the coherence bandwidth of the channel, then the channel is said to be frequency selective. This means that different paths can be resolved by the receiver. The maximum number of resolvable paths L is given by

$$L = \left\lfloor \frac{T_m}{T_c} \right\rfloor + 1 = \lfloor N R_b T_m \rfloor + 1 \tag{3}$$

where T_m denotes the rms delay spread, T_c the chip time, N the spreading code length (N_q for Q-PSK, N_b for B-PSK, and N_d for D-PSK), and R_b the channel bit rate.

B. Channel Model Using the Pico Cellular Measured Power Delay Profiles

The power delay profile measurements have been carried out in the frequency bands 2.4, 4.75, and 11.5 GHz in an indoor environment [11]. In the analytical channel model we have described the path gains by a Rician pdf, while in the measured pico cellular channel model we have used the true measured values of the path gains for the calculations. This is a more practical situation than assuming the path gains to be Rician distributed. An measured power delay (MPD) profile is shown in Fig. 1. The peaks of the profile shown in Fig. 1 are estimated. The estimation is based on the fact that each local maximum in the original profile represents a ray. Due to the inherent diversity of the direct sequence spread spectrum (DS/SS), the delayed versions of the initial pulse can be grouped in clusters which can be resolved independently. Two signals having a time difference less than the chip time T_c are treated as a single signal. These two signals can be combined by means of a vectorial summation. For example, if we assume that the delay, amplitude and phase of ray 1 are respectively denoted as τ_1, β_1, and ϕ_1, and the delay, amplitude, and phase of ray 2 are respectively denoted as τ_2, β_2, and ϕ_2, then the delay, amplitude, and phase of the resulting signal are given by

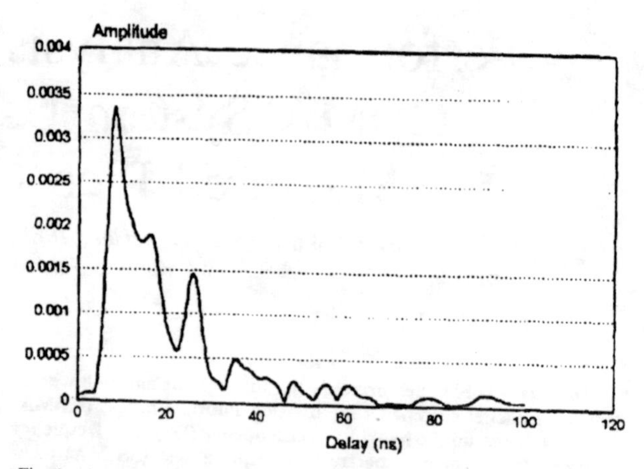

Fig. 1. Measured MDP.

$$\tau_r = \frac{\beta_1\tau_1 + \beta_2\tau_2}{\beta_1 + \beta_2} \tag{4}$$

$$\beta_r = \sqrt{(\beta_1 \cos\phi_1 + \beta_2 \cos\phi_2)^2 + (\beta_1 \sin\phi_1 + \beta_2 \sin\phi_2)^2} \tag{5}$$

and

$$\phi_r = \arctan\left(\frac{\beta_1 \sin\phi_1 + \beta_2 \sin\phi_2}{\beta_1 \cos\phi_1 + \beta_2 \cos\phi_2}\right). \tag{6}$$

If we assume selection diversity, we lock on the strongest received path. If no diversity is being used, we lock on the first received path.

III. Packet Success Probability

The packet success probability from a particular transmitter's viewpoint given $(k - 1)$ other packets in the channel is represented by $P_k(k)$. We can interpret the packet success probability as the probability that the packet is received without any bit errors. This is the probability which incorporates the physical aspects, such as modulation type and the channel statistics. Although the bit errors in a packet are often highly correlated, we assume independent bit errors in a packet for the sake of mathematical convenience. When channel coding is used, this is a reasonable assumption, because in that case the information bits are interleaved by coding bits which creates less correlation among the information bits. For independent bit errors in a packet we can write [4]

$$P_k(k) = [1 - P_e(k)]^{N_p} \tag{7}$$

where $P_e(k)$ denotes the average bit error probability as a function of the number simultaneously transmitting users k.

The expression for $P_e(k)$ is determined by the modulation type.

In order to determine the packet success probability, we first have to consider the average bit error probability, according to (7). We have to keep in mind that we consider the channel to be of the slow fading type, which means that the channel statistics do not change considerably for the duration of one packet. First the packet success probability is derived for D-PSK, Q-PSK, and B-PSK modulation using the analytical model and then it is derived for B-PSK modulation using the measured delay profile.

A. Packet Success Probability and Bit Error Probability in Case of D-PSK

In case of hybrid DS/SFH with D-PSK modulation and selection diversity the conditional bit error probability as a function of the number of simultaneously transmitting users is given by

$$P_e(k \mid \beta_{\max}, \tau_{kl}, L) = \sum_{n_i=0}^{k-1} P_e(n_i \mid \beta_{\max}, \tau_{kl}, L) P(n_i). \quad (8)$$

Here $P(n_i)$ represents the probability of having n_i active interferers out of $k-1$ active transmitters and is given by

$$P(n_i) = \binom{k-1}{n_i}\left(\frac{1}{q}\right)^{n_i}\left(1-\frac{1}{q}\right)^{k-1-n_i} \quad (9)$$

where q is the number of frequencies in the hopping pattern. In case of selection diversity and additive white Gaussian noise (AWGN) the conditional bit error probability $P_e(n_i \mid \beta_m, \tau_{kl}, L)$ is given by [6]

$$P_e(n_i \mid \beta_m, \{\tau_{kl}\}, L) = Q(a,b) - \frac{1}{2}\left(1 + \frac{\mu}{\sqrt{\mu_0 \mu_{-1}}}\right) \times \exp\left(-\frac{a^2+b^2}{2}\right) I_0(ab) \quad (10)$$

where $Q()$ denotes the Marcum Q-function. Here we have assumed fixed delays, phase angles bits, and considered that the maximum of the envelope has been found (all other signals are seen as noise). According to [6] the μ-parameters in (10), representing the variances and the covariance, respectively, of the matched filter output of the receiver, are assumed to be random Gaussian variables, which are characterized by a Gaussian distribution. The expressions for the μ-parameters are given here below

$$\mu_0 = \frac{P}{8}\sum_{k=1}^{n_i}\left[E(X_k^2) + E(\hat{X}_k^2) + E(Y_k^2) + E(\hat{Y}_k^2)\right] + \frac{P}{4}\left[E(X_1\hat{X}_1) + E(Y_1\hat{Y}_1)\right] + 2\sigma_N^2 \quad (11a)$$

$$\mu_{-1} = \frac{P}{8}\sum_{k=1}^{n_i}\left[E(X_k^2) + E(\hat{X}_k^2) + E(Y_k^2) + E(\hat{Y}_k^2)\right] + 2\sigma_N^2 \quad (11b)$$

$$\mu = \frac{P}{8}\sum_{k=1}^{n_i}\left[E(X_k\hat{X}_k) + E(Y_k\hat{Y}_k)\right] + \frac{P}{8}\left[E(\hat{X}_k) + E(\hat{Y}_k)\right] \quad (11c)$$

where X_1, X_k, $\hat{X}_1$, and $\hat{X}_k$ are given by

$$X_1 = \sum_{l, l\neq j} \beta_{1l}\cos(\psi_{1l})R_{1l}(\tau_{1l}) \quad (12a)$$

$$X_k = \sum_{l} \beta_{kl}\cos(\psi_{kl})R_{kl}(\tau_{kl}) \quad (12b)$$

$$\hat{X}_1 = \sum_{l, l\neq j} \beta_{1l}\cos(\psi_{1l})\hat{R}_{1l}(\tau_{1l}) \quad (12c)$$

$$\hat{X}_k = \sum_{l} \beta_{kl}\cos(\psi_{kl})\hat{R}_{kl}(\tau_{kl}) \quad (12d)$$

respectively (similar expressions for Y). The partial correlation functions R_{1k} and $\hat{R}_{1k}$ as well as a and b are defined in [5]. Substitution of the conditional bit error probability (10) in (7) yields the conditional packet success probability. The conditioning on the random variables is now removed by weighting the conditional success probability with the probability density functions of the μ-parameters and with the Rician fading statistics, i.e., the probability density function of the maximal path gain β_m. For an order of diversity equal to M and assuming that all path gains are equally distributed independent Rician variables, this pdf becomes

$$f_{\beta_m}(\beta_m) = M\left[\int_0^{\beta_m} \frac{z}{\sigma_r^2}\exp\left(-\frac{A^2+z^2}{2\sigma_r^2}\right)\cdot I_0\left(\frac{A_z}{\sigma_r^2}\right)dz\right]^{(M-1)} \cdot \frac{\beta_m}{\sigma_r^2}\exp\left(-\frac{A^2+\beta_m^2}{2\sigma_r^2}\right)\cdot I_0\left(\frac{A\beta_m}{\sigma_r^2}\right). \quad (13)$$

The unconditional packet success probability is then given by

$$P_k'(k) = \int_{\mu}\int_{\mu_{-1}}\int_{\mu}\int_{\beta_m}\left[1 - P_e(k \mid \beta_m, \mu_0, \mu_{-1}, \mu)\right]^{N_p} \cdot f_{\beta_m}(\beta_m) f_{\mu_0}(\mu_0) f_{\mu_{-1}}(\mu_{-1}) \cdot f_{\mu}(\mu) d\beta_m d\mu d_{\mu_{-1}} d_{\mu_0}. \quad (14)$$

B. Packet Success Probability and Bit Error Rate in Case of B- and Q-PSK

In case of B- and Q-PSK modulation, the conditional bit error probability is given by

$$P_e(k \mid \alpha) = \sum_{n_i=0}^{k-1} P_e^{B,Q}(n_i \mid \alpha)P(n_i) \quad (15)$$

where P_e^B and P_e^Q are the bit error probabilities for B-PSK and Q-PSK modulation, respectively. We have derived these bit error probabilities for both selection diversity and maximal ratio combining.

1) Selection Diversity: Selection diversity means that the largest of a group of M signals carrying the same information is selected. The order of diversity M equals the number of antennas times the number of paths. The bit error probabilities in case of selection diversity are given as

$$P_e^B(n_i \mid \alpha_{\max}) = \frac{1}{2}\text{erfc}\left\{\left[\left(\frac{E_b(2\sigma_r^2 + A^2)}{N_0}\right)^{-1}\frac{1}{\alpha_{1j}^2} + \frac{2L}{3N\alpha_{1j}^2}(1+n_i)\right]^{-0,5}\right\} \quad (16)$$

and

$$P_e^Q(n_i \mid \alpha_{\max}) = \frac{1}{2}\text{erfc}\left\{\left[\left(\frac{E_b(2\sigma_r^2 + A^2)}{N_0}\right)^{-1}\frac{1}{\alpha_{1j}^2} + \frac{L}{3N\alpha_{1j}^2}(1+n_i)\right]^{-0,5}\right\} \quad (17)$$

for B-PSK and Q-PSK, respectively. Substitution of the conditional bit error probabilities given in (16) or (17) in (7) yields the conditional packet success probability for B- or Q-PSK modulation. Similar to D-PSK, the conditioning is removed by weighting the conditional bit error probability, given in (15), with the pdf of the normalized maximal path gain α_m. This pdf is obtained by a change of variables in (11) in the following way: $\alpha = \beta/\sqrt{A^2 + 2\sigma_r^2}$ and $y = z/\sqrt{A^2 + 2\sigma_r^2}$. For the pdf of the normalized maximum path gain we then find

$$f_{\alpha_{\max}}(\alpha) = M\left[\int_0^{\alpha} 2y(1+R)\exp(-R-(1+R)y^2) \times I_0\left(2\sqrt{R(1+R)}y\right)dy\right]^{(M-1)} \cdot 2\alpha(1+R)\exp\left(-\alpha^2(1+R)-R\right) \times I_0\left(2\sqrt{R(1+R)}\alpha\right). \quad (18)$$

The unconditional packet success probability is now given by

$$P_k'(k) = \int_{\alpha_m}[1 - P_e(k \mid \alpha_m)]^{N_p} f_{\alpha_m}(\alpha_m)\, d\alpha_m. \quad (19)$$

2) Maximal Ratio Combining: In case of maximal ratio combining, the contributions of the several resoluted paths are added together. There is a compensation for the phase shift and a weighting associated with the signal strength. The bit error probability for B-PSK and Q-PSK modulation are given, respectively, as

$$P_e^B(n_i \mid \alpha_{\max}) = \frac{1}{2}\text{erfc}\left\{\left[\left(\frac{E_b(2\sigma_r^2 + A^2)}{N_0}\right)^{-1}\frac{1}{Mv} + \frac{2L}{3NMv}(1+n_i)\right]^{-0,5}\right\} \quad (20)$$

and

$$P_e^Q(n_i \mid \alpha_{\max}) = \frac{1}{2}\text{erfc}\left\{\left[\left(\frac{E_b(2\sigma_r^2 + A^2)}{N_0}\right)^{-1}\frac{1}{Mv} + \frac{L}{3NMv}(1+n_i)\right]^{-0,5}\right\} \quad (21)$$

where v is a normalized variable and is defined as $v = t_n/M(2\sigma_r^2 + A^2)$ with $t_n = \sum_{m=1}^{M}\beta_{1m}^2$. In order to remove the channel statistics we have to average these bit error probabilities with the pdf of v, which is given by

$$f(v) = M[1+R]\left[v\left(1+\frac{1}{R}\right)\right]^{M-1/2} \times \exp(-MR - Mv(1+R)) \cdot I_{M-1}\left(2M\sqrt{vR(1+R)}\right). \quad (22)$$

The unconditional packet success probability is now given by

$$P_k'(k) = \int_v [1 - Pe(k \mid v)]^{N_p} f_v(v)dv. \quad (23)$$

C. Packet Success Probability and Bit Error Rate in Case of the Measured Power Delay Profile for B-PSK Modulation

In case of the measured pico cellular channel model we do not approximate the path gains, delays and phase shifts by random variables. In fact we use the true measured values for the calculations. The average bit error probability for the measured pico cellular channel model is given as (24) (shown at the bottom of the page). If we look at the downlink, the channel is the same for all k users. Then the subscript k in (24) disappears. The packet success probability can now be calculated by using (7).

IV. Performance Results

The performance has been investigated in terms of throughput and delay. We have used the definition for the throughput S, [4], [10], as the expected number of successfully received packets per time slot, given by

$$S = \sum_{k=1}^{C} kp_k P_k'(k) \quad (25)$$

where p_k is the probability density function of the composite arrivals in a given time slot. For identical, independent, and a finite number of users, this pdf is a binomial distribution given by

$$p_k = \binom{C}{k}\left[\frac{G}{C}\right]^k\left[1-\frac{G}{C}\right]^{C-k} \quad (26)$$

where the offered traffic G is defined as the average number of transmissions (new plus retransmitted packets) per time slot by C users. The average normalized delay D of the system is defined as the average number of time slots between the generation and the successful reception of a packet. According to [4] the normalized delay is given by

$$D = 1.5 + \left[\frac{G}{S} - 1\right](\lfloor\delta + 1\rfloor + 1) \quad (27)$$

where $(G/S - 1)$ represents the average number of retransmissions needed for a packet to be received successfully and δ denotes the mean of the retransmission delay. The round trip propagation delay can be neglected in a pico cellular environment.

$$P_e(k) = \sum_{n_i=0}^{k-1}\frac{1}{2}\text{erfc}\left\{\sqrt{\frac{\frac{PT^2}{8}\beta_{1j}^2}{\frac{PT^2}{6N}\sum_{l,l\neq j}\beta_{1l}^2\cos^2(\Psi_{1l}) + n_i\frac{PT^2}{6N}\sum_{l=1}^{L}\beta_{kl}^2\cos^2(\Psi_{kl}) + \frac{N_0T}{8}}}\right\} \cdot P(n_i). \quad (24)$$

The throughput S and delay D are computed using (25)–(27) and the analytical expressions of Section III for hybrid DS/SFH CDMA. The throughput and delay are also evaluated for DS and SFH CDMA by adapting the hybrid DS/SFH analytical expressions. In the case of direct sequence, all users use the same carrier frequency; hence, the number of hopping frequencies q then becomes equal to one. The number of active interferers n_i is then equal to the number of active transmitters k. For the probability of having n_i active interferers we then have

$$P(n_i) = 1. \tag{28}$$

In case of SFH, the direct sequence part has to be left out of the model by taking $N = 1$. The number of resolvable paths, L in this case, is equal to one, because there is no inherent spread-spectrum anymore.

We have considered two types of FEC codes, viz. the (7, 4) Hamming code and the (23, 12) Golay code. The figures between brackets (n, k) means that k source bits are transformed into a block of n channel bits by coding. The conditional bit error probability at the hard decision output is given by

$$P_{ecl} = \frac{1}{n} \sum_{m=t+1}^{n} m \binom{n}{m} P_e^m (1 - P_e)^{n-m} \tag{29}$$

where t denotes the number of errors the code can correct and P_e the conditional bit error probability at the hard decision output.

A. Analytical Channel Model Assuming Identically Distributed Path Gains

For the comparison of the three modulation types D-PSK, Q-PSK, and B-PSK we have kept the bandwidth, information bit rate and the rms delay spread fixed. This means that the ratio Nq/T has to be the same for the three systems. Unless stated otherwise we have chosen the parameters such that the ratio Nq/T, which is proportional to the transmission bandwidth, equals approximately 200 MHz. This is a practical value for spread-spectrum communications systems.

Since the bit rates for B-PSK and D-PSK are equal to $1/T$ and for Q-PSK equal to $1/T_q = 1/2T$, the bit rate constraint leads to $T_q = 2T$, where T_q is the bit duration in case of Q-PSK and where T is the bit duration in case of B- and D-PSK. Assuming the same number of hopping frequencies, q for the three systems, we finally have $N_q = 2N_b = 2N_d$, where N_q, N_b, and N_d denote the spreading code lengths in case of Q-PSK, B-PSK, and D-PSK, respectively.

When FEC coding is used, we would like to keep the information bit rate R_i the same. For the channel bit rate R_b we can write $R_b = (n/k)R_i$. Since the ratio n/k approximately equals two for the (7, 4) Hamming and (23, 12) Golay code, the channel bit rate should be doubled in case of FEC coding in order to have the information bit rate fixed. In all cases R_i is taken 64 kb/s. We have considered a signal-to-noise (SNR) atio of 20 dB (in case of FEC coding this becomes 17 dB). The packet length N_p is taken equal to 1024 b. In the analytical model we have assumed the rms delay spread T_m to be equal to

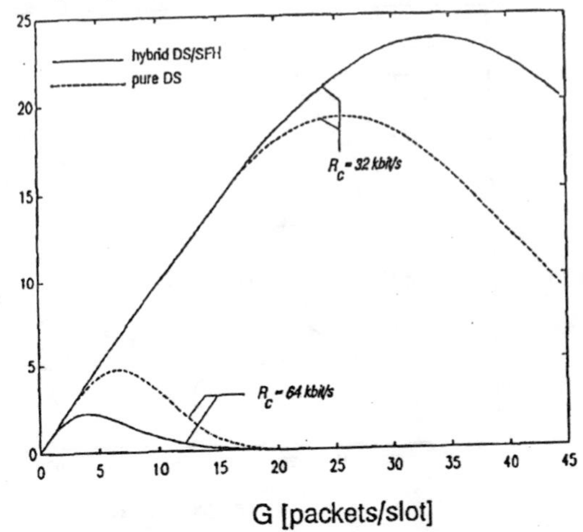

Fig. 2. Comparison of the throughput of hybrid DS/SFH CDMA and DS CDMA with D-PSK, with the bit rate as a parameter, using the analytical channel model. R = 6, 8 dB; M = 4; q = 2; $N_{DS/S_{SFH}}$ = 127; N_{DS} = 255.

250 ns. The Rice parameter R, which characterizes the indoor environment is taken equal to 6.8 dB.

In Table I we have presented the throughput S normalized on the offered traffic G for Q-, B-, and D-PSK modulation with the order of selection diversity M as a parameter. It is seen that Q-PSK yields relatively the largest throughput and D-PSK relatively the poorest. It is obvious that selection diversity enhances the performance in comparison with nondiversity (M = 1).

Figure 2 depicts the throughput comparison for hybrid DS/SFH CDMA and DS CDMA for D-PSK modulation at fixed bandwidth of approximately 16 MHz for two bit rates, viz. 32 kb/s and 64 kb/s. It is seen that for a relatively low bit rate, which corresponds with a relatively small number of resolvable paths according to (3), the hybrid system yields a larger throughput than the DS system. For a higher bit rate and consequently a larger number of resolvable paths, pure DS performs better than hybrid DS/SFH with this parameter setting and the chosen bandwidth. The number of hopping frequencies should be increased in order to fully explore the capabilities of hybrid DS/SFH. In this case, where the number of hopping frequencies only equals two, DS outperformes hybrid DS/SFH. We also see that an increase in bit rate decreases the performance for both hybrid DS/SFH and pure DS.

Figure 3 shows the effect of FEC coding on the delay for hybrid DS/SFH with Q- and B-PSK modulation at fixed bandwidth for the same number of hopping frequencies, but with varying N. Q-PSK modulation and (23, 12) Golay coding yields the best performance. In this case, the (7, 4) Hamming code does not improve the performance for both modulation techniques. Since the bandwidth is kept fixed in the noncoding and coding situation, the spreading code length N is reduced by a half in case of coding. This means that the multi-user and multipath interference terms have a larger influence in

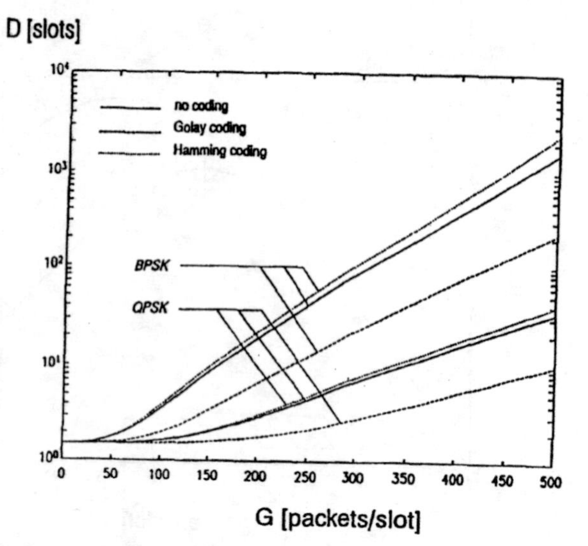

Fig. 3. Effect of FEC coding on the delay of hybrid DS/SFH CDMA with B- and Q-PSK, using the analytical channel model. $M = 4$; $q = 10$; $L = 5$; $N_q = 255$ (coding); $N_q = 511$ (no coding); $N_b = 127$ (coding); $N_b = 255$ (no coding).

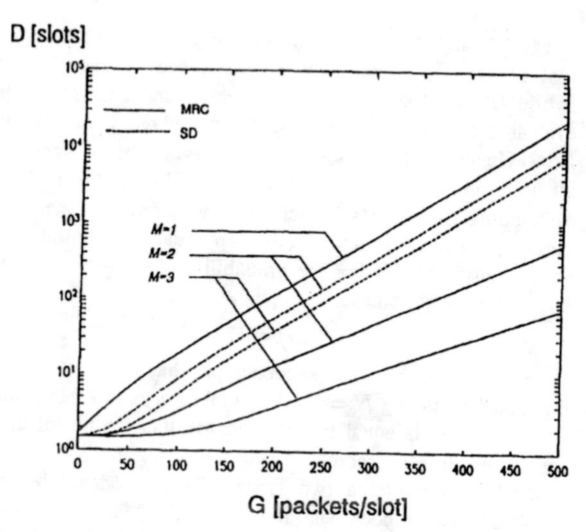

Fig. 4. Effect of diversity on the delay performance of hybrid DS/SFH CDMA with B-PSK with M as a parameter, using the analytical channel model. $N_b = 127$; $L = 3$.

that case [see (16) and (17)], which has to be overcome by the FEC codes. The Hamming code has a poorer coding gain than the Golay code and so the Hamming code isn't able to improve the performance in this case.

Figure 4 show the delay performance of hybrid DS/SFH CDMA with B-PSK modulation, selection diversity, maximal ratio combining, and order of diversity M as a parameter. It is seen that maximal ratio combining is superior over selection diversity, which becomes more obvious for larger values of M. Maximal ratio combining uses all available paths to construct a decision variable, while selection diversity only uses the strongest path to come to a decision variable. In the case of maximal ratio combining, a more reliable decision can be made than in the case of selection diversity.

Figure 5 depicts the throughput comparison of hybrid DS/SFH CDMA with the throughput of the pure DS CDMA, for B-PSK and Q-PSK modulation without FEC coding. We can see that hybrid DS/SFH CDMA has a higher throughput than the pure DS CDMA. In this situation the number of hopping frequencies is large enough to outperform DS as opposed to Fig. 2. Although the multi-user and multipath interference terms are slightly larger with hybrid DS/SFH than in case of DS [according to (16) and (17)], the number of hopping frequencies is large enough to overcome the effect of the interference.

In Fig. 6 the throughput using hybrid DS/SFH is compared with the throughput using pure DS and SFH for B- and Q-PSK modulation, with FEC coding. For the pure SFH technique the number of resolvable paths L is equal to one. So there is no inherent diversity which is the case with pure DS. We can see that hybrid DS/SFH has the highest throughput. Besides, we see that the maximum throughput for both SFH and DS is almost equal, however the throughput for SFH does not vary much over the large range of G. In the case of SFH with this parameter setting, the number of hopping frequencies is

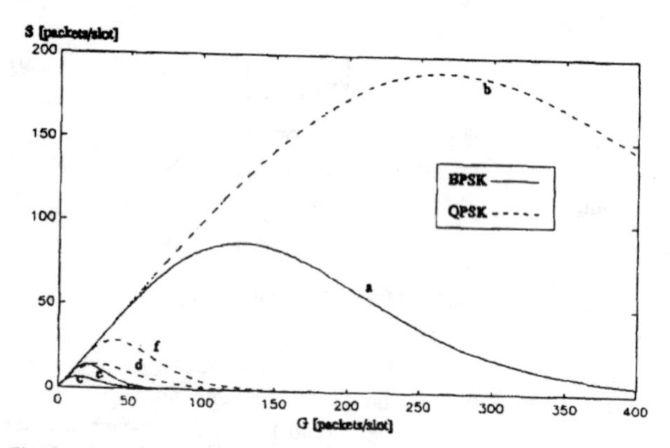

Fig. 5. Throughput of hybrid DS/SFH CDMA versus the DS CDMA without FEC coding, using the analytical channel model: (a) Hybrid: $L = 3, M = 4, q = 10, N = 127$; (b) Hybrid: $L = 3, M = 4, q = 10, N = 255$; (c) $L = 21, M = 1, q = 1, N = 1270$, DS; (d) $L = 21, M = 1, q = 1, N = 2550$, DS; (e) $L = 21, M = 4, q = 1, N = 1270$, DS; (f) $L = 21, M = 4, q = 1, N = 2550$, DS.

large enough to have a slightly better performance than DS, although the interference terms according to (16) and (17) are larger than in case of DS. The results obtained here are such that the possibilities of SFH will be investigated further in future work.

B. Channel Model Using the Pico Cellular Measured Power Delay Profiles

Results are presented in the following figures for the measured pico cellular channel model. In Figs. 7 and 8, respectively, the throughput and delay of hybrid DS/SFH CDMA has been compared with the throughput of DS CDMA for a fixed bandwidth (80 MHz) in the 2.4 GHz band using the measured pico cellular channel model. It is seen that hybrid DS/SFH CDMA is superior to DS CDMA.

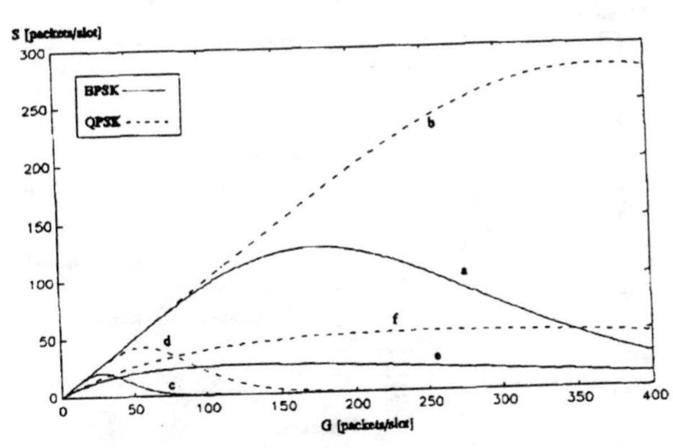

Fig. 6. Throughput of hybrid DS/SFH CDMA versus the DS CDMA and SFH CDMA with FEC coding, using the analytical channel model: (a) Golay coding, Hybrid: $L = 3, M = 4, q = 10, N = 63, (23, 12)$; (b) Golay coding, Hybrid: $L = 3, M = 4, q = 10, N = 127, (23, 12)$; (c) Golay coding, DS: $L = 21, M = 4, q = 1, N = 630, (23, 12)$; (d) Golay coding, DS: $L = 21, M = 4, q = 1, N = 1270, (23, 12)$; (e) Golay coding, SFH: $L = 1, M = 4, q = 630, N = 1, (23, 12)$; (f) Golay coding, SFH: $L = 1, M = 4, q = 1270, N = 1, (23, 12)$.

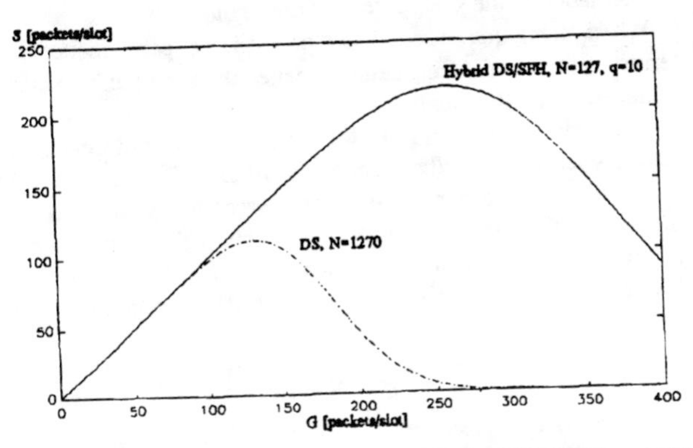

Fig. 7. Throughput comparison between hybrid DS/SFH CDMA and DS CDMA with B-PSK modulation, for a bandwidth of 80 MHz in the 2.4 GHz band, using the measured pico cellular channel model. $K = 400$, bandwidth = 80 MHz, $R_b = 64$ Kb/s, and SNR = 20 dB.

Figures 9 and 10, respectively, depict the throughput and delay performance of a hybrid DS/SFH CDMA system for a fixed number of hopping frequencies ($q = 6$) and different values for the bit rate R_b and N. We can see that the highest throughput is obtained with $R_b = 125$ kb/s and $N = 255$. A high bit rate combined with a smaller spreading sequence length yields a worse throughput and higher delays.

In Figs. 11 and 12, respectively, the throughput and delay of hybrid DS/SFH CDMA, using the measured pico cellular channel model has been compared with the performance obtained using the analytical model. We can see that the maximum throughput obtained using the MPD channel model is higher than in case of the analytical IID channel model. This is due to the fact that in the IID channel model it is assumed that all paths arrive with the same power at the receiver, as opposed to the MPD channel where delayed paths arrive with less power at the receiver than the first arriving path. This means that in case of the MPD channel the multipath interference is less than in case of the IID channel model,

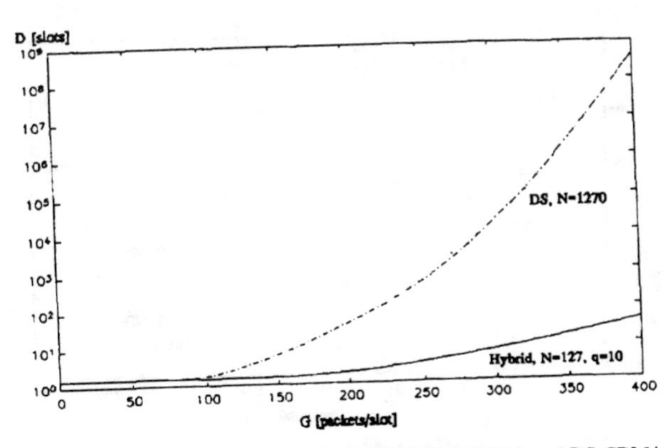

Fig. 8. Delay comparison between hybrid DS/SFH CDMA and DS CDMA with B-PSK modulation, for a bandwidth of 80 MHz in the 2.4 GHz band, using the measured pico cellular channel model. $K = 400$, bandwidth = 80 MHz, $R_b = 64$ Kb/s, and SNR = 20 dB.

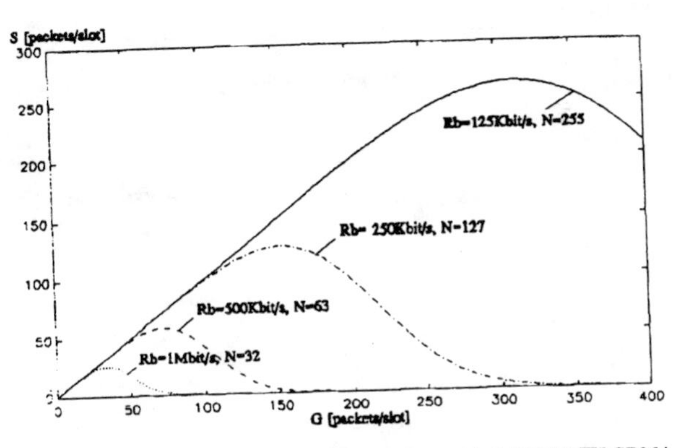

Fig. 9. Influence of R_b and N on the throughput of hybrid DS/SFH CDMA with B-PSK modulation, at a fixed number of hopping frequencies q, using the measured pico cellular channel model. $K = 400$, bandwidth = 200 MHz, SNR = 20 dB, and $q = 6$.

which yields a better performance than the case when using the IID model.

V. Conclusion

A performance analysis of hybrid DS/SFH, DS, SFH CDMA using analytical and measured pico cellular channels has been presented. The throughput and delay characteristics of hybrid DS/SFH CDMA has been compared with DS CDMA and SFH CDMA for analytical channels. The performance of hybrid DS/SFH CDMA has been compared for Q-PSK, B-PSK, and D-PSK modulation, using the analytical channel model. The performance of hybrid DS/SFH CDMA with B-PSK modulation has been compared for analytical and measured pico cellular channels.

The comparison of hybrid DS/SFH with D-, B-, and Q-PSK under the constraint of a fixed bandwidth has shown that Q-PSK yields the best and D-PSK the worst performance of the three modulation methods. With the demodulation of D-PSK modulated signals more noise is involved than demodulating B- and Q-PSK modulated signals, which degrades the performance. In case of Q-PSK, the spreading code sequence

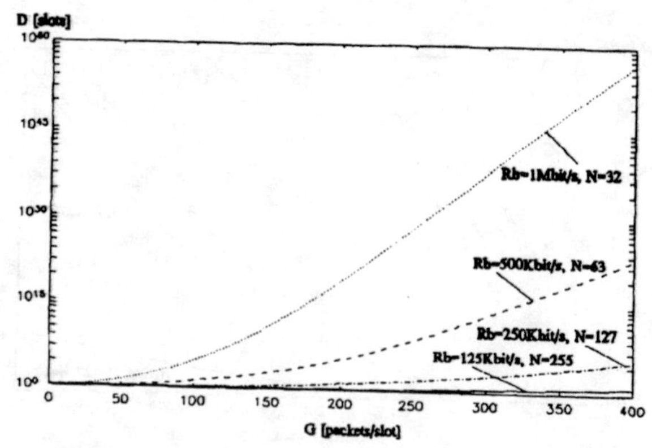

Fig. 10. Influence of R_b and N on the delay of hybrid DS/SFH CDMA with B-PSK modulation, at a fixed number of hopping frequencies q, using the measured pico cellular channel model. $K = 400$, bandwidth = 200 MHz, SNR = 20 dB, and $q = 6$.

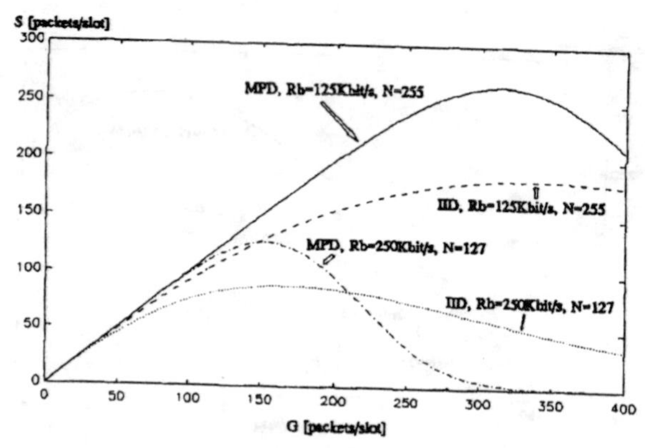

Fig. 11. Throughput comparison of hybrid DS/SFH CDMA with B-PSK modulation, using the measured pico cellular channel model (MPD) and the analytical channel model (IID) for two different bit rates. $K = 400$, bandwidth = 200 MHz, SNR = 20 dB, and $q = 6$.

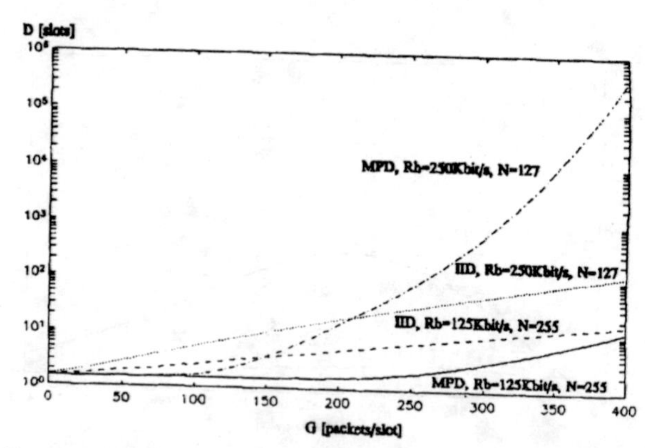

Fig. 12. Delay comparison of hybrid DS/SFH CDMA with B-PSK modulation, using the measured pico cellular channel model (MPD) and the analytical channel model (IID) for two different bit rates. $K = 400$, bandwidth = 200 MHz, SNR = 20 dB, and $q = 6$.

N is twice as large as in case of B-PSK in order to fulfill the bandwidth constraint. This yield a better performance for systems using Q-PSK than systems using B-PSK.

With the constraint of a fixed bandwidth and bit rate a trade-off can be made between the number of frequencies in the hopping pattern q and the spreading code length N. Hybrid DS/SFH outperformes DS and SFH only with the combination of a relatively large number of hopping frequencies and a relatively small spreading code sequence. If we use relativly small bandwidths and high bitrates, then DS outperformes hybrid DS/SFH. This is due to the fact that at a small bandwidth the capabilities of hybrid DS/SFH, such as a relatively large number of hopping frequencies can not be fully explored. Additionally higher bitrates involve a higher number of resolvable paths, which increases the multipath interference. In this case, the relatively high length of the spreading code of DS in that case yield a better performance than the combination of a relatively low number of hopping frequencies and a relatively low spreading code length.

The use of FEC coding can enhance the performance in some cases. If the multi-user and multipath interference is relatively large, which is the case for a large number of resolvable paths, the use of FEC coding only improves the performance if the code rate is large enough.

When using measured delay profiles the performance in terms of throughput and delay is better than the performance obtained with the analytical model. The difference performance using the two models is due to the difference in interference power. When using the MPD channel model the paths arrive with different power at the receiver, as opposed to the case of the analytical model. The analytical channel model in this case describes the worst case performance and yields a lower bound of the throughput and delay performance.

REFERENCES

[1] R. Prasad, H. S. Misser, and A. Kegel, "Performance analysis of direct-sequence spread-spectrum multiple-access communication in an indoor Rician fading channel with D-PSK modulation," *Electron. Lett.*, vol. 26, pp. 1366–1367, Aug. 16, 1990.

[2] H. S. Misser, C. A. F. J. Wijffels, and R. Prasad, "Throughput analysis of CDMA with D-PSK modulation and diversity in indoor Rician fading radio channels," *Electron. Lett.*, Vol. 27, no. 7, pp. 601–602, Mar. 28, 1991.

[3] R. Prasad, C. A. F. J. Wijffels, and K. L. A. Sastry, "Performance analysis of slotted CDMA with D-PSK modulation, diversity and BCH coding in indoor radio channels," *AEÚ*, vol. 46, no. 6, pp. 375–382, 1992.

[4] C. A. F. J Wijffels, H. S. Misser, and R. Prasad, "A micro-cellular CDMA system over slow and fast Rician fading radio channels with forward error correction coding and diversity," *IEEE Trans. Veh. Technol.*, vol. 42, pp. 570–580, Nov. 1993.

[5] R. Prasad, H. S. Misser, and A. Kegel, "Performance evaluation of direct-sequence spread-spectrum multiple-access for indoor wireless communications in a Rician fading channel," *IEEE Trans. Commun.*, vol. 43, pp. 581–592, Feb./Mar./Apr. 1995.

[6] M. Kavehrad and B. Ramamurthi, "Direct-sequence spread spectrum with D-PSK modulation and diversity for indoor wireless communications," *IEEE Trans. Commun.*, vol. COM-35, pp. 224–236, Feb. 1987.

[7] J. Wang and M. Moeneclay, "Hybrid DS/SFH spread-spectrum multiple-access with predetection diversity and coding for indoor radio," *IEEE J. Select. Areas Commun.*, vol. 10, no. 4, pp. 705–713, May 1992.

[8] ——, "Hybrid DS/SFH SSMA with predetection diversity and coding over indoor radio multipath Rician-fading channels," *IEEE Trans. Commun.*, vol. 40, no. 10, pp. 1654–1662, Oct. 1992.

[9] E. A. Geraniotis, "Noncoherent hybrid DS/SFH spread-spectrum multiple access communications," *IEEE Trans. Commun.*, vol. COM-34, pp. 862–872, Sept. 1986.

[10] A. Polydoros and J. Silvester, "Slotted random access spread-spectrum networks: An analytical framework," *IEEE J. Select. Areas Commun.*, vol. SAC-5, no. 6, pp. 989–1002, July 1987.
[11] G. J. M. Janssen and R. Prasad, "Propagation measurements in an indoor radio environment at 2.4 GHz, 4.75 GHz and 11.5 GHz," in *42nd Veh. Technol. Society VTS Conference Frontiers of Technology*, Denver, Colorado, May 10–13, 1992, pp. 617–620.
[12] J. G. Proakis, *Digital Communications*, 2nd edn. New York: McGraw-Hill, 1989.

Fevzi Çakmak (S'96) was born on June 3, 1970, in Istanbul, Turkey. He received the M.Sc. degree from the faculty of electrical engineering of Delft University of Technology, The Netherlands, in 1995.

During his study, he was a trainee at the International Cooperation Department of Teléfonica de España. In 1995, he joined the Network and Service Control Department of KPN Research. His activities and interests there were in the areas of mobile communications, B-ISDN signaling using SVC's, B-ISDN standardization, ATM, and TINA. In 1996, he joined CMG Telecommunications and Utilities BV, where he is working on GSM.

René G. A. Rooimans (M'93) was born in Rotterdam, The Netherlands, on May 3, 1967. He received the B.Sc.E.E. degree from Delft University of Technology, Delft, The Netherlands, in 1989 and 1993, respectively.

He worked as a Research Fellow at the Telecommunications and Traffic-Control Systems Group, Delft University of Technology, Delft. He is currently with the Telecommunications and Information Systems Department of EZH, one of the four Power Utility Companies in The Netherlands. His research interests are in the field of radio and mobile communications and spread-spectrum communications.

Ramjee Prasad (M'88–SM'90) was born in Babhnaur (Gaya), Bihar, India, on July 1, 1946. He received the B.Sc.(Eng) degree from Bihar Institute of Technology, Sindri, India, and the M.Sc.(End.) and Ph.D. degrees from Birla Institute of Technology (BIT), Ranchi, India, in 1968, 1970, and 1979, respectively.

In 1970, he joined BIT as a Senior Research Fellow and became an Assistant Professor in 1980. While he was with BIT, he supervised many research projects in the area of microwave and plasma engineering. From 1983 to 1988, he was with the University of Dar es Salaam (UDSM), Tanzania, where he became a Professor in Telecommunications with the Department of Electrical Engineering, in 1986. At UDSM, he was responsible for the collaborative project "Satellite Communications for Rural Zones" with the Eindhoven University of Technology, The Netherlands. Since February 1988, he has been with the Telecommunications and Traffic-Control Systems Group, Delft University of Technology (DUT), The Netherlands, where he is actively involved in the area of personal indoor and mobile radio communications (PIMRC). His current research interests lies in wireless networks, packet communications, multiple access protocols, adaptive equalizers, spread-spectrum CDMA systems, and multimedia communications. He is currently involved in the European ACTS project FRAMES (Future Radio Wireband Multiple Access System) as a Project Leader of DUT, and is a member of a working group of European cooperation in the field of scientific and technical research (Cost-231) project dealing with "Evolution of Land Mobile Radio (including personal) Communications." He has served as a member of advisory and program committees of several IEEE international conferences. He has also presented keynote speeches, invited papers, and tutorials on PIMRC at various universities, technical institutions, and IEEE Conferences. He has published over 200 technical papers, and is the coordinating editor and one of the Editors-in-Chief of a Kluwer International Journal on *Wireless Personal Communications*; he is also a member of the editorial boards of other international journals including the *IEEE Communications Magazine* and the *IEE Electronics and Communications Engineering Journal*.

Dr. Prasad is listed in the US *Who's Who in the World*. He was Organizer and Interim Chairman of the IEEE Vehicular Technology/Communications Society Joint Chapter, Benelux Section. He is now the Elected Chairman of the Joint Chapter. He is also founder of the IEEE Symposium on Communications and Vehicular Technology (SCVT) in the Benelux and he was the Symposium Chairman of SCVT'93 and SCVT'95. He is a Fellow of the IEE, a Fellow of the Institution of Electronics & Telecommunications Engineers, a Member of the New York Academy of Sciences, and a Member of The Netherlands Electronics and Radio Society (NERG).

A CDMA-Distributed Antenna System for In-Building Personal Communications Services

Howard H. Xia, *Senior Member, IEEE*, Angel B. Herrera, Steve Kim, and Fernando S. Rico

***Abstract*— To investigate applications of spread spectrum code division multiple access (CDMA) technology to in-building personal communications services (PCS), comprehensive studies have been conducted for a CDMA PCS distributed antenna system in the 1.8 GHz band. The CDMA PCS distributed antenna system was set up with three nodes, each having two time-delayed elements, in a Qualcomm two story office building in San Diego. This paper presents measurement and modeling results on coverage, voice quality (frame error rate), reduction of transmit power, and path diversity for the in-building CDMA PCS distributed antenna system. Wideband CDMA signal coverage was predicted by using a ray tracing tool to find optimum placement of the distributed antennas. Using three nodes mounted in the ceiling space between the first and second floors, with each active element transmitting at −5 dBm in the system, the ray-tracing prediction shows good signal coverage in both floors of the building. The prediction results are confirmed by measurements at numerous discrete points with a standard deviation of 3.3 dB. Measurements using various combinations of number of nodes and delay elements showed significant time and path diversity advantages for the CDMA-distributed antenna system in indoor radio environments. Trade-offs between diversity gain and self-interference due to uncaptured finger energy in fringe areas are discussed.**

I. Introduction

THE commercial success of personal communications services (PCS) will rely on advanced radio technologies to make efficient use of the PCS frequency spectrum, and to provide high quality performance and low system complexity. The direct sequence code division multiple access (DS-CDMA) technology has been accepted as a digital cellular standard (TIA/EIA/IS-95) [1], and has been developed as a PCS standard by the Joint Technical Committee (JTC) on Wireless Access in the United States [2]. Some of the unique features of CDMA, such as high system capacity, low transmit power, multipath mitigation, soft hand-off, uniform coverage and path diversity due to RAKE receiver, show promise for PCS applications. In this paper, we study the application of a CDMA-distributed antenna system to in-building PCS. The study involves measurement and modeling on coverage, voice quality (frame error rate), reduction of transmit power, and path diversity for the in-building CDMA PCS distributed antenna system.

Distributed antenna systems are commonly used in remote areas as a way to provide coverage extension, or in hard to reach spots to combat shadowing. However, in practice, the signals simulcast by antennas at different locations may result in deep multipath fading in overlapping regions. Taking advantage of the ability of DS-CDMA to distinguish multipath signals, a distributed antenna system has been proposed by Qualcomm for CDMA PCS applications [3]. The system is able to capture the simulcast signals before the occurrence of multipath fading, and then uses the captured simulcast signals to obtain path diversity gain. In this system, a radio signal is distributed through a coaxial cable to a string of antenna elements with time delays greater than one over the CDMA channel bandwidth. On forward link, using coherent combination at the RAKE receiver of the delayed version of a signal transmitted through multiply placed antennas, the mobile station is expected to experience path diversity gain against multipath and shadow fading. Similarly, on reverse link, the base station demodulators can capture up to four independently fading signals received by antenna elements placed at different locations so as to achieve path diversity.

Such a distributed antenna system is extremely useful for in-building applications. In contrast to outdoor mobile users, indoor users are more likely to expect indoor wireless services with quality comparable to that of wireline networks. Therefore, more stringent system performance is required for in-building wireless systems. However, the indoor propagation environment is harsh, in which frequency flat fading results from multipaths with very short time delays and shadowing results from wall and floor blockage. The short multipaths generated by reflections and diffusions within a building generally cannot be resolved by the CDMA receiver with RF bandwidth of 1.23 MHz. By introducing time delay elements between the antennas in the distributed antenna system, deliberate multipaths with >1 μs time delays are created which can be processed by the CDMA RAKE receiver. Since multipath signals from different paths at different times generally fade independently, significant time and path diversity gain against multipath fading and shadowing is therefore expected to be achieved in this distributed antenna system under indoor propagation environments. With no signal processing in the nodes, and no switching process during handoffs from element to element in the same distributed antenna string, the time and path diversity advantage can be exploited by using the three finger CDMA RAKE receiver. This makes a simple and inexpensive distributed antenna system possible for in-building PCS.

To investigate the impact of the distributed antenna system in indoor environments, a distributed antenna system with

Manuscript revised October 23, 1995. This paper was presented at the IEEE Vehicular Technology Conference, Stockholm, Sweden, June, 1994.

The authors are with AirTouch Communications, Walnut Creek, CA 94598 USA.

Publisher Item Identifier S 0733-8716(96)01949-X.

0733–8716/96$05.00 © 1996 IEEE

three nodes, each having two time-delayed antenna elements, was set up in a Qualcomm two story office building in San Diego. Wideband CDMA signal coverage was predicted by using a ray-tracing tool to find optimum placement of the nodes. Time and path diversity advantages of the distributed antenna system were studied by using various combinations of number of nodes and number of delay elements. Trade-offs between diversity gain and self-interference due to uncaptured finger energy in fringe areas are also discussed based on the pilot channel signal-to-interference ratio E_c/I_o measurements for up to eight delay elements in a single node.

II. Distributed Antenna System Description

The core CDMA PCS system proposed to the Joint Technical Committee (JTC) maintains the 1.23 MHz RF channel bandwidth of TIA/EIA/IS-95 and provides up to 14.4 kbps transmission rate [2]. A wider RF bandwidth (2.46 MHz) and higher data rates (up to 76.8 kbps) have also been proposed for enhanced speech quality and advanced data services in the extended CDMA PCS systems [4]. The CDMA PCS system tested in San Diego was a 1.8 GHz, upbanded version of the 800 MHz cellular CDMA system (TIA/EIA/IS-95), which has an RF channel bandwidth of 1.23 MHz and a transmission rate up to 9.6 kbps. For this test system, the CDMA receiver can resolve multipath signals with time delays greater than about 1 μs.

The in-building distributed antenna test system, as shown in Fig. 1, was set up in a Qualcomm building in San Diego. In order to provide more stringent system performance, which is generally expected by in-building users, the test system was designed with two transmit/receive elements in each node to provide path diversity even within the coverage area of the node. A string of three nodes was used for coverage in the 300 ft long by 100 ft wide two-story office building. The node antennas were placed above the dropped ceiling tiles of the first floor with the dipole pointing up in order to cover both floors of the building. No filtering, up and down conversion, or signal processing occur in the nodes. Each node consists only of two separate simple transmit/receive paired antenna elements. The antenna element has an amplifier and a surface acoustic wave (SAW) delay element. The SAW delay element inserts ~2 μs time delay between the nodes and 8 μs time delay between the two radiating elements within the node. A pair of coaxial cables was used to distribute the transmitted signals to the radiating elements in the nodes. Another pair of coaxial cables was used to receive signals from all antenna elements in the nodes. All signal processing is done in the base station located within the building.

III. Coverage

It is extremely critical for an in-building distributed antenna system to provide adequate signal coverage and diversity without sacrificing system economics. In order to obtain optimum transmit power, optimum number of nodes, and optimum antenna placement locations, wideband CDMA signal strength was predicted for both floors of the building by using a ray-tracing tool developed at Polytechnic University [5], [6]. In this tool, the specular reflection and transmission at interior and exterior walls are taken into account by using 2-D ray-tracing procedures, while signal reduction due to irregular scattering of furniture and ceiling fixtures in the vertical dimension is calculated relative to Fresnel zone clearance. As opposed to time intensive 3-D ray tracing, the 2-D ray-tracing approach can account for up to 19 levels of reflection or transmission for each transmit site so as to assure prediction accuracy. The ray-tracing prediction has shown good agreement with measurements made in a number of Pacific Telesis buildings [7].

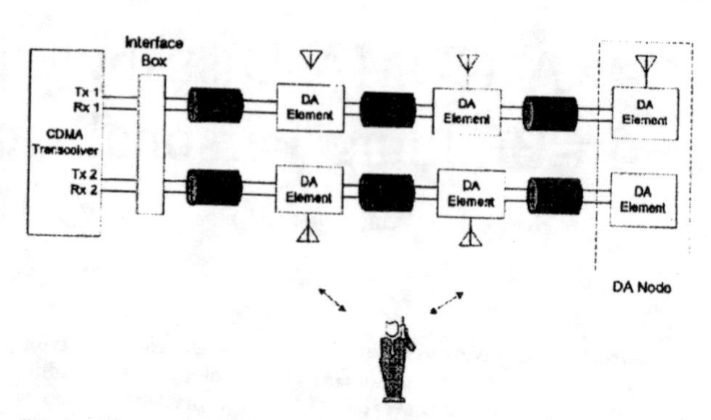

Fig. 1. Distributed antenna system configuration.

In order to design the most economical distributed antenna system, ray-tracing predictions were made for various configurations of transmit power, number of nodes, and node placements in the Qualcomm building. In Fig. 2, we depict the floor plan for the first floor of the Qualcomm building. As shown by the floor plan, this building consists of walled offices, labs, and machine shops. A similar configuration was used for the second floor of this two-story building. Based on prediction, three nodes were finally chosen to be placed in equally spaced locations to provide coverage in this rectangular building. Their locations are shown in Fig. 2 as asterisks. The nodes were mounted in the ceiling space between the first and second floors so as to provide signal coverage on both floors of the building. In Fig. 3, we show the predicted first floor signal coverage using the three nodes. In the prediction, the transmit power from each of the radiating elements was specified at −5 dBm, which was the actual transmit power level used in the measurements.

To achieve confidence with the theoretical prediction, signal strength measurements were also made at numerous discrete points throughout the building. The test points are shown in Fig. 2 by numbered dots associated with the locations. A comparison between the measured and predicted first floor signal strengths is shown in Fig. 4 for all of the measurement locations. The prediction results, with a standard deviation of 3.3 dB, agree very well with the measurements. Similar comparison results were obtained for the second floor. These comparisons lend credence to the following discussion relative to the prediction results.

As shown in Fig. 3, good signal coverage is provided in the first floor of the building using only three two-element nodes with each radiating element transmitting only −5 dBm.

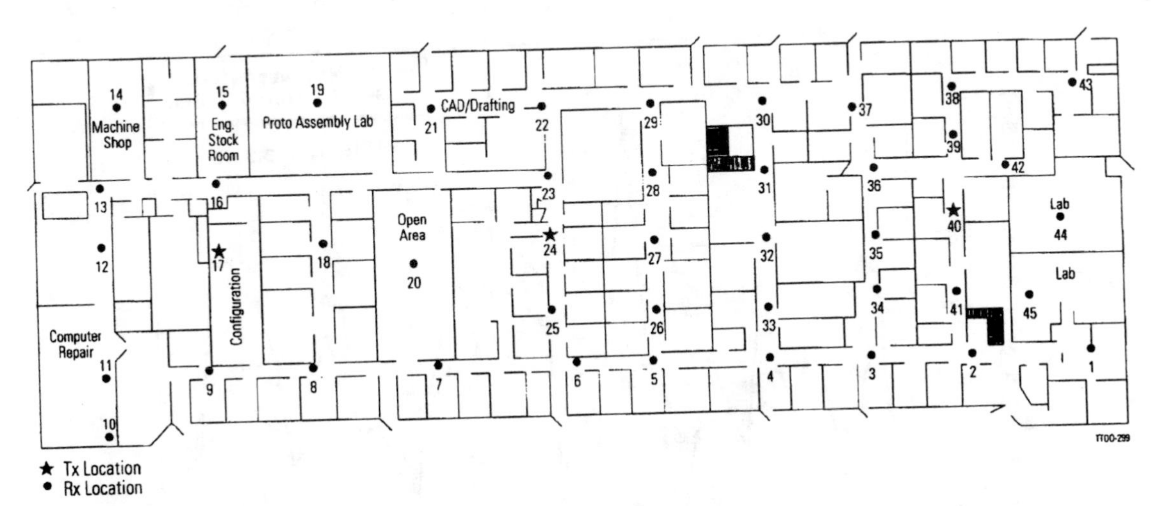

Fig. 2. Floor plan and test locations.

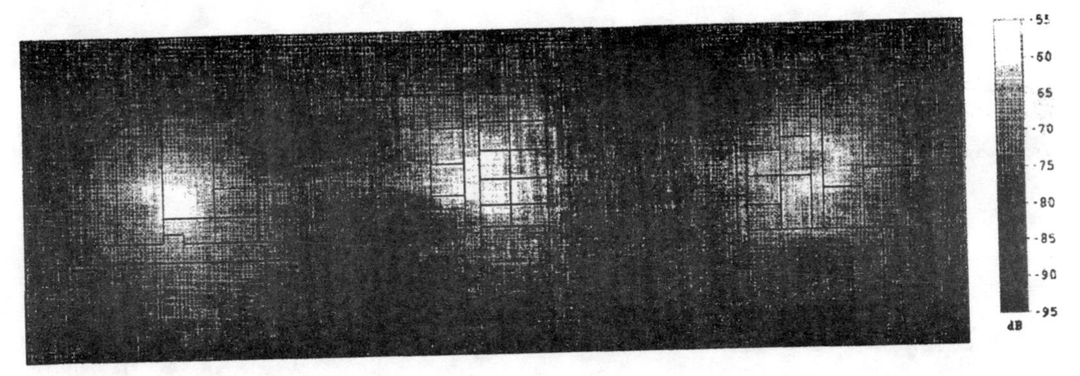

Fig. 3. Signal coverage predicted by a ray-tracing model.

Even though the received signal level in the second floor, as opposed to that in the first floor, is about 10 dB lower due to floor penetration loss, the three nodes provide sufficient signal coverage on the second floor. This implies significant transmit power reduction in the mobile stations, since in the CDMA system the mobile station transmit power is proportional to the base station transmit power as a result of power control. In fact, compared to the case with a single base station antenna mounted on top of the building, it is observed that the mobile transmit power reduced more than 10 dB on average in the three node distributed antenna system. In other words, this implies that 10 times more simultaneous mobile users can be contained in the distributed antenna system for a DS-CDMA system whose capacity is interference limited.

IV. Path Diversity

To study the time and path diversity advantages of the in-building distributed antenna system, measurements were made for two configurations: one with two radiating elements in the nodes and one with a single radiating element in the nodes. The latter case was made possible by disabling one string of radiating elements in the system. In Fig. 5, we show the mobile receive power for both configurations measured at a large number of locations throughout the first floor of the building. These test locations are coincident with those shown in Fig. 2 for the coverage measurements. Similar path diversity results were obtained for the measurements made on the second floor.

While the two element system provides path diversity between nodes and within the coverage areas of each node, the single-element system diversity only exists in the overlapping areas of the nodes. As observed in Fig. 5, diversity advantage is gained for the system with two radiating elements in all but two measurement locations, points 5 and 27. An average 4.8 dB improvement was obtained for the mobile received power in the system with two radiating elements over the system with a single radiating element. Generally, only an average improvement of 3 dB in the received power is expected as a result of doubling the transmit power through two radiating elements. The excess 1.8 dB improvement is therefore considered as an average time and path diversity advantage in the two-element distributed antenna system due to the resolution of additional independently fading signals from different paths at different times. A reduction of an equivalent amount of transmitted energy is expected in the reverse direction as a result of power control.

V. Self-Interference Due to Uncaptured Multipath Energy

There is concern that the multipath energy may not be totally captured by the CDMA receiver due to limited number

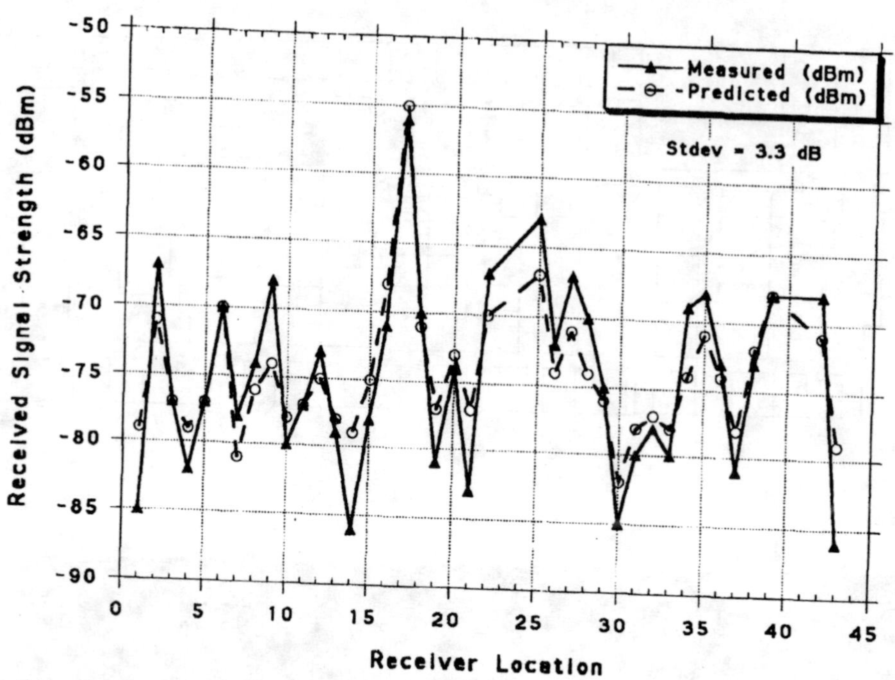

Fig. 4. Comparison of measured and predicted signal strength.

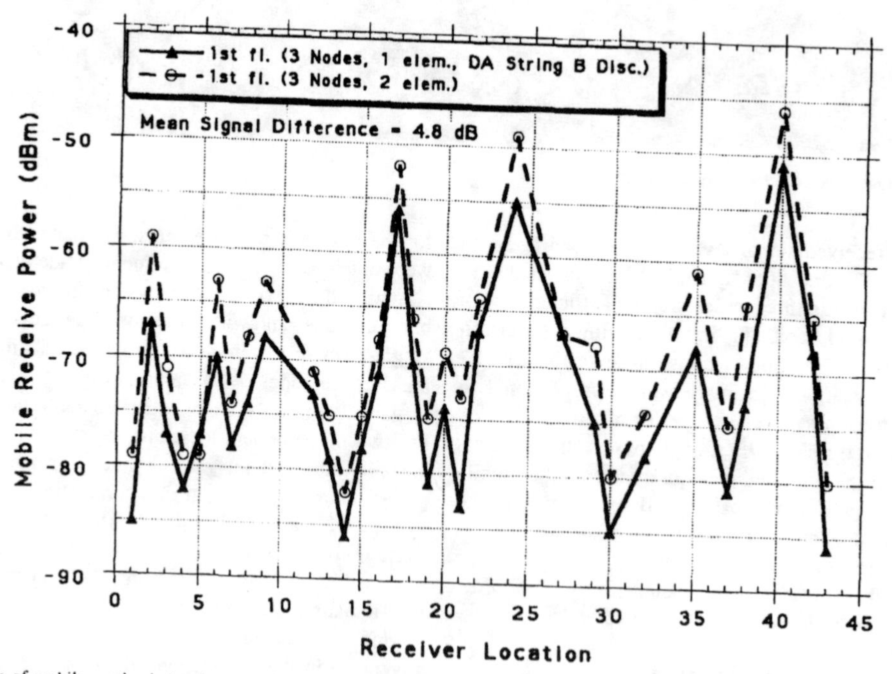

Fig. 5. Comparison of mobile received signal strength for system configurations with two radiating elements and with a single radiating element.

of correlators. Instead of contributing to the desired signal, the uncaptured multipath energy acts as self-interference at the CDMA receiver during the processing of the captured multipath signals. To study the significance of self-interference due to uncaptured multipath energy, the pilot E_c/I_o, the forward link frame error rate (FER), and mobile receive and transmit power are tested at the CDMA receiver under a system configuration with up to eight radiating elements in a single node.

In Fig. 6, we plot the mobile receive power (shown as solid symbols) versus the number of radiating elements in the node for both cases with no other simulated users in the system

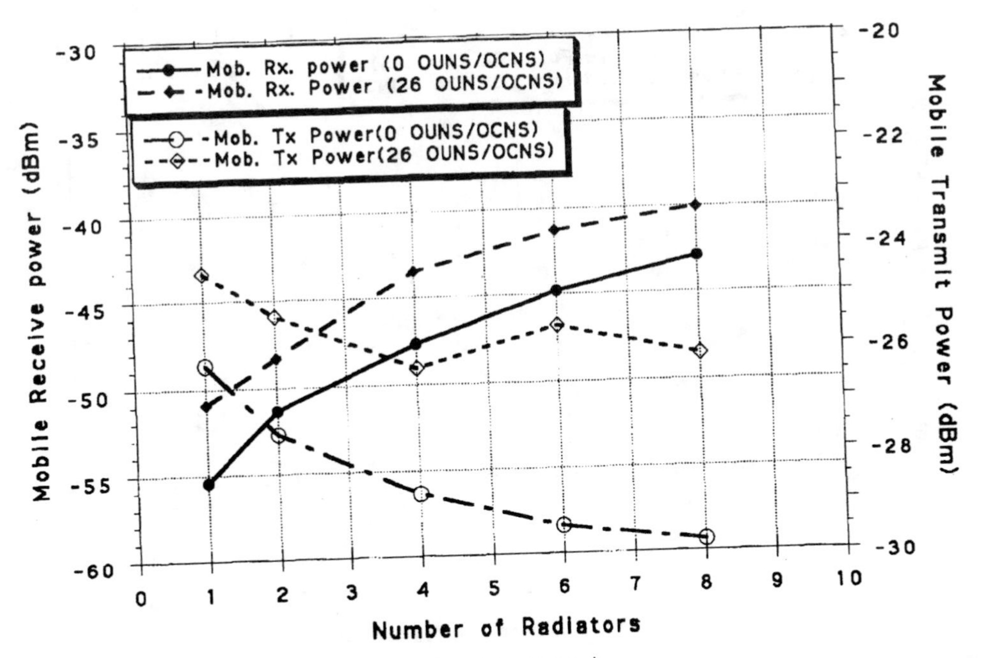

Fig. 6. Mobile receive and transmit power for a various number of radiators in a node.

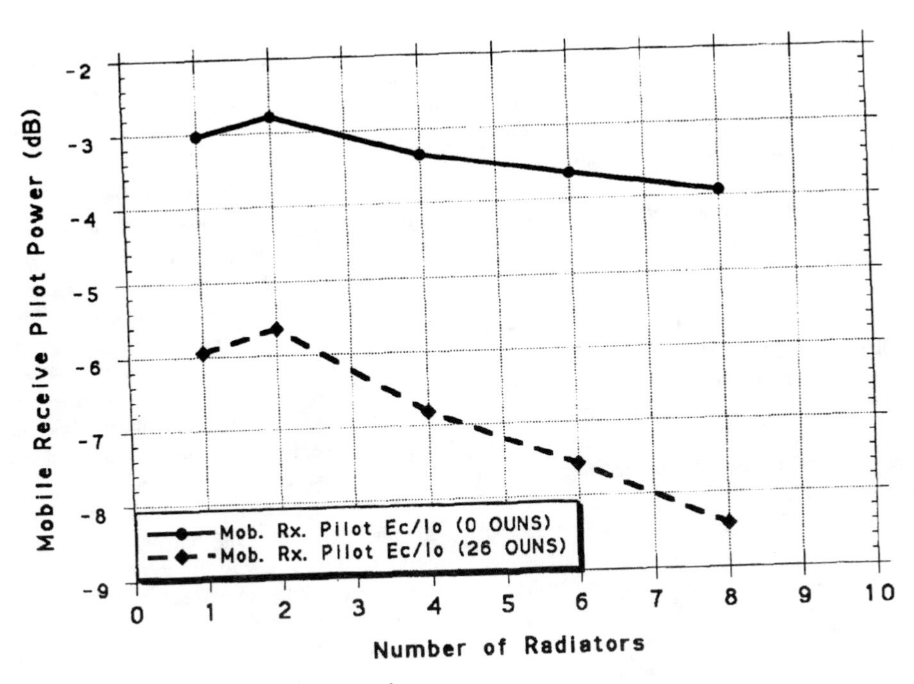

Fig. 7. Pilot E_c/I_o for a various number of radiators in a node.

(0 OUNS/OCNS) and with 26 other simulated users in the system (26 OUNS/OCNS). The mobile transmit power (shown by open symbols) is also plotted in this figure whose scale is marked in the right-hand side vertical axis. As expected, the mobile receives more power as the number of radiating elements increases. However, the RAKE receiver employing three correlators can only process three time-delayed multipath signals. The remaining multipath energy received by the mobile user then becomes self-interference. Generally, the mobile transmit power is inversely proportional to the mobile receive power as a result of open-loop power control in the CDMA system. The mobile transmit power is further adjusted according to the reverse link performance at the base station during closed-loop power control. As shown in Fig. 6, for the case with 26 other simulated users in the system (26 OUNS/OCNS), the mobile transmit power does not decrease

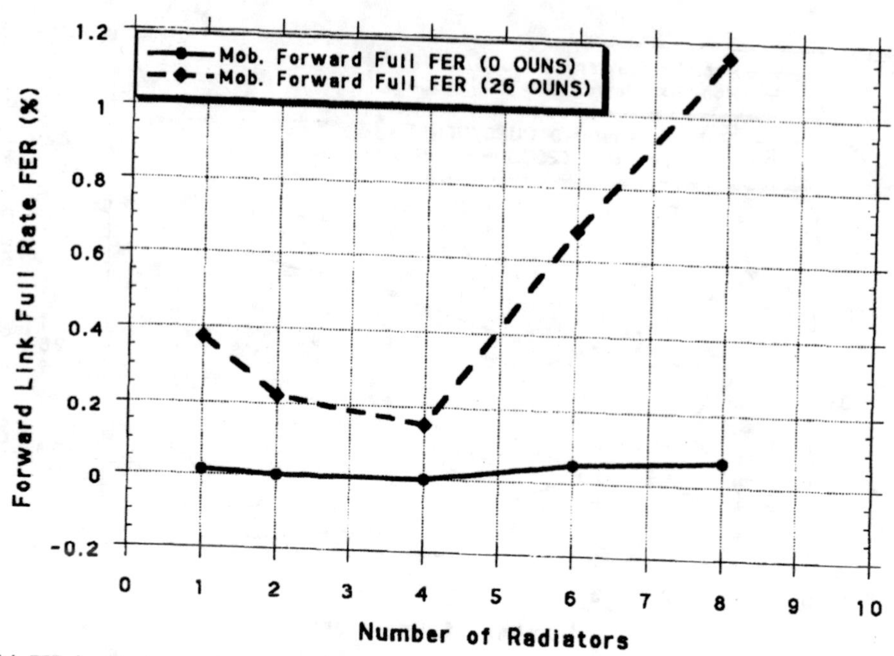

Fig. 8. Forward link FER for a various number of radiators in a node.

as the mobile receive power increases when more than four radiating elements are used. This implies that for a heavily loaded system (26 OUNS/OCNS) more power is transmitted by the mobile user to overcome the self-interference due to uncaptured multipath signals when the number of radiators exceeds four.

The self-interference due to uncaptured multipath energy can be measured by the pilot E_c/I_o. In Fig. 7, we depict the pilot E_c/I_o versus the number of radiating elements in the node for both cases with no other simulated users in the system (0 OUNS/OCNS) and with 26 other simulated users in the system (26 OUNS/OCNS). As shown in Fig. 7, the pilot E_c/I_o increases as the number of radiators increases from one to two, but decreases as the number of radiators increase beyond four.

To lend further credence to the above observation, we plot in Fig. 8 the forward link FER, which is a more direct measure of voice quality in CDMA systems, versus the number of radiating elements in the node. Again, the results are shown for both cases with no other simulated users in the system (0 OUNS/OCNS) and with 26 other simulated users in the system (26 OUNS/OCNS). While the system is resilient enough to maintain good FER performance when no other users are on the system, the FER performance is further degraded in a heavily loaded system (26 OUNS/OCNS) when more than four artificial multipaths are presented.

While significant time and path diversity advantages are shown in Section IV for the in-building distributed antenna system with two radiating elements in each node, more radiators in the same node may not provide additional diversity gain but degrade system performance as a result of self-interference due to uncaptured multipath energy. Trade-offs between diversity gain and self-interference need to be carefully justified. Proper system engineering is required for the distributed antenna system to minimize overlapping areas with more than three artificial multipaths.

VI. Conclusion

Measurements on performance of a distributed antenna system in a two story office building shows great promise for applications of the CDMA PCS technology in indoor environments. Good signal coverage is provided on both floors of the building using three nodes with two radiating elements, each element transmitting only −5 dBm. The mobile transmit power reduced more than 10 dB on average for the three node distributed antenna system. Significant time and path diversity advantages were observed for the distributed antenna system employing time delay elements to create deliberate multipath signals. More radiators in one distributed node may not provide additional diversity gain, but may result in self-interference due to multipath energy uncaptured by the CDMA receiver with limited number of correlators. Proper system engineering is required for the distributed antenna system to minimize overlapping areas with more than three artificial multipaths.

References

[1] Telecommunication Industry Association, TIA/EIA/IS-95, "Mobile station-base station compatibility standard for dual-mode wideband spread spectrum cellular system," July 1993.

[2] Joint Technical Committee (JTC) on Wireless Access, PN3384, "Personal station-base station compatibility requirements for 1.8 to 2.0 GHz code division multiple access (CDMA) personal communications systems," *Ballot Version*, 1994.

[3] A. Salmasi and K. S. Gilhousen, "On the system design aspects of code division multiple access (CDMA) applied to digital cellular and personal communications networks," in *Proc. 41st IEEE Veh. Technol. Conf.*, St. Louis, MO, May 1991, pp. 57–62.

[4] Motorola Inc. and Qualcomm Inc., "The CDMA PCS system common air interface proposal," to the *Joint Technical Committee (JTC) on Wireless Access*, Nov. 1993.

[5] W. Honcharenko, H. L. Bertoni, J. Dailing, J. Qian, and H. D. Yee. "Mechanism governing propagation on single floors in modern office buildings," *IEEE Trans. Veh. Technol.*, vol. 41, pp. 496–504, 1992.

[6] H. L. Bertoni, W. Honcharenko, L. R. Maciel, and H. H. Xia, "UHF propagation prediction for wireless personal communications," invited paper, in *Proc. IEEE, Wireless Networks for Mobile and Personal Communications*, vol. 82, no. 9, pp. 1333–1359, 1994.

[7] W. Honcharenko, H. H. Xia, S. Kim, and H. L. Bertoni, "Measurements of fundamental propagation characteristics inside buildings in the 900 and 1900 MHz bands," in *Proc. 43rd IEEE Veh. Technol. Conf.*, Secaucus, NJ, May 1993.

Howard H. Xia (S'88–M'91–SM'95) was born in Canton, China, on August 16, 1960. He received the B.S. degree in physics from South China Normal University, Canton, China, in 1982. He received the M.S. degree in physics, the M.S. degree in electrical engineering, and the Ph.D. degree in electrophysics, all from Polytechnic University, Brooklyn, NY, in 1986, 1988, and 1990, respectively.

Since 1994, he has been with AirTouch Communications as a principal engineer, where he is responsible for systems design and optimization of analog and digital cellular networks, and systems deployment of personal communications systems. Prior to AirTouch, he was a Senior Staff Engineer with Telesis Technologies Laboratory from 1991 to 1994, where he was engaged in research on various aspects of personal communications service (PCS). From 1990 to 1991, he was a Senior Engineer with PacTel Cellular and conducted research and development on analog and digital cellular networks. Since 1993, he has served as a member of the United States delegation to participate in ITU-R (formerly CCIR) activities on standardization of the third generation international mobile systems-FPLMTS/IMT-2000. He has published numerous articles on subjects of indoor and outdoor radio propagation, spectrum sharing, and CDMA systems design.

Dr. Xia was co-recipient of the IEEE Vehicular Technology Society's 1993 Neal Shepherd Award for best paper on propagation.

Angel B. Herrera received a B.S. degree in electrical engineering from California State University, Los Angeles, in 1988.

In 1990, he joined PacTel, Irvine, CA, where he was involved in the design of cellular systems. Also, he worked in the development and testing of CDMA systems. In 1995, he joined Cox California PCS, Incorporated, Irvine, CA, and is now Manager of System Performance. He is responsible for CDMA system optimization and performance.

Steve Kim was born in Seoul, Korea. He received the B.S. degree from the University of Southern California, Los Angeles, CA, in 1991.

Since 1991, he has been with AirTouch Communications (formerly PacTel Corporation) and has been involved in indoor and outdoor propagation prediction, Korea CDMA system planning and deployment, and Spain GSM deployment.

Fernando S. Rico was born in Orange, CA on March 6, 1957. He received the B.S. degree in engineering technology from California State Polytechnic University, Pomona, in 1981.

He has worked in various telecommunications engineering positions, starting at Pacific Telephone in February, 1982. He worked in the AirTouch Cellular Los Angeles engineering department between 1984 and 1988 in the RF design and traffic engineering groups. He then worked at Pacific Bell's Science and Technology department between 1988 and 1992 helping develop early PCS technical strategies. He worked at Telesis Technologies Laboratory between 1992 and 1994, testing the feasability of CDMA at PCS frequencies. Since 1994, he has been involved in design, development, and implementation of digital and analog cellular networks for the AirTouch Communications' domestic and international markets as the Director of RF Planning and Assessment in Walnut Creek, CA.

IEEE JOURNAL ON SELECTED AREAS IN COMMUNICATIONS, VOL. 12, NO. 4, MAY 1994

Traffic Handling Capability of a Broadband Indoor Wireless Network Using CDMA Multiple Access

Chang G. Zhang, H. M. Hafez and David D. Falconer

***Abstract*—CDMA (code division multiple access) may be an attractive technique for wireless access to broadband services because of its multiple access simplicity and other appealing features. In order to investigate traffic handling capabilities of a future network providing a variety of integrated services, this paper presents a study of a broadband indoor wireless network supporting high-speed traffic using CDMA multiple access. The results are obtained through the simulation of an indoor environment and the traffic capabilities of the wireless access to broadband 155.5 MHz ATM-SONET networks using the mm-wave band. A distributed system architecture is employed and the system performance is measured in terms of call blocking probability and dropping probability, The impacts of the base station density, traffic load, average holding time, and variable traffic sources on the system performance are examined. The improvement of system performance by implementing various techniques such as handoff, admission control, power control and sectorization are also investigated.**

I. Introduction

WIRELESS communications is a rapidly expanding field with many potential applications. Future wireless communication systems built for indoor environments are expected to handle integrated voice, data, and video services. ATM (Asynchronous Transfer Mode) is selected as the basis for future broadband ISDN facilities [1]. A broadband indoor wireless system can serve as an access to wired backbone SONET LAN's. The design of wireless broadband networks will mainly depend on the traffic characteristics, user density, channel bandwidth, access method, and performance measures. As far as access methods are concerned, a CDMA scheme is one of the most attractive multiple access schemes for mobile and personal communications [2], [3]. Its advantages include soft handoff capacity and voice activation capability. It is also effective in combating multipath fading and interference. However, it may require tight adaptive power control to compensate for the near–far problem [4], [5]. In order to provide a better understanding of a broadband indoor wireless system and CDMA access protocol, it is important to evaluate its traffic handling capabilities. This paper presents a simulation study of such a system.

Manuscript received June 30, 1993; revised November 14, 1993. This research was supported by a grant from the Canadian Institute for Telecommunications Research (CITR) under the NCE program of the Government of Canada. This paper was presented at conference Wireless '93 under the title "Traffic Characteristics of an Indoor Wireless Broadband Network Using CDMA Multiple Access," July 13, Calgary, Canada.

The authors are with the Department of Systems & Computer Engineering, Carleton University, Ottawa, Ontario, Canada K1S 5B6.

IEEE Log Number 9215407.

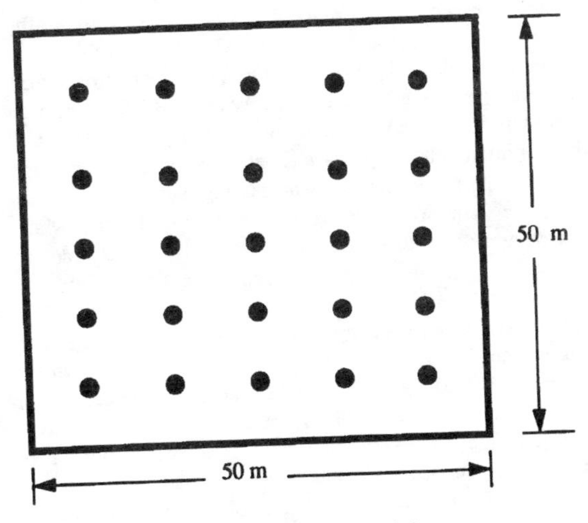

Fig. 1. Illustrative coverage area in traffic simulation.

II. CDMA System Model

We consider an indoor wireless network model which consists of a number of base stations uniformly distributed in the service area of 50 × 50 square meters as shown in Fig. 1. The number of base stations is variable. This floor design represents a typical indoor signal propagation environment such as a large conference room. In a conventional centralized architecture in which the building area is divided into microcells, different radio frequencies are assigned to adjacent cells to minimize interference; each radio channel is re-used at cells separated by several cell diameters. In contrast to this approach, we adopted a distributed network architecture proposed by Leung [6] in which the same radio frequency band is used over the entire coverage area, with multiple base stations interconnecting each portable to the backbone LAN. This architecture eliminates the need for frequency coordination, and greatly eases network configuration and management. For instance, adding portables in the existing network coverage area is straightforward, with no more involvement of network management functions than is necessary for fixed portables.

We adopt direct sequence spread spectrum multiple access operating in a random access mode as a wireless channel access method in which channel coding permits the simultaneous transmission of multi-users. This means that the system is modeled as an ALOHA system where unsucessful transmissions are caused entirely by multiple access interference [7], [8]. Both uplink and downlink radio channels are configured as

0733–8716/94$04.00 © 1994 IEEE

multipoint/broadcast channels shared by all portables and base stations, which access the respective channels under regulation of the access protocol. However, CDMA capacity is generally limited by the uplink capacity [9]; hence, only the uplink is modeled in this simulation.

Perfect code synchronization is assumed, implying that the intended receiving base station always has knowledge of the code of the transmitting portable. Since high rate video transmission has the most significant impact on the system performance, we focused most of our efforts on video traffic of data rate 20 Mbps, transmitted as ATM cells, in this work. The transmitted line bit rate is a SONET OC-3 rate of 155.52 Mbps and thus the corresponding duty cycle is 7.75, i.e., each active terminal is only transmitting about 1/8 of the time. Co-channel interference is partly alleviated by this duty cycle and partly by the spread spectrum processing gain.

Service requests are generated uniformly in the whole service area. Due to technology limitations, we assume the achievable chip rate of spread spectrum is 1.55 Gchip/s. In other words, we have a processing gain of only about 10. Because of the large required bandwidth even if no spreading is used, the system is assumed to use radio frequencies in the mm-wave (20-60 GHz). To simplify the signal processing of base stations, an assumption is also made that a single long code is assigned to every user which uses its own portion of the code. Direct sequence spread spectrum signaling using a single pseudonoise spreading code employed over the channel can also take advantage of its multipath rejection and signal capturing properties. By using spread spectrum signaling, multipath and multiaccess interference signals are attenuated by the processing gain when they are out of phase by more than one chip with the signal to which the receiver is in synchronization. Such a low processing gain may not provide enough protection against interference, but it is reinforced by the use of asynchronous random access within the 7.75 duty cycle.

The signaling bandwidth for the uplink is about 2 GHz. The same bandwidth at a different frequency is also needed for the downlink. For the mm-wave band, the bandwidth resource is abundant. In this work, we assumed and used radio propagation parameters in the frequency range of 20–30 GHz. A common channel is shared by all base stations and portables. To make the wireless link easily compatible with the backbone OC-3 ATM–SONET network, the same ATM cell format, i.e., 54 bytes per cell including overhead and redundant code bits (48 data + 5 ATM header + 1 wireless header) is adopted in the wireless link. The wireless header contains signaling bits which take care of the realization of call setup, admission control, power control, and multi-media traffic handling. The general system block diagram is shown in Figure 2. The wide-band system can support multimedia traffic, i.e., base stations operating at a OC-3 rate receive transmissions from various portables and forward them to a wired backbone LAN. Portables can be mobile voice phone sets, data servers, and video desktops. They generate various types of traffic that is packetized into a common ATM cell format. Note that the simulation does not deal with the details of ATM frame structures,

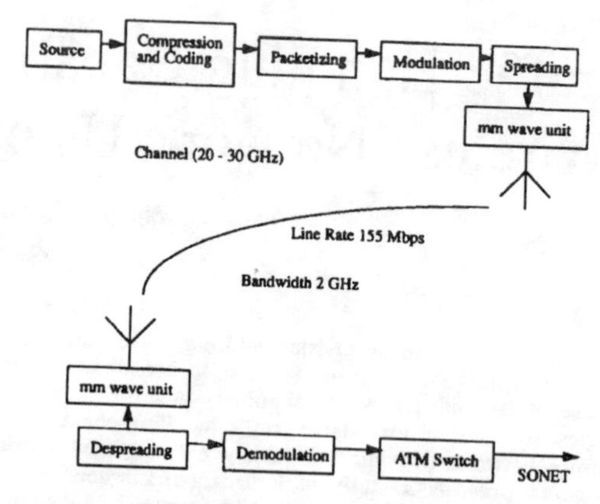

Fig. 2. Simulation system block diagram with associated functions. The mm wave unit performs the conversion from IF to mm waveband in the transmitter and from mm waveband to IF in the receiver.

it enters into simulation only as a duty cycle mentioned earlier.

In the indoor environment, the signal suffers from fading, shadowing loss, and also intersymbol interference (ISI) which is a main impairment of an indoor multipath channel. A combination of antenna sectorization, selection diversity, and bandwidth spreading modulation may be effective in combatting ISI problems [10] [11] for this type of application. We can reasonably assume that the multipath effect and ISI could be greatly reduced by employing the commutation signaling and RAKE type receiver with diversity combining. The simulation model does not contain the implementation details of transceiver. We consider co-channel interference to be the major source of communication impairment. At a base station, the detection signal to interference ratio of $-12\,\mathrm{dB}$ (equivalent to $E_b/N_O = 7\mathrm{dB}$) is taken as a reception threshold which corresponds to the bit error rate less than 10^{-3} [9]. Although this criterion was proposed by Qualcomm to qualify the voice communication, we assume that this threshold is also suitable for high data rate traffic when powerful long constraint-length channel coding and interleaving are exploited. The weakness of the low processing gain can be compensated to some extent by taking advantage of microdiversity. In our system, we used a relatively high base station density. Since the number of portables within a certain area may be limited by the available network throughput capacity per unit area, depending upon the traffic load, the arrangement can be made to provide adequate radio coverage over a single floor. As a result any portable can access the four closest base stations, and the best one is chosen to forward signal to the backbone LAN. The service requests arrive according to a Poisson pattern.

III. Radio Signal Characteristics

Radio signal propagation in the 20-60 GHz band depends mainly on: attenuation and multipath effects by walls, space dividers, doors, windows, furniture, people, etc. [12], [13]. Mm-wave propagation inside a building is ray-like and the attenuation behaves as $R^{-\gamma}$ where R is the distance from the portable to the base station and γ is the path exponent that varies with frequency. In the frequency range of interest, i.e., 20–30 GHz band, the γ is chosen to be in the range of 1.9–2.1 [14]. Other effects are ignored in this study. The base station antenna is assumed to be mounted at a height of approximately 10 feet above the floor.

At a base station j, the received local mean signal power from the desired portable i is

$$S_{ji} = S_{TX} + G_{RX} + G_{TX} - L(R_{ji}) \quad (1)$$

where all quantities are expressed in dB or dBm, and $i \leq /, \mathrm{m}$ (the number of active portables in the system), and $j \leq N$ (the number of base stations. R_{ji} (meter) is the distance between the ith portable and the jth base station. S_{TX} is the portable transmitter power (0 dBm). G_{RX} and G_{TX} are the antenna gains for a portable and a base station, respectively. We assume initially that all portables transmit at the same power level, and their antennas have the same omni-directional patterns in azimuth and the same gain. Two types of base station antennas are studied: (1) omni-directional antenna with zero gain; sectorized antenna with angular distribution of antenna gain. The path loss in dB is given by

$$L(R_{ji}) = \left(10\gamma \log(R_{ji}) + 10 \log \frac{P_r(1m)}{P_t}\right) \quad (2)$$

where $P_r(1m)/P_t$ is the normalized received power at a separation distance of 1 meter. According to wide-band measurements of indoor radio channel, the worst value of the second term is approximately 80 dB [15]. In our case, we take 62 dB. In this study we consider a worst-case "open environment" with no walls between the base stations that would provide isolation between microcells.

In our spread spectrum system, the total radio frequency spectrum is shared by all portables. Hence, a desired transmission is interfered with by all currently active portables in the system. The local mean signal power in dBm at the lth base station due to the nth co-channel interfering portable takes the same expression as the desired signal

$$I_{ln} = S_{TX} + G_{RX} + G_{TX} - L(R_{ln}) \quad (3)$$

where R_{ln} is the distance from the nth co-channel portable to the lth base station. Exact modeling of the multiuser interference especially in a fast multipath fading environment still remains an open problem. In this study, the multiuser interference is modeled approximately as an uncorrelated Gaussian random process so that its constituent power spectral densities add. The total local mean interfering power in dBm at the lth basestation is the power sum of all co-channel interfering portables.

$$I_{tot} = 10\left(\log \sum_{n=1, n\neq i}^{m} 10^{(I_{ln}/10)}\right) \quad (4)$$

where m denotes all active co-channel portables at the instant of calculation. Note that $n = i$ is excluded from the sum. The signal to interference plus noise ratio in dB is

$$\mathrm{SIN}R_{ji} = (S_{ji} - 10\log[10^{I_{tot}/10} + 10^{(N_{RX})/10}]) \quad (5)$$

where N_{RX} (dBm) is the thermal noise level at the receiver which is assumed to be approximately $-82\,\mathrm{dBm}$. The ith portable measures $\mathrm{SIN}R$ and selects four serving base stations

$$\Phi_i = \max(\mathrm{SIN}R_{i1}, \mathrm{SIN}R_{i2}, \mathrm{SIN}R_{i3}, \mathrm{SIN}R_{i4}) \quad (6)$$

If the $\mathrm{SIN}R$ measured at one or more base stations exceeds a prespecified detection threshold $\mathrm{SIN}R_{Th}$, the call can be successfully set up, and a reliable communication is said to be established. The value of the $\mathrm{SIN}R_{Th}$ is chosen to provide some level of acceptable communication performance. In a spread system, $\mathrm{SIN}R_{Th}$ is a function of i) E_b/N_0 required to achieve a vocoder bit error rate specification, and ii) the available processing gain, which is determined by the spreading bandwidth, W, and the information bit rate R_b which takes different value depending on the traffic sources. $\mathrm{SIN}R_{Th}$ is expressed as

$$\mathrm{SIN}R_{Th} = \frac{E_b/N_0}{(\alpha W)/R_b} \quad (7)$$

where α is the duty cycle, and $\mathrm{SIN}R_{Th}$ is assumed to be at least $-12\,\mathrm{dB}$ for establishment of a reliable communication. This value of $\mathrm{SIN}R_{Th}$ translates to an E_b/N_0 of 7 dB.

IV. Traffic Model

A discrete even driven simulation was chosen for the traffic model representation which is described in this section. There are M (250) portables in the user set and they transmit or receive 20 Mbps video calls. In general, by a call we mean a transmission of a succession of cells for a duration of time. Portables' positions are random in a 50 meter by 50 meter square area. Each idle user in the user set independently generates access attempts with an average rate λ calls/sec. λ is varied to account for different traffic loads. The interarrival time between access attempts is exponentially distributed with mean $1/\lambda$, and the call arriving process approaches Poisson when the user set is large. The portables are uniformly distributed in the service area. The mobility of portables is not considered in this study. The whole traffic load for video source has an exponential distribution with holding time of 30 minutes mean per call. In the case of multi-sources, considered later, it is a combination of different types of traffic with corresponding traffic characteristics.

The access scheme is CDMA. Any given portable selects a set of four base stations based on average signal strength calculation and requests transmission. Upon receiving the request, the four base stations measure the average signal strength to be received relative to the total interference signal strength. If this ratio at one or more base stations is above the detection threshold, the transmission is authorized, and the portable will take the best link and transmit for the duration of the call. During the process of transmission, new arrivals may come into the system and cause interference. This may lead the signal to interference ratio of the transmitting portable to drop below the detection threshold, forcing this call to drop. Since the state of each base station is updated in terms of signal to interference ratio, whenever there is an arrival or a departure, it is always possible that some calls have to be dropped before their completion as the result of SINR variations.

V. Simulation Methods

The computation intensive call-by-call stochastic simulation is written using Simscript II.5 and C. Base stations are distributed in a regular uniform pattern in the service area. In the simulations the number of base stations is varied. A portable is placed at a random position in the service area and access is attempted. On a successful access, a new user record is generated and the system state is updated for an additional co-channel interference. By updating the system state, we mean every active portable's signal to interference ratio at its serving base is updated. Whenever the updated SINR goes below the threshold, the call has to be dropped and then returns to the user set. A failed access is due to lesser SINR than threshold. The blocking and dropping statistics are updated and the user is returned to the user set. The readings are recorded after the system reaches steady state.

VI. Simulation Results and Discussions

The objective of this simulation is to investigate the system performance. The selected criteria for evaluation of the system in an indoor environment are blocking probability and dropping probability under various traffic conditions. By positioning portables at many uniformly separated locations in the service area, a measure of performance over the service area is achieved. These performance statistics are also referred to as area averaged statistics. The simulation results are grouped under a few different headings. The analyses of various results on the broadband CDM system are associated.

A. Performance of the Basic System

In CDMA, the blocking probability depends on power level, base station density, traffic condition, transmission channel, traffic type, etc. All those system parameters are controlled in the simulation such that the blocking will be caused mainly by multiuser interference, i.e., the traffic load relative to the number of base stations in the system. The random access character of CDMA/ALOHA makes the system relatively simple. On the other hand, ongoing transmissions are not well protected. During the service period of time, the quality of the service may deteriorate as the result of new subsequent users. When the interference is serious enough to cause the SINR of that ongoing call to fall below the treshold, then the call is dropped. The dropping probability is primarily decided by multiuser interference and base station density.

Fig. 3 illustrates blocking probability and dropping probability vs. the offered traffic loads under a fixed base station density of 32 per 1000 square meters. For simplicity, power control is not implemented in the basic system. The traffic load is defined as the total number of active users at a given instant of time, on an average basis, multiplied by user rate, divided by the service area. Because of long holding time (average 30 minutes) and access scheme, this system suffers from a high dropping probability. Increasing the traffic load

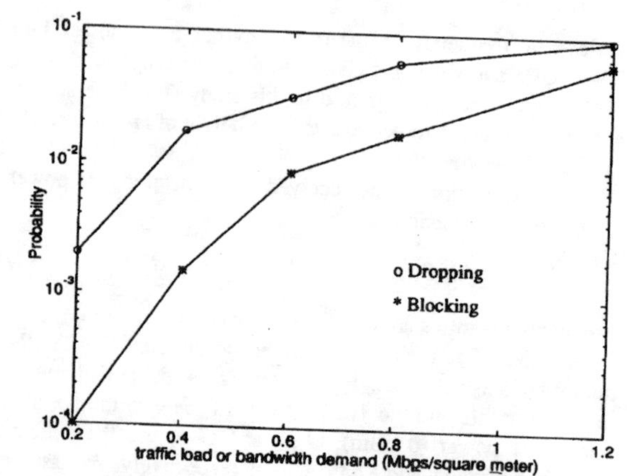

Fig. 3. Blocking and dropping probabilities vs. offered traffic load for an indoor basic system under fixed base station density of 32/1000 square meters.

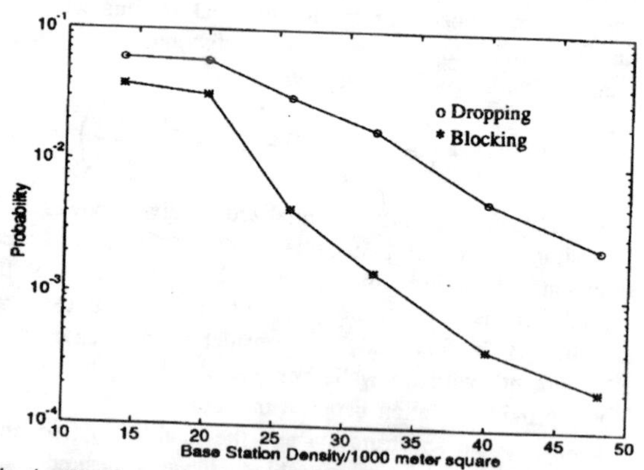

Fig. 4. Blocking and dropping probabilities vs. base station density for an indoor basic system under a fixed traffic load of 0.4 Mbps/square meters.

causes both blocking and dropping to increase. It seems that the basic system can support a traffic load no higher than about 0.4 Mbps/square meter or 12 Mbps per base station.

Fig. 4 presents the blocking and dropping probability versus base station density for a given traffic load at 0.4 Mbps/square meter. It is understood that more base stations can support more users or improve system performance accordingly. The simulation results in this figure indicate that significant improvement can be achieved when the base station density exceeds 20/1000 square meter.

Fig. 5 illustrates how traffic load varies with base station density for a fixed value of the blocking probability of 1% and that of the dropping probability of 3%. As seen from the figure, below a certain value of base station density, 20/1000 square meter in our case, increasing the base station density only improves system capability a little. However, a linear relation is observed beyond that value between the increase of the system capacity and the increase of the base station density. Note that there seems a performance limit for the system. As the base station density increases beyond 40/1000 square meters, the capacity tends to saturate. This agrees with

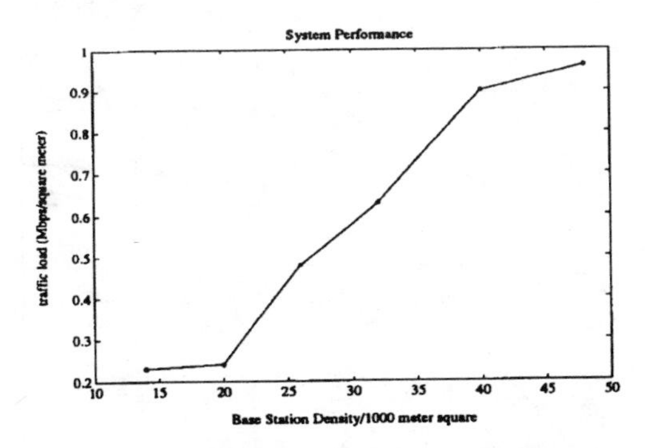

Fig. 5. Offered traffic load vs. base station density for an indoor basic sytem under fixed blocking probability of 1% and dropping probability of 3%.

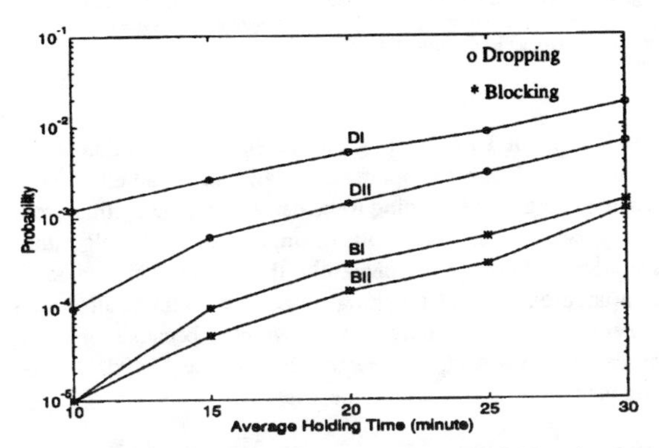

Fig. 6. Blocking and dropping probabilities vs. average holding time for a given traffic load of 0.4 Mbps/meter square, and a fixed base station density of 32/1000 square meters. The curves BI, DI, BII, and DII stand for blocking and dropping probabilities of systems without or with handoff, respectively.

a previous study of infrared used as a transmission link and CSMA employed as a multiple access protocol [16]. The important message here is that the operation of system should be restricted to the linear range, i.e. the base station density be limited from 20-40/1000 meter square, provided that 1% blocking and 3% dropping are acceptable performance.

To this point, we have assumed that the high rate video traffic has an exponentially distributed holding time with 30 minutes mean. Different mean holding times were also examined in this work, but it was limited in a reasonable range. In our simulation program, the mean holding time is an adjustable system parameter. Therefore, the variations of the average holding time could be investigated. By fixing the traffic load at 0.4 Mb/s/square meter, and the base station density at 32/1000 square meters, we ran simulations at various average holding times. The results (BI and DI) are shown in Fig. 6. The interesting conclusion drawn from this figure is that (1) the system performance in terms of blocking and dropping probabilities improves as the mean holding time decreases; (2) the higher dropping than blocking is inherent and it is apparently not directly caused by longer holding time.

It is understood that a longer holding time could cause a performance degradation, but the higher dropping probability seems unusual, even for relatively short holding time. One might attribute it to higher taffic load. Our previous results have indicated that this is not the case. As a matter of fact, this is due to two reasons: the access scheme and the lack of power control. We are using CDMA with very small processing gain. So the protection of the desired signal is rather fragile. In addition, without power control, any ongoing transmission could be wiped by nearby newcomers. Certainly higher dropping probability should always be avoided for a practical system. It seems that some other effective mechanisms have to be implemented to achieve this goal.

B. Performance of the System With Handoff

In order to alleviate the effect of higher dropping probability, we implement a handoff mechanism according to which whenever an active portable's transmission is jeopardized by other interferers, it starts a search algorithm. All base stations around will be checked. If there exists any other base station providing better service, then it switches over. In this discrete event-driven simulation model each portable updates its state when a new arrival or departure occurs. Therefore each portable is tracing its own status during the course of simulation. A search algorithm allows any portable to change from one base station to another when it detects the current link is not acceptable. Therefore, a local higher load can be shared by nearby available resources. To what extent the performance can be improved with this handoff mechanism is of interest.

Fig. 7 presents the simulation results of this system in terms of blocking and dropping probabilities under a fixed base station density of 32/1000 square meters. To compare with the basic system, the simulation results without handoff are also given in the same figure. It shows that the improvement of blocking probability is rather small, but the dropping probability improvement is more significant. Similar to Fig. 4, the effect of system performance on base station density is also investigated. Under a given traffic load, both blocking and dropping probabilities decrease as the base station density increases. The curves in Fig. 8 look like the ones in Fig. 4, but they are shifted down. From this graph, we see that the dropping is significantly reduced, especially for larger base station density. Handoff actually provides site diversity which enhances the probability of successful reception and therefore increases the network throughput. For traffic with different mean holding time, the handoff improvement is different too. In Fig. 6, this effect is presented as curves BII and DII representing blocking and dropping probabilities. In contrast to the basic system (BI and DI), the handoff is more effective for traffic with shorter mean holding time.

C. Performance of the System with Admission Control

As we have seen previously the system exhibits higher dropping probability than blocking probability. This is certainly an unwanted behavior. We have attributed this to small processing gain and the absence of power control. On the other hand, it may be useful to look at an extereme case,

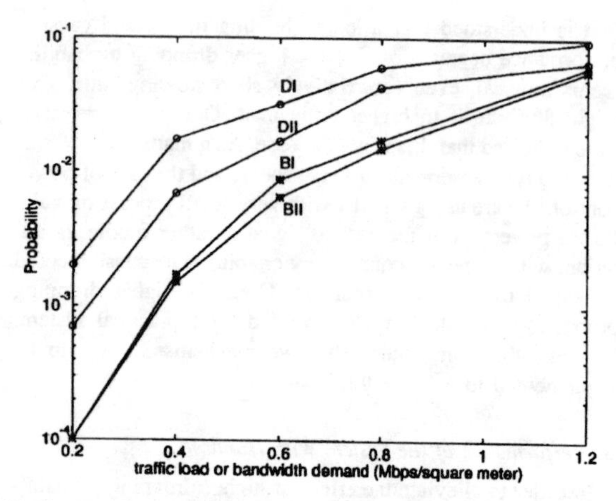

Fig. 7. Blocking and dropping probabilities vs. traffic load for a basic system with (BII and DII) or without (BI and DI) handoff under a fixed base station density of 32/1000 square meters.

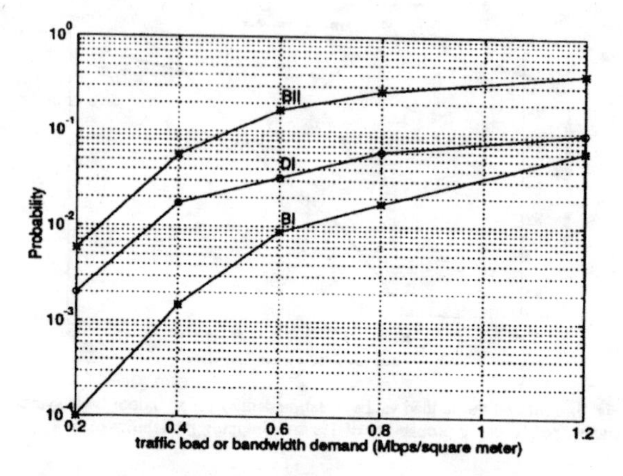

Fig. 9. Blocking and dropping probabilities vs. base station density under a given base station density of 32/1000 square meters for a system with (BII) or without (BI and DI) admission control.

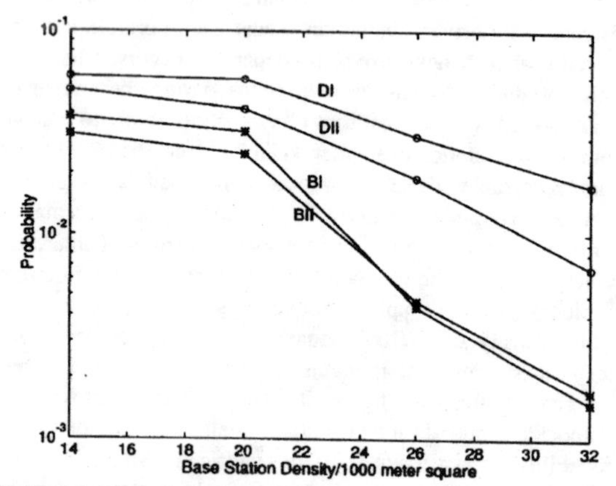

Fig. 8. Blocking and dropping probabilities vs. base station density under a given traffic load of 0.4 Mbps/square meter for a system with (BII and DII) or without (BI and DI) handoff.

i.e., admission control by which the dropping probability can be completely removed. This can be realized by employing a certain controller which regulates each service request. A portable intending to transmit has to communicate with the central controller first. Upon receiving the request the controller then communicates with each base station to see the effect of incoming transmission on all currently served portables. If none of the active portables' transmission quality is degraded to an unacceptable threshold due to this incoming portable, then a permission is granted. Otherwise, it is blocked. In fact, base stations and all active portables are tracing their link quality individually when the system state is updated. They do not have global knowledge about the impact of upcoming calls. For instance, if a call is accepted by a local base station, it may cause other ongoing calls to be dropped. In this model, all portables are transmitting at the same power level. An admission control algorithm estimates the interference of an upcoming call at each base station based on the line-of-sight component. A call can be admitted only when no single call is going to be caused to drop by this event at any base station. The results of implementing this algorithm are presented in Fig. 9. Apparently, the dropping is eliminated at a large expense of blocking probability. Technically, such a central controller may not be viable, because of extra processing overhead. But some performance bounds can be obtained from such an extreme case.

D. Performance of the Basic System with Different Types of Traffic

So far, we have only investigated system behavior under an extreme traffic condition, i.e., a single type of high rate video traffic. We have collected useful information on such a worst case. But in the real world, a broadband system will support integrated multi-media traffic. To see the effect of different types of traffic, a simulation was conducted for five types of traffic. The traffic statistics are

Type I: 45% of traffic with 64 Kbps data rate, and an exponentially distributed holding time of 3 minutes mean, corresponding to voice users.

Type II: 5% of traffic with 128 Kbps data rate, and an exponentially distributed holding time of 4 minutes mean, corresponding to low bit rate data users.

Type III: 15% of traffic with 0.5 Mbps data rate, and an exponentially distributed holding time of 5 minutes mean, corresponding to bursty data users.

Type IV: 15% of traffic with 5 Mbps data rate, and an exponentially distributed holding time of 30 minutes mean, corresponding to interactive data users.

Type V: 20% of traffic with 20 Mbps data rate, and an exponentially distributed holding time of 60 minutes mean, corresponding to video users.

Corresponding to those different source rates, the different duty cycles are associated accordingly in simulation, i.e. their values are 2421.88, 1210.94, 310, 31, and 7.75, respectively.

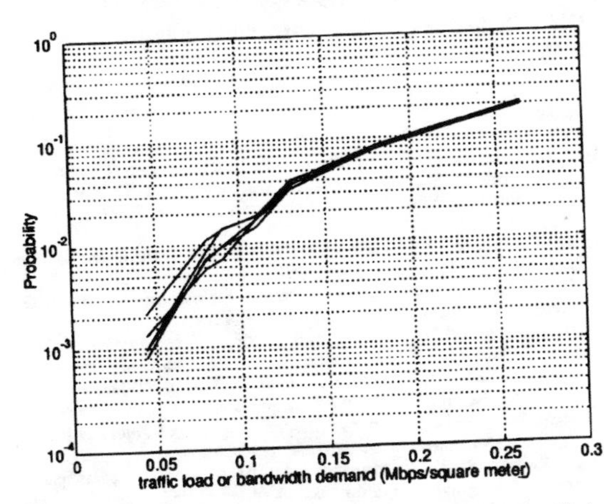

Fig. 10. Blocking probabilities of 5 types of traffic vs. offered traffic load for a fixed base station density of 14.4/1000 square meters.

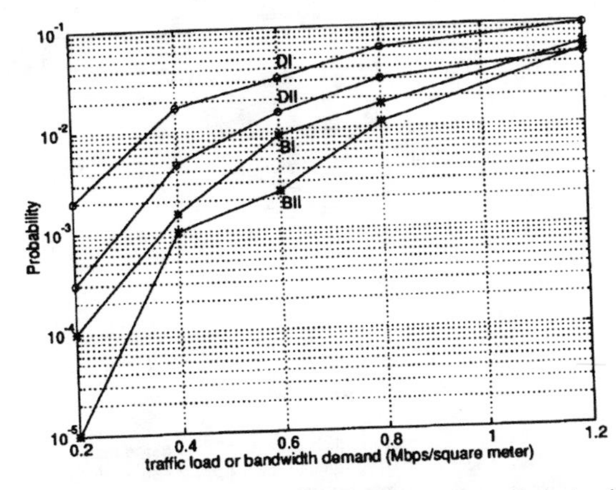

Fig. 12. Blocking and dropping probabilities vs. offered traffic load under a given base station density of 32/1000 square meters for a system with (BII and DII) or without (BI and DI) power control.

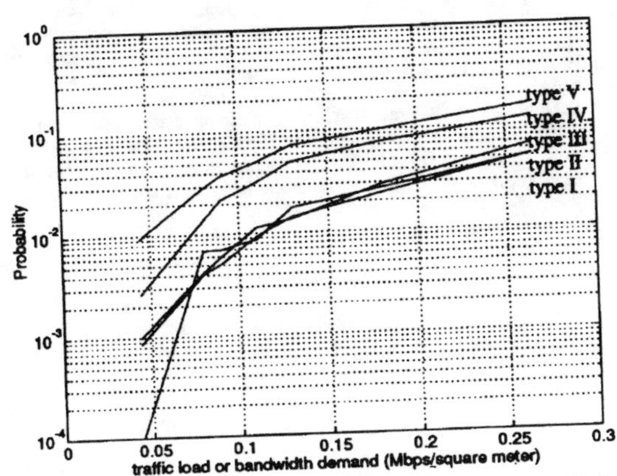

Fig. 11. Dropping probabilities of 5 types of traffic vs. offered traffic load for a fixed base station density of 14.4/1000 square meters.

Simulation results are shown in Fig. 10 and Fig. 11. The interesting thing is that the blocking probabilities are about the same for different types of traffic, but the dropping probabilities are quite distinct only for the two high rate traffic types which are significantly larger than the other three types of traffic.

E. Performance with Power Control

In a CDMA system all users transmit in the same frequency channel and hence the received spectrum at the base station shows a quite uneven distribution. The differences in power levels cause each communication link to have different degrees of communication quality. Also the effect of the stronger signal on the weaker one due to cross correlation could cause unacceptable degradation for the weaker signal. The high dropping probability of our system seen previously is also attributed to the absence of power control. Hence power control is necessary to equalize the received power at the base station and maintain an acceptable communication quality, although the implementation of power control may increase the system's complexity. For an indoor communication system, it may not be that hard due to a lower speed of portables. We implemented a power control scheme in our system in order to see the impact of power control on system performance. The power control is based on each microcell. Each base station covers a certain service area. All active portables appearing in the same microcell have their power adjusted to the same signal level. Note that every portable selects its own link in this system, and base stations always know how many portables are being served. Power control algorithm forces all portables sharing a common base station to adjust their power based on the line-of-sight component so that the base station receives equal power from these portables. Interference from portables controlled by other base stations may be higher or lower. By implementing such a algorithm, the simulation results are displayed in Fig. 12. Compared to the performance of a basic system represented by BI and DI, both blocking and dropping probabilities are improved significantly. The dropping probability is reduced more than blocking probability. It is not suprising to see the performance improvement introduced by power control. For a CDMA system, power control is an effective way to combat the near/far problem and reduce the dropping probability accordingly. On the other hand, implementation of power control leads to higher data processing overhead, and complexity.

F. Performance of the Basic System With Sectorized Base Antennas

The utilization of sectored antennas tends to increase the system capacity. Our previous results indicate that a system with relatively small processing gain may have difficulty in supporting a high traffic load. By implementing sectored antennas for the base stations, the system performance should be improved. In our model system. we still keep omnidirectional antennas for the portables, and antennas with three sectors are

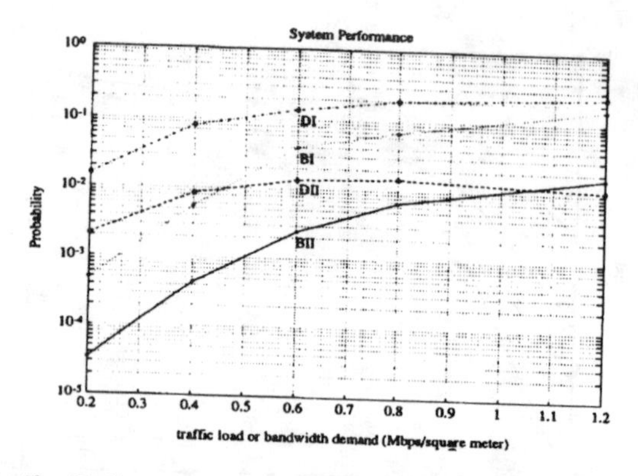

Fig. 13. Blocking and dropping probabilities vs. offered traffic load under a given base station density of 1/100 square meters for a system with (BII and DII) or without (BI and DI) three sectored base-site antennas.

used for the base stations. For systems using CDMA multiple access, using sectored antennas increases the resources or reduces the microcell area. Each sector is still operating at the same frequency, but it senses less interference, because only those active portables located inside the same section will be sources of interference.

The system performance is simulated with a single type of traffic source, i.e. 20Mbps data rate and 30 minute mean holding time. The results are shown in Fig. 13. It is seen that the improvement in terms of blocking and dropping probabilities, compared to the basic system, is significant. Even at a higher offered load of 1.2 Mbps/square meters, the blocking and dropping probabilities are well below 10^{-2}. The interesting thing is that as the traffic load increases, the dropping probability increases at first, then it turns flat showing some kind of saturation. Compared to the blocking probability, the dropping probability is still bigger. Once again, we see that higher dropping probability is associated with the absence of power control. Even with sectorization, the problem does not seem to be solved, but it has become very much less serious. At a load of 1.1 Mbps/square meters, the blocking probability approaches the dropping probability. The results here imply that sectorization should be a promising approach to increase the system capacity.

VII. Conclusions

We have investigated performance of an indoor wireless system using CDMA access. A broadband communication system using CDMA access with small processing gain appears not to have great potential to support higher traffic of large bandwidth demand in terms of acceptable performance. However, employing power control and sectored antennas techniques would make the system competitive with other systems.

Acknowledgment

The authors are grateful for fruitful discussions with G. Stamatelos. They also acknowledge the suggestions of the anonymous reviewers, which have led to improvements in the presentation.

References

[1] J. B. Lyles and C. D. Swinehart, "The Emerging gigabit environment and the role of local ATM," *IEEE Commun. Mag.*, p. 52, Apr. 1992.
[2] A. J. Viterbi, "When to spread spectrum-A sequel," *IEEE Commun. Mag.*, p. 12, Apr. 1985.
[3] W. C. Y. Lee, "Overview of cellular CDMA," *IEEE Trans. Vehic. Technol.*, vol. 40, p. 291, May 1991.
[4] D. L. Schilling *et al.*, "Spread spectrum for commercial communications," *IEEE Commun. Mag.*, p. 66, Apr. 1991.
[5] A. M. Saleh and L. J. Cimini, Jr., "Indoor radio communications using time-division multiple access with cyclical slow frequency hopping and coding," *IEEE J. Select. Areas Commun.*, vol. 7, p. 59, Jan. 1989.
[6] V. C. M. Leung, "Internetworking wireless terminals to local area networks via radio bridges," *IEEE, Int. Conf. Select. Topics in Wireless Commun.*, Vancouver, Canada, June 1992.
[7] D. Raychaudhuri, "Performance analysis of random access packet-switched code division multiple access systems," *IEEE Trans. Commun.*, vol. COM-23, no. 6, p. 895, June 1981.
[8] R. K. Morrow and J. S. Lehnert, "Packet throughput in slotted ALOHA DS/SSMA radio systems with random signature sequences," *IEEE Trans. Commun.*, vol. 40, no. 7, p. 1223, June 1992.
[9] K. S. Gilhousen *et al.*, "On the capacity of a cellular CDMA system," *IEEE Trans. Vehic. Technol.*, vol. 40, p. 303, May 1991.
[10] G. L. Turin, "Communication signaling—An anti-mulitpath technique," *IEEE J. Select. Areas Commun.*, vol. SAC-2, no. 4, p. 548, July 1984.
[11] H. Leib, "A digital transmission approach for indoor millimeter wave-band systems," in *Proc. ICUPC'93*, Oct. 1993, p. 880.
[12] L. Golding and A. Livne, "RALAN-A radio local area network for voice and data communications," in *Proc.IEEE GLOBECOM*, 1988, p. 48.1.1.
[13] A. R. Tharek and J. P. McGeehan, "Indoor propagation and bit error rate measurements at 60 GHz using phase-locked ocillator," in *Proc.IEEE Vehic. Technol. Conf.*, 1988, p. 127.
[14] G. A. Kalivas, M. El. Tanany, and S. A. Mahmoud, "Millimeter-wave channel measurements for indoor wireless communications," in *Proc. IEEE Vehic. Technol. Conf.*, 1992, p. 609.
[15] P. F. M. Smulders and A. G. Wagemans, "Wide-band measurements of MM-wave indoor radio channels," in *Proc. Third Int. IEEE Symp. on PIMRC*, Boston, MA, Oct. 1992, p. 10.5.1.
[16] A. L. M. Gerla, "Wireless communications in the automated factory environment," *IEEE Network*, vol. 2, p. 64, May 1988.

Chang G. Zhang was born in Xinjang, China. He received the M.Sc. degree in statistical physics from Liaoning Normal University, Dalian, China, in 1985, the Ph.D. degree in solid state physics from the University of Ottawa, Canada, in 1992 and the M.Sc. degree in system engineering from Carleton University, Canada. in 1993.

Since 1992, he has been involved microcellular architecture of broadband indoor wireless communication systems research with the Canadian Institute of Telecommunication Research. In November 1993, he joined Bell-Northern Research as a System Engineer engaged in wireless access technology research.

H. M. Hafez received the B.Sc. and M.Sc. degrees in electrical engineering from the University of Alexandria, Egypt, in 1971 and 1974, respectively, and the Ph.D. degree in electrical engineering from Carleton University, Ottawa, Canada, in 1980.

Between 1980 and 1981, he was with the Federal Department of Communications, Canada, as a consultant in systems engineering in the area of mobile communications and spectrum management. In 1981, he joined the Department of Systems and Computer Engineering at Carleton University, where he is currently an Associate Professor. He has acted as a consultant to several industrial and government organizations, including Bell Canada, Bell Northern Research, NovaTel, and Bell Cellular.

David D. Falconer (M'68–SM'83–F'86) was born in Moose Jaw, Saskatchewan, Canada on August 15, 1940. He received the B.A.Sc. degree in engineering physics from the University of Toronto in 1962, and the S.M. and Ph.D. degrees in electrical engineering from M.I.T. in 1963 and 1967, respectively.

After a year as a Postdoctoral Fellow at the Royal Institute of Technology, Stockholm, Sweden, he was with Bell Laboratories, Holmdel, NJ, from 1967 to 1980 as a Member of the Technical Staff and later as Group Supervisor. During 1976–1977, he was a Visiting Professor at Linköping University, Linköping, Sweden. Since 1980, he has been at Carleton University, Ottawa, Canada, where he is a Professor in the Department of Systems and Computer Engineering. His interests are in digital communications and communications theory, with particular application to wireless communications systems. He was Editor for Digital Communications for the IEEE TRANSACTIONS ON COMMUNICATIONS from 1981 to 1987. He is a member of the Association of Professional Engineers of Ontario. He was awarded the Communications Society Prize Paper Award in Communications Circuits and Techniques in 1983 and again in 1986. He was a co-recipient of the *IEEE Vehicular Technology Transactions* Best Paper of the Year award in 1992. He was a consultant to Bell-Northern Research, working on ISDN access in 1986–1987 and to Codex/Motorola, working on cellular CDMA techniques, in 1990–1991 during sabbaticals. He is currently leading a research project on broadband indoor wireless communication, involving several universities, sponsored by CITR (Canadian Institute for Telecommunications Research).

UNIT 5
OTHER APPLICATIONS OF SPREAD SPECTRUM

There are many applications of spread-spectrum techniques. We have selected three papers for this unit. The first one is "*The Global Positioning System*" by Getting, which describes the Global Positioning System (GPS). Direct-sequence spread- spectrum formats are used for GPS signals. The second paper, "*Mobile Access to an ATM Network Using a CDMA Air Interface*" by McTiffin et al., discusses the use of direct-sequence spread-spectrum CDMA for the radio access of mobiles to an ATM network. The final paper, "*The Application of a Novel Two-Way Mobile Satellite Communications and Vehicle Tracking System to the Transportation Industry*" by Jacobs et al., describes the service features and technical characteristics of a satellite vehicle tracking system called OmniTRACS. Both direct-sequence and frequency-hopping techniques are used by the system.

The Global Positioning System

Originated 30 years ago for military missions, the satellite-based GPS has grown to have commercial and scientific applications as well

"The most significant development for safe and efficient navigation and surveillance of air and spacecraft since the introduction of radio navigation 50 years ago"—that is how the National Aeronautical Association described the Global Positioning System on Feb. 10, when it bestowed its coveted Robert J. Collier Trophy on the system's designers: Aerospace Corp., the Global Positioning Team, IBM Federal Systems, Rockwell International, the U.S. Air Force, and the U.S. Naval Research Laboratory. The trophy has been awarded annually since 1912 "for the greatest achievement in aeronautics or astronautics in America...the value of which has been thoroughly demonstrated in actual use during the preceding year."

Although the Global Positioning System, or GPS as it is commonly called, was not fully operational at the outbreak of the Persian Gulf War in January 1991, its performance there under fire evoked an explosive demand in the military for this "force multiplier," which allows accurate location of fighting units. The use of GPS during that war also supported the U.S. Defense Mapping Agency's World Grid System established in 1984—the WGS-84.

Today civilian use of the GPS has been growing: for ship and aircraft navigation, for precise surveying and geological studies, and for establishing precise worldwide time that could aid in synchronizing digital communications. Such promise in new areas is striking for a system that had its beginning some 30 years ago, when Aerospace Corp., El Segundo, CA, started studies on how to improve radio navigation systems.

THE PROBLEM WITH STARS. For centuries, the heavenly bodies were used to set time standards and latterly to accurately locate astronomical observatories. The stars are so far away that, for all practical purposes, they appear to stand still as the earth rotates on its axis under them once a day. The planets, however, are seen to move relative to the stars, and some (such as Jupiter) have moons whose orbits can be used as chronometers, albeit with some practical problems.

Ivan A. Getting Fellow, The IEEE

Often positions on the earth were measured relative to astronomical observatories, called datum points. Nations set up their own map grids, which they base on datum observatories within their boundaries, and sometimes even agree to standardize those grids. Still, even now, national map grids may be mismatched at their edges by half a kilometer or more, in effect dethroning the prime meridian at the Greenwich Observatory as their zero longitude reference!

But if cloud, rain, or fog obscure the stars and planets at night, the astronomical approach to position location and navigation fails. The classical navigator must then fall back on dead-reckoning or inertial navigation, which depends on his knowing exactly where he started or being close enough to an identifiable landmark whose position is known to him. If, in addition, he can measure his direction of travel, elapsed time, and true speed, he can compute his present position.

Many devices have been developed to aid dead reckoning. The most sophisticated is the inertial navigation system, which relies on accelerometers to determine changes in velocity and gyroscopes to determine changes in direction and thus subsequent position. In the mid-20th century, Charles Stark Draper at the Massachusetts Institute of Technology (MIT) in Cambridge was the outstanding contributor to inertial navigation.

Unfortunately for the classical navigator, all dead-reckoning systems—including inertial navigation—suffer from increasing position error with elapsed time. After World War II, nonetheless, inertial navigation systems were refined so expertly that they could guide ballistic missiles from launch to target—if the position and target of the launch point were known and the time of powered flight was relatively short (say a few minutes).

Today, inertial navigation systems are also used on aircraft. The pilot enters his initial location at a surveyed airport. Because of drifts in the inertial system, though, accuracies of location better than 1 or 2 kilometers per hour of flight are difficult to achieve. In short, dead-reckoning systems are aids to navigation, but cannot replace accurate position location.

LORAN: FIRST RADIONAVIGATION. The first electrotechnology used to combat the effects of poor visibility on position finding was radio waves. Between 1912 and 1915, Reginald Fessenden in Boston first tackled fog-bound lighthouses, from which he transmitted spark-gap–generated radio waves simultaneously with acoustic underwater signals for measuring the intervening distance. Also, time signals were sent by radio from designated land-based stations to correct the chronometers of ships. So that a mariner could sense the direction, the so-called direction-finding radio beacons were installed along the shore. But taking the bearing of radio beacons was not accurate and was limited to the shores of industrialized nations. That situation lasted until World War II.

The second electrotechnology for improving position finding was a byproduct of the development of pulsed radar: the realization that short time intervals could be measured accurately by repeated pulses and, when combined with known characteristics of electromagnetic propagation, could yield accurate measurements of range (distance). In short, range is the product of the velocity of radio propagation and the time interval between transmission and reception.

In October 1940, Alfred Loomis, then chief of Section D-1 (microwave) of the National Defense Research Council (NDRC) in Washington, DC, suggested combining radio technology with the then-new technology of accurate measurement of time intervals. If two radio transmitters were located at carefully surveyed positions, and if each transmitted a short pulse in synchronization, then a receiver on a ship would receive both pulses [Fig. 2]. The separation in the time of arrival of the two pulses would determine a line of position—actually a hyperbola—characteristic of the time interval between the received pulses. If two sets of transmitting stations each provided its own line of position, then the unique location at their intersection was where the ship had to be [Fig. 3].

These pulsed radio beacons formed a true navigation system. The observer was passive, there could be any number of ships, there was no interference between users, and the position was known only to the user. To meet the operational coverage area (at first the North Atlantic), the chosen frequency was low enough (1.95 MHz) to provide both ground-wave propagation out to 1200 km at sea during the day as well as skywaves to about 2000 km at night.

This system, called Loran for LOng RAnge Navigation, was developed at MIT's Radiation Laboratory during World War II to support convoys in stormy winter weather in the Atlantic Ocean. With an accuracy of

0018-9235/93/$3.00©1993 IEEE

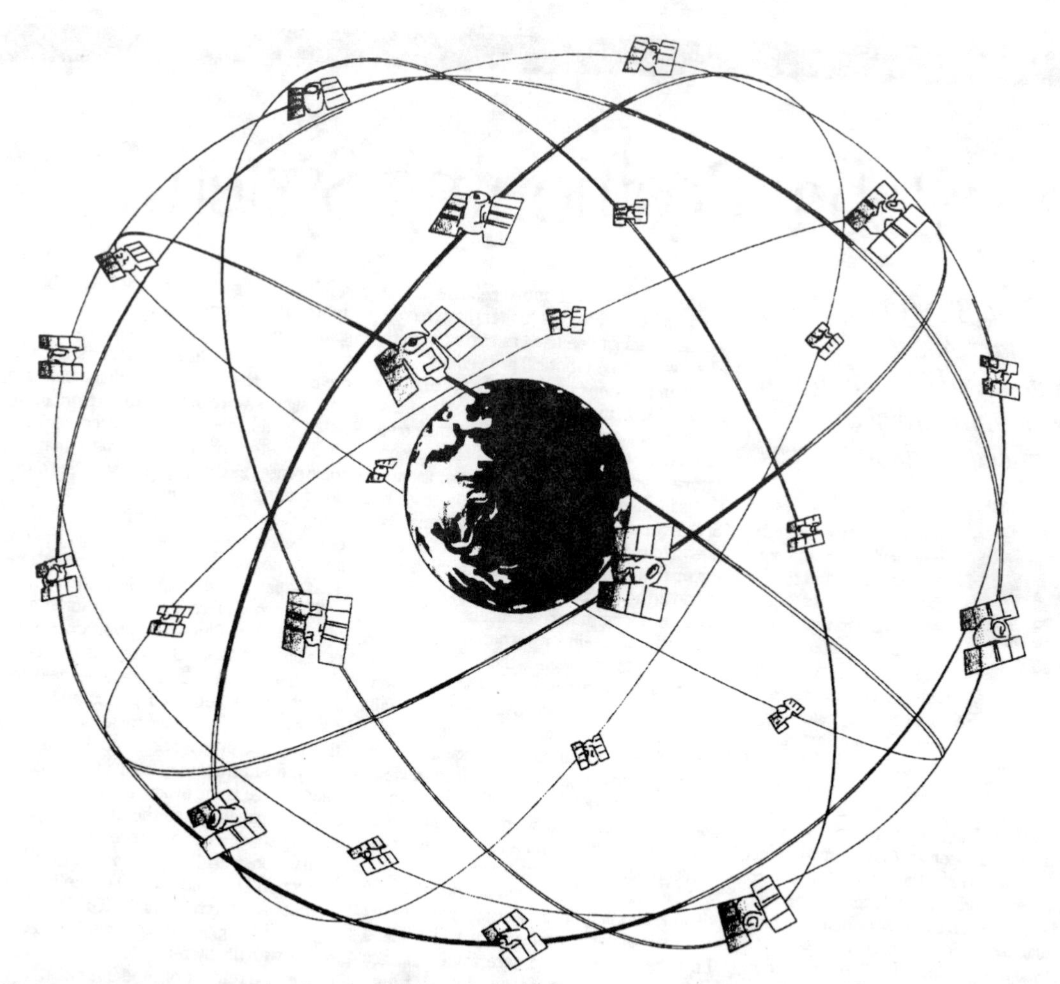

[1] The 24 satellites of the Global Positioning System (GPS) circle the earth every 12 hours at a height of 20 200 km. Four satellites are located in each of six planes inclined at 55 degrees to the plane of the earth's equator. Each satellite continuously transmits pseudorandom codes at two frequencies in the 1000-MHz band, and covers half the earth with accurately synchronized time signals and data about its own position.

about 1.5 km, Loran was the first true all-weather position-finding system. After the war, and operating at somewhat different frequencies, the system was taken over by the U.S. Coast Guard as Loran-A, which was considered a major technological breakthrough at the time.

Following World War II, a number of variants based upon the Loran principle of using differences in the time of arrival of radio signals were designed. To provide better world coverage with fewer transmitters, a low-frequency (10–14 kHz) system called Omega was established, although its accuracy was limited [Table I]. While Omega used continuous-wave radiation rather than pulses as did the original Loran, in effect it gauged the difference in time of arrival from ground stations by measuring the relative phase angles of transmissions from pairs of stations.

To increase accuracy, Loran-C, using ground wave propagation (100 kHz), was developed for tactical aircraft. Loran-C was accurate to about 100 meters, depending on local calibration, the moisture of the soil, local topography, and other factors. It was widely used during the Vietnam War.

3-D NAVIGATION. All the early Loran systems and Omega were two-dimensional; that is, they helped locate position in latitude and longitude only. All depended on the synchronization of the ground stations.

The first concept for a three-dimensional type of Loran system—which would include altitude as well as latitude and longitude—was proposed by Raytheon Co., Lexington, MA, in 1960. The proposal was a response to U.S. concern about the vulnerability of its fixed-based Intercontinental Ballistic Missiles (ICBMs), whose geographical locations were well known to the Soviet Union, making them vulnerable in a preemptive attack. To counter that liability, the U.S. Air Force embarked upon a mobile version of the Minuteman ICBMs mounted on railroad cars, which could be moved at will and fired from sites that need not have been accurately surveyed beforehand.

Such a mobile system tremendously complicated the guidance of the missiles. An ICBM's conventional inertial navigation system required knowledge of not only the position coordinates of the launch point, but also of the local orientation of the zenith, the azimuth of the launch, and measurements of gravitational anomalies determined out to several hundred kilometers.

The navigation system proposed by Raytheon to the Air Force on Feb. 15, 1960, was called Mosaic (Mobile System for Accurate ICBM Control). In its simplest version, Mosaic used four 3000-MHz (S-band) continuous-wave transmitters at somewhat different frequencies, with their modulation all locked to atomic clocks and synchronized through communications links.

Although during flight the missile would measure arrival time differences of the signals in a manner similar to Omega, much greater accuracy was possible at the 3000-MHz band because of its large bandwidth and its immunity to atmospheric propagation variations. The missile would continuously compute its position during powered flight, correcting its heading and engine cut-off, thus solving the navigation problem for arrival at the target. By this technique, the guidance of the missile would be independent of the coordinates of the launch point.

Before plans for Mosaic could be put into operation, though, Secretary of Defense Robert McNamara canceled the Mobile Minuteman program early in 1961. Whether Mosaic would have been developed if that program had been continued remains moot, but a key offshoot of Mosaic's concept was its introduction to several of the planning people at Space Technology Laboratories (STL), in El Segundo and Los Angeles, many of whom transferred to Aerospace Corp. at its formation in 1960.

A third electrotechnology critical to the GPS was the introduction of artificial satellites. In the United States, as in the Soviet Union, the basic technology for launching satellites evolved from the ballistic missile programs. In 1960, the Department of Defense (DOD) assigned space launching to the Air Force, which selected Aerospace Corp. as its system engineer.

It should be noted that in 1960 the Minuteman guidance and control and computer package represented the most advanced application of electronic technology, including speed, reliability, and compactness. STL and North American Aviation Co. (at several sites in the Los Angeles area) had made major contributions; but so had Bell Telephone Laboratories, MIT's Lincoln Laboratory, and many others for their efforts in solid-state electronics. Without these advances as well as progress in atomic clocks, the GPS would not have been possible.

TRANSIT DEBUTS. Satellites were first used in position finding in a simple but reliable two-dimensional system called Transit, developed by the Navy and its affiliated Applied Physics Laboratory (APL) run by Johns Hopkins University, Baltimore, MD.

In 1957, George Weiffenback and William Guier of APL demonstrated that they could establish the ephemeris (the almanac) of the original Soviet satellite, Sputnik 1, by carefully measuring the Doppler shift in the frequency of its continuous-wave transmitter. The next year, Frank McClure suggested that the inverse was also true: if you knew the ephemeris of a satellite, then you could determine your position on the earth by observing the Doppler shift of a stable transmitter on the satellite [Fig. 4].

It was the simplest conceivable satellite position-finding system. Specifically, it was designed to meet the Navy's requirement for accurately locating ballistic missile submarines and other ships at the ocean's surface. Altitude was assumed to be zero and was referred to the earth's geoid (the shape of the earth as determined by the constant of gravity—essentially, sea level). The first Transit satellite was launched in 1959, and the Transit system was operational in 1968. By 1990, Transit included seven operational satellites and six spares stored in orbit.

This satellite system was the first to introduce corrections in the velocity of propagation of radio waves through the ionosphere by transmitting a signal at two frequencies, 150 MHz and 400 MHz. The Transit receiver on the earth's surface measured the difference in the time of arrival of the two signals superimposed on both frequencies and estimated the delay introduced by the free-electron density in the ionosphere.

When a Transit satellite is in view of the user, successive fixes can be taken every 30 seconds, and the satellite may be in view for up to 20 minutes. Sightings of the satellites are intermittent, depending on the user's latitude: every 30 minutes at 80 degrees north or south latitude to every 110 minutes at the equator. The accuracy of a single fix can be as good as 50–200 meters for a stationary or slowly moving observer. Where the altitude is accurately known, repeated measurements at a fixed land position can be averaged over many observations, requiring long periods of time—up to days—to give accuracies approaching a few meters.

Thus, Transit had many shortcomings: it was slow, intermittent, and two-dimensional, and it was subject to errors with even the slightest motion of the observer. In short, Transit, while a big step forward in radio position location, is impractical for use on aircraft or missiles.

BIRTH OF NAVSTAR. Aerospace Corp. was established in July 1960 to apply, as the Secretary of the Air Force stated, "the full resources of modern science and technology to the problem of achieving those continued advances in ballistic missiles and space systems which are basic to national security." In the company's first six months, it hired some 200 veterans from the Space Technology Laboratories as well as more than 100 other scientists and engineers; within two years it had nearly 1500.

Consistent with the company's mission, two sets of studies on national security were begun. The first was Project 75, which was to define ballistic missile systems that would be in use by around 1975, 15 years into the future. Defining military space systems was the focus of the second, which was called Project 57 (the inverse of 75).

Among the space systems studies, navigation of aircraft using radio signals from satellites was a key area of interest. It was in the navigation study, directed by Phil Diamond, that the Global Positioning System concept was born. In 1963, the Space Division of the Air Force began its support of GPS and named it Project 621B, but the initials GPS stuck.

The system was designed to meet the

(Continued on p. 43)

Navigation systems compared

System	Position accuracy, meters	Velocity accuracy, m/s	Range of operation	Comments
Global Positioning System, GPS	16 (SEP)	≥0.1 (rms per axis) (c)	Worldwide	24-hour all-weather coverage; specified position accuracy available to authorized users
LOng-RAnge Navigation, Loran C (a)	180 (CEP)	No velocity data	U.S. coast and continental, selected overseas areas	Localized coverage; limited by skywave interference
Omega	2200 (CEP)	No velocity data	Worldwide	24-hour coverage; subject to VLF propagation anomalies
Standard inertial navigation systems, Std INS (b)	≤1500 after 1st hour (CEP)	0.8 after 2 hours (rms per axis)	Worldwide	24-hour all-weather coverage; degraded performance in polar areas
TACtical Air Navigation, Tacan (a)	400 (CEP)	No velocity data	Line of sight (present air routes)	Position accuracy is degraded mainly by azimuth uncertainty, which is typically on the order of 1.0 degree
Transit (a)	200 (CEP)	No velocity data	Worldwide	90-minute interval between position fixes suits slow vehicles (better accuracy available with dual frequency measurements)

SEP, CEP = spherical and circular probable error (linear probable error in three and two dimensions).
(a) Federal Radionavigation Plan, December 1984.
(b) SNU-84-1 *Specification for USAF Standard Form Fit and Function* (F^3) *Medium Accuracy Inertial Navigation Set/Unit*, October 1984.
(c) Dependent on integration concept and platform dynamics.

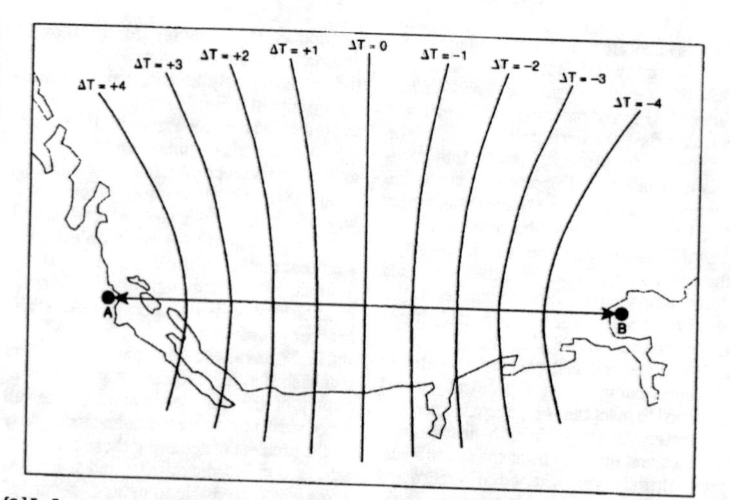

[2] In Loran-A, the first radio-based LOng-RAnge Navigation system, a pair of ground-based radio stations (A and B) simultaneously transmit a pulsed signal. To compute position, the navigator notes the difference in the arrival times of the signals (ΔT). If the pulses arrived at different times, the time difference established a line of position—actually, a hyperbola.

(Continued from p. 38)
requirements of tactical aircraft; they need to be able to determine their position continuously anywhere in the world, with 3-D accuracy adequate for bomb delivery. Since GPS was set up to serve an unlimited number of users who would not be required to radiate signals (which might expose their locations), the user equipment had to be "unattended and cheap" and thus available in large quantities.

All Aerospace Corp. GPS studies recommended a system with three features in common. First, the system had to pinpoint accurately the positions of the satellites—in effect, replacing stars with satellites. Second, it had to be able to replace light with radio waves that penetrate the ionosphere and clouds—in short, providing an all-weather service. Finally, it should be able to measure relative time using the Loran-type technique that compares relative time of arrival at the user with the radio signals from time-synchronized satellites.

To synchronize the time to a common standard, each transmitter in the satellites included atomic clocks, with updates from a ground master station. The ground control segment would consist of a master control station plus a number of monitor stations at accurately located geographical positions [Fig. 5]. These monitor stations would track the satellites, determine their ephemerides, and check the satellite clocks.

In a Loran-type navigation system, the user measures the time difference of arrival from pairs of synchronized transmitters. Since conventional crystal oscillators are adequate for this purpose, Hideyoshi Nakamura and his colleagues at Aerospace Corp. suggested the use of signals from four satellites, measuring four independent time differences of arrival and adding these to a clock connected to the same user's quartz oscillator. Of course, the user's clock has an unknown bias relative to the satellite system time.

There are thus four unknowns: the user's 3-D location and the user's clock bias from system time. Since the four measurements of differences in time of arrival are independent, the equations can be solved for the four unknowns. The clock bias can lock the user's clock to the satellite system time; and if the system ground station is synchronized with universal standard time (UTC), then every user throughout the world is automatically linked to UTC as well. Measuring a signal's time of arrival from four time-synchronized satellites became an important characteristic of the GPS system [Fig. 6].

It is, of course, necessary to relate the GPS longitude and latitude to the various map systems, and the GPS altitude to the height above the accepted geoid at those coordinates. If the ephemerides are stated in terms of a well-defined earth gravitational model, a worldwide mapping grid results.

The new aspect, however, is the introduction of a worldwide time standard. Studies made at Aerospace Corp. in 1963 and 1964 indicated that the then-current technology would permit instantaneous and continual measurement of the three spatial coordinates to an accuracy of about 10–20 meters and of time to about 30–60 ns. Detailed studies on military applications were made; and the possibilities for civilian use seemed boundless.

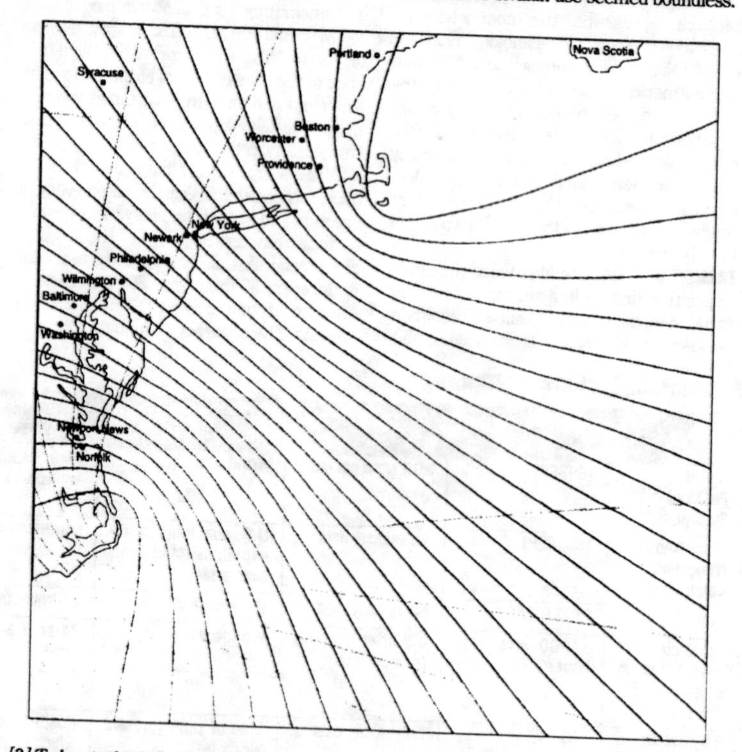

[3] To locate the definitive position of an ocean-going ship using Loran-A, the navigator must establish the intersection of two lines of position. That intersection could be determined either from three synchronized ground stations [shown] or from two sets of two synchronized stations [not shown]. The navigator's device for measuring the time difference of arrival need not be synchronized with the transmitters. Through the 1950s, the hyperbolic lines of position from pairs of ground stations were identified on nautical charts by the use of different colors [shown here as green, red, and purple].

The cost of a satellite in orbit is roughly proportional to its weight. The GPS studies indicated that a medium-sized launch vehicle such as the Atlas would be necessary for the GPS satellites. The satellite would therefore have primary power derived from solar cells of about 1 kW. Allowing for all other requirements from systems on the satellite for power, approximately 200 W were available for each of the two GPS transmitters, providing roughly 20 W of radiated power for each.

Moreover, to provide worldwide coverage, each satellite antenna had to be designed so that the radio signal was spread evenly over an entire hemisphere of the earth—a large area indeed! The result was that the signal strength reaching the user's equipment was some 30 decibels (a factor of 1000) below ambient noise from all the other sources that a receiver sees when it is pointed toward the sky.

BANDWIDTH COMPRESSION. The solution to this vexing problem was found in bandwidth compression. Suppose that the position data are needed only once each second. If the transmitted signal has a bandwidth of, say, 10 MHz, information-processing theory requires the signal-to-noise ratio to be improved by 60 dB (a factor of 1 million) if the signal is compressed to 10 Hz. In other words, compressing the bandwidth brings the range-measurement information to some 30 dB above the ambient noise in the 10-cycle bandwidth of the user's position computation.

In 1965, James B. Woodford and Pete Soule conceived the use of wide-band pseudorandom sequences of binary digits modulated on a continuous carrier employing code-division multiplexing. With this approach, all satellites can transmit on the same frequency without interference. All can also transmit characteristic identifying codes while bandwidth-compression techniques within the receiver extract the accurate time of arrival.

By 1965, the system concept had been developed in sufficient detail for presentation at technical meetings. Based on this concept, beginning in 1968, the Air Force placed contracts with industry to develop user equipment. In 1970, satellite experiments were performed, and in 1971 and 1972, using balloon-carried transmitters simulating satellites, a number of tests on receivers in helicopters and airplanes were conducted at White Sands Proving Ground in New Mexico. To everyone's delight, the technique pinpointed the positions of the aircraft to within an accuracy of 15 meters.

At this time, Aerospace Corp.'s satellite system design placed 16 satellites into inclined geosynchronous orbits whose ground tracks formed four oval-shaped clusters extending 30 degrees north and south of the equator. Such a scheme permitted an orderly evolution of the system since it required only four satellites at first to demonstrate its operational capabilities.

That demonstration was important for funding. Air Force research and development offices had continued their support of the GPS's development, but without a showing of a working satellite system, support from the DOD's operational elements would not have been available.

COORDINATING FOR DEFENSE. The GPS work by Aerospace Corp. for the Air Force stimulated additional satellite-based position-location and navigation work by the Navy. These extra undertakings were done both at APL, based on the Transit, and at the Naval Research Laboratory, based on a system called Timation that involved measuring one-way transit time and precise atomic clocks at the user. To obtain world coverage with its Timation system, NRL proposed using 21–27 satellites in 8-hour orbits. The Army also proposed using its system called Secor (Sequential Correlation of Range).

In 1968, a three-service committee, later called Navseg (Navigation Satellite Executive Committee), was organized to coordinate the effort of the various satellite navigation groups. While this committee did provide for the exchange of information and made constructive suggestions for various field tests, it had no authority to enforce its recommendations. The Navy, however, did proceed with its tests of atomic clocks in satellites.

On Aug. 17, 1974, the Deputy Secretary of Defense determined that a joint three-service program based on the GPS concept be established and that the Air Force should be its program manager. The program was renamed Navstar, but the term GPS continues as well as GPS/Navstar.

General Kenneth Schultz, then Commander of the Air Force Space Division, had appointed Colonel Brad Parkinson as the Air Force GPS program manager. His counterpart at Aerospace Corp. was engineering manager Walt Melton (who was succeeded by Bruce Adams, then by Ed Lassiter, by Allen Boardman in 1979, and later by David Nelson).

Once the DOD decided on a joint program, a system configuration was rapidly developed. The satellites were to be placed in 12-hour inclined orbits, a compromise between the Air Force advocacy of 24-hour orbits and the Navy's 8-hour orbits. This was a poor choice from the standpoint of the Van Allen belts of trapped radiation around the earth and their effects on long-life satellites. But the Navy was concerned that if a cluster of satellites in 24-hour orbits was first installed over the Americas, Congressional and DOD budget considerations might not provide coverage of, say, the Indian Ocean.

Besides, they argued, 8- or 12-hour orbits would necessarily provide for overflight of U.S. land, where system ground stations would be free of foreign intervention. While there were valid arguments on both sides, compromise and cooperation were the order of the day, and the present six-plane, 12-hour, 24-satellite constellation was selected [Fig. 1].

The initial funding of the GPS/Navstar was about US $150 million. By full operational capability, the costs would be in the

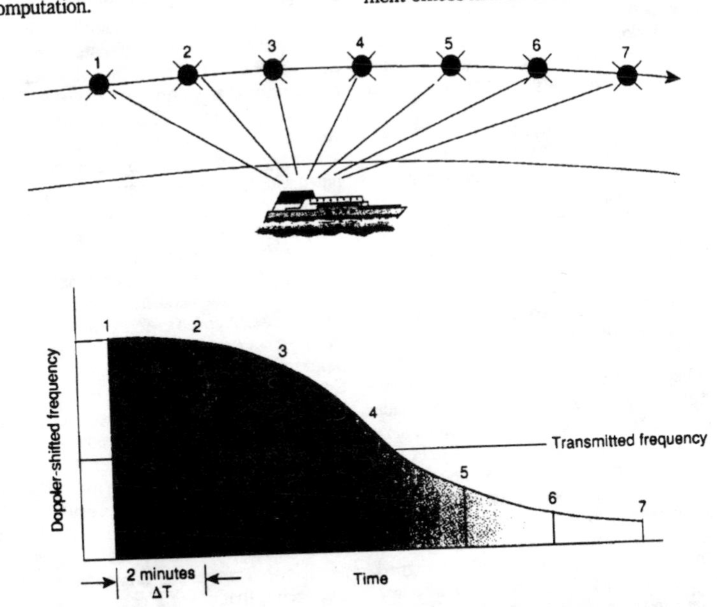

[4] In the U.S. Navy's Transit system the first two-dimensional radionavigation position-finding system using a satellite, the satellite is in a polar orbit at the relatively low altitude of 1100 km and transmits its position data every 2 minutes. It transmits at a fixed frequency and uses the varying time difference of signal arrival (Doppler effect) over the satellite's path during the observation time. The ship's position is calculated from the shape and amount of the Doppler-shifted curve [purple]. To reduce the average intervals between position fixes to 90 minutes, seven satellites are used.

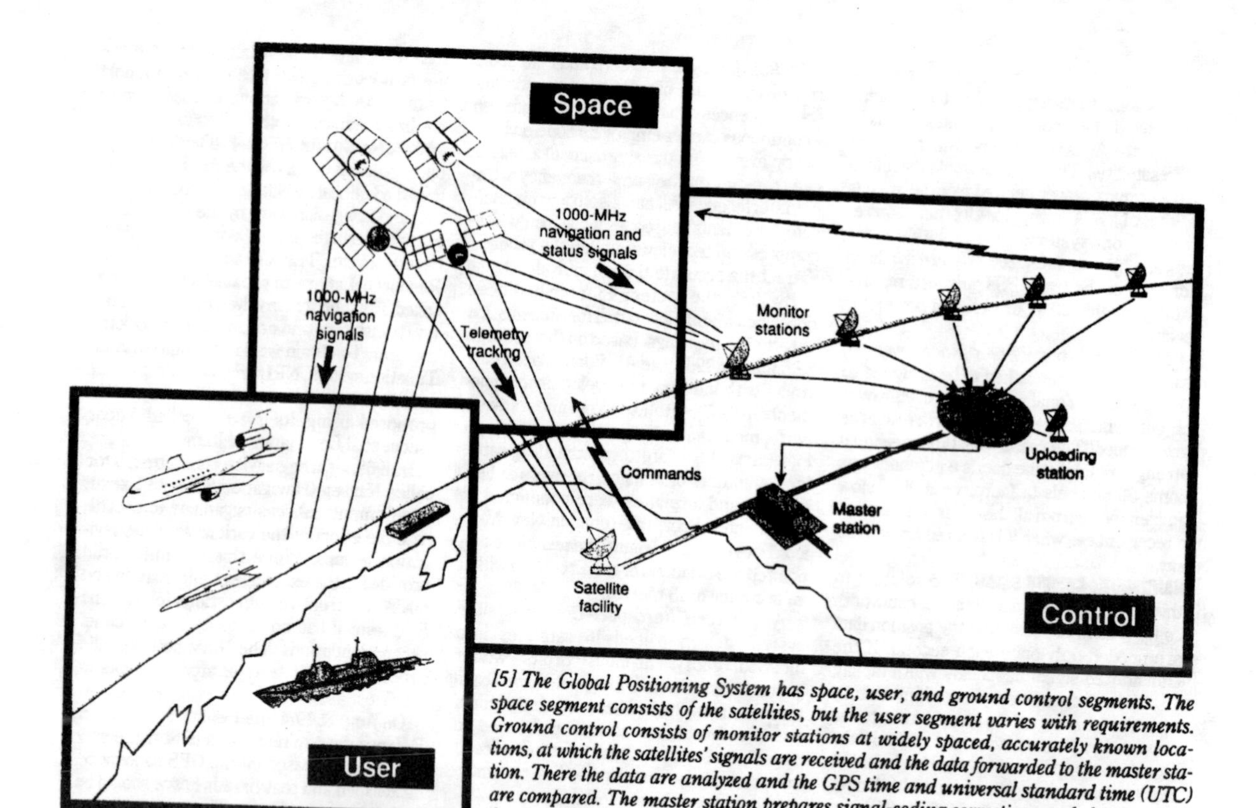

[5] The Global Positioning System has space, user, and ground control segments. The space segment consists of the satellites, but the user segment varies with requirements. Ground control consists of monitor stations at widely spaced, accurately known locations, at which the satellites' signals are received and the data forwarded to the master station. There the data are analyzed and the GPS time and universal standard time (UTC) are compared. The master station prepares signal-coding corrections and change orders for the satellite control facility, which uploads the data to the satellites.

range of $10 billion. Since the GPS/Navstar is a support system (as opposed to a weapons system), finding money within the DOD during the development phase became an increasingly difficult task.

In addition, skeptics abounded at each corner: the satellites were vulnerable; the user receivers could be jammed; and there was concern over what measures could be taken to prevent an enemy from using the GPS against the United States.

To overcome this last objection, the Air Force Ground Control Station deliberately introduced satellite timing and position errors into the satellite transmissions. This move reduced the accuracy of user-horizontal position finding of civilian and "unauthorized" users to 100 meters 90 percent of the time—adequate for navigation but not useful for weapon delivery. During war and as stipulated by the President of the United States, these artificially imposed errors would be removed for authorized U.S. and allied users. This concept is dubbed selective availability.

HOW GPS WORKS. Most radionavigation systems in use in 1965 used discrete carrier frequencies to identify signals from different transmitters. Those using continuous-wave transmission modulated the carrier at multiples of subfrequencies whose phase measurement at the receiver provided accurate measurement of arrival time.

In Aerospace Corp.'s approach of using the same frequencies for all GPS satellite transmitters, code signals are used for transmitter identification and time ticks, as well as transmission of system data. Code modulation in addition provides bandwidth compression for accurate measurement and signal enhancement.

All GPS satellites transmit the same two navigation frequencies, designated L1 and L2, as well as a UHF intrasatellite communications link, and S-band links (SGLS) to ground stations. The L1 and L2 transmissions are in an internationally assigned navigation frequency band: L1 at 1575.42 MHz, and L2 at 1227.6 MHz.

These frequencies are generated by multiplying the atomic clock output of 10.23 MHz by 154 and 128 respectively. Both frequencies are needed at the receiver for precise measurements when corrections must be made for ionospheric propagation in a manner similar to that of the Transit system. Either frequency can be used in a stand-alone mode if satellite broadcast coefficients for the ionospheric corrections are processed in the user's equipment.

A precision code (the P code) is designed to repeat every 280 days or until a new key is sent to the code generators. To a casual observer, the P code appears to be random noise. User receivers also generate the same P code, and compare their code with the code received from the satellite. The receiver matches its self-generated code with the received satellite signal by introducing a delay. When the signals match (correlate), the receiver delay is actually the time of arrival as measured by the user's clock.

Since the GPS system consists of 24 satellites, the P code is divided into 40 seven-day segments that are continually repeated. Each satellite has its own assigned segment. Even so, there are 6.5 trillion 1s and 0s in the seven-day code. The length of this code, combined with the weekly repetition rate, makes acquisition of the P-code signal slow unless the user can accurately initialize his or her receiver in position and/or time.

To assist in acquisition, each satellite is assigned a coarse acquisition code (the C/A code). The C/A codes are chosen from a family of orthogonal (that is, statistically independent) codes of a class called Gold Codes—one for each satellite. There are 1023 chips in the code, which repeats a thousand times a second. The C/A code is locked into the P code for rapid transfer. It can, however, be used to compute the user position for users who do not have access to the P code and for civilian receivers that do not require a high level of accuracy (though practice has shown that the C/A code accuracy approaches that of the P code).

Since the P code and the C/A code are chosen not to interfere with each other, they are modulated on the same carrier, one phase shifted 90 degrees and then combined, a process called phase quadrature [Fig. 6].

Finally, the user must have the daily ephemeris and time corrections that each satellite has received from the ground mas-

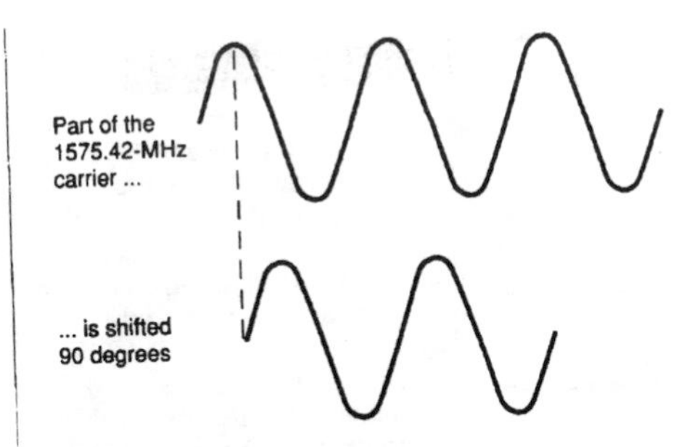

[6] Each GPS satellite transmits two sets of codes: the precision code (or P code) for accurately determining the signal's arrival time, and the coarse acquisition code (or C/A code) for less accurate tracking and for assisting the receiver in acquiring the P code. To conserve power in the satellite, the carrier is divided into two parts before transmission. One part is modulated by the P code, while the other is phase-shifted 90 degrees and modulated by the C/A code. The two are then combined for amplification and transmission, a process known as phase quadrature.

ter control station. Each satellite transmits the system data on both L1 and L2. If a GPS receiver is not already loaded with an almanac describing the ephemerides of all the satellites, it must establish an almanac by acquiring any satellite in view. Since the satellite configuration can change over long periods of time, the almanac needs to be updated about every six months.

While all satellites broadcast at the same frequencies, because of their rapid motion in orbit as well as the motion of the observer, the frequencies at the user equipment receiver are shifted by the Doppler effect. Thus, each user receiver must "tune in" on the satellites in view at the nominal frequencies of L1, L2, or both.

In addition to the frequency tuning, each receiver must acquire the C/A code and then the P code, if so equipped, and read the navigation message to get current satellite data. These steps may require substantial delays if the initializing coordinates, including time, inserted into the receiver are in gross error. This delay is called "time to first fix."

CIVILIAN APPLICATIONS. The GPS user equipment comes in many variants, depending upon the application. For example, a hand-held single-channel portable unit, if initialized with approximate position and time, requires about 2 minutes to get the first position and time readings. After that, it computes the data every half-second. It uses only the C/A code, but that code is entirely adequate for marine navigation and land surface vehicles.

Thousands of these modified commercial units were used by the Army in the Persian Gulf War—for position finding, navigation, artillery fire control, and so on. In the meantime, manufacturers all over the world embarked on the production and sale of multiple-channel receivers integrated with computers and displays, which made the navigation of yachts match video games for entertainment.

User equipment for high dynamic vehicles, such as supersonic aircraft and missiles, are more complex, especially if in addition, they are designed to operate against sophisticated enemies equipped with jammers. The GPS user equipment, because of its spread-spectrum design, provides an initial 60 dB of signal enhancement. If an automatic self-nulling antenna is used, some additional 30 dB of jamming resistance is achieved. The overall jamming margin depends, of course, on the type of jamming and the specific design of the user's equipment. An average overall jamming margin over ambient noise of 45–60 dB is often quoted.

Every piece of user equipment in highly maneuverable aircraft should be integrated with an inertial navigation unit. In a fully integrated design, the GPS helps calibrate the inertial navigation system drifts; and when the GPS becomes temporarily inoperative for whatever reason, the inertial navigation system carries on until the GPS reacquires. Such a combination GPS-inertial navigation system contributes to the accuracy of a fast-moving maneuverable vehicle and it adds antijam capabilities to, say, cruise missiles.

In the early Aerospace Corp. studies, it was shown that the relative position accuracy of two user receivers locked on the same satellites and not far removed from each other—say zero to tens of kilometers—was extremely high. The largest error contributors in the stand-alone GPS receiver are those associated with the satellite positions and the ionospheric-induced errors—each about 2–3 meters in the pseudo-range error budget. These errors, however, are identical for adjacent users and hence cancel out in relative measurements. This is true even if selective availability is on.

So, for example, if a GPS master station receiver is used at a harbor on a carefully surveyed spot, all ships in that harbor with corrections from the master station will enjoy better absolute accuracy (1 meter), even in the presence of selective availability, than an independent GPS receiver operating in the P code without selective availability. This procedure, called Differential GPS, is being installed by the Coast Guard in all major U.S. harbors.

Differential GPS expands the use of GPS into many new areas: refueling in air under conditions of almost zero visibility becomes possible, stand-off missiles can be controlled from mother planes, sonobuoys equipped with receivers can be accurately located, and so forth. As demonstrated in Sweden and at O'Hare Airport in Chicago, aircraft and service vehicles at airports can be located to 1 meter.

There are other special classes of user equipment, especially for fixed users who

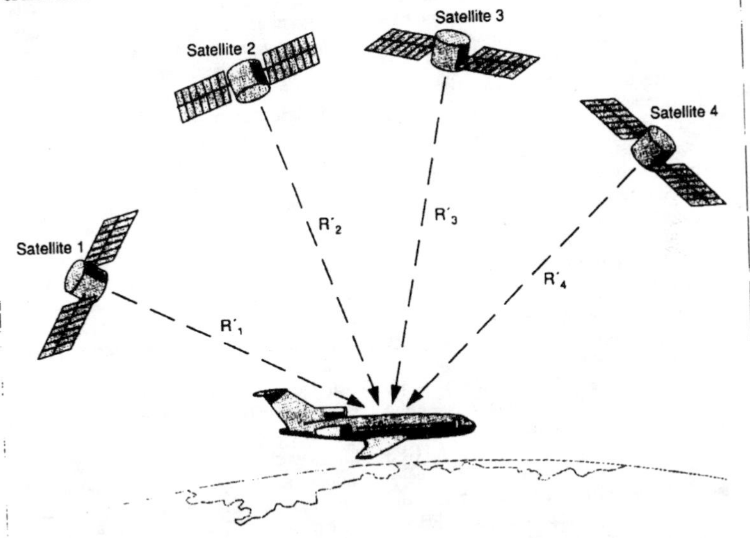

[7] With the GPS system, a user can determine his precise position in three dimensions from four satellites (Satellites 1, 2, 3, and 4). The user's clock determines a pseudo-range R′ to each satellite by noting the arrival time of the signal. The four R′ distances each include an unknown error due to the inaccuracy of the user's inexpensive clock. There are four unknowns: the user's longitude, latitude, altitude as measured from the center of the earth, and the user's clock error to bring his or her clock into synchronization with system time. Solving four range equations yields the user position information to an accuracy of 15 meters and the time to an accuracy of 100 ns.

seek extreme position accuracy and/or time synchronization. Essentially, in addition to locking on the P code, the receiver may also reconstruct the carrier, of which there are 154 cycles per time chip of the P code. This allows the measurement in principle of the time of reception to a fraction of one cycle of the carrier. Such special receivers are called codeless or interferometric receivers and have been applied in geodesy and time synchronization.

Scientists, too, have espoused the GPS/Navstar system for geodesy and weather research. In 1983–85, the Air Force Geophysics Laboratory, next door to the Lincoln Laboratory, working with associates at MIT, performed a Three-Dimensional Geodetic Control experiment on a well-documented 35-station survey field in Germany. They concluded that the relative positions of the monuments in the surveyed area, up to 10 km apart, were confirmed by the GPS survey to 10 mm plus 1 part per million horizontally and 1.6 parts vertically.

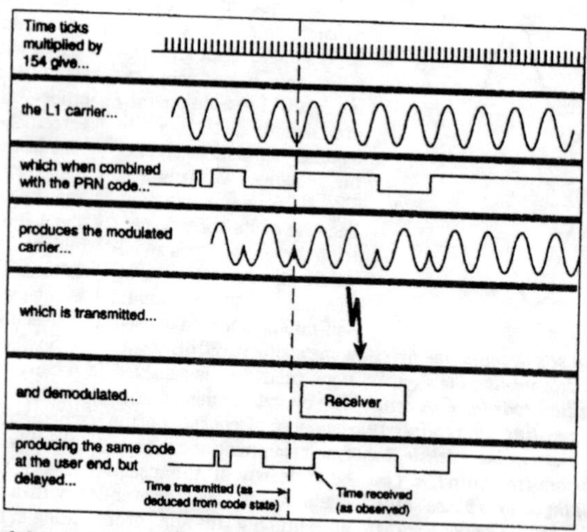

[8] Aboard the satellite is a cesium atomic clock generating time ticks at a frequency of 10.23 MHz. The time ticks are multiplied by 154 for the L1 carrier. The precision P code consists of a series of chips of 1s and 0s at the atomic clock chipping rate of 10.23 MHz; thus each chip contains 154 cycles of the carrier. Because the P code appears to the casual observer to be like random noise, it is called pseudo-random noise (PRN). User receivers generate the same P code. When the signals match (correlate), the receiver code delay is the time of arrival as measured by the user's clock.

Over the past five years, scientists have organized a worldwide geodetic effort centered on the use of the GPS/Navstar system. The University Navstar Consortium (Unavco), with headquarters in Boulder, CO, has supported more than 70 domestic and international campaigns. Its researchers have included in their studies independent GPS ephemeris determinations, specialized data reduction software, and so on. Currently, Unavco has 23 international university and research institution members. The U.S. National Aeronautics and Space Administration (NASA) has sponsored some of these programs under its Dynamics of the Solid Earth (DOSE) program.

The U.S. National Geodetic Survey publishes very accurate GPS ephemerides. GPS is the major means used by the International Bureau of Weights and Measures in France to support International Atomic Time.

Advanced equipment for commercial GPS users has followed the trend of scientific application. Such equipment can perform all the normal positioning functions of the GPS system as well as accurate ground surveys.

WHAT LIES AHEAD? To recover some of the costs of building and operating the GPS, the U.S. Congress considered charging civilians for using the system. Fortunately, Congressional members had the wisdom instead to follow the U.S. tradition that provides safety aids to navigation to the public of all nations without charge.

On the other side of the coin, the U.S. Department of Transportation (DOT) and the DOD are leaning toward reducing or canceling other forms of radionavigation aids, including Transit, OMNI-DME, Omega, and Loran-C. Also being studied is the use of GPS for civilian international aeronautical navigation by the International Civil Aeronautics Organization, better known as ICAO.

Recent tests by the U.S. Federal Aviation Administration (FAA) and others have demonstrated successfully all three FAA categories of blind landings of aircraft, opening up for all-weather operation the use of small airfields and commercial fields surrounded by rough terrain and other obstructions. The ICAO and the FAA are looking for "integrity" in system operation—that is, a system that offers essentially 100 percent reliability and provides an immediate warning of failure.

It has taken 30 years to bring the GPS to its current level of military appreciation—as demonstrated in the Persian Gulf War. Yet, like many defense-oriented developments, it is finding a place of importance in the civil sector as well.

In fact, there has been an explosive civilian growth in the manufacture of many types of user equipment (by some 160 manufacturers worldwide), new opportunities in the employment of such equipment, and even the establishment of trade journals and dedicated slick magazines. The FAA and ICAO are testing the GPS for likely adoption in air travel; and the Maritime Commission may approve its use as the primary navigation system. Using GPS, scientists will continue to study in greater depth our wobbly, less than perfect, and somewhat deformable world prone to earthquakes and other natural disasters.

Other current plans call for the DOD, through the Air Force, to maintain a 24-satellite GPS system, which is scheduled to start operating next year. Additional replacement satellites are on order.

In about the year 2003, a new production satellite replenishment generation (Block IIF) that meets even higher levels of performance will be required. With continued research on the ionosphere, on the anomalies affecting the motion of the satellites, and on other factors, improved accuracy of the information to the users' receivers will be attained. Advances in receiver designs incorporating high dynamic range, massive parallel correlators, and filters, and improved integration with host ancillary equipment will substantially improve overall performance. The changes will much enhance the system's overall use and help to ensure its integrity.

TO PROBE FURTHER. The history of the GPS system are described in Chapter 7, "Navigation Satellites," in a commemorative book, *The Aerospace Corporation, Its Work: 1960–1980*, by Everett T. Welmers (The Aerospace Corp., El Segundo, CA, 1980). Much of the background information is expanded in a book by Ivan A. Getting, *All in a Lifetime* (Vantage Press, New York, 1989).

The original Loran program and its development during World War II are described in great detail in Henry E. Guerlac's book, *Radar in World War II* (Tomash Publishers and the American Physical Society, 1987). GPS technical specifications and system descriptions are available through the GPS Joint Program Office, U.S. Air Force, Box 92960, Worldway Postal Center, Los Angeles, CA 90009.

There are at least two publications devoted to news about GPS. One is *GPS World*, a monthly from Advanstar Communications, 195 Main St., Metuchen, NJ 08840. Another is *GPS Report*, published biweekly by Phillips Publishing, 7811 Montrose Rd., Potomac, MD 20854.

Current policy and future plans for all U.S. radionavigation systems, civil and military, are described in a recent government publication, *1992 Federal Radionavigation Plan*, available through the National Technical Information Service, Springfield, VA 22161, DOT-VNTSC-RSPA-92-2/DOD-4650.5. ◆

ABOUT THE AUTHOR. Ivan A. Getting (F) is a consultant and a member of the U.S. Air Force Scientific Advisory Board and the Board of Trustees of the Environmental Research Institute of Michigan (ERIM). As the founding president of Aerospace Corp., El Segundo, CA (1960–77), and earlier as vice president in charge of research and development at Raytheon Co., Lexington, MA (1951–60), he was the force behind the establishment of the Navstar Global Positioning System.

Mobile Access to an ATM Network Using a CDMA Air Interface

M. J. McTiffin, A. P. Hulbert, T. J. Ketseoglou, W. Heimsch, and G. Crisp

Abstract—**The paper presents a possible integrated system concept for a Direct Sequence Spread Spectrum CDMA radio access system suitable for third-generation mobile radio. The system has been conceived to take account of such diverse services as low bit rate voice and quasi-broadband services at rates of up to 256 kb/s. Broadband services imply the use of the ATM transmission technique, and particular attention is paid to the mutual impact of CDMA and ATM. An efficient Automatic Repeat Request technique is described which gives a suitably low overall error rate and a soft capacity limit. The proposed solution represents a quantum advance on today's CDMA solutions and integrates well with the ATM fixed network.**

I. Introduction

VARIOUS bodies worldwide are currently working towards the specification of the next generation of mobile telecommunications systems. In particular, the International Telecommunication Union (ITU) and the European Telecommunications Standards Institute (ETSI) are developing standards for Future Public Land Mobile Telecommunication Systems (FPLMTS) and Universal Mobile Telecommunications System (UMTS), respectively. In the shorter term, the services which will be offered by third-generation mobile systems are likely to be telephony and those supported by Narrowband Integrated Services Digital Networks (N-ISDNs), and indeed voice services can be expected to predominate for many years to come. However, in the longer term it is anticipated that there will be a demand for higher bandwidth services and this will require third-generation mobile systems to interface to broadband networks using Asynchronous Transfer Mode (ATM) transmission. Thus, the future air interface must carry narrowband services effectively while providing the flexibility to carry higher bandwidth services as the demand arises. A radio access technique well suited to this diverse requirement is Direct Sequence Spread Spectrum Code Division Multiple Access (CDMA).

CDMA and ATM have characteristics which separately and in combination can offer significant advantages in the cellular mobile radio and cordless office environments, especially when a wide range of service rates must be carried. Both allow a given transmission link to support a number of simultaneous virtual connections which can be used on demand and this can simplify routing and reduce overheads. With certain CDMA handoff regimes, the statistical multiplexing properties of ATM enable the most efficient use of the access network. In addition, mobile digital radio techniques, including CDMA, use burst mode transmission resulting in packetization delays. The process of filling ATM cells with speech also involves a packetization delay and the harmonization of mobile radio, and ATM standards can result in systems where the overall delay is less than the sum of the individual delays of the two packetization processes.

Within the fixed network, broadband transmission systems offer data rates in excess of 100 Mb/s. However, the inherent nature of radio communications, in terms of transmitter power constraints and limited spectrum availability, restricts the maximum data rate which is possible over the air interface. The maximum rate will also depend on the radio cell size. Thus, in the context of an air interface operating in the region of 2 GHz, broadband services must be regarded as being similar to those provided in the fixed network but operating at lower bit rates. Radio transmission is also significantly more error prone than broadband networks, tending to reduce capacity further due to the error control overheads.

In the following, Section II considers the requirements imposed by the services likely to be carried, and Section III outlines the relevant characteristics of CDMA and ATM. Sections IV and V go on to present the air interface and network issues which should be considered, with particular emphasis on the interaction of the CDMA and ATM techniques. An integrated system concept built on this analysis is then outlined in Section VI. The reliable transmission of data depends on a suitable Automatic Repeat Request (ARQ) strategy, and Section VII discusses such strategies.

II. Nature of Services Provided over the Air Interface

A number of key services have been identified as being representative of the requirements to be met. The services seen as having the most impact are those with a high demand (e.g., telephony), which are delay-critical (e.g., telephony and videotelephony), requiring a high bandwidth (e.g., high-speed data, videotelephony), and those intolerant of errors (e.g., data and videotelephony). It is generally recognized that speech should be of "toll" quality and that data should be carried with a low overall Bit Error Rate (BER) [1], [2]. In addition, in the broadband environment there is a need to handle bursty services efficiently.

Based on the expected advances in technology, the constraints imposed by the radio environment, and the need to use spectrum efficiently, the following basic service characteristics

Manuscript received May 2, 1993; revised November 6, 1993.
M. J. McTiffin and A. P. Hulbert are with Roke Manor Research Ltd., Hants, S051 OZN, U.K.
T. J. Ketseoglou and W. Heimsch are with Siemens AG, D-81359 Munich, Germany.
G. Crisp is with GPT Ltd., Nottingham, NG9 1LA, U.K.
IEEE Log Number 9215586.

0733-8716/94$04.00 © 1994 IEEE

have been assumed. Speech is encoded at 8 kb/s, using coding with a block structure, and is able to tolerate a 10^{-3} BER. With regard to current trials being conducted within ITU [3], a frame time of 10 ms is assumed. The target for the additional delay due to the mobile system is 25 ms one way and includes coding, transcoding, packetization, and transmission delays. Data services are to be carried at rates in the range of 8–256 kb/s, with an additional one-way delay of up to 200 ms and an overall BER of 10^{-6} or better, which can be achieved with ARQ. Although the design has been based around these values, lower data rates can be carried but with reduced efficiency and other speech rates are possible. It should also be possible to carry delay-sensitive Constant Bit Rate (CBR) services, such as videoconferencing, to current standards.

III. CDMA and ATM

ATM, in combination with CDMA, can offer certain advantages. The essential characteristics of the two techniques are outlined.

CDMA allows many users to share the same radio frequency spectrum simultaneously through the use of spread spectrum [4]. Each individual connection across the radio interface can be distinguished by the CDMA code allocated to that connection. Since there is a relatively large number of codes, they can be allocated to new connections as they are set up or when a mobile affiliates to a new base station (BS). User data can then be transmitted over the air interface without the need for additional channel assignment. The CDMA code, thus, represents a virtual connection over the air interface.

Statistical multiplexing of user traffic occurs on the air interface itself as user data is generated and sent at essentially random times. When the system capacity is exceeded, the interference between users increases above a threshold resulting in excessive error rates. In a power controlled system, this can lead to instability so traffic must be carefully managed.

For the most efficient use of the radio environment, a mobile must communicate through the base station which requires it to use the least transmit power. Any departure from this rule will reduce the capacity of the system. This is particularly true for a CDMA system, and one consequence is that in CDMA systems, mobiles tend to change frequently the base stations through which they communicate.

ATM subdivides data for transmission into small fixed-size packets called ATM cells [5]. These consist of a 48 octet (byte) information field for carrying user data and 5 octets of header containing control information. The control information includes an address which indicates the connection of which the ATM cell forms a part. The address is only significant on an individual network link and translation occurs at each switching node. Addresses are assigned at connection establishment and exist for the duration of the connection. The header address information thus identifies a virtual connection over the fixed network.

Unlike traditional transmission systems where network bandwidth is assigned even though a user has no data to send, ATM only uses network capacity when there is data to be transmitted. Traffic from different users is statistically multiplexed onto common high-speed transmission systems. As the total traffic approaches the bandwidth of the transmission system, queues build up leading to delay. When buffers start overflowing, ATM cells are discarded and data is lost.

IV. Radio Access Design Issues

The radio access specification involves the reconciliation of many, often conflicting, requirements. The design issues for any digital radio interface include the bit rate, error performance requirements, delay characteristics, and overall efficiency. For the system considered here, a set of different trade-offs of these issues must be made for each service and integrated into a cohesive whole. Moreover, the design must lead to a cost-effective solution, implying a high degree of commonality of implementation across the different requirements.

The most demanding requirement is probably the wide range of bit rates to be supported. For a spread spectrum system, the service bit rate can be varied either by varying the spreading factor while maintaining a constant spread bandwidth or by varying the number of codes transmitted in parallel, each with a fixed spreading factor, which carry the data between them. Since the arguments are different for the downlink and the uplink, these are considered separately.

- **On the downlink**, a variable spreading factor is intuitively the more obvious approach, requiring considerably less hardware than multiple codes. However, a variable spreading factor does not readily permit the use of orthogonal codes since orthogonal code sets usually comprise codes of the same length. By fixing a base bit rate corresponding to the minimum requirement and providing the higher rates as multiples of this, code orthogonality can be achieved with multiple bit rates.
  The complexity of a receiver for multiple parallel codes need not be excessive if a new architecture known as precombining rake is employed. In this architecture, the path parameters for the received signal are derived from a downlink pilot signal and used for phase, amplitude, and time alignment of the various multipath components which are combined prior to despreading. Despreading (and, hence, demodulation) of the various codes is then simply achieved by a single real correlator per code. On this basis, parallel codes appear preferable.
- **On the uplink**, code orthogonality cannot practically be achieved since there is no common pilot for channel parameter extraction, and the transmission of parallel codes from one mobile would greatly increase the mobile transmitter's peak to mean transmit power ratio requirements. In this case, the use of a variable spreading factor is the preferred option.

There are two aspects to the provision of variable bit rates:

- **Service specific bit rates**—The bit rate of the air interface connection is established at the time the connection is set up. If the connection is CBR, the bit rate allocated will be the actual bit rate. If the connection is Variable Bit Rate (VBR), the bit rate allocated will be the peak bit rate.

- **Variable bit rate services**—Having set up a VBR connection, the air interface will carry any data which arrives for transportation at the peak bit rate allocated at connection set up. When there is no data to transmit, there will be no transmission. Thus, VBR services are implemented using discontinuous transmission (DTX). One such service is speech which, over the air interface, is treated as a VBR service .

The simplest approach to satisfying the different error performance requirement of the various services would be to provide that required by the most demanding service for all services. However, more efficient use of the spectrum would result from tailoring the error performance of each service to its requirement. Several approaches are available for controlling error rates, notably choice of forward error correction code, choice of the ratio of energy per bit to noise (or interference) spectral density (E_b/N_o), and the use of ARQ.

Uniquely for CDMA, there is little (if any) compromise between error performance and capacity for the forward error correction code [6]. The code chosen should provide the highest coding gain at the output BER of interest for the service. The BER curves of the better codes are such that the best code for one BER is likely to be best, or nearly the best, for another. Convolutional codes work well on these channels, being particularly well suited to exploiting the available soft decision information. The most convenient means of varying the error rates is by varying the E_b/N_o. However, for very demanding BER requirements on non–delay-critical services, improvements in capacity can be achieved through the use of optimized ARQ, as is explored in Section VII.

A vital issue for a universal communications system is the handling of delay-critical services, notably voice. Low transmission delay implies a short transmission frame. This conflicts with the provision of time diversity, which is required to combat the effects of flat multipath fading. An alternative approach to handling the effects of fading is fast power control. This is required anyway in a CDMA system to mitigate the effects of the "Near/Far" problem. It is shown in [7] that fast power control can greatly reduce the required E_b/N_o at low speeds. Very short transmission frames can be transmitted with excellent E_b/N_o performance at all speeds, provided sufficient power control updates are transmitted in every frame. Simulations show that, for practical purposes, sixteen ± 1 dB updates per frame effectively combat fading irrespective of the frame size. Of course, the power control transmission overhead increases for shorter frames, but the available flexibility allows optimization to be performed. It is logical to use the same frame time on the air interface as is used by the speech coder, since it allows continuous transmission, but this results in a delay of twice the frame time. Correspondingly shorter transmission frame times could be included to match the reduced frame times of higher bit rate voice coders. However, here the 10–ms frame time is assumed to apply to all services.

Restricting the particular bit rates carried to 8×2^n kb/s $(0 \leq n \leq 5)$ allows the uplink spreading factor to be varied simply in factors of 2. In practice, every transmission frame contains fixed overheads (e.g., CRC and payload header) so that, for example, doubling the spreading factor would more than halve the available payload. This effect can be compensated by introducing a different degree of puncturing of the convolutional code for each of the bit rates.

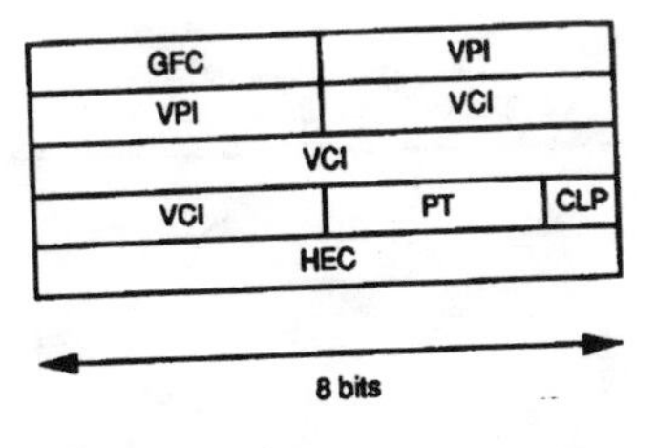

Fig. 1. ATM cell header format at the user network interface (UNI).

A. Transporting ATM Cell Header Information (Fig. 1)

In general, a terminal will require a number of simultaneous connections; for example, a voice terminal will have signaling and speech connections and a multimedia terminal may have video and data connections as well. However, the number of simultaneous connections per mobile will be limited. Thus, although ATM connections could be identified by transmitting the ATM address, the overhead can be reduced by providing translation at the mobile and in the network between an air interface connection identity and the ATM address. Connection identity information can be removed altogether if each connection is carried on a separate CDMA code. This would also allow the basic error rate of the link to be tailored to the service as discussed earlier.

An alternative would be to use one CDMA code per mobile, leading to all connections to a particular mobile being treated identically over the air interface. Only when data is received without errors can the service be determined reliably. Low error rates require ARQ with a consequent increase in transmission delay, which would be unacceptable for voice traffic. Also, undetected errors in the ATM address or connection identity could cause ATM cells to be misrouted.

Given that there are sufficient CDMA codes and the increase in complexity due to additional codes is marginal, one code (or, at high data rates, a number of codes) per connection has been used in the system concept. This provides a simple method of minimizing the payload overhead and the chances of misrouting.

In addition to the ATM address, other ATM cell header fields can be removed over the air interface. The Generic Flow Control (GFC) field is only significant at the User Network Interface (UNI), where it is used to control access from a number of terminals. The Header Error Control (HEC) field is redundant, since the error detection and correction provided by digital radio systems are generally more powerful. Within the fixed network, the HEC is also used for ATM cell delineation, which can be provided here by radio system framing. The Payload Type (PT) and Cell Loss Priority (CLP)

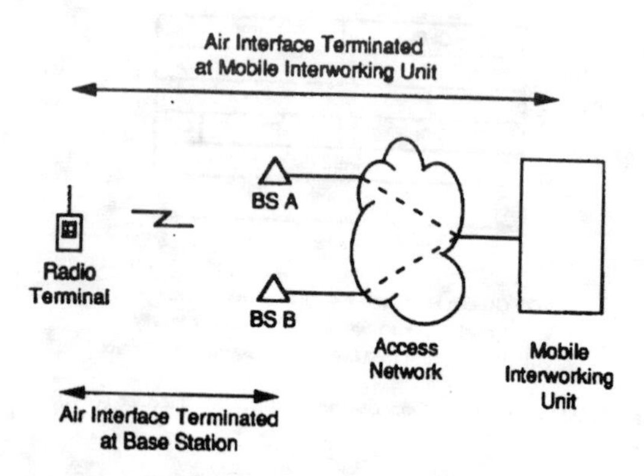

Fig. 2. Network configuration for diversity options.

fields will still have to be carried as they have network-wide significance, but the payload overhead due to ATM is still reduced to some 1%. At the receiving side of the interface, the original ATM cell less GFC can be reconstructed and a default value inserted in the GFC field before a new HEC is calculated.

V. Access Network Design Issues

CDMA systems operate best if every mobile communicates through its optimum base station. Thus, a mobile may be affiliated to a number of base stations and communicate via one or more depending upon the instantaneous radio conditions and the handoff regime imposed by the system. Use of a nonoptimum base station will lead to a reduction in system capacity, but this may be acceptable if there are compensatory advantages such as a significantly reduced infrastructure cost.

In considering where the air interface should be logically terminated, the base station and the Mobile Interworking Unit (MIU) represent the two extremes in a range of options (Fig. 2). In one case, no air interface control information is passed over the access network and, in the other, it is passed between all base stations and the serving MIU. Intermediate solutions are possible but, for simplicity, these alternatives are first compared. From the network point of view, a number of possible diversity techniques can been distinguished:

- **Parallel transmission**—depending upon the radio conditions, a mobile may communicate through a number of base stations in parallel. These paths are combined at the mobile and a point within the network. Additional network transmission capacity will be required in proportion to the number of mobiles with parallel transmission.
- **Selection diversity**—a mobile may communicate through one of a number of base stations on a frame-by-frame basis. There will be no overall increase in traffic on the uplink, but the route taken by any one frame is indeterminate. Statistical multiplexing of the uplink information from a number of mobiles onto the access network link allows this characteristic to be exploited. ATM transmission is a good way of achieving this. On the downlink, the parallel transmission of information is still required since it is not known in advance which base station will be used for a particular frame.
- **Soft hard handoff**—a mobile may be affiliated to more than one base station but will communicate through a particular base station for many frames at a time. Thus, as long as the network can respond quickly when a mobile wishes to handoff to another base station, no additional traffic paths are required. To prevent breaks in transmission, it may be necessary to transmit downlink data in parallel for short periods.
- **Hard handoff**—a mobile is only affiliated to a single base station, and handoff to a new base station will be a relatively long procedure–probably leading to a break in transmission.
- **Mixture of regimes**—there is no reason why the uplink and downlink should use the same handoff regimes and hybrid schemes may have advantages.

If data is transmitted over diverse paths, the original data can only be recovered with confidence once the paths are combined. On the uplink, ARQ, signaling termination, transcoding, echo control, data encryption, etc. should all be performed on the network side of this point. Table I (with reference to Fig. 2) illustrates the impact of the diversity techniques on the network. Dual diversity is assumed for simplicity, but the principles apply to higher degrees of diversity. Generalizing, it can be seen that most techniques imply some transmission over the access network between base station and the air interface logical end point, wherever this point may be (e.g., another base station).

The preferred option in the first instance is to terminate the air interface at the MIU, since this allows a greater number of diversity techniques to be considered. This centralization of functions will also tend to lead to simpler and more easily maintained base stations. However, where parallel transmission is used, centralization will result in a greater use of network transmission capacity and in this case devolution of the combiner/splitter function may be desirable but with consequent control complexity. Similar considerations apply to the downlink when selection diversity is used.

There is a number of methods by which information could be transmitted over the access network. When transmitting voice, each 10-ms speech coder frame could be carried in a separate ATM cell but the occupancy will be low. However, to keep within the delay budget it is not possible to wait for another frame for the same connection. The occupancy can be increased by including speech frames from a number of connections in one ATM cell, but this can add complexity and delay. In fact, although a single speech code frame per ATM cell appears inefficient, such a cell does represent 10 ms of speech, which in the broadband network with 64 kb/s coding would be carried in a little under two ATM cells.

When transmitting ATM cells, the use of ARQ requires a payload overhead to be passed with the user ATM cell between the logical end points of the air interface. Other overhead information may also be carried, depending on the diversity technique employed. There is, thus, more information to be transmitted over the access network per user ATM cell than

TABLE I
DIVERSITY OPTIONS AND AIR INTERFACE LOGICAL END POINTS

RADIO SYSTEM DIVERSITY	AFFILIATED TO BS	BS USED FOR TRANSMISSION	WHEN	INTERFACE END POINT
Parallel Transmission	A & B	A & B	Simultaneously	MIU
Selection Diversity	A & B	A or B	Per Frame	MIU
Soft Hard Handoff	A & B	A or B	For Many Frames	BS? or MIU
Hard Handoff	A or B	A or B	Long Term	BS or MIU
Up and Down Links Different	A & B	A &/or B		BS? or MIU

can be carried within a single access network ATM cell. This can be overcome in a number of ways:

- **Use two access network ATM cells per user ATM cell**—This leads to the access network ATM cell being half filled with user data (50% "payload efficiency"), but allows a simple encapsulation of air interface frames. Depending on the diversity technique, the remaining space could be used to report on radio signal quality. The access network ATM cell address can relate directly to the user connection, and this can be exploited to simplify routing.
- **Generate a stream of user ATM cells plus overheads**—The stream is segmented into ATM cells using standard ATM Adaptation Layer (AAL) processes [8], [9] before being transported to the MIU in ATM cells. Although the payload efficiency rises to some 80%, the remaining space is taken up with AAL overheads and cannot be used for radio quality data.
- **Send the overhead in separate ATM cells**—ARQ sequence numbers would still be transmitted over the air interface but only sent between base station and MIU periodically, at connection initiation and at discontinuities. A parallel access network connection could not be used, since ATM cell sequence is important but is not guaranteed between ATM connections. If user cells and overhead information used the same connection, a means of differentiating between the two would be required. A special pattern would be susceptible to imitation, and modification to the header would affect standards. Thus, although the cell efficiency could approach 99% there could be significant difficulties in implementing this approach.
- **Avoid the need to transmit overhead information**—This could be achieved by logically terminating the air interface at the base station, but it would limit the diversity options available, probably to hard handoff and possibly soft/hard handoff.

Given the complexity of the other techniques and despite its potential inefficiency, the simple mapping of user ATM cells into two access network ATM cells is assumed. For commonality, each speech coder frame is assumed to be carried in a separate ATM cell.

VI. SYSTEM CONCEPT

A variety of user terminals is foreseen, ranging from service specific (voice terminals, pagers, notebook computers) to standard fixed network N-ISDN and broadband terminals interfacing via special radio terminals (Fig. 3). N-ISDN traffic is assumed to be transported in ATM cells. Terminals gain access via the common air interface to base stations which are connected over a broadband access network to an MIU. The following describes uplink transmission: the reverse process occurs on the downlink.

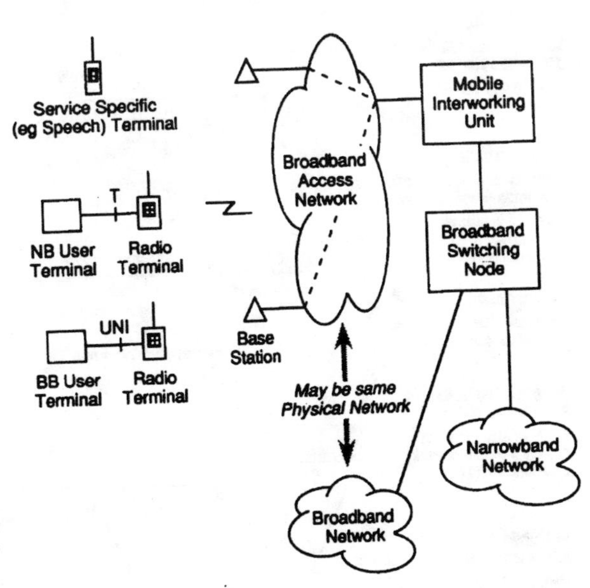

Fig. 3. Outline system architecture.

Fig. 4 shows the proposed method of transporting ATM cells. On the uplink, a standard ATM cell is passed from the user terminal to the radio terminal where the ATM header is removed. The user information field is split into two, and a payload overhead (containing, e.g., ARQ control data plus the PT and CLP) attached to each half to form air interface half cells. These are transmitted on a CDMA code, which is uniquely related to the connection address within the user's ATM cell (higher data rates can use multiple unique codes). At the base station, the air interface half cells are encapsulated into standard ATM cells and transmitted to the MIU over the broadband access network which is logically separate from, but may be physically integrated with, the main broadband network. The ATM cell header address used for the access network connection is also unique to the connection and is mapped from the CDMA code.

At the MIU, a standard ATM cell which contains the original user information field is formed from the two air interface half cells and forwarded to the Broadband Switching Node. The network ATM header address is translated from the connection identity, as denoted by the access network header address, and the PT and CLP values are taken from the payload overhead. The Header Error Control (HEC) field is calculated from the contents of the rest of the header. This interface is assumed to be the Node Network Interface (NNI), which does not include the GFC.

Signaling is assumed to be carried in ATM cells. To prevent an inordinate number of CDMA codes from being required, all signaling is sent on a common code with individual signaling

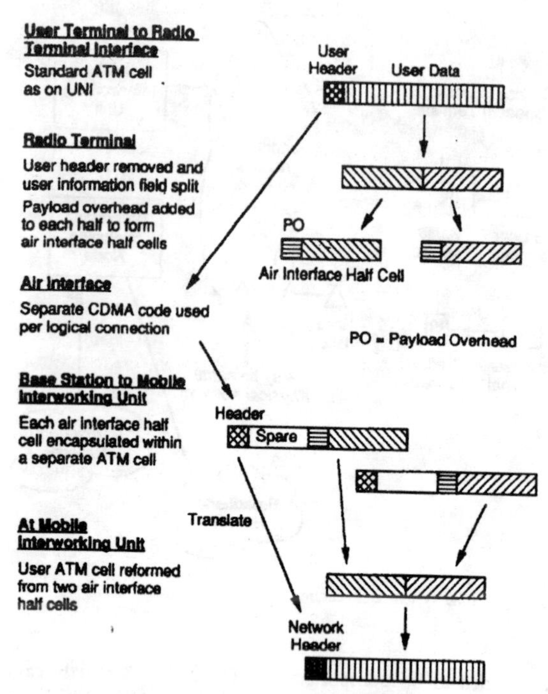

Fig. 4. Transmission of ATM cells from mobile terminal to fixed network.

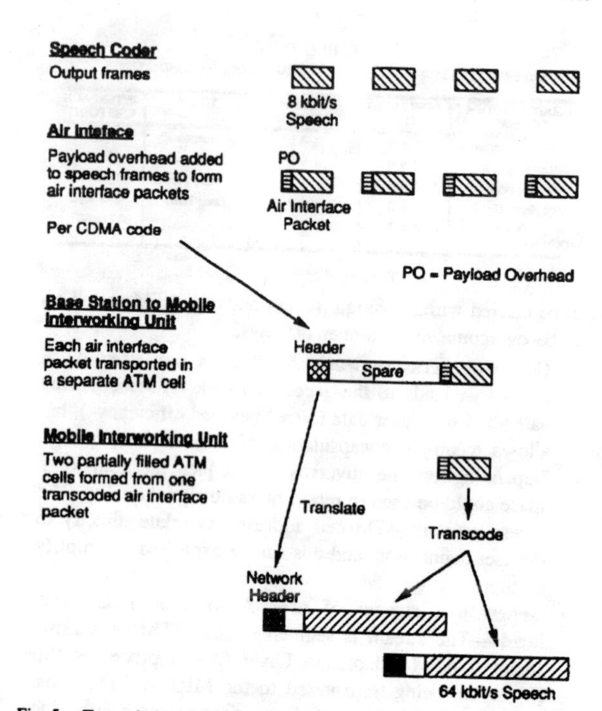

Fig. 5. Transmission of 8 kb/s speech from mobile terminal to fixed network.

relationships being distinguished by an address in the payload overhead.

Voice services are carried in a similar manner to data (Fig. 5). On the uplink, the output of the speech coder frame is combined with a payload overhead to form an air interface packet. A separate CDMA code is used for each voice connection over the air interface. At the base station, the air interface packet is encapsulated into an ATM cell for transmission to the MIU where transcoding is performed.

Here, it is assumed that to minimize interworking delays on connections involving mobiles, each ATM cell within the main broadband network will carry 5 ms of speech. This is different from the likely broadband network standard, where an ATM cell would contain 5.875 ms of PCM encoded speech [8], [10] and presupposes that the amount of speech per ATM cell can be different for calls involving mobiles. If this is not done, the overall delay will be significantly increased. For example, if the speech coder and transcoder are considered in isolation, a 10-ms speech coder frame interworking with an ATM network with cells containing 5.875 ms of speech gives an overall delay of at least 15.875 ms. If the ATM cell were to carry 5 ms of speech, the minimum overall delay reduces to 10 ms. Other speech coder frame times may require ATM cells to be filled to a different extent. To be fully effective, speech terminals and signaling throughout the broadband network should support such partially filled cells.

This approach will lead to to an increase in ATM cell traffic for voice connections to mobiles, but this increase only applies to voice traffic between mobiles and fixed network users: traffic between fixed network terminations will be unaffected. In addition, voice traffic between mobiles need not be transcoded, and individual speech coder frames could be carried across the fixed network in partially filled ATM cells. This would reduce the network bandwidth occupied by such connections and would result in a lower end-to-end delay than might otherwise be expected.

VII. Error Control

The proposed system provides a multiservice capability incorporating both non–delay-critical and delay-critical services. In this section, the two requirements are considered separately.

A. Non–Delay-Critical Services

Errors in non–delay-critical services can be most effectively handled by inclusion of a feedback protocol (ARQ). In general, the data services will be bursty so will require some form of Medium Access Control (MAC). Owing to its simplicity, a slotted ALOHA protocol based on CDMA [11] is proposed. This ALOHA protocol will employ hybrid ARQ [12] techniques to achieve and maintain the required quality of service. To understand better the associated trade-offs in the system concept which has been developed, a number of error control schemes were simulated.

The simulation was of a 8 kb/s data service using the 8 kb/s voice connection in a data-oriented fashion. This is the lowest rate service and was taken as a worst case since it represents the greatest channel variation per bit transmitted. The simulation environment was based on Differential Binary Phase Shift Keying (DBPSK) modulation with multisymbol detection on transmitted symbols to improve performance [13]. A special bit reliability metric for the fading environment was used to provide soft decision information for the decoding process. A Rayleigh fading environment was assumed for the channel model with two antennas and two multipath components per antenna. Closed loop power control was implemented and, in all simulations, a mobile speed of 30 mph was assumed. The carrier was taken to be at 2 GHz for the purposes of Doppler spectrum calculations, while a bandwidth of 5 MHz was assumed.

Investigations centered around two ARQ techniques. The first was based on type-I hybrid ARQ [12], while the second was based on type-II ARQ [12] with incremental redundancy transmission. A description of the two techniques follows.

Type-I Hybrid ARQ: The Type-I hybrid ARQ employed half-rate convolutional coding of constraint length 9 with generator polynomials 561 and 753 in octal [14]. The details of the simulation were slightly different from the overall proposal in that the data service was simulated with a frame length of 12 ms rather than 10 ms. The data packet included 8 tail bits so that, for an air interface packet of 112 bits, 240 encoded bits were transmitted per packet. For a radio packet of 12 ms, this corresponds roughly to an 8 kb/s data rate during activity. Standard block interleaving techniques were also employed to mitigate the channel memory effects (due to Doppler).

The decoding algorithm was based on a soft decision input Viterbi Algorithm (VA). Two different overall decoding approaches were simulated. The first was the conventional decoding technique in which packets are decoded when they arrive and discarded if errored. As an alternative, code combining [15], [16] was also investigated as a means to achieve adaptive operation. Here, a simplified version of code combining was used in which packets were combined pairwise. Untruncated operation (unlimited number of transmissions) was assumed.

Type-II Hybrid ARQ: In this scheme, the inner code was based on the same convolutional code as for the Type I hybrid ARQ scheme while the outer code was based on Reed-Solomon (R-S) coding. The R-S code operates in an incremental redundancy fashion, so that redundancy is sent on demand [17]. Due to the short packet size, shortening of the R-S code was necessary. In the simulations, it was found that the (36, 18) shortened R-S code defined over Galois Field (GF) GF(2^6) gave the optimum performance. The operation of this scheme is as follows: in the first transmission, there is no redundancy in the R-S code so decoding according to the inner convolutional code takes place; if this decoding trial fails, then the next 18 symbols are sent in the next transmission. Now, due to the redundancy of the code, e_2 errors and s_2 erasures can be corrected in the total 36 symbols provided the following inequality is satisfied:-

$$2.e_2 + s_2 \leq N - K = 18 \tag{1}$$

For improved performance, a first decoding trial takes place based on the second transmission alone (assigning erasures to the whole first transmission). If this fails, then Generalized Minimum Distance (GMD) decoding [18] takes place, in which a sequence of decoding trials is exploited to improve the performance. Information available out of the VA was used to transfer reliability information from the inner code to the outer code. If the second stage decoding fails, then the algorithm restarts by sending the initial part of the R-S code.

Numerical Evaluation: The following simplifying assumptions were made in evaluating the performance of ARQ:

— The probabilities of frame error in successive repeats of any one frame transmission are uncorrelated. For a practical ARQ scheme, successive repeats of a single frame will not take place contiguously, so the assumption should be reasonably well justified.
— There is no overhead in applying the ARQ. This assumption is clearly unrealistic. At the very least, some additional redundancy will be required for reliably detecting errored frames. This assumption is justified on the basis that, when comparing different error control schemes, similar additional redundancy will be required in each case.
— Perfect detection of errored frames is achieved.
— All signaling of acknowledgments or repeat requests is achieved with 100% reliability and consumes no capacity.
— ARQ repeats will continue until the frame is successfully delivered.

In the work described, the performance measure taken was the average number of transmissions for a packet success, $\overline{L}$. Let $P_s(1)$ and $P_s(2)$ denote the probabilities of successfully decoding a packet in exactly 1 and 2 transmissions, respectively. Then, it can be shown that for type-I ARQ:

$$\overline{L} = \frac{1}{Ps(1)} \tag{2}$$

while, for code combining and type-II ARQ,

$$\overline{L} = \frac{2 - P_s(1)}{Ps(1) + Ps(2)} \tag{3}$$

Fig. 6 presents the results for the performance of the three schemes investigated. In the figure, the horizontal axis corresponds to the E_b/N_o delivered by the closed loop power control loop (this corresponds to actual measurements taken upon adding up all antenna and multipath contributions). The vertical axis takes account of the average number of transmissions for successful packet delivery by weighting the E_b/N_o according to the number of transmissions required ($\overline{L}$). Thus, the vertical axis reflects the overall average interference generated by the modeled link per bit successfully delivered. For low E_b/N_o values, the Protocol E_b/N_o is large, reflecting the very large number of transmissions required per delivered frame. For high E_b/N_o values, the Protocol E_b/N_o is equal to the average measured data E_b/N_o reflecting the fact that every message is delivered at the first attempt.

It can be clearly seen that code combining ARQ outperforms the other methods considered. This is an interesting result because combining gives a relatively simple means for achieving this type of robust performance, in comparison with both conventional type-I hybrid ARQ (no code combining) and type-II hybrid ARQ. The latter system, although highly complex (concatenated coding scheme, GMD decoding), does not achieve a performance comparable to the code combining scheme. This results from many factors, including the slow

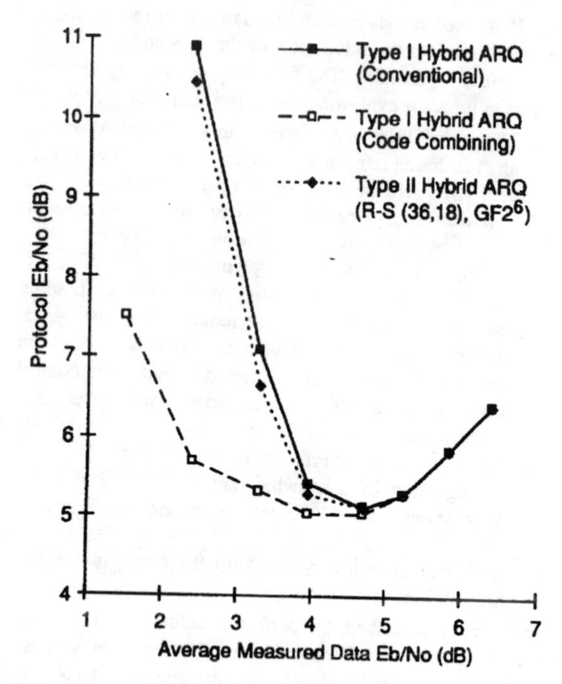

Fig. 6. Summary of ARQ performance.

fading rate of the channel and the system operation in a relatively low E_b/N_o. The main conclusion here is that code combining with type-I hybrid ARQ is the preferable technique to use, both due to its simplicity and robustness under fluctuating conditions.

It is also noteworthy that under the condition of optimum E_b/N_o, the improvement through code combining is very modest. For a network operating below capacity, power control can maintain the E_b/N_o at the optimum level at all times. However, as the network loading approaches its capacity limit, the E_b/N_o cannot be maintained and there is a danger of overall power control instability. The use of code combining allows detection of the overload condition through measurement of the repeat request rate and a backing off in the offered traffic without loss of capacity during the overload condition. Thus, the use of code combining affords a form of soft capacity limit which is applicable to data requiring guaranteed delivery.

B. Delay-Critical Services

The error control requirements for the delay-critical services can only reasonably be met by means of forward error correction coding. The evaluation of the optimum inner code for data services also applies to the main code for delay-critical services. Thus, for delay-critical services, convolutional coding of the type specified, Type-I hybrid ARQ, is proposed.

VIII. Conclusion

A system concept based on marrying the best features of CDMA and ATM has been developed. It has been shown that there is very good match between the attributes of the two, leading to a highly flexible interface between the mobile connection and the broadband network. High bandwidth services, up to 256 kb/s, can be supported and can coexist with good spectral efficiency with voice users operating on the same RF carrier.

The target overall delay of 25 ms can probably just be met with a broadband access network, assuming the total speech coder/decoder delay is similar to the speech coder frame time and that ATM cells carrying 5 ms of speech are supported by the broadband network. Lower bandwidth transmission could be used within the access network to reduce costs but lower transmission rates lead to longer delays. Total infrastructure bandwidth requirements and, thus, cost can also be significantly affected by the diversity technique used.

A code-combining ARQ protocol has been presented which allows the data capacity of the system to be maintained even under overload conditions. The stability margin of the system is thereby increased, which in turn eases the problem of radio network management.

The development of concepts for third-generation mobile telecommunications systems provides a unique opportunity to migrate telephony services onto ATM networks. By developing harmonized standards for the transport of information across radio access systems, through ATM networks, and into other networks, e.g., PSTN and ISDN, the overall transmission delay can be constrained to acceptable levels.

Acknowledgment

The work described in this paper was performed as part of the Siemens and GPT advanced mobility program. The authors wish to thank the directors of Siemens AG, GPT Ltd., and Roke Manor Research Ltd. for permission to publish the paper. The authors also wish to thank Dr. L. Bömer of Siemens ÖN ZL for his continuing support of the work described in the paper.

References

[1] ITU-T, "Operational and service provisions for future public land mobile telecommunication systems," *Draft Recommend. F.115*, COM 1-R7, May 1993.

[2] ETSI, "Framework for services to be supported by universal mobile telecommunications system," Draft ETR/SMG-50201, Aug.1993.

[3] ITU-T, "Rapporteurs's report on question Q.L/15 (Encoding of speech signals at bit rates around 8 kbit/s)," TD SG15/2, Sept. 1993.

[4] G. R. Cooper and R. W. Nettleton, "A spread spectrum technique for high-capacity mobile communications," *IEEE Trans. Vehic. Tech.*, pp. 264–275, Nov. 1978.

[5] CCITT, "B-ISDN ATM layer specification," *Draft Revised Recommend. I.361*, COM XVIII-R116, July 1992.

[6] A. J. Viterbi, "Spread spectrum communications—Myths and realities," *IEEE Commun. Mag.*, pp. 11–18, May 1979.

[7] A. P. Hulbert, "Myths and realities of power control," *IEE Colloq. Spread Spectrum Techn. Radio Commun. Syst.*, Savoy Place, London, Apr. 1993.

[8] CCITT, "B-ISDN ATM adaptation layer (AAL) specification," *Draft Revised Recommend.* I.363, COM XVIII-R116, July 1992.

[9] ITU-T, "Recommendation I.363," (Section 6–AAL5), COM 13-9, Mar. 1993.

[10] CCITT, "Pulse code modulation (PCM) of voice frequencies," Recommend. G.711, CCITT Blue Book, 1989.

[11] A. Polydoros and J. Silvester, "Slotted random access spread-spectrum networks: An analytical framework," *IEEE J. Select. Areas Commun.*, vol. SAC-5, pp. 989–1002, July 1987.

[12] S. Lin and D. J. Costello, *Error Control Coding: Fundamentals and Applications*. Englewood Cliffs, NJ: Prentice Hall, 1983.

[13] D. Divsalar and K. M. Simon, "Multiple-symbol differential detection of MPSK," *IEEE Trans. Commun.*, vol. 38, no. 3, pp. 300–308, Mar. 1990.
[14] J. G. Proakis, *Digital Communications.* New York: McGraw Hill, 1989.
[15] D. Chase, "Code combining: A maximum likelihood decoding approach for combining an arbitrary number of noisy packets," *IEEE Trans. Commun.*, vol. 33, pp. 385–393, May 1985.
[16] T. Ketseoglou, "Coding adaptivity issues in spread-spectrum random-access networks," *Ph.D. Dissertation*, Dept. EE-Syst., Univ. Southern California, Aug. 1990.
[17] D. M. Mandelbaum, "An adaptive-feedback coding scheme using incremental-redundancy," *IEEE Trans. Inform. Theory*, vol. IT-20, pp. 388–389, May 1974.
[18] G. D. Forney, "Generalized minimum distance decoding," *IEEE Trans. Inform. Theory*, vol. IT-9, pp. 64–74, Apr. 1963.

Thomas J. Ketseoglou was born in Athens, Greece on March 26, 1960. He received the B.S. degree from the University of Patras, Patras, Greece, in 1982, the M.S. degree from the University of Maryland, College Park, in 1986, and the Ph.D. degree from the University of Southern California, Los Angeles in 1990, all in electrical engineering.

Since January 1991, he has been with Siemens AG, Munich, Germany, involved in various projects focusing on wireless communications. His research interests are in the area of wireless communications with the emphasis on modulation and coding, service integration, and networking aspects. He is a member of the Technical Chamber of Greece.

Michael McTiffin received the B.Sc. degree in electrical and electronic engineering from the University of Bristol, England in 1969.

He joined the then Plessey Company and for many years has worked in telecommunications research and development, initially in the area of digital switching but later specializing in system aspects. He contributed to the standardization of the GSM digital cellular radio system and more recently participated in the European RACE project. BLNT, which produced a broadband switch coupled to an optical customer loop. He is now a member of a team looking at concepts for third-generation mobile radio systems with responsibility for system and network aspects.

Wolfgang Heimsch was born in Stuttgart, Germany. He received the Ph. D. degree in electrical engineering from the Technical University, Munich, Germany in 1986.

In the same year, he joined the Siemens Research and Development Department, where he worked on BICMOS microprocessors. He is now in charge of the Development Department of Siemens Mobile Radio Terminals.

Pete Hulbert received the degree in electronic engineering from the University of Southampton, England in 1975.

The greater part of his career since that time has been spent working in radio communications research at Roke Manor Research in Romsey, England. Initially working on radio equipment design both in the RF and baseband areas, he has moved on to work at the systems level, particularly on digital radio communications for the civil market. Simulation has played a key role in many of his activities, the most recent being in the area of CDMA cellular mobile radio. He is a member of the UK Institution of Electrical Engineers and is a Chartered engineer. He served as Honorary Editor for the UK IEE Proceedings Part F (Communications Radar and Signal Processing) between 1983 and 1987.

Graham Crisp was born in Nottingham, UK in 1942.

Over the years, he has held a variety of posts concerned with the development and evolution of public, private, and mobile telecommunications networks and systems architectures. In 1988, when GPT Limited was formed by the amalgamation of GEC's and Plessey's telecommunication interests, he transferred from Plessey into GPT's Corporate Central Engineering Research function. In GPT, as a Senior Manager, he has a special responsibility for telecommunications user mobility. Among other tasks during this period with GPT, he has chaired the ETSI committee responsible for the standardization of data and telematic services on the GSM digital cellular system.

The Application of a Novel Two-Way Mobile Satellite Communications and Vehicle Tracking System to the Transportation Industry

Irwin M. Jacobs, *Fellow, IEEE*, Allen Salmasi, *Member, IEEE*, and Thomas J. Bernard

***Abstract*—A novel two-way mobile satellite communications and vehicle position reporting system currently operational in the United States and Europe is described. The system characteristics and service operations are described in detail, and technical descriptions of the equipment and signal processing techniques are provided. Additionally, the application of this technology to the land transportation industry, and specifically the over-the-road truckload motor carrier segment, with the type of benefits being derived by the current users through experience is also described.**

I. Introduction

THE motor carrier industry has suffered since its inception from a lack of accurate information as to the whereabouts of millions of dollars in assets and the inability to monitor and redirect drivers and equipment in order to take advantage of changing business conditions and opportunities. This has been especially true in the industry segment known as "Truckload" (TL), which is characterized by acceptance and delivery of full truckload freight in almost random fashion throughout the country, creating the necessity for frequent communications with home base and a continuing operational frustration at headquarters due to a lack of specific knowledge as to the whereabouts of the majority of the company assets. The problem is not as acute in the "less than truckload" (LTL) industry segment where they engage in local pick-up and delivery, consolidation at local terminals followed by relatively constant short to medium hauls between known locations. Until recently, the only method available to a motor carrier's headquarters for communication with trucks on the road has been by requiring drivers to make frequent stops (two to four, typically) each day to telephone the dispatch function with the latest information with regard to location and driving status. This system has bred: 1) inaccuracies, 2) overloaded dispatch personnel who spend the majority of their time talking with drivers on the telephone, and 3) unhappy drivers who find themselves wrestling large rigs in and out of truck stops in order to conform to the policies of their employer only to find themselves placed on hold for extended periods of time. Each stop can take from 45 to 90 minutes and the need for frequent check calls can substantially encroach upon the driver's weekly maximum on-duty time of 70 hours, costing him money and at the expense of what should be adequate rest time. This, in turn, contributes to driver turnover, which represents one of the major problems in this industry, currently running at 100% per year or more at most irregular route truckload carriers.

Manuscript received June 1990.

The authors are with QUALCOMM Inc., 10555 Sorrento Valley Road, San Diego, CA 92121.

IEEE Log Number 9040542.

The lack of accurate information as to truck routing and adherence to schedule has resulted in service failures (late deliveries), which the motor carrier has not been in a position to handle proactively by informing its customer of such failures in advance and arranging for revised unloading appointments. Also, fuel tax record keeping has been hampered by the awkward way in which information is manually collected and occasionally subjecting the carrier to major tax fines as a result of state audits. Out-of-route mileage has remained high because of the lack of accurate information, and deadhead miles are too great because of the inability to redirect vehicles to new loads or because of a lack of a total picture as to the locations of all of the equipment in a given area. Emergency conditions experienced en route cannot be handled efficiently because then invariably take place in remote areas where the driver has great difficulty in contacting headquarters.

Within the past two years, a unique two-way mobile satellite communications and tracking system has been introduced to the trucking industry in the United States and Europe, providing significant improvements and solutions to the above-noted problems. In this paper, we describe the service features and the technical characteristics of such a mobile satellite system, OmniTRACS®. This system operates on a secondary basis in the 12/14 GHz band, (Ku-band), which has been allocated to the Fixed Satellite Services (FSS) on a primary basis. Developed by QUALCOMM, Inc., the system is the first operational domestic mobile Ku-band satellite service to provide two-way messaging and position reporting services to mobile users throughout the United States. Blanket authority ("license") to construct and operate a Ku-band network of mobile and transportable earth stations and a Hub earth station was granted for the United States Federal Communications Commission (FCC) in February 1989 for an application filed in December 1987 [1]. Using a pair of Ku-band transponders on a domestic FSS satellite as described below, the system is capable of serving a population of user terminals ranging between 40 000–80 000 users depending on the average length of the messages being transmitted and the frequency of transmissions. As demand for the system grows, additional Ku-band transponders can be leased from the satellite operators(s) and modularly added to the system to increase its overall capacity.

The system began operational tests in January 1988 in which a mobile terminal was driven from the west coast to the east coast of the United States and back in constant communication with a Hub facility. Fully operational system providing for commercial services began in August 1989. Terminals have since been installed in vehicles ranging from 18-wheel tractor-trailers to minivans, marine vessels and automobiles to bicycles.

Operation has been very successful in all kinds of environments from wide open Western freeways to the concrete canyons of New York City. Initially, a position reporting capability integrated into the mobile terminal utilized LORAN-C derived position information to be transmitted automatically on a scheduled basis or on-demand to a central location monitoring and communicating with the mobile units. A satellite-based position reporting system using the direct range measurements for the forward link signal of the primary satellite and a second "beacon" signal from a second satellite has been introduced.

The 1979 World Administrative Radio Conference allocated the 14.0–14.5 GHz frequency band to the land mobile satellite service for earth-to-space transmissions on a secondary basis. No companion downlink in the 11.7–12.2 GHz frequency band for space-to-earth transmissions was allocated because a downlink transmission from an FSS satellite appears identical regardless of whether it is being received by fixed or mobile terminals. Hence, the land mobile satellite service transmissions in the earth-to-space direction do not cause unacceptable interference into the primary FSS services, while the system tolerates interference from the primary services. In the space-to-earth direction, the land mobile satellite transmissions (the forward link carrier) are coordinated in the U.S. in a manner similar to the way video carriers are coordinated by the FSS operators. Outside the U.S., the system may be operated at Ku-band frequencies above 10.7 GHz where applicable and desirable. It is also feasible to extend the range of services through this system to provide aeronautical and maritime mobile satellite services on a secondary basis in the same FSS bands in a manner similar to the way in which land mobile satellite services are provided.

As previously noted the mobile-to-hub link operates on a secondary basis, which means that the signal must not cause interference to primary users and must be sufficiently robust to withstand interference from primary users to provide an economically viable and reliable service. A key aspect in the design of the system is the completely successful implementation of sophisticated signal processing techniques in a low-cost mobile terminal. A second key feature is the design (patented) of a low cost, highly reliable 19 dB gain mechanically steered antenna that constantly tracks the satellite as the mobile vehicle changes direction.

As of April 1990, in the U.S., over 8000 mobile terminals are in operation using a pair of existing Ku-band transponders aboard GTE Spacenet Corporation's GSTAR-I satellite with another backlog of approximately 6000 mobile terminals scheduled for shipments. A demonstration system, EutelTRACS®, is currently operational in Europe using a pair of Ku-band transponders aboard a EUTELSAT satellite. Typical users include those involved in public safety, transportation, public utilities, resource extraction, construction, agriculture, national maintenance organizations, private fleets, and others who have a need to send and receive information to vehicles, marine vessels, or aircraft enroute. As vehicles travel across the U.S., they move out of range of conventional land-based communication systems. This satellite-based system eliminates the range problems inherent with land-based systems in the U.S., creating a true nationwide network enabling users to manage mobile resources efficiently and economically.

II. System and Service Description

Fig. 1 shows the end-to-end system overview. The Ku-band mobile satellite communications network has three major components:

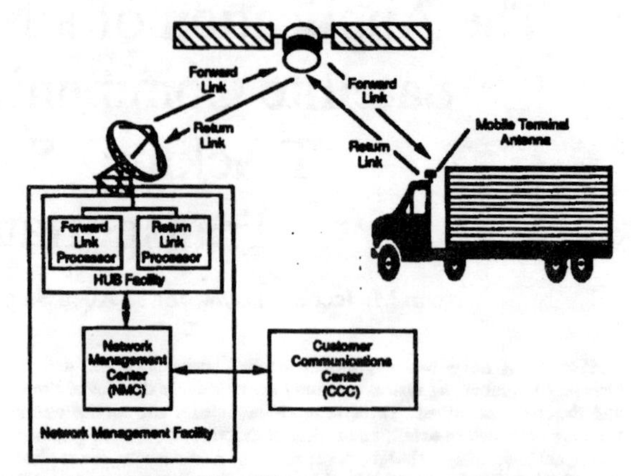

Fig. 1. Ku-band mobile satellite communications system block diagram.

1) A Network Management Facility (NMF) for controlling and monitoring the network.
2) Two Ku-band transponders, aboard a U.S. domestic satellite located at 103° West Longitude.
3) Two-way data-communication and position-reporting mobile and transportable terminals.

All message traffic passes through the NMF. The NMF contains a 7.6 m earth station including modems (the "Hub") for communication with the mobile terminals via the satellite, a Network Management Center (NMC) for network monitoring and control as well as message formatting, processing, management and billing, and land-line modems for connection with Customer Communications Centers (CCC's) [1]–[3].

The mobile and transportable satellite communication network features two-way data messaging, position reporting, fleet broadcasting, call accounting and message confirmation.

A. Service Features

Ku-band satellite communications are commonplace in the commercial market and, in fact, the fixed very small aperture terminals (VSAT) network applications is one of the major growing markets for the satellite communications industry. Telephony, television, and private data networks are already extensive users of Ku-band satellites, plus Ku-band components are mass produced, making terminal costs relatively inexpensive. This provides for very reasonably priced communication services and equipment at Ku-band frequencies. In the U.S., other potential mobile satellite service providers must rely on tentative satellite launches years into the future at a cost of hundreds of millions of dollars or on securing frequency spectrum not yet sufficiently available for such use. This system utilizes existing Ku-band hardware components and facilities, both on the ground and in space, with an innovative and proprietary system design that takes advantage of novel signal processing techniques to deliver the mobile and transportable network service.

Nationwide—The current operational system covers the continental U.S., including metropolitan, rural and coastal areas, so long as a direct-line-of-sight to the satellite from the mobile antenna is available.

Existing Satellites—Uses Ku-band transponders on existing geostationary satellites which have been operational for many years, substantially reducing the amount of initial capital investment required and the launch risks.

Two-Way—Allows dispatcher or driver to initiate or respond to preformatted or free-form messages. Emergency, group and fleet-wide messages are also available. Most importantly, it provides CRC error checking and positive acknowledgment of each unit addressed and message sent, ensuring the sender of a successful error-free transmission.

Data Communication—Dispatchers use computer terminals for message creation, transmission, and response. Unlike traditional voice systems, messages are received by a computer terminal in a fraction of the time it takes for diction of the same information, and can be stored for convenient viewing or later recall.

Demand for data communication systems has increased as the needs of mobile users have expanded and changed. Greater speed and accuracy in mobile communications, access to information stored in computer databases, improvements in response time from the field, system security and privacy, mobile fleet management and control, and use of preformatted and storable messages are some of the advantages that the satellite data communication system provides.

Position Reporting—Initially for U.S. operations, LORAN-C derived position information in the mobile terminal was made available to the dispatcher or driver on a scheduled basis or on-demand. Alternatively, and for operations in other parts of the world, radio-determination providing for radio-location and radio-navigation satellite services may be provided either through the use of the Global Positioning System (GPS) or by direct range measurements through two or more satellites depending on the cost of GPS receivers and availability/coverage provided by GPS as well as other Ku-band satellites. In April 1990, QUALCOMM introduced the first provision of vehicle position location through the use of two Ku-band satellites and direct range measurements, while eliminating LORAN-C receiver circuit card for the mobile terminal, to obtain navigation accuracy of less than 1000 ft throughout the continental United States and without the error conditions prevalent in the existing LORAN-C system due to atmospheric interference and the lack of radio signal coverage.

This position reporting capability is beneficial in vehicle management, such as *ad hoc* dispatching, scheduling of shipments, vehicle arrival time management, accident location, recovery in the event of hijacking, development of accurate state fuel tax reporting records and many other emergency and nonemergency situations.

B. System Description

The system, as used in the U.S., uses two transponders in a single Ku-band satellite. One transponder is used for a moderate rate (5–15 kb/s, with the higher data rates presently being implemented) continuous data stream from the Hub to all mobile terminals in the system. The system users can also use the terminals in a transportable or fixed mode. Messages are addressed to individual mobile terminals or to groups of mobile terminals on this channel. The antenna utilized for the mobile terminal is vertically polarized on both receive and transmit. This necessitates the use of two transponders, one horizontally polarized and the other vertically polarized, for the mobile service. Because transmission from a mobile at Ku-band in the U.S. falls under a secondary allocation, the mobile units must not interfere with fixed services and at the same time must tolerate interference from those services. Accordingly, to achieve this objective for the return link (mobile terminal to the Hub) a combination of frequency hopping and direct sequence spread spectrum waveforms is utilized, together with low power and low data rate transmissions.

For the forward link, the system uses a triangular FM dispersal waveform similar to that used by satellite video carriers, resulting in interference properties similar to television signals but with substantially less energy outside of a 2 MHz bandwidth.

A second transponder on the same satellite is used by the return link. Each mobile terminal has a low transmit power level (+19 dBW EIRP). This power level allows data rates on the return link ranging from 55 to 165 b/s, which is dynamically adjusted depending on available link margin for each individual return link transmission from the terminal. The antenna pattern of the mobile terminal is rather broad (approximately 6° beamwidth along the orbital arc), and therefore the potential exists to cause interference to users in adjacent satellites. To mitigate this interference, several techniques are used.

1) Direct-sequence spread-spectrum techniques are used to spread the instantaneous power spectral density of each mobile uplink over a bandwidth of 1.0 MHz.
2) Frequency hopping and FDM techniques are used to ensure that the power spectral density produced by the combination of all active mobile terminals is uniformly spread over a bandwidth of up to 54 MHz, which can be adjusted to a narrower or wider bandwidth to optimize the return link throughput capacity.
3) The transmissions of the mobile terminals are very carefully controlled. A mobile terminal will not transmit unless commanded to do so, either as a direct request (acknowledgment, report, etc.) or as a response to a carefully defined—and limited—group poll. This polling technique controls the number and frequency location of mobile transmitters at all times so that the level of interference can be tightly regulated. Furthermore, reception of the command through the forward link at appropriate signal levels implies that the antenna is correctly oriented for transmission.
4) Back-up satellite frequency plans and bandwidth can be downloaded "over the air" to the mobile terminal providing for automatic switch-over to back-up transponders and/or satellites.

As a result of the preceding techniques, a network consisting of tens of thousands of mobile terminals causes no unacceptable interference to adjacent satellites.

III. The Mobile Satellite Terminal

Fig. 2 shows a functional block diagram of the mobile terminal. A microprocessor implements all of the signal processing, acquisition and demodulation functions. The antenna has a asymmetric pattern optimized for operations in the U.S. (approximately 40° 3 dB beamwidth in elevation and approximately 6° beamwidth in azimuth). It is steerable in azimuth only. A low-noise amplifier and conventional down-conversion chain provide a signal to the microprocessor for acquisition, tracking and demodulation. During transmission, an up-conversion and spreading chain provide a signal in the 14–14.5 GHz band to the 1.0 W power amplifier. This signal is transmitted via the steerable antenna that has a maximum gain of +19 dBi for a total transmit power of +19 dBW.

Whenever the mobile unit is not in receive synchronization, it executes a receive acquisition algorithm until data from the satellite can be demodulated. At this point, the antenna is pointed toward the satellite and messages can be received from

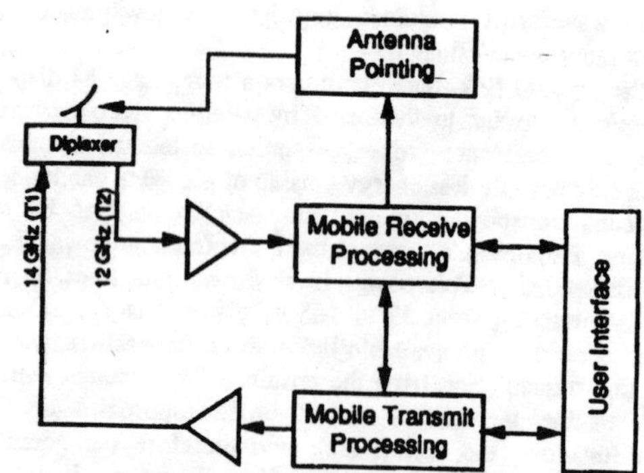

Fig. 2. Mobile terminal block diagram.

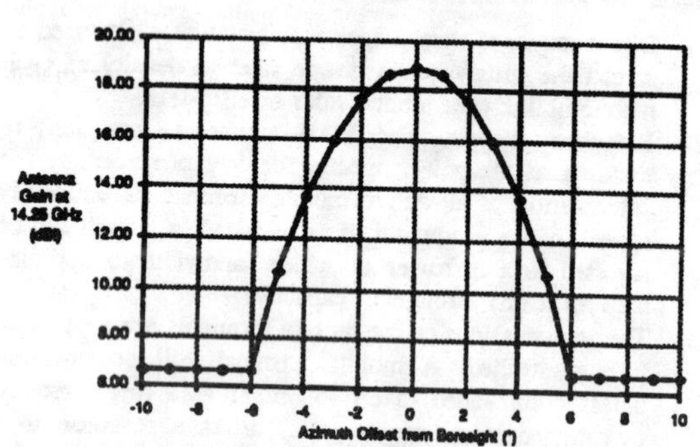

Fig. 3. Mobile terminal transmit antenna gain (azimuth cut).

the Hub. When commanded by the Hub, the mobile may start transmission of a message. The terminal is half-duplex, and transmissions are done at a 50% duty cycle to allow for continued antenna tracking of the received downlink signal. If at any time during a transmission the receive signal is lost, the terminal ceases transmission to prevent interference from being generated.

A. Description of Return Link Modulating Signal

At the lowest return-link data rate, binary data at 55.1 b/s is rate 1/3 convolutionally encoded to produce code symbols at a rate of 165.4 symbols per second. These code symbols are used five at a time to drive a 32-ary FSK modulator at a rate of 33.1 FSK Bd. A 50% transmit duty cycle produces an FSK symbol period of 15.1 μs. The tones out of the FSK modulator are direct-sequence spread at a rate of 1.0 Megachip per second for an instantaneous bandwidth of 1.0 MHz. This 1.0 MHz bandwidth signal is then frequency hopped over up to a 54 MHz bandwidth. To maximize system capacity in areas with good satellite G/T, a 3.0x data rate of 165.4 b/s is also provided. This is implemented by a three times FSK symbol rate.

B. Return Link Power Density

Fig. 3 shows the mobile terminal main lobe antenna pattern in azimuth keeping elevation at maximum gain. This Fig. assumes a nominal boresight gain of 19 dBi. The sidelobes of this antenna are asymmetric and nonuniform, but stay below −12.0 dB relative to boresight gain.

Table I shows the maximum transmit power density link budget for the return link with 250 units transmitting simultaneously. Table I combined with the transmit antenna pattern of Fig. 3 produces the EIRP power density shown in Fig. 4. Also

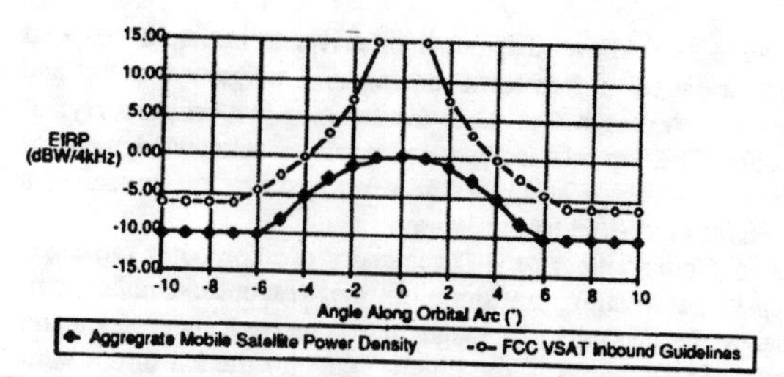

Fig. 4. Return link power density compared to the U.S. VSAT in-bound power density limits.

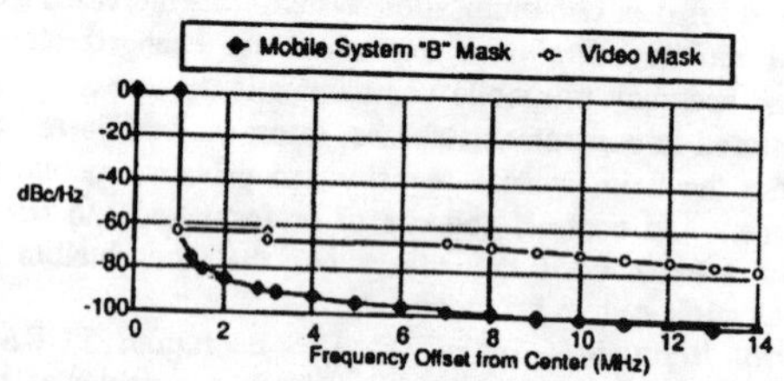

Fig. 5. Suggested coordination mask for the system forward link compared with video coordination mask.

Fig. 6. Satellite mobile communication terminal.

TABLE I
SYSTEM RETURN LINK POWER DENSITY

Maximum Tx Power	1.26 W	1.0 dBW
Maximum Tx Antenna Gain		19.0 dBi
Occupied BW	48 000 000 Hz	−76.8 dB/Hz
FCC Reference BW	4000 Hz	36.0 dB-Hz/4kHz
Number of Uplinks	250	24.0 dB
Tx Duty Cycle	50%	−3.0 dB
System EIRP Density		0.2 dBW/4kHz

TABLE II
FORWARD LINK POWER DENSITY

Satellite Transmit EIRP		44.0 dBW
Occupied Bandwidth	2 000 000 Hz	−63.0 dB/Hz
Reference FCC Bandwidth	4000 Hz	36.0 dB-Hz/4kHz
Transmit Power Density		17.0 dBW/4kHz

shown in Fig. 4 is the U.S. Ku-band guideline for inbound transmissions out of VSAT's, which indicates that the aggregate EIRP (measured in dBW/4 KHz) generated by 250 mobile units transmitting simultaneously is below the U.S. inbound transmission guideline for one VSAT.

C. Description of Forward Link Modulating Signal

Binary data at variable rates of 4960.3–14 880.90 b/s rate 1/2 block encoded to produce code symbols at rates of 9920.6 to 29 761.8 symbols per second, respectively. These code symbols are used to drive a BPSK modulator at rates of 9920.6 to 29 761.8 PSK symbols, respectively. A triangle wave FM dispersal waveform is then applied, resulting in similar coordination properties to video signals for coordinating with adjacent satellite transponders or cross-polarized co-channel transponder on the same satellite.

Coordination for the forward link waveform is easier than coordination of a video signal. The frequency band in the range of ±1.0 MHz about the center of the dispersal waveform will contain relatively high instantaneous power densities, but the frequency bands outside of this ±1.0 MHz range will drop off rapidly—even faster than video signals. Fig. 5 shows the suggested coordination mask for the forward link signal compared with the coordination mask suggested for video signals. The power density signals are shown in Table II.

D. Description of Mobile Terminal Hardware

The mobile terminal, shown in Fig. 6, consists of three components: the outdoor unit, the communication unit, and the display unit. The mobile terminal is compact and user-friendly, and provides for multiple message memory. The presently operational system in the U.S. supports a variety of preformatted and free-form text messages of up to 1900 characters in either direction. Each terminal is individually addressable with minimal power requirements and is capable of performing remote diagnostics. The mobile terminal can operate in a temperature range between −30° and 70°C, and provisions are made for quick installation, unattended service operations, and easy maintenance.

The outdoor unit contains the antenna assembly and the front-end RF electronics. The outdoor unit (including antenna) is approximately 11-in in diameter, 7 in high, and weighs approximately 10 lb. The outdoor unit can be mounted on the roof of a vehicle, marine vessel, or a mast (for a truck or vessel cab). The communication unit contains an analog section, digital electronics and the LORAN-C receiver, which are contained on four circuit card assemblies. The communication unit is contained in a single 4.4-in H × 9.22-in W × 12.5-in L unit weighing 16 lb and requires 12 VDC power, 35 W maximum power rating, and it can be mounted anywhere in the vehicle since it does not require operator access.

The display unit consists of a 40-character by four-line display and an ABCD or QWERTY keyboard. The display unit also provides several function keys for preprogrammed user functions. The display contains indicators for message waiting, satellite synchronization, and power indication. It also provides a maintenance mode for in-field troubleshooting and mobile unit initialization. The display unit is 11 in × 7.5 in × 2.75 in and can be located on the dashboard or any other convenient location. Cost of the mobile unit is in the U.S. $3900–$4500 range, depending on the volume ordered.

IV. The Network Management Facility (NMF)

The NMF is composed of a Hub and a Network Management Center. The Hub consists of a Ku-band earth station, the Forward Link Processor (FLP) and Return Link Processor (RLP) components, which are hardware/firmware subsystems designed and developed specifically for this system. The above equipment is used with a standard "off-the-shelf" Ku-band earth station with an interface at a 140 MHz IF frequency. For the U.S. operations, the primary earth station contains an antenna of 7.6 m in size. The NMC consists of a fully redundant midrange computer configuration and custom designed software components.

The users at various fixed locations can communicate with the NMC through conventional phone lines, packet switched networks, fixed satellite service VSAT's or USAT's, which in turn broadcasts the messages over a satellite link to the designated mobile terminals. The mobile terminals automatically acknowledge the receipt of messages allowing the system to detect and retransmit unreceived messages. The mobile terminal user can create and transmit messages back to the NMC and their respective user centers either as a reply to forward-link transmissions or at any time desired. The Hub is intended to provide service to a multiplicity of users through sharing of the satellite channel as well as the NMF/Hub facility.

The NMC is the focal point for the entire system providing the following network features:

- NMC to mobile (forward messages) and mobile to NMC (return messages) communication paths management and control;
- user fixed location interface with the NMC over dial-in land-lines, packet switched networks or fixed satellite service VSAT's;
- user validation to prevent unauthorized access;
- validation to prevent transmission of messages to unauthorized destinations;
- validation to ensure the destination mobile equipment configuration supports the transaction;
- ensured user message delivery;
- automatically updated mobile position reporting information;
- unique message identification through the use of forward and return message numbers;
- capability to transmit individual, group, and system addressed messages;
- user definable macro message formats;
- periodic system messages to ensure synchronization and to determine system operational status;
- system wide channel usage accounting data (user name, account number, etc.) written to disk files and available for billing purposes;
- operator console providing system status and control;
- system administrative interface providing data base editing and report capabilities;
- capability for upgrade of the NMC computer to a more powerful computer configuration.

Users are required to perform a sign-on process prior to transmitting their message(s) to the NMC. Multiple messages may be sent to the NMC during each session. Users are notified at the time of sign-on of pending mobile unit messages and, if present, these messages can be transmitted back to the user.

System security features require user identification prior to using the system to prevent unauthorized system users and invalid transmissions. Users sign on by means of a special access code that includes their account number and password. The NMC validates the sign on by comparing the user supplied values with those stored in the User File. If the user supplied account number and password match those in the User File, the user is allowed access to the NMC. Three unsuccessful sign on attempts result in a line disconnect. Attempts to use the system by unauthorized individuals are recorded and the statistics on such attempts are available to the operator through the NMC monitor. Only after validation are user messages accepted.

When a user transmits a message to the NMC, the NMC acknowledges that the message has been accepted and queued for transmission. When the message is transmitted over the satellite link and received successfully by the mobile unit, it automatically acknowledges the message (only messages addressed to specific mobile units are acknowledged; group addressed messages are never acknowledged). This message acknowledgement ensures the message was delivered and was received correctly by the mobile unit. If a user message is not acknowledged within the time out period, the NMC automatically retransmits the message N times.

The user is notified, via a message confirmation, that the message has either been received by the mobile unit or that the message is undeliverable. In addition to ensuring message delivery, the NMC and mobile unit perform duplicate message detection to prevent a single message from being reported more than once.

The NMC supports individual, group, and system messages. Macro messages are the commonly used formatted or "fill-in-the-blanks" messages by the driver and the dispatcher that are partially pre-stored in the terminal for quick data entry and transmissions. These types of messages, sometimes referred to as canned messages or preformatted messages, allow the user dispatch centers and mobile terminals to send prestored/preformatted messages, allow the user dispatch centers and mobile terminals to send prestored/preformatted messages. Each individual user may have its own predefined messages. All mobile units in a user's fleet must use the same macros. Each user fleet may define up to 63 forward macros and 63 return macros, which are down-loaded from the NMC on a regular basis to provide for updates of macros by each user.

VI. System Enhancements

Since initial implementation of the system, the need for a number of enhancements has been recognized and the following have been implemented:

1) TrailerTRACS™—a nominal hardware addition which provides for automatic identification and position reporting whenever a trailer is attached or dropped from a power unit (e.g., tractor) equipped with the system.

2) Driver Pager—a local paging unit which is operated by the system and which notifies the driver (up to 1000 yards from the vehicle) when an important message is waiting for him in the truck—can also be used as an alarm for "wake-up."

3) Message Return Receipt—provides the ability to have the system notify central operations or dispatch whenever a message has actually been read—analogous to certified mail with a return receipt.

4) Panic Button—provides the ability for a driver to hit one key or button, thereby initiating a "panic" message which is intercepted by the Network Management Center, causing a telephone call to be initiated to appropriate personnel to inform them of serious trouble at the location of the truck so that authorities can be dispatched. This feature is a requirement of the Military Traffic Management Command (MTMC) in connection with certain categories of defense ammunition shipments.

VII. System Benefits

In January 1989, Alex, Brown & Sons Incorporated Transportation Group reported that two-way mobile satellite communications and vehicle tracking services could be cost justified by most carriers due to such benefits as: enhanced equipment and driver productivity, better customer service, elimination of most truck-stops, improvements in driver satisfaction, reduction in clerical activities, fuel tax data reporting, integration with Electronic Data Interchange (EDI), trailer's "last-position" reporting, efficient handling of sensitive shipments and others.

The transportation industry has an increasingly acute need for mobile communications on a nationwide basis. Companies involved in public transit, trucking, railroads, marine transport, and aviation require communication with and tracking of their mobile fleets. Trucking operations are subject to frequent changes in routing and scheduling. Trucks are more vulnerable than other types of transportation to mechanical failure, accidents, local traffic delays and adverse road conditions, increasing the need for mobile communications. Because the OmniTRACS system uses data messages to communicate information, the cost per message is significantly less than voice transmissions. The time spent sending and receiving messages is decreased, thus saving money at both the driver and dispatch levels. In addition, the nationwide transmission range of the system allows for messages to be sent en route, which eliminates unnecessary stops.

Of particular interest is the ability of the OmniTRACS system to track vehicles, specifically those carrying hazardous materials, weapons, or petroleum. The cargo's value, coupled with the risks involved, requires constant surveillance of the truck, marine vessel, train or aircraft. The sheer number of government and private industry vehicles traveling across the country make it difficult to monitor the location of each vehicle at a specific time. The OmniTRACS position reporting system allows a dispatcher to locate any vehicle at any time with the desired level of accuracy throughout most of the continental U.S. This feature is beneficial for day-to-day monitoring of a fleet, as well as locating a vehicle that has been stolen or involved in an accident. Recovery of valuable merchandise and equipment is possible by using the OmniTRACS system.

As of August 1, 1990, the OmniTRACS system has been operating for over 24 months, is processing over 300 000 transactions per day, and the mobile terminals have nearly 1 billion miles of road experience on U.S. highways. Experience thus far reflects averages of seven messages per day, per vehicle, at 70 characters per message. Forward link messages represent one-third of the traffic and two-thirds of the transmitted characters. Return link messages (from the driver) tend to rely heavily on macros. More than 50 truckload motor carriers have implemented such systems in part or all of their fleet and approximately 100 firms are currently evaluating such systems. System benefits currently being experienced are as follows.

1) Reduced telephone expenses—A tangible early benefit directly related to the reduction in telephone contacts being made by drivers. In some cases, drivers who previously called several times each day are making contact less than once per week.

2) Increased revenue miles—One recent carrier reported an increase of 5.95% in overall revenue miles achieved in comparing single truckload units in his fleet equipped with the OmniTRACS system as compared to the rest of the fleet. This study covered a period of 20 weeks. For teams of truckload units, the increase averaged to a more dramatic 7%. These increases are due to a higher incidence of pre-dispatching due to improved load planning and increased miles by drivers due to savings of one to three hours per day in eliminating long stops for check calls.

3) Reduced "deadhead" miles—reductions experienced by as much as 10% (example: from 7% to 6.4%).

4) Reduced driver turnover—early reports from users indicate that driver turnover is reduced by 10% or more in fleets equipped with the system.

5) Reduced accidents—Accidents per million miles and accident dollar per million miles have been reduced due primarily to the reduced frequency of diverting to truck stops.

6) Increased dispatch efficiency—Most carriers find that dispatchers and driver managers can handle anywhere from 50 to 100% more drivers because of reduced telephone traffic and the ability to locate trucks without contacting the driver.

References

[1] Memorandum Opinion, Order and Authorization in the Matter of QUALCOMM, INC. Application for Blanket Authority to Construct and Operate a Network of 12/14 GHz Transmit/Receive Mobile and Transportable Earth Stations a Hub Earth Station, Feb. 14, 1989, the U.S. Federal Communications Commission.

[2] F. P. Antonio, K. S. Gilhousen, I. M. Jacobs, L. A. Weaver, Jr., "Technical characteristics of the OmniTRACS—The first operational mobile Ku-band satellite communications system," in *Proc. Mobile Satellite Conf.*, JPL Pub. 88-9, pp. 203–208, May 3–5, 1988, Pasadena, CA.

[3] A. Salmasi, "An overview of the OmniTRACS—the first operational mobile Ku-band satellite communications system," in *Proc. Mobile Satellite Conf.*, JPL Pub. 88-9, pp. 63–68, 1988, Pasadena, CA.

[4] T. J. Bernard, "Logistics benefits from two-way satellite tracking," *Defense Transport. J.*, Feb. 1990.

Irwin M. Jacobs (S'55–M'60–F'74) received the B.E.E. degree from Cornell University, Ithaca, NY, in 1956, and the M.S. and Sc.D. degrees in electrical engineering from the Massachusetts Institute of Technology (MIT), Cambridge, MA, in 1957 and 1959, respectively.

From 1959 to 1966 he was an Assistant/Associate Professor of Electrical Engineering at MIT and a staff member of the Research Laboratory of Electronics. During the academic year 1964–65, he was a NASA Resident Research Fellow at the Jet Propulsion Laboratory, California Institute of Technology, Pasadena. In 1966, he joined the newly formed Department of Applied Electrophysics, now the Department of Electrical Engineering and Computer Science, at the University of California, San Diego. In 1972, he resigned as Professor of Information and Computer Science to devote full time to LINKABIT Corporation. As Co-Founder, President, Chief Executive Officer and Chairman, he guided the growth of LINKABIT from a few part-time employees in 1969 to over 1400 employees in 1985 located in San Diego, Boston, and Washington, DC. In August 1980, LINKABIT merged with M/A-COM Incorporated, and the following February, he became a member of the M/A-COM Board Of Directors. In February 1983, he became Executive Vice President. While at LINKABIT and M/A-COM, he led the development of the Air Force DUAL Modem, the first commercial VSATS, and VideoCipher™ among other products. In April 1985, he resigned from his position at M/A-COM. On July 1, 1985, he became a Founder and the Chairman and President of QUALCOMM Inc., San Diego, CA.

Allen Salmasi (S'77–M'79) received the B.S. degree in electrical engineering and applied mathematics and the M.S. degree in electrical engineering and management economics from Purdue University, West Lafayette, IN, in 1977 and 1979, respectively, the M.S. degree in applied mathematics and an Engineer's degree in electrical engineering from the University of Southern California, Los Angeles, in 1983 and 1984, respectively. He has completed the doctoral candidacy requirements at the University of Southern California, for the Ph.D. degree in electrical engineering.

From 1974 to 1984, he was employed by the National Aeronautics and Space Administration (NASA) and the Jet Propulsion Laboratory, Pasadena, CA, where he held various technical and management positions involved in the development of NASA's Land Mobile Satellite Service program. He found Omninet Corporation in 1984, at various times holding offices of Chairman of the Board, President, and CEO, through August 1988. He joined QUALCOMM Inc., San Diego, CA, as a member of the Board of Directors and Vice President of Planning in 1988. He also serves as General Manager of the Digital Cellular System and Personal Communication Network development.

Thomas J. Bernard studied business administration at Los Angeles City College, Los Angeles, CA, and at San Fernando State University, San Fernando, CA.

He spent 21 years with ROHR Industries, a California Aerospace subcontractor with sales in excess of $500 million, where he held various positions. His last appointment with ROHR was in 1979, when he was named Vice President and General Manager of the ROHR Flxible Bus Co., a subsidiary with 2000 employees where a significant turn around in profitability was achieved. In 1978, Mr. Bernard was named President of ROHR Flxible. From 1982 to 1986 he was with M/A-COM LINKABIT where from 1982 to 1984, he served as Vice President, Communication Networks, responsible for management of the development, implementation and market production of the IDX 3000 series digital PABX and the MX24A Multiplexer. In 1985, he was appointed Executive Vice President and General Manager, M/A-COM Telecommunications Division, Western Operations. During this period, extensive development and introduction of new products and technologies were achieved with particular emphasis on very small aperture earth station (VSAT), digital cross-connect systems for the TELCO Market and a new radio-telephone (bandwidth efficient) product called "Ultraphone." He joined QUALCOMM Inc., San Diego, CA, in September 1986, as Vice President of Commercial Programs and was appointed Vice President and General Manager, OmniTRACS, in 1988.